ORTHOPAEDIC SPORTS MEDICINE

PRINCIPLES AND PRACTICE

ORTHOPAEDIC SPORTS MEDICINE

PRINCIPLES AND PRACTICE

Volume 1

Jesse C. DeLee, M.D.
Associate Clinical Professor of Orthopaedics and
Director, University of Texas Health Science Center
at San Antonio Sports Medicine Fellowship Program
San Antonio, Texas

David Drez, Jr., M.D.
Clinical Professor of Orthopaedics and
Head, Louisiana State University Knee and
Sports Medicine Fellowship Program
Lake Charles, Louisiana

W.B. SAUNDERS COMPANY
A Division of Harcourt Brace & Company
Philadelphia London Toronto Montreal Sydney Tokyo

W.B. SAUNDERS COMPANY
A Division of
Harcourt Brace & Company

The Curtis Center
Independence Square West
Philadelphia, Pennsylvania 19106

Library of Congress Cataloging-in-Publication Data

Orthopaedic sports medicine : principles and practice / [edited by]
Jesse C. DeLee, David Drez, Jr.—1st ed.

     p.   cm.

Includes index.

ISBN 0–7216–2836–2

1. Sports—Accidents and injuries.   2. Musculoskeletal system—
   Wounds and injuries.   3. Sports medicine.   I. DeLee, Jesse.
   II. Drez, David.

[DNLM: 1. Sports Medicine. 2. Athletic Injuries. 3. Orthopedics.
QT 260 077 1994]

RD97.078   1994

617.3′008′8796—dc20

DNLM/DLC   93–28370

Orthopaedic Sports Medicine: Principles and Practice

ISBN Volume I 0–7216–2834–6
Volume II 0–7216–2835–4
Two Volume Set 0–7216–2836–2

Printed in the United States of America.

Last digit is the print number: 9   8   7   6   5   4   3   2   1

*We dedicate these volumes to our wives,
Ann and Judy, to our families who support
us, and to all our teachers past and present*

CONTRIBUTORS

John P. Albright, M.D.
Professor and Director of Sports Medicine Services, Department of Orthopaedic Surgery, University of Iowa College of Medicine, Iowa City, Iowa

David B. Allen, M.D.
Associate Professor of Pediatrics, University of Wisconsin Medical School; Director of Pediatric Endocrinology, University of Wisconsin Children's Hospital, Madison, Wisconsin

Bob Anderson, B.S.
President of Stretching, Inc., Palmer Lake, Colorado

Bruce E. Andrea, M.D.
Fellow, Division of Cardiology, University of Colorado Health Sciences Center, Denver, Colorado

Jack T. Andrish, M.D.
Orthopaedic Surgeon, Section of Sports Medicine, Section of Pediatric Orthopaedics, Department of Orthopaedics, Cleveland Clinic Foundation, Cleveland, Ohio

Mark E. Baratz, M.D.
Clinical Instructor, Department of Orthopaedics, Medical College of Pennsylvania, Philadelphia; Orthopaedic Surgeon, Allegheny General Hospital, St. Clair Hospital, and Mercy Hospital, Pittsburgh, Pennsylvania

Martin H. Beck, M.D.
Bern, Switzerland

James B. Bennett, M.D.
Clinical Professor, Department of Orthopedic Surgery, and Chief of Hand and Upper Extremity Service, Baylor College of Medicine; Senior Active Staff, The Methodist Hospital, Houston, Texas

John A. Bergfeld, M.D.
Head of Section of Sports Medicine, Department of Orthopaedic Surgery, The Cleveland Clinic Foundation, Cleveland, Ohio

Thomas M. Best, M.D., Ph.D.
Associate Research Professor, Department of Biomedical Engineering, Duke University, and Department of Orthopaedic Surgery, Duke University Medical Center, Durham, North Carolina

Bruce D. Beynnon, Ph.D.
Assistant Professor, Department of Orthopaedics, University of Vermont College of Medicine, Burlington, Vermont

Joseph B. Blanda, M.D.
Instructor of Orthopaedic Surgery, Northeastern Ohio Universities College of Medicine; Active Staff, Akron City Hospital, Akron, Ohio

R. Luke Bordelon, M.D.
Clinical Professor, Louisiana State University School of Medicine; Attending, Children's Hospital, New Orleans; Director of Foot Clinic, Doctors' Hospital of Opelousas, Opelousas, Louisiana

James F. Bruce, Jr., M.D.
Consultant, Hand Surgery, University of Virginia School of Medicine, Charlottesville, Virginia; Orthopaedic Surgery Staff, West Georgia Medical Center, La Grange, Georgia; Consulting Staff, Roosevelt Warm Springs Institute for Rehabilitation, Warm Springs, Georgia

Michael E. Brunet, M.D.
Professor of Orthopaedic Surgery and Director of Sports Medicine, Tulane University Medical Center and Tulane University School of Medicine; Active Staff, Tulane Medical Center, The Medical Center of Louisiana, Veterans Administration Medical Center of New Orleans, Touro Infirmary, and Children's Hospital, New Orleans, Louisiana

Joseph A. Buckwalter, M.S., M.D.
Professor of Orthopaedic Surgery, University of Iowa and University of Iowa Hospitals and Clinics, Iowa City, Iowa

Edmund R. Burke, Ph.D.
Associate Professor, Biology Department, University of Colorado at Colorado Springs, Colorado Springs, Colorado

Kenneth P. Butters, M.D.
Clinical Instructor, University of Oregon Health Sciences University; Orthopedic Staff, Sacred Heart Hospital, Eugene, Oregon

S. Terry Canale, M.D.
Professor, Department of Orthopaedic Surgery, University of Tennessee; Chief of Pediatric Orthopaedics, Le

Bonheur Children's Medical Center; Staff, The Campbell Clinic, Inc., Memphis, Tennessee

Thomas O. Clanton, M.D.
Clinical Associate Professor, Department of Orthopaedic Surgery, The University of Texas Health Science Center at Houston, and Chief, Foot and Ankle Section, Hermann Hospital, Houston, Texas

Robert O. Cone III, M.D.
Clinical Assistant Professor, The University of Texas Health Science Center at San Antonio; Attending, St. Luke's Lutheran Hospital, South Texas Methodist Hospital, Nix Medical Center, Humana Women's Hospital, Rehabilitation Institute of San Antonio, and South Texas Ambulatory Surgical Hospital, San Antonio, Texas

Michael J. Coughlin, M.D.
Staff, Orthopaedic Surgery, St. Alphonsus Regional Medical Center, Boise, Idaho

Jay S. Cox, M.D.
Professor of Orthopedic Surgery, Pennsylvania State University School of Medicine; Orthopedic Staff, Centre Community Hospital, State College, Pennsylvania

Ralph J. Curtis, Jr., M.D.
Clinical Assistant Professor of Orthopaedic Surgery, The University of Texas Health Science Center at San Antonio; Orthopaedic Surgery Associates, San Antonio, Texas

Sandra L. Curwin, B.Sc., M.Sc., Ph.D.
School of Physiotherapy, Halifax, Nova Scotia, Canada

Samuel D. D'Agata, M.D.
Team Physician, Gettysburg College, Gettysburg; Member, Hanover Orthopaedic Associates; Orthopaedic Surgeon, Hanover General Hospital, Hanover, Pennsylvania

Robert D. D'Ambrosia, M.D.
Professor and Chairman, Department of Orthopaedics, Louisiana State University Medical Center, New Orleans, Louisiana

Dale M. Daniel, M.D.
Associate Clinical Professor, Department of Orthopedics, University of California at San Diego; Staff Orthopedist, Kaiser Hospital, San Diego, California

Kenneth E. DeHaven, M.D.
Professor of Orthopaedic Surgery and Vice Chairman, Department of Orthopaedics, University of Rochester School of Medicine; Attending, Strong Memorial Hospital and Monroe Community Hospital, Rochester, New York

Jesse C. DeLee, M.D.
Associate Clinical Professor of Orthopaedics and Director, University of Texas Health Science Center at San Antonio Sports Medicine Fellowship Program, San Antonio, Texas

David Drez, Jr., M.D.
Clinical Professor of Orthopaedics and Head, Louisiana State University Knee and Sports Medicine Fellowship Program, Lake Charles, Louisiana

Frank J. Eismont, M.D.
Professor of Orthopedic Surgery and Director of Orthopedic Spine Service, University of Miami School of Medicine; Co-Director of the Spinal Cord Injury Unit, Jackson Memorial Hospital, Miami, Florida

Richard D. Ferkel, M.D.
Clinical Instructor, Orthopedic Surgery, UCLA Center for the Health Sciences; Chief of Arthroscopy, Wadsworth Veterans Hospital; Attending, Southern California Orthopedic Institute, Van Nuys, California

Gerald A.M. Finerman, M.D.
Professor and Acting Chief, Division of Orthopedic Surgery, University of California at Los Angeles, Los Angeles, California

Timothy P. Finney, M.D.
Clinical Assistant Professor of Orthopaedic Surgery, Louisiana State University Medical Center, New Orleans, Louisiana

Marc J. Friedman, M.D.
Assistant Clinical Professor of Orthopedic Surgery, UCLA School of Medicine; Attending Physician, Southern California Orthopedic Institute, Van Nuys, California

Daniel Fritschy, M.D.
Assistant Professor, Hôpital Cantonal Universitaire Faculty of Medicine, Geneva, Switzerland

Freddie H. Fu, M.D.
Blue Cross of Western Pennsylvania Professor of Orthopaedic Surgery and Vice Chairman/Clinical Department of Orthopaedic Surgery, University of Pittsburgh, Pittsburgh, Pennsylvania

Brian J. Galinat, M.D.
Orthopaedic Surgeon, Delaware Orthopaedic Center, Wilmington, Delaware

William E. Garrett, Jr., M.D., Ph.D.
Associate Professor, Duke University Medical Center, Durham, North Carolina

Thomas A. Gennarelli, M.D.
Professor of Neurosurgery, University of Pennsylvania School of Medicine; Staff, Hospital of the University of Pennsylvania and Children's Hospital of Philadelphia, Philadelphia, Pennsylvania

Christian Gerber, M.D.
Professor of Orthopedics, University of Bern; Chairman, Department of Orthopedics, Kantonspital, Fribourg, Switzerland

William A. Grana, M.D.
Clinical Professor, Department of Orthopedic Surgery and Rehabilitation, University of Oklahoma Health Sciences Center; Attending, Presbyterian Hospital, Oklahoma City, Oklahoma

J. David Grauer, M.D.
Orthopedic Fellow, The Orthopedic Specialty Hospital, Salt Lake City, Utah

David P. Green, M.D.
Clinical Professor, Department of Orthopaedics, University of Texas Health Science Center at San Antonio; President, The Hand Center of San Antonio, San Antonio, Texas

Letha Y. Griffin, M.D., Ph.D.
Team Physician, Georgia State University; Clinical Instructor, Emory University School of Medicine; Staff Physician, Peachtree Orthopaedic Clinic, Atlanta, Georgia

Michael L. Gross, M.D.
Private Practice, Orthopaedic and Sports Medicine Associates, Emerson; Attending Physician, Pascack Valley Hospital, Westwood, New Jersey

Peter Hanson, M.S., M.D.
Professor of Medicine, Cardiology Section, University of Wisconsin Medical School; Co-Director, Preventive Cardiology and Cardiac Rehabilitation Programs; Staff, University Hospital, Madison, Wisconsin

Richard J. Hawkins, M.D., FRCS(C)
Clinical Professor, University of Colorado, Department of Orthopedics, Denver; Consultant, Steadman-Hawkins Clinic, Vail, Colorado

Edward G. Hixson, M.D., FACS, FACSM
Expedition Physician, China-Everest 1982, 1983; Seven Summits Everest Expedition, China-Everest 1983; Director, Medical Supervisory Team, U.S. Nordic Ski Team 1976–1986; General Surgeon, Adirondack Medical Center, Saranac Lake, New York

Roch B. Hontas, M.D.
Assistant Professor, Department of Orthopaedic Surgery, Tulane University School of Medicine; Active Staff, Tulane Medical Center, 7th Ward General Hospital, and St. Tammany Parish Hospital, New Orleans, Louisiana

Cheryl Hubley-Kozey, B.Sc., M.Sc., Ph.D.
Associate Professor, School of Physiotherapy, Halifax, Nova Scotia, Canada

Peter A. Indelicato, M.D.
Professor, University of Florida; Chief, Sports Medicine Service at Hampton Oaks/Shands Hospital, Gainesville, Florida

Roland P. Jakob, M.D.
Associate Professor and Vice-Chairman, Department of Orthopaedic Surgery, Inselspital, University of Bern, Bern, Switzerland

Nizar N. Jarjour, M.D.
Assistant Professor of Medicine, Department of Medicine, Section of Pulmonary and Critical Care, University of Wisconsin School of Medicine, Madison, Wisconsin

Frank W. Jobe, M.D.
Clinical Professor, Department of Orthopaedics, University of Southern California School of Medicine, Los Angeles; Associate, Kerlan-Jobe Orthopaedic Clinic, Inglewood; Medical Director, Biomechanics Laboratory, Centinela Medical Center, Inglewood, California

Robert J. Johnson, M.D.
Professor, Department of Orthopaedics, University of Vermont College of Medicine; Attending in Orthopaedics, Medical Center Hospital of Burlington, Vermont, and Fanny Allen Hospital, Colchester, Vermont

Pekka Kannus, M.D., Ph.D.
Docent (Associate Professor) of Sports Medicine, University of Jyväskylä, Jyväskylä; Chief Physician, Accident and Trauma Research Center, The UKK-Institute, Tampere, Finland

Ronald P. Karzel, M.D.
Attending Orthopedic Surgeon, Southern California Orthopedic Institute, Van Nuys, California

James S. Keene, M.D.
Professor of Orthopedic Surgery and Head Team Physician, University of Wisconsin Athletic Teams, University of Wisconsin, Madison, Wisconsin

Scott H. Kitchel, M.D.
Clinical Instructor, Division of Orthopedic Surgery, University of Oregon; Staff Surgeon, Sacred Heart Hospital, Eugene; Staff Surgeon, Shriners Hospitals for Crippled Children, Portland Unit, Portland, Oregon

Robert F. Lemanske, Jr., M.D.
Associate Professor of Medicine and Pediatrics and Head Director of Pediatric Allergy, University of Wisconsin School of Medicine, Madison, Wisconsin

Russell C. Linton, M.D.
Team Orthopaedic Surgeon, Mississippi State University; Staff Orthopaedic Surgeon, Golden Triangle Regional Medical Center, Columbus, Mississippi

Francis R. Lyons, M.D.
Associate, St. Vincent's Hospital Clinical School, University of Melbourne; Orthopaedic Surgeon, St. Vincent's Hospital, Melbourne, Australia

Roger A. Mann, M.D.
Director, Foot Fellowship, Oakland; Associate Clinical Professor of Orthopedic Surgery, University of California School of Medicine, San Francisco; Director, Foot Surgery, Summit Hospital, Oakland, California

Angus McBryde, Jr., M.D.
Professor and Chairman, Department of Orthopaedics, University of South Alabama, Mobile, Alabama

Frank C. McCue, III, M.D.
Alfred R. Shands Professor of Orthopaedic Surgery and Plastic Surgery of the Hand and Director, Division of Sports Medicine and Hand Surgery, University of Virginia School of Medicine; Professor of Education, University of Virginia, Curry School of Education, Charlottesville, Virginia

Thomas L. Mehlhoff, M.D.
Clinical Assistant Professor of Orthopedic Surgery, Baylor College of Medicine; Active Staff, The Methodist Hospital, Houston, Texas

Morris B. Mellion, M.D.
Clinical Associate Professor, Departments of Family Practice and Orthopaedic Surgery (Sports Medicine), University of Nebraska Medical Center; Associate Professor, School of Health, Physical Education, and Recreation, University of Nebraska at Omaha; Consulting, University of Nebraska Medical Center and Children's Memorial Hospital; Associate Staff, Immanuel Medical Center; Courtesy Staff, Methodist Hospital, Omaha, Nebraska

Lyle J. Micheli, M.D.
Associate Clinical Professor of Orthopaedic Surgery, Harvard Medical School; Director, Division of Sports Medicine, Children's Hospital, Boston, Massachusetts

Chad W. Millet, M.D.
Clinical Instructor, Louisiana State University School of Medicine, Department of Orthopaedics, New Orleans, Louisiana

Nicholas G.H. Mohtadi, M.D., M.Sc., FRCS(C)
Clinical Assistant Professor, University of Calgary Sport Medicine Centre; Attending, Foothills Hospital, Alberta Children's Hospital, Calgary, Alberta, Canada

Michael A. Monmouth, M.D.
Clinical Assistant Professor, University of Texas Medical School, Houston; Attending Orthopaedic Surgeon, St. John Hospital, Nassau Bay, Clear Lake Regional Medical Center, Webster, and Hermann Hospital, Houston, Texas

Bernard F. Morrey, M.D.
Professor of Orthopedics, Mayo Medical School; Chairman, Department of Orthopedics, Mayo Clinic, Mayo Foundation, Rochester, Minnesota

Van C. Mow, Ph.D.
Professor of Mechanical Engineering and Orthopaedic Bioengineering, College of Physicians and Surgeons, Columbia University, New York, New York

Sam Nassar, M.D.
Associate Professor, Department of Orthopaedic Surgery, Wayne State University School of Medicine, Detroit, Michigan

Frank R. Noyes, M.D.
Clinical Professor, Department of Orthopaedic Surgery, University of Cincinnati Medical Center; Director, Cincinnati Sportsmedicine Research and Education Foundation, Cincinnati, Ohio

Michael J. Pagnani, M.D.
Orthopaedic Surgeon, The Lipscomb Clinic, Nashville, Tennessee

Andrew W. Parker, M.D.
Fellow, Louisiana State University Knee and Sports Medicine Fellowship, Knee, Shoulder, and Sports Medicine Center, Lake Charles, Louisiana

Robert Patek, M.D.
Assistant Clinical Professor, Department of Orthopaedics, Northwestern University, Chicago; Attending Physician, Lutheran General Hospital, Park Ridge, Illinois

Lonnie E. Paulos, M.D.
The Orthopedic Specialty Hospital, Medical Director, Orthopedic Biomechanics Institute, Salt Lake City, Utah

Jacquelin Perry, M.D.
Professor of Orthopaedics and Professor of Biokinesiological and Physical Therapy, University of Southern California, Los Angeles; Chief of Pathokinesiology, Rancho Los Amigos Hospital, Downey; Consultant to Biomechanics Laboratory, Centinela Hospital Medical Center, Inglewood, California

Marilyn Pink, M.S., R.P.T.
Professor, Mount St. Mary's College; Assistant Administrator of Biomechanics Laboratory, Centinela Hospital Medical Center, Inglewood, California

John W. Powell, Ph.D., ATC
Research Associate, Department of Orthopaedic Surgery, University of Iowa Hospitals and Clinics, Iowa City, Iowa

John A. Racanelli, M.D.
Fellow, Louisiana State University Knee and Sports Medicine Fellowship, Knee, Shoulder, and Sports Medicine Center, Lake Charles, Louisiana

William D. Regan, M.D., FRCS(C)
Clinical Assistant Professor of Orthopaedic Surgery, University of British Columbia; Active Staff, University Hospital, University of British Columbia, and Shaughnessy Sites, Vancouver, British Columbia, Canada

Bruce Reider, M.D.
Associate Professor of Surgery, Section of Orthopaedic Surgery and Rehabilitation Medicine, University of Chicago; Director of Sports Medicine, University of Chicago Hospitals, Chicago, Illinois

Per A.F.H. Renstrom, M.D.
Professor, Department of Orthopaedics and Rehabilitation, University of Vermont; Attending, Medical Center Hospital of Vermont, Burlington, and Fanny Allen Hospital, Colchester, Vermont

Thomas S. Roberts, M.D.
Assistant Clinical Professor of Orthopaedic Surgery, University of Arkansas for Medical Sciences, Little Rock, Arkansas

James B. Robinson, M.D.
Assistant Clinical Professor in Family Practice and Sports Medicine, Tuscaloosa Family Practice Program; Medical Director, DCH Sports Medicine, DCH Medical Center, Tuscaloosa; Chairman of Medicine Department, Northport Hospital, Northport, Alabama

Charles A. Rockwood, Jr., M.D.
Professor and Chairman Emeritus, Department of Orthopaedics, University of Texas Health Science Center at San Antonio, San Antonio, Texas

Alberto G. Schneeberger, M.D.
Bern, Switzerland

Mario Schootman, M.S.
Injury Prevention Research Center, The University of Iowa, Department of Preventive Medicine and Environmental Health, Iowa City, Iowa

Wayne J. Sebastianelli, M.D.
Director of Athletic Medicine and Assistant Professor of Orthopaedic Surgery, The Pennsylvania State University, Milton S. Hershey Medical Center; Attending, Milton S. Hershey Medical Center and Centre Community Hospital, Hershey, Pennsylvania

Richard Simon, M.D.
Director of Sportsmedicine, Universal Medical Center, Plantation, Florida

Kenneth M. Singer, M.D.
Clinical Assistant Professor of Surgery, Division of Orthopedic Surgery and Rehabilitation, Oregon Health Sciences University; Chief-of-Staff, Sacred Heart General Hospital, Eugene, Oregon

T. David Sisk, M.D.
Professor and Chairman, Department of Orthopaedic Surgery, University of Tennessee-Campbell Clinic, Memphis, Tennessee

William D. Stanish, M.D.
Professor of Surgery, Dalhousie University, and Director of the Orthopaedic and Sports Medicine Clinic of Nova Scotia, Halifax, Nova Scotia, Canada

Carl L. Stanitski, M.D.
Professor of Orthopaedic Surgery, Wayne State University; Chief, Department of Orthopaedic Surgery, Children's Hospital of Michigan, Detroit, Michigan

James W. Strickland, M.D.
Clinical Professor, Department of Orthopaedic Surgery, Indiana University School of Medicine; Chairman, Department of Hand Surgery, St. Vincent Hospital and Health Care Center, Indianapolis, Indiana

Roy Terry, M.D.
Resident, University of Mississippi Medical Center, Jackson, Mississippi

James E. Tibone, M.D.
Associate Clinical Professor, Department of Orthopaedics, University of Southern California, Los Angeles; Associate, Kerlan-Jobe Orthopaedic Clinic, Inglewood, California

Joseph S. Torg, M.D.
Professor of Orthopaedic Surgery, University of Pennsylvania School of Medicine; Staff, Hospital of the University of Pennsylvania and Children's Hospital of Philadelphia, Philadelphia, Pennsylvania

Peter A. Torzilli, Ph.D.
Associate Professor, Cornell University Medical College; Director and Senior Scientist, Laboratory for Soft Tissue Research, Hospital for Special Surgery, New York, New York

W. Michael Walsh, M.D.
Adjunct Graduate Associate Professor, School of Health, Physical Education and Recreation, University of Nebraska at Omaha; Clinical Associate Professor of Orthopaedic Surgery, University of Nebraska Medical Center; Courtesy Staff, Children's Memorial Hospital, Methodist Hospital, Bishop Clarkson Memorial Hospital, Omaha Surgical Center, and Immanuel Medical Center; All-Team Orthopaedist, University of Nebraska at Omaha, Omaha, Nebraska

Jon J.P. Warner, M.D.
Assistant Professor of Orthopaedic Surgery and Assistant Director of the Sports Medicine Institute, University of Pittsburgh, Pittsburgh, Pennsylvania

Russell F. Warren, M.D.
Professor of Orthopaedic Surgery, Cornell University Medical Center; Chief, Sports Medicine and Shoulder Service, The Hospital for Special Surgery, New York, New York

Gerald R. Williams, Jr., M.D.
Assistant Professor, Department of Orthopaedic Surgery, University of Pennsylvania; Attending Surgeon, Shoulder Service, Hospital of the University of Pennsylvania and Philadelphia Veterans Administration Hospital, Philadelphia, Pennsylvania

Savio L.-Y. Woo, Ph.D.
Professor and Vice Chairman for Research, Musculoskeletal Research Center, Department of Orthopaedic Surgery, University of Pittsburgh Medical Center; Professor of Mechanical Engineering, University of Pittsburgh, Pittsburgh, Pennsylvania

Dale Christopher Young, M.D.
Assistant Instructor of Orthopaedics, Medical College of Virginia; Attending, Chippenham Medical Center and Johnston-Willis Hospital, Richmond, Virginia

Eleanor A. Young, Ph.D., R.D., L.D.
Professor, Department of Medicine, Division of Gastroenterology and Nutrition, University of Texas Health Science Center at San Antonio; Consultant, Medical Center Hospital and Audie Murphy Veterans Administration Hospital; Governing Board (Vice-Chair), Santa Rosa Northwest, Santa Rosa Rehabilitation Hospital, Villa Rosa Hospital, San Antonio, Texas

Mary L. Zupanc, M.D.
Associate Professor, Department of Pediatrics and Neurology, University of Wisconsin School of Medicine; Staff, University of Wisconsin Hospitals and Clinics, Madison, Wisconsin

PREFACE

One might ask, "Why publish a book on sports medicine when the subspecialty simply treats athletes with orthopaedic injuries?" There is no question the orthopaedic sports medicine specialist must be soundly schooled in orthopaedic knowledge; however, the athletes often present special considerations that are not part of the everyday practice of orthopaedics. The intense desire and determination of the athlete to return to sports are not commonly encountered in the day-to-day practice of orthopaedics. The various stress syndromes of the musculoskeletal system that occur secondary to training and athletic competition and the specialized rehabilitation techniques also present special problems not seen in other orthopaedic patients.

The contributors to this two-volume treatise on orthopaedic sports medicine are leaders in their subspecialty. They have shared with us their experience and perspectives by addressing specific problems of the athlete in their areas of expertise. Each contributor gives an excellent review of the particular topic and completes the contribution with his preferred treatment.

The early chapters address certain nonorthopaedic conditions that the orthopaedic sports medicine specialist must address when caring for the athlete. Nutrition, heat illness, biomechanics, and sports psychology are areas that must be part of the orthopaedic sports medicine physician's data base.

In the later chapters, contributors discuss the diagnosis and various methods of treatment, and present their chosen method of treatment of specific athletic injuries. When to return an athlete to activity is an important responsibility the orthopaedic sports medicine specialist must assume. Each contributor has addressed this issue in his particular anatomic area.

It is our hope that this work will prove useful to our fellow orthopaedic surgeons. If so, it will be a tribute to the contributors who have shared with us their expertise. We are indebted to them for their time and effort.

We also offer our sincere appreciation to Marti Daigle, and to the staff at the W.B. Saunders Company for their patience and untiring efforts in bringing these volumes to publication.

JESSE C. DeLEE, M.D. DAVID DREZ, JR., M.D.

CONTENTS

Volume 1

Volume 2

BASIC SCIENCE OF SOFT TISSUE

Muscle and Tendon

Thomas M. Best, M.D.
William E. Garrett, Jr., M.D., Ph.D.

SKELETAL MUSCLE

Structure

Skeletal muscle constitutes the single largest tissue mass in the body, comprising 40% to 45% of the total body weight in the average individual. Its structure is closely related to its function, which is to generate force resulting in joint and limb locomotion and movement. The system can be thought of as a series of mechanical levers that move the body and its limbs. Consequently, the moment arm is the most important parameter because it influences the magnitude of the muscle forces. To resist a given externally applied moment, the larger the moment arm, the less the muscle force that is required. The gross structure of skeletal muscle is highly variable, as is obvious from a study of surface anatomy and deep dissection. In general, muscle originates from bone or dense connective tissue, either directly or from a tendon of origin. The muscle fibers themselves pass distally, usually to a tendon of insertion, which connects with bone. This structural framework is necessary to support the unit against injury and to organize the individual units into tissues and organs (Fig. 1A–1).

The muscle-tendon unit can cross one, two, or more joints. Generally speaking, muscles that cross one joint are located close to bone and are frequently more involved in postural or tonic activity (e.g., soleus). Morphologically they are broad and flat, possessing a decreased speed of contraction but increased strength (force output) compared with the two-joint muscles. Two-joint or phasic muscles lie more superficially within a compartment. Examples include the gastrocnemius, which is more superficial than the soleus, and the rectus femoris, which is the most superficial of the quadriceps group. Compared with one-joint muscles, these muscles have a greater speed of shortening and capacity for length change; however, they are less effective in producing tension over the full range of motion.

Architecture

Architecture is a major consideration in an analysis of muscle function. The arrangement and organization of fibers within the muscles and of mus-

This work was supported in part by a grant from the Medical Research Council of Canada.

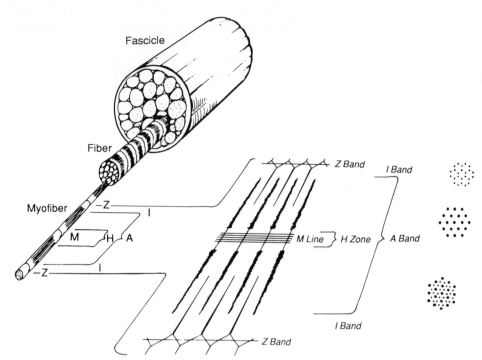

FIGURE 1A–1
Schematic drawing of the structure of striated muscle showing the organizational framework necessary for effective function (see text for further explanation of structures).

cles in the body play a substantial role in the way in which an organism functions. The muscle fiber is the basic structural unit of skeletal muscle, and the sarcomere is the smallest contractile unit of the fiber. These fibers are grouped into small bundles known as fascicles, which are usually oriented obliquely to the longitudinal axis. Consequently, fiber arrangement within the muscle is quite variable, and a large number of configurations can be seen (Fig. 1A–2), including fusiform, parallel, unipennate, bipennate, and multipennate. In general, fusiform muscles permit greater range of motion. Pennate muscles are generally more powerful (force of contraction) than parallel-fibered muscles of the same weight because their organization allows a larger number of fibers to work in parallel. Moreover, because pennate muscles contain short fibers, the maximal velocity of shortening is lower and the work performed can be considerably less [149].

Consequently, the pennate muscle resists elongation and has a steep passive length-tension diagram.

Due to their oblique arrangement, muscle fibers seldom extend the entire length of the muscle belly, and there is no location within the muscle belly that is crossed by all fibers. It is because of this arrangement that the cross-sectional area of muscle is difficult to measure. In most muscles the fiber length is less than half the muscle length. Fiber length is an important determinant of the amount of shortening possible in the muscle. Also, the force produced is proportional to the cross-sectional area of the muscle and the orientation of the muscle fibers. More important, it has been shown that force production is independent of fiber type when cross-sectional area differences are taken into consideration. The role of muscle architecture in force development and change in length has been frequently recognized recently (Fig. 1A–3).

In addition to the arrangement of muscle fibers, the fibrous connective tissue network within the muscle is important in a discussion of architecture. Tendons often are spread out on the surface or within the muscle substance, providing a wide area for attachment of muscle fibers. Connective tissue surrounds the whole muscle (epimysium), each bundle of fibers (perimysium), and the individual fibers themselves (endomysium). This connective tissue framework is continuous within the muscle and attaches to the tendon of insertion. It is this continuous network that is essential to the efficiency with which force is generated and movement results. The

FIGURE 1A–2
Muscle fiber architecture. *A,* Parallel; *B,* unipennate; *C,* bipennate; *D,* fusiform.

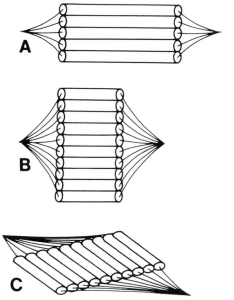

FIGURE 1A–3
The role of muscle architecture in force development and length change. Length of *A* is twice that of *B;* cross-sectional area of *A* equals that of *B*, whereas maximum force of *A* is one-half that of *B*, whereas maximum length change of *A* is twice that of *B*. In *C* the force is diminished by only a small factor when fibers are arranged in pennate fashion.

area of the muscle-tendon junction is a specialized region connecting the fibers with the tendon. Recently, interest has developed in examining this region in stretch-induced skeletal muscle injury.

The Myofibrillar Proteins

Muscle fibers are composed of four major proteins: myosin, actin, tropomyosin, and troponin. Together with the remaining structural proteins (Table 1A–1) they comprise about 12% of the total weight of vertebrate striated muscle.

Myosin is a hexameric molecule composed of two high-molecular-weight (200,000) heavy chains (HC) and four low-molecular-weight light chain (LC) subunits termed A-1, A-2, and DTNB (two units). Figure 1A–4 shows a typical myosin molecule. This molecule can be cleaved by trypsin to yield two fragments, heavy meromyosin (HMM) and light meromyosin (LMM). Papain cleaves the LMM fragment into a globular protein, S-1, and a helical protein, S-2. The ATPase activity and the actin-combining property of myosin are completely associated with the HMM component, whereas the solubility properties of the molecule are associated with the LMM fraction. Functionally, it is the HC component that possesses ATPase activity, whereas the LC component appears to regulate this action and therefore is not essential for ATPase activity.

The other three major structural proteins are incorporated into the thin filaments. Of these, actin is present in the largest amount. Actin molecules are small, roughly spherical structures arranged in the thin filaments as if to form a twisted strand of beads. If a muscle extracted for myosin is treated with acetone, dried, and resuspended in distilled water, the remaining molecule is known as G actin. This molecule polymerizes in the presence of $MgCl_2$ or NaCl to give F actin, which is a double-stranded helical polymer of the G actin subunits [75]. The process of polymerization requires energy, which is provided as the bound adenosine 5′-triphosphate (ATP) hydrolyzes to adenosine diphosphate (ADP) during the transformation from the G to the F form of actin. It is important to remember that both actin and myosin molecules have a recognizable polarity that is essential to muscular contraction.

The remaining two major structural proteins, tropomyosin and troponin, constitute a protein complex that enables calcium to regulate the contraction-relaxation cycle of actomyosin. Tropomyosin molecules (molecular weight, approximately 65,000 daltons) are long, thin proteins that attach end to end, forming a very thin filament on the surface of an actin strand. Each strand carries its own filament, which lies near the groove between the paired strands. Together with the troponin myofibrillar protein they collectively form "native tropomyosin." The two main subunits are termed the alpha and beta polypeptide chains with molecular weights of 34,000 and 35,000 daltons, respectively. These two chains differ mainly in their cysteine content and electrophoretic mobility, and it is the ratio of these two subunits that varies among fiber types [41, 174]. The primary effect of native tropomyosin on F actin is to make actomyosin highly sensitive to the calcium concentration of the system. In other words, the tropomyosin molecule helps to confer calcium sensitivity on the ATPase activity of skeletal muscle

TABLE 1A–1
Relative Proportions of Myofibrillar Proteins in Rabbit Skeletal Muscle

Protein	Percentage of Total Structural Protein
Myosin	55
Actin	20
Tropomyosin	7
Troponin	2
C protein	2
M proteins	<2
Alpha-actinin	10
Beta-actinin	2

Adapted from Carlson, F. D., and Wilkie, D. R. *Muscle Physiology.* Englewood Cliffs, NJ, Prentice-Hall, 1974.

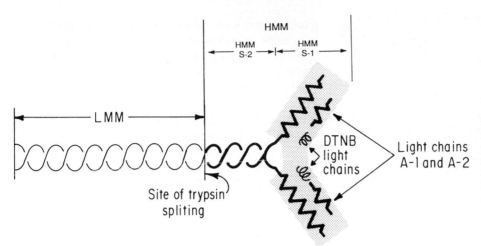

FIGURE 1A–4
Schematic drawing of myosin molecule showing subunit structure. The heavy chain component possesses the ATPase activity, and the light chain confers the solubility properties of the molecule. HMM, heavy meromyosin; LMM, light meromyosin. (Adapted from Carlson, F.D., and Wilkie, D.R.: *Muscle Physiology.* Englewood Cliffs, NJ: Prentice-Hall, 1974.)

contraction. At low calcium concentrations the tropomyosin threads move out of the groove along actin and cover the region on the actin surfaces where the myosin crossbridges attach. When calcium concentrations approach 10^{-5} M, tropomyosin binds to its target protein, troponin-C, and allosteric conformational changes with the entire troponin-tropomyosin complex result, with movement of the tropomyosin further back into the grooves [185]. This permits actin and myosin to bind, leading to subsequent ATP hydrolysis and initiation of skeletal muscle contraction. Troponin has a more or less globular shape and sits adjacent to the tropomyosin molecule. It is a noncovalent complex of three subunits, troponin T, C, and I, each of which has a distinct physiologic function in muscle. Thin filaments are commonly 1 micron (μ) in length, and each troponin-tropomyosin complex is associated with seven actin monomers.

The C protein has been shown to be located in the crossbridge-bearing region of the thick filament. The M protein is associated with the enzyme creatine phosphokinase. It is localized to the M line, which occurs in the middle of the thick filament.

Ultrastructure

Figure 1A–5 is a typical electron micrograph of skeletal muscle. The banded arrangement is due to the repetition of dark and light bands. On electron microscopy the dark bands are composed of thick filaments with small projections or crossbridges extending from the filament. The primary constituent protein of the thick filament is myosin. The light band is composed of thin filaments that are smaller than thick filaments. The chief protein of the thin filaments is actin. This banded appearance persists down to the level of individual fibrils. The A band gets its name from the fact that it is anisotropic to polarized light, whereas the I band is isotropic. The

A band may be made to appear either lighter or darker than the alternating I bands depending on the setting of the polarizing microscope used. The basic functional unit of skeletal muscle is the sarcomere, which extends from one Z band to the next and is further divided into I bands and A bands. The Z band is composed of at least four proteins: alpha actinin, desmin, filamin, and zeugmatin. The I band contains actin, tropomyosin, and troponin; the A band consists of myosin and the actin-tropomyosin-troponin complex. With contraction and muscle shortening the light bands get narrower and the dark bands do not change in length. Finally, there is a region near the center of the A band termed the H zone. The M line is in the center of the H zone.

The Sliding Filament Model

As with so many discoveries in science, it is the simultaneous work of several groups that leads to important findings. In 1954 Hugh E. Huxley and Jean Hanson of the University of Cambridge wrote a paper describing a model for the configuration of muscle proteins. At the same time, another group headed by Andrew F. Huxley at University College, London, proposed the same model, though in less detail. Consequently, the two papers were published consecutively in the same journal [112, 122]. The essential features of the model remain the same today. Due to the presence of the A band and its high index of refraction, it was suggested that an arrangement of rods or filaments is stacked parallel to the long axis of the muscle. These rods are composed of thick filaments of myosin. At the same time another set of filaments, the thin filaments, extends from the Z line through the I band, and part way into the A band, stopping short of the H zone. During contraction there is no change in the length of the A band or in the actin and myosin

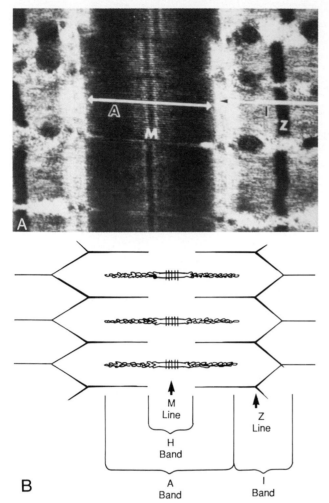

FIGURE 1A–5
Electron micrograph *(A)* and schematic drawing *(B)* of skeletal muscle.

molecules. In addition, the length of the I band decreases with sarcomere shortening. If resting muscle is lengthened or shortened there is somehow a change in the sarcomere pattern that conforms with the expected mechanical behavior of the two sets of filaments sliding past one another. During the next few years advances in electron microscopy greatly enhanced the ability of researchers to view skeletal muscle, and in 1971 the now famous "sliding filament" theory of skeletal muscle contraction was proposed by Huxley and Huxley [114, 120].

Activation of Contraction: The Sarcoplasmic Reticulum

Regulation of skeletal muscle contraction is accomplished primarily by calcium ions, which are stored in the membrane-bound areas of the cyto-plasm known as the sarcoplasmic reticulum. The exact role of calcium is not well defined because it is known that rigor complexes can be formed in the presence or absence of calcium [185]. The sarcoplasmic reticulum has two distinct units, the longitudinal tubules and the transverse tubular system. The latter bisects the longitudinal tubules, which expand outward to form two large lateral sacs in the region of the Z line. These two lateral sacs and their corresponding transverse tubule are collectively referred to as a triad. Mammalian skeletal muscle has two triads per sarcomere (Fig. 1A–6).

An action potential passes over the sarcolemma and into the transverse tubular system as well as longitudinally into the sacs, resulting in an increase in calcium permeability and a release of calcium ions into the sarcoplasm. This free calcium binds to the regulatory proteins, allowing for interaction of the thick and thin filaments. Crossbridge cycling continues as long as the free calcium concentration is maintained. Following completion of the electrical event, relaxation of the muscle occurs by active transport of the calcium into the longitudinal tubules of the sarcoplasmic reticulum forcing the calcium to dissociate from troponin and allowing the tropomyosin molecule to snap back into the groove, where it prevents any further crossbridge attachment.

Fiber-Type Differences

Different striated muscles exhibit significant variations in structure at both the histologic and ultrastructural levels. There are also differences in innervation, physiology, biochemistry, and circulation (Table 1A–2). Currently, two major classes of fiber types are recognized as being structurally, physiologically, and metabolically distinct [75]. The type I, or slow-twitch oxidative fibers (SO), have the slowest contraction time and the lowest content of glycogen and glycolytic enzymes. They are rich in mitochondria and myoglobin and are quite fatigue resistant. Morphologically, their sarcomeres contain a wide Z band. The type IIa, or fast-twitch oxidative fibers (FOG), have a faster contraction time than type I fibers. In addition, they have a higher content of mitochondria and myoglobin than type IIb fibers. Physiologically, they are more fatigue resistant than type IIb fibers. As well, they have high levels of myosin ATPase, oxidative enzymes, and glycogen. Morphologically, their Z bands are slightly narrower than those in type I fibers. The type IIa fiber is termed an intermediate fiber because it possesses both forms of myosin that are present in the type I

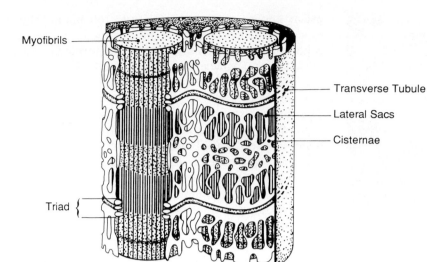

FIGURE 1A–6
Details of the sarcoplasmic reticulum, the system of membranes responsible for transmission of the electrical signal from one muscle cell to the next. Each electrical signal passes inward along the transverse tubule, causing release of calcium from the lateral sacs.

and IIb fibers. Most mammalian carnivores, including humans, also possess a type IIc or superfast fiber. This fiber is most prominent in the jaw muscles. It possesses a unique form of myosin that can be distinguished from both the slow and fast types characteristic of type I and type II fibers. At birth up to 10% of muscle fibers may be classified as type IIc; this declines to approximately 2% after the first year of life. Therefore, the type IIc fiber is often thought of as being undifferentiated [46]. Interestingly enough, during physical training as many as 10% of these fibers may be present in some muscles of endurance athletes [183]. Their presence has yet to be explained, although it is felt by some that this fiber type is a transitional form between type I and type IIa fibers.

Biochemically, the structural proteins in the sarcomere are also distinct. Myosin, tropomyosin, and troponin have distinct structural isomers in the different fiber types [41, 143]. Recent investigations have shown more heterogeneity in the structural proteins than had been previously expected. Rather than three distinct fiber types there appear to be a discrete number of basic sets of structural protein isomers for the fast and slow muscles [80].

Histochemical analysis of human muscle shows

TABLE 1A–2
Histochemical Reactions to Human Skeletal Muscle

	Muscle Fiber Type			
	I	**IIa**	**IIb**	**IIc**
		Fast Twitch		
Example	**Slow Twitch**	**Fatigue Resistant**	**Fatiguing**	**Superfast Twitch**
Routine ATP-ase	○	●	●	●
ATP-ase preincubated pH 4.6	●	○	●	●
ATP-ase preincubated pH 4.3	●	○	○	○
NADH-TR	●	○	○	○
SDH	●	○	○	○
Glycerophosphate-menadione-linked	○	○	○	○
PAS	○ + ○	●	○	○
Phosphorylase	○ + ○	●	●	●

○ = 0   ○ = 1+   ○ = 2+   ● = 3+

Abbreviations: NADH-TR = nicotinamide adenine dinucleotide trireductase; SDH = succinate dehydrogenase; PAS = periodic acid-Schiff stain.
From Dubowitz, V., and Brooke, M. H. *Muscle Biopsy: A Modern Approach.* London, W. B. Saunders, 1973.

that there are discrete differences between muscles in regard to the composition of their fiber type. The difference in the sensitivity of myosin to retaining or losing ATPase after exposure to either high or low pH is a reliable method for classifying muscle fibers histochemically [55, 163]. This work has been done by several groups [25, 26, 49, 51, 127, 183]. With the exception of Saltin and associates [183], who obtained their results from needle biopsies, these data were derived primarily from specimens obtained at autopsy or during surgery. In summary, these studies showed that muscles are composed of mixtures of fiber types and that those with similar function have similar fiber populations. Although most muscles in humans have a mean fiber composition of 50% slow-twitch (ST) and 50% fast-twitch (FT) fibers, some muscles, such as the soleus and triceps brachii, have a predominance of either ST or FT fibers. The tonic or postural muscles (e.g., soleus) are usually situated closer to the bony skeleton and have a greater proportion of type I fibers. In contrast, the phasic or faster contracting muscles lie closer to the surface and have a higher proportion of type II fibers [127].

In 1972 the first study examining the relationship between the fiber composition of humans and their athletic abilities was published [78]. Muscle biopsies were obtained from the vastus lateralis and deltoid muscles of 74 male subjects of various athletic abilities. Findings from this study suggested that high-performance athletes tended to have fiber compositions that would be advantageous for their particular event. That is, endurance athletes had high percentages of type I fibers, whereas those in non-endurance events (e.g., sprinters) tended to possess higher percentages of FT fibers in their muscles. In 1976 Thorstensson and colleagues found that people with higher percentages of type II muscle fibers in the quadriceps were able to generate more knee extensor force and torque at higher velocities than their counterparts with fewer type II fibers [199]. Similarly, athletes who demonstrated excellence in sprint events had a relatively higher percentage of type II fibers, whereas elite distance runners had predominantly type I fibers [132, 183].

It should be stressed here that considerable variation exists in the fiber composition of skeletal muscle within each of the subgroups of athletic ability. What this fact probably indicates is that factors other than fiber type composition also contribute to performance. Additionally, it points out that a single biopsy, in many cases, probably does not allow a good estimate of the entire muscle.

Physiology

Central Nervous System Control of Force Production

It was first demonstrated in the seventeenth century that muscle contracts without changing its volume. Experiments by Jan Swammerdam in the mid-1600s used an isolated preparation of frog muscle, and the idea was extended to human muscle by Francis Glisson in 1677 [157]. Since that time, much of what we know today about skeletal muscle has had its foundations in the work of A. V. Hill and his colleagues at University College, London.

There are two basic strategies that allow the central nervous system (CNS) to control force production. Either the rate of discharge of firing neurons can be increased or the number of active motor units can be increased, which is known as recruitment.

Mechanical Events: Twitch and Tetanus

Following electrical excitation of the muscle, the first mechanical event recorded is not the development of force but rather a latency period lasting about 15 msec during which the muscle held isometrically produces no force. Abbot and Ritchie showed that in an isometrically contracting muscle there can be a short-lived fall in tension before positive tension is developed [4]. In response to a single stimulus, a single transient rise in tension, known as a twitch, is produced. Similarly, two twitches separated by an appropriate time interval will result in two identical force recordings. If the frequency of stimulation is increased, a summation of the force recordings begins known as unfused tetanus (Fig. 1A–7). Eventually, a level is reached at which an increase in frequency results in no further increase in force. This is usually referred to as tetanic fusion and commonly occurs at a frequency of 50 to 60 Hz in mammalian skeletal muscle at body temperature.

Classic experiments in muscle physiology have looked at the relationship between the length of a muscle and the tension produced at that length. In 1951, Aubert and co-workers developed the tension-length curves for frog sartorius muscle [2]. The passive curve was measured on the resting muscle at a series of different lengths, whereas tetanized curve was measured at a series of constant lengths as the muscle was maintained in isometric contraction. From Figure 1A–8, it can be seen that the tension is greater when the muscle is tetanized compared with that when the muscle is passively stretched to the same length.

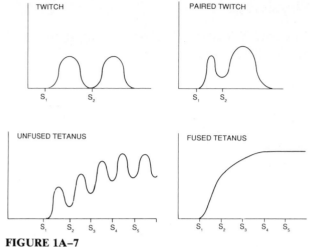

FIGURE 1A–7
Twitch and tetanus. As the frequency of stimulation is increased, muscle force rises to an eventual plateau level known as fused tetanus.

The Motor Unit

The second mechanism for controlling force production involves the concepts of the motor unit and recruitment. Much of our understanding of skeletal muscle physiology is based on an understanding of the properties and organization of the motor unit. The motor unit is classically defined as the alpha-motoneuron and the muscle fibers it innervates [32]. The topic of motor unit properties has been reviewed in depth by Close [40] and Buchthal and Schmalbruch [30].

A single alpha-motoneuron can innervate from 10 to 2000 muscle fibers [57]. General characteristics of the motor unit include the following. First, within a motor unit all fibers are homogeneous with regard to histochemically identifiable contractile and met-

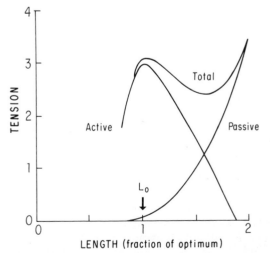

FIGURE 1A–8
Tension-length curve of skeletal muscle. L_0 = rest length.

abolic properties [65, 138, 140]. Second, the fibers in a motor unit are generally distributed throughout the muscle [48]. Third, fibers within a given motor unit are rarely located adjacent to each other. Finally, the motor end plate has a characteristic location on the muscle fiber at approximately the midpoint of the length of the fiber, and the density of the fibers decreases as a function of the distance from the motor end plate [50, 139].

One of the important areas of research in skeletal muscle physiology involves the systematic response of the different motor units to distinct physiologic tasks. The existence of an orderly procedure for motor unit recruitment by the central nervous system is based on the work of Henneman and coworkers [89–93] and their discovery of the "size principle." Basically, there appears to be a continuum of thresholds for motor unit activation based on the size of the motor unit. Slow-twitch motor units are innervated by small, low-threshold alpha-motoneurons, whereas fast-twitch motor units are innervated by larger, higher threshold alpha-motoneurons. It is this size principle that helps to account for the smoothness and uniqueness of skeletal muscle movement and locomotion.

The relative importance of increased firing rate and recruitment as mechanisms for increasing the force of voluntary contraction has been investigated. In a study using human subjects [152] it was shown that only at low levels of force output is recruitment the primary mechanism for increased force production. Increased firing rate becomes the more important mechanism at intermediate (greater than 50% maximum force) force levels and thereafter accounts for the large majority of increased force production (Fig. 1A–9). By analyzing electromyographic (EMG) signals of contracting motor units, DeLuca and co-workers [43] identified the common drive principle that accounts for the behavior of the firing rates of motor units and provides a rather simple explanation for the control of motor unit activation. This property of common drive explains the fact that the nervous system does not control the firing rates of motor units individually; rather, it acts on the pool of motoneurons in a uniform fashion.

Contractile Properties

The basic feature that differentiates motor units is their contractile properties, which are often expressed as the time to peak tension in a twitch and the closely associated one-half relaxation time. A slow-twitch motor unit possesses a relatively long time to peak tension, whereas a short time to peak tension is characteristic of the fast-twitch motor unit. A prime determinant of the twitch property of muscle is the rate at which myosin splits ATP into

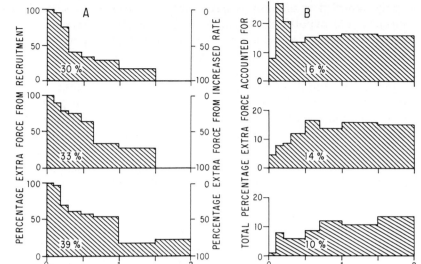

FIGURE 1A–9
A, Calculated percentage of increased force due to recruitment and increased firing rate for three subjects. *B,* Calculated total percentages of force accounted for by the units studied at various force levels. (From Milner-Brown, H.S., Stein, R.B., and Yemmon, R. Changes in firing rate of human motor units during linearly changing voluntary contractions. *J Physiol (Lond)* 230:371–390, 1973.)

ADP + Pi. This enzyme is referred to as the actin-activated myosin ATPase or, more commonly, the ATPase [16, 17, 184, 189]. It appears that under normal physiologic conditions of high concentrations of magnesium, the enzymatic activity of myosin is greatly reduced and that the high rates of ATP hydrolysis that occur with muscle contraction are due to interactions that remove the inhibitory effects of the divalent cations, magnesium and calcium.

Differences in the specific ATPase activities of myosin are due to the existence of several forms of the protein [27, 28, 163]. Traditionally, myosin isoenzymes have been identified based on their susceptibility to loss of ATPase activity in response to an alteration in pH [27, 28]. Myosin from slow-twitch muscles is acid stable but alkaline labile, whereas the opposite is true for fast-twitch muscles. The sliding filament hypothesis of skeletal muscle contraction argues that myosin is the most important contractile protein because it hydrolyzes ATP to yield energy for formation of the actomyosin complex. Barany and others were the first to demonstrate a relationship between the contractile and biochemical properties of skeletal muscle when they found that the activity of the ATPase of myosin correlated closely with the intrinsic speed of muscle shortening (Fig. 1A–10). It has been argued that since the rate-limiting step of muscle contraction may be the rate of energy delivery from ATP hydrolysis, one can examine changes in the ATPase of myosin in an effort to look at alterations in contractile characteristics. It is now known that much of the regulation of skeletal muscle involves myosin, and it therefore comes as no surprise that physiologic adaptation to a given stimulus often involves the myosin molecule [167].

The Adaptability of Mammalian Skeletal Muscle

It has been known for some time now that physiologic overload of skeletal muscle can result in adaptation of all components of the motor unit including the muscle itself, the neuromuscular junction, and the corresponding alpha-motoneuron [32]. This "plasticity" of skeletal muscle has been studied by numerous groups under different conditions to demonstrate that the different fiber types within a muscle adapt to various forms of overload in several ways [11, 31, 58, 123, 124, 142, 146, 160, 172, 180, 205]. Interestingly enough, each of these adaptations results in a logical alteration of both the morphology

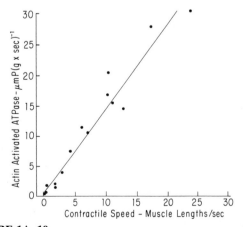

FIGURE 1A–10
Relationship between maximal speed of shortening and actin-activated myosin ATPase from a variety of animal species. (Reproduced from *The Journal of General Physiology*, 1967, 50:197–218, by permission of the Rockefeller University Press.)

and the function of the muscle fiber. Central to this concept is the hypothesis that muscle adapts to the function it performs, which in turn implies that specificity becomes an important factor in any type of functional overload, including exercise training. A variety of stimuli including cross-reinnervation, electrical stimulation, hypergravity stress, thyrotoxicosis, compensatory hypertrophy, and exercise have been used to study and elucidate mechanisms for this adaptive response.

Cross-Reinnervation and Electrical Stimulation. The cross-reinnervation experiments of Buller and colleagues [31] demonstrated that the isometric twitch speed of cat skeletal muscle is largely determined by the motor innervation it receives. Salmons and Vrbova [181] showed that transformation of a fast-twitch into a slow-twitch muscle can be brought about with unaltered innervation by transmitting to the nerve (via implanted electrodes) a frequency pattern that is normally delivered to a slow-twitch muscle. It has also been shown that neural influences are largely responsible for the reciprocal changes that occur in both functional [205] and biochemical [16] properties. However, neither the influence of a chemical trophic substance released from a nerve ending nor the pattern of electrical stimulation could be ruled out as the mechanism affecting the expression of various fiber types.

It is well known that in adult mammals the motoneurons innervating slow-twitch skeletal muscle generate a sustained, low-frequency pattern of activity, whereas motoneurons supplying fast-twitch muscles induce intermittent bursts of more intense activity [32]. Attempts have been made to rule out the existence of a chemotrophic substance affecting the expression of different fiber types by several workers including Salmons and Sreter [180], who were able to show that long-term electrical stimulation can be used both to reproduce and to oppose the effects of cross-reinnervation. In the absence of an actual change in the pattern of impulse activity reaching a muscle, cross-reinnervation has no significant effect on force-velocity properties of skeletal muscle. As a result, these observations discounted the theory that a chemotrophic substance affects the fundamental physiologic and biochemical differences between fast and slow muscle. From this, it was therefore concluded that the pattern of muscle stimulation plays an important role in determining the functional properties of skeletal muscle [142]. Specific trophic factors, yet to be determined, may still exist.

Compensatory Hypertrophy. Several researchers have induced functional overload in skeletal muscle by employing a technique of ablating synergistic muscles to study the remaining intact system. This type of overload, referred to as compensatory hypertrophy, has been used to induce both histochemical and biochemical alterations in skeletal muscle fibers. Typical experiments [11, 123, 160, 172] involve compensatory hypertrophy of the fast-twitch rat plantaris muscle to show a decrease in calcium-activated ATPase activity, an increase in the number of histochemically determined alkaline labile fibers, and an altered myosin light chain pattern. These changes are not as complete as those observed with cross-reinnervation and electrical stimulation. This method of inducing functional overload is unique in that there is no direct manipulation of the innervation of the intact muscle. This method represents a major departure from the cross-reinnervation and electrical stimulation models and offers new insight into skeletal muscle adaptation to functional overload. Because compensatory hypertrophy does not directly interfere with the intact nerve, it more closely represents an ideal physiologic situation. There is no reason to suspect that the factors operating here to affect fiber type transformations are different from those associated with hypergravity stress because neural input is not manipulated in either model. Changes in the soleus and plantaris muscle of rats maintained for 6 months under hypergravity stress reinforce Lomo's conjecture that the pattern of electrical stimulation determines the contractile properties of skeletal muscle [142]. Other models have shown that muscle is capable of responding to a hormonal stimulus [58, 124].

Exercise. Training programs are based on the principles of overload, specificity, and reversibility [56]. Overload simply means that a certain level of stimulus is necessary for adaptation to occur. Specificity of training implies that a specific stimulus for adaptation results in structural and functional changes in specific elements of skeletal muscle. Finally, reversibility simply implies that the effects of training can be reversed. Discontinuation of a training stimulus can result in detraining and a concomitant decrease in the adaptive changes that have previously occurred.

The adaptive changes of skeletal muscle to endurance exercise are well documented [77, 182]. Typically, endurance exercise is characterized by activation of large muscle groups that generate high metabolic loads resulting in adaptation of the respiratory and circulatory transport systems as well as the enzymatic capacity of the muscle. In certain circumstances this type of training can double the oxidative capacity of skeletal muscle. Saltin and Gollnick [182] were the first to quantify the effects of an endurance training program on the oxidative enzymes of muscle in man. In response to a 5-month bicycle ergometer program of 1 hour/day 4 days a week at a load requiring approximately 75% of maximal oxygen power, the succinyl dehydrogenase (SDH) and phosphofructokinase (PFK) enzymatic activities were doubled. The magnitude of the increased SDH activity is similar to the results shown by Holloszy [100] for endurance-trained rats. In contrast, anaerobic capacity, as measured histochemically by alpha-glycerophosphate dehydrogenase activity, was increased in the fast-twitch fibers only [182]. Mean oxygen uptake increased 13%, and muscle glycogen was 2.5 times higher than the level prior to training. In neither of these studies did a change occur in the percentages of slow- or fast-twitch fibers as identified from myosin ATPase activity. These studies and others demonstrate the tremendous capacity for adaptation in the oxidative potential of skeletal muscle in animals and humans.

High-force, low-repetition training results in an increase in muscle strength that is proportional to the cross-sectional area of the muscle. For years now it has been debated whether this is due to muscle hypertrophy (i.e., an increase in the size of the muscle fibers themselves) or to muscle hyperplasia (an increase in the number of muscle fibers). Both arguments have been presented using a number of models [81, 82, 101]. Although these studies have been limited by lack of an animal model to simulate the human condition, it appears likely that the majority of the change is due to muscle hypertrophy. Along with an increase in the size of the fibers an increase in the amount of contractile proteins, particularly myosin, also occurs. The contributions of hyperplasia, fiber splitting, fiber branching, and fiber fusion to the increase in muscle mass are yet to be determined. It is also important to remember that there is a strong neurologic component in strength training. In an untrained person it has been estimated that approximately 80% of the motor units can be stimulated voluntarily. Training results in an alteration in CNS firing of motor neurons to achieve better synchronization of muscle activation.

In response to prolonged endurance exercise, athletes have been found to possess increased skeletal muscle capillary density [8]. The predominant effect of endurance training on skeletal muscle structure is a marked increase in volume density of the mitochondria. Tesch and co-workers [197] investigated a similar response in weight lifters and found that the adaptive response to long-term weight and power lifting includes fast-twitch fiber hypertrophy, which results in reduced capillary density.

There is tremendous variability in the fiber types composing human skeletal muscle. The percentages of this composition vary considerably among muscles and among individuals. It is logical to think that certain compositions of fiber types would be quite advantageous for particular athletic events. Saltin and others [183] were the first to show the conversion of type IIb to type IIa fibers; however, the ratio of type I to type II fibers remained constant. Previous reports in the literature of increases in the percentage of "red" compared with "white" fibers employed oxidative capacity as measured by SDH or DPNH-diaphorase activity to classify fibers. In no instance had a change been demonstrated in fiber characteristics as determined histochemically by myosin ATPase.

Green and co-workers [84, 85] were the first to demonstrate in animals that a transition from fast-twitch to slow-twitch fibers can be brought about by increased contractile activity. Similar to chronic nerve stimulation, high-intensity endurance running led to transitions from fast- to slow-twitch fibers in the sarcoplasmic reticulum with a reduction in parvalbumin content. In addition, there was a decrease in fast-type and in increase in slow-type myosin light chains. These studies demonstrated that increased contractile activity may induce changes that are qualitatively similar to those seen in chronic nerve stimulation and that endurance training affects not only the metabolic properties of the muscle fiber but also causes fast to slow transitions in the Ca^{+2}-handling system. As judged by conventional histochemical techniques, a decrease in type IIb fibers with concomitant increases in type IIa and I fibers in the plantaris, extensor digitorum longus (EDL), and vastus lateralis muscles occurred. It was concluded that increased contractile activity, as brought about in a physiologic manner, is capable of inducing fiber type transitions [84].

A similar change was identified in humans by Howald and others [104], who showed, in a group of sedentary subjects, transformations of type IIb into type IIa fibers, and of type IIa into type I fibers. The fiber type changes were depicted by histochemical staining of myofibrillar ATPase and were accompanied by an enhancement of the oxidative capacity in all fiber types. A similar alteration in fiber types with exercise has yet to be shown in highly trained athletes.

Under a variety of conditions skeletal muscle fibers possess the potential for synthesis of all types of myofibrillar proteins. Based on the results of exercise studies and the various types of overload presently discussed, it appears logical to suggest that skeletal muscle fiber transformations occur only when a departure from normal function is both severe and sustained. The time course and amount of stimulus required remain uncertain at present. This hypothesis is supported by the intrinsic properties of motor units, including the recruitment principle, which states that in order to recruit large motor units the stimulus must be very great. One must keep in mind that the other forms of overload discussed here are rather nonphysiologic conditions and are not characteristic of physical training. It may be that in well-trained athletes exercise does not meet the necessary demands of this principle, and therefore the fiber type transformations depicted histochemically with other forms of overload are not observed in athletes. Such a question presents a considerable challenge to all who are interested in the field of sports medicine.

Muscle Injury and Repair

Muscle injury results from several mechanisms governed by separate pathologic processes. Not enough basic research regarding muscle injury has been done to elucidate the precise changes that occur with injury, and experimental and scientific data regarding prevention and rehabilitation are often lacking.

Muscle Laceration

Laceration of muscle is much more common in trauma than in athletics, where most injuries are the result of nonpenetrating direct trauma or indirect trauma such as muscle strains. The pathologic changes and recovery of muscle following laceration have been investigated recently [72]. Following laceration and repair, the two segments of the muscle heal by dense scar formation. More importantly, muscle does not regenerate across the scar, and functional continuity is not restored. The muscle segment isolated from the motor point loses its innervation. Following healing, the segment isolated from the motor point develops the histologic picture of denervated muscle. Electrical mapping studies show that muscle activation does not cross the scar created within the muscle [72]. Consequently, the muscle loses a significant proportion of its ability to produce tension. Partial lacerations also decrease the ability of the muscle to generate tension, but the decrease is less than that in muscles that are completely transected. The isolated segment may be able to transmit force and to shorten, but the active contractile function of the muscle remains in the portion with an intact nerve supply (Fig. 1A–11).

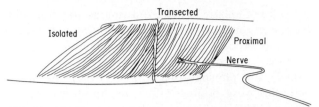

FIGURE 1A–11
Schematic drawing of lacerated muscle. The laceration leaves fibers intact proximally and distally while dividing the central fibers. Scar tissue isolates the distal segment from its nerve supply. (From Garrett, W.E., Jr., Seaber, A.V., Bokswich, J., et al. Recovery of skeletal muscle following laceration and repair. *J Hand Surg* 9a(5):683–692, 1984.)

Treatment should, therefore, stress the repair or reconstruction of the muscle using the long tendons of origin and insertion to anchor the repair; muscle tissue alone is inadequate for suture repair. Reattachment of the functioning muscle segment to the tendon of the isolated muscle by connective tissue could, therefore, preserve some function in the lacerated muscle.

Muscle Cramps

Ordinary muscle cramps are common during and following athletic exercise and are frequent even in young healthy people not involved in athletics. They occur most frequently in the gastrocnemius complex and can arise during exercise, at rest, or while asleep. There is still considerable uncertainty about the etiology of muscle cramps. It is well known that their onset frequently follows contraction of shortened muscles. The cramp often originates as fasciculations from a single focus or several distinct foci within the muscle and then spreads throughout the muscle in an irregular pattern. Electromyographic studies reveal fascicular twitching in a single focus, followed by very high frequency discharges within the muscle fibers [44]. The electrical activity reveals that the entire motor unit is involved, and therefore the initiating source is within the motor nerve fiber rather than within the individual muscle fibers themselves. Specifically, the focus is thought to be located in the terminal arborizations of the motor nerve fibers. Layzer [141] supported these findings on peripheral motor nerve involvement and suggested that the disturbance could arise from hyperexcitable motoneurons in the spinal cord.

Ordinary muscle cramps are associated with a variety of conditions unrelated to exercise. Excessive sweating or diuresis can cause saline loss and may produce cramps. Renal failure patients on chronic hemodialysis often have muscle cramps.

These conditions may be related to an alteration in sodium concentration, and administration of a saline solution is sometimes helpful [191]. Lowered levels of serum calcium or magnesium have also been implicated [44]. However, neither of these ionic disturbances is necessarily present in muscle cramps following exercise, the precise mechanism of which remains uncertain.

The cramp can sometimes be interrupted by forceful stretching of the involved muscle or activation of the antagonistic muscle. Following resolution of the knotted and painful contraction, the muscle shows evidence of altered excitability and fasciculations for many minutes following the cramp. The muscle may also be painful for several days after the event. Correction of electrolyte and water disturbances is thought to be helpful in preventing cramps, and adequate water and supplemental sodium may be given empirically, although the value of this treatment has not been proved. Drugs have been more helpful in treating cramps occurring in nonathletic individuals. Quinine sulfate and chloroquine phosphate have been beneficial, particularly for night cramps [153, 164]. The use of these medications to prevent or control exercise-induced muscle cramps has not been studied extensively, and their efficacy is therefore questionable at this time.

Despite their common occurrence, the etiology of muscle cramps during exercise remains poorly understood. Many of the studies to date have been conducted in ultraendurance athletes to investigate the proposed mechanisms, which include dehydration, electrolyte disturbances, and muscle fatigue. Maughan [147] followed 90 competitors at the 1982 Aberdeen marathon and found no correlation between hydration status and electrolyte balance and the incidence of muscle cramps. In an unpublished report, Kantorowsk and others [130] studied athletes participating in the 1989 Ironman Championship in Hawaii. Despite more extreme environmental conditions than in Maughan's study, no correlation between dehydration and the incidence of muscle cramping was found. Both studies used the percentage of body weight loss as a means of assessing hydration status.

Delayed-Onset Muscle Soreness

Muscle pain following unaccustomed vigorous exercise is a common phenomenon in athletes. It is especially marked following the initiation or resumption of training after a period of time without training. This pain should be distinguished from discomfort occurring during exercise, which is often associated with muscle fatigue and is metabolic in origin. Typically, delayed-onset muscle soreness begins a number of hours after exercise and is quite prominent on the first and second days after activity. The painful areas are noted particularly along the tendon or fascial connections within the muscle.

Several different pathologic mechanisms have been proposed to explain delayed-onset muscle soreness. Many of the concepts relating to delayed muscle soreness were introduced by Hough at the beginning of this century [103]. He distinguished the pain associated with the immediate fatigue of exercise from the pain noted in the muscle 1 to 2 days following exercise. He documented the fact that delayed pain did not necessarily follow more fatiguing work. On the contrary, he found that exercise routines that produced considerable fatigue did not produce as much delayed-onset muscle pain. However, rhythmic contractions marked by high intensity and relatively little fatigue were much more likely to be associated with delayed soreness. He concluded that delayed soreness was associated more with the amount of tension developed in the muscle than with its fatigue. Exercises associated with a sudden contraction or jerk were most likely to cause some pain. He proposed that the diminution in the ability of the muscle to produce tension as well as the soreness could be explained by small ruptures within the muscle. The exact site of the rupture either within the muscle fibers or within the connective tissue was uncertain.

Asmussen gave some support to the studies of Hough [13]. Negative or eccentric work was shown to produce more delayed-onset muscle soreness than positive work, in spite of the much greater fatigue induced by positive work. Asmussen concluded that the pain was due primarily to mechanical stress rather than fatigue and metabolic waste products. He also felt that the connective tissue within the muscle rather than the muscle fibers themselves might be the location of the injury. Abraham investigated the hypothesis that connective tissue breakdown might be associated with delayed-onset muscle soreness by monitoring hydroxyproline levels [5]. It has been shown that excretion of hydroxyproline in the urine is an index of the rate of collagen degradation [168]. Hydroxyproline is a modified amino acid found almost exclusively in collagen. Following a weight-lifting program a significant increase in urinary hydroxyproline occurred in subjects experiencing delayed muscle soreness. Elevated levels of myoglobin excretion were also noted, but the elevated levels occurred both in subjects who developed pain and in those subjects who did not have pain. Thus, there was a correlation between muscle soreness and collagen breakdown. In contrast, it has been shown that blood lactic acid concentration measured from serum was not related to exercise-induced delayed muscle soreness [187]. Exercise does create a significant increase in serum muscular enzymes and myoglobin, but these values usually remain within normal limits and are no different in subjects with soreness and subjects without soreness [19].

An alternative theory of muscle soreness implicates muscle spasm and electrical activity as the cause of pain rather than connective tissue breakdown [45, 135]. DeVries proposed that exercise produces ischemia that subsequently causes pain [45]. Pain initiates reflex tonic muscle contraction, prolonging the ischemia and promoting a vicious cycle. Using quantitative electromyography, he demonstrated that muscular activity was present when pain was present. Stretching of the muscle diminished the pain and likewise the electromyographic activity. Kraus [136] advocated the use of surface anesthesia to interrupt the pain spasm. Abraham [5] reinvestigated the electromyographic data and was unable to show significant changes in subjects with and without muscle soreness. The weight of the evidence, therefore, seems to be with the advocates of the "torn tissue" theory of muscle damage as a cause of delayed muscle soreness. However, it is still possible that electromyographic changes may accompany the tears in the tissue, and treatment that alters the muscle spasm or electromyographic manifestations may be of benefit in treating delayed-onset muscle soreness.

Recent studies of delayed muscle soreness have investigated the changes that occur on an ultrastructural level (Fig. 1A–12). Electron microscopy of muscle in subjects with pain in the vastus lateralis following cycling showed that significant changes had occurred in the sarcomere and the cross-striated pattern [61]. Three days following heavy exercise 50% of the muscle fibers displayed disorganization of the myofibrillar material. Armstrong [10] disputes these findings, demonstrating histologic evidence of injury in less than 5% of muscle fibers active during exercise. Change occurred predominantly in the type II fibers, and pathologic changes were noted particularly at the Z-band level as mild broadening to complete disruption [61]. At the same time, muscle strength (force output) was diminished during maximal knee extension exercises.

Delayed myofibrillar damage is probably the best established injury resulting from eccentric action. Muscle-specific enzymes, such as creatine kinase and lactate dehydrogenase, are released following completion of exercise, and this release often continues for several days [128].

In summary, much has been learned about the entity of delayed-onset muscle soreness. Several

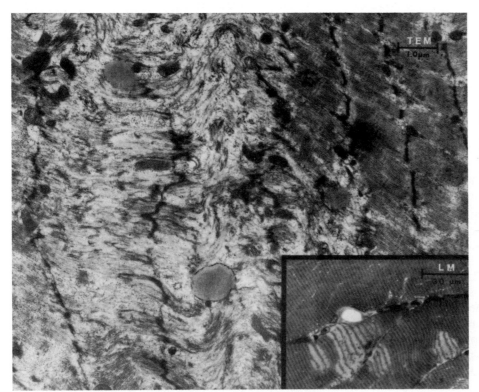

FIGURE 1A–12
Transmission electron micrograph (TEM) and light microscope (LM) views of injured rat soleus muscle fibers immediately after downhill walking. Note the A-band disruption and Z-band damage. (TEM micrograph reproduced by permission from Dr. R.W. Ogilvie. LM micrograph reproduced with permission from Armstrong, R.B., Marum, P., Tullson, P., et al. Acute hypertrophic response of skeletal muscle to the removal of synergists. *J Appl Physiol* 46:835–842, 1975.)

questions remain including questions about the mechanism of injury from repeated eccentric actions. The amount and type of stimulus needed to produce muscle damage is one important area that needs to be explored. Following muscle damage the regions that swell include the endomysium and perimysium where the pain receptors are located. What is the role of connective tissue damage in eccentric exercise? In chronically trained animals and humans, there is evidence to support the idea that a modification of connective tissue occurs around muscles subjected to eccentric loads [134, 145].

Muscle Contusions

Direct trauma to muscle is a common athletic injury, particularly in contact sports. Damage and partial disruption of muscle fibers occur, and intramuscular hematoma is frequently associated with these injuries. Direct trauma may affect any muscle, but the quadriceps ("charley horse") and gastrocnemius muscles are more prone to these injuries. The injuries are characterized by tenderness, diffuse swelling or a discrete hematoma, and limitation of motion and strength.

Quadriceps Contusions. Adequate acute treatment of muscle injuries is important to limit hematoma formation and inflammation. Jackson and Feagin [126] reviewed quadriceps injuries occurring in military cadets and found them to be a significant cause of athletic and occupational disability. Similar results were reported by Ryan [178]. Both authors stress the importance of grading the severity of the injury initially. The initial grade correlates well with the amount and duration of the disability. The possible pathologic mechanisms were described by Ryan [178], but there are few scientific data demonstrating the pathologic processes involved. The treatment regimen generally includes rest and ice with an early return to gentle motion [61]. Prolonged immobilization has been associated with longer periods of disability than shorter delays in restoration of motion. Active and passive motion should be emphasized, and care is necessary in therapy to avoid reinjury. A recent study [12] has offered promise in treatment of these injuries by instituting early immobilization in 120 degrees of knee flexion for the first 24 hours following the injury. The average time needed for return to full athletic activity was reduced from an average of 18 to 3.5 days. Although this study offers much needed insight into the treatment of acute muscular contusions, the long-term efficacy of such treatment remains to be evaluated.

Myositis Ossificans. An unfortunate complication of muscle contusions is the occurrence of myositis ossificans, the calcification or actual ossification of the tissue at the site of injury. The pathogenesis of heterotopic bone formation is poorly understood. It usually becomes radiologically evident 2 to 4 weeks following a severe contusion and is often connected to the underlying bone [126]. In one study heterotopic bone was present in approximately 20% of patients with a quadriceps hematoma [176].

The mass may enlarge or may be symptomatic for several months before stabilization occurs. It is important to be aware of the association of the mass and its roentgenographic appearance with a previous contusion because the condition can also mimic osteogenic sarcoma (Fig. 1A–13). The histologic features may also be similar if a biopsy is performed early in the course of myositis ossificans.

Heterotopic bone may resorb with time. Recovery of normal function is possible even in the presence of myositis ossificans, but the recovery period is longer than that associated with an uncomplicated contusion. No specific treatment is recommended in addition to the treatment of contusions. Specifically, early surgery is to be avoided because it may exacerbate heterotopic bone formation and prolong disability. Surgery may be considered late in the course of the disease to remove the heterotopic bone if it is causing symptoms [106]. In general, surgery should be considered only if the presence of the bone mass is causing symptoms; this does not occur frequently. Surgery should be considered only after the heterotopic bone is mature and no changes are occurring in the orthopaedic and radiologic evaluation of the patient.

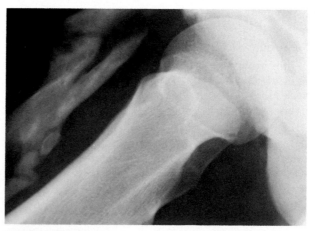

FIGURE 1A–13
X-ray of myositis ossificans of the rectus femoris. This 22-year-old patient suffered a quadriceps contusion that resulted in the condition shown. The resultant heterotopic bone gradually resorbed over time.

Compartment Syndromes

Much interest has been directed toward the diagnosis, management, and pathophysiologic basis of compartment syndromes in recent years. This syndrome is a pathologic condition of skeletal muscle characterized by increased interstitial pressure within an anatomically confined muscle compartment that interferes with the circulation and function of the muscle and neurovascular components of the compartment. The elevated pressure within the muscle compartment may be due to many different factors. Most frequently, transient ischemia is the inciting factor because it causes muscle edema once some circulation has been restored. Hemorrhage within the compartment or direct trauma to muscle can also cause pressure elevation or a compartment syndrome. Much of the current emphasis on compartment syndromes is due to the clinical observation that if the entity is recognized early, the elevated pressure can be relieved (usually by incising the investing fascia), and the circulation and function of the compartmental muscles and neurovascular components can be restored.

The pathophysiology of compartment syndromes has been investigated by a number of laboratories [87]. Increased pressure in the compartment results from increased fluid. The fluid increase can originate from hemorrhage, intracellular edema, or extracellular edema. The pressure within the compartment can be measured by several techniques including needle manometer [206], wick catheter [154], solid state transducer [148], and noninvasive auscultation. Threshold pressures above which significant muscle damage can occur have been proposed to be 30 to 40 mm Hg [78] or within 10 to 30 mm Hg of diastolic blood pressure [206].

When the intracompartmental pressure is elevated, it is postulated that capillary perfusion is compromised, and the skeletal muscle within the compartment is subject to ischemic injury. The level of pressure necessary to interfere with the capillary circulation will not necessarily occlude the major arteries running through the compartment. Therefore, the presence of a pulse distal to the compartment does not rule out the presence of a compartment syndrome. Clinical evaluation relies on evidence of pain, particularly with active extension of the limb, increased compartment pressure noted with palpation, and altered nervous function as noted by paresthesias in the sensory distribution of nerves within the compartment. Abnormalities in nerve function with increasing intracompartmental pressure have been demonstrated. The various objective methods of measuring tissue pressure listed above have been used extensively in clinical situations to increase the reliability of the diagnosis.

Acute Compartment Syndromes. Acute compartment syndromes have been associated with a variety of injuries common in athletics [131]. Direct trauma to bone or soft tissue is the most frequently noted injury. A review by Mubarak [154] stresses the association of compartment syndromes and fractures. Tibial shaft fractures comprise a large proportion of fractures leading to compartment syndromes. However, direct soft tissue injury and muscle trauma can also result in elevated pressure and compromise tissue perfusion. As stated earlier, the cause of the increased pressure can be edema, hemorrhage, or a combination of the two.

In addition to direct trauma, indirect injury due to exertion is well recognized as a cause of compartment syndromes. Indirect injuries can be acute or chronic; they have been reviewed well recently by Veith [202] and Mubarak [154]. The acute syndromes are not well understood. Several factors deserve mention. Intense muscular activity alone causes a large rise in interstitial pressure that might prevent normal capillary perfusion. Intermittent pressure levels of greater than 100 mm Hg are common during some forms of exercise [148]. As a result, muscle perfusion during such exercise is possible only intermittently when the pressure falls between muscular contractions. Increasing exercise causes a muscle volume increase that may be as large as 20% associated with increased blood content and intracompartmental fluid accumulation. The increased fluid component probably raises tissue pressure measurements at rest as well. Therefore, the combination of intermittent high interstitial pressures associated with muscle activity and elevated rest pressure due to compartment fluid expansion can predispose an athlete to an acute compartment syndrome.

Acute exertional compartment syndromes are not common [154]. They are usually associated with intense muscular activity, particularly in individuals unaccustomed to such activity such as military recruits. Confirmation of compartment pressures by one of the techniques mentioned earlier is preferable. Treatment should consist of decompression of the muscle and neurovascular components by fascial release.

Chronic Compartment Syndromes. Chronic exertional compartment syndromes are more frequent clinically than acute forms and have received considerable attention in the recent literature. The presenting complaints are usually those of pain or a deep ache over the anterior or lateral compartment. The discomfort usually occurs after a relatively long exercise period and is usually severe enough to cause the athlete to either stop his activity or reduce the intensity of the exercise. The symptoms are often bilateral [96, 98]. Sensory changes may also be present. Occasionally, muscle hernias may be present, and the hernias may be near the fascial opening through which the distal branch of the superficial peroneal nerve passes to reach the subcutaneous tissue (Fig. 1A–14).

Chronic or recurrent compartment syndrome is difficult to diagnose clinically. Corroboration with objective pressure measurements is desirable. Resting pressure values measured by several techniques may be slightly higher in patients with the chronic syndrome. However, the primary characteristic distinguishing the chronic condition is pressure elevation above normal during exercise and a slower return to resting value at the end of exercise [202]. These findings have consistently identified chronic compartment syndromes.

Treatment of chronic or recurrent compartment syndrome has consisted of elective fasciotomy of a single compartment if conservative measures and activity alteration are unsuccessful [154, 202]. Postoperative results in a relatively small number of

cases have been gratifying subjectively, and pressure measurements have returned toward normal. One should be aware that fascial release adversely affects the strength of a muscle, and, therefore, these procedures should not be advocated without accurate diagnosis and counseling [64].

Medial Tibial Syndrome. Some mention should be made of the condition commonly termed "medial tibial syndrome" or "shin splints." Previously, this syndrome of exercise-related pain localized to the medial aspect of the distal third of the tibia had been ascribed to a recurrent deep posterior compartment syndrome [29, 169]. However, objective measurement of pressure within the anterior and posterior compartments has not shown any pressure elevation [65]. This entity is most likely due to a stress reaction of the bone or muscle originating from the bone in response to repetitive use. The condition is characterized by pain along the medial aspect of the tibia coursing across the junction of muscle and tendon to bone. It is frequently found in athletes running long distances on hard surfaces. It is also more common in athletes with significant hindfoot valgus and midfoot pronation, often called "flatfoot."

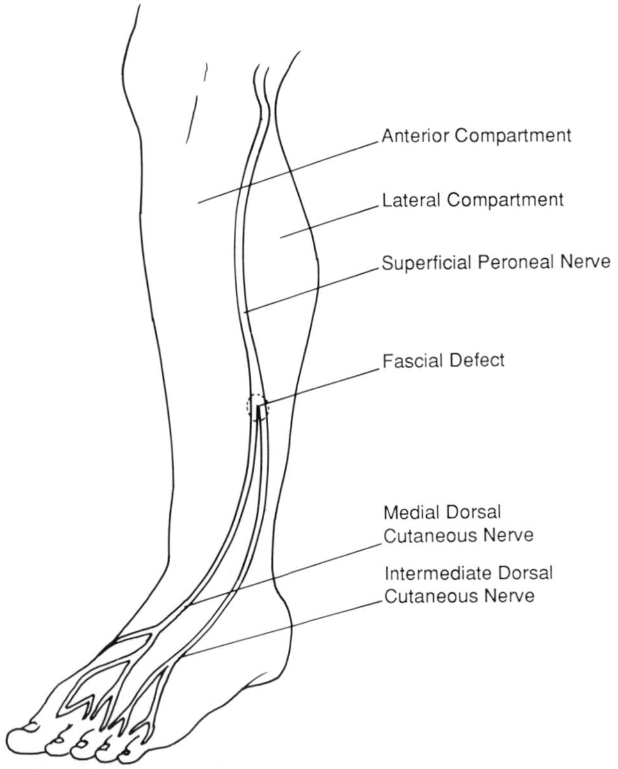

Muscle Strain Injuries

The clinical significance of stretch-induced muscle injuries or "strains" is readily evident to those treating occupational or sport-related injuries. This type of injury is usually cited as the most frequent injury in sports [76, 137, 165, 178]. It is somewhat surprising that only recently has there been increased interest in developing an understanding of the pathophysiology and biomechanics of these common problems.

Clinical Studies of Muscle Strain Injury

Mechanism of Injury. The literature on the clinical aspects of muscle injury is voluminous, but there is a relatively small amount of supporting scientific data. Indirect strain injuries are caused by stretching or a combination of muscle activation and stretching. It is widely felt clinically that muscle strain injury occurs in response to forcible stretching of a muscle either passively or, more often, when the muscle is activated [137, 170, 210]. Muscle strain injuries most often occur during eccentric contractions [76, 165, 210]. With eccentric contractions muscle may be more prone to injury for several reasons. If muscle is injured by excessive force developing in the muscle-tendon unit, higher forces are possible with eccentric contraction [190]. It has been shown that active muscle force production can

FIGURE 1A–14
Schematic drawing of the relationship of the branches of the superficial peroneal nerve to the fascial defect. (Modified from Garfin, S.R., Mubarak, S.J., and Owen, C.A. Exertional anterolateral compartment syndrome. *J Bone Joint Surg* 59-AP:404, 1977.)

Labels on figure:
Anterior Compartment
Lateral Compartment
Superficial Peroneal Nerve
Fascial Defect
Medial Dorsal Cutaneous Nerve
Intermediate Dorsal Cutaneous Nerve

be significantly higher when muscle is stretched while activated than when it is held at the same length or allowed to shorten. In addition, more force is produced by the passive or connective tissue element of muscle as it is being stretched [54]. It is believed that passive force provides little resistance until enough stretch is applied to the muscle. However, passive force in a muscle-tendon unit actually limits the range of motion of some joints. If excessive strain causes muscle injury, eccentric muscle function is associated with muscle stretching by definition.

Sports medicine personnel and athletes are well aware that certain muscles are more prone to injury than others. Muscles at risk for injury usually include the "two-joint" muscles (i.e., muscles that cross two or more joints and are therefore subject to stretch at more than one joint) [23]. A frequent characteristic of the injured muscles is their ability to limit the range of motion of a joint because of the intrinsic tightness in the muscle (e.g., hamstring muscles can limit knee extension when the hip is flexed). Similarly, the gastrocnemius can limit ankle dorsiflexion when the knee is extended. With these muscles, physiologic joint motion can place the muscles in positions of increased passive tension within the joint. Another characteristic of muscles at risk for injury is that they often function in an eccentric manner as they are used in sport. With eccentric contractions the muscle can be considered to be controlling or regulating motion as a function of energy absorption. It is clear that much of the muscle action involved with running or sprinting is eccentric [21, 22, 36, 37]. For instance, the hamstrings act not so much to flex the knee as to decelerate knee extension during running. Similarly, the quadriceps act as much to prevent knee flexion as to power knee extension in running [38, 125, 144]. These muscles are acting to control joint motion or to decelerate the joint and are therefore acting eccentrically.

Epidemiologic studies reveal that muscle strain injuries occur most often in sprinters or athletes engaging in activities of high velocity. They are more common in sports that require bursts of speed or rapid acceleration such as track and field, football, basketball, rugby, or soccer [165]. Another characteristic of muscles most likely to be injured is their relatively high percentage of type II or fast-twitch muscle fibers [68]. These muscles are generally more superficial in the extremities and cross two or more joints. That these muscles have higher percentages of fast-twitch fibers suggests that the body must require faster contractions in these muscles for kinesiologic reasons. The higher speeds of contraction may be a factor predisposing to injury.

Structural Changes with Muscle Strain Injury. Clinical data about the exact nature of the changes occurring in muscle following strain injury are relatively sparse. The injury may be partial or complete depending on whether the muscle-tendon unit is grossly disrupted [137]. Complete tears are characterized by muscle asymmetry at rest compared to the contralateral contour. With contraction, a torn muscle will show a bulge toward the side of the muscle-tendon unit that is still attached to bone.

Muscle strain injuries can be distinguished clinically from exercise-induced muscle soreness. A strain injury is an acute and usually painful event that is recognized by the patient as an injury. Muscle soreness is a condition characterized by muscle pain often occurring 12 to 24 hours after exercise and usually without a single identifiable injury [10, 88]. The two conditions are alike in that they are more prone to occur with eccentric exercise [13, 60, 62, 186]. Incomplete injuries are characterized more by focal pain and swelling. In both injuries, passive stretching and active contraction of the affected region cause discomfort.

Direct muscle injury or contusion causes injury to muscle at the place of contact. However, the location of pathologic changes in a muscle following strain injury has not been so well defined until recently. The vulnerable site in an indirect strain injury appears to be a site near the musculotendinous junction or the tendon-bone junction [66]. Injuries involving the muscle belly itself appear to have a strong predilection to occur near the junction but not precisely at the true histologic junction. Although surgical exploration of muscle injuries has not been common, a number of references to the surgical findings exist. These studies confirm the existence of tears near the muscle-tendon junction in (1) the gastrocnemius medial head (often incorrectly called a plantaris rupture) [47, 151], (2) the rectus femoris muscle [171], (3) the triceps brachii muscle [14], (4) the adductor longus muscle [193], (5) the pectoralis major muscle [166], and (6) the semimembranosus muscle [156]. More recently, high-resolution imaging studies have localized acute hamstring injuries to the region of the muscle-tendon junction (Fig. 1A–15). The size of the area of the muscle-tendon junction in humans is often surprising to clinicians. The hamstring muscles, for example, have an extended musculotendinous junction in the posterior thigh. The proximal tendon and muscle-tendon junctions of the long head of the biceps femoris and the semimembranosus extend for well over half the total length of these muscles.

Bleeding often occurs following muscle injury; however, it often takes one or more days following

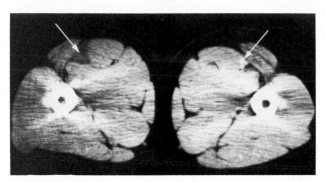

FIGURE 1A–15
Acute left biceps femoris muscle strain and chronic right hamstring injury. A prone axial computed tomographic (CT) image of the proximal thighs in this patient demonstrates an area of low density in the region of the long head of the left biceps femoris muscle, typical of an acute muscle strain. Calcifications are noted in the comparable muscle group on the right side probably due to an old injury in this patient. (From Garrett, W.E., Jr., Rich, F.R., Nikolaou, P., et al. Computed tomography of hamstring muscle strains. *Med Sci Sports Exerc* 21(5):510. © The American College of Sports Medicine, 1989.)

injury to detect subcutaneous ecchymosis. Blood collections within the subcutaneous tissue show that bleeding is not confined to the muscle proper; rather it escapes through the perimysium and fascia to the subcutaneous space. Computed tomography has shown that there is an inflammatory or edematous response within the muscle tissue itself [70]. In certain instances it appears that bleeding can occur and a hematoma can form between the muscle tissue and the surrounding fascial compartment, as shown by ultrasonography [59]. This finding is in contrast to that often seen with direct muscular contusion, with which bleeding often occurs within the midsubstance [175, 176].

Prevention and Treatment. Without a firm understanding of the mechanisms and pathophysiology of muscle strain injury, it has been difficult in the past to find agreement on the best methods of preventing or treating these often debilitating problems. A number of factors have often been cited as relevant in a predisposition to muscle-strain injury. Most athletes routinely practice stretching largely because it is felt to prevent muscle injury [18, 105, 208]. Adequate warm-up is also cited as a way of preventing muscle injury [137, 208]. There are indications that training programs employing adequate stretching and warm-up exercises can help to decrease muscle injuries; however, the training programs employed a number of other variables, and the individual effect of single factors has not yet been determined.

In addition to these preventive measures of stretching and warm-up, there are some risk factors that might be avoided to prevent injury. Fatigue is

thought to predispose muscle to injury [137]. Previous incomplete injury is thought to predispose muscle to a subsequent injury [165]. Although these factors are widely felt to be important risk factors for muscle injury, there have been few solid clinical or laboratory studies to support these hypotheses.

Authors' Preferred Method of Treatment of Muscle Strain Injuries

Treatment of muscle strain injury has varied considerably [137, 165]. Immediate treatment usually involves relative rest, ice, and compression. Further treatment modalities usually include physical therapy to improve range of motion as well as functional strengthening exercises, bandaging, and medications. Medications have included topical anesthetics, analgesics, muscle relaxants, and anti-inflammatory agents (steroidal and nonsteroidal). Occasionally, surgical intervention has been advocated in persons with complete dissociation of the muscle-tendon unit [151, 162]. These treatment regimens are empirically adapted from clinical practice, and few studies have ever been performed experimentally or clinically to demonstrate the effects of the different forms of treatment.

Based on available laboratory and clinical studies, we have devised the following treatment regimen for muscle strain injuries. Initial management includes ice and anti-inflammatory medications, usually nonsteroidal agents. We avoid immobilization and prefer to begin active stretching and muscle activation as soon as these exercises can be performed without great discomfort. Following the initial injury, the tensile strength of the muscle-tendon unit is weaker than normal, and large forces should be avoided. However, forces large enough to disrupt the muscle are likely to occur only during the return to uncontrolled activity, not in a controlled rehabilitation setting. We stress full recovery of muscle length and joint range of motion. Strengthening exercises are resumed early, and progressive resistance is emphasized. Controlled exercise and running are resumed when the patient is comfortable in doing so. The time recommended for return to sport is the most critical decision in rehabilitation because most reinjuries occur at this time. In general, only when range of motion is complete, strength is equal to that on the opposite side, and practice with mild pain is possible can the athlete return to competition.

We find that application of ice during the acute phase of the injury is helpful. Heat is quite helpful prior to performing stretching exercises after the acute phase. Therapeutic exercise should be of relatively high intensity to impart a strengthening effect. Isokinetic devices are particularly useful because the resistance is accommodating, and the injured athlete can work at a comfortable level through a full range of motion.

Laboratory Studies of Muscle Strain Injury

Until recently relatively few laboratory studies investigating the pathophysiology and mechanisms of muscle strain injury have been performed. One of the first investigations was a 1933 study by McMaster demonstrating that normal tendon did not rupture when the gastrocnemius muscle-tendon unit of rabbits was pulled to failure [150]. Interestingly, the healthy tendon did not fail even after it had been partially transected. Failure occurred at the bone-tendon junction, the myotendinous junction, or within the muscle. Our laboratory had previously studied muscle recovery in rabbit hindlimb muscle following muscle laceration and repair [72]. More recently, this model was adapted to the study of muscle strain injury using standard techniques of electrophysiology and biomechanics. Initial experiments demonstrated that activation of normal muscle by nerve stimulation alone did not cause either complete or incomplete disruption [67]. There was a reduction of force, and failure of excitation occurred, but no disruption of the muscle-tendon unit resulted. To obtain gross or microscopic muscle injury, stretch of the muscle was required. The forces produced at the time of muscle failure even without muscle activation were several times the maximum isometric force produced by the activated muscles [69]. This result demonstrated that passive forces within the muscle might be as important as the active forces involved with muscle strain injury. Stretch-induced injury was studied in response to passive stretch (i.e., stretch without muscle activation). In all cases, the neurovascular supply was left intact. The rabbit hindlimb was immobilized by skeletal fixation in a special frame. The tendon could be attached to an Instron materials testing device or to a force transducer. Muscles were activated by peroneal nerve stimulation when required.

Passive Stretch. Passive stretch was evaluated in five muscles with varied fiber architectural arrangements [69]. Muscles were stretched from the proximal or distal tendon without preconditioning or muscle activation. Strain rates of 1, 10, and 100 cm/ minute were tested. Muscles consistently demonstrated injury near the muscle-tendon junction (usually distally). The tendons were broad and flat with large areas of fiber attachment. The muscle fibers ran obliquely from the tendons of origin to the tendons of insertion, and the muscle-tendon junction region extended well into the muscle belly. A small and variable amount of muscle fiber was left attached to the tendon, usually 0.1 to 1.0 mm in

length. These experiments demonstrated that disruption occurred predictably near the muscle-tendon junction within the strain rates tested and for all muscles tested regardless of architectural features or direction of strain. Subsequent studies have confirmed similar findings at strain rates of up to 50 cm/ second.

Classic electrophysiologic studies of muscle have demonstrated that the active force production of muscle is proportional to the cross-sectional area of the muscle fibers, whereas shortening ability is proportional to the length of the muscle fiber. These concepts have been applied to human muscle performance [207]. It was felt that the biomechanical factors of muscle in response to stretch might also be related to fiber length. However, the amount of strain based on fiber lengths varied widely from 75% to 225% of the resting fiber length. It is apparent that strain injury does not occur after a relatively constant fiber strain. It is also interesting that previous studies have shown that the ends of the muscle fibers near the muscle-tendon junction do not strain as much as the more central areas of the fibers [83]. At this time it is not certain why the ends of the fibers near the muscle-tendon junction behave differently and are more susceptible to injury.

Active Stretch. Experiments were performed to measure the amount of force needed to produce failure, energy absorption prior to failure, and muscle length prior to failure in passive and active muscles [71]. Because it is felt clinically that the majority of muscle injuries occur during powerful eccentric contractions, conditions were devised to evaluate this situation. Muscles were stretched to the point of failure under three conditions of motor nerve activation: (1) tetanically stimulated, (2) submaximally stimulated, and (3) unstimulated. The results were somewhat surprising. The total amount of strain prior to failure did not differ among the three groups. The force generated at failure was only about 15% higher in stimulated muscles. The location of failure near the myotendinous junction did not change. However, the energy absorbed was approximately 100% higher in muscles stretched to failure while activated (Fig. 1A–16). These data confirm the importance of considering muscles as energy absorbers. The passive components of stretched muscle have the ability to absorb energy, but the potential to absorb energy is greatly increased by concomitant active contraction of the muscle.

This concept helps to explain the ability of muscles to prevent injury to themselves as well as the supporting joint structures. Muscles can be injured when they are incapable of withstanding a certain

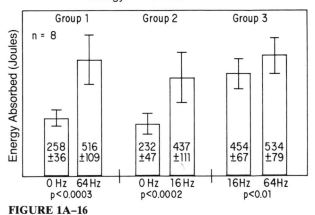

FIGURE 1A–16
Average relative energy absorbed by the muscle-tendon unit prior to failure in groups 1 through 3. All values −/+ SD. 0 Hz, no stimulation; 16 Hz, wave-summated stimulation; 64 Hz, tetanic stimulation. (From Garrett, W.E., Jr., Safran, M.R., Seaber, A.V., et al. Biomechanical comparison of stimulated and nonstimulated skeletal muscle pulled to failure. *Am J Sports Med* 15(5):451, 1987.)

force or strain. The ability of a muscle to withstand force and strain is a measure of the energy absorbed by the muscle prior to failure. In engineering terms, strain energy is the area under the curve relating stress to strain. It can be considered that there are two components of the ability to absorb energy. The passive component is not dependent on muscle activation and is a property that is due to the connective tissue elements within the muscle including the muscle fibers themselves and the connective tissue associated with the cell surface as well as between fibers. In addition, there is an extra ability to absorb energy based on the contractile mechanism of the muscle. In the experiments previously noted it is clear that the active component can double the ability of muscle to absorb energy. Therefore, conditions that diminish the contractile ability of the muscle might also diminish the ability of muscle to absorb energy. Muscle fatigue and muscle weakness are often considered as factors predisposing muscle to injury. Fatigue and weakness both imply that the active ability to absorb energy is diminished.

Just as the ability of the muscle to absorb energy can protect a muscle, it can also protect associated bone and joint structures [170]. It should also be noted that at low levels of strain most energy absorption is due to the active rather than the passive elements. Most physiologic activity in eccentrically contracting muscle occurs at relatively small levels of muscle strain, and energy absorption is due more to active than to passive force in the muscle.

Nondisruptive Injury. The studies cited above all evaluated the biomechanics of muscle strained to failure or complete muscle disruption. Previous work has documented the physiologic and histologic recovery of muscle after nondisruptive stretch-induced injury [159]. Nondisruptive injury was created by stretching unstimulated muscle and observing the force-displacement relationship. When the slope of the curve was no longer linear, the muscle was considered to have undergone a "plastic" deformation with a resulting alteration in the material structure. Muscle recovery was monitored physiologically and histologically. Nondisruptive injuries were created by stretching the muscles using 80% of the force necessary to disrupt the contralateral muscle.

Histologic studies show that injuries that are nondisruptive to whole muscle do result in disruption of a small number of muscle fibers near the muscle-tendon junction. The fibers do not tear at the actual junction of the muscle fiber and the tendon; rather, the tears occur within fibers that are a short distance from the tendon. These experiments have rarely resulted in disruption near the central portion of the muscle fibers. In the acute phase, the injuries are marked by disruption and some hemorrhage within the muscle (Fig. 1A–17). By 24 to 48 hours following the injury an inflammatory reaction becomes pronounced. Invading inflammatory cells and edema are present. By the seventh day the inflammatory reaction begins to be replaced by fibrous tissue near the site of injury. Although some regenerating muscle fibers are present, normal histology is not restored and scar tissue is persistent.

Functional recovery of the muscle was determined by physiologic testing of maximal force production in response to nerve stimulation. Immediately following the injury, muscle produced 70% of normal force. By 24 hours only 50% of normal force production was recorded. Recovery then followed and by 7 days was 90% complete (Fig. 1A–18). These results were confirmed in a later study evaluating the effect of nonsteroidal anti-inflammatory drugs on healing muscle injuries [161]. These studies demonstrated that the recovery of contractile ability is relatively rapid. It may be that the initial loss of function can be attributed to the hemorrhage and edema at the site of injury.

The tensile strength of muscle following a nondisruptive strain has also been evaluated in a preliminary study [161]. Using anterior tibialis muscles in rabbits, the muscles were stretched as in the studies cited above. The tensile strength of unstimulated muscle returned to only 77% of normal by 7 days, in contrast to 90% recovery of active force-generating ability. Tensile strength may be a reasonable indicator of the susceptibility of muscle to injury. A

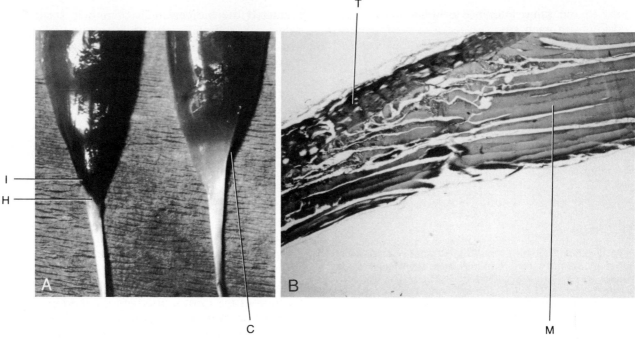

FIGURE 1A–17
A, Gross appearance of tibialis anterior muscle following controlled passive strain injury. A small hemorrhage (H) is visible at the distal tip of injured muscle at 24 hours. I, injured; C, control. *B,* Histologic appearance of TA muscle immediately after passive strain injury. Note the rupture of fibers at the distal muscle-tendon junction, along with hemorrhage. T, tendon; M, intact muscle fibers. Masson's stain (100×). (From Nikolaou, P.K., Macdonald, B.L., Glisson, R.R., Seaber, A.V., and Garrett, W.E., Jr. Biomechanical and histological evaluation of muscle after controlled strain injury. *Am J Sports Med* 15(1):13, 1987.)

previous injury may therefore predispose a muscle to further injury. Additional studies are needed to examine these possibilities.

Studies of nondisruptive strain injury show that a strain injury can indeed be produced characterized by disruption of fibers near the muscle-tendon junction with a subsequent inflammatory reaction and fibrosis in the healing phase. Recovery of contractile and tensile strength can be followed. Nondisruptive injury results in an injury that can recover physiologically. Tensile strength recovery is slower, indicating that injured muscles may be more likely to sustain a second strain injury.

Viscoelastic Behavior of Muscle

Among the factors felt to be important in the behavior of muscle and the prevention of injury are innate flexibility, warm-up, and stretching before exercise. Usually the response of muscle to stretching has been explained on a neurophysiologic basis with reference to stretch reflexes [102, 188]. However, many properties of muscle in response to stress may be explained with reference to the viscoelastic properties common to biologic tissues in general. Our laboratory has investigated viscoelasticity in muscle tissue and compared the response of muscle-tendon units and tendon alone [194]. Many investigators have examined the viscoelastic response of tendon and ligament [34, 204]. These studies have demonstrated that, when a ligament or tendon is stretched and held at a constant length, the tension at that length gradually decreases. This decrease in tension over time is known as stress relaxation. Additionally, cyclic stretching of ligaments and ten-

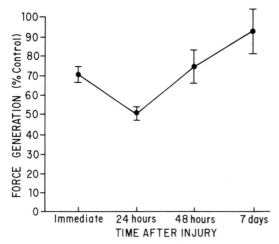

FIGURE 1A–18
Percentage of control force generation over range of frequencies versus time after controlled passive injury. Immediately after injury, N = 30; 24 hours, N = 7; 48 hours, N = 8; 7 days, N = 8. All values +/− SEM. (From Nikolaou, P.K., Macdonald, B.L. Glisson, R.R., et al. Biomechanical and histologic evaluation of muscle after controlled strain injury. *Am J Sports Med* 15(1):10, 1987.)

dons to the same length results in a decrease in tension with each stretch. There are a number of classic studies of viscoelasticity in muscle physiology [3]. However, most of these studies pertain to the viscoelastic behavior of active force production. Much less is known about the viscoelastic behavior of the stretching of muscle in a manner pertaining to current athletic and rehabilitation regimens of muscle stretch.

Another series of experiments was performed to determine whether muscle-tendon units also respond to repeated stretching with a reduction in tension. The study evaluated the biomechanical changes in the EDL muscle-tendon unit using two types of cyclic repetitive stretching. EDL units were stretched to 78.4 Newtons (N) from an initial tension of 1.96 N, held at this tension for 30 seconds, and returned to their initial length [194]. This cycle was repeated ten times. Successive cyclic stretches required an increase in length to attain the predetermined tension. The total length increase after ten stretching cycles averaged 3.45%. Eighty percent of this total length increase occurred during the first four stretches. The stress relaxation curve following the first stretch was significantly different from that following the second stretch. Similarly, the third and fourth cycles were different. The final six relaxation curves showed no significant difference from each other (Fig. 1A–19).

The second testing regimen involved stretching EDL units 10% beyond their resting length and then returning them to resting length. This stretching cycle was completed ten times to the same length for each specimen. The peak tensions were found to decrease with each stretch, with an overall drop

in peak tension of 16.6% from the first to the tenth stretch. The decrease in peak tension during the first four stretches was statistically significant; however, peak tensions in cycles five through ten did not differ statistically.

These data suggest that repetitive stretching will lead to a reduction of load on the muscle-tendon unit at a given length. The stretching regimen was chosen to mimic the stretching routines commonly practiced in sports. Stretching clearly led to a reduction in tension for a given length of muscle. It has been shown that this effect is independent of any effect from reflexes or other influences mediated by the central nervous system [194]. Of course, reflex effects and central nervous system influences may be involved in addition to viscoelastic responses.

Effect of Repetitive Stretching on Failure Properties. The studies cited above demonstrate that viscoelastic behavior in muscle-tendon units is quite separate from reflex effects in the muscle. It is apparent that these effects can have great significance in flexibility and performance factors during muscle use in sports and exercise. It is not clear whether these effects have any bearing on the prevention of muscle injury. Stretching has often been advocated as a means of preventing injury. This hypothesis has been tested in two different rabbit muscles subjected to cyclic stretching [20]. The muscles were cyclically stretched using 50% or 70% of the force needed to produce failure in the contralateral leg. It was found that ten cycles at 50% of maximum force resulted in a significant increase in the length of muscle stretched at failure without a change in force at failure or energy absorption. Muscles that were cyclically stretched to 70% of maximum force behaved differently. Many of them demonstrated macroscopic evidence of disruption before completing the ten cycles. Length and force to failure were unaffected in the remaining specimens. These data show a protective effect of cyclic stretching in some cases and demonstrate that muscles subjected to stretch without any preconditioning are more subject to injury. It is also clear that stretch that produces more than 70% of maximum sustainable force in the muscle can make the muscle more likely rather than less likely to be injured. These data also demonstrate the effects of architecture on the response of the muscle-tendon unit.

It appears that a cyclic stretching routine may make muscle less likely to be injured because it increases the length to which a muscle can stretch before failure occurs. Stretching therefore produces significant effects on muscle, both at physiologic lengths where viscoelastic effects produce stress relaxation and less stiff muscles and at highly stretched lengths at which the failure properties of muscle are affected.

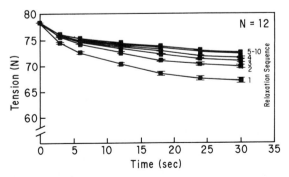

FIGURE 1A–19

Relaxation curves for extensor digitorum longus (EDL) muscle-tendon units subjected to repeated stretch to the same tension. There was a statistically significant (p < 0.05) difference between the first relaxation curve and the subsequent nine curves. The second relaxation curve also showed a statistically significant difference from the other nine curves (p < 0.05). There were no differences in curves 3 through 10. (From Taylor, D.C., Dalton, J.D., Seaber, A.V., et al. Viscoelastic properties of muscle-tendon units. The biomechanical effects of stretching. *Am J Sports Med* 18(3):304, 1990.)

Effects of Warm-Up. The value of warm-up is emphasized in the prevention of muscle injuries. The effects of muscular activity may be totally different than the viscoelastic effects associated with stretching. However, it is known that viscoelastic materials are sensitive to temperature, which changes within the muscle with activation. Studies evaluating these changes in failure properties of muscle have been performed. Muscle was held isometrically and stimulated for one single tetanic contraction lasting 10 to 15 seconds [71, 179]. This contraction was associated with a temperature rise of approximately 1°C within the muscle. A single contraction did lead to changes in muscle failure properties. More stretch was possible prior to failure, and more force was produced.

The changes may be due in part to the small temperature change within the muscle. However, the changes may also be due to viscoelastic or stretch effects even though the muscle was held isometrically. The muscle belly and fibers shorten minimally, and the tendon and muscle-tendon junction undergo slight stretch in response to muscle activation. Although the total length of the muscle-tendon unit does not change, it is quite likely that the area susceptible to injury undergoes some degree of strain in response to the force of contraction, and the resulting change in failure properties may still be due to stretch effects rather than to the temperature change.

Summary of Basic Studies on Muscle Injury

Laboratory studies of muscle injury support many of our present notions about muscle injury and its prevention. The etiology of most muscle injuries involves powerful eccentric contractions. Both disruptive and nondisruptive injuries show pathologic changes near the muscle-tendon junction. Laboratory studies have emphasized the importance of active muscular contractions in the absorption of energy.

Basic studies are also demonstrating factors that can prevent muscle injuries or substantially alter muscle properties. The separate effects of stretch, activation, and temperature are being evaluated with respect to manipulations that may improve performance or prevent injury. In addition, controlled injuries can be created in the muscle, and different methods of treatment can be evaluated as muscle recovery is followed.

Laboratory studies have several implications regarding the prevention of muscle injuries. Stretch is always required to injure normal muscle; specifically, strong active contractions involving shortening of the muscle do not appear to create injury. Therefore, the ability of muscle to withstand stretch should be important in preventing injury. Increasing the extensibility of muscle can be done in several ways. Preliminary stretching increases extensibility primarily because of the viscoelastic nature of muscle. The stretch should be held or slowly increased over time to allow the time-dependent stress relaxation and creep to occur in the muscle. Several stretches that are each held for a period of time seem to be beneficial. Ballistic stretching should be avoided because high velocities result in increased forces for the same stretch, and the quick shortening does not allow time-dependent or viscous changes to occur in the muscle. Empirically, three to four stretches held for 5 to 10 seconds or more seem useful.

Warm-up also increases muscle extensibility. The usual athletic warm-up involves both increasing the temperature in the muscle by metabolic activity and stretching the muscles and tendons by active muscle force production. Because extensibility of connective tissue increases with temperature, a warm-up period before a stretching routine may be effective.

Longer term adaptations in muscle are also very important in the prevention of acute injury. Often the physiologic requirement of a muscle in sports is the control or deceleration of a joint or limb. Therefore, the muscle is required to absorb the kinetic energy of the limb. The ability of the muscle to absorb energy can protect it from injury. Strong muscles can absorb more energy than weak muscles; strong muscles undergo less deformation or stretch than weaker muscles. Therefore, strengthening can help prevent strain injury.

Conditioning has a similar effect. Fatigue is a situation in which the ability of muscle to generate active force is declining. Fatigued muscles therefore can absorb less energy than nonfatigued muscles. Muscle strength and conditioning seem to be valuable components of an injury prevention program, particularly in people in whom muscle injury is most likely.

Conclusions

More clinical and basic laboratory studies of muscle injuries have become available in recent years. Clinical studies emphasizing the imaging of muscle injuries provide information about the location and nature of the initial injury and the clinical course. Few studies exist that provide good information on effective means of preventing injury. Of course, clinical studies may be difficult to perform because of the large numbers of subjects who must be followed to obtain reliable epidemiologic data. However, such studies are necessary.

Basic laboratory studies can be quite helpful in the practical management and prevention of muscle strain injuries. Model studies have demonstrated the pathophysiology of these injuries, and the findings are quite consistent with the clinical findings. As the process of injury is becoming better understood, more emphasis is being placed on the treatment and prevention of these injuries and on methods of improving performance.

TENDON

Tendons are dense, regularly arranged collagenous structures connecting muscle to bone. On microscopic examination, tendons consist of a network of interlacing fibers with various shaped cells and ground substance. Up to 85% of the dry weight of this structure is collagen, and it is therefore not surprising that the mechanical and physiologic behavior of collagen is the most important factor in determining the properties of this tissue [34]. Tendons must be capable of resisting large tensile stresses in order to perform their primary function, which is to transmit forces from muscle to bone. In addition, tendon permits the muscle belly to remain at an appropriate distance from the joint over which it acts by controlling the length of the moment arm. In addition to this load-transmitting role, tendons satisfy both kinematic requirements (they must be flexible enough to bend at joints) and damping requirements (they must absorb sudden shock to limit damage to muscle).

Structure

For many years tendons and ligaments were classified similarly as dense, regularly arranged connective tissues [18], and at times the terms were used interchangeably [3, 6, 120]. However, there are significant differences between the two with respect to structure, and histologic and biochemical properties [7]. A tendon is composed primarily of densely packed collagen fibers that are arranged more parallel to their longitudinal axis than is the case in ligament. These collagen fibers run the entire length of the structure. Each collagen fiber in turn is composed of thinner fibrils. Fibroblasts, which are few in number, are located more centrally in the tendon between the collagen bundles or fibrils. Our present knowledge of tendon morphology is outlined in Figure 1A–20.

The surface of the tendon is enveloped in a white, glistening, synovial-like membrane, the epitenon. It is continuous on its inner surface with the endotenon, a thin layer of connective tissue that binds collagen fibers and also contains lymphatics, blood vessels, and nerves [41]. In some tendons the epitenon is surrounded by a loose areolar tissue called the paratenon. Typically, the paratenon surrounds tendons that move in a straight line and are capable of great elongation owing to the presence of elastic fibers. This paratenon functions as an elastic sheath permitting free movement of the tendon against the surrounding tissue. Together the epitenon and paratenon compose the peritendon (Fig. 1A–21).

In some tendons the paratenon is replaced by a true synovial sheath or bursa consisting of two layers lined by synovial cells. This double-layered sheath, which is lined by synovium, is referred to as a tenosynovium. Within this synovial sheath the mesotendon carries important blood vessels to the tendon [117]. The flexor tendons of the forearm and

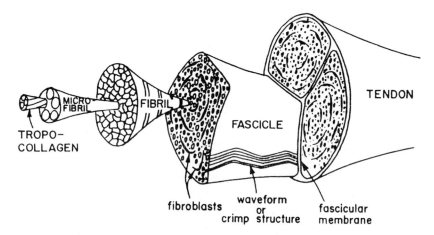

FIGURE 1A–20
Basic tendon morphology. (Adapted from Kastelic, J., Galeski, A., and Baer, E. The multicomposite structure of tendon. *Connect Tissue Res* 6:11–23, 1978.)

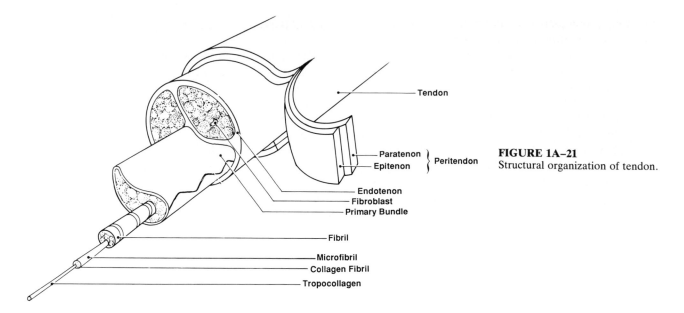

FIGURE 1A–21
Structural organization of tendon.

hand are surrounded by this well-defined sheath lined with synovial cells. Consequently, this sheath (paratenon) has been termed a tenosynovium. In the absence of a synovial lining, the paratenon is often called a tenovagium. In contrast, the Achilles tendon sheath (paratenon) has a synovial lining and therefore is termed a tenosynovium.

The perimysium becomes continuous with the endotenon at the musculotendinous junction. The tendon-bone interface marks the site where collagen fibers enter bone as fibers of Sharpey and the endotenon becomes continuous with the periosteum. The insertion of tendon into bone is generally classified into two types. The simpler, termed direct insertion, occurs when the tendon fibrils pass directly into bone through zones of fibrocartilage with little interdigitation into the surrounding periosteum. As described by Cooper and Misol [31], in this type of insertion the tendon inserts into a zone of fibrocartilage and then into a layer of mineralized fibrocartilage and finally into bone. The periosteum is therefore continuous with the endotendon. Dissipation of force is effectively achieved through this gradual transition from tendon to fibrocartilage to bone. The second type of insertion is more complex and involves the periosteum as well as the underlying bone. Here the superficial fibrils insert into the periosteum while the deeper fibrils fan out into bone directly.

Tendon is well-vascularized tissue, although less so than muscle. The blood supply to tendon has several sources including the perimysium, periosteal attachments, and surrounding tissues. Blood supplied through the surrounding tissues reaches the tendon through the paratenon, mesotenon, or vin-

cula. A distinction between "vascular" and the so-called "avascular" tendons has been made to denote differences in blood supply. Vascular tendons are surrounded by a paratenon and receive vessels along their borders; these vessels then coalesce within the tendon. The relatively avascular tendons are contained within tendinous sheaths, and the mesotenons within these sheaths function as vascularized conduits called vincula. Thus, the muscle-tendon and tendon-bone junctions along with the mesotenon are the three types of vascular supply to the tendon inside the sheath. Other sources of nutrition [91, 92] include diffusional pathways from the synovial fluid, which provide a considerable supply of nutrients for the flexor tendons of the hand.

The nervous supply to a tendon is sensory in nature. The proprioceptive information supplied to the central nervous system by these nerves is usually picked up through mechanoreceptors located near the musculotendinous junction.

Biochemistry

The cellular component of tendon is the tenocyte, which is responsible for the production of collagen and the matrix proteoglycans. Like all types of connective tissue, tendons consist of relatively few cells (fibroblasts) and an abundant extracellular matrix. In both tendons and ligaments the main constituent is collagen, along with small amounts of elastin, ground substance, and water. Collagen constitutes approximately one-third of the total protein in the body and is present in large amounts in specialized connective tissues such as tendon, ligament, skin, joint capsule, and cartilage.

Collagen

Modern research on the molecular structure of collagen began in the 1950s with the use of the electron microscope (Fig. 1A–22). Ramachandran and Kartha [150] and Rich and Crick [152] were the first to develop models depicting the triple-helical structure of collagen. Today, 12 different but homologous collagen types are recognized (for a review, see ref. 130). It is often convenient to think of two major classes of collagen, those that are fiber-forming and those that are not. Collagen types I, II, and III are known as fibril-forming collagens. After being secreted into the extracellular space, these collagens assemble into collagen fibrils. Collagen types I and III are the main forms comprising normal connective tissue (e.g., skin), and of these,

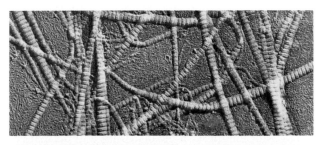
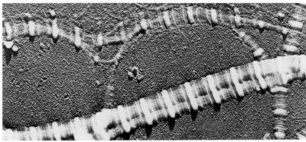

FIGURE 1A–22
Electron micrograph of collagen demonstrating the periodicity and regularity of the molecule. Precipitated from collagen solution by dialysis against 1% NaCl. (Bloom, W., and Fawcett, D.W. *A Textbook of Histology,* 8th ed. Philadelphia, W.B. Saunders, 1962, p. 105. Original investigators: J. Gross, F.O. Schmitt, and J.H. Highberger.)

type I is much more common, constituting 90% of the collagen in the body. The remaining nine or ten collagen types constitute the second major group, of which types IV and V are the basement membrane collagens.

The fibroblast is a spindle-shaped, contractile cell that is mainly responsible for the synthesis of connective tissue matrix precursors including collagen, elastin, and proteoglycans [20, 136]. Collagen is produced within the fibroblast as a large precursor molecule (procollagen), which is then secreted and cleaved extracellularly to form tropocollagen. Soluble tropocollagen molecules then form noncovalent crosslinks, resulting in insoluble collagen molecules that aggregate to form collagen fibrils. After collagen fibrils have been synthesized in the extracellular space, they are greatly strengthened by the formation of covalent crosslinks within and between the constituent collagen molecules. In its normal triple helical state, mature collagen can be degraded only by collagenase, whereas ruptured collagen fibrils are also susceptible to digestion by trypsin. When isolated collagen fibrils are viewed in an electron microscope, they exhibit cross-striations every 64 to 68 nm. This pattern reflects the packing arrangement of the individual collagen molecules in the fibril.

Collagen is the strongest fibrous protein in the body. These fibers are arranged parallel to their longitudinal axis, which results in tendon having one of the highest tensile strengths of all soft tissues. All types of collagen have in common a triple helical domain, which is combined differently with globular and nonhelical structural elements (Fig. 1A–23). The triple helical collagen molecule is quite stiff compared with a single polypeptide chain. The most common collagen molecule, type I collagen (also found in skin and bone), is composed of three alpha-peptide chains, each with about 1000 amino acids, resulting in a total molecular weight of about 340,000 daltons [149, 153]. The alpha chains exist in several different isomeric forms (see Table 1A–3). Type I collagen contains two alpha-1 chains and one alpha-2 chain [131]. These three alpha chains are wound around each other in a regular helix to generate a rodlike collagen molecule about 300 nm long and 1.5 nm in diameter. Normal human adult flexor tendons are composed largely of type I collagen (more than 95%); the remaining 5% consists of type III and type IV collagen [71, 135]. The amino acid sequence of the collagen molecule has been studied extensively to understand the crosslinking mechanism of these structures. They are arranged in a characteristic triple helical pattern that gives the molecule both its rodlike form and its rigid properties. Every third amino acid in the alpha chain is glycine; other amino acids commonly pres-

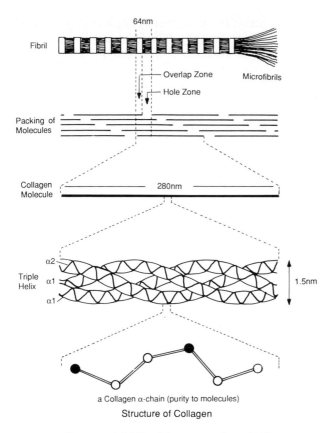

FIGURE 1A–23

Microstructure of collagen showing the three alpha chains of the triple helix. Three separate alpha chains are wrapped around each other to form a ropelike triple-stranded, helical rod. Every third amino acid of the native molecule is glycine.

Physical Properties of Collagen. The mechanical properties of soft collagenous tissues are highly dependent on their structural integrity, which is determined primarily by the architecture and properties of the collagen fibers as well as by the amount of elastin present in the tissue. In addition, the physical properties of collagen are highly dependent on covalent crosslinks within and between the molecules. The triple helix conformation of collagen is stabilized mainly by hydrogen bonds. The three chains are held together strongly by hydrogen bonds between glycine residues and between hydroxyl groups of hydroxyproline. This helical conformation is reinforced by hydroxyproline- and proline-forming hydrogen bonds to the other two chains. The degree of crosslinking is a key to both the tensile strength of collagen and its resistance to enzymatic and chemical breakdown. This conclusion is evident clinically in the condition of lathyrism, in which an absence of crosslinking produces a collagen that is significantly weakened, resulting in an increased incidence of aortic rupture, abnormal curvature of the spine, and problems with skin breakdown and wound dehiscence.

Elastin

Elastin is a protein found in connective tissues that permits these structures to undergo great changes in length without incurring any permanent change in structure while expending the least amount of energy in the process. It is elastin that is responsible for the wavy pattern of the tendon when viewed in a light microscope. Tendons of the extremities possess small amounts of this structural protein, whereas elastic ligaments such as the ligamentum flavum and ligamentum nuchae have greater proportions of elastin. The greatest amount

ent are proline (15%) and hydroxyproline (15%) [149]. Consequently, two-thirds of the collagen molecule consists of three amino acids. Hydroxyproline is derived from proline and is almost unique to collagen, and another amino acid, hydroxylysine, is unique to collagen.

TABLE 1A–3
Principal Collagen Types and Their Properties

Type	Molecular Formula	Polymerized Form	Distinctive Features	Tissue Distribution
I	[alpha-1 (I)]₂ alpha-2 (I)	Fibril	Low hydroxylysine Low carbohydrate	Skin, tendon, bone, ligaments, cornea
II	[alpha-1 (II)]₃	Fibril	High hydroxylysine High carbohydrate	Articular cartilage, intervertebral disk, notochord, vitreous body of eye, fetal collagen
III	[alpha-1 (III)])₃	Fibril	High hydroxyproline Low hydroxylysine Low carbohydrate	Skin, blood vessels, internal organs
IV	[alpha-1 (IV)₂ alpha-2 (IV)	Basal lamina	Very high hydroxylysine High carbohydrate	Basal laminae

of elastin in most tendons is found at the fascicle surface [158]; it usually comprises less than 1% by dry weight. On the other hand, the elastin content of the aorta can be as high as 30% to 60% of dry weight. Elastin, like collagen, has lysine-derived crosslinks. The amino acids desmosine and isodesmosine are unique to elastin. Their formation is dependent on the presence of copper. The elastic potential of elastin is due primarily to the crosslinking of lysine residues via desmosine, isodesmosine, and lysin-onorleucine.

Ground Substance

About 1% of the total dry weight of tendon is composed of ground substance, which consists of proteoglycans, glycosaminoglycans, structural glycoproteins, plasma proteins, and a variety of small molecules. The physiologic importance of these structures is due in large part to their waterbinding capacity, which helps to account for the viscoelastic properties of tendinous materials. It is currently theorized that proteoglycans and glycosaminoglycans are important for stabilizing the collagenous skeleton of connective tissue. Proteoglycans are high-molecular-weight macromolecules consisting of a protein core to which glycosaminoglycan side chains are attached. Glycosaminoglycans (GAGs) are macromolecules containing repeated disaccharides composed of a hexosamine residue and a uronic acid residue. Glycosaminoglycans that are abundant in mammalian tissues include hyaluronic acid, chondroitin sulfate, dermatan sulfate, keratan sulfate, and heparin-heparan sulfate. Except for hyaluronic acid, the GAGs are negatively charged owing to the presence of sulfate or carboxyl groups, and this confers predictable mechanical and chemical properties on the connective tissue. It has been shown that regions of tendon that experience primarily tensile forces have a lower proteoglycan content and higher rates of collagen synthesis than areas that experience frictional and compressive forces as well as tensile forces [14].

Mechanical Properties of Tendon

The primary function of tendon is to transmit muscle forces to the skeletal system to provide joint and limb locomotion and movement. To do this effectively tendons must be capable of resisting high tensile forces with limited elongation. Tensile strength as high as 98 N/mm^2 has been reported [42]. The densely packed collagen fiber bundles

arranged parallel to the length of the tissue provide efficient resistance to tensile loading. On the other hand, tendons have weak resistance against shear and compression forces. From a functional point of view, tendons are designed to transmit loads with minimal energy loss and deformation.

Biomechanical studies of tendon have revealed that the stress-strain relationship is similar to other collagenous tissues such as ligament and skin. As with most biologic tissues, tendons demonstrate complex time-dependent and history-dependent nonlinear viscoelastic properties [1, 185, 186]. These include stress relaxation (decreased stress with time under constant deformation) and creep (increased deformation with time under constant load) (Fig. 1A–24). Because of the history-dependent properties of the structure, the shape of the load-deformation curve is dependent on the previous loading history. Clinically we recognize these history- and time-dependent characteristics; for example, increased tendinous and ligamentous laxity occurs following exercise. Several models have been developed to describe and predict the mechanical behavior of tendons and other biologic tissues [49, 172, 185].

Figure 1A–25 represents a typical load-elongation curve for tendon. Abrahams described three distinct regions of such a curve prior to rupture of the structure [1]. Under tension the fibers straighten, and the system becomes stiffer. Different components of the structure take up loads at different levels, resulting in a nonlinear, concave, upward

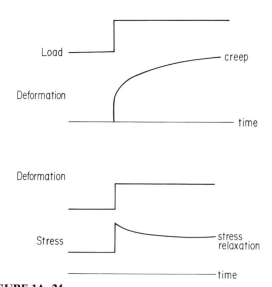

FIGURE 1A–24
Biomechanical properties of collagen. Under a constant load the tendon will undergo time-dependent relaxation (creep), whereas under a constant deformation the structure will undergo stress relaxation (i.e., reduction in load over time).

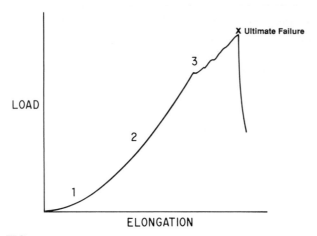

FIGURE 1A–25
Typical load-elongation curve to failure showing (1) primary or "toe" region, (2) secondary or "linear" region, (3) end of secondary region.

load-deflection or stress-strain curve. The initial "toe" region represents the alignment of fibers in the direction of stress. This toe portion of the curve results when the fiber bundle waviness straightens out. In this region little force is required initially to elongate the tissue, which may protect the tendon from damage during rapid force development. The toe region is followed by a steep linear portion of the curve in which the majority of the fibers are aligned parallel to the longitudinal axis of tension. The slope of the curve in this linear region is often referred to as the elastic stiffness of the tendon. This linear region results from elongation of the helical structure on the fibrillar and macromolecular levels. In the end of this region small force reductions are sometimes observed, which can usually be attributed to failure of a few fiber bundles. After the second or linear region, major failure of the tendon usually occurs in an unpredictable fashion. Upon failure, the fibers recoil and blossom into a tangled bud at the ruptured end. At strains of 4% to 8%, collagen fibers begin to slide past one another, resulting in disruption of their crosslinked structure. Although normal physiologic forces or loads are reported to cause strain of less than 4% [5, 42], certain activities such as sports can occasionally cause stronger loads. Consequently, there is a large margin of safety because the maximum isometric contractile force of a muscle is usually about one-third of the maximum load of the tendon. Perhaps more important is the fact that repetitive loading at submaximal failure loads can result in fatigue and eventual failure of the structure.

Large variations in ultimate tensile strength and maximum strain are usually attributable to differences in species, type, and age of the tendons studied. Testing conditions such as temperature and humidity are also important factors in the results obtained. It is important to keep these variables in mind when comparing the results of different studies. Preconditioning or "mechanical stabilization" of the specimen prior to testing will help to eliminate some of this variation. One must also distinguish between the structural properties of the tendon-bone complex and the mechanical or material properties of the tendon itself. Structural properties (i.e., ultimate load, ultimate deformation, linear stiffness, and energy absorbed at failure) describe the tensile properties of the tendon-bone complex. They are obtained directly from the load-deformation curve. On the other hand, mechanical or material properties are represented by the stress-strain relationship and are properties of the tendon itself (Fig. 1A–26).

Adaptability of Collagen

Aging. After collagen maturation, the mechanical properties of tendon reach a plateau followed

A-STRUCTURAL PROPERTIES OF BONE-TENDON COMPLEX (LOAD-DEFORMATION CURVE)

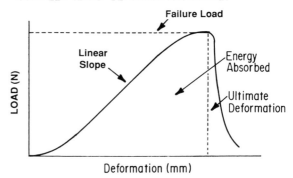

B-MECHANICAL PROPERTIES OF TENDON SUBSTANCE (STRESS-STRAIN CURVE)

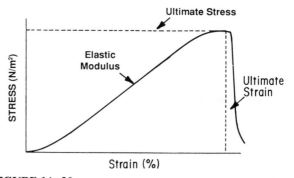

FIGURE 1A–26
Representative plots of tensile testing to failure. The structural properties are obtained from the load-deflection curve *(A)*, and the mechanical-material properties are obtained from the stress-strain curve *(B)*.

shortly thereafter by a decrease in tensile strength. This decrease in tensile strength is correlated with decreases in both the amount of insoluble collagen and the total amount of collagen present [179]. There is also a concomitant increase in stiffness [177], which is likely due to a marked increase in collagen crosslinking [12, 45]. Other extracellular changes include a decrease in the content of mucopolysaccharides and water.

Training. Studies of exercise-related changes in tendon properties are inconclusive. Most studies have shown that training results in increased tensile strength of tendons, measured as maximal failure load [169, 173, 192]. Increased tensile strength is also true of the ligament-bone interface [169, 192]. However, a recent study [188] showed that following 1 year of moderate exercise flexor tendons in swine showed no difference in mechanical properties, cross-sectional area, or collagen content compared to control animals. In similar studies increases in strength, size, and collagen content were found in extensor tendons [192]. In a study using the peroneus brevis tendons of rabbits, Viidik [173] found that the ultimate load was higher for trained than for nontrained animals but the weight, water, and collagen content of these tendons were no different. There may be a fundamental difference in the responses of flexor and extensor tendons to exercise, flexor tendons having a limited capacity for adaptability and extensor tendons having a greater training potential. These differences need to be more carefully documented.

Ultrastructural investigations have demonstrated that exercise leads to an increased number of collagen fibrils that are thinner in diameter compared with controls [107, 108]. Studies of exercised rabbits revealed that the collagen fiber crimp angle is increased while crimp length and elastic modulus are lowered [193]. Interestingly, anabolic steroids accentuate these changes and can lead to increased collagen dysplasia [5, 193]. Further research is needed in this area.

Immobilization. Several studies have demonstrated a decrease in the tensile strength of tendon [8, 119] and an increased collagen turnover [8] following immobilization. Like ligament, the mechanical properties of tendon show a decrease in stiffness with immobilization. One must keep in mind that many of the differences in these studies may be attributable to differences in age of the animals studied as well as to duration of immobilization.

Based on the information currently available, Woo and associates [189] have developed a hypothetical curve that predicts the mechanical response of tendons and ligaments to various time periods of exercise and immobilization (Fig. 1A–27). The relationship between stress and motion and the mechanical properties of the tissue is highly nonlinear. This diagram suggests that for tendons and ligaments within the normal range of physiologic activity, immobilization results in profound shifts in deformation properties when subjected to increasing forces. On the other hand, short-term training has little or no observable effect on these properties, and even long-term training has a minimal effect. The clinical significance of these animal results suggests that connective tissue is more responsive to a decrease in mechanical stimuli than to progressively higher loads.

Corticosteroid Treatment. No area of tendon research has received as much attention in the orthopaedic and sports medicine literature as that of corticosteroid injections into or around tendinous tissue. The marked anti-inflammatory effect of glucocorticoids is well recognized and often implemented in the treatment of athletic injuries. It is also widely accepted that the biosynthesis of collagen is inhibited by glucocorticoids [114]. There are case reports of local injections around the Achilles tendon [84] and patellar tendon [126] resulting in rupture of the tendinous tissue. Bilateral rupture of the Achilles tendon in patients receiving oral glucocorticoid therapy has been reported by numerous

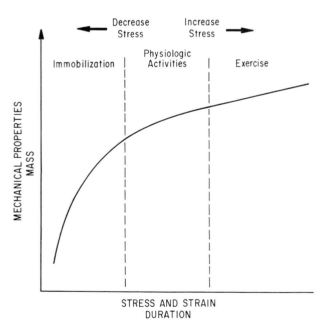

FIGURE 1A–27
A hypothetical curve showing the nonlinear properties of collagenous tissues and the effects of stress and motion on the equilibrium responses of soft connective tissues. (Reprinted from Woo, SL-Y. Mechanical properties of tendons and ligaments: 1. Quasistatic and nonlinear viscoelastic properties. *Biorheology* 19:397, 1982. With permission of Pergamon Press Ltd., Headington Hill Hall, Oxford OX3 OBW, UK.)

groups [85, 106]. Thus, the effects of corticosteroids can be systemic as well as local.

Laboratory studies of the effects of corticosteroid injections have produced somewhat confusing results. Important variables include the duration of the study, amount of steroid injected, and site of injection. Oxlund [124] showed that local administration of hydrocortisone acetate, 20 mg/kg every third day for 24 days, around the peroneal tendons of rats produced an increase in the tensile strength and stiffness of muscle tendons with no change in collagen content. In the same study another group of animals received injections into both knee joint cavities that resulted in decreased tensile strength of the posterior cruciate ligament-bone interface. Systemic effects of this local cortisol treatment included decreased thickness and fat content of the skin. A similar study showed that intramuscular injections of prednisolone increased the strength of muscle tendons in rabbits [128]. Following daily injections of prednisolone (2 mg/kg/body weight for 14 days the mechanical properties and collagen content of the peroneus tertius and longus tendons were evaluated. This treatment resulted in increased maximum load, maximum stress, and energy absorption for the muscle tendons. Elastic stiffness, measured after exhaustion of the viscous properties of the tendons, was also increased. It was concluded that 2 weeks of prednisolone treatment increased the strength and stiffness of muscle tendons. Increased stiffness was felt to be due to an increased stabilization of the collagen crosslinking pattern.

To evaluate the long-term effects of corticosteroid injections, Oxlund [125] injected 10 mg/kg of cortisone around the peroneal tendons every third day for 55 days. He found that although the mechanical properties of the tendons were not altered, their dry weight and hydroxyproline content were reduced. The thickness and collagen content of skin remote from the injection site were reduced although the strength of skin specimens was increased.

As a result of these studies and others it is currently felt that corticosteroids act on collagenous tissues in two ways [125]. Initially, during the first 1 to 2 weeks, corticosteroids induce a relatively fast increase in the mechanical and structural stability of the injected tissues. This increase is believed to be due to a change in the crosslinking pattern of the collagen. Eventually, progressive thinning and a reduction in collagen occur mainly secondary to an inhibition of collagen synthesis; these ultimately lead to a reduction in collagen content if the treatment is continued over a longer period of time.

Nonsteroidal Anti-inflammatory Drugs. Vogel was the first to show that indomethacin treatment resulted in increased tensile strength in rat tail tendons [180]. Increases in the proportion of both the insoluble fraction and the total collagen content were also noted. Other studies have confirmed similar findings [121]. Carlstedt and co-workers [25] examined the influence of indomethacin on the biomechanical and biochemical properties of rabbit tendons and found increased strength in tendon repair following indomethacin treatment. There was a slight decrease in the amount of soluble collagen, which may have been due to increased crosslinking following indomethacin treatment. It was concluded that the increased tendon strength resulted from the increased crosslinkage in the collagen.

Tendon Healing

Mechanisms of Tendon Injury

For the most part, tendons are subjected to axial loads. Special anatomic situations arise, for instance, in the flexor tendons of the hand when these tendons wrap around articular surfaces, resulting in compressive loading of the tendon. During active flexion the pressure between the pulley and the flexor tendon may be as high as 700 mm Hg [11]. These compressive loads are capable of altering the histologic structure of the tendon [122]. Injury to tendons can result from either acute trauma (e.g., laceration) or repetitive loading (e.g., overuse injury). The former will be discussed first with respect to flexor tendon injuries of the hand, and overuse tendinous injuries incurred in sports will be covered last.

Considerable scientific data are available regarding the acutely traumatized tendon. Much of this work has been done on tendons of the hand owing to their propensity for injury. Injury to tendon can occur in numerous ways; in the hand, avulsion directly from bone and midsubstance transection of the tendon itself are the two major mechanisms. These injuries often occur following crush injuries to the hand, of which 75% involve associated injury to the surrounding soft and hard tissues.

Primary Tendon Healing

As in other areas in the body, tendon healing proceeds in three phases: an inflammatory stage, a reparative or collagen-producing stage, and a remodeling phase. There are different opinions about the exact mechanisms occurring in primary tendon healing or in the healing of two divided tendon ends brought into apposition by sutures. There are two different theories of tendon healing. One theory suggests that healing depends on the surrounding tissues and that the tendon itself plays no significant role [86, 89, 130, 131, 138–141]. This theory holds that the tendon is an inert, almost avascular structure whose cells are incapable of contributing to the healing process. Using a canine model, Potenza [138, 140] showed that the tendon is invaded by fibrovascular tissue at the location of suture placement. At 28 days the collagen produced by these fibroblasts is immature, but by 128 days it is indistinguishable from that of normal tendon. In contrast, several recent studies [55, 57, 87, 91, 92, 97, 98, 102] have suggested that the inflammatory response is nonessential to the healing process and that tendons do possess an intrinsic capacity for repair. Lundborg and Rank [91, 92] demonstrated an intrinsic tendon repair response consisting of proliferation of tendon cells and production of mature collagen in rabbit flexor tendons. Efforts to demonstrate the intrinsic capacity of tendon healing previously failed owing to an inability experimentally to isolate the tendon itself from the inflammatory response discussed above. Lindsay and Thomson [87] were the first to show (in chickens) that an experimental tendon suture zone could be isolated from the perisheath tissues and that healing progressed at the same rate when the perisheath tissues were intact. Later it was demonstrated that in isolated segments of profundus tendon in rabbits these researchers were able to show that an active metabolic process existed in the experimentally free tendons by the presence of both anabolic and catabolic enzymes [102]. It was further demonstrated that sutured free tendon grafts of rabbit flexor tendons healed without adhesions within a vascular synovial environment of the suprapatellar bursa [90]. Consequently, it is now accepted that tendons may possess both intrinsic and extrinsic capabilities for healing and that the contribution of each of these two mechanisms probably depends on the location, extent, and mechanism of injury.

Tendon healing begins with the formation of a blood clot and an inflammatory reaction that includes an outpouring of fibrin and inflammatory cells. The degree of inflammation is related to the size of the wound and the amount and type of trauma that has occurred. A clot forms between the two tendon ends that is then invaded by cells resembling fibroblasts and migratory capillary buds. This process occurs during the first 3 days after injury, and it is believed that the fibroblasts arise from locally resident cells in the perivascular tissue [61]. This inflammatory phase is very much evident until the eighth to tenth day following injury. Collagen synthesis begins within the first week and reaches its maximum level after about 4 weeks. By 3 months this process still proceeds at a rate three to four times normal. Type I collagen is synthesized and extruded into the extracellular space as procollagen, which is converted to type I collagen by the enzyme procollagenase. Initially the collagen fibrils are oriented perpendicular to the long axis of the tendon, but by 2 months these fibrils are usually oriented parallel to the axis of tensile loading. Restoration of the gliding function of the tendon depends upon the dissolution and reformation of the collagen fibers during the scar remodeling phase. This starts about the fifteenth day, and by 28 days most of the fibroblasts and collagen between the tendon stumps are oriented longitudinally. Collagenase is present in the wound on the second day after injury, and somewhere between the fourth and sixth weeks collagen synthesis and collagen degradation reach an equilibrium. Collagen maturation and remodeling begin in the third week and can continue for up to 1 year following injury [59, 101]. The strength of the tendon repair results from the organization of collagen fibrils at the bone site; these fibrils crosslink with each other and with those of the tendon on each side of the wound.

Biomechanics of Tendon Healing

Paget [129] was the first to study the biomechanics of tendon healing when he examined the breaking strength of two healing Achilles tendons in rabbits. Mason and Allen [101] performed the classic experimental study on tendon healing using the extensor carpi radialis and flexor carpi ulnaris tendons in dogs to show that tensile strength progressed through three phases parallel with the phases of healing: (1) rapid decrease in tensile strength as a result of wound edema, which lasts about 5 days; at this time tensile strength depends primarily on the suture; (2) increase in tensile strength, reaching a plateau about the sixteenth day; (3) second increase in tensile strength, beginning between the nineteenth and twenty-first days and continuing for an undetermined period of time (length of study was 72 days). The histologic processes of repair corresponding to the three phases were: (1) exudation and fibrous union; (2) fibroplasia; (3) maturation, organization, and differentiation.

It was concluded from the work of Mason and Allen that the mechanical strength of the healing tendon is closely related to the three histologic phases of the healing process. According to Mason and Allen, function and motion during the first two phases of healing resulted in increased cellular reaction and separation at the suture lines. Also, active unprotected use even after 3 weeks of immobilization may be associated with stretching of the suture line and always leads to an increased cellular response. This work encouraged clinicians for the next 40 years to immobilize patients with injured tendons until the third phase of the healing process, when range of motion was encouraged to stimulate increased tendon strength and gliding.

The strength of an injured tendon that has been properly sutured increases rapidly during the fibroplastic phase, when granulation tissue is produced to repair the defect. Quantitative changes in acid mucopolysaccharides (hydroxyproline and hexosamine) accompany collagen production and increased tensile strength [39]. Bryant and Weeks showed that the ratio of wound collagen to mucopolysaccharide content is a direct measure of increased tensile strength [21]. In addition, the strength of the healing tendon increases as the collagen becomes stabilized by crosslinks and the fibrils are assembled into fibers. During the maturation phase, the mechanical strength of healing tendon increases owing to remodeling and reorganization of the fiber architecture. A gradual shift of collagen production from type III to type I may also contribute to increased mechanical strength.

Factors Affecting Healing

As emphasized by Akeson and co-workers [5], the factors affecting tendon repair are different from those associated with the healing of ligament.

Active Mobilization. Mason and Allen [101] showed that a tendon that was immobilized for 3 weeks and then mobilized tripled its tensile strength in just 2 weeks. Their study suggested that mobilization may stimulate both the healing rate and the strength of repaired flexor tendons. In other studies tendons undergoing minimal tension at the site of repair were weaker at 3 weeks than those with significant tension, but this strength increased rapidly during the following 3 weeks [77]. Early active mobilization did not improve tendon gliding; on the contrary, it increased tension across the suture line, which led to gap formation and tendon ischemia.

From the work of Mason and Allen [101] in 1941 it was concluded that immobilization was necessary after tendon surgery to prevent suture problems and wound insufficiency. This concept has been challenged several times over the years. Early active mobilization (less than 3 weeks following surgery) by contraction of the attached muscle is contraindicated according to the results of every study on primary tendon healing. Poor results are probably due to increased tension on the suture line with resultant ischemia, tenomalacia, and possible tendon rupture or gap formation between the tendon ends [76]. Early movement by active flexion of the involved muscle is likely to produce a gap leading to a weaker repair. Therefore, active mobilization may have a deleterious effect on tendon healing.

Stress. Tendons that undergo minimal tension at the site of repair are weaker at 3 weeks than those with significant tension. Mechanical stress promotes orientation of the collagen fibrils [133]. Remodeling of collagen scar tissue into mature tendon tissue depends upon the presence of tensile forces [138, 174]. How much tension is necessary to promote an acceptable clinical response is uncertain at this time.

Controlled Passive Mobilization. The concept of immediate passive mobilization was introduced by Kleinert and co-workers [80], who demonstrated that during limited active extension there is reciprocal relaxation of the flexor tendons, thereby allowing passive extension of the repaired tendon. This limited motion was found to be effective both experimentally and clinically in decreasing the tethering effect of adhesions. The biomechanical effects of controlled passive mobilization have been investigated [187]. Rates of both tendon repair and gliding function were significantly improved with early passive mobilization. Repaired tendons at 12 weeks postrepair had regained over one-third the strength of intact control tendons. These tendons maintained good gliding function within the sheath during the repair process. Tensile strength during tendon healing increased faster with controlled passive mobilization than with immobilization. From a biomechanical point of view, the optimal procedure appears to be to use strong suture repair to reduce gap formation and scar tissue in the early phase of healing. Following the initial healing phase, controlled passive mobilization and tensile stress placed on the tendon appear to promote earlier reorganization and remodeling of the collagen, leading to achievement of higher tensile strength.

Corticosteroids. Previously in this chapter we have discussed the effects of steroid injections on healthy tendons. It is well accepted that large doses of corticosteroids inhibit wound healing [44, 163]. Similar studies on the effects of small and moderate doses are conflicting. There is a lack of experimental data on the healing of injured tendons. Studies done on other fibroblastic structures have shown that corticosteroids suppress the formation of adhesions [26, 195] but may lower the tensile strength of sutured tendons with subsequent spontaneous rupture [26, 38]. Oxlund injected rats with cortisone and noted increased stiffness in 10-day wounds but decreased energy absorption prior to tendon failure and 20 days of healing in skin wounds [127]. These studies were confirmed by Gottrup [56], who found decreased strength prior to failure and decreased energy absorption in healing stomach incisions of rats following long-term cortisone treatment.

Overuse Tendon Injuries

In contrast to traumatic tendinous injury, sport-related injuries most often involve repetitive sub-maximal loading of the tissues, resulting in what is commonly referred to as "overuse injury." In response to cyclic loading dysfunction may result from either changes due to fatigue within tendons and their connections or inflammatory changes in surrounding tissue. It has been estimated that as many as 50% of all sports injuries are due to overuse and that the tissue most often affected is the musculotendinous unit [63]. The most frequent problem with overuse in the upper extremity involves the supraspinatus tendon with or without associated biceps tendinitis in swimmers, throwers, and weight lifters. Tennis players frequently have problems in the proximal wrist extensor tendon (i.e., "tennis elbow"). Other upper extremity overuse syndromes include "golfers elbow," or tendinitis of the flexor or pronator tendon, and "crossover" tendinitis of the adductor pollicis longus and extensor pollicis brevis tendons in rowers. Overuse problems in lower extremity tendons are frequently seen in runners and quite often involve the Achilles and posterior tibial tendons. Ballet dancers typically present with flexor hallucis longus tendinitis, and sports involving jumping (e.g., basketball, volleyball) often cause symptoms in the patellar tendon.

Classification of Overuse Injuries. There is still some confusion in the literature about a universal classification of overuse tendon injuries and the pathologic entities responsible for them. A classification of Achilles tendon disorders [167] provides a guide to the structural manifestations of overuse injury as follows: (1) peritendinitis, or inflammation of the peritendon; (2) tendinosis with peritendinitis; (3) tendinosis without peritendinitis; (4) partial rupture; and (5) total rupture. Others have added a sixth category, tendinitis, in which the primary site of injury is the tendon and there is an associated reactive peritendinitis [28]. Unfortunately, this classification is not universal because some tendons lack a paratenon and have synovial sheaths. The term peritendinitis also includes the terms paratenonitis, tenosynovitis, and tenovaginitis if a single layer of areolar connective tissue or a double-layered tendon sheath invaginates the tendon.

Significant tendon degeneration in the presence of inflammation of the peritendinous tissue (i.e., the paratenon or tendon sheath) is referred to as tendinosis with peritendinitis. This has been documented in competitive runners [29]. Tendinosis without peritendinitis describes the clinical condition of tendon degeneration without an accompanying inflammatory response. It has been shown histologically that tendon degeneration can occur without clinical symptoms [28].

Spontaneous tendon rupture during sporting activities is not infrequent. It occurs despite laboratory studies demonstrating that under normal circumstances healthy tendon is not the weak link of the musculotendinous unit. Various other studies have shown that healthy tendon is stronger than its muscle or muscle-tendon junction. These results have led researchers and clinicians to suggest that there must be some underlying pathology that results in spontaneous tendon rupture during sport-related activity.

Recent studies have shown that in cases of chronic tendon pain the pathologic lesion is typical of a degenerative process rather than an inflammatory one and that this degeneration occurs in areas of diminished blood flow. Several authors have documented the existence of areas of marked degeneration without acute or chronic inflammation in the majority of these cases [9, 10, 29, 36, 47, 165]. These changes are separate and distinct from the site of rupture. A review of patients with "chronic tendinitis syndrome" revealed similar findings of tendon degeneration [116, 167]. Nirschl has described the pathology of chronic tendinitis as angiofibroblastic hyperplasia [167]. Microscopically, one sees a characteristic pattern of fibroblasts and vascular, atypical, granulationlike tissue [115, 116]. Cells characteristic of acute inflammation are virtually absent. These observations have led many to suggest that lack of blood flow in certain areas may predispose a person to tendinous rupture or chronic tendinitis. Areas of decreased vascular supply have been demonstrated in both the supraspinatus and Achilles tendons [105, 164]. Interestingly enough, rupture of these tendons occurs regularly in the areas of diminished blood supply [28]. Biopsy specimens of young patients with chronic tendinitis have revealed a change in the morphology of tenocytes adjacent to areas of collagen degeneration [28]. A more recent study [68] of a 25-year-old with spontaneous rupture of the Achilles tendon showed an area of degeneration marked by loss of collagen and an increase in mucopolysaccharide as well as water and glycoprotein content.

Clinical Considerations

Diagnosis. There are many common clinical presentations of acute and chronic pathologic processes involving tendons. These entities are common in the general population and in particular among athletes.

Trauma-Induced Conditions. Some conditions involve trauma-induced or chronic overload changes within the tendon substance. These conditions often involve degenerative changes without significant accompanying inflammatory changes and may be considered a form of tendinosis. These conditions cause symptoms of localized pain, especially when the tendon is placed under tension by muscle action or stretch. Often the degeneration occurs near the bony origin or insertion of the tendon. The following tendons are often involved:

1. *Wrist extensors.* The extensor carpi radialis brevis is involved near its origin on the lateral epicondyle. The condition is frequently termed lateral epicondylitis or tennis elbow.

2. *Patellar tendon.* Pathologic changes occur near the attachment of this tendon on the patella that lead to a painful condition called "jumper's knee" or infrapatellar tendinitis. It is particularly painful during periods of high tension in the tendon such as in jumping or landing from a jump.

3. *Posterior tibial tendon.* Injury and complete rupture may occur near the tarsal attachment. This injury is common in runners and dancers.

4. *Supraspinatus tendon.* This frequent tendinopathy occurs near the junction of the rotator cuff and the greater tuberosity. Changes can occur within the tendon causing pain in response to eccentric loading. This injury is common in eccentric loading during exercises typically performed by throwing athletes.

Peritendinitis. Conditions characterized by reactive and often inflammatory changes in the peritendinous structures were identified above as peritendinitis. Peritendinitis may occur with or without tendinosis. The following tendons are examples of those that may be associated with a peritendinitis:

1. *Achilles tendon.* The peritendinous tissue around the tendon may demonstrate significant fibrosis and inflammatory change.

2. *Rotator cuff tendons.* In addition to the tendinosis noted earlier under Classification of Overuse Injuries, significant involvement of the peritendinous tissue may occur with or without degenerative changes in the tendon. The subacromial bursa may be considered part of the peritendon of the rotator cuff, and its involvement is common in the well-known impingement syndrome. In this condition the symptoms often emanate from inflamed peritendinous tissue rather than from traction or tension within the tendon. For example, the shoulder pain experienced in impingement syndrome often occurs during complete rest and at night rather than in response to tendon loading.

3. *Flexor hallucis longus.* A common injury in dancers, peritendinitis of this tendon is characterized by pain from the peritendon at a site of injury near the tendon sheath posterior to the medial malleolus.

4. *Abductor pollicis longus.* Inflammation within the peritendon here is caused by a tight sheath in the first extensor compartment of the wrist.

The above examples demonstrate that changes can occur within the tendon, the peritendon, or both. Usually these conditions are characterized by pain. They may also lead to complete failure of a tendon with resulting loss of function. Complete failure of the Achilles tendon and the rotator cuff are perhaps the best known conditions of tendon failure. As discussed earlier, there is considerable evidence that tendon failure usually involves a tendon that has a preexisting degenerative change or a tenuous area of vascular supply [28, 105, 164].

Treatment of Problems of the Tendon and Peritendon. Most pathologic conditions involving tendinous and peritendinous structures are mild and will resolve spontaneously. In these conditions, time and changes in the routine of physical activity are often successful. Restriction of physical activity can include anything from rest to maintenance of vigorous exercise with avoidance of motions that cause pain. Nonsteroidal anti-inflammatory agents are useful for both their analgesic and anti-inflammatory effects.

Therapeutic exercises are also frequently prescribed as treatment for painful conditions of tendons. The added stimulus of training may help to strengthen the injured area or the surrounding normal tendon. Recently, eccentric exercises have been widely used with promising results. These exercises are usually designed to increase the strength of the tendon and are performed at a highly intense level. For example, traction injuries to the rotator cuff are often treated by vigorous strengthening of the shoulder abductors and external rotators. Athletes using overhead motions such as baseball pitchers and tennis players may benefit from a program of preventive exercise.

Corticosteroid injections also have a place in the management of tendon problems. Injections are best used to treat inflammatory involvement of the peritendinous structures. Direct injection into the tendons is usually avoided because glucocorticoids inhibit collagen biosynthesis and because the nature of the intratendinous pathology often involves no inflammation. When inflamed peritendinous structures are responsible for the pain, injections may be quite helpful. Injections of the subacromial bursa, the retrocalcaneal bursa, and the first dorsal compartment of the wrist (for DeQuervain's tenosynovitis) can be very useful clinically.

Complete rupture of tendons is not infrequent. In injuries occurring in athletes, these conditions are usually best treated by surgical repair. Rotator cuff tears are best treated acutely. Achilles tendon ruptures can be treated conservatively [23, 82, 118]. However, strength recovery may be better with surgery, and the risk of reinjury is less with surgery [19, 67, 74]. Good surgical repair will also diminish the need for immobilization and restricted activity.

Certain conditions involving the tendons create significant pain and may be treated surgically when they do not respond to conservative management. As discussed previously, it has recently been noted that cases of failed response to treatment of chronic overuse injuries usually involve pathology of the tendon. This pathology appears to involve a chronic degenerative process secondary to diminished vascular supply rather than acute inflammation. These conditions may involve pathology within the tendons; examples are tennis elbow and jumper's knee. Tennis elbow can be successfully treated by surgical removal of the abnormal portion of the tendon of origin of the wrist extensors. Similarly, jumper's knee may respond to excision of the abnormal portion of the patellar tendon near its attachment to the patella.

Summary

Pathologic involvement of tendons and peritendinous tissues is very common in sports medicine. Certain tendons and certain locations are involved frequently. An attempt should be made to understand the pathologic processes, at least insofar as which tissues are involved and how. This understanding will provide a more rational approach to treatment and prevention.

References

Skeletal Muscle

1. Abbot, B.C., and Aubert, X.M. Changes of energy in a muscle during very slow stretches. *Proc R Soc Lond [Biol]* 139:104–117, 1951.
2. Abbot, B.C., and Aubert, X.M. The force exerted by active striated muscle during and after change of length. *J Physiol* 117:77–86, 1952.
3. Abbot, B.C., and Lowy, J. Stress relaxation in muscle. *Proc R Soc Lond [Biol]* 146:281–288, 1956.
4. Abbot, B.C., and Ritchie, J.M. The onset of shortening in striated muscle. *J Physiol* 113:336–345, 1951.
5. Abraham, W.M. Factors in delayed muscle soreness. *Med Sci Sports Exerc* 9:11–20, 1977.
6. Almekinders, L.C., Garrett, W.E., and Seaber, A.V. Histopathology of muscle tears in stretching injuries. *Orthopaedic Transactions.* Presented at the Orthopaedic Research Society, Atlanta, GA, 1984.
7. Almekinders, L.C., Garrett, W.E., and Seaber, A.V. Pathophysiologic response to muscle tears in stretching injuries. *Orthopaedic Transactions.* Presented at the Orthopaedic Research Society, Atlanta, GA, 1984.
8. Andersen, P. Capillary density in skeletal muscle of man. *Acta Physiol Scand* 95:203–205, 1975.
9. Apple, D.V., O'Toole, J., and Annis, C. Professional basketball injuries. *Physician Sports Med* 10(11):81–86, 1982.
10. Armstrong, R.B. Mechanisms of exercise-induced delayed onset muscular soreness: A brief review. *Med Sci Sports Exerc* 16 (6):529–538, 1984.
11. Armstrong, R.B., Marum, P., Tullson, P., and Saubert C.W. Acute hypertrophic response of skeletal muscle to the removal of synergists. *J Appl Physiol* 46:835–842, 1975.
12. Aronen, J.A., Chronister, R., Ove, P., and McDevitt, E.R. Thigh contusions: Minimizing the length of time before return to full athletic activities with early immobilization in 120 cases of knee flexion. Presented at the 16th Annual Meeting of the American Orthopaedic Society for Sports Medicine, Sun Valley, Idaho, 16–19 July, 1990.
13. Asmussen, E. Observations on experimental muscular soreness. *Acta Rheum Scand* 2:109–116, 1956.
14. Bach, B.R., Warren, R.F., and Wickiewicz, T.L. Triceps rupture: A case report and literature review. *Am J Sports Med* 15(3):285–289, 1987.
15. Baldwin, K.M., Winder, W.W., and Holloszy, J.O. Adaptation of actomyosin ATPase in different types of muscle to endurance exercise. *Am J Physiol* 229:422–426, 1975.
16. Barany, M. ATPase activity of myosin correlated with speed of muscle shortening. *J Gen Physiol* (Suppl. Pt 20), 50:197–218, 1967.
17. Barany, M., and Close, R.I. The transformation of myosin in cross-innervated rat muscles. *J Physiol (Lond)* 213:455–474, 1971.
18. Beaulieu, J.E. Developing a stretching program. *Physician Sports Med* 9(11):59–65, 1981.
19. Besson, C., Rochcongar, P., Beauverger, Y., Dassonvill, J., Aubree, M., Catherine, M. Study of the valuations of serum muscular enzymes and myoglobin after maximal exercise test and during the next 24 hours. *Eur J Appl Physiol* 47:47–56, 1981.
20. Best, T.M., Glisson, R.R., Seaber, A.V., Garrett, W.E. The response of muscle-tendon units of varying architecture to cyclic passive stretching. Presented at the 35th Annual Meeting of the Orthopaedic Research Society, Las Vegas, 6–9 February, 1989.
21. Bosco, C., Montanari, G., Tarkka, I., Latteri, F., Cozzi, M., Iachelli, G., Faina, M., Colli, R., Dal Monte, A., LaRosa, M., Ribacchi, R.T., Giovenali, P., and Cortili, G. The effect of prestretch on mechanical efficiency of human skeletal muscle. *Acta Physiol Scand* 131:323–329, 1987.
22. Bosco, C., Vhtasalo, J.T., Komi, P.V., and Luhtanen, P. Combined effect of elastic energy and myoelectrical potentiation during stretch-shortening cycle exercise. *Acta Physiol Scand* 114:557–565, 1982.
23. Brewer, B.J. Mechanism of injury to the musculotendinous unit. Instr Course Lec 17:354–358, 1960.
24. Bronson, D.D., and Schachat, F.H. Heterogeneity of contractile proteins. *J Biol Chem* 257(7):3937–3944, 1982.
25. Brooke, M.H., and Engel, W.K. The histographic analysis of human biopsies with regard to fiber types. 1. Adult male and female. *Neurology* 19:221–223, 1969.

26. Brooke, M.H., and Engel, W.K. The histographic analysis of human biopsies with regard to fiber types. 2. Children's biopsies. *Neurology* 19:591–605, 1969.

27. Brooke, M.H., and Kaiser, K. Three "myosin adenosine triphosphatase" systems: The nature of their pH lability and sulfhydryl dependence. *J Histochem Cytochem* 18:670–672, 1970.

28. Brooke, M.H., and Kaiser, K. The use and abuse of muscle histochemistry. *Ann NY Acad Sci* 228:121–144, 1974.

29. Bryk, E., and Grantham, S.A. Shin splints: A chronic deep posterior ischemic compartmental syndrome of the leg? *Ortho Rev* 12(4):29–40, 1983.

30. Buchthal F., and Schmalbruch H. Motor unit of mammalian muscle. *Physiol Rev* 60:90–142, 1980.

31. Buller, A.J., Eccles, J.C., and Eccles, R.M. Interactions between motorneurons and muscles in respect of the characteristic speeds of their responses. *J Physiol (Lond)* 150:417–439, 1960.

32. Burke, R.E., and Edgerton, V.E. Motor unit properties and selective involvement in movement. *Exerc Sport Sci Rev* 3:31–81, 1975.

33. Burkett, L.N. Causative factors in hamstring strains. *Med Sci Sports Exerc* 2(1):39–45, 1970.

34. Butler, D.L., Noyes, F.R., and Grood, E.S. Measurement of the biomechanical properties of ligaments. *In*: Bahniuk, G., and Burstein, A. *CRC Handbook of Engineering and Biology,* Vol. 1B. Boca Raton, FL, CRC Press, 1978, pp. 279–314.

35. Carlson, F.D., and Wilkie, D.R. *Muscle Physiology.* Englewood Cliffs, NJ, Prentice Hall, 1974.

36. Cavagna, G.A. Storage and utilization of elastic energy in skeletal muscle. *Exerc Sports Sci Rev* 5:89–129, 1977.

37. Cavagna, G.A., Heglund, N.C., and Taylor, C.R. Mechanical work in terrestrial locomotion: Two basic mechanisms of minimizing energy expenditure. *Am J Physiol* 233(2):R243–R261, 1977.

38. Cavagna, G.A., Komaek, L., and Mazzoleni, S. The mechanics of sprint running. *J Physiol* 217:23–35, 1971.

39. Christensen, C.S., and Wiseman, D.C. Strength, the common variable in hamstring strain. *Athletic Trainer* 7(2):36–40, 1972.

40. Close, R.I. Dynamic properties of mammalian skeletal muscles. *Physiol Rev* 52:129–197, 1972.

41. Cummins, P., and Perry, S.V. Troponin I from human skeletal and cardiac muscles. *J Biochem* 171:251–259, 1978.

42. D'Ambrosia, R.D., Zelis, R.F., Chuinard, R.G., and Wilmore, J. Interstitial pressure measurements in the anterior and posterior compartments in athletes with shin splints. *Am J Sports Med* 5(3):127–131, 1977.

43. DeLuca, C.J. Control properties of motor units. *J Exp Biol* 115:125–136, 1985.

44. Denny-Brown, D. Seminars on neuromuscular physiology. *Am J Med* (Sept): 368–390, 1953.

45. DeVries, H.A. Quantitative electromyographic investigation of the spasm theory of muscle pain. *Am J Phys Med* 45(3):119–134, 1966.

46. Dubowitz, V., and Brooke, M.H. Muscle Biopsy: A modern approach. (*Major Problems in Neurology,* Vol.2.). London, W.B. Saunders, 1973.

47. Durig, M., Schuppisser, J.P., Gauer, E.F., and Muller, W. Spontaneous rupture of the gastrocnemius muscle. *Injury* 9:143–145, 1977–78.

48. Eisen, A., Karpati, G., Carpenter S., and Danton, J. The motor unit profile of the rat soleus in experimental myopathy and reinnervation. *Neurology* 24:878–884, 1974.

49. Edgerton, V.R., Smith, J.L., and Simpson, D.R. Muscle fibre type populations of human muscles. *Histochem J* 19:257–266, 1975.

50. Edstrom, L., and Kugelber E. Histochemical composition, distribution of fibers, and fatigueability of single motor units. *J Neurol Neurosurg Psychiatr* 31:424–433, 1968.

51. Edstrom, L., and Nystrom, B. Histochemical types and sizes of fibres in normal human muscles. *Acta Neurol Scand* 45:257–269, 1969.

52. Ekstrand, J., and Gillquist, J. The frequency of muscle tightness and injuries in soccer players. *Am J Sports Med* 10:75–78, 1982.

53. Ekstrand, J., Gillquist, J., and Liljedahl, S.O. Prevention of soccer injuries: Supervision by doctors and physiotherapists. *Am J Sports Med* 11:116–120, 1983.

54. Elftman, H. Biomechanics of muscle. *J Bone Joint Surg* 48A:363, 1966.

55. Engel, W.K. The essentiality of histo- and cytochemical studies of skeletal muscle in the investigation of neuromuscular disease. *Neurology* 12:778–794, 1962.

56. Faulkner, J.A. New perspectives in training for maximum performance. *JAMA* 205:741–746, 1986.

57. Feinstein, B., Lindegard, B., Nyman, E., and Wohlfart, G. Morphologic studies of motor units in normal human muscles. *Acta Anat* 23:127–142, 1955.

58. Fitts, R.H., Winder, W.W., Brooke, M.H., Kaiser, K.K., and Holloszy, J.O. Contractile, biochemical, and histochemical properties in thyrotoxic rat soleus muscle. *Am J Physiol* 238 (*Cell Physiol* 7) C15–C20, 1980.

59. Fornage, B.D., Tokuche, D.H., Segal, P., and Rifkin, M.D. Ultrasonography in the evaluation of muscular trauma. *J Ultrasound Med* 2:549–554, 1983.

60. Friden, J., Seger, J., and Ekblom, B. Sublethal muscle fibre injuries after high-tension anaerobic exercise. *Eur J Appl Physiol* 57:360–368, 1988.

61. Friden, J., Sjostrom, M., and Ekblom, B. A morphological study of delayed muscle soreness. *Experientia* 37:506, 1981.

62. Friden, J., Sjostrom, J., and Ekblom, B. Myofibrillar damage following intense eccentric exercise in man. *Int J Sports Med* 4:170–176, 1983.

63. Garfin, S.R., Mubarak, S.J., and Owen, C.A. Exertional anterolateral compartment syndrome. *J Bone Joint Surg.* 59A:404, 1977.

64. Garfin, S.R., Tipton, C.M., Mubarak, S.J., Woo, S. L-Y., Hargens, A.R., and Akeson, W.H. Role of fascia in maintenance of muscle tension and pressure. *J Appl Physiol* 51(2):317–320, 1981.

65. Garnett R.A.F., O'Donovan, M.J., Stephens J. A., and Taylor A. Motor unit organization of human medial gastrocnemius. *J Physiol (Lond)* 287:33–43, 1978.

66. Garrett, W.E., Jr. Injuries to the muscle-tendon unit. *Instr Course Lect* 37:275–282, 1988.

67. Garrett, W.E., Jr., Almekinders, L., and Seaber, A.V. Biomechanics of muscle tears in stretching injuries. *Trans Ortho Res Soc* 9:384, 1984.

68. Garrett, W.E., Jr., Califf, J.C., and Bassett, F.H. Histochemical correlates of hamstring injuries. *Am J Sports Med* 12(2):98–103, 1984.

69. Garrett, W.E., Jr., Nikolaou, P.K., Ribbeck, B.M., Glisson, R.R., and Seaber, A.V. The effect of muscle architecture on the biomechanical failure properties of skeletal muscle under passive extension. *Am J Sports Med* 16(1):7–12, 1988.

70. Garrett, W.E., Jr., Rich, F.R., Nikolaou, P., and Vogler, J.B., III. Computed tomography of hamstring muscle strains. *Med Sci Sports Exerc* 21(5):506–514, 1989.

71. Garrett, W.E., Jr., Safran, M.R., Seaber, A.V., Glisson, R.R., and Ribbeck, B.M. Biomechanical comparison of stimulated and nonstimulated skeletal muscle pulled to failure. *Am J Sports Med* 15(5):448–454, 1988.

72. Garrett, W.E., Jr., Seaber, A.V., Bokswich, J., Urbaniak, J.R., and Goldner, J.L. Recovery of skeletal muscle following laceration and repair. *J Hand Surg* 9a(5):683–692, 1984.

73. Garrett, W.E., Jr., and Tidball, J. Myotendinous junction: Structure and failure. *In* Woo, S.L.-Y, and Buckwalter, J.A. (Eds.), *Injury and Repair of Musculoskeletal Soft Tissues.* American Academy of Orthopaedic Surgeons and National Institute of Arthritis and Musculoskeletal and Skin Diseases, 1987.

74. Garrick, J.G.K, and Requa, R.K. Epidemiology of women's gymnastics injuries. *Am J Sports Med* 8(4):261–264, 1980.

75. Gauthier, G.F. Skeletal muscle fiber types. *In* Engel, A.G., and Banker, B.Q. (Eds.), *Myology,* Vol. 1. New York, McGraw-Hill, 1986, pp. 255–284.

76. Glick, J.M. Muscle strains: Prevention and treatment. *Physician Sports Med* 8:73–77, 1980.

77. Gollnick, P.D., Armstrong, R.B., Saltin, B., Saubert, C.W. IV., Sembrowich, W.L., and Shepherd, R.E. Effect of training on enzyme activity and fiber composition of human skeletal muscle. *J Appl Physiol* 34(1):107–111, 1973.

78. Gollnick, P.D., Armstrong, R.B., Saubert C.W. IV, Piehl, K., and Saltin, B. Enzyme activity and fiber composition in skeletal muscle of untrained and trained men. *J Appl Physiol* 33:312–319, 1972.

79. Gollnick, P.D., and Matoba, H. The muscle fiber composition of skeletal muscle as a predictor of athletic success. *Am J Sports Med* 12(3):212–217, 1984.

80. Gollnick, P.D., and Saltin, B. Skeletal muscle physiology. *In* Teitz, C.C. (Ed.), *Scientific Foundations in Sports Medicine.* Philadelphia, B.C. Decker, 1989.

81. Gollnick, P.D., Timson, B.F., Moore, R.L., et al: Muscular enlargement and number of fibers in skeletal muscle of rats. *J Appl Physiol* 50:936–943, 1981.

82. Gonyea, W., Ericson, G.C., and Bonde-Peterson, F. Skeletal muscle fiber splitting induced by weight-lifting exercise in cats. *Acta Physiol Scand* 99:105–109, 1980.

83. Grana, W.A., and Schelberg-Karnes, E. How I manage deep muscle bruises. *Physician Sports Med* 11(16):123–125, 1983.

84. Green, H.J., Klug, G.A., Reichmann, H., Seedorf, U., Wiehrer, W., and Pette, D. Exercise-induced fibre type transitions with regard to myosin, parvalbumin, and sarcoplasmic reticulum in muscles of the rat. *Pflugers Arch* 400:432–438, 1984.

85. Green, H.J., Reichmann, H., and Pette, D. Fibre type specific transformations in the enzyme activity pattern of rat vastus lateralis muscle by prolonged endurance training. *Pflugers Arch* 399:216–222, 1983.

86. Gross, J., Highberger, J.H., and Schmitt, F.O. Extraction of collagen from connective tissue by neutral salt solutions. *Proc Nat Acad Sci USA* 41:1, 1955.

87. Hargens, A.R., and Akeson, W.H. *Pathophysiology of the Compartment Syndrome.* Philadelphia, W.B. Saunders, 1981.

88. Heiser, T.M., Weber, J., Sullivan, G., Clare, P., and Jacobs, R.R. Prophylaxis and management of hamstring muscle injuries in intercollegiate football. Presented at the American Orthopaedic Society of Sports Medicine, Atlanta, GA, 8 February, 1984.

89. Henneman, E. Relation between size of neurons and their susceptibility to discharge. *Science* 126:1345–1347, 1957.

90. Henneman, E., Clamann, H.P., Gillies, J.D., and Skinner, R.D. Rank order of motoneurons within a pool: Law of combination. *J Neurophysiol* 37:1338–1349, 1974.

91. Henneman, E., and Olson, C.B. Relations between structure and function in the design of skeletal muscles. *J Neurophysiol* 28:581–598, 1965.

92. Henneman, E., Somjen, G., and Carpenter, D.O. Excitability and inhibitability of motoneurons of different sizes. *J Neurophysiol* 28:599–620, 1965.

93. Henneman, E., Somjen, G., and Carpenter, D.O. Functional significance of cell size in spinal motoneurons. *J Neurophysiol* 28:560–580, 1965.

94. Hill, A.V. The heat of shortening and the dynamic constants of muscle. *Proc R Soc Lond [Biol]* 126:136–195, 1938.

95. Hill, A.V. The onset of contraction. *Proc R Soc Lond [Biol]* 136:242–254, 1949.

96. Hill, A.V. The transition from rest to full activity in muscle: The velocity of shortening. *Proc R Soc Lond [Biol]* 138:329–338, 1951.

97. Hill, A.V. The mechanics of active muscle. *Proc R Soc Lond [Biol]* 141:104–117, 1953a.

98. Hill, A.V. Chemical change and mechanical response in stimulated muscle. *Proc R Soc Lond [Biol]* 141:314–320, 1953b.

99. Hill, A.V. The "instantaneous" elasticity of active muscle. *Proc R Soc Lond [Biol]* 141:161–178, 1953c.

100. Holloszy, J.O. Biochemical adaptations in muscle. Effects of exercise on mitochondrial oxygen uptake and respiratory enzyme activity in skeletal muscle. *J Biol Chem* 242:2278–2282, 1967.

101. Holly, R.G., Barnett, J.R., Ashmorn, C.R., Taylor, R.G., and Mulé, P.A. Stretch-induced growth in chicken wing muscles: A new model for stretch hypertrophy. *Am J Physiol* 238:C62–C71, 1980.

102. Holt, L.E. *Scientific Stretching for Sport (3-S).* Halifax, Nova Scotia, Sport Research, 1981.

103. Hough, T. Ergographic studies in muscular soreness. *Am J Physiol* 7:76–92, 1902.

104. Howald, H., Hoppeler, H., Claasen, H., Mathieu, O., and Straub, R. Influences of endurance training on the ultrastructural composition of the different muscle fiber types in humans. *Pflugers Arch* 403:369–376, 1985.

105. Hubley-Kozey, C.L., and Standish, W.D. Separating fact from fiction about a common sports activity: Can stretching prevent athletic injuries? *J Musc Skel Med* 25–32, 1984.

106. Hughston, J.C., Whatley, G.S., and Stone, M.M. Myositis ossificans traumatica (myo-osteosis). *South Med J* 55:1167–1170, 1962.

107. Huxley, A.F. Muscle structure and theories of contraction. *Prog Biophys Biophys Chem* 7:255–318, 1957.

108. Huxley, A.F. A note suggesting that the cross-bridge attachment during muscle contraction may take place in two stages. *Proc R Soc Lond [Biol]* 183:83–86, 1973.

109. Huxley, A.F. Muscular contraction. *J Physiol* 243:1–43, 1974.

110. Huxley, A.F. *Reflections of Muscle.* Liverpool, Liverpool University Press, 1979.

111. Huxley, A.F., and Julian, F.J. Speed of unloaded shortening in frog striated muscle fibres. *J Physiol* 177:60P–61P, 1964.

112. Huxley, A.F., and Niedergerke, R. Structural changes in muscle during contraction: Interference microscopy of living muscle fibres. *Nature* 173:971–973, 1954.

113. Huxley, A.F., and Peachey, L.D. The maximum length for contraction in vertebrate striated muscle. *J Physiol* 156:150–165, 1961.

114. Huxley, A.F., and Simmons, R.M. Proposed mechanism of force generation in striated muscle. *Nature* 233:533–538, 1971.

115. Huxley, A.F., and Taylor, R.E. Local activation of striated muscle fibres. *J Physiol* 144:426–441, 1958.

116. Huxley, H.E. Electron microscope studies of the organization of the filaments in striated muscle. *Biochim Biophys Acta* 12:387–394, 1953.

117. Huxley, H.E. Electron microscope studies on the structure of natural and synthetic protein filaments from striated muscle. *J Mol Biol* 7:281–308, 1963.

118. Huxley, H.E. Structural arrangements and the contraction mechanism in striated muscle. *Proc R Soc Lond [Biol]* 160:442–448, 1964.

119. Huxley, H.E. The mechanism of muscular contraction. *Science* 164:1356–1366, 1969.

120. Huxley, H.E. The structural basis of muscular contraction. *Proc R Soc Lond* 178:131–149, 1971.

121. Huxley, H.E., and Brown, W. The low-angle X-ray diagram of vertebrate striated muscle and its behavior during contraction and rigor. *J Mol Biol* 30:383–434, 1967.

122. Huxley, H.E., and Hanson, J. Changes in the cross-striations of muscle during contraction and stretch and their structural interpretation. *Nature* 173:973–976, 1954.

123. Ianuzzo, C.D., and Chen, V. Metabolic character of hypertrophied rat muscle. *J Appl Physiol* 46:738–742, 1979.

124. Ianuzzo, D., Patel, P., Chen, V., O'Brien, P., and Williams, C. Thyroidal trophic influence on skeletal muscle myosin. *Nature* 270:74–76, 1977.

125. Inman, V.T., Ralston, H.J., and Todd, F. *Human Walking.* Baltimore, Williams & Wilkins, 1981.

126. Jackson, D.W., and Feagin, J.A. Quadriceps contusions in young athletes. *J Bone Joint Surg* 55A:95–101, 1973.

127. Johnson, M.A., Polgar, J., Weightman, D., and Appleton, D. Data on the distribution of fibre types in thirty-six

human muscles: An autopsy study. *J Neurol Sci* 18:111–129, 1973.

128. Jones, D.A., Newham, D.J., Round, J.M., and Tolfree, S.E.J. Experimental human muscle damage: Morphologic changes in relation to other indices of damage. *J Physiol* 375:435–448, 1986.

129. Julian, F.J., and Morgan, D.L. The effect on tension of non-uniform distribution of length changes applied to frog muscle fibres. *J Physiol* 293:379–392, 1979.

130. Kantorowsk, P.G., Hiller, W.D.B., Garrett, W.E., Smith, R., and O'Toole, M. Cramping studies in 2600 endurance athletes. Presented at the 37th Annual Meeting of the American College of Sports Medicine, Salt Lake City, Utah, 24 May, 1990.

131. Kennedy, J.C., and Roth, J.H. Major tibial compartment syndromes following minor athletic trauma. *Am J Sports Med* 7(3):201–203, 1979.

132. Komi, P.V., Rusko, H., Vos, J., Vihko, V. Anaerobic performance capacity in athletes. *Acta Physiol Scand* 100:107–114, 1977.

133. Korn, E.D. Actin polymerization and its regulation by proteins from nonmuscle cells. *Physiol Rev* 62:672–737, 1982.

134. Kovanen, V., Suominen, H., and Heikkinen, E. Connective tissue of "fast" and "slow" skeletal muscle in rats—effects of endurance training. *Acta Physiol Scand* 108:173–180, 1980.

135. Kraus, H. The use of surface anesthesia in the treatment of painful motion. *JAMA* 116:2582–2583, 1941.

136. Kraus, H. Evaluation and treatment of muscle function in athletic injury. *Am J Surg* 98:353–362, 1959.

137. Krejci, V., and Koch, P. *Muscle and Tendon Injuries in Athletes.* Chicago, Year Book, 1979.

138. Kugelberg, E. Histochemical composition, contraction speed, and fatiguability of rat soleus motor units. *J Neurol Sci* 20:177–198, 1973.

139. Kugelberg, E., Edstrom, L., and Abruzzese, M. Mapping of motor units in experimentally reinnervated rat muscle. *J Neurol Neurosurg Psychiat* 33:319–329, 1970.

140. Kugelberg, E., and Lindegren, B. Transmission and contraction fatigue of rat motor units in relation to succinate dehydrogenase activity of motor unit fibres. *J Physiol (Lond)* 288:285–300, 1979.

141. Layzer, O. *In* Vinken, P.J., and Bruyn, G.W. (Eds.), *Handbook of Clinical Neurology.* New York, Elsevier, 1986, pp. 295–316.

142. Lomo, *In* Pette, D. (Ed.), *Plasticity of Muscle.* New York, Walter de Gruyter, 1980, pp. 297–309.

143. Lowey, S., and Risby, D. Light chains from fast and slow muscle myosins. *Nature* 234:81–85, 1971.

144. Mann, R.A., and Hagy, J. Biomechanics of walking, running, and sprinting. *Am J Sports Med* 8(5):345–350, 1980.

145. Margaria, R. Positive and negative work performances and their efficiencies in human locomotion. *In* Cumming, G.R., Snidal, D., Taylor, A.W. (Eds.), *Environmental Effects on Work Performance.* Edmonton, Canadian Association of Sports Sciences, 1972, pp. 215–228.

146. Martin, W.D., and Romond, E.M. Effects of chronic rotation and hypergravity on muscle fibers of soleus and plantaris muscles of the rat. *Exp Neurol* 49:758–771, 1975.

147. Maughan, R.J. Exercise-induced muscle cramp: A prospective biochemical study in marathon runners. *J Sport Sci* 4(1):31–34, 1986.

148. McDermott, A.G.P., Marble, A.E., Yabsley, R.H., and Phillips, D. Monitoring dynamic anterior compartment pressures during exercise: A new technique using the STIC catheter. *Am J Sports Med* 10(2):83–89, 1982.

149. McMahon, T.A. *Muscles, Reflexes, and Locomotion.* Princeton, NJ, New University Press, 1984.

150. McMaster, P.E. Tendon and muscle ruptures: Clinical and experimental studies on the causes and location of subcutaneous ruptures. *J Bone Joint Surg* 15:705–722, 1933.

151. Miller, W.A. Rupture of the musculotendinous junction of the medial head of the gastrocnemius muscle. *Am J Sports Med* 5(5):191–193, 1977.

152. Milner-Brown, H.S., Stein, R.B., and Yemmon, R. Changes in firing rate of human motor units during linearly changing voluntary contractions. *J Physiol (Lond)* 230:371–390, 1973.

153. Moss, H.K., and Hermann, L.G. Night cramps in human extremities. *Am Heart J* 35:403–408, 1948.

154. Mubarak, S.J. Etiologies of compartment syndromes. *In* Mubarak, S.J., and Hargens, A.R. (Eds.), *Compartment Syndromes and Volkmann's Contracture.* Philadelphia, W.B. Saunders, 1981.

155. Mubarak, S., and Hargens, A. Exertional compartment syndromes. *In* American Academy of Orthopaedic Surgeons Symposium, *The Foot and Leg in Running Sports.* St. Louis, C.V. Mosby, 1982.

156. Muller, W. *The Knee: Form, Function, and Ligament Reconstruction.* New York, Springer-Verlag, 1982.

157. Needham, D.M. *Machina Carnis, The Biochemistry of Muscular Contraction in Its Historical Development.* Cambridge, Cambridge University Press, 1971.

158. Nicholas, J.A. Injuries to knee ligaments: Relationship to looseness and tightness in football players. *JAMA* 212(13):2236–2239, 1970.

159. Nikolaou, P.K., Macdonald, B.L., Glisson, R.R., Seaber, A.V., Garrett, W.E., Jr. Biomechanical and histological evaluation of muscle after controlled strain injury. *Am J Sports Med* 15(1):9–14, 1987.

160. Noble, E.G., Dabrowski, B.L., and Ianuzzo, C.D. Myosin transformation in hypertrophied rat muscle. *Pflugers Arch* 396:260–262, 1983.

161. Obremskey, W.T., Seaber, A.V., Ribbeck, B.M., and Garrett, W.E., Jr. Biomechanical and histological assessment of a controlled muscle strain injury treated with piroxicam. *Trans Orthop Res Soc* 13:338, 1988.

162. O'Donoghue, D.H. *Treatment of Athletic Injuries.* Philadelphia, W.B. Saunders, 1984, pp. 51–63.

163. Padykula, H.A., and Herman, E. The specificity of the histochemical method for adenosine triphosphatase. *J Histochem Cytochem* 3:170–183, 1955.

164. Parrow, A., and Samuelsson, S.M. Use of chloroquine phosphate—A new treatment for spontaneous leg cramps. *Acta Med Scand* 181(2):237–244, 1967.

165. Peterson, L., and Renstrom, P. Preventive measures. *In* Grana, W.A. (Ed.), *Sports Injuries: Their Prevention and Treatment.* Chicago, Year Book, 1986, pp. 86–104.

166. Peterson, L., and Stener, B. Old rupture of the adductor longus muscle. *Acta Orthop Scand* 47:653–657, 1976.

167. Pette, D. (Ed.). *Plasticity of Muscle.* New York, Walter de Gruyter, 1980.

168. Prockop, D.J., and Sjoerdsma, A. Significance of urinary hydroxyproline in man. *J Clin Invest* 40:843–849, 1961.

169. Puranen, J. The medial tibial syndrome: Exercise ischaemia in the medial fascial compartment of the leg. *J Bone Joint Surg* 56B:712–715, 1974.

170. Radin, E.L., Simon, S.R., Rose, R.M., and Paul, I.L. *Practical Biomechanics for the Orthopaedic Surgeon.* New York, Wiley, 1979.

171. Rask, M.R., and Lattig, G.J. Traumatic fibrosis of the rectus femoris muscle. *JAMA* 221(3):368–369, 1972.

172. Reichmann, H., Srihari, T., and Pette, D. Ipsi- and contralateral fibre transformations by cross-reinnervation: A principle of symmetry. *Pflugers Arch* 397:202–208, 1983.

173. Reneman, R.S. The anterior and the lateral compartment syndrome of the leg due to intensive use of muscles. *Clin Orthop* 113:69–80, 1975.

174. Romero-Herrera, A.E., Nasser, S., and Lieska, N.G. Heterogeneity of adult human striated muscle tropomyosin. *Muscle Nerve* 5:713–718, 1982.

175. Rooser, B. Quadriceps contusion with compartment syndrome: Evacuation of hematoma in 2 cases. *In* Anderson, J.L., George, F.J., Shephard, R.J., Torg, J.S., and Eichner, E.R. *Year Book of Sports Medicine.* Chicago, Year Book, 1988.

176. Rothwell, A.G. Quadriceps hematoma: A prospective clinical study. *Clin Orthop Rel Res* 171:97–103, 1982.

177. Roy, R.R., Meadows, I.D., Baldwin, K.M., and Edgerton, V.R. Functional significance of compensatory overloaded rat fast muscle. *J Appl Physiol* 52(2):473–478, 1982.

178. Ryan, A.J. Quadriceps strain, rupture, and charlie horse. *Med Sci Sports Exerc* 1(2):106–111, 1969.

179. Safran, M.R., Garrett, W.E., Jr., Seaber, A.V., Glisson, R.R., and Ribbeck, B.M. The role of warmup in muscular injury prevention. *Am J Sports Med* 16(2):123–129, 1988.

180. Salmons, S., and Sreter, F.A. Significance of impulse activity in the transformation of skeletal muscle type. *Nature* 263:30–34, 1976.

181. Salmons, S., and Vrbova, G. The influence of activity on some contractile characteristics of mammalian fast and slow muscles. *J Physiol (Lond)* 210:L38–L49, 1969.

182. Saltin, B., and Gollnick, P.D. Skeletal muscle adaptability: Significance for metabolism and performance. *In* Peachy, L.D., Adrian, R. and Geiger, S.R. (Eds.), *Handbook of Physiology. Section 10: Skeletal Muscle.* Baltimore, Williams & Wilkins, 1983, pp. 555–631.

183. Saltin, B., Henricksson, J., Nygaard, E., and Anderson, P. Fiber types and metabolic potentials of skeletal muscles in sedentary man and endurance runners. *Ann NY Acad Sci* 310:3–29, 1977.

184. Samaha, F.J., Guth, L., and Albers, R.W. Differences between slow and fast muscle myosin. Adenosine triphosphatase activity and release of associated proteins by chloromercuriphenylsulfonate. *J Biol Chem* 245:219–224, 1970.

185. Schaub, M.C., and Watterson, J.G. Control of the contractile process in muscle. *Trends Pharmacol Sci* 2(10):279–282, 1981.

186. Schwane, J.A., Johnson, S.R., Vandenakker, C.B., and Armstrong, R.B. Delayed-onset muscular soreness and plasma CPK and LDK activities after downhill running. *Med Sci Sports Exerc* 15(1):51–56, 1983.

187. Schwane, J.A., Watrous, B.G., Johnson, S.R., and Armstrong, R.B. *Physician Sports Med* 11(3):124–131, 1983.

188. Solveborn, S-A. *The Book About Stretching.* New York, Japan Publications, 1983.

189. Sreter, F.A., Seidel, J.C., and Gergely, J. Studies on myosin from red and white skeletal muscles of the rabbit. I. Adenosine triphosphatase activity. *J Biol Chem* 241:5772–5776, 1966.

190. Stauber, W.T. Eccentric action of muscles: Physiology, injury and adaptation. *Exerc Sport Sci Rev* 17:157–185, 1989.

191. Stewart, W.K., Fleming, L.W., and Manuel, M. Muscle cramps during maintenance hemodialysis. *Lancet* 1:1049–1051, 1972.

192. Sudmann, E. The painful chronic anterior lower leg syndrome. *Acta Orthop Scand* 50:573–581, 1979.

193. Symeonides, P.P. Isolated traumatic rupture of the adductor longus muscle of the thigh. *Clin Orthop Rel Res* 88:64–66, 1972.

194. Taylor, D.C., Dalton, J.D., Seaber, A.V., and Garrett, W.E., Jr. Viscoelastic properties of muscle-tendon units. The biomechanical effects of stretching. *Am J Sports Med* 18(3):300–309, 1990.

195. Teitz, C.C. *Scientific Foundations of Sports Medicine.* Philadelphia, B.C. Decker, 1989.

196. Tesch, P., and Karlsson, J. Isometric strength performance and muscle fibre type distribution in man. *Acta Physiol Scand* 103:47–51, 1978.

197. Tesch, P.A., Thorsson, A., and Kaiser, P. Muscle capillary supply and fiber type characteristics in weight and power lifters. *Am J Physiol* 35–38, 1984.

198. Thorstensson, A. Observations on strength training and detraining. *Acta Physiol Scand* 100:491–493, 1977.

199. Thorstensson, A., Grimby, G., and Karlsson, J. Force-velocity relations and fiber composition in human knee extensor muscles. *J Appl Physiol* 40(1):12–16, 1976.

200. Thorstensson, A., Hutten, B., vonDobelin, W., and Karlsson, J. Effect of strength training on enzyme activities and fibre characteristics in human skeletal muscle. *Acta Physiol Scand* 96:392–399, 1976.

201. Tipton, C.M., Schild, R.J., and Tomanek, R.J. Influence of physical activity on the strength of knee ligaments in rats. *Am J Physiol* 212:283–287, 1966.

202. Veith, R.G. Recurrent compartmental syndromes due to intensive use of muscles. *In* Matsen, F.A. (Ed.), *Compartmental Syndromes.* New York. Grune & Stratton, 1980.

203. Viidik, A. The effect of training on the tensile strength of isolated rabbit tendons. *Scand Plast Reconstr Surg* 1:141–147, 1967.

204. Viidik, A. Functional properties of collagenous tissues. *Int Rev Connect Tissue Res* 6:127–215, 1973.

205. Weeds, A. *In* Pette, D. (Ed.), *Plasticity of Muscle.* New York, Walter de Guyter 1980. pp. 607–615.

206. Whitesides, T.E., Haney, T.C., Morimoto, K., and Harada, H. Tissue pressure measurements as a determinant for the need of fasciotomy. *Clin Orthop Rel Res* 113:43–51, 1975.

207. Wickiewicz, T.L., Roy, R.R., Powell, P.L., and Edgerton, V.R. Muscle architecture of the human lower limb. *Clin Orthop Rel Res* 179:275–283, 1983.

208. Wiktorsson-Moller, M., Oberg, B., Ekstrand, J., and Gillquist, J. Effects of warming up, massage, and stretching on range of motion and muscle strength in the lower extremity. *Am J Sports Med* 11(4):249–252, 1983.

209. Woo, S.L-Y., and Buckwalter, J.S. (Eds.). *Injury and Repair of the Musculoskeletal Soft Tissues.* Park Ridge, Ill, American Academy of Orthopaedic Surgeons, 1988.

210. Zarins, B., and Ciullo, J.V. Acute muscle and tendon injuries in athletes. *Clin Sports Med* 2(1):167–182, 1983.

Tendon

1. Abrahams, M. Mechanical behaviour of tendon in vitro: A preliminary report. *Med Biol Eng* 5:433–443, 1967.

2. Akeson, W.H. An experimental study of joint stiffness. *J Bone Joint Surg* 43A:1022–1034, 1961.

3. Akeson, W.H., Wor, S.-L.-Y., and Frank, C.B. The biology of ligaments. *In* Hunter, L.Y., and Funk, F.J. (Eds.), *Rehabilitation of the Injured Knee,* Vol. 3. St. Louis, C.V. Mosby, 1984.

4. Akeson, W.H., Amiel, D., and LaViolette, D. The connective-tissue response to immobility: A study of the chondroitin-4 and -6 sulfate and dermatan sulfate changes in periarticular connective tissue of control and immobilized knees of dogs. *Clin Orthop* 51:183–197, 1967.

5. Akeson, W.H., Frank, C.B., Amiel, D., and Woo, SL-Y. Ligament biology and biomechanics. *In* Finnerman, G. (Ed.), *American Academy of Orthopaedic Surgeons Symposium on Sports Medicine.* St. Louis, C.V. Mosby, 1985, pp 11–51.

6. Amiel, D., Abel, M.F., Kleiner, J.B., and Akeson, W.H. Synovial fluid nutrient delivery in the diarthrial joint: An analysis of rabbit knee ligaments. *J Orthop Res* 4:90–95, 1986.

7. Amiel, D., Frank, C.B., Harwood, F.L., Fronek, J., and Akeson, W.H. Tendons and ligaments: A morphological and biochemical comparison. *J Orthop Res* 1(3):257, 1984.

8. Amiel, D., Woo, SL-Y, Harwood, F.L., and Akeson, W.H. The effect of immobilization on collagen turnover in connective tissue: A biochemical-biomechanical correlation. *Acta Orthop Scand* 53:325–332, 1982.

9. Arner, O., and Lindholm, A. Subcutaneous rupture of the Achilles tendon. *Acta Chir Scand* Supp 239, 1959.

10. Arner, O., Lindholm, A., and Orell, S.R. Histologic changes with subcutaneous rupture of the Achilles tendon. *Acta Chir Scand* 116:484–490, 1958.

11. Azar, C.A., Fleegler, E.J., and Culver, J.E. Dynamic anatomy of the flexor pulley system of the fingers and thumb [abstract]. *J Hand Surg* 9A:595, 1984.

12. Bailey, A.J., Robins, S.P., and Balian, G. Biological significance of the intermolecular crosslinks of collagen. *Nature* 251:105–109, 1974.

13. Banes, A.J., Enterline, D., Bevin, A.G., Salisbury, R.E.: Effects of trauma and partial devascularization on protein

synthesis in the avian flexor profundus tendon. *J Trauma* 21:505–512, 1981.

14. Banes, A.J., Link, G.W., Bevin, A.G., Peterson, H.D., Gillespie, Y., Bynum, D., Watts, S., and Dahners, L. Tendon synovial cells secrete fibronectin in vivo and in vitro. *J Orthop Res* 6:73–82, 1988.

15. Banes, A.J., Link, G.W., Peterson, H.D., Yamauchi, M., and Mechanic, G.L.: Temporal changes in collagen cross-link formation at the focus of trauma and at sites distant to a wound. *In* Gruber, D., Walker, R.I., Macvittie, T.J., and Conklin, J.J. Eds., *The Pathophysiology of Combined Injury and Trauma.* New York, Academic Press, 1987, pp. 257–273.

16. Barfred, T. Experimental rupture of the Achilles tendon: Comparison of experimental sutures in rats of different ages and living under different conditions. *Acta Orthop Scand* 42:406–428, 1971.

17. Becker, H. Primary repair of flexor tendons in the hand without immobilization: Preliminary report. *Hand* 10:37–47, 1978.

18. Bloom, W., and Fawcett, D.W. *A Textbook of Histology* (8th ed.) Philadelphia, W.B. Saunders, 1962, p. 105.

19. Bradley, J.P., and Tibone, J.E. Percutaneous and open surgical repairs of Achilles tendon ruptures. A comparative study. *Am J Sports Med* 18(2):188–195, 1990.

20. Branwood, A.W. The fibroblast. *Int Rev Connect Tissue Res* 1:1, 1963.

21. Bryant, W.M., and Weeks, P.M. Secondary wound tensile strength gain: A function of collagen and mucopolysaccharide interaction. *Plast Reconstr Surg* 39:84–91, 1967.

22. Bunnell, S. Repair of tendons in the fingers and description of two new instruments. *Surg Gynecol Obstet* 26:103–110, 1918.

23. Carden, D.G., Noble, J., Chalmers, J., Lunn, P., and Ellis, J. Rupture of the calcaneal tendon. Early and late management. *J Bone Joint Surg* 69B:416–420, 1987.

24. Carlstedt, C.A. Mechanical and chemical factors in tendon healing. Effects of indomethacin and surgery in the rabbit. *Acta Orthop Scand* Suppl 224:1–75, 1987.

25. Carlstedt, C.A., Madsen, K., and Wredmark, T. The influence of indomethacin on tendon healing. A biomechanical and biochemical study. *Arch Orthop Trauma Surg* 105(6):332–336, 1986.

26. Carstam, N. The effect of cortisone on the formation of tendon adhesions on tendon healing. *Acta Chir Scand* Suppl. 182, 1953.

27. Chaplin, D.M., and Greenlee, T.K., Jr. The development of human digital tendons. *J Anat* 120:253–274, 1975.

28. Clancy, W.G. Tendon trauma and overuse injuries. *In* Leadbetter, W.B., Buckwalter, J.A., and Gordon, S.C., Eds., *Sports Induced Inflammation.* Park Ridge, IL, American Academy of Orthopaedic Surgeons, 1989, pp. 609–618.

29. Clancy, W.G., Neidhart, D., and Brand, R.L. Achilles tendonitis in runners: A report of five cases. *Am J Sports Med* 4:46–57, 1976.

30. Cooney, W.P., An, K.N., and Chao, E.Y.S. Direct measurement of tendon forces in the hand. *Trans Orthop Res Soc* 11:53, 1986.

31. Cooper, R.R., and Misol, S. Tendon and ligament insertion. *J Bone Joint Surg* 52A:1–19, 1970.

32. Cowan, M.A., and Alexander, S. Simultaneous bilateral rupture of Achilles tendons due to triamcinolone. *Br Med J* 1:1658, 1961.

33. Cronkite, A.E. The tendon strength of human tendons. *Anat Rec* 64:173–186, 1936.

34. Dale, W.C. A composite materials analysis of the structure, mechanical properties, and aging of collagenous tissue. Ph.D. thesis, Case Western Reserve University, Cleveland, Ohio, 1981.

35. Davidson, J.M., Klagsbrun, M., Hill, K.E., Buckley, A., Sullivan, R., Brewer, P.S., and Woodward, S.C. Accelerated wound repair, cell proliferation, and collagen accumulation are produced by a cartilage-derived growth factor. *J Cell Biol* 100:1219–1227, 1985.

36. Davidsson, L., and Salo, M. Pathogenesis of subcutaneous tendon ruptures. *Acta Chir Scand* 135:209–212, 1969.

37. Distefanco, V.J., and Nixon, J.E. Achilles tendon rupture: Pathogenesis, diagnosis and treatment by a modified pullout wire technique. *J Trauma* 12:671–677, 1972.

38. Douglas, L.G., Jackson, S.H., and Lindsay, W.K. The effects of dexamethasone, norethandrole, promethazine and a tension-relieving procedure on collagen synthesis in healing flexor tendons as estimated by tritiated proline uptake studies. *Can J Surg* 40:36–46, 1967.

39. Dunphy, J.E., and Udupa, K.N. Chemical and histochemical sequences in the normal healing of wounds. *N Engl J Med* 253:847–851, 1955.

40. Duran, R.J., and Houser, R.G. Controlled passive motion following flexor tendon repair in zones 2 and 3. *In American Academy of Orthopaedic Surgeons Symposium on Tendon Surgery in the Hand.* St. Louis, C.V. Mosby, 1975, pp. 105–114.

41. Edwards, D.A.W. The blood supply and lymphatic drainage of tendons. *J Anat (Lond)* 80:147, 1946.

42. Elliott, D.H. Structure and function of mammalian tendon. *Biol Rev* 40:392–421, 1965.

43. Enwemeka, C.S., Spielholz, N.I., and Nelson, A.J. The effect of early functional activities on experimentally tenotomized Achilles tendons in rats. *Am J Phys Med Rehabil* 67(6):264–269, 1988.

44. Erlich, H.P., and Hunt, T.K. Effects of cortisone and vitamin A on wound healing. *Ann Surg* 167:324, 1968.

45. Eyre, D.R. Cross-linking in collagen and elastin. *Ann Rev Biochem* 53:717–748, 1984.

46. Forrester, J.C., Zederfeldt, B.H., Hayes, T.L., and Hunt, T.K. Wolff's law in relation to the healing skin wound. *J Trauma* 10:770–779, 1970.

47. Fox, J.M., Blazena, M.E., Jobe, F.W., Kirlan, R.E., Carter, V.S., Shields, C.L., and Carlson, G.J. Degeneration and rupture of the Achilles tendon. *Clin Orthop* 107:221–224, 1975.

48. Frank, C., Woo, SL-Y., Amiel, D., Harwood, F., Gomez, M., and Akeson, W. Medial collateral ligament healing: A multidisciplinary assessment in rabbits. *Am J Sports Med* 11:379–389, 1983.

49. Fung, Y.C. Stress-strain history. Relations of soft tissues in simple elongation. *In* Fung, Y.C., Perrone, N., and Anliker, M. (Eds.), *Biomechanics: Its Foundations and Objectives.* Englewood Cliffs, NJ, Prentice-Hall, 1972.

50. Gelberman, R.H., Botte, M.J., Spiegelman, J.J., and Akeson, W.H. The excursion and deformation of repaired flexor tendons treated with protected early motion. *J Hand Surg* 11A:106–110, 1986.

51. Gelberman, R.H., Manske, P.R., Vande Berg, J.S., Lesker, P.A., and Akeson, W.H. Flexor tendon repair in vitro: A comparative histologic study of the rabbit, chicken, dog and monkey. *J Orthop Res* 2:39–48, 1984.

52. Gelberman, R.H., Posch, J.L., and Jurist, J.M. High-pressure injection injuries of the hand. *J Bone Joint Surg* 57A:935–937, 1975.

53. Gelberman, R.H., Vande Berg, J.S., Lundborg, G.N., and Akeson, W.H. Flexor tendon healing and restoration of the gliding surface: An ultrastructural study in dogs. *J Bone Joint Surg* 65A:70–80, 1983.

54. Gelberman, R.H., Woo, SL-Y, Lothringer, K., Akeson, W.H., and Amiel, D. Effects of early intermittent passive mobilization on healing canine flexor tendons. *J Hand Surg* 7:170–175, 1982.

55. Gelberman, R.H., Woo, SL-Y, Lothringer, K., Akeson, W.H., and Amiel, D. Effects of early intermittent passive mobilization on healing canine flexor tendons. *J Hand Surg* 7:170–175, 1982.

56. Gottrup F. Healing of incisional wounds in the stomach and duodenum. Ph.D. Thesis, Institute of Anatomy, University of Aarhus, Aarhus, Denmark, 1983.

57. Graham, M.F., Becker, H., Cohen, I.K., Merritt, W., and Diegelmann, R.F. Intrinsic tendon fibroplasia: Documentation by in vitro studies. *J Orthop Res* 1:251–256, 1984.

58. Greenlee, T. K., Jr., Beckham, C., and Pike, D. A fine structural study of the development of the chick flexor digital tendon: A model for synovial sheathed tendon healing. *Am J Anat* 143:303–314, 1975.

59. Greenlee, T.K., Jr., and Pike, D. Studies of tendon healing in the rat. *Plast Reconstr Surg* 48:260, 1971.

60. Greenlee, T.K., Jr., and Ross, R. The development of the rat flexor digital tendon: A fine structure study. *J Ultrastruct Res* 18:354–376, 1967.

61. Grillo, H.C. Origin of fibroblasts in wound healing: An autoradiographic study of inhibition of cellular proliferation by local X-irradiation. *Ann Surg* 157:453, 1963.

62. Haut, R.C., and Little, R.W. A constitutive equation for collagen fibers. *J Biomech* 5:423–430, 1972.

63. Herring, S., and Nilson, K. Introduction to overuse injuries. *Clin Sports Med* 6(2):225–239, 1987.

64. Hitchcock, T.F., Light, T.R., Bunch, W.H., Knight, G.W., Patwardhan, A.G., and Hollyfield, R.L. The effect of immediate controlled mobilization on the strength of flexor tendon repairs. *Trans Orthop Res Soc* 11:216, 1986.

65. Hunt, T.K., and Van Winkle, W., Jr. Normal repair. *In* Hunt, T.K., and Dunphy, J.E. (Eds.), *Fundamentals of Wound Management.* New York, Appleton-Century-Crofts, 1979, pp. 2–67.

66. Hutton, P., and Ferris, B. Tendons. *In* Bucknall, T.E., and Ellis, H. (Eds.), *Wound Healing for Surgeons.* London, Balliere Tindall, 1984, pp. 286–296.

67. Inglis, A.E., and Sculco, T.P. Surgical repair of ruptures of the tendo Achilles. *Clin Orthop* 156:160–169, 1981.

68. Ippolito, E., Natali, P.G., Postacchini, F., Accinni, L., and Demartino, C. Morphological immunochemical and biochemical study of rabbit achilles tendon at various ages. *J Bone Joint Surg* 62A:583–598, 1980.

69. Ismail, A.M., Balakrishnan, R., and Rajakumar, M.K. Rupture of patellar tendon after steroid infiltration. *J Bone Joint Surg* 51B:503–505, 1969.

70. Jenkins, R.B., and Little, R.W. A constitutive equation for parallel-fibered elastic tissue. *J Biomech* 7:397–402, 1974.

71. Jimenez, S.A., Yankowski, R., and Bashey, R.I. Identification of 2 new collagen alpha chains in extracts of lathyritic chick embryo tendons. *Biochem Biophys Res Commun* 81:1298–1306, 1978.

72. Kastelic, J., and Baer, E. Reformation in tendon collagen: The mechanical properties of biological materials. Presented at the 34th Symposium of the Society for Experimental Biology, 1980.

73. Kastelic J., Galeski, A., and Baer, E. The multicomposite structure of tendon. *Connect Tissue Res* 6:11–23, 1978.

74. Kellam, J.F., Hunter, G.A., and McElwain, J.P. Review of the operative treatment of Achilles tendon ruptures. *Clin Orthop* 201:80–83, 1985.

75. Kessler, I. The 'grasping' technique for tendon repair. *Hand* 5:253–255, 1973.

76. Ketchum, L.D. Primary tendon healing: A review. *J Hand Surg* 2:428–435, 1977.

77. Ketchum, L.D., Martin, N.L., and Kappel, D.A. Experimental evaluation of factors affecting the strength of tendon repairs. *Plast Reconstr Surg* 59:708–719, 1977.

78. Kiririkko, K.I., Laitinen, O., Aer, J., and Halme, J. Studies with C-proline on the action of cortisone on the metabolism of collagen in the rat. *Biochem Pharmacol* 14:1445–1451, 1965.

79. Kiiskinen, A. Physical training and connective tissues in young mice—physical properties of Achilles tendons and long bones. *Growth* 41:123–137, 1977.

80. Kleinhart, H.E., Kutz, J.E., Atasoy, E., and Stormo, A. Primary repair of flexor tendons. *Orthop Clin North Am* 4:865, 1973.

81. Kuhn, K. The classical collagens: Types I, II, and III. *In* Mayne, R., and Burgeson, R.E. (Eds.), *Structure and Function of Collagen Types.* New York, Academic Press, 1987.

82. Lea, R.B., and Smith, L. Rupture of the Achilles tendon: Nonsurgical treatment. *Clin Orthop* 60:115–118, 1968.

83. Leddy, J.P., and Packer, J.W. Avulsion of the profundus insertion in athletes. *J Hand Surg* 2:66–69, 1977.

84. Lee, H.B. Avulsion and rupture of the tendo calcaneus after injection of hydrocortisone. *Br Med J* 2:395, 1957.

85. Lee, M.L.H. Bilateral rupture of Achilles tendon. *Br Med J* 1:1829–1830, 1961.

86. Lindsay, W.K., and Birch, J.R. The fibroblast in flexor tendon healing. *Plast Reconstr Surg* 34:223–232, 1964.

87. Lindsay, W.K., and Thomson, H.G. Digital flexor tendons: An experimental study. Part 1. The significance of each component of the flexor mechanism in tendon healing. *Br J Plast Surg* 12:289–316, 1959.

88. Lindsay, W.K., Thomson, H.G., and Walker, F.G. Digital flexor tendons: An experimental study: Part II. The significance of a gap occurring at the line of suture. *Br J Plast Surg* 13:1–9, 1960.

89. Lipscomb, P.R., and Wakim, K.G. Regeneration of severed tendons: An experimental study. *Mayo Clinic Proc* 36:271–276, 1961.

90. Lundborg, G. Experimental flexor tendon healing without adhesion formation. A new concept of tendon nutrition and intrinsic healing mechanism. *Hand* 3:235–238, 1976.

91. Lundborg, G., and Rank, F. Experimental intrinsic healing of flexor tendons based upon synovial fluid nutrition. *J Hand Surg* 3:21–31, 1978.

92. Lundborg, G., and Rank, F. Experimental studies on cellular mechanisms involved in healing of animal and human flexor tendon in synovial environment. *Hand* 12:3–11, 1980.

93. Manske, P.R., Bridwell, K., and Lesker, P.A. Nutrient pathways to flexor tendons of chickens using tritiated proline. *J Hand Surg* 3:352–357, 1978.

94. Manske, P.R., Gelberman, R.H., Vande Berg, J.S., and Lesker, P.A. Intrinsic flexor tendon repair: Morphological study in vitro *J Bone Joint Surg* 66A:385–396, 1984.

95. Manske, P.R., and Lesker, P.A. Avulsion of ring finger flexor digitorum profundus tendon: An experimental study. *Hand* 10:52–55, 1978.

96. Manske, P.R., and Lesker, P.A. Nutrient pathways of flexor tendons in primates. *J Hand Surg* 7:436–444, 1982.

97. Manske, P.R., and Lesker, P.A. Biochemical evidence of flexor tendon participation in the repair process: An in vitro study. *J Hand Surg* 9B:117–120, 1984.

98. Manske, P.R., and Lesker, P.A. Histological evidence of intrinsic flexor tendon repair in various experimental animals: An in vitro study. *Clin Orthop* 182:297–304, 1984.

99. Manthorpe, R., Helin, G., Kofod, B., and Lorenzen, I. Effects of glucocorticoid on connective tissue of aorta and skin in rabbits. *Acta Endocrinol (KBH)* 77:310–324, 1974.

100. Martin, G.R., Timpl, R., Muller, P.I.K., and Kuhn, K. *Trends Biochem Sci* 10:285–287, 1985.

101. Mason, M.L., and Allen, H.S. The rate of healing of tendons: An experimental study of tensile strength. *Ann Surg* 113:424–459, 1941.

102. Matthews, P., and Richards, H. The repair potential of digital flexor tendons: An experimental study. *J Bone Joint Surg* 56B:618–625, 1974.

103. Matthews, P., and Richards, H. Factors in the adherence of flexor tendon after repair: An experimental study in the rabbit. *J Bone Joint Surg* 58B:230–236, 1976.

104. McKenzie, A.R. An experimental multiple barbed suture for the long flexor tendons of the palm and fingers: Preliminary report. *J Bone Joint Surg* 49B:440–447, 1967.

105. McNab, I. Rotation cuff tendinitis. *Ann R Coll Surg Engl* 53:271–287, 1973.

106. Melmed, E.P. Spontaneous bilateral rupture of the calcaneal tendon during steroid therapy *J Bone Joint Surg* 47B:104–105, 1965.

107. Michna, H. Morphometric analysis of loading-induced changes in collagen-fibril populations in young tendons. *Cell Tissue Res* 236:465–470, 1984.

108. Michna, H. Tendon injuries induced by exercise and anabolic steroids. *Intern Orthop* 11:157–162, 1987.

109. Munro, I.R., Lindsay, W.K., and Jackson, S.H. A syn-

chronous study of collagen and mucopolysaccharide in healing flexor tendons of chickens. *Plast Reconstr Surg* 45:493–501, 1970.

110. Murray, G. A method of tendon repair. *Am J Surg* 99:334–335, 1960.

111. Nessler, J.P., and Mass, D.P. Direct-current electrical stimulation of tendon healing in vitro. *Clin Orthop* 217:303–312, 1987.

112. Neuberger, A., and Slack, H.G.B. Metabolism of collagen from liver, bone, skin and tendon in normal rat. *Biochem J* 53:47–52, 1953.

113. Nicolandoni, C. Ein Orchlag zur Schnennaht. *Wien Klin Wochenschr* 52:1413–1417, 1980.

114. Nimni, M.E., and Bavetta, L.A. Collagen synthesis and turnover in the growing rat under the influence of methylprednisolone. *Proc Soc Exp Biol Med* 117:618–623, 1964.

115. Nirschl, R.P. Prevention and treatment of elbow and shoulder injuries in the tennis player. *Clin Sports Med* 7(2):289–308, 1988.

116. Nirschl, R.P. Rotator cuff tendinitis: Basic concepts of pathoetiology. *Instr Course Lect* 38, 1989.

117. Nisbet, N.W. Anatomy of the calcanea tendon of the rabbit. *J Bone Joint Surg* 42B:360, 1960.

118. Nistor, L. Surgical and non-surgical treatment of Achilles tendon rupture. A prospective randomized study. *J Bone Joint Surg* 63A:394–399, 1981.

119. Noyes, F.R. Functional properties of knee ligaments and alterations induced by immobilization. *Clin Orthop* 123:210–242, 1977.

120. O'Donoghue, D.H., Rockwood, C.A., Zaricznyj, B., and Kenyon, R. Repair of knee ligaments in dogs. I. The lateral collateral ligament. *J Bone Joint Surg* 43A:1167, 1961.

121. Ohkawa, S. Effects of orthodontic forces and anti-inflammatory drugs on the mechanical strength of the periodontium in the rat mandibular first molar. *Am J Orthod* 81:498–502, 1982.

122. Okuda, Y., Gorski, J.P., An, K-N, and Amadio, P.C. Biochemical, histological, and biomechanical analyses of canine tendon. *J Orthop Res* 5:60–68, 1987.

123. Owoeye, I., Spielholz, N.I., Fetto, J., and Nelson, A.J. Low-intensity pulsed galvanic current and the healing of tenotomized rat Achilles tendons: Preliminary report using load-to-breaking measurements. *Arch Phys Med Rehabil* 68(7):415–418, 1987.

124. Oxlund, H. The influence of a local injection of cortisol on the mechanical properties of tendons and ligaments and the indirect effect on skin. *Acta Orthop Scand* 51:231–238, 1980.

125. Oxlund, H. Long-term local cortisol treatment of tendons and the indirect effect on skin. *Scand J Plast Reconstr Surg* 16:61–66, 1982.

126. Oxlund, H. Changes in connective tissues during corticotropin and corticosteroid treatment. Ph.D. Thesis, Institute of Anatomy, University of Aarchus, Aarchus, Denmark, 1983.

127. Oxlund, H., Fogdestam, I., and Viidik, A. The influence of cortisol on wound healing of the skin and distant connective tissue response. *Surg Gynecol Obstet* 148:876–880, 1979.

128. Oxlund, H., Manthorpe, R., and Viidik, A. The biomechanical properties of connective tissue in rabbits as influenced by short-term glucocorticoid treatment. *J Biomech* 14:129–133, 1981.

129. Paget, J. Healing of injuries in various tissue. *Lect Surg Pathol* 1:262–274, 1853.

130. Peacock, E.E., Jr. A study of the circulation in normal tendons. *Ann Surg* 45:415–423, 1959.

131. Peacock, E.E., Jr. Some problems in flexor tendon healing. *Surgery* 45:415–423, 1959.

132. Peacock, E.E., Jr. Fundamental aspects of wound healing relating to the restoration of gliding function after tendon repair. *Surg Gynecol Obstet* 119:241–250, 1964.

133. Peacock, E.E., Jr. Biological principles in the healing of long tendons. *Surg Clin N Am* 45:461–476, 1965.

134. Piez, K.A. *Ann Rev Biochem* 37:547–570, 1968.

135. Piez, K.A., Miller, E.J., Lane, J.M., et al: The order of the CNBr peptides from the alpha 1 chain of collagen. *Biochem Biophys Res Commun* 37:801–805, 1969.

136. Porter, K.R. Cell fine structure and biosynthesis of intercellular macromolecules. *In Connective Tissue: Intercellular Macromolecules. Biophys J* 4:2–167, 1964.

137. Postlethwaite, A.E., Keski-Oja, J., Moses, H.L., et al: Stimulation of the chemotactic migration of human fibroblasts by transforming growth factor *B J Exp Med* 165:251–256, 1987.

138. Potenza, A.D. Tendon healing within the flexor digital sheath in the dog. *J Bone Joint Surg* 44A:49–64, 1962.

139. Potenza, A.D. Critical evaluation of flexor tendon healing and adhesion formation within artificial digital sheaths. *J Bone Joint Surg* 45A:1217–1233, 1963.

140. Potenza, A.D. The healing of autogenous tendon grafts within the flexor digital sheath in dogs. *J Bone Joint Surg* 46A:1462–1484, 1964.

141. Potenza, A.D. Prevention of adhesions to healing digital flexor tendons. *JAMA* 187:187–191, 1964.

142. Potenza, A.D. Tendon healing within the flexor digital sheath in the dog. *J Bone Joint Surg* 44A:49–64, 1964.

143. Potenza, A.D. Concepts of tendon healing and repair. *In American Academy of Orthopaedic Surgeons Symposium on Tendon Surgery in the Hand.* St. Louis, C.V. Mosby, 1975, pp. 18–47.

144. Potenza, A.D. Tendon and ligament healing. *In Owen, R., Goodfellow, J., Bullough, P. (Eds.), Scientific Foundations of Orthopaedics and Traumatology.* London, Heinemann, 1980, pp. 300–305.

145. Potenza, A.D., and Herte, M.C. The synovial cavity as a "tissue culture in situ"—Science or nonsense? *J Hand Surg* 7:196–199, 1982.

146. Prockop, D.J., and Guzman, N.A. Collagen diseases and the biosynthesis of collagen. *Hosp Pract* December, 61–68, 1977.

147. Puddu, G., Ippolito, E., and Postacchini, F. A classification of achilles tendon disease. *Am J Sports Med* 4:145–150, 1976.

148. Pulvertaft, R.G. Suture materials and tendon junctures. *Am J Surg* 109:346–352, 1965.

149. Ramachandran, G.N. Molecular structure of collagen. *Int Rev Connect Tissue Res* 1:127–182, 1963.

150. Ramachandran, G.N., and Kartha, G. *Nature (Lond)* 174:269–270, 1954.

151. Renstrom, P., and Johnson, R. Overuse injuries in sports: A review. *Sports Med* 2:316–333, 1985.

152. Rich, A., and Crick, F.H.C. *Nature (Lond)* 176:915–916, 1955.

153. Rich, A., and Crick, F.H.C. The molecular structure of collagen. *J Mol Biol* 3:483, 1961.

154. Rigby, C.B.J., Hirai, N., Spikes, J.D., and Erying, H. The mechanical properties of rat tail tendon. *J Gen Physiol* 43:265–283, 1959.

155. Roberts, J.M., Goldstrohm, G.L., Brown, T.D., and Mears, D.C. Comparison of unrepaired, primarily repaired, and polyglactin mesh reinforced Achilles tendon lacerations in rabbits. *Clin Orthop* 181:244–249, 1983.

156. Ross, R. The fibroblast and wound repair. *Biol Rev* 43:51–96, 1968.

157. Rowe R.W.D. The structure of rat tail tendon. *Connect Tissue Res* 14:9–20, 1985.

158. Rowe, R.W.E. The structure of rat tail tendon fascicles. *Connect Tissue Res* 14:21–30, 1985.

159. Salter, R.B., Bell, R.S., and Keeley, F.W. The protective effect of continuous passive motion on living articular cartilage in acute septic arthritis: An experimental investigation in the rabbit. *Clin Orthop* 159:223–247, 1981.

160. Salter, R.B., Minister, R.R., Bell, R.S., et al: Continuous passive motion and the repair of full-thickness articular cartilage defects: A one-year follow-up. *Trans Orthop Res Soc* 7:167, 1982.

161. Salter, R.B., and Ogilvie-Harris, D.J. Healing of intra-articular fractures with continuous passive motion. *Instr Course Lect,* 28:102–117, 1979.

162. Salter, R.B., Simmonds, D.F., Malcolm, B.W., Rumble, E.J., Macmichael, D., and Clements, N.D. The biological effect of continuous passive motion on the healing of full-thickness defects in articular cartilage: An experimental investigation in the rabbit. *J Bone Joint Surg* 62A:1232–1251, 1980.
163. Sandberg, N. The relationship between administration of cortisone and wound healing in rats. *Acta Chir Scand* 127:446, 1964.
164. Schatzker, J., and Branemark, P. Intravital observation on the microvascular anatomy and microcirculation of the tendon. *Acta Orthop Scand* Suppl 126, 1969.
165. Skeoch, D. Spontaneous partial subcutaneous ruptures of the tendoachilles. *Am J Sports Med* 9:20–22, 1981.
166. Smaill, G.B. Bilateral rupture of Achilles tendons. *Br Med J* 1:1657, 1961.
167. Smart, G.W., Taunton, J.E., and Clement, D.B. Achilles tendon disorders in runners: A review. *Med Sci Sports Exerc* 12(4):231–243, 1980.
168. Suominen, H., Kiiskinen, A., and Heikkinen, E. Effects of physical training on metabolism of connective tissues in young mice. *Acta Physiol Scand* 108:17–22, 1980.
169. Tipton, C.M., Schild, R.J., and Tomanek, R.J. Influence of physical activity on the strength of knee ligaments in rats. *Am J Physiol* 212:783–787, 1967.
170. Urbaniak, J.R., Cahill, J.D., Jr., and Mortenson, R.A. Tendon suturing materials: Analysis of tensile strengths. *In American Academy of Orthopaedic Surgeons Symposium on Tendon Surgery in the Hand.* St. Louis, C.V. Mosby, 1975, pp. 70–80.
171. Verdan, C.E. Half a century of flexor-tendon surgery: Current status and changing philosophies. *J Bone Joint Surg* 54A:472–491, 1972.
172. Viidik, A. Biomechanical behaviour of soft connective tissues. *In* Akkas, N. (Ed.), *Progress in Biomechanics.* Amsterdam, Sijthoff and Noordhoff, 1979, pp. 75–113.
173. Viidik, A. The effects of training on the tensile strength of isolated rabbit tendons. *Scand J Plast Reconstr Surg* 1:141–147, 1967.
174. Viidik, A. Experimental evaluation of the tensile strength of isolated rabbit tendons. *Bio-Med Eng* 2:64–67, 1967.
175. Viidik, A. Mechanical properties of parallel-fibred collagenous tissues. *In* Viidik, A., and Vuust, J.(Eds.), *Biology of Collagen.* London, Academic Press, 1972, pp. 237–255.
176. Viidik, A. Tensile strength properties of Achilles tendon systems in trained and untrained rabbits. *Acta Orthop Scand* 40:261–272, 1969.
177. Viidik, A., Danielsen, C.C., and Oxlund, H. Fourth International Congress of Biorheology Symposium on Mechanical Properties of Living Tissues: On fundamental and phenomenological models, structure and mechanical properties of collagen, elastin and glycosaminoglycan complexes. *Biorheology* 19:437–451, 1982.
178. Viidik, A.V., and Gottrup, F. Mechanics of healing soft tissue wounds. *In* Schmid-Schonbein, G.W., Woo, SL-Y., and Zweifach, B.W. (Eds.), *Frontiers in Biomechanics.* New York, Springer-Verlag, 1986, pp. 263–279.
179. Vogel, H.C. Influence of maturation and age on mechanical and biochemical parameters of connective tissue of various organs in the rat. *Connect Tissue Res* 6:161–166, 1978.
180. Vogel, H.C. Mechanical and chemical properties of various connective tissue organs in rats as influenced by non-steroidal antirheumatic drugs. *Connect Tissue Res* 5:91–95, 1977.
181. Wade, P.J.F., Muir, I.F.K., and Hutcheon, I.L. Primary flexor tendon repair: The mechanical limitations of the modified Kessler technique. *J Hand Surg* 11B:71–76, 1986.
182. Weber, E.R. Nutritional pathways for flexor tendons in the digital theca. *In* Hunter, J.M., Schneider, L.H., and Mackin, E.J. (Eds.), *Tendon Surgery in the Hand.* St. Louis, C.V. Mosby, 1987, pp. 91–99.
183. Wertheim, H.G. Memoire sur l'elasticite et al cohesion des principeaus tissus du corps humain. *Chim Phys* 21:385–414, 1847.
184. White, A., Handler, P., and Smith, E.L. *Principles of Biochemistry.* New York, McGraw-Hill, 1973.
185. Woo, SL-Y. Mechanical properties of tendons and ligaments: 1. Quasistatic and nonlinear viscoelastic properties. *Biorheology* 19:385–396, 1982.
186. Woo, SL-Y. Biomechanics of tendons and ligaments. *In* Schmid-Schonbein, G.W., Woo, SL-Y, and Zweifach, B.W. (Eds.), *Frontiers in Biomechanics.* New York, Springer-Verlag, 1986, pp. 180–195.
187. Woo, SL-Y, Gelberman, R.H., Cobb, N.G., Amiel, D., Lothringer, K., and Akeson, W.H. The importance of controlled passive mobilization of flexor tendon healing. *Acta Orthop Scand* 52:615–622, 1981.
188. Woo, SL-Y., Gomez, M.A., Amiel, D., Ritter, M.A., Gelberman, R.H., and Akeson, W.H. The effects of exercise on the biomechanical and biochemical properties of swine digital flexor tendons. *J Biomech Eng* 103:51–56, 1981.
189. Woo, SL-Y., Gomez, M.A., Woo, Y-K, and Akeson, W.H. Mechanical properties of tendons and ligaments: II. The relationships of immobilization and exercise on tissue remodeling. *Biorheology* 19:397–408, 1982.
190. Woo, SL-Y., Lee, T.Q., Gomez, M.A., Sato, S., and Field, F.P. Temperature dependent behaviour of canine medial collateral ligament. *J Biomech Eng* 109:68–71, 1987.
191. Woo, SL-Y., Orlando, C.O., Camp, J., and Akeson, W.H. Effects of postmortem storage by freezing on ligament tensile behaviour. *J Biomech* 19:399–404, 1986.
192. Woo, SL-Y., Ritter, M.A., Amiel, D., Sanders, T.M., Gomez, M.A., Kuei, S.C., Garfin, S.R., and Akeson, W.H. The biomechanical and biochemical properties of swine tendons. Long-term effects of exercise on the digital extensions. *Connect Tissue Res* 7:177–183, 1980.
193. Wood, T.O., Cooke, P.H., and Goodship, A.E. The effect of exercise and anabolic steroids on the mechanical properties and crimp morphology of the rat tendon. *Am J Sports Med* 16(2):153–158, 1988.
194. Wray, R.C., and Weeks, P.M. Experimental comparison of technics of tendon repair. *J Hand Surg* 5:144–148, 1980.
195. Wrenn, R.N., Goldner, J.L., and Markee, J.L. An experimental study of the effect of cortisone on the healing process and tensile strength of tendons. *J Bone Joint Surg* 36A:588–601, 1954.

Ligaments

Joseph A. Buckwalter, M.D.
Savio L.-Y. Woo, Ph.D.

The specialized dense fibrous tissue structures, ligaments, tendons, and joint capsules, form a major component of the musculoskeletal system [13]. These tissues consist primarily of highly oriented, tightly packed collagen fibrils that give them flexibility and great tensile strength. Despite their similarities, the dense fibrous tissues differ in structure and function. Skeletal ligaments and joint capsules have much lower length-to-width ratios than tendons. Unlike tendons, they connect adjacent bones to each other rather than connecting muscle to bone, and they more often form layered sheets or lamellae of collagen fibrils rather than cords. This chapter does not include a discussion of the ligaments that connect soft tissues to supported or suspended organs such as the thyroid and liver because these structures do not form part of the skeletal system.

Skeletal ligaments and, to a lesser degree, joint capsules help to stabilize synovial and nonsynovial joints by guiding normal joint motion and preventing abnormal motion. They also may have sensory functions in providing joint proprioception and initiating protective reflexes [7, 13]. Sports activities frequently result in ligament sprains or tears that compromise ligament function and thereby decrease joint stability. Significant joint instability due to loss of ligament function eliminates participation in competitive sports and may lead to degeneration of the affected joint [36, 63].

This section first outlines the types of ligaments and their structure and composition. Next the response of ligaments to injury is reviewed, including inflammation, repair, and remodeling of ligaments, and finally, ligament autografts, allografts, xeno-grafts, and grafts formed from reconstituted collagen are reviewed.

TYPES OF LIGAMENTS

Surgeons and anatomists have identified hundreds of skeletal ligaments. Generally, they have named ligaments by their location and bony attachments (e.g., anterior glenohumeral ligament, anterior talofibular ligament) or by their relationship to other ligaments (e.g., medial collateral ligament of the knee, posterior cruciate ligament of the knee). Many ligaments, like the anterior cruciate ligament, form discrete, easily identifiable structures with distinct mechanical functions, but others, like the hip joint capsular ligaments, blend with the joint capsule, making it difficult to define their exact structure and function.

The anatomic relationships of ligaments to the joints they help stabilize vary. They may be intracapsular, capsular, or extracapsular. A thin layer of synovium covers intracapsular ligaments including the anterior cruciate ligament of the knee and the ligamentum teres of the hip. The areolar connective tissue that surrounds the joints covers capsular ligaments including the capsular ligaments of the hip and glenohumeral joint. Fat and areolar connective tissue separate extracapsular ligaments, including the coracoacromial ligament, the coracoclavicular ligaments, and the costoclavicular ligaments from the underlying joint capsule and capsular ligaments. Despite these differences, the presumed primary functions of intracapsular, capsular, and extracapsular ligaments remain the same—stabilizing the relationship between adjacent bones and stabilizing and guiding joint motion.

STRUCTURE OF LIGAMENTS

Skeletal ligaments vary in length, shape, and thickness. They cross joints with wide ranges of motion such as the hip and glenohumeral joint as well as those with little or no normal motion such as the sacroiliac joint and the proximal tibiofibular joint. Grossly, they appear as firm, white fibrous bands, sheets, or thickened strips of joint capsule securely anchored to bone. They consist of a proximal bone insertion, the substance of the ligament or capsule, and a distal bone insertion. Most insertions are no more than a millimeter thick, so they contribute only a small amount to the volume and length of the ligament.

SUBSTANCE OF LIGAMENTS

Bundles of collagen fibrils form the bulk of the ligament substance [13, 14, 30, 94]. Some ligaments, including the anterior cruciate ligament of the knee, consist of more than one band of collagen fibril bundles; as the joint moves, different bands become taut [15]. The alignment of collagen fiber bundles within the ligament substance generally follows the lines of tension applied to the ligament during normal activities, but light microscopic examination shows that the collagen bundles also may have a wave or crimp pattern. The crimp pattern of matrix organization may allow slight elongation of the ligament without incurring damage to the tissue [94]. In some regions ligamentary cells align themselves in rows between collagen fiber bundles, but in other regions the cells lack apparent orientation relative to the alignment of the matrix collagen fibers. Scattered blood vessels penetrate the ligament substance forming small-diameter, longitudinal vascular channels that lie parallel to the collagen bundles. Nerve fibers lie next to some vessels, and nerve endings with the structure of mechanoreceptors have been found in some ligaments [7, 13, 14].

INSERTIONS

Insertions of ligaments attach the flexible ligament substance to the rigid bone and allow motion between the bone and the ligament without damage to the ligament. Despite their small size, ligament insertions have a more variable and elaborate structure than the ligament substance [10, 11, 22, 32, 98], and they may have different mechanical properties. Measurement of ligament mechanical properties shows that during activity insertions or the ligament regions near insertions deform more than the ligament substance [101].

Ligament insertions vary in size, strength, angle of the ligament collagen fiber bundles relative to the bone, and proportion of ligament collagen fibers that penetrate directly into bone [13, 22, 32]. Based on the angle between the collagen fibrils and the bone and the proportion of the collagen fibers that penetrate directly into bone, investigators group ligament insertions into two types: direct and indirect.

Direct Insertions

Direct ligament insertions into bone, like the femoral insertion of the medial collateral ligament of the knee or the tibial insertion of the anterior cruciate ligament of the knee, consist of sharply defined regions where the ligament appears to pass directly into the cortex of the bone [10, 11, 22, 52, 98]. Attempts to separate a ligament with a direct insertion from the bone usually fail unless the surgeon cuts the substance of the ligament through the region next to the insertion.

The thin layer of superficial ligament collagen fibers in direct insertions joins the fibrous layer of the periosteum. Most of the ligament insertion consists of deeper fibers that directly penetrate the cortex, often at a right angle to the bone surface. In ligaments that approach the bone surface at a right angle, the ligament collagen fibers follow a straight line into the bone, but in ligaments that approach the bone surface at an oblique angle, the ligament collagen fibers make a sharp turn to enter the insertion. The angle of ligament collagen fibril insertion may not be apparent from gross examination. For example, the medial collateral ligament of the femur approaches the surface of the femur obliquely; grossly, it appears to insert at a 50- to 70-degree angle, but microscopic study shows that the collagen fibers of the ligament enter the bone at about a 90-degree angle.

The deeper collagen fibers pass through four zones with increasing stiffness [10, 11, 22, 98]: ligament substance, fibrocartilage, mineralized fibrocartilage, and bone. In the fibrocartilage zone the cells become larger and more spherical than the cells in most regions of the ligament substance. A sharp border of mineralized and unmineralized matrix separates the fibrocartilage zone from the mineralized fibrocartilage zone. From this latter zone, the ligament collagen fibers pass into the bone and blend with the bone collagen fibers.

Indirect Insertions

Indirect or oblique ligament insertions into bone [10, 11, 22, 52, 54, 98], like the tibial insertion of the medial collateral ligament of knee or the femoral insertion of the lateral collateral ligament, are less common than direct insertions. They usually cover more bone surface area than direct insertions, and their boundaries cannot be easily defined because the ligament passes obliquely along the bone surface rather than directly into the cortex. Ligaments with indirect insertions into bone can often be elevated from the bone without cutting the ligament substance and may not have a zone of fibrocartilage.

Like direct insertions, indirect insertions have superficial and deep collagen fibers, but the superficial ligament collagen fibers passing into the fibrous layer of the periosteum form most of the substance of indirect insertions. The deeper collagen fibers of indirect insertions approach the bone cortex at oblique angles and do not pass through well-defined fibrocartilage zones. The structure of indirect insertions, particularly the distribution of ligament fibers between bone and periosteum, may change with skeletal development and alter the mechanical properties of the insertions. A study of the maturation of the rabbit medial collateral ligament insertion into the tibia showed that with skeletal growth more of the ligament collagen fibers entered the bone, decreasing the contribution of the periosteal component to the insertion [54]. Changes occurring with age in the periosteum may also alter the structure and mechanical properties of the insertions [98].

COMPOSITION OF LIGAMENTS

Different ligaments and different regions of the same ligament may vary slightly in matrix composition and in cell shape and density [1, 28, 30]. Yet they all consist of fibroblasts surrounded by an extracellular matrix formed by two components: a solid, highly ordered arrangement of macromolecules, primarily type I collagen, and water that fills the macromolecular framework [13, 30]. The composition of the matrix, the organization of the matrix macromolecules, and the interaction between the matrix macromolecules and the tissue water determine the material properties of the tissue. The

biochemical composition of ligament insertions has not been extensively studied because of the difficulties of isolating and analyzing small volumes of tissue, but, like ligament substance, the insertions consist of a small number of cells surrounded by an abundant extracellular matrix formed mainly by type I collagen fibrils.

Cells

Fibroblasts are the dominant cell of ligaments, although endothelial cells of small vessels and nerve cell processes also exist within the substance of ligaments [11, 13, 22, 30, 32, 71]. Fibroblasts form and maintain the extracellular matrix. They vary in shape, activity, and density among ligaments, among regions of the same ligament, and with the age of the tissue [13, 22, 30, 32, 71]. Many fibroblasts have long, small-diameter cell processes that extend between collagen fibrils [22]. Younger ligaments have a high concentration of cells that synthesize new matrix. These cells frequently have a large volume of cytoplasm containing substantial amounts of endoplasmic reticulum. With increasing age the cells become less active, but at any age they synthesize the matrix macromolecules necessary to maintain the structure and function of the tissue.

Matrix

Water

Tissue fluid contributes 60% or more of the wet weight of most ligaments. Because many ligament cells lie at some distance from vessels, these cells must depend on diffusion of nutrients and metabolites through the tissue fluid. In addition, the interaction of the tissue fluid and the matrix macromolecules influences the material properties of the tissue.

Matrix Macromolecules

Four classes of molecules (collagens, elastin, proteoglycans, and noncollagenous proteins) form the molecular framework of the ligament matrix and constitutes about 40% of the wet weight of most ligaments [13, 30, 111].

Collagens

Fibrillar collagens have the form of cylindrical cross-banded fibrils when examined by electron microscopy. These fibrils give ligaments their form and tensile strength and constitute 70% to 80% of the dry weight of the ligament. Type I collagen is the major component of the molecular framework. It comprises more than 90% of the collagen content of ligaments. Type III collagen constitutes about 10% of the collagen, and other collagen types may be present as well. In contrast to ligaments, tendons have less type III collagen [1].

Elastin

Most ligaments have little elastin (usually less than 5%), but a few, such as the nuchal ligament and the ligamentum flavum, have high concentrations of elastin (up to 75%). Elastin forms protein fibrils or sheets, but elastin fibrils lack the cross-banding pattern of fibrillar collagens and differ in amino acid composition, including two amino acids not found in collagen (desmosine and isodesmosine). Also unlike collagen, elastin amino acid chains assume random coils when the molecules are not stressed. This conformation of the amino acid chains makes it possible for elastin to undergo some deformation without rupturing or tearing and then, when unloaded, return to its original size and shape.

Proteoglycans

Proteoglycans form only a small portion of the macromolecular framework of the ligament, usually less than 1% of the dry weight [30]. Nonetheless, they may have important roles in organizing the extracellular matrix and interacting with the tissue fluid [8, 9, 11, 13, 38, 39, 64, 74]. Most ligaments have a higher concentration of glycosaminoglycans than do tendons [1].

Like tendon, meniscus, and articular cartilage, ligaments contain two known classes of proteoglycans: large articular cartilage-type proteoglycans containing long, negatively charged chains of chondroitin and keratan sulfate, and smaller proteoglycans that contain dermatan sulfate. Because of their long chains of negative charges, the large articular cartilage-type proteoglycans tend to expand to their maximum domain in solution until restrained by the collagen fibril network. As a result, they maintain water within the tissue and exert a swelling pressure, thereby contributing to the material properties of the tissue and filling the regions between the collagen fibrils. The small dermatan sulfate proteoglycans usually lie directly on the surface of collagen fibrils and appear to affect the formation, organization, and stability of the extracellular matrix, including collagen fibril formation and diameter, rather than playing a direct role in providing its mechanical properties [74, 75].

Noncollagenous Proteins

These molecules consist of protein, and many of them also contain a few monosaccharides and oligosaccharides [11, 13, 30]. Although noncollagenous proteins contribute only a small percentage of the dry weight of dense fibrous tissues, they appear to help organize and maintain the macromolecular framework of the collagen matrix, aid the adherence of cells to the framework, and possibly influence cell function. One specific noncollagenous protein, fibronectin, has been identified in the extracellular matrix of ligaments. It may be associated with several matrix component molecules as well as with blood vessels. Other noncollagenous proteins undoubtedly exist within the ligament matrix, but their identity and functions have not yet been defined.

HEALING OF LIGAMENTS

Ligament healing (i.e., the restoration of the structural integrity of a strained or torn ligament following injury) depends on the response of the tissue to injury [12]. Strains and tears disrupt the matrix, damage blood vessels, and injure or kill cells. Damage to the cells, matrices, and blood vessels and the resulting hemorrhage start a response that includes inflammation, repair, and remodeling [12, 102]. These events do not occur separately but form a continuous sequence of cell, matrix, and vascular changes that begin with the release of inflammatory mediators and end when remodeling of the repaired tissue ceases [12].

Inflammation

Inflammation, the cellular and vascular response to injury, includes release of inflammatory mediators, vasodilatation, increased blood flow, exudation of plasma, and migration of inflammatory cells [12]. These tissue events cause swelling, erythema, increased temperature, pain, and impaired function. Acute inflammation lasts 48 to 72 hours after most ligament injuries and then gradually resolves as repair progresses. Some of the events that occur during inflammation, including the release of cytokines or growth factors (small soluble molecules that promote cell migration, proliferation, and differentiation), may help stimulate tissue repair [12].

Inflammation begins immediately after injury with the release of inflammatory mediators from damaged cells. These mediators promote vascular dilatation and increase vascular permeability. Exudation of fluid from vessels in the injured region causes tissue edema. Ligamentous tissue immediately surrounding the injury becomes swollen and increasingly friable. Uninjured ligamentous tissue at a distance from the injury also becomes edematous owing to exudation of fluid from the dilated vessels. Blood escaping from the damaged vessels forms a hematoma that temporarily fills the injured site. Fibrin accumulates within the hematoma, and platelets bind to fibrillar collagen, thereby achieving hemostasis and forming a clot consisting of fibrin, platelets, red cells, and cell and matrix debris. The clot provides a framework for vascular and fibroblast cell invasion. As they participate in clot formation, platelets release vasoactive mediators and the cytokines or growth factors transforming growth factor-beta (TGF-beta) and platelet-derived growth factor (PDGF).

Within hours after injury, polymorphonuclear leukocytes appear in the damaged tissue and the clot, followed shortly by monocytes that increase in number until they become the predominant cell type. Enzymes released from the inflammatory cells help to digest necrotic tissue, and monocytes phagocytose small particles of necrotic tissue and cell debris. Endothelial cells near the injury site begin to proliferate, creating new capillaries that grow toward the region of tissue damage. Release of chemotactic factors and cytokines from endothelial cells, monocytes, and other inflammatory cells help stimulate the migration and proliferation of the fibroblasts that begin the repair process [12].

Repair

Ligament repair, the replacement of necrotic or damaged tissue by cell proliferation and synthesis of new matrix [12], depends on the fibroblasts that migrate into the injured tissue and clot. Within 2 to 3 days after injury, fibroblasts within the wound begin to proliferate rapidly and synthesize new matrix. They replace the clot and necrotic tissue with a soft, loose fibrous matrix containing high concentrations of water, glycosaminoglycans, and type III collagen. Inflammatory cells and fibroblasts fill this initial repair tissue. Within 3 to 4 days, vascular buds from the surrounding tissue grow into the repair tissue and then canalize to allow blood flow to the injured tissue and across small tissue defects. This friable vascular granulation tissue fills the tissue defect and extends for a short distance into the surrounding tissue but has little tensile strength.

During the next several weeks, as repair progresses, the composition of the granulation tissue changes. Water, glycosaminoglycan, and type III collagen concentrations decline, the inflammatory cells disappear, and the concentration of type I collagen increases. Newly synthesized collagen fi-brils increase in size and begin to form tightly packed bundles, and the density of fibroblasts decreases. Matrix organization increases [19, 20, 29, 60] as the fibrils begin to align along the lines of stress, the number of blood vessels decreases, and small amounts of elastin may appear within the site of injury. The tensile strength of the repair tissue increases as the collagen content increases.

Remodeling

Repair of many ligament injuries results in an excessive volume of highly cellular tissue with limited tensile strength and a poorly organized matrix. Remodeling reshapes and strengthens this tissue by removing, reorganizing, and replacing cells and matrix [12]. In most ligament injuries, evidence of remodeling appears within several weeks of injury as the numbers of fibroblasts and macrophages decrease, fibroblast synthetic activity decreases, and the fibroblasts and collagen fibrils assume a more organized appearance. As these changes occur, collagen fibrils grow in diameter, the concentration of collagen and the ratio of type I collagen to type III collagen increase, and the water and proteoglycan concentrations of the repair tissue decline.

In the months following injury the volume of the repair tissue decreases and the degree of repair tissue orientation continues to increase, presumably partially in response to loads applied to the repair tissue. The most apparent signs of remodeling disappear within 4 to 6 months of injury: The cell density and the number of small blood vessels decline to near normal levels, and the collagen concentration increases nearly to normal, but removal, replacement, and reorganization of repair tissue continue to some extent for years [29]. Although mature repair tissue restores the structural integrity of the ligament, it can usually be distinguished from normal tissue and may differ in composition. Most important, mature ligament repair tissue lacks the tensile strength of normal ligament tissue (usually it has 50% to 70% of normal tensile strength) [4, 27, 31, 102]. Generally, this strength deficit does not cause clinically apparent disturbances of joint function, at least partially because the volume (or cross-sectional area) of the repaired tissue remains greater than the volume of uninjured ligament.

Variables That Influence Ligament Healing

Many variables influence ligament healing [12]. Among the most important are type of ligament, size of the tissue defect, and the amount of loading applied to the ligament repair tissue. Injuries to capsular and extracapsular ligaments stimulate production of repair tissue that will fill most defects, but injuries to intracapsular ligaments, such as the anterior cruciate ligament, often fail to produce a successful repair response. This may be due to the synovial environment, limited vascular ingrowth, and fibroblast migration from surrounding tissues or other factors. The extent of the injury and the type of treatment can influence the volume of necrotic tissue and the size of the tissue defect. Treatments that achieve or maintain apposition of torn ligament tissue and stabilize the injury site decrease the volume of repair tissue necessary to heal the injury. They may also minimize scarring and help provide near normal tissue length. For these reasons, it is generally desirable to avoid wide separation of ruptured ligament ends and to select treatments that maintain some stability of the injured site during the initial stages of repair. Early controlled loading of ligament repair tissue can promote healing, but excessive loading will disrupt repair tissue and delay or prevent healing [4, 12, 13, 34, 53, 88, 90] (see also Chap. 1C for a more detailed discussion of the effects of loading on repair tissues).

LIGAMENT GRAFTS

Most ligament sprains and tears heal satisfactorily (i.e., the response of the ligament and the surrounding tissues to the injury restores the structural integrity of the ligament), but ruptures of some ligaments (like the anterior cruciate ligament) or severe or untreated injuries of other ligaments may fail to heal [12]. Failure to heal or loss of ligamentous tissue can leave the patient with an unstable joint. For this reason, surgeons have used autogenous and allogenic dense fibrous tissues to reconstruct ligaments that have failed or are likely to fail to heal. Currently, autografts and allografts are most commonly used in reconstruction of a deficient anterior cruciate ligament, although other ligaments have also been reconstructed with dense fibrous tissue grafts. Xenografts have not been widely used for ligament reconstruction, and artificial ligaments composed of reconstituted collagen remain experimental.

Following transplantation, dense fibrous tissue autografts and allografts heal to the host tissues and undergo remodeling that includes revascularization, repopulation with cells from the recipient site, and synthesis of new matrix by these cells [3, 5, 83, 77, 65]. Grafts also undergo a change in biomechanical properties when placed in the recipient site that adversely affects their ability to withstand large tensile loads [65, 99]. For example, the central and medial portions of the patellar tendon are stronger than the anterior cruciate ligament [68], but transplantation of these portions of the patellar tendon to reconstruct the anterior cruciate ligament reduces their strength to 10% to 30% of the strength of the anterior cruciate ligament [15, 65, 93]. As the graft heals to the recipient tissues and undergoes remodeling, its mechanical properties improve, but it never approaches the stiffness and strength of the graft tissue before transplantation. Despite the failure of grafts to restore the biomechanical properties of ligaments to normal in experimental studies, clinical evaluations show that they can significantly improve the stability of joints with ligamentous insufficiency [33, 42, 44, 80, 85]. (Clinical testing may fail to show the weakness of grafts detected by experimental studies because clinical tests of ligament function use small loads and do not attempt to test the ultimate strength of the reconstruction.)

Healing of the site of the graft attachment may be responsible for most of the increase in graft strength observed following transplantation. A study of rabbit patellar tendon autografts supported this possibility; it showed that the site of failure of the grafts changed with time [6]. Immediately after transplantation most grafts failed at the femoral fixation site; 6 weeks later the grafts tore through the graft substance within the tibial tunnel; and 30 weeks after transplantation five of six grafts ruptured in the intra-articular part of the graft, and one failed within the femoral bone tunnel. Fifty-two weeks after transplantation all grafts tested failed in the intra-articular part of the graft substance. These results show that initially the weakest regions of the reconstruction are the bone attachment sites and the parts of the graft within the bone tunnels. Healing of these regions with bone presumably increases their strength, making the intra-articular parts of the graft the weakest regions.

Autografts

Autografts can be transplanted with a vascular pedicle that maintains their blood supply or as free tissue. Tissues used for ligament autografts include fascia lata, portions of the patellar tendon, other tendons, and meniscus [33]. Clinically, these types of autografts heal and can reconstruct deficient ligaments [33], but they may differ in structure and mechanical properties.

Nonvascularized Autografts

Harvesting an autograft leaves the matrix intact, but disruption of the blood supply causes ischemic necrosis or injury to some graft cells [50]. Studies of cell function in nonvascularized grafts suggest that graft cells can survive transplantation [35], but the proportion of cells that survive has not been clearly established. Soon after transplantation, inflammatory cells, fibroblasts, and vessels at the recipient site invade the graft and increase the cell density [6, 50, 91]. While the new cells and vessels penetrate the substance of the matrix, the graft usually swells. The increased water content correlates closely with decreased graft stiffness and strength [56].

Once the new cells establish themselves within the graft, they begin the remodeling process. The collagen fibril alignment within the graft may appear disorganized during the initial stages of remodeling, but progressive realignment occurs, and edema and cell density decrease until the histologic appearance of the graft approaches that of normal tissue [6]. Changes in the biochemical composition of the graft matrix accompany the changes in histologic appearance. Patellar tendon normally lacks type III collagen and has a low concentration of glycosaminoglycans. After patellar tendon transplantation to reconstruct the anterior cruciate ligament, type III collagen and glycosaminoglycan concentrations of the graft increase until they reach levels normally found in the anterior cruciate ligament, and the pattern of collagen cross-links changes to resemble that found in the anterior cruciate ligament [1, 2]. Despite their normal appearance and the changes in their matrix composition, autografts generally have significantly less stiffness and strength than normal ligaments. Most studies show that they reach only about 20% to 40% of normal values 6 months to a year after implantation [6, 56, 100].

The variables that affect the healing and ultimate strength of autografts have not been extensively studied, but loading of the grafts appears to influence graft healing and remodeling. Excessive or insufficient tension on the graft may adversely affect revascularization and remodeling. A study of patellar tendon autograft reconstructions of anterior cruciate ligaments in dogs showed that grafts fixed under high tension (39 Newtons) had poor revascularization and myxoid degeneration 3 months after surgery [103]. Grafts fixed under low tension (1 N) showed more extensive revascularization, no degeneration, and greater strength and stiffness. The presence of a Dacron prosthesis (placed roughly parallel to an autograft) increased the stiffness and strength of the reconstruction in the immediate postoperative period, presumably because of the better immediate fixation of the prosthetic material [104]. Three months after surgery the strength of the augmented grafts had decreased, but the strength of the grafts without Dacron augmentation had increased. The investigators suggested that the Dacron augmentation delayed revascularization and remodeling of the grafts and that the declining strength of the augmented grafts resulted from the delay in remodeling combined with degradation of the Dacron's material properties. The reason for the apparent inhibition of graft remodeling associated with augmentation remains uncertain. It may result from decreased loading of the graft because augmentation of an autograft with a prosthesis did not delay revascularization and remodeling when the synthetic graft was not parallel to the autograft [55].

The extent to which a graft restores the normal anatomy of a ligament and is subjected to normal loading and the presence of well-vascularized surrounding tissues that can supply cells and vessels to invade the graft may also influence the results. A study of fresh medial collateral ligament autografts in rabbits showed that under ideal conditions these autografts can achieve near normal strength [78]. The investigators removed the bone medial collateral ligament complexes completely and then replaced them using internal fixation. In the autograft complexes the ligament substance and insertions showed early weakening, but graft strength gradually increased. At 24 weeks the graft complexes showed their lowest load at failure, 65% of control values. At 48 weeks the load at failure had increased to about 90% of control values, and the material properties of the grafts could not be distinguished from the material properties of control samples. The authors emphasized that immediate anatomic replacement of a ligament represents the optimal circumstance for recovery of function and that their report represents the best possible mechanical recovery of any dense fibrous tissue graft. Rapid cell and vascular invasion of the graft from surrounding soft tissues may also have contributed to recovery of mechanical properties of the graft. The results also suggest that even under optimal circumstances, healing and incorporation of the autograft proceed slowly; whether an autograft transplanted under optimal biologic and mechanical conditions can eventually recover normal stiffness and strength remains uncertain.

Vascularized Autografts

To decrease ischemic necrosis and avoid the necessity for recipient site cells to revascularize and repopulate the graft, surgeons have developed methods of maintaining the vascularity of autografts [16, 15, 24, 51, 69, 72]. If vascularized grafts healed more rapidly and maintained or achieved greater ultimate strength, they would have significant advantages, but studies of the healing and biomechanical properties of vascularized and nonvascularized patellar tendon graft reconstructions of the anterior cruciate ligament in dogs and primates do not show significant advantages of vascularized autografts [16, 24]. Vascularized grafts in dogs initially had a better blood supply and matrix organization than nonvascularized grafts [24], but at 12 weeks or more after transplantation little or no difference in matrix organization, cell population, and blood supply was evident between the two types of graft [24]. Testing of graft strength showed that vascularized grafts had greater strength 6 weeks postoperatively, equal strength 12 weeks postoperatively, and slightly less strength 20 weeks postoperatively. In cynomologus monkeys, vascularized and nonvascularized grafts showed low levels of stiffness and strength 7 weeks postoperatively (24% and 16% of control values, respectively) [16]; stiffness and strength increased to 57% and 39% of control values; respectively, by 1 year postoperatively.

Specific Tissues Used for Autografts

Iliotibial Band/Fascia Lata

The ease and low morbidity of harvesting the distal portion of the fascia lata, the iliotibial band, make it an attractive source of tissue for ligament autografts [43], but experimental studies show that these grafts fail to restore the normal biomechanical properties of the anterior cruciate ligament complex. One group of investigators used the distal portion of the fascia lata, the iliotibal tract, for reconstruction of the anterior cruciate ligament in dogs [70]. Four years after reconstruction the grafts remained considerably weaker than normal anterior cruciate ligament complexes, and the treated knees showed evidence of instability and arthritic change. Another group found that the mean ultimate load to failure of iliotibial tract grafts used to reconstruct anterior cruciate ligaments was about 40% of control values [86]; a similar study found that the stiffness and ultimate load values of such grafts were 45% and 40%, respectively, of control values [91]. Fascia lata grafts used to replace the anterior cruciate ligament

in goats produced similar results: Two months postoperatively mean graft stiffness was no more than 10% of control value for the anterior cruciate ligament, and mean graft ultimate load strength was no more than 15% of control value [40]. Eight weeks after reconstruction the anterior posterior translation of the treated knees was about seven times greater than that of control knees.

Patellar Tendon

Because of their size, strength, and availability, autografts consisting of the central or medial third of patellar tendon with their bony insertion sites have been widely used for reconstruction of the anterior cruciate ligament [47, 69, 72, 81]. However, experimental studies show that despite the great strength of these grafts before transplantation, after transplantation the material properties of the grafts fail to reach those of the normal anterior cruciate ligament complex [76]. In most studies, the grafts had approximately one-third of the ultimate load strength of controls 24 months following reconstruction, and in at least some studies, the reconstructed joints showed increased anterior posterior translation and evidence of joint degeneration.

Multiple investigations of patellar tendon autografts in dogs have confirmed that these grafts fail to restore normal stiffness and strength to the anterior cruciate ligament. One group of investigators found that in dogs patellar tendon autografts and anterior cruciate ligaments initially had similar stiffness; the ultimate load values of the grafts were about 72% of control values [103, 104]. After implantation the ultimate load values of the femoral patellar tendon graft tibia complex declined to about 10% of control values, and graft stiffness declined to about 11% of control values. Three months after surgery the ultimate load values of the grafts had increased to 20% of control values, and graft stiffness had increased to 22% of control values, but the grafted knees had nearly three times as much anterior posterior instability as control knees. Twenty months after surgery the anterior translation of the grafted knees had decreased to a level only 1 mm greater than that of controls, but the ultimate load values for the grafts had reached only 32% of control values. Other studies show that 4 to 6 months after reconstruction the stiffness, ultimate loads to failure, and energy absorbed to failure of patellar tendon grafts are commonly less than half normal values, and joints with patellar tendon grafts have increased laxity [59, 76, 84].

Studies of patellar tendon autograft reconstructions of anterior cruciate ligaments in goats, rabbits, and rhesus monkeys also show that grafts do not develop the properties of normal anterior cruciate ligaments. Patellar tendon reconstructions of the anterior cruciate ligament in goat knees had a mean ultimate load value of about 45% of control value, and mean graft stiffness was about 36% of control value [59]. All knees developed patellofemoral degenerative changes and increased anterior posterior joint translation 2 to 24 months after surgery [59]. Twelve months after surgery patellar tendon autograft reconstructions of the anterior cruciate ligament in rabbits had ultimate loads and stiffness that were 15% and 24%, respectively, of those of the control anterior cruciate ligaments, and the knees had twice as much anterior laxity as normal knees [6]. Rhesus monkey knees with patellar tendon reconstructions of the anterior cruciate ligament had less than a 1-mm increase in anterior tibial translation and no evidence of degenerative changes 3 to 12 months after surgery [18]. Stiffness of the graft complex increased from 39% at 3 months to 47% of control at 12 months, and ultimate load values increased from 26% at 3 months to 52% of control at 12 months [18].

Semitendinosus Tendon

Although investigators have not directed as much attention toward semitendinosus tendon reconstructions of the anterior cruciate ligament as they have toward patellar tendon reconstructions, the semitendinosus tendon provides a clinically useful alternative dense fibrous tissue for anterior cruciate ligament reconstruction or augmentation of anterior cruciate ligament repairs [17, 33, 62, 80, 105]. One study of rabbits showed that the ultimate load strength of semitendonosis tendon anterior cruciate ligament reconstructions was less than 30% of controls and that 26 weeks after reconstruction the grafts failed at loads of less than 15% of control values [49]. Evaluation of the reconstructed knees showed degenerative changes.

Meniscus

Menisci have also been used to reconstruct anterior cruciate ligaments [26, 44, 87, 95]. The reported clinical experience shows that meniscal reconstructions of the anterior cruciate ligament restores knee stability in most patients [26, 44, 87], but one report described failure of meniscal grafts in all 13 patients in the series [95].

Experimental studies show that menisci have a tensile strength similar to that of other dense fibrous tissue grafts. Medial menisci used for anterior cruciate ligament reconstructions in dogs failed at an ultimate load of about 30% of controls 6 months after surgery [21], and meniscal grafts in rabbits failed at an ultimate load of about 25% of controls 1 year postoperatively [61]. Histologic examination of meniscal grafts from both humans and rabbits

showed that the grafts did not develop the appearance of ligamentous tissue [26, 61]. In the human biopsy specimens the grafts retained the woven pattern of collagen fibers characteristic of meniscal tissue [26]. (See Chapter 2, Section B.)

Allografts

Dense fibrous tissue allografts acquired from carefully selected donors and processed and stored according to established standards and guidelines, such as those of the American Association of Tissue Banks, offer a currently accepted alternative to autografts [67, 82, 89]. Dense fibrous tissue allografts in humans have not shown evidence of clinically detectable rejection or a higher rate of infection than would be expected in autografts [67, 82, 89]. Because allograft reconstruction of a ligament does not require harvest of an autograft, thereby decreasing surgical time and eliminating complications at the donor site, many surgeons use this technique to reconstruct the anterior cruciate ligament [42, 82, 85]. Less frequently, they use allografts to reconstruct other ligaments, including the medial collateral ligament of the knee.

Dense fibrous tissue allografts may consist of the substance of donor fascia, tendon, or ligament, or they may consist of tendon or ligament and their bone insertions. Grafts that include the bone insertions usually allow better immediate fixation to the recipient site bone and retain the structure of the insertion. Although dense fibrous tissue allografts that lack bone insertions heal to the recipient site [82], they may present greater problems with initial secure fixation to the host bone and possibly greater laxity after healing [67]. Because they do not include allograft bone, allografts consisting only of dense fibrous tissue have the potential for fewer problems with host immunologic reactions.

Examination of allografts at intervals after transplantation shows that host cells and vessels invade the allografts and that allografts heal to the host tissue [5, 82–84]. Although some allografts may remodel more slowly than autografts [77], most studies show that incorporation and remodeling of allografts closely resemble the incorporation and remodeling characteristics of autografts [66, 82, 84, 93, 97]. Following healing and remodeling, comparable types of allografts and autografts show similar strength and stiffness [66, 84, 93, 97].

Fresh allografts can incite a prominent inflammatory reaction including lymphocytic infiltration and synovitis [5]. Freezing the grafts decreases this reaction, and in animal experiments, frozen and freeze-dried allografts have not caused immunologic reactions that lead to destruction of the grafts [5, 46, 66, 84, 93], but animal and human studies show that they may elicit humeral and cellular immune responses [73, 86, 93]. Currently, most ligament allografts are preserved by deep freezing or freeze-drying. The available evidence suggests that these methods of preservation do not adversely alter the biomechanical properties of the grafts [66]. However, allograft sterilization may adversely affect graft mechanical properties.

Aging may also adversely affect the mechanical properties of allografts. With increasing age the stiffness and strength of human anterior cruciate ligaments and other dense fibrous tissues decline [99]. Therefore, older donors may not provide optimal allografts for ligament reconstruction.

Anterior Cruciate Ligament Allografts

Several groups of investigators have examined the results of bone–anterior cruciate ligament–bone allograft reconstructions of the anterior cruciate ligament in animals. One group determined the ultimate loads at failure of fresh frozen graft complexes in dogs [93]. Nine months after surgery the loads at graft failure were found to be 15% of control values. Another group performed a similar study but found that loads at graft failure were 89% of control values 9 months postoperatively [66]. A possible explanation for this difference in allograft strength relative to control values is that the first group's control values (mean maximum load 1160 N) [93] were about one order of magnitude greater than the control values reported by the second group (mean maximum load 115 N) [66]. The higher control values agree more closely with those reported by other investigators [23, 56, 100, 104]. The studies may also have differed in technique and stability of graft fixation.

A study of freeze-dried anterior cruciate ligament bone allograft reconstructions in goats agreed closely with the first study of allograft reconstructions in dogs. It showed that reconstructed knees in goats had significantly greater anterior posterior translation than did control knees 1 year after reconstruction and that stiffness and ultimate load values for the graft complexes were 35% and 25%, respectively, of control values [45, 46].

Other experiments have measured the strength of dense fibrous tissue allografts that do not include bone insertions. Thirty weeks after reconstruction of dog anterior cruciate ligaments with fresh frozen

patellar tendon allografts the ultimate load strength of the grafts was 29% of control values [84]. Reconstruction of dog anterior cruciate ligaments with freeze-dried flexor tendons produced similar results 9 months after surgery [96, 97]. Freeze-dried fascia lata reconstructions of dog anterior cruciate ligaments had ultimate load strengths of less than 20% of control values at 6 weeks, but their ultimate load strength increased to 67% of control values at 24 weeks [23].

Medial Collateral Ligament Allografts

Allograft reconstruction of the medial collateral ligaments has not been studied as extensively as allograft reconstruction of the anterior cruciate ligament. Because vascularized soft tissues surround medial collateral ligament allografts, cells and vessels can potentially invade these grafts more rapidly than is possible with anterior cruciate ligament allografts [41]. Medial collateral ligament allografts may remodel more rapidly and achieve greater average strength relative to control ligaments than anterior cruciate ligament allografts [41]. But the reported experiments in dogs and rabbits suggest that, as with allograft reconstruction of the anterior cruciate ligament, allograft reconstruction of the medial collateral ligament restores only part of the normal strength of the ligament [41, 77, 79].

Three weeks following reconstruction of the medial collateral ligament in dogs with fresh frozen dog patellar tendon allografts, new vessels and cells had invaded the peripheral regions, but the central region of the grafts lacked cells, and the collagen fiber bundles appeared fragmented [41]. Fifteen weeks following surgery, the grafts showed evidence of hypervascularity and longitudinal alignment of collagen fibrils. Thirty and 52 weeks following surgery, the hypervascularity remained, and alignment of the collagen fiber bundles resembled the alignment of fiber bundles in normal ligaments. One year following surgery the mean stiffness of the grafts did not differ from that of normal ligaments, ultimate load at graft failure was about 75% of control values, and the average peak stress of the grafts was about 40% of the value for normal ligaments.

Bone–medial collateral ligament–bone allografts in rabbits were significantly tighter than controls 3 weeks following surgery, and they remained tight, but the allografts had inferior structural and material properties [79]. Twelve and 48 weeks following surgery the grafts had only about 60% of the strength of controls. The investigators concluded that although the biomechanical properties of allografts slowly improve, they may never achieve normal values. These results suggest that when the allografts have healed to the recipient site tissues and have been revascularized by host blood vessels and repopulated with new cells, they may reach a state of equilibrium in which their material properties neither improve nor deteriorate. This equilibrium state may provide sufficient mechanical strength to withstand the loads associated with normal activities but may not be sufficient to withstand unusually high loads.

Xenografts

The availability of xenografts makes them a potentially attractive tissue for reconstruction of ligaments. The small number of appropriate human donors and the technical difficulties and expense of acquiring ideal allografts limit the availability of these tissues. Bovine xenografts from young animals could be acquired, prepared for use, and stored in large quantities, thereby reducing the cost of the tissue relative to allografts.

Experimental studies suggest that xenograft ligament reconstructions partially restore ligament function and that the host tissues invade and begin to incorporate and replace the xenograft collagen [48, 57, 58]. Bovine xenograft reconstructions of dog medial collateral ligaments had near normal loads to failure 4 months postoperatively, and the host fibroblasts and vessels progressively invaded the grafts [60]. The host cells appeared to synthesize new matrix within the grafts.

Early results of ligament reconstruction with xenografts in humans showed that the grafts could restore anterior cruciate ligament function, but a clinical follow-up study of 40 knees that had been treated for anterior cruciate deficiency with bovine tendon xenografts preserved in gluteraldyhide showed a high incidence of graft failure and synovitis [92]. About half the grafts ruptured between 12 and 20 months after the operation. Six of the first 30 knees developed severe synovitis within 8 months of surgery. Treatment of these knees included total synovectomy and graft removal. The authors then changed their graft-rinsing procedure for the last 10 knees and found no evidence of synovitis following this change.

Reconstituted Collagen Implants

Ligament prostheses formed from reconstituted collagen offer a possible alternative to xenografts

[4, 25, 37]. Preliminary experiments show that host cells do invade these implants and that they can develop strength similar to that of autogenous dense fibrous tissue grafts. In the future these biologic implants could be created in various sizes and shapes and with different concentrations and orientations of collagen fibrils, and they could incorporate cytokines or other growth factors that might stimulate cell migration, proliferation, and differentiation [4].

References

1. Amiel, D., Frank, C., Harwood, F., Fronek, J., and Akeson, W. Tendons and ligaments: a morphological and biochemical comparison. *J Orthop Res* 1:257–265, 1984.
2. Amiel, D., Kleiner, J. B., Roux, R. D., Harwood, F. L., and Akeson, W. H. The phenomenon of "ligamentization": anterior cruciate ligament reconstruction with autogenous patellar tendon. *J Orthop Res* 4:162–172, 1986.
3. Amiel, D., and Kuiper, S. Experimental studies on anterior cruciate ligament grafts. *In* Daniel, D., Akeson, W. H., and O'Connor, D. D. (Eds.), *Knee Ligaments: Structure, Function, Injury, and Repair.* New York, Raven Press, 1990.
4. Andriacchi, T., Sabiston, P., DeHaven, K., Dahners, L., Woo, S., Frank, C., Oakes, B., Brand, R., and Lewis, J. Ligament: Injury and repair. *In* Woo, S. L.-Y., and Buckwalter, J. D. (Ed.), *Injury and Repair of the Musculoskeletal Soft Tissues.* Park Ridge, IL, American Academy of Orthopaedic Surgeons, 1988.
5. Arnoczky, S. P., Warren, R. F., and Ashlock, M. A. Replacement of the anterior cruciate ligament using a patellar tendon allograft: An experimental study. *J Bone Joint Surg* 68-A:376–385, 1986.
6. Ballock, R. T., Woo, S. L.-Y., Lyon, R. M., Hollis, J. M., and Akeson, W. H. Use of patellar tendon autograft for anterior cruciate ligament reconstruction in the rabbit: A long-term histologic and biomechanical study. *J Orthop Res* 7:474–485, 1989.
7. Barrack, R. L., and Skinner, H. B. The sensory function of knee ligaments. *In* Daniel, D., Akeson, W. H., and O'Connor, D. D. (Eds.), *Knee Ligaments: Structure, Function, Injury, and Repair.* New York, Raven Press, 1990.
8. Bray, D. F., Frank, C. B., and Bray, R. C. Cytochemical evidence for a proteoglycan-associated filamentous network in ligament extracellular matrix. *J Orthop Res* 8:1–12, 1990.
9. Buckwalter, J. A. Cartilage. *In* Delbecco, R. (Ed.), *Encyclopedia of Human Biology.* San Diego, Academic Press, 1990.
10. Buckwalter, J. A., and Cooper, R. R. Bone structure and function. *Instr Course Lect* 36:27–48, 1987.
11. Buckwalter, J. A., and Cooper, R. R. The cells and matrices of skeletal connective tissues. *In* Albright, J. A., and Brand, R. A. (Eds.), *The Scientific Basis of Orthopaedics.* Norwalk, Appleton & Lange, 1987.
12. Buckwalter, J. A., and Cruess, R. Healing of musculoskeletal tissues. *In* Rockwood, C. A., and Green, D. P. (Eds.), *Fractures.* Philadelphia, J. B. Lippincott, 1991.
13. Buckwalter, J. A., Maynard, J. A., and Vailas, A. C. Skeletal fibrous tissues: Tendon, joint capsule, and ligament. *In* Albright, J. A., and Brand, R. A. (Eds.), *The Scientific Basis of Orthopaedics.* Norwalk, Appleton & Lange, 1987.
14. Burks, R. T. Gross anatomy. *In* Daniel, D., Akeson, W. H., and O'Connor, J. (Eds.), *Knee Ligaments: Structure, Function, Injury, and Repair.* New York.
15. Butler, D. L. Anterior cruciate ligament: Its normal response and replacement. *J Orthop Res* 7:910–921, 1989.

16. Butler, D. L., Grood, E. S., Noyes, F. R., Olmstead, M. L., Hohn, R. B., Arnoczky, S. P., and Siegel, M. G. Mechanical properties of primate vascularized vs. nonvascularized patellar tendon grafts; changes over time. *J Orthop Res* 7:68–79, 1989.

17. Cho, K. O. Reconstruction of the anterior cruciate ligament by semitendinosis tenodesis. *J Bone Joint Surg* 57-A:608–613, 1975.

18. Clancy, W. G., Narechania, R. G., Rosenberg, T. D., Gmeiner, J. G., Wisnefske, D. D., and Lange, T. A. Anterior and posterior cruciate ligament reconstruction in rhesus monkeys. *J Bone Joint Surg* 63-A:1270–1284, 1981.

19. Clayton, M. L., Miles, J. S., and Abdulla, M. Experimental investigations of ligamentous healing. *Clin Orthop Rel Res* 61:146–153, 1968.

20. Clayton, M. L., and Weir, G. J. Experimental investigations of ligamentous healing. *Am J Surg* 98:373–378, 1959.

21. Collins, H. R., Hughston, J. C., DeHaven, K. E., Bergfeld, J. A., and Evarts, C. M. The meniscus as a cruciate ligament substitute. *Am J Sports Med* 2(1):11–21, 1974.

22. Cooper, R. R., and Misol, S. Tendon and ligament insertion: A light and electron microscopic study. *J Bone Joint Surg* 52-A:1–21, 1970.

23. Curtis, R. J., Delee, J. C., and Drez, D. J. Reconstruction of the anterior cruciate ligament with freeze dried fascia lata allografts in dogs: A preliminary report. *Am J Sports Med* 13(6):408–414, 1985.

24. deKorompay, V. L., and Dessouki, E. Vascularized versus nonvascularized patellar tendon for intraarticular cruciate reconstruction: Histological and biomechanical analysis in dogs. *Am J Sports Med* 15(4):399, 1987.

25. Dunn, M. G., Tria, A. J., Bechler, J. R., Zawadsky, J. P., Kato, P., and Silver, G. H. Development of a reconstituted collagen anterior cruciate ligament prosthesis: Preliminary animal results. *Am J Sports Med* 18(5):554–555, 1990.

26. Ferkel, R. D., Fox, J. M., Pizzo, W. D., Friedman, M. J., Snyder, S. J., Dorey, F., and Kasimian, D. Reconstruction of the anterior cruciate ligament using a torn meniscus. *J Bone Joint Surg* 70-A:715–723, 1988.

27. Frank, C., Amiel, D., Woo, S. L.-Y., Akeson, W. Normal ligament properties and ligament healing. *Clin Orthop Rel Res* 196:15–25, 1985.

28. Frank, C., McDonald, D., Lieber, R. L., and Sabiston, P. Biochemical heterogeneity within the maturing rabbit medial collateral ligament. *Clin Orthop Rel Res* 236:279–285, 1988.

29. Frank, C., Schachar, N., and Dittrich, D. Natural history of healing in the repaired medial collateral ligament. *J Orthop Res* 1:179–188, 1983.

30. Frank, C., Woo, S., Andriacchi, T., Brand, R., Oakes, B., Dahners, L., Dehaven, K., Lewis, J., and Sabiston, P. Normal ligament: structure, function, and composition. *In* Woo, S. L.-Y., and Buckwalter, J. H. (Eds.), *Injury and Repair of the Musculoskeletal Soft Tissues.* Park Ridge, IL, American Academy of Orthopaedic Surgeons, 1988.

31. Frank, C., Woo, S. L.-Y., Amiel, D., Harwood, T., Gomez, M., and Akeson, W. Medial collateral ligament healing: A multidisciplinary assessment in rabbits. *Am J Sports Med* 11:379–389, 1983.

32. Frank, C. B. and Hart, D. A. The biology of tendons and ligaments. *In* Mow, V. C., Ratcliff, A., and Woo, S. L.-Y. (Eds.), *Biomechanics of Diarthrodial Joints.* New York, Springer-Verlag, 1990.

33. Friedman, M. J., Sherman, O. H., Fox, J. M., Pizzo, W. D., Snyder, S. J., and Ferkel, R. J. Autogenic anterior cruciate ligament (ACL) anterior reconstruction of the knee: A review. *Clin Orthop Rel Res* 196:9–14, 1985.

34. Fronek, J., Frank, C., Amiel, D., Woo, S. L.-Y., Cootts, R. D., and Akeson, W. H. The effects of intermittent passive motion (IPM) in the healing of medial collateral ligament. *Trans Orthop Res Soc.* 8:31, 1983.

35. Fulkerson, J. P., Berke, A., and Parthasarathy, N. Collagen biosynthesis in rabbit intraarticular patellar tendon transplants. *Am J Sports Med* 18(3):249–253, 1990.

36. Gillquist, J. Knee stability: Its effect on articular cartilage. *In* Ewing, D. D. (Ed.), *Articular Cartilage and Knee Joint Function: Basic Science and Arthoroscopy.* New York, Raven Press, 1990.

37. Goldstein, J. D., Tria, A. J., Zawadsky, J. P., Kato, Y. P., Christiansen, D., and Silver, F. H. Development of a reconstituted collagen tendon prosthesis. *J Bone Joint Surg* 71-A:1183–1191, 1989.

38. Hardingham, T. E. Proteoglycans: Their structure, interactions and molecular organization in cartilage. *Biochem Soc Trans* 9:489–497, 1981.

39. Hascall, V C. Interactions of cartilage proteoglycans with hyaluronic acid. *J Supramol Structure* 7:101–120, 1977.

40. Holden, J. P., Grood, E. S., Butler, D. L., Noyes, F. R., Mendenhall, H. V., Kampen C. L. V., and Neidich, R. L. Biomechanics of fascia lata ligament replacements: Early postoperative changes in the goat. *J Orthop Res* 6:639–647, 1988.

41. Horibe, S., Shino, K., Nagano, J., Nakamura, H., Tanaka, M., and Ono, K. Replacing the medial collateral ligament with an allogenic tendon graft: An experimental canine study. *J Bone Joint Surg* 72-B:1044–1049, 1990.

42. Indelicato, P. A., Bittar, E. S., Prevot, T. J., Woods, G. A., Branch, T. P., and Hugel, M. Clinical comparison of freeze-dried and fresh frozen patellar tendon allografts for anterior cruciate ligament reconstruction of the knee. *Am J Sports Med* 18(4):335–342, 1990.

43. Insall, J., Joseph, D. M., Aglietti, P., and Campbell, R. D. Bone-block iliotibial-band transfer for anterior cruciate insufficiency. *J Bone Joint Surg* 63-A:560–569, 1981.

44. Ivey, F. M., Blazina, M. E., Fox, J. M., and Pizzo, W. D. Intraarticular substitution for anterior cruciate insufficiency. *Am J Sports Med* 8(6):405–410, 1980.

45. Jackson, D. W., Grood, E. S., Arnoczky, S. P., Butler, D. L., and Simon, T. M. Cruciate reconstruction using freeze dried anterior cruciate ligament allograft and a ligament augmentation device (LAD): An experimental study in a goat model. *Am J Sports Med* 15(6):528–538, 1987.

46. Jackson, D. W., Grood, E. S., Arnoczky, S. P., Butler, D. L., and Simon, T. M. Freeze dried anterior cruciate ligament allografts: Preliminary studies in a goat model. *Am J Sports Med* 15(4):295–303, 1987.

47. Jones, K. C. Reconstruction of the anterior cruciate ligament: A technique using the central one-third of the patellar ligament. *J Bone Joint Surg* 45-A:925–932, 1963.

48. Jurgutis, J., Gambardella, R., Nimni, M., Marshall, G. J., Gendler, E., and Sarmiento, G. The replacement of anterior cruciate ligaments with heterologous implants. *Orthop Trans* 6:99–100, 1982.

49. Kennedy, J. C. Intraarticular replacement in the anterior cruciate ligament-deficient knee. *Am J Sports Med* 8(1):1–8, 1980.

50. Kleiner, J. B., Amiel, D., Harwood, F. L., and Akeson, W. H. Early histologic, metabolic, and vascular assessment of anterior cruciate ligament autografts. *J Orthop Res* 7:235–242, 1989.

51. Lambert, K. L. Vascularized patellar tendon graft with rigid internal fixation for anterior cruciate ligament insufficiency. *Clin Orthop Rel Res* 172:85–89, 1983.

52. Laros, G. S., Tipton, C. M., and Cooper, R. R. Influence of physical activity on ligament insertions in the knees of dogs. *J Bone Joint Surg* 53-A:275–286, 1971.

53. Long, M., Frank, C., Schachar, N., Dittrich, D., and Edwards, G. E. The effects of motion on normal and healing ligaments. *Trans Orthop Res Soc* 7:43, 1982.

54. Matyas, J. R., Brodie, D., Anderson, M., and Frank, C. B. The developmental morphology of a "periosteal" ligament insertion: Growth and maturation of the tibial insertion of the rabbit medial collateral ligament. *J Orthop Res* 8:412–424, 1990.

55. McCarthy, J. A., Steadman, J. R., Dunlap, J., Shively, R., and Stonebrook, S. A nonparallel, nonisometric synthetic graft augmentation of a patellar tendon anterior cruciate ligament reconstruction: A model for assessment of stress shielding. *Am J Sports Med* 18(1):43–49, 1990.

56. McFarland, E. G., Morrey, B. F., An, K. N., and Wood, M. B. The relationship of vascularity and water content to tensile strength in a patellar tendon replacement of the anterior cruciate in dogs. *Am J Sports Med* 14:436–448, 1986.

57. McMaster, W. C. Bovine xenograft collateral ligament replacement in the dog. *J Orthop Res* 3:492–498, 1985.

58. McMaster, W. C. A histologic assessment of canine anterior cruciate substitution with bovine xenograft. *Clin Orthop Rel Res* 196:196–201, 1985.

59. McPherson, G. K., Mendenhall, H. V., Gibbons, D. F., Plenk, H., and Roth, J. H. Experimental, mechanical and histologic evaluation of the Kennedy ligament augmentation device. *Clin Orthop Rel Res* 196:186–195, 1985.

60. Mello, M. S., Godo, C., Vidal, B. C., and Abujadi, J. M. Changes in macromolecular orientation on collagen fibers during the process of tendon repair in the rat. *Ann Histochem* 20:145–152, 1975.

61. Misou, A., Vallianatos, P., Piskopakis, N., and Nicolaou, P. Cruciate ligament replacement using a meniscus. *J Bone Joint Surg* 70-B:784–786, 1988.

62. Mott, W. H. Semitendinosis anatomic reconstruction for cruciate ligament insufficiency. *Clin Orthop Rel Res* 172:90–92, 1983.

63. Mow, V. C., Setton, L. A., Ratcliff, A., Howell, D. S., and Buckwalter, J. A. Structure-function relationship of articular cartilage and the effects of joint instability and trauma on cartilage. *In* Brandt, K. D. (Ed.), *Cartilage Changes in Osteoarthritis*. Indianapolis, Indiana University School of Medicine, 1990.

64. Muir, H. Proteoglycans as organizers of the extracellular matrix. *Biochem Soc Trans* 11:613–622, 1983.

65. Newton, P. O., Horibe, S., and Woo, S. L.-Y. Experimental studies on anterior cruciate ligament autografts and allografts. *In* Daniel, D., Akeson, W. H., and O'Connor, J. (Eds.), *Knee Ligaments: Structure, Function, Injury, and Repair*. New York, Raven Press, 1990.

66. Nikolaou, P. K., Seaber, A. V., Glisson, R. R., Ribbeck, B. M., and Bassett, F. H. Anterior cruciate ligament allograft transplantation: Long-term function, histology, revascularization, and operative technique. *Am J Sports Med* 14(5):348–360, 1986.

67. Noyes, F. R., Barber, S. D., and Mangine, R. E. Bone-patellar ligament-bone and fascia lata allografts for reconstruction of the anterior cruciate ligament. *J Bone Joint Surg* 72-A:1125–1136, 1990.

68. Noyes, F. R., Butler, D. L., Grood, E. S., Zernicke, R. G. and Hefzy, M. S. Biomechanical analysis of human ligament grafts used in knee-ligament repairs and reconstructions. *J Bone Joint Surg* 66-A:344–352, 1984.

69. Noyes, F. R., Butler, D. L., Paulos, L. E., and Grood, E. S. Intra-articular cruciate reconstruction: I. Perspectives on graft strength, vascularization, and immediate motion after replacement. *Clin Orthop Rel Res* 172:71–77, 1983.

70. O'Donoghue, D. H., Frank, G. R., Jeter, G. L., Johnson, W., Zeiders, J. W., and Kenyon, R. Repair and reconstruction of the anterior cruciate ligament in dogs: Factors influencing long-term results. *J Bone Joint Surg* 53-A:710–718, 1971.

71. Oakes, B. W. Experimental studies on the development, structure and properties of elastic fibers and other components of connective tissues. Ph.D. Thesis. Clayton, Australia, 1978.

72. Paulos, L. E., Butler, D. L., Noyes, F. R., and Grood, E. S. Intra-articular cruciate reconstruction: I. Replacement with vascularized patellar tendon. *Clin Orthop Rel Res* 172:78–84, 1983.

73. Pinkowski, J. L., Reiman, P. R., and Chen, S.-L. Human lymphocyte reaction to freeze-dried allograft and xenograft ligamentous tissue. *Am J Sports Med* 17(5):595–600, 1989.

74. Poole, A. R., Webber, C., Pidoux, I., Choi, H., and Rosenberg, L. C. Localization of a dermatan sulfate proteoglycan (DS-PGII) in cartilage and the presence of an immunologically related species in other tissues. *J Histochem Cytochem* 34(5):619–625, 1986.

75. Rosenberg, L., Choi, H. U., Neame, P. J., Sasse, J., Roughley, P. J., and Poole, A. R. Proteoglycans of soft connective tissue. *In* Leadbetter, W. B., Buckwalter, J. A., and Gordon, S. L. (Eds.), *Sports Induced Inflammation—Basic Science and Clinical Concepts*. Park Ridge, IL, American Academy of Orthopaedic Surgeons, 1990.

76. Ryan, J. R., and Drompp, B. W. Evaluation of tensile strength of reconstructions of the anterior cruciate ligament using the patellar tendon in dogs. *So Med J* 59:129–134, 1966.

77. Sabiston, P., Frank, C., Lam, T., and Shrive, N. Allograft transplantation: A morphological and biochemical evaluation of a medial collateral ligament complex in a rabbit model. *Am J Sports Med* 18(2):160–168, 1990.

78. Sabiston, P., Frank, C., Lam, T., and Shrive, N. Transplantation of the rabbit medial collateral ligament. I. Biomechanical evaluation of fresh autografts. *J Orthop Res* 8:35–45, 1990.

79. Sabiston, P., Frank, C., Lam, T., and Shrive, N. Transplantation of the rabbit medial collateral ligament. II. Biomechanical evaluation of frozen/thawed allografts. *J Orthop Res* 8:46–56, 1990.

80. Sgaglione, N. A., Warren, R. F., Wickiewicz, T. L., Gold, D. A., and Panariello, R. A. Primary repair with semitendinosus tendon augmentation of acute anterior cruciate ligament injuries. *Am J Sports Med* 18(1):64–73, 1990.

81. Shelbourne, K. D., Whitaker, H. J., McCarroll, J. R., Rettig, A. C., and Hirschman, L. D. Anterior cruciate ligament injury: Evaluation of intraarticular reconstruction of acute tears without repair: two to seven year followup of 155 athletes. *Am J Sports Med* 18(5):484–489, 1990.

82. Shino, K., Inoue, M., Horibe, S., Hamada, M., and Ono, K. Reconstruction of the anterior cruciate ligament using allogeneic tendon: Long-term followup. *Am J Sports Med* 15(5):457–465, 1990.

83. Shino, K., Inque, M., Horibe, S., Nagano, J., and Ono, K. Maturation of allograft tendons transplanted into the knee. *J Bone Joint Surg* 70-B:556–560, 1988.

84. Shino, K., Kawasaki, T., Hirose, H., Gotoh, I., Inque, M., and Ono, K. Replacement of the anterior cruciate ligament by an allogeneic tendon graft: An experimental study in the dog. *J Bone Joint Surg* 66-B:672–681, 1984.

85. Shino, K., Kimura, T., Hirose, H., Inoue, M., and Ono, K. Reconstruction of the anterior cruciate ligament by allogeneic tendon graft. *J Bone Joint Surg* 68-B:739–746, 1986.

86. Thorson, E. P., Rodrigo, J. J., Vasseur, P. B., Sharkey, N. A., and Hietter, D. O. Comparison of frozen allograft versus fresh autogenous anterior cruciate ligament replacement in the dog. *Orthop Trans* 12:65, 1987.

87. Tillberg, B. The late repair of torn cruciate ligaments using menisci. *J Bone Joint Surg* 59-B:15–19, 1977.

88. Tipton, C. M., James, S. L., Mergner, W., and Tcheng, T.-K. Influence of exercise on strength of medial collateral ligaments of dogs. *Am J Physiol* 218:894–901, 1970.

89. Tomford, W. W., Thongphasuk, J., Mankin, H. J., and Ferraro, M. J. Frozen musculoskeletal allografts: a study of the clinical incidence and causes of infection associated with their use. *J Bone Joint Surg* 72-A:1137–1143, 1990.

90. Vailas, A. C., Tipton, C. M., Matthes, R. D., and Gart, M. Physical activity and its influence on the repair process of medial collateral ligaments. *Connect Tissue Res* 9:25–31, 1981.

91. van-Rens, T. G., Berg, A. F. v. d., Huiskes, R., and Kuypers, W. Substitution of the anterior cruciate ligament: A long-term histologic and biomechanical study with autogenous pedicled grafts of the iliotibial band in dogs. *Arthroscopy* 2(3):139–154, 1986.

92. Van-Steensel, C. J., Schreuder, O., Bosch, B. F. V. D., Paassen, H. C. V., Menke, H. E., Voorhorst, G., and Gratama, S. Failure of anterior cruciate-ligament reconstruction using tendon xenograft. *J Bone Joint Surg* 69-A:860–864, 1987.

93. Vasseur, P. B., Rodrigo, J. J., Stevenson, S., Clark, G.,

and Sharkey, W. Replacement of the anterior cruciate ligament with a bone-ligament-bone anterior cruciate ligament allograft in dogs. *Clin Orthop Rel Res* 219:268–277, 1987.

94. Viidik, A. Simultaneous mechanical and light microscopic studies of collagen fibers. *Z Anat Entwickl-Ges.* 136:204–212, 1972.

95. Walsh, J. J. Meniscal reconstruction of the anterior cruciate ligament. *Clin Orthop Rel Res* 89:171–177, 1972.

96. Webster, D. A. Freeze-dried flexor tendons in anterior cruciate ligament reconstruction. *Clin Orthop Rel Res* 181:238–243, 1983.

97. Webster, D. A., and Werner, F. W. Mechanical and functional properties of implanted freeze-dried flexor tendons. *Clin Orthop Rel Res* 180:301–309, 1983.

98. Woo, S., Maynard, J., Butler, D., Lyon, R., Torzilli, P., Akeson, W., Cooper, R., and Oakes, B. Ligament, tendon, and joint capsule insertions into bone. *In* Woo, S. L-Y., and Buckwalter, J. A. (Eds.), *Injury and Repair of the Musculoskeletal Soft Tissues.* Park Ridge, IL, American Academy of Orthopaedic Surgeons, 1988.

99. Woo, S. L.-Y., and Adams, D. J. The tensile properties of human anterior cruciate ligament (ACL) and ACL graft tissues. *In* Daniel, D., Akeson, W. H., and O'Connor, J. (Eds.), *Knee Ligaments: Structure, Function, Injury, and Repair.* New York, Raven Press, 1990.

100. Woo, S. L.-Y., and Buckwalter, J. A. Ligament and tendon autografts and allografts. *In* Friedlander, G. E., and Goldberg (Eds.), *Biologic Restoration of Bone and Articular Surfaces.* Park Ridge, IL, American Academy of Orthopaedic Surgeons, 1991.

101. Woo, S. L.-Y., Gomez, M. A., Seguchi, Y., Endo, C. M., and Akeson, W. H. Measurement of mechanical properties of ligament substance from a bone-ligament-bone preparation. *J Orthop Res* 1:22–29, 1983.

102. Woo, S. L.-Y., Horibe, S., Ohland, K. J., and Amiel, D. The response of ligaments to injury: Healing of the collateral ligaments. *In* Daniel, D., Akeson, W. H., and O'Connor, J. (Eds.), *Knee Ligaments: Structure, Function, Injury, and Repair.* New York, Raven Press, 1990.

103. Yoshiya, S., Andrish, J. T., Manley, M. T., and Bauer, T. W. Graft tension in anterior cruciate ligament reconstruction: An in vivo study in dogs. *Am J Sports Med* 15(5):464–470, 1987.

104. Yoshiya, S., Andrish, J. T., Manley, M. T., and Kurosaka, M. Augmentation of anterior cruciate ligament reconstruction in dogs with prostheses of different stiffness. *J Orthop Res* 4:475–485, 1986.

105. Zaricznyj, B. Reconstruction of the anterior cruciate ligament using free tendon graft. *Am J Sports Med* 11:164–176, 1983.

Effects of Repetitive Loading and Motion on the Musculoskeletal Tissues

Joseph A. Buckwalter, M.D.
Savio L.-Y. Woo, Ph.D.

Few subjects are as important for the practice of sports medicine as the effects of loading and motion on the musculoskeletal tissues. Maintenance of the structural integrity and function of the musculoskeletal system depends on the responses of the tissues to repetitive loading. An understanding of the effects of training and the results of decreased training and immobilization is required to understand overuse injuries, and treatment of these injuries requires a knowledge of the responses the musculoskeletal tissues to repetitive loading. Changes in the patterns of tissue loading can strengthen or weaken normal tissues, and controlled loading and motion can accelerate rehabilitation and repair of injured tissues [22]. Excessive, premature, or uncontrolled loading and motion of repaired tissue can inhibit or stop repair [22].

Although the muscle, bone, cartilage, tendon, and ligament of skeletally mature individuals and the specific structures formed from these tissues may not appear to change over time, the cells in these tissues are degraded and replaced throughout life [20]. If replacement balances degradation, in both rate and type of tissue produced, the composition and volume of the tissue remain constant. A persistent imbalance between replacement and degradation progressively changes the composition and sometimes the volume of the tissue forming a specific structure.

Cells can also alter the organization of a tissue by changing their shape and the alignment of matrix macromolecules [25, 53, 55, 63, 71, 79, 124, 133]. Cells change the tissue organization extensively during growth and development and during remodeling of repair tissue, but they also alter normal mature tissues in response to changes in repetitive loading. Changes in the orientation and organization of the matrix macromolecules of bone, cartilage, ligament, and tendon can alter the mechanical properties of the tissues without altering their volume or composition.

The rate of tissue turnover and the composition, structure, and mechanical properties of these tissues depend not only on the genome of the cells but also on the repetitive loading of the tissue. The cells respond to persistent changes in tissue stresses and strains due to external loadings by changing the balance between their synthetic and degradative activities and—in at least some tissues—between their proliferative activity and the organization of the tissue. In general, decreased stresses and strains decrease cell synthetic function and the organization of the matrix, whereas increased stresses and strains, up to a critical level, increase cell synthetic function and matrix organization. Stresses and strains above the critical level will disrupt the tissue or cause tissue degeneration. Immature tissues may be more responsive to repetitive loading than mature tissues [30]. During formation and growth of the musculoskeletal system repetitive loading may influence not only the matrix composition and organization and tissue volume but also the type of tissue formed and the shape of tissue structures like bones and joints [32, 33]. For these reasons, the loading history of a tissue influences the state of the tissue, and an alteration in the pattern of loadings can alter this state.

Adjustment or adaptation of tissues to persistent

increases or decreases in stresses and strains in the tissue can influence the function of the musculoskeletal system. Adaptation of tissues to increased repetitive loading makes possible the increases in tissue strength necessary for athletic activities and rehabilitation of tissues following injury. Failure of the tissues to adapt to increased loading can cause disruption or degeneration of the tissue. Adaptation of tissues to decreased loading decreases tissue strength and increases the probability of mechanical failure.

The first part of this section reviews the current concepts of the relationships between repetitive loading and maintenance of normal tissue composition, structure, and mechanical properties. Subsequent parts discuss the adaptation of musculoskeletal tissues to persistent changes in loading, the mechanisms of the adaptive response, and the consequences of failure of the adaptive response. The last part summarizes the adaptive responses of normal muscle, bone, dense fibrous tissues, and cartilage to increased and decreased repetitive loading and the effects of repetitive loading on repair tissue.

MAINTENANCE OF TISSUE COMPOSITION, STRUCTURE, AND MECHANICAL PROPERTIES

The type, intensity and frequency, of loadings necessary to maintain the normal composition, structure, and mechanical properties of most tissues vary over a broad range. Variation of loading within this range does not appear to alter the tissue. Only when the intensity and frequency of loading exceeds or falls below the level necessary to maintain the tissue does the balance between synthesis and degradation change, causing a change in the mechanical properties of the tissue. For this reason, moderate increases or decreases in exercise patterns do not cause detectable alterations in the tissues.

Although experimental studies of musculoskeletal tissues show that maintenance of tissue integrity requires repetitive loading to balance degradation and synthesis, these studies have not defined the range of specific loading patterns nor the levels necessary to produce this balance. They have shown that in the absence of loading, tissue degradation exceeds replacement and the tissue atrophies and that because tissues vary in their sensitivity to loading, the same loadings produce different responses in different tissues. The response of a tissue to loading, and therefore the range of loading that will maintain the tissue integrity, also varies with the

age of the individual and probably with a variety of other systemic factors including nutritional state and hormonal balance.

Maintenance of specific structures such as articular cartilage, periarticular dense fibrous tissues (including joint capsules and some ligaments), and tendons requires more than a balance between degradation and synthesis. In addition to the volume and composition of the tissue, its internal cellular and molecular organization, the shape of the structure, and the relationships with surrounding tissues must be maintained. Cyclic compression stimulates proteoglycan synthesis by chondrocytes, and cyclic tension stimulates synthesis of type I collagen by tendon fibroblasts, but cyclic compression alone will not maintain the integrity and function of patellofemoral joint articular cartilage, and cyclic tension alone will not maintain the integrity and function of the flexor pollicis longus tendon. These tissue structures require repetitive motion and complex patterns of loading to maintain their internal organization, shape, and normal relationships with surrounding tissues.

TISSUE ADAPTATION TO REPETITIVE LOADING

When repetitive loading and, for some tissues, motion exceed or fall below the range necessary to maintain the tissue, the cells alter the composition and organization and thus the mechanical properties of the tissue [45, 129]. That is, they adjust or adapt the tissue to the pattern of repetitive loading. Like the range of loading necessary to maintain the integrity of the tissue, the degree of change in loading pattern necessary to cause an adaptive response varies according to the tissue involved, the individual, age, and other systemic variables. Some tissues such as muscle and bone respond quickly to a change in the pattern of repetitive loading, but others show little or no adaptive response except with prolonged unloading and immobilization. The variability of the response can cause an imbalance in the strength of different tissues in an individual and makes it difficult to predict the results of a training program or a period of decreased activity for a specific individual.

Decreased Loading

When loading falls below the frequency and intensity necessary to maintain the tissue, synthesis fails to keep pace with degradation, and the organization of the matrix macromolecules decreases. These changes weaken the tissues. For example, immobilization of a leg in a long leg nonweight-bearing cast for 6 weeks decreases the strength of the bone and the tendon and ligament insertions. Immobilization also alters the composition, matrix molecular organization, and mechanical properties of the articular cartilage. Less extreme decreases in loadings and motion can also alter the composition, structure, and mechanical properties of the tissues. These alterations increase the probability of injury. For example, if a competitive athlete stops training for several months or more and then attempts to resume abruptly his or her former maximum level of activity, the weakened tissues may fail to withstand the loads.

Increased Loading

When loading exceeds the frequency and intensity necessary to maintain the tissue, cell synthetic function may exceed degradation, and the organization of the matrix macromolecules may change to accommodate the increased loads. This adaptive response to repetitive increased loading can increase the strength and sometimes the volume of the tissue, thereby improving the performance of the musculoskeletal system. Conditioning or training programs that progressively increase the frequency and intensity of exercise rely on this adaptive response. If it fails to occur or if the intensity of training exceeds the capacity of the adaptive response, the strength of the tissues will not increase to meet the demands imposed by training, and the individual may suffer an injury.

MECHANISMS OF TISSUE ADAPTATION

Physicians have recognized that musculoskeletal tissues adapt to changes in repetitive loading for at least 250 years. In his thesis presented in 1723, Nicholas André noted that moderate exercise strengthens and shapes the musculoskeletal tissues and that carefully applied persistent loading will correct bony deformities [10]. In the later part of the nineteenth century, Just Lucas-Championniére taught that rest injured cartilage, ligaments, and muscles whereas mobility accelerated repair of these tissues following injury [68]. During the same period, Julius Wolff stressed the ability of bone to adapt to changes in repetitive loading [68]. Subsequently, other physicians and scientists documented

the validity of these clinical observations, but the cellular mechanisms of musculoskeletal tissue adaptation remained unexplained.

Experimental studies reported during the last 25 years have documented the effects of exercise on bone, tendon, ligament, and cartilage, and recent investigations of isolated tissues and cells have shown that repetitive loading of various types and intensities influences cell shape and cell synthetic and proliferative functions [11, 16, 19, 38, 71]. Furthermore, they have shown that loading alters the alignment of the matrix [88]; the matrix can transmit loads to cells, and the cells can realign the matrix macromolecules in response to the loads [55, 71]. These recent studies have helped define repetitive loading more precisely in terms of its intensity, duration, frequency, and pattern, shown that cells modify tissues in response to loads, and measured the tissue responses in terms of cell synthesis of specific matrix molecules, cell proliferation, matrix organization, and tissue mechanical properties. Despite these advances in ability to show the results of cyclic mechanical forces on cells and tissues, the systemic and local mechanisms of tissue responses to repetitive loading remain poorly understood.

Systemic Mechanisms

Regular exercise causes systemic changes including alterations in vascular perfusion, metabolism, collagen turnover [61], and hormonal balance. These changes may influence the adaptive response of specific tissues to repetitive loading, possibly by altering the sensitivity of the tissues to cyclic loads. The influence of systemic changes on local tissue responses to repetitive loading has not been studied, partially because of difficulties in isolating the potential effects of different systemic changes associated with training.

Local Mechanisms

The local mechanism of adaptation to loading consists of detection by cells of repetitive tissue strains and their response to these strains by modification of tissue. Experimental studies of the local effects of tissue loading show that cells may detect tissue strain through either deformation of the cells or alterations in the matrix due to deformation of the tissue. Cyclic stretching or compression of mesenchymal cells can align intracellular microfilaments along the axis of tension, change cell shape [5, 74], and alter synthesis of DNA, matrix molecules, and prostaglandins [16, 54, 57, 77, 91, 99, 109, 115, 149]. Deformation of the matrix can alter matrix macromolecular organization, fluid flow, streaming currents, pressure gradients, or electrical fields and these matrix alterations can influence cell function [54, 56, 83, 88, 91, 105, 114, 138]. Loading of a tissue also can affect vascular perfusion and diffusion through the matrix, thereby affecting the flow of nutrients and metabolites [12]. Cells may also sense and respond to deformation or damage to matrix macromolecules that result from repetitive loading. The function of the nervous system in the adaptation of bone, muscle, and dense fibrous tissue remains uncertain. Unlike other tissues, cartilage lacks innervation, and thus its local mechanisms of tissue adaptation must not depend on the direct influence of peripheral nerves.

Local Mechanisms of Adaptation in Skeletal Muscle

At least some local mechanisms of skeletal muscle adaptation may differ significantly from mechanisms used in other tissues [17, 31, 41]. Musculoskeletal tissue cells other than muscle cells respond to tension, compression, or shear forces applied to the tissue by producing changes in matrix synthesis, cell proliferation, and matrix organization. Muscle also can respond to repetitive passive stretching, and muscle cells can alter their extracellular matrix [31]. Yet, unlike the cells of other tissues, skeletal muscle cells actively generate force and change shape as they contract, and they respond to persistent changes in the pattern of repetitive contractions primarily by initiating changes in cell structure, volume, and function.

FAILURE OF TISSUE ADAPTATION

Overuse injuries—tissue disruption and inflammation or tissue degeneration resulting from repetitive loading—occur frequently in association with conditioning programs [34]. These injuries cause pain and impair tissue function. In many instances, they result from a failure of tissue adaptation [20]. The adaptive response may fail to meet the increased demands imposed by a training program because the repetitive loadings exceed the capacity of the adaptive response to modify the tissue. Loadings that exceed the tissue's strength disrupt the tissue, causing inflammation and repair. Not all tissue disruptions produce macroscopically apparent injury: Grossly intact tissue may nevertheless have

suffered extensive damage to the macromolecular framework of the tissue matrix that leaves the tissue significantly weakened [69]. Levels of loading lower than those that disrupt the tissue and cause inflammation can disturb tissue turnover or the adaptive response and cause degeneration of the tissue. Degeneration may weaken a tissue but does not necessarily result in disruption.

Tissue Disruption

Patients and physicians readily recognize tissue disruption due to acute single loads. They often have more difficulty in recognizing injuries caused by repetitive loading. Acute forced rotation and translation of the tibia relative to the femur can rupture the anterior cruciate ligament, and a blow to the forearm can fracture the ulna. These acute injuries cause mechanical failure of the tissue and initiate the injury and repair processes. Repetitive loading that exceeds the strength of the tissue also can rupture or fracture the matrix, damage or kill cells, and tear vessels. A muscle tendon junction may fail as a result of repetitive loading, and increased exercise may cause a stress fracture. Tissue damage initiates inflammation in vascularized tissues, including release of inflammatory mediators and growth factors and migration of inflammatory cells to the site of injury. Repair (i.e., replacement of damaged or lost cells and matrix with new cells and matrix) follows inflammation.

Tissue Degeneration

In dense fibrous tissues such as tendon, disturbances in tissue turnover or in the tissue adaptive response can cause degeneration. Degeneration alters the organization and composition of the matrix and probably cell metabolism as well [34]. The tissue remains grossly intact with little or no clinically apparent inflammation. Light microscopic examination of the tissue usually shows edema with minimal vascular proliferation or inflammatory cell infiltration. Electron microscopic examination often shows disruption of collagen microfibrils and splitting or fragmentation of collagen fibrils. In some areas the collagen fibrils lose their orientation relative to the long axis of the tendon. In other areas a loose myxoid matrix accumulates between regions of dense collagen fibrils. These degenerative changes in the tissue decrease the strength of the tissue and allow increased deformation of the tissue with loading. The weakened tissue may rupture more easily, and loading of degenerated tissue may cause pain.

THE RESPONSES OF SPECIFIC TISSUES TO REPETITIVE LOADING

Although all musculoskeletal tissues can respond to repetitive loading, they vary in magnitude and type of response to specific patterns of loading. Skeletal muscle and bone provide the most easily demonstrated and extensively studied examples of the adaptive response. Cartilage and dense fibrous tissues also respond to loading, but the responses are more difficult to measure. Repair tissue, like immature normal tissues [102], may be more sensitive to cyclic loading and motion than mature normal tissues. The available evidence shows that optimal remodeling of musculoskeletal repair tissue requires controlled loading of progressively increasing intensity and frequency, but excessive or uncontrolled loading or loading applied too early following injury may interfere with repair [22, 127]. The responses of tissues can be separated into those resulting from decreased loading of normal tissues, increased loading of normal tissues, and loading of repair tissue.

Skeletal Muscle

Decreased Use of Normal Skeletal Muscle

Decreased use of skeletal muscle due to a change in training or injury causes easily detectable changes in muscle volume, structure, and function. Within weeks of a decrease in training frequency or intensity, muscle mass decreases, individual muscles lose definition, and strength declines. Myofiber and myofibril volume decreases, and the oxidative capacity of the muscle declines.

Immobilization produces more severe loss of muscle structure, volume, and function. Two weeks of immobilization decreases muscle fiber size and causes loss of myofibrils [35]. As the duration of immobilization increases, the mitochondria enlarge, lose their cristae, and disintegrate. Eventually the muscle cells contain only amorphous protein, vesicles, and fragments of membranes. As the myofibers degenerate, fibrous tissue and fat begin to fill the interfiber spaces. Changes in muscle volume and function accompany these structural alterations [35]. Six weeks of cast immobilization decreased the weight of cat muscles nearly 25%, and 22 weeks of cast immobilization decreased muscle weight nearly 70%. The ability of muscles to generate tension decreased as muscle weight decreased.

If the cell membranes, blood supply, and nerve supply remain intact, immobilized muscle has the potential to regain normal structure, volume, and function [31, 35]. Within days of resumption of regular muscle use, the myofibers begin to form new myofibrils and mitochondria, and cell volume and strength increase.

Increased Use of Normal Skeletal Muscle

Persistent increases in use also change the structure, functional capacity, and often the volume of skeletal muscle [17, 86]. The specific adaptive changes that occur in muscle depend on the pattern of increased use. Three patterns that produce different muscle responses are [31] (1) low-tension, high-repetition use, which primarily increases muscle endurance; (2) high-tension, low-repetition use which primarily increases muscle strength; and (3) stretching, which primarily increases muscle strength. Initially, muscle may respond to a training program with rapid changes in structure and function, but as adaptation occurs the rate of change decreases, and eventually the muscle reaches a stable state.

Low-tension high-repetition exercise like walking, running, cycling, or swimming, performed for 30 to 60 minutes at a time, increases the capacity of muscle cells for sustained effort. This type of endurance training increases the number and size of muscle cell mitochondria, muscle glycogen concentration, and the proportion of muscle cells identified as having high oxidative capacity [31, 87]. These changes can double the oxidative capacity of the muscle [31, 87].

Strength training programs usually consist of high-tension, low-repetition muscle use. These programs increase muscle strength and usually volume, primarily by causing cell hypertrophy, that is, increasing the number of myofibrils. Strength training programs generally do not increase muscle oxidative capacity. Stretching of muscles also increases strength [31]. The tension generated by stretching accelerates muscle protein turnover and can cause hypertrophy.

Use of Injured Muscle

Although the effects of loading injured muscle have not been extensively studied, controlled loading and motion may promote muscle repair and regeneration [22]. A study of crushing injury to rat muscle showed that early mobilization accelerated resolution of the hematoma and inflammation, muscle regeneration, and increases in muscle strength [65, 66].

Bone

Decreased Loading of Normal Bone

Persistent decreases in cyclic loading of bone cause removal of bone cells to exceed replacement. Osteoclasts resorb trabecular and cortical bone, and osteoblasts fail to replace enough bone to maintain the mass and strength of the tissue [21]. Cessation of training may not produce readily detectable changes in bone volume, shape, and strength like those seen in skeletal muscle, but prolonged immobilization of a limb causes radiographic changes including decreased density of cancellous bone, loss of trabeculae, and thinning and increased porosity of cortical bone. Regular contraction of the muscles of an immobilized limb may decrease the loss of bone [28], but without regular loading due to weight bearing bone mass declines to less than half the normal value after 12 weeks or more of immobilization [73, 126]. These alterations decrease bone strength and increase the probability of fracture.

Increased Loading of Normal Bone

Persistent increases in cyclic loading of bone cause formation of bone to exceed bone removal and can result in dramatic increases in bone volume and strength [21, 108]. Experimental removal of the ulnar diaphysis in young pigs showed the capacity of bone to adapt rapidly to persistent increased loading [52]. Three months after surgery, the bone volume of the remaining overloaded radius approached that of the normal ulna and radius combined. Athletes who repetitively use one arm more than the other, such as tennis players or baseball pitchers, increase the bone mass in the arm they use most frequently.

Loading of Injured Bone

Optimal fracture healing appears to require at least some cyclic loading of the repaired tissue; it may stimulate formation and mineralization of fracture callus and promote remodeling of the repair

tissue [22]. Experimental studies show that denervation retards fracture healing [104], possibly by decreasing loading of the repair tissue or by inhibiting the function of growth factors that require activation by neurotransmitters. Previous exercise [60] and loading following injury can increase the rate of fracture healing in animals [51, 113], and clinical experience shows that early or even immediate loading of long bone fractures does not impair, and may even promote, fracture healing [22, 37, 85, 112].

Dense Fibrous Tissues

Decreased Loading of Normal Dense Fibrous Tissue

Decreases in loading and motion of dense fibrous tissues change their matrix composition, morphology, and mechanical properties. Unloading dense fibrous tissues that normally resist tensile stresses alters matrix turnover so that with time, matrix degradation exceeds formation, the newly synthesized matrix is less well organized or aligned along the lines of stress, and tissue stiffness and strength decline. Animal studies show that prolonged immobilization (usually 6 weeks or more) decreases the glycosaminoglycan and water content of ligamentous, capsular, and tendonous tissue, decreases the degree of orientation of the matrix collagen fibrils, and may increase collagen crosslinking [2, 4, 7, 8, 23, 119–123, 136, 145, 148]. With prolonged immobilization both total ligament collagen mass and ligament stiffness decline. Decreased use of rat limbs due to denervation increased matrix turnover and caused a net loss of collagen mass; collagen synthesis and degradation both increased so that the proportion of new collagen relative to old collagen increased, but collagen synthesis did not balance degradation, resulting in a net loss of collagen [72]. Allowing active joint motion in dogs while preventing weight bearing for 8 weeks decreased bone density but did not cause resorption or weakening of the knee ligaments [73]. These results suggest that maintenance of bone structure and mechanical properties requires weight bearing, but loading due to muscle contractions and active motion will maintain the composition and mechanical properties of the periarticular dense fibrous tissues for at least 2 months.

Decreased loading also alters ligament insertions into bone [23, 76, 93, 140, 142]. The extent and severity of the alterations depend to some extent on the type of ligament insertion. In some tendon, ligament, and joint capsule insertions (called direct insertions; see Section B of this chapter), most collagen fibrils pass directly into the bone matrix through a series of well-defined zones that include the substance of the tendon, ligament, or joint capsule, a zone of fibrocartilage, a zone of calcified cartilage, and finally bone [140]. In other insertions (called indirect or periosteal insertions), many of the collagen fibrils join the periosteum, and relatively few fibrils pass obliquely into the substance of the bone matrix [140].

Decreased ligament loading due to immobilization usually produces more extensive changes in the periosteal type of insertion. In these insertions, subperiosteal osteoclasts resorb much of the bony insertion of ligaments subjected to prolonged immobilization, leaving the ligament attached primarily to periosteum [140]. In direct insertions, resorption occurs around the insertion, but relatively little resorptive activity occurs within the insertion. The cruciate ligaments are an example of a direct ligament insertion, and the tibial insertion of the medial collateral ligament is an example of the periosteal type of insertion. Prolonged immobilization causes bone resorption around the periphery of the cruciate ligament insertion but only limited resorptive activity beneath the insertion site and in the zone of mineralized fibrocartilage [92, 93]. In contrast, prolonged immobilization causes significant diffuse resorption of the bony part of the tibial insertion of the medial collateral ligament [76, 140, 142]. These changes, particularly those in the periosteal insertions, weaken the bone ligament junction significantly within 6 to 8 weeks.

Following resumption of normal joint loading, the cells in the insertion site begin to form new bone, restoring the structure and mechanical properties of the insertions toward normal, but complete restoration of the structure and strength of the insertions site following 6 to 8 weeks of immobilization requires a longer period of active loading [7, 148]. Six to eight weeks of activity following immobilization of dog knees left ligament insertions significantly weaker than normal insertions, and the available evidence suggests that complete restoration of normal ligament insertion structure and mechanical properties requires up to 1 year of activity [76, 93, 142]. Muscle contractions alone did not prevent changes due to decreased ligament loading because isotonic exercises during immobilization did not prevent weakening of ligament insertions [93].

Increased Loading of Normal Dense Fibrous Tissues

Experimental studies show that repetitive exercise can increase the strength, size, matrix organization, and possibly collagen content of tendons and ligaments and their insertions into bone [1, 7, 22, 83, 84, 88, 119–121, 130–133, 140, 141, 146–148, 150, 151]. These studies cannot separate the systemic effects of exercise from the local effects of repetitive loading on the tissues, but investigations of the response of dense fibrous tissues to loading show that the tissues respond directly to increased loadings. Application of tension to cultured tendons increased protein and DNA synthesis [115], and a recent in vivo study shows that increased loading alone can cause adaptation of dense fibrous tissue [147]. Insertion of a pin underneath rabbit medial collateral ligaments increased the load on the ligaments by 200% to 350%. Over a 12-week period the presence of the pin significantly increased the strength of the bone ligament complex.

Aging decreases ligament stiffness and strength and may decrease the adaptive response to repetitive loading. In one experiment a life-long training program did not improve the mechanical properties of canine medial collateral ligaments or flexor tendons or prevent age-related deterioration of the mechanical properties of the dense fibrous tissue [137]. From this work, it appears that age-related changes in ligaments may eventually negate the potential benefits of training.

Dense fibrous tissues respond not only to changes in intensity and frequency of loading but also to changes in the type of loading. Tendon regions regularly subjected to tension during normal activities differ from regions regularly subjected to compression in tissue structure, matrix composition, and cell synthetic activity [6, 49, 82]. Tendon regions subjected primarily to tension consist of linearly arranged dense collagen fibrils and elongated cells and have a lower proteoglycan content, different proportions of proteoglycan types, and a higher rate of collagen synthesis than regions subjected to compression as well as tension [49, 75, 82, 134, 135]. Tendon regions subjected primarily to compression consist of a network of collagen fibrils separated by a proteoglycan-containing matrix and more rounded cells than those found in tension-bearing regions, and the cells of these regions synthesize larger proteoglycans than the cells of the regions subjected primarily to tension [82]. These differences may be caused, at least in part, by differences in the type of loading. Subjecting tendons to compression increased the hyaluronic acid and chondroitin sulfate content, whereas applying tension to the same tendon decreased the glycosaminoglycan content [49]. Like tendons, different ligaments and even regions of the same ligament vary in thickness, matrix composition, and water content [44]. Presumably at least some of these differences result from adaptation to differences in loading.

Loading of Injured Dense Fibrous Tissue

Controlled loading applied at the optimal time during repair of an injured dense fibrous tissue can promote healing [9, 22, 23, 27, 29, 43, 47, 64, 80, 81, 103, 119–121, 127, 128, 143]. Tensile loading of tendon repair tissue appears to cause the repair cells and matrix collagen fibrils to line up parallel to the line of tension [11, 13]. Lack of tension leaves the cells and fibers disoriented. Loading may also alter the rate of tendon repair [47]. Three weeks following injury, surgically repaired tendons treated with early mobilization had twice the strength of repaired tendons treated with immobilization [48]. Twelve weeks following injury, repaired tendons treated with early mobilization still had greater strength than repaired tendons treated with an initial period of immobilization. Controlled loading and motion soon after injury also can accelerate ligament repair by increasing the wet and dry weight of injured ligaments, improving matrix organization, and inducing more rapid return of normal DNA content and normal collagen synthesis and strength [9, 46, 79, 128].

Excessive or uncontrolled loading of injured tissues disrupts the repair tissue, causes further damage, and may delay or prevent repair [22]. In a study of medial collateral ligament healing in rats, forced exercise increased the strength of ligament repair tissue in stable knees [29], but in unstable knees such exercise did not increase the stiffness and strength of the repair tissue and did increase joint instability. Another study of the effects of anterior cruciate ligament transection showed that temporary immobilization of the knee prevented the development of osteophytes, suggesting that early motion following injury may have increased instability [98].

Cartilage

Decreased Loading and Motion of Normal Cartilage

Persistent decreases in loading and motion of a synovial joint cause changes in articular cartilage that parallel the changes seen in the periarticular dense fibrous tissues: Chondrocytes change their synthetic activity, cartilage proteoglycan concentration decreases, tissue volume may decrease, and mechanical properties deteriorate [1, 14, 15, 26, 39, 40, 42, 50, 58, 59, 62, 67, 70, 96, 97, 101, 106, 107, 116–118, 125]. Although loading of an immobilized

joint by repetitive muscle contraction may help preserve the cartilage [28], maintenance of normal synovial joint structure, composition, and function requires both loading and motion [100].

Cartilage alterations occur soon after a persistent decrease in joint use. Forty days following experimental tibial fractures in dogs the articular cartilage of the operated limb had a significantly lower glycosaminoglycan concentration [96]. Presumably, the decrease in gylcosaminoglycan concentration resulted from decreased loading and motion of the joints due to the fracture. Cast immobilization of dog limbs also damaged articular cartilage [67, 70, 97, 101]. Six weeks or more of cast immobilization decreased cartilage thickness, uronic acid content, and proteoglycan synthesis and diminished the ability of proteoglycans to form aggregates. Cessation of immobilization followed by ambulation in a pen for 3 weeks reversed the changes. Treadmill exercise after cessation of immobilization prevented the reversal, indicating that intense and frequent loading of the damaged cartilage impeded repair [97].

A study comparing the effects of 6 weeks of immobilization of dog knees using external fixators with the effects of 6 weeks of immobilization in long leg casts suggests that the rigidity of joint immobilization influences the severity of damage to the joint [15]. The external fixators rigidly immobilized the knees, but the long leg cast immobilization allowed 8 to 15 degrees of motion. Cartilage water content increased 7% in both groups of knees. Hexouronate concentration decreased 23% in the joints treated with casts and 28% in joints immobilized by rigid external fixators. The rigid fixators also produced more severe depression of proteoglycan synthesis and proteoglycan loss and impaired cartilage recovery. Within a week after remobilization, joints treated with cast immobilization had recovered near normal hexouronate content, but rigidly immobilized joints showed little or no evidence of recovery of hexouronate content.

Continued immobilization of joints eventually causes irreversible damage including contracture of periarticular dense fibrous tissues and muscles that act across the joint, loss of articular cartilage, and obliteration of the joint cavity by fibrofatty tissue [39, 42]. Once fibrofatty tissue fills the joint, attempts to restore motion forcefully tear the intra-articular tissue, often in a different plane from that of the original joint cavity, and avulse fragments of the remaining articular cartilage [39, 40].

The duration of immobilization necessary to damage a synovial joint irreversibly probably varies among joints and among species. Most animal studies suggest that controlled remobilization can reverse the damage caused by several months of immobilization. In rat joints, remobilization could reverse changes due to 30 days of immobilization, but 60 days of immobilization caused irreversible changes [40]. Two weeks of immobilization of rabbit knees did not cause any detectable permanent changes, but after 6 weeks some joints had developed contractures [42]. In dog knees, 6 days of immobilization decreased proteoglycan synthesis 41%, and 3 weeks of immobilization caused loss of proteoglycan aggregation; 2 weeks of motion restored proteoglycan aggregation to normal [101]. Another study of dog joints showed that 15 weeks of remobilization improved but did not completely restore the mechanical properties of dog articular cartilage subjected to 11 weeks of cast immobilization [67].

Increased Loading of Normal Cartilage

Increased loading and motion of articular cartilage, up to a certain level, may increase matrix synthesis relative to matrix degradation. In dogs, increased loading of a limb instituted by cast immobilization of the opposite limb or moderate running exercise (4 km/day), increased cartilage glycosaminoglycan concentration and thickness [70]. Strenuous running (20 km/day) reduced cartilage thickness and glycosaminoglycan concentration in normal joints, suggesting that loading and motion above a certain level may adversely affect articular cartilage.

Joint instability presumably increases articular cartilage loading and as a result can damage cartilage [89, 90], but some experimental studies suggest that it initially stimulates matrix synthesis. Transection of dog anterior cruciate ligaments causes knee instability, but in one set of experiments it also increased cartilage proteoglycan synthesis and concentration. Cartilage thickness, as measured by magnetic resonance imaging 3 years after surgery, also increased [18]. Although these responses of cartilage to increased loading may have beneficial effects, the mechanical properties and durability of articular cartilage subjected to prolonged increases in loading and motion have not been described.

Loading and Motion of Injured Cartilage

Differences in type of tissue damage and repair response separate acute cartilage injuries into three types [22, 26]: (1) loss or abnormalities of the matrix macromolecules, primarily proteoglycans, without disruption of tissue; (2) disruption of cartilage without injury to subchondral bone; and (3) disruption of cartilage and subchondral bone. Chondrocytes can restore lost proteoglycans if the collagen matrix of the articular cartilage remains intact and if enough chondrocytes remain viable. Chondrocytes cannot repair disruptions of tissue such as chondral fractures or cartilage lacerations. Following injury to cartilage, chondrocytes briefly increase their synthetic and proliferative activity, but they do not migrate to the site of injury or produce new cells and matrix that fill the tissue defect. Disruption of subchondral bone along with cartilage causes hemorrhage and initiates inflammation and repair by cells from the bone and the bone blood vessels. A fibrin clot forms in the tissue defect. Mesenchymal cells migrate into the clot and produce repair tissue that usually fills the bone defect and most of the chondral defect [26]. Cells in the chondral portion of the defect then produce repair tissue that usually has a matrix with a composition intermediate between that of fibrocartilage and articular cartilage [26].

The effects of loading and motion on injured cartilage vary with the type of injury [24, 26]. The observations that prolonged joint immobilization and unloading injure articular cartilage, that cyclic loading increases chondrocyte synthetic activity, and that resumption of use following joint immobilization improves articular cartilage composition and mechanical properties show that controlled loading and motion can stimulate repair of cartilage damage if the damage is limited to loss of matrix proteoglycans. There is no evidence that joint use promotes repair of injuries that disrupt the cartilage alone. Multiple studies of the effects of early passive motion treatment on osteochondral injuries report that passive motion improves the initial quality of cartilage repair tissue and the generation of cartilage by periosteal and perichondrial grafts [36, 86, 94, 95, 110, 111, 144]. Yet early motion alone does not predictably restore a normal articular surface [24].

Despite the necessity of loading and motion to maintain the health of articular cartilage and their potentially beneficial effects following many joint injuries, excessive or uncontrolled early loading and motion may delay healing or disrupt or prevent formation of cartilage repair tissue [3, 22, 24, 26, 97, 139]. Intense loading soon after injury to cartilage may interfere with restoration of lost matrix proteoglycan. Loss of matrix proteoglycan decreases cartilage stiffness and increases cartilage permeability. Repetitive intense loading of cartilage that has suffered loss of matrix proteoglycan, before the chondrocytes restore the matrix proteoglycan content, may cause further damage [22, 24, 26]. Guinea pig knees subjected to chemical injury developed cartilage fibrillation and osteophytes after 3 weeks of unrestrained active joint use, whereas immobilization for 3 weeks prevented fibrillation and osteophyte formation [139]. In contrast, running on a treadmill for 3 weeks following prolonged immobilization of dog knees prevented reversal of the immobilization-induced changes in cartilage proteoglycans [97]. These observations suggest that brief protection from loading and motion may allow chondrocytes to repair some types of damage to matrix. Excessive loading and motion may also damage the repair tissue that forms following osteochondral injury. Examination of the effect of abrading the femoral heads of dogs showed that protection of an abraded surface from loading allowed formation of cartilage repair tissue [3]. Areas subjected to heavy loading formed little or no repair tissue.

References

1. Akeson, W. H. The response of ligaments to stress modulation and overview of the ligament healing response. *In* David, D., Akeson, W., and O'Connor, J. (Eds.), *Knee Ligaments: Structure, Function, Injury, and Repair.* New York, Raven Press, 1990.
2. Akeson, W. H., Amiel, E., Mechanic, G. L., Woo, S. L.-Y., Harwood, F. L., and Hammer, M. L. Collagen cross-linking alterations in joint contractures: Changes in the reducible cross-links in periarticular connective tissue collagen after nine weeks of immobilization. *Connect Tissue Res* 5:15–19, 1977.
3. Akeson, W. H., Miyashita, C., Taylor, T. K., LaViolette, D., and Amiel, D. Experimental cup arthroplasty of the canine hip. *J Bone Joint Surg* 51-A:149–164, 1969.
4. Akeson, W. H., Woo, S. L.-Y., Amiel, D., Coutts, R. D., and Daniel, D. The connective tissue response to immobility: biochemical changes in periarticular connective tissue of the immobilized rabbit knee. *Clin Orthop Rel Res* 93:356–362, 1973.
5. Albrecht-Buehler, G. Role of cortical tension in fibroblast shape and movement. *Cell Motil Cytoskeleton* 7:54–67, 1987.
6. Amadio, P. C., Bergland, L. J., and Ahn, K. N. Tensile properties of biochemically discrete zones of canine flexor tendon. *Trans Orthop Res Soc* 14:251, 1989.
7. Amiel, D., Schroeder, H. v., and Akeson, W. H. The response of ligaments to stress deprivation and stress enhancement. *In* Daniel, D., Akeson, W., and O'Connor, J. (Eds.), *Knee Ligaments: Structure, Function, Injury, and Repair.* New York, Raven Press, 1990.
8. Amiel, D., Woo, S. L.-Y., Harwood, F. L., and Akeson, W. H. The effect of immobilization on collagen turnover in connective tissue: A biochemical-biomechanical correlation. *Acta Orthop Scand* 53:325–332, 1982.
9. Andriacchi, T., Sabiston, P., DeHaven, K., Dahners, L., Woo, S., Frank, C., Oakes, B., Brand, R. and Lewis, J. Ligament: Injury and repair. *In* Woo, S. L.-Y., and Buckwalter, J. A. (Eds.), *Injury and Repair of the Musculoskeletal Soft Tissues.* Park Ridge, IL, American Academy of Orthopaedic Surgeons, 1988.
10. Andrè, N. *Orthopaedia: or, The Art of Correcting and Preventing Deformities in Children.* London, A. Millar, 1743.
11. Arem, A. J., and Madden, J. W. Effects of stress on healing wounds: I. Intermittent noncyclical tension. *J Surg Res* 20:93–102, 1976.
12. Bader, D. L., Barnhill, R. L., and Ryan, T. J. Effect of externally applied skin surface forces on tissue vasculature. *Arch Phys Med Rehabil* 67:807–811, 1986.
13. Bair, G. R. The effect of early mobilization versus casting on anterior cruciate ligament reconstruction. *Trans Orthop Res Soc* 5:108, 1980.
14. Baker, W. C., Thomas, T. G., and Kirkaldy-Willis, W. H. Changes in the cartilage of the posterior intervertebral joints after anterior fusion. *J Bone Joint Surg* 51-B:736–746, 1969.
15. Behrens, F., Kraft, E. L., and Oegema, T. R. Biochemical changes in articular cartilage after joint immobilization by casting or external fixation. *J Orthop Res* 7:335–343, 1989.
16. Binderman, I., Shimshoni, Z., and Somjen, D. Biochemical pathways involved in the translation of physical stimulus into biological message. *Calcif Tissue Int* 36:S82–S85, 1984.
17. Booth, F. W. Perspectives on molecular and cellular exercise physiology. *J Appl Physiol* 65:1461–1471, 1988.
18. Braunstein, E. M., Brandt, K. D., and Albrecht, M. MRI demonstration of hypertophic articular cartilage repair in osteoarthritis. *Skeletal Radiol* 19:335–339, 1990.
19. Brunette, D. M. Mechanical stretching increases the number of epithelial cells synthesizing DNA in culture. *J Cell Sci* 69:35–45, 1984.
20. Buckwalter, J. A. Overview of pathophysiologic mechanisms in sports inflammation. *In* Leadbetter, W. B., Buckwalter, J. A., and Gordon, S. L. (Eds.), *Sports Induced Inflammation—Basic Science and Clinical Concepts.* Park Ridge, IL, American Academy of Orthopaedic Surgeons, 1990.
21. Buckwalter, J. A., and Cooper, R. R. Bone structure and function. *Instr Course Lect* 36:27–48, 1987.
22. Buckwalter, J. A., and Cruess, R. Healing of musculoskeletal tissues. *In* Rockwood, C. A., and Green, D. (Eds.), *Fractures.* Philadelphia, J. B. Lippincott, 1991.
23. Buckwalter, J. A., Maynard, J. A., and Vailas, A. C. Skeletal fibrous tissues: Tendon, joint capsule, and ligament. *In* Albright, J. A., and Brandt, R. A. (Eds.), The Scientific Basis of Orthopaedics. Norwalk, CT, Appleton & Lange, 1987.
24. Buckwalter, J. A., and Mow, V. C. Cartilage repair as treatment of osteoarthritis. *In* Moskowitz, R. W., Howell, D. S., Goldberg, V. M., and Mankin, H. J. (Eds.), *Osteoarthritis: Diagnosis and Management.* Philadelphia, W. B. Saunders, 1991.
25. Buckwalter, J. A., Mower, D., Ungar, R., Schaeffer, J., and Ginsberg, B. Growth plate chondrocyte profiles and their orientation. *J Bone Joint Surg* 67-A:942–955, 1985.
26. Buckwalter, J. A., Rosenberg, L. C., and Hunziker, E. Articular cartilage: Composition, structure, response to injury and methods of facilitating repair. *In* Ewing, J. W. (Ed.), *The Science of Arthroscopy.* New York, Raven Press, 1990.
27. Burks, R., Daniel, D., and Losse, G. The effect of continuous passive motion on anterior cruciate ligament reconstruction stability. *Am J Sports Med* 12:323–327, 1984.
28. Burr, D. B., Frederickson, R. G., Pavlinch, C., Sickles, M., and Burkart, S. Intracast muscle stimulation prevents bone and cartilage deterioration in cast-immobilized rabbits. *Clin Orthop Rel Res* 189:264–278, 1984.
29. Burroughs, P., and Dahners, L. E. The effect of enforced exercise on the healing of ligament injuries. *Am J Sports Med* 18(4):376–378, 1990.
30. Butler, D. L., and Siegel, A. Alterations in tissue response: Conditioning effects at different ages. *In* Leadbetter, W. B., Buckwalter, J. A., and Gordon, S. L. (Eds.), *Sports Induced Inflammation—Basic Science and Clinical Concepts.* Park Ridge, IL, American Academy of Orthopaedic Surgeons, 1990.

31. Caplan, A., Carlson, B., Faulkner, J., Fishman, D., and Garrett, W. Skeletal muscle. *In* Woo, S. L.-Y., and Buckwalter, J. A. (Eds.), *Injury and Repair of the Musculoskeletal Soft Tissues.* Park Ridge, IL, American Academy of Orthopaedic Surgeons, 1988.

32. Carter, D. R. Mechanical loading histories and cortical bone remodeling. *Calcif Tissue Int* 36:S19–S24, 1984.

33. Carter, D. R., and Wong, M. The role of mechanical loading histories in the development of diarthrodial joints. *J Orthop Res* 6:804–816, 1988.

34. Clancy, W. G. Tendon trauma and overuse injuries. *In* Leadbetter, W. B., Buckwalter, J. A., and Gordon, S. L. (Eds.), *Sports Induced Inflammation—Basic Science and Clinical Concepts.* Park Ridge, IL, American Academy of Orthopaedic Surgeons, 1990.

35. Cooper, R. R. Alterations during immobilization and regeneration of skeletal muscle in cats. *J Bone Joint Surg* 54-A:919–953, 1972.

36. Coutts, R. D., Woo, S. L.-Y., Amiel, D., van Schroeder, H. P. and Kwan, M. K. A histological, biochemical and biomechanical evaluation of rib perichondrial autografts in full thickness articular cartilage defects: A one year study. *Clin Orthop Rel Res* in press, 1991.

37. Dehne, E., Metz, C. W., Deffer, P. A., and Hall, R. M. Non-operative treatment of the fractured tibia by immediate weight bearing. *J Trauma* 1:514–535, 1961.

38. DeWitt, M. T., Handley, C. J., Oakes, B. W., and Lowther, D. A. In vitro response of chondrocytes to mechanical loading. The effect of short term mechanical tension. *Connect Tissue Res* 12:97–109, 1984.

39. Enneking, W. F., and Horowitz, M. The intra-articular effects of immobilization on the human knee. *J Bone Joint Surg* 54-A:973–985, 1972.

40. Evans, E. B., Eggers, G. W. N., Butler, J. K., and Blumel, J. Experimental immobilization and remobilization of rat knee joints. *J Bone Joint Surg* 42-A:737–758, 1960.

41. Faulkner, J. A. New perspectives in training for maximum performance. *JAMA* 205:741–746, 1968.

42. Finsterbush, A., and Friedman, B. Reversibility of joint changes produced by immobilization in rabbits. *Clin Orthop Rel Res* 111:290–298, 1975.

43. Frank, C., Akeson, W. H., Woo, S. L.-Y., Amiel, D., and Coutts, R. D. Physiology and therapeutic value of passive joint motion. *Clin Orthop Rel Res* 185:113–125, 1984.

44. Frank, C., McDonald, D., Lieber, R. L., et al. Biochemical heterogeneity within the maturing rabbit medial collateral ligament. *Clin Orthop Rel Res* 236:279–285, 1988.

45. Frank, C. B., and Hart, D. A. Cellular responses to loading. *In* Leadbetter, W. B., Buckwalter, J. A., and Gordon, S. L. (Eds.), *Sports Induced Inflammation—Basic Science and Clinical Concepts.* Park Ridge, IL, American Academy of Orthopaedic Surgeons, 1990.

46. Fronek, J., Frank, C., Amiel, D., Woo, S. L.-Y., Coutts, R. D. and Akeson, W. H. The effects of intermittent passive motion (IPM) in the healing of medial collateral ligament. *Trans Orthop Res Soc* 8:31, 1983.

47. Gelberman, R., Goldberg, V., An, K.-N., and Banes, A. Tendon. *In* Woo, S. L.-Y., and Buckwalter, J. A. (Eds.), Park Ridge, IL, American Academy of Orthopaedic Surgeons, 1988.

48. Gelberman, R. H., Woo, S. L.-Y., Lothringer, K., Akeson, W. H., and Amiel, D. Effects of early intermittent passive mobilization on healing canine flexor tendons. *J Hand Surg* 7(2):170–175, 1982.

49. Gillard, G. C., Reilly, H. C., Bell-Booth, P. G., and Flint, M. H. The influence of mechanical forces on the glycosaminoglycan content of the rabbit flexor digitorum profundus tendon. *Connect Tissue Res* 7:37–46, 1979.

50. Good, S. C. A study of the effects of experimental immobilization on rabbit articular cartilage. *J Anat* 108:497–507, 1971.

51. Goodship, A. E., and Kenwright, J. The influence of induced micromovement upon the healing of experimental tibial fractures. *J Bone Joint Surg* 67-B:650, 1985.

52. Goodship, A. E., Lanyon, L. E., and McFie, H. Functional adaptation of bone to increased stress. *J Bone Joint Surg* 61-A:539–546, 1979.

53. Gould, R. P., Selwood, L., Day, A., and Wolpert, L. The mechanism of cellular orientation during early cartilage formation in the chick limb and regenerating amphibian limb. *Exp Cell Res* 83:287–296, 1974.

54. Gray, M. L., Pizzanelli, A. M., Godzinsky, A. J., and Lee, R. C. Mechanical and physicochemical determinants of chondrocyte biosynthetic response. *J Orthop Res* 6:777–792, 1988.

55. Grinnell, F., and Lamke, C. R. Reorganization of hydrated collagen lattices by human skin fibroblasts. *J Cell Sci* 66:51–63, 1984.

56. Grodzinsky, A. J. Electromechanical and physicochemical properties of connective tissue. *CRC Rev Biomed Eng* 9(2):133–199, 1983.

57. Hall, A. C., and Urban, J. P. G. Responses of articular chondrocytes and cartilage to high hydrostatic pressure. *Trans Orthop Res Soc* 14:49, 1989.

58. Hall, M. C. Cartilage changes after experimental immobilization of the knee joint of the young rat. *J Bone Joint Surg* 45-A:36–44, 1963.

59. Hall, M. C. Cartilage changes after experimental relief of contact in the knee joint of the mature rat. *Clin Orthop Rel Res* 64:64–76, 1969.

60. Heikkinen, E., Vihersaari, T., and Penttinen, R. Effect of previous exercise on fracture healing: A biochemical study with mice. *Acta Orthop Scand* 45:481–489, 1974.

61. Heikkinen, E., and Vuori, I. Effect of physical activity on the metabolism of collagen in aged mice. *Acta Physiol Scand* 84:543–549, 1972.

62. Helminen, H. J., Jurvelin, J., Kuusela, T., Heikkila, R., Kiviranta, I., and Tammi, M. Effects of immobilization for 6 weeks on rabbit knee articular surfaces as assessed by the semiquantitative steromicroscopic method. *Acta Anat* 115:327–335, 1983.

63. Holmes, L. B., and Trelstad, R. L. Cell polarity in precartilage mouse limb mesenchyme cells. *Dev Biol* 78:511–520, 1980.

64. Indelicato, P. A., Hermansdorfer, J., and Huegel, M. Nonoperative management of complete tears of the medial collateral ligament of the knee in intercollegiate football players. *Clin Orthop Rel Res* 256:174–177, 1990.

65. Jarvinen, M. Healing of crush injury in rat striated muscle. 2. Histological study of the effect of early mobilization and immobilization on the repair process. *Acta Pathol Microbiol Immunol Scand* 83:269–282, 1975.

66. Jarvinen, M. Healing of a crush injury in rat striated muscle. 4. Effect of early mobilization and immobilization of the tensile properties of gastrocnemius muscle. *Acta Chir Scand* 142:47–56, 1976.

67. Jurvelin, J., Kiviranta, I., Saamanen, A.-M., Tammi, M., and Helminen, H. J. Partial restoration of immobilization-induced softening of canine articular cartilage after remobilization of the knee (stifle) joint. *J Orthop Res* 7:352–358, 1989.

68. Keith, A. *Menders of the Maimed: The Anatomical and Physiological Principles Underlying the Treatment of Injuries to Muscles, Nerves, Bones, and Joints.* London, Oxford University Press, 1919.

69. Kennedy, J. C., Hawkins, R. J., Willis, R. B., and Danylchuk, K. D. Tension studies of human knee ligaments: Yield point, ultimate failure, and disruption of the cruciate and tibial collateral ligaments. *J Bone Joint Surg* 58-A:350–355, 1976.

70. Kiviranta, I. *Joint Loading Influences on the Articular Cartilage of Young Dogs.* Thesis. Kuopio, University of Kuopio, 1987.

71. Klebe, R. J., Caldwell, H., and Milam, S. Cells transmit spatial information by orienting collagen fibers. *Matrix* 9:451–458, 1989.

72. Klein, L., Dawson, M. H., and Heiple, K. G. Turnover of collagen in the adult rat after denervation. *J Bone Joint Surg* 59-A:1065–1067, 1977.

73. Klein, L., Heiple, K. G., Torzilli, P. A., Golberg, V. M., and Burstein, A. H. Prevention of ligament and meniscus atrophy by active joint motion in a non-weight bearing model. *J Orthop Res* 7:80–85, 1989.
74. Kolega, J. Effects of mechanical tension on protrusive activity of and microfilament and intermediate filament organization in an epidermal epithelium moving in culture. *J Cell Biol* 102:1400–1411, 1986.
75. Koob, T. J., and Vogel, K. G. Proteoglycan synthesis in organ cultures from regions of tendon subjected to different mechanical forces. *Biochem J* 246:589–598, 1987.
76. Laros, G. S., Tipton, C. M., and Cooper, R. R. Influence of physical activity on ligament insertions in the knees of dogs. *J Bone Joint Surg* 53-A:275–286, 1971.
77. Leung, D. Y. M., Glagov, S., and Mathews, M. B. A new in vitro system for studying cell response to mechanical stimulation: Different effects of cyclic stretching and agitation on smooth muscle cell biosynthesis. *Exp Cell Res* 109:285–298, 1977.
78. Lewis, L. B., and Trelstad, R. L. Patterns of cell polarity in the developing mouse limb. *Dev Biol* 59:164–173, 1977.
79. Long, M., Frank, C., Schachar, N., Dittrich, D., and Edwards, G. E. The effects of motion on normal and healing ligaments. *Trans Orthop Res Soc* 7:43, 1982.
80. Mason, M. L., and Allen, H. S. The rate of healing of tendons: An experimental study of tensile strength. *Ann Surg* 113:424–459, 1941.
81. Mason, M. L., and Shearon, C. G. The process of tendon repair—An experimental study of tendon suture and tendon graft. *Arch Surg* 25:615–692, 1932.
82. Merrilees, M. J., and Flint, M. H. Ultrastructural study of tension and pressure zones in a rabbit flexor tendon. *Am J Anat* 157:87–106, 1980.
83. Michna, H. Morphometric analysis of loading-induced changes in collagen-fibril populations in young tendons. *Cell Tissue Res* 236:465–470, 1984.
84. Michna, H. Tendon injuries induced by exercise and anabolic steroids in experimental mice. *Int Orthop* 11:157–162, 1987.
85. Mooney, V., Nickel, V., Harvey, J. P., and Snelson, R. Cast brace treatment for fractures of the distal part of the femur. *J Bone Joint Surg* 52-A:1563–1578, 1970.
86. Moran, M. E., Kreder, H. J., Salter, R. B., and Keeley, F. W. Biological resurfacing of major full thickness defects in joint surfaces by neochondrogenesis with cryopreserved allogenic periosteum stimulated by continuous passive motion. *Trans Orthop Res Soc* 14:542, 1989.
87. Morgan, T. E., Cobb, L. A., Short, F. A., Ross, R., and Gunn, D. R. Effects of long-term exercise on human muscle mitochondria. *Adv Exp Med Biol* 11:87–95, 1971.
88. Mosler, E., Folkhard, W., Knorzer, E., Nemetschek-Gansler, H., Nemetschek, T. H., and Kock, M. H. J. Stress-induced molecular rearrangement in tendon collagen. *J Mol Biol* 182:589–596, 1985.
89. Mow, V. C., Bigliani, L. U., Flatow, E. L., Pollock, R. G., Parisien, M. V., Soslowsky, L. J., Guilak, F., and Pawluk, R. J. The role of joint instability in joint inflammation and cartilage deterioration: a study of the glenohumeral joint. *In* Leadbetter, W. B., Buckwalter, J. A., and Gordon, S. L. (Eds.), *Sports Induced Inflammation—Basic Science and Clinical Concepts.* Park Ridge, IL, American Academy of Orthopaedic Surgeons, 1990.
90. Mow, V. C., Setton, L. A., Ratcliff, A., Howell, D. S., and Buckwalter, J. A. Structure-function relationship of articular cartilage and the effects of joint instability and trauma on cartilage. *In* Brandt, K. D. (Ed.), *Cartilage Changes in Osteoarthritis.* Indianapolis, Indiana University School of Medicine, 1990.
91. Norton, L. A., Rodan, G. A., and Bourret, L. A. Epiphyseal cartilage cAMP changes produced by electrical and mechanical perturbations. *Clin Orthop Rel Res* 124:59–68, 1977.
92. Noyes, F. R., DeLucas, J. L., and Torvik, P. J. Biomechanics of anterior cruciate ligament failure: An analysis of strain-rate sensitivity and mechanisms of failure in primates. *J Bone Joint Surg* 56-A:236–253, 1974.
93. Noyes, F. R., Torvik, P. J., Hyde, W. B., and DeLucas, J. L. Biomechanics of ligament failure. II. An analysis of immobilization, exercise, and reconditioning effects in primates. *J Bone Joint Surg* 56-A:1406–1418, 1974.
94. O'Driscoll, S. W., Keeley, F. W., and Salter, R. B. The chondrogenic potential of free autogenous periosteal grafts for biological resurfacing of major full-thickness defects in joint surfaces under the influence of continuous passive motion: An experimental study in the rabbit. *J Bone Joint Surg* 68-A:1017–1035, 1986.
95. O'Driscoll, S. W., Keeley, F. W., and Salter, R. B. Durability of regenerated articular cartilage produced by free autogenous periosteal grafts in major full-thickness defects in joint surfaces under the influence of continuous passive motion: A follow-up report at one year. *J Bone Joint Surg* 70-A:595, 1989.
96. Olah, E. H., and Kostensky, K. S. Effect of altered functional demand on the glycosaminoglycan content of the articular cartilage of dogs. *Acta Biol Acad Sci Hung* 23(2):195–200, 1972.
97. Palmoski, M. J., and Brandt, K. D. Aspirin aggravates the degeneration of canine joint cartilage caused by immobilization. *Arthritis Rheum* 25:1333–1342, 1982.
98. Palmoski, M. J., and Brandt, K. D. Immobilization of the knee prevents osteoarthritis after anterior cruciate ligament resection. *Arthritis Rheum* 25:1201–1208, 1982.
99. Palmoski, M. J., and Brandt, K. D. Effects of static and cyclic compressive loading on articular cartilage plugs in vitro. *Arthritis Rheum* 27(6):675–681, 1984.
100. Palmoski, M. J., Colyer, R. A., and Brandt, K. D. Joint motion in the absence of normal loading does not maintain normal articular cartilage. *Arthritis Rheum* 23(3):325–334, 1980.
101. Palmoski, M. J., Perricone, E., and Brandt, K. D. Development and reversal of a proteoglycan aggregation defect in normal canine knee cartilage after immobilization. *Arthritis Rheum* 22:508–517, 1979.
102. Pedrini-Mille, A., Pedrini, V. A., Maynard, J. A., and Vailas, A. C. Response of immature chicken meniscus to strenuous excercise: Biochemical studies of proteoglycan and collagen. *J Orthop Res* 6:196–204, 1988.
103. Piper, T. L., and Whiteside, L. A. Early mobilization after knee ligament repair in dogs: An experimental study. *Clin Orthop Rel Res* 150:277–282, 1980.
104. Retief, D. H., and Dreyer, C. H. Effects of neural damage on the repair of bony defects in the rat. *Arch Oral Biol* 12:1035–1039, 1967.
105. Rodan, G. A., Bourret, L. A., and Norton, L. A. DNA synthesis in cartilage cells is stimulated by oscillating electric fields. *Science* 199(10):690–692, 1978.
106. Roth, J. H., Mendenhall, H. V., and McPherson, G. K. The effect of immobilization on goat knees following reconstruction of the anterior cruciate ligament. *Clin Orthop Rel Res* 229:278–282, 1988.
107. Roy, S. Ultrastructure of articular cartilage in experimental immobilization. *Ann Rheum Dis* 29:634–642, 1970.
108. Rubin, C. T. Skeletal strain and the functional significance of bone architecture. *Calcif Tissue Int* 36:S11–S18, 1984.
109. Sah, R. L.-Y., Kim, Y.-J., Doong, J.-Y. H., Grodzinsky, A. J., Plaas, A. H. K., and Sandy, J. D. Biosynthetic response of cartilage explants to dynamic compression. *J Orthop Res* 7:619–636, 1989.
110. Salter, R. B., Minster, R. R., Bell, R. S., Wong, D. A., and Bogoch, E. R. Continuous passive motion and the repair of full-thickness articular cartilage defects: A one-year follow-up. *Trans Orthop Res Soc* 7:167, 1982.
111. Salter, R. B., Simmonds, D. F., Malcolm, B. W., Wong, D. A., and Bogoch, E. R. The biological effect of continuous passive motion on healing of full-thickness defects in articular cartilage: an experimental study in the rabbit. *J Bone Joint Surg* 62-A:1232–1251, 1980.
112. Sarmiento, A. A Functional below-the-knee cast for tibial fractures. *J Bone Joint Surg* 49-A:855–875, 1967.

113. Sarmiento, A., Schaeffer, J. F., Beckerman, L., Latta, L. L. and Enis, J. E. Fracture healing in rat femora as affected by functional weight-bearing. *J Bone Joint Surg* 59-A:369, 1977.

114. Schwartz, E. R., Kirkpatrick, P. R., and Thompson, R. C. The effect of environmental pH on glycosaminoglycan metabolism by normal human chondrocytes. *J Lab Clin Med* 87(2):198–205, 1976.

115. Slack, C., Flint, M. H., and Thompson, B. M. The effect of tensional load on isolated embryonic chick tendons in organ culture. *Connect Tissue Res* 12:229–247, 1984.

116. Tammi, M., Kiviranta, I., Peltonen, L., Jukka, J., and Helminen, H. J. Effects of joint loading on articular cartilage collagen metabolism: Assay of procollagen prolyl 4-hydroxylase and galactosylhydroxylysyl glucosyltransferase. *Connect Tiss Res* 17:199–206, 1988.

117. Tammi, M., Saamanen, A.-M., Jauhiainen, A., Malminen, O., Kiviranta, I., and Helminen, H. Proteoglycan alterations in rabbit knee articular cartilage following physical exercise and immobilization. *Connect Tissue Res.* 11:45–55, 1983.

118. Thaxter, T. H., Mann, R. A., and Anderson, C. E. Degeneration of immobilized knee joints in rats. *J Bone Joint Surg* 47-A:567–585, 1965.

119. Tipton, C. M. Ligamentous strength measurements from hypophysectomized rats. *Am J Physiol* 221:1144–1150, 1971.

120. Tipton, C. M., James, S. L., Mergner, W., and Tcheng, T.-K. Influence of exercise on strength of medial collateral ligaments of dogs. *Am J Physiol* 218:894–901, 1970.

121. Tipton, C. M., Matthes, R. D., Maynard, J. A., and Carey, R. A. The influence of physical activity on ligaments and tendons. *Med Sci Sports* 7(3):165–175, 1975.

122. Tipton, C. M., Matthes, R. D., and Sandage, D. S. In situ measurement of junction strength and ligament elongation in rats. *J Appl Physiol* 37(5):758–761, 1974.

123. Tipton, C. M., Schild, R. J., and Tomanek, R. J. Influence of physical activity on the strength of knee ligaments in rats. *Am J Physiol* 212(4):783–787, 1967.

124. Trelstad, R. L. Mesenchymal cell polarity and morphogenesis of chick cartilage. *Dev Biol* 59:153–163, 1977.

125. Troyer, H. The effect of short-term immobilization on the rabbit knee joint cartilage. *Clin Orthop Rel Res* 107:249–257, 1975.

126. Uhthoff, H. K., and Jaworski, Z. F. G. Bone loss in response to long-term immobilization. *J Bone Joint Surg* 60-B:420–429, 1978.

127. Urschel, J. D., Scott, P. G., and Williams, H. T. G. The effect of mechanical stress on soft and hard tissue repair; A review. *Br J Plast Surg* 41:182–186, 1988.

128. Vailas, A. C., Tipton, C. M., Matthes, R. D., and Gart, M. Physical activity and its influence on the repair process of medial collateral ligaments. *Connect Tissue Res* 9:25–31, 1981.

129. Videman, T. Connective tissue and immobilization: Key factors in musculoskeletal degeneration? *Clin Orthop Rel Res* 221:26–32, 1987.

130. Viidik, A. The effect of training on the tensile strength of isolated rabbit tendons. *Scand J Plast Reconstr Surg* 1:141–147, 1967.

131. Viidik, A. The elasticity and tensile strength of the anterior cruciate ligament in rabbits as influenced by training. *Acta Physiol Scand* 74:372–380, 1968.

132. Viidik, A. Tensile strength properties of achilles tendon systems in trained and untrained rabbits. *Acta Orthop Scand* 40:261–272, 1969.

133. Vilart, R., and Vidal, B. D.-C. Anisotropic and biomechanical properties of tendons modified by exercise and denervation: Aggregation and macromolecular order in collagen bundles. *Matrix* 9:55–61, 1989.

134. Vogel, K. G., Keller, E. J., Lenhoff, R. J., Campbell, K., and Koob, T. J. Proteoglycan synthesis by fibroblast cultures initiated from regions of adult bovine tendon subjected to different mechanical forces. *Eur J Cell Biol* 41:102–112, 1986.

135. Vogel, K. G., and Thonar, E. J.-M. Keratan sulfate is a component of proteoglycans in the compressed region of adult bovine flexor tendon. *J Orthop Res* 6:434–442, 1988.

136. Walsh, S., Frank, C., and Hart, D. Immobilization alters cell function in growing rabbit ligaments. *Trans Orthop Res Soc* 13:57, 1988.

137. Wang, C. W., Weiss, J. A., Albright, J., Buckwalter, J. A., Martin, R., and Woo, S. L.-Y. Life-long exercise and aging effects on the canine medial collateral ligament. *Orthop Trans* 14:488–489, 1990.

138. Watson, J., Haas, W. G. D., and Hauser, S. S. Effect of electric fields on growth rate of embryonic chick tibiae in vitro. *Nature* 254:331–332, 1975.

139. Williams, J. M., and Brandt, K. D. Immobilization ameliorates chemically induced articular cartilage damage. *Arthritis Rheum* 27:208–216, 1984.

140. Woo, S., Maynard, J., Butler, D., Lyon, R., Torzilli, P., Akeson, W., Cooper, R., and Oakes, B. Ligament, tendon, and joint capsule insertions into bone. *In* Woo, S. L-Y., and Buckwalter, J. A. (Eds.), *Injury and Repair of the Musculoskeletal Soft Tissues.* Park Ridge, IL, American Academy of Orthopaedic Surgeons, 1988.

141. Woo, S. L., Gomez, M. A., and Amiel, D. The effects of exercise on the biomechanical and biochemical properties of swine digital flexor tendons. *J Biomech Eng.* 103:51–56, 1981.

142. Woo, S. L.-Y., Gomez, M. A., Sites, T. J., Newton, P. O., Orlando, C. A., and Akeson, W. H. The biomechanical and morphological changes in the medial collateral ligament of the rabbit after immobilization and remobilization. *J Bone Joint Surg.* 69-A:1200–1211, 1987.

143. Woo, S. L.-Y., Horibe, S., Ohland, K. J., and Amiel, D. The response of ligaments to injury: Healing of the collateral ligaments. *In* Daniel, D., Akeson, W. H., and O'Connor, J. (Eds.), *Knee Ligaments: Structure, Function, Injury, and Repair.* New York, Raven Press, 1990.

144. Woo, S. L.-Y., Kwan, M. K., Lee, T. Q., Field, F. P., Kleiner, J. B., and Coutts, R. D. Perichondrial autograft for articular cartilage: Shear modulus of neocartilage studied in rabbit cartilage autografts. *Acta Orthop Scand* 58:510–518, 1987.

145. Woo, S. L.-Y., Matthews, J. V., Akeson, W. H., Amiel, D., and Convery, F. R. Connective tissue response to immobility. *Arthritis Rheum* 18:257–264, 1975.

146. Woo, S. L.-Y., Ritter, M. A., Amiel, D., Sanders, T. M., Gomez, M. A., Kuei, S. C., Garfin, S. R., and Akeson, W. H. The biomechanical and biochemical properties of swine tendons—long term effects of exercise on the digital extensors. *Connect Tissue Res* 7:177–183, 1980.

147. Woo, S. L.-Y., and Tkach, L. V. The cellular and matrix response of ligaments and tendons to mechanical injury. *In* Leadbetter, W. H., Buckwalter, J. A., and Gordon, S. L. (Eds.), *Sports Induced Inflammation—Basic Science and Clinical Concepts.* Park Ridge, IL, American Academy of Orthopaedic Surgeons, 1990.

148. Woo, S. L.-Y., Wand, C. W., Newton, P. O., and Lyon, R. M. The response of ligaments to stress deprivation and stress enhancement: biomechanical studies. *In* Daniel, D., Akeson, W. H., and O'Connor, J. (Eds.), *Knee Ligaments: Structure, Function, Injury, and Repair.* New York, Raven Press, 1990.

149. Yeh, C.-K., and Rodan, G. A. Tensile forces enhance prostaglandin E synthesis in osteoblastic cells grown on collagen ribbons. *Calcif Tissue Int* 36:S67–S71, 1984.

150. Zuckerman, J., and Stull, G. A. Effects of exercise on knee ligament separation force in rats. *J Appl Physiol* 26(6):716–719, 1969.

151. Zuckerman, J., and Stull, G. A. Ligamentous separation force in rats as influenced by training, detraining, and cage restriction. *Med Sci Sports* 5(1):44–49, 1973.

Tissue Effects of Medications in Sports Injuries

Joseph A. Buckwalter, M.D.
Savio L.-Y. Woo, Ph.D.

Physicians treating sports injuries to tendons, muscle tendon junctions, ligaments, and synovial joints commonly prescribe anti-inflammatory medications, primarily oral nonsteroidal anti-inflammatory drugs and injectable corticosteroids, with the intent of decreasing pain and promoting recovery [1, 20, 33, 48]. Some physicians also use short courses of oral corticosteroids [12, 30]. Ideally, partial suppression of inflammation could reduce secondary tissue damage resulting from release of degradative enzymes and other events during inflammation, limit disuse changes and allow earlier rehabilitation by decreasing pain and swelling, and accelerate healing by shortening the duration of acute inflammation [49]. Nonsteroidal and steroidal anti-inflammatory drugs suppress inflammation and often provide analgesia, but their efficacy in minimizing tissue damage and accelerating a return to normal function after injury have not been proved in controlled studies [1, 48, 88], and repeated use of some of these medications can adversely alter the properties of normal musculoskeletal tissues and impair healing. For these reasons, physicians treating patients with sports injuries should understand the potential adverse effects as well as the potential benefits of these medications.

Besides the accepted anti-inflammatory medications, some athletes use other drugs without the advice of a physician in an attempt to relieve pain and promote recovery from injury. These medications include anabolic steroids and dimethyl sulfoxide [2, 36, 37, 44, 49, 58, 72, 84, 89]. Since these drugs may have detrimental effects on the tissues, they are also discussed in this chapter.

The first part of the chapter surveys the types of anti-inflammatory drugs and drugs used to promote recovery from injury. The next section reviews the effects of these drugs in sports injuries, and the last section discusses the specific effects of these drugs on the dense fibrous tissues and cartilage.

DRUGS USED TO TREAT SPORTS INJURIES

Nonsteroidal Anti-inflammatory Drugs

Commonly used nonsteroidal anti-inflammatory drugs include aspirin, diflunisal, fenoprofen calcium, ibuprofen, indomethacin, naproxen, piroxicam, phenylbutazone, sulindac, and tolmetin sodium. These chemically heterogeneous drugs differ in their specific actions, but they share important clinical and tissue effects [1, 20, 88]. They all have some analgesic, antipyretic, and anti-inflammatory activity and a variable potential for causing gastric or intestinal ulceration and interfering with platelet function. They all inhibit the synthesis and release of prostaglandins. Damaged cells release prostaglandins, and the available evidence shows that prostaglandins act as mediators of inflammation. Presumably, nonsteroidal anti-inflammatory drugs suppress inflammation primarily by inhibiting prostaglandin synthesis, although they may also affect inflammation by other mechanisms.

Corticosteroids

Corticosteroids used for their anti-inflammatory activity include cortisone, hydrocortisone, prednisone, methylprednisolone, triamcinolone, and dexamethasone. They suppress or prevent the initial events in inflammation by inhibiting capillary dilation, migration of inflammatory cells, and tissue edema. Once the inflammatory process starts, they also inhibit capillary and fibroblast proliferation and collagen synthesis. These later effects can compromise healing [8, 12, 21, 38, 86].

These medications also influence carbohydrate, lipid, nucleic acid, and protein metabolism, fluid and electrolyte balance, and the function of cells in many organs and organ systems including the kidney, liver, skeletal muscle, bone, cartilage, cardiovascular system, immune system, and nervous system [12]. Individual drugs vary considerably in their spectrum of activity. For example, at equivalent dose levels, triamcinolone has five times the antiinflammatory activity of hydrocortisone, and dexamethasone has 25 times the anti-inflammatory activity of hydrocortisone, but hydrocortisone causes sodium retention whereas triamcinolone and dexamethasone have little or no effect on sodium retention.

Along with the intended therapeutic results, administration of corticosteroids can cause tissue damage and disturb the function of a variety of tissues and organ systems. Despite multiple reports of the deleterious consequences of corticosteroids [8, 12, 49], the relationships between corticosteroid dose and specific complications remain unclear, and the reports have not established a safe dose of these drugs [5]. Short-term moderate or low-dose oral corticosteroid therapy has not been proved to cause significant complications in normal people, but multiple case reports describe bone necrosis associated with short-term high-dose therapy [5, 83]. Reported complications of prolonged or repeated use of oral corticosteroids include fluid and electrolyte imbalance, glucose intolerance, hypertension, increased susceptibility to infections, impaired wound healing, bone necrosis, tendon ruptures, gastrointestinal ulceration, myopathy, behavioral disturbances, osteoporosis, and, in children, inhibition of growth [8, 12, 38]. Corticosteroid injections generally cause less serious adverse effects, although one case report described a patient who developed bone necrosis following multiple corticosteroid injections [76]. Surveys of problems associated with corticosteroid injections show that subcutaneous fat necrosis and loss of skin pigmentation are the most common complications [19, 47, 48]. Tendon ruptures and accelerated joint destruction occur less frequently, and few systemic complications of local corticosteroid injections have been reported [19, 47, 48].

Anabolic Steroids

Examples of anabolic steroids (i.e., steroids derived from testosterone with anabolic activity) include methyltestosterone, testosterone propionate, methandrostenolone, oxandrolone, and stanozolol. They all have androgenic and anabolic activity but vary in the ratio of anabolic to androgenic activity. For example, testosterone propionate has an anabolic-androgenic ratio of 1:1, but stanozolol has a ratio of 100:1. These drugs have two generally accepted medical uses—treatment of selected types of anemia and treatment of hypogonadal males [36, 37]. Although the predominant activities of these medications are anabolic and androgenic, like other steroids, they influence cell function in many tissues and organ systems including muscle, liver, reproductive organs, the immune system, the central nervous system, and the hematopoietic system.

Some athletes use oral and injectable anabolic steroids with the intent of improving performance through gains in strength and ability to endure increased training [36, 37, 45]. The efficacy of anabolic steroids for these purposes is questionable. Reviews of the results of anabolic steroid use by athletes show that these medications do not predictably improve athletic performance [36, 37, 45]; they do not improve aerobic capacity, and they have inconsistent effects on strength. They may help to increase strength, as measured by a once repeated maximum weight lift, in athletes who have been training intensively in a weight-lifting program before the start of steroid use and continue to train intensively and maintain a high-protein diet [36, 37]. Using other measures of strength, and in athletes who do not meet these criteria, anabolic steroids have not been shown to increase strength predictably, and these drugs have not been shown to improve performance in specific sports.

Complications of anabolic steroid use occur frequently. More than 30% of athletes taking anabolic steroids reported subjective side effects including changes in libido, aggressiveness, and muscle spasm [36, 37]. Use of these drugs also causes abnormalities in liver function tests, decreased serum testosterone levels, and decreased spermatogenesis. In addition, benign and malignant liver tumors have been associated with anabolic steroid use [36, 37].

Dimethyl Sulfoxide

The physical and chemical characteristics of dimethyl sulfoxide, a clear, colorless liquid, make it an exceptional solvent, better than water for many substances. Because it lowers the freezing point of fluids and protects cells against damage due to freezing, it has an important role in preserving tissues and cells such as erythrocytes, platelets, and bone marrow elements [72, 84]. When applied topically, it easily penetrates the skin and appears in the blood within minutes. It probably has local anesthetic activity and may have a central analgesic effect [84]. In some experiments it appears to have anti-inflammatory activity, and several studies suggest that it may reduce collagen synthesis or enhance collagen degradation [84]. Currently accepted uses include preservation of cells and treatment of interstitial cystitis of the bladder, gastrointestinal amyloidosis, and dermatologic lesions of scleroderma [72].

EFFECTS OF DRUGS IN SPORTS INJURIES

Nonsteroidal Anti-inflammatory Drugs

Nonsteroidal drugs have established roles in the treatment of chronic inflammatory diseases involving the musculoskeletal system, including rheumatoid arthritis and other rheumatologic disorders, and they form the primary medical therapy for osteoarthritis. The efficacy of nonsteroidal anti-inflammatory drugs in providing symptomatic improvement and their tissue effects have been examined in patients with these chronic conditions and in animal experiments designed to assess the effects of repeated use of these medications on normal tissues.

Despite their widespread use for treatment of acute sports injuries, such as ligament and joint capsule sprains, and chronic sports injuries, such as patellar or Achilles tendonitis, the efficacy of nonsteroidal anti-inflammatory medications following these injuries has not been clearly defined [1, 20, 88]. Currently, physicians must base their use of these drugs for the treatment of sports injuries primarily on clinical experience and an understanding of their general analgesic and anti-inflammatory activities.

Nonsteroidal anti-inflammatory agents decrease acute soft tissue inflammation, and clinical experience suggests that they decrease the pain associated with tissue injury and possibly joint and soft tissue stiffness as well [1, 20, 32, 33]. There is less evidence that these drugs can promote restoration of normal tissue function following injury or hasten a return to participation in sports. A study of ligament repair in rats showed that piroxicam increased the strength of healing rat ligaments 14 days after injury if the drug was administered for the first 6 days after injury [31]. It did not affect the ultimate strength of the healed ligaments or the strength of normal ligaments. An experimental study suggested that a nonsteroidal anti-inflammatory drug promoted return of function following muscle strains [3], but other experimental and clinical studies have not clearly shown that nonsteroidal anti-inflammatory drugs promote more rapid return to full function or improve performance following injury [88].

Corticosteroids

The anti-inflammatory potency of corticosteroids far exceeds that of the available nonsteroidal medications, but the frequency of serious complications associated with their use also exceeds that of nonsteroidal agents. Like the nonsteroidal anti-inflammatory drugs, corticosteroids have a generally accepted role in the treatment of chronic inflammatory diseases of the musculoskeletal system including rheumatoid arthritis. Appropriate use of these medications for treatment of sports injuries is less clear.

Injections

Despite limited documentation of their efficacy in the treatment of sports injuries, many physicians use corticosteroid injections for these injuries [19, 48, 49]. One investigator reported that the symptoms of bursitis and tendonitis responded more frequently to corticosteroid injections than other conditions, including knee synovitis associated with internal derangements and acromioclavicular joint arthritis [48], but in many patients the symptoms eventually returned following the injections. Experimental studies show that corticosteroids decrease scar tissue adhesions following tendon injuries [40] and decrease joint stiffness following fractures [35]. They have not shown that these medications accelerate healing or a return to normal function [48].

The author of a review of corticosteroid injections for treatment of acute and chronic sports injuries stressed that these injections should be used with caution [47, 48]. He advised physicians to consider corticosteroid injections only after other nonsurgical treatments including exercise and rest have failed and when the physician can identify a discrete, palpable source of the patient's symptoms. No more than three injections spaced weeks apart should be given, and a second or third injection should be given only if the first injection decreased the symptoms. Following the injection, the patient should have a period of rest or protection from further injury. Corticosteroid injections should not be used immediately after an acute injury, immediately before a competitive event, or in the presence of infection. Corticosteroids should not be injected into tendons or ligaments.

Oral Corticosteroids

Although oral corticosteroids have potent anti-inflammatory effects, few physicians use them for the treatment of sports injuries [12]. Lack of any peer-reviewed studies of the use of these drugs for treatment of sports injuries during the last 10 years [12, 49] makes it difficult to assess either their efficacy in improving function or accelerating a return to activity following injury or their safety. Because of the difficulty of showing the efficacy of oral corticosteroids and their potential complications, some authors recommend that physicians not use these medications for the routine treatment of sports injuries [12, 49].

Anabolic Steroids

Among some groups of athletes, anabolic steroids have a reputation for expediting recovery from injury [49]. The available objective evidence does not confirm this reputed effect [36, 37, 45]. No reported studies have shown accelerated healing of ligament, tendon, or joint injuries due to anabolic steroids [36, 37], and several reports suggest that these drugs may increase the probability of certain injuries [36, 44, 58, 89].

Dimethyl Sulfoxide

Some athletes have reported excellent results from topical dimethyl sulfoxide treatment of musculoskeletal soft tissue injuries, but attempts to prove the efficacy of this treatment in controlled trials have produced conflicting results [17, 18, 55, 72, 73, 77, 78, 84]. Application of 60% to 95% dimethyl sulfoxide reportedly relieved the symptoms of acute bursitis within 30 minutes in about 90% of patients [77, 78]. A study of acute sprains, strains, bursitis, and tendonitis treated with 80% dimethyl sulfoxide found significantly better results with dimethyl sulfoxide than with placebo [17, 18]; but another study found 13 treatment failures among 20 patients with acute bursitis or tendonitis [55]. A double-blind trial of dimethyl sulfoxide treatment of

rotator cuff tendonitis and tennis elbow did not find any significant benefit from the drug [73].

EFFECTS OF ANTI-INFLAMMATORY DRUGS ON DENSE FIBROUS TISSUES

Nonsteroidal Anti-inflammatory Drugs

Thus far, despite the extensive use of nonsteroidal anti-inflammatory medications, no evidence of damage to normal dense fibrous tissues caused by these drugs has been reported [1, 31, 88]. Under some experimental circumstances they may increase tissue strength [87]. Because some events occurring during inflammation may stimulate tissue repair [21, 23], suppression of inflammation could delay cell activity necessary for formation of new tissue. The anti-inflammatory activity of nonsteroidal anti-inflammatory medications may have an effect on the early stages of dense fibrous tissue repair, but clinically significant inhibition of healing has not been documented, and one study showed that these drugs may temporarily increase the strength of healing dense fibrous tissues [31].

Corticosteroids

In contrast to nonsteroidal anti-inflammatory medications, steroids have been reported to cause harmful effects in normal and injured dense fibrous tissues [48, 63]. They alter the metabolism of normal tissues, and many authors have reported spontaneous tendon and plantar fascial ruptures following corticosteroid injections or systemic corticosteroid use [9, 19, 29, 34, 39, 43, 46, 50, 57, 79]. It is difficult to decide whether inflammation or injury weakened the tissues before steroid injections in these patients, but animal experiments show that steroids inhibit matrix synthesis by normal mesenchymal cells [4], and clinical experience shows that multiple steroid injections may cause tissue atrophy [48].

The mechanism of apparent spontaneous rupture of tendons following corticosteroid use remains uncertain. Normal composition, structure, and mechanical properties of these tissues depend on matrix turnover (i.e., steady replacement of degraded matrix molecules). Conceivably, corticosteroids suppress synthesis of the matrix macromolecules, thereby preventing replacement of matrix lost during normal turnover. With time, this negative balance would weaken the tissue. It is also possible that corticosteroid injections might directly disrupt matrix organization [41, 85]. Injection causes hemorrhage and cell and matrix damage within tendons [6], and the damage associated with steroid injections may be more severe than that associated with saline injections [85]. Some authors have also found hyaline material in the region of a corticosteroid injection into dense fibrous tissue and have suggested that the injection may have caused necrosis [6, 41, 85].

Studies that have examined the strength of normal tendons after steroid injections have yielded inconsistent results. Some of the inconsistencies may result from differences in doses of corticosteroid, location of the injection (into the tissues surrounding the tendon or into the tendon itself), time of testing after injection, and methods of measuring tendon strength. Two groups of investigators found that corticosteroid injections did not weaken normal rabbit tendons [51, 56, 74], but repeated intraarticular injections of large doses of corticosteroids decreased the strength and stiffness of monkey anterior cruciate ligament-bone ligament-bone units [63]. Two other studies showed decreases in tendon strength following injections directly into the tendon substance [41, 85]. In one of the latter studies, 48 hours after corticosteroid injection into normal rabbit Achilles tendons, the ultimate failure strength of the tendons had decreased 35% [41]. Microscopic examination of the injected tendons showed evidence of disruption of the normal collagen fibril arrangement and the appearance of clefts within the matrix. Two weeks after injection the failure strength of the injected tendons had improved to near normal, an amorphous eosinophilic staining material had appeared within the substance of the matrix, and the collagen fibril arrangement appeared to be nearly normal. These results showed that injection of corticosteroids directly into dense fibrous tissue weakens the tissue, but following a single injection the cells can restore the matrix toward normal. Because of the potential increased risk of tendon rupture demonstrated by experimental studies and the clinical reports of spontaneous tendon ruptures associated with corticosteroid injections, several authors have advised physicians either to use extreme caution in selecting corticosteroid injections for treatment of tenosynovitis or to avoid using this treatment [48, 49, 82, 85].

Corticosteroids also alter the healing of dense fibrous tissues. Steroid-mediated inhibition of fibroblast proliferation and synthesis of new matrix has the benefit of decreasing adhesions between injured dense fibrous tissues and the surrounding tissues [42, 90], but it also causes wounds to gain strength more slowly [38, 40]. In experimental studies corticosteroid injections of transected tendons decreased tendon weight, load to failure, and energy to failure [40, 90]. Presumably, these consequences of corticosteroid injections result from inhibition of the synthetic function of tendon cells. They may also prolong healing and increase the probability of complications, including failure of healing and wound disruption [8, 38, 48, 86].

Anabolic Steroids

Anabolic steroids may also weaken normal dense fibrous tissues. Several reports have described tendon ruptures or tears in association with the use of anabolic steroids [36, 44]. Experimental studies also suggest that these drugs damage dense fibrous tissues [58, 89]. Administration of an anabolic steroid to mice subjected to an endurance training program caused degenerative changes in muscle tendon junctions including increased variation in collagen fibril diameter and organization, disruption of collagen fibrils, and calcification [58]. A study of rat tendons showed that exercise in animals given anabolic steroids caused tendons to reach breaking strains earlier and supported the argument that anabolic steroids may predispose dense fibrous tissues to injury [89].

Dimethyl Sulfoxide

Like corticosteroids and anabolic steroids, dimethyl sulfoxide may weaken dense fibrous tissues. In tissue cultures it inhibits fibroblast proliferation [14] and decreases collagen synthesis by at least one cell line [7]. If the drug has the same consequences in normal tissues in vivo, it could increase the probability of tendon, ligament, and joint injuries. One experimental study supports this suggestion [2]. Investigators washed the skin area of mice Achilles tendons with a 70% solution of dimethyl sulfoxide and then measured the strength of the Achilles tendons. They found a variable effect on the force required to separate the tendons. In the first week of treatment it decreased 20.2%. Over the next 2 weeks it increased and then decreased again. The investigators concluded that the decreased separation force due to dimethyl sulfoxide treatment made the tendons more susceptible to injury.

EFFECTS OF ANTI-INFLAMMATORY DRUGS ON CARTILAGE

Nonsteroidal Anti-inflammatory Drugs

Selected nonsteroidal anti-inflammatory agents alter the synthetic activity of normal chondrocytes and possibly the degradation of proteoglycans. As a result, they change the composition and possibly the mechanical properties of the cartilage matrix. A series of studies shows that prolonged administration of salicylates and several other nonsteroidal anti-inflammatory agents suppresses proteoglycan synthesis in normal cartilage and sometimes alters the organization of the cartilage matrix by interfering with proteoglycan aggregate formation [15, 16, 65, 67, 69, 70, 71]. The clinical significance of these effects of nonsteroidal anti-inflammatory drugs on normal cartilage in vivo remains uncertain, but a significant decrease in cartilage proteoglycan concentration decreases cartilage stiffness and increases cartilage permeability [22, 25, 26, 61, 62]. These changes might make the tissue more vulnerable to injury [21]. However, none of the reported studies show that nonsteroidal anti-inflammatory medications cause progressive degeneration of normal cartilage, and one study found that a nonsteroidal anti-inflammatory drug decreased cartilage proteoglycan turnover and increased the stiffness of normal cartilage [75].

Nonsteroidal anti-inflammatory agents also affect chondrocyte function in injured or degenerating cartilage [15, 16, 66, 68, 71], and several investigations show that aspirin suppresses proteoglycan synthesis more severely in osteoarthritic cartilage than in normal cartilage [68, 69, 71]. Prolonged oral administration of aspirin aggravated the degeneration of canine articular cartilage caused by immobilization [66] and exacerbated degeneration of articular cartilage in unstable joints [68]. In dogs with knee instability due to transection of the anterior cruciate ligament, prolonged administration of aspirin decreased articular cartilage thickness, cartilage proteoglycan content, and proteoglycan synthesis compared with dogs with knee instability that did not receive aspirin. Although these studies show that nonsteroidal anti-inflammatory agents, particularly aspirin, alter chondrocyte synthetic function, the clinical significance of these observations, like the effects of nonsteroidal agents on normal cartilage, has not been defined.

Corticosteroids

Many experimental studies show that repeated intra-articular injections of corticosteroids cause progressive deterioration of normal articular cartilage and that increasing amounts of corticosteroids increase the severity of damage to cartilage [10, 52, 53, 60]. Following intra-articular corticosteroid injections, chondrocyte synthesis of collagen and proteoglycans decreases rapidly and profoundly [10, 52, 53, 64]. Then matrix proteoglycan concentration drops, decreasing cartilage stiffness and increasing permeability [62]. Acute or repetitive loading of the damaged articular cartilage may cause mechanical disruption of the weakened matrix and progressive loss of the articular cartilage [10, 11, 21, 24]. Systemic corticosteroid administration also depresses chondrocyte synthetic activity [54].

In otherwise normal joints, if the damage to articular cartilage due to corticosteroids leaves the tissue physically intact with enough viable chondrocytes, the cells will attempt to repair the damage. Following cessation of intra-articular steroid injections, chondrocytes increase their rates of proteoglycan and collagen synthesis by up to 900% [11]. The increase results from accelerated activity by existing cells and an apparent increase in the number of cells. Under favorable circumstances, the increased matrix synthesis will return the matrix proteoglycan concentration toward normal [11].

Corticosteroid injections of synovial joints damaged by rheumatoid or osteoarthritis often provide rapid relief from pain [28]. In patients with advanced joint disease, in whom synovial inflammation contributes to the symptoms and to progression of the disease, suppression of inflammation by corticosteroids might help to maintain joint function. Unfortunately, other effects of the corticosteroids may more than offset these potential benefits. Corticosteroid injections presumably suppress chondrocyte synthetic activity in injured or degenerated cartilage at least as effectively as they do in normal cartilage. Suppression of chondrocyte synthetic activity may prevent the cells from repairing matrix defects due to injury or disease, thereby hastening the deterioration of the articular cartilage. Many clinical descriptions of rapid joint disintegration in patients with rheumatoid arthritis and osteoarthritis following intra-articular steroid injections show that this may indeed occur [13, 27, 59, 80, 81, 82, 91]. Although steroid-induced inhibition of chondrocyte synthetic activity may have accelerated the joint destruction in these patients, decreased pain may have also had a role. Relief of pain following the injections allows the patients to increase their activity, and the increased loading may contribute to loss of the damaged articular cartilage [27]. The apparent frequency of this problem and the results of clinical and experimental studies have led several physicians to recommend abandonment of the practice of multiple joint injections with corticosteroids and to insist on strong justification for single injections [82].

References

1. Abramson, S. B. Nonsteroidal anti-inflammatory drugs: mechanisms of action and theraputic considerations. *In* Leadbetter, W. B., Buckwalter, J. A., and Gordon, S. L. (Eds.), *Sports Induced Inflammation—Basic Science and Clinical Concepts.* Park Ridge, IL, American Academy of Orthopaedic Surgeons, 1990.
2. Albrechtsen, S. J., and Harvey, J. S. Dimethyl sulfoxide: Biomechanical effects on tendons. *Am J Sports Med* 10:177–179, 1982.
3. Almekinders, L. C., and Gilbert, J. A. Healing of experimental muscle strains and the effects of nonsteroidal antiinflammatory medication. *Am J Sports Med* 14:303–308, 1986.
4. Anastassiades, T., and Dziewiatkowski, D. The effect of cortisone on the metabolism of connective tissues in the rat. *J Lab Clin Med* 75(5):826–839, 1970.
5. Archer, A. G., Nelson, M. C., Abbondanzo, S. L., and Bogumil, G. P. Case report 554. *Skeletal Radiol* 18:380–384, 1989.
6. Balasubramaniam, P., and Prathap, K. The effect of injection of hydrocortisone into rabbit calcaneal tendons. *J Bone Joint Surg.* 54-B:729–734, 1972.
7. Banes, A. J., Mebes, S., Smith, R., et al. DMSO normalizes collagen synthesis in MAV-Z(O) infected chick embryo cells. *Gen Pharmacol* 10:521–523, 1979.
8. Baxter, J. D., and Forsham, P. H. Tissue effects of glucocorticoids. *Am J Med* 53:573–589, 1972.
9. Bedie, S. S., and Ellis, W. Spontaneous rupture of the calcaneal tendon in rheumatoid arthritis after local steroid injection. *Ann Rheum Dis* 29:494–495, 1970.
10. Behrens, F., Shepard, N., and Mitchell, N. Alteration of rabbit articular cartilage by intra-articular injections of glucocorticoids. *J Bone Joint Surg* 57-A:70–76, 1975.
11. Behrens, F., Shepard, N., and Mitchell, N. Metaboic recovery of articular cartilage after intra-articular injections of glucocorticoid. *J Bone Joint Surg* 58-A:1157–1160, 1976.
12. Behrens, T. W., and Goodwin, J. S. Oral corticosteroids. *In* Leadbetter, W. B., Buckwalter, J. A., and Gordon, S. L. (Eds.), *Sports Induced Inflammation—Basic Science and Clinical Concepts.* Park Ridge, IL, American Academy of Orthopaedic Surgeons, 1990.
13. Bentley, G. and Goodfellow, J. W. Disorganization of the knees following intra-articular hydrocortisone injections. *J Bone Joint Surg* 51-B:498–502, 1969.
14. Berliner, D. L., and Ruhman, A. G. The influence of DMSO on fibroblast proliferation. *Ann NY Acad Sci* 141:159–164, 1967.
15. Brandt, K. D., and Palmoski, M. J. Proteoglycan content determines the susceptibility of articular cartilage to salicylate-induced suppression of proteoglycan synthesis. *J Rheum* 10 (Suppl. 9):78–80, 1983.
16. Brandt, K. D., and Palmoski, M. J. Effects of salicylates and other nonsteroidal anti-inflammatory drugs on articular cartilage. *Am J Med* 77:65–69, 1984.
17. Brown, J. H. Clinical experience with DMSO in acute musculoskeletal conditions comparing a noncontrolled series with a controlled double-blind study. *Ann NY Acad Sci* 141:496–505, 1967.
18. Brown, J. H. A soluble blind clinical study of DMSO for acute injuries and inflammations compared to accepted standard therapy. *Curr Ther Res* 13:536–540, 1971.
19. Bruno, L. P., and Clarke, R. P. The use of local corticosteroid injections in orthopaedic surgery. 56th annual meeting of the American Academy of Orthopaedic Surgeons, Las Vegas, February, 1989.
20. Buchanan, W. W. Aspirin and nonacetylated salicylates: Use in inflammatory injuries incurred during sporting activities. *In* Leadbetter, W. B., Buckwalter, J. A., and Gordon, S. L. (Eds.), *Sports Induced Inflammation—Basic Science and Clinical Concepts.* Park Ridge, IL, American Academy of Orthopaedic Surgeons, 1990.
21. Buckwalter, J. A., and Cruess, R. Healing of musculoskeletal tissues. *In* Rockwood, C. A., and Green, D. P. (Eds.), *Fractures.* Philadelphia, J. B. Lippincott, 1991.
22. Buckwalter, J. A., Hunziker, E. B., Rosenberg, L. C., Coutts, R. D., Adams, M. E., and Eyre, D. R. Articular cartilage: Composition and structure. *In* Woo, S. L.-Y., and Buckwalter, J. A. (Eds.), *Injury and Repair of the Musculoskeletal Soft Tissues.* Park Ridge, IL, American Academy of Orthopaedic Surgeons, 1988.
23. Buckwalter, J. A., Maynard, J. A., and Vailas, A. C. Skeletal fibrous tissues: Tendon, joint capsule, and ligament. *In* Albright, J. A., and Brand, R. A. (Eds.), *The Scientific Basis of Orthopaedics.* Norwalk, Appleton & Lange, 1987.
24. Buckwalter, J. A., and Mow, V. C. Cartilage repair in the treatment of osteoarthritis. *In* Moskowitz, R. W., Howell, D. S., Mankin, H. J., and Goldberg, J. W. (Eds.), *Osteoarthritis.* Philadelphia, W.B. Saunders, 1991.
25. Buckwalter, J. A., Rosenberg, L. C., and Hunziker, E. B. Articular cartilage: Composition, structure, response to injury, and methods of facilitating repair. *In* Ewing, J. W. (Ed.), *Articular Cartilage and Knee Joint Function: Basic Science and Arthroscopy.* New York, Raven Press, 1990.
26. Buckwalter, J. A., Rosenberg, L. C., Coutts, R., Hunziker, E., Reddi, A. H., and Mow, V. C. Articular cartilage: Injury and repair. *In* Woo, S. L.-Y., and Buckwalter, J. A. (Eds.), *Injury and Repair of the Musculoskeletal Soft Tissues.* Park Ridge, IL, American Academy of Orthopaedic Surgeons, 1988.
27. Chandler, G. N., and Wright, V. Deleterious effect of intra-articular hydrocortisone. *Lancet* 2:661–663, 1958.
28. Chandler, G. N., Wright, V., and Hartfall, S. J. Intra-articular therapy in rheumatoid arthritis: Comparison of hydrocortisone acetate tertiary butyl acetate and hydrocortisone acetate. *Lancet* 2:659–661, 1958.
29. Cowan, M. A., and Alexander, S. Simultaneous bilateral rupture of achilles tendons due to triamcinolone. *Br Med J* 2:1658, 1961.
30. Cox, J. S. Current concepts in the role of steroids in the treatment of sprains and strains. *Med Sci Sports Exerc* 16:216–218, 1984.
31. Dahners, L. E., Gilbert, J. A., Lester, G. E., Taft, T. N., and Payne, L. Z. The effect of nonsteroidal antiinflammatory drug on the healing of ligaments. *Am J Sports Med* 16:641–646, 1988.
32. Ducan, J. J., and Farr, J. E. Comparison of declofenac sodium and aspirin in the treatment of acute sports injuries. *Am J Sports Med* 16:656–659, 1988.
33. Dupont, M., Beliveau, P. and Theriault, G. The efficacy of antiinflammatory medication in the treatment of the acutely sprained ankle. *Am J Sports Med* 15:41–45, 1987.
34. Ford, L. T., and DeBender, J. Tendon rupture after local steroid injection. *South Med J* 72:827–830, 1979.
35. Grauer, J. D., Kabo, J. M., Dorey, F. J., and Meals, R. A. The effects of dexamethasone on periarticular swelling and joint stiffness following fracture in a rabbit hindlimb model. *Clin Orthop Rel Res* 242:277–284, 1989.
36. Haupt, H. A. The role of anabolic steroids as modifiers of sports-induced inflammation. *In* Leadbetter, W. B., Buckwalter, J. A., and Gordon, S. L. (Eds.), *Sports Induced Inflammation—Basic Science and Clinical Concepts.* Park Ridge, IL, American Academy of Orthopaedic Surgeons, 1990.

37. Haupt, H. A., and Rovere, G. D. Anabolic steroids: A review of the literature. *Am J Sports Med* 12:469–484, 1984.

38. Howes, E. L., Plotz, C. M., Blunt, J. W., and Ragan, C. Retardation of wound healing by cortisone. *Surgery* 28(2):177–181, 1950.

39. Ismail, A. M., Balakrishnan, R., and Rajakumar, M. K. Rupture of patellar ligament after steroid injection. *J Bone Joint Surg* 51-A:503–505, 1969.

40. Kapetanos, G. The effect of local corticosteroids on the healing and biomechanical properties of the partially injured tendon. *Clin Orthop Rel Res* 163:170–179, 1982.

41. Kennedy, J. C., and Willis, R. B. The effects of local steroid injections on tendons: A biomechanical and microscopic correlative study. *Am J Sports Med* 4(1):11–21, 1976.

42. Ketchum, L. D. Effects of triamcinolone on tendon healing and function. *Plast Reconstr Surg* 47(5):471–482, 1971.

43. Kleinman, M., and Gross, A. E. Achilles tendon rupture following steroid injection: Report of three cases. *J Bone Joint Surg* 65-A:1345–1347, 1983.

44. Kramhoft, M., and Solgaard, S. Spontaneous rupture of the extensor pollicus longus tendon after anabolic steroids. *J Hand Surg* 11-B:87, 1986.

45. Lamb, D. R. Anabolic steroids in athletics: How well do they work and how dangerous are they? *Am J Sports Med* 12:31–38, 1984.

46. Leach, R., Jones, R., and Silva, T. Rupture of the plantar fascia in athletes. *J Bone Joint Surg* 60-A:537–539, 1978.

47. Leadbetter, W. B. Corticosteroid injection for the treatment of athletic injury. *Med Sci Sports Exerc* 15:103, 1983.

48. Leadbetter, W. B. Corticosteroid injection therapy in sports injuries. *In* Leadbetter, W. B., Buckwalter, J. A., and Gordon, S. L. (Eds.), *Sports Induced Inflammation—Basic Science and Clinical Concepts.* Park Ridge, IL, American Academy of Orthopaedic Surgeons, 1990.

49. Leadbetter, W. B. Overview of modifiers of inflammation. *In* Leadbetter, W. B., Buckwalter, J. A., and Gordon, S. L. (Eds.), *Sports Induced Inflammation—Basic Science and Clinical Concepts.* Park Ridge, IL, American Academy of Orthopaedic Surgeons, 1990.

50. Lee, H. B. Avulsion and rupture of the tendo calcaneus after injection of hydrocortisone. *Br Med J* 2:395, 1957.

51. Mackie, J. W., Goldin, B., Foss, M. L., Cockrell, J. L. Mechanical properties of rabbit tendons after repeated antiinflammatory steroid injections. *Med Sci Sports* 6(3):198–202, 1974.

52. Mankin, H. J., and Conger, K. A. The acute effects of intra-articular hydrocortisone on articular cartilage in rabbits. *J Bone Joint Surg* 48-A:1383–1388, 1966.

53. Mankin, H. J., and Conger, K. A. The effect of cortisol on articular cartilage of rabbits. *Lab Invest* 15(4):794–800, 1966.

54. Mankin, H. J., Zarins, A., and Jaffe, W. L. The effect of systemic corticosteroids on rabbit articular cartilage. *Arth Rheum* 15:593–599, 1972.

55. Marmor, L., and Wilke, B. Experience with DMSO. *Calif Med* 105:28–30, 1966.

56. Matthews, L. S., Sonstegard, D. A., and Phelps, D. B. A biomechanical study of rabbit patellar tendon: Effects of steroid injections. *J Sports Med* 2:349–357, 1975.

57. Melmed, E. P. Spontaneous bilateral rupture of the calcaneal tendon during steroid therapy. *J Bone Joint Surg* 47-B:104–105, 1965.

58. Michna, H., and Stang-Voss, C. The predisposition to tendon rupture after doping with anabolic steroids. *J Sports Med* 4:59, 1983.

59. Miller, W. T., and Restifo, R. A. Steroid arthropathy. *Radiology* 86:652–657, 1966.

60. Moskowitz, R. W., Davis, W., Sammarco, J., Mast, W., and Chase, S. W. Experimentally induced corticosteroid arthropathy. *Arth Rheum* 13:236–243, 1970.

61. Mow, V. C., Proctor, C. S., and Kelly, M. A. Biomechanics of articular cartilage. Nordin, M. (Ed.), *Basic Biomechanics of the Musculoskeletal System.* Philadelphia, Lea & Febiger, 1989.

62. Mow, V. C., and Rosenwasser, M. P. Articular cartilage: Biomechanics. *In* Woo, S. L.-Y., and Buckwalter, J. A. (Eds.), *Injury and Repair of the Musculoskeletal Soft Tissues.* Park Ridge, IL, American Academy of Orthopaedic Surgeons, 1988.

63. Noyes, F. R., Grood, E. S., Nussbaum, N. S., and Cooper, S. M. Effect of intra-articular corticosteroids on ligament properties. *Clin Orthop Rel Res* 123:197–209, 1977.

64. Oegema, T. R., and Behrens, F. Proteoglycan aggregate synthesis in normal and chronically hydrocortisone-suppressed rabbit articular cartilage. *Arch Biochem Biophys* 206(2):277–284, 1981.

65. Palmoski, M. J., and Brandt, K. D. Effects of some nonsteroidal antiinflammatory drugs on proteoglycan metabolism and organization in canine articular cartilage. *Arth Rheum* 23:1010–1020, 1980.

66. Palmoski, M. J., and Brandt, K. D. Aspirin aggravates the degeneration of canine joint cartilage caused by immobilization. *Arth Rheum* 25:1333–1342, 1982.

67. Palmoski, M. J., and Brandt, K. D. Benoxaprofen stimulates proteoglycan synthesis in normal canine knee cartilage in vitro. *Arth Rheum* 26:771–774, 1983.

68. Palmoski, M. J., and Brandt, K. D. In vivo effect of aspirin on canine osteoarthritic cartilage. *Arth Rheum* 26:994–1001, 1983.

69. Palmoski, M. J., and Brandt, K. D. Relationship between matrix proteoglycan content and the effects of salicylate and indomethacin on articular cartilage. *Arth Rheum* 26:528–531, 1983.

70. Palmoski, M. J., and Brandt, K. D. Effects of salicylate and indomethacin on glycosaminoglycan and prostaglandin E2 synthesis in intact canine knee cartilage ex vivo. *Arth Rheum* 27:398–403, 1984.

71. Palmoski, M. J., Colyer, R. A., and Brandt, K. D. Marked suppression by salicylate of the augmented proteoglycan synthesis in osteoarthritic cartilage. *Arth Rheum* 23:83–91, 1980.

72. Percy, E. C. Dimethyl sulfoxide: Its role as an anti-inflammatory agent in athletic injuries. *In* Leadbetter, W. B., Buckwalter, J. A., and Gordon, S. L. (Eds.), *Sports Induced Inflammation—Basic Science and Clinical Concepts.* Park Ridge, IL, American Academy of Orthopaedic Surgeons, 1990.

73. Percy, E. C., and Corson, J. D. The use of DMSO in tennis elbow and rotator cuff tendinitis: A double blind study. *Med Sci Sports Exerc* 13:215–219, 1981.

74. Phelps, D., Sonstegard, D. A., and Mattews, L. A. Corticosteroid injection effects on the biomechanical properties of rabbit patellar tendons. *Clin Orthop Rel Res* 100:345–348, 1974.

75. Ratcliffe, A., and Mow, V. C. Structural and functional relationships of articular cartilage: The effect of naproxen. *In* Moskowitz, R. W., *Effects of NSAIDs on Bone and Joint Disease: New Insights.* Fair Lawn, NJ, Medical Publishing Enterprises, 1990.

76. Roseff, R., and Canoso, J. J. Femoral osteonecrosis following soft tissue corticosteroid infiltration. *Am J Med* 77:1119–1120, 1984.

77. Rosenbaum, E. E., Herschler, R. J., and Jacob, S. W. DMSO in musculoskeletal disorders. *JAMA* 192:309–313, 1965.

78. Rosenbaum, E. E., and Jacob, S. W. DMSO in acute musculoskeletal injuries and inflammations. *Northwest Med* 63:167–168, 1964.

79. Smaill, G. B. Bilateral rupture of Achilles tendons. *Br Med J* 1:1657–1658, 1961.

80. Steinberg, C. L., Duthie, R. B., and Piva, A. E. Charcot-like arthropathy following intra-articular hydrocortisone. *JAMA* 181(10):851–854, 1962.

81. Sweetnam, D. R., Mason, R. M., and Murray, R. O. Steroid arthropathy of the hip. *Br Med J* 1:1392–1394, 1960.

82. Sweetnam, R. Corticosteroid arthropathy and tendon rupture. *J Bone Joint Surg* 51-B:397–398, 1969.

83. Taylor, L. J. Multifocal avascular necrosis after short-term high-dose steroid therapy. *J Bone Joint Surg* 66-B:431–433, 1984.

84. Trice, J. M., and Pinals, R. S. Dimethyl Sulfoxide: A review of its use in the rheumatic disorders. *Sem Arth Rheum* 15(1):45–60, 1985.

85. Unverferth, L. J., and Olix, M. L. The effect of local steroid injections on tendon. *J Sports Med* 1:31–37, 1973.

86. Vogel, H. G. Tensile strength of skin wounds in rats after treatment with corticosteroids. *Acta Endocrinol* 64:295–303, 1970.

87. Vogel, H. G. Mechanical and chemical properties of various connective tissue organs in rats as influenced by nonsteroidal antirheumatic drugs. *Connect Tissue Res* 5:91–95, 1977.

88. Weiler, J. M., Albright, J. P., and Buckwalter, J. A. Nonsteroidal anti-inflammatory drugs in sports medicine. *In* Lewis, A. J., and Furst, D. E. (Eds.), *Nonsteroidal Anti-Inflammatory Drugs*. New York, Marcel Dekker, 1987, pp. 71–88.

89. Wood, T. O., Cooke, P. H., and Goodship, A. E. The effect of exercise and anabolic steroids on the mechanical properties and crimp morphology of the rat tendon. *Am J Sports Med* 16(2):153–158, 1988.

90. Wrenn, R. N., Goldner, J. L., and Markee, J. L. An experimental study of the effect of cortisone on the healing process and tensile strength of tendons. *J Bone Joint Surg* 36-A:588–601, 1954.

CHAPTER TWO

INJURIES TO CARTILAGE AND MENISCUS

Sports Injuries to Articular Cartilage

Joseph A. Buckwalter, M.D.
Van C. Mow, Ph.D.

Synovial joints make possible the rapid controlled movements necessary for sports. Normal function of these complex diarthrodial structures depends on the structural integrity and macromolecular composition of articular cartilage. Sports-related traumatic disruptions of cartilage structure or alterations in cartilage macromolecular composition or organization change the biomechanical properties of the tissue and compromise joint function. Even more important, they can lead to progressive lifelong pain and disability. Sports injuries to articular cartilage present far more difficult diagnostic and treatment problems than injuries to ligament, tendon, or bone, partly because less is known about these injuries and partly because of the unique structure and function of articular cartilage.

The specialized composition and organization of articular cartilage [34, 35] make the diagnosis of many injuries difficult, but they also provide the unique biomechanical properties that make possible normal synovial joint function. Articular cartilage is very soft and yields its interstitial water easily when compressed [6, 106, 140, 143], yet it is very stiff in tension along planes parallel to the articular surface [98, 205]. Intact cartilage provides a smooth, lubricated gliding surface with a coefficient of friction better than most manmade bearing materials [56, 84, 142, 144]. In the joint, cartilage distributes the loads of articulation, thereby minimizing peak stresses acting on the subchondral bone while the tensile strength of the tissue maintains its structural integrity under loading. These biomechanical properties make the tissue remarkably durable and wear resistant [115], enabling it to last many decades even under high and repetitive stresses.

Alterations in the mechanical properties of cartilage due to injury or disease or associated with increasing age have not been well defined, but the available information shows that these properties change with age and loss of structural integrity. Cartilage from skeletally immature joints (bones

This work was sponsored in part by grants from the National Institutes of Health: grant AR34758 on articular cartilage healing and passive motion (JB) and grant AR5-35542 on articular cartilage injuries and response due to anterior cruciate resection and joint immobilization (VCM).

with open growth plates) is much stiffer than cartilage from skeletally mature joints (those with closed growth plates) [173]. Older cartilage and fibrillated cartilage have much lower tensile stiffness and strength [1, 99]. Declines in stiffness and strength may increase the probability of injury to cartilage.

Participation in sports often subjects the articular cartilage to intense repetitive compressive forces that can cause injury and deterioration of the tissue. Falls or other high-energy impacts can damage the articular cartilage without disrupting the articular surface. These abnormally large forces generate high shear stresses at the cartilage-subchondral bone junction, causing matrix lesions [7, 55] that may lead to clinically significant cartilage deterioration and joint dysfunction [39, 55, 132, 168]. Furthermore, repetitive trauma that leaves the articular surface intact can cause other injuries, including subtle damage to the matrix macromolecular framework and cartilage cells (chondrocytes) [55, 57, 168, 169]. These injuries disrupt the well-organized macromolecular fabric of the matrix, alter cell function, and disturb normal cell-matrix interactions, thereby adversely altering the cell activities required to maintain cartilage mechanical properties [30, 36, 85, 120–123, 148]. Since cartilage lacks nerves and blood vessels, damage limited to cartilage is not likely to be detected at the time of injury, but cartilage with this type of damage may become fatigued and fail more easily when subjected to acute or repetitive trauma. Less frequent but more severe sports injuries can acutely disrupt the articular surface by fracturing both cartilage and the underlying bone [25, 86]. Therefore, both acute traumatic injuries and repetitive excessive loadings resulting from participation in sports can lead to cartilage deterioration and loss of synovial joint function.

Despite their importance, sports injuries to articular cartilage remain poorly understood. Because injuries limited to cartilage do not cause pain or inflammation, patients and physicians rarely suspect cartilage damage following excessive acute or repetitive joint loading. Even when the physician suspects that cartilage injury exists, making a precise diagnosis of many types of cartilage injuries is difficult or impossible at present. Partly because of the difficulty in making a diagnosis, the natural history of many types of cartilage injury remains unknown, and it seems likely that some of the deterioration of articular surfaces now attributed to sports-related ligamentous or meniscal injuries is due to undetected cartilage damage.

This chapter first reviews the important aspects of articular cartilage composition, organization, and biomechanical properties that make possible normal synovial joint function. The next sections summarize the response of cartilage to different types of injury and the results of cartilage repair, cartilage shaving, and abrasion of subchondral bone. The last section describes the use of cartilage grafts to replace lost or damaged articular surfaces.

COMPOSITION OF ARTICULAR CARTILAGE

Like the dense fibrous tissues and meniscus, articular cartilage consists of cells, matrix water, and a matrix macromolecular framework [8, 9, 27, 28, 34, 35, 66, 146, 148, 151]. However, unlike the most dense fibrous tissues, cartilage lacks nerves, blood vessels, and a lymphatic system. These conditions are responsible for its unusual physiologic requirements, cell behavior, and responses to injury.

Chondrocytes

Only one type of cell exists within normal cartilage—the highly specialized chondrocyte [34, 35, 191, 208]. These cells contribute relatively little to the total volume of mature human articular cartilage, usually 5% or less. Like other mesenchymal cells [27, 191], chondrocytes surround themselves with their extracellular matrix and rarely form cell to cell contacts. In normal cartilage they are isolated in the extracellular matrix, and, because the tissue lacks blood vessels, the cells depend on diffusion through the matrix for their nutrition and rely primarily on anaerobic metabolism.

Figures 2A–1A and B show a histologic section of normal adult articular cartilage with the chondrocytes embedded in the matrix and a schematic representation of chondrocyte morphology. Three distinct zones of chondrocytes are seen: (1) the superficial tangential zone (STZ) with ellipsoidal cells with their long axes aligned parallel to the surface; (2) the middle zone with spherical cells randomly distributed throughout the region; and (3) the deep zone with similar spherical cells forming columns aligned perpendicular to the tidemark and the calcified zone. More details on this arrangement will be presented later in the chapter.

Articular cartilage chondrocytes contain the organelles necessary for matrix synthesis, including endoplasmic reticulum and Golgi membranes. Also, they frequently contain intracytoplasmic filaments and glycogen, and at least some chondrocytes have a cilium that extends from the cell into the extracellular collagen-proteoglycan matrix. These struc-

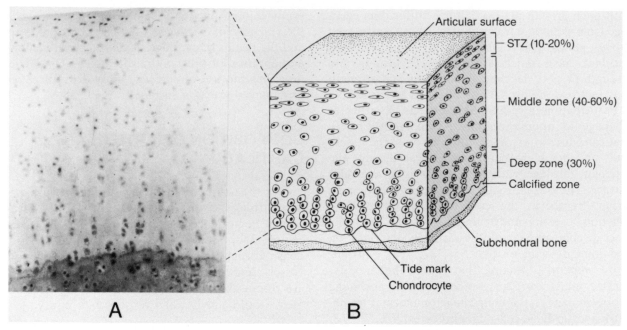

FIGURE 2A–1

Normal articular cartilage structure. Histologic *(A)* and schematic *(B)* views of a section of normal articular cartilage. The tissue consists of four zones: the superficial tangential zone (STZ), the middle zone, the deep zone, and the calcified zone. Notice the differences in cell alignment among zones. The cells of the superficial zone have an ellipsoidal shape and lie with their long axes parallel to the articular surface. The cells of the other zones have a more spheroidal shape. In the deep zone they tend to align themselves in columns perpendicular to the joint surface. (From M. Nordin and V.H. Frankel: *Basic Biomechanics of the Musculoskeletal System,* 2nd ed. Philadelphia, Lea & Febiger, 1989, pp. 31–57. Used with permission.)

tures may have a role in sensing mechanical changes in the matrix. After completion of skeletal growth, chondrocytes rarely divide, but throughout life they synthesize and maintain the extracellular matrix that gives cartilage its essential material properties. Synthesis and turnover of proteoglycans are relatively fast, whereas collagen synthesis and turnover are very slow [151, 154, 191].*

Extracellular Matrix

Tissue Fluid

Water contributes up to 80% of the wet weight of articular cartilage, and the interaction of water with the matrix macromolecules significantly influences the material properties of the tissue [59, 92, 113, 114, 122, 125, 127, 140, 151, 197]. This tissue fluid contains gases, small proteins, metabolites, and a high concentration of cations to balance the negatively charged proteoglycans [125, 127, 152, 198]. The volume, concentration, and behavior of the

*For more details concerning turnover of cartilage matrix macromolecules, see the excellent review by Lohmander on cartilage proteoglycan turnover 117.

tissue water depend on its interaction with the structural macromolecules. In particular, large aggregating proteoglycans organize the tissue water and impede its flow through the matrix.

The large proteoglycans also help to maintain the fluid within the matrix and the fluid electrolyte concentrations. These matrix macromolecules have large numbers of negative charges that attract positively charged ions and repel negatively charged ions. This increases the concentration of positive ions like sodium and decreases the concentration of negative ions like chloride. The increase in the total inorganic ion concentration increases the tissue osmolarity, that is, it creates a Donnan effect. The collagen network resists the Donnan osmotic pressure caused by the inorganic ions associated with proteoglycans [125–127]. This interaction between proteoglycans and tissue fluid significantly influences the compressive stiffness and resilience of articular cartilage [109, 110, 127, 140, 153].

Structural Macromolecules

The structural macromolecules that provide 20% to 40% of the wet weight of cartilage include collagens, proteoglycans, and noncollagenous proteins

[28, 117, 151]. Chondrocytes synthesize all three types of molecules from amino acids and sugars, but differences in the types and organization of the amino acids and sugars give each type of molecule a different form and function [27, 28, 117, 151]. Abnormalities in these molecules or in their organization can adversely affect the durability and mechanical properties of the cartilage and may lead to deterioration of the articular surface [1, 31, 78, 84, 127, 140–142, 147, 151].

The three classes of macromolecules differ in their concentrations within the tissue and in their contributions to tissue properties. Collagens contribute about 60% of the dry weight of cartilage, proteoglycans contribute 25% to 35%, and the noncollagenous proteins and glycoproteins contribute 15% to 20%. Collagens are distributed relatively uniformly throughout the depth of the cartilage except for a collagen-rich region near the surface [114]. The collagen fibrillar meshwork and crosslinking among collagen fibers give cartilage its form and tensile strength [99, 173, 183, 205]. Figure 2A–2A and B show a schematic representation of the fibrillar collagen ultrastructure throughout the depth of the

tissue, along with three scanning electron micrographs that show the appearance of the fibrillar collagen network in the three zones of uncalcified cartilage [145]. Proteoglycans and noncollagenous proteins bind to the collagenous meshwork or become mechanically entrapped within it, and water fills this molecular framework. Proteoglycans give cartilage its stiffness in compression and its resilience. Figure 2A–3 shows a proteoglycan monomer attached to a hyaluronate chain and a linking protein *(A),* a proteoglycan aggregate *(B),* and an electron micrograph of a proteoglycan aggregate *(C).* Some noncollagenous proteins help organize and stabilize the matrix macromolecular framework, whereas others help chondrocytes bind to the macromolecules of the matrix.*

Collagens. Collagens, a family of 13 or more protein molecules produced by more than 20 distinct genes, form a critical part of every tissue [65, 154]. They have a region consisting of three amino acid chains wound into a triple helix. This triple helical

---

*The reader is referred to references 28, 29, 34, 35, 117, and 151 for more detailed reviews of cartilage molecular organization.

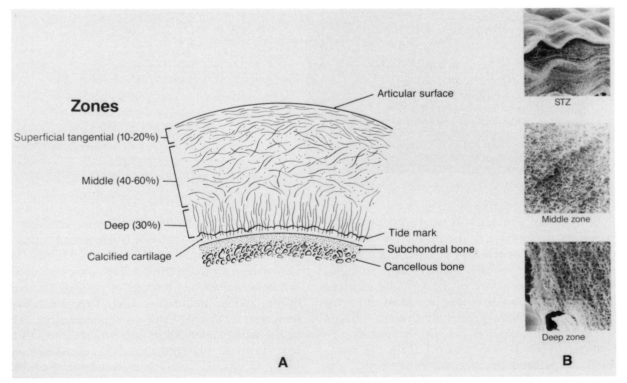

FIGURE 2A–2
Interterritorial matrix collagen fibril orientation in normal articular cartilage. Schematic *(A)* representation and scanning electron micrographs *(B)* of the interterritorial matrix collagen fibril orientation and organization in normal articular cartilage. In the superficial zone the fibrils lie roughly parallel to the articular surface. In the middle zone they assume a more random alignment, and in the deep zone they lie roughly perpendicular to the articular surface. (From M. Nordin and V.H. Frankel: *Basic Biomechanics of the Musculoskeletal System,* 2nd ed. Philadelphia, Lea & Febiger, 1989, pp. 31–57. Used with permission.)

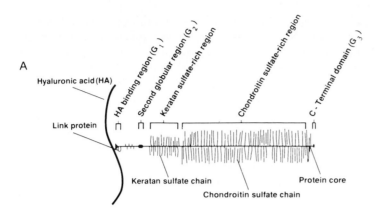

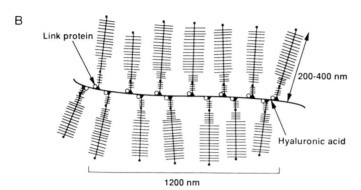

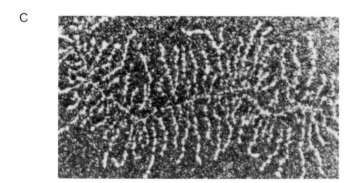

FIGURE 2A–3

Proteoglycan structure. *A,* Details of proteoglycan monomer structure showing chondroitin sulfate and keratan sulfate chains and the interaction of the monomer with hyaluronate chain and link protein. *B,* Molecular conformation of a typical proteoglycan aggregate showing size of the molecule. *C,* An electron micrograph of a proteoglycan aggregate.

region makes up a major part of all collagen molecules and provides some of the distinctive structural and mechanical properties of collagens. Collagens, by virtue of their tensile stiffness and strength, contribute to the structure of the articular cartilage matrix, and articular cartilage, like most other tissues, contains multiple genetically distinct collagen types, specifically, types II, VI, IX, X, and XI.

Collagen types II, IX, and XI form the crossbanded fibrils seen on electron microscopy. The organization of these fibrils into a tight meshwork that extends throughout the tissue provides the tensile stiffness and strength of articular cartilage [1, 98, 99, 173, 205]. This meshwork also contributes to the cohesiveness of the tissue by mechanically entrapping the large proteoglycans. The principal

articular cartilage collagen, type II, accounts for 90% to 95% of the cartilage collagen and forms the primary component of the crossbanded fibrils. Type IX collagen contains both collagenous and noncollagenous regions and has one or possibly two chondroitin sulfate chains [24, 199]. Type IX collagen molecules bind covalently to the superficial layers of the crossbanded fibrils and project into the matrix. Type XI collagen molecules also bind covalently to type II collagen molecules and probably form part of the interior structure of the crossbanded fibrils. The functions of type IX and type XI collagens remain uncertain, but presumably, acting together they help form and stabilize the collagen fibrils assembled primarily from type II collagen. The position of type IX collagen molecules on the surface of the fibrils suggests that they may influence the diameter and stability of fibrils and interact with other matrix macromolecules. The position of type XI collagen molecules within the fibrils suggests that they may contribute to the formation and influence the diameter of fibrils.

The functions of type VI and type X collagen remain unknown. Type VI collagen is often described as an adhesion protein and in some regions reaches its highest concentration in the immediate vicinity of the chondrocytes. Type X collagen appears near the cells of the calcified cartilage zone of articular cartilage and the hypertrophic zone of the growth plate where the longitudinal cartilage septa begin to mineralize. This pattern of distribution suggests that it contributes to mineralization of cartilage.

Proteoglycans

Proteoglycans consist of a protein core and one or more glycosaminoglycan chains (long, unbranched polysaccharide chains consisting of repeating disaccharides that contain an amino sugar). Each disaccharide unit has at least one negatively charged carboxylate or sulfate group so that the glycosaminoglycans form long strings of negative charges that repel other negatively charged molecules and attract cations. These molecules vary in the sugars included in their polysaccharide chains, but they all consist of repeating disaccharide units containing a derivative of either glucosamine or galactosamine [117, 172]. Glycosaminoglycans found in cartilage include hyaluronic acid, chondroitin sulfate, keratan sulfate, and dermatan sulfate.

Articular cartilage contains at least three types of proteoglycans: a large aggregating proteoglycan called aggrecan that contains large numbers of chondroitin sulfate and keratan sulfate chains, and two small dermatan sulfate-containing, nonaggregating proteoglycans called biglycan and decorin. Other matrix molecules that have glycosaminoglycan chains and therefore could be classified as proteoglycans include type IX collagen (discussed in the section on collagens) and fibromodulin (discussed in the section on noncollagenous proteins and glycoproteins). The tissue probably also contains other small proteoglycans and some large nonaggregating proteoglycans. The aggrecan molecules fill most of the interfibrillar space of the cartilage matrix [28, 34, 35, 79, 117, 151]. They contribute about 90% of the total cartilage matrix proteoglycan, while large nonaggregating proteoglycans contribute 10% or less and small nonaggregating proteoglycans contribute about 3%. The large nonaggregating proteoglycans resemble the large aggregating proteoglycans in structure and composition [117] and may represent degraded aggregating proteoglycans.

The large aggregating proteoglycans consist of protein core filaments with many covalently bound chondroitin sulfate and keratan sulfate chains (see Fig. 2A–3) [29, 117, 151, 172]. Chondroitin sulfate and keratan sulfate form about 95% of the molecule, and protein forms about 5%. The protein cores consist of five domains: three globular domains and two extended domains (see Fig. 2A–3). A short extended region of the protein core separates the G1 (first globular domain) region and the G2 (second globular domain) region. A longer extended region of the protein core contains covalently bound keratan sulfate and chondroitin sulfate chains and separates the G2 and G3 (third globular domain) regions. Keratan sulfate chains cluster together in a region near the G2 domain, the keratan sulfate-rich region, and the chondroitin sulfate chains cluster together in a region between the keratan sulfate-rich regions and the G3 domain, the chondroitin sulfate-rich region. The G1 domain binds noncovalently to hyaluronic acid filaments and small proteins called link proteins. The functions of the G2 and G3 domains remain unknown. Since each keratan sulfate and chondroitin sulfate chain contains many negative charges, adjacent chains repel each other and tend to maintain aggrecan molecules in an expanded form. This conformation promotes the trapping of proteoglycans within the fine collagen meshwork [79, 151].

In the articular cartilage matrix, most aggregating proteoglycan monomers noncovalently associate with hyaluronic acid filaments and link proteins to form proteoglycan aggregates (see Fig. 2A–3). These large molecules have a central hyaluronic acid backbone that can vary in length from several hundred nanometers to more than 10,000 nm [29, 31, 117, 151]. Large aggregates may have more than 300 associated monomers. Link proteins and other small noncollagenous proteins stabilize the association between monomers and hyaluronic acid. They appear to have a role in directing the assembly of aggregates and help to maintain the stability of the matrix and provide added mechanical strength [29, 147].

Aggregate formation helps to anchor proteoglycans within the matrix, preventing their displacement during deformation of the tissue, and also helps to organize and stabilize the relationship between proteoglycans and collagen fibrils. Recently, it has been found that proteoglycans at physiologic concentrations form elastic networks capable of storing energy [78, 141]. Link protein greatly increases the stiffness and strength of these proteoglycan networks [147]. The interaction between the collagen and proteoglycan networks provides the strength and cohesiveness of the articular cartilage extracellular matrix [140, 148].

The interaction of large proteoglycans with the tissue fluid contributes significantly to the compressive stiffness and resilience of cartilage. These proteoglycans have a structure that fills a large volume with negatively charged glycosaminoglycan chains (see Fig. 2A–3) that interact with water and cations. By repelling each other, these charged chains hold the monomers stiffly extended, thereby inflating the collagen fibril meshwork with water. Compression of the intact matrix drives the glycosaminoglycan chains closer together, increasing resistance to further compression and forcing water out of the molecular domain. Release of compression allows the molecules to reexpand and imbibe the lost fluid.

Comparison of the maximum volume that can be occupied by proteoglycans in solution with their concentration in articular cartilage matrix shows that if the cartilage matrix proteoglycans expanded fully they would fill a volume many times larger than the tissue that contains them [29]. In the matrix, their domains must overlap or be collapsed, and the repulsive forces from charges on these molecules and the osmotic pressure generated by counterions associated with these charges exert a constant pressure to expand. Only the collagen fibril meshwork restrains the expansion of the proteoglycans [58, 109, 110, 125–127, 152, 153]. Disruption of this collagen meshwork releases the matrix proteoglycans to extend their protein cores and glycosaminoglycan chains, thereby increasing the concentration of water and decreasing the proteoglycan concentration within the extracellular matrix. These molecular changes increase the permeability and decrease the stiffness of the matrix. Thus, loss of proteoglycans, or collagen network disruption due to acute or repetitive trauma, will have significant effects on the mechanical properties of articular cartilage [1, 6, 99, 127, 140, 148].

Small nonaggregating proteoglycans have shorter protein cores than aggrecan molecules, and two of them, biglycan and decorin, contain a different type of glycosaminoglycan, dermatan sulfate, as well as other glycosaminoglycans [172]. They also have far fewer glycosaminoglycan chains than aggrecan molecules; biglycan has two glycosaminoglycan chains, and decorin has one. In adult articular cartilage, at least some of these smaller proteoglycans form close associations with collagen fibrils. Unlike the large aggregating molecules, they do not fill a large volume of the tissue or contribute directly to the mechanical behavior of the tissue. Instead, they bind to other macromolecules and probably influence cell function. For example, the dermatan sulfate proteoglycans appear to inhibit cartilage repair.

Noncollagenous Proteins and Glycoproteins

The noncollagenous proteins and glycoproteins are not as well understood as the collagens and proteoglycans. They consist primarily of protein and have a few attached monosaccharides and oligosaccharides. At least some of these molecules appear to help organize and maintain the macromolecular structure of the matrix, whereas others may help stabilize the relationship between chondrocytes and other matrix macromolecules.

Link proteins help organize and stabilize the matrix through their effects on proteoglycan aggregation [117, 147, 151]. Other noncollagenous proteins found in articular cartilage may help mediate the adhesion of chondrocytes to the matrix and possibly stabilize the relationship between the chondrocytes and the matrix [117]. Fibromodulin, a glycoprotein that contains keratan sulfate, also has been found within cartilage. Because of its keratan sulfate content, it may be considered a form of proteoglycan as well as a glycoprotein. It appears to be associated with cartilage collagen fibrils and may influence collagen turnover.

Cell-Matrix Interactions

Maintenance of cartilage depends on continual complex interactions between chondrocytes and the matrix they synthesize. Normal degradation of matrix macromolecules, especially proteoglycans, requires that the cells continually synthesize new molecules [117, 151, 172]. The cells can sense the content of the matrix and respond appropriately. If they did not, the tissue would lose its biomechanical properties. For example, experimental depletion of matrix proteoglycan with papain stimulates proteoglycan synthesis [117]. If the cells did not replace the lost proteoglycans, the tissue would deteriorate. Mechanical loading also affects cartilage homeostasis [97, 103, 160, 161, 194]. These interactions between the cells and their matrix have considerable importance for sports injuries to articular cartilage because mechanical injury that interferes with the ability of chondrocytes to replace matrix macromolecules or lack of appropriate mechanical stimulation will lead to deterioration of the tissue. However, at present, the mechanism or mechanisms through which chondrocyte synthetic function is stimulated or suppressed remain unknown.

Although the details of the mechanisms of chondrocyte control and modulation are unknown, it is known that, in general, chondrocytes sense and respond to changes in patterns of matrix deformation due to persistent changes in joint use. It is likely that both mechanical and physicochemical events occurring during matrix deformation play significant roles in stimulating chondrocytes. When cartilage is loaded, it is deformed. Figure 2A–4 shows diagrams of a chondrocyte embedded in the charged extracellular matrix in an undeformed state and a deformed state. Deformation alters the charge density around the cells and induces a streaming potential throughout the tissue during compression. These physicochemical effects vary throughout the different zones of the charged collagen-proteoglycan extracellular matrix owing to varying proteoglycan concentrations according to depth from the surface [28, 117, 127, 151, 172].

Results from recent compression studies have suggested that these physicochemical events are important in modulating chondrocyte proteoglycan biosynthesis [72, 177]. Additionally, the increase in charge density within the extracellular matrix increases the interstitial Donnan osmotic pressure and osmotic pressure gradients and produces streaming polarization and electro-osmosis effects around the chondrocytes [110, 125]. These electrical events may also be important in stimulating chondrocyte biosynthetic activities. The stress-strain environment and the strain-energy density around the cell may also play important roles in stimulating chondrocytes [76, 77].

In addition to these mechanical, electrical, and physicochemical events, biochemical agents such as growth factors, cytokines, and enzymes have been shown to be potent stimulators of chondrocytes. Indeed, each of these stimulators may play a role in modulating chondrocyte activities, and their effect may be synergistic. How the mechanical, electrical, and physicochemical forces work, how each of the chemical agents modulates cartilage homeostasis, and how they influence each other are, at present, entirely unknown. Studies addressing these important questions offer great challenges in the future.

ORGANIZATION OF ARTICULAR CARTILAGE

To form articular cartilage, chondrocytes organize the collagens, proteoglycans, and noncollagenous proteins into a unique, highly ordered structure (see Figs. 2A–1 and 2A–2). The composition, organization, and mechanical properties of the matrix, cell morphology and probably cell function vary according to the depth of these elements from the articular surface [28, 34, 117, 127, 151, 172]. Matrix composition, organization, and function also vary with distance from the cell [34, 35, 181].

Zones of Articular Cartilage

Morphologic changes in articular cartilage cells and matrix from the articular surface to the subchondral bone make it possible to identify four zones or layers of articular cartilage: the superficial tangential zone, the middle or transitional zone, the deep or radial zone, and the zone of calcified cartilage (see Figs. 2A–1 and 2A–2). Although each zone has distinct morphologic features, the boundaries between zones cannot be sharply defined. Recent biologic and mechanical studies have shown that this morphologic zonal organization has functional significance. Cells in the different zones differ in shape, size, and orientation relative to the articular surface and appear to differ in synthetic activity. They may also respond differently to mechanical loading, suggesting that development and maintenance of normal articular cartilage depends in part on differentiation of phenotypically distinct populations of chondrocytes. The heterogeneity of chondrocytes presumably is responsible for the differences in matrix composition and organization that result in different mechanical properties in each zone.

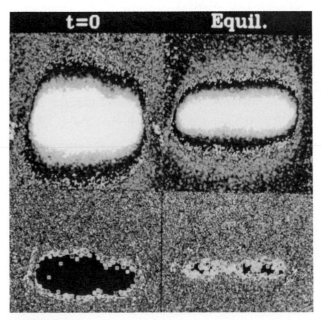

FIGURE 2A–4
Confocal laser microscopic view of a chondrocyte from the middle zone of articular cartilage embedded in the extracellular matrix in the unloaded state (t = 0) and the compressed state at equilibrium (equil.)

Superficial Zone

The thinnest zone, the superficial tangential zone, has two layers. A sheet of fine fibrils with little polysaccharide and no cells covers the joint surface (see Fig. 2A–1). This layer presumably corresponds to the clear film that can be stripped from the articular surface in some regions. By phase-contrast microscopy it appears as a narrow bright line, the "lamina splendens." In the next layer of the superficial zone, flattened ellipsoid-shaped chondrocytes are arranged so that their major axes parallel the articular surface (see Fig. 2A–1). They synthesize a matrix that has a high collagen concentration and a low proteoglycan concentration relative to the other cartilage zones. Water content is the highest in this zone, averaging 80% [114, 127, 151].

Transitional Zone

The transitional, or middle, zone has several times the volume of the superficial zone (see Fig. 2A–1). The cells of this zone have a higher concentration of synthetic organelles, endoplasmic reticulum, and Golgi membranes than the cells of the superficial zone. They assume a spheroidal shape and synthesize a matrix with collagen fibrils of a larger diameter and a higher concentration of proteoglycans than that found in the superficial zone. In this zone the proteoglycan concentration is higher than that in the superficial zone, but the water and collagen concentrations are lower [114, 127, 197].

Deep Zone

The chondrocytes in the deep zone resemble those of the middle zone, but they tend to align themselves in columns perpendicular to the joint surface (see Fig. 2A–1). This zone contains the collagen fibrils with the largest diameter, the highest concentration of proteoglycans, and the lowest concentration of water. The collagen fibers of this zone pass through the "tidemark" (a thin basophilic line seen on light microscopic sections of decalcified articular cartilage that marks the boundary between calcified and uncalcified cartilage) [39, 171] into the calcified zone anchoring adult articular cartilage to the subchondral bone [171].

Zone of Calcified Cartilage

A zone of calcified cartilage lies between the deep zone of uncalcified cartilage and the subchondral bone (see Fig. 2A–2). The cells of the calcified cartilage zone have a smaller volume per cell than the cells of the deep zone and contain only small amounts of endoplasmic reticulum and Golgi membranes. During aging, the tidemark advances, causing thinning of the uncalcified cartilage [38, 39]. This remodeling process may be due to repetitive microtrauma in the deep zone of cartilage [7, 55, 168]. Results of stress-strain analysis suggest that cartilage thinning is detrimental to the tissue [10, 168].

Matrix Regions

Variations in the matrix within zones have been described by dividing the matrix into regions or compartments called the pericellular region, the territorial region, and the interterritorial region (Figs. 2A–5 and 2A–6) [28, 34, 164, 181]. The pericellular and territorial regions appear to serve the needs of chondrocytes, that is, they bind the cell membranes to the matrix macromolecules and protect the cells from damage during loading and deformation of the tissue [77]. They may also help transmit mechanical signals to the chondrocytes. The primary function of the interterritorial matrix is to provide the mechanical properties of the tissue [127, 140, 148].

Pericellular Matrix

Chondrocyte cell membranes appear to attach to the thin rim of the pericellular matrix that covers the cell surface (see Figs. 2A–5 and 2A–6). This matrix region probably contains noncollagenous proteins and possibly nonfibrillar collagens and is rich in proteoglycans [28, 181]. It contains little or no fibrillar collagen.

Territorial Matrix

An envelope of territorial matrix surrounds the pericellular matrix of individual chondrocytes and, in some locations, pairs or clusters of chondrocytes and their pericellular matrices (see Figs. 2A–5 and 2A–6). In the deep zone, a territorial matrix surrounds each chondrocyte column. The thin collagen fibrils of the territorial matrix nearest to the cell appear to adhere to the pericellular matrix. At a distance from the cell they decussate and intersect at various angles, forming a fibrillar basket around the cells [164]. This collagenous basket may provide

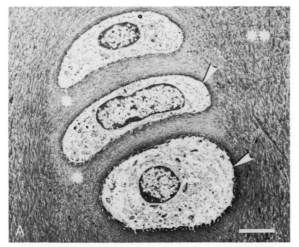

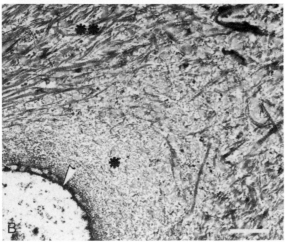

FIGURE 2A–5
A and *B,* Electron micrographs showing chondrocytes and the regions of articular cartilage matrix. In both electron micrographs, short cell processes protrude from the chondrocytes through the pericellular matrix (arrow heads) to the border between the pericellular matrix and the territorial matrix (*). The interterritorial matrix (**) contains larger collagen fibrils and surrounds the territorial matrices, pericellular matrices, and cells. (From Buckwalter, J. A., Hunziker, E. B., Rosenberg, R. C., et al. Articular cartilage: Composition and structure. *In* Woo, S. L.-Y., and Buckwalter, J. A. (Eds.), *Injury and Repair of the Musculoskeletal Soft Tissues.* Park Ridge, IL, American Academy of Orthopaedic Surgeons, 1988, pp. 405–426.)

mechanical protection for the chondrocytes during loading and deformation of the tissue. The results of a recent stress-strain analysis support this concept; it showed that the matrix regions surrounding the cells function as a protective buffer against the development of high stresses and strains in the chondrocyte when the tissue is loaded [77]. An abrupt increase in collagen fibril diameter and a transition from the basket-like orientation of the collagen fibrils to a more parallel arrangement marks the boundary between the territorial and interterritorial matrices. However, many collagen fibrils con-

nect the two regions, making it difficult to identify the precise division between these two regions [28, 181].

Interterritorial Matrix

The interterritorial matrix (see Figs. 2A–5 and 2A–6) makes up most of the volume of mature articular cartilage. It contains the largest diameter collagen fibrils. Unlike the collagen fibrils of the territorial matrix, these fibrils are not organized to surround the chondrocytes, and they change their orientation relative to the joint surface 90 degrees from the superficial zone to the deep zone (see Fig. 2A–2) [46, 111, 150, 170].

In the superficial zone, the fibril diameters are relatively small, and the fibrils generally lie parallel to the articular surface. Creation of pinhole defects in some articular surfaces produces split lines that show a tendency to extend parallel to the plane of joint motion [17, 88], suggesting that collagen fibrils in the interterritorial matrix of the superficial zone also may be oriented relative to the motion of the joint. In the middle zone interterritorial matrix collagen fibrils assume more oblique angles relative to the articular surface, and in the deep zone they generally lie perpendicular to the joint surface. Since the interterritorial matrix forms most of the volume of cartilage and since collagen provides the tensile stiffness and strength of articular cartilage, these biomechanical properties should vary with changes in collagen fibril orientation and organization in the interterritorial matrix (Fig. 2A–2), and experimental studies show that cartilage tensile stiffness and strength do vary among cartilage zones [99, 173, 205].

MECHANICAL PROPERTIES OF CARTILAGE

The behavior of cartilage when subjected to compression, tension, or shear depends on the concentration, properties, and organization of the matrix macromolecules, the water content, and the physical and electrical interactions between the water and the macromolecular framework [98, 144, 148]. Because cartilage has a solid phase (the macromolecular framework of collagens, proteoglycans, and noncollagenous proteins) and a fluid phase (the tissue water), it behaves as a biphasic (two-phase) viscoelastic material; that is, its response to loading combines viscosity, a property characteristic of

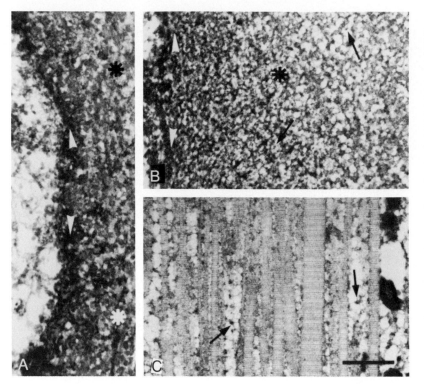

FIGURE 2A–6
Electron micrographs showing the matrix compartments of articular cartilage. *A,* The pericellular matrix (white arrowheads) consists of a narrow, dense coat of proteoglycans and possibly glycoproteins as well as other nonfibrillar molecules. The territorial matrix (*) surrounds the pericellular matrix. Notice the intimate contact between the pericellular matrix and the cell membrane. *B,* The territorial matrix (*) consists of a dense network of fine collagen that may extend around one cell or a group of cells (a territorium or chondron). Notice the absence of a sharp border between the pericellular and territorial matrices. *C,* The interterritorial matrix consists of parallel large-diameter collagen fibrils and fibers with matrix granules (dark arrows in *B* and *C*) interspersed between fibrils. These granules consist of proteoglycans precipitated with ruthenium hexamine trichloride. (From Buckwalter, J. A., Hunziker, E. B., Rosenberg, R. C., et al. Articular cartilage: Composition and structure. *In* Woo, S. L.-Y., and Buckwalter, J. A. (Eds.), *Injury and Repair of the Musculoskeletal Soft Tissues.* Park Ridge, IL, American Academy of Orthopaedic Surgeons, 1988, pp. 405–426.)

fluids, with elasticity, a property characteristic of solids.

When subjected to a constant load or a constant deformation, the response of a viscoelastic material varies with time. A viscoelastic material subjected to a constant load responds with a rapid initial deformation followed by further slow, progressive deformations until it reaches an equilibrium state, a behavior called creep. A viscoelastic material subjected to a constant deformation responds with a high initial stress followed by a slowly progressive decrease in the stress required to maintain the deformation, a behavior called stress relaxation.

In articular cartilage, because it is a porous-permeable hydrated soft tissue, creep and stress relaxation in compression are predominantly caused by fluid flow through the matrix [106, 140, 148]. In shear, when no interstitial fluid flow occurs, creep and stress relaxation occur because the macromolecular framework is altered [82, 83, 140, 185, 188]. An important characteristic of articular cartilage is its ability to undergo deformation and then revert to its original form by reversing fluid flow. This feature is important in joint lubrication and tissue fluid circulation, particularly because of the load-bearing function of cartilage and because of its lack of a vascular supply and lymphatic system [142, 144].

Changes in the composition or organization of the matrix macromolecular framework can cause a de-

terioration of these essential mechanical properties. Collagen fibrils provide tensile strength but little resistance to compression. Interaction of the proteoglycans with water provides resistance to compression, swelling pressure, and resilience, but little tensile strength. For these reasons, disruption of the collagen fibril meshwork allows the proteoglycans to expand, increasing the water concentration and decreasing the proteoglycan concentration [110, 126, 148, 152, 153]. Increased water concentration and decreased proteoglycan concentration decrease cartilage stiffness and increase matrix permeability [6, 140, 148]. These changes make the tissue less capable of supporting the loads of articulation and more vulnerable to further injury from loading, which may cause progressive mechanical failure of the matrix and clinical osteoarthritis [36, 37, 148].

CARTILAGE INJURY AND REPAIR

The response of cartilage to trauma and the potential for repair of cartilage depend to a large extent on the type of injury sustained and whether the injury involves the subchondral bone. Sports-related synovial joint trauma can fracture or rupture the cartilage matrix, causing visible splits in the articular surface, or it can damage the macromolecular framework and alter cell function without disrupting the surface of the tissue [7, 23, 34, 51, 55, 96, 100, 128, 148, 169, 210].

Since cartilage lacks blood vessels, damage to the cartilage alone does not cause inflammation. If an injury disrupts both the cartilage and the subchondral bone, the bone blood vessels participate in the inflammation that initiates the fracture healing [36]. The clot and repair tissue from bone can then fill the articular cartilage defect and follow the sequence of inflammation, repair and remodeling as in the repair of other tissues such as ligament [36]. However, unlike repair tissue in ligaments, the repair tissue that fills cartilage defects differentiates initially toward articular cartilage rather than toward dense fibrous tissue [33, 34, 36, 37].

Differences in the potential for cartilage repair separate acute articular cartilage injuries into three general types [30, 34, 36, 37]: (1) loss of matrix macromolecules or disruption of the macromolecular framework without visible tissue disruption; (2) mechanical disruption of articular cartilage alone; and (3) mechanical disruption of articular cartilage and subchondral bone. Each of these types of cartilage damage presents a different problem for repair, although the categories overlap. Progressive loss of matrix macromolecules or disruption of the organization of the matrix macromolecular framework eventually results in mechanical disruption of

the tissue, and mechanical disruption of the cartilage may release tissue factors that stimulate matrix degradation and loss of matrix macromolecules.

Table 2A–1 lists the types of acute sports-related injuries to cartilage caused by direct blunt trauma, high-energy joint loading, and forceful joint twisting and summarizes the current understanding of the clinical presentation, tissue response, and potential for healing of these injuries. The exact mechanisms of injury and natural history of these injuries remain poorly understood. Injuries limited to the articular cartilage do not cause pain or inflammation, but these injuries commonly occur in association with injuries to other tissues that have nerves and blood vessels, including synovium, ligament, joint capsule, meniscus, and bone. Therefore, people with injuries of the articular cartilage may have symptoms and signs that result from damage to innervated vascularized tissues.

Matrix Damage Without Visible Tissue Disruption

Acute or repetitive direct blunt trauma and acute or repetitive high-energy joint loading can cause cartilage damage without visible tissue disruption (see Table 2A–1). Because of the demands of com-

TABLE 2A–1
Acute Sports Injuries to Articular Cartilage

Injury Type	Clinical Presentation	Tissue Response	Potential for Healing
Damage to matrix or cells without visible disruption of articular surface	No known symptoms Direct inspection of the articular surface and current clinical imaging methods cannot detect this type of injury	Synthesis of new matrix macromolecules Cell proliferation?	If basic matrix structure remains intact and a suffcent number of viable cells remain, the cells can restore the normal tissue composition If the matrix or cell population sustains significant damage or if the tissue sustains further damage, the lesion may progress
Cartilage disruption (cartilage fractures or ruptures)	May cause mechanical symptoms, synovitis, and joint effusions	No fibrin clot formation or inflammation. Synthesis of new matrix macromolecules and cell proliferation, but new tissue does not fill cartilage defect	Depending on the location and size of the lesion and the structural integrity, stability, and alignment of the joint, the lesion may or may not progress
Cartilage and bone disruption (osteochondral fractures)	May cause mechanical symptoms, synovitis, and joint effusions	Formation of a fibrin clot, inflammation, invasion, of new cells, and production of new tissue	Depending on the location and size of the lesion and the structural integrity, stability, and alignment of the joint, the lesion may or may not progress

petitive or recreational sports, these injuries probably occur frequently in athletes. Presumably they are often associated with other joint injuries such as ligament and meniscal tears, but lack of a tissue inflammatory response, nerves in cartilage, or visible disruption of the articular surface prevents detection of these injuries even by direct examination of the articular surface or magnetic resonance imaging.

The intensity and type of cartilage loading that can cause matrix damage without gross tissue disruption has not been well defined [7, 148, 169]. Physiologic levels of joint loading do not appear to cause cartilage injury, but impact loading above that associated with normal activities and less than that necessary to produce cartilage disruption can cause tissue alterations. Other causes of this type of cartilage injury include traumatic or surgical disruption of the synovial membrane, prolonged joint immobilization, some medications, joint irrigation, and synovial inflammation [7, 30, 34, 36, 37, 55].*

Cartilage damage without tissue disruption has not been studied as extensively as other types of cartilage injury, but the available evidence shows that loss of proteoglycans or disruption of their organization occurs before other signs of tissue injury [3, 129, 192]. The loss of proteoglycans and alteration of their molecular structure may be due to either increased degradation or altered synthesis of the molecules [30, 34, 36, 44]. Loss of matrix proteoglycans decreases cartilage stiffness and increases its hydraulic permeability [6, 140, 148]. These alterations may cause greater loading of the remaining macromolecular framework, including the collagen fibrils, increasing the vulnerability of the tissue to damage from further impact loading [37].

These injuries may cause matrix abnormalities other than loss of proteoglycans, such as ruptures or distortions of the collagen fibril meshwork or disruptions of the collagen fibril-proteoglycan relationships, and they may alter chondrocyte function or even damage the chondrocytes. For example, impact loading of dog articular cartilage caused cartilage swelling, collagen fibril swelling, and disturbances in the relationships between collagen fibrils and proteoglycans [55]. Presumably these changes represent more severe matrix damage than a decrease in proteoglycan concentration. Mildly fibrillated cartilage from human femoral condyles shows a significant decrease in tensile stiffness and an increase in swelling, whereas cartilage specimens obtained adjacent to focal lesions show a dramatic loss of tensile properties and an even greater increase in swelling [1, 2].

Lack of a reliable method of detecting cartilage injuries that do not cause visible disruption of the matrix in humans has made it impossible to define the natural history of this type of damage. Probing the intact articular surface may show soft regions, but the significance of softening remains uncertain. Methods that may eventually help to identify these injuries in humans include imaging techniques that show the state of water within the cartilage and devices that measure cartilage stiffness.

The ability of chondrocytes to sense changes in matrix composition and to synthesize new molecules makes it possible for them to repair damage to the macromolecular framework. The available evidence indicates that following a loss of proteoglycans, the cells increase synthesis of these macromolecules and begin to restore the matrix concentration of proteoglycans toward normal. As a result, the material properties of the matrix return toward normal. Following significant depletion of proteoglycans, the process of repairing the matrix may require many weeks or possibly months to complete [36]. If the cells do not repair significant matrix macromolecular abnormalities, or if the loss of matrix molecules progresses, the tissue will deteriorate [37]. It is not clear at what point this type of injury becomes irreversible and leads to progressive loss of articular cartilage. Presumably, if the basic collagen meshwork remains intact and if enough chondrocytes remain viable, the chondrocytes can restore the matrix as long as the loss of matrix proteoglycan does not exceed the amount that can be rapidly produced by the cells [30, 34, 37]. When these conditions are not met, the cells will not succeed in their attempt to restore the matrix, the chondrocytes will be exposed to excessive loads, and the tissue will degenerate [37].

For these reasons, insults that cause this type of articular cartilage injury, including immobilization, exposure of articular cartilage, and inflammation, should be minimized. Since this type of matrix macromolecular injury may temporarily increase the vulnerability of cartilage to mechanical injury, it appears advisable to minimize the intensity of impact loading of cartilage following severe blunt trauma, prolonged immobilization, or inflammation.

Articular Cartilage Injuries That Disrupt the Tissue

Severe blunt trauma, penetrating injuries, and fractures can cause visible disruptions of the articu-

---

*The chapter on Loading and Motion discusses the effects of decreased loading and motion on articular cartilage and the chapter on Medications discusses the effects of medications on articular cartilage.

lar cartilage matrix. These injuries rupture, lacerate, or fracture the matrix macromolecular framework and kill chondrocytes at the site of injury without directly damaging the subchondral bone. Penetrating joint injuries that lacerate cartilage rarely occur as a result of sports, but study of experimental lacerations has provided most of our current understanding of the potential for healing of injuries limited to cartilage.

The lack of blood vessels and cells that can repair significant tissue defects limits the response of cartilage to injury [30, 34, 36, 37, 55, 120, 121, 123]. Because cartilage lacks blood vessels, these injuries do not cause hemorrhage, fibrin clot formation, or inflammation, and the local response to injury depends entirely on chondrocytes. Undifferentiated mesenchymal cells cannot migrate from blood vessels to the site of injury, proliferate, differentiate, and synthesize a new matrix. Chondrocytes are tightly encased in the collagen-proteoglycan matrix, and therefore they cannot migrate through the matrix to the site of injury. However, they do respond to tissue injury by proliferating and increasing the synthesis of matrix macromolecules near the injury site [30, 34, 36, 37, 120, 121, 123]. The newly synthesized matrix and proliferating cells do not fill the tissue defect, and soon after injury the increased proliferative and synthetic activity ceases.

When cartilage injury is associated with damage to the synovial membrane, blood may fill the joint. Even when this happens, fibrin clots do not form in the cartilage injury, and cells from the synovium and blood vessels do not migrate into the cartilage defect. The failure of clot formation and cell migration and adhesion may be due to proteoglycans, specifically, dermatan sulfate proteoglycans, which inhibit cell adhesion to the cartilage matrix.

Cartilage Lacerations

Penetrating injuries to synovial joints and surgical instruments passed across an articular surface can cut or abrade articular cartilage without damaging subchondral bone. Because this type of injury has been studied more extensively than blunt trauma to cartilage, more is known about the response of the tissue in such injuries [30, 34, 36, 120, 121, 123]. Lacerations perpendicular to the articular surface kill chondrocytes at the site of injury and create matrix defects [30, 34, 130]. Because these lesions cannot cause hemorrhage or initiate an inflammatory response and because fibrin clots rarely form on exposed damaged articular cartilage, platelets do not bind to the damaged cartilage, and a fibrin clot

does not appear. Inflammatory cells, capillaries, and undifferentiated mesenchymal cells do not migrate to the site of injury. Chondrocytes near the site of injury proliferate and form clusters, or clones, and synthesize new matrix. They do not migrate to the site of the lesion, perhaps because the matrix restrains their movement or perhaps because of their limited ability to respond to injury. The new matrix they produce remains near the cells and therefore does not repair the damage. Shortly after the injury, these cell proliferative and synthetic activities cease. Unlike similar osteoarthritic lesions, experimental lacerations of articular cartilage do not show evidence of progression.

Superficial lacerations or abrasions of cartilage made tangential or parallel to the articular surface also do not stimulate a successful repair response [34, 36, 71]. Some cells next to the site of injury may die, while others show evidence of increased proliferation and matrix synthesis. A thin layer of new, acellular matrix may form over the surface. The available evidence shows that, as in cartilage defects perpendicular to the surface, the remaining normal tissue does not deteriorate.

Blunt Trauma That Disrupts Cartilage Alone

During sports activities, impact loading, twisting, and direct blows to synovial joints occur frequently. The resulting compression of an articular surface can rupture the cartilage matrix, producing chondral fissures, flaps, or fractures without bone injury. If the compressive force is sufficiently high, the uncalcified cartilage may shear off the calcified cartilage [7]. The mechanisms of these injuries have not been extensively studied, but the available evidence shows that impact loading of the articular surface can rupture the matrix [169].

Disruption of normal articular cartilage by a single impact requires substantial force. A study of the response of human articular cartilage to blunt trauma showed that articular cartilage could withstand impact loads of up to 25 Newtons/sq mm (25 MPa) without apparent damage. Impact loads exceeding this level caused chondrocyte death and cartilage fissures [169], and the authors suggested that a stress level that could cause acute cartilage disruption required a force greater than that necessary to fracture the femur. Another study measured the pressure on human patellofemoral articular cartilage during impact loading and found that impact loads less than the level necessary to fracture bone caused stresses greater than 25 MPa in some regions of the articular surface. With the knee flexed 90 degrees, 50% of the load necessary to cause a bone fracture produced joint pressures greater than 25 MPa for nearly 20% of the patellofemoral joint. At 70% of the bone fracture load nearly 35% of the contact area of the patellofemoral joint pressures exceeded 25 MPa, and at 100% of the bone fracture load 60% of the patellofemoral joint pressures exceeded 25 MPa [81]. These results suggest that impact loads can disrupt cartilage without fracturing bone.

Other experimental investigations show that repetitive impact loads can split the articular cartilage matrix and initiate progressive cartilage degeneration [51, 202]. Cyclic loading of human cartilage samples in vitro caused surface fibrillation [202], and periodic impact loading of bovine metacarpal phalangeal joints in vitro combined with joint motion caused rapid degeneration of articular cartilage [166]. Repeated overuse of rabbit joints in vivo combined with peak overloading caused articular cartilage damage including formation of chondrocyte clusters, fibrillation of the matrix, thickening of subchondral bone, and penetration of subchondral capillaries into the calcified zone of the articular cartilage [51, 167]. The extent of cartilage damage appeared to increase with longer periods of repetitive overloading, and deterioration of the cartilage continued following cessation of excessive loading.

An investigation of the effects of repetitive loading of cartilage plugs also showed that repetitive loading disrupts the tissue and that the severity of the damage increases as the load and the number of loading cycles increase [210]. Two hundred and fifty cycles of a 1000 pound/sq in. compression load caused surface abrasions. Five hundred cycles produced primary fissures penetrating to calcified cartilage, and 1000 cycles produced secondary fissures extending from the primary fissures. After 8000 cycles the fissures coalesced and undermined the cartilage fragments. Higher loads caused similar changes with fewer cycles. The experiments suggested that repetitive loading can cause propagation of vertical cartilage fissures from the joint surface to calcified cartilage and extension of oblique fissures into areas of intact cartilage, extending the damage and creating cartilage flaps and free fragments.

Clinical studies have identified articular cartilage fissures, flaps, and free fragments similar to those produced experimentally by single and repetitive impact loads [23, 96, 100, 128]. In at least some patients, acute impact loading of the articular surface or twisting movements of the joint apparently caused these injuries. In other patients, cartilage damage may have resulted from repetitive loading. Frequently, other joint injuries including rupture of the anterior cruciate ligament and meniscal tears occur in association with cartilage damage.

Taken together, the clinical and experimental studies suggest that closed injuries to synovial joints, including direct blows and loading combined with torsion, can split articular cartilage matrix without causing bone fractures. These injuries disrupt the cartilage matrix macromolecular framework and kill chondrocytes near the injury. Since chondrocytes cannot repair these matrix injuries, either the fissures remain unchanged, or they progress. The experimental studies suggest that excessive repetitive loading weakens the cartilage macromolecular framework before visible matrix disruption occurs. Presumably the chondrocytes could repair at least some of this molecular damage before cartilage fissures developed if the tissue were protected from further injury.

Osteochondral Fractures and Osteochondral Defects

Sports injuries can cause fractures that extend through cartilage into the subchondral bone. Severe osteochondral fractures may result in loss of part of the articular surface. Unlike injuries limited to cartilage, fractures that extend into subchondral bone cause pain, hemorrhage, and fibrin clot formation and activate the inflammatory response (see Table 2A-1) [16, 30, 33, 34, 37, 41, 43, 49, 53, 69, 120, 121, 123, 131, 135]. Because undifferentiated mesenchymal cells migrate into the region of the fibrin clot, proliferate, and synthesize a new matrix, most osteochondral defects fill with new cells and matrix.

Soon after an osteochondral injury, blood escaping from the damaged bone blood vessels forms a hematoma that temporarily fills the injury site. Fibrin forms within the hematoma, and platelets bind to fibrillar collagen and establish hemostasis. A continuous fibrin clot fills the bone defect and extends for a variable distance into the cartilage defect. Platelets within the clot release potent vasoactive mediators including serotonin, histamine, and thromboxane A_2 and growth factors or cytokines (small proteins that influence multiple cell functions including migration, proliferation, differentiation, and matrix synthesis) including transforming growth factor-beta and platelet-derived growth factor [36]. Bone matrix also contains growth factors including transforming growth factor-beta, bone morphogenic protein, platelet-derived growth factor, insulin-like growth factor I, insulin-like growth factor II, and possibly others. Release of these growth factors may play an important role in the repair of osteochondral defects. In particular, they probably stimulate vascular invasion and migration of undifferentiated cells into the clot and influence the proliferative and synthetic activities of the cells.

Shortly after entering the tissue defect the undifferentiated mesenchymal cells proliferate and synthesize a new matrix. Within 2 weeks of injury, some of the mesenchymal cells assume the rounded form of chondrocytes and begin to synthesize a matrix that contains type II collagen and a relatively high concentration of proteoglycans [30, 34, 36, 37]. These cells produce regions of hyaline-like cartilage in the chondral and bone portions of the defect. In many osteochondral defects the regions of hyaline-like cartilage first appear next to the exposed bone matrix, leaving the central region of the defect filled with more fibrous tissue [33].

Six to eight weeks following injury, the repair tissue within the chondral region of most defects contains many chondrocyte-like cells in a matrix consisting of type II collagen, proteoglycans, some type I collagen, and noncollagenous proteins [33]. Unlike the cells in the chondral portion of the defect, the repair cells in the bone portion of the defect produce immature bone, fibrous tissue, and hyaline-like cartilage [30, 33, 37]. They soon restore the original level of subchondral bone. Capillaries that approached or entered the chondral portion of the defect recede.

Six months after injury, the mesenchymal cells have repaired the bone defect with a tissue consisting primarily of bone but also containing some regions of fibrous tissue, small blood vessels, and hyaline cartilage [33]. In contrast, the chondral portions of large osteochondral defects rarely fill completely with repair tissue [30, 33, 37]. In animal experiments repair tissue filled about two-thirds of the total volume of the chondral portion of large osteochondral defects and more than 95% of the total volume of the bone portion of the defects [34, 36], and the tissue in the chondral portion of the defect differed significantly in composition from that in the bone portion of the same defect [33]. The chondral repair tissue did not contain bone or blood vessels and had a significantly higher proportion of hyaline-like cartilage. In most regions of the chondral defects, it had a composition and structure intermediate between hyaline cartilage and fibrocartilage, and it rarely replicated the elaborate structure of normal articular cartilage. The differences in the differentiation of the repair tissue in the chondral and bony parts of the same defect show that the environment in the two regions causes the same

repair cells to produce different types of tissue. It is not clear whether the important differences in environment are mechanical, biologic, electrical, or other unknown factors.

Occasionally, the cartilage repair tissue persists unchanged or progressively remodels to form a functional joint surface, but in most large osteochondral injuries the chondral repair tissue begins to show evidence of depletion of matrix proteoglycans, fragmentation, and fibrillation, increasing collagen content, and loss of cells with the appearance of chondrocytes within a year or less [30, 34, 36, 37, 69, 135]. The remaining cells often assume the appearance of fibroblasts as the surrounding matrix comes to consist primarily of densely packed collagen fibrils. This fibrous tissue usually fragments and often disintegrates, leaving areas of exposed bone [30, 34, 37, 135].

The inferior mechanical properties of cartilage repair tissue may be responsible for its frequent deterioration [37]. Several experimental studies show that even repair tissue that successfully fills osteochondral defects lacks the stiffness of normal articular cartilage [12, 48, 203]. Cartilage repair tissue formed in rabbit metatarsophalangeal joint arthroplasties deformed more easily and took longer to recover from deformation than normal articular cartilage [48]. Repair cartilage formed in pig joints swelled in Ringer's solution more than normal cartilage and had greater permeability and less stiffness on compression than normal cartilage [203]. Detailed study of chondral repair cartilage in primate osteochondral defects also showed that the repair tissue was more permeable and less stiff on compression than normal articular cartilage [12–14].

Differences in matrix composition and organization may explain the differences between the mechanical properties of repair cartilage and the mechanical properties of normal cartilage [37]. The increased swelling of repair cartilage indicates a lack of organization or weakness of the collagen fibril meshwork. Microscopic studies of repair cartilage support this suggestion. They show that the orientation of the collagen fibrils in even the most hyaline-like cartilage repair tissue does not follow the pattern seen in normal articular cartilage [33]. In addition, the repair tissue cells may fail to establish the normal relationships between matrix macromolecules, in particular, the relationship between cartilage proteoglycans and the collagen fibril network. This may occur because of lack of organization of the macromolecules, insufficient concentrations of some macromolecules, or the presence of molecules that interfere with the assembly of a normal cartilage matrix. For example, the presence of type I collagen or high concentrations of dermatan sulfate proteoglycans might interfere with the establishment of normal articular cartilage collagen proteoglycan relationships.

The decreased stiffness and increased permeability of repair cartilage matrix subjects the macromolecular framework to increased strain fields during joint use, resulting in progressive structural damage to the matrix collagen and proteoglycans [37]. Mechanical failure of the matrix may expose the repair chondrocytes to excessive loads, further compromising their ability to restore the matrix. Thus, cartilage repair tissue may initially have a composition and structure that closely resembles normal articular cartilage, but defects in organization of the matrix macromolecular framework could compromise the function and durability of the repair tissue.

Clinical experience and experimental studies suggest that the success of chondral repair in osteochondral injuries may depend to some extent on the severity of the injury as measured by the volume of tissue or surface area of cartilage injured, the stability of the injury site, and the age of the individual [36]. Experimental studies suggest that smaller osteochondral defects heal more predictably and successfully than larger defects [33, 34, 36, 49, 179]. A study of intra-articular fractures of the distal femur in rabbits showed that anatomically reduced cartilage fractures stabilized by compression fixation of the bone fracture healed with apparently normal articular cartilage [136]. Inadequately reduced and adequately reduced fractures that were not stabilized by compression fixation healed with fibrocartilage.

The age of the individual may also influence the success of osteochondral repair. Potential age-related differences in repair have not been thoroughly investigated [37], but bone heals more rapidly in children than in adults, and the articular cartilage chondrocytes in skeletally immature animals show a better proliferative response to injury and synthesize larger proteoglycan molecules than those from mature animals [26, 29, 31, 195].

SURGICAL TREATMENT OF ARTICULAR CARTILAGE DAMAGE

Because localized cartilage disruption can compromise joint function and may lead to progressive cartilage deterioration, surgeons have sought methods of treating localized cartilage damage. Currently, shaving damaged cartilage and abrading exposed subchondral bone are the most common surgical treatments for damaged cartilage.

Shaving Fibrillated Articular Cartilage

Many surgeons shave fibrillated cartilage with the intent of decreasing joint symptoms and leaving a smoother articular surface. Shaving degenerating articular surfaces can remove frayed and fibrillated superficial cartilage, but the efficacy of this procedure in decreasing pain, improving joint function, or stimulating restoration of a damaged articular surface has not been established [32, 37].

Several reports describe decreased symptoms following arthroscopic débridement of loose cartilage fragments and flaps, torn menisci, osteophytes, and proliferative synovium in osteoarthritic knee joints [15, 187], but experimental and clinical studies of shaving fibrillated cartilage have not shown clear benefits from the procedure. Shaving normal rabbit patellar cartilage did not stimulate significant repair but also did not cause progressive deterioration [137]. In one series of patients, shaving fibrillated patellar cartilage produced unpredictable results [18–20]. Only 25% of patients had satisfactory results, and the investigators concluded that the procedure of shaving areas of affected patellar cartilage "is disappointing and ineffective" [18–20]. Another group studied regions of human femoral articular cartilage after arthroscopic shaving for treatment of cartilage damage. They did not find evidence of restoration of a smooth articular surface and suggested that shaving may have increased fibrillation and chondrocyte necrosis in and adjacent to the region of the original defect [182].

Despite these observations, removing fragments of degenerating cartilage and joint irrigation may decrease symptoms of mechanical catching and pain in some patients. Experimental injection of cartilage fragments into rabbit knees produced an inflammatory arthritis that included joint effusions, increased levels of synovial enzymes, and articular cartilage friability, pitting, and discoloration [64]. This report suggests that cartilage fragments can contribute to synovitis. Presumably débridement of fibrillated cartilage and the associated joint irrigation would temporarily decrease this synovial inflammation by removing cartilage particles, degradative enzymes, and inflammatory mediators.

Abrasion of Subchondral Bone

Arthroscopic abrasion offers a potentially attractive treatment of articular surfaces that have small regions of full-thickness cartilage loss [32]. The surgeon removes the most superficial layers of subchondral bone, usually 1 to 3 mm, to disrupt intraosseous vessels [93–95]. The resulting hemorrhagic exudate forms a fibrin clot, and undifferentiated cells invade the clot, forming repair tissue over the abraded bone. Protection of the joint from excessive loading allows the repair tissue to remodel and form a new articular surface.

Controlled trials of arthroscopic abrasive treatment of full-thickness cartilage loss have not been published, but several authors have reported that arthroscopic abrasion of the knee can decrease pain in 60% or more of patients [68, 93–95]. Abrasion of regions of exposed subchondral bone has resulted in formation of a fibrocartilaginous repair tissue that, in at least some patients, persists for years [93–95]. Examination of the repair tissue shows that, like the repair tissue formed in the chondral regions of experimental osteochondral defects [30, 33, 34], it has a fibrocartilaginous appearance and contains variable concentrations of type II collagen [94, 95]. No studies of the tissue formed following abrasion have shown regeneration of normal articular cartilage.

Despite the ability of arthroscopic abrasion to stimulate formation of cartilage repair tissue and the encouraging reports of symptomatic improvement, it is difficult to assess the value of this procedure. The available evidence suggests that it can provide temporary symptomatic improvement in selected patients [32, 37].

CARTILAGE GRAFTS

Because of the inferior mechanical properties of most cartilage repair tissue that forms following osteochondral injury or surgical treatment of cartilage defects, investigators and surgeons have explored the use of a variety of cartilage grafts, including osteochondral autografts and allografts and periosteal and perichondrial grafts, to replace regions of damaged or lost articular surface. More recently, several groups of investigators have developed methods of isolating chondrocytes or undifferentiated mesenchymal cells, growing them in culture and then implanting them in a gel or other artificial matrix to replace articular cartilage.

Cartilage Autografts

Lack of donor sites limits use of cartilage autografts. Animal experiments show that articular car-

tilage autografts transplanted with a thin shell of bone heal to the recipient site tissue [42, 54, 112, 162, 165]. The chondrocytes in most adequately stabilized autografts remain viable, and the matrix remains intact for a year or more. Because of the limited sources of cartilage for grafting, surgeons rarely use cartilage autografts in humans. Sources of possible cartilage autografts include the proximal tibiofibular joint, the sternum, and the patella. Animal experiments show that sternal osteochondral autografts can replace segments of articular cartilage [193], and surgeons have used osteochondral patellar grafts to replace severely damaged portions of the tibial articular surface in humans [204]. Radiographs show that the bone of the graft heals with the recipient site bone, and clinical evaluation shows that grafts can provide satisfactory joint function without knee effusions or evidence of degeneration for 18 years or more [91].

Periosteal and Perichondrial Autografts

Periosteum and perichondrium provide other sources of tissue to repair articular cartilage defects. Cells from both tissues can synthesize the necessary matrix macromolecules to form hyaline cartilage and survive transplantation without a vascular pedicle [4, 5, 40, 47, 60, 155, 156, 175, 176, 209]. In experimental and clinical studies, most grafts have been harvested from the recipient, but allografts also can restore a joint surface [139].

Experimental work shows that periosteal grafts will fill large defects in rabbit articular surfaces with hyaline-like cartilage and that this tissue remains intact for as long as 1 year after transplantation [155, 156, 157, 159, 175, 176]. Perichondrial grafts have produced similar results in rabbits and dogs [4, 5, 60–63, 104, 105, 186, 206]. The graft cells produce a matrix that contains type II collagen [4, 104, 156], and the concentration of type II collagen may increase with time [5]. The mechanical properties of these grafts have not been thoroughly examined, but one study showed that the viscoelastic properties of perichondrial grafts improved following surgery and approached the properties of normal cartilage 26 weeks after surgery [206]. These observations suggest that the grafts remodel so that their composition and mechanical properties more closely resemble those of articular cartilage.

Although these experimental studies show that periosteal and perichondrial cells can survive transplantation and form a new articular surface, the results vary among animals and possibly among regions of the joints. An investigation of rabbit perichondrial grafts found that 50% produced unacceptable results due to fractures, failures of graft attachment, or infection [4]. A subsequent examination of rabbit perichondrial grafts showed that 38% of the grafts produced unacceptable results [5]. A study of periosteal grafts in rabbits showed that these grafts frequently succeeded in rabbit femoral condyles and patellar grooves but usually failed in the patella [158].

Despite the evidence that grafts can improve their composition and mechanical properties after transplantation [5, 206], other studies show that at least some of the grafts deteriorate. In one set of experiments, 2 to 8 months after surgery dog perichondrial grafts had formed new cartilage with smooth articular surfaces, but by 12 to 17 months the grafts had degenerated, leaving exposed bone in some regions and fragments of graft tissue in others [61, 62]. A study of periosteal grafts in sheep also found that the grafts deteriorated with use of the joint [174]. One year following placement of periosteal grafts, moderately well differentiated fibrocartilage covered the previously exposed bone, but 2 years after transplantation the graft tissue had degenerated.

The age of the graft may influence the results. Periosteum and perichondrium become thinner and less cellular with age, and the potential of the cells to proliferate and synthesize new matrix decreases [36]. The possible differences in results with grafts of different ages have not been thoroughly studied experimentally, but cryopreserved periosteal allografts from young rabbits produced better results than grafts from older animals [139].

Early motion of the joint following grafting may influence the results. Although it is not clear how motion affects cell function in cartilage repair tissue, motion stimulates fibrocartilage formation following joint resection [138], and treatment with passive motion may promote cartilage formation by periosteal grafts [52, 155–157, 159, 209]. Grafts placed in immobilized joints also form cartilage, but they form less cartilage than grafts treated with passive motion [52]. Despite these encouraging results, the long-term benefits of passive motion treatment remain uncertain [36]. One group of investigators found that the apparent beneficial effect of early motion treatment of rabbit perichondrial grafts did not persist [104, 108, 206]. Passive joint movement appeared to improve the quality of the grafts initially, but 1 year after transplantation there were no apparent benefits of the early motion. Thus, it appears that passive joint motion soon after graft transplantation may affect the initial behavior of the graft cells, but the mechanism of the effect, the optimal timing and duration of motion, and the long-term benefits have not yet been established.

In addition to these experimental studies, surgeons have replaced damaged or lost articular cartilage in human osteoarthritic and rheumatoid joints with perichondrial grafts. After removing degenerated or damaged cartilage, the surgeon places a rib perichondrial autograft in the defect [63, 87, 163, 186, 207]. Most reports of this procedure have described results for joints of the upper extremity. Some patients have experienced improved range of motion and decreased pain, but the results have not been predictable [63].

One clinical series showed that the results of rib perichondrial allograft arthroplasties in metacarpophalangeal and proximal interphalangeal joints depended to a large extent on the age of the patient [184]. One hundred percent of patients in their twenties had good results with metacarpophalangeal joint arthroplasties. Only 75% of the patients in their thirties treated with similar procedures had good results. Seventy-five percent of patients in their teens had good results with proximal interphalangeal joint arthroplasties compared with 66% of patients in their twenties, and none of the patients over 40 years old. These results agree with the concept that the ability of perichondrium or periosteum to produce new cells and matrix declines with age and that the best clinical results with these procedures may be expected in skeletally immature patients or those who have recently reached skeletal maturity.

Cartilage Allografts

Cartilage allografts have the advantage of providing osteochondral segments of any size or shape without donor site morbidity. Surgeons have used allografts to replace segments of articular surfaces, entire articular surfaces, and entire synovial joints. Biopsies of the allograft cartilage show that many chondrocytes remain viable years after transplantation [50]. Large grafts have been used for joint reconstruction following tumor resections or major trauma [70, 124]. Smaller grafts, usually consisting of articular cartilage and a thin shell of subchondral bone, have been used to replace damaged regions of articular cartilage in young, physically active patients. Generally, the larger grafts have caused more frequent and more severe surgical and postoperative complications, including infection and mechanical failure [118, 124, 134].

Fresh and cryopreserved grafts have been used experimentally and clinically. Fresh grafts presumably have the advantage of maintaining the maximum viability of the chondrocytes [50], and a study of large dog osteochondral allografts showed that fresh grafts produced better results than frozen grafts [190]. Use of preserved grafts makes it possible to accumulate a bank of grafts of different sizes and shapes, and freezing has the advantage of decreasing graft immunogenicity [67, 190]. Recent studies report that cryopreservation and storage for up to 28 days does not alter the mechanical properties or structure of cartilage [101, 107]. Chondrocytes can survive freezing and thawing [180, 196], but in one study only a few chondrocytes survived freezing. Frozen grafts had more evidence of structural deterioration and lower concentrations of glycosaminoglycans than fresh grafts [190].

Cartilage allografts can survive transplantation and heal to the recipient site tissues [80, 162]. A study of articular allografts used to replace the tibial articular surface in skeletally mature rabbits showed that articular cartilage, growth plate, and cultured chondrocyte allografts resulted in significantly better repair than did the natural repair response [11]. The authors concluded that correctly positioned and secured cartilage allografts could repair limited articular defects in mature rabbits. Other studies have shown that at least some allograft cartilage degenerates when subjected to loading [42, 178, 190], and one recent study of dog osteochondral allografts showed that allograft cartilage became thin, dull, and roughened [190].

The host response to osteochondral allografts may depend to some extent on the size of the graft, including the amount of bone transplanted with the cartilage, freezing of the graft, and antigen matching. Small cartilage allografts do not cause an apparent inflammatory reaction [80, 162], but large osteochondral allografts can cause synovial inflammation [189, 190]. To some extent, this difference may depend on the amount of bone in the graft. Large antigen-mismatched osteochondral allografts stimulate systemic humoral, cell-mediated, and antibody-dependent cell-mediated immune responses and local immune responses [189, 190], but even large antigen-matched osteochondral grafts can cause synovitis [190].

A study of large osteochondral allografts in dogs suggests that host immune responses adversely affect cartilage grafts [190]. The authors compared the results of leukocyte antigen-mismatched frozen allografts, leukocyte antigen-mismatched fresh allografts, leukocyte antigen-matched fresh allografts, and leukocyte antigen-matched frozen allografts. Leukocyte antigen-mismatched fresh allografts stimulated the most severe inflammatory response. In-

vasive pannus appeared more frequently in joints with fresh grafts, especially those joints with leukocyte antigen-mismatched grafts. In some of these joints, the pannus eroded the cartilage to the subchondral bone. Antigen mismatching increased cartilage deterioration and exacerbated the damage due to freezing. Fresh antigen-matched grafts produced results similar to those seen with autogenous grafts. The results of this study showed that, at least for large segmental osteochondral allografts, fresh tissue-matched grafts produced the best results.

Clinical studies show that fresh osteochondral allografts can replace localized regions of damaged articular cartilage in humans [116, 133, 134]. Fresh osteochondral grafts used to replace portions of damaged tibial plateaus decreased pain and improved function in 10 of 12 patients followed for more than 2 years [116]. Evaluation of 40 knees 2 to 10 years after transplantation of fresh osteochondral allografts for localized degeneration of the articular surface showed that 31 of the grafts had healed and 9 had failed [134]. Of the 31 successful transplants, 13 had an excellent result, 14 had a good result, and 4 had a fair result. The authors recommended use of fresh osteochondral shell allografts for treatment of post-traumatic degenerative arthritis of the patella, for post-traumatic arthritis and traumatic defects of the tibial plateau, and for traumatic defects, osteochondritis dissecans, and avascular necrosis of the femoral condyle. They advised against use of grafts for unicompartmental degenerative arthritis of the knee involving both the femur and the tibia. Only 3 of 10 of these procedures succeeded. These studies suggest that fresh allografts can provide at least temporary improvement for selected patients with disabling symptoms due to isolated regions of degenerated or damaged cartilage.

Chondrocyte and Artificial Matrix Grafts

Synthetic matrix grafts, with or without cells and growth factors that stimulate cartilage formation, offer another method of replacing regions of damaged or lost articular cartilage [21, 22, 30, 34, 37, 45, 73–75, 89, 90, 200, 201]. The creation of synthetic matrices that vary in size and shape makes it possible to fill any chondral defect precisely. A synthetic matrix provides a framework for cell migration and attachment and may give the cells some protection from excessive loading. The cells included in these grafts may be autografts or allografts. Chondrocytes or mesenchymal cells harvested from the intended recipient or from another individual can be grown and maintained in culture and then reimplanted.

Most synthetic matrices used to replace articular cartilage and to implant growth factors or cells consist of reconstituted collagen, but one group of investigators has reported improved cartilage repair with carbon fiber pads [149]. The matrix composition and organization can influence cell migration, proliferation, and differentiation [102, 119]. For example, an in vitro study showed that a collagenous matrix promoted formation of cartilage by mesenchymal cells [119]. Several experimental studies have shown the feasibility of implanting chondrocytes or mesenchymal cells in cartilage defects [34, 37]. The implanted cells survive and synthesize a collagenous matrix that often resembles that of normal articular cartilage [89, 90, 200, 201]. This tissue appears to resemble hyaline cartilage more closely than the tissue that forms in defects not treated with chondrocyte artificial matrix and synthetic matrix transplants. In one series of experiments the investigators created 4-mm diameter osteochondral defects in rabbit articular surfaces and then placed collagen gels containing allograft chondrocytes in the defects [200, 201]. Eighty percent of the treated defects showed successful healing 24 weeks later. Other groups have also reported improved cartilage healing using similar methods [73, 74, 89, 90].

Thus far, investigators have not identified the optimal type of synthetic matrix or defined the benefits of implanting various cell types and growth factors. Nor have they shown that implantation of synthetic matrices containing chondrocytes or undifferentiated mesenchymal cells or growth factors can predictably restore a durable articular surface in large cartilage defects. Yet these studies show that this approach has the potential to improve repair of limited osteochondral defects.

CONCLUSIONS

The specialized composition and organization of articular cartilage provide its unique biomechanical properties that make possible normal synovial joint function. Blunt trauma, high-energy joint loading, and forceful joint loading and twisting can damage cartilage without causing visible disruption of the matrix, fracture cartilage without damaging the subchondral bone, or fracture both cartilage and subchondral bone. Injuries that alter the biomechanical properties of cartilage compromise the smooth pain-free joint motion necessary for participation in sports and may lead to progressive deterioration of the articular surface. Damage to cartilage that does not cause visible matrix disruption has been studied in animals, but this type of injury is difficult to detect in humans. Improved imaging methods or devices that measure in vivo cartilage mechanical properties may solve this problem. Currently, the natural history of these injuries remains unknown, but the experimental evidence shows that if the basic structure of matrix remains intact and if the cartilage is spared further injury, the chondrocytes can repair damage. Chondrocytes do not heal injuries that disrupt cartilage, but injuries that disrupt cartilage and bone stimulate an inflammatory response and migration of undifferentiated cells into the injury site. These cells proliferate, differentiate into chondrocyte-like cells, and synthesize a new matrix, but the tissue they produce usually fails to restore the normal volume of articular cartilage and the repair tissue lacks the biomechanical properties of articular cartilage. Because cartilage has a limited capacity for restoring normal tissue after significant injury, surgeons and other investigators have sought reliable methods of repairing or replacing damaged articular surfaces. Many diverse approaches have been taken with various success rates both in animal studies and in limited clinical use. Thus far, all of these approaches have important limitations, but the knowledge gained from these studies provides a basis for developing better methods of treating injuries of articular cartilage and preventing progression of cartilage damage.

References

1. Akizuki, S., Mow, V. C., Muller, F., Pita, J. C., Howell, D. S., and Manicourt, D. H. The tensile properties of human knee joint cartilage I: Influence of ionic conditions, weight bearing, and fibrillation on the tensile modulus. *J Orthop Res* 4:379–392, 1986.
2. Akizuki, S., Mow, V. C., Muller, F., Pita, J. C., and Howell, D. S. The tensile properties of human knee joint cartilage II: The influence of weight bearing, and tissue pathology on the kinetics of swelling. *J Orthop Res* 5:173–186, 1987.
3. Altman, R. D., Tenenbaum, J., Latta, L., Riskin, W., Blanco, L. N., and Howell, D. S. Biomechanical and biochemical properties of dog cartilage in experimentally induced osteoarthritis. *Ann Rheum Dis* 43:83–90, 1984.
4. Amiel, D., Coutts, R. D., Abel, M., Stewart, W., Harwood, F., and Akeson, W. H. Rib perichondrial grafts for the repair of full thickness articular cartilage defects. *J Bone Joint Surg* 67-A:911–920, 1985.
5. Amiel, D., Coutts, R. D., Harwood, F. L., Ishizue, K. K., and Kleiner, J. B. The chondrogenesis of rib perichondrial grafts for repair of full thickness articular cartilage defects in a rabbit model: A one year post-operative assessment. *Connect Tissue Res* 18:27–39, 1988.
6. Armstrong, C. G., and Mow, V. C. Variations in the intrinsic mechanical properties of human cartilage with age,

degeneration and water content. *J Bone Joint Surg* 64-A:88–94, 1982.

7. Armstrong, C. G, Mow, V. C., and Wirth, C. R. Biomechanics of impact-induced microdamage to articular cartilage: A possible genesis for chondromalacia patella. *In* Finerman, G. M. (Ed.), *Symposium on Sports Medicine: The Knee.* St. Louis, C. V. Mosby, 1985, pp. 70–84.
8. Arnoczky, S. P., and Warren, R. F. Microvasculature of the human meniscus. *Am J Sports Med* 10:990–995, 1982.
9. Arnoczky, S. P., and Warren, R. F. The microvasculature of the meniscus and its response to injury—An experimental study in the dog. *Am J Sports Med* 11:131–141, 1983.
10. Askew, M. J., and Mow, V. C. The Biomechanical function of the collagen ultrastructure of articular cartilage. *J Biomech Eng*, 100:105–115, 1978.
11. Aston, J. E., and Bentley, G. Repair of articular surfaces by allografts of articular and growth-plate cartilage. *J Bone Joint Surg* 68-B:29–35, 1986.
12. Athanasiou, K. A. Biomechanical assessment of articular cartilage healing and interspecies variability. Ph. D. Thesis, Columbia University, New York, December 1988, 15:95–96.
13. Athanasiou, K. A., Spilker, R. L., Buckwalter, J. A., Rosenwasser, M. P., and Mow, V. C. Finite element biphasic modeling of repair articular cartilage. *Adv Bioeng* 95–96, 1989.
14. Athanasiou, K. A., Rosenwasser, M. P., Spilker, R. L., Buckwalter, J. A., and Mow, V. C. Effects of passive motion on the material properties of healing articular cartilage. *Trans Orthop Res Soc* 15:156, 1990.
15. Baumgaertner, M. R., Cannon, W. D., Vittori, J. M., Schmidt, E. S., and Maurer, R. C. Arthroscopic debridement of the arthritic knee. *Clin Orthop Rel Res* 253:197–202, 1990.
16. Bennett, G. A., and Bauer, W. Further studies concerning the repair of articular cartilage in dog joints. *J Bone Joint Surg* 17:141–150, 1935.
17. Benninghoff, A. Form und Bau der Gelenkknorpel in ihren Beziehungen zur Funktion: Zweiter Teil. Der Aufbau des Gelenkknorpels in Seinen Beziehungen zur Funktion. *Z Zellforsch Mikrosk Anat* 2:783–862, 1925.
18. Bentley, G. The surgical treatment of chondromalacia patellae. *J Bone Joint Surg* 60-B:74–81, 1978.
19. Bentley, G. Chondromalacia patellae. *J Bone Joint Surg* 52-A:221–232, 1980.
20. Bentley, G., and Dowd, G. Current concepts of etiology and treatment of chondromalacia patellae. *Clin Orthop Rel Res* 189:209–228, 1984.
21. Bentley, G., and Greer, R. B. Homotransplantation of isolated epiphyseal and articular cartilage chondrocytes into joint surfaces of rabbits. *Nature* 230:385–388, 1971.
22. Bentley, G., Smith, A. U., and Mukerjhee, R. Isolated epiphyseal chondrocyte allografts into joint surfaces. An experimental study in rabbits. *Ann Rheum Dis* 37:449–458, 1978.
23. Bradley, J., and Dandy, D. J. Osteochondritis dissecans and other lesions of the femoral condyles. *J Bone Joint Surg* 71-B:518–522, 1989.
24. Bruckner, P., Vaughn, L., and Winthalter, K. H. Type IX collagen from sternal cartilage of chicken embryo contains covalently bound glycosaminoglycans. *Proc Natl Acad Sci USA* 82:2608–2615, 1985.
25. Brown, T. D., Anderson, D. D., Nepola, J. V., Singerman, R. J., Pedersen, D. R., and Brand, R. A. Contact stress aberrations following imprecise reduction of simple tibial plateau fracture. *J Orthop Res* 6:851–862, 1988.
26. Buckwalter, J. A., Kuettner, K. E., and Thonar, E. J.-M. Age-related changes in articular cartilage proteoglycans: Electron microscopic studies. *J Orthop Res* 3:251–257, 1985.
27. Buckwalter, J. A., and Cooper, R. R. The cells and matrices of skeletal connective tissues. *In* Albright, J. A. and Brand, R. A. (Eds.), *The Scientific Basis of Orthopaedics.* Norwalk, CT, Appleton & Lange, 1987, pp. 1–30.
28. Buckwalter, J. A., Hunziker, E. B., Rosenberg, R. C.,

Coutts, R. D., Adams, M. E., and Eyre, D. R. Articular cartilage: Composition and structure. *In* Woo, S. L-Y., and Buckwalter, J. A. (Eds.), *Injury and Repair of the Musculoskeletal Soft Tissues.* Park Ridge, IL, American Academy of Orthopaedic Surgeons, 1988, pp. 405–426.
29. Buckwalter, J. A., and Rosenberg, L. C. Electron microscopic studies of cartilage proteoglycans. *Elec Microsc Rev* 1:87–112, 1988.
30. Buckwalter, J. A., Rosenberg, L. C., Coutts, R., Hunziker, E., Reddi, A. H., and Mow, V. C. Articular cartilage: Injury and repair. *In* Woo, S.L.-Y., and Buckwalter, J. A. (Eds.), *Injury and Repair of the Musculoskeletal Soft Tissues.* Park Ridge, IL, American Academy of Orthopaedic Surgeons, 1988, pp. 465–482.
31. Buckwalter, J. A., Rosenberg, L. C., and Ungar, R. Age related changes in link protein function. *Orthop Trans* 13:258, 1989.
32. Buckwalter, J. A. Arthroscopic treatment of osteoarthritic knee joints. *In* Brandt, K. D. (Ed.), *Cartilage Changes in Osteoarthritis.* Indianapolis, University of Indiana Press, 1990, pp. 137–141.
33. Buckwalter, J. A., and Olmstead, M. Cartilage repair in primates and rabbits. Unpublished observations, 1990.
34. Buckwalter, J. A., Rosenberg, L. A., and Hunziker, E. B. Articular cartilage: Composition, structure, response to injury, and methods of facilitation repair. *In* Ewing, J. W. (Ed.), *Articular Cartilage and Knee Joint Function: Basic Science and Arthroscopy.* New York, Raven Press, 1990, pp. 19–56.
35. Buckwalter, J. A. Cartilage. *In* Dulbecco, R. (Ed.) *Encyclopedia of Human Biology,* Vol. 2. San Diego, Academic Press, 1991, pp. 201–215.
36. Buckwalter, J. A., and Cruess, R. Healing of musculoskeletal tissues. *In* Rockwood, C. A., and Green, D. (Eds.), *Fractures.* Philadelphia, J. B. Lippincott, in press, 1991.
37. Buckwalter, J. A., and Mow, V. C. Cartilage repair in osteoarthritis. *In* Moskowitz, R. W., Howell, D. S., Goldberg, V. M. and Mankin, H. J. (Eds.), *Osteoarthritis: Diagnosis and Management* (2nd ed.). Philadelphia, W. B. Saunders, 1992, pp. 71–107.
38. Bullough, P. G. The geometry of diarthrodial joints, its physiologic maintenance, and the possible significance of age-related changes in geometry-to-load distribution and the development of osteoarthritis. *Clin Orthop Rel Res* 156:61–66, 1981.
39. Bullough, P. G., and Jagannath, A. The morphology of the calcification front in articular cartilage. *J Bone Joint Surg* [Br] 65:72–78, 1983.
40. Bulstra, S. K., Homminga, G. N., Buurman, W. A., Terwindt-Rouwenhorst, E., and van der Linden, A. J. The potential of adult human perichondrium to form hyaline cartilage in vitro. *J Orthop Res* 8:328–335, 1990.
41. Calandruccio, R. A., and Gilmer, W. S., Jr. Proliferation, regeneration, and repair of articular cartilage of immature animals. *J Bone Joint Surg* 44-A:431–455, 1962.
42. Campbell, C. J., Ishida, H., Takahashi, H., and Kelly, F. The transplantation of articular cartilage. *J Bone Joint Surg* 45-A:1579–1592, 1963.
43. Campbell, C. J. The healing of cartilage defects. *Clin Orthop Rel Res* 64:45–63, 1969.
44. Carney, S. L., Billingham, M. E. J., Muir, H., and Sandy, J. D. Structure of newly synthesized 35S-proteoglycans and 35S-proteoglycan turnover products of cartilage explant culture from dogs with experimental osteoarthritis. *J Orthop Res* 3:40–47, 1985.
45. Chesterman, P. J., and Smith, A. U. Homotransplantation of articular cartilage and isolated chondrocytes. An experimental study in rabbits. *J Bone Joint Surg* 50-B:184–197, 1968.
46. Clarke, I. C. Articular cartilage: A review and scanning electron microscope study, 1. The interterritorial fibrillar architecture. *J Bone Joint Surg* 53B:732–750, 1971.
47. Cohen, J., and Lacroix, D. Bone and cartilage formation by periosteum. *J Bone Joint Surg* 37-A:717–730, 1955.

48. Coletti, J. M., Akeson, W. H., and Woo, S. L.-Y. A comparison of the physical behavior of normal articular cartilage and arthroplasty surface. *J Bone Joint Surg* 54-A:147–160, 1972.

49. Convery, F. R., Akeson, W. H., and Keown, G. H. The repair of large osteochondral defects: An experimental study in horses. *Clin Orthop Rel Res* 82:253–262, 1972.

50. Czitrom, A. A., Keating, S., and Gross, A. E. The viability of articular cartilage in fresh allografts after clinical transplantation. *J Bone Joint Surg* 72-A:574–581, 1990.

51. Dekel, S., and Weissman, S. L. Joint changes after overuse and peak overloading of rabbit knees in vivo. *Acta Orthop Scand* 49:519–528, 1978.

52. Delaney, J. P., O'Driscoll, S. W., and Salter, R. B. Neochondrogenesis in free intraarticular periosteal autografts in an immobilized and paralyzed limb. *Clin Orthop Rel Res* 248:278–282, 1989.

53. DePalma, A. F., McKeever, C. D., and Subin, D. K. Process of repair of articular cartilage demonstrated by histology and autoradiography with tritiated thymidine. *Clin Orthop Rel Res* 48:229–242, 1966.

54. Depalma, A. F., Tsaltas, T. T., and Mauler, G. G. Viability of osteochondral grafts as determined by uptake of S35. *J Bone Joint Surg* 45-A:1565–1578, 1963.

55. Donohue, J. M., Buss, D., Oegema, T. R., and Thompson, R. C. The effects of indirect blunt trauma on adult canine articular cartilage. *J Bone Joint Surg* 65-A:948–956, 1983.

56. Dowson, D., Unsworth, A., Cooke, A. F., and Gvozdanovic, D. Lubrication of joints. *In* Dowson, D., and Wright, V. (Eds.), *An Introduction to the Biomechanics of Joints and Joint Replacements.* London, Institute of Mechanical Engineers, 1981, pp. 120–145.

57. Emery, I. H., and Meachim, G. Surface morphology and topography of patello-femoral cartilage fibrillation in Liverpool necropsies. *J Anat* 116:103–120, 1973.

58. Eisenberg, S. R., and Grodzinsky, A. J. The kinetics of chemically induced nonequilibrium swelling of articular cartilage and corneal stroma. *J Biomech Eng* 109:79–89, 1987.

59. Edwards, J. Physical characteristics of articular cartilage. *Proc Inst Mech Eng* 181:16–24, 1967.

60. Engkvist, O., Johansson, S. H., Ohlsen, L., and Skoog, T. Reconstruction of articular cartilage using autologous perichondrial grafts. *Scand J Plast Reconstr Surg* 9:203–206, 1975.

61. Engkvist, O. Reconstruction of patellar articular cartilage with free autologous perichondrial grafts. *Scand J Plast Reconstr Surg* 13:361–369, 1979.

62. Engkvist, O., and Ohlsen, L. Reconstruction of articular cartilage with free autologous perichondrial grafts. *Scand J Plast Reconstr Surg* 13:269–279, 1979.

63. Engkvist, O., and Johansson, S. H. Perichondrial arthroplasty: A clinical study in twenty-six patients. *Scand J Plast Reconstr Surg* 14:71–87, 1980.

64. Evans, C. H., Mazzocchi, R. A., Nelson, D. D., and Rubash, H. E. Experimental arthritis induced by intraarticular injection of allogenic cartilagienous particles into rabbit knees. *Arthritis Rheum* 27:200–215, 1984.

65. Eyre, D. R. The collagens of musculoskeletal soft tissues. *In* Leadbetter, W. B. Buckwalter, J. A., and Gordon, S. L. (Eds.), *Sports Induced Inflammation—Basic Science and Clinical Concepts.* Park Ridge, IL, American Academy of Orthopaedic Surgeons, 1990, pp. 161–170.

66. Fithian, D. C., Kelly, M. A., and Mow, V. C. Material properties and structure-function relationships in the menisci. *Clin Orthop Rel Res* 252:19–30, 1990.

67. Friedlaender, G. E. Immune responses to osteochondral allografts. *Clin Orthop Rel Res* 174:58–68, 1983.

68. Friedman, M. J., Berasi, D. O., Fox, J. M., Pizzo, W. D., Snyder, S. J., and Ferkel, R. D. Preliminary results with abrasion arthroplasty in the osteoarthritic knee. *Clin Orthop Rel Res* 182:200–205, 1984.

69. Furukawa, T., Eyre, D. R., Koide, S., and Glimcher, M. J. Biochemical studies on repair cartilage resurfacing experimental defects in the rabbit knee. *J Bone Joint Surg* 62A:79–89, 1980.

70. Gebhardt, M. C., Roth, Y. F., and Mankin, H. J. Osteoarticular allografts for reconstruction in the proximal part of the humerus after excision of a musculoskeletal tumor. *J Bone Joint Surg* 72-A:334–345, 1990.

71. Ghadially, F. N., Thomas, I., Oryschak, A. F., and Lalonde, J.-M. Long-term results of superficial defects in articular cartilage: A scanning electron-microscope study. *J Pathol* 121:213–217, 1977.

72. Gray, M., Pizzanelli, A. M., Grodzinsky, A. J., and Lee, R. C. Mechanical and physiochemical determinants of the chondrocyte biosynthetic response. *J Orthop Res* 6:777–792, 1988.

73. Grande, D. A., Pitman, M. I., Peterson, L., Menche, D., and Lein, M. The repair of experimentally produced defects in rabbit articular cartilage by autologous chondrocyte transplantation. *J Orthop Res* 7:208–218, 1989.

74. Grande, D. A., Singh, I., and Pugh, J. Healing of experimentally produced lesions in articular cartilage following chondrocyte transplantation. *Anat Rec* 218:142–148, 1987.

75. Green, W. T. Articular cartilage repair. Behavior of rabbit chondrocytes during tissue culture and subsequent allografting. *Clin Orthop Rel Res* 124:237–250, 1977.

76. Guilak, F., Meyer, B. C., Ratcliffe, A., and Mow, V. C. The effect of static loading on proteoglycan biosynthesis and turnover in articular cartilage explants. *Trans Orthop Res Soc* 16:50, 1991.

77. Guilak, F., Ratcliffe, A., Hunziker, E. B., and Mow, V. C. Finite element modeling of articular cartilage chondrocytes under physiological loading. *Trans Orthop Res Soc* 16:366, 1991.

78. Hardingham, T. E., Muir, H., Kwan, M. K., Lai, W. M., and Mow, V. C. Viscoelastic properties of proteoglycan solutions with varying proportions present as aggregates. *J Orthop Res* 5:36–46, 1987.

79. Hascall, V. C. Interactions of cartilage proteoglycans with hyaluronic acid. *J Supramol Structure* 7:101–120, 1977.

80. Hamilton, J. A., Barnes, R., and Gibson, T. Experimental homografting of articular cartilage. *J Bone Joint Surg* 51-B:566–567, 1969.

81. Haut, R. C. Contact pressures in the patellofemoral joint during impact loading on the human flex knee. *J Orthop Res* 7:272–280, 1989.

82. Hayes, W. C., and Mockros, L. F. Viscoelastic properties of human articular cartilage. *J Appl Physiol* 31:562–568, 1971.

83. Hayes, W. C., and Bodine, A. J. Flow-independent viscoelastic properties of articular cartilage matrix. *J Biomech* 11:407–420, 1978.

84. Hou, J. S., Lai, W. M., Holmes, M. H., and Mow, V. C. Squeeze film lubrication for articular cartilage with synovial fluid. *In* Mow, V. C., Ratcliffe, A., and Woo, S. L-Y. (Eds.), *Biomechanics of Diarthrodial Joints,* Vol. 2. New York, Springer Verlag, 1990, pp. 347–368.

85. Howell, D. S. Etiopathogenesis of osteoarthritis. *In* Moskowitz, R. M., Howell, D. S., Goldberg, V. M., and Mankin, H. J. (Eds.), *Osteoarthritis: Diagnosis and Management.* Philadelphia, W. B. Saunders, 1984, pp. 129–146.

86. Huberti, H. H., and Hayes, W. C. Patello-femoral contact pressures: The influence of Q-angle and tendo-femoral contact. *J Bone Joint Surg* 66-A:715–724, 1984.

87. Huid, I., and Anderson, L. I. Perichondrial autograft in traumatic chondromalacia patellae. *Acta Orthop Scand* 52:91–93, 1981.

88. Hultkrantz, W. Ueber die Spaltrichtungen der Gelenkknorpel. *Verh Anat Ges* 12:248–256, 1898.

89. Itay, S., Abramovici, A., and Nevo, Z. Use of cultured embryonal chick epiphyseal chondrocytes as grafts for defects in chick articular cartilage. *Clin Orthop Rel Res* 220:284–303, 1987.

90. Itay, S., Abramovici, A., Ysipovitch, Z., and Nevo, Z. Correction of defects in articular cartilage by implants of cultures of embryonic chondrocytes. *Trans Orthop Res Soc* 13:112, 1988.

91. Jacobs, J. E. Follow-up notes on articles previously published in the journal: Patellar graft for severely depressed comminuted fractures of the lateral tibial condyle. *J Bone Joint Surg* 47-A:842–847, 1965.
92. Jaffe, F. F., Mankin, H. J., Weiss, C., and Zarins, A. Water binding in the articular cartilage of rabbits. *J Bone Joint Surg* 56-A:1031–1039, 1974.
93. Johnson, L. L. *Diagnostic and Surgical Arthroscopy*. St. Louis, C. V. Mosby, 1980.
94. Johnson, L. L. Arthroscopic abrasion arthroplasty. Historical and pathologic perspective: Present status. *Arthroscopy* 2:54–59, 1986.
95. Johnson, L. L. The sclerotic lesion: Pathology and the clinical response to arthroscopic abrasion arthroplasty. *In* Ewing, J. W. (Ed.), *Articular Cartilage and Knee Joint Function. Basic Science and Arthroscopy*. New York, Raven Press, 1990, pp. 319–334.
96. Johnson-Nurse, C., and Dandy, D. J. Fracture-separation of articular cartilage in the adult knee. *J Bone Joint Surg* [Br] 67-B:42–43, 1985.
97. Jurvelin, J., Kiviranta, I., Tammi, M., and Helminen, H. J. Softening of canine articular cartilage after immobilization of the knee joint. *Clin Orthop Rel Res* 207:246–252, 1986.
98. Kempson, G. E., Muir, H., Pollard, C., and Tuke, M. The tensile properties of the cartilage of human femoral condyles related to the content of collagen and glycosaminoglycans. *Biochim Biophys Acta* 297:456–472, 1973.
99. Kempson, G. E. Mechanical properties of articular cartilage. *In* Freeman, M. A. R. (Ed.), *Adult Articular Cartilage*. Tunbridge Wells, Kent, England, Pitman Medical, 1979, pp. 333–414.
100. Kennedy, J. C., Grainger, R. W., and McGraw, R. W. Osteochondral fractures of the femoral condyles. *J Bone Joint Surg* 48-B:436–440, 1966.
101. Kiefer, G. N., Sundby, K., McAllister, D., Shrive, N. G., Frank, C. B., Lam, T., and Schachar, N. S. The effect of cryopreservation on the biomechanical behavior of bovine articular cartilage. *J Orthop Res* 7:494–501, 1989.
102. Kimura, T., Yasui, N., Ohsawa, S., and Ono, K. Chondrocytes embedded in collagen gels maintain cartilage phenotype during long-term cultures. *Clin Orthop Rel Res* 186:231–239, 1984.
103. Kiviranta, I., Jurvelin, J., Tammi, M., Saamanen, A.-M., and Helminin, H. J. Weight-bearing controls glycosaminoglycan concentration and articular cartilage thickness in the knee joints of young beagle dogs. *Arthritis Rheum* 30:801–809, 1987.
104. Kleiner, J. B., Coutts, R. D., Woo, S. L.-Y., Amiel, D., Lee, T. O., Rosenstein, A. D., Harwood, F. L., and Field, F. P. The short-term evaluation of different treatment modalities upon full-thickness articular cartilage defects: A study of rib perichondrial chondrogenesis. *Trans Orthop Res Soc* 11:282, 1986.
105. Kulick, M. I., Brent, B., and Ross, J. Free perichondrial graft from the ear to the knee in rabbits. *J Hand Surg* 9A:213–215, 1984.
106. Kwan, M. K., Lai, W. M., and Mow, V. C. Fundamentals of fluid transport through cartilage in compression. *Ann Biomed Eng* 12:537–558, 1984.
107. Kwan, M. K., Wayne, J. S., Woo, S. L.-Y., Field, F. P., Hoover, J., and Meyers, M. Histological and biomechanical assessment of articular cartilage from stored osteochondral shell allografts. *J Orthop Res* 7:637–644, 1989.
108. Kwan, M. K., Woo, S. L.-Y., Amiel, D., Kleiner, J. B., Field, F. P., and Coutts, R. D. Neocartilage generated from rib perichondrium: A long-term multidisciplinary evaluation. *Trans Orthop Res Soc* 12:277, 1987.
109. Lai, W. M., Hou, J. S., and Mow, V. C. A triphasic theory for articular cartilage swelling. *Proc Biomech Symp ASME* AMD 98:33–36, 1989.
110. Lai, W. M., Hou, J. S., and Mow, V. C. A triphasic theory for the swelling and deformational behavior of cartilage. *J Biomech Eng* in press, 1991.
111. Lane, J. M., and Weiss, C. Review of articular cartilage collagen research. *Arthritis Rheum* 18:553–562, 1975.
112. Lane, J. M., Brighton, C. T., Ottens, H. R., and Lipton, M. Joint resurfacing in the rabbit using an autologous osteochondral graft. *J Bone Joint Surg* 59-A:218–222, 1977.
113. Linn, F. C., and Sokoloff, L. Movement and composition of interstitial fluid of cartilage. *Arthritis Rheum* 8:481–494, 1965.
114. Lipshitz, H., Etheredge, R., and Glimcher, M. J. Changes in the hexosamine content and swelling ratio of articular cartilage as functions of depth from the surface. *J Bone Joint Surg* 58-A:1149–1153, 1976.
115. Lipshitz, H., and Glimcher, M. J. In vitro studies of the wear of articular cartilage. II. Characteristics of the wear of articular cartilage when worn against stainless steel plates having characterized surface. *Wear* 52:297–339, 1979.
116. Locht, R. C., Gross, A. E., and Langer, F. Late osteochondral allograft resurfacing for tibial plateau fractures. *J Bone Joint Surg* 66-A:328–335, 1984.
117. Lohmander, S. Proteoglycans of joint cartilage. Structure, function, turnover and role as markers of joint disease. *Bailliere's Clin Rheumatol* 2:37–62, 1988.
118. Lord, C. F., Gebhardt, M. C., Tomford, W. W., and Mankin, H. J. Infection in bone allografts: Incidence, nature and treatment. *J Bone Joint Surg* 70-A:369–376, 1988.
119. Major, G., DerMark, K. V., Reddi, H., Heinegard, D., Franzen, A., and Silberman, M. Acceleration of cartilage and bone differentiation on collagenous substrata. *Collagen Rel Res* 7:351–370, 1987.
120. Mankin, H. J. The reaction of articular cartilage to injury and osteoarthritis: Part II. *N Engl J Med* 291:1335–1340, 1974.
121. Mankin, H. J. The reaction of articular cartilage to injury and osteoarthritis: Part I. *N Engl J Med* 291:1285–1292, 1974.
122. Mankin, H. J., and Thrasher, A. Z. Water content and binding in normal and osteoarthritic human cartilage. *J Bone Joint Surg* 57-A:76–80, 1975.
123. Mankin, H. J. The response of articular cartilage to mechanical injury. *J Bone Joint Surg* 64A:460–466, 1982.
124. Mankin, H. J., Deppelt, S. H., Sullivan, T. R., and Tomford, W. W. Osteoarticular and intercalary allograft transplantation in the management of malignant tumors of bone. *Cancer* 50:613–630, 1982.
125. Maroudas, A. Biophysical chemistry of cartilaginous tissues with special reference to solute and fluid transport. *Biorheology* 12:233–248, 1975.
126. Maroudas, A. Balance between swelling pressure and collagen tension in normal and degenerate cartilage. *Nature* 260:808–809, 1976.
127. Maroudas, A. Physicochemical properties of articular cartilage. *In* Freeman, M. A. R. (Ed.), *Adult Articular Cartilage*. Tunbridge Wells, Pitman Medical, 1979, pp. 215–290.
128. Matthewson, M. H., and Dandy, D. J. Osteochondral fractures of the lateral femoral condyle. *J Bone Joint Surg* [Br] 60-B:199–202, 1978.
129. McDevitt, C. A., Gilbertson, E. M. M., and Muir, H. An experimental model of osteoarthrosis: Early morphological and chemical changes. *J Bone Joint Surg* 59-B:24–35, 1977.
130. Meachim, G. The effect of scarification on articular cartilage in the rabbit. *J Bone Joint Surg* 45B:150–161, 1963.
131. Meachim, G., and Roberts, C. Repair of the joint surface from subarticular tissue in the rabbit knee. *J Anat* 109:317–327, 1971.
132. Meachim, G. Morphological pattern of cartilage fibrillation. *J Pathol* 115:231–238, 1975.
133. Meyers, M. H. Resurfacing of the femoral head with fresh osteochondral allografts: Long-term results. *Clin Orthop Rel Res* 197:111–114, 1985.
134. Meyers, M. H. Akeson, W., and Convery, F. R. Resurfacing the knee with fresh osteochondral allograft. *J Bone Joint Surg* 71-A:704–713, 1989.
135. Mitchell, N., and Shepard, N. The resurfacing of adult rabbit articular cartilage by multiple perforations through

the subchondral bone. *J Bone Joint Surg* 58-A:230–233, 1976.

136. Mitchell, N., and Shepard, N. Healing of articular cartilage in intra-articular fractures in rabbits. *J Bone Joint Surg* 62-A:628–634, 1980.

137. Mitchell, N., and Shepard, N. Effect of patellar shaving in the rabbit. *J Orthop Res* 5:388–392, 1987.

138. Mooney, V., and Fergurson, A. B. The influence of immobilization and motion on the formation of fibrocartilage in the repair granuloma after joint resection in the rabbit. *J Bone Joint Surg* 48-A:1145–1155, 1966.

139. Moran, M. E., Kreder, H. J., Salter, R. B., and Keeley, F. W. Biological resurfacing of major full thickness defects in joint surfaces by neochondrogenesis with cryopreserved allogenic periosteum stimulated by continuous passive motion. *Trans Orthop Res Soc* 14:542, 1989.

140. Mow, V. C., Holmes, M. H., and Lai, W. M. Fluid transport and mechanical properties of articular cartilage. *J Biomech* 17:377–394, 1984.

141. Mow, V. C., Mak, A. F., Lai, W. M., Rosenberg, L. C., and Tang, L. H. Viscoelastic properties of proteoglycan subunits and aggregates in varying solution concentrations. *J Biomech* 17:325–338, 1984.

142. Mow, V. C., and Mak, A. F. Lubrication of diarthrodial joints. *In* Skalak, R., and Chien, S. (Eds.), *Bioenginering Handbook*. New York, McGraw Hill, 1986, pp. 5.1–5.34.

143. Mow, V. C., Kwan, M. K., Lai, W. M., and Holmes, M. H. A finite deformation theory for nonlinearly permeable soft hydrated biological tissues. *In* Woo, S.L-Y., Schmid-Schonbein, G., and Zweifach, B. (Eds.), *Frontiers in Biomechanics*. New York, Springer-Verlag, 1986, pp. 153–179.

144. Mow, V. C., and Rosenwasser, M. P. Articular cartilage: Biomechanics. *In* Woo, S. L-Y., and Buckwalter, J. A. (Eds.), *Injury and Repair of the Musculoskeletal Soft Tissues*. Park Ridge, IL American Academy of Orthopaedic Surgeons, 1988, pp. 427–464.

145. Mow, V. C., Proctor, C. S., Kelly, M. A. Biomechanics of articular cartilage. *In* Nordin, M., and Frankel, V. H. (Eds.), *Basic Biomechanics of the Musculoskeletal System* (2nd ed.). Philadelphia, Lea & Febiger, 1989, 31–57.

146. Mow, V. C., Fithian, D. C., and Kelly, M. A. Fundamentals of articular cartilage and meniscus biomechanics. *In* Ewing, J. W. (Ed.), *Articular Cartilage and Knee Joint Function*. New York, Raven Press, 1989, pp. 1–18.

147. Mow, V. C., Zhu, W., Lai, W. M., Hardingham, T. E., Hughes, C., and Muir, H. The influence of link protein stabilization on the viscometric properties of proteoglycan aggregate solutions. *Biochim Biophys Acta* 112:201–208, 1989.

148. Mow, V. C., Setton, L. A., Ratcliff, A., Howell, D. S., and Buckwalter, J. A. Structure-function relationships of articular cartilage and the effects of joint instability and trauma on cartilage. *In* Brandt, K. D. (Ed.), *Cartilage Changes in Osteoarthritis*. Indianapolis, University of Indiana Press, 1990, pp. 22–42.

149. Muckle, D. S., and Minns, R. J. Biological response to woven carbon fiber pads in the knee: A clinical and experimental study. *J Bone Joint Surg* 72B:60–62, 1990.

150. Muir, H., Bullough, P., and Maroudas, A. The distribution of collagen in human articular cartilage with some of its physiological implications. *J Bone Joint Surg* 52B:554–563, 1970.

151. Muir, H. Proteoglycans as organizers of the extracellular matrix. *Biochem Soc Trans* 11:613–622, 1983.

152. Myers, E. R., Lai, W. M., and Mow, V. C. A continuum theory and an experiment for the ion-induced swelling behavior of articular cartilage. *J Biomech Eng* 106:151–158, 1984.

153. Myers, E. R., Armstrong, C. G., and Mow, V. C. Swelling pressure and collagen tension. *In* Hukins, D. W. L. (Ed.), *Connective Tissue Matrix*. New York, Macmillan Press, 1984, pp. 161–168.

154. Nimni, M. E. Collagen: Structure, function and metabolism in normal and bibrotic tissues. *Semin Arthritis Rheum* 13:1–86, 1983.

155. O'Driscoll, S. W., and Salter, R. B. The induction of neochondrogenesis in free intra-articular periosteal autografts under the influence of continuous passive motion: An experimental study in the rabbit. *J Bone Joint Surg* 66A:1248–1257, 1984.

156. O'Driscoll, S. W., Keely, F. W., and Salter, R. B. The chondrogenic potential of free autogenous periosteal grafts for biological resurfacing of major full-thickness defects in joint surfaces under the influence of continuous passive motion: An experimental study in the rabbit. *J Bone Joint Surg* 68A:1017–1035, 1986.

157. O'Driscoll, S. W., and Salter, R. B. The repair of major osteochondral defects in joint surfaces by neochondrogenesis with autogenous osteoperiosteal grafts stimulated by continuous passive motion: An experimental investigation in the rabbit. *Clin Orthop Rel Res* 208:131, 1986.

158. O'Driscoll, S. W., Delaney, J. P., and Salter, R. B. Failure of experimental patellar resurfacing using free periosteal autografts. *Orthop Trans* 13:294–295, 1989.

159. O'Driscoll, S. W., Kelley, F. W., and Salter, R. B. Durability of regenerated articular cartilage produced by free autogenous periosteal grafts in major full-thickness defects in joint surfaces under the influence of continuous passive motion: A follow-up report at one year. *J Bone Joint Surg* 70A:595, 1989.

160. Palmoski, M., Perricone, E., and Brandt, K. D. Development and reversal of a proteoglycan aggregation defect in normal canine knee cartilage after immobilization. *Arthritis Rheum* 22:508–517, 1979.

161. Palmoski, M. J., and Brandt, K. D. Running inhibits the reversal of atrophic changes in canine knee cartilage after removal of a leg cast. *Arthritis Rheum* 24:1329–1337, 1981.

162. Pap, K., and Krompecher, S. Arthroplasty of the knee: Experimental and clinical experience. *J Bone Joint Surg* 43-A:523–537, 1961.

163. Pastacaldi, P., and Engkvist, O. Perichondrial wrist arthroplasty in rheumatoid patients. *Hand* 11:184–190, 1979.

164. Poole, C. A., Flint, M. H., and Beaumont, B. W. Chondrons extracted from canine tibial cartilage: Preliminary report on their isolation and structure. *J Orthop Res* 6:408–419, 1988.

165. Porter, B. P., and Lance, E. M. Limb and joint transplantation. *Clin Orthop Rel Res* 104:249–274, 1974.

166. Radin, E. L., and Paul, I. L. Response of joints to impact loading. *Arthritis Rheum* 14:356–362, 1971.

167. Radin, E. L., Martin, R. B., Burr, D. B., Caterson, B., Boyd, R. D., and Goodwin, C. Effects of mechanical loading on the tissues of the rabbit knee. *J Orthop Res* 2:221–234, 1984.

168. Radin, E. L., Burr, D. B., Fyhrie, D., Brown, T. D., and Boyd, R. D. Characteristics of joint loading as it applies to osteoarthrosis. *In* Mow, V. C., Ratcliffe, A., and Woo, S. L.-Y. (Eds.), *Biomechanics of Diarthrodial Joints*, Vol 1. New York, Springer-Verlag, 1990, pp. 437–451.

169. Repo, R. U., and Finlay, J. B. Survival of articular cartilage after controlled impact. *J Bone Joint Surg* 59-A:1068–1075, 1977.

170. Redler, I., and Zimmy, M. L. Scanning electron microscopy of normal and abnormal articular cartilage and synovium. *J Bone Joint Surg* 52A:1395–1404, 1970.

171. Redler, I., Zimny, M. L., Mansell, J., and Mow, V. C. The ultrastructure and biomechanical significance of the tidemark of articular cartilage. *Clin Orthop Rel Res* 112:357–362, 1975.

172. Rosenberg, L., Choi, H. U., Naeme, P. J., Sasse, J., Roughley, P. J., and Poole, A. R. Proteoglycans of soft connective tissues. *In* Leadbetter, W. B., Buckwalter, J. A., and Gordon, S. L. (Eds.), *Sports Induced Inflammation—Basic Science and Clinical Concepts*. Park Ridge, IL, American Academy of Orthopaedic Surgeons, 1990, pp. 171–188.

173. Roth, V., and Mow, V. C. The intrinsic tensile behavior of the matrix of bovine articular cartilage and its variation with age. *J Bone Joint Surg* 62-A:1102–1117, 1980.

174. Rothwell, A. G. Synovium transplantation onto the cartilage denuded patellar groove of the sheep knee joint. *Orthopaedics* 13:433–442, 1990.

175. Rubak, J. M. Reconstruction of articular cartilage defects with free periosteal grafts: An experimental study. *Acta Orthop Scand* 53:175–180, 1982.

176. Rubak, J. M., Poussa, M., and Ritsila, V. Chondrogenesis in repair of articular cartilage defects by free periosteal grafts in rabbits. *Acta Orthop Scand* 53:181–186, 1982.

177. Sah, L. Y., Kim, Y. O., Doong, J. Y. H., Grodzinsky, A. J., Plass, A. H. K., and Sandy, J. D. Biosynthetic response of cartilage explants to dynamic compression. *J Orthop Res* 7:619–636, 1989.

178. Salenius, P., Holmstrom, T., Koskinen, E. V. S., and Alho, A. Histological changes in clinical half-joint allograft replacements. *Acta Orthop Scand* 53:295–299, 1982.

179. Salter, R. B., Simmonds, D. F., Malcolm, B. W., Rumble, E. J., MacMichael, D., and Clements, N. D. The biological effect of continuous passive motion on healing of full-thickness defects in articular cartilage: An experimental study in the rabbit. *J Bone Joint Surg* 62A:1232–1251, 1980.

180. Schachar, N., Nagao, M., Matsuyama, T., MacAllister, D., and Ishii, S. Cryopreserved articular chondrocytes grow in culture, maintain cartilage phenotype, and synthesize matrix components. *J Orthop Res* 7:344–351, 1989.

181. Schenk, R. K., Eggli, P. S., and Hunziker, E. B. Articular cartilage morphology. *In* Kuettner, K. E., Schleyerbach, R., and Hascall, V. C. (Eds.), *Articular Cartilage Biochemistry*. New York, Raven Press, 1986, pp. 3–22.

182. Schmid, A., and Schmid, F. Results after cartilage shaving studied by electron microscopy. *Am J Sports Med* 15:386–387, 1987.

183. Schmidt, M. B., Chun, L. E., Eyre, D. R., and Mow, V. C. The relationship between collagen cross-linking and tensile properties of articular cartilage. *Trans Orthop Res Soc* 12:134, 1987.

184. Seradge, H., Kutz, J. A., Kleinert, H. E., Lister, G. D., Wolff, T. W., and Atasoy, E. Perichondrial resurfacing arthroplasty in the hand. *J Hand Surg* 9A:880–886, 1984.

185. Simon, W. H., Mak, A. F., and Sprit, A. A. The effect of shear fatigue on bovine articular cartilage. *J Orthop Res* 8:86–93, 1990.

186. Skoog, T., and Johansson, S. H. The formation of articular cartilage from free perichondrial grafts. *Plast Reconstr Surg* 57:1–6, 1976.

187. Sprague, N. F. Arthroscopic debridement for degenerative knee joint disease. *Clin Orthop* 160:118–123, 1981.

188. Sprit, A. A., Mak, A. F., and Wassell, R. P. Nonlinear viscoelastic properties of articular cartilage in shear. *J Orthop Res* 7:43–49, 1988.

189. Stevenson, S. The immune response to osteochondral allografts in dogs. *J Bone Joint Surg* 69-A:573–582, 1987.

190. Stevenson, S., Dannucci, G. A., Sharkey, N. A., and Pool, R. R. The fate of articular cartilage after transplantation of fresh and cryopreserved tissue-antigen-matched and mismatched osteochondral allografts in dogs. *J Bone Joint Surg* 71-A:1297–1307, 1989.

191. Stockwell, R. S. *Biology of Cartilage Cells*. Cambridge, Cambridge University Press, 1979.

192. Stockwell, R. A., and Billingham, M. E. J. Early response of cartilage to abnormal factors as seen in the meniscus of the dog knee after cruciate ligament section. *Acta Biol Hung* 35:281–291, 1984.

193. Stover, S. M., Pool, R. R., and Lloyd, K. C. K. Repair of surgically created osteochondral defects with autogenous sternal osteochondral grafts. *Trans Orthop Res Soc* 14:543, 1989.

194. Tammi, M., Paukkonen, K., Kiviranta, I., Jurvelin, J., Saamanen, A. M., and Helminen, H. J. Joint loading-induced alterations in articular cartilage. *In* Helminen, H. J., Kiviranta, J., Saamanen, A.-M., Tammi, M., Paukkonen, K., and Juruelin, J. (Eds.), *Joint Loading*. Bristol, Wright, 1987, pp. 64–88.

195. Thonar, E. J., Buckwalter, J. A., and Kuettner, K. E. Maturation-related differences in the structure and composition of proteoglycans synthesized by chondrocytes from bovine articular cartilage. *J Biol Chem* 261:2467–2474, 1986.

196. Tomford, W. W., Fredricks, G. R., and Mankin, H. J. Studies on cryopreservation of articular cartilage chondrocytes. *J Bone Joint Surg* 66-A:253–259, 1984.

197. Torzilli, P. A., Rose, D. E., and Dethmers, D. A. Equilibrium water partition in articular cartilage. *Biorheology* 19:519–537, 1982.

198. Torzilli, P. A. The influence of cartilage conformation on its equilibrium water partition. *J Orthop Res* 3:473–483, 1985.

199. Van der Rest, M., and Mayne, R. Type IX collagen-proteoglycan from cartilage is covalently cross-linked to type II collagen. *J Biol Chem* 263:1615–1618, 1988.

200. Wakitani, S., Kimura, T., Hirooka, A., Ochi, T., Yoneda, M., Yasui, N., and Ono, K. Repair of rabbits' articular surfaces by allograft of chondrocytes embedded in collagen gels. *Trans Orthop Res Soc* 13:440, 1988.

201. Wakitani, S., Kimura, T., Hirooka, A., Ochi, T., Yoneda, M., Natsuo, N., Owaki, H., and Ono, K. Repair of rabbit articular surfaces with allograft chondrocytes embedded in collagen gel. *J Bone Joint Surg* 71B:74–80, 1989.

202. Weightman, B. O., Freeman, M. A. R., and Swanson, S. A. V. Fatigue of articular cartilage. *Nature* 244:303–304, 1973.

203. Whipple, R. R., Gibbs, M. C., Lai, W. M., Mow, V. C., Mak, A. F., Wirth, C. R. Biphasic properties of repaired cartilage at the articular surface. *Trans Orthop Res Soc* 10:340, 1985.

204. Wilson, W. J., and Jacobs, J. E. Patellar graft for severely depressed comminuted fractures of the lateral tibial condyle. *J Bone Joint Surg* 34-A:436–442, 1952.

205. Woo, S. L-Y., Akeson, W. H., and Jemmott, G. F. Measurements of nonhomogeneous directional properties of articular cartilage. *J Biomech* 9:785–791, 1976.

206. Woo, S. L.-Y., Kwan, K. K., Lee, T. Q., Field, F. P., Kleiner, J. B., and Coutts, R. D. Perichondrial autograft for articular cartilage: Shear modulus of neocartilage studied in rabbits. *Acta Orthop Scand* 58:510–515, 1987.

207. Wu, G., and Johnson, D. E. Perichondrial arthroplasy in the hand: A case report. *J Hand Surg* 8:445–453, 1983.

208. Zanetti, M., Ratcliffe, A., and Watt, F. M. Two subpopulations of differentiated chondrocytes identified with a monoclonal antibody to keratan sulfate. *J Biol Chem* 101:53–59, 1985.

209. Zarnett, R., Delaney, J. P., O'Driscoll, S. W., and Salter, R. B. Cellular origin and evolution of neochondrogenesis in major full-thickness defects of a joint surface treated by free autogenous periosteal grafts and subjected to continuous passive motion in rabbits. *Clin Orthop Rel Res* 222:267–274, 1987.

210. Zimmerman, N. B., Smith, D. G., Pottenger, L. A., and Cooperman, D. R. Mechanical disruption of human patellar cartilage by repetitive loading in vitro. *Clin Orthop Rel Res* 229:302–307, 1988.

Injuries to Meniscus

Joseph A. Buckwalter, M.D.
Van C. Mow, Ph.D.

As recently as the 1970s some surgeons assumed that knee menisci were vestigial structures that had little effect on joint function, and they treated small meniscal tears by total menisectomy. Since then, understanding of meniscal function and common treatments of meniscal injuries have changed considerably. Advances in our understanding of meniscal function have come from comparisons of meniscal structure and function among species, investigations of the relationships between the biomechanical function and the composition and organization of menisci, and studies showing that total menisectomy adversely affects joint function and increases the probability of joint degeneration. Appreciation of the adverse effects of loss of menisci on joint function has encouraged surgeons and other investigators to seek methods of promoting meniscal healing and replacing menisici.

Study of menisci in other species illustrates their importance in the specialized synovial joints. They exist in many animals other than man [9, 61]. Monkeys, some forms of bat, and amphibians such as crocodiles and bullfrogs have menisci, although in the latter two animals the menisci are discoid in shape. In birds and chickens the menisci are C-shaped, similar to the knee joints of humans and the larger bovine species. Bovine knee menisci are closer to discoid in shape than those of human knees, which are often described as semilunar-shaped [9, 55, 56, 90]. There is also a discoid-shaped fibrocartilaginous disc in the temporomandibular joint of man* and rodents [26, 61], and a meniscus is present in the ankle joint of kangaroos (a joint subjected to large forces when the animal jumps) [61]. From these and other comparative anatomic studies, investigators have recognized that a firm, intra-articular, fibrocartilaginous structure with great tensile strength is necessary in joints in which rotation and translation occur because these joints require excellent lubrication characteristics and probably additional mechanical stability, and in joints in which high forces are transmitted across the joint [44, 61, 84, 98].

Recent investigations utilizing microscopy, biochemistry, and bioengineering have significantly advanced understanding of the relationships between meniscal function and meniscal composition and organization. Like articular cartilage, knee meniscal tissue [9, 16, 17, 19] performs important mechanical functions including load bearing [4, 14, 15, 34, 38, 53, 58, 62, 91, 93, 94, 97, 107], shock absorption [57, 90, 106], improving joint stability [57, 50, 60, 96, 108], and possibly participating in joint lubrication [27, 61, 76, 78, 90]. These functions depend on a highly organized extracellular matrix consisting of fluid and a macromolecular framework formed of collagen (types I, II, III, V, VI), proteoglycans, elastin, and noncollagenous proteins. Unlike articular cartilage, some regions of meniscus have a blood supply (the peripheral 25% to 30% of the lateral meniscus and the peripheral 30% of the medial meniscus [6, 7, 9, 24, 55, 65]) and a nerve supply (the peripheral regions of the meniscus and especially the meniscal horns [9, 83, 114, 118]).

Since loss of menisci alters the loading of articular cartilage in ways that may increase the probability and severity of degenerative joint disease [4, 9, 14, 25, 27, 34, 57, 72, 73, 95], surgeons and investigators have studied the response of menisci to injury and sought methods of preserving, repairing, and re-

---

*The composition and structure of the temporomandibular joint disc are very similar to knee meniscus. Details on this topic may be found in the following references [26, 68, 81, and 82].

This work was sponsored in part by NIH grant AR 5-37950 for research on knee meniscus (VCM). We thank Ms. Diane MacKinnon for her editorial assistance.

placing menisci [31, 41–43, 79, 89, 103, 104]. Experimental and clinical studies show that tears through the vascular regions of the meniscus can heal, but tears through the avascular regions do not undergo a repair process that can heal a significant tissue defect [7–9, 55]. Recent research suggests that methods of stimulating healing of tears in avascular meniscal tissue and of replacing meniscal tissue exist and may help to maintain or restore meniscal function [8–12, 49, 79, 103, 104, 109–111].

This chapter first summarizes the current understanding of meniscal composition, structure, mechanical properties, blood supply, and nerve supply.* Subsequent sections review the response of menisci to injury and methods of stimulating meniscal healing. The final section discusses meniscal grafts.

COMPOSITION

Like bone, cartilage, and ligament, meniscus consists of scattered cells surrounded by an abundant extracellular matrix [9, 16, 17, 41, 65, 79, 92]. Continued structural integrity and function of the tissue depend on interactions between the cells and their surrounding matrix. The material properties of the tissue result from the composition and organization of the matrix macromolecules and the interactions between these solid components of the matrix and the tissue fluid.

Cells

Based on morphologic characteristics, there are two major types of meniscal cells [40, 65]. Near the surface, the cells have flattened ellipsoid or fusiform shapes, and in the deep zone the cells are spherical or polygonal. These differences in cell shape and size between the superficial and deep regions of the tissue resemble the changes in cell morphology seen between the superficial and deep regions of articular cartilage (see Chapter 2A). Like the cells from the superficial and deep zones of articular cartilage, the superficial and deep meniscal cells appear to have different synthetic functions or perhaps different responses to loading [110].

Within the meniscus, these cells produce and maintain the macromolecular framework of the tissue. They contain the synthetic organelles, endo-plasmic reticulum, and Golgi membranes, which are necessary to accomplish their primary function of synthesizing matrix macromolecules and, like most other mesenchymal cells, lack cell-to-cell contacts [9, 16]. Because most of them lie at a distance from blood vessels, they rely on diffusion through the matrix for transport of nutrients and metabolites.

Meniscal cells attach their membranes to matrix macromolecules via adhesion proteins (fibronectin, thrombospondin, type VI collagen [39, 65, 66, 87], and the matrix, particularly the pericellular region, protects the cells from damage due to physiologic loading of the tissue [46, 47]. Deformation of the macromolecular framework of the matrix causes fluid flow through the matrix [35, 78, 80, 90] and influences meniscal cell function [46, 47]. In all these features, the meniscal cells resemble articular cartilage chondrocytes. However, because meniscal tissue is much more fibrous than hyaline cartilage (see Chapter 2A), some authors have proposed that meniscal cells be called fibrochondrocytes [65, 109].

Extracellular Matrix

Water

Water contributes 65% to 75% of the total weight of meniscus [1, 35, 88, 90]. Table 2B–1 summarizes the water, S-glycosaminoglycan, and hydroxyproline content of meniscus [35]. Although some portion of the water may reside within the intrafibrillar space of the collagen fibers [51, 64, 105], most of it is retained within the tissue in the solvent domains of the proteoglycans by their strong hydrophilic tendencies and by Donnan osmotic pressure exerted by the counter ions associated with the negative charge groups on the proteoglycans [28, 52, 59, 63, 75, 90].* Because the pore size of the tissue is extremely small (<60 Å [75]), very large hydraulic pressures are required to overcome the drag of frictional resistance in forcing fluid flow through the tissue [75, 90]. Thus, interactions between water and the macromolecular framework of the matrix significantly influence the viscoelastic properties of the tissue (this subject will be discussed in more detail below).

Matrix Macromolecules

Collagens

The macromolecular framework of the meniscal matrix consists primarily of collagens, which may

---

*For a comprehensive reference on the knee meniscus, the reader is referred to the monograph by Mow and co-workers [79].

*Section A in this chapter includes a more extensive discussion of the interaction of aggregating proteoglycans and the tissue fluid.

TABLE 2B–1
Composition of Meniscus by Region

Region	n	Sulfated Glycosaminoglycan (% Dry Weight)	Water Content	Hydroxyproline (% Dry Weight)
LA	18	1.80 + 0.50	75.02 + 2.14	14.3 + 3.7
LC	18	1.68 + 0.56	72.99 + 2.40	13.2 + 2.0
LP	18	1.75 + 0.45	73.39 + 2.44	15.2 + 3.1
MA	12	2.20 + 1.01	72.12 + 9.73	13.2 + 3.6
MC	14	2.06 + 0.68	76.77 + 2.68	13.9 + 3.4
MP	18	1.94 + 0.83	74.88 + 7.32	13.9 + 3.6

Abbreviations: LA = lateral anterior; LC = lateral central; LP = lateral posterior; MA = medial anterior; MC = medial central; MP = medial posterior.

From Fithian, D. C., Kelly, M. A., and Mow, V. C. Material properties and structure-function relationships in the menisci. *Clin Orthop Rel Res* 252:19–31, 1990.

contribute up to 95% of the dry weight of the tissue (Table 2B–1) [35]. Most of this collagen is type I [9, 32, 65]. Types II, III, V, and VI collagen each may contribute from 1% to 2% of the total tissue collagen, although, strictly speaking, type VI may be classified as a matrix glycoprotein [65, 66, 117].

The large-diameter type I collagen fibrils lie mostly in the outer radial two-thirds of the meniscus and give this tissue its ultrastructural arrangement and tensile stiffness and strength [13, 21, 32, 33, 35, 40, 78, 90]. The type II collagen fibrils have smaller diameters and are located in the inner one-third, nonvascularized region of the meniscus near the surface of the tissue [23]. This inner region is also rich in proteoglycans and has a hyaline appearance [3, 9]. Little or no information exists about type III and type V collagen in the meniscus. However, type VI collagen may have a role in stabilizing the type I and II collagen framework of the meniscus and in maintaining fibrochondrocyte adhesion to the matrix [9, 65, 117].

Proteoglycans

Some meniscal regions have a proteoglycan concentration of up to 3% of their dry weight [1, 9, 35, 65, 67, 92]. Like proteoglycans from other dense fibrous tissues, including tendon, ligament, and hyaline cartilage, meniscus proteoglycans can be divided into two general types: large aggregating proteoglycans that expand to fill large volumes of matrix and contribute to tissue hydration and the mechanical properties of the tissue, and smaller nonaggregating proteoglycans that usually have a close relationship with fibrillar collagen [2, 18, 65].*

Electron microscopic studies of the large aggregating proteoglycans from meniscus show that they have the same structure as the large aggregating proteoglycans from articular cartilage [18, 92]. Although the large aggregating proteoglycans influence the material properties of meniscus, their low concentration suggests that they probably contribute less to the properties of meniscus than to the properties of articular cartilage [35, 63, 59, 75, 90].* As with the quantitatively minor collagens, the smaller nonaggregating meniscal proteoglycans may help organize and stabilize the matrix, but at present, the function of these small meniscal proteoglycans remains unknown.

Noncollagenous Proteins

Noncollagenous proteins also form part of the macromolecular framework of meniscus and may contribute as much as 10% of the dry weight of the tissue in some regions [3, 65]. Two specific noncollagenous proteins, link protein and fibronectin, have been identified in meniscus [65]. Link protein is required for the formation of stable proteoglycan aggregates capable of forming strong networks [48, 77]. Fibronectin serves as an attachment protein for cells in the extracellular matrix [39, 87]. Other noncollagenous proteins such as thrombospondin [37] also exist and may serve as adhesive proteins in the tissue, thus contributing to the structure and the mechanical strength of the matrix. However, the exact details of their composition and function in the meniscus remain largely unknown.

Elastin

Most dense fibrous tissues, including meniscus, contain elastin, but it contributes less than 1% of the meniscus dry weight [9, 16, 17, 65]. The contribution of elastin to the mechanical properties of these tissues is uncertain, but it seems unlikely that sparsely distributed elastic fibers have a significant role in the organization of the matrix or in determining the mechanical properties of the tissue.

---

*Section A in this chapter contains a more extensive discussion of the proteoglycans found in cartilage.

STRUCTURE

Within the meniscus, the diameter and orientation of the collagen fibrils and cell morphology vary from the surface to the deeper regions [9, 13, 21, 40, 65, 52, 78, 90, 113]. The highly ordered arrangement of collagen fibrils within the tissue correlates closely with the biomechanical properties of meniscus. As in articular cartilage, the mechanisms responsible for organizing the collagen fibrils within meniscus remain unknown, but a study of meniscal development suggests that weight bearing may influence meniscal collagen fibril organization [24].

A thin layer, approximately 200 mm thick, forms the meniscal surface. It is rich in type II collagen and consists of a randomly woven mesh of fine collagen fibrils that lie parallel to the surface. Below this surface layer, large, circumferentially arranged collagen fiber bundles (mostly type I collagen) course throughout the entire body of each meniscus [9, 13, 21, 35, 90]. These circumferential collagen bundles give meniscus great tensile stiffness and strength parallel to their orientation [21, 35, 90, 112, 113]. They insert into the anterior and posterior meniscal attachment sites on the tibial plateau, and large forces are transmitted through these attachment sites. Figure 2B–1A is a schematic illustration of these large-fiber bundles and the thin superficial surface layer. Figure 2B–1B is a macrophotograph of a bovine medial meniscus with the surface layer removed, showing the large collagen bundles of the deep zone.

Any radial section of meniscus (Fig. 2B–2) shows radially oriented bundles of collagen fibrils or "radial tie fibers" weaving among the circumferential collagen fibril bundles from the periphery of the meniscus to the inner region [21, 52, 99]. Their prevalence and the tensile properties of specimens prepared from radial sections argue that these radial tie fibers form sheaths that provide a secure "wrap" around the loosely arranged large circumferential collagen bundles [52, 99]. Presumably, they help to increase the stiffness and strength of the tissue in a radial direction [99], thereby resisting longitudinal splitting of the collagen framework. Recent finite element computer predictions of meniscal function show that large radial stresses and strains are indeed developed in the central region of the tissue that are perhaps capable of producing longitudinal splits similar to those common in bucket handle tears [86, 101, 102]. In cross-section these tie fibers appear to be more abundant in the middle and posterior sections than in the anterior sections of the meniscus. This arrangement may be important for meniscal function, since load transmission studies [4] have shown that most loads of tibiofemoral articulation

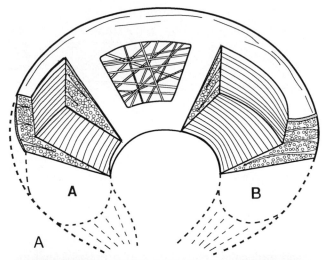

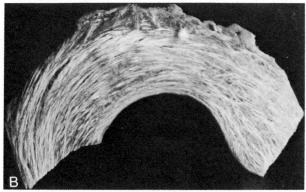

FIGURE 2B–1

The collagen fibril organization in the meniscus. *A,* Diagram of collagen fiber architecture throughout the meniscus. Collagen fibers of the thin superficial sheet are randomly distributed in the plane of the surface and are predominantly arranged in a circumferential fashion deep in the substance of the tissue. A indicates the anterior insertion of the meniscus into the tibia and B indicates the posterior insertion. (Adapted from Bullough, P. G., Munuera, L., Murphy, J., et al. *J Bone Joint Surg.* 52B:564–570, 1970.)

B, Macrophotograph of bovine medial meniscus with the surface layer removed showing the large circumferentially arranged collagen bundles of the deep zone. (From Proctor, C. S., Schmidt, M. B., Whipple, R. R., Kelly, M. A., and Mow, V. C. Material properties of the normal medial bovine meniscus. *J Orthop Res* 7:771–782, 1989.)

are transmitted through the middle and posterior portions of the meniscus.

MECHANICAL PROPERTIES

Mechanical functions of the meniscus include distributing loads over a broad area of articular cartilage, absorbing shock during dynamic loading, improving joint stability and, possibly, participating in joint lubrication [4, 14, 15, 27, 34, 38, 50, 53, 57, 58, 60–76, 78, 90, 91, 93, 94, 96, 97, 106–108]. These functions depend on the solid phase of the meniscal matrix (primarily the two major matrix

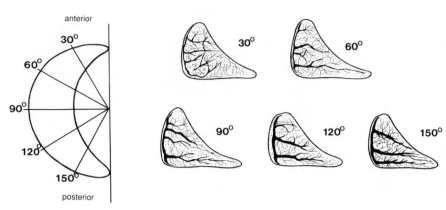

FIGURE 2B–2
Radial collagen fiber bundles of the meniscus. Radial tie fibers consisting of branching bundles of collagen fibrils extend from the periphery of the meniscus to the inner rim in every radial section through the meniscus. They are more abundant in the posterior sections and gradually diminish as the sections progress toward the anterior region of the meniscus. (From Kelly, M. A., Fithian, D. C., Chern, K. Y., and Mow, V. C. Structure and function of meniscus: Basic and clinical implications. *In* Mow, V. C., Ratcliffe, A., and Woo, S. L.-Y. (Eds.), Biomechanics of Diarthrodial Joints, Vol. 1. New York, Springer-Verlag, 1990, pp. 191–211.)

macromolecules, collagens and proteoglycans) and on the tissue fluid. In biomechanical terms, the collagen network and proteoglycans form a cohesive porous-permeable solid matrix [75, 90]. The interaction of the tissue fluid with the macromolecular solid matrix and, in particular, the flow of water through the framework, make important contributions to the mechanical properties of the tissue.

Figure 2B–3 shows the load-carrying mechanisms for tendons and ligaments, articular cartilage, and meniscus. These three tissues have strong viscoelastic tendencies manifested by creep and stress relaxation behavior in response to loading and deformation. Creep is the increasing deformation that occurs with time when a viscoelastic material or structure is subjected to a constant load, whereas stress relaxation is the decreasing stress that occurs with time when a viscoelastic material or structure is subject to a constant deformation. How creep and stress

relaxation occur depends on the viscoelastic properties of the macromolecules and their organization within the individual tissues.

Tendons and ligaments are linear fibrous structures that transmit loads along the length of their fibers. Type I collagen fibers predominate in these materials, and they form large undulating fiber bundles with high levels of tensile stiffness and strength [22, 36, 115]. The viscoelastic behavior of tendons and ligaments results from the uncoiling of the collagen fibrils within the substance of these tissues [36, 115] and has significant implications for warm-up and stretching exercises before participation in sports. When tendons and ligaments are loaded in tension, they will creep, thus loosening the joint, and when tendons and ligaments are held at a constant stretch, less load will be required to maintain the stretched position with time, thus resulting in a less stiff joint.

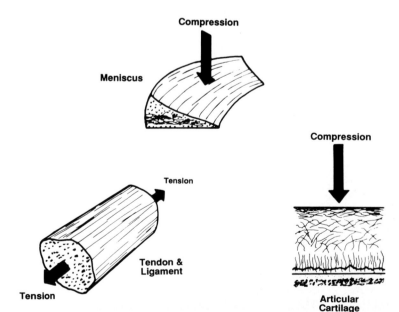

FIGURE 2B–3
Direction of primary loading for three major types of connective tissues (tendons and ligaments, articular cartilage, and meniscus). (From Mow, V. C., Fithian, D. C., and Kelly, M. A. Fundamentals of articular cartilage and meniscus biomechanics. *In* Ewing, J. W. (Ed.), *Articular Cartilage and Knee Joint Function: Basic Science and Arthroscopy.* New York, Raven Press, 1990, pp. 1–78.)

The loading pattern on articular cartilage is complex, both temporally and spatially. Diarthrodial joints are subject to very high stresses, and these stresses are subject to fast fluctuations during gait or rapid sports movements [4, 14, 15, 38, 45, 58, 85, 106, 107]. Thus, the direction of loading and the distribution or pattern of stresses and strains are highly variable within articular cartilage [75, 116]. However, there are specific structure-function relationships: (1) at the articular surface, large tensile stresses are produced, and hence there is a strong collagen-rich surface zone; (2) in the middle zone, tensile stresses may be produced in any direction, depending on the loading pattern and motion; hence, a randomly distributed collagen network in this zone would seem to function best; and (3) in the deep zone, tensile and shear stresses are produced, and hence, theoretically, a radial fiber orientation would be optimal [74, 116].

High compressive stresses are also developed within articular cartilage due to joint loading. They are resisted by both physicochemical and mechanical forces. A large Donnan osmotic swelling pressure exists, arising from the counter ion concentration associated with the negatively charged groups on the proteoglycan molecules [28, 63, 64]. This accounts for 30% to 50% of the total equilibrium compressive stiffness of cartilage [59, 63, 64]. The stiffness of the collagen-proteoglycan matrix (i.e., without ionic effects) accounts for the remainder of the equilibrium compressive stiffness [74, 75].

Articular cartilage also exhibits pronounced viscoelastic effects (see Section A in this chapter). This viscoelasticity is mainly due to the high frictional drag exerted on the microporous matrix by interstitial fluid flow [74, 75]. Indeed, very high interstitial pressures are required to force fluid to flow through the permeable solid matrix. When cartilage is suddenly loaded, the interstitial hydraulic pressure is largely responsible for its initial compressive stiffness. With time, creep occurs as the interstitial fluid is forced to flow away from the high pressure regions and exudes from the tissue. This process slowly relieves the high hydraulic pressure, and the load acting on the tissue is slowly transferred to the solid matrix. Load sharing (ratio of loads carried between the fluid pressure and the solid matrix stress) during normal physiologic function is estimated to be 22:1. Disturbances of the cartilage matrix, like cartilage degeneration, increase permeability and decrease compressivity. These changes compromise the fluid pressure mechanism of load carriage and the biphasic viscoelastic behavior (creep and stress relaxation). It is important to emphasize that in articular cartilage viscoelasticity is mainly due to interstitial fluid flow, whereas in tendons and ligaments viscoelasticity is mainly due to movement of matrix macromolecules, primarily fibrillar collagen.

The material properties of meniscal tissue differ from those of tendons and ligaments and from those of articular cartilage [35, 75, 116]. Figure 2B–3 shows that the meniscus is loaded in compression in a direction perpendicular to the predominant collagen fiber direction. However, because of the triangular shape of the meniscal tissue and its location along the periphery of the tibiofemoral articulation, the compressive force tends to extrude the meniscus outward toward the joint margins. A high tensile stress must be developed in the circumferential collagen fibers of the tissue to resist this extrusion effect [97]. This circumferential tensile stress is often referred to as a hoop stress (a term derived from the hoops of a barrel). Thus, the geometric configuration of the meniscus and its nearly frictionless articulations with the femoral and tibial surfaces provide an efficient mechanism for converting the compressive loadings in the knee into tensile loads running parallel to the circumferentially arranged, strong collagen fibers. In this sense, the meniscus behaves like ligament.

Recent studies of the tensile stiffness and strength of bovine and human menisci have shown that meniscal tissue is anisotropic and inhomogeneous [35, 90, 112, 113]. Figure 2B–4A shows that, for the bovine medial meniscus, the posterior specimens are significantly stiffer in tension than the anterior specimens except at the surface [90, 112, 113]. Figure 2B–4B shows that, for both posterior and anterior specimens, the circumferential specimens are stiffer in tension than the radial specimens except at the surface [90, 112, 113]. The surface of meniscal tissue is more than five times stiffer than the surface zone of articular cartilage [35]. In general, the tensile stiffness of meniscal specimens harvested from the circumferential direction may be as much as 100 times greater than the stiffness of specimens obtained from the radial direction. Also, compared with articular cartilage, both posterior and circumferential meniscal deep zone tissues are more than 20 times stiffer [35, 75, 116]. A recent study of human medial and lateral menisci (Fig. 2B–5 [35]) showed that the central and posterior parts of human medial meniscus have less tensile stiffness than the anterior part of the medial meniscus or all parts of the lateral meniscus. Nevertheless, the tensile stiffness of the human meniscus is far greater than that of human articular cartilage [5]. This high tensile stiffness and strength is consistent with the ligament-like function of the meniscus.

Along with the large hoop stresses that can de-

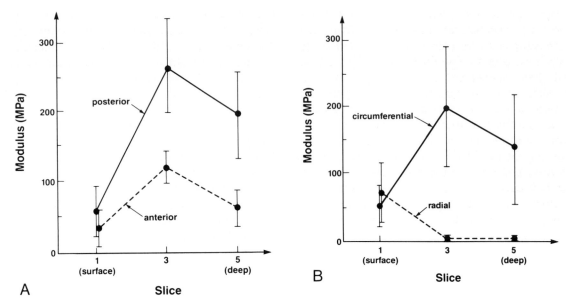

FIGURE 2B–4

Graphs showing that menisci are inhomogeneous and anisotropic in terms of their tensile stiffness. *A,* Tensile stiffness of bovine medial meniscus varies between anatomic regions (posterior and anterior) and among zones (depth). This type of variation indicates that the meniscus has inhomogeneous material properties. *B,* Tensile stiffness of bovine medial meniscus varies with direction (circumferential and radial). This type of variation indicates that the meniscus has anisotropic material properties. (*A* and *B,* from Proctor, C. S., Schmidt, M. B., Whipple, R. R., Kelly, M. A., and Mow, V. C. Material properties of the normal medial bovine meniscus. *J Orthop Res* 7:771–782, 1989.)

velop along the circumferential fibers as a result of the extrusive forces on the meniscus, large radial stresses and strains may also develop deep within the substance of the tissue. If they develop, they are resisted only by the relatively weak radial tie fibers [90, 99]. Indeed, recent finite element models have predicted the existence of large radial stresses and strains deep within the midsection of the meniscus [101, 102]. The location and pattern of the high radial stresses and strains suggest that longitudinal lesions similar to bucket handle tears of the meniscus are caused by these mechanical forces.

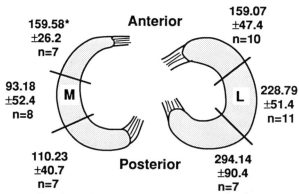

FIGURE 2B–5

Variation in circumferential tensile stiffness of human medial and lateral menisci. Note that the medial central and medial posterior menisci are weakest as measured by tensile stiffness. (From Fithian, D. C., Kelly, M. A., and Mow, V. C. Material properties and structure-function relationships in the menisci. *Clin Orthop Rel Res* 252:19–31, 1990.)

In compression, meniscus closely resembles articular cartilage except that the high Donnan osmotic swelling pressure component is not present in meniscus owing to its low proteoglycan concentration. Nevertheless, the collagen-proteoglycan framework does form a porous-permeable solid matrix. Like articular cartilage, it exhibits pronounced biphasic viscoelastic effects in compression (i.e., its viscoelasticity is mainly due to the high frictional drag exerted on the microporous collagen-proteoglycan matrix by interstitial fluid flow) [75, 90]. Indeed, even higher interstitial pressures are required to force fluid to flow through the porous-permeable meniscal solid matrix because its permeability is one-sixth that of articular cartilage [75, 90, 116]. This means that the frictional drag of fluid flow through meniscus is six times that of articular cartilage. Also, meniscal tissue has a compressive stiffness one-half that of articular cartilage [75, 90, 116]. Thus, when the soft meniscal tissue is loaded in compression, it can be deformed easily, forcing interstitial fluid flow. By virtue of its high frictional drag, a large amount of energy (per unit volume of tissue) is dissipated when the tissue is compressed. By virtue of its mass within the knee, the meniscus provides an excellent energy absorption mechanism for the joint, damping out shocks experienced by the knee from normal daily activities and sports [106]. These properties of meniscal tissue may help protect the cartilage and subchondral bone from damage resulting from excessive physiologic loads.

The meniscus also has unique shear properties owing to its ultrastructural construction. The large type I collagen fiber bundles are held together by the radial tie sheaths very much like timber stacks in lumber yards. When sheared in planes containing the circumferential collagen fibers (Fig. 2B–6 [inset]), resistance to shear is provided only by the sparse radial ties (as shown). Consequently, the shear modulus of the meniscus is very low. In addition, Figure 2B–6 shows that the shear modulus decreases with increasing shear strain, a phenomenon not previously found in connective tissues. This characteristic is important because it means that meniscal tissues can be easily shaped to conform to the anatomic forms of the mating femoral and tibial articulating surfaces in the axial plane. This may be an important consideration in selecting meniscal allografts.

The studies summarized above show that the function and tensile properties of meniscus are very similar to those of tendons and ligaments, and its function and material properties in compression are very similar to those of articular cartilage. Figure 2B–7 provides a summary of the mechanical properties of articular cartilage, meniscus, and ligament in tension, compression, and shear, showing the similarities and differences among these tissues. This figure also shows that meniscal shear properties span a broad range and occupy a unique place in the spectrum of connective tissue mechanical properties.

BLOOD SUPPLY

Branches from the geniculate arteries form a capillary plexus along the peripheral borders of the

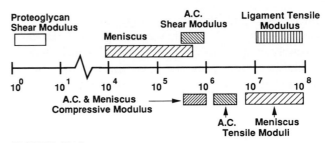

FIGURE 2B–7
Comparison of the material properties of articular cartilage, meniscus, and ligament in terms of tension, compression, and shear. Note that meniscus differs from both articular cartilage and ligament.

menisci [6, 7, 9, 65]. Small radial branches project from these circumferential parameniscal vessels into the meniscal substance. During the fetal period, blood vessels have been observed throughout the meniscus, the greatest number occurring in the peripheral one-third of the tissue [24]. From birth through adolescence the densities of meniscal cells and blood vessels decrease. In the adult, the penetrating vessels extend into only 10% to 30% of the medial meniscus and 10% to 25% of the lateral meniscus, leaving the more central regions without blood vessels.

NERVE SUPPLY

Nerves enter the joint capsule, the knee ligaments, and the periphery of the menisci and the meniscal horns. They do not enter the central regions of the menisci [9, 83, 114, 118]. As with other intra-articular connective tissues, the functions of nerve endings in these tissues have not been clearly defined, but some authors have suggested that meniscal nerves contribute to joint proprioception [83, 118].

INJURY AND REPAIR

Traumatic meniscal tears occur frequently in young, active people. A sudden change in direction while running, forceful squatting, twisting the knee, or external forces applied to the knee such as rotation, varus, valgus, or hyperextension subject the meniscus to tension, compression, and shear. Tension, compression, or shear forces that exceed the strength of the meniscal matrix in any direction (i.e., circumferential or radial) tear the tissue. Acute traumatic injuries of normal meniscal substance usually produce longitudinal or transverse tears, although the morphology of tears can be quite com-

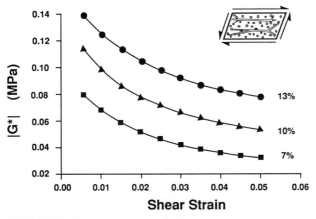

FIGURE 2B–6
Graph showing that the shear stiffness of meniscal specimens containing circumferential collagen fibers (inset) decreases with increasing shear strain. This is known as a shear-softening effect; it enhances the ability of the tissue to conform to its mating articulating surfaces. Note that shear stiffness increases with increasing compressive strain (7%, 10%, 13%). This behavior promotes knee joint stability.

plex [86], and the morphology of tears due to overloading of normal meniscal tissue depends strongly on the direction and rate of stretch [90].

Unlike acute traumatic tears through apparently normal meniscal tissue, degenerative meniscal tears occur in association with age-related degenerative changes in the tissue. These degenerative tears are most common in individuals over 40 years of age. Often these people do not recall a specific injury, or they recall only a minor load applied to the knee. Degenerative tears often have complex shapes or may appear as horizontal clefts or flaps as though they were produced by shear failure (see Fig. 2B–6 [inset]). Multiple degenerative tears often occur within the same meniscus. These features of degenerative meniscal tears suggest that, at least sometimes, they result more from age-related changes in the collagen-proteoglycan solid matrix than from specific acute trauma.

The response of meniscal tissue to tears depends on whether the tear occurs through a vascular or an avascular portion of the meniscus [7, 9]. The vascular regions respond to injury like other vascularized dense fibrous tissues. The tissue damage initiates a sequence of cellular and vascular events recognized as inflammation, repair, and remodeling [20]. This sequence of events can result in healing of a meniscal injury and restoration of the tissue structure and function. The avascular regions of meniscal tissue, like articular cartilage, cannot repair significant tissue defects.

Repair in Vascular Regions of the Meniscus

When a tear occurs through the vascular regions of the meniscus, the damage to blood vessels causes hemorrhage and fibrin clot formation. Injury to meniscal cells and clot formation activate inflammation. Platelets within the clot and inflammatory cells release mediators that stimulate cell migration, proliferation, and differentiation [20]. The nature of the response of meniscal cells to injury and their contribution to repair of vascularized regions of the meniscus are uncertain.

As in other vascularized dense fibrous tissues such as tendon and ligament, mesenchymal cells and vascular buds invade the fibrin clot. If the injury site is sufficiently stable to allow repair, granulation tissue replaces the fibrin clot, and vessels soon cross the site of the defect. Repeated loading and motion early during repair may prevent successful healing by disrupting the immature repair tissue. Successful repair results in replacement of damaged or lost

tissue with new tissue consisting of a high concentration of fibroblasts and small blood vessels surrounded by a newly formed, poorly organized matrix.

Following successful repair, the newly formed tissue begins to remodel [20]. Cell density and vascularity decline, excess tissue is resorbed, and the collagen fibrils at the injury site assume a higher degree of orientation. Loading of the meniscus presumably influences remodeling of meniscal repair tissue in the same manner that it may influence collagen fibril organization in immature tissue [24], but the relationships between loading and remodeling have not been defined. Clinical and experimental studies show that repair and subsequent remodeling of the repair tissue in the vascular regions of the meniscus can restore the structural integrity of the tissue and presumably at least a significant portion of its material properties [9, 49, 55, 104]. However, very little basic scientific information exists on the composition, remodeling, structure, and material properties of meniscal repair tissue.

Meniscal Regeneration

Resection of a portion of the meniscus through the peripheral vascularized region or resection of the entire meniscus initiates production of repair tissue that can extend from the remaining peripheral tissue into the joint [29, 30, 54, 56, 69–71, 100]. Although the repair cells usually fail to produce tissue identical in composition and structure to normal meniscal tissue, many authors have referred to this phenomenon as meniscal regeneration [29, 54, 56]. Although some repaired menisci grossly resemble normal menisci, it is not certain whether or not they are capable of normal meniscal function, and the mechanical properties of "regenerated" meniscal tissue have not been studied.

Surgeons have reported meniscal regeneration in many clinical situations and investigators have examined the tissue produced by meniscal regeneration in animals, but the mechanisms and conditions that promote this type of repair and the functional importance of meniscal regeneration remain poorly understood. It can occur repeatedly in the same knee [30] and occasionally occurs following total knee replacement [29]. In rabbits, meniscal regeneration occurs more frequently on the medial side of the knee than on the lateral side, and development of degenerative changes in articular cartilage following meniscectomy is inversely correlated with the extent of formation of meniscal repair tissue

[69–71]. Synovectomy appears to prevent meniscal regeneration, suggesting that synovial cells contribute to formation of meniscal repair tissue [54], but the predictability, frequency, and factors responsible for meniscal regeneration remain unknown.

Repair in Avascular Portions of the Meniscus

The response of meniscal tissue to tears in the avascular portion resembles the response of articular cartilage to lacerations in many respects [20]. Experimental studies show that a penetrating injury to the avascular region of the meniscus does not cause an apparent repair or inflammatory reaction (see Section A in this chapter). Cells in the region of the injury, like chondrocytes in the region of an injury limited to the articular cartilage, may proliferate and synthesize new matrix, but there is no evidence that they can migrate into the site of the defect or produce new matrix that can fill it.

Improving Repair in the Avascular Portion of the Meniscus

Because of the ineffective repair response of meniscal cells in the avascular region of the meniscus, investigators have developed several methods to stimulate repair. The more promising approaches include creation of a vascular access channel to the injury site and stimulation of cell migration to the avascular region using implantation of a fibrin clot, an artificial matrix, or growth factors [9, 49].

Attempts to extend vascular channels into the avascular zone of the meniscus have shown some promise. Creation of full-thickness channels from the periphery of the meniscus to experimentally created longitudinal bucket handle-like lesions in the avascular portion of the meniscus stimulated healing. Blood vessels from the peripheral meniscal tissue and presumably mesenchymal cells migrated through the channel and healed the meniscal lesion. This approach also healed lesions in the avascular portions of sheep and rabbit menisci, but it may compromise the biomechanical function of the meniscus by destroying the integrity of the peripheral meniscal rim and circumferential collagen fiber bundles. To avoid this problem, some investigators have used trephines to create vascular tunnels that connect the vascular peripheral portion of the meniscus with a lesion in the avascular region [9].

The synovium can provide a source of vessels and cells for meniscal healing. Suturing a flap of vascular synovial tissue into a longitudinal lesion in the avascular portion of the meniscus brings a blood supply and new cells to the injury site [9]. Abrasion of synovium to stimulate proliferation of the synovial fringe into the meniscus also allows blood vessels to enter the avascular regions. Although the results of these attempts to increase the blood supply appear promising, the quality of the repair tissue, its biomechanical properties, and the long-term results of these methods have not been evaluated.

Investigators have also attempted to stimulate meniscal repair without participation of blood vessels. This approach assumes that either the meniscal cells can repair defects in the avascular meniscal regions if they are appropriately stimulated, or that mesenchymal cells from the vascular region can migrate into the avascular region. Several experimental studies support these assumptions. They have shown that cultured meniscal fibrochondrocytes can proliferate and synthesize matrix when exposed to chemotactic and mitogenic factors found in wound hematomas [109, 110].

Methods that might stimulate meniscal repair without participation of blood vessels include implantation of a fibrin clot that presumably contains platelet-derived growth factor and possibly other growth factors that stimulate mesenchymal cell migration, proliferation, and matrix synthesis. A fibrin clot would also act as a scaffolding for the migration of cells and could serve as a vehicle for implantation of specific growth factors. An experimental study showed that implanting an exogenous fibrin clot in defects in the avascular regions of dog menisci stimulated proliferation of fibrous connective tissue that eventually assumed the appearance of fibrocartilaginous tissue, although it differed from normal meniscal tissue histologically and grossly [8]. The source of the repair cells was not identified, but they may have arisen from adjacent synovium and from meniscal tissue. Clinical experience with injection of fibrin clots into meniscal defects also suggests that these clots stimulate repair. In one study, 92% of isolated meniscal tears treated with fibrin clots healed, compared with 59% of isolated meniscal tears treated without fibrin clots [49]. Implantation of a synthetic matrix [103], possibly containing growth factors, might also stimulate repair.

MENISCAL GRAFTS

To prevent the degeneration of the articular cartilage that may result from loss of a meniscus [14, 34, 57, 70, 72, 73, 95], many surgeons currently attempt to preserve or repair menisci when possible.

Because of the limitations of intrinsic meniscal repair in the avascular portions of the meniscus and because some meniscal injuries or degenerative diseases prevent successful repair, even in the vascularized regions of the meniscus, surgeons cannot always preserve or repair damaged menisci. In these circumstances, partial or total meniscectomy may be necessary. To protect knee joint cartilage from degeneration after these procedures, investigators and surgeons have been exploring several approaches to meniscal tissue replacement.

Meniscal Allografts

Experimental studies have shown that allograft menisci will heal to host tissues. Studies of dog meniscal allografts [10, 11] showed that cryopreservation and short-term storage did not alter the morphology or mechanical properties of the menisci, but only about 10% of the meniscal cells remained metabolically active, meniscal cell synthetic activity decreased to less than 50% of normal, and total metabolic activity of the tissue declined with increasing storage time. Transplanted cryopreserved dog allograft menisci did not cause apparent rejection reactions, healed to the host tissues, and appeared to function normally after transplantation [11, 12]. Transplantation of lyophilized and deep frozen menisci in sheep also showed that grafts can heal to the recipient site tissues [111].

Experimental studies have shown that meniscal grafts remodel and may help to decrease the probability of degenerative joint disease following removal of the original meniscus. With time, the cell density of dog meniscal grafts increased; 3 months after transplantation they had a cell numerical density and level of cell metabolic activity similar to those of normal menisci. Six months following transplantation small blood vessels penetrated the grafts to about one-third of their width from the periphery. The articular cartilage underlying the menisci remained intact 6 months following transplantation, but the exposed tibial cartilage had fissures and degenerative changes. These changes appeared less severe than those found in dog knees subjected to total meniscectomy [25], suggesting that the allografts provide some protection for the articular cartilage. The long-term results of experimental meniscal allografts have not been reported, so the efficacy of allografts for decreasing probability or severity of degenerative joint disease remains uncertain [79].

Limited clinical studies confirm that meniscal allografts heal with host tissues [111, 119]. One group

of investigators used fresh meniscal allografts [119]. They found that "most" of the grafts appeared structurally sound and functional at follow-up evaluation and reported survival of grafts for as long as 8½ years, although no biomechanical tests were performed to assess their mechanical properties. Another group used lyophilized and deep frozen allografts [111]. They reported that operative complications and rejection reactions did not occur, but the grafts decreased in size. Although these clinical results show the potential for allograft meniscal replacement, other studies are needed to determine whether meniscal allografts decrease the frequency and severity of degenerative joint disease after loss of menisci.

Synthetic Matrix Meniscal Grafts

Synthetic matrices, created from reconstituted collagen, fibrin, or other materials and shaped to fit specific meniscal defects, can replace lost or damaged meniscal tissue [79, 103]. Recipient site cells and blood vessels might grow into these synthetic matrices and remodel them to resemble meniscal tissue. Initial experimental investigations suggest that synthetic collagen matrices may have the potential to replace menisci [79, 103]. These synthetic matrices could also serve as vehicles for implantation of growth factors that promote host cell migration, proliferation, and differentiation in the graft or for implantation of cultured mesenchymal cells that could synthesize a new matrix.

CONCLUSIONS

The meniscus is a specialized intra-articular fibrocartilaginous structure of the knee. By virtue of its composition, ultrastructure, and mechanical properties, it functions as a load-bearing and shock-absorbing structure. Along with the knee ligaments, the menisci help stabilize normal knees, and they may also assist in lubrication of the knee. Unlike ticular cartilage, some regions of adult meniscus have blood vessels and a nerve supply, especially the meniscal horns. Loss of one or both menisci alters the loading of articular cartilage and has been shown to increase the probability and severity of degenerative joint disease. Experimental and clinical studies show that tears through the vascular regions of the meniscus can heal, but tears through the avascular regions do not undergo a significant repair process. The vascular regions respond to injury like other vascularized dense fibrous tissues.

The avascular regions of the meniscus, like articular cartilage, cannot repair significant tissue defects. Clinical and experimental studies show that repair and subsequent remodeling of the repair tissue in the vascular regions of the meniscus can restore the structural integrity of the tissue. To prevent the degeneration of articular cartilage that may result from loss of a meniscus, surgeons have attempted to preserve or repair the menisci when possible. Initial studies of meniscal allografts show that these grafts can heal to the host tissues and may help restore meniscal function. Synthetic matrices, created from reconstituted collagen, fibrin, and other materials and shaped to fit specific meniscal defects, also have shown promise as a method of replacing severely damaged or absent menisci. The addition of growth factors or mesenchymal cells to these synthetic matrices may produce even better results.

References

1. Adams, M. E., and Muir, H. The glycosaminoglycans of canine menisci. *J Biochem* 197:385–389, 1981.
2. Adams, M. E., McDevitt, C. A., Ho, A., and Muir, H. Isolation and characterization of high-buoyant-density proteoglycans from semilunar menisci. *J Bone Joint Surg* 68-A:55–64, 1986.
3. Adams, M. E., and Hukins, D. W. L. The extracellular matrix of the meniscus. *In* Mow, V. C., Arnoczky, S. P. and Jackson, D. W. (Eds.), *Knee Meniscus—Basic and Clinical Foundations.* New York, Raven Press, 1991.
4. Ahmed, A. M., Burke, D. L., and Yu, A. In-vitro measurement of static pressure distribution in synovial joints—part I. Tibial surface of the knee. *J Biomech Eng* 105:216–225, 1983.
5. Akizuki, S., Mow, V. C., Muller, F., Pita, J. C., Howell, D. S., and Manicourt, D. H. The tensile properties of human knee joint cartilage I: Influence of ionic conditions, weight bearing, and fibrillation on the tensile modulus. *J Orthop Res* 4:379–392, 1986.
6. Arnoczky, S. P., and Warren, R. F. Microvasculature of the human meniscus. *Am J Sports Med* 10:90–95, 1982.
7. Arnoczky, S. P., and Warren, R. F. The microvasculature of the meniscus and its response to injury: An experimental study in the dog. *Am J Sports Med* 11:131–141, 1983.
8. Arnoczky, S. P., McDevitt, C. A., Warren, R. F., Spivak, J., and Allen, A. Meniscal repair using and exogenous fibrin clot: An experimental study in the dog. *Trans Orthop Res Soc* 11:452, 1986.
9. Arnoczky, S. P., Adams, M., DeHaven, K., Eyre, D. R., and Mow, V. C. Meniscus. *In* Woo, S. L.-Y., and Buckwalter, J. A. (Eds.), *Injury and Repair of the Musculoskeletal Soft Tissues.* Park Ridge, IL, American Academy of Orthopaedic Surgeons, 1988, pp. 487–537.
10. Arnoczky, S. P., McDevitt, C. A., Schmidt, M. B., Mow, V. C., and Warren, R. F. The effect of cryopreservation on canine menisci: A biochemical, morphologic, and biomechanical evaluation. *J Orthop Res* 6:1–12, 1988.
11. Arnoczky, S. P., and Milachowski, K. A. Meniscal allografts: Where do we stand? *In* Ewing, J. W. (Ed.), *Articular Cartilage and Knee Joint Function: Basic Science and Arthroscopy.* New York, Raven Press, 1990, pp. 129–136.
12. Arnoczky, S. P., Warren, R. F., and McDevitt, C. A. Meniscal replacement using a cryopreserved allograft. *Clin Orthop Rel Res* 252:121–128, 1990.
13. Aspden, R. M., Yarker, Y. E., and Hukins, D. W. L. Collagen orientations in the meniscus of the knee joint. *J Anat* 140:371–380, 1985.
14. Bourne, R. B., Finaly, J. B., Papadopoulos, P., and Andreae, P. The effect of medial meniscectomy on strain distribution in the proximal part of the tibia. *J Bone Joint Surg* 66-A:1431–1437, 1984.
15. Brown, T. D., and Shaw, D. T. In vitro contact stress distribution on the femoral condyles. *J Orthop Res* 2:190–199, 1984.
16. Buckwalter, J. A., and Cooper, R. R. The cells and matrices of skeletal connective tissues. *In* Albright, J. A., and Brand, R. A. (Eds.), *The Scientific Basis of Orthopaedics* (2nd ed.). Norwalk, CT, Appleton & Lange, 1987, pp. 1–30.
17. Buckwalter, J. A., Maynard, J. A., and Vailas, A. C. Skeletal fibrous tissues: Tendon, joint capsule, and ligament. *In* Albright, J. A., and Brand, R. A. (Eds.), *The Scientific Basis of Orthopaedics* (2nd ed.). Norwalk, CT, Appleton & Lange, 1987, pp. 387–406.
18. Buckwalter, J. A., and Rosenberg, L. C. Electron microscopic studies of cartilage proteoglycans. *Elec Microsc Rev* 1:87–112, 1988.
19. Buckwalter, J. A. Cartilage. *In* Dulbecco, R. (Ed.), *Encyclopedia of Human Biology,* Vol. 2. San Diego, Academic Press, 1991, pp. 201–215.
20. Buckwalter, J. A., and Cruess, R. L. Healing of the musculoskeletal tissues. *In* Rockwood, C. A., and Green, D. (Eds.), *Fractures.* Philadelphia, J. B. Lippincott, in press, 1991.
21. Bullough, P. G., Munuera, L., Murphy, J., and Weinstein, A. M. The strength of the menisci of the knee as it relates to their fine structure. *J Bone Joint Surg* 52-B:564–570, 1970.
22. Butler, D. L., Kay, M. D., and Stouffer, D. C. Comparison of material properties in fascicle-bone units from human patellar tendon and knee ligaments. *J Biomech* 19:425–432, 1986.
23. Cheung, H. S. Distribution of type I, II, III and V in the pepsin solubilized collagen in bovine menisci. *Connect Tissue Res* 16:343–356, 1987.
24. Clark, C. R., and Ogden, J. A. Development of the human knee joint. *J Bone Joint Surg* 65-A:538–547, 1983.
25. Cox, J. S., Nye, C. E., Schaefer, W. W., and Woodstein, I. J. The degenerative effects of partial and total resection of the medial meniscus in dog's knees. *Clin Orthop Rel Res* 109:178–183, 1975.
26. DeBont, L. G. M. Temporomandibular joint. Articular cartilage structure and function. Ph.D Thesis, Department of Oral and Maxillofacial Surgery, University Hospital, Groningen, The Netherlands, 1985, pp. 1–82.
27. DeHaven, K. E. The role of the meniscus. *In* Ewing, J. W. (Ed.), *Articular Cartilage and Knee Joint Function: Basic Science and Arthroscopy.* New York, Raven Press, 1990, pp. 103–116.
28. Donnan, F. G. The theory of membrane equilibria. *Chem Rev* 1:73–90, 1924.
29. Espley, A. J., and Waugh, W. Regeneration of menisci after total knee replacement: A report of five cases. *J Bone Joint Surg* 63-B:387–390, 1981.
30. Evans, D. K. Repeated regeneration of a meniscus in the knee. *J Bone Joint Surg* 45-B:748–749, 1963.
31. Ewing, J. W. Arthroscopic treatment of degenerative meniscal lesions and early degeneration of arthritis of the knee. *In* Ewing, J. W. (Ed.), *Articular Cartilage and Knee Joint Function: Basic Science and Arthroscopy.* New York, Raven Press, 1990, pp. 137–146.
32. Eyre, D. R., and Muir, H. The distribution of different molecular species of collagen in fibrous, elastic and hyaline cartilages of the pig. *Biochem J* 151:595–601, 1975.
33. Eyre, D. R., and Wu, J. J. Collagen of fibrocartilage: A distinctive molecular phenotype in bovine meniscus. *FEBS Lett* 158:265–270, 1983.
34. Fairbank, T. J. Knee joint changes after meniscectomy. *J Bone Joint Surg* 30-B:664–670, 1948.

35. Fithian, D. C., Kelly, M. A., and Mow, V. C. Material properties and structure-function relationships in the menisci. *Clin Orthop Rel Res* 252:19–31, 1990.

36. Frank, C., Woo, S. L.-Y., Andriacchi, T. P., Brand, R., Oakes, B., Dahners, L., DeHaven, K., Lewis, J., and Sabiston, P. Normal ligament: Structure, function and composition. *In* Woo, S. L.Y., and Buckwalter, J. A. (Eds.), *Injury and Repair of the Musculoskeletal Soft Tissues*. Park Ridge, IL, American Academy of Orthopaedic Surgeons, 1988, pp. 45–100.

37. Frazier, W. A. Thrombospondin: A modular adhesive glycoprotein of platelets and nucleated cells. *J Cell Biol* 105:625–632, 1987.

38. Fukubayashi, T., and Kurosawa, H. The contact area and pressure distribution pattern of the knee. *Acta Orthop Scand* 51:871–879, 1980.

39. Furcht, L. T. Structure and function of the adhesive glycoprotein fibronectin. *Mod Cell Biol* 95:369–377, 1982.

40. Ghadially, F. N., Thomas, I., Yong, N. K., and LaLonde, J. M. A. Ultrastructure of rabbit semilunar cartilages. *J Anat* 125:499–517, 1978.

41. Ghosh, P., and Taylor, T. K. F. The knee joint meniscus. A fibrocartilage of some distinction. *Clin Orthop Rel Res* 224:52–63, 1987.

42. Gilquist, J., and Oretorp, N. Arthroscopic partial meniscectomy. *Clin Orthop Rel Res* 167:29–33, 1982.

43. Goodfellow, J. He who hesitates is saved. *J Bone Joint Surg* 62-B:1, 1982.

44. Goodsir, J. Mechanism of the knee joint. *In Anatomical Memoirs of John Goodsir*, Vol. 2. Edinburgh, Adam and Charles Black, 1858, pp. 220–231.

45. Greenwald, A. S., and O'Connor, J. J. The transmission of load through the human hip joint. *J Biomech* 4:507–528, 1971.

46. Guilak, F., Ratcliffe, A., and Mow, V. C. The stress-strain environment around a chondrocyte: A finite element analysis of cell-matrix interactions. Goldstein, S. A. (Ed.). New York, Adv Biong ASME 17:395–398, 1990.

47. Guilak, F., Ratcliffe, A., Hunziker, E. B., and Mow, V. C. Finite element modeling of articular cartilage chondrocytes under physiological loading conditions. *Trans Orthop Res Soc* 16:366, 1991.

48. Hardingham, T. E. The role of link protein in the structure of cartilage proteoglycan aggregates. *Biochem J* 177:237–247, 1979.

49. Henning, C. E., Lynch, M. A., Yearout, K. M., Vequist, S. W., Stallbaumer, R. J., and Decker, K. A. Arthroscopic meniscal repair using an exogenous fibrin clot. *Clin Orthop Rel Res* 252:64–72, 1990.

50. Hsieh, H. H., and Walker, P. S. Stabilizing mechanisms of the loaded and unloaded knee joint. *J Bone Joint Surg* 58-A:87–93, 1976.

51. Katz, E. P., Wachtel, E. J., and Maroudas, A. Extrafibrillar proteoglycans osmotically regulate the molecular packing of collagen in cartilage. *Biochim Biophys Acta* 882:136–139, 1986.

52. Kelly, M. A., Fithian, D. C., Chern, K. Y., and Mow, V. C. Structure and function of the meniscus: Basic and clinical implications. *In* Mow, V. C., Ratcliffe, A., and Woo, S. L-Y. (Eds.), *Biomechanics of Diarthrodial Joints*, Vol. 1. New York, Springer-Verlag, 1990, 191–211.

53. Kettelkamp, D. B., and Jacobs, A. W. Tibiofemoral contact area—determination and implications. *J Bone Joint Surg* 54-A:349–356, 1972.

54. Kim, J.-M., and Moon, M. S. Effect of synovectomy upon regeneration of meniscus in rabbits. *Clin Orthop Rel Res* 141:287–294, 1979.

55. King, D. The healing of semilunar cartilages. *J Bone Joint Surg* 18:333–342, 1936.

56. King, D. Regeneration of semilunar cartilage. *Surg Gynecol Obstet* 62:167–170, 1936.

57. Krause, W. R., Pope, M. H., Johnson, R. J., and Wilder, D. G. Mechanical changes in the knee after meniscectomy. *J Bone Joint Surg* 58-A:599–604, 1976.

58. Kurosawa, H., Fukubayashi, T., and Nakajima, H. Load-bearing mode of the knee joint. *Clin Orthop Rel Res* 144:283–290, 1980.

59. Lai, W. M., Hou, J. S., and Mow, V. C. Triphasic theory for the swelling properties of hydrated charged soft biological tissues. *In* Mow, V. C., Ratcliffe, A., and Woo, S. L-Y. (Eds.), *Biomechanics of Diarthrodial Joints*, Vol. 1. New York, Springer-Verlag, 1990, pp. 283–312.

60. Levy, I. M., Torzilli, P. A., and Warren, R. F. The effect of medial meniscectomy on anterior-posterior motion of the knee. *J Bone Joint Surg* 64-A:883–888, 1982.

61. MacConaill, M. A. The function of intra-articular fibrocartilages, with special reference to the knee and the inferior radio-ulnar joints. *J Anat* 66:210–227, 1932.

62. Maquet, P. G., Van de Berg, A. J., and Simonet, J. C. Femorotibial weight-bearing areas. *J Bone Joint Surg* 57-A:766–771, 1975.

63. Maroudas, A. Physicochemical properties of articular cartilage. *In* Freeman, M. A. R. (Ed.), *Adult Articular Cartilage*. Tunbridge Wells, Kent, England, Pitman Medical, 1979, pp. 215–290.

64. Maroudas, A., and Bannon, C. Measurement of swelling pressure in cartilage and comparison with the osmotic pressure of constituent proteoglycans. *Biorheology* 18:619–632, 1981.

65. McDevitt, C. A., and Webber, R. J. The ultrastructure and biochemistry of meniscal cartilage. *Clin Orthop Rel Res* 252:8–18, 1990.

66. McDevitt, C. A., Miller, R. R., and Spindler, K. P. The cells and cell matrix interactions. *In* Mow, V. C., Arnoczky, S. P., and Jackson, D. W. (Eds.), *Knee Meniscus—Basic and Clinical Foundations*. New York, Raven Press, 1991.

67. McNicol, D., and Roughley, P. J. Extraction and characterization of proteoglycan from human meniscus. *Biochem J* 185:705–713, 1980.

68. Mills, D. K., Daniel, J. C., and Scapino, R. Histological features and in-vitro proteoglycan synthesis in the rabbit craniomandibular joint disc. *Arch Oral Biol* 33:195–202, 1988.

69. Moon, M. S., Kim, J.-M., and OK, I. Y. The normal and regenerated meniscus in rabbits: Morphologic and histologic studies. *Clin Orthop Rel Res* 182:264–269, 1984.

70. Moon, M. S., and Chung, I. S. Degenerative changes after meniscectomy and meniscal regeneration. *Int Orthop* 12:17–19, 1988.

71. Moon, M. S., Woo, Y. K., and Kim, Y. I. Meniscal regeneration and its effects on articular cartilage in rabbit knees. *Clin Orthop Rel Res* 227:298–304, 1988.

72. Moskowitz, R. W., Davis, W., Sammarco, J., Martens, M., Baker, J., Mayor, M., Burstein, A. H., and Frankel, V. H. Experimentally induced degenerative joint lesions following partial meniscectomy in the rabbit. *Arthritis Rheum* 16:397–405, 1973.

73. Moskowitz, R. W., Howell, D. S., Goldberg, V. M., Muniz, O., and Pita, J. C. Cartilage proteoglycan alterations in an experimental model of rabbit osteoarthritis. *Arthritis Rheum* 22:155–163, 1979.

74. Mow, V. C., and Lai, W. M. Recent developments in synovial joint biomechanics. *Soc Ind Appl Math Rev* 22:275–317, 1980.

75. Mow, V. C., Holmes, M. H., and Lai, W. M. Fluid transport and mechanical properties of articular cartilage: A review. *J Biomech* 377–394, 1984.

76. Mow, V. C., and Mak, A. F. Lubrication of diarthrodial joints. *In* Skalak, R., and Chien, S. (Eds.), *Handbook of Bioengineering*, New York, McGraw-Hill, 1986, pp. 5.1–5.34.

77. Mow, V. C., Zhu, W. B., Lai, W. M., Hardingham, T. E., Hughes, C., and Muir, H. The influence of link protein stabilization on the viscometric properties of proteoglycan aggregate solutions. *Biochim Biophys Acta* 992:201–208, 1989.

78. Mow, V. C., Fithian, D. C., and Kelly, M. A. Fundamentals of articular cartilage and meniscus biomechanics. *In* Ewing,

J. W. (Ed.), *Articular Cartilage and Knee Joint Function: Basic Science and Arthroscopy.* New York, Raven Press, 1990, pp. 1–78.

79. Mow, V. C., Arnoczky, S. P., and Jackson, D. W. *Knee Meniscus—Basic and Clinical Foundations.* New York, Raven Press, 1991.

80. Myers, E. R., Zhu, W., and Mow, V. C. Viscoelastic properties of articular cartilage and meniscus. *In* Nimni, M. E. (Ed.), *Collagen: Chemistry, Biology and Biotechnology,* Vol. 2. Boca Raton, FL, CRC Press, 1988, pp. 268–288.

81. Nakano, T., and Scott, P. G. Proteoglycans of the articular disc of the bovine temporomandibular joint. I. High molecular weight chondroitin sulfate proteoglycan. *Matrix* 9:277–283, 1989.

82. Nakano, T., and Scott, P. G. A quantitative chemical study of glycosaminoglycans in the articular disc of the bovine temporomandibular joint. *Arch Oral Biol* 9:749–757, 1989.

83. O'Connor, B. L., and McConnaughey, J. S. The structure and innervation of cat knee menisci, and their relation to a "sensory hypothesis" of meniscal function. *Am J Anat* 153:431–442, 1978.

84. Parsons, F. G. The joints of mammals compared with those of man. Part II. Joints of the hind limb. *J Anat Physiol* 34:301–323, 1900.

85. Paul, J. P. Force action transmitted by joints in the human body. *Proc R Soc Lond* 192B:163–172, 1976.

86. Pavlov, H., Ghelman, B., and Vigorita, V. J. *Atlas of the Knee Menisci.* New York, Appleton-Century-Croft, 1983.

87. Pierschbacher, M. D., Hayman, E. G., and Ruoslahti, E. The cell attachment determinant in fibronectin. *J Cell Biochem* 28:115–126, 1985.

88. Peters, T. J., and Smillie, I. S. Studies on the chemical composition of the menisci of the knee joint with special reference to the horizontal cleavage lesion. *Clin Orthop Rel Res* 86:245–252, 1972.

89. Price, C. T., and Allen, W. C. Ligament repair in the knee with preservation of the meniscus. *J Bone Joint Surg* 60-A:61–65, 1978.

90. Proctor, C. S., Schmidt, M. B., Whipple, R. R., Kelly, M. A., and Mow, V. C. Material properties of the normal medial bovine meniscus. *J Orthop Res* 7:771–782, 1989.

91. Radin, E. L., DeLamotte, F., and Maquet, P. Role of the menisci in the distribution of stress in the knee. *Clin Orthop Rel Res* 185:290–293, 1984.

92. Roughley, P. J., McNicol, D., Santer, V., and Buckwalter, J. A. The presence of a cartilage-like proteoglycan in the adult human meniscus. *Biochem J* 197:77–83, 1981.

93. Seedhom, B. B. Transmission of the load in the knee joint with special reference to the role of the menisci. Part I. *Eng Med* 8:207–218, 1979.

94. Seedhom, B. B., and Hargreaves, D. J. Transmission of the load in the knee joint with special reference to the role of the menisci. Part II. *Eng Med* 8:220–228, 1979.

95. Shapiro, F., and Glimcher, M. J. Induction of osteoarthritis in the rabbit knee joint. Histologic changes following meniscectomy and meniscal lesion. *Clin Orthop Rel Res* 147:287–295, 1980.

96. Shoemaker, S. C., and Markolf, K. L. The role of the meniscus in the anterior-posterior stability of the loaded anterior cruciate-deficient knee. *J Bone Joint Surg* 68-A:71–79, 1986.

97. Shrive, N. G., O'Connor, J. J., and Goodfellow, J. W. Load bearing in the knee joint. *Clin Orthop Rel Res* 131:279–287, 1978.

98. Sisson, S. *The Anatomy of the Domestic Animal.* Philadelphia, W. B. Saunders, 1917.

99. Skaggs, D. L., and Mow, V. C. Function of radial tie fibers in the meniscus. *Trans Orthop Res Soc* 15:248, 1990.

100. Smillie, I. S. Observations on the regeneration of the semilunar cartilages in man. *Br J Surg* 31:398–401, 1944.

101. Spilker, R. L., Donzelli, P. S., and Mow, V. C. Finite element model of meniscus response to a sudden overload. *Trans Orthop Res Soc* 16:293, 1991.

102. Spilker, R. L., and Donzelli, P. S. A biphasic finite element model of the meniscus for stress-strain analysis. *In* Mow, V. C., Arnoczky, S. P., and Jackson, D. W. (Eds.), *Knee Meniscus: Basic and Clinical Foundations.* New York, Raven Press, 1991.

103. Stone, K. R., Rodkey, W. G., Webber, R. J., McKinney, L., and Steadman, J. R. Future directions: Collagen-based prostheses for meniscal regeneration. *Clin Orthop Rel Res* 252:129–135, 1990.

104. Stone, R. G., Spears, T. D., and Bean, J. W. A 2- to 9-year review of meniscal repair (open and arthroscopic repair). *In* Ewing, J. W. (Ed.), *Articular Cartilage and Knee Joint Function: Basic Science and Arthroscopy.* New York, Raven Press, 1990, pp. 117–128.

105. Torzilli, P. A. Influence of cartilage conformation on its equilibrium water partition. *J Orthop Res* 3:473–483, 1985.

106. Voloshin, A. S., and Wosk, J. Shock absorption of meniscectomised and painful knees. A comparative in vivo study. *J Biomed Eng* 5:157–161, 1983.

107. Walker, P. S., and Erkman, M. J. The role of the meniscus in force transmission across the knee. *Clin Orthop Rel Res* 109:184–192, 1975.

108. Wang, C. J., and Walker, P. S. Rotary laxity of the human knee. *J Bone Joint Surg* 56-A:161–170, 1974.

109. Webber, R. J., Harris, M. G., and Hough, A. J. Cell culture of rabbit meniscal fibrochondrocytes: Proliferative and synthetic response to growth factors and ascorbate. *J Orthop Res* 3:36–42, 1985.

110. Webber, R. J. In vitro culture of meniscal tissue. *Clin Orthop Rel Res* 252:114–120, 1990.

111. Weismeier, K. G., Milachowski, K., Wirth, C. J., and Kohn, D. Meniscus transplantation: An experimental study and clinical results. *J Bone Joint Surg* 71-B:711, 1989.

112. Whipple, R. R., Wirth, C. R., and Mow, V. C. Mechanical properties of the meniscus. Spilker, R. L. (Ed.). *Adv Bioeng* 11:32–33, 1984.

113. Whipple, R. R., Wirth, C. R., and Mow, V. C. Anisotropic and zonal variations in the tensile properties of the meniscus. *Trans Orthop Res Soc* 10:367, 1985.

114. Wilson, A. S., Legg, P. G., and McNeur, P. G. Studies on the innervation of the medial meniscus in the human knee joint. *Anat Rec* 165:485–492, 1969.

115. Woo, S. L.-Y., Gomez, M. A., and Akeson, W. H. The time dependent viscoelastic properties of canine medial collateral ligament. *J Biomech Eng* 103:293–298, 1981.

116. Woo, S. L.-Y., Mow, V. C., and Lai, W. M. Biomechanical properties of articular cartilage. *In* Skalak, R., and Chien, S. (Eds.), *Handbook of Bioengineering.* New York, McGraw-Hill, 1986, pp. 4.1–4.44.

117. Wu, J. J., Eyre, D. R., and Slayter, H. S. Type VI collagen of the intervertebral disc. Biochemical and electron microscopic characterization of the native protein. *Biochem J* 248:373–381, 1987.

118. Zimny, M. L., Albright, D. H., and Dabezies, E. J. Mechanoreceptors in the human medial meniscus. *Acta Anat* 133:35–40, 1988.

119. Zukor, D. J., Cameron, J. C., Brooks, P. J., Oakeshott, R. D., Farine, I., Rudan, J. F., and Gross, A. E. The fate of human meniscal allografts. *In* Ewing, J. W. (Ed.), *Articular Cartilage and Knee Joint Function: Basic Science and Arthroscopy.* New York, Raven Press, 1990, 147–152.

BASIC BIOMECHANICAL PRINCIPLES: FORCES, MOMENTS, AND EQUILIBRIUM

Peter A. Torzilli, Ph.D.

Engineering, physics, and mathematical principles have long been used to study the mechanics of the human body. During everyday activity the human body is subjected to a variety of mechanical loads. These loads are composed of external forces, most notably gravity and ground reaction forces, and internal forces generated by muscles, bone, and soft tissue deformation. A delicate balance between external and internal forces is necessary to provide the body with stability during movement and while at rest. The study of these forces and their effects on body movement and stability is termed *biomechanics*. Biomechanics is based on the principles of engineering mechanics, the study of forces and their effects on stability and motion. Mechanics broadly covers two engineering areas: *Statics* is the study of bodies in *equilibrium* or at rest as a result of the combination of forces acting upon the bodies, and *dynamics* is the study of the *motion* of bodies as a result of the forces acting upon them. Furthermore, in the study of the *kinematics* of body motion the forces producing the motion are ignored, and only the motion itself (translations and rotations) are of interest.

When the system of forces acting on a body is balanced, the system has no external effect on the body, that is, the body is in equilibrium or at rest, and the problem is one of statics. However, when the force system has a resultant different from zero, the body will move or be accelerated, and the problem becomes one of dynamics. Finally, when the internal effects of an external force system are considered or when the changes in shape or deformation of the body are important, the problem becomes one of mechanics of materials.

The fields of statics, dynamics, and mechanics of materials is too broad to cover adequately in this chapter. Only the basic concepts of statics will be discussed. The concepts of forces, moments, and free-body diagrams and the principles of static equilibrium will be reviewed. There are several excellent biomechanics textbooks that deal with these and other areas of biomechanics [2, 3, 7, 10, 14, 15]. The reader is referred to these publications for additional information. This chapter will deal only with the basic principles governing the analysis of problems in static equilibrium and will attempt to provide the reader with the appropriate concepts and methodology needed to analyze these types of biomechanics problems.

UNITS OF MEASURE

The basic units of mechanics are *length* (L), *mass* (M), and *time* (t). All other quantities can be represented in these basic units or dimensions. For

The material in this chapter is based in part on material developed by the author while lecturing at The Hospital for Special Surgery. The author wishes to thank The Hospital for Special Surgery for the support provided him in the development of this material.

TABLE 3–1
Basic Units of Mechanics

	CGS	MKS	United Kingdom	United States
Mass	Gram	Kilogram	Pound mass (lbm)	Slug or pound mass
Length	Centimeter	Meter	Foot	Foot
Time	Second	Second	Second	Second
Force	Dyne	Newton (N)	Poundal	Pound force (lbf)

instance, velocity units = L/t, acceleration units = L/t², *and force units = ML/t²*. Four common systems of units currently exist. These are the CGS system (gram, centimeter, second), the MKS system (kilogram, meter, second), the English system (pound, foot, second), and the U.S. system (slug, foot, second) (see Table 3–1). Units of one system are easily converted into units of another (Table 3–2). Although all four systems are currently used, the MKS system is now considered the standard system of units used in scientific publications.

FORCES

Probably the most basic unit in mechanics is force. *Force* is defined as the action of one object or body on another body. The action can be associated with direct or indirect contact, for example, magnetic, electrostatic, and gravitational actions. Force is a *vector* quantity, having both *magnitude* and *direction* (line of action). A *scaler* quantity has only magnitude, for example, length, mass, and time.

To define a force adequately four properties must be reported—its magnitude, line of action, direction or sense, and point of application (Fig. 3–1). The magnitude of the force describes the size or amount of the force, whereas the line of action defines its orientation. The manner in which the force is directed, for example, a push or pull, is indicated by its direction and the location of the point of application (where the force is applied). This is illustrated in Figure 3–2. A change in any one of these four properties will change the definition of the force.

When several forces act in a given situation, they are called a *system of forces* or a *force system*. Force systems can be classified according to the arrangement of the lines of action of the forces of the system. The forces may be coplaner (in the same plane) or noncoplaner (parallel or nonparallel), concurrent (acting at the same point) or nonconcurrent, and colinear (along the same line of action) or noncolinear. The most general system of forces is one in which the forces are noncoplaner, nonparallel, and nonconcurrent. In other words, they do not lie in a common plane, they are not all parallel to each other, and they do not all intersect at a common point.

The *resultant* of a force system is the simplest force system that can replace the original system without changing its external effect on the body. The resultant of a force system can be a single force, a single moment or torque, or a force and moment. A *resultant force* is a *single force* representing the combined effect of two or more forces (Fig. 3–3). Likewise, any single force may be replaced or resolved into two or more equivalent force vectors.

A simple method of determining the resultant force is achieved by using the *parallelogram law*, which states that the resultant of *two concurrent forces* (acting at the same point) is equal to the diagonal of the parallelogram whose sides are formed by the two forces (Fig. 3–4). Another technique of obtaining the resultant force is by using the *polygon* method, in which the forces are drawn sequentially tip to tip, tip to tip, and so on, the resultant force being the line from the original point of application of the system of forces to the final tip (Fig. 3–5).

TABLE 3–2
Unit Conversion Factors

Length	Mass	Force
1 in. = 2.54 cm	1 slug = 32.2 lbm	1 lbf = 445,000 dynes
1 ft = 30.5 cm	1 g = 2.205 × 10⁻³ lbm	1 lbf = 32.2 poundals
1 ft = 0.305 m	1 g = 0.685 × 10⁻⁴ slugs	1 lbf = 4.448 Newtons
		1 N = 10⁵ dynes
		1 N = 0.2248 lbf

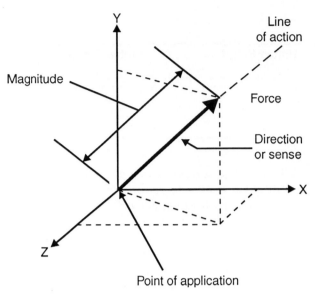

FIGURE 3–1
A force has four properties—a magnitude, line of action, direction, and point of application. When referenced to a cartesian coordinate system, the force is easily defined.

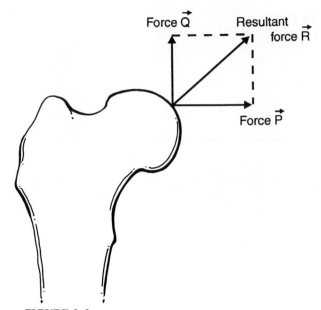

FIGURE 3–3
Resultant force R is equivalent to the two forces P and Q.

Complicated force systems can thus be resolved by using either the parallelogram method on two forces at the same time, or the polygon method. This process is illustrated in the following example.

Example: The resultant single quadriceps muscle force can be found by resolving the four individual component muscle forces (Fig. 3–6*A*). Using either

method the muscle forces (1) and (2) can be resolved into force (1,2), and forces (3) and (4) can be resolved into force (3,4). The resultant force is then easily determined as the combination of forces (1,2) and (3,4) (Fig. 3–6*B*).

One important aspect of determining the resultant force is the fact that it enables one to predict joint motion. If the resultant force of a system of forces passes through the point of joint contact between two bodies, no motion will result. This is illustrated in Figure 3–7*A*, showing the resultant muscle force passing through the point of joint contact between the humerus and the glenoid. No motion will result, assuming all forces are coplaner. However, in Figure 3–7*B* the resultant force passes medially to the joint contact point, causing an adduction motion of the humerus relative to the scapula.

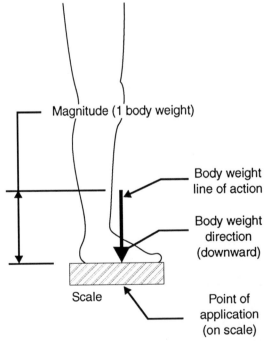

FIGURE 3–2
Example of the four properties of a force illustrated when standing on a scale.

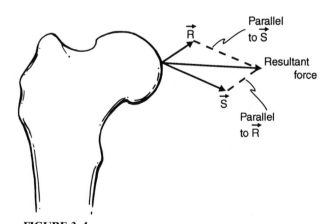

FIGURE 3–4
The resultant force R is equivalent to the two forces R and S.

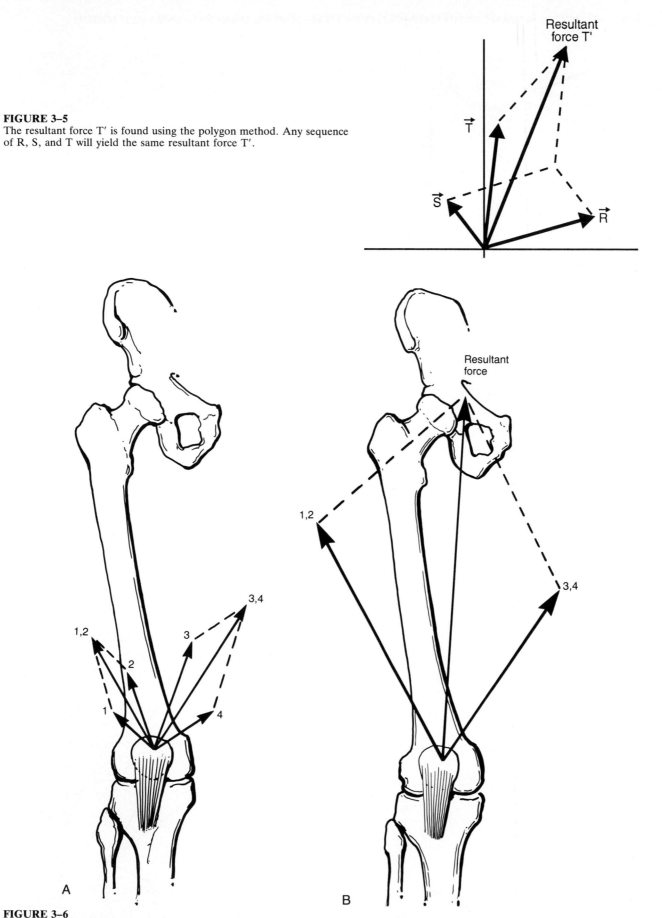

FIGURE 3–5
The resultant force T′ is found using the polygon method. Any sequence
of R, S, and T will yield the same resultant force T′.

FIGURE 3–6
A, The four individual components of the quadriceps muscle force 1, 2, 3 and 4 can be resolved into two components, 1,2 and 3,4.
B, A single resultant force (1,2,3,4) can be found by combining the two forces 1,2 and 3,4.

125

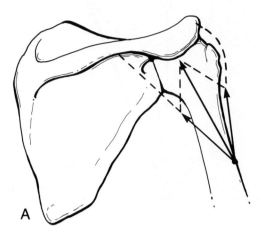

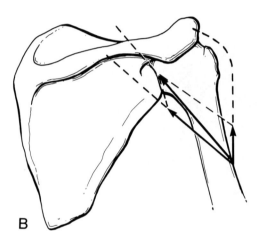

FIGURE 3–7

A, The resultant muscle force is directed through the joint contact point, resulting in no rotational motion. *B,* Since the resultant force is directed medial to the joint contact point, adduction rotation will result.

FIGURE 3–8

A, The quadriceps and hamstring muscles act to stabilize the knee joint. *B,* Each force (Q,H) can be resolved into compressive (C), flexion (F), and extension (E) components.

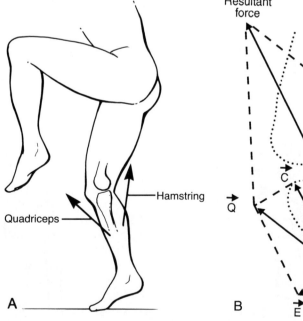

Another use for the *resolution of forces* (replacing a system of forces with a single resultant force) is to resolve or partition a single force into two or more components. This is especially useful when it is desirable to determine the *compression and shear components* of a resultant force acting at a joint and whether the resultant force will cause joint motion.

Example: The quadriceps (Q) and hamstring (H) muscle forces shown in Figure 3–8*A* can each be resolved into compressive and flexion (shear) components as shown in Figure 3–8*B*. The resolved components of force parallel to the long axis of the tibia will act to compress the joint surfaces, while forces perpendicular to the long axis will result in either flexion or extension. Whether the tibia will rotate, and in which direction (flexion or extension), will depend on whether one shear component is greater than the other. A single resultant joint reaction force can easily be found by extending the quadriceps (Q) and hamstrings (H) force to a common point of intersection (O) and then using the parallelogram method (see Fig. 3–8*B*). In this case, the resultant joint reaction force passes through the point of joint contact (d), indicating no motion or static equilibrium.

Example: Shown in Figures 3–9*A* and *B* are two positions of the elbow, each with similar magnitudes of biceps force. However, the biceps force in Figure 3–9*B* acts perpendicular to the long axis of the forearm. Resolution of the biceps force in Figure 3–9*A* into components parallel and perpendicular to the long axis of the forearm indicates that a smaller force component is present for elbow flexion. Another way of stating this is that for equal magnitudes of biceps force, a smaller component of the biceps force is available to resist elbow extension, say from gravity or an externally applied force, than that shown in Figure 3–9*B*.

MOMENTS

The *moment* of a force (torque) is the action of a force that causes rotation about some axis (Fig. 3–10). The moment is equal to the product of the magnitude of the force and the perpendicular distance from the line of action of the force to the chosen axis. The units of moments are Newton meters. Moments can either cause rotation or restrict rotation. Like a force, a moment is a vector quantity. Its line of action is along the line of rotation about which the force acts.

The magnitude of a moment can change significantly according to the line of action of the force vector. This is illustrated in Figure 3–11*A* for three

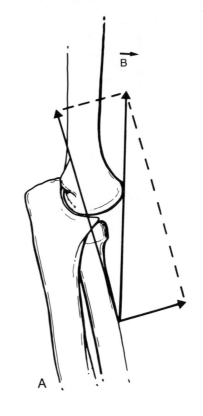

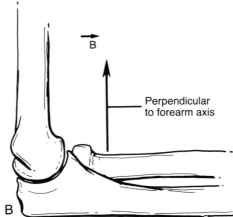

FIGURE 3–9
A, The biceps force (B) can be resolved into two components, one perpendicular to the forearm and one parallel with the forearm. The perpendicular component will act to resist rotation, whereas the parallel component will cause joint compression. *B,* The biceps force (B) is perpendicular to the forearm and greater than the perpendicular component shown in Figure 3–9*A*.

joint reaction forces of equal magnitude but different lines of action, F_a, F_b, and F_c. As the joint reaction force moves laterally from F_a to F_b, the magnitude of the (bending) moments along the stem increases as the distance from the line of action of the force to the stem increases, as shown in Figure 3–11*B*.

Force F_a produces the greatest magnitude of the bending moment at point 2. In this case, the lateral

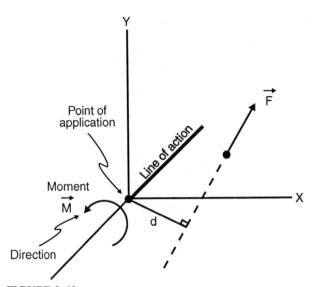

FIGURE 3–10
The moment (M) is equal to the magnitude of the force (F) times the perpendicular distance (d) to the force.

aspect of the stem will be in tension. Rotating the force laterally (F_a to F_b) produces increased bending moments at all locations along the stem, with the greatest moment now occurring at point 5. However, rotating the force medially (F_a to F_c) results not only in reduced bending moments but also in a zero bending moment at point 4 and a reversal of the direction of the moment at point 5. This latter moment will result in compression on the lateral aspect at point 5.

Moments can be summed in a manner similar to that used for forces, producing a single resultant moment equivalent to the original system of moments (Fig. 3–12). Each moment, that is, the force times the perpendicular distance about the axis of rotation, can be added or subtracted depending upon the direction of the moment. If, after summing all individual moments, the resultant moment M_R is equal to zero, no motion or rotation will result, and the body will be in static equilibrium. On the other hand, if the resultant moment is not zero, either motion will result or an additional moment of equal magnitude but opposite direction will be required to balance the system of forces.

FREE BODY DIAGRAMS

A *free body diagram* is a sketch of a body, a portion of a body, or two or more bodies completely isolated or free from all other bodies, showing all

the forces exerted by all other bodies on the one being considered (Fig. 3–13). A free body diagram has three essential characteristics: (1) it is a diagram or sketch of the body; (2) the body is shown completely separated (isolated, cut free) from all other bodies, including foundations, supports, and so on; and (3) the action on the body of each body removed in the isolating process is shown as a force or forces

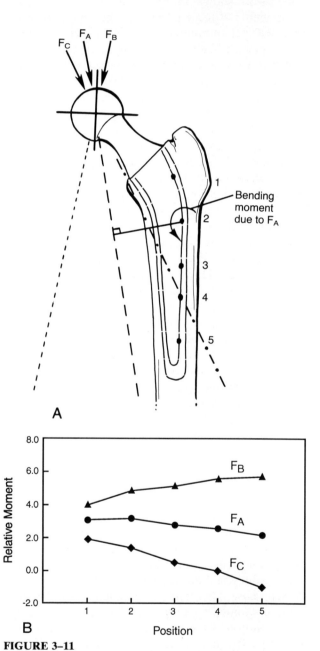

FIGURE 3–11
A, Bending moments about the lateral aspect of the hip stem resulting from the joint reaction forces shown in *B*. *B*, Relative moments along the stem for the three forces (F_A, F_B, F_C) shown in *A*.

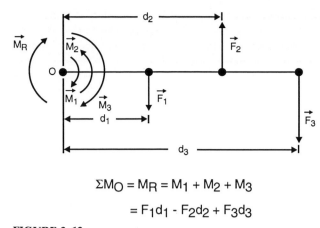

$$\Sigma M_O = M_R = M_1 + M_2 + M_3$$

$$= F_1 d_1 - F_2 d_2 + F_3 d_3$$

FIGURE 3–12
The resultant moment (M_R) is the sum of the three moments M_1, M_2, and M_3.

on the diagram. Typical free body diagrams and reaction forces are given in Figure 3–14.

Understanding and developing an ability to draw free body diagrams are probably the most important techniques needed for analyzing problems in biomechanics. Learn to draw free body diagrams and you will learn biomechanics. Unfortunately, the negative corollary is also true.

Each force in a complete free body diagram should be labeled either with its known magnitude or a symbol used when it is unknown. The direction of the unknown forces, when not obvious at a glance, may be assumed and corrected later if found to be incorrect. The slope or angle of inclination of all forces (their line of action) not obviously horizontal or vertical should also be indicated.

Example: The following free body diagrams of the lower extremity will illustrate the concepts involved. Shown in Figure 3–15A is the lower half of an individual in a one-legged stance (equilibrium). The upper body has been removed and its force component indicated. The ground reaction force is shown at the foot, and the weight (gravity) of the upper and lower legs is also indicated. A free body diagram can be drawn for each lower extremity, as shown in Figure 3–15B and C.

Whenever two bodies are separated, the appropriate forces must be inserted, as shown at the hip and foot in Figure 3–15B. Note that a muscle and quadriceps force have also been shown. These two may be redundant but are necessary to equilibrate the lower extremity fully. Additional muscle forces can be added as needed for completeness.

To further illustrate the technique of free body diagrams, a free body diagram of the knee joint

shown in Figure 3–15C is shown in Figure 3–15D. The femur, tibia, patella, and menisci can be further separated and a free body diagram constructed for each individual body. Each individual force is drawn on each body, and the corresponding forces between two bodies are drawn with similar lines of action but in opposite directions. Thus, when the bodies shown in Figure 3–15E are reassembled (moved into contact), similar forces will cancel, resulting in the free body diagram shown in Figure 3–15D.

When bodies have broad areas of contact, as between the meniscus, tibia, and femur (Fig. 3–15E), the resultant force can be distributed over the area of contact between the two bodies. Of course, this distributed force system can be resolved into a single resultant (joint contact) force for purposes of analysis. This illustrates the concept of a *contact stress* or pressure, whereby a resultant joint reaction force is distributed over a contact area, that is, a force per unit area or *stress*.

STATIC EQUILIBRIUM

Static equilibrium is the condition occurring when a body is in equilibrium and remains at rest with zero velocity. For this condition to exist, the forces and moments existing within and about the body must be balanced. For every free body diagram (system of forces and moments) it is possible to replace the forces and moments by a single force

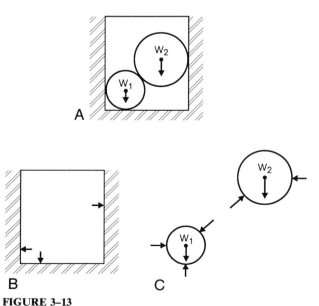

FIGURE 3–13
The free body diagram of each component in *A* is shown in *B* and *C*. Each ball has a unique set of forces acting on it to keep it in equilibrium.

Name of Body to be Removed	Sketch of Reacting Bodies	Action of Body Removed	Description

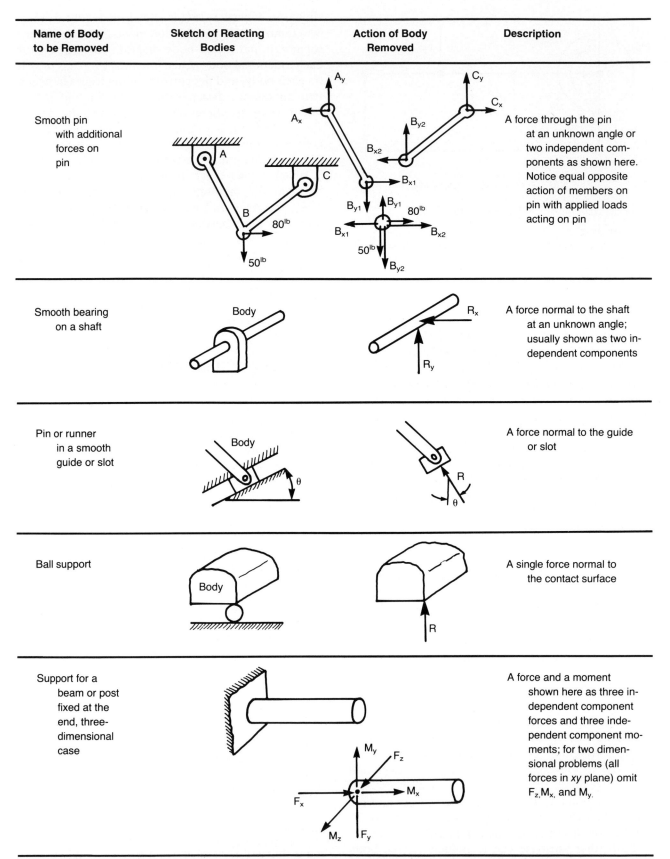

Smooth pin with additional forces on pin			A force through the pin at an unknown angle or two independent components as shown here. Notice equal opposite action of members on pin with applied loads acting on pin
Smooth bearing on a shaft			A force normal to the shaft at an unknown angle; usually shown as two independent components
Pin or runner in a smooth guide or slot			A force normal to the guide or slot
Ball support			A single force normal to the contact surface
Support for a beam or post fixed at the end, three-dimensional case			A force and a moment shown here as three independent component forces and three independent component moments; for two dimensional problems (all forces in xy plane) omit F_z, M_x, and M_y.

FIGURE 3–14
Typical free body diagrams and reaction forces. (From Higdon/Stiles/Davis/Evces, *ENGINEERING MECHANICS: Statics and Dynamics*, 2nd Vector Edition, © 1976, p. 115. Adapted by permission of Prentice Hall, Englewood Cliffs, New Jersey.)

Name of Body to be Removed	Sketch of Reacting Bodies	Action of Body Removed	Description
Earth			Always a vertical force equal to the weight and passing through the center of gravity of the body
Flexible cord, rope, cable (weight neglected)			Always a single force (tension) along the cord
Smooth surface			Always a single force perpendicular to the smooth surface
Roller			Always a force perpendicular to the surface on which roller can roll
Smooth pin			A force through the pin at an unknown angle; usually shown as two independent components

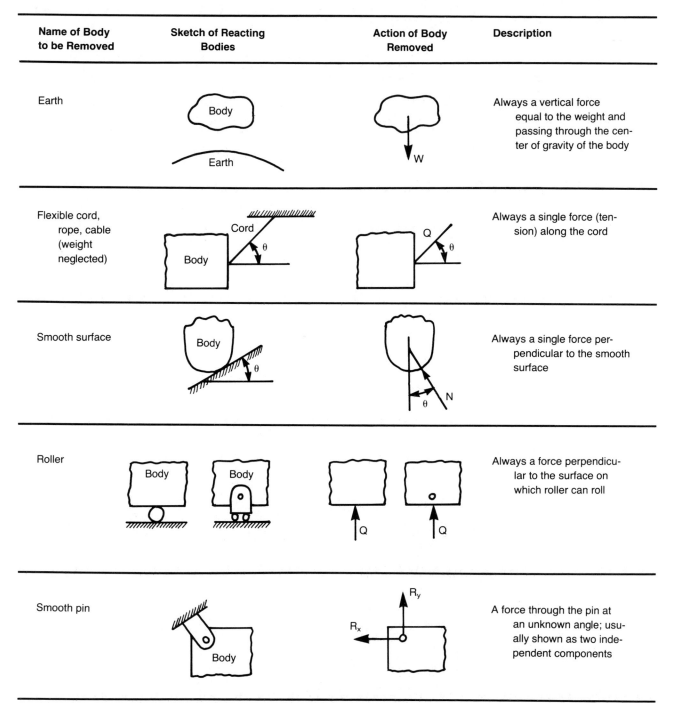

FIGURE 3–14 *Continued*

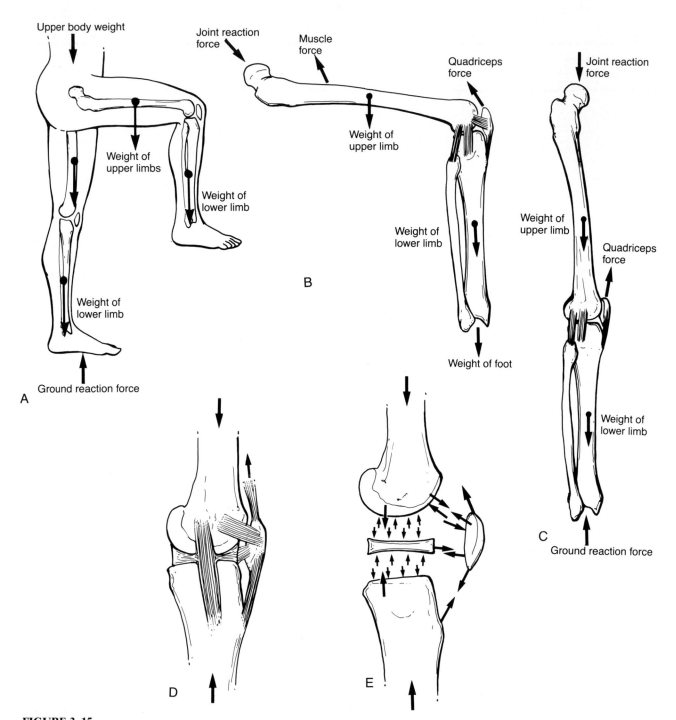

FIGURE 3–15

*A,*Illustration of the technique of free body diagrams. The free body diagrams for each lower extremity are shown in *B* and *C. D,* Free body diagram of the knee joint. *E,* Free body diagram for each component of the knee joint shown in *D.* When reassembled, the forces of equal magnitude but opposite direction cancel each other out, resulting in the free body diagram of shown in *D.*

and a single moment acting at some specific point. This technique (force and moment summation) has been illustrated previously. The resultant force will be independent of the specific point chosen, that is, having the same magnitude, direction, line of action, and point of application. However, the moment will always depend on the chosen point. For any system of forces and moments the single force and moment are the *resultant force* and the *resultant moment* for that system.

Example: The resultant force for the body segment weights shown in Figure 3–16 is simply the sum of the individual forces (weights) acting along the same line of action. In this case, all forces are in the direction of gravity. The resultant body segment force is also in this direction and is of equal magnitude but opposite in direction to the ground reaction force W. The resultant body segment force is independent of any point chosen to determine the resultant moment. The resultant moment about

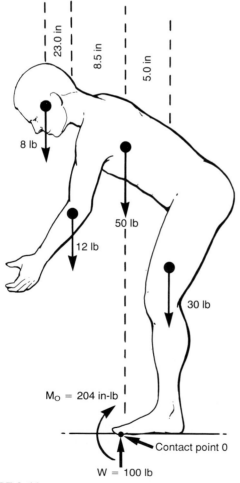

FIGURE 3–16
If the sum of all forces and moments about a specific point equal zero, the body is in static equilibrium.

point 0 (ground contact point) is calculated as follows:

$$M_R = (8 \text{ lb}) (31.5 \text{ in.}) + (12 \text{ lb}) (8.5 \text{ in.})$$
$$+ (50 \text{ lb}) (0 \text{ in.}) - (30 \text{ lb}) (5.0 \text{ in.})$$
$$= 204 \text{ in.-lb (counterclockwise)}$$

where counterclockwise moments are considered positive, and the perpendicular distances from each force (8, 12, 50 and 30 lb) to point 0 have been used (31.5, 8.5, 0 and 5.0 in.).

Since the resultant body segment moment M_R about point 0 is different from zero, equilibrium will not be achieved and the body will move unless an equal and opposite moment exists to balance M_R. This latter moment M_0 is shown in Figure 3–16 being exerted by the foot at point 0. Now, the resultant force and resultant moment for both the body segments and the floor reaction sum to zero, indicating a state of static equilibrium. A similar result would be found if any other point had been chosen to sum forces and moments.

The above example illustrates the fundamental principle for determining static equilibrium. The necessary and sufficient conditions for a rigid body to be in a condition of static equilibrium are that the resultant force F_R and the resultant moment M_R about any point equal zero. That is,

$$F_R = \Sigma F_i = 0$$
$$M_R = \Sigma M_i = 0$$

The concept of static equilibrium can be used to find internal muscle and joint reaction forces as illustrated in Figure 3–17 for a one-legged stance. A free body diagram of the loaded hip is shown to the right with unknown forces drawn in for the abductor muscle F_A and the hip reaction force F_H. Setting the resultant force equal to zero yields

$$F_R = F_H - F_A - 136 \text{ lb} = 0$$
$$F_H - F_A = 136 \text{ lb}$$

This equation yields one relationship between the two unknown forces. A second relationship can be found by calculating the resultant moment M_R about any point and setting this equal to zero. The resulting system of two equations can then be solved simultaneously for the unknown forces. However, by studying the free body diagram of the hip, a point may be found whereby one of the unknown forces can be eliminated (moment arm distance equal to zero). This would leave one unknown force that could be solved directly. For instance, summing moments about point H yields

$$M_H = F_A(2 \text{ in.}) - (136 \text{ lb}) (5 \text{ in.}) = 0$$
$$F_A = 340 \text{ lb}$$

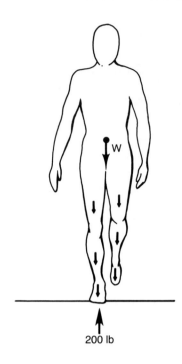

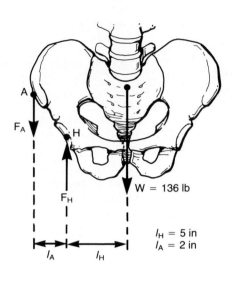

$W = 136$ lb

$l_H = 5$ in
$l_A = 2$ in

FIGURE 3–17
The forces F_A and F_H can be determined by finding the resultant force and resultant moment about any point and setting them equal to zero.

200 lb

The hip joint reaction force can be determined either by using the first equation or by summing moments about point A,

$$M_A = F_H(2 \text{ in.}) - (136 \text{ lb}) (7 \text{ in.})$$
$$F_H = 476 \text{ lb}$$

Inserting the unknown forces into the first equation provides a check on the calculations.

As a help in determining the unknown forces the following step-by-step procedure is provided:

1. Carefully determine what data are given and what results are required.

2. Draw a free body diagram of the member or group of members on which some or all of the unknown forces are acting.

3. Observe the type of force system that acts on the free body diagram.

4. Determine the number of independent equations of equilibrium available for the type of force system involved. At any particular point six equations can be written. When reference is made to a cartesian coordinate system this will yield three force equations along each axis and three moment equations about each axis.

5. Compare the number of unknown forces on the free body diagram with the number of independent equations of equilibrium available for the force system.

6a. If there are as many independent equations of equilibrium as unknowns, proceed with the solution by writing and solving the system of equations of equilibrium.

6b. If there are more unknowns to be evaluated than independent equations of equilibrium, draw a free body diagram of another body and repeat steps 3, 4, and 5 for the second free body diagram.

7a. If there are as many independent equations of equilibrium as unknowns for the second free body diagram, proceed with the solution by writing and solving the necessary equations of equilibrium.

7b. If there are still more unknowns to be evaluated than independent equations of equilibrium for the second free body diagram, compare the total number of unknowns on both free body diagrams with the total number of independent equations of equilibrium available for both diagrams.

8. If there are as many independent equations of equilibrium as unknowns for both diagrams, proceed to solve the problem by writing and solving the equations of equilibrium. If there are still more unknowns than independent equations, repeat steps 6b and 7. If there are still too many unknowns after as many free body diagrams have been drawn as there are individual bodies in the problem, then the problem is *statically indeterminate*—that is, not all the unknowns can be evaluated by statics alone.*

These step-by-step procedures are illustrated in the following examples.

Example: The subject squatting in Figure 3–18*A* weighs 200 lb. To achieve static equilibrium a muscle force F_G must be exerted at the hip to balance the

---

*Special cases may arise in which certain unknowns (but never all of them) may be evaluated even though there are more unknowns than independent equations of equilibrium.

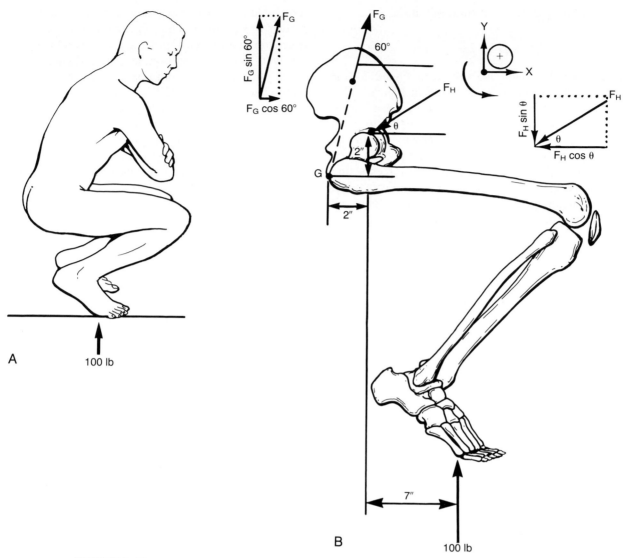

FIGURE 3–18
A and *B*, Free body diagrams used to determine the hip joint reaction force F_H and muscle force F_G.

body weight. A free body diagram of the lower extremity is shown in Figure 3–18*B*, in which the weight of the lower leg has been neglected. The two unknown forces for the muscle force F_G and the hip reaction force F_H are included. The muscle force line of action is known from its attachments (origin and insertion), whereas F_H acts at an unknown angle Θ with respect to the X axis.

The forces F_G and F_H can be easily found from the equilibrium conditions applied to the free body diagram. The resultant forces in the X and Y directions yield

$$F_{R_x} = F_G \cos 60° - F_H \cos \Theta = 0$$
$$F_{R_y} = F_G \sin 60° - F_H \sin \Theta + 100 \text{ lb} = 0$$

where the components of F_G and F_H are resolved along the X and Y axes using trigonometric relationships, and the positive directions are as shown. These two equations contain three unknowns (F_G, F_H, and Θ), requiring a third equilibrium relationship (moment equilibrium). Choosing the resultant moment at the hip or point G will eliminate either F_H or F_G, respectively. However, choosing the hip will also eliminate the unknown angle Θ and directly yield F_G. Resolving F_G into its X and Y components and then taking moments about the hip yields

$$M_{RH} = F_G \cos 60° \, (2 \text{ in.}) - F_G \sin 60° \, (2 \text{ in.})$$
$$+ \, 100 \text{ lb} \, (7 \text{ in.}) = 0$$
$$F_G = 956 \text{ lb}$$

From the two force equilibrium equations

$$F_H = 1133 \text{ lb}$$
$$\Theta = 60°$$

Example: The lower leg shown in Figure 3–18B can be further separated as shown in Figure 3–19. The unknown patellar tendon force F_P has a line of action as shown whereas the joint reaction force F_J and its line of action Θ are unknown. Again the limb weight is neglected. In this example the joint reaction force F_J is the resultant of all forces acting on the tibia, including contact forces and soft tissue forces (cruciates, meniscus, etc.). Resolving all forces along the coordinate axes yields

$$F_{R_x} = F_P \cos 30° - F_J \cos \Theta = 0$$
$$F_{R_y} = F_P \sin 30° - F_J \sin \Theta = 0$$

Choosing the joint contact point as the point about which to resolve the moments yields F_P directly.

$$M_J = F_P \,(2 \text{ in.}) - 100 \text{ lb} \,(12 \text{ in.}) = 0$$
$$F_P = 600 \text{ lb}$$

The two force equilibrium equations can now be used to solve for F_J, yielding

$$F_J = 656 \text{ lb}$$
$$\Theta = 38°$$

Example: Even though these two examples are idealizations of the actual forces acting at the hip and knee, they do demonstrate that high muscle and

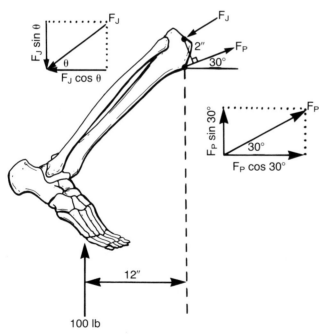

FIGURE 3–19
Free body diagram of the lower extremity shown in Figure 3–18A.

joint contact forces can result from the simple maneuver of squatting. The joint reaction force, which is greater than three times the body weight, can be resolved into one component perpendicular to the tibial joint surface (compressive) and another one parallel to it (shear). If the joint surfaces are considered frictionless or nearly so, the shear component can be assumed to be negligible and set equal to zero. Thus, only a compressive component will exist, with a line of action normal to the tangent to the articular surfaces at the point of contact. This simple assumption now makes it possible to calculate the cruciate ligament forces.

Arbitrarily cutting the lower leg in Figure 3–19 at midshaft, the free body diagram for each half is shown in Figure 3–20. If the anterior cruciate ligament is assumed to be the only soft tissue acting to restrain anterior tibial translation (resulting from the anterior shear component of the patellar force F_P), a system of equilibrium force and moment equations can be determined to find the cruciate force F_{ACL}. However, in this three-force system, it is possible to calculate F_{ACL} directly by resolving the moments about an axis (point) located at the intersection of the lines of action of F_J and F_P. Using radiographic or anatomic measurements, the point of application and line of action of F_J, F_P, and F_{ACL} can be determined and the intersection of F_J and F_P calculated from geometry, as shown in Figure 3–20.

Once the intersection location has been found, the resultant equilibrium force F_C and moment M_C at this point (the cut midshaft section) are easily determined from the free body diagram of the lower half. This yields $F_C = 100 \text{ lb}$ and $M_C = 817 \text{ in.-lb}$. Resolving (summing) the moments about the intersection gives

$$M_C = F_{ACL} \cos 15° \,(5 \text{ in.}) + F_{ACL} \sin 15° \,(0.5 \text{ in.}) - 817 \text{ lb}$$

from which the force in the anterior cruciate ligament is found to be

$$F_{ACL} = 165 \text{ lb}$$

The remaining joint and patellar forces can now be easily found from the force equilibrium conditions.

Example: As a further example of the use of these techniques, the amount of anterior tibial translation in the above example can be estimated. Given in Table 3–3 are typical geometric and mechanical (material and structural) properties of the anterior cruciate ligament and commonly used substitutions [8]. Using both the material and structural properties listed in Table 3–3 will yield a range (bounds) for the amount of anterior tibial translation that results when the ligament stretches as a result of the

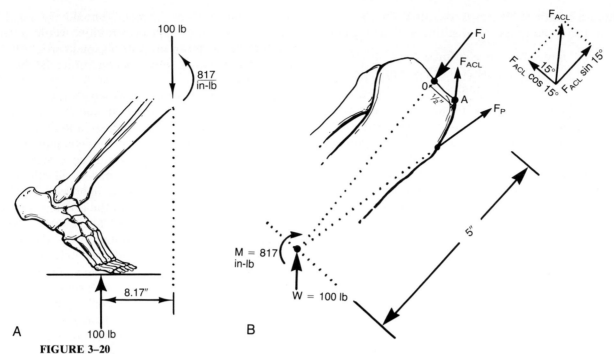

FIGURE 3–20
A and *B*, Detailed free body diagrams of the tibia used to determine the force in the anterior cruciate ligament.

static 734 N (165 lb) force. Based on material properties, the stress in the anterior cruciate ligament is calculated from

$$\text{stress} = \text{force/area}$$
$$= (734\ \text{N})/(44\ \text{mm}^2) = 16.7\ \text{N/mm}^2$$

The amount of ligament stretch (change in length) can be found from the stain (percentage of change in length) and the modulus,

$$\text{strain} = \text{stress/modulus}$$
$$= (16.7\ \text{N})/(110\ \text{N/mm}^2) = 15.2\%$$

and

$$\text{stretch} = \text{strain} \times \text{length}$$
$$= (0.152)\ (50\ \text{mm})$$
$$= 7.6\ \text{mm}$$

TABLE 3–3
Material and Structural Properties of the Anterior Cruciate Ligament

Tissue	Width (mm)	Area (mm²)	Max Load (N)	Stiffness (kN/M)	Modulus (MPa)	Max Stress (MPa)	Elongation[a] (%)
Anterior cruciate ligament	—	44	1730	182	110	40	44
Patellar tendon (one-third)	15	50	2900	700	340	60	20–25
Semitendinosus	—	14	1300	180	350	90	35
Fascia latae	16	8	620	120	400	80	27
	45[b]		1770	340			
Gracilis	—	8	850	170	600	110	27
Iliotibial band	18	37	660	110	—	—	—
	25[2]		920	160			
Quadriceps tendon							
Medial third	14	29	370	50			
Central third	16	17	270	40			
Lateral third	14	29	250	30			

[a]Bone-ligament-bone
[b]Normal width available for transfer
Abbreviations: MPa = 10^6 N/M² = N/mm².

Adapted from Noyes, F. R., Butter, D. L., Grood, E. S., et al. Biomechanical analysis of human ligament grafts used in knee-ligament repairs and reconstructions. *J. Bone Joint Surg* 66-A:344, 1984.

where a length of 50 mm is assumed. On the other hand, using the property of stiffness directly yields

$$\begin{aligned} \text{stretch} &= \text{force/stiffness} \\ &= (734 \text{ N})/(182 \text{ N/mm}) \\ &= 4.0 \text{ mm} \end{aligned}$$

The amount of anterior translation can now be approximated as the component of stretch along the anteroposterior direction (cos 15°), or 3.9 mm and 7.3 mm for the stiffness and material methods, respectively. These two amounts represent the low and high bounds. If the patellar tendon is used as a substitute, the bounds become 1.0 mm and 2.1 mm, respectively. These smaller bounds are obvious from Table 3–3 because the patellar tendon is 3.9 times stiffer, has a modulus 3.1 times higher, and has half the elongation-to-failure capability compared to the anterior cruciate ligament.

However, this static equilibrium analysis should not constitute the only criterion for clinical decisions. The dynamic and kinematic consequences of using a particular tissue as a substitute are important factors. In this case, although the patellar tendon is obviously stronger, it may also limit motion and could adversely affect knee kinematics. Thus, even though the knee may be *stable,* the altered kinematics could cause damage to the surrounding soft tissues (articular cartilage, meniscus).

Example: As a final example, the results of an idealized static equilibrium analysis of stair climbing are given in Table 3–4. In this case, the quadriceps force, the resultant joint reaction force (compression and shear), the patellar force, and the forces in the anterior or posterior cruciate ligaments were calculated for flexion angles of 20 and 90 degrees. Also included are the values of the forces when the patella has been removed, effectively reducing the moment arm of the patellar tendon force and increasing all other forces significantly.

Note that as the quadriceps act, the tibia is translated anteriorly, resulting in a resisting force in the anterior cruciate ligament. Should the quadriceps not act, the opposite motion would result because the posterior cruciate ligament would have to exert a resisting force to equilibrate (stabilize) the knee.

More important is the significant increase in ligament forces that would result should the patella be removed. Compared to the maximum loads possible (Table 3–3), it is obvious that the ligaments would rupture. How can this be? It will be remembered that in this static equilibrium analysis all other soft tissues and muscle forces were ignored. Furthermore, the constraining effect of the articular surface geometry was also neglected. These components will exert considerable restraining forces and lessen the force in the cruciates. Unfortunately, inclusion of all possible forces makes a simple analysis impossible and results in an *indeterminate* solution, one in which not all forces can be found. However, with the use of more sophisticated engineering and mathematical techniques, good approximations for all forces can be realized [11, 15].

SUMMARY

Study of the mechanics of the human body is a complex problem often requiring sophisticated and advanced engineering and mathematical techniques. However, in many instances a good understanding of mechanics can be obtained by performing a basic biomechanical analysis. This chapter has attempted to provide the basic principles of biomechanics governing forces, moments, and static equilibrium and to show the reader how to apply these principles. The principles developed in this chapter are not new, as any sophomore engineering student can attest. Furthermore, their application to the human body is also not unique to this chapter. Any of the numerous textbooks available on biomechanics will provide more detail and additional examples. What

TABLE 3–4
Static Equilibrium Analysis of One-Legged Stance During Stair Climbing

Flexion Angle	Quadriceps Force	Percentage of Body Weight[a]		Patellar Force	ACL	PCL[b]
		Joint Force				
		Compression	Shear			
20	0.9	1.8	0.2	0.8	0.2	0.2
90	4.0	4.3	0.7	7.7	0.7	2.6
90 (w/o patella)	7.7	10.9	7.1	15.3	7.1	7.5

[a]For purposes of calculation, body weight is assumed to be equal to 200 lb.
[b]Force in the PCL only when quadriceps are not active and ACL force is zero.
Abbreviations: ACL, anterior cruciate ligament; PCL, posterior cruciate ligament.

was attempted here was the presentation of these principles in terminology that can be easily understood by the individual without an engineering or technical background. This chapter is based in part on basic science lectures given at The Hospital for Special Surgery. It became obvious during these lectures that a strong understanding of forces, moments, and static equilibrium is required before study of the more advanced aspects of human body mechanics is possible. Mechanics, and specifically static equilibrium, is the basis upon which more sophisticated analyses must be built. The techniques presented in this chapter have been simplified to provide a conceptual basis of the subject through the analysis of human body mechanics.

One of the most important aspects of this chapter is the use of free body diagrams. This technique is vital to an understanding of biomechanics. Free body diagrams, which are usually absent or used incorrectly, have led to misconceptions or errors about the mechanics of the human body. A better and more detailed understanding of how each musculoskeletal component interacts to provide equilibrium and motion will be more easily attained through the careful use of free body diagrams and the subsequent inclusion of the forces and moments acting on each body. Even without sophisticated mathematical analyses the use of free body diagrams will yield an abundance of information about what forces are acting, how they act, and where they act. The importance of free body diagram analyses cannot be overemphasized.

References

1. Arnoczky, S. P., and Torzilli, P. A. Biomechanical analysis of forces acting about the canine hip. *Am J Vet Med* 42:1581–1585, 1981.
2. Black, J., and Dumbleton, J. H. *Clinical Biomechanics. A Case History Approach.* Edinburgh, Churchill Livingstone, 1981.
3. Dumbleton, J. H., and Black, J. *An Introduction to Orthopaedic Materials.* Springfield, IL, Charles C Thomas, 1975.
4. Higdon, A., and Stiles, W. B. *Engineering Mechanics: Statics and Dynamics.* Englewood Cliffs, NJ, Prentice Hall, 1962.
5. Kapandji, I. A. *The Physiology of the Joints.* Edinburgh, Churchill Livingstone, 1976.
6. Morris J. M. Biomechanical aspects of the hip joint. *Orthop Clin North Am* 2:33–54, 1971.
7. Nordin, M., and Frankel, V. H. *Basic Biomechanics of the Musculoskeletal System.* (2nd ed.). Philadelphia, Lea & Febiger, 1989.
8. Noyes, F. R., Butler, D. L., Grood, E. S., Zernicke R. F., and Hefzy, M. S. Biomechanical analysis of human ligament grafts used in knee-ligament repairs and reconstructions. *J Bone Joint Surg* 66A:344, 1984.
9. Perry, J., and Noyes, F. R. Analysis of knee joint forces during flexed knee stance. *J Bone Joint Surg* 57A:961–967, 1975.
10. Radin, E. L., Simon, S. R., Rose, R. M., and Paul, I. L. *Practical Biomechanics for the Orthopaedic Surgeon.* New York, Wiley, 1979.
11. Rydell, N. Biomechanics of the hip joint. *Clin Orthop Rel Res* 92:6–15, 1973.
12. Smidt, G. L. Biomechanical analysis of knee flexion and extension, *Biomechanics* 6:79–92, 1973.
13. Torzilli, P. A. Biomechanics of the elbow. *In* Inglis, A. E. (Ed.), *Symposium on Total Joint Replacement of the Upper Extremity.* St. Louis, C. V. Mosby, 1982, pp. 150–168.
14. Walker, P. S. *Human Joints and Their Artificial Replacements.* Springfield, IL, Charles C Thomas, 1977.
15. Williams, R., and Lissner, H. *Biomechanics of Human Motion* (2nd ed, edited by B. Le Veau). Philadelphia, W. B. Saunders, 1977.

SCIENTIFIC PRINCIPLES OF SURGICAL TREATMENT

Christian Gerber, M.D.
Martin H. Beck, M.D.
Alberto G. Schneeberger, M.D.

SUTURE MATERIAL*

Essentially no orthopaedic procedure is carried out without the use of sutures. The selection of the type and size of the suture material is more closely related to what a surgeon has learned empirically during his training than to logical conclusions based on material properties and imposed demands. The material and size of the sutures used in reconstructive procedures such as rotator cuff repairs or anterior cruciate ligament reconstructions have poorly known mechanical characteristics, and the role of the suture material and the conceivable surgical suturing techniques in the development of fixation failures are not well established. The technique of tying knots is not given particular attention because most knots are thought to be safe for tissues that are not under tension. It is established, however, that success in orthopaedic operations such as tendon or ligament repairs depends on the type of suture material used and the technique of using the suture [9, 21, 22, 23, 24, 26, 27, 40]. Alternatives to suturing techniques, especially for soft tissue fixation to bone, have been studied more recently and have been found to be successful [2]. Thus, a more precise analysis of the role of suture and suturing techniques is needed now to allow a more scientifically based selection of suture material, suturing and knotting technique, and potential alternatives.

Imposed Demands on Suture Materials

Different demands are imposed on suture materials. Their importance or priority may change with the specific application of the suture material considered. A variety of mechanical, biomechanical, and biologic properties should be considered when a specific suture material is selected. Such properties include the following:

1. Biological characteristics
   a. Biocompatibility
2. Mechanical characteristics
   a. Ultimate tensile strength
   b. Elasticity, deformation under load (gap formation under tensile load)
   c. Adequate maintenance of mechanical properties during healing (absorbable sutures)
   d. Knotting properties (ease of knotting, loss of strength after tying knots, number of knots necessary for stable knot)
3. Handling characteristics
   a. Ease in practical use

This chapter summarizes the current knowledge of suture materials and attempts to compare potential advantages and disadvantages of the sutures used in daily practice. Table 4–1 is a compilation of the most important data from the literature and from our own studies and is intended to be an easily

---

*Copyright for this section (Suture Material) is held by Christian Gerber, M.D.

accessible reference for a rational selection of the optimal suture for a specific application.

Biologic Characteristics, Biocompatibility

A number of studies have assessed the compatibility of different suture materials [4, 5, 15, 17, 20, 26, 30, 31, 34, 35, 39, 45] either by semiquantitative analysis of the histologic foreign body reaction to the implanted material or by experimental or clinical testing of the healing properties of the sutured tissues. Biocompatibility depends on the type of suture material, its structure (braided versus monofilament [1], the amount of material implanted, and the site of implantation.

In general the implantation of suture material causes an initial inflammatory reaction characterized by the presence of polymorphonuclear cells, lymphocytes, and monocytes. This acute inflammatory foreign body reaction peaks between days 2 and 7. By day 4, mononuclear cells start to predominate, and fibroblasts appear. By day 7, mature fibroblasts are present, and the foreign material becomes encapsulated in a fibrous mantle by day 10 [1]. No further tissue reaction is then to be expected if the implanted material is nonresorbable. Most currently used nonresorbable materials are quite inert and are therefore extremely well tolerated. Resorbable materials elicit a "second" boost of inflammatory reaction at the time of their resorption. The intensity of this reaction depends on the specific chemical process that leads to resorption and on the amount of the material to be resorbed. Suture materials such as polyglactin (Vicryl), polyglycolic acid (Dexon Plus), polygluconate (Maxon), and polydioxanone (PDS) are resorbed by simple hydrolysis, whereas catgut requires an enzymatic degradation that tends to provoke a much more intense soft tissue reaction.

Of the commonly used materials, monofilament stainless steel provokes the least foreign body reaction in skin and other musculoskeletal tissues. Almost no foreign body reaction is seen after implantation of nylon, polypropylene (Prolene), polyester (Ethibond, Tevdek, Ticron, and so on), and polybutester (Novafil), and also none to the absorbable materials polygluconate (Maxon) and polydioxanone (PDS). Polyglycolic acid (Dexon Plus) and polyglactin (Vicryl), which are also dissolved by simple hydrolysis, also elicit a minimal foreign body response (Table 4–1) [18, 30, 31, 34] but are probably tolerated somewhat less well than polydioxanone (PDS) and polygluconate (Maxon). The tissue response may be more pronounced if these materials

are used in the skin. Catgut and, at the time of its resorption, also chromic catgut cause a moderate to intense inflammatory response [17, 31]. Silk, which used to be the standard for skin closure, is probably the material that is least well tolerated of all suture materials still in use [34], and its use experimentally has been proved to compromise the results of intra-articular ligament repairs [26]. Optimal selection of materials for skin closure has also been shown clinically to reduce the incidence of wound infection (polydioxanone [PDS] plus polypropylene versus chromic catgut plus silk [45]).

A monofilament structure has a variety of biologic advantages. For one thing, there is less surface area exposed to the body, so that the foreign body reaction is less intense than that seen with multifilament sutures [1]. Because less suture material is exposed to hydrolysis, monofilament sutures retain their mechanical properties longer [5]. There is increasing concern [1, 4] that braided capillary materials may favor the propagation of infection, whereas noncapillary materials or monofilament sutures do not. It appears that bacteria can colonize these materials, not so much on their surface as within the suture, where immunocompetent cells have insufficient access. This accounts for the lack of support for the use of multifilament sutures in potentially contaminated situations [1]. Because the implantation of braided suture material very close to the skin incites a much more intense and lasting reaction than when the suture material is buried in well-vascularized tissue [15], and because contamination very close under the epidermis is always possible, we feel that the use of braided suture materials immediately under the epidermis should be avoided.

Clearly, large amounts of suture material incite more intense foreign body reactions. Therefore, sutures with optimal strength and knotting characteristics are needed so that small sutures requiring few throws for a stable knot can be utilized.

Mechanical Characteristics

Ultimate Tensile Strength

From an engineering standpoint, ultimate tensile strength should be measured in relation to the square area of the material tested, and material properties become system properties if the suture material is knotted. For practical purposes, however, the orthopaedic surgeon selects the suture size exclusively by the USP (United States Pharmacopeia) size indicated in numbers ranging from 11–0 to 5. When comparing data from different suture

TABLE 4–1

Characteristics of Commonly Used Suture Materials

Suture Materials	Maximum Tensile Strength Knotted (N)	Elongation Under 40-N Load (%)	Knots Required (n)	Tensile Strength/ After n Weeks (%/n)	Location of Implant	Tissue Reaction
Ethibond 3	106 ± 9	4.5 ± 0.3	1=1=1=1/2=1=1			
Surgilon 3	88 ± 6	12.8 ± 0.5				
Mono. steel 2	134 [37]		1=1/2=1/1=2 [37]			
PDS II 2	109 ± 15	14.7 ± 2.2				
Mult. steel 2	102 [37]		1=1/2=1/1=2 [37]			
Dexon Plus 2	101 ± 7	6.1 ± 0.6				
Vicryl 2	90 ± 7	7.5 ± 0.4				
Tevdek II 2	84 ± 7	5.5 ± 0.4				
Mersilene 2	83 ± 7; 71 [37]	5.4 ± 0.5				
Ethibond 2	82 ± 3	5.6 ± 0.4				
Surgilon 2	82 ± 5	13.9 ± 0.5				
Novafil 2	77 ± 14	20.9 ± 2.0				
Dacron 2	73 ± 8	6.2 ± 0.5				
Dermalon 2	69 ± 9; 62 ± 4 [44]	12.3 ± 2.0				
PDS II 1	85 ± 8	20.2 ± 2.5	1=1=1=1=1 [35]			
Maxon 1	85 ± 19; 77 [41]	14.8 ± 3.0	2=1=1=1/(1=1=1)			
Dexon Plus 1	71 ± 7	8.5 ± 0.5	2=1=1=1 [36]			
Tevdek II 1	69 ± 5	5.9 ± 0.6	1=1=1=1/2=1=1			
Vicryl 1	66 ± 6	8.3 ± 1.0	2=1=1=1 [36]			
Ticron 1	68 ± 4	9.3 ± 0.7				
Ethibond 1	65 ± 4	7.0 ± 0.5	1=1=1=1/2=1=1			
Surgilon 1	65 ± 9	16.0 ± 1.3				
Novafil 1	57 ± 8; 48 [41]	25.7 ± 2.5				
Prolene 1	55 [41]		1=1=1=1 [35]			
Suturamid 1	52 ± 7	15.5 ± 1.5				
Dermalon 1	43 ± 5 [44]					
Ethilon 1	40 [41]					
Mono. steel 0	83 [37]	20.4 ± 3.8	1=1/2=1/1=2 [37]	50%/3 [5]	Rabbit: subcut. fat [5]	
Maxon 0	67 ± 15; 68 [41]		1=1/2=1/1=2 [37]	50%/6 [5]; 50%/5 [35]; 0%/8 [35]	Rabbit: under trapezius [35]; fat [5]	
Mult. steel 0	65 [37]		1=1=1=1 [35]			
PDS II 0	59 ± 6	26.7 ± 2.6				
Ethibond 0	54 ± 7	7.8 ± 0.7	1=1=1=1 [30];	50%/2 [5, 36]	Rabbit: under trapezius [36]; fat [5]	+ to ++ [36]
Dexon Plus 0	53 ± 6	12.1 ± 2.7	2=1=1=1 [36]			
Vicryl 0	52 ± 8	12.0 ± 2.8	1=1=1=1=1 [30]; 2=1=1=1 [36]	50%/2 [5, 36]	Rabbit: under trapezius [36]; fat [5]	+ to ++ [36]
Tevdek II 0	48 ± 3	8.3 ± 0.8				
Ticron 0	46 ± 5	11.5 ± 2.5				
Surgilon 0	44 ± 3	25.5 ± 3.6	1=1=1=1 [35];			
Prolene 0	42 ± 2; 36 [41, 37]	32.6 ± 4.2	1=1=1/1=2 [37]			
Novafil 0	41 ± 6; 33 [41]	28.8 ± 1.9				
Ethilon 0	41 ± 3; 29 [41]	34.6 ± 4.7				
Mersilene 0	39 [37]					
Dermalone 0	32 ± 5 [44]					
Mono. steel 2/0	43 [41]		1=1 [19]			
Maxon 2/0			1=1=1=1 [35]			
PDS 2/0			1=1 [19]			
Mult. steel 2/0						

Material						
Dexon Plus 2/0			1=1=1=1 [30]; 2=1=1=1 [36]			
Dexon 2/0			1=1=1=1=1 [30]; 2=1=1=1 [36]	60%/2 [10]; 2%/4 [10]	Rat: subcut. fat [10]	(+) to + 10
Vicryl 2/0			1=1=1 [35]	60%/2 [10]; 4%/4 [10]	Rat: subcut. fat [10]	(+) to + 10
Dermalon 2/0	28 ± 3 [44]					
Prolene 2/0	27 [41]					
Novafil 2/0	25 [41]					
Ethilon 2/0	20 [41]					
Mono. steel 3/0	33 [38]		1=1 [19, 20, 38]/2 = 1 [20, 36]/1=2 [38]			
Maxon 3/0	32 [41]		1=1=1 [35]			
PDS 3/0			1=1 [19, 20, 38]/2 = 1 [20, 38]/1=2 [38]			
Mult. steel 3/0	24 [38]		1=1=1=1 [30]; 2=1=1=1 [36]			
Dexon Plus 3/0			1=1=1=1 [30]; 2=1=1=1 [36]			
Vicryl 3/0			1=1=1/1=2 [20, 36]; 1=1=1 [35]			
Prolene 3/0	14 [38]; 17 [41]		1=1/1=2/2=1 [20]			
Mersilene 3/0	16 [38]		1=1=1 [20]			
Tevdek 3/0			1=1=1/1=2 [20]; 1=1=1=1 [19]			
Dacron 3/0			1=1=1=1/1=2 [20]			
Dermalon 3/0	14 ± 1 [44]		1=1=1=1/1=2 [20]			
Ethilon 3/0	14 [41]					
Novafil 3/0	13 [41]					
Mono. steel 4/0	19 ± 2 [39]	28.7 ± 20 [39]	1=1 [19, 39]	50%/3 [31]	Rat: peritoneum [31]	(+) to +++ [31]
Maxon 4/0	17 ± 2 [39]; 21 [41]	65.8 ± 16 [39]	1=1=1 [39]	50%/4 [31]	Rat: peritoneum [31]	(+) to +++ [31]
PDS 4/0	16 ± 4 [39]		1=1 [39]			
Mult. steel 4/0			1=1=1=1 [30]			
Dexon Plus 4/0	13 ± 2 [39]	30.5 ± 6 [39]	1=1=1=1 [39]	50%/2 [10]; 5%/4 [10]	Rat: subcut. fat [10]	(+) to + [10]
Vicryl 4/0			1=1=1=1 [30]	55 [31]–65 to %/2; 0 [31]–8 to %/4	Rat: subcut. fat [10] peritoneum [31]	(+) to + [10]; ++ to +++ [31]
Prolene 4/0	11 [41]		1=1=1 to 1=1=1=1=1 [39]			
Ticron 4/0	10 ± 2 [39]	27.6 ± 9 [39]				
Dermalon 4/0	9 ± 1 [39]; 10 ± 1 [44]	78.9 ± 22 [39]	1=1=1 to 1=1=1=1=1 [39]			
Novafil 4/0	8 ± 1 [39]; 9 [41]	130.9 ± 6 [39]	1=1=1=1 [39]			
Ethilon 4/0	8 [41]					
Chromic catgut 4/0				50%/1 [31]; 25%/2 [31]	Rat: peritoneum [31]	+++ [31]
Mono. steel 5/0					Dog: tendon [34]	+ [34]
Mult. steel 5/0					Dog: tendon [34]	+ [34]
Dexon 5/0					Dog: tendon [34]	+ [34]
Ticron 5/0					Dog: tendon [34]	+ [34]
Nylon mo. 5/0					Dog: tendon [34]	(+) [34]
Chromic catgut 5/0					Dog: tendon [34]	+ [34]
Silk 5/0					Dog: tendon [34]	+++ [34]
Sil. silk 5/0					Dog: tendon [34]	+++ [34]

Tissue reaction: (+) very mild, + mild, ++ moderate, +++ intense, ++++ very intense materials of different sizes, it should be understood that the square area (or diameter) of a specific USP number of one material may be different from the square area (diameter) of the same USP number of another material [44]. In addition, we never use unknotted sutures in daily practice. The mechanical properties of suture materials presented in Table 4–1 refer, therefore, to knotted sutures (unless indicated otherwise) of commercially available sizes expressed in USP numbers. This appears reasonable because the surgeon is not interested in verifying the diameter of a certain suture and because essentially all sutures invariably fail at the site of the knot [32, 38]. Until recently, only thin suture materials had been widely tested [5, 18, 19, 29, 35, 36, 41, 42], and the mechanical properties of heavier suture materials (e.g., sizes 0 to 3) were only sparsely documented. Repairs of large musculotendinous units are, however, performed with the use of thick sutures, which have so far not been proved to be adequate let alone optimal for their respective applications. We recently tested the mechanical in vitro properties of the heavier sutures (gauges 0 to 3) [32]. Not all suture materials were available in all sizes. The mechanical in vitro properties known of the suture materials commonly used are summarized in Table 4–1.

Maximum tensile in vitro strength of comparable USP-size sutures was found for monofilament stainless steel and the absorbable monofilament polygluconate (Maxon) and polydioxanone (PDS). Braided absorbable polyglactin (Vicryl) and polyglycolic acid (Dexon Plus) as well as braided polyester (Mersilene, Ethibond, Tevdek) also showed excellent ultimate tensile strength, whereas nylon (Dermalon, Prolene) was somewhat weaker. Our own data are in rough agreement with those of the literature. In contrast to the study of Bourne and colleagues [5], we did not find a decrease in ultimate tensile strength for wet sutures compared to the dry material.

Deformation Under Tensile Load

Although monofilament sutures tend to be more compliant, they are favored in newer arthroscopic techniques [7] because they can be passed more easily through arthroscopic instruments. Suture elongation under load and thread or knot failure, however, may lead to gap formation and impair successful healing and functional recovery [22, 33].

Table 4–1 shows that the very strong, resorbable monofilament polygluconate (Maxon) and polydioxanone (PDS II) sutures are very compliant under tensile loads, as opposed to the also very strong, resorbable but braided polyglycolic acid (Dexon) and polyglactin 910 (Vicryl) sutures, which are very

stiff. Among the commonly used nonabsorbable sutures, only monofilament stainless steel is stiff. The other monofilament sutures (polypropylene [Prolene] and nylon) are very compliant. The most commonly used nonabsorbable braided suture material is polyester; it is characterized by high tensile strength and low compliance.

Clinically, it may be advantageous to have a very compliant suture, especially in a running-type suture, because it may yield rather than break [28]. For tendons, strong suture repair techniques that prevent gapping are optimal [22, 45] because scar and adhesion formation is reduced and early functional treatment, which promotes remodeling, can be undertaken. An optimal tendon-to-bone repair should not allow gap formation under tensile load, but moderate extensibility may be beneficial for healing. If a rotator cuff tendon is sutured to a trough in the greater tuberosity, transosseous suture loops of roughly 7 cm are tied over the proximal humeral cortex. If such repairs are brought under a tension of, for example, 200 N, which is possible when the arm is lowered, the elasticity of the suture material alone may allow gapping of the repair. Under a load of 200 N, the suture material properties alone would allow a gap of 2.2 mm in the case of Ethibond No. 3, of 3.2 mm for Ethibond No. 1, of 6.7 mm for Surgilon No. 1, and of 9.1 mm for PDS II No. 1 [32]. Thus, suture material alone may prevent stable anchorage of a tendon in bone if the repair is subjected to load, and the selection of appropriate suture material therefore is critical. Also, it is of critical importance to know the loads to which a repair will be subjected in order to determine the optimal suture material.

Maintenance of Mechanical Properties in Vivo

Although poorly documented, it is commonly accepted that nonresorbable sutures maintain their strength throughout wound healing. This is not the case for absorbable suture materials. Simple soaking in Ringer's lactate has reduced ultimate tensile strength by 4% to 13% in one study [5], but has not impaired mechanical properties in other studies [32]. The implantation of these resorbable sutures alters their tensile strength throughout the healing process; catgut loses its strength essentially within 1 week. Polyglactin 910 (Vicryl) loses about 50% of its initial ultimate tensile strength after 2 weeks, 75% by 3 weeks, and all of its strength by 4 weeks [5]. The material is resorbed after 56 [17] to 90 [10] days. Polyglycolic acid (Dexon Plus) has similar mechanical properties in vivo; 20% of its initial strength is lost within the first week [39], 50% by 2 weeks, and essentially 100% by 4 weeks. Whereas Vicryl is stronger than Dexon in the first 35 days, the latter remains longer in the wound than the former [10] and is only resorbed after about 120 days. Polygluconate (Maxon) loses its strength distinctly less rapidly; its very high initial strength is reduced to roughly 50% after 3 weeks and is lost by 6 weeks [5]. Polydioxanone (PDS) is initially somewhat less strong, but because it is hydrolyzed more slowly than Maxon it is already distinctly stronger at 3 weeks after implantation and maintains about 50% of its initial breaking strength for 6 weeks [5]. Its resorption requires roughly half a year [17].

The selection of the appropriate suture material depends on the expected type and rate of wound healing. For simple adaptation of subcutaneous tissue that is not under tension, the rapid resorption of polyglycolic acid or of polyglactin 910 may be desirable. Because the fibroblastic response dictates that a healing wound rapidly regains strength between days 5 and 14 and because collagen content increases until day 42 [1] with subsequent remodeling of the wound, a suture material such as polygluconate (Maxon) may have optimal resorption characteristics for a wound that is under slight tension. If a tendon or fascia is repaired under moderate tension and if a longer period of protection is desired, a material such as polydioxanone (PDS) may currently be optimal. If, however, prolonged holding power is required and if gapping is to be prevented, braided polyester (Ethibond, Tevdek, Mersilene) may be the best alternative.

Knotting Properties

Tera and Åberg [38] have introduced an internationally accepted terminology for knotting techniques. They distinguished between parallel and crossed knots and established that, in general, parallel knots (Fig. 4–1) are more stable than crossed knots. Knotting properties appear to be similar for different sizes of the same material [19, 37]. Knots are indeed of great importance because suture material almost invariably fails at the site of the knot [20, 32, 38]. Whereas stainless steel loses little strength by knotting, catgut and silk lose a large part of their strength [37, 38]. Indeed, the failure strength of suture material may be reduced by 10% to 70% by tying a knot [20]. Slipping of the knot depends on the number of throws and on the material. In general, any knot with six throws should not slip [5]. We were, however, able to show that the 2=1=1 configuration leads to stable knots for all tested sutures [32], and this is our preferred suture technique, although some sutures may require even fewer throws. Zechner and colleagues [47] have shown that the reduction in tensile strength can be partly prevented if knots are tied in the horizontal branch of the suture rather than in the

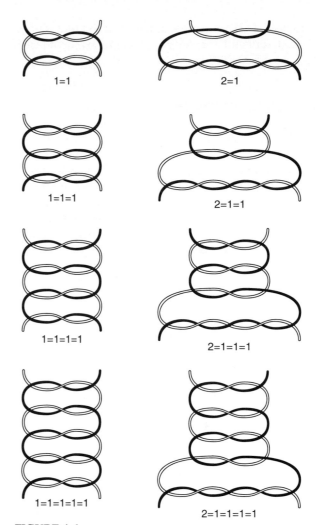

FIGURE 4–1
Parallel knots and the nomenclature of Tera and Åberg [37].

region where linear stress is applied. Although one tends to believe that knots of monofilament sutures tend to slide easily and therefore need multiple knots [39], this is not true in reality; polygluconate has excellent knotting properties [5, 31, 32], as have other monofilament sutures. The braided resorbable sutures tend to slip less if the suture material is wet [5]. Knotting is more delicate in coated materials such as Ticron (braided polyester) because coating reduces friction and therefore favors slipping of the knot. Coated and particularly stiff sutures are better knotted with double throws [20, 39].

Surgically, the sliding knot [25] is the most convenient knot; it corresponds to a parallel square knot and has the corresponding properties. It should be noted, however, that a double throw for the first knot increases tensile strength in a statistically significant manner [32]. If such a double throw is pinched with a ribbed needle holder, the tensile strength of sutures of size 0 to 3 is not significantly impaired [32], whereas pinching a smaller size of thread reduced the tensile strength by up to 30% [5].

Handling Properties

Modern suture materials are available with swedged-on precision needles, which are adapted to the intended use of the suture and thus essentially solve needle problems. Among suture materials, stainless steel has little popularity despite its superb soft tissue tolerance because of its stiffness [9] and its potential for breaking if the suture is kinked. Handling of the other sutures is mostly a matter of personal preference, and there are no clear-cut advantages in the handling properties of one suture as opposed to another.

Personal Approach to the Selection of Sutures

To fix large tendons to bone or to secure a bone block of a patellar ligament–bone complex to the bone, we prefer heavy (No. 2 to No. 3), nonabsorbable, braided polyester (Ethibond) sutures. They are inert, have a very high tensile strength, and are extremely stiff, so that they prevent gapping [13, 14]. For tendon sutures we use fine polypropylene (Prolene) or polydioxanone (PDS) sutures. They are somewhat elastic but have extremely high tensile strength, and polydioxanone (PDS) maintains its mechanical properties sufficiently long [3] to allow tendon healing. If strength needs to be maintained, as in closures of aponeuroses (fascia lata), we prefer to use heavier elastic running sutures with polydioxanone (PDS). In subcuticular tissue the breaking strength of the suture can be lost rapidly, so 4–0 polygluconate (Maxon) appears optimal. For sutures placed very close to the skin and in questionably contaminated situations we try to avoid the use of braided suture materials. For closure of the skin we prefer polypropylene (Prolene) 4–0 or 3–0. We do not use very rapidly absorbed polyglactin (Vicryl Rapid) or polyglycolic acid (Dexon) [8], which have been recommended for skin closure [12], but they can be used for approximation of subcuticular tissues [11]. Within the skin these materials tend to cause irritation and may serve as a wick, promoting contamination. Although it is still recommended [43], we have no more use for any form of catgut.

The selection of suture remains personal in every field of surgery [16], but selection should be done on a rational basis.

ANTIBIOTIC PROPHYLAXIS

Postoperative wound infection is the most worrysome and possibly the most serious local complication of almost any orthopaedic procedure, and modern orthopaedics would be inconceivable without effective measures for its prevention and treatment. The discovery of the role of airborne bacteria in the development of sepsis by Pasteur [112] was the basis for Lister's formulation of the three basic principles for the prevention and treatment of septic complications [100, 114]:

First, germs must be prevented from getting into the wound, during or after the operation.
Second, germs should be killed (eliminated) outside the wound.
Third, if germs are present in the wound, they must be killed before they can spread.

These three principles are valid to this day— indeed, they summarize the basic goals of one of the major efforts of modern orthopaedic surgery, namely, to reduce infection by minimizing bacterial contamination of the operative field. In addition to the psychological and medical dimension, prevention of sepsis also has an economic dimension because the total cost of an orthopaedic operation rises dramatically if infection ensues [82, 109, 110].

Lister started the concept of disinfection. Hygienic measures for preventing contamination of wounds through the use of antiseptic soaps and solutions to disinfect the skin [59, 62, 77, 79, 90, 101, 120] and sterile surgeons' clothes and masks to prevent intrusion of airborne bacteria into the wound [48, 81] allow the surgeons to reduce the rate of infection to 3% to 4% of all major procedures [69, 110]. Newer developments such as the introduction of ultraclean air [49, 57, 97] by Charnley [57] and the use of prophylactic antibiotics [74, 75, 76, 85, 91] again reduced the infection rate significantly, so that in the 1990s we operate with infection rates of below 1%.

Some patients are at higher risk, and they need particular attention and preventive measures. Established risk factors for the development of postoperative sepsis include [53, 56, 126]:

Malnutrition including dehydration
Immune deficiency and immunosuppression
Chronic hypoxia
Malignancy
Diabetes
Extremes of age
Renal and liver failure

Local skin problems (psoriasis, scarring, radiation fibrosis)

Lymphedema, venostasis, major vessel disease

Specific procedures carry a higher risk of postoperative sepsis than others, namely, those that involve the use of large implants such as total joints or allografts [67, 83, 116, 128] and operations that last a long time (more than 90 minutes) [62, 80, 94, 109]. There are also environmental factors that probably play a role in the frequency of infection (e.g., hygienic conditions in the region, hospital, or operating room; immunologic competence of certain populations toward certain infective agents); these are *not* or only partly under the influence of the surgeon. There are, however, a large number of factors that are under the influence of the surgeon, and scrupulous attention to them will reduce the incidence of infection. Of particular relevance for the orthopaedic surgeon are (1) preoperative assessment and preparation of the patient; (2) preoperative antibiotic prophylaxis; (3) intraoperative prevention of septic complications; and (4) postoperative prevention of septic complications.

The focus of this chapter is on antibiotic prophylaxis, which appears to be the most controversial topic and the one most subject to change. I feel nonetheless that other considerations are as important and deserve to be alluded to.

Preoperative Assessment

The history and clinical examination should determine whether there is any increased risk due to a general health problem such as decreased immunocompetence (HIV, recent infectious foci [genitourinary tract infection, dental problems, etc.]), malnutrition, obesity, chronic hypoxia, or medical treatment [steroids], or due to a local condition such as skin problems (e.g., infectious dermatitis, acne, skin erosion, psoriasis, ingrown toenails), arterial insufficiency, venous stasis, lymphedema, prior surgery at the operative site with relevant scar formation, or prior history of local infection.

If such risk factors are present, the physician should determine whether he can treat the condition with an acceptable delay to reduce the risk of the elective procedure to a minimum. As long as medical or surgical management of a significant risk factor is possible, it must be undertaken *before* elective surgery. With medical treatment it should be considered, for example, that prolonged preoperative hospitalization increases the incidence of infection significantly [62]. If the risk has to be accepted because of either the urgency of the indication or the lack of an alternative treatment, the indication must be reconsidered. If the risk/benefit ratio is such that the indication is maintained, additional specific preventive measures may be indicated.

The skin should be prepared for surgery. Shaving is not of proven benefit. In fact, if it is performed the day before surgery, it may result in nicks of the skin that can become contaminated and may be a source of infection [62]. It is generally accepted that treatment of the operative region with antiseptic soap creates a protective film on the skin that is effective in reducing the number of germs at the operative site [62, 81]. Our patients receive a prescription for Betadine soap, and are instructed to scrub the operative site once a day for 5 minutes with this liquid soap for the 4 days before admission. They are also encouraged to use the soap for showering during these 4 days. A study of compliance with and efficacy of this prescription is not yet available. Upon admission, the patient showers using Betadine soap. The operative site is thoroughly scrubbed, a depilating cream is applied at the exact site of the incision, and the hair is removed.

The skin is then disinfected with Betadine solution and covered with sterile dressings. Theoretically it is preferable to remove the hair immediately before surgery. In practice, however, many operative theaters do not allow one to shave the operative site because of fear of possible contamination of the room; in addition, preoperative positioning and preparation may need to be performed by relatively poorly trained personnel, so that optimal hygienic handling of the operative site in the preparation room is not ascertained. If depilation is performed on the ward, disinfection with Betadine solution and sterile dressings are considered indispensable.

Preparation and draping in the operating room have become well standardized. Impermeable fabrics have largely replaced the more conventional materials for the gowns of the surgical team [48]. The use of plastic adhesive drapes is controversial [103]. Their value in the prevention of infection has not been proved. We feel that an adhesive that does not stick perfectly to the skin and allows tissue fluid to collect is probably detrimental; conversely, a perfectly adapted drape may be advantageous because it permits minimal contact with the skin of the patient, which should always be considered contaminated [34]. We use a polydione-impregnated adhesive (Ioban) for major procedures.

Antibiotic Prophylaxis

Systemic Antibiotic Prophylaxis

After successful treatment of infections with antibiotics, it appeared logical to use antibiotics to prevent the development of septic complications. At first these drugs were used without any scientific background and failed to show any beneficial effect. In fact, the rate of infection seemed to be increased following the use of "prophylactic" antibiotics [86, 87, 91, 113, 122, 125, 129].

With an unacceptably high infection rate and the devastating sequalae of septic complications, the absolute need to find an efficient means of preventing infection prompted basic research [51, 52, 67, 78, 83, 115, 116, 121, 127, 128]. Better designed clinical [69, 75, 76, 85] studies led us to the current understanding of the role of antibiotic prophylaxis. The most influential studies were those of Burke [52], which clearly established that in order to be effective, antibiotics in sufficient concentration must be at the operative site by the time of the skin incision. This scientific evidence, which has been amply confirmed [136], explained a number of pessimistic results, which had been obtained with too late administration of the drugs [72, 87, 129]. Further studies have provided pharmacologic data such as bone and soft tissue penetration and concentrations of different antibiotics [89, 127, 135]. Other studies have identified the germs to be expected in orthopaedic infections [48, 53, 69, 93, 96, 99, 104, 133], thereby providing the basis on which effective prophylactic antibiotics can be selected [119].

Is Antibiotic Prophylaxis Necessary?

Traditionally, the infection rate in orthopaedic procedures has been between 3% and 24% [50, 55, 69, 72, 76, 85, 104, 109] with an average rate of about 4% for major fractures or total joint replacement surgery. Because postoperative sepsis has serious somatic, psychological, and economic implications [82, 92, 110, 130], all effective prophylactic measures are clearly necessary.

Based on the extensive hygienic precautions now used, such as meticulous disinfection and draping of the skin of the patient, careful preparation of the surgeon including optimal protection with masks, gloves, and clothing [66], and the use of laminar air flow systems, bacterial contamination of the operative field seems virtually impossible in a clean (aseptic) procedure. There is conclusive evidence, however, that however clean the surgery may be and under whatever condition the operation may have taken place, each surgical wound is contaminated with bacteria by the time of wound closure [49, 51, 63, 88, 92, 97].

All prophylactic antibiotics, systemic [52, 71, 136] as well as local [73, 115], are most effective if they are present at the time of inoculation of the bacteria and lose their efficacy with time elapsed after inoculation [115, 136]. The surgical trauma, which starts with the skin incision, causes additional soft tissue trauma that compromises the host defense mechanisms [53], including local oxygen delivery, thereby promoting infection [56, 95]. Given the facts that infection is a relevant risk, that each surgical wound is contaminated, and that timely administered antibiotics can prevent the multiplication and spread of germs, it appears clearly necessary to eradicate the contaminating germs by administration of adequate doses of effective antibiotics at the time of inoculation of bacteria.

Is Antibiotic Prophylaxis Efficient?

Several reviews have pointed out the difficulties inherent in designing worthwhile studies to determine whether antibiotic prophylaxis actually lowers the infection rate [48, 58, 65, 84, 86, 92, 111]. The series of patients must be enormous, so that only multicenter studies can be used; the antibiotics must have been administered *before* skin incision in every case; the protocol must be prospective and truly randomized; because the definition of infection is not easy, clean and clean-contaminated surgeries must be distinguished; and finally, follow-up of at least 2 years must be available.

Although very few studies are methodologically impeccable, such studies do exist [85], and their results have been confirmed in 90% of all the more recent clinical or experimental studies [50, 54, 55, 65, 68, 69, 72, 84, 85, 104, 105, 124, 130, 136]. Therefore, the clinical efficiency of correctly administered, systemic prophylactic antibiotics in major orthopaedic procedures (major fracture surgery, total joint and allograft replacement) is beyond any doubt. There is, conversely, no evidence that antibiotic prophylaxis is necessary or beneficial for *minor* clean procedures such as diagnostic arthroscopies.

What Kind of Antibiotic Should Be Administered?

The ideal antibiotic should fulfill the following criteria [98].

1. It should be highly active against the bacteria expected to be present.
2. It should reach sufficient concentrations in bone and soft tissue.
3. It should be bactericidal.
4. It should have a sufficiently long half-life.
5. It should not be toxic.

6. It should not create resistant strains of bacteria.

A series of antibiotics are suboptimal for prophylaxis on the basis of these considerations. For example, the aminoglycosides are relatively toxic, the penicillins (ampicillin, amoxycillin), and the tetracyclines induce plasmide resistance, and the macrolides or sulfonamides are essentially bacteriostatic drugs. The two main groups of antibiotics with sufficient documentation of efficacy are therefore the isoxazolylpenicillins and the cephalosporins.

The isoxazolylpenicillins (cloxacillin, dicloxacillin, methicillin) have been widely used in orthopaedics and have proved effective in preventing *Staphylococcus aureus* infection [50, 55, 68, 72, 91, 107]. They continue to be used by authors [49] who rarely see gram-negative infections and who do not see methicillin-resistant staphylococci. The current epidemiologic data in most large orthopaedic centers, however, show that coagulase-negative, methicillin-resistant staphylococci as well as gram-negative infections are a reality [69, 76, 104], so that the isoxazolylpenicillins have been abandoned in favor of the cephalosporins.

The cephalosporins are inhibitors of cell wall synthesis. Three generations are distinguished. The prototypes of the three generations are first-generation, cefazolin; second-generation, cephamandole; and third-generation, cefotaxime. Cephalosporins have very low toxicity and good bone and soft tissue penetration [127, 135], and they easily reach the desired fourfold minimal inhibitory concentration for any sensitive organisms [98] at the desired site. Third-generation cephalosporins are optimal against gram-negative organisms and should not be used for prophylaxis in orthopaedic surgery. The second-generation cephalosporins have better activity against gram-negative organisms (*Pseudomonas aeruginosa* excluded) and against *staphylococcus epidermidis* than first-generation drugs. However, they are less effective against *Staphylococcus aureus* than first-generation drugs. The latter are effective against streptococci and staphylococci (although not methicillin-resistant staphylococci) but are only relatively effective against S. *epidermidis* [71, 98, 108]. In institutions where gram-negative or S. *epidermidis* infections are prevalent, a second-generation cephalosporin may therefore be indicated as routine prophylaxis [74, 75, 76, 104]. For the most common infections in orthopaedics, however, first-generation cephalosporins such as cefazolin are still the drugs of choice [64, 71, 89, 102, 127]. Finally, drugs with a long half-life are preferred [118].

How Much, How Long?
The low toxicity of the drugs and their short duration of action direct the use of relatively high doses (2 g of cefazolin or 2 g of cefamandole) at the induction of anesthesia, at least 10 minutes before inflation of the tourniquet [98].

The correct length of time of administration of antibiotics is controversial. Whereas there is still no justification for recommending a single-dose regimen, there is no study proving that antibiotic prophylaxis of more than 1 day is superior to a 24-hour regimen [60, 65, 74, 98, 117, 130, 134]. Some authors recommend administration of antibiotics until aspiration drains are removed [69, 85], whereas others explicitly state there is no justification for awaiting removal of drains [133]. A single study in the literature has concluded that single-dose antibiotic prophylaxis is less efficient than longer administration of the drug [75]; this report has been expanded by a subsequent study by the same authors showing that a single dose of a longer acting drug is just as efficient as a multiple dose regimen [74]. A number of studies have shown that the value of antibiotic prophylaxis is lost if the procedures last a relatively long time [62, 80, 94]. It appears, therefore, logical to give a second dose intraoperatively if an operation lasts longer than 2 hours.

Authors' Preferred Method of Treatment
We use 2 g of cefazolin at the induction of anesthesia or at least 10 minutes before inflation of the tourniquet. We add 1 g of cefazolin after 2 hours of surgery or at the time of closure of the wound and then continue with 1 g of cefazolin every 8 hours for a total of 24 hours. In extremely high risk cases we continue this regimen until the drains are removed.

There are specific indications for the use of other antibiotics such as vancomycin [70, 98] or even gentamicin. We use a combination of these two drugs (1 g of vancomycin before induction of anesthesia and 500 mg every 12 hours for 24 hours plus three doses of 80 mg of gentamicin in 24 hours) in healthy individuals with normal renal function, in high-risk patients (e.g., immunosuppressed patients with total joint replacement), and in patients who have had severe immediate hypersensitivity reactions to penicillin. Those patients with pure skin reactions to penicillin are treated with cephalosporins [92].

Problems Related to the Use of Systemic Antibiotic Prophylaxis
Complications of systemic antibiotic prophylaxis are rarely reported. Allergic reactions are rare with cephalosporins. Pseudomembranous colitis may occur but is usually associated with prolonged use of the drug. However, symptoms of diarrhea and abdominal cramps should suggest the diagnosis and demand evaluation for the clostridium toxin.

Drug resistance so far has been a manageable problem. It is noteworthy, however, that cefazolin-resistant gram-negative bacteria were found in 59% of the positive swabs in a study at one institution using cefazolin prophylaxis routinely. This could be reduced to 31% within 2 years by changing the antibiotic to a different drug [49]. It seems, therefore, prudent to monitor drug resistance within an institution and adapt the prophylactic antibiotics accordingly [48].

A particular problem with antibiotic prophylaxis is that it may be exceedingly difficult to obtain compliance. The antibiotics are usually administered by the anesthesiology team at the time of induction of anesthesia. A study by Crossley and Garner [61] demonstrated that only 49% of the patients did receive prophylaxis prior to surgery, 10% received it more than 4 hours before surgery, and 17% received it after surgery.

Local Antibiotic Prophylaxis

Irrigation of the wound with solutions containing antibiotics has not been common practice in our country and is not generally recommended in Europe [98]. Petty [115] has established that the same principles apply to the use of these drugs as to the use of systemic antibiotics. The wound must be irrigated immediately after the skin is incised because irrigation with a 0.1% kanamycin and 0.1% neomycin solution within 1 minute of inoculation eradicates the bacteria, whereas a delay of 5 minutes renders irrigation with antibiotic solution useless. Rosenstein and colleagues [121] experimentally confirmed the value of irrigation with a bacitracin and polymyxin solution, but they also demonstrated that irrigation alone was as effective. Although clinical studies [73, 78, 106, 130] are controversial, it is believed that we should remain open to future advances. At present we still use Ringer's solution without admixture of antibiotics.

There is no doubt that implanted cement favors the development of wound infection [83, 116]. We believe that the value of antibiotics mixed with polymethylmethacrylate has been proved [91, 98]. Although we are aware of the impaired mechanical quality of cement to which antibiotics have been added, we use only cement to which antibiotics have been added. It should be noted that bactericidal concentrations of antibiotics are obtained in the wound after total hip replacement if cement impregnated with antibiotics is used without systemic antibiotics [98].

Intraoperative Prevention of Septic Complications

Very meticulous surgery without destruction of muscle and soft tissues as well as prevention of postoperative hematoma is of paramount importance. Gloves should be changed no later than after 2 hours of surgery [65, 123], and masks should never be worn for more than one operation. We irrigate the wounds copiously and take utmost care to close the wound as atraumatically as possible. The wound edges should be adapted perfectly to permit closure of the skin barrier within 24 to 48 hours.

Postoperative Prevention of Septic Complications

Occlusive wound dressings are used for 48 hours [103, 131]. They are then removed using sterile gloves and instruments. Should a hematoma develop, it is surgically evacuated under optimal intraoperative conditions.

It should be recognized that antibiotic prophylaxis is only one factor in preventing infection. It cannot make up for inadequate surgical technique with soft tissue damage and postoperative hematoma formation, or for poor draping or contamination of the air through excessive conversation during surgery. Only strict adherence to all the rules of aseptic technique can lead to the goal of zero percent wound infection.

TOURNIQUET

History

Temporary arrest of the circulation via compression in cases of traumatic amputation with acute hemorrhage is an instinctive move. Ligation of a limb was first reported by Ambroise Paré, who controlled hemorrhage while performing an amputation of a leg in 1552. This method became used more widely, and in 1674 Morel suggested the use of a "garrot" for circular compression of a limb before amputation or major surgery. In 1712 Jean Louis Petit described a screw device that applied pressure transcutaneously directly over an artery. He named this device a "tourniquet" (from the French *tourner*, to turn).

The next step in the development of bloodless surgery was suggested by Lister in 1864 and consisted of elevation of the extremity prior to application of the tourniquet. His aim was to reduce the blood pool, particularly the venous pool, as much as possible prior to application of the tourniquet. In 1873 von Esmarch popularized the concept of exsanguination prior to application of the tourniquet. His original technique consisted of exsanguination of the limb with a tight bandage followed by application of a rubber band, which served as a tourniquet. Esmarch gave credit to Grandesso-Sylvestry for this concept. Later, Von Langenbeck introduced a flat rubber bandage for exsanguination that has not been improved significantly to date. Martin proposed a softer latex bandage that is nowadays preferred by some surgeons because of its better handling properties.

The tourniquet itself was developed from simple cords to rubber tubes to flat rubber belts and finally to pneumatic devices derived from the Riva-Rocci blood pressure apparatus. In 1904 Cushing introduced a pneumatic tourniquet linked to a pump that inflated rapidly to above systolic pressures. He observed that slowly inflating pumps produced venous pooling by cutting off venous return before arterial inflow. Further refinements, particularly the introduction of manometers, led to the currently used devices.

In 1983 Neimkin and Smith introduced a double-cuff tourniquet. By inflating these cuffs alternately the authors prolonged tourniquet time to 3.5 hours without introducing clinically apparent sequelae [144, 171]. This is an established method of achieving a bloodless field but is rarely used because of the fear of damaging nerves and muscles through too prolonged oxygen deprivation. It is nonetheless one established possibility of extending tourniquet time in well-selected circumstances.

"Could a jeweler repair a watch immersed in ink?" asked Bunnel, and indeed, modern high-standard surgical techniques in reconstructive surgery of the extremities can be achieved only in a bloodless field. It is necessary that such a device be used correctly without jeopardizing the structure and function of the musculoskeletal tissues of the extremity.

Indications and Practical Application

If a tourniquet is used, it should be inflated only after the incision is definitely planned, draping is completed, and electrocautery and aspiration devices are in place and working. Prophylactic antibiotics must be administered 10 but at least 5 minutes before the cuff is inflated [140]. Katz [161]

recorded peak serum concentrations of antibiotics within 20 minutes after intravenous injection. Administration after cuff inflation showed low antibiotic levels with questionable antibiotic activity.

Cuff Circumference, Width, and Pressure

Crenshaw and co-workers [143] investigated the influence of cuff widths of 8, 12, and 18 cm on tissue fluid pressures. Tissue fluid pressures were maximal at midcuff; however, pressure distribution under the cuff formed a wider plateau in a 12-cm than in an 8-cm cuff. In addition, no decrease in tissue fluid pressure was found. Compared with a pressure transmission to tissues near bone of 64% in an 8-cm cuff, a 12- or 18-cm cuff provided pressure transmission of 95%. Limb circumference was found to have no influence on transmission of pressure to the deep tissues. It was concluded that with a wider cuff less pressure is required to achieve a bloodless field. Tissue pressures are highest in the midcuff area and fall to near zero at 1 to 2 cm outside the edges of the cuff [154, 176]. Pressure also decreases as a function of the depth of the soft tissue. Tissue displacement away from the midcuff area, suggested by the steep fall of pressures near the edges of the cuff, is maximal in tissue regions beneath the cuff edges [176]. Accumulation of shear stress and greater tissue deformation at this site probably cause nerve lesions beneath the tourniquet edges. In conclusion, it seems clear that the largest cuff that can be easily applied should be utilized.

Although it is widely accepted that suitable pressures for the upper extremity are 250 to 300 mm Hg and 350 to 400 mm Hg for the lower extremity, impressive pressures such as 750 mm Hg [142] or even 1150 mm Hg [166] on the thigh are occasionally recommended. In order to minimize the risk of soft tissue injury, the lowest tourniquet pressure that maintains a bloodless field should be used [182]. Shaw [182] found that pneumatic tourniquet pressures do not reflect underlying soft tissue pressures. The tissue pressure is consistently lower and decreases with soft tissue depth. This tendency becomes more pronounced as the circumference of the limb increases. In a normotensive individual of normal habitus, pressures higher than 300 to 350 mm Hg should rarely be required. However, in obese patients with significant hypertension, correspondingly higher pressures are required. The generally recommended pressures are derived from clinical experience. Fletcher [147] recommends pressures of 250 to 300 mm Hg for the upper limb and

500 to 550 mm Hg for the lower limb, regardless of the actual blood pressure. Rorabeck [178] considers 100 to 150 mm Hg above systolic pressure to be sufficient. Reid and colleagues stated that the sole predominant factor in many institutions seems to be a desire to avoid any significant bleeding distal to the tourniquet [177]. Based on his clinical investigations, in which he used an average pressure of 189.9 ± 24.1 mm Hg for the upper extremity and 231 ± 26.5 mm Hg for the lower extremity, he recommends adding 50 mm Hg to the occlusion pressure in the upper and 75 mm Hg to the occlusion pressure in the lower limb (as measured by Doppler). He also showed that there is a significant correlation between occlusion pressure, tourniquet pressure, and thigh circumference in the lower limb. Circumference of the arm did not correlate with occlusion pressure. Egkher and colleagues [145] showed also that the cuff pressure only had to be slightly higher than systolic blood pressure. The difference between systolic blood pressure and cuff pressure depends, however, on the circumference of the limb, the age of the patient, and the tissue turgor. The authors constructed a unit with a minicomputer that provided a permanent pressure control in the pneumatic cuff depending on the varying systolic blood pressures during the operation. The average difference in pressure was 10 to 70 mm Hg above systolic blood pressure, which was clinically sufficient.

In clinical practice we have found that pressures that are too close to the systolic pressure may cause bleeding, which then requires opening and reinflating the tourniquet, so that we use pressures of 150 mm Hg above systolic pressure for the upper limb and 200 mm Hg above the systolic value for the lower extremity.

Duration

The maximum safe duration of ischemia is not well established. Damage to soft tissues depends on the ischemic tolerance of these tissues, e.g., muscle and nerve. In man, the onset of irreversible muscle damage is considered to be 6 hours. Pedowitz considered 2 hours to be the threshold for skeletal muscle injury induced by tourniquet compression at clinically relevant inflation pressures [176], and indeed, clinical problems occur when tourniquet times of more than 2 hours are necessary. Many investigations have been carried out to determine the optimal duration of tourniquet ischemia. Green [152] reported periods ranging from 45 minutes to 4 hours. Like most others, he considered a tourni-

quet time of 2 hours to be most widely acceptable. Although the tolerance to ischemia is longer, it seems well established that greater muscle injury is induced *beneath* the tourniquet than distal to it. This can be diminished if, for example, for a 4-hour total tourniquet time hourly 10-minute reperfusion times are allowed. This results in significantly less muscle injury beneath the cuff than is possible without intermittent reperfusion.

The effect of ischemia on muscle metabolism has been widely investigated to determine optimal tourniquet duration. Post-tourniquet serum level of myoglobins has been found to be either initially increased with a subsequent decrease occurring over 6 to 8 hours [160] or unchanged after 2 hours of tourniquet time [163], indicating no detectable muscle cell damage. Based on the normalization of intravascular pH, pO_2, and pCO_2, Wilgis [186] recommended the release of the tourniquet after 90 minutes for 10 minutes and then reinflation for another 90 minutes, which allows a tourniquet time of 3 hours. However, he observed that in the post-tourniquet time little or no oxygen diffusion took place across the capillary bed and that many capillaries were bypassed via meta-arteriolar shunts. Sapega and associates [180] recommended an initial period of 1½ hours with a release for a period of 5 minutes and an additional 90 minutes of ischemia because a normal intracellular high-energy phosphate profile had been reestablished within 5 minutes. Newman [172] showed by phosphorus-31 nuclear magnetic resonance spectroscopy that hourly release periods of 10 minutes prevented ATP depletion and maintained rapid metabolic recovery. Release periods of 5 minutes did not prevent ATP depletion and in addition caused deterioration in tissue pH and prolonged recovery time. Rorabeck [178] showed in an experiment on the sciatic nerve of dogs that the slowing in conduction velocity and the time required for it to return to normal after release of the tourniquet varied directly with the amount and duration of applied pressure. A tourniquet applied for 1 hour was always followed by a complete return of conduction velocity to normal, but a tourniquet inflated for 3 hours to 500 mm Hg caused permanent slowing in the conduction velocity. Mechanical interference with arterial supply of the nerve is the most important cause of tourniquet palsy, which is due to impairment and recovery of conduction velocity in the portion of the nerve beneath the tourniquet.

The disadvantages of temporary release of the tourniquet are swollen and bloodstained tissues, so that many structural details become obscured [144]. This problem can be bypassed with the use of a double-cuff tourniquet, as described by Neimkin and colleagues, which allows ischemia times of up to 4 hours for the upper extremity [171]. By sequential inflation of the cuffs at hourly intervals, nerve compression at one anatomic location can be decreased. Since the limiting factor of the conventional cuff is the onset of neural dysfunction, this seems an acceptable approach. The main disadvantage of the double cuff, which is not discussed by the authors, is the need for smaller cuffs, which in turn sets the underlying tissues under higher absolute pressures and shear forces. This could explain why 70% of the patients in the series of Dreyfuss and colleagues [144] showed a transient decrease in subjective sensibility.

If tourniquet duration is longer than 2 hours, a "breathing" period of 10 minutes is often recommended after 90 minutes before the cuff is inflated for another 90 minutes. Once removed, the tourniquet should not be reapplied until the pH, pO_2, and pCO_2 have returned to normal. Some authors (Gershuni, personal communication) have experimental evidence, however, that this approach is no more favorable than a continued application of 3 hours' duration.

For training purposes, deflating the tourniquet before wound closure should be the standard approach. It is preferable to literally remove the tourniquet from the extremity to prevent venous stasis. In the post-tourniquet phase, it seems important to wait out the recovery phase before attempting to cauterize bleeders [186]. Application of a circumferential cast prior to tourniquet release may be hazardous because limb swelling immediately after release is approximately 10% [183]. It is our clinical practice to release the tourniquet after 2 hours and allow a breathing period of at least 30 minutes. Antibiotic prophylaxis is repeated, and the tourniquet then reinflated.

Contraindications to Tourniquet Use

There are contraindications to the use of the tourniquet and contraindications to the use of exsanguination. Contraindications for the use of tourniquets are:

Chronic ischemia of the limb because the chronically hypoxic limb tolerates complete anoxia very poorly.

Calcification of major feeding vessels: They may be incompressible, and the tourniquet may lead to increased bleeding due to venous stasis.

Severe crush injury with tissue hypoxia (to prevent additional damage).

Sickle cell anemia may constitute a relative contraindication. Hypoxia and acidosis are ideal conditions for inducing sickling of erythrocytes with the risk of thrombosis, infarction, and hemolysis. There are reports, however, that provided the limb is carefully exsanguinated, the patient is at no particular risk [147].

In cases of proved or suspected deep venous thrombosis no tourniquet should be used because of the danger of mobilization of a thrombus leading to potential pulmonary embolism [139, 156, 165, 168, 170].

Exsanguination is contraindicated

In infective states
In tumor surgery (including biopsies)
In cases of proved or suspected deep venous thrombosis no tourniquet should be used because of the danger of mobilization of a thrombus leading to subsequent pulmonary embolism [139, 156, 165, 168, 170].

Fletcher reports a case of cardiac arrest following bilateral leg exsanguination [147]. Exsanguination of more than one limb at a time results in an effective increase in circulating blood volume of up to 800 ml. This should be taken into account in patients with poor cardiac reserve.

Fractures are also considered contraindications except for the hand, where more proximally an exsanguination is debatable.

Extreme care must be taken in patients receiving steroid treatment (e.g., rheumatoid arthritis) because their skin may dissociate during the application of the Esmarch band, resulting in an exfoliative dermatitis, which may be a devastating complication.

Complications

Although the tourniquet has become universally accepted and reported complications are rare [175], its use is not without dangers [152]. The tourniquet is potentially harmful, but awareness of the potential problems and meticulous attention to detail can prevent or at least minimize the known complications. Most complications are related either to the duration of ischemia, which is easily controlled, or to the pressure exerted by the pneumatic cuff. Other complications are related to concomitant vascular or metabolic disorders.

Today post-tourniquet paralysis is essentially related to accidentally high pressures caused by defective pressure gauges [137, 162]. Ischemic damage to the muscle fibers seems not to be a problem because the tolerance of muscle ischemia is about 6 hours, a duration that is never reached in clinical practice. It is known, however, that hypothermia slows the muscular metabolism, enhances resistance to anoxia, and hastens postanoxic recovery [158]. As opposed to hypothermia, the effect of steroids is controversial. Newman [172] denies any effect of steroids or heparin on metabolic function based on his study using phosphorus-31 NMR spectroscopy. In contrast, Goto and colleagues [151] found that changes in acidosis, oxygen saturation, and enzyme release were more favorable in the group treated with methylprednisolone, although the mechanism for these effects remains unclear. Measurements of myoglobin serum levels after tourniquet release are somewhat controversial [157, 163], but it appears that with the usual application times, significant muscle fiber destruction, as evidenced by increased myoglobin serum levels, is not observed. Compartment syndromes have occurred only rarely after the use of pneumatic tourniquets [153, 175]. O'Neil reported a transient compartment syndrome of the forearm resulting from faulty tourniquet function [174]. Patients with McArdle's disease, a muscle phosphorylase deficiency disease, develop compartmentlike symptoms with vigorous activity [164], and the use of a pneumatic tourniquet for only a short time may lead to compartment syndrome in patients with this rare disease [175].

Common also is damage to nerve conduction, ranging from mild neurapraxia to complete nerve destruction [175] due to ischemia and concentration of pressure at the edges of the pneumatic cuff. Nitz et al. reported a 60% to 85% incidence of electromyographically detectable muscle denervation after tourniquet use [173]. Complete paralysis following tourniquet application in humans, however, is rare (1 in 8000 cases according to Middleton [167]). Since the introduction of the pneumatic cuff and the subsequent decrease in the use of the Esmarch bandage, complete nerve palsies have become even more rare [162]. This is explained by an experiment performed by McLaren and colleagues, which showed in mongrel dogs that pressure concentrations under the rubber bandage of Esmarch were constant and relevant, whereas pressure concentration could not be found under pneumatic tourniquets [169]. In a well-arranged experiment with rats, Nitz and associates [173] tried to determine the role of pressure and ischemia on neural function. They compared tourniquet-induced alterations of nerve conduction with vascular manipulation of the sciatic nerve by occluding the extrinsic and the intrinsic blood supply to the nerve. They showed that blood flow during tourniquet application was totally oc-

cluded in the intraneural vessels. Electromyographic (EMG) and compound muscle action potential (CMAP) abnormalities identified in muscles of animals subjected to tourniquet compression were similar to those treated by vascular occlusion and were attributed to nerve injury. These findings suggest that ischemia most likely causes loss of nerve function in those animals treated with vascular occlusion. However, no direct evidence is available that relates the compression component associated with tourniquet application to pressure.

Degenerative alterations of the main arteries may give rise to other problems. In patients with Monckeberg's calcinosis [159] or simple arteriosclerosis the tourniquet may fail to produce or maintain a bloodless field owing to incompressibility of the arterial vessels. Thrombosis of an artery due to disruption of an atheromatous plaque in a superficial femoral artery under tourniquet application was reported by Giannestras [149] but must be extremely rare. The formation of venous thrombi seems not to be affected by a pneumatic tourniquet [138].

Pulmonary embolism, however, has been described following exsanguination with an Esmarch bandage and application of a pneumatic tourniquet [181]. Embolism has occurred during exsanguination and application of the tourniquet [139, 156, 165] and after deflation [168, 170] owing to mobilization of a preexisting thrombus.

Pressure sores as a result of soft tissue kinking beneath the tourniquet are uncommon and usually occur in relatively obese patients with abundant skin and subcutaneous tissue in relation to muscle mass. Carefully applying the tourniquet over a padding prevents this complication. A similar problem is burning injuries, which are not well documented in the literature but seem to be rather frequent in clinical practice. If the tourniquet is applied and then alcoholic disinfectants are applied to an elevated extremity, the disinfectant will flow under the tourniquet, placed under pressure, and then cause burns, which may be very painful and occasionally disfiguring. It is therefore crucial that the surgeon ensure that the disinfectant cannot flow under the tourniquet.

Post-tourniquet swelling is more pronounced after prolonged tourniquet time or initially insufficient pressure, which allows arterial inflow [175]. Silver showed that an exsanguinated limb will swell by approximately 10% of its original volume after release of the tourniquet, of which 5% is contributed by the exsanguination [183].

Two specific complications have been reported with the use of digital tourniquets. One is digital artery spasm due to excessive pressure beneath the tourniquet, and the second is digital gangrene, which results from inadvertent failure to remove the tourniquet from the base of the finger [175].

It is obvious that almost all these complications can be avoided if careful handling of the tourniquet is ensured.

Authors' Preferred Method

For reconstructive surgery of the extremities we use the largest possible tourniquet. We prefer to use a sterile tourniquet because this prevents the problems of burning injuries under the cuff, gives easy access in case of dysfunction, and facilitates removal at the end of the operation for hemostasis. The cuff is tightly applied over a padding as proximally as possible at the root of the extremity 10 minutes after administration of antibiotic prophylaxis. Exsanguination is used when not contraindicated but never in patients with fractures. The pressures used are approximately 150 mm Hg above systolic pressure for the upper extremity and 200 mm Hg above systolic pressure for the lower limb. We never extend tourniquet time beyond 2 hours. After this time the tourniquet is released, antibiotic prophylaxis is repeated, and the tourniquet is not reinflated sooner than approximately 30 minutes. The tourniquet is always released before skin closure so that optimal hemostasis can be obtained.

References

Suture Material

1. Aston, S. J. The choice of suture material for skin closure. *J Derm Surg* 2:57–61, 1976.
2. Beck, M. Fixation der Infraspinatussehne in einer Knochennute am Humeruskopf des Schafs. Thesis, University of Bern, Switzerland, 1992.
3. Birdsell, D. C., Tustanoff, E. R., and Lindsay, W. K. Collagen production in regenerating tendon. *Plast Reconstr Surg* 37:504–511, 1966.
4. Blomstedt, B., Osterberg, B., and Bergstrand, A. Suture material and bacterial transport. *Acta Chir Scand* 143:71–73, 1977.
5. Bourne, R. B., Bitar H., and Andreae, P. R. In vivo comparison of four absorbable sutures: Vicryl, Dexon Plus, Maxon and PDS. *Can J Surg* 31:43–45, 1988.
6. Breuninger, H., and Haueisen, S. Anwendung von resorbierbarem Nahtmaterial in der operativen Dermatologie. *Z Hautkr* 60:453–457, 1984.
7. Caspari, R. B. Current development of instrumentation for arthroscopy. *Clin Sports Med* 6:619–636, 1987.
8. Clough, J. V., and Alexander-Williams, J. Surgical and economic advantages of polyglycolic acid suture material in skin closure. *Lancet* 1:194–195, 1975.
9. Cowan, R. J., and Courtemanche, A.D. An experimental study of tendon suturing techniques. *Can J Surg* 2:373–379, 1959.
10. Craig, P.H., Williams, J. A., Davis, K. W., Magoun, A.D.,
11. Levy, A. J., Bogdansky, S., and Jones, J. P. A biologic comparison of polyglactin 910 and polyglycolic acid synthetic absorbable sutures. *Surg Gynecol Obstet* 141:1–10, 1975.
12. Deutsch, H. L. Observations on a new absorbable suture material. *J Derm Surg* 1:49–51, 1975.
13. Duprez, K., Bilweis, J., Duprez, A., and Merle, M. Experimental and clinical study of fast absorption cutaneous suture material. *Ann Chir Main* 7:91–96, 1988.
14. Forward, A.D., and Cowan, R. J. Tendon suture to bone. An experimental investigation in rabbits. *J Bone Joint Surg* 45A:807–823, 1963.
15. Gerber, C., Beck, M., Schlegel, U., and Schneeberger, A.G. The in vitro stability of different rotator cuff repair techniques. Presented at the Annual Meeting of the American Shoulder And Elbow Surgeons, Washington, D.C., February, 1992.
16. Haaf, U., and Breuninger, H. Resorbierbares Nahtmaterial in der menschlichen Haut: Gewebereaktion und modifizierte Nahttechnik. *Hautarzt* 39:23–27, 1988.
17. Hartko, W.J., Ghanekar, G., and Kemmann, E. Suture materials currently used in obstetric-gynecologic surgery in the United States: A questionnaire survey. *Obstet Gynecol* 59:241–246, 1982.
18. Henke, R., and Kairies U. Neues synthetisches resorbierbares Nahtmaterial im Tierexperiment und in der ersten klinischen Testung. *Z Chirurg* 109:641–650, 1984.
19. Herrmann, J. B., Kelly, R.J., and Higgins, G. A. Polyglycolic acid sutures. *Arch Surg* 100:486–490, 1970.
20. Herrmann, J. B. Tensile strength and knot security of surgical suture materials. *Am Surg* 7:209–217, 1971.
21. Holmlund, D. E. W. Knot properties of surgical suture materials. A model study. *Acta Chir Scand* 140:355–362, 1974.
22. Krackow, K. A., Thomas, S. C., and Jones, L. C. Ligament–tendon fixation: An analysis of a new stitch and comparison with standard techniques. *Orthopedics* 11:909–917, 1988.
23. Lindsay, W. K., Thomson, H. G., and Walker, F. G. Digital flexor tendons: An experimental study. Part II: The significance of a gap occuring at the line of suture. *Br J Plast Surg* 13:1–9, 1960.
24. Mason, M. L., and Shearon, C. G. Process of tendon repair: Experimental study of tendon suture and tendon graft. *Arch Surg* 25:615–692, 1932.
25. Mason, M. L. Primary and secondary tendon suture: A discussion of the significance of the technique in tendon surgery. *Surg Gynecol Obstet* 70:392–402, 1940.
26. Mast, J. W., Jakob, R. P., and Ganz, R. Planning and reduction technique in fracture surgery. Berlin, Springer-Verlag, 1989, p. 250.
27. O'Donoghue, D. H., Frank, G. R., Jeter,G. L., Johnson, W., Zeiders, J. W., and Kenyon, R. Repair and reconstruction of the anterior cruciate ligament in dogs. Factors affecting long term results. *J Bone Joint Surg* 53A:710–718, 1971.
28. Peacock, E. E. Some technical aspects and results of flexor tendon repair. *Surgery* 58:330–342, 1965.
29. Roedeheaver, G. T., Nesbit, W.S., and Edlich, R. F. Novafil, a dynamic suture for wound closure. *Ann Surg* 204:193–199, 1986.
30. Roedeheaver, G.T., Thacker, J. G., and Edlich, R. F. Mechanical performance of polyglycolic acid and polyglactin 910 synthetic absorbable sutures. *Surg Gynecol Obstet* 153:835–841, 1981.
31. Roedeheaver, G.T., Thacker, J. G., and Owen, B.S. Knotting and handling characteristics of coated synthetic absorbable sutures. *J Surg Res* 35:525–530, 1983.
32. Sanz, L. E., Patterson, J. A., Kamath, R., Willett, G., Ahmed, S. W., and Butterfield, A.B. Comparison of Maxon suture with Vicryl, chromic catgut, and PDS sutures in fascial closures in rats. *Obstet Gynecol* 71:418–422, 1988.
33. Schneeberger, A. G., Schlegel, U., and Gerber C. Mechanical properties of suture materials commonly used in

orthopaedic surgery. An in vitro study of suture sizes 0 to 3. In press, 1992.

33. Seradge, H. Elongation of the repair configuration following flexor tendon repair. *J Hand Surg* 8:182–185, 1986.

34. Srugi, S., and Adamson, J. E. A comparative study of tendon suture materials in dogs. *Plast Reconstr Surg* 50:31–35, 1972.

35. Stone, I. K., Masterson, B. J., and von Frauenhofer, J. A. Knot stability and tensile strength of an absorbable suture material. *Surface and Coatings Technology* 27:287–293, 1986.

36. Stone, I. K., von Fraunhofer, J. A., and Masterson, B. J. Mechanical properties of coated absorbable multifilament suture materials. *Obstet Gynecol* 67:737–740, 1986.

37. Tera, H., and Åberg, C. Strength of knots in surgery in relation to type of knot, type of suture material and dimension of suture thread. *Acta Chir Scand* 143:75–83, 1977.

38. Tera, H., and Åberg, C. Tensile strengths of twelve types of knot employed in surgery, using different suture materials. *Acta Chir Scand* 142:1–7, 1976.

39. Trail, I. A., Powell, E. S., and Noble, J. An evaluation of suture materials used in tendon surgery. *J Hand Surg* 14B:422–427, 1989.

40. Urbaniak, J. R., Cahill, J. D., and Mortensen, R. A. Tendon suturing methods: Analysis of tensile strengths. *In* American Academy of Orthopaedic Surgeons. *Symposium on Tendon Surgery in the Hand.* St. Louis, C.V. Mosby, 1975, pp. 70–80.

41. von Frauenhofer, J. A., Storey, R. J., and Masterson, B. J. Tensile properties of suture materials. *Biomater* 9:324–327, 1988.

42. von Frauenhofer, J. A., Storey, R. S., Stone, I. K., and Masterson, B. J. Tensile strength of suture materials. *J Biomed Mater Res* 19:595–600, 1985.

43. Webster, R. C., McCullough, E. G., Giandello, P. R., and Smith, R.C. Skin wound approximation with new absorbable suture material. *Arch Otolaryngol* 111:517–519, 1985.

44. Whitley, J. Q., Prewitt, M. J., and Kusy, R. P. Relationship of the diameter and tensile strength of nylon sutures to the USP specification and the effect of preconditioning. *J Appl Biomat* 1:315–320, 1990.

45. Willat, D. J., Durham, L., Ramadan, M. F., and Bark-Jones, N. A prospective randomized trial of suture material in aural wound closure. *J Laryngol Otol* 102:788–790, 1988.

46. Woo, S. L., and Buckwalter, J. A. (Eds.) *Injury and Repair of the Musculoskeletal Soft Tissues.*

47. Zechner, W., Buck-Gramcko, D., Lohmann, H., and Stock, W. Ueberlegungen zur Verbesserung der Nahttechnik bei Beugesehnen-verletzungen, klinische und experimentelle Studie. *Handchir Mikrochir Plast Chir* 17:8–13, 1985.

Antibiotic Prophylaxis

48. Aebi, B., Gerber, C., and Ganz, R. Infektprophylaxe bei orthopädischen Wahleingriffen mit besonderer Berücksichtigung des alloplastischen Gelenkersatzes. *Helv Chir Acta* 56:387–397, 1989.

49. Aglietti, P., Salvati, E. A., Wilson, P. D., and Kutner, L. J. Effect of a surgical unidirectional filtered air flow unit on wound bacterial contamination and wound healing. *Clin Orthop* 101:99–104, 1974.

50. Boyd, R. J., Burke, J. F., and Colton, T. A. A double-blind clinical trial of prophylactic antibiotics in hip fractures. *J Bone Joint Surg* 55A:1251–1258, 1973.

51. Burke, J. F. Identification of the sources of staphylococci contaminating the surgical wound during operation. *Ann Surg* 158:898–904, 1963.

52. Burke, J. F. The effective period of preventive antibiotic action in experimental incisions and dermal lesions. *Surgery* 50:161–168, 1961.

53. Burke, J. F. Use of preventive antibiotics in clinical surgery. *Am Surg* 39:6–11, 1973.

54. Burnett, J. W., Gustilo, R. B., Williams, D. N., and Kind, A. C. Prophylactic antibiotics in hip fractures. *J Bone Joint Surg* 62A:457–462, 1974.

55. Carlsson, A. S., Lindgren, L., and Lindberg, L. Prophylactic antibiotics against early and late deep infections after total hip replacement. *Acta Orthop Scand* 48:405, 1977.

56. Chang, N., Goodson, W. H., Gottrup, F., and Hunt, T. K. Direct measurement of wound and tissue oxygen tension in postoperative patients. *Ann Surg* 197:470–478, 1983.

57. Charnley, J. Postoperative infection after total hip replacement with special reference to air contamination in the operating room. *Clin Orthop* 87:167–187, 1972.

58. Chodak, G. W., and Plaut, M. E. Use of systemic antibiotics for prophylaxis in surgery: A critical review. *Arch Surg* 112:326–334, 1977.

59. Connell, J. F., and Rousselot, L. M. Povidone iodine: Extensive surgical evaluation of a new antiseptic agent. *Am J Surg* 108:849–855, 1964.

60. Conte, J. E., Cohen, S. N., Roe, B. B., and Elasthoff, R. M. Antibiotic prophylaxis and cardiac surgery, a prospective double blind comparison of single dose versus multiple dose regimens. *Ann Intern Med* 76:943–949, 1972.

61. Crossley, K. B., and Gardner, L. C. Antimicrobial prophylaxis in surgical patients. *JAMA* 245:722, 1981.

62. Cruse, P. J. E., and Foord, R. A Five-year prospective study of 23,649 surgical wounds. *Arch Surg* 107:206–210, 1973.

63. Culbertson, W. R., Altmeier, W. A., Gonzales, L. L. and Hill, E. O. Studies on the epidemiology of clean operative wounds. *Ann Surg* 154:599–610, 1961.

64. DiPiro, J. T., Bowden, T. A., and Hooks, V. H. Prophylactic parenteral cephalosporins in surgery: Are the newer agents better? *JAMA* 252:3277–3279, 1984.

65. Doyon, F., Evrard, J., and Mazas, F. Evaluation des essais thérapeutiques publiés sur antibioprophylaxie en chirurgie orthopédique. *Rev Chir Orthop* 75:72–76, 1989.

66. Ducel, G. La préparation du chirurgien. *Cahiers d'enseignement de la SOFCOT 37* Paris, Expansion Scientifique, 1990, pp. 61–65.

67. Elek, S. D., and Conen, P. E. The virulence of *Staphylococcus pyogenes* for man. A study of the problem of wound infection. *Br J Exp Pathol* 38:573, 1957.

68. Ericson, C., Lidgren, L., and Lindberg, L. Cloxacillin in the prophylaxis of postoperative infections of the hip. *J Bone Joint Surg* 55A:808–813, 1973.

69. Evrard, J., Doyon, F., Acar, J. F., Salord, J. C., Mazas, F., and Flamant, R. Two-day cefamandole versus five-day cephazolin prophylaxis in 965 total hip replacements. Report of a multicentre double-blind randomised trial. *Int Orthop* 12:69–73, 1988.

70. Farber, B. F., Karchmer, A. E., Buckley, M. J., and Moellering, R. C. Vancomycin prophylaxis in cardiac operations: Determinations of an optimal dosage regimen. *J Thorac Cardiovasc Surg* 933–935, 1983.

71. Fitzgerald, R. H., and Thompson, R. L. Cephalosporin antibiotics in the prevention and treatment of musculoskeletal sepsis. *J Bone Joint Surg* 65A:1201–1205, 1983.

72. Fogelberg, E. V., Zetzman, E. K., and Stinchfield, F. E. Prophylactic penicillin in orthopaedic surgery. *J Bone Joint Surg* 52A:95–98, 1970.

73. Forbes, G. B. Staphylococcal infection of operation wounds with special reference to topical antibiotic prophylaxis. *Lancet* 2:505–509, 1961.

74. Garcia, S., Lozano, M. L., Gatell, J. M., Soriano, E., Ramon, R., and Sanmiguel, J. G. Prophylaxis against infection. Single dose cefonicid compared with multiple-dose cefamandole. *J Bone Joint Surg* 73A:1044–1048, 1991.

75. Gatell, J. M., Garcia, S., Lozano, L., Soriano, E., Ramon, R., and Sanmiguel, J. G. Perioperative cefamandole prophylaxis against infections. *J Bone Joint Surg* 69A:1189–1193, 1987.

76. Gatell, J. M., Riba, J., Lozano, M. L., Mana, J., Ramon, R., and Sanmiguel, J. G. Prophylactic cephamandole in orthopaedic surgery. *J Bone Joint Surg* 66A:1219–1222, 1984.

77. Ghosh, J., Maisels, D.O., and Woodcock, A.S. Preoperative skin disinfection. *Br J Surg* 54:551–553, 1967.
78. Gingrass, R. P., Close, A. S., and Ellison, E. H. The effect of various topical and parenteral agents on the prevention of infection in experimental and contaminated wounds. *J Trauma* 4:763–783, 1964.
79. Goldblum, S. E., Ulrich, J. A., Goldman, R. S., Reed, W. P., and Avasthi, P. S. Comparison of 4% chlorhehidene gluconate in a detergent base (Hibiclens) and povidone-iodine (Betadine) for the skin preparation of hemodialysis patients and personnel. *Am J Kidney Dis* 2:548–554, 1983.
80. Goldmann, D. A., Hopkins, C. C., Karchmer, A. W., et al. Cephalotin prophylaxis in cardiac valve surgery: A prospective double-blind comparison of two-day and six-day regimens. *J Thorac Cardiovasc Surg* 73:470–479, 1977.
81. Green, D. P. General principles. *In* Green, D. P. *Operative Hand Surgery,* 2nd ed. New York, Churchill Livingstone, 1988, pp. 3–7.
82. Green, J. W., and Wenzel, R. P. Postoperative wound infection: A controlled study of the increased duration of hospital stay and direct cost of hospitalization. *Ann Surg* 185:264–268, 1977.
83. Gristina, A. G., Rovere, G. D., Shoji, J., and Nicastro, J. P. An in vitro study of bacterial response to inert and reactive materials and to methyl-methacrylate. *J Biomed Mat Res* 10:273–281, 1984.
84. Guglielmo, B. J., Hohn, D. C., Koo, P. J., Hunt, T. K., Sweet, R. L., and Conte, J. E. Antibiotic prophylaxis in surgical procedures: A critical analysis of the literature. *Arch Surg* 118:943–955, 1983.
85. Hill, C., Mazas, F., Flamant, R., and Evrard, J. Prophylactic cephazolin versus placebo in total hip replacement. *Lancet* 1:795–797, 1981.
86. Hirschmann, J. V., and Inui, T. S. Antimicrobial prophylaxis: Critique of recent trials. *Rev Infect Dis* 2:1, 1982.
87. Howe, C. W. Postoperative wound infection due to *Staphylococcus aureus*. *N Engl J Med* 251:411–417, 1954.
88. Howe, C. W., and Marston, A. T. A study of sources of postoperative staphylococcal infection. *Surg Gynecol Obstet* 115:266–275, 1962.
89. Jones, S., DiPiro, J. T., Nix, D. E., and Bhatti, N. A. Cephalosporins for prophylaxis in operative repair of femoral fractures: Levels in serum, muscle, and hematoma. *J Bone Joint Surg* 67A:921–924, 1985.
90. Joress, S. M. A study of disinfection of the skin. A comparison of povidone-iodine with other agents used for surgical scrubs. *Ann Surg* 155:296–304, 1962.
91. Josefsson, G., Lindberg, L., and Wiklander, B. Systemic antibiotics and gentamicin-containing bone cement in the prophylaxis of post-operative infections in total hip arthroplasty. *Clin Orthop* 159:194–200, 1981.
92. Kaiser, A. B. Antimicrobial prophylaxis in surgery. *N Engl J Med* 315:1129–1138, 1986.
93. Kaiser, A. B. Surgical wound infections. *N Engl J Med* 324:123–124, 1991.
94. Kaiser, A. B., Herrington, J. L., Jacobs, J. K., Mulherin, J. L., Roach, A.C., and Sawyers, J. L. Cefoxitin versus erythromycin, neomyin and cefazolin in colorectal operations: Importance of the duration of the surgical procedure. *Ann Surg* 198:525–530, 1983.
95. Knighton, D. R., Halliday, B., and Hunt, T. K. Oxygen as an antibiotic: The effect of inspired oxygen on infection. *Arch Surg* 119:199–204, 1984.
96. Krieger, J. N., Kaiser, D. L., and Wenzel, R. P. Nosocomial urinary tract infections cause wound infections postoperatively in surgical patients. *Surg Gynecol Obstet* 156:313–318, 1983.
97. Lidwell, O. M., Lowbury E. J. L., Whyte, W., Blowers, R., Stanley, S. J., and Lowe, D. Effect of ultraclean air in operating rooms on deep sepsis in the joint after total hip or knee replacement: A randomised study. *Br Med J* 285:10–14, 1982.
98. Lindberg, L. Antibiothérapie prophylactique en chirurgie orthopédique. *Cahiers d'enseignement de la SOFCOT 37.* Paris, Expansion Scientifique, 1990, pp. 66–75.

99. Lindberg, L. Evaluation bactériologiques des infection orthopédiques postopératoires. *Cahiers d'enseignement de la SOFCOT 37.* Paris, Expansion Scientifique, 1990, pp. 80–83.
100. Lister, J. On the antiseptic principles in the practice of surgery. *Lancet* 2:353, 1867.
101. Lowbury, E. J. L., Lilly, H. A., and Bull, J. P. Disinfection of the skin of operation sites. *Br Med J* 2:1039–1044, 1960.
102. Mader, J. T., and Wilson, K. J. Comparative evaluation of cephamandole versus cephalotin in the treatment of experimental *Staphylococcus aureus* osteomyelitis in rabbits. *J Bone Joint Surg* 65A:507–513, 1983.
103. Matter, P. Pre-, Intra-, and postoperative guidelines. *In* Müller, M. E., Allgöwer, M., Schneider, R., and Willenegger, H. *Manual of Internal Fixation,* 3rd ed. Heidelberg, Springer-Verlag, 1991, pp. 413–415.
104. Marotte, J. H., Frottier, J., Cazalet, G., Lord, G., Blanchard, J. P., and Guillamon, J. L. Antibiothérapie préventive et infection post-opératoire en chirurgie orthopédique. 1983 prothéses totales de hanche. *Rev Chir Orthop* 71:79–86, 1985.
105. Nach, D. C., and Keim, H. A. Prophylactic antibiotics in spinal surgery. *Orthop Rev* 2:27–30, 1973.
106. Nachamie, B. A., Siffert, R. S., and Bryer, M. S. A study of neomycin instillation into orthopaedic surgical wounds. *JAMA* 204:139–141, 1968.
107. Nelson, C. L., Green, T. G. , Porter, R. A., and Warren, R. D. One day versus seven days of preventive antibiotic therapy in orthopaedic surgery. *Clin Orthop* 176:258–263, 1983.
108. Neu, H. Cephalosporin antibiotics as applied in surgery of bones and joints. *Clin Orthop* 190:50–64, 1984.
109. Nichols, R. L. Postoperative wound infection. *N Engl J Med* 307:1701–1702, 1982.
110. Nichols, R. L. Techniques known to prevent postoperative wound infection. *Infect Control* 3:34–37, 1982.
111. Norden, C.W. A critical review of antibiotic prophylaxis in orthopaedic surgery. *Rev Infect Dis* 5:928–932, 1983.
112. Pasteur, L. Mémoire sur les corpuscules organisées qui existent dans l'atmosphère. *Ann Sci Nat* 16:5, 1861. Quoted in Peltier, L. F. *Fractures, A History and Iconography of Their Treatment.* San Francisco, Norman Publishing, 1990.
113. Pavel, A., Smith, R. I., Ballard, A., and Larsen, I. J. Prophylactic antibiotics in clean orthopaedic surgery. *J Bone Joint Surg* 56A:777–782, 1974.
114. Peltier, L. F. *Fractures, A History and Iconography of Their Treatment.* San Francisco, Norman Publishing, 1990.
115. Petty, W. Quantitative determination of the effect of antimicrobial irrigating solutions on bacterial contamination of experimental wounds. *Trans Orthop Res Soc* p. 9, 1979.
116. Petty, W., Spanier, S., Shuster, J. J., and Silverthorne, C. The influence of skeletal implants on incidence of infection. *J Bone Joint Surg* 67A:1236–1244, 1985.
117. Pollard, J. P., Hughes, S. P. F., Scott, J. E., Evans, M. J., and Benson, M. K. D. Antibiotic prophylaxis in total hip replacement. *Br Med J* 1:707–709, 1979.
118. Polk, H. C., Trachtenberg, L., and Finn, M. P. Antibiotic activity in surgical incisions: The basis for prophylaxis in selected operations. *JAMA* 244:1353–1135, 1980.
119. Quintiliani, R., and Nightingale, C. Principles of antibiotic usage. *Clin Orthop* 190:31–35, 1984.
120. Ritter, M. A., French, M. L. V., Eitzen, H. E., and Gioe, T. J. The antimicrobial effectiveness of operative site preparative agents. *J Bone Joint Surg* 62A:826–828, 1980.
121. Rosenstein, B. D., Wilson, F. C., and Funderburk, C. H. The use of bacitracin irrigation to prevent infection in postoperative skeletal wounds. An experimental study. *J Bone Joint Surg* 71A:427–430, 1989.
122. Sanchez-Ubeda, R., Fernand, E., and Rousselot, L. M. Complication rate in general surgical cases: The value of penicillin and streptomycin as postoperative prophylaxis—a study of 511 cases. *N Engl J Med* 259:1054–1050, 1958.
123. Sanders, R., Fortin, P., Ross, E., and Helfet, D. Outer gloves in orthopaedic procedures. Cloth compared with latex. *J Bone Joint Surg* 72A:914–917, 1990.

124. Scales, J. T., Towers, A. G., and Roantree, B. M. The influence of antibiotic therapy on wound inflammation and sepsis associated with orthopaedic implants: A long-term clinical survey. *Acta Orthop Scand* 43:85–100, 1972.

125. Schonholtz, G. J., Borgia, C. A., and Blair, J. D. Wound sepsis in orthopaedic surgery. *J Bone Joint Surg* 44A:1548–1552, 1962.

126. Sheftel, T. G., Mader, J. T., and Pennick, J. J. Methicillin-resistant *Staphylococcus aureus* osteomyelitis. *Clin Orthop* 198:231–239, 1985.

127. Smith, B. R., Rolston, K. V., LeFrock, J. L., and Bayliff, C. Bone penetration of antibiotics. *Orthopaedics* 6:187–193, 1983.

128. Southwood, R. J., Rice, J. L., McDonald, P. J., Hakendorf, P. H., and Rozenbilds, M. A. Infection in experimental hip arthroplasties. *J Bone Joint Surg* 67B:229–231, 1985.

129. Stevens, D. P. Postoperative orthopaedic infections. A study of etiological mechanisms. *J Bone Joint Surg* 46A:96–102, 1964.

130. Stone, H. H., Haney, B. B., Kolb, L. D., Geheber, C. E., and Hooper, C. A. Prophylactic and preventive antibiotherapy: Timing, duration and economics. *Ann Surg* 189:691–699, 1979.

131. Vecsei, V. Prévention de l'infection après l'opération. *Cahiers d'enseignement de la SOFCOT 37.* Paris, Expansion Scientifique, 1990, pp. 84–93.

132. Waterman, N. G., and Pollard, N. T. Local antibiotic treatment of wounds. *In* Maibach, H. I., and Rovee, D. T. (Eds.), *Epidermal Wound Healing.* Chicago, Year Book Medical Publishers, 1972, pp. 267–280.

133. Williams, D. N. Antibiotic prophylaxis in bone and joint surgery. *In* Gustilo, R. B., Grueninger, R. P., Tsukayama, D. T. (Eds.), *Orthopaedic Infection.* Philadelphia, W. B. Saunders, 1989, pp. 60–65.

134. Williams, D. N., and Gustilo, R. B. The use of preventive antibiotics in orthopaedic surgery. *Clin Orthop* 190:83, 1984.

135. Williams, D. N., Gustilo, R. B., Beverly, R., and Kind, A. C. Bone and serum concentrations in five cephalosporin drugs: Relevance to prophylaxis and treatment in orthopaedic surgery. *Clin Orthop* 179:253, 1983.

136. Worlock, P., Slack, R., Harvey, L., and Mawhinney, R. The prevention of infection in open fractures. An experimental study of the effect of antibiotic therapy. *J Bone Joint Surg* 70A:1341–1347, 1988.

Tourniquet

137. Aho, K., et al. Pneumatic tourniquet paralysis. Case report. *J Bone Joint Surg* 65B:441–443, 1983.

138. Angus, P. D., et al. The pneumatic tourniquet and deep venous thrombosis. *J Bone Joint Surg* 65B:336–339, 1983.

139. Araki, S., et al. Fatal pulmonary embolism following tourniquet inflation. A case report. *Acta Orthop Scand* 62:488, 1991.

140. Bannister, G. C., et al. The timing of tourniquet application in relation to prophylactic antibiotic administration. *J Bone Joint Surg* 70B:322–324, 1988.

141. Benzon, H. T., et al. Changes in venous blood lactate, venous blood gases and somatosensory evoked potentials after tourniquet application. *Anesthesiology* 69:677–682, 1988.

142. Campbell, W. C., et al. A pneumatic tourniquet. *J Bone Joint Surg* 19:832–833, 1937.

143. Crenshaw, A. G., et al. Wide tourniquet cuffs more effective at lower inflation pressures. *Acta Orthop Scand* 59:447–451, 1988.

144. Dreyfuss, U. Y., et al. Sensory changes with prolonged double-cuff tourniquet time in hand surgery. *J Hand Surg* 13A:736–740, 1988.

145. Egkher, E., et al. Blutdruckabhängige, prozessgesteuerte Blutsperre zur Minimierung des Tourniquet-Syndroms. Unfallchirurgie 12:200–203, 1986.

146. Estrera, A. S., et al. Massive pulmonary embolism: A complication of the tourniquet ischemia. *J Trauma* 22:60–62, 1982.

147. Fletcher, I. R. The arterial tourniquet. Review article. *Ann R Coll Surg* 65:409–417, 1983.

148. Gardner, V. O., et al. Contractile properties of slow and fast muscle following tourniquet ischemia. *Am J Sports Med* 12:417–423, 1984.

149. Giannestras, N. J. Occlusion of the tibial artery after a foot operation under tourniquet. A case report. *J Bone Joint Surg* 59A:682–683, 1977.

150. Gidlöf, A., et al. The effect of prolonged total ischemia on the ultrastructure of human skeletal muscle capillaries. A morphometric analysis. *Int J Microcirc Clin Exp* 7:67–86, 1987.

151. Goto, H., et al. Effect of high-dose of methylprednisolone on tourniquet ischaemia. *Can J Anaesth* 35:484–488, 1988.

152. Green, D. P. The tourniquet. In *Operative Hand Surgery*, 2nd ed. New York, Churchill Livingstone, 1988.

153. Greene, T. L., et al. Compartment-syndrome of the arm—a complication of the pneumatic tourniquet. A case report. *J Bone Joint Surg* 65A:270–273, 1983.

154. Hargens, A. R., et al. Local compression patterns beneath pneumatic tourniquets applied to arms and thighs of human cadavers. *J Orthop Res* 5:247–252, 1987.

155. Hrl, M., et al. Effect of tourniquet ischaemia on carbohydrate metabolism of dog skeletal muscle. *Eur Surg Res* 17:53–60, 1985.

156. Hofmann, A. A., et al. Fatal pulmonary embolism following tourniquet inflation. A case report. *J Bone Joint Surg* 67A:633–634, 1985.

157. Ikemoto, Y., et al. Changes in serum myoglobin levels caused by tourniquet ischemia under normothermic and hypothermic conditions. *Clin Orthop* 234:296–302, 1988.

158. Irving, G. A., et al. The protective role of local hypothermia in tourniquet-induced ischaemia of muscle. *J Bone Joint Surg* 67B:297–301, 1985.

159. Jeyaseelan, S., et al. Tourniquet failure and arterial calcification. Case report and theoretical dangers. *Anaesthesia* 36:48–50, 1981.

160. Jorgensen, H. R. J. Myoglobin release after tourniquet ischemia. *Acta Orthop Scand* 58:554–556, 1987.

161. Katz, J. F., et al. Tissue antibiotic levels with tourniquet use in orthopedic surgery. *Clin Orthop* 165:261–264, 1982.

162. Klenerman, L. Tourniquet paralysis. *J Bone Joint Surg* 65B:374–375, 1983.

163. Laurence, A. S., et al. Serum myoglobin following tourniquet release under anaesthesia. *Eur J Anaesth* 5:143–150, 1988.

164. Layzer, R. B. McArdle's disease in the 1980s. *N Engl J Med* 312:370–371, 1985.

165. Lecoutre, D., et al.: Embolie pulmonaire fatale par manoevre d'exsanguination. *Can Anaesthesiol* 37:289–291, 1989.

166. McElvenny, R. T. The tourniquet—its clinical application. *Am J Surg* 69:94–106, 1945.

167. Middleton, R. W. D., et al. Tourniquet paralysis. *Aust N Z J Surg* 44:124–128, 1974.

168. McGrath, B. J., et al. Massive pulmonary embolism following tourniquet deflation. *Anesthesiology* 74:618–620, 1991.

169. McLaren, A. C., et al. The pressure distribution under tourniquets. *J Bone Joint Surg* 67A:433–438, 1985.

170. Messahel, F. M. Incidence of pulmonary embolism in total knee arthroplasty. *Middle East J Anesthesiol* 11:187–192, 1991.

171. Neimkin, R. J., et al. Double tourniquet with linked mercury manometers for hand surgery. *J Hand Surg* 8:938–941, 1983.

172. Newman, R. J. Metabolic effects of tourniquet ischaema studied by nuclear magnetic resonance spectroscopy. *J Bone Joint Surg* 66B:434–440, 1984.

173. Nitz, A. J., et al. Pneumatic tourniquet application and nerve integrity: Motor function and electrophysiology. *Exp Neurol* 94:264–279, 1986.

174. O'Neil, O., et al. Transient compartment syndrome of the forearm resulting from venous congestion from a tourniquet. *J Hand Surg* 14A:894–896, 1989.

175. Palmer, A. K. Complications from tourniquet use. *Hand Clinics* 2:301–305, 1986.
176. Pedowitz, R. A. Tourniquet-induced neuromuscular injury. A recent review of rabbit and clinical experiments. *Acta Orthop Scand* 62(Suppl 245), 1991.
177. Reid, H. S., et al. Tourniquet hemostasis. A clinical study. *Clin Orthop* 177:230–234, 1983.
178. Rorabeck, C. H. Tourniquet-induced nerve ischemia: An experimental investigation. *J Trauma* 20:280–286, 1980.
179. Rudge, P. Tourniquet paralysis with prolonged conduction block. *J Bone Joint Surg* 56B:716–720, 1974.
180. Sapega, A. A., et al. Optimizing tourniquet application and release times in extremity surgery. *J Bone Joint Surg* 67A:303–314, 1985.
181. Sermeus, L., et al. Pulmonary embolism confirmed by transoesophageal echocardiography. *Anaesthesia* 47:28–29, 1992.
182. Shaw, J. A., et al. The relationship between tourniquet pressure and underlying soft-tissue pressure in the thigh. *J Bone Joint Surg* 64A:1148–1152, 1982.
183. Silver, R. Limb swelling after release of a tourniquet. *Clin Orthop* 206:86–89, 1986.
184. Valli, H., et al. Arterial hypertension associated with the use of a tourniquet with either general or regional anaesthesia. *Acta Anaesth Scand* 31:279–283, 1987.
185. Van Roekel, H. E., et al. Tourniquet pressure: The effect of limb circumference and systolic blood pressure. *J Hand Surg* 10B:142–144, 1985.
186. Wilgis, E. F. S. Observations on the effects of tourniquet ischemia. *J Bone Joint Surg* 53A:1343–1348, 1971.

STATISTICS IN SPORTS INJURY RESEARCH

Mario Schootman, M.S.
John W. Powell, Ph.D., ATC
John P. Albright, M.D.

Beyond case reports there is an extraordinary degree of difficulty involved in the execution of successful investigations into the cause, treatment, and prevention of sports injuries. As noted by Noyes and Albright [38], ". . . sports medicine research requires consideration of dimensions which other clinical research projects do not require." This implies that much of the more important work to be done is difficult, time-consuming, and expensive.

As sports medicine practitioners, our combined medical and sports-oriented background allows us to conceptualize the practical significance of study findings. However, our limitations are never more apparent than when it is time to determine the best method by which the data will be collected and analyzed. It is clear that we must also develop a working relationship with our comrades in the field of preventive medicine and biostatistics. However, to function at a most efficient level with our collaborators, we must become better acquainted with the uses and limitations of the most commonly used study designs.

As participation in sports and emphasis on sports injury prevention continue to grow, researchers are beginning to take up the challenge of investigating the associated injury patterns. The American Orthopaedic Society for Sports Medicine (AOSSM) Research Committee has been sponsoring workshops to assist clinicians in upgrading their knowledge of research techniques, procedures, analyses and presentation. Although the medical community has been conducting research into a wide variety of disease areas, they have only recently begun the task of examining sports injury prevention programs. Noyes and Albright indicate that the study

of sports injuries (sports traumatology) combines epidemiology and clinical medicine to analyze the frequency and determinants of injuries sustained by athletes, to prevent injuries, or to alter patterns of participation which contribute to injury [38].

A recent document published by the AOSSM as a supplement to the *American Journal of Sports Medicine* entitled Sports Injury Research describes the difficulties in conducting investigations in sports-related injuries. This supplement is intended to provide entry-level instruction to the unsophisticated practitioner. The various papers in the AOSSM supplement offer specific discussions to assist the researcher in the development of a basic program of systematic investigation. The task of the current chapter is to extend the researcher's ability to incorporate statistical analysis of research data into the design of the project. It will address a variety of research designs, both epidemiologic and clinical, and will discuss the pros and cons of different statistical applications inherent in each of these designs. This chapter does not aim for completeness on all topics discussed. It is an overview and a brief description of the possibilities of epidemiologic tools applicable to sports injury research. Rather, in most instances, the reader is referred to more elaborate texts for further details on the specific subject.

EPIDEMIOLOGY

Epidemiology is defined as the medical science dealing with the occurrence, causes, and prevention of disease. It is most often thought of as a scientific method used in public health programs to investigate the outbreak of diseases. By identifying the patterns of disease occurrence, the epidemiologist seeks to find factors associated with the onset of the disease and to make recommendations for control and prevention. The techniques of epidemiology have been used and refined for decades.

The use of epidemiologic techniques to study injury patterns, or *injury epidemiology* as it is called, is a very young science compared with disease epidemiology. In fact, it was not until the early 1960s that the principles of disease epidemiology became popular in the study of injuries. The first modern study of the occurrence of sports injuries was initiated by LaCava [28]. Since that time, numerous projects addressing the patterns of sports-related injuries have been conducted.

Although injury research shows similarities with disease epidemiology (i.e., both study health-related risks), several distinct differences exist. First, although most diseases have an insidious onset, most injuries have a sudden onset. Cancer develops over a period of time and is often difficult to diagnose. On the other hand, a fracture occurs in a split second and is very easy to recognize. Second, a disease may result from exposure to a specific set of conditions that at first may be unknown to the clinician and could have been present in a wide variety of circumstances. On the other hand, injuries occurring under specific conditions (e.g., traffic-related injuries) can occur only when exposure to that situation can be generated.

Sports injuries are a unique case of the general idea of an injury. They occur only when an athlete is exposed to sports. Unlike injuries that occur to the general population, sports injuries occur under specific conditions of participation, generally conducted at a known time and place. As participants, athletes also contribute unique qualities to the injury picture. They have an eagerness to participate and a determination to play regardless of the playing conditions. Their eagerness to play despite the somewhat hazardous conditions of sports is quite different from a worker's lack of desire to be exposed to another health-related risk, such as asbestos. This quality of the athlete creates a unique problem when trying to assess the impact of the injury on the individual's performance. Additionally, the athlete who has sustained an injury will challenge the healing of the injury by wanting to return to full participation in sports before the injury has been fully resolved. On the contrary, an individual who has contracted a specific disease will generally return to active participation in society only after he has fully recovered from that disease.

Despite these differences, epidemiologic principles have been frequently applied to sports injury research. At present it is still unclear which techniques can be applied successfully considering the differences between the two disciplines. Except for a few studies [23, 60, 62], techniques for studying sports injuries have not been investigated extensively. This may be due in part to a lack of awareness of available epidemiologic and statistical methods by investigators studying risk factors associated with sports injuries. As more work is done by sports medicine professionals, the use of more complex study designs and analyses will emerge. The purpose of this chapter is to describe the applicability of several epidemiologic techniques and statistical methods to the study of sports injuries. Throughout this chapter, the effect of muscle flexibility on the occurrence of muscle strains is used to illustrate the applicability of epidemiologic concepts to the study of sports injuries.

Important Epidemiologic Concepts

Before entering the main discussion of research designs and statistical analyses, several basic concepts will be briefly described for the unfamiliar reader. They deal specifically with the types of study designs, injury definition, injury recording, population at risk, statistical significance, sampling, bias, and estimates of the frequency of injury occurrence (rate and risk).

Types of Study Designs

The central issue in studying factors affecting the occurrence of sports injuries is the need to determine the mechanisms by which data will be collected and analyzed. Before searching for risk factors, descriptive data have to be collected regarding the frequency, type, and severity of the injuries occurring in specific populations. The study of the procedures used to identify this information is generally referred to as *descriptive epidemiology*. After the injury has been described, the search for risk factors affecting the occurrence of the injury can be initiated. The study of this process is called *analytic epidemiology*.

Within the latter type of epidemiology, several study designs can be distinguished depending on the use of the time factor. The first design, which uses time in a prospective way, is called a *cohort* study. In this type of study design the research team has identified and developed a plan for the recording of risk factors associated with the occurrence of the injury in members of the study population prior to the actual occurrence of the injuries (Table 5–1).

When, for example, epidemiologists are looking at the occurrence of muscle strains during participation in football, they identify and record risk factors (e.g., muscle flexibility) prior to the occurrence of the muscle strain. In a cohort study players are tracked over time to determine who did and who did not develop a muscle strain. This design is especially valuable for estimating the frequency and severity of injuries occurring in specified populations.

As in cohort studies, data in *surveillance* studies are collected using time in a prospective manner. This design is characterized by the routine and orderly collection of information about risk factors and their association with the occurrence of injuries. In most instances, many risk factors are continuously monitored by the research team. One of the strengths of this type of design is its ability to identify specific cases as they occur and to relate those cases to the many risk factors.

Another type of analytic study, the *case-control*

design, emerges when time is used in a retrospective manner. In this type of study, the injuries have already occurred, and the researcher is looking back in time (retrospectively) to determine who did and who did not have a risk factor present before or at the time of the onset of injury (see Table 5–1). Using the muscle strain example, the researcher knows which players sustained muscle strains and looks back in time to determine who among these were considered to have flexible muscles and who were not. When using a case-control design it is assumed that the data about risk factors are available, either from the study subjects or existing records, which may not always be the case. Furthermore, the data may not be of sufficient quality or quantity to make valid inferences about the study findings because they were not collected for this specific study. The case-control design is unsuitable for estimating the frequency and severity of injuries because it is not known how many players were at risk for developing the observed injuries. This design is, however, especially suitable for studying rare injuries (e.g., abdominal muscle strains), contrary to cohort studies, which are more suitable for studying more frequent injuries (e.g., hamstring strains).

In the fourth basic type of analytic study design neither a prospective nor a retrospective manner of data collection is used, but a single point in time. This type of design is called a *cross-sectional* study. When using this design the researcher determines at a single point in time whether or not a player has the risk factor (muscle flexibility or lack of flexibility) and whether or not the player is injured (has a muscle strain). This type of study design can also be used to estimate the frequency and severity of injuries but only at a single point in time. In many instances cross-sectional designs cannot be used to determine the effect of a risk factor on the occurrence of injuries. Special consideration must be given to the temporal relationship between the risk factor and the occurrence of injuries (i.e., it may be difficult to determine whether the risk factor preceded the injury or the injury preceded the risk factor). For example, did the lack of flexibility produce the muscle strain, or was flexibility reduced because of the strain?

Injury Definitions

As with any research project, the operational definitions assigned to various aspects of the project will dictate the logical utility of the findings. The most critical definition in the epidemiology of sports

TABLE 5–1
Graphic Presentation of Three Analytic Study Designs

Cohort Study

Flexibility		*Muscle Strain*
Present	→	Yes
	→	No
Absent	→	Yes
	→	No
Time frame: Present		Future

Case-Control Study

Flexibility		*Muscle Strain*
Present ←		
		Yes
Absent ←		
Present ←		
		No
Absent ←		
Time frame: Past		Present

Cross-Sectional Study

Flexibility	*Muscle Strain*
Present	Present
Present	Absent
Absent	Present
Absent	Absent
Time frame:	Point in time

injuries is the definition of an injury. The specific nature of the operational definition of injury is extremely important for selecting the types of procedures that are to be used to interpret the findings and for making comparisons between studies. Several different definitions have been used, including professional determination of the extent of tissue damage, a physician or emergency room visit, filing of an insurance claim, and loss of participation time. Separate definitions can also be combined into one operational definition of injury. For example, a sports injury may be considered reportable when a player seeks the expertise of an available medical professional, is unable to participate in the next session, and is found to have specific tissue damage. Such a definition means that each and every one of these criteria must be met before an injury is considered reportable. This may limit the number of observed injuries compared to studies using only one of the components of this combined operational definition.

Because of the existence of a wide variety of operational definitions describing the occurrence of sports injuries, comparisons between studies are difficult and sometimes impossible. Therefore, in injury epidemiology there is a need for a uniform operational definition of the word injury.

Injury Recording (Numerator)

Once the definition of injury has been decided, it is important to determine the number of variables in the study that will be recorded. Variations can include simply recording the injury as an event or as an event with a long list of risk factors possibly associated with the occurrence of the injury. Generally, risk factors are divided into two main groups—internal or personal factors, such as height and weight of the player, and external or environmental factors, such as type of playing surface and weather conditions [30, 33].

The quality of the recorded risk factors and injuries, and therefore of the entire study, depends heavily on the thoroughness of the data recorder. If the data have not been collected with accuracy and commitment by the data collector, the study findings may not be of sufficiently high quality to make valid inferences. The quality as well as the quantity of the collected data has implications for the study findings. It is more difficult to make valid inferences based on only a few injuries. The number of injuries in the study strongly depends on the operational

definition and the population under consideration as discussed in the next section.

Population at Risk (Denominator)

In epidemiology the population at risk is defined as those individuals having a chance of developing a specified type of disease. By extending this approach to sports injury research, the population at risk can be defined as those athletes who have a chance of sustaining an injury during their sports participation. The population at risk has to be known to make valid inferences about the magnitude of the problem of injuries across studies. If 25 people with muscle strains are observed during a study period, it is essential to know whether these injuries occurred in a population of 100 or 1000 players.

During the study period, players may or may not participate in sports activities. If they do not participate they are consequently not at risk of sustaining an injury. There are numerous reasons for player nonparticipation. For example, if a player is listed on the third-string team and the question at hand concerns game exposure, such a player is generally not at risk because he did not actually play in the game. This player will participate in practice sessions and will be at risk for practice-related injuries. It is very important to be able to account for these variations in participation if players at risk are to be associated with the injury patterns.

In some studies the population at risk from which the injuries were accumulated may be unknown. In studying injuries that are referred to a physician's office or clinic, the population at risk from which these referrals came is likely to be unknown because only a portion of the injured players are referred. Different referral patterns may exist, for example, based on the perceived professional capability of the office or clinic. Use of this type of analysis will provide descriptors for the local physician or clinic but will not allow comparisons among various sites unless the referral patterns and the population from which the cases arose are the same.

When the population at risk is known, the frequency of an individual's participation in sports activities may also be known. In most instances the population time, that is, the number of participants at risk multiplied by the time they were at risk (e.g., 1000 player-games or 100,000 player-games), must be available to compare findings across studies. Knowing how many games (or other measure of exposure) generated, for example, 25 muscle strains as described earlier is essential for comparisons with other muscle strain data.

Statistical Significance

The concept of statistical significance is commonly used in all forms of research. It is used to estimate the potential for obtaining differences in the data that are not attributable to random variation. Statistics can be used, for example, to make inferences about college athletes regarding the relationship between muscle flexibility and the occurrence of a muscle strain.

Statistical significance refers to the probability of the observed effect (difference between two groups) being created by a prescribed set of mathematical procedures. It is designed to determine whether the observed differences are a function of random occurrence and is most often expressed as a probability (p value). This p value is the probability of observing, when the null hypothesis is true, a value of the test statistic at least as extreme as the value actually observed. In the case previously cited, the null hypothesis is an explicit statement of the absence of association between the dependent and independent variables, that is, there is no association between muscle flexibility and the occurrence of muscle strains. An alternative hypothesis states that there is an association between the dependent and independent variables. This hypothesis can be either one-sided or two-sided in its statement. In a one-sided statement, the influence of the independent variable on the dependent variable is described as either positive or negative. A two-sided hypothesis describes a relationship between the dependent and independent variables but makes no statement about the positive or negative influence of the independent variable on the dependent variable. In the example of muscle strains, a two-sided hypothesis might be that an association between flexibility and muscle strains exists, but no statement is made about the injury-reducing or -producing quality of flexibility. In a one-sided hypothesis, the directionality of the statement is made explicit. For example, muscle inflexibility increases the risk (or decreases the risk) of sustaining a muscle strain.

In addition to the testing of a hypothesis, a *confidence interval* can be calculated to determine the statistical usefulness of the findings. A confidence interval is a tool that will aid in describing the variation that may be present in the data. The greater the variation in the data, the wider the confidence interval. The application and use of confidence intervals in data analysis associated with epidemiologic studies are discussed later in the section on analytical epidemiology.

Before moving on to the design and analysis of specific study designs, it is important to comment on the difference between the concepts of statistical significance and clinical importance. Statistical significance has been described as a set of mathematical procedures designed to estimate the effect of one variable on another. Clinical importance, on the other hand, refers to the magnitude of the observed effect during the specific conditions of the study and its relative ability to make a difference in the risk of injury. The terms statistical significance and clinical importance should not be used interchangeably. Even when a statistical test results in a significant finding, there is no assurance that the observed estimate, when applied to the real world, is of clinical importance. Statistical significance is a function of the nature of the data, the study design, and the mathematical tools used to look for significance. For example, when using a large enough sample size, even the most clinically unimportant finding (a very small effect size) can become statistically significant. Clinical importance deals directly with whether the observed effect actually makes a difference in the frequency of injury.

Sampling

After selecting the appropriate study design and the method of data collection the next step is to determine the size of the sample to estimate the effect of the study factor. There are various methods to determine the size of the sample needed in observational and experimental studies. All have the following steps in common [22].

First, the null hypothesis and either a one- or a two-sided alternative hypothesis has to be stated. Second, the researcher must select the appropriate statistical test based on the type of dependent and independent variables involved. Dependent variables are variables whose values are hypothesized to be influenced by the study factors. Independent variables are factors that are considered to influence the prediction of the frequencies associated with the dependent variable. In our case, the dependent variable is the occurrence of muscle strains, whereas one of the independent variables possibly predicting this occurrence is muscle flexibility. Statistical tests are based on specific assumptions about the type of dependent and independent variables. Therefore, it is essential to select appropriate statistical tests before the data are collected.

Third, a reasonable effect size has to be chosen.

Effect size refers to the magnitude of the effect that the investigators want to find in order to obtain clinical importance. The magnitude of the effect size and the sample size required to observe this effect are inversely related—that is, the larger the effect size, the smaller the sample size can be. Conversely, the smaller the effect size, the larger the sample size required. For example, the research team may want to determine what decrease in production of injuries would warrant the use of preventive knee braces in college football.

The fourth step needed to determine sample size is to set an acceptable level of probability that the null hypothesis will be rejected when it is actually true (alpha), usually a probability of 0.05. Also, the researcher must consider the probability that the study results will fail to reject the null hypothesis (no effect from the use of preventive knee braces) when it is actually true (there is an effect of using braces). This probability is referred to as beta, and it may vary more than alpha. A beta of between 0.20 and 0.10 is quite often used.

The final step needed in calculating the sample size is to use the appropriate formula. The formula by which the sample size is calculated depends heavily on the type of study intended. Different formulas have to be used for different designs [41, 52]. For example, when alpha and beta equal 0.05 and 0.10, respectively, and we want to find that players with decreased muscle flexibility are 50% more likely to sustain a muscle strain, we need 89 players in each group (one group comprises players with decreased flexibility and the other group consists of players with increased flexibility). This calculation is based on the chi-square test* and use of a cohort study design. Although the formula used to calculate this sample size is beyond the scope of this chapter, the reader is referred to Breslow and Day [2], Hulley and Cummings [22], and Schlesselman [52] for further detail on calculation of sample sizes.

Bias

After selecting the appropriate study design and estimating the sample size, the study is conducted by using the appropriate method of data collection. However, during data collection bias may occur. Bias is defined as any systematic error in the design, conduct, or analysis of a study that results in a mistaken estimate of a risk factor's effect on the risk of injury. Sources of bias include the way in

which subjects are selected for the study (selection bias), measurement error or misclassification of subjects on one or more variables (information bias), and failure to adjust for variables that distort the relationship between the study factor and the dependent variable (confounding). Such flaws in the research design or analysis will affect the validity of the estimate of the effect of the study factor and may lead to spurious conclusions.

Different types of studies are susceptible to different types of bias. Case-control studies, for example, may obtain information retrospectively about risk factors by using interviews or questionnaires. Because injured players may know more or have different information about their exposure to an alleged risk factor than uninjured players, recall bias may occur. Because information about risk factors is collected prospectively when using cohort studies, they do not suffer from recall bias. However, prospective studies may suffer from what is called follow-up bias. That is, players that leave the study are not representative of players that remain in the study population. They may, for example, be older or have had previous injuries. Recall and follow-up bias are only two of many biases that can distort the relationship between the study factor and the dependent variable.

It is common practice to determine the efficacy of preventive devices by using "historical controls" [53]. These types of controls can be found when a study design is used that consists of two sequential periods. During the first part of the study no preventive devices were worn by the players, but during the second study period almost all players wore preventive devices. In many instances, if players were not randomized to both conditions (preventive knee braces worn or no preventive braces worn) and the research team used two sequential study periods, a bias may have occurred. There may have been differences in players' participation that may have affected their risk of injury; e.g., a player may have been a linebacker in one study period and a defensive tackle in the other. This difference may result in an inability to compare both groups to determine the effect of the use of the preventive device.

It is important to note that many biases such as recall bias, follow-up bias, and the use of historical controls occur in the design of the study and can, only with great difficulty, be corrected in the analysis. It is therefore essential to consider the presence of bias before the study is actually implemented. (See Kleinbaum and co-workers [25] and Schlesselman [52] for additional information on bias detection and control.)

*See formula 5 for calculation of the chi-square test.

Frequency of Injury Occurrence

In descriptive epidemiology there are two separate indicators describing the frequency of the occurrence of sports injuries—*rate* and *risk*. Although the risk of injury can be calculated from the rate of injury, these indicators are not equivalent and should be clearly distinguished from each other. It is important in the presentation of the study findings for the research team to be fully aware of both the differences between the concepts of rate and risk and the specific ways in which these differences should be interpreted.

Rate

The concept of rate can be described as a ratio that is associated with the instantaneous change of one phenomenon (a sports injury) with another variable, usually a unit of time [15, 35, 36]. When using rate as a measure of injury occurrence, a case (the numerator) may be referred to either as an injury or as an injured player, and it is important for the investigator to differentiate clearly between the two. A rate can be expressed as either an absolute or a relative measure depending on whether reference is made to the size of the population at risk (number of players) from which the cases arose [15, 19]. The number of muscle strains per football game is an example of an absolute rate; no reference is made to the population at risk. When the number of players participating in each game is included in the denominator, a relative rate is created, for example, the number of muscle strains per player-game. These rate calculations are generally considered estimates of the potential for injury under the study conditions.

Risk

Risk is the probability that a player will incur a certain type of injury over a specified period of time [36]. Since risk is stated as a probability, its value ranges from 0 to 1.0, it is dimensionless and must make reference to the study period. When a player's risk is 1.0, every player can expect to be injured. When it is 0.0, no injuries are expected. The most important thing to remember in establishing risk is that because it is a proportion, the items in the numerator must be of the same type (players injured) as those in the denominator (players at risk). For example, the number of athletes with a muscle strain divided by the total number of athletes at risk will establish a risk statement. The number of muscle strains divided by the population at risk does not create a risk because a player may have multiple injuries. This is a statement of injuries divided by players, not players divided by players.

DESCRIPTIVE EPIDEMIOLOGY

The two most important concepts associated with the area of descriptive epidemiology are *incidence,* the occurrence of new cases within a specific time period, and *prevalence,* the number of cases that exist at a specific time or during a period of time [35]. An example of incidence is the occurrence of muscle strains during a single season. Prevalence, on the other hand, is the number of muscle strains existing at a specific date or period (September 10 or the month of September), which would include both new injuries and those left from a previous time period. Both concepts play a role in the study of sports injuries and will be described in detail in the following sections.

Incidence

As in disease epidemiology, assessing incidence for specified populations is fundamental to both descriptive and etiologic investigations. Incidence is defined as the number of cases that appear during a given study period and can be presented in different ways. For example, at a certain clinic 15 new muscle strains were reported in female gymnasts during a 3-month period. In this example, a "case" refers to a type of sports injury sustained by sports participants. Case may also be used to refer to an individual (an injured player). Using the latter definition of a case, 11 newly injured soccer players were evaluated at a hospital clinic during a 3-month period. The research team should clearly specify the definition of a case during the presentation of their findings. Neither definition for estimating incidence refers to the size of the population from which cases arose or the amount of exposure accumulated by the population at risk. Lack of reference to population at risk or exposure makes comparisons between studies impossible unless the population at risk and their exposure are equal in both studies. For example, a quality assurance committee informs a surgeon that the number of patients with numbness and tingling from use of a tourniquet during surgery is twice as great as that reported by his peers. Without consideration of the number of patients at risk, comparison of quality may be very misleading. The literature provides several incidence measures that have been used to describe the occurrence of sports injuries. For example, the number of injuries [42, 56] and the number of injured players [30, 48] that occurred over a specified study period, and the number of injuries per unit of time [61] represent only a few of the measures used.

Two concepts of incidence can and should be distinguished—risk and rate. Risk is defined as the probability that an individual will be injured and is measured at the level of the individual player, that is, the risk a specific player has of sustaining an injury during the study period. Rate is defined as the frequency of injuries occurring during a specific time period and is measured at the level of the population (players at risk). The injury rate equals the number of injuries per opportunity (exposure). These two concepts are especially valuable for assessing individual risk of participation and for describing the importance of specific risk factors associated with the occurrence of injuries.

The following discussion will examine these two concepts in greater detail. We will use a series of hypothetical data to show the specific steps needed to develop risk and rate estimates and their respective interpretations.

Hypothetical Test Data

Table 5–2 displays the hypothetical participation of 8 football players during a study period of 9 games and the injury frequency associated with their participation. This table will be used throughout the discussion of incidence and prevalence to illustrate the theoretical concepts involved. A "+" in Table 5–2 represents a player at risk for sustaining a specific type of injury during that particular game. There are two ways by which players are not at risk for sustaining an injury: They are injured, represented by an O, or they do not play because of administrative reasons (×), such as an examination or personal circumstances. For example, player

number 6 participated in games 1 and 2, was injured in game 3, returned to participation in game 5, and did not participate in games 8 and 9. This results in 6 exposures for player number 6. In Table 5–2 exposure is assessed at the level of a game, that is, a player sustaining an injury during a game still contributes an entire game to the total amount of exposure. Thus, player number 6 had 6, not 5, exposures out of a total potential 9 exposures. Note that player number 7 did not participate in game number 1 due to the sequela of an injury sustained before the start of the 9-game study period.

Rate of Injury

In descriptive epidemiology there are a variety of ways to create an incidence rate. The quality of the various calculations will depend on the number of injuries considered in the numerator and the ability of the research team to develop a reasonable denominator. Our discussion will center on two different techniques for estimating the rate of injury, absolute and relative incidence rates.

Absolute Incidence Rate
The number of new cases per unit of time is known as the absolute incidence rate [15], or instantaneous rate of development [36]. The size of the population from which these cases were accumulated is unknown. Examples are the number of fatalities per year due to playing football [37], the number of players per year treated by physicians, and the number of emergency room visits for sports injuries for a specific time period [49]. The number of injuries, or injured participants, per parachute jump is also considered an absolute incidence rate.

When referring to Table 5–2, the absolute incidence rate equals 6 injuries (including 2 injuries for

TABLE 5–2
Hypothetical Follow-Up of Eight Football Players Over a Study Period of Nine Games

Player Number										Number of Games at Risk
8	+	×	×	+	+	⊕	O	O	O	4
7	O	+	+	+	+	+	+	+	+	8
6	+	+	⊕	O	+	+	+	×	×	6
5	+	+	+	+	×	×	×	×	×	4
4	+	⊕	O	O	+	+	⊕	O	O	5
3	+	+	+	⊕	O	O	O	O	O	4
2	+	+	⊕	O	O	+	+	+	+	7
1	+	+	+	+	+	+	+	+	+	9
Game Number	1	2	3	4	5	6	7	8	9	**Total = 47**

+ = player is at risk; O = player is injured; × = player is not at risk due to administrative reasons; ⊕ = player injured during game.

player number 4) in 9 games, or 0.67 injuries per game. The absolute incidence rate simply addresses the number of injuries (numerator) divided by the number of exposure-events (e.g., games [denominator]). No reference is made to any of the characteristics of the games or the size of the study population from which these data were accumulated. If we use the number of injured players as the numerator rather than the number of injuries, the absolute incidence rate equals 5 injured players in 9 games, or 0.56 players per game. The two absolute incidence rates differentiate between the rate of injury for the study group (0.67) and the rate of injury per player (0.56).

The lack of reference to the population at risk has been described by Walter and colleagues [62] as a limitation of the use of absolute incidence rates. This limitation may be evident in the way in which the injuries are selected for inclusion in the study, that is, the recorded injuries may not be a reflection of all injuries in the population. Clinicians, for example, may not see a representative sample of all injuries because only specific types of injuries are actually referred to the clinic. These referral patterns may be quite different among clinics and are based on local and professional reputations. This limitation results in an inability to study the risk factors associated with the occurrence of sports injuries except for the specific types of injuries being recorded. This lack of population descriptors severely limits the ability to identify any high-risk groups that may exist. Finally, comparisons among the findings from different studies are very difficult when the characteristics and size of the population at risk are unknown.

Relative Incidence Rate

An alternative to the absolute incidence rate is the *relative incidence rate*, which takes into consideration the size of the population from which the cases were accumulated [15]. Various studies have used terms such as frequency rate [11], frequency ratio [30], and case rate frequency [39] to describe their data. These terms are equivalent to the term relative incidence rate.

The concept of relative incidence rate uses the number of new cases divided by the population-time during which they were accumulated. In this procedure, a case can be referred to as either an injury or an injured player, and the number of injuries or injured players per unit of time is then estimated. When using the relative incidence rate, the researcher assumes that the population under study reflects a constant relative incidence rate during the study period and within each age category. This is referred to as population stability [34, 36].

Depending on the design of the study, population-time (denominator) can be expressed at various levels of detail, such as the number of seconds [10], minutes [4], games [55], or even seasons. Selecting the level of exposure to be recorded will determine the sensitivity of the incidence measure created. As the detail of the accumulated exposure increases, the practical difficulties encountered in the recording of the data also increase, specifically, the workload of the data collectors. These considerations also apply to the exposure assessment needed for calculating the absolute incidence rate.

When relative incidence rates are used, the research team should keep in mind that this procedure makes no distinction between a group of a few players who accumulated a large amount of population-time (exposure) and a larger group of players who accumulated the same amount of population-time [18, 57]. For example, 100 players participating in 5 games create the same population-time as 50 players participating in 10 games (i.e., 500 player-games). It can be hypothesized that both groups differ with regard to internal risk factors such as physical fitness that may predispose the athlete to injury. Furthermore, the relative incidence rate is not sensitive to factors outside the athlete's control that affect his or her participation time. For example, a player may not participate because another player has more ability or because he is held out of the game for administrative or personal reasons. Under these conditions, it has been suggested that the relative incidence rate should be calculated separately for players with a high average amount of population-time and those with a low average amount of population-time [36]. The important thing to remember is to interpret the results based on these considerations.

For practical purposes, the three items of information necessary to give epidemiologic meaning to a relative incidence rate are (1) the numerator (the number of injured players or injuries), (2) the denominator (the population of players from which the cases are derived), and (3) a specific time period. One of three methods can be employed to determine the amount of population-time (denominator) used in the calculation of the relative incidence rate; the exact population-time, a uniform population-time estimate, and an average population-time estimate.

Exact Population-Time

By examining the data in Table 5–2, we can determine the number of players who participated in each of the 9 games and can establish an exposure (population-time) component of 47 player-games for this example. Since six injuries occurred during the 47 player-games, a relative incidence rate of 0.13

injuries (6/47) per player-game results. If we count only the number of injured players, the relative incidence rate becomes 0.11 injured players (5/47) per player-game. The exposure that results from assessing the participation of every player in every game may be referred to as the exact population-time. Under some conditions, the number of recorded injuries may be small relative to the total number of player-exposures. The relative incidence rate, in this situation, is often expressed in terms of a reference population, for example, the number of injuries or injured players per 100 or 1000 player-exposures.

An advantage of using the exact population-time is that it takes into account the fact that subjects are often not at risk during the entire study period. Individual athletes may enter and exit the population at risk at any point in time during the study. This loss of exposure may be due to injury or other administrative reasons. By calculating the exact population-time, the denominator contains only the number of players and the actual time they were at risk.

An important consideration when studying sports injuries is the knowledge that athletes may sustain multiple injuries during a specific study period. These multiple injury situations may be represented in three ways: (1) an athlete sustains a different type of injury to the same body part; (2) the athlete received the same type of injury to a different body part; or (3) the athlete has the same type of injury to the same body part (a re-injury). If multiple injuries are included in the data base, a relative incidence rate can be calculated for conditions in which the player was injured more than once. Calculation of two relative incidence rates, one for new injuries and one for re-injuries, should be preferred to a relative incidence rate that does not discriminate between players with one and those with more than one injury [31].

To be able to differentiate between the rate of new injuries and that of recurrent injuries, specific considerations must be used in calculating the population-time. For example, the population-time for each player from the beginning of the study until the onset of the first injury is part of the population-time for players who sustain a first injury. The time from the recovery date of the first injury until the onset of the second injury is part of the population-time for players who sustain a second injury, and so on. Players who have not had a first injury will not contribute to the exposure-time for the calculation of the relative incidence rate for the second injury. From Table 5–2 it can be seen that 5 players sustained one injury, and 1 player sustained a second

injury. The relative incidence rate for players sustaining a first injury equals 0.14, which reflects the 5 injured players (numerator) and the 37 exposures (5 injured and 2 uninjured). The relative incidence rate for a second injury equals 0.10. Three players (2, 4, and 6) sustained a first injury and are consequently at risk for sustaining a second injury; they account for 10 player-games. Because only player number 4 sustained a second injury, the relative incidence rate for a second injury equals 0.10. This type of differentiation requires accumulation of the exact population-time for all athletes during the study period.

Uniform Population-Time Estimate

In addition to monitoring each individual player in the study population to determine the exact population-time, an estimate of the population-time can be made by considering the number of injured players and the duration of the study period [2, 15].

Suppose that all injured players sustained only one injury and that the injuries were uniformly distributed among the players and occurred at the midpoint of the study period. This situation implies that each injured player contributed one-half of a study period to the population-time. Thus, to obtain an estimate of the population-time, the number of injured players multiplied by half of the study period must be added to the population-time accumulated by the uninjured players. For example, among a group of 100 uninjured players at the beginning of a 2-year study period 30 sustained a muscle strain. The estimated population-time is calculated as follows: Of the 100 uninjured players at the beginning of the study, 70 remained uninjured while 30 sustained an injury. Thus, each of the 70 players contributed 2 years to the population-time, while the 30 injured players each contributed 1 year. Therefore, the total population-time is the sum of 140 and 30 player-years, which equals 170. The relative incidence rate is 30 injured players divided by 170 population-years, which equals 0.18 injured players per player-year.

Using this approach and the data given in Table 5–2, the relative incidence rate equals 0.09 injured players per player-game and is calculated as follows: Seven players were uninjured at the beginning of the 9 games, of whom 5 sustained an injury. Thus, each of the 2 uninjured players contributed 9 games to the population-time, resulting in 18 player-games. The 5 injured players each contributed one-half of the duration of the study period (i.e., 4.5 games), resulting in an additional 22.5 player-games being included in the population-time. The relative incidence rate is calculated as before and equals 0.12 injured players per player-game (5 injured players divided by 40.5 player-games), which is only slightly different from the relative incidence rate (0.11) calculated by using the exact population-time.

One advantage of using the uniform population-time is the ease with which it is obtained because it is based on consideration of the population rather than on the individual player. Although the exact population-time is more accurate and allows for a variety of interpretive levels, it is very difficult to obtain because of the burden placed on the data recorder. A disadvantage of using the uniform population-time is the assumption of only one injury per injured player. This negates the ability to examine incidence rates for subsequent injuries.

Average Population-Time Estimate

The third mechanism for obtaining the data needed for the denominator of the relative incidence rate is to record the average number of players at risk during each exposure and multiply this by the amount of exposure that occurred during the study period [47]. If an average of 13 players are at risk during a soccer game and 15 games were played, during which 7 injuries occurred, the relative incidence rate is 0.04 injuries per player-game (7/(13*15)). This method is similar to that used in the exact population-time approach but allows the recorder to maintain a team exposure record rather than a player-by-player exposure record. For very large teams, such as those used in college football, this approach is more acceptable, especially if multiple sites are to be used for data collection [44, 45, 46]. The average number of players at risk during an exposure can, for example, be calculated from a sample of the exposures in the study period. The approximation of the population-time is simply the average number of players at risk in the sample multiplied by the total number of games derived from a roster. This eliminates the extensive detail necessary for calculating the exact population-time.

In a sample of three of the games listed in Table 5–2 (games 4, 7, and 8 were selected by random occurrence), the average number of players at risk equals 4.3. Players injured in these games were also counted as being at risk. The approximation of the population-time equals the average number of players at risk (4.3) multiplied by the duration of the study period (9 games), resulting in 39.0 player-games. Thus, the relative incidence rate by this method equals 0.13 injured players per player-game (5 injured players divided by 39.0 player-games), which is similar to the relative incidence rate obtained by calculating the exact population-time.

Risk of Injury

A question always asked by individuals who participate in sports is, "What is the risk that I will be injured?" To answer this question, the risk or probability that an individual will be injured must be calculated. Using epidemiologic techniques, there are three different ways in which to approach the assessment of the risk of sustaining an injury—estimation of the cumulative incidence, use of an actuarial method, and use of the relative incidence rate [25]. Each method has advantages and disadvantages that affect the ability of the research team to assess the risk of injury that faces each participant. It should be noted that the appropriate method for establishing the risk of injury will be affected by the design and overall operation of the study.

Cumulative Incidence

The most commonly used method for assessing risk of injury during a study period is calculation of the *cumulative incidence*. This measure is defined as the proportion of subjects injured during the study period of the total number of players uninjured at the beginning of the study period [15, 36]. For example, of 1000 uninjured soccer players at the beginning of a two-season study period, 24 players developed a muscle strain. In this case, the cumulative incidence equals 0.024, that is, the risk a player has of sustaining a muscle strain during two consecutive seasons.

The cumulative incidence is used predominantly in situations in which each member of the study group is followed until the injury occurs or until the risk period ends. This method does not account for attrition from the study group for reasons other than injury [35, 36]. Players who quit the team, for example, might have developed an injury if they had remained in the study group and were followed over the entire study period. From the data in Table 5–2, if players are removed from the population at risk if they sustain an injury and if it is assumed that player number 7 is not at risk at the beginning of the study period due to an injury, a cumulative incidence of 0.71 (5/7) is calculated. This reflects 5 injured players, not 6 injuries, sustained among the 7 uninjured players who began the study.

In the study of sports injuries, the cumulative incidence may not be the most accurate estimate of risk unless all subjects in the observed population are followed for the entire follow-up period or are known to develop an injury during the interval [25]. Additionally, the concept of competing causes may interfere with the estimate of risk using the cumulative incidence. This is particularly important when the injury being considered is relatively rare compared to other types of injuries. For example, injuries to the face in football players are less common than injuries to the knee [45]. Therefore, when studying facial injuries, football players have different periods of participation not because of facial injuries but because of the occurrence of knee injuries. The cumulative incidence is not suited to deal with different periods of participation for different players in the study group.

In the literature one other measure of incidence has frequently been used to describe the occurrence of sports injuries. Some studies have used the total number of injuries that occurred divided by the total number of players at risk over the study period [30, 40, 56]. There are three disadvantages to using this type of ratio for estimating the risk of sports injuries. First and most important, the outcome of this ratio is not the risk of injury incurred by the player during the study period. Since the cumulative incidence is a ratio, it must reflect a proportion in which the numerator includes the same items as the denominator (players must be divided by players).

Second, some studies wrongly multiply this calculated ratio by 100, which produces an incidence percentage [27]. The result is uninterpretable because in some conditions an "incidence" of more than a 100% could result. In addition, the term percentage indicates a proportion, and the question would then be, "a percentage of what?"

Finally, most studies using this ratio (number of injuries divided by number of players at risk), have not discriminated between new and subsequent injuries to the same player, thereby complicating the interpretation. The number of injuries per player at risk, when calculated using this procedure, has little epidemiologic meaning and should be avoided in describing the risk of sports injuries.

Another ratio useful in sports injury research is the number of injuries per injured player. This ratio can be used to obtain an indication of the number of multiple injuries sustained by players. This ratio is always greater than or equal to 1.0. If the ratio equals 1.0, each player sustained only one injury. When this ratio is greater than 1.0, some players have sustained multiple injuries. If, for example, 50 muscle strains were sustained by 40 players, the injury:injured player ratio equals 1.25. In most cases, existing studies did not use this ratio, yet many provided the number of injuries and injured players from which this ratio can be calculated.

Actuarial Method

The *actuarial method* of estimating risk is especially valuable when the durations of the individual follow-up periods for uninjured players vary substantially [25]. The following formula can be used to estimate the risk (R) a player has of sustaining an injury during the study period (t_0 until t_1), where t_0 equals the starting time and t_1 the ending time of the study period, I represents the number of new cases (injured players) accumulated during the study

period, N_0 equals the number of uninjured players at the beginning of the study period, and W represents the number of withdrawals from the study population during the study period:

$$R(t_0, t_1) = \frac{I}{N_0 - (W/2)}$$

Formula 1

Formula 1 represents the number of uninjured players that would be expected to produce I new cases if all players could be followed for the entire period (the cumulative incidence discussed earlier is a special use of this actuarial method).

The actuarial method assumes that the mean withdrawal time occurs at the midpoint of the study period. That is, players sustain one injury during the study period, and these injuries occur at the midpoint of the study, after which the player is removed from the population at risk. Referring to Table 5–2 and using formula 1, the risk equals 0.67 ($4/(7 - 1)$). Four players sustained an injury during the 9 games, after which they were removed from the population at risk. Seven players were uninjured at the beginning of the study period, assuming player number 7 was injured before the start of the study period. Two players (numbers 5 and 8) were removed from the population at risk before they sustained an injury. Even though these players eventually returned to play and were subsequently injured, they are not counted in this calculation. This results in a risk of 0.67 for players to sustain an injury during the 9 games. However, even when all withdrawals actually occurred at the midpoint of the study period, the actuarial method results in a biased estimate of risk [9, 17]. Several alternatives for this problem have been proposed [16] but will not be discussed here.

Using the Relative Incidence Rate

The third method for estimating risk is based on the calculation of the relative incidence rate. When it is assumed that the relative incidence rate is constant over age and time, a relationship (see Formula 2) between relative incidence rate and cumulative incidence ($CI_{\Delta t}$) exists [34, 36]. Since $CI_{\Delta t}$ is a proportion, only the first injury sustained can be used to calculate risk.

The risk, i.e., $CI_{\Delta t}$, that a player will sustain an injury during the study period (Δt) is related to the relative incidence rate (RIR) as follows:

$$CI_{\Delta t} = 1 - \exp^{[-RIR^*\Delta t]}$$

Formula 2

For example, the relative incidence rate for a certain sport equals 3.1 per 1000 player-games. The risk incurred by a player of sustaining a muscle strain during 20 games of participation in that sport equals 0.06 ($1 - 2.718^{-0.0031*20}$). When RIR is small (less than 0.10), the $CI_{\Delta t}$ is approximately equal to $RIR^*\Delta t$ based on formula 2.

Referring to Table 5–2, the $CI_{\Delta t}$ based on the relative incidence rate ($=0.1064$) equals 0.62 ($1 - 2.718^{-0.1064*9}$). This represents the risk incurred by a player of sustaining an injury during 9 games of sports participation and is similar to the value of the CI calculated directly (0.71).

Prevalence

Miettinen [35] defined prevalence as "the existence of a particular state among members of a population." Based on the time during which the subjects affected by the specific disease (injury) are assessed, two types of prevalence can be distinguished.

First, *point prevalence* is the proportion of a defined population affected by the injury in question at a specified point in time [2]. The numerator consists of the number of existing cases at that point in time, while the denominator consists of the total population (injured and uninjured) from which the injuries are ascertained. For example, of the total number of 150 soccer players on October 24, 19 are injured. Thus, the point prevalence equals 0.13 (19/150). No reference is made to the time elapsed between the onset of the injury and the point in time at which the prevalence is assessed. Therefore, injuries with a longer duration will be overrepresented in the point prevalence compared to injuries with a shorter duration [19].

Point prevalence can also be calculated from Table 5–2. For example, the prevalence at game 4 equals 0.50. This figure was calculated as follows: Four players were injured at game 4, whereas 4 were not. Thus, the point prevalence at game 4 equals 4 divided by 8 ($= 0.50$).

The second type of prevalence that can be estimated is the period prevalence, which is also a proportion of injured players. However, cases are observed over a period of time unlike point prevalence in which a single point in time is used. The period prevalence consists of the number of injured players and the number of players that are injured during the period within which the disease (injury) is ascertained. For example, during the month starting October 24, 1985, 11 players were found to be injured. The number of uninjured players was 150. Therefore, the period prevalence equals 0.07 (11/161). It can be seen from Table 5–2 that the 3-game period prevalence from game 4 until game 6 equals 0.63 (5/8). Of the total of 8 players, 5 were found to be injured during that period.

Period prevalence is of limited usefulness because epidemiologists generally require knowledge of whether an injury is new or old. Also, an injury is more likely to be included in a prevalence study when the time needed for recovery is long [19]. Period and point prevalence are typically assessed by a cross-sectional study design, which will be described in the next section, along with other designs. Because of its limited applications to the study of sports injuries, period prevalence will not be discussed here in further detail.

ANALYTIC EPIDEMIOLOGY

So far we have been discussing the methods and tools available from the area of descriptive epidemiology for describing and interpreting the injury frequencies associated with sports. Once various patterns of injury rates and individual risks have been identified, it is time to move on to investigation of the factors associated with the potential for injury. Analytic, or etiologic epidemiology, is an area of study that refers to those projects that search for factors associated with the occurrence of sports injuries. Two general types of designs are commonly used to study etiologic factors in epidemiology—observational and experimental designs.

Observational Studies

The strength of epidemiologic research comes from the characteristic inclusion of information associated with the populations at risk of sustaining injury. This systematic observation of the characteristics of different populations sets epidemiologic research apart from other types of biomedical research. *Observational* studies, which are the most frequently used studies in epidemiology, are characterized by the ability of the research team to observe events in a target population without manipulating the influential factors under investigation. Since there is no manipulation of study factors among the subjects included in the study, observational studies do not suffer from the ethical problem of withholding prevention under conditions of personal risk (i.e., who should receive the preventive conditions?). Observational studies require a specific amount of very accurate recording of data relating to the risk factors under consideration. One of the limitations of the observational study design centers around the difficulty arising from the tendency to attribute the findings of the study to a specific risk factor because of the potential influence of uncontrollable factors.

Observational study designs for investigating risk factors affecting the occurrence of sports injuries can be categorized into three basic types: cohort, case-control, and cross-sectional designs. We will examine the basic description of these models and the types of analytic procedures that are appropriate for each design.

Cohort Designs

Of the various types of observational studies used in sports injury research, one of the most popular is the cohort design. Using this technique, a specified number of athletes with and without a certain risk factor (e.g., flexibility) are monitored over a specified study period. During this study period the number of injuries and factors associated with them are recorded. For example, of a population of 350 football players who were followed for one season, 200 were considered flexible according to a set of predetermined criteria, and 150 were not. During this study period, some of the athletes developed a muscle strain whereas others did not. Table 5–3 presents a hypothetical example of the structure of data obtained by using this type of design. It can be seen from this table that of the 200 football players considered flexible (the index group), 27 sustained muscle strains. In the control group, consisting of 150 football players not considered flexible, 35 players sustained a muscle strain.

Under ideal study conditions, researchers using the cohort design monitor and record information about each athlete along with the athlete's exposure to the study factors. If the study design calls for a player-by-player analysis, detailed information about the exposure of each player is required. If the analysis is to be conducted while considering a group

level of exposure, exact player exposures are not required. Under these conditions, a variety of procedures can be used to estimate the total exposure figures for this type of study population. Analytic techniques such as estimates of the *relative risk* and *incidence density ratios* are only two examples of procedures used to analyze cohort studies. The following sections will describe applications of some of these techniques.

Surveillance Design

Another type of study that is similar to the general cohort approach is called the surveillance design. This design also defines and collects data using prospective procedures. Surveillance studies continually monitor a group of athletes and the characteristics of their participation in a given sport. The monitoring of the activities of the study group allows excellent documentation of the amount of opportunity that exists for injury. Examples of successful use of this type of study design in sports injury research are the National Football League Injury Surveillance System [45], the National Athletic Injury/Illness Reporting System [5], and the National Athletic Trainers' Association's High School Injury Study [46].

Survival Design

A special type of the cohort design has been called the survival design. It requires in-depth data collection on individuals and their exposure to specific risk situations. It is frequently used in cancer epidemiologic [7] and orthopaedic procedures [6, 13] but is relatively uncommon in sports injury research. The key issue in survival analysis is the ability to determine who is at risk for each kind of event as a function of time. The basic tool used to interpret these findings is called a survival curve. The survival curve represents the reduction in the proportion of uninjured persons compared to the number of persons being injured during the study period. The basic idea is to compare the survival curve for those exposed to the study factor with the survival curve for those not exposed to the study factor.

TABLE 5–3
Hypothetical Data Obtained by Using a Cohort or Case-Control Study Design for Analyzing the Effect of Flexibility on the Occurrence of Muscle Strains

Flexibility	Injured	Not Injured	Total
Yes	27 (a)	173 (b)	200 (m_1)
No	35 (c)	115 (d)	150 (m_0)
Total	62 (n_1)	288 (n_0)	350 (n)

Cohort Analysis

The data in Table 5–3 will be used to describe the various analyses associated with cohort studies. For purposes of clarification, specific identities have been assigned to the various cells in the frequency table. Using these data, we can determine the measure of association known as the relative risk. The relative risk (RR) will help to interpret the association between exposure for the index group (players who are flexible) and its associated injuries and exposure of the control group (players who are not flexible) and its injuries. The formula for calculating relative risk is:

$$RR = \frac{a/m_1}{c/m_0}$$

Formula 3

It represents the number of times injuries are more or less likely to occur in the index group compared to the control group. If the relative risk equals unity (RR = 1), no association between exposure (flexibility) and the occurrence of injury (muscle strains) exists. When the relative risk is greater than 1, the risk factor is associated with a higher frequency of injuries. If the relative risk is less than 1, the risk factor is associated with fewer injuries. When applying the data from Table 5–3, the relative risk equals 0.58 (27/200 [35/150]). This means that the risk of sustaining a muscle strain for players who are flexible is 0.58 times less than the risk for players who are not. To more clearly understand the meaning of the 0.58 figure, a 95% confidence interval is established. This procedure is used to determine an upper and a lower number, in which 95% of the estimates of relative risk (RR) would exist. The 95% Taylor series confidence interval (CI) is calculated according to formula 4 [25].

$$95\% \ CI = RR \ exp\left[\pm 1.96 \sqrt{\frac{1-a/m_1}{m_1(a/m_1)} + \frac{1-c/m_0}{m_0(c/m_0)}} \right]$$

Formula 4

Using these data, the lower limit of the confidence interval = 0.37 ($0.58*2.718^{-0.46}$) and the upper limit = 0.92 ($0.58*2.718^{0.46}$) and does not include 1. This finding indicates that the number of muscle strains that occurred in the flexible group differ statistically from those that occurred in the group considered to be inflexible.

The relative risk of 0.58 can also be tested to determine whether it differs statistically from 1 (no relationship between flexibility and occurrence of muscle strains) by conducting a chi-square test on the frequency data given in Table 5–3 (formula 5):

$$Chi\text{-}square = \frac{(n-1)(ad-bc)^2}{n_1 n_0 m_1 m_0}$$

Formula 5

Using this test, it was found that there is a statistical relationship between flexibility and the occurrence of muscle strains (chi-square = 5.67; degrees of freedom = 1; p < .05). This finding, coupled with the finding that 1 was not included in the 95%

confidence interval, indicates that the relative risk of 0.58 demonstrates the preventive effect of the presence of flexibility on the occurrence of muscle strains.

Surveillance Analysis

This type of design is particularly well suited for the interpretation of findings using relative incidence rates. A ratio of two relative incidence rates, known as the incidence density ratio, can be calculated [47]. This ratio can be used to assess the effect of etiologic factors on the occurrence of sports injuries. The incidence density ratio represents the relative risk of the group represented in the numerator compared to the group represented in the denominator. See Table 5–4 for an example of data collected by surveillance systems. We will use the incidence density ratio to assess these hypothetical data in relation to their potential risk of injury.

When using the data given in Table 5–4 the RIR for flexible players and players not considered flexible are as follows:

Flexible: (240/4700)*1000 =
51.1 per 1000 player-games

Not flexible: (276/5100)*1000 =
54.1 per 1000 player-games

Thus, the incidence density ratio = 51.1/54.1 = 0.94.

Formula 6 can be used to test whether both relative incidence rates differ statistically from each other [25]:

$$Z = \frac{(A - m_1 p_0)}{\sqrt{(m_1 p_0 q_0)}}$$

Formula 6

where A represents the number of flexible players with a muscle strain, p_0 represents the proportion of player-games played by flexible players (L_1/L),

TABLE 5–4

Hypothetical Data Obtained by Using a Surveillance Study Design to Determine the Effect of Flexibility on the Occurrence of Muscle Strains

	Flexible	Not Flexible	Total
Number of players with muscle strain	240 (a)	276 (b)	516 (m_1)
Number of player-games	4700 (L_1)	5100 (L_0)	9800 (L)

and q_0 equals $1 - p_0$. When using the data listed in Table 5–4, the Z statistic equals -0.658, indicating that the incidence density ratio presented is not statistically different from 1.

The 95% confidence interval of the incidence density ratio (IDR) can be used, as described earlier, to determine if the condition of "no association" exists between these two rates. The confidence interval is determined as shown in Formula 7 [25]:

$$95\% \text{ CI} = \text{IDR}^{(1 \pm 1.96/Z)}$$

Formula 7

where Z is the test statistic (Z score) displayed in formula 6. The 95% confidence interval equals (0.78 − 1.12). These data do include 1, which indicates a condition of no association or no statistically significant difference between the rate of injury for flexible players and that for players not considered flexible.

Survival Analysis

A type of analytic procedure used for survival analysis involves comparison of the survival curves described above. Of the different prospective studies, survival design and its analysis require the greatest amount of detail in recording the daily exposures for the subjects. This is a practical issue that needs to be addressed, especially as the size of the population at risk increases. Although survival techniques may be applicable to studies designed to determine the length of time prior to an injury for a particular player, using them to study specific injuries to the athlete (e.g., anterior cruciate ligament sprains) may be very difficult. In general, survival analysis works best in the study of conditions that are relatively common, a situation that does not always exist in investigating specific types of sports injuries. An extensive amount of literature is available on the design and analysis of survival techniques in a variety of disciplines [8, 29], which use the concept of life tables extensively. The reader is directed to these sources for thorough discussions of the techniques and applications of this type of analysis.

Case-Control Design

The second type of observational design, which is especially well suited for studying relatively rare occurrences, is the *case-control* study, often referred to as case-referent or retrospective study. It involves comparison of a group of identified cases with one or more groups of noncases (controls). In a cohort design, as described earlier, the data are recorded prior to the event to be studied (i.e., a prospective method of collection). Although Table 5–3 has been used to illustrate the analysis of cohort studies, we will use the same data to reflect the analytic procedures used for the case-control design.

In a case-control design, the cases (players with a muscle strain) and controls (players without muscle strain) are asked to recall whether or not they were exposed to certain agents or conditions prior to their injury, for example, flexibility. In addition to asking for information from cases and controls, data regarding exposure to risk factors may be obtained from existing records. Using the data from Table 5–3, of 62 football players with a muscle strain, 27 (44%) were considered to be flexible, whereas of the 288 players without muscle strain, 173 (56%) were considered to be flexible. Although it was easy for players to recall the data needed for this specific question, more complex questions about weight, body composition, specific activity at the time of injury, and so on, might be recalled with considerably less ease. Much of the quality of this type of retrospective data collection depends on the existing recording systems used and the time elapsed between the injury and the questioning period.

Case-control studies are especially useful for studying injuries that occur infrequently. They are usually less costly and require less time to conduct than cohort studies. For example, the study of eye injuries in football would require an extremely large population or an extremely long study period if a cohort design were used. A case-control study can select a number of cases from a large population rather than waiting for them to occur.

Although case-control techniques are widely used in epidemiologic research, some specific limitations should be noted. One of these limitations centers on the difficulties encountered in subject selection or case identification procedures. Knowing exactly where to go to get cases may not be obvious. Also, the variability that can exist in retrospective data collection can present special difficulties in the analysis and reliability of the data.

Case-Control Analysis

As with cohort analysis, the task is to determine whether the findings in the index group are different from those represented by the control group. The relative risk in case-control studies can be approximated by using the concept of the *odds ratio*.

Using the data given in Table 5–3, an odds ratio or risk estimate can be created for the relationship between muscle flexibility and the occurrence of muscle strains. The question at hand is, "Do flexible players have an equal, greater, or lesser risk of sustaining a muscle strain compared with players who are not flexible?'' Like relative risk, an odds ratio of 1 indicates no association between flexibility and the occurrence of muscle strains. A finding of less than 1 indicates that the study factor is associated with fewer injuries, and an odds ratio of greater than 1 indicates a greater frequency of injury associated with the study factor. In this example, the odds ratio (OR) is calculated as follows:

$$OR = \frac{a * d}{b * c} = \frac{27 * 115}{173 * 35} = 0.51$$

Formula 8

This means that the risk of sustaining an injury for those who are considered flexible is almost twice as low as that for players considered not flexible. Approximate confidence intervals (95% CI) can be calculated by using Taylor's series of confidence intervals according to formula 9 [25]:

$$95\% \ CI = OR * \exp[\pm 1.96 * \sqrt{(1/a + 1/b + 1/c + 1/d)}]$$

Formula 9

The 95% confidence interval of the odds ratio given above has a lower limit of 0.29 ($0.51*2.718^{-0.55}$) and an upper limit of 0.89 ($0.51*2.718^{0.55}$).

The odds ratio can be tested to determine whether it has a statistically significant variation from 1 by using the Mantel-Haenszel chi-square statistic with 1 degree of freedom (formula 5). The chi-square statistic, under these conditions, equals 5.67 and is statistically significant at $p < .05$. This is consistent with the finding that 1 is not included in the 95% confidence interval of the odds ratio. Based on these data, it would appear that athletes who are flexible have a substantially lower risk of sustaining a muscle strain than those who are not flexible.

Cross-Sectional Design

By using the third type of observational study design, *cross-sectional* studies, the method of data collection is neither prospective, as in cohort studies, nor retrospective, as in case-control studies. In cross-sectional studies the investigators determine, at one point in time, the presence or absence of an injury (muscle strain) and the presence or absence of the study factor (flexibility). This type of design is useful in describing injuries and their basic distribution patterns among subgroups within the population. Cross-sectional studies can be used to investigate associations provided the study factor in question preceded the occurrence of injury, which is often impossible to determine.

Cross-sectional designs by definition provide information about the existence of injuries at a specific point in time; that is, they provide information about prevalence. These types of studies are particularly useful in planning for health care facilities and to determine the patterns of existing injury frequencies. Prevalence is of limited value in studying etiologic hypotheses because of the frequent difficulty of determining the sequence between the study factor and the identification of the sports injury. Cross-sectional designs are impractical when an injury is considered infrequent because a very large population would be required to identify cases. The cross-sectional design has very limited utility in the study of sports injuries because injuries with longer rehabilitation times would be overrepresented in the findings. The reader is referred to Kleinbaum and colleagues [25] for a more thorough discussion of the cross-sectional design and its implications for research on prevalence.

Multivariate Approach

Much of the data that appear in the literature on etiologic factors associated with injury frequency attempt to determine the risk of injury solely by examining the study factor and its relationship to injury occurrence. It is common practice to use this single factor, or univariate, approach in searching for factors that affect the occurrence of sports injuries. For example, in reviewing numerous studies that examined the influence of preventive knee braces on the occurrence of knee injuries in football, the univariate approach was frequently used to address the issue of the efficacy of these devices [53]. Many of these studies did not address other related factors such as type of session (game or practice), position of the player, type of shoes worn, and amount of population-time involved.

Epidemiologists have recognized that numerous factors are associated with the occurrence of diseases and injuries. The multi-factor, or multivariate, approach is consistently described as superior to the univariate approach because it provides an opportunity to consider more of the large number of variables that may be associated with the event. The multivariate approach is becoming more common in sports injury research as the limitations associated with the univariate approach become better documented [53] and study designs become more sophisticated.

Extraneous Factors

The multivariate approach carries with it some special considerations in the selection of variables to be considered in the study. The research team should keep in mind that some variables will be considered primary foci, study factors, but other variables that affect the relationship between the study factors and the injury event also exist. This latter type of variable is referred to as an *extraneous factor,* and they must be carefully evaluated to assess the "true" effect of the study factor. The majority of these extraneous factors can be divided into three groups that are of specific concern for observational studies. The three categories are confounders, co-variables, and effect modifiers.

Confounders are factors associated with both the study factor (e.g., knee brace) and the occurrence of injuries (e.g., medial collateral ligament [MCL] injuries). For example, when preventive knee braces are worn only during one type of session (game or practice) and the risk of sustaining an MCL injury is different during each session, the type of session is a confounding variable for the relationship between the use of a preventive knee brace and the occurrence of an MCL injury.

The second group of extraneous variables, called *covariables,* is associated only with the occurrence of injuries and not with the study factor. For example, the type of session is classified as a covariable when the probability of sustaining an MCL injury is different during practices and games but all players in the study group wear preventive knee braces during both sessions.

The third group of extraneous factors, called effect modifiers, are factors that affect the relationship between the study factor and injury occurrence. For example, type of session is an effect modifier if a relationship exists between the use of preventive knee braces and MCL injuries for injuries sustained during games but not during practice.

Controlling for extraneous factors can be done either in the design of the study or in the analysis of the findings, or in both phases. Controlling for extraneous variables in the design of a project could be done by, for example, matching players with certain characteristics and the presence of the study factor (players who wear preventive knee braces) to players with the same characteristics but without the study factor (players who do not wear preventive knee braces).

When controlling for extraneous variables during the analysis of findings, generally two methods are used, stratified analysis and mathematical modeling. Although both methods are effective, the simpler stratified analysis technique becomes more difficult to interpret as the number of extraneous variables increases. The strength of mathematical modeling is its capability of considering numerous variables.

Stratified Analysis

Stratification is accomplished by making cross-classification tables based on the values of the identified extraneous variables. For example, in a hypothetical study of soccer players the investigators want to determine the risks of playing with and without shin guards. Because the type of session (game or practice) will likely affect the relative incidence rate and use of shin guards, this factor has to be controlled. In this case, a separate table showing the association between shin guards and the occurrence of injuries within each type of session (one for games and one for practices) must be made. Stratification is not limited to one factor but can be extended to incorporate multiple factors.

Stratified analysis is particularly worthwhile when the data under examination address the following concerns [25]. First, there are a sufficient number of injuries in all levels of the data table. All cells of all tables have to be filled with more than a few injured players to make valid inferences about the study findings. Second, there must be a specific rationale for the selection of the control variables that are considered important to the overall project. Third, an appropriate scheme of categorization for each variable must be identified prior to analyzing the data. There are several ways to form strata, and the exact nature of the data partitions that create the table will have an important effect on the findings.

When using stratified analysis, there are various statistical estimates that can be used to address the risk of injury (e.g., the odds ratio, the relative risk, and the incidence density ratio [25], to name a few). In the final analysis, a risk estimate for the relationship between the study factor and the occurrence of injuries without controlling for any extraneous factors can be compared with an appropriate adjusted risk estimate. Several methods are available for making this comparison. The exact nature of each of these methods is beyond the scope of this chapter but is discussed in the work of Kleinbaum and colleagues [25].

Mathematical Modeling

An alternative and more complex method of controlling for extraneous variables is called mathematical modeling. There are several advantages of using this strategy compared to stratified analysis [25]. The biggest advantage is that it is a feasible choice when there are small numbers of injured players in the cells of the cross-classification table. Examples of mathematical modeling strategies used in sports injury research are logistic regression [1, 54] and loglinear analysis [3, 43].

Logistic regression is used to estimate the probability that an injury-event will occur. It includes the ability to consider both continuous and categorical variables that will predict an event in a dichotomous dependent variable (e.g., the presence or absence of injuries). Schootman and Powell [54] determined, by using logistic regression, factors potentially associated with the severity of high school football injuries. Using logistic regression, you would set the dependent variable (time lost) into two groups (less than 8 days and 8 days or more) and then determine whether height (continuous variable) or type of surface (categorical variable) could predict the time loss category. With most calculations a variety of measures of association can be developed, with the odds ratio being used most often.

Another mathematical modeling technique applicable to the study of sports injuries is loglinear analysis. This technique is an extension of the chi-square test previously described and is used to determine the simultaneous relationship between two or more variables. It is classified as an independence model and uses only categorical data. These independence models can be used to determine in football, for example, whether the type of surface (grass or Astroturf), the period of a game, and the type of play (running or passing) are independent of each other when one is establishing the likelihood of injury. The model then searches for the interactive effects among the variables included in the table. Estimates of the likelihood of injury in the various cells can then be established and interpreted.

Other mathematical modeling procedures, such as multiple regression analysis, discriminant analysis, and multiple analysis of variance may be appropriate for predicting the occurrence of an event. The applicability of these procedures in epidemiologic research relating to sports injuries has yet to be determined.

Experimental Designs

Experimental designs, or intervention studies, are characterized by randomly allocating subjects to certain predetermined conditions according to the investigator's choice prior to the beginning of the study. Sitler and associates [55] used this type of design to study the efficacy of preventive knee braces. Ekstrand and Gillquist [14] also used a random allocation procedure to investigate the use of a preventive program to reduce injuries in soccer players. In most cases, random allocation ensures that player factors are equally represented among both control and experimental groups, although this fact needs to be double-checked. It is sometimes

difficult to justify a random allocation procedure if the device or procedure being studied is perceived by the subjects as preventive. To withhold such a preventive feature from a group exposed to risk presents a variety of ethical issues that need to be overcome. If the preventive feature is a device or piece of equipment, the costs of purchasing these materials, standardization of the devices, and verification that variations in the device are not extraneous variables can be very costly. These limitations are generally responsible for the small number of intervention studies done in the field of sports injuries. Practically speaking, strong intervention studies are very difficult to accomplish in etiologic sports injury research.

DIAGNOSTIC TESTS

Studies of diagnostic tests are designed to determine how well a new test or testing procedure can discriminate between two conditions, such as the presence or absence of laxity of a ligament. In preparing a research project to examine the differences between testing protocols, several steps have to be taken to determine the appropriateness of a new diagnostic test [22]:

1. Ascertain the need for a new diagnostic test.
2. Select the subjects.
3. Determine the availability of a golden standard.
4. Establish a standardized and blinded study design.
5. Estimate the sample size.
6. Find a sufficient number of willing subjects.
7. Report the results in terms of sensitivity and specificity.

The key issue is the golden standard against which the new test is rated. A golden standard is a reference procedure or test that is always positive in patients in whom the condition is (injury) present and always negative in patients without the condition. Table 5–5 is constructed to assist in the discussion and is based on the previous statements.

TABLE 5–5
Typical Data Table for the Evaluation of a New Diagnostic Test

Test Result	Condition	
	Present	*Absent*
Positive	True-positive (tp)	False-positive (fp)
Negative	False-negative (fn)	True-negative (tn)

It can be seen from this table that four distinct situations exist. The condition is present in patients and the test diagnosed these patients as having the condition (true-positive); the patients have the disease but the test indicated otherwise (false-negative); the patients do not have the condition but the test indicated that they did (false-positive); and the patients did not have the condition and this was indicated correctly by the test (true-negative). The ability of the test to identify subjects with and without the condition (injury) can be determined by calculating sensitivity and specificity, respectively.

Sensitivity is defined as the proportion of subjects with the condition who have a positive test (tp/ (tp + fn)), whereas *specificity* is defined as the proportion of subjects without the condition who have a negative test (tn/(fp + tn)). For the strongest possible test, the investigators want both indices (sensitivity and specificity) to be as high as possible, preferably near 100%. An example of a new diagnostic test is magnetic resonance imaging (MRI) to detect neurologic diseases [21]. The diagnostic capabilities of MRI can be compared to the capabilities of computed tomography or plain x-ray film by calculating sensitivity and specificity parameters.

Another example is a comparison of tests for knee instability, such as the Lachman test and the anterior drawer test, with arthroscopy as a more objective standard for determining the presence of anterior cruciate ligament (ACL) injuries [24]. See Table 5–6 for a hypothetical example of the relationship between diagnosing an ACL injury by the Lachman test and by arthroscopy. In this case, arthroscopy is considered the golden standard.

Based on Table 5–6 the sensitivity of the Lachman test equals 81% (121/150), whereas the specificity equals 87% (109/125). Thus, of the total number of subjects with an ACL injury (as diagnosed by arthroscopy), 81% were correctly diagnosed by the Lachman test. Conversely, of the total number of subjects without an ACL injury as diagnosed by arthroscopy, 87% were diagnosed as such by the Lachman test.

TABLE 5–6
Hypothetical Data to Determine the Relationship Between the Lachman Test and Arthroscopy in Detecting the Presence of ACL Injuries

Lachman Test	Arthroscopy		Total
	Present	*Absent*	
Present	121	16	137
Absent	29	109	138
Total	150	125	275

TABLE 5–7
Hypothetical Data to Estimate the Interobserver Reliability of a Test Used to Assess Flexibility*

Observer B	Observer A		Total
	Present	*Absent*	
Present	64 (a) [39.44]	11 (b) [35.55]	75
Absent	7 (c) [31.55]	53 (d) [28.44]	60
Total	71	64	135

*Expected values for each cell, based on a calculation similar to that for the chi-square test, are listed in brackets.

Like sensitivity and specificity, measures of the predictive values *positive* and *negative* of the test can be determined. Predictive values are used to determine the ability of the test to classify subjects with and without the condition. *Predictive value positive* is defined as the probability that a subject who has a positive test actually has the condition (tp/(tp + fp)). *Predictive value negative* is defined as the probability that a subject who has a negative test does not have the condition (tn/ (fn + tn)). As with sensitivity and specificity, both predictive values must be high for the test to predict the presence or absence of the condition accurately. In Table 5–6, the predictive values positive and negative equal 80% (109/138) and 88% (121/137), respectively. Thus, of the 137 players with a positive test (indicating the presence of an ACL injury), 80% actually had an ACL injury. Similarly, of the 138 players with a negative test, 88% did not actually have an ACL injury.

In addition to assessing the association between two diagnostic procedures, the extent of the agreement between two persons performing the same test can be determined, also known as the *interobserver reliability*. In assessing interobserver reliability the same test is replicated by two observers to determine the presence of a condition. Interobserver reliability is designed to determine how consistent both recorders are in observing the presence or absence of the condition (e.g., flexibility or injury). Table 5–7 lists hypothetical data observed by the two investigators for estimating interobserver reliability.

The interobserver reliability can be estimated by using the kappa coefficient [58]. Kappa is defined according to formula 10:

$$\text{Kappa} = \frac{\text{observed} - \text{expected proportion in agreement}}{1 - \text{expected proportion in agreement}}$$

Formula 10

Kappa can be negative, but its value is 0.0 when

agreement occurs only at the chance-expected level, and its value is 1.0 when agreement is perfect. Based on Table 5–7, the observed proportion of agreement is defined as (a + d)/N, and the expected proportion of agreement is based on the expected values calculated like those used in a chi-square test. For example, the expected value for d equals (60*64)/135. The expected values are displayed between brackets in Table 5–7. Using the data from this table, kappa is found to equal 0.73. This means that there is a high level of agreement between both observers about the presence of the studied condition.

Another type of observer reliability is intraobserver reliability, which is based on the same formula for kappa. The intraobserver reliability is the degree of consistency scored by one observer in multiple trials. In our example, the interobserver and the intraobserver reliability have the same value. Other measures can be used to calculate interobserver and intraobserver reliability but will not be discussed here.

DESIGN OF CLINICAL TRIALS

In the final part of this chapter we will discuss briefly the design of studies intended to identify the effectiveness of treatments or interventions of injuries. The use of a valid study design is of crucial importance in making valid inferences about the study findings. In this particular approach, the use of randomized clinical trials is extremely suitable for comparing the effectiveness of treatments and should be employed whenever possible [50]. A randomized clinical trial contains four main elements: the selection of patients for the trial, the random allocation of treatments to patients, the treatment period, and the statistical analysis [59]. An example of a random clinical trial is a study done by Sandberg and colleagues [51]. In this study patients with knee injuries were randomly assigned to operative and nonoperative groups. During follow-up, factors such as range of motion, swelling, and muscle force in both groups were compared.

Before admission into a clinical trial each patient should fulfill all of the inclusion criteria adopted and none of the exclusion criteria. Depending on these criteria, the number of eligible patients, the ability to generalize the study's findings to other populations, and internal validity will be determined. For example, the use of very tight inclusion criteria will limit the number of study subjects available. Also, the findings can be generalized only to patients with the same characteristics as those in the study population. On the other hand, use of a homogeneous study group enhances the internal validity of the findings. The use of less tight inclusion criteria will increase the number of eligible patients, and the findings can be generalized to a more heterogeneous population, but under these conditions the internal validity may be reduced.

After obtaining a sufficient number of eligible patients, the next step is to randomize them to two (or more) treatment regimens. Randomization of eligible patients is preferred for the design of the study and enhances the statistical analysis of a clinical trial. It also ensures that comparable groups are formed, not only with regard to known variables but also with regard to unknown relevant variables. All patients fulfilling the inclusion criteria should be randomized (after obtaining their consent) to one of the treatment regimens.

Three fundamental designs are available for assessing the effect of treatments [59]. The first is called a group comparison and encompasses different treatments that are randomly allocated to two (or more) groups of patients. Second, the matched pairs trial uses pairs of patients identical with respect to all relevant factors except the type of treatment. And third, the crossover trial includes patients who all receive both treatments in sequential order.

As with other types of studies done in injury research, the effect of different biases needs to be considered during design and analysis. One of the more common biases can be controlled by using a double-blind protocol (i.e., neither the patient nor the investigator knows to which treatment group the patient has been assigned). In some instances double-blind designs are impossible, for instance, when the principal investigator or the patient knows the patient's treatment assignment. A physician who is investigating the effectiveness of a surgical procedure would be an example of a program where double-blind protocols would be impractical.

A variety of different statistical tests are available for interpreting data obtained by a randomized clinical trial. The following are merely examples of the tests available. When the outcome variable is dichotomous the Mantel-Haenszel chi-square test, the odds ratio, and logistic regression can be used, all of which were described previously. If the outcome variable is continuous, analysis of variance can be employed [26]. Analysis of variance is an extension of the Student's t-test and is used to test three or more means statistically. In addition to the statistical techniques mentioned in this section, other statistical tools can be used, but because of limited space they will not be discussed in this chapter. The interested reader is referred to Kleinbaum and Kupper [26] for more information.

SUMMARY

In the study of factors affecting the occurrence of sports injuries, the first issue to be tackled is the assessment of the problem by using incidence and prevalence. Two measures of incidence have been described, rate and risk. Rate of injury, either absolute or relative, is defined as the number of injuries or injured players per unit of time (e.g., game, practice, parachute jump, and so on). The injury rate, or incidence density, is especially valuable in testing etiologic hypotheses. Different ways of estimating this rate are available depending on the level of detail with which the exposure has been assessed. The counterpart of the use of rates is the estimation of risk, defined as the probability a player has of sustaining an injury during the study period. Unlike rate, risk is estimated at the level of the individual. Three methods can be employed to estimate risk—cumulative incidence, the actuarial method, and incidence densities. Several advantages and disadvantages of each method have been mentioned.

Prevalence is less suitable for testing hypotheses in etiologic sports injury research. Two types of prevalence can be distinguished, point and period prevalence. The main difference between both concepts is the use of the time factor. Point prevalence estimates the proportion of injured players at a point in time, whereas period prevalence uses a period of time to estimate the proportion.

Next, a study design has to be chosen to determine the effect of an etiologic factor on the occurrence of sports injuries. The available study designs can be separated into two main groups depending on whether or not randomization is used to allocate the subjects to various conditions. Studies that use randomization are called experimental studies, whereas observational studies do not use random allocation of subjects. Three observational study designs have been discussed—cohort, case-control, and cross-sectional studies. The main difference between these designs is the use of the time factor. Cohort studies start with one group of exposed players and compare it to a control group (unexposed players) in regard to the development of injuries during subsequent follow-up. Case-control studies start with a group of injured players and compare it to a group of uninjured players with regard to their previous exposure. Cross-sectional studies use a single point in time to estimate the presence of the study factor and the existence of sports injuries.

After selecting the appropriate study design, the size of the sample has to addressed. The method used to estimate the size of the sample depends on many factors, such as alpha, beta, and type of study.

During design and analysis of the study the presence of bias has to be determined. Different types of study designs are vulnerable to different types of bias. Case-control studies are especially vulnerable to recall bias, whereas cohort studies may suffer from follow-up bias.

The next phase in conducting a study of sports injuries is the analysis of the obtained data. Depending on the type of study design, different types of information with different levels of detail are obtained, which strongly limit the applicable statistics. The relative risk and the odds ratio with their respective confidence intervals and statistical tests have all been described. Although the univariate approach is common practice in etiologic sports injury research, a plea was made for the use of the multivariate approach. Analysis of multivariate models can be accomplished by using stratified analysis or mathematical modeling. Both have been described and both have specific advantages and disadvantages in studying the risk of sports injuries.

Diagnostic tests are also part of sports injury research. The development of a new and better diagnostic test is fully warranted. The value of the new diagnostic test can be described by calculating its sensitivity, specificity, and predictive values positive and negative. The existence of a gold standard is a necessity, however. Interobserver and intraobserver reliabilities are also valuable for using diagnostic tests. Kappa is used to assess both reliabilities.

The final section of this chapter consists of ways to evaluate the effectiveness of treatments. Randomized clinical trials should be used whenever possible. The main advantage of the use of these trials is the random allocation of subjects to two or more treatments. This secures the random distribution of known and unknown factors across treatments. Several statistical techniques can be used to estimate the effectiveness of the treatments depending on the scale of the dependent variables and the design of the study.

Finally, sports medicine and sports injury research are young disciplines compared with other aspects of medical research. These developing disciplines have led to an increasing interest in the development of scientific methods applicable to the study of sports injuries. Because of the many variations that exist in sports medicine in regard to injuries and exposures, special consideration has to be given to the applicability of methods derived from disciplines such as epidemiology and biostatistics. To develop methods that are specifically applicable to sports medicine research, sports medicine professionals must work closely with professionals in these two disciplines. The initiation of this working relationship may be difficult at first because practitioners in neither field have sufficient knowledge to communicate in both disciplines. As time passes, however, more and more sports medicine professionals are learning to apply scientifically sound research methods, and epidemiologists are becoming more interested in sports medicine. The strength and the future of the continuing development of the scientific foundations needed for preventing sports injuries are dependent on the interdisciplinary cooperation achieved by individuals in many professional disciplines.

References

1. Backx, F. J. G., Erich, W. B. N., Kemper, A. B. A., and Verbeek, A. L. M. Sports injuries in school-aged children. An epidemiologic study. *Am J Sports Med* 17(2):234–240, 1989.
2. Breslow, N. E., and Day, N. E. *Statistical Methods in Cancer Research*. Vol. 1. *The Analysis of Case-Control Studies*. Lyon, IARC Scientific Publications, no. 32, 1980.
3. Buckley, W. E. A multivariate analysis of conditions attendant to concussion injuries in college football 1975–1982. Thesis, Pennsylvania State University, 1985.
4. Cahill, B. R., and Griffith, E. H. Exposure to injury in major college football. *Am J Sports Med* 7:183–185, 1979.
5. Clarke, K. S., Alles, W. F., and Powell, J. W. An epidemiological examination of the association of selected products with related injuries in football 1975–1977. Contract #SPSC-C-77-0039. Washington, D.C., U.S. Consumer Product Safety Commission, 1979.
6. Cornell, C. N., and Ranawat, C. S. Survivorship analysis of total hip replacements. Results in a series of active patients who were less than fifty-five years old. *J Bone Joint Surg* 68-A(9):1430–1434, 1986.
7. Cox, D. R., and Oakes, D. *Analysis of Survival Data*. London, Chapman and Hall, 1984.
8. Crowley, J. Statistical analysis of survival data. *Ann Rev Public Health* 5:385–411, 1984.
9. Cutler, S. J., and Ederer, F. Maximum utilization of the life table method in analyzing survival. *J Chron Dis* 8:699–712, 1958.
10. Dagiau, R. P., Dillman, C. J., and Milner, E. K. Relationship between exposure time and injury in football. *Am J Sports Med* 8:257–260, 1980.
11. Damron, F. C. Injury surveillance systems for sports. *In* Vinger, P. F., and Hoerner, E. F. (Eds.), *Sports Injuries: The Unthwarted Epidemic*. Littleton, MA, PSG Publishing, 1981.
12. Daniel, W. W. *Applied Nonparametric Statistics*. Boston, Houghton Mifflin, 1978.
13. Dobbs, H. S. Survivorship of total hip replacement. *J Bone Joint Surg* 62-B(2):68–173, 1980.
14. Ekstrand, J., and Gillquist, J. Prevention of sports injuries in football players. *Int J Sports Med* 5(Suppl.):140–144, 1984.
15. Elandt-Johnson, R. C. Definition of rates: Some remarks on their use and misuse. *Am J Epidemiol* 102:267–271, 1975.
16. Elandt-Johnson, R. C. Various estimators of conditional probabilities of death in follow-up studies: Summary of results. *J Chron Dis* 30:247–256, 1977.
17. Fleiss, J. L., Dunner, D. L., Stallone, F., and Fieve, R. R. The life table: A method for analyzing longitudinal studies. *Arch Gen Psychiat* 33:107–112, 1976.
18. Fletcher, R. H., Fletcher, S. W., and Wagner, E. H. *Clinical Epidemiology: The Essentials*. Baltimore, Williams & Wilkins, 1982.
19. Freeman, J., and Hutchinson, G. B. Incidence, prevalence, and duration. *Am J Epidemiol* 112:707–723, 1980.
20. Greenberg, R. S., and Kleinbaum, D. G. Mathematical modelling strategies for the analysis of epidemiologic research. *Ann Rev Public Health* 6:223–245, 1985.
21. Hayes, R. G., and Nagle, C. E. Diagnostic imaging of intracranial trauma. *Physician Sports Med* 18(2):69–79, 1990.
22. Hulley, S. R., and Cummings, S. R. (Eds.): *Designing Clinical Research*. Baltimore, Williams & Wilkins, 1988.
23. Hunter, R. E., and Levy, I. M. Vignettes. *Am J Sports Med* 16(Suppl.):25–36, 1987.
24. King, S., Butterwick, D. J., and Cuerrier, J. P. The anterior cruciate ligament: A review of recent concepts. *J Orthop Sports Phys Ther* 8(3):110–114, 1986.
25. Kleinbaum, D. G., Kupper, L. L., and Morgenstern, H. *Epidemiologic Research: Principles and Quantitative Methods*. Belmont, CA, Lifetime Learning Publications, 1982.
26. Kleinbaum, D. G., and Kupper, L. L. *Applied Regression Analysis and Other Multivariable Methods* (2nd ed.). Boston, PWS-Kent Publishing, 1988.
27. Kranenborg, N. Sports participation and injuries. *Geneeskunde Sport* 13:89–93, 1980 (in Dutch).
28. La Cava, G. A clinical and statistical investigation of traumatic lesions due to sport. *J Sports Med Phys Fitness* 1:8–19, 1961.
29. Lawless, J. E. *Statistical Models and Methods for Lifetime Data*. New York, Wiley, 1982.
30. Lysens, R. J., Ostyn, M. S., Auweele, Y. V., Lefevre, J., Vuylsteke, M., and Renson, L. The accident-prone and overuse-prone profiles of the young athlete. *Am J Sports Med* 17(2):612–619, 1989.
31. MacMahon, B., and Pugh, T. F. *Epidemiology: Principles and Methods*. Boston, Little, Brown, 1970.
32. Mantel, N., and Haenszel, W. Statistical aspects of the analysis of data from retrospective studies. *J Nat Cancer Inst* 22(4):719–748, 1959.
33. Mechelen, W. van, Hlobil, H., and Kemper, H. C. G. *How Can Sports Injuries Be Prevented?* Oosterbeek, The Netherlands National Institute for Sports Health Care, 1987.
34. Miettinen, O. S. Estimability and estimation in case-referent studies. *Am J Epidemiol* 103:226–235, 1976.
35. Miettinen, O. S. *Theoretical Epidemiology*. New York, Wiley & Sons, 1985.
36. Morgenstern, H., Kleinbaum, D. G., and Kupper, L. L. Measures of disease incidence used in epidemiologic research. *Int J Epidemiol* 9:97–104, 1980.
37. Mueller, F. O., and Blyth, C. S. Fatalities and catastrophic injuries in football. *Physician Sports Med* 10(10):135–139, 1982.
38. Noyes, F. R., and Albright, J. P. Foreword to sports injury research. *Am J Sports Med* 16(Suppl.):vi–x, 1988.
39. Olson, O. C. The Spokane study: High school football injuries. *Physician Sports Med* 7(12):75–82, 1979.
40. O'Toole, M. L., Hiller, D. B., Smith, R. A., and Sisk, T. D. Overuse injuries in ultraendurance triathletes. *Am J Sports Med* 17(2):514–518, 1989.
41. Phillips, A. N., and Pocock, S. J. Sample size requirements for prospective studies with examples for coronary heart disease. *J Clin Epidemiol* 42:639–648, 1989.
42. Pino, E. C., and Colville, M. R. Snowboard injuries. *Am J Sports Med* 17(2):778–781, 1989.
43. Powell, J. W. A multivariate epidemiological model applied to injuries to the knee in college football 1975–1978. Thesis, Pennsylvania State University, 1980.
44. Powell, J. W. Pattern of knee injuries associated with college football 1975–1983. *Athl Training* 20:104–109, 1985.
45. Powell, J. W. Incidence of injury associated with playing surfaces in the National Football League 1980–1985. *Athl Training* 22:202–206, 1987.
46. Powell, J. W. 636,000 injuries annually in high school football. *Athl Training* 22:19–22, 1987.
47. Powell, J. W., and Schootman, M. Knee injuries associated with playing surfaces in the National Football League: A multifactor approach. In press, 1991.

48. Prager, B. I., Fitton, W. L., Cahill, B. R., and Olson, G. H. High school football injuries: A prospective study and pitfalls of data collection. *Am J Sports Med* 17(2):561–685, 1989.
49. Pritchett, J. W. A statistical study of physician care pattern in high school football injuries. *Am J Sports Med* 10:96–99, 1982.
50. Raskob, G. E., Lofthouse, R. N., and Hull, R. D. Methodological guidelines for clinical trials evaluating new therapeutic approaches in bone and joint surgery. *J Bone Joint Surg* 67-A(8):1294–1297, 1985.
51. Sandberg, R., Balkfors, B., Nilsson, B., and Westlin, N. Operative versus non-operative treatment of recent injuries to the ligaments of the knee. *J Bone Joint Surg* 69-A(8):1120–1126, 1987.
52. Schlesselman, J. J. *Case-Control Studies.* Oxford, Oxford University Press, 1982.
53. Schootman, M. The effect of preventive knee braces on the occurrence and severity of knee injuries in American football. Master's thesis. Free University, Amsterdam, 1990 (in Dutch).
54. Schootman, M., and Powell, J. P. Severity of high school football injuries. Submitted 1992.
55. Sitler, M., Ryan, J., Hopkinson, W., Wheeler, J., Santomier, J., Kolb, R., and Polley, D. The efficacy of a prophylactic knee brace to reduce knee injuries in football. *Am J Sports Med* 18:310–315, 1990.
56. Smith, A. D., and Ludington, R. Injuries in elite pair skaters and ice dancers. *Am J Sports Med* 17(2):484–488, 1989.
57. Szklo, M. Design and conduct of epidemiologic studies. *Prev Med* 16:142–149, 1987.
58. Thompson, W. D., and Walter, S. D. A reappraisal of the kappa coefficient. *J Clin Epidemiol* 41:949–959, 1988.
59. Tygstrup, N., Lachin, J. M., and Juhl, E. (Eds.). *The Randomized Clinical Trial and Therapeutic Decisions.* New York, Marcel Dekker, 1982.
60. Wallace, R. B. Application of epidemiologic principles to sports injury research. *Am J Sports Med* 15(Suppl.):22–24, 1987.
61. Wallace, R. B., and Clark, W. R. The numerator, denominator, and the population at-risk. *Am J Sports Med* 15:55–56, 1987.
62. Walter, S. D., Sutton, J. R., McIntosh, J. M., and Connolly, C. The aetiology of sports injuries. *Sports Med* 2:47–58, 1985.

REHABILITATION PRINCIPLES AND TECHNIQUES

Cheryl Hubley-Kozey, B.Sc., M.Sc., Ph.D.
William D. Stanish, M.D.
Sandra L. Curwin, B.Sc., M.Sc., Ph.D.

BASIC PRINCIPLES OF REHABILITATION

Rehabilitation refers to an active process that takes place to normalize variables altered by disease, injury, or surgery, implying that an improvement associated with reconditioning will take place. This definition of rehabilitation may pose some problems for the athletic population, especially with the determination of "normal" values for variables assessed during the rehabilitation process. For example, strength is an important performance variable in many sports, and therefore is trained to an extent that may place the athlete at an extreme range with respect to measures from a normal population (Fig.

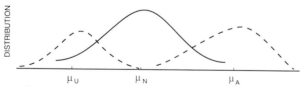

FIGURE 6–1
For a given performance variable such as knee extensor muscle strength, the mean value for a normally healthy population (μ_N) may be much lower than the mean value for a population of elite sprint cyclists (μ_A). Although there is overlap between the two curves, values obtained from nonathletic populations should not be considered the standard for comparison. Note also that following an injury an athlete may fall within the unhealthy curve (μ_U) for a given performance variable. This suggests that a great deal of retraining may be required to move him or her back within the population range, more training than that necessary to make such an athlete comparable to the normal population.

6–1). Consequently, normative values derived from a nonathletic population may not suffice as comparative measures for an athletic group.

The first part of this chapter defines the variables most often used to assess the effectiveness of rehabilitation programs for orthopaedic problems. This provides a common basis for the discussion of rehabilitation objectives and techniques that follows. Special consideration is given to the unique characteristics of athletes, including suggestions for training and conditioning for strength, endurance, and flexibility.

Muscle Strength and Power

Muscle strength is considered an important variable to be assessed during rehabilitation because many injuries, diseases, or other traumas have associated muscle strength deficits. The cause of the deficit may be damage to the muscle-tendon unit, neurologic inhibition from swelling or pain, disuse atrophy, or one of a number of other factors discussed in the literature [80, 81]. The quantitative measurement of muscular strength and its changes during rehabilitation provides an evaluation of the integrity and function of the neuromuscular unit.

There are several definitions of muscle strength in the literature, with most relating to the ability of the muscle to generate a tensile force [52, 53, 71,

77, 78]. This intrinsic force-generating capability produces and resists movements essential to the performance of many athletic skills. Therefore, muscular strength should be evaluated by assessing the force production capabilities of the muscle. Manual muscle testing and circumference measurement are often used clinically and provide relevant information for the clinician. However, they do not relate directly to muscle force and should not be considered alternative methods of assessing strength because they lack the specificity required by the above definition. Manual muscle testing may even be inappropriate for many members of the athletic population because the maximum measure is limited by the ability of the tester to resist the movement. Differentiating the grades of strength with manual muscle testing is also subjective. The cross-sectional area of a muscle, known to be closely related to muscle force production [1], is the variable that circumferential muscle measurement is meant to assess. However, there are many confounding factors that make a simple relationship between muscle force and limb circumference difficult [78]. Therefore, in practical terms, the fundamental consideration in defining muscle strength uses the concept of force generation.

The tensile force produced by a muscle is influenced by several factors, including the cross-sectional area of the muscle; its fiber-type composition; the geometry of fiber orientation; the number, order, and frequency of motor units recruited; and many of the other factors discussed in Chapter 1. The length of the muscle and the velocity of lengthening or shortening also have an effect on the force transmitted from the muscle to its attached tendon [14, 36, 37, 100]. Direct in vivo measurement of muscle force requires the placement of a force transducer directly on the tendon, which requires surgical intervention. This is not a clinically viable procedure for practical, technical, and ethical reasons. The electromyographic signal has been related to muscle force output, but further research is required before this relationship can be exploited clinically [38, 49]. The most commonly used clinical estimate of muscle strength is therefore the moment produced by a muscle or muscle group. Use of this externally measured variable relies on knowledge of the biomechanical principles of the human musculoskeletal system.

The moment of force produced about a joint axis of rotation has been used as a measure of the strength of a muscle (or muscle group) by many authors [51, 71, 72, 75, 77]. The SI (Système Internationale) unit for measuring a moment of force is the Newton (N), whereas in the British system it is the foot pound (ft-lb). The knee extensor muscle group, shown in Figure 6–2, illustrates the relationship between muscle force and the muscle moment. The latter is the product of two components: the force generated by the muscle (measured in Newtons) and the moment arm length of the muscle (measured in meters).

The muscle force (F_{KE}) in Figure 6–2A is a single vector representing all the muscles acting to produce a specific movement. Thus, the knee extensor force vector represents the vector addition of the four components of the quadriceps, vastus lateralis, vastus medialis, vastus intermedius, and rectus femoris (Fig. 6–2B). An understanding of the functional anatomy of these muscles can be used to partially isolate their action and evaluate their individual contribution to force production. The contribution of the rectus femoris to knee extension, for example, can be affected by altering its length by changing the angle of the hip joint. Although complete isolation of a muscle within a group is difficult to achieve, the positions shown in muscle testing manuals may be used if an assessment of an individual muscle contribution is deemed necessary [45].

The second component of the moment of force is the moment arm length of the muscle. This is the perpendicular distance between the joint axis of rotation and the vector representing muscle force. The muscle moment arm length can be altered by either changing the distance of the muscle's insertion point from the joint center or changing the joint angle, or by the presence of bony structures, such as the patella, which can increase the moment arm length of the knee extensors at some joint angles [77, 78]. Because the muscle moment of force is simply the product of the force vector multiplied by the moment arm, it is easy to see that muscle force can be obtained by dividing the muscle moment of force by the moment arm length of the muscle; this relates to our original definition of muscle strength. Although we know how moment arm lengths vary with changes in joint angle (for some of the major muscle groups), it is unusual in the clinical setting to perform an actual calculation of muscle force. This division would require either the use of standardized moment arm lengths and their associated assumptions derived from the literature, or the development of a system such as radiography to measure accurately the moment arm lengths during a testing session. This is unnecessary unless there has been a major change in moment arm length due to an intervention, for example, a patellectomy. Such an alteration in moment arm length would affect the relationship between muscle force and muscle moment of force, and direct pre- and post-

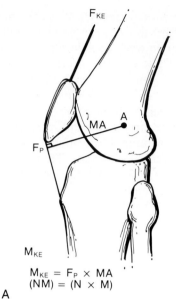

F_{KE} = force vector of the
knee extensor muscle
group
F_P = force in the patellar
ligament, which is equal to
F_{KE}
A = knee joint axis of rotation
MA = moment arm length
of knee extensors
M_{KE} = moment of force of knee
extensor

$$M_{KE} = F_P \times MA$$
$$(NM) = (N \times M)$$

A

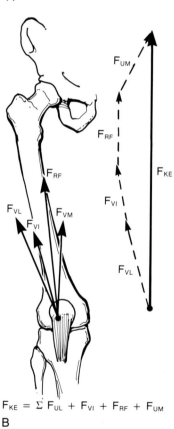

F_{VL} = vastus lateralis force vector
F_{VI} = vastus intermedius force vector
F_{RF} = rectus femoris force vector
F_{VM} = vastus medialis force vector

$$F_{KE} = \Sigma F_{UL} + F_{VI} + F_{RF} + F_{UM}$$

B

FIGURE 6–2
A, The moment of force produced by the knee extensor muscle group (M_{KE}) is the product of the knee extensor force (F_{KE}) times its moment arm length (MA). The MA length is the distance from the line of force F_P to the axis of rotation A. Note that F_P is the force in the patellar ligament; it is approximately equal to F_{KE}. *B,* The force vector for the knee extensor muscle group is the vector sum of the four quadriceps force vectors. All four vectors act at their insertion, which in this case is the patellar ligament on the tibia.

treatment comparisons could not be made. Because such extreme changes are uncommon, muscle moment of force suffices as a measure of muscle strength. Any alterations in the muscle moment of force are assumed, therefore, to be changes in muscle force production because the moment arm length is not normally modified.

The muscle moment of force is considered a valid

measurement of muscle strength and can be used to assess the effects of muscle strength-altering programs. The use of the muscle moment of force also has practical implications because movements usually take place around a joint axis. Thus the moment of force, although a derived value, can provide information concerning the ability of an individual to perform a specific task or movement when the kinetics of the movement are known. A knowledge of the basic concept of muscle moment of force provides the foundation for understanding and interpreting other derived variables, such as mechanical work and power, which are calculated using the muscle moment of force.

Several devices have been used to measure muscle strength, including strain gauges, cable tensiometers, weights, and torque transducers [51, 52, 53, 71, 77]. These devices provide an indirect measure

of the moment of force produced by a muscle or muscle group. A static situation is presented to demonstrate the relationship between the measured moment of force and muscle strength. The equilibrium equations illustrated in Figure 6–3 show that the knee extensors' moment is the vector addition of the moment recorded by the device, the moment due to the weight of the limb, and a flexor muscle moment, all acting about the knee joint axis of rotation. These components should be assessed separately because their contributions to the muscle moment can change substantially depending on the measuring device used, the position of the limb, and the direction of the movement with respect to gravity.

The moment due to the limb's weight can vary with limb position, limb mass, or limb length; therefore, these values should enter into the calculation

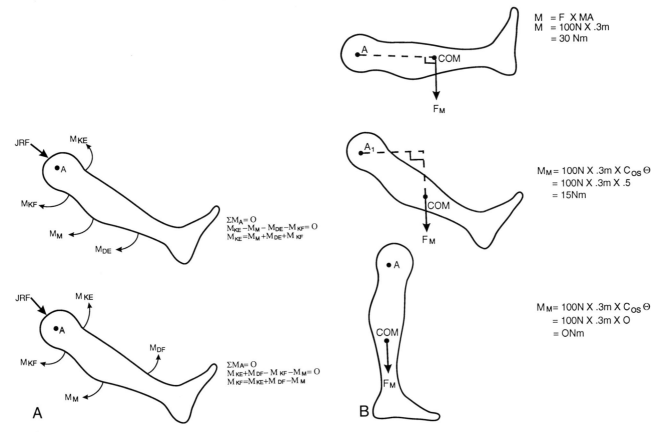

FIGURE 6–3

A, In the static situation, the sums of all the moments of force and forces acting on the system about A are equal to zero. In this example, the axis of rotation is the knee joint center and the free body diagram has four moments of force represented. M_{KE} represents the knee extensor muscle moment (as illustrated in Fig. 6–2); M_{KF} represents the knee flexor moment; M_M represents the mass of the lower leg and foot; M_{DE} and M_{FE} represent the moment of the measuring device for extension and flexion, respectively. The upper diagram and equations are for calculation of the knee extensor moment, whereas the lower diagram is for the knee flexor group. *B*, This diagram illustrates how the moment of the mass of the lower leg and foot (M_M) varies with limb position. M_M is the weight of the leg and foot, F_M, times the perpendicular distance between the line of this force and the axis of rotation A. If the knee were flexed to 90 degrees, the line of force of F_M would act straight through the joint axis of rotation, and M_M would equal zero. M_M would also equal zero if the movement were performed in the horizontal plane with gravity eliminated because F_M represents the mass of the leg and foot times the acceleration due to gravity ($F = MA$). When movements are performed in the sagittal plane, the maximum M_M occurs when the limb is horizontal and the moment arm is longest.

of the muscle moment. This is particularly important when dealing with large body segments, such as the trunk, total lower limb, or even the lower leg. Large segment masses and long segment lengths result in a large limb moment of force if movements are against, or in the direction of, gravitational forces (see Fig. 6–3). Figure 6–3B illustrates how this limb moment changes with change in limb position. From this diagram, the moment due to the weight of the limb, if not considered, causes an underestimation of the knee extensor (KE) moment of force and an overestimation of the knee flexor (KF) moment of force. Movement in a horizontal plane would result in a limb moment equal to that shown in Figure 6–3A. Several studies have addressed the issue of gravitational errors in the calculation of muscle moment of force [27, 28, 104].

The antagonist muscle moment can alter the moment of force recorded on a measuring device even if the agonist muscle moment is constant (see Fig. 6–3A). Such antagonist muscle moments can be the result of high levels of coactivation associated with spasm or other neuromuscular problems, or from the passive resistance of the antagonist muscle's noncontractile elements as a result of adaptive shortening. The antagonist muscle moments, because they are difficult to measure, are usually assumed to be zero, but this does not justify ignoring them completely. An evaluation of the contribution of antagonist muscle forces due to spasm or passive shortening should be made, or their presence at least noted for future reference. Estimates of muscle activation can be obtained through electromyography or, crudely, through palpation. One factor associated with increases in antagonistic activity has been velocity of movement [63]. Knowledge of the resistance to passive range of motion can be useful in particular when a shortening contraction of the agonist is present. A recent study by Vrahas and colleagues [95] gave estimates for values of passive moments at the hip joint at various ranges of motion. Determining the total antagonist muscle's contribution to a moment of force presents a difficult and still unresolved problem for the biomechanist. The validity of the muscle moment is, however, associated with the magnitude of error resulting from these antagonist moments of force. If present, these antagonistic moments of force result in an underestimation of the moment of force produced by the agonist.

Lastly, the measuring device is an important component of the muscle moment equation. If a force-measuring device is used, the angle and point of application must be recorded so that the force can be translated into a moment of force around the

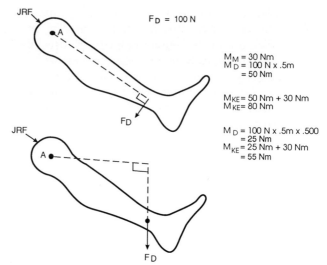

FIGURE 6–4
The same magnitude of force applied at the same distance from the axis of rotation may result in very different moments of force about the joint axis of rotation. Thus, the M_{KE} calculated for each would differ. *A,* The force of the measuring device, F_D, is applied perpendicular to the long axis of the limb; therefore, the moment arm length is 0.5 meters. *B,* F_D represents the force of a free weight, which always acts vertically downward. M_D decreases with increasing knee flexion because the moment arm is related to the cosine of the angle at the knee (0.500 for the 45-degree angle shown).

joint axis of rotation [78]. Torque (moment of force) measuring devices require alignment of the joint and apparatus axes of rotation, but point and angle of force application are not required. Figure 6–4 illustrates the differences in muscle moment of force calculated when two different applications of the same force, at the same distance from the joint axis of rotation, are used.

In summary, to achieve a valid measure of the muscle moment of force, all the components discussed must be entered as accurately as possible into the equilibrium equation. This moment can then be used as a measure of the strength of the muscle or muscle group being evaluated.

Types of Strength

Static (isometric) muscle strength has been assessed by measuring of the moment of force produced during a maximum isometric effort; that is, there is no external change in muscle length (Fig. 6–5A). This can be achieved by stabilizing the joint angle. Studies indicate that reproducibility can be improved by asking the patient to hold the activation for a short time interval, repeating the measurement, and allowing appropriate rest periods between

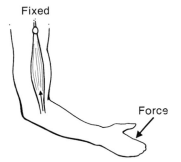

Joint angle fixed
No external muscle length change
Tendon loaded (in tension)
Isometric activation
No external work (no movement)
Metabolic cost

A

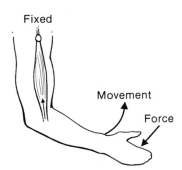

Joint angle changes
External muscle length decreases
Tendon loaded (in tension)
Concentric activation
Positive external work (force and
    movement are in
    same direction)
Metabolic cost

B

FIGURE 6–5
A, Isometric, *B,* concentric, and *C,* eccentric activations are the result of different changes in muscle length during activation. In each type of activation, the tendon is being loaded, and the force produced by the muscle is being transmitted to the bone via the tendon.

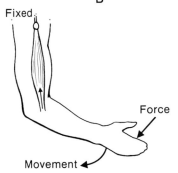

Joint angle changes
External muscle length increases
Tendon loaded (in tension)
Eccentric activation
Negative external work (force and movement are
    in opposite directions)
Metabolic cost

C

activations [12, 52, 71, 72, 78]. Measurements made at one joint angle (muscle length) do not necessarily reflect the muscle's performance at other joint angles [75, 77] because muscle force is related to the length of the muscle (Fig. 6–6). Therefore, strength measurements are often taken at several joint angles within the working range of the muscle. Problems associated with various injuries may manifest themselves as strength deficits at specific angles within this working range; for example, the differences in knee extensor moment of force in the flexed versus extended knee may be greater than normally expected for an individual suffering from a patellofemoral disorder. High patellofemoral compressive forces in the flexed position may result in an inhibition of muscle excitation and a reduction in force output. This finding might be masked if measurements were not made at differing joint angles.

Dynamic strength, as the name implies, refers to the force-generating capabilities of the muscle dur-

ing movement. Again, the muscle moment of force is considered to reflect this capability. Muscle work and, to some extent, power are also considered to reflect muscle strength, although the latter is usually discussed separately. The mechanical work of a muscle group is calculated by taking the integral of the product of the moment of force and the angular displacement [104]. Power is simply the product of the moment of force and the angular velocity [103]. Therefore, both muscle work and power are derived from the moment of force. Dynamic movements may involve lengthening (eccentric) or shortening (concentric) activations of the muscle. Although both types of activation transmit, via tendon, the tension produced by the muscle to its bony attachments, total muscle length increases during eccentric activation and decreases during concentric activation [49]. Figure 6–5 illustrates isometric, concentric, and eccentric muscle activations. The muscle's ability to generate tension, for a given level of motor unit

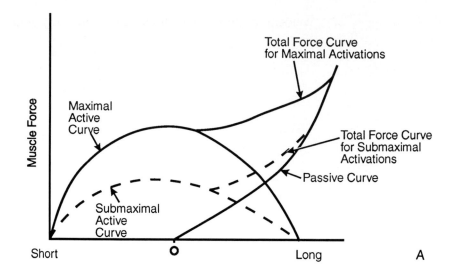

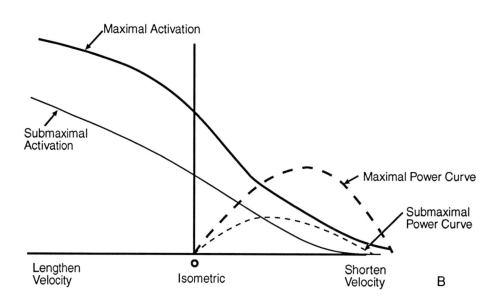

FIGURE 6–6
A graphic depiction of the classic *A,* force–length and *B,* force–velocity relationships for skeletal muscle. The *solid curves* reflect the assumption that the activation level of the muscle is maximal throughout the length or velocity changes. The *dotted line* represents a submaximal activation. The *power curve* illustrates that power is equal to the force times the velocity of muscle length change.

activation, differs with the type of activation. Figure 6–6 illustrates the force–velocity relationship for maximal and submaximal activations. Winter [103] presents a series of hypothetical curves for different activation levels, although these need to be substantiated empirically.

The peak value of the muscle moment has commonly been used as a measure of dynamic strength [75, 87]. This value can occur at any angle within the range of movement, although ranges have been published for some of the major muscle groups [77]. As with static muscle strength, this single measure may not adequately reflect the total muscle capability, particularly when assessing abnormal conditions.

The mechanics of the dynamic situation are more complex than those of the static situation and will not be presented in the same detail. The major reason for the increase in complexity is the addition of acceleration (both linear and angular) forces of the limb and the measuring tool. Devices such as the Cybex (Lumex, Ronkonkoma, NY) or KinCom (Chattecx, Chattanooga, TN) can help simplify the problem by controlling maximal angular velocity of the limb, resulting in a constant velocity movement once this maximal velocity has been attained. Acceleration forces are minimized, and the moment measured by the device can then be related to the muscle moment of force in a manner similar to that described for the static situation. These types of constant angular velocity movements are referred to as isokinetic.

As mentioned previously, the physiologic significance of the peak torque as a reflection of the muscle's capability throughout the range over which the muscle would be required to work has been questioned. It has, however, been shown to be a highly reliable measure with both the Cybex and KinCom dynamometers [26]. Other measures used to assess dynamic strength include average moments of force and mechanical work over specified joint ranges as well as the moment of force at specific angles within the range of motion [27, 43, 53, 71]. Data collection via a computer system makes these values easy to calculate. The peak torque value may suffice as a very general assessment of the muscle's dynamic capability, whereas the latter three measures may be more valuable in assessing the overall dynamic performance of the muscle. Measurements of moment of force and mechanical work calculated during the performance of various athletic tasks may also provide a valuable assessment of muscular strength capabilities [13, 41, 53, 71]. These assessments require sophisticated kinematic and kinetic analysis procedures.

Muscle power reflects the ability of the muscle to perform work in a given time interval. The average muscle power is simply the mechanical work divided by the time required to perform that work. Obviously, a shorter time requirement for a given amount of work would result in a higher power output. Peak power is the maximal power produced during a movement. This value is easily obtained using an isokinetic device by multiplying the moment of force by the angular velocity. Thus, power depends on both the force produced and the speed of movement. The optimum power curve, shown in Figure 6–6, is related to the force–velocity characteristics of the muscle and its anatomic configuration [31, 36, 87, 100]. A good overview of muscle strength and power measurements is provided in the text *Physiological Testing of the High Performance Athlete* [71]. Muscle power may be important in athletic muscle strength evaluation because many sports require bursts of muscle activation, such as sprinting and jumping. For an athlete involved in these sports, an inability to produce a powerful muscle activation may be indicative of an abnormal state or an ineffective retraining program.

Figure 6–6 provides a basis for understanding the relationships between muscle strength and power. The curves shown distinguish the differences between isometric, concentric, and eccentric activations, and show the association of force output and velocity of muscle length change to power production. Little work has been done on power absorption [62], although it may be a significant factor in understanding injury mechanisms because eccentric muscle activations have been implicated in many soft tissue injuries [18, 84]. The fact that many of these dynamic relationships from human skeletal muscles have relied on isokinetic evaluations has raised questions concerning their direct comparison with early work using isolated muscle preparations [36, 37] and the extrapolation to a varying velocity environment [14]. It is hoped that further research will provide better insight into these relationships.

Muscular Endurance

The reconditioning of endurance should be an important part of the rehabilitation process for athletes who are involved in long-duration, repetitive sports. Susceptibility to fatigue can be indicative of central or peripheral nervous system problems, neuromuscular junction dysfunction, limited substrate availability, and numerous other problems.

Muscular endurance relates to the muscle's ability to resist fatigue. Physiologically, muscles with a higher aerobic capacity (using oxygen) are best at resisting fatigue [46, 71, 88]. There are, however, many factors that affect this ability, including neurologic mechanisms, substrate availability, type of work, and subject motivation. Many evaluations of endurance focus on the sustained repetitive activation of a muscle or muscle group against a resistance [15, 58]. These tests are usually stopped when the individual can no longer produce a percentage of the maximal voluntary measure, or when a preset number of repetitions has been completed. The number of repetitions (to, say, 50%), the time from the start to the end of the test, or the decrease in torque output are often recorded. These assessment models provide an overall indication of muscular endurance; however, they do not distinguish physiologic from psychological mechanisms.

Over the past few decades, a great deal of work has been done with electromyographic (EMG) signals, their relationship to muscle force production, and the associated changes in the signal with fatigue [38]. The EMG signal reflects the spatial and temporal summation of the motor unit action potentials of skeletal muscle and thus provides a link to the muscle's physiology. For a more detailed discussion of EMG, the reader is referred to the work of Basmajian and Deluca [8]. In general, the EMG signal becomes more synchronous as the muscle fatigues. If a maximal effort is being resisted by the muscle, the EMG amplitude remains constant and the force generated by the muscle decreases as it fatigues. If a submaximal level of the muscle's voluntary output is being resisted, the force remains constant and the amplitude of the EMG signal increases as the muscle becomes fatigued. Both situations reflect the muscle's inability to effectively produce a force. It is possible that tests using EMG signals could be used to assess the endurance capabilities of a muscle. Such tests provide a basis for separating physiologic and psychological mechanisms of fatigue. The changes in the electromyogram have been shown to be quite repeatable, and although they currently do not distinguish between the various physiologic mechanisms of fatigue, investigations are being directed at this area [7, 10]. Before clinical applications of this approach are used, these questions must be thoroughly addressed.

Flexibility

The term flexibility refers to the ability of the muscle to elongate within the physical limitations imposed by the joint. Linear measurements of muscle length require numerous calculations and assumptions [39] so the measurement often used to reflect flexibility is the angular displacement between two adjacent limb segments (joint range of motion) [42]. For normal, healthy joints, this measurement is usually restricted by the bony configuration of the joint (e.g., elbow extension) or by the length of the muscle–tendon unit (e.g., hamstrings and hip flexion). With an abnormal joint, swelling, pain, and other physical limitations may prevent large ranges of motion. These limitations are addressed during rehabilitation, and many techniques are used to reduce pain and swelling, including physical modalities and drug therapy. The muscle–tendon unit's contribution to restriction of motion is the focus, however, of flexibility reconditioning.

Factors limiting muscle extensibility include both passive and active resistance [82]. Both produce a resistance force, as illustrated by the force–length relationship in Figure 6–6. Passive resistance reflects compliance of the noncontractile elements, whereas active resistance is due to activation of the muscle contractile mechanism. The reader is referred to Alter [3] for an in-depth discussion of these two factors. Passive resistance is altered by muscle temperature, immobilization, and tissue loading [85, 97, 98]. Figure 6–7 demonstrates how passive tension, generated by stretching the muscle's elastic components, affects the muscle's ability to elongate. Low-load, long-duration applications of force to a heated

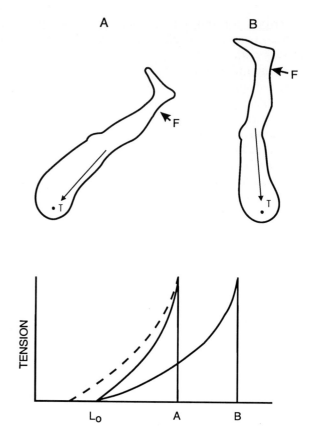

FIGURE 6–7
The passive length–tension curve for skeletal muscle can alter the ability of the muscle to elongate. Here the hip extensor muscle group (including the hamstrings if the knee is extended) resists the hip flexor moment that is being applied to the lower limb (F × D). Muscle length can be limited by high resistive forces of the muscle in a relaxed state. A would be less flexible than B, which would be reflected in the shorter muscle length and smaller joint range of motion. The rupture point of the muscle can be reached at a shorter length in two different ways: (1) by an overall shift of the curve to the left to a new resting length (L_0) due to adaptive shortening of the muscle (*dotted line*); or (2) by reduced compliance of the muscle, for example, as a result of increased amounts of connective tissue following muscle injury (*solid line, A*). The curve represented by B shows both increased compliance compared to A and a passive adaptation (increased length).

muscle have been shown to improve this tissue compliance and to produce the best plastic deformation and long-term gains in muscle extensibility [50, 73, 97, 98]. In contrast, active resistance relates to the ability of the muscle to relax during elongation. An inability to turn off the active force-generating mechanism resists, and thus prevents, lengthening. Neurologic factors can play an important role in this ability. Most of the proprioceptive neuromuscular facilitation techniques used to enhance muscle flexibility are based on reducing this active resistance through reciprocal inhibition, although there is some debate about their effectiveness [34, 59, 64].

Evaluation of flexibility, therefore, should focus on measuring the resistance of the muscle to elongation both statically and dynamically. Few studies have addressed this issue, and standardized protocols for measuring the resistance and length have not been clearly established. The measure most often used is the relative angular displacement between two adjacent segments, as mentioned above. Because the first resistive mechanism within the constraints of the joint structure itself is often the muscle tendon unit, this measure indirectly evaluates the extensibility of muscles crossing the joint [42]. Standardized protocols are available for measuring joint range of motion and provide a basis for evaluating static flexibility [24, 42, 62]. Various devices, such as goniometers and flexometers, have been used for these assessments [62].

Dynamic measurements of flexibility have been proposed, but they have not been standardized. It would seem to be important for athletes to be able to move dynamically through a range of motion without stresses on the muscle tendon unit. The development of standardized protocols is required to provide an operational definition of dynamic flexibility and its measurement. Future research measuring the joint range of motion required to complete various athletic skills and the time involved should prove fruitful when comparing the injured athlete's capabilities with the demands of the sport. Motion analysis techniques undoubtedly provide the means by which such evaluations will be performed. Until these techniques can be used routinely, or the body of research data increases, the nature of the sport must be subjectively assessed with respect to the maximal movement required at a joint and the athlete's movement capability. The notion that extreme flexibility is necessary for athletic participation lacks both common sense and empiric data to support its validity. Although it has not been proved empirically, some authors agree that the performance of motor tasks without placing undue stresses on the muscle–tendon unit is a reasonable goal of flexibility conditioning [3, 44].

The above discussion provides a basis for delineating rehabilitation objectives and training factors associated with muscle strength, power, endurance, and flexibility. The remainder of this chapter focuses on the importance of these variables in the rehabilitation of the athlete.

TRAINING AND CONDITIONING

Rehabilitation after injury or surgery is an extremely important facet of achieving and maintaining wellness. A great deal of interest and research has surrounded rehabilitation, particularly as related to knee and shoulder disorders [19, 20, 21, 25, 57]. Much of these data therefore form the cornerstone of the remainder of this chapter. In the desire to offer useful information to the reader, an attempt has been made to distill the scientific literature and construct a practical guide to current techniques for rehabilitation.

Historically, injuries to the musculoskeletal system were treated with prolonged immobility and the tardy introduction of therapeutic exercise [4]. These cautious tactics respected neither the biology of healing nor the contributions of Julius Wolffe, or even Hippocrates. Indeed, the modern thrust in rehabilitation after surgery or injury is toward the rapid introduction of stress [11, 90, 92, 93]. This controlled stress may take one of many forms, including continuous passive motion (CPM) [30], electrical stimulation [17, 25, 83], or simply passive manual movement [99]. The common thread in these techniques is the knowledge that injured tissues within the musculoskeletal system respond in a most positive fashion to progressive stress. Challenging the tissue progressively, within its inherent tensile strength, improves the biomechanical profile of the injured structure [61, 91].

Rehabilitation techniques that promote early movement must be tailored to the age and physical state of the patient [29, 60, 76, 92, 94]. The severity of the injury must, of course, be appreciated so that overly zealous rehabilitation is avoided. Therapeutic techniques that promote stress beyond the tensile strength of the healing tissue invariably promote further injury and retard recovery. Unfortunately, many "soft" data surround rehabilitation. How can one be certain that a specific training technique is appropriate and not too aggressive, actually disturbing healing?

The discomfort experienced by a patient frequently dictates the progression of a training program. Monitoring such a response has significant frailties because it is a subjective index. Furthermore, the results of animal studies frequently do not parallel our findings in humans, making extrapolation from such studies difficult [31, 61, 70]. Despite these recognizable uncertainties, one must push forward in rehabilitating the injured athlete. Early and progressive stress must remain one of the fundamentals.

Certain principles must be accepted to create a successful program of rehabilitation:

1. Understand and grade the severity of the injury
2. Understand and respect the biology of healing

3. Understand the biomechanics of the athlete's activity

4. Set the goals or objectives for recovery based on 1, 2, and 3 above

A word of caution: the injured athlete is a very difficult patient. There are consistently both intrinsic and extrinsic pressures to return the athlete to action before rehabilitation is complete. This temptation must be resisted with scientific principles.

STRENGTH AND ENDURANCE TRAINING

The muscle–tendon unit is extremely vulnerable to injury in athletics and, following injury or surgery, must be reconditioned to achieve a normalized state [18, 19, 23, 47]. This objective, shared by therapist, physician, and athlete, must be tailored to the individual. Factors that may affect the rehabilitation process include age, preinjury fitness level, and postinjury expectations. A thorough knowledge of the sporting activity and its mechanical and physiologic demands is essential in order to return the athlete to his or her activity safely and effectively [13].

Any injury to the muscle–tendon unit or other parts of the musculoskeletal system (e.g., stress fracture, ligament sprain) triggers a series of neurohormonal interactions, which provoke an immediate "shutdown" of the affected extremity [55, 56, 81]. Atrophy, demineralization, and trophic changes rapidly ensue, particularly if the injured part is not treated with expeditious management of the pain component, followed by judicious reconditioning. Intrinsic to all injuries of the musculoskeletal system is a loss of strength [51]. Rehabilitation, as the task of normalization, must be directed toward thwarting further loss of strength and promoting strength retraining to achieve a preinjury milieu. Some of the guidelines to be followed during a reconditioning program are:

1. Determine clinically the severity of the injury.

2. Following injury, measure the muscle–tendon performance statically and dynamically.

3. Monitor and measure the improvement in isometric and dynamic performance.

4. Progressively introduce sport-specific skills, progressing from low-velocity to high-velocity tasks.

5. Return to athletic activities with careful monitoring.

6. Measure muscle–tendon function 2 to 3 months after return to sports to delineate residual deficits in order to prevent future injury.

The difficulty of assessing muscle strength in the rehabilitation of athletes lies in the difficulty of determining what is normal. Although there are numerous strength-testing methods and devices available, usually the uninjured limb is considered representative of the normal strength of a muscle group. This may not be appropriate for many athletes. Tennis players, baseball players, and quarterbacks may be expected to have increased strength in their dominant limbs and, indeed, this has been shown to be true [53]. It becomes very difficult, then, to determine the end point of rehabilitation after injury when no history of the preinjury strength levels exists. This, of course, is a good argument for the need to pretest athletes, preferably when they are fully conditioned. Alternatively, normative data for athletes in various sports may be used to provide a reference for the relative strength (ratio) of the dominant or nondominant limb, and this can be used as a guideline to ending rehabilitation. It should be noted that it is often impractical to continue to carry out rehabilitation in the health care setting once improvement has slowed. This is certainly true in athletic rehabilitation, where increments in strength, although important, may be small. It then becomes very important to design appropriate exercise programs that the athlete may carry out with little or no supervision.

The design of a tailored exercise program depends heavily on accurate assessment of muscle strength and endurance [51, 53]. There are numerous methods available, but it is highly desirable, in testing the athletic population, to have a device that (1) allows testing of multiple muscle groups, (2) allows testing at multiple speeds, (3) provides quantitative results, (4) is sensitive to small changes in performance, and (5) allows test results to be accurately reproduced. The need for these criteria eliminates traditional manual muscle testing as anything but a gross screening method because the tester is unable to determine small changes in strength [45, 52]. Small, handheld dynamometers have greatly improved manual muscle testing by allowing quantitation of the results, and this method may have a place in testing muscle groups in unusual positions that are representative of postures encountered during athletic participation but cannot be reproduced on large strength-testing devices.

Accurate assessment of muscle strength usually requires an isokinetic device such as a Cybex, KinCom, Biodex, or Lido. These machines have become increasingly sophisticated and allow accurate testing and storage of test results [26, 58, 71, 72, 75]. Their chief drawbacks are uniplanar movement, placement of the limb in a relatively static position, and firm fixation of the body in space—all of which are excellent for reproducing test results but not very representative of sports participation. Also, it should be stressed again that it is the moment of force rather than muscle force production itself that is being measured.

Alternatively, a gross motor skill, such as vertical jump, may be evaluated [6]. An outcome measure (height jumped) can be recorded, and, if a force platform is available, quantitative data on task performance can be calculated. Such methods provide an overall evaluation of lower limb performance but do not pinpoint the source of any deficiencies that may be noted because more than one joint is involved in the movement [39]. These tasks may be made more representative of sports-related movements, and isolated testing can be performed late if deemed necessary. The addition of kinematic data and more sophisticated analytic techniques may also provide estimates of individual joint contributions during such gross motor tasks [39, 41].

Techniques of Strength Retraining

The basic science of muscle behavior has been discussed previously in the text, as have the definitions of various types of muscle contractions. The practical application of these issues follows.

Static Strength (Isometric) Training

Isometrics have been used as a foundation of several rehabilitation protocols for strengthening [6, 15, 20, 25]. As discussed previously, the muscle is activated by the athlete or patient, resulting in agonist activation but no movement of the adjacent joints. No obvious muscle shortening takes place. Originally described by Hettinger and Muller in 1953 [35], this type of muscle activity has long been employed as a technique in muscle rehabilitation. Some controversy has surrounded the therapeutic value of this type of muscle behavior in restoring an injured or atrophic muscle–tendon unit [16, 25, 75].

In practical terms, isometric muscle activations can be developed with or without applied resistance. For example, isometric quadriceps exercise can be performed with the patient in the supine position, voluntarily firing the quadriceps muscle mass. This can be accomplished with the knee in full extension or at varying joint angles. Fixed resistance can be applied at any site below the tibial tubercle to prevent a change in knee or hip angle during the isometric contraction. Traditionally, the muscle (in

this case, the extensor mechanism) activation is maintained for 3 to 5 seconds, depending on the fitness level of the individual. Fatigue or pain may ensue, which will thwart an optimal, that is, maximal, contraction. Controversy surrounds the relative value of maximal versus submaximal isometric contractions.

Isometric strength conditioning has distinct limitations. Obviously, the activity is being performed in a static posture with the adjacent joints firmly fixed at a particular angle. This limits any strength improvements to these specific joint angles [71, 75]. This muscle performance can hardly be transferred to the dynamic demands placed on the muscle–tendon unit during athletic activity. Ideally, muscle retraining is conducted throughout the entire range of motion required for a specific athletic skill. Furthermore, enhancement of motor skill, which is fundamental to athletics, cannot be duplicated by isometric exercise. There is little if any biofeedback and thus poor carryover to the sporting activity.

Isometric exercises are most important in the early phases of rehabilitation because they are relatively painless and are simple to perform. Joints can be positioned so that joint contact forces are reduced while maximal muscle contraction is achieved. This avoids placing excessive stress on adjacent joints, which may have been the site of injury or surgery. The patellofemoral joint is a particularly good example of this situation. As shown in Figure 6–8, quadriceps forces remain constant at various degrees of knee flexion, whereas patellofemoral contact forces become smaller as the knee approaches full extension. The simplicity of isometric exercise means that major expenditures in terms of equipment purchase are not necessary. Gadgets such as rubber tubing or bands are widely used by therapists in the design of isometric exercise programs and can be used by the patient in both the clinical setting and the home environment. Other forms of simple instrumentation, or even manual resistance, can be used in the clinic to provide stationary resistance.

Dynamic Strength Training

Dynamic exercises are introduced as the injured tissue matures, initially through short arcs of motion and with minimal manual resistance. The tissue must be rendered stronger and sufficiently flexible to absorb the forces intrinsic to the activity without loss of integrity; thus, the degree of muscle strengthening must be qualitatively directed toward the normative values for athletes in that specific sport.

FIGURE 6–8
A comparison of patellofemoral compression forces and quadriceps tensile force as the quadriceps resists a constant load at varying degrees of knee flexion. If the aim of the exercise is maximum effort by the quadriceps and minimum compressive force on the painful joint, clearly it is preferable to do the exercise in full extension. This method of strengthening is most useful in the early stages of rehabilitation and can be supplanted by more resistance training as pain subsides.

Progression is governed by the limits of pain and the biology of healing.

The injured athlete must not only regain the skills, balance, and coordination that are basic to his or her sport but must also reach satisfactory thresholds of muscle performance in order to return to action safely and successfully. The muscle training program should parallel, as closely as possible, the forces applied to the muscle–tendon unit in the specific sporting activity [18, 71]. Because there is no consensus on how this is best achieved, strength reconditioning typically involves one or more of the following types of dynamic exercise, performed either eccentrically or concentrically: (1) isotonic, (2) isokinetic, or (3) variable resistance.

Isotonic Strength Training

Isotonic exercise is defined as constant tension. The most common example of this form of exercise is the lifting of free weight that, on the surface, would be assumed to provide constant tension. Basic biomechanics reveals, however, that because of alterations in muscle moment arm length and the muscle force–length relationship, the demands

placed on the muscle by the same weight at different parts in the range of motion may vary considerably (see Fig. 6–4). The ability to develop force is also affected by the muscle force–velocity relationship, as described earlier [14, 100]. Free weights tend to be employed at relatively slow velocities, which enhances force production but infrequently duplicates the muscle activity necessary for the sporting challenge [5, 66].

Because the amount of weight lifted is limited by the weakest point in the range of motion for a particular movement, free weights may not provide sufficient resistance for optimum strength gains. Various modifications have been introduced to commercial strength-training devices to eliminate this shortcoming, such as a cam system that alters the moment arm of the applied weight so that a larger resisting moment is produced at points in the range where the muscle is able to produce a larger muscle moment of force because of an increase in its moment arm length [5]. Basically, the expectation of the use of such machines is the development of a technique that provides constant muscle tension throughout a full range of motion, thus providing a training experience that challenges the muscle–tendon unit in a more physiologic fashion. Expense and availability may limit their use.

Free weights are readily available in virtually all rehabilitation units and most exercise establishments. They must be used cautiously because unrealistic expectations may predispose to injury. An assistant, or spotter, is essential to ensure safety. Gains in strength as a result of using free weights have been well chronicled [21] and, if the exercises are done progressively, very little pain is experienced. Motor performance is said to be superior following isotonic training compared to isometric exercises; however, this area remains controversial [16, 66].

Caution is essential when introducing free weights during the early period of recovery from injury. In our experience, manual resistance provided by a skilled therapist or team member is far more effective and safe. Different directions, velocities, and intensities of training can readily be produced, and the therapist can accommodate any pain encountered during movement by immediately changing the resistance.

Isokinetic Strength Training

Exercise at a constant velocity is referred to as isokinetic work [58, 66]. This form of exercise is usually provided by machinery that controls the speed of the movement. The level of effort of a contraction by the injured athlete may be submaximal or maximal, based on the stage of recovery for that particular injury. As the injury matures, assuming that no major discomfort is produced, the athlete achieves a more maximal contraction, thus increasing the isokinetic challenge [65, 68]. The velocity of the activity may be varied to improve the recruitment of various muscle fiber types [32, 63, 68]; however, the maximum velocities available for training with these devices is far lower than angular velocities achieved during athletic involvement, which may limit their utility in the final stages of rehabilitation.

As discussed above, these machines are extremely useful for assessing muscle strength within the limitations described. In the clinical setting, athletes are very compliant with these forms of exercise and particularly enjoy the graphic feedback on performance that is provided by the computer display that most devices employ. Instruction from a therapist is imperative, but constant supervision is rarely necessary for the familiar user. There is very little muscle soreness if concentric activity alone is performed (e.g., with Cybex), and the dividends in terms of strength gains have been well documented [58, 65, 66, 72, 75]. This equipment is very expensive and not nearly as hardy as free weights.

Manual Resistance

Exercises performed against manual resistance have many advantages. The athlete is usually uncomfortable in the early stages of rehabilitation and frequently timid. Manual resistance allows the therapist to control not only the quantity of resistance but also the velocity of effort and the range of motion. In addition, the limb is supported by the therapist, which reassures the apprehensive patient. The controlling factors for this type of exercise are the discomfort experienced by the patient coupled with objective evidence of renewed inflammation, such as heat, swelling, and tenderness. Following the successful performance of an adequate range of motion against manual resistance, the injured patient or athlete may then be introduced to free

weights and isokinetic exercise if the machinery is available.

Once dynamic strengthening has been accomplished to a satisfactory threshold, progressive introduction of sports-specific skills should be encouraged. The timing for the reintroduction of these skills is highly controversial; however, the practitioner is always safe in monitoring and combining the dynamometer performance with the performance on the practice field.

Eccentric Strength Training

Eccentric activation refers to the situation in which the muscle–tendon unit is lengthened while active. The movement that takes place during this activity, called negative work, is fundamental to most athletic activities, particularly those in which rapid deceleration is involved [48, 49]. Eccentric forces are developed on reversal of direction during athletic performance, for example, jumping and throwing.

Eccentric training or conditioning is introduced during the later stages of rehabilitation, once healing has advanced. Because eccentric loading may be painful and causes muscle soreness if conducted too aggressively [49], increasing pain, swelling, or muscle tenderness should encourage the therapist to reduce the activity. This training is gradually inculcated in the rehabilitation program as part of the motor skill training for a specific sporting activity, that is, the exercise is designed to simulate movements occurring during the athlete's sport [18]. For the jumping athlete with an injury to the extensor mechanism, the following program has proved to be successful:

1. General warm-up
2. Stretch quadriceps and hamstrings
3. Three sets of 10 repetitions of semisquat exercise (Figure 6–9). Progress is made from a slow,

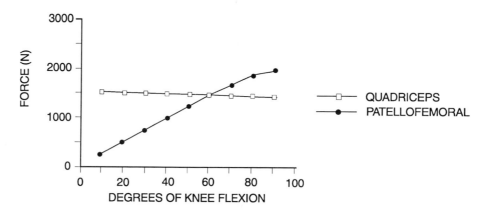

FIGURE 6–9
The motion performed to load the patellar tendon using the eccentric exercise program. Initially, movement is done slowly and carefully, but the patient soon progresses to moving rapidly downward and stopping, the so-called "drop and stop." The aim is to increase the load on the patellar tendon. If this type of exercise is used for performance training, the downward movement should be followed by an immediate upward jumping motion.

controlled movement to a rapid "drop and stop" (or a drop followed by an upward jump) as discomfort lessens. Some pain should be felt by the end of the exercise session

4. Repeat the muscle stretches

5. Apply ice for 5 to 10 minutes to the affected area.

RANGE OF MOTION AND FLEXIBILITY TRAINING

Although range of motion and flexibility are often considered synonymous in the athletic setting, these movements really involve two separate variables. One is joint motion, which may be limited by capsular tightness, the presence of a torn or displaced structure within the joint, bony limitations, or pain. The second variable to be considered is the extensibility of soft tissues that are not part of the joint, such as muscles and tendons. These two variables are sometimes referred to as static and active range of motion [42, 43, 82] because of the nature of the tissues restricting motion.

It is extremely important in designing a rehabilitation program to determine which structures are responsible for any limitation of motion that may be present. For example, after surgery, knee joint motion is often limited by capsular tightness, which results from the immobilization position of the joint, but it may also be limited by hamstring muscle tightness. Stretching exercises performed in a manner that reaches the limits of muscle flexibility before the end of joint motion leaves the capsular tightness unaddressed.

The concept of *active flexibility* is extremely important in the athletic population. Joint motion measured under static conditions may be entirely normal, but when other joints are placed in a position that creates tension in a two-joint muscle, motion may become limited. Ankle joint motion after a sprain may seem normal compared to the uninjured side, but when knee extension is maintained, range of motion (ROM) becomes restricted because of soft tissue tightness in the gastrocnemius muscle. This tightness alters the mechanics of movement by causing early knee flexion or heel-off during running and may predispose the athlete to reinjury.

The specific ROM and flexibility needs of an athlete are largely determined by his or her sports involvement. The mechanical demands of certain sports require, and thus induce, flexibility in some muscle groups and tightness in others [23, 47, 54, 67]. It is not yet known whether specific flexibility training reduces the likelihood of athletic injury, although anecdotal evidence suggests that such is the case. The concept of muscle imbalance contributing to injury production has also been suggested [3, 19, 23]. This theory holds that tight muscle groups must be stretched, whereas lengthened muscles should be strengthened, thus recreating "balance" between opposing muscle groups. Applications of this theory include (1) stretching the calf muscles and strengthening the dorsiflexors (runners), (2) stretching the pectoral muscles and strengthening the lateral rotators of the shoulder (pitchers and swimmers), and (3) stretching the iliopsoas and quadriceps and strengthening the hamstrings (gymnasts).

Flexibility has long been considered an important component of physical fitness, and various stretching techniques have found their way into athletic training and rehabilitation [3, 9, 33]. Unfortunately, the utility of flexibility training in injury prevention has not been proved [44]. Scientific evidence to support flexibility training in injury prevention is lacking, but anecdotal reports of enhanced performance and reduced risk of soft tissue injury have ensured a place for muscle stretching in the training programs of athletes of all skill levels. The design of stretching exercises for rehabilitation and prevention is usually based on the injury patterns occurring in a particular sport, and the theory that stretching tight muscles makes them less likely to be injured in the future.

Flexibility is an easily trainable skill when a stretching program is regularly followed [9, 22, 34, 40, 99, 102]. Improvement in flexibility takes place only when the joint is moved through a range of motion that places the surrounding muscles under tension, and studies of athletes have shown that their flexibility patterns are quite sport-specific [54, 67]. The value of improving flexibility has been subjectively recognized by both the therapeutic and sports communities, and stretching exercises have become an integral part of rehabilitation and fitness programs.

Various stretching techniques have been incorporated into training programs. The three most commonly used classes of stretching exercises are (1) ballistic or active stretching, (2) static stretching, and (3) proprioceptive neuromuscular facilitation (PNF). Each of these techniques has been shown in clinical trials to be effective in increasing joint range of motion when practiced regularly.

Ballistic Stretching

This stretching technique, a common element in many calisthenics programs, involves quick, forceful, repetitive movements through the joint's range of motion. The force generated by the active movement of the athlete's body or limb is used to overcome the resistance provided by those muscles opposing the movement, that is, the muscles to be stretched. These rapid movements, however, initiate a stretch reflex by activating the muscle spindles, causing activation of the muscle group being stretched [56, 59].

Static Stretching

A static stretch places the muscle–tendon unit under a slow, gentle stretch that is maintained for a period of 20 to 60 seconds. This steady stretch has been shown to reduce the intensity of the stretch reflex and depress the tone of the stretched muscle group [56, 82]. This type of stretching is preferable to ballistic stretching because it expends less energy, provides some prevention or relief of muscular distress, and reduces the risk of injury caused by exceeding the extensibility of the tissue being stretched [22].

Proprioceptive Neuromuscular Facilitation Stretching

PNF stretching techniques also attempt to reduce muscle tone by stimulating the Golgi tendon organs. Although a number of PNF techniques are in use, they rely on the same basic principles. One common PNF technique [86] uses a three-step sequence:

1. The muscle–tendon unit is slowly stretched to the end of its range and held for several seconds
2. With a partner providing support, a maximal isometric contraction against resistance is performed and held for a period of 5 to 10 seconds
3. The muscle–tendon unit is then relaxed and slowly stretched further by the partner.

It is believed that an isometric contraction with the muscle–tendon unit at its greatest length will produce the greatest tension in the unit and therefore maximally stimulate the Golgi tendon organs. This causes reciprocal inhibition of the muscle under tension and allows it to be stretched further upon relaxation [55, 56, 73, 82].

Which of the stretching techniques provides the greatest improvement in flexibility is a source of some debate. Both ballistic and static stretching have been shown to improve flexibility equally [9, 96]. However, the risk of injury with ballistic stretching probably restricts its use to the final stages of

preparticipation, after other stretching techniques have been used. Some investigators have found that PNF techniques yield greater increases in joint range of motion than either ballistic or static stretches [69, 86, 101], although the superiority of PNF has been questioned [59, 64].

In determining which stretching technique will provide the best results, one must remember that the collagenous elements of the muscle–tendon unit must be lengthened along with its contractile component. Greater gains in flexibility are achieved when a low load is applied for a long duration than when a larger load is applied rapidly, as is the case in ballistic stretching [97, 98]. In vitro studies using collagenous tissue confirmed the benefit of a low-load, long-duration technique in creating a lasting increase in tissue length [50, 97, 98]. This result indicates either the static or PNF technique of stretching as the method of choice.

A potential disadvantage of PNF stretching is that a partner is required. Both the subject and the partner must be familiar with the stretching technique to avoid risk of injury. Also, there is some evidence to suggest that muscle activity may actually be increased during PNF stretching and that subjects often report pain during the stretch phase [59, 64]. Although valuable in the clinical setting in the hands of a skilled therapist, these techniques should be used carefully.

SUMMARY

Modern rehabilitation techniques have evolved rapidly with the recognition of the deleterious influences of prolonged immobilization and the benefits of early, controlled exercise. The use of exercise early after injury demands a thorough understanding of the mechanical and biologic principles involved in exercise and healing. Application of these principles to the rehabilitation of athletes requires sport-specific knowledge concerning motor patterns, strength and flexibility requirements, and metabolic demands. There is still, however, a great deal to be learned about optimization of rehabilitation strategies.

References

1. Alexander, R. M., and Vernon, A. The dimensions of knee and ankle muscles and the forces they exert. *J Hum Movement Studies* 1:115–123, 1975.
2. Alnaqeeb, M. A., Al Zaid, N. S., and Goldspink, G. Connective tissue changes and physical properties of developing and aging skeletal muscle. *J Anat* 139:677–689, 1984.

3. Alter, M. J. *Science of Stretching.* Champaign, IL, Human Kinetics Books, 1988.

4. Amiel, D., Akeson, W. H., Harwood, F. L., and Frank, C. B. Stress deprivation effect on metabolic turnover of the medial collateral ligament collagen: A comparison between nine and twelve week immobilization. *Clin Orthop* 172:265–270, 1983.

5. Ariel, G. Barbell versus dynamic variable resistance. *US Sports Assoc News* 1:7, 1977.

6. Ball, J. R., Rich, R. Q., and Wallis, E. L. Effects of isometric training on vertical jumping. *Res Q* 35:231–235, 1964.

7. Barnes, W. S., and Williams, J. H. Effects of ischemia on myoelectrical signal characteristics during rest and recovery from static work. *Am J Phys Med Rehabil* 66:249–263, 1987.

8. Basmajian, J. V., and DeLuca, C. J. *Muscles Alive: Their Function Revealed Through Electromyography* (5th ed). Baltimore, Williams & Wilkins, 1985.

9. Beaulieu, J. E. Developing a stretching program. *Physician Sportsmed* 9:59–69, 1981.

10. Boubrit, M., Duchene, J., and Goubel, F. Is lactic acid responsible for EMG power spectrum changes observed during exercise? *J Physiol* 345:141p, 1983.

11. Butler, D. L., Grood, E. S., Noyes, F. R., and Zernicke, R. F. Biomechanics of ligaments and tendons. *Exerc Sport Sci Rev* 6:125–182, 1978.

12. Caldwell, L. S., Chaffin, D. B., Dukes, F. N., Kroemer, K. H. E., Laubach, L. L., Snook, S. H., and Wasserman, D. E. A proposed standard procedure for static muscle strength testing. *Am Ind Hyg Assoc J* 35:201–206, 1974.

13. Cavagna, G. A., Komarek, L., and Mazzoleni, S. The mechanics of sprint running. *J Physiol* 217:709–721, 1971.

14. Chapman, A. E. The mechanical properties of human muscle. *Exerc Sports Sci Rev* 13:443–501, 1985.

15. Cotten, D. Relationship of the duration of sustained voluntary isometric contraction to the changes in endurance and strength. *Res Q* 38:274–366, 1967.

16. Counsilman, J. The importance of speed and exercise. *Athletic Journal* 57:72–75, 1976.

17. Currier, D. P., and Mann, R. Muscle strength development by electrical stimulation in healthy individuals. *Physiotherapy* 63:915–921, 1983.

18. Curwin, S., and Stanish, W. D. *Tendinitis: Etiology and Treatment.* Lexington, MA, Collamore Press, 1984.

19. DeHaven, K. E., and Lintner, D. M. Athlete injuries: Comparison by age, sport and gender. *Am J Sports Med* 14:218–224, 1986.

20. Delitto, A., and Lehman, R. D. Rehabilitation of the athlete with a knee injury. *Clin Sport Med* 8:805–840, 1989.

21. Delorme, T. L. Restoration of muscle power by heavy resistance exercises. *J Bone Joint Surg* 27:645–667, 1945.

22. DeVries, H. A. Evaluation of static stretching procedures for improvement of flexibility. *Res Q* 33:222–229, 1962.

23. Ekstrand, J. and Gillquist, J. The frequency of muscle tightness and injuries in soccer players. *Am J Sports Med* 10:75–78, 1982.

24. Ekstrand, J., Wiktorsson, M., Oberg, B., and Gillquist, J. Lower extremity goniometric measurements: A study to determine their reliability. *Arch Phys Med Rehabil* 63:171–175, 1982.

25. Eriksson, E., and Haggmark, T. Comparison of isometric muscle training and electrical stimulation: Supplementing isometric muscle training in the recovery after major knee surgery. *Am J Sports Med* 7:169–171, 1979.

26. Farrell, M., and Richards, J. G. Analysis of the reliability and validity of the kinetic communicator exercise device. *Med Sci Sports Exerc* 18:44–49, 1986.

27. Figoni, S. F., Christ, C., Morris, A. F., and Massey, B. H. Effects of speed, hip and knee angle and gravity on hamstrings to quadriceps torque ratios. *Med Sci Sports Exerc* 17:126, 1985.

28. Fillyaw, M., Bevins, T., and Fernandez, L. Importance of correcting isokinetic peak torque for the effect of gravity when calculating knee flexor to extensor muscle ratios. *Phys Ther* 66:23–29, 1986.

29. Florid, A., Ippolito, E., and Postacchini, F. Age-related changes in the metabolism of tendon cells. *Connect Tissue Res* 9:95–97, 1981.

30. Frank, C., and Akeson, W. H. Physiology and therapeutic value of passive joint motion. *Clin Orthop* 185:113–125, 1984.

31. Garrett, W. E., Nikolaov, P. K., Ribbeck, B. M., Glisson, R. R., and Seaber, A. V. The effect of muscle architecture on the biomechanical failure properties of skeletal muscle under passive extension. *Am J Sports Med* 16:7–12, 1988.

32. Gollnick, P. D., Armstrong, R. B., Sembrowich, W. L., et al. Glycogen depletion pattern in human skeletal muscle fibers after heavy exercise. *J Appl Physiol* 34:615–618, 1973.

33. Harris, M. L. Flexibility. *Phys Ther* 49:591–601, 1962.

34. Hartley-O'Brien, S. J. Six mobilization exercises for active range of hip flexion. *Res Q Exerc Sport* 51:625–635, 1980.

35. Hettinger, T., and Muller, E. Muskeleistung and Muskeltraining. *J Arb Physiol* 15:111–126, 1953.

36. Hill, A. V. The heat shortening and the dynamic constants of muscle. *Proc R Soc B Lond [Biol]* 126:136–195, 1938.

37. Hill, A. V. The series elastic component of muscle. *Proc R Soc Lond [Biol]* 137:270–280, 1950.

38. Hof, A. L., and Van den Berg, J. EMG to force processing I: an electrical analogue of the Hill muscle model. *J Biomech* 14:747–748, 1981.

39. Hubley, C. L. Kinematic and electrical activity of five lower limb muscles during a fundamental movement pattern. *In* Whittle, M., and Harris, D. (Eds), *Biomechanical Measurement in Orthopaedic Practice.* Oxford, Claredon Press, 1985, pp. 260–265.

40. Hubley, C. L., Kozey, J. W., and Stanish, W. D. The effects of static stretching and stationary cycling on range of motion at the hip joint. *J Orthop Sports Phys Ther* 6:104–109, 1984.

41. Hubley, C. L., and Wells, R. P. A work-energy approach to determine individual contributions to vertical jump performance. *Eur J Appl Physiol* 50:247–254, 1983.

42. Hubley-Kozey, C. L. Testing flexibility. *In* MacDougall, J. D., Wenger, H. A. and Green, H. J. (Eds.), *Physiological Testing of the High Performance Athlete.* Champaign, IL, Human Kinetics Publishers, 1991, pp. 305–359.

43. Hubley-Kozey, C. L., and Grainger, J. The effects of increasing angular velocities on mechanical work of the knee flexors and extensors. *Can J Appl Sport Sci* 10:14, 1985.

44. Hubley-Kozey, C. L., and Stanish, W. D. Can stretching prevent athletic injuries. *J Musculoskeletal Med* 1:25–32, 1984.

45. Kendall, H. D., Kendall, F. P., and Wadsworth, G. *Muscle Testing and Function* (2nd ed.). Baltimore, Williams & Wilkins, 1971.

46. Kernell, D., Donselaar, Y., and Eebeek, D. Effects of physiological amounts of high and low rate chronic stimulation on fast muscles of the cat hindlimb. II: Endurance related properties. *J Neurophysiol* 58:614, 1987.

47. Kirby, R. L., Simms, F. C., Symington, V. J., and Garner, J. B. Flexibility and musculoskeletal symptomatolgy in female gymnasts and age-matched controls. *Am J Sports Med* 9:160–164, 1981.

48. Komi, P. V. Physiological and biomechanical correlates of muscle function: Effects of muscle structure and stretch-shortening cycle on force and speed. *Exerc Sports Sci Rev* 12:81–121, 1984.

49. Komi, P. V., and Buskirk, E. R. The effect of eccentric and concentric muscle conditioning on tension and electrical activity of human muscle. *J Ergonomics* 15:417–434, 1972.

50. Kotke, F. J., Pauley, D. L., and Ptak, R. A. The rationale for prolonged stretching for correction of shortening of connective tissue. *Arch Phys Med Rehabil* 47:345–352, 1966.

51. Kozey, C. L., and Stanish, W. D. Strength deficiencies between muscle groups of individuals with chronic ankle sprains. *Can J Appl Sports Sci* 9:24, 1984.

52. Kroemer, K. H. E., and Marres, W. S. Towards an objective assessment of the maximal voluntary contraction com-

ponent in routine muscle strength measurements. *Eur J Appl Physiol* 45:1–9, 1980.

53. Kulig, K., Andrews, J. G., and Hay, J. G. Human strength curves. *Exerc Sports Sci Rev* 12:417–466, 1984.

54. Leighton, J. R. Flexibility characteristics of three specialized skill groups of champion athletes. *Arch Phys Med Rehabil* 38:580–583, 1957.

55. McArdle, W. D., and Katch, V. L. Nerve control of human movement. *In Exercise Physiology: Energy, Nutrition and Human Performance* (2nd ed.). Philadelphia, Lea & Febiger, 1986, pp. 305–319.

56. McCrea, D. A. Spinal cord circuitry and motor reflexes. *Exerc Sports Sci Rev* 14:105–141, 1986.

57. Minkoff, J., and Sherman, O. H. Considerations pursuant to the rehabilitation of the anterior cruciate injured knee. *Exerc Sports Sci Rev* 15:297–349, 1987.

58. Moffroid, M., Whipple, R., Hofkosh, J., Lowman, E., and Thistle, H. A. Study of isokinetic exercise. *Phys Ther* 49:735–749, 1969.

59. Moore, M. A., and Hutton, R. S. Electromyographic investigation of muscle stretching techniques. *Med Sci Sports Exerc* 12:322–329, 1980.

60. Nicholas, J. A., and Friedman, M. J. Orthopaedic problems in middle-aged athletes. *Physician Sportsmed* 7:39–46, 1979.

61. Nikolaov, P. K., MacDonald, B. L., Glisson, R. R., Seaber, A. V., and Garrett, W. E. Biomechanical and histological evaluation after controlled strain injury. *Am J Sports Med* 15:9–14, 1987.

62. Norkin, C. C., and White, D. J. *Measurement of Joint Motion: A Guide to Goniometry*. Philadelphia, F. A. Davis, 1987.

63. Osternig, L. R., Hamill, J., Corlos, D. M., and Lader, J. Electromyographic patterns accompanying isokinetic exercise under varying speed and sequencing conditions. *Am J Phys Med* 63:289–297, 1984.

64. Osternig, L. R., Robertson, R., Troxel, R., and Hansen, P. Muscle activation during proprioceptive neuromuscular facilitation stretching techniques. *Am J Phys Med* 66:298–307, 1987.

65. Pipes, T. V. Isokinetics versus isotonic strength training in adult man. *Med Sci Sports* 7:262–264, 1975.

66. Pipes, T. V. Strength training modes: What is the difference? *J Scholastic Coach* 46:36–37, 1977.

67. Reid, D. C., Burnham, R. S., Saboe, L. A., and Kushner, S. F. Lower extremity flexibility patterns in classical ballet dancers and their correlation to lateral hip and knee injuries. *Am J Sports Med* 15:347–352, 1987.

68. Rose, S. J., and Rothstein, J. M. Muscle mutability: General concepts and adaptations to altered patterns of use. *Physiotherapy* 62:1773–1787, 1982.

69. Sady, S. P., Wortman, M., and Blanke, D. Flexibility training: Static or proprioceptive neuromuscular facilitation? *Arch Phys Med Rehabil* 63:261–263, 1982.

70. Safran, M. R., Garrett, W. E., Seaber, A. V., Glisson, R. R., and Ribbeck, B. M. The role of warm-up in muscular injury prevention. *Am J Sports Med* 16:123–129, 1988.

71. Sale, D. G., and Norman, R. W. Testing strength and power. *In* MacDougall, J. D., Wenger, H. A., and Green, H. J. (Eds.), *Physiological Testing of the High Performance Athlete*. Champaign, IL, Human Kinetics Publishers, 1991, pp 7–106.

72. Sanderson, D. J., Musgrove, T. P., and Ward, D. A. Muscle imbalance between hamstring and quadriceps during isokinetic exercise. *Aust J Physiotherapy* 30:107–110, 1984.

73. Sapega, A. A., Quendenfeld, T. C., Moyer, R. A., and Butler, R. A. Biophysical factors in range of motion exercise. *Physician Sportsmed* 9:57–65, 1981.

74. Schultz, P. Flexibility: Day of the static stretch. *Physician Sportsmed* 7:109–117, 1979.

75. Scudder, G. N. Torque curves produced at the knee during isometric and isokinetic exercises. *Arch Phys Med Rehabil* 61:68–73, 1980.

76. Sheppard, R. J. *Physical Activity in Aging* (2nd ed.). Rockville, MD, Aspen Publishers, 1987.

77. Smidt, G. L. Biomechanical analysis of knee flexion and extension. *J Biomech* 6:79–92, 1973.

78. Smidt, G. L., and Rogers, M. W. Factors contributing to the regulation and clinical assessment of muscular strength. *Phys Ther* 62:1283–1290, 1982.

79. Smith, E. L., and Gilligan, C. Exercise, sport and physical activity for the elderly: Principles and problems of programming. *In* McPherson, B. D. (Ed.), *Sport and Aging*. Champaign, IL, Human Kinetics Publishers, 1986.

80. Soderberg, G. L. Kinesiology. Muscle mechanics and pathomechanics: Their clinical relevance. *Phys Ther* 63:216–220, 1983.

81. Spencer, J. D., Hayes, K. C., and Alexander, I. J. Knee joint effusion and quadriceps reflex inhibition in man. *Arch Phys Med Rehabil* 65:171–177, 1984.

82. Stanish, W. D., and Hubley-Kozey, C. L. Neurophysiology of stretching. *In* D'Ambrosia, R., and Drez, D. (Eds.), *Prevention and Treatment of Running Injuries*. Thorofare, NJ, Slack Inc., 1989, pp. 209–219.

83. Stanish, W. D., Rubinovich, M., Kozey, J., and MacGillivary, G. The use of electricity in ligament and tendon repair. *Physician Sportsmed* 13:109–116, 1985.

84. Stanton, P., and Purdam, C. Hamstring injuries in sprinting—the role of eccentric exercises. *J Orthop Sports Phys Ther* 10:343–349, 1989.

85. Stolov, W. C., and Thompson, S. C. Soleus immobilization contracture in young and old rats. *Arch Phys Med Rehabil* 60:555–557, 1979.

86. Tanigawa, M. C. Comparison of the hold-relax procedure and passive mobilization on increasing muscle length. *Phys Ther* 52:725–735, 1972.

87. Thorstensson, A., Grimby, G., and Karlsson, J. Force velocity relations and fiber composition in human knee extensor muscles. *J Appl Physiol* 40:12–16, 1976.

88. Thorstensson, A., and Karlsson, J. Fatiguability and fibre composition in human knee extensor muscles. *Acta Physiol Scand* 98:318–322, 1976.

89. Thys, H., Faraggiona, T., and Margaria, R. Utilization of muscle elasticity in exercise. *J Appl Physiol* 32:491–494, 1972.

90. Tipton, C. M., Vailas, A. C., and Matthes, R. D. Experimental studies on the influences of physical activity on ligaments, tendons and joints: A brief review. *Acta Med Scand (Suppl.)* 711:157–168, 1986.

91. Urschel, J. D., Scott, P. G., and Williams, H. T. G. The effect of mechanical stress on hard and soft tissue repair: A review. *Br J Plast Surg* 41:182–186, 1988.

92. Vailas, A. C., Pedrini, V. A., Pedrini-Mille, A., and Holloszy, J. O. Patellar matrix changes associated with aging and voluntary exercise. *J Appl Physiol* 58:1572–1576, 1985.

93. Vailas, A. C., Tipton, C. M., Matthes, R. D., and Gart, M. Physical activity and its influence on the repair process of medial collateral ligaments. *Connect Tissue Res* 9:25–37, 1981.

94. Viidik, A.: Connective tissues—possible implications of the temporal changes for the aging process. *Mech Ageing Dev* 9:267–285, 1979.

95. Vrahas, M. S., Brand, R. A., Brown, T. D., and Andrews, J. G. Contributions of passive tissue to the intersegmental moment at the hip. *J Biomech* 23:357–362, 1990.

96. Wallin, D., Ekblom, B., Graham, R., and Nordenborg, T. Improvement of muscle flexibility: A comparison between two techniques. *Am J Sports Med* 13:263–268, 1985.

97. Warren, C. G., Lehmann, J. F., and Koblanski, J. N. Elongation of rat tail tendon: Effect of load and temperature. *Arch Phys Med Rehabil* 52:465–474, 1971.

98. Warren, C. G., Lehmann, J. F., and Koblanski, J. N. Heat and stretch procedures: An evaluation using rat tail tendon. *Arch Phys Med Rehabil* 57:122–126, 1976.

99. Wiktorsson-Moller, M., Oberg, B., Ekstrand, J., and Gillquist, J. Effects of warming up, massage, and stretching on range of motion and muscle strength in the lower extremity. *Am J Sports Med* 11:249–252, 1983.

100. Wilkie, D. R. The relation between force and velocity in human muscle. *J Physiol (Lond)* 110:249–280, 1950.
101. Williford, H. N. A comparison of proprioceptive neuromuscular facilitation and static stretching techniques. *Am Correct Ther J* 39:30–33, 1985.
102. Williford, H. N., East, J. B., Smith, F. H., and Burry, L. A. Evaluation of warm-up for improvement in flexibility. *Am J Sports Med* 14:316–319, 1986.
103. Winter, D. A. *Biomechanics of Human Movement.* New York, John Wiley, 1979.
104. Winter, D. A., Wells, R. P., and Orr, G. W. Errors in the use of isokinetic dynamometers. *Eur J Appl Physiol* 46:397–408, 1981.
105. Wood, T. O., Cooke, P. H., and Goodship, A. E. The effect of exercise and anabolic steroids on the mechanical properties and crimp morphology of the rat tendon. *Am J Sports Med* 16:153–158, 1988.
106. Wood, T. O., Viidik, A., Urcell, J. D., and DeLorme, T. L. Restoration of muscle power by heavy resistance exercise. *J Bone Joint Surg* 27:645–652, 1945.

THERAPEUTIC MODALITIES

Cold Modalities

William A. Grana, M.D.

Cryotherapy is the application of cold for the treatment of injury or disease. This section deals with the beneficial, therapeutic effects of cold on soft tissue injury. Today, cold is a widely used physical agent based primarily on both clinical and empirical evidence [2, 11, 12, 14, 26]. As such, cold is used for both the immediate care and the rehabilitation of soft tissue injury [17, 24, 30]. Immediate care is defined as that portion of treatment that occurs during the first 48 hours following injury. Most clinicians feel that cold is a standard treatment for soft tissue injury because of its efficacy and its cost effectiveness. The purpose of this discussion is to examine the mechanism of action of cold in soft tissue injury, the methods of application, and the indications and contraindications for its use.

HISTORICAL REVIEW

All of the ancient physicians including Hippocrates, Galen, Celsus, Avecenna, and Seneca recognized the beneficial effects of cold [19]. Throughout history, cold in the form of snow, ice, or cold drinks and water has been used to treat various illnesses in which inflammation, fever, or burning pain is the main clinical manifestation. The simple principle has been to apply an agent that produces a physical sensation opposite to the clinical symptom. More scientifically, ice has been used to produce hemostasis and analgesia and to decrease the inflammation in infection [19]. In the eighteenth century, ice was first produced in quantity artificially, but it was not until more than 100 years later that it became commercially available. This technologic advance allowed greater clinical application for the management of inflammation and for analgesia [19].

CURRENT CONCEPTS AND CONTROVERSIES

Cold affects the inflammatory response that follows soft tissue injury through its effects on metabolism and circulation. Cold reduces enzymatic function and therefore decreases metabolism [5]. Cold also produces vasoconstriction, which may affect the extension of hematoma formation, but it does not occur quickly enough to prevent local hemorrhage following an injury [5, 6, 16]. An excellent review by Kowal of the physiologic effects of cryotherapy on inflammation summarizes these findings [18].

Inflammation is delayed by the application of cold but is not eliminated [18]. Similarly, edema is not prevented in the animal model by the application of cold [23]. Investigation of the use of cold to decrease soft tissue swelling or edema in the animal model has yielded either conflicting or inconclusive information. Models of crush injury or other trauma demonstrate a decrease or no change in the occurrence of swelling when cold is applied. Cold application *does not prevent* edema or swelling [22, 23].

The vasoconstriction that occurs following cold application is an effect of cold on alpha-2 adrenoreceptors in the terminal arterioles and overcomes the mild *depressant* effect of cold on smooth muscle [10]. Blood flow decreases to a body part as measured by radioactive xenon flow studies as the temperature decreases. Following the surface application of cold, this change in blood flow occurs to a depth of 2 to 4 cm [17, 21]. In addition, the vascular response to cold is delayed by 5 to 10 minutes. Therefore, cold will produce minimal benefit on the occurrence of acute hemorrhage, but it may theoretically prevent the extension of a hematoma [17, 21, 28, 29].

In addition to these specific effects of cold, a circulatory phenomenon occurs commonly in the periphery, particularly in the fingers and toes, called cold-induced vasodilatation (CIVD). It explains the erythema that occurs after cold application. This response is also known as the "hunting response" and occurs following cold exposure. This vasodilatation seems to be limited to the periphery, particularly the fingers and toes, and does not occur in deeper tissues such as large muscle groups. There is a lessening of the vasoconstrictive effect of cold in deeper tissues in response to long application, but no vasodilatation occurs in these deeper tissues. The significance of this phenomenon is unknown but may be protective in nature. In addition, this phenomenon has no effect on core temperature, and, in general, core temperature is not affected by the therapeutic use of cold [17, 20].

Cold decreases pain by blocking the sensory transmission of pain impulses as a result of the slowing or elimination of nerve conduction. In addition, cold reduces muscle spasm through its effect on the muscle spindle and therefore relieves pain [1, 4, 17].

Finally, cold increases the stiffness of collagen and therefore decreases ligament elasticity and muscular flexibility. Cold application affects function through complex processes that are mediated by both neural and mechanical means. Although cold decreases connective tissue extensibility, its immediate effect is to improve muscle contraction. However, in studies of muscle strength, at the end of a normal treatment period of 20 to 30 minutes, strength declined by 2% to 15% [3, 18, 29]. Strength then increased for up to 3 hours after treatment. However, both the loss of extensibility and the

decreased strength are reasons to avoid cold prior to vigorous exercise or sports competition. In the normal subject, brief cold application does not improve passive joint motion. In addition, these effects occur only with cold application that promotes muscle cooling, not with skin cooling alone as in local ice application to a specific anatomic area [3, 8, 18, 27, 29].

The effects of cold depend upon the temperature of the applied agent, the duration of its application, the surface area over which it is applied, and the body mass to which the cold is applied. Based on experimental work, the optimal duration of application is 20 to 30 minutes because any longer time does not increase the rewarming time required for the body part [17]. Cold will penetrate to a depth of approximately 2 to 4 cm, again, depending on the mass of the tissue involved and the constancy of the cold applied [17].

COMPLICATIONS AND CONTRAINDICATIONS

Cold application has been associated with nerve palsy at both the elbow and the knee and therefore must be done cautiously, after the patient has been apprised of any detrimental symptoms to allow recognition [7, 9].

The management of a neural injury due to cold application follows the usual management of a closed injury to a motor nerve from any other cause. The injured part is splinted in a functional position and observed. If the injury persists beyond 24 to 48 hours, passive range of motion of the joints involved is begun. After 2 weeks, if there is still persistent motor loss, a baseline electromyogram (EMG) is done to document the injury. This should be repeated in 6 weeks. If there is no change in the physical or EMG findings, in all likelihood the injury is permanent [7, 9]. There is no support for or report in the literature of surgical exploration or neurolysis of the nerve involved following such an injury.

On the other hand, it is not completely clear whether these effects are due to cold alone or to the combination of cold and compression, since these two modalities are frequently used together. Sprays such as ethyl chloride are used for local anesthesia for soft tissue injections or aspiration of a joint, but these pose a significant risk with improper prolonged use that includes burns and frostbite [7, 9].

Individuals who are sensitive to cold for a variety of reasons may risk local burns or systemic complications as a result of cold application. For this reason, in certain patients relative or absolute contraindications to the use of cold may exist. Contraindications to the use of cold include the presence of Raynaud's phenomenon, cold allergy, cryoglobulinemia, and paroxysmal cold hemoglobinuria. Relative contraindications include arthritic conditions, pheochromocytoma, anesthetic skin, or a patient unresponsive from cardiac disease. These recommendations are based on the fact that cold produces vasoconstriction and decreases metabolism and neurofunction [17, 19].

CLINICAL EVALUATION OF COLD MODALITIES

Cold may be applied by the use of ice in a variety of forms or by such means as gel refrigerant packs, chemical mixtures that produce an exothermic reaction when mixed, coolant sprays such as ethylchloride or ethylfluoride that produce local skin anesthesia, or electromechanical devices that produce a combination of refrigeration and compression. In terms of efficacy, ice in one form or another is the best method. Following this, in descending order of effectiveness, are the refrigerant gel packs, chemical packs, and refrigerant compression systems [17, 25].

Although there is much information about the physiologic effects of cold on inflammation, metabolism, circulation, and nerve conduction as noted earlier, little information has been obtained from *controlled* investigations on the specific clinical uses of cold. The use of cold for acute and chronic management of soft tissue injury is based on these physiologic effects, but the clinical information available does not categorize or standardize the severity of injury, the type of injury, the treatment regimen, the modalities used for treatment, and the criteria used to classify results. Most of the clinical information available relates to common soft tissue injuries such as ankle sprains and thigh contusions. These studies are qualitative, involve a small number of cases, and use return to sport as the criterion for success [13, 15, 31]. Nonetheless, the available data do seem to support the beneficial effects of cold applied during the first 72 hours for the acute symptoms and as well as a more rapid return to "full" activity after recovery from soft tissue injury.

TREATMENT USES

Cold is beneficial when applied immediately to the area of a soft tissue injury as well as during the rehabilitation phase of an injury because of its effects in reducing pain and muscle spasm. Cold should be used for treatment of soft tissue injury within the first 48 hours because of its beneficial effect on the inflammatory response. In addition, pain relief occurs in both the acute and rehabilitative phases of a soft tissue injury because cold decreases spasm and slows nerve conduction. Finally, cold is used in the rehabilitation phase to allow therapeutic exercise (termed "cryokinetics") and to facilitate mobilization primarily by reducing muscle spasm and providing pain relief. On the other hand, cold should not be used prior to athletic activity because it produces decreased muscular flexibility due to increased collagen stiffness.

AUTHOR'S RECOMMENDED USE OF COLD

I recommend the use of cold in both the acute and rehabilitative phases of soft tissue injury. To a certain extent, the specific usage will depend upon the anatomic area involved in the injury. However, some generalizations can be made, in that my preferred method of cold application is an ice pack, ice bath, or ice cup (massage). This type of application is cost effective and safe and should be available in every clinic or training room. If crushed ice is not available, and one wishes to use a conforming type of ice pack, a "poor man's" refrigerant gel pack can be created by using a large bag of frozen peas, and this can be reused at least several times. I do not care for chemical packs or refrigerant gel packs because these can break, and the contents are caustic and may produce dermal injury.

Also, I do not use coolant sprays or refrigerant compression apparatus at all because I do not feel that these have an advantage with regard to clinical efficacy, convenience, cost, safety, or duration of effect [24, 25].

The method of application will depend upon the site of injury. I am cautious about wrapping, particularly about bony prominences, because cold application plus compression may produce injury to the subcutaneous nerves, such as the peroneal or ulnar nerve. The choice of device (ice pack, ice bath, or ice massage) will depend upon the area to be covered—the larger the area, the more likely one is to use an ice bath; the smaller the area, the more likely one is to use ice massage.

Acute Phase

The duration of application is 20 to 30 minutes, depending upon how superficial the area involved is. Large muscular areas such as the thigh or the calf may require a longer duration of application. Cold is applied every 2 hours while the patient is awake. This allows sufficient time for rewarming [17, 18]. The frequency of application is also a matter of convenience, and, if it is needed and can be accomplished with supervision, then round-the-clock application is used. The general rule of thumb for duration of application is that if the patient complains that the area of application seems hot, it is time to stop because this may indicate the presence of cold-induced vasodilatation of the subcutaneous tissues. This is particularly true of the hands and feet [17].

In summary, during the acute phase I apply ice for 20 to 30 minutes every 2 hours for the first 48 to 72 hours [17, 18].

Rehabilitation Phase

During the rehabilitation phase, I use ice along with active range of motion exercise to control pain and inflammation. This treatment may be combined with high-voltage muscle stimulation to assist in return of motion. Once swelling and pain have diminished sufficiently, ice is combined with active and passive range of motion to the tolerance of any pain produced. The ice is used both prior to and during the therapeutic exercise bout. This is done for 20 to 30 minutes twice a day in the training or therapy room setting. In between, the injury site is protected. For example, in the case of an ankle spasm, a tape job or air cast with or without crutch ambulation is used. In the nontraining room setting for the same ankle spasm, the patient may be placed in a functional, prefabricated short leg cast or a bivalved plaster or fiberglass cast that can be removed for treatment.

When a larger surface area is to be covered by the cold application, an ice bath can be used; the duration of treatment is then 5 to 10 minutes prior to the exercise bout. If the exercise is not supervised by the trainer or therapist, the bath is used during exercise. If the a lower extremity is immersed in the ice bath, one may wish to use a "toe guard" to protect the toes from the effects of the cold. Once the effects of swelling and pain have subsided and the patient has progressed to more vigorous exercise and functional activity, the ice is used for 20 to 30 minutes *after* the therapeutic exercise bout. One exception to this is the use of cold prior to contract-relax exercises, which are done when the range of motion is close to normal.

RESEARCH NEEDS

There is a need for controlled, standardized, clinical research to document the effectiveness of cold for the management of soft tissue injury. Much of the data that are available do not document the severity of an injury. Therefore, the effects of a specific regimen or treatment are not documented either. For example, it would be useful and important to know whether ice adds any benefit to the use of compression and elevation that comprise usual treatment of most soft tissue injuries of the extremity. Such data could be acquired only by documenting the severity of the injury and then measuring the effect as a percentage of improvement. This would allow a more standardized approach to the use of cold and also document the effects of cold. Cold remains a mainstay in the management of soft tissue injury now and will continue to do so in the future, but, it is to be hoped, with better scientific documentation of its effectiveness.

References

1. Abramson. Effect of tissue temperature in blood motor nerve conduction velocity. *JAMA* 198:1082–1088, 1966.
2. Barnes, L. Cryotherapy—Putting injury on ice. *Physician Sports Med* 3:130–136, 1979.
3. Barnes, W. S., and Larson, M. R. Effects of localized hyper- and hypothermia on maximal isometric grip strength. *Am J Phys Med* 64(6):306–314.
4. Bell, K. R., and Lehmann, J. F. Effect of cooling on H- and T-reflexes in normal subjects. *Arch Phys Med Rehabil* 68:490–493, 1987.
5. Brooks and Duncan. The influence of temperature on wounds. *Ann Surg* 114:1069–1075, 1971.
6. Clarke, R. S. J., Hellon, R. F., and Lind, A. R. Vascular reactions of human forearm to cold. *Clin Sci* 17:165–179, 1958.
7. Collins, K., Storey, M., and Peterson, K. Peroneal nerve palsy after cryotherapy. *Physician Sports Med* 14(5):105–108, 1986.
8. Cornelius and Jackson. The effects of cryotherapy and PNF on hip extensor flexibility. *Athletic Training* 19:183, 1984.
9. Drez, D., Faust, D. C., and Evans, J. P. Cryotherapy and nerve palsy. *Am J Sports Med* 9:256–257, 1981.
10. Faber, J. E. Effect of local tissue cooling on microvascular smooth muscle and postjunctional α_2-adrenoceptors. *Am J Physiol* July:255, 1988.
11. Grana, W. A., Karr, J., and Stafford, M. Rehabilitation techniques for athletic injury. *Instruct Course Lect* 34:393–400, 1985.
12. Hocutt, J. E. Cryotherapy. *Am Fam Physician* 23(3):141–144, 1981.
13. Hocutt, J. E., et al. Cryotherapy in ankle sprains. *Am J Sports Med* 10(5):316–319, 1982.
14. Kalenak, A., et al. Athletic injuries: Heat vs. cold. *Am Fam Physician* 12(5):131–134, 1975.
15. Kalenak, A, et al. Treating thigh contusions with ice. *Physician Sports Med*:65–67, 1975.
16. Kellett, J. Acute soft tissue injuries—A review of the literature. *Med Sci Sports Exerc* 18(5):489–500.

17. Knight, K. L. *Cryotherapy: Theory, Technique and Physiology.* Chattanooga, Chattanooga Corporation, 1985.
18. Kowal, M. A. Review of physiological effects of cryotherapy. *J Orthop Sports Phys Ther* 66–73, 1983.
19. Lehman, D. D. *Therapeutic Heat and Cold* (3rd ed.). Baltimore, Williams & Wilkins, 1982.
20. Lewis, T. Observations upon the reactions of vessels of the human skin to cold. *Heart* 15:177–208, 1930.
21. Lowdon and Moore. Determinants and nature of intramuscular temperature changes during cold therapy. *Am J Phys Med* 54:223, 1975.
22. Matsen, F. A. 3d, Questad K., Matsen, A. L., et al. The effect of local cooling on postfracture swelling. *Clin Orthop* 109:201, 1975.
23. McMaster, W. C., and Liddle, S. Cryotherapy: Influence on posttraumatic limb edema. *Clin Orthop* 150, 1980.
24. McMaster, W. C. A literary review on ice therapy in injuries. *Am J Sports Med* 5:124–126, 1977.
25. McMaster, W. C., Liddle, S., and Waugh, T. R. Laboratory evaluation of various cold therapy modalities. *Am J Sports Med* 6:291–294, 1978.
26. Meausen, R., and Lievans, P. The use of cryotherapy in sports injuries. *Sports Med* 3:398–414.
27. Newton, R. A. Effects of vapocoolants on passive hip flexion in healthy subjects. *Phys Ther* 65(7):1034–1036, 1985.
28. Pappenheimer, S. L., Eversole, S. L., and Soto-Rivera, A. Vascular responses to temperature in the isolated perfused hindlimb of the cat (Abstract). *Am J Physiol* 155:458, 1948.
29. Thorsson, O., et al. The effect of local cold application on intramuscular blood flow at rest and after running. *Med Sci Sports Exerc* 17(6):710–713, 1985.
30. Wayloms, G. W., et al. The physiological effects of ice massage. *Arch Phys Med Rehab* 1967.
31. Yackzaw, L., Adams, C., and Francis, H. T. The effects of ice massage on delayed muscle soreness. *Am J Sports Med* 12:159, 1984.

Heat Modalities

Jay S. Cox, M.D.

Thermotherapy is the use of heat to increase the temperature in various tissues and has been used for the treatment of a wide spectrum of orthopaedic conditions. Until the twentieth century heat therapy was limited to superficial heat modalities such as hot towels and warm baths. These treatments were inexpensive and were frequently successful in eliminating pain and discomfort for a variety of conditions. Later, the development of other superficial heat modalities such as whirlpool baths, hydrocollator packs, paraffin baths, and infrared lamps allowed the process of heat transfer to be more efficient and scientifically controlled.

Diathermy was developed in the 1920s as a means of delivering heat to deeper tissues. Diathermy is the use of high-frequency electromagnetic currents to induce heating of biologic tissues by vibration and distortion of tissue molecules.

The use of ultrasound for medical purposes was developed in the early 1950s. High-frequency acoustic vibrations at frequencies too high to be perceived by the human ear are used to impart thermal energy to deep tissues. Because of the prompt thermal effect, which can occur at great depth, and also the ease of application, ultrasound has replaced diathermy as the preferred technique for deep heating.

THE MECHANISM OF HEAT TRANSFER

Transmission of heat to tissue occurs by one of four mechanisms: conduction, convection, radiation, and conversion. Conduction and radiation describe the transfer of energy entirely as a result of temperature differences. Conduction is the exchange of thermal energy between two surfaces in physical contact with each other; the hydrocollator pack is an example of this type of heat transfer.

Radiation is the transfer of heat from a warm source to a cooler source through a conducting medium such as air. Infrared lamps deliver heat by this mechanism. Radiation techniques have the advantage of not allowing the heat source to contact the patient's skin, and they also allow treatment over a large area. Convection is a more rapid process of heat transfer than conduction or radiation and depends on both temperature difference and mass transport. Convection occurs when air or water molecules move across the body, creating temperature variations. Whirlpool baths are an example of heat application by convection. Conversion is the transformation of a form of nonthermal energy, such as electric, electromagnetic, or acoustic energy, into heat energy. These techniques, consisting of diathermy and ultrasound, allow far greater tissue penetration than superficial heat modalities.

BASIC SCIENCE

For any heat modality to be effective, the target tissue temperature must reach a therapeutic range. This range is usually considered to be 104° to 113°F (40° to 45°C). The beneficial effect usually begins at 3 minutes and lasts for several minutes. The upward limit of duration of the beneficial effect seems to be 30 minutes, but a rapid rate of temperature elevation may have a mild additional benefit [11]. The application of heat results in hyperemia, and local blood flow is increased. This effect occurs because heat inhibits the sympathetic vasoconstrictive nerve fibers in the skin, resulting in vasodilatation. Both local lymphatic and venous drainage increase, and removal of metabolic waste due to the inflammatory process is facilitated. The supply of oxygen, antibodies, leukocytes, and phagocytes to the target tissues is vastly increased because of this hyperemia.

The thermotherapeutic response also results in an increase in overall metabolic rate. Increased metabolic rate creates additional heat, thus causing further vasodilatation and increased capillary permeability. One of the undesirable side effects of this increased capillary permeability is increased tissue edema.

The analgesic effect of heat is not fully understood but is thought to be the result of several mechanisms. Heat is believed to act selectively on free nerve endings and peripheral nerve fibers to increase the pain threshold and may also decrease sensory nerve conduction velocity. Pain caused by muscle spasm is relieved by heat owing to a direct effect on the gamma fibers of the muscle spindles, which demonstrate decreased activity and sensitivity to stretch. Muscle spasm is also reduced because heat-induced vasodilatation results in the removal of breakdown products from injured tissues. These byproducts, which include prostaglandin, histamine, and bradykinin, stimulate the nerve fibers that cause the pain-spasm-pain cycle. Heat alters the viscoelastic properties of collagen tissue, resulting in an increase in extensibility or elongation of this tissue. For the best therapeutic benefit, stretching should be used in conjunction with heat. When heated, the collagen in the capsule of joints also elongates, resulting in capsular laxity and relief of joint stiffness. Recent studies indicate that heating articular tissue to 108°F (42°C) may inhibit the articular enzymes responsible for the inflammatory cycle in joint inflammations [3, 12]. The intra-articular temperature is normally lower than that of the surrounding tissues, even when an inflammatory arthritis exists [7, 8, 9]. Local hyperemia in joints may inhibit articular enzymes and help the action of anti-inflammatory medications [2, 12, 14].

In summary, the physiologic effects of thermotherapy include altered pain sensation, vasodilatation, increased collagen extensibility, enhanced nutrition and metabolism, and decreased sensitivity to muscle stretch.

CLINICAL APPLICATIONS

Thermotherapy is used for chronic conditions that can benefit most from its known physiologic effects. Heat is effective in treating chronic inflammation, subacute or organizing hematoma, joint stiffness, muscle spasm, and other various pain syndromes. It is effective in some subacute inflammatory conditions. One must be careful in using heat therapy in patients with acute inflammatory conditions because the inflammation may be aggravated. Another contraindication to the use of heat is acute trauma with resulting hemorrhage or hematoma formation. Heat exacerbates hemorrhage by causing vasodilatation in the acutely ruptured vessels and would also cause increased edema formation by increasing the permeability of the capillary walls. Other contraindications include impaired skin sensation for pain or temperature, poor thermal regulation, impaired skin circulation, scar areas, and malignancy.

SUPERFICIAL HEAT MODALITIES

Superficial modalities transfer heat by conduction, convection, and radiation. The term superficial is indeed appropriate because there is general agreement that none of these modalities can have a depth of penetration greater than 1 cm [1].

Warm Whirlpool

For whirlpool treatment of an upper extremity problem, temperatures should be controlled at 98° to 110° F (37° to 45°C), and for treatment of a lower extremity problem the temperature should be between 98° and 104°F (37° to 40°C). If full-body treatment is indicated, temperatures should be limited to a range of 90° to 102°F (37° to 39°C) to avoid general hyperthermia. Treatment should continue for 15 to 20 minutes, which is usually long enough to reduce muscle spasm and stimulate vasodilatation. Range of motion exercises for affected areas can be instituted during the whirlpool treatment. Mild exercise may increase additional blood flow to the deeper structures.

Contrast Baths

In patients in whom maximal vasodilatation is desirable or when subacute or gravity-dependent swelling is present, contrast baths should be considered. Temperature for the warm cycle should be 104° to 110°F (40° to 43°C), and temperature for the cold cycle should be 50° to 59°F (10° to 15°C). The warm cycle should last for approximately 5 minutes and the cold cycle 1 or 2 minutes. Treatment should continue for 30 minutes and conclude with the cold cycle. Contrast baths have the theoretical advantage of producing alternate vasodilatation and vasoconstriction, which can be effective in controlling residual edema formation [13].

Hot Packs

Commercially available hydrocollator packs consist of canvas packs of petroleum distillate. Packs are heated in special units controlled by a thermostat. This temperature is carefully controlled to a range of 122° to 158°F (50° to 70°C). Skin burns are avoided by wrapping the packs in towels, which should be 1 inch in thickness. The duration of treatment is between 15 and 20 minutes. Hot packs are best indicated for treatment of muscle spasm and may also be used as a preliminary measure with specific exercise programs.

Paraffin Baths

Paraffin baths are ideal for the treatment of hand and foot problems. The extremities should be well cleansed prior to being placed in a paraffin bath because paraffin is an excellent growth medium for skin bacteria. The bath, which consists of 1 gallon of mineral oil to 2 pounds of paraffin in a ratio of 7:1, is heated to 126°F (51°C). One of two techniques is used. "Dipping" involves immersing the extremity in the paraffin for a few seconds and then removing it, allowing the paraffin to harden. This procedure is repeated until several layers have accumulated. With "wrapping," the extremity is dipped into the bath and then wrapped in a plastic bag surrounded by several layers of towels to act as an insulator. Treatment should continue for 20 to 30 minutes. The risk of skin burns is substantial, so this treatment should be monitored closely. This is particularly true in patients with various collagen or vascular disorders who may have more sensitive skin.

Infrared Heat

Heat transfer with infrared heat occurs by radiation. The heat is transferred from a warmer source to a cooler source through a conductive medium such as air. Effectiveness can be improved by placing warm, moist towels on the area to be treated. The infrared lamp should be placed approximately 20 inches from the body, and treatment should last 20 minutes. Dry towels should be used to drape off areas that do not require treatment. Continuous monitoring of the skin is necessary to prevent burning. The primary advantage of infrared lamps compared with hot packs is that a minimum of equipment is needed and the required maintenance is minimal.

DEEP HEAT MODALITIES

Diathermy

Diathermy is the application of high-frequency electromagnetic currents to induce deep heating in biologic tissues by means of vibration and distortion of the tissue molecules. The heat is caused by the resistance of the tissue to the passage of the energy. Its use requires a very high degree of skill and expertise because diathermy doses are not precisely controlled, and the amount of heating the patient receives cannot be accurately prescribed or directly measured. Diathermy as a therapeutic agent is subdivided into two distinct entities: shortwave diathermy and microwave diathermy. The shortwave diathermy unit is basically a radio transmitter. The power source is capable of generating a radio frequency, and electrode placement will have a significant effect on how much energy is delivered to the patient. The primary benefits of diathermy are the same as those of heat in general—tissue temperature rise, dilatation of the blood vessels, increased filtration and diffusion, increased capillary permeability, increased blood flow, and increased metabolic rate.

Some of the indications for the use of shortwave diathermy are chronic inflammatory conditions such as rheumatoid arthritis, bursitis, tendinitis, osteoarthrosis, sprains, strains, and neuritis. Contraindications for the use of shortwave diathermy are metal implants, fresh hemorrhage in tissues, pacemakers, intrauterine devices, malignancies, arteriosclerosis, phlebitis, wet dressings, pregnancy, and infection. It should not be used near the gonads or epiphyseal plates in children, nor over the pelvic area in menstruating women. It should not be used around the eyes for any prolonged periods of time nor over contact lenses.

Microwave diathermy also heats tissues by electromagnetic radiation but differs from shortwave diathermy in that it utilizes higher frequencies and shorter wave lengths. This gives an advantage in that less than 10% of the energy is lost from the machine as it is applied to the patient. The two frequencies that are currently available in this country are 2456 megahertz and 915 megahertz. The lower frequency is more effective in penetrating fat, but with either frequency, subcutaneous fat should be no more than 1 cm in thickness to avoid overheating. The heating effect is caused by the intramolecular vibration of the molecules that are high in polarity [13]. If subcutaneous fat is greater than 1 cm thick, the fat temperature will rise to a level that is too uncomfortable before there is a rise in temperature in the deeper tissues.

The same precautions and contraindications listed for shortwave diathermy apply to microwave diathermy. In general, microwave diathermy has been very effective in the treatment of fibrous muscular contractions and also in the treatment of tendinitis and chronic tenosynovitis.

In summary, the beneficial effects of diathermy are those of heat in general. However, with diathermy, tissues that are 3 to 5 cm deep can be heated, but to what degree has not yet been specifically determined. The physiologic actions are well known, but more information is necessary about the actual clinical effects. One would have to determine how much the placebo effect in itself is a factor in the response of some patients to this therapy.

Ultrasound

Ultrasound is acoustic energy that causes molecular motion as it passes through physical substances, producing thermal and nonthermal effects. Ultrasonic waves are those composed of higher frequencies than are detectable by the human ear. Audible sound ranges between 16,000 and 20,000 hertz (Hz), so frequencies above this level are defined as ultrasound. Standard therapeutic systems operate at a frequency of between 750,000 and 1,000,000 decibels/second (Hz). The penetration of ultrasonic waves is inversely related to their frequency. Because higher frequency ultrasound has a higher rate of absorption in the superficial tissues, less energy is available to the deeper tissues. Therefore, higher frequencies are used to treat superficial and cutaneous lesions, and lower frequencies are used to treat deeper areas with thicker soft tissue such as the back, abdomen, and other muscle areas. Fat has a relatively low absorption rate, and muscle absorbs considerably more than fat. Bone absorbs more ultrasonic energy than any other tissue, and this is attributed to the high protein content and density of bone compared with other tissues.

Thermal Effects

As the ultrasonic beam travels through tissue, the energy is either conducted, absorbed, or reflected. Absorption of the ultrasonic energy produces the thermal or heating effect. Ultrasonic energy passes easily through subcutaneous fat, so it can produce temperature increases in the deep tissues. Low-frequency ultrasound can heat tissues to a depth of 10 cm, and temperature rises can be produced in a short span of time. Since metal consists of dense, homogeneous material, most ultrasonic energy is conducted through metal implants and is not reflected or absorbed. Thus, there is no increased thermal effect when ultrasound is used over metallic devices.

Nonthermal Effects

A newer development in ultrasound therapy is the use of pulsed ultrasound, which uses a lower intensity of sound and produces nonthermal responses without appreciable generation of heat. These pulsed beams allow mechanical effects rather than increases in tissue temperature. It is thought that these nonthermal effects may reduce inflammation, and because they do not increase tissue temperature, they may be of use in the treatment of acute traumatic injuries.

Cavitation may occur during the use of ultrasound and is a potentially harmful effect. This is the phenomenon in which gas bubbles are trapped within the tissues. These bubbles enlarge and vibrate in response to the ultrasonic waves. The sudden expansion and collapse of cavities result in focal high pressures and temperatures in the tissues, which can cause local tissue damage [4].

Therapeutic Use of Ultrasound

Sound waves move poorly through air, so the sound head is applied directly to the skin with the use of a coupling agent. This coupling agent acts as a conductive medium at the air-transducer interface. The most effective coupling agents are glycerin and mineral oil. Only the area under the sound head will receive the energy transfer, so the sound head must be moved in circular motions over the area to be treated. The sound head must not remain stationary because this can result in a rapid rise in tissue temperature, creating damage to the endothelial cells and blood vessels. Areas that have small or irregular surfaces, such as the hands and feet, can be effectively treated by submerging them in water and positioning the sound head 1 to 2 cm from the body part. Treatment should last for 5 to 10 minutes, depending on the size of the area to be treated. Patients often report feeling a sense of warmth during treatment, but if pain is elicited, the intensity should be lowered. Pain generally indicates overheating of the bone periosteum.

Clinical Indications

The prime indication for the use of ultrasound is any condition in which deep and prompt tissue heating is desired. One effective use is the reduction of muscle spasm. This ability of ultrasound to reduce the spasm is believed to derive from a direct alteration in the skeletal muscle contractile process, reduced muscle spindle activity, reduced pain sensation, and increased blood flow to the involved musculature. Ultrasound is used to reduce the pain of joint contractures in that the heat increases the extensibility of the collagen fibers by altering the properties of the viscoelastic components of the connective tissue. As mentioned with diathermy, stretching must be incorporated along with the application of heat. Ultrasound has been said to have an effect on calcium deposits in various tendons and tissues, but the ability of ultrasound to remove these deposits is not well proven. The pain in these conditions is relieved by reducing the inflammation and edema in the tendons and bursae. One of the most recent developments in the use of ultrasound is in the treatment of fractures. Two recent reports have indicated that fracture healing is accelerated with the use of ultrasound [6, 10].

Contraindications to the use of ultrasound include areas of suspected malignancy or infection, open growth plates in children, the thoracic region in patients who have a cardiac pacemaker, and any area of acute hemorrhage or inflammation. It should not be used over the abdomen in pregnant women because of the potential for damage to the fetus nor over the abdomen in women who are menstruating because of the possibility of increased bleeding. It should not be used over the eyes, heart, or gonads.

Phonophoresis

Phonophoresis is the technique of using the mechanical energy of ultrasound to introduce medications through the skin to deeper tissues. Medications such as corticosteroids and salicylates can be used in patients with a variety of inflammatory conditions. The most common medication used in phonophoresis is 10% hydrocortisone cream. Griffin and associates [5] conducted a double-blind study to compare the effects of hydrocortisone phonophoresis with the use of ultrasound alone. There were 102 patients in this series with chronic inflammatory conditions, and 68% of those receiving phonophoresis with hydrocortisone cream experienced relief from pain and improved range of motion in joints compared with only 28% of the patients treated with ultrasound alone. Medications delivered by ultrasound have been recovered as deep as 10 cm

within tissues after 5 minutes of treatment. However, it is not known how long this medication may remain in the area of application and in what concentration.

References

1. Abramson, D. I., Tuck, S., Lee, S. W., Richardson, G., and Chu, L. S. W. Vascular basis for pain due to cold. *Arch Phys Med Rehabil* 47:300–305, 1966.
2. Benz, R., Beckers, F., and Zimmermann, U. Reversible electrical breakdown of lipid bilayer membranes: A charge-pulse relaxation study. *J Membrane Biol* 48:181–204, 1979.
3. Castor, C. W., and Yaron, M. Connective tissue activation: VIII. The effects of temperature studies in vitro. *Arch Phys Med Rehabil* 57:5–9, 1976.
4. Dyson, M. Mechanisms involved in therapeutic ultrasound. *Physiotherapy* 73:116–120, 1987.
5. Griffin, J. E., Echternach, J. L., Price, R. E., et al. Patients treated with ultrasonic driven hydrocortisone and with ultrasound alone. *Phys Ther* 47:594–601, 1967.
6. Heckman, J. D. Acceleration of tibial fracture healing by noninvasive low intensity pulsed ultrasound. Paper presented at the annual meeting of the American Academy of Orthopaedic Surgeons, Anaheim, CA, March, 1991.
7. Hollander, J. L., and Horvath, S. M. The influence of physical therapy procedures on the intra-articular temperature of normal and arthritic subjects. *Am J Med Sci* 218:543–548, 1949.
8. Hollander, J. L., Stoner, E. K., Brown, E. M., Jr., et al. Joint temperature measurement in the evaluation of anti-arthritic agents. *J Clin Invest* 30:701–706, 1951.
9. Horvath, S. M., and Hollander, J. L. Intra-articular temperature as a measure of joint reaction. *J Clin Invest* 28:469–473, 1949.
10. Kristiansen, T. K. Acceleratory effect of ultrasound on healing time of Colles' fractures. Paper presented at the annual meeting of the American Academy of Orthopaedic Surgeons, Anaheim, CA, March, 1991.
11. Lehman, J. F., Warren, C. G., and Scham, S. M. Therapeutic heat and cold. *Clin Orthop* 99:207–244, 1974.
12. McGhie, J. B., Wold, E., Pettersen, E. O., et al. Combined electron radiation and hyperthermia: Repair of DNA strand breaks in NHIK 3025 cells irradiated and incubated at 37°, 42.5°, or 45°C. *Radiat Res* 96:31–40, 1983.
13. Prentice, W. E. *Therapeutic Modalities in Sports Medicine.* St. Louis, Times Mirror, 1986, pp. 102, 149.
14. Song, C. W. Effect of hyperthermia on vascular functions of normal tissues and experimental tumors: Brief communication. *J Natl Cancer Inst* 60:711–713, 1978.

Electrical Stimulation in Sports Medicine

Kenneth M. Singer, M.D.

The use of electrical modalities in the treatment of athletes has become quite common in the last two decades [52]. At present a wide variety of devices that can apply electricity in differing ways are available for clinical use. However, the scientific basis that forms the rationale for the use of electricity has, unfortunately, lagged behind its clinical application, and therefore the clinician is easily confused when attempting to decide when to use electrical stimulation and how to choose among the commercially available products. This section will attempt to describe the current knowledge of the application of electrical principles in sports medicine situations.

In essence, there are three commonly recognized applications of electrotherapy for the treatment of soft tissue injuries. The first is the attempt to modify pain by means of transcutaneous stimulation of sensory nerves. This is accomplished by using a transcutaneous electric nerve stimulation (TENS) unit. The second application is the use of electrotherapy to strengthen muscles, and electrical stimulation (ES) used for this purpose following surgery has become a common rehabilitation technique. It has been recently suggested that normal muscles may be strengthened by ES, thus raising the question of whether ES could be used in conditioning programs. The third application of electrotherapeutics is to enhance healing, and at present there is very little objective evidence in the literature about its efficacy in such clinical situations.

This section will review the current literature on these uses of electrical stimulation and offer some suggestions for its use based upon this review and on personal observation. However, a disclaimer is necessary. This discussion is meant to be generic in nature. For purposes of illustration certain commercially available devices may be used as examples, but nothing in this presentation is intended to either endorse or detract from any specific product.

HISTORY

The history of the use of electricity for medical reasons is long and interesting [39, 56]. Scribonius Largus, a Roman physician, described the use of electric eels, rays, and torpedo fish in the treatment of gout and headaches. The animal was placed on the painful area and the foot or hand was inserted into a container with the animal; after the animal was irritated it emitted a retaliatory jolt of electricity that relieved the patient's symptoms. Dioscorides, a Greek physician, expressed doubt about the value of the torpedo fish in curing headache but encouraged its use for the treatment of hemorrhoids!

The use of live animals was not very dependable, but the appeal of electricity as a therapy remained, and in the 1700s and 1800s numerous prototype generators were developed so that electricity could be produced at the convenience of the physician rather than at the whim of the fish. The development of the Leyden jar, which is simply a glass container lined with tinfoil on both surfaces, solved the problem of how the electricity could be stored for later use.

Even politicians and theologians became involved. Ben Franklin used an electric generator to treat paralytics and epileptics. John Wesley, the founder of Methodism, was an ardent advocate of electrotherapeutics and described its use for sciatica, kidney stones, and even angina.

Alessandro Volta and Luigi Galvani discovered in the late 1700s that electricity could cause muscles

213

to contract but feuded about the source of the electricity. Aldini, Galvani's nephew, became interested in the possibility of using electricity to restart the heart. Since this interest coincided with the French Revolution, there was an abundance of freshly decapitated bodies available for his experiments.

Throughout the 1800s and early 1900s a variety of battery-operated faradic devices were marketed and advocated for a variety of conditions. They were promoted for treating acne, abscesses, alopecia, anemia, asthma, constipation, corns, goiter, gout, headaches, hemorrhoids, lumbago, urinary incontinence, shingles, and many other ailments.

In 1965 Melzack and Wall [40] postulated their gate theory of pain control, lending scientific plausibility to the long-standing claims of symptomatic benefit derived from electricity.

In the past 25 years the use of electrical stimulation modalities in sports medicine as well as in many other areas in medicine has become much more sophisticated and very widely used, yet there is still no strong consensus among the various disciplines involved about the most realistic place for the use of this technique in clinical practice.

MECHANISM OF ACTION

The neurophysiologic basis of nerve stimulation and muscle contraction is quite complex, and the reader is referred to basic texts for more detailed information. However, a brief basic description will be useful for the purposes of this section. The therapeutic use of electrical stimulation depends upon the ability of muscle and nerve tissue to respond to externally applied electrical stimuli. The nerve fiber, be it motor or sensory, responds to externally applied electrical impulses much as it would if the stimulus had come from the spinal cord. This stimulus causes the membrane to depolarize, generating an action potential that is transmitted along the nerve.

Nerve fibers function in an "all or none" manner, that is, a stimulus, if strong enough to exceed the threshold of the nerve fiber, will generate an action potential in that fiber, and the width and amplitude of the action potential are always the same regardless of the size of the stimulus. As the action potential travels along the nerve fiber, the nerve membrane must repolarize to return to its normal state. During this refractory period, additional stimuli will not be transmitted; until repolarization is complete, no stimulus will have an effect. Thus, the action potential in an individual nerve fiber cannot

be changed by altering the size of the stimulus; either the nerve fires or it does not. Continued stimulation results in repetitive action potentials of equal amplitude.

However, there are different types and sizes of fibers. In general, the larger the diameter of the fiber, the lower the stimulus required to begin propagation of the wave, the higher the conduction velocity, the larger the electrical response, and the shorter the duration of the response. Additionally, some fibers are myelinated and some are not, conferring different electrical characteristics. Thus, larger, myelinated fibers are easier to excite and have a faster conduction velocity than either smaller fibers or unmyelinated fibers of the same size (Fig. 7C–1).

There are three main types of nerve fibers, types A, B, and C, with subgroups of each. Types A and B fibers are covered with myelin, and their conduction velocity is increased. Type A fibers are larger than type B fibers and may be either motor or sensory. Type B fibers are smaller, slower, and autonomic. Type C fibers are the smallest and transmit cutaneous pain.

Most peripheral nerves contain more than one type of fiber, and all have different thresholds. Therefore, when a nerve is stimulated externally, any combination of motor, sensory, or autonomic

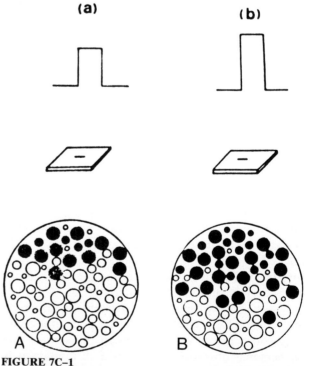

FIGURE 7C–1
Cross-section of a typical nerve showing the mixture of different fiber types.

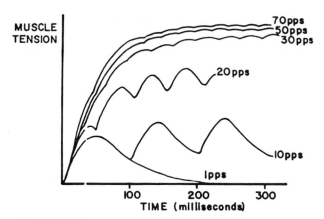

FIGURE 7C–2
Stimulation of a muscle demonstrating how increasing the frequency and strength of the stimulus can increase the strength of the muscle contraction.

fibers may be stimulated. Although each fiber functions in an all or none manner, the different fibers traversing the same nerve will be recruited at different rates, thus explaining why varying the stimulus can produce a different physiologic response.

Muscle tissue responds to electrical stimulation differently than nerve tissue. If an impulse travels down the motor nerve fiber to the end plate of an innervated muscle fiber, it induces a chemically mediated, all or none contraction in a single muscle fiber. However, the magnitude of the contraction can be altered by the timing of subsequent electrical stimuli. Unlike a nerve fiber, the contractile mechanism does not experience a refractory period to depolarization, and therefore subsequent closely spaced stimuli can cause increased strength of contraction of the individual muscle fibers (Fig. 7C–2).

As muscle stimuli increase in *frequency,* summation of contractions occurs, the final result being a tetanic, or sustained, contraction. As the repetitive stimuli accumulate, both the strength and the duration of contractions increase compared with individual twitches.

If the *intensity* of stimulation is increased, the muscle contraction increases in strength but by a different mechanism. The functional unit of a muscle is the motor unit. This consists of one motoneuron and the muscle fibers innervated by the motoneuron. In different muscles different numbers of fibers are innervated by a single neuron. For example, the gastrocnemius may have 1000 or more fibers per motoneuron, whereas the extraocular muscles contain only five to ten fibers per motoneuron. In muscles with multiple muscle fibers, stimuli of greater intensity will recruit more motor units and thus cause a stronger contractile response.

Electrical devices may be powered by common household current or by batteries. However, from a physiologic standpoint, what is important is the *output* of the electrical device, not what powers it. The power source does not matter; the electronics of the device can alter it to whatever output is desired. The electrical stimulator can therefore emit either alternating current, such as that found in common household current, or direct current, which is like that produced by a battery. Alternating current can have different wave forms, as can direct current, and these can be square, sinusoidal, spiked, intermittent, and so on. Similarly, direct current can be continuous or intermittent (Fig. 7C–3). Galvanic current is direct current.

The *wave form* is the shape of the electrical pulse, and many of its factors can vary. The rise time, the amplitude (in amperes), and the duration (in milliseconds) can all affect the physiologic response to stimulation.

Frequency is the number of cycles or pulses per second, and it can be uniform or modulated, with

FIGURE 7C–3
Different types of electrical impulses.

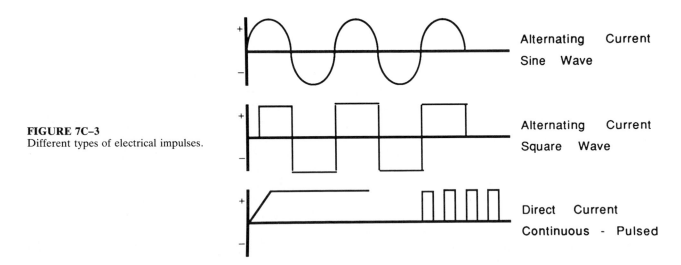

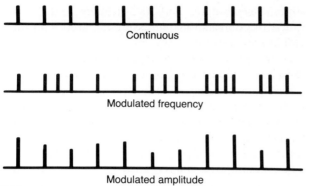

FIGURE 7C–4
The frequency of stimulation can be uniform or modulated, and the amplitude can also be modulated or varied.

varying time intervals between each pulse (Fig. 7C–4).

USE OF ELECTRICAL STIMULATION TO RELIEVE PAIN

Electrical stimulation was first used clinically in the 1960s as a therapeutic trial to decide which patients would be suitable for surgical implantation of dorsal column stimulators. However, many patients responded so well to the use of transcutaneous electric nerve stimulation (TENS) that an impetus began to develop TENS stimulators with wider applicability and portability. The small TENS units as we now know them were introduced in the early 1970s, and more than 100 different commercially available units are manufactured today [25].

Mechanism of Action of TENS

The perception of pain is not well understood; however, there are mechanisms of operation that must be understood in order to understand how TENS works. The first is the gate control theory, described and popularized by Melzack and Wall [40] in 1965. They described the existence of a gating mechanism in the dorsal horn of the spinal cord and postulated that pain perception could be modulated by altering or "closing the gate."

Pain sensation is carried to the spinal cord by small, slow, unmyelinated (type C) nerve fibers. Light touch and proprioception sensations are carried by way of fast, larger myelinated (type A) fibers. Both types of fibers pass through the substantia gelatinosa of the spinal cord, where they synapse with transmission cells, which then carry the information to the brain. These transmission

cells form junctions with both type A and type C fibers.

Transmission cells are inhibited by impulses from either cell type, and therefore either type can inhibit the transmission of messages to the brain of the other. The gate theory postulates that if the larger, more easily excited type A sensory fibers are overstimulated, that stimulus will reach the transmission junctions first and effectively flood the pathways to the brain, thereby "closing the gate" to transmission of pain by the slower type C fibers and effectively blocking reception of the painful stimuli. On the other hand, input from the type C fiber afferents inhibits the transmission cells in the substantia gelatinosa, thus "opening the gate" and allowing painful stimuli to be carried further up the neuraxis.

The system is in fact more complex than this description implies because there are also inputs descending from higher areas in the brain and spinal cord that modify the transmission of pain in the substantia gelatinosa. Thus, although the theory initially postulated by Melzack and Wall is an oversimplification, it brought to the attention of scientists and clinicians the fact that pain perception could be modulated somewhere within the neuraxis if the appropriate stimulus could be delivered and the appropriate neural substrate on which such stimuli might act could be found, and it has provided considerable insight into the mechanism of pain perception [3].

A second mechanism by which TENS can affect pain perception is chemically mediated. Stimulation of sensory nerves can also stimulate the release of endogenous morphinelike opiates, either beta-endorphins from the pituitary gland or enkephalins from the spinal cord. Since opiate receptor sites are located at numerous locations throughout the central nervous system, stimuli that cause the release of either of these substances will decrease pain perception [3].

A third pain mechanism in the midbrain modulates central transmission of pain. Under normal circumstances, sensory input from the periphery to the part of the reticular formation that acts as the central biasing mechanism diminishes pain perception. In situations when the normal sensory input is lost, the central biasing mechanism is less inhibitory, allowing augmentation of pain transmission to higher centers. TENS may substitute normal afferent input and trigger the inhibitory influence of the central biasing mechanism [61].

Conventional TENS consists of the administration of high-frequency (approximately 100 Hz) stimuli of low intensity and theoretically acts by means of the gate control theory. It increases large fiber afferent transmission and inhibits the small fiber (painful) stimuli. Low-frequency TENS (approximately 5 to 25 Hz), often referred to as acupuncturelike TENS, has been shown to be reversible through administration of an opiate antagonist, naloxone, suggesting that the effect is probably due to the release or activation of endogenous pain-relieving opiates [11]. The gate control theory does not explain why pain relief may persist after TENS stimulation has stopped, but the activation of endogenous endorphins or enkephalins explains the more prolonged analgesic effect [16].

TENS units (Fig. 7C–5) may take advantage of these concepts in their electronic composition. Use of a phasic stimulation pattern will apparently select the A fibers as opposed to the C fibers, which seem to respond to continuous stimulation patterns. Small, unmyelinated C fibers are unexcitable at pulse widths of less than 200 μsec, whereas the A fibers can be depolarized with stimuli as short as 2 μsec. This allows selective activation of cutaneous afferents, resulting in gate closure. Also, stimulation of sensory nerves with frequencies above the 50-Hz range will increase the pain threshold by depolarizing type A fibers without concomitant excitation of C fibers but does not release endorphins, implicating the gate control theory, whereas stimulation with lower frequencies will stimulate endorphin release. Therefore, from a theoretical point of view, a TENS unit might use one or the other stimulation pattern, depending on the desired effect. Or it may be that use of both frequencies might give the best results.

FIGURE 7C–5
A standard portable TENS unit.

One should be cognizant of the difference between high-frequency and low-frequency TENS.

The many stimulators on the market vary with respect to their stimulation parameters such as amplitude, pulse width, wave form, and frequency. The amplitude may vary from 30 to 120 milliamperes and the pulse width from 9 to 350 μsec. In general, stimulators that produce a stimulus of high amplitude have a lower pulse width and vice versa. Much attention has been given to the wave form, but very little is known about its effect.

Although considerable controversy exists about the correlation of stimulation parameters with pain relief, there is no evidence that any specific technique or stimulator is consistently better than any other [25]. In fact, there may be no correlation between pain-relieving effects and specific pulse widths, rates, or stimulus intensities [51, 61, 66].

Effectiveness of TENS

Confusion about the effectiveness of TENS exists for several reasons. Patients have varying perspectives. Some come with a preestablished belief that it will not work, whereas others view it optimistically, much like any other form of treatment. TENS units are easily available and are often obtained by patients "just to try" from nonprescription sources such as friends or family, and when obtained in this manner they may not be used correctly. Moreover, pain is so subjective that it is difficult to determine the effectiveness of any pain-relieving modality or medication. Pain is difficult to quantify, pain-grading techniques are crude at best, and therefore study design problems abound.

In addition, subjective testimony about the effectiveness of any treatment or medication is difficult to evaluate because of the well-described placebo effect. Most studies evaluating pain relief fail to take into account the clearly beneficial placebo contributions, which can be responsible for up to 30% of pain relief. The placebo effect is maximal when the treatment is applied directly to the site of the pain and often decreases with time. Both the direct application and decrease in effect over time are characteristic of the use of TENS units, and therefore it is often difficult to determine if the placebo effect is responsible for the benefit.

There are other important variables. It is possible that the beneficial value of the relationship between the patient and the therapist may affect the results, and there is wide variability in the motivation and willingness of patients to participate. Because of all these reasons, it is difficult to design studies in clinical practice that will control the variables sufficiently to warrant a clear picture of the efficacy of TENS [45].

Objective Evaluation of TENS

There are many reports describing pain control in a variety of orthopaedic conditions, the majority of the reports being subjective testimonials. We will attempt to limit this review to the objective data available in the literature.

One of the most successful uses for TENS is in the management of postoperative pain [59]. In postoperative situations the location and description of acute pain are usually precise and allow use of a specific treatment approach. Homogeneous groups of patients and matched control groups are available, unlike the situation with other painful conditions. When a single surgical procedure is studied, reasonably objective comparative data can be obtained. In such situations, studies seem to show that TENS units may be effective in relieving pain, decreasing narcotic usage, decreasing the length of the hospital stay, and improving mobilization time after surgery. For example, there appear to be lower medication requirements and shorter hospital stays following laminectomy [25, 51] when the electrodes are placed parallel to the incision and TENS is used for 24 to 48 hours after surgery.

Arvidsson and Eriksson [2] compared the results of high-frequency TENS with placebo TENS used to modify pain after knee surgery. They found that placebo TENS had no significant effect on pain perception or the ability to contract the quadriceps muscle (as measured by integrated electromyography [EMG]), whereas high-frequency TENS decreased pain perception by 50% at rest and 11% after quadriceps contraction and increased quadriceps muscle contraction ability considerably.

Jensen and colleagues [24] reported that a TENS-treated group experienced less pain, required less narcotics, and regained strength more quickly following arthroscopic meniscectomy compared with a group of patients treated without TENS and another group treated with a placebo TENS unit. Harvie [23] studied patients who had undergone a variety of surgical procedures on the knee and noted that narcotic usage decreased by 75% to 100% and strength and range of motion returned faster when TENS was used. Smith and co-workers [58] studied two groups of patients. Following single-incision arthrotomy and meniscectomy, those treated with TENS had a shorter average hospital stay, initiated straight leg raising sooner, and began ambulation

sooner than a comparable group that did not use TENS. These researchers also noted a decrease in the amount of narcotics used. Similarly, following total knee reconstructions, the group treated with TENS had a shorter hospital stay, could perform straight leg raise sooner, and used less narcotic analgesia. Unfortunately, no placebo TENS unit was used in their control groups.

However, studies exist that show that TENS is not effective when the same parameters are evaluated.

In an experimentally produced pain setting, comparison of blood beta-endorphin levels following stimulation with conventional TENS, low-frequency TENS, and TENS without batteries failed to show any appreciable differences [46], and similarly, no differences were found in pain threshold in that study.

Those studies that do purport to show that TENS is effective do have common features. TENS is most effective in drug-naive patients and in those who had not used narcotics preoperatively for more than 2 weeks in the 6 months before surgery. Poor pain relief was invariably found in drug-experienced patients, perhaps suggesting a cross-tolerance between narcotics and TENS [58, 59]. In a pain questionnaire study performed in an attempt to predict which patients with chronic pain would be more likely to respond favorably to TENS treatment, it was found that older, retired patients who had had pain for less than 1 year, had undergone limited or no surgery, and used non-narcotic analgesics were more likely to experience pain relief [26]. Johansson suggested that patients with neurogenic pain responded more favorably that did patients with somatogenic or psychogenic pain. Patients with pain in the extremities did better that those with axial pain. In other studies, age, sex, and intensity of pain did not correspond with the response to treatment [27].

It thus appears that many patients will obtain short-term relief, which will decrease to minimal relief as time progresses. TENS is usually effective in patients with acute injury, postoperative pain, or exacerbation of chronic musculoskeletal syndromes and does not help much after the acute stage is over.

There are numerous articles that claim to show subjective control of pain in a variety of chronic conditions, but objective data again are lacking, and most of these reports are deficient in that under such circumstances pain-grading techniques are extremely difficult to design and interpret. TENS is not very effective in providing relief in many chronic pain syndromes [15, 37]. In addition, in chronic pain conditions there are many more variables of prognostic significance, including duration of pain, correlation of pain with amount of soft tissue damage, other pain-relieving modalities previously tried, and a variety of psychologic and physiologic factors that make any comparison studies extremely difficult to design and evaluate [10].

Indications for Use of TENS

TENS is commonly used in a variety of acute and chronic pain conditions; however, indications for the use of TENS in all these situations cannot be supported by the current literature. TENS may be beneficial in any acute and some chronic pain situations in some individuals. The difficulty for the clinician arises in deciding when to use it. This decision must be made on an individual basis, taking into account the nature of the patient, the type of pain-causing condition, and other factors such as compliance, cost, and availability. Its routine use in any specific situation is not recommended; however, when difficulties arise it may be helpful, specifically, in acute pain conditions caused by either trauma or surgery, in individuals who have not been exposed to narcotics, and in patients with somatic as opposed to psychogenic pain.

Some specific uses of TENS in sports medicine have been advocated aside from those mentioned above. In certain situations, such as severe patellofemoral pain syndromes or adhesive capsulitis, the pain may prevent the patient from accomplishing a potentially beneficial exercise program. TENS may diminish the pain sufficiently to allow participation in physical therapy. TENS has been advocated in the treatment of acute shoulder dislocations, acromioclavicular separations, lateral epicondylitis, hip

pointers, ankle sprains, and Osgood-Schlatter's disease, among others [25]. In the author's experience, low-dose analgesics, when effective and not contraindicated, are equally effective in many of these situations and much less costly.

Practical Aspects of Use of TENS

For TENS to be effective, electrode placement must be appropriate. Effective stimulation can occur when the electrodes are placed over the peripheral nerve, the dermatome, or the trigger points but is usually best when situated over the area of maximal pain. Much controversy exists about electrode placement sites, and probably there are no exact best sites. Sites should be selected on an empirical basis, sometimes locally and sometimes over an appropriate dermatome or trigger area. An experienced physical therapist will be very effective in selecting the correct placement in the individual patient (Fig. 7C–6) [61].

There are different types of electrodes. Most are held against the skin with straps, using saline or gels to improve conduction. Electrodes used with portable units are usually held in place with adhesives and changed every 2 to 4 days. Although electrode placement is very important, and even electrode size might be, the actual type of electrode used does not seem to affect the outcome.

Virtually all TENS units allow the operator to exert some control over the stimulation parameters, particularly frequency and amplitude. The usually recommended method is to turn up the amplitude until there is a barely perceptible awareness of stimulation. In some instances, increasing the current or frequency until tingling and muscle fascicu-

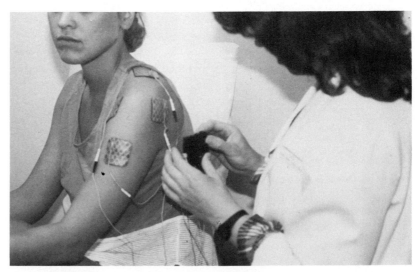

FIGURE 7C–6
Use of a conventional portable TENS unit in a clinical setting.

lations occur and then decreasing it until the discomfort is gone seems effective. As with electrode placement, variation of the stimulation parameters is usually done on an empirical basis, and no relationship has been demonstrated between pain relief and any given stimulation parameter [61].

If gate control (high-frequency) stimulation is desired, settings of between 50 and 100 Hz should be used [25, 61]. If endorphin release is desired, one selects intensities that are nonpainful in the frequency range of 1.5 to 10 Hz. Low-frequency stimulation treatments may of necessity be shorter because higher intensities are used, whereas high-frequency stimulation may achieve better results if it is used for longer periods.

Conventional TENS comprises a continuous stream of pulses of current at a constaant rate. Burst TENS comprises a burst of current pulses with the bursts repeating at a constant rate. In one study in which patients with chronic pain in a spine rehabilitation program were given a choice between burst TENS and conventional TENS, all chose burst TENS [5]. However, most objective studies comparing the various forms of TENS—low-frequency, high-frequency, burst frequency, hyperstimulation, TENS with galvanic stimulator—do not lead to the conclusion that any one form is better than any other. Clearly, further study is needed in this area [26]. For more details about the actual use of TENS, the reader is referred to the monograph by Kahn [29].

Contraindications and Complications in the Use of TENS

There are very few contraindications to the use of TENS. Stimulation should not occur in the area of the carotid sinus because it may trigger a vasovagal reflex. TENS units should not be used with demand types of cardiac pacemakers, although they are safe with fixed rate pacemakers. Use of TENS to treat undiagnosed pain in athletes should be discouraged because symptoms may be masked and further injury may accrue. No electrical device should be used during at least the early stages of pregnancy, and we would not recommend their use in patients with epilepsy or mental retardation.

Complications from TENS use are extremely rare. There are, however, certain situations in which the use of TENS, although not contraindicated, is known to be associated with a poor response and should probably not be used. These include patients with pain of psychogenic origin and drug-experienced individuals with pain from any source. Patients who have wound drainage do not seem to respond well. Improper electrode placement is also associated with a poor response.

Neuroprobe

The neuroprobe is a type of electrical stimulation unit that can be used to search out tender areas or trigger points and may be effective in reducing pain. It may also be used for locating ideal sites for electrode placement with the conventional TENS unit (Fig. 7C–7).

Interferential

Another type of stimulation establishes a crossing pattern of electrical fields such that electrical signals from two sets of electrodes with the same wave

FIGURE 7C–7
Use of the neuroprobe.

form arrive at a point from two different directions, thereby achieving accumulated intensity. This area is called the interference pattern, and it can often be used as another way of modifying pain.

Summary of TENS

In summary, TENS units can be effective in controlling pain in some individuals and in some conditions. The response is variable and unpredictable; there are few complications and few contraindications. The results are usually not long-lasting, and it is often difficult to determine how much of the result is related to the placebo effect.

TENS is most likely to be effective in patients with acute pain or postoperative pain who have not had extensive narcotic or drug exposure. It is important for someone who is familiar with the device and its use to instruct the patient how most effectively to apply the electrodes and to adjust the settings properly. In some situations TENS can be effective in moderating pain sufficiently to allow better exercise programs. Whether or not it is cost effective must be determined on an individual basis.

Much of the information provided by TENS manufacturers used in marketing, as well as that in the literature, makes unclear the distinction between theory or hypothesis and fact. Many questions concerning the efficacy of TENS for pain control are difficult to answer. Nonetheless, TENS has been applied almost indiscriminately to heterogeneous populations of patients with pain. Many of these pain conditions are poorly defined, and the results may vary depending on whether TENS is used early in the course of treatment or is used as a last resort. In addition, changes have occurred in the types of units available, as conventional TENS has been replaced by burst, modulated, brief-intense, or acupuncturelike TENS, yet there has been insufficient research to determine whether the newer forms are any more effective than conventional TENS. It often appears that the amount of money spent in marketing is greater than that spent in research.

Many questions exist about the basic mechanisms of electrical stimulation-induced effects on pain perception, and many fundamental questions about how TENS alters the pain experience remain unanswered. It is important for clinicians to assume greater responsibility and become more critical in their evaluation of the literature on TENS [45].

For additional information, the reader is referred to a 658-article bibliography on pain and all aspects of its mechanism and control [48].

ELECTRICAL STIMULATION OF MUSCLE

The question of improvement in strength with electrical stimulation arises in two completely separate circumstances: (1) the issue of whether electrical stimulation (ES) can enhance the strength of a normal muscle, and if so how much compared with other strengthening modalities, and (2) the possible benefits of ES used during rehabilitation to strengthen muscles weakened by injury or surgery or to decrease the muscle atrophy that results from injury or surgery. To improve strength, maximal or nearly maximal muscle contractions must occur to the point of fatigue. Therefore, we must seek information about whether maximum contractile forces can be produced by ES as well as or better than those produced by voluntary effort.

Strengthening of Normal Muscle

The use of electrical stimulation to enhance the strength of normal muscle is a relatively new concept. It originates from claims made by a scientist named Kots from the Soviet Union [31], who reported at a 1977 symposium at Concordia University that use of a specific stimulation program was able to increase greatly the strength of normal muscles. His program consisted of using a 2500-Hz frequency current in 10 msec bursts separated by 10 msec of rest. He used a duty cycle consisting of 10 seconds of stimulation applied to the muscle followed by 50 seconds of rest (Fig. 7C–8). He reported that after only 3 weeks of this program there were major gains in strength and endurance, a decrease in subcutaneous fat, and increased limb girth, much greater than could have been produced by an exercise program. Apparently this information, though widely quoted, has never been published, and no data have been published that I am aware of that can substantiate these results [33, 47, 63].

Many studies describe the effects of ES on muscle strengthening, most of them involving the quadriceps. A more detailed description of these data can be found in the excellent review article by Selkowitz [55].

Certain well-known principles of muscle strengthening apply to any effects that may be achieved by electrical stimulation. Specificity of training exists, so that when a muscle is trained isometrically it responds isometrically, and there is not very much crossover between an isometrically strengthened muscle and sustained isokinetic or isotonic gains. Although some work has been done with isotonic

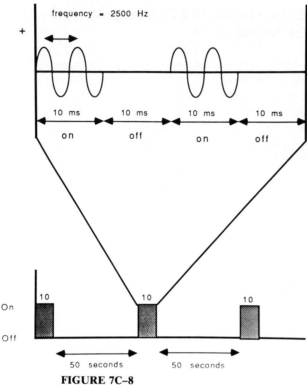

FIGURE 7C–8
Stimulation pattern advocated by Kots.

and isokinetic strengthening by means of electrical stimulation, most of the time the stimulator is used isometrically.

Selkowitz [54] found that in normal volunteers isometric training with ES alone improved isometric strength of the quadriceps significantly compared with controls subjected to no training. The relative increase in isometric strength was determined by the ability of the subject to tolerate a longer and more forceful maximum voluntary isometric contraction (MVIC). Although others [4, 7, 60] have achieved similar results qualitatively, large quantitative differences exist between various studies using different methods of stimulation. Studies attempting to reproduce Kots' techniques with the commercially available Electro-Stim 180 device do not show appreciably better results than those achieved with other techniques, and in Boutelle's [4] study, the standard deviations are so large that they allow overlap between the pretest and post-test results.

The evidence is less convincing that isometric training using ES can increase isokinetic strength, with some studies showing that some strength gains can be made [4, 53], and others showing that the effects are not statistically valid [4, 7, 35]. No strength gains have been found at isokinetic speeds of greater than 120 degrees/second [55].

Halbach and Strauss reported an increased iso-

metric quadriceps femoris muscle strength of 22% with ES alone, but strength gains recorded from subjects exercising isokinetically were even more marked, with a 42% increase in strength [19]. Similarly, Kubiak [35] showed that isometric exercises increased strength more than ES.

Strength training depends on progressively increasing the strength of muscle contractions, and therefore, whatever method can produce the strongest contractions should be able to effect the greatest strength increases. In normal individuals, maximal voluntary muscle contractions appear to be able to generate more knee extension torque than can electrical stimulation of the quadriceps muscle [64].

When groups training in isometric exercise programs are compared with groups using ES in an effort to strengthen the quadriceps, there appear to be no appreciable differences in strength gains [7, 36, 38]. In studies in which training groups using both isometric exercises and electrical stimulation are compared with groups using only isometric exercise, again no statistically valid differences have been found [6, 7, 22, 34, 35]. It is interesting to note, however, that the impression of the subjects being tested was quite different. In a post-test questionnaire, 60% of those undergoing electrical stimulation along with isometric exercises felt that ES had increased the muscle response compared with maximum voluntary contractions, and in those who underwent ES in addition to maximum contractions, 87% thought that ES increased the strength of the contractions despite test data indicating that it did not [32]. This subjective response explains why athletes using such devices are often willing to give testimonial support for their efficacy and also why such claims, in the absence of more objective data, are not valid. Similarly, ES combined with an isokinetic exercise program produced no greater strength gains than isokinetic exercises alone [44].

Direct current has also been used in an attempt to strengthen muscles. However, when comparing high-voltage galvanic (HVG) stimulation in normal subjects with an exercise group and a control group, no increase in strength was found in either the control or the HVG-stimulated group. There was, however, a significant increase in strength in the isometric exercise group, indicating that, like other forms of stimulation, HVG stimulation is not as effective as isometric exercise in increasing strength [41].

Theoretically, if ES could produce contractions greater than those produced by maximum voluntary contractions, strength gains could be achieved. The rationale behind the strength enhancement is that electrical stimulation will either increase the maximum contractile force in each muscle fiber or that it will recruit more fibers to contract with a given stimulus. At present it has not been demonstrated that ES units can accomplish that, nor is it known whether the athlete could tolerate stronger contractions than those produced by a MVIC [55].

Strengthening Muscle After Injury or Surgery

Under some circumstances a strong voluntary muscle contraction is not possible. Muscles paralyzed by upper motorneuron lesions can be strengthened using electrical stimulation [49]. In athletes, muscles may be adversely affected by injury, surgery of an adjacent joint (such as the quadriceps with knee surgery), or immobilization following injury or surgery [12]. If individuals are rendered incapable of sustaining a maximal voluntary contraction because of such conditions, it is theoretically possible that ES may be beneficial in restoring normal function.

However, the literature gives conflicting opinions on this issue. Gould and co-workers [17], in a study of patients undergoing open meniscectomy, found that neither ES nor exercise prevented quadriceps strength loss after immobilization, but that a group undergoing electrical stimulation lost less muscle volume and less strength, were able to discontinue crutches earlier, and used less pain medication than a similar group treated with a conventional isometric training program. Eriksson and Haggmark [13] found that electrical stimulation combined with exercise was more effective in preventing muscle atrophy than exercise alone in patients following surgery.

Morrissy and co-workers [42] studied the loss of strength in patients undergoing anterior cruciate ligament (ACL) reconstruction followed by 6 weeks of cast immobilization by comparing two groups, one utilizing ES on the immobilized quadriceps and the other receiving no ES. Neither group was engaged in an exercise program during that period. The authors showed that muscle stimulation can decrease the significant strength loss that results from immobilization at 6 weeks, but when examined 12 weeks after surgery or 6 weeks after immobilization was discontinued, the strength losses were identical for the stimulated and nonstimulated group. They also showed, as have others, that stimulation does not increase the girth of the leg and that it seemed to have no effect on the pain that occurred during maximal quadriceps contraction.

Sisk and his group [57] at the Campbell Clinic were unable to demonstrate any beneficial effect of ES following ACL reconstruction and cast immobilization. They compared two groups, one of which received ES and an isometric exercise program, and the other the exercise program without ES. Isometric knee strength was measured using maximum isometric contractions at the seventh, eighth, and ninth postoperative weeks, and no significant differences were found between their two groups.

Wigerstad-Lossing and colleagues [65] studied two groups of patients similar to those of Sisk's group. They compared two groups of patients after undergoing ACL surgery, one receiving ES and voluntary exercises and the other undergoing voluntary exercises only. They noted significant differences 6 weeks after surgery, with the group receiving ES exhibiting less strength loss, less reduction in cross-sectional area of the quadriceps as measured by computed tomography, and less reduction in oxidative and glycolytic muscle enzyme activity during the immobilization period. It would have been interesting to see whether the differences still existed at a later time.

DeLitto and colleagues [9] compared two groups of patients early after ACL surgery. One group was placed on a program of simultaneous contractions of the quadriceps and hamstrings, and the other group underwent simultaneous ES of the quadriceps and hamstrings 5 days a week for 3 weeks during the first 6 weeks after ACL reconstruction. Those undergoing ES finished the program with higher percentages of both extension and flexion torque and significantly greater isometric strength gains, which were statistically significant. They concluded that during the early phase of postoperative rehabilitation ES seems to be effective. This conclusion differs from that reported in other studies [7, 36, 54], and again, it would have been interesting to see whether the strength gains persisted.

Arvidsson and associates [1], in a study following ACL surgery, found differences in females but not in males in measuring quadriceps wasting by computed tomography. There were no significant differences between the two groups in the activity of oxidative or glycolytic enzymes on percutaneous muscle biopsies.

Six weeks of immobilization following ligament injury or surgery causes changes in the muscle fibers of the vastus lateralis and decreased oxidative capacity as indicated by succinate dehydrogenase activity [18, 20, 21]. Use of ES with or without exercise programs has been found by some to be effective in retarding the biochemical changes accompanying immobilization [12, 13, 20, 21, 62], whereas others found no differences [20, 21].

Johnson and colleagues [28] observed a strength increase of 25% in patients with mild chondromalacia patellae compared with an increase of 200% in patients with severe cases of chondromalacia. This suggests that the greater the atrophy, the more effective the electrical stimulation.

It may be that the benefits are achieved by reeducating the muscle, and that ES functions by facilitating early voluntary muscle contractions or by achieving a certain amount of pain relief. In a questionnaire study Currier and Mann [8] showed that healthy subjects training with ES experienced less muscle soreness than those training isometrically despite similar gains in strength.

Practical Aspects of Use of Electrical Stimulation

Electrical stimulation is performed by stimulating the motor nerve to a muscle, so it is very much like TENS. If a TENS unit were used and the electrodes were placed over the motor nerve, a muscle-stimulating effect could be obtained if the stimulating current were increased. Although there are variations in the stimulation parameters used, such as frequency, amplitude, and wave form, it appears that the body has difficulty in distinguishing them qualitatively; electricity is electricity, whether it originates from a TENS unit, a muscle stimulating unit, or a spark plug. If a sensory nerve is stimulated, pain modification occurs. If a motor nerve is stimulated, contraction of the muscle occurs. This is not to say that stimulation parameters are not important—they are. We just do not know very much about them. No standardization exists in the choice of stimulation parameters at present.

The application of muscle stimulation is straightforward (Fig. 7C–9). The electrodes are placed so that stimulation can reach the motor nerve to the muscle involved. Because the electrical impulse will travel through tissue, the exact site of placement is not critical. The frequency and amplitude are increased until a tetanic contraction can be obtained without significant discomfort. The length of stimulation is controversial, as is the frequency, but probably 7- to 10-second stimulations are needed many times throughout the day. It was formerly thought that the more stimulation was performed, the more the gain in strength, but more recent evidence, including the paper by Sisk and colleagues [57] indicates that strength gains may not accrue even with long periods of stimulation. It is clear that because exercise is probably at least as effective as muscle stimulation, if not more so, the individual

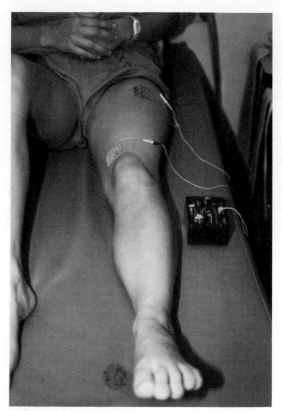

FIGURE 7C–9
Use of electrical stimulation to improve quadriceps strength.

should be engaged in an active exercise program while undergoing stimulation.

Contraindications

Contraindications to the use of ES for strengthening muscle are essentially the same as those described earlier for TENS.

Summary of Electrical Stimulation to Strengthen Muscle

Despite numerous attempts, no published study in the western countries has been able to reproduce the Soviet data as reported by Kots. At the present time, there seems to be no evidence indicating that strength gains in normal muscle can be enhanced by ES above the gains that can be achieved with voluntary contractions, whether ES is used alone or in conjunction with traditional maximum voluntary exercise programs. It must be remembered that this is such a highly politicized issue that all of the data and information known may not be available for review.

In the rehabilitation setting, there is evidence, although it is not conclusive, that suggests that ES may retard muscle strength loss or increase strength in muscles that are weakened by injury, surgery, or disuse, and when used in conjunction with voluntary exercise, it may be more effective than exercise programs alone in increasing muscle strength. This presupposes a muscle that is partially denervated because of injury, surgery, or disuse from immobilization. The parameters appear to be the same in partially denervated muscles and those subjected to disuse. It is generally agreed that ES is effective in maintaining and improving muscle strength in weakened musculature, but how much more effective it is in relation to voluntary exercise cannot be stated conclusively. It may be that its benefits are a result of its reeducative role in facilitating early voluntary muscle contractions and perhaps also the result of pain relief accompanying the nerve stimulation.

It appears that the effect is not long lasting and may well be no better than gains that could be produced by a good exercise program. Some data suggest that ES may be suited to the initial management of motor reeducation because active exercise is the eventual objective in rehabilitation programs.

Research in the area of strengthening weakened muscles is urgently needed. The optimal stimulus parameters, which include pulse shape, charge, duration, frequency, and intensity, that are required to produce the desired motor response need to be established. At present there are no systematic studies that relate all of these stimulus parameters to force production in different muscles. For example, at the present time no consistent differences have been demonstrated between the ability of monophasic or pulsating direct current and biphasic or alternating outputs to induce muscle contractions. Each type of current possibly has its advantages and disadvantages, but these have yet to be demonstrated.

ELECTRICAL STIMULATION TO PROMOTE HEALING

Recently, several devices have been advocated for use in acute injury situations to speed soft tissue healing. High-voltage galvanic stimulation has long been advocated for this, and there is sporadic information in the literature to support, in selected instances, the claims that it will increase local circulation, decrease edema, and facilitate the healing of skin ulcers [30]. In tissue culture studies, direct current electrical stimulation was shown to enhance tenoblastic repair [43], but there are no in vivo studies to confirm this. Good studies are difficult to perform, and it is difficult to substantiate these claims.

More recently, other forms of stimulation, sometimes known as microelectrical nerve stimulation (MENS), have purported to enhance healing of musculotendinous or ligamentous injuries. An early device was the Electroacuscope, and a more recent one is the Myomatic. We do not have good data to support their efficacy at this time, but there is much testimonial and promotional support in the lay press.

SUMMARY

As with all modalities in sports medicine, electrical stimulation does not offer any magical solutions. Electrical stimulation to control pain certainly has some applications. Its use for speeding rehabilitation by augmenting muscle strength immediately following injury or surgery is perhaps beneficial but is less clear based upon the available literature. In my opinion, there is not sufficient evidence in the literature to warrant its use in routine postoperative care.

This is another area in medicine in which randomized prospective evaluation of these forms of treatment would have been of great value before they became so widely established in therapeutic practice. There is so much testimonial and anecdotal support for the use of these modalities that one must be vigilant to separate the science from the marketing strategy. Research in all areas of electrical stimulation is badly needed, and it is to be hoped that answers will be forthcoming to delineate its role more clearly.

References

1. Arvidsson, I., Arvidsson, H., Eriksson, E., and Jansson, E. Prevention of quadriceps wasting after immobilization: An evaluation of the effect of electrical stimulation. *Orthopedics* 9:1519–1528, 1986.
2. Arvidsson, I., and Eriksson, E. Postoperative TENS pain relief after knee surgery: Objective evaluation. *Orthopedics* 9:1346–1351, 1986.
3. Bishop, B. Pain: Its physiology and rationale for management. *Phys Ther* 60:13–37, 1980.
4. Boutelle, D., Smith, B., and Malone, T. A strength study utilizing the Electro-Stim 180. *J Orthop Sports Phys Ther* 7:50–53, 1985.
5. Brill, M. M., and Whiffen, J. R. Application of 24-hour burst TENS in a back school. *Phys Ther* 65:1355–1357, 1985.
6. Currier, D. P., Lehman, D., and Lightfoot, P. Electrical stimulation in exercise of the quadriceps femoris muscle. *Phys Ther* 59:1508–1512, 1979.
7. Currier, D. P., and Mann, R. Muscular strength development by electrical stimulation in healthy individuals. *Phys Ther* 63:915–921, 1983.

8. Currier, D. P., and Mann, R. Pain complaint: Comparison of electrical stimulation with conventional isometric exercise. *J Orthop Sports Phys Ther* 5:318–323, 1984.

9. DeLitto, A., Rose, S. J., McKowen, J. M., Lehman, R. C., Thomas, J. A., and Shively, R. A. Electrical stimulation versus voluntary exercise in strengthening thigh musculature after anterior cruciate ligament surgery. *Phys Ther* 68:660–663, 1988.

10. Deyo, R. A., Bass, J. E., Walsh, N. E., Schoenfeld, L. S., and Ramamurthy, S. Prognostic variability among chronic pain patients: Implications for study design, interpretation, and reporting. *Arch Phys Med Rehab* 69:174–178, 1988.

11. Ericksson, M. B. E., Sjolund, B. H., and Nielzen, S. Long term results of peripheral conditioning stimulation as an analgesic measure in chronic pain. *Pain* 6:335–347, 1979.

12. Eriksson, E. Rehabilitation of muscle function after sport injury—major problem in sports medicine. *Int J Sports Med* 2:1–6, 1981.

13. Eriksson, E., and Haggmark, T. Comparison of isometric muscle training and electrical stimulation supplementing isometric muscle training in the recovery after major knee ligament surgery: A preliminary report. *Am J Sports Med* 7:169–171, 1979.

14. Fahey, T. D., Harvey, M., Schroeder, R. V., and Ferguson, F. Influence of sex differences and knee joint position on electrical stimulation-modulated strength increases. *Med Sci Sports Exerc* 17:144–147, 1985.

15. Fried, T., Johnson, R., and McCracken, W. Transcutaneous electrical nerve stimulation: Its role in the control of chronic pain. *Arch Phys Med Rehabil* 65:228–231, 1984.

16. Gersh, M. R., and Wolf, S. L. Applications of transcutaneous electrical nerve stimulation in the management of patients with pain. State-of-the-art update. *Phys Ther* 65:314–322, 1985.

17. Gould, N., Donnermeyer, D., Gammon, G. G., Pope, M., and Ashikaga, T. Transcutaneous muscle stimulation to retard disuse atrophy after open meniscectomy. *Clin Orthop* 178:190–195, 1983.

18. Haggmark, T., Jansson, E., and Eriksson, E. Fiber type area and metabolic potential of the thigh muscles in man after knee surgery and immobilization. *Int J Sports Med* 2:12–17, 1981.

19. Halbach, J. W., and Straus, D. Comparison of electro-myo stimulation to isokinetic training in increasing power of the knee extensor mechanism. *J Orthop Sports Phys Ther* 2:20–24, 1980.

20. Halkjaer-Kristensen, J., and Ingemann-Hansen, T. Wasting of the human quadriceps muscle after knee ligament injuries: Muscle fiber morphology. *Scand J Rehabil Med Suppl* 13:12–20, 1985.

21. Halkjaer-Kristensen, J., and Ingemann-Hansen, T. Wasting of the human quadriceps muscle after knee ligament injuries: Dynamic and static muscle function. *Scand J Rehabil Med Suppl* 13:29–37, 1985.

22. Hartsell, H. D. Electrical muscle stimulation and isometric exercise effects on selected quadriceps parameters. *J Orthop Sports Phys Ther* 8:203–209, 1986.

23. Harvie, K. W. A major advance in the control of postoperative knee pain. *Orthopedics* 2:129–131, 1979.

24. Jensen, J. E., Conn, R. R., Hazelrigg, G., and Hewett, J. E. The use of transcutaneous neural stimulation and isokinetic testing in arthroscopic knee surgery. *Am J Sports Med* 13:27–33, 1985.

25. Jensen, J. E., Etheridge, G. L., and Hazelrigg, G. Effectiveness of transcutaneous electrical neural stimulation in the treatment of pain. Recommendations for use in the treatment of sports injuries. *Sports Med* 3:79–88, 1986.

26. Jette, D. U. Effect of Different forms of transcutaneous electrical nerve stimulation on experimental pain. *Phys Ther* 66:187–193, 1986.

27. Johansson, F., Almay, B. G. L., Von Knorring, L., and Terenius, L. Predictors for the outcome of treatment with high frequency transcutaneous electrical nerve stimulation in patients with chronic pain. *Pain* 9:55–61, 1980.

28. Johnson, D. H., Thurston, P., and Ashcroft, P. J. The Russian technique of faradism in the treatment of chondromalacia patellae. *Physiotherapy* (Canada) 29:266–268, 1977.

29. Kahn, J. *Principles and Practice of Electrotherapy.* New York, Churchill Livingstone, 1987.

30. Kloth, L. C., and Feedar, J. A. Acceleration of wound healing with high voltage, monophasic pulsed current. *Phys Ther* 68:503–508, 1988.

31. Kots, Y. M. Electrostimulation. Notes from lectures presented at Canadian-Soviet Exchange Symposium on Electrostimulation of Skeletal Muscles, Concordia University, Montreal, Canada, December, 1977. Notes from *Electrostim USA* (transl. by D. Babkin and N. Temtesenko).

32. Kramer, J., Lindsay, D., Magee, D., Mendryk, S., and Wall, T. Comparison of voluntary and electrical stimulation contraction torques. *J Orthop Sports Phys Ther* 5:324–331, 1984.

33. Kramer, J. F., and Mendryk, S. W. Electrical stimulation as a strength improvement technique: A review. *J Orthop Sports Phys Ther* 4:91–98, 1982.

34. Kramer, J. F., and Semple, J. E. Comparison of selected strengthening techniques for normal quadriceps. *Physiotherapy* (Canada) 35:300–304, 1983.

35. Kubiak, R. J., Whitman, K. M., and Johnston, R. M. Changes in quadriceps femoris muscle strength using isometric exercise versus electrical stimulation. *J Orthop Sports Phys Ther* 8:537–541, 1987.

36. Laughman, R. K., Youdas, J. W., Garrett, T. R., and Chao, E. Y. S. Strength changes in the normal quadriceps femoris muscle as a result of electrical stimulation. *Phys Ther* 63:494–499, 1983.

37. Lehmann, T. R., Russell, D. D. Efficacy of electroacupuncture and TENS in the rehabilitation of chronic low back pain patients. *Pain* 26:277–290, 1986.

38. McMiken, D. F., Todd-Smith, M., and Thompson, C. Strength of human quadriceps femoris muscles by cutaneous electrical stimulation. *Scand J Rehabil Med* 15:25–28, 1983.

39. McNeal, D. R. 2000 years of electrical stimulation. *In* Hamercht, F. T., and Reswick, J. B. (Eds.). *Functional Electrical Stimulation—Application Neuroprosthesis,* New York, Marcel Dekker, 1977.

40. Melzack, R., and Wall, P. D. Pain mechanisms, a new theory. *Science* 150:971, 1965.

41. Mohr, T., Carlson, B., Sulentic, C., and Landry, R. Comparison of isometric exercise and high volt galvanic stimulation on quadriceps femoris muscle strength. *Phys Ther* 65:606–609, 1985.

42. Morrissy, M. C., Brewster, C. E., Shields, C. L., and Brown, M. The effects of electrical stimulation on the quadriceps during postoperative knee immobilization. *Am J Sports Med* 13:40–45, 1985.

43. Nessler, J. P., and Mass, D. P. Direct-current electrical stimulation of tendon healing in vitro. *Clin Orthop Rel Res* 217:303–312, 1987.

44. Nobbs, L. A., and Rhodes, E. C. The effect of electrical stimulation and isokinetic exercises on muscular power of the quadriceps femoris. *J Orthop Sports Phys Ther* 8:260–268, 1986.

45. Nolan, M. F. Selected problems in the use of transcutaneous electrical nerve stimulation for pain control—an appraisal with proposed solutions. A special communication. *Phys Ther* 68:1694–1698, 1988.

46. O'Brien, W. J., Rutan, F. M., Sanborn, C., and Omer, G. E. Effect of transcutaneous electrical nerve stimulation on human blood beta-endorphin levels. *Phys Ther* 64:1367–1374, 1984.

47. Owens, J., and Malone, T. Treatment parameters of high frequency electrical stimulation as established on the Electro-Stim 180. *J Orthop Sports Phys Ther* 4:162–168, 1983.

48. *Phys Ther* (Bibliography) 65:322–336, 1985.

49. Ragnarsson, K. T. Physiologic effects of functional electrical stimulation-induced exercises in spinal cord injured individuals. *Clin Orthop Rel Res* 223:53–63, 1988.

50. Reynolds, A. C., Abram, S. E., Anderson, R. A., Vasudevan, S. V., and Lynch, M. T. Chronic pain therapy with

TENS: Predictive value of questionnaires. *Arch Phys Med Rehabil* 64:311–313, 1983.

51. Richardson, R. R., and Siquiera, E. B. Transcutaneous electrical neurostimulation in postlaminectomy pain. *Spine* 5:361–365, 1980.

52. Robinson, A. J., and Snyder-Mackler, L. Clinical application of electrotherapeutic modalities. *Phys Ther* 8:1235–1238, 1988.

53. Romero, J. A., Sanford, T. L., Schroeder, R. V., and Fahey, T. D. The effects of electrical stimulation of normal quadriceps on strength and girth. *Med Sci Sports Exerc* 14:194–197, 1982.

54. Selkowitz, D. M. Improvement in isometric strength of the quadriceps femoris muscle after training with electrical stimulation. *Phys Ther* 65:186–196, 1985.

55. Selkowitz, D. M. High frequency electrical stimulation in muscle strengthening. A review and discussion. *Am J Sports Med* 17:103–111, 1989.

56. Sheon, R. P. Transcutaneous electrical nerve stimulation: From electric eels to electrodes. *Postgrad Med* 75:71–74, 1984.

57. Sisk, T. D., Stalka, S. W., Deering, M. B., and Griffin, J. W. Effect of electrical stimulation on quadriceps strength after reconstructive surgery of the anterior cruciate ligament. *Am J Sports Med* 15:215–220, 1985.

58. Smith, M. J., Hutchins, R. C., and Hehenberger, D. Transcutaneous neural stimulation use in postoperative knee rehabilitation. *Am J Sports Med* 11:75–82, 1983.

59. Solomon, R. A., Vienstein, M. C., and Long, D. M. Reduction of postoperative pain and narcotic use by transcutaneous electrical nerve stimulation. *Surgery* 87:142–146, 1980.

60. Soo, C-L., Currier, D. P., and Threlkeld, A. J. Augmenting voluntary torque of healthy muscle by optimization of electrical stimulation. *Phys Ther* 68:333–337, 1988.

61. Soric, R., and Devlin, M. Transcutaneous electrical nerve stimulation: Practical aspects and applications. *Postgrad Med* 78(4):101–107, 1985.

62. Stanish, W., Valiant, G., Bonen, A., and Belcastro, A. N. Effects of immobilization and electrical stimulation on muscle glycogen and myofibrillar ATPase. *Can J Appl Sport Sci* 7:267–271, 1982.

63. St. Pierre, D., Taylor, A. W., Lavoie, M., Sellers, W., and Kots, Y. M. Effects of 2500 Hz sinusoidal current on fibre area and strength of the quadriceps femoris. *J Sports Med* 26:60–66, 1986.

64. Walmsley, R. P., Letts, G., and Vooys, J. A comparison of torque generated by knee extension with a maximal voluntary muscle contraction vis-a-vis electrical stimulation. *J Orthop Sports Phys Ther* 6:10–17, 1984.

65. Wigerstad-Lossing, I., Grimby, G., Jonsson, T., Morelli, B., Peterson, L., and Renström, P. Effects of electrical muscle stimulation combined with voluntary contractions after knee ligament surgery. *Med Sci Sports Exerc* 20:93–98, 1988.

66. Wolf, S. L., Gersh, M. R., and Rao, V. R. Examination of electrode placements and stimulating parameters in treating chronic pain with conventional transcutaneous electrical nerve stimulation. *Pain* 11:37–47, 1981.

Exercise

Lonnie E. Paulos, M.D.
J. David Grauer, M.D.

Physical activity is a primary stimulus for the growth and repair of musculoskeletal tissues [30]. "Therapeutic exercise" is defined as a form of therapy that utilizes physical activity to facilitate the return of normal musculoskeletal function [30]. It is used to correct an impairment following injury or surgery, to improve musculoskeletal function as in training, or to maintain a healthy level of physical fitness [68]. Exercise is a complex interaction involving the cardiovascular system to provide endurance, the metabolic and structural demands of skeletal muscle, the mobility of joints, the stability of ligaments and tendons, proprioceptive feedback, and the mental state of the individual. All of these factors must be kept in mind when prescribing exercise as therapy.

RESPONSE TO IMMOBILIZATION

Immobilization is often prescribed following injury or surgery. The effects of immobilization on skeletal muscle, tendon, ligament, joint capsule, and articular cartilage are dramatic. Mechanical and biomechanical changes occur in these tissues, producing joint stiffness and contracture [5, 123]. Decreases in water and total hexosamine content occur, collagen crosslinking increases, the underlying bone becomes osteoporotic, and the joint space is replaced by fibrofatty connective tissue [1–3, 4, 35, 123].

Tendon and Ligament

Immobilization leads to a change in the passive tension of the entire muscle-tendon unit, which may cause shortening and contracture of the structure [110, 118]. Stress-induced changes occur at the myotendinous junction [123]. Maturation and remodeling of fibroblasts following laceration of the rabbit Achilles tendon have been demonstrated to be faster in the mobilized than in the immobilized limb, leading to the appearance of larger, more mature collagen fibrils in the mobilized Achilles tendon [116]. Controlled passive motion has been shown to enhance the tensile strength of healing tendon following laceration to a greater degree than does immobilization [40, 78, 124]. Thus, immobilization may adversely affect the muscle-tendon unit, and mobilization in the presence of tendon injury and repair has been shown to enhance tendon healing and strength.

Ligament is also adversely affected by immobilization. Alterations in collagen crosslinking, synthesis, and degradation occur in response to immobilization [1, 128, 129]. These changes lead to decreased strength and stiffness of the ligament [119]. Ligament contraction and shortening also occur, contributing to joint stiffness [130, 131]. The strength of soft tissue-bone junctions declines with joint immobilization [90, 123]. Many of these processes are reversible with increased activity levels (remobilization) [70, 123, 125]. Also, ligament-bone units and bone-tendon complexes become stronger with exercise [123]. This strength increase appears to be related to the type and duration of the activity [123]. Prolonged immobilization also decreases proprioceptive feedback from the joint and surrounding tendons and ligaments [30]. This loss of feedback control may contribute to reinjury.

Skeletal Muscle

The effects of immobilization on skeletal muscle are also well known. Long-term immobilization results in a readjustment of sarcomere number, depending on the position of immobilization [49, 117]. The removal of tension results in measurable decrements of strength and produces atrophy of muscle structure as demonstrated by decreases in oxidative capacity and the cross-sectional area [18, 19, 30, 59, 103]. Bed rest of healthy subjects for 21 days has been shown to reduce muscle strength by 26% [31]. Knee immobilization by a long leg cast and crutches for 14 days in normal young adults caused a 25% loss of strength in the quadriceps, hamstrings, and plantar flexors [47]. Significant loss of strength also occurs after surgical procedures. Extensor muscle displays a 50% decrease in strength 4 weeks following open meniscectomy [48].

Although the loss of muscle strength is predictable, the quantity is variable. Furthermore, strength limitations are often underestimated by manual testing [96]. Strength loss in the upper extremity has been shown to vary between 1% and 5% per day following cast immobilization of the elbow [87]. On the other hand, the quadriceps must experience a 50% loss of strength before weakness affects gait patterns [58].

Inactivity leads to cardiopulmonary deconditioning, resulting in decreased efficiency in delivery of oxygen to skeletal muscle [96], as energy costs for comparable activities increase. This response results in decreased activity levels of the individual and a progressive loss in cardiopulmonary conditioning.

Immobilization and inactivity thus have dramatic and total body effects. Joints become stiff and contracted, bone and ligament insertions become weakened, skeletal muscle loses strength and atrophies, and proprioceptive feedback and cardiopulmonary conditioning are compromised. If injury or surgical trauma is also involved, the loss of connective tissue integrity and the inflammatory response of healing tissues will have additional detrimental effects. The goal of exercise is to prevent or reverse these changes while protecting healing tissues.

RESPONSE TO EXERCISE

Skeletal Muscle

Histology and Physiology

Systems that provide energy for muscle contraction and exercise can be divided into anaerobic and aerobic systems (Table 7D–1) [107]. The anaerobic system is a very rapid source of energy whereby muscles metabolize glucose and produce lactic acid. Its primary use is in high-intensity, short-duration

TABLE 7D–1
Anaerobic vs. Aerobic Exercise

Anaerobic	Aerobic
Muscle metabolizes glucose to form lactic acid	Muscle metabolizes glycogen to form CO_2
Oxygen not required	Oxygen required
Lactic acid buildup; fatigue is limiting factor	Oxygen availability to muscle cell is limiting factor
Used for high-intensity, short-duration exercise	Used for prolonged endurance exercise
Example: 400 meter sprint or less, shot put	Example: Marathon run

activities because fatigue is the primary limiting factor. The aerobic system requires the presence of oxygen to metabolize glycogen. The availability of oxygen is the primary limiting factor. Aerobic metabolism is utilized for prolonged endurance activities.

Different striated muscles have different histologic, physiologic, and biochemical characteristics [123]. It is unusual to identify a muscle of homogeneous fiber types, although a certain fiber type often predominates in any given muscle. Striated muscle fibers are either slow-twitch (type I) or fast-twitch (type II). Type I fibers are slow-twitch fibers that have slow contraction times, low concentrations of glycogen and glycolytic enzymes, large stores of triglycerides, and rich concentrations of mitochondria and myoglobin. They are very resistent to fatigue. Type I fibers contain slow-type myosin heavy chains, and the sarcomeres contain a wide Z band [123].

Type II fast-twitch fibers have been subdivided into four subtypes, IIA, IIB, IIC, and IIM, according to a somewhat variable content of mitochondria and myoglobin [39]. In general, type II fast-twitch fibers have a lower content of mitochondria and myoglobin and rely more on glycolic pathways for energy production. They generally contain more fast-type myosin heavy chains, have faster contraction times, and are less fatigue resistant than type I fibers. Type II fibers can generate maximal tension in one-third the time needed by type I slow-twitch fibers and generally participate in high-intensity, short-duration activities that are dependent upon anaerobic energy. Type I slow-twitch fibers have a much greater aerobic capacity and generally participate in endurance activities requiring oxidative metabolic pathways. Muscle tissue displays active and passive behavior [33]. The total tension developed in a muscle is the sum of the voluntary active tension plus the passive tension, as demonstrated in the Blix curve (Fig. 7D–1) [33].

Physiology and Biochemistry

The transport of oxygen by the respiratory and circulatory systems to the muscle mitochondria is essential for muscular function. Pulmonary ventilation is responsible for bringing oxygen to the bloodstream. Increased oxygen requirements during exercise are met by an increase in breathing rate, which does not limit physical performance in healthy individuals [107]. Oxygen must diffuse across the alveoli to reach the bloodstream. Conditioning can increase the number of alveoli available by improving lung aeration and the total surface area available for oxygen diffusion, thereby improving oxygen delivery to the bloodstream [107].

The distribution of oxygen through the bloodstream to the skeletal muscle depends upon the cardiac output. Cardiac output is the amount of blood pumped by the heart per minute and is determined by stroke volume and heart rate. Both stroke volume and heart rate increase in response to exercise. Endurance training can improve stroke volume, which will require a lower heart rate at any given activity level [107]. Cardiac output multiplied by the amount of oxygen extracted by skeletal muscle (arterial-venous oxygen difference) is called the VO_2, which represents the quantity of oxygen transported to and used by skeletal muscle. Under conditions of maximal exercise, this figure is known as the VO_2 max; it indicates the maximum amount of oxygen capable of being transported to and used by skeletal muscle. VO_2 max can be increased by 15% to 20% with training [107].

The Effect of Strength Training on Skeletal Muscle

In general, continuous submaximal training has the most effect in improving the oxygen transport system, whereas high-intensity, interval-type training causes changes in the structural and biochemical properties of muscle [107]. The distribution of specific muscle fiber types is thought to be genetically determined. Although training does not seem to change the proportion of fast- and slow-twitch fibers, selective recruitment probably occurs with specific types of exercise [107]. A 5-month bicycle training program caused a selective enlargement in slow-twitch fibers but no change in the percentages of slow- or fast-twitch fibers in the vastus lateralis [42].

Metabolic changes as well as increased muscle mass and strength have been documented following exercise [113]. An 8-week "sprint training" program produced increases in the creatine phosphokinase

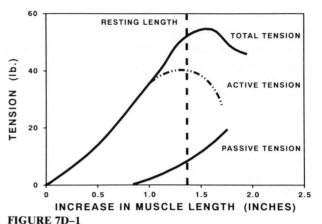

FIGURE 7D–1
Blix curve for muscle demonstrating the passive tension secondary to passive muscle stretching and the active tension secondary to contracting myofibrils in relation to muscle length.

and total phosphagen levels in the vastus lateralis in humans [113]. This may represent an enhanced ability to replenish adenosine triphosphate (ATP) from adenosine diphosphate (ADP). These male subjects showed an increase in thigh circumference and an increase in total body weight. They showed an increase in maximal voluntary isometric contraction as well as improved performance in dynamic strength tests. There was no change, however, in the distribution of muscle fiber types.

The effect of a progressive, dynamic exercise program for the leg extensors has been demonstrated in human subjects [112]. Significant increases in dynamic and isometric strength were documented. Single-repetition maximum squat increased by 67%, Sargent jump increased by 22%, and maximum voluntary isometric contraction increased by 13%. The program resulted in increases in lean body mass and leg muscle mass but no change in body weight. There were no changes in muscle fiber composition or in concentration of creatine phosphokinase. An increase in myokinase was documented, reflecting a potential increased capacity to metabolize ATP.

Weight-lifting exercises in cats have been shown to cause muscle hypertrophy, to increase isometric twitch and tetanic tensions, and to alter the subtype of fast-twitch fibers from fast-twitch oxidative (slower contraction, more oxygen-dependent) to fast-twitch glycolytic (faster contraction, less oxygen-dependent) [44]. The overall proportion of fast- to slow-twitch fibers remained constant, however. Increases in muscle weight and fiber diameter after exercise are well documented. In addition, an increase in the total number of muscle fibers secondary to muscle fiber splitting has been reported, whereby one fiber was seen histologically to have "split" into two fibers [45]. The mechanism by which this may occur in response to exercise is unclear. This finding has been demonstrated by only one researcher, and thus it has yet to be determined whether fiber splitting truly occurs.

The effect of training on muscle metabolism is substantial. A 6-week training program induces a higher rate of lipid utilization, a decreased rate of carbohydrate utilization, a slower increase in lactate levels, and an increased utilization of exogenous glucose [69]. A training program is also capable of improving the serum lipid profile by decreasing total cholesterol, low-density lipoproteins (LDL), and triglyceride levels; increasing serum high-density lipoproteins (HDL); and increasing the content of mitochondria, oxidative enzymes, and lactate threshold, thereby increasing oxidative potential and the ability to perform endurance exercise [43, 51, 79]. Studies have also shown an increase in the capillary bed of muscles in response to training, although this is controversial [43, 123]. Training can also increase lipid oxidation, thereby decreasing the rate of glycolysis and lactate formation [69]. Endurance training has also been shown to affect neuromuscular synaptic function. Synaptic efficiency improved in cats trained on a treadmill [34].

Adaptive changes occur during a training program. As the program continues, unless there is a progressive increase in running speed or incline, animals become capable of performing daily exercise tasks with relatively less participation of fast-twitch fibers [10]. These fibers are recruited only when excitatory input into the motor neuron pad increases to sufficiently high levels. This is one reason why progressive exercises are useful and implies that to improve the greatest spectrum of fiber types, one must combine "steady state" exercise with repeated intervals at high intensity.

Thus, strength training produces selective recruitment in muscle fiber type, increased muscle weight and fiber diameter, and possibly increased number of fibers. An increase in muscle strength results. Endurance training improves oxygen utilization of muscle through increases in oxidative enzymes, in number and size of mitochondria, and in lipid utilization, thereby sparing endogenous carbohydrate stores and decreasing lactate formation.

Thus, skeletal muscle specifically adapts to imposed demands. Stimuli required to produce desirable changes in skeletal muscle can be organized into a training program [123]. These programs are based on the principles of overload, specificity, and reversibility [36].

Exercise of skeletal muscle beyond a critical level is required to increase its structural or functional capability. Thus, muscle must be "overloaded" and subjected to loads greater than those encountered during normal activities [30]. This critical threshold is necessary whether the objective is to increase contractile strength or muscular endurance. To elicit strength gains, contractions of at least 60% of maximum intensity are necessary [30]. The threshold necessary to increase muscular endurance is unclear. Specificity of training means that the structural and functional adaptation of muscle is specific to the applied stimulus [123]. The principle of specific adaptation to imposed demands (SAID) means that the training program must adapt the individual to the specific demands that will be placed upon him.

High-tension, low-frequency contractions cause muscle hypertrophy and strength gains [36]. The tension developed in contracting muscle is defined as the torque or strength of the muscle contraction. Muscle mass is increased predominantly by increasing the cross-sectional area of muscle fibers [123]. Fast-twitch type II fibers undergo hypertrophy to a greater degree than do slow-twitch type I fibers [123]. The role of muscle fiber splitting in the increase of muscle mass remains controversial. Programs designed to increase muscle strength do not improve the oxidative capacity of the muscle. Endurance training (low-tension, high repetition) significantly improves the oxidative capacity of muscle, however. Although the percentages of type I and type II fibers are not changed, an increase in the percentage of high-oxidative type II fibers relative to other type II fibers probably does occur [106]. Adaptation occurs in response to a particular training program, and the rate of structural and functional change of the muscle will reach a plateau unless the training stimulus is progressively increased (progressive resistance exercise, or PRE) [123].

In addition, all of these changes in response to training are reversible. Discontinuation of the stimulus will result in regression of the adaptive changes [36]. These changes also occur after surgical intervention. Exercise may be prescribed as rehabilitation following surgery or trauma, for training purposes in athletic endeavors, or as a general conditioning program. The goal of rehabilitation is balanced bilateral muscular strength and full antagonistic muscle balance [6]. Restoration of muscle strength following surgery or trauma is very important to protect the bone, ligaments, and joints during and after the healing phase. The distinction between rehabilitation and training is an important one, and efforts to restore an injured limb to normal function must utilize a different approach from that used for training of athletes or conditioning of a healthy limb.

TABLE 7D–2
Modes of Application: Vocabulary

Strength	The ability to perform work or move a load
Power	The ability to accelerate a load (work/time); dependent upon level of strength and velocity of contraction
Endurance	The ability to perform repetitive submaximal contractions
Isometric contraction	Muscular contraction producing tension without a significant change in fiber length
Isotonic contraction	Muscular contraction that moves a load at constant resistance through an arc of motion
Static resistance	Isotonic contraction moving a load against a constant unchanging resistance through a range of motion
Variable resistance	Isotonic contraction moving a load against a resistance that changes through the range of motion (via cams or pulleys); differs from isokinetic contraction in that velocity is not constant
Isokinetic contraction	Muscular contraction performed through an arc of motion at a constant velocity against variable resistance
Concentric contraction	Muscular contraction performed while the muscle itself is shortening
Eccentric contraction	Muscular contraction performed while the muscle itself is lengthening
Torque	A rotary force

The most effective way to increase muscular strength is through the use of voluntary maximal muscular contractions [14, 21, 25, 32, 37, 61, 104, 114]. To obtain a voluntary maximal contraction when lifting a weight, the muscle must repetitively lift the weight until it can no longer do so. A voluntary maximal muscular contraction is thus obtained only during the final repetition [37]. Training specificity means that muscular response and adaptation are very specific to the type of exercise performed. This applies not only to the type of exercise performed (i.e., endurance vs. strength) but also to the specific range of motion and plane of movement of the limb and joint [102]. Exercises performed at one particular range of motion or plane of movement will enhance performance only through that particular arc. It seems that only the speed at which exercises are performed does not follow training specificity [30]. Thus, strengthening programs must include exercise at the full arc of motion as well as in different planes of movement. Motor performance is defined as the ability to perform specific motor tests such as the 50-yard dash or vertical jump. For athletes, it is important that training specifically enhance motor performance through improved endurance or muscle strength.

There are many variables in strength training. These include the accessibility of training equipment and facilities and the cost of a particular training regimen. After trauma or reconstructive surgery, exercises must be kept within the tolerance range of healing tissues. This requirement often demands limitations in the type, intensity, or range of motion of particular exercises as well as protective bracing or splinting. Thus, many factors must be considered when prescribing an exercise program. The program must be specifically tailored to the needs and ultimate goals as well as the limitations of the individual. Types of training include isometric, isotonic (concentric vs. eccentric), variable resistance, and isokinetic (Table 7D–2) [37]. These different types have been scientifically evaluated and compared, and each has advantages and disadvantages (Table 7D–3).

TYPES OF STRENGTH TRAINING

Isometric Exercise

Isometric contraction is the development of tension within a muscle without significant change in the fiber length [37]. No joint motion or work is accomplished. When maximal tension is achieved, an insignificant amount of shortening does occur, however [56, 63, 86]. This amount is less than 10% of the original length. Isometric contractions are usually performed against an immovable object or against too much resistance for the muscle to overcome. The resistance can be maximal, allowing for maximal voluntary contraction because it is a function of the applied force [100]. Hettinger and Muller first introduced the concept of isometric training in 1953, claiming that a 6-second contraction once a day 5 days a week could increase muscle strength by 5% per week [57]. Subsequent studies have confirmed that isometric exercises can increase muscle strength at variable rates [24, 56, 76, 86, 88, 89]. In addition, maximal isometric contractions produce greater strength gains than submaximal contractions [115]. Strength gains achieved through isometric training are also seen when the muscles are tested

TABLE 7D–3
Advantages and Disadvantages of Different Types of Exercise

Exercise Type	Advantages	Disadvantages
Isometric	Easily performed Can be performed when immobilization is required	Static: does not improve performance Strength improvement is specific to joint angle Lack of feedback Difficult to determine maximal contraction
Isotonic	Improved motor performance Excellent feedback	Difficult to load muscle maximally through entire arc of motion (improved with variable resistance or DAPRE techniques)
Isokinetic	Improved motor performance Allows for accurate strength testing Voluntary maximal contraction through entire arc of motion	Requires expensive equipment Nonphysiologic: may cause patellofemoral overload
Functional	Easily performed Optimal improvement in agility and motor performance More physiologic than isokinetic exercise	

dynamically and are more prevalent at low rather than high velocities [74]. Maximal strength gains utilizing isometric exercise can take 8 weeks to develop [56, 86].

Increases in limb circumference attributed to muscular hypertrophy have been reported after isometric training [82, 102]. Muscular endurance, however, is not increased with isometric training [50]. Isometric exercises have been shown to decrease atrophy of muscles that have been immobilized in plaster [105].

Strength gains through isometric training are very specific to the joint angle at which the exercises are performed [12, 13, 38, 74, 82, 119]. Strength improvement will not be as great at angles other than those at which the exercises are performed. Thus, if isometric quadriceps training is performed only in full extension, isometric strength increases for the quadriceps will occur only if the limb is in terminal knee extension. No strength improvement through the remaining arc of knee motion can be expected to occur. It is important to perform isometric contractions at several angles of joint motion and to choose the specific angles based upon the functional requirements of the patient to maximize strength gains. Because isometrics are static in nature, they do not improve motor performance [9, 24, 73, 122]. This result has been attributed to the fact that static exercises can actually reduce the maximal speed of limb movement, and motor performance often relies on movement at maximal velocities [37, 109]. An advantage of isometrics is the ease of performing the exercises. They can be performed virtually anywhere and do not require expensive equipment. A

disadvantage is the lack of feedback to the individual. There is no object to move, and thus a lack of sense of accomplishment may be experienced [37]. This disadvantage also makes it more difficult to determine whether a maximal contraction is being performed.

The recommended frequency and duration of isometric exercises vary. Most authorities recommend performing maximal contractions for approximately 5 seconds, five to ten times per day, three to five days per week at three different joint angles [37, 107]. To be most effective, these guidelines should be used as a minimum, and exercises should be performed as frequently as possible throughout the day. To improve the muscular response, isometric exercises may be augmented with electrical neuromuscular stimulation at 2-hour intervals three to four times a day [93]. Neuromuscular stimulation is valuable because it facilitates initiation of isometric contractions in the painful postoperative or post-injury period.

If progression to isotonic exercises requiring joint motion is prohibited and prolonged use of isometrics is required, the use of spectrum isometrics is important. Spectrum isometrics is a program employing isometric exercises at a variety of joint positions to improve strength through a greater arc of motion.

Isotonic Exercise

Isotonic exercise is classically defined as the movement of a load at constant resistance through an arc of motion [37, 93]. There are two types of isotonic exercise: concentric and eccentric. Concentric exercise is defined as exercise performed by muscle contraction while the muscle itself is shortening [37, 93]. Eccentric exercise is defined as exercise performed by muscle contraction while the muscle itself is lengthening [37, 93]. Resistance in isotonic exercise can be static or variable.

Isotonic exercises have been the foundation of limb rehabilitation since their introduction and popularization by DeLorme in 1945 [27]. DeLorme introduced the concept of "heavy resistance exercises" and reported that because the rate and extent of muscle hypertrophy were proportional to resistance, maximal resistance exercises should be used. He emphasized that low-repetition, high-resistance exercises produce power, high-repetition, low-resistance exercises produce endurance, and each is incapable of producing results obtained by the other [27]. He felt that weakened, atrophied muscles should be built up with heavy resistance exercises until normal power is restored prior to initiating endurance training [27].

In 1948 DeLorme adopted the term progressive resistance exercise (PRE) to replace the older term heavy resistance exercise [28]. The amount of resistance recommended is based on the weight that can be moved through a range of motion, not to exceed 10 consecutive repetitions. In addition, he employed the principle of counterbalancing to make possible the administration of PREs to muscles so weak that they cannot complete an arc of motion against gravity. The technique of counterbalancing uses weight applied via a pulley to assist in the support of the limb against gravity. The amount of load the muscle must overcome to move against gravity is thus actually less than the weight of the limb. He defined load-resisting exercises as those in which the exercise load resists the muscle, and load-assisting exercises as those in which the exercise load assists the muscle as in counterbalancing. The increases in muscle loads for load-resisting exercises are based upon a 10-repetition maximum, which is the most weight a muscle can lift through an arc of motion 10 times. The increases in muscle loads for load-assisting exercises are based on a 10-repetition minimum, which is the least amount of assistance required by the muscle to help it lift the extremity 10 times through the arc. This load is determined by decreasing the assisting weight until the muscle is able to move the limb only 10 times against gravity. DeLorme advocated performing three sets of 10 repetitions, using 50% of a 10-repetition maximum for the first set, 75% of a 10-repetition maximum for the second set, and the full 10-repetition maximum for the third set. When a 10-repetition

maximum can be lifted 15 times, a new repetition maximum is established (Table 7D–4).

In 1950 DeLorme demonstrated the clinical usefulness of PREs in treating patients with femoral fractures [29]. Subsequently, many reports have documented the efficacy of PREs in producing muscle hypertrophy and strength gains. A 12-week program of elbow extension PREs produced significant gains in arm circumference and strength of elbow extension [81]. These gains were subsequently lost if the exercises were terminated. A 10-week PRE program demonstrated gains in strength and muscle hypertrophy in both men and women [120]. In addition, a PRE program can elicit these desired effects without concomitant masculinizing effects or marked weight changes in women [80]. Studies comparing different PRE programs have demonstrated similar improvements in static and dynamic strength but somewhat variable hypertrophy [11, 91]. Improvement in motor performance has also been demonstrated [32]. Increases of lean body mass of 5 to 6 pounds and decreases in body fat of 5 to 6 pounds during 8 weeks of training with PREs have been reported [37, 98, 120].

A major disadvantage of isotonic exercise is that a muscle cannot be loaded to maximum capacity because muscle strength varies with joint position. If a particular exercise is performed with constant weight at a particular velocity through an arc of motion, only a single voluntary maximal contraction is performed per set. This would be the last repetition before the weight could no longer be lifted and would occur only at the muscle's weakest point through the arc of motion. Maximum muscle tension

TABLE 7D–4
Load-Resisting and Load-Assisting Exercises

Load-Resisting Exercises

First set of 10 repetitions: Use half of 10-repetition maximum

Second set of 10 repetitions: Use three-fourths of 10-repetition maximum

Third set of 10 repetitions: Use 10-repetition maximum

The 10-repetition maximum is the most weight that can be lifted correctly through a full arc of motion for 10 repetitions.

Load-Assisting Exercises

First set of 10 repetitions: Use twice the 10-repetition minimum

Second set of 10 repetitions: Use one and a half times the 10-repetition minimum

Third set of 10 repetitions: Use 10-repetition minimum

The 10-repetition minimum is the least amount of assistance required by the muscle to help it lift the extremity 10 times through the arc.

cannot be developed for the stronger joint positions. This can be compensated for partially if the maximal force is applied through the arc by either accelerating or decelerating the weight. Thus, if the weight is lifted as quickly as possible, it is possible to perform a voluntary maximal contraction through the entire arc [37]. Many exercises do not accommodate themselves to such explosive movements. Because muscle strength varies with joint position and the velocity at which the exercise is performed varies, total muscle development is unpredictable with isotonic exercises.

To compensate for these limitations, variable resistance strength training modalities were developed (Nautilus, Universal, MARCY, Eagle, and so on). Variable resistance exercise is defined as exercise in which the amount of resistance offered to the muscle is varied through cams or pulleys in an attempt to match the strength curve of the individual [37]. Thus, muscle tension is increased or decreased throughout the range of motion depending upon joint position. These weight machines may not completely accomplish this goal, however. Muscle length-tension ratios vary among individuals, so all individual strength curves cannot be ideally matched with a single apparatus. Nevertheless, these systems do emphasize flexibility and strength through the full range of motion, and this is important. Increases in strength with as few as 10 to 12 repetitions of variable resistance isotonic exercises have been reported [97]. Motor performance has been improved with this type of training as well [97].

In 1979 the daily adjustable progressive resistive exercise (DAPRE) technique was described in an attempt to ensure that a patient works near his optimal capacity [65]. This program reportedly allows for individual differences in the rate at which muscle strength is regained and provides an objective method for increasing resistance in accordance with strength increases. During the first and second sets, 10 repetitions against one-half the working weight and six repetitions against three-fourths the working weight are performed. During the third and fourth sets, the maximum number of all repetitions is performed. The number of repetitions performed during the fourth set is used to determine the working weight for the next session (Table 7D–5).

The number of sets and repetitions for isotonic exercises should be somewhat individualized to allow an individual to use maximal effort [37]. A study comparing different weight-training programs found that three sets of six repetitions per set were best for improving strength [14]. Three sets of 6 to 10 repetitions of maximum exercises performed at least three times per week are generally recom-

TABLE 7D–5
DAPRE Technique

Set	Weight	Repetitions
1	Half working weight	10
2	Three-forths working weight	6
3	Full working weight	Maximum[a]
4	Adjusted working weight	Maximum[b]

General guidelines for adjustment of working weight:

Adjusted Working Weight

Repetitions in Each Set	Fourth Set[c]	Next Session[d]
0–2	Decrease 5—10 lb	Decrease 5–10 lb
3–4	Decrease 0–5 lb	Keep same
5–6	Keep same	Increase 5–10 lb
7–10	Increase 5–10 lb	Increase 5–15 lb
11–	Increase 10–15 lb	Increase 10–20 lb

[a]The number of repetitions performed during the third set is used to determine the working weight for the fourth set according to the guidelines above.
[b]The number of repetitions performed during the fourth set is used to determine the working weight for the next session.
[c]Repetitions performed during the third set are used to determine the adjusted working weight for the fourth set.
[d]Repetitions performed during the fourth set are used to determine the working weight for the next session (usually the next day).

mended [14, 27, 28, 30, 37]. These exercises can be performed with free weights or machines. Free weights are less expensive and more accessible but require a spotter and are somewhat more dangerous than weight machines. A resistance program should be designed to "overload" but not "overwhelm" [30]. The combination of atrophy, an aggressive exercise program, and a highly motivated patient can trigger the "overwork syndrome" described by Knowlton [66]. This syndrome is characterized by exercise that is progressed too rapidly, resulting in inflammation and synovitis. Contractile ability can actually decline regardless of the exercises employed. After a program is under way, a gradual and controlled increase in exercise intensity is necessary to obtain maximal strength gains [30]. When working with weak, atrophied muscles, 10 repetitions of an initial resistance of 25% maximum should be used [23]. This should be gradually increased to 50% and then to 75% of maximum.

If during isotonic exercise the muscle is contracting as the limb returns to the starting position, then eccentric, or negative exercise is being utilized as well (as in the DeLorme knee extension PRE when the knee returns to flexion slowly). Eccentric contraction occurs when a muscle lengthens while resisting load. When this type of exercise was popularized initially, it was felt that because a greater amount of weight could be used, greater strength gains would be possible [37]. Although significant gains in strength have been reported with eccentric training, most reports show no significant difference compared with concentric isotonic training [61, 62, 77]. One study did show greater strength gains with eccentric compared with concentric training programs [67]. It may be that equal strength gains can be achieved with less effort with eccentric contractions [62]. A group trained eccentrically with two sets of six repetitions at 120% of the single repetition maximum achieved strength gains equal to those achieved by a group trained concentrically with two sets of 10 repetitions at 80% of the single repetition maximum [62]. The tension developed in a muscle is directly related to the electrical activity as measured by electromyography (EMG) [37]. When performing eccentric and concentric contractions with equal weights, more EMG activity is present during concentric contractions [8]. Thus, to cause equal muscle tension to develop equal strength, more weight should be necessary during eccentric training than with concentric exercise. Further research in this area is clearly necessary.

Strength gains with concentric and eccentric exercises probably occur independently at different rates, and should be addressed separately [17]. Muscle develops selectively with either concentric or eccentric training, so a training program should include both types of exercise in the isotonic phase. Disadvantages of eccentric training include muscle soreness during the first few weeks, use of heavier weights, requiring an assistant and perhaps causing an increased safety hazard, and the need for longer sessions due to the slow nature of eccentric contractions [37].

Isokinetic Exercise

Isokinetic exercise is exercise performed at a constant velocity of speed, which is predetermined [37, 94]. To maintain a constant velocity of contractions, the resistance offered is variable throughout the arc of motion. The muscle group is placed under more uniform tension. Because the speed is regulated, resistance occurs in proportion to the force applied throughout the entire range of motion [64]. Any muscle exertion encounters a corresponding counteractive resistant force to maintain the predetermined velocity. Increased muscular effort will meet increased resistance rather than increased velocity, as occurs with isotonic contractions [99]. This allows voluntary maximal contractions to occur throughout the entire range of motion, depending upon the motivation of the individual. If voluntary

maximal contractions do occur through the entire range of motion, maximal strength gains should result [37].

Significant gains in muscle strength have been reported using isokinetic training at a variety of repetition and set frequencies as well as at slow or fast velocities (22.5 to 180 degrees/second) [72, 83, 98]. Strength gained at fast velocities will carry over to all velocities slower than the training velocity, whereas strength gained at slow velocities will not carry over to velocities faster than the training velocity [84, 98]. Thus, to compete at a fast velocity of movement, an athlete should train isokinetically at fast velocities.

Isokinetic training has been shown to increase motor performance [95, 98]. Isokinetic training at fast velocities has been reported to increase motor performance to a greater degree than does training at slow velocities [98]. Decreases in percentage of body fat and increases in lean body weight have been reported with isokinetic training, and these changes are greater when training is performed at faster velocities [98].

Advantages of isokinetic exercises include lack of muscle soreness [98], the potential to achieve the greatest increases in strength throughout the entire arc of motion, a greater potential for improving motor performance activities, no requirement for a weight stack that must be moved, and decreased workout time due to the high speed of movement [37].

Disadvantages include the high expense of isokinetic equipment (Cybex, Merac, etc.), difficulty of assessing the effort made, and the lack of positive feedback because no weights are visibly moved [37]. In addition, isokinetic exercises are nonphysiologic and may frequently cause overload of the patellofemoral joint.

Optimum positioning and limb stabilization are required for isokinetic training or testing [6]. The optimum axis of joint motion must also be determined. Selection of appropriate velocities must take into account the functional limitations and ultimate goals of the individual. After appropriate warm-up and stretching, 5 to 10 submaximal repetitions at each velocity should be performed to confirm proper alignment and patient tolerance [6]. Isokinetic machines allow accurate strength testing, and a baseline test should be performed prior to institution of the training program as well as at regular intervals for retesting. Total peak torque (rotary force) as a percentage of the contralateral limb, torque curve analysis, and power endurance capacity should be determined [6].

To perform an isokinetic test, the machine must

be accurately adjusted to the individual limb. The agonist and antagonist muscle groups are generally tested on both limbs at a predetermined velocity (degrees/second). A curve is generated as torque versus position in degrees (Fig. 7D–2). Strength comparisons then can be made between agonist and antagonist muscle groups and between one limb and the other. This comparison is generally expressed as a percentage of the injured limb to the normal contralateral limb for each muscle. Isokinetic sessions can vary from 10 to 20 minutes each and should be performed three times a week [6]. Adequate supervision is recommended during isokinetic training, particularly during the postoperative phase, because high speed training can have detrimental effects on healing tissues [6]. These effects may include production of excessive patellofemoral joint reaction forces and unpredictable forces on healing anterior cruciate ligament grafts.

COMPARISON STUDIES

Isometric and Isotonic Exercise

The literature contains numerous studies comparing different modes of exercise. Comparisons between isometric and isotonic exercises are somewhat controversial. Generally, static training improves static strength, and dynamic training improves dynamic strength [15]. Static strength testing is not as accurate as dynamic strength testing for measuring strength increases produced by dynamic exercises, and conversely, dynamic strength testing is not as accurate as static strength testing for measuring strength increases produced by static exercises. Thus, if one wishes to measure strength gains, the measurement technique must coincide with the training technique [37, 83, 112]. Comparison studies between isometric and isotonic exercises therefore produce conflicting results. Some reports of isotonic versus isometric exercise have shown no difference in muscle hypertrophy [71, 75], whereas other reports have demonstrated greater increases in strength and hypertrophy with isotonic exercises [102, 115]. Motor performance improves to a greater extent with isotonic than with isometric exercises [16, 20, 37]. On the other hand, isometric training has an advantage in that it can be performed virtually anywhere and requires no special equipment.

Isometric training was very popular from 1955 to the early 1970s. This popularity waned when it was demonstrated that strength gains from isometric training was specific to the joint angle at which it was performed [64]. Isotonic exercises probably produce a more uniform increase in strength over the entire range of motion, improving strength and endurance more effectively [107]. As a result, physical educators, coaches, and athletes adopted the principles of progressive resistance exercises to improve athletic and occupational skills most effectively [64]. Isometric training continues to have an important role, especially in postinjury or postop-

FIGURE 7D–2
Example of a typical curve, generated by an isokinetic machine, of muscle torque versus position in degrees.

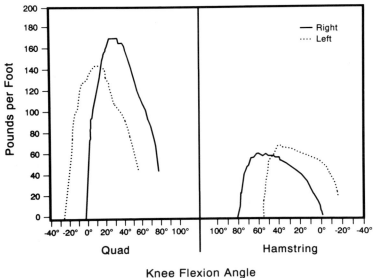

erative rehabilitation, when joint motion must be restricted.

Isotonic and Isokinetic Exercise

Several studies comparing isotonic and isokinetic training have demonstrated that isokinetic exercises produce greater strength increases [83, 98, 111]. Pipes and Wilmore concluded that isokinetic exercises were superior to isotonic exercises in producing muscular strength, increased anthropometric measurements, and superior performance on motor performance tasks [98]. In addition, isokinetic exercises performed at high speeds showed the greatest overall gains in strength. Isokinetic training is as effective as isotonic training in effecting decreases in body fat percentages and increases in lean body mass [98]. Isokinetic training is reported to cause less muscle soreness than isotonic training [98].

Isometric and Isokinetic Exercise

Isokinetic training increases isometric and isokinetic strength to a greater degree than does isometric training [37, 83, 111]. Electromyographic comparisons have demonstrated that isokinetic training is preferred over isometric training for strength development [22, 37]. Muscular endurance is improved to a greater degree with isokinetic than with isometric training [107].

Isotonic Fixed and Variable Resistance Exercise

Comparisons of isotonic fixed resistance and variable resistance exercises have produced conflicting results [37]. Variable resistance has been reported to be superior to isotonic training in producing strength gains [7], whereas another study reported no significant difference between the two [12].

Specific Modalities

Isometric training requires no weights or special equipment. The patient is instructed in techniques of specific muscle contraction without adjacent joint motion. An example of this is the "quad set" in which a quadriceps contraction is performed and maintained for several seconds while the hip and knee remain immobile.

Isotonic training can be performed with free weights or exercise machines such as the Universal or Nautilus system. Isokinetic training requires more specialized and expensive equipment such as the Cybex or Merac system. These machines allow the performance of exercises as well as quantitative strength testing.

Considerable interest has developed in the technique of circuit weight training (CWT) to improve overall fitness, including endurance as well as strength. CWT refers to a series of weight-training exercises of 12 to 15 repetitions using from 40% to 60% of a one-repetition maximum weight [41]. Minimal rest (15 to 30 seconds) is allowed between each station. Gettman and Pollock showed that CWT can increase aerobic capacity by 5% and lean body mass by 1 to 3.2 kg and can decrease fat by 0.8% to 2.9% [41]. Strength improved 32%. Hempel and Wells demonstrated only minimal aerobic benefit from the Nautilus Express Circuit, a form of CWT [55]. Thus, CWT does not develop high levels of aerobic fitness but does produce strength gains and can help to maintain overall fitness.

Weight training supplemented by endurance, flexibility, or motor skill activities has been used to augment strength increases with generalized conditioning [64]. Because motor performance is dependent upon proprioceptive feedback from the joints, tendons, and ligaments, proprioceptive conditioning should accompany any training or rehabilitation program [30]. It has been shown that immobilization decreases proprioceptive feedback [30]. Kinesthetic exercises such as the balance board can assist proprioceptive rehabilitation.

The use of proprioceptive neuromuscular facilitation (PNF) will also enhance recovery. PNF is defined as an exercise technique utilizing proprioceptive, cutaneous, and auditory input to produce functional improvement in motor output [101]. These exercises involve stimulation of proprioceptors to elicit a neuromuscular response and are frequently performed with the aid of a physical therapist. For example, manual stretching of the muscle during a contraction will initiate the stretch reflex, thereby increasing motor nerve input and facilitating the contraction. PNF is best utilized to correct deficiencies in strength, flexibility, and coordination in response to neuromuscular demands [101].

FUNCTIONAL EXERCISES

The use of functional exercises for rehabilitation is gaining popularity. Functional exercise is defined as the restoration of strength and agility through dynamic exercise [129]. Functional exercises frequently employ several muscle groups simultaneously and tend to be sport specific. They are separate from and have several advantages over isokinetic exercises. Functional exercises do not require the use of expensive equipment and can be performed virtually anywhere. Functional rehabilitation can effect optimal improvements in agility and motor performance [60, 85, 108, 129]. Skip ropes, agility drills, and paddleball, basketball, and badminton equipment have been successfully used for knee rehabilitation by the United States military services [129]. The use of an unstable balance board (for dynamic joint control training) caused a greater improvement in muscle reaction time than did a simple muscle training in one study [60]. Bicycling and swimming have proved valuable in knee rehabilitation [85, 98, 108]. Currently used modalities for rehabilitation following knee ligament reconstruction include leg presses, mini squats, stretch cords, trampoline, and pogo sticks. These modalities are much more practical than expensive isokinetic machines for most patients and produce excellent results. In many cases, functional exercises can be used instead of isokinetic training. Total body fitness must not be overlooked during any rehabilitation program to prevent deconditioning, minimize the loss of skilled movements, and enhance the mental satisfaction of the patient [30].

INDICATIONS AND CONTRAINDICATIONS

Therapeutic exercise is exercise that is used for rehabilitation following injury or surgery to improve musculoskeletal function or to maintain conditioning [30]. Maintenance of the patient's current level of aerobic fitness, initiation of range of motion and strengthening exercises at appropriate phases of tissue healing, protection of healing tissue from disruptive biomechanical stresses, and patient education about the treatment plan are all important factors in prescribing an exercise program [30]. For example, knee rehabilitation following anterior cruciate ligament reconstruction must be performed in such a way that force on the healing graft is minimized. This entails limiting active terminal knee extension against resistance until adequate graft healing strength has been achieved because anterior tibiofemoral shear forces are high at this position [92, 93]. Such limitation can be accomplished by using a hinged brace with an extension stop. Thus, each situation must be evaluated in terms of bio-

mechanics prior to prescribing an exercise program. A warm-up period should precede each exercise session. This will increase both local and general circulation, thereby preparing muscles, tendons, and joints for activity [230]. Warm-up can be accomplished through the use of heat or several unloaded repetitions prior to the beginning of the session. The lack of warm-up could result in soft tissue injury.

Isometric Exercise

Isometric exercises should be initiated immediately in the postoperative or postinjury period. Specific indications for isometric exercises include the need to increase or preserve muscle strength when the associated joint cannot be mobilized or loaded to protect healing tissues. There are no contraindications to isometric exercise except the inability to strengthen the muscle throughout a full range of motion compared with isotonic or isokinetic exercise. Electrical neuromuscular stimulation can be an important adjunct in the prevention of atrophy during the isometric phase because it facilitates muscle contraction during the painful postoperative or postinjury phase.

The effect of isometric exercise can be further enhanced by the "crossover effect" [93]. Strengthening of the opposite limb by as much as 30% can occur through vigorous exercise of one limb [52–54]. Therefore, during the isometric phase the contralateral uninjured limb should undergo isotonic or isokinetic strengthening exercises to take advantage of this effect and to effect maximal strengthening of the injured limb. Stationary cycling with the contralateral uninvolved leg should also be performed to maintain cardiovascular and neuromuscular endurance [93]. If prolonged isometrics are required and the joint can be mobilized but not loaded, then spectrum isometrics are indicated [93].

Isotonic Exercise

Once the patient can perform active voluntary contractions and joint motion begins, isometrics and neuromuscular stimulation are replaced with isotonic progressive resistance exercises and range of motion exercises [93]. Following soft tissue injury, active assisted range of motion is generally preferred to passive stretching. Indications for isotonic exercise include the need to increase muscle strength once active range of motion is allowed. These exercises are contraindicated if soft tissue healing is not adequate to sustain cycling loading or in patients with patellofemoral disorders in which high joint reactive forces must be avoided. Initial resistance should be low and should be gradually increased through isotonic phases. For example, following anterior cruciate ligament reconstruction, quadriceps and hamstring exercises are started at less than 10% body weight [93]. Because of the specificity of strength training, both concentric and eccentric isotonic exercises should be performed. Healing tissues require protection during the isotonic phase. This may take the form of taping, splinting, or use of an orthosis. Frequently, range of motion must be restricted at first and gradually increased throughout the rehabilitation period. Protected weight-bearing is often necessary and must be individualized to the particular diagnosis. General fitness activities must continue to prevent generalized deconditioning. Flexibility, range of motion, and strengthening of the other extremities should be maintained through stretching and resistive exercises [48].

Isokinetic Exercise

Because of the potential for excellent strength gains, isokinetic exercises can be an integral component of rehabilitation if they are available. Isokinetic exercises can be performed when isotonic exercises can be performed pain free with 10% to 20% body weight resistance. A contraindication to isokinetic quadriceps training is patellofemoral pain or abnormal tracking.

A disadvantage of the use of isokinetic exercises is the expense of and limited access to the specialized equipment required. Optimal limb positioning and stabilization are required. Selection of velocity, range of motion, and pad placement must be individualized and must take into account the postinjury or postsurgical status of the limb. For example, following anterior cruciate ligament reconstruction, recommendations include high pad placement and initiation of exercises at high velocity to minimize resistance, and to reduce patellofemoral joint reactive forces. Once full activity has been reached by the patient, maintenance of strength is best accomplished through continued use of isotonic exercises because of the expense of and limited access to isokinetic machines.

Exercises needed to build muscular endurance must begin early in the rehabilitation period. The endurance capabilities of muscle are generally slow to return. Endurance requires repetitive high-speed exercises, which can be accomplished through the use of high-velocity isokinetic exercises or, more conveniently, use of a stationary bicycle.

Functional Exercise

Cycling can be initiated when knee flexion reaches 110 degrees and at least 10% of body weight is being lifted isotonically [93]. Cycling should be initiated with no resistance. Adjustment in seat height can be made to compensate for limitations in range of motion. Swimming is a valuable exercise for building muscular endurance and can also be initiated early in the rehabilitation period, concomitant with initiation of joint motion [85, 93, 108]. After adequate soft tissue healing has occurred, outdoor cycling, stationary track, or rowing exercises can be added. Cardiovascular endurance will also require recovery and maintenance following a period of inactivity. The exercises prescribed for muscular endurance will also improve cardiovascular endurance.

After adequate gains in strength and endurance have been achieved (e.g., a hamstrings-quadriceps ratio of at least 70% to 80% [127] and quadriceps strength equal to 75% of body weight), a progressive running program can be initiated. Return to sports can occur in the absence of joint arthrosis or effusion and in the presence of joint stability. We suggest that involved to noninvolved limb strength ratios should be at least 90% as tested isokinetically. Neuromuscular coordination and enhancement are facilitated through the use of skill drills specific to the sport that the patient is returning to. Lifetime conditioning should continue through maintenance sports such as bicycling, swimming, weight training, track skiing, jogging, and rowing.

SUMMARY

Prescription of exercise is a necessary and valuable tool in the rehabilitation of a limb following injury or surgery, in training for athletics, and in general body conditioning. Each of the various forms of exercise has advantages and disadvantages, but all are valuable when utilized correctly and prescribed individually for a particular patient. The response to training is very specific to the particular type and application of exercise, so the use of a variety of exercise modalities is helpful. A specifically designed exercise program implemented at the appropriate stage of healing will correct physical impairment, improve musculoskeletal function, and promote total body fitness, thus facilitating return to normal activities and reducing the chance of reinjury [30]. The treating physician must design and supervise the exercise program based upon the pathology present and the type of surgery performed.

References

1. Akeson, W. H. An experimental study of joint stiffness. *J Bone Joint Surg* 43A:1022–1034, 1961.
2. Akeson, W. H., Amiel, D., and La Violette, D. The connective tissue response to immobility: A study of the chondroitin-4 and -6 sulfate and dermatan sulfate changes in periarticular connective tissue of control and immobilized knees of dogs. *Clin Orthop* 51:183–197, 1967.
3. Akeson, W. H., Woo, S. L.-Y., Amiel, D., Coutts, R. D., and Daniel, D. The connective tissue response to immobility: Biochemical changes in periarticular connective tissue of the immobilized rabbit knee. *Clin Orthop* 93:356–362, 1973.
4. Akeson, W. H., Woo, S. L.-Y., Amiel, D., and Doty, D. H. Rapid recovery from contracture in rabbit hindlimb: A correlative biomechanical and biochemical study. *Clin Orthop* 122:359–365, 1977.
5. Akeson, W. H., Woo, S. Y.-L., Amiel, D., and Matthews, J. V. Biomechanical and biochemical changes in the periarticular connective tissue during contracture development in the immobilized rabbit knee. *Connect Tissue Res* 2:315–323, 1974.
6. Allman, F. L. Rehabilitative exercises in sports medicine. *Instruc Course Lect* 34:389–392, 1985.
7. Ariel, G. Barbell vs. dynamic variable resistance. *U.S. Sports Association News* I:7, 1977.
8. Asmussen, E. Positive and negative work. *Acta Physiol Scand* 28:364–382, 1953.
9. Ball, J. R., Rich, R. Q., and Wallis, E. L. Effects of isometric training on vertical jumping. *Res Q Am Assoc Health Phys Educ* 35:231–235, 1964.
10. Baldwin, K. M., and Winder, W. W. Adaptive responses in different types of muscle fibers to endurance exercise. *Ann NY Acad Sci* 3:411–428, 1978.
11. Barney, V. S., and Bangerter, B. L. Comparison of three programs of progressive resistance exercise. *Res Q* 32:138–146, 1961.
12. Belka, D. Comparison of dynamic, static, and combination training on dominant wrist flexor muscles. *Res Q Am Assoc Health Phys Educ* 39:244–250, 1968.
13. Bender, J. A., and Kaplan, H. M. The multiple angle testing method for the evaluation of muscle strength. *J Bone Joint Surg* 45A:135–140, 1963.
14. Berger, R. Effect of varied weight training programs on strength. *Res Q Am Assoc Health Phys Educ* 37:259–333, 1962.
15. Berger, R. A. Comparison of static and dynamic strength increases. *Res Q* 33:329–333, 1962.
16. Berger, R. Effects of dynamic and static training on vertical jumping ability. *Res Q* 34:419–424, 1963.
17. Bonde-Peterson, F., Knuttgen, H. G., and Nendrikisson, J. Muscle metabolism during exercise with concentric and eccentric contractions. *J Appl Physiol* 33:792–795, 1972.
18. Booth, F. W. Time course of muscular atrophy during immobilization of hind limbs in rats. *J Appl Physiol* 43:656–661, 1977.
19. Booth, F. W., and Seider, M. J. Effects of disuse by limb immobilization on different muscle fiber types. *In* Pette, D. (Ed.), *Plasticity of Muscle.* New York, de Gruyter, 1980.
20. Campbell, R. C. Effects of supplemental weight training on the physical fitness of athletic squads. *Res Q* 33:343–348, 1962.
21. Chu, E. The effect of systematic weight training on athletic power. *Res Q Am Assoc Health Phys Educ* 21:188–194, 1950.
22. Clarke, D. H. Adaptations in strength and muscular endurance resulting from exercise. *In* Wilmore J. (Ed.), *Exercise and Sports Reviews.* New York, Academic Press, 1973.
23. Colson, J. H. C., and Armour, W. J. *Sports Injuries and Their Treatment.* Stanley Paul, Anchor Press.
24. Cotton, D. Relationship of the duration of sustained voluntary isometric contraction to changes in endurance and strength. *Res Q Am Assoc Health Phys Educ* 38:366–374, 1967.
25. Counsilman, J. The importance of speed in exercise. *Athletic J* 57:72–75, 1976.
26. Dahners, L. E. Ligament contraction—A correlation with cellularity and actin staining. *Trans Orthop Res Soc* 11:56, 1986.
27. DeLorme, T. L. Restoration of muscle power by heavy resistance exercises. *J Bone Joint Surg* 27:645–667, 1945.
28. DeLorme, T. L., and Watkins, A. L. Techniques of progressive resistance exercise. *Arch Phys Med* 29:263–273, 1948.
29. DeLorme, T. L., West, F. E., and Shriber, W. J. Influence of progressive-resistance exercises on knee function following femoral fractures. *J Bone Joint Surg* 32A:910–924, 1950.
30. Dickinson, A., and Bennet, K. M. Therapeutic exercise. *Clin Sports Med* 4(3), 1985.
31. Dietrich, J. E., Whedon, G. D., and Shorr, E. Effects of immobilization upon various metabolic and physiologic functions of normal men. *Am J Med* 4:3, 1944.
32. Dintiman, G. B. Effect of various training programs on running speed. *Res Q Am Assoc Health Phys Educ* 35:456, 1964.
33. Dumbleton, J. H., and Black, J. Principles of mechanics. *In* Black, J., and Dumbleton, J. H. (Eds.), *Clinical Biomechanics.* Churchill Livingstone, New York, 1981, pp. 359–399.
34. Eccles, R. M., and Westerman, R. A. Enhanced synaptic function due to excess use. *Nature* 184:460–461, 1959.
35. Enneking, W. F., and Horowitz, M. The intra-articular effects of immobilization on the human knee. *J Bone Joint Surg* 54A:973–985, 1972.
36. Faulkner, J. A. New perspectives in training for maximum performance. *JAMA* 205:741–746, 1986.
37. Fleck, S. J., and Schutt, R. C. Types of strength training. *Clin Sports Med* 4(1):159–168, 1985 (Jan).
38. Gardner, G. Specificity of strength changes of the exercised and nonexercised limb following isometric training. *Res Q Am Assoc Health Phys Educ* 34:98–101, 1963.
39. Gauthier, G. F. Skeletal muscle fiber types. *In* Engel, A. G., and Banker, B. Q. (Eds.), *Myology,* Vol. 1. New York, McGraw-Hill, 1986, pp. 255–284.
40. Gelberman, R. H., Woo, S. L.-Y., Lathringer, K., Akeson, W. H., and Amiel, D. Effects of early intermittent passive mobilization on healing canine flexor tendons. *J Hand Surg* 7:170–175, 1982.
41. Gettman, L. R., and Pollock, M. L. Circuit weight training: A critical review of its physiological benefits. *Physician Sports Med* 9:44–60, 1981.
42. Gollnick, P. D., Armstrong, R. B., Saltin, B., Saubert, C. W. IV, Sembrowich, W. L., and Shepherd, R. E. Effect of training on enzyme activity and fiber composition of human skeletal muscle. *J Appl Physiol* 34(1):107–111, 1973.
43. Gollnick, P. D., Hermansen, L., and Saltin, B. The muscle biopsy: Still a research tool. *Physician Sports Med* 8:50:55, 1980.
44. Gonyea, W., and Bonde-Peterson, F. Alterations in muscle contractile properties and fiber composition after weightlifting exercise in cats. *Exp Neurol* 59:75–84, 1978.
45. Gonyea, W., Ericson, G. C., and Bonde-Petersen, F. Skeletal muscle fiber splitting induced by weight-lifting exercise in cats. *Acta Physiol Scand* 99:105–109, 1977.
46. Gould, N., Donnermeyer, D., Pope, M., Alvarez, R., and Ashikaga, T. Transcutaneous muscle stimulation to retard

disuse atrophy after open meniscectomy. *Clin Orthop* 178:190–197, 1983.

47. Gould, N., Donnermeyer, D., Pope, M., and Ashikaga, T. Transcutaneous muscle stimulation as a method to retard disuse atrophy. *Clin Orthop* 164:215–220, 1982.

48. Grana, W. A., Karr, J., and Stafford, M. Rehabilitation techniques for athletic injury. *Instruct Course Lect* 34:393–400, 1985.

49. Griffin, G. E., Williams, P. E., and Goldspink, G. Region of longitudinal growth in striated muscle fibres. *Nature* 232:28–29, 1971.

50. Grimby, G., von Heijne, C., Hook, O., and Wedel, H. Muscle strength and endurance after training with repeated maximal isometric contractions. *Scand J Rehab Med* 5:118–123, 1973.

51. Heath, G. W., Ehsani, A. A., Hagberg, J. M., et al. Exercise training improves lipoprotein lipid profiles in patients with coronary artery disease. *Am Heart J* 105:889–895, 1983.

52. Hellebrandt, F. A., Houtz, S. J., and Kirkorian, A. M. Influence of bimanual exercise on unilateral work capacity. *J Appl Physiol* 2:452–466, 1950.

53. Hellebrandt, F. A., Houtz, S. J., Parrish, A. M., et al. Tonic neck reflexes in exercises of stress in man. *Am J Phys Med* 35:144–159, 1956.

54. Hellebrandt, F. A., and Waterland, J. C. Indirect learning: The influence of unimanual exercise on related muscle groups of the same and the opposite side. *Am J Phys Med* 41:45–55, 1962.

55. Hempel, L. S., and Wells, C. L. Cardiorespiratory cost of Nautilus express circuit. *Physician Sports Med* 13:82–97, 1985.

56. Hettinger, T. *Physiology of Strength.* Springfield, IL, Charles C Thomas, 1961, p. 82.

57. Hettinger, T., and Muller, E. Muskelleistung und Muskeltraining. *Arb Physiol* 15:111–126, 1953.

58. Hill, J., Moynes, D. R., Yocum, L. A., Perry, J., and Jobe, F. W. Gait and functional analysis of patients following patellectomy. *Orthopaedics* 6:724–728, 1983.

59. Hogan, E. L., Dawson, D. M., and Romahul, F. C. A. Enzymatic changes in denervated muscle. *Arch Neurol* 13:274–282, 1965.

60. Ihara, H., and Nakayoma, A. Dynamic joint control training for knee ligament injuries. *Am J Sports Med* 14(4):309–315, 1986.

61. Johnson, B. L. Eccentric vs. concentric muscle training for strength development. *Med Sci Sports* 4:111–115, 1972.

62. Johnson, B. L., Adamczyk, J. W., Tennoe, K. O., and Stromme, S. B. A comparison of concentric and eccentric muscle training. *Med Sci Sports* 8:35–38, 1976.

63. Karpovich, P. V. *Physiology of Muscular Activity.* Philadelphia, W. B. Saunders, 1959.

64. Katch, F. I., and Drumm, S. S. Effects of different modes of strength training on body composition and anthropometry. *Clin Sports Med* 5(3):413–459, 1986.

65. Knight, K. L. Knee rehabilitation by the daily adjustable progressive resistive exercise technique. *Am J Sports Med* 7(6):336–337, 1979.

66. Knowlton, G. C., and Bennett, R. L. Overwork. *Arch Phys Med* 38:118, 1957.

67. Komi, P. N., and Buskirk, E. R. Effect of eccentric and concentric muscle conditioning on tension and electrical activity of human muscle. *Ergonomics* 15:417–434, 1972.

68. Kottke, F. J. Therapeutic exercise. *In* Kottke, F. J., Stillwell, G. K., and Lehmann, J. F. *Handbook of Physical Medicine and Rehabilitation* (3rd ed.). Philadelphia, W. B., Saunders, 1982.

69. Krzentowski, G., Pirnay, F., Luycky, A. S., Lacroix, M., Mosora, F., and Lefebure, P. J. Effect of physical training on utilization of a glucose load given orally during exercise. *Am J Physiol* 246:P412–E417, 1984.

70. Laros, G. S., Tipton, C. M., and Cooper, R. R. Influence of physical activity on ligament insertions in the knees of dogs. *J Bone Joint Surg* 53A:275–286, 1971.

71. Leach, R. E., Stryker, W. S., and Zohn, D. A. A comparative study of isometric and isotonic quadriceps exercise programs. *J Bone Joint Surg* 47A:1421–1426, 1965.

72. Lesmes, G. R., Costill, D. L., Coyle, E. F., and Fink, W. J. Muscle strength and power changes during maximal isokinetic training. *Med Sci Sports* 10:266–269, 1978.

73. Lindeburg, F. A., Edwards, P. K., and Heath, W. D. Effect of isometric exercise on standing broad jumping ability. *Res Q Am Assoc Health Phys Educ* 34:478–483, 1963.

74. Lindh, M. Increase of muscle strength from isometric quadriceps exercises at different knee angles. *Scand J Rehab Med* 22:33–36, 1979.

75. Lorback, M. M. A study comparing the effectiveness of short periods of static contractions to standard weight training procedures in the development of strength and muscle girth. Thesis (Master's), Pennsylvania State University, 1955.

76. MacConaill, M. A., and Basmajian, J. V. *Muscles and Movements: A Basis for Human Kinesiology.* Baltimore, Williams & Wilkins, 1967, p. 421.

77. Mannheimer, J. D. A comparison of strength gain between concentric and eccentric contractions. *Phys Ther* 49:1201–1207, 1969.

78. Mason, M. L., and Allen, H. S. The rate of healing of tendons: An experimental study on tensile strength. *Ann Surg* 194:424–459, 1964.

79. Matsen, F. A., and Teitz, C. C. *Health Maintenance: Orthopaedic Knowledge Update II.* Chicago, American Association of Orthopaedic Surgeons, 1987.

80. Mayhew, J. L., and Gross, P. M. Body composition changes in young women with high resistance weight training. *Res Q* 45:433–440, 1974.

81. McMorris, R. O., and Elkins, E. C. A study of production and evaluation of muscle hypertrophy. *Arch Phys Med Rehabil* 3:420–426, 1954.

82. Meyers, C. R. Effects of two isometric routines on strength, size, and endurance in exercised and non-exercised arms. *Res Q Am Assoc Health Phys Educ* 38:430–440, 1967.

83. Moffroid, M. T., Whipple, B. A., Hofkosh, J., Leuman, E., and Thistle, H. A study of isokinetic exercise. *Phys Ther Rev* 49:734–747, 1969.

84. Moffroid, M. T., and Whipple, R. H. Specificity of speed exercise. *Phys Ther* 50:1692–1699, 1970.

85. Montgomery, J. B., and Steadman, J. R. Rehabilitation of the injured knee. *Clin Sports Med* 4(2):333–343, 1985.

86. Moore, J. C. Active resistive stretch and isometric exercise in strengthening wrist flexion in normal adults. *Arch Phys Med Rehabil* 2:246–269, 1971.

87. Müller, E. A. Influence of training and of inactivity on muscle strength. *Arch Phys Med Rehabil* 51:449–462, 1970.

88. Müller, E. A. Training muscle strength. *Ergonomics* 2:216–222, 1959.

89. Müller, E. A. Physiology of muscle training. *Rev Can Biol* 21:303–313, 1962.

90. Noyes, F. R., De Lucas, J. L., and Torvik, P. J. Biomechanics of anterior cruciate ligament failure: An analysis of strain-rate sensitivity and mechanisms of failure in primates. *J Bone Joint Surg* 56A:236–253, 1974.

91. O'Shea, P. Effects of selected weight training programs on the development of strength and muscle hypertrophy. *Res Q* 37:95–102, 1966.

92. Paulos, L. E., Noyes, F. R., Grood, E., and Butler, D. L. Knee rehabilitation after anterior cruciate ligament reconstruction and repair. *Am J Sports Med* 9:140–149, 1981.

93. Paulos, L. E., Payne, F. C., III, and Rosenberg, T. D. Rehabilitation after anterior cruciate ligament surgery. *In* Jackson, D., and Drez, D. (Eds.), *The Anterior Cruciate Knee.* St. Louis, C. V. Mosby, 1987, pp. 291–314.

94. Perrine, J. J., and Edgerton, V. R. Muscle force, velocity, and power relationships under isokinetic loading. *Med Sci Sports Exerc* 10:159–166, 1978.

95. Perrine, J. J., and Edgerton, V. R. Isokinetic anaerobic ergometry presentation. *Med Sci Sports* 7:78, 1975.

96. Perry, J. Scientific basis of rehabilitation. *Instruct Course Lect* 34:385–388, 1985.

97. Peterson, J. A. Total conditioning: A case study. *Athletic J* 1:40, 1975.

98. Pipes, T. V. Isokinetic vs. isotonic strength training in adult men. *Med Sci Sports* 7:262–274, 1975.

99. Pipes, T. V. Strength training modes: What's the difference? *Scholastic Coach* 46:96–124, 1977.

100. Pipes, T. V., and Wilmore, J. H. Isokinetic vs. isotonic strength training in adult men. *Med Sci Sports* 7:262–274, 1975.

101. Prentice, W. E., and Kooima, E. F. The use of proprioceptive neuromuscular facilitation techniques in the rehabilitation of sports-related injury. *Athletic Training* Spring, 1:26–31, 1986.

102. Rasch, P. J., and Morehouse, L. E. Effect of static and dynamic exercises on muscular strength and hypertrophy. *J Appl Physiol* 11:29, 1957.

103. Rilensberick, D. H., Gamble, J. G., and Max, S. R. Response of mitochondrial enzymes to decreased muscular activity. *Am J Physiol* 225:1295, 1973.

104. Rosentwig, J., and Hinson, M. M. Comparison of isometric, isotonic, and isokinetic exercises by electromyography. *Arch Phys Med Rehabil* 53:249–252, 1972.

105. Rozier, C. K., Elder, J. D., and Brown, M. Prevention of atrophy by isometric exercise of a casted leg. *J Sports Med* 19:191–194, 1979.

106. Saltin, B., and Gollnick, P. Skeletal muscle adaptability: Significance for metabolism and performance. *In* Peachy, L. D., Adrian, R. H., and Geiger, S. R. (Eds.), *Handbook of Physiology:* Section 10. *Skeletal Muscle.* Bethesda, American Physiological Society, 1983, pp. 555–631.

107. Singer, K. M. Health maintenance of the musculoskeletal system. *In Orthopaedic Knowledge Update I.* Chicago, American Association of Orthopaedic Surgeons, 1984.

108. Steadman, J. R. Rehabilitation after knee ligament surgery. *Am J Sports Med* (8)4:294–296, 1980.

109. Swegan, D. B. The comparison of static contraction with standard weight training in effect on certain movement speeds and endurance. Thesis (Ph.D., Microcard), Pennsylvania State University, 1957.

110. Tardieu, C., Tabavy, J.-C., Tabavy, C., and Tardieu, G. Adaptation of connective tissue length to immobilization in the lengthened and shortened positions in rat soleus muscle. *J Physiol [Paris]* 78:214–220, 1982.

111. Thistle, H. G., Hislop, H. J., Moffroid, M., and Lowman, E. W. Isokinetic contractions: A new concept of resistance exercise. *Arch Phys Med* 48:279–282, 1967.

112. Thorstensson, A., Hulten, B., von Dobeln, W., and Karlsson, J. Effect of strength training on enzyme activities and fibre characteristics in human skeletal muscle. *Acta Physiol Scand* 96:392–398, 1976.

113. Thorstensson, A., Sjodin, B., and Karlsson, J. Enzyme activities and muscle strength after "sprint training" in man. *Acta Physiol Scand* 94:313–318, 1975.

114. Walters, L., Stewart, C. L., and LeClaire, J. F. Effect of short bouts of isometric and isotonic contractions on muscular strength and endurance. *Am J Phys Med* 39:131–141, 1960.

115. Ward, P. E. The effects of isometric and isotonic exercises on strength, endurance, anthropometric measurements. Thesis (Master's) University of Washington, 1963.

116. Williams, I. F., Craig, A. S., Parry, D. A. D., Goodship, A. E., Shah, J., and Silver, I. A. Development of collagen fibril organization and collagen crimp patterns during tendon healing. *Int J Biol Macromol* 7:275–282, 1985.

117. Williams, P. E., and Goldspink, G. The effect of immobilization on the longitudinal growth of striated muscle fibers. *J Anat* 116:45–55, 1973.

118. Williams, P. E., and Goldspink, G. Connective tissue changes in immobilized muscle. *J Anat* 138:343–350, 1984.

119. Williams, M., and Stutzman, L. Strength variation throughout the range of joint motion. *Phys Ther Rev* 39:145–152, 1959.

120. Wilmore, J. H. Alterations in strength, body composition, and anthropometric measurements consequent to 10-week weight training program. *Med Sci Sports* 6:133–138, 1974.

121. Wilson, C. J., and Dahners, L. E. An examination of the mechanism of ligament contracture. *Med Sci Sports* 7:286–291, 1988.

122. Wolbers, C. P., and Sills, F. P. Development of strength in high school boys by static muscle contractions. *Res Q Am Assoc Health Phys Educ* 27:446–450, 1956.

123. Woo, S. L.-Y., and Buckwalter, J. A. *Injury and Repair of the Musculoskeletal Soft Tissues.* Chicago, American Association of Orthopaedic Surgeons, 1988.

124. Woo, S. L.-Y., Gelberman, R. H., Cobb, N. G., Amiel, D., Lothringer, K., and Akeson, W. H. The importance of controlled passive mobilization on flexor tendon healing: A biomechanical study. *Acta Orthop Scand* 52:615–622, 1981.

125. Woo, S. L.-Y., Gomez, M. A., Sites, T. J., Newton, P. O., Orlando, C. A., and Akeson, W. H. The biomechanical and morphological changes in the medial collateral ligament of the rabbit after immobilization and remobilization. *J Bone Joint Surg* 69A:1200–1211, 1987.

126. Woo, S. L.-Y., Matthews, J. V., Akeson, W. H., Amiel, D., and Convery, E. R. Connective tissue response to immobility: Correlative study of biomechanical and biochemical measurements of normal and immobilized rabbit knees. *Arthritis Rheum* 18:257–264, 1975.

127. Wyatt, M. P., and Edwards, A. M. Comparison of quadriceps and hamstring torque values during isokinetic exercise. *J Orthop Sports Phys Ther* 3:48–56, 1981.

128. Wyke, B. D. *Principles of General Neurology.* Amsterdam, Elsevier, 1969.

129. Yamamoto, S. K., Hartman, C. W., Feagin, J. A., and Kimball, G. Functional rehabilitation of the knee: A preliminary study. *J Sports Med* 3(6):228–291, 1976.

S E C T I O N E

Scientific Basis of Therapeutic Modalities

Lyle J. Micheli, M.D.
Michael A. Monmouth, M.D.

STATIONARY BICYCLE
ELASTIC RESISTANCE
CONTINUOUS PASSIVE MOTION
DIMETHYL SULFOXIDE
ANABOLIC STEROIDS

STATIONARY BICYCLE

The stationary bicycle, or bicycle ergometer, is one of the most widely available mechanisms in fitness and health clubs, physical therapy facilities, and gymnasiums. Even the simplest of these devices usually has some measurement instrument for calculating work output, adjustable seat and handlebar heights, and, of course, variable speed of operation (Fig. 7E–1). The lowest possible setting is 25 watts on most machines, with the typical training range from 70 to 110 watts of work.

The earliest work on measuring and enhancing aerobic fitness using the maximum oxygen uptake (Vo_2 max) was done by Astrand and Salton [16] with cycle ergometers. Although the apparatus has largely been supplanted by treadmill testing in many research situations now, it still is used in cardiac rehabilitation, in testing or training individuals who cannot run, and, with modifications, in training and testing upper extremities.

Most protocols in cardiac rehabilitation simply use the pulse rate, which has been found to be a safe and consistent parameter. Training is usually done in the midrange between resting pulse and maximal pulse rate. It has been found that this will result in an equivalent work output of approximately 70% Vo_2 max. This appears to be independent of patient weight, height, and other parameters.

Maintaining or attaining enhanced aerobic fitness is not specific to the rehabilitation of the cardiac patient. Improved aerobic fitness is the foundation of most sports training programs. In addition, Blair and associates have recently proved a positive correlation between aerobic fitness and longevity [20].

Describing aerobic fitness as cardiovascular fitness is incorrect, since many of the adaptive changes to increased aerobic work occur in the skeletal muscle bed; cardiac or pulmonary changes are secondary and certainly not limiting factors.

A second and growing use of the stationary cycle is for rehabilitation of the lower extremity following injury or surgery [49–52, 106]. One of the most obvious advantages of stationary cycling as a rehabilitative tool is the widespread availability of the device.

Indications for Use

First, use of this device is indicated for maintenance of aerobic fitness following illness or injury or surgery. Second, use of the stationary cycle provides well leg exercise when the opposite extremity is immobilized. It has been shown that exercise of the well leg also has beneficial effects on the immobilized leg. The third indication is rehabilitation of the injured extremity itself, specifically to increase the range of motion, endurance, strength, and power.

Use of the stationary cycle in rehabilitation is contraindicated for a patient with limited motion at the knee or hip who has severe pain associated with the use of the cycle. This restriction includes patients with patellofemoral arthrosis. Of course, use is also contraindicated with patients who have associated injuries or illnesses that render their use of the cycle unsafe.

FIGURE 7E–1
The stationary bicycle may be used for cardiovascular fitness enhancement or for lower extremity rehabilitation.

Clinical Applications

1. Clinical applications include the rehabilitation of a patient with an injured or painful knee following knee surgery or for knee overuse injury.

2. The cycle ergometer or stationary cycle may be specifically indicated for rehabilitation of a patient after anterior cruciate ligament surgery because of the low level of anterior shear associated with the use of this machine.

3. The stationary cycle may have particular use in patients who cannot run or swim for aerobic or rehabilitative exercise because they cannot absorb the impact of running or cannot swim.

4. The stationary cycle may be particularly useful following ankle injury or surgery.

5. The stationary cycle may be especially useful following back injury or surgery. Care must be taken to adjust the cycle so that no pain is sustained while performing the cycling activity.

6. Certain patients with disease states such as rheumatoid arthritis or multiple sclerosis or exces-

sively weak or cachectic patients who cannot tolerate other aerobic and endurance training devices can still be rehabilitated effectively using the stationary cycle.

Need for Further Research

First, additional research is needed in cycle design. It is possible to develop cycles that employ different body positions, such as semireclining or reclining; cycles with variable crank lengths, including a decreased length to accommodate patients with limited ranges of motion; and cycles with pedals and handgrips that can be used interchangeably for either upper or lower extremity rehabilitation.

Second, additional research is required regarding attainable strength, power, and endurance in the injured extremity when the cycle is used as a rehabilitation device. Although it is commonly believed that restoration of complete strength after injury can be attained using resistive weight training or strength training, the ability of patients using the cycle to attain similar levels of strength, endurance, power, and range of motion following injury has not been well documented.

Further study is also needed in the specific mechanics of cycling as they relate to the biomechanics of both uninjured and injured extremities.

Cycling Validation

One study of importance for the aerobic use of the stationary bicycle was done by Costill and co-workers [33] in 1977. This study demonstrated that stationary cycling enhanced restoration of aerobic enzymes of muscles in the postoperative period. In this study, patients were immobilized with casts for 2 weeks after meniscectomies. After 2 weeks, the patients were divided into two groups. One group actively exercised the knee using a stationary cycle in addition to performing knee extension exercises. The other group used only resistive exercises with a 20-pound weight. At the end of 6 weeks, the patients in the cycling group had restored the aerobic enzyme succinic acid dehydrogenase (SDH) to normal levels. The enzyme had been reduced 26% to 42% in the first 2 weeks of casting. In the group of patients using resistive exercise only, SDH activity fell an additional 10% to 16%. In the group who had had aerobic training in addition to injured leg cycling, resulting SDH levels were in most cases higher than the SDH levels of the normal leg. In addition, the maximum oxygen uptake (VO_2 max) was also maintained by cycling with the injured leg.

In 1985 a study by Henning and colleagues [77] of cruciate ligament injury demonstrated the lack of deformation of the cruciate ligament during cycling. In this study, measurement devices were implanted in the anterior cruciate ligament of patients with first-degree anterior cruciate ligament (ACL) strains. These patients were then tested while performing various activities. The amount of deformation caused by each activity was compared with the deformation resulting from an 80-pound Lachman test. Cycling was found to produce 7% as much elongation as an 80-pound Lachman test. Furthermore, a one-leg half-squat at 21% plus the use of a 20-pound boot from full extension to 22 degrees flexion created an 87% to 121% elongation in the 80-pound Lachman test.

This study supports the use of cycling for rehabilitation in patients with injured ACLs or in those who had undergone cruciate repair.

In 1980 McLeod and Blackburn [106] published a study of the biomechanics of knee rehabilitation with cycling that also provided information useful for rehabilitation in knee ligament injuries. They demonstrated that the quadriceps can be effectively rehabilitated with cycling while placing minimal stresses on the anterior cruciate ligament, joint capsule, and posterior structures of the knee. By modifying the seat position and pedal positions, they could further decrease these forces. These researchers found that 102 degrees of knee flexion was necessary for cycling. They suggested that a starting speed of 40 rpm for 10 minutes would be appropriate. The goal was to progress to a level of 60 rpm and a work duration of 60 minutes. They suggested that sprints be added once this level was reached.

It was noted that the lower the seat position, the more it decreased gastrocnemius function and increased hamstring function, providing protection for the cruciate ligament. The researchers thought also that toe pedaling was more effective than heel pedaling in protecting the anterior cruciate ligament in capsule structures.

A study by Ericson and Nisell [51] in 1986 lends further support to the usefulness of cycling for rehabilitation and refines its parameters. They studied six healthy subjects and made several important observations. First, the anterior pedal position does create less anterior shear force on the ACL. They identified a midposition on the cycle in which the tibiofibular shear force was low (37 degrees) and then compared the compressive and shear forces of cycling with other activities. The comparison revealed that cycling had only a 1.2 times body weight (BW) compression force compared with two to four times BW for normal walking, four times BW for stair climbing, three to seven times BW for rising from a normal chair, and two to three times BW for lifting a 12-kilogram burden.

Also, the vastus medialis and vastus lateralis groups appeared to be the muscle groups most involved in cycling. These two muscle groups had 54% and 50%, respectively, of the maximal electromyographic (EMG) isometric activity, whereas the more central rectus femoris muscle activity was lower (12% maximum EMG).

Ericson and Nisell postulated that this difference in muscle use was due to the fact that the rectus femoris is a two-joint muscle. Therefore it may be necessary to address this in a later phase of rehabilitation. The researchers noted that the anterior tibial shear force increased significantly when the posterior instead of the anterior foot position was used. They thought that this was most likely due to the increased extension of the knee. Finally, the compressive force was reduced by a decrease in workload or an increase in saddle weight.

ELASTIC RESISTANCE

Use of elastic resistance in rehabilitation is well known and well accepted. The elastic resistance devices currently suggested include a door attachment, handles, and foot attachments that allow lower extremity exercise. Indications for this modality include rehabilitation of the knee, ankle, shoulder, elbow, and wrist. There is some potential for use in patients with problems of the hip and lower back as well (Fig. 7E–2).

The major advantage of this modality is that it is simple, inexpensive, and versatile. It can be used both in the therapy clinic and at home. Additionally, it can be used to supplement more sophisticated techniques. It can be used to enhance or to supply resistance in both concentric and eccentric exercises. It can also be used for tethered running and tethered side-to-side exercise and is valuable in allowing motions needed in a specific sport—for example, a side push-off exercise using the tethered cord simulates the turn in skiing. In addition, elastic resistance is particularly useful for strengthening the ankle, for which very few free-weight or machine techniques can be used effectively.

Contraindications for the use of elastic resistance include a lack of appropriate instruction and a lack of appropriate measures after ligament surgery or joint injury. Care should be taken to avoid ranges

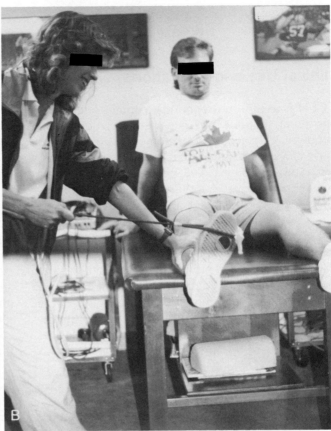

FIGURE 7E–2
A, Elastic bands may be used by the patient for dynamic progressive resistive exercise (PRE). *B,* Elastic bands facilitate active PRE with a therapist.

of motion that create either pain or cramping in the patellar area and that might place ligament procedures at risk.

Recommendations for further research include the following areas:

1. Quantification of the amount of resistance delivered by increases in the length or size of cord.

2. EMG evaluation of the muscles involved during specific exercises related to a specific sport would provide important information. The timing of specific exercises and the duration of given movements should be quantified and patterned.

3. Finally, ranges of motion in the weight-bearing position that create stress on ligaments and the patellofemoral mechanism should be evaluated.

The validation of elastic resistance has not been done at this time.

CONTINUOUS PASSIVE MOTION

Continuous passive motion (CPM) is a rehabilitation technique that involves, in its simplest form, passive motion of an extremity by an externally applied force. There are at least 13 devices now being used to mobilize injured or postoperative extremities in this way. These were first introduced clinically in Canada in 1978 and in the United States in 1982. Their clinical use is based on the work of Salter and co-workers [117, 118, 135–139], which began in 1970. Based on animal work, primarily with rabbits, they demonstrated an enhanced rate of healing of articular cartilage, prevention of cartilage degeneration after septic arthritis, enhanced healing of patellar tendon and knee ligaments, and accelerated clearing of hemarthrosis. Subsequent clinical studies by Salter and Ogilvie-Harris [138] and others [56, 101], including Coutts and associates [36], have shown a decreased incidence of deep venous thrombosis, decreased postoperative swelling, significant decrease in postoperative medication, and, following total knee replacement, a statistically increased range of motion of the knee.

The mechanisms by which CPM accomplishes these results are unknown. They may include direct cellular response to external stress or a generalized enhanced circulatory increase in the extremity.

Clinical Use and Indications

CPM is used in postoperative patients to enhance pain relief, improve the general circulation of the extremity, enhance the nutrition of the joint, particularly the articular cartilage, and prevent contractions and adhesions. The clinical indications also include postoperative rehabilitation and management following ligament repair [26, 125, 143] or other intra-articular surgery; internal fixation of fractures, particularly articular fractures [138]; synovectomy; joint manipulation for contracture; operations on the extremities or joints for infection; and total joint replacement [36, 57, 131]. There are continuous passive motion devices available for both the upper and lower extremities; although primarily used for elbow and knee disorders, certain devices can also be used to mobilize the ankle and shoulder.

Contraindications

There are few contraindications for the use of CPM. Examples include a noncompliant patient or a clinical situation in which lack of information or experience by nursing personnel or therapists would render use of the device a risk to the patient.

Risks

The risks in the use of CPM devices include the following: disruption of the surgical measures such as ligament repair or fracture fixation, the possibility of hemorrhage in the immediate postoperative period, and malfunction of the device.

Further Research

Further research with the modality is needed to refine the dose-response curve and to standardize protocols for use. Much of this research could be done with animal models, and it would be appropriate for clinically controlled studies in humans. Additional research is needed to differentiate protocols for the various clinical conditions for which it is now used, including status following ligament repair, status following joint infection, and status following joint replacement. In particular, additional research is needed to develop equipment that

delivers the basic clinical requirements of passive motion at an efficient but cost-effective level. Investigation of the relative interaction between neuromuscular stimulation and CPM is a necessity. Presently, two devices on the market incorporate additional neuromuscular stimulation, including possible stimulation of antagonistic muscle groups. There is little information and no clinical research at this time on which to base the indications for or relative benefits of this combination. Human clinical studies—in particular, controlled studies of the relative decrease in postoperative swelling, rate of wound healing, synovial health as identified by synovial clearance studies, and articular cartilage healing—are needed to supplement the extensive laboratory animal work done by Dr. Salter and colleagues. An additional need is identification of structures that are at risk of stress from these devices and from CPM at the biomechanical level. This identification could be accomplished with further cadaver studies or human clinical studies employing implanted strain gauges.

A Rationale for Passive Motion Programs

The effect of stress deprivation on connective tissue has been documented in numerous animal studies [1–7, 9, 23, 55, 74, 109, 159, 172]. The consistent feature of the gross appearance of the periarticular and synovial tissues of immobilized joints is fibrofatty connective tissue proliferation within the joint space. In a rat model studied by Evans and colleagues [55], there was significant growth within 30 days of fibrofatty tissue over the cartilage surface and destruction of the cartilage in 60 days. Ulceration develops at cartilage contact points, with rate and severity depending on the rigidity of the mobilization and the degree of joint compression. There were secondary changes in subchondral bone.

The biomechanical and biochemical response to stress alteration has been studied by Akeson and colleagues [1–7] in San Diego in both dog and rabbit models. Casts and internal fixation were used to immobilize the hind limb. Biomechanical changes of the knee composite were evaluated with an arthrograph. Joint stiffness in the knee was measured in terms of torque-angular deformation. Following immobilization of 2 weeks, increased joint stiffness was seen as an increase in torque needed to move the joint and an increase in the area of hysteresis. Progressive stiffness was seen between the second and twelfth weeks of the study.

Biomechanical changes in ligaments were noted

with testing of the isolated medial collateral ligament (MCL)–bone complex following 9 weeks of immobilization. Load to failure under tension revealed a decrease in linear slope, in ultimate load to failure, and in energy absorbed to about one-third that of the control limb. The point of failure changed from a midsubstance ligament tear in controls to an avulsion at the fibular attachment site.

In another study, animals that were mobilized after 9 weeks of immobilization with cage activity for 12 weeks were studied [9]. The MCL recovery was near normal; however, the attachment site at the tibia remained weak, and failure occurred at two-thirds of the load applied in the control limb.

Still another study on the conditioning effect of the connective tissue was performed in San Diego using miniature swine [173]. Pigs were randomly divided into two groups. One group's activity consisted of eating and sleeping, whereas the second group of animals ran 6 miles a day, 5 days a week. After 3 months of exercise, there was no difference in the connective tissue in the control and experimental animals. After 1 year of exercise the nonexercised animals had hypertrophy of the distal tibia, the area of the extensor tendon was increased 21%, and load to failure of the extensor tendon was increased 62%. Thus numerous animal studies verify that stress deprivation results in rapid changes in joint mechanics, whereas exercise results in a slow increase in the strength of connective tissues.

Other studies have evaluated the biomechanical effects of immobilization [48, 74, 127, 159, 168, 169, 172]. The water content of connective tissue fibers, which is normally 65% to 75%, decreases. It is believed that water serves as a spacer between individual collagen fibers, permitting discrete movement of one fiber past another. The largest change in stress-deprived tissues is a drop in the glycosaminoglycan (GAG) content that reduces normal connective tissue pliability. The normal turnover of GAG is short, with a half-life in terms of days, as opposed to collagen, which has a turnover period of months to 1 to 2 years. In the immobilized limb, there is a decrease in the rate of GAG production. A loss of GAG and water results in a decrease in fiber-to-fiber distance and friction between fibers with motion. Studies of collagen metabolism reveal that collagen turnover continues at a reduced level. Randomly oriented fibers are produced that result in cross-linking between preexisting fibers.

Beneficial Effects of Motion

Animal studies have revealed that there is an increased clearance rate in the synovial joint with joint motion: hemarthrosis is cleared more rapidly [117], as are large particles injected into the joint. Septic joints recover with less cartilage destruction [136]; cartilage lesions have an improved healing rate [139]; and there is improved healing of sutured tendons, as shown in patellar tendon laceration of a rabbit model [135].

Gelberman and co-workers [63, 65], using canine paw models, have demonstrated that 5 minutes of passive motion markedly improved the healing of the sutured flexor tendon. The healing of the moved tendon was more intrinsic as opposed to the extrinsic healing of the completely immobilized paw. The tensile strength of the healed tendon was greater in the joint that was moved than in the one that was immobilized. Inoue and associates [85] reported improved healing in a dog model of the MCL tendon in the mobilized animal as opposed to that in the immobilized animal.

In conclusion, there is a wealth of literature supporting the thesis that motion and stress are important for the maintenance of normal connective tissue and the healing of injured connective tissue. The literature also supports the hypothesis that motion enhances blood flow, decreases pain, and may be a factor in neuromuscular retraining programs.

DIMETHYL SULFOXIDE

The biologic and anti-inflammatory properties of dimethyl sulfoxide (DMSO) were first recognized in 1964 [134]. Since then, clinical trials have yielded conflicting reports of the efficacy of DMSO, and it is not presently approved by the Food and Drug Administration for general use [146].

However, it is approved in various concentrations for veterinary use as an irrigant for interstitial cystitis and as a topical solution for scleroderma. It is also readily available over the counter as a chemical solvent. Thus it can easily be obtained by athletes to treat acute sprains and strains. Recent chemical analysis of random samples of DMSO found that all such samples were without impurities except for traces of water [133].

Adverse reactions include a localized dermatitis and a characteristic garlic odor of the breath, which have made it difficult to perform true double-blind studies [92, 146]. Further studies are needed to better determine the effectiveness of DMSO in the treatment of musculoskeletal injuries.

ANABOLIC STEROIDS

The use of anabolic steroids by competitive athletes appears to have grown in recent years. Among elite athletes, a number of well-publicized scandals have highlighted the abuse of these drugs. During the 1988 Olympic Games in Seoul, South Korea, a popular Canadian sprinter was stripped of his gold medal after a world record–breaking performance because he later tested positive for the anabolic steroid stonozolol [105]. Of greater concern are recent reports that the use of anabolics by adolescents may be increasing [24].

Epidemiology of Use

Anabolic steroids have become popular among athletes because of their purported ability to increase muscle mass and improve strength. The first use of anabolic steroids by humans reportedly occurred during World War II, when they were given to German troops to enhance their aggressiveness. The first reported use of the agents to enhance athletic performance occurred in 1954, when Russian and Bulgarian weight lifters dramatically improved their performance. The use of the drugs by other athletes in other countries and in a great variety of sports soon followed [162].

The use of anabolic steroids has certainly been most visible among athletes. Estimates from a number of informal surveys in the past suggested that as many as 80% to 90% of athletes involved in power sports such as weight lifting used them [31, 60, 98, 100, 103, 110, 162]. Recent surveys have attempted to document more accurately the extent of anabolic steroid use among other groups. Their use among intercollegiate males has been estimated to range from 2% to 20% [130]. In two surveys, use among adolescent males has been determined to be 6.6% in one study [24] and as high as 11% in the other [90].

Anabolic steroids have been used primarily by weight lifters and body builders and by participants in field events such as the shot put and discus throw. However, their use has spread to encompass football players, sprinters, distance runners, wrestlers, and swimmers, including women athletes [158].

The direct effect of anabolic steroids on athletic performance remains controversial. Nevertheless, there is presently a consensus that these agents probably can help to increase muscle size and strength but only under specific conditions of prior and concurrent strength training and not without the risk of potentially serious side effects.

Pharmacology and Physiology of Testosterone [29, 76, 91]

Anabolic steroids are derivatives of the hormone testosterone. Testosterone is the principal biologically active androgen in both men and women. It is a 19-carbon derivative of cholesterol. In males it is secreted primarily by the Leydig cells of the testes. In women the adrenal secretion of androstenedione and dehydroepiandrosterone (DHEA) provides the major physiologic supply of androgenic hormones, probably exerting their effect through conversion to testosterone. Androgens are also metabolized to form the feminizing hormone estradiol.

Testosterone and other androgens in the circulation are bound to plasma proteins, including sex hormone–binding globulin (SHBG). However, this binding is completely reversible and provides a rapid supply of active hormone. Testosterone and other androgens pass freely through cell membranes and interact with cytoplasmic receptors. These complexes subsequently bind to nuclear chromatin to activate transcription of messenger RNA from DNA and induction of protein synthesis.

In males secretion of testosterone by the Leydig cells is stimulated by luteinizing hormone (LH) from the anterior pituitary, which in turn is stimulated by gonadotropin-releasing hormone (GnRH) secreted from the hypothalamus. Testosterone as the end product acts as a negative-feedback inhibitor on GnRH.

Testosterone has both androgenic and anabolic effects. The androgenic effects include sexual differentiation in utero, maturation at puberty, and development of male secondary sex characteristics such as deepening of the voice, facial and body hair distribution, development of a normal libido, and spermatogenesis. The anabolic effects of testosterone include the stimulation of muscle development and erythropoiesis.

Pharmacology and Physiology of Anabolic Steroids [29, 45, 96, 144, 166]

Anabolic steroids were developed in an attempt to maximize the anabolic, as opposed to the androgenic, effects of testosterone. Historically, anabolic steroids have been used to treat specific types of anemias because of their ability to stimulate erythropoiesis. They have also been used in hypogonadal males to stimulate sexual development.

When accompanied by adequate protein and caloric intake, anabolic steroids have been shown to be effective in reversing a negative nitrogen balance produced by glucocorticosteroids under conditions of stress such as trauma, burns, or infection. However, they are not thought to offer any significant advantage over parenteral nutrition in these situations in improving nitrogen balance.

Anabolic steroids are broadly classified into two different groups: those that are active when taken orally, and those that are active when given parenterally (Table 7E–1).

The orally active anabolic steroids are produced by adding an alkyl group at the C-17 position of the D-ring of testosterone. This alkylation at C-17 affords increased resistance to hepatic inactivation, thus allowing oral administration. Methandrostenolone (Dianabol), one of the more frequently used oral agents, has a double bond between C-1 and C-2 that also contributes to resistance to hepatic metabolism.

The parenterally active anabolic steroids are 17-beta esterified derivatives of testosterone. They are given intramuscularly. They are highly lipid soluble and thus have a prolonged duration of action, being slowly absorbed from their sites of injection. Nandrolone decanoate (Deca-Durabolin) is one of the more commonly used injectable steroids, employed when a sustained effect is desired. The synthetic anabolic steroids have a considerably lower affinity for SHBG than does testosterone. Consequently, more unbound substrate is available to interact with cytoplasmic receptors of target cells. This may help to explain the enhanced biologic activity of anabolic steroids compared with naturally occurring testosterone.

Both the androgenic and anabolic effects of synthetic and naturally occurring androgens are thought to be mediated through a common receptor, leaving unexplained the apparent increase of anabolic-to-androgenic ratio of synthetic anabolic steroids compared with testosterone. This increase might be best explained by differences in the enzymatic metabolism of the target tissue.

For example, the enzyme 5-alpha-reductase converts testosterone into an even more potent derivative. The same is true of the 5-alpha-reduced derivative of methandrostenolone. However, in target tissue methandrostenolone resists reduction because of its structure. Thus, in sex accessory tissue in which 5-alpha reduction is high, methandrostenolone has less androgenic activity compared with testosterone, which can be reduced to a more active compound. However, in muscle, in which there is little 5-alpha-reductase activity, there is no appreciable difference in the potency of testosterone compared with methandrostenolone. Therefore, compared with testosterone, methandrostenolone has a greater anabolic-to-androgenic activity ratio.

In the case of nandrolone decanoate, its 5-alpha-reduced product has a lower affinity than the parent compound for the androgen receptor. Thus in sex accessory tissue, in which 5-alpha-reductase is high, nandrolone decanoate is reduced to a less active compound. Therefore it also has a greater anabolic-to-androgenic activity ratio compared with testosterone [29, 91]. However, all synthetic anabolic steroids maintain some androgenic activity, which can lead to untoward side effects.

Effects of Anabolic Steroids on Athletic Performance

The majority of studies on the effects of anabolic steroids on sports performance have concentrated on the increases in overall strength and aerobic performance in athletes. Haupt and Rovere [75] performed an extensive review of the literature regarding these issues in 1984. They found that studies were consistent in reporting no improvement in aerobic performance in athletes treated with anabolic steroids. However, of 24 studies investigating the effects of anabolic steroids on human strength, 14 reported significant strength increases with steroids [13, 14, 22, 60, 78, 87, 88, 119–121, 145, 156, 163, 167], whereas 10 reported no increases in strength [40, 56, 58, 62, 67, 79, 89, 102, 140, 154]. In an attempt to explain this inconsistency in the literature, they conducted a critical statistical analysis of the materials and data used in all these studies. Haupt and Rovere [75] concluded that it was the combination of previous training in weight lifting followed by continuous training in weight lifting during the steroid regimen that was most consistently associated with significant increases in

TABLE 7E–1
Characteristics of Anabolic Steroids

Oral Steroids
Alkylation at C-17
Increased resistance to hepatic inactivation, allowing oral uptake
Shorter half-life
Shorter duration of action
Drug detection more difficult

Injectable Steroids
Esterified derivation of testosterone
Highly lipid soluble
Slower absorption from injection sites
Prolonged duration of action
More easily detected

strength. Several earlier investigators also contended that weight lifting had to be combined with a high-protein diet to achieve a significant increase in strength while taking anabolic steroids.

These findings can also be explained in part by the anticatabolic effect of anabolic steroids. Anabolic steroids may improve the utilization of ingested protein and convert a negative nitrogen balance into a positive one [25, 56, 75, 96, 157] by reversing the catabolic effects of glucocorticosteroids released during periods of stress. It has been noted, however, that anabolic steroids may fail to induce significant nitrogen retention in athletes ingesting a diet deficient in calories and protein (Table 7E–2).

In the normal nonstressed state, the anticatabolic effects of anabolic steroids are not a factor, and the purely anabolic effects that result in a positive nitrogen balance are short-lived because the body strives to maintain homeostasis. However, during intensive weight lifting the athlete can be pushed into a catabolic state by stressing the body, resulting in release of glucocorticosteroids. Nitrogen retention is subsequently decreased, as is muscle protein formation. Thus the intensely trained weight lifter is thought to be in a chronic autobolic state. This may explain why many weight lifters reach a plateau in their training. When these athletes are given anabolic steroids that are profoundly antiautobolic,

nitrogen retention is improved, decreasing the loss of muscle protein, which may contribute to an overall increase in strength. Consequently, an athlete not previously trained in weight lifting would not be in an autobolic state and thus would not be expected to gain any benefits from the antiautobolic effects of anabolic steroids.

Anabolic steroids also have been reported to provide a placebo effect and motivational effects that induce a state of euphoria and diminish fatigue [129]. This would enhance training and might also lead to gains in strength.

Side Effects of Anabolic Steroids

Most adverse reactions to anabolic steroids are related to prolonged administration and excessive doses (see Table 7E–2). In men the use of anabolic steroids leads to inhibition of GnRH by the hypothalamus. This in turn inhibits the secretion of follicle-stimulating hormone (FSH) and LH, resulting in decreased spermatogenesis and testosterone production as well as testicular atrophy [82, 150]. These effects appear to be reversible when steroids are stopped. However, abnormal sperm counts may persist for several months or longer.

Gynecomastia has been reported in mature males taking anabolic steroids and results from the peripheral conversion of androgens to estrogenic substances. Breast tissue hypertrophy usually diminishes after cessation of anabolic steroids; however, it does not completely disappear [29]. Further adverse effects in men include worsening acne and accelerated baldness in those who are genetically predisposed to it.

In women, the use of anabolic steroids can result in severe virilizing effects. These include increased facial hair, baldness, deepening of the voice, and enlargement of the clitoris. These effects appear to be irreversible. Menstrual irregularity or cessation, increased libido, and aggressiveness have also been observed but seem to be reversible once the steroids have been stopped [41, 83, 97, 151].

In children, anabolic steroid use can cause gynecomastia in both sexes. It can also lead to premature closure of the epiphyses of long bones, resulting in stunted growth [42, 44, 166]. Other side effects include acne and oily skin. Acne occurs commonly from stimulation of the sebaceous glands (see Table 7E–2) [94].

Anabolic steroids are also thought to increase the risk of musculotendinous injury because of the discrepancy between muscle and tendon strength [98, 101].

TABLE 7E–2
Effects and Side Effects of Anabolic Steroids

Anabolic Effects
Improved nitrogen retention (with a diet sufficient in calories and protein)
Decreased loss of muscle protein in catabolic state

Androgenic Side Effects
  Men
    Accelerated baldness
    Gynecomastia
    Decreased spermatogenesis
    Testicular atrophy
    Acne

  Women
    Increased facial hair
    Baldness, deepening of voice
    Menstrual irregularities
    Increased libido
    Increased aggressiveness

Other Side Effects
Hepatic dysfunction (with oral steroids only)
Musculotendinous injury
Aggressive behavior
Mood elevation
Decreased HDL, increased cardiovascular risk
Physeal growth arrest in children

HDL = high-density lipoprotein.

Aggressive behavior, mood elevation, and irritability as well as frank psychotic and manic symptoms are some of the psychological changes associated with steroid use [129, 152, 158, 175].

Levels of high-density lipoprotein (HDL) cholesterol, thought to be protective in reducing atherosclerosis, have been shown to be reduced with anabolic steroid use. The occurrence of myocardial infarctions in a few young men known to use anabolic steroids suggests that this reduction in HDLs leads to an increased risk of cardiovascular disease [28, 84, 107, 150, 164].

Hepatic dysfunction seems to be confined to the use of oral alkylated anabolic steroids only. Elevation of liver function test (LFT) results, including serum glutamic pyruvic transaminase (SGPT), serum glutamic oxaloacetic transaminase (SGOT), alkaline phosphatase, and L-lactate dehydrogenase (LDH), has been noted. These changes are reversible if steroid treatment ceases [75, 166, 175]. It has been emphasized that SGOT, SGPT, and LDH are found in significant concentrations in other body tissues, including skeletal muscle. Thus, these levels may be increased owing to muscle activity alone and are not necessarily related to hepatic dysfunction. Two enzymes are thought to be more specific in detecting hepatic dysfunction from anabolic steroids [75]. These are a liver-specific isoenzyme of LDH and alkaline phosphatase, found primarily in liver and bone.

Pelosis hepatitis is the formation of multiple small blood-filled cystic lesions within the liver. It is a rare disorder, historically associated with tuberculosis. However, recent clinical reports suggest an association with oral anabolic use. These cysts can rupture, causing severe hemorrhage, liver failure, and death [18, 166]. Both benign and malignant liver tumors have been reported in conjunction with the use of anabolic steroids, including two cases of hepatocellular carcinoma in healthy young adults [28, 123].

Current Trends

Most anabolic steroids are thought to be obtained on the black market from "gym dealers." However, some athletes have also obtained them from pharmacists and physicians. Anabolic steroids can be purchased in many countries without prescription; thus, athletes training abroad have ready access to them and can serve as suppliers to other athletes.

Anabolic steroids are usually taken in cycles; that is, they are taken for a period of time and then are taken in progressively lower doses or not at all. In addition, several types of anabolic steroids may be taken during a cycle. This is called "stacking" and is an attempt to maximize the effects of anabolic steroids as close as possible to the time of competition while minimizing undue risk of drug detection.

Injectable forms of anabolic steroids were popular in recent years because they are associated with less toxicity to the liver. However, most users are now aware that injectable anabolic steroids can be detected in the urine months after use, whereas oral anabolic steroids usually are not detectable 2 to 14 days after use. Some researchers believe that this awareness has led to a switch from the longer-lasting, more easily detectable, injectable steroids to the oral preparations, which are more hepatotoxic but can be cleared from the system in less than 1 month [38].

More recently, testosterone has also been used to maintain the effects of previously taken anabolic steroids. Testosterone is difficult to detect in drug screening programs and thus may be used near the end of a cycle before a competition. However, the ratio of testosterone to epitestosterone can be determined and is used to help detect exogenous use of testosterone [39].

There have been reports of the use of human chorionic gonadotropin hormone, which is taken to permit a gradual secretion of natural testosterone [98]. Human growth hormone has even been tried by some athletes, even though its anabolic effects are questionable [34].

Traditionally, drug testing has been preannounced or is limited to the period immediately before or after major competitions. However, it is generally agreed that one of the most effective ways of discouraging the use of anabolic steroids is by random drug testing throughout the year. This technique has already been challenged as a violation of civil liberties and remains controversial [38].

Most of the studies documenting significant gains in strength in athletes taking anabolic steroids have involved weight lifters. However, similar gains in strength may also be possible in other athletes involved in power sports such as football or sprinting if the athlete is simultaneously involved in a vigorous strengthening program. This possibility deserves further study.

Future studies of the effects of anabolic steroids may prove difficult because of the ethics of giving higher than therapeutic doses to subjects involved in prospective trials. In addition, because of the motivational effects of anabolic steroids, the subject can usually distinguish them from placebos, making double-blind studies difficult.

Most of the studies of the side effects of anabolic

steroids have been in patients undergoing therapeutic treatment for medical conditions. However, an increasing number of complications are also being reported in athletes taking anabolic steroids. Further investigation into the long-term health effects of anabolic steroids in athletes is anticipated [37].

Although it is important from a scientific point of view to continue to study the potential benefits as well as the adverse effects of anabolic steroids, it is the opinion of most sports-governing bodies and medical organizations that the use of anabolic steroids is contrary to the ethics of sport, which includes equitable competition and prohibits the use of foreign substances to gain an unfair advantage [34]. From this aspect alone, it is believed that anabolic steroids have no place in athletics.

References

1. Akeson, W. H. An experimental study of joint stiffness. *J Bone Joint Surg* 43A:1022–1034, 1961.
2. Akeson, W. H., Amiel, D., and LaViolette, D. The connective-tissue response to immobility. A study of chondroitin-4 and 6-sulfate and dermatan sulfate changes, in periarticular connective tissue of control and immobilized knees of dogs. *Clin Orthop* 51:183–197, 1967.
3. Akeson, W. H., Amiel, D., Mechanic, G. L., Woo, S. L. Y., and Harmond, F. L. Collagen cross-linking alterations in joint contractures: Changes in the reducible crosslinks in periarticular connective tissue collagen after nine weeks of immobilization. *Connect Tissue Res* 5:15, 1977.
4. Akeson, W. H., Amiel, D., and Woo, S. L. Y. Immobility effects on synovial joints: The pathomechanics of joint contracture. *Biorheology* 17:95, 1980.
5. Akeson, W. H., Amiel, D., Woo, S. L. Y., and Harwood, F. L. Mechanical imperatives for synovial joint homeostasis: The present potential for their therapeutic manipulation. *Proceedings of the 3rd International Congress on Biorheology.* 1978, p. 47.
6. Akeson, W. H., Woo, S. L. Y., Amiel, D., Coutts, R. D., and Daniel, D. The connective tissue response to immobility: Biomechanical changes in periarticular connective tissue of the immobilized rabbit knee. *Clin Orthop* 93:356–362, 1973.
7. Akeson, W. Y., Woo, S. L. Y., Amiel, D., and Frank, C. B. The biology of ligaments. *In* Hunter, L.Y., and Funk, F. H. (Eds.), *Rehabilitation of the Injured Knee.* St. Louis, C.V. Mosby, 1984, pp. 93–148.
8. American College of Sports Medicine. Position stand on the use of anabolic steroids in sports. *Med Sci Sports Exerc* 19:534–539, 1987.
9. Amiel, D., Akeson, W. H., Harwood, F. L., and Mechanic, G. L. Effect on nine-week immobilization of the types of collagen synthesized in periarticular connective tissue from rabbit knees. *Trans Orthop Res Soc* 5, 1980.
10. Amiel, D., Frank, C., Harwood, F., Fronek, J., and Akeson, W. H. Tendons and ligaments: A morphological and biochemical comparison. *J Orthop Res* 1:257, 1984.
11. Amiel, D., Frey, C., Woo, S. L. Y., Harwood, F., and Akeson, W. Value of hyaluronic acid in the prevention of contracture formation. *Clin Orthop* 196:306–311, 1985.
12. Amiel, D., Kleiner, J., and Akeson, W. H. The natural history of the anterior cruciate ligament autograft of patellar tendon origin. *Am J Sports Med* (in press, 1992).
13. Ariel, G. The effect of anabolic steroid (methandrostenolone) upon selected physiological parameters. *Athletic Training* 7:190–200, 1972.
14. Ariel, G. Residual effect of an anabolic steroid upon isotonic muscular force. *J Sports Med Phys Fitness* 14:103–111, 1974.
15. Arm, S. W., Pope, M. H., Johnson, R. H., Fischer, R. A., Arvidsson, T., and Eriksson, E. The biomechanics of anterior cruciate ligament rehabilitation and reconstruction. *Am J Sports Med* 12:8–18, 1984.
16. Astrand, P. O., and Saltin, B. Maximal oxygen uptake and heart rate in various types of muscle activity. *J Appl Physiol* 16:977, 1961.
17. Bagheri, S. A., and Boyer, J. Pelosis hepatitis associated with androgenic-anabolic steroid therapy. *Ann Intern Med* 81:610–618, 1974.
18. Bailey, A. J., and Robins, S. P. Development and maturation of the cross-links in the collagen fibers of skin. *Front Matrix Biol* 1:130, 1973.
19. Baker, W. C., Thomas, T. G., and Kirkaldy-Willis, W. H. Changes in the cartilage of the posterior intervertebral joints after anterior fusion. *J Bone Joint Surg* 51B:737, 1969.
20. Blair, S. N., Kohl, H. W., Paffenbarger, R. S., Jr., et al. Physical fitness and all-cause mortality. A prospective study of healthy men and women. *JAMA* 202:2395–2401, 1989.
21. Blokker, C. P., Rorabeck, C. H., and Bourne, R. B. Tibial plateau fractures. *Clin Orthop* 182:193–199, 1984.
22. Bowers, R. W., and Reardon, J. P. Effects of methandrostenolone (Dianabol) on strength development and aerobic capacity. *Med Sci Sports* 4:54, 1972.
23. Brooke, J. S., and Slack, H. G. B. Metabolism of connective tissue in limb atrophy in the rabbit. *Ann Rheum* 18:129, 1959.
24. Buckley, W. E., Yesalis, C. E., Friedl, K. E., et al. Estimated prevalence of anabolic steroid use among male high school seniors. *JAMA* 260:3441–3445, 1989.
25. Bullock, G., White, A. M., and Worthington, J. The effect of catabolic and anabolic steroids on amino acid incorporation by skeletal muscle ribosomes. *Biochem J* 108:417–425, 1968.
26. Burks, R., Daniel, D., and Losse, G. The effect of continuous passive motion on anterior cruciate ligament reconstruction stability. *Am J Sports Med* 12:323–327, 1984.
27. Cabaud, H. E., Chatty, A., Gildengorin, V., and Feltman, R. J. Exercise effects on the strength of the rat anterior cruciate ligament. *Am J Sports Med* 8:79–85, 1980.
28. Cohen, J. C., Noakes, T. D., and Bernade, A. J. Hypercholesterolemia in male powerlifters using anabolic androgenic steroids. *Physician Sportsmed* 16:49–56, 1988.
29. Colby, H. D., and Longhurst, P. A. Fate of anabolic steroids in the body. *In* Thomas, J. (Ed.), *Drugs, Athletes and Physical Performance.* New York, Plenum, 1988, pp. 11–30.
30. Convery, F. R., Akeson, W. H., and Keown, G. H. The repair of large osteochondral defects. *Clin Orthop* 82:253–262, 1972.
31. Cooper, D. L. Drugs and the athlete. *JAMA* 221:1007–1011, 1972.
32. Cooper, R. R., and Misel, S. Tendon and ligament insertion. *J Bone Joint Surg* 52A:1, 1970.
33. Costill, D. L., Fink, W. J., and Habansky, A. J. Muscle rehabilitation after knee surgery. *Physician Sportsmed* 5:71–74, 1977.
34. Council on Scientific Affairs. Drug abuse in athletes: Anabolic steroids and human growth hormone. *JAMA* 259:1703–1705, 1988.
35. Coutts, R. D., Amiel, D., Woo, S. L. Y., Woo, Y. K., and Akeson, W. H. Establishment of an appropriate model for the growth of perichondrium in a rabbit joint milieu. *Trans Orthop Res Soc* 196, 1983.
36. Coutts, R. D., Toth, C., and Kaita, J. H. The role of continuous passive motion in the rehabilitation of the total knee patient. *In* Hungerford, D. (Ed.), *Total Knee Arthroplasty—A Comprehensive Approach.* Baltimore, Williams & Wilkins, 1984, pp. 126–134.

37. Cowart, V. S. Study proposes to examine football players, power lifters for possible long-term sequelae from anabolic steroid use in 1970s competition. *JAMA* 257:3021–3025, 1987.

38. Cowart, V. S. Some predict increased steroid use in sports despite drug testing, crackdown on suppliers. *JAMA* 257:3025–3029, 1987.

39. Cowart, V. S. Athlete drug testing receiving more attention than ever before in history of competition. *JAMA* 261:3510–3516, 1989.

40. Crist, D. M., Stackpole, P. J., and Peake, G. T. Effects of androgenic-anabolic steroids on neuromuscular power and body composition. *J Appl Physiol* 54:366–370, 1983.

41. Damste, P. H. Voice change in adult women caused by virilizing agents. *J Speech Hear Disord* 32:126–132, 1967.

42. Daniel, W. A., Jr., and Benett, D. L. The use of anabolic-androgenic steroids in childhood and adolescence. *In* Kochakian, C. D. (Ed.), *Anabolic-Androgenic Steroids.* New York, Springer-Verlag, 1976.

43. Dose, D. B. Arthroscopy pain management: A new approach. Presented at the Free-Standing Ambulatory Surgical Association Meeting, New Orleans, May 1985.

44. Dyment, P. G. The adolescent athlete and ergogenic aids. *J Adolesc Health Care* 8:68–73, 1987.

45. Eikelboom, F. A., and Van der Vies, J. (Eds.), Anabolics in the 80s. *Acta Endocrinol* (Suppl.)271, 1985.

46. Engkvist, O. Reconstruction of patellar articular cartilage with free autologous perichondrial grafts. An experimental study in dogs. *Scand J Plast Reconstr Surg* 13:361, 1979.

47. Engkvist, O., and Johansson, S. H. Perichondrial arthroplasty. A clinical study in twenty-six patients. *Scand J Plast Reconstr Surg* 14–71, 1980.

48. Enneking, W. F., and Horowitz, M. The intra-articular effects of immobilization on the human knee. *J Bone Joint Surg* 54A:973, 1972.

49. Ericson, M. O., Bratt, A., and Nisell, R. Muscular activity during ergometer cycling. *Scand J Rehab Med* 17:53–61, 1985.

50. Ericson, M. O., Ikholm, J., and Svensson, O., et al. The forces on ankle joint structures during ergometer cycling. *Foot Ankle* 6:135–142, 1985.

51. Ericson, M. O., and Nisell, R. Tibiofemoral joint forces during ergometer cycling. *Am J Sports Med* 14:285–293, 1986.

52. Ericson, M. O., Nisell, R., and Ekholm, J. Varus and valgus loads on the knee joint during ergometer cycling. *Scand J Sports Med* 6:39–45, 1984.

53. Eriksson, E. Sports injuries of the knee ligaments: Their diagnosis, treatment, rehabilitation, and prevention. *Med Sci Sports* 3:133–144, 1976.

54. Eriksson, E., and Haggmark, T. Comparison of isometric muscle training and electrical stimulation supplementation of isometric muscle training in recovery after major knee ligament surgery. *Am J Sports Med* 7:169–171, 1979.

55. Evans, E. B., Eggers, G. W. N., Butler, J. K., and Blumel, J. Experimental immobilization and remobilization of rat knee joints. *J Bone Joint Surg* 42A:737–758, 1960.

56. Fahey, T. D., and Brown, C. H. The effects of an anabolic steroid on the strength, body composition, and endurance of college males when accompanied by a weight-training program. *Med Sci Sports* 5:272–276, 1973.

57. Fisher, R. L., Kloten, K., Bzdyra, B., and Cooper, J. A. Continuous passive motion (CPM) following total knee replacement. *Conn Med* 49:498, 1985.

58. Fowler, W. M., Jr., Gardner, G. W., and Egstrom, G. H. Effect of an anabolic steroid on physical performance of young men. *J Appl Physiol* 20:1038–1040, 1965.

59. Frank, C., Akeson, W. H., Woo, S. L. Y., and Coutts, R. D. Physiology and therapeutic value of passive joint motion. *Clin Orthop* 185:113–125, 1984.

60. Freed, D. L. J., Banks, A. J., Longson, D., and et al. Anabolic steroids in athletics: Crossover double blind trial on weightlifters. *Br Med J* 2:471–473, 1975.

61. Fronek, H., Frank, C., Amiel, D., Woo, S. L. Y., Coutts, R. D., and Akeson, W. H. The effect of intermittent passive motion (IPM) on the healing of the medial collateral ligament. *Trans Orthop Res Soc* 31, 1983.

62. Gasner, S. W., Jr., Early, R. G., and Carlson, B. R. Anabolic steroid effects on body composition in normal young men. *J Sports Med Phys Fitness* 11:98–103, 1971.

63. Gelberman, R. H., Menon, J., Gonsalves, M., and Akeson, W. H. The effects of mobilization on the vascularization of healing flexor tendons in dogs. *Clin Orthop* 153:283–289, 1980.

64. Gelberman, R. H., Van deBorg, J. S., Lundberg, G. N., and Akeson, W. H. Flexor tendon healing and restoration of the gliding surface. *J Bone Joint Surg* 65A:70–80, 1983.

65. Gelberman, R. H., Woo, S. L. Y., Lothringer, K., Akeson, W. H., and Amiel, D. Effects of early intermittent passive mobilization on healing canine flexor tendons. *J Hand Surg* 7:170, 1982.

66. Godfrey, C. M., Jayawardena, H., Quance, T. A., and Welsh, P. Comparison of electro-stimulation and isometric exercise in strengthening the quadriceps muscle. *Physiother Can* 31:2–4, 1979.

67. Golding, L. A., Freydinger, J. E., and Fishel, S. S. Weight, size, and strength unchanged with steroids. *Physician Sportsmed* 2:39–43, 1974.

68. Goldfuss, A. J., Moorehouse, C. A., and LeVeau, B. F. Effect of muscular tension on knee instability. *Med Sci Sports* 4:267–271, 1973.

69. Goldman, B. Liver carcinoma in an athlete taking anabolic steroids. *J Am Osteopath Assoc* 5:25, 1985.

70. Goodship, A. E., Lanyon, L. E., and McFie, H. Functional adaptation of bone to increased stress. *J Bone Joint Surg* 61A:539–546, 1979.

71. Gould, N., Donnermeyer, D., Pope, M., and Ashikage, T. Transcutaneous muscle stimulation as a method to retard disuse atrophy. *Clin Orthop* 164:215–220, 1982.

72. Greene, W. B. Use of continuous passive slow motion in the postoperative rehabilitation of difficult pediatric knee and elbow problems. *J Pediatr Orthop* 3:419–423, 1983.

73. Guyton, A. C., Barber, B. J., and Moffatt, D. S. Theory of intestinal pressures. *In* Hargens A (Ed.), *Tissue Fluid Pressure and Composition.* Baltimore, Williams & Wilkins, 1980.

74. Hall, M. C. Cartilage changes after experimental immobilization of the knee joint of the young rat. *J Bone Joint Surg* 45A:36–44, 1963.

75. Haupt, H. A., and Rovere, G. D. Anabolic steroids: a review of the literature. *Am J Sports Med* 12:469–484, 1984.

76. Hedge, G. A., Colby, H. D., and Goodman, R. L. (Eds.), *Clinical Endocrine Physiology.* Philadelphia, W.B. Saunders, 1987.

77. Henning, C. E., Lynch, M. A., and Glick, K. R. An in vivo strain gauge study of elongation of the anterior cruciate ligament. *Am J Sports Med* 13:22–26, 1985.

78. Hervey, G. R., Hutchinson, I., Knibbs, A. V., et al. Anabolic effects of methandienone in men undergoing athletic training. *Lancet* 2:699–702, 1976.

79. Hervey, G. R., Knibbs, A. V., Burkinaw, L., et al. Effects of methandienone on the performance and body composition of men undergoing athletic training. *Clin Sci* 60:457–461, 1981.

80. Hohl, M. Tibial condylar fractures. *J Bone Joint Surg* 49A:455–467, 1967.

81. Hohl, M., and Luck, J. V. Fractures of the tibial condyle. *J Bone Joint Surg* 38A:1001–1018, 1956.

82. Holma, P., and Aldercreutz, H. Effect of an anabolic steroid (methandienone) on plasma LH–FSH and testosterone and on the response to intravenous administration of LRH. *Acta Endocrinol* 83:856–864, 1976.

83. Houssay, A. B. Effects of anabolic-androgenic steroids on the skin, including hair and sebaceous glands. *In* Kochakian, C. D. (Ed.), *Anabolic-Androgenic Steroids.* New York, Springer-Verlag, 1976.

84. Hurley, B. F., Seals, D. R., Hugberg, J. M., et al. High-density lipoprotein cholesterol in bodybuilders vs. power-

lifters—negative effects of androgen use. *JAMA* 252:507–513, 1984.

85. Inoue, M., Gomez, M., Hollis, V., et al. Medial collateral ligament healing: Repair vs. nonrepair. *Trans Orthop Res Soc* 11:72, 1986.
86. Jackson, D. S., and Mechanic, G. Cross-link patterns of collagens synthesized by cultures of 3T6 and 3T3 fibroplasts and by fibroplasts of various granulation tissues. *Biochem Biophys Acta* 336, 1974.
87. Johnson, L. C., Fisher, G., Silvester, L. J., et al. Anabolic steroid: Effects on strength, body weight, oxygen uptake, and spermatogenesis upon mature males. *Med Sci Sports* 4:43–45, 1972.
88. Johnson, L. C., and O'Shea, J. P. Anabolic steroid: Effects on strength development. *Science* 164:957–959, 1969.
89. Johnson, L. C., Roundy, E. S., Allsen, P. E., et al. Effect of anabolic steroid treatment on endurance. *Med Sci Sports* 7:287–289, 1975.
90. Johnson, M. D., Jay, M. S., Shoup, B., et al. Anabolic steroid use by male adolescents. *Pediatrics* 83:921–924, 1989.
91. Keenan, E. J. Anabolic and androgenic steroids. *In* Thomas, J. (Ed.), *Drugs, Athletes and Physical Performance.* New York, Plenum, 1988, pp. 11–30.
92. Kellett, J. Acute soft tissue injuries—a review of the literature. *Med Sci Sports Exerc* 18:489–500, 1986.
93. King, S., Butterwick, D. J., and Cuerrier, J. P. The anterior cruciate ligament: A review of recent concepts. *J Orthop Sports Phys Ther* 8:110–122.
94. Kiraly, C. L., Alen, M., Rahkila, P., and Horsmanheimo, M. Effect of androgenic and anabolic steroids on the sebaceous glands in power athletes. *Acta Derm Venereal (Stockh)* 67:36–40, 1987.
95. Klein, L., Dawson, M. H., and Heiple, K. S. Turnover of collagen in the adult cat after denervation. *J Bone Joint Surg* 39A:1065, 1977.
96. Kochakian, C. D. (Ed.). *Anabolic-Androgenic Steroids.* Handbook of Experimental Pharmacology, New Series. New York, Springer-Verlag, 1976.
97. Kruskemper, H. L. *Anabolic Steroids.* New York, Academic Press, 1968.
98. Lamb, D. R. Anabolic steroids in athletics: How well do they work and how dangerous are they? *Am J Sports Med* 12:31–38, 1984.
99. Laros, G. S., Tipton, C. M., and Cooper, R. R. Influence of physical activity on ligament insertions in the knees of dogs. *J Bone Joint Surg* 53A:275–286, 1971.
100. Ljungqvist, A. The use of anabolic steroids in top Swedish athletes. *Br Med J* 9:82, 1975.
101. Lombardo, J. A., Longscope, C., and Boy, R. D. Recognizing anabolic steroid abuse. *Patient Care* 19:28–47, 1985.
102. Loughton, S. J., and Ruhling, R. O. Human strength and endurance responses to anabolic steroids and training. *J Sports Med Phys Fitness* 17:285–296, 1977.
103. Leuking, M. T. Steroid hormones in sports, special reference: Sex hormones and their derivatives. *Int J Sports Med (Suppl.)* 3:65–67, 1982.
104. Lynch, J. A., Baker, P. L., Polly, R. E., McCoy, M. T., Sund, K., and Roudybush, D. Continuous passive motion: A prophylaxis for deep venous thrombosis following total knee replacement. Presented at 1984 Orthopaedic Research Society Meeting.
105. Marshall, E. The drug of champions. *Science* 242:183–184, 1988.
106. McLeod, W. D., and Blackburn, T. A. Biomechanics of knee rehabilitation with cycling. *Am J Sports Med* 8:175–182, 1980.
107. McNutt, R. A., Ferenchick, J. S., and Kirlin, P. C. Acute MI in a 22-year-old world class lifter using anabolic steroids. *Am J Cardiol* 62:164, 1988.
108. Melzak, R., and Wall, P. D. Psychophysiology of pain. Evolution of pain theories. *Int Anesthesiol Clin* 8:3–34, 1970.
109. Mooney, V., and Ferguson, A. B., Jr. The influence of immobilization and motion on the formation of fibrocartilage in the repair granuloma after joint resection in the rabbit. *J Bone Joint Surg* 48A:1145–1155, 1966.
110. Morey, S. W., and Passariello, K. *Steroids—A Comprehensive and Factual Report.* Tampa, FL, S.W. Morey and K. Passariello Publishers, 1982.
111. Morrisey, M. C., Brewster, C. E., Shields, C. L., and Brown, M. The effects of electrical stimulation on the quadriceps during post-operative knee immobilization. *Am J Sports Med* 13:40–45, 1985.
112. Neuberger, A., and Slack, H. G. B. The metabolism of collagen from liver, bones, skin, and tendon in the normal rat. *Biochem J* 53:47, 1953.
113. Nisell, R., Ericson, M. O., Nemeth, G., and Ekholm, J. Tibiofemoral joint forces during isokinetic knee extension. *Am J Sports Med* 17:49–54, 1989.
114. Noyes, F. R. Functional properties of knee ligaments and alterations induced by immobilization. *Clin Orthop* 123:210–242, 1977.
115. Noyes, F. R., Butler, D. L., Paulos, L. E., and Grood, E. S. Intraarticular cruciate reconstruction. I. Perspectives on graft strength, vascularization, and immediate motion after replacement. *Clin Orthop* 172–177, 1983.
116. Noyes, F. R., Towik, P. J., Hyde, W. B., and DeLucas, J. L. Biomechanics of ligament failure. *J Bone Joint Surg* 56A:1406–1418, 1974.
117. O'Driscoll, S. W., Kumar, A., and Salter, R. B. The effect of continuous passive motion on the clearance of a hemarthrosis from a synovial joint. *Clin Orthop* 176:305–311, 1983.
118. O'Driscoll, S. W., and Salter, R. B. The induction of neochondrogenesis in free intraarticular periosteal autografts under the influence of continuous passive motion. *J Bone Joint Surg* 66A:1248–1257, 1984.
119. O'Shea, J. P. The effects of anabolic steroid on dynamic strength levels of weight lifters. *Nutr Rep Int* 4:363–370, 1971.
120. O'Shea, J. P. A biochemical evaluation of the effects of stanozolol on adrenal, liver, and muscle function in humans. *Nutr Rep Int* 2:351–362, 1970.
121. O'Shea, J. P., and Winkler, W. Biochemical and physiological effects of an anabolic steroid in competitive swimmers and weightlifters. *Nutr Rep Int* 2:351–362, 1970.
122. Ohlsen, L. Cartilage regeneration from perichondrium. Experimental and clinical applications. *Plast Reconstr Surg* 62:507, 1978.
123. Overly, W. L., Dankoff, J. A., Wang, B. K., and Sing, U. D. Androgens and hepatocellular carcinoma in an athlete. *Ann Intern Med* 100:158–159, 1984.
124. Parker, F., and Keefer, C. S. Gross and histologic changes in the knee joint in rheumatoid arthritis. *Arthritis Pathol* 20:507, 1935.
125. Paulos, L. Rehabilitation and mobilization techniques of early motion. Presented at Anterior Cruciate Ligament—New Concepts Course. Long Beach, CA, October 1985.
126. Paulos, L., Noyes, F. R., Grood, E., and Butler, D. L. Knee rehabilitation after anterior cruciate ligament reconstruction and repair. *Am J Sports Med* 9:140–149, 1981.
127. Peacock, E. E. Comparison of collagenous tissue surrounding normal and immobilized joints. *Surg Forum* 14:440, 1963.
128. Perry, C. R., Evans, L. S., Rice, S., Fobarty, J., and Burdge, R. E. A new surgical approach to fractures of the lateral tibial plateau. *J Bone Joint Surg* 66A:1236–1240, 1984.
129. Pope, H. G., and Katz, D. L. Affective and psychotic symptoms associated with anabolic steroids use. *Am J Psychiatr* 145:487–490, 1988.
130. Pope, H. G., Katz, D. L., and Champoux, R. Anabolic-androgenic steroid use among 1010 college men. *Physician Sportsmed* 16:75–81, 1988.
131. Porter, B. Crush fractures of the lateral tibial tubercle. *J Bone Joint Surg* 52B:676–687, 1970.
132. Poussa, M., Rubak, J., and Ritxila, V. Differentiation of

the osteochondrogenic cells of the periosteum in chondrotrophic environment. *Acta Orthop Scand* 52:235, 1981.

133. Reeves, W. P., Creswell, M., and Squires, W. B. Analysis of commercial samples of dimethyl sulfoxide. *Med Sci Sports Exerc* 21:348–349, 1989.

134. Rosenbaum, E. E., and Jacob, S. W. Dimethyl sulfoxide (DMSO) in acute subdeltoid bursitis. *Northwest Med* 63:156–158, 1964.

135. Salter, R. B., and Bell, R. S. The effect of continuous passive motion on the healing of the partial thickness lacerations of the patellar tendon in the rabbit. *Ann R Coll Phys Surg Can* 14:209, 1981.

136. Salter, R. B., Bell, R. S., and Kelley, R. W. The protective effect of continuous passive motion on living articular cartilage in acute septic arthritis. *Clin Orthop* 159:223–247, 1981.

137. Salter, R. B., and Field, P. The effects of continuous compression on living articular cartilage. *J Bone Joint Surg* 42A:31–49, 1960.

138. Salter, R. B., and Ogilvie-Harris, D. J. Fractures involving joints: Part II. Healing of intraarticular fractures with continuous passive motion. *AAOS Instr Course Lect* 28:102–117, 1979.

139. Salter, R. B., Simmonds, D. F., Malcolm, B. W., Runble, E. J., MacMichael, D., and Clements, N. O. The biological effect of continuous passive motion on the healing of full thickness defects in articular cartilage. *J Bone Joint Surg* 62A:1232–1251, 1980.

140. Samuels, L. T., Henschel, A. F., and Keys, A. Influence of methyl testosterone on muscular work and creatinine metabolism in normal young men. *J Clin Endocrinol Metab* 2:649–654, 1942.

141. Schiller, S., Matthews, M. D., Cifonelli, J., and Dorfman, A. A metabolism of mucopolysaccharides in animals. Further studies on skin utilizing C-14 glucose, C-14 acetate, and S-35 sodium sulfate. *J Biol Chem* 218:139, 1956.

142. Silfverskiold, J. P., Steadman, J. R., Higgins, R. M., Hagerman, T., and Atkins, J. A. Rehabilitation of the anterior cruciate ligament in the athlete. *Sports Med* 6:308–319, 1988.

143. Skyhar, M. J., Danzig, L. A., Margens, A. R., and Akeson, W. H. Nutrition of the anterior cruciate ligament. Effects of continuous passive motion. *Am J Sports Med* 13:415, 1985.

144. Snyder, P. J. Clinical use of androgens. *Annu Rev Med* 35:207–217, 1984.

145. Stamford, B. A., and Moffatt, R. Anabolic steroids: effectiveness as an ergogenic aid to experienced weight trainers. *J Sports Med* 14:191–197, 1974.

146. Stanitski, C. L. Pharmacological management of musculoskeletal injuries. *In* Strauss, R. H. (Ed.), *Drugs and Performance in Sports.* Philadelphia, W.B. Saunders, 1987, pp. 180–181.

147. Steadman, J. R. Nonoperative measures for patellofemoral problems. *Am J Sports Med* 7:374–375, 1979.

148. Steadman, J. R. Rehabilitation of acute injuries of the anterior cruciate ligament. *Clin Orthop* 172:129–132, 1983.

149. Steadman, J. R., Forster, R. S., and Silfverskiold, J. P. Rehabilitation of the knee. *Clin Sports Med* 8:605–627, 1989.

150. Strauss, R. H. Anabolic steroids. *In* Strauss, R. H. (Ed.), *Drugs and Performance in Sports.* Philadelphia, W.B. Saunders, 1987, pp. 59–67.

151. Strauss, R. H., Liggett, M. I., and Lanese, R. R. Anabolic steroid use and perceived effects in ten weight-trained women athletes. *JAMA* 253:2871–2873, 1985.

152. Strauss, R. H., Wright, J. E., and Finerman, G. A. M. Anabolic steroid use and health status among forty-two weight-trained male athletes. *Med Sci Sports Exerc* 14:19, 1982.

153. Strauss, R. H., Wright, J. E., and Finerman, G. A. M. Side effects of anabolic steroids on weight-trained men. *Physician Sportsmed* 11:87–96, 1983.

154. Stromme, S. B., Meen, H. D., and Aakvaag, A. Effects of an androgenic-anabolic steroid on strength development and plasma testosterone levels in normal males. *Med Sci Sports* 6:203–208, 1974.

155. Swann, D. A., Radin, E. L., and Nazimiec, M. Role of hyaluronic acid in joint lubrication. *Ann Rheum Dis* 33:318, 1974.

156. Tahmindjis, A. J. The use of anabolic steroids by athletes to increase body weight and strength. *Med J Aust* 1:991–993, 1976.

157. Tainter, M. L., Arnold, A., Beyler, A. L., et al. Synthetic anabolic steroids in geriatrics. *Med Times* 91:747–759, 1963.

158. Taylor, W. N. *Anabolic Steroids and the Athlete.* Jefferson, NC, McFarland & Co., 1982.

159. Thaxter, T. H., Mann, R. A., and Anderson, C. E. Degeneration of immobilized knee joints in rats. *J Bone Joint Surg* 42A:567, 1960.

160. Tipton, C. M., Matthes, R. D., and Martion, R. R. Influence of age and sex on the strength of bone-ligament junctions in knee joints in rats. *J Bone Joint Surg* 60A:230, 1978.

161. Von Reyhar, C. On the cartilage and synovial membranes of the joints. *J Anat Physiol* 8:261, 1974.

162. Wade, N. Anabolic steroids—doctors denounce them, but athletes aren't listening (News). *Science* 176:1399–1403, 1972.

163. Ward, P. The effect of anabolic steroid on strength and lean body mass. *Med Sci Sports* 5:277–282, 1973.

164. Webb, O. L., Laskarzewski, P. M., and Glueck, C. J. Severe depression of high-density lipoprotein cholesterol levels in weight lifters and body builders by self-administered exogenous testosterone and anabolic-androgenic steroids. *Metabolism* 33:971–975, 1984.

165. Weichselbaum, A. Die feineren Verandungen des gelenk Knorpels bei fungoser Synovits and Caries der Gelenkenden. *Virchows Arch* 73:461, 1978.

166. Wilson, J. D., and Griffin, J. E. The use and misuse of androgens. *Metabolism* 29:1278–1295, 1980.

167. Win-May, M., and Mya-Tu, M. The effect of anabolic steroids on physical fitness. *J Sports Med Phys Fitness* 15:266–271, 1975.

168. Woo, S. L. Y., et al. Effect of immobilization and exercise on strength characteristics of bone—medial collateral ligament-bone complex. *Am Soc Mech Eng Symp* 32:62, 1979.

169. Woo, S. L. Y., Gomez, M. A., Amiel, D., Newton, P. O., Orlando, C. A., Sites, T., and Akeson, W. H. The biomechanical and biochemical changes of the MCL following immobilization and remobilization. *J Bone Joint Surg* (in press, 1992).

170. Woo, S. L. Y., Gomez, M. A., Sequichi, Y., Endo, C. M., and Akeson, W. H. Measurement of mechanical properties of ligament substance from a bone-ligament preparation. *J Orthop Res* 1:22, 1983.

171. Woo, S. L. Y., Kuie, S. C., and Amiel, D. The response to cortical long bone secondary to exercise training. Transactions of the 26th Annual Meeting of the Orthopedic Research Society 5:256, 1980.

172. Woo, S. L. Y., Matthews, J. V., Akeson, W. H., Amiel, D., and Convery, R. Connective tissue response to immobility. Correlative study of biomechanical and biochemical measurements of normal and immobilized rabbit knees. *Arthritis Rheum* 18:257, 1975.

173. Woo, S. L. Y., Ritter, M. A., Amiel, D., Sanders, T. M., Gomez, M. A., Kuei, S. C., Garfin, S. T., and Akeson, W. H. The biomechanical and biochemical properties of swine tendons—long-term effects of exercise on the digital extensors. *Connect Tissue Res* 7:117–183, 1980.

174. Woo, S. L. Y., Ritter, M. A., Gomez, M. A., Kuei, S. C., and Akeson, W. H. The biomechanical and structural properties of swine digital flexor tendons secondary to running exercise. *Orthop Trans* 4:165–166, 1980.

175. Wright, J. E. Anabolic steroids and athletics. *Exerc Sports Sci Rev* 8:149–202, 1980.

Scientific, Medical, and Practical Aspects of Stretching

Bob Anderson, B.S.
Edmund R. Burke, Ph.D.

PHYSIOLOGY OF FLEXIBILITY AND STRETCHING

Many physical therapists, athletic trainers, and physicians consider flexibility to be one of the most important objectives in the conditioning programs of athletes. *Flexibility* can be defined as the range of motion of a joint or a series of joints that are influenced by muscles, tendons, ligaments, bones, and bony structures.

Flexibility is influenced by a number of factors. These include the level and type of activity performed, with full range of motion promoting improved flexibility and limited range of motion leading to reductions in flexibility. Gender and age are other factors, with females having greater flexibility than males, and with flexibility increasing up to young adulthood and then decreasing with age. Temperature is also a factor, with flexibility increasing with heat and decreasing with cold. Flexibility also appears to be highly specific to the joint being evaluated. In other words, one can be highly flexible in one joint and have limited range of motion in another [29].

Muscle tissue, with its great contractile properties, has the additional and unique ability to lengthen greatly under certain conditions. The Golgi tendon organs (GTOs) are sensory receptors that provide the muscle with certain information and are located at the muscle-tendon junctures. The purpose of the GTOs is protection. When a muscle is subjected to sustained stretch, it receives sensory information as to the force and duration of the stretching action. Given enough information, the receptors trigger an inhibiting response that causes the muscle to relax and, therefore, achieve greater length than it originally possessed. If a joint such as the knee or elbow refuses to "stretch" out, it is possible the cause is related to other soft tissues of the joint (such as ligaments) and not the muscles surrounding the joint [12].

Static Versus Dynamic Flexibility

Two kinds of flexibility are often described in the literature: dynamic and static. It is obvious that the ability to flex and extend a joint through a wide range of motion (which is measured virtually in the static position) is not necessarily a good criterion of the stiffness or looseness of that same joint as it applies to the ability to move the joint quickly with little resistance to movement. Range of motion is one factor, the only one that has been widely investigated to date. How easily the joint can be moved in the middle of the range of motion, where the speed is necessarily greatest, is quite another factor [19].

Dynamic flexibility refers to the forces that resist throughout the range of motion of a joint, or joint "stiffness," to put it loosely. It is difficult to measure dynamic flexibility either quantitatively or qualitatively. The major force resisting joint mobility is the elasticity of the soft tissues, such as muscles, tendons, fascia, ligaments, and skin.

Static flexibility describes the range of motion about a joint. It is relatively easy to measure, compared with dynamic flexibility.

This chapter was published originally in *Clinics in Sports Medicine* 10(1):63–86, 1991. Reprinted with permission of W. B. Saunders Co.

Measuring Flexibility

Because flexibility is not a general measurement, but rather specific to each joint or group of joints, it can be seen that a simple test such as the sit-and-reach test provides very limited information about the flexibility of the trunk and hips while bending forward. Although general tests of this nature are important, the following are more accurate ways to determine the flexibility of a joint:

1. Goniometry: A goniometer consists of a 180-degree protractor, which may have two extended arms, one fixed at the zero line and one mobile, or just one mobile arm that can be locked in any position. The center point of the goniometer is aligned with the center of the joint, and readings are taken in extreme positions.

2. Flexometry: The Leighton Flexometer [25] contains a rotating circular dial marked off in degrees and a pointer counterbalanced to ensure that it always points vertically. It is strapped on the appropriate body segment, and the range of motion is determined with respect to the perpendicular [23].

3. Electrogoniometry: The ELGON is a protractor-like device in which the protractor has been replaced by a potentiometer. The potentiometer provides an electrical signal that is directly proportional to the angle of the joint. This device can give continuous recordings during a variety of activities. The versatility of this unit allows a much more accurate and realistic assessment of functional flexibility, or that degree of flexibility exhibited during actual physical activity [29].

Benefits Derived from Flexibility Exercise

1. *Injury prevention.* Muscles possessing greater extensibility are less likely to be overstretched during vigorous activity, lessening the likelihood of injury [11].

2. *Reduced muscle soreness.* Stretching, especially after exercise, can help reduce the next-day muscle soreness that often results from a strenuous workout [11].

3. *Skill enhancement.* Optimal flexibility aids athletic performance. Sufficient shoulder flexibility is necessary before the serve in tennis can be properly mastered. Proficient golf skills require flexibility throughout the hips, trunk, and shoulder regions [13].

4. *Muscle relaxation.* Stiff, tight muscles are relaxed by easy, gentle stretching. After sitting in a

chair for several hours, we look forward to a full, hearty stretch [20].

Flexibility Training

There are two methods used to stretch muscles and other soft tissues that limit flexibility.

1. Ballistic (bouncing) stretching is a rapid, jerky movement in which a body part is put into motion and momentum carries it through the range of motion until the muscles are stretched to the limits. As the athlete bounces, the muscle responds by contracting, to protect itself from overstretching. Thus, internal tension develops in the muscle and prevents it from being fully stretched.

2. Static stretching involves placing the muscles at their greatest possible length and holding them in that position for a minimum of 15 to 30 seconds. Golgi tendon organs act as tension sensors and can be responsible for initiating sensory impulses, resulting in reduced resistance to stretched soft tissue. These proprioceptors simply serve to inhibit muscle contraction in the stretched tissue. This relaxation phenomenon does not result when a stretch is performed quickly [9, 22].

STRETCHING VERSUS WARM-UP

Warm-up and stretching are not the same thing. Warm-up is an activity that raises the total body temperature, as well as temperature of the muscles, to prepare the body for vigorous exercise.

Physiology of Warm-Up

Theoretically, the following physiologic changes take place during warm-up and should enhance performance:

1. *Increased muscle temperature/reduced injury.* The temperature increases within the muscles that are used during the warm-up. A warmed muscle both contracts more forcefully and relaxes more quickly. Therefore, both speed and strength should be enhanced during exercise [4, 26].

2. *Increased blood temperature.* The temperature of the blood as it travels through the muscle increases. It is an established fact that as blood temperature rises, the amount of oxygen it can hold becomes reduced (especially at the partial pressures in the muscle). This makes available more oxygen to the working muscles [3, 5, 6].

3. *Improved range of motion.* The range of motion around joints is increased, especially if flexibility exercises are part of the program [27, 30].

4. *Reduced risk from sudden, strenuous exercise.* Barnard and colleagues [7, 8] have shown that proper warm-up will limit the amount of subendocardial ischemia when people participate in sports or any sudden, vigorous activity.

Types of Warm-Up

There are two basic types of warm-up that may be used to prepare for strenuous activity, and each is effective provided there is an increase in muscle temperature.

1. Related warm-up. When the specific skills of an event are performed during the warm-up, it is referred to as related warm-up. For a cyclist, this may mean getting on a bicycle, riding for a few miles, and adding a few short sprints to ensure that all the muscle fibers are warmed up.

2. Unrelated warm-up. In the second type of warm-up, the movements performed, i.e., calisthenics or flexibility exercises, are different from the actual skills of the activity. Track and field athletes begin their warm-up with some form of unrelated exercise.

Which type of warm-up is preferred? If immediate participation in the actual activity would likely result in muscle or joint injuries, then unrelated warm-up is preferred. For example, if a long jumper demands great flexibility in his event, and if he were to start jumping immediately, he may injure his muscles. He would warm up with calisthenics, light running, and stretching and then work into a routine of related warm-up of actual jumping.

When related warm-up is used it starts slowly and progresses into more intense activity. Basketball players, generally, begin their warm-up with slow-paced lay-up drills, followed by dribbling and shooting drills.

Duration of Warm-Up

Whatever warm-up you choose, it should be intense enough to increase body temperature but not so intense as to cause fatigue. When an athlete begins to sweat, it means that the internal temperature has been raised to a desired level. Obviously, the intensity and duration of the warm-up must be adjusted to the individual athlete. Better performance results when a 15- to 30-minute warm-up of unrelated (stretching) and related exercises (sports related) are used with a few minutes of high-intensity exercise.

The effects of the warming-up may last up to 45 minutes. However, the closer a warm-up is to the event, the more beneficial it will be in terms of effective performance. The warm-up should begin to taper off 10 to 15 minutes prior to training or competition and end 5 to 10 minutes before the race starts, if possible. This will allow recovery from any slight fatigue without losing the effects of the warm-up.

PROPRIOCEPTIVE NEUROMUSCULAR FACILITATION

Proprioceptive neuromuscular facilitation (PNF) may be defined as a method "promoting or hastening the neuromuscular mechanism through stimulation of the proprioceptors" [24]. It was developed after World War II as a physical therapy procedure for the rehabilitation of patients. Techniques had to be developed to restore men with neurologic problems to a functioning state.

PNF involves a number of techniques of bodily manipulation in coordination with the subjects' own movement that can improve strength, endurance, range of motion, and joint function of the musculoskeletal system. The words are defined as follows: "proprioceptive: receiving stimulation from within the tissue of the body; neuromuscular: pertains to the nerves and the muscles; facilitation: the promotion or hastening of any natural processes" [24]. Hence, PNF refers to the improvement of flexibility through stimulation of the nerves and muscles internally. This technique involves the use of the principles of reciprocal innervation and the stretch reflex.

As a muscle is passively or actively stretched, it is brought to a point of limitation before pain will develop. This is the point at which the proprioceptive organs send a message to the brain to terminate the movement before further elongation occurs. At this point the muscles being stretched (antagonist) are contracted for 6 to 8 seconds at an intensity of 50% to 100% of maximum. This contraction allows for the inhibition of the muscle spindles and Golgi tendon organs and subsequent stretch reflex of the stretch receptors. After releasing the contraction, the muscle is stretched to a new point of limitation and held for 6 to 8 seconds.

There are numerous techniques that combine isotonic and isometric contractions in different combinations. Alter [1], in his book *Science of Stretching,* offers over eight PNF techniques and exercises.

Two of the most commonly used PNF stretches in rehabilitation and athletics are the contract-relax technique and the agonist contract-relax technique. In the former, the muscle is passively taken through a range of motion and then contracted for 6 to 8 seconds, then relaxed, and then taken once again passively to an increased pain-free range. This is repeated three to six times (Fig. 7F–1A). In the agonist contract-relax technique, the individual maximally contracts the muscle opposite the muscle to be stretched against a resistance for 6 to 8 seconds. Then, the agonist muscle is relaxed and the antagonist muscle is stretched (Fig. 7F–1B).

PNF is most often used with athletes and individuals who have less-than-normal range of motion or who have lost normal range of motion. For example, the preseason physical of a baseball pitcher may reveal less-than-normal range of motion in the shoulder. PNF exercises that stretch the rotator muscles of the shoulder may be prescribed.

Although PNF techniques offer many potential benefits, there are also some disadvantages. PNF stretching needs to be closely monitored if chance of soft tissue injury is to be minimized. Individuals with heart disease or high blood pressure and those individuals not aware of the possibility of the Valsalva phenomenon happening should be cautioned in the use of PNF [1].

PNF Contract-Relax Technique for the Hamstring and Spinal Muscles

Stretch No. 1. While sitting on the floor with both legs together, the individual leans forward to the point of limitation while holding the legs straight and toes pointed upward. The athlete pushes back against his or her partner (contracting the spinal muscles) and pushes the legs against the floor (contracting the hamstrings) for a 6- to 8-second isometric contraction. The partner resists the movement. After the release of the contraction, the athlete

stretches to a new point of limitation and once again holds the stretch for 6 to 8 seconds while the partner puts light pressure on the back (see Fig. 7F–1A).

PNF Agonist Contract-Relax Technique for the Hamstring and Spinal Muscles

Stretch No. 2. This exercise can be performed individually and is good for individuals with poor flexibility in the lower back. In the first part of the exercise, the athlete (on the left) places both hands on the thighs above the knees, the legs slightly bent and extended upward. The athlete isometrically contracts the hip flexors and quadriceps of the legs against the hands for 6 to 8 seconds. He or she then releases the contraction and places both hands behind the thigh (athlete on the right). He or she then pulls the legs toward the chest and holds the stretch for 6 to 8 seconds (see Fig. 7F–1B).

CLINICAL EVALUATION OF FLEXIBILITY

Measuring Flexibility

Stretch No. 1. Shoulder: One method of assessing flexibility of the shoulder requires nothing more than a measuring stick. The right elbow is raised and the right hand reaches down between the shoulder blades. The left hand is placed in the small of the back with the palm facing away from the back. The person being tested attempts to overlap the fingers of the two hands. The score is determined by the distance between the hands, if not overlapped, or the amount of finger overlap. Repeat, reversing the positions of the hands [1]. This method checks external rotation of upper arm and inward rotation of lower arm (Fig. 7F–2A).

Stretch No. 2. Another good check for tight

FIGURE 7F–1
A and *B*, Proprioceptive neuromuscular facial (PNF) contract-relaxation techniques for hamstring and spinal muscles. Dotted areas indicate those most likely to feel the stretch.

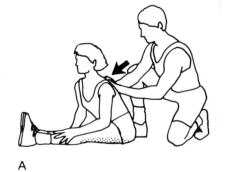

A

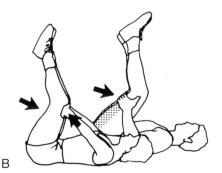

B

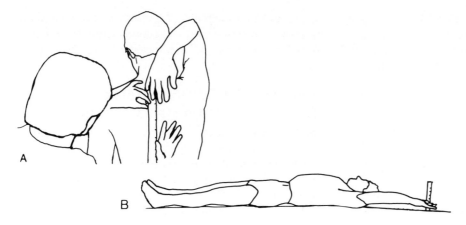

shoulders (especially anterior tightness, in weight-lifters, poor posture and round shoulders, and so on) is to have the patient lie on his or her back and extend the arm like the Statue of Liberty—straight up with elbow straight. The shoulder, elbow, and wrist should all contact the table easily. To show progress, as athletes improve with stretching, measure the distance of the hand from the table (N. A. DiNubile, personal communication, 1990) (Fig. 7F–2B).

Stretch No. 3. Hip flexibility: the iliopsoas test (Thomas test) [27]. Anterior hip tightness is common in runners and sedentary persons (Fig. 7F–3).

Starting position is supine, with the patient positioned at the end of the table. The knees are flexed over the end of the table and the hips are extended. Then both knees are brought to the chest and held firmly with both hands. The side being tested is then relaxed down toward the tabletop toward hip extension, and the knee should remain flexed to 90 degrees as it drops off the end of the table.

Flexion of the hip indicates tightness of the iliopsoas, and extension of the knee indicates tightness of the rectus femoris. Furthermore, if the patient's anterior hip and iliopsoas are tight, the thigh would not touch the table and the entire pelvis would need to rotate down for the thigh to contact the table. The number of degrees away from the table measures the degree of flexion contracture of the hip.

Stretch No. 4. Hamstring test [14]. Start supine, with the hip and knee flexed to 90 degrees. The examiner slowly extends the knee until muscle resistance is felt, being careful to avoid a change of hip position or quadriceps assistance by the patient. Keeping the lower back flat is a prerequisite to accurate testing [1]. The knee should be fully extended as the hip is flexed to 90 degrees; if tight, the knee will remain flexed (Fig. 7F–4).

Stretch No. 5. Quadriceps test (Ely test). Start prone. The examiner slowly flexes the knee until muscle resistance is felt. The knee should flex to 135 degrees freely, without springing back when pressure is released. The hip should not flex. The heel should touch buttocks, but if tight, it either does not reach or it reaches buttock and the pelvis (on the same side) rises up off the table (Fig. 7F–5).

Stretch No. 6. Iliotibial band test (Ober's test) [28]. Start in straight lateral side-lying position. The

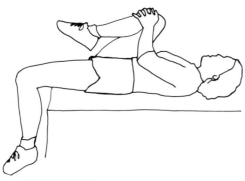

FIGURE 7F–3
Clinical evaluation of flexibility in the hip.

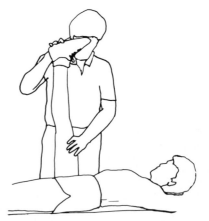

FIGURE 7F–4
Clinical evaluation of flexibility in the hamstrings.

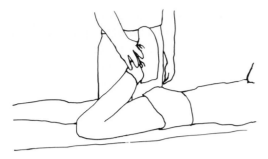

FIGURE 7F–5
Clinical evaluation of flexibility in the quadriceps (Ely Test).

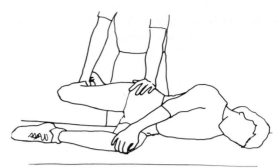

FIGURE 7F–6
Clinical evaluation of flexibility in the iliotibial band (Ober's Test).

knee closest to the table is flexed at 90 degrees and held by the patient with both hands. The examiner stands behind the patient, steadies the pelvis with one hand, and grasps the ankle of the upper leg with the other hand. The knee is flexed to 90 degrees, and the entire leg is pulled posteriorly into full hip extension. The examiner should not attempt to lift the thigh or depress it. In full hip extension the thigh should be adducted at least 15 degrees below the horizontal plane. The rear leg (knee) should be able to drop to table level. If it stays up, the iliotibial band across the lateral aspect of the knee is tight (Fig. 7F–6).

Stretch No. 7. The Gastrocnemius test [14]. Start supine with the knee fully extended. The examiner inverts the foot to lock the subtalar joint, then dorsiflexes the ankle. Twenty degrees of ankle dorsiflexion should be present. If the ankle cannot get to neutral or better, then the calf is tight (Fig. 7F–7).

Obviously, there are many other areas of the body in which flexibility is important. Details for different tests of flexibility, including standards for evaluation, are included in several sources [10, 15–17, 21].

STRETCHES FOR THE MAJOR MUSCLE GROUPS

How to Stretch

There are many methods of stretching, some more complex than others, but if practiced regularly and with sensitivity, all seem to yield results.

The method I prefer is probably the simplest of all, the static stretch method. It is easy to learn and do. But regardless of what method you use, there are certain criteria that apply to all.

Stretching is definitely not a contest. What someone else can do has nothing to do with you. It is very important to understand this point so an indi-

vidual does not strain or tear tissue trying to be like someone else.

Other than the ballistic method (bouncing), a certain amount of time, varying from 5 to 60 seconds or more, will be spent in a sustained stretch in almost all forms of stretching, whether various forms of static stretching or PNF.

The key to proper stretching, regardless of the method, lies in the feeling created when you are sustaining a stretch. The feeling of the stretch tells you whether you are stretching correctly or not.

When you are stretching correctly, the feeling is mild and comfortable, not painful. When you stretch to a point at which you feel a *slight* tension ("the easy stretch" [2]), then stop and hold for 10 to 20 seconds. As you hold the stretch, the feeling should subside somewhat. This is a relaxation response; it indicates that an accommodation has taken place. If the feeling of stretch grows in intensity as the position is held, then the stretch is too great and the person should ease off a bit into a more comfortable feeling. To develop increased flexibility, move a *slight bit further* into the stretch ("the developmental stretch" [2]) after you have done the

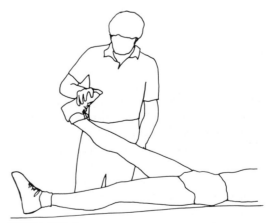

FIGURE 7F–7
Clinical evaluation of flexibility in the gastrocnemius.

easy stretch, until you feel an increase in stretch tension. Now stop and hold for 10 to 20 seconds. This feeling should either stay the same or become slightly less intense as you hold it. Remember, any feeling of stretch that builds in intensity as the stretch is held indicates that you are stretching too far.

Staying relaxed is very important. Be certain the extremities are not held tight. The jaw (the most important part of the body to keep relaxed), shoulders (think "shoulders hang, shoulders down"), hands, and feet should all be kept relaxed as you stretch.

Breathing is slow, deep, and rhythmic, with mental concentration on relaxation and the area being stretched.

Stretching is not exercise per se. Exercise is extension and contraction done rhythmically. So when you stretch you do not have to accept the old exercise adage "no gain without pain" or "if it hurts, I must be doing it right."

A little bit less is better than a little bit more. If there are to be lasting results from stretching, they will come from doing it mildly and regularly and not from straining. Then a person can learn to stretch easy and relaxed without the pressure to push too hard.

Remember the importance of the feeling of the stretch as an excellent way to tell whether you are stretching correctly.

Stretches for various areas of the body are described in detail in the following sections.*

Stretches for the Shoulders, Back, Neck, and Arms

Stretch No. 1. In a standing or sitting position, interlace your fingers above your head. Now, with your palms facing upward, push your arms slightly back and up. Feel the stretch in arms, shoulders, and upper back. Hold stretch for 15 seconds. Do not hold your breath. This stretch can be done anywhere, anytime, and is excellent for slumping shoulders. Keep knees slightly flexed (Fig. 7F–8A).

Stretch No. 2. With arms overhead, hold the elbow of one arm with the hand of the other arm. Keeping knees slightly bent (1 inch), gently pull your elbow behind your head as you bend from your hips to the side. Hold an easy stretch for 10 seconds. Do both sides. Keeping your knees slightly bent will give you better balance (Fig. 7F–8B).

Stretch No. 3. Shoulder shrug: Raise the top of your shoulders toward your ears until you feel slight tension in your neck and shoulders. Hold this feeling of tension for 3 to 5 seconds, then relax your shoulders downward into their normal position. Do this two or three times. This stretch is good to use at the first signs of tightness or tension in the shoulder and neck area (Fig. 7F–8C).

Stretch No. 4. From a stable, aligned sitting position, turn your chin toward your left shoulder to create a stretch on the right side of your neck. Hold correct stretch tensions for 10 to 15 seconds. Do each side twice (Fig. 7F–8D).

Stretch No. 5. To stretch the side of your neck and top of your shoulder, lean your head sideways toward your left shoulder as your left hand pulls your right arm down and across, behind your back. Hold an easy stretch for 10 seconds. Do both sides (Fig. 7F–8E).

Stretch No. 6. The next stretch is done with your fingers interlaced behind your back. Slowly turn your elbows inward while straightening your arms. This is an excellent stretch for shoulders and arms. Hold for 5 to 15 seconds. Do twice (Fig. 7F–8F).

Stretch No. 7. Hold a towel near both ends so that you can move it with arms straight up, over your head and down behind your back. Do not strain or force it. Your hands should be far enough apart to allow for relatively free movement up, over, and down. To isolate and add further stretch to the muscles of a particular area, hold the stretch at any place during this movement for 10 to 20 seconds (Fig. 7F–8G).

Stretch No. 8. Place both hands a shoulder width apart on a fence or ledge and let your upper body drop down as you keep your knees slightly bent (1 inch). Your hips should be directly above your feet. To change the area of the stretch, bend your knees just a bit more or place your hands at different heights. Find a stretch that you can hold for at least 30 seconds. This will take some of the kinks out of a tired upper back. The top of the refrigerator, a file cabinet, or a chain-linked fence is good to use for this stretch. *Remember to always bend your knees when coming out of this stretch* (Fig. 7F–8H).

Stretch No. 9. From the position illustrated in Figure 7F–8L, with your palms flat and fingers pointed back toward your knees, slowly lean backward to stretch the forearms and wrists. Be certain to keep your palms flat. Hold a comfortable stretch for 20 to 25 seconds. *Do not overstretch* (Fig. 7F–8L).

Stretch No. 10. With your legs bent under you, reach forward with one arm and grab the end of the mat, carpet, or anything you can hold onto. If you

---

*These stretches have been excerpted from Anderson, B. A.: *Stretching.* Bolinas, Shelter Publications, 1980; with permission.

FIGURE 7F–8

A–O, Stretches for the major muscle groups: shoulders, back, neck, and arms. Dotted areas indicate those most likely to feel the stretch.

cannot grab onto something, just pull with your hand. Do likewise pulling on end of mat. Hold stretch for 20 seconds. Stretch each side and do not strain. You should feel the stretch in your shoulders, arms, sides, upper back, and even in your lower back (Fig. 7F–8J).

Stretch No. 11. With arms extended overhead and palms together as shown, stretch arms upward and slightly backward. Breathe in as you stretch upward, holding the stretch for 5 to 8 seconds. This is a stretch for the muscles of the outer portions of the arms, shoulders, and ribs (Fig. 7F–8K).

Stretch No. 12. Here is a simple stretch for your triceps and the top of your shoulders. With arms overhead, hold the elbow of one arm with the hand of the other arm. Gently pull the elbow behind your head, creating a stretch, then stop. Now move the arm away from your body and down as you resist the movement with your opposite hand. Hold this isometric contraction for 6 seconds (PNF technique). Then relax and gently pull the arm over until you feel the right stretch. Do it slowly and hold for 15 seconds. Do not use drastic force to limber up. Stretch both sides (Fig. 7F–8I).

Stretch No. 13. Interlace your fingers out in front of you at shoulder height. Turn your palms outward as you extend your arms forward to feel a stretch in your shoulders, middle of upper back, arms, hands, fingers, and wrists. Hold an easy stretch for 15 seconds, then relax and repeat (Fig. 7F–8M).

Stretch No. 14. To stretch your shoulder and the middle of your upper back, gently pull your elbow across your chest toward your opposite shoulder. Hold stretch for 10 seconds (Fig. 7F–8N).

Stretch No. 15. This stretch for the upper body stretches the muscles laterally along the spine. Stand about 12 to 24 inches away from a fence or wall with your back toward it. With your feet about shoulder-width apart and toes pointed straight ahead, slowly turn your upper body around until you can easily place your hands on the fence or wall at about shoulder height. Turn in one direction and touch the wall, return to the starting position, and then turn in the opposite direction and touch the wall. Do not force yourself to turn any farther than is fairly comfortable. If you have a knee problem, do this stretch very slowly and cautiously. Be relaxed and do not force. Hold for 10 to 20 seconds. Keep knees slightly bent 1 inch (Fig. 7F–8O).

Stretches for the Upper and Lower Back

Most of these stretches are done while lying on your back.

Stretch No. 1. With your knees flexed, interlace your fingers behind your head at about ear level and relax. Then use the power of your arms to slowly pull your head forward until you feel a slight stretch in the back of the neck. Hold for 5 to 10 seconds, then slowly return to the original starting position. Do this three or four times to gradually loosen up the upper spine and neck (Fig. 7F–9A).

Stretch No. 2. Variation: Gently pull your head and chin toward your left knee and hold for 5 seconds. Relax and lower your head back down to the floor, then pull your head gently toward your right knee. Repeat two or three times (Fig. 7F–9K).

Stretch No. 3. With the back of your head on the floor, turn your chin toward your shoulder (as you keep your head resting on the floor). Turn chin only as far as needed to get an easy stretch in the side of your neck. Hold 5 seconds, then stretch to the other side. Repeat two or three times (Fig. 7F–9C).

Stretch No. 4. Shoulder blade pinch. From a bent-knee position, with your fingers interlaced behind your head, pull your shoulder blades together to create tension in the upper back area. (As you do this your chest should move upward.) Hold this controlled tension for 4 to 5 seconds, then relax and gently pull your head forward as shown in Figure 7F–9A. This will help release tension and allow the neck to be stretched effectively (Fig. 7F–9D).

Stretch No. 5. Lower back flattener. To relieve tension in lower back area, tighten your gluteus muscles and, at the same time, tighten your abdominal muscles to flatten your lower back. Hold this tension for 5 to 8 seconds, then relax. Repeat two or three times. Concentrate on maintaining constant muscle contraction. This pelvic tilting exercise will strengthen the gluteus and abdominal muscles so that you are able to sit and stand with better posture. Use these tension controls when sitting and standing (Fig. 7F–9E).

Stretch No. 6. Shoulder blade pinch and gluteus tightener. Simultaneously do the shoulder blade pinch, flatten your lower back, and tighten your gluteus muscles. Hold 5 seconds, then relax and pull your head forward to stretch the back of your neck and upper back. Repeat three or four times (Fig. 7F–9F).

Stretch No. 7. From a bent-knee position, with your head resting on the floor, put one arm above your head (palm up) and the other arm down along your side (palm down). Now reach in opposite directions at the same time to create a controlled stretch in your shoulders and back. Hold stretch for 6 to 8 seconds. Do both sides at least twice. Keep your lower back relaxed and flat (Fig. 7F–9G).

Stretch No. 8. Elongation stretch: Extend your

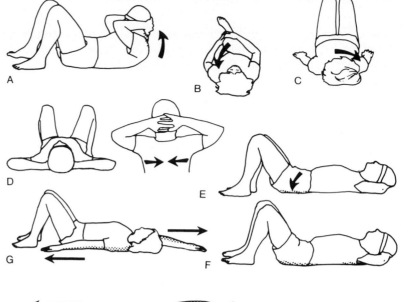

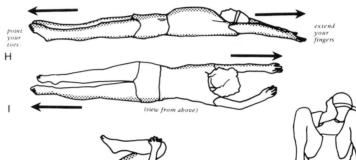

FIGURE 7F–9
A–P, Stretches for the major muscle groups in the upper and lower back. Dotted areas indicate those most likely to feel the stretch.

arms overhead and straighten out your legs. Now reach as far as is comfortable in opposite directions with your arms and legs. Stretch for 5 seconds and then relax (Fig. 7F–9*H*).

Stretch No. 9. Now stretch diagonally. Point the toes of your left foot as you extend your right arm. Stretch as far as is comfortable. Hold 5 seconds and then relax. Usually, doing this stretch three times is sufficient for reducing tension and tightness (Fig. 7F–9*I*).

Stretch No. 10. Pull your right leg toward your chest. For this stretch keep the back of your head on the floor or mat if possible, but do not strain. Hold an easy stretch for 30 seconds. Repeat, pulling your right leg toward your chest. Be certain to keep your lower back flat. This is a very good position for the legs, feet, and back (Fig. 7F–9*J*).

Stretch No. 11. Variation. Pull your knee to your chest, then gently pull the right knee across your body toward your opposite shoulder to create a stretch on the outside of your right hip. Hold an easy stretch for 20 seconds. Do both sides (Fig. 7F–9*B*).

Stretch No. 12. Stretch for the lower back and side of hip. Bend one knee at 90 degrees and, with your opposite hand, pull that bent leg up and over your other leg as shown in the illustration. Turn your head to look toward the hand of the arm that is straight (head should be resting on floor, not held up). Now, using the hand on your thigh (resting just above knee), pull your bent leg down toward the floor until you get the right stretch feeling in your lower back and side of hip. Keep feet and ankles relaxed. Be certain the backs of your shoulders are flat on the floor. If not, the angle changes between the shoulders and the hips, and it is more difficult to create a proper stretch. Hold an easy stretch for 30 seconds on each side (Fig. 7F–9*L*).

Stretch No. 13. From a bent-knee position move your chin downward to your "Adam's apple" in an attempt to flatten the cervical spine to the floor or mat. Hold for 3 to 5 seconds and repeat two or three times (Fig. 7F–9*M*).

Stretch No. 14. A stretch for the lower back, side, and top of hip. Keep knees almost together and rest your feet on the floor. Interlace your fingers behind your head and rest your arms on the floor. Now lift the left leg over the right leg (left drawing). From here, use left leg to pull right leg toward floor (right drawing) until you feel a good stretch along the side of the hip or in the lower back. Stretch and be relaxed. Keep the upper back, back of head, shoulders, and elbows flat on the floor. Hold for 30 seconds. The idea is not to touch the floor with your right knee, but to stretch within *your* limits. Repeat

stretch for other side, crossing right over left leg and pulling down to the right (Fig. 7F–9*N*).

Stretch No. 15. Lie prone with your hands just to the outside of your shoulders. Now slowly extend your arms until you feel a comfortable stretch. Be certain to keep the front of the hips in contact with the floor as you do this stretch. Also, keep your eyes and head forward and not up so your back is not put into hyperextension. Hold for 4 to 6 seconds. Repeat several times (Fig. 7F–9*O*).

Stretch No. 16. Standing with knees slightly bent, place palms on lower back just above hips, fingers pointing downward. Gently push your palms forward to create an extension in the lower back. Hold comfortable pressure for 10 to 12 seconds. Repeat twice. Use this stretch after sitting for an extended period of time (Fig. 7F–9*P*).

Series of Stretches for Groin, Hips, and Back

Stretch No. 1. Relax with knees bent and soles of your feet together. This comfortable position will stretch your groin. Hold for 30 seconds (Fig. 7F–10*A*).

Stretch No. 2. Variation. From this lying groin stretch, gently rock your legs as one unit (see illustration) back and forth about 10 to 12 times. These are easy movements of no more than 1 inch in either direction. Initiate movements from top of hips. This will gently limber up your groin and hips (Fig. 7F–10*B*).

Stretch No. 3. Put the soles of your feet together and hold on to your feet. Now contract the abdominals as you gently pull yourself forward, bending at the hips, until you feel a mild stretch in your groin. You may also feel a stretch in the back. Do not make initial movement for stretch from head and shoulders. Move from the hips. Hold for 20 to 40 seconds (Fig. 7F–10*C*).

Stretch No. 4. With hands supplying slight resistance on insides of opposite thighs, try to bring knees together, just enough to contract the muscles in the groin. Hold this stabilized tension for 5 to 8 seconds, then relax and stretch the groin as in the preceding stretch (see Fig. 7F–10*C*). This will help relax a tight groin area. This technique of tension-relax-stretch is valuable for athletes who have had groin problems (Fig. 7F–10*H*).

Stretch No. 5. Pull your knee across your body toward your opposite shoulder until an easy stretch is felt on the side of the hip. Hold for 30 seconds. Do both sides. This is a good stretch for runners and sedentary persons (Fig. 7F–10*F*).

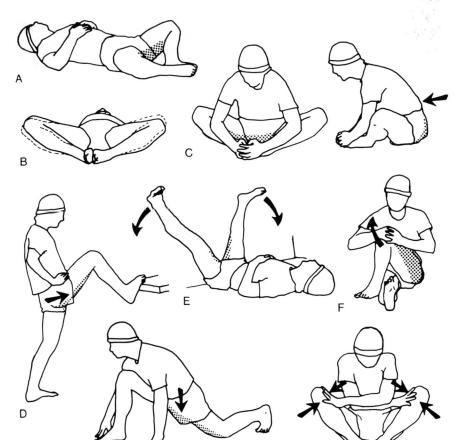

FIGURE 7F-10
A–H, Stretches for the major muscle groups in the groin, hips, and back. Dotted areas indicate those most likely to feel the stretch.

Stretch No. 6. It is possible to stretch your groin from this position by slowly separating your legs, with your heels resting on the wall, until you feel an easy stretch. Hold the stretch 30 seconds and relax (Fig. 7F–10*E*).

Stretch No. 7. Place one leg forward until the knee of the forward leg is directly over the ankle. Your other knee should be resting on the floor. Now, without changing the position of the knee on the floor or the forward foot, lower the front of your hip downward to create an easy stretch. This stretch should be felt in front of the hip and possibly in your hamstrings and groin. This will help relieve tension in the lower back. Hold the stretch for 30 seconds (Fig. 7F–10*G*).

Stretch No. 8. Place the ball of your foot up on a secure support of some kind (wall, fence, table). Keep the down leg pointed straight ahead. Now bend the knee of the up leg as you move your hips forward. This should stretch your groin, hamstrings, and front of hip. Hold for 30 seconds. This stretch will make it easier to lift your knees. If possible, for balance and control, use your hands to hold onto the support. Do both legs (Fig. 7F–10*D*).

Stretches for Hamstring Muscles

Stretch No. 1. Begin in the bent-knee position shown. This position contracts the quadriceps and relaxes the hamstrings. Hold for 30 seconds. The primary function of the quadriceps is to straighten the leg. The basic function of the hamstrings is to bend the knee. Because these muscles have opposing actions, tightening the quadriceps will relax the hamstrings. Now, as you hold this bent-knee position, feel the difference between the front of the thigh and the back of the thigh. The quadriceps should feel hard and tight while the hamstrings should feel soft and relaxed. It is easier to stretch the hamstrings, as in the next stretch, if they have been relaxed first (Fig. 7F–11*A*).

Stretch No. 2. Sit down and straighten your right leg. The sole of your left foot will be resting next to the inside of your straightened leg. Lean slightly forward from the hips and stretch the hamstrings of your right leg. Find an easy stretch and relax. If you cannot touch your toes comfortably, use a towel to help you stretch. Hold for 30 seconds. Do not lock your knee. Your right quadriceps should be soft and

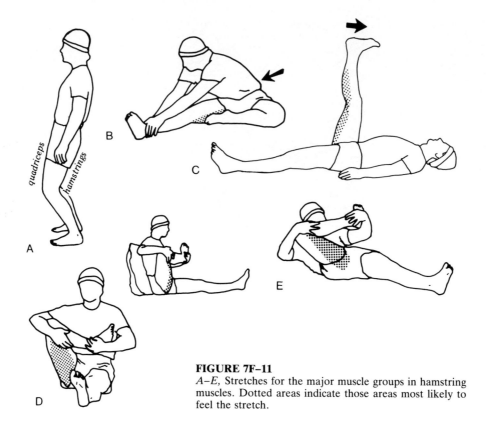

FIGURE 7F–11

A–E, Stretches for the major muscle groups in hamstring muscles. Dotted areas indicate those areas most likely to feel the stretch.

relaxed during the stretch. Keep your right foot upright with the ankle and toes relaxed. Then stretch the left leg (Fig. 7F–11*B*).

Stretch No. 3. Lie on your back and lift your leg up toward a 90-degree angle at the thigh joint. Keep the low back flat against the floor during the stretch. Hold stretch for 15 to 20 seconds. Do both legs (Fig. 7F–11*C*).

Stretch No. 4. To stretch the upper hamstrings and hip, hold on to the outside of your ankle with one hand, with your other hand and forearm around your bent knee. Gently pull the leg as one unit toward your chest until you feel an easy stretch in the back of the upper leg. You may want to do this stretch while you rest your back against something for support. Hold for 15 to 30 seconds. Be certain the leg is pulled as one unit so that no stress is felt in the knee (Fig. 7F–11*D*).

Stretch No. 5. Begin this stretch lying down, then lean forward to hold onto your leg as described in the previous stretch. Gently pull leg as one unit toward your chest until you feel an easy stretch in the buttocks and upper hamstring. Hold for 20 seconds. Doing this stretch in a prone position will increase the stretch in the hamstrings for people who are relatively flexible in this area. Do both legs and compare (Fig. 7F–11*E*).

Series of Stretches for the Knee/ Quadriceps Area

Stretch No. 1. Opposite hand to opposite foot— quadriceps and knee stretch. Hold top of left foot (from inside of foot) with right hand and gently pull, heel moving toward buttocks. The knee bends at a natural angle in this position and creates a good stretch in knee and quads. This is especially good to do if you have trouble or feel pain stretching in the hurdle stretch position leaning back, or when pulling the right heel to buttocks with the right (same) hand. Pulling the opposite hand to the opposite foot does not create any adverse angles in the knee and is especially good in knee rehabilitation and with problem knees. Hold for 30 seconds. Do both legs (Fig. 7F–12*A*).

Stretch No. 2. Extend your foot in back of you, setting the top of it on a table, fence, or bar behind you at a comfortable height. Think of pulling your leg through (moving your leg forward) from the front of your hip to create a stretch for the front of the hip (iliopsoas) and quadriceps. Flex your gluteus muscles as you do this stretch. Keep the down knee slightly bent (1 inch) and upper body vertical. The foot on the ground should be pointed straight ahead.

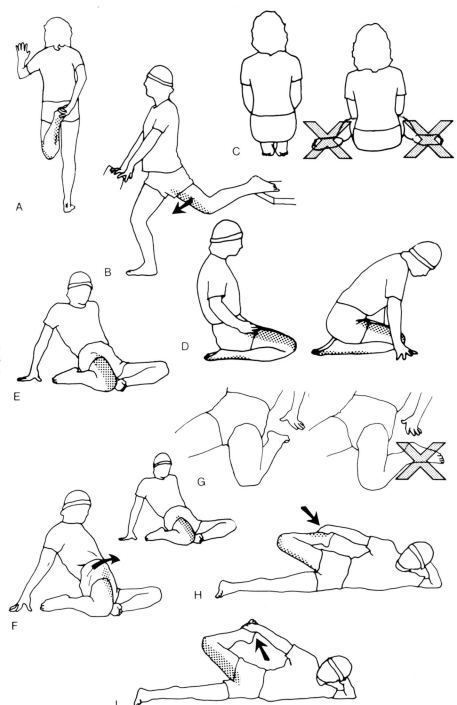

FIGURE 7F–12
A–I, Stretches for the muscles in the knee/quadriceps area. Dotted areas indicate those most likely to feel the stretch.

You can change the stretch by slightly bending the knee of the supporting leg a little more. Hold an easy stretch for 20 seconds. Learn to feel balanced and comfortable in this stretch through relaxed practice. This is a very safe way to stretch the knee-quadriceps area (Fig. 7F–12*B*).

Stretch No. 3. Do not let your feet flare out to the sides when doing a weight-bearing knee-quadriceps stretch. A flared-out position of the lower legs and feet may cause overstretching of the inside (medial collateral) ligaments of the knee (Fig. 7F–12*C*).

Stretch No. 4. Sit on your feet, your toes pointed behind you. Do not let your feet flare to the outside. If your ankles are tight, put your hands on the outside of your legs on the floor and use your hands for support to help you maintain an easy stretch. Do not strain. Hold for 15 to 30 seconds. *Be careful if you have had any knee problems. If pain is present, discontinue this stretch* (Fig. 7F–12*D*).

Stretch No. 5. Sit with your right leg bent, with your right heel just to the outside of your right hip. The left leg is bent and the sole of your left foot is next to the inside of your upper right leg. (Do not let your right foot flare out to the side in this position.) Now slowly lean straight back until you feel an easy stretch in your right quadriceps. Use hands for balance and support. Hold an easy stretch for 15 to 30 seconds. *Do not hold any stretches that are painful to the knee* (Fig. 7F–12*E*).

Stretch No. 6. After stretching your quadriceps, practice tightening the buttocks on the side of the bent leg as you turn the hip over. This will help stretch the front of your hip and give a better overall stretch to the upper thigh area. After contracting the buttocks muscles for 5 to 8 seconds, let the buttocks relax. Then continue to stretch quadriceps for another 10 to 15 seconds (Fig. 7F–12*F*).

Stretch No. 7. In this stretch position your foot should be extended back with the ankle flexed. If your ankle is tight and restricts the stretch, move your foot just enough to the side to lessen the tension in your ankle. Try not to let your foot flare out to the side in this position. By keeping your foot pointed straight back, you take the stress off the inside of your knee. The more your foot flares to the side, the more stress there is on your knee (Fig. 7F–12*G*).

Stretch No. 8. Lie on your left side and rest the side of your head in the palm of your left hand. Hold the top of your right foot with your right hand between the toes and ankle joint. Gently pull the right heel toward the right buttock to stretch the ankle and quadriceps (front of thigh). Hold an easy stretch for 10 seconds. *Never stretch the knee to the point of pain. Always be in control* (Fig. 7F–12*H*).

Stretch No. 9. Move the front of your right hip forward by contracting the right buttocks (gluteus) muscles as you push your right foot into your right hand. This should stretch the front of your thigh. Hold a comfortable stretch for 10 seconds. Keep the body in a straight line. Now stretch the left leg in the same way. You may also get a good stretch in the front of the shoulder (Fig. 7F–12*I*).

Stretches for the Gastrocnemius/Soleus Area

Stretch No. 1. To stretch your calf, stand close to a solid support and lean on it with your forearms, head resting on hands. Bend one leg and place your foot on the ground in front of you, with the other leg straight behind. Slowly move your hips forward, keeping your lower back flat. Be certain to keep the heel of the straight leg on the ground, with toes pointed straight ahead or slightly turned in as you hold the stretch. Hold an easy stretch for 15 to 30 seconds and do not bounce. Stretch the other leg. If you are an extreme pronator, use Stretch No. 4 (see Fig. 7F–13*D*) instead (Fig. 7F–13*A*).

Stretch No. 2. To create a stretch for the calf and Achilles tendon, lower your hips downward as you slightly bend your knee. Be certain to keep your back flat. Your back foot should be slightly toes-in or straight ahead during the stretch. Keep your heel down. This stretch is good for developing ankle flexibility. Hold stretch for 25 seconds. The Achilles tendon area needs only a *slight feeling of stretch* (Fig. 7F–13*B*).

Stretch No. 3. Assume a bent-knee position with your heels flat, toes pointed straight ahead and feet about shoulder-width apart. Hold this position for 30 seconds. In this position you will be able to stretch the gastrocnemius/soleus area (Fig. 7F–13*C*).

Stretch No. 4. Slowly pull the toes back toward the shin until you cannot go any further, then stop and hold the foot dorsiflexed. Next, slowly bend forward from the thigh joint until you feel a stretch in the back of the lower leg. Now hold this stretch for approximately 10 to 15 seconds as you keep the foot dorsiflexed. This is an excellent stretch for the rear of the lower leg (Fig. 7F–13*D*).

SPECIAL NEEDS

The hypermobile athlete who engages in activities that require hypermobility needs strength training. Any joint that is hypermobile will benefit from increased tone and strength from the surrounding

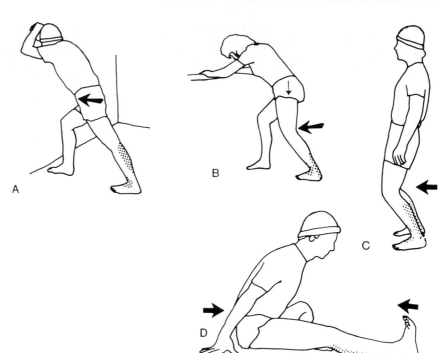

FIGURE 7F–13
A–D, Stretches for the muscles in the gastrocnemius/soleus area. Dotted areas indicate those most likely to feel the stretch.

muscles. It is recommended that the hypermobile athlete concentrate on developing increased strength through resistive weight training for the area(s) of hypermobility.

As increased strength is developed in the areas of hypermobility, mild, light stretching could possibly be used to reduce excess muscle tension. However, the athlete should be careful not to use stretching to promote hypermobility by stretching too far.

As for the tight athlete with repeated strains, an overall stretching program with the emphasis on relaxation may be helpful. Many tight athletes may not like to stretch because they are not very flexible. It is important to de-emphasize flexibility at first so the athlete can relax and concentrate on the *feeling of the stretch* and not how far he or she can stretch. First an athlete learns how to stretch to reduce tension and then, after using stretching regularly, flexibility gains will be experienced [18]. Patience is necessary.

Generally, if a joint is hypermobile, strengthen it; if the musculature is "tight," stretch it.

References

1. Alter, M. J. *Science of Stretching.* Champaign, IL, Human Kinetics, 1988, pp. 90, 120.
2. Anderson, B. A. *Stretching.* Bolinas, Shelter Publications, 1980.
3. Asmussen, E., and Boje, O. Body temperature and capacity for work. *Acta Physiol Scand* 10:1, 1945.
4. Astrand, P. O., and Rodahl, K. *Textbook of Work Physiology.* New York, McGraw-Hill, 1977.
5. Barcroft, J., and Edholm, O. G. The effect of temperature on blood flow and deep temperature in the human forearm. *J Physiol* 102:5, 1943.
6. Barcroft, J., and King, W. R. The effect of temperature on the dissociative curve of blood. *J Physiol* 39:374, 1909.
7. Barnard, R. L., Gardner, G. W., Diaco, N. V., et al: Cardiovascular responses to sudden strenuous exercise. Heart rate, blood pressure, and ECG. *J Appl Physiol* 34:833, 1973.
8. Barnard, R. J., MacAlpin, R., Kattus, A. A., et al: Ischemic response to sudden strenuous exercise in healthy men. *Circulation* 48:936, 1973.
9. Beaulieu, J. E. Developing a stretching program. *Physician Sportsmed* 9(11):59, 1981.
10. Beck, J. L., and Day, W. R. *Overuse Injuries.*
11. Bryant, S. Flexibility and stretching. *Physician Sportsmed* 12(2):171, 1984.
12. Chu, D. Developing flexibility. *In* Cooper, P. (Ed.), *Aerobics Theory and Practice.* Sherman Oaks, Aerobics and Fitness Association of America, 1985, p. 129.
13. Ciullo, J. V., and Zarins, B. Biomechanics of the musculotendinous unit. *Clin Sports Med* 2:71, 1983.
14. Cooper, D. L., and Fair, J. Developing and testing flexibility. *Physician Sportsmed* 6:137–138, 1978.
15. Corbin, C. B., Dowell, L. J., Lindsey, R., et al: *Concepts in Physical Education* (4th ed.). Dubuque, Iowa, William C. Brown, 1981.
16. Corbin, C. B., and Lindsey, R. *Fitness for Life* (2nd ed.). Glenview, IL, Scott, Foresman, 1983.
17. Cureton, T. K. *Physical Fitness of Champion Athletes.* Urbana, IL, University of Illinois Press, 1951.
18. deVries, H. A. Electromyographic observation of the effect of static stretching upon muscular distress. *Res Q* 32:468–479, 1961.
19. deVries, H. A. Flexibility. *In Physiology of Exercise* (3rd ed.). Dubuque, IA, William C. Brown, 1980, p. 463.
20. deVries, H. A., Wiswell, R. A., Bulbulion, R., et al: Tranquilizer effect of exercise. *Am J Physical Med* 60(2):57, 1981.
21. Fleischman, E. A. *The Structure and Measurement of Physical Fitness.* Englewood Cliffs, NJ, Prentice-Hall, 1964.

22. Guyton, A. C. *Textbook of Medical Physiology* (6th ed.). Philadelphia, W. B. Saunders, 1981, p. 631.
23. Hebbelinck, M. Flexibility. *In* Dirix, A., Knuttgen, H. G., and Tittel, K. (Eds.), *The Olympic Sports Medicine Book.* Oxford, Blackwell Scientific Publications, 1988, p. 216.
24. Knott, M., and Voss, D. E. *Proprioceptive Neuromuscular Facilitation.* New York, Harper & Row, 1968, p. 3.
25. Leighton, J. R. The Leighton flexometer and flexibility test. *J Assoc Phys Ment Rehabil* 20:86, 1966.
26. Martin, B. J., Robinson, S., Wiegman, D. L., et al: Effect of warm-up on metabolic responses to strenuous exercise. *Med Sci Sports* 7:146, 1975.
27. Sapega, A. A., Quendenfeld, T. C., Moyer, R. A., et al: Biophysical factors in range-of-motion exercise. *Physician Sportsmed* 9(12):57, 1981.
28. Tachdjian, M. A. *Pediatric Orthopaedics.* Philadelphia, W. B. Saunders, 1972.
29. Wilmore, J. H., and Costill, D. L. *Training for Sport and Activity* (3rd ed.). Dubuque, IA, William C. Brown, 1988, p. 373.
30. Wright, V., and Johns, R. J. Physical factors concerned with the stiffness of normal and diseased joints. *Bull Johns Hopkins Hosp* 106:215, 1960.

NUTRITION IN SPORTS

Eleanor A. Young, Ph.D., R.D., L.D.

NUTRITION AND PHYSICAL PERFORMANCE INTERRELATIONSHIPS

The role of diet and exercise as an integral aspect of life, particularly of physical fitness and performance, has been recorded in ancient Greek and Roman history. Concepts of positive health and a lifestyle that strongly encouraged diet, exercise, and physical training for athletes were very much a vital aspect of life. Persistence of these concepts led eventually to formal establishment of the Olympic Games in 1896. It was thought that physical beauty, strength, and health, when combined with moral and spiritual values promoted through exercise, particularly competitive exercise, would lead to balance and perfection of the human person [60].

Considerable scientific knowledge has been gained during recent years to confirm the essential role of diet and nutrition in athletic performance [1, 30, 47, 62, 69] as well as in health and prevention of disease [3, 20, 31, 34, 48, 49, 58, 61, 63, 64]. A great deal of research has resulted in a broader understanding of metabolic fuel homeostasis before, during, and after physical exercise and the regulation of this homeostasis by the intensity and duration of exercise, the effect of physical training, and the effects of altered nutrition and obesity. These factors, in addition to genetics, age, sex, and body composition, influence physical performance, as do education and the behavioral characteristics of certain lifestyles (e.g., use of smoking, alcohol, or drugs).

From a medical perspective, a major goal in considering the role of nutrition in athletic performance is to encourage proper nutrition, regular exercise, and physical fitness throughout the life cycle. Physicians, particularly orthopaedic surgeons, may need to consider nutritional factors in the prevention or treatment of athletic injuries. They may also want to emphasize patient education and encourage patients to make changes leading to optimal nutrition and physical fitness as related to health and disease.

Strong support for the coupling of nutrition and physical performance has recently been published by the U.S. Department of Health and Human Services in *Healthy People 2000: National Health Promotion and Disease Prevention Objectives* [67]. These fitness objectives include those listed in Table 8–1 and bring into sharp focus the role of both nutrition and exercise in moving the population toward health and physical fitness by the year 2000.

EXERCISE FUEL: TRANSFORMATION, TRANSFER, AND OXIDATION OF FOOD FUELS

It is from the foods we eat and the beverages we consume that we obtain the essential nutrients that are required to meet the body's need for fuel. These nutrients include carbohydrates, fats, proteins, vitamins, minerals, and water (Table 8–2). The human body has the unique and marvelous ability to transform the energy derived from foods into energy forms that are then used to perform work. By a series of complex reactions, energy obtained from dietary proteins, carbohydrates, and lipids is shunted into a common metabolic pathway to yield compounds that in turn produce energy that the body uses for metabolic and physical work. Carbohydrates are converted into glucose, fats into glycerol and fatty acids, and proteins into amino acids. All of these units can enter the metabolic pathways of acetyl co-enzyme A and then the tricarboxylic acid (TCA) or Krebs cycle to yield energy. This energy

TABLE 8–1
Nutrition-Exercise Objectives for the Year 2000

1. Reduce overweight to a prevalence of no more than 20% among people aged 20 and older and no more than 15% among adolescents aged 12 to 19 years.
2. Increase to at least 30% the proportion of people aged 6 and older who engage regularly, preferably daily, in light to moderate physical activity for at least 30 minutes/day.
3. Increase to at least 20% the proportion of people aged 18 and older and to at least 75% the proportion of children and adolescents aged 6 to 17 who engage in vigorous physical activity that promotes the development and maintenance of cardiorespiratory fitness 3 or more days per week for 20 or more minutes per occasion.
4. Reduce to no more than 15% the proportion of people aged 6 and older who engage in no leisure-time physical activity.
5. Increase to at least 50% the proportion of overweight people aged 12 and older who have adopted sound dietary practices combined with regular physical activity to attain an appropriate body weight.
6. Increase to at least 50% the proportion of school physical education class time that students spend being physically active, preferably engaged in lifetime physical activities.
7. Increase to at least 50% the proportion of children and adolescents in grades 1 to 12 who participate in daily school physical education.
8. Increase the proportion of worksites offering employer-sponsored physical activity and fitness programs.
9. Increase community availability and accessibility of physical activity and fitness facilities.
10. Increase to at least 50% the proportion of primary care providers who routinely assess and counsel their patients about the frequency, duration, type, and intensity of each patient's physical activity practices.
11. Increase to at least 75% the proportion of primary care providers who provide nutrition assessment and counseling or referral to qualified nutritionists or dietitians.

Data from U.S. Department of Health and Human Services, Public Health Service. *Healthy People 2000: National Health Promotion and Disease Prevention Objectives.* Washington, D.C., U.S. Dept. of Health and Human Services, 1990.

is then used for basal metabolism, the specific dynamic action of metabolizing foods, and for physical activity.

This intriguing and complex transformation of nutrients into energy is accomplished through the action of high-energy compounds, adenosine triphosphate (ATP) molecules, and creatine phosphate (CP) molecules. Food energy is shuttled primarily via high-energy ATP compounds and is supported by a reservoir of high-energy CP compounds. This process, often referred to as phosphorylation, is generated by oxidation of dietary carbohydrates, fats, and proteins. ATP, the energy currency of the body, enables biologic work to be done. This energy is available for immediate use by the body but is stored in only small amounts within the cells. It can provide enough energy for maximal exercise for only a few seconds. However, a decrease in the ATP concentration due to exercise almost immediately stimulates a conversion of energy stored in nutrients to provide fuel to resynthesize ATP, and this requires oxygen. The cell concentration of CP is some three to five times that of ATP and is sometimes referred to as a "reserve store" of energy or high-phosphate bonds. The breakdown of glucose to pyruvic acid (glycolysis) does not require oxygen, and these reactions are called *anaerobic*. However, the metabolic reactions that take place in the conversion of pyruvate to acetyl co-enzyme A (Krebs cycle), during which this compound is degraded to CO^2 and H_2O, do require oxygen and are referred to as *aerobic*. Thus, food nutrients provide the potential energy that is converted into metabolic energy. Protein, carbohydrate, and lipids enter into this dynamic, interrelated process of providing fuel for exercise.

Recent excellent reviews of fuel liberation and utilization during physical performance are available [12, 13, 38, 39, 50, 68, 70]. In addition, it is now known that the release and utilization of metabolic fuels during exercise are tightly regulated by a complex, integrated neural and hormonal response. Insulin, glucagon, epinephrine, norepinephrine, growth hormone, and cortisol are involved in the dynamic interrelated release and use of energy fuels [28, 50, 68].

Energy used for exercise of short duration and high intensity, such as weight-lifting, 25-yard swim, shot put, or pole vault, utilizes mostly high-energy ATP and CP and is largely anaerobic. In contrast, endurance types of exercise such as marathon running, distance cycling, swimming, or hiking require oxygen for continual conversion of stored energy to provide fuel for aerobic energy. These demands are modified by many factors, such as training and dietary pattern.

Components of energy expenditure include carbohydrates, lipids, and protein available in the body. Fuel stores for the average well-fed male are shown in Table 8–3 [5]. It is well known that the major sources of fuel for exercise are triglycerides stored in the adipose tissue. Carbohydrate stores consist primarily of glycogen in the muscle and liver, with a small percentage from plasma glucose. Protein has only a limited role as energy fuel during exercise. It is not known whether diet modification can enhance

TABLE 8–2
Useful Definitions

Adipose tissue:	Made up of body cells that store fat. Serves as concentrated store of available fuel for energy production. Excessive storage leads to excess body fat (obesity). This is associated with several major diseases (e.g., coronary heart disease, diabetes) and may decrease physical performance.
Aerobic:	Metabolic process in energy transfer that requires oxygen.
Amino acids:	Compounds or units that make up specific proteins.
Essential amino acids:	Nine amino acids that are dietary essential compounds and cannot be synthesized by the human body. They are methionine, threonine, tryptophan, isoleucine, leucine, lysine, valine, phenylalanine, and histidine (infants only).
Anaerobic:	Metabolic process in energy transfer that does not require oxygen.
Basal metabolism:	Basal metabolic rate (BMR) is the amount of energy needed to maintain the internal activities of the body while at digestive, physical, and emotional rest. It is the measure of energy required for body processes such as heartbeat, breathing, kidney function, brain activity, etc.
Carbohydrates:	Hydrates of carbon consisting of carbon (C), hydrogen (H) and oxygen (O), in the general formula $C_n(H_2O)_n$. Usually referred to as sugars and starches. Energy yield: 4 kcal/g.
Cholesterol:	A fat-related compound, a sterol. Normal constituent of bile acids, cell membranes, brain and nervous tissues, steroid hormones. Elevated blood levels associated with increased vascular disease (e.g., atherosclerosis).
Dietary fiber:	Naturally occurring woodlike fibrous materials found mostly in plants that are not digested or are only partially digested by digestive enzymes. Most common dietary fibers: cellulose, hemicellulose, lignin, pectins, gums.
Disaccharides:	Sometimes called double sugars. Two single carbohydrate units linked together. Most common double sugars: maltose, sucrose, lactose.
Energy:	The power or force or capacity for work. Enables the body to maintain life-sustaining activities.
Total energy requirements:	The sum of energy needed for BMR, physical activities, and the thermogenic effect of food intake (about 10% of the body's total energy needs for metabolism).
Fat-soluble vitamins:	Vitamins A, D, E, K
Linolenic acid:	An essential dietary fatty acid that must be obtained in the diet because it cannot be synthesized in the human body. Most common sources: vegetable oils.
Lipids:	A group of organic substances of a fatty nature, including fats, oils, waxes. One gram yields 9 kilocalories. Most common dietary lipids: butter, margarine, vegetable oils, salad dressings, cream, bacon.
Metabolism:	All of the chemical processes in the body by which substances in foods (proteins, carbohydrates, fats) are converted into energy forms that the body uses to do work.
Minerals:	Mineral elements required in the diet. Yield no calories but serve as building blocks, activators, regulators, transmitters, and controllers of numerous essential body functions.
Major minerals:	Calcium, phosphorus, magnesium, sodium, potassium, chloride, sulfur.
Trace minerals:	Iron, iodine, zinc, copper, manganese, chromium, cobalt, selenium, molybedenum, fluoride.
Monosaccharides:	Sometimes called simple sugars. Single units of carbohydrates. Most common simple sugars: glucose, galactose, fructose.
Omega-3 fatty acids:	Long-chain fatty acids of 20 carbons with five double bonds. Common source: fish oils. May be beneficial in combating coronary heart disease.
Polysaccharides:	Sometimes called complex carbohydrates. Most common polysaccharides: starch, glycogen, dextrins.
Proteins:	Complex organic units composed of building blocks or units called amino acids. Contain C, H, O, and N. Energy yield: 4 kcal/g. Most common examples: casein (milk), albumin (egg white), gluten (wheat), myosin (muscle), hemoglobin (red blood cells).
Vitamins:	A group of diverse essential nutrients characterized as follows: (1) organic dietary substances; (2) necessary to perform special metabolic functions in the body; (3) cannot be synthesized by the body; (4) must be supplied in the diet; (5) required only in very small amounts; (6) lack leads to a specific deficiency disease; (7) excessive amounts may be toxic. Yield no calories, but some are essential components of numerous metabolic functions.
Water-soluble vitamins:	Thiamin, riboflavin, niacin, ascorbic acid, pyridoxine, pantothenic acid, biotin, folic acid, cobalamin.

TABLE 8–3
Body Fuel Stores in Adult Males

	Kg	Amount kcal
Fat from adipose triglycerides	12	100,000
Muscle protein	6	24,000
Glycogen: liver	0.07	280
muscle	0.40	1600
Glucose from body fluids	0.02	80
Free fatty acids from body fluids	0.004	4
Total		135,964

or decrease specific body protein stores for fuel during exercise. At most, protein may contribute only 5% to 10% of the total energy turnover.

Carbohydrates as Fuel

The utilization of fuels for exercise is now known to be modified by several factors: exercise intensity, duration of exercise, endurance training, repetition of exercise over time, and diet. Although a greater amount of fuel from fat is available for energy, carbohydrate fuel is more versatile and important in that it can be utilized anaerobically, its combustion is more efficient in liberating energy, and it can provide an aerobic substrate at a rate double that of fat [39]. However, the total capacity for energy release from carbohydrate is much less than that from fat stores. It is for this reason that dietary intake of carbohydrate prior to and during exercise is of critical importance. This need has led to manipulation of the diet to maximize the muscle glycogen stores needed for specific forms of exercise, particularly prolonged endurance types of exercise. These manipulations have been referred to as carbohydrate or glycogen loading, or supercompensation. Various modifications of carbohydrate loading have been proposed, and all involve some degree of muscle glycogen depletion followed by ingestion of a high-carbohydrate diet just prior to the exercise event [13, 39]. Liver glycogen stores are also important in maximizing exercise performance. To maintain high levels of liver glycogen, a carbohydrate-rich diet is essential because a high-fat, high-protein diet cannot prevent a drop in liver glycogen levels [46].

It is clear from many studies now published that maximum glycogen stores in the muscles and liver are essential to delay hypoglycemia, nausea, dizziness, and exhaustion during endurance exercise. Clearly, the diet is crucial in this manipulation. A

50% drop in liver glycogen can occur even after an overnight fast, and early morning endurance-type exercise can be maximized by consuming adequate carbohydrates 2 to 4 hours prior to the event to replenish liver glycogen stores [13, 28, 30, 68, 70].

Carbohydrate intake immediately prior to exercise or during exercise has been studied in an effort to maximize the amount of readily available fuel and to prevent exhaustion of glycogen stores, leading to subsequent fatigue and exhaustion. Exercise intensity and duration, nutritional status, exercise training status, and the type of fuel substrate used may influence the result [12, 28, 30, 68]. The impact of carbohydrate ingestion immediately prior to exercise is controversial. Fructose may be a more effective carbohydrate fuel than glucose, but it should be used only after trial tests in specific circumstances. Intake of a sugar load within 30 minutes of activity is likely to precipitate an overshoot in insulin release from the liver with a subsequent decrease in blood sugar. The rise in serum insulin will inhibit the release of fatty acids for energy. This has a negative impact on glycogen stores that may cause glycogen depletion sooner, lead to fatigue, and thus compromise optimal energy performance. Water alone is the optimal drink immediately preceding exercise.

Carbohydrate ingestion during prolonged exercise has been shown to enhance performance and delay fatigue [13, 28, 30, 68, 70]. In practice, a beneficial effect may depend on ensuring an isotonic solution taken in small amounts at 10- to 15-minute intervals not to exceed 1 liter/hour [20]. Sodium chloride may be added to replace electrolytes, and the fluid may be chilled to enhance gastric emptying and to lower core body temperature if extreme heat exists [8].

Pregame Meal

The utilization of energy fuels both before and during exercise depends on effective carbohydrate management. The pregame meal should be one that provides adequate carbohydrate energy and that ensures adequate hydration. Generally, the pregame meal is high in carbohydrates and relatively low in protein and fat. Carbohydrates are more readily digested and absorbed and can more effectively replenish muscle glycogen that may have decreased during an overnight fast. On the other hand, high-fat and high-protein foods, such as steak and eggs, have not been shown to have any significant beneficial effect. Studies have indicated that a high-carbohydrate pregame meal may contribute to improved performance. A sample pregame meal is

TABLE 8–4
Typical Pregame Meal

Pregame Meal, 500 Kilocalories	
Milk, skim	8 oz.
Bread	2 slices
Cheese, mozzarella	2 slices
Margarine spread	1 tsp
V-8 juice	8 oz.
Pregame Meal, 900 Kilocalories	
Milk, skim	8 oz.
Broiled/baked chicken	¼ cup diced chicken
Potatoes, mashed	1 cup
Bread	2 slices
Margarine spread	1 tsp
Applesauce	1 cup
Angel food cake	1 slice

given in Table 8–4 and should be modified in total calories depending on the age and sex of the athlete and the type of exercise performed.

The time of the pregame meal is also important; it should allow at least a 3-hour period for digestion and absorption before the event. It is important to remember that the pregame meal cannot "make up" for lack of attention and compliance to a well-balanced diet during training.

It is recommended that familiar foods always be used for the pregame meal to avoid any unsuspected adverse food reactions. Commercial liquid meals for athletes have been found to be satisfactory for many, and they provide additional liquid for hydration. Low-fiber foods within 48 hours of game time may lessen the bulk of gastrointestinal residue remaining in the bowel. The use of cooked rather than raw fruits and vegetables, avoidance of nuts and seeds, and use of plain sugar cookies instead of oatmeal-raisin cookies may be beneficial.

MEETING HUMAN NUTRIENT REQUIREMENTS

Recommended Dietary Allowances

The recommended dietary allowances (RDA) [65] represent the most reliable scientific standards for essential energy and nutrient needs of healthy persons living in the United States. The RDAs are updated periodically to reflect the most recent scientific evidence. For most nutrients, the allowances are based on the average physiologic requirements for absorption of each nutrient plus an adjustment factor to compensate for incomplete utilization, variation in requirements among individuals, and differences in bioavailability among food sources of

each nutrient. The RDAs provide a safety factor appropriate for each nutrient and therefore exceed the basal requirements for most people. Exceptions are the RDAs for energy, which reflect mean population requirements for each age and sex group because an added allowance factor for individuals could be excessive and contribute to obesity in a large segment of the population [65].

Derivation of the RDAs is discussed in detail by the Subcommittee on the 10th Edition of the RDAs, Food and Nutrition Board, [65] and equations for predicting resting energy expenditure, energy expenditure for various physical activities, and estimates of daily energy allowances at different levels of physical activity for men and women are provided [65]. Estimated energy needs can be reasonably calculated from Tables 8–5 through 8–8. These tables may be very helpful to the physician, the athletic coach or trainer, and the athlete in determining energy needs in a scientific manner. Use of these tables allows one to calculate energy needs on an individual basis according to age and sex if the resting metabolic rate, type of activity, and time spent in all activities over a 24-hour period are known. Table 8–5 provides the equation for predicting basal metabolic rate or resting energy expen-

TABLE 8–5
Equations for Predicting Resting Energy Expenditure from Body Weight*

Sex and Age Range (Years)	Equation to Derive REE in kcal/day	R†	SD†
Males			
0–3	$(60.9 \times wt‡) - 54$	0.97	53
3–10	$(22.7 \times wt) + 495$	0.86	62
10–18	$(17.5 \times wt) + 651$	0.90	100
18–30	$(15.3 \times wt) + 679$	0.65	151
30–60	$(11.6 \times wt) + 879$	0.60	164
>60	$(13.5 \times wt) + 487$	0.79	148
Females			
0–3	$(61.0 \times wt) - 51$	0.97	61
3–10	$(22.5 \times wt) + 499$	0.85	63
10–18	$(12.2 \times wt) + 746$	0.75	117
18–30	$(14.7 \times wt) + 496$	0.72	121
30–60	$(8.7 \times wt) + 829$	0.70	108
>60	$(10.5 \times wt) + 596$	0.74	108

*From WHO. *Energy and Protein Requirement Technical Report.* Series 724. Geneva, WHO, 1985.
†Correlation coefficient (R) of reported BMRs and predicted values, and standard deviation (SD) of the differences between actual and computed values.
‡Weight in kg of the person.
Reprinted by permission from *Recommended Dietary Allowances,* 10th edition. Copyright 1989 by the National Academy of Sciences. Courtesy of the National Academy Press, Washington, D.C.

diture (REE) by sex and age. For example, a 30-year-old woman weighing 55 kg can estimate her REE as follows: 55 × 8.7 + 829 = 1307.5 kcal/day. Table 8–6 provides the approximate energy expenditure needed for various activities for males and females. This information is used in the example shown in Table 8–7 to estimate the total daily energy allowance for a very sedentary or very active day for a 70-kg male and a 58-kg female. Table 8–8 provides median heights and weights and recommended energy intake at various ages for men and women. These data can be used to provide a rough estimate of energy requirements if the exact proportion of time spent for different activities is not known.

The athlete can utilize these tables to gain infor-

TABLE 8–6
Approximate Energy Expenditure for Various Activities in Relation to Resting Needs for Males and Females of Average Size*

Activity Category†	Representative Value for Activity Factor per Unit Time of Activity
Resting Sleeping, reclining	REE × 1.0
Very light Seated and standing activities, painting trades, driving, laboratory work, typing, sewing, ironing, cooking, playing cards, playing a musical instrument	REE × 1.5
Light Walking on a level surface at 2.5 to 3 mph, garage work, electrical trades, carpentry, restaurant trades, housecleaning, child care, golf, sailing, table tennis	REE × 2.5
Moderate Walking 3.5 to 4 mph, weeding and hoeing, carrying a load, cycling, skiing, tennis, dancing	REE × 5.0
Heavy Walking with load uphill, tree felling, heavy manual digging, basketball, climbing, football, soccer	REE × 7.0

*Based on values reported by Durin, J. U. G. A., and Passmore, R. *Energy, Work and Leisure.* London, Heinemann, 1967, and WHO. *Energy and Protein Requirement Technical Report.* Series 724. Geneva, WHO, 1985.

†When reported as multiples of basal needs, the expenditures of males and females are similar.

Reprinted by permission from *Recommended Dietary Allowances,* 10th edition. Copyright 1989 by the National Academy of Sciences. Courtesy of the National Academy Press, Washington, D.C.

TABLE 8–7
Example of Calculation of Estimated Daily Energy Allowances for Exceptionally Active and Inactive 23-Year-Old Adults

Step 1: Derivation of Activity Factor*

Activity as Multiples of REE	Very Sedentary Day		Very Active Day	
	Duration (hr)	Weighted REE Factor	Duration (hr)	Weighted REE Factor
Resting 1.0	10	10.0	8	8.0
Very light 1.5	12	18.0	8	12.0
Light 2.5	2	5.0	4	10.0
Moderate 5.0	0	0	2	10.0
Heavy 7.0	0	0	2	14.0
TOTAL	24	33.0	24	54.0
MEAN		1.375		2.25

Step 2: Calculation of Energy Requirement, Kcal per Day

Gender	Resting Energy Expenditure	Very Sedentary Day (REE × 1.375)	Very Active Day (REE × 2.25)
Male, 70 kg	1,750	2,406	3,938
Female, 58 kg	1,350	1,856	3,038

*Activity patterns are hypothetical. As an example of use of the ranges within a class of activity, very light activity is divided between sitting and standing activities.

Reprinted by permission from *Recommended Dietary Allowances,* 10th edition. Copyright 1989 by the National Academy of Sciences. Courtesy of the National Academy Press, Washington, D.C.

mation about REE and the energy needed for specific kinds of exercise multiplied by the time spent in a specific exercise. This can provide a fairly good estimate of overall energy needs during a 24-hour period. A better estimate will be obtained if these data are calculated over several consecutive days and an average value obtained. It is definitely known that energy requirements are influenced by age, sex, growth status, body size, climate (temperature, humidity), pregnancy, and lactation. These factors must be taken into consideration in determining energy needs as related to energy expenditure.

Protein Contributions to Energy Needs

The amount of dietary protein that should contribute to the total energy intake is given in Table 8–9 [65]. Approximately 14% to 18% of total food energy intake in the United States is derived from proteins, and 65% of protein intake is from animal sources. This may be of concern to athletes who wish to curtail fat intake because animal sources of protein tend to be high in lipids. Several recent studies confirm that the amount of protein required

TABLE 8–8
Median Heights and Weights and Recommended Energy Intake

Category	Age (Years) or Condition	Weight (kg)	Weight (lb)	Height (cm)	Height (in)	REE* (kcal/day)	Average Energy Allowance (kcal)† Multiples of REE	Per kg	Per day‡
Infants	0.0–0.5	6	13	60	24	320		108	650
	0.5–1.0	9	20	71	28	500		98	850
Children	1–3	13	29	90	35	740		102	1,300
	1–6	20	44	112	44	950		90	1,800
	7–10	28	62	132	52	1,130		70	2,000
Males	11–14	45	99	157	62	1,440	1.70	55	2,500
	15–18	66	145	176	69	1,760	1.67	45	3,000
	19–24	72	160	177	70	1,780	1.67	40	2,900
	25–50	79	174	176	70	1,800	1.60	37	2,900
	51+	77	170	173	68	1,530	1.50	30	2,300
Females	11–14	46	101	157	62	1,310	1.67	47	2,200
	15–18	55	120	163	64	1,370	1.60	40	2,200
	19–24	58	128	164	65	1,350	1.60	38	2,200
	25–50	63	138	163	64	1,380	1.55	36	2,200
	51+	65	143	160	63	1,280	1.50	30	1,900
Pregnant	1st trimester								+0
	2nd trimester								+300
	3rd trimester								+300
Lactating	1st 6 months								+500
	2nd 6 months								+500

*Calculation based on FAO equations.
†In the range of light to moderate activity, the coefficient of variation is ±20%.
‡Figure is rounded.
Reprinted with permission from *Recommended Dietary Allowances*, 10th edition. Copyright 1989 by the National Academy of Sciences. Courtesy of the National Academy Press, Washington, D.C.

by the athlete may be higher than the RDA level set for nonathletes. This may be particularly important for endurance athletes [15, 17, 32, 41, 71]. Some of the findings that support the need for a protein intake higher than that given in the RDAs (0.8 g/kg/day) include (1) decreased synthesis of protein during and for some time after exercise; (2) increased protein catabolism during and following exercise; and (3) increased output of alanine from the exercising muscle. Because the protein intake of most adults in the United States is about 100 g/day, an amount significantly above the RDA levels, this is usually more than sufficient for most athletes. The amount of protein consumed by athletes may be as high as 150 g/kg/day.

Many athletes adhere to a belief that very large amounts of protein, often in the form of liquid protein supplements, will contribute to increased muscle size and strength. This is a fallacy that should be corrected. Protein in excess of 12% to 15% of total calories is not needed. This is the percentage of calories from dietary protein currently recommended (RDA) for the healthy adult. A typical example follows for a sedentary male and an athletic male, both weighing 70 kg. The sedentary male may consume around 2500 kcal/day, and the athlete 5000 kcal/day. If 12% to 15% of the total energy intake

is calculated, it is clear that the athlete consumes more than sufficient protein to cover his basal protein needs as well as any additional need for protein due to exercise.

Sedentary 70-kg adult
2500 kcal/day
12% protein = 75.4 g protein/day
= 1.08 g protein/kg/day

Athlete 70-kg adult
5000 kcal/day
12% protein = 150.9 g protein/day
= 2.16 g protein/kg/day

Most athletes, especially endurance athletes, consume this amount of calories and protein or more, and this amount above the current RDAs is easily met by most diets consumed by athletes. There is no need for "super" amounts of dietary protein above this level. Because excess protein cannot be stored in the body, nitrogen from excess protein is cleaved from amino acids and excreted through the kidneys. The remainder of the amino acid fragment is shunted into the energy source for fuel. This places an added burden on the kidneys to excrete

TABLE 8–9
Recommended Allowances of Reference Protein and U.S. Dietary Protein

Category	Age (Years) or Condition	Weight (kg)	Derived Allowance of Reference Protein*		Recommended Dietary Allowance	
			(g/kg)	(g/day)	(g/kg)†	(g/day)
Both sexes	0–0.5	6	2.20‡		2.2	13
	0.5–1	9	1.56		1.6	14
	1–3	13	1.14		1.2	16
	4–6	20	1.03		1.1	24
	7–10	28	1.00		1.0	28
Males	11–14	45	0.98		1.0	45
	15–18	66	0.86		0.9	59
	19–24	72	0.75		0.8	58
	25–50	79	0.75		0.8	63
	51+	77	0.75		0.8	63
Females	11–14	46	0.94		1.0	46
	15–18	55	0.81		0.8	44
	19–24	58	0.75		0.8	46
	25–50	63	0.75		0.8	50
	51+	65	0.75		0.8	50
Pregnancy	1st trimester			+1.3		+10
	2nd trimester			+6.1		+10
	3rd trimester			+10.7		+10
Lactation	1st 6 months			+14.7		+15
	2nd 6 months			+11.8		+12

*Data from WHO. *Energy and Protein Requirement Technical Report.* Series 724. Geneva, WHO, 1985.

†Amino acid score of typical U.S. diet is 100 for all age groups, except young infants. Digestibility is equal to reference proteins. Values have been rounded upward to 0.1 g/kg.

‡For infants 0 to 3 months of age, breastfeeding that meets energy needs also meets protein needs. Formula substitutes should have the same amount of amino acid composition as human milk, corrected for digestibility if appropriate.

Reprinted with permission from *Recommended Dietary Allowances,* 10th edition. Copyright 1989 by the National Academy of Sciences. Courtesy of the National Academy Press, Washington, D.C.

excess nitrogen and requires increased water intake. Moreover, protein is an expensive source of calories, and protein supplements added to the diet represent an uneconomical approach to muscle development. Although many athletes may not maintain nitrogen balance on 1.0 to 1.5 g/kg body weight/day, there is insufficient evidence to support a protein intake of more than 2.5 to 3.0 g/kg/day.

Some athletes may be at risk for too low a protein intake that could compromise athletic performance. Athletes at risk include (1) vegetarians with ill-designed vegetable protein diets that include an array of amino acids that do not complement one another to provide adequate essential amino acids for protein synthesis; (2) athletes with so low a calorie intake protein must be utilized as fuel (jockeys, gymnasts, dancers, figure skaters, and so on); (3) athletes with imbalanced diets comprising excessive carbohydrate intake and insufficient protein intake; and (4) athletes who perform multiple daily training sessions that do not allow recovery of glycogen stores between sessions, thus necessitating utilization of proteins for energy.

Athletes at risk for too high a protein intake include (1) athletes with hepatic or renal disease;

(2) athletes who fail to take in adequate water to handle a high protein load, thus placing themselves at risk for dehydration; (3) athletes with excessive protein intake that is subsequently converted to fat and stored in adipose tissue; and (4) athletes with high protein intake that precipitates increased urinary calcium loss, thus increasing the need for greater calcium intake and possibly leading to loss of calcium from bone.

Carbohydrate Contributions to Energy Needs

Carbohydrates form the bulk of caloric intake, and for endurance athletes carbohydrates may represent 60% to 70% of total dietary calories. Dietary carbohydrates should be complex, with a limited intake of refined simple sugars. The bulk of dietary carbohydrates in the form of vegetables, fruits, pastas, and grain products also provide an array of required vitamins and minerals. Dietary simple sugars may be beneficial for pregame and postgame and easy-to-carry carbohydrate-rich snacks, such as raisins, dried apricots or prunes, fruit juices, and pretzels.

Dietary carbohydrates are usually digested and absorbed rapidly and almost completely, eventually shunting glucose through the circulatory system to the liver and muscle glycogen stores. Carbohydrate is the limiting substrate for endurance exercise, and for this reason dietary measures that optimize glycogen stores can enhance exercise endurance. This can be done by (1) increasing glycogen stores before exercising, primarily by ingesting a high-carbohydrate diet; (2) replacing carbohydrates utilized during exercise; and (3) decreasing the rate of carbohydrate use by enhancing fat utilization for energy, possibly by means of caffeine ingestion for some athletes [30] or by means of endurance training [21].

Lipid Contributions to Energy

Dietary fat for the athlete should be limited to less than 30% of the total energy intake, with saturated fats contributing less than 10% of total calories. Dietary cholesterol should be less than 300 mg/day. Dietary fat intake above these levels has been associated with decreased exercise performance and an increased risk of cardiovascular disease [36, 66]. In addition to dietary fats, body stores of fat are also derived through lipogenesis from excess dietary carbohydrates or by protein in the diet. Lipid synthesis occurs in the liver and adipose tissue and is influenced by total caloric intake. The body's enormous capacity to store fat can account for 5% to over 50% of total body weight. Fat stores may exceed carbohydrate stores by 30- to 60-fold [21]. Fat stores are dynamic and fluctuate with the balance between energy intake and energy expenditure. Fat is usually transported in the form of albumin-bound fatty acids, which also exist in a dynamic equilibrium between adipocyte uptake and release of free fatty acids [21]. Fatty acids can be activated to acetyl coenzyme A through acyl CoA synthetase and subsequently undergo beta-oxidation with entry of the acetyl units into the Krebs cycle. The contribution of fat to energy production for exercise is nearly a mirror image of the contribution from carbohydrates. The major dietary fat, triglyceride, is hydrolyzed into glycerol and three fatty acid molecules. The glycerol molecule can enter into the glucose metabolic pathway leading into the Krebs cycle, while the fatty acids undergo beta-oxidation and can also enter into the Krebs cycle via the acetyl CoA pathway. Thus, both glucose, the primary carbohydrate fuel, and glycerol and the fatty acids, the primary fuel from fats, are metabolized through the Krebs cycle. The roles of triglycerides in muscle fibers and adipocytes in relation to energy needs for exercise are not well established. At rest and during light activity, fat is the preferred fuel for many tissues; however, during heavy exercise, mobilization of free fatty acids may be too slow. Endurance training can enhance fat utilization during submaximal exercise, and this can exert a glycogen-sparing effect, thus enhancing the capacity for endurance during exercise [21].

Alcohol, a Nonenergy Source for Exercise

Alcohol yields 7 calories/g and is thus almost as "rich" in energy as fat. Although alcohol consumption can provide a potent source of calories, alcohol cannot be utilized directly or indirectly by the muscle [57]. Moreover, several well-established adverse effects of alcohol for the athlete include (1) malabsorption of thiamine, folic acid, vitamin B_{12}, riboflavin, niacin, ascorbic acid, and pyridoxine; (2) erosive gastritis; (3) decrease in hepatic release of glucose; (4) inhibition of free fatty acid oxidation; and (5) myocardial depression [57].

VITAMINS AND MINERALS FOR THE ATHLETE

Vitamins

Thirteen vitamins have been isolated, analyzed, and synthesized. Table 8–10 lists the RDA for most of these, and estimated safe intakes for biotin and pantothenic acid are given in Table 8–11. Vitamins are by definition essential for life and must be obtained from the diet. A well-balanced diet including foods from a variety of sources should provide adequate amounts of all essential vitamins. This is especially true for athletes because the total energy intake for most athletes far exceeds that of sedentary nonathletes, and consequently a greater amount and variety of vitamins are also available to the athlete through dietary intake. Only under circumstances of known deficiency or in persons known to be at risk of deficiency are vitamin supplements necessary. There is no evidence that the intake of supplemental vitamins can enhance normal health status, improve exercise performance, or increase one's ability in athletic training except when a deficiency of vitamins exists. Supplemental intake of vitamins does not necessarily increase the blood levels of the water-soluble vitamins because excess amounts of these vitamins are usually excreted in the urine. There are reports in the medical literature of the harmful effects of excessive intake of several of the water-soluble vitamins (e.g., pyridoxine, niacin, and riboflavin). However, excess intake of the fat-soluble vitamins A, D, E, and K may be stored in the body and is therefore much more likely to become toxic. Fat-soluble vitamins should not be taken in excessive amounts without medical supervision [25]. Thus, there is no medical reason for megavitamin therapy for the healthy athlete who eats a varied diet. Megavitamin dosing at 10 times the RDA will not lead to a supercharged athletic performance and may prove to be harmful, thus compromising performance.

It should be recognized that several of the B-complex vitamins—thiamine, riboflavin, niacin, pyridoxine, cobalamin (B_{12}), folic acid, biotin, and pantothenic acid—all play crucial roles as coenzyme factors in energy metabolism and therefore are very much involved in the release of energy from carbohydrates, fats, and proteins during metabolism. A deficiency in any of these vitamins could compromise release of energy for exercise [2, 25].

Minerals

Fifteen minerals are listed in Tables 8–10 through 8–12, and these essential nutrients have a variety of metabolic roles. They serve as builders, regulators, activators, transmitters, and controllers. Sodium, potassium, and chloride play crucial roles in the shifts of body fluids. Calcium and phosphorus play a primarily structural role. Oxygen is part of the heme core in hemoglobin, cobalt is part of the vitamin B_{12} core, and iodine serves as a constituent in thyroxine. The variety of functions of these nutrients in reference to specific aspects of athletic performance and health are evident. Some minerals have specific additional functions.

Calcium. Calcium is the major mineral of the human body and, together with phosphorus, contributes to the formation of the bones and teeth. In an ionized form calcium plays important roles in muscular contraction and in transmission of nerve impulses. Calcium catalyzes the actions of myosin and actin filaments that permit sliding contractions to take place in muscle, including cardiac muscle. Calcium is required to trigger neural signals from one neuron to another and eventually to the target

TABLE 8–10
Food and Nutrition Board, National Academy of Sciences—National Research Council Recommended Dietary Allowances,* Revised 1989

Designed for the maintenance of good nutrition of practically all healthy people in the United States

Category	Age (Years) or Condition	Weight† (kg)	Weight† (lb)	Height† (cm)	Height† (in)	Protein (g)	Fat-Soluble Vitamins Vitamin A (μg RE)‡	Vitamin D (μg)¶	Vitamin E (mg α-TE)§	Vitamin K (μg)
Infants	0.0–0.5	6	13	60	24	13	375	7.5	3	5
	0.5–1.0	9	20	71	28	14	375	10	4	10
Children	1–3	13	29	90	35	16	400	10	6	15
	4–6	20	44	112	44	24	500	10	7	20
	7–10	28	62	132	52	28	700	10	7	30
Males	11–14	45	99	157	62	45	1,000	10	10	45
	15–18	66	145	176	69	59	1,000	10	10	65
	19–24	72	160	177	70	58	1,000	10	10	70
	25–50	79	174	176	70	63	1,000	5	10	80
	51 +	77	170	173	68	63	1,000	5	10	80
Females	11–14	46	101	157	62	46	800	10	8	45
	15–18	55	120	163	64	44	800	10	8	55
	19–24	58	128	164	65	46	800	10	8	60
	25–50	63	138	163	64	50	800	5	8	65
	51 +	65	143	160	63	50	800	5	8	65
Pregnant						60	800	10	10	65
Lactating	1st 6 months					65	1,300	10	12	65
	2nd 6 months					62	1,200	10	11	65

*The allowances, expressed as average daily intakes over time, are intended to provide for individual variations among most normal persons as they live in the United States under usual environmental stresses. Diets should be based on a variety of common foods in order to provide other nutrients for which human requirements have been less well defined. See text for detailed discussion of allowances and of nutrients not tabulated.

†Weights and heights of Reference Adults are actual medians for the U.S. population of the designated age, as reported by NHANES II. The median weights and heights of those under 19 years of age were taken from Hamill et al. (1979)

‡Retinol equivalents. 1 retinol equivalent = 1 μg retinol or 6 μg β-carotene. See text for calculation of vitamin A activity of diets as retinol equivalents.

muscle cells. Calcium is an important activator of certain enzymes, such as ATPase, in the release of energy for muscle contraction. In addition, calcium is required in the crosslinking of fibrin, thus allowing blood clotting to occur. Calcium also controls the passage of fluids through cell membranes by regulating cell permeability. Thus, the essential functions of calcium have a direct impact on muscle, nerve, and bone structure, all of which are essential for athletic performance.

Yet, in the United States calcium is the mineral that is most frequently found to be deficient in the diet. Some 75% of all adults in the United States consume less than the RDA, and about 25% of all females in this country consume less than 300 mg of calcium per day, whereas the recommended intake for most adult females is 1000 to 1200 mg/day [27]. This deficiency is considered a major contributor to decreased bone mass leading to osteoporosis, which affects about 24 million people in the United States. This disease accounts for more than 1.2 million fractures/year, including 500,000 spinal fractures and 230,000 hip fractures. Although a decrease in estrogen in older women contributes to osteoporosis, two primary defenses are (1) adequate dietary calcium

from early life on, and (2) regular exercise that delays the rate of bone loss with age [42].

Secondary amenorrhea associated with excessive exercise, low calcium intake, and reduced body mass is thought to be a contributor to hormonal imbalance and has a possible negative effect on bone mass [10, 22, 42, 44, 59]. This condition appears to be a paradox, but it is a very real problem that is sometimes difficult to identify, especially if it is associated with specific eating disorders. This abnormality can seriously compromise athletic performance in the young female athlete and can lead to the serious problem of osteoporosis in later life.

Iron. Iron is a functionally active component of hemoglobin in the red blood cells, and in this strategic position it has an influence on physical exercise. Hemoglobin carries about 80% of the total 3 to 5 g of iron normally found in the human body and also significantly increases the oxygen-carrying capacity of blood by 65-fold. Iron is a structural component of myoglobin stored in the muscle and aids in the storage and transport of oxygen in the muscle cell. Cytochromes, which serve as catalysts in intracellular energy transfer, also contain iron. Iron reserves, such as hemosiderin and ferritin, are

TABLE 8–10 *Continued*
Food and Nutrition Board, National Academy of Sciences—National Research Council Recommended Dietary Allowances,* Revised 1989
Designed for the maintenance of good nutrition of practically all healthy people in the United States

	Water-Soluble Vitamins						Minerals						
Vitamin C (mg)	Thiamin (mg)	Riboflavin (mg)	Niacin (mg NE)#	Vitamin B$_6$ (mg)	Folate (µg)	Vitamin B$_{12}$ (µg)	Calcium (mg)	Phosphorus (mg)	Magnesium (mg)	Iron (mg)	Zinc (mg)	Iodine (µg)	Selenium (µg)
30	0.3	0.4	5	0.3	25	0.3	400	300	40	6	5	40	10
35	0.4	0.5	6	0.6	35	0.5	600	500	60	10	5	50	15
40	0.7	0.8	9	1.0	50	0.7	800	800	80	10	10	70	20
45	0.9	1.1	12	1.1	75	1.0	800	800	120	10	10	90	20
45	1.0	1.2	13	1.4	100	1.4	800	800	170	10	10	120	30
50	1.3	1.5	17	1.7	150	2.0	1,200	1,200	270	12	15	150	40
60	1.5	1.8	20	2.0	200	2.0	1,200	1,200	400	12	15	150	50
60	1.5	1.7	19	2.0	200	2.0	1,200	1,200	350	10	15	150	70
60	1.5	1.7	19	2.0	200	2.0	800	800	350	10	15	150	70
50	1.2	1.4	15	2.0	200	2.0	1,200	1,200	280	15	12	150	45
60	1.1	1.3	15	1.4	150	2.0	1,200	1,200	300	15	12	150	50
60	1.1	1.3	15	1.5	180	2.0	1,200	1,200	280	15	12	150	55
60	1.1	1.3	15	1.6	180	2.0	800	800	280	15	12	150	55
60	1.0	1.2	13	1.6	180	2.0	800	800	280	10	12	150	55
70	1.5	1.6	17	2.2	400	2.2	1,200	1,200	320	30	15	175	65
95	1.6	1.8	20	2.1	280	2.6	1,200	1,200	355	15	19	200	75
90	1.6	1.7	20	2.1	260	2.6	1,200	1,200	340	15	16	200	75

§As cholecalciferol. 10 µg cholecalciferol = 400 IU of vitamin D.

¶α-Tocopherol equivalents. 1 mg d-α tocopherol = 1 α-TE. See text for variation in allowances and calculation of vitamin E activity of the diet as α-tocopherol equivalents.

#1 NE (niacin equivalent) is equal to 1 mg of niacin or 60 mg of dietary tryptophan.

Reprinted with permission from *Recommended Dietary Allowances*, 10th edition. Copyright 1989 by the National Academy of Sciences. Courtesy of the National Academy Press, Washington, D.C.

TABLE 8–11
Estimated Safe and Adequate Daily Dietary Intakes of Selected Vitamins and Minerals*

Category	Age (Years)	Vitamins	
		Biotin (μg)	Pantothenic Acid (mg)
Infants	0–0.5	10	2
	0.5–1	15	3
Children and adolescents	1–3	20	3
	4–6	25	3–4
	7–10	30	4–5
	11+	30–100	4–7
Adults		30–100	4–7

Category	Age (Years)	Trace Elements†				
		Copper (mg)	Manganese (mg)	Fluoride (mg)	Chromium (μg)	Molybdenum (μg)
Infants	0–0.5	0.4–0.6	0.3–0.6	0.1–0.5	10–40	15–30
	0.5–1	0.6–0.7	0.6–1.0	0.2–1.0	20–60	20–40
Children and adolescents	1–3	0.7–1.0	1.0–1.5	0.5–1.5	20–80	25–50
	4–6	1.0–1.5	1.5–2.0	1.0–2.5	30–120	30–75
	7–10	1.0–2.0	2.0–3.0	1.5–2.5	50–200	50–150
	11+	1.5–2.5	2.0–5.0	1.5–2.5	50–200	75–250
Adults		1.5–3.0	2.0–5.0	1.5–4.0	50–200	75–250

*Because there is less information on which to base allowances, these figures are not given in the main table of RDA and are provided here in the form of ranges of recommended intakes.

†Since the toxic levels for many trace elements may be only several times usual intakes, the upper levels for the trace elements given in this table should not be habitually exceeded.

Reprinted with permission from *Recommended Dietary Allowances,* 10th edition. Copyright 1989 by the National Academy of Sciences. Courtesy of the National Academy Press, Washington, D.C.

TABLE 8–12
Estimated Sodium, Chloride, and Potassium Minimum Requirements of Healthy Persons

Age	Weight (kg)*	Sodium (mg)*†	Chloride (mg)*†	Potassium (mg)‡
Months				
0–5	4.5	120	180	500
6–11	8.9	200	300	700
Years				
1	11.0	225	350	1,000
2–5	16.0	300	500	1,400
6–9	25.0	400	600	1,600
10–18	50.0	500	750	2,000
>18§	70.0	500	750	2,000

*No allowance has been included for large, prolonged losses from the skin through sweat.

†There is no evidence that higher intakes confer any health benefit.

‡Desirable intakes of potassium may considerably exceed these values (~3,500 mg for adults).

§No allowance included for growth. Values for those below 18 years assume a growth rate at the 50th percentile reported by the National Center for Health Statistics and averaged for males and females.

Reprinted with permission from *Recommended Dietary Allowances,* 10th edition. Copyright 1989 by the National Academy of Sciences. Courtesy of the National Academy Press, Washington, D.C.

found in the liver, spleen, and bone marrow and can replenish the loss of iron from the functional compounds. These iron-containing compounds clearly play essential roles associated with exercise: oxygen transport in the red blood cell; storage and transport of oxygen to muscle; and catalysts for energy transfer reactions within cells. Iron deficiency anemia is associated with a dampened capacity for even mild exercise [37, 56].

As much as 60% of female endurance athletes have been reported to be iron deficient [16, 52]. It is highly recommended [52] that athletes be screened for anemia, suboptimal hemoglobin levels, and risk factors that may lead to anemia. Risk factors include (1) a diet low in iron, protein, ascorbic acid, folate, or vitamin B_{12}; (2) high iron losses, primarily through menstruation; and (3) excessive intense training. Screening could identify non-iron-associated anemia and help to initiate appropriate interventions.

There is considerable confusion about iron deficiency, and investigators have tried to distinguish iron deficiency without anemia from iron deficiency with anemia. Terms used include "sports anemia," "pseudoanemia," "mild anemia," and "exercise induced anemia." Currently, it is thought that training generates increased serum levels of renin, aldosterone, vasopressin, and albumin, which tend to increase the intravascular plasma volume, leading to a dilutional effect. Hemoglobin levels of 12 to 13 g/100 mL of blood in women and men, respectively, indicate a situation approaching clinical anemia. Although iron losses through the gastrointestinal tract and sweat and trauma to the foot pads of runners have been implicated in "sports anemia," most recent studies suggest that a transient dilutional effect occurs in the early periods of training, followed by a return to pretraining levels after 6 to 8 weeks. Thus, although total hemoglobin may actually increase with training, a dilutional effect occurs as plasma volume expands. Despite this effect, aerobic capacity and exercise performance during training consistently increase [52, 53, 55]. Careful screening of athletes for iron deficiency anemia will help to sort out a true iron deficiency anemia from a pseudoanemia. True iron deficiency should be treated aggressively because failure or delay to attend to this will compromise athletic performance.

WATER AND ELECTROLYTES

Water is perhaps the most crucial of all the essential nutrients. Water is necessary for transport of all essential nutrients into the body cells for metabolism, and also for excretion of metabolic waste products from the body through the kidneys, skin, lungs, and gastrointestinal system. An imbalance of body water can quickly and seriously compromise physical performance. Meeting the human need for a continuous supply of water and maintaining water balance constitute major nutritional and physiologic functions. Table 8–10 [65] provides the estimated minimum levels for sodium, chloride, and potassium, the major electrolytes. Water and these electrolytes are essential dietary components.

Approximately 55% to 60% of the human body is water. Water has three major functions: (1) it gives structure and form to the body through tissue turgor; (2) it provides an aqueous environment essential for cell metabolism; and (3) it provides for a stable body temperature. Dehydration compromises these functions and in turn has a negative impact on physical performance. Electrolytes play prominent roles in controlling water balance and normal body hydration. Several organ systems, namely, the kidneys, lungs, skin, and gastrointestinal tract, play important and specific roles in helping to maintain normal body water balance and adequate hydration.

Physical performance, especially for endurance athletes, can impose significant stress on body water and electrolyte balance; however, the body has a remarkable ability to adjust to varying circumstances associated with physical performance. The normal requirement for 2.5 to 3.10 liters of water intake/day may be increased five- to six-fold by exercise. Excretion of water and electrolytes is also increased to maintain body water and electrolyte balance. Body water loss leading to dehydration can severely compromise cardiovascular function and work capacity.

The athlete needs to be knowledgeable about factors that can modify the maintenance of water and electrolyte balance: appropriate clothing to permit transfer of heat; environmental humidity; acclimatization techniques; continual maintenance of adequate fluid intake; personal capacity for workload; appropriate training; age; and degree of fatness. The frequent practices of using diuretics or laxatives to effect water loss from the body, restricting fluid intake, and inducing excessive sweating to "make weight" are all practices that will compromise appropriate water and electrolyte balance and athletic performance.

Replacement of water and electrolytes is at the top of the list of maneuvers to maintain optimal nutritional status. Adequate water intake before, during, and after endurance exercise is an absolute must. Replacement of electrolytes may not be needed during short-term events but may be beneficial in endurance events [6, 14, 24, 54]. The osmolality of replacement drinks is an important consideration because moderate concentrations of glucose or electrolytes (7% to 8%) may inhibit gastric emptying and could compromise thermoregulation during prolonged exercise, especially when temperatures are warm [51]. Recent studies in the use of glucose polymers suggest that these compounds have possible benefits because they maximize carbohydrate while minimizing osmolality [6, 14, 24, 54].

BASIC FOOD GROUPS

A simplified, systematic way to ensure an intake of adequate calories and all essential nutrients is to use a basic food group plan instead of calculating exact amounts of protein, vitamins, minerals, and so on that are needed each day. Foods have been grouped into four major groups according to their nutrient composition (Table 8–13). By selecting the

TABLE 8–13
Basic Food Groups

Milk Group: 2 or more servings a day
   Choices: milk
            yogurt
            cheese
            cottage cheese
            buttermilk
Meat and Protein Group: 2 or more servings a day
   Choices: fish   dry peas
           poultry   beans
           eggs   nuts
           beef   pork
           lamb
Fruit and Vegetable Group: 4 or more servings a day
   Choices: raw fruits
            green and yellow vegetables
            citrus juice
            berries
Bread and Cereal Group: 4 or more servings a day
   Choices: whole-grain enriched breads
            whole-grain enriched cereals
            pasta
            rice
Supplemental Group: use sparingly
   Choices: sugar   oils
           honey   fats
           syrup   bacon
           preserves sauces
           candy   mayonnaise
           desserts gravies
           baked foods
           deep-fat fried foods

number of recommended servings of each of the four food groups daily, all essential nutrients will be received. Depending on energy expenditure, additional servings can be added when energy needs are high. The basic four groups include (1) milk and dairy foods, (2) meats and protein foods, (3) fruits and vegetables, and (4) breads and cereals. Fats, oils, and sugar are placed in a separate supplemental group and should be selected sparingly. It is much easier to remember these food groups as a guide to ingesting essential nutrients in the daily diet than to count grams of protein, milligrams of B vitamins, or International Units of vitamins A and D.

WEIGHT CONTROL AND EXERCISE

Although there have been strong promotional efforts in the United States to develop regular exercise patterns and to maintain a desirable body weight range throughout life, the prevalence of obesity has not decreased. In fact, there has been an increase in the overall prevalence of obesity in this country, primarily among children and adoles-

cents [23, 43]. Most recent data indicate that 34 million adults aged 20 to 74 are obese. Obesity, especially severe obesity, is associated with increased morbidity and mortality [33, 43]. Severe obesity increases the risks of coronary heart disease, high blood cholesterol levels, high blood pressure, stroke, diabetes, kidney disease, pulmonary disease, anesthetic risk, gallbladder disease, and certain forms of cancer [33, 43]. Because of the very important role of energy expenditure in preventing and treating obesity, a strong commitment to a regular lifelong exercise pattern—or lack thereof—is a crucial issue.

The disorder of obesity is much more complex than was formerly believed. Certainly today, obesity is still not well understood, and many questions remain unanswered. Obesity is probably not a single, simple disorder but is probably a constellation of several interrelated disorders. Numerous approaches to the treatment of obesity have been used, and the results of many studies show a general failure to modify the body composition of most obese persons. A dropout rate in these studies ranging from 20% to 80% attests to the poor success rate. Energy output and energy intake both need to be considered to bring energy balance into equilibrium. Excess weight gain throughout life is very often more closely associated with reduced physical activity than with increased calorie intake. As a nation, the calorie intake per person has steadily decreased in the United States while the prevalence of obesity has increased. These observations have led to stronger support for the role of exercise for long-term weight control [9, 18, 40], and recent studies confirm the effectiveness of regular aerobic exercise [4, 11, 19]. A combination of reduced caloric intake plus enhanced energy expenditure provides the most effective approach to weight reduction. Table 8–7 illustrates clearly the significant difference in the energy requirements for a 70-kg male on a very sedentary day (2406 kcal/day) and on a very active day (3938 kcal/day). Based on the fact that 1 pound of body fat is approximately equal to 3500 kcal, the calorie expenditure during the sedentary day would be less than the energy value of 1 pound of body fat (0.68 lb). The calorie expenditure during the very active day would equal the calorie value of 1.13 lb. On the more active day with a greater energy expenditure, greater body weight loss could occur compared with the less active day with a lower energy expenditure. This assumes, of course, that all other factors that could influence energy expenditure, such as energy intake, were held constant. Using this example, the very active daily energy expenditure over the course of time could equal the energy value of 7.91 lb/week or 31.64 lb/month. How much weight loss would occur realistically would have to consider the actual amount of *energy intake* during this same period of time.

To lose 2 pounds of body fat during a 1-week period, one would have to decrease the usual energy intake by 7000 kcal or increase the energy expenditure by 7000 kcal. A combination of both (i.e., decrease the energy intake by 3500 kcal *and* increase the energy expenditure by 3500 kcal) could effect a body weight loss of 2 pounds. A decrease of 1000 kcal per day could effect a 2-pound weight loss per week; an increase of 1000 kcal per day could add 2 pounds of body weight per week.

Energy expenditures for selected sports activities (in kcal/min^{-1}/kg^{-1}) are listed in Table 8–14. The estimated energy used for various sports can serve as a guideline for the effective coupling of regular exercise *and* appropriate calorie intake to maintain desirable body weight.

Perhaps an even more basic problem is the prevention of obesity from early life onward [3]. This requires considerable education about the compo-

nents of a healthful diet that meets essential nutrient needs as well as a commitment to a regular exercise program. Many people do not have a clear understanding of what they should weigh, what their calorie intake is or should be, or how to balance energy intake with energy expenditure. A simple nomogram of Bray's body mass index (Fig. 8–1) may be very helpful in monitoring energy balance to achieve desirable body weight [7, 33]. A great deal more attention as well as support for marketing these few strategic maneuvers could have a nationwide impact on reducing the prevalence of obesity, especially in children, and could help millions of Americans achieve a more healthful lifestyle. Such an effort would be accompanied by a significant reduction in the risk factors associated with obesity, a lessening of the disorders caused by or associated with obesity, and a reduction in the national health care cost that is precipitated by these diseases. Successful education to effect a more realistic energy balance could also lead to a more positive self-image for millions of Americans and a more productive and satisfying lifestyle.

NUTRITION SUPPLEMENTS AND ERGOGENIC AIDS

In the search for substances and practices that might enhance athletic performance, many nutrients as well as non-nutritive substances have been explored by athletes. Ergogenic aids can be defined as nutritional, physical, pharmacologic, or psychologic substances or practices thought to enhance physical performance. Surveys indicate that the prevalence of use of ergogenic aids by athletes ranges from 2% to 68%, with nutrient supplements high on the list of various substances [26, 28, 35, 45]. Most athletes use ergogenic aids with the intention of improving their strength, skill, and speed.

Anabolic steroids in sports were banned by the International Olympic Committee in 1968. Accurate records of use are difficult to discover, but current estimates suggest that anabolic steroids are used by almost 100% of athletes involved in body building, power-lifting, weight-lifting, shot put, and discus throw. Among male high school seniors, the usage rate has been estimated to be 6.6%, accounting for some 500,000 users, and it may be 5% among intercollegiate athletes [29]. The adverse effects of these drugs have been widely publicized and will not be further commented on here.

Human growth hormone (HGH), a polypeptide secreted by the anterior pituitary gland, has gained widespread use more recently. This hormone may

TABLE 8–14
Energy Expenditure in Selected Sports Activities

Activity	(kcal/min^{-1}/kg^{-1})
Archery	0.065
Badminton	0.097
Basketball	0.138
Boxing, in ring	0.222
Canoeing, racing	0.103
Cricket, batting	0.083
Cycling, racing	0.169
Football	0.132
Golf	0.085
Gymnastics	0.066
Judo	0.195
Racquetball	0.178
Running, cross-country	0.163
Running, horizontal	
9 min/mile	0.193
6 min/mile	0.252
Skiing, hard snow, level, moderate	0.119
speed uphill, maximum speed	0.274
Skindiving, considerable motion	0.276
Squash	0.212
Swimming, backstroke	0.169
breaststroke	0.162
crawl, fast	0.156
side stroke	0.122
Tennis	0.109
Volleyball	0.050
Walking, normal pace	0.080

From McArdle, W.D., Katch, R.I., and Katch, V.L. *Exercise Physiology*. Philadelphia, Lea & Febiger, 1991, pp. 804–811.

NOMOGRAM FOR BODY MASS INDEX

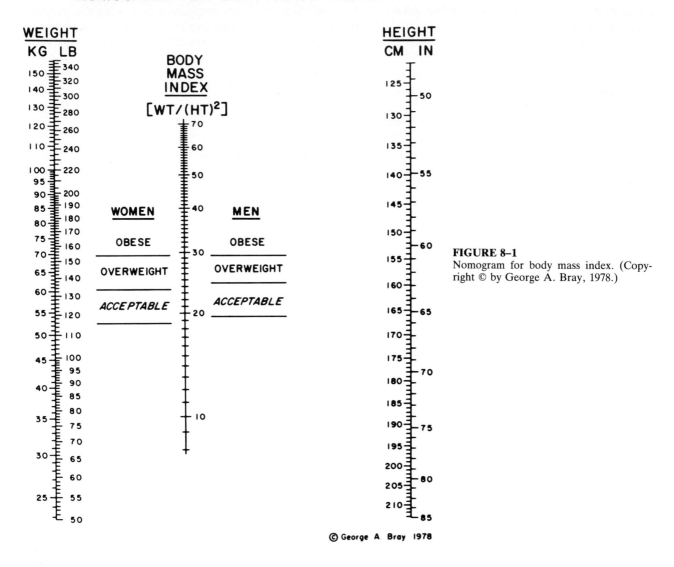

FIGURE 8–1
Nomogram for body mass index. (Copyright © by George A. Bray, 1978.)

© George A. Bray 1978

stimulate body growth and increase strength potential. It is expensive, and access to it is limited; however it is undetectable by routine procedures for drug testing. HGH has a potential for abuse, and its use should be monitored cautiously [29].

Although there are numerous claims that megavitamin and megamineral therapies increase energy and endurance, the only well-documented effects of most vitamin or mineral therapy is related to the return of performance ability if a true deficiency exists. Adverse reactions related to large intakes of either vitamins or minerals in well-nourished persons can precipitate a variety of effects including liver or kidney damage, hair loss, anorexia, diabetes, diarrhea, and muscle pain [29].

Massive doses of protein or of single amino acids have been popular ergogenic aids despite the fact that such practices have not been shown to increase endurance or muscle strength. High doses of amino acids (e.g., tryptophan) may have very serious side effects, particularly in light of several recent deaths that have been associated with high intakes of tryptophan.

Extracts from the pancreas, liver, lungs, and testicles have been advertised with claims of improving the function of the corresponding human organs. The quality of these compounds is highly questionable. It is likely that proteins and amino acids in these products will be digested and absorbed like any other dietary protein, and they are not likely to provide any superior athletic advantage.

Most investigators of the effectiveness of ergogenic substances have found little substantial evidence that "super" amounts of nutrients will provide super power or efficiency to athletes . . . if athletes are well nourished and do not have any nutrient deficiencies. Supplemental nutrients have a positive role only when need to prevent or treat a deficiency of a specific nutrient exists. These persistent findings strongly suggest the need to provide accurate information about nutrient supplementation to all athletes, particularly serious professional athletes.

NUTRITIONAL SUPPORT IN CRITICAL CARE OF INJURED ATHLETES

The preoperative, injured, or acutely ill patient may require extraordinary nutritional support. The rate of weight loss with acute illness or injury may seriously compromise body water and lean body mass. Catabolic breakdown of body tissue after injury is accompanied by significant losses of nitrogen and large increases in resting energy expenditure.

Careful nutritional assessment and continued monitoring of critical care patients are crucial. Nutritional support by means of enteral or parenteral nutrition will be imperative for appropriate treatment and even for survival in many instances. The orthopaedic surgeon may frequently be the physician called on to provide care for injured athletes. It is essential to be familiar with the metabolic utilization of fuel substrates and how this use may be influenced by individual variability, type and severity of injury, and opportunities of access to administration of appropriate nutritional support. The urgent and often comprehensive management of critical care patients can be among the most complicated and challenging problems encountered in the medical-surgical care of injured athletes. Central to this care is the nutritional support of cellular metabolism that will require the considerable nutrition knowledge and expertise of the caring physician.

SUMMARY

The role of nutrition in orthopaedic sports medicine has several dimensions. An understanding of how food energy is transformed and transferred to work energy (sports) is essential in learning how to maximize physical performance. Numerous factors may modify the fuel efficiency of the body, and a knowledge of how to modify these factors to provide a positive advantage may place an athlete on the "cutting edge" of performance.

Establishing a lifelong healthful connection between diet and exercise may be the most important way to reduce the morbidity and mortality that is now known to be associated with obesity and a sedentary lifestyle. Ergogenic aids that do not live up to their name or claims are unlikely to be helpful. An understanding of basic scientific nutrition will serve all sportsmen and sportswomen well, especially if it is practiced from early life onward. Even in the presence of unfortunate injury, nutritional support may be the most important key to survival . . . and sports once again.

References

1. Bannister, R. Health, fitness, and sport. *Am J Clin Nutr* 49(5):927–930, 1989.
2. Belko, A.Z. Vitamins and exercise—an update. *Med Sci Sports Exerc* 19(5):S191–S196, 1987.
3. Blair, S.N., Kohl, H.W., III, Paffenbarger, R.S., Jr., Clark, D.G., Cooper, K.H., and Gibbons, L.W. Physical fitness and all-cause mortality. A prospective study of healthy men and women. *JAMA* 262(17):2395–2401, 1989.
4. Bouchard, C., Tremblay, A., Nadeau, A., Dussault, J., Despres, J.P., Githeriault, Lupien, P.J., Serresse, O., Boulay, M.R., and Fournier, G. Long-Term exercise training with constant energy intake. 1. Effect on body composition and selected metabolic variables. *Int J Obesity* 14:57–73, 1990.
5. Cahill, G.F., Aoki, T.T., and Rosini, A.A. Metabolism in obesity and anorexia. *Nutr Brain* 3:1–70, 1979.
6. Carter, J.E., and Gisolfi, C.V. Fluid replacement during and after exercise in the heat. *Med Sci Sports Exerc* 21(5):532–539, 1989.
7. Chaine, G., Cormier, L., Moutillet, M., Noreau, L., Leblanc, C., and Landry, F. Body Mass Index as a discriminant function among health-related variable and risk factors. *J Sports Med Phys Fit* 29(3):253–261, 1989.
8. Costill, D.L., and Saltin, B. Factors limiting gastric emptying during rest and exercise. *J Appl Physiol* 37:679–683, 1974.
9. Dahlkoetter, J.A., Callahan, E.J., and Linton, J. Obesity and the unbalanced energy equation: Exercise versus eating habit change. *J Consult Clin Psychiatr* 47:898–905, 1979.
10. Dalsky, G.P. Effect of exercise on bone: Permissive influence of estrogen and calcium. *Med Sci Sports Exerc* 22:281–285, 1990.
11. Despres, J.P., Bouchard, C., Tremblay, A., Savard, R., and Marcotte, M. Effects of aerobic training on fat distribution in male subjects. *Med Sci Sports Exerc* 17:113–118, 1985.
12. Devlin, J.T., and Horton, E.S. Exercise. *In* Kinney, J.M., Jeejeebhoy, K.N., Hill, G.L., and Owen, O.E. (Eds.), *Nutrition and Metabolism in Patient Care*. Philadelphia, W.B. Saunders, 1988, pp. 225–233.
13. Devlin, J.T., and Horton, E.S. Metabolic fuel utilization during postexercise recovery. *Am J Clin Nutr* 49(5):944–948, 1989.
14. Edwards, T.L., Santeusanio, D., and Wheeler, K.B. Endurance cyclists given carbohydrate solutions during moderate-intensity rides. *Tex Med* 82:29–31, 1986.
15. Evans, W.J., Fisher, E.C., Hoerr, R.A., and Young, V.R. Protein metabolism and endurance exercise. *Phys Sports Med* 11:63–72, 1983.
16. Expert Scientific Working Group Summary of a report on assessment of the iron nutritional status of the U.S. population. *Am J Clin Nutr* 42(6):1318–1330, 1985.
17. Felig, P., and Wahren, J. Fuel homeostasis in exercise. *N Engl J Med* 293:1078–1084, 1975.

18. Foreyt, J.P., Scott, L.W., and Gotto, A.M. Weight control and nutrition education programs in occupational settings. *Pub Health Rep* 95:127–136, 1980.

19. Franklin, B.A., and Rubenfire, M. Losing weight through exercise. *JAMA* 244:377–379, 1980.

20. Goldfine, A., Ward, A., Taylor, P., Carlucci, D., and Rippe, J.M. Exercising to health. *Phys Sports Med* 19(6):81–93, 1991.

21. Gollnick, P.D., and Saltin, B. Fuel for muscular exercise: Role of fat. *In* Horton, E.S. and Terjung, R.L. (Eds.), *Exercise, Nutrition and Energy Metabolism.* New York, Macmillan, 1988, pp. 72–88.

22. Gonzalez, E.R. Premature bone loss found in some nonmenstruating sports-women. *JAMA* 248:513–514, 1982.

23. Gortmaker, S.L., Dietz, W.H., Jr., Sobol, A.M., and Wehler, C.A. Increasing pediatric obesity in the United States. *Am J Dis Child* 141:535–540, 1987.

24. Grassi, M., Frajoli, A., Messina, B., Mammucari, S., and Mennuni, G. Mineral waters in treatment of metabolic changes from fatigue in sportsmen. *J Sports Med Phys Fit* 30(4):441–449, 1990.

25. Guilland, J.C., Penaranda, T., Gallet, C., Boggio, V., Fuchs, F., and Klepping, J. Vitamin status of young athletes including the effects of supplementation. *Med Sci Sports Exerc* 21(4):441–449, 1989.

26. Haymes, E.M. Proteins, vitamins, and iron. *In* Williams, M.H. (Ed.), *Ergogenic Aids in Sports.* Champaign, IL, Human Kinetics, 1983, pp. 27–55.

27. Heaney, R.P., Gallagher, S.C., Johnston, C.C., Neer, R., Parfitt, A.M., and Whedon, G.D. Calcium nutrition and bone health in the elderly. *Am J Clin Nutr* 36:986–1013, 1982.

28. Horton, E.S. Metabolic fuels, utilization, and exercise. *Am J Clin Nutr* 49(5):931–937, 1989.

29. Hough, D.O. Anabolic steroids and ergogenic aids. *Am Fam Pract* 41(4):1157–1164, 1990.

30. Hultman, E. Nutritional effects on work performance. *Am J Clin Nutr* 49(5):949–957, 1989.

31. International Federation of Sports Medicine Position Statement. Physical exercise: An important factor for health. *Phys Sports Med* 18(3):155–156, 1990.

32. Kendrick, Z.W., and Lowenthal, D.T. Drug-nutrient interactions. *In* Horton, E.S., and Terjung, R.L. (Eds.), *Exercise, Nutrition and Energy Metabolism.* New York, Macmillan, 1988, pp. 196–212.

33. Koop, C.E. *The Surgeon General's Report on Nutrition and Health.* Public Health Service Publ. No. 88-50210. Washington, D.C., U.S. Dept. Health and Human Services, 1988, pp. 275–309.

34. Kottke, T.E., Caspesren, C.J., and Hill, C.S. Exercise in the management and rehabilitation of selected chronic disease. *Prev Med* 13:47–65, 1984.

35. Krowchuk, D.P., Anglin, T.M., Goodfellow, D.B., Stancin, T., Williams, P., and Zimet, G.D. High school athletes and the use of ergogenic aids. *Am J Dis Child* 143:486–489, 1989.

36. Lemon, P.W.R. Protein and exercise: Update 1987. *Med Sci Sports Med* 19(5):S179–S190, 1987.

37. Manore, M.M., Besenfelder, P.D., Wells, C.L., Carroll, S.S., and Hooker, S.P. Nutrient intakes and iron status in female long-distance runners during training. *J Am Diet Assoc* 89(2):257–259, 1989.

38. McArdle, W.D., and Katch, F.I. *Exercise Physiology. Energy, Nutrition, and Human Performance.* Philadelphia, Lea & Febiger, 1991, pp. 85–144.

39. McGilvery, R.W. The use of fuels for muscular work. *In* Howard, H. and Poortmans, J.R. (Eds.), *Metabolic Adaptations to Prolonged Physical Exercise.* Basel, Birkhauser Verlag, 1975, pp. 12–30.

40. Miller, P.M., and Sims, K.L. Evaluation and component analysis of a comprehensive weight control program. *Int J Obesity* 5:57–65, 1981.

41. Millward, D.J., Davies, C.T.M., Halliday, D., Wordone, S.L., Matthews, D., and Rennie, M. Effect of exercise on protein metabolism in humans as explored with stable isotopes. *Fed Proc* 41:2686–2691, 1982.

42. National Institutes of Health Consensus Conference. Osteoporosis. *JAMA* 252:799–802, 1984.

43. National Institutes of Health Consensus Development Conference Statement. Health implications of obesity. *Ann Intern Med* 103:1073–1077, 1985.

44. Nelson, M.E., Fisher, E.C., Catsos, P.D., Meredith, C.N., Turksoy, R.N., and Evans, W.J. Diet and bone status in amenorrheic runners. *Am J Clin Nutr* 43:910–916, 1986.

45. Nieman, D.C., Gates, J.R., Butler, J.V., Pollett, L.M., Dietrich, S.J., and Lutz, R.D. Supplementation patterns in marathon runners. *J Am Diet Assoc* 89:1615–1619, 1989.

46. Nilsson, L.H., and Hultman, E. Liver glycogen in man: The effect of total starvation or a carbohydrate-poor diet followed by carbohydrate refeeding. *Scand J Clin Lab Invest* 32:325–330, 1973.

47. Parizkova, J. Age-dependent changes in dietary intake related to work output, physician fitness, and body composition. *Am J Clin Nutr* 49(5):962–967, 1989.

48. Powell, K.E., Caspersen, C.J., Koplan, J.P., and Ford, E.S. Physical activity and chronic diseases. *Am J Clin Nutr* 49(5):999–1006, 1989.

49. Prokop, L. International Olympic Committee Medical Commission's policies and programs in nutrition and physical fitness. *Am J Clin Nutr* 49(5):1065, 1989.

50. Ravussin, E., and Bogardus, C. Relationship of genetics, age, and physical fitness to daily energy expenditure and fuel utilization. *Am J Clin Nutr* 49(5):968–975, 1989.

51. Rehrer, N.J., Beckers, E., Brouns, F., TenHoor, R., and Saris, W.H.M. Exercise and training effects on gastric emptying of carbohydrate beverages. *Med Sci Sports Exerc* 21(5):540–549, 1989.

52. Risser, W.L., Lee, E.J., Poindexter, H.B., West, M.S., Pivarnik, J.M., Risser, J.M.H., and Hickson, J.F. Iron deficiency in female athletes: Its prevalence and impact on performance. *Med Sci Sports Exerc* 20(2):116–121, 1988.

53. Risser, W.L., and Risser, J.M.H. Iron deficiency in adolescents and young adults. *Phys Sports Med* 18(12):87–101, 1990.

54. Seidman, D.S., Ashkenazi, I., Arnon, R., Shapiro, Y., and Epstein, Y. The effects of glucose polymer beverage ingestion during prolonged outdoor exercise in the heat. *Med Sci Sports Exerc* 23(4):458–462, 1991.

55. Selby, G.B. When does an athlete need iron? *Phys Sports Med* 19(4):96–102, 1991.

56. Schoene, R.B., Escorrou, P., Robertson, H.T., Nilson, K.L., Parsons, J.R., and Smith, N.J. Iron repletion decreases maximal exercise lactate concentrations in female athletes with minimal iron-deficiency anemia. *J Lab Clin Med* 102(2):306–312, 1983.

57. Schurch, P.M., Radimsky, J., Iffland, R., and Hollman, W. The influence of moderate prolonged exercise and a low carbohydrate diet on ethanol elimination and on metabolism. *Eur J Appl Physiol* 48:407, 1982.

58. Serfass, R.C., and Gerberich, S.G. Exercise for optimal health: Strategies and motivational considerations. *Prev Med* 13:79–99, 1984.

59. Shangold, M., Rebar, R.W., Colston, A., and Schiff, I. Evaluation and management of menstrual dysfunction in athletes. *JAMA* 263(12):1665–1669, 1990.

60. Simopoulos, A.P. Nutrition and fitness from the first Olympiad in 776 BC to 393 AD and the concept of positive health. *Am J Clin Nutr* 49(5):921–926, 1989.

61. Smith, C.W. A practical guide for helping the patient achieve a healthy lifestyle. *J Am Board Fam Pract* 2:238–246, 1989.

62. Smith, M. Position of the American Dietetic Association. Nutrition for physical fitness and athletics performance. *J Am Diet Assoc* 87(7):933–938, 1987.

63. Strasser, A.L. Exercise should play an important role in corporate fitness, stress reduction. *Occup Health Safety* 58(11):6, 1989.

64. Stroot, P. A priority for the World Health Organization: promoting healthy ways of life. *Am J Clin Nutr* 49(5):1063–1064, 1989.

65. Subcommittee on the Tenth Edition of the RDAs, Food and Nutrition Board, National Research Council. *Recommended Dietary Allowances.* Washington, D.C., National Academy Press, 1989.

66. Thompson, P.D., Culliname, E.M., Eshelman, R., Sady, S.P., and Herbert, P.N. The effects of high-carbohydrate and high fat diets on the serum lipid and lipoprotein concentrations of endurance athletes. *Metabolism* 33:1003–1010, 1984.

67. U.S. Department of Health and Human Services, Public Health Service. *Healthy People 2000: National Health Promotion and Disease Prevention Objectives.* Washington, D.C., U.S. Dept. of Health and Human Services, 1990, pp. 93–110.

68. Williams, C. Diet and endurance fitness. *Am J Clin Nutr* 49(5):1077–1083, 1989.

69. Wilmore, J.H., and Freund, B.J. Nutritional enhancement of athletes' performance. *Nutr Abst Rev* 54(1):1–16, 1984.

70. Wright, E.D. Nutrition and exercise. *In* Paige, D.M. (Ed.), *Clinical Nutrition.* St. Louis, C.V. Mosby, 1988, pp. 677–717.

71. Young, V.R. Protein and amino acid metabolism in relation to physical exercise. *In* Winick, M. (Ed.), *Nutrition and Exercise.* New York: John Wiley and Sons, 1986, pp. 9–32.

SPORTS PHARMACOLOGY

Ergogenic Drugs in Sports

James B. Robinson, M.D.

Athletic performance is influenced by a multitude of variables, including the genetic component of the athlete, which is probably the most important, fitness level, skill level, diet, quality of the opponent, environment, health, coaching, location of competition, sleep, motivation, psychology, and luck. Also among the factors that may influence athletic performance are drugs. Unfortunately, athletes today have turned to drugs in an attempt to enhance their performance to achieve better results. Drugs in sports can be divided into three categories. The first category consists of therapeutic agents used by the team physician or trainer to enhance the healing process or to treat illness. The second category is "recreational drugs," or drugs of abuse commonly found in the community such as cocaine, marijuana, alcohol, and narcotics. Third, the category called performance-enhancing drugs consists of drugs or chemicals used by athletes to try to gain a competitive edge over other athletes.

Ergogenic is a word derived from the Greek words *ergo,* meaning to work, and *gennan,* meaning to produce, and is defined as any substance that has a tendency to increase work output [22]. Individuals and athletes have used substances to improve performance for thousands of years, dating back to ancient Greek times in the third century BC [66].

Substances including mushrooms, herbs, oils, potions, concoctions, and even blood and body parts of animals were used by athletes to enhance their performance. Although drugs are only a small part of the numerous factors that influence performance, differences in performance among elite athletes may be so small that any variation may make a large difference in competitive results. Numerous performance-enhancing aids are used by athletes today, and the present chapter will discuss some of the more common substances. Each drug will be described in regard to its pharmacologic mechanisms, its desired effect, and its adverse effects on the individuals who use it. Hundreds of other drugs used by athletes to enhance performance are not discussed here for reasons of space and because their use changes so rapidly.

ANABOLIC-ANDROGENIC STEROIDS

Doubtless the most common and most publicized ergogenic drugs abused by athletes are anabolic-androgenic steroids. From the time athletes first began using anabolic steroids in the 1950s [1, 42, 78] until the recent loss by Ben Johnson of his gold medal, steroids have been a topic of much discussion and controversy. Anabolic steroids are synthetic forms of the male hormone testosterone. The anabolic effects refer to the tissue-building effects, and the androgenic effects refer to the masculinizing properties possessed by these medications. In 1849, Berthold [8] first demonstrated the androgenic effect of testes implanted in castrated roosters. After that time the hormone was isolated and was first produced in 1935 [1, 55]. It was reported in the 1940s that German SS troops used anabolic-androgenic steroids to increase their strength and aggressiveness [42, 59]. The first report of use in athletes occurred in 1954 with the use by Russian power lifters in the Olympic Games [42, 59, 78]. It was not until 1956 that anabolic steroids were first used in this country at the York Barbell Club in Pennsylvania with the help of Dr. John Zeigler and the CIBA Drug Company in their development of Dianabol [38, 41, 78]. Anabolic steroids were not tested in athletes until 1976 [41], and the first report of athletes being banned for steroid use occurred in 1983 [42] at the Pan American games in Caracas. Since that time, numerous athletes have been banned from competition because they tested positive for steroids. Several different preparations are available for both oral and parenteral use. Some of them are produced by pharmaceutical companies for therapeutic reasons, and others are made through various other means illegitimately. The drugs were first developed for several therapeutic uses, including treatment of certain anemias, hereditary angioneurotic edema, osteoporosis, catabolic states, and, of course, replacement therapy in male children who lack testosterone for one reason or another.

The incidence of use varies from study to study. It is generally estimated that usage in high school is anywhere from 1.1% to 18% [2, 14]; one of the more recent studies shows that 6.6% of high school seniors have used anabolic steroids during their senior year, and as many as 30% of these are nonathletes [14]. At the college level, usage ranges from 5.8% to 20% [2, 25, 63], and in the National Football League, up to 6% of players may use these drugs [75]. It has been estimated that up to 100% of body builders have used anabolic-androgenic steroids at some time in their training; usage in females varies between 2.5% and 7% [43, 75].

Several mechanisms of action of anabolic steroids have been proposed, the first of which is that they increase protein synthesis [1, 68]. Anabolic steroids are taken up by cell receptors and are brought into the cellular structure, producing an increase in DNA-dependent RNA polymerase, which increases protein synthesis. They also decrease the catabolic

effect of the glucocorticoids that are released with high stress activities such as high-intensity workouts [1, 73]. They are also believed to enhance aggressive behavior, which may be a key to their success as perceived by athletes [1, 12]. A placebo effect may play a large role also. There are numerous conflicting studies in the literature at present about the efficacy of anabolic steroids in athletes. However, after careful review of the literature, it is evident that anabolic-androgenic steroids can cause an increase in muscle mass, an increase in strength, and, theoretically, an enhancement of performance in sports in which strength and muscle size are important [1]. These effects are largely determined by the intensity of the workout, the weight-training experience of the individual, diet, the amount of drug used, the length of time the drug was used, the number of different drugs used, and other variables.

The desired effects of anabolic steroids include:

1. Increase in strength.
2. Increase in weight.
3. Increase in aggressiveness.
4. Capability of sustaining repetitive, high-intensity workouts.
5. Enhanced performance.

Most of these desired effects have been documented in the literature by one source or another [1, 22, 42]. Some athletes have stated that one of the effects desired from the use of anabolic steroids is improved aerobic capacity; however, the studies indicate that no improvement in aerobic capacity occurs in individuals taking these medications [1].

The desired benefits of these medications are often outweighed by their adverse effects. Most of the adverse effects documented in the literature have been demonstrated in individuals taking the medication for therapeutic reasons. Athletes differ from these individuals on several counts. First, the dosage of the medication taken by the athlete can be up to 40 to 100 times the therapeutic dosage. Second, athletes may use several different drugs in combination (termed "stacking") comprising one or two injectables combined with one or two oral medications at the same time. Third, athletes generally do not take anabolic steroids continuously for long periods of time. They use drugs in "cycles," taking the medications for a period of 6 to 8 weeks and having anywhere from two to three cycles a year. Most cycles are undertaken during the off-season training program and are not used during the competitive year. It is, therefore, difficult to extrapolate the possible adverse effects seen in individuals using the drug for therapeutic reasons with usage in athletes; however, with more attention paid to some of the adverse effects and with better testing, we are beginning to see some of these effects in athletes also.

Adverse effects can be grouped into several different categories [1, 42]. Adverse effects on the liver include cholestasis, an alteration in liver function tests, especially with oral anabolic steroids, and tumors, both benign and malignant. Approximately 38 cases of hepatic tumors have been reported in the literature, and only three of these have occurred in athletes [42]. Peliosis hepatitis, a condition often associated with chronic diseases such as tuberculosis, consists of blood-filled cysts in the liver that can occur with prolonged steroid use; however, none have been reported in athletes thus far [42]. Effects on the cardiovascular system include alterations in the blood lipids with a marked decrease in high-density lipoproteins (HDL) and a possible increase in the total cholesterol level also [1, 42]. Blood pressure increases, and there is occasional glucose intolerance [1, 42]. There have been reports of ultrastructural changes in the heart at the electron microscopic level. The significance of these changes is unknown at this time; however, there have been case reports of athletes developing cardiomyopathy when using androgenic steroids. Whether this condition is actually due to these medications or not can only be surmised. The effects on the male reproductive system include oligospermia and azoospermia, testicular atrophy, gynecomastia, prostatic hypertrophy, and decreases in follicle-stimulating hormone (FSH) and luteinizing hormone (LH) levels [1, 42]. Effects on the female reproductive system include decreases in FSH, LH, estrogen, and progesterone, which can lead to menstrual irregularities [1, 42]. Also, masculinizing effects do occur and are irreversible. Clitoromegaly has also been reported. The effects on the psychological system are gaining more and more attention. These include changes in libido, mood swings, an increase in aggressiveness, depression, possible addiction, and certain pathologic states including drug-induced insanity, affective disorders, and psychosis [62]. Withdrawal has also been a concern. Effects are also seen in children including early epiphyseal closure, possibly leading to a decrease in stature; however, the full potential for growth of the individual is not known [1, 42]. Early maturation is also seen, but the effects on the developing hormonal system in these children cannot be determined. Miscellaneous effects include acne, alopecia, headaches, and tumors. Anabolic steroids have also been postulated to decrease immune function. Studies in rats have demonstrated weakness of the collagen structures and possibly an increased damage to ligaments and muscles; however, direct evidence of these effects in athletes has not been proved. Acquired immunodeficiency syndrome (AIDS) [70] and hepatitis have also been reported in athletes who share needles, and there are other effects that are just unknown at this time. Another problem that arises in athletes taking anabolic steroids is the numerous other drugs they take to offset the adverse effects of the steroids. These include tamoxifen to block estrogenic effects, human chorionic gonadotropin (HCG) to block oligospermia, stimulants, antibiotics, human growth hormone, diuretics, probenecid, amino acids, caffeine, clomiphene, thyroid hormones, anti-inflammatories, L-dopa, insulin, vitamins, hyaluronidase, periactin, and numerous recreational drugs.

Legal Considerations

The use of anabolic steroids is banned by numerous athletic organizations including the NFL, National Collegiate Athletic Association (NCAA), the International Olympic Committee (IOC), and others. Detection of these drugs in the urine or blood can result in sanctions and possible fines. The medications are prescription medications, and therefore their distribution and use without a prescription is considered illegal. These medications are controlled substances (Class II) and thus require physician accountability for discharge. In the past 5 years at least 150 individuals have been indicted at the federal level for distribution of anabolic steroids. This number will continue to rise as awareness and concern about the use of anabolic steroids increases. The National Institute of Drug Abuse is currently considering treating people who use these drugs the same as people who abuse other drugs. It has been estimated that approximately 80% of anabolic steroids are obtained from the black market; the other 20% come from medical professionals including doctors, pharmacists, and veterinarians [20]. Although tighter controls may decrease the illegal dispensing of these drugs by the medical profession, it will only increase the amount obtained by other means.

HUMAN GROWTH HORMONE

Because of all the attention that has been paid to anabolic steroids in sports recently, athletes have attempted to find other ways to obtain the same results given by steroids without the adverse effects or the possibility of being banned from competition for the use of steroids. One of the newest drugs being used by athletes is human growth hormone [21]. The emergence of human growth hormone as a potential ergogenic aid is of grave concern to the physician taking care of athletes.

History

Until 1985 human growth hormone was obtained from the pituitary gland of cadavers, purified, and used for the treatment of growth hormone–deficient patients. It was a very expensive process, and supplies were limited for obvious reasons. Four documented cases of Creutzfeldt-Jakob disease caused by a slow virus limited the use of cadaveric supplies of human growth hormone [34]. However, in 1985 with the aid of recombinant DNA techniques from bacteria, synthetic human growth hormone was developed [51]. Because the supply now easily meets the demand and the price of the hormone has markedly decreased, athletes are starting to use growth hormones on a regular basis. The first suspected use of growth hormone in an athlete occurred in 1983 [51, 76]. Many athletes experimented with "monkey juice," which was derived from rhesus monkey pituitary extract. It was felt that this extract could have the same beneficial effects as human growth hormone. However, animal growth hormone is completely without physiologic effects in humans, and the price of a rhesus monkey was higher than the amount of pituitary extract that could be obtained and sold.

Pharmacology

Growth hormone is a polypeptide secreted by the anterior pituitary that has numerous physiologic effects in different areas of the body. An increase in amino acid uptake and protein synthesis occurs at the muscular level. The hormone increases lipolysis by increasing mobilization of free fatty acids and by increasing sensitivity of the lipolytic effects of catecholamines. It also inhibits glucose uptake at the tissue level and increases the amount of somatomedin C and insulinlike growth factor, both of which are important in the growth of skeletal tissues. Because of this effect, the hormone increases somatic growth in patients who have deficient growth hormone.

Numerous factors affect growth hormone secretion [48, 51, 54]. Factors that stimulate growth hormone release include sleep, the largest surge occurring approximately 60 to 90 minutes after the onset of sleep, stress, hypoglycemia, exercise of high intensity, duration, and frequency (depending on the age, sex, and fitness level of the individual), and amino acids, including lysine, arginine, and ornithine. Other factors include various hormones such as estrogen, glucagon, vasopressin, and growth-hormone–releasing hormone and medications including clonidine, propranolol, L-dopa, and bromocriptine. Other stimulants for release include protein depletion, starvation, anorexia, and chronic renal failure.

Use of human growth hormone is limited for therapeutic purposes. It is most commonly used for replacement in growth-hormone–deficient individuals. There are approximately 4,000 documented cases of growth-hormone–deficient individuals in the United States in whom this medication is used [57]. Other therapeutic uses include treatment of osteoporosis and obesity; its use in trauma patients is still investigational.

Desired Effects

Athletes take human growth hormone for several reasons, including:

1. To increase body size
2. To increase strength
3. To increase lean body mass and decrease fat
4. To gain anabolic effects without the use of steroids and possible detection

Human growth hormone is also used to increase the height of a prepubescent growing child.

Studies on the effects of growth hormone in athletes are limited for obvious reasons, including the unethical and possible legal ramifications of prospective studies using growth hormone in athletes. Studies that have been done thus far are animal studies. It has been shown that growth hormone will increase the size and weight of muscle, but only in atrophied muscle, not normal tissue [9, 37]. In rats, atrophied muscle size and strength were both increased with the use of growth hormone [37, 52]. Whether these effects are seen in humans is yet to be determined. Because of their increased lipolytic effect, growth hormones could theoretically increase lean body mass and decrease fat, which would be desirable in certain athletes such as body builders [51]. It is not uncommon for a physician to receive a phone call from a concerned parent asking the physician to prescribe growth hormone for a child in an attempt to increase the child's height, thereby increasing his chances of excelling in the sport of basketball. Parents seem to think that if a child is taller, he or she will obviously play basketball better and thus become financially secure through a professional career. Although this result may seem attractive to the parents, the long-term effect of growth hormone on the child could possibly be devastating, aside from the fact that natural ability plays a greater role in success than actual height.

Adverse Effects

The adverse effects of exogenous growth hormone in athletes have yet to be fully determined. Our only model so far for the long-term consequences of excess growth hormone is acromegaly [54]. Growth hormone–secreting tumors of the pituitary that cause acromegaly have a fatality rate of up to 89% by the age of 60. Complications of this disease in these patients include myopathy of skeletal and cardiac muscle, bony hypertrophy, thickening of the skin, peripheral neuropathy, and visceromegaly. Advanced coronary artery disease and hypertension are also seen. Almost all acromegalic patients become diabetic because of the anti-insulin effect of growth hormones. Other miscellaneous effects include acne, an increase in the amount of sweat, hirsutism, colonic polyps, and giantism. As with the early studies on anabolic steroids, we can only speculate about the possible long-term effects of growth hormone use in athletes and therefore should be cautious when advising athletes about the effects. As with other medications, the amount, frequency, and duration of drug use all determine the possible long-term consequences.

AMPHETAMINES

Another group of ergogenic aids often used by athletes is stimulants [50]. In this category, amphetamines are one of the most popular agents. Amphetamine compounds were first synthesized in 1887; however, it was not until 1930 that the pressor effects of these medications were identified [50, 60]. The drugs were first used in 1935 for the treatment of narcolepsy, and abuse soon followed, peak incidence in the United States occurring in the 1960s [65]. Athletes in various sports abuse these drugs for their power to mask fatigue and enhance performance [50]. In the 1960s Mandell and colleagues reported widespread use of amphetamines by professional football players in the NFL [52]. It was estimated that approximately 66% of players in the NFL used amphetamines at one time, most of them

regularly. The incidence of usage today varies from study to study. In 1984 in a study of college athletes it was estimated that 8% of college athletes had used amphetamines in the prior year [2]. However, only 3% of them used the drugs to enhance performance. In 1960 the cyclist Kurt Jensen died during the Rome Olympics from the apparent adverse effects of amphetamines. The same thing happened to the English cyclist Tommy Simpson during the 1967 Tour de France, which drew attention to the possible grave consequences of the use of these drugs.

Pharmacology

Amphetamines are sympathomimetic amines. Their effects are similar to those of the endogenous catecholamines epinephrine, norepinephrine, and dopamine. Amphetamines have several proposed mechanisms of action. First, they stimulate the release of catecholamines from the nerve cells. Second, they displace catecholamines from the receptor sites, allowing an increased amount of these substances to remain at the synaptic cleft and enhancing transmission. Third, they inhibit reuptake of catecholamines, thus increasing the amount of catecholamines in the synaptic cleft. Fourth, they are a partial catecholamine agonist. Finally, they inhibit the breakdown of catecholamines in the cleft.

These medications have both peripheral and central sympathetic effects. Peripheral effects include an increase in systolic and diastolic blood pressure, increased heart rate, dilatation of pupils, decreased salivation, and an increase in fatty acid release, and bronchial smooth muscle relaxation. Central nervous system effects include increased alertness, decrease in sensation of fatigue, elevation of mood, respiratory stimulation, and appetite suppression.

These medications were initially developed to treat narcolepsy; however, they have also been used to treat hyperactivity, morbid obesity, depression, and chronic pain syndrome.

Desired Effects

Athletes use these preparations for several reasons [50], including:

1. Increased endurance
2. Increase in energy substrates by releasing fatty acids
3. Increased reaction time
4. Masking of fatigue
5. Appetite suppression

There are a few good studies on the effects of amphetamines on athletes; however, only two warrant mention. The first, by Smith and Beecher [72], demonstrates improved performance in over 75% of athletes (runners, swimmers, and throwing athletes) with high doses of these medications. Although the increments of improvement were small, theoretically they could make a difference in each of these sports. The other study, by Chandler and Blair [16], demonstrated an increase in time to exhaustion following ingestion of amphetamines. In this study, lactic acid production was increased, thus demonstrating that amphetamines do not actually prevent fatigue but only mask the symptoms. There are several other studies that show no improvement in athletic performance with the use of these medications; however, they are difficult to interpret owing to their poor design and number of subjects [6, 27, 32]. Based on the evidence available, amphetamines can improve athletic performance by masking fatigue and thus increasing reaction time, increasing concentration, and increasing fine motor coordination [50].

There is another group of athletes who use amphetamines for a different reason. This group includes athletes in whom body composition may affect the judging of their sport such as gymnasts and divers. These medications are used as anorectics to decrease appetite, thus enabling these athletes to maintain or lose body weight to appear lean.

Adverse Effects

The adverse effects of these medications are seen in many body systems [74]. Central nervous system effects include tremulousness, anxiety, insomnia, agitation, dizziness, irritability, headaches, psychosis, seizures, and possibly even cerebrovascular accidents. Effects on the cardiovascular system include arrhythmias, hypertension, and angina pectoris. Effects on the gastrointestinal system include nausea, vomiting, anorexia, diarrhea, and dry mouth. Other adverse effects include an increased risk of hyperthermia in athletes exercising in the heat, vasculitis, withdrawal, tolerance, addiction, and possibly death.

In summary, amphetamines are sympathomimetic amines that increase the amount of catecholamine response. This increase in catecholamine action can be beneficial to athletes because it masks the symptoms of fatigue and increases reaction time, endurance, and energy substrates. The use of these medications has been limited by some athletes because of their adverse effects including tremulousness, agitation, anxiety, insomnia, nausea, vomiting, and others.

CAFFEINE

One stimulant used by athletes and found in numerous over-the-counter medications and food is caffeine [23, 50]. The use of caffeine dates back to Paleolithic times [71]. It is a plant alkaloid that is found most commonly in two plants, *Coffea arabica* and *Cola acuminata*. It belongs to the class of agents known as methylxanthines; theophylline and theobromine are also methylxanthines. Caffeine is thought to exert its effects by three mechanisms:

1. Translocation of intracellular calcium, which frees up more calcium for muscular contraction.
2. Inhibition of phosphodiesterase, which leads to an increase in cyclic AMP. This increased cyclic AMP increases glycogenolysis and lipolysis, thus allowing utilization of free fatty acids for energy substrate and allowing glycogen sparing at the muscle level.
3. Blockage of adenosine receptors, which blocks the sedative properties of adenosine.

Desired Effects

The desired effects of caffeine are several [23, 64, 71]. First, athletes use it to decrease fatigue and thus increase concentration and alertness. Second, the increased lipolysis that then yields to glycogen sparing at the muscle level theoretically produces improved enhancement of endurance by allowing more energy substrate at the muscle level for long periods of time. Third, by allowing more free calcium to be available at the muscle level, an increase in strength of muscle contraction is postulated. And fourth, a combination of the above leads to increased endurance and performance.

Several studies have been done on the effectiveness of caffeine to enhance performance [19, 61, 64]. Most of the studies seem to indicate that performance in endurance events benefits by caffeine more than short-term, high-intensity workouts. Doses of up to 5 mg per kilogram were needed to improve athletic performance in endurance events.

Adverse Effects

Several adverse effects can be attributed to caffeine use [23, 50, 64]. As with other central nervous system stimulants, nervousness, irritability, insomnia, and anxiety have been reported. Cardiovascular effects of tachycardia and arrhythmias have been reported. Caffeine acts as a diuretic and can also produce cystitis in some people. There have also been reports of a withdrawal phenomenon from the use of caffeine.

In summary, caffeine is a stimulant in the class of methylxanthines that works at the cellular level by increasing extracellular calcium, and thus allowing more calcium for muscle contraction, and by inhibiting phosphodiesterase, thereby increasing cyclic AMP, which increases glycogenesis and lipolysis and allows glycogen sparing, thus possibly improving endurance. It also decreases sedation by blocking adenosine receptors.

Like other central nervous system stimulants, caffeine has several adverse effects that may be detrimental to the athlete's performance, including restlessness, anxiety, irritability, insomnia, tachycardia, arrhythmias, and then possibly death.

Caffeine is banned by both the IOC and the NCAA in large doses [56, 77]. A urine concentration of greater than 12 μg/mL or 15 μg/mL is defined as abuse by the IOC and the NCAA, respectively. This amount is equivalent to approximately four to five cups of coffee.

BLOOD DOPING

Blood doping, or induced erythrocythemia, is defined as an increase in hemoglobin following reinfusion of an athlete's own blood to try to improve the oxygen-carrying capacity of hemoglobin, thus improving aerobic performance [7]. Induced erythrocythemia was first reported in 1947 [58]; however, it was not until 1984, when seven members of the U.S. cycling team, including four medalists, admitted to blood doping in their hotel rooms prior to the event, that it became controversial in sports [29, 47]. In 1985, blood doping was banned by the IOC.

Physiology

Aerobic metabolism is defined as the breakdown of glycogen in the presence of oxygen to form adenosine triphosphate (ATP), which is the energy source for muscle contraction. Aerobic metabolism has a higher yield of ATP formation than does anaerobic metabolism. By increasing both the hemoglobin concentration in the blood and the amount of oxygen that can be transported to the muscle,

improved oxygen delivery may result so that more ATP can be produced theoretically enhancing activities dependent on aerobic metabolism. It has been estimated that one unit of whole blood, or 275 mL of packed red blood cells, can add 100 mL of oxygen to the total oxygen-carrying capacity of blood. Because an athlete's blood volume can circulate through the body up to five times a minute, the potential oxygen-carrying capability is about 500 mL/minute [53].

Procedure

There are two basic methods of inducing erythrocythemia [36]. One is to infuse the athlete's own blood back into his system (autologous reinfusion), and the other is to infuse heterologous blood from the blood bank. In the autologous reinfusion method, blood is donated by the athlete, stored, and then reinfused at a later time when the athlete's hemoglobin and hematocrit have returned to a near-normal state. There are two methods of storage. One is refrigeration at 4°C; however, the blood is good for only 4 to 5 weeks, which does not allow enough time for normalization of the athlete's hemoglobin and hematocrit. The other method of storage is freezing with glycerol. Blood stored in this way can last for years, and this is the method of choice. In general, an athlete removes two units of blood about 4 to 8 weeks prior to an event. The stored packed cells are mixed with glycerol and frozen, then thawed and reconstituted with saline and reinfused approximately 1 to 7 days prior to the event. The postinfusion hypovolemia leads to a shift in protein-free plasma from the intravascular to the interstitial compartment. This results in normal volemia with a greater hemoglobin concentration, and thus has the potential for increased oxygen-carrying capacity.

Desired Effects

The basic desired effect of induced erythrocythemia is to increase the hemoglobin level, thus increasing oxygen-carrying capacity and, theoretically, the endurance of the athlete [56, 77]. Early reports did not support the concept of blood doping to improve endurance [80]; however, the technique has been perfected, and recent studies have indicated that there is a definite beneficial effect [11, 81]. Differences in results in these studies are related to improper designs and controls and differences in time to transfusion and amount of blood transfused.

Recent studies in which the athlete himself is used as his own control have indicated that blood doping causes an increase of up to 25% in endurance capacity [69]. This increase is limited by the athlete's initial aerobic fitness, athletes in moderate physical fitness showing the greatest improvement in endurance.

Adverse Effects

Numerous adverse effects are related to induced erythrocythemia [7, 36]. If heterologous blood is used, a transfusion reaction may occur in 3% to 10% of individuals. Hepatitis may occur in up to 10% of people, along with the risk of transmission of AIDS. Other hemolytic anemias and immune sensitizations can occur also. When autologous blood is used, bacterial infections and problems related to mislabeling and mishandling of blood can occur. Debate arises about many aspects of the adverse effects of polycythemia. The possibility of increased viscosity exists, leading to congestive heart failure, hypertension, and cerebrovascular accidents. However, most young, healthy athletes are able to handle the extra viscosity without difficulty.

Erythropoietin

Erythropoietin is a hormone released from the kidneys in response to low hematocrit to stimulate red blood cell production from the bone marrow. Recently (in 1987), erythropoietin was manufactured by recombinant DNA technique [31]. Although it is difficult to obtain and is used mostly in patients with chronic renal failure, erythropoietin has been shown to increase the hematocrit of patients with chronic renal failure by up to 35%, and the results last up to 7 months [31]. Erythropoietin used by athletes to try to induce erythrocythemia has already been reported in the literature. Cyclists and other aerobic athletes, mostly in Europe, have been using the drug for the last several years; however, there have been some severe adverse effects, including possible death. Further study on the use of this substance in athletes is warranted.

Final Comment

Although blood doping and induced erythrocythemia with erythropoietin have been banned by the IOC, detection of these maneuvers is very difficult by present means. There is currently no blood or urine test that can accurately detect blood doping [7]. However, several tests have been proposed. One is to measure an athlete's hemoglobin in two separate samples 1 to 2 weeks apart prior to competition. The other is to measure the serum iron and bilirubin concentrations to detect the hemolysis that occurs with transfusion of frozen blood cells. A test is also being developed to measure erythropoietin levels [7]; however, all of these methods involve invasive procedures that are very difficult to enforce in athletes. Other methods of detecting and preventing blood doping are being investigated; however, this has proved very challenging to the sports medicine physician who deals with the abuse of ergogenic aids.

BICARBONATE LOADING

Bicarbonate loading is defined as the use of sodium bicarbonate (common baking soda) prior to competition to neutralize the acid (lactic acid) produced by anaerobic metabolism. In anaerobic activity such as intermediate distance running—for instance, 800 meters—lactic acid is produced by the anaerobic breakdown of glycogen. As lactic acid increases at the tissue level, pH decreases, causing fatigue. It was first realized that lactic acid causes fatigue in 1907. However, it was not until 1931 that the effects of reducing lactic acidosis were noted [35]. Several studies on the use of bicarbonate ingestion to buffer lactic acid to decrease fatigue and improve anaerobic endurance have been done [18, 44, 67, 79]. Some studies have shown a decrease in running time in athletes who have ingested sodium bicarbonate approximately 30 minutes before an event. A total of 300 mg/kg is necessary to achieve a beneficial effect [35]. The effect is not seen in events such as short duration sprints; however, in intermediate events, such as 800- to 1500-meter runs, a decrease in running time has been noted.

Adverse Effects

Although no serious adverse effects have been reported, the use of such large amounts of sodium bicarbonate (300 mg/kg) may result in such irritating effects as diarrhea, which may occur in 50% of athletes after 30 minutes, belching, water retention, and a possible increase in blood pressure in hypertensive athletes. Although rare, the possibility of severe metabolic alkalosis exists and may lead to confusion, apathy, and even tetany.

Although the use of bicarbonate loading is limited to only certain events, and studies are not all conclusive, it does deserve mention here. There is no direct policy on the use of bicarbonate loading. There is also no testing that can be performed for bicarbonate loading; however, in theory, it is an ergogenic aid and is morally and ethically unjust.

MISCELLANEOUS ERGOGENIC AIDS

Amino Acids and Proteins

Athletes have used amino acid supplements and protein supplements in conjunction with other ergogenic aids to try to enhance their benefit [4]. Amino acids are divided into essential and nonessential. Essential amino acids are those not being produced by the body and that therefore must be taken in by diet. Essential amino acids include histidine, isoleucine, leucine, lysine, methionine/cysteine, phenylalanine/tyrosine, threonine, tryptophan, and valine. A well-balanced diet containing the recommended daily allowances for proteins will contain all of the essential amino acids needed for proper nutrition. However, athletes who are vegetarians or who do not take in milk or dairy products may be deficient in some of these. The recommended daily allowance for protein intake is 0.8 g per kilogram, but some studies have indicated that athletes may require up to 1.4 g per kilogram and weight lifters and endurance athletes may require as much as 2 to 2.4 g per kilogram [4]. Most athletes eating a normal balanced diet will meet these protein requirements without any difficulty. However, one must be wary of athletes on an inadequate diet, such as some endurance athletes and lactovegetarians.

Desired Effects

Athletes take amino acids and protein supplements as ergogenic aids for several reasons [13, 30, 49]. Increases in body size and strength and prevention of protein wasting are very important for endurance athletes. Most athletes utilize either glycogen or fatty acids for energy; however, in some endurance athletes up to 10% of energy expenditure comes from the breakdown of proteins. These athletes have a higher need for added proteins. Athletes also take amino acids and protein supplements to stimulate the body's natural release of growth hormone [45]. Three amino acids, lysine, arginine, and ornithine, have been studied in relation to their effect on growth hormone release. These studies are very inconclusive and actually contradictory on some points, and there is no good evidence to suggest that ingestion of these amino acids alone will increase the release of growth hormone.

Very few adverse effects are associated with ingestion of a large amount of amino acids and protein supplements; however, dehydration along with renal damage and hepatic damage is possible [10, 30]. Eosinophilic myositis syndrome with tryptophan ingestion has also been described [17]. There is no good evidence that amino acids are harmful; however, no conclusive studies have been done.

Other protein supplements are often used by athletes including spirulina [28], a protein substance derived from algae that has been found to have no ergogenic capabilities; however, it may contain toxic bacteria and therefore may be detrimental to health. Gelatin has also been used but is actually a very poor source of protein that is deficient in numerous amino acids. It has also not been demonstrated to be ergogenic [33]. Athletes use gelatin in the hope of increasing phosphocreatine to enhance their anaerobic capacity; however, this result has not been demonstrated in studies.

Amino acids and protein supplements are often used by athletes to enhance performance. However, there have been no good studies demonstrating that athletes maintaining a normal diet and normal protein intake will benefit from the use of these substances. Although no severe adverse effects or long-term complications of such use have been reported, further study is needed.

VITAMINS AND MINERALS

Vitamins and minerals are substances that are used in daily metabolism at the cellular level but are not synthesized by the body and must be consumed in the diet [5, 39, 82]. It has been estimated that between 30% and 80% of all athletes take multivitamins and iron in the belief that they improve athletic performance. Vitamins and minerals can improve performance in athletes who have vitamin deficiencies; however, these are rare if the athlete maintains a well-balanced diet. There are some vitamins that in megadoses can be toxic, and no ergogenic effects have been proved.

Vitamins most commonly used by athletes include thiamine (vitamin B_1). Thiamine is a coenzyme used in carbohydrate metabolism. When a deficiency exists, a decline in muscle endurance may occur; however, no improvement in endurance is associated with megadoses. Riboflavin (vitamin B_2) is a coenzyme used in the electron transport chain; no increase in energy utilization occurs with megadoses of this vitamin. Nicotinic acid (niacin) is utilized for ATP production during glycolysis and fat synthesis. Excess niacin may inhibit free fatty acid mobilization and facilitate glycogen utilization, which may decrease aerobic capacity. Pyridoxine (vitamin B_6) is a coenzyme used in several chemical reactions including gluconeogenesis, glycogenolysis, and lipid synthesis. It may actually cause a more rapid depletion of glycogen stores and thus may have an adverse effect on aerobic capacity. Cyanocobalamin (vitamin B_{12}) is a coenzyme utilized in the synthesis of red blood cells. No evidence of increased aerobic endurance has been seen in athletes taking vitamin B_{12} supplements. Ascorbic acid (vitamin C) is essential for collagen synthesis and epinephrine and corticosteroid synthesis and may actually help facilitate iron absorption in deficiency states, along with iron supplementation. It has been touted as an "antioxidant," but no change in VO_2 max or aerobic endurance has been demonstrated in studies. Calciferol (vitamin D) promotes calcium absorption but has no effect on athletic performance. Alpha-tocopherol (vitamin E) is an antioxidant and may protect red blood cell membrane phospholipids from oxidation. It has been used to help improve aerobic capacity during training at high altitudes. It has produced no demonstrated changes in aerobic or anaerobic performance or any effect on lactate levels in normal athletes.

There are also "pseudovitamins" including paraminobenzoic acid (PABA) [28] and choline [39], but these have no known nutritional value. Other substances touted as ergogenic aids also have no nutritional value, including vitamin P (the bioflavins) [28, 39] and vitamin B_{15} (pangamic acid) [28, 39], which is actually not a vitamin at all. Its chemical formula has numerous variations, none of which are regulated.

Megadoses of vitamins often have adverse effects [39], especially the fat-soluble vitamins, including vitamins A, D, E, and K. Some of the more common toxic effects include cirrhosis, hypercalcemia, and bone reabsorption, seen with megadoses of vitamin A; hypercalcemia, hypertension, and arrhythmias, associated with megadoses of vitamin D; weakness, fatigue, nausea, and vomiting, seen with megadoses of vitamin E; sensory neuropathy, associated with toxic doses of vitamin B_6; diarrhea, nausea, and decreased B_{12} levels with toxic doses of vitamin C.

Vitamins and minerals are essential compounds for normal metabolic activities. Supplements of these compounds have not demonstrated any ergo-genic potential, and large doses may actually produce harmful effects. Although most athletes take a multivitamin with iron on a daily basis, a well-balanced diet will more than meet the daily vitamin requirements of these athletes.

PHOSPHATE LOADING

Ingestion of phosphate has been used by athletes to improve their aerobic performance by increasing oxygen delivery through the increased production of 2,3-diphosphoglyceride (2,3-DPG) [24, 46]. An increase in 2,3-DPG at the red blood cell level will shift the oxygen dissociation curve to the right, resulting in increased oxygen unloading at the tissue level, thus improving oxygenation and, theoretically, aerobic performance. Ingestion of phosphate has been shown to increase the level of DPG, which is the same result that occurs with high-altitude training [46]. However, the effects of this increased 2,3-DPG are offset at the lung by decreased oxygen absorption. The studies that have been done have been limited and criticized for their methodology and demonstrated improvement only at high altitudes.

Although adverse effects of phosphate ingestion are not clear, hyperphosphatemia may increase the effects of metastatic calcification in the soft tissues.

CARNITINE

Carnitine is an amino acid that is used as a substrate for enzymes. One of its primary functions is the transport of long-chain fatty acids into the mitochondria to facilitate fatty acid metabolism, which in theory may improve aerobic endurance by decreasing glycogen utilization at the muscle level [15, 26]. There are no studies demonstrating the ergogenic effects of carnitine in a person ingesting a normal well-balanced diet; however, people who eat diets that are poor in red meat and dairy products, such as lactovegetarian diets, may have a deficiency of carnitine and may benefit from carnitine substitution. Athletes ingest numerous other substances that theoretically enhance performance, but studies and information on these substances, which change from day to day, are very limited [28]. These substances include cyproheptadine (periactin), which is used to stimulate appetite; ginseng, which is used as a stimulant; octacosanol, found in wheat germ, which is used by athletes in an attempt to improve stamina, strength, and reaction time; and gamma-oryzonal, which is used to stimulate growth hormone release and also as an antioxidant to neutralize free radicals released during exercise. Athletes believe that the use of antioxidants decreases free radicals and thus eliminates cell damage and increases recovery time after exertion.

SUMMARY

In summary, there are numerous factors that affect athletic performance. One of these factors is performance-enhancing drugs. Although not all athletes use performance-enhancing drugs, a large and growing number of individuals are turning toward these chemicals to achieve goals that they may not have been able to achieve through training alone.

The drugs most commonly used by athletes are anabolic steroids, which are synthetic analogues of the male hormone testosterone. Athletes use anabolic-androgenic steroids to increase strength, body weight, lean body mass, and aggressiveness, to allow high-intensity workouts, and to enhance performance. These medications are associated with some adverse effects including changes in the liver, cardiovascular system, male and female reproductive systems, and psychological system, and they have adverse effects on children. They are banned by the International Olympic Committee and most other governing bodies of the various sports. They are a controlled substance by the Drug Enforcement Agency (DEA); however, they are mostly obtained through illegal means. Human growth hormone (somatotropin) is also used by athletes to increase size and strength, increase lean body mass by decreasing fat tissue, enhance performance, and to try to enhance the growing potential in children as well. This synthetic hormone can have some serious adverse effects, including cardiovascular changes, diabetes, and acromegaly. Although human growth hormone is also banned by most governing bodies, there is still no test to detect its use, and athletes have turned to using this performance-enhancing drug instead of anabolic steroids to keep from being caught and disqualified.

Amphetamines are sympathomimetic chemicals used by athletes to increase endurance, energy substrates, and reaction time, to mask fatigue, and to decrease appetite in some athletes who desire a lean body appearance. Amphetamines stimulate catecholamine release, displace catecholamine from the receptor site, or inhibit the breakdown and reuptake of catecholamines, thus allowing potentiation of the effect of catecholamine at the receptor sites. The adverse effects of amphetamines include changes in the central nervous system including tremulousness,

anxiety, agitation, and even possibly stroke. Changes in the cardiovascular system and gastrointestinal system are possible, as are heat illness and even death. They, too, are banned by most governing bodies in sports and are tested for routinely. Caffeine, a compound found in everything from medications to soft drinks, is sometimes taken by athletes to decrease fatigue, increase the breakdown of fatty acids, thus sparing glycogens at the muscle level to decrease their consumption, increase muscle contraction, and improve endurance. It causes a translocation of intracellular calcium, thus providing more calcium for uptake, and can inhibit the enzyme phosphodiesterase, thus increasing cyclic AMP and enhancing glycogenolysis, allowing the utilization of fatty acids for energy substrate. The adverse effects of large doses of caffeine include symptoms similar to those of other stimulants, including restlessness, anxiety, insomnia, cardiovascular effects, and the possibility of heat illness due to its diuretic effect.

Induced erythrocythemia, or blood doping, is used by athletes to improve aerobic performance. Infusion of the athlete's own blood, or stored blood prior to an event, will increase the hemoglobin level and thus the oxygen-carrying capacity of blood to the muscle, thus increasing aerobic endurance. Adverse effects depend upon the method used, either heterologous or autologous infusion, and include the transmission of hepatitis, AIDS, immune sensitizations, and other infections. There is also the possibility of increased viscosity and with it congestive heart failure, hypertension, and stroke. Erythropoietin is a hormone released by the kidneys to stimulate red blood cell production. Although this hormone is now produced by recombinant DNA techniques to treat patients with anemia due to chronic renal failure and other illnesses, athletes have begun to use this drug to increase their own hemoglobin and hematocrit levels and improve their aerobic performance. Adverse effects are just becoming known; however, reports of death from hyperviscosity states have occurred.

Other ergogenic aids include bicarbonate loading—the ingestion of sodium bicarbonate to neutralize lactic acid production during anaerobic metabolism, thus decreasing fatigue and enhancing performance in anaerobic events. Amino acids and proteins are taken by athletes to increase size and strength and prevent protein wasting in endurance athletes. However, no studies have indicated the actual effectiveness of these compounds. Vitamins and minerals are also ingested for different reasons, but there is no actual proof of their effectiveness.

The use of ergogenic aids continues to be a problem that physicians must face on a daily basis when they deal with competitive athletes. Although much is known about most of these compounds, further study is warranted to truly define their effectiveness and their adverse effects in the athletes who use them. Athletes will continue to use performance-enhancing drugs to gain the competitive edge that may allow them to attain goals that they believe would be unattainable on their own. Society plays a role in the perpetuation of this practice by placing such a high value on sports and the success of athletes. The axiom "winning equals success" is often taken to extremes by both athletes and their followers. Athletes will continue to try to gain success by all means possible, and performance-enhancing drugs are just one aspect of this quest.

References

1. American College of Sports Medicine. Position stand on the use of anabolic-androgenous steroids in sports. *Med Sci Sports Exerc* 19:534–539, 1987.
2. Anderson, W. A., and McKeag, D. B. The substance use and abuse habits of college student-athletes. Presented to National Collegiate Athletic Association Council by the College of Human Medicine, Michigan State University. June, 1985.
3. Apostolakis, M., Deligiannis, A., and Madena-Pyrgaki, A. The effects of human growth hormone administration on the fractional status of rat atrophied muscle following immobilization. *Physiologist* 23 (Suppl.):S111, 1980.
4. Aronson, V. Protein and miscellaneous ergogenic aids. *Physician Sportsmed* 14(5):199–202, 1986.
5. Aronson, V. Vitamins and minerals as ergogenic aids. *Physician Sportsmed* 14(3):209–212, 1986.
6. Belleville, J., and Dorey, F. Effect of nefopam on visual tracking. *Clin Pharmacol Ther* 26:457–463, 1979.
7. Berglund, B. O. Development of techniques for the detection of blood doping in sport. *Sports Med* 5:127–135, 1988.
8. Berthold, A. A. Transplantation der Hoden. *Arch Anat Physiol Wiss Med* 16:42, 1849.
9. Bigland, B., and Jehring, B. Muscle performance in rats, normal and treated with growth hormone. *J Physiol* 116:129, 1952.
10. Brenner, B. M., Meyer T. W., and Hosteter, T. H.: Mechanisms of disease. *N Engl J Med* 307(11):652–659, 1982.
11. Brien, A. J., and Simon T. L. The effects of red blood cell infusion on 10-km race time. *JAMA* 257:2761, 1987.
12. Brooks, R. V. Anabolic steroids and athletes. *Phys Sportsmed* 8(3):161–163, 1980.
13. Brotherhood, J. R. Nutrition and physical performance. *Sports Med* 1:350, 1984.
14. Buckley, W. E., Yesalis, C. E. III, Friedl, K. E., Anderson, W. A., Streit, A. L., and Wright, J. E. Estimated prevalence of anabolic steroid use among male high school seniors. *JAMA* 260(23):3441–3445, 1988.
15. Carnitine. *Medical Letter* 28:88, 1986.
16. Chandler, J., and Blair, S. The effect of amphetamines on selected physiological components related to athletic success. *Med Sci Sports Exerc* 12:65–69, 1980.
17. Clauw, D. J., Nashel, D. J., Yumhau, A., and Katz, P. Tryptophan-associated eosinophilic connective tissue disease. *JAMA* 263(11):1502–1506, 1990.
18. Costill, D. L. Exercise physiology: Is sodium bicarbonate an aid to sprint performance? *Sports Med* 10:4, 1988.
19. Costill, D. L., Dalsky, G. P., and Fink W. J. Effects of caffeine ingestion on metabolism and exercise performance. *Med Sci Sports* 10:155, 1978.
20. Cowart, V. S. Some predict increased steroid use in sports despite drug testing, crackdown on suppliers. *JAMA* 257:3025, 1987.
21. Cowart, V. S. Human growth hormone: The latest ergogenic aid? *Physician Sportsmed* 16(3):175, 1988.
22. Coyle, E. F. Ergogenic aids. *Clin Sports Med* 3(3):731–742, 1984.
23. Delbeke, F. T., and Debackere, M. Caffeine: Use and abuse in sports. *Int J Sports Med* 5:179–182, 1984.
24. Dempsey, J. A., et al. Muscular exercise, 2,3-DPG and oxyhemoglobin affinity. *Int J Physiol* 30:34, 1971.
25. Dezelsky, T. L., Toohey, J. V., and Shaw, R. S. Nonmedical drug use behavior at five United States universities: A 15-year study. *Bull Narc* 37:49, 1985.
26. Dipalma, J. R. L-Carnitine: Its therapeutic potential. *Am Fam Physician* 34:127, 1986.
27. Domino, E., Albers, J., et al. Effects of d-amphetamine on quantitative measures of motor performance. *Clin Pharmacol Ther* 13:251–257, 1972.
28. Dubick, M. A. Dietary supplements and health aids—A critical evaluation, Part 3. Natural and miscellaneous products. *J Nutri Education* 15(4):123–129, 1983.
29. Eichner, E. R: Blood doping: Results and consequences from the laboratory and the field. *Physician Sportsmed* 15(1):121–129, 1987.
30. Ergogenic aid of the month: Protein supplements. *Sports Med Digest* 9:6, 1987.
31. Eschbach, J. W., Egrie, J. C., Downing, M. R., Browne, J. K., and Adamson, J. W. Correction of anemia of end-stage renal disease with recombinant human erythropoietin. Results of a combined phase I and II clinical trial. *N Engl J Med* 316:73, 1987.
32. Evans, M., Martz, R., et al. Effects of dextroamphetamine on psychomotor skills. *Clin Pharmacol Ther* 19:777–781, 1976.
33. Fad of the month: Gelatin and glycine. *Sports Med Digest* 10:6, 1988.
34. Fradkin, J. E., Schonberger, L. B., Mills, J. L., Gunn, W. J., Piper, J. M., Wysowski, D. K., Thomson, R., Durako, S., and Brown, P. Creutzfeldt-Jakob disease in pituitary growth hormone recipients in the United States. *JAMA* 265(7):880–884, 1991.
35. Gao, J., Costill, D. L., Horswill, C. A., and Parle, S. H. Sodium bicarbonate improves performance in interval swimming. *Eur J Appl Physiol* 58:171–174, 1988.
36. Gledhill, N. Blood doping and related issues: A brief review. *Med Sci Sports Exerc* 14(3):183–189, 1982.
37. Goldberg, A. L., and Goodman, H. M. Relationship between growth hormone and muscle work in determining muscle size. *J Physiol* 200:655, 1969.
38. Goldman, B. *Death in the Locker Room: Steroids, Cocaine and Sports.* The Body Press, Tucson, 1987.
39. Grandjean, A. C. Vitamins, diet, and the athlete. *Clin Sports Med* 2(1):105–114, 1983.
40. Hallagan, J. B., Hallagan, L. F., and Snyder M. B. Anabolic androgenic steroid use by athletes. *N Engl J Med* 321(15):1042–1045, 1900.
41. Hatton, C. K., and Catlin, D. H. Detection of androgenic anabolic steroids in urine. *Clin Lab Med* 7(3):655–665, 1987.
42. Haupt, H. A., and Rovere, G. D. Anabolic steroids: A review of the literature. *Am J Sports Med* 12(6):469–484, 1984.
43. Hazelden Health Promotion Services. *Elite Women Athletes Survey.* Hazelden Health Promotion Services, January, 1987.
44. Hooker, S., Morgan, C., and Wells, C. Effect of sodium bicarbonate ingestion on time to exhaustion and blood lactate of 10K runners. *Med Sci Sports Exerc* 19(Suppl.):S67, 1987.
45. Isidori, A., Monaco, A. L., and Cappa, M. A study of growth hormone release in man after oral administration of amion acids. *Curr Med Res Opinion* 7(7), 1981.
46. Jain, S. C., et al. Effect of phosphate supplementation on oxygen delivery at high altitude. *Int J Biometeorol* 31:249, 1987.

47. Klein, H. G. Blood transfusions and athletes. Games people play. *N Engl J Med* 312:854, 1985.
48. Karagiorgos, A., Garcia, J. F., and Brooks, G. A. Growth hormone response to continuous and intermittent exercise. *Med Sci Sports* 11(3):302–307, 1979.
49. Lemon, P. W., Yarasheski, K. E., and Dolny, D. G. The importance of protein for athletes. *Sports Med* 1:474, 1984.
50. Lombardo, J. A. Stimulants and athletic performance (Part 1 of 2): Amphetamines and caffeine. *Physician Sportsmed* 14(11):129–140, 1986.
51. Macintyre, J. D. Growth hormone and athletes. *Sports Med* 4:129–142, 1987.
52. Mandell, A. J., Stewart, K. D., and Russo, P. V. The Sunday syndrome: From kinetics to altered consciousness. *Fed Proc* 40:2693, 1981.
53. McCardle, W. D., Katch, F. I. and Katch V. L. *Exercise Physiology—Energy, nutrition, and human performance.* Philadelphia, Lea & Febiger, 1981, p. 305.
54. Melmed, S. Acromegaly. *N Engl J Med* 322(14):966–977, 1990.
55. Murad, F., and Haynes, R. C., Jr. Androgens. *In* Gilman, A. G., et al. (Eds.), *Goodman and Gilman's The Pharmacologic Basis of Therapeutics* (7th ed). New York, Macmillan, 1986, p. 1440.
56. National Collegiate Athletic Association. *The 1990–1991 NCAA Drug Testing Program Mission.* 4, Overland Park, KS, NCAA Publishing, 1990.
57. A new biosynthetic human growth hormone. *Med Lett* 29(745):73, 1987.
58. Pace, N., Lozner, E. L., Consolatio, W. V., Pitts, G. C., and Pecora, L. S. The increase in hypoxia tolerance of normal men accompanying the polycythemia induced by transfusion of erythrocytes. *Am J Physiol* 148:152, 1947.
59. Perlmutter, G., and Lowenthal, D. T. Use of anabolic steroids by athletes. *Am Fam Physician* 32:208, 1985.
60. Piness, G., Miller, H., and Alles, G. A. Clinical observations on phenylaminoethanol sulfate. *JAMA* 94:790, 1930.
61. Pohlman, E. T., Despres, J. P., Besseite, H., Fontaine, E., Tremblay, A., and Bouchard, C. Influence of caffeine on the resting metabolic rate of exercise-trained and inactive subjects. *Med Sci Sports Exerc* 17:689, 1985.
62. Pope, H. G., and Katz, D. L. Affective and psychotic symptoms associated with anabolic steroid use. *Am J Psychiatry* 145:487, 1988.
63. Pope, H. G., Katz, D. L., and Champoux, R. Anabolic-androgenic steroid use among 1,010 college men. *Physician Sportsmed* 16(7):75, 1988.
64. Powers, S. K., and Dodd, S. Caffeine and endurance performance. *Sports Med* 2:165, 1985.
65. Prinzmetal, M., and Bloomberg, W. The use of benzedrine for the treatment of narcolepsy. *JAMA* 105:2051, 1935.
66. Puffer, J. The use of drugs in swimming. *Clin Sports Med* 5:77, 1986.
67. Robinson, K., and Verity, L. S. Effect of induced alkalosis on rowing ergometer performance (rep) during repeated 1-mile work-outs. *Med Sci Sports Exerc* 19 (Suppl.):S67, 1987.
68. Rogozkin, V. A. Anabolic steroid metabolism in skeletal muscle. *J Steroid Biochem* 11:923–926, 1979.
69. Sawka, M. N., et al. Erythrocyte reinfusion and maximal aerobic power: An examination of modifying factors. *JAMA* 257:1496, 1987.
70. Sklarek, H. M., Mantovani, R. P., Erens, E., Heisler, D., Niederman, M. S., and Fein, A. M. AIDS in a bodybuilder using anabolic steroid. *N Engl J Med* 311:1701, 1984.
71. Slavin, J. L., and Joensen, D.J. Caffeine and sports performance. *Phys Sportsmed* 13(5):191, 1985.
72. Smith, G. M., and Beecher, H. K. Amphetamine sulfate and athletic performance: 1. Objective effects. *JAMA* 170:542–557, 1959.
73. Snochowski, M., Dahlberg, E., Eriksson, E., and Gustafsson, J. A. Androgen and glucocorticoid receptors in human skeletal muscle cytosol. *J Steroid Biochem* 14:765–771, 1981.
74. Strauss, R. H. Drugs in sports. *In* Strauss, R.H. (Ed.), *Sports Medicine.* Philadelphia, W. B. Saunders, 1984, p. 481.
75. Terney, R., and McLain, L. G. The use of anabolic steroids in high school students. *Am J Dis Child* 144:99–103, 1990.
76. Underwood, L. E. Occasional notes—Report of the Conference on Uses and Possible Abuses of Biosynthetic Human Growth Hormone. *N Engl J Med* 311(9):606–608, 1984.
77. U. S. Olympic Committee, Division of Sports Medicine and Science. Colorado Springs, Drug Education and Control Policy, 1988.
78. Wade, N. Anabolic steroids: Doctors denounce them, but athletes aren't listening. *Science* 176:1399–1403, 1972.
79. Wilkes, K., Gledhill, N., and Smyth, R. Effect of acute induced metabolic alkalosis on 800-m racing time. *Med Sci Sports Exerc* 15:277, 1983.
80. Williams, M. H., Goodwin, A. R., Perkins, R., and Bocrie, J. Effect of blood reinjection upon endurance capacity and heart rate. *Med Sci Sports* 5:181, 1973.
81. Williams, M. H., Wesserdine, S., and Somma, T. The effect of induced erythrocythemia upon 5-mile treadmill run time. *Med Sci Sports Exerc* 13:169, 1981.
82. Williams, M. H. Vitamin and mineral supplements to athletes: Do they help? *Clin Sports Med* 3(3):623–637, 1984.

Treatment of Hypertension in Athletes

Peter Hanson, M.D.
Bruce E. Andrea, M.D.

Athletes are usually considered to be free of cardiovascular disease because of their apparent high level of physical fitness. However, hypertension commonly begins in young adulthood and may occur in 5% to 10% of persons in the age group of 20 to 30 years [19]. The incidence of hypertension continues to increase with age and is found in 20% to 25% of middle-aged adults [15, 19]. Therefore, a significant number of athletes, from recreational to elite, will have some degree of hypertension [28].

The treatment of hypertension requires appropriate consideration of the potential influence of exercise and training on hypertension and the potential side effects of antihypertensive medications on athletic performance. This section will outline the current clinical strategy for the evaluation and treatment of hypertension in adult (over 18 years) athletes.

EPIDEMIOLOGIC CONSIDERATIONS

Hypertension is defined according to population mean values of repeated resting blood pressure (Table 9B–1). Normal blood pressure includes all values below 140/85. Hypertension is classified as mild (borderline), moderate, or severe. The absolute cut-off values for these classes may vary somewhat according to different sources [15, 19, 30].

The majority of hypertensive patients fall into the mild-to-moderate range, and this includes most competitive or recreational athletes. The overall incidence of mild-to-moderate hypertension in athletes was found in a recent study to be approximately 50% less than the baseline frequency in a general clinic population [28].

A variety of risk factors contribute to the probability of development of hypertension (Table 9B–2). These include genetic, metabolic, and behavioral factors. In general, a family history of hypertension increases the probability of future development of hypertension [19]. Overall, the black population shows a higher incidence of hypertension than white or Asian groups. The underlying reason for this is not understood. Finally, men exhibit hypertension earlier than women [19].

Metabolic risk factors include obesity and glucose intolerance [42]. These conditions produce an abnormal regulation of vascular volume and peripheral resistance, which mediates the increase in blood pressure.

Behavioral factors include repeated exposure to excessive environmental and social stress leading to chronic (neurogenic) activation of the sympathetic nervous system [11]. Other behavioral factors include excessive alcohol consumption and recreational or ergogenic drug abuse including stimulants

TABLE 9B–1
Classification of Hypertension

	Blood Pressure (mm Hg)	
	Diastolic	*Systolic**
Normal	<85	<135
High normal	85–90	135–140
Mild hypertension (borderline)	90–104	140–159
Moderate hypertension	105–114	>159
Severe hypertension	>114	>159

*Hypertension may be predominantly systolic. *Isolated* systolic hypertension is defined as greater than 160 with normal or borderline diastolic values.

TABLE 9B–2
Risk Factors for Hypertension

Genetic Factors
   Family history
   Males more than females
   Blacks more than whites; Asians least
Metabolic Factors
   Obesity
   Glucose intolerance
   Endocrine disorders
Behavioral Factors
   High sodium intake
   Alcohol consumption
   Drug abuse
      Stimulants
      Anabolic steroids
      Tobacco (?)

---

such as amphetamines and cocaine, and possibly anabolic steroids [13]. The relationship of these ergogenic agents to the development of sustained hypertension is unclear but should be considered in competitive athletes. While tobacco use has not been shown to result in sustained hypertension, it can raise blood pressure acutely. This may have significance in athletes who chew tobacco.

CLINICAL PATHOPHYSIOLOGY OF HYPERTENSION

For clinical purposes, hypertension is divided into two categories, primary hypertension and secondary hypertension. Primary hypertension accounts for more than 95% of cases of sustained high blood pressure. Multiple regulatory mechanisms contrib-

ute to the evolution of primary hypertension. These include (1) abnormal neuroreflexes and sympathetic control of peripheral resistance, (2) abnormal renal and metabolic control of vascular volume and compliance, and (3) abnormal local smooth muscle and endothelial control of vascular resistance. The interaction among these factors is illustrated in Figure 9B–1.

A detailed discussion of the altered control mechanisms operative in primary hypertension is beyond the scope of this chapter. Briefly, the combined neurohumoral and metabolic abnormalities contribute to a gradual increase in systemic vascular resistance that is characteristic of primary hypertension. Activation of sympathetic nervous system and other pressor hormones such as angiotensin probably mediates the increase in vascular resistance. However, a variety of studies have shown only small or moderate increases in circulating catecholamines [12] and variable levels of renin and angiotensin [27]. More recent studies have implicated a primary defect in calcium and sodium balance in vascular smooth muscle as an underlying common defect in primary hypertension [26]. Abnormalities in endothelial-derived vascular control factors may also be important [35]. Structural changes within vascular smooth muscle ultimately develop, so that the vascular resistance eventually is self-maintained. Thus, the multiple factors mediating the development of hypertension ultimately produce a sustained elevation of systemic vascular resistance. Finally, there is an alteration in arterial baroreceptor function that allows the baroreflexes to be "reset" to maintain a higher systemic arterial pressure level [1].

The hemodynamic patterns associated with hypertension vary with the stage or severity of hyper-

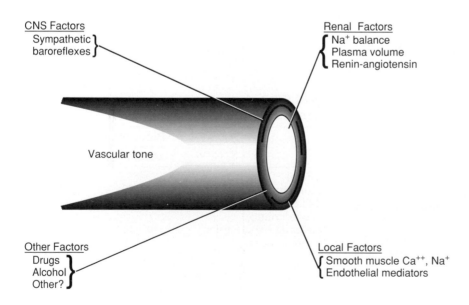

FIGURE 9B–1
Summary of physiologic factors that may contribute to abnormal arteriolar vascular tone in hypertension.

tension. The early phase of mild or borderline hypertension is frequently characterized by increased resting heart rate and cardiac output. Although vascular resistance is in the normal range for this group, it is inappropriately high for the corresponding cardiac output. In moderate hypertension cardiac output and heart rate are normal, and vascular resistance is increased. With severe hypertension, there is a further increase in vascular resistance and varying depression of cardiac output due to increased vascular afterload [32].

Secondary hypertension accounts for approximately 5% of cases of sustained hypertension. The major causes of secondary hypertension are summarized in Table 9B–3. They include renal, endocrine, and vascular abnormalities. Secondary causes should be considered when hypertension develops either in younger patients or rapidly in adult patients with no prior history of hypertension. In addition, secondary hypertension should be considered when elevated blood pressure is poorly responsive to routine antihypertensive therapy.

Renal vascular disease causes an increased release of renin, which stimulates the enzymatic conversion of plasma angiotensin I to angiotensin II. Angiotensin II is a potent peripheral vasoconstrictor and also stimulates the release of aldosterone, which promotes renal retention of sodium and water. Tumors of the adrenal medulla and cortex are less common causes of secondary hypertension. Pheochromocytoma is caused by a catecholamine-releasing adenoma within the adrenal medulla. Adrenal cortical adenomas may release cortisol or aldosterone. These renal-endocrine causes of secondary hypertension are successfully treatable by surgery or medical management.

TABLE 9B–3
Causes of Secondary Hypertension

Renal
Vascular obstruction
Parenchymal disease
Renin-secreting tumor
Endocrine
Adrenal medullary
Pheochromocytoma
Adrenal cortex
Aldosterone-producing tumor
Cortisol-producing tumor
Thyroid
Hyperthyroidism
Hypothyroidism
Hyperparathyroidism
Oral contraceptives
Other
Coarctation of aorta

Regardless of cause, increased blood pressure levels produce a predictable pattern of end-organ pathology within the cardiac chambers and vascular tree. Left ventricular hypertrophy is nearly a universal response and may be seen by echocardiographic measurements in patients with borderline and mild hypertension [45]. With sustained hypertension concentric left ventricular hypertrophy increases, and abnormal diastolic relaxation is seen. Vascular changes include progressive degeneration of medium and small arterial vessels (arteriolosclerosis). The retinal and renal glomerular arterioles are exquisitely sensitive to hypertensive vascular degeneration. The cerebral circulation is also altered, and the risk of stroke increases significantly if hypertension remains untreated [38].

The end-organ damage from hypertension usually evolves over a period of years. However, rapid increases in blood pressure that may occur in secondary hypertension may be poorly tolerated, whereas gradual increases in blood pressure and isolated systolic hypertension in older patients may be surprisingly well tolerated for long periods without morbid events.

Blood Pressure Response to Exercise in Hypertensive Patients

With dynamic exercise, cardiac output increases owing to increased heart rate, stroke volume, and cardiac contractility. Blood flow to working muscle is augmented by local vasodilation, whereas blood flow to nonworking muscle and visceral organs is diminished by sympathetic vasoconstriction. The net result is a rise in systolic blood pressure with little change in diastolic pressure and a decrease in vascular resistance. In mild-to-moderate hypertension, cardiac output increases normally; however, systolic and diastolic pressure and vascular resistance are higher at all levels of exercise compared with normotensive subjects [32]. In patients with severe hypertension, cardiac output is lower than in age-matched controls because stroke volume is decreased. Systolic and diastolic pressure and vascular resistance are markedly increased [32].

The normal response to isometric exercise is a combined rise in systolic and diastolic pressure, commonly referred to as a *pressor* response. The pressor response is mediated by reflex increases in cardiac output with little or no change in vascular resistance [17]. The magnitude of the blood pressure rise is proportional to the combined size of the muscle mass and the percentage of maximal effort used in performing the isometric contractions [46].

The peak systolic and diastolic pressures achieved during dynamic and isometric exercise is substantially increased in hypertensive patients. However, the *relative* increase in blood pressure (from resting values) is similar to that seen in normotensive control subjects. These findings suggest that blood pressure is "reset" and maintained at higher levels throughout the spectrum of activity from rest to peak exercise. Figure 9B–2 illustrates the general pattern of blood pressure responses to exercise stress in mild hypertensive subjects. Note that hypertensive subjects exhibit greater systolic and diastolic blood pressures under conditions of treadmill exercise or isometric handgrip.

Recent studies suggest that exercise blood pressure responses may provide additional criteria for diagnosis and management of hypertension [3]. Blood pressure values exceeding 180/120 mm Hg at 50% of maximum handgrip (90 seconds) and treadmill blood pressure values exceeding 180/90 at 50% intensity or 225/90 mm Hg at 100% intensity are usually defined as hypertensive responses (Table 9B–4). Some patients with mild hypertension may show an early rise in systolic pressure to levels of

TABLE 9B–4
Blood Pressure Responses to Isometric Handgrip and Treadmill Exercise in Normal and Hypertensive Young Adults

	Isometric Handgrip (50% max)	Treadmill Exercise	
		50%	**100%**
Normal	<180/120	<180/80	<220/80
Mild hypertension	180–190 / 120–130	180–190 / 80–90	210–220 / 80–90
Moderate hypertension	>190 / 130	>190 / 90	>220 / 90

Unpublished data from Ward, A., and Hanson, P.

180 to 200 mm Hg during submaximal exercise, and these values are maintained at maximal exercise. Such patients may exhibit an excessively high systolic blood pressure with activities of daily living, although resting blood pressures and peak exercise blood pressures are within the normal range. Studies comparing echocardiographic measurements of cardiac hypertrophy with exercise blood pressures and the levels seen on ambulatory blood pressure monitoring indicate that patients with elevated blood pressure responses during exercise or activity also show evidence of left ventricular hypertrophy (LVH) [8] and are at greater risk of development of sustained hypertension [3, 52].

CLINICAL EVALUATION

Confirmation of Hypertension

Hypertension in athletes is frequently detected during routine physical examinations for sports participation or during casual blood pressure monitoring by athletic trainers. In these instances, blood pressure measurement techniques and the surrounding conditions may vary considerably. Some athletes may exhibit "clinic" or "white coat" hypertension [36]. Therefore, it is important to document apparent hypertension carefully under standardized measurement conditions [22].

Blood pressure should be measured in the upright seated position with the measured arm maintained at midchest (right atrial) level. Measurements should be made after a 10-minute period of rest and during abstinence from stimulants such as coffee and other caffeinated drinks. Repeated measurements should be obtained during the initial evaluation, with follow-up confirmatory readings and ini-

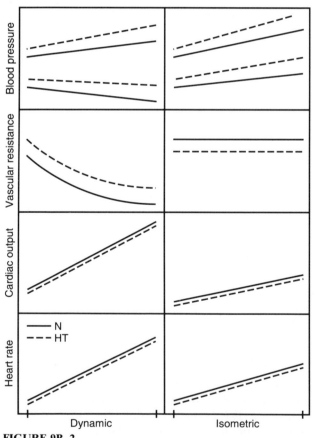

FIGURE 9B–2
Comparison of hemodynamic responses to dynamic and isometric exercise in normotensive (——) and hypertensive (– – –) subjects.

tiation of therapy being based on the degree of hypertension observed [19].

It is important to use the appropriate blood pressure cuff size. Many athletes are engaged in heavy weight-training, and their arm size exceeds the upper limit of the standard adult blood pressure cuff. Hypertensive blood pressure values will occur if the cuff size is relatively small compared to arm size. The recommended cuff diameters and dimensions for adult pressure measurements are summarized in Table 9B–5.

Another potential measurement error may occur in athletes with relative bradycardia. If cuff pressures are released too quickly, the large pressure decrement between pulses may lead to significant errors in blood pressure determinations. It is recommended that the cuff pressure be released at a rate of 3 to 5 mm Hg/second to avoid this error.

Athletes who show hypertensive readings in the seated position should not be allowed to lie supine in order to "rest." Increased venous return in the supine position activates cardiopulmonary baroreceptors that reflexively reduce blood pressure, which may falsely correct true hypertension [1]. In some cases, self-measured blood pressures at home two to three times daily for 1 week are helpful in confirming mild or borderline hypertension [6].

Ambulatory blood pressure measurements using automated recorders have recently been used to distinguish clinic hypertension from true hypertension [43]. Continuous ambulatory blood pressure monitoring is especially useful for patients with unreliable home blood pressure and clinic blood pressure values. Ambulatory blood pressure monitoring is also valuable in confirming the efficacy of antihypertensive therapy as well as alerting one to the possibility of an abnormal hypertensive response to normal daily activities or exercise [53].

Further History and Physical Examination

Once hypertension is suspected, a more thorough history and physical examination should be performed, with an emphasis on discerning the coexistence of other cardiovascular risk factors, the possibility of secondary hypertension, and the existence of end-organ pathology. As such, the history should assess the patient's habits (salt intake, tobacco use, coffee and alcohol consumption), and a personal or family history of hypercholesterolism, diabetes, hypertension, heart disease, premature death, kidney disease, pheochromocytoma, hypoaldosteronism, or hypoparathyroidism. Legitimate and illicit drug use should also be assessed, including the use of nonsteroidal anti-inflammatory drugs, decongestants, stimulants (including diet pills), oral contraceptives, and anabolic steroids.

A review of systems approach can alert one to the presence of secondary hypertension as well as to the presence of end-organ damage. In parallel, the examination should concentrate on endocrine, renal, and circulatory function. This includes palpation of the thyroid and possibly enlarged polycystic kidneys, auscultation for carotid, subclavian, renal, or femoral artery bruits, and assessment of peripheral pulses. The presence of arteriolar end-organ damage can readily be assessed with a thorough retinal examination.

Determination of cardiac hypertrophy may be a confusing problem in athletes owing to physiologic hypertrophy resulting from athletic training. The cardiac silhouette may be at the upper limits of normal on chest x-ray, and routine 12-lead electrocardiograms (ECGs) may show evidence of left ventricular hypertrophy by accepted voltage criteria. Even echocardiographic studies may show hypertrophy that is indistinguishable from LVH [8, 40, 45].

Laboratory studies should include a urinalysis and measurements of electrolytes, blood urea nitrogen (BUN), and creatinine. Specialized laboratory studies are usually unnecessary in patients with primary hypertension. However, renal- or endocrine-mediated secondary hypertension should be suspected if hypertension is new, rapid in onset, or resistant to routine therapy.

Specialized studies include plasma catecholamines, thyroid function, and aldosterone and renin levels. If renin-mediated renal vascular disease is suspected, the captopril "inhibition" test is highly sensitive in establishing the diagnosis of renal vascular disease [41]. Measurement of plasma renin 60 minutes after an oral dose of captopril (50 mg) will

TABLE 9B–5
Recommended Bladder Dimensions for Adult Blood Pressure Cuff

Arm Circumference (cm)	Cuff Name	Bladder Width (cm)	Circumference (cm)
17–26	Small adult	11	17
24–32	Adult	13	24
32–42	Large adult	17	32
42–50	Thigh	20	42

Adapted from Kirkendall, W. M., Feinleib, M., Freis, E. D., et al. Recommendations for human blood pressure determinations by sphygmomanometers. *Circulation* 62:1146A–1155A, 1980.

show significant increases in plasma renin (150%) from baseline levels. The presence of renal vascular hypertension is confirmed by use of a renal scan or arteriogram. Detailed diagnosis of suspected secondary hypertension should be done in a referral center.

TREATMENT

Nonpharmacologic Initial Approach to Mild Hypertension

Modification of all known contributory nutritional and behavioral factors should be pursued. These include inappropriately high dietary sodium intake, licorice consumption, alcohol abuse, tobacco in any form, and recreational or surreptitious ergogenic aids (amphetamines, cocaine, anabolic steroids, growth hormone, blood doping, and erythropoietin). Additionally, contraceptive agents used by women may contribute to their elevated blood pressure.

Restriction of sodium chloride (salt) and high dietary fat intake requires dietary consultation because many athletes are unaware of "hidden" salt sources in convenience foods. It should be emphasized that both sodium and chloride are now implicated in salt-mediated hypertension (not just sodium alone) [25]. It is not possible to predict accurately which patients have salt-sensitive hypertension; thus, a trial of restricted salt intake is usually warranted as an initial therapeutic intervention. Approximately 50% of young adults with hypertension will respond to this intervention [51]. Restriction of sodium chloride intake to 5 g daily ($= 2$ g Na^+ or 88 mmol) is recommended as a general measure unless sweat loss of sodium chloride is excessive due to warm weather training [24]. Increasing the intake of calcium and potassium has been recommended as a possible form of therapy for hypertension. The blood pressure reduction in most studies of these therapies is modest [21, 37].

Obesity is an important factor in most young adults with hypertension [42]. This is usually not a problem in most athletes, although some "power event" athletes may have excessive subcutaneous fat stores. Ideal weight goals should be based on subcutaneous fat measurements obtained with skin calipers or underwater weighing, if available.

Aerobic exercise is also recommended for treatment of mild hypertension [2, 10, 18]. Since most athletes exercise in excess of the usually recommended levels, additional exercise therapy may be irrelevant. However, for some power event athletes,

additional low-level aerobic training such as jogging or cycling for 30 minutes daily may provide some benefit [18].

Pharmacologic Therapy

The choice of drug therapy for treatment of hypertension in athletes must be individualized and carefully monitored for potential side effects. Recommendations for hypertensive therapy developed in the late 1970s emphasized a "stepped-care" approach in which all patients were routinely treated with a diuretic followed by a beta-blocking agent or alpha-blocking agent as necessary. This approach has now been replaced by individualized monotherapy based on the physiologic characteristics of the patient's hypertension. Routine use of diuretics as initial therapy for hypertension is now discouraged because of the multiple metabolic side effects of these agents.

Table 9B–6 provides a summary of the five major classes of antihypertensive agents along with comments about their hemodynamic and metabolic actions and their potential adverse side effects on sports training or competition. The major emphasis is on appropriate choice of drug class. Individual preparations will be discussed briefly according to their specific advantages or disadvantages for hypertensive therapy in athletes. Recent reviews on this topic are available in references 4, 23, and 34.

Diuretics

Diuretics have been a mainstay of antihypertensive therapy for several decades. Their mechanism of action consists of inhibiting sodium and water uptake in the proximal kidney tubules. This results in decreased plasma volume and decreased cardiac output with a lowering of blood pressure. Systemic vascular resistance is unchanged acutely, but with long-term treatment it gradually declines while plasma volume may return toward normal. Diuretics are most effective in volume- and salt-dependent hypertension and may be used in low doses to augment the antihypertensive effect of other agents such as vasodilators and angiotensin-converting enzyme (ACE) inhibitors. It should be noted that diuretics are the least effective class of antihypertensive agents for control of exercise-induced hypertension [23, 34]. In addition, control of blood pressure with diuretics does not reverse ventricular hypertrophy associated with hypertension [9].

Diuretic therapy also increases the urinary loss of

TABLE 9B–6
Summary of Major Classes of Antihypertensive Agents and Potential Effects on Exercise and Sports Participation

Drug Classes	Hemodynamic Effects	Metabolic and CNS	Effects on Exercise
1. Diuretics Thiazide (hydrochlorothiazide)		Urinary loss of K^+, Mg^{++}	Hypovolemia, orthostatic hypotension
Loop inhibitors (Lasix)	↓ Plasma volume ↓ Cardiac output ↓ SVR (long-term) Note: little or no control of exercise hypertension	Increases in plasma cholesterol, glucose, uric acid	*Long-term:* Hypokalemia; hypomagnesemia, muscle weakness, cramps, possible rhabdomyolysis, possible arrhythmias
Potassium-sparing (triamterene, amelioride)		Reduced K^+ loss	
2. Beta-Adrenergic Blocking Agents Nonselective (beta-1 or beta-2) (propranolol)	↓ HR (20% to 30%) ↓ Contractility ↑ SVR (muscle and skin)	Inhibition of lipolysis and glycogenolysis; increased cholesterol; reduced HDL cholesterol; CNS depression (lipophilic beta blockers)	Significant loss of VO_2 max due to decreased cardiac output and skeletal muscle flow; impairment of substrate mobilization; earlier fatigue and lactate threshold; exercise bronchospasm may occur
Selective (beta-1) (atenolol, bisoprolol)	Less effect on beta-2 vasodilation	Less impairment of glycogenolysis and lipolysis	Less effect on bronchial smooth muscle
Combined (beta/alpha-1) (labetalol)	↓ SVR; less impairment of muscle blood flow		Combined beta/alpha-1 drug may be best choice if beta blockade is necessary
3. Alpha-Adrenergic Blocking Agents Peripheral (alpha-1) (prazosin, terazosin)	↓ SVR; marked orthostatic BP decrease after oral dose; only limited control of exercise hypertension	No major changes in energy metabolism	VO_2 max preserved; no major effect on training or sports performance
Central (alpha-2) (clonidine, guanabenz)	↓ SVR ↓ HR (minor)	Same; mild-moderate drowsiness, dry mouth (minimized with transcutaneous preparations)	Same as above
4. Vasodilators Direct (hydralazine, minoxidil)	↓ SVR ↑ HR, ↑ Q (reflex)	Headaches, flushing, fluid retention secondary to activation of renin-angiotensin system; lupus erythematosus reaction possible	Potential for competitive "steal" of muscle blood flow due to generalized vasodilation
Calcium channel blockers (nifedipine, diltiazem, verapamil)	↓ SVR ↑ HR (nifedipine) ↓ HR (verapamil)	Headaches, flushing, fluid retention (nifedipine), constipation (verapamil)	VO_2 max generally preserved; potential for competitive "steal" of muscle blood flow due to generalized vasodilation
5. Angiotensin-Converting Enzyme Inhibitors (captopril, enalapril, fosinopril)	↓ SVR No ↑ in HR	No major effect on energy metabolism; potential for ↑ K^+	No impairment of VO_2 max, training, or competition

Abbreviations: HR, heart rate; SVR, systemic vascular resistance; VO_2 max, maximal oxygen uptake; Q, cardiac output.

potassium and magnesium, leading to the undesirable side effects of hypokalemia and hypomagnesemia, both of which are associated with muscle weakness, cramps, and possible rhabdomyolysis. These side effects may occur during warm weather exercise and training such as football practice [24]. Another complication of potassium and magnesium depletion is cardiac arrhythmia [17]. The extent of this problem in sports is not known but could potentially contribute to unexplained sudden death during exercise.

If diuretics are selected for use as an antihypertensive agent, they should be used in low doses (hydrochlorothiazide 12.5 to 25 mg daily) and should be combined with a potassium-sparing agent such as triamterene. Loop diuretics such as furosemide (Lasix) are inappropriate for antihypertensive therapy and should be avoided.

Beta-Adrenergic Blocking Agents

Beta-adrenergic blocking agents reduce blood pressure by inhibiting cardiac contractility and heart rate. Systemic vascular resistance does not decrease and may thus contribute to impairment of regional blood flow to muscle and viscera [48]. These effects are most noticeable in nonselective beta-blocking agents such as propranolol, which inhibit both beta-1 and beta-2 adrenergic receptors. Beta-blocking agents also have a variety of undesirable effects on metabolism, including inhibition of glycogenolysis and lipolysis [31]. In addition, they cause an increase in serum cholesterol and a reduction in the high-density lipoprotein (HDL) cholesterol fraction.

Many studies have shown that acute and chronic beta blockade produces a significant loss of VO_2 max due to the combined effects of diminished heart rate, decreased cardiac output, and impaired muscle blood flow [20, 34, 48]. In addition, inhibition of normal substrate mobilization can lead to earlier fatigue and lactate accumulation with submaximal and maximal exercise [20, 31]. Beta blockade may also induce or worsen bronchospasm at rest and during exercise. Finally, lipophilic beta-blocking agents (propranolol, metoprolol) commonly produce depression, fatigue, and drowsiness by their direct effect on the central nervous system.

Some of the undesirable metabolic side effects of beta blockade are lessened with beta-1 selective beta-blocking agents (metoprolol, atenolol, bisoprolol). However, impairment of cardiac output and VO_2 max is similar in both selective and nonselective beta-blocking agents [34]. Thus, beta-adrenergic blocking agents are not recommended for the treatment of hypertension in athletes unless the patient shows evidence of hyperadrenergic function (tachycardia at rest, during postural change, or during submaximal exercise).

Antihypertensive agents that combine beta blockade and alpha blockade (beta/alpha-1; labetalol) offer a potentially useful combination that may be more appropriately applied to athletes. This class of agents produces beta blockade with effective reduction in systemic vascular resistance and less impairment of muscle blood flow and VO_2 max [33].

Alpha-Adrenergic Blocking Agents

Alpha-adrenergic blocking agents include those causing peripheral (alpha-1) blockade and those causing combined central and peripheral (alpha-2) blockade. Prazosin and doxazosin are competitive inhibitors of peripheral alpha-1 receptors in vascular smooth muscle. These drugs result in a decrease in systemic vascular resistance with little or no reflex increase in heart rate or cardiac output. Alpha I agents may produce a marked orthostatic blood pressure decrease after an initial oral dose. This response gradually attenuates after prolonged use. Minor fluid retention may develop due to decreased renal perfusion.

Treatment with alpha-1 blockade may enhance the release of free fatty acids and glucose during exercise. VO_2 max is maintained, and there appear to be no reported effects on exercise performance [4].

The centrally acting alpha-2 adrenergic blocking agents (clonidine, guanabenz) also reduce systemic vascular resistance by modulating sympathetic outflow from medullary nuclei [44]. Clonidine also has no major effects on energy metabolism. Oral clonidine commonly produces mild-to-moderate drowsiness, dry mouth, and impotence. In addition, abrupt discontinuation of oral clonidine may produce a transient "rebound" increase in sympathetic activity. These effects are significantly minimized with transcutaneous clonidine. Several studies have reported satisfactory control of blood pressure responses to exercise with no impairment of VO_2 max in patients treated with weekly patches of transcutaneous clonidine [7].

Vasodilators

Direct-acting vasodilators (hydralazine, minoxidil) are occasionally used for management of an acute hypertensive emergency. They reduce systemic vascular resistance by causing direct relaxation of vascular smooth muscle. There is a secondary increase in heart rate and cardiac output due to activation of baroreflexes. Activation of the renin-angiotensin system and fluid retention occur during longer-term management. Hydralazine may also cause a vasculitis-arthritis syndrome similar to that seen in lupus erythematosus. These wide-ranging side effects limit the usefulness of direct vasodilators for long-term management of hypertension. Finally, the direct-acting vasodilators do not reverse ventricular hypertrophy associated with hypertension.

Calcium Channel Blockers

Calcium channel blockers include a family of agents with different molecular configurations. They all act to inhibit the calcium slow-channel conductance in vascular smooth muscle. This produces a generalized vasodilation and corresponding decrease in blood pressure. Calcium channel blockers are useful for acute and long-term management of hypertension. Satisfactory blood pressure control at rest and during exercise has been documented for all preparations currently available [4, 23, 34]. These include nifedipine, diltiazem, and verapamil. There are some minor differences in hemodynamic patterns that may be important in the treatment of athletes. Nifedipine is the most powerful vasodilator and causes some reflex increase in heart rate at rest and during submaximal exercise. Nifedipine also produces mild lower extremity edema and vascular headaches. Verapamil and diltiazem produce mild sinoatrial and atrioventricular junctional conduction block in addition to vasodilation, so that reflex increases in heart rate are minimal. Minor impairment of maximal heart rate may occur in some individuals treated with verapamil or diltiazem [4, 23, 34].

Calcium channel blockers have no major metabolic or renal side effects. However, there is a potential for a competitive "steal" of muscle blood flow during continued exercise due to vasodilation of nonexercising muscle and other tissues. Several studies have indicated an earlier onset of lactate threshold with a corresponding reduction in VO_2 max, which was attributed to competition between blood flow to exercising and nonexercising regions [5]. Calcium channel blockers are effective in reversing ventricular hypertrophy induced by hypertension [39].

Angiotensin-Converting Enzyme Inhibitors

ACE inhibitors reduce blood pressure by causing competitive inhibition of angiotensin-converting enzyme in circulating plasma and also in the vascular smooth muscle wall. ACE inhibitors are most effective in patients with hypertension associated with activation of the renin-angiotensin axis. However, accumulated experience has shown that ACE inhibitors may also be effective in patients with normal plasma renin levels. ACE inhibitors are also effective in reversing ventricular hypertrophy associated with hypertension [39].

ACE inhibitors produce a prompt decrease in blood pressure with little or no reflex increase in heart rate. There is a potential for significant orthostatic hypotension after an initial dose of an ACE inhibitor, particularly in sodium-depleted states. Therefore, blood pressure must be monitored frequently after an initial dose.

ACE inhibitors have no major effect on energy metabolism. However, serum potassium may increase owing to a decrease in aldosterone action on the distal kidney tubule. Thus, potassium supplements should be avoided in patients treated with ACE inhibitors.

Nonproductive coughing and exacerbation of bronchospasm may occur in some individuals treated with ACE inhibitors. (This response may be related to simultaneous blockade of bradykininase by ACE inhibitors, which allows bradykinin levels to increase in the lungs.)

Available studies have shown that ACE inhibitors produce adequate control of exercise blood pressure and no impairment of VO_2 max [14]. Thus, in selected athletes ACE inhibitors provide an excellent pharmacologic profile for treatment of mild to moderate hypertension. The effectiveness of ACE inhibitors may be improved with the addition of low-dose diuretic therapy.

Summary

Mild to moderate hypertension in athletes should be treated initially with nonpharmacologic intervention for 3 to 6 months. Clinic and home blood pressures should be monitored frequently during this trial period. Treatment with low-dose antihypertensive agents should be considered if the blood pressure fails to respond to nondrug therapy.

The initial choice of monodrug therapy should be a calcium channel blocker, an alpha-1 or alpha-2 blocker, or an ACE inhibitor. Treatment should be initiated at a low dose and observed for effectiveness over a period of 6 to 8 weeks. Diuretic therapy may be used as an initial treatment in patients in whom salt sensitivity or volume-dependent hypertension is suspected. Low-dose diuretics may also facilitate the antihypertensive effects of calcium channel blockers, alpha-adrenergic blockers, and ACE inhibitors. Serum potassium levels should be evaluated at intervals of 2 to 4 months in patients on ACE inhibitors, diuretics, or potassium-sparing diuretics.

In some instances, beta blockers may be indicated if there is evidence of hyperadrenergic activity (resting tachycardia, and so on). A combined beta/alpha-1 agent such as labetalol is recommended in these cases.

When choosing therapy in collegiate or Olympic caliber athletes, one should be familiar with the substances banned by the National Collegiate Athletic Association (NCAA) and the International Olympic Committee [49]. In general, diuretics are banned given their potential to mask the urinary detection of other banned substances and their potential use in helping the athlete artificially "make weight." Beta-adrenergic blocking drugs are banned only in precision events (all archery and shooting events including biathlon, skating, luge, bobsled, ski jumping, equestrian events, fencing, gymnastics, sailing, diving, and synchronized swimming).

Antihypertensive therapy may be withdrawn or reduced after a period of 6 months if satisfactory control of blood pressure is attained. Several studies have indicated that a small but significant number of patients will remain normotensive after blood pressure is adequately controlled [29]. The reason for this normalization is not fully understood but may involve reestablishment of baroreceptor control mechanisms.

CRITERIA FOR SPORTS PARTICIPATION

Guidelines for participation by hypertensive patients in individual or team sports is currently based on accumulated clinical experience and objective criteria of blood pressure control and the presence or absence of end-organ involvement. In 1984, a Bethesda Conference Task Force [14] developed guidelines for sports participation by athletes with arterial hypertension; these are summarized in Table 9B–7. The Task Force classified sports according to intensity (low, moderate, high) and type of exercise performed (dynamic and isometric). In general, patients with well-controlled hypertension (blood pressure less than 140/90 mm Hg) who show no evidence of end-organ involvement may participate in moderate- to high-intensity dynamic sports. Participation in sports with a high isometric exercise component is recommended only after careful evaluation and discussion of the potential risks with the patient. Recommendations for additional testing such as treadmill exercise testing or isometric handgrip evaluation were not specifically recommended but would be helpful in such cases.

Patients with mild to moderate hypertension that is not well controlled (blood pressure over 140/90) or who show evidence of end-organ involvement should be limited to low-intensity sports activity. Some patients in this group may also engage in moderate- to high-intensity dynamic sports that have a low isometric component.

Some patients with identified secondary hyperten-

TABLE 9B–7
Guidelines for Sports Participation in Athletes with Hypertension

Clinical Category	Blood Pressure Treatment Status	End-Organ Involvement	Recommendation for Sports Participation
Borderline (labile)	Nonpharmacologic (with follow-up)	None	No restriction
Mild to moderate	Controlled (≤140/90)	None	No restriction for moderate- to high-intensity sports Should limit isometric training or sports
	Not controlled (>140/90) or	None	Limit sports participation to low-intensity dynamic activities
Moderate to severe (previously)	Controlled	Present	Avoid isometric training or sports
	Currently controlled	None	Low-intensity dynamic sports participation (individual patients may participate in moderate- to high-demand aerobic sports)
Secondary hypertension (renal origin)	Controlled	Renal	Low-intensity sports participation Avoid "body collision" sports

Low-intensity sports include bowling, golf.

Moderate- to high-intensity dynamic sports include baseball, basketball, field hockey, distance running, soccer, swimming, tennis, volleyball.

Moderate- to high-intensity isometric sports include field (throwing) events, gymnastics, weight lifting, wrestling.

Moderate- to high-intensity dynamic sports with significant isometric component include rowing at race-pace, sprinting, football, downhill skiing.

Data from Frohlich, E. D., Lowenthal, D. T., Miller, H. S., et al. Task force IV. Systemic arterial hypertension. *J Am Coll Cardiol* 6:1218–1221, 1985.

sion due to kidney disease (polycystic kidneys, renal parenchymal disease, or a single kidney) should be limited to low-intensity competitive sports in which there is no danger of body collision or sudden impact.

The presence of end-organ involvement is based on physical examination and laboratory data outlined earlier under Clinical Evaluation. Three major criteria for end-organ involvement include (1) arteriolar narrowing with arterial-venous nicking and evidence of retinal hemorrhage, (2) the presence of left ventricular hypertrophy on electrocardiographic or echocardiographic evaluation, and (3) abnormal renal function as evidenced by proteinuria or diminished renal clearance studies. The presence of any one or a combination of these factors requires careful consideration for participation in sports activities.

It should be noted that the Task Force emphasized that there is no available evidence to indicate that there is an immediate risk related to participation in more intensive competitive sports by patients with target-organ involvement whose blood pressure at rest is well controlled by therapy. One may want to further document control of blood pressure during the usual sport participation, either in a simulated laboratory setting or in the field employing an ambulatory monitor. Close medical follow-up is crucial.

AUTHORS' PREFERRED METHOD OF TREATMENT

A schematic diagram for the treatment of suspected hypertension in athletes is summarized in Figure 9B–3.

1. Confirmation of hypertension is the first and most important step in this process. Blood pressure should be measured both in the clinic and at home over a period of 2 weeks or more. Selection of proper cuff size is essential to avoid inaccurate measurement and false hypertensive readings. Ambulatory blood pressure monitoring should be considered (when available) if clinic and home blood pressure readings seem unreliable.

2. Physical examination should be performed with emphasis on potential end-organ involvement. A detailed examination of the retinal vessels and auscultation of the carotid, aortic-renal, and lower extremity vessels should be included. Determining the presence of left ventricular hypertrophy may be difficult in highly trained athletes. Chest x-rays and

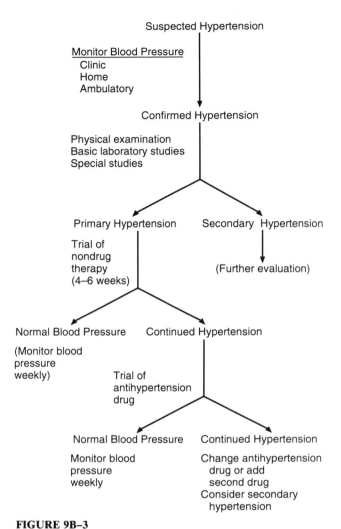

FIGURE 9B–3
Clinical approach to diagnosis and treatment of hypertension in athletes.

echocardiographic studies are usually not warranted during the initial evaluation.

3. Basic laboratory work should include urinalysis, renal function studies (BUN and creatinine), and electrolytes. Specialized studies of catecholamines, renin, aldosterone, thyroxin, or other hormones should be reserved for patients with possible secondary hypertension. In these cases, athletes should be referred to a specialist for evaluation.

4. Nondrug therapy should be utilized initially for 4 to 6 weeks. These measures include moderate restriction of salt intake, weight loss based on percentage of body fat, and avoidance of stimulants and alcohol. Blood pressure should be monitored twice daily at home and during weekly clinic visits. If nondrug treatment is successful, blood pressure self-monitoring should continue at a reduced frequency.

5. Antihypertensive drug therapy should be ini-

tiated in athletes who fail to respond to a trial of nondrug therapy. The details of antihypertensive therapy have been outlined earlier under Treatment. Initial drug dosage should be in the low range of the recommended adult dosage. Home blood pressure self-monitoring and clinic blood pressure monitoring should be continued as outlined for nondrug therapy. Patients who show normalization of blood pressure may be continued on maintenance antihypertensive therapy with periodic monitoring of blood pressure and pertinent laboratory studies in those on diuretic therapy. Withdrawal of antihypertensive therapy may be attempted to see if normalization of blood pressure is maintained.

Athletes showing continued evidence of hypertension while on antihypertensive therapy may require an alternate drug or an additional medication. Failure to respond to appropriate changes in antihypertensive therapy should raise the question of possible secondary causes of hypertension.

It should be emphasized that many athletes fail to receive proper follow-up care for hypertension after participation in sports has ended. Athletes with hypertension should be advised of the need for long-term follow-up and management within their local community.

References

1. Abboud, F. The sympathetic system in hypertension. *Hypertension* 4(Suppl 2):II208–II225, 1982.
2. Cade, R., Mars, D., Wagemaker, H., et al. Effect of aerobic exercise training on patients with systemic arterial hypertension. *Am J Med* 77:785–790, 1984.
3. Chaney, R. H., and Eyman, R. K. Blood pressure at rest and during maximal dynamic and isometric exercise as predictors of systemic hypertension. *Am J Cardiol* 62:1058–1061, 1988.
4. Chick, T. W., Halperin, A. K., and Gacek, E. M. The effect of antihypertensive medications on exercise performance: A review. *Med Sci Sports Exerc* 20:447–454, 1988.
5. Choong, C. Y. P., Roubin, G. S., Wei-Feng, S., Harris, P. J., and Kelly, D. T. Effects of nifedipine on systemic and regional oxygen transport and metabolism at rest and during exercise. *Circulation* 71:787–796, 1985.
6. Cottier, C., Julius, S., Gajendragadkar, S. V., and Schork, A. Usefulness of home BP determination in treating borderline hypertension. *JAMA* 248:555–558, 1982.
7. Davies, S. F., Graif, J. L., Husebye, D. G., Maddy, M. M., McArthur, C. D., Path, M. J., O'Connell, M., Iber, C., and Davidman, M. Comparative effects of transdermal clonidine and oral atenolol on acute exercise performance and response to aerobic conditioning in subjects with hypertension. *Arch Intern Med* 49:1551–1556, 1989.
8. Devereux, R. B., Pickering, T. G., Harshfield, G. A., Kleinert, H. D., Denby, L., Clark, L., Pregibon, D., Jasm, M., Kleiner, B., Borer, J. S., and Laragh, J. H. Left ventricular hypertrophy in patients with hypertension: Importance of blood pressure response to regularly recurring stress. *Circulation* 68:470–476, 1983.
9. Drayer, J. I. M., Gardin, J. M., Weber, M. A., and Aronow, W. S. Changes in ventricular septal thickness during diuretic therapy. *Clin Pharmacol Ther* 75 (Suppl. 3A):283–288, 1982.
10. Duncan, J. J., Farr, J. E., Upton, S. J., Hagan, R. D., Oglesby, M. E., and Blair, S. N. The effects of aerobic exercise on plasma catecholamines and blood pressure in patients with mild essential hypertension. *JAMA* 254:2609–2613, 1985.
11. Egan, B. Neurogenic mechanisms initiating essential hypertension. *Am J Hypertension* 2:357S–362S, 1989.
12. Esler, M., Jennings, G., Korner, P., Willett, I., Dudley, F., Hasking, G., Anderson, W., and Lambert, G. Assessment of human sympathetic nervous system activity from measurements of norepinephrine turnover. *Hypertension* 11:3–20, 1988.
13. Freed, D. L., Banks, A. J., Longson, D., and Burley, D. M. Anabolic stenosis in athletes: Crossover double-blind trial on weightlifters. *Br Med J* 2:471–473, 1975.
14. Frohlich, E. D., Lowenthal, D. T., Miller, H. S., Pickering, T., and Strong, W. B. Task force IV. Systemic arterial hypertension. *J Am Coll Cardiol* 6:1218–1221, 1985.
15. Gifford, R. W., Kirkendall, W., O'Connor, D. T., and Weidman, W. AHA Scientific Council Special Report: Office evaluation of hypertension. *Circulation* 79:721–731, 1989.
16. Hanson, P., and Nagle, F. Isometric exercise, cardiovascular responses in normal and cardiac populations. *Cardiology Clin* 5:157–170, 1987.
17. Hollifield, J. W., and Slaton, P. E. Thiazide diuretics, hypokalemia and cardiac arrhythmias. *Acta Med Scand* Suppl. 647:67–73, 1986.
18. Jennings, G., Nelson, L., Nestel, P., Esler, M., Korner, P., Burton, D., and Bazelmans, J. The effects of changes in physical activity on major cardiovascular risk factors, hemodynamics, sympathetic function, and glucose utilization in man. *Circulation* 73:30–40, 1986.
19. Joint National Committee on the Detection, Evaluation, and Treatment of High Blood Pressure. Report. *Arch Intern Med* 148:1023–1038, 1988.
20. Kaiser, P. Physical performance and muscle metabolism during beta adrenergic blockade in man. *Acta Physiol Scand* Suppl. 536:1–44, 1984.
21. Kaplan, N. M., Carnegie, A., Raskin, P., Heller, J. A., and Simmons, M. Potassium supplementation in hypertensive patients with diuretic-induced hypokalemia. *N Engl J Med* 312:746–749, 1985.
22. Kirkendall, W. M., Feinleib, M., Freis, E. D., and Mark, A. L. Recommendations for human blood pressure determinations by sphygmomanometers. *Circulation* 62:1146A–1155A, 1980.
23. Klaus, D. Management of hypertension in actively exercising patients: Implications for drug selection. *Drugs* 37:212–213, 1989.
24. Knochel, J. P., Dolin, L. N., and Hamburger, R. J. Pathophysiology of intense physical conditioning in a hot climate: Mechanisms of potassium depletion. *J Clin Invest* 51:242–255, 1972.
25. Kurtz, T. W., Hamoudi, A. A-B., and Morris, R. C. Salt-sensitive essential hypertension in men. Is the sodium ion alone important? *N Engl J Med* 317:1043–1048, 1987.
26. Kwan, C. Y. Dysfunction of calcium handling by smooth muscle in hypertension. *Can J Physiol Pharmacol* 63:366–374, 1985.
27. Laragh, J. H., et al. The renin axis and vasoconstriction volume analysis for understanding and treating renal hypertension. *Am J Med* 68:4–13, 1975.
28. Lehmann, M., Durr, H., Meikelbach, H., and Schmid, A. Hypertension and sports activity: Institutional experience. *Clin Cardiol* 13:197–208, 1990.
29. Levinson, P. D., Khatri, I. M., and Freis, E. D. Persistence of normal blood pressure after withdrawal of drug treatment in mild hypertension. *Arch Intern Med* 142:2265–2268, 1982.
30. Lewin, A., Blaufox, M. D., Castle, H., Entwisle, G., and Langford, H. Apparent prevalence of curable hypertension in the Hypertension Detection and Follow-up Program. *Arch Intern Med* 145:424–427, 1985.

31. Lundborg, P. H., Astrom, C., Bengtsson, C., Fellenius, E., and Von Schenck, H. Effect of β-adrenoreceptor blockade on exercise performance and metabolism. *Clin Sci* 61:299–305, 1981.
32. Lund-Johansen, P. Hemodynamics in essential hypertension. *Clin Sci* 59:3435–3545, 1980.
33. Lund-Johansen, P. Short and long-term hemodynamic effect of labetalol in essential hypertension. *Am J Med* 75:24–31, 1983.
34. Lund-Johansen, P. Exercise and antihypertensive therapy. *Am J Cardiol* 59:98A–107A, 1987.
35. Luscher, T. F., Diederich, D., Buhler, F. R., and Vanhoutte, P. M. Interactions between platelets and vessel walls. Role of endothelium-derived vasoactive substances. *In* Laragh, J. G., and Brenner, B. M. (Eds.), *Hypertension: Pathophysiology, Diagnosis and Management.* New York, Raven Press, 1990, pp. 637–648.
36. Mancia, G., Bertinieri, G., Grassi, G., Parati, G., Pomidossi, B., Ferrari, A., Gregorini, L., and Zanchetti, A. Effects of blood pressure measurements by the doctor on patient's blood pressure and heart rate. *Lancet* 2:695–698, 1983.
37. McCarron, D. A., and Morris, C. D. Blood pressure response to oral calcium in persons with mild or moderate hypertension. *Ann Intern Med* 103:825–831, 1985.
38. McMahon, S. W., Cutler, J. A., and Furberg, C. D. The effects of drug treatment for hypertension on morbidity and mortality from cardiovascular disease. A review of randomized controlled trials. *Prog Cardiovasc Dis* 2(Suppl.):99–118, 1986.
39. Messerli, F. H. Reduction of left ventricular hypertrophy: A goal of antihypertensive therapy. *Cardiovasc Rev Rep* 10(8)(Suppl.):14–18, 1989.
40. Morganroth, J., Maron, B., and Henry, W. L. Comparative left ventricular dimensions in trained athletes. *Ann Intern Med* 82:521–524, 1975.
41. Muller, F. B., Sealy, J. B., Case, D. B., Atlas, S. A., Pickering, T. G., Pecker, M. S., Preibisz, J. J., and Laragh, J. H. The captopril test for identifying renovascular disease in hypertensive patients. *Am J Med* 80:635–637, 1986.
42. Peterson, H. R., Rothschild, M., Weinberg, C. R., Fell, R. D., McLeish, K. R., and Pfeifer, M. Body fat and the activity of the autonomic nervous system. *N Engl J Med* 318:1077–1082, 1988.
43. Pickering, T. G., Harshfield, G. A., Kleinert, H. D., Blank, S., and Laragh, J. H. Blood pressure during normal daily activities, sleep, and exercise: Comparison of values in normal and hypertensive subjects. *JAMA* 247:992–996, 1982.
44. Punnen, S., Urbanski, R., Krieger, A. J., and Sapru, H. N. Ventrolateral medullary pressor area: Site of hypotensive action of clonidine. *Brain Res* 422:336–346, 1987.
45. Reichek, N., and Devereux, R. B. Left ventricular hypertrophy: Relationship of anatomic, echocardiographic and electrocardiographic findings. *Circulation* 63:623–632, 1981.
46. Seals, D. R., Washburn, R. H., and Hanson, P. G. Increased cardiovascular response to static contraction of larger muscle groups. *J Appl Physiol* 54:434–437, 1983.
47. Tesch, P. A., and Kaiser, P. Effects of β-blockade on O_2 uptake during submaximal and maximal exercise. *J Appl Physiol* 54:901–905, 1983.
48. Trap-Jensen, J., Causen, J. P., Noer, I., et al. The effects of beta adrenoreceptor blockade on cardiac output, liver blood flow and skeletal muscle flow in hypertensive patients. *Acta Physiol Scand* Suppl. 440:30, 1976.
49. United States Olympic Committee. Guide to banned medications. Sportsmediscope: USOC Sports Medicine and Science Division Newsletter 7:1–5 1988 (Hotline 1-800-233-0393, Colorado Springs, CO).
50. Walther, R. J., and Tifft, C. P. High blood pressure and the competitive athlete: Guidelines and recommendations. *Physician Sports Med* 13:93–113, 1985.
51. Weinberger, M. H., Miller, J. D., Luft, F. C., Grim, C. E., and Fineberg, N. S. Definitions and characteristics of sodium sensitivity and blood pressure resistance. *Hypertension* Suppl. 8:II-127–II-134, 1986.
52. Wilson, N. V., and Meyer, B. M. Early prediction of hypertension using exercise blood pressure. *Prev Med* 10:62–68, 1981.
53. Zachariah, P. K., Sheps, S. G., and Smith, R. L. Clinical use of home and ambulatory blood pressure monitoring. *Mayo Clin Proc* 64:1436–1446, 1989.

Management of Exercise-Induced Asthma

Nizar N. Jarjour, M.D.
Robert F. Lemanske, Jr., M.D.

Asthma is a very common disease affecting about 7% of Americans [21] and has a prevalence and morbidity that, for unclear reasons, seem to have increased over the past few years. It is generally accepted that asthma is an airway disease characterized by an increased responsiveness of the tracheobronchial tree to a variety of stimuli resulting in a widespread narrowing of the airways that changes in severity either spontaneously or as a result of therapy. A variety of stimuli are known to provoke asthmatic symptoms including allergens, viral infections, cigarette smoke, air pollutants, cold air, and exercise. Asthma attacks are characterized by wheezing, cough, chest tightness, and shortness of breath. Occasionally, cough can be the only manifestation of an asthma attack.

Exercise-induced asthma (EIA) is the occurrence of an asthma attack during or shortly after exercise. In many patients, exercise is just one of many factors triggering asthma; in others it may be the only mechanism provoking asthmatic attacks. Although EIA can be troublesome, especially when misdiagnosed, once it has been identified, the treatment is usually effective in controlling the symptoms and enabling patients to achieve their full potential in activities ranging from weekend jogging to Olympic competition.

CHARACTERISTICS

The Immediate Response

Exercise-induced asthmatic symptoms most commonly occur within the first 15 minutes after exercise cessation. In addition to this immediate response,

symptoms occurring hours after resolution of the immediate response have also been reported. In general, the immediate or early exercise-induced asthmatic response starts during or shortly after cessation of vigorous exercise, peaks in 5 to 10 minutes, and improves gradually with complete resolution in 10 to 30 minutes (Fig. 9C–1). In some patients, however, EIA may become gradually

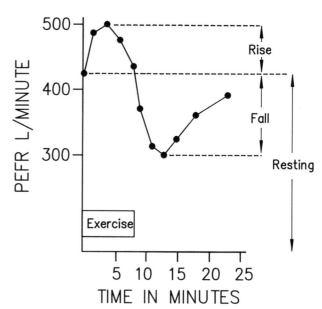

FIGURE 9C–1
Changes in peak expiratory flow rate (PEFR) recorded during and after 8 minutes of treadmill running in an asthmatic patient. Note that PEFR increased during the first minutes of exercise, then decreased gradually to reach the nadir 5 to 10 minutes after cessation of exercise, and rose back to near baseline level 20 minutes later. (From Anderson, S., Seale, J.P., Ferris, L., et al. An evaluation of exercise-induced asthma. *J Allergy Clin Immunol* 64:612–624, 1979.)

worse for 30 to 60 minutes before it starts to improve. The degree of airway obstruction can be quantified using a peak flow meter, a simple hand-held instrument that measures the greatest expiratory flow, which is called the peak expiratory flow rate (PEFR). Alternatively, a spirometer can be used; in addition to PEFR, spirometry will provide such data as forced expiratory volume in the first second of expiration (FEV_1) and forced vital capacity (FVC). The reduction in PEFR, FEV_1 and the FEV_1/FVC ratio are used to assess the intensity of airway obstruction that may occur following exercise.

When PEFRs are serially evaluated in asthmatics during exercise, they are found to increase with the start of exercise, probably secondary to the release of endogenous catecholamines, and subsequently decrease, reaching a low point 5 to 10 minutes following exercise (Fig. 9C–1). The incidence of EIA varies depending on the type of exercise, its intensity and duration, the condition of the inspired air, and the baseline pulmonary function [30, 63].

The athlete's response to increasing duration and intensity of exercise includes an increased heart rate (HR) and increased oxygen consumption ($\dot{V}O_2$), the latter measurement representing the amount of oxygen (in milliliters) used by the body metabolism (per minute). Both of these parameters can be used to assess the magnitude of the exercise response. Heart rate is easier to measure and simpler to follow continuously during exercise. Generally, in patients without heart disease, the maximum heart rate reached during exercise is related to age. The expected maximum heart rate for a subject is derived from a simple formula (220 − age [in years]); thus, maximum heart rate for a healthy 25-year-old person is 195 beats/minute.

It is frequently noted that a critical level of exercise intensity is needed to cause EIA; generally, to result in EIA, a given exercise must cause an increase in heart rate above 80% of the maximum predicted heart rate. Interrupted or mild exercise is less likely to cause EIA than continued or high-intensity exercise [24]. Although the incidence of EIA among subjects with asthma varies from 50% to 80% in published reports, the majority (up to 90%) of patients with asthma will experience EIA following exercise under sufficiently stressful conditions [43]. Other airway diseases that are associated with airway hyperresponsiveness, such as cystic fibrosis, viral upper respiratory tract infections, and chronic bronchitis, are also associated with wheezing during and following exercise.

The Late Response

Late asthmatic responses have been described following allergen challenge, they consist of airway obstruction that develops 3 to 8 hours after resolution of the immediate response following allergen exposure. The pathophysiology of the late response is thought to involve pulmonary mast cell activation with subsequent release of potent chemoattractants that induce an influx of inflammatory cells to the airways [41]. In addition to allergen challenge, late responses have been reported to occur following exercise [6, 7]. Some investigators have found that up to 30% of subjects with EIA have both immediate and late responses [9], whereas others have considered the late response merely a reflection of diurnal variations in pulmonary function as opposed to a manifestation induced by the exercise stimulus itself [50]. When observed, the occurrence of the late response is not related to disease severity [9, 42], the magnitude of the early response [9], or the degree of histamine bronchial responsiveness [9]. In general, it is more common in children than adults [40], and some investigators have found that a slower rate of recovery from the early response may predict development of the late response in children with EIA [9, 42].

Serum levels of neutrophil chemotactic activity (NCA) have been shown to rise in parallel with changes in pulmonary function during the early and late asthmatic response in children [40]. Because NCA may be a mast cell–derived mediator, these findings have led to the assumption that mast cells are involved in the pathogenesis of EIA. The ability of cromolyn sodium, a drug that may inhibit mast cell mediator release, to attenuate EIA, has provided additional support for the potential participation of mast cells in the pathogenesis of EIA.

The Refractory Period

In some patients it has been noted that repeated periods of exercise over short periods of time are associated with a declining incidence and intensity of asthmatic symptoms. The interval of time in which this occurs has been called the refractory period. It is defined as the time after the occurrence of EIA during which less than half of the initial airway response is provoked by a second exercise challenge [53]. That is, when exercise is repeated during this period, the subject has less than a 50% fall in postexercise pulmonary function (FEV_1, and PEFR) compared to that recorded following the first exercise challenge. Approximately half of all patients with EIA have such a refractory period of about 1 hour [2].

The mechanism inducing the refractory period is unestablished. Interestingly, patients do not show a change in histamine airway responsiveness during the refractory period [28] and can still develop airway obstruction if exposed to an appropriate allergen [68]. The latter observation suggests that degranulation of mast cells following exercise is not a factor in inducing a refractory state.

It is not always necessary to induce a perceptible asthmatic attack to initiate a refractory period. Indeed, small multiple sprints of 30 seconds in duration, performed 2 minutes apart for 30 minutes prior to exercise, can result in a reduction in the severity of EIA [52]. This capacity for induction of a refractory period can be used to benefit the asthmatic athlete. The "prescription" for warm-up should be individualized based on the patient's age, the level of fitness, and the type of activity he or she is engaged in. If a successful warm-up routine can be established, it can be used prior to the start of exercise in addition to the prophylactic use of medication for optimal control of EIA.

PATHOGENESIS

The pathophysiology of EIA is not yet fully established. During the past two decades, however, several hypotheses have emerged including heat and water loss, airway rewarming, and the release of mediators of immediate hypersensitivity.

Heat and Water Loss

It has been recognized that certain types of exercise, such as jogging and cycling, are more likely to induce EIA than others, such as swimming, and that EIA is usually more severe when asthmatics perform physical activity while breathing cold air. Chen and associates [10] reported that EIA could be prevented if subjects breathed fully saturated air at body temperature and suggested that EIA was related to heightened water and heat loss secondary to hyperventilation during exercise. Strauss and colleagues [63] found that exercise and breathing cold air are additive in producing limitation of air flow in subjects with asthma. Later, by directly recording the air stream temperature in the tracheobronchial tree in normal subjects, McFadden and associates [44] demonstrated a significant fall in temperature during hyperventilation of cold air that was not localized

to one region of the respiratory tract but varied with the rate and depth of respiration and the conditions of inspired air. As the inspired air passes from the nose to the lungs, airways perform important tasks that include warming the inspired air to body temperature and fully saturating it with water. Thus, airways lose heat and water to condition inspired air. Although the upper airway performs the majority of this task, as the requirements for conditioning increase with hyperpnea, larger segments of the tracheobronchial tree contribute significantly as well. From these and other observations, McFadden and his colleagues proposed that respiratory heat loss, which is the total amount of energy lost to warm inspired air to body temperature, was the main stimulus for EIA. Deal and co-workers [12] found that the magnitude of EIA was proportional to the total thermal load that was placed on the airways, which is a combination of energy lost directly in heating inspired air and indirectly in humidifying it.

Other investigators have felt that water loss may be a more important factor than heat loss in inducing EIA during the air conditioning process. Hahn and colleagues [27] demonstrated the presence of significant EIA in asthmatic subjects who exercised while breathing warm dry air; these patients had airway water loss by evaporation with little or no direct heat loss during exercise. Hahn and his associates calculated that a critical respiratory water loss rate of 0.41 ± 0.09 mL/minute was needed to induce EIA. Anderson [1] suggested that evaporative water loss was the pivotal mechanism for EIA and proposed that it acted as a stimulus for EIA by causing a transient increase in the osmolarity of the epithelial lining fluid that in turn caused airway obstruction. This hypothesis was further supported by the findings of Schoeffel and co-workers [54], who studied the airway response to breathing nebulized water and hypotonic, isotonic, and hypertonic saline. Marked airway obstruction was induced by hyper- and hypo-osmolar solutions in asthmatic subjects but not in normal volunteers, and no significant changes were seen with isotonic saline. Similar results were found by Eschenbacher and colleagues [19]. Furthermore, in an animal model, Boucher and his associates [8] demonstrated that tracheal epithelial fluid is hyperosmolar whereas distal airway bronchial epithelial fluid is iso- or hypo-osmolar, adding support to the water loss and airway drying hypothesis [1].

Although the relative importance of heat or water loss in provoking EIA remains controversial, it is obvious that the two occur simultaneously under most circumstances. When inspired air is colder, it is usually drier, since cold air has less capacity to carry water vapor. Thus, when breathing cold air, the airways have the dual task of warming up the inspired air to body temperature while saturating it fully with water vapor, in essence losing water and heat simultaneously. The amount of heat and water lost by the airways in the process of conditioning inspired air is related to both the volume of inspired air per breath and the breathing rate (i.e., minute ventilation). Thus, both water and heat loss are directly related to minute ventilation.

The biologic or physiologic pathways that transform airway heat or water loss into airway obstruction are not well established. It has been proposed that heat and water loss lead to reflex bronchospasm directly; however, this hypothesis does not explain why the peak of airway obstruction usually occurs 5 to 10 minutes after cessation of exercise, a time when heat and water loss have returned to baseline levels. Furthermore, heat and water loss cannot solely explain the refractory period or the late phase response that are known to occur in some patients with EIA. Thus, despite the temporal association of heat and water loss in triggering EIA, additional factors seem to be necessary to produce the clinical manifestations of EIA.

Airway Rewarming

Recent studies have shown that conditions of the inspired air during the recovery period immediately after exercise can affect the severity of EIA. McFadden and colleagues noted that breathing warm air, ambient or body temperature, during the recovery period following vigorous exercise produced greater airway obstruction immediately after exercise bronchoprovocation than breathing cold air in both normal and asthmatic adult subjects [45]. By directly recording the tracheobronchial tree temperature before, during, and after exercise in asthmatics and normals, they demonstrated that the magnitude of heat loss and airway cooling was similar in both groups. Asthmatic subjects, however, had a more rapid rise in air stream temperature after exercise had ceased, suggesting that they had a rebound hyperemia of the bronchial vascular bed, which may be responsible for EIA [22]. Smith and associates, using similar postexercise recovery conditions in asthmatic adolescent patients, were unable to reproduce these findings consistently [62].

Mediators of Hypersensitivity

Mediators of hypersensitivity are a group of potent cell products that are released by a variety of inflammatory cells (eosinophils, mast cells, and so on) in response to a variety of stimuli. Many of these mediators (e.g., histamine) are stored inside some of these inflammatory cells (e.g., histamine is stored in mast cells and basophils) and are released from these granules to the outside of the cell (degranulation). Once released, histamine can cause smooth muscle contraction, vasodilation, and local edema. Other mediators that are felt to be important in the pathogenesis of asthma include prostaglandins and leukotrienes.

It is well known that drugs that are thought to inhibit mast cell degranulation, such as cromolyn sodium, can attenuate EIA. Furthermore, mediator release from mast cells can be induced by changes in temperature or tonicity [16]. These facts led to the hypothesis that physical stimuli (airway dryness or cooling) produce EIA by inducing airway mast cells to release mediators (e.g., histamine) that are capable of causing airway obstruction [38]. Peripheral blood histamine levels have been found to be elevated [39] or unchanged from baseline [13] when measured during EIA. Interpretation of these assays is complicated by the presence of histamine-rich basophils in the peripheral blood that can lead to markedly elevated plasma histamine levels. A putative mast cell mediator, neutrophil chemotactic factor (NCF), has been found to be increased during EIA [39] but not after eucapnic hyperventilation-induced asthma. These findings suggest that EIA and hyperventilation-induced asthma may be caused by different mechanisms or that NCF release is merely related to exercise and is not important for the development of EIA. Therefore, despite some evidence of mast cell mediator release into peripheral blood, it cannot be clearly established that pulmonary mast cells are the source of these mediators nor that they are the pathway leading to EIA. Indeed, Broide and colleagues recently performed bronchoalveolar lavage in a group of seven asthmatics with EIA before and after exercise challenge and then measured the levels of mast cell mediators found in the lavage fluid. They found no increase in the levels of these mediators following exercise challenge; therefore, they concluded that mast cells and their mediators are not involved in the pathogenesis of EIA [7a].

DIAGNOSIS

Although the majority of patients with EIA have a known history of asthma and EIA, a small but significant group are not aware of their symptoms and are even missed on questionnaires specifically designed to elicit their identity [66]. A high degree of suspicion by teachers, trainers, coaches, and parents is needed to make the initial important step in identifying children and young adults with exercise-induced difficulties. Coughing, wheezing, chest tightness, shortness of breath, and inability to keep pace with peers all indicate the possibility of asthma as opposed to a general lack of physical fitness or poor motivation. As previously discussed, one should keep in mind that other airway diseases can enhance airway responsiveness and lead to exercise-induced airway obstruction, both on an acute (e.g., viral upper respiratory tract infections) basis and in chronic (e.g., cystic fibrosis) conditions.

Exercise challenge testing is of value in confirming the diagnosis, assessing the severity, and evaluating the efficacy of a given treatment in controlling EIA [15, 58]. An evaluation by a physician is strongly recommended prior to exercise challenge testing. It is advisable to obtain a pertinent history and perform a physical examination, placing special emphasis on the cardiopulmonary and musculoskeletal problems that may limit the performance of exercise testing or make it unsafe. When detection of EIA or assessment of its severity is the primary goal of the challenge, medications that may attenuate EIA should be withheld for a varying period of time depending on the preparation. In general, 24 to 48 hours for long-acting theophylline, 8 to 12 hours for inhaled beta agonists and cromolyn, and 24 to 48 hours for antihistamines are sufficient. Steroids can be continued provided that the dose has been stable for 3 to 4 weeks and with the understanding that they may attenuate the response to exercise. No patient should undergo exercise challenge during an exacerbation or following a recent worsening of his or her asthma.

Obviously, when the goal of exercise testing is to assess the efficacy of a given therapy, it should be continued with special emphasis on patient compliance. When the primary purpose is to detect latent asthma, methacholine bronchoprovocation may be a more sensitive test than exercise bronchoprovocation [56]. Methacholine bronchoprovocation (or challenge) is done by giving the subject who is suspected of having asthma a very dilute methacholine solution (0.04 mg/mL) by nebulizer (five full breaths) and checking pulmonary function 5 minutes after the dose. If there is no change in pulmonary function compared to prechallenge baseline values, the next dose of methacholine (higher concentration) is given, and pulmonary function is again checked 5 minutes later. This procedure is repeated until a predetermined fall in FEV_1 (usually 20% from baseline) is reached. The concentration of methacholine required to produce this effect is termed the PD_{20} (provocative dose required to drop a predetermined parameter of pulmonary function by 20%).

Several protocols for exercise bronchoprovocation have been adopted. Some investigators have used free running as a method of exercise challenge [57]; however, most laboratories utilize a treadmill or a cycle ergometer and conduct a protocol that consists of 6 to 8 minutes of steady-state exercise at high intensity (80% to 90% of maximum predicted heart rate) following a 2- to 4-minute warm-up at 50% to 70% maximum predicted heart rate [15, 17]. Continuous electrocardiographic (ECG) monitoring is strongly recommended for following heart rate response to exercise and for detecting potential arrhythmias. Monitoring oxygen consumption ($\dot{V}O_2$) can also be used to assess exercise intensity.

Measurements of pulmonary function should be made shortly before exercise to serve as a baseline, then 3 to 5 minutes after cessation of exercise and every 5 minutes thereafter for 20 to 30 minutes, and finally 1 hour after exercise termination. Spirometry is the most commonly used test; a drop in FEV_1 greater than 12% from baseline is considered an indication of EIA, and a drop of greater than 20% is more diagnostic and less variable when the test is done on multiple occasions under similar conditions [14]. Peak expiratory flow rates have also been used. The advantage of this measurement is that it can be taken using a portable meter that can be used in field studies by coaches and athletes themselves; however, it is more variable than the FEV_1. On the other hand, field studies have the added advantage of permitting the inclusion of other factors in addition to exercise intensity such as outdoor air, type of exercise, and preexercise warm-up [15]. Finally, plethysmography, which allows measuring airway resistance independent of effort, can be used to enhance sensitivity and permit evaluation in the individual who has difficulty in performing satisfactory spirometric maneuvers or who has short-lived bronchospasm that improves with the deep inspiration that is required to obtain spirometric or peak flow measurements.

TREATMENT

As reviewed, multiple mechanisms have been proposed for the pathogenesis of EIA. Based on these observations, treatment recommendations are summarized in Figure 9C–2. As diagrammed, the type of exercise, the use of warm-up exercises, the climatic conditions during and following exercise, and the medical therapy being used (see below) are all important considerations. Exercise performed in warm moist air with an adequate warm-up period would be the least likely to provoke EIA. Thus, swimming preceded by multiple short laps in the pool as a warm-up would be advisable for individuals who have a history of difficulty with running events. If a child or young adult prefers running sports, longer distances preceded by short sprints as a warm-up would be preferable to events such as the half-mile or mile run. For small children playing outdoors in winter, the use of a scarf or face mask during playtime would diminish respiratory heat and water loss [25]. Finally, as discussed below, the use of medications *prior to* exercise can also be very effective in attenuating or preventing EIA.

Factors relevant in evaluating the efficacy of various medications used to treat EIA include the drug dosage used, its timing and route of delivery, and the definition of an acceptable protective effect by a particular medication. Because normal individuals will not experience a drop in FEV_1 of 10% or more after exercise, some studies designate a drug as effective in controlling EIA if treatment is associated with at least a $\leq 10\%$ decrease in the postexercise FEV_1. Others consider a 50% reduction in the postexercise FEV_1 fall compared to pretherapy or no therapy to be significant protection against EIA. A statistically significant protective effect found in a given study should be interpreted in light of its clinical importance and appropriateness to the conditions of the athlete in question. In general, drugs that can be used in the management of EIA include beta-receptor agonists, theophylline, cromolyn sodium, anticholinergics, and corticosteroids. The overall pharmaceutical approach to the treatment of EIA is summarized in Figure 9C–3.

Beta-Receptor Agonists

Beta-receptor agonists given by the inhaled route prior to exercise are the treatment of choice for EIA [23] because they are effective in reducing or completely preventing EIA in 80% of patients. Beta-receptor agonists are bronchodilators as well as an effective treatment for EIA, so they are an optimal choice in individuals with preexisting airway obstruction at the time they plan to exercise. Newer preparations that have longer durations of action and more beta-2 receptor selectivity (beta-2 receptors are more numerous in the lung compared to beta-1 receptors) are generally recommended (e.g., albuterol, terbutaline, bitolterol, pirbuterol, fenoterol). Beta agonists have systemic effects outside

EIA TREATMENT

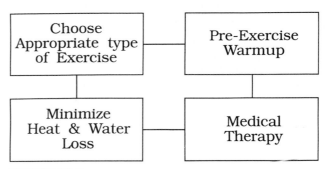

FIGURE 9C–2
Overall treatment approaches to individuals with exercise-induced asthma.

the lungs. Through interactions with cardiac beta-1 receptors, they can increase heart rate and contractility. Stimulation of similar receptors in the central nervous system can lead to respiratory stimulation, increased wakefulness, and increased psychomotor activity. These extrapulmonary effects can influence the performance of an athlete. Side effects can be minimized if the medications are inhaled rather than taken orally. Older, nonselective beta-1 agonists are banned in competitive sports, whereas selective beta-2 agonists are not [20a].

The ability to suppress EIA is not entirely related to the bronchodilatory effect of the drug [55]. Furthermore, the duration of protection from EIA achieved by a beta agonist appears to be shorter than the bronchodilatory effect [60, 61]. The activity of various beta-2 receptor agonists in preventing

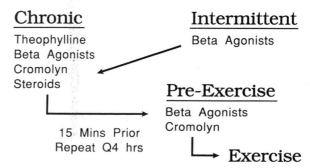

FIGURE 9C–3
Exercise-induced asthma. Step-wise medical therapy. The first step in medical therapy for EIA is optimal control of chronic asthma symptoms by the regular or intermittent use of bronchodilators with or without cromolyn or steroids. Preexercise prophylaxis is best achieved by using a beta-agonist inhaler (and cromolyn if needed) 10 to 15 minutes prior to exercise, repeating it every 2 to 4 hours.

EIA is summarized in Table 9C–1. Mechanisms by which beta agonists attenuate EIA include bronchodilation prior to exercise, enhanced exercise-associated bronchodilation, and suppression of mediator release or blockage of their effects at the smooth muscle level [59].

The inhalation route offers distinct advantages including a rapid onset of action (minutes), sustained bronchodilation (4 to 6 hours with albuterol or similar drugs), and a relatively low incidence of side effects (tremor, palpitations) compared to oral preparations. To achieve maximum deposition of the inhaled drug in the lungs, it is important to inhale properly from the aerosol device. The patient should be advised to shake the canister well, place the mouthpiece about 1.5 inches in front of the mouth, maximally exhale to functional residual capacity, and discharge the inhaler while taking a slow (5 to 10 seconds) deep breath to total lung capacity. A brief period of breath holding (10 seconds) after full inspiration will further enhance drug deposition in the lungs. Many surveys have found that major errors exist in the technique of inhaler use in a significant proportion (up to 50%) of patients using metered-dose inhalers, which can be related to lack of knowledge of the proper technique or difficulty in coordination. In those who fail to master an acceptable technique despite adequate teaching, a spacer device between the actuator and the mouth can enhance drug delivery.

Isoproterenol. Isoproterenol is a nonselective beta-receptor agonist with a rapid onset (2 to 5 minutes) and a short duration of action (1 to 2 hours). Due to its beta-1 properties, cardiac side effects are more common. Isoproterenol is infrequently used for the prophylaxis or therapy of EIA owing to its more frequent side effects (palpitations, tremors) and its short duration of action.

Metaproterenol. Metaproterenol has both beta-1 and beta-2 effects and a longer duration of action than that of isoproterenol but shorter than that of

albuterol [4]. The protective effect of metaproterenol starts within 5 minutes, peaks at 15 minutes, and has a 2-hour duration. Interestingly, its protective effect against EIA is shorter than its bronchodilatory effect [37]. Oral preparations of metaproterenol are also capable of reducing the severity of EIA after 1 hour of use. Some oral forms are associated with more side effects (e.g., tremors, agitation), and they should be limited to patients who are unable to use a metered-dose inhaler effectively (e.g., children under 7 years) [37].

Albuterol. Albuterol is a selective beta-2 receptor agonist that has a significant EIA protective effect beginning from 10 to 15 minutes and lasting up to 4 hours after inhaler use (two puffs or 180 mg). Its efficacy (80%) has been demonstrated to be better than that of cromolyn sodium (50%) [49] and is of longer duration than that of metaproterenol [4] in the prophylaxis of EIA. Albuterol has also been found to be more protective than ipratropium bromide, a cholinergic antagonist, in children with EIA [65]. Compared to theophylline, atropine, cromolyn, and placebo, albuterol has been found to provide maximum protection against EIA. As with metaproterenol, oral preparations of albuterol are available and also provide protection against EIA.

Other selective beta-2 receptor agonists that can be used for the prevention and therapy of EIA include terbutaline sulfate, fenoterol (not available in the United States), and pirbuterol (recently released in the United States), all of which have bronchodilator properties similar to those of albuterol [2a].

In summary, beta-agonist inhalers are the drugs of choice for the prevention and therapy of EIA. Selective beta-2 receptor agonists provide a longer duration of action with fewer side effects than nonselective drugs. Optimal dosage, usually 1 to 2 inhalations taken 10 to 15 minutes before exercise, should be prescribed on an individual basis, and appropriate attention should be paid to the technique of inhaler use. For individuals who find it difficult to master the optimum technique, especially children, the use of a spacer device can greatly enhance proper delivery of the medication.

TABLE 9C–1
Activity of Several Beta-2 Agonists in Preventing Exercise-Induced Asthma

| Drug | Route of Administration | Activity | |
		Onset (min)	Duration (h)
Isoproterenol	Inhaled	5	½–1
Metaproterenol	Inhaled	15	2–4
Albuterol	Inhaled	15	4–6
Terbutaline	Inhaled	15	1*
	Oral	15	2–5

*Evaluated out to 1 hour postdose only.

Theophylline

Although its precise mechanism of action is still not clearly established, theophylline has been used for more than 50 years as a bronchodilator. Despite its success as a bronchodilator for the management of asthma in general, its role in the prevention of EIA is much less prominent [18]. Theophylline appears to be most helpful in preventing EIA in patients who are taking the drug on a chronic basis for asthma. Slow-release forms of theophylline make possible a convenient dosing schedule (once or twice a day) with less variation in serum concentrations. Several drugs are known to interact with the metabolism or clearance of theophylline leading to either higher (e.g., oral contraceptives, propranolol, cimetidine, erythromycin) or lower (e.g., phenobarbital, phenytoin, rifampin) serum theophylline levels. Theophylline elimination is increased in smokers and decreased in elderly patients and those with liver or heart disease; thus, dosing should be chosen accordingly and adjusted based on blood levels. Interestingly, although its effectiveness as a bronchodilator is related to its serum concentration, some [48] but not all [5] investigators have been able to establish a similar relationship when the drug is evaluated in the context of EIA. Theophylline in general has been found to be more effective than placebo but is less potent and more inconsistent than beta agonists [34, 47, 55]. In addition to bronchodilation, other potential beneficial effects of theophylline include its positive cardiac inotropic effect, improved conductivity of the diaphragm, and improved skeletal muscle function. The significance of these effects in otherwise healthy individuals, however, is unclear, and the well-known side effects of theophylline (nausea, palpitations, arrhythmias) in many instances outweigh these theoretical benefits. Therefore, the effects of theophylline on sports performance are minimal. Theophylline is approved for use by the United States Olympic Committee.

In summary, although theophylline is less effective than beta agonists in the prophylaxis of EIA, it can be helpful in the overall management of asthma, thereby providing additional bronchodilator support during exercise for these individuals.

Anticholinergics

The use of anticholinergic drugs in diseases characterized by airway obstruction is based on the importance of the cholinergic nervous system regulation of airway caliber. Atropine, a cholinergic antagonist that has been used by many physicians as a bronchodilator to treat exacerbations of asthma, is limited by its systemic absorption and resultant side effects (e.g., dry mouth, tachycardia) when it is given by the aerosol route. Ipratropium bromide, available in a metered-dose inhaler for the treatment of chronic obstructive pulmonary disease, retains the beneficial antimuscarinic effects of atropine at the level of the airway but lacks the systemic side effects because of its poor systemic absorption. Because ipratropium bromide is not associated with systemic side effects, it is unlikely to affect athletic performance. Although ipratropium bromide has found its main use in chronic bronchitis, it is also useful in selected patients with asthma. Ipratropium bromide is a slower and longer acting bronchodilator than beta agonists and, in general, is less effective in the prevention of EIA. The addition of ipratropium bromide to a beta agonist does not provide any additional protective effect against EIA compared to the beta agonist alone [51]. Some studies suggest that the efficacy of ipratropium bromide can be enhanced by using higher doses [29] or by giving it to a subgroup of EIA subjects characterized by the absence of the refractory period [69] or to those with predominantly large airway obstruction (as shown by helium response). In summary, ipratropium bromide is a less favorable choice than beta agonists for the prevention or therapy of EIA.

Cromolyn Sodium

Cromolyn sodium is a nonbronchodilator medication used in the treatment of asthma. It has been shown to protect against the development of asthma when given prior to exposure to a variety of stimuli such as sulfur dioxide, cold air, industrial chemicals, and allergens. Its exact mechanism of action is not established but may involve suppression of mast cell mediator release, suppression of inflammatory cells, or interference with heat or water loss across the airway. Inhaled cromolyn has also been found to offer effective prophylaxis against EIA; significant protection is demonstrable 15 to 30 minutes after inhalation and lasts up to 2 hours. Its relative efficacy (50%) is dose dependent [46, 67] and is less than that of albuterol (80%) [65] with possibly some additive benefit in those not completely protected by a beta agonist [35]. Cromolyn is available for inhaled use in three forms: powder from a capsule, a metered-dose inhaler, and a solution for nebulization. The required dose in the metered-dose inhaler (2 to 4 mg) is much less than that needed in the powder form (20 mg). Cromolyn's advantages are its very low incidence of side effects, its lack of effect on athletic performance, and its ability to block the late response that may occur in some patients with EIA. Its disadvantage is that, unlike beta-receptor agonists, it is not a bronchodilator, and it should not be used in the presence of baseline airway obstruction. Thus, its usefulness lies in the prophylaxis rather than in the treatment of EIA.

In summary, cromolyn should be considered for

the prophylactic management of EIA, especially in those who experience side effects from beta agonists or in whom the beneficial effects of beta agonists are incomplete, in which case the combination of beta agonist and cromolyn may have a synergistic effect.

Corticosteroids

Because airway obstruction in asthma can be due to inflammation and edema as well as to bronchospasm, corticosteroid therapy has been found to be extremely valuable in the overall treatment of asthmatic symptoms that are not controlled by bronchodilator therapy alone. Although very effective, corticosteroids have many side effects that mandate careful consideration prior to prescribing them for an asthmatic. These include peptic ulcer exacerbation or bleeding, hypertension, cataract formation, weight gain, diabetes mellitus, adrenal suppression, and osteoporosis. High-dose corticosteroids can cause mood alteration and may therefore affect athletic performance. Side effects of oral corticosteroids are seen mainly with prolonged use (months to years) of high-dose steroids (more than 10 mg prednisone/day). Patients who require corticosteroids for adequate control of airway obstruction should be informed about the side effects associated with their use and carefully monitored to ensure early detection and appropriate management of any side effects should they develop over time.

Inhaled forms of corticosteroids, when used in the recommended doses, have the advantage of delivering the drug directly to the target organ while not producing the clinically significant side effects seen with the oral forms [36]. Although corticosteroids are not "first-line" drugs for EIA treatment, regular use of the inhaled forms can decrease the severity of EIA [31] and enhance the protective effect of beta agonists when they are administered routinely for at least 4 weeks [32]. Inhaled corticosteroids may be associated with oral thrush (candidiasis), dysphonia (vocal cord dysfunction), and cough. The occurrence of oral thrush can be completely prevented by rinsing the mouth after each use of the inhaler. When it occurs, it is easily treated with oral nystatin and temporary discontinuation of the inhaled steroid. Vocal cord dysfunction appears to be a local form of steroid myopathy that improves following the withdrawal or lowering of the dose of inhaled steroid. Coughing while inhaling the medicine is occasionally reported and is thought to be related to the propellant in the inhaled steroid.

Coughing may be less common with some preparations (triamcinolone) than others (beclomethasone).

Although systemic side effects are occasionally seen with inhaled steroids, they are very infrequent and are seen mainly when steroids are given in doses higher than those recommended by the manufacturer. Reported side effects include mild suppression of the hypothalamic-pituitary-adrenal axis, delayed growth in children, accelerated osteoporosis, and cataract formation [36].

Thus, asthmatic patients whose EIA is refractory to bronchodilators or cromolyn prophylaxis should be given a trial of routine inhaled corticosteroids. Most important, such patients should be more carefully evaluated for their overall level of symptom control and pulmonary function because such situations suggest suboptimal treatment. Commonly used steroid inhalers in the United States include beclomethasone dipropionate, flunisolide, and triamcinolone acetate.

In summary, aerosolized steroids are not beneficial in the prophylaxis of EIA when administered immediately prior to exercise; however, long-term use can modify EIA and reduce the required beta-agonist inhaler dose needed to block it.

EFFECTS OF PHYSICAL TRAINING ON THE SEVERITY OF EIA

Exercise is a common stimulus for asthma, and thus asthmatics may tend to avoid physical activities that frequently provoke an attack. This may have a significant influence on a patient's lifestyle and sense of well-being, especially among children. The effects of physical training on EIA have been studied with encouraging but occasionally conflicting results. Most of the discrepancies can be explained by differences in study design, method of analysis, or interpretation of results. In most studies, patients experience improvement in self-confidence and the ability to cope with asthma after training; however, objective evidence of improvement in EIA has been demonstrated by some but not all investigators [3, 20, 26, 33, 64]. Since well-structured physical training programs may help to decrease the severity of EIA in addition to improving work capacity and self-confidence, they should be encouraged. Patients should be advised to use prophylactic medications properly to maximize their participation and enjoyment in a full range of activities.

TREATMENT OF EIA IN ATHLETES PARTICIPATING IN NATIONAL AND INTERNATIONAL COMPETITIONS

A substance that is banned by a national or international official sports organization remains a banned substance, even when prescribed by a physician for a clinically justifiable purpose. Approved antiasthma medications include theophylline, cromolyn, anticholinergics, corticosteroids, and specific beta-2 agonists. The National Collegiate Athletic Association (NCAA) publishes a list of banned drugs that was modified recently. The United States Olympic Committee makes available a drug list and a hotline (1-800-233-0393) for questions relating to these issues [11]. Due to possible changes in the policies of various athletic organizations, it is important always to obtain the most recent guidelines of the organization involved at least a month prior to any scheduled competition. This should allow time to adjust therapeutic strategies well in advance of the actual sporting event.

AUTHORS' RECOMMENDED METHOD OF TREATMENT

EIA can be controlled in the majority of asthmatic patients and athletes by proper use of physical training, preexercise warm-up, correct choice of exercise, use of antiasthma medications, and use of methods that minimize airway heat and water loss during exercise. Physical training improves cardiopulmonary fitness, sense of well-being and self-confidence. Furthermore, because improved fitness permits the athlete to perform the same exercise with less minute ventilation (i.e., less heat and water loss), the frequency and severity of EIA may decrease. The choice of exercise type is most important in the amateur athlete. Certain types of exercise, such as biking, jogging, and skiing, are more likely to cause EIA than swimming or aerobic exercises. A preexercise warm-up can help to decrease the severity of EIA by inducing a refractory period. Because up to half of asthmatics may develop a refractory state, athletes should be encouraged to perform a routine warm-up prior to exercise to determine whether such an activity may attenuate their asthmatic response.

The use of antiasthma medications will help to control EIA in the majority of asthmatics. For those with asthmatic symptoms associated only with exercise, the prophylactic use of a beta-2 agonist inhaler prior to exercise will prevent EIA in most

($\approx 80\%$) cases. In a small group of asthmatic patients, the protection provided by a beta-2 agonist may be incomplete, and in these athletes the addition of inhaled cromolyn sodium can provide a second level of prophylaxis. In asthmatic patients in whom symptoms are provoked by exercise as well as other stimuli, drug therapy can be enhanced by the regular use of a steroid inhaler, and possibly oral theophylline to ensure overall control of asthma.

Because the pivotal stimuli for EIA appear to be airway heat and water loss, any measures taken to minimize these losses can help to decrease the incidence and severity of EIA. Measures include avoiding outdoor exercise during cold, dry winter days and using a scarf over the nose and mouth when exposure to these conditions is unavoidable.

SUMMARY

Exercise-induced asthma is a common problem in both asthmatic children and adults. It is characterized by airway obstruction that occurs shortly after vigorous physical activity; it improves gradually and is usually completely reversed within 30 to 60 minutes postexercise. Half of the patients with EIA may experience a refractory period for at least 1 hour following their initial exercise-induced asthmatic response. The pathogenesis of EIA is not clearly established but may be influenced by heat and water loss across the airway wall during exercise, mast cell mediator release, the rate of airway rewarming, or the intensity of the exercise stimulus itself. Despite the lack of precise knowledge about the pathogenesis of EIA, its occurrence can be prevented in most cases. Proper choice of a suitable exercise type and exercise conditions and use of a prophylactic beta-2 agonist should help to achieve optimal control of EIA in the majority of athletes suffering from asthma and allow them to engage in a full range of sports activities.

The outstanding record of asthmatic athletes on the 1984 and 1988 U.S. Olympic teams should be an encouragement to patients with asthma to pursue excellence in physical fitness and sports unrestricted by their illness.

References

1. Anderson, S. D. Is there a unifying hypothesis for exercise-induced asthma? *J Allerg Clin Immunol* 73:660–665, 1984.
2. Anderson, S. D. Exercise-induced asthma: The state of the art. *Chest* (Suppl.)87:191S–195S, 1985.
2a. Anderson, S. D., Seale, J. P., Ferris, L., Schoeffel, R.,

and Lindsay, D. A. An evaluation of pharmacotherapy for exercise-induced asthma. *J Allerg Clin Immunol* 73:612–624, 1979.

3. Arborelius, M., Jr., and Svenonius, E. Decrease of exercise-induced asthma after physical training. *Eur J Respir Dis* (Suppl.)65:25–31, 1984.

4. Berkowitz, R., Schwartz, E., Bukstien, D., Grunstein, M., and Chai, H. Albuterol protects against exercise-induced asthma longer than metaproterenol sulfate. *Pediatrics* 77:173–178, 1986.

5. Bierman, C. W., Shapiro, G. G., Pierson, W. E., and Dorsett, C. S. Acute and chronic theophylline therapy in exercise-induced bronchospasm. *Pediatrics* 80:845–849, 1977.

6. Bierman, C. W., Spiro, S. G., and Petheram, I. Characterization of the late response in exercise-induced asthma. *J Allerg Clin Immunol* 74:701–706, 1984.

7. Bierman, C. W., Spiro, S. G., and Petheram, I. Late response in exercise-induced asthma (EIA) (Abstract). *J Allerg Clin Immunol* 65:206, 1980.

7a. Broide, D. H., Eisman, S., Ramsdell, J. W., Ferguson, P., Schwartz, L. B., and Wasserman, S. I. Airway levels of mast cell-derived mediators in exercise-induced asthma. *Am Rev Respir Dis* 141: 563–568, 1990.

8. Boucher, R. C., Stutts, M. J., Bromberg, P. A., and Gatzy, J. T. Regional differences in airway surface liquid composition. *J Appl Physiol* 50:613–620, 1981.

9. Boulet, L. P., Legris, C., Turcotte, H., and Hebert, J. Prevalence and characteristics of late asthmatic responses to exercise. *J Allerg Clin Immunol* 80:655–662, 1987.

10. Chen, W. Y., and Horton, D. J. Heat and water loss from the airways and exercise induced asthma. *Respiration* 34:305–310, 1977.

11. Clarke, K. S. Sports medicine and drug control programs of the U.S. Olympic Committee. *J Allerg Clin Immunol* 73:740–744, 1984.

12. Deal, E. C., McFadden, E. R., Jr., Ingram, R. H., Jr., Strauss, R. H., and Jaeger, J. J. Role of respiratory heat exchange in production of exercise-induced asthma. *J Appl Physiol* 46:467–475, 1979.

13. Deal, E. C., Jr., Wasserman, S. I., Soter, N. A., Ingram, R. H., Jr., and McFadden, E. R., Jr. Evaluation of role played by mediators of immediate hypersensitivity in exercise-induced asthma. *J Clin Invest* 65:659–665, 1980.

14. Eggleston, P. A. Exercise challenge: Laboratory evaluation of exercise-induced asthma: Methodologic consideration. *J Allerg Clin Immunol* 64:604–608, 1979.

15. Eggleston, P. A. Methods of exercise challenge. *J Allerg Clin Immunol* 73:666–669, 1984.

16. Eggleston, P. A., Kagey-Sobotka, A., Schleimer, R. P., and Lichtenstein, L. M. Interaction between hyperosmolar and IgE-mediated histamine release from basophils and mast cells. *Am Rev Respir Dis* 130:86–91, 1984.

17. Eggleston, P. A., Rosenthal, R. R., Anderson, S. A., Anderson, R., Bierman, C. W., Bleecker, E. R., Chai, H., Cropp, G. J. A., Johnson, J. D., Konig, P., Morse, J., Smith, L. J., Summers, R. J., and Trautlein, J. J. Study group on exercise challenge, bronchoprovocation committee, American Academy of Allergy. Guidelines for the methodology of exercise challenge testing of asthmatics. *J Allerg Clin Immunol* 64:642–645, 1979.

18. Ellis, E. Inhibition of exercise-induced asthma by theophylline. *J Allerg Clin Immunol* 73:690–692, 1984.

19. Eschenbacher, W. L., Boushey, H. A., and Sheppard, D. Alteration in osmolarity of inhaled aerosols cause bronchoconstriction and cough, but absence of a permeant anion causes cough alone. *Am Rev Respir Dis* 129:211–215, 1984.

20. Fitch, K. D., Blitvich, J. D., and Morton, A. R. The effect of running training on exercise-induced asthma. *Ann Allerg* 57:90–94, 1986.

20a. Fitch, K. D. The use of anti-asthmatic drugs. Do they affect sports performance? *Sports Med* 3:136–150, 1986.

21. Gergen, P. J., Mullally, D. I., and Evans, R. National survey of prevalence of asthma among children in the United States, 1977–1980. *Pediatrics* 81:1–7, 1988.

22. Gilbert, I. A., Fouke, J. M., and McFadden, E. R., Jr. Heat and water flux in the intrathoracic airways and exercise-induced asthma. *J Appl Physiol* 63:1681–1691, 1987.

23. Godfrey, S., and Konig, P. Inhibition of exercise-induced asthma by different pharmacological pathways. *Thorax* 31:137–140, 1976.

24. Godfrey, S. Symposium on special problems and management of allergic athletes. *J Allerg Clin Immunol* 73:630–633, 1984.

25. Gravelyn, T. R., Capper, M., and Eschenbacher, W. L. Effectiveness of heat and moisture exchange in preventing hyperpnea-induced bronchoconstriction in subjects with asthma. *Thorax* 42:877–880, 1987.

26. Haas, F., Pasierski, S., Levine, N., Bishop, M., Axen, K., Pineda, H., and Haas, A. Effect of aerobic training on forced expiratory airflow in exercising asthmatic humans. *J Appl Physiol* 63:1230–1235, 1987.

27. Hahn, A., Anderson, S. D., Morton, A. R., Black, J. L., and Fitch, K. D. A reinterpretation of the effects of temperature and water content of inspired air in exercise-induced asthma. *Am Rev Respir Dis* 130:575–579, 1984.

28. Hahn, A. G., Nogrady, S. G., Tumilty, D. M., Lawrence, S. R., and Morton, A. R. Histamine reactivity during the refractory period after exercise-induced asthma. *Thorax* 39:919–923, 1984.

29. Hartley, J. P., and Davies, P. M. Cholinergic blockade in prevention of exercise-induced asthma. *Thorax* 35:680–685, 1980.

30. Haynes, R. L., Ingram, R. H., and McFadden, E. R. An assessment of the pulmonary response to exercise in asthma and an analysis of the factors influencing it. *Am Rev Respir Dis* 114:739–752, 1976.

31. Henriksen, J. M. Effect of inhalation of corticosteroids on exercise induced asthma: Randomized double blind crossover study of budesonide in asthmatic children. *Br Med J* 291:248–249, 1985.

32. Henriksen, J. M., and Dahl, R. Effects of inhaled budesonide alone and in combination with low-dose terbutaline in children with exercise-induced asthma. *Am Rev Respir Dis* 128:993–997, 1983.

33. Henriksen, J. M., and Nielsen, T. T. Effect of physical training on exercise-induced bronchoconstriction. *Acta Paediatr Scand* 72:31–36, 1983.

34. Ioli, F., Donner, C. F., Fracchia, C., Patessio, A., and Aprile, C. Sustained release anhydrous theophylline in preventing exercise-induced asthma. *Respiration* 46:105–113, 1984.

35. Konig, P. The use of cromolyn in the management of hyperreactive airways and exercise. *J Allerg Clin Immunol* 73:686–689, 1984.

36. Konig, P. Inhaled corticosteroids, their present and future role in the management of asthma. *J Allerg Clin Immunol* 82:297–306, 1988.

37. Konig, P., Eggleston, P. A., and Serby, C. W. Comparison of oral and inhaled metaproterenol for prevention of exercise-induced asthma. *Clin Allerg* 11:597–604, 1981.

38. Lee, T. H., Assoufi, B. K., and Day, A. B. The link between exercise, respiratory heat exchange and the mast cell in bronchial asthma. *Lancet* 1:520–522, 1983.

39. Lee, T. H., Brown, M. J., Nagy, L., Causon, R., Walport, M. J., and Kay, A. B. Exercise-induced release of histamine and neutrophil chemotactic factor in atopic asthmatics. *J Allerg Clin Immunol* 70:73–81, 1982.

40. Lee, T. H., Nagakura, T., Papageorgiou, N., Iikura, Y., and Kay, A. B. Exercise-induced late asthmatic reactions with neutrophil chemotactic activity. *N Engl J Med* 308:1502–1505, 1983.

41. Lemanske, R. F., and Kaliner, M. Late phase allergic reactions. *In* Middleton, E., Jr., Reed, C. E., Ellis, E. F., Adkinson, N. F., Jr., and Yuninger, J. W. (Eds.), *Allergy: Principles and Practice,* Vol. 1. St. Louis, C.V. Mosby, 1988.

42. Likura, Y., Inui, H., Nagakura, T., and Lee, T. H. Factors predisposing to exercise-induced late asthmatic responses. *J Allerg Clin Immunol* 75:285–289, 1985.

43. McFadden, E. R., Jr. Exercise-induced asthma: Assessment of current etiologic concepts. *Chest* 91 (Suppl.):151S–157S, 1987.

44. McFadden, E. R., Jr., Denison, D. M., Waller, J. F., Assoufi, B., Peacock, A., and Sopwith, T. Direct recordings of the temperatures in the tracheobronchial tree in normal man. *J Clin Invest* 69:700–705, 1982.

45. McFadden, E. R., Jr., Lenner, A. M., and Strohl, K. P. Postexertional airway rewarming and thermally induced asthma. New insights into pathophysiology and possible pathogenesis. *J Clin Invest* 78:18–25, 1986.

46. Patel, K. R., Berkin, R. E., and Kerr, J. W. Dose-response study of sodium cromoglycate in exercise-induced asthma. *Thorax* 37:663–666, 1982.

47. Philips, M. J., Ollier, S., Trembath, P. W., Boobis, S. W., and Davies, R. J. The effects of sustained-release aminophylline on exercise-induced asthma. *Br J Dis Chest* 75:181–189, 1981.

48. Pollock, J., Keichel, F., Cooper, D., and Weinberger, M. Relationship of theophylline concentration to inhibition of exercise-induced bronchospasm and comparison with cromolyn. *Pediatrics* 60:840–844, 1977.

49. Rohr, A. S., Siegel, S. C., Katz, R. M., Rachelefsky, G. S., Spector, S. L., and Lanier, R. A comparison of inhaled albuterol and cromolyn in the prophylaxis of exercise-induced bronchospasm. *Ann Allerg* 59:107–109, 1987.

50. Rubinstein, I., Levison, H., Slutsky, A. S., Hak, H., Wells, J., Zamel, N., and Rebuck, A. S. Immediate and delayed bronchoconstriction after exercise in patients with asthma. *N Engl J Med* 317:482–485, 1987.

51. Sanguinetti, C. M., DeLuca, S., Gasparini, S., and Massei, V. Evaluation of Duovent in the prevention of exercise-induced asthma. *Respiration* 50 (Suppl. 2):181–185, 1986.

52. Schnall, R. P., and Landau, L. I. Protective effects of repeated short sprints in exercise-induced asthma. *Thorax* 35:828–832, 1980.

53. Schoeffel, R. E., Anderson, S. D., Gillam, I., and Lindsay, D. A. Multiple exercise and histamine challenges in asthmatic patients. *Thorax* 35:164–170, 1980.

54. Schoeffel, R. E., Anderson, S. D., and Altounyan, R. E. C. Bronchial hyper-reactivity in response to inhalation of ultrasonically nebulized solutions of distilled water and saline. *Br Med J* 283:1285–1287, 1981.

55. Seale, J. P., Anderson, S. D., and Lindsay, D. A. A comparison of oral theophylline and oral salbutamol in exercise-induced asthma. *Aust NZ J Med* 7:270–275, 1977.

56. Shapiro, G. G. Methacholine challenge, relevance for the allergic athlete. *J Allerg Clin Immunol* 73:670–675, 1984.

57. Shapiro, G. G., Pierson, W. E., Fukurawa, C. T., and Bierman, C. W. A comparison of the effectiveness of free running and treadmill exercise for assessing exercise-induced bronchospasm in clinical practice. *J Allerg Clin Immunol* 64:609–611, 1979.

58. Silverman, M., Konig, P., and Godfrey, S. Use of serial exercise tests to assess the efficacy and duration of action of drugs for asthma. *Thorax* 28:574–578, 1973.

59. Sly, R. M. Beta-adrenergic drugs in the management of asthma in athletes. *J Allerg Clin Immunol* 73:680–685, 1984.

60. Sly, R. M., Heimlich, E. M., Busser, R. J., and Strick, L. Exercise-induced bronchospasm. Evaluation of isoproterenol, phenylephrine and the combination. *Ann Allerg* 26:253–258, 1967.

61. Sly, R. M., Heimlich, E. M., Ginsburg, J., Busser, R. J., and Stick, L. Exercise induced bronchospasm: Evaluation of metaproterenol. *Ann Allerg* 26:253–258, 1968.

62. Smith, C. M., Anderson, S. D., Walsh, S., and Mcelrea, M. S. An investigation of the effects of heat and water exchange in the recovery period after exercise in children with asthma. *Am Rev Respir Dis* 140:598–605, 1989.

63. Strauss, R. H., McFadden, E. R., Ingram, R. H., and Jaeger, J. J. Enhancement of exercise-induced asthma by cold air. *N Engl J Med* 297:743–747, 1977.

64. Svenonius, E., Kautto, R., and Arborelius, M., Jr. Improvement after training of children with exercise-induced asthma. *Acta Paediatr Scand* 72:23–30, 1983.

65. Svenonius, E., Arborelius, M., Jr., Wieberg, R., and Ekberg, P. Prevention of exercise induced asthma by drugs inhaled from metered aerosols. *Allergy* 43:252–257, 1988.

66. Tsankas, J. N., Milner, R. D. G., Bannister, O. M., and Boon, A. W. Free running asthma screening test. *Arch Dis Childh* 63:261–265, 1988.

67. Tullett, W. M., Tan, K. M., Wall, R. T., and Patel, K. R. Dose-response effect of sodium cromoglycate pressurized aerosol in exercise-induced asthma. *Thorax* 40:41–44, 1985.

68. Weiler-Ravell, D., and Godfrey, S. Do exercise and antigen-induced asthma utilize the same pathways? *J Allerg Clin Immunol* 67:391–397, 1981.

69. Wilson, N. M., Barnes, P. J., Vickers, H., and Silverman, M. Hyperventilation-induced asthma, evidence for two mechanisms. *Thorax* 37:657–662, 1982.

Management of the Young Athlete With Type I Diabetes Mellitus

David B. Allen, M.D.

Diabetes mellitus is a chronic metabolic disorder caused by an absolute or relative deficiency of insulin; it affects almost 1 in 500 children under the age of 18. Worldwide incidence rates vary from 1 in 100,000 in Japan to 30 in 100,000 in Scandinavia, and the disorder is approximately four times more common in whites than in blacks. The predominant form of diabetes affecting children is type I diabetes mellitus, a disorder in which the insulin-producing islet cells of the pancreas are selectively destroyed by the body's immune system. The resulting insulin *deficiency* invariably requires treatment with exogenous insulin and is the hallmark of type I insulin-dependent diabetes mellitus (IDDM). In contrast, type II diabetes mellitus occurs almost exclusively during adulthood as a result of increasing *resistance* to the action of insulin that may or may not require treatment with exogenous insulin. Children and adolescents may also manifest either type I or type II diabetes mellitus in association with genetic syndromes (e.g., Down's syndrome), drug treatment (e.g., glucocorticoids), or other systemic diseases (e.g., cystic fibrosis). Other features differentiating type I diabetes mellitus from the more common adult-onset type II diabetes mellitus are described in Table 9D–1.

Physical exercise is traditionally considered beneficial in the treatment of IDDM [28]. Attainment and maintenance of ideal body weight, improvement in self-image, and decreases in hypertension and lipid-related cardiovascular risk factors [12, 18] can all be achieved by the diabetic patient who exercises. Observations of exercise-associated reductions in blood sugar [14] have been substantiated by in vivo and in vitro [27] findings of increased insulin sensitivity due to enhanced insulin receptor-binding after physical training; these effects can also be seen after a single exercise session in an inactive individual [9]. Although the value of exercise in improving long-term metabolic control remains controversial [20], athletic participation by young individuals with IDDM is encouraged to achieve the same health benefits enjoyed by exercising nondiabetic individuals. Management of these athletes requires knowledge of the metabolic responses to exercise in diabetic individuals. Careful monitoring and adjustment of insulin doses and nutrition plans can then make possible the safe and successful participation of patients with IDDM in virtually any athletic activity.

PERTINENT PHYSIOLOGY AND PATHOPHYSIOLOGY

Glucose Metabolism During Exercise

Maintenance of a normal plasma glucose level during exercise depends upon a precise balance between mobilization of fuel sources and glucose utilization by tissues. The exercising muscle fiber can increase its metabolic rate and production of adenosine triphosphate (ATP) tremendously, with oxidative processes 50 times, and glucose uptake 35 times, higher than resting levels [30]. Primary sources of energy include fat and carbohydrate present in muscle and glucose released from the liver as a result of breakdown and mobilization of stored glycogen (glycogenolysis). The relative contributions of these sources vary with the duration and intensity of exercise. At the onset of muscular activity, energy is derived predominantly from anaerobic breakdown of muscle glycogen to form

TABLE 9D–1
Distinguishing Characteristics of the Two Major Types of Diabetes Mellitus

Characteristic	Type I	Type II
Synonym	Juvenile onset, insulin-dependent (IDDM)	Adult-onset, noninsulin-dependent (NIDDM)
Age of onset	Most often during growing years, occasionally during adulthood	Adulthood; rarely seen in children/adolescents
Contributing causative factors	Immunologic destruction of insulin-producing cells of the pancreas	Age, obesity, pregnancy
Predominant insulin abnormality	Insulin deficiency	Insulin resistance
Response to stress/illness/failure to administer insulin	Hyperglycemia and ketoacidosis	Hyperglycemia without ketoacidosis
Treatment with diet/weight reduction alone	Negligible efficacy	Improvement expected

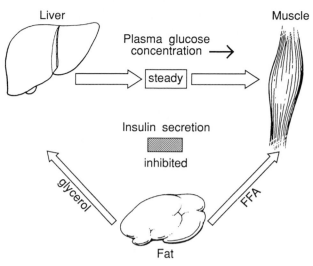

FIGURE 9D–1
The response to exercise in healthy individuals and in insulin-dependent diabetics. When plasma insulin is normal or slightly diminished, hepatic glucose production increases markedly, as does skeletal muscle utilization of glucose, whereas blood glucose remains unchanged. FFA, free fatty acids. (J. M. Ekoe, Overview of diabetes mellitus and exercise. *Medicine and Science in Sports and Exercise* 21(4):353–368, 1989, © The American College of Sports Medicine.)

lactate. Aerobic metabolism of a combination of hepatic-derived glucose and free fatty acids from adipose tissue contributes the majority of energy after 20 to 30 minutes of activity. With prolonged exercise (60 to 90 minutes), free fatty acids constitute the principal energy source, although glycogen remains an important fuel, and depletion of muscle glycogen has been shown to coincide with the time of exhaustion [10]. Intensity of exercise is measured by the percentage of the individual's maximum oxygen consumption (VO₂ max) required for fuel utilization. During exercise at high intensity, oxidation of glucose for energy predominates; with less intense exercise, fat utilization is preferred. Physical training enables the athlete to perform the same work at a lower percentage of VO₂ max) and therefore to conserve glucose and improve endurance by utilizing a greater proportion of free fatty acids.

Insulin is an anabolic hormone that has important effects on carbohydrate, protein, and fat metabolism. A commonly held misconception is that insulin exerts its major effect on blood glucose by increasing the uptake of glucose into tissue such as muscle. In truth, the major means by which insulin decreases blood glucose is by suppressing the *production* of glucose by the liver. Insulin inhibits protein degradation, thus encouraging growth and preventing weight loss and tissue breakdown. Utilization of fat for energy (lipolysis) is highly sensitive to the inhibitory effects of insulin, and the appearance of fat-derived ketones in blood or urine signifies marked insulin deficiency in the diabetic patient.

Fuel supply during exercise is orchestrated by a variety of hormonal responses. In exercising nondiabetic individuals (Fig. 9D–1), insulin secretion is inhibited by increased sympathetic (alpha-adrenergic) nervous system activity. This suppression of endogenous insulin release and increased secretion of hormones that oppose the actions of insulin (counterregulatory hormones), such as glucagon, epinephrine, and norepinephrine, allows hepatic glucose production to increase, satisfying the demands of exercising muscle (Table 9D–2). Lowered

TABLE 9D–2
Actions of Major Counterregulatory Hormones

Hormone	Mechanism of Hyperglycemic Effect
Glucagon*	Activates hepatic glycogenolysis and gluconeogenesis
Epinephrine*	Stimulates hepatic glucose production, limits peripheral glucose utilization, suppresses insulin secretion
Growth hormone	After initial glucose-lowering effect, limits glucose transport into cells, mobilizes fat, and provides gluconeogenic substrate (glycerol)
Cortisol	Initially inhibits glucose utilization; with time, mobilizes substrate (amino acids and glycerol) for gluconeogenesis

*Hormones important in recovery from acute hypoglycemia

serum insulin concentrations also facilitate lipolysis and the liberation of free fatty acids and glycerol (utilized with lactate, pyruvate, and alanine as precursors for gluconeogenesis). Stimulation of insulin-independent glucose uptake by skeletal muscle through exercise compensates for declining insulin concentrations and facilitates delivery of substrate to exercising tissues. The exact mechanism by which exercise independently stimulates glucose transport into muscle remains unknown.

Physical training increases the athlete's sensitivity to insulin [17], an effect that predominates in skeletal muscle rather than liver. An increase in insulin binding [26] due to augmentation of both insulin receptor number and affinity has been demonstrated following even short periods of physical training. In addition, reductions in body adipose tissue contribute to a greater insulin effect after prolonged training. Changes in insulin sensitivity correlate directly with improvements in VO_2 max [17], although this may primarily reflect frequency of exercise rather than fitness per se.

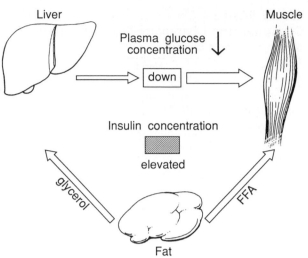

FIGURE 9D–2
The response to exercise in hyperinsulinemic insulin-dependent diabetics. When plasma insulin is increased, skeletal muscle utilization of glucose during exercise increases markedly, but the increase in hepatic glucose production is smaller than normal; thus, blood glucose levels fall. (J. M. Ekoe, Overview of diabetes mellitus and exercise. *Medicine and Science in Sports and Exercise* 21(4):353–368, 1989, © The American College of Sports Medicine.)

Glucose Regulation During Exercise in Athletes With IDDM

Soon after the implementation of insulin therapy, it was observed that physical activity could reduce the insulin requirements of patients with IDDM and that the decrease in blood glucose following an insulin injection was magnified by subsequent exercise [16]. In contrast to the effect of insulin suppression observed in nondiabetic athletes, absorption of exogenously administered insulin is enhanced by exercise in the individual with IDDM (Fig. 9D–2) [6]. This effect is pronounced when insulin is administered less than 60 minutes before exercise and, for most activities, is augmented when a leg (versus an arm) injection site is used [29]. Increased serum insulin levels would inhibit hepatic glucose production and peripheral lipolysis. At the same time, continued insulin-independent glucose uptake by exercising muscles depletes energy stores. In addition, deficiencies of counterregulatory hormones such as glucagon (common) and epinephrine (less common) develop in many individuals with IDDM and further limit fuel availability during exertion. Thus, the balance between energy supply and utilization may be disrupted in the exercising diabetic athlete by an unregulated (either excessive or insufficient) insulin effect and a variable antagonism of insulin by counterregulatory hormones.

Consequences of Excessive Insulin Effect During Exercise

Suppression of glycogenolysis by excessive insulin action combined with insulin-independent glucose uptake by working muscles can result in hypoglycemia during, immediately after, or several hours after exercise. Failure to anticipate the occurrence or intensity of exercise and to make appropriate adjustments in insulin dosage and caloric intake account for most instances of hypoglycemia during activity. Most diabetic patients are aware of the effect of exercise on blood glucose and alter their daily insulin and nutrition plan accordingly. This caution, combined with some immediate hyperglycemic effects of strenuous exercise [29], explains the relative rarity of hypoglycemia during exercise or 1 to 2 hours after exercise.

A more common but perhaps less widely recognized occurrence is the development of hypoglycemia 6 to 15 hours after strenuous activity [2]. A 2-year prospective case study of 300 children and adolescents with IDDM revealed that 48 (16%) experienced such postexercise late-onset (PEL) hypoglycemia [19]. Distinctive characteristics of this phenomenon are summarized in Table 9D–3. In addition to its delayed occurrence, PEL hypoglycemia is distinguished by its severity. Stupor, coma, and/or seizures during the episode are relatively common, and warnings of impending hypoglycemia are often absent. Patient age, duration of IDDM, and "tightness" of metabolic control are unrelated to the likelihood of experiencing PEL hypoglycemia. On the other hand, a clear predisposing factor is the associated intensity or duration of exercise, considered to be exceptional for essentially all patients affected. Diabetic athletes making the transition from an untrained to a trained state are more likely to experience delayed hypoglycemia.

Intuitively, one would expect the hypoglycemic effect of exercise to disappear within a few hours after the activity. In fact, the occurrence of nocturnal hypoglycemia following an intervening meal and bedtime snack, and occasionally in the absence of nocturnal exogenous insulin, indicates that the effects of acute exercise on glucose metabolism persist for several hours. Enhanced glucose uptake and glycogen synthesis by muscle [7], increased sensitivity of muscle to insulin [23], and increased binding of insulin to monocytes [15] combine to improve glucose tolerance in untrained subjects for at least 18 hours after exercise [9]. This increased sensitivity to insulin, combined with the avid extraction of glucose from the circulation for the repletion of muscle and liver glycogen stores depleted during exercise, explains the occurrence of nocturnal PEL hypoglycemia in most cases. Diabetic individuals with diminished glucagon, epinephrine, and cortisol responses to hypoglycemia have an additional risk of experiencing delayed hypoglycemia. Deficient counterregulation contributes to the lack of warning

signs and increased severity of PEL hypoglycemia and may explain its occurrence in patients not using insulin with peak activity in the night or early morning.

Consequences of Insufficient Insulin Effect During Exercise

High-intensity, short-term exercise is normally associated with transient rises in plasma glucose levels, which peak 5 to 15 minutes after exercise is stopped and return to baseline within 40 to 60 minutes [21]. Suppression of insulin secretion, stimulation of the sympathetic nervous system and release of counterregulatory hormones, and stimulation of hepatic glucose production in excess of peripheral glucose uptake combine to produce this glycemic response. Following the termination of exercise, insulin secretion is promptly stimulated in the presence of elevated plasma glucose levels and declining epinephrine concentrations. Consequently, transiently elevated plasma glucose levels return rapidly to normal [4].

In the athlete with IDDM, the postexercise rise in plasma insulin levels does not occur, and hyperglycemia following intense exertion may be sustained and of greater magnitude [3]. This effect is exaggerated when a preexercise insulin deficiency is present and the plasma glucose concentration is elevated (Fig. 9D–3). Severe insulin deficiency results in accelerated lipolysis and ketogenesis and is heralded (in the nonfasting state) by ketonuria. Exercise in this setting leads to further impairment of glucose utilization, enhancement of lipolysis, and stimulation of hepatic glucose production and ketogenesis. In addition to prompt elevation in blood glucose concentrations, defective peripheral clearance of ketones leads to a rapid worsening of an already compromised metabolic state [8].

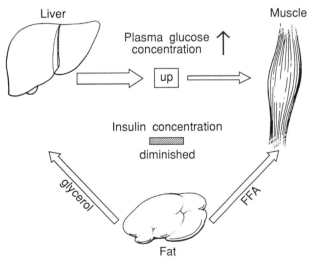

FIGURE 9D–3
The response to exercise in insulin-deficient insulin-dependent diabetics. When plasma insulin is markedly diminished, hepatic glucose production is markedly increased during exercise, but the increase in skeletal muscle utilization of glucose is smaller than under normal circumstances; thus, blood glucose levels rise. (J. M. Ekoe, Overview of diabetes mellitus and exercise. *Medicine and Science in Sports and Exercise* 21(4):353–368, 1989, © The American College of Sports Medicine.)

TABLE 9D–3
Clinical Characteristics of Postexercise Late-Onset Hypoglycemia

Occurs 6 to 15 hours following unusually strenuous or prolonged activity (frequently nocturnal)
Frequently severe (stupor, coma, and/or seizure) with few warning signs
Unrelated to age of patient or duration of diabetes
Unrelated to strict metabolic control or excessive evening insulin administration

CLINICAL EVALUATION

Safe and successful athletic participation by individuals with IDDM is achievable and desirable but requires a commitment to a program of frequent monitoring of blood sugar values and careful nutritional planning. The physician's initial evaluation should emphasize this point and assess the individual's ability and willingness to embark on such a program (Table 9D–4). The patient's prior history of metabolic control and attention to diabetic care is the most informative indicator of future performance. Measurements of glycosylated hemoglobin (a long-term indicator of blood glucose control) should reflect (at least) a satisfactory metabolic status according to the physician caring for the individual. A well-established routine of self-monitoring of blood glucose levels and, when indicated, urine ketones should be evident.

The athlete's awareness and knowledge of the possibility of hypoglycemia should be explored, and strategies for prevention (presented later) should be

TABLE 9D–4
Preparticipation Evaluation of the Diabetic Athlete

I. Has a routine of stable blood glucose control been established?
   Requirements for the athlete:
   A. Recent satisfactory measurement of glycosylated hemoglobin levels
   B. Habitual (two to three times/day) self-monitoring of blood glucose levels
   C. Understanding of indications for or interpretation of urine ketone measurement
II. Has the possibility of hypoglycemia been anticipated?
   Requirements for the athlete:
   A. Recognition of early warning signs of hypoglycemia
   B. Knowledge of treatment strategies for mild hypoglycemia (e.g., hard candy, glucose tablets)
   C. Knowledge of strategies to be followed in case of pre- and postexercise alterations in insulin dosage
   D. Medic-Alert bracelet or necklace indicating diabetic state
   E. Provision of glucagon (1 mg for subcutaneous or IM injection by trained personnel) for treatment of severe hypoglycemia
III. Are complications of diabetes present?
   Requirements for the athlete:
   A. Recent evaluation of blood pressure, neurologic function, joint mobility, and skin condition
   B. Recent retinal examination by licensed ophthalmologist
   C. Screening laboratory evaluation for blood lipid abnormalities and diabetic nephropathy

discussed. Because alterations in insulin sensitivity (and complications from hypoglycemia) are more frequently observed during the transition from the untrained to the trained state, the individual's current level of physical activity and fitness should influence the caution with which new activity is undertaken. An important historical point in the history that should be clarified is the frequency of previous hypoglycemic reactions and, in particular, the individual's ability to detect the symptoms that herald their onset. Lack of warning signs of hypoglycemia indicates a likely deficiency in both the glucagon and epinephrine responses and increases the probability of occurrence of potentially dangerous hypoglycemia during and after exercise.

Diabetic individuals are predisposed to the development of coronary artery disease, although manifestations of this during adolescence are extraordinarily rare. Elevations in serum total cholesterol and triglycerides and reductions in high-density lipoprotein cholesterol (together comprising a "high risk" lipid profile) are frequently seen when diabetes management is suboptimal. With sustained correction of hyperglycemia, these lipid derangements usually, but not always, return to normal. A detailed family history of premature coronary artery disease, stroke, or hypertension should be elicited and, if positive, should prompt a thorough evaluation of the athlete's lipoprotein profile. Detection of familial lipid disorders and marked abnormalities in serum total cholesterol levels (e.g., more than 300 mg/dL) should *not* interfere with sports participation but should focus attention on the need for improvement in blood glucose control and referral for additional lipid-lowering therapy.

Most adolescents with IDDM do not manifest its long-term effects. Nevertheless, a careful physical examination should focus on detecting any microvascular, neurologic, or musculoskeletal complications of IDDM, several of which can be worsened by exercise. Particular attention should be directed toward individuals beginning a fitness program who have poor control of long-standing diabetes. Both diabetic retinopathy [5] and painful neuropathy [1] can become worse abruptly with rapid correction of hyperglycemia. Patients with proliferative retinopathy may develop retinal or vitreous hemorrhages when blood pressure becomes transiently elevated during vigorous exertion. Athletic activities requiring heavy lifting or straining should be avoided by these individuals. Rarely, retinal detachment may occur [13]. Urinary albumin excretion, one of the manifestations of diabetic nephropathy, can become worse with exercise [22]. This may represent merely a transient hemodynamic response, and it has not been shown that exercise has any deleterious effect on the progression of renal disease. Hypertension, on the other hand, is known to accelerate the progression of diabetic nephropathy and should be carefully screened out. Peripheral neuropathy associated with even minor degrees of anesthesia will predispose the athlete to soft tissue and joint injuries. Limited joint mobility, a sign of poor metabolic control and increased risk of microvascular disease [24], can also impair performance and increase the risk of injury. Autonomic neuropathy may lead to a decreased cardiovascular response to exercise [11], a lower maximal aerobic capacity [25], and a blunted awareness of hypoglycemia and response to dehydration, each of which can impair physical work capacity.

Because significant vascular and neurologic complications usually do not become apparent during the first 10 years of IDDM, participation in virtually any sport is possible for the diabetic child. Only in extremely rare circumstances does retinopathy or nephropathy contraindicate participation in activities likely to result in trauma (e.g., football) or in increases in intracranial pressure and blood pressure (e.g., heavy weight-lifting). However, although participation in any chosen sport is rarely prescribed, not all sports activities are equally *beneficial* to the prevention of long-term diabetic complications. Longevity and freedom from vascular and neurologic complications are enhanced by the attainment of ideal body weight and increased lean body mass (which increases sensitivity to insulin), relatively low dietary protein intake (which delays progression of renal disease), and avoidance of hypertension (which, if present, accelerates retinopathy and nephropathy). Sports that emphasize aerobic conditioning, muscle tone, and endurance and that have the potential for lifelong participation are more likely to contribute to achieving these goals. Consequently, counseling the young diabetic patient about prudent choices of sports activities is a valuable role for the health care professional.

TREATMENT

Guidelines for Preventive Management

The challenging responsibility of the physician supervising the athletic activities of a diabetic patient is to anticipate and prevent untoward events, not to treat them. Preparation begins with the establishment of at least reasonably good metabolic control through a consistent routine of insulin administration and caloric intake. Glycemic control should be documented by obtaining measurements of glycosylated hemoglobin levels. Records of home blood sugar monitoring alone should not be relied upon because these are notoriously unreliable. Gradual introduction of a fitness program should precede athletic participation by several weeks. If possible, and paying careful attention to preventing injury, daily exercise (rather than three or four sessions per week) should be encouraged to avoid the need for frequent readjustment of the daily insulin and nutrition regimen. Attainment of fitness *prior to* the athletic season allows insulin sensitivity to increase gradually and reduces the risk of hypoglycemia during enforced practice and competition.

Strategies for preventing hypoglycemia and hyperglycemia during and after each training session or competition are summarized in Table 9D–5. The athlete should incorporate these procedures into an individualized regimen that is modified according to an estimation of the duration and intensity of exer-

TABLE 9D–5
Prevention of Hypoglycemia or Hyperglycemia with Exercise

Before Exercise
1. Estimate intensity, duration, and energy expenditure needed for exercise
2. Eat a meal 1 to 3 hours before exercise
3. Insulin:
   a. Administer insulin more than 1 hour before exercise
   b. Decrease insulin that has peak activity coinciding with exercise period
4. Assess metabolic control:
   a. If blood sugar is less than 100 mg/dL, take supplemental preexercise snack
   b. If blood sugar is more than 250 mg/dL, delay exercise and measure urine ketones
   c. If urine ketones are positive, take insulin and delay exercise until negative

During Exercise
1. Supplement calories with carbohydrate feedings (30 to 40 g for adults, 15 to 25 g for children) every 30 minutes during extended, strenuous activity
2. Replace fluid losses adequately
3. Monitor blood glucose during exercise of long duration

After Exercise
1. Monitor blood sugar, including overnight if amount of exercise is not habitual
2. Increase calorie intake for 12 to 24 hours after activity, according to intensity and duration of exercise
3. Reduce insulin dosage that reaches peak effect in evening and night, according to intensity and duration of exercise.

Adapted with permission from Horton, E. S. Role and management of exercise in diabetes mellitus. *Diabetes Care* 11:201–211, 1988. Copyright © 1988 by American Diabetes Association, Inc.

cise. In anticipation of heightened insulin sensitivity and insulin-independent glucose uptake, administration of insulin should be timed to avoid the occurrence of the peak activity of short-acting (regular) insulin during exercise; and reductions in dosage of longer-acting insulin (e.g., NPH) may also be needed. Because exercise can cause deterioration of metabolic control that is already poor, elevated blood glucose levels and ketonuria should delay the start of an exercise program until adequate glycemic control and absence of ketosis have been achieved.

In the diabetic athlete, a lack of exercise-induced reduction in serum insulin levels may cause inappropriate suppression of hepatic glycogenolysis and gluconeogenesis during periods of increased energy utilization. Therefore, provision of adequate calories during prolonged exercise is important. A rough estimate of caloric needs assumes an energy expenditure (for most aerobic sports) of 600 kcal/hour or 10 kcal/minute for an adult individual. At an intensity of approximately 50% maximum aerobic capacity, glucose oxidation might be expected to supply approximately 50% of the energy, or 5 kcal/minute (1.25 g carbohydrate). For each 30 minutes of exercise, therefore, supplementary feedings of 30 to 40 g of carbohydrate (15 to 25 g for school-aged children) will aid in preventing hypoglycemia. (Note that the failure of exercising IDDM patients to lose weight and gain improved metabolic control is also due largely to this need for supplemental caloric intake) [32]. Because fluid losses are increased by glycosuria, and an awareness of and response to dehydration can be impaired by autonomic dysfunction, adequate fluid replacement is crucial for the athlete with IDDM.

Awareness of the possibility of postexercise late-onset (PEL) hypoglycemia is the most important step in preventing it. Education of diabetic athletes about the need to distinguish between extraordinary and ordinary physical activity is paramount. The lack of warning signs of impending PEL hypoglycemia emphasizes the need to implement preventive measures in the hours following unusually strenuous or prolonged activity. Reductions in both short- and long-acting insulin dosages as well as increases in caloric intake are appropriate in proportion to the degree to which exertion has exceeded the daily routine. The frequency of blood glucose monitoring should be increased so that appropriate readjustments can be made to avoid worsening of metabolic control. In general, athletes with IDDM who experience a significant increase in appetite or who feel weaker and more exhausted than usual in the evening following exercise can minimize or avoid hypoglycemia by increasing caloric intake, taking less

insulin that peaks in the evening and overnight, and checking blood glucose levels more frequently during the night.

An occasional highly motivated diabetic athlete receives continuous subcutaneous insulin through an insulin pump. These devices utilize only short-acting (regular) insulin delivered through exposed tubing and a needle temporarily implanted most often in the skin of the abdomen. Continuous low-dose regular insulin is supplemented by additional premeal insulin also delivered through the pump setup. The pump can be detached temporarily, allowing for participation in water sports. Although this method of insulin administration does not always improve blood glucose control, it does allow greater flexibility in timing of meals and activity. When appropriate adjustments in insulin dosage (especially during the evening and night following intense exercise) and caloric intake detailed above are made, the insulin pump can be used safely by the athlete. However, the following potential hazards of the insulin pump deserve precautionary attention: (1) Suitable protection of the device should be provided, which may be difficult during contact sports; (2) proper needle placement must be monitored to prevent cessation of insulin delivery and development of ketoacidosis during exercise; (3) conversely, the patient must be aware that continuous insulin administration is occurring and should be interrupted if hypoglycemia requiring treatment occurs.

Guidelines for Acute Management of Hypoglycemia

Hypoglycemia during exercise may occur in spite of preventive efforts, particularly when the duration of activity is more prolonged than expected or the degree of exertion is exceptional. Warning signs of impending hypoglycemia tend to be idiosyncratic and reproducible for each diabetic individual, so clinical symptoms are reliable diagnostic indicators of declining blood glucose levels (typically in the range of 50 to 70 mg/dL) and are usually readily identified. Symptoms of mild hypoglycemia (e.g., dizziness, fatigue, extreme hunger, headache) are best treated by readily available and easily absorbed sources of carbohydrate. Fruit juices, oral glucose tablets, and candy usually suffice. These should be supplemented by a food containing complex carbohydrate and protein, which will produce a sustained glycemic response.

An athlete who is unconscious or impaired to such an extent that protection of the airway is uncertain should *not* be given oral preparations. The blood glucose level in these circumstances is usually less than 40 mg/dL. Although confirmation of hypoglycemia using a blood glucose test strip or meter is desirable, treatment should not be delayed if these diagnostic tools are not available; that is, temporary elevation of the blood glucose level is not hazardous to a (nonhypoglycemic) diabetic athlete and should be attempted based upon the presumptive diagnosis of hypoglycemia. Parenteral glucagon (1 mg given subcutaneously or intramuscularly) should be available to properly instructed trainers or other responsible persons and is the treatment of choice in these circumstances. Because the effect of glucagon is not long-lasting, supplemental carbohydrate and protein should also be given when the mental status has improved. Occasional gastrointestinal upset and (less frequently) vomiting are unavoidable side effects of glucagon administration but should not deter its use when indicated.

CRITERIA FOR SPORTS PARTICIPATION

An individual with IDDM should meet the following criteria before participation in an athletic or conditioning program is permitted:

1. Knowledge, technical mastery, and consistent application of home blood sugar monitoring techniques.

2. Achievement of reasonable metabolic control as documented by measurements of glycosylated hemoglobin levels, blood sugar levels of less than 250 mg/dL, and the absence of ketonuria.

3. Evidence of passing a recent screening for diabetic microvascular complications; if such complications are present, the patient should demonstrate awareness of compensatory alterations in exercise choice or intensity needed for safe participation.

4. Knowledge of preventive strategies needed to avoid hypoglycemia during and after exercise.

5. If possible, prior arrangements with the team trainer or other responsible person should be made to ensure emergency treatment of hypoglycemia.

Consideration of these guidelines can help to fulfill the common desire of young patients with IDDM to participate in sports. In addition to the likely, although unproved, long-term benefits of improved glycemic control and reduced microvascular complications, athletic activity provides an opportunity to pursue a meaningful personal goal. Participating with peers in sports promotes feelings of mastery, control, and individuality that significantly enrich the psychological and physical well-being of young persons with IDDM.

References

1. Allen, D. B., and MacDonald, M. J. Ice-water addiction complicating acute painful diabetic neuropathy. *Diabetes Care* 10:796–797, 1987.
2. Allen, D. B., and MacDonald, M. J. Preventing postexercise late-onset hypoglycemia. *Pract Diabetology* 8(1):1–9, 1989.
3. Berk, M. A., Clutter, W. E., Skor, D. A., Shah, S. D., Gingerich, R. P., Parvin, C. A., and Cryer, P. E. Enhanced glycemic responsiveness to epinephrine in insulin dependent diabetes mellitus is the result of the inability to secrete insulin. *J Clin Invest* 75:1842–1851, 1985.
4. Calles, J., Cunningham, J. J., Nelson, L., Brown, N., Nadel, E., Sherwin, R. S., and Felig, P. Glucose turnover during recovery from intensive exercise. *Diabetes* 32:734–738, 1983.
5. Daneman, D., Drash, A. L., Lobes, L. A., Becker, D. J., Baker, L. M., and Travis, L. B. Progressive retinopathy with improved control in diabetic dwarfism (Mauriac's syndrome). *Diabetes Care* 4:360–365, 1981.
6. Farrannini, E., Linde, B., and Faber, O. Effect of bicycle exercise on insulin absorption and subcutaneous blood flow in the normal subjects. *Clin Physiol* 2:59–70, 1982.
7. Fell, R. D., Terblanche, S. E., Ivy, J. L., Young, J. C., and Holloszy, J. O. Effect of muscle glycogen content of glucose uptake following exercise. *J Appl Physiol* 52:434, 1982.
8. Fery, F., de Mairtelaer, V., and Balasse, E. O. Mechanism of the hyperketonaemic effect of prolonged exercise in insulin-deprived type 1 (insulin-dependent) diabetic patients. *Diabetologia* 30:298–304, 1987.
9. Heath, G. W., Gavin, J. R., III, Hinderliter, J. M., Hagberg, J. M., Bloomfield, S. A., and Holloszy, J. O. Effects of exercise and lack of exercise on glucose tolerance and insulin sensitivity. *J Appl Physiol* 55:512, 1983.
10. Hermansen, L., Hultman, R., and Saltin, B. Muscle glycogen during prolonged severe exercise. *Acta Physiol Scand* 71:129–139, 1967.
11. Hilsted, J., Galbo, H., and Christensen, N. J. Impaired cardiovascular responses to graded exercise in diabetic autonomic neuropathy. *Diabetes* 28:313–319, 1979.
12. Horton, E. S. The role of exercise in the treatment of hypertension in obesity. *Int J Obes* 5(Suppl. 1):165, 1981.
13. Horton, E. S. Role and management of exercise in diabetes mellitus. *Diabetes Care* 11:201–211, 1988.
14. Kemmer, F. W., Berchtold, P., Berger, M., Starke, A., Cüppers, H. J., Gries, F. A., and Zimmerman, H. Exercise-induced fall of blood glucose in insulin-treated diabetics unrelated to alteration of insulin mobilization. *Diabetes* 28:1131, 1979.
15. Koivisto, V. A., Soman, V., Conrad, P., Hendler, R., Nadel, E., and Felig, P. Insulin binding to monocytes in trained athletes. *J Clin Invest* 64:1011, 1979.
16. Lawrence, R. D. The effect of exercise on insulin action in diabetes. *Br Med J* 1:648, 1926.
17. LeBlanc, J., Nadeau, A., Boulay, M., and Rousseau-Migneron, S. Effects of physical training and adiposity on glucose metabolism and ^{125}I-insulin binding. *J Appl Physiol: Repirat Environ Exerc Physiol* 46:235, 1979.
18. Lipson, L. C., Bonow, R. W., Schaefer, E. J., Brewer, H., and Lindgren, F. T. Effect of exercise conditioning on plasma high-density lipoprotein and other lipoproteins. *Atherosclerosis* 37:529, 1980.
19. MacDonald, M. J. Postexercise late-onset hypoglycemia in insulin-dependent diabetic patients. *Diabetes Care* 10:584–588, 1987.
20. Merrero, D. G., Fremion, A. S., and Golden, M. P. Improving compliance with exercise in adolescents with insulin-

dependent diabetes mellitus: Results of a self-motivated home exercise program. *Pediatrics* 81:519, 1988.

21. Mitchell, T. H., Abraham, G., Schiffrin, A., Leiter, L. A., and Marliss, E. B. Hyperglycemia after intense exercise in IDDM subjects during continuous subcutaneous insulin infusion. *Diabetes Care* 11:311–317, 1988.

22. Mogensen, C. E., and Vittinghus, E. Urinary albumin excretion during exercise in juvenile diabetes. *Scand J Clin Lab Invest* 35:295–300, 1975.

23. Richter, E. A., Baretto, L. P., Goodman, M. N., and Ruderman, N. B. Muscle glucose metabolism following exercise in the rat. Increased sensitivity to insulin. *J Clin Invest* 69:785, 1982.

24. Rosenbloom, A. L., Silverstein, J. H., Lezotte, D. C., Richardson, K., and McCallum, M. Limited joint mobility in childhood diabetes mellitus indicates increased risk for microvascular disease. *N Engl J Med* 305:191–194, 1981.

25. Rubler, S. Asymptomatic diabetic females: Exercise testing. *NY State J Med* 81:1185–1191, 1981.

26. Soman, V. R., Koivisto, V. A., Brantham, P., and Felig, P. Increased insulin binding to monocytes after acute exercise in normal man. *J Clin Endocrinol Metab* 47:216, 1978.

27. Soman, V. J., Koivisto, V. A., Deibert, D., Felig, P., and DeFronza, R. A. Increased insulin sensitivity and insulin binding to monocytes after physical training. *N Engl J Med* 301:1200, 1979.

28. Vranic, M., and Berger, M. Exercise and diabetes mellitus. *Diabetes* 28:147, 1979.

29. Wahren, J., Felig, P., and Hagenfeldt, L. Physical exercise and fuel homeostasis in diabetes mellitus. *Diabetologia* 14:213, 1978.

30. Wallberg-Henriksson, H. Acute exercise: Fuel homeostasis and glucose transport in insulin-dependent diabetes mellitus. *Med Sci Sports Exerc* 21(4):356–361, 1989.

31. Zinman, B., Murray, F. T., Vranic, M., Albisser, A. M., Leibel, B. S., McClean, P. A., and Marliss, E. B. Glucoregulation during moderate exercise in insulin treated diabetics. *J Clin Endocrinol Metab* 45:641, 1977.

32. Zinman, B., Zuniga-Guajardo, S., and Kelly, D. Comparison of the acute and long-term effects of exercise on glucose control in Type I diabetes. *Diabetes Care* 7:515–519, 1984.

Therapeutic Drug Use and Epilepsy in Sports

Mary L. Zupanc, M.D.

In the United States and elsewhere, millions of children and adults participate in organized and unorganized sports. Because 0.5% of the general population has epilepsy, persons with seizures are often sports participants [3, 9, 12, 16]. The diagnosis of epilepsy per se does not preclude participation in sports. As health care providers who treat patients with epilepsy, it is our job to ensure that the epileptic patient *can* participate within safe guidelines.

TERMINOLOGY

Epilepsy, by definition, refers to chronic recurrent seizures. An epileptic seizure per se is the clinical manifestation of an abnormal synchronous discharge of a group of neurons in the cerebral cortex. The features of a clinical epileptic seizure differ according to the anatomic location of the neurons in the cortex. In this way, an epileptic seizure can cause an altered state of consciousness, tonic-clonic movements, stereotypic or repetitive movements, loss of muscle tone, sensory or psychic experiences, or autonomic dysfunction. These disturbances usually, but not always, are accompanied by electrographic epileptogenic activity during scalp electroencephalography (EEG). There are many different types of epileptic seizures. Unfortunately, in the past, the classification of epileptic seizures has been confusing, and many redundant terms have been used. In 1970 and again in 1981 (a revision), the International League Against Epilepsy published the International Classification of Epileptic Seizures [6]. According to this classification, epileptic seizures are divided into two categories—generalized and partial. Generalized seizures are characterized by bilat-

erally synchronous discharges on the EEG. They can be further subdivided into nonconvulsive and convulsive seizures. Nonconvulsive seizures include absence seizures ("petit mal") and myoclonic seizures. Convulsive seizures are the more familiar tonic-clonic seizures ("grand mal"). Partial seizures are seizures that begin in one part of the brain. Partial seizures can also be subdivided into simple partial seizures, *without* impairment of consciousness, and complex partial seizures, *with* impairment of consciousness. These partial seizures can encompass a whole range of symptoms including motor, sensory, autonomic, or psychic phenomena. For example, a typical partial seizure may include staring, diminished responsiveness, and stereotypic hand movements. If a partial seizure evolves into a generalized tonic-clonic seizure, it is called a secondary generalized seizure.

Epileptic seizures, if recurrent, should be termed epilepsy. Epileptic syndromes have also been identified and classified [5]. The three big subdivisions are partial epilepsies, generalized epilepsies, and unclassified epilepsies. Simplistically, generalized epilepsies are often idiopathic, contain a genetic component, are associated with bilateral synchronous discharges on EEG, and often have a good prognosis. Partial epileptic syndromes are called the localization-related epilepsies. They can be further divided into idiopathic, age-related onset, and symptomatic. The idiopathic category implies that a specific etiology remains unknown, and all diagnostic studies are negative. In the age-related category, the most common localization-related epilepsy is benign rolandic epilepsy. This epilepsy is a self-limited epilepsy that is 100% benign and is outgrown by the time the child reaches puberty. It is associated with temporal-central spikes seen during drowsiness on the EEG; the seizures can be either partial or secondarily generalized tonic-clonic seizures. The symptomatic category comprises epileptic syndromes caused by focal brain abnormalities such as malformations, trauma, and tumor. Infections and toxic, metabolic, and neurodegenerative causes are also categorized under this heading; these causes usually result in unifocal or multifocal damage to the brain. Unfortunately, symptomatic localization-related epilepsies tend to have a poorer prognosis for good control or cure.

Various medications are commonly used to treat epilepsy. They include phenobarbital, carbamazepine, phenytoin, valproate/valproic acid, ethosuximide, clonazepam, and primidone. In general, carbamazepine, phenytoin, and phenobarbital/primidone are commonly used to treat partial seizures. Valproate can also be prescribed for partial seizures. However, valproate is most commonly prescribed for generalized epilepsies, convulsive or nonconvulsive. Absence epilepsy is usually treated with valproate, ethosuximide, or clonazepam. Myoclonic seizures are most commonly treated with valproate or clonazepam. Most epileptic patients, probably close to 75% to 80%, are easily treated and remain seizure-free on monotherapy. Some patients will require polytherapy, although rarely with more than two drugs.

Unfortunately, these medications also carry side effects, many of which can affect performance in an athletic event. Phenobarbital, phenytoin, and primidone have all been reported to produce significant cognitive and behavioral side effects. Phenobarbital has been reported to reduce IQ scores in children, an effect that outlasts administration of the drug [10]. In addition, phenobarbital and primidone have been reported to produce sedation in adults, irritability, poor attention and concentration in children, depression in adolescents, mood lability, and sleeping disorders. Phenytoin has also been reported to depress cognitive function, slow overall performance, and produce sedation [24].

Valproate and carbamazepine are more newly developed drugs and have not been reported to produce severe cognitive and sedative side effects. However, carbamazepine can cause sedation and diplopia, especially at high doses. It can also suppress the bone marrow, in particular, the white blood cells, resulting in an increased susceptibility to infection [24]. Valproate rarely causes sedation but frequently is associated with weight gain, transient hair loss, and easy bruising-bleeding (as a result of low numbers of platelets, i.e., thrombocytopenia) [24]. The liver toxicity reported with valproate is exceedingly rare in late childhood (over 10 years),

adolescent, and adult populations. In children less than 2 years of age who are taking multiple anticonvulsants and have major cerebral malformations, however, the risk of idiopathic fatal liver toxicity with valproate is 1 in 500 [8]. Ethosuximide is another drug that is commonly used to treat generalized epilepsies, primarily absence epilepsy. Its main side effect is nausea and vomiting. It can also cause sedation and suppression of the bone marrow [24].

Fortunately, most of these drugs are well tolerated by the epileptic population, resulting in good seizure control and few side effects.

MANAGEMENT OF SEIZURES

Although epilepsy in the vast majority of epileptic individuals who participate in sports is well controlled, occasionally an individual does have a seizure while engaging in the sport of his choosing. This can be embarrassing for the individual, frightening to those who observe it, and at times dangerous, particularly if it occurs during a swimming event. For the team physician, coach, and other supervisors of the event, the main tenet to remember is to stay calm and keep others calm. A seizure is almost always a self-limited event that requires *minimal* intervention. The next recommendation is to prevent the individual with epilepsy from hurting himself. This goal can be accomplished by removing objects close to him, particularly during a generalized tonic-clonic seizure. If the individual has been swimming, a skilled lifeguard should remove the person from the water as quickly as possible. A tongue blade or other object should *never* be inserted between the teeth of an individual who is having a seizure; he will not swallow his tongue, and chances are that he will bite the object in two or injure the person who is attempting to insert the object. If a mouthpiece is present, it should be removed if this can be accomplished easily; otherwise, forget it. Once a seizure is over, usually within minutes, the individual is often tired. He should be allowed to lie down. If the episode has been a convulsive (i.e., tonic-clonic) seizure, the individual should be turned onto his side to prevent aspiration of vomitus. If the episode has been a nonconvulsive seizure (e.g., partial complex seizure consisting of stereotypic movements or bizarre behaviors), this is usually not necessary, although the individual is often lethargic and semi-responsive following the seizure. Needless to say, if a seizure occurs during a sporting event, the individual should not participate further that day. The one exception to this rule may be the epileptic who has absence epilepsy ("petit mal") and experiences occasional break-through seizures.

Once the physical needs of the individual who has experienced a seizure have been met, attention should be turned to the emotional needs of this individual and others. The embarrassment that many people feel following a seizure should be acknowledged and addressed. This implies listening to the individual and his feelings. A simple "How are you feeling?" may be enough to elicit a conversation. Alternatively, the person who had the seizure may not want to discuss it at all. However, the significant adults in this person's life may want to let him know that they are willing to listen and talk at any time in the future. The same thing applies to other people who witnessed the event. These people need to know that seizures are not harmful, that the person will recover, and that most individuals with epilepsy lead perfectly normal lives with only rare or no breakthrough seizures. Unfortunately, the lay public still has many misconceptions about epilepsy.

CLINICAL EVALUATION OF THE SPORTS PARTICIPANT

Evaluation of an individual with epilepsy who wishes to participate in sports is relatively easy. There are only a few questions that need to be answered in assessing the ability of an individual with epilepsy to participate in sports: (1) How well controlled are his or her seizures? (2) Is the individual experiencing significant side effects from the anticonvulsant medication? (3) In what sport activity does the person wish to participate? (4) How great is the person's desire to participate?

TREATMENT OPTIONS AND CRITERIA FOR SPORTS PARTICIPATION

Individuals with epilepsy, even those whose seizures are not under total control, should be allowed to participate in a variety of sports. Severely limiting sports participation may impose unduly harsh restrictions on the individual with epilepsy, resulting in a feeling of being stigmatized as an "outsider." As stated by the American Academy of Pediatrics in their Committee Report on this subject, "in today's culture, sports and athletics are extremely important, and unnecessarily strict interpretations of medical conditions may do more harm than good" [12].

Most authors agree that individuals with epilepsy can participate in the great majority of sports even if incomplete control of seizures exists [3, 11, 14, 16]. There is also widespread agreement that some sports by their very nature pose situations that could be dangerous even if only one breakthrough seizure were to occur. These sports include rock-climbing, rope-climbing, activity on parallel bars, high diving, and prolonged underwater swimming [12]. In addition, if it is known that hyperventilation will increase one's propensity to have a seizure, other sports activities *may* have to be modified, for example, track, swimming, basketball, and so on. In the same vein, if fatigue is a major precipitant of an individual's seizures, reevaluation of the appropriateness of a particular sports activity may be necessary. In most circumstances, the sport can be allowed but with certain restrictions.

There are two areas that spark controversy among the various authors: body contact sports and swimming. Individuals with epilepsy used to be universally denied participation in body contact sports such as football. It was hypothesized that repeated head trauma would result in further brain injury with eventual aggravation of the epileptic condition. There are no studies in the medical literature that prove that chronic head trauma increases the frequency of epileptic seizures [1, 7, 14]. In fact, recent authors have argued against restriction, stating that it does not happen in reality [3, 14]. Their conclusion is that the decision to participate in body contact sports must be made on an individual basis. Frequent seizures (occurring daily or weekly) are considered a relative contraindication by these authors because of the possibility that a seizure will incapacitate the patient at an inopportune moment, for example, a football lineman experiencing a seizure right after the beginning of a play.

Swimming represents a special case. Fatalities of individuals with epilepsy in swimming pools, ocean, and bathtubs are well-documented but relatively rare events. However, an individual with epilepsy who swims is four times more likely to drown than the normal individual who swims [16, 19–21]. Various contraindications for swimming have been proposed by different authors [2, 4, 15, 19–21, 23], and include

1. Frequent seizures, more than once a month
2. A seizure-free status that has lasted for less than 3 months
3. A patient who is going through a period of adjustment to anticonvulsant drug therapy
4. Noncompliance with drug regimen
5. Unstable blood levels of drug
6. Lack of one-to-one supervision

7. Murky water in oceans and lakes
8. Mental retardation

Frequent seizures pose a danger to the individual with epilepsy who swims. Oceans and lakes with murky water are particularly dangerous because it may be impossible to find a drowning individual before the onset of disastrous consequences. Mental retardation increases the risk of swimming even further, according to some studies [2, 19, 23].

Most authors concur that participation in swimming can be allowed if an individual with epilepsy has been seizure-free for 6 months, is mentally normal, has stable and therapeutic drug levels, and is closely supervised by a lifeguard who is aware of the individual's condition and knows cardiopulmonary resuscitation (CPR) [11, 16, 19]. The "buddy system" should also be used. If these requirements are met, swimming poses little risk.

In reality, bathing is a much *bigger* risk [16–18]. Several studies have shown that taking a bath presents a more serious risk of drowning than swimming [13, 22]. In fact, it is the older adolescent and adult who run the highest risk. They tend to bathe alone, whereas younger children are closely supervised during bath time by their parents.

AUTHOR'S PREFERRED METHOD OF TREATMENT

In my practice, I allow the great majority of patients to participate in athletics and sports. The opportunities to participate are much greater than they used to be. Even individuals who are physically or mentally disabled and who have epilepsy can become active in sports, for example, in the Special Olympics.

I allow my patients to participate in contact sports such as football. The rationale behind my recommendation is always based on striking a balance between the patient's medical condition and his or her perceived need to participate in contact sports. I am still somewhat reluctant to be enthusiastic about contact sports for people in the general population as well as for individuals with epilepsy: Why expose oneself unnecessarily to possible head trauma, such as a concussion? Even mild head trauma can result in attention or concentration problems, memory problems, headache, and other somatic complaints. Additionally, moderate to severe head trauma can result in further brain injury and a decreased seizure threshold. Individuals with epilepsy, like the general population, should be outfitted with the proper protective equipment for the particular contact sport involved, including a mouthpiece. An individual with epilepsy will not choke on the mouthpiece in the event of a seizure.

My recommendations for swimming are quite generous. If an individual with epilepsy can be closely supervised (one on one) by a competent lifeguard who is trained in CPR and who is aware of the individual's condition, I usually allow swimming if the individual has been seizure-free for more than 3 months. Modifications of this recommendation occur if (1) the individual's usual seizure is very brief and results in no or only brief alteration of consciousness, and (2) the individual's desire to swim is great.

If an individual with epilepsy has chosen a sport that I deem unsafe (based on the criteria outlined previously), I attempt to steer him or her into an alternative sport or to set limitations or modifications on the sport chosen.

SUMMARY

In summary, I encourage my patients with epilepsy to participate in sports and athletic events. These individuals can participate in the majority of sports activities with no or only a few restrictions or modifications. The sense of satisfaction and the growth in self-esteem that result from participation usually far outweigh the risks.

References

1. Aisenson, M. R. Accidental injuries in epileptic children. *Pediatrics* 2:85, 1948.
2. Berggreen, S. M. Accidents and surgical emergencies in a population of mentally retarded children. *Acta Paediatr Scand* 62:289–296, 1972.
3. Berman, W. Sports and the child with epilepsy. *Pediatrics* 74(2):320, 1984.
4. Bowerman, D. L., Levisky, J. A., and Urich, R. W.: Premature deaths in persons with seizure disorders: Subtherapeutic levels of anticonvulsant drugs in postmortem blood specimens. *J Forensic Sci* 23:522–526, 1978.
5. Commission on Classification and Terminology of the International League Against Epilepsy. Proposal for classification of epilepsies and epileptic syndromes. *Epilepsia* 26:268–278, 1985.
6. Commission on Classification and Terminology of the International League Against Epilepsy. Proposal for revised clinical and electroencephalographic classification of epileptic seizures. *Epilepsia* 22:489–501, 1981.
7. Corbitt, R. W. Epileptics and contact sports. AMA Committee on the Medical Aspects of Sports. *JAMA* 229(7):820–821, 1974.
8. Dreifuss, F. E., Santilli, N., Langer, D. H. Valproic acid hepatic fatalities: A retrospective review. *Neurology* 37:379–385, 1987.
9. Engel, J., Jr. *Seizures and Epilepsy.* (Contemporary Neurology Series.) Philadelphia, F. A. Davis, 1989.

10. Farwell, J. R., Young, J. L., Hirtz, D. G. Phenobarbital for febrile seizures—Effects on intelligence and on seizure recurrence. *N Engl J Med* 322:364–369, 1990.

11. Freeman, J. M. Epilepsy and swimming. *Pediatrics* 76(1):139, 1985.

12. Fremont, A. C., et al. Sports and the child with epilepsy. Committee on Children with Handicaps and Committee on Sports Medicine. *Pediatrics* 72(6):884–885, 1983.

13. Krohn, W. Causes of death among epileptics. *Epilepsia* 4:315, 1963.

14. Livingston, S., and Berman, W. Participation of the epileptic child in contact sports. *J Sports Med* 2(3):170–174, 1974.

15. Livingston, S., Pauli, L. L., and Pruce, I. Epilepsy and drowning in childhood. *Br Med J* 3:515–516, 1977.

16. O'Donahoe, N. V. What should the child with epilepsy be allowed to do? *Arch Dis Child* 58:934–937, 1983.

17. Orlowski, J. P. Prognostic factors in drowning and the post-submersion syndrome. *Crit Care Med* 6:94, 1978.

18. Orlowski, J. P. Prognostic factors in pediatric cases of drowning and near-drowning. *Ann Emerg Med* 8:176–179, 1979.

19. Orlowski, J. P., Rothner, D., and Lueders, H. Submersion accidents in children with epilepsy. *Am J Dis Child* 136:777–780, 1982.

20. Pearn, J. H. Epilepsy and drowning in childhood. *Br Med J* 1:1510–1511, 1977.

21. Pearn, J., Part, R., and Yamaoka, R. Drowning risks to epileptic children: A study from Hawaii. *Br Med J* 2:1284–1285, 1978.

22. Sonnen, A. E. H. Epilepsy and swimming. Epilepsy: A clinical and experimental research. *Monograph Neurol Sci* 5:265–270, 1980.

23. Williams, C. E. Accidents in mentally retarded children. *Dev Med Child Neurol* 15:660–662, 1973.

24. Woodbury, D. M., Penry, J. K., and Pippenger, C. E. (Eds.). *Antiepileptic Drugs* (2nd ed.). New York, Raven Press, 1982.

THE TEAM PHYSICIAN

W. Michael Walsh, M.D.
Morris B. Mellion, M.D.

PHILOSOPHY

In the June 18, 1989, *Philadelphia Inquirer,* a headline blares "Team Doctors: A Crisis in Ethics." In the article that follows, staff writers Angelo Cataldi and Glen Macnow point out that "under pressure from coaches and owners to keep players on the field, doctors often must choose between winning and the health of their patients. Former National Football League (NFL) star lineman Charlie Krueger puts it more bluntly: 'The doctor works for the team, and he doesn't know what the Hippocratic oath means as it pertains to you.' " The same article quotes one well-known team physician as saying, "Being a team physician is a bum deal." Another says, "It's a lousy job." Malpractice insurance carriers often offer high rates or refuse to insure physicians with team doctor roles. Even at the college level, doctors are open to such legal charges as those brought in the widely publicized Buonicanti v. Citadel team physician case in which the team doctor was accused of contributing to a disastrous cervical spine injury. Grim reading for the young orthopaedist or family physician who is contemplating taking on the care of an athletic team!

Is that all there is? Headaches, ethical conflicts, and threats of lawsuits? Fortunately not. Especially at the grassroots level, few positions offer as much enjoyment, satisfaction, and stimulation as that of team physician. Being a team physician may be a privilege, a challenge, or a threat. Usually, the experience is a mixture of the three. In ideal circumstances, the privileges and challenges far outweigh the threatening aspects. Caring for an athletic team adds new dimensions to the practice of medicine— the pressures of time, society, and money on the athlete. At all levels of competition, the element of time is paramount. The season's schedule marches inexorably on. If the athlete is not fit to play, the coach must use others in his or her place. The challenge for the physician is to find the fastest safe treatment, thus allowing the athlete to return to play quickly and with small chance of re-injury.

The importance placed by our competitive society on "winning" further complicates the decision making of the team physician. The value of an individual player to the competitive success of the team is often blown out of proportion. This is particularly true at the lower levels of competition, where the welfare of the athlete must come before any thought of team victory. At higher levels of play, the margins of safety often become obscure. What would be an easy decision about a junior high school quarterback may be extremely difficult when it involves an NCAA Division I competitor leading a team into postseason play.

Medical decisions may influence not only the athlete and his or her teammates but also the coach and the institution. When injuries plague a team and key athletes are sidelined, their loss may affect the coach's ability to keep his job, the school's opportunity to participate in lucrative postseason play, and the professional team's success in selling tickets.

Our purpose in this chapter is to discuss the various *functions, relationships,* and *responsibilities* of the team physician to maximize the pleasures of that job while minimizing the frustrations.

ROLES AND FUNCTIONS OF THE TEAM PHYSICIAN

Who Serves As Team Physician?

A variety of doctors serve as team physicians. The 1987 Physician and Sports Medicine survey of 29,000 team physicians revealed that 46% were family physicians and general practitioners, and 17% were orthopaedists. Other specialties with 4% or more representatives included osteopathy, internal medicine, general surgery, pediatrics, and obstetrics/gynecology [12].

A good team physician must address the physical, emotional, and spiritual needs of the athletes. The job requires a comprehensive approach [4, 7, 8, 14]. It is performed within the context of the sport and the needs of the team. The real success of the specialist as team physician depends on his ability to meet the athletes' varied medical and psychosocial needs. This ability depends on a broad knowledge base that includes athletics as well as medicine. Caring for athletes requires a detailed knowledge not only of the musculoskeletal system but also of the processes of growth and development, cardiorespiratory function, gynecology, dermatology, and neurology. The physician must be well versed in psychology and human behavior, particularly as they relate to issues of performance anxiety, motivation, and group interactions. To achieve a high level of success, the physician must understand and appreciate the role of rehabilitation in the care of the athlete. Three areas of pharmacology are also critical: therapeutics, performance-enhancing substances, and recreational drugs. Finally, the team physician must be well grounded in exercise physiology, biomechanics, and the physical demands of specific sports.

Often an orthopaedist and a family physician or other generalist share the responsibility for an athletic team. This may be an ideal situation because although the majority of injury problems are musculoskeletal, they constitute only a fraction of the athlete's health care needs [13]. Lombardo has referred to the primary care physician as the "captain of the ship, director of the symphony, and jack of all trades" [7].

Availability

Availability is a cornerstone for success as a team physician [4, 10, 13]. Personal presence and a well-organized coverage system are essential. The most common service team physicians provide is game coverage. "Sidelines" are the front lines of sports medicine, especially in contact and collision sports. A physician who covers a team solely from the stands or the office does not truly deserve the title team physician.

A second venue of coverage is the athletic training room. The physician can demonstrate interest in the team by seeing athletes in their own environment rather than in the office. Athletes are more relaxed in their familiar setting and generally provide a better history and a higher level of compliance with recommendations given there. On the other hand, team physicians often make special accommodations in their office schedules to squeeze in athletes from their teams who have urgent problems. Most athletic activity goes on during the normal work day, and it is not reasonable for the team physician to be continuously available on site. The team physician or other appropriate back-up should always be available to the coach or athletic trainer.

An additional but extremely important element of an effective relationship is some unstructured time spent with the athletic trainer or coach. Often, it is the informal conversations that cement a good working relationship.

Qualifying Athletes to Play

Decisions about qualifications to play are always difficult. With youngsters in community leagues, a very conservative approach may be warranted. Sometimes, however, coaches and parents may have an expanded notion of why it is important for a particular player to be on the field on a given day. At the high school level, the opportunity to compete often influences scholarship opportunities. For a few performers at the highest level, a scholarship may mean an opportunity to attend NCAA Division I schools. But for the majority of these high school athletes, athletic scholarships may represent their only opportunity to afford an education at Division II or Division III NCAA schools as well as universities, colleges, and junior colleges with many other affiliations. For these athletes, the ability to play well may truly be their only option for financing a college education.

College level participation carries an even larger set of economic and social rewards. Although only a tiny percentage of college athletes advance to the professional level, many more gain employment and social opportunities from their achievements in athletics. Finally, in professional sports, there is a phalanx of young recruits knocking at the door to replace any injured player. Medical decisions are very difficult in this context.

Preparticipation Evaluation

Team physicians are now involved in a variety of preventive aspects of athletic medical care. Foremost among these is the preparticipation evaluation.

Even before the physician-athlete relationship exists, the prospective participant must be accepted as a member of the team. In most circumstances this means undergoing some type of preparticipation physical evaluation. The completeness, complexity, and sophistication of this evaluation vary widely with the level of competition and the nature of the sport. At the junior high school level this examination may require only a general health evaluation to rule out conditions that contraindicate participation in certain sports. The major challenges at this level are the wide range of musculoskeletal maturation that exists in peripubertal athletes and the possibility of discovering previously unidentified congenital defects that could pose serious health threats [2, 6, 11, 15].

By the time the athlete reaches senior high school and college level participation, natural selection processes have been at work, and the likelihood of finding a serious congenital problem is much less. Here the usual finding is a previous orthopaedic injury, so examiners must be extremely adept at performing musculoskeletal examinations and appropriate preparticipation rehabilitation techniques.

By the time professional level competition has been reached, much has usually occurred in the athlete's past. Examinations are usually rigorous and sophisticated because of the economic aspects of professional sports. Extensive performance testing, routine x-ray screening, and the use of computerized musculoskeletal testing techniques are commonplace. These topics are beyond the scope of this chapter, which is oriented more toward the volunteer team physician at the high school level.

The role of the team physician in the preparticipation evaluation is to determine the qualifications of athletes, both general and sport-specific, in order to counsel them on appropriate sports and to treat and rehabilitate deficits. An excellent policy statement, guide, and form for use in the preparticipation physical evaluation is currently being formulated jointly by the American Academy of Family Physicians, the American Academy of Pediatrics, the American Osteopathic Academy of Sports Medicine, and the American Orthopaedic Society for Sports Medicine. Until this document appears, many other sources contain a history and evaluation form that can be used as a model [2, 9, 15].

Overall, the primary goal of any preparticipation physical examination is to help maintain the health and safety of the athlete. Beyond that, its mandatory objectives are to:

1. Detect conditions that might limit competition.
2. Detect conditions that might predispose the athlete to injury.

3. Meet certain legal and insurance requirements.

In an ideal situation these objectives can be expanded to include the following:

1. Determination of general health status.
2. Physician counseling.
3. Maturity assessment.
4. Fitness evaluation.

The timing of the preparticipation examination is always something of a dilemma. Ideally, it should take place approximately 6 weeks prior to the beginning of the sports season to allow for correction or rehabilitation of any defects that are identified. However, this often means the middle of the summer when few athletes are available and when the physician's schedule may already be overcommitted.

The frequency of the preparticipation examination has been the subject of much controversy. Untold manhours have been wasted in performing routine full examinations on the same group of healthy high school athletes year after year. We believe that, ideally, a complete screening examination should be performed on entry into any given level of school, that is, junior high school, senior high school, and college. After the initial full examination, in subsequent years an interim health questionnaire would be filled out and a limited physical examination performed focusing on the new areas of illness or injury disclosed by the questionnaire. This approach has been adopted by the NCAA, but at the moment few states have seen fit to follow suit. Consequently, in many locales a full screening examination must be done annually to fulfill state requirements.

The venue for the preseason examination has also been the subject of much discussion. Undoubtedly, there are great advantages to having this examination done by the athlete's primary care physician in his own office. A physician-patient relationship already exists. Past records are available. There is a great measure of privacy for counseling. Overall, such a setting greatly enhances continuity of health care. On the other hand, many athletes may not have a primary care physician. Many physicians may have extremely limited appointment time, especially during the time of year when every athlete needs an examination. There are certainly varying interest and skill levels among primary care physicians in dealing with athletes and sports. Such examinations also tend to be significantly more costly.

Because of these disadvantages, the preseason screening examination is often carried out as a multistation examination utilizing a medical team. The advantages are those of personnel with specialized interests in sports, efficiency in handling large numbers of athletes, the ability to move into nontraditional areas such as performance testing, and an immediate line of communication open to coaches, trainers, and school nurses. The disadvantages of such a setting are apparent. This is usually a noisy environment with fairly hurried examinations. It is almost impossible to counsel athletes on delicate issues in such a setting. Follow-up is often spotty. A word should also be said about what appears to be the worst of both worlds, that is, one overworked physician trying to examine a gymnasium or locker-room full of athletes on one day after office hours. This setting tends to promote the "warm body" type of evaluation.

The ideal situation appears to be a private office setting in which the examination is done by a knowledgeable team physician who is also the athlete's family physician or pediatrician. This is seldom practical. When it is not, a well-organized station-to-station examination is a good alternative. If one chooses to organize such a screening program, the following information may be of benefit.

Adequate manpower is a must. Team members can include physicians, dentists, house officers, medical students, physician assistants, nurse practitioners, athletic trainers, physical therapists, exercise physiologists, and dietitians. Special history and physical examination forms are extremely valuable. The best arrangement is to have the history form on one side and the physical examination form on the other. Many different examples of such forms are available [2, 6, 15]. Room may be left on the same form for parental consent to emergency treatment and for the team physician's final evaluation and sign-off at the end of the examination.

A variable number of stations can be used depending on the space and manpower available. A typical set-up might include the stations shown in Table 10–1. Some additional clarification is relevant. Station 1 should not include actually filling out the medical history form; this should have been done previously at home by parents or guardians along with the athlete. Station 2 is one of the most important. Here the validity of the written health history is ascertained. All athletes have a tendency to "pencil whip" the history form. Unless it is double-checked by a health care professional, major points in the history may go overlooked. Station 6 may well be divided into more stations, at each of which only a portion of the medical examination is done. If the entire examination is done at one station, plenty of help should be available because otherwise this will undoubtedly become a bottleneck.

TABLE 10–1
Stations for Preparticipation Examination

		Required Personnel
Station 1	Sign-in	Athletic director or coach
Station 2	Review history	Physician, physician assistant, or nurse
Station 3	Height and weight	Coach, trainer, other volunteer
Station 4	Visual acuity	Trainer or nurse
Station 5	Vital signs	Trainer, physician assistant, or nurse
Station 6	Medical examination	Physician(s)
Station 7	Orthopaedic examination	Physician(s)
Station 8	Review-reassessment	Team physician
Station 9	Body composition	Physiologist, trainer, or therapist
Station 10	Flexibility	Therapist or trainer

Station No. 7, the orthopaedic examination, may take various forms. If nonorthopaedic surgeons are performing the examination, a quick 12-point orthopaedic screening examination may be used [2]. If the evaluation is actually done by orthopaedic surgeons, they may wish to go into much more detail, especially regarding the examination of ankles, knees, and shoulders.

Station No. 8 is undoubtedly the most difficult. Here a physician, preferably the team physician, must review all of the data generated during the examination and decide whether the athlete may safely participate or not. In looking back over the examination, important information can be found in the answers to questions on the health history form about medications and the potential presence of cardiac disease. Medications may impair the athlete's ability to compete and may also tip off the physician to the presence of other illnesses that may not otherwise have come to light. Questions about passing out or becoming dizzy during exercise as well as a family history of premature heart disease or sudden death may alert the physician to the possibility of potentially lethal cardiac lesions. In the physical examination itself, blood pressure will be the most frequent abnormality. If the blood pressure falls outside the accepted limits for the age group (Table 10–2), it should be repeated, taking care to use a large enough cuff for the athlete's arm size. Disturbances of cardiac rhythm or cardiac murmurs not meeting the characteristics of physiologic murmurs deserve further investigation before the athlete is approved for competition [11, 15].

TABLE 10–2
Classification of Blood Pressure

Children	
Age (Years)	**Normal Pressure (mm Hg)**
<6	120/70
6 to 9	130/75
10 to 13	135/80
14 to 17	140/85

Adults	
Classification	**Diastolic**
Normal blood pressure	<85
High normal blood pressure	85–89
Mild hypertension	90–104
Moderate hypertension	105–114
Severe hypertension	>115
	Systolic (When Diastolic Is <90 mm Hg)
Normal blood pressure	<140
Borderline isolated systolic hypertension	150–159
Isolated systolic hypertension	>160

Skin infections demand appropriate treatment before the athlete is allowed to participate in sports involving body contact, either with another competitor or with mats as in wrestling. Liver, spleen, or kidney enlargement will be a limiting factor in contact and collision sports. The presence of a hernia by itself is not necessarily a disqualifying factor but may require further discussion with the athlete and parents.

The loss of any one of a pair of organs (eye, kidney, testicle) has traditionally been a disqualifying feature for contact or collision sports or other activities in which the second of the paired organs may be in jeopardy. However, this traditional conservatism has been challenged repeatedly in courts of law and has always been decided in favor of the athlete who wishes to participate. In this case, it is recommended that a formal document be drafted by the school or its lawyer to be signed by the athlete and parents or guardians indicating that they are completely informed about the potential consequences.

The most common abnormal findings at Station 8 will be evidence of previous orthopaedic injuries, particularly as the age level of the group screened goes up. Consequently, the most important recommendation stemming from the orthopaedic examination is that of full rehabilitation of previous orthopaedic injuries prior to the start of a new season.

These recommendations are provided to the athletic trainer, school nurse, or coaching staff for action and may be invaluable.

The final decision, as evidenced by a licensed physician's signature, is that the athlete may

1. Participate without restriction.
2. Participate only after completing further evaluations or rehabilitation.
3. Not participate in specific sports due to certain disqualifying factors.

Shaffer [14] has warned against making guarantees or assurances that it will be safe for a young athlete to participate in sports activities. He suggests simply stating that a review of the health history and the medical examination have not disclosed conditions that would make participation in supervised activities inadvisable. Lists of disqualifying conditions are available from different sources [2, 15]. Again, these lists have become guidelines rather than rigid edicts because of the repeatedly reaffirmed right of youngsters to participate in sports.

Medical Supervision of Athletes

The traditional function of a team physician has been medical supervision of the athlete. This role, which started small, has now been greatly expanded. The physician is responsible for on-site coverage at the field, gym, arena, pool, or other venue. Obviously, the presence of a physician is often limited to high-risk situations and high-risk sports. When a physician is not available for coverage, certified athletic trainers or other personnel trained in the prevention of injuries and in early evaluation and care of injured athletes should be present. When physicians and athletic trainers are not available at high school or higher levels of competitions, coaches should be certified in American Red Cross advanced first aid and cardiopulmonary resuscitation or the equivalent.

In addition to personnel, proper equipment is also important. Many lists of suggested equipment, for both the physician and the sponsoring organization, have been published [1, 3, 14].

Supervision at tournaments often involves coordination of a larger medical team to ensure that skilled care is available at all times and venues. The pure volunteer who mans the sidelines of contact and collision sports on Fridays or Saturdays must be a well-trained clinician. The majority of injuries require extreme skill in musculoskeletal evaluation. Nonmusculoskeletal injuries certainly occur as well. For example, head and facial injuries, eye trauma, heat illness, and abdominal injuries may also test the physician's evaluating and decision-making abilities. In athletes with traumatic conditions, many of the most difficult decisions are related to return to participation. Should the athlete be allowed back during the same contest? Should he or she be held out of competition? If so, for how long? What are the parameters for allowing return to play?

Common medical conditions occur in athletes. Most of these are nontraumatic and nonmusculoskeletal. The team physician often treats streptococcal throat infections, infectious mononucleosis, cystitis, and skin infections. The big difference is that the team physician must relate the medical disorder and treatment to the athlete's ability to return to play. For example, the splenomegaly accompanying infectious mononucleosis in the football player or the herpes dermatitis in a wrestler may present significant dilemmas for the athlete wishing to compete.

The team physician often treats conditions that are not physical but psychological. A complete team physician must be willing and able to counsel athletes and to refer them to consultants when necessary. Problems of alcohol and drug abuse, performance anxiety, failed emotional relationships, unwanted pregnancy, eating disorders, and deaths in the family are all common. The death of a team member, particularly as a result of competition, calls for the greatest degree of sensitivity and skill in dealing with the grieving process and with media attention and in helping the team back into competition [5]. The team physician should be able and willing to evaluate the circumstances and either treat or refer.

The team physician is frequently asked for advice on the proper conditioning techniques to be used to prevent or rehabilitate injuries. The requests may be for both preseason and in-season conditioning as well as for general and sport-specific techniques. The advice may relate to endurance, strength, flexibility, agility, or nutrition. It often concerns issues of protective equipment. It is important to understand the scientific merits and liabilities of the equipment worn by athletes for practice and play. Optimally, the physician understands the considerations involved in the selection and fit of the equipment and in the prevention of injuries and reinjuries.

Education

The team physician often serves as an educator or resource person at many levels and for many audiences in sports medicine (Table 10-3). The

TABLE 10–3
The Team Physician as Educator

1. Audiences
   a. Athletes
   b. Coaches
   c. Athletic trainers
   d. Administration, especially athletic directors
   e. Medical personnel
      i. Medical students
      ii. House officers
      iii. Colleagues and consultants
   f. Paramedical personnel
   g. Parents
   h. General public
2. Methods
   a. One-on-one instruction and precepting
   b. In-service training
   c. Lectures, workshops, seminars
   d. Formal instructional courses
   e. Newsletters
   f. Audio and video instructional tapes
   g. Written books and articles in sports and professional journals

Mellion, M.B., and Walsh, W.M. The team physician. *In* Mellion, M.B., Walsh, W.M., and Shelton, G.L. (Eds.), *The Team Physician's Handbook.* Philadelphia, Hanley & Belfus, 1990, p. 6.

teaching is often intrinsically linked with consultation. Sometimes a medical opinion is sought, and at other times issues arise involving conditioning, protective equipment, rehabilitation, or virtually anything in the range of medicine and sport. Team physicians commonly use a variety of educational techniques varying from one-on-one instruction and precepting to workshops, lectures, formal instructional courses, newsletters, audiotapes and videotapes, and books and articles in sports and professional journals.

Part of the physician's role is to educate the coach about improvements that can be made in medical and preventive care. Continuing education of the coaching and athletic training staff is important to eliminate outdated and possibly harmful techniques. Archaic attitudes and misconceptions have a way of persisting in sports as in other aspects of life. When they increase the risk of harm to athletes, the team physician is responsible for dispelling them. A secondary benefit is often protection of the coach and team from possible future liability.

The other side of education for the team physician is his or her own education. Medicine and sports change, and the physician must stay current. The team physician also has the opportunity to balance formal learning with the lessons about life, sport, and medicine that come from observing highly motivated athletes and coaches perform.

Administration

The team physician may be expected to perform a variety of administrative functions. These include developing a general system of care for the team and establishing guidelines for consultation with the team physician and other consultants. The physician is responsible for coordinating medical supervision among medical personnel, athletic trainers, and other resources. This role includes planning and organizing the preparticipation evaluation. The physician, in concert with the athletic trainer and often the coach, must determine the content and standards of the evaluation, establish the guidelines for participation, arrange the location, and secure and coordinate the activities of the necessary personnel.

The team physician should develop a prearranged system of emergency care. Both general medical supplies and emergency equipment should be readily available in the training room and the athletic venue. At high-risk events, an ambulance should be present to facilitate rapid emergency transportation. If a patient needs to be transferred for medical care, a system should be in place to notify the hospital in advance while the patient is en route. This often involves establishing a communication system with a telephone on the sidelines or using radio or cellular phone communication.

Healer

Even in this highly technical era of medicine, the role of the physician as healer is preeminent. The intangible effect of the "laying on of hands" should not be underestimated. Supportive care by a good team physician is often the key to an athlete's recovery and successful return to play.

Communication and Liaison

The ideal team physician is a skilled communicator who can often resolve conflicts or enhance cooperation among members of the athletic and sports medicine teams. Certain relationships often require delicate communications for the benefit of the athlete and the team:

Athlete – coach
Athlete – parents
Team physician – parents
Team physician – athlete's family doctor
Athlete – medical colleagues and consultants
Athletic trainer – coach
Injured athlete – press

Often the tact and skill of an experienced team physician can be a healing influence in these relationships as well as for the athlete individually. This leads to a discussion of the relationships involving the team physician.

RELATIONSHIPS INVOLVING THE TEAM PHYSICIAN

The Sports Medicine Team

The team physician is not alone in caring for the athletes on the team. He is the key "player" on a sports medicine team that consists of the physician, the athlete, the coach, and the athletic trainer, when one is available. The care of athletes is a team effort in which members of the sports medicine team support each other for the benefit of the athlete and the athletic team. This team concept leads to the best possible care for the athlete [4, 7, 8, 13, 14].

Each of the key players on the sports medicine team has a support system (Fig. 10–1). The athlete's support system includes teammates, family and significant others, friends, teachers, and the athletic trainer. The coach's support system includes the athletic director, the school or league administration, the coaching staff and other coach colleagues, the equipment manager, and the athletic trainer. The team physician's support staff is much more complex. It includes clinical support, research support, and the athletic trainer. Elements of clinical support include other medical specialists, physical therapists, sports psychologists or psychiatrists, nutritionists, dentists, podiatrists, equipment managers, and health educators. The research support for a team physician includes medical researchers, exercise physiologists, sports psychologists, kinesiologists, nutritionists, physical educators, sociologists, and the athletic equipment industry.

The athletic trainer occupies a unique position at the center of athletic health care. The athletic trainer is a therapist, counselor, and confidant for the athlete, an advisor and friend to the coach, and the "eyes and ears" for the team physician. In this last capacity, the athletic trainer provides triage and screening, supervises conditioning, care, and rehabilitation, and provides continuous functional evaluation of the athlete. Every high school and college athletic program should have a certified athletic trainer. Happy is the team physician who finds himself working with a talented, full-time, certified athletic trainer, who thereby lightens the physician's burden immeasurably. The physician and athletic trainer can develop an excellent working relation-

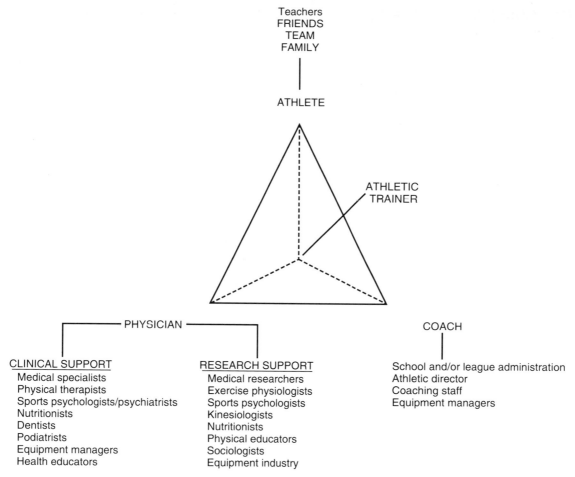

FIGURE 10–1

The sports medicine team. (Modified from Mellion, M.B. *Office Management of Sports Injuries and Athletic Problems*. Philadelphia, Hanley & Belfus, 1988.)

ship by establishing well-delineated lines of authority, responsibilities, procedures, and routines.

Special relationships exist between physicians and coaches. These are as varied as human beings can make them and, ideally, are cordial and filled with mutual respect. The physician recognizes the goals and tasks confronting the coach, and the coach appreciates that the team physician is ultimately responsible for the health and safety of the athlete. In other words, both should work toward the common goal of safe participation and team success. One of the worst situations that can arise is an adversarial relationship between the coach and the medical staff.

Relationship of Team Physician to Institution

It is important for the physician to establish an explicit formal relationship with the school, league, or team [14]. This agreement should include the job description, any fiscal arrangements, and a statement of expectations. When possible, especially if monetary arrangements are involved, the agreement should be in writing. If a formal contract is not appropriate, it is even more critical to discuss these items before the season begins. The team physician may be hired or solicited by a variety of people including the athletic director, coach, athletic trainer, or business manager [13]. Consequently, it is extremely important for the job description of the team physician to identify explicitly the person to whom the physician reports. The job description should also indicate what services are to be provided by the team physician both at home and away. Remuneration for services or travel and other benefits should be delineated. The job description should include all of the expectations that the institution has of the physician as well as those that the physician has of the institution.

RESPONSIBILITIES OF THE TEAM PHYSICIAN

The ethical responsibilities of the team physician reflect the many relationships involved in the care of the team. Perhaps the foremost responsibility is to allow athletes to participate. In public and school-based programs, athletes have a right to participate if there is no valid medical contraindication against it. They even have the right to insist on participation by waiving the liability of the school or institution if they are injured or have a medical condition that increases risk. Consequently, the team physician should know both the sports medicine literature and the sport well enough to avoid disqualifying athletes from participating for insignificant or outdated reasons. On the other hand, the physician has an equally important responsibility to protect the athlete from injury, reinjury, and permanent disability. When there are valid reasons contraindicating participation, the athlete must be counseled and thoroughly informed about the risks and dangers. Sometimes it is necessary to protect the athlete from himself or herself. For example, it may be especially difficult to reason with an athlete who has a "participate at any cost" attitude.

It is the team physician's responsibility to provide optimum health care for the athlete. Superficial encounters are not acceptable. Good record-keeping is basic. Health care starts with the preparticipation physical examination, which should be thorough and timely. The physician is responsible for putting together a team of competent examiners who can meet with the athletes early enough before the competitive season to allow treatment and rehabilitation of many defects that may be found.

The physician's responsibility for the athlete's confidentiality is often a difficult problem in the sports setting. Generally, the physician has a relationship to the school or professional club that inhibits the normal strict rules governing doctor-patient confidentiality. The athlete should be given advance notice of any potential sharing of medical information, and it is best that this notice be in writing with the athlete signing an acknowledgment. Even with such an acknowledged release of information, the team physician must be sensitive to the issue of how widely medical information is disseminated. The physician, however, has a broader responsibility to contribute to the success of the team. The players, the coach, and the athletic trainer have all dedicated their time and effort to the sport. They place their trust in the physician to provide expert and compassionate care.

Institutional responsibilities are also important for the team physician. The leagues, schools, and teams that have team physicians usually have made large commitments in personnel and finances. They may reasonably expect the team physician to be a contributor to the overall success of the athletic program. The primary job of the physician is to provide optimal care for the athletes. At the college and professional level, the physician is expected to prescreen potential scholarship and professional athletes to protect the institution from funding an athlete who is not definitely qualified to participate.

Another major institutional responsibility of the physician is to protect the institution from unnecessary liability. Provision of high-quality, timely medical care is basic in this regard. Additionally, the preparticipation evaluation may identify athletes who are at high risk of illness or injury and can then either provide successful treatment and rehabilitation or disqualify them from playing. Often the physician can provide safety information about the field of play or about dangerous athletic techniques—information that may help to prevent unnecessary injuries.

Legal and Medicolegal Considerations

The team physician faces a variety of legal and medicolegal considerations. First among these is the issue of institutional and professional liability. When a physician functions as a team physician for pay or other tangible remuneration, the normal rules of medical liability apply. However, when the team physician acts as a volunteer and treats an athlete without compensation, "good samaritan" laws apply. Almost every state has a good samaritan law, and more than three-fourths have sports-specific laws. It is important to note that many good samaritan laws require that the physician conform to the standard of care in the community. Preparticipation evaluations are not covered by good samaritan laws even if they are performed without charge.

Difficult issues occur when physicians travel with athletic teams outside of the states or countries in which they are licensed. For major athletic events, the host state or country generally passes legislation granting licenses to visiting team physicians. For routine competitions and tournaments, it is recommended that the traveling team physician work through the host team or tournament physician or a local physician in the host town. Good samaritan laws apply in states that have them if the physician is not compensated. No suit has yet been brought against a team physician traveling with a team to another state over the issue of practicing without a license.

Another item of legal significance is that of providing care to minors. Permission to provide care should be obtained from the responsible adults prior to the season, and appropriate forms should be carried with the team at all times [4, 14]. These considerations are magnified in the context of decision-making about a return to competition after an injury has occurred. One must act prudently. On the sideline in the heat of battle it is far better to err on the side of conservatism than to risk permanent injury. McKeag [8] has suggested guidelines to help in making decisions in these return-to-play situations on the sidelines of sporting events:

1. A definite diagnosis has been made.
2. The injury cannot be worsened by continued play.
3. The athlete can compete fairly and protect himself.

One other thorny legal issue is the athlete's right to participate. During the past two decades, several judicial decisions have affirmed that an athlete has a right to participate in junior high school, high school, and college sports in spite of a variety of disqualifying conditions. All of these cases have been decided in favor of the athlete, if he is willing to release the appropriate institution from liability. A recent case of national note involved a 17-year-old basketball player with hypertrophic cardiomyopathy who wanted to continue to play in spite of the fact that his older brother had died from the same disease in an exercise-related death. This case raises the question of taking reasonable judicial rulings to an unreasonable extreme.

TEAM PHYSICIAN REWARDS

Team physicians derive a variety of rewards from their service. Most find that the greatest reward is the immense personal satisfaction derived from providing a service to the community while working with young, highly motivated people. Rarely are tangible rewards able to match this element. Serving as a team physician is a labor of love. At anything less than the professional team level, most of the time spent by the physician will be as a "volunteer." Above the high school level, some compensation may be part of the agreement. However, this will be extremely variable, ranging from pure volunteerism in some collegiate environments to a considerable retainer or fee-for-service arrangement with a professional club. Some colleges and universities provide team physician services through the student health organization or an affiliated medical school, but most of these arrangements include large amounts of "volunteer" time as well. Surgeons may receive fees for surgical procedures performed, but often these services are provided on a discount basis as well.

The less tangible but equally important benefit of being a team physician is credibility, both in medicine and in the sport. Undoubtedly, an affiliation with a team, from high school to the professional level, enhances a physician's prestige in the community and may contribute to practice building as well.

References

1. Dick, A.D. Chalktalk for the team physician. *Am Fam Phys* 28(3):231–236, 1983.
2. Harvey, J. The pre-participation examination of the child athlete. *Clin Sports Med* 1(3):353–369, 1982.
3. Hershman, E.B., and Andrish, J.T. The team physician's bag: What to take to the game. *J Musculoskel Med* September, 23–27, 1985.
4. Howe W.B. Primary care sports medicine: A part-timer's perspective. *Physician Sports Med* 16(1):103–114, 1988.
5. Karofsky, P.S. Death of a high school hockey player. *Physician Sports Med* 18(2):99–103, 1990.
6. Linder, C.W., Durant, R.H., Seklecki, R.M., and Strong, W.B. Pre-participation health screening of young athletes: Results of 1268 examinations. *Am J Sports Med* 9(3):187–193, 1981.
7. Lombardo, J.A. Sports medicine: A team effort. *Physician Sports Med* 13(4):72–81, 1985.
8. McKeag, D.B. The role and responsibility of the team physician. Presented at the American College of Sports Medicine, The Team Physician Course, Hilton at Walt Disney World Village, Florida, March 1989.
9. Mellion, M.B., Walsh, W.M., and Shelton, G.L. (Eds). *The Team Physician's Handbook*. Philadelphia, Hanley & Belfus, 1990.
10. Redfearn, R.W. The physician's role in school sports programs. *Physician Sports Med* 8(9):67–71, 1980.
11. Salem, D.N. and Isner, J.M. Cardiac screening for athletes. *Orthop Clin North Am* 11(4):687–695, 1980.
12. Samples, P. The team physician: No official job description. *Physician Sports Med* 16(1):169–175, 1988.
13. Scheiderer, L.L. A survey of the practices of appointing and utilizing intercollegiate athletic team physicians. *Athletic Training* 22(3):211–218, 1987.
14. Shaffer, T.E. So you've been asked to be the team physician. *Physician Sports Med* 4:57–63, 1976.
15. Smith, N.J. (Ed.). *Sports Medicine: Health Care for Young Athletes*. Evanston, IL, American Academy of Pediatrics, 1983.

THE FEMALE ATHLETE

Letha Y. Griffin, M.D., Ph.D.

Athletics are still largely viewed as a world . . . in which women are intruders but not rightful heirs [130].

HISTORICAL CONSIDERATIONS

The historical background of any subject is usually fascinating. Such is the case with women's sports. Although this book is primarily a text of practical knowledge written to aid the reader in the diagnosis, treatment, and rehabilitation of sports injuries and illnesses, a slight divergence into the development of women's sports is relevant because it provides a needed perspective into the woman athlete's attitudes and responses to sport and injury.

The shorter, lighter, and less muscular structure of women has since early time been interpreted to mean that women are weaker, softer, less durable, and more vulnerable than men [130, pp. xv–xi]. This idea of women's physical limitations was reflected in early Greek society when only men were allowed to participate in the Olympic Games. Women were not even permitted to watch the competition; instead, beautiful women (weak and timid) were awarded as prizes to the victorious men [130, pp. xv–xi]. Only in Sparta were girls trained in physical sport, and even this training was not intended to render them more competitive and aggressive as role leaders in Spartan society but merely to make them better "breeders" of future Spartan male warriors [64].

Nowhere was feminine physical frailty more exaggerated than in our country (Table 11–1). In midnineteenth century America, both popular and scientific opinion propagated the notion that physical exertion in women would induce nervous shock, threaten reproduction (the only true reason for the existence of women), and result in all manner of physical maladies [130, pp. 15–41]. So closely allied were the qualities of weakness, frailness, paleness, softness, and inertia with women that these traits became synonymous with femininity [55]. Under the guise of protecting women, society shielded them, and in so doing contained and restricted them greatly, not only in sport but also in all ways of living, making them dependent on the "stronger" male [12]. So as to not sound chauvinistic, let us quickly point out that the prejudice was bisexual. Men were thought to be incapable of softness, tenderness, child cuddling, and caring because these qualities were not associated with masculinity. Masculinity had through the ages become equated with strength, quickness, skill, agility, and a certain degree of insensitivity [11, 138].

The concept of female fragility continued to be propagated by physical educators throughout the early twentieth century, retarding the development of sport for women [54]. Although many women in the early 1900s maintained a lifestyle that incorporated their ideals of femininity, there was still a large segment of the female population who had to work for a living in the mills and factories, where they were expected to perform all manner of physical labor [54]. Management became concerned about the physical fitness of these women employees, and thus encouraged the organization of competitive sports for them. Their efforts sparked the development of the first era of female athletics (approximately 1917 to 1936) [81, 88].

Industry's efforts were aided by educators who began to promote sports for women in colleges and universities [15]. Softball emerged, and indoor courts were built for volleyball and basketball. Increasing numbers of women participated in the existing sports of croquet, archery, bowling, tennis, and golf. These activities now developed a competitive nature, losing their aspect of being merely leisure activities of the rich [63]. Large gymnasiums and pools were built to accommodate sport participation. Organized competition with national and international female women's tournaments devel-

TABLE 11–1
Overview of the Historical Development of Women's Sports in the United States

1880–1917 (The Beginning)
Primarily social events
Individual participation, not teams
Principally upper class
Primarily outdoors
Never involved hard effort
No attempt made to adapt clothing for sport
Sample sports: croquet, archery, bowling, golf, tennis

1917–1936 (First Rise of Women's Athletics)
Increased number of sports
Increased participation of women in sports in colleges and universities
Indoor facilities constructed
Sport activities organized by working classes
Sport facilities supported by industry and community
Activity by YWCA
Gymnasiums built
Sample sports: racquetball, volleyball, basketball, softball, croquet, archery, bowling, tennis, golf, badminton, fencing, swimming, diving, skiing, figure skating, speed skating, bobsledding, field hockey, lacrosse, polo

1936–1960 (Split and First Decline)
Working Class
  Sport still flourishes among small numbers of the working class
  Increased number of organized tournaments in this sequence of popularity
  Participants begin to play competitively
  Clothing adapted to sport
Upper Classes, Colleges, and Universities
  Sports deemphasized in colleges and universities by physical educators
  Only a few activities like mild calisthenics allowed
  Educators think it unfeminine for women to engage in vigorous physical activity

1960–Present (Second Rise of Women's Athletics)
Rapid increase in numbers of female athletes in colleges and universities
Industrial sport continues to flourish
Government and towns organize sport facilities for women
Women become more involved in recreational sports in the community
Motivation for sport participation broadens to include play, competition, and socialization

oped, and clothing modifications for sports were seen [54].

During the early 1930s, when competitive sports for women appeared to be flourishing, significant resistance developed among some college and university leaders [72]. These educators felt that women participating in sports would become masculinized. They felt it their obligation, therefore, to use their influence to stop the development of competitive sports for women [15, 43, 130, pp. xv–xli]. Only a

few sports, the most popular being gymnastics (then really a form of mild calisthenics) were encouraged [15, 44]. Students themselves had virtually no say in the matter. The organization of informal games by women students during their free time was prohibited by authorities. Unfortunately, the concept that sport was harmful for women dominated the thinking of college and university educators until the 1960s [64].

Middle-class working women were less subject to the influence of physical educators. Hence, organized sports for women continued in local communities, but the intensity of participation was less than it had been during the flourishing period of women's sports in the 1920s [15]. It was from this working class in the thirties and forties that athletes developed who were capable of competing in national and international events.

The number of working women increased markedly in the 1940s when women were needed to replace men called to war [81]. The increase in the number of working women was probably a primary cause of the rise of the "new femininity" that began in the 1960s [15]. Women started fighting for equality of pay and equal career opportunities; equality in sports was a natural extension of this process. Colleges were forced to develop women's sports by the Title IX ruling of the Educational Assistance Act of 1972, which required institutions receiving federal funds to offer women equal opportunities in all programs. In 1972, fewer than 10,000 college women participated in sports, and no scholarships were available to them. By 1991, 91,000 women participated in National Collegiate Athletic Association (NCAA) sports; they were awarded 10,000 scholarships worth $51,000,000 [6].

In the 1960s, the Commission on Intercollegiate Athletics for Women was organized to encourage women's intercollegiate competitive events. National championships for women were organized as well. This represented a radical departure from a 1957 statement of college and university educators in which they proclaimed that extramural competition should not lead to county, state, district, or national championships [43]. Women's sports events no longer had to be social gatherings played only for fun but could be competitive. This was a major step for college sports.

On an international level, women's ability to participate effectively in rigorous sports has continued to be recognized by the inclusion of additional women's events each Olympic year. In the 1984 Summer Olympics, marathon and long distance cycling events were opened to women. In the 1988 Olympics, archery, cycle sprints, 10,000-meter run,

50-meter freestyle swimming, air pistol shooting, 470-class yachting, and the Alpine combined and super giant slalom were opened to women [132].

ANATOMIC AND PHYSIOLOGIC DIFFERENCES BETWEEN MALES AND FEMALES THAT AFFECT PERFORMANCE

An understanding of the anatomic and physiologic differences between the sexes is important because these differences form the basis of the uniqueness of each sex in athletic ability (Table 11–2). One should realize that these differences represent average values only and that there is significant variability within each sex. Moreover, knowledge of these differences should not be used to define male and female sport dominions nor to catalogue the abilities of each sex. Rather, it should be used to design training and rehabilitation programs that maximize the genetic potential of each sex.

Anatomic Considerations

In physique, women have smaller bones and less articular surfaces [64, pp. 15–23]. Males have longer legs, their leg length being 56% of their total height, compared to 51.2% in females [46, 71, 128]. The heavier, larger, more rugged structure of the male may be a mechanical and structural advantage in some athletic activities. His longer bones act as greater levers, producing more force for striking, hitting, or kicking [53]. Females have narrower shoulders, wider pelvises, and greater valgus of the elbow, varus of the hip, and valgus of the knee [46, 128]. The greater varus angle of the hip and valgus angle of the knee in women have been blamed by

TABLE 11–2
Female Anatomy Compared with Male Anatomy

Females have:
  Shorter legs
  Lower centers of gravity
  Smaller body frames
  Narrower shoulders
  Wider pelvises
  Smaller thoracic cavities
  Smaller lungs
  Smaller heart sizes
  Greater valgus angle of elbows, knees
  Greater varus angles of hips

TABLE 11–3
Female Physiology Compared with Male Physiology

Females have:
  Smaller aerobic and anaerobic capacities
  Lower basal metabolic rates
  Equal sordific responses
  Less muscle mass per body weight
  Greater percentage of fat per body weight

some for the increased number of overuse syndromes seen about the hip and the knee in the unconditioned female athlete [47, 104]. The female's increased valgus of the elbow is felt to give her a unique throwing style. The wider pelvis and shorter lower extremities of the female result in a lower center of gravity (56.1% of her total height compared to 56.7% in the male) [1]. This gives the female a distinct advantage in balance sports such as gymnastics. In fact, the balance beam, one of the four major events of women's gymnastic competition, is too difficult for most males and is not included in their competition.

Physiologic Differences

For the same body weight, female athletes have smaller heart sizes, lower diastolic and systolic blood pressures, and smaller lungs and thoracic cavities than equally trained male athletes (Table 11–3) [104, 128]. These differences decrease the female athlete's effectiveness in both anaerobic and aerobic activities. Anaerobic activities are the quick burst-type activities that result from contracting muscle fibers that utilize nonoxygen-requiring metabolic pathways, whereas aerobic activities are those slow endurance activities produced by muscle fibers that utilize oxygen-requiring metabolic pathways for energy.

Aerobic capacity is dependent upon one's genetic ability to consume oxygen maximally. Maximum oxygen consumption (termed VO_2 max) reflects the body's ability to extract and utilize oxygen maximally for aerobic metabolism [92]. One's VO_2 max measures the lung's ability to extract oxygen from the environment, the heart's ability to deliver this oxygen to the muscle, and the muscle's ability to assimilate oxygen maximally in energy pathways. Because of her smaller heart size, stroke volume, and lung and muscle mass, a female of the same weight and conditioned state as a male has a lower baseline VO_2 max upon which to build her aerobic capacity. Again, let us not forget that within each of these categories, there is a continuum for each sex. Some females at the upper end of the VO_2 max capacity have a greater genetic baseline than some males on the lower end of the masculine profile. Furthermore, training can and does enhance one's maximal oxygen consumption, but the baseline on which the athlete builds is genetically determined [71, 140].

With regard to body composition, normal females of college age have approximately 25% fat per body weight, whereas normal college males have about 15% fat per body weight [52]. In studies analyzing body fat in conditioned athletes, great variability among athletes participating in different sports has been found [17, 75, 140]. In track, the average conditioned female has about 10% to 15% body fat, whereas the conditioned male generally has less than 7% [17, 140, 141]. As noted earlier, the female has less muscle mass per body weight (approximately 23%) than an equally trained male (approximately 40%) [64, pp. 15–46]; therefore, it is more difficult for her to achieve the same power and speed as an equally trained male of the same body weight [53]. The greater percentage of fat per body weight in the female adds to her disability in power and speed.

The female's increased percentage of body fat per body weight is still a disadvantage even if she is matched with a male of equal muscle mass because the female must use the same muscle mass to "energize" her extra body fat. If the male is made to perform in a weighted vest, the weight being equal to the excess body fat of the female, his VO_2 max is lowered to her range. The extra load of fat women must carry has a direct negative effect on their work capacity and stamina [108, 135].

However, fat does insulate and give one buoyancy [36]—an important feature for swimmers, especially in natural water. Women hold the record for swimming the English Channel with the best one-way time of 7 hours, 40 minutes set in 1978 by Penny Dean [80]. Fat may also provide an energy source when glycogen is depleted. Ernst van Aaken proposed that women had the ability to convert from carbohydrate metabolism to fatty acid metabolism more readily than males and hence had a physiologic edge in ultra-endurance events [131]. His work, however, has not been substantiated by others [84, 135].

At one time females were reported to be more prone to heat exhaustion than males because they needed higher core body temperatures to increase their sordific (sweating) response [29]. However, recent evidence has demonstrated that conditioned males and females are equivalent in their thermal regulation [37, 50, 91].

Basal metabolic rate (BMR) is lower in females [109], and therefore for the same amount of activity, the female needs fewer calories. Caloric expenditure of women (18 to 19 years of age) is reported to be about 37 kcal/m² body surface area/hour, whereas caloric expenditure of males in this age group is about 40 kcal/m² body surface area/hour [64, pp. 15–46]. This difference in BMR is important when planning training tables and pregame meals for female athletes. One must plan nutritionally balanced meals with fewer total calories than those in meals intended for male athletes [48]. This fact, although it seems so obvious, can easily be overlooked. For example, when females first entered the military academies, the diet of the cadets was not altered, and the women gained weight eating the high-calorie foods provided for the male cadets. The nutritionists at the academies learned that they had to provide nutritious meals with fewer total calories (lean meat, fewer sauces, more fruits and vegetables) for their female cadets [4].

Coordination and dexterity are difficult parameters to measure but probably are equal between the sexes.

The above anatomic and physiologic comparisons were made between hormonally mature males and females. During the prepubertal years the female best approximates her male counterpart in height, weight, heart size, and aerobic ability [64, pp. 15–46, 128]. Puberty in the male is a time of maturation of fitness, but in the female it is a period of great alteration in physical characteristics and abilities, making her no longer equal to her male counterpart in size or strength. The female must adjust her timing and performance techniques to accommodate her new increases in height and weight without the help of a parallel increase in muscle mass. One needs only to recall the prepubertal Nadia Comaneci of the 1976 Olympics, the skillful, crowd-pleasing gymnast who performed so well. By the time of the next Olympics, puberty had exacted its toll with an increased percentage of body fat, curves replacing her previously straight lines, and an alteration in her center of gravity. Several years were required to modify her performance technique to allow for her new body dimensions.

PSYCHOLOGICAL CONSIDERATIONS IN THE FEMALE ATHLETE

As discussed under historical considerations, we have for years considered independence, fortitude, aggressiveness, achievement, and the desire to win or conquer as masculine or nonfeminine qualities, yet these characteristics are important in the development of the successful athletic competitor [126]. The winning male athlete has just proved his masculinity, whereas the winning female athlete often feels a need to justify her femininity [11]. This attitude is slowly changing, but it will be years before the prejudices arbitrarily qualifying masculinity and femininity are completely changed in our minds. The female athlete may face times of depression because she feels she is not living up to the perceived expectations of her sex. Those involved in women's sports must be aware that such feelings can develop and be ready to help the athlete cope with them.

Moreover, in studies of personality characteristics of athletes, women are found to be lower in dominance and confidence than men and higher in impulsiveness, tension, and general anxiety [9, 126]. In surveys done in the past, female athletes stated that they participated in sports because they enjoyed interacting and socializing with other players; males reported greater concern with the competitive aspect of sport, that is, their motivation to play was "to win" [11]. More recent studies do report that motivational drives in female athletes are changing as participation in sports increases. Winning is becoming more important. Perhaps many of the previously reported differences in personality characteristics in male and female athletes were more cultural than biologic. Still, unique personality traits of each sex probably exist despite cultural influences. Because motivation is intimately related to performance (some feel that performance is merely skill plus motivation), those working with athletes (trainers, coaches, and physicians) must be attuned to their psychological needs as well as their physiologic requirements.

GYNECOLOGIC CONSIDERATIONS

Effects of Exercise on Menstruation

Normal ovulation and menstruation depend upon proper hormonal secretion by the hypothalamus, the pituitary, and the ovaries [82]. A delicate balance of positive and negative feedback channels exists within this system. The hypothalamus secretes a factor that acts on the pituitary, causing it to release the gonadotropins, follicle-stimulating hormone (FSH) and luteinizing hormone (LH). These protein hormones promote the development of ovarian follicles and the production of estrogen and progesterone by the ovary.

Ovarian follicles, which are composed of the egg or ovum and its encasing cell, have two basic stages: the follicular stage, which precedes ovulation, and the luteal phase, which follows ovulation. Follicle-stimulating hormone causes follicles within the ovary to mature and produce estrogen. As more estrogen is produced by the generating follicles, the pituitary responds by producing a surge of LH. This surge of LH, combined with the FSH produced by the pituitary, acts on the ovary to stimulate ovulation in the follicle that is most mature at that time. After ovulation, the follicle that produced the ovum involutes, forming the corpus luteum, which secretes progesterone as well as estrogen, thus beginning the luteal or second phase of the monthly menstrual cycle [138]. The normal menstrual cycle averages 28 days, with a range of 25 days to 32 days [82]. The length of the follicular and luteal phases of the cycle are equal (Fig. 11–1).

Delayed Menarche

The onset of menstrual function, called menarche, has been reported to be delayed in some highly competitive athletes who train intensively. Data on the degree of delay in menarche in female athletes vary. The average age of menarche in girls in the United States is 12.5 to 12.8 years (Table 11–4) [82]. Studies of competitive runners have reported menarche delayed to approximately 15.1 years, compared with controls of 12.9 years [97]. A study of college swimmers and runners reported that each year of training before menarche delayed menarche by 5 months [40]. In a study of ballerinas, the mean age of menarche was reported to be 14.3 years, and in a retrospective study of college athletes, the reported age of menarche was 13.4 years compared with controls of 12.1 to 12.8 years [74].

In contrast, Erdelyi in 1976 reported no change in the onset of menarche in Hungarian athletes (the average age of menarche in these women was 13.5 years) [35]. It is interesting that the age for menarche in controls in this Hungarian study was similar to the age at onset of menarche in some studies of U.S. athletes but is delayed compared with U.S. controls.

Initiation of menarche is dependent on the maturation of the hypothalamus-pituitary-ovarian axis. It has been stated that 17% body fat is needed for the onset of menses [64, pp. 15–46]. Others believe that not only decreased body fat in some competitive athletes (most notably ballerinas and long distance runners) but also the altered nutritional states resulting from fad diets begun in the early teenage

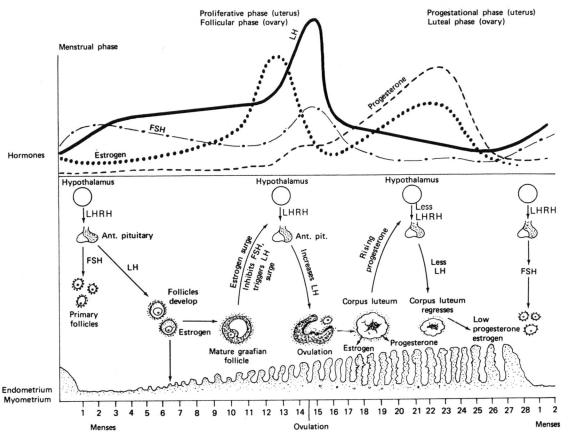

FIGURE 11-1
Hormone levels and events of the ovaries and endometrial cycles in the absence of fertilization. (Reproduced by permission BIOLOGY OF WOMEN, by Ethel Sloane. Delmar Publishers Inc., Albany, NY; copyright 1980 [ISBN: 0-8273-4938-6].)

years result in the delay of menarche. However, not all athletes with delayed menses have decreased body fat, leading some to theorize that the emotional and physical stresses of vigorous training and intensive competition may influence endocrine function at the level of hypothalamic control.

Amenorrhea

Many of the same factors that have been reported to delay menarche also appear to be involved in the development of secondary amenorrhea in athletes. Primary amenorrhea is the delay of menarche past

TABLE 11-4
Gynecologic Facts

Average age of menarche	Nonathletes, 12.5 years
	Athletes, 13.5–15.5 years
Normal menstrual cycle	28 days (range 25–35 days)
Primary amenorrhea	No menses by age 16
Secondary amenorrhea	Lack of normal menses for 3 months

the age of 16 [119]. Secondary amenorrhea is the cessation of periods for more than 3 consecutive months following menarche [82]. Amenorrhea has been reported to occur in 10% to 30% of competitive athletes [119, 145]. An even greater number of female athletes report alterations in their periods or fewer than the normal number of menstrual periods (oligomenorrhea) [27, 73, 97].

Surprisingly, this trend was not reported prior to 1970. In the Tokyo Olympics in 1964, 66 women participated, and the incidence of amenorrhea was reported to be only 1.5% [145]. Whether this sample population was too small to reflect the true incidence of menstrual irregularities in competitive athletes at this time or whether other variables existed is not known.

Predisposing factors in the development of secondary amenorrhea in the athlete have been reported to be nulliparity, young age of the athlete, intensive training prior to the onset of menses, stress (psychological and physiologic), intensity of training, increased endorphin levels, and altered body fat [14, 18, 28, 73, 79, 82, 117, 119, 136].

Genetic predisposition is probably also a factor

because many athletes with secondary amenorrhea report irregular periods prior to the beginning of their athletic careers and therefore appear to have a more "sensitive" hormonal balance [113, 119].

Perhaps the greatest proportion of the literature on secondary amenorrhea concerns its relationship with body weight. Some claim that 22% body fat is needed for normal menstrual function [41]. Others state that a minimal weight (115 pounds) is important despite body height [124]. Still others report that the rapidity of weight loss is the crucial factor [26]. Others believe that it is not the number of pounds lost or the rate of loss that is crucial but the percentage of total body weight lost that is important because a 10% to 15% total body weight loss reflects the loss of approximately one-third of the normal body fat [26].

Different reasons have been proposed to explain why the percentage of body fat may be critical in maintaining normal menstrual function. Some feel that because fat is one of the extragonadal sources of estrogen, a certain critical amount of body fat is needed for sufficient aromatization of androgens to estrogens [14]. Others argue that this extragonadal estrogen cannot be that vital in menstrual control because there are other extragonadal sources of estrogen that can compensate for decreased fat production (e.g., muscle). Some authors claim that the hypothalamus interprets decreased body fat as decreased energy stores needed to support reproduction and lactation and therefore "shuts off" the reproductive system [124]. Another theory is that decreased fat stores alter thermoregulation, which influences normal hormonal cycling [124].

It is known that secondary amenorrhea is common in individuals with anorexia nervosa. Some data suggest, however, that psychological (emotional) factors, not merely weight loss in anorexics, triggers secondary amenorrhea. This idea has led some to theorize that perhaps in athletes the physical and psychological stresses of sport result in both weight loss and secondary amenorrhea [39].

Stress secondary to changes in employment or lifestyle can result in periods of amenorrhea in women. Anderson reported that 86% of the women freshmen in the 1976 entering class at the U.S. Military Academy were amenorrheic, even though they averaged 17% body fat [4]. Psychological and physiologic stresses increase opiates, which increase secretion of prolactin. Increased secretion of prolactin may block the action of the gonadotrophins on the ovary [28].

In summary then, the etiology of secondary amenorrhea in athletes is still not clear. Furthermore, it is not even certain when and how extensively ath-

letes with secondary amenorrhea should be evaluated to make certain that their loss of normal periods is truly secondary to their athletic activity and is not related to ovarian failure, hypothyroidism, or pituitary tumors. Most authorities now state that athletes who experience secondary amenorrhea should have a routine history and physical examination by their physician. Laboratory studies should include a complete blood count, urinalysis, SMA-18, and thyroid screen. No additional hormonal screening needs to be done unless the athlete's amenorrhea persists for more than 1 year or unless abnormalities are found on her history or physical examination [28].

Most athletes with secondary amenorrhea have low estrogen levels. Some clinicians believe that these women should be treated with low-dose estrogen for 25 days each month. Progesterone is typically added during the last 10 days of estrogen administration to stimulate withdrawal bleeding [28, 115, 116]. Estrogen is recommended to maintain the normal urogenital epithelium and stimulate the endometrium as well as for its beneficial effects on the cardiovascular system and its reported effects in the prevention of osteoporosis.

Amenorrhea and Osteoporosis. Osteoporosis in the hypoestrogenic athlete has recently gained much attention [19, 21, 31, 45, 94]. Although other diminished estrogen states such as hyperprolactinemia and premature menopause have been linked with osteoporosis, the potential risk of osteoporosis in athletes with secondary hypoestrogenic amenorrhea was not realized until the early 1980s. A study done at the University of California in San Francisco on six hypoestrogenic amenorrheic athletes reported decreased bone mineral content in vertebrae as measured by the quantitative computed tomographic technique [21]. These initial data have since been substantiated by others [31]. However, bone mineral loss has been found only in the cancellous vertebrae and not in the cortical bone of the appendicular skeleton [76, 86]. Moreover, Snyder reported that elite amenorrheic and oligomenorrheic oarswomen had vertebral bone densities approximately equal to those of controls [123], and Bilanin and colleagues reported a lower vertebral bone mineral density in male long distance runners compared with controls [10]. These data suggest that perhaps the decreased bone mineral content seen in amenorrheic runners is related more to their running than to their amenorrhea. More data need to be obtained before meaningful conclusions can be drawn.

The mechanism of the protective effect of estrogen on bone mineral content is still controversial. It may be related to calcium balance because lack of estrogen seems to increase daily calcium requirements by decreasing calcium absorption and increasing calcium excretion. Although many still do not advocate the administration of exogenous estrogen to hypoestrogenic amenorrheic athletes because of the association between exogenous estrogen replacement and endometrial carcinoma, they do recommend calcium supplementation in doses of 1200 to 1500 mg daily, especially for athletes whose diets do not contain sufficient dairy products and other foods rich in calcium.

Secondary amenorrhea in athletes appears to be reversible [105, 125]. Gaining weight, decreasing the intensity of training, or otherwise altering the physical and psychological stresses on the athlete usually restores normal cycling [28]. In fact, as stated previously, most athletes with secondary amenorrhea experience a return of normal menstrual function within 1 year even without overtly altering their training schedules. Moreover, this temporary amenorrheic state has not been found to interfere with future fertility [119].

Interestingly, in studies of hormonal cycling in high-intensity athletes, investigators have found that many athletes without secondary amenorrhea still have altered cycles with prolonged follicular phases and shortened luteal phases during their peak training months [118]. Again, this alteration in normal hormonal balance appears to be reversible and does not affect fertility. More research is needed in this area.

Athletes should be warned that secondary amenorrhea is not a guarantee against pregnancy. It is a temporary interruption in ovulation. The athlete could begin ovulating in any month and hence could become pregnant if she engages in unprotected intercourse at that time [120].

Contraception

Contraception is of concern to many athletes. No sport contraindicates the use of any contraceptive method [114]. Personal preference dictates the athlete's choice. Some athletes have found that oral contraceptive agents cause increased water retention and a feeling of sluggishness premenstrually [114]. Diaphragms, if worn during training episodes, have been reported by some to cause cramping and abdominal pain. Others object to the inconvenience of spermicidal jellies, foams, suppositories, or prophylactics. Information and advice on contraception should be available to those athletes who request it.

The Effects of Menstruation on Performance

Prior to the reports of osteoporosis associated with amenorrhea, many athletes felt that the amenorrheic state was an advantage because they did not need to worry about the effects of menstruation on performance. Some athletes have noted no change in athletic performance during the menstrual cycle. In fact, Olympian medals have been won by athletes in every phase of the menstrual cycle [127]. However, some women, particularly those who retain fluid premenstrually, have a perceived limitation of performance at this time. They claim that beginning about 3 days prior to menses and including the first 2 days of their periods they are more easily tired, lack stamina, and have increased emotional lability. Others have found that performance in short-duration events (track, field, gymnastics, and so on) is not affected by hormonal cycling, but performance in long-duration events (cycling, rowing, and so on) is affected [16, 145].

One study involving high-performance athletes reported no change in heart rate, anaerobic threshold, work capacity, grip strength, treadmill times at 90% and 100% VO_2 max, and all-out running when measured premenstrually, on day 1 of the menstrual cycle, at midcycle, and postmenstrually [115]. It may be that conditioning minimizes the effect of menstruation on performance.

Vaginitis

Vaginitis (an infection of the vagina) occurs not uncommonly in women and typically results in a purulent, pruritic discharge. In the female athlete symptoms of vaginitis can be irritating enough to be a distraction from performance. Therefore, vaginitis should be recognized and treated promptly.

Yeast and the sexually transmitted diseases (STDs), including *Trichomonas, Chlamydia,* and herpes infections, are the most common causes of vaginitis. These infections may be easily diagnosed by examining a smear of the vaginal discharge [120]. Mycostatin vaginal suppositories or Monistat cream may be used to treat monilial vaginitis. Oral Flagyl is effective in treating *Trichomonas* infections, and oral or topical Acyclovir is prescribed for genital herpes.

Bacterial pelvic inflammatory disease (PID) is the least common infection producing a vaginal discharge. Women with PID typically experience more pain than those with yeast infection or the more common sexually transmitted types of vaginitis.

They may be febrile and systemically ill. Diagnosis is confirmed and the appropriate antibiotics prescribed on the basis of a Gram stain and culture of the vaginal discharge obtained on pelvic examination.

Dysmenorrhea

Dysmenorrhea, or pain with menstruation, is thought to be secondary to the release of prostaglandins by the lining of the uterus. It is less frequently seen in athletes than in nonathletes [4, 114]. Prostaglandin inhibitors such as ibuprofen (Motrin, Advil, Nuprin) and sodium naproxen (Naprosyn) may be helpful in controlling dysmenorrhea.

Exercise During Pregnancy

In years past it was assumed that pregnancy should be a time of inactivity. Many felt that any bouncing or strain on a gravid uterus would certainly result in a miscarriage and might also result in stretching the uterine broad ligaments, leading to uterine prolapse [93, 135, pp. 127–158]. These and other similar myths have been disproved during the last 10 years. We know that the skin, muscles, and fascia of the abdominal wall, the thick uterine muscle lined by membranes, and the amniotic fluid that bathes the fetus in a shock-absorbing medium will act to maximum advantage to disperse forces encountered by direct blows or abrupt movements. Hence, normal physical activity, including well-structured exercise programs, will not predispose a woman to spontaneous abortion in a normal pregnancy.

However, certain physiologic changes do occur during pregnancy that may influence sport activity because exercise imposes additional requirements on the altered physiology of pregnancy. The systems primarily affected by these changes are the cardiovascular system, the respiratory system, the musculoskeletal system, and the metabolic and endocrine systems. The most dramatic changes are seen in the cardiovascular system [30, 59, 145].

During pregnancy, the circulatory system is hyperdynamic, that is, the total blood volume is increased by 35% [85]. Cardiac stroke volume increases, heart rate increases, and hence cardiac output is elevated. Cardiac output is increased by approximately one-third compared with the nonpregnant state. Exercise elevates this cardiac output even higher until near term, when the increase in cardiac output becomes progressively smaller with exercise owing to the decrease in cardiac reserve brought about by the obstruction of the vena cava by the enlarged uterus [59, 85].

Increases in progesterone during pregnancy cause increased sensitivity of the respiratory center in the medulla to carbon dioxide, resulting in a decrease of carbon dioxide in maternal blood by about 25% [85]. The breathlessness experienced by pregnant women during mild exercise is probably a manifestation of increased sensitivity to carbon dioxide. Tidal volume increases in pregnancy, causing a rise in minute volume to about 40% above the levels in nonpregnancy [30, 59]. The respiratory rate is not altered. Minute ventilation during exercise also increases during the later months of pregnancy. The rate of increase is proportional to the workload. As a result of increased body weight, more oxygen is required during exercise in pregnancy, and hence, maximum exercise capacity is reached at a lower level of work than in the nonpregnant state.

Pulmonary ventilation during submaximal running at 6 miles per hour dramatically increases as pregnancy progresses. This increase in the respiratory exchange ratio is out of proportion to body weight during the last trimester. It may result from the fact that the work rate exceeds the anaerobic threshold, precipitating the need to blow off carbon dioxide in order to decrease metabolic acidosis caused by the increase in serum lactic acid [85].

Exercise may induce an initial bradycardia in the fetus, but the fetal heart rate rapidly returns to normal within 3 minutes after exercise is begun [30, 59]. It was originally thought that because exercise imposes increased circulatory demands on muscle, blood would be shunted away from the placenta, causing circulatory embarrassment to the fetus. If such shunting does occur, it must be minimal and is compensated for by enhanced arterial venous oxygen extraction, which maintains fetal oxygenation; hence, fetal heart rate is not increased.

Increased body heat brought about by exercise, especially in the first trimester, is of concern [32, 100]. Elevations in temperature above certain critical levels can kill rapidly proliferating cells. Hyperthermia can result in fetal abnormalities, including central nervous system defects (microcephaly, seizures, mental deficiency), facial dysmorphogenesis, cataracts, arthrogryposis, hernias, and tooth defects. It is felt that a critical level for inducing a teratogenic effect is a core body temperature of 40° to 41°C (104° to 106°F) [32]. Therefore, it is recommended that during exercise, core body temperature should remain below 38.9°C (102°F). In summer, running should be done in the cool early morning or late evening hours. Plenty of water should be consumed to avoid heat stress. Furthermore, during pregnancy, women are cautioned to restrict hot tub bath exposure at 39°C (102°F) to less than 15 minutes. The duration of sauna bathing can be somewhat longer because evaporation from the steam will help to lower core body temperature more effectively than with hot tub baths.

Elevated estrogen and relaxin hormone levels during pregnancy cause softening of ligaments, especially of the pelvic ring. Such relaxation at the sacroiliac joints and pubic symphysis can cause gait abnormalities (the waddling gait of pregnancy). Pain with running has been attributed to this pelvic relaxation [30, 85]. However, this laxity has not been reported to cause an increased rate of joint dislocations or ligament sprains if proper conditioning programs accompany sport participation.

At one time it was felt that perineal exercises would make the muscles of the pelvic floor and perineum too tight and hence delay labor. The belief now is that strengthening of the perineal and abdominal muscles results in less fatigue and backache during pregnancy and may aid the mother during the second stage of labor [101, 143].

Athletic women appear to regain their prepregnant weight sooner after pregnancy [34]. However, weight gain during pregnancy has been reported to be slightly less in athletic than in nonathletic women [22]. Moreover, exercise may have a beneficial psychological effect on the mother's self-image and should be encouraged throughout most pregnancies. Exercise restrictions exist in mothers with diabetes, hypertension, cardiovascular abnormalities, cervical defects, previous miscarriages, and other circumstances considered to be high-risk pregnancy states. Mothers in this category should seek counsel from their obstetricians before exercising.

OTHER MEDICAL ISSUES

Iron Deficiency

Because a female can lose up to 1.2 to 2 mg of iron/day during menstruation, menstruating females have a higher incidence of iron deficiency than does the general population. If this iron deficiency is significant, it may result in anemia [96]. However, an athlete can be iron deficient without being anemic. Iron deficiency is best described as the absence of iron in storage areas and can be quantitated by measuring the saturation of transferrin. A saturation of transferrin of between 15% and 20% is considered borderline iron deficiency [87].

Iron is the mineral present in hemoglobin molecules within the red cell that is necessary for oxygen transport. Hence, a normal iron level is especially important in athletes, in whom maximal oxygen-carrying capacity is needed for maximal energy output [24, 49, 95]. Iron may also participate as a catalyst in removing the lactate formed during aerobic exercise. This role of iron has not yet been clearly elucidated, but we do know that women who are iron deficient but not anemic have increased lactate levels with submaximal exercise compared with women with normal iron levels [87].

Strenuous exercise during early conditioning programs has been reported to lower plasma iron levels in both sexes, perhaps because of increased destruction of red cells with vigorous activity. This has been termed sports anemia [96].

Iron supplementation for the menstruating woman is advised if her diet does not contain adequate amounts of iron-rich foods such as liver, oysters, beef, turkey, dried apricots, and prune juice. The recommended daily allowance of iron for menstruating women is 18 mg/day [96]. One must remember that only 10% of all ingested iron is absorbed through the gut. Ferrous sulfate in doses of 960 to 1920 mg/day (providing 200 to 400 mg of elemental iron) may be used to supplement dietary iron sources in the anemic athlete. A multivitamin pill plus iron may be used instead of ferrous sulfate as an iron supplement for the nonanemic athletic female. However, most multivitamin pills contain only 12 mg of iron. Megadoses of iron are not helpful; one cannot make blood "iron rich" and thereby improve oxygen-carrying capacity. In fact, overdoses of iron can be toxic.

The symptoms of iron deficiency reflect not only the level of iron but also the rate of development of the deficiency. Generally, no symptoms will be noted unless the iron deficiency is severe enough to produce a detectable anemia. Symptoms of severe anemia include fatigue, irritability, and headaches.

Urinary Tract Infections

Infections of the urinary tract, and particularly of the bladder, are more frequent in women than in men. To minimize the risks of infection, female athletes should be encouraged to drink plenty of fluids, especially water, which increases urine flow, promoting bacterial "washout" [5]. If urinary symptoms of itching, burning, frequency, urgency, or blood in the urine develop, a urinalysis should be done with a urine culture to follow if indicated so that the cause of the symptoms can be ascertained and treatment instituted as quickly as possible to minimize time lost from participation.

Use and Abuse of Anabolic Steroids

Some women athletes, like some men athletes, have experimented with exogenous androgens to increase strength. Certainly women will note significant increases in muscle size if they take androgens, exercise to fatigue, and eat a balanced diet [61, 144]. Some dispute still exists about whether "androgen-augmented" muscle is as strong per unit size as normal muscle. Numerous articles that argue both sides of this issue are available [68, 144].

The risk of anabolic steroids in women include the possibly irreversible changes of increase in size of the clitoris, lowering of the voice, increased facial and body hair, male-patterned baldness, fluid retention, increased blood pressure, and increased risks of heart attack, stroke, and liver abnormalities [67, 68, 144]. In addition, the use of anabolic steroids by athletes in national and international competition is illegal and may result in suspension of the athlete.

CONDITIONING

Conditioning is as important for female athletes as it is for male athletes. This fact seems obvious today, but during the rapid rise of organized athletics for women in the early 1970s, sport participation for women was not paralleled by sport preparedness. Women learned the skills of sports, but often they were not provided with proper instruction in programs designed to develop the strength, flexibility, and cardiovascular endurance needed for optimum performance.

Early studies of injury rates of women athletes reported increased numbers of minor injuries compared with men participating in similar sports [33, 65]. Some interpreted this information as indicating that women were physically and physiologically inferior to men and were unable to participate in vigorous sports, whereas in fact, these data merely reflected the fact that women were not receiving adequate off-season, preseason, and in-season conditioning.

Conditioning programs for women do not markedly differ from those designed for men [65]. Off-season conditioning should focus on cardiovascular endurance. Aerobic activities like swimming, biking, or running are excellent. Although women on the whole have lower maximal oxygen consumption than men, women athletes can increase their cardio-vascular endurance by proper aerobic conditioning [8]. As recommended for male athletes, women should perform their aerobic activity at approximately 70% to 80% maximum heart rate for at least 20 minutes three to four times per week [2, 3]. Even in anaerobic sports like ice skating, gymnastics, and basketball, athletes find cardiovascular stamina allows them to minimize fatigue, thereby enhancing performance and reducing the risk of injury.

Several weeks before the start of the season, the athlete should begin to emphasize strengthening programs designed to condition major muscle groups, placing particular emphasis on those muscle groups specific for her sport. For example, volleyball players need strong quadriceps for jumping as well as strong deltoids, pectorals, and biceps for striking the ball, whereas soccer players may concentrate primarily on lower extremity strength. Women can strengthen muscles by isometric, isotonic, and isokinetic techniques [99, 121, 142].

Weight training with constant, variable, or accommodating resistance will result in increased strength in women as well as in men, but will typically not result in as large increases in muscle bulk as it does in men. In fact, it has been stated that women can increase their strength by 40% without any increase in muscle size [139]. Muscle size is a hormonally regulated characteristic. It is the higher androgen levels in the male that result in increased muscle bulk when muscles are maximally challenged.

Initially, it was not felt that stretching was important in women's conditioning programs because women were reported to be more flexible than men. It may be true that the bulkier, tighter physique is less common in women, but it is not true that women, especially those involved in weight-training programs, should not stretch before sport activity and before workouts. Stretching guidelines for women are similar to those established for men, that is, they should warm up adequately before stretching and avoid ballistic stretching. Muscles are stretched by repetitive, slow, gentle elongation of the muscle fibers [69].

In-season conditioning programs are generally structured toward developing skills needed for a particular sport, and only short periods of time are devoted to maintaining the level of cardiovascular fitness achieved during the off season and maintaining the strength and flexibility achieved during preseason training [1].

Women interested in maintaining fitness but not in performing any particular single sport activity may well achieve the tone they want while maintaining the cardiovascular fitness they need through aerobic conditioning only. Recreational sport participants such as those engaged in softball or tennis will find it advantageous to develop strength and flexibility in the needed muscle groups as well as learning sport-specific skills. In addition, these individuals will find that developing cardiovascular fitness will enable them to perform their activity with less fatigue.

Because of the higher incidence of patellofemoral stress syndrome in women, many authorities do not advocate stair or hill climbing as a conditioning drill for women [57]. These activities do build quadriceps strength, but they also result in increased patellofemoral forces. Similarly, women should approach knee extension with free or accommodating weights with caution, especially if they try to lift through an arc of 90 degrees to full extension. Many women will do better if they lift from a beginning position of 60 degrees of flexion to full extension because in this range patellofemoral forces are less than if one initiates extension with the knee flexed to 90 degrees [8]. To decrease patellofemoral forces while biking indoors, women should raise their bike seat so that the knee is fully extended when the pedal is on the downstroke.

In summary, the adage "the poorly conditioned athlete is an injury-prone athlete" is as true for both competitive and recreational female athletes as it is for male athletes. Generally, conditioning programs for women, like those for men, should include muscle strengthening and flexibility drills as well as exercises to improve cardiovascular endurance. Moreover, conditioning principles, including overload progression and specificity of training response, are as applicable to women as they are to men.

INJURIES

With the growth of women's athletics, many thought there would be an increase in the numbers and types of injuries as women became more aggressive and competitive in sports. Many clinicians felt that women athletes would present with an entirely new range of injuries. In fact, early studies of injuries in female athletes did report increases in the number of injuries sustained [33, 65]. This high injury rate probably reflected the lack of adequate conditioning in the women rather than a true physiologic weakness and predisposition to injury. Later studies surveying injury rates in conditioned women athletes demonstrated no increases in injuries compared with male athletes [23, 137]. Instead, as women have become more serious in their sport participation, training and conditioning techniques have improved, and the numbers of injuries have decreased. Injuries appear to be more sport specific than sex specific; that is, injury types and rates are similar for men and women in the same sport but differ for women participating in different sports.

Because anatomic and physiologic differences do exist between the sexes (see earlier section), even though injury rates and types are similar for both sexes, some injuries will occur with a slightly greater frequency in women. These include paronychia, overuse injuries of the shoulder [62], patellofemoral stress syndrome [57], anterior cruciate ligament injuries [77], anterior and posterior ankle impingement [56], bunions [66], spondylolysis [58], lumbar apophysitis [83, 122], and stress fractures [107].

Paronychia

Paronychia, secondary to improper trimming of cuticles, may be somewhat more common in women. Creams applied to the hands around the cuticles to prevent dried cuticles from developing hangnails may be helpful. Fingernails should be kept short to prevent an athlete's nails from injuring other players or herself.

Shoulder Problems

Women engaged in sports requiring the arm to be either forcefully or repetitively placed in an overhead pronated position may develop overuse syndromes related to rotator cuff strain (shoulder subluxation, impingement, or a combination of these two entities). Diagnosis and treatment of these entities are similar in both sexes [1] (see Chap. 15D). The incidence of overuse problems was reportedly increased in women in injury reports written in the 1970s [122, 137]. Some felt that because of the significant lack of muscle strength measured in the upper extremity even in athletic women [51], the rate of such injuries would remain high. However, they failed to consider that most of these women, even though they were performing sporting activities requiring upper extremity strength, had not done much to condition or strengthen the upper body during their training. As women began incorporating such strengthening exercises into their conditioning routines, the maximal strength obtained in the upper extremity approached that of males, with deficits similar to those in the lower extremity and based on genetic differences and muscle mass, not on a genetically weak upper-to-lower extremity strength ratio [25].

In the military academies, women were initially

required to do only flexed arm hangs instead of pull-ups because it was felt that their upper body strength would not permit the latter [70, 106]. However, after women had been present at the academies about a year, most of them were able to do complete pull-ups without difficulty [129].

Patellofemoral Stress Syndrome

Patellofemoral stress syndrome (retropatellar pain) is not uncommon in female athletes [56]. As in males, the onset of pain typically results from the performance of repetitive knee flexion-extension activities (running, mountain climbing, soccer, and so on). The term retropatellar pain or patellofemoral stress syndrome [57] or "patellalgia" [98] has been reserved to describe a symptom complex of a painful but stable patella, whereas chondromalacia is used to describe pathologic wear of the articular cartilage of the patella, which generally results either from direct or indirect trauma to the articular surface or from aging [38].

Athletes with the patellofemoral stress syndrome may complain that their knee "hurts," or they may localize the pain to the front of the knee or along the medial border of the patella. The pain is generally made worse with downhill activities. It is intensified by prolonged sitting with the knee in an acutely flexed position or by frequent kicking, as in soccer. Initially, the pain may be vague and may follow the activity, but if the athlete continues to perform despite the pain, the pain may soon become present with activity and may prevent completion of the activity. "Give way" weakness is not uncommon. The athlete states than when going downstairs or down a hill or after getting up from a seated position, her knee will just "give way."

The patellofemoral stress syndrome is thought to result from altered patellofemoral mechanics, and the pain is felt to be due to either a stretching of the medial retinaculum or increased or altered forces on the patellar hyaline cartilage and the bone beneath by the patella tracking laterally [57].

The increased incidence of patellofemoral stress syndrome in females compared to males [57, 102] is felt to be related to the higher rate of occurrence in women of the factors that result in a laterally sitting and tracking patella, that is, wider pelvis, greater valgus angle of the knee, increased femoral anteversion, and a tendency toward a poorly developed vastus medialis obliquis portion of the quadriceps [13, 47]. A more complete discussion of the history, physical examination, and management of this complex overuse injury is found in Chapter 22D.

Anterior Cruciate Ligament Injuries

Although increased ligamentous laxity in the female was initially thought to contribute to an increased incidence of anterior cruciate ligament injuries in female athletes [77], especially those engaged in high-risk sports such as snow skiing, volleyball, and basketball, studies measuring ligamentous laxity using a knee arthrometer have not demonstrated increased laxity of the anterior cruciate ligament or posterior cruciate ligament in women [134].

If one then carefully reviews earlier published reports, one finds that they measured joint flexibility or overall soft tissue laxity about the joint, not isolated ligamentous laxity [78, 103, 134]. However, since the surrounding soft tissues of the joint (muscle, capsule, and so on) are felt to be the secondary restraints of the anterior and posterior cruciate ligaments, the observed overall joint flexibility may be a contributing etiologic factor in injuries to these knee ligaments [47]. Moreover, because the hamstrings and the quadriceps are the dynamic restraints for static cruciates [90], lack of sufficient strength of these muscles may contribute to an increased incidence of anterior and posterior cruciate ligament injuries. Hence, proper off-season, preseason, and in-season conditioning programs to strengthen these muscle groups are encouraged.

Some have suggested that if a cruciate ligament injury occurs in a woman with hyperextension laxity of the knee, a ligament reconstruction is advisable because this individual will have greater difficulty compensating for the increased laxity caused by the cruciate ligament. However, one must consider all parameters in these women. A woman with increased flexibility of the knee who sustains a cruciate injury may be at greater risk of reinjury if a cruciate reconstruction is not done. On the other hand, because patellofemoral tracking problems are more common in women, and because one of the most common complications of an anterior cruciate ligament reconstruction, which utilizes the autologous bone-patella tendon-bone for the cruciate substitution [89], is patellofemoral pain [110], one must consider the increased risk of such problems in women who undergo this procedure and advise patients accordingly. Also, the incidence of postoperative stiffness in women undergoing ligamentous reconstruction has been reported to be greater than that occurring in males.

Ankle Impingement

Ankle impingement syndromes are seen in sports in which extreme ankle motion is desired, that is, gymnastics, dancing, and diving [56, 58]. Anterior capsular impingement may occur when a gymnast "lands short" during dismounts or tumbling routines, causing the foot to be hyperflexed [122]. Pain is typically described as an "aching discomfort" intermittently felt over the anterior aspect of the ankle. It increases with activity and improves with rest. On physical examination, the joint typically is not swollen but is tender to palpation across the entire anterior aspect of the ankle (tibial-talar joint). Rest, oral nonsteroidal anti-inflammatory drugs, ice, and protection from ankle hyperflexion (proper landing techniques, pads placed anteriorly over the ankle, and so on) may be helpful in decreasing symptoms.

Forced ankle plantar flexion in gymnastics, ballet, and diving may result in irritation of the posterior ankle capsule in women [42]. Posteriorly, the talus may have associated with it a separate ossicle (the os trigonum) or an elongated posterior process that can become traumatized by hyperflexion activity, resulting in pain over the entire posterior ankle capsule. Occasionally, repetitive trauma can cause a stress fracture of this process. Rest from those activities that aggravate symptoms while maintaining the conditioned state is recommended and often results in relief of symptoms. As with anterior capsular impingement, oral anti-inflammatory agents, physical therapy modalities (phonophoresis, iontophoresis, electrical stimulation, and so on) may be helpful in decreasing inflammation and hence symptoms.

Foot Disorders

The incidence of some foot disorders appears to be related at least in part to improper shoe wear and is more common in female than in male athletes [102]. These disorders include bunions, bunionettes, corns, and calluses. Although the tendency toward increased spreading of the metatarsals distally, resulting in a wide forefoot and prominence of the first and fifth metatarsal heads, has a genetic basis, shoes that are too narrow cause pressure over the bursa that lies over the flare, resulting in inflammation and pain in the bursa. Moreover, constant pressure from a too narrow shoe rubbing over this bony flare results in thickening of the metatarsal head. If shoes are too short or too wide in the heel (allowing the foot to slide forward), the longer toes, typically the second and third toes, become jammed into the end of the shoe, and a hammertoe deformity with a subsequent callus developing over the flexed

proximal interphalangeal joint may occur (see Chap. 24J).

Bunions, bunionettes, and hammertoes can be painful and disabling [66, pp. 536–564]. Athletes with a tendency toward these foot disorders should be cautioned during preseason examinations to be careful about shoe selection. Once symptoms from bunions, bunionettes, or hammertoes develop, they should be treated promptly with shoe alterations, anti-inflammatory creams, ice, and proper padding.

Ballet dancers, because of the fit of their toe shoes, have a high incidence of foot disorders. As with other athletes, prompt, aggressive, conservative treatment is indicated [112]. Caution should be used when contemplating surgical relief for these foot disorders because surgical procedures may alter the mechanics of the foot disadvantageously, resulting in metatarsalgia, diffuse foot pain, or stress fractures of the midmetatarsals [56, 111].

Spondylolysis and Vertebral Apophysitis

Spondylolysis as an acute stress fracture of the pars interarticularis may occur in female gymnasts, divers, and skaters, in whom maximum repetitive flexion-extension motions of the spine are common. Frequently, with the acute injury routine films do not show the defect, but a bone scan of the area may be positive. Rest is recommended to allow for complete healing and prevent the development of chronic spondylolysis. A more complete discussion of this entity appears elsewhere in this text.

Another cause of back pain in the skeletally immature athlete participating in sports that require repetitive hyperextension and hyperflexion of the spine is vertebral apophysitis, or irritation of the anterior superior and inferior apophyseal plates of the vertebral bodies, which is believed to result from traction on these structures by the anterior longitudinal ligament. An increased frequency of this condition has been reported in very young gymnasts who practice their sport 4 to 5 hours a day [60]. Gymnastics requires tremendous back flexibility because the routines include repetitive flexion-extension maneuvers. Unlike Scheuermann's disease, apophysitis does not seem to be associated with Schmorl's nodes and may result in increased lumbar lordosis rather than an increase in thoracic kyphosis.

Rest frequently relieves symptoms, but bony changes may persist. Prior to returning to their sport, these youngsters should begin a program to increase the strength and flexibility of the back and abdominal musculature. Symptoms determine when the child may resume full participation in sports. The long-range prognosis of multiple vertebral apophysites is not known. One probably should encourage a child with multiple vertebral apophysites to select a sport that does not require repetitive extremes of flexion or extension of the thoracic or lumbar spine [133].

Stress Fractures

An increased number of stress fractures in female Army cadets compared with males (10% versus 1%) was reported in 1976 when females first entered the military academy [107]. The initial explanation was that this increased incidence of stress fractures was probably attributable to a lack of proper conditioning. However, after Cann and associates in 1984 reported decreased bone mass in young amenorrheic women with various endocrine disorders [20], the question arose as to whether all female athletes with low estrogenic amenorrhea might have a deficiency in bone mass that would make them more prone to stress fractures, similar to postmenopausal women. Much debate has been raised on this issue [7, 31], but no conclusive evidence is yet available.

SUMMARY

During the last two decades, a clearer understanding of the anatomic and physiologic factors that influence women's participation in sports has resulted not only in a greater number of women participating in a wider variety of sports but also in better structured conditioning programs for women athletes, which have helped to decrease the number of injuries they incur. However, many questions still exist that need to be addressed if we are to see our women reach peak performance capabilities while incurring a minimal number of risks. Ongoing research will enable us to work toward the realization of these goals.

References

1. American Academy of Orthopaedic Surgeons. *Athletic Training and Sports Medicine* (2nd ed). Park Ridge, IL, American Academy of Orthopaedic Surgeons, 1991.
2. American College of Sports Medicine. Position statement on the recommended quantity and quality of exercise for developing and maintaining fitness in healthy adults. *Med Sci Sports* 10(3):vii, 1978.
3. American College of Sports Medicine. *Guidelines for Graded Exercise Testing and Exercise Prescription* (3rd ed.) Philadelphia, Lea & Febiger, 1986.
4. Anderson, J. Women's sports and fitness programs of the U.S. Military Academy. *Physician Sports Med* 7(4):72–78, 1979.
5. Andriole, U. Urinary tract infections and pyelonephritis. *In* Wyngaarden, J. (Ed.), *Cecil Textbook of Medicine* (17th ed.). Philadelphia, W.B. Saunders, 1985, pp. 619–623.
6. Baker, M. D. Personal communication, 1991.
7. Barrow, G., and Saha, S. Menstrual irregularity and stress fractures in collegiate female distance runners. *Am J Sports Med* 16(3):209–215, 1988.
8. Benas, D. Special considerations in women's rehabilitation programs. *In* Hunter, L., and Funk, F. (Eds.), *Rehabilitation of the Injured Knee*. St. Louis, C.V. Mosby, 1984, pp. 393–405.
9. Berlin, P. The woman athlete. *In* Gerber, E., Felshin, J., Berlin, P., and Wyrick, W. (Eds.), *The American Woman in Sport*. Reading, Addison Wesley, 1974, pp. 283–400.
10. Bilanin, J., Blanchard, M., and Russek-Cohen, E. Lower vertebral bone density in male long distance runners. *Med Sci Sports Exerc* 21(1):66–70, 1989.
11. Birrell, S. The psychological dimensions of female athletic participation. *In* Boutilier, M., and SanGiovanni, L. (Eds.), *The Sporting Woman*. Champaign, IL, Human Kinetics, 1983, pp. 49–91.
12. Birrell, S. Discourses on the gender/sport relationship: From women in sport to gender relations. *In* Pandolf, K. (ed.), *Exercise and Sport Sciences Reviews*. New York, Macmillan, 1988, pp. 459–502.
13. Bloom, M. Differentiating between meniscal and patellar pain. *Physician Sports Med* 17(8):95–108, 1989.
14. Bonen, A., and Keizer, H. Athletic menstrual cycle irregularity: Endocrine response to exercise and training. *Physician Sports Med* 12(8):78–94, 1984.
15. Boutilier, M., and SanGiovanni, L. *The Sporting Woman*. Champaign, IL, Human Kinetics, 1983, pp. 23–47.
16. Brooks-Gunn, J., Gargiulo, J., and Warren, M. The menstrual cycle and athletic performance. *In* Puhl, J., and Brown, C. (Eds.), *The Menstrual Cycle and Physical Activity*. Champaign, IL, Human Kinetics, 1986, pp. 13–28.
17. Butts, N. Physiological profile of high school female cross country runners. *Physician Sports Med* 10(11):103–111, 1982.
18. Calabrese, L., Kirkendall, D., Floyd, M., Rapoport, S., Williams, G., Weiker, G., and Bergfeld, J. Menstrual abnormalities, nutritional patterns, and body composition in female classical ballet dancers. *Physician Sports Med* 11(2):86–98, 1983.
19. Caldwell, F. Light boned and lean athletes: Does the penalty outweigh the reward? *Physician Sports Med* 12(9):139–149, 1984.
20. Cann, C., Martin, M., Genant, H., and Jaffe, R. Decreased spinal mineral content in amenorrheic women. *JAMA* 251:626–629, 1984.
21. Carr, C., Gevant, H., Ettinger, G., and Gordon, G. Spinal mineral loss in oophorectomized women. *JAMA* 244(18):2050–2059, 1980.
22. Clapp, J., and Dickstein, S. Endurance exercise and pregnancy outcome. *Med Sci Sports Exerc* 16:556–562, 1984.
23. Clarke, K., and Buckley, W. Women's injuries in collegiate sports. *Am J Sports Med* 8(3):187–191, 1980.
24. Cooter, G., and Moribray, K. Effect of iron supplementation and activity on serum iron depletion and hemoglobin levels in female athletes. *Research* 49:114–117, 1978.
25. Cox, J., and Lenz, H. Women in sports: The naval academy experience. *Am J Sports Med* 7(6):355–357, 1979.
26. Dale, E. Exercise and the menstrual cycle. *Emory University Department of Obstetrics and Gynecology Bulletin* 6(1):48–53, 1984.
27. Dale, E., Gerlach, D., and Wilhote, A. Menstrual dysfunction in distance runners. *Obstet Gynecol* 54:47, 1979.
28. Diddle, A. Athletic activity and menstruation. *Sportsmed* 76(5):619–624, 1984.
29. Dill, D., Yousef, M., and Nelson, J. Responses of men and women to two-hour walks in desert heat. *J Appl Physiol* 35:231–235, 1973.

30. Dresslendorfier, R. Physical training during pregnancy and lactation. *Physician Sportsmed* 6(2):76–80, 1978.

31. Drinkwater, B., Nilson, K., Chesnut, C., Bremner, W., Shainholtz, S., and Southworth, M. Bone mineral content of amenorrheic and eumenorrheic athletes. *N Engl J Med* 311(5):277–280, 1984.

32. Edward, M. Is hyperthermia a human teratogen? *Am Heart J* 98(3):277–279, 1979.

33. Eisenberg, T., and Allen, W. Injuries in a women's varsity athletic program. *Physician Sportsmed* 6(3):112–116, 1978.

34. Erdelyi, G. Gynecological survey of female athletes. Presented at the Second National Conference on Medical Aspects of Sports Sponsored by the American Medical Association, Washington, D.C., November 27, 1960, pp. 174–178.

35. Erdelyi, G. Effects of exercise on the menstrual cycle. *Physician Sportsmed* 4(3):79–81, 1976.

36. Fahey, T. Endurance training. *In* Shangold, M., and Mirkin, G. (Eds.), *Women and Exercise: Physiology and Sports Medicine.* Philadelphia, F.A. Davis, 1988, pp. 65–78.

37. Ferstle, J. Christine Wells: Asking the right questions. *Physician Sportsmed* 10(7):157–160, 1982.

38. Ficat, R., and Hungerford, D. *Disorders of the Patellofemoral Joint.* Baltimore, Williams & Wilkins, 1977, pp. 170–179.

39. Frisch, R. Food intake, fatness, and reproductive ability. *In* Vigersky, R. (Ed.), *Anorexia Nervosa.* New York, Raven Press, 1977, pp. 149–160.

40. Frisch, R. Delayed menarche and amenorrhea of college athletes in relation to age of onset of training. *JAMA,* 246:1559, 1982.

41. Frisch, R., and McArthur, J. Menstrual cycle: Fitness as a determinant of minimal weight for height necessary for maintenance or onset. *Science* 185(9):949–961, 1976.

42. Gelabert, R. Preventing dancers' injuries. *Physician Sportsmed* 8:69–76, 1980.

43. Gerber, E. Historical survey. *In* Gerber, E., Felshin, J., Berlin, P., and Wyrick, W. (Eds.), *The American Woman in Sport.* Reading, Addison Wesley, 1974, pp. 3–47.

44. Gerber, E. Sport in society. *In* Gerber, E., Felshin, J., Berlin, P., and Wyrick, W. (Eds.), *The American Woman in Sport.* Reading, Addison Wesley, 1974, pp. 86–135.

45. Gonzalez, E. Premature bone loss found in some nonmenstruating sportswomen. *JAMA* 24(5):513–514, 1983.

46. Hale, R. Factors important to women engaged in vigorous physical activity. *In* Strauss, R. (Ed.), *Sports Medicine.* Philadelphia, W.B. Saunders, 1984, pp. 250–269.

47. Haycock, C., and Gillette, G. Susceptibility of women athletes to injury: Myths versus reality. *JAMA* 236:163–165, 1976.

48. Haymes, E. Iron supplementation. *In* Stull, G. (Ed.), *Encyclopedia of Physical Education, Fitness, and Sports,* Vol. 2. Salt Lake City, Brighton, 1980, pp. 335–344.

49. Haymes, E. Iron deficiency and the active woman. American Alliance for Health, Physical Education and Recreation Research Report 2:91–97, 1983.

50. Haymes, E. Physiological response of female athletes to heat stress: A review. *Physician Sportsmed* 12(3):45–59, 1984.

51. Heyward, V., Johannes-Ellis, S., and Romer, J. Gender differences in strength. *Res Q Exerc* 57:154–158, 1986.

52. Higdon, R., and Higdon, H. What sports for girls? *Today's Health,* 10:21, 1967.

53. Hoffman, T., Stauffer, R., and Jackson, A. Sex differences in strength. *Am J Sports Med* 7(4):265–267, 1979.

54. Howell, M., and Howell, R. Women in sport and physical education in the United States 1900–1914. *In* Howell, R. (Ed.), *Her Story in Sport: A Historical Anthology of Women in Sports.* West Point, NY, Leisure Press, 1982, pp. 154–164.

55. Howell, R. American Women, 1800–1860: Recreational pursuits and exercise. *In* Howell, R. (Ed.), *Her Story in Sport: A Historical Anthology of Women in Sports.* West Point, Leisure Press, 1982, pp. 70–79.

56. Hunter, L. Women's athletics: The orthopaedic surgeon's viewpoint. *Clin Sports Med* 3(4):809–827, 1984.

57. Hunter, L., Andrews, J., Clancy, W., and Funk, F. Common orthopaedic problems of the female athlete. *Instr Course Lect* 31:126–152, 1982.

58. Hunter-Griffin, L. Orthopedic concerns. *In* Shangold, M., and Mirkin, G. (Eds.), *Women and Exercise: Physiology and Sports Medicine.* Philadelphia, F.A. Davis, 1988, pp. 195–219.

59. Hutchinson, P., Cureton, K., and Sparling, P. Metabolic and circulatory responses to running during pregnancy. *Physician Sportsmed* 9(8):55–61, 1981.

60. Jackson, D., and Wiltse, L. Low back pain in young athletes. *Physician Sportsmed* 11:53–60, 1974.

61. Johnson, L., and O'Shea, J. Anabolic steroids: Effects on strength development. *Science* 164:957–959, 1969.

62. Kennedy, J., and Hawkins, R. Swimmer's shoulder. *Physician Sportsmed* 4:34–38, 1974.

63. Kenney, K. The realm of sports and the athletic woman 1850–1900. *In* Howell, R. (Ed.), *Her Story in Sport: A Historical Anthology of Women in Sports.* West Point, NY, Leisure Press, 1982, pp. 107–140.

64. Klafs, C., and Lyon, J. *The Female Athlete* (2nd ed.). St. Louis, C.V. Mosby, 1978, pp. 3–12.

65. Kosek, S. Nature and incidence of traumatic injury to women in sports. *In* Proceedings, National Sports Safety Congress, Cincinnati, Ohio, 1973, pp. 50–52.

66. Kulund, D. *The Injured Athlete* (2nd ed.). Philadelphia, J.B. Lippincott, 1988, pp. 46–47.

67. Lamb, D. Anabolic steroids and athletic performance. *Sportsmed Dig* 6(7):1, 1984.

68. Lamb, D. Anabolic steroids in athletes: How well do they work and how dangerous are they? *Am J Sports Med* 12(1):31–38, 1984.

69. Lamb, D. *Physiology of Exercise* (2nd ed.). New York, Macmillan, 1984, pp. 366–393.

70. Lenz, H. Women's sports and fitness programs at the U.S. Naval Academy. *Physician Sportsmed* 7(4):42–50, 1979.

71. Lesmes, G., Fox, E., Stevens, C., and Otto, R. Metabolic responses of females to high intensity interval training of different frequencies. *Med Sci Sports Exerc* 10(3):229–232, 1978.

72. Lucas, J., and Smith, R. Women's sport: A trial of equality. *In* Howell, R. (Ed.), *Her Story in Sport: A Historical Anthology of Women in Sports.* West Point, NY, Leisure Press, 1982, pp. 239–265.

73. Lutter, J., and Cushman, S. Menstrual patterns in female runners. *Physician Sportsmed* 10(9):60–72, 1982.

74. Malena, R. Age at menarche in athletes and nonathletes. *Med Sci Sports* 5(1):11, 1973.

75. Malena, R., Harper, H., and Avent, H. Physique of female track and field athletes. *J Am Coll Sports Med* 3:32, 1982.

76. Marcus, R., Cann, C., Madvig, P., Minkoff, J., Goddard, M., Bayer, M., Martin, M., Gaudiani, L., Haskell, W., and Genant, H. Menstrual function and bone mass in elite women distance runners: Endocrine and metabolic features. *Ann Intern Med* 102:158–163, 1985.

77. Marshall, J., and Barbash, H. *The Sports Doctor's Fitness Book for Women.* New York, Delacorte Press, 1981, pp. 15–33.

78. Marshall, J., Johanson, N., Wickiewicz, T., Tischler, H., Koslin, B., Zeno, S., and Meyers, A. Joint looseness: A function of the person and the joint. *Med Sci Sports Exerc* 12:189–194, 1980.

79. McArthur, J., Bullen, B., and Beitens, I. Hypothalamus amenorrhea in runners of normal body composition. *Endocr Res Commun* 7:13–27, 1980.

80. McWhirter, N. *Guiness Book of Women's Sports Records.* New York, Sterling Publishing, 1979, p. 146.

81. Melpomene Institute for Women's Health Research. *The Bodywise Woman.* Englewood Cliffs, NJ, Prentice-Hall, 1990, pp. 10–28.

82. Menstrual Changes in Athletes: A Round Table. *Physician Sportsmed* 9(11):99–112, 1981.

83. Micheli, L. Low back pain in the adolescent: Differential diagnosis. *Am J Sports Med* 7(6):362–364, 1979.
84. Mirkin, G. Nutrition for sports. *In* Shangold, M., and Mirkin, G. *Women and Exercise: Physiology and Sports Medicine.* Philadelphia, F.A. Davis, 1988, pp. 91–93.
85. Mullinax, K., and Bryan, D. Exercise and pregnancy. *Emory University Department of Obstetrics and Gynecology Bulletin* 6(1):61–73, 1984.
86. Nelson, M., Fisher, E., Castos, P., Meredith, C., Turksoy, R., and Evans, W. Diet and bone status in amenorrheic runners. *Am J Clin Nutr* 43:910–916, 1986.
87. Nilson, K., and Schoene, R. Iron repletion decreases maximal exercise lactate concentration in female athletes with minimal iron deficiency anemia. *J Lab Clin Med* 102:306–312, 1983.
88. Noonkester, B. The american sportswoman from 1900–1920. *In* Howell, R. (Ed.), *Her Story in Sport: A Historical Anthology of Women in Sports.* West Point, NY, Leisure Press, 1982, pp. 178–222.
89. Noyes, F., Butler, D., Paulos, L., and Grood, E. Intraarticular cruciate reconstruction I: Perspectives on graft strength, vascularization, and immediate motion after replacement. *Clin Orthop Rel Res* 172:71–77, 1983.
90. Noyes, F., and McGinniss, G. Controversy about treatment of the knee with anterior cruciate ligament laxity. *Clin Orthop Rel Res* 198:61–76, 1985.
91. Nunnaley, S. Physiological response of women to thermal stress: A review. *Med Sci Sports* 10(4):250–255, 1978.
92. O'Toole, M., and Douglas, P. Fitness: Definition and development. *In* Shangold, M., and Mirkin, G. (Eds.), *Women and Exercise: Physiology and Sports Medicine.* Philadelphia, F.A. Davis, 1988, pp. 3–22.
93. Paisley, J., and Mellion, M. Exercise during pregnancy. *Am Fam Physician* 38:143–150, 1988.
94. Parrish, M. Exercising to the bone. *Women's Sports* April, 5(4):25–32, 1983.
95. Pate, R. Sports anemia: A review of the current research literature. *Physician Sportsmed* 11(2):115–126, 1983.
96. Pate, R., Maguire, M., and Wyk, J. Dietary iron supplementation in women athletes. *Physician Sportsmed* 7(9):81–86, 1979.
97. Paul, B. Running before menarche. *Melpomene* February, 1:5–6, 1983.
98. Percy, E., and Strother, R. Patellalgia. *Physician Sportsmed* 13(7):43–59, 1985.
99. Pipes, T. Strength training modes: What's the difference? *Scholastic Coach* 46(10):96, 1977.
100. Pleet, H., Graham, J., and Smith, D. Central nervous system and facial defects associated with maternal hyperthermia. *Pediatrics* 67(6):785–789, 1981.
101. Pomerance, J., Gluck, L., and Lynch, V. Physical fitness in pregnancy: Its effects on pregnancy outcome. *Am J Obstet Gynecol* 119:867–876, 1974.
102. Potera, C. Women in sports: The price of participation. *Physician Sportsmed* 14:149–153, 1986.
103. Powers, J. Characteristic features of injuries in the knee in women. *Clin Orthop Rel Res* 143:120–124, 1979.
104. Powers, J. Title IX: Knee. *In* American Academy of Orthopaedic Surgeons Symposium. *The Athlete's Knee: Surgical Repair and Reconstruction.* St. Louis, C.V. Mosby, 1980, p. 125.
105. Prior, J., Yuen, B., Clement, P., Bowie, L., and Thomas, J. Reversible luteal phase changes and infertility associated with marathon training. *Lancet* 2:269–270, 1982.
106. Protzman, R. Physiologic performance of women compared to men. Observations of cadets at the United States Military Academy. *Am J Sports Med* 7(3):191–194, 1979.
107. Protzman, R., and Griffis, C. Stress fractures in men and women undergoing military training. *J Bone Joint Surg* 59A:825, 1977.
108. Puhl, J.. Women and endurance: Some factors influencing performance. *In* Drinkwater, B. (Ed.), *Female Endurance Athletes.* Champaign, IL, Human Kinetics, 1986, pp. 41–58.
109. Roy, S., and Irvin, R. *Sports Medicine.* Englewood Cliffs, NJ, Prentice-Hall, 1983, pp. 457–467.
110. Sachs, R., Daniel, D., Stone, M., and Garfein, R. Patellofemoral problems after anterior cruciate ligament reconstruction. *Am J Sports Med* 17(6):760–765, 1989.
111. Sammarco, G. Dance injuries. *In* Nicholas, J., and Hershman, E. (Eds.), *The Lower Extremity and Spine in Sports Medicine.* St. Louis, C.V. Mosby, 1986, pp. 1406–1439.
112. Sammarco, G. Diagnosis and treatment in dancers. *Clin Orthop Rel Res* 187:176–187, 1987.
113. Schwartz, B., Cumming, D., Riordan, E., Selye, M., Yen, S., and Rebar, R. Exercise-associated amenorrhea: A distinct entity? *Am J Obstet Gynecol* 141:662–670, 1981.
114. Shangold, M. Gynecologic concerns in exercise and training. *In* Shangold, M. and Mirkin, G. (Eds.), *Women and Exercise: Physiology and Sports Medicine.* Philadelphia, F.A. Davis, 1988, pp. 186–194.
115. Shangold, M. Sports and menstrual function. *Physician Sportsmed* 8(8):66–70, 1980.
116. Shangold, M. Evaluating menstrual irregularity in athletes. *Physician Sportsmed* 10(2):21–24, 1982.
117. Shangold, M. Exercise and the adult female: Hormonal and endocrine effects. *Exerc Sport Sci Rev* 12:53, 1984.
118. Shangold, M., Freeman, R., Thysen, B., and Gatz, M. The relationship between long distance running, plasma progesterone, and luteal phase length. *Fertil Steril* 31:130–133, 1979.
119. Shangold, M., and Levine, H. The effect of marathon training upon menstrual function. *Am J Obstet Gynecol* 143:862, 1982.
120. Shangold, M., and Mirkin, G. *The Complete Sports Medicine Book for Women.* New York, Miller Press, 1985, pp. 111–124.
121. Shlerman, G. Conditioning the athlete. *In* Haycock, C. (Ed.), *Sports Medicine for the Athletic Female.* Oradell, NJ, Medical Economics, 1980, pp. 49–59.
122. Snook, G. Injuries in women's gymnastics. *Am J Sports Med* 7(4):242–244, 1979.
123. Kaiserauer, S., Snyder, A., Sleeper, M., and Zierath, J. Nutritional, physiological, and menstrual status of distance runners. *Med Sci Sports Exerc* 21(2):120–125, 1989.
124. Speroff, L., and Redwine, D. Exercise and menstrual function. *Physician Sportsmed* 8(5):42–52, 1980.
125. Stager, J., Ritchie-Flanagan, G., and Robert-Shaw, D. Reversibility of amenorrhea in athletes. *N Engl J Med* 310:51–52, 1984.
126. Straub, W. Psychology of the athlete. *In* Haycock, C. (Ed.), *Sports Medicine for the Athletic Female.* Oradell, NJ, Medical Economics, 1980, pp. 7–17.
127. Thomas, C. Special problems of the female athlete. *In* Ryan, A. and Allman, F. (Eds.), *Sports Medicine.* New York, Academic Press, 1974, pp. 347–373.
128. Thomas, C. Factors important to women participants in vigorous athletics. *In* Strauss, R. (Ed.), *Sports Medicine and Physiology.* Philadelphia, W.B. Saunders, 1979, pp. 304–319.
129. Thomas, J. Women's sports and fitness programs at the U.S. Air Force Academy. *Physician Sportsmed* 7(4):59–68, 1979.
130. Twin, S. *Out of the Bleachers.* Old Westbury, NY, Feminist Press, 1979, p. xxxvii.
131. Ullyot, J. *Women's Running.* Mountain View, CA, World Publications, 1976, pp. 91–92.
132. United States Olympic Committee. *The USA in the Olympic Movement.* Colorado Springs, The United States Olympic Committee, 1988.
133. Walsh, W., Huurman, W., and Shelton, G. Overuse injuries of the knee and spine in girls' gymnastics. *Clin Sports Med* 3(4):839–850, 1984.
134. Weesner, C., Albohm, M., and Ritter, M. A comparison of anterior and posterior cruciate ligament laxity between female and male basketball players. *Physician Sportsmed* 14:149–154, 1986.
135. Wells, C. *Women, Sport, and Performance.* Champaign, IL, Human Kinetics, 1985, pp. 19–33.

136. Wentz, Al. Psychogenic amenorrhea and anorexia nervosa. *In* Givens, R. (Ed.), *Endocrine Causes of Menstrual Disorders.* Chicago, Year Book, 1977.
137. Whiteside, P. Men's and women's injuries in comparable sports. *Physician Sportsmed* 8(3):130–140, 1980.
138. Williams, J., and Sperryn, P. *Sports Medicine* (2nd ed.). London, Edward Arnold, 1976, pp. 210–225.
139. Wilmore, J. Alteration in strength, body composition and anthropometric measure consequent to a 10-week training program. *Med Sci Sport* 6(2):133–138, 1974.
140. Wilmore, J., and Brown, C. Physiological profile of women distance runners. *Med Sci Sports Exerc* 6(3):178–181, 1974.
141. Wilmore, J., Brown, C., and Davis, J. Body physique and composition of the female distance runner. *In* Milvy, P. (Ed.), *The Marathon: Physiological, Medical, Epidemiological, and Psychological Studies.* New York, New York Academy of Sciences, 1977, pp. 764–776.
142. Wilson, H. Rehabilitation of the injured athlete. *In* Haycock, C. (Ed.), *Sports Medicine for the Athletic Female.* Oradell, NJ, Medical Economics, 1980, pp. 207–244.
143. Woodward, S. How does strenuous maternal exercise affect the fetus? A review. *Birth* 8:17–27, 1981.
144. Wright, J. Anabolic steroids and athletes. *Exerc Sports Sci Rev* 8:149–202, 1980.
145. Zaharieva, E. Survey of sportswomen at the Tokyo Olympics. *J Sports Med Phys Fitness* 5:215–219, 1965.

ENVIRONMENTAL STRESS

Heat Intolerance Problems

Carl L. Stanitski, M.D.

Illness resulting from a thermal challenge during exercise presents a spectrum of conditions ranging from benign to fatal. Heat syncope, cramps, exhaustion, and stroke are manifestations of the body's inability to respond to an increased thermal load. This chapter will review thermoregulatory physiology, recognition of heat intolerance disorders, and treatment of these conditions with emphasis on prevention.

The outdated term "sunstroke" should not be used for heat intolerance conditions because these conditions may occur in the absence of sun and have been known to occur indoors in gymnasia with high heat and excessive humidity.

During exercise, an athlete is surrounded by his or her own microclimate. Endogenous heat is produced by muscle work, the amount of heat produced being proportional to the intensity and duration of the exercise. The body's ability to handle this thermal load requires elaborate endocrine, exocrine, circulatory, and neuromuscular regulatory mechanisms [4, 9, 42, 43, 53, 58, 64, 70, 74].

Heat intolerance problems are associated with a wide variety of athletic activities [1, 2, 6, 38, 67, 75, 76]. Endurance races (10K, marathon, triathlon), preseason football, soccer or field hockey sessions, and various hot weather sports camps are sources of exercise-induced heat intolerance disorders. The number of endurance athletes participating in distance runs, marathons, and triathlons has significantly increased during the last decade. This phenomenon is not isolated to the United States. In Auckland's "Round the Bays" race, upward of 70,000 runners take part annually [68]. The London marathon commonly has 15,000 participants [69].

The true incidence of heat intolerance problems in the athletic population is unknown. Complications from these conditions are usually underreported. During late summer preseason practices many deaths reported as cardiac in origin may be secondary to heat stroke, cardiovascular collapse being the end stage of this circumstance [8, 59, 60].

BASIC THERMOREGULATORY PHYSIOLOGY

The two major factors involved in heat intolerance problems are endogenous heat production and the effect of the athlete's environment in terms of temperature and humidity. Thermohomeostasis, the body's complex response to its heat production and the external environment, creates a balanced system in which the heat produced is equivalent to the heat lost. Exercise in a hot environment magnifies the body's ordinary physiologic responses to work [1, 7, 11, 20, 21, 24, 29, 43, 44, 48, 56, 58, 64, 73]. These responses include increases in body core (absolute central) temperature and skin (surface) temperature, metabolic heat production, pulse rate, blood pressure, and sweat rate. Sweat rate is proportional to metabolic rate, which is proportional to the level of exertion and the attendant environmental temperature [12, 15, 25, 32, 52].

Normal heat production is the result of thyroid thermogenesis (i.e., the body's metabolic response to circulating thyroid hormones) and the action of adenosine triphosphate (ATP) on the sodium pump of cell membranes. Seventy-five to 80% of the chemical energy involved in muscle contraction is changed to heat, which is then dispersed via the bloodstream [1, 5, 16, 50]. During vigorous exercise, large muscle mass activity increases the normal resting metabolic demands (which are usually minimal) by factors of 8 to 10 [43]. Hypothalamic neural control appears to be the seat of action of the body's coordinated efforts to withstand both endogenous and exogenous heat challenges [13, 18, 56, 74]. Thermal-sensitive cells in the preoptic and anterior hypothalamic neurons are affected by temperature changes in circulating blood.

Heat production is proportional to body weight. Heat loss is proportional to body surface area. For each 1°F rise in core temperature, metabolic demands for oxygen consumption increase by 6% to 7% of normal. Metabolic heat production is also proportional to the intensity of exercise and the function of exercise efficiency. Marked differences will occur between an experienced athlete with very efficient motion and a clumsy novice, and heat production can vary between 5000 and 15,000 kcal/hour [22].

To prevent excessive rise in body core temperature, two physiologic mechanisms exist: increased skin blood flow and increased sweating. Evaporation is a primary source of heat dissipation. During exercise, approximately 80% of the total heat removed is removed by means of the sweat evaporation mechanism. For a 70-kg man, every 100 mL of sweat evaporated decreases a rise in mean core body temperature by 1°C [13, 15, 25, 32, 52, 70]. Normal 70-kg adult male sweat production is approximately 1 liter per hour of exercise. A diminution in sweat production is noted with an increase in age, even in VO_2 max-matched controls. Arteriosclerotic cardiovascular disease and diabetes mellitus both cause diminution in skin blood flow.

Significant differences exist between children and adults in ability to handle a thermal load during exercise [3, 9, 10, 11, 12]. Children have a relatively increased body surface area to body mass and are slower to become acclimatized to a heat challenge. Children's sweating capacity is also reduced. They have a higher threshold for sweating and a decreased sweat production. In children, the amount of electrolytes lost during sweating is less than that lost in adults. Their sweat is even more hypotonic than that of adults. Curiously, even though children have a greater density of sweat glands than adults, the adult gland is able to produce two and one-half times as much sweat as a child's. At puberty, the adult form of sweat production becomes manifest.

Female children have a 10% to 12% higher body surface area to mass ratio than boys and have higher amounts of heat loss in response to mild heat stress. In general, boys and girls are equal in response to heat challenges at rest. Boys have higher sweating rates than girls. Because girls have a higher surface area-mass ratio, their metabolic rate of heat production is diminished.

Previously discussed adult male-female differences in response to thermal stress may be related to different fitness levels. Studies correlating male-female differences in response to thermal stress in patients matched for fitness (VO_2 max) found no differences between the sexes [16, 17, 39, 58].

With heavy exercise, muscle demands are increased and require increased cardiac output. This elective vascular shunting causes a diminution in blood flow to the skin with secondary compromise of skin sweating and evaporation as a mechanism for heat dissipation. It has been questioned whether a selective brain cooling mechanism is present because of the enhancement of scalp and facial sweating.

Muscle blood flow tends to be unchanged with changes in temperature. By contrast, skin blood flow is significantly changed by thermal stress [44, 52, 53], a reflection of the increase in heart rate and, if blood volume is adequate, increased circulating volume. Sunburn will cause a diminution in both skin blood flow and response to stress.

No change of rate of heat removal from the body's core and no change in metabolic heat production (as measured by oxygen consumption) were noted when the body was sprayed during exercise [12]. Cutaneous vasoconstriction secondary to the spray may negate the effects of evaporative cooling of one's heat transfer from the core to the periphery. Experiments using skin water spraying in an attempt to diminish thermal stress show no change in rectal temperature, sweat loss, plasma volume, or total heat production effects [14, 29, 55].

Heat intolerance problems reflect the body's inability to handle a heat load, producing subsequent cumulative deficits that affect heat transport, sweat production, and failure of temperature regulatory mechanisms. A vicious cycle ensues of decreased circulating blood volume secondary to hypohydration, diminished sweat production, increased body heat production, and inability to dissipate this heat within a fixed thermal environmental challenge [7, 20, 31, 36, 39, 42].

During exercise, the muscle, visceral, and skin vascular beds are in complex competition to maintain blood pressure and demands to support muscle metabolism and thermal regulation. Traditionally it has been thought that splanchnic bed vasoconstriction caused a diminution in renal and visceral blood flow in an attempt to maintain blood pressure and shunt blood to the extremities and skin to provide enough oxygen and nutrients for muscle work and vascular supply for sweat production and heat dissipation [23, 24, 27, 32, 44]. This shunting may be a source of the renal, hepatic, and other organ failures that occur in heat stroke [45, 46, 59, 60]. With progressive cardiovascular decompensation owing to hypovolemia, cardiovascular failure occurs with a variety of the collapse states seen with heat intolerance problems.

These heat intolerance problems comprise a continuum of physiologic inability of the body to respond to the heat challenge of exercise in a hostile thermal environment. Heat intolerance problems are seen in athletes in circumstances of intense exercise carried out within rigid systems, either self-imposed or enforced by coaches or peers. Without such pressures, most athletes stop before problems occur or at the first significant sign of problems. Only those who attempt to continue despite such warning signs later develop significant heat intolerance sequelae.

There are large individual variances in tolerance to an ambient or exercise heat challenge. Those at risk include the unfit, obese, dehydrated, very young, very old, unacclimatized people, and people with a past history of thermal intolerance to exercise. It is vital to recognize the athlete who is predisposed to an inadequate response to a heat challenge during exercise [2, 19, 20, 31, 33, 47, 51].

ENVIRONMENTAL EFFECTS

The environment must be considered in terms of ambient temperature, humidity, air movement velocity, and radiant heat sources. The general temperature, humidity, and wind speed reported for a general area by the local weather service reflects broad guidelines and may be highly inaccurate for specific exercise locales. Measurement of specific environmental factors should be done at the exercise site so that appropriate assessment of exercise risks in a potentially hazardous thermal setting can be made (Fig. 12A–1).

Important ambient factors controlling the amount of heat that can be lost by the body include air temperature, humidity, radiant heat (solar, surface [e.g., synthetic turfs]), and amount of wind. If skin temperature is greater than the surrounding environmental temperature, cooling occurs by both conduction and radiation. If the reverse is true, i.e., environmental temperature is greater than skin temperature, the only heat dissipated is lost by sweat evaporation. Humidity (percentage of water vapor in the air) is a major determinant of sweat evaporation rate and volume.

Solar radiation may account for as much as 150 kcal/hour of heat gain. Such a radiant source is then augmented if an athlete is exposed to a heat sink such as artificial turf, whose surface temperature can often be measured at 105° to 110°F.

A wet bulb thermometer reading compared with a dry bulb thermometer reading yields a measure of relative humidity when each is exposed to a constant rate of air movement. An increase in wet bulb thermometer value equals no evaporation secondary to atmospheric conditions that produce a high skin residual water vapor. A relative humidity of 60% or above means that sweat is not evaporated unless there is significant air movement. As a guide, if the wet bulb temperature alone is over 75°F (24°C), rest periods should be given every 30 minutes. At wet bulb temperatures above 76°F (24.4°C), only brief light exercise should be done until climate conditions improve.

A widely used method of assessing temperature, humidity, air movement, and radiant heat is the wet bulb globe temperature (WBGT). WBGT is calculated accordingly:

WBGT = (0.1 × ambient dry bulb
temperature) +
(0.2 × black globe temperature) +
(0.7 × wet bulb temperature)

A black globe (e.g., a toilet tank float painted black plus a thermometer) measures radiant heat. It is evident that the wet bulb temperature is the major factor in the equation (70%), a measure of the importance of humidity in thermal stress. A WBGT below 65°F (18°C) indicates a low risk for heat illness; between 65° and 73°F (18° to 23°C), a moderate risk; and 74° to 82°F (23° to 28°C), a high risk. Prudence calls for modification of exercise intensity

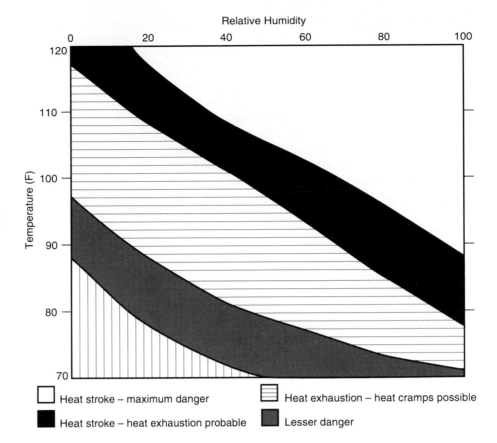

FIGURE 12A–1
Assessment of exercise risks in potentially dangerous thermal environments.

and duration as well as changes in clothing and equipment to eliminate as many risks of heat intolerance as possible.

THERMAL ACCLIMATIZATION

Acclimatization (i.e., the body's adaptation to an increased external heat load), usually requires 2 to 3 weeks of graduated exposure to increased environmental temperature and humidity. Physiologically, acclimatization to this type of thermal challenge for a given exercise load is demonstrated by decreased core and skin temperature, decreased pulse rate, increased ability (in amount of time) to withstand heat stress, increased sweating rate, and decreased electrolytes present within the sweat [13, 16, 28, 34, 36, 62, 66, 73].

Proposed mechanisms of acclimatization include stimulating sweat and cutaneous blood flow, increasing evaporative cooling, increasing cardiovascular stability, improving body fluid dynamics, stimulating earlier onset of sweating, and increasing sweat sensitivity [13, 16, 28, 34].

Heat acclimatization causes a 10% to 25% expansion in plasma volume. This increase in circulating plasma volume improves heat tolerance and en-

hances one's ability to make a physical effort. The exact mechanism of producing this hypervolemia is unknown and may be related to exogenous thermal stresses, associated endurance training, or a combination of both [24, 28, 34, 66].

All of these measures are designed to improve the body's response to the increased environmental thermal stress. Unfortunately, most athletes are unable to spend time adjusting to the heat challenge because many challenges occur unpredictably. Although most athletes at the high school and college level are not able to become acclimated to the hot conditions that often exist in late summer preseason practices, acclimatization methods should be kept in mind. If at all possible, exercise in progressively increasingly hotter circumstances over increasingly longer time periods should be done as part of precamp conditioning programs (e.g., exercise indoors in a warmer environment). Optimization of acclimatization is best achieved by training in an environment comparable to the one in which the competition will occur.

Trained, unacclimatized runners are more easily acclimatized or adapted to heat stress than those who are untrained. Endurance trained, acclimatized athletes undergo less cardiovascular strain because of their improved sweat mechanism, which requires diminished peripheral blood flow resistance compared with those who are untrained.

Most tests of exercise thermal adjustment and acclimatization have been done in a limited number of very fit, competitive or elite male and some female athletes in extremely controlled (treadmill, chamber) environments at a high intensity of effort [13, 16, 22, 32]. Whether such data are applicable to ordinary runners or novice, overweight, or unconditioned athletes is debatable because this latter group tends to be out on the endurance trails longer, are less fit, and have fewer coping mechanisms. Previous studies on acclimatization may not have been carried out long enough because more than 2 hours of testing are required to increase core temperatures to preacclimatization levels for a fixed exercise stress.

PREVENTION OF HEAT ILLNESS

There has fortunately been a heightening of awareness of exercise-related thermal problems in the athletic population. Only the most backward coaching staffs still believe that water deprivation during practice is a means to and source of mental toughness. Despite this awareness, vigilance must be continued to identify participants at risk because *heat intolerance problems are not a disease—they are an accident waiting to happen, and in most cases they are predictable and therefore preventable.* Judicious assessment of the thermal risk existing during exercise must be done by coaching staffs, race and tournament directors, and other individuals using guidelines such as the WBGT index. All athletes and coaches must be aware that heat intolerance problems exist and may occur in the best of athletes. In addition to being a health hazard, hyperthermia may also be a limiting factor in endurance sports performance [31].

The competitive macho nature of sports and the feeling of invulnerability among athletes must be balanced by the realization that body fluid loss does occur and must be made up to allow peak performance in both routine and excessive thermal environments [23, 24, 26, 32, 37, 40, 41, 43, 51, 56, 64, 76]. Deliberate loss of sweat to "make weight" so that the individual can play on teams that have body weight limits is to be vigorously condemned [9, 10]. Children are not designed to be placed in sweat boxes to participate in football leagues.

The well-conditioned athlete has a more effective sweat mechanism and produces more sweat in response to the environmental challenge. Athletes in poor physical condition are at high risk for heat intolerance problems [8, 22, 34, 47, 65, 66, 74]. Their general poor condition produces inefficient muscle use with a rapid rise in core temperature secondary to excessive muscle activity. Because of their poorer physical conditioning, the workload is less well tolerated. Coaches who demand excessive work as a punitive measure only put the athlete (who is ill conditioned to begin with) at higher risk for a heat intolerance disorder. Obesity causes a greater rise in core temperature for a given level of hypohydration and a reduced tolerance for prolonged exercise [8]. The group of athletes most commonly afflicted with fatal heat stroke are obese high school football middle linemen [8].

Appropriate equipment must be chosen to allow maximum surface area exposure for heat dissipation [8, 33, 51]. Since 50% of heat loss occurs through the hypervascular scalp and neck area, protective headgear should be worn only at times of direct challenges [20, 33]. During the remainder of the time, helmets should be removed to allow body cooling to occur. Practices in sports requiring protective armor (e.g., football and lacrosse) should be done in shorts and shirts without pads until reasonable acclimatization has occurred. Then, periods requiring full equipment contact should be relatively short (25 to 30 minutes) and scheduled for a time of day when the thermal challenge is least. Mesh cut-off jerseys with short sleeves should be worn to increase the amount of surface area exposed for thermal exchange.

Athletes who have had fluid losses because of gastrointestinal disorders with vomiting and diarrhea are likewise at risk for heat intolerance problems because of their relative state of hypohydration. They should not be involved in vigorous exercise until normal body weight has been restored. Blood donors are transiently relatively hypovolemic, and fluid needs to be replaced prior to accepting an exercise challenge. Patients with chronic diseases that may affect oxygen transport (anemia) or sweating (cystic fibrosis) are also at risk for heat-induced disorders.

An athlete who is in the midst of a febrile illness must be considered at risk for heat intolerance problems. A rise in body core temperature of 1°F produces a 6% to 7% increase in metabolic demand. Any athlete with a fever should be held out of vigorous exercises until he or she is afebrile.

A variety of drugs (recreational, over-the-counter, and prescription) may have a negative effect on athletic performance in heat [19, 35, 47]. Medications that can compromise the body's thermal response include diuretics (diminution of blood volume), phenothiazines, antihistamines, calcium blockers, antidepressants, antiseizure medications, anti-Parkinson's medications, caffeine, ethanol, and soft drinks (that contain caffeine). The effect of anabolic steroids and the water retention associated with them are not known in relation to their users' ability to respond to an exercise-induced thermal challenge.

Patients with a previous history of heat intolerance disorders need to be observed carefully because they tend to be at higher risk for recurrence of these disorders than patients who have never had such an experience [47, 63].

Hydration is the key to preventing disorders of heat intolerance. Hydration needs to be started early in practice and continued throughout the practice or athletic event. An athlete who has sustained cumulative fluid losses due to exercise or illness is at significant risk for heat intolerance problems. Athletes who are deliberately hypohydrated in an attempt to "make weight" also fall into this category. A loss of greater than 3% body weight from sweating causes an increase in core temperature proportionate to the amount of dehydration. Greater than 5% body weight sweat loss causes a

marked risk of heat intolerance disorders. Those with more than 7% body weight sweat loss are susceptible to potentially fatal heat stroke because of the cumulative effect of this hypohydration [10, 23, 26, 32, 42, 51, 64, 76].

To counteract the risk of cumulative hypohydration, all athletes engaged in vigorous successive daily exercise should be weighed nude before and after each practice session. Those with a body weight loss of more than 2% per practice should not be allowed to return to practice until the fluid loss is made up. Unfortunately, thirst is an inadequate mechanism for the drive to replace lost fluid. Usually only 50% of body fluid loss is made up in response to thirst. Fluid intake must be encouraged between sessions and during sessions to overcome hypohydration (Fig. 12A–2).

A rough guide of 1 pint of fluid per pound of sweat loss can be used in replacement therapy. The primary source of fluid replacement should be Adam's ale-water. Commercially available exercise electrolyte replacement drinks may have excessive amounts of glucose and may be hyperosmolar. Because of high concentrations of glucose, gastric

NAME	DATE										
	TIME										
	DRESS										
		WEIGHT		WEIGHT		WEIGHT		WEIGHT		WEIGHT	
		IN	OUT	IN	OUT	IN	OUT	IN	OUT	IN	OUT

FIGURE 12A–2
Fluid measurements to counteract hypohydration in athletes exercising in heat.

emptying time is delayed even when these fluids are taken chilled. It is suggested that, if such drinks are used, their concentrations be diluted by factors of two to three to improve ease of ingestion and decrease gastric emptying time [49, 61]. Drinks should be cold because cold stimulates thirst and increases rapidity of gastric emptying. No adverse effects have been reported on the use of cold drinks before, during, or after exercise.

Once significant hypohydration or dehydration has occurred, it cannot be corrected during exercise because fluid ingested at that point will be excreted through the urine [1, 23, 24]. Dehydration causes decreased total plasma volume, diminished stroke volume, increased heart rate, and diminished cardiac output with decreased glomerular filtration rate and decreased renal and skin blood flow. Liver glycogen is depleted, electrolytes are lost, and muscle strength and endurance are reduced. Mental acuity falls, and compromised thermoregulatory mechanisms result.

Salt tablets and salt replacement should not be encouraged because aldosterone is increased with exercise, and sodium is conserved by both the kidneys and sweat glands. Concentration of blood sodium is increased secondary to serum fluid loss [26, 32, 40, 56, 74]. Additional ingestion of sodium can cause further depletion of the intracellular space by hypernatremic dehydration. Excessive water inges-

tion in marathoners may lead to hyponatremia with secondary CNS effects [57]. Salt is not required in the form of salt tablets because sweat is hypotonic, and sweat loss is primarily water and not electrolyte. The average diet of high school and college students is rich in salt from ingestion of fast foods and junk foods, and excessive table salt added at mealtime is usually unnecessary and potentially harmful.

Acclimatization, fluid replacement, clothing modifications, and adaptation of activity to the environmental thermal challenge are the keys to the prevention of heat intolerance problems.

HEAT INTOLERANCE ILLNESSES

The Medical Research Council of Great Britain characterized heat intolerance problems into three types—heat stroke, heat exhaustion, and heat cramps [72]. The World Health Organization includes those three categories and, in addition, heat syncope [71] (Table 12A–1).

Heat syncope occurs while a person is standing following exercise in the heat [71]. The resultant venous pooling causes decreased cardiac output and stroke volume, and a transient fainting episode ensues. The core body temperature is normal. Re-

TABLE 12A–1
Forms of Heat Illness

Heat Illness	Prevention	Clinical Findings	Treatment
Heat syncope	Avoid standing about after vigorous exercise	Normal temperature Transient fainting episode	Recumbent position with legs elevated Oral hydration if previously underhydrated
Heat cramps	Avoid salt deficiency Ensure acclimatization Ensure adequate hydration Ensure appropriate conditioning	Normal temperature Painful contraction of large muscle groups (gastrocnemius, soleus; hamstrings)	Stop exercise Eliminate water and salt deficits
Heat exhaustion	See under Heat cramps Ensure adequate intra-event hydration Recognize environmental condition risk	Temperature elevated but <104°F (40°C) Fatigue Sweaty, flushed skin Vertigo or syncope Vomiting	See under Heat cramps Monitor core temperature IV fluids usually required (1 to 2 liters over 4 hours)
Heat stroke	See under Heat cramps Identify athletes at risk	Temperature usually >104°F (40°C) Syncope Coma Shock Multisystem failure Dry, flushed skin (usually but not necessarily)	Emergent condition Remove from thermal environment Cooling bath IV hydration Treat specific organ system compromise

covery is rapid following a brief period of recumbency and then mild activity. Oral fluids are usually not required unless a hypohydrated state has been present prior to the onset of syncope.

Heat cramps are extremely painful contractions in large muscle groups, particularly the hip extensors and hamstrings [1, 21, 25, 40, 43, 48, 56, 67]. Because the cramp affects only a few muscle bundles at a time and then moves on throughout the muscle, it appears to "wander" within that large muscle group. These cramps are due to decreased hydration as well as a diminution in serum sodium and chloride. Massaging the muscle usually does not relieve the symptoms. Treatment consists of fluid replacement with an intravenous infusion of 1 to 1½ liters of normal saline solution or ingestion of 1 liter of a 1% sodium chloride solution. This solution can easily be made by dissolving two 10-grain sodium chloride tablets in a liter of water. The intravenous route of fluid administration results in more rapid resolution of symptoms. Oral ingestion of a salty solution is usually not very palatable, and athletes commonly refuse to ingest the amount necessary to restore normal balance. Such muscle cramps can be recurrent if cumulative deficits of fluid and electrolytes are present. The best way to prevent such cramps is to ensure adequate hydration prior to the onset of exercise. Activity can be resumed after hydration and electrolyte balance have become normal.

Heat exhaustion is a more severe form of heat intolerance than either heat syncope or heat cramps and is the most common form of heat intolerance in the athletic population. The water depletion type of heat exhaustion, manifest by thirst and oliguria and presenting as lethargy, confusion, nausea, vomiting, and fatigue [7, 21, 24, 27, 42, 43, 48, 56, 74], is the more common type. It may be combined with the less common salt depletion type, which results from deficiencies of sodium chloride in the diet or because of high sodium chloride losses in body fluids with subsequent losses of extracellular fluid and decreased plasma volume and cardiac output. Decreases in urine and serum sodium chloride are noted [20, 26, 27]. This type of heat exhaustion is not characterized by thirst or relieved by fluid ingestion alone.

The spectrum of presentation of heat exhaustion is wide. The 1984 Olympics women's marathon showed dramatic effects of dehydration leading to heat exhaustion in a Swiss marathoner, Gabriella Anderson-Schiess. It is important to institute treatment for this because continued participation in a hostile thermal environment can rapidly lead to heat stroke. Heat exhaustion in athletes should not be considered merely as poor conditioning or excessive fatigue during or following a workout. Treatment should consist of fluid replacement until normal body weight is obtained and polyuria begins. The athlete should not be allowed to resume vigorous activity until normal body weight is reached, and, in severe cases, until normal serum electrolytes are restored. There is no universal agreement on the ideal fluid to be given intravenously for heat exhaustion therapy. The most important oral fluid should be water. Controversy exists about whether the addition of glucose helps or hurts [49, 61]. The addition of glucose may be of value if accumulated muscle and liver glycogen depletion has occurred. Such depletion is often noted after the third or fourth day of intense workouts twice a day.

Unconditioned and unacclimatized military reservists had significant frequency of exercise heat exhaustion following an exercise challenge and an 8.5% hospital readmission rate within 48 to 72 hours of the initial occurrence. Of those studied, 40 of 42 had an elevated serum glutamic oxaloacetic transaminase (SGOT) level, which was felt to indicate cellular injury [63]. Measurement of this enzyme may be useful in assessing cellular damage and as a guideline for return to endurance activity.

Heat Stroke

Exercise-associated heat stroke is the second most common cause of death in athletes in the United States. Animals refuse to continue running if their core temperature exceeds a certain amount. This built-in protective mechanism is overridden in humans, who commonly ignore warning signs prior to the onset of heat stroke.

Heat stroke is a disaster waiting to happen. Excessive, prolonged, vigorous exercise raises body core temperature by a considerable amount. Adequate compensatory mechanisms may not be available in a hostile thermal environment, and heat stroke will then ensue. Initial heat stroke symptoms may be muscle ache, gastrointestinal upset, headache, and nausea, which appear to be flulike symptoms and most likely will produce a misdiagnosis. Acute mental breakdown from cerebral involvement may result in hospitalization in a psychiatric unit [1, 4, 8, 18, 20, 27, 38, 42, 43, 47, 56, 59, 75]. This condition has a 50% to 70% fatality rate, with 80% of the fatalities being due to circulatory failure. The magnitude of heat stroke is a response to progressive hypohydration with decreased circulatory volume, further compromising the body's ability to dissipate heat. It may yield paradoxical shivering, which can cause increased body temperature.

The body must be understood in a relative sense because a pulse of 100 to 110 beats/minute in a highly conditioned athlete indicates significant tachycardia in a person who has a "normal" resting pulse of 40 to 45. Oral and axillary temperatures are inadequate measures of core temperature. Temperature is best ascertained by rectal, esophageal, or tympanic membrane probes. Monitoring of the core temperature during treatment will provide data indicating the efficacy of the therapeutic regimen.

The clinical presentation of heat stroke usually involves a high body temperature in a person in eminent danger of vascular collapse. Although sweating may be diminished or absent (as a sign of loss of normal thermal homeostasis), patients may have heat stroke and still be sweating. Heat stroke may occur with cataclysmic suddenness. Its sequelae are often related to the accumulated time spent at an elevated body temperature of between 41° and 42°C. A rapid rise in core body temperature to over 106°F causes protein denaturation.

Heat stroke may cause a spectrum of signs and symptoms, including tachycardia, hypotension, electrocardiographic changes, arrhythmias, acute renal failure (which can lead to chronic renal disease), myoglobinuria, hyperkalemia, increased serum uric acid, coagulopathy, hepatic impairment, nausea, vomiting, and diarrhea, irritability, coma, and delirium.

Metabolic acidosis secondary to excessive lactic acid buildup may be coupled with a respiratory alkalosis secondary to hyperventilation. Any blood gas results need to be evaluated in terms of corrected values for elevated body temperature.

Treatment of Heat Stroke

The basic principles of treatment of heat stroke include removal (if possible) of the patient from the site of increased temperature, cooling of the patient, management of dehydration, and specific supports for multi-organ failure. Such support includes correction of acid base imbalance, cardiac arrhythmias, disseminated intravascular clotting, renal failure, and similar conditions [20, 45, 46, 59]. In patients with exercise-associated heat intolerance problems with temperatures between 41° and 43°C, a 50% to 70% mortality exists.

Multisystem reduced organ function occurs with heat stroke. Central nervous system abnormalities include weakness, dizziness, confusion, headaches, delirium, and coma. A coma lasting more than 24 hours, especially one associated with seizures, bodes ominously for full recovery. In approximately 10% to 35% of patients with heat stroke, acute renal tubular necrosis results secondary to three mechanisms: direct tubular injury from heat, myoglobin precipitation in the tubules, and tubular damage resulting from diminished renal blood flow [45]. Vascular endothelial damage leads to further hemorrhage within the vascular tree and progressive disseminated intravascular coagulopathy [45]. Liver involvement is usually associated with central lobular necrosis and marked cholestasis [46]. Because of ischemia in the gastrointestinal tract, heat stroke victims may experience a transient malabsorption syndrome. Gastrointestinal function returns to normal in about 3 months following small bowel mucosal regeneration [60]. The endocrine stress response is high in patients with exercise-involved heat stroke [4]. These endocrine stress responses return to normal after recovery.

No specific pharmaceutical measures (other than hydration) have been shown to prevent heat intolerance problems. Certain medications may be of value in heat stroke management. Niacin may be used to produce a vascular cutaneous flush in an attempt to decrease vasoconstriction secondary to the use of ice for cooling, which causes shivering. Shivering may also be controlled by diazepam (Valium). Blocking of beta-endorphins by opiate antagonists may offer an effective means of controlling heat stroke [30]. The role of opiates (specifically naloxone) has been suggested in the treatment of exercise-induced heat intolerance [30]. The rise in exercise-induced temperature has been abolished using such opiates. The mechanism of this reduction is not understood but may be due to stabilizing sympathetic and parasympathetic tone.

SUMMARY

Exercise in normothermic conditions is stressful to the body's homeostatic mechanisms. When the challenge of an increased thermal load is superimposed, many athletic participants cannot cope. Compromise of the thermoregulatory mechanism may be manifest in a variety of ways reflecting a spectrum of involvement from rapidly resolving and benign to catastrophic fatal collapse. Responsibility for prevention of heat illness lies with the athlete, coaches, activity directors, and others because the circumstances in which these conditions occur are usually predictable. Appropriate modification of these circumstances will benefit all concerned with the athlete's welfare as the major goal.

References

1. Adams, W.C., Fox, R.H., Fry, A.J., and McDonald, I.C. Thermoregulation during marathon running in cool, moder-

ate, and hot environments. *J Appl Physiol* 38:1030–1037, 1975.

2. American College of Sports Medicine. Position statement on prevention of heat injuries during distance running. *Med Sci Sports* 7:7, 1975.

3. Anderson, R.K., and Kennedy, W.L. Effect of age on heat-activated sweat gland density and flow during exercise in dry heat. *J Appl Physiol* 63:1089–1094, 1987.

4. Appenzeller, O., Khogali, M., Carr, D.B., Gumaa, K., Mustafa, M.K.Y., Jamjoom, A., and Skipper, B. Makkah Hajj: Heat stroke and endocrine responses. *Ann Sports Med* 3:30–32, 1986.

5. Araki, T., Toda, Y., Matsushita, K., and Tsujino, A. Age differences in sweating during muscular exercise. *Jap J Phys Fitness Sports Med* 28:239, 1979.

6. Armstrong, L.E., Hubbard, R.W., DeLuca, J.P., and Christensen, E.L. Heat acclimatization during summer running in the northeastern United States. *Med Sci Sports Exerc* 19(2):131–136, 1987.

7. Armstrong, L.E., Hubbard, R.W., Kraemer, W.J., et al. Signs and symptoms of heat exhaustion during strenuous exercise. *Ann Sports Med* 3:182–189, 1987.

8. Barcenas, C., Hoeffler, H.P., and Lie, J.T. Obesity, football, dog days and siriasis: A deadly combination. *Am Heart J* 92:237, 1976.

9. Bar-Or, O. Thermoregulation and fluid and electrolyte needs. *In* Smith, N.J. (Ed.), *Sports Medicine: Health Care for Young Athletes.* Evanston, IL, The American Academy of Pediatrics, 1983.

10. Bar-Or, O., Dotan, R., Inbar, O., Rothstein, A., and Zonder, H. Voluntary hypohydration in 10- to 12-year-old boys. *J Appl Physiol* 48:104, 1980.

11. Bar-Or, O. Climate and the exercising child: A review. *Int J Sports Med* 1:53, 1980.

12. Bassett, D.R., Jr., Nagle, F.J., Mookerjee, S., Darr, K.C., Ng, A.V., Voss, S.G., and Napp, J.P. Thermoregulatory responses to skin wetting during prolonged treadmill running. *Med Sci Sports Exerc* 19(1):28–32, 1987.

13. Baum, E., Bruck, K., and Schwennicke, H.P. Adaptive modifications in the thermoregulatory system of long-distance runners. *J Appl Physiol* 40:404–410, 1976.

14. Brebner, D.F., and Kerslake, D.M. The time course of the decline in sweating produced by wetting the skin. *J Physiol* 175:295–302, 1964.

15. Brengelmann, G.L., Wyss, C., and Rowell, L.B. Control of forearm blood flow during periods of steadily increasing skin temperature. *J Appl Physiol* 35:77–84, 1973.

16. Brown, C.H., and Wilmore, J.H. The effects of maximal resistance training on the strength and body composition of female athletes. *Med Sci Sports* 6:174, 1974.

17. Burke, E.J. Physiological effects of similar training programs in males and females. *Res Q* 48:510, 1977.

18. Cabanac, M. Face fanning: A possible way to prevent or cure brain hyperthermia. *In* Khogali, M., and Hales, J.R.S. (Eds.), *Heat Stroke and Temperature Regulation.* Sydney, Academic Press, 1983, pp. 213–221.

19. Claremont, A.D., Costill, D.L., Fink, W., and Van Handel, P. Heat tolerance following diuretic induced dehydration. *Med Sci Sports* 8:239–243, 1976.

20. Clowes, G.H.A., Jr., and O'Donnell, T.F., Jr. Heat stroke. *N Engl J Med* 291:564, 1974.

21. Committee on Sports Medicine. Climatic heat stress and the exercising child. *Pediatrics* 69:808, 1982.

22. Conley, D.L., and Krahenbuhl, G.S. Running economy and distance running performance of highly trained athletes. *Med Sci Sports* 12:357–360, 1980.

23. Convertino, V.A., et al. Role of thermal and exercise factors in the mechanism of hypovolemia. *J Appl Physiol* 48:657–664, 1980.

24. Costill, D.L. Sweating: Its composition and effects on body fluids. The marathon: Physiological medical, epidemiological and psychological studies. *Ann NY Acad Sci* 301:160–174, 1977.

25. Costill D.L. Water and electrolyte requirements during exercise. *Clin Sports Med* 3(3):639–648, 1984.

26. Costill, D.L., and Fink, W.J. Plasma volume changes following exercise and thermal dehydration. *J Appl Physiol* 37:521–525, 1974.

27. Costrini, A.M., Pitt, H.A., Gustafson, A.B., et al. Cardiovascular and metabolic manifestations of heat stroke and severe heat exhaustion. *Am J Med* 66:296–302, 1979.

28. Davies, C.T.M. Effect of acclimatization to heat on the regulation of sweating during moderate and severe exercise. *J Appl Physiol* 50:741–746, 1981.

29. Davies, C.T.M., Brotherhood, J.R., and Zeidifard, E. Temperature regulation during severe exercise with some observations on skin wetting. *J Appl Physiol* 41:772–776, 1976.

30. DeMeirler, K., Arentz, T., et al. The role of endogenous opiates in thermal regulation of the body during exercise. *Br Med J* 290:739–740, 1985.

31. Drinkwater, B.L. Heat as a limiting factor in endurance sports. *Am Acad Phys Educ* 18:93–100, 1984.

32. Fortney, S.M., Nadel, E.R., Wenger, C.B., and Bove, J.R. Effect of blood volume on sweating rate and body fluids in exercising humans. *J Appl Physiol* 51:1594–1600, 1981.

33. Fox, E.L., Mathews, D.K., Kaufman, W.S., and Bowers, R.W. Effects of football equipment on thermal balance and energy cost during exercise. *Res Q Am Assoc Health Phys Ed* 37:332, 1966.

34. Gisolfi, C.V., and Robinson, S. Relations between physical training, acclimatization and heat tolerance. *J Appl Physiol* 26:530–534, 1969.

35. Gordon, N.F., van Rensburg, J.P., Russell, H.M.S., Kielblock, A.J., and Myburgh, D.P. Effect of beta-adrenoceptor blockade and calcium antagonism, alone and in combination, on thermoregulation during prolonged exercise. *Int J Sports Med* 8:1–5, 1987.

36. Greenleaf, J.E. and Greenleaf, C.J. Human acclimation and acclimatization to heat: A compendium of research. NASA Technical Report No. TMX-62008. Moffett Field, CA, Ames Research Center, 1970, pp. 1–188.

37. Greenleaf, J.E., and Castle, B.L. Exercise temperature regulation in man during hypohydration and hyperhydration. *J Appl Physiol* 30:847–853, 1971.

38. Hanson, P.G., and Zimmerman, S.W. Exertion heatstroke in novice runners. *JAMA* 242:154, 1979.

39. Haymes, E.M. Physiological responses of female athletes to heat stress: A review. *Phys Sportsmed* 12:45–54, 1984.

40. Hiller, W.D.B., O'Toole, M.L., Massimino, F., et al. Plasma electrolyte and glucose changes during the Hawaiian ironman triathalon (abstract). *Med Sci Sports Exerc* 17:219, 1985.

41. Hiller W.D.B., O'Toole, M.L., Fortress, E.E., Laird, R.H., Imbert, P.C., and Osk, T.D. Medical and physiological considerations in triathalons. *Am J Sports Med* 15(2):164–167, 1987.

42. Hubbard, R.W. Effects of exercise in the heat on predisposition to heatstroke. *Med Sci Sports Exerc* 11:66–71, 1979.

43. Hubbard, R.W., and Armstrong, L.E. The heat illness: Biochemical, ultrastructural and fluid-electrolyte considerations. *In* Pandolf, K.B., Sawka, M.N., and Gonzalez, R.R. (Eds.), *Human Performance, Physiology and Environmental Medicine at Terrestrial Extremes.* Indianapolis, Benchmark Press, 1988, pp. 305–359.

44. Johnson, J.M. Regulation of skin circulation during prolonged exercise. *Ann NY Acad Sci* 301:195–212, 1977.

45. Kew, M.C., Abrahams, C., and Seftel, H.C. Chronic interstitial nephritis as a consequence of heatstroke. *Q J Med* 39:189, 1970.

46. Kew, M.C., Berson, I., et al. Liver damage in heatstroke. *Am J Med* 49:192, 1970.

47. Kilbourne, E.M., Choi, K. Risk factors for heat stroke. *JAMA* 247:3332–3336, 1982.

48. MacDougall, J.D., Reddan, W.G., Layton, C.R., and Dempsey, J.A. Effects of metabolic hyperthermia on performance during prolonged exercise. *J Appl Physiol* 36:538–544, 1974.

49. Mallard, D., Owen, K.C., Kregel, P., Wall, T., and Gisolfi, C.V. The year book of sports medicine. Chapter 1. Exercise physiology and medicine: Effects in ingesting carbohydrate

beverages during exercise in the heat. *Exerc Med Sci Sports* 18:568–575, 1986.

50. Mitchell, J.W. Energy exchanges during exercise. *In* E.R. Nadel (Ed.), *Problems with Temperature Regulation.* New York, Academic Press, 1977, pp. 11–26.

51. Murphy, R.J. Heat illness and athletics. *In* Strauss, R.H. (Ed.), *Sports Medicine and Physiology.* Philadelphia, W.B. Saunders, 1979.

52. Nadel, E.R. Control of sweating rate while exercising in the heat. *Med Sci Sports* 11:31–35, 1979.

53. Nadel, E.R., et al. Circulatory regulation during exercise in different ambient temperatures. *J Appl Physiol* 46:430–437, 1979.

54. Nadel, E.R., and Stolwijk J.A.J. Effect of skin weekenders on sweat gland response. *J Appl Physiol* 35:689–694, 1973.

55. Nadel, E.R., Wenger, B.C., Roberts, M.F., Stolwijk, J.A., and Catarell, E. Physiological defenses against hyperthermia of exercise. *Ann NY Acad Sci* 301:98–109, 1977.

56. Nelson, P.B., Robinson, A.G., Kapoor, W., and Rinaldo, J. Hyponatremia in a marathoner. *Phys Sportsmed* 18(10):78–87, 1988.

57. Nunneley, S.A. Physiological responses of women to thermal stress: A review. *Med Sci Sports* 10:250, 1978.

58. O'Donnell, T.F., Jr. Acute heatstroke: Epidemiologic, biochemical, renal and coagulation studies. *JAMA* 234:824, 1975.

59. O'Donnell, T.F., Jr., and Clowes, G.H., Jr. The circulatory abnormalities of heat stroke. *N Engl J Med* 287:734–737, 1972.

60. Owen, M.D., Kregel, K.C., Wall, P.T. and Gisolfi, C.V. Effects of ingesting carbohydrate beverages during exercise in the heat. *Med Sci Sports Exerc* 18:568–575, 1986.

61. Piwonka, R.W., Robinson, S., Gay, V.L., and Manalis, R.S. Preacclimatization of men to heat by training. *J Appl Physiol* 20:379–384, 1965.

62. Roberts, P., Hubbard, R.W., and Kerstein, M.D. Serum glutamic-oxaloacetic transaminase (SGOT) as a predictor of recurrent heat illness. *Mil Med* 152:408–410, 1987.

63. Senay, L.C. Effects of exercise in the heat on body fluid distribution. *Med Sci Sports Exerc* 11:42–48, 1979.

64. Strydom, N.B., Wyndham, C.H., Williams, C.G., et al. Acclimatization to humid heat and the role of physical conditioning. *J Appl Physiol* 21:636–642, 1966.

65. Strydom, N.B., and Williams, C.G. Effects of physical conditioning on the state of heat acclimatization of Bantu laborers. *J Appl Physiol* 27:262–265, 1969.

66. Sutton, J.R., and Bar-Or, O. Thermal illness in fun running. *Am Heart J* 100:778, 1980.

67. Sutton, J., Coleman, M.K., Millar, A.P., Lazarus, L., and Russo, P. The medical problems of mass participation in athletic competition: The "city-to-surf" race. *Med J Aust* 2:127, 1972.

68. Sutton, J.R. Community participation in long distance athletic events—The London Marathon. *In* Sergeant, A.J., and Young, A. *Proc Conference Community Health and Fitness,* The London Polytechnic Institute, 1982.

69. Taylor, N.A.S. Eccrine sweat glands: Adaptations to physical training and heat acclimation. *Sports Med* 3:387–397, 1986.

70. U.S. Department of Health, Education and Welfare. *WHO International Classification of Diseases.* PHS No. 992. Washington, D.C., U.S. Government Printing Service, 1963.

71. Weiner, J.S., and Horne, G.O. A classification of heat illness. *Br Med J* 1:1533, 1958.

72. Wells, C.L., Constable, S.H., and Haan, A.L. Training and acclimatization: Effects on responses to exercise in a desert environment. *Aviat Space Environ Med* 51:105–112, 1980.

73. Wyndham, C.H. The physiology of exercise under heat stress. *Ann Rev Physiol* 35:193–220, 1973.

74. Wyndham, C.H. Heatstroke and hyperthermia in marathon runners. *Ann NY Acad Sci* 301:128–138, 1977.

75. Wyndham, C.H., and Strydom, N.B. The danger of inadequate water intake during marathon running. *S Afr Med J* 43:893–896, 1969.

Cold Injury

Edward G. Hixson, M.D.

Cold injury has historically been most significant to the military. Many armies have suffered more casualties from cold than enemy fire. In 218 B.C. Hannibal lost 20,000 of 46,000 men crossing the Alps [1]. Napoleon invaded Russia with 385,000 troops and returned with 3000. There were 250,000 deaths from cold injury [26], and cold injury accounted for 10% of U.S. casualties in World War II and Korea (90,000 in the former, 9000 in Korea). More than 7.5 million U.S. troop days were lost in World War II [10]. The German and especially Russian losses were much greater. Appropriately, much of the available information on cold injury comes from the military.

The effect of cold on athletic activity is less dramatic. For athletic events in a controlled environment (e.g., swimming, or basketball) there is no effect. Athletic activities performed in cold weather are affected, particularly endurance events requiring submaximal exercise for long durations (e.g., cross-country skiing, biathlon). Endurance events are now more frequent in wilderness environments (e.g., marathons, ultramarathons, "Iron Man," triathlon, speed hikes, "survival of the fittest," and so on). Cold injury is not only more likely but the risk is greater. Nowhere, however, is the significance of cold injury greater than for wilderness recreational sports. Most peacetime cold injury casualties are seen here (e.g., in mountaineering, ski touring, back packing, trekking, snowmobiling, hunting, and so on).

PHYSIOLOGY OF COLD INJURY

Man is an animal who evolved in the tropics. He can survive in a cold environment only by increased heat production and decreased heat loss achieved by insulation. When thus protected, man can toler-ate extremes of ambient temperature from $-50°C$ to $100°C$ ($-58°F$ to $212°F$). Core temperature may vary only through a range of 4°C without impairment of function. Core temperature is the temperature of the internal organs, especially the heart, lungs, and brain, in contrast to surface temperature, which is skin temperature. Core temperature is relatively constant at 99°F. Skin temperature fluctuates greatly. The living cell can tolerate extremes from $-1°C$ to 45°C (31°F to 113°F) without cell death. The temperature range from 28°C to 34°C (82°F to 93°F) is the thermoneutral zone. If ambient temperatures are in this range the nude human body will not cool and can easily dissipate metabolic heat.

Heat Production

Man must maintain his core temperature (37.6°C or 99.6°F) within a narrow range (4°C or about 12°F). The body is able to maintain a core temperature only by constant heat production. Under basal conditions of rest, about 1 kilocalorie (kcal) per kilogram per hour is produced by the body's metabolic activity. Maximal exercise may increase this by roughly a factor of 10. The body is about 25% efficient—it utilizes about 25% of its energy for work, 75% being released as heat. The source of the energy is food—carbohydrate, fat, and protein. Carbohydrates produce about 4 kcal/g, protein slightly less, and fats 9 kcal/g. The average individual requires roughly 2400 kcal daily for nonstrenuous activity. Many athletes require a significantly higher caloric intake to sustain increased muscle activity, 6600 to 10,000 kcal daily. The military recommends 4000 kcal daily for troops training in a cold environment [25]. High-carbohydrate diets are generally preferred for athletes, with carbohydrate accounting for 60%, protein for 15%, and fats for 25% of calories.

Heat Loss

Heat production must be balanced by heat loss. The body loses most of its heat through the skin (approximately 90%). This loss is variable by a factor of about 100 from maximum vasodilation to vasoconstriction. Blood flow to the skin is involuntarily controlled. Areas of high blood flow, such as the head and neck, lose heat fastest. Hence the old climbers' dictum, "When your feet are cold, put your hat on." Another area of significant heat loss is the lungs. This loss is quite variable and is under both voluntary and involuntary control; respiratory rate changes dramatically with the level of exercise, response to altitude, and other factors.

The body loses heat in four ways, by (1) conduction, (2) convection, (3) evaporation, and (4) radiation (Fig. 12B–1). Conduction is direct loss from contact. Because air conducts heat poorly, trapping layers of warm air next to the body is the basic principle of insulation. Contact with water, rock, steel, and other solids that conduct heat better increase conduction losses. Sleeping on the ground is accompanied by increased conduction losses, emphasizing the need for more insulation under the sleeper. Increased conduction losses are most dramatic when the body is wet. Immersion in water increases the rate of cooling 100 times faster than air at the same temperature [18]. As water temperature decreases, cooling increases in a linear fashion [11]. The wet athlete is at great risk, especially if he has been accidentally immersed in cold water, such as in a sailing accident, or his own sweat such as a marathon skier or runner.

Convection loss is caused by the motion of air or water across the body surface. Heat is removed relative to the velocity of the air or water. "Wind chill" is a means of expressing increased heat loss by wind on the human body. Wind chill is most often expressed as the "equivalent" temperature in motionless air [22]. The rate of heat loss varies directly with the difference in temperature between the body and the air [27]. Heat loss varies directly also with the square of the velocity of the wind. Doubling the velocity increases heat loss fourfold. However, at very high wind speeds, the loss is not this great because there is not enough contact time to warm the air.

Evaporation is the body's major means of controlling heat loss. Whether through perspiration or external wetting, when water evaporates, heat is absorbed. One gram of water converted to vapor absorbs about 500 calories (heat of vaporization) in addition to increased conduction losses from wetting. Roughly one-third of the body's evaporation losses come from the lungs, whereas two-thirds come from the skin. The relative humidity of the ambient air is important. In dry air, evaporation is faster. The respiratory rate increases as altitude increases. Evaporative heat loss is greatly increased

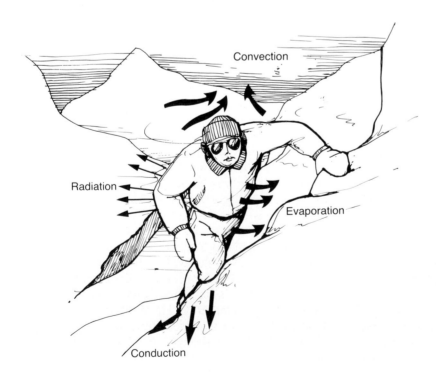

Convection

Radiation

Evaporation

Conduction

FIGURE 12B–1
The four ways in which heat is lost.

as one ascends to high altitude, as is evaporative water loss from the lungs. This loss may require an additional 3 or 4 liters of water intake daily to prevent dehydration, and 1500 to 2000 kcal of heat may be lost [27]. Heavy exercise, even in a cold environment, may raise the body temperature perhaps as high as 41.1°C or 106°F, producing as much perspiration as 1.8 liters hourly [20]. It is easy to see why the athlete, especially the high altitude climber, may lose tremendous amounts of heat and water, especially from the skin and lungs. When the climber stops climbing, he is wet, dehydrated, cold, and exhausted. He may spend the night in a forced bivouac with limited food and water in subzero windy conditions at an extreme altitude.

The greatest amount of heat (approximately 55% to 65% of the body's heat loss) is lost by radiation. A warm body gives off heat as infrared radiation to a colder body or surroundings. The amount of radiation loss is relatively constant and is related to the mass of the body and its surface area. Radiation losses are not affected greatly by clothing. Special "reflective" surfaces on clothing do not change radiation losses significantly, contrary to the boasts of manufacturers. Children cool faster than adults because they have a greater surface area to mass ratio [14]. (This is also true for conduction and convection losses.) The infrared radiation given off is the basis for the use of infrared radiation detectors to find lost individuals.

Control

Maintenance of a balance between heat loss and heat production is essential to maintain a constant core temperature. With variable climatic conditions, the body is able to control heat loss and heat production through both voluntary and involuntary mechanisms. The body senses ambient temperature through the skin, and skin temperature thus may vary greatly. The "thermostat" is located in the hypothalamus of the brain and regulates heat loss and production. Voluntary muscular activity produces heat, as does involuntary shivering. Shivering can increase heat production as much as fivefold [15]. The amount of heat production is limited by the body's store of muscle glycogen.

Heat loss is largely controlled by vasoconstriction. Cutaneous blood flow may vary by as much as 100 times in the range from maximum vasoconstriction to maximum vasodilation. The body loses 90% to 95% of its heat through the skin. Piloerection, important to animals when they "fluff their fur," improves insulation by trapping a layer of motionless

air close to the skin. This response is present in nearly hairless man as "goose bumps," which do not appreciably increase insulation.

The body may also increase heat production involuntarily through nonshivering thermogenesis. The mechanism is complex. Cold causes increased output of epinephrine and nonepinephrine. Increased levels of epinephrine and norepinephrine result in increased metabolic work and therefore heat production. The amount of heat produced by this mechanism is insignificant with respect to that produced by shivering.

FACTORS INFLUENCING COLD INJURY

Clothing

Clothing design and materials must make use of physiologic principles to be effective as insulation. In our high-technology age there is no lack of variety. The basic principle is to maintain layers of warm, motionless air close to the body. For exercise in a cold environment, the use of "layered clothing" is important. The elevation of body temperature with exercise should be anticipated. To avoid sweating and soaking insulation by removing layers as overheating occurs, the athlete should dress down. After stopping, he dresses up, adding layers to prevent chilling.

The use of "warm when wet" materials such as wool, which retains 80% of its insulating value when wet, is important. Fiber pile is similar and lighter. It does not absorb water and may be easily shaken or wrung out. Material that wicks water away from the body (polypropylene for underwear) is beneficial and dry on the body. Down has been the gold standard for loft and insulating ability. Many climbers replace this with newer fibers of materials such as polyester (e.g., Qualofil and Polarguard), which shed water. Down clumps and is useless when wet. Often down is reserved for situations in high, dry, extreme cold.

Wind protection is important and fabrics such as Gortex and Entrant provide wind protection (wool and fiber pile offer little wind protection) while allowing water vapor to escape. These materials provide a reasonable barrier to water droplets in rain or snow. Nylon or plastic rainwear, although waterproof, will cause the exercising athlete to become drenched by perspiration because water vapor cannot escape and condenses.

Mittens are warmer than gloves. Newton's law states that cooling is proportional to surface area.

The lower surface area to mass ratio of the mittened hand causes the hand to lose heat more slowly. Adding insulation to fingers of gloves may be counterproductive by increasing surface area and heat loss. Plastic boots using closed-cell foam liners keep and maintain dry insulation. Being rigid, they also prevent the constricting effect of crampon straps and snowshoe bindings. For cold weather, all clothing should be loose, avoiding constriction of circulation. Dry is best. The military has recommended spraying the feet daily with antiperspirant to keep the feet and socks dry.

The vapor barrier concept is applicable for extreme cold. The concept is that insensible water loss may be suppressed by a feedback mechanism, thereby preserving heat and avoiding insensible evaporative heat loss. Vapor barriers (an impervious layer of nylon or plastic) have generally been used as liners for socks and sleeping bags.

The head and neck area must be protected. Hat, balaclavas, scarves, and so on, are important.

Nutrition and Hydration

For athletic activity of any kind, nutrition and hydration are extremely important. These two needs are especially crucial in a cold environment. Stored muscle glycogen is the limiting factor of muscle activity, whether shivering or athletic. Activity of longer duration burns an increased amount of stored fats. In response to cold, urinary output is increased, leading to "cold diuresis." Vasoconstriction of the skin increases central blood volume and renal blood flow. The usual requirement for about 2 quarts of fluid daily may be significantly increased in the cold.

Heredity

There is abundant anecdotal evidence about cold tolerance among certain ethnic and racial groups, Eskimos, Sherpa, aborigines, and so on. Much adaptation to and tolerance of cold is cultural and learned. A physiologic response, cold-induced vasodilation (CIVD), is measurable. Cold produces vasoconstriction, especially in the hands. After 5 to 10 minutes of constriction these vessels dilate briefly, avoiding ischemic damage. Inhabitants of cold regions such as Eskimos and aborigines have increased CIVD. Blacks have decreased CIVD and thus an increased risk of frostbite.

The Mind

Cold injury often accompanies severe adverse environmental conditions compounded by a human catastrophe. It has been shown over and over that a positive attitude, particularly in a leadership position, is the best defense. Those afflicted by injury, altitude, starvation, or hypothermia are often apathetic. A positive constructive attitude in one member of the group may mobilize everyone's efforts and enable the group to survive.

Another occasionally life-saving quality is "knowing when to quit." Mountaineers in particular strive to reach their summit or goal. When reaching this goal will risk life and limb, it is time to turn back. This decision is best made early in the face of deteriorating weather. This recommendation may seem simplistic and self-evident, but it is often a very hard decision for a highly motivated mountaineering team to make.

TYPES OF COLD INJURY

When heat loss exceeds heat production, cold injury results. Cold injury may be one of two types, generalized and localized. Hypothermia is generalized cold injury. Localized injury may be either freezing injury (frostbite) or nonfreezing injury. There are also several syndromes that are triggered by cold exposure.

HYPOTHERMIA

Hypothermia occurs when the body core temperature drops below 35°C (95°F). There are two classes, primary and secondary. Primary hypothermia occurs accidentally to otherwise healthy individuals. Secondary hypothermia occurs in the course of systemic disease. The former, primary hypothermia, is the subject of this discussion.

Hypothermia has also been classified as acute and chronic. Acute hypothermia results from sudden exposure to extreme cold. An example is a sudden immersion in icy water (usually less than 10°C or 50°F). For persons with acute hypothermia such as an immersion, exposure times are measured in minutes or a few hours. Chronic hypothermia occurs more slowly. A climber or hiker lost in bad weather may slowly develop chronic hypothermia over a period of many hours. It may take many hours for avalanche victims to develop hypothermia (if the snow is dry). Victims of chronic hypothermia such as those who become lost in a blizzard are often described by the press as succumbing to death from "exposure." The incidence of hypothermia is unknown; it is both a wilderness and an urban problem and has occurred in every state of the union from Alaska to Florida. Mortality is significant and may be as high as 70% for people with a core temperature of below 30°C (86°F) [8]. For all degrees of hypothermia, an overall mortality of 17% has been reported [3].

Diagnosis

The sine qua non of a diagnosis of hypothermia is the determination of core body temperature. This requires the use of rectal or esophageal probes, not oral or axillary measurements. Decreased core temperature establishes the diagnosis and indicates the severity of hypothermia. Core temperature correlates with mortality and is a guide for treatment. An increased mortality of 1.8% has been observed for each degree centigrade of temperature drop [16]. The average clinical thermometer does not read below 34°C (93°F to 94°F).* Any core temperature decrease represents hypothermia, but most individuals with a temperature above 35°C (95°F) are alert and rewarm spontaneously if protected. At this level, field treatment by one's companions is successful. Most victims will be actively shivering at this level.

Mild hypothermia is characterized by a core temperature in the range of 32°C to 35°C (90°F to 95°F). The victim is usually pale and cool, maximally vasoconstricted. Uncontrollable shivering is present for maximum thermogenesis. There are varying degrees of confusion, disorientation, and incoherence. The patient is usually ambulatory but may be ataxic. Fine hand movements are usually difficult, but large muscle activity is possible. There is cardiovascular stability. This group can usually be rewarmed by external means satisfactorily. Mild hypothermics may be rewarmed by the use of passive external rewarming (PER). In previously healthy people with mild hypothermia the body has the ability to generate enough heat to rewarm itself if it is supported and protected (endogenous thermogenesis) [4]. The victim is dried, placed at rest, and insulated (with blankets or similar covering). Spontaneous rewarming with complete recovery is the rule. It is important to monitor the patient to make sure this desired

result occurs. Simple adjunctive measures such as heating pads, prewarmed blankets, warm IV solutions, and similar measures may be helpful. These measures are technically "active" methods. However, they have limited ability to add significant amounts of heat and are considered adjunctive to passive external rewarming. A warm environmental temperature is desirable and should be at least 32°C (90°F).

Moderate hypothermia, in which the core temperature is 28°C to 32°C (82°F to 90°F) is more serious. Vital signs are affected, and hypotension, bradycardia, hypoventilation, and cardiac irregularities such as atrial fibrillation are common. Mental status progresses to stupor or coma. Shivering ceases, the muscles become rigid with areflexia, and the pupils dilate. At approximately 30°C the body becomes poikilothermic; there is no ability to generate heat. The body will cool to the ambient temperature. Death is inevitable if heat is not added. Active rewarming is needed because PER would be obviously ineffective. Active methods fall into two categories, active external rewarming (AER) and active core rewarming (ACR). AER methods include heating pads, heating blankets, and hot water bottles applied to high blood flow areas. Most effective is immersion in a water bath (40°C). The main disadvantage of this method is rewarming shock. If maximally vasoconstricted skin is warmed, vasodilation is produced, and the blood volume may be decreased centrally. Cardiac resuscitation in a water bath is impossible, and monitoring is difficult. If a water bath is used the extremities should be left out and only the trunk immersed. This complication of rewarming shock and "after drop" has led some to advise against its use [4]. On the other hand, this may be the only method available in rural hospitals. Satisfactory results may be obtained with previously healthy young individuals and those with acute hypothermia; however, most experts agree that core rewarming methods are preferable if they are available.

Severe hypothermia exists when the core temperature drops below 28°C (82°F). Victims in this group have the highest mortality. In truth, they appear dead. Blood pressure is unobtainable. An extremely slow respiratory rate is present. Profound coma, dilated pupils, and rigid and areflexic muscles are present. An electrocardiogram is mandatory. Such an individual should not be pronounced dead until the core has been warmed to at least 30°C (82.2°F); then and only then should resuscitation attempts cease if there has been no response [29]. For patients with moderate and severe hypothermia, ACR methods are preferred. Airway rewarming is the most

---

*For general use, low reading thermometers are available from relatively few sources; Dynamed, Inc., 6200 Yarrow Drive, Carlsbad, California, 92008; Carolian Biological Products, P.O. Box 187, Gladstone, Oregon, 97027; and Zeal G.H., 8 Lombard Street, Loudon, S.W. 19, England.

common and most available means of ACR. Most victims of severe hypothermia require endotracheal intubation for ventilation and oxygenation. The respirator is equipped with a heated nebulizer set at 40°C to 45°C, which will rewarm most hypothermics satisfactorily. For awake patients, a face mask can be used, but it is less efficient than intubation. All IV fluids should be heated to 40°C to 45°C (a microwave oven is an effective method of heating IV bags easily). Intravenous fluids are not very effective unless massive amounts are given, and this is not always desirable. (One liter of fluid at 42°C (107°F) supplies only 14 kcal, which would raise the temperature of a 70-kg patient only one-third of a degree centigrade) [17]. Operating room blood warmers are effective in warming IV fluid. Heated nasogastric, colonic, mediastinal, and thoracic irrigations have been used, but these are not very effective or widely employed.

Another rewarming method is the use of warmed peritoneal dialysis. Standard dialysis fluid is warmed to 40°C to 45°C (104°F to 113°F) and given through a standard percutaneous peritoneal dialysis catheter. An exchange rate of 6 liters/hour is effective [4].

Cardiopulmonary bypass is an effective means of ACR. Cardiopulmonary bypass teams and equipment are available in medical centers with cardiac surgery programs. There is an additional benefit in that circulatory support is also provided. With this method the blood is warmed to 38°C to 40°C (100°F to 104°F) [4].

When circulatory support is not needed, hemodialysis equipment may be used (the patient must have sufficient cardiac output to circulate his blood through the apparatus). The standard shunt is used, and the blood is circulated through a heating device. A single-vessel technique is available as well, using a subclavian catheter [4, 19]. Arteriovenous shunts with heat exchangers have been previously suggested for treatment of hypothermia [9].

The Chinese have used an interesting method. Twenty-eight profoundly hypothermic infants were rewarmed using a microwave oven with a marked decrease in mortality rate of 29.6% [13].

Medical Problems Unique to Hypothermia

Regardless of the rewarming technique used, many problems occur that are unique to hypothermia. Monitoring and continuous assessment of the patient's status by practitioners with knowledge and experience in the treatment of hypothermia are essential to reduce mortality.

Cardiac Complications. As the victim becomes hypothermic, cardiac activity and heart rate are slowed. The conduction system in particular is slowed. Irregularities, notably atrial fibrillation, occur frequently. The J wave may be seen on the electrocardiogram, which, in the absence of other systemic disease, is virtually pathognomonic of hypothermia, although it is not always present. Myocardial function is decreased. Below 28°C (82°F), ventricular fibrillation is common and is easily triggered. Of great significance is the fact that the cold heart is refractory to electrical defibrillation and cardiac drugs [29]. The use of Swan-Ganz catheters in patients with hypothermia may be hazardous because fibrillation may be precipitated by contact with the heart. Hypothermics should be handled gently. The question of the advisability of chest compression for cardiopulmonary resuscitation (CPR) in the wilderness situation often arises. For victims of acute hypothermia (immersion) when CPR can be administered on the scene and continued on the way to the hospital, there is no question, the answer is yes. If the hypothermic shows no signs of life whatsoever, CPR should be done according to the guidelines published by the American Heart Association (AHA). It has been shown that CPR does provide adequate perfusion to sustain life in the hypothermic state [23]. Signs of life in people with severe hypothermia, however, are hard to determine. CPR should not be withheld on the basis of muscle rigidity, fixed dilated pupils, absent pulse, absent respiration, or because the victim appears to have been dead for an unknown period of time. Controversy arises when the victim has extreme bradycardia, hypoventilation, and signs of hypoperfusion caused by the extreme vasoconstriction. For normothermics, CPR would be indicated in this situation to augment perfusion. For hypothermics, however, the Wilderness Medical Society (WMS) has recommended modification of the AHA guidelines. If there is *any sign of life,* CPR is withheld [4]. The reason for this is that bradycardia, hypofusion, and hypoventilation may provide adequate oxygenation of the hypothermic person when the demands for oxygen are reduced. CPR in this instance may convert extreme bradycardia to ventricular fibrillation. For hospital care of such patients modification of the advanced cardiac life support (ACLS) standards is necessary based upon knowledge and experience with hypothermia.

In the wilderness situation, one is often faced with a situation in which providing continuous CPR is difficult, and interruptions are inevitable. This is not a reason to withhold CPR [4]. Actually, hypothermics may tolerate interruptions in CPR better than normothermics.

Another problem is that in the field it is difficult to detect the weak signs of cardiac or pulmonary activity that are able to sustain the life of a hypothermic patient. In normothermic patients, CPR is used when there is inadequate perfusion. In people with chronic hypothermia, CPR should not be used if there is any sign of cardiac or respiratory activity. The minimal cardiac output and ventilation sufficient to perfuse a hypothermic patient may be difficult to detect. Cardiac compression in this instance may cause fibrillation. When there is no cardiac or respiratory activity, CPR may adequately perfuse hypothermic patients [23]. This area of controversy needs continual reexamination. The American Heart Association's standards for CPR as refined by the Wilderness Medical Society are the best guidelines [4]. Modification of ACLS and basic life support (BLS) techniques must keep pace with increased knowledge.

The Metabolic Icebox. The hypothermic patient with no apparent signs of life may be resuscitated and may completely recover. Core temperatures of 15.2°C (59°F) for infants and 16°C (61°F) for adults have been recorded with complete recovery [4]. Cold water and hypothermia account for the increased survival time of near-drowning victims. The brain death that occurs when perfusion is decreased in normothermia is markedly delayed in hypothermia. The caveat not to accept "cold and dead," only "warm and dead," is appropriate. It is important to note that the hypothermic patient develops many electrolyte and acid-base problems. It is best to remember that many of these problems resolve with rewarming; medical treatment should be given in light of this statement to avoid iatrogenic problems. Many complications occur such as pneumonia, gastrointestinal bleeding, pancreatitis, renal failure, pulmonary failure, and disseminated intravascular coagulation, to name a few. These are treatable, and complete recovery without sequelae is possible even for victims of severe hypothermia.

Paradoxical Disrobing. A bizarre situation occasionally arises in patients with hypothermia. The confused victim may completely disrobe in a cold windy environment, hastening his demise. Nude dead hypothermics have been mistaken for victims of sexual assault.

Prevention

Hypothermia is best prevented. For athletic events at risk the event should be cancelled or postponed in cold, wet, windy weather. Proper use of clothing to decrease heat loss can be life-saving.

The mnemonic VIP, familiar to mountaineers, is important. Clothing should allow *ventilation* to allow escape of water vapor during exercise. It should *insulate* to conserve heat, especially when resting. One should *protect* oneself from the wind and rain with shelter or clothing as appropriate. Insulation should employ the COLD principle—*clean,* avoid *overheating,* and keep it *layered,* and loose, and stay *dry.* Nutrition and hydration are important. To produce heat one must have enough food (fuel). This is more important prior to exercise than during it. Hydration is important, particularly during exercise lasting longer than 1 hour.

If accidental catastrophe occurs such as sudden immersion (falling overboard in cold water from a yacht), remember the mnemonic HELP (heat escape lessening position). The victim assumes a curled-up fetal position and avoids activity. More than one victim should huddle together (Fig. 12B–2). Personal flotation devices (PFA) are of great advantage because "staying afloat" uses energy and speeds heat loss. Use of the "drownproofing" technique will increase heat loss.

Cold and Athletic Performance

For athletic events held in a controlled environment there is no cold effect (e.g., swimming, basketball, wrestling). Cold weather at outdoor sporting events does affect performance. For hockey players, an increase in maximum oxygen uptake (VO_2 max) in a cold environment has been noted. There is an associated decrease in plasma lactate [21, 24], but there is no change in power. It is suggested that this response is accounted for by the increased need for thermogenesis [21]. For endurance athletes engaging in longer-duration, submaximal exercise, significantly reduced endurance capacity is noted. This effect has been noted in cross-country skiers at very low temperatures ($-28°C$) [6]. An increased VO_2 max is noted if core temperature is decreased [2] or if there is direct cooling of muscle [5]. It is apparent that endurance athletes must use additional energy for thermogenesis. For aerobic endurance athletes, decreased performance may result under cold conditions. The strategy for competition in the cold should include heat conservation. There may be some advantage to training in the cold to obtain maximum benefit in competition.

Athletes sometimes ask whether, if accidentally immersed in cold water, will the increased heat resulting from muscle activity allow one to swim to shore, protected against hypothermia? The answer is unquestionably NO! The average swimmer in 50°F

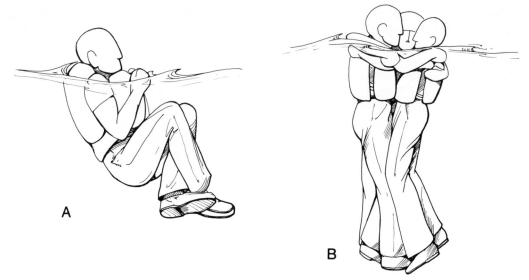

FIGURE 12B–2
Heat exchange lessening positions (HELP).

(10°C) water can swim only about 1 km before hypothermia overtakes him [12]. This concept applies to all endurance events in a cold environment. During exercise, one is protected by the increase in body temperature. When reserves run out, the exhausted athlete is at great risk for hypothermia. In 1964 during the British 73-km speed hike, the Four Inns Walk, wet and windy weather accounted for seven cases of hypothermia with three deaths. Only 22 of 240 starters finished [28]. It is important for race officials to anticipate environmental factors and the risk of hypothermia. Athletes and coaches must also consider this as well as strategies for conserving heat for maximum performance.

On an international level, the Federation Internationale de Ski (FIS) has addressed this problem. Rule 303.4 states, "If the temperature is below −20°C measured at the coldest point of the course, a competition will be postponed or cancelled by the jury. With difficult weather conditions (e.g., strong wind, high humidity, heavy snowfall or high temperature) the jury may, in consultation with the doctors responsible, postpone or cancel the competition" [7].

LOCALIZED COLD INJURY

Freezing Cold Injury, Frostbite

Frostbite has been a major source of morbidity in the military. Allied forces had over 1,000,000 cases in World War II and Korea. During November and December of 1942, the Germans performed 15,000 amputations for cold injury [45]. According to W. J. Mills in the *Encyclopaedia Britannica*, "frostbite, a literal freezing of living tissue, superficial or deep, occurs whenever heat loss from a tissue is sufficient to permit ice formation." Frostbite requires an ambient temperature below 0°C or 32°F. Duration of exposure, temperature, and wind chill are the major variables to be considered. The hypoxia of altitude is also significant as a predisposing and aggravating factor.

Pathophysiology

As skin temperature drops, vasoconstriction occurs. Cold is the strongest stimulus for vasoconstriction. Local peripheral frostbite may be the price of preserving core temperature through this mechanism. If the skin is cooled to 10°C (50°F), maximum vasoconstriction occurs and is reversed every 5 to 10 minutes by "the hunting response," or CIVD [45]. With respect to susceptibility to frostbite particularly, as well as cold injury in general, CIVD explains racial and ethnic patterns of cold adaptation. Smoking causes vasoconstriction and may aggravate or predispose the smoker to frostbite, as does vascular disease. Alcohol causes vasodilation, which theoretically would benefit the drinker, although in cold conditions the cold stimulus is so great it overcomes this vasodilating effect. As further heat is slowly lost, ice crystals form in the extracellular space (cooling over hours). If rapid cooling occurs (over minutes), intracellular crystals form, and these are more damaging [40]. Extracellular ice crystals take up water, causing cellular dehydration, and a toxic concentration of electrolytes builds up in the cells. At below −20°F (−4°C) approximately 90% of the water within the cell freezes [45]. The cell is further damaged by secondary vascular injury. Stasis, sludging, and thrombosis occur in the microcirculation. With thawing there is capillary leakage and marked edema. The cell is further injured by ischemia. Tissues are affected differently; nervous, vascular, and muscle tissues are affected most. Skin, fascia, and connective tissue suffer less, and tendon and bone are quite resistant. Skin loses its sensation at about 10°C (50°F) [36]. Slow freezing usually occurs in frostbite, and much of the tissue loss is due to the secondary vascular injury that occurs. Rapid cooling, such as that resulting from contact with supercooled liquids or at extreme altitudes, produces more immediate cell death by freezing. The freeze, thaw, refreeze sequence is particularly damaging.

Clinical Presentation

Frostbite is classified as superficial or deep. With superficial frostbite there is little tissue loss, whereas with deep frostbite significant tissue loss occurs [41]. Prior to thawing, the frostbitten extremity is not swollen. It is hard, cold, and feels "frozen." It may be white, bluish, or mottled. Sensation is absent. Depth of injury is difficult to determine. Superficial injury has a more rubbery consistency than deep injury, which is rock hard. With deep injury, soft tissue may be frozen to bone over bony prominences. With superficial frostbite the skin is usually mobile over bony prominences such as joints and malleoli. Thawing is painful, with a marked erythemic blush extending out to the tips of the digits as they thaw. Sensation usually returns with hyperemia. This is followed by swelling and blister formation (in 1 hour to several days). Large blisters that extend to the tips of the digits and contain a straw-colored fluid indicate more superficial injury. Less prominent proximal blisters with dark fluid indicate deeper injury. Blisters break spontaneously in 7 to 10 days. Over the next few weeks the skin becomes mummified, as a dry eschar forms and tissues shrink. With deep injury this process occurs more rapidly. With rapid freezing at extreme altitude an eschar forms in days without swelling. Three to six weeks after injury the eschar demarcates, the dead tissue forming a definite line with the proximal viable tissue. With superficial injury the eschar separates, leaving healed new epithelium beneath. With deep cold injury, autoamputation occurs.

Treatment

The preferred method of treatment has been best described by William J. Mills, Jr. [41]. The basic principle is rapid rewarming. The frozen extremity is placed in a water bath at 40°C to 42°C (104°F to 108°F). This is best done in the hospital. Approximately 30 minutes or less is usually sufficient to thaw an extremity, as evidenced by a hyperemic blush progressing to the tips of the digits. This is very painful, and narcotics are often necessary.

Following thawing, open treatment is used. Sterile technique is important. Lambswool is used to separate the fingers or toes to avoid maceration. Blebs are left intact (the skin is its own best dressing). A cradle keeps bedclothes off the injury. Motion is encouraged; however, weight bearing is prohibited. Tetanus toxoid is given when indicated. Antibiotics are not given routinely. Some advocate penicillin for 24 to 48 hours to prevent streptococcal infection. Ibuprofen is given.

Hydrotherapy is begun twice daily for about 15 to 20 minutes at 90°F to 98°F in a whirlpool. An antiseptic solution is used (pHisohex or Betadine). Motion is easier in the bath and is encouraged. The agitation of the whirlpool is adequate for any debridement needed. Formal surgical debridement is avoided, and only dead skin from ruptured blebs is removed.

When an eschar forms, escharotomy may be beneficial to allow better motion. Demarcation is allowed to occur naturally. Amputation is delayed and is indicated only when the soft tissue is clearly demarcated and begins to separate from the living. Autoamputation is appropriate and may give optimum results.

Surgical amputation is done only after demarcation occurs. If infection develops (wet gangrene) it will be evident by odor, purulent drainage, fever, leukocytosis, pain, occasional proximal erythema, swelling, and similar signs. In such a case, antibiotics and early amputation are indicated. Guillotine or modified guillotine amputations are indicated in frostbite. Revision, skin grafting, and other reconstructive techniques are best done as secondary procedures. Major amputations (e.g., below knee, above knee, forearm) are uncommon in cases of frostbite and usually are needed only when infection develops or when there is accompanying fracture or other injury. Primary amputations with immediate postsurgical prosthetics fitting (IPPF) may be done in clean cases under optimum conditions with good results. IPPF is most likely to give good results as a secondary procedure.

Other modalities have been advocated as adjunctive treatment. Low-molecular-weight dextran has been recommended but is unlikely to be beneficial [45]. Heparin and other anticoagulants are of no benefit [44]. Surgical sympathectomy has been advocated, but most practitioners now feel that it carries no significant benefit and a risk of morbidity. Medical sympathectomy with vasodilators does not alter the outcome [45]. Antibiotics are indicated only for infection. Ibuprofen has been shown to be of benefit by helping to relieve pain as well as inflammation. It also inhibits prostaglandin synthesis. Blister fluid in burn and frostbite victims has been shown to contain prostaglandinlike substance such as thromboxane at elevated levels [43]. These substances encourage vasoconstriction, platelet aggregation, and leukocyte adherence. Inhibition of these effects through ibuprofen offers potential benefit [34]. Good hydration and nutrition are necessary for recovery. Psychological support to obtain a positive attitude is beneficial. Exercise is encouraged; it is often easily done and can be supervised at the time of the whirlpool bath. Patients are encouraged to exercise as frequently as possible on their own.

Prediction of Injury

Prediction of the extent of injury on a clinical basis alone is difficult. Thermography, arteriography, Doppler ultrasound, and other studies have been used to predict the extent of injury. The technique that holds the most promise is triple-phase bone scanning. Scanning is done for three phases: (1) flow, (2) blood pool, and (3) delayed. Meticulous technique is needed, and the scan is repeated in 7 days. In one such study, an accurate level of demarcation was predicted after 48 hours and was later confirmed at amputation in five patients. This technique has been used thus far to monitor and prognosticate, not to determine an amputation site [38]. The concept of early prediction of level of injury with early primary amputation is attractive but not yet acceptable.

Sequelae

Hyperhidrosis, pain, decreased sensation, and cold sensitivity are common sequelae of frostbite. Fat pad atrophy of the terminal phalanges is common. A particularly devastating problem in children with frostbite is premature epiphyseal closure with resulting growth disturbance. This most commonly takes the form of shortening with arrested growth

of the terminal phalanges, occasionally with malalignment [42].

Treatment of Frostbite in the Field

Prehospital treatment of frostbite may be the major determining factor in the degree of injury that results. Frostbite is best prevented. Proper clothing and proper use of clothing are essential. Clothing should be loose and should not constrict the circulation. Cross-country ski pole straps may constrict circulation to the fingers. Mittens are better than gloves. Snowmobile boots with felt liners carry a risk because the boots are very warm when dry. When wet by perspiration or snow over the top wicking down, the felt swells. When it is cold enough to freeze, a severe cold injury may result, demarcated at the top of the boot! Mountaineers prefer plastic boots with closed-cell foam liners and overboots or supergaiters. Frostbite is unlikely with the military Mickey Mouse or vapor barrier boot. However, it is soft and is not satisfactory with crampons, snowshoes, or skis.

Extreme altitude, wind, and extreme cold in combination produce immediate deep injury. In 1960 Chu Yin-hua enabled the Chinese Everest Team to reach the summit by removing his boots to climb a difficult rock pitch, the second step. They were successful in reaching the summit (29,028 feet); however, Chu lost the toes of both feet at the midmetatarsal level [35].

Corneal injury can occur with extreme cold and high wind. Most cases resolve when treated as snow blindness. Rare cases are severe enough to require corneal transplant.

There are anecdotal reports of ingestion of supercooled alcohol producing deep esophageal frostbite [32].

When frostbite occurs in the field, the major decision is when to evacuate and when to rewarm in the field. One can walk many miles on frozen feet without causing significant increase in the injury. Once frozen feet have been thawed, the victim is a litter case. Unless spontaneous rewarming is inevitable, rewarming should not be done in the field. When rewarming is done, litter evacuation should be available along with the ability to prevent refreezing. In 1982, Larry Nielson developed frostbite of his fingers and toes at 27,500 feet on Mt. Everest. He was able to descend to 23,000 feet with frozen extremities. After spontaneous rewarming, litter evacuation was required to base camp at 17,000 feet.

One should never thaw by an open fire, automobile exhaust, heaters, etc. The insensitive frostbitten tissues will often be further damaged by thermal burn. In 1812 Napoleon's chief surgeon, Baron DeLarrey, noted that soldiers with cold injury who rested closest to the fire had the highest mortality.

Nonfreezing Cold Injury

Trench Foot (Immersion Foot)

Trench foot occurs at ambient temperatures of between 0°C and 10°C (32°F to 50°F). The injury requires a long duration of exposure to cold water (days to weeks). The intense, prolonged vasoconstriction results in ischemic injury. In World War I the British had over 115,000 casualties from trench foot [28]. The injury is initially hyperemic and is followed by cyanotic mottling of the skin and intense swelling. Complete resolution with open treatment is usual. Severe injury may result in gangrene, requiring amputation. The injury is basically a neurovascular injury, and painful debilitating paresthesias may result.

Chilblains (Pernio)

"Cold sores" or typical skin lesions occur following cold exposure, usually in exposed areas, at temperatures of 0°C to 15°C (32°F to 60°F) [47]. The lesions are localized, erythematous, or cyanotic and are often raised as plaques or nodes. Blistering or ulceration may occur but is seldom a serious problem. The use of topical fluorinated corticosteroids may aggravate the problem [31].

ILLNESS STIMULATED BY COLD

There are some disease entities that are stimulated by cold but not caused by it. Cold urticaria is one example. In cold urticaria, hives are produced by cold exposure. The diagnosis is made by a simple provocative test. Applying an ice cube to the skin for 5 minutes produces a wheal. The problem is rarely severe. Life-threatening hypotension is possible if such an individual is immersed in cold water. Systemic symptoms such as headaches, fever, and arthralgia can occur [46]. Antihistamines may be useful.

Another disease for which cold is the trigger is Raynaud's phenomenon, which is a well-known vasospasm of the fingers in response to cold. It is intermittent and usually occurs in young women. It is occasionally severe enough to cause ulceration and tissue loss and has been treated with medication and even sympathectomy. Pavlovian conditioning at home is simple, inexpensive, and effective and carries no morbidity; it is the treatment of choice [33]. The protocol for this program is well described in reference 33.

Cold may often trigger asthma. Cold-induced bronchoconstriction (CIB) is a variant of exercise-induced bronchoconstriction and is brought on by the hyperventilation of cold air in susceptible individuals. It is a direct airway effect [30]. Theophylline is minimally effective in blocking bronchoconstriction as a bronchodilator [39]. Cromolyn sodium has been shown to be effective, as have metaproterenol, albuterol, and terbutaline. However, metaproterenol is banned by the International Olympic Committee. Theophylline and cromolyn are acceptable, as are terbutaline and albuterol. Vigorous training and good warm-up may benefit the patient and make pharmacologic agents unnecessary.

SUMMARY

Whenever athletic activity interacts with a cold environment, cold injury is possible. With knowledge of cold physiology and cold injury, most of the problems presented can be avoided. When accidents occur, they must be treated properly.

References

Hypothermia and Physiology

1. Bangs, C., and Hamlet, M. D. Hypothermia and cold injury. *In* Auerbach, P. S., and Geehr, E. C. (Eds.), *Management of Wilderness and Environmental Emergencies* (2nd ed.). St. Louis, C. V. Mosby, 1989.
2. Bergh, V., and Ekbloom, B. Physical performance and peak aerobic power at different body temperatures. *J Appl Physiol* 46:885–889, 1978.
3. Danzyl, D. F., and Pozos, R. S. Multicenter hypothermia survey. *Ann Emerg Med* 16:9, 1987.
4. Danzyl, D. F., Pozos, R. S., and Hamlet, M. D. Accidental hypothermia. *In* Auerbach, P. S., and Geehr, E. C. (Eds.), *Management of Wilderness and Environmental Emergencies* (2nd ed.). St. Louis, C. V. Mosby, 1989.
5. Davies, C., and Young, K. Effect of temperature on the contractile properties and muscle power of triceps surae in humans. *J Appl Physiol* 55:191–195, 1983.
6. Faulkner, J. A., White, T. P., and Markley, J. M. The 1979 Canadian ski marathon: A natural experiment in hypothermia. *In* Nagle, F. J., and Montage, J. H. (Eds.), *Exercise in Health and Disease—Balke Symposium.* Springfield, IL, Charles C Thomas, 1981.
7. Federation Internationale de Ski. *The International Ski Competition Rules (I.C.R.): Book II, Cross-Country and Nordic Combined.* Istanbul, 36th International Ski Congress, 1968.
8. Fox, R. H., Woodward, P. M., and Fry, A. S. Diagnosis of hypothermia in the elderly. *Lancet* I:424–427, 1971.

9. Gregory, R. T. Unpublished data. Arctic Medical Research Laboratory, Department of the Army, Fairbanks, Alaska, 1972.

10. Hamlet, M. D. An overview of medically related problems in the cold environment. *Milit Med* 152:393–396, 1987.

11. Hayward, J. S. The physiology of immersion hypothermia. *In* Pozos, R. S., and Wilttmers, L. E. (Eds.), *The Nature and Treatment of Hypothermia*. Minneapolis, University of Minnesota Press, 1958.

12. Hayward, J. S. Immersion hypothermia. *In* Wilkerson, J. A., Bangs, C. C., and Hayward, J. S. (Eds.), *Hypothermia, Frostbite, and Other Cold Injuries*. Seattle, The Mountaineers, 1986.

13. Huang, Z., Sun, Q., and Shen, M. Rewarming with microwave irradiation for severe cold injury syndrome. *Chin Med J* 93(2):119–120, 1980.

14. Keating, W. R. Accidental immersion hypothermia and drowning. *Practitioner* 219:183, 1977.

15. Lampietro, P. F. Heat production from shivering. *J Appl Physiol* 15:632, 1960.

16. Miller, J. W., Danzyl, D. F., and Thomas, D. M. Urban accidental hypothermia: 135 cases. *Ann Emerg Med* 9:456–461, 1980.

17. Myers, R. A., Britten, J. S., and Crowly, R. A. Hypothermia, quantitative aspects. *J Am Coll Emerg Physicians* 8:523, 1979.

18. Nadel, E. R. Energy exchanges in water. *Underseas Biomed Res* 2(2):149, 1984.

19. O'Keefe, K. M. Treatment of accidental hypothermia and rewarming techniques. *In* Robert, J. R., and Hedger, J. R. (Eds.), *Clinical Procedures in Emergency Medicine*. Philadelphia, W. B. Saunders, 1985.

20. Pugh, L. G., Corbett, C. E., and Johnson, R. H. Rectal temperatures, weight losses, and sweat rates in marathon running. *J Appl Physiol* 23:347, 1967.

21. Quiron, A., Therminarios, A., Pellerei, E., Melthor, D., Lawrencelle, L., and Tauche, M. Aerobic capacity, anaerobic threshold, and cold exposure with speed skaters. *J Sportsmed Phys Fitness* 28(1):27–34, 1988.

22. Siple, P. *Windchill in the Northern Hemisphere*. Technical Report Edition 1982. Washington, D.C., Falkwaki and Hastings, Quartermaster Research and Engineering Command, 1958.

23. Standards and guidelines for cardiopulmonary resuscitation (CPR) and emergency cardiac care (ECC). JAMA 255:2905–2984, 1986.

24. Therminaros, A., Quiron, A., Pellerei, E., and Laurencelle, L. Effects of cold exposure on physiologic responses obtained during a short exhaustive exercise. *J Sportsmed Phys Fitness* 28(1):27–34, 1988.

25. U.S. Army Medical Research and Development Command. Report No. T1/78: Behavioral evaluation of a winter warfare training exercise. Natick, MA, U.S. Army Medical Research and Development Command, 1977.

26. Whayne, T. F., and DeBakey, M. E. *Cold Injury, Ground Type*. Washington, D.C., Office of the Surgeon General, Department of the Army, 1958.

27. Wilkerson, J. A. Avoiding hypothermia. *In* Wilkerson, J. A., Bangs, C. C., and Hayward, J. S. (Eds.), *Hypothermia, Frostbite, and Other Cold Injuries*. Seattle, The Mountaineers, 1986, chap. 2.

28. Wilkerson, J. A., Bangs, C. C., and Hayward, J. S. *Hypothermia, Frostbite, and Other Cold Injuries*. Seattle, The Mountaineers, 1986.

29. Wilmore, D. W. Fever, hyperpyrexia, hypothermia. *In* Wilmore, D. W., Brennan, M. F., Harken, A. H., Holcroft, J. W., and Meakeas, J. L. (Eds.), *Scientific American, Medical Care of the Surgical Patient*, Vol. 1. New York, Scientific American, 1988, chap. 10.

Frostbite

30. Berk, J. L., Lenner, K. A., and McFadden, E. D. Cold-induced bronchoconstriction: Role of cutaneous reflexes vs. direct airway effects. *J Appl Physiol* 63(2):659–664, 1982.

31. Burry, J. H. Adverse effects of topical fluorinated corticosteroids on chilblains (Letter). *Med J Aust* 146:451, 1987.

32. Hamlet, M. P. Frostbite and hypothermia. Presented at the 7th Annual Sportsmedicine Meeting, Lake Placid Sports Medicine Society, Lake Placid, New York, 1987.

33. Hamlet, M. D. Cold injuries, Raynaud's disease, and hypothermia. *In* Casey, M. J., Foster, C., and Hixson, E. G. (Eds.), *Winter Sports Medicine*. Philadelphia, F. A. Davis, 1990, chap. 18.

34. Heggars, J. P. Experimental and clinical observations on frostbite. *Ann Emerg Med* 16:1056, 1987.

35. Hixson, E. G. Personal communication after examination of Chu Yin Hua, 1984.

36. Holm, P. C., and Vangaard, L. Frostbite. *Plast Reconstr Surg* 54:544, 1974.

37. Israel, R. H., Kohan, J. M., Poe, R. H., Kallag, M. C., Greenblath, D. W., and Rathbun, S. Inhaled metaproterenol is superior to inhaled cromolyn in protecting against cold air induced bronchoconstriction. *Respiration* 53:225–231, 1988.

38. Mehta, R. C., and Wilson, M. A. Frostbite injury: Prediction of tissue viability with triple phase bone scanning. *Radiology* 120:511–514, 1989.

39. Merland, N., Cartier, A., L'Archeveque, J., Ghezzo, H., and Malo, J. Theophylline minimally inhibits bronchoconstriction induced by dry cold air inhalation in asthmatic subjects. *Am Rev Respir Dis* 137:1304–1308, 1988.

40. Merryman, H. T. Mechanics of freezing in living cells and tissues. *Science* 123:515, 1956.

41. Mills, W. J., Jr. Frostbite. *Alaska Med* 15(2):35, 1973.

42. Reed, M. H. Growth disturbances in the hands following thermal injuries in children. *J Can Assoc Radiol* 39:95, 1989.

43. Robson, M. C., and Heggars, J. P. Evaluation of hand frostbite blister fluid as a clue to pathogenesis. *J Hand Surg* 6:43, 1981.

44. Schumaker, H. B. Studies in experimental frostbite: The effect of heparin in preventing gangrene. *Surgery* 22:900, 1947.

45. Smith, D. J., Robson, M. C., and Heggars, J. P. Frostbite. *In* Auerbach, P. S., and Geehr, E. C. (Eds.), *Management of Wilderness and Environmental Emergencies* (2nd ed.). St. Louis, C. V. Mosby, 1989.

46. Wagner, W. O. Urticaria: A challenge in diagnosis and treatment. *Postgrad Med* 83(5), 321–329, 1985.

47. Wilkerson, J. A. Other cold injuries. *In* Wilkerson, J. A., Bangs, C. C., and Hayward, J. S. (Eds.), *Hypothermia, Frostbite, and Other Cold Injuries*. Seattle, The Mountaineers, 1986.

Altitude Stress

Edward G. Hixson, M.D.

High altitude is best defined as altitude ranging from 1500 to 3500 m (5000 to 11,000 feet). Below this level healthy individuals suffer few effects, but above this level some physiologic effects are noted. Very high altitude ranges from 3500 to 5800 m (10,000 to 19,000 feet). In this zone, man is able to acclimatize himself to the physiologic stress of altitude and exercise. Above 5800 m (19,000 feet), which is considered extreme altitude, man can survive only by acclimatization [13]. No human habitation has ever persisted above 18,000 feet. This region has been referred to as "the death zone" (Fig. 12C–1).

Rate of ascent is important. Rapid ascent refers to an ascent of minutes or hours in duration; a fast ascent takes place over days, and a slow or gradual ascent is made over weeks [18]. Commercial airlines are required to pressurize cabins to 436 mm Hg above the outside pressure. For most heights, this is equivalent to experiencing a rapid ascent to 6000 or 7000 feet and is easily tolerated by most people. Rapid ascent to over 25,000 feet may be fatal. In 1875, the balloon Zenith made such an epic ascent, and only one of the three occupants survived [18]. If one breathes 100% oxygen, ascent to roughly 45,000 feet is possible. Blood boils at 65,000 feet. The Federal Aviation Administration (FAA) requires pilots and passengers in unpressurized cabins to use oxygen between 10,000 and 12,000 feet for exposures of more than 30 minutes. Above 12,000 feet, oxygen is required at all times [7]. The summit of Mt. Everest, 29,028 feet (8848 m), is the highest altitude achieved by man without supplemental oxygen (Reinhold Messner and Peter Habler, 1978 Austrian Everest Expedition). Such a feat requires maximum acclimatization and is attainable by only a very few individuals who are physiologically able to tolerate this altitude, perhaps on the basis of genetic advantage.

ATHLETIC EVENTS AT ALTITUDE

High, very high, and extreme altitudes have been the exclusive realm of the mountaineer. Balloonists, hang gliders, glider pilots, parachutists, and aviators ascend to altitude equipped with oxygen and pressurized suits and cabins when appropriate. Competitive athletic events, such as speed hikes, marathons, and ultramarathon races, are becoming more frequent at high and very high altitude. Many treks now go to high altitude, and guided climbs to extreme altitude are now available to many fit individuals. Approximately 40 million people live at elevations above 8000 feet, and an equal number visit yearly [25]. Hundreds of thousands visit high altitude regions of the world yearly [25]. There are over 100,000 active mountain climbers in the United States and many more in the world [23]. Thousands of mountaineers ascend extreme altitude yearly, and more than 100 climbers have reached the summit of Mt. Everest.

Competitive athletic events are usually not held above 2290 m (8700 feet). The International Federation of Sportsmedicine (FIMS) banned competition above 3050 m (10,000 feet) in 1974. The International Ski Federation (FIS) restricts the highest elevation of a cross-country course to 1800 m (6000 feet) [8]. Biathlon has a similar rule. For competitive athletes the concern is not only for altitude illness but also for the physiologic effects of altitude on the athlete. Concern was great prior to the 1968 Mexico City Olympics (7500 feet, or 2300 m).

PHYSIOLOGIC CHANGES AT ALTITUDE

Air is 20.93% oxygen, and this remains constant regardless of altitude. As altitude increases, baro-

Suggested Altitudes

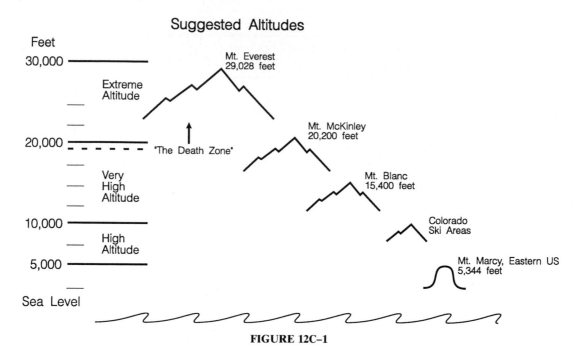

FIGURE 12C–1

metric pressure decreases in linear fashion [9]. As barometric pressure decreases, the partial pressure of oxygen (PO_2) decreases in direct proportion. As the partial pressure of oxygen of ambient air decreases, the partial pressure of oxygen in the blood (PaO_2) also decreases.

Barometric pressure at sea level is 760 mm Hg, normal PaO_2 is 90 to 95 mm Hg, $PaCO_2$ is 40 mm Hg, and oxygen saturation of blood is 96%. As one ascends to 3000 m (about 10,000 feet), barometric pressure decreases to 560 mm Hg, and PaO_2 is 60 mm Hg, roughly the level at which most people develop symptoms of altitude stress. As one ascends to 5800 m (19,000 feet), a very high altitude range, barometric pressure decreases to 360 mm Hg, PaO_2 is roughly 46 mm Hg, $PaCO_2$ is 20 mm Hg, and oxygen saturation of blood is 80%. Up to this point man can become acclimatized without deterioration in performance. At the extreme altitude of the summit of Mt. Everest (29,028 feet [8848 m]), barometric pressure is 253 mm Hg [30]. From measurement of alveolar air samples, PaO_2 at this altitude is computed to be 28 mm Hg, and $PaCO_2$ is 7.5 mm Hg [32]. A corresponding oxygen saturation would be 60%. Under these circumstances, man is near death. Maximum oxygen uptake is calculated to be in the range of 5 mL/minute/kg [31]. This is barely enough to maintain metabolism. In 1924, Norton reached 28,000 feet on Mt. Everest without supplemental oxygen, but it was not until 1978 that Messner and Habler reached the summit without supplemental oxygen. It took 54 years to climb the last 1000 feet without supplemental oxygen.

Acclimatization

Man must become acclimatized to very high and extreme altitudes by ascending slowly to be able to exercise, to avoid altitude illness, and to survive. The stimulus to acclimatization is hypoxia. The result of acclimatization is maximization of oxygen delivery to the cell for a given altitude. Maximum oxygen uptake decreases roughly 10% for each 1000 m of gain in altitude over 1500 m [13]. This is not changed by acclimatization. Maximum oxygen uptake is limited by the oxygen available in the ambient air. Acclimatization occurs by a multitude of complex changes taking place on a systemic as well as a cellular level. Hyperventilation is an important aspect of acclimatization, as shown dramatically by a $PaCO_2$ of only 7.5 mm Hg on the summit of Everest. That this occurs with a relatively normal pH indicates the acid-base compensation that occurs. Hematopoiesis is stimulated by altitude. An increased number of red blood cells accounts for hematocrits in the range of 50% to 70% [15]. Maximum benefit from increased hematocrit occurs at about 57%; above this level the decreased blood flow resulting from decreased viscosity may outweigh the benefits of more hemoglobin [18]. Hematocrits over 75% are characteristic of chronic mountain sickness, or Monge's disease [24]. This degree of polycythemia is maladaptive. Increased erythropoiesis at altitude has a parallel in illicit "blood doping"; athletes may take autologous transfusions to enhance their performance in endurance sports. (See later discussion in section on Training.)

It is generally agreed that an altitude of above 3000 m (10,000 feet) is needed to stimulate acclimatization. Gradual ascent allows sufficient time. Maximum acclimatization occurs at 3 to 6 weeks. One should allow 1 day for each 1000 feet of ascent over 5000 feet, and 1 day for each 500 feet above 14,000 feet. The old climbers' dictum, "climb high but sleep low" is appropriate.

Deterioration

At extreme altitude (above 5800 m [19,000 feet]) the body deteriorates rapidly. This is the "death zone." Here the beneficial effects of acclimatization are exceeded by the effects of deterioration. This deterioration is observed as rapid weight loss with attendant loss of strength and vigor. Anecdotally, some climbers have reported a 25% body weight loss during a 3-month expedition. For the China-Everest 1982 expedition, an average 16% body weight loss was observed [16]. Weight is lost as fat and muscle in roughly equal amounts. There is a myth that if climbers "fatten up" prior to an expedition, this fat will be lost, sparing muscle. However, a high percentage of body fat does not protect against muscle loss. Sherpas, who have low body fat (10% to 15%), generally do not lose weight at high and very high altitudes [2]. Anorexia occurs at high altitude, demanding increased caloric needs to avoid negative nitrogen balance [14]. These needs would obviously be greatly increased at extreme altitude. There is preliminary evidence that increased caloric intake in a fit climber may avoid weight loss during a short exposure to extreme altitude. (G. Roach reached the summit of Everest and returned with a gain of 2 pounds at the end of the expedition [17].) A high-carbohydrate diet will have the same effect as lowering the summit 2000 feet [19]. Carbohydrates use less oxygen when oxidized to produce energy. The deterioration that occurs at altitude has been considered altitude illness and termed high altitude cachexia (HAC) by the author. High altitude deterioration is a generalized deterioration of mind and body. Symptoms are weakness, lethargy, anorexia, and loss of vigor. The most obvious sign is loss of weight and decreased muscle mass as well as fat. Methods of minimizing the effects of deterioration are of obvious benefit to the high altitude climber, who must function at the highest level of strength, endurance, and mental activity to reach the summit or even survive.

HIGH ALTITUDE ILLNESS

All altitude illness shares a common etiology, the hypoxia attendant on decreased barometric pressure, although, for discussion, each syndrome of altitude illness is treated separately here. Often, the victim of altitude illness has some symptoms of many of the syndromes discussed, although one may predominate.

High Altitude Acute Mountain Sickness

Acute mountain sickness (AMS) is usually mild and self-limited. It can be minimized or avoided completely by gradual ascent and is uncommon in the acclimatized climber. Symptoms are headache, weakness, shortness of breath, anorexia, nausea, and vomiting. Sleeplessness and Cheyne-Stokes breathing are common. At elevations of 2400 m (8000 feet) in Colorado, 12% of newcomers were noted to be affected [20]. Rapid ascents to high altitude of 3000 to 4392 m (14,400 feet) produced symptoms of AMS in 67% [12]. It is of major significance that some cases do not resolve and progress to more severe altitude illness. At 4243 m, 8% of cases of AMS progress to more severe illness [11]. Symptoms are usually most severe after 48 to 72 hours and are gone by 4 or 5 days.

AMS is treated and can often be prevented by acetazolamide. The mechanism of action of acetazolamide is complex. It is a carbonic inhibitor and blocks the enzyme that facilitates the combination of CO_2 and water to form bicarbonate in the blood. The rate of breathing is stimulated, and blood pH is lowered. All these changes make acclimatization to altitude easier, and therefore, altitude illness, especially AMS, is alleviated. One breathes more easily, receives more oxygen without raising the blood pH, and gets rid of some retained fluid by diuresis because the drug is a mild diuretic. It is approved for treatment and prevention of AMS by the Food and Drug Administration (FDA).

Acetazolamide for prophylaxis or treatment is given at a dosage of 250 mg twice daily beginning 1 day prior to ascent. Descent or oxygen administration is also effective treatment but is rarely necessary. Rest is important. Exercise exacerbates the symptoms. The best prophylaxis is gradual ascent, 1 day for each 1000 feet above 5000 feet, and 1 day for each 500 feet above 14,000 feet. Hydration, fitness, and a high-carbohydrate diet are helpful; however, physical fitness does not protect one from any high altitude illness. Many young, fit climbers go "too high, too hard, too fast," and AMS may be the price.

High Altitude Pulmonary Edema

High altitude pulmonary edema (HAPE) is a severe form of altitude illness that is most likely to cause death. It is usually seen 1 to 4 days after arrival at an altitude of 2400 m (8000 feet) or above. A too high, too hard, too fast ascent often precedes HAPE. Fitness is not protective. The disease is most common in young males. It is also noted as a reentry phenomenon when acclimatized individuals or high altitude natives descend for a few days and return. The most significant symptoms are dyspnea and cough. Cyanosis is common, as are elevated heart and respiratory rates. Pulmonary rales lead to frank edema and a copious pink, frothy sputum. Symptoms of AMS may or may not occur concomitantly or may predate HAPE. The incidence of HAPE varies from less than 1 in 10,000 skiers in Colorado to 2% of climbers on Mt. McKinley, to 15% of Indian recruits flown to altitude [13]. Of those who go rapidly to 9000 feet, 1% will get HAPE. Of those who get HAPE, 10% of cases will be fatal [19]!

Pulmonary hypertension is present [22]. The "edema" is a permeability edema that results from capillary leakage with normal left heart function [27]. In many ways it is similar to adult respiratory distress syndrome (ARDS). Prevention is achieved by ascending gradually. HAPE does, however, strike fit, well-acclimatized mountaineers. In these cases it may be related to excessive exercise at extreme altitude.

Treatment is early descent with the first symptom (a descent of only 1000 feet may be sufficient). Oxygen is indicated. Diuretics such as furosemide may be useful; however, there is a risk of hypotension with their use. Almost all sick climbers are severely dehydrated, and diuretic therapy has a high risk of hypotension. Morphine has been recommended as with pulmonary edema of cardiac origin. Respiratory depression from morphine is the main hazard of its use. Again, descent is the most beneficial treatment that can be done in the field. Many cases resolve as descent is made to the hospital. For severe disease in hospitalized patients, endotracheal intubation with positive end-expiratory pressure (PEEP) ventilation has been beneficial. Some cases will progress to a fatal outcome even with the most aggressive respiratory support. However, most cases resolve within 24 to 48 hours of descent. Radiographs are diagnostic.

High Altitude Cerebral Edema

AMS is a neurologic disorder, mild and self-limited. High altitude cerebral edema (HACE) is a severe and often fatal neurologic disorder that may result from a progression of AMS. It does occur de novo in acclimatized climbers at extreme altitude. In both AMS and HACE there is increased intracranial pressure which in the latter progresses to frank edema. Cerebrospinal fluid (CSF) pressure may be over 300 mm of water in HACE [21]. Edema is noted at autopsy and on computed tomographic scans.

The hallmark symptom is severe headache. Ataxia or mental changes such as confusion and hallucinations are often present; these may progress to stupor or coma. A variety of focal neurologic signs may occur (i.e., hemiparesis and seizures). Papilledema is often present. HACE rarely occurs below 12,000 feet. It takes 2 to 3 days to develop. It is the most severe form of altitude illness and has the highest mortality, morbidity, and likelihood of permanent sequelae. No permanent neurologic effects have been observed to result from altitude exposure alone without altitude illness [4].

The treatment is descent. Dexamethasone, 4 to 8 mg given every 6 hours, is indicated, as is oxygen. For the hospitalized patient, intubation and controlled hyperventilation may be helpful in lowering the CSF pressure. Diuretics may be of benefit but may aggravate injury if, due to dehydration, hypotension were to occur, decreasing cerebral blood flow.

High Altitude Retinal Hemorrhage

The eye is the window to the brain. In HACE retinal vessels are engorged with hemorrhage, leading to frank papilledema. Many climbers experience benign asymptomatic retinal hemorrhages. High altitude retinal hemorrhage (HARH) is seen in 50% of climbers ascending to 5000 m and in virtually 100% of those ascending above 6500 m [33]. HARH usually resolves after 1 to 2 weeks without sequelae. If hemorrhage involves the macula, a blind spot or scotoma results.

Other High Altitude Problems

Thrombotic and embolic disease (e.g., phlebitis, pulmonary emboli, cerebral thrombosis, etc.) are more common at altitude. As a result of increased hematopoiesis, hematocrits as high as 70% to 76% occur. In HAPE a hypercoagulability has been noted [6]. Dehydration aggravates this. Neurologic problems ranging from transient ischemic attacks to frank stroke occur. Monge's disease, or chronic mountain sickness, is associated wiith polycythemia and pulmonary disease in high altitude natives.

High altitude is marked by intense ultraviolet radiation because there is less atmosphere available to filter out these rays. Reflection from snow increases the problems. Severe sunburn or snow blindness may result from only 1 hour of exposure if the eyes or skin is unprotected. Glacier glasses with side panels and maximum ultraviolet block sunscreen are needed.

On a glacier at high altitude, the temperature can go from subzero at night to 80°F or 90°F in a matter of hours. Heat injury is possible.

As we have stated, most altitude illness is decreased in incidence and severity by acclimatization. General advice to mountaineers may be summarized as follows:

1. Gradual ascent—1000 feet daily when over 5000 feet and 500 feet daily when over 14,000 feet.

2. Decreased work load, especially anaerobic, following arrival at altitude.

3. Good hydration.

4. High-carbohydrate diet.

5. Low salt intake. Most people with altitude illness and many new arrivals at altitude note generalized edema. Increased blood pressure may be noted. Many backpack dehydrated foods have a high salt content.

6. In patient with illness worse than AMS, or when AMS does not resolve, descend.

ATHLETIC TRAINING AND COMPETITION AT ALTITUDE

Competitive athletic events, even when restricted to moderate altitude, are significantly affected by altitude. Altitude illness is rarely a factor. The main concern is the physiologic effect on performance.

In 1968 the games of the nineteenth Olympiad were held in Mexico City at 2300 m (7349 feet). Since that time, there has been considerable increased interest in the effects of moderate altitude on athletic training and performance. The lowered barometric pressure and its attendant decrease in partial pressure of oxygen are the major factors of significance. At 3000 m there is 31% less oxygen in the air; at Mexico City there is a 24% reduction. Maximum oxygen uptake (VO_2 max) decreases roughly 10% for each 1000 m over 1500 m [13]. For endurance events at the altitude of Mexico City, one would have expected time increases of 15% to 20% to occur. However, increases of only 7% to 10% were observed [29]. Although VO_2 max cannot

be changed for a given altitude by acclimatization, performance, which is related to many variables, can.

Temperature

With increasing altitude, ambient temperature decreases 2°C for each 300 m [28]. Concern about cold injury at moderate altitude was not significant for the latitude of Mexico City and its usual weather. Weather and season are more significant factors than altitude for athletic competition.

Gravity

Acceleration due to gravity is decreased 0.3 cm/sec^2 for each 1000 m increase in altitude. Conceivably, this fact benefits jumping, pole vault, and ballistic events. However, its actual significance is negligible (0.13% at 4000 m) [28].

Air Resistance

The "thin" air at altitude decreases wind resistance, and therefore athletes must expend less energy to overcome it. Work done against wind resistance is proportional to the third power of the velocity [28]. The faster the event, the greater the significance. At sea level, roughly 11% of energy is expended against wind resistance. For the 5000-m run at Mexico City, a 3.4% improved performance would be anticipated based on these data. For 100 m, a 14.38% improvement would be anticipated [28]. Usually in Olympic events times are 2.9% slower than those in world records. At Mexico City, times were only 0.99% off world record times [5], and more world records were broken for short events (less than 1 minute) than in previous Olympic competitions. Bob Beamon's dramatic long jump exceeded the world record by 80 cm and was a dramatic achievement not completely explained by thin air. For short anaerobic events in which wind resistance is a factor, performance will be enhanced at altitude.

Acclimatization

To allow an athlete to compete at maximum performance, an acclimatization period of 3 to 6 weeks has been recommended [10]. For aerobic events at an altitude of 2000 m, 10 to 20 days of acclimatization has been found necessary for maximum performance [1]. For short anaerobic (less than 1 minute) events, little or no benefit from acclimatization is to be expected, but it is reasonable to allow several days to allow any symptoms of AMS to resolve. For short-duration events, competition after 72 hours of acclimatization might be optimal. Cardiac output has been observed to decrease with acclimatization, but at 72 hours, this has not had time to occur [28]. There is marked individual variation, but certainly the benefits of acclimatization are less clear for sports of brief duration.

Training

It is tempting to hypothesize that the benefits of acclimatization gained from training at altitude would enhance sea level performance. For endurance events, autologous transfusions have been found to increase performance. They are effective, but the International Olympic Committee (IOC) has banned "blood doping." The practice involves drawing several units of blood from an athlete several weeks prior to competition. The athlete's own hematopoietic system is thus stimulated to make up the loss. The stored autologous blood is then given back to the athlete 1 day before competition. Hemoglobin and oxygen-carrying capacity are increased, as is the athlete's performance in endurance events. This effect persists for about 1 week. The advantage gained from altitude acclimatization would be similar. However, there are many deleterious effects that detract from this benefit such as slower pace, among others. Altitude training has not had wide acceptance. Increased aerobic power, plasma volume, and hemoglobin concentration have been noted in athletes training 10 weeks at 2000 m. Training at 2400 m produced no benefit [3]. Many athletes find that the hyperventilation achieved from altitude acclimatization gives a sensation of breathlessness in sea level performance [13]. Logistic, financial, environmental, and other factors are involved as well. Skiers may have to train at high altitude to find snow whether or not they ascribe to the benefits of altitude training. An understanding of the effects of altitude is important for athletes and coaches regardless of their training philosophy.

The Aerobic Exerciser

Intermittent use of the aerobic exerciser (a backpack apparatus that combines expired and inspired air to simulate a 7500-foot elevation)* has been advocated as a training aid. This device has been shown to produce increased VO_2 max for cyclists. Other parameters were not changed (i.e., ventilation, maximum heart rate, anaerobic threshold, and hemoglobin) [26]. This improvement may simply be due to increased training intensity produced by hypoxic stress. Whether or not this device allows athletes to become acclimatized to altitude by sea level training or whether it improves sea level performance has yet to be proved.

Nature and Nurture

The 1968 Mexico City Olympics demonstrated the advantage held by natives of high altitude in competing in submaximal exercise endurance events held at altitude. In the 1500-m run, Kipcharge Keino of Kenya beat Jim Ryan, the world record holder, who had previously beaten Keino twice in sea level competition. In the 10,000-m competition, the first five places were taken by residents of 1500 to 2000 meters (Temu of Kenya was the gold medalist). The sea level world record holder, Ron Clarke, finished sixth and collapsed. There are numerous other examples of the advantage held by altitude natives in athletic competitions performed at moderate altitude. They have no advantage in sea level competition. Growing up at altitude has been stressed as a major factor. A high level of activity by altitude natives during youth probably produces a benefit. It is certainly evident that a 3- to 6-week acclimatization period will not enable a sea level athlete to achieve the performance of the altitude resident in competition at altitude.

SUMMARY

For the mountaineer, an understanding of the stress imposed by altitude is important to optimize his chances for a successful summit, if not for survival itself, and to avoid altitude illness. For the athlete, the effects of altitude stress must be considered in relation to the competition site, the acclimatization required, and training. The effects of cold stress are intimately related to those of altitude. Nowhere else in sports medicine is knowledge more effective in preventing injury than in regard to cold and altitude.

---

*PO_2 Aerobic Exerciser. The Inspired Air Corporation, Westlake Village, California.

References

1. Astrand, P. O., and Rodall, K. *Textbook of Work Physiology*. New York, McGraw-Hill, 1977.
2. Bager, S., and Blume, F. D. Weight loss and changes in body composition at high altitude. *J Appl Physiol* 57(5):1580–1585, 1985.
3. Blake, B., Nagle, F. J., and Davies, A. Altitude and maximum performance in work and sports activity. *JAMA* 194(6):176, 1965.
4. Clark, C. F., Healon, R. K., and Wiens, A. M. Neuropsychological functioning after prolonged high altitude exposure in mountaineering. *Aviation Sci Exper Med* 54(3):202–207, 1983.
5. Craig, A. B. Olympics, 1968, postmortem. *Med Sci Sports* 1(4):177–180, 1969.
6. Dickerson, J. G. Altitude related deaths in seven trekkers in the Himalaya. *Thorax* 38:646, 1983.
7. Federal Aviation Administration. *Federal Aviation Regulations* (11th ed.). Regulation 135. Washington, D.C., 1987.
8. Federation Internationale de Ski. *International Ski Competition Rules*. 36th International Ski Congress, Istanbul, 1968. Berne, FIS, 1988.
9. Frisancho, A. R. Functional adaptation to high altitude hypoxia. *Science* 187:313, 1975.
10. Goddard, R. F. Recommendations to The International Olympic Committee. *In International Symposium on the Effects of Altitude on Physical Performance*. Albuquerque, The Athletic Institute, 1966.
11. Hackett, P. H., Rennie, D., and Levine, H. D. The incidence, importance and prophylaxis of acute mountain sickness. *Lancet* 2:1149, 1976.
12. Hackett, P., and Roach, R. C. Medical therapy of altitude illness. *Ann Emerg Med* 12:980–986, 1987.
13. Hackett, P. H., Roach, R. C., and Sutton, J. R. High altitude medicine. *In* Auerbach, P. S., and Geehr, E. C. (Eds.), *Management of Wilderness and Environmental Emergencies* (2nd ed.). St. Louis: C. V. Mosby, 1986, p. 17.
14. Hartman, G., and Oberli, H. Nutrition and heavy alpine physical performance. *Bibl Nutr Dieta* 27:126–132, 1979.
15. Heath, D., and Williams, D. R. *Man at High Altitudes*. New York, Churchill-Livingstone, 1981.
16. Hixson, E. G. Unpublished observations, China-Everest '82 expedition.
17. Hixson, E. G. Unpublished observations, Expedition physician 1983 Seven Summits Everest Expedition.
18. Houston, C. S. *Going Higher*. Burlington, VT, C. S. Houston, 1983.
19. Houston, C. S. *Going Higher* (3rd ed.). Boston, Little, Brown, 1987.
20. Houston, C. S. Incidence of acute mountain sickness. *Am Alpine J* 27(59):162, 1985.
21. Houston, C. S., and Dickenson, J. Cerebral form of high altitude illness. *Lancet* 2:758, 1975.
22. Hultgren, H. N., Lopex, C. E., and Lundberg, E. Physiologic studies of pulmonary edema at high altitude. *Circulation* 29:393, 1964.
23. McLennan, J. G., and Ungersman, J. Mountaineering accidents in the Sierra Nevada. *Am J Sportsmed* 2:160–163, 1983.
24. Monge, C. C., and Whittenburg, J. Increased hemoglobin oxygen affinity at extreme altitude. *Science* 186:843, 1974.
25. Moore, L. G. Altitude aggravated illness, examples from pregnancy and pre-natal life. *Ann Emerg Med* 16:965–973, 1983.
26. Paul, J. S., Vacaro, P., and Ben-Ezra, V. An evaluation of the PO_2 aerobic exerciser as an ergogenic aid in training competitive cyclists. *J Sportsmed* 25:104–110, 1985.
27. Schoene, R. B. Pulmonary edema at high altitude. *Clin Chest Med* 6(3):491–507, 1983.
28. Shephard, R. J. Altitude. *In* Dirix, A., Knottgren, H. O., and Tittel, K. (Eds.), *The Olympic Book of Sportsmedicine*. Oxford, Blackwell Scientific Publications, 1988.
29. Stiles, M. H. What medicine learned from the 1968 Olympics. *Medical Times* 102(8):123–128, 1984.
30. West, J. B. Barometric pressure at extreme altitude on Mt. Everest: Physiologic significance. *J Appl Physiol* 54:1187–1194, 1988.
31. West, J. B. Man at extreme altitude. *J Appl Physiol* 52(6):1393–1399, 1982.
32. West, J. B., Lahiri, S., Maret, K. H., Peters, R. M., and Pizzo, C. J. Pulmonary gas exchange on the summit of Mt. Everest. *J Appl Physiol* 55:678–689, 1983.
33. Wiedman, M., and Tabin, G. High altitude retinal hemorrhage as a prognostic indicator in altitude illness. *Int Ophthalmol Clin* 26(2), 1986.

SUDDEN CARDIAC DEATH IN AN ATHLETE

Timothy P. Finney, M.D.
Robert D. D'Ambrosia, M.D.

The last 10 years have seen an explosion in the number of people participating in some type of athletic activity. People of all ages are now becoming more health conscious, maintain better nutritional practices, and dedicate more time to vigorous physical activities. The obvious benefits of vigorous habitual exercise are well documented and include improved hemodynamic and skeletal muscle adaptation, weight control, and improved cardiovascular function (Table 13–1). Less known are the possible risks of vigorous exercise, which can range from a minor overuse syndrome to catastrophic sudden cardiac death (Table 13–2).

The purpose of this chapter is to examine exercise-related sudden death, its etiology, and its underlying pathologic process. Exercise-induced sudden death in the young athlete is due to an entirely different pathologic lesion from that seen in the athlete older than 30 years (Table 13–3). Autopsy studies have shown multiple structural cardiovascular abnormalities underlying sudden death in athletes less than 30 years of age [61, 102, 105, 108]. The various abnormalities include myocardial, coronary vessel, and conduction system disorders. In athletes older than 30 years of age, the underlying pathologic lesion is almost always related to severe coronary artery disease [60, 101, 114, 115]. By studying the various causes of sudden death in the athlete one should be better equipped to deal with this problem, find those at risk, and prevent occurrence of this catastrophe.

EXERCISE-RELATED SUDDEN DEATH IN THE YOUNG (LESS THAN 30 YEARS OF AGE)

Sudden cardiac death of a young, seemingly healthy athlete occurs relatively infrequently, but when a case does occur, it is highly publicized in the media. It is estimated that approximately 10 to 25 sudden deaths per year occur in the United States among the millions of young athletes participating in recreational sports [34, 108]. In the young athlete less than 30 years of age, sudden death is usually due to structural cardiovascular disease, with hypertrophic cardiomyopathy being one of the most fre-

TABLE 13–1
Benefits of Vigorous Exercise

Cardiovascular (decreased heart rate, decreased blood pressure, increased maximal oxygen consumption)
Lipid levels (decreased triglycerides, increased high-density lipoprotein/cholesterol)
Psychologic (decreased stress, increased self-esteem)
Health awareness (decreased tobacco use)
Longevity (decreased aging)

TABLE 13–2
Risks of Running

Orthopaedic ("overuse syndrome"—knee, tendon, stress fracture)
Heart (exhaustion, stroke, dehydration)
Cold (frostbite, dehydration)
Trauma (auto accidents, dog bites)
Endocrine (delayed menarche, amenorrhea)
Psychosocial ("addiction," increased type A behavior)
Sudden death (coronary artery disease, cardiomyopathy, arrhythmia)

From Ludmerer, K. M., and Kissane, J. M. (Eds.). Sudden death in a 47-year-old Marathon runner. Clinopathologic Conference. *Am J Med* 76:517–526, 1984.

TABLE 13–3
Sudden Cardiac Death in Athletes

Young Athletes (Age 30 Years and Younger)
Myocardial
    Hypertrophic cardiomyopathy
    Idiopathic concentric left ventricular hypertrophy
    Myocarditis
    Sarcoidosis
    Lipomatous infiltration of the heart
Coronary artery disorders
    Anomalous left coronary artery from right sinus of
        Valsalva
    Anomalous right coronary artery from left sinus of
        Valsalva
    Single coronary artery
    Coronary artery hypoplasia
    Coronary artery from pulmonary artery
Aortic disorders
    Marfan's cystic medial necrosis
Valvular disorders
    Congenital aortic stenosis
    Mitral valve prolapse
Cardiac conduction system disorders
    Idiopathic long QT syndrome
    Histologic abnormalities of sinoatrial or
        atrioventricular nodes

Older Athletes (Older Than 30 Years)
    Coronary artery disease
    Systemic causes of cardiovascular disease

quently documented causes of sudden death (Fig. 13–1*A*). Most studies show that sudden death in the young athlete who is free from any coronary artery disease can be the result of a myocardial, congenital coronary artery, aortic, cardiac conduction, or valvular disorder [61]. Each of these will be discussed separately.

Myocardial Disorders

The most commonly reported cause of sudden death in the young athlete is myocardial pathology (Table 13–4). Maron and colleagues [61] reported a series of 29 highly conditioned competitive athletes aged 13 to 30 who suffered sudden death. They showed that 22 of 29, or 67%, had definite structural myocardial lesions causing death. The majority of the deaths were due to hypertrophic cardiomyopathy (HCM). HCM has been found to be inherited as a autosomal dominant disorder with incomplete penetrance and results in a hypertrophic nondilated left ventricle [61]. The cardiac muscle fibers in the ventricular wall as well as the septal fibers become hypertrophied, which impedes blood flow into the aorta. This ventricular enlargement is also associated with a decrease in volume of the left ventricle, which does not contract normally or as forcefully.

All of this leads to a decrease in cardiac output. The severity of symptoms experienced by the patient can be affected by anything that alters myocardial contractility, ventricular volume (e.g., exercise), drugs, or body position. Autopsy studies of patients with HCM show a disorganization of the myocardial fibers and septal wall, which may also lead to the conduction disorders that may be the cause of sudden death. All cases of HCM show no sign of other cardiac or systemic disease such as aortic stenosis or systemic hypertension that could cause left ventricular hypertrophy.

In most cases of sudden death in the young athlete there are no prodromal signs or symptoms. In the study conducted by Maron and colleagues, 21 of 29 athletes had been entirely asymptomatic even upon retrospective questioning of family members [61]. The remaining eight patients had transient cardiac-related symptoms including syncope in three, presyncope in one, chest pain in two, periodic mild fatigue in one, and mild fatigue, presyncope, and palpitations in one. Even with these prodromal symptoms, the actual cardiac abnormality was never correctly diagnosed. This makes it imperative to maintain a high index of suspicion when attempting to diagnose those at risk for sudden death.

Even seemingly minor signs and symptoms should raise the suspicion of more serious cardiac abnormalities. The most important screening test is a careful history and physical examination. Any history of syncopal episodes, chest pain, or shortness of breath should be evaluated further with the help of a cardiologist. Since hypertrophic cardiomyopathy is inherited as an autosomal dominant disorder [60], any family history of sudden death should raise the suspicion of HCM or other congenital heart problems, and careful evaluation is then warranted. Physical examination revealing any heart murmur or abnormal cardiac rate or rhythm should be further evaluated prior to starting vigorous exercise or activities. Even something as insignificant as increased fatigue can be a sign of an underlying cardiac problem. After a careful history and physical examination, anyone in whom there is a suspicion of an underlying cardiac problem should be further evaluated according to symptoms. A baseline he-

TABLE 13–4
Myocardial Disorders

Hypertrophic cardiomyopathy (HCM)
Idiopathic concentric left ventricular hypertrophy
    (ICLVH)
Myocarditis
Sarcoidosis

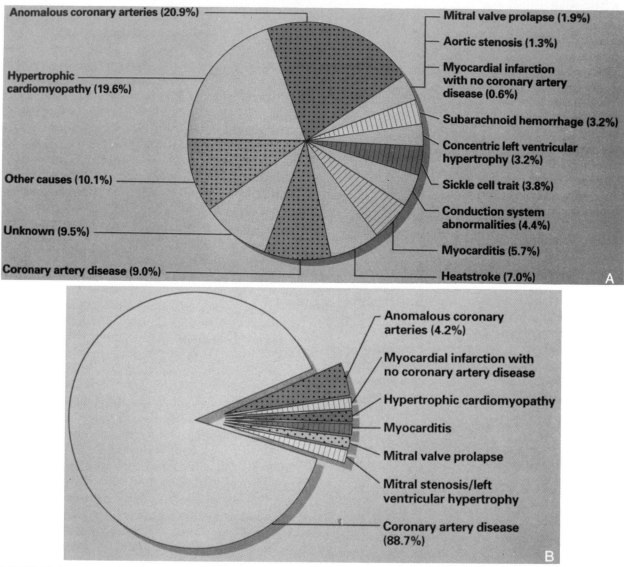

FIGURE 13–1

A, Causes of exercise-related sudden death in persons aged 30 and younger. *B,* Causes of exercise-related sudden death in persons aged 40 and older. (From Chillag, S., Bates, M., and Voltin, R. Sudden death: Myocardial infarction in a runner with normal coronary arteries. *Physician Sportsmed* 18(3):89–94, 1990. Reproduced with permission of McGraw-Hill, Inc.)

moglobin and hematocrit should be obtained, and a lipid profile can aid in the diagnosis of occult hypercholesterolemia. A routine chest x-ray and electrocardiogram (ECG) are probably indicated in anyone with exercise-induced syncope, chest pain, or shortness of breath (Fig. 13–2). The debate over whether a two-dimensional echocardiogram should be used in the routine screening of young patients is controversial. Although it is expensive, it is still the most sensitive and specific test for HCM and other cardiac abnormalities. Still, it should be employed only by a cardiologist who feels that the patient's signs and symptoms warrant structural evaluation of the cardiovascular system.

The most important point for the primary physician is to determine when further work-up is needed for a young patient in whom a cardiovascular problem is suspected. Once a prodromal sign is suspected, the patient should be evaluated with the above-noted work-up. All the disorders noted in Table 3–3 can have similar presenting signs and symptoms, so it is vital that the preparticipation screening examination begin with a thorough history and physical examination.

Another myocardial condition that has been linked to sudden death is idiopathic concentric left ventricular hypertrophy [42, 57, 62, 68, 94, 105, 107]. In this disorder, the entire left ventricle is concentrically enlarged, but there is no cellular disarray or genetic inheritance. The disorder results in abnormal left ventricular contractility but does not affect the coronary arteries. Signs and symptoms may be silent or may have the same presenting signs as HCM or any cardiac disorder (Table 13–5). The diagnosis is confirmed by two-dimensional echocardiography [61].

In Maron and associates' study [61] preexisting cardiovascular disease was suspected clinically in only 7 of 29 athletes. This fact is alarming, and the condition may often be missed on a routine preparticipation physical examination. In five athletes in whom cardiac disease was suspected, the ECGs were all abnormal, and subsequent echocardiograms may have diagnosed the pathology.

Exercise-Induced Versus Pathologic Cardiac Hypertrophy

How does one diagnose physiologic exercise-induced left ventricular hypertrophy from pathologic hypertrophy? Screening of athletes by history and physical examination, family history, and ECG may identify the most obvious cardiovascular abnormalities such as systemic hypertension or aortic valvular stenosis [1]. However, HCM is usually not associated with a loud murmur and may produce a normal tracing on ECG. If there is any suspicion of HCM such as prodromal symptoms or an abnormal chest x-ray, an echocardiogram would be the most sensitive screening test to obtain in these athletes. Because an echocardiogram is the most sensitive test for ruling out structural cardiovascular abnormalities [57], should it be used routinely as a screening test in the athlete? Braden and Strong [11] feel strongly that the routine use of a thorough preparticipation history and physical examination can either detect the most frequent causes of sudden death or raise enough suspicion to justify more elaborate tests. Including the echocardiogram routinely in preparticipation evaluations is financially, logistically, and legally impractical. It is important to remember that if a patient dies from HCM, the patient's entire family should be screened for HCM because of its autosomal dominant inheritance pattern [59].

Coronary Artery Disorders

In the young athlete, congenital coronary artery anomalies cause exercise-induced sudden cardiac death more commonly than atherosclerotic coronary artery disease [61, 64, 86, 88]. In Maron and colleagues' series [61] four patients were found to have an anomalous origin of the left coronary artery from the right anterior sinus Valsalva. This was the most frequent type of anomaly seen in the coronary arteries. The oblique course of the anomalous coronary artery (Fig. 13–3) between the aorta and the pulmonary artery results in kinking, and an ischemic condition develops, especially during strenuous activity and high oxygen demands by the myocardium. The developing ischemia may result in death by an infarct or a deadly arrhythmia. This condition could be heralded by exertional chest pain and is confirmed by stress ECG and coronary angiogram.

TABLE 13–5
Signs and Symptoms of Hypertrophic Cardiomyopathy and Idiopathic Concentric Left Ventricular Hypertrophy

HCM
  Hypertrophied nondilated left ventricle
  Myocardial cellular disarray
  Coronary arteries with narrow lumens and thick walls
ICLVH
  Unexplained ventricular hypertrophy
  Ventricle enlarged concentrically
  No bizarre cellular disarray
  Normal coronary arteries

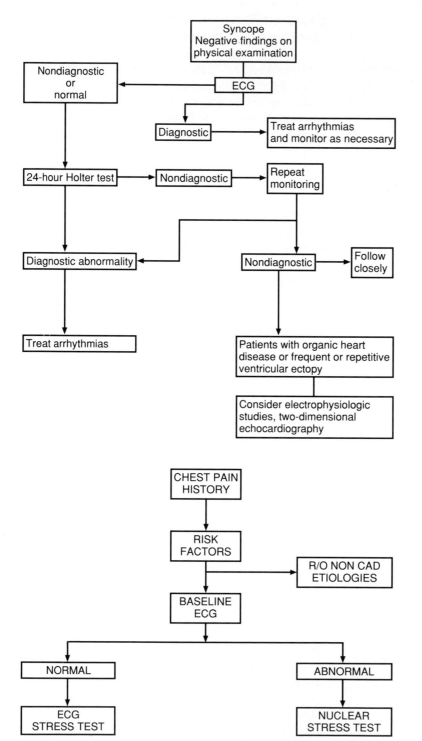

FIGURE 13–2
Approach to the patient with syncope or stable angina. PE, physical examination; R/O, rule out; CAD, coronary artery disease; ECG, electrocardiogram.

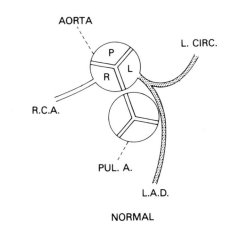

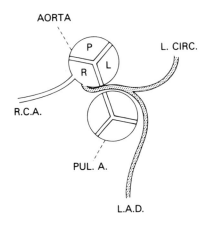

FIGURE 13–3
A, Normal origins of left coronary artery from left sinus of Valsalva and of right coronary artery from right sinus of Valsalva. *B,* Anomalous origin of left coronary artery from right anterior sinus of Valsalva. Other congenital coronary artery anomalies recognized to be responsible for some exercise-related deaths in young victims of sudden death include anomalous origin of right coronary artery from left sinus of Valsalva, single coronary artery, origin of a coronary artery from the pulmonary artery, and coronary artery hypoplasia. (Reproduced with permission from Maron, B.J., Roberts, W.C., McAllister, H.A., et al. Sudden death in young athletes. *Circulation* 62:218–229. Copyright 1980 American Heart Association.)

Another notable coronary artery anomaly that has been reported but is extremely rare resulted in the death of basketball star Pistol Pete Maravich [105, 108]. Absence of a left coronary artery with a single right coronary artery and marked cardiac hypertrophy were found at autopsy in this case. Other conditions such as coronary artery spasm could result in exercise-induced sudden death, but in the majority of these cases some prodromal symptoms are reported by the patient, most notably exertional chest pain or syncopal episodes. A discussion of preparticipation screening for sudden cardiac death is presented later in this chapter.

Aortic Disorders

With the sudden death of U.S. Olympic volleyball star Flo Hyaman in January, 1986, aortic rupture secondary to Marfan's syndrome has received much attention [27]. Marfan's syndrome has an autosomal dominant inheritance pattern and is thought to result in abnormal crosslinking of collagen and elastin [102, 105]. Affected individuals are characteristically tall with long thin extremities and have a weakness of the joint capsules, ligaments, tendons, and fasciae that results in joint dislocations, hernias, and kyphoscoliosis. In the cardiovascular system this results in abnormalities of the aortic and heart valves. In patients dying from acute aortic dissection, pathologic changes involve cystic medial necrosis and moderate degeneration of the elastic elements with disorganization of the smooth muscle bundles in the tunica media of the aorta. Valvular abnormalities such as ballooning, redundant mitral cusps, and elongated and thin chordae tendineae may be present. The diagnosis of Marfan's syndrome is usually suggested by physical examination, but echocardiography is required to evaluate the degree of aortic dilatation before a decision is made about sports participation. The Sixteenth Bethesda Conference made two recommendations for patients with Marfan's syndrome pursuing sports participation: (1) Patients with Marfan's syndrome should not participate in sports in which there is a danger of body collision (class II sports); (2) patients with aortic root dilation or mitral regurgitation may participate in sports of low intensity (class Ib), and selected cases may engage in some sports with high dynamic and low static demands (class Ia.2) (see Table 13–6) [27, 72].

Of the 257 patients with Marfan's syndrome observed between 1939 and 1970, the average age at death of the 72 deceased patients was 32 years. Of the 72 patients who died, the cause of death in 52 was cardiovascular complications, with aortic dilation rupture or dissection accounting for 80% of these [12].

Valvular Disorders

Various valvular disorders have been reported to cause sudden death in young athletes. Topaz and colleagues [102] studied 50 deaths in children, adolescents, and young adults and found two cases of sudden death due to congenital aortic stenosis.

An alarming number of sudden deaths (12) was attributed to mitral valve prolapse (MVP) [29]. Mitral valve prolapse is a common disorder with a

TABLE 13–6
Classification of Sports

I. Intensity and type of exercise performed
  A. High to moderate intensity
    1. High to moderate dynamic and static demands
      Boxing
      Crew/rowing
      Cross-country skiing
      Cycling
      Downhill skiing
      Fencing
      Football
      Ice hockey
      Rugby
      Running (sprint)
      Speed skating
      Water polo
      Wrestling
    2. High to moderate dynamic and low static demands
      Badminton
      Baseball
      Basketball
      Field hockey
      Lacrosse
      Orienteering
      Ping-pong
      Race walking
      Racquetball
      Running (distance)
      Soccer
      Squash
      Swimming
      Tennis
      Volleyball
    3. High to moderate static and low dynamic demands
      Archery
      Auto racing
      Diving
      Equestrian
      Field events (jumping)
      Field events (throwing)
      Gymnastics
      Karate or judo
      Motorcycling
      Rodeoing
      Sailing
      Ski jumping
      Water skiing
      Weight lifting

  B. Low intensity (low dynamic and low static demands)
    Bowling
    Cricket
    Curling
    Golf
    Riflery

II. Danger of body collision
    Auto racing*
    Bicycling*
    Boxing
    Diving*
    Downhill skiing*
    Equestrian*
    Football
    Gymnastics*
    Ice hockey
    Karate or judo
    Lacrosse
    Motorcycling*
    Polo*
    Rodeoing*
    Rugby
    Ski jumping*
    Soccer
    Water polo*
    Water skiing*
    Weight lifting*
    Wrestling

*Increased risk if syncope occurs.
Reprinted with permission from the American College of Cardiology (Journal of the American College of Cardiology, 1986, Vol. 53, pp. 481–494).

References
Asmussen, E. Similarities and dissimilarities between static and dynamic exercise. *Circ Res* 48(6)(Suppl. I):3–10, 1981.
Mitchell, J. H., Hefner, L. L., and Monroe, R. G. Performance of the left ventricle. *Am J Med* 53:481–494, 1972.
Mitchell, J. H., and Wildenthal, K. Static (isometric) exercise and the heart: Physiological and clinical considerations. *Annu Rev Med* 25:369–381, 1974.
Shaffer, T. E. The health examination for participation in sports. *Pediatr Ann* 7:27–40, 1978.
Sonnenblick, E. H., Ross, J., Jr., and Braunwald, E. Oxygen consumption of the heart. Newer concepts of its multifactorial determination. *Am J Cardiol* 22:328–336, 1968.
Strong, W. B., and Alpert, B. S. The child with heart disease: Play, recreation and sports. *Curr Prob Cardiol* 6:1038, 1981.

prevalence of approximately 5% in the general population [29]. Although it is usually a benign condition, it can be a rare cause of sudden death. The hallmark finding on physical examination is a systolic (nonejection) click with or without a late systolic murmur. Some patients may be symptom-free with the condition, whereas others may demonstrate atypical chest pain, palpitations, dizziness, abnormal electrocardiographic findings, or arrhythmias [75, 114].

The etiology of sudden death in patients with mitral valve prolapse is possibly related to the friction exerted by the chordae upon the left ventricular wall [29, 114]. This may result in ventricular ectopy. A familial inheritance of mitral valve prolapse has been recognized and sudden death has been reported in relatives of involved subjects [29, 114], a fact also substantiated by Topaz and associates [102], who reported three subjects who had mitral valve prolapse and a family history of sudden death. It has been recommended that a postexercise electrocardiogram be done in all patients with a late systolic murmur or a nonejection click. Noninvasive tests may help in the diagnosis of MVP including two-dimensional echocardiography and ECG. In patients with MVP and symptoms suggestive of arrhythmias, an exercise ECG may be needed to bring out the associated arrhythmia. The Sixteenth Bethesda Conference made the following recommendations for sports participation in patients with MVP: Patients with documented mitral valve prolapse and the following criteria may participate only in low-intensity class Ib sports: (1) history of syncope, (2) family history of sudden death due to mitral valve prolapse, (3) chest pain that is made worse by exercise, (4) repetitive forms of ventricular ectopic activity or sustained supraventricular tachycardia, particularly if made worse by exercise, (5) moderate or marked mitral regurgitation, and (6) association with Marfan's syndrome. Patients without any of these findings may engage in all competitive sports [72].

Cardiac Conduction System Disorders

The terminal event in most cases of sudden cardiac death is usually a fatal arrhythmia, and most cases of sudden death occur in patients with some type of structural defect. However, in most series, 8% to 10% of postmortem examinations are unable to demonstrate any structural heart defect [61]. Sudden death in some patients with normal hearts may be accounted for by metabolic and electrolyte abnormalities, exercise-induced arrhythmias, or possibly neurally mediated arrhythmias [22, 53, 55].

A disorder called idiopathic long QT syndrome as reported by Schwartz and colleagues [93] may result in an exercise-induced fatal arrhythmia. Van Camp [105, 108] found that although some patients had no structural cardiac abnormality, they usually had increased cardiac mass. This could result in areas of cardiac conduction ectopy. Establishing criteria for sports participation in a patient with an arrhythmia is very difficult. There are no concrete studies to date that show whether or not a particular arrhythmia predisposes an athlete to sudden death. It is the responsibility of the physician to diagnose these problems and determine whether the patient has symptoms due to the arrhythmia. Certain arrhythmias by themselves are dangerous no matter in what clinical situation they are seen. Generally, these arrhythmias are characterized by either a very rapid or a very slow ventricular rate that can compromise the cardiac output [72, 45]. Atrial flutter or atrial fibrillation with ventricular rates of 200 to 300 beats per minute, sustained ventricular tachycardia, and atrioventricular (AV) block or sinus node disease with very slow ventricular rates can significantly decrease cardiac output and cause sudden death. Although most cases are benign, the guidelines set forth by the Sixteenth Bethesda Conference provide an excellent framework for determining sports participation. The reader is referred to the Sixteenth Bethesda Conference, which provides a complete list of such arrhythmias with basic guidelines and recommendations regarding sports participation [72].

SUDDEN CARDIAC DEATH IN OLDER ATHLETES (GREATER THAN 30 YEARS OF AGE)

Although the majority of sudden cardiac deaths in young athletes are due to structural, often congenital, disorders, it is clear that the majority of sudden deaths in athletes over 30 years of age are due to preexisting coronary artery disease [58, 76, 101, 111, 114, 115] (Fig. 13–1C). Thompson and colleagues [101] studied 18 cases of death occurring during running or jogging and found that 13 were due to coronary artery disease. Waller and Roberts [115] studied the cases of five white male runners who died while running. These men ran 22 to 176 km/week for 1 to 10 years, and at autopsy all showed severe atherosclerotic luminal narrowing of the major epicardial coronary arteries. Vermani and associates [111] studied a series of 30 sudden deaths that occurred during jogging. The patients ran 7 to 105 miles/week with a mean of 33 miles and had been running the distance for a mean of 10 years. Twenty-two of the 30 patients (75%) showed evidence of severe coronary artery atherosclerosis.

Although most of the patients mentioned here were clinically asymptomatic, the majority had some risk factors or abnormal laboratory test result that could have given a warning signal. Many patients had hypercholesterolemia, systemic hypertension, or a strong family history of coronary artery disease.

The patient is considered to have "known" coronary artery disease if either greater than 50% narrowing of one or more coronary arteries is evident on angiography or, more often, if the history and noninvasive studies indicate a high probability of atherosclerotic coronary vascular disease [101, 114]. The best way to determine the prognosis for the patient is to assess the degree of left ventricular dysfunction by stress ECG or stress nuclear studies [12, 72]. This method allows the cardiologist to stratify three levels of risk by noninvasive studies. High-risk patients have decreased left ventricular function at rest, whereas those with low risk have normal or near normal left ventricular function at rest and no evidence of ischemia or ventricular tachycardia after exercise tolerance testing. Those with moderate risk fall between these extremes as outlined by the Sixteenth Bethesda Conference [72].

Low-risk patients can participate in low-intensity sports (class Ib), but it is the recommendation of the task force that these patients refrain from moderate- to high-intensity sports. Some exceptions can be made on a case by case basis. Patients with moderate to high risk should be discouraged from participating in all competitive sports but should participate in low-intensity cardiovascular fitness programs as outlined in postoperative coronary bypass protocols. These patients should be carefully monitored by a physician during these activities [72].

Even marathon runners are not immune from severe, life-threatening coronary artery disease according to Bossler [7], although the prevalence of coronary artery disease in these runners is probably low.

PREPARTICIPATION EVALUATION

With more and more weekend athletes participating in strenuous activities, what evaluation process should be followed for these athletes? Should all athletes over 30 years of age undergo an exercise ECG? What is the role of echocardiography?

It is generally agreed that anyone of any age who

TABLE 13–7
Warning Signals of Cardiovascular Abnormalities

Symptoms
Syncope or near syncope
Exercise intolerance
Exertional chest pain
Dyspnea
Family History
Coronary artery disease
Sudden death
Heart murmur
Hypertension
Physical Examination
Hypertension
Cardiac arrhythmias
Heart murmur
Marfan's habitus
Echocardiogram

begins a new program of vigorous exercise should undergo a thorough preparticipation history and physical examination [3, 34]. Focusing on any possible cardiovascular warning signal is important (Table 13–7). If the patient history, family history, or physical findings warrant, laboratory tests should be obtained to rule out any other cardiac risk factors (Table 13–8).

The role of exercise ECG is not as clear. The exercise stress test may produce ischemia or malignant ventricular arrhythmias and is probably the most important screening test for detecting occult coronary disease [9, 13, 15, 28, 56, 85, 98]. In the asymptomatic patient with occult coronary artery disease the exercise ECG looses its sensitivity. Used in combination with a stress thalium scan, the predictive value of real coronary artery disease is increased [104], (see Fig. 13–2).

As recommended by the Bethesda Conference [72], before beginning any vigorous exercise program, an exercise ECG should be considered in (1) men 45 years of age or older, (2) women 55 years old or older, (3) men under 45 and women under 55 with significant risk factors for coronary heart disease, such as abnormal serum lipid levels, total cholesterol level greater than 250 mg/100 mL, or exercise intolerance or ventricular arrhythmias.

TABLE 13–8
Cardiac Risk Factors

Hypercholesterolemia	High fat diet
Tobacco Smoking	Family history
Hypertension	High stress
Obesity	Alcohol intake

There are some reported risk factors involved with exercise ECG, but even in those prone to coronary heart disease, the incidence of myocardial infarction and death is very low.

Use of echocardiography is useful in ruling out a structural cardiovascular defect, which causes sudden death more often in young athletes [57]. Its use in the diagnosis of coronary heart disease is probably not as important. If there is any suggestion of a structural defect such as a valvular disorder or left ventricular hypertrophy, the use of echocardiography may be helpful.

By no means is a person doomed to a sedentary life if he or she is found to have coronary heart disease. Sensible, well-supervised chronic exercise does appear to lower the overall risk of cardiac arrest. Therefore, Van Camp suggests that it is important to screen those at risk and make proper exercise suggestions while encouraging those free from coronary heart disease to be physically active [105, 108].

SUMMARY

Sudden cardiac death in athletes, although rare, is a significant problem, and health care professionals should be educated about the reasons for its occurrence and methods of prevention. The causes have been found to be entirely different in young athletes less than 30 years old and older athletes above 30 years of age. By far the most common cause of sudden cardiac death in the older athlete, even one who has been active for many years, is coronary artery disease. In younger athletes various structural cardiac abnormalities result in sudden death.

Although the occurrence of sudden cardiac death may never be totally eliminated, the incidence could be decreased if sensible preparticipation goals were set. Life-threatening problems can be discovered by performing a routine history and physical examination but one needs to be aware of the various causes. If warranted, further laboratory studies such as stress ECG or echocardiography may discover a problem that could result in exercise-induced sudden cardiac death.

By no means is this chapter meant to deter anyone from participating in an athletic endeavor. There is no medical evidence that physical activity, even strenuous exercise, is harmful to a healthy cardiovascular system. It is a hopeful sign that health care professsionals can discover those who are at risk and prescribe safe but helpful physical activity regimens for them.

References

1. Alpert, S.A., et al., Athletic heart syndrome *Physician Sportsmed* 12(7):103–107, 1989.
2. Ambrose, J.A., Winters, S.L., Arora R.R., Haft, J.L., Goldstein, J., Rentron, K.P., Gorlin, R., and Fuster, V. Coronary angiographic morphology in myocardial infarction: A link between the pathogenesis of unstable angina and myocardial infarction. *J Am Coll Cardiol* 6:1233–1238, 1985.
3. American College of Sports Medicine: Position statement on prevention of heat injuries during distance running. *Med Sci Sports* 7:7–9, 1975.
4. American College of Sports Medicine. *Guidelines for Graded Exercise Testing and Exercise Prescription.* Philadelphia, Lea & Febiger, 1980.
5. American Heart Association, Committee on Exercise. *Exercise Testing and Training of Apparently Healthy Individuals.* Dallas, American Heart Association, 1972.
6. Balady, G.J., Weiner, D.A., McCabe, C.H., and Ryan, T.J. Value of arm testing in detecting coronary artery disease. *Am J Cardiol* 55:37–39, 1985.
7. Bossler, T.J. Athletic activity and longevity (letter to the editor). *Lancet* 1:712–713, 1972.
8. Beller, G.A., and Gibson, R.S. Sensitivity, specificity, and prognostic significance of noninvasive testing for occult or known coronary disease. *Prog Cardiovasc Dis* 24:241–270, 1987.
9. Blumenthal, D.S., Weiss, J.I., Mellits, E.D., and Gerstenblith, G. The predictive value of a strongly positive stress test in patients with minimal symptoms. *Am J Med* 70:1005–1010, 1981.
10. Bonow, R.O., Ket, K.M., Rosing, D.R., Lan, K.K., Lakatos, E., Borer, J.S., Bacharach, S.L., Green, M.V., and Epstein, S.E. Exercise-induced ischemia in mildly symptomatic patients with coronary-artery disease and preserved left ventricular function. *N Engl J Med* 311:1339–1345, 1984.
11. Braden, D.S., and Strong W.B. Preparticipation screening for sudden cardiac death in high school and college athletes. *Physician Sports Med* 16(10):128–140, 1988.
12. Braunwald, E.E. *Heart Disease: A Textbook of Cardiovascular Medicine* (3rd ed.). Philadelphia, W.B. Saunders, 1988.
13. Bruce, R.A., Hossack, K.F., DeRouen, T.A., and Hofer, V. Enhanced risk assessment of primary coronary heart disease events by maximal exercise testing. *J Am Coll Cardiol* 2:565–573, 1983.
14. Caiozzo, V.J., Davis, D.A., Ellis, J.F., Azus, J.L., Vandagriff, R., Prietto, C.A., and McMaster, W.C. A comparison of gas exchange indices used to detect the anaerobic threshold. *J Appl Physiol* 53:1184–1189, 1982.
15. Chaitman, B.R. Changing role of the exercise electrocardiogram as a diagnostic and prognostic test for chronic ischemic heart disease. *J Am Coll Cardiol* 8:1195–1210, 1986.
16. Cheitlin, M.D. The intramural coronary artery: Another cause for sudden death with exercise? *Circulation* 62:238–239, 1980.
17. Chillag, S., Bates, M., and Voltin, R. Sudden death: Myocardial infarction in a runner with normal coronary arteries. *Physician Sportsmed* 18(3):89–94, 1990.
18. Chung, E.K. Exercise ECG testing: Is it indicated for asymptomatic individuals before engaging in any exercise program? *Arch Intern Med* 140:895–896, 1980.
19. Clausen, J.P. Circulatory adjustments to dynamic exercise and effect of physical training in normal subjects and in patients with coronary artery disease. *Prog Cardiovasc Dis* 18:459–495, 1976.
20. Coplan, N.L., Gleim, G.W., and Nicholas, J.A. Principles of exercise prescription for patients with coronary artery disease. *Am Heart J* 112:145–149, 1986.

21. Coplan, N.L., Gleim, G.W., and Nicholas, J.A. Using exercise respiratory measurements to compare methods of exercise prescription. *Am J Cardiol* 58:832–836, 1986.

22. Coplan, N.L., Gleim, G.W., Scandura, M., Fisher, E., and Nicholas, J. Post-exercise potassium flux and exercise intensity. *Med Sci Sports Exerc* 19:S83, 1987.

23. Davis, J.A. Anaerobic threshold: Review of the concept and directions for future research. *Med Sci Sports Exerc* 17:6–18, 1985.

24. Davies, M.J., and Thomas, A.C. Plaque fissuring—the cause of acute myocardial infarction, sudden ischaemic death, and crescendo angina. *Br Heart J* 53:363–373, 1985.

25. Davies, M.J., and Thomas, A. Thrombosis and acute coronary artery lesions in sudden cardiac ischemic death. *N Engl J Med* 310:1137–1140, 1984.

26. DeBusk, R., Pitts, W., Haskell, W., and Huston, N. Comparison of cardiovascular responses to static-dynamic effort and dynamic effort alone in patients with chronic ischemic heart disease. *Circulation* 477–483, 1974.

27. Demak, R. Marfan syndrome: A silent killer. *Sports Illustrated* 64 (Feb. 17):30–35, 1986.

28. Doyle, J.T. Epidemiologic aspects of the asymptomatic positive exercise test. *Circulation* 75(Suppl. II):II-12–II-13, 1987.

29. Duren, D.R., Becker, A.E., and Dunning A.J. Long-term follow-up of idiopathic mitral valve prolapse in 300 patients: A prospective study. *J Am Coll Cardiol* 11:42–47, 1988.

30. Eichner, E.R. Sickle cell trait and risk of exercise induced death. *Physician Sportsmed* 15(12):41–43, 1987.

31. Epstein, S.E. Values and limitations of the electrocardiographic response to exercise—the assessment of patients with coronary artery disease. *Am J Cardiol* 42:667–673, 1978.

32. Epstein, S.E. Implications of probability analysis on the strategy used for noninvasive detection of coronary artery disease. *Am J Cardiol* 42:667–673, 1978.

33. Epstein, S.E. Implications of probability analysis on the strategy used for noninvasive detection of coronary artery disease. *Am J Cardiol* 46:491–499, 1980.

34. Epstein, S.E., and Maron, B.J. Sudden death and the competitive athlete: Perspectives on preparticipation screening studies. *J Am Coll Cardiol* 7220–7230, 1986.

35. Falk, E. Unstable angina with fatal outcome: Dynamic coronary thrombosis leading to infarction and/or sudden death. *Circulation* 71:699–708, 1985.

36. Froelicher, V.F., Longo, M.R., Triebwasser, J.H., and Lancaster, M.C. Value of exercise testing for screening asymptomatic men for latent coronary artery disease. *Prog Cardiovasc Dis* 18:265–276, 1976.

37. Fuster, V., Steele, P.M., and Chesebro, J.H. Role of platelets and thrombosis in coronary atherosclerotic disease and sudden death. *J Am Coll Cardiol* 5:175B–184B, 1985.

38. Goldschlager, N., Selzer, A., and Cohn, K. Treadmill stress tests as indicators of presence and severity of coronary artery disease. *Ann Intern Med* 85:277–286, 1976.

39. Gleim, G.W., Coplan, N.L., Scandura, M., Holly, T., and Nicholas, J.A. Myocardial oxygen demand at equivalent systemic oxygen consumption for different exercise modes. *Med Sci Sports Exerc* 18S:82, 1986

40. Gleim, G.W., Zabetakis, P.M., DePasquale, E.E., Michelis, M.F., and Nicholas, J.A. Plasma osmolality, volume, and renin activity at the "anaerobic threshold." *J Appl Physiol* 56:57–63, 1984.

41. Graves, J.E., Pollock, M.L., Montain, S.J., Jackson, A.S., O'Keefe, J.M. The effect of hand-held weights on the physiological responses to walking exercise. *Med Sci Sports Exerc* 19:260–265, 1987.

42. Hardarson, T., Curiel, R., and De La Calzada, C.S. Prognosis and mortality of hypertrophic obstructive cardiomyopathy. *Am J Cardiol* 53:902–907, 1984.

43. Hopkirk, J.A.C., Uhl, G.S., Hickman, J.R., Fischer, J., and Medina, A. Discriminant value of clinical and exercise variables in detecting significant coronary artery disease in asymptomatic men. *J Am Coll Cardiol* 3:887–894, 1984.

44. Hossack, K.F., and Hartwig, R. Cardiac arrest associated with supervised cardiac rehabilitation. *J Cardiac Rehabil* 2:402–408, 1982.

45. James, T.N., Froggatt, P., and Marshall, T.K. Sudden death in young athletes. *Ann Intern Med* 67:1013–1021, 1967.

46. Jokl, E. Sudden death after exercise due to myocarditis. *Medicine and Sports* 5:115–165, 1971.

47. Kannel, W., Belanger, A., D'Agostino, R., and Israel, I. Physical activity and physical demand on the job and risk of cardiovascular disease and death: The Framingham Study. *Am Heart J* 112:802–805, 1986.

48. Kark, J.A., Posey, D.M., Schumacher, H.R., and Ruehle, C.J. Sickle-cell trait as a risk factor for sudden death in physical training. *N Engl J Med* 317:781–787, 1987.

49. Kent, J.A., Rosing, D.R., Ewels, C.J., Lipson, L., Bonow, R., and Epstein, S.E. Prognosis in asymptomatic or mildly symptomatic patients with coronary artery disease. *Am J Cardiol* 49:1823–1831, 1982.

50. Koplan, J.P., Powell, K.E., Sikes, R.K., Shirley, R.W., and Campbell, C.C. An epidemiologic study of the benefits and risks of running. *JAMA* 248:3118–3121, 1982.

51. Lambert, E.C., Menon, V.A., Wagner, H.R., and Vlad, P. Sudden unexpected death from cardiovascular disease in children. A cooperative international study. *Am J Cardiol* 34:189–196, 1974.

52. Lehman, M., Keul, J., Huber, J., and De Prada, M. Plasma catecholamines in trained and untrained volunteers during graduated exercise. *Int J Sports Med* 2:143–147, 1981.

53. Lehman, M., Schmid, P., and Jeul, J. Plasma catecholamines and blood lactate accumulation during incremental exhaustive exercise. *Int J Sports Med* 6:78–81, 1985.

54. Liberthson, R.R., Dinsmore, R.E., and Fallon, J.T. Aberrant coronary artery origin from the aorta. *Circulation* 59:748–754, 1979.

55. Lown, B., Verrier, R.L., and Rabinowitz, S.H. Neural and psychologic mechanisms and the problem of sudden cardiac death. *Am J Cardiol* 39:890–902, 1972.

56. Malinow, M.R., McGarry, D.L., and Muehl, K.S. Is exercise testing indicated for asymptomatic active people? *J Cardiac Rehabil* 4:376–380, 1984.

57. Maron, B.J. Structural features of the athlete's heart as defined by echocardiography. *J Am Coll Cardiol* 7:190–203, 1986.

58. Maron, B.J., Epstein, S.E., and Roberts, W.C. Causes of sudden death in competitive athletes. *J Am Coll Cardiol* 7:204–214, 1986.

59. Maron, B.J., Lipson, L.C., Roberts, W.C., Savage D.D., and Epstein S.E. Malignant hypertrophic cardiomyopathy: Identification of a subgroup of families with unusually frequent premature deaths. *Am J Cardiol* 41:1133–1140, 1978.

60. Maron, B.J., Roberts, W.C., and Epstein, S.E. Sudden death in hypertrophic cardiomyopathy: A profile of 78 patients. *Circulation* 65:1388–1394, 1982.

61. Maron, B.J., Roberts, W.C., McAllister, H.A., Rosing, D.R., and Epstein, S.E. Sudden death in young athletes. *Circulation* 62:218–229, 1980.

62. Maron, B.J., Roberts, W.C., Edwards, J.E., McAllister, H.A., Foley, D.D., and Epstein, S.E. Sudden death in patients with hypertrophic cardiomyopathy: Characterization of 26 patients without functional limitation. *Am J Cardiol* 41:803–810, 1978.

63. Maron, B.J., Savage, D.D., Wolfson, J.K., and Epstein, S.E. Prognostic significance of 24 hour ambulatory electrocardiographic monitoring in patients with hypertrophic cardiomyopathy: A prospective study. *Am J Cardiol* 48:252–257, 1981.

64. McClellan, J.T., and Jokl, E. Congenital anomalies of coronary arteries as cause of sudden death associated with physical exertion. *Medicine and Sport* 5:91–98, 1971.

65. McHenry, M.M. Medical screening of patients with coronary artery disease. *Am J Cardiol* 33:752–756, 1974.

66. McHenry, P.L., O'Donnell, J., Morris, S.N., and Jordan, J.J. The abnormal exercise electrocardiogram in apparently

healthy men: A predictor of angina pectoris as an initial coronary event during long-term follow-up. *Circulation* 70:547–551, 1984.

67. McKenna, W.J., Deanfield, J., and Faruqui, A. Prognosis in hypertrophic cardiomyopathy. *Am J Cardiol* 47:532–538, 1981.

68. McKenna, W.I., England, D., and Doi, Y.C. Arrhythmia in hypertrophic cardiomyopathy. I. Influence prognosis. *Br Heart J* 46:168–172, 1981.

69. McManus, B.M., Waller, B.F., Graboys, T.B., Mitchell, J.H., Siegel, R.J., Miller, H.S. Jr., Froelich, V.F., and Roberts, W.C. Exercise and sudden death. Part I. *Curr Probl Cardiol* 6:1–89, 1981.

70. McNeer, J.F., Margolis, J.R., Lee, K.L., Kisslo, J.A., Peter, R., Kong, Y., Behar, V.S., Wallace, A.G., McCants, C.B., and Rosati, R.A. The role of the exercise test in the evaluation of patients for ischemic heart disease. *Circulation* 57:64–70, 1978.

71. Mead, W.F., Pyfer, H.R., Trombold, J.C, and Frederick, R.C. Successful resuscitation of two near simultaneous cases of cardiac arrest with a review of fifteen cases occurring during supervised exercise. *Circulation* 53:187–189, 1976.

72. Mitchell, J.H., Maron, B.J., and Epstein, S.E. (Eds.). Sixteenth Bethesda Conference. Cardiovascular abnormalities in the athlete: Recommendations regarding eligibility of competition. *J Am Coll Cardiol* 6:1186–1232, 1986.

73. Morales, A.R., Romanelli, R., and Boucek, R.J. The mural left anterior descending coronary artery, strenuous exercise and sudden death. *Circulation* 62:230–237, 1980.

74. Morris, S.M., and McHenry, P.L. Role of exercise stress testing in healthy subjects and patients with coronary heart disease. *Am J Cardiol* 42:659–666, 1978.

75. Neuspiel, D.R., and Kuller, L.H. Sudden and unexpected death in childhood and adolescence. *JAMA* 254:1321–1325, 1985.

76. Noakes, T.D. Heart disease in marathon runners: A review. *Med Sci Sports Exerc* 19:187–189, 1987.

77. Northcote, R.J., and Ballantyne, D. Sudden cardiac death in sport. *Br Med J* 287:1357–1359, 1983.

78. Opie, L.H. Sudden death and sport. *Lancet* 1:263–266, 1975.

79. Paffenbarger, R.S., Jr., Hyde, R.T., Wing, A.L., and Steinmetz, C.H. A natural history of athleticism and cardiovascular health. *JAMA* 252:491–495, 1984.

80. Paffenbarger, R.S., Jr., Wing, A.L., Hyde, R.T., and Jung, D.L. Physical activity and incidence of hypertension in college alumni. *Am J Epidemiol* 117:245–257, 1983.

81. Park, R.C., and Crawford, M.H. Heart of the athlete. *Curr Probl Cardiol* 10:41–52, 1985.

82. Ragosta, M., Crabtree, J., Sturner, W.Q., and Thompson, P.D. Death during recreational exercise in the state of Rhode Island. *Med Sci Sports* 16:339–342, 1984.

83. Redwood, D.R., Borer, J.S, and Epstein, S.E. Whither the ST segment during exercise? *Circulation* 54:703–706, 1976.

84. Richter, E.A., Ruderman, N.B., and Schneider, S.H. Diabetes and exercise. *Am J Med* 70:201–209, 1981.

85. Rifkin, R.D., and Hood, W.B. Bayesian analysis of electrocardiographic exercise stress testing. *N Engl J Med* 297:681–686, 1977.

86. Roberts, W.C. Major anomalies of coronary arterial origin seen in adulthood. *Am Heart J* 111:941–963, 1986.

87. Roberts, W.C., McAllister, H.A., Jr., and Ferrans, V.J. Sarcoidosis of the heart. A clinico-pathologic study of 35 necropsy patients (Group I) and review of 78 previously described necropsy patients (Group II). *Am J Med* 63:86–108, 1977.

88. Roberts, W.C., Siegel, R.J., and Zipes, D.P. Origin of the right coronary artery from the left sinus of Valsalva and its functional consequences: Analysis of 10 necropsy patients. *Am J Cardiol* 49:863–868, 1982.

89. Sanmarco, M.E., Ponius, S., and Selvester, R.H. Abnormal blood pressure response and marked ischemic ST segment depression as predictors of severe coronary artery disease. *Circulation* 61:572–578, 1980.

90. Savage, D.D., Seides, S.F., and Clark, C.E. Electrocardiographic findings in patients with obstructive and nonobstructive hypertrophic cardiomyopathy. *Circulation* 58:402–408, 1978.

91. Scheuer, J., and Tipton, C.M. Cardiovascular adaptations to physical training. *Annu Rev Physiol* 39:221–251, 1977.

92. Schwade, J., Blomquist, C.G., and Shapiro, W. A comparison of the response to arm and leg work in patients with ischemic heart disease. *Am Heart J* 94:203–208, 1977.

93. Schwartz, P.J., Periti, M., and Malliani, A. The long Q-T syndrome. *Am Heart J* 89:378–390, 1975.

94. Shaw, P.M., Adelman, A.G., and Wigle, E.D. The natural (and unnatural) course of hypertrophic obstructive cardiomyopathy. A multicenter study. *Circ Res* 34(Suppl. II):II-179–II-195, 1973.

95. Sheffield, T., Haskell, W., Heiss, G., Kioschos, M., Leon, A., Roitman, D., and Schrott, H. Safety of exercise testing volunteer subjects: The Lipid Research Clinics' Prevalence Study Experience. *J Cardiac Rehab* 2:395–400, 1982.

96. Siscovick, D.S., Weiss, N.S, Fletch, R.H., and Lasky, T. The incidence of primary cardiac arrest during vigorous exercise. *N Engl J Med* 311:874–877, 1984.

97. Stuart, R.J., and Ellestad, M.H. National survey of exercise stress testing facilities. *Chest* 77:94–97, 1980.

98. Subcommittee on Exercise Testing, American College of Cardiology/American Heart Association Task Force on Assessment of Cardiovascular Procedures. Guidelines for exercise testing. *J Am Coll Cardiol* 8:725–738, 1986.

99. Theine, G., Nava, A., Corrado, D., and Pennelli, R.L. Right ventricular cardiomyopathy and sudden death in young people. *N Engl J Med* 318:129–133, 1988.

100. Thompson, P.D., Funk, E., Carleton, R.A., and Sturner, W.Q. Incidence of death during jogging in Rhode Island from 1975 through 1980. *JAMA* 247:2535–2538, 1982.

101. Thompson, P.D., Stern, M.P., Williams, P., Duncan, K., Hashell, W.L., and Wood, P.D. Death during jogging or running. A study of 18 cases. *JAMA* 242:1265–1267, 1979.

102. Topaz, O., and Edwards, J.E. Pathologic features of sudden death in children, adolescents and young adults. *Chest* 87:476–482, 1985.

103. Uhl, G.S., and Froelicher, V. Screening for asymptomatic coronary artery disease. *J Am Coll Cardiol* 1:946–955, 1983.

104. Uhl, G.S., Kay, T.N., and Hickman, J.R. Computer enhanced thallium scintigrams in asymptomatic men with abnormal exercise test. *Am J Cardiol* 48:1037–1043, 1981.

105. Van Camp, S.P. Exercise-related sudden deaths: Risks and causes (Part 1 of 2). *Physician Sportsmed* 16(5):97–112, 1988.

106. Van Camp, S.P. The Fixx tragedy: A cardiologist's perspective. *Physician Sportsmed* 12(9):153–157, 1984.

107. Van Camp, S.P., and Bloor, C.M. Cardiovascular pathology and cardiac hypertrophy in sudden death in young athletes (Abstract). *Med Sci Sports Exerc* 19(Suppl. 2):S91, 1987.

108. Van Camp, S.P., and Choi, J.H. Exercise and sudden death. *Physician Sportsmed* 16(3):49–52, 1988.

109. Van Camp, S.P., and Peterson, R.A. Cardiovascular complications of outpatient cardiac rehabilitation programs. *JAMA* 256:1160–1163, 1986.

110. Virmani, R., Robinowitz, M., Clark, M.A., and McAllister, H.A. Jr. Sudden death and partial absence of the right ventricular myocardium: A report of three cases and a review of the literature. *Arch Pathol Lab Med* 106:163–167, 1982.

111. Virmani, R., Robinowitz, M., and McAllister, H.A., Jr. Nontraumatic death in joggers. A series of 30 patients at autopsy. *Am J Med* 72:874–882, 1982.

112. Voigt, J., and Agdal, N. Lipomatous infiltration of the head. An uncommon cause of sudden, unexpected death in a young man. *Arch Pathol Lab Med* 106:497–498, 1982.

113. Wahren, J., and Bygdeman, S. Onset of angina pectoris in relation to circulatory adaption during arm and leg exercise. *Circulation* 44:432–444, 1971.

114. Waller, B.F. Exercise-related sudden death in young (age <30 years) and old (age >30 years) conditioned subjects.

In Wenger, N.K. (Ed.), *Exercise and the Heart.* Philadelphia, F.A. Davis, 1985, pp. 9–74.

115. Waller, B.F., and Roberts, W.C. Sudden death while running in conditioned runners aged 40 years or over. *Am J Cardiol* 45:1292–1300, 1980.

116. Wasserman, K. The anaerobic threshold measurement to evaluate exercise performance. *Am Rev Respir Dis* 129:S35–S40, 1984.

117. Weaver, W.D., Cobb, L.A., and Hallstrom, A.P. Characteristics of survivors of exertion and nonexertion-related cardiac arrest: Value of subsequent exercise testing. *Am J Cardiol* 50:671–676, 1982.

118. Weiner, D.A., McCabe, C.H., and Ryan, R.J. Prognostic assessment of patients with coronary disease by exercise testing. *Am Heart J* 105:749–755, 1983.

119. Weiner, D.A., McCabe, C.H., and Ryan, T.J. Identification of patients with left main and three vessel coronary disease with clinical and exercise test variables. *Am J Cardiol* 49:21–27, 1980.

120. Weiner, D.A., Ryan, T.L., McCabe, C.H., Chaitman, B.R., Sheffield, L.T., Ferguson, J.C., Fischer, L.D., and Tristani, F. Prognostic importance of a clinical profile and exercise test in medically treated patients with coronary artery disease. *J Am Coll Cardiol* 3:772–779, 1984.

121. Williams, P.T., Wood, .D., Haskell, W.L., and Vranizan, K. The effects of running mileage and duration on plasma lipoprotein levels. *JAMA* 247:2674–2679, 1982.

122. Zabetakis, P.M., Gleim, G.W., Coplan, N.L., Michelis, M.F., and Nicholas, J.A. Serum potassium and plasma aldosterone during dynamic exercise above the lactate threshold. *Med Sci Sports Exerc* 18:S40, 1986.

HEAD AND CERVICAL SPINE INJURIES

Joseph S. Torg, M.D.
Thomas A. Gennarelli, M.D.

The purpose of this chapter is to present clear, concise guidelines for the classification, evaluation, and emergency management of injuries that occur to the head and neck as a result of participation in competitive and recreational activities. Although all athletic injuries require careful attention, the evaluation and management of injuries to the head and neck should proceed with particular consideration. The actual or potential involvement of the nervous system creates a high-risk situation in which the margin for error is low. An accurate diagnosis is imperative, but the clinical picture is not always representative of the seriousness of the injury at hand. An intracranial hemorrhage may present initially with minimal symptoms yet follow a precipitous downhill course, whereas a less severe injury, such as neurapraxia of the brachial plexus that is associated with alarming paresthesias and paralysis, will resolve swiftly and allow a quick return to activity. Although the more severe injuries are rather infrequent, this low incidence coincidentally results in little, if any, management experience for the on-site medical staff.

EMERGENCY MANAGEMENT

There are several principles that should be considered by individuals responsible for athletes who may sustain injuries to the head and neck [93, 94].

1. The team physician or trainer should be designated as the person responsible for supervising on-the-field management of the potentially serious injury. This person is the "captain" of the medical team.

2. Prior planning must ensure the availability of all necessary emergency equipment at the site of potential injury. At a minimum, this should include a spine board, stretcher, and equipment necessary for the initiation and maintenance of cardiopulmonary resuscitation (CPR).

3. Prior planning must ensure the availability of a properly equipped ambulance as well as a hospital equipped and staffed to handle emergency neurologic problems.

4. Prior planning must ensure the immediate availability of a telephone for communicating with the hospital emergency room, ambulance, and other responsible individuals in case of an emergency.

Managing the unconscious or spine-injured athlete is a process that should not be done hastily or haphazardly. Being prepared to handle this situation is the best way to prevent actions that could convert a repairable injury into a catastrophe. Be sure that all the necessary equipment is readily accessible and in good operating condition and that all assisting personnel have been trained to use it properly. On-the-job training in an emergency situation is inefficient at best. Everyone should know what must be done beforehand so that on a signal the game plan can be put into effect.

A means of transporting the athlete must be immediately available in a high-risk sport such as football and "on call" in other sports. The medical facility must be alerted to the athlete's condition and estimated time of arrival so that adequate preparation can be made.

The availability of the proper equipment is essential! A spine board is necessary and is the best means of supporting the body in a rigid position. It is essentially a full body splint. By splinting the body, the risk of aggravating a spinal cord injury, which must always be suspected in the unconscious athlete, is reduced. In football, bolt cutters and a sharp knife or scalpel are also essential if it becomes necessary to remove the face mask. A telephone must be available to call for assistance and to notify the medical facility. Oxygen should be available and is usually carried by ambulance and rescue squads, although it is rarely required in an athletic setting. Rigid cervical collars and other external immobilization devices can be helpful if properly used. However, manual stabilization of the head and neck is recommended even if other means are available.

Properly trained personnel must know, first of all, who is in charge. Everyone should know how to perform CPR and how to move and transport the athlete. They should know where emergency equipment is located, how to use it, and the procedure for activating the emergency support system. Individuals should be assigned specific tasks beforehand, if possible, to prevent duplication of effort. Being well prepared helps to alleviate indecisiveness and second-guessing.

Prevention of further injury is the single most important objective. Do not take any action that could possibly cause further injury. The first step should be to immobilize the head and neck by supporting them in a stable position (Fig. 14–1A, B). Then, in the following order, check for breathing, pulse, and level of consciousness.

If the victim is breathing, simply remove the mouth guard, if present, and maintain the airway. It is necessary to remove the face mask only if the respiratory situation is threatened or unstable or if the athlete remains unconscious for a prolonged period. Leave the chin strap on.

Once it is established that the athlete is breathing and has a pulse, evaluate the neurologic status. The level of consciousness, response to pain, pupillary response, and any unusual posturing, flaccidity, rigidity, or weakness should be noted.

At this point, simply maintain the situation until transportation is available or until the athlete regains consciousness. If the athlete is face down when the ambulance arrives, change his position to face up by logrolling him onto a spine board (see Figs. 14–1 to 14–4). Gentle longitudinal traction should be exerted to support the head without attempting to correct alignment. Make no attempt to move the injured person except to transport him or to perform CPR if it becomes necessary.

If the athlete is not breathing or stops breathing, the airway must be established. If he is face down, he must be turned to a face-up position. The safest and easiest way to accomplish this is to logroll the athlete into a face-up position. In an ideal situation the medical support team is made up of five members: the leader, who controls the head and gives the commands only; three members to roll; and another to help lift and carry when it becomes necessary. If time permits and the spine board is on the scene, the athlete should be rolled directly onto it. However, breathing and circulation are much more important at this point.

With all medical support team members in position, the athlete is rolled toward the assistants—one at the shoulders, one at the hips, and one at the

FIGURE 14–1
A, Athlete with suspected cervical spine injury may or may not be unconscious. However, all who are unconscious should be managed as though they had a significant neck injury. *B*, Immediate manual immobilization of the head and neck unit. First check for breathing. (From Torg, J. S. (Ed.), *Athletic Injuries to the Head, Neck and Face.* Philadelphia, Lea & Febiger, 1982.)

FIGURE 14–2

A, Logroll to a spine board. This maneuver requires four individuals: the leader to immobilize the head and neck and command the medical-support team, and the remaining three individuals positioned at the shoulders, hips, and lower legs. *B*, Logroll. The leader uses the crossed-arm technique to immobilize the head. This technique allows the leader's arms to "unwind" as the three assistants roll the athlete onto the spine board. *C*, Logroll. The three assistants maintain body alignment during the roll. (From Torg, J. S. (Ed.), *Athletic Injuries to the Head, Neck and Face.* Philadelphia, Lea & Febiger, 1982.)

knees. They must maintain the body in line with the head and spine during the roll. The leader maintains immobilization of the head by applying slight traction and by using the crossed-arm technique. This technique allows the arms to unwind during the roll (Fig. 14–2*A–C*).

The face mask must be removed from the helmet before rescue breathing can be initiated. The type of mask that is attached to the helmet determines the method of removal. Bolt cutters are used with the older single- and double-bar masks. The newer masks that are attached with plastic loops should be removed by cutting the loops with a sharp knife or scalpel. Remove the entire mask so that it does not interfere with further rescue efforts (Fig. 14–3*A–C*). Once the mask has been removed, initiate rescue breathing following the current standards of the American Heart Association.

Once the athlete has been moved to a face-up position, quickly evaluate breathing and pulse. If there is still no breathing or if breathing has stopped, the airway must be established. The jaw thrust technique is the safest first approach to opening the airway of a victim who has a suspected neck injury because in most cases it can be accomplished by the rescuer grasping the angles of the victim's lower jaw and lifting with both hands, one on each side, displacing the mandible forward while tilting the head backward. The rescuer's elbows should rest on the surface on which the victim is lying (Fig. 14–4).

If the jaw thrust is not adequate, the head tilt–jaw lift should be substituted. Care must be exercised not to overextend the neck. The fingers of one hand are placed under the lower jaw on the bony part near the chin and lifted to bring the chin forward, supporting the jaw and helping to tilt the

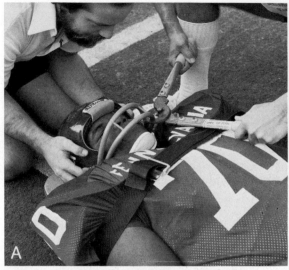

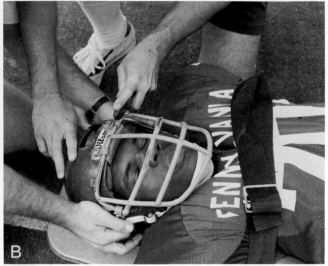

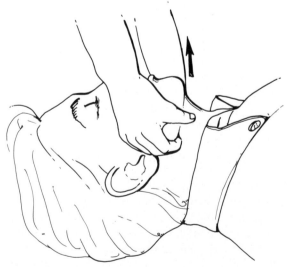

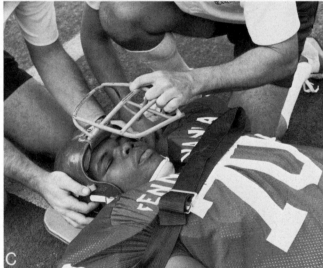

FIGURE 14–3

A, Remove double and single masks with bolt cutters. Head and helmet must be securely immobilized. *B*, Remove "cage"-type masks by cutting the plastic loops with a utility knife. Make the cut on the side of the loop away from the face. *C*, Remove the entire mask from the helmet so that it does not interfere with further resuscitation efforts. (From Torg, J. S. (Ed.), *Athletic Injuries to the Head, Neck and Face.* Philadelphia, Lea & Febiger, 1982.)

FIGURE 14–4

Jaw-thrust maneuver for opening the airway of a victim with a suspected cervical spine injury.

head back. The fingers must not compress the soft tissue under the chin, which might obstruct the airway. The other hand presses on the victim's forehead to tilt the head back (Fig. 14–5).

The transportation team should be familiar with handling a victim with a cervical spine injury, and they should be receptive to taking orders from the team physician or trainer. It is extremely important not to lose control of the care of the athlete; therefore, be familiar with the transportation crew that is used. In an athletic situation, prior arrangements with an ambulance service should be made.

Lifting and carrying the athlete require five individuals; four to lift, and the leader to maintain immobilization of the head. The leader initiates all actions with clear, loud verbal commands (Fig. 14–6A, B).

The same guidelines apply to the choice of a medical facility as to the choice of an ambulance:

FIGURE 14–5
Head tilt–jaw lift maneuver for opening the airway. Used if jaw thrust is inadequate or if a helmet is being worn.

Be sure it is equipped and staffed to handle an emergency head or neck injury. There should be a neurosurgeon and an orthopaedic surgeon to meet the athlete upon arrival. Roentgenographic facilities should be standing by.

Once the athlete is in a medical facility and permanent immobilization measures have been instituted, the helmet can be removed. The chin strap may now be unfastened and discarded. The athlete's head is supported at the occiput by one person while the leader spreads the earflaps and pulls the helmet off in a straight line with the spine (Fig. 14–7*A, B*).

Despite the advent of such "high-tech" imaging modalities as computed axial tomography (CT) and magnetic resonance imaging (MRI), the initial radiographic examination of a patient with suspected or actual cervical spine trauma remains a routine roentgenographic examination. The preliminary study, while immobilization of the head, neck, and trunk is maintained, includes an anteroposterior (AP) and lateral examination of vertebrae C1–C7. If a major fracture, subluxation, dislocation, or evidence of instability is not evident, the remainder of the routine examination, including open mouth and oblique views, should be obtained. Depending on the neurologic and comfort status of the patient, lateral flexion and extension views should be obtained at some point. CT and MRI may provide more detailed information; however, horizontally oriented fractures and subtle subluxations are best identified on the routine radiographs. The choice of imaging technique will depend on the results of the

FIGURE 14–6
A, Four members of the medical support team lift the athlete on the command of the leader. *B,* The leader maintains manual immobilization of the head. The spine board is not recommended as a stretcher. An additional stretcher should be used for transporting the patient over long distances. (From Torg, J. S. (Ed.), *Athletic Injuries to the Head, Neck and Face.* Philadelphia, Lea & Febiger, 1982.)

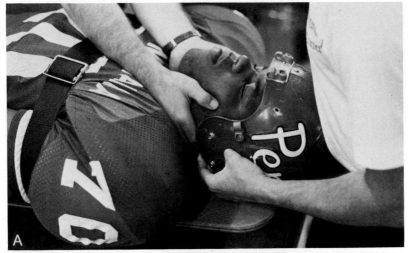

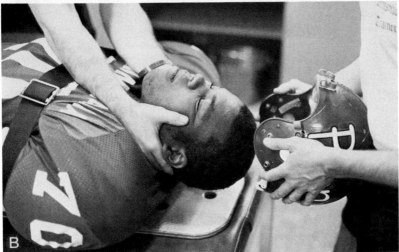

FIGURE 14–7
A, The helmet should be removed only when permanent immobilization can be instituted. The helmet may be removed by detaching the chin strap, spreading the earflaps, and gently pulling the helmet off in a straight line with the cervical spine. *B,* The head must be supported under the occiput during and after removal of the helmet. (From Torg, J. S. (Ed.), *Athletic Injuries to the Head, Neck and Face.* Philadelphia, Lea & Febiger, 1982.)

routine examination, the neurologic status of the patient, the preference of the responsible physician, and the availability of the imaging modalities.

HEAD INJURIES

The athlete who receives a blow to the head, or a sudden jolt to the body that results in a sudden acceleration-deceleration force to the head, should be carefully evaluated. If the individual is ambulatory and conscious, the entire spectrum of intracranial damage, ranging from a grade 1 concussion to a more severe intracranial condition, must be considered (Table 14–1). Initial on-field examination should include an evaluation of the following:

1. Facial expression
2. Orientation to time, place, and person
3. Presence of post-traumatic amnesia
4. Presence of retrograde amnesia
5. Abnormal gait

Traumatic injuries to the brain can be classified as diffuse or focal. The immediate and definitive management of athletically induced trauma to the brain depends on the nature and severity of the injury. Those responsible for managing such injuries must understand the problems from the standpoint of basic pathomechanics.

Diffuse Brain Injuries

Diffuse brain injuries are associated with widespread or global disruption of neurologic function and are not usually associated with macroscopically visible brain lesions. Diffuse brain injuries result from shaking of the brain within the skull and thus are lesions caused by the inertial or acceleration effects of a mechanical input to the head. Both theoretic and experimental evidence points to rotational acceleration as the primary mechanism of injury in diffuse brain injuries [23, 24, 25, 27, 77].

Since diffuse brain injuries, for the most part, are

TABLE 14–1
Guidelines for Return to Play After Concussion

	First Concussion	Second Concussion	Third Concussion
Grade 1 (mild)	May return to play if asymptomatic*	Return to play in second week if asymptomatic at that time for 1 week	Terminate season; may return to play next season if asymptomatic
Grade 2 (moderate)	Return to play after being asymptomatic for 1 week	Minimum of 1 month; may return to play then if asymptomatic for 1 week; consider terminating season	Terminate season; may return to play next season if asymptomatic
Grade 3 (severe)	Minimum of 1 month; may then return to play if asymptomatic for 1 week	Terminate season; may return to play next season if asymptomatic	

*No headache, dizziness, or impaired orientation, concentration, or memory during rest or exertion.

Grade 1 (mild): No loss of consciousness; post-traumatic amnesia <30 minutes; Grade 2 (moderate): Loss of consciousness; <5 minutes or post-traumatic amnesia >30 minutes; Grade 3 (severe): Loss of consciousness; >5 minutes or post-traumatic amnesia >24 hours.

From Cantu, R. *In* Torg J. S. (Ed.), *Athletic Injuries to the Head, Neck and Face* (2nd ed.) St. Louis, Mosby-Year Book, 1991.

not associated with visible macroscopic lesions, they have historically been lumped together to include all injuries not associated with focal lesions. More recently, however, diagnostic information has been gained from CT scans and MRI, as well as from neurophysiologic studies, that make it possible to define more clearly several categories within this broad group of diffuse brain injuries [5].

Three categories of diffuse brain injury are recognized:

1. Mild concussion. Several specific concussion syndromes involve temporary disturbances of neurologic function without loss of consciousness.
2. Classic cerebral concussion. This is a temporary, reversible neurologic deficiency caused by trauma that results in temporary loss of consciousness.
3. Diffuse axonal injury. This takes the form of prolonged traumatic brain coma with loss of consciousness lasting more than 6 hours. Residual neurologic, psychologic, or personality deficits often result because of structural disruption of numerous axons in the white matter of the cerebral hemispheres and brain stem.

Mild Cerebral Concussion

The syndromes of mild cerebral concussion are included in the continuum of diffuse brain injuries; they represent the mildest form of injury in this spectrum. Mild concussion syndromes are those in which consciousness is preserved but some degree of noticeable temporary neurologic dysfunction occurs. These injuries are exceedingly common and, because of their mild degree, often are not brought

to medical attention; however, they are the most common brain injuries encountered in sports medicine [26].

A grade 1 mild concussion, the mildest form of head injury, results in confusion and disorientation unaccompanied by amnesia. This temporary confusion, without loss of consciousness, lasts only momentarily after the injury. This concussion syndrome is completely reversible, and there are no associated sequelae. An individual with a grade 1 mild concussion is confused, has a dazed look, and may exhibit mild unsteadiness of gait. However, post-traumatic amnesia (forgetting events after the injury) and retrograde amnesia (forgetting the events before the injury) are not prominent features. This clinical picture is best described by the athletes themselves who say, "I had my bell rung." Usually the state of confusion is short-lived, and the athlete is completely lucid in 5 to 15 minutes. When his mind is clear, he may return to his former activity under the watchful supervision of the team physician or trainer. However, associated symptoms such as vertigo, headaches, photophobia, and labile emotions should preclude returning to the game [29].

A grade 2 mild concussion is characterized by confusion associated with retrograde amnesia that develops after 5 to 10 minutes. Again, this is an extremely frequent event. Athletes may experience a "ding," and, although confused, they may continue coordinated sensorimotor activities after the injury [98]. If examined immediately, these players have total recall of the events immediately before impact. However, retrograde amnesia develops 5 to 10 minutes later, and thereafter they do not remember the impact or events immediately before the impact. The amnesia usually covers only several minutes before the injury; it may diminish somewhat, but players always have some degree of permanent, though brief, retrograde amnesia despite resumption of completely normal consciousness. The confusion and disorientation completely resolve [29].

Individuals manifesting amnesia should not be permitted to return to play that day. These athletes require careful postinjury evaluation. They may develop the "postconcussion syndrome," characterized by persistent headaches, inability to concentrate, and irritability. In some instances, these symptoms may last for several weeks after the injury, and participation in the sport is precluded as long as symptoms are present.

As the mechanical stresses to the brain increase in grade 3 mild concussion, confusion and amnesia are present from the time of impact. Athletes can usually continue to play although they have no recollection of previous events. By this stage, some degree of post-traumatic amnesia occurs in addition to retrograde amnesia. The confusion may last many minutes, but then the level of consciousness returns to normal, usually with some permanent degree of retrograde and post-traumatic amnesia [29].

These three syndromes of mild cerebral concussion have been frequently witnessed and described in detail [49, 50]. Although consciousness is preserved, it is clear that some degree of cerebral dysfunction has occurred. The fact that memory mechanisms appear to be the most sensitive to trauma suggests that the cerebral hemispheres rather than the brain stem are the location of the mild injury forces. The degree of cerebral cortical dysfunction, however, is not sufficient to disconnect the influence of the cerebral hemispheres from the brain stem activating system, and therefore consciousness is preserved. Few cortical functions except memory seem to be in jeopardy, and the only residual deficit that patients with mild concussion syndromes have is the brief retrograde or post-traumatic amnesia. However, because definite alteration of brain function has occurred, athletes who sustain grades 2 and 3 mild cerebral concussions should not be permitted to participate in the remainder of the contest.

Classic Cerebral Concussion

Classic cerebral concussion is seen in the "knocked-out" player. This individual is in a paralytic coma, usually recovers after a few seconds or minutes, and then passes through stages of stupor, confusion with or without delirium, and finally an almost lucid state with automatism before becoming fully alert. Such an individual will most certainly have retrograde and post-traumatic amnesia. If the loss of consciousness lasts for more than several minutes or if there are other signs of a deteriorating neurologic state, the patient should be immediately transported to a hospital.

The athlete who has been rendered unconscious should be initially evaluated to determine whether he is breathing, whether there is a pulse, and the level of consciousness. If unobstructed respirations and an adequate pulse are present, there is no immediate need to do anything except keep in mind that head and neck injuries are frequently associated. Therefore, the player should be protected from injudicious manipulation or movement.

Such patients frequently remain semistuporous for more than several minutes. They should be removed from the field on a spine board or stretcher and should not be permitted to stagger off. An athlete who has been rendered unconscious for any length of time should not be allowed to return to contact activity that day, even if he is mentally clear. Overnight observation in a hospital should be considered for those who experience more than a transient loss of consciousness.

Insufficient attention has been given to the precise stages of recovery from classic cerebral concussion. Although, by definition, loss of consciousness is transient and reversible, the sequelae of concussion are commonplace. Some sequelae such as headache or tinnitus may reflect injuries to the head, the inner ear, or other noncerebral structures. However, subtle changes in personality and in psychologic or memory functioning have been documented and must have a cerebrocortical origin. Thus, although most patients with classic cerebral concussion experience no sequelae other than amnesia for the events of impact, some individuals may have other long-lasting, although subtle, neurologic deficiencies that must be investigated further [31, 33]. Cantu [13], employing a somewhat more simplified classification of cerebral concussion, has developed criteria for return to play following concussion (see Table 14–1).

Second Impact Syndrome

Delayed brain swelling may occur following a concussion complicated by the post-concussion syndrome, that is, persistence of any symptoms such as of headache, vertigo, inability to concentrate, irritability, tenderness, or memory disturbance. Return to contact activity before complete resolution of a concussion and post-concussion syndrome can result in the second impact syndrome. The effects of delayed brain swelling associated with this phenomenon can be disastrous. The implications in making clinical decisions regarding the return of an individual to a contact activity are quite apparent.

Focal Brain Syndromes

In discussing the occurrence of intracranial hematoma resulting from athletic injury, two major points must be emphasized. First, owing to recent developments in the clinical evaluation of patients and correlated animal research, there is now a satisfactory understanding of the mechanism of occurrence of focal intracranial hematoma, which is somewhat different from the older concepts of head injuries [32]. Second, management of such patients has advanced rapidly and has changed dramatically during the last decade from what was accepted medical practice in the past.

The entire spectrum of traumatic intracranial hematomas occurs in sports injuries [44]. These include cerebral contusions, intracerebral hematomas, epidural hematomas, and acute subdural hematomas. The presentation of athletes with head injuries who have had serious trauma is similar in most instances. Management depends on definitive diagnosis and varies according to the underlying pathologic process.

Intracerebral Hematoma and Contusion

These injuries occur in patients with a significant intracerebral pathologic condition who may not have suffered loss of consciousness or focal neurologic deficit but who do have persistent headache or periods of confusion after a head injury and posttraumatic amnesia. As with any patient who has suffered a head injury, athletes with such symptoms should undergo a CT scan to permit early differentiation between solid intracerebral hematoma and hemorrhagic contusion with surrounding edema.

Epidural Hematoma

Epidural hematoma results when the middle meningeal artery, which is imbedded in a bony groove in the skull, tears as a result of a skull fracture crossing this groove. Because the bleeding in this instance is arterial, accumulation of clot continues under high pressure, leading to a potentially serious brain injury.

The classic description of an epidural hematoma is that of loss of consciousness at the time of injury, followed by recovery of consciousness in a variable period, after which the patient is lucid. This is followed by the onset of increasingly severe headache, decreased level of consciousness, dilation of one pupil, usually on the same side as the clot, and decerebrate posturing and weakness, usually on the side opposite the hematoma. In our experience, however, only one-third of patients with epidural hematoma present with this classic history. Another third of patients do not become unconscious until late in the course of illness, and the remaining third are unconscious from the time of injury and remain unconscious throughout their course. The absence of a classic clinical picture of epidural hematoma cannot be relied on to rule out this diagnosis, and the best diagnostic test for evaluating these patients is a CT scan [12].

Acute Subdural Hematoma

Athletic head injuries result from inertial loading at a lower level than that associated with serious head injuries caused by vehicular accidents or falling from heights. Also, acute subdural hematomas occur much more frequently than epidural hematomas in athletes. In patients with head injuries in general, approximately three times as many acute subdural hematomas occur as do epidural hematomas.

Two main types of acute subdural hematomas have been clearly identified: (1) those with a collection of blood in the subdural space, apparently not associated with underlying cerebral contusion or edema, and (2) those with collections of blood in the subdural space but associated with an obvious contusion on the surface of the brain and hemispheric brain injury with swelling. The mortality for simple subdural hematomas is approximately 20%, but this increases to more than 50% for subdural hematomas with an underlying brain injury [12].

Patients with an acute subdural hematoma typically are unconscious, may or may not have a history of deterioration, and frequently display focal neurologic findings. Patients with simple subdural hematomas are more likely to have had a lucid interval following their injury and are less likely to be unconscious at admission than patients with hemispheric injury and brain swelling. It is necessary to obtain a CT or MRI scan to diagnose an acute subdural hematoma. The size of the subdural clot relative to the size of the midline shift of the brain structures can be evaluated best by CT scan. Of patients with acute subdural hematoma, 84% also have an associated hemorrhagic contusion or intracerebral hematoma with associated brain swelling [40].

The term acute subdural hematoma raises the image of a large collection of clotted blood in the intracranial cavity, compressing the brain substance and causing brain compromise due to the space occupied by the hematoma. This is not an infrequent consequence of closed head trauma, but this type of subdural hematoma is more common in adults who have a degree of cortical atrophy.

Young athletes, and especially children, frequently develop only minimal subdural hematomas with underlying cerebral hemispheric swelling. This type of brain injury is not the result of a space-occupying mass resulting from clotted blood causing brain compression but rather swollen brain tissue causing a consequent rise in intracranial pressure. The advent of CT and MRI scanning permits an accurate differential diagnosis between these two conditions, which frequently cause similar clinical pictures. The modalities of treatment for these two distinct types of acute subdural hematomas are quite different [11].

Brain Swelling

Brain swelling is a poorly understood phenomenon that can accompany any type of head injury. Swelling is not synonymous with cerebral edema, which refers to a specific increase in brain water. Such an increase in water content may not occur in brain swelling, and current evidence favors the concept that brain swelling is due in part to increased intravascular blood within the brain. This is caused by a vascular reaction to head injury that leads to vasodilation and increased cerebral blood volume. If this increased cerebral blood volume continues long enough, vascular permeability may increase, and true edema may result [40].

Although brain swelling may occur in any type of head injury, the magnitude of the swelling does not correlate well with the severity of the injury. Thus, both severe and minor head injuries may be complicated by brain swelling. The effects of brain swelling are thus additive to those of primary brain injury and may in certain instances be more severe than the primary injury itself.

Despite our lack of knowledge about the precise mechanism that causes brain swelling, it can be conceptualized in two general forms. It should be remembered that many different types of brain swelling exist and that acute and delayed brain swelling represent phenomenologic rather than mechanistic entities.

Acute brain swelling occurs in several circumstances. Swelling that accompanies focal brain lesions tends to be localized, whereas diffuse brain injuries are associated with generalized swelling. Focal swelling is usually present beneath contusions but does not often contribute additional deleterious effects. On the other hand, the swelling that occurs with acute subdural hematomas, although principally hemispheric in distribution, may cause more mass effect than the hematoma itself. In such circumstances, the small amount of blood in the subdural space may not be the entire reason for the patient's neurologic state. If the hematoma is removed, the acute brain swelling may progress so rapidly that the brain protrudes through the craniotomy opening. Every neurosurgeon is all too familiar with external herniation of the brain, which, when it occurs, is difficult to treat.

The more serious types of diffuse brain injuries are associated with generalized rather than focal acute brain swelling. Although not all patients with diffuse axonal injury have brain swelling, the incidence of swelling is higher than that in patients with either classic cerebral concussion or one of the mild concussion syndromes. Because of the serious nature of the underlying injury, it is difficult to determine the extent of swelling in these patients. The swelling, although widespread throughout the brain, may not cause a rise in intracranial pressure for several days. This late rise in pressure probably reflects the formation of true cerebral edema, and it may be that diffuse swelling associated with severe diffuse brain injuries is harmful because it produces edema. In any event, this type of swelling is different from the type of swelling associated with acute subdural hematomas.

Delayed brain swelling may occur minutes to hours after a head injury. It is usually diffuse and is often associated with the milder forms of diffuse brain injuries. Whether delayed swelling is the same as or a different phenomenon from the acute swelling of the more serious diffuse injuries is unknown. However, in less severe diffuse injuries there is a distinct time interval before delayed swelling becomes manifest, thus confirming that the primary insult to the brain was not serious. Considering the high frequency of mild concussions, the incidence of delayed swelling must be low. However, when it occurs, delayed swelling can cause profound neurologic changes or even death [59].

In its most severe form, severe delayed swelling can cause deep coma. The usual history is that of an injury associated with a mild concussion or a classic cerebral concussion from which the patient recovers. Minutes to hours later the patient becomes lethargic, then stuporous, and finally lapses into a coma. The coma may be either a light coma with appropriate motor responses to painful stimuli, or a deep coma associated with decorticate or decerebrate posturing.

The key differences between these patients and those with diffuse axonal injury is that in the latter the coma and abnormal motor signs are present from the moment of injury, whereas with delayed cerebral swelling there is a time interval without these signs. This distinction is significant, however, because with diffuse axonal injury a certain amount of primary structural damage has occurred at the moment of impact, but this is not present in cases of pure delayed swelling. Therefore, the deleterious effects of delayed swelling should be potentially reversible, and if these effects are controlled the outcome should be good. However, such control may be difficult. Vigorous monitoring of and attention to intracranial pressure is necessary, and prompt and vigorous treatment of raised intracranial pressure is required to control brain swelling. If this is accomplished successfully, the mortality from increased intracranial pressure associated with diffuse brain swelling should be low.

Principles of Management

As knowledge of physiology and pathophysiology has increased, so has the ability to resuscitate seriously ill or severely injured people successfully. The 1950s saw the start of successful treatment of acute respiratory and postoperative problems, followed by satisfactory cardiac resuscitation and emergency cardiac care in the 1960s. Innovations in critical care medicine were extended in the form of brain resuscitation in the 1970s. Such care is based on the concept that the degree of permanent neurologic, intellectual, and psychologic deficit after brain trauma with coma is only partly the result of the initial injury and is certainly in part due to secondary changes, which can be worsened or improved by the quality of the supportive care received. Head injuries, by their very nature, require resuscitation, that is, therapy initiated after the insult. The proper care of patients with head injuries, athletic or otherwise, depends on a full appreciation and use of brain resuscitation measures in an intensive care setting.

Treatment for focal intracranial hematoma consists of removal of the hematoma and recognition of and treatment for the underlying brain injury. Included in this concept is that of resuscitation of the brain, which is therapy designed to have specific neuron-saving potential once general resuscitation methods and supportive care have begun [10].

First aid should consist of getting the patient safely into a supine position and determining the vital signs and the significance of any associated injuries. Initial treatment should be to establish an adequate and useful airway and begin hyperventilation maneuvers. This can be accomplished by using a manual resuscitation bag with supplemental oxygen, if available. The patient should then be transferred as quickly as possible to a medical facility where diagnosis and treatment of brain injury can begin. Although these measures are important for all patients who have suffered concussion, they are vital for patients who remain comatose after trauma. Also, consideration must be given to the possibility of a concomitant cervical spine injury and appropriate measures taken. Once patients arrive in the emergency room and their cardiorespiratory status is determined to be stable, endotracheal intubation is immediately performed on comatose patients. A CT or MRI scan is obtained as soon as possible to provide an immediate diagnosis of the intracranial condition. Patients are then categorized as either surgical or nonsurgical cases, depending on the size of the intracranial hematoma.

CERVICAL SPINE INJURIES

Athletic injuries to the cervical spine may involve the bony vertebrae, the intervertebral discs, the ligamentous supporting structures, the spinal cord, roots, and peripheral nerves, or any combination of these structures. The panorama of injuries observed runs the spectrum from the "cervical sprain syndrome" to fracture-dislocations with permanent quadriplegia [4]. Fortunately, severe injuries with neural involvement occur infrequently. However, those responsible for the emergency and subsequent care of the athlete with a cervical spine injury should possess a basic understanding of the variety of problems that can occur.

The various athletic injuries to the cervical spine and related structures are:

1. Nerve root–brachial plexus injury
2. Stable cervical sprain
3. Muscular strain
4. Nerve root–brachial plexus axonotmesis
5. Intervertebral disc injury (narrowing-herniation) without neurologic deficit
6. Stable cervical fractures without neurologic deficit
7. Subluxations without neurologic deficit
8. Unstable fractures without neurologic deficit
9. Dislocations without neurologic deficit

10. Intervertebral disc herniation with neurologic deficit
11. Unstable fracture with neurologic deficit
12. Dislocation with neurologic deficit
13. Quadriplegia
14. Death

Criteria for return to contact activities following congenital and traumatic problems of the cervical spine are included at the end of this chapter.

Nerve Root–Brachial Plexus Injury

The most common and poorly understood cervical injuries are the pinch-stretch neurapraxias of the nerve roots and brachial plexus. Typically, following contact with head, neck, or shoulder, a sharp burning pain is experienced in the neck on the involved side that may radiate into the shoulder and down the arm to the hand. There may be associated weakness and paresthesia in the involved extremity lasting from several seconds to several minutes. Characteristically, there is weakness of shoulder abduction (deltoid), elbow flexion (biceps), and external humeral rotation (spinatis). The key to the nature of this lesion is its short duration and the presence of a full, pain-free range of neck motion. Although the majority of these injuries are short-lived, they are worrisome because of the occasional plexus axonotmesis that occurs. However, the youngster whose paresthesia completely abates, who demonstrates full muscle strength in the intrinsic muscles of the shoulder and upper extremities, and who, most importantly, has a full, pain-free range of cervical motion may return to his activity.

Persistence of paresthesia, weakness, or limitation of cervical motion requires that the individual be protected from further exposure and that he undergo neurologic, electromyographic, and roentgenographic evaluation. Persistent or recurrent episodes require a complete neurologic and radiographic-imaging work-up. If routine roentgenographic films of the cervical spine are negative and a preganglionic root lesion is suspected, MRI, plain myelography, or CT myelography should be considered. Disc herniation, foraminal narrowing, and extradural intraspinal masses should be considered in the differential diagnosis. A complete electromyographic (EMG) examination, including both nerve conduction studies and a needle electrode examination, may be helpful. These studies should be delayed for 3 to 4 weeks from the time of the initial injury. Nerve conduction studies should include both routine conduction as well as sensory nerve action potential (SNAP) evaluations [35].

Electrode evaluation of the cervical spine musculature will differentiate between preganglionic root injuries and plexus pathologies [35].

Brachial plexus injuries can be classified by the staging system described by Seddon [15]. Neurapraxia is the mildest lesion and corresponds to a reversible aberration in axonal function without intrinsic axonal disruption. The episode is transient and completely reversible with resolution occurring almost immediately or within 2 weeks. The next level of nerve injury is axonotmesis, in which injury results in disruption of the axon and myelin sheath. The epineurium, however, remains intact. Following injury, Wallerian degeneration occurs from the point of injury distally, and complete regeneration must occur for function to return. The most severe injury, neurotmesis, results from an irreversible insult resulting in nerve laceration, crushing, or stretching such that the endoneurium, epineurium, and perineurium are all disrupted. Recovery does not occur. It should be noted that in the absence of frank shoulder dislocation or a penetrating injury, this lesion is rarely if ever seen as a result of athletic activity.

The "burner" syndrome characteristically seen in contact activities such as football is generally believed to be a traction injury of the brachial plexus involving the upper trunk (C5–C6) [65]. The mechanism results from depression of the ipsilateral shoulder with forceful deviation of the head and cervical region to the opposite side. However, it is also quite apparent that a mechanism involving neck extension, ipsilateral rotation, and compression in individuals with intervertebral foraminal narrowing can result in root compression. This mechanism can be demonstrated by the Spurling maneuver, in which the head is extended, rotated toward the involved arm, and compressed. A positive test will reproduce the pain or symptoms.

With regard to management on the field of the athlete with a burner syndrome, generally, demonstrable weakness and anesthesia associated with burning pain are transient. If there is no associated neck pain or limitation of cervical motion, an asymptomatic individual may return to his or her activity. However, if examination reveals decreased neck motion and pain, the athlete should be removed from competition pending further evaluation. In individuals with persistent weakness appropriate studies should be done as mentioned to rule out other significant cervical spine pathology. In most individuals, EMG will show an upper trunk plexopathy, and the athlete's progress can be used as a guideline for return to activities. Although a mild degree of weakness may be tolerable if strength has been documented to be improving, weakness that will render him unable to perform and protect himself should preclude further activity. Those with recurrent burners may continue to participate as long as they have normal, pain-free motion and full muscle strength. Prevention of recurrent plexus neurapraxia is based on a year-round neck and shoulder muscle strengthening program and appropriate "horse-collar" type neck rolls to prevent extreme neck deviation. Straps that bind the helmet to the shoulder pads should be prohibited [35].

Acute Cervical Sprain Syndrome

An acute cervical sprain is a collision injury frequently seen in contact sports. The patient complains of having "jammed" his neck with subsequent pain localized to the cervical area. Characteristically, the patient presents with limitation of cervical spine motion but without radiation of pain or paresthesia. Neurologic examination is negative, and roentgenograms are normal.

Stable cervical sprains and strains eventually resolve with or without treatment. Initially, the presence of a serious injury should be ruled out by performing a thorough neurologic examination and determining the range of cervical motion. Range of motion is evaluated by having the athlete perform the following actions: actively nod his head, touch his chin to his chest, extend his neck maximally, touch his chin to his left shoulder, touch his chin to his right shoulder, touch his left ear to his left shoulder, and touch his right ear to his right shoulder. If the patient is unwilling or unable to perform these maneuvers actively while standing erect, proceed no further. The athlete with less than a full, pain-free range of cervical motion, persistent paresthesia, or weakness should be protected and excluded from activity. Subsequent evaluation should include appropriate roentgenographic studies, including flexion and extension views to demonstrate fractures or instability.

In general, treatment of athletes with "cervical sprains" should be tailored to the degree of severity of the injury. Immobilizing the neck in a soft collar and using analgesics and anti-inflammatory agents until there is a full, spasm-free range of neck motion is appropriate. It should be emphasized that individuals with a history of collision injury, pain, and limited cervical motion should have routine cervical spine x-rays. Also, lateral flexion and extension roentgenograms are indicated after the acute symptoms subside. Marked limitation of cervical motion, persistent pain, or radicular symptoms or findings may require MRI to rule out intervertebral disc injury.

Intervertebral Disc Injuries

Acute herniation of a cervical intervertebral disc associated with neurologic findings and occurring as an isolated entity is rare in the athlete. However, an acute onset of transient quadriplegia in an athlete who has sustained head impact but has negative cervical spine roentgenograms should prompt consideration of an acute rupture of a cervical intervertebral disc. The syndrome of acute anterior spinal cord injury, as described by Schneider [66–68], may be observed in individuals with instability associated with acute disc herniation: "The acute anterior cervical spinal cord injury syndrome may be characterized as an immediate acute paralysis of all four extremities with a loss of pain and temperature to the level of the lesion, but with preservation of posterior column sensation of motion, position, vibration and part of touch." The pressure of the disc is exerted on the anterior and lateral columns, whereas the posterior columns are protected by the denticulate ligaments. Magnetic resonance imaging or a CT myelogram should be performed to substantiate the diagnosis. Anterior discectomy and interbody fusion for a patient with neurologic involvement or persistent disability because of pain should be considered.

Albright and colleagues [3] studied 75 University of Iowa freshmen football recruits who had had roentgenograms of their cervical spines after playing football in high school but before playing in college. Of this group, 32% had one or more of the following: "occult" fracture, vertebral body compression fractures, intervertebral disc-space narrowing, or other degenerative changes. Of this group, only 13% admitted to a positive history of neck symptoms. The development of early degenerative changes or intervertebral disc-space narrowing in this group was attributed to the effect of repetitive loading on the cervical spine as a result of head impact from blocking and tackling.

Acute and chronic cervical intervertebral disc injury without frank herniation or neurologic findings occurs with considerable frequency in the athlete. Associated with a history of injury are neck pain and limited cervical spine motion. Roentgenograms may demonstrate disc-space narrowing and marginal osteophytes. Magnetic resonance imaging frequently demonstrates disc bulge without herniation. In general, management is conservative, withholding permission to engage in activity until the youngster is asymptomatic and has a full range of cervical spine motion.

Cervical Vertebral Subluxation Without Fracture

Axial compression-flexion injuries incurred by striking an object with the top of the helmet can result in disruption of the posterior soft tissue supporting elements with angulation and anterior translation of the superior cervical vertebrae. Fractures of the bony elements are not demonstrated on roentgenograms, and the patient has no neurologic deficit. Flexion-extension roentgenograms demonstrate instability of the cervical spine at the involved level, manifested by motion, anterior intervertebral disc-space narrowing, anterior angulation and displacement of the vertebral body, and fanning of the spinous processes. Demonstrable instability on lateral flexion-extension roentgenograms in young, vigorous individuals requires aggressive treatment. When soft tissue disruption occurs without an associated fracture, it is likely that instability will result despite conservative treatment. When anterior subluxation greater than 20% of the vertebral body is due to disruption of the posterior supporting structures, a posterior cervical fusion is recommended.

Cervical Fractures or Dislocations: General Principles

Fractures or dislocations of the cervical spine may be stable or unstable and may or may not be associated with neurologic deficit. When fracture or disruption of the soft tissue supporting structure immediately violates or threatens to violate the integrity of the spinal cord, implementation of certain management and treatment principles is imperative. These include the following:

1. Protection of the cord from further injury
2. Expeditious reduction
3. Attainment of rapid and secure stability
4. Implementation of an early rehabilitation program

The first goal is to protect the spinal cord and nerve roots from injury through mismanagement. It has been estimated that many neurologic deficits occur after the initial injury. That is, if a patient with an unstable lesion is carelessly manipulated during transportation to a medical facility or subsequently managed inappropriately, further encroachment on the spinal cord can occur.

Second, once appropriate roentgenograms have been obtained and qualified orthopaedic and neurosurgical personnel are available, the malaligned cervical spine should be reduced as quickly and gently as possible. This will effectively decompress the spinal cord. When dislocation or anterior angulation and translation are demonstrated roentgenographically, immediate reduction is attempted with skull traction utilizing Gardner-Wells tongs. These tongs can be easily and rapidly applied under local anesthesia without shaving the head in the emergency room or in the patient's bed. Because these tongs are spring-loaded, it is unnecessary to drill the outer table of the skull for their application. The tongs are attached to a cervical-traction pulley, and weight is added at a rate of 5 lb/disc space or 25 to 40 lb for a lower cervical injury. Reduction is attempted by adding 5 lb every 15 to 20 minutes and is monitored by lateral roentgenograms.

Unilateral facet dislocations, particularly at the C3–C4 level, are not always reducible using skeletal traction. In such instances, closed skeletal or manipulative reduction under nasotracheal anesthesia may be necessary. The expediency of early reduction of cervical dislocations must be emphasized.

It has been proposed that the presence of a bulbocavernous reflex indicates that spinal shock has worn off and that, except for recovery of an occasional root at the site of the injury, neither motor or sensory paralysis will be resolved regardless of treatment. The bulbocavernous reflex is produced by pulling on the urethral catheter. This stimulates the trigone of the bladder, producing a reflex contraction of the anal sphincter around the examiner's gloved finger. Although the presence of a bulbocavernous reflex is generally a sign that there will be no further neurologic recovery below the level of injury, this is not always true. The presence of this reflex does not give the clinician license to handle the situation in an elective fashion. The malalignments and dislocations of the cervical spine associated with quadriparesis should be reduced as quickly as possible, by whatever means necessary, if maximum recovery is to be expected.

In most instances in which a vertebral body burst fracture is associated with anterior compression of the cord, decompression is logically effected through an anterior approach with an interbody fusion. Likewise, intervertebral disc herniation with cord involvement is best managed through an anterior discectomy and interbody fusion. In patients with cervical fractures and dislocations, posterior cervical laminectomy is indicated only rarely when excision of foreign bodies or bony alignment of the spine is the most effective method for decompression of the cervical cord.

Indications for surgical decompression of the spinal cord have been delineated. A documented increase in neurologic signs is the clearest mandate for surgical decompression. Further observation, expectancy, and procrastination in this situation are contraindicated. Persistent partial cord or root signs, with objective evidence of mechanical compression, are also an indication for surgical intervention.

The use of parenteral corticosteroids to decrease the inflammatory reactions of the injured cord and surrounding soft tissue structures is indicated in the management of acute cervical spinal cord injuries. The efficacy of methyl prednisolone in improving neurologic recovery when given in the first 8 hours has been recently demonstrated. The recommended regimen is a bolus of 30 mg/kg body weight of methyl prednisolone administered intravenously followed by an infusion of 5.4 mg/kg/hour for 23 hours [9].

The third goal in managing fractures and dislocations of the cervical spine is to effect rapid and secure stability in order to prevent residual deformity and instability with associated pain, and the possibility of further trauma to the neural elements. White and associates recognized that the literature is neither always clear nor consistent in describing what constitutes an unstable cervical spine [95, 96]. Using fresh cadaver specimens, they performed load displacement studies on sectioned and unsectioned two-level cervical spine segments to determine the horizontal translation and rotation that occurred in the sagittal plane after each ligament was transected. The experiments constituted a quantitative biomechanical analysis of the effects of destroying ligaments and facets on the stability of the cervical spine below C2 in an attempt to determine cervical stability. The express purpose of the study was to establish indications for treatment methods to stabilize the spine. Although the intent of the study was to define clinical instability to formulate treatment standards and was not intended to establish criteria for a return to contact athletics, it appears that their findings are relevant to this latter issue.

White and his colleagues described clinical stability as the ability of the spine to limit its patterns of displacement of physiologic loads to prevent damage or irritation of the spinal cord or the nerve roots [95, 96]. They delineated four important findings. First, in sectioning the ligaments, small increments of change in stability occur followed without warning by sudden, complete disruption of the spine under stress. Second, removal of the facets alters the motion segment such that in flexion there is less angular displacement and more horizontal displacement. Third, the anterior ligaments contribute more to stability in extension than the posterior ligaments,

and in flexion the posterior ligaments contribute more than the anterior ligaments. The fourth and most relevant finding from the standpoint of criteria for return to contact sports is as follows: The adult cervical spine is unstable, or on the brink of instability, when one or more of the following conditions is present: (1) all of the anterior or all of the posterior elements are destroyed or are unable to function; (2) more than 3.5 mm horizontal displacement of one vertebra exists in relation to an adjacent vertebra measured on lateral roentgenograms (resting or flexion-extension) (see Fig. 14–34); or (3) there is more than 11 degrees of rotation difference compared to that of either adjacent vertebra measured on a resting lateral or flexion-extension roentgenogram (see Fig. 14–35).

The method of immobilization depends on the postreduction status of the injury. Thompson and colleagues have concisely delineated the indications for using nonsurgical and surgical methods for achieving stability [78]. These concepts for managing cervical spine fractures and dislocations may be summarized as follows:

1. Patients with stable compression fractures of the vertebral body, undisplaced fractures of the lamina or lateral masses, or soft tissue injuries without detectable neurologic deficit can be adequately treated with traction and subsequent protection with a cervical brace until healing occurs.

2. Stable, reduced facet dislocation without neurologic deficit can also be treated conservatively in a halo jacket brace until healing has been demonstrated by negative lateral flexion-extension roentgenograms.

3. Unstable cervical spine fractures or fracture-dislocations without neurologic deficit may require either surgical or nonsurgical methods to ensure stability.

4. Absolute indications for surgical stabilization of an unstable injury without neurologic deficits are late instability following closed treatment and flexion-rotation injuries with unreduced locked facets.

5. Relative indications for surgical stabilization in patients with unstable injuries without neurologic deficit are anterior subluxation greater than 20%, certain atlantoaxial fractures or dislocations, and unreduced vertical compression injuries with neck flexion.

6. Cervical spine fractures with complete cord lesions require reduction followed by stabilization by closed or open means, as indicated.

7. Cervical spine fractures with incomplete cord lesions require reduction followed by careful evaluation for surgical intervention.

The fourth and final goal of treatment is rapid and effective rehabilitation started early in the treatment process.

A more specific categorization of athletic injuries to the cervical spine can be made. Specifically, these injuries can be divided into those that occur in the upper cervical spine, the midcervical spine, and the lower cervical spine.

Upper Cervical Spine Fractures and Dislocations

Upper cervical spine lesions involve C1 through C3. Although they rarely occur in sports, several specific injuries that can occur to the upper cervical vertebrae deserve mention [30]. The transverse and alar ligaments are responsible for atlantoaxial stability (Fig. 14–8A). If these structures are ruptured from a flexion injury with translation of C1 anteriorly, the spinal cord can be impinged between a posterior aspect of the odontoid process and the posterior rim of C1 (Fig. 14–8B). The patient gives a history of head trauma and complains of neck pain, particularly with nodding, and may or may

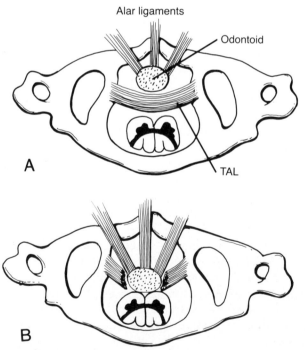

FIGURE 14–8
The atlanto-axial complex as seen from above *(A)*. The disruption of the transverse ligament (TAL) with intact alar ligaments results in C1–C2 instability without cord compression *(B)*. (Redrawn from Hensinger, R. N. Congenital anomalies of the atlantoaxial joint. *In* The Cervical Spine Research Society Editorial Committee. *The Cervical Spine* (2nd ed., p. 242). Philadelphia, J. B. Lippincott, 1989.)

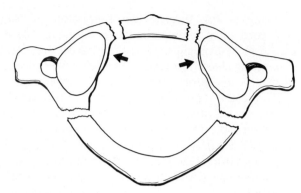

FIGURE 14–9
Schematic representation of the four-part comminuted burst fracture of the atlas as seen from above. (Redrawn from Jefferson, G. Fracture of the Atlas vertebra. *Br J Surg* 7:407–422, 1919. By permission of the Publishers Butterworth-Heinemann Ltd.)

not present with cord signs. Roentgenographically, lateral views of the C1–C2 articulation demonstrate increase of the atlantodens interval (ADI). This interval is normally 3 mm in the adult. With transverse ligament rupture, it may increase up to 10 to 12 mm depending on the status of the alar and accessory ligaments. Note that increase in the ADI may only be seen when the neck is flexed. Fielding states that atlantoaxial fusion may be the "conservative" treatment for this lesion. He recommends posterior C1–C2 fusion using wire fixation and an iliac bone graft [20].

Fractures of the atlas were described by Jefferson in 1920 [38]. These may be of two types: posterior arch fractures and burst fractures. Posterior arch fractures are the more common of these two types, and with a brace support they go on to satisfactory fibrous or bony union. Burst fractures result from an axial load transmitted to the occipital condyles, which then disrupt the integrity of both the anterior and posterior arches of the atlas (Fig. 14–9). Roentgenograms demonstrate bilateral symmetric overhang of the lateral masses of the atlas in relation to the axis, with an increase in the paraodontoid space on the open mouth view. Clinically, the patient characteristically has pain and imitates the nodding motion. These fractures are considered stable when the combined lateral overhang of the atlas measures less than 7 mm. When the transverse diameter of the atlas is 7 mm greater than that of the axis, a transverse ligament rupture should be suspected (Fig. 14–10). Treatment, as recommended by Fielding, includes head-halter traction until muscle spasm resolves, followed by a brace support [20]. If flexion-extension roentgenograms subsequently demonstrate significant instability, fusion may be indicated.

Fractures of the odontoid have been classified into three types by Anderson and D'Alonzo [6].

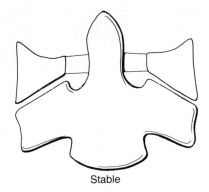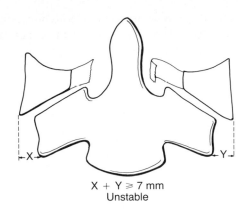

FIGURE 14–10
Illustration of a comminuted Jefferson fracture with both the transverse ligament intact (stable configuration) and a transverse ligament rupture (unstable configuration). (Redrawn from White, A. A., and Panjabi, M. M. *Clinical Biomechanics of the Spine.* Philadelphia, J. B. Lippincott, 1978, p. 204.)

Stable

$X + Y \geq 7$ mm
Unstable

Type I is an avulsion of the tip of the odontoid at the site of the attachment of the alar ligament and is a rare and stable lesion. Type II is a fracture through the base at or just below the level of the superior articular processes. Type III involves a fracture of the body of the axis (Fig. 14–11). When the odontoid is not displaced, planograms may be required to identify the lesion.

The mechanism of odontoid fractures has not been clearly delineated. However, they appear to be due to head impact. All routine cervical spine roentgenographic studies should include the open

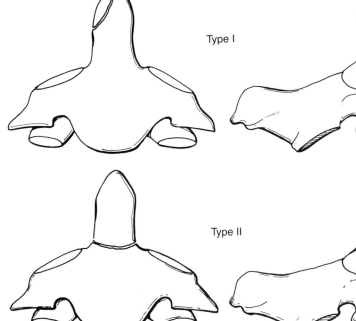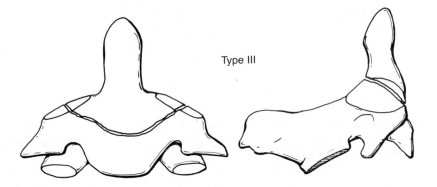

Type I

Type II

Type III

FIGURE 14–11
Illustration of the three types of odontoid fractures in both the AP and lateral planes. Type I is an oblique avulsion fracture from the upper portion of the odontoid. Type II is a fracture of the odontoid process at its base. Type III is an odontoid fracture through the body of C2. (Redrawn from Anderson, L. D., and D'Alonzo, R. T. Fractures of the odontoid process of the axis. *J Bone Joint Surg* 56A:1663–1674, 1974.)

mouth view to identify lesions involving the odontoid as well as the atlas. If these are negative, and if a lesion in this area is suspected, planograms or bending films may further delineate pathologic changes in this area.

Managing type II fractures is a problem. It has been reported that 36% to 50% of these lesions treated initially with plaster casts or reinforced cervical braces fail to unite. Cloward has reported that 85% of his patients heal within 3 months when treated with the halo brace [16].

It is necessary to stabilize fibrous unions or nonunited fractures of the odontoid surgically if they are demonstrated to be unstable on flexion and extension views. Stabilization may be effected either through posterior C1–C2 wire fixation and fusion or anterior fusion of C1–C2 by a dowel graft through the articular facets, as described by Cloward [16].

Fractures through the arch of the axis are also known as traumatic spondylolistheses of C2 or hangman's fractures. These are relatively rare lesions. The mechanism of injury is generally recognized to be hyperextension. This injury is inherently unstable, but it has been shown to heal with predictable regularity without surgical intervention.

Midcervical Spine Fractures and Dislocations

Acute traumatic lesions of the cervical spine at the C3–C4 level are rare and are generally not associated with fractures. These lesions are classified as follows: (1) acute rupture of the C3–C4 intervertebral disc; (2) anterior subluxation of C3 on C4; (3) unilateral dislocation of the joint between the articular processes; and (4) bilateral dislocation of the joint between the articular processes [82, 83, 89].

An episode of transient quadriplegia in an athlete who has sustained head impact but has a negative cervical spine roentgenogram suggests acute rupture of the C3–C4 intervertebral disc. The syndrome of acute anterior spinal cord injury, as described by Schneider and associates [67, 68], may be observed. A cervical myelogram or MRI will substantiate the diagnosis. Anterior discectomy and interbody fusion may be the most effective treatment of this lesion.

Anterior subluxation of C3 on C4 is a result of a shearing force through the intervertebral disc space that disrupts the interspinous ligament as well as the posterior supporting structure. Roentgenograms demonstrate narrowing of the intervertebral disc space, anterior angulation and translation of C3 on C4, an increase in the distance between the spinous processes of the two vertebrae, and instability without fracture of the bony elements (Fig. 14–12).

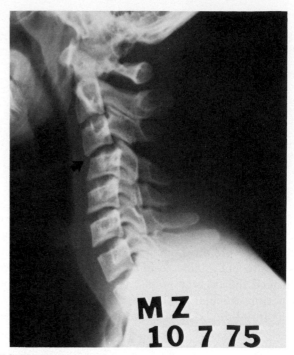

FIGURE 14–12
Roentgenograph demonstrates C3–C4 subluxation as manifested by anterior intervertebral disc space narrowing, anterior angulation, displacement of the superior vertebral body, and fanning of the spinous processes. (From Torg, J. S., Sennett, B., Vegso, J. J., et al. Axial loading injuries to the middle cervical spine: Analysis and classification. *Am J Sports Med* 19(1):17–25, 1991.)

Spinal fusion may be necessary for adequate stabilization in such cases, in contrast to cervical spine instability caused by fracture, in which adequate reduction and subsequent bony healing result in stability. When the patient has posterior instability, posterior fusion is preferable to an anterior interbody fusion.

Unilateral facet dislocation at C3–C4 may result in immediate quadriparesis. This injury involves the intervertebral disc space, the interspinous ligament, the posterior ligamentous supporting structures, and the one facet with resulting rotatory dislocation of C3 on C4 without fracture (Fig. 14–13). At this level, strong skeletal traction does not usually yield a successful reduction, and closed manipulation under general anesthesia is necessary to disengage the locked joint between the articular processes.

Bilateral facet dislocation at the C3–C4 level is a grave lesion (Fig. 14–14). Skeletal traction may not reduce the lesion, and the prognosis for this injury is poor.

Lower Cervical Spine Fractures and Dislocations

Lower cervical spine fractures or dislocations are those involving C4 through C7. In injuries resulting

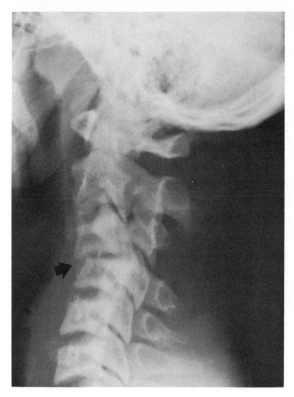

FIGURE 14–13
Unilateral C3–C4 facet dislocation resulting in complete motor and sensory deficit distal to the lesion. There is fanning of the spinous processes of C3 and C4 and more than 20% anterior displacement of the body of C3 on C4. (From Torg, J. S., Sennett, B., Vegso, J. J., et al. Axial loading injuries to the middle cervical spine: Analysis and classification. *Am J Sports Med* 19(1):17–25, 1991.)

from various athletic endeavors, the majority of fractures or dislocations of the cervical spine, with or without neurologic involvement, involve this segment. Although unilateral and bilateral facet dislocations occur, they are relatively rare. The vast majority of severe, athletically incurred cervical spine injuries are fractures of the vertebral body with varying degrees of compression or comminution [43].

Unilateral Facet Dislocations. Unilateral facet dislocations are the result of axial loading, flexion-rotation types of mechanisms. The lesion may be truly ligamentous without any associated vertebral fracture. In such instances, the facet dislocation is stable and is usually associated with neurologic involvement. Roentgenograms demonstrate less than 50% anterior shift of the superior vertebra on the inferior vertebra. Attempts should be made to reduce the facet dislocation by skeletal traction. However, as with similar lesions described at the C3–C4 level, it may not be possible to effect a closed reduction. In this instance, open reduction under direct vision through a posterior approach

with supplemental posterior element bone grafting should be performed.

Bilateral Facet Dislocations. Bilateral facet dislocations are unstable and are almost always associated with neurologic involvement. These injuries are associated with a high incidence of quadriplegia. Lateral roentgenograms demonstrate greater than 50% anterior displacement of the superior vertebral body on the inferior vertebral body. Immediate treatment, as previously described, consists of closed reduction with skeletal traction. Such lesions are generally reducible by skeletal traction and are then treated by halo-brace stabilization and posterior fusion. It should be noted that instability is directly related to the ease with which the lesion is reduced, since the easier it is to reduce, the easier it is to redislocate. If skeletal traction is unsuccessful, either manipulative reduction under sedation or general anesthesia or open reduction under direct vision is recommended. When the dislocation is reduced closed and the reduction is maintained, immobilization should be effected by use of the halo brace for 8 to 12 weeks. Corrective bracing should continue for an additional 4 weeks.

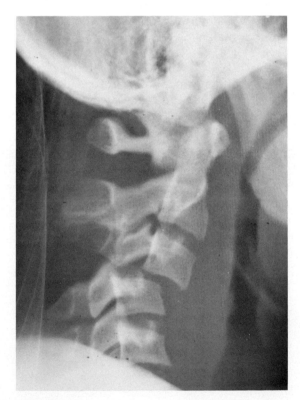

FIGURE 14–14
Bilateral facet dislocation at the C3–C4 level demonstrates anterior angulation as well as translation greater than 50% of the width of the vertebral body associated with spinous fanning. The lesion resulted in quadriplegia. (From Torg, J. S., Sennett, B., Vegso, J. J., et al. Axial loading injuries to the middle cervical spine: Analysis and classification. *Am J Sports Med* 19(1):17–25, 1991.)

Vertebral Body Compression Fractures. Compression fractures of the vertebral body are a result of axial loading. Vertebral body fractures of the cervical spine can be classified into five types [87].

Type I. Simple wedge or vertebral end plate compression fractures of the cervical vertebrae are common injuries that respond to conservative management and rarely if ever are associated with neurologic involvement (Fig. 14–15*A, B*). It is important to differentiate these lesions from compression fractures that are associated with disruption of the posterior element soft tissue supporting structures. The latter lesions are unstable and are frequently associated with neurologic involvement, including quadriplegia.

Type II. An isolated anterior-inferior vertebral body or "teardrop" fracture is without displacement, has intact posterior elements, and is not associated with neurologic involvement (Fig. 14–16*A, B*). This is a relatively stable fracture and may be treated conservatively [41, 81, 90].

Type III. Comminuted burst vertebral body fractures have intact posterior elements, but displacement of bony fragments into the vertebral canal may place the cord in jeopardy. Late settling of the fracture with deformity can occur. Surgical stabilization is recommended (Fig. 14–17).

Type IV. The axial load three part–two plane vertebral body fracture consists of three fracture parts: (1) an anteroinferior teardrop; (2) a sagittal vertebral body fracture; and (3) disruption of the posterior neural arch [32]. This lesion is unstable

and is almost always associated with quadriplegia. Careful evaluation of the routine anteroposterior roentgenogram or CT scan is necessary to appreciate the sagittal vertebral body fracture, a finding that portends a grave prognosis (Fig. 14–18*A–D*).

Type V. Vertebral body three part–two plane compression fracture associated with disruption of posterior elements of an adjacent vertebra. This is an extremely unstable fracture (Fig. 14–19*A, B*).

Cervical Spinal Stenosis With Cord Neurapraxia and Transient Quadriplegia

Characteristically, the clinical picture of cervical spinal cord neurapraxia with transient quadriplegia involves an athlete who sustains an acute transient neurologic episode of cervical cord origin with sensory changes that may be associated with motor paresis involving both arms, both legs, or all four extremities after forced hyperextension, hyperflexion, or axial loading of the cervical spine [80, 91]. Sensory changes include burning pain, numbness, tingling, or loss of sensation; motor changes consist of weakness or complete paralysis. The episodes are transient, and complete recovery usually occurs in 10 to 15 minutes, although in some cases gradual resolution does not occur for 36 to 48 hours. Except for burning paresthesia, neck pain is not present at the time of injury. There is complete return of motor function and full, pain-free cervical motion.

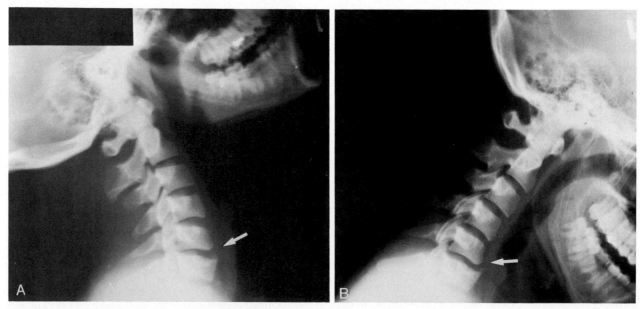

FIGURE 14–15

A and *B*, Type I vertebral body end plate compression fracture involving the superior aspect of C6 *(arrow)*. Extension and flexion views demonstrate absence of evidence of instability. (From Torg, J. S. (Ed.), *Athletic Injuries to the Head, Neck and Face.* Philadelphia, Lea & Febiger, 1982.)

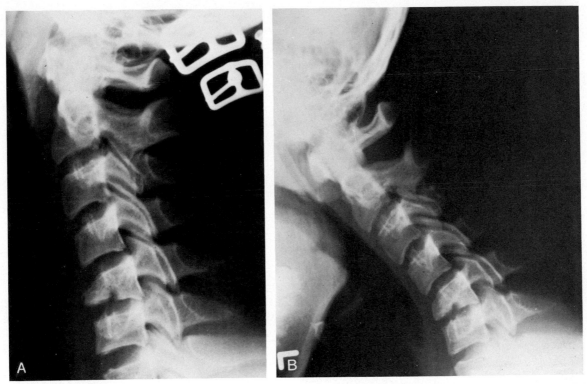

FIGURE 14–16

A, Type II anterior-inferior vertebral body fracture demonstrates the characteristic isolated teardrop fracture, without displacement into the vertebral canal. *B*, Lateral flexion roentgenograms of the lesion demonstrate maintenance of adjacent disc space height as well as a lack of subluxation or spinous process fanning. If there is no disruption of the posterior elements, this is a relatively stable lesion. (From Torg, J. S. (Ed.), *Athletic Injuries to the Head, Neck and Face.* Philadelphia, Lea & Febiger, 1982.)

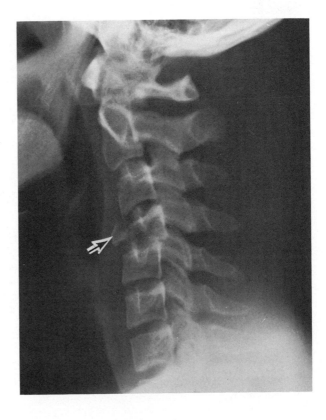

FIGURE 14–17

Type III comminuted burst fracture of C4 with displacement of fragments into the vertebral canal *(arrow).* (From Torg, J. S. (Ed.), *Athletic Injuries to the Head, Neck and Face.* Philadelphia, Lea & Febiger, 1982.)

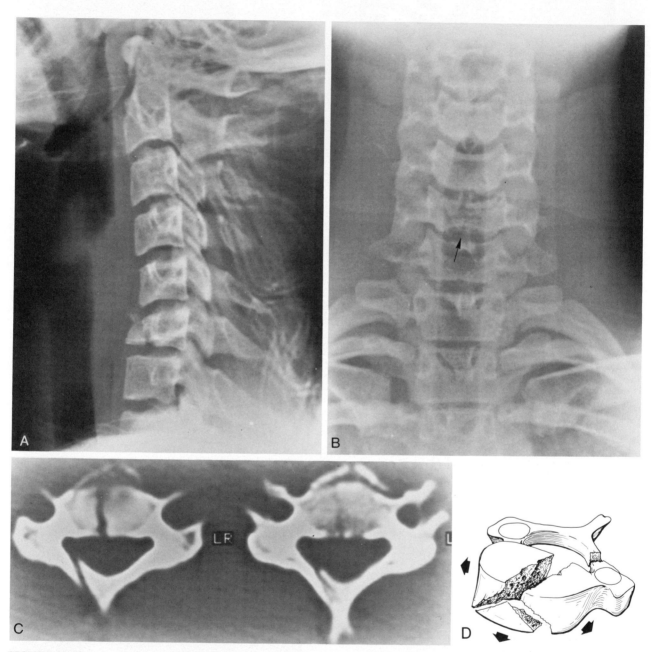

FIGURE 14–18

A, A three part–two plane fracture of C6. The patient had no permanent neurologic sequelae. Lateral view demonstrates prevertebral soft tissue swelling and an anteroinferior fracture fragment of C6 involving the entire vertebral body height (VBH) and one-third of the vertebral body width (VBW). There is approximately 1 mm of posterior displacement of the inferior aspect of the posterior vertebral body. The C6–C7 intervertebral disc space is minimally narrowed posteriorly with associated capsular disruption and "fanning." (From Torg, J. S., et al. The axial load teardrop fracture. *Am J Sports Med* 19(4), 1991.)

B, Frontal view demonstrates a faint, linear radiolucency through the C6 vertebral body indicating a sagittal vertebral body fracture (arrow). There is mild lateral mass displacement. *C*, CT examination demonstrates the sagittal fracture extending completely through the vertebral body with disruption of the lamina on the right. *D*, Diagrammatic representation of the three part–two plane vertebral body compression fracture demonstrates the anteroinferior teardrop as well as the sagittal vertebral body fractures and associated fracture through the lamina.

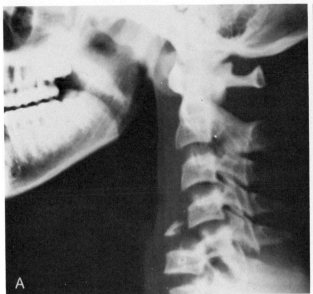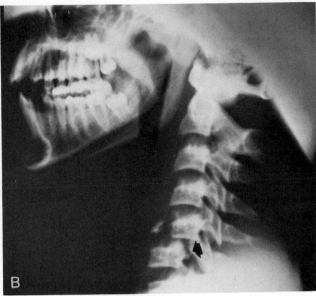

FIGURE 14–19
A, Type V compression fracture of the vertebral body associated with fractures of the neural arch of an adjacent vertebra. Settling and posterior displacement of the superior vertebral segment occur. *B*, Distraction of the superior vertebral segment with skeletal traction permits visualization of fractures through the pedicles (arrow) of C6 in addition to the three part–two plane fracture of the body of C5. (From Torg, J. S. (Ed.), *Athletic Injuries to the Head, Neck and Face.* Philadelphia, Lea & Febiger, 1982.)

Routine x-ray films of the cervical spine show no evidence of fracture or dislocation. However, a demonstrable degree of cervical spinal stenosis is present.

Determination of Spinal Stenosis: Method of Measurement

In order to identify cervical stenosis, a method of measurement is needed. The standard method, the one most commonly employed for determining the sagittal diameter of the spinal canal, involves measuring the distance between the middle of the posterior surface of the vertebral body and the nearest point on the spinolaminar line. Using this technique, Boijsen [8] reported that the average sagittal diameter of the spinal canal from the fourth to the sixth cervical vertebra in 200 healthy individuals was 18.5 mm (range, 14.2 to 23 mm). The target distance he used was 1.4 m. Others have noted that values of less than 14 mm are uncommon and fall below the standard deviation for any cervical segment. Other measurements reported in the literature vary greatly. It is the variations in the landmarks and the methods used to determine the sagittal distance as well as the use of different target distances for roentgenography that have resulted in inconsistencies in the so-called normal values. Therefore, the standard method of measurement for spinal stenosis is a questionable one.

The Ratio Method. An alternative way to determine the sagittal diameter of the spinal canal was devised by Pavlov and colleagues and is called the ratio method [52]. It compares the standard method of measurement of the canal with the anteroposterior width of the vertebral body at the midpoint of the corresponding vertebral body (Fig. 14–20). The actual measurement of the sagittal diameter in millimeters, as determined by the conventional method, is misleading both as reported in the literature and in actual practice because of variations in the target distances used for roentgenography and in the landmarks used for obtaining the measurement. The ratio method compensates for variations in roentgenographic technique because the sagittal diameter of both the canal and the vertebral body is affected similarly by magnification factors. The ratio method is independent of variations in technique, and the results are statistically significant. Using the ratio method of determining the dimension of the canal, a ratio of the spinal canal to the vertebral body of less than 0.80 is indicative of cervical stenosis. We believe that the ratio of the anteroposterior diameter of the spinal canal to that of the vertebral body is a consistent and reliable way to determine cervical stenosis (Fig. 14–21*A*, *B*) in those individuals who have experienced episodes of cervical cord neurapraxia. However, the ratio has a very low predictive value and should not be used as a screening tool.

On the basis of these observations, it may be concluded that the factor that explains the described

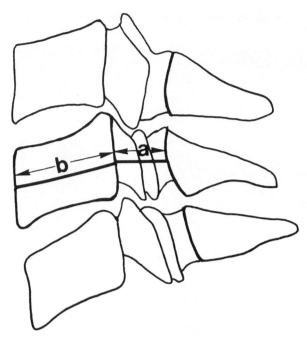

$$ratio = \frac{a}{b}$$

FIGURE 14–20

The ratio of the spinal canal to the vertebral body is the distance from the midpoint of the posterior aspect of the vertebral body to the nearest point on the corresponding spinolaminar line (a) divided by the anteroposterior width of the vertebral body (b). (From Torg, J. S., Pavlov, H., Gennario, S. E., et al. Neurapraxia of the cervical spinal cord with transient quadriplegia. *J Bone Joint Surg* 68A(9):1354–1370, 1986.)

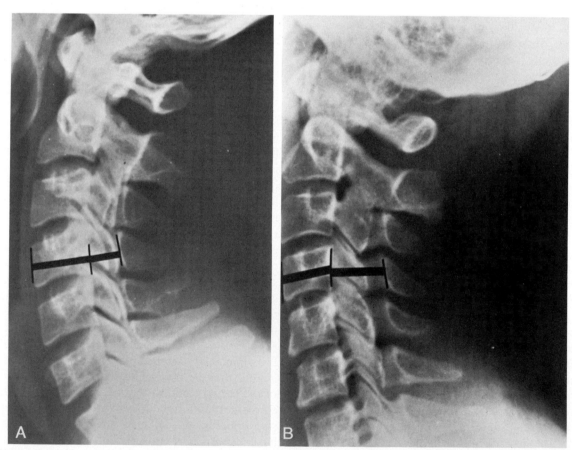

FIGURE 14–21

A and *B*, A comparison between the ratio of the spinal canal to the vertebral body of a stenotic patient versus that of a control subject is demonstrated on lateral roentgenograms of the cervical spine. The ratio is approximately 1:2 (0.50) in the stenotic patient compared with 1:1 (1.00) in the control subject. (From Torg, J. S., Pavlov, H., Gennario, S. E., et al. Neurapraxia of the cervical spinal cord with transient quadriplegia. *J Bone Joint Surg* 68A(9):1354–1370, 1986.)

neurologic picture of cervical spinal cord neurapraxia is diminution of the anteroposterior diameter of the spinal canal, either as an isolated observation or in association with intervertebral disk herniation, degenerative changes, post-traumatic instability, or congenital anomalies. In instances of developmental cervical stenosis, forced hyperflexion, or hyperextension of the cervical spine, further decreases the caliber of an already narrow canal, as explained by the pincer mechanism of Penning [53] (Fig. 14–22). In patients whose stenosis is associated with osteophytes or a herniated disc, direct pressure can occur, again when the spine is forced in the extremes of flexion and extension. It is further postulated that with an abrupt but brief decrease in the anteroposterior diameter of the spinal canal, the cervical cord is mechanically compressed, causing transient interruption of either motor or sensory function, or both, distal to the lesion. The neurologic aberration that results is transient and completely reversible.

A review of the literature has revealed few reported cases of transient quadriplegia occurring in athletes. Attempts to establish the incidence indicate that the problem is more prevalent than may be expected. Specifically, in a population of 39,377 exposed participants, the reported incidence of transient paresthesia in all four extremities was 6 per 10,000, whereas the reported incidence of paresthesia associated with transient quadriplegia was 1.3 per 10,000 in the one football season surveyed. From these data, it may be concluded that the prevalence of this problem is relatively high and that an awareness of the etiology, manifestations, and appropriate principles of management is warranted.

Characteristically, after an episode of cervical spinal cord neurapraxia with or without transient quadriplegia, the first question raised concerns the advisability of restricting activity. In an attempt to address this problem, 117 young athletes have been interviewed who sustained cervical spine injuries associated with complete permanent quadriplegia while playing football between the years 1971 and 1984. None of these patients recalled a prodromal experience of transient motor paresis. Conversely, none of the patients in this series who had experienced transient neurologic episodes subsequently sustained an injury that resulted in permanent neurologic injury. On the basis of these data, it is concluded that a young patient who has had an episode of cervical spinal cord neurapraxia with or without an episode of transient quadriplegia is not predisposed to permanent neurologic injury because of it.

With regard to restrictions on activity, no definite recurrence patterns have been identified to establish firm principles in this area. However, athletes who have this syndrome associated with demonstrable cervical spinal instability or with acute or chronic degenerative changes should not be allowed further participation in contact sports. Athletes with developmental spinal stenosis or spinal stenosis associated with congenital abnormalities should be treated on an individual basis. Of the six youngsters with obvious cervical stenosis who returned to football, three had a second episode and withdrew from the activity, and three returned without any problems at 2-year follow-up. The data clearly indicate that individuals with developmental spinal stenosis, with or without associated symptoms, are not predisposed to more severe injuries with associated permanent neurologic sequelae.

PREVENTION

Athletic injuries to the cervical spine that result in injury to the spinal cord are infrequent but

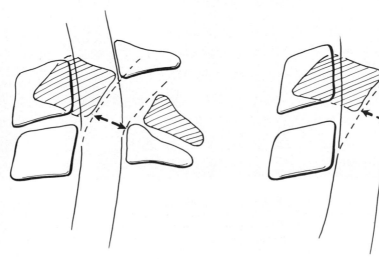

FIGURE 14–22
The pincers mechanism, as described by Penning, occurs when the distance between the posteroinferior margin of the superior vertebral body and the anterosuperior aspect of the spinolaminar line of the subjacent vertebra decreases with hyperextension, resulting in compression of the cord. With hyperflexion, the anterosuperior aspect of the spinolaminar line of the superior vertebra and the posterosuperior margin of the inferior vertebra would be the "pincers." (Redrawn from Torg, J. S., Pavlov, H., Gennario, S. E., et al: Neurapraxia of the cervical spinal cord with transient quadriplegia. *J Bone Joint Surg* 68A(9):1354–1370, 1986.)

catastrophic events. Accurate descriptions of the mechanism or mechanisms responsible for a particular injury transcend simple academic interest. Before preventive measures can be developed and implemented, identification of the mechanisms involved in the production of the particular injury is necessary. Because the nervous system is unable to recover significant function after severe trauma, prevention assumes a most important role when considering these injuries.

Injuries resulting in spinal cord damage have been associated with football [3, 18, 22, 84–88], water sports [1, 2, 17, 39, 47, 63, 75], wrestling [97], rugby [14, 45, 60, 62, 64, 70, 71], trampolining [21, 34, 73, 79], and ice hockey [74, 76]. The use of epidemiologic data, biomechanical evidence, and cine-matographic analysis has (1) defined and supported the involvement of axial load forces in cervical spine injuries occurring in football; (2) demonstrated the success of appropriate rule changes in the prevention of these injuries; and (3) emphasized the need for employment of epidemiologic methods to prevent cervical spine and similar severe injuries in other high-risk athletic activities.

Data on cervical spine injuries resulting from participation in football have been compiled by a national registry since 1971 [84–86]. Analysis of the epidemiologic data and cinematographic documentation clearly demonstrates that the majority of cervical fractures and dislocations were due to axial loading (Fig. 14–23). On the basis of these observations, rule changes banning both deliberate

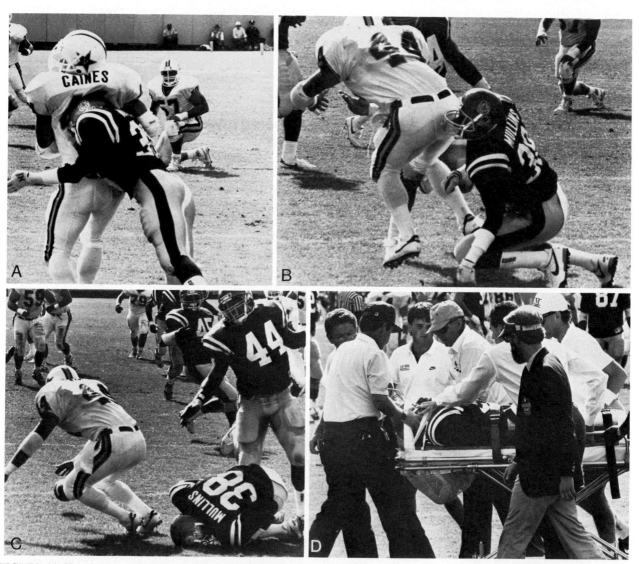

FIGURE 14–23
A, With the advent of the polycarbonate helmet–face mask protective device, use of the top or crown of the helmet as the initial point of contact in blocking and tackling became prevalent. Contact is made, the head abruptly stops, the momentum of the body continues, and the cervical spine is literally crushed between the two. In this instance the fracture-dislocation transected the spinal cord. *B,* The injured player collapses, having been rendered quadriplegic. *C,* Further collapse is noted. *D,* The player is evacuated on a spine board and stretcher. (Photos by Randy Green, Vanderbilt Stundent Communications.)

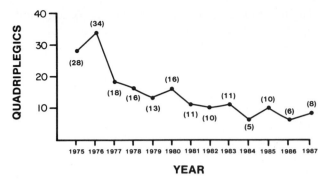

FIGURE 14–24

The yearly incidence of permanent cervical quadriplegia for all levels of participation (1975 to 1987) decreased dramatically in 1977 following initiation of the rule changes prohibiting the use of head-first tackling and blocking techniques.

"spearing" and the use of the top of the helmet as the initial point of contact in making a tackle were implemented at the high school and college levels. Subsequently, a marked decrease in cervical spine injury rates has occurred. The occurrence of permanent cervical quadriplegia decreased from 34 in 1976 to five in the 1984 season (Fig. 14–24).

Identification of the cause of cervical quadriplegia and prevention of cervical quadriplegia resulting from football involve four areas: (1) the role of the helmet–face mask protective system; (2) the concept of the axial loading mechanism of injury; (3) the effect of the 1976 rule changes banning spearing and the use of the top of the helmet as the initial point of contact in tackling; and (4) the necessity for continued research, education, and enforcement of rules.

The protective capabilities provided by the modern football helmet have resulted in the advent of playing techniques that have placed the cervical spine at risk of injury with associated catastrophic neurologic sequelae. Available cinematographic and epidemiologic data clearly indicate that cervical spine injuries associated with quadriplegia occurring as a result of football are not hyperflexion accidents [36]. Instead, they are due to purposeful axial loading of the cervical spine as a result of spearing and head-first playing techniques (Fig. 14–25A–D). As an etiologic factor, the modern helmet–face mask system is secondary, contributing to these injuries

FIGURE 14–25

A, Subject (No. 37, foreground) lines up in front of ball carrier in preparation for tackling. B, Preimpact position shows tackler about to ram ball carrier with the crown of his helmet. C, At impact, contact is made with the top of the helmet. Although the neck is slightly flexed, it is clearly not hyperflexed. The major force vector is transmitted along the axial alignment of the cervical spine. D, The tackler recoils following impact.

because of its protective capabilities that have permitted the head to be used as a battering ram, thus exposing the cervical spine to injury.

Classically, the role of hyperflexion has been emphasized in cervical spine trauma whether the injury was due to a diving accident, trampolining, rugby, or American football [69]. Epidemiologic and cinematographic analyses have established that most cases of cervical spine quadriplegia that occur in football have resulted from axial loading [84–86, 88]. Far from being an accident or an untoward event, techniques are deliberately used that place the cervical spine at risk of catastrophic injury. Recent laboratory observations also indicate that athletically induced cervical spine trauma results from axial loading [7, 37, 48, 56, 57].

In the course of a collision activity, such as tackle football, most energy inputs to the cervical spine are effectively dissipated by the energy-absorbing capabilities of the cervical musculature through controlled lateral bending, flexion, or extension motion. However, the bones, discs, and ligamentous structures can be injured when contact occurs on the top of the helmet when the head, neck, and trunk are positioned in such a way that forces are transmitted along the longitudinal axis of the cervical spine.

When the neck is in the anatomic position, the cervical spine is extended due to normal cervical lordosis. When the neck is flexed to 30 degrees, the cervical spine straightens (Fig. 14–26*A,B*). In axial

loading injuries, the neck is slightly flexed, and normal cervical lordosis is eliminated, thereby converting the spine into a straight segmented column. Assuming the head, neck, and trunk components to be in motion, rapid deceleration of the head occurs when it strikes another object, such as another player, trampoline bed, or lake bottom. This results in the cervical spine being compressed between the rapidly deaccelerated head and the force of the oncoming trunk. When the maximum amount of vertical compression is reached, the straightened cervical spine fails in a flexion mode, and fracture, subluxation, or unilateral or bilateral facet dislocation can occur (Fig. 14–27*A–E*) [51].

Refutation of the "freak accident" concept with the more logical principle of cause and effect has been most rewarding in dealing with problems of football-induced cervical quadriplegia. Definition of the axial loading mechanism in which a football player, usually a defensive back, makes a tackle by striking an opponent with the top of his helmet has been a key element in this process. Implementation of rules changes and coaching techniques eliminating the use of the head as a battering ram have resulted in a dramatic reduction in the incidence of quadriplegia since 1976. We believe that the majority of athletic injuries to the cervical spine associated with quadriplegia also occur as a result of axial loading [51, 87].

Tator and associates [75] studied 38 acute spinal

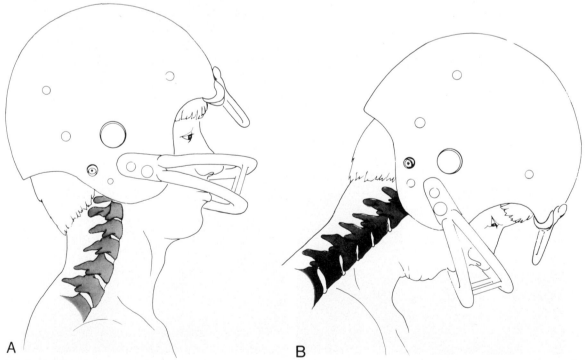

A   B

FIGURE 14–26
A, When the neck is in a normal upright anatomic position, the cervical spine is slightly extended because of the natural cervical lordosis. *B*, When the neck is flexed slightly, to approximately 30 degrees, the cervical spine is straightened and converted into a segmented column

FIGURE 14–27
A–E, Biomechanically, a straight cervical spine responds to an axial load force like a segmented column. Axial loading of the cervical spine first results in compressive deformation of the intervertebral discs *(A* and *B).* As the energy input continues and maximum compressive deformation is reached, angular deformation and buckling occur. The spine fails in a flexion mode *(C)* with resulting fracture, subluxation, or dislocation *(D* and *E).* Compressive deformation to failure with a resultant fracture, dislocation, or subluxation occurs in as little as 8.4 msec [12]. (From Torg, J. S., Glasgow, S. G. *Clin J Sports Med* 1:12–27, 1991.)

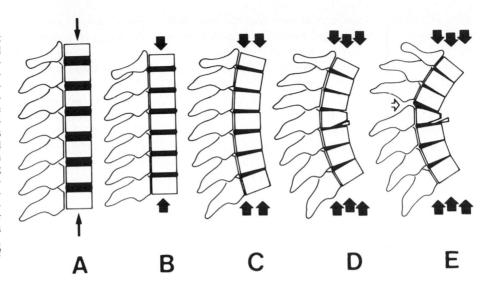

A   B   C   D   E

cord injuries due to diving accidents and observed that, "In most cases the cervical spine was fractured and the spinal cord crushed. The top of the head struck the bottom of the lake or pool." Scher [62], reporting on vertex impact and cervical dislocation in rugby players, observed that, "When the neck is slightly flexed, the spine is straight. If significant force is applied to the vertex when the spine is straight, the force is transmitted down the long axis of the spine. When the force exceeds the energy-absorbing capacity of the structures involved, cervical spine flexion and dislocation will result." Tator and Edmonds [74] reported on the results of a national questionnaire survey by the Canadian Committee on the Prevention of Spinal Injuries due to Hockey, which recorded 28 injuries involving the spinal cord, 17 of which resulted in complete paralysis. They noted that in this series the most common mechanism involved was a check in which the injured players struck the boards "with the top of their heads, while their necks were slightly flexed." Reports in the recent literature that deal with the mechanism of injury involved in cervical spine injuries resulting from water sports (diving), rugby, and ice hockey support our thesis [1–3, 14, 17, 18, 21, 22, 34, 39, 45, 47, 60, 62–64, 70, 71, 73–76, 79, 84–88, 97].

CRITERIA USED TO GAUGE RETURN TO CONTACT ACTIVITIES FOLLOWING CERVICAL SPINE INJURY

Injury to the cervical spine and associated structures as a result of participation in competitive athletic and recreational activities is not uncommon.

It appears that the frequency of these various injuries is inversely proportional to their severity. Whereas Albright [3] has reported that 32% of college football recruits sustained "moderate" injuries while in high school, catastrophic injuries with associated quadriplegia occur in less than 1 in 100,000 participants per season at the high school level. As indicated, the variety of possible lesions is considerable and the severity variable. The literature dealing with diagnosis and treatment of these problems is considerable. However, conspicuously absent is a comprehensive set of standards or guidelines for establishing criteria for permitting or prohibiting return to contact sports (boxing, football, ice hockey, lacrosse, rugby, wrestling) following injury to the cervical spinal structures. The explanation for this void appears to be twofold. First, the combination of a litigious society and the potential for great harm should things go wrong makes "no" the easiest and perhaps most reasonable advice. Second and perhaps most important, with the exception of the matter of transient quadriplegia, there is a lack of credible data pertaining to postinjury risk factors. Despite a lack of credible data, this chapter will attempt to establish guidelines to assist the clinician as well as the patient and his parents in the decision-making process [92].

Cervical spine conditions requiring a decision as to whether or not participation in contact activities is advisable and safe can be divided into two categories: (1) congenital or developmental conditions, and (2) post-traumatic conditions. Each condition has been determined to present either no contraindication, relative contraindications, or an absolute contraindication on the basis of a variety of parameters. Information compiled from over 1200 cervical spine injuries documented by the National Football Head and Neck Injury Registry has provided insight into whether various conditions may or may not predispose to more serious injury [84–86]. A review of the literature in several instances provides significant data for a limited number of specific conditions. Analysis of many conditions predicated on an understanding of recognized injury mechanisms has permitted categorization on the basis of "educated" conjecture. And last, much reliance has been placed on personal experience that must be regarded as anecdotal.

The structure and mechanics of the cervical spine enable it to perform three important functions. First, it supports the head as well as the variety of soft tissue structures of the neck. Second, by virtue of segmentation and configuration, it permits multiplanar motion of the head. Third, and most important, it serves as a protective conduit for the spinal cord and cervical nerve roots. A condition that impedes or prevents the performance of any of the three functions in a pain-free manner either immediately or in the future is unacceptable and constitutes a contraindication to participation in contact sports.

The following proposed criteria for return to contact activities in the presence of cervical spine abnormalities or following injury are intended only as guidelines. It is fully acknowledged that for the most part they are at best predicated on anecdotal experience, and no responsibility can be assumed for their implementation.

Critical to the application of these guidelines is the implementation of coaching and playing techniques that preclude the use of the head as the initial point of contact in a collision situation. Exposure of the cervical spine to axial loading is an invitation to disaster and makes all safety standards meaningless.

Congenital Conditions

Odontoid Anomalies. Hensinger [36] has stated that "patients with congenital anomalies of the odontoid are leading a precarious existence. The concern is that a trivial insult superimposed on already weakened or compromised structure may be catastrophic." This concern became a reality during the 1989 football season when an 18-year-old high school player was rendered a respiratory-dependent quadriplegic while making a head tackle that was vividly demonstrated on the game video. Postinjury roentgenograms revealed an os odontoidium with marked C1–C2 instability (Fig. 14–28*A,B*). Thus, the presence of odontoid agenesis, odontoid hypo-

plasia, or os odontoidium are all absolute contraindications to participation in contact activities.

Spina Bifida Occulta. This is a rare, incidental roentgenographic finding that presents no contraindication.

Atlanto-occipital Fusion. This rare condition is characterized by partial or complete congenital fusion of the bony ring of the atlas to the base of the occiput. Signs and symptoms are referable to the posterior columns due to cord compression by the posterior lip of the foramen magnum and usually occur in the third or fourth decade. They usually begin insidiously and progress slowly, but sudden onset or instant death has been reported. Atlanto-occipital fusion as an isolated entity or coexisting with other abnormalities constitutes an absolute contraindication to participation in contact activities.

Klippel-Feil Anomaly. This eponym is applied to congenital fusion of two or more cervical vertebrae. For purposes of this discussion, the variety of abnormalities can be divided into two groups: type I—mass fusion of the cervical and upper thoracic vertebrae (Fig. 14–29); and Type II—fusion of only one or two interspaces (Fig. 14–30). To be noted, a variety of associated congenital problems have been associated with congenital fusion of the cervical vertebrae and include pulmonary, cardiovascular, and urogenital problems. Pizzutillo [54] has pointed out that "children with congenital fusion of the cervical spine rarely develop neurologic problems or signs of instability." However, he further states that "the literature reveals more than 90 cases of neurologic problems . . . that developed as a consequence of occipital cervical anomalies, late instability, disc disease, or degenerative joint disease." These reports included cervical radiculopathy, spasticity, pain, quadriplegia and sudden death. Also, "more than two thirds of the neurologically involved patients had single level fusion of the upper area, whereas many cervical patients with extension fusions of five to seven levels had no associated neurologic loss." Despite this, a type I lesion, a mass fusion, constitutes an absolute contraindication to participation in contact sports. A type II lesion with fusion of one or two interspaces with associated limited motion or associated occipitocervical anomalies, involvement of C2, instability, disc disease, or degenerative changes also constitutes an absolute contraindication to participation. On the other hand, a type II lesion involving fusion of one or two interspaces at C3 and below in an individual with a full cervical range of motion and an absence of occipitocervical anomalies, instability, disc disease,

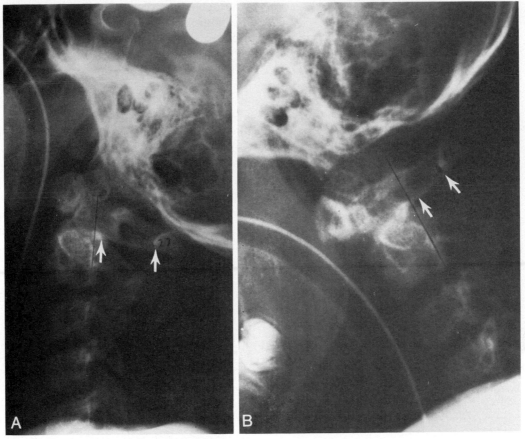

FIGURE 14–28
Inherent instability at C1 in a patient with an os odontoidium. This condition resulted in respiratory-dependent quadriplegia following a spear tackle by this 18-year-old high school football player. The reduction in the space available for the cord is vividly demonstrated by the *(A)* lateral extension and *(B)* flexion views postinjury. (From Torg, J. S., Glasgow, S. G. *Clin J Sports Med* 1:12–27, 1991.)

FIGURE 14–29
Type I Klippel-Feil deformity with multiple level fusions and deformities as demonstrated on the lateral roentgenogram. (From Torg, J. S., Glasgow, S. G. *Clin J Sports Med* 1:12–27, 1991.)

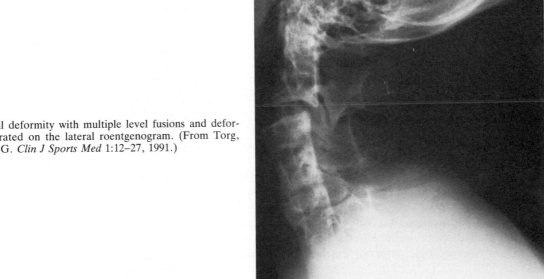

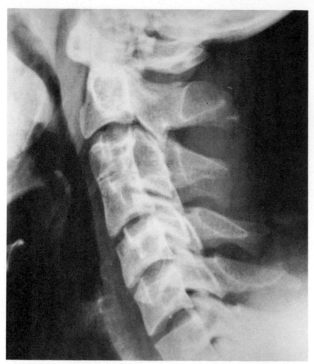

FIGURE 14–30
Type II Klippel-Feil deformity with a one-level congenital fusion at C3–C4 involving both the vertebral bodies and the lateral masses. (From Torg, J. S., Glasgow, S. G. *Clin J Sports Med* 1:12–27, 1991.)

or degenerative changes should present no contraindication.

Developmental Conditions

Developmental Narrowing (Stenosis) of the Cervical Spinal Canal. This condition and its association with cervical cord neurapraxia and transient quadriplegia has been well defined [80, 91]. The definition of narrowing or stenosis as a cervical segment with one or more vertebrae that have a canal-vertebral body ratio of 0.8 or less is predicated on the fact that 100% of all reported clinical cases have fallen below this value at one or more levels. To be noted, 12% of asymptomatic controls also fell below the 0.8 level, as did 32% of asymptomatic professional and 34% of asymptomatic college players. In the group of reported symptomatic players, there was in every instance complete return of neurologic function, and in those who continued with contact activities recurrence was not predictable.

Clearly, the presence of developmental narrowing of the cervical spinal canal does not predispose to permanent neurologic injury. Eismont and colleagues [19] have indicated, on the basis of experience of cervical fractures or dislocations resulting from automobile accidents, that the degree of neurologic impairment was inversely related to the anteroposterior diameter of the canal [19]. Due to the all or nothing pattern of axial load football spine injuries, this phenomenon has not been observed in sports-related injuries.

The presence of a canal-vertebral body ratio of 0.8 or less is not a contraindication to participation in contact activities in asymptomatic individuals. We further recommend against preparticipation screening roentgenograms in asymptomatic players. Such studies will not contribute to safety, are not cost-effective, and will only contribute to the hysteria surrounding this issue.

In individuals with a ratio of 0.8 or less who experience either motor or sensory manifestations of cervical cord neurapraxia there is a relative contraindication to return to contact activities. In these

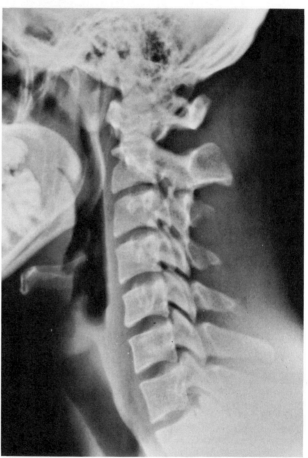

FIGURE 14–31
The ratio of the spinal canal to the vertebral body is the distance from the midpoint of the posterior aspect of the vertebral body to the nearest point on the corresponding spinolaminar line divided by the anteroposterior width of the vertebral body. A ratio of less than 0.8 indicates the presence of developmental narrowing (stenosis). Lateral roentgenogram of a 20-year-old intercollegiate football player who had one episode of transient quadriplegia that lasted 10 minutes following a hyperflexion injury. The canal–vertebral body ratios are narrow from C3 through C7. Specifically, the ratio at C4 measures 0.6. This player returned to active playing for two seasons without a recurrence. (From Torg, J. S., Glasgow, S. G. *Clin J Sports Med* 1:12–27, 1991.)

instances, each case must be determined on an individual basis depending on the understanding of the player and his parents and their willingness to accept any presumed theoretical risk (Fig. 14–31).

An absolute contraindication to continued participation applies to those individuals who experience a documented episode of cervical cord neurapraxia associated with any of the following:

1. Ligamentous instability
2. Intervertebral disc disease
3. Degenerative changes
4. MRI evidence of cord defects or swelling
5. Symptoms or positive neurologic findings lasting more than 36 hours
6. More than one recurrence

"Spear Tackler's Spine." Analysis of material recently received by the National Football Head and Neck Injury Registry has identified a subset of football players with "spear tackler's spine." The entity consists of (1) developmental narrowing (stenosis) of the cervical canal; (2) persistent straightening or reversal of the normal cervical lordotic curve on an erect lateral roentgenogram obtained in the neutral position; (3) concomitant preexisting post-traumatic roentgenographic abnormalities of the cervical spine; and (4) documentation of the individual employing spear tackling technique (Fig. 14–32*A, B*). In two instances in which preinjury

roentgenograms as well as video documentation of axial loading of the spine due to spear tackling were available, a C3–C4 bilateral facet dislocation resulted in one instance and C4–C5 fracture dislocation in the other, both players being rendered quadriplegic. It is postulated that the straightened "segmented column" alignment of the cervical spine combined with head-first tackling techniques predisposed these individuals to an axial loading injury of the cervical segment. Thus, this combination of factors constitutes an absolute contraindication of further participation in collision sports.

Traumatic Conditions of the Upper Cervical Spine (C1–C2)

The anatomy and mechanics of the C1–C2 segments of the cervical spine differ markedly from those of the middle or lower segments. Lesions with any degree of occipital or atlantoaxial instability portend a potentially grave prognosis. Thus, almost all injuries involving the upper cervical segment that involve a fracture or ligamentous laxity are an absolute contraindication to further participation in contact activities (Fig. 14–33*A,B*). Healed, nondisplaced Jefferson fractures, healed type I and type II odontoid fractures, and healed lateral mass fractures of C2 constitute relative contraindications providing

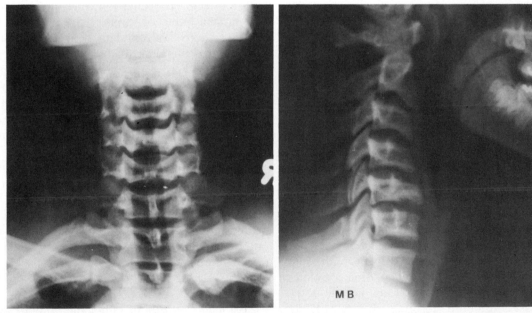

FIGURE 14–32
Roentgenograms of a 19-year-old intercollegiate linebacker with spear tackler's spine. On the anteroposterior view *(A)* the cervical spine is noted to be tilted toward his left. This represents a wry neck attitude frequently seen in those with either acute or chronic cervical injury. The lateral view *(B)* demonstrates several manifestations of spear tackler's spine: (1) a cervical kyphosis, (2) developmental narrowing of the cervical canal, and (3) an old compression injury of C5. The kyphotic deformity was fixed in both flexion and extension. He subsequently sustained a bilateral C3–4 facet dislocation and was rendered quadriplegic as a result of spear tackling.

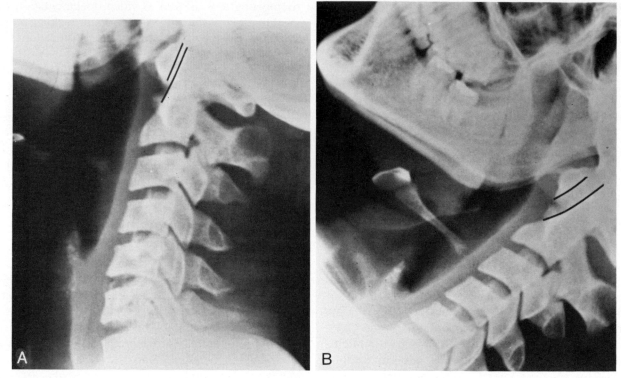

FIGURE 14–33
The atlas-dens interval (ADI) is the distance on the lateral roentgenogram between the anterior aspect of the dens and the posterior aspect of the anterior ring of the atlas. In children, the ADI should not exceed 4.0 mm, whereas the upper limit in the normal adult is less than 3.0 mm. C1–C2 instability is vividly demonstrated in the above extension *(A)* and flexion *(B)* views. (From Torg, J. S., Glasgow, S. G. *Clin J Sports Med* 1:12–27, 1991.)

the patient is pain-free, has a full range of cervical motion, and has no neurologic findings.

Because of the uncertainty of the results of cervical fusion, the gracile configuration of C1, and the importance of the alar and transverse odontoid ligaments, fusion for instability of the upper segment constitutes an absolute contraindication regardless of how successful the fusion appears roentgenographically.

Traumatic Conditions of the Middle and Lower Cervical Spine

Ligamentous Injuries. The criteria of White and Panjabi for defining clinical instability were intended to help establish indications for surgical stabilization (Figs. 14–34 and 14–35) [95, 96]. However, although the limits of displacement and angulation correlated with disruption of known structures, no one determinant was considered absolute. In view of the observations of Albright and colleagues [3] that 10% (7 of 75) of the college freshmen in their study demonstrated "abnormal motion," as well as on the basis of our own experience, it appears that in many instances some degree of "minor instability" exists in populations of both high school and college

football players without apparently leading to adverse effects. The question, of course, is what are the upper limits of "minor" instability? Unfortunately, there are no data available relating this question to the clinical situation that allow reliable standards. Clearly, however, lateral roentgenograms that demonstrate more than 3.5 mm of horizontal displacement of either vertebra in relation to another or more than 11 degrees of rotation than either adjacent vertebra represent an absolute contraindication to further participation in contact activities (Fig. 14–36*A*,*B*). With regard to lesser degrees of displacement or rotation, further participation in sports enters the realm of "trial by battle," and such situations can be considered a relative contraindication depending on such factors as level of performance, physical habitus, position played (i.e., interior lineman vs. defensive back), and so on.

Fractures

An *acute fracture* of either the body or posterior elements with or without associated ligamentous laxity constitutes an absolute contraindication to participation.

The following healed stable fractures in an asymptomatic patient who is neurologically normal and has a full range of cervical motion can be considered

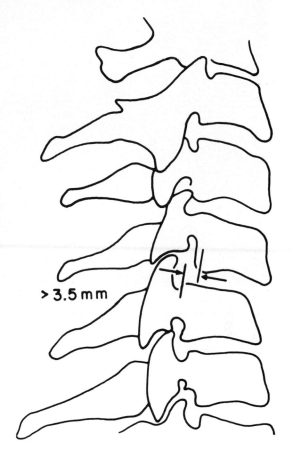

FIGURE 14–34

The method for determining translatory displacement, as described by White and colleagues [95]. Using the posteroinferior angle of the superior vertebral body as one point of reference and the posterosuperior angle of the vertebral body below, the distance between the two in the sagittal plane is measured. A distance of 3.5 mm or greater is suggestive of clinical instability. (From White, A. A., Johnson, R. M., Panjabi, M. M. et al. *Clin Orthop* 109:85, 1975.)

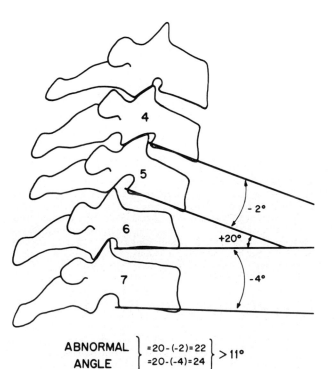

FIGURE 14–35

Abnormal angulation between two vertebrae at any one interspace is determined by comparing the angle formed by the projection of the inferior vertebral body borders with that of either the vertebral body above or the vertebral body below. If the angle at the interspace in question is 11 degrees or greater than either adjacent interspace, it is considered by White and associates [95] as clinical instability. (From White, A. A., Johnson, R. M., Panjabi, M. M. et al. *Clin Orthop* 109:85, 1975.)

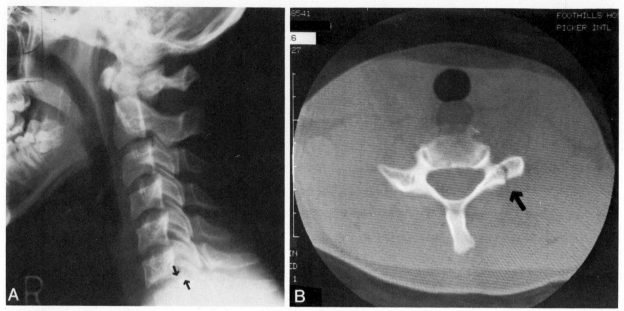

FIGURE 14–36

A, Lateral roentgenogram of the cervical spine in the erect neutral position of a 21-year-old college football player demonstrates anterior translation of C6 on C7 of greater than 3.5 mm (arrows). *B*, A CT scan of C6 in the sagittal plane demonstrates a fracture through the lateral mass (arrow). Persistent displacement despite healing of the fracture constitutes an absolute contraindication to further participation in contact sports. (From Torg, J. S., Glasgow, S. G. *Clin J Sports Med* 1:12–27, 1991.)

to present *no contraindication* to participation in contact activities:

1. Stable compression fractures of the vertebral body without a sagittal component on anteroposterior roentgenograms and without involvement of either the ligamentous or the posterior bony structures (Fig. 14–37).

2. A healed stable end plate fracture without a sagittal component on anteroposterior roentgenograms or involvement of the posterior or bony ligamentous structures (Fig. 14–38*A,B*).

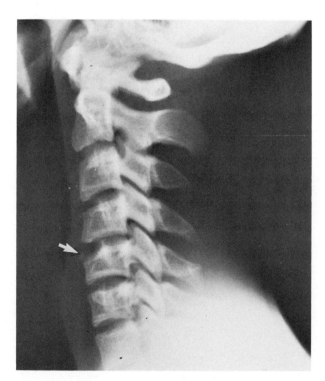

FIGURE 14–37

Lateral roentgenogram of the cervical spine taken in the erect neutral position demonstrates an anterosuperior compression defect in the vertebral body of C5 (arrow). This limbus deformity resulted from a prior compression injury to the ring epiphysis. There is no evidence of angulation, displacement, or indication of any adherent stability of the spine. Such a radiographic finding would not constitute a contraindication to further participation. (From Torg, J. S., Glasgow, S. G. *Clin J Sports Med* 1:12–27, 1991.)

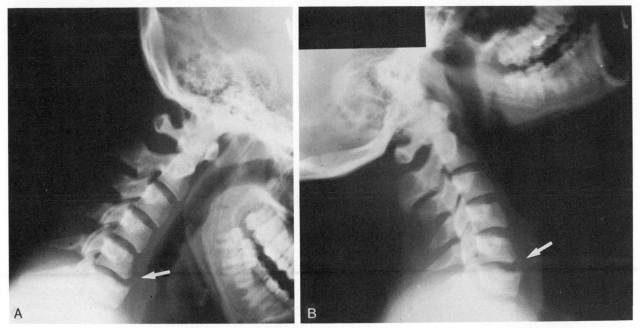

FIGURE 14–38
Lateral flexion *(A)* and extension *(B)* views of a healed, stable end plate fracture involving the superior aspect of C6 in a 22-year-old intercollegiate football player. The injury had occurred 4 years earlier and consisted of a sore neck; he missed two games but did not have a radiograph taken. There were no subsequent problems despite participation in high school and college varsity football. (From Torg, J. S., Glasgow, S. G. *Clin J Sports Med* 1:12–27, 1991.)

3. Healed spinous process "clay shoveler" fractures.

Relative contraindications apply to the following healed stable fractures in individuals who are asymptomatic and neurologically normal and have a full pain-free range of cervical motion:

1. Stable displaced vertebral body compression fractures without a sagittal component on antero-posterior roentgenograms. The propensity for these fractures to settle causing increased deformity must be considered and carefully followed (Fig. 14–39A,B).

2. Healed stable fractures involving the elements of the posterior neural ring in individuals who are asymptomatic, neurologically normal, and have a full pain-free range of cervical motion (Fig. 14–40A,B). In evaluating radiographic and imaging studies to find the location and subsequent healing of a posterior neural ring fracture it is important to understand that a rigid ring cannot break in one location [72]. Thus, healing of paired fractures of the ring must be demonstrated.

Absolute contraindications to further participation in contact activities exist in the presence of the following fractures:

1. Vertebral body fracture with a sagittal component (Fig. 14–41A–C and 14–42)

2. Fracture of the vertebral body with or without displacement with associated posterior arch fractures or ligamentous laxity (Fig. 14–43)

3. Comminuted fractures of the vertebral body with displacement into the spinal canal

4. Any healed fracture of either the vertebral body or the posterior components with associated pain, neurologic findings, and limitation of normal cervical motion

5. Healed displaced fractures involving the lateral masses with resulting facet incongruity

Intervertebral Disc Injury

There is no contraindication to participation in contact activities in individuals with a healed anterior or lateral disc herniation treated conservatively (Fig. 14–44) or in those requiring an intervertebral discectomy and interbody fusion for a lateral or central herniation who have a solid fusion, are asymptomatic and neurologically negative, and have a full pain-free range of motion (Fig. 14–45A,B).

A *relative contraindication* exists in individuals with either conservatively or surgically treated disc disease with residual facet instability. An *absolute contraindication* exists in the following situations:

1. Acute central disc herniation (Fig. 14–46)

Text continued on page 458

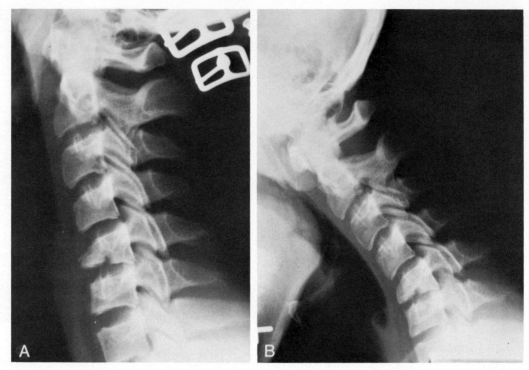

FIGURE 14–39

A, Lateral roentgenogram of the cervical spine taken while in a cervical brace demonstrates a displaced compression fracture of the vertebral body of C5. Notable is the fact that there is no associated angulation, displacement, intervertebral disc space narrowing, facet incongruity, or fanning of the spinous processes. *B*, Lateral flexion view demonstrates pathologic angulation as defined by White and associates [95]. There is no translation, disc space narrowing, facet incongruity, or fanning of the spinous processes, suggesting a stable lesion. The increased angulation is attributed to the deformity of the vertebral body. Assuming that progression of the deformity or evidence of instability occurred and that the patient had a pain-free neck with a normal range of motion, this situation would constitute a relative contraindication to participation in contact activities depending on the player's level, position, and willingness to accept risk of reinjury. (From Torg, J. S., Glasgow, S. G. *Clin J Sports Med* 1:12–27, 1991.)

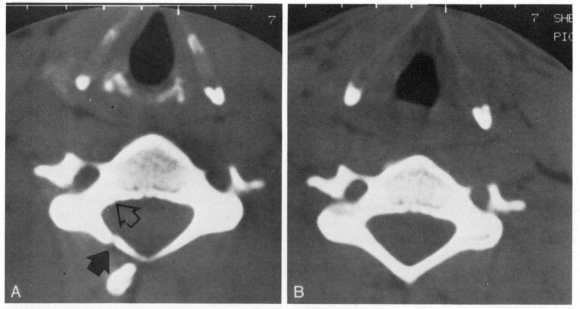

FIGURE 14–40

A, CT scan of a vertebral neural arch in the transverse plane demonstrating a hairline fracture through the lateral mass (open arrow) as well as a more evident nondisplaced fracture through the ipsilateral lamina (closed arrow). *B*, The patient was treated in a halo brace with satisfactory evidence of healing as demonstrated on CT scan. Following immobilization and the return of normal pain-free motion, he was permitted to return to contact activity after rehabilitation was fully effected and pain-free cervical range of motion and paravertebral muscle strength returned. (From Torg, J. S., Glasgow, S. G. *Clin J Sports Med* 1:12–27, 1991.)

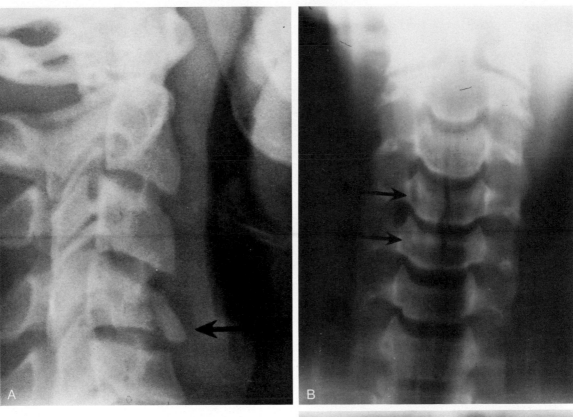

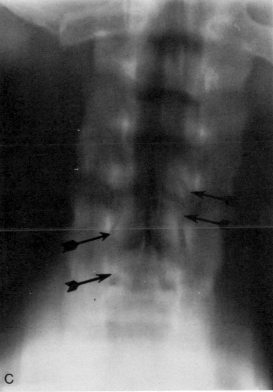

FIGURE 14–41

A, Lateral view of the cervical spine of a 17-year-old high school football player who was struck on the top of the head with a spring-loaded tackling device demonstrates a teardrop fracture of C4 (arrow). *B*, Anteroposterior views demonstrate sagittal fractures through the body of C4 and C5. *C*, Laminagrams in the anteroposterior projection through the neural arch demonstrate concomitant fractures through the posterior structures. Although the athlete remained neurologically intact and successful healing of the fractures occurred, return to contact activities was absolutely contraindicated because of the involvement of both anterior and posterior elements. In keeping with Steel's "rule of the ring," a sagittal fracture through the vertebral body is associated with a disruption of the neural arch. (From Torg, J. S., Glasgow, S. G. *Clin J Sports Med* 1:12–27, 1991.)

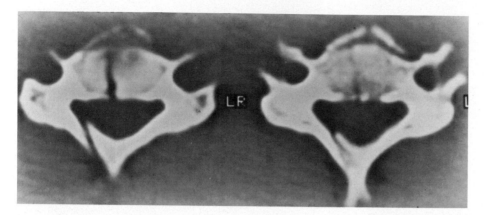

FIGURE 14–42
CT scans in the transverse plane of a cervical vertebra of a high school football player injured while spear tackling. Although there was no neurologic involvement, this axial load teardrop fracture vividly demonstrates involvement of both the anterior and posterior vertebral structures and constitutes an absolute contraindication to further participation in contact activities. (From Torg, J. S., Glasgow, S. G. *Clin J Sports Med* 1:12–27, 1991.)

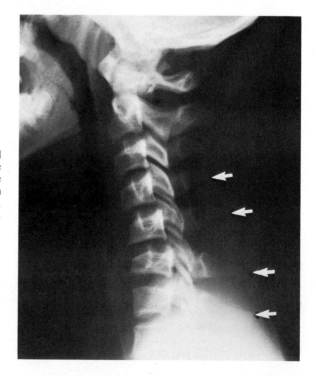

FIGURE 14–43
Lateral roentgenograms of the cervical spine in the erect neutral position demonstrate an anterosuperior compression defect in the vertebral body of C6 (large arrow). In addition, there is fanning of the C5–C6 spinous process indicating posterior instability due to disruption of the intraspinous and posterior longitudinal ligaments (small arrows). This situation constitutes an absolute contraindication to contact sports. (From Torg, J. S., Glasgow, S. G. *Clin J Sports Med* 1:12–27, 1991.)

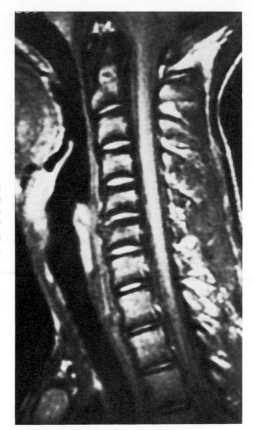

FIGURE 14–44
Magnetic resonance sagittal image of the cervical spine in a 17-year-old high school football player with a history of prior neck injury. An anterior intervertebral disc herniation with disc space changes at the C5–C6 level is seen (arrow). At the time of follow-up examination, the youngster was asymptomatic and neurologically normal and had a pain-free range of cervical motion. He was permitted to return to contact activities. (From Torg, J. S., Glasgow, S. G. *Clin J Sports Med* 1:12–27, 1991.)

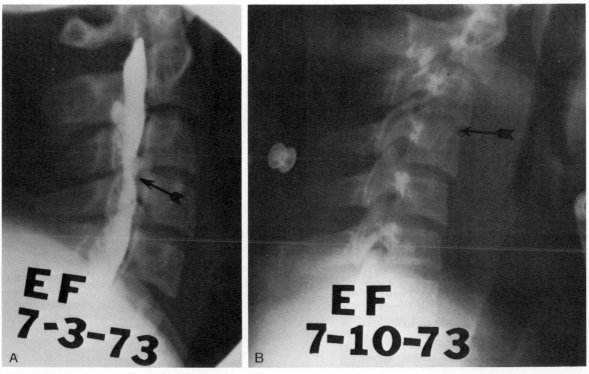

FIGURE 14–45
A, A myelogram demonstrates a herniated nucleus pulposus at C3–C4 (arrow) in a 21-year-old college football player who was injured when he struck a blocking dummy with his head. *B*, A C3–C4 anterior discectomy and interbody fusion was performed because of persistent symptoms. Lateral roentgenograms 1 week postsurgery demonstrated excellent graft placement. Normally, a solid one-level interbody fusion in an individual who is asymptomatic and neurologically negative and has a full range of motion will allow the individual to return to contact activities. In this particular case, however, the athlete had a congenital narrowing (stenosis) of the cervical canal, had had an episode of transient quadriplegia, and demonstrated reversal of the normal cervical lordosis. Because of this, further contact activities were absolutely contraindicated. (From Torg, J. S., Glasgow, S. G. *Clin J Sports Med* 1:12–27, 1991.)

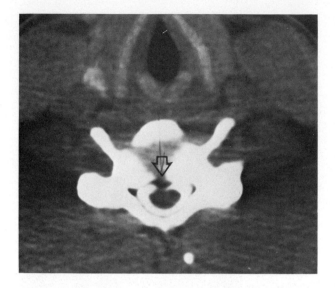

FIGURE 14–46
Transverse section of a CT myelogram through the C5–C6 interspace demonstrates a small central herniation without pressure on the spinal cord. The patient, a high school football player, had an episode of cervical cord neurapraxia associated with congenital narrowing (stenosis) of the cervical canal. Lateral roentgenograms demonstrated reversal of the normal cervical lordosis. In addition, he had a wry neck attitude and decreased neck motion. His clinical situation represents an absolute contraindication to participation in contact activities. (From Torg, J. S., Glasgow, S. G. *Clin J Sports Med* 1:12–27, 1991.)

2. Acute or chronic "hard disc" herniation with associated neurologic findings, pain, or significant limitation of cervical motion (Fig. 14–47)

3. Acute or chronic hard disc herniation with associated symptoms of cord neurapraxia due to concomitant congenital narrowing (stenosis) of the cervical canal.

Status Following Cervical Spine Fusion

A stable one-level anterior or posterior fusion in a patient who is asymptomatic, neurologically negative, and pain-free and has a normal range of cervical motion presents no contraindication to continued participation in contact activities (Fig. 14–48).

Individuals with a stable two- or three-level fusion who are asymptomatic and neurologically negative and have a pain-free full range of cervical motion have a relative contraindication. Because of the presumed increased stresses at the articulations of the adjacent uninvolved vertebrae and the propensity for development of degenerative changes at these levels, it appears to be the rare exception who should be permitted to continue contact activities (Fig. 14–49).

In individuals with more than a three-level anterior or posterior fusion, an absolute contraindication exists as far as continued participation in contact activities (Fig. 14–50).

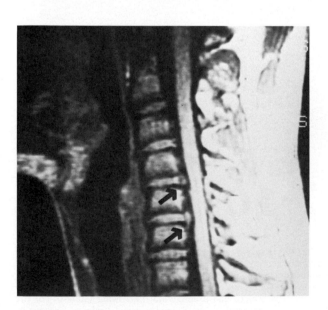

FIGURE 14–47
A sagittal MRI scan of the cervical spine of a 17-year-old high school football player who complained of posterior neck pain while butt blocking as well as a right unilateral transient radiculopathy or "burner." Visualized are intervertebral disc herniations at C4–C5 and C5–C6 that are indenting the spinal cord at both levels (arrows). Although the neurologic examination was normal, the presence of a wry neck attitude, limited neck extension, congenital canal narrowing (stenosis), and reversal of a normal cervical lordosis on roentgenogram precluded the individual from participation in contact sports. (From Torg, J. S., Glasgow, S. G. *Clin J Sports Med* 1:12–27, 1991.)

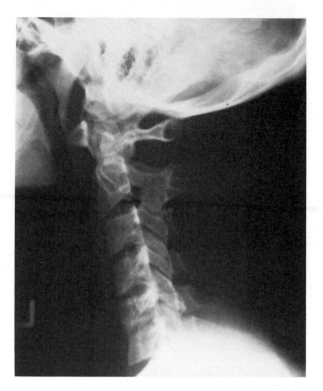

FIGURE 14–48
Lateral roentgenogram of 28-year-old professional ice hockey player who underwent a successful one-level interbody fusion at C5–C6 for instability. He subsequently played 2 years without a problem. (From Torg, J. S., Glasgow, S. G. *Clin J Sports Med* 1:12–27, 1991.)

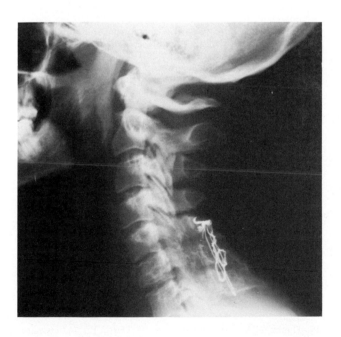

FIGURE 14–49
Lateral roentgenogram of the cervical spine of a 28-year-old former professional football player who had undergone a C4–C5–C6 posterior fusion for a post-traumatic instability. He subsequently returned to play 2 years of professional football; however, he developed stiffness, neck discomfort, and limited motion. The individual who elects to return to contact activities following more than a two-level fusion must understand that the probability of symptoms resulting from degenerative changes at the articulations above and below the fusion is increased. (From Torg, J. S., Glasgow, S. G. *Clin J Sports Med* 1:12–27, 1991.)

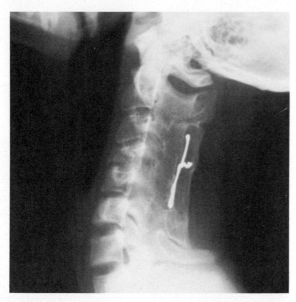

FIGURE 14–50

Lateral roentgenograms of an 18-year-old who had injured his neck playing football when he was 13 years old. At that time, a three-level posterior fusion and wiring was performed; however, it appears that periosteal stripping of adjacent vertebrae above and below resulted in a five-level fusion. Such a situation is an absolute contraindication to participation in contact activities. (From Torg, J. S., Glasgow, S. G. *Clin J Sports Med* 1:12–27, 1991.)

References

1. Albrand, O. W., and Walter, J. Underwater deceleration curves in relation to injuries from diving. *Surg Neurol* 4:461–465, 1975.
2. Albrand, O. W., and Corkill, G. Broken necks from diving accidents: A summer epidemic in young men. *Am J Sports Med* 4:107–110, 1976.
3. Albright, J. P., Moses, J. M., Feldich, H. G., et al. Non-fatal cervical spine injuries in interscholastic football. *JAMA* 236:1243–1245, 1976.
4. Allen, B. L., Jr., Ferguson, R. L., Lehman, T. R., and O'Brien, R. P. A mechanistic classification of closed, indirect fractures and dislocations of the lower cervical spine. *Spine* 7:1–27, 1982.
5. Alves, W. M. Football-induced mild head injury. *In* Torg, J. S. (Ed.), *Athletics Injuries to the Head, Neck and Face* (2nd ed.). St. Louis, Mosby–Year Book, 1991.
6. Anderson, L. D., and D'Alonzo, R. T. Fractures of the odontoid process of the axis. *J Bone Joint Surg* 56A:1663, 1974.
7. Bauze, R. J., and Ardran, G. M. Experimental production of forward dislocation of the cervical spine. *J Bone Joint Surg* 60B:239–245, 1978.
8. Boijsen, E. The cervical spinal canal in intraspinal expansive processes. *Acta Radiol* 42:101–115, 1954.
9. Bracken, M. D., Shepard, M. J., Collins, W. F., et al. A randomized, controlled trial of methylprednisolone or naloxone in the treatment of acute spinal cord injury. *N Engl J Med* 322(20):1405, 1990.
10. Bruce, D. A., Gennarelli, T. A., and Langfitt, T. W. Resuscitation from coma due to head injury. *Crit Care Med* 6:254, 1978.
11. Bruce, D. A., Schut, L., Bruno, L. A., et al. Outcome following severe head injuries in children. *J Neurosurg* 48:679–688, 1978.
12. Bruno, L. A. Focal intracranial hematoma. *In* Torg, J. S. (Ed.), *Athletic Injuries to the Head, Neck and Face* (2nd ed.). St. Louis, Mosby–Year Book, 1991.
13. Cantu, R. C. Criteria for return to competition after a closed head injury. *In* Torg, J. S. (Ed.), *Athletic Injuries to the Head, Neck and Face* (2nd ed.). St. Louis, Mosby-Year Book, 1991.
14. Carvell, J. E., Fuller, D. J., Duthrie, R. B., and Cockin, J. Rugby football injuries to the cervical spine. *Br Med J* 286:49–50, 1983.
15. Clancy, W. G. Brachial plexus and upper extremity peripheral nerve injuries. *In* Torg, J. S. (Ed.), *Athletic Injuries to the Head, Neck, and Face*. Philadelphia, Lea & Febiger, 1982, pp. 215–220.
16. Cloward, R. B. Acute cervical spine injuries. *Clin Symp* 32:2–32, 1980.
17. Coin, C. G., Pennink, M., Ahmad, W. D., and Keranen, V. J. Diving-type injury of the cervical spine: Contribution of computed tomography to management. *J Comput Assist Tomog* 3:362–372, 1979.
18. Dolan, K. D., Feldick, H. G., Albright, J. P., and Moses, J. M. Neck injuries in football players. *Am Fam Phys* 12:86–91, 1975.
19. Eismont, F. J., Clifford, S., Goldberg, M., et al. Cervical sagittal spinal canal size in spine injuries. *Spine* 9:663–666, 1984.
20. Fielding, J. W., et al. Athletic injuries to the atlanto-axial articulation. *Am J Sports Med* 6:226, 1978.
21. Frykman, G., and Hilding, S. Hop pa studsmatta kan orska allvarliga skador (Trampoline jumping can cause serious injury). *Lakartidningen* 67:5862–5864, 1970.
22. Funk, F. J., Jr., and Wells, R. E. Injuries of the cervical spine in football. *Clin Orthop Rel Res* 109:50–58, 1975.
23. Gennarelli, T. A., Segawa, H., Wald, U., et al. Physiological response to angular acceleration of the head. *In* Grossman, R. G., and Gildenberg, P. L. (Eds.), *Head Injury: Basic and Clinical Aspects*. New York, Raven Press, 1982, pp. 129–140.
24. Gennarelli, T. A. The state of the art of head injury biomechanics. *In Proceedings of the American Association of Automotive Medicine* 29:447–463, 1985.
25. Gennarelli, T. A. Mechanisms and pathophysiology of cerebral concussion. *J Head Trauma Rehabil* 1:23–29, 1986.
26. Gennarelli, T. A. Cerebral concussion and diffuse brain injuries. *In* Cooper, P. R. (Ed.), *Head Injury*. Baltimore, Williams & Wilkins, 1987, pp. 108–124.
27. Gennarelli, T. A. Mechanisms of cerebral concussion, contusion and other effects of head injury. *In* Youmans, J. (Ed.), *Neurological Surgery*. Philadelphia, W. B. Saunders, 1990, pp. 1953–1964.
28. Gennarelli, T. A. Head injury mechanisms. *In* Torg, J. S. (Ed.), *Athletic Injuries to the Head, Neck and Face* (2nd ed.). St. Louis, Mosby-Year Book, 1991.
29. Gennarelli, T. A. Cerebral concussion and diffuse brain injuries. *In* Torg, J. S. (Ed.), *Athletic Injuries to the Head, Neck and Face* (2nd ed.). St. Louis, Mosby-Year Book, 1991.
30. Glasgow, S. G. Upper cervical spine injuries (C1 and C2). *In* Torg, J. S. (Ed.), *Athletic Injuries to the Head, Neck and Face* (2nd ed.). St. Louis, Mosby-Year Book, 1991.
31. Gronwall, D., and Wrightson, P. Memory and information processing capacity after closed head injury. *J Neurol Neurosurg Psychiatry* 44:889–895, 1981.
32. Gurdjian, E. E., Webster, J. E., and Lissner, H. R. Observations on the mechanisms of brain concussion, contusion, and laceration. *Surg Gynecol Obstet* 101:684, 1955.
33. Guthkelch, A. N. Posttraumatic amnesia, postconcussional symptoms and accident neurosis. *Eur Neurol* 19:91–102, 1980.
34. Hammer, A., Schwartzbach, A. L., Darre, E., and Osgaard, O. Svaere neurologiske skader some folge af trampolinspring (Severe neurologic damage resulting from trampolining). *Ugeskrift Laeger* 143:2970–2974, 1981.
35. Hershman, E. B. Injuries to the brachial plexus. *In* Torg, J.

S. (Ed.), *Athletic Injuries to the Head, Neck and Face* (2nd ed.). St. Louis, Mosby-Year Book, 1991.

36. Hensinger, R. N. Congenital anomalies of the odontoid. *In* The Cervical Spine Research Society Editorial Committee. *The Cervical Spine* (2nd ed.). Philadelphia, J. B. Lippincott, 1989, pp. 248–257.

37. Hodgson, V. R., and Thomas, L. M. Mechanisms of cervical spine injury during impact to the protected head. *Twenty-fourth Stapp Car Crash Conference*, 1980, pp. 15–42.

38. Jefferson, G. Fracture of the atlas vertebra. *Br J Surg* 7:407, 1920.

39. Kewalramani, L. S., and Taylor, R. G. Injuries to the cervical spine from diving accidents. *J Trauma* 15:130–142, 1975.

40. Langfitt, T. W., Tannenbaum, H. M., and Kassell, N. F. The etiology of acute brain swelling following experimental head injury. *J Neurosurg* 24:47–56, 1966.

41. Lee, C., Kim, K. S., and Rogers, L. F. Triangular cervical body fragments: Diagnostic significance. *Am J Roentgenol* 138:1123–1132, 1982.

42. Lee, C., Kim, C. S., and Rogers, L. F. Sagittal fracture of the cervical vertebral body. *Am J Roentgenol* 139:55–60, 1982.

43. Leventhal, M. R. Management of lower cervical spine injuries (C4–C7). *In* Torg, J. S. (Ed.), *Athletic Injuries to the Head, Neck and Face* (2nd ed.). St. Louis, Mosby-Year Book, 1991.

44. Lindsay, K. W., McLatchie, G., and Jennett, B. Serious head injuries in sports. *Br Med J* 281:789–791, 1980.

45. McCoy, G. F., Piggot, J., Macafee, A. L., and Adair, I. Injuries of the cervical spine in schoolboy rugby football. *J Bone Joint Surg* 66B:500–503, 1984.

46. McCoy, S. H., and Johnson, K. A. Sagittal fracture of the cervical spine. *J Trauma* 16:310–312, 1976.

47. Mennen, U. Survey of spinal injuries from diving: A study of patients in Pretoria and Cape Town. *South Afr Med J* 59:788–790, 1981.

48. Mertz, H. J., Hodgson, V. R., Murray, T. L., and Nyquist, G. W. An assessment of compressive neck loads under injury-producing conditions. *Physician Sportsmed* 6(11):95–106, 1978.

49. Ommaya, A. K., and Hirsch, A. E. Tolerance for cerebral concussion from head impact and whiplash in primates. *J Biomech* 4:13, 1971.

50. Ommaya, A. K., and Gennarelli, T. A. Cerebral concussion and traumatic unconsciousness. Correlation of experimental and clinical observations on blunt head injuries. *Brain* 97:633, 1974.

51. Otis, J. C., Burstein, A. H., and Torg, J. S. Mechanisms and pathomechanics of athletic injuries. *In* Torg, J. S. (Ed.), *Athletic Injuries to the Head, Neck and Face* (2nd ed.). St. Louis, Mosby-Year Book, 1991.

52. Pavlov, H., Torg, J. S., Robie, B., and Jahre, C. Cervical spinal stenosis: Determination with vertebral body ratio method. *Radiology* 164:771–775, 1987.

53. Penning, L. Some aspects of plain radiography of the cervical spine in chronic myelopathy. *Neurology* 12:513–519, 1962.

54. Pizzutillo, P. D. Kipple-Feil syndrome. *In* The Cervical Spine Research Society Editorial Committee. *The Cervical Spine* (2nd ed.). Philadelphia, J. B. Lippincott, 1987, pp. 258–271.

55. Richman, S., and Friedman, R. Vertical fracture of cervical vertebral bodies. *Radiology* 62:536–542, 1954.

56. Roaf, R. Experimental investigations of spinal injuries. *J Bone Joint Surg* 41B:855, 1959.

57. Roaf, R. A study of the mechanics of spinal injuries. *J Bone Joint Surg* 42B:810–823, 1960.

58. Roaf, R. International classification of spinal injuries. *Paraplegia* 10:78–84, 1972.

59. Saunders, R. L., and Harbaugh, R. E. The second impact in catastrophic contact sports head trauma. *JAMA* 252:538–539, 1984.

60. Scher, A. T. The high rugby tackle—An avoidable cause of cervical spinal injury? *South Afr Med J* 53:1015–1018, 1978.

61. Scher, A. T. Diversity of radiological features in hyperextension injury of the cervical spine. *South Afr Med J* 58:27–30, 1980.

62. Scher, A. T. Vertex impact and cervical dislocation in rugby players. *South Afr Med J* 59:227–228, 1981.

63. Scher, A. T. Diving injuries to the cervical spinal cord. *South Afr Med J* 59:603–605, 1981.

64. Scher, A. T. 'Crashing' the rugby scrum—An avoidable cause of cervical spinal injury. *South Afr Med J* 61:919–920, 1982.

65. Schneck, C. D. Anatomy of the innervation of the upper extremity. *In* Torg, J. S. (Ed.), *Athletic Injuries to the Head, Neck and Face* (2nd ed.). St. Louis, Mosby-Year Book, 1991.

66. Schneider, R. C. A syndrome in acute cervical spine injuries for which early operation is indicated. *J Neurosurg* 8:360–367, 1951.

67. Schneider, R. C., Charie, G., and Pantek, H. The syndrome of acute central cervical spinal cord injury. *J Neurosurg* 11:546–577, 1954.

68. Schneider, R. C. The syndrome of acute anterior spinal cord injury. *J Neurosurg* 12:95–123, 1955.

69. Schneider, R. C., and Kahn, E. A. Chronic neurologic sequelae of acute trauma to the spine and spinal cord. Part I. The significance of the acute-flexion or "tear-drop" fracture dislocation of the cervical spine. *J Bone Joint Surg* 38A:985–997, 1956.

70. Silver, J. R. Rugby injuries to the cervical cord. *Br Med J* 1:192, 1979.

71. Silver, J. R. Injuries of the spine sustained in rugby. *Br Med J* 288:37–43, 1984.

72. Steel, H. H. Personal communication, 1965.

73. Steinbruck, J., and Paseslack, V. Trampolinspringen—ein gefahrlicher Sport? (Is trampolining a dangerous sport?) *Münch Med Wochenschrift* 120:985–988, 1978.

74. Tator, C. H., and Edmonds, V. E. National survey of spinal injuries in hockey players. *Can Med Assoc J* 130:875–880, 1984.

75. Tator, C. H., Edmonds, V. E., and New, M. L. Diving: A frequent and potentially preventable cause of spinal cord injury. *Can Med Assoc J* 124:1323–1324, 1981.

76. Tator, C. H., Ekong, C. E. U., Rowed, D. A., Schwartz, M. L., Edmonds, V. E., and Cooper, P. W. Spinal injuries due to hockey. *Can J Neurol Sci* 11:34–41, 1984.

77. Thibault, L. E., and Gennarelli, T. A. Biomechanics and craniocerebral trauma. *In* Povlishock, J., and Becker, D. (Eds.), *Central Nervous System Trauma Status Report.* NINCDS, 1985, pp. 370–390.

78. Thompson, R. C., et al. Current concepts in management of cervical spine fractures and dislocations. *Am J Sports Med* 3:159, 1975.

79. Torg, J. S., and Das, M. Trampoline-related quadriplegia: Review of the literature and reflections on the American Academy of Pediatrics' position statement. *Pediatrics* 74:804–812, 1984.

80. Torg, J. S., Pavlov, H., Genuario, S. E., Sennett, B., Robie, B., and Jahre, C. Neurapraxia of the cervical spinal cord with transient quadriplegia. *J Bone Joint Surg* 68A:1354–1370, 1986.

81. Torg, J. S., Pavlov, H., O'Neill, M. J., Nichols, C. E., and Sennett, B. The axial load teardrop fracture: A biomechanical, clinical, and roentgenographic analysis. *Am J Sports Med* 19(4), 1991.

82. Torg, J. S., Sennett, B., Vegso, J. J., and Pavlov, H. Axial loading injuries to the middle cervical spine segment: An analysis and classification of twenty-five cases. *Am J Sports Med* 19(1):6–20, 1991.

83. Torg, J. S., Truex, R. C., Marshall, J., Hodgson, V. R., Quedenfeld, T. C., Spealman, A. D., and Nichols, C. E. Spinal injury at the level of the third and fourth cervical vertebrae from football. *J Bone Joint Surg* 59A:1015, 1977.

84. Torg, J. S., Truex, R., and Quedenfeld, T. C. The National Football Head and Neck Injury Registry: Report and conclusions. *JAMA* 241:1477–1479, 1979.

85. Torg, J. S., Vegso, J. J., O'Neill, J., and Sennett, B. The epidemiologic, pathologic, biomechanical and cinematographic analysis of football-induced cervical spine trauma. *Am J Sports Med* 18:50–57, 1990.

86. Torg, J. S., Vegso, J. J., and Sennett, B. The National Football Head and Neck Injury Registry: 14-year report on cervical quadriplegia, 1971 through 1985. *JAMA* 254:3439–3443, 1985.

87. Torg, J. S., Wiesel, S. W., and Rothman, R. H. Diagnosis and management of cervical spine injuries. *In* Torg, J. S. (Ed.), *Athletic Injuries to the Head, Neck and Face.* Philadelphia, Lea & Febiger, 1982.

88. Torg, J. S., Vegso, J. J., and Torg, E. Cervical quadriplegia resulting from axial loading injuries: Cinematographic, radiographic, kinetic and pathologic analysis. Philadelphia, American Academy of Orthopaedic Surgeons Audio-Visual Library, University of Pennsylvania, 1987.

89. Torg, J. S., and Pavlov, H. Middle cervical spine injuries. *In* Torg, J. S. (Ed.), *Athletic Injuries to the Head, Neck and Face* (2nd ed.). St. Louis, Mosby-Year Book, 1991.

90. Torg, J. S. Axial load "teardrop" fracture. *In* Torg, J. S. (Ed.), *Athletic Injuries to the Head, Neck and Face* (2nd ed.). St. Louis, Mosby-Year Book, 1991.

91. Torg, J. S., and Fay, C. M. Cervical spinal stenosis with cord neurapraxia and transient quadriplegia. *In* Torg, J. S. (Ed.), *Athletic Injuries to the Head, Neck and Face* (2nd ed.). St. Louis, Mosby-Year Book, 1991.

92. Torg, J. S., and Glasgow, S. G. Criteria for return to contact activities after cervical spine injury. *In* Torg, J. S. (Ed.), *Athletic Injuries to the Head, Neck and Face* (2nd ed.). St. Louis, Mosby-Year Book, 1991.

93. Vegso, J. J., and Torg, J. S. Field evaluation and management of intracranial injuries. *In* Torg, J. S. (Ed.), *Athletic Injuries to the Head, Neck and Face* (2nd ed.). St. Louis, Mosby-Year Book, 1991.

94. Vegso, J. J., and Torg, J. S. Field evaluation and management of cervical spine injuries. *In* Torg, J. S. (Ed.), *Athletic Injuries to the Head, Neck and Face* (2nd ed.). St. Louis, Mosby-Year Book, 1991.

95. White, A. A., Johnson, R. M., and Panjabi, M. M. Biomechanical analysis of clinical stability in the cervical spine. *Clin Orthop* 109:85–93, 1975.

96. White, A. A., and Panjabi, M. M. *Clinical Biomechanics of the Spine.* Philadelphia, J. B. Lippincott, 1978.

97. Wu, W. Q., and Lewis, R. C. Injuries of the cervical spine in high school wrestling. *Surg Neurol* 23:143–147, 1985.

98. Yarnall, P. R., and Lynch, S. The "ding" amnestic states in football trauma. *Neurology* 23:196–197, 1973.

THE SHOULDER

Functional Anatomy, Biomechanics, and Kinesiology

James Tibone, M.D.
Robert Patek, M.D.
Frank W. Jobe, M.D.
Jacquelin Perry, M.D.
Marilyn Pink, M.S., R.P.T.

Normal shoulder function is the product of a complex anatomic arrangement that allows for significant joint mobility limited only by intact passive and finally coordinated dynamic restraints. Although this arrangement provides for a tremendous spectrum of functional capabilities, it also has a high potential for dysfunction if the physiologic limits of any one of the integral component structures is either exceeded or impaired. This situation, although not uncommon in the general population, occurs with even greater frequency in competitive athletes. When one considers that simple arm elevation requires motion at three separate joints, movement at the scapulothoracic articulation and coordinated recruitment of the deltoid, rotator cuff and scapular muscles, one can readily appreciate the level of demand placed on individual shoulder structures during the performance of more rigorous athletic activities.

Because a high degree of functional interplay exists, injury to even a single isolated shoulder structure can lead not only to a functional deficit but, more important, can create an altered biomechanical state that either taxes, limits, or potentiates injury to other structures. The sports medicine specialist must therefore be able not only to recognize the functional deficit but also to identify and treat the site of primary or secondary pathology. If necessary, he or she should also assist the athlete in restoring and maintaining normal shoulder mechanics. The ability to do so requires a thorough understanding of basic shoulder anatomy and biomechanics.

Previous authors have studied the anatomy and biomechanics of the various individual structures that comprise the shoulder. Just as these data have been applied to the general population, they can also be applied to athletic individuals. The subtle differences that do exist are for the most part quantitative and only rarely have qualitative differences been appreciated [47]. The present chapter reviews the functional anatomy, biomechanics, and kinematics of the shoulder. The chapter has been arranged by anatomic site, and biomechanical data have been included where appropriate. The final section of the chapter presents complex kinematic and electromyographic data that apply to the athlete's shoulder.

FUNCTIONAL ANATOMY

Motion occurs at four separate sites within the shoulder complex. These areas are the sternoclavicular joint, the acromioclavicular joint, the glenohumeral joint, and the scapulothoracic articulation [71, 74, 110]. Although the structures are functionally integrated, they will be presented separately for the purposes of anatomic discussion (Fig. 15A–1).

Sternoclavicular Joint

The sternoclavicular joint is the only true synovial joint connecting the upper extremity and the axial skeleton. Although it lacks inherent bony stability, joint apposition is secured and maintained by the interarticular disc and the anterior sternoclavicular, posterior sternoclavicular, costoclavicular, and infraclavicular ligaments [33, 135]. Of these, the anterior sternoclavicular ligament and the interarticular disc are the two components with the greatest structural significance. The anterior sternoclavicular ligament provides the primary restraining force to upward displacement, and the interarticular disc provides the primary restraining force to medial displacement of the proximal end of the clavicle [12]. Although stability is enhanced by these structures, they still allow 35 degrees of elevation, 35 degrees of translation, and 50 degrees of rotation at the sternoclavicular joint [71, 109]. This motion plays an important permissive role in the achieve-

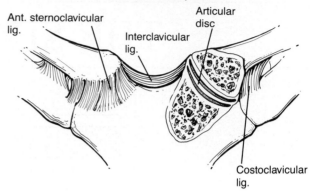

FIGURE 15A–2
The demonstration of the anatomy of the sternoclavicular joint.

ment of global shoulder motion [31, 57] (Fig. 15A–2).

Acromioclavicular Joint

The acromioclavicular joint is a plane joint that demonstrates 20 to 50 degrees of lateral inclination in the frontal plane [33, 135]. Unlike the sternoclavicular joint, little structural support is provided by the surrounding capsule, the variable interarticular disc, or the relatively weak acromioclavicular ligament [74].

Biomechanical data have shown that the acromioclavicular ligament resists axial rotation and posterior translation, whereas joint stability is provided by the stout coracoclavicular ligaments [42]. The conoid posteromedially and the trapezoid anterolaterally are anatomically distinct and serve separate biomechanical functions. The conoid resists anterior and superior translation, whereas the trapezoid resists axial compression of the distal end of the clavicle. In addition to providing structural support, the coracoclavicular ligaments bind the scapula to the clavicle, prevent scapular tilt, and limit potential excess rotation at the acromioclavicular joint [33, 71].

Motion at the acromioclavicular joint is facilitated by rotation of the S-shaped clavicle. This bone serves functionally to lengthen the coracoclavicular ligaments, allowing a potential 20 to 30 degrees of motion in the vertical, horizontal, and frontal planes [71, 74]. This motion is necessary to achieve full overhead shoulder elevation [57] (Fig. 15A–3).

Glenohumeral Joint

Although movement at the sternoclavicular and acromioclavicular joints extends the field of motion,

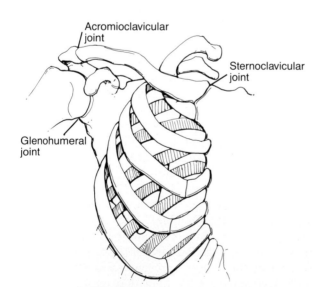

FIGURE 15A–1
The articulations of the shoulder joint.

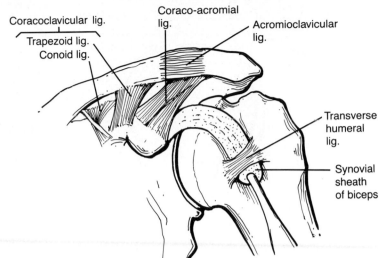

FIGURE 15A–3
The demonstration of the anatomy of the acromio-clavicular joint.

the central core of shoulder motion occurs at the glenohumeral joint [31, 104, 116]. The great three-dimensional mobility exhibited by this joint is a product of the relatively incongruous articular surfaces and a surrounding soft tissue envelope that provides both dynamic active and varying degrees of passive stability without restricting joint motion. This joint must maintain a fine balance between functional mobility and adequate stability. The static stabilizers are the joint surfaces and the capsulolabral complex. The dynamic stabilizers are the rotator cuff and the scapular rotators (trapezius, serratus anterior, rhomboids, and levator scapula).

Humeral Head

The articular portion of the humeral head approximates one-third of a sphere [15] and has an average diameter of 45 ± 10 mm [74, 116]. Relative to the shaft, the humeral head is medially angulated 45 degrees and retroverted 30 degrees [57]. This position not only affords the arm a more forward and lateral position, but, more important, aligns the humeral articular surface with the opposing glenoid fossa [100] (Fig. 15A–4).

Glenoid Fossa

The glenoid fossa is slender, shallow, and pear-shaped and has an average size of 41 by 25 mm [100]. This represents 50% of the contour and 33% of the surface area of the humeral head. Relative

to the plane of the scapula, the fossa is retroverted 7 degrees [115] and angled downward 5 degrees [41, 103], a position offering little anterior and no inferior bony support (Fig. 15A–5). Because bony contact is minimal, stability of the glenohumeral joint is provided largely by the surrounding soft tissue envelope made up of the glenoid labrum, the joint capsule, the glenohumeral ligaments, and the rotator cuff (Fig. 15A–6).

Glenoid Labrum

The fibrocartilaginous labrum acts to expand the depth and increase the area of the glenoid [100]. It is triangular on cross-section with a thin, free intra-articular apex and three sides that face the humeral head, the joint capsule, and the glenoid surface, respectively [135]. Firm peripheral attachments are

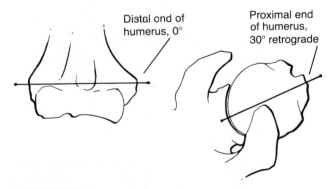

FIGURE 15A–4
The glenohumeral articulation with 30° of retroversion is demonstrated.

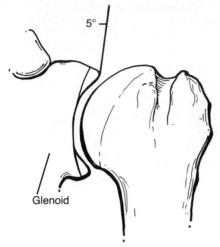

FIGURE 15A–5
The 5° downward tilt of the glenoid is demonstrated.

normally maintained at the glenoid margin and act to anchor the labrum to the glenoid. The presence of an intact labrum increases humeral contact area by up to 75% in the vertical direction and by 56% in the transverse direction [100, 114]. This additional contact area improves glenohumeral joint stability without compromising joint motion.

Joint Capsule

Mobility is further enhanced by a thin redundant joint capsule that has almost twice the surface area of the humeral head [33]. This arrangement allows a tremendous range of joint motion, passive stability being provided only by selective tightening of various portions of the joint capsule depending on arm

position. At rest, with the arm in the dependent position, the superior portion of the capsule is taut, and the inferior region is lax [71]. With overhead elevation, this relationship is reversed. Similarly, external rotation tightens the anterior and relaxes the posterior capsule, whereas horizontal flexion does just the opposite [100].

Glenohumeral Ligaments

Structural reinforcement of the anterior capsule is provided by folds in its inner wall that have been designated as the superior, middle, and inferior glenohumeral ligaments [35, 83]. Like the capsule, these ligaments are called upon differentially based upon arm position and rotation. The superior glenohumeral ligament parallels the biceps tendon and assists in supporting the dependent arm [100]. The middle glenohumeral ligament is located in a plane between the anterior superior band of the inferior glenohumeral ligament and the subscapularis tendon. The middle glenohumeral ligament, along with the inferior glenohumeral ligament and subscapularis tendon have been shown to resist external rotation at 45 degrees of shoulder abduction [130].

Anatomically and biomechanically, the most significant structure is the inferior glenohumeral ligament [100, 116]. It is composed of an axillary pouch, an anterior superior band, and a posterior superior band [83]. Unlike the superior and middle glenohumeral ligaments, which play important but less significant functional roles, the inferior glenohumeral ligament has been shown to be the primary restraint to external humeral rotation at 90 degrees of shoulder abduction [130]. Overhead athletes

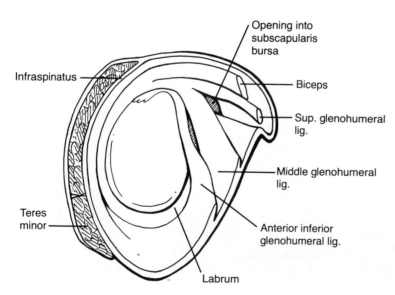

FIGURE 15A–6
The static stabilizers of the ligamentous labral complex as well as the dynamic stabilizers of the rotator cuff are demonstrated.

often place their arm in this position and appear to be at risk for injury to the inferior glenohumeral ligament. Loss of the integrity of the inferior glenohumeral ligament has been implicated as a major cause of anterior glenohumeral instability in this population.

Rotator Cuff

Like the glenoid labrum, the joint capsule and the glenohumeral ligament complex, the rotator cuff contributes to both motion and stability in the glenohumeral joint [18, 57]. The rotator cuff is made up of the supraspinatus, infraspinatus, teres minor, and subscapularis. These muscles originate on the scapula and terminate in short, flat, broad tendons that fuse intimately with the capsule of the shoulder forming a musculotendinous cuff [71, 135] (Fig. 15A–7).

These muscles balance the upward pull of the deltoid, creating a smooth rotational movement of the humeral head on the glenoid surface during shoulder elevation [57]. Individually, owing to their location relative to the glenohumeral joint, the

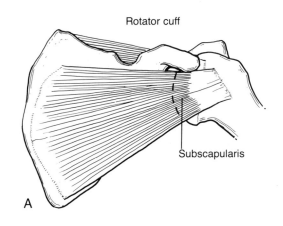

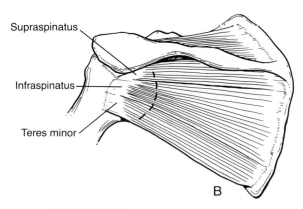

FIGURE 15A–7
The four rotator cuff muscles are demonstrated.

subscapularis facilitates internal rotation, and the infraspinatus and teres minor assist in external rotation of the humerus.

In addition to their vital role in shoulder elevation, the individual rotator cuff muscles have been shown to be capable of independent passive and dynamic functions [47]. The subscapularis acts as a passive restraint to external rotation during the initial stages of abduction [130]. It also acts as a forceful dynamic internal rotator of the humerus [100]. Similarly, the infraspinatus and teres minor provide passive restraint to internal rotation, and force applied through these tendons has been shown to facilitate as well as limit external rotation of the humerus and also decrease strain in the inferior glenohumeral ligament [18].

Scapula Rotators

The scapular stabilizers are made up of the upper, middle, and lower trapezius, the levator scapulae, the serratus anterior, and the rhomboid musculature. Postural support is provided by the levator scapulae and upper trapezius. Upward rotation is facilitated by the trapezius and serratus anterior. Retraction is provided by the middle trapezius and rhomboids, protraction by the serratus anterior, and depression by the inferior portion of the serratus anterior and lower trapezius [110]. These muscles help stabilize the glenohumeral joint by positioning the scapula for optimum bony stability in any position.

Scapulothoracic Articulation

The scapulothoracic articulation is not a true joint but instead represents the riding of the scapula on the posterior surface of the thoracic cage [33, 71]. The scapula is a mobile structure owing to its lack of bony continuity with the axial skeleton. Other than its articulation at the acromioclavicular joint, the scapula is supported only by the coracoclavicular ligaments and the surrounding muscular attachments [135].

Although scapular motion can occur independently, it normally occurs as an integral component of overall shoulder motion. Motion of the scapula not only contributes to global mobility but also provides bony support to the glenohumeral joint and increases muscle effectiveness because optimal leverage and effective muscle fiber length are maintained throughout a large range of shoulder motion [110].

BIOMECHANICAL PRINCIPLES

Though anatomically distinct, the various component structures about the shoulder are functionally interrelated. This complex interplay has been well studied, and pertinent basic biomechanical principles are presented in the following section.

Shoulder Motion

Perry has described three patterns of shoulder motion: elevation, internal and external rotation, and horizontal flexion and extension [100, 110].

Elevation

Shoulder elevation can be defined in three planes: sagittal (flexion), scapular (neutral), and coronal (abduction) [100]. The scapular plane is approximately 30 to 45 degrees anterior to the coronal plane [100, 103]. Although theoretically the shoulder is capable of a 180-degree arc of elevation, studies [15, 41] have shown that men and women on average are limited to 167 and 171 degrees of elevation, respectively. Posterior elevation (extension) is limited to 60 degrees. Sagittal plane elevation is restricted by capsular torsion [68] and coronal plane elevation by impingement of the greater tuberosity on the acromion [100]. Scapular or neutral plane elevation is therefore recommended both for examination and for rehabilitation of the shoulder [110] (Fig. 15A–8).

Internal and External Rotation

Internal and external rotation occur largely at the glenohumeral joint. The amount of rotation varies depending on the degree of elevation. With the arm at the side the maximum arc of motion is 180 degrees, external rotation being greater than internal rotation (108 degrees vs. 71 degrees) [15, 100]. This difference is due to the apposition of the forearm against the trunk. At 90 degrees of abduction the maximum arc is 120 degrees with internal rotation being the greatest component [13]. This relationship does not hold true in the overhead athlete because external rotation is often increased and internal rotation proportionately decreased in such athletes, preserving the normal arc. Overhead there is very little rotation at the glenohumeral joint [13].

Horizontal Flexion and Extension

Horizontal flexion and extension also occurs at the glenohumeral joint. The maximum arc is 180 degrees, 135 degrees being anterior and 45 degrees being posterior. Motion of this plane is limited by the humeral articular surface [100, 110].

Scapulohumeral Rhythm

Total arm elevation is the sum of motion at the glenohumeral joint and the scapulothoracic articulation [30]. While the range is variable over the entire arc the average reported ratio is 3 degrees of glenohumeral motion to every 2 degrees of scapulothoracic motion [41, 57, 103, 110, 115]. The difference is largely due to variable participation of the scapula in shoulder elevation with scapular motion lagging during the first 30 and during the last 60 degrees of shoulder elevation [57, 110].

Instant Center Analysis

Glenohumeral Motion

Poppen and Walker [103] performed a radiographic analysis of glenohumeral joint motion during shoulder elevation in the scapular plane. They found that, unlike the knee joint, which exhibits a combination of rolling and gliding, gliding or rotation is the predominant motion in the glenohumeral joint. They found that after a 3-mm upward shift at the initiation of abduction, the contact point of the humeral head on the glenoid surface remained within 1 mm of its initial contact point. The instant center of rotation of the humeral head also stayed within 6 ± 2 mm of the initial center of rotation.

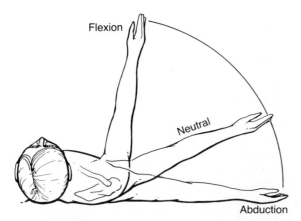

FIGURE 15A–8
The planes of shoulder elevation are illustrated.

These findings indicate that almost pure rotational movement occurs in the glenohumeral joint. This normal rotation is absent in some patients with pain or pathologic processes relative to the shoulder and can be measured radiographically.

Howell and colleagues have performed a similar analysis of glenohumeral joint mechanics in the horizontal plane [55]. They found that the humeral head was centered in the glenoid cavity throughout the horizontal plane of motion except when the arm was in maximum extension and external rotation. In this position, the center of the humeral head rests approximately 4 mm posterior to the center of the glenoid cavity. When the arm was then flexed or rotated from this position, the humeral head glided anteriorly, producing a shear stress on the articular surface of the glenoid and labrum. Analysis of their patients with documented anterior glenohumeral instability revealed abnormal anterior translation of the humeral head with this technique.

Scapular Motion

Poppen and Walker also studied scapular motion during shoulder elevation [103]. They found that during the first 60 degrees of elevation, scapular motion was quite variable. They refer to this as the setting stage. At between 60 and 120 degrees of elevation, rotation of the scapula was centered at the scapular spine, and between 120 and 180 degrees rotation was centered at the base of the glenoid. This variability of scapular motion is related to the motion that occurs at the sternoclavicular and acromioclavicular joint. At between 0 and 120 degrees of shoulder elevation, the clavicle elevates at the sternoclavicular joint. Between 120 and 180 degrees, the clavicle rotates 40 degrees and then moves upward and also rotates at the acromioclavicular joint [110]. During this motion, the coracoclavicular ligaments are functionally lengthened by rotation of the S-shaped clavicle [57].

Muscle Forces

The demand torque experienced by the shoulder musculature is a product of force times lever length (T = F × L). As the arm is raised, the center of gravity of the upper extremity moves away from the body. This creates a moment arm or lever that proportionately increases demand torque at the shoulder. Accordingly, an increase in load will also increase this torque value [110].

To balance the increased torque created by arm elevation or changing load, significant force must be generated by the shoulder musculature. Muscle force is dependent on several factors including the number of motor cells, muscle fiber alignment, muscle fiber length, type and speed of contraction, and lever length [100, 110]. Deltoid function in shoulder elevation is improved by optimizing several of these factors. Its multipennate structure increases the number of cells per cross-sectional unit. Oblique fiber alignment increases muscle cell availability and therefore power. Optimal fiber length is preserved by scapular rotation, and muscle lever length is increased as the deltoid's line of pull moves away from the center of humeral rotation during shoulder elevation [110].

Joint Forces

Muscles crossing a joint create joint forces when they are active. Because of their oblique orientation, the deltoid and rotator cuff muscles cause both compressive and shear forces at the glenohumeral joint [102]. Compressive forces tend to stabilize and shear forces cause displacement of the humeral head [110]. The amount of these forces is determined by the relative intensity of muscle contraction, and the ratio varies with muscle alignment and the angle of shoulder abduction [102].

Force Couple

Owing to their downward and medial orientation, the pull of the rotator cuff tendons tends to balance the upward and outward pull of the deltoid during shoulder elevation. These forces acting in opposite directions create a "force couple" that causes the humeral head to turn about its axis of rotation, creating a gliding motion at a point of contact on the glenoid surface. Clearly, unopposed action of either the deltoid or the rotator cuff changes this normal relationship and could lead potentially to subluxation of the humeral head and obviously impaired shoulder function. Instability causes abnormal joint loading patterns that can produce deleterious changes in the hyaline cartilage, which can lead to osteoarthritis [84].

The force couple mechanism that functions in casual shoulder elevation does not always apply in the rapid and precise movements that occur in sports, as the next section on kinematics will demonstrate.

KINEMATICS AND MUSCLE FUNCTION OF THE SHOULDER

Inman and associates [57] provided the first comprehensive biomechanical analysis of shoulder function in 1944. They created a dynamic model of the shoulder using anatomic, roentgenographic, and electromyographic systems of analysis. The ability to analyze motion with electromyography yielded several tenets of shoulder mechanics that are now accepted as common knowledge. Inman and colleages proposed that the supraspinatus and the deltoid muscles act as a single unit throughout elevation of the shoulder. They also proposed that the subscapularis, infraspinatus, and teres minor acted as a functional unit to depress the humerus continuously throughout this motion.

An understanding of shoulder biomechanics is of paramount importance to sports medicine physicians because athletic injuries to the shoulder are common. It was Jobe's [62] observation that athletes often experience a selective weakness of specific rotator cuff muscles rather than the generalized muscle impairment that Inman's model predicts. This led to laboratory investigation of whether Inman's conclusions regarding single plane motion analysis could be applied to sport-specific activities.

The Centinela Hospital Biomechanics Laboratory in Inglewood, California, began to study shoulder motion during baseball pitching and other overhead sports activities with the goal of providing a basis for optimal treatment and prophylaxis of athletic injuries. In addition, the effects of selected common shoulder pathologies were studied.

Methods

Dynamic electromyography and high-speed film analysis were used to identify the functions of the major muscles that control the shoulder. Electromyographic (EMG) data were collected using the Basmajian single-needle technique [7]. Dual 50-micron insulated wires with 2-mm bared tips were inserted into the muscle using a 25-gauge needle as a cannula. The accuracy of placement was determined by a manual muscle examination and signal display on the oscilloscope.* The subject wore an FM transmitter belt pack, and the telemetered EMG information was recorded on multichannel tape for later retrieval and review.

Motion analysis was conducted with the aid of 16-mm cameras, which were set at speeds varying from 50 to 200 frames/second. The subjects were photographed from two or three points of view (anterior, posterior, lateral, or overhead) depending on the needs of the study. The cameras were pin-registered to ensure accurate simultaneous recording. An electronic pulse marked the film and EMG records simultaneously to synchronize the data. After processing, the films were reviewed on a stop-action projector,* which permitted single-frame viewing.

The EMG data were converted from analog to digital signals and quantitated by computer integration with sampling at 2500 times/second. After excluding the low-level activity identified as the resting signal, the greatest 1-second electromyographic signal during the manual muscle strength test (MMT) was selected as the normalizing value or 100% MMT. The activity patterns were assessed every 20 msec and were expressed as a percentage of the normalization base. The MMT percentages were then averaged during the different stages of the sport.

Baseball—Pitching

The "normal" group of subjects consisted of skilled male pitchers who had no known shoulder pathology [66, 67]. Amateur pitchers and skilled pitchers with shoulder pathology were compared to this group [48]. Their ages ranged from 21 to 34 years. The muscles monitored included the anterior, middle, and posterior deltoids, supraspinatus, infraspinatus, teres minor, subscapularis, triceps, pectoralis major, latissimus dorsi, serratus anterior, brachialis, biceps brachii, and trapezius. The baseball pitch was divided into five stages (Fig. 15A–9).

Stage I. The *wind-up* is the period of preparation that ends when the ball leaves the gloved hand.

Stage II. *Early cocking* is a period of shoulder abduction and external rotation that begins as the ball is released from the nondominant hand and ends with foot-ground contact.

Stage III. In *late cocking*, the arm is maximally abducted and externally rotated. This stage ends with the first forward motion.

Stage IV. *Acceleration* starts with internal rotation of the humerus and ends with ball release.

Stage V. *Follow-through* starts with ball release and ends when motion ceases.

The effect of glenohumeral instability on shoulder activity was analyzed in 15 skilled throwers [48].

---

*Tektronix, model SIIIA, Tektronics, Beaverton, OR.

*L.W. International, model 110, Photo-Optical Digitizer, Woodland Hills, CA.

Stages of Pitching

FIGURE 15A–9
The five phases of pitching. *Phase 1:* wind-up. *Phase 2:* early cocking. *Phase 3:* late cocking. *Phase 4:* acceleration. *Phase 5:* follow-through.

The average age was 23 years. Eleven subjects had a diagnosis of chronic recurrent anterior dislocation. The symptoms of instability were isolated with no evidence of impingement. All diagnoses were confirmed at the time of surgery. Eight throwers were also evaluated who had a diagnosis of impingement syndrome [80]. These athletes did not have instability and had similar EMG test results.

During a baseball pitch, the deltoid muscle is responsible for arm elevation with active forward flexion and abduction of the humerus [5, 73]. The three heads of the deltoid, the anterior, middle, and posterior, exhibit peak activity during early cocking when the arm is elevated to 90 degrees. Subsequently, in late cocking, the activity of the deltoid decreases with a concurrent increase in rotator cuff muscle activation. This sequential pattern of muscle activity, beginning with the deltoid and ending with the rotator cuff, contradicts the forced couple proposed by Inman and associates [57], in which the deltoid and rotator cuff contract simultaneously. This pattern also demonstrates the importance of the rotator cuff in throwing and identifies these muscles as generating the primary force in late cocking.

The supraspinatus has been thought to play an important role in humeral abduction. Its activity, however, peaks during late cocking when the arm is already abducted and is most prone to anterior subluxation [130]. Therefore, the supraspinatus contributes to the stability of the joint by drawing the humeral head toward the glenoid [6, 75, 102]. This finding correlates with the compressive force pattern noted by Poppen and Walker at the glenohumeral joint in abduction [102]. The greater use of the supraspinatus muscle by the amateur pitchers compared to the professionals suggests the need for specific muscle conditioning and proper shoulder mechanics. In amateur pitchers the supraspinatus can easily become fatigued because of this increased activity, which can lead to tendinitis. This "overuse syndrome" may then cause a subtle compromise of glenohumeral stability. When the muscles become conditioned and fire in a synchronous manner, the athlete can use them more efficiently and economically, thus reducing the chance of overuse and injury.

The infraspinatus and teres minor are responsible for external rotation of the shoulder [6, 75]. Their action contributes to stability by drawing the head toward the glenoid fossa. The activity patterns are similar for both muscles with peak activity occurring in late cocking and follow-through, although both lagged behind the supraspinatus in timing.

The peak activity of the subscapularis occurs in late cocking when it contracts eccentrically to protect the anterior joint, which is under extreme tension at the position of maximal abduction and external rotation. The subscapularis continues to function as an internal rotator to help carry the arm across the chest during acceleration and follow-through [66, 67, 69].

During the pitch, the arm is cocked at 90 degrees of abduction with horizontal extension and maximum external rotation. This motion tightens the anterior shoulder capsule and ligaments. Relief from the strain on the anterior structures is provided by the rotator cuff muscles. The infraspinatus and teres minor contract to actively rotate the arm externally. At the same time, they pull the humeral head posteriorly and relieve the stress on the anterior capsule and labrum. The humeral head during shoulder motion usually remains centered in the glenoid fossa. However, recent research studies have shown that in the cocking position, the humeral head is displaced 3 to 4 mm posteriorly in the glenoid [55]. This is a direct result of the action of the infraspinatus and teres minor. These muscles have a short lever arm based on their insertion site on the humeral head. Therefore, to meet the demands of pitching or overhead activity, greater than average strength in these muscles may be needed to reduce the strain on the anterior shoulder capsule [100]. It is therefore very important in a patient with anterior subluxation to rehabilitate the posterior rotator cuff, namely, the infraspinatus and teres minor, in addition to performing the usual internal rotation exercises to strengthen the subscapularis. The subscapularis does protect the anterior shoulder, however, by preventing excessive external rotation and by contracting eccentrically to permit deceleration of the externally rotating shoulder. Precise timing and sequential contraction of the rotator cuff muscles protect the shoulder from injury. When the muscles become fatigued from overuse, this sequential activation pattern is disturbed, and injury with resultant inflammation results.

Professional throwing athletes demonstrated selective use of individual rotator cuff muscles. They were able to use the subscapularis exclusively among the other rotator cuff muscles during the acceleration phase. In contrast, amateurs tended to use all the rotator cuff and biceps brachii muscles. Proper coordinated motion of the trunk, shoulder, and elbow by professional pitchers makes use of the supraspinatus, infraspinatus, teres minor, and biceps brachii less necessary for acceleration. Because repetitive, overhead activities such as pitching can lead to muscle strains and tendinitis, overuse syndromes may occur less often in the athlete who uses his muscles economically [107]. Efficient muscle use learned through training improves endurance and reduces the chance of injury secondary to overuse.

The concept of separate and independent actions of the deltoid and rotator cuff helps to explain the selective muscle weakness seen in the throwing athlete. Therefore, a rehabilitation program was designed to strengthen specifically the individual rotator cuff muscles in the throwing athlete. This program is effective for injury reduction and postsurgical rehabilitation, and it is used by both amateur and professional athletes [85].

The role of the biceps brachii muscle at the shoulder is controversial [8, 43]. EMG analysis demonstrated that the biceps functions predominantly at the elbow rather than at the shoulder during pitching. The similarity of the biceps brachii and the brachialis firing patterns supports this finding. Biceps activity correlates with the motion of

the elbow as it flexes during late cocking and as it decelerates during follow-through. The heads of the triceps work together to produce elbow motion.

The pectoralis major and latissimus dorsi muscles act as internal rotators and contract eccentrically to protect the joint, along with the subscapularis, during late cocking. Increasing activity during acceleration indicates that intense internal rotation and forceful arm depression provide the principal propulsive force.

The serratus anterior moves the scapula to place the glenoid in the optimal position for stability [7, 57]. The glenoid is a stable platform on which the humeral head articulates. Serratus anterior activity is high during late cocking due to scapular protraction and upward rotation, which allows the scapula to keep pace with the horizontally flexing and externally rotating humerus. If the serratus becomes fatigued or does not contract at the proper time, shoulder impingement or subluxation can occur. The relatively low level of trapezius activity during the cocking and acceleration phases implies that this muscle provides primarily supplementary scapular stabilization to enhance the rotational action of the serratus anterior during pitching. During follow-through, the adductive action of the trapezius decelerates scapular protraction.

Analysis of the shoulder in throwers with subacromial impingement demonstrates differences in the pattern of muscle use between these subjects and normals. During late cocking, deltoid activity continued in the impingers while decreasing in the normals. The lower activity of the supraspinatus means that it is less able to assist the deltoid during cocking. This imbalance between the deltoid and the supraspinatus may contribute to the "impingement syndrome."

The internal rotators (subscapularis, pectoralis major, and latissimus dorsi) and serratus anterior also showed differences during early and late cocking in impinged and normal shoulders. They showed reduced activity that may contribute to increased external rotation, thus predisposing or aggravating the impingement syndrome. This increased external rotation may then lead to shoulder subluxation.

Differences in neuromuscular activities may account for initial or persistent impingement problems in the throwing athlete. Conditioning and retraining of these muscles must be accomplished as part of a preventive or rehabilitative program.

Evaluation of patients with an isolated diagnosis of glenohumeral instability also revealed differences from normals [47]. In these subjects, the increased activity of the biceps brachii during acceleration was mild, but this may represent a compensatory mechanism to help stabilize the humeral head against the glenoid. This is consistent with the difference observed in late cocking and acceleration when the arm is most prone to subluxation.

The mild enhancement of supraspinatus activity observed throughout cocking and acceleration may be an attempt to stabilize the joint in the unstable patient by drawing the humeral head toward the glenoid.

The pectoralis major, subscapularis, and latissimus dorsi demonstrated markedly decreased activity during the pitch in patients with instability. Inhibition of the synergistic activity of these muscles allows persistent or accentuated external rotation. This neuromuscular difference is postulated to be a factor in producing or maintaining chronic anterior instability.

Decreased serratus anterior activity in the patient with instability diminishes horizontal protraction of the scapula, which normally begins during late cocking. Early fatigue of the serratus anterior will then add to the stress on the anterior restraints.

An overhead athlete, in particular a thrower, should strengthen and retrain the muscles about the shoulder as part of a conservative or postsurgical rehabilitation program. Laboratory data support the concept that the internal rotator and scapular protractor muscles are the key areas in patients with instability.

After careful clinical evaluation of throwing athletes with impingement findings, many were found to have underlying anterior instability. These EMG findings support the premise that the primary problem, instability, leads to secondary stress on the rotator cuff tendons.

Swimming

Shoulder action during swimming was studied in seven skilled swimmers in a pool and in five on dry land in the Biomechanics Laboratory [90]. Their ages ranged from 19 to 35 years. The muscles that were monitored were the pectoralis major (clavicular head), biceps brachii, subscapularis, supraspinatus, infraspinatus, latissimus dorsi, serratus anterior, and middle deltoid. The EMG apparatus worn by the aquatic subjects was enclosed in a waterproof pouch and taped to the back where it would allow unencumbered movement of the upper extremities. Three strokes were performed in the water: freestyle, butterfly, and breaststroke. The dry land subjects performed the freestyle and butterfly strokes while lying prone on an elevated platform that allowed unimpeded upper extremity motion.

The swimming strokes were broken down into the pull-through and recovery phases as described by Richardson and colleagues [108] (Figs. 15A–10, 15A–11, and 15A–12). The different strokes in swimming have two phases, a pull-through phase and a recovery phase. In the freestyle stroke, the pull-through phase is subdivided into hand entry, midpull-through, and end of pull-through. During hand entry, the shoulder is internally rotated and abducted and the body roll begins. In midpull-through, the shoulder is at 90 degrees abduction and neutral rotation. Body roll occurs at a maximum of 40 to 60 degrees from horizontal. With the end of pull-through, the shoulder is internally rotated and fully adducted. The body has returned to horizontal. The recovery phase is subdivided into elbow lift, midrecovery, and hand entry. In elbow lift, the shoulder begins to abduct and rotate externally. The body roll begins in the opposite direction from pull-through. In midrecovery, the shoulder is abducted to 90 degrees and externally rotated beyond neutral. Body roll reaches a maximum of 40 to 60 degrees. Breathing occurs by turning the head to the side. In hand entry, the shoulder is externally rotated and maximally abducted, and the body is returned to neutral roll (see Fig. 15A–10).

FIGURE 15A–10
The phases of the free stroke. *Phase 1:* pull-through phase; hand entry, mid pull-through, end pull-through. *Phase 2:* recovery phase, elbow lift, mid-recovery, and hand entry. (From Pink, M., Perry, J., et al. The normal shoulder during freestyle swimming: An electromyographic and cinematographic analysis of twelve muscles. *Am J Sports Med* 19(6), 1991.)

Swimming—Freestyle

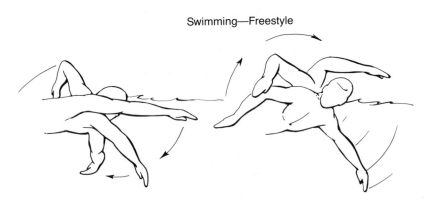

Swimming—Backstroke

FIGURE 15A–11
Pull-through and recovery phases of the backstroke.

In the backstroke pull-through phase at hand entry, the shoulder is externally rotated and abducted with the beginning of body roll. In midpull-through, the shoulder is abducted 90 degrees in neutral rotation with maximum body roll. At the end of pull-through, the shoulder is internally rotated and adducted, and body roll is horizontal. In the recovery phase of the backstroke, there is hand lift rather than elbow lift. In hand lift, the shoulder begins abduction and external rotation, and the body roll allows the arm to clear the water. In midrecovery, the shoulder is 90 degrees abducted, and body roll is maximum. In hand entry, the shoulder is at maximum abduction, and body roll is neutral (see Fig. 15A–11).

In the butterfly stroke, the pull-through phase is the same as in the freestyle, but there is absence of body roll in all stages. To avoid shoulder flexion or extension, the hands are spread apart at the midpull-through stage. The recovery phase is again similar to that in the freestyle with an absence of body roll. Body lift allows both arms to clear the water. Shoulder flexion and extension do not occur (see Fig. 15A–12).

Shoulder activity during swimming also empha-

sizes the importance of the rotator cuff. The supraspinatus, infraspinatus, and middle deltoid are predominant in the recovery phase. They abduct and externally rotate the extremity in preparation for a new pull-through. This position, similar to the cocking phase of throwing, places the shoulder at risk for subacromial impingement [50]. The serratus anterior also has an important function during recovery. It allows the acromion to rotate clear of the abducting humerus and provides a stable glenoid on which the humeral head may rotate as in throwing. The dry land data indicated that the serratus anterior works at nearly maximal levels to accomplish this. If this muscle becomes fatigued during the course of a number of cycles, scapular rotation may not coincide with humeral abduction. As a result, impingement may occur.

The biceps brachii muscle exhibited erratic activity during all of the strokes and functioned primarily at the elbow, which is similar to its role in pitching. The latissimus dorsi and pectoralis major were found to be propulsive muscles with a resulting action similar to that of the acceleration phase of throwing.

The swimming study indicated that particular attention must be paid to conditioning the rotator cuff

Swimming—Butterfly

 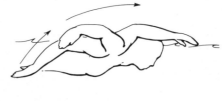

FIGURE 15A–12
Pull-through and recovery phases of the butterfly stroke.

Tennis—Serve

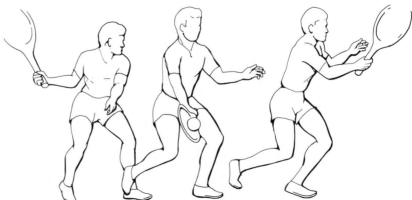

FIGURE 15A–13
The five stages of a tennis serve. *Stage 1:* preparation. *Stage 2:* early cocking. *Stage 3:* late cocking. *Stage 4:* acceleration. *Stage 5:* follow-through. (From Morris, M., Jobe, F. W., et al. Electromyographic analysis of elbow function in tennis players. *Am J Sports Med* 17(2), 1989.)

and serratus anterior muscles in an effort to decrease the common problem of swimmer's shoulder impingement syndrome [19].

Tennis

The shoulder was analyzed during the serve, forehand, and backhand in six male college tennis players who were free of orthopaedic shoulder disorders [78]. Their ages ranged from 18 to 21 years. The testing was done on an outdoor court. All subjects used a single extremity forehand stroke, and two of the six used the two-handed backhand technique. Electromyographic signals were recorded from the middle deltoid, supraspinatus, subscapularis, pectoralis major, biceps brachii, latissimus dorsi, and serratus anterior muscles. The tennis

serve was divided into the same five stages as a baseball pitch. The forehand and backhand ground strokes were divided into three stages (Figs. 15A–13, 15A–14, and 15A–15).

Stage I. *Racquet preparation* begins with shoulder turn and ends with the initiation of weight transfer to the front foot.

Stage II. *Acceleration* begins with weight transfer to the front foot accompanied by forward racquet movement and culminates at ball impact.

Stage III. *Follow-through* begins at ball impact and ends with completion of the stroke.

Analysis of the forehand ground stroke revealed a relatively passive wind-up sequence. Trunk rotation provided some of the force for shoulder motion. In follow-through, there was a marked decrease in activity among the accelerating muscles and a con-

Tennis—Forehand

FIGURE 15A–14
The three stages of a tennis forehand. *Stage 1:* racquet preparation. *Stage 2:* acceleration. *Stage 3:* follow-through. (From Ryu, R, K., Jobe, F. W., et al. An electromyographic analysis of shoulder function in tennis players. *Am J Sports Med* 16(5), 1988.)

Tennis—Backhand

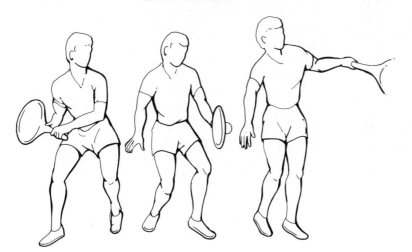

FIGURE 15A–15
The three stages of a tennis backhand. *Stage 1:* preparation. *Stage 2:* acceleration. *Stage 3:* follow-through. (From Ryu, R. K., Jobe, F. W., et al. An electromyographic analysis of shoulder function in tennis players. *Am J Sports Med* 16(5), 1988.)

comitant increase in the external rotators responsible for deceleration. The backhand ground stroke was similar in concept but opposite in muscle activity to the forehand. Follow-through demonstrated deceleration with increased activity of the internal rotators.

The tennis serve requires a complex sequence of muscle activity with phases similar to those observed in baseball pitching. The deltoid muscle function is low during cocking compared to pitching because trunk rotation contributes to shoulder abduction. The acceleration and follow-through stages in the tennis serve demonstrated muscle patterns and activity that were similar to those observed in throwing in baseball. Because the motions for serving the tennis ball were similar to those for pitching, tennis players may benefit from the same conditioning program as that outlined for pitchers. Likewise, emphasis should be placed on rehabilitating the rotator cuff and serratus anterior muscles.

Golf

Normal shoulder activity during a golf swing was evaluated in seven professional golfers [65]. All of the golfers were right-handed; therefore, all references to muscle activity refer specifically to a right-handed golf swing. The muscles monitored included the anterior, middle, and posterior deltoids, supraspinatus, infraspinatus, subscapularis, pectoralis major, and latissimus dorsi on both the right and left sides. The phases of the golf swing were divided into four stages (Fig. 15A–16):

Stage I. *Take away* begins with the initiation of motion in the address position and ends with the backswing.

Stage II. *Forward swing* starts at the termination of the backswing and continues until the club becomes horizontal with the ground.

Stage III. *Acceleration* begins at the horizontal position of the club and ends with ball contact.

Golf

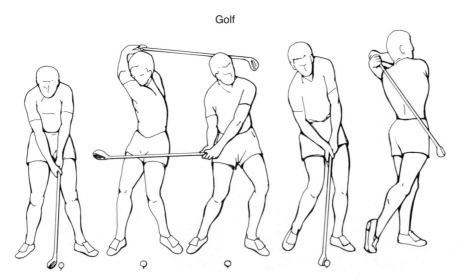

FIGURE 15A–16
Four stages of the golf swing. *Stage 1:* take away. *Stage 2:* forward swing. *Stage 3:* acceleration. *Stage 4:* follow-through. (From Jobe, F. W. Electromyographic shoulder activity in men and women professional golfers. *Am J Sports Med* 17(6), 1989.)

Stage IV. *Follow-through* starts at ball contact and continues until motion ceases.

Analysis of the golf swing revealed a different pattern of activity in its four stages from that of the overhand pitch, the tennis serve, and the swimming stroke. A relative quiescence of the deltoid and dominance of rotator cuff muscle activity was observed. This may be explained by the limited elevation of the arm in the golf swing compared to that in pitching, swimming, or tennis. However, like these other sports, rotator cuff activity in golf was dictated by the demands of motion and by obligatory synergy with the deltoid. The subscapularis was more active than any other muscle throughout the golf swing. In addition, the rotator cuff muscles showed as much activity on the right shoulder as in the left shoulder. The latissimus dorsi and pectoralis major provided power bilaterally and showed marked activity during acceleration. Therefore, the golfer should emphasize conditioning of the rotator cuff, pectoralis major, and latissimus dorsi muscles on both sides of the body.

SUMMARY

These data correlate well with the original work of Inman and colleagues [57]. Slow elevation of the arm in any of the basic planes presents a prolonged three-dimensional challenge to stabilize the arm in space. Consequently, the humeral rotator muscles act in synchrony with the deltoid, which is raising the arm. The EMG data, however, have demonstrated that the rapid, precise motion characterizing individual sports stimulates selective muscle activity and specific periods of great intensity. Such information has provided a physiologic basis for prophylactic and therapeutic exercise protocols [64]. Expansion of this research to include subjects with shoulder pathology has strengthened therapeutic guidelines.

Although each of the sports studied has its unique phasic pattern, there also is a similarity among the four [100]. With the exception of swimming, each arm cycle begins with a gentle approach to the appropriate starting position. Subsequently, the shoulder structures are "cocked" to provide a tense, forceful unit ready for an accelerated release [128]. Once the critical effect has been accomplished, the muscles decelerate the limb so that residual forces do not cause injury. Baseball pitching provides the clearest model, but the same pattern is evident in other sports under different phasic terms.

These data demonstrate the importance of the rotator cuff for shoulder stabilization during athletic demands. The rotator cuff muscles do not always act in synergy with the deltoid. They act according to their mechanical qualities and are function- or sport-specific. Studies that compared professional pitchers with amateurs further refined this principle and demonstrated that individual muscles of the rotator cuff act independently. Evaluation of the different sports revealed that although the rotator cuff function is important in all, the emphasis and role of the individual muscles varied. The importance of serratus anterior muscle activity in stabilizing and protracting the scapula was a consistent finding. Thus, it is critical to have an understanding of the demands placed upon the shoulder in a particular sport or activity to be able to understand and treat pathologic conditions.

References

1. Alderink, G. J., and Kuck, D. J. Isokinetic shoulder strength of high school and college-aged pitchers. *J Orthop Sports Phys Ther* 7(4):163–172, 1986.
2. Anderson, M. Comparison of muscle patterning in the overarm throw and tennis serve. *Res Q* 50:541–553, 1979.
3. Arborelius, U. P., Ekholm, O., Gunner, N., Svennson, O., and Nissel, R. Shoulder joint load and muscular activity during lifting. *Scand J Rehab Med* 18:71–82, 1986.
4. Atwater, A. E. Biomechanics of overarm throwing movements and of throwing injuries. *Exerc Sport Sci Rev* 7:43–85, 1979.
5. Barnes, D. A., and Tullos, H. S. An analysis of 100 symptomatic baseball players. *Am J Sports Med* 6(2):62–67, 1978.
6. Basmajian, J. V., and Bazant, F. J. Factors preventing downward dislocation of the adducted shoulder joint. An electromyographic and morphological study. *J Bone Joint Surg* 41A:1182–1186, 1959.
7. Basmajian, J. V., and DeLuca, C. J. *Muscles Alive: Their Function Revealed by Electromyography.* Baltimore, Williams & Wilkins, 1985, pp. 265–289.
8. Basmajian, J. V., and Latif, A. Integrated actions and functions of the chief flexors of the elbow: A detailed electromyographic analysis. *J Bone Joint Surg* 39A:1106–1118, 1957.
9. Bateman, J. E. Cuff tears in athletes. *Orthop Clin North Am* 4(3):721–745, 1973.
10. Bateman, J. E., and Fornasier, V. L. *The Shoulder and Neck* (2nd ed.). Philadelphia, W. B. Saunders, 1978.
11. Batterman, C. Mechanics of the crawl arm stroke. *Swimming World* 7:4–5, 1966.
12. Bearn, J. G. Direct observation of the function of the capsule of the sternoclavicular joint in clavicular support. *J Anat* 101:159–170, 1967.
13. Bechtol, C. O. Biomechanics of the shoulder. *Clin Orthop* 146:37–41, 1980.
14. Bonci, C. M., Hensal, F. J., and Torg, J. S. A preliminary study on the measurement of static and dynamic motion at the glenohumeral joint. *Am J Sports Med* 14(1):12–17, 1986.
15. Boone, D. C., and Azen, S. P. Normal range of motion of joints in male subjects. *J Bone Joint Surg* 61A:756–759, 1979.
16. Bost, F. C. The pathological changes in recurrent dislocation of the shoulder. *J Bone Joint Surg* 26:595–613, 1942.
17. Brown, L. P., Niehues, S. L., Harrah, A., Yavorsky, P., and Hirshman, P. H. Upper extremity range of motion and

isokinetic strength of the internal and external shoulder rotators in major league baseball players. *Am J Sports Med* 16(6):577–585, 1988.

18. Cain, P. R., Mutschler, T. A., Fu, F. H., and Lee, S. K. Anterior stability of the glenohumeral joint. A dynamic model. *Am J Sports Med* 15(2):144–148, 1987.

19. Clancy, W. G. Shoulder problems in overhead-overuse sports. *Am J Sports Med* 7(2):138–140, 1979.

20. Clarys, J. P. A review of EMG in swimming: Explanation of facts and/or feedback information. *In* Hollander, A. P., Huijing, P., and de Goat, G. (Eds.), *Biomechanics and Medicine in Swimming*. Champaign, IL, Human Kinetics, 1983, pp. 123–135.

21. Clarys, J. P., Kiscott, J., and Lewillie, L. A cinematographical, electromyographic and resistive study of water polo and competition front crawl. *In* Cerguighlini, S., Verano, A., and Wastenmueller, J. (Eds.), *Biomechanics III*. Basel, S. Karger, 1972, pp. 446–452.

22. Codman, E. A. *The Shoulder*. Brooklyn, G. Miller, 1934.

23. Colachis, S. C., Jr., Strohm, B. R., and Brecher, V. L. Effects of axillary nerve block on muscle force in the upper extremity. *Arch Phys Med Rehabil* 50:647–654, 1969.

24. Colachis, S. C., Jr., and Strohm, B. R. Effects of suprascapular and axillary nerve blocks on muscle force in upper extremity. *Arch Phys Med Rehabil* 52:22–29, 1971.

25. Cotton, R. E., and Rideout, D. F. Tears of the rotator cuff: A radiological and pathological necropsy survey. *J Bone Joint Surg* 46B:314–328, 1964.

26. Counsilman, J. E. Forces in swimming two types of crawl stroke. *Res Q* 26:127–139, 1955.

27. Cyprian, J. M., Vasey, H. M., Burdet, A., Bonvin, J. C., Kritsikis, N., and Vaugnat, P. Humeral retrotorsion and glenohumeral relationship in the normal shoulder and in recurrent anterior dislocation (scapulometry). *Gen Orthop* 175:8–17, 1983.

28. deAndrade, M. S., Grant, C., and Dixon, A. St. J. Joint distension and reflex muscle inhibition in the knee. *J Bone Joint Surg* 47A:313–322, 1965.

29. DeHaven, K. E., and Evarts, C. M. Throwing injuries of the elbow in athletes. *Orthop Clin North Am* 4(3):801–808, 1973.

30. DeLuca, C. J., and Forrest, W. J. Force analysis of individual muscles acting simultaneously on the shoulder joint during isometric abduction. *J Biomech* 6:385–393, 1973.

31. Dempster, W. T. Mechanism of shoulder movement. *Arch Phys Med Rehabil* 46A:49, 1965.

32. Dempster, W. T. *Space Requirements of the Seated Operator*. WADC Technical Report 55-159. Washington, D.C., Office of Technical Services, U.S. Dept of Commerce, 1955.

33. DePalma, A. F. *Surgery of the Shoulder* (3rd ed.). Philadelphia, J. B. Lippincott, 1983.

34. DePalma, A. F., Cooke, A. J., and Prabhaker, M. The role of the subscapularis in recurrent anterior dislocations of the shoulder. *Clin Orthop* 54:35–49, 1967.

35. Detrisac, D. A., and Johnson, L. L. *Arthroscopic Shoulder Anatomy. Pathologic and Surgical Implications*. Thorofare, NJ, Slack, 1986.

36. DiStefano, V. Functional anatomy and biomechanics of the shoulder joints. *Athlete Training* 12:141–144, 1977.

37. Doody, S. G., Freedman, L., and Waterland, J. C. Shoulder movements during abduction in the scapular plane. *Arch Phys Med Rehabil* 51:595–604, 1970.

38. Dupuis, R., Adrian, M., Yoneda, Y., and Jack, M. Forces acting on the hand during swimming and their relationships to muscular, spatial, and temporal factors. *In* Teraud, J., Bodingfield, E. W., Nelson, R. C., and Morehouse C. A. (Eds.), *Swimming III*. International Series on Sport Sciences 8. Baltimore, University Park Press, 1979, pp. 110–116.

39. Ekholm, J., Arborelius, U. P., Hillered, L., and Ortqvist, A. Shoulder muscle EMG and resisting moment during diagonal exercise movements resisted by weight and pulley circuit. *Scand J Rehabil Med* 10:179–185, 1978.

40. Elftman, H. Biomechanics of muscle with particular application to studies of gait. *J Bone Joint Surg* 48A:363–377, 1966.

41. Freedman, L., and Munro, R. R. Abduction of the arm in the scapular plane: Scapular and glenohumeral movements. *J Bone Joint Surg* 48A:1503–1510, 1966.

42. Fukuda, K., Craig, E. V., An, K., Cofield, R. H., and Chao, E. Y. Biomechanical study of the ligamentous system of the acromioclavicular joint. *J Bone Joint Surg* 68(3):434–440, 1986.

43. Furlani, J. Electromyographic study of the biceps brachii in movements of the glenohumeral joint. *Acta Anat* 96:270–284, 1976.

44. Galinat, B. J., and Howell, S. M. The containment mechanism: The primary stabilizer of the glenohumeral joint in the horizontal plane. *J Bone Joint Surg* 70(2):227–232, 1988.

45. Galinat, B. J., Howell, S. M., Marone, P. J., and Renzi, A. J. Normal and abnormal mechanics of the glenohumeral joint in the coronal plane. *Am Orthop Assoc Resident Conf* (Abstract) pg. 553.

46. Gainor, B. J., Piotrowski, G., Puhl, J., Allen, W. C., and Hagen, R. The throw: Biomechanics and acute injury. *Am J Sports Med* 8(2):114–118, 1980.

47. Glousman, R., Jobe, F., Tibone, J., Moynes, D., Antonelli, D., and Perry, J. Dynamic electromyographic analysis of the throwing shoulder with glenohumeral instability. *J Bone Joint Surg* 70A:220–225, 1988.

48. Gowan, I. D., Jobe, F. W., Tibone, J. E., Perry, J., and Moynes, D. R. A comparative electromyographic analysis of the shoulder during pitching. *Am J Sports Med* 15(6):586–590, 1987.

49. Hagberg, M. Electromyographic signs of shoulder muscular fatigue in two elevated arm positions. *Am J Sports Med* 60(3):111–121, 1981.

50. Hawkins, R. J., and Hobeika, P. E. Impingement syndrome in the athletic shoulder. *Clin Sports Med* 2:391–405, 1983.

51. Hawkins, R. J., and Kennedy, J. C. Impingement syndrome in athletes. *Am J Sports Med* 8(3):151–158, 1980.

52. Haxton, H. A. Absolute muscle force in the ankle flexors of man. *J Physiol* (Lond) 103:267–273, 1944.

53. Hitchcock, H. H., and Bechtol, C. O. Painful shoulder. *J Bone Joint Surg* 30A:263–273, 1978.

54. Howell, S. M., and Galinat, B. J. The glenoid-labral socket. A constrained articular surface. *Clin Orthop* 243:122–125, 1989.

55. Howell, S. M., Galinat, B. J., Renzi, A. J., and Marone, P. J. Normal and abnormal mechanics of the glenohumeral joint in the horizontal plane. *J Bone Joint Surg* 70A:227–232, 1988.

56. Howell, S. M., Imobersteg, A. M., Seger, S. H., and Marone, P. J. Clarification of the role of the supraspinatus muscle in shoulder function. *J Bone Joint Surg* 68A:398–404, 1986.

57. Inman, V. T., Saunder, M., and Abbott, L. C. Observations of the function of the shoulder joint. *J Bone Joint Surg* 26(1):1–30, 1944.

58. Ivey, F. M., Calhoun, J. H., Rusche, K., and Bierschenk, J. Normal values for isokinetic testing of shoulder strength. *J Bone Joint Surg* 9(1):47, 1985.

59. Jensen, R. K., and Blanksby, B. A model for upper extremity forces during the underwater phase of the front crawl. *In* Lewillie, L., and Clarys, J. P. (Eds.), *Proceedings of the Second International Symposium on Biomechanics in Swimming*. Baltimore, University Park Press, 1974, pp. 145–153.

60. Jiang, C. C., Otis, J. C., Warren, R. F., and Wickiewicz, T. L. Muscle excursion measurements and moment arm determinations of rotator cuff muscles. *Orthop Res Soc* (Abstract) pg. 96.

61. Jobe, F. W. Serious rotator cuff injuries in sports medicine. *Clin Sports Med* 2(2):407–412, 1983.

62. Jobe, F. W. Personal communication, 1980.

63. Jobe, F. W., and Bradley, J. P. Rotator cuff injuries in

baseball. Prevention and rehabilitation. *Sports Med* 6:378–387, 1988.

64. Jobe, F. W., and Moynes, D. R. Delineation of diagnostic criteria and a rehabilitation program for rotator cuff injuries. *Am J Sports Med* 10:336–339, 1982.

65. Jobe, F. W., Moynes, D. R., and Antonelli, D. J. Rotator cuff function during a golf swing. *Am J Sports Med* 14(5):388–392, 1986.

66. Jobe, F. W., Moynes, D. R., Tibone, J. E., and Perry, J. An EMG analysis of the shoulder in pitching. A second report. *Am J Sports Med* 12(3):218–220, 1984.

67. Jobe, F. W., Tibone, J. E., Perry, J., and Moynes, D. R. An EMG analysis of the shoulder in throwing and pitching: A preliminary report. *Am J Sports Med* 11(1):3–5, 1983.

68. Johnston, T. B. The movements of the shoulder joint. *Br J Surg* 25:252, 1937.

69. Jones, D. W., Jr. The role of shoulder muscles in the control of humeral position (an EMG study) (Master thesis). Cleveland, Case Western Reserve University, August 3, 1970.

70. Kaltsas, D. S. Comparative study of the properties of the shoulder joint capsule with those of other joint capsules. *Clin Orthop* 173:20–26, 1983.

71. Kent, B. E. Functional anatomy of the shoulder complex. A review. *Phys Ther* 8:867–888, 1971.

72. King, J. W., Brelsford, H. J., and Tullos, H. S. Analysis of the pitching arm of the professional baseball pitcher. *Clin Orthop* 67:116–123, 1969.

73. Kuland, D. N., McCue, F. C., Rockwell, D. A., and Fiek, J. H. Tennis injuries: Prevention and treatment. *Am J Sports Med* 7(4):249–253, 1979.

74. Lucas, D. B. Biomechanics of the shoulder. *Arch Surg* 107:425–432, 1973.

75. MacCowaill, M. A., and Basmajian, J. V. *Muscles and Movements: A Basis for Human Kinesiology.* New York, Robert Krieger, 1977.

76. Maki, S., and Gruen, T. Anthropometric study of the glenohumeral joint. *Transactions, 22nd Annual Meeting of the American Orthopaedic Research Society* 1:173, 1976.

77. Matsen, F. A., III. Biomechanics of the shoulder. *In* Frankel, V. H., and Nordin, M. (Eds.), *Basic Biomechanics of the Skeletal System.* Philadelphia, Lea & Febiger, 1980.

78. McCormick, J., Ruy, R. K. N., Jobe, F. W., Moynes, D. R., and Antonelli, D. J. An EMG analysis of shoulder function in tennis players. Unpublished study. Inglewood, CA, Biomechanics Laboratory, Centinela Hospital, 1985.

79. McLaughlin, H. L. Rupture of the rotator cuff. *J Bone Joint Surg* 44A:979–983, 1982.

80. Miller, L., Jobe, F. W., Moynes, D. R., Antonelli, D. J., and Perry, J. EMG analysis of shoulders in throwers with subacromial impingement. Unpublished study. Inglewood, CA, Biomechanics Laboratory, Centinela Hospital, 1985.

81. Miyashita, M. Arm action in the crawl stroke. *In* Lewillie, L., and Clarys, J. P. (Eds.), *Proceedings of the Second International Symposium on Biomechanics in Swimming.* Baltimore, University Park Press, 1974, pp. 167–173.

82. Morrey, B. F., and Chao, E. Y. Recurrent anterior dislocation of the shoulder. *In* Black, J., and Dumbleton, J. H. (Eds.), *Clinical Biomechanics. A Case History Approach.* New York, Churchill Livingstone, 1981, pp. 24–46.

83. Moseley, H. F., and Overgaard, B. The anterior capsular mechanism in recurrent anterior dislocation of the shoulder. *J Bone Joint Surg* 44B:913–927, 1962.

84. Mow, V. C., Bigliane, L. U., and Flatow, E. L. Pathophysiology of anterior shoulder instability and cartilage degeneration. 63rd Annual Meeting, New York Orthopaedic Hospital Alumni Association, New York, April, 1989.

85. Moynes, D. R. Prevention of injury to the shoulder through exercises and therapy. Symposium on injuries to the shoulder in the athlete. *Clin Sports Med* 2:413–422, 1983.

86. Moynes, D. R., Perry, J., Antonelli, D. J., and Jobe, F. W. Electromyography and motion analysis of the upper extremity in sports. *Phys Ther* 66(12):1905–1911, 1986.

87. Murray, M. P., Gore, D. R., Gardner, G. M., and Mollinger, L. A. Shoulder motion and muscle strength of normal men and woman in two age groups. *Clin Orthop* 192:268–273, 1985.

88. Nelson, R. C., and Pike, N. L. Analysis and comparison of swimming starts and strokes. *In* Morehouse, C., and Nelson, R. C. (Eds.), *Swimming Medicine IV.* Baltimore, University Park Press, 1977, pp. 347–360.

89. Nuber, G. W., Gowan, I. D., Perry, J., Moynes, D. R., and Jobe, F. W. EMG analysis of classical shoulder motion. Unpublished study. Inglewood, CA, Biomechanics Laboratory, Centinela Hospital, 1984.

90. Nuber, G. W., Jobe, F. W., Perry, J., Moynes, D. R., and Antonelli, D. Fine wire electromyography analysis of muscles of the shoulder during swimming. *Am J Sports Med* 14(1):7–11, 1986.

91. Ohwovoriole, E. N., and Mekow, C. A technique for studying the kinematics of human joints. Part II: The humeroscapular joint. *Orthopaedics* 10(3):457–462, 1987.

92. Ovensen, J., and Nielsen, S. Experimental distal subluxation in the glenohumeral joint. *Arch Orthop Trauma Surg* 104:78–81, 1985.

93. Ovensen, J., and Nielson, S. Anterior and posterior shoulder instability. A cadaver study. *Acta Orthop Scand* 57:324–327, 1986.

94. Ovensen, J., and Nielson, S. Posterior instability of the shoulder. *Acta Orthop Scand* 57:436–439, 1986.

95. Pappas, A. M., Zawacki, R. M., and McCarthy, C. F. Rehabilitation of the pitching shoulder. *Am J Sports Med* 13(4):223–235, 1985.

96. Pappas, A. M., Zawacki, R. M., and Sullivan, T. J. Biomechanics of baseball pitching. A preliminary report. *Am J Sports Med* 13(4):216–222, 1985.

97. Paulos, L. E., France, E. P., Harner, C. D., and Straight, C. B. Biomechanical evaluation of rotator cuff fixation methods. *Orthop Res Soc* (Abstract) pg. 447.

98. Pedegana, L. R., Elsner, R. C., Robert, D., Lang, J., and Farewell, V. The relationship of upper extremity strength of throwing speed. *Am J Sports Med* 10(6):352–354, 1982.

99. Penny, J. N., and Welsh, R. P. Shoulder impingement syndrome in athletes and their surgical management. *Am J Sports Med* 9:11–15, 1981.

100. Perry, J. Anatomy and biomechanics of the shoulder in throwing, swimming, gymnastics, and tennis. *Clin Sports Med* 2(2):247–270, 1983.

101. Perry, J., Antonelli, D., and Gronley, J. Relationships between EMG and force; final report. In press.

102. Poppen, N. K., and Walker, P. S. Forces at the glenohumeral joint in abduction. *Clin Orthop* 135:165–170, 1978.

103. Poppen, N. K., and Walker, P. S. Normal and abnormal motion of the shoulder. *J Bone Joint Surg* 58A:195–201, 1976.

104. Post, M. *The Shoulder: Surgical and Nonsurgical Management* (2nd ed.). Philadelphia, Lea & Febiger, 1988.

105. Priest, J. D., and Nagel, D. A. Tennis shoulder. *Am J Sports Med* 1:28–42, 1976.

106. Reeves, B. Experiments of the tensile strength of the anterior capsular structures of the shoulder in man. *J Bone Joint Surg* 50B:858–865, 1968.

107. Richardson, A. B. Overuse syndromes in baseball, tennis, gymnastics, and swimming. *Clin Sports Med* 2(2):379–390, 1983.

108. Richardson, A. B., Jobe, F. W., and Collins, H. R. The shoulder in competitive swimming. *Am J Sports Med* 8(3):159–163, 1980.

109. Rockwood, C. A., and Odor, J. M. Spontaneous atraumatic anterior subluxation of the sternoclavicular joint. *J Bone Joint Surg* 71A:1280–1288, 1989.

110. Rowe, C. *The Shoulder.* New York, Churchill Livingstone, 1988.

111. Rowe, C. R., and Zarins, B. Recurrent transient subluxation of the shoulder. *J Bone Joint Surg* 63A:863–871, 1981.

112. Ryu, R. K., McCormick, J., Jobe, F. W., Moynes, D. R., and Antonelli, D. J. An electromyographic analysis of

shoulder function in tennis players. *Am J Sports Med* 16(5):481–485, 1988.

113. Saha, A. K. Anterior recurrent dislocation of the shoulder. *Acta Orthop Scand* 68:479–493, 1967.

114. Saha, A. K. Dynamic stability of the glenohumeral joint. *Acta Orthop Scand* 42:491–505, 1971.

115. Saha, A. K. Mechanism of shoulder movements and a plea for the recognition of "zero position" of glenohumeral joint. *Clin Orthop* 173:3–10, 1983.

116. Sarrafian, S. K. Gross and functional anatomy of the shoulder. *Clin Orthop* 173:11–19, 1983.

117. Schleihauf, R. E. A hydrodynamic analysis of swimming propulsion. *In* Terauds, J., and Bedingfield, E. W. (Eds.), *Swimming III. Proceedings of the Third International Symposium on Biomechanics in Swimming.* Baltimore, University Park Press, 1978, pp. 70–109.

118. Seirig, A., and Baz, A. A mathematical model for swimming mechanics. *In* Lewillie, L., and Clarys, U. P. (Eds.), *Swimming, Water Polo and Diving.* Brussels, University Libre, 1970.

119. Shevlin, M. G., Lehmann, J. F., and Lucci, J. A. Electromyographic study of the function of some muscles crossing the glenohumeral joint. *Arch Phys Med Rehabil* 50:264–270, 1969.

120. Slater-Hammel, A. T. An action current study of contraction-movement relationships in the tennis stroke. *Res Q* 20:424–431, 1949.

121. Slocum, D. B. The mechanics of some common injuries to the shoulder in sports. *Am J Surg* 98:394–400, 1959.

122. Staples, O. S., and Watkins, A. L. Full active abduction in traumatic paralysis of the deltoid. *J Bone Joint Surg* 25:85–89, 1943.

123. Sugahara, R. Electromyographic study of shoulder movements. *Jpn J Rehabil Med* 11:41–52, 1974.

124. Sullivan, P. E., and Portney, L. G. Electromyographic activity of shoulder muscles during unilateral upper extremity proprioceptive neuromuscular facilitation patterns. *Phys Ther* 60:283–288, 1980.

125. Tarbell, T. Some biomechanical aspects of the overhead throw. *In* Cooper, J. M. (Eds.), *C.I.C. Symposium on Biomechanics,* Indiana University, 1970. Chicago, Athletic Institute, 1971.

126. Ting, A., Barto, P., Ling, B., Jobe, F. W., and Moynes, D. R. EMG analysis of lateral biceps muscle action in shoulders with rotator cuff tears. Unpublished study. Inglewood, CA, Biomechanics Laboratory, Centinela Hospital, 1986.

127. Townley, C. O. The capsular mechanism in recurrent dislocation of the shoulder. *J Bone Joint Surg* 32A:370–380, 1950.

128. Tullos, H. S., and King, J. W. Throwing mechanism in sports. *Orthop Clin North Am* 4(3):709–720, 1973.

129. Tullos, H. S., Wendell, D. E., Woods, G. W., Wukasch, D. C., Cookley, D. A., and King, J. W. Unusual lesions of the pitching arm. *Clin Orthop* 88:169–182, 1972.

130. Turkel, S. J., Panio, N. W., Marshall, J. L., and Girgis, F. G. Stabilizing mechanisms preventing anterior dislocation of the glenohumeral joint. *J Bone Joint Surg* 63A:1208–1217, 1981.

131. Van Linge, B., and Mulder, J. D. Function of the supraspinatus syndrome. *J Bone Joint Surg* 45B:750–754, 1963.

132. Warren, R. F. Subluxation of the shoulder in athletes. *Clin Sports Med* 2(2):339–355, 1983.

133. Weber, E. F. Ueber die Langenverhaltnisse der Fleischfasen der Muskeln im Allgemeinen. Berichte ueber die Verhandlungen der Saechsischen Gesellschaft der Wissenschaften zu Leipzig, Mathematisch-Physische Klasse, 1951, p. 63.

134. Weiner, D. S., and MacNab, I. Superior migration of the humeral head. *J Bone Joint Surg* 52B:524–527, 1970.

135. Williams, P. L., Warwick, R. *Gray's Anatomy* (36th ed.). Baltimore, Williams & Wilkins, 1980, pp. 453–473.

136. Yoshizawa, M., Ikamoto, T., Kumamoto, M., and Oka, H. Electromyographic study of two styles in the breaststroke as performed by the swimmers. *In* Asmussen, E., and Jorgensen, K. (Eds.), *Biomechanics VI-B.* International Series on Biomechanics 2B. Baltimore, University Park Press, 1978.

Injuries to the Acromioclavicular Joint, the Sternoclavicular Joint, Clavicle, Scapula, Coracoid, Sternum, and Ribs

INJURIES TO THE ACROMIOCLAVICULAR JOINT

Gerald R. Williams, M.D.
Charles A. Rockwood, Jr., M.D.

Acromioclavicular dislocation was recognized as early as 400 B.C. by the father of medicine, Hippocrates (460–377 B.C.) [2]. He not only described the injury but also cautioned against mistaking it for glenohumeral dislocation. Hippocrates wrote, "Physicians are particularly liable to be deceived in this accident (for as the separated bone protrudes, the top of the shoulder appears low and hollow), so that they may prepare for a dislocation of the shoulder; for I have known many physicians otherwise not expert at the art who have done much mischief by attempting to reduce shoulders, thus supposing it as a case of dislocation" [2]. Hippocrates' recommended treatment consisted of compressive bandaging in an attempt to hold the distal end of the clavicle in a reduced position. Nearly 600 years later, the famous Greco-Roman physician Galen (A.D. 129–199) [2] diagnosed his own acromioclavicular dislocation, which he sustained while wrestling in the Palestra. He attempted to treat his dislocation after the method of Hippocrates but soon abandoned this treatment because it was so uncomfortable. It is appropriate that one of the earliest reported cases of acromioclavicular dislocation was related to sports, because certainly today sports participation is one of the most common causes of acromioclavicular dislocations.

The reported incidence of acromioclavicular injury is variable. Rowe and Marble [112] retrospectively reviewed the medical records of the Massachusetts General Hospital and found 52 acromioclavicular joint injuries among 603 shoulder girdle injuries. Most occurred in the second decade of life. Thorndike [128] reported acromioclavicular joint involvement in 223 of 578 athletes with shoulder injuries. In one 5-year period, our institution recorded 520 acromioclavicular injuries [96]. More than 300 of the injuries occurred in the first three decades of life (Table 15B–1). Acromioclavicular dislocation is more common in males (5:1 to 10:1) and is more often incomplete than complete (approximately 2:1) [111].

In 1917, Cadenat [29] emphasized that the treatment of acromioclavicular injuries depends on whether the dislocation is complete or incomplete. Although some authors report surgical repair for selected incomplete dislocations [14], the great majority of type I and type II injuries do not require surgical intervention [19, 23, 47, 54, 60, 95, 100]. The treatment of complete dislocations, however, is much more controversial. Many surgeons adamantly

TABLE 15B–1
Frequency of Acromioclavicular Injury Among 520 Injuries with Respect to Age (by Decade)

1st	2nd	3rd	4th	5th	6th	7th	8th	9th
4	112	226	117	29	5	8	16	3

Age range: 9 to 87
From Rockwood, C. A., and Green, D. P. (Eds.), *Fractures in Adults,* 3rd ed. Philadelphia, J. B. Lippincott, 1990.

recommend operative reduction and repair of all acute, complete acromioclavicular dislocations. Other authors, with equal vigor, advocate routine nonoperative treatment in all cases of complete dislocation. Still other surgeons recommend operative repair in selected circumstances.

The confusion surrounding the treatment of acromioclavicular injuries is not clarified by a review of the literature. Although many articles have been written about acromioclavicular injuries, there are few if any prospective studies comparing the results of operative and nonoperative treatment in two well-matched groups that are large enough to make statistically significant conclusions.

Cadenat [29] accurately described the pathology associated with acromioclavicular dislocation in 1917. The acromioclavicular ligaments initially give way, resulting in a minor sprain. As the injuring force is continued, the coracoclavicular ligament is disrupted, and a complete acromioclavicular dislocation occurs. Some years later Tossy and associates [129] and Allman [7] formally classified acromioclavicular dislocations into three types: (1) a type I injury in which the acromioclavicular ligaments are partially torn, (2) a type II injury in which the acromioclavicular ligaments are completely torn and the coracoclavicular ligament is stretched, and (3) a type III dislocation in which both the acromioclavicular and the coracoclavicular ligaments are completely disrupted.

In 1984, Rockwood [109] expanded the classification to include three additional types of acromioclavicular dislocation. In a type IV injury the clavicle is grossly displaced posteriorly into or through the trapezius muscle. A type V injury is an exaggeration of a type III injury, with severe vertical separation of the clavicle from the scapula. In a type VI injury the clavicle is dislocated inferiorly into either a subacromial or a subcoracoid position. The salient features of this expanded classification are summarized in Figure 15B–1.

ANATOMY

The acromioclavicular joint is a diarthrodial joint located between the lateral end of the clavicle and the medial margin of the acromion process of the scapula (Fig. 15B–2). According to Tyurina [131], the articular surfaces are initially hyaline cartilage. At age 17, on the acromial side of the joint, and at age 24, on the clavicular side, the hyaline cartilage becomes fibrocartilage. Bosworth [22] states that the average size of the adult acromioclavicular joint is 9 mm by 19 mm.

The acromioclavicular joint contains an intra-articular, fibrocartilaginous disc that may be complete or partial (meniscoid). The disc has great variation in size and shape. DePalma [38] has demonstrated that with age the meniscus undergoes rapid degeneration until it is essentially no longer functional beyond the fourth decade. Recent work by Peterson [104] and Salter and colleagues [114] substantiates these findings.

The acromioclavicular joint is surrounded by a thin capsule that is reinforced above, below, anteriorly, and posteriorly by the superior, inferior, anterior, and posterior acromioclavicular ligaments, respectively (Fig. 15B–3). The fibers of the superior acromioclavicular ligament, which is the strongest of the capsular ligaments, blend with the fibers of the deltoid and trapezius muscles. By virtue of their attachments on the superior aspect of the clavicle and the acromion process, the deltoid and trapezius are important dynamic stabilizers of the acromioclavicular joint.

The coracoclavicular ligament is the strongest of the static stabilizers of the acromioclavicular joint. Its fibers run from the outer inferior surface of the clavicle to the base of the coracoid process of the scapula (Figs. 15B–2 to 15B–4). The coracoclavicular ligament has two components, the conoid and the trapezoid ligaments. A bursa may separate the two ligaments. The conoid ligament [68] is cone-shaped, with the apex of the cone attaching to the posteromedial side of the base of the coracoid process. The base of the cone attaches to the conoid tubercle on the posterior undersurface of the clavicle. The conoid tubercle is located at the apex of the posterior clavicular curve, which is at the junction of the lateral third of the flattened clavicle with the middle two-thirds of the triangularly shaped shaft. The trapezoid ligament [68] arises anterior and lateral to the conoid ligament on the coracoid process. This is just posterior to the attachment of the pectoralis minor tendon. The trapezoid ligament extends superiorly to a rough line on the undersurface of the clavicle. This line extends anteriorly and laterally from the conoid tubercle.

Salter and colleagues [114] measured the components of the coracoclavicular ligament in 20 cadavers. They found the trapezoid ligament to vary from 0.80 to 2.50 cm in length and from 0.80 to 2.50 cm in width. The conoid ligament varied from 0.70 to 2.5 cm in length and from 0.40 to 0.95 cm in width. According to Bosworth [22], the average space between the clavicle and the coracoid process is 1.3 cm. Bearden and co-workers [15] reported a range of values for the coracoclavicular interspace of 1.1 to 1.3 cm.

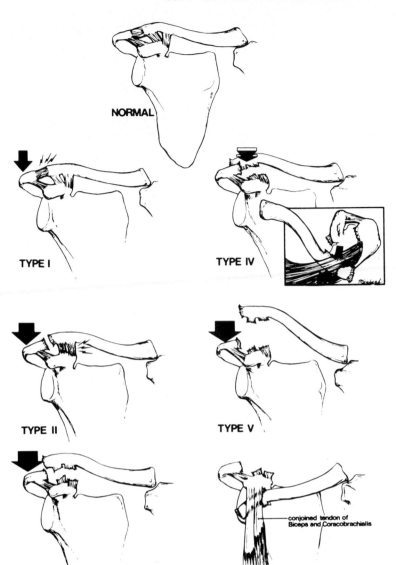

FIGURE 15B–1

Schematic drawings of the classification of ligamentous injuries that can occur to the acromioclavicular joint. *(Type I)* In a type I injury a mild force applied to the point of the shoulder does not disrupt either the acromioclavicular or the coracoclavicular ligament. *(Type II)* A moderate to heavy force applied to the point of the shoulder disrupts the acromioclavicular ligaments, but the coracoclavicular ligaments remain intact. *(Type III)* When a severe force is applied to the point of the shoulder both the acromioclavicular and the coracoclavicular ligaments are disrupted. *(Type IV)* In a type IV injury not only are the ligaments disrupted but the distal end of the clavicle is also displaced posteriorly into or through the trapezius muscle. *(Type V)* A violent force applied to the point of the shoulder not only ruptures the acromioclavicular and coracoclavicular ligaments but also disrupts the muscle attachments and creates a major separation between the clavicle and the acromion. *(Type VI)* This is an inferior dislocation of the distal clavicle in which the clavicle is inferior to the coracoid process and posterior to the biceps and coracobrachialis tendons. The acromioclavicular and coracoclavicular ligaments have also been disrupted. (From Rockwood, C. A., and Green, D. P. (Eds.), *Fractures in Adults*, 2nd ed. Philadelphia, J. B. Lippincott, 1984.)

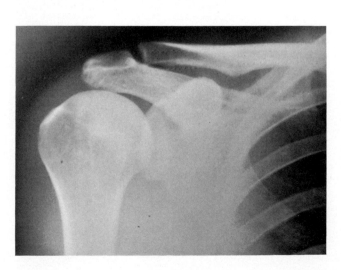

FIGURE 15B–2

Anteroposterior view of the normal shoulder. Note the acromioclavicular joint, the coracoid process, and the coracoclavicular interspace. (From Rockwood, C. A., and Green, D. P. (Eds.), *Fractures in Adults*, 2nd ed. Philadelphia, J. B. Lippincott, 1984.)

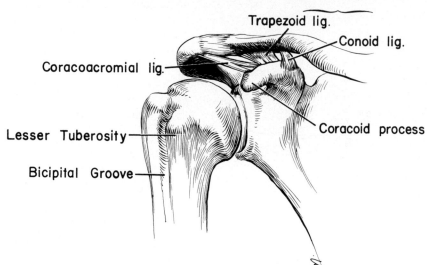

Coracoclavicular lig.

Trapezoid lig.

Conoid lig.

Coracoacromial lig.

Coracoid process

Lesser Tuberosity

Bicipital Groove

FIGURE 15B–3
Normal anatomy of the acromioclavicular joint. (From Rockwood, C. A., and Green, D. P. (Eds.), *Fractures in Adults*, 2nd ed. Philadelphia, J. B. Lippincott, 1984.)

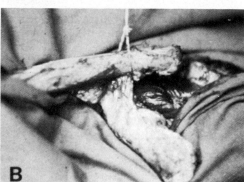

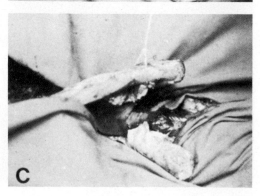

FIGURE 15B–4
The importance of the acromioclavicular and coracoclavicular ligaments for stability of the acromioclavicular joint is shown, using a fresh cadaver. *A*, With the muscles and acromioclavicular capsule and ligaments resected and the coracoclavicular ligaments intact, the clavicle can be displaced anteriorly, as shown, or posteriorly from the articular surface of the acromion. *B*, However, since the coracoclavicular ligaments are intact, the clavicle cannot be displaced significantly upward. *C*, Following the transection of the coracoclavicular ligaments, the clavicle can be displaced completely above the acromion process. This suggests that the horizontal stability of the acromioclavicular joint is accomplished by the acromioclavicular ligaments, and vertical stability is obtained through the coracoclavicular ligaments. (From Rockwood, C. A., and Green, D. P. (Eds.), *Fractures in Adults,* 2nd ed. Philadelphia, J. B. Lippincott, 1984.)

BIOMECHANICS

The coracoclavicular ligament helps to couple glenohumeral abduction and flexion to scapular rotation on the thorax. Full overhead elevation cannot be accomplished without combined and synchronous glenohumeral and scapulothoracic motion [31, 66, 71]. Inmann and co-workers [66] noted that the clavicle rotates about its longitudinal axis through an arc of 40 to 50 degrees during full abduction. As the clavicle rotates upward, it dictates scapulothoracic rotation by virtue of its attachment to the scapula—by means of the conoid and trapezoid ligaments. This "synchronous scapuloclavicular" motion was originally described by Codman [30] and more recently referred to by Kennedy and Cameron [73, 74].

In addition to mediating synchronous scapuloclavicular motion, the coracoclavicular ligament also strengthens and stabilizes the acromioclavicular articulation [1, 7, 14–16, 21, 29, 62, 63, 67, 74, 88, 99, 111, 113, 129]. Cadenat [29] in 1917 emphasized the importance of the coracoclavicular ligament in stabilizing the acromioclavicular joint. He concluded that a moderate blow to the acromion process would rupture the acromioclavicular ligaments, producing an incomplete acromioclavicular dislocation. A heavier blow would then rupture the acromioclavicular and coracoclavicular ligaments and produce a complete dislocation. He agreed with the studies of Poirier and Rieffel (1891) [105], Delbet [as cited by Cadenat, 29], and Mocquot [29], which concluded that the trapezoid and conoid portions of the coracoclavicular ligament must be divided to produce a complete dislocation of the acromioclavicular joint.

Alternatively, Urist [133] concluded that complete acromioclavicular dislocation could occur without rupture of the coracoclavicular ligament. As a result of a series of cadaver experiments, he concluded that division of only the acromioclavicular ligaments and joint capsule, combined with detachment of the deltoid and trapezius muscles, was sufficient to allow "complete dislocation" of the clavicle anteriorly and posteriorly away from the acromion (i.e., in a horizontal plane). Vertical displacement of the clavicle under these circumstances was only minimal. Only after the coracoclavicular ligament was transsected did a complete vertical dislocation of the acromioclavicular joint occur.

Rockwood [109, 110] has repeated some of the cadaver studies of Urist [133] and agrees with his anatomic findings. Indeed, with the muscles and acromioclavicular ligaments detached, the clavicle can be displaced in a horizontal direction, either anterior or posterior to the acromion process (Fig. 15B–4 A). However, only very slight vertical displacement is noted (Fig. 15B–4B). Only when the conoid and trapezoid ligaments have been divided can the lateral end of the clavicle be vertically and totally dislocated above the acromion process (Fig. 15B–4C).

Fukuda and colleagues [49] recently studied the individual ligament contributions to acromioclavicular joint stability. They performed load displacement tests with fixed displacement after sequential ligament sectioning. The contributions of the acromioclavicular, trapezoid, and conoid ligaments were determined at small and large displacements. At small displacements, the acromioclavicular ligaments were the primary restraint to both posterior (89%) and superior (68%) translation of the clavicle. At large displacements, the conoid ligament provided the primary restraint (62%) to superior translation while the acromioclavicular ligaments continued to be the primary restraint (90%) to posterior translation. The trapezoid ligaments served as the primary restraint to acromioclavicular joint compression at both large and small displacements.

These experiments have led to the following conclusions: (1) the horizontal stability of the acromioclavicular joint is controlled by the acromioclavicular ligaments, and (2) vertical stability is controlled by the coracoclavicular ligament.

CLINICAL EVALUATION

History

The athlete who sustains an acromioclavicular dislocation commonly reports either one of two mechanisms of injury—direct or indirect.

Direct Force. Injury by direct force occurs when the athlete falls onto the point of the shoulder with the arm at the side in an adducted position (Fig. 15B–5). This is by far the most common cause of acromioclavicular injury. Other types of collisions sustained in sports such as hockey, football, and karate can also cause acromioclavicular joint injury. The end result of a downward force applied to the superior aspect of the acromion (assuming that there is no sternoclavicular joint injury or clavicle fracture) is injury to either or both of the acromioclavicular and coracoclavicular ligaments.

Indirect Force. Less commonly, the acromioclavicular joint can be injured as the result of an indirectly applied force. The force from a fall on the outstretched hand is transmitted up the arm, through the shaft of the humerus, into the humeral

FIGURE 15B–5
The most common mechanism of injury is a direct force that occurs from a fall on the point of the shoulder. (From Rockwood, C. A., and Green, D. P. (Eds.), *Fractures in Adults*, 2nd ed. Philadelphia, J. B. Lippincott, 1984.)

head, and into the acromion process. The resultant strain is referred only to the acromioclavicular ligaments and not to the coracoclavicular ligament because the coracoclavicular interspace is actually

decreasing during loading (Fig. 15B–6). Acromioclavicular joint injury as the result of a distally directed force applied to the upper extremity has been reported by Liberson [76]. He reported a case in which the scapula was forcibly drawn inferiorly and anteriorly by a sudden change in the position of a heavy burden carried by the patient.

Physical Examination

When an acromioclavicular joint injury is suspected, the patient should be examined in the standing or sitting position whenever possible. The weight of the arm stresses the acromioclavicular joint and exaggerates any existing deformity.

In a type I injury, minimal to moderate tenderness and swelling are present over the acromioclavicular joint without palpable displacement of the joint. Usually there is only minimal pain with arm movement. Tenderness is not present in the coracoclavicular interspace.

Acromioclavicular subluxation, or a type II injury, is characterized by moderate to severe pain at the acromioclavicular joint. If the patient is examined shortly after injury, the outer end of the clavicle may be noted to be slightly superior to the acromion. Shoulder motion produces pain in the acromioclavicular joint. The distal end of the clavicle is unstable in the horizontal plane (i.e., the distal clavicle can be grasped and "rocked" anteriorly and posteriorly). Tenderness can also be elicited in the coracoclavicular interspace.

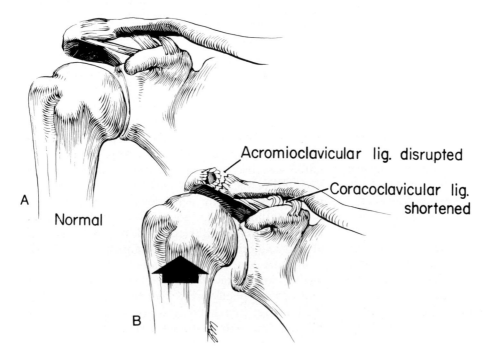

Acromioclavicular lig. disrupted

Coracoclavicular lig. shortened

A Normal

B

FIGURE 15B–6
An indirect force applied up through the upper extremity (e.g., a fall onto the outstretched hand) may displace the acromion superiorly from the clavicle, producing injury to the acromioclavicular ligaments. However, stress is not placed on the coracoclavicular ligaments. (From Rockwood, C. A., and Green, D. P. (Eds.), *Fractures in Adults*, 2nd ed. Philadelphia, J. B. Lippincott, 1984.)

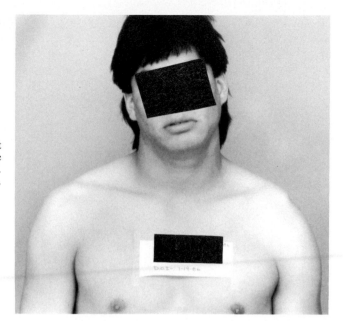

FIGURE 15B–7
This patient has a complete type III dislocation of the left acromioclavicular joint. The left shoulder is drooping, and there is prominence of the left distal clavicle. (From Rockwood, C. A., and Matsen, F. A. (Eds.), *The Shoulder*. Philadelphia, W. B. Saunders, 1990.)

The patient with a type III injury, a complete acromioclavicular dislocation, characteristically presents with the upper extremity held adducted close to his body and supported in an elevated position to relieve the pain in the acromioclavicular joint. The clavicle appears "high-riding" as the shoulder complex droops inferiorly and medially. The clavicle may be prominent enough to tent the skin (Fig. 15B–7). Moderate pain is the rule, and any motion of the arm, particularly abduction, increases the pain. Tenderness is noted at the acromioclavicular joint, the coracoclavicular interspace, and along the superior aspect of the lateral one-fourth of the clavicle. The lateral clavicle is unstable in both the horizontal and vertical planes. According to Cadenat [29], Delbet noted that the clavicle can be so prominent and unstable in this situation that it can be depressed like a piano key.

The hallmark of a type IV acromioclavicular dislocation is posterior displacement of the distal end of the clavicle. The patient exhibits all of the clinical findings of a type III acromioclavicular injury. In addition, examination of the seated patient from above reveals that the outline of the displaced clavicle is inclined posteriorly compared to the contralateral uninjured shoulder (Fig. 15B–8). Occasionally the clavicle is displaced so severely posteriorly that it becomes "button-holed" through the trapezius muscle and tents the posterior skin. Consequently, motion of the shoulder is more painful than that with a type III injury.

The type V acromioclavicular injury is an exaggeration of the type III injury in which the distal end of the clavicle appears to be grossly displaced superiorly toward the base of the neck (Fig. 15B–9). This apparent upward displacement is the result of a downward displacement of the upper extremity and produces a grossly disfigured shoulder. The patient has more pain than with a type III injury, particularly over the distal half of the clavicle. This is secondary to the extensive muscle and soft tissue disruption that occurs with this injury.

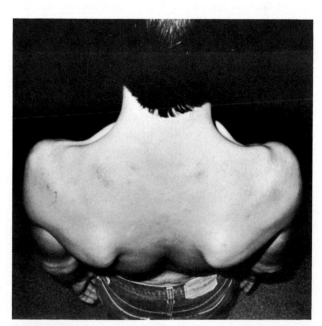

FIGURE 15B–8
Superior view of a seated patient who has a type IV acromioclavicular injury of the right shoulder. The right clavicle has a more posterior inclination than the normal left clavicle. Note the prominent bump on the dorsal posterior aspect of the right shoulder. (From Rockwood, C. A., and Green, D. P. (Eds.), *Fractures in Adults*, 2nd ed. Philadelphia, J. B. Lippincott, 1984.)

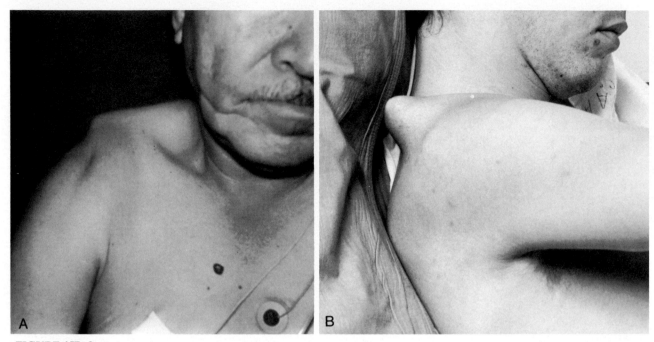

FIGURE 15B–9
Clinical photographs of patient with type V acromioclavicular dislocation. *A,* A severe upward displacement of the right clavicle has occurred into the base of this patient's neck. *B,* Note the severe upward displacement of the clavicle in this patient's right shoulder. There was so much tension on the skin that it was becoming necrotic. (From Rockwood, C. A., and Green, D. P. (Eds.), *Fractures in Adults,* 2nd ed. Philadelphia, J. B. Lippincott, 1984.)

In a type VI, or inferior, acromioclavicular dislocation, the superior aspect of the shoulder has a flat appearance compared to the rounded contour of the normal shoulder. With the subcoracoid dislocation, the acromion is extremely prominent. There is a palpable inferior step-off to the superior surface of the coracoid process. Since subcoracoid dislocation of the clavicle is often the result of severe trauma, there may be associated fractures of the clavicle and upper ribs or injury to the upper roots of the brachial plexus. These associated injuries may produce so much swelling of the shoulder that the disruption of the acromioclavicular joint may not be recognized initially. Accompanying vascular injuries were not present in the patients presented by Patterson [103], McPhee [84], Schwartz and associates [117], and Gerber and Rockwood [50]. However, all of the adult cases of subcoracoid dislocation of the clavicle reported by Patterson [103], McPhee [84], and Gerber and Rockwood [50] had transient paresthesias prior to reduction of the dislocation. Following the reduction, the neurologic deficit cleared.

Radiographic Evaluation

The acromioclavicular joint requires one-third to one-half of the x-ray penetration needed to image the denser glenohumeral joint for good quality radiographs. Radiographs taken of the acromioclavicular joint using routine shoulder technique will be overpenetrated (i.e., dark), and small fractures may be overlooked (Fig. 15B–10 *A*). Therefore, the x-ray technician must be specifically requested to take radiographs of the acromioclavicular joint rather than of the shoulder (Fig. 15B–10*B*).

Anteroposterior Views. Routine anteroposterior views should be taken with the patient standing or sitting, with his back placed against the x-ray cassette and his arms hanging unsupported at his sides. Because of significant individual variation in acromioclavicular joint anatomy and because the coracoclavicular interspace will vary with the angle of the x-ray beam and with the distance between the beam and the patient, both acromioclavicular joints should ideally be imaged simultaneously on one large (14-inch by 17-inch) cassette. Large patients with broad shoulders may be more conveniently studied using two smaller (10-inch by 12-inch) cassettes and identical x-ray technique.

Zanca [142] reviewed 1000 radiographs of patients with shoulder pain and noted that on a true anteroposterior view of the acromioclavicular joint the distal clavicle and acromion were superimposed on the spine of the scapula. Therefore, he recommended a 10- to 15-degree cephalic tilt in order to project an unobscured image of the joint (Figs.

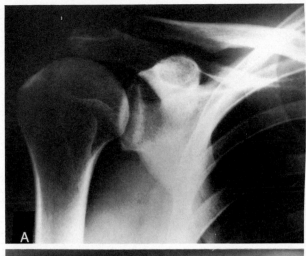

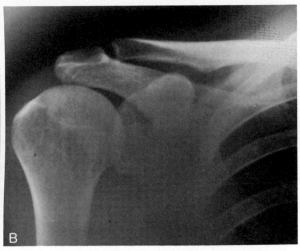

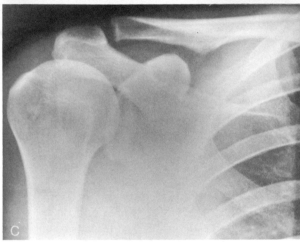

FIGURE 15B–10

Explanation of why the acromioclavicular joint is poorly visualized on routine shoulder x-rays. *A*, This routine anteroposterior view of the shoulder shows the glenohumeral joint well. However, the acromioclavicular joint is too dark to be interpreted because that area of the anatomy has been overpenetrated by the x-rays. *B*, When the exposure usually used to take the shoulder films is decreased by two-thirds, the acromioclavicular joint is well visualized. However, the inferior corner of the acromioclavicular joint is superimposed on the acromion process. *C*, Tilting the tube 15 degrees upward provides a clear view of the acromioclavicular joint. (From Rockwood, C. A., and Green, D. P. (Eds.), *Fractures in Adults*, 2nd ed. Philadelphia, J. B. Lippincott, 1984.)

15B–10*C* and 15B–11). This view is useful, particularly when there is suspicion of a small fracture or loose body on routine views (Fig. 15B–12).

Lateral Views. As with any musculoskeletal injury, a radiograph in one plane is not sufficient to classify an acromioclavicular injury. An axillary lateral view should be taken of the injured as well as the normal shoulder when an acromioclavicular dislocation is suspected. The cassettes should be placed on the superior aspect of the shoulder and should be medial enough to expose as much of the lateral third of the clavicle as possible. This will reveal any existing posterior displacement of the clavicle as well as any small fractures that may have been missed on the anteroposterior view.

Stress Radiographs. Bossart and colleagues [20] reviewed the stress x-rays of 82 patients with suspected acromioclavicular joint injury. Five patients with type III acromioclavicular injury were identified as a result of these stress radiographs. Because of this low percentage yield, Bossart and colleagues did not believe that stress radiographs of the acro-

mioclavicular joint were justified. Patients who present with a clinically obvious acromioclavicular joint injury and typical deformity suggestive of complete dislocation (i.e., types III, IV, or V) often demonstrate maximal coracoclavicular interspace widening on routine anteroposterior views alone. Occasionally, however, the dislocation will not be apparent on routine radiographs because the patient has supported his injured shoulder with the opposite arm and reduced the acromioclavicular joint (Fig. 15B–13). In addition, some cases of type II injury (i.e., subluxation) can be difficult to differentiate from type III (complete) dislocation on routine views. Therefore, stress films of both shoulders to test the integrity of the coracoclavicular ligament should be routinely performed when acromioclavicular dislocation is suspected.

Anteroposterior Stress Views. The technique for anteroposterior stress views is similar to that described above for routine anteroposterior radiographs except that weights (10 to 15 pounds) are suspended from each arm with wrist straps (Fig.

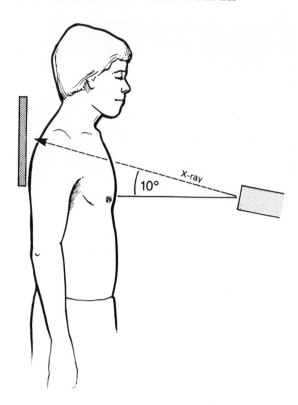

FIGURE 15B–11
Position of the patient for the Zanca view—a 10- to 15-degree cephalic tilt of the x-ray tube for visualizing the acromioclavicular joint. (From Rockwood, C. A., and Matsen, F. A. (Eds.), *The Shoulder*. Philadelphia, W. B. Saunders, 1990.)

15B–14*A*). The weights should be hanging from the wrists rather than held by the patient to encourage complete muscle relaxation (Fig. 15B–14*B*).

Bannister and colleagues [12] recently described "weight-lifting" in addition to "weight-bearing" stress views. An anteroposterior radiograph is taken with a weight suspended from the arm. After infiltration of the acromioclavicular joint with local anesthetic, the patient is asked to lift the weight by shrugging his shoulders. If, during this weight-lifting view, the coracoclavicular distance becomes wider, the patient is thought to have disrupted not only the acromioclavicular and coracoclavicular ligaments but also the entire deltoid and trapezius muscle attachments. Bannister and associates recom-

mended early surgical repair in this group of patients. However, this recommendation was based on only six patients, three of whom were treated nonoperatively. Therefore, the significance of this weight-lifting radiograph is still unclear.

Lateral Stress Views. Alexander [5] described a lateral stress view of the acromioclavicular joint to help identify acromioclavicular dislocation. He called it the "shoulder forward view." The patient is positioned as if a true scapular lateral radiograph is to be taken. The patient is then asked to thrust both shoulders forward. Radiographs are taken of both the injured and the normal side for comparison. The acromioclavicular joint on the normal side will maintain its reduced position. However, on the

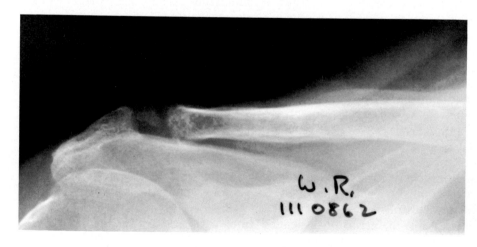

FIGURE 15B–12
The Zanca x-ray view demonstrates a loose body in the acromioclavicular joint. (From Rockwood, C. A., and Matsen, F. A. (Eds.), *The Shoulder*. Philadelphia, W. B. Saunders, 1990.)

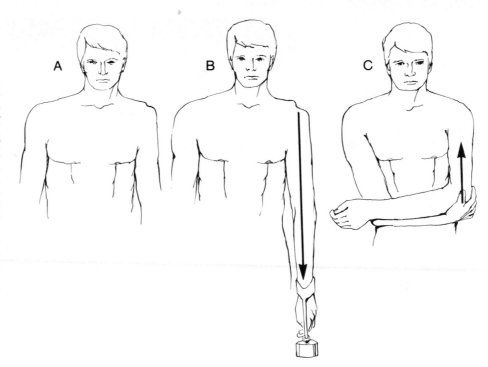

FIGURE 15B–13
Importance of obtaining stress x-rays of the acromioclavicular joint. *A*, The patient presents only a mild deformity in the vicinity of the acromioclavicular joint. *B*, With weight added to the wrist of the upper extremity, the deformity is exaggerated. *C*, If the radiograph is taken of the acromioclavicular joint without weights attached to the upper extremity, the patient may, in an effort to relieve the pain, lift the shoulder upward, presenting a view that shows minimal separation of the acromioclavicular joint. (From Rockwood, C. A., and Green, D. P. (Eds.), *Fractures in Adults*, 2nd ed. Philadelphia, J. B. Lippincott, 1984.)

injured side, the acromion will be displaced anteriorly and inferiorly with respect to the distal clavicle (Fig. 15B–15). Waldrop and co-workers [138] recommended this view for routine evaluation of all acromioclavicular injuries. Some cases of type II injury with significant posterior instability may be difficult to differentiate from complete dislocation using this view.

Radiographic Findings in a Type I Injury. In a type I injury of the acromioclavicular joint the radiographs are normal compared to the uninjured

shoulder except for mild soft tissue swelling. There is no widening, separation, or deformity of the acromioclavicular joint.

Radiographic Findings in a Type II Injury. Radiographs in a patient with a type II injury may reveal a slight elevation of the lateral end of the clavicle. Compared with the normal side, the acromioclavicular joint may appear to be widened. This widening is probably the result of a slight medial rotation of the scapula and slight posterior displacement of the clavicle by the pull of the trapezius muscle. Stress

FIGURE 15B–14
Technique of obtaining stress x-rays of the acromioclavicular joint. *A*, Anteroposterior x-ray films are made of both acromioclavicular joints with 10 to 15 pounds of weight hanging from the wrists. *B*, The distance between the superior aspect of the coracoid and the undersurface of the clavicle is measured to determine whether or not the coracoclavicular ligaments have been disrupted. One large horizontally positioned 14- by 17-inch x-ray cassette can be used in small patients to visualize both shoulders on the same film. In large patients it is better to use two horizontally placed smaller cassettes and take two separate films to obtain the measurements. (From Rockwood, C. A., and Green, D. P. (Eds.), *Fractures in Adults*, 2nd ed. Philadelphia, J. B. Lippincott, 1984.)

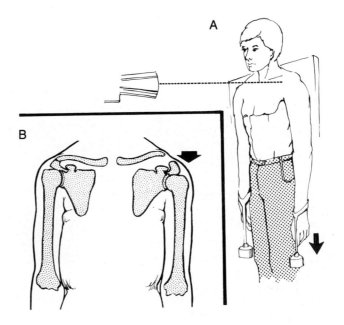

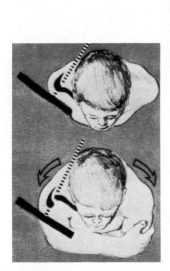
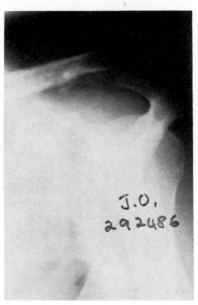
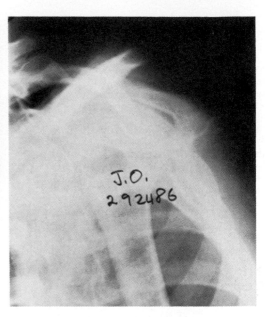

FIGURE 15B–15

Technique of obtaining the Alexander or scapular lateral view to evaluate injuries of the acromioclavicular joint. *(Left),* Schematic drawing illustrating how the shoulders are thrust forward at the time the radiograph is taken. *(Center),* Alexander view taken with the shoulder in the relaxed position. Note that the acromioclavicular joint is only minimally displaced. *(Right),* With the shoulders thrust forward, there is gross displacement of the acromioclavicular joint, and the acromion is displaced anteriorly and inferiorly under the distal end of the clavicle. (From Rockwood, C. A., and Green, D. P. (Eds.), *Fractures in Adults*, 3rd ed. Philadelphia, J. B. Lippincott, 1991.)

films of both shoulders reveal that the coracoclavicular interspace of the injured shoulder is the same as that of the normal shoulder (Fig. 15B–16).

Radiographic Findings in a Type III Injury. In

obvious cases of complete acromioclavicular dislocation the joint is totally displaced. The lateral end of the clavicle is displaced completely above the superior border of the acromion, and the coracocla-

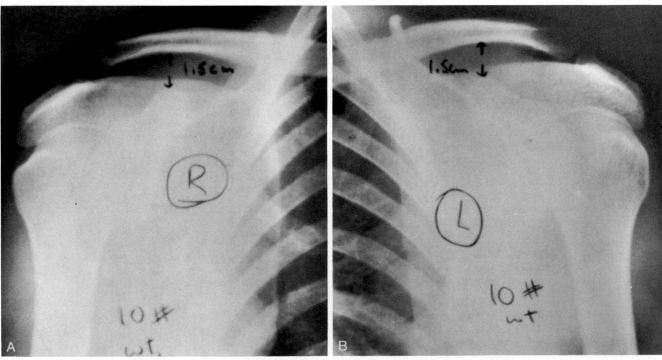

FIGURE 15B–16

X-ray appearance of a type II acromioclavicular joint injury to the right shoulder. With stress, the coracoclavicular distance in both shoulders measures 1.5 cm. However, in the injured right shoulder the acromioclavicular joint is widened compared with the normal left shoulder. (From Rockwood, C. A., and Green, D. P. (Eds.), *Fractures in Adults*, 2nd ed. Philadelphia, J. B. Lippincott, 1984.)

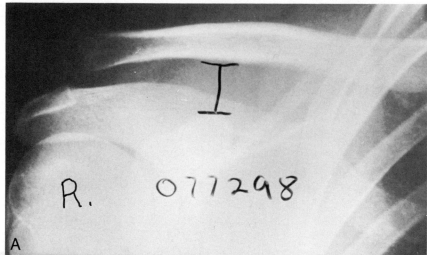

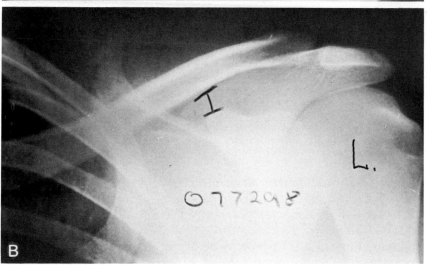

FIGURE 15B–17
X-ray appearance of a type III injury to the right shoulder. Stress x-ray films were made to compare the right with the left shoulder. Not only is the right acromioclavicular joint displaced compared with the left, but, more significantly, there is a great increase in the coracoclavicular interspace in the injured right shoulder compared with the normal left shoulder. (From Rockwood, C. A., and Green, D. P. (Eds.), *Fractures in Adults*, 2nd ed. Philadelphia, J. B. Lippincott, 1984.)

vicular interspace is significantly greater than that in the normal shoulder (Fig. 15B–17). When the case is questionable, which it often is, stress films comparing the injured to the normal shoulder will reveal a major discrepancy in the coracoclavicular distances (25% to 100%).

Rarely, complete acromioclavicular dislocation will be accompanied by a fracture of the coracoid process rather than by disruption of the coracoclavicular ligaments. Although a fracture of the coracoid process is frequently difficult to visualize on routine radiographs, its existence should be suspected when there is complete acromioclavicular separation in association with a normal coracoclavicular interspace compared to the uninjured shoulder. The best "special view" for visualizing the coracoid fracture is the Stryker notch view (Fig. 15B–18). This view is taken with the patient supine, the involved arm flexed, and the palm placed on top of the head. The x-ray beam is angled 10 degrees cephalically (see Fig. 15B–18).

Children and adolescents occasionally sustain a variant of complete acromioclavicular dislocation in which the clavicle is relatively displaced superiorly through a longitudinal rent in the periosteal sleeve [45, 58, 71]. In spite of the apparent superior displacement of the clavicle, the acromioclavicular ligaments, the coracoclavicular ligament, and the confluent deltotrapezius fascia are not torn and remain attached to the intact periosteal sleeve. Falstie-Jensen and Mikkelsen [45], as well as Katznelson and colleagues [71] have reported this injury in young adults as well as children. Havranek [58] reported on 10 children with a relative superior displacement of the clavicle associated with a Salter-Harris II fracture of the distal clavicular physis rather than an acromioclavicular dislocation.

Radiographic Findings in a Type IV Injury. Although a type IV injury can be associated with relative upward displacement of the clavicle and an increase in the coracoclavicular interspace, the most striking feature of this injury is posterior displace-

FIGURE 15B–18
Technique for taking the Stryker notch view to demonstrate fractures of the base of the coracoid. The patient is supine with a cassette placed posterior to the shoulder. The humerus is flexed approximately 120 degrees so that the patient's hand can be placed on top of the head. The x-ray beam is directed 10 degrees superiorly. (From Rockwood, C. A., and Matsen, F. A. (Eds.), *The Shoulder*. Philadelphia, W. B. Saunders, 1990.)

ment of the distal clavicle as seen on the axillary lateral x-ray (Fig. 15B–19*A*, *B*). In patients in whom an axillary lateral radiograph cannot be taken (i.e., patients with heavy, thick shoulders or patients with multiple injuries), a CT scan may be of great value in confirming clinical suspicions of a posteriorly dislocated acromioclavicular joint (Fig. 15B–19*C*, *D*).

Radiographic Findings in a Type V Injury. The characteristic radiographic feature of a type V acromioclavicular dislocation is a marked (2 to 3 times normal) increase in the coracoclavicular interspace (Fig. 15B–20). The clavicle appears to be grossly displaced superiorly away from the acromion. However, radiographs reveal that the clavicle on the injured side is actually at approximately the same level as the clavicle on the normal side and that the scapula is displaced inferiorly (Fig. 15B–21).

Radiographic Findings in a Type VI Injury. There are two basic types of inferior acromioclavicular dislocation: subacromial and subcoracoid. Radiographic findings in the subacromial type include a decreased coracoclavicular distance (i.e., less than that on the uninjured side) and subacromial displacement of the distal end of the clavicle. A subcoracoid dislocation is characterized by a reversed coracoclavicular distance, the clavicle being displaced inferior to the coracoid process. Since this

injury is usually the result of severe trauma, it is often accompanied by multiple fractures of the clavicle and ribs.

Other Modalities

Schmid and colleagues [116] reported the use of ultrasonography for the diagnosis of 22 cases of type III acromioclavicular dislocation. Ultrasound demonstrated visible instability of the distal clavicle, incongruity of the joint, hematoma formation, or visible ligament remnants in all cases. However, in spite of the advent of such sophisticated imaging modalities as ultrasound, CT scan, and magnetic resonance imaging, plain radiography continues to be the most readily available and cost-effective method for routine investigation of injuries to the acromioclavicular joint.

TREATMENT OPTIONS

Types I and II Injuries

Most authors [19, 23, 47, 54, 60, 95, 100] recommend nonsurgical treatment for incomplete acromioclavicular dislocations (i.e., type I and type II injuries). However, a report by Bergfeld and colleagues [17] and a study by Cox [32] suggest that untreated type I and type II injuries may lead to more chronic disability than was previously recognized.

Many types of nonoperative treatment have been reported for type II acromioclavicular dislocations. Some authors attempt to reduce the subluxation through the use of compressive bandages and slings, adhesive tape strapping, braces, harnesses, traction techniques, and many types of plaster casts. Urist [133–135] reviewed the literature extensively and summarized more than 35 forms of nonoperative management for acromioclavicular dislocations. He advocated the use of a plaster cast device that incorporated an elastic strap to support the arm and depress the clavicle. Allman [7] recommended the use of a Kenny-Howard sling or harness (Fig. 15B–22) for 3 weeks. Regardless of the method of reduction or immobilization chosen, 3 to 6 weeks of continuous uninterrupted pressure on the superior aspect of the clavicle are required for the ligaments to heal and to maintain the reduction. Alternatively, many authors recommend a sling for 10 to 14 days, or until symptoms subside. This is followed by an early and gradual rehabilitation program.

Persistent pain following a type II acromioclavicular injury has been reported by Bateman [14].

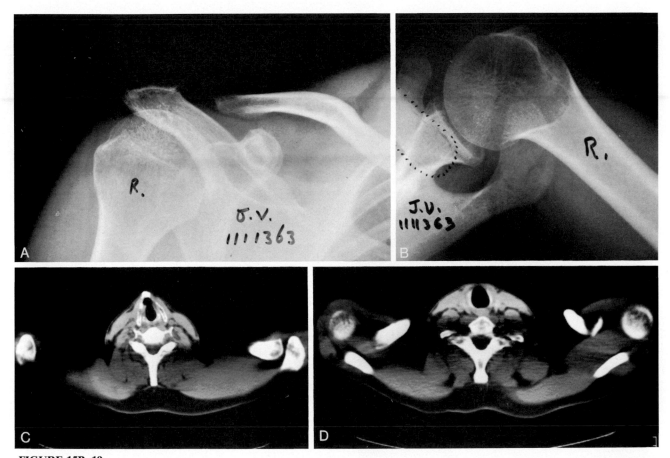

FIGURE 15B–19
Type IV dislocation of the acromioclavicular joint. *A*, The anteroposterior x-ray reveals obvious deformity of the acromioclavicular joint. The distal end of the clavicle appears to be inferior to the acromion. *B*, The axillary view confirms that the clavicle is displaced posteriorly away from the acromion process. *C*, CT scans reveal that the left clavicle is in its normal position adjacent to the acromion. Note that the right clavicle is completely absent. *D*, A lower CT cut demonstrates the posterior displacement and also confirms the fact that the clavicle is inferior to the acromion. (From Rockwood, C. A., and Matsen, F. A. (Eds.), *The Shoulder*. Philadelphia, W. B. Saunders, 1990.)

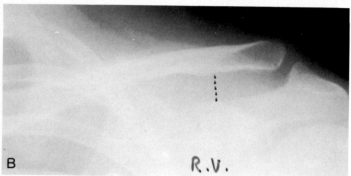

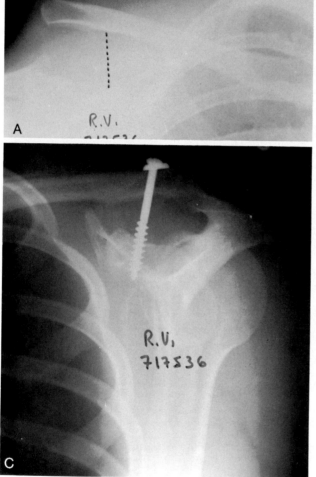

FIGURE 15B–20

X-ray appearance of a type V acromioclavicular joint injury. *A*, Note on the soft tissue radiograph the severe displacement of the right clavicle compared with the normal left shoulder. *B*, The coracoclavicular interspace in the right shoulder was almost three times that of the left shoulder. *C*, At the time of operative repair there was considerable soft tissue injury with stripping of the trapezius and deltoid muscles off the clavicle. The clavicle was reduced down to its normal articulation with the acromion and temporarily held in place with a coracoclavicular lag screw while the repaired coracoclavicular ligaments healed. (From Rockwood, C. A., and Green, D. P. (Eds.), *Fractures in Adults*, 2nd ed. Philadelphia, J. B. Lippincott, 1984.)

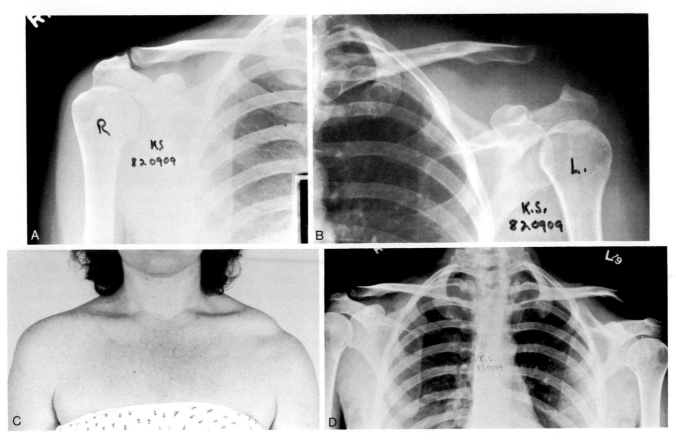

FIGURE 15B–21

A–D, Comparison of the clinical deformity with the x-ray deformity in a patient with a type V injury. Note that a line drawn through both clavicles would indicate that they are essentially at the same level. The deformity is due to the inferior displacement of the upper extremity, which is secondary to the loss of the coracoclavicular suspensory ligaments. (From Rockwood, C. A., and Matsen, F. A. (Eds.), *The Shoulder.* Philadelphia, W. B. Saunders, 1990.)

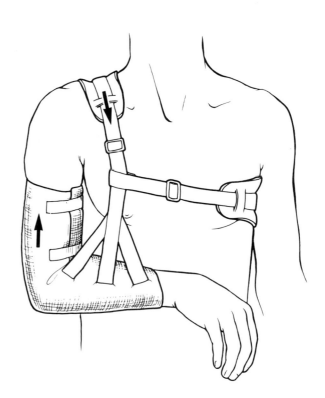

FIGURE 15B–22

Schematic drawing of shoulder harness (Kenny-Howard type) that is used in the nonoperative treatment of injuries of the acromioclavicular joint. The strap that runs over the top of the shoulder and under the elbow is tightened sufficiently to reduce the clavicle to the acromion. A halter strap around the trunk keeps the harness from slipping off the top of the shoulder. (From Rockwood, C. A., and Green, D. P. (Eds.), *Fractures in Adults,* 2nd ed. Philadelphia, J. B. Lippincott, 1984.)

Causes of persistent symptoms include post-traumatic osteolysis of the clavicle, torn capsular ligaments trapped within the joint, loose pieces of articular cartilage, and a detached intra-articular meniscus that is displaced in and out of the joint like a torn meniscus in the knee. Bateman [14] refers to these persistent problems as "internal derangement" of the acromioclavicular joint. Brosgol [24] reported on the use of an arthrogram of the acromioclavicular joint to delineate the pathology associated with persistent acromioclavicular symptoms.

Acromioclavicular joint arthroplasty may be required to relieve persistent pain following a type II acromioclavicular injury. If the articular surface of the clavicle is degenerative, excision of the distal 2 cm of the clavicle, in addition to joint debridement and meniscectomy, is required. Because the coracoclavicular ligament is intact, the scapula is not displaced inferiorly and the coracoclavicular interspace is preserved. This procedure has been described by Mumford [90] and Gurd [57].

Type III Injuries

The treatment of a type III acromioclavicular dislocation is extremely controversial. The earliest writings on the subject by Hippocrates, Galen, and Paul of Aegina recommended closed reduction and the use of compressive bandages to maintain the clavicle in the reduced position [2]. In 1974, Powers and Bach [106] polled all chairmen of approved residency training programs in the United States on this subject and reported the following findings: (1) The majority of program chairmen treated type III acromioclavicular injuries by open reduction, (2) 60% of the chairmen used temporary acromioclavicular fixation, and 35% used coracoclavicular fixation, and (3) nonoperative treatment was rarely advocated and was often inadequate.

Several recent articles seem to indicate that nonoperative treatment is enjoying a resurgence. Anzel and Streitz [8] reported good results in 31 patients with type II and type III acromioclavicular injuries treated with a dynamic splint. Walsh and associates [139], Glick and associates [52], Sellers and colleagues [119], and MacDonald and colleagues [81] could document little significant strength deficit in patients treated nonoperatively for complete acromioclavicular dislocation. Jakobsen [66] recommended nonoperative treatment over operative treatment for complete acromioclavicular dislocations because he noted similar functional results in both groups, shorter convalescent time in the non-

operatively treated group, and the association of more easily managed complications with the nonoperatively treated group.

In 1987, Larson and Hede [75] prospectively compared three methods of treatment for acromioclavicular dislocation: (1) reduction of the joint, repair of the coracoclavicular ligament, imbrication of the deltotrapezius fascia, and acromioclavicular fixation; (2) short-term immobilization followed by early range of motion; and (3) closed reduction and bandaging. Although their patient groups were not specifically matched, the results are informative. Two of twenty-five patients in group I required reoperation because of a persistently painful acromioclavicular joint; both underwent distal clavicle excision with excellent results. Osteolysis of the clavicle and persistent pain in association with acromioclavicular fixation have been reported by other authors [44]. Three of twenty-nine patients in group II required late operation, two because of pain and one because the distal clavicle was tenting the skin. All three patients underwent distal clavicle excision combined with coracoclavicular ligament reconstruction, leading to two excellent results and one good result. Larson and Hede [75] recommended early surgical repair in patients with extreme prominence of the clavicle and in patients whose occupation demanded prolonged heavy lifting or repeated use of the arm with the shoulder abducted or flexed to 90 degrees.

Nonoperative Treatment

Numerous methods of nonoperative treatment of complete acromioclavicular dislocation have been reported (Table 15B–2). The two most commonly used methods are (1) closed reduction and application of a sling and harness device to maintain the reduction of the clavicle and (2) short-term sling support followed by early range of motion. Multiple studies have reported good functional results in spite of residual deformity in patients treated in this manner of "skillful neglect" [59, 64, 66, 81]. Although plaster casts are occasionally still used today, they are no longer popular. This is undoubtedly due in part to the reported good results with other, more convenient forms of nonoperative management.

The prototype of the sling and harness immobilization device is the Kenny-Howard sling. The sling supports the forearm and arm. A strap or harness over the top of the shoulder and down around the elbow is tightened so that it depresses the clavicle (Fig. 15B–23). A separate halter strap around the trunk keeps the harness from slipping off the top of

TABLE 15B–2
Nonoperative Treatment for Acromioclavicular Dislocation Reported in the Literature

Form of Treatment	Authors*
Adhesive strapping	Rawlings, Thorndike and Quigley, Benson, Bakalim and Wilppula
Sling or bandage	Jones, Watson-Jones, Hawkins
Brace and harness	Giannestras, Warner, Currie, Anderson and Burgess, Anzel and Streitz
Crotch loops, stocking, garter, and strap	Darrow and associates, Spigelman, Varney and associates
Figure-of-eight bandage	Usadel
Sling and pressure dressing	Goldberg
Abduction traction and suspension in bed	Caldwell
Casts	Urist, Howard, Shaar, Hart, Trynin, Stubbins and McGaw, Dillehunt, Key and Cornwell, Gibbens

*See original for complete reference data.
From Rockwood, C. A., and Green, D. P. (eds.), Fractures in Adults, 3rd ed. Philadelphia, J. B. Lipppincott, 1991, p 884.

the shoulder. To maintain the reduction, the sling and harness must maintain continuous pressure under the elbow and on top of the clavicle for 6 weeks. Additional methods of reduction and immobilization have been described. These include crotch loops, stockings, garter straps, and special pads [36, 126, 137].

Convenience, shorter rehabilitation times, dissatisfaction with the results of surgical repairs, and reports of good functional outcomes have led some authors to favor "skillful neglect" over other forms of treatment. Nicoll [97] states that operative procedures for acromioclavicular dislocation are unwarranted and that maintenance of reduction by closed means is unattainable. He recommends a sling for 1 to 2 weeks followed by exercises. Hawkins [59], Imatani and colleagues [64], Bjerneld and colleagues [19], Dias and associates [40], Sleeswijk-Visser and co-workers [124], and Schwarz and Leixnering [118] all reported 90% to 100% satisfactory results with 5- to 7-year follow-up in patients with type III acromioclavicular dislocations treated by skillful neglect.

This form of treatment is especially popular among physicians who manage athletic injuries. Athletes who sustain an injury to the acromioclavicular joint are very likely to be back in competition within a few days or weeks, subjecting themselves to the same type of violent contact. In addition, the shorter rehabilitation time associated with this form of treatment allows the athlete to return to competition sooner. Glick [51] especially favors skillful neglect in athletes because it allows rapid and safe return to competition. Glick and associates [52] reported on 35 athletes with chronic, unreduced type III acromioclavicular dislocations. The results were good, and no athlete was disabled. They recommend skillful neglect and emphasize the role of aggressive postinjury strengthening.

Operative Treatment

A review of over 300 articles pertaining to the operative treatment of acromioclavicular dislocation reveals that approximately half have contributed only a new surgical technique or a new twist to an old technique. Although many specific techniques have been reported, only four basic types of acromioclavicular repairs are currently being performed: (1) acromioclavicular repair, (2) coracoclavicular repair, (3) distal clavicle excision, and (4) dynamic

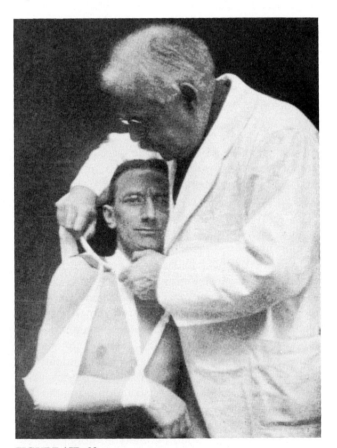

FIGURE 15B–23
Closed treatment of dislocations of the acromioclavicular joint. Sir Robert Jones is seen applying a sling and bandage that holds the arm elevated while depressing the lateral end of the clavicle. (From Jones, R. *Injuries of Joints*. London, Oxford University Press, 1917. By permission of Oxford University Press.)

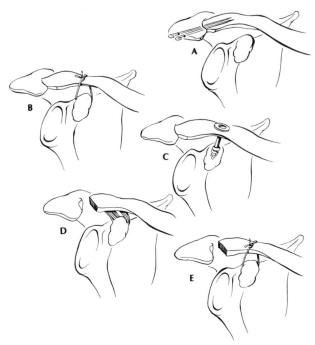

FIGURE 15B–24

Various operative procedures used for treatment of injuries to the acromioclavicular joint. *A,* Steinmann pins across the acromioclavicular joint. *B,* Suture between the clavicle and the coracoid process. *C,* A lag screw between the clavicle and the coracoid process. *D,* Resection of the distal clavicle when the coracoclavicular ligaments are intact. *E,* Resection of the distal clavicle with suture, fascia, or ligament between the clavicle and the coracoid process when the coracoclavicular ligaments are missing. (From Rockwood, C. A., and Green, D. P. (Eds.), *Fractures in Adults,* 3rd ed. Philadelphia, J. B. Lippincott, 1991.)

muscle transfers. Many of the specific procedures being used today are combinations, modifications, or modifications of modifications of previously described procedures (Fig. 15B–24).

Intra-articular Acromioclavicular Repair

Sage and Salvatore [113] advocated acromioclavicular ligament repair and reinforcement of the superior acromioclavicular ligament with the intra-articular, fibrocartilaginous meniscus. Neviaser [92–94] introduced superior acromioclavicular ligament reconstruction through transfer of the coracoid attachment of the coracoacromial ligament to the superior aspect of the clavicle. He did not recommend repair of the coracoclavicular ligament. Ho and colleagues [61] also advocated superior acromioclavicular ligament reconstruction using the coracoacromial ligament. Various other methods of acromioclavicular ligament repair or reconstruction have been recommended by many authors [3, 9, 14, 29, 65, 89, 99, 125]. Bundance and Cook [27] as well as Bartonicek and colleagues [13] emphasized the importance of imbricating the deltotrapezius fascia to reinforce the acromioclavicular repair.

Many methods of acromioclavicular fixation have been reported in conjunction with acromioclavicular repair or reconstruction. Smooth pins of one type or another have been recommended by numerous authors [9, 13, 14, 27, 35, 46, 65, 85, 89, 92–94, 99, 113, 143]. A Stuck nail was advocated by Stephens [126A]. Ahstrom [3] advocated the use of a 9/64-inch, fully threaded pin. Linke and Moschinski [79] supplemented two smooth pins with a superior acromioclavicular wire that acted like a tension band. A partially threaded screw was introduced by Simmons and Martin [122]. Vainionpaa and colleagues [136] and Paavolainen and associates [101] recommended AO cortical and malleolar screws. Acromioclavicular fixation using the Balsar hook plate has been reported by Albrecht [4], Dittel and colleagues [41], Dittmer and co-workers [42], Kaiser and associates [69], Mlasowski and associates [86], and Shindler and colleagues [115].

A comparison of acromioclavicular methods of fixation was reported by Eskola and co-workers [44]. They performed a prospective, randomized trial of 100 patients using three types of acromioclavicular fixation: (1) smooth pins, (2) threaded pins, and (3) a cortical screw. Thirteen of the eighty-six patients who were available for review developed symptomatic osteolysis of the distal clavicle. Eight of the thirteen cases of osteolysis occurred in 25 patients who underwent cortical screw fixation. Smith and Stewart [125] advocated distal clavicle excision in conjunction with acromioclavicular fixation as a means of decreasing degenerative changes in the acromioclavicular joint.

Pin migration is a serious complication that has been reported with acromioclavicular fixation. Mazet [82] reported migration of a Kirschner wire into the lung 76 days after its insertion into the right acromioclavicular joint. Norrell and Llewellyn [98] reported migration of a Steinmann pin from the right acromioclavicular joint into the spinal canal. The pin was found in the subarachnoid space anterior to the spinal cord. It extended transversely through the spinal cord at the level of the first thoracic vertebra. It was easily removed in the direction from which it came. Lindsey and Gutowski [77] reported migration of a Kirschner wire into the neck posterior to the carotid sheath. Eaton and Serletti [43] removed a pin from the pleura of the lung that had migrated across the midline of the body. Urban and Jaskiewicz [132] encountered a case of pin migration into the ipsilateral pleural cavity 3 months after surgery. Sethi and Scott [120] reported laceration of the subclavian artery from a Hagie pin that had migrated from the ipsilateral acromioclavicular joint. Retief and Meintjes [108] reported a case in which a Kirschner wire placed in the acromioclavicular joint had migrated through the thoracic cavity and was lodged behind the liver. During the 6 weeks that elapsed between the Kirschner wire's insertion and its removal, the patient had suffered a pneumothorax secondary to the wire's penetration of the lung. Grauthoff and Klammer [56] reported five cases of pin migration into the aorta, subclavian artery, or lung. Although the incidence of pin migration can be diminished by bending the portion of the pin that protrudes from the skin, migration can still occur if the pin breaks. Patients should be warned of this complication and informed that the pins must be removed. Patients should be followed radiographically on a regular basis until the pin has been removed. The pin should be removed prematurely should any evidence of migration be apparent radiographically.

Extra-articular Coracoclavicular Repairs

The use of a coracoclavicular lag screw in the treatment of complete acromioclavicular dislocations was introduced by Bosworth [21] in 1941. The screw was placed percutaneously, using local anesthesia and fluoroscopic guidance. Bosworth did not recommend either repair of the coracoclavicular ligaments or exploration of the acromioclavicular joint. He referred to the procedure as a screw suspension operation, not a fixation, because the screw suspended the scapula from the clavicle rather than fixed it to it.

Percutaneous insertion of a cannulated coracoclavicular screw was also reported by Tsou [130] in 1989. Tsou fluoroscopically placed a guide pin from the clavicle to the coracoid process. After adequate positioning of the pin within the coracoid had been confirmed radiographically, a cannulated drill bit and screw were sequentially passed over the guide pin. Tsou reported a 32% technical failure rate in 53 patients with complete acromioclavicular dislocation using this technique.

Several authors have reported good results using coracoclavicular fixation and various modifications of Bosworth's original technique. In 1968, Kennedy and Cameron [74] and Kennedy [73] reported on repair of acromioclavicular dislocations using a coracoclavicular technique. A Bosworth screw was used to bring the clavicle into contact with the top of the coracoid process. The acromioclavicular joint was thoroughly debrided, the deltoid and trapezius muscles were repaired back to the clavicle, and the coracoclavicular ligament was preserved but not repaired. In acute cases, bone fragments created by drilling the hole in the clavicle were placed in the coracoclavicular interspace in an effort to gain permanent bone fixation between the clavicle and the coracoid. Weitzman [141] reported using a Bosworth screw inserted open, under general anesthesia. Although he did not expose or repair the coracoclavicular ligament, he did expose and debride the acromioclavicular joint and imbricate the deltoid and trapezius muscle attachments.

Jay and Monnet [67] reported a modification of Bosworth's technique that they learned from Amspatcher, at the University of Oklahoma Medical Center. In a series of 31 cases of acromioclavicular joint dislocation in which the acromioclavicular joint was debrided, the coracoclavicular ligaments were repaired, a Bosworth screw was used to hold the acromioclavicular joint reduced temporarily, and the trapezius and deltoid muscles were repaired back to the clavicle (Fig. 15B–25). Jay and Monnet [67] recommended intraoperative radiography to ensure proper placement of the screw. Lowe and Fogarty [80] used a similar technique in 21 patients.

Other methods of coracoclavicular fixation have been reported. Beardon and co-workers [15] as well as Albrecht [12] recommended the use of two loops of stainless steel wire around the coracoid process and clavicle. Synthetic Dacron, arterial graft, or velour Dacron graft between the clavicle and the coracoid process were reported by Tagliabue and Riva [127], Park and colleagues [102], Nelson [91], Kappakas and McMaster [70], Fleming and associates [48], and Goldberg and colleagues [55]. Bunnell [28] in 1928 and Lom [79] in 1988 reported the use of fascia lata for reconstruction of the coracoclavicular ligaments.

Coracoclavicular fixation using grafts or synthetic material has been associated with a variety of complications. Goldberg and colleagues [55] recognized

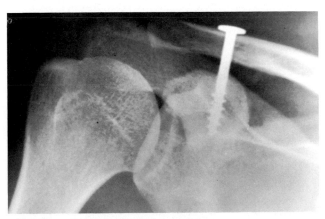

FIGURE 15B–25
Postoperative anteroposterior x-ray of the shoulder with Bosworth screw in place. Note that the acromioclavicular joint has been reduced and the coarse lag threads of the screw are well seated into the coracoid process. (From Rockwood, C. A., and Green, D. P. (Eds.), *Fractures in Adults*, 2nd ed. Philadelphia, J. B. Lippincott, 1984.)

erosion of the Dacron graft through the distal clavicle in some cases. Dahl [33, 34] has reported on the use of a synthetic velour or Dacron coracoclavicular loop for fixation. He described marked clavicular erosion caused by the loop. In one case the prosthesis had eroded through the entire clavicle. In all seven adult cases, there was significant evidence of clavicular erosion (Fig. 15B–26). Moneim and Balduini [87] recognized a case of coracoid fracture following a reconstruction of the coracoclavicular ligaments using a Dacron graft passed under the coracoid and through two drill holes in the clavicle. Park and associates [102] found superior results and no resorption in patients treated with double velour Dacron grafts compared with the older knitted Dacron vascular grafts.

Excision of the Distal Clavicle

McLaughlin [83] credits Facassini (1902) with the first distal clavicle excision but gives no reference. Cadenat [29] states that distal clavicle excision was first performed by Morestin but cites no reference or date. In 1941 Gurd [57] in Montreal and Mumford [90] in Indiana independently described their experiences with excision of the distal end of the clavicle.

Mumford [90] performed distal clavicle excision for chronic acromioclavicular subluxation (i.e., type II injury) in patients with symptomatic degenerative changes and an intact coracoclavicular ligament. Cook and Tibone [31] reported on 17 athletes who underwent distal clavicle excision for chronic pain following a type II acromioclavicular injury. Sixteen of the seventeen patients returned to their preinjury performance level. The most common complaint was a decrease in maximum bench press strength.

Cybex testing demonstrated some weakness at slow speeds but little or no weakness at faster speeds.

Gurd [57] recommended distal clavicle excision for chronic symptomatic, unreduced complete acromioclavicular dislocation (i.e., type III injury). Mumford [90] emphasized that simple distal clavicle excision in the presence of coracoclavicular ligament disruption and an unstable tender clavicle is not sufficient because it will result in a shorter, unstable, tender distal clavicle. In this situation, as originally pointed out by Mumford, distal clavicle excision must be combined with coracoclavicular ligament reconstruction.

Weaver and Dunn [140] in 1972 reported on 15 patients with type III acromioclavicular joint dislocations who underwent excision of the distal 2 cm of the clavicle combined with coracoclavicular ligament reconstruction using the coracoacromial ligament. Twelve of the fifteen patients had acute dislocations and three had chronic conditions. Weaver and Dunn transferred the acromial attachment of the coracoacromial ligament into the intramedullary canal of the remaining clavicle. They did not recommend the use of supplementary internal fixation. Although this procedure has been called the Weaver-Dunn procedure, it was originally described in detail by Cadenat in 1917 [29]. Rauschning and co-workers [107] reported stable and painless shoulders in 18 patients who underwent distal clavicle excision and coracoclavicular ligament reconstruction using the coracoacromial ligament.

Kawabe and associates [72] and Shoji and co-workers [121] reported transfer of the coracoacromial ligament using a small bone block from the

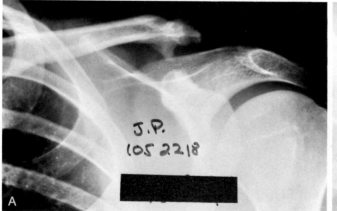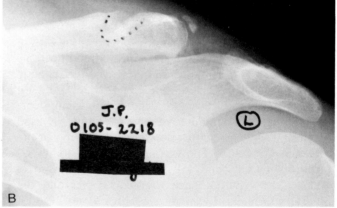

FIGURE 15B–26

A and *B*, X-rays of patient with a type III injury who was previously treated with a Dacron graft around the coracoid and over the top of the clavicle. At the time of secondary reconstruction, the Dacron graft was noted to be partially eroded down through the clavicle. (From Rockwood, C. A., and Matsen, F. A. (Eds.), *The Shoulder.* Philadelphia, W. B. Saunders, 1990.)

acromion. Kawabe and associates fixed the coracoacromial ligament with its attached bone block to the distal clavicle using a screw. This procedure was performed without distal clavicle excision. In some cases supplementation with acromioclavicular fixation using smooth wires was also done. Shoji and colleagues also transferred the coracoacromial ligament with a bone block in 12 acute cases and three chronic cases of complete acromioclavicular dislocation.

Dynamic Muscle Transfer

Transfer of the short head of the biceps muscle with or without the accompanying tendon of the coracobrachialis for repair of complete acromioclavicular dislocation has been described by several authors [10, 11, 25, 26, 39, 53, 123]. Brunelli [26] in 1956 originally reported transfer of the coracoid process with the attached short head of the biceps to the lateral end of the distal clavicle through a deltopectoral incision. No attempt was made to reduce the acromioclavicular joint intraoperatively. Brunelli noted that reduction occurred during the first 2 to 3 days postoperatively as the dynamic transfer pulled the distal clavicle inferiorly.

Because of a "critical evaluation of the results," Brunelli [25] modified his original technique. He emphasized the importance of obtaining reduction intraoperatively prior to transferring the coracoid. If closed reduction was not possible, he recommended open reduction of the acromioclavicular joint. In addition, he moved the attachment site of the coracoid process on the clavicle to a more medial position so that it was directly superior to the base of the coracoid process.

Bailey [11] in 1964 and Bailey and associates [10] in 1972 reported good results with transfer of the coracoid process to the clavicle for complete acromioclavicular dislocation. In 1965 Dewar and Barrington [39] reported a similar procedure for treatment of chronic cases of complete acromioclavicular joint dislocation. Berson and co-workers [18] reported satisfactory results in 23 acute and 6 chronic acromioclavicular dislocations. Glorian and Delplace [53] reported a modification of the procedure in which fixation was supplemented by two temporary acromioclavicular joint pins.

Skjeldal and co-workers [123] reported a high incidence of complications associated with repair of complete acromioclavicular dislocation using coracoid process transfer. They reported 17 patients with a 7½-year follow-up. Complications included fragmentation of the tip of the coracoid process in three patients, deep infection in two patients, pain at rest in two patients, and pain with motion in three patients. As a result of this high complication rate, Skjeldal and co-workers did not recommend this procedure.

POSTOPERATIVE CARE

Patients with Acromioclavicular Fixation

Patients are encouraged to move the hand and elbow but are discouraged from abducting the shoulder. Motion must be limited to prevent breakage or migration of the pins across the acromioclavicular joint. Rowe [111] recommends that abduction motion be limited to 40 degrees, and DePalma [37] recommends no abduction until the pins are removed. Most authors recommend removal of the pins after 6 to 8 weeks. After pin removal, range of motion and strengthening exercises can be initiated.

Patients with Coracoclavicular Fixation

Bosworth [21] recommended a sling until the soft tissues had healed. However, he allowed the patient to perform activities of daily living such as bathing, dressing, and feeding on the first postoperative day. In addition, he encouraged daily removal of the sling to perform pendulum exercises and active assisted range of motion. The patient was restricted from heavy work for 8 weeks. Kennedy [73] used no form of external splintage and encouraged a gradual use of range of motion exercises. He anticipated full abduction 7 to 10 days postoperatively. The patient was allowed to return to vigorous athletic activities 6 to 8 weeks postoperatively.

Other authors have advocated a slower progression to regular activity to protect the coracoclavicular fixation. Alldredge [6] recommended no postoperative immobilization beyond the second postoperative day. However, he limited the patient's activities for 5 to 6 weeks. Bearden and colleagues [15] supported the arm in a sling for 10 to 14 days following surgery. The patient was then instructed to avoid strenuous activity, such as lifting weights. Jay and Monnet [67] recommended a sling for 4 weeks postoperatively, at which time active exercises were started. Weitzman [141] recommended a sling and swathe immediately after surgery. However, on the first postoperative day a plaster shoulder Velpeau cast was applied. This cast was worn for 4 weeks and was then removed to allow active exercises.

The recommendations concerning removal of the fixation device vary considerably. Neither Bosworth [21] nor Kennedy [73] recommended removal of the coracoclavicular lag screw. Weitzman [141], on the other hand, recommended coracoclavicular screw removal under local anesthesia after 3 months. Jay and Monnet [67] removed the screw under local anesthesia at 8 weeks. Bearden and colleagues [15] removed the coracoclavicular wire loops 6 to 8 weeks postoperatively. This procedure was performed under local anesthesia if possible or general anesthesia if necessary. They recommended removing all fragments of wire, even if they were broken. Alldredge [6], however, did not remove the coracoclavicular wires if they were broken and the shoulder had a full range of motion. Otherwise, he recommended wire loop removal 6 to 8 weeks postoperatively.

Patients Undergoing Excision of the Distal Clavicle

Patients undergoing distal clavicle excision in the presence of an intact coracoclavicular ligament are encouraged to begin passive range of motion exercises immediately after the operation. Active range of motion and strengthening exercises are begun after complete healing of the deltoid and trapezius has occurred (2 to 3 weeks).

Patients undergoing distal clavicle excision in conjunction with coracoclavicular ligament reconstruction require a longer period of protection from heavy lifting or contact to allow healing of the reconstructed coracoclavicular ligament. Weaver and Dunn [140] recommended a Velpeau dressing or a sling postoperatively. Circumduction exercises were begun on the first postoperative day. Full active use of the shoulder was not allowed until 5 weeks postoperatively.

Patients Undergoing Coracoid Process Transfer

Brunelli and Brunelli [25] recommended that the elbow be kept in 90 degrees of flexion for 3 to 5 days postoperatively. It was then progressively extended to increase the pull on the transferred tendon in order to reduce the distal clavicle. Six to eight weeks should be allowed for adequate healing of the coracoid process to the clavicle before any heavy lifting or contact sports are allowed.

Type IV, V, and VI Injuries

Operative measures are frequently required for type IV, V, and VI injuries. Closed reduction is very difficult if not impossible in type IV and VI dislocations. The tremendous amount of soft tissue injury associated with type V injuries dictates early operative repair in the great majority of cases.

AUTHORS' PREFERRED METHOD OF TREATMENT

Type I Injury

As emphasized by Cadenat in 1917 [29], the treatment of acromioclavicular joint injuries depends upon whether the injury is an incomplete or a complete dislocation. Patients with a type I acromioclavicular injury are given a sling for convenience and encouraged to use ice for the first 12 hours. Immediate active range of motion exercises of the shoulder are encouraged.

Criteria for Return to Athletics. The athlete is withheld from competition, particularly contact sports, until there is a full painless range of motion and the joint is nontender to palpation. This usually takes approximately 1 to 2 weeks.

Type II Injury

The treatment of a type II acromioclavicular injury involves rest, ice for the first 12 hours, and a sling for support. The patient should be encouraged to begin gentle range of motion exercises and activities of daily living as soon as symptoms permit. This usually coincides with approximately the seventh postinjury day.

Criteria for Return to Athletics. Since the coracoclavicular ligament has undergone plastic deformation, it should be protected from recurrent injury for approximately 6 to 8 weeks. Assuming that the athlete has regained a pain-free range of motion, return to sport may be attempted prior to this with the use of a protective pad placed on the superior aspect of the acromioclavicular joint to guard against a superior blow.

Type III Injury

In an athlete with a type III acromioclavicular dislocation, our preferred treatment varies depending upon the type of sport. We prefer to repair an acute type III injury in many athletes because we believe that this will result in a stronger, more durable shoulder. However, nonoperative treatment is preferred in athletes who participate in sports involving repeated violent trauma to the shoulder (e.g., hockey, football, rugby, soccer, and so on).

The high-caliber throwing athlete represents a special case of complete acromioclavicular dislocation. Although several authors have documented little significant strength deficit in patients treated nonoperatively for type III acromioclavicular dislocation, the issue of fatigability has not been adequately addressed in the literature [52, 81, 119, 139]. In addition, the effect of complete disruption of the coracoclavicular ligaments on normal "synchronous scapuloclavicular" motion is unknown. However, scapulothoracic rotation during overhead elevation is linked to clavicular rotation by virtue of the intact coracoclavicular ligament. In addition, the importance of synchronous glenohumeral and scapulothoracic motion in normal overhead function has been well documented [30, 65, 70]. Therefore, as a general rule, we prefer to perform operative reduction and coracoclavicular ligament repair in a high-caliber throwing athlete with a type III acromioclavicular dislocation.

Nonoperative Treatment

If nonoperative treatment is indicated, the shoulder is placed in a sling, and ice is applied during the first 12 hours. With the assistance of oral analgesics, the patient is encouraged to perform active range of motion exercises and activities of daily living within the first 3 to 4 days following injury. A functional range of motion is normally achieved within 7 days postinjury. Within 2 to 3 weeks, the patient should have a full range of motion and little if any discomfort (Fig. 15B–27).

Criteria for Return to Athletics. In general, heavy lifting and contact are avoided for 6 to 8 weeks. However, an earlier return to sport is sometimes possible through the use of a protective pad as mentioned in the section on type II injury.

Operative Treatment

The operative technique used by the senior author during the past 27 years was learned from Dr. James Amspatcher at the University of Oklahoma Medical Center. The current technique encompasses some of the best ideas of several previously described procedures. The acromioclavicular joint is explored and debrided, the coracoclavicular ligament is reapproximated, and a coracoclavicular lag screw is used for temporary internal fixation. The deltoid and trapezius muscles are imbricated and repaired back to the clavicle.

Preoperative Preparation. A 10- by 12-inch x-ray cassette is placed on the operating table beneath the patient's shoulder. In addition, a small roll or pad is placed beneath the scapula to elevate the injured shoulder. The operation is performed on a regular operating room table in the "beach-chair position." A special, narrow headrest is used to allow complete access to the superior aspect of the injured shoulder. The head should be slightly deviated toward the normal shoulder. The anesthesiologist, along with his or her equipment, is moved toward the opposite shoulder to allow both the surgeon and an assistant to stand at the top of the table. Care must be exercised during draping to include the coracoid process, the top of the clavicle, and the base of the neck in the operative field.

Skin Incision. A straplike incision, approximately 3 inches in length, is made on the superior aspect of the clavicle in the lines of Langer. It begins 1 inch posterior to the clavicle, crosses the clavicle approximately 1 inch medial to the acromioclavicular joint, and ends just medial to the tip of the coracoid process (Fig. 15B–28A). Full-thickness skin flaps are created to expose the distal 2 inches of the clavicle, the acromioclavicular joint, and the anterior deltoid.

The Anterior Deltoid. If the deltoid and trapezius muscle attachments have not been stripped entirely off the distal clavicle as a result of the injury, this interval must be longitudinally divided to allow the clavicle to be dislocated superiorly. Retraction of the deltoid muscle anteriorly and distally allows visualization of the torn ends of the coracoclavicular ligament and the base of the coracoid process. Occasionally, the deltoid will have been stripped off the distal clavicle with an intact periosteal sleeve. In this situation, it is necessary to split the deltoid distally in line with its fibers for approximately 2 inches. This will allow exposure of the coracoclavicular ligaments and the base of the coracoid process (see Fig. 15B–28B). After the distal end of the clavicle has been dislocated superiorly using a towel clip, Lawin clamp, or bone hook, the acromioclavicular joint is thoroughly debrided. This includes the intra-articular disc and any loose fragments of the acromioclavicular joint capsule or ligaments (see Fig. 15B–28B).

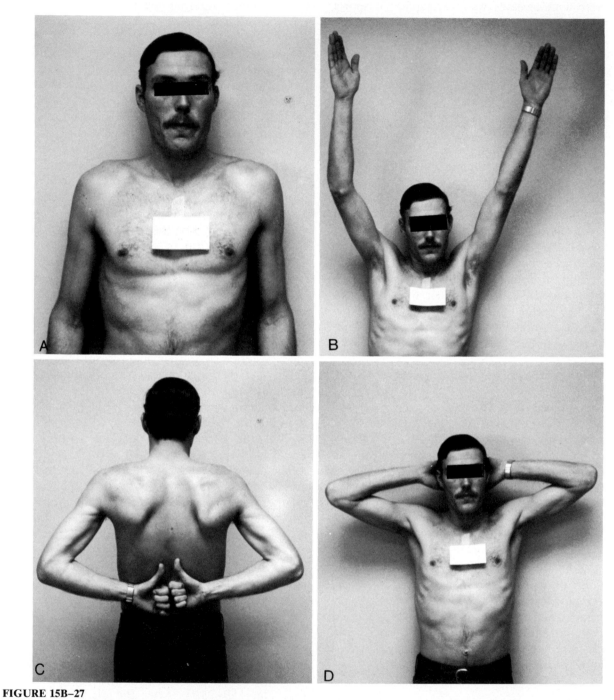

FIGURE 15B–27

A–D, This patient had suffered an acute type III acromioclavicular dislocation of his right shoulder on the previous day. He was initially treated for 4 days in a sling and then allowed to use the arm for everyday living activities. He still has a bump over the top of the right shoulder; however, he has full range of motion and essentially no pain. He is not involved in heavy physical labor. (From Rockwood, C. A., and Green, D. P. (Eds.), *Fractures in Adults,* 2nd ed. Philadelphia, J. B. Lippincott, 1984.)

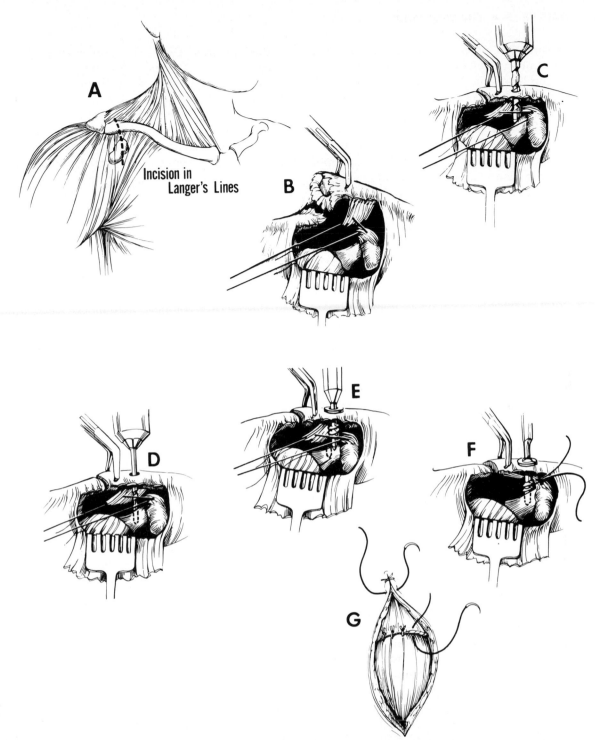

FIGURE 15B–28

The senior author's repair for a complete acromioclavicular dislocation. *A*, The skin incision is about 3 inches in length and extends from the posterior edge of the clavicle, 1 inch medial to the acromioclavicular joint and then down in Langer's lines to a point just medial to the tip of the coracoid process. *B*, The deltoid secondary to the injury is usually subperiosteally stripped away from the distal clavicle. It may have to be surgically detached to aid in identification and reapproximation of the coracoclavicular ligaments and the base of the coracoid process. The distal end of the clavicle can be lifted up with a towel clip or a bone hook to aid in the placement of the sutures in the coracoclavicular ligament. The acromioclavicular joint is thoroughly debrided of the meniscus. If the acromioclavicular ligaments are amenable to repair, they are preserved and repaired later. *C*, The distal end of the clavicle is held reduced adjacent to the acromion with a towel clip. A ³⁄₁₆-inch drill bit is used to make a hole in the clavicle directly above the base of the coracoid. *D*, Through the ³⁄₁₆-inch hole in the clavicle, a ⁹⁄₆₄-inch drill bit is used to create a hole through both cortices of the base of the coracoid. *E*, The specially designed lag screw of appropriate length is then placed through the clavicle until the smooth shank of the screw lies in the clavicle. *F*, The nonthreaded nipple end of the screw is then passed into the hole of the coracoid, and the screw is tightened home to depress the clavicle down to the level of the acromion. The stay sutures in the coracoclavicular ligaments are then tied, and the screw is tightened another half-turn to take any tension off the reapproximated ligaments. *G*, The muscle attachments of the deltoid and trapezius are carefully repaired and, if possible, are imbricated over the top of the clavicle and the acromioclavicular joint. (From Rockwood, C. A., and Green, D. P. (Eds.), *Fractures in Adults*, 2nd ed. Philadelphia, J. B. Lippincott, 1984.)

507

Reapproximation of the Coracoclavicular Ligament. The coracoclavicular ligament is a dense, heavy, strong ligament. It has a very definite purpose in the shoulder: to support the scapula and the upper extremity from the clavicle. If surgical repair of acromioclavicular joint dislocation is to be undertaken, the coracoclavicular ligament should be reapproximated, as opposed to depending upon unreliable scar tissue to provide stability. The torn ends of the ligament are isolated and "tagged" with two or three nonabsorbable sutures. The sutures are not tied until the clavicle is reduced (see Fig. 15B–28B).

Temporary Internal Fixation. A special coracoclavicular lag screw* is used to hold the clavicle temporarily in a reduced position. With the acromioclavicular joint reduced, a ³⁄₁₆-inch hole is drilled through the distal clavicle, directly over the base of the coracoid (see Fig. 15B–28C). A ⁹⁄₆₄-inch drill bit is then used to drill holes in both cortices of the base of the coracoid process. This is done under direct vision by retracting the deltoid anteriorly and distally (see Fig. 15B–28D). A depth gauge is used to determine the length of the screw. The threads of the screw are then passed through the clavicle until the smooth portion of the shank is within the drill hole in the clavicle. The nipplelike prominence on the end of the screw is then placed within the drill hole in the superior cortex of the coracoid process (see Fig. 15B–28E). The screw is advanced until the acromioclavicular joint is reduced and the threads engage both cortices. We prefer to use this special coracoclavicular lag screw instead of a standard 6.5-mm AO cancellous screw for several reasons. First, the thread-to-shaft diameter ratio is larger and should result in better pull-out strength. Second, the smooth blunt tip may be used to engage the hole in the coracoid process without interference from the threads. And finally, protrusion of the blunt tip beneath the coracoid process is less likely to injure the vital structures inferior to the coracoid process than the sharp tip on the cancellous screw.

The sutures in the coracoclavicular ligament are then tied. The screw is advanced an additional one-quarter turn to relieve tension on the repair of the coracoclavicular ligament (see Fig. 15B–28F). The deltoid and trapezius muscle attachments are then imbricated over the acromioclavicular joint and distal clavicle (see Fig. 15B–28G). A radiograph is taken to ensure that the screw is vertical and that the threads engage both cortices of the base of the coracoid process.

*Rockwood screw, Howmedica Special Product Division.

Postoperative Care. The arm is supported in a sling for 1 to 2 weeks. However, the patient is allowed to use the arm for activities of daily living. Lifting, pushing, or pulling is prohibited for 8 weeks after surgery. The screw is routinely removed 6 to 8 weeks postoperatively under local anesthesia.

Criteria for Return to Athletics. Following screw removal, the patient is instructed not to perform any heavy lifting, pushing, or pulling and to avoid contact sports for 10 to 12 weeks from the initial operative repair. Athletes are not permitted to return to contact sports or to place undue stress on the shoulder until 12 weeks postoperatively, and only after they have recovered full strength of the shoulder and a full range of motion.

Chronic Dislocation. Some patients treated nonoperatively for complete acromioclavicular dislocation develop persistent pain in the acromioclavicular joint. These patients require excision of the distal 2 cm of the clavicle and reconstruction of the coracoclavicular ligament. The skin incision and much of the surgical approach are identical to that described for acute repair of a type III injury (Fig. 15B–29A). After the distal 2 cm of the clavicle have been excised, the medullary canal is drilled and curetted (see Fig. 15B–29B, C). The proximal attachment of the coracoacromial ligament is then removed from the acromion (see Fig. 15B–29D, E). Additional length of ligament can be obtained by removing the ligament from the undersurface of the acromion rather than from its anterior surface. As Cadenat [29] noted in 1917, incising the anterior fasciculus of the coracoid attachment of the ligament also yields additional length. The clavicle is then reduced to its anatomic location using the special coracoclavicular lag screw as described in the section on acute injury. The coracoacromial ligament is then transferred into the medullary canal of the clavicle (see Fig. 15B–29F). The sutures in the ligament are brought through two superior drill holes in the clavicle and tied over the top of the clavicle. The lag screw is then advanced a quarter turn to relieve tension on the transferred coracoacromial ligament. The deltoid and trapezius muscles are then imbricated as described in the section on acute repair.

Postoperative Care. Postoperatively, the patient's arm is supported in a sling for 4 weeks. At 7 to 10 days after surgery the patient can use the arm for everyday living activities but is instructed to avoid any heavy lifting, pushing, or pulling. After 10 to 12 weeks, the screw should be removed under local anesthesia. Heavy lifting and contact should be avoided until full strength and full range of motion have returned.

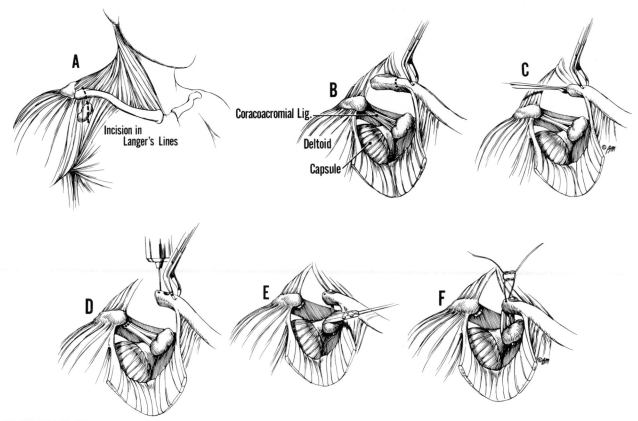

FIGURE 15B–29
The senior author's method of reconstructing a chronic type III, IV, V, or VI acromioclavicular dislocation. *A*, The incision is made in Langer's lines. *B*, The distal end of the clavicle is excised. *C*, The medullary canal is drilled out and curetted to receive the transferred coracoacromial ligament. *D*, Two small drill holes are made through the superior cortex of the distal clavicle. The coracoacromial ligament is carefully detached from the acromion process. *E*, With the coracoacromial ligament detached from the acromion, a heavy nonabsorbable suture is woven through the ligament. *F*, The ends of the suture are passed out through the two small drill holes in the distal end of the clavicle. The coracoclavicular lag screw is inserted, and when the clavicle is reduced down to its normal position, the sutures used to pull the ligament snugly up into the canal are tied. (From Rockwood, C. A., and Green, D. P. (Eds.), *Fractures in Adults,* 2nd ed. Philadelphia, J. B. Lippincott, 1984.)

Type IV, V, and VI Injuries. Type IV, V, and VI injuries most often require surgery. Closed reduction should be attempted in type IV and type VI injuries. If this is successful, indications for operative repair are similar to those outlined for the type III injury. Should closed reduction be unsuccessful, open reduction and surgical repair as described for the acute type III injury should be performed. The deformity and soft tissue injury associated with type V acromioclavicular dislocation are so severe that surgery is almost mandatory (see Fig. 15B–20).

References

1. Abbott, L. C., Saunders, J. B., Hagey, J., and Jones, E. W. Surgical approaches to the shoulder joint. *J Bone Joint Surg* 31A(2):235–255, 1949.
2. Adams, F. L. *The Genuine Works of Hippocrates* (Vols. 1, 2). New York, William Wood, 1886.
3. Ahstrom, J. P., Jr. Surgical repair of complete acromioclavicular separation. *JAMA* 217(6):785–789, 1971.
4. Albrecht, F. The Balser plate for acromioclavicular fixation. *Chirurg* 53(11):732–734, 1983.
5. Alexander, O. M. Dislocation of the acromioclavicular joint. *Radiography* 15:260, 1949.
6. Alldredge, R. H. Surgical treatment of acromioclavicular dislocations (proceedings). *J Bone Joint Surg* 47(A):1278, 1965.
7. Allman, F. L., Jr. Fractures and ligamentous injuries of the clavicle and its articulation. *J Bone Joint Surg* 49A:774–784, 1967.
8. Anzel, S. H., and Streitz, W. L. Acute acromioclavicular injuries: a report of 19 cases treated nonoperatively employing dynamic splint immobilization. *Clin Orthop* 103:143–149, 1974.
9. Augereau, B., Robert, H., and Apoil, A. [Treatment of severe acromioclavicular dislocation: A coracoclavicular ligamentoplasty technique derived from Cadenat's procedure.] *Ann Chir* 35(9):720–722, 1981.
10. Bailey, R. W., O'Connor, G. A., Tilus, P. D., and Baril, J. D. A dynamic repair for acute and chronic injuries of the acromioclavicular area (proceedings). *J Bone Joint Surg* 54A:1802, 1972.
11. Bailey, R. W. A dynamic repair for complete acromioclavicular joint dislocation (proceedings). *J Bone Joint Surg* 47A:858, 1965.
12. Bannister, G., Stableforth, P., and Hutson, M. The management of acute acromioclavicular dislocation. Presented

at the 4th International Conference on Surgery of the Shoulder, New York, October 4–7, 1989.

13. Bartonicek, J., Jehlicka, D., and Bezvoda, Z. [Surgical treatment of acromioclavicular luxation.] *Acta Chir Orthop Traumatol Cech* 55(4):289–309, 1988.

14. Bateman, J. E. *The Shoulder and Neck.* Philadelphia, W. B. Saunders, 1972.

15. Bearden, J. M., Hughston, J. C., and Whatley, G. S. Acromioclavicular dislocation: Method of treatment. *Am J Sports Med* 1(4):5–17, 1973.

16. Behling, F. Treatment of acromioclavicular separations. *Orthop Clin North Am* 4(3):747–757, 1973.

17. Bergfeld, J. A., Andrish, J. T., and Clancy, W. G. Evaluation of the acromioclavicular joint following first- and second-degree sprains. *Am J Sports Med* 6(4):153–159, 1978.

18. Berson, B. L., Gilbert, M. S., and Green, S. Acromioclavicular dislocations: Treatment by transfer of the conjoined tendon and distal end of the coracoid process to the clavicle. *Clin Orthop* 135:157–164, 1978.

19. Bjerneld, H., Hovelius, L., and Thorling, J. Acromioclavicular separations treated conservatively: A five-year follow-up study. *Acta Orthop Scand* 54:743–745, 1983.

20. Bossart, P. J., Joyce, S. M., Manaster, B. J., and Packer, S. M. Lack of efficacy of "weighted" radiographs in diagnosing acute acromioclavicular separation. *Ann Emerg Med* 17(1):47–51, 1988.

21. Bosworth, B. M. Acromioclavicular separation: New method of repair. *Surg Gynecol Obstet* 73:866–871, 1941.

22. Bosworth, B. M. Complete acromioclavicular dislocation. *N Engl J Med* 2 41:221–225, 1949.

23. Bowers, K. D. Treatment of acromioclavicular sprains in athletes. *Phys Sportsmed* 11(1):79–89, 1983.

24. Brosgol, M. Traumatic acromioclavicular sprains and subluxations. *Clin Orthop* 20:95–98, 1956.

25. Brunelli, G., and Brunelli, F. The treatment of acromioclavicular dislocation by transfer of the short head of biceps. *Int Orthop* 12(2):105–108, 1988.

26. Brunelli, G. Proposta de un nuovo methodo de correzione chirurgia del la lussazione acromion clavicolare. *Bull Soc Med Chir Bresciana* 10:95–98, 1956.

27. Bundens, W. D., Jr., and Cook, J. I. Repair of acromioclavicular separations by deltoid-trapezius imbrication. *Clin Orthop* 20:109–114, 1961.

28. Bunnell, S. Fascial graft for dislocation of the acromioclavicular joint. *Surg Gynecol Obstet* 46:563–564, 1928.

29. Cadenat, F. M. The treatment of dislocations and fractures of the outer end of the clavicle. *Int Clin* 1:145–169, 1917.

30. Codman, E. A. Rupture of the supraspinatus tendon and other lesions in or about the subacromial bursa. *In* Codman, E. A. *The Shoulder.* Boston, Thomas Todd, 1934.

31. Cook, F. F., and Tibone, J. E. The Mumford procedure in athletes: An objective analysis of function. *Am J Sports Med* 16(2):97–100, 1988.

32. Cox, J. S. The fate of the acromioclavicular joint in athletic injuries. *Am J Sports Med* 9(1):50–53, 1981.

33. Dahl, E. Follow-up after coracoclavicular ligament prosthesis for acromioclavicular joint dislocation (proceedings). *Acta Chir Scand* (Suppl.) 506:96, 1981.

34. Dahl, E. [Velour prosthesis in fractures and dislocations in the clavicular region.] *Chirurg* 53:120–122, 1982.

35. Dannohl, C. H. [Angulation osteotomy at the clavicle in old dislocations of the acromioclavicular joint.] *Aktuel Traumatol* 14(6):282–284, 1984.

36. Darrow, J. C., Smith, J. A., and Lockwood, R. C. A new conservative method for treatment of type III acromioclavicular separations. *Orthop Clin North Am* 11(4):727–733, 1980.

37. DePalma, A. F., Callery, G., and Bennett, G. A. Variational anatomy and degenerative lesions of the shoulder joint. *Instr Course Lect* 6:255–281, 1949.

38. DePalma, A. F. Surgical anatomy of the acromioclavicular joints. *Clin Orthop* 13:7–12, 1959.

39. Dewar, F. P., and Barrington, T. W. The treatment of chronic acromioclavicular dislocation. *J Bone Joint Surg* 47B(1):32–35, 1965.

40. Dias, J. J., Steingold, R. A., Richardson, R. A., Tesfayohannes, B., and Gregg, P. J. The conservative treatment of acromioclavicular dislocation: Review after five years. *J Bone Joint Surg* 69B(5):719–722, 1987.

41. Dittel, K. K., Pfaff, G., and Metzger, H. [Results after operative treatment of complete acromioclavicular separation (Tossy III injury).] *Aktuel Traumatol* 17(1):16–22, 1987.

42. Dittmer, H., Jauch, K. W., and Wening, V. [Treatment of acromioclavicular separations with Balser's hookplate.] *Unfallheilkunde* 87(5):216–222, 1984.

43. Eaton, R., and Serletti, J. Computerized axial tomography—a method of localizing Steinmann pin migration: A case report. *Orthopedics* 4(12):1357–1360, 1981.

44. Eskola, A., Vainionpaa, S., Korkala, O., and Rookanen, P. Acute complete acromioclavicular dislocation. A prospective randomized trial of fixation with smooth or threaded Kirschner wires or cortical screw. *Ann Chir Gynaecol* 76(6):323–326, 1987.

45. Falstie-Jensen, S., and Mikkelsen, P. Pseudodislocation of the acromioclavicular joint. *J Bone Joint Surg* 64B(3):369, 1982.

46. Fama, G., and Bonaga, S. [Safety pin synthesis in the cure of acromioclavicular luxation.] *Chir Organi Mov* 73(3):227–235, 1988.

47. Ferguson, A. B., Jr., and Bender, J. *The ABC's of Athletic Injuries and Conditioning.* Baltimore, Williams & Wilkins, 1964.

48. Fleming, R. E., Tomberg, D. N., and Kiernan, H. A. An operative repair of acromioclavicular separation. *J Trauma* 18(10):709–712, 1978.

49. Fukuda, K., Craig, E. V., An, K. N., Cofield, R. H., and Chao, E. Y. S. Biomechanical study of the ligamentous system of the acromioclavicular joint. *J Bone Joint Surg* 68A(3):434–439, 1986.

50. Gerber, C., and Rockwood, C. A., Jr. Subcoracoid dislocation of the lateral end of the clavicle: A report of three cases. *J Bone Joint Surg* 69A(6):924–927, 1987.

51. Glick, J. Acromioclavicular dislocation in athletes: Autoarthroplasty of the joint. *Orthop Rev* 1(4):31–34, 1972.

52. Glick, J. M., Milburn, L. J., Haggerty, J. F., and Nishimoto, D. Dislocated acromioclavicular joint: Follow-up study of 35 unreduced acromioclavicular dislocations. *Am J Sports Med* 5(6):264–270, 1977.

53. Glorian, B., and Delplace, J. [Dislocations of the acromioclavicular joint treated by transplant of the coracoid process.] *Rev Chir Orthop* 59:667–679, 1973.

54. Goldberg, D. Acromioclavicular joint injuries: A modified conservative form of treatment. *Am J Surg* 71(4):529–531, 1946.

55. Goldberg, J. A., Viglione, W., Cumming, W. J., Waddell, F. S., and Ruz, P. A. Review of coracoclavicular ligament reconstruction using Dacron graft material. *Aust NZ J Surg* 57(7):441–445, 1987.

56. Grauthoff, V. H., and Kalmmer, H. L. [Complications due to migration of a Kirschner wire from the clavicle.] *Fortschr Röntgenstr* 128(5):591–594, 1978.

57. Gurd, U. F., Leutenegger, A., and Ruedi, T. [Results of Bosworth procedures for Tossy III acromioclavicular luxations.] *Z Unfallchir Versicherungsmed Berufskr* 79(3):171–174, 1986.

58. Havranek, P. Injuries of distal clavicular physis in children. *J Pediatr Orthop* 9(2):213–215, 1989.

59. Hawkins, R. J. *The Acromioclavicular Joint.* Paper prepared for AAOS Summer Institute, Chicago, July 10–11, 1980.

60. Hill, J. A. Acromioclavicular separations need conservative treatment: Same results achieved with surgical care. *Orthopedics Today* 6(9):25, 1986.

61. Ho, W. P., Chen, J. Y., and Shih, C. H. The surgical

treatment of complete acromioclavicular joint dislocation. *Orthop Rev* 17(11):1116–1120, 1988.

62. Horn, J. S. The traumatic anatomy and treatment of acute acromioclavicular joint in the elderly. *Arch Gerontol Geriatr* 3:259–265, 1984.
63. Hoyt, W. A., Jr. Etiology of shoulder injuries in athletes. *J Bone Joint Surg* 49A(4):755–766, 1967.
64. Imatani, R. J., Hanlon, J. J., and Cady, G. W. Acute complete acromioclavicular separation. *J Bone Joint Surg* 57A(3):328–332, 1975.
65. Inman, V. T., McLaughlin, H. D., Neviaser, J., and Rose, C. Treatment of complete acromioclavicular dislocation. *J Bone Joint Surg* 44A:1008–1011, 1962.
66. Jakobsen, B. W. [Acromioclavicular dislocation. Conservative or surgical treatment?] *Ugeskr Laeger* 151(4):235–238, 1989.
67. Jay, G. R., and Monnet, J. C. The Bosworth screw in acute dislocations of the acromioclavicular joint. Presented at Clinical Conference, University of Oklahoma Medical Center, April, 1969.
68. Johnston, T. B., Davies, D. V., and Davies, F. (Eds.). *Gray's Anatomy* (32nd ed.). London, Longmans, Green, 1958.
69. Kaiser, W., Ziemer, G., and Heymann, H. [Treatment of acromioclavicular luxations with the Balser hookplate and ligament suture.] *Chirurg* 55(11):721–724, 1984.
70. Kappakas, G. S., and McMaster, J. H. Repair of acromioclavicular separation using a Dacron prosthesis graft. *Clin Orthop* 131:247–251, 1978.
71. Katznelson, A., Nerubay, J., and Oliver, S. Dynamic fixation of the avulsed clavicle. *J Trauma* 16(10):841–844, 1986.
72. Kawabe, N., Watanabe, R., and Sato, M. Treatment of complete acromioclavicular separation by coracoacromial ligament transfer. *Clin Orthop* 185:222–227, 1984.
73. Kennedy, J. C. Complete dislocation of the acromioclavicular joint: 14 years later. *J Trauma* 8(3):311–318, 1968.
74. Kennedy, J. C., and Cameron, H. Complete dislocation of the acromioclavicular joint. *J Bone Joint Surg* 36B(2):202–208, 1954.
75. Larsen, E., and Hede, A. Treatment of acute acromioclavicular dislocation: three different methods of treatment prospectively studied. *Acta Orthop Belg* 53(4):480–484, 1987.
76. Liberson, F. The role of the coracoclavicular ligaments in affections of the shoulder girdle. *Am J Surg* 44(1):145–157, 1939.
77. Lindsey, R. W., and Gutowski, W. T. The migration of a broken pin following fixation of the acromioclavicular joint: A case report and review of the literature. *Orthopedics* 9(3):413–416, 1986.
78. Linke, R., and Moschinski, D. [Combined method of operative treatment of ruptures of the acromioclavicular joint.] *Unfallheilkunde* 87:223–225, 1984.
79. Lom, P. [Acromioclavicular disjunction: I. Diagnosis and classification; II. Surgical treatment—the author's modification.] *Rozhl Chir* 67(4):253–270, 1988.
80. Lowe, G. P., and Fogarty, M. J. P. Acute acromioclavicular joint dislocation: Results of operative treatment with the Bosworth screw. *Aust NZ J Surg* 47(5):664–667, 1977.
81. MacDonald, P. B., Alexander, M. J., Frejuk, J., and Johnson, G. Comprehensive functional analysis of shoulders following complete acromioclavicular separation. *Am J Sports Med* 16(5):475–480, 1988.
82. Mazet, R. J. Migration of a Kirschner wire from the shoulder region into the lung: Report of two cases. *J Bone Joint Surg* 25A(2):477–483, 1943.
83. McLaughlin, H. L. *Trauma*. Philadelphia, W. B. Saunders, 1959.
84. McPhee, I. B. Inferior dislocation of the outer end of the clavicle. *J Trauma* 20(8):709–710, 1980.
85. Mikusev, I. E., Zainulli, R. V., and Skvortso, A. P. [Treatment of dislocations of the acromial end of the clavicle.] *Vestn Khir* 139(8):69–71, 1987.
86. Mlasowski, B., Brenner, P., Duben, W., and Heymann, H. Repair of complete acromioclavicular dislocation (Tossy stage III) using Balser's hookplate combined with ligament sutures. *Injury* 19(4):227–232, 1988.
87. Moneim, M. S., and Balduini, F. C. Coracoid fractures as a complication of surgical treatment by coracoclavicular tape fixation. *Clin Orthop* 168:133–135, 1982.
88. Moseley, H. F. Athletic injuries to the shoulder region. *Am J Surg* 98:401–422, 1959.
89. Moshein, J., and Elconin, K. F. Repair of acute acromioclavicular dislocation, utilizing the coracoacromial ligament (proceedings). *J Bone Joint Surg* 51A:812, 1969.
90. Mumford, E. B. Acromioclavicular dislocation. *J Bone Joint Surg* 23:799–802, 1941.
91. Nelson, C. L. Repair of acromioclavicular separations with knitted Dacron graft. *Clin Orthop* 143:289, 1979.
92. Neviaser, J. S. Acromioclavicular dislocation treated by transference of the coracoacromial ligament. *Bull Hosp Jt Dis Orthop Int* 12(1):46–54, 1951.
93. Neviaser, J. S. Acromioclavicular dislocation treated by transference of the coracoacromial ligament. *Arch Surg* 64:292–297, 1952.
94. Neviaser, J. S. Acromioclavicular dislocation treated by transference of the coracoacromial ligament: A long-term follow-up in a series of 112 cases. *Clin Orthop* 58:57–68, 1968.
95. Neviaser, R. J. Injuries to the clavicle and acromioclavicular joint. *Orthop Clin North Am* 18(3):433–438, 1987.
96. Nguyen, V. Personal communication, 1989.
97. Nicoll, E. E. Annotation: Miners and mannequins (editorial). *J Bone Joint Surg* 36B(2):171–172, 1954.
98. Norrell, H., and Llewellyn, R. C. Migration of a threaded Steinmann pin from an acromioclavicular joint into the spinal canal: A case report. *J Bone Joint Surg* 47A:1024–1026, 1965.
99. O'Donoghue, D. H. *Treatment of Injuries of Athletes*. Philadelphia, W. B. Saunders, 1970.
100. Oh, W. H., and Garvin, W. Subluxation of the distal clavicle. *Orthop Clin North Am* 11(4):813–818, 1980.
101. Paavolainen, P., Bjorkenheim, J. M., Paukku, P., and Slatis, P. Surgical treatment of acromioclavicular dislocation: A review of 39 patients. *Injury* 14:415–420, 1983.
102. Park, J. P., Arnold, J. A., Coker, T. P., Harris, W. D., and Becker, D. A. Treatment of acromioclavicular separations: A retrospective study. *Am J Sports Med* 8(4):251–256, 1980.
103. Patterson, W. R. Inferior dislocation of the distal end of the clavicle. *J Bone Joint Surg* 49A:1184–1186, 1967.
104. Petersson, C. J. Degeneration of the acromioclavicular joint: A morphological study. *Acta Orthop Scand* 54:434–438, 1983.
105. Poirier, P., and Rieffel, H. Mechanisme des luxations sur acromiales de la clavicule. *Arch Gen Med* 1:396–422, 1981.
106. Powers, J. A., and Bach, P. J. Acromioclavicular separations—closed or open treatment. *Clin Orthop* 104:213–223, 1974.
107. Rauschning, W., Nordesjo, L. O., Nordgren, B., Sahlstedt, B., and Wigren, A. Resection arthroplasty for repair of complete acromioclavicular separations. *Arch Orthop Traumatol Surg* 97:161–164, 1980.
108. Retief, P. J., and Meintjes, F. A. Migration of a Kirschner wire in a body: A case report. *S Afr Med J* 53:557–558, 1978.
109. Rockwood, C. A., Jr. Injuries to the acromioclavicular joint. *In* Rockwood, C. A., Jr., and Green, D. P. *Fractures in Adults* (Vol. 1). Philadelphia, J. B. Lippincott, 1984, pp. 860–910.
110. Rockwood, C. A., Jr., and Young, D. C. Disorders of the acromioclavicular joint. *In* Rockwood, C. A., Jr., and Matsen, F. A. (Eds.), *The Shoulder*. Philadelphia, W. B. Saunders, 1990.
111. Rowe, C. R. Symposium on surgical lesions of the shoulder: Acute and recurrent dislocation of the shoulder. *J Bone Joint Surg* 44A:977–1012, 1962.

112. Rowe, C. R., and Marble, H. C. Shoulder girdle injuries. *In* Cave, E. F. (Ed.), *Fractures and Other Injuries*. Chicago, Year Book, 1961.

113. Sage, F. P., and Salvatore, J. E. Injuries of acromioclavicular joint: Study of results in 96 patients. *South Med J* 56:486–495, 1963.

114. Salter, E. G., Nasca, R. J., and Shelley, B. S. Anatomical observations on the acromioclavicular joint and supporting ligaments. *Am J Sports Med* 15(3):199–206, 1987.

115. Schindler, A., Schmid, J. P., and Heyse, C. [Hookplate fixation for repair of acute acromioclavicular separation: Review of 41 patients.] *Unfallchirurg* 88(12):533–540, 1985.

116. Schmid, A., and Schmid, F. [Use of arthrosonography in diagnosis of Tossy III lesions of acromioclavicular joints.] *Aktuel Traumatol* 18(3):134–138, 1988.

117. Schwarz, N., and Kuderna, H. Inferior acromioclavicular separation: Report of an unusual case. *Clin Orthop* 234:28–30, 1988.

118. Schwarz, N., and Leixnering, M. [Results of nonreduced acromioclavicular Tossy III separations.] *Unfallchirurg* 89:248–252, 1986.

119. Sellers, R., Tibone, J., Tonino, P. M., and Moynes, D. Strength testing of third degree acromioclavicular dislocations. Personal communication, 1988.

120. Sethi, G. K., and Scott, S. M. Subclavian artery laceration due to migration of a Hagie pin. *Surgery* 80(5):644–646, 1976.

121. Shoji, H., Roth, C., and Chuinard, R. Bone block transfer of coracoacromial ligament in acromioclavicular injury. *Clin Orthop* 208:272–277, 1986.

122. Simmons, E. H., and Martin, R. F. Acute dislocation of the acromioclavicular joint. *Can J Surg* 11:473–479, 1968.

123. Skjeldal, S., Lundblad, R., and Dullerud, R. Coracoid process transfer for acromioclavicular dislocation. *Acta Orthop Scand* 59(2):180–182, 1988.

124. Sleeswijk-Visser, S. V., Haarsma, S. M., and Speeckaert, M. T. C. Conservative treatment of acromioclavicular dislocation: Jones strap vs. Mitella (proceedings). *Acta Orthop Scand* 55(4):483, 1984.

125. Smith, M. J., and Stewart, M. J. Acute acromioclavicular separations. *Am J Sports Med* 7(1):62–71, 1979.

126. Spigelman, L. A harness for acromioclavicular separation. *J Bone Joint Surg* 51A(3):585–586, 1969.

126A. Stephens, H. E. G. Stuck nail fixation for acute dislocation of the acromioclavicular joint (proceedings). *J Bone Joint Surg* 51B(1):197, 1969.

127. Tagliabue, D., and Riva, A. [Current approaches to the treatment of acromioclavicular joint separation in athletes.] *Ital J Sports Traumatol* 3(1):15–24, 1981.

128. Thorndike, A., Jr., and Quigley, T. B. Injuries to the acromioclavicular joint: A plea for conservative treatment. *Am J Surg* 55:250–261, 1942.

129. Tossy, J. D., Mead, N. C., and Sigmond, H. M. Acromioclavicular separations: useful and practical classification for treatment. *Clin Orthop* 28:111–119, 1963.

130. Tsou, P. M. Percutaneous cannulated screw coracoclavicular fixation for acute acromioclavicular dislocations. *Clin Orthop* 243:112–121, 1989.

131. Tyurina, T. V. [Age-related characteristics of the human acromioclavicular joint.] *Arkh Anat Giston Embriol* 89(11):75–81, 1985.

132. Urban, J., and Jaskiewicz, A. [Idiopathic displacement of Kirschner wire to the thoracic cavity after the osteosynthesis of acromioclavicular joint.] *Chir Narzadow Ruchu Ortop Pol* 49(4):399–402, 1984.

133. Urist, M. R. Complete dislocation of the acromioclavicular joint: The nature of the traumatic lesion and effective methods of treatment with an analysis of 41 cases. *J Bone Joint Surg* 28:813–837, 1946.

134. Urist, M. R. Complete dislocation of the acromioclavicular joint (follow-up notes). *J Bone Joint Surg* 45A:1750–1753, 1963.

135. Urist, M. R. The treatment of dislocation of the acromioclavicular joint: a survey of the past decade. *Am J Surg* 98:423–431, 1959.

136. Vainionpaa, S., Kirves, P., and Laike, E. Acromioclavicular joint dislocation—surgical results in 36 patients. *Ann Chir Gynecol* 70:120–123, 1981.

137. Varney, J. H., Coker, J. K., Cawley, J. J. Treatment of acromioclavicular dislocation by means of a harness. *J Bone Joint Surg* 34A(1):232–233, 1952.

138. Waldrop, J. I., Norwood, L. A., and Alvarez, R. G. Lateral roentgenographic projections of the acromioclavicular joint. *Am J Sports Med* 9(5):337–341, 1981.

139. Walsh, W. M., Peterson, D. A., Shelton, G., and Newmann, R. D. Shoulder strength following acromioclavicular injury. *Am J Sports Med* 13(3):153–158, 1985.

140. Weaver, J. K., and Dunn, H. K. Treatment of acromioclavicular injuries, especially complete acromioclavicular separation. *J Bone Joint Surg* 54A(6):1187–1197, 1972.

141. Weitzman, G. Treatment of acute acromioclavicular joint dislocation by a modified Bosworth method: Report on 24 cases. *J Bone Joint Surg* 49A(6):1167–1178, 1967.

142. Zanca, P. Shoulder pain: Involvement of the acromioclavicular joint: Analysis of 1000 cases. *Am J Roentgenol* 112(3):493–506, 1971.

143. Zaricznyj, B. Late reconstruction of the ligaments following acromioclavicular separation. *J Bone Joint Surg* 58A(6):792–795, 1976.

INJURIES TO THE STERNOCLAVICULAR JOINT

Gerald R. Williams, M.D.
Charles A. Rockwood, Jr., M.D.

The sternoclavicular joint is one of the least commonly injured joints in the body. The majority of the literature consists only of small series involving three or four cases. Rowe and Marble reported an incidence of 3% of sternoclavicular dislocations among 1603 shoulder girdle injuries [76A]. In comparison, there was an 85% incidence of glenohumeral dislocations and a 12% incidence of acromioclavicular dislocations. In this series, as well as in Rockwood's experience, dislocation of the sternoclavicular joint was not as rare as posterior dislocation of the glenohumeral joint [74].

Injuries to the sternoclavicular joint are classified according to the anatomic location of the dislocated

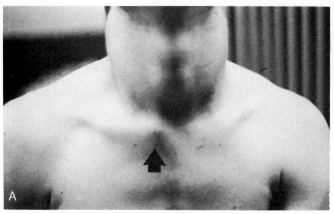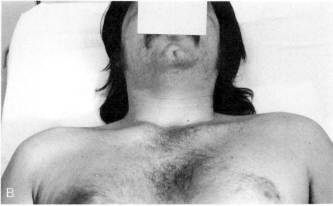

FIGURE 15B–30
A, Clinically there is an evident anterior dislocation of the right sternoclavicular joint (arrow). *B*, When the clavicles are viewed from around the level of the patient's knees, it is apparent that the right clavicle is dislocated anteriorly. (From Rockwood, C. A., and Green, D. P. (Eds.), *Fractures in Adults*, 2nd ed. Philadelphia, J. B. Lippincott, 1984.)

medial clavicle and to the etiology of the dislocation. Anterior sternoclavicular dislocation, the most common type, occurs when the medial end of the clavicle is displaced anteriorly or anterosuperiorly with respect to the anterior margin of the sternum (Fig. 15B–30) [74]. Posterior sternoclavicular dislocation is uncommon and is characterized by posterior or posterosuperior displacement of the medial clavicle with respect to the posterior margin of the sternum (Figs. 15B–31 and 15B–32) [75]. Etiologically, sternoclavicular injuries are classified as either traumatic or atraumatic [74].

CLASSIFICATION

Anatomic

Although anterior dislocation of the sternoclavicular joint is much more common than posterior dislocation, the ratio of anterior to posterior dislocations has only rarely been reported. A review of nearly 250 references dealing with injuries of the sternoclavicular joint reveals that more than 60% discuss only the rare posterior dislocation. This is undoubtedly because of the severe complications

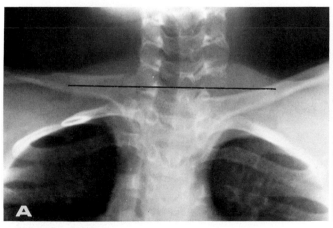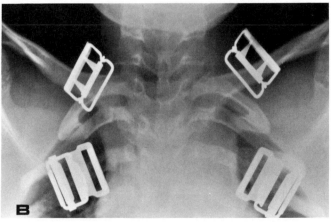

FIGURE 15B–31
A, Posterior dislocation of the left sternoclavicular joint as seen on the 40-degree cephalic tilt x-ray projection in a 12-year-old boy. The left clavicle is displaced inferiorly to a line drawn through the normal right clavicle. *B*, Following closed reduction, the medial ends of both clavicles are in the same horizontal position. The buckles are part of the figure-of-eight clavicular harness that is used to hold the shoulders back after reduction. (From Rockwood, C. A., and Green, D. P. (Eds.), *Fractures in Adults*, 2nd ed. Philadelphia, J. B. Lippincott, 1984.)

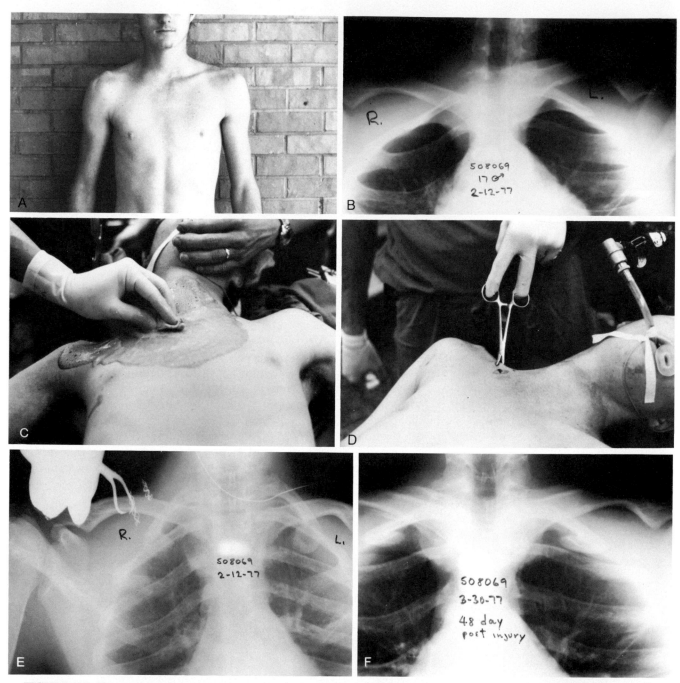

FIGURE 15B–32

Posterior dislocation of the right sternoclavicular joint. *A*, A 16-year-old boy with a 48-hour-old posterior displacement of the right medial clavicle that resulted from direct trauma to the anterior right clavicle. He noted immediate onset of difficulty in swallowing and some hoarseness in his voice. *B*, The 40-degree cephalic tilt x-ray projection confirmed the presence of the posterior displacement of the right medial clavicle compared with the left clavicle. Because of the patient's age, this was considered most likely to be a physeal injury of the right medial clavicle. *C*, Because the injury was 48 hours old, we were unable to reduce the dislocation with simple traction on the arm. The right shoulder was surgically cleansed to allow use of a sterile towel clip. *D*, With the towel clip attached securely around the clavicle and using continued lateral traction, a visible and audible reduction was obtained. *E*, Postreduction x-rays showed that the medial clavicle had been restored to its normal position. The reduction was quite stable, and the patient's shoulders were held back with a figure-of-eight strap. *F*, The right clavicle has remained reduced. Note particularly the periosteal new bone formation along the superior and inferior borders of the right clavicle. This results from a physeal injury in which the epiphysis remains adjacent to the manubrium while the clavicle is displaced out of a split in the periosteal tube. (From Rockwood, C. A., and Green, D. P. (Eds.), *Fractures in Adults*, 2nd ed. Philadelphia, J. B. Lippincott, 1984.)

that can be associated with it. The largest series from a single institution is that of Nettles and Linscheid [63], who reported 60 patients with sternoclavicular dislocations—57 anterior and 3 posterior. The ratio of anterior to posterior dislocations in this series was approximately 20:1. Waskowitz [82] reviewed 18 cases of sternoclavicular dislocations; none of these dislocations was posterior. In Rockwood's series of 273 traumatic injuries of the sternoclavicular joint, there were 121 patients with anterior dislocation and 41 patients with posterior dislocation [74].

Etiologic

Traumatic Injury. There are three grades of traumatic sternoclavicular injury. A mild sprain, or grade I injury, is characterized by intact ligaments and a stable joint. In a moderate sprain, or grade II injury, there is subluxation of the sternoclavicular joint, and the capsule and surrounding ligaments may be partially disrupted. In a severe sprain (grade III injury), the sternoclavicular ligaments and capsule are completely disrupted, and the joint is dislocated either anteriorly or posteriorly.

The most common cause of dislocation of the sternoclavicular joint is vehicular accidents; sports participation is the second most common cause [63, 66, 82]. Omer [66] in his review of patients from 14 military hospitals, accumulated 82 cases of dislocation of the sternoclavicular joint. He reported that almost 80% of these cases occurred as a result of vehicular accidents (47%) and athletic competition (31%).

The youngest reported patient to have sustained a sternoclavicular dislocation is a 7-month-old infant girl. According to Wheeler and associates [84], the injury occurred when the infant was lying on her left side and her older brother accidentally fell on her, compressing her shoulders together. The closed reduction was unstable, and the child was immobilized in a figure-of-eight bandage for 5 weeks. At the end of 10 weeks the child had a full range of motion, and there was no evidence of sternoclavicular instability. Rockwood has encountered a case of anterior injury in a 3-year-old patient that occurred as a result of an automobile accident [74] (Fig. 15B–33).

Atraumatic Injury. Atraumatic instability of the sternoclavicular joint is either acquired or congenital. Spontaneous subluxation or dislocation of one or both sternoclavicular joints may occur during overhead motion, particularly in patients with systemic ligamentous laxity. This atraumatic instability is usually nonpainful (Fig. 15B–34). Congenital defects of the sternoclavicular joint, with loss of bone substance on either side of the joint, can predispose to subluxation or dislocation. Guerin [35] first reported congenital luxations of the sternoclavicular

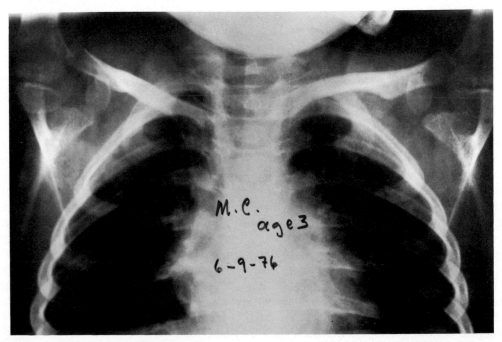

FIGURE 15B–33
X-ray of a 3-year-old child with traumatic anterior dislocation of the left sternoclavicular joint. The chest film demonstrates that the left clavicle is superior to the right, suggesting an anterior displacement of the left medial clavicle. (From Rockwood, C. A., and Green, D. P. (Eds.), *Fractures in Adults*, 2nd ed. Philadelphia, J. B. Lippincott, 1984.)

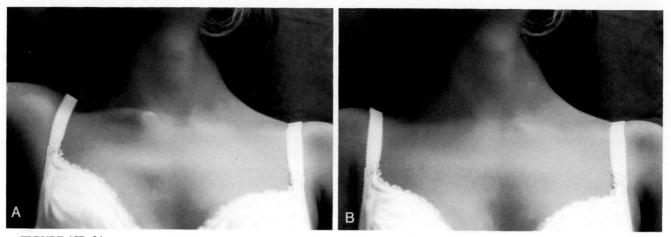

FIGURE 15B–34
Spontaneous anterior subluxation of the sternoclavicular joint. *A*, With the arms in the overhead position, the medial end of the right clavicle dislocates spontaneously anteriorly with no trauma. *B*, When the arm is brought back down to the side, the medial end of the clavicle reduces spontaneously, usually with no significant discomfort. (From Rockwood, C. A., and Matsen, F. A. (Eds.), *The Shoulder*. Philadelphia, W. B. Saunders, 1990.)

joint in 1841. Cooper [20] described a patient with scoliosis so severe that the shoulder was displaced forward enough to dislocate the medial clavicle posteriorly. Newlin [64] reported a case of bilateral congenital posterior dislocation of the sternoclavicular joint in a 25-year-old man. The posteriorly dislocated medial clavicles simulated an intrathoracic mass.

HISTORY

A review of the early literature on this subject indicates that, in the nineteenth century, dislocations of the sternoclavicular joint were managed essentially the same way as fractures of the medial clavicle [20, 45]. Sir Astley Cooper [20], in his 1824 text, recommended that the injury be treated not only with a clavicle bandage but also with a sling "which through the medium of the os humeri and scapula supports it and prevents the clavicle from being drawn down by the weight of the arm."

Although Cooper did not report having seen an isolated traumatic posterior dislocation of the sternoclavicular joint, he did describe a posterior dislocation of the sternoclavicular joint in a patient who had severe scoliosis [20]. As the severity of the scoliosis progressed, the scapula advanced laterally around the chest wall, forcing the medial end of the clavicle posterior to the sternum. The patient eventually developed significant dysphagia because of extrinsic pressure on the esophagus from the posteriorly displaced medial clavicle. Davey, a surgeon in Suffolk, resected the medial end of the clavicle. Davey must have been an excellent surgeon, for in

1824 he resected 1 inch of the medial clavicle using a saw! He protected the vital structures behind the sternoclavicular joint by introducing "a piece of well beaten sole leather under the bone while he divided it." The patient recovered and had no more difficulty with swallowing. This case probably represents the first surgical resection of the medial end of the clavicle [20].

The first case of traumatic posterior dislocation of the sternoclavicular joint to appear in the literature was probably that reported by Rodrigues [76] in 1843. He reported a "case of dislocation inward of the internal end of the clavicle." The patient's left shoulder was against a wall when the right side of the chest and thorax were compressed almost to the midline by a cart. The patient experienced immediate onset of dyspnea, which persisted for 3 weeks. The physician noted that the patient appeared to be suffocating because his face was blue. The left shoulder was swollen and painful, and there was "a depression on the left side of the superior extremity of the sternum." Posteriorly directed manual pressure on this depression greatly increased the sensation of suffocation. Rodrigues noted that posterior displacement of the shoulder and the lateral end of the clavicle resulted in forward displacement of the medial end of the clavicle and a diminution in the sensation of asphyxia. Therefore, his treatment consisted of binding the left shoulder backward with a cushion between the two scapulae, but only after the patient had been "bled" twice within the first 24 hours. Rodrigues may have seen other cases of posterior sternoclavicular dislocation, since he stated that the patient "retained a slight depression of the internal extremity of the clavicle; such, however, is the ordinary fate of the patients who present this form of dislocation."

In the late nineteenth century a number of articles concerning sternoclavicular injury appeared in England, Germany, and France. It was not until the 1930s, however, that articles pertaining to sternoclavicular joint injury appeared in the American literature. These early American authors included Duggan [25], Shafer [43], and Loman [53].

ANATOMY

Although the clavicle is the first long bone of the body to ossify (fifth intrauterine week), its medial epiphysis is the last of the long bone epiphyses to appear and to close (Fig. 15B–35) [33, 34, 70]. The medial clavicular epiphysis ossifies between the eighteenth and twentieth years and fuses with the shaft of the clavicle between the twenty-third and twenty-fifth years [33, 34, 70]. Webb and Suchey [83] concluded, after an autopsy study of the medial end of the clavicle in 605 males and 254 females, that complete union of the medial clavicular epiphysis may not be present until 31 years of age. With this knowledge in mind, it is conceivable that many of the so-called sternoclavicular dislocations are

actually fractures through the unfused medial physeal plate.

The medial end of the clavicle and the sternum articulate to form a diarthrodial joint. Because the upper extremity is suspended from the lateral clavicle, the sternoclavicular joint is the only true articulation between the upper extremity and the axial skeleton (Fig. 15B–36). The articular surface of the clavicle is much larger than that of the sternum, and in the adult, both are covered with fibrocartilage. The enlarged, bulbous medial clavicle presents a saddle-shaped (i.e., concave from anterior to posterior and convex from superior to inferior) articular surface to the notch of the sternum [33, 34]. Because the clavicular notch of the sternum is concave, the joint surface is not congruent. Cave [17] has demonstrated that, in 2.5% of patients, the inferior aspect of the medial clavicle articulates with the superior aspect of the first rib through a small facet located at the synchondral junction between the first rib and the sternum.

The sternoclavicular joint has the distinction of having the least amount of bony stability of any major joint in the body. This lack of stability results from the fact that less than half of the medial clavicle articulates with the upper angle of the sternum. Grant [33] remarked that "the two (make) an ill

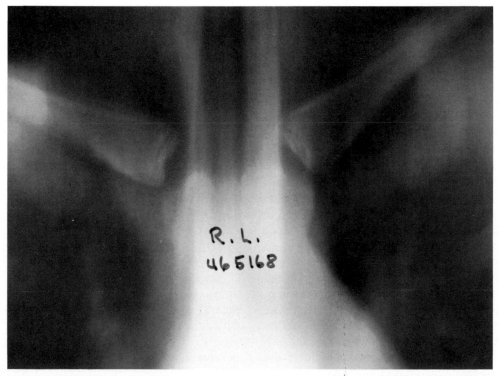

FIGURE 15B–35
Tomogram demonstrating the thin, waferlike disc of the epiphysis of the medial clavicle. (From Rockwood, C. A., and Green, D. P. (Eds.), *Fractures in Adults*, 2nd ed. Philadelphia, J. B. Lippincott, 1984.)

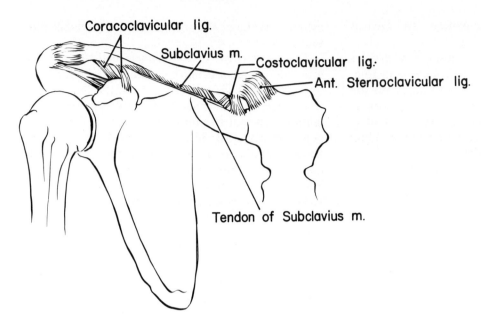

Coracoclavicular lig.

Subclavius m.

Costoclavicular lig.

Ant. Sternoclavicular lig.

Tendon of Subclavius m.

FIGURE 15B–36
Normal anatomy around the sternoclavicular and acromioclavicular joints. Note that the tendon of the subclavius muscle arises in the vicinity of the costoclavicular ligament from the first rib and has a long tendon structure. (From Rockwood, C. A., and Green, D. P. (Eds.), *Fractures in Adults*, 2nd ed. Philadelphia, J. B. Lippincott, 1984.)

fit.'' Digital palpation in the superior sternal notch reveals that, with motion of the upper extremity, a large part of the medial clavicle is completely above the superior margin of the sternum.

Because of the relative bony incongruity, the sternoclavicular joint must rely on its surrounding ligaments for most of its stability. These ligaments include the intra-articular disc ligament, the extra-articular costoclavicular ligament (i.e., the rhomboid ligament), the interclavicular ligament, and the capsular ligament. The intra-articular disc ligament is a very dense, fibrous structure that arises from the synchondral junction of the first rib with the sternum, passes through the sternoclavicular joint, and attaches on the superior and posterior aspects of the medial clavicle. Anteriorly and posteriorly the disc blends into the fibers of the capsular ligament. The disc divides the joint into separate medial and lateral compartments (Fig. 15B–37) [33, 34]. DePalma [23] has shown that, rarely, the two joint compartments communicate through a perforation in the disc. The intra-articular disc ligament acts as a check rein against medial displacement of the medial clavicle (Fig. 15B–37).

The short and strong costoclavicular or rhomboid ligament arises from the superior surface of the first rib and the adjacent synchondral junction with the sternum and passes superiorly to attach to the undersurface of the medial end of the clavicle. The site of attachment on the medial clavicle has often been referred to as the rhomboid fossa [33, 34]. However, Cave [17] studied the attachment of the costoclavicular ligament on 153 clavicles and could demonstrate a true depression, or fossa, in only 30% of the specimens. The clavicle was flat (60%)

or elevated (10%) in the remainder of the specimens. According to Cave, the costoclavicular ligament averages 1.3 cm in length, 1.9 cm in maximum width, and 1.3 cm in thickness. The costoclavicular ligament consists of two components, an anterior and a posterior fasciculus (see Fig. 15B–36) [7, 17,

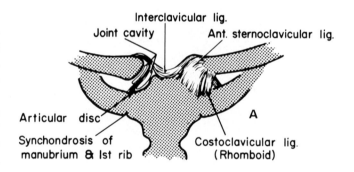

Interclavicular lig.

Joint cavity

Ant. sternoclavicular lig.

Articular disc

Synchondrosis of manubrium & 1st rib

Costoclavicular lig. (Rhomboid)

A

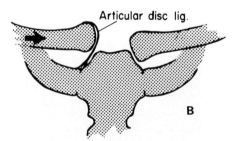

Articular disc lig.

B

FIGURE 15B–37
A, Normal anatomy around the sternoclavicular joint. Note that the intra-articular disc ligament divides the sternoclavicular joint cavity into two separate spaces and inserts onto the superior and posterior aspects of the medial clavicle. *B*, The intra-articular disc ligament acts as a check rein against medial displacement of the medial clavicle. (From Rockwood, C. A., and Green, D. P. (Eds.), *Fractures in Adults*, 2nd ed. Philadelphia, J. B. Lippincott, 1984.)

34]. Bearn [7] has demonstrated the presence of a consistent bursa between these two fasciculi.

The anterior fasciculus of the costoclavicular ligament arises anteromedially on the first rib and courses superolaterally to the clavicle. The fibers of the posterior fasciculus arise posterolaterally on the first rib and course superomedially to the clavicle. This crossing configuration provides rotational stability for the sternoclavicular joint during clavicular rotation, which occurs during full overhead elevation of the arm [34]. The two-part costoclavicular ligament is, in many ways, similar in configuration to the two-part coracoclavicular ligament at the distal end of the clavicle.

The interclavicular ligament connects the superomedial aspect of each clavicle with the capsular ligaments and the upper sternum (see Fig. 15B–37). According to Grant [33], this band may be homologous with the wishbone of birds. This ligament assists the capsular ligaments in maintaining "shoulder poise" (i.e., holding the clavicle and shoulder up). It can be tested by palpating in the superior sternal notch during overhead elevation. With elevation of the arm, the ligament is lax, but as the arms return to the side, the ligament becomes taught.

The components of the capsular ligament represent thickenings in the actual joint capsule (see Figs. 15B–36 15B–37). The ligament covers and reinforces the anterosuperior and posterior aspects of the joint. The anterior portion of the capsular ligament is heavier and stronger than the posterior portion. The capsular ligament attaches primarily on to the medial clavicular epiphysis. However, there is some secondary blending of the fibers of the ligament into the metaphysis [10, 21, 70].

BIOMECHANICS

Each of the sternoclavicular ligaments imparts a certain amount of stability to the joint. The intra-articular disc ligament acts as a check against medial displacement of the inner clavicle (see Fig. 15B–37). Bearn [7] has shown experimentally that the anterior fibers of the costoclavicular ligament resist excessive upward rotation of the medial clavicle and that the posterior fibers resist excessive downward rotation. In addition, the anterior fibers resist lateral displacement, and the posterior fibers resist medial displacement of the clavicle.

"Shoulder poise" is the result of resistance to downward displacement of the lateral end of the clavicle by the ligaments surrounding the sternoclavicular joint. Bearn [7] has shown that the capsular ligament is the strongest and most important structure in preventing upward displacement of the medial clavicle (i.e., downward displacement of the lateral clavicle). He studied the individual contributions of each of the sternoclavicular ligaments to resistance against downward displacement of the lateral clavicle in postmortem specimens. He determined that sectioning of the costoclavicular, intra-articular disc, and interclavicular ligaments had little or no effect on resistance to displacement of the lateral clavicle. Division of the capsular ligament alone resulted in downward depression of the distal end of the clavicle. Once the capsular ligament had been cut, the intra-articular disc ligament failed with only 5 pounds of weight at the lateral end of the clavicle. Bearn's findings have many clinical implications for the mechanisms of injury of the sternoclavicular joint.

The sternoclavicular joint is probably the most frequently moved joint of long bones in the body. Almost any motion of the upper extremity is transferred proximally to the sternoclavicular joint. The joint is freely movable and functions as a ball and socket. It has motion in almost all planes, including rotation [44, 54]. The clavicle is capable of 30 to 35 degrees of upward elevation, 35 degrees of combined forward and backward movement, and 45 to 50 degrees of rotation around its longitudinal axis through the sternoclavicular joint (Fig. 15B–38).

CLINICAL EVALUATION OF STERNOCLAVICULAR INJURY

History

The sternoclavicular joint is one of the least commonly dislocated joints in the body. This might seem surprising because it is also one of the most incongruent and mobile joints in the body. However, the rarity with which dislocation of this joint occurs is a testament to the extreme strength of the surrounding ligamentous structures. Traumatic dislocation of the sternoclavicular joint usually occurs only after tremendous forces, either direct or indirect, have been applied to the shoulder.

A direct force applied to the anteromedial aspect of the clavicle displaces the medial clavicle posteriorly behind the sternum and into the mediastinum (Fig. 15B–39). This may occur in a variety of ways: a knee landing directly on the medial end of the clavicle, a kick being delivered to the front of the medial clavicle, or a vehicle running over a person's chest or pinning a person to a wall. Anterior sternoclavicular dislocation as a result of direct trauma

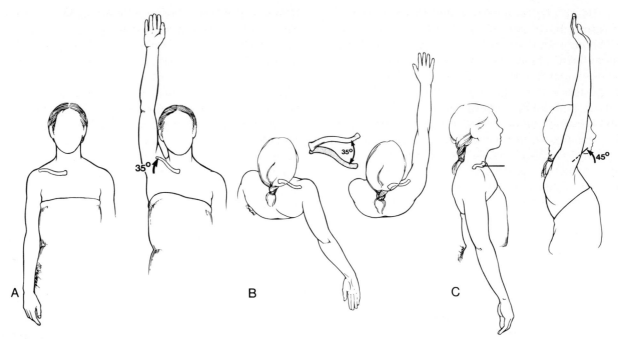

FIGURE 15B–38
Motions of the clavicle and the sternoclavicular joint. *A*, With full overhead elevation the clavicle is elevated 35 degrees. *B*, With adduction and extension the clavicle displaces anteriorly and posteriorly 35 degrees. *C*, The clavicle rotates on its long axis 45 degrees as the arm is elevated to the full overhead position. (From Rockwood, C. A., and Green, D. P. (Eds.), *Fractures in Adults*, 2nd ed. Philadelphia, J. B. Lippincott, 1984.)

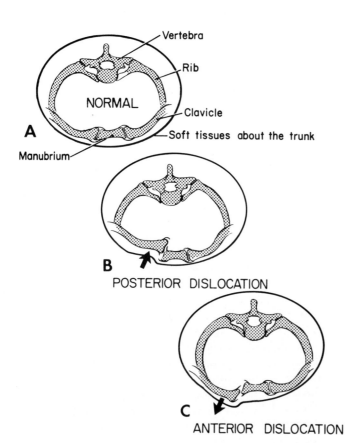

FIGURE 15B–39
A cross-section through the thorax at the level of the sternoclavicular joint. *A*, Normal anatomic relationships. *B*, Posterior dislocation of the sternoclavicular joint. *C*, Anterior dislocation of the sternoclavicular joint. (From Rockwood, C. A., and Green, D. P. (Eds.), *Fractures in Adults*, 2nd ed. Philadelphia, J. B. Lippincott, 1984.)

is most unusual because it requires the sternum and the entire thorax to be driven posteriorly while the clavicle remains in place.

The sternoclavicular joint is most commonly injured by indirect forces applied at the anterolateral or posterolateral aspect of the shoulder. If the shoulder is compressed and rolled forward, an ipsilateral posterior sternoclavicular dislocation results. An ipsilateral anterior sternoclavicular dislocation occurs when the shoulder is compressed and rolled backward (Fig. 15B–40). One of the most common causes, according to Rockwood [74], is a pile-up in a football game. In this instance, a player falls to the ground, landing on the lateral aspect of his shoulder. Before the athlete can get out of the way, several other players pile on top of his opposite shoulder. This imparts significant compressive force

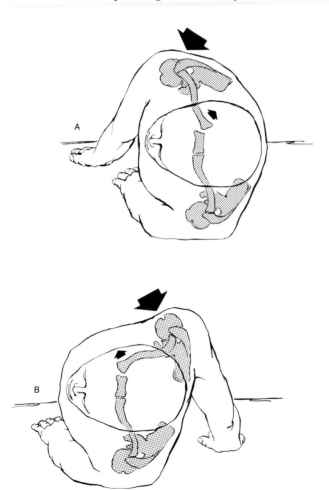

FIGURE 15B–40
Mechanisms that produce anterior or posterior dislocations of the sternoclavicular joint. *A*, If the patient is lying on the ground and a compression force is applied to the posterior lateral aspect of the shoulder, the medial end of the clavicle will be displaced posteriorly. *B*, When the lateral compression force is directed from the anterior position, the medial end of the clavicle is dislocated anteriorly. (From Rockwood, C. A., and Green, D. P. (Eds.), *Fractures in Adults*, 2nd ed. Philadelphia, J. B. Lippincott, 1984.)

that is transmitted down the shaft of the clavicle to the sternoclavicular joint. Mehta and co-workers [59] reported that three of four posterior sternoclavicular dislocations were produced by indirect force. Heinig [40] reported injury by indirect force in eight of nine cases of posterior sternoclavicular dislocations. Indirect force was the most common mechanism of injury in Rockwood's series of 168 patients [74].

Physical Examination

In a mild sprain, or type I injury, the ligaments of the joint are intact. The patient complains of mild to moderate pain, particularly with movement of the upper extremity. The joint may be slightly swollen and tender to palpation, but instability is absent. A moderate sprain results in a subluxation, or type II injury, of the sternoclavicular joint. The ligaments are elongated but intact. Swelling is noted and pain is marked, particularly with any movement of the arm. Anterior or posterior subluxation may be apparent upon inspection of the joint when compared to the other side. In addition, the medial clavicle may be unstable anteroposteriorly when manually stressed.

A severe sprain results in a complete dislocation, or type III injury. The dislocation may be either anterior or posterior. Certain signs and symptoms are common to both anterior and posterior dislocations. The patient has severe pain, which is increased with any movement of the arm. Axial compression of the shoulders is particularly painful. The patient usually supports the injured arm across the trunk with the normal arm. The affected shoulder appears to be shortened and thrust forward compared with the normal shoulder. In addition, the head may be tilted toward the side of the dislocated joint. When the patient is placed in a supine position on the examination table, the discomfort is exaggerated, and the involved shoulder will not lie flat on the table.

In an anterior dislocation, the medial end of the clavicle is visibly and palpably prominent anterior to the sternum (see Fig. 15B–30). It may be fixed in this position, or it may be quite mobile. The symptoms associated with posterior dislocation of the sternoclavicular joint are more severe than those associated with anterior dislocation. Venous congestion may be present in the neck or upper extremity. Stankler [79] reported on two patients with unrecognized posterior dislocations who developed venous engorgement of the ipsilateral arm. Breathing difficulties, shortness of breath, or a choking sensation may be noted. The patient may also complain of difficulty in swallowing or a tight feeling in the throat. In addition, arterial circulation to the upper extremity may be diminished. In severe cases, complete shock or a pneumothorax may be present.

Inspection of a patient with a posterior dislocation of the sternoclavicular joint reveals that the medial clavicle, which normally presents an anterior-superior fullness over the sternum, is less prominent compared to the normal side. The medial end of the clavicle, which is usually palpable, is displaced posteriorly. The articular corner of the sternum is more easily palpated compared to the normal side.

Although differentiation between an anterior and a posterior sternoclavicular dislocation is usually possible with physical examination, radiography should be used to verify the clinical impression. The senior author has encountered six patients who were thought clinically to have an anterior dislocation but turned out, on radiographic evaluation, to have a posterior dislocation.

Radiographic Evaluation

Occasionally, routine anteroposterior or posteroanterior radiographs of the chest or sternoclavicular joint suggest that "something is wrong" with one of the sternoclavicular joints. One or the other of the medial clavicles may appear to be displaced. Ideally, a second view in a plane orthogonal to the anteroposterior view should be taken. However, normal anatomic constraints do not allow this.

Several authors have recommended special projections to determine the position of the medial clavicle with respect to the sternum accurately [28, 29, 41, 46, 49, 71, 78]. Hobbs [41] recommended a view that approximates a 90-degree cephalocaudal lateral view of the sternoclavicular joint. The patient is seated high enough to lean forward over the x-ray table. With the cassette on the table, the patient leans forward so that the nape of his flexed neck is almost parallel to the table and his anterior rib cage is against the cassette (Fig. 15B–41). The flexed elbows straddle the cassette and support the head and neck. The x-ray source, which is located above the nape of the neck, projects a beam that passes through the cervical spine, projecting the sternoclavicular joints onto the cassette.

Rockwood [74] described a 40-degree cephalic tilt view of both sternoclavicular joints. He called this view the "serendipity" view because of the manner in which it was discovered. The patient is positioned supine in the center of the x-ray table with an 11-by 14-inch, nongrid cassette beneath his shoulders and neck. The tube is tilted 40 degrees cephalically

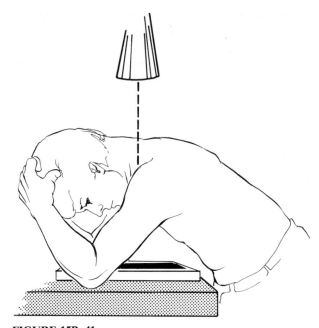

FIGURE 15B–41
Positioning of the patient for x-ray evaluation of the sternoclavicular joint as recommended by Hobbs. (Modified from Hobbs, D. W. The sternoclavicular joint: A new axial radiographic view. *Radiology* 90:801, 1968. Reproduced with permission from Rockwood, C. A., and Green, D. P. (Eds.), *Fractures in Adults*, 2nd ed. Philadelphia, J. B. Lippincott, 1984.)

and centered directly on the sternum. The distance from the beam to the patient's sternoclavicular joint varies with the thickness of the thorax to project both clavicles onto the film. In children, the tube should be 45 inches from the sternoclavicular joint. In adults, whose anteroposterior chest diameter is greater, the distance should be 60 inches (Fig. 15B–42).

Tomograms are helpful for distinguishing among sternoclavicular dislocation, medial clavicular fracture, and medial clavicular physeal injury (Fig. 15B–43). In 1959 Baker [3] stated that tomography was far more valuable than either routine films or the fingertips of the examining physician. Morag and Shahin [62] used tomography to evaluate the sternoclavicular joint in 20 patients. They recommended its routine use in the evaluation of problems of the sternoclavicular joint.

Because of its ability to visualize transverse sections of the body, the CT scan is the best technique for evaluating injuries to the sternoclavicular joint (Fig. 15B–44). The patient should be positioned supine in the CT scanner rather than being placed in a lateral decubitus position. In this way, a scan can be obtained of both sternoclavicular joints to compare the normal with the abnormal joint. Hartman and Dunnagan [39] reported on the use of CT arthrography of the sternoclavicular joint. They

FIGURE 15B–42
Positioning of the patient for the "serendipity" view of the sterno-clavicular joints. The x-ray tube is tilted 40 degrees from the vertical position and is aimed directly at the manubrium. The nongrid cassette should be large enough to receive the projected images of the medial halves of both clavicles. In children the tube distance from the patient should be 45 inches; in thicker chested adults the distance should be 60 inches. (From Rockwood, C. A., and Green, D. P. (Eds.), *Fractures in Adults*, 2nd ed. Philadelphia, J. B. Lippincott, 1984.)

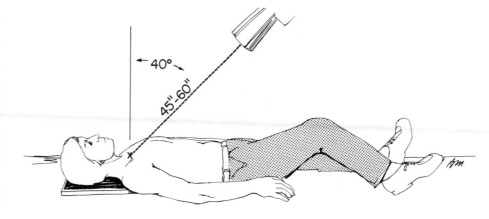

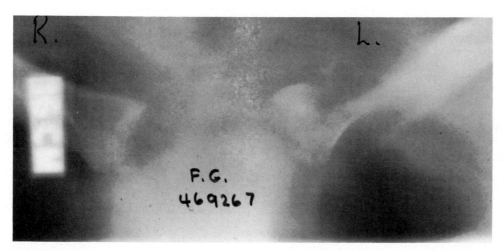

FIGURE 15B–43
Tomogram demonstrating a fracture of the left medial clavicle. The clinical pre-x-ray diagnosis was an anterior dislocation of the left sternoclavicular joint. (From Rockwood, C. A., and Green, D. P. (Eds.), *Fractures in Adults*, 2nd ed. Philadelphia, J. B. Lippincott, 1984.)

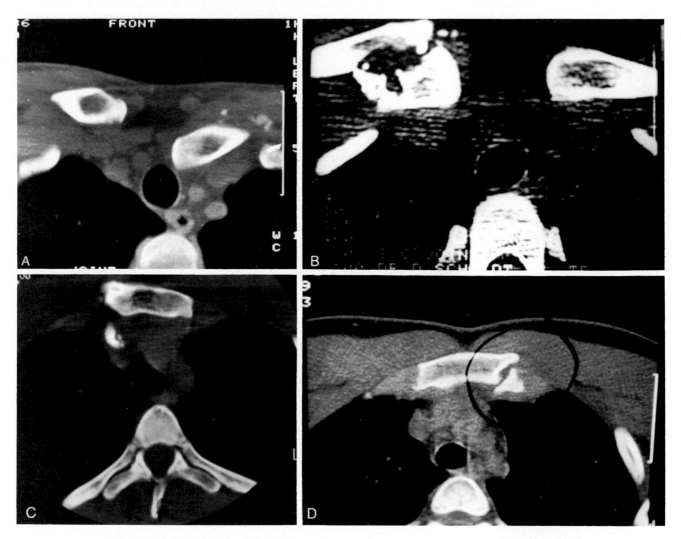

FIGURE 15B–44

CT scans of the sternoclavicular joint demonstrating various types of injuries. *A*, A posterior dislocation of the left clavicle compresses the great vessels and produces swelling of the left arm. *B*, A fracture of the medial clavicle does not involve the articular surface in this instance. *C*, A fragment of bone is displaced posteriorly into the great vessel. *D*, A fracture of the medial clavicle is displaced into the sternoclavicular joint. (From Rockwood, C. A., and Matsen, F. A. (Eds.), *The Shoulder.* Philadelphia, W. B. Saunders, 1990.)

were able to demonstrate a capsular disruption in a patient following a traumatic injury to the joint. The clinical significance of this, however, is unclear.

TREATMENT OPTIONS

Traumatic Injuries

Type I and II Injuries

Mild to moderate sprains (i.e., type I and type II injuries) of the sternoclavicular joint are most often treated nonoperatively. A type I injury, in which there is only minor injury to the ligaments and the joint remains reduced, requires only a sling and a gradual return to everyday activities. In sternoclavicular subluxation, or type II injury, the joint may be reduced by drawing the shoulders backward as if reducing and holding a fracture of the clavicle. This is true whether the subluxation is anterior or posterior. DePalma [22] suggests a plaster figure-of-eight dressing. McLaughlin [58] emphasizes the use of a sling to support the arm as well as a splint to hold the shoulder backward. Allman [2] prefers the use of a soft figure-of-eight bandage combined with a sling. Occasionally, he uses adhesive strapping over the medial end of the clavicle.

Some authors [6, 22] have recommended open reduction, ligament repair, and temporary internal fixation of the sternoclavicular joint with pins drilled from the clavicle into the sternum for subluxations in which closed reduction has failed. DePalma [24] applies a figure-of-eight plaster cast postoperatively. In addition, he supports the arm with a sling and swathe. The pins and cast are removed after 6 weeks. The placement of pins across the sternoclavicular joint, however, is associated with many serious complications. There have been six reported deaths [18, 32, 51, 63, 77] and three near deaths [11, 67, 85] from complications of transfixing the sternoclavicular joint with Kirschner wires or Steinmann pins. The pins, either intact or broken, migrated into the heart, pulmonary artery, innominate artery, or aorta. In 1984 Gerlach and colleagues [32] from West Germany reported two deaths that resulted from migrating nails from the sternoclavicular joint. The nails migrated to the heart and caused cardiac tamponade and death. The physicians were charged with manslaughter because of negligence. Because of the association of such serious consequences with pinning the sternoclavicular joint, one should use extreme caution when selecting this form of treatment.

Persistent pain following nonoperative treatment of a type II injury may require joint exploration.

Bateman [6] has emphasized the significance of torn intra-articular discs in the pathogenesis of these persistent symptoms. Duggan [25] reported a case in which, several weeks after an injury to the sternoclavicular joint, the patient still experienced popping in the joint. He incised the capsule and found a completely loose intra-articular disc. He commented that this looked like "an avulsed fingernail." Following repair of the capsule and excision of the disc, the patient was symptom free.

Type III Injuries or Complete Dislocation

Anterior Dislocation. Anterior dislocation of the sternoclavicular joint is generally easy to reduce by closed means. With the patient in a supine position and with a 3- to 4-inch thick pad between the shoulders, gentle direct pressure is applied over the anteriorly displaced medial clavicle. Although the reduction is frequently easy to obtain, the clavicle usually redislocates when the pressure is released. If the sternoclavicular joint is stable following reduction, it can be immobilized for 6 weeks using a soft figure-of-eight dressing, a commercial clavicle strap or harness, or a plaster figure-of-eight cast. Although some authors have recommended operative repair of unstable anterior sternoclavicular dislocations, most patients do well when treated nonoperatively [74]. In addition, Rockwood and Odor [74] reported a high rate of complications and poor results in patients treated operatively for anterior sternoclavicular dislocations.

Posterior Dislocation. Posterior sternoclavicular dislocation can be a life-threatening injury. Only after the history, physical examination, and appropriate consultation have ruled out injury to the pulmonary and vascular systems should further work-up and treatment of the sternoclavicular injury be embarked on. If, on the other hand, significant injury to the vital structures posterior to the sternoclavicular joint has occurred, it should be appropriately addressed prior to instituting treatment for the posterior sternoclavicular dislocation. Worman and Leagus [85] reported a posterior dislocation of the sternoclavicular joint in which the medial end of the clavicle had perforated the right pulmonary artery. At operation, the clavicle was found to be impaled within the pulmonary artery, thus preventing exsanguination. Had the injury not been suspected and a closed reduction been performed in the emergency department, the result could have been disastrous.

Although many early authors [15, 19, 30, 40, 57, 58, 61, 68, 77, 80] advocated open reduction and fixation for posterior sternoclavicular dislocation, most current authors recommend closed reduction and external immobilization [12, 15, 23, 27, 30, 40, 50, 57, 61, 72, 77, 80]. Some authors [68, 80] even reported changing from open reduction to closed reduction because of the observation that the dislocation had reduced very easily under direct vision intraoperatively. Many techniques for closed reduction of posterior sternoclavicular dislocation have been reported. The use of traction with the arm in abduction has been reported by DePalma [23], Fery and colleagues [30], McKenzie [57], Mitchell and Cobey [61], Rockwood [72], and Salvatore [77]. Under general anesthesia, the patient is placed supine on the operating table with a 3- to 4-inch thick sandbag placed between the shoulders and with the involved shoulder near the edge of the table (Fig. 15B–45). Lateral traction is applied to the abducted arm, which is then gradually extended. The clavicle may then reduce with an audible "snap" or "pop." The reduction is almost always stable. Overzealous extension may cause the medial end of

the clavicle to bind against the back of the manubrium. In some cases, the medial clavicle must be manipulated to dislodge it from behind the sternum. Initially, this should be attempted by manually grasping the medial shaft of the clavicle. Should this fail, the skin can be sterilely prepared, and a sterile towel clip can be used to grasp the medial clavicle to apply lateral and anterior traction (see Fig. 15B–32).

Buckerfield and Castle [12] reported the use of traction with the arm in adduction for closed reduction of posterior sternoclavicular dislocations. With this technique [12], the patient is placed supine on the operating table with a 3- to 4-inch bolster between the shoulders. Distal traction is then applied to the adducted arm while a downward force is exerted on the shoulders. Reduction occurs when the clavicle is levered over the first rib into its normal position. Other authors have reported other similar traction techniques for closed reduction [15, 27, 40, 80].

Following reduction, the joint is usually stable. A

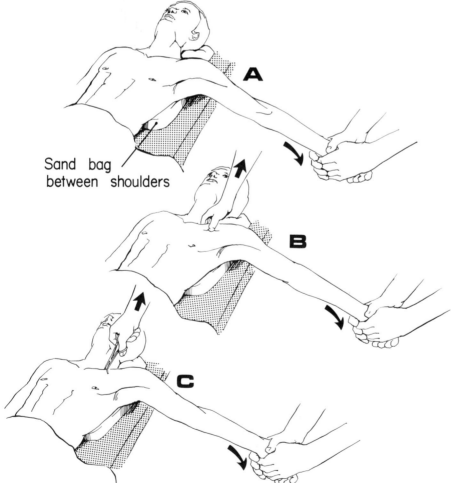

Sand bag between shoulders

FIGURE 15B–45
Technique of closed reduction of the sternoclavicular joint. *A*, The patient is positioned supine with a sandbag placed between the two shoulders. Traction is then applied to the arm against countertraction in an abducted and slightly extended position. In patients with anterior dislocations, direct pressure over the medial end of the clavicle may reduce the joint. *B*, In patients with posterior dislocations, in addition to the traction it may be necessary to manipulate the medial end of the clavicle with the fingers to dislodge the clavicle from behind the manubrium. *C*, In stubborn cases of posterior dislocation, it may be necessary to prepare the medial end of the clavicle surgically and place a towel clip *around* the medial clavicle to lift it back into position. (From Rockwood, C. A., and Green, D. P. (Eds.), *Fractures in Adults*, 2nd ed. Philadelphia, J. B. Lippincott, 1984.)

figure-of-eight dressing or a commercially available figure-of-eight strap should be used to hold the shoulders backward for 4 to 6 weeks to allow ligament healing.

Most adult patients do not tolerate significant posterior displacement of the clavicle [9, 31, 42, 52]. Therefore, operative intervention is indicated should closed reduction fail. If open reduction can be performed without damaging the intact anterior sternoclavicular ligaments, the joint will remain stable with the shoulders held back in a figure-of-eight dressing. However, if all the ligamentous supports of the sternoclavicular joint are disrupted, reduction must be combined with ligament repair or reconstruction. Damage to the articular surface may necessitate resection of the medial 1 to 1½ inches of the clavicle. The remaining clavicle should be stabilized to the first rib.

Many techniques for restoring sternoclavicular stability have been reported. Elting [27], Denham and Dingley [21], Brooks and Henning [10], DePalma [24], and Brown [11] recommended fixation of the sternoclavicular joint using Kirschner wires or Steinmann pins in addition to ligament repair or reconstruction. As mentioned previously, the complications associated with this type of fixation are life-threatening.

In an effort to avoid the use of pins across the sternoclavicular joint, many other authors have reported alternative techniques for sternoclavicular fixation. Habernek and Hertz [38], Nutz [65], Pfister and Weller [69], Kennedy [47], Tagliabue and Riva [81], Hartman and Dunnagan [39], Bankart [4], Ecke [26], and Stein [80] have all advocated the use of various types of wire sutures. Burri and Neugebauer [13] recommended a figure-of-eight loop of carbon fiber. Haug [39A] reported the use of a specially designed plate to stabilize the sternoclavicular joint.

Sternoclavicular ligament reconstruction has been reported using various tendons. Elting [27] supplemented ligament repair with a short toe extensor. Maguire [56], Booth and Roper [8], Barth and Hagen [5], Lunseth and colleagues [55], and Burrows [14] reconstructed the sternoclavicular ligaments using the tendons of the sternocleidomastoid, subclavius, or pectoralis major muscles. In addition, the use of fascia lata has been reported by Bankart [4], Milch [60], Lowman [53], Speed [79B], Key and Conwell [48], and Allen [1].

The treatment of physeal injuries to the medial clavicle is similar to that described for anterior and posterior dislocations. However, remodeling as a result of physeal activity may reduce or eliminate any residual deformity. Therefore, physeal injuries

with anterior and asymptomatic posterior displacement may not require reduction. Posterior displacement associated with significant symptoms, however, should be treated with the same urgency as a simple posterior sternoclavicular dislocation.

Atraumatic Injuries

Acquired spontaneous subluxation or dislocation of the sternoclavicular joint usually occurs in young people under the age of 20; there is a female preponderance. As the arm is flexed or abducted, one or both clavicles spontaneously subluxate anteriorly (see Fig. 15B–34). Spontaneous reduction occurs when the arm is returned to the side. The hallmark of this condition is a lack of significant trauma. It is usually associated with laxity in other joints. Rockwood, in a review of 37 cases, found it to be a self-limiting condition [75]. He recommended that surgical reconstruction not be attempted because of increased pain, an unsightly scar, and persistent subluxation or dislocation (Fig. 15B–46).

Congenital subluxation or dislocation of the sternoclavicular joint may result from the absence or partial absence of bone or muscles around the sternoclavicular joint. Specific rehabilitation or surgical procedures are usually not necessary because this condition also responds well to simple nonoperative management [35, 64].

AUTHORS' PREFERRED METHOD OF TREATMENT

Type I Injuries

A simple sling, rest, ice, and analgesics are normally adequate for type I injuries or mild sprains. The patient is encouraged to use the arm for everyday living activities within the first 5 to 7 days.

Criteria for Return to Athletics. In general the athlete may return to his sport when he has regained a full pain-free range of motion and the joint is nontender. This normally occurs within 7 to 14 days.

Type II Injuries

A figure-of-eight bandage or clavicle strap is helpful in patients with type II injuries to hold the shoulders back and the sternoclavicular joint reduced. Symptoms normally resolve enough to allow removal of the figure-of-eight harness within the

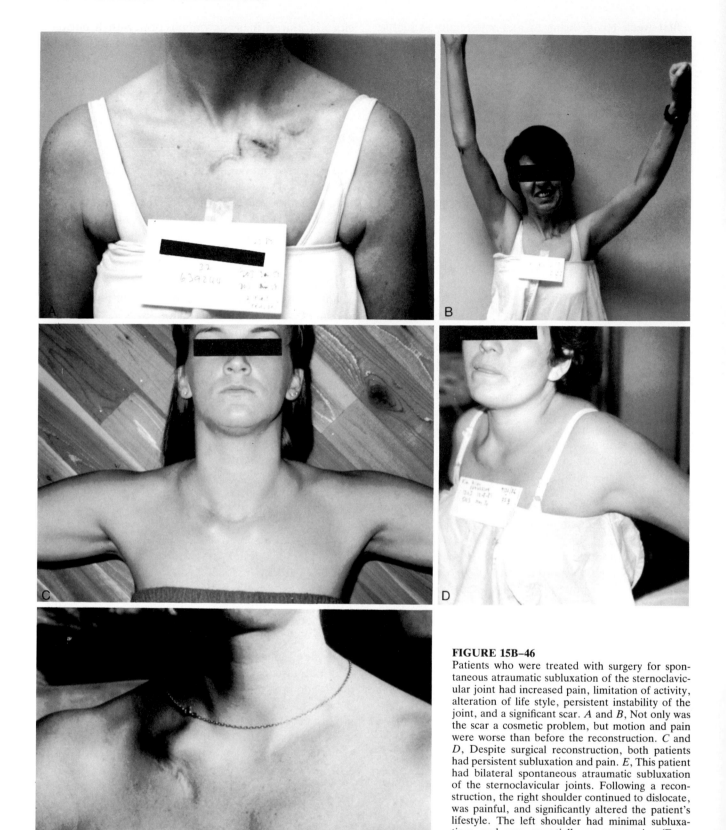

FIGURE 15B–46

Patients who were treated with surgery for spontaneous atraumatic subluxation of the sternoclavicular joint had increased pain, limitation of activity, alteration of life style, persistent instability of the joint, and a significant scar. *A* and *B*, Not only was the scar a cosmetic problem, but motion and pain were worse than before the reconstruction. *C* and *D*, Despite surgical reconstruction, both patients had persistent subluxation and pain. *E*, This patient had bilateral spontaneous atraumatic subluxation of the sternoclavicular joints. Following a reconstruction, the right shoulder continued to dislocate, was painful, and significantly altered the patient's lifestyle. The left shoulder had minimal subluxations and was essentially asymptomatic. (From Rockwood, C. A., and Matsen, F. A. (Eds.), *The Shoulder*. Philadelphia, W. B. Saunders, 1990.)

first week to 10 days. At this point, the patient is encouraged to return gradually to everyday living activities.

Criteria for Return to Athletics. Since this type of injury has been characterized by plastic deformation of the surrounding ligaments of the sternoclavicular joint, full participation in contact sports should be avoided for a total of 6 weeks from the time of injury. Return to noncontact sports may be attempted when the athlete has regained a full, pain-free range of motion. Contact sports should be avoided for an additional 4 to 6 weeks to allow ligament healing.

Type III Injuries

Nearly all type III, or complete, sternoclavicular dislocations are managed with initial closed reduction. Acute traumatic posterior dislocations are stable after closed reduction and may be maintained in a reduced position using a figure-of-eight dressing. Most anterior dislocations, however, are unstable following reduction. However, the deformity is usually well tolerated, and operative repair with internal fixation is associated with significant complications.

Anterior Dislocations. Reduction of an anterior sternoclavicular dislocation should be performed with the assistance of muscle relaxants and narcotics administered intravenously. The patient is placed supine on the table with a stack of three or four towels between the shoulder blades. Direct backward pressure on the medial end of the clavicle is applied while an assistant applies gentle traction on the arm. If the joint is unstable following reduction, as frequently occurs, a sling is prescribed for 2 weeks, and the patient is encouraged to begin using the arm as soon as the discomfort is gone. The patient should be informed that the sternoclavicular joint is unstable but that operative repair is hazardous and unnecessary. If the joint is stable following reduction, which rarely occurs, the patient is placed in a figure-of-eight dressing for 4 to 6 weeks to allow ligament healing.

Criteria for Return to Athletics. Regardless of whether or not the reduction is stable, athletes involved in contact sports should not be allowed to participate until 6 to 8 weeks after injury to allow soft tissue healing. Return to noncontact sports may be attempted after the athlete has recovered a full range of motion and is pain-free. This normally occurs around the 4-week mark.

Posterior Dislocations. Acute posterior dislocations should be reduced only after injury to the mediastinal structures has been ruled out. The pa-

tient is placed in a supine position with a bolster between the scapulae. The affected arm and shoulder should extend beyond the edge of the table so that the shoulder can be abducted and extended during reduction. In some cases, intravenous narcotics, muscle relaxants, or tranquilizers will be adequate to accomplish reduction. However, if the patient is having extreme discomfort and is overly anxious, general anesthesia will be necessary. The reduction is performed by placing gentle traction longitudinally on the abducted arm while countertraction is applied by an assistant from the opposite side of the table. The traction is gradually increased while the shoulder is brought into extension. Reduction is usually accompanied by an audible "pop" or "snap." If this method is unsuccessful, the skin should be surgically prepared, and a sterile towel clip should be used to gain purchase on the medial clavicle percutaneously (see Fig. 15B–32). The towel clip should pass completely around the medial clavicle because the dense cortical bone in the clavicle is not easily penetrated by the towel clip. Longitudinal traction is applied through the arm while the medial clavicle is manipulated superiorly and anteriorly by lifting on the towel clip.

If closed reduction fails by either of the above two methods, an open reduction should be performed. The patient is placed supine on the operating table with a sandbag or three or four towels between the scapulae. The entire upper extremity should be surgically prepared so that lateral traction can be applied during the procedure. In addition, a folded sheet should be placed around the patient's thorax so that countertraction can be applied during the procedure. An anterior incision, which parallels the superior border of the medial 3 to 4 inches of the clavicle and extends caudally over the sternum just medial to the involved sternoclavicular joint, is utilized. The procedure should be done with the assistance of a cardiothoracic surgeon. The sternoclavicular joint is incised superiorly, and care is taken to preserve the anterior sternoclavicular ligaments. The medial clavicle is reduced through the use of traction and countertraction combined with direct manipulation of the medial clavicle using a clamp or an elevator. If the anterior ligaments have been preserved, the joint will be stable following reduction. If the anterior capsule has been damaged or is insufficient to prevent instability of the joint following reduction, the medial 1 to 1½ inches of the clavicle is excised. The residual clavicle is stabilized to the first rib using 1 mm Dacron tape.

Following reduction, either open or closed, of an acute posterior sternoclavicular dislocation, the shoulder should be held back in a well-padded figure-of-eight clavicle strap or dressing. With closed reduction, the period of immobilization is approximately 3 to 4 weeks. With open reduction, the period of immobilization is slightly longer (i.e., 4 to 6 weeks) to allow soft tissue healing.

Criteria for Return to Athletics. Following either closed or open reduction, return to sports should be postponed for a total of 10 to 12 weeks postreduction. The complications associated with chronic or recurrent posterior dislocations are so severe that complete ligament healing should be ensured prior to return to competition.

Physeal Injuries

The treatment of physeal injuries of the medial clavicle is slightly different from the management of true dislocations. Although the remodeling potential of the medial clavicular physis is good, closed reduction should be attempted whenever significant displacement has occurred. Residual deformity can be accepted with anterior displacement and with posterior displacement that is asymptomatic. Very rarely, residual posterior displacement will be accompanied by compression of the vital structures within the mediastinum. In this instance, open reduction of the physeal injury is indicated. Following reduction, the shoulder should be held backward with a figure-of-eight strap or dressing for 3 to 4 weeks.

Criteria for Return to Athletics. Return to non-contact sports can be attempted at 6 weeks postinjury or when the athlete has regained full, painless range of motion and the fracture site and medial clavicle are nontender. Contact sports should be avoided for an additional 4 to 6 weeks to allow remodeling and strengthening of the medial clavicle to occur.

The true incidence of medial clavicular physeal injury is unknown. Because the medial clavicular epiphysis does not ossify until age 18, the diagnosis is difficult to verify radiographically. However, because the capsule attaches mainly onto the epiphysis, it is conceivable that the majority of "dislocations" in patients under the age of 25 are actually physeal separations rather than true dislocations. This distinction is clinically relevant because residual displacement in this injury can potentially diminish with time as remodeling occurs.

References

1. Allen, A. W. Living suture grafts in the repair of fractures and dislocation. *Arch Surg* 6:1007–1020, 1928.
2. Allman, F. L. Fracture and ligamentous injuries of the clavicle and its articulations. *J Bone Joint Surg* 49A:774–784, 1967.
3. Baker, E. C. Tomography of the sternoclavicular joint. *Ohio State Med J* 55:60, 1959.
4. Bankart, A. S. B. An operation for recurrent dislocation (subluxation) of the sternoclavicular joint. *Br J Surg* 26:320–323, 1938.
5. Barth, E., and Hagen, R. Surgical treatment of dislocations of the sternoclavicular joint. *Acta Orthop Scand* 54:746–747, 1983.
6. Bateman, J. E. *The Shoulder and Neck.* Philadelphia, W. B. Saunders, 1972.
7. Bearn, J. G. Direct observations on the function of the capsule of the sternoclavicular joint in clavicular support. *J Anat* 101(1):159–170, 1967.
8. Booth, C. M., and Roper, B. A. Chronic dislocation of the sternoclavicular joint: An operative repair. *Clin Orthop* 140:17–20, 1979.
9. Borrero, E. Traumatic posterior displacement of the left clavicular head causing chronic extrinsic compression of the subclavian artery. *Phys Sportsmed* 15(7):87–89, 1987.
10. Brooks, A. L., and Henning, G. D. Injury to the proximal clavicular epiphysis (proceedings). *J Bone Joint Surg* 54A(6):1347–1348, 1972.
11. Brown, J. E. Anterior sternoclavicular dislocation—a method of repair. *Am J Orthop* 31:184–189, 1961.
12. Buckerfield, C. T., and Castle, M. E. Acute traumatic retrosternal dislocation of the clavicle. *J Bone Joint Surg* 66A(3):379–384, 1984.
13. Burri, C., and Neugebauer, R. Carbon fiber replacement of the ligaments of the shoulder girdle and the treatments of lateral instability of the ankle joint. *Clin Orthop* 196:112–117, 1985.
14. Burrows, H. J. Tenodesis of subclavius in the treatment of recurrent dislocation of the sternoclavicular joint. *J Bone Joint Surg* 33B(2):240–243, 1951.
15. Butterworth, R. D., and Kirk, A. A. Fracture dislocation sternoclavicular joint. A case report. *Va Med* 79:98–100, 1952.
16. Cadenat, F. M. The treatment of dislocation and fractures of the outer end of the clavicle. *Int Clin* 1:145–169, 1917.
17. Cave, A. J. E. The nature and morphology of the costoclavicular ligament. *J Anat* 95:170–179, 1961.
18. Clark, R. L., Milgram, J. W., and Yawn, D. H. Fatal aortic perforation and cardiac tamponade due to a Kirschner wire migrating from the right sternoclavicular joint. *South Med J* 67(3):316–318, 1974.
19. Collins, J. J. Retrosternal dislocation of the clavicle (proceedings). *J Bone Joint Surg* 54B(1):203, 1972.
20. Cooper, A. *A Treatise on Dislocations and Fractures of the Joints* (2nd Am. ed. from 6th London ed.). Boston, Lilly & Wait and Carter & Hendee, 1832.
21. Denham, R. H., Jr., and Dingley, A. F., Jr. Epiphyseal separation of the medial end of the clavicle. *J Bone Joint Surg* 49A:1541–1550, 1963.
22. DePalma, A. F. The role of the disks of the sternoclavicular and the acromioclavicular joints. *Clin Orthop* 13:222–233, 1959.
23. DePalma, A. F. Surgical anatomy of acromioclavicular and sternoclavicular joints. *Surg Clin North Am* 43:1541–1550, 1963.
24. DePalma, A. F. *Surgery of the Shoulder* (2nd ed.). Philadelphia, J. B. Lippincott, 1973.
25. Duggan, N. Recurrent dislocation of sternoclavicular cartilage. *J Bone Joint Surg* 13:365, 1931.
26. Ecke, H. Sternoclavicular dislocations. Personal communication, 1984.
27. Elting, J. J. Restrosternal dislocation of the clavicle. *Arch Surg* 104:35–37, 1972.
28. Fedoseev, V. A. [Method of radiographic study of the sternoclavicular joint.] *Vestn Rentgenol Radiol* 3:88–91, 1977.
29. Fery, A. M., Rook, F. W., and Masterson, J. H. Retrosternal

dislocation of the clavicle. *J Bone Joint Surg* 39A(4):905–910, 1957.

30. Fery, A., and Leonard, A. [Transsternal sternoclavicular projection. Diagnostic value in sternoclavicular dislocations.] *J Radiol* 62(3):167–170, 1981.

31. Gangahar, D. M., and Flogaites, T. Retrosternal dislocation of the clavicle producing thoracic outlet syndrome. *J Trauma* 18(5):369–372, 1978.

32. Gerlach, D., Wemhoner, S. R., and Ogbuihi, S. [On two cases of fatal heart tamponade due to migration of fracture nails from the sternoclavicular joint.] *Z Rechtsmed* 93(1):53–60, 1984.

33. Grant, J. C. B. *Method of Anatomy* (7th ed.). Baltimore, Williams & Wilkins, 1965.

34. Gray, H., and Goss, C. M. (Eds.). *Anatomy of the Human Body* (28th ed.). Philadelphia, Lea & Febiger, 1966, pp. 324–326.

35. Guerin, J. Recherches sur les luxations congenitales. *Gaz Med Paris* 9:97, 1841.

36. Gunson, E. F. Radiography of sternoclavicular articulation. *Radiog Clin Photog* 19(1):20–24, 1943.

37. Gunther, W. A. Posterior dislocation of the sternoclavicular joint. *J Bone Joint Surg* 31A(4):878–879, 1949.

38. Habernek, H., and Hertz, H. [Origin, diagnosis and treatment of traumatic dislocation of sternoclavicular joint.] *Aktuel Traumatol* 17(1):223–228, 1987.

39. Hartman, T. J., and Dunnagan, W. A. Cinearthrography of the sternoclavicular joint (abstract). *Personal communication.* 1979.

39A. Haug, W. Retention einer seltenen Sterno-clavicular-hexations Fraktur mittels modifizierter Y-Platte der AO. *Aktuel Traumatol* 16:39–40, 1986.

40. Heining, C. F. Retrosternal dislocation of the clavicle: Early recognition, x-ray diagnosis, and management. *J Bone Joint Surg* 50A(4):830, 1968.

41. Hobbs, D. W. Sternoclavicular joint: New axial radiographic view. *Radiology* 90:801–802, 1968.

42. Holmdahl, H. C. A case of posterior sternoclavicular dislocation. *Acta Orthop Scand* 23:218–222, 1953–1954.

43. Howard, F. M., and Shafer, S. J. Injuries to the clavicle with neurovascular complications: A study of fourteen cases. *J Bone Joint Surg* 47A:1335–1346, 1965.

44. Inman, V. T., Saunders, J. B., and Abbott, L. C. Observations on the function of the shoulder joint. *J Bone Joint Surg* 26:1–30, 1944.

45. Jones, R. *Injuries of Joints* (2nd ed.). London, Oxford University Press, 1917, pp. 53–55.

46. Kattan, K. R. Modified view for use in roentgen examination of the sternoclavicular joints. *Radiology* 108(3):8, 1973.

47. Kennedy, J. C. Retrosternal dislocation of the clavicle. *J Bone Joint Surg* 31B(1):74–75, 1949.

48. Key, J. A., and Conwell, H. E. *The Management of Fractures, Dislocations, and Sprains* (5th ed.). St. Louis, C. V. Mosby, 1951, pp. 458–461.

49. Kurzbauer, R. The lateral projection in roentgenography of the sternoclavicular articulation. *Am J Roentgenol* 56(1):104–105, 1946.

50. Leighton, R. K., Buhr, A. J., and Sinclair, A. M. Posterior sternoclavicular dislocations. *Can J Surg* 29(2):104–106, 1989.

51. Leonard, J. W., and Gifford, R. W. Migration of a Kirschner wire from the clavicle into pulmonary artery. *Am J Cardiol* 16:598–600, 1965.

52. Louw, J. A., and Louw, J. A. Posterior dislocation of the sternoclavicular joint associated with major spinal injury. *S Afr Med J* 71(12):791–792, 1987.

53. Lowman, C. L. Operative correction of old sternoclavicular dislocation. *J Bone Joint Surg* 10(3):740–741, 1928.

54. Lucas, D. B. Biomechanics of the shoulder joint. *Arch Surg* 107:425–432, 1973.

55. Lunseth, P. A., Chapman, K. W., and Frankel, V. H. Surgical treatment of chronic dislocation of the sterno-clavicular joint. *J Bone Joint Surg* 57B(2):193–196, 1975.

56. Maguire, W. B. Safe and simple method of repair of recurrent dislocation of the sternoclavicular joint (abstract). *J Bone Joint Surg* 68B(2):332, 1986.

57. McKenzie, J. M. M. Retrosternal dislocation of the clavicle: A report of two cases. *J Bone Joint Surg* 45B(1):138–141, 1963.

58. McLaughlin, H. *Trauma.* Philadelphia, W. B. Saunders, 1959.

59. Mehta, J. C., Sachdev, A., and Collins, J. J. Retrosternal dislocation of the clavicle. *Injury* 5(1):79–83, 1973.

60. Milch, H. The rhomboid ligament in surgery of the sternoclavicular joint. *J Int Coll Surg* 17(1):41–51, 1952.

61. Mitchell, W. J., and Cobey, M. D. Retrosternal dislocation of clavicle. *Med Ann DC* 29(10):546–549, 1960.

62. Morag, B., and Shahin, N. The value of tomography of the sternoclavicular region. *Clin Radiol* 26:57–62, 1975.

63. Nettles, J. L., and Linscheid, R. Sternoclavicular dislocations. *J Trauma* 8(2):158–164, 1968.

64. Newlin, N. S. Congenital retrosternal subluxation of the clavicle simulating an intrathoracic mass. *Am J Roentgenol* 130:1184–1185, 1978.

65. Nutz, V. [Fracture dislocation of the sternoclavicular joint.] *Unfallchirurg* 89(3):145–148, 1986.

66. Omer, G. E. Osteotomy of the clavicle in surgical reduction of anterior sternoclavicular dislocation. *J Trauma* 7(4):584–590, 1967.

67. Pate, J. W., and Wilhite, J. Migration of a foreign body from the sternoclavicular joint to the heart: A case report. *Am Surg* 35(6):448–449, 1969.

68. Peacock, H. K., Brandon, J. R., and Jones, O. L. Retrosternal dislocation of the clavicle. *South Med J* 63(1):1324–1328, 1970.

69. Pfister, U., and Weller, S. [Luxation of the sternoclavicular joint.] *Unfallchirurg* 8(2):81–87, 1982.

70. Poland, J. Separation of the epiphyses of the clavicle. *In Traumatic Separation of Epiphyses of the Upper Extremity.* London, Smith, Elder, 1898.

71. Ritvo, M., and Ritvo, M. Roentgen study of the sternoclavicular region. *Am J Roentgenol* 53(5):644–650, 1947.

72. Rockwood, C. A., Jr. Dislocation of the sternoclavicular joint. *In* Rockwood, C. A., Jr., and Green, D. P. (Eds.), *Fractures* (Vol. 1). Philadelphia, J. B. Lippincott, 1975, pp. 756–787.

73. Rockwood, C. A., Jr. Injuries to the sternoclavicular joint. *In* Rockwood, C. A., Jr., and Green, D. P. (Eds.), *Fractures in Adults* (Vol. 1, 2nd ed.). Philadelphia, J. B. Lippincott, 1984, pp. 910–948.

74. Rockwood, C. A., Jr. Disorders of the sternoclavicular joint. *In* Rockwood, C. A., Jr., and Matsen, F. A. (Eds.), *The Shoulder.* Philadelphia, W. B. Saunders, 1990.

75. Rockwood, C. A., Jr., and Odor, J. M. Spontaneous atraumatic anterior subluxations of the sternoclavicular joint in young adults. Report of 37 cases. *Orthop Trans* 12(3):557, 1988.

76. Rodrigues, H. Case of dislocation inwards of the internal extremity of the clavicle. *Lancet* 1:309–310, 1843.

76A. Rowe, C. R., and Marble, H. C. Shoulder girdle injuries. *In* Cave, E. F. (Ed.), *Fractures and Other Injuries.* Chicago, Year Book, 1958.

77. Salvatore, J. E. Sternoclavicular joint dislocation. *Clin Orthop* 58:51–54, 1968.

78. Schmitt, W. G. H. Articulatis Sternoclavicularis: Darstellung in einer zweiter Ebene. *Roentgenpraxis* 34:262–267, 1981.

78A. Spar, I. Psoriatic arthritis of the sternoclavicular joint. *Conn Med* 42(4):225–226, 1978.

78B. Speed, K. *A Textbook of Fractures and Dislocations* (4th ed.). Philadelphia, Lea & Febiger, 1942, pp 282–290.

79. Stankler, L. Posterior dislocation of clavicle: A report of 2 cases. *Br J Surg* 50:164–168, 1962.

80. Stein, A. H. Retrosternal dislocation of the clavicle. *J Bone Joint Surg* 39A(3):656–660, 1957.

81. Tagliabue, D., and Riva, A. Le lussazioni sterno-claveari. *Minerva Ortoped* 36(11):876–871, 1985.

82. Waskowitz, W. J. Disruption of the sternoclavicular joint: An analysis and review. *Am J Orthop* 3:176–179, 1961.
83. Webb, P. A. O., and Suchey, J. M. M. Epiphyseal union of the anterior iliac crest and medial clavicle in a modern multiracial sample of American males and females. *Am J Phys Anthropol* 68:457–466, 1985.
84. Wheeler, M. E., Laaveg, S. J., and Sprague, B. L. S-C joint disruptions in an infant. *Clin Orthop* 139:68–69, 1979.
85. Worman, L. W., and Leagus, C. Intrathoracic injury following retrosternal dislocation of the clavicle. *J Trauma* 7(3):416–423, 1967.

FRACTURES OF THE CLAVICLE

Dale Christopher Young, M.D.
Charles A. Rockwood, Jr., M.D.

Fractures of the clavicle are relatively common sports injuries. Because of their frequency and the subcutaneous nature of the clavicle, fractures of the clavicle are usually readily identified. Clavicular fractures also usually unite regardless of treatment method. Nonetheless, the potential complications of clavicular fractures and the difficulty inherent in their management attest to the importance of these injuries.

HISTORICAL REVIEW

Although descriptions of the management of clavicular fractures date back to Hippocrates, the earliest sports-related case dates to 1702, when William III died following an equestrian accident in which he sustained a clavicular fracture and a false aneurysm of the subclavian artery [6]. His death provides evidence of the occasional serious nature of clavicular fractures. Numerous historical treatment modalities for the reduction of clavicular fractures have been described, including the popular figure-of-eight dressing that was originally described by Lucas-Championneire in the late 1860s [6].

EPIDEMIOLOGY

Clavicular fractures are common. They occur with a frequency of about 1/1000 people per year. According to statistics of the American Academy of Orthopaedic Surgeons, clavicular fractures have a bimodal type of incidence by both age and sex (Fig. 15B–47) [11]. Most clavicular fractures occur in men before the age of 25; then they become progressively uncommon and are very uncommon in men between 35 and 55 years, at which time their frequency increases again. A similar pattern is present in women, but it shows equal frequency between women younger than 25 and those over 75.

CURRENT CONCEPTS AND CONTROVERSIES

1. Treatment of acute totally displaced midshaft clavicular fractures is controversial.
2. Treatment of displaced medial and distal clavicular fractures may be difficult.

PERTINENT ANATOMY

The clavicle ossifies through a combination of intramembranous and echondral processes. The medial clavicular epiphysis, which is most important for longitudinal growth, ossifies between the ages of 12 and 19, and the physis fuses between the ages of 22 and 25 [17]. Thus, many medial sternoclavicular injuries in young adults are growth plate injuries.

The clavicle is stoutly anchored medially to the sternum and laterally to the scapula by a combination of extra-articular and capsular ligaments (Fig. 15B–48) [7]. These structures are discussed in greater detail in the sections specific to injuries of these joints.

The tubular medial clavicle resists axial loading and protects the costoclavicular space where the medial cord and origin of the ulnar nerve place the ulnar nerve at risk in patients with medial clavicular fractures, clavicular nonunions, and healed fractures with exuberant callus [2, 4, 19, 26, 45]. The weaker junction of the medial tubular clavicle and flattened lateral clavicle places the middle clavicle at risk for fractures [25].

BIOMECHANICS

The clavicle functions as a yardarm that maintains the width of shoulders against gravity and muscle pull. Excision of the clavicle allows the shoulder to collapse medially under the pull of the proximal

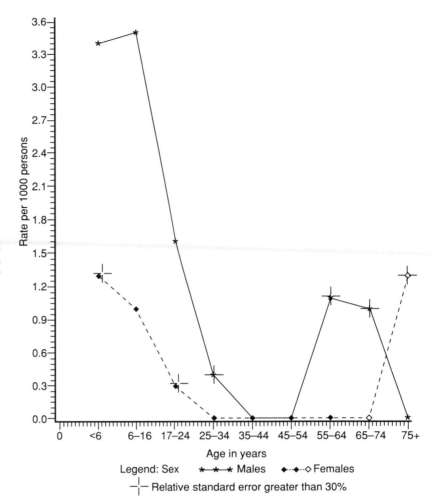

FIGURE 15B–47
Estimated annual incidence rates by age and sex for fractures of the clavicle. (Redrawn through the courtesy of the American Academy of Orthopaedic Surgeons.)

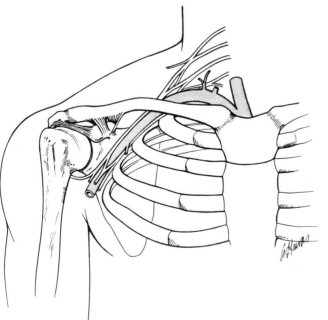

FIGURE 15B–48
The clavicle is bound securely by ligaments at both the sternoclavicular and acromioclavicular joints. It is the only bony strut from the torso to the extremity. In addition, the brachial plexus and greater vessels are seen posterior to the medial third of the clavicle between the clavicle and the first rib. (From Rockwood, C. A., and Matsen, F. A. (Eds.), *The Shoulder*. Philadelphia, W. B. Saunders, 1990.)

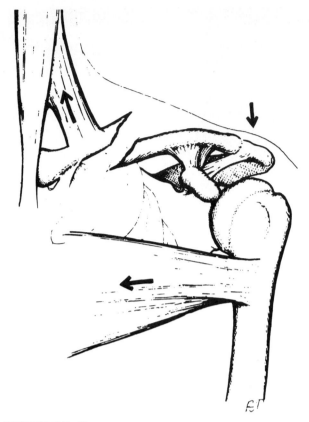

FIGURE 15B–49
When displacement occurs, the proximal fragment is drawn superiorly and posteriorly by the pull of the sternocleidomastoid, while the distal segment droops forward due to the weight of gravity and the pull of the pectoralis. (From Rockwood, C. A., and Matsen, F. A. (Eds.), *The Shoulder.* Philadelphia, W. B. Saunders, 1990.)

musculature with motion and can be painful and disabling. When displacement occurs in clavicular fractures, the sternocleidomastoid pulls the proximal fragment superiorly and posteriorly whereas the distal fragment droops inferiorly under the forces of gravity and the pull of the pectoralis major (Fig. 15B–49).

CLASSIFICATION

There is no generally accepted classification for clavicle fractures. Craig's classification is as follows [6]:

Group I	Fracture of the middle third
Group II	Fracture of the distal third
Type I	Minimal displacement
Type II	Displaced secondary to a fracture medial to the coracoclavicular ligaments
A	Conoid and trapezoid ligaments remain attached to the distal fragment
B	Conoid ligament is torn, trapezoid ligament remains attached to the distal fragments
Type III	Fractures of the articular surface
Type IV	Ligaments intact to the periosteum (children), with displacement of the proximal fragment (the so-called pseudodislocation injury)
Type V	Comminuted, with ligaments attached neither proximally nor distally but to an inferior, comminuted fragment
Group III	Fracture of the proximal third
Type I	Minimal displacement
Type II	Displaced (ligaments ruptured)
Type III	Intra-articular
Type IV	Epiphyseal separation (children and young adults)
Type V	Comminuted

This classification further subdivides Neer's classification of distal clavicular fractures [36, 37] and provides a classification for medial clavicular fractures.

Thompson's classification system [49, 50] is based on location, fracture displacement, and the presence or absence of neurovascular compromise. Thompson's classification is as follows:

Type	Location
I	Acromioclavicular joint
II	Lateral third
III	Middle third
IV	Medial third
V	Sternoclavicular joint
	Fracture Pattern
A	Bony contact
B	No bony contact
	With or without comminuted fragment
C	Neurovascular compromise

Thompson has emphasized the clinical significance of the type IIIB, middle clavicle, 100% displaced fracture. He believes that this fracture pattern is at such high risk for nonunion that primary open reduction and internal fixation of an acute type IIIB fracture should be considered if it cannot be converted to a type IIIA fracture by closed reduction. He has found uniformly that the trapezius muscle is entrapped in the fracture site in such cases.

CLINICAL EVALUATION

History

Clavicular fractures may result from direct or indirect trauma. Patients involved in stick sports

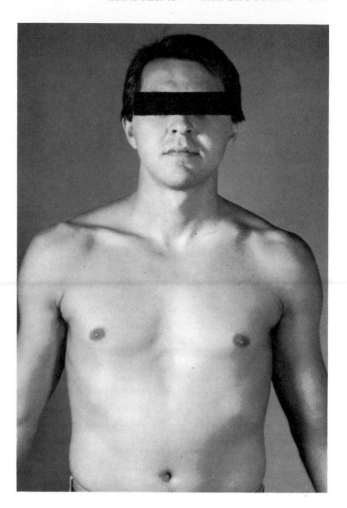

FIGURE 15B–50
With an adult clavicle fracture, the arm may droop forward and downward to varying degrees. (From Rockwood, C. A., and Matsen, F. A. (Eds.), *The Shoulder*. Philadelphia, W. B. Saunders, 1990.)

such as lacrosse and hockey frequently sustain a direct blow to the clavicle. In addition, a fall directly onto the bone itself can result in fracture. Clavicular injuries may also occur from a fall onto the outstretched arm, but this is probably a less common mechanism of injury [46]. Medial clavicular fractures are most commonly secondary to indirect trauma from a force directed to the lateral arm.

The history is usually clear in clavicular fractures. The patient complains of pain at the fracture site and supports the arm on the affected side. Frequently the deformity is obvious (Fig. 15B–50). A careful history and examination can be essential, however, because posterior sternoclavicular fracture dislocations and posteriorly displaced lateral clavicular fractures are not infrequently missed.

Physical Examination

Careful evaluation of clavicular fractures is important, particularly to ensure that the appropriate subsequent radiographs and special studies are ob-

tained. A complete neurovascular examination must be performed to identify uncommonly associated brachial plexus or vascular injuries. Chest auscultation is performed to evaluate for pulmonary injury.

Diagnostic Studies

Regardless of what portion of the clavicle is fractured, one must first address the associated injuries, which are potentially more serious. A chest x-ray should be performed in all patients with clavicular fractures who are comatose, who have sustained multiple trauma, or whose findings on physical examination suggest a thoracic or pulmonary injury [8, 15, 27, 47, 48, 53]. The chest x-ray is useful in identifying a hemothorax or pneumothorax and in screening for associated rib fractures [14]. Some authors recommend spine films to rule out a first rib fracture [52]. Abnormal vascular findings or an ipsilateral asymmetric bruit warrants further vascular evaluation, usually arteriography. Both arterial and venous injuries may occur in patients with

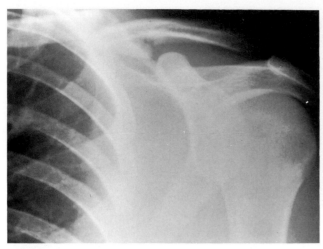

FIGURE 15B–51
An anteroposterior view of this left comminuted clavicular fracture poorly defines and identifies the fracture fragments owing to overlying bone in the area of the fracture fragments. (From Rockwood, C. A., and Matsen, F. A. (Eds.), *The Shoulder.* Philadelphia, W. B. Saunders, 1990.)

clavicular trauma [12, 16, 22, 24, 29, 51]. For shaft fractures of the clavicle, an anteroposterior view and a 45-degree cephalic tilt anteroposterior view are recommended (Figs. 15B–51 and 15B–52). Rowe has suggested that the films include the proximal humerus, scapular, and upper lung regions [43].

For fractures of the distal third of the clavicle, a trauma series may be adequate. Neer recommends an additional three views when indicated: (1) an anteroposterior view of both shoulders with 10 pounds of weight suspended from each wrist to assess concomitant ligamentous injury (Figs. 15B–53 and 15B–54) [2], (2) an anterior 45-degree oblique view, and (3) a posterior 45-degree oblique

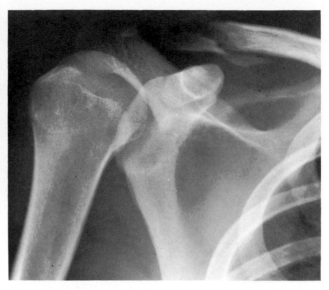

FIGURE 15B–53
A fracture of the right distal clavicle. In an anteroposterior view, the fracture location suggests ligamentous involvement, with the ligaments attached to the distal fragment. (From Rockwood, C. A., and Matsen, F. A. (Eds.), *The Shoulder.* Philadelphia, W. B. Saunders, 1990.)

view—the latter two views delineating fracture displacement [36, 37]. Articular fractures of the distal clavicle may require tomography or a CT scan (Fig. 15B–55).

Fractures of the medial clavicle may be particularly difficult to detect on plain x-rays. A 45-degree cephalic tilt view may be helpful [6, 36]. When

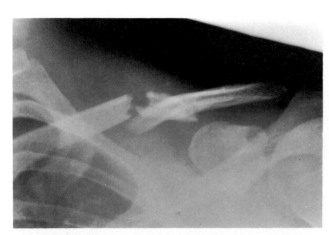

FIGURE 15B–52
However, when a 20- to 45-degree cephalic tilt view is obtained, the fracture anatomy is more clearly delineated. (From Rockwood, C. A., and Matsen, F. A. (Eds.), *The Shoulder.* Philadelphia, W. B. Saunders, 1990.)

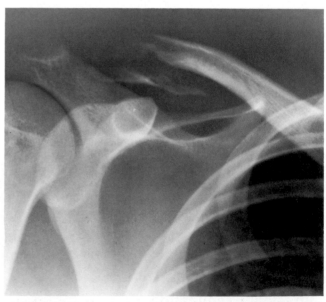

FIGURE 15B–54
The extent of ligamentous involvement is confirmed on a weighted view, which shows separation of the coracoclavicular distance. The coracoclavicular ligaments are attached to the distal clavicular segment. (From Rockwood, C. A., and Matsen, F. A. (Eds.), *The Shoulder.* Philadelphia, W. B. Saunders, 1990.)

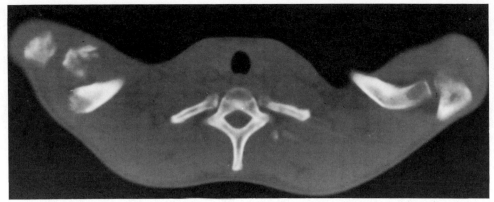

FIGURE 15B–55
CT scan of a right clavicular fracture. Not only does it confirm the site of the fracture as the distal clavicle, it also identifies a previously unsuspected intra-articular extension of the distal clavicular fracture. (From Rockwood, C. A., and Matsen, F. A. (Eds.), *The Shoulder*. Philadelphia, W. B. Saunders, 1990.)

clinical suspicion of medial clavicular injury is high, a CT scan is usually indicated.

TREATMENT OPTIONS

The basic principles governing the treatment of clavicular fractures are to (1) achieve a satisfactory reduction, (2) maintain a satisfactory reduction, and (3) minimize immobilization of the ipsilateral shoulder and arm. Craig has recently outlined the various methods of treatment [6]:

1. Simple support of the arm. This method may be utilized when the fracture alignment is acceptable and does not warrant a closed reduction. Either a sling, a sling-and-swathe, a Sayre bandage, or a Velpeau bandage may be employed. Anderson and colleagues [1] recently found no differences in healing, functional, or cosmetic results in comparing treatment with a simple sling with treatment with a figure-of-eight bandage.

2. Reduction. Reduction is usually achieved by bringing the distal fragment to a superior and posterior position. This can be achieved using a figure-of-eight bandage (Fig. 15B–56), either alone [10, 42] or combined with plaster reinforcement [5, 54] or by using various forms of Spica cast immobilization [21, 36, 39, 43]. Few studies have addressed the question of whether a vigorous effort to achieve reduction provides functional or cosmetic benefit or whether it minimizes late complications. Thompson believes that the 100% displaced middle third fracture is at particular risk for nonunion unless bony contact can be achieved and maintained [50].

3. Open or closed reduction with internal fixation. Techniques for internal fixation include intra-

medullary devices [3, 23, 32, 35, 38, 40, 41, 44, 55] and plate fixation [20, 33].

The complications of clavicular fractures include nonunion, malunion, neurovascular injury, and post-traumatic arthritis. Nonunion is the most frequent complication. Several factors may contribute to nonunion, including treatment modality, inadequate immobilization, severe trauma, previous fracture through the same site, fracture location, and fracture displacement. Some of these factors are beyond the control of the surgeon, but the surgeon can affect treatment modality, duration of immobilization, and extent of acceptable fracture displacement. It is generally agreed that displaced distal clavicular fractures are at high risk for nonunion, but in view of recent studies [18, 28, 50], we need to better define the acceptability of displacement in middle third fractures as well. Jupiter and Leffert pointed out that the degree of displacement was the most significant factor in nonunion [18]. Manske and Szabo demonstrated interposition of the trapezius muscle in the fracture site, predisposing to nonunion [28].

The duration of treatment is variable depending on age as well as fracture location and displacement, but, in general, in young adults these fractures will heal in 4 to 6 weeks and in adults in 6 weeks or longer [43].

Nonunited clavicular fractures may be asymptomatic, but a return to contact sports may be inadvisable even in this situation. Clavicular nonunions have been successfully treated using a variety of surgical techniques including open reduction and internal fixation with dynamic compression plates and intramedullary devices. Bone grafting is usually performed and is particularly indicated in the patient with atrophic nonunion [18].

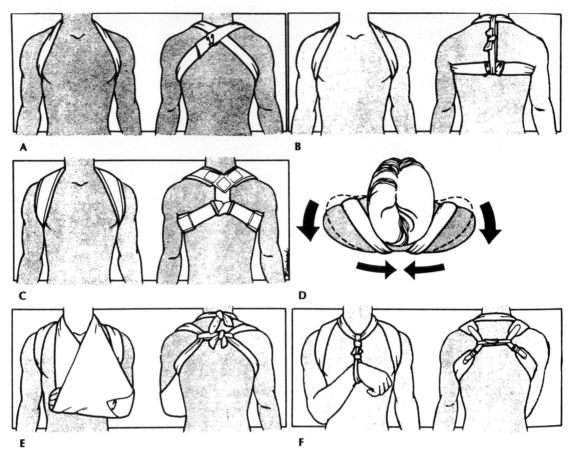

FIGURE 15B–56

Various types of figure-of-eight bandages. The figure-of-eight method is intended to maintain a reduction that has been achieved by closed means. *A,* Stockinette that has been padded with three layers of sheet wadding and held in place with safety pins. *B,* Padded stockinette that is not crossed in back. The upper and lower borders are tied to each other and tightened daily, increasing the tension to maintain the reduction. *C,* A commercial figure-of-eight support. *D,* Superior view of the patient showing how the figure-of-eight support pulls the shoulder up and backward. *E,* A modified figure-of-eight bandage with a sling. *F,* A figure-of-eight support used with a collar and cuff. (From Rockwood, C. A., and Green, D. P. (Eds.), *Fractures in Adults* (2nd ed.). Philadelphia, J. B. Lippincott, 1984.)

Neurovascular injuries associated with clavicular fractures are uncommon but do occur, both early and late [2, 4, 9, 12, 13, 16, 19, 22, 24, 26, 30, 31, 45, 47, 51]. Post-traumatic arthritis of the distal clavicle following treatment of a clavicular fracture may respond to nonoperative measures, but in certain circumstances and when the coracoclavicular ligaments are intact, the lateral 2 cm of the clavicle may be excised [34]. Excision of the medial clavicle is rarely indicated. Malunion of clavicular fractures is rarely troublesome.

Complications may be avoided by judicious use of early surgery. Indications for primary operative treatment of acute clavicular fractures include:

1. Open fractures
2. Multiple trauma, when closed means are impossible or undesirable
3. Neurovascular compromise requiring immediate exploration

4. Severe displacement with tenting of the skin that fails to respond to closed reduction
5. Completely displaced fractures of the middle third of the clavicle in selected patients [49]
6. Patients who are unable to tolerate closed treatment (neurologic disorders and so on)
7. Distal clavicular fractures
   a. Completely displaced, irreducible fractures of the distal clavicle in which the coracoclavicular ligaments are intact
   b. Displaced distal clavicular fractures with rupture of the coracoclavicular ligaments

REHABILITATION

Rehabilitation measures for clavicular fracture are usually similar regardless of the treatment modality. As soon as the acute pain has resolved, the patient may begin gentle pendulum exercises of the ipsilateral shoulder, active use of the shoulder up to 40 degrees of flexion, isometric deltoid and rotator cuff strengthening, and normal use of the ipsilateral elbow and hand. The patient should not start a program of range of motion exercises beyond 40 to 45 degrees of flexion until there is clinical evidence that healing is occurring, usually at 4 to 6 weeks. As union progresses, range of motion may be increased and resistive exercises instituted. When radiographic union is present, full active use of the arm is allowed.

CRITERIA FOR RETURN TO SPORTS PARTICIPATION

The patient should not return to noncontact sports until a painless full range of motion is present, the fracture is healed, and near normal strength is restored. This usually requires more than 6 weeks after the injury. Contact sports place the clavicle at even greater risk for refracture and should be delayed until the union is solid, probably for at least 4 to 6 months. No special bracing is required, but donut pads may be fabricated to protect the clavicle from reinjury.

AUTHORS' PREFERRED METHOD OF TREATMENT

Fractures of the Shaft

We manage most fractures of the shaft with a well-padded commercial figure-of-eight splint. Occasionally a closed reduction is necessary. A postreduction radiograph is utilized to evaluate the reduction. A sling is rarely needed to support the extremity. Rehabilitation as outlined previously is started as soon as the patient can participate comfortably.

It is rare that a fracture of the shaft of the clavicle requires an open reduction and internal fixation. If a completely displaced fracture of the midshaft of the clavicle cannot be reduced with closed manipulation, we observe the patient for 2 to 3 weeks in the figure-of-eight splint. If after observation the *complete* displacement persists, we recommend an open reduction and internal fixation as described by Thompson [49]. We prefer to use an intramedullary modified Hagie pin instead of a plate and screws for fixation.

Fractures of the Distal Clavicle

Nondisplaced fractures are treated symptomatically with a sling and early rehabilitation. In patients with displaced fractures, a portion of the coracoclavicular ligament remains attached to the distal fragment, and the distal fragment remains adjacent to the acromion. The weight of the arm pulls the distal fragment inferiorly, and the trapezius muscle pulls the proximal fragment superiorly, creating a gross displacement of the fracture. In most instances, we have been able to reduce such a fracture with a figure-of-eight splint. The splint is applied so that it pulls the distal fragment back and up. More recently, we have used a modified figure-of-eight splint.* The modified device has better padding and, most important, a rigid vertical upright that prevents superior displacement of the device. Reduction of the fracture can be confirmed by using a scapular-lateral x-ray.

If the figure-of-eight splint does not reduce the fracture, an open reduction is used. Depending on the exact position of the fracture, the method of internal fixation may include intramedullary pins or reduction and stabilization of the fracture with a coracoclavicular lag screw inserted between the medial fragment and the coracoid process.

Fractures of the Medial Clavicle

This is a very rare fracture, and unless there is significant posterior displacement with potential or actual injuries to the adjacent neurovascular structures, medial clavicular fractures are treated symptomatically in a figure-of-eight splint or with a sling. Reduction of posterior fracture dislocations of the clavicle requires great care and usually the presence in the operating room of a thoracic surgeon scrubbed and ready to operate. Fixation of the fracture should not be done with pins across the fracture and into the sternum because there is a danger of migration of the pins into the mediastinum. Plate and screws or nonmetallic heavy sutures can be used.

References

1. Anderson, K., Jensen, P., and Lauritzen, J. Treatment of clavicular fractures. Figure-of-eight bandage vs. a simple sling. *Acta Orthop Scand* 57:71–74, 1987.
2. Berkheiser, E. J. Old ununited clavicular fractures in the adult. *Surg Gynecol Obstet* 64:1064–1072, 1937.
3. Breck, L. Partially threaded round pins with oversized

*Mesa Products, San Antonio, TX.

threads for intramedullary fixation of the clavicle and the forearm bones. *Clin Orthop* 11:227–229, 1958.

4. Campbell, E., Howard, W. B., and Breklund, C. W. Delayed brachial plexus palsy due to ununited fracture of the clavicle. *JAMA* 139:91–92, 1949.
5. Cook, T. Reduction and external fixation of fractures of the clavicle in recumbency. *J Bone Joint Surg* 36A:878–880, 1954.
6. Craig, E. V. Fractures of the clavicle. *In* Rockwood, C. A., Jr., and Matsen, F. A. (Eds.), *The Shoulder*. Philadelphia, W. B. Saunders, 1990.
7. DePalma, A. *Surgery of the Shoulder* (3rd ed.). Philadelphia, J. B. Lippincott, 1983.
8. Dugdale, T. W., and Fulkerson, J. B. Pneumothorax complicating a closed fracture of the clavicle. A case report. *Clin Orthop* 221:212–214, 1987.
9. Enker, S. H., and Murthy, K. K. Brachial plexus compression by excessive callus formation secondary to a fractured clavicle. A case report. *Mt. Sinai (NY) J Med* 37(6):678–682, 1970.
10. Giancecchi, F., Bugli, G., Camurri, G. B., et al. Fratture composte de clavicola. Considerazioni su 160 casi trattati con bendaggio soffice a otto. *Minerva Ortop* 35:795–801, 1984.
11. Grazier, K. L., Holbrook, T. L., Kelsey, J. L., and Stauffer, R. N. *The Frequency of Occurrence, Impact and Cost of Musculoskeletal Conditions in the United States*. Chicago, American Academy of Orthopaedic Surgeons, 1984.
12. Howard, F. M., and Schafer, S. J. Injuries to the clavicle with neurovascular complications. A study of fourteen cases. *J Bone Joint Surg* 47A:1335–1346, 1965.
13. Iqbal, Q. M. Axillary artery thrombosis associated with fracture of the clavicle. *Med J Malaysia* 26:68–70, 1972.
14. Iqbal, Q. M. Fracture of first rib with associated fracture of the clavicle. *Med J Malaysia* 25:223–225, 1971.
15. Jackson, W. J. Clavicle fractures. Therapy is dictated by the patient's age. *Consultant* 177, 1982.
16. Javid, H. Vascular injuries of the neck. *Clin Orthop* 28:70–78, 1963.
17. Jit, I., and Kulkrani, M. Times of appearance and fusion of epiphysis at the medial end of the clavicle. *Indian J Med Res* 64(5):773–792, 1976.
18. Jupiter, J. B., and Leffert, R. D. Nonunion of the clavicle. *J Bone Joint Surg* 69A:753–760, 1987.
19. Kay, S. P., and Eckardt, J. J. Brachial plexus palsy secondary to clavicular nonunion. A case report and literature survey. *Clin Orthop* 206:219–222, 1986.
20. Khan, M. A. A., and Lucas, H. K. Plating of fractures of the middle third of the clavicle. *Injury* 9:263–267, 1978.
21. Kini, M. G. A simple method of ambulatory treatment of fractures of the clavicle. *J Bone Joint Surg* 23:795–798, 1941.
22. Klier, I., and Mayor, P. B. Laceration of the innominate internal jugular venous junction. Rare complication of fracture of the clavicle. *Orthop Rev* 10:81–82, 1981.
23. Lengua, F., Nuss, J. M., Lechner, R., Baruthio, J., and Veillon, F. [The treatment of fracture of the clavicle by closed-medio-lateral pinning.] *Rev Chir Orthop* 73(5):377–380, 1987.
24. Lim, E., and Day, L. J. Subclavian vein thrombosis following fracture of the clavicle. A case report. *Orthopedics* 10(2):349–351, 1987.
25. Ljunggren, A. E. Clavicular function. *Acta Orthop Scand* 50:261–268, 1979.
26. Lusskin, R., Weiss, C. A., and Winer, J. The role of the subclavius muscle in the subclavian vein syndrome (costoclavicular syndrome) following fracture of the clavicle. *Clin Orthop* 54:75–84, 1967.
27. Malcolm, B. W., Ameli, F. N., and Simmons, E. H. Pneumothorax complicating a fracture of the clavicle. *Can J Surg* 22(1):84, 1979.
28. Manske, D. J., and Szabo, R. M. The operative treatment of midshaft clavicular nonunions. *J Bone Joint Surg* 67A:1367–1371, 1985.

29. Matry, C. Fracture de la clavicule gauche au tiers interne. Blessure de la vein sour-claviere. *Osteosynthese Bull Mem Soc Nat Chir* 58:75–78, 1932.
30. Miller, D. S., and Boswick, J. A. Lesions of the brachial plexus associated with fractures of the clavicle. *Clin Orthop* 64:144–149, 1969.
31. Mital, M. A., and Aufranc, O. E. Venous occlusion following greenstick fracture clavicle. *JAMA* 206:1301–1302, 1968.
32. Moore, T. O. Internal pin fixation for fracture of the clavicle. *Am Surg* 17:580–583, 1951.
33. Mueller, M. E., Allgower, N., and Willenegger, H. *Manual of Internal Fixation*. New York, Springer-Verlag, 1970.
34. Mumford, E. B. Acromioclavicular dislocation. *J Bone Joint Surg* 23:799–802, 1941.
35. Murray, G. A method of fixation for fracture of the clavicle. *J Bone Joint Surg* 22:616–620, 1940.
36. Neer, C. S., II. Fractures of the clavicle. *In* Rockwood, C. A., Jr., and Green, D. P. (Eds.), *Fractures in Adults* (2nd ed.). Philadelphia, J. B. Lippincott, 1984, pp. 707–713.
37. Neer, C. S., II. Fractures of the distal third of the clavicle. *Clin Orthop* 58:43–50, 1968.
38. Neviaser, R. J., Neviaser, J. S., and Neviaser, T. J. A simple technique for internal fixation of the clavicle. *Clin Orthop* 109:103–107, 1975.
39. Packer, B. D. Conservative treatment of fracture of the clavicle. *J Bone Joint Surg* 26:770–774, 1944.
40. Paffen, P. J., and Jansen, E. W. Surgical treatment of clavicular fractures with Kirshner wires. A comparative study. *Arch Chir Neerl* 30:43–53, 1978.
41. Perry, B. An improved clavicular pin. *Am J Surg* 112:142–144, 1966.
42. Quigley, T. B. The management of simple fracture of the clavicle in adults. *N Engl J Med* 243:286–290, 1950.
43. Rowe, C. R. An atlas of anatomy and treatment of mid-clavicular fractures. *Clin Orthop* 58:29–42, 1968.
44. Rush, L. V., and Rush, H. L. Technique of longitudinal pin fixation in fractures of the clavicle and jaw. *Mississippi Doctor* 27:332, 1949.
45. Sakellarides, H. Pseudoarthrosis of the clavicle. *J Bone Joint Surg* 43A:130–138, 1961.
46. Stanley, D., Trowbridge, E. A., and Norris, S. H. The mechanism of clavicular fracture. *J Bone Joint Surg* 70B:461–464, 1988.
47. Steenburg, R. W., and Ravitch, M. M. Cervico-thoracic approach for subclavian vessel injury from compound fracture of the clavicle: Considerations of subclavian axillary exposures. *Ann Surg* 1(57):839, 1963.
48. Stimson, L. A. *A Treatise on Fractures*. Philadelphia, Henry A. Lea's Son and Company, 1883, p. 332.
49. Thompson, J. S. Classification of clavicle fractures has an impact on operative indications (abstract). *Proceedings of the Fourth International Conference on Surgery of the Shoulder*, New York, 1989.
50. Thompson, J. S. Shoulder girdle injuries. *In* Champion, H. R., Robbs, J. V., and Trunkey, D. D. (Eds.), *Rob and Smith's Operative Surgery: Trauma Surgery* (4th ed.). London, Butterworth, 1989.
51. Tse, D. H. W., Slabaugh, P. B., and Carlson, P. A. Injury to the axillary artery by a closed fracture of the clavicle. *J Bone Joint Surg* 62A:1372–1373, 1980.
52. Weiner, D. S., and O'Dell, H. W. Fractures of the first rib associated with injuries to the clavicle. *J Trauma* 9(5):412–422, 1969.
53. Yates, D. W. Complications of fractures of the clavicle. *Injury* 7(3):189–193, 1986.
54. Young, C. S. The mechanics of ambulatory treatment of fractures of the clavicle. *J Bone Joint Surg* 13:299–310, 1931.
55. Zenni, E. J., Jr., Krieg, J. K., and Rosen, M. J. Open reduction and internal fixation of clavicular fractures. *J Bone Joint Surg* 63A(1):147–151, 1981.

OSTEOLYSIS OF THE CLAVICLE

Francis R. Lyons, M.D.
Charles A. Rockwood, Jr., M.D.

This condition, peculiar to the lateral end of the clavicle and characterized by symptomatic resorption of bone over a period of weeks to many months, remains uncertain in origin. Common to most theories of etiology is the role of trauma—hence the frequent appearance of the condition under the term post-traumatic osteolysis of the clavicle. However, it is increasingly recognized that trauma sufficient to cause overt disruption of the acromioclavicular joint is not a prerequisite for the onset of the condition [3]. The condition may be precipitated by a single, apparently minor injury to the acromioclavicular joint [5, 17] or even by the repeated microtrauma of strenuous physical activity. This has prompted Brunet and colleagues [2] to coin the term atraumatic osteolysis of the clavicle.

Historical review of this entity indicates that osteolysis of the lateral end of the clavicle was little recognized prior to a report by Madsen [10] in 1963 of nine cases. Before his report the condition received scant attention and probably went largely unrecognized. Several authors [6, 10] credit the first description of the condition to Werder [20] in 1950. Still others [2, 9] credit the earliest reference to Dupas in 1936. In actual fact, that paper referred to osteolysis of the metacarpal heads rather than to osteolysis of the lateral end of the clavicle. However, Urist [19] in 1946 in the English literature did refer to the condition in relation to acromioclavicular dislocation. He described two cases of what he termed "avascular necrosis or osteochondritis" of the lateral clavicle following that injury. Gross and radiologic examination of the excised specimens in these cases revealed sclerotic bone, irregular areas of resorption, and new subperiosteal bone formation "typical of the healing process of crushed spongiosa." He also noted atrophy or rarefaction at the acromial end of the clavicle with virtually all severe dislocations of the acromioclavicular joint.

Stahl [18] in 1954 reported two cases of osteolysis related to acromioclavicular dislocation or subluxation and made one of the earliest references to the progression of typical radiologic changes over a period of a year, describing the "wooly" outline of the acromial end of the clavicle. He also mentioned briefly that this condition could occur after only "slight" traumatic injury to the acromioclavicular joint and surmised that it may result from a hyper-reaction to trauma or possibly absorption relative to "fine fracture lines" in the acromial end of the clavicle. This is one of the earliest references to the fact that the condition readily follows apparently trivial trauma without frank acromioclavicular subluxation or dislocation. This phenomenon was also alluded to in 1958 by Ehricht [4] in a report of osteolysis of the clavicle in a worker who used a vibrating pneumatic tool.

Throughout the past two decades, reports have continued to emerge of this condition following either trivial or significant acromioclavicular trauma. The reports of Murphy and associates [12] in 1975 and Kaplan and Resnick [7] in 1986 described the occurrence of osteolysis that resulted from repetitive overhead lifting of heavy objects in an occupational setting. It was the report of Cahill [3] in 1982 that focused attention upon the relevance of this "atraumatic" phenomenon to the athlete. He reported osteolysis of the distal clavicle in 46 male athletes, none of whom had suffered an acute injury to the acromioclavicular joint. Of these 46 men, 45 were competitive weight lifters or lifted weights as part of their training. Pain and local tenderness at the acromioclavicular joint associated with osteoporosis, loss of subchondral bone detail, and cystic changes in the distal part of the clavicle were present in all to varying degrees. Brunet and colleagues [2] in 1986 reported the same condition in a baseball player who supplemented his training with a weight-lifting program. The condition was also identified by Seymour [16] in a softball pitcher who used a 360-degree windmill-type action. All of these reports testify to the occurrence of osteolysis with minor repetitive trauma. The condition has also received attention in the field of sporting endeavors with single or repeated episodes of more significant trauma to the shoulder. The sport of judo has been the subject of several reports as a potential cause of osteolysis when the participant suffers repeated falls onto the point of the shoulder [14, 17]. Sports involving repeated shoulder-to-shoulder or shoulder-to-wall impact, such as football, rugby, hockey, and ice hockey have also been identified as precipitating causes of osteolysis [6, 8, 11, 13, 15]. Cycling accidents involving falls onto the point of the shoulder also have been reported as causes of this condition in the athlete [14, 15].

There appears to be a predilection for this condition to occur in males. Reports of osteolysis of the clavicle in females following single significant traumatic events but without acromioclavicular subluxation have been made [1, 10, 12]. However, there are apparently no reports of its occurrence in female athletes in the setting of repetitive microtrauma.

The etiology of this curious condition remains obscure. Aseptic necrosis of bone is among the more widely supported theories of its cause. Reference has already been made to Urist's description of two cases of "avascular necrosis." Murphy and associates [12] added histologic support to the theory of aseptic necrosis. Their operative findings revealed destruction of the clavicular articular cartilage and preservation of the acromial side of the joint. The remaining lateral clavicular bone was sclerotic and difficult to cut. Intraoperative cultures were negative. Microscopy confirmed the impression of sclerosis and underlying aseptic necrosis. Orava and colleagues [14] lent support to this avascular necrosis etiology on the basis of reduced isotopic uptake found 4 months following injury in a patient with radiologically proven osteolysis. However, their findings were in sharp contrast to those of Cahill [3], who found that isotopic uptake in the lateral end of the clavicle was always enhanced in the presence of plain x-ray changes of osteolysis. Cahill interpreted this finding of positive scintigraphy as suggesting that the lesion in the distal part of the clavicle could have resulted from microfracture of subchondral bone and subsequent attempts at repair. He postulated that during weight lifting excessive stresses were concentrated on the articular surface of the distal clavicle and that the initiating event in the production of symptoms could be subchondral stress fracture. He claimed support for his theory by the finding of microfractures of the subchondral bone in "about" 50% of the surgical specimens from 21 operatively treated patients. Histologic sections of the resected lateral clavicle were typical of those seen in patients with degenerative joint disease and paralleled the radiographic findings. What remains unclear about the pathology of osteolysis is whether stress fracture is a consequence of avascular necrosis, whether it precipitates avascular necrosis, or indeed, whether either process may lead to osteolysis independently. The conflicting bone scan findings of Cahill [3] and Orava and colleagues [14] additionally confuse the issue. Isotopic bone scanning is generally performed only after symptoms become chronic, and it may be that the frequent finding of the increased isotope uptake may represent the reparative phase of aseptic necrosis.

The histologic finding of vascular proliferation

and villous hypertrophy of the synovium as well as the occasional involvement of the acromial side of the joint led Levine and colleagues [9] to suggest that synovial pathology played a role in the etiology of this condition. Their suspicion was heightened by the complete reconstitution of the acromion after synovectomy and lateral clavicular excision. Brunet and associates [2] also gave support to the synovial theory. An operative specimen was described as showing inflamed synovial tissue extending through the articular cartilage of the clavicle and invading the subarticular bone. The synovium was morphologically villous and hypertrophic with occasional multinuclear giant cells. The resulting direct communication between the lesion and the acromioclavicular joint was cited as a feature distinguishing this process from intra-osseous ganglion. As well as the synovial changes, Brunet and co-workers noted the presence of trabecular microfractures. Morrison [11] also noted synovial pathology in the form of fibrinous changes. Understanding the actual etiology of the condition is further complicated by the fact that Murphy and co-workers [12] observed no synovial changes and reported that the acromion and intra-articular meniscus remained normal. Madsen [10] had originally suggested that an autonomic nervous mechanism might cause clavicular osteolysis on the basis that a discrepancy in people's size was a frequent finding. However, this theory seems to have gained little support from subsequent authors.

CLINICAL FEATURES OF OSTEOLYSIS OF THE CLAVICLE

History

The history provided by the patient almost invariably indicates that the process began with some form of shoulder trauma. The role of trauma is especially apparent when the patient provides a history of a single significant injury. This is most commonly a direct blow to the point of the shoulder, in the form of either a fall or body contact. Except in cases of frank dislocation of the acromioclavicular joint, the patient does not notice deformity but may report some minor local swelling. Typically, patients who have experienced a single traumatic event notice moderate to severe initial pain. The pain often recedes, only to recur as an aching pain some weeks to months later as the osteolytic process becomes established. An intermediate group of patients may give a history of repeated, less severe, but still readily recognizable traumatic episodes to the shoulder. Examples of this type of patient might include a judo expert who is repeatedly thrown to the mat or a hockey player who habitually blocks with the shoulder or collides with the wall. More insidious is the history provided by the patients characterized in Cahill's [3] report, who suffered repetitive microtrauma to the shoulders. Although his patients were involved in a wide variety of sports, the common underlying theme was that all but one were involved in a rigorous weight-training routine. These patients noted the onset of symptoms after bench pressing, performing dips on parallel bars, or doing push-ups, often to the point where these activities became intolerable due to pain. All of these athletes were men, and all were doing weight-training exercises on a regular basis, usually at least 3 times per week. Pain may lead to the complaint of restricted motion of the shoulder, and painful crepitus at the lateral end of the clavicle may also be noted.

Physical Examination

The findings of deformity in the region of the acromioclavicular joint depend upon the degree of severity of the initial injury. If it has been sufficient to result in subluxation or even dislocation of the acromioclavicular joint, the clinical deformity will be readily apparent. However, many patients do experience significant trauma to the point of the shoulder but have no obvious derangement of the acromioclavicular joint. It may be that these patients have experienced the less readily diagnosed type I acromioclavicular injury. If these patients go on to develop acromioclavicular osteolysis, they will not have an obvious clinical deformity, nor will patients with osteolysis secondary to repetitive microtrauma. As mentioned, swelling in the region of the acromioclavicular joint may be a complaint, and indeed this may be detected clinically. Murphy and associates [12] was able to aspirate a 4-mL acromioclavicular joint effusion 1 month after a specific injury had occurred in a patient who at that time had a normal acromioclavicular x-ray and subsequently went on to develop osteolysis. This effusion may be the source of swelling, or, alternatively, swelling may be a soft tissue response to partial disruption of the acromioclavicular ligaments.

As symptoms become established over a period of weeks to months, localized point tenderness over the acromioclavicular joint becomes evident. Glenohumeral motion may be secondarily restricted as pain worsens. Specific tests designed to stress the acromioclavicular joint will aggravate the discomfort of this condition. These tests include forward flexing the shoulder to 90 degrees and then forcibly ad-ducting the arm across the chest. Similar aggravation of acromioclavicular symptoms will be produced by flexing the shoulder to 180 degrees and then forcibly adducting the overhead arm toward the head. Taking the patient in a bear hug and axially loading the acromioclavicular joint will similarly elicit discomfort. If any doubt remains about the localization of symptoms to the acromioclavicular joint, a selective injection using local anesthetic alone will almost invariably bring swift and complete relief of symptoms in this condition. If this does not happen, the diagnosis is in doubt. The injection test is best administered using a 25-gauge needle that is just one-half inch long. By entering the joint from above with the short needle, one can ensure that the injection enters the acromioclavicular joint rather than the subacromial space. Entry to the joint is additionally confirmed by the easy passage of just 2 to 3 mL of local anesthetic and the ease with which this fluid can be withdrawn.

Diagnostic Studies

Plain x-ray examination of the acromioclavicular joint remains the mainstay of the diagnosis of lateral clavicular osteolysis. Levine and associates [9] have arbitrarily divided the radiographic changes of osteolysis into three phases, lytic, reparative, and "burnt out." They noted that the lytic phase could begin as early as 2½ weeks after injury but could be delayed in onset for up to 9 months. This phase is characterized by soft tissue swelling in the region of the acromioclavicular joint and demineralization and loss of subarticular cortex in the lateral clavicle. Focal cystic erosions may also be seen. Levine and his colleagues documented bone loss from the lateral clavicle that could range from 0.5 to 3 cm. Calcification in the soft tissues and subperiosteal reaction, especially where there was initial subluxation or dislocation of the acromioclavicular joint, were also seen. As bone loss from the clavicle continued an apparent widening of the acromioclavicular joint space was seen (Fig. 15B–57). Levine and colleagues [9] and Cahill [3] also reported a number of cases in which there were accompanying lytic changes in the acromial side of the joint. However, this phenomenon was apparently quite rare and should not be considered a typical finding. The initial lytic phase may continue for a period of 12 to 18 months. A reparative phase then begins, and typically during the next 4 to 6 months the eroded lateral end of the clavicle is reconstituted to a varying extent. However, invariably a residual widening of the acromioclavicular joint occurs along with a tapered appear-

FIGURE 15B–57
Post-traumatic osteolysis of the clavicle in a 35-year-old weight lifter.

ance of the lateral clavicle. Small isolated bone remnants may be detected in the widened acromioclavicular interval. At the completion of the reparative phase the patient is left in the final "burnt out" phase, by which time symptoms have usually subsided, but the radiologic hallmarks of the preceding osteolytic process remain permanently.

Radiographs taken soon after acute trauma to the shoulder show no evidence of the osteolytic process. But of course if trauma has been sufficient there may be features of acromioclavicular subluxation or dislocation. It is appropriate to radiograph both shoulders to detect acromioclavicular derangement as well as to assess lateral clavicular demineralization. Bearing in mind that the symptoms of osteolysis may precede radiologic changes and also that these changes may be delayed for many months after injury, it may be necessary to take x-rays of the acromioclavicular joint on a serial basis until either the symptoms subside or the radiologic diagnosis of osteolysis is established.

Radiographic technique is very important. Standard anteroposterior views of the shoulder are inadequate for visualization of the acromioclavicular joint; if these are relied upon, osteolysis may be missed. There are two explanations for this. First, adequate penetration of the glenohumeral joint will result in an overpenetrated view of the acromioclavicular joint. Second, a straight shoulder anteroposterior film will show the acromioclavicular joint superimposed upon the spine of the scapula, making interpretation difficult. Optimal visualization of the acromioclavicular joint is provided by taking an anteroposterior film with the x-ray beam tilted in a 25- to 30-degree cephalic direction. A soft tissue technique is essential to avoid overpenetration and to best reveal dystrophic calcification in the soft tissue and periosteal reaction.

Isotopic bone scanning is also a useful investigative tool but is by no means diagnostic. Reference has already been made to this procedure in the section on etiology of osteolysis. Cahill [3] has reported the greatest experience with bone scanning. In all patients with clear-cut radiologic changes typical of osteolysis, bone scanning was positive. In fact, he demanded a positive scan to confirm the diagnosis. He also noted that occasionally positive scintigraphic findings were seen in the acromion as well. His paper suggests that scanning was performed in his patients after the onset of plain radiographic changes. Thus, no statement could be made about the bone scan findings early in the evolution of the condition. Although no false-negative scan occurred in his series, there were three false-positive bone scans, each of these being in patients with bilaterally positive bone scans but only unilateral changes on plain x-rays. One must also bear in mind that both Beraneck and associates [1] and Orava and associates [14] reported single cases of negative bone scans in association with positive x-ray changes of osteolysis.

Several authors have reported that laboratory tests including serum calcium, phosphorus, and alkaline phosphatase determinations are negative in patients with clavicular osteolysis [10, 11, 12].

Prior to arriving at a definitive diagnosis of post-traumatic osteolysis of the clavicle on the basis of the preceding clinical and diagnostic studies, one must of course consider alternative diagnoses. In fit young athletes these are probably largely theoretical considerations. However, conditions that do exhibit signs resembling those of clavicular osteolysis include intra-osseous ganglion, rheumatoid arthritis, scleroderma, massive osteolysis, secondary hyperparathyroidism, acromioclavicular sepsis, and multiple myeloma.

TREATMENT OPTIONS

This is a condition in which treatment options are few, and there is little scope for individual preference of management. The decision essentially lies between conservative treatment in the form of rest and avoidance of aggravating activities, and surgery consisting of lateral clavicular excision. At first glance, the decision between the two courses of action might seem self-explanatory. However, the situation is complicated by a number of factors, not the least of which is that, apart from the report of Cahill [3], most papers discuss individual case reports or very small series. The point has been made that the natural history of this condition probably runs a self-limiting course of 1 to 2 years. What cannot be ascertained from the literature is how many of these patients who are followed in a strictly conservative fashion will be left with residual symptoms in some form. If we knew for a fact that all patients recovered fully with time, then we might have stronger grounds for recommending a lengthy period of conservative treatment. The main objective of conservative treatment is to rest the shoulder or, as a compromise, to modify training activities or sports participation so that the condition is not aggravated. A number of authors [8, 11, 15] have suggested that early immobilization even to the point of using a sling will lead to rapid relief of symptoms and may even reduce the extent of bone loss, leading to early cessation of resorption. However, there appears to be no sound scientific basis for this claim.

Cahill [3] mentioned the use of intra-articular steroid injections. He used this measure in only three patients and made no comment upon its efficacy. The fact that the injections were administered in only the first three cases out of a series of 46 might imply that he found injection to be of little value. Morrison [11] claimed lasting relief from steroid injection in a single case. If steroids are to be used, it must be upon an empirical basis without scientific foundation, and indeed, one may wonder whether the catabolic effect of local steroid may accelerate the osteolytic process.

Cahill also referred in passing to the use of nonsteroidal anti-inflammatory drugs, but again no comment was made upon their efficacy. Kaplan and Resnick [7] and Morrison [11] each claimed good response to nonsteroidal anti-inflammatory drugs in isolated cases.

Apart from the unknown factors of the time needed for recovery and the completeness of recovery with conservative treatment, one must bear in mind the unique difficulty of pursuing a lengthy conservative course in the athletic population. Cahill has addressed this problem well. He found that alteration between the training program and rest resulted in relief for most patients with moderate symptoms. A number of patients with severe symptoms refused surgery and ceased weight training altogether. However, this latter approach is probably the exception in the athletic population. Cahill was faced with a group whose indication for surgery was their refusal to give up their training program or modify it. The more orthodox indication for surgery was, of course, a lack of response to conservative treatment in those who had been willing to give it a fair trial. Twenty-one of Cahill's 46 patients eventually required surgery in the form of lateral clavicular excision. This is a much higher

operative rate than that reported in other smaller series, where good results could be expected with lengthy conservative treatment and operative treatment was the exception. Again, this high rate of surgery probably reflects the reluctance of the athlete to refrain from sport for any period of time. Cahill found the results of surgery satisfying. Fourteen of the nineteen patients treated with surgery returned to their previous levels of sporting activity. Of the five who did not, two had withdrawn voluntarily, and three withdrew because they could not compete at their previous levels. Pain was not a determining factor in withdrawal from sports after surgery.

POSTOPERATIVE MANAGEMENT AND REHABILITATION

Should lateral clavicular excision be undertaken, postoperative management ought to include complete rest in a sling for the first 1 to 2 weeks to allow soft tissue healing and to protect the deltotrapezius fascia repair. During this period the sling may be removed to perform pendulum exercises and assisted passive forward flexion of the shoulder. There is no reason why the patient cannot continue to pursue a weight-lifting program designed for the unaffected extremities. When the priority of healing has been achieved, the patient may resume a stretching program to maximize range of motion of the shoulder prior to undertaking a formal strengthening schedule. This should address individual strengthening of the rotator cuff muscles and the three parts of the deltoid. Full recovery and return to preinjury strength may be anticipated about 6 to 8 weeks after surgery. Training from then on would be individually tailored to avoid or modify any aggravating maneuvers until discomfort eases.

AUTHORS' PREFERRED METHOD OF TREATMENT

The authors prefer to encourage a conservative course of treatment for as long as the patient will comply. Should the symptoms of osteolysis begin after an identifiable acute injury or recur following repetitive microtrauma, a short period of complete rest in a sling may be appropriate. A course of nonsteroidal anti-inflammatory drugs is prescribed when pain is particularly troublesome. Similarly, an acute flare of symptoms can occasionally be brought under control by a single steroid injection. Bearing in mind the long natural history of the condition, one needs to encourage the patient to persist with the conservative line and avoid aggravating activities even to the point of staying out of sports to ensure complete rest. The authors persist with these measures as long as the patient tolerates his discomfort and keeps away from sports. The patient will inevitably indicate when these factors become intolerable. At that point, the only alternative is excision of the lateral 2.5 cm of the clavicle.

The authors employ a 7-cm shoulder strap incision in the lines of Langer over the distal end of the clavicle. The deltotrapezius fascia is incised longitudinally over the distal clavicle and reflected anteriorly and posteriorly in a subperiosteal fashion. The precise location of the coracoclavicular ligaments must be determined, and the 2.5 cm of clavicle lateral to the coracoclavicular ligaments is resected after sectioning it with a small oscillating saw. The bone is removed, and the competence of the coracoclavicular ligaments is then confirmed and any sharp bone edges at the osteotomy site bevelled. The acromioclavicular joint space is debrided by excising the intra-articular meniscus and any residual scar tissue. A secure repair of the deltotrapezius fascia is obtained using nonabsorbable Dacron suture. Subcutaneous tissue is approximated with 2.0 Dacron, and a running subcuticular skin closure is achieved. Postoperative management is as previously outlined with special emphasis placed on protection of the deltotrapezius fascia repair as well as specific strengthening of the cuff muscles and the three parts of the deltoid in an isometric fashion, using either a pulley system of weights or rubber-resistant material.

References

1. Beraneck, L., Pere, G., and Crouzet, J. Osteolyse post-traumatique de la clavicule: A propos d'un cas avec hypofix-ation scintigraphique. *Rev Rheum Mal Osteo-artic* 49:699–700, 1982.
2. Brunet, M. E., Reynolds, M. C., Cook, S. D., and Brown, T. W. Atraumatic osteolysis of the distal clavicle: Histologic evidence of synovial pathogenesis: A case report. *Orthopedics* 9:557–559, 1986.
3. Cahill, B. R. Osteolysis of the distal part of the clavicle in male athletes. *J Bone Joint Surg* 64A:1053–1058, 1982.
4. Ehricht, H. G. Die Osteolyse im lateralen Claviculaencle nach Pressluftschaden. *Arch Orthop Trauma Surg* 50:576–582, 1959.
5. Griffiths, C. J., and Glucksman, E. Post-traumatic osteolysis of the clavicle: A case report. *Arch Emerg Med* 3:129–132, 1986.
6. Jacobs, P. Post-traumatic osteolysis of the outer end of the clavicle. *J Bone Joint Surg* 46B:705–707, 1964.
7. Kaplan, P. A., and Resnick, D. Stress-induced osteolysis of the clavicle. *Radiology* 158:139–140, 1968.
8. Lamont, M. K. Osteolysis of the outer end of the clavicle. *NZ Med J* 95:241–242, 1982.
9. Levine, A. H., Pais, M. J., and Schwartz, E. E. Post-traumatic osteolysis of the distal clavicle with emphasis on early radiologic changes. *Am J Roentgenol* 127:781–784, 1976.
10. Madsen, B. Osteolysis of the acromial end of the clavicle following trauma. *Br J Radiol* 36:822–828, 1963.
11. Morrison, I. S. Post traumatic osteolysis of the acromial end of the clavicle. *Australas Radiol* 22:183–186, 1978.
12. Murphy, O. B., Bellamy, R., Wheeler, W., and Brower, T. D. Post-traumatic osteolysis of the distal clavicle. *Clin Orthop* 109:108–114, 1975.
13. Norfray, J. F., Tremaine, M. J., Groves, H. C., and Bachman, D. C. The clavicle in hockey. *Am J Sports Med* 5:275–280, 1977.
14. Orava, S., Virtanen, K., Holopainen, Y. V. O. Post-traumatic osteolysis of the distal end of the clavicle. Report of three cases. *Ann Chir Gynaecol* 73:83–86, 1984.
15. Quinn, S. F., Glass, T. A. Post-traumatic osteolysis of the clavicle. *South Med J* 76:307–308, 1983.
16. Seymour, E. Q. Osteolysis of the clavicular tip associated with repeated minor trauma to the shoulder. *Radiology* 123:56, 1977.
17. Smart, M. J. Traumatic osteolysis of the distal ends of the clavicles. *J Can Assoc Radiol* 23:264–266, 1972.
18. Stahl, F. Considerations on post-traumatic absorption of the outer end of the clavicle. *Acta Orthop Scand* 23:9–13, 1954.
19. Urist, M. R. Complete dislocations of the acromioclavicular joint. *J Bone Joint Surg* 28:813–837, 1946.
20. Werder, H. Post-traumatische des Schlüssel bienendes. *Schweiz Med Wochenschr* 34:912–913, 1950.

FRACTURES OF THE SCAPULA

Gerald R. Williams, M.D.

Charles A. Rockwood, Jr., M.D.

The incidence of scapular fracture has been reported to be 3% to 5% of shoulder girdle injuries [8, 28] and 0.4% to 1% of all fractures [21, 34]. Although any area of the scapula may be fractured, the neck (10% to 60%) and body (49% to 89%) are most commonly involved [1, 12, 15, 33].

When scapular fracture does occur, it is often the result of severe trauma. This is evidenced by the fact that in most series motor vehicle or motorcycle accidents are reported as the cause of injury in over 50% of the cases [12, 15, 33]. Consequently, associated injuries are common. Rowe reported that 71% of the patients in his series of scapular fractures had other associated injuries: 45% had fracture of other bones, including the ribs, sternum, and spine; 3% of patients sustained a pneumothorax and exhibited subcutaneous emphysema; 4% sustained brachial plexus injuries; and 19% sustained other shoulder girdle dislocations [29].

Several classification systems for scapular fractures have been reported in the literature. Zdravkovic and Damholt [35] divided scapular fractures into three types: type I fractures, or fractures of the body; type II fractures, or fractures of the apophyses (including the coracoid and acromion); and type III fractures, or fractures of the superior lateral angle (i.e., scapular neck and glenoid). Zdravkovic and Damholt considered the type III fracture to be the most difficult to treat; these represented only 6% of their series.

Thompson and co-workers [32] presented a classification system that divided these fractures according to the likelihood that associated injuries would be present. Their cases all resulted from blunt trauma. Class I fractures included fractures of the coracoid and acromion process and small fractures of the body. Class II fractures comprised glenoid and scapular neck fractures. Class III fractures included major scapular body fractures. Thompson and colleagues [32] reported that class II and III fractures were much more likely to have associated injuries.

Wilbur and Evans [33] described 40 patients with 52 scapular fractures. The patients were divided into two groups based upon fracture location: group I, which included patients with fractures of the scapular body, neck, and spine, and group II, which included fractures of the acromion process, coracoid process, or glenoid. They reported unsatisfactory results of treatment in patients in group II because of residual pain and loss of glenohumeral motion.

Intra-articular glenoid fractures have been classified by Ideberg [10, 11] into five types: type I, avulsion of the anterior margin; type II, transverse fractures of the glenoid fossa with inferior subluxation of the humeral head and glenoid fragment; type III, oblique fracture through the glenoid extending medially through the superior border of the scapula (often associated with acromioclavicular fracture or acromioclavicular dislocation); type IV, a horizontal fracture beginning in the midportion of the glenoid and extending medially directly across the scapula to exit the medial border; and type V, a combination of the horizontal or type IV fracture with a fracture separating the inferior half of the glenoid (Fig. 15B–58).

ANATOMY

The scapula is enveloped by multiple layers of muscles. The anterior surface provides attachment for the subscapularis, the serratus anterior, the omohyoid, the pectoralis minor, the conjoined tendon of the coracobrachialis and short head of the biceps, the long head of the biceps, and the long head of the triceps (Fig. 15B–59). The posterior surface of the scapula provides muscular attachment sites for the levator scapulae, rhomboid major, rhomboid minor, latissimus dorsi, teres major, teres minor, a portion of the long head of the triceps, the deltoid, the trapezius, the supraspinatus, the infraspinatus, and a portion of the omohyoid (Fig. 15B–60). The intramuscular position of the scapula provides it with great mobility and a protective cushion that are no doubt responsible for the low incidence of scapular injury.

The osseous components of the scapula, which consist of the body and spine, the coracoid process, the acromion process, the glenoid, and the inferior angle, arise from several ossification centers [16, 23, 31]. At birth, the body and spine form one ossified mass. However, the coracoid process, the acromion process, the glenoid, and the inferior angle are all cartilaginous. The coracoid process is a coalescence of four or five centers of ossification. The center of ossification for the midportion of the coracoid appears at age 3 to 18 months and may be bipolar. The ossification center for the base of the coracoid, which includes the upper third of the glenoid, appears at 7 to 10 years. Two ossification centers appear at age 14 to 16 years: a center for the tip and a shell-like center at the medial apex of the coracoid process. The ossification centers for the base and the midportion of the coracoid coalesce during adolescence at age 14 to 16 years. The other ossification centers fuse at the age of 18 to 25 years (Figs. 15B–61 and 15B–62).

The acromion is a coalescence of two or three centers of ossification that appear between the ages of 14 and 16 years, coalesce at the age of 19, and fuse to the spine at the age of 20 to 25 years. Failure of the anterior acromion ossification center to fuse to the spine gives rise to the os acromiale. This unfused apophysis is present in 2.7% of random patients and is bilateral in 60% of cases [13]. The size of the os acromiale depends upon which of the four ossification centers of the acromion have failed to fuse (Fig. 15B–63). The most common site of nonunion is between the mesoacromion and the meta-acromion, which corresponds to the midacromioclavicular joint level. An axillary lateral radiograph clearly demonstrates the lesion (Fig. 15B–64). Norris [22] has reported that the os acromiale has been mistaken for fracture and that there is an association between the os acromiale and a rotator cuff tear.

The inferior angle of the scapula arises from an

Text continued on page 552

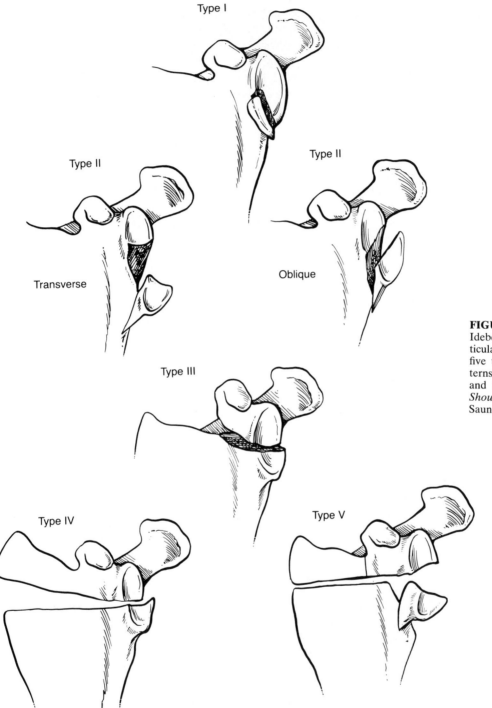

FIGURE 15B–58
Ideberg's classification of intra-articular fracture of the glenoid into five types based on fracture patterns. (From Rockwood, C. A., and Matsen, F. A. (Eds.), *The Shoulder*. Philadelphia, W. B. Saunders, 1990.)

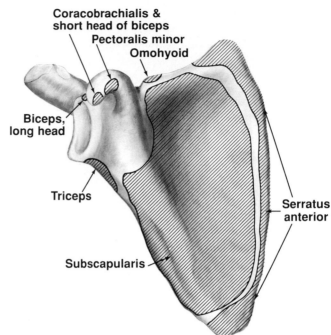

FIGURE 15B–59
The muscle attachments to the anterior surface of the scapula. (From Rockwood, C. A., and Matsen, F. A. (Eds.), *The Shoulder*. Philadelphia, W. B. Saunders, 1990.)

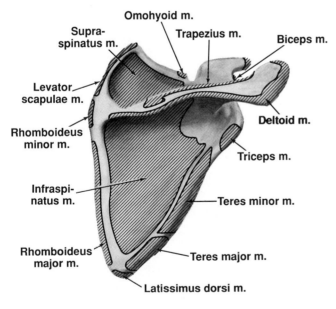

FIGURE 15B–60
The muscle attachments to the posterior surface of the scapula. (From Rockwood, C. A., and Matsen, F. A. (Eds.), *The Shoulder*. Philadelphia, W. B. Saunders, 1990.)

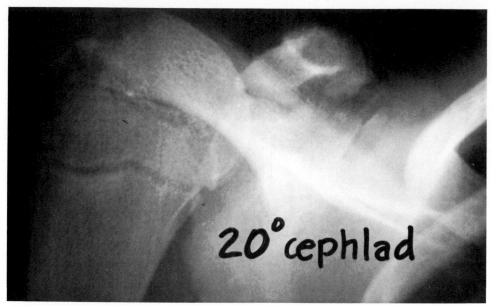

FIGURE 15B–61
A normal ossification pattern at the base of the coracoid. A crescent-shaped center is seen at the apex of the coracoid. (From Rockwood, C. A., and Matsen, F. A. (Eds.), *The Shoulder*. Philadelphia, W. B. Saunders, 1990.)

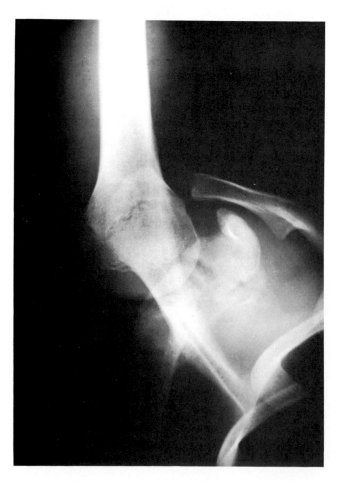

FIGURE 15B–62
An epiphyseal line is seen across the upper third of the glenoid because this portion of the glenoid ossifies in common with the base of the coracoid. This may be confused with a fracture and is the precise location of most type III glenoid fractures. (From Rockwood, C. A., and Matsen, F. A. (Eds.), *The Shoulder*. Philadelphia, W. B. Saunders, 1990.)

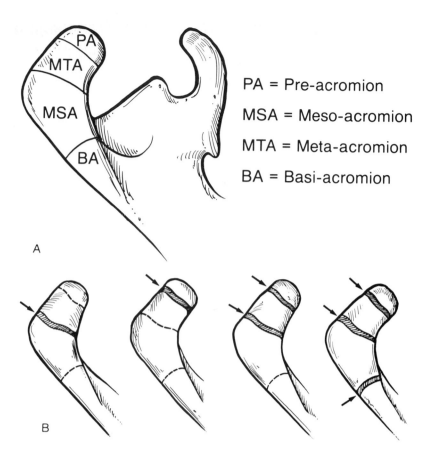

PA = Pre-acromion

MSA = Meso-acromion

MTA = Meta-acromion

BA = Basi-acromion

FIGURE 15B–63

A, The diagram represents the ossification centers of the acromion. *B*, The most common site of failure of ossification lies between the meso-acromion and the meta-acromion. (From Rockwood, C. A., and Matsen, F. A. (Eds.), *The Shoulder.* Philadelphia, W. B. Saunders, 1990.)

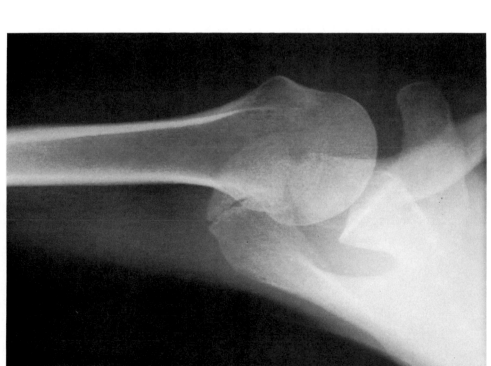

FIGURE 15B–64

Os acromiale. (From Rockwood, C. A., and Matsen, F. A. (Eds.), *The Shoulder.* Philadelphia, W. B. Saunders, 1990.)

ossification center that appears at age 15 and fuses with the remainder of the scapula at age 20. The vertebral border arises from an ossification center that appears at age 16 to 18 and fuses by the twenty-fifth year.

The glenoid fossa ossifies from four sources: (1) the coracoid base (including the upper third of the glenoid), (2) the deep portion of the coracoid process, (3) the body, and (4) the lower pole, which joins with the remainder of the body of the scapula at age 20 to 25.

Because many athletes are adolescents and because many of the apophyses do not fuse until age 25, caution must be exercised in interpreting radiographs of the scapula. The os acromiale is the most frequently quoted unfused apophysis and can be confused with fracture [13, 22]. In addition, the physes at the base of the coracoid and the tip of the coracoid process can be difficult to distinguish from fracture. In the appropriate setting, a radiograph of the contralateral scapula is useful in determining whether or not a radiographic "line" is truly a fracture or an unfused apophysis.

CLINICAL EVALUATION

History

Scapular fractures in athletes can result from either direct or indirect mechanisms of injury. Although most athletes are not subjected to the high-energy trauma associated with motor vehicular accidents, direct blows to the scapula can occur with enough force to cause fracture in contact sports such as hockey and football. Direct blows to the acromion can cause either acromion fracture or acromioclavicular separation. In addition, direct blows to the scapula or to the lateral aspect of the shoulder can cause scapular body fractures or glenoid fractures. Alternatively, glenoid fractures can be the result of indirect trauma incurred during a violent glenohumeral dislocation.

Physical Examination

The athlete with a scapular fracture typically presents with his arm adducted and protected from all movements. Abduction is especially painful. Although ecchymosis is less than expected from the degree of bony injury present, severe local tenderness is a reliable finding [3]. Athletes with scapular body fractures or coracoid process fractures fre-

quently complain of increasing pain with deep inspiration secondary to the pull of the pectoralis minor or serratus anterior muscles. Frequently, rotator cuff function is extremely painful and weak secondary to inhibition from intramuscular hemorrhage. This has been described as a "pseudo-rupture" of the rotator cuff [20] and frequently resolves within a few weeks.

It is important to reemphasize that scapular fracture is often associated with other injuries that need more urgent treatment. Significant associated injuries have been reported to occur in 35% to 98% of all patients with scapular fractures [3]. The highest incidence of serious associated injuries occurs in fractures sustained during high-speed motor vehicle accidents [1, 5, 12, 15, 32, 33]. McLennen and Ungersma [18] reported 16 pneumothoraces in 30 patients who presented with fractured scapulae. Ten of the sixteen pneumothoraces were delayed in onset from 1 to 3 days. The authors recommended a follow-up chest x-ray, physical examination, and blood gas determination in all patients with scapular fractures. Other series have reported an overall incidence of pneumothorax associated with scapular fracture of between 11% and 38% [1, 5, 32]. Ipsilateral rib fractures [32], pulmonary contusion [5, 32], and brachial plexus injury [1, 5, 12, 15, 32] have also been reported in association with scapular fractures. Physical examination should be directed toward detecting any of these possible associated injuries.

Radiographic Evaluation

Most scapular fractures can be adequately visualized using routine radiographic views. A true anteroposterior view of the scapula combined with an axillary or true scapular lateral view will demonstrate most scapular body or spine fractures, glenoid neck fractures, and acromion fractures (Figs. 15B–65 to 15B–67). "Special" views may be required in selected circumstances. The Stryker notch view, as described later in this chapter, is useful for coracoid fractures (Fig. 15B–68) [3]. The apical oblique view described by Garth [7] and the West Point lateral view [26, 27] are useful views for evaluating anterior glenoid rim fractures. The CT scan is a useful adjunct when evaluating intra-articular glenoid fractures. The contralateral normal shoulder should be scanned as well as the involved shoulder to provide a means for comparison of the pathologic findings noted in the involved shoulder [3].

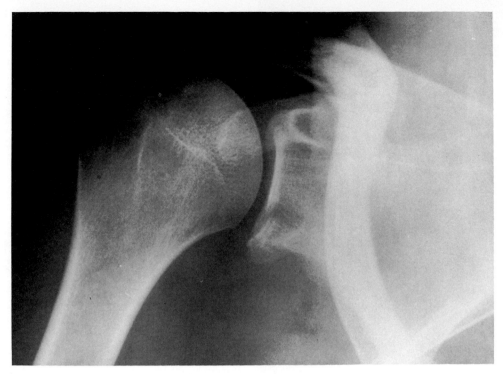

FIGURE 15B–65
A true anteroposterior view of the glenoid showing an anterior-inferior glenoid fracture. (From Rockwood, C. A., and Matsen, F. A. (Eds.), *The Shoulder*. Philadelphia, W. B. Saunders, 1990.)

Glenoid rim fractures associated with glenohumeral instability pose perhaps the most difficult decisions for treatment of fractures of the scapula among athletes. The athlete does not always relay a history of glenohumeral dislocation in association with his initial injury. Since the decision regarding operative or nonoperative treatment for these glenoid rim fractures depends upon whether or not they are associated with instability [3], the physician should make every attempt possible to verify the presence or absence of instability. In this regard, stress views with or without fluoroscopic control or an examination under anesthesia may be helpful.

TREATMENT OPTIONS

The recommended treatment for specific types of scapular fractures varies depending upon whether the fracture is intra- or extra-articular. The great majority of extra-articular fractures (i.e., glenoid neck, scapular body or spine, acromion, and coracoid fractures) are managed nonoperatively [2, 4, 14, 17, 25]. Intra-articular fractures, particularly those associated with glenohumeral instability, are managed operatively [4, 10, 24, 29, 30].

Extra-articular Fractures

Glenoid Neck

Fracture of the neck of the scapula is the second most common scapular fracture [3]. The glenoid articular surface is intact, and the fracture line extends from the suprascapular notch area across the neck of the scapula to its lateral border inferior to the glenoid. The glenoid and coracoid may be comminuted or may remain as an intact unit. Although glenoid neck fractures are often impacted, their displacement is limited by an intact clavicle and by the acromioclavicular and coracoclavicular ligaments [8].

Most authors recommend closed treatment for glenoid neck fractures [2, 4, 14, 25]. For displaced fractures, DePalma [4] recommends closed reduction and olecranon pin traction for 3 weeks followed by a sling. Bateman [2] favors closed reduction and a shoulder spica cast for 6 to 8 weeks in cases in which "shortening of the neck is sufficient to favor subluxation or interfere with abduction." McLaughlin [17] casts doubt on the usefulness of closed reduction in these fractures because most of them are impacted and difficult to move. Lindblom and

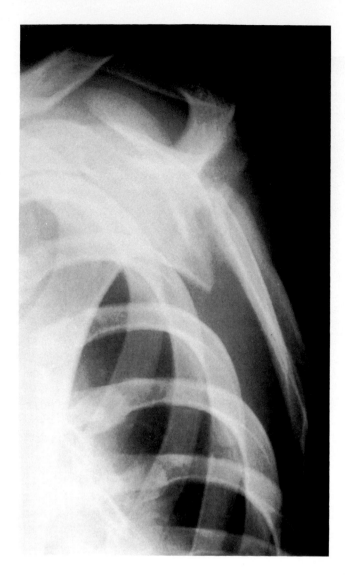

FIGURE 15B–66
A tangential scapular lateral view (trauma series lateral view) showing a displaced scapular body fracture with a bayonet position. (From Rockwood, C. A., and Matsen, F. A. (Eds.), *The Shoulder*. Philadelphia, W. B. Saunders, 1990.)

FIGURE 15B–67
The fractured base of the acromion is well seen on a tangential scapular lateral view. (From Rockwood, C. A., and Matsen, F. A. (Eds.), *The Shoulder*. Philadelphia, W. B. Saunders, 1990.)

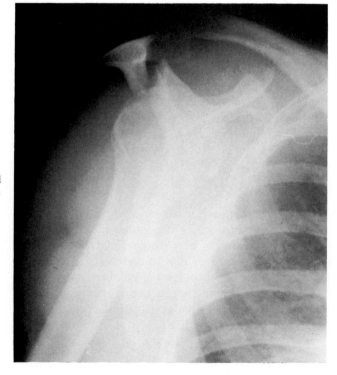

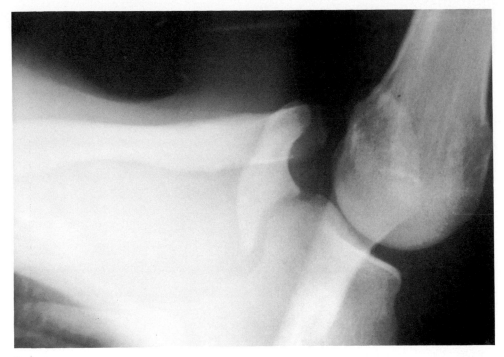

FIGURE 15B–68
A fracture of the base of the coracoid, seen best on a Stryker notch view. (From Rockwood, C. A., and Matsen, F. A. (Eds.), *The Shoulder*. Philadelphia, W. B. Saunders, 1990.)

Levin [14] studied a series of scapular neck and body fractures and concluded that, if untreated, all fractures healed in the position displayed at the time of the original injury (i.e., without additional displacement).

Other authors [6, 8] have recommended open reduction of glenoid neck fractures. Gagney and colleagues [6] reported only one good result among 12 displaced glenoid neck fractures treated closed. They theorized that healing of the glenoid neck in the displaced position would "disorganize the coracoacromial arch." Therefore, they recommended open reduction and internal fixation. Hardegger and co-workers [8] recommended open reduction and scapular fixation for displaced glenoid neck fractures associated with a fracture of the clavicle or disruption of the coracoclavicular ligaments. They postulated that a severe displacement in these injuries would result in "functional imbalance" of the shoulder mechanism.

Most series report good functional results in patients with glenoid neck fractures regardless of the method of treatment [9, 11, 14, 33] (Fig. 15B–69). Hitzrot and Bolling [9] in 1916 stated that manipulation and traction had no effect on displaced glenoid neck fractures and that the results were so satisfactory without reduction that attempts to achieve reduction were unnecessary. Armstrong and Vanderspuy [1] reported that six of seven of their patients with glenoid neck fractures had some residual stiffness, but no patient had a functional disability. Zdravkovic and Damholt [35] came to the same conclusion in their report, in which patients had an average of 9 years of follow-up.

Scapular Body Fracture

Fracture of the body of the scapula is the most common type of scapular fracture and is correlated with the highest incidence of associated injury [3]. These fractures may be quite comminuted and displaced. However, malunion is rarely associated with clinical symptoms [4, 25, 28, 29]. Consequently, most authors favor a sling, ice, and supportive measures until the initial pain subsides [17, 25, 28, 29]. Neer and Bateman [2] reported immobilization using cross-strapping with adhesive moleskin in a nonambulatory patient with a scapular body fracture. This type of immobilization, however, has been associated with residual shoulder stiffness [17].

Acromion Fracture

Although fracture of the acromion is rare, when it does occur it is usually the result of one of two mechanisms. First, an acromion fracture can result

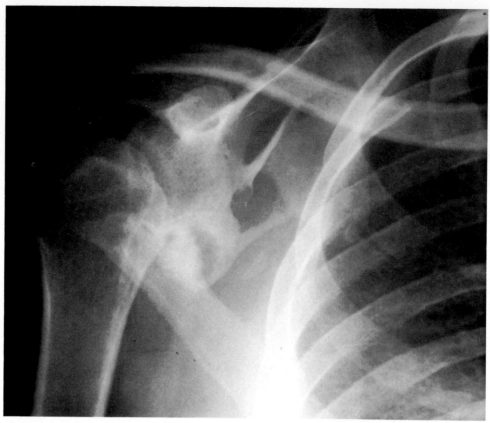

FIGURE 15B–69
A healed glenoid neck fracture with marked medial displacement and full range of motion. (From Rockwood, C. A., and Matsen, F. A. (Eds.), *The Shoulder*. Philadelphia, W. B. Saunders, 1990.)

from a downward blow directly applied to the superior aspect of the acromion. Second, acromion fracture can result from superior displacement or dislocation of the humeral head. When the injury is a result of a downward blow to the acromion, acromioclavicular dislocation is much more common than acromion fracture. However, when a fracture does occur, it is usually minimally displaced. Caution should be used in distinguishing this minimally displaced fracture from an os acromiale. In questionable cases, a radiograph of the contralateral side may be helpful because the os acromiale is bilateral in 60% of the cases [13]. The supraspinatus outlet view may be useful in estimating the amount of displacement if any is present [19A]. Significant displacement of an acromion fracture resulting from a downward blow to the acromion should alert the clinician to possible associated brachial plexus avulsions [3, 19]. Significant superior displacement of the acromion associated with superior displacement or dislocation of the humeral head should alert the clinician to possible injury to the rotator cuff [19].

The majority of acromion fractures, because they are minimally displaced, should be treated closed

[19, 25, 28, 29, 33]. McLaughlin [17] stated that "bony union is the rule, despite the presence or absence of immobilization, provided the fragments are in apposition." Neer [19] recommended symptomatic treatment only. Wilbur and Evans [33], on the other hand, reported residual stiffness in patients with acromion fractures. They recommended cast immobilization in 60 degrees of abduction, 25 degrees of flexion, and 25 degrees of external rotation for 6 weeks. Most authors recommend open reduction and internal fixation for markedly displaced acromion fractures to prevent nonunion or malunion and secondary impingement [18A, 19, 29].

Glenoid (Intra-articular) Fractures

Although intra-articular fractures of the glenoid have been classified by Ideburg into five types [10], most authors base their recommendations for treatment upon the presence or absence of associated instability rather than on fracture type [3, 4, 10, 24, 29, 30]. Anatomic reduction of the articular surface is not as critical as it is in lower extremity weight-

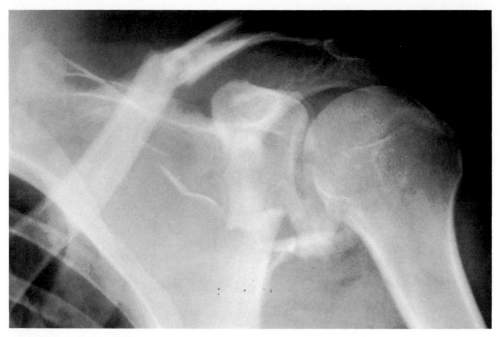

FIGURE 15B–70
A combined glenoid articular surface fracture with satisfactory position. (From Rockwood, C. A., and Matsen, F. A. (Eds.), *The Shoulder*. Philadelphia, W. B. Saunders, 1990.)

bearing joints [29]. Many authors have reported good functional results in patients with intra-articular glenoid fractures that were not associated with instability [10, 24, 25, 29] (Figs. 15B–70 and 15B–71). When instability is associated with an intra-articular glenoid fracture, however, open reduction is indicated [4, 10, 24]. Ideburg [10] does not believe that the size of the fragment is a prognostic indicator for further instability. DePalma, on the other hand [4], thinks that glenohumeral instability will result if the fragment is greater than one-fourth of the glenoid or if it is displaced more than 10 mm. He recommends immediate open reduction in this situation [4]. Rockwood [24] recommends open reduction of the fragment with screw fixation if the fracture involves one-fourth or more of the glenoid and is associated with instability. Rowe and co-workers [29, 30], however, have reported a 2% recurrence rate of instability when the fragment was excised and the capsule was repaired back to the remainder of the glenoid, even if the fragment involved one-fourth or one-third of the glenoid.

AUTHORS' PREFERRED METHOD OF TREATMENT

Glenoid Neck Fracture

The majority of glenoid neck fractures are impacted and stable and do not require any reduction to obtain a good clinical result. Symptomatic local care, followed by passive exercises, will result in a rapid return of motion. Strengthening exercises may be instituted at 4 to 6 weeks.

Criteria for Return to Athletics. Healing is normally complete after approximately 6 weeks. However, return to sports should be delayed until range of motion has returned to normal, and the strength of the shoulder is 90% of that of the uninvolved extremity. This normally takes 3 to 4 months.

Body and Spine Fractures

Assuming that serious associated injuries have been ruled out, symptomatic treatment is indicated for virtually all patients with this type of fracture. Ice and sling immobilization are used initially. Within 2 to 3 weeks, passive range of motion and stretching exercises can be instituted. As pain and swelling subside, active range of motion and strengthening exercises can be instituted.

Criteria for Return to Athletics. The athlete should be withheld from competition until the fracture has healed and there is a full range of motion. Foam padding over the posterior aspect of the scapula helps to cushion blows that may be encountered when the athlete returns to contact competition.

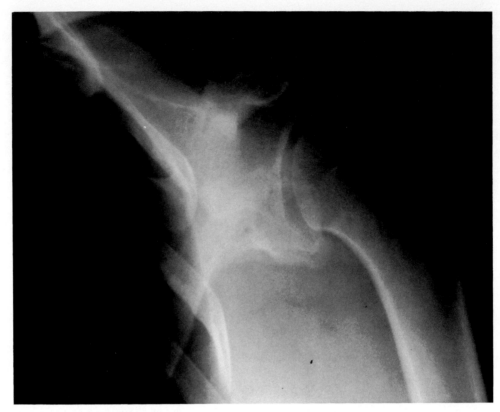

FIGURE 15B–71
This fracture healed well without problems and with good preservation of the joint surface. (From Rockwood, C. A., and Matsen, F. A. (Eds.), *The Shoulder*. Philadelphia, W. B. Saunders, 1990.)

Acromion Fractures

Most acromion fractures, as previously stated, are stable and are minimally displaced. Therefore, a sling is required for only 3 to 5 days. When pain diminishes enough to permit exercise, active and passive range of motion exercises are begun. Resisted deltoid exercises are avoided for 6 weeks to allow fracture union. In the rare instance of a displaced acromion fracture, open reduction and internal fixation are performed. Caution should be exercised to rule out the presence of an os acromiale, rotator cuff tear, or brachial plexus injury.

Criteria for Return to Athletics. The athlete with an acromion fracture, regardless of whether it was displaced and required fixation, should be withheld from competition until fracture union is complete and range of motion is pain-free. This normally requires between 6 and 12 weeks. If the fracture has been accompanied by a complication such as a rotator cuff tear or brachial plexus injury, appropriate treatment should be instituted and return to sport delayed accordingly.

Glenoid Fractures

Stable Glenohumeral Joint. Those glenoid fractures that are not associated with instability are treated symptomatically, using a sling for immobilization, until pain permits range of motion exercises (7 to 10 days). Occasionally, olecranon pin traction is useful to reduce a stellate or comminuted fracture through "ligamentotaxis." Strengthening exercises are added as range of motion is restored and absence of pain permits.

Criteria for Return to Athletics. Return to sport is possible after fracture union has occurred, range of motion has reached its maximum level, and strength has returned to within 90% of that of the opposite extremity. The athlete should be warned about the possibility of developing glenohumeral arthritis, particularly if he or she is involved in a sport that places a large demand on the shoulder.

Unstable Glenohumeral Joint. Intra-articular glenoid fractures associated with glenohumeral instability are best treated with surgical repair. Most often the fracture occurs at the anteroinferior glenoid rim and is accompanied by anterior instability. If the fragment involves a significant portion of the glenoid (i.e., 25%) and is large enough to accept a small fragment screw, open reduction and internal fixation through an anterior deltopectoral approach are performed. If the fragment is too small to accommodate a screw, it is excised, and the anteroinferior capsule is repaired to the raw surface of the remaining glenoid. Pendulum exercises are begun immediately postoperatively. At 2 to 3 weeks postoperatively, the patient is encouraged to use the arm for everyday living activities, and gentle passive flexion and external rotation exercises are begun. At 6 to 8 weeks, stretching and strengthening exercises are instituted.

Posterior glenoid rim fracture, when accompanied by posterior glenohumeral instability, is approached in a manner similar to that used for its anterior counterpart. If the fragment comprises 25% or more of the articular surface, it is reduced and stabilized with a screw. Smaller fragments are excised, and the capsule is reattached to the remaining glenoid.

Criteria for Return to Athletics. Regardless of whether the fragment has been excised or fixed, the presence of glenohumeral instability dictates postponement of a return to competition for 6 months to a year.

References

1. Armstrong, C. P., and Vanderspuy, J. The fractured scapula: importance in management based on a series of 62 patients. *Injury* 15:324–329, 1984.
2. Bateman, J. E. *The Shoulder and Neck* (2nd ed.). Philadelphia, W. B. Saunders, 1978.
3. Butters, K. P. The scapula. *In* Rockwood, C. A., Jr., and Matsen, F. A. (Eds.), *The Shoulder.* Philadelphia, W. B. Saunders, 1990.
4. DePalma, A. F. *Surgery of the Shoulder* (3rd ed.). Philadelphia, J. B. Lippincott, 1983, pp. 366–367.
5. Fischer, R. P., Flynn, T. C., Miller, P. W., and Thompson, D. A. Scapular fractures and associated major ipsilateral upper-torso injuries. *Curr Concepts Trauma Care* 1:14–16, 1985.
6. Gagney, O., Carey, J. P., and Mazas, F. Les Fractures recentes de l'omoplate a propos de 43 cas. *Rev Chir Orthop* 70:443–447, 1984.
7. Garth, W. P., Jr., Shappey, C. E., and Ochs, C. W. Roentgenographic demonstration of instability of the shoulder: The apical oblique projection—a technical note. *J Bone Joint Surg* 66A(9):1450–1453, 1984.
8. Hardegger, F. H., Simpson, L. A., and Weber, B. G. The operative treatment of scapular fractures. *J Bone Joint Surg* 66B:725–731, 1984.
9. Hitzrot, T., and Bolling, R. W. Fracture of the neck of the scapula. *Ann Surg* 63:215–234, 1916.
10. Ideberg, R. Unusual glenoid fractures: A report on 92 cases. *Acta Orthop Scand* 58:191–192, 1987.
11. Ideberg, R. Fractures of the scapula involving the glenoid fossa. *In* Bateman, J. E., and Welsh, R. P. *Surgery of the Shoulder.* Philadelphia, B. C. Decker, 1984, pp. 63–66.
12. Imatani, R. J. Fractures of the scapula. A review of 53 fractures. *J Trauma* 15:473–478, 1975.
13. Liberson, F. Os acromiale—a contested anomaly. *J Bone Joint Surg* 19:683–689, 1937.
14. Lindblom, A., and Leven, H. Prognosis in fractures of the body and neck of the scapula. *Acta Chir Scand* 140:33, 1974.
15. McGahan, J. P., Rab, G. T., and Dublin, A. Fractures of the scapula. *J Trauma* 20:880–883, 1980.
16. McClure, J. G., and Raney, R. B. Anomalies of the scapula. *Clin Orthop* 110:22–31, 1975.
17. McLaughlin, H. L. *Trauma.* Philadelphia, W. B. Saunders, 1959.
18. McLennen, J. G., and Ungersma, J. Pneumothorax complicating fractures of the scapula. *J Bone Joint Surg* 64A:598–599, 1982.
18A. Neer, C. S., II. Fractures about the shoulder. *In* Rockwood, C. A., Jr., and Green, D. P. (Eds.), *Fractures.* Philadelphia, J. B. Lippincott, 1984, pp. 713–721.
19. Neer, C. S., II. *Shoulder Reconstruction.* Philadelphia, W. B. Saunders, 1990.
19A. Neer, C. S., II, and Poppen, N. K. Supraspinatus outlet. *Orthop Trans* 11:234, 1987.
20. Neviaser, J. Traumatic lesions: Injuries in and about the shoulder joint. *Instr Course Lect* 13:187–216, 1956.
21. Newell, E. D. Review of over 2,000 fractures in the past seven years. *South Med J* 20:644–648, 1927.
22. Norris, T. R. Unfused epiphysis mistaken for acromion fracture. *Orthopedics Today* 3(10):12–13, 1983.
23. Ogden, J. A., and Phillips, S. B. Radiology of postnatal skeletal development. *Skel Radiol* 9:157–169, 1983.
24. Rockwood, C. A., Jr. Disorders of the glenohumeral joint. *In* Rockwood, C. A., Jr., Green, D. P. (Eds.), *Fractures in Adults* (2nd ed.). Philadelphia, J. B. Lippincott, 1984.
25. Rockwood, C. A., Jr. Management of fractures of the scapula. *J Bone Joint Surg* 10:219, 1986.
26. Rockwood, C. A., Jr. Personal communication, 1989.
27. Rokous, J. R., Feagin, J. A., and Abbott, H. G. Modified axillary roentgenogram. *Clin Orthop* 82:84–86, 1972.
28. Rowe, C. R. Fractures of the scapula. *Surg Clin North Am* 43:1565–1571, 1963.
29. Rowe, C. R. *The Shoulder.* New York, Churchill Livingstone, 1987, pp. 373–381.
30. Rowe, C. R., Patel, D., Southmayd, W. W. The Bankart procedure—a long-term end-result study. *J Bone Joint Surg* 60A:1–16, 1978.
31. Tachdjian, M. O. *Pediatric Orthopedics.* Philadelphia, W. B. Saunders, 1972, pp. 1553–1555.
32. Thompson, D. A., Flynn, T. C., Miller, P. W., Fischer, R. P. The significance of scapular fractures. *J Trauma* 25:974–977, 1985.
33. Wilbur, M. C., and Evans, E. B. Fractures of the scapula—an analysis of forty cases and review of literature. *J Bone Joint Surg* 59A:358–362, 1977.
34. Wilson, P. D. *Experience in the Management of Fractures and Dislocations (Based on an Analysis of 4390 Cases) by the Staff of the Fracture Service Massachusetts General Hospital, Boston.* Philadelphia, J. B. Lippincott, 1938.
35. Zdravkovic, D., and Damholt, V. V. Comminuted and severely displaced fractures of the scapula. *Acta Orthop Scand* 45:60–65, 1974.

FRACTURES OF THE CORACOID

Gerald R. Williams, M.D.
Charles A. Rockwood, Jr., M.D.

Fractures of the coracoid process of the scapula are uncommon and have received especially little attention in the sports medicine literature. Injury may occur as an apparently isolated phenomenon or in association with other injuries about the shoulder girdle. The coracoid process of the scapula serves as an anchoring point for the attachment of multiple ligaments and muscles. Ligaments include the coracohumeral and coracoacromial ligaments and the conoid and trapezoid components of the coracoclavicular ligaments. These last two ligaments perform an essentially suspensory function, exerting a static upward force upon the scapula through the coracoid process [5]. On the contrary, the muscular attachments exert a dynamic, active, and largely inferior force upon the coracoid. The muscular origins comprise the pectoralis minor from the body and tip of the coracoid and the conjoined tendon from the tip incorporating the coracobrachialis and the short head of the biceps brachii. Consideration of these ligamentous and muscular attachments to the coracoid will give some insight into the proposed mechanisms of coracoid fracture. The location of the coracoid fracture (i.e., tip, body or base in relation to the musculoligamentous structures) will also determine the stability of the fracture and hence the propensity for displacement.

MECHANISM OF INJURY

Mariani [13] has suggested that direct and indirect mechanisms cause acute coracoid fracture. The *direct* type of injury appears to be a relatively rare phenomenon. A direct blow to the coracoid from the exterior, culminating in fracture because of the coracoid's deep-seated, sheltered anatomic location, must involve massive trauma more common to motor vehicle accidents than to sporting endeavors. However, direct trauma to the coracoid from the interior may arise in two circumstances. Anterior translation of the humeral head in subcoracoid glenohumeral dislocations may result in a direct coracoid impact that is sufficient to cause fracture [1, 7, 22, 23]. This too must be considered an uncommon injury, but it has been proposed that the combination of glenohumeral dislocation with coracoid fracture may be underdiagnosed. As will be discussed later, this shortcoming may be related to

the difficulty of obtaining a good axillary lateral radiograph in an acutely painful shoulder or to the widespread practice of relying on the more difficult to interpret and less readily reproducible scapular lateral radiograph. McLaughlin [12] considered glenohumeral dislocation the commonest cause of coracoid fracture. It has also been suggested that an undetected coracoid fracture might account for occasional cases of prolonged convalescence after glenohumeral dislocation [1] and may conceivably be confused with anterior instability or rotator cuff pathology. The other direct mechanism of coracoid fracture would in theory involve a blow to the lateral clavicle causing inferior displacement and impact with the coracoid [13]. This would result in acromioclavicular ligamentous disruption but would preserve the coracoclavicular ligaments. Although this scenario appears not to have received specific attention in the literature, it is possible that some apparently isolated, undisplaced coracoid fractures might arise in this manner. The stress radiograph, to be discussed later, would be of particular relevance in this situation.

So-called *indirect* mechanisms probably account for the majority of fractures of the coracoid process. This is the mechanism that is probably most often responsible for isolated coracoid fractures [4, 13]. Smith [20] described this mechanism in terms of a sudden, violent, and resisted contraction of the conjoined tendon and pectoralis minor. Mariani [13] concluded that the coracoid is especially vulnerable to the stress of muscular action when the arm is in the position of abduction and extension. Benton and Nelson [2] also drew attention to the stress placed on the coracoid when the arm is in this position. Another indirect mechanism of coracoid fracture might involve a direct blow or fall onto the point of the shoulder [16], causing superior subluxation or dislocation of the lateral clavicle. Rather than the more common rupture of the coracoclavicular ligaments, the coracoid (proximal to the coracoclavicular ligaments) may fail [3, 11, 14, 15, 20, 21]. The pain accompanying the coracoid fracture may overshadow the acromioclavicular disruption so that, again, this injury may be misinterpreted as an isolated coracoid fracture if stress radiographs are not obtained [13, 20].

In addition to these direct and indirect mechanisms involving significant trauma, repetitive forces of a lesser degree have been reported by a number of authors to be a source of coracoid stress fracture. Boyer [4] and Sandrock [19] in separate reports (apparently of the same patient) described a fracture of the coracoid base in a young female trapshooter. The position of the gun butt directly over the coracoid tip was confirmed radiologically. Symptoms resolved and the fracture healed when shooting was stopped. This appears to be an example of repetitive direct trauma resulting in stress fracture. A case of indirect trauma resulting from repetitive muscular action and leading to coracoid stress fracture in its distal half was reported by Benton and Nelson [2]. They described a 19-year-old tennis player with a 4-year history of shoulder pain. It had been of insidious onset and was aggravated by the service position. This patient eventually required excision of the distal fragment and reattachment of the conjoined tendon.

PATTERN OF CORACOID FRACTURE

The pattern of coracoid fracture is variable. Most such fractures occur through the base of the process [1, 4, 10, 13, 17], and this is almost invariably the case with associated acromioclavicular injuries. One apparent exception to this is found in the report of Montgomery and Loyd [14] of two adolescents with coracoid apophyseal avulsions at the site of attachment of the coracoclavicular ligament. Basal coracoid fractures may very rarely involve a significant portion of the superior portion of the glenoid articular surface. Fractures of the body or tip of the

coracoid process of the scapula without acromioclavicular injury seem to be related more to violent muscular action [2, 5, 7, 24] but have been associated with anterior glenohumeral dislocation [23]. These more distal fractures also appear to be more troublesome in terms of delayed union or nonunion and the related problem of displacement [7]. Both of these problems are well demonstrated in the previously mentioned case report by Benton and Nelson [2] of a distal stress fracture in a young tennis player. These problems of union and displacement and hence the possibility of persistent symptoms seem to be related to the location of the fracture in relation to the coracoclavicular ligaments [5]. A fracture within the broad area of the attachment of these ligaments is likely to be splinted and minimally displaced. However, as the fracture line moves toward the tip and hence beyond the attachments of the conoid and trapezoid ligaments, the coracoid tip becomes increasingly subject to the displacing action of the pectoralis minor and the conjoined tendon [2, 7].

We have described the connection of coracoid fracture with acromioclavicular separation and glenohumeral dislocation. Certain other associations have also been recognized. Zilberman and Rajovitsky [24] encountered coracoid fractures in conjunction with clavicular shaft and acromion fractures. Wolf and colleagues [22] described a combination of coracoid base fracture and avulsion of a thin spicule from the superior border of the scapula medial to the coracoid. The authors have also seen this pattern of injury, and it may be that the fragments are connected by the suprascapular ligament (Fig. 15B–72).

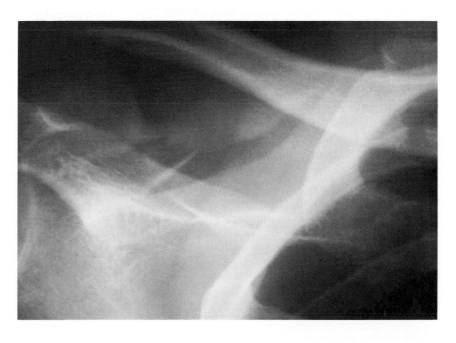

FIGURE 15B–72
An anteroposterior radiograph of a patient with a fracture through the base of the coracoid process that includes the superior border of the scapula.

Neurologic injuries may also be seen with coracoid fractures [16]. The brachial plexus deep to the coracoid and pectoralis minor may be contused, resulting in either specific or subtle patchy neurologic deficits. Basal coracoid fractures may especially result in suprascapular nerve entrapment and may be confused with rotator cuff tears [16].

CLINICAL FEATURES OF CORACOID FRACTURES

History

The history of coracoid fractures caused by shoulder injury is not particularly specific. Approximately one-third of reports of coracoid fractures attribute the injury to a motor vehicle accident [3]. The next most common history obtained is that of a fall onto the point of the shoulder or a direct blow to the shoulder. It is of interest that football injuries involving both of these mechanisms appear regularly in the literature. Typical football injuries have included a fall onto the shoulder, a direct blow from running into a goal post, and apparently the shoulder impact sustained during a rugby scrum [13]. A fall backward onto the extended abducted and externally rotated arm also appears to be a well-established mechanism [5, 7] and may be brought out in the history.

As previously discussed, coracoid stress fracture appears to be a real entity in sportsmen [2, 4, 19]. A history of acute injury is conspicuously absent, and recalcitrant symptoms of insidious onset may pose a diagnostic dilemma. Pain is an invariable complaint with acute fracture of the coracoid but may be poorly localized at the front of the shoulder. Pain may be aggravated by arm movements that exert muscular forces upon the coracoid. The patient may recognize these particular movements, which include elbow flexion [6], shoulder flexion with the elbow extended [2, 13], and combined shoulder abduction, extension, and external rotation [6]. Similarly, the patient may volunteer the fact that the pain is aggravated by deep inspiration due to pectoralis minor activation [2, 13].

As with all shoulder girdle injuries, neurologic symptoms [16] may be prominent, especially in the form of transient paresthesias. This is not surprising in view of the intimate relationship of the major neurovascular structures to the coracoid and pectoralis minor.

Physical Examination

Unless there has been an associated injury to the acromioclavicular joint or a glenohumeral disloca-tion, there will usually be no striking external abnormality. Falls onto or direct blows to the shoulder may, of course, result in localized areas of contusion or abrasion. Despite the deep-seated location of the process, swelling or loss of definition in the deltopectoral interval may be detected [6]. Marked localized tenderness upon palpation is a key finding. Specific stress tests [2, 6, 13] such as resisted elbow flexion, resisted straight arm raising, and coughing may also sharply localize discomfort to the coracoid region. Whenever suspicion of a coracoid injury is raised, attention should be specifically directed to the acromioclavicular joint and vise versa. Local acromioclavicular tenderness, swelling, subluxation, or obvious superior dislocation of the lateral clavicle may be apparent. As always, comparison with the normal shoulder may be of considerable assistance. Neurologic examination is mandatory in cases of coracoid fracture. Because deficits may be patchy and subtle, a thorough brachial plexus assessment is essential. Special emphasis should be placed on suprascapular nerve evaluation because of the risk of entrapment [16].

Diagnostic Studies

Although it is possible to diagnose some coracoid fractures with a plain anteroposterior radiograph [6], it is likely that most will be overlooked without additional views [2] (Fig. 15B–73) due to the fact that the coracoid process is foreshortened and projected over the acromion and spine of the scapula in this view [9]. Of course, the anteroposterior view should be part of the routine shoulder series and is of particular relevance in detecting associated acromioclavicular injuries or fractures [22, 24] (Fig. 15B–74).

Many authors have stressed the value of the axillary lateral view in diagnosing coracoid fractures [1–4, 16, 23], but even this view may fail to demonstrate a basal coracoid fracture [9] (Fig. 15B–75 A,B). Froimson [6] has also pointed out that the abduction required for a good axillary lateral view may be difficult to obtain because of the pain it provokes in patients with acute coracoid fractures. A much better profile of the coracoid, including the base, can be obtained by tilting the x-ray beam in a cephalic direction. This obviates the need to move the patient's arm. Most authors recommend a supine position with a 30- to 35-degree cephalic tilt [9, 13, 15, 23, 24]. Froimson found that a cephalic tilt of as much as 45 to 60 degrees was quite useful. Although the Stryker notch view was originally intended to identify the Hill Sachs lesion characteristic of anterior glenohumeral dislocations, the au-

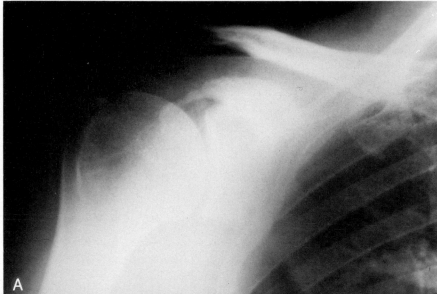

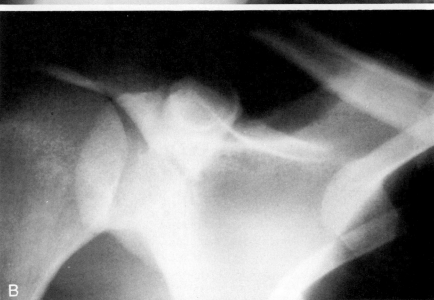

FIGURE 15B–73
A, A coracoid process fracture that is well visualized on a routine anteroposterior radiograph. *B*, Fractures of the base of the coracoid process can be difficult to visualize on routine anteroposterior views.

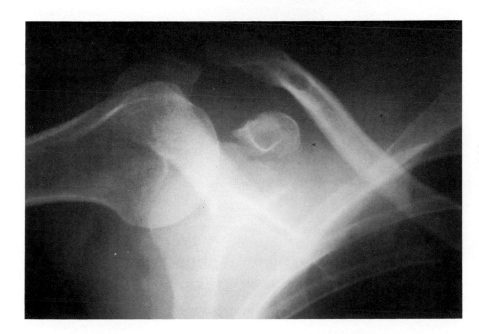

FIGURE 15B–74
Combined coracoid process fracture and acromioclavicular dislocation as demonstrated on routine anteroposterior radiograph.

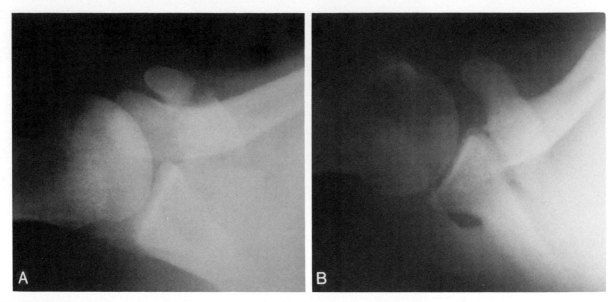

FIGURE 15B–75

A, Fractures of the tip of the coracoid process are frequently well visualized on an axillary lateral radiograph. *B*, Basal fractures of the coracoid process can be difficult to demonstrate, even on an axillary lateral radiograph.

thors have found this technique especially useful in studying coracoid fractures [8] (Fig. 15B–76). Kopecky and colleagues [10] reported on the value of CT scanning when doubt exists. The authors have also found that this modality is helpful in clarifying coracoid fracture morphology (Fig. 15B–77).

Specific attention to the acromioclavicular joint in the presence of coracoid fracture should be paid to the extent that stress views of the acromioclavic-

ular joint should be a *routine* part of the assessment of coracoid fractures. The coracoclavicular distance will be maintained regardless of the degree of acromioclavicular separation that exists [3, 13, 20] (Fig. 15B–78). Erect anteroposterior films of both shoulders with and without weights using a softer technique appropriate to the acromioclavicular joint are necessary.

Normal coracoid epiphyses or apophyses should not be confused with fractures [2, 8]. The basal epiphyseal plate fuses at puberty. Smaller accessory

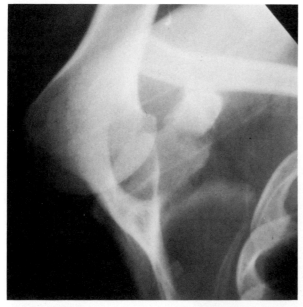

FIGURE 15B–76

The Stryker notch view is very helpful in demonstrating basal coracoid process fractures that are difficult to visualize with other routine views.

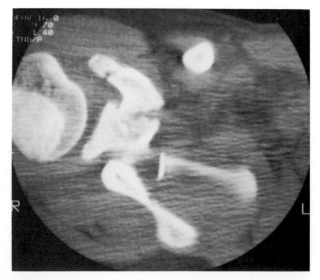

FIGURE 15B–77

CT scanning is occasionally useful for clarifying the morphology of certain coracoid fractures.

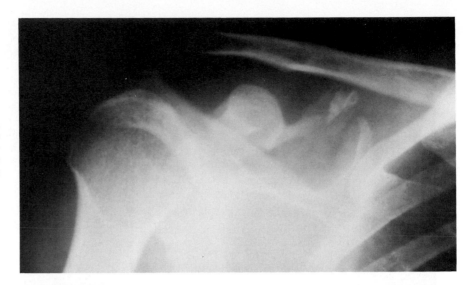

FIGURE 15B–78
An anteroposterior stress radiograph of this patient with a combined coracoid process fracture and acromioclavicular dislocation reveals that the coracoclavicular interspace has been maintained.

ossification centers that are shell-like and rounded may be seen medial to the coracoid base or at its very tip (Fig. 15B–79 A, B). The only other diagnostic study apart from radiography that may be necessary is electromyography, which is indicated if suprascapular nerve entrapment is suspected [16].

TREATMENT OPTIONS

Acute isolated fracture of the coracoid base is almost invariably treated conservatively with the expectation of a good result [5]. If the acromioclavicular joint is sound, the basal fracture is splinted

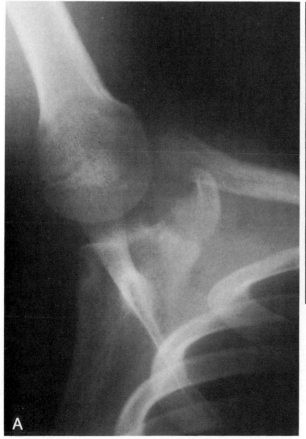

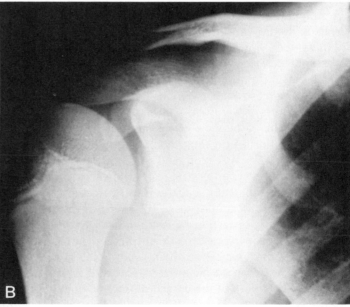

FIGURE 15B–79
A, The normal basal coracoid physis as demonstrated on the Stryker notch view. B, The normal ossification center at the tip of the coracoid process.

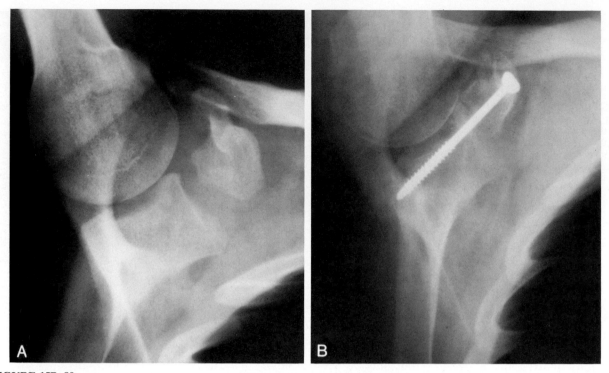

FIGURE 15B–80

A, Stryker notch view demonstrating a nonunion of the base of the coracoid process in an 18-year-old football player. *B*, The nonunion was fixed through an anterior deltopectoral approach with an interfragmentary compression screw and bone graft.

by the coracoclavicular ligaments, and displacement is minimal. Fracture surfaces are relatively large and predominantly cancellous. Prompt union is generally anticipated; however, the senior author has had personal experience with a case of nonunion requiring bone grafting and screw fixation (Fig. 15B–80 *A,B*). This case may have been related to a premature return to vigorous activities. Essentially, the basic treatment ought to be symptomatic, resting the affected arm in a sling, administering analgesia for the initially severe pain, and gradually mobilizing the shoulder as symptoms regress and healing, monitored radiographically, occurs. Basal coracoid fracture with suprascapular nerve palsy is a rare indication for early operative exploration. The prognosis for recovery of suprascapular entrapment appears to be poor once cancellous bone has formed [16].

As the location of an isolated coracoid fracture approaches the tip of the coracoid process, opinions about treatment diverge. The closer the fracture gets to the tip of the coracoid process, the smaller is the stabilizing effect of the coracoclavicular ligaments and the greater is the propensity for displacement or nonunion exerted by the muscular attachments at the tip. Rowe [18] recommends simple approximation of fragments with nonresorbable sutures in this situation, whereas McLaughlin [12] considers that fibrous union is not uncommon and

is rarely accompanied by any residual symptoms. However, there is no doubt that marked displacement and delayed union of these more distal fractures may significantly delay recovery [2, 5, 7] and that satisfactory results will follow surgical treatment of selected fractures. Essentially, the choice is between screw fixation of the fragment or, if it is especially distal, excision of the fragment and reattachment of the pectoralis minor and conjoined tendon to the residual coracoid stump. Similarly, for a combined coracoid fracture and acromioclavicular dislocation there appears to be no single best line of management. Bernard and associates [3], in a comprehensive review of this dual injury, found that surgical and nonsurgical methods of treatment appear to offer equally favorable results. It is also of interest that coracoid nonunion appeared to be no more common with this injury. Should nonunion arise, its combination with complete acromioclavicular dislocation appears to be compatible with a functional pain-free result [11]. However, when dealing with athletes who have unusual physical demands and are seeking the best functional result, one may be swayed toward an anatomic restoration. If this course is taken, the fracture location itself may preclude coracoclavicular fixation techniques, and transarticular pins may become necessary [20] (Fig. 15B–81). Although screw fixation of the cor-

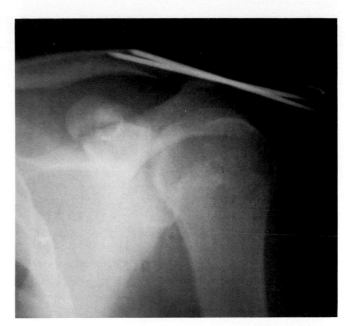

FIGURE 15B–81
Combined coracoid process fracture and acromioclavicular dislocation treated with reduction and acromioclavicular wires.

acoid back to the body of the scapula is technically difficult, it is another alternative.

AUTHORS' PREFERRED METHOD OF TREATMENT OF CORACOID FRACTURES

Without Acromioclavicular Dislocation

Because the coracoclavicular and acromioclavicular ligaments remain intact in this injury, the fracture is stable. Therefore, treatment with a sling for comfort is sufficient. Pendulum exercises should be encouraged to prevent loss of motion to the shoulder. However, overhead elevation is restricted for 4 to 6 weeks to allow healing to occur at the base of the coracoid process.

Criteria for Return to Athletics. The athlete may return to competition following complete healing of the fracture and return of a full, painless range of motion. This usually requires in the neighborhood of 6 to 10 weeks.

With Acromioclavicular Dislocation

When a coracoid process fracture is accompanied by a severely displaced acromioclavicular disloca-

tion, open reduction with internal fixation is indicated. Fixation options are limited because of the fracture of the coracoid process. Therefore, in spite of the small risk of acromioclavicular arthritis, fixation in the form of transarticular smooth pins is indicated. The pins are removed at 6 to 8 weeks when x-rays reveal healing of the fracture. Before the pins are removed, the patient is not permitted to raise his arm overhead.

Criteria for Return to Athletics. Following pin removal, fracture healing, and restoration of a full, pain-free range of motion, the athlete may return to competition. This normally requires 8 to 12 weeks.

Fractures of the Distal Coracoid Process

Treatment for these fractures is symptomatic. The majority of patients become asymptomatic in 8 to 10 weeks regardless of the presence or absence of nonunion. Should a symptomatic nonunion result, excision of the fragment and reattachment of the pectoralis minor or conjoined tendons to the residual coracoid stump is curative.

Criteria for Return to Athletics. The athlete can return to competition when pain allows a full range of motion. This normally occurs approximately 4 to 6 weeks following injury.

POSTOPERATIVE MANAGEMENT AND REHABILITATION

Because of the important strong muscular and ligamentous attachments to the coracoid process, rehabilitation even after conservative treatment needs to follow a relatively delayed course. To do otherwise risks aggravating the symptoms and incurring an even lengthier delay in returning to sport as well as the distinct possibility of a symptomatic coracoid nonunion. A general conditioning program may be reasonably maintained, but specific shoulder mobilization, particularly strengthening, must be resisted according to the individual's progress until symptoms show continuing resolution and radiologic progress is advanced. This is a condition that does not allow great variance in individual treatment preferences. As previously discussed, perhaps the greatest controversy concerns the place of surgery for coracoid fracture associated with acromioclavicular dislocation. The authors are unable to make a single firm recommendation on this point because the decision is strongly based on the athlete's likely future demands on the shoulder.

References

1. Benchetrit, E., and Friedman, B. Fracture of the coracoid process associated with subglenoid dislocation of the shoulder: A case report. *J Bone Joint Surg* 61A:295–296, 1979.
2. Benton, J., and Nelson, C. Avulsion of the coracoid process in an athlete: Report of a case. *J Bone Joint Surg* 53A:356–358, 1971.
3. Bernard, T. N., Jr., Brunet, M. E., and Haddad, R. J., Jr. Fractured coracoid process in acromioclavicular dislocations. Report of four cases and review of the literature. *Clin Orthop* 175:227–232, 1983.
4. Boyer, D. W., Jr. Trapshooter's shoulder: Stress fracture of the coracoid process. Case report. *J Bone Joint Surg* 57A:862, 1975.
5. Derosa, G. P., and Kettlekamp, D. B. Fracture of the coracoid process of the scapula. Case report. *J Bone Joint Surg* 59A:696–697, 1977.
6. Froimson, A. I. Fracture of the coracoid process of the scapula. *J Bone Joint Surg* 60A:710–711, 1978.
7. Garcia-Elias, M., and Salo, J. M. Nonunion of a fractured coracoid process after dislocation of the shoulder. A case report. *J Bone Joint Surg* 67B:722, 1985.
8. Hall, R. H., Isaac, F., and Booth, C. R. Dislocations of the shoulder with special reference to accompanying small fractures. *J Bone Joint Surg* 41A:489–494, 1959.
9. Kohler, A., and Zimmer, E. A. *Borderlands of the Normal and Early Pathologic in Skeletal Roentgenology* (3rd Am. ed.). New York, Grune & Stratton, 1968, pp. 156–159.
10. Kopecky, K. K., Bies, J. R., and Ellis, J. H. CT diagnosis of the coracoid process of the scapula. *Comput Radiol* 8:325–327, 1984.
11. Lasda, N. A., and Murray, D. G. Fracture-separation of the coracoid process associated with acromioclavicular dislocation: Conservative treatment—a case report and review of the literature. *Clin Orthop* 134:222–224, 1978.
12. McLaughlin, H. L. *Trauma*. Philadelphia, W. B. Saunders, 1959, p. 239.
13. Mariani, P. P. Isolated fracture of the coracoid process in an athlete. *Am J Sports Med* 8:129–130, 1980.
14. Montgomery, S. P., and Loyd, R. D. Avulsion fracture of the coracoid epiphysis with acromioclavicular separation: Report of two cases in adolescents and review of the literature. *J Bone Joint Surg* 59A:963–965, 1977.
15. Protass, J. J., Stampfli, F. V., and Osmer, J. C. Coracoid process fracture diagnosis in acromioclavicular separation. *Radiology* 116:61–64, 1975.
16. Rockwood, C. A., Jr. Dislocations about the shoulder. *In* Rockwood, C. A., Jr., and Green, D. P. (Eds.), *Fractures in Adults* (2nd ed.). Philadelphia, J. B. Lippincott, 1984, pp. 719–721.
17. Rounds, R. C. Isolated fracture of the coracoid process. *J Bone Joint Surg* 31A:662–663, 1949.
18. Rowe, C. R. Fractures of the scapula. *Surg Clin North Am* 43:1565–1571, 1963.
19. Sandrock, A. R. Another sports fatigue fracture. Stress fracture of the coracoid process of the scapula. *Radiology* 117:274, 1975.
20. Smith, D. M. Coracoid fracture associated with acromioclavicular dislocation. A case report. *Clin Orthop* 108:105–167, 1975.
21. Urist, M. R. Complete dislocations of the acromioclavicular joint. *J Bone Joint Surg* 28:813–837, 1946.
22. Wolf, A. W., Shoji, H., Chuinard, R. G. Unusual fracture of the coracoid process. A case report and review of the literature. *J Bone Joint Surg* 58A:423–424, 1976.
23. Wong-Pack, W. K., Bobechko, P. E., and Becker, E. J. Fractured coracoid with anterior shoulder dislocation. *J Can Assoc Radiol* 31:278–279, 1980.
24. Zilberman, Z., and Rejovitzky, R. Fracture of the coracoid process of the scapula. *Injury* 13:203–206, 1981.

SNAPPING SCAPULA

Frances R. Lyons, M.D.
Charles A. Rockwood, Jr., M.D.

Snapping scapula refers to a tactile-acoustic phenomenon originating from the scapulothoracic articulation. It may be symptomatic or not. The term probably encompasses a number of entities ranging from physiologic phenomena through ill-defined muscular, ligamentous, or bursal conditions to distinct entities leading to bony incongruity at the scapulothoracic articulation. In the latter setting, the sequence of identifiable architectural abnormalities leading to pathologic scapulothoracic biomechanics is easily understood, and in general management of the problem is undisputed. However, the snapping scapula in the absence of skeletal problems remains a perplexing and controversial problem. In this situation, diagnosis may be imprecise and treatment largely empirical.

The number of case reports theorizing about snapping scapula that appeared in the late nineteenth and early twentieth century French, German, and Italian literature suggests that it is a condition that provoked a good deal of curiosity. Milch and Burman [8] in 1933 collated and summarized most of those reports. Apart from a single case report of snapping scapula secondary to an inferior angle exostosis by McWilliams [3] in 1914, their paper was the first to address the subject in depth in the English literature. Since then, Milch has made ongoing reports of his experience with snapping scapula [5–7]. Other than these, discussion of the topic since the paper of Milch and Burman has been scant [2, 13].

Milch and Burman's original paper provides interesting historical insights into the interest snapping scapula aroused prior to that time. It is apparent that this puzzling condition was addressed with zeal by some investigators. They indicate that no fewer than 38 reports on the topic appeared between the first report of Boinet in 1867 and theirs. Their report also indicates some skepticism on the part of some investigators about the significance of the condition, and, indeed, this attitude may persist in the orthopaedic community today. For example, they mention Lotheissen (1908) as follows: "Indeed, Lotheissen, [who] was interested himself in the mechanism of production of these sounds and who, by practice, was able to develop them in himself, naively remarked that 'with time, one acquires a certain virtuosity in producing noises.' "

Milch and Burman acknowledged a variety of audible or palpable phenomena arising from the scapulothoracic articulation. These included a soft friction sound arising from the normal muscle action that should be considered physiologic. They considered a loud, often painful snapping sound a clear indication of underlying mechanical abnormalities. Between these extremes is an intermediate group of less severe palpable or audible clicking, popping, snapping, or grating noises. They suggested that this group of sounds is probably significant, but in the absence of demonstrable pathology and especially in the absence of pain, this conclusion remains controversial.

Many mechanisms of scapula snapping have been proposed with most fitting into the two broad categories proposed by Milch [6]. These categories are (1) changes in the congruence of the anterior scapular surface and the underlying chest wall, and (2) changes in the interposed soft tissues, muscles, or bursae between the scapula and the chest wall.

The classic mechanism of snapping in the first category is the subscapular exostosis or osteochondroma [10]. These osteochondromas are generally sessile with an overlying fluid-filled bursa and snap over the underlying ribs in a ratchetlike fashion. Exostoses less commonly arise from the ribs. Rib deformity after fracture or associated with scoliosis has also been suggested as a cause of snapping scapula. Malunited fractures of the body of the scapula are also a consideration.

It was Milch who proposed abnormal anterior angulation of the superomedial angle of the scapula as the source of incongruence, and it was this concept that formed the basis of the superomedial excision of the scapula that he popularized. It is likely that anterior angulation of the superomedial angle is a normal variation. However, distinctly abnormal shapes of this region of the scapula have been associated with the Sprengel deformity and the tubercle of Luschka. This latter condition has been thought to be an anatomic variant, but Milch considered it a form of osteochondroma that happens to arise on the anterior surface of the superomedial scapular angle [8]. Bateman [1] suggested that bony spurring of the superomedial angle may be a response to forceful repetitive shoulder action. He theorized that minute periosteal tears lead to the proliferation of a small beaklike process, which in turn leads to snapping. Michele and colleagues [4] suggested that the scapular-thoracic incongruity may result from round shoulders or poor posture leading to scapular sagging and what they termed the scapulo-costal syndrome.

The role of pathology in the interposed soft tissues, muscles, or bursae in the causation of snapping scapula is more difficult to assess. Again Milch has been the major contributor to this area. He suggested that this group of entities may actually be the most common cause of snapping scapula. He postulated that the pathology may be an "interstitial myofibrosis" leading to inadequate scapular cushioning. He supported this theory with the finding of "interfascicular fibrosis" in operative specimens following resection of the superomedial angle of the scapula [5]. Along similar lines, Rockwood [11] found a fibrotic flap that rode up and down under the vertebral border of the scapula with motion. The patient had suffered a fall and may have avulsed part of the rhomboid major insertion. Resection of the flap eliminated the snapping.

"Bursitis" in the absence of bony pathology has been advanced as either a cause or an entity associated with snapping scapula and may be a cause of scapulothoracic problems in the athlete. Milch [6] described the presence of three normal adventitious bursae about the scapula. These are located at the superior angle between the serratus anterior and subscapularis, between the serratus anterior and thoracic wall, and deep to the inferior angle of the scapula. O'Donoghue [9] suggested that inflammation of a bursa, especially that between the serratus anterior and the chest wall, may be a source of pain and crepitus in athletes. More recently, Sisto and Jobe [12] reported on "scapulothoracic bursitis" in professional baseball pitchers. Their series comprised four athletes said to have tender palpable masses at the inferior angle of the scapula. In each case, the excised mass proved to be a chronically inflamed bursa with a dense fibrous wall. One of these patients had an audible click at the inferior angle of the scapula that was attributed to the bursal inflammation.

CLINICAL FEATURES OF SNAPPING SCAPULA

History

Bearing in mind the mixed etiology of snapping scapula, there may be great diversity in the history and clinical findings. The condition may arise in any age group but is most commonly encountered in adolescents and young adults. The onset of snapping may be spontaneous or may follow a specific injury, major or trivial. A relationship to repetitive activities involving forcible setting of the scapula to the chest wall or wide scapulothoracic excursion such as occurs in pitching may be evident. The patient is often able to reproduce the snapping or grating phenomenon at will with particular maneuvers.

Snapping may or may not be painful, but in athletes in whom an etiology of bursal inflammation is most likely, pain is usual. The series of Sisto and Jobe [12] indicated that pain was experienced during the early and late cocking phases and during the acceleration phase of pitching. Pain was alleviated with the follow-through. Scapular prominence is a rare complaint and is invariably associated with an underlying bony mass.

Physical Examination

The source of symptoms, be it pain or an audible or palpable phenomenon, can usually be sharply localized to a discrete area about the scapula. In searching for this site, the patient should be in-structed to perform the maneuver that reproduces the discomfort. This will also give the examiner an idea of the character of the snapping and thus whether a bony incongruity or soft tissue etiology is more likely.

Rarely, there will be a noticeable prominence of part of the scapula, sometimes to the point where the scapula appears winged. Long thoracic nerve function will be intact. This deformity is almost invariably associated with a coarse grating noise or sensation and an underlying subscapular exostosis or osteochondroma.

If the snapping is originating from the periscapular or subscapular soft tissue or bursal inflammation, then physical findings may be few and subtle. Milch's "interfascicular fibrotic" phenomenon was most often found at the superomedial angle. Scapulothoracic bursitis is also encountered in this location as well as the inferior angle. Sisto and Jobe [12] reported that the swelling of the bursitis in this latter location was palpable. The authors suggest that palpation of such a swelling may be difficult in the muscular athlete.

Diagnostic Studies

Studies are directed toward establishing whether or not there is a bony congruency problem. If this is excluded, one may reasonably assume a soft tissue etiology. The plain anteroposterior scapular x-ray is generally normal. Even large subscapular exostoses may be overlooked on this view. The scapular lateral x-ray is crucial. Particular attention should be directed toward the inferior angle, a common site for exostosis (Fig. 15B–82), and the superior angle, where abnormal anterior angulation, exostosis, or spurring due to repetitive muscular microtrauma (Fig. 15B–83) may be seen. If doubt exists about the significance of the findings, comparison with the scapular lateral view of the asymptomatic shoulder may aid interpretation.

CT scanning may assist in further defining any bony lesions noted on plain radiographs (Fig. 15B–84); however, isotopic bone scanning has been found to be of no assistance [2].

TREATMENT OPTIONS

If a bone lesion such as a scapular or costal exostosis is diagnosed and symptoms are sufficiently disturbing, excision of the offending mass is generally simple and curative [9]. If a discrete bony problem is not evident, then diagnosis, to say noth-

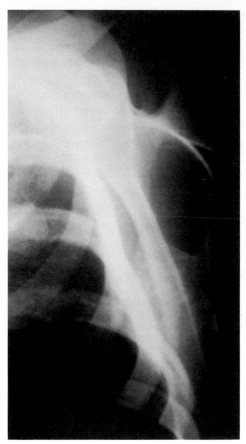

FIGURE 15B–82
Scapular lateral radiograph reveals an exostosis of the inferior angle of the scapula in this patient with a snapping scapula.

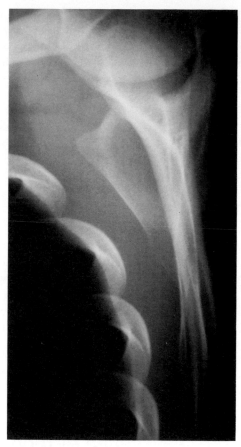

FIGURE 15B–83
Prominence of the superomedial angle of the scapula is demonstrated on the scapular lateral radiograph of this patient with a snapping scapula.

ing of treatment, is much more difficult. One needs to know the patient well to gauge how much of a problem scapular snapping really is. Caution is needed, especially in the patient with painless, seemingly habitual snapping and particularly when minor trauma, worker's compensation, or litigation is involved [10]. Emphasis in these cases should be upon rehabilitation by way of strengthening the periscapular stabilizing muscles.

When discomfort or disability is more severe and no bony problem can be identified, several authors [2, 6] have advocated empirical resection of that portion of the scapula related to the snapping, grating, or grinding. This is most often the superomedial angle or the vertebral border. In these cases the excised bone is normal [2], but Milch has reported "interfascicular fibrosis" in the associated muscle insertions [5]. Periscapular symptoms in the athlete are more commonly bursal in origin secondary to forceful repetitive stresses occurring between the anterior surface of the scapula and the underlying thoracic wall. Rest, ice, local heat, stretching, and strengthening shoulder exercises, anti-inflam-

matory medications, and cortisone injections have all been recommended for this condition [6, 9, 12]. O'Donoghue [9] found bursal infiltration with local anesthetic to be of diagnostic assistance. The need for surgical treatment of scapulothoracic bursitis is rare, but Sisto and Jobe [12] found it necessary in

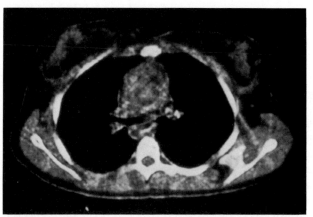

FIGURE 15B–84
CT scan of both shoulders of a patient with a left snapping scapula. Notice the exostosis of the inferior angle.

four professional pitchers after conservative measures failed. Duration of symptoms averaged 18 months prior to surgery, and all patients responded well to surgical excision of a chronically inflamed bursa at the inferior angle of the scapula.

Excision of bony or bursal lesions is generally achieved simply using a muscle-splitting approach. However, particular care should be exercised in the region of the supramedial angle of the scapula to avoid injury to the spinal accessory nerve.

POSTOPERATIVE MANAGEMENT AND REHABILITATION

Postoperative care should include a specific program to maintain full glenohumeral and elbow motion as well as early restoration of scapulothoracic mobility. Specific exercises addressing the periscapular stabilizers are also important. Sisto and Jobe [12] recommend that this program begin 1 week after bursal excision, and they allow the throwing athlete to resume gentle tossing 6 weeks postoperatively and then to return to full speed as tolerated.

AUTHORS' PREFERRED METHOD OF TREATMENT

In the absence of a clearly demonstrable exostosis arising off the scapula or underlying ribs, the author would urge caution and conservatism in undertaking treatment of the snapping scapula, especially when snapping develops spontaneously or after a minor injury. The author supports conservative treatment that emphasizes scapular stabilizing exercises. These include isometric shoulder shrugging exercises using a weight appropriate to the patient, wall push-ups, knee push-ups, and, eventually, conventional push-ups. When snapping or local scapular pain is attributed to bursitis in the throwing athlete, the author again strongly recommends conservative treatment. This includes the scapula stabilizing exercises, rest from aggravating activities, local moist heat, nonsteroidal anti-inflammatory medications, and, rarely, local infiltration with steroid preparations.

References

1. Bateman, J. E. *The Shoulder and Neck*. Philadelphia, W. B. Saunders, 1972, pp. 187–188.
2. Cameron, H. U. Snapping scapulae: A report of three cases. *Eur J Inflam* 7:66–67, 1984.
3. McWilliams, C. A. Subscapular exostosis with adventitious bursa. *JAMA* 63:1473–1474, 1914.
4. Michele, A. A., Davies, J. J., Krueser, F. J., and Lichtor, J. M. Scapulocostal syndrome (fatigue-postural paradox). *NY State J Med* 50:1353–1356, 1950.
5. Milch, H. Partial scapulectomy for snapping of the scapula. *J Bone Joint Surg* 32A:561–566, 1950.
6. Milch, H. Snapping scapula. *Clin Orthop* 20:139–150, 1961.
7. Milch, H. Partial scapulectomy for snapping of the scapula. *J Bone Joint Surg* 44A:1696–1697, 1962.
8. Milch, H., and Burman, M. S. Snapping scapula and humerus varus: Report of six cases. *Arch Surg* 26:570–588, 1933.
9. O'Donoghue, D. H. *Treatment of Injuries to Athletes*. Philadelphia, W. B. Saunders, 1962, pp. 14–141.
10. Parsons, T. A. The snapping scapula and subscapular exostoses. *J Bone Joint Surg* 55B:345–349, 1973.
11. Rockwood, C. A., Jr. The snapping scapula. *Instr Course Lect*, American Academy of Orthopaedic Surgeons, 1986.
12. Sisto, D. J., and Jobe, F. W. The operative treatment of scapulothoracic bursitis in professional pitchers. *Am J Sports Med* 14:192–194, 1986.
13. Strizak, A. M., and Cowen, M. H. The snapping scapula syndrome: A case report. *J Bone Joint Surg* 64A:941–942, 1982.

FRACTURES OF THE STERNUM

Francis R. Lyons, M.D.
Charles A. Rockwood, Jr., M.D.

Sternal injuries other than contusion resulting from direct injury are rare in sportsmen. Prior to the advent of the automobile, sternal fracture was considered an unusual injury. But recently sternal fracture or fracture-dislocation has become relatively common in automobile accidents. Several authors have documented an almost one-to-one relationship between fractures of the sternum and the number of automobiles on the road [3, 5]. Despite the rarity of sternal fractures, De Tarnowsky [2] in 1905 proposed a mechanistic classification of the injury that was reconsidered in depth by Fowler [4] in 1957 and remains valid today (Fig. 15B–85). The mechanisms proposed were (1) direct, (2) indirect, and (3) muscular.

The last mechanism refers to violent muscular action that essentially tears the sternum into proximal and distal fragments through the action of opposing muscle groups. This type of injury is extremely rare but has been reported secondary to

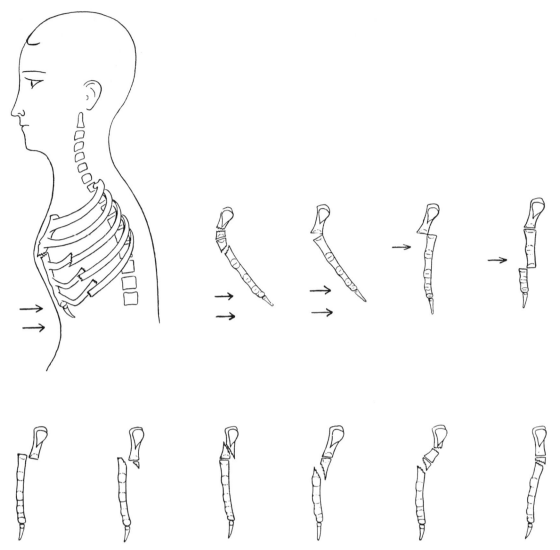

FIGURE 15B–85
Mechanistic classification of sternal fractures. (From Fowler, A. W. Flexion compression injury of the sternum. *J Bone Joint Surg* 39B:487, 1957.)

convulsions and tetanus as well as in a clown performing a back spring [2]! For our purposes this mechanism must be considered a curiosity. The first two mechanisms of injury may have more relevance to the athlete.

As the term suggests, direct sternal injury results from a direct force applied to the sternum leading to contusion or fracture (Fig. 15B–86). Most often the blow is to the body of the sternum in its lower part. The relatively flexible lower portion is displaced inward, and a fracture or sternomanubrial disruption occurs in the stiffer segment cephalad to the applied force. If a direct (and greater) force is applied at a more superior level, the sternum or sternomanubrial junction might be expected to fail at the same level as the applied direct force. The direct mechanism is not infrequently associated with

rib fractures, and in cases of severe trauma the anterior chest wall may be flail.

The indirect or contrecoup mechanism of sternomanubrial injury results from flexion-compression of the cervicothoracic spine. Almost invariably this results in fracture-dislocation of the sternomanubrial joint with posterior and inferior displacement of the manubrium relative to the body of the sternum. The initiating event may be a forced flexion of the cervical spine in the form of a heavy blow to the back of the head or a fall onto the same area. A heavy fall on the buttocks could result in the same forceful flexion of the upper spine. As the upper thoracic spine flexes, the relatively fixed upper two ribs carry the manubrium posteriorly. This posterior displacement may be aided by the chin striking the manubrium as the neck reaches extreme flexion

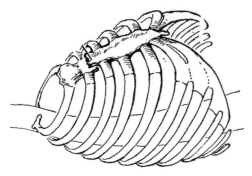

FIGURE 15B–86
A direct force applied to the sternum or manubrium can result in contusion or fracture. (From DePalma, A. F. *The Management of Fractures and Dislocations—An Atlas*, 3rd ed. Philadelphia, W. B. Saunders, 1981.)

(Fig. 15B–87). The body of the sternum, being attached to the more mobile lower ribs, may actually be driven forward by the sudden rise in intra-abdominal and intra-thoracic pressure that accompanies such an injury. This injury is commonly associated with a wedge fracture of the upper thoracic vertebrae [7].

As previously mentioned, sternomanubrial fractures, dislocations, or combinations thereof are rare in athletes. Reports in the literature in this context are extremely scarce, with nearly all discussions being devoted to road trauma. However, Johnson and co-workers [8] recently reported a manubriosternal dislocation in a football halfback. He had sustained multiple direct helmet-chest contacts, which may have been a direct mechanism of injury. Fortuitously, a game film clearly revealed an acute cervical spine flexion with the chin contacting the manubriosternal junction. This evidence, as well as the typical posterior manubrial displacement relative to the body, favored an indirect mechanism (Fig. 15B–88). De Tarnowsky [2] reported cases of an acrobat and a gymnast, each of whom fell onto the occiput, suffering similar "indirect" injuries. This mechanism may also have accounted for the sternal fracture that occurred in a diver entering shallow water with the arms outstretched, the force being transmitted along the clavicles and driving the manubrium posteriorly. Santos [10] mentioned a cyclist who sustained a transverse fracture of the middle third of the sternum after a fall. However, further details of the mechanism of injury were not given.

The author has been able to locate only one report of a sternal fracture resulting from a probable direct mechanism in a remotely athletic endeavor [6]. This concerned a professional rodeo bullrider who was stomped in the chest by a bull! It was of interest that this fracture of the body of the sternum progressed to pseudoarthrosis that required bone grafting and wire fixation.

CLINICAL FEATURES OF FRACTURES OF THE STERNUM

History

The mechanism of a patient's injury such as a direct blow to the anterior chest wall or forced flexion of the cervical thoracic junction may arouse suspicion about the possibility of a sternal injury. There will be a complaint of anterior chest wall pain that is aggravated by inspiration. This may be a feature distinguishing this injury from simple sternal contusion. Crepitus with respiratory excursion or deformity of the anterior chest wall may be a complaint in displaced fractures. Upper thoracic pain could also be a symptom with injuries involving a contrecoup mechanism. Shortness of breath immediately following injury is the usual response to a severe blow to the chest. However, this symptom rarely persists more than a few minutes.

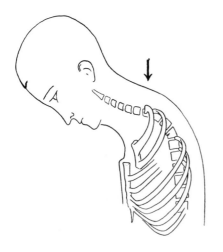

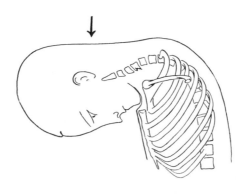

FIGURE 15B–87
A fall or blow to the upper part of the spine *(A)* or to the back of the head and neck *(B)* can result in sternal fracture. Posterior displacement of the manubrium can be exaggerated by a blow from the chin as the neck is hyperflexed *(B)*. (From Fowler, A. W. Flexion compression injury of the sternum. *J Bone Joint Surg* 39B:487, 1957.)

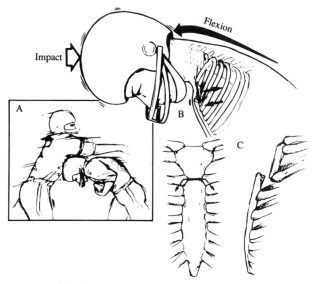

FIGURE 15B–88
Sternomanubrial dislocation caused by hyperflexion of the cervical spine combined with axial load. (From Johnson, C. D., Mackenzie, J. W., and Zawadsky, J. P. Manubriosternal dislocation in a football athlete. *Surg Rounds Orthop* 2:45–50, 1989.)

Physical Examination

Invariably there is localized tenderness at the level of the injury, and a step-off may be palpable or even visible with severe displacements. There may be palpable crepitus at the level of the injury with respiratory movements. Intrathoracic injury is not uncommonly encountered with the violent injuries associated with road trauma. This problem is thought to be nonexistent in athletic injuries. Nevertheless, the patient should be screened for the presence of associated rib fractures and clinical evidence of underlying pulmonary contusion, pneumothorax, or vascular injury.

Diagnostic Studies

A properly penetrated lateral radiograph of the chest is the key to a definitive diagnosis (Fig. 15B–89). The fracture will be overlooked in more than 80% of cases if the anteroposterior chest x-ray alone is relied upon [1]. All the same, this view must be obtained with regard to possible pneumothorax or mediastinal injury. Because in athletic injuries we are usually dealing with young individuals, it is of considerable importance to recognize the common variations in the time of fusion of the four segments of the body of the sternum [9]. At about the seventh year the third and fourth segments fuse. The second and third fuse at about age 14, but the first and second segments remain unfused until about age 21.

Fusion of these segments may be delayed at any stage, which may easily lead to a wrong diagnosis of fracture.

Electrocardiographic evidence of myocardial contusion has been reported in as many as 50% of patients with sternal fractures [1]. Again, such studies refer almost exclusively to patients with road trauma. Even so, this investigation seems a wise precaution in the athlete with such an injury.

TREATMENT OPTIONS

Sternal contusions and most sternal fractures may be treated conservatively by means of rest and analgesia until symptoms subside. Contact sports should be avoided until this occurs. It is unusual for symptoms to persist beyond 6 to 12 weeks, even in patients with more severe injuries.

If the fracture or manubriosternal dislocation is markedly displaced, typically with the body fragment lying anterior to the manubrium, a postural reduction is usually readily obtained. The patient is positioned supine with a sandbag placed transversely under the shoulders, slightly below the level of the spines of the scapulae (Fig. 15B–90). Reductions so obtained are generally stable, and treatment may be continued in the same way as that used for undisplaced fractures. Cases requiring reduction under anesthesia, or even open reduction, internal

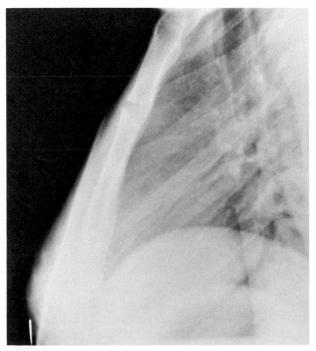

FIGURE 15B–89
Lateral radiograph of the chest revealing a sternomanubrial dislocation.

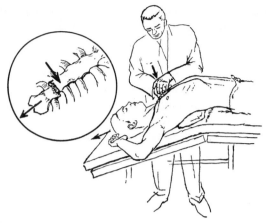

FIGURE 15B–90
Method of reduction of fractured sternum. If necessary, an assistant may apply traction by grasping the patient's arm at the axilla and pulling cephalad. Insert shows detail of the forces applied in reduction. (From DePalma, A. F. *The Management of Fractures and Dislocations–An Atlas*, 3rd ed. Philadelphia, W. B. Saunders, 1981.)

fixation, and bone grafting as in the case of a footballer with an unstable sternomanubrial dislocation reported by Johnson and colleagues [8], must be considered exceptional in the athlete.

AUTHORS' PREFERRED METHOD OF TREATMENT

The authors have no personal experience with sternal injury in the athletic setting other than simple contusion, thus reflecting the rarity of the injury.

There appear to be no further special recommendations to be made with regard to rehabilitation after this injury other than avoidance of aggravating activities and contact sport. There is probably no reason why a general conditioning or training program could not be continued within the limits of discomfort posed by the injury.

References

1. Buckman, R., Trooskin, S. Z., Flancbaum, L., and Chandler, J. The significance of stable patients with sternal fractures. *Surg Gynaecol Obstet* 164:261–265, 1987.
1A. DePalma, A. F. *The Management of Fractures and Dislocations—An Atlas* (3rd ed.). Philadelphia, W. B. Saunders, 1981.
2. DeTarnowsky, G. Contrecoup fracture of the sternum. *Ann Surg* 41:253–264, 1905.
3. Foley, N. T., and Mattox, K. L. Fractures of the sternum. *Curr Concepts Trauma Care* Fall:9–11, 1985.
4. Fowler, A. W. Flexion-compression injury of the sternum. *J Bone Joint Surg* 39B:487–497, 1957.
5. Helal, B. Fracture of the manubrium sterni. *J Bone Joint Surg* 46B:602–607, 1964.
6. Hensinger, R. N., and Berkoff, H. A. Traumatic nonunion of the sternum: Report of a case. *J Trauma* 15:159–162, 1975.
7. Jenyo, M. S. Post-traumatic fracture-dislocation of the manubriosternal joint with a wedge fracture of the body of the fourth thoracic vertebra. *J Trauma* 25:274–275, 1985.
8. Johnson, C. D., Mackenzie, J. W., and Zawadsky, J. P. Manubriosternal dislocation in a football athlete. *Surg Rounds Orthop* 2:45–50, 1989.
9. O'Donoghue, D. H. *Treatment of Injuries to Athletes.* Philadelphia, W. B. Saunders, 1962, pp. 313–318.
10. Santos, G. H. Treatment of displaced fractures of the sternum. *Surg Gynaecol Obstet* 166:272–274, 1988.

FRACTURES OF THE RIB

Francis R. Lyons, M.D.
Charles A. Rockwood, Jr., M.D.

Rib fractures in the athlete are not particularly common. Essentially they can be divided into two broad categories. Those arising from a direct blow to the thoracic wall are largely self-explanatory. The blow, if from a blunt object, may result in a fracture of one or perhaps two ribs at the exact site of impact. Alternatively, the blow may be applied over a broader area, causing compression of the chest in one of its dimensions. This compression occurs most commonly in the anteroposterior plane, leading to lateral and often multiple rib fractures. Direct trauma may less commonly disrupt the costochondral junction rather than the rib proper [5].

The second category of rib fracture consists of what might be termed indirect injuries. These are the rib fractures that arise from violent or repetitive muscular action. This mixed group includes the so-called stress fracture of the first rib and avulsion fractures of the anterior parts of the lower or "floating" ribs. The latter condition has been reported in baseball pitchers by Tullos and colleagues [8]. They theorized that this injury may arise from the opposing pull of the latissimus dorsi, internal oblique, and serratus posterior inferior, all of which are attached to the lower ribs (Fig. 15B–91). This theory is supported by the observation of similar spontaneous fractures in the lower ribs of heavy laborers [3].

Fractures of the first rib have provoked consid-

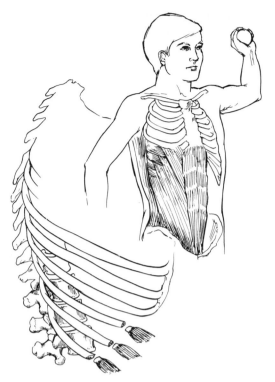

FIGURE 15B–91
Avulsion fractures of the floating ribs may result from the opposing pulls of the latissimus dorsi, internal obliques, and serratus posterior inferior. (From Tullos, H. S., Erwin, W. D., Woods, G. W., et al. Unusual lesions of the pitching arm. *Clin Orthop Rel Res* 88:169–182, 1972.)

erable interest in the literature. In general orthopaedics this injury is considered an indicator of major trauma and the possibility of serious underlying pulmonary or neurovascular damage. However, in athletes the injury is relatively benign in that massive direct trauma is not the initiating event. Rather, it may follow apparently trivial injury or repetitive activity such as pitching. That it may occur with or without preceding symptoms suggests that this injury may be a form of stress fracture (Fig. 15B–92). It has also been reported secondary to sudden muscular exertion, suggesting that fracture of the first rib in the athlete may comprise a spectrum of stress fracture, indirect acute trauma due to opposing muscular forces, or a combination of both.

Whatever the mechanism, a fracture of the first rib appears invariably to pass through the broad, flat, thin portion of the first rib in the region of the subclavian groove. Curran and Kelly [1] have proposed that the action of the scaleni opposed to the intercostals and upper part of the serratus anterior leads to failure of the rib at this point. Such first-rib fractures have most often been reported in baseball pitchers [1, 2, 4, 8]. The rib associated with either the pitching or the nonpitching extremity may be involved. It has also been encountered as a sudden

event without preceding symptoms in basketball players reaching for rebounds, hence the term rebound rib [7].

A similar, probably stress-related, phenomenon has been observed in fractures of the lower rib in golfers. Rasad [6] reported three cases of spontaneous rib fracture in novice golfers. The fractures were near the posterolateral angle and occurred in pairs ranging from the fourth through the seventh ribs. They were always on the side of the chest wall opposite the patient's handedness. He suggested that the mechanism may involve excessive muscular forces secondary to faulty technique.

CLINICAL FEATURES OF RIB FRACTURES

History

If rib fracture is secondary to a single blow, the patient invariably can recall the event causing it such as a heavy tackle or being crushed in a player pile-up. His "wind is knocked out," and initially he struggles for breath. Attempts at deep inspiration provoke severe local pain and muscle spasm, which is a feature that distinguishes this injury from simple

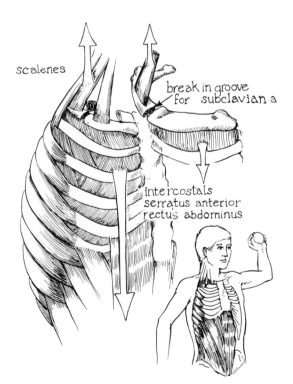

FIGURE 15B–92
Activities such as pitching may result in a stress fracture of the first rib. (From Tullos, H. S., Erwin, W. D., Woods, G. W., et al. Unusual lesions of the pitching arm. *Clin Orthop Rel Res* 88:169–182, 1972.)

rib contusion. The stress type of fracture typically seen in the first rib will have a somewhat more variable history. Direct trauma is not involved. There may be an insidious onset of aching, but more often there are no premonitory symptoms, and fracture occurs quite suddenly during sporting activity, frequently with a snap, sometimes audible, in the shoulder or root of the neck [2, 4, 7]. The accompanying pain is then severe and is aggravated by deep inspiration and upper limb movements. The complaint of sudden shoulder pain and a snap must be distinguished from symptoms typical of glenohumeral subluxation.

Physical Examination

In cases of direct trauma to the ribs there is sharply localized tenderness at the site of fracture. Compression of the chest in an anteroposterior plane may aggravate pain in patients with a lateral rib fracture. Pneumothorax as the result of a rib fracture is rare in sports. Still, the clinical features of pneumothorax should be sought, including palpation for subcutaneous emphysema. A displaced rib fracture may injure the intercostal vessels running along the inferior border of the rib, in which case there may be marked local swelling and hematoma formation. If the costochondral junction has fractured anteriorly, it is occasionally possible to

palpate a step if the injury is displaced or feel a click with inspiration. Hepatic, splenic, or renal injury should always be kept in mind in patients with injury to the lower ribs.

Physical findings accompanying stress fracture of the first rib are few. There may be some discomfort during deepest inspiration or with palpation deep in the root of the neck. Because shoulder pain is a common complaint with this injury, glenohumeral assessment is important.

Diagnostic Studies

Clinical suspicion of rib fracture is generally confirmed by plain radiographs, the one exception being fracture of the costochondral junction. Fractures in the region of the anterior or posterior portions of the ribs are best seen on plain anteroposterior chest x-ray. Clear visualization of fractures at the lateral angle may require oblique films. The stress-type fracture of the first rib is also usually visualized readily on radiographs. The fracture may be partially obscured by the clavicle or the superimposed posterior part of the second rib. This may necessitate repeating the films using a slight caudal or cephalic tilt of the x-ray beam as appropriate. Should any doubt remain, isotopic bone scanning of the symptomatic first rib fracture is confirmatory [2, 4].

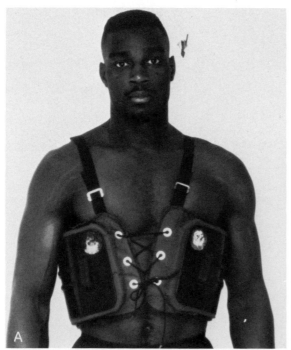

FIGURE 15B–93
"Flak" jacket used to protect the athlete with rib fractures.

TREATMENT OPTIONS

Rib fracture from direct trauma in the athlete generally requires symptomatic treatment alone. Even a small pneumothorax or hemothorax may be managed expectantly. However, patients with hemothorax should be admitted to the hospital for clinical and radiologic monitoring of the condition. Otherwise, the priority is pain relief. Intercostal nerve block may be useful if pain and associated muscle spasm are severe. A long-acting agent such as bupivacaine, infiltrating not only the intercostal nerve of the fractured rib but also those immediately above and below it, is most effective. The value of a chest binder for rib fracture seems to depend upon the individual preference of the physician. A chest binder may provide a splinting effect, especially in lower rib fractures. Nonunion is not a complication of this type of traumatic fracture, and malunion is of little importance. Severe pain usually settles within about a week of the fracture, and light training may be resumed as discomfort allows.

Displaced fractures of the costochondral junction anteriorly may be more troublesome. Union of this injury is slow, and symptoms are correspondingly more persistent. A pseudoarthrosis may be the disabling outcome, and the occasional need for its excision has been reported [5].

Fracture of the first rib is different in that the pain produced by the injury is generally significant only if the patient attempts to resume sporting activity. Between training periods there may be little or no discomfort. The only treatment needed is rest or perhaps even a sling if shoulder movement aggravates the discomfort. There is much individual variation in the time needed for resolution of symptoms. This period demands much patience from the athlete. Pseudoarthrosis is relatively common with this fracture but is consistently reported as painless [2]. Even so, symptoms may recur if return to sport is premature. Rare complications such as Horner's syndrome or subclavian aneurysm have been seen after fractures of the first rib involving major direct trauma. However, these complications are unheard of in the athletic setting.

AUTHORS' PREFERRED METHOD OF TREATMENT

Assuming that the rib fracture has not been accompanied by a concomitant injury such as a pneumothorax, hemothorax, Horner's syndrome, or subclavian artery injury, fractures of the rib are treated symptomatically with oral analgesics.

Criteria for Return to Athletics. Frequently, rib fractures become less sensitive to stress at approximately 2 weeks postinjury. Normally, the athlete can return to play at approximately the 2-week mark using a rib belt or chest binder. Intercostal blocks are a useful adjunct at this point. The rib cage is routinely protected for 6 to 8 weeks using a "flak" jacket (Fig. 15B–93).

References

1. Curran, J. P., and Kelly, D. A. Stress fracture of the first rib. *Am J Orthop* 8:16–18, 1966.
2. Gurtler, R., Pavlov, H., and Torg, J. S. Stress fracture of the ipsilateral first rib in a pitcher. *Am J Sports Med* 13:277–279, 1985.
3. Horner, D. B. Lumbar back pain arising from stress. Fracture of the lower ribs. *J Bone Joint Surg* 46A:1553–1556, 1964.
4. Lankenner, P. A., Lyle, J. M. Stress fracture of the first rib. *J Bone Joint Surg* 67A:159–160, 1985.
5. O'Donoghue, D. H. *Treatment of Injuries to Athletes.* Philadelphia, W. B. Saunders, 1962, pp. 307–313.
6. Rasad, S. Golfer's fractures of the ribs. Report of three cases. *Am J Roentgenol* 120:901–903, 1974.
7. Sacchetti, A. D., Beswick, D. R., and Morse, S. D. Rebound rib: Stress-induced first rib fractures. *Ann Emerg Med* 12:177–179, 1983.
8. Tullos, H. S., Erwin, W. D., Woods, G. W., Wukasch, D. C., Cooley, D. A., and King, J. W. Unusual lesions of the pitching arm. *Clin Orthop* 88:169–182, 1972.

Glenohumeral Instability

Michael J. Pagnani, M.D.
Brian J. Galinat, M.D.
Russell F. Warren, M.D.

Shoulder pain is common in the athletic population. The high loads generated at the glenohumeral joint in throwing, swimming, and racquet sports frequently lead to pathologic changes in athletes who repetitively perform these activities. The large forces directed upon the shoulder in contact sports are also a prominent cause of injury. Although abnormal anterior laxity of the glenohumeral joint is the most common type of instability, posterior and multidirectional instabilities are increasingly recognized as important causes of shoulder disability. The signs and symptoms associated with these types of instability are often more subtle than those associated with the unidirectional anterior type, and the results of treatment have often been less satisfactory. Recent advances in understanding of the pathomechanics of the unstable shoulder have confirmed that no single "essential lesion" is responsible for glenohumeral instability. Pathology varies with the degree and type of instability, and different anatomic structures play stabilizing roles as the position of the shoulder changes. Attempts at treatment should specifically address the pathology encountered in a particular case, and stability should not be obtained at the expense of function. An approach to the diagnosis and treatment of shoulder instability based on our current knowledge of the pathomechanics of the shoulder is presented in this section.

BASIC SCIENCE OF SHOULDER STABILITY

The glenohumeral joint is notable for its freedom from constraint, which endows it with considerable mobility. The shoulder has the greatest range of motion of all the joints in the human body [61]. Bony restraints to motion are minimal. Only a small portion of the relatively large humeral head contacts the glenoid fossa in any shoulder position [16, 122]. Although the articulating portions of the humeral head and glenoid are quite congruous [40], the shoulder joint does not act in a strictly ball-and-socket fashion. When the arm is adducted, the area of contact between the articular surfaces is represented by the "bare area" of the glenoid. Articular contact becomes more peripheral at 90 degrees of abduction. At the extremes of motion, rotation of the humeral head is coupled with its translation upon the glenoid [53, 68, 69, 125, 171, 172]. For instance, with extension and external rotation, the head translates posteriorly. This translation may be lost in the unstable shoulder. Two factors contribute to the "rollback" phenomenon: the development of tension in the periarticular soft tissues and the fact that the articular surfaces of the humeral head and the glenoid fossa are not perfectly congruous. Normally, translation of the head on the glenoid is confined to only a few millimeters [20, 53, 69, 125, 126, 171, 174]. The mobility of the scapula is, to some degree, related to this remarkable stability. Motion of the scapula allows the glenoid to adjust to changes in arm position.

It is the surrounding soft tissue envelope that primarily confers stability, within a finite area, to the normal glenohumeral joint. This stability is due both to the static effect of ligaments and tendons and to dynamic mechanisms associated with muscular contraction. It is helpful to consider the surrounding soft tissues as a four-layered entity [33] (Fig. 15C–1). The most superficial layer is composed

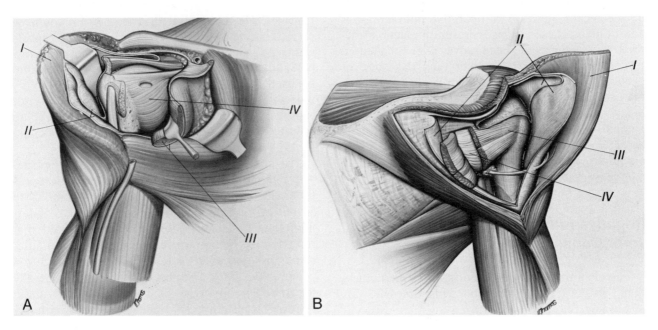

FIGURE 15C–1

Supporting layers of the glenohumeral joint. *A*, Anterior structures: I, deltoid, II, clavipectoral fascia, III, subscapularis, IV, anterior capsule. The defect in layer IV represents the "rotator interval." *B*, Posterior structures: I, deltoid, II, posterior scapular fascia, III, infraspinatus, teres minor, IV, posterior capsule. (From Cooper, D. E., O'Brien, S. J., and Warren, R. F. Supporting layers of the glenohumeral joint. *Clin Orthop* (in press.)

of the deltoid and pectoralis major muscle bellies and their corresponding fascias. The second layer is composed anteriorly of the clavipectoral fascia, the conjoined tendon of the coracobrachialis and the short head of the biceps, and the coracoacromial ligament. The posterior scapular fascia is the posterior component of the second layer. The third layer is formed by the rotator cuff, and the deepest layer is represented by the shoulder capsule (including the glenohumeral ligaments and the glenoid labrum). All four layers operate in a complex interaction to stabilize the joint, and pathology that interferes with this intricate system leads to instability.

Static Mechanisms

Bony Joint Conformity

The corresponding articular surfaces of the humeral head and glenoid are very nearly congruous in the normal situation [40]. The humeral head is retroverted an average of 30 degrees with respect to the distal humeral condyles. The scapular plane lies in approximately 35 degrees of anteversion in relation to the coronal plane of the body, and the glenoid is retroverted an average of 7 degrees with respect to the scapular plane [140]. Saha's influential work stimulated investigation of the effect of varia-

tion in glenoid version, size, shape, and tilt on shoulder stability [140]. Patients with a relatively "shallow" glenoid fossa or with an abnormal size or tilt may be at increased risk for instability. Basmajian and Bazant [13] implicated the upward vertical tilt of the inferior portion of the glenoid in limiting inferior translation. Brewer and associates [22] found increased glenoid retroversion in adolescents with posterior instability. On the other hand, in a study that used the posterior border of the acromion as the reference axis for glenoid version, Galinat and colleagues [45, 46] found no association between version and anterior or posterior instability. Theories that attribute instability to abnormal torsion also remain questionable. Most North American surgeons have avoided glenoid osteotomy [14, 148] and rotational humeral osteotomy [141, 179] in the initial surgical treatment of shoulder instability because of a general feeling that the primary disturbance in most cases is unrelated to bony alignment.

A bony defect in the humeral head is commonly associated with glenohumeral instability. With anterior instability, the defect is noted in the posterolateral portion of the head (Hill-Sachs lesion) [42, 65]. In contrast, an anteromedial lesion is often noted with posterior instability (reverse Hill-Sachs lesion). Bony defects of the anterior (or posterior) glenoid rim may also be seen. These bony deficiencies alter the normal mechanics of the joint, and some authors [30, 98, 99, 179] recommend that they be specifically addressed if operative intervention is entertained. These lesions represent impaction fractures that occur as the head moves over the glenoid rim. They are the effect rather than the cause of instability [65]. Townley [161], in cadaveric experiments, showed that the Hill-Sachs lesion did not result in anterior dislocation unless it was accompanied by capsular disruption. Stability can usually be achieved by correcting the associated soft tissue pathology without additional measures designed to compensate for bone loss [136].

Negative Intra-articular Pressure and Joint Cohesion

The glenohumeral joint is normally bathed in less than 1 mL of free synovial fluid. This joint fluid aids in holding the articular surfaces together with viscous and intermolecular forces [95, 150]. Additionally, the normal intra-articular pressure is negative, creating a relative vacuum that resists translation [25, 72, 87, 172]. If these properties are disrupted by venting the capsule with a needle and introducing air or fluid, subluxation tends to occur [87, 172]. Habermeyer and associates [51] found that traction on the arm caused an increase in negative pressure in normal shoulders but that no increase occurred in unstable shoulders; this suggests that the vacuum effect is somehow lost in the unstable shoulder.

Glenoid Labrum

The labrum is a fibrous rather than a fibrocartilaginous structure that is intimately attached to the glenoid rim [34, 104]. The superior attachment of the labrum is loose and "meniscal-like," whereas the inferior attachment is firm (Fig. 15C–2A). Mobility of the labrum above the transverse equator of the glenoid is normal, not pathologic; in contrast, inferior mobility is abnormal. The fibers of the biceps tendon intermingle with the superior labrum, and the inferior glenohumeral ligament blends into the inferior labrum.

Several authors [16, 104] have interpreted Bankart's original [9] description of the "essential lesion" in recurrent anterior instability as being detachment of the glenoid labrum. As discussed later in this chapter, Bankart also emphasized the importance of the anterior *capsular* detachment in recurrent instability and stated (in 1948) that "it may be that in the past I have laid too much stress upon the role of the fibrocartilage or glenoid labrum. This may be torn from bone or the capsule may be torn

from it." He also noted that the labrum could be excised during operative repair as long as the capsule was reattached to bone [11]. Other investigators [104, 161] have since shown that labral detachment by itself is insufficient to cause anterior instability unless it is accompanied by disruption of the anterior capsule. The term Bankart lesion has evolved to refer to capsular-periosteal separation at the glenoid neck.

The labrum does have some role in anteroposterior glenohumeral stability; the average anteroposterior depth of the glenoid is doubled, from 2.5 to 5 mm, by the presence of the labrum [68]. Vanderhooft and colleagues [166] recently reported that resection of the labrum reduced resistance to anterior translation by 20% in shoulders subjected to a compressive load. The posterior labrum is generally felt to be less important than the anterior labrum in resisting translation [57, 58, 105]. Superoinferior translation may be partially impeded by the presence of the labrum as well. Patients with a relatively small labrum may be at increased risk for instability. Separation of the labrum from the glenoid rim decreases the concavity of the socket. In addition, anteroinferior labral detachment is usually associated with some degree of capsular disruption from the glenoid neck. This capsular-periosteal separation creates laxity in an important stabilizer, the inferior glenohumeral ligament, which is intimately connected to the labrum [101].

Ligamentous and Capsular Restraints

The shoulder capsule is large, loose, and redundant to allow for the large range of shoulder motion. The capsule contains discrete capsular ligaments, which are important in understanding the pathomechanics of shoulder instability. The ligaments come under tension when the joint is placed at the extremes of motion, and they represent the final checkrein to instability when all other mechanisms have been overwhelmed [95].

Rupture of the capsule was felt to be the primary lesion in the unstable shoulder in early investigations into the pathologic basis of shoulder instability [31, 92, 130] and has been noted in some cases clinically [127, 133]. Intra-capsular dislocation, characterized by capsular-periosteal disruption at the anterior glenoid neck, was noted by Broca and Hartmann [23] in 1890 and by Perthes [123] in 1906. The importance of anterior capsular stripping in relation to the labrum and the bony glenoid was stressed in the English literature by Bankart beginning in 1923. In

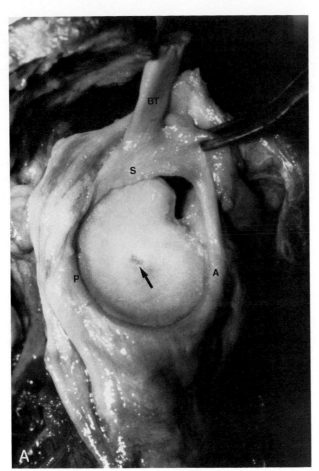
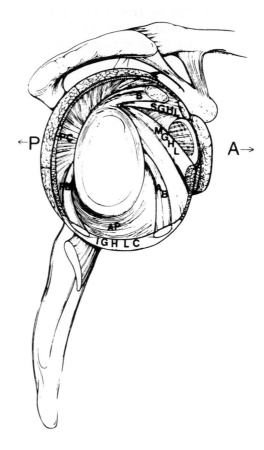

FIGURE 15C–2

Capsular and labral anatomy of the shoulder. *A*, View of the glenoid and labrum after disarticulation of the shoulder. Note the normal mobility of the superior labrum (S). The biceps tendon (BT) is intimately associated with this portion of the labrum. Note also the "bare area" of the glenoid, demonstrated by the arrow (A = anterior, P = posterior).

B, *Capsuloligamentous complex of the shoulder. A, anterior, P, posterior, B, biceps tendon, SGHL, superior glenohumeral ligament, MGHL, middle glenohumeral ligament, AB, anterior band of inferior glenohumeral ligament complex (IGHLC), PB, posterior band of IGHLC, AP, axillary pouch, PC, posterior capsule. (From O'Brien, S. J., Neves, M. C., Arnoczky, S. P., et al. The anatomy and histology of the inferior glenohumeral ligament complex of the shoulder.* Am J Sports Med *18:449–456, 1990.)*

his original [4] description of recurrent anterior dislocation, he stated, "The essential feature is the detachment of the capsule from the fibro-cartilaginous glenoid ligament." In this context, he appeared to refer to the labrum as the "glenoid ligament," and, contrary to the interpretations of earlier writers, he noted that capsular detachment *from* the labrum was the essential defect. He later [11] reiterated, "The only abnormal laxity which is encountered is due to the fact that the capsule is detached from the glenoid margin . . ." and that "the only rational treatment is to reattach the glenoid ligament (or the capsule) to the bone from which it has been torn" [10].

Disruption of the *lateral* capsule from the humeral neck has been noted in some cases of anterior dislocation [8, 107].

Three anterior glenohumeral ligaments have been described: the superior glenohumeral ligament (SGHL), the middle glenohumeral ligament (MGHL), and the inferior glenohumeral ligament complex (IGHLC) [36, 41, 144]. The SGHL originates from the anterosuperior labrum anteriorly to the biceps tendon and inserts superior to the lesser tuberosity near the bicipital groove. The MGHL arises adjacent to the SGHL and extends medially to attach on the lesser tuberosity in intimate association with the subscapularis tendon. The IGHLC extends from the anteroinferior labrum to insert just inferior to the MGHL. The IGHLC is actually composed of three functionally disparate parts: an anterior band, a posterior band, and an interposed axillary pouch [110] (Fig. 15C–2*B*). DePalma and colleagues [36] noted several variations in the composition of the glenohumeral ligaments. The MGHL is the most variable structure; it may be robust in some cases and completely absent in others.

The posterior capsule includes the area posterior to the biceps and superior to the posterior band of the IGHLC. This area is the thinnest part of the capsule; there is no direct posterior ligamentous reinforcement.

The role of the anterior glenohumeral ligaments in preventing instability is complex and varies with shoulder position and with the direction of the translating force. In 1910, Delorme examined the function of the individual ligaments and described increased tightening of the IGHL with external rotation and abduction [35]. Many years later, Turkel and colleagues [164] showed that the IGHLC is the primary check against *anterior* instability with the arm abducted 90 degrees. Subsequently, Warren and colleagues [176] examined the restraints to *posterior* translation with the arm flexed, adducted, and internally rotated. They were unable to create posterior dislocation after excising the infraspinatus, teres minor, and entire posterior capsule. If in addition the anterosuperior capsule (including the SGHL) was transected, posterior instability did occur. Conversely, incision of the anterosuperior capsule alone did not cause instability unless the posterior structures were also disrupted. These findings led to the concept that the glenohumeral joint capsule should be considered a "circle" in which abnormal motion in one direction requires capsular damage on the opposite side of the joint. This concept has been supported by the work of Oveson and Nielson [117, 118] and Terry and associates [159].

O'Brien and associates and Schwartz and colleagues [110, 112, 146, 147] later found that the IGHLC (with the posteroinferior capsule) is also the primary static stabilizer against *posterior* instability when the arm is abducted 90 degrees. The relative contribution of the components of the IGHLC changes with flexion-extension or rotation of the arm. When the shoulder is in 90 degrees of abduction and 30 degrees of extension, the anterior band of the IGHLC becomes the prime stabilizer against both anterior and posterior translation. When the 90-degree abducted arm is forward flexed to 30 degrees, the posterior band of the IGHLC becomes the prime anteroposterior stabilizer. The IGHLC is also the primary anteroposterior stabilizer with the arm in 45 degrees of abduction. The role of the IGHLC in preventing *inferior* translation increases with increasing abduction of the arm [20, 113, 159, 171, 172]. At 90 and 45 degrees of abduction, the IGHLC is the primary stabilizer against inferior motion.

Recently, the functional relationship of the three components of the IGHLC has been better defined by Warner, Bowen, and their associates [20, 110, 112, 171, 172]. With internal rotation and abduction, the anterior band of the IGHLC moves inferiorly to resist inferior translation as the posterior band shifts posteriorly to prevent anteroposterior motion (Fig. 15C–3). With external rotation and abduction, the posterior band moves inferior to the head, and the anterior band shifts anteriorly. These reciprocal movements of the different portions of the inferior glenohumeral ligament are thought to contribute to "rollback" of the head upon the glenoid.

In the *adducted* arm, the SGHL has been found to have a significant role in limiting inferior translation [20, 63, 171, 172] as well as anterior translation [113]. Inferior translation was increased in specimens in which the SGHL was not well formed [171, 172]. The SGHL has also been shown to have a secondary role in limiting posterior translation in the flexed, adducted, and internally rotated shoul-

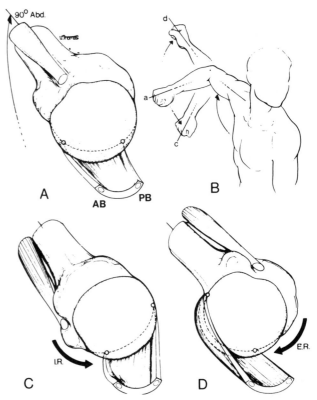

FIGURE 15C–3

A–D, Functional relationship of the components of the IGHLC in the abducted shoulder. With internal rotation, the anterior band (AB) moves inferiorly while the posterior band (PB) shifts posteriorly. With external rotation, the posterior band assumes an inferior position, and the anterior band is anterior to the humeral head. (From O'Brien, S. J., Neves, M. C., Arnoczky, S. P., et al. The anatomy and histology of the inferior glenohumeral ligament complex of the shoulder. *Am J Sports Med* 18:449–456, 1990.)

der. The SGHL does not appear to restrain posterior translation in the abducted arm [110, 112, 146, 147].

The MGHL has a role in limiting anterior translation in the midrange of abduction and serves as a secondary restraint in the lower ranges of abduction [110, 112, 113]. Some authors feel that the presence of a stout MGHL cord with a small foramen between it and a normal IGHLC may protect against anterior instability [100, 104].

The coracohumeral ligament (CHL) extends from the base of the coracoid to the transverse humeral ligament, which bridges the greater and lesser tuberosities. Although some authors [13, 63, 116, 159] feel that the CHL has a significant role in controlling translation (especially inferiorly), this view is not supported by the ligament cutting studies that have been performed at our institution [20, 32, 171, 172]. Cooper and colleagues [32] found that the coracohumeral ligament usually represents a fold in the capsule between the subscapularis and supraspinatus tendons and that it does not have a suspensory role. Warner and associates [172] reported that cutting the CHL in isolation while maintaining the SGHL did not affect inferior translation. In contrast, Helmig and co-workers [63] noted a significant increase in inferior translation after sectioning the CHL and a further significant increase in translation if the anterosuperior capsule (presumably including the SGHL) was cut as well. Further sectioning of the posterior capsule did not significantly alter translation.

The space between the superior border of the subscapularis and the anterior margin of the supraspinatus has been termed the rotator interval [108]. The SGHL and the CHL are located in this region of the capsule. A relatively large interval has been associated clinically with inferior instability [108] as well as anterior instability [137]. Harryman and associates [54] recently found that this portion of the capsule plays a significant role in preventing inferior subluxation of the adducted shoulder and acts as a secondary restraint against posterior translation. These findings correlate with the function of the SGHL. The presence and size of the rotator interval should be addressed during operative repair.

The capsule also contains numerous synovial recesses, and increased capsular redundancy is an important factor in the pathogenesis of shoulder instability, particularly instability of the multidirectional type [106, 157, 165]. Pollock and associates [124] have shown that the IGHLC undergoes considerable plastic deformation prior to failure. These data support the concept that gradual stretching of the capsule can occur with repetitive microtrauma, which is a common etiologic factor in anterior and posterior subluxation as well as in multidirectional instability. The idea of post-traumatic capsular laxity was strongly presented by Thomas in the early part of this century [157, 158]. Capsular laxity may also occur in some patients due to inherent soft tissue laxity.

Dynamic Mechanisms

The rotator cuff tendons and the tendon of the long head of the biceps may have an important role in controlling glenohumeral translation. Joessel [82], in 1880, originally proposed that damage to the rotator cuff could allow increased excursion of the humeral head upon the glenoid. The effect of passive musculotendinous tension within the rotator cuff appears to have some static role in preventing translation. Jens [75] implicated an abnormally proximal insertion site of the subscapularis as the primary lesion in selected cases of anterior instability. Symeonides [156] felt that lengthening and laxity of the subscapularis muscle was the dominant factor in most cases of anterior instability. It appears that the subscapularis plays a role in limiting anterior translation in the lower ranges of abduction and that the teres minor and infraspinatus have a similar function with regard to posterior stability [117, 118, 164]. As Rowe and associates [136] have pointed out, however, significant lesions of the subscapularis are uncommon surgical findings. Instead, the primary mechanism by which these muscles affect glenohumeral stability appears to be a dynamic one, and is associated with a coordinated system of selective muscular contraction.

It is likely that the muscles of the rotator cuff serve a complementary function in order to adjust tension in the capsuloligamentous system. Terry and associates [159] have theorized that stretch receptors within the capsular ligaments could be activated by tension to induce selective contraction of the surrounding musculature to protect the ligaments at the extremes of motion. A recent study has documented the presence of Ruffini end-organs and Pacinian corpuscles in the capsulolabral system [167]. An abnormality in the muscular response to neuroreceptor input may be present in patients with voluntary instability.

The contraction of the rotator cuff and biceps muscles causes the humeral head to be compressed into the glenoid and increases the load needed to translate the head [27, 68, 71]. This function has been the basis for conservative treatment of shoulder instability that incorporates strengthening of the rotators. McKernan and associates [96], in cadaveric experiments, showed that simulated maximal contraction of the posterior rotator cuff muscles reduced anterior ligamentous strain. Thus, the rotator cuff may function as a secondary stabilizer in cases of capsuloligamentous deficiency. In a related study, McKernan and colleagues [97] proposed that the ability of the cuff muscles to act as secondary stabilizers may be inadequate to counteract completely the degree of capsuloligamentous damage seen with the classic Bankart lesion. The stabilizing effect of a particular muscle is dependent on the position of the shoulder; a change in position may render a secondary stabilizer nonfunctional. Interestingly, Howell and Kraft [70] noted a persistence of normal glenohumeral translation patterns in patients with instability who underwent a selective nerve block that paralyzed the supraspinatus and infraspinatus muscles. This finding suggests that rotator cuff compressive forces are not the only factors that provide secondary stability in the unstable shoulder.

A second group of muscles affects glenohumeral stability. The scapular rotators (trapezius, rhomboids, latissimus dorsi, serratus anterior, and levator scapulae) position the scapula to provide a stable "platform" beneath the humeral head. This allows the glenoid to adjust to changes in arm position. We have observed that winging of the scapula is associated with posterior instability (Fig. 15C–4). This phenomenon may be a purposeful action that is designed to control translation. The periscapular muscles should not be neglected during shoulder rehabilitation.

The activities of various muscles during the act of throwing have been investigated [21, 76, 77]. There are five basic phases in throwing:

1. Wind-up: Begins with the initiation of movement and ends when the ball leaves the gloved hand.
2. Early cocking: Begins with abduction and external rotation of the shoulder and ends when the stride foot strikes the ground.
3. Late cocking: Continues to maximal external rotation of the shoulder.
4. Acceleration: Begins with internal rotation of the humerus and continues until ball release.
5. Follow-through: Extends from ball release until all motion is complete.

The muscles surrounding the glenohumeral joint perform two important functions during the act of pitching. They contract concentrically to generate ball velocity, and they protect the joint from the translatory forces by eccentric or isometric contraction [119]. In the wind-up phase, there is minimal muscular activity. As cocking proceeds, the deltoid is activated as the shoulder begins to elevate. The supraspinatus, infraspinatus, and teres minor are then brought into action, followed by activation of the subscapularis. The rotator cuff is thought to "fine tune" shoulder elevation. An increase in biceps activity in late cocking is thought to be primarily due to its role as an elbow flexor [50]. Cocking is limited by the eccentric actions of the subscapularis, latissimus dorsi, and pectoralis major. These muscles theoretically act to protect the anterior joint structures against the developing tension.

During the acceleration phase, the rotator cuff, pectoralis major, and latissimus dorsi cause rapid internal rotation of the shoulder while the serratus anterior maintains the scapula against the chest wall.

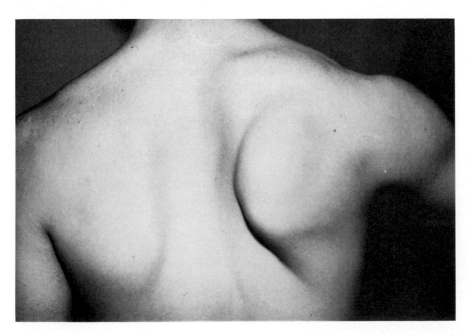

FIGURE 15C–4
Winging of the scapula in a patient with posterior instability.

Whereas amateur throwers tend to fire all of the muscles of the rotator cuff as well as the biceps to generate this acceleration force, professional pitchers selectively fire the subscapularis. The increased muscular economy and efficiency of professional pitchers is felt to provide relative protection against fatigue and overuse. In the final phase, all of the muscles surrounding the shoulder contract to decelerate the arm and forearm. The rotator cuff in particular may function to decelerate anterior translation of the humeral head upon the glenoid.

In a study of throwers with anteriorly unstable shoulders, an imbalance in the usual pattern of muscle recruitment was seen [49]. In these patients, selective weakness of specific muscles was noted rather than a generalized impairment. Internal rotator activity was decreased in all phases, and serratus anterior activity was diminished in the late cocking and acceleration phases. An increase in biceps activity was also noted and was felt to represent an attempt to stabilize the shoulder through the compressive action of the biceps tendon. Rodosky and associates [131] recently simulated biceps contraction during the cocking phase and noted an increase in IGHLC strain after the biceps tendon was sectioned. Increased activity of the biceps may help to explain the so-called SLAP lesion that has been noted recently in throwers [131, 152, 153].

An understanding of the complex act of throwing, asynchronous firing patterns, and specific muscular weakness in the unstable shoulder may provide the basis for specific goals in rehabilitation. Cain and associates [27] simulated rotator cuff contraction during the cocking phase of throwing and found that contraction of the infraspinatus and teres minor reduced the strain on the IGHLC. This result has direct implications for strengthening the external rotators in patients with anterior subluxation. Rehabilitative measures that stress eccentric muscle control and scapular control may be of prime importance [149]. The scapula must rotate upward (horizontally protract) in synchrony with arm elevation. Dysfunction of the serratus anterior, which is primarily responsible for this motion, may predispose to increased stress and subsequent injury.

The high loads [120] delivered to the shoulder area in the throwing athlete can damage the rotator cuff itself as well as the capsular structures. Since the rotator cuff muscles appear to function as secondary restraints on glenohumeral motion, capsuloligamentous disruption could subject them to overuse, fatigue, and injury. Abnormal anterior translation may also result in compression of the cuff against the glenoid. These factors may explain the common association of impingement symptoms and instability in throwers.

Summary

Discrete anatomic deficiencies in the capsular structures of the glenohumeral joint appear to be responsible for the great majority of unstable shoulders. It appears that the IGHLC is the primary static restraint against anterior, posterior, and inferior translation when the arm is abducted between 45 and 90 degrees. In the midrange of abduction, the MGHL and subscapularis assist the IGHLC in resisting anterior translation, whereas the teres minor and infraspinatus help prevent posterior translation. When the arm is adducted, the SGHL and the MGHL stabilize against anterior movement, the posterior capsule and the SGHL resist posterior motion, and the SGHL and IGHLC are the primary restraints against inferior translation. The role of the CHL in limiting inferior translation is controversial. The classic Bankart lesion (capsular-periosteal separation at the anteroinferior glenoid neck) renders the important IGHLC nonfunctional. Inferior capsular rupture has a similar effect.

Enlargement of the rotator interval appears to result in abnormal inferior translation and may also be related to increases in anteroposterior motion. Laxity of the capsule may occur on a genetic basis or may be due to plastic deformation resulting from extrinsic forces; the resultant enlargement of the capsule could lead to multidirectional instability.

The labrum deepens the glenoid socket and may predispose to instability when it is damaged or inherently small. The labrum is intimately associated with the IGHLC, and labral detachment reduces the restraining effect of the IGHLC.

The rotator cuff, biceps, and scapular rotators also play important roles in the stability of the glenohumeral joint. Contraction of the cuff and biceps appears to increase the load needed to translate the humeral head, and selective contraction of the individual muscles may adjust tension within the capsuloligamentous structures. The complex function of the shoulder in athletic activities requires a coordinated, synchronous interaction of all these muscles. Muscular dysfunction may predispose to instability, and, conversely, instability may result in musculotendinous pathology as secondary mechanisms of restraint become overwhelmed. Abnormalities in bony anatomy and loss of the normal vacuum effect within the glenohumeral joint may be related to some cases of instability.

With further investigation, the specific location of

the anatomic lesion or lesions associated with a particular pattern of instability will become better defined. Nonoperative treatment and rehabilitative measures should avoid stress to damaged structures and should be aimed at specific strengthening regimens for dysfunctional or inadequate muscular activity. Operative treatment should be directed at restoring the pathologic capsulolabral disruption or reducing abnormal capsular laxity. Such an approach will maintain a high level of function and will stabilize the shoulder without causing undue restriction of motion.

CLASSIFICATION OF SHOULDER INSTABILITY

Shoulder instability may be classified on the basis of frequency and chronology, direction and degree of instability, etiology, and the presence or absence of voluntary control. With regard to the temporal background of the patient's complaints, the condition may be described as acute, recurrent, or chronic (i.e., fixed or locked). The instability may occur anteriorly, posteriorly, or inferiorly, or it may be multidirectional. The articular surfaces may become completely separated (dislocation), or symptoms may result from abnormal translation without complete separation (subluxation). Complete dislocation may be said to occur when anterior or posterior translation exceeds one-half of the sum of the diameters of the glenoid and humeral head [112].

An episode of macrotrauma is less common in patients with posterior or multidirectional instability than in those with anterior dislocation. On the other hand, microtrauma associated with repetitive use is common in those with anterior subluxation as well as in those with posterior and multidirectional instability, especially in throwers and swimmers. Some patients are unable to relate the onset of their symptoms to either trauma or repetitive use; this atraumatic subgroup is more commonly found in those with multidirectional instability. (However, multidirectional instability may also result when traumatic injury occurs in a patient with laxity of the capsule or a large rotator interval.)

Finally, the ability of the patient to display instability voluntarily should be determined. A certain number of these patients are psychologically disturbed. Although classic teaching has been to avoid operation in all patients with voluntary instability [138], there is recent evidence that there is a subgroup of patients who can voluntarily demonstrate their instability but do not have underlying psychiatric disease. Generally, these patients have a more positional type of instability in which the head slides out posteriorly with flexion, adduction, and internal rotation of the arm [44].

PHYSICAL EXAMINATION OF THE SHOULDER

A careful history and physical examination remain the cornerstone of diagnosis in shoulder instability. Examination of the glenohumeral joint begins with a visual inspection for deformity, asymmetry, and atrophy. The cervical spine is assessed for range of motion, pain, and radicular symptoms. A routine neurovascular assessment should be made. The shoulder region is palpated for evidence of local tenderness. Passive and active arcs of motion in each shoulder and the presence or absence of impingement signs are noted. The impingement sign is positive if maximum passive forward flexion of the shoulder produces pain in the terminal 10 to 15 degrees of motion [177]. Muscular strength, particularly in abduction and external rotation of the shoulder, is evaluated. Tests that address the issue of instability more specifically are then performed.

Drawer tests are designed to assess translation of the humeral head on the glenoid. The degree of translation is noted in the anterior, posterior, and inferior directions. Normally, translation is equal anteriorly and posteriorly and is greatest in neutral flexion-extension and neutral rotation. These tests are best performed with the patient supine and the arm in 90 degrees of abduction and neutral rotation. Other positions of abduction and rotation may also be examined in an attempt to correlate pathology with the existing basic science data presented earlier. Translation is graded as 1+ if there is increased translation compared to the opposite shoulder but neither subluxation nor dislocation occurs. If the head can be subluxated over the glenoid rim but then spontaneously reduces, translation is graded as 2+. Frank dislocation without spontaneous reduction constitutes 3+ translation [1]. It is essential that the opposite shoulder be tested for comparison.

Grading of the sulcus sign (Fig. 15C–5) is based on the distance between the inferior margin of the lateral acromion and the humeral head when a downward traction force is applied to the adducted arm. Less than 1 cm of distance represents a 1+ sulcus, 1 to 2 cm indicates a 2+ sulcus, and more than 2 cm requires a grade of 3+ [1]. An abnormal sulcus sign reflects laxity of the SGHL and IGHLC and is indicative of inferior instability. It is pathognomonic of multidirectional instability. Assessment of inferior translation with the arm abducted more

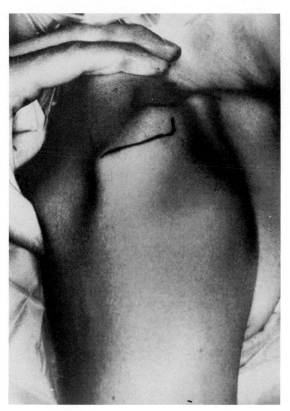

FIGURE 15C–5
Sulcus sign. A downward traction force is applied to the adducted arm and the distance between the lateral acromion and the humeral head is noted. (From Altchek, D. W., Warren, R. F., Skyhar, M. J., et al. T-plasty modification of the Bankart procedure for multidirectional instability of the anterior and inferior types. *J Bone Joint Surg* 73A:105–112, 1991.)

FIGURE 15C–6
Posterior apprehension sign. The arm is forward flexed and internally rotated as the humerus is loaded with posteriorly directed force and adducted across the chest.

than 45 degrees may, however, more accurately reflect tension on the inferior capsule [19, 20, 171].

Apprehension tests are designed to induce anxiety and protective muscular contraction as the shoulder is brought to a position associated with instability. The anterior apprehension test is performed with the arm abducted and externally rotated. The examiner progressively increases the degree of external rotation and notes the development of patient apprehension. The posterior stress test is performed with the arm internally rotated and forward flexed to 90 degrees. Pain is noted as the humerus is loaded in an anteroposterior direction and progressively adducted across the chest (Fig. 15C–6).

The relocation test [21, 78, 79] is also helpful in evaluating anterior instability. The examiner's hand is placed over the anterior shoulder of the supine patient. A posteriorly directed force is applied with the hand to prevent anterior translation of the head. The shoulder is then abducted and externally rotated as it is in the anterior apprehension test. A positive result in this test is obtained when this anterior pressure allows increased external rotation and diminishes associated pain and apprehension.

Ligamentous laxity is commonly associated with shoulder instability and can be measured objectively on physical examination. The degree of thumb hyperabduction with the wrist volar flexed can be estimated by noting the distance between the thumb and the volar forearm. If the thumb reaches the forearm, the test is considered positive. An assessment is also made for index metacarpophalangeal extension in excess of 90 degrees, elbow hyperextension, and knee hyperextension (Fig. 15C–7).

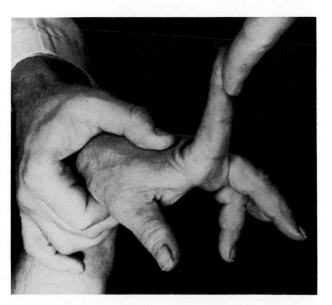

FIGURE 15C–7
Signs of ligamentous laxity should be evaluated in patients with shoulder instability. Index metacarpophalangeal extension in excess of 90 degrees is one indicator of generalized laxity.

Examination under anesthesia (EUA) is a valuable tool because the awake patient often guards against vigorous attempts to evaluate translation. Under anesthesia, the shoulder is stressed inferiorly at 0 and 90 degrees of elevation with the arm in neutral rotation. To assess anteroposterior laxity at 90 degrees of elevation, one of the examiner's hands is used to deliver a translatory load to the humerus. The opposite hand is used to sense the degree of translation or the presence of crepitation. If the anterior and posterior bands of the IGHLC are intact, internal and external rotation of the arm will compress the joint. By placing the arm in neutral rotation, translation is maximized. One must be careful in interpreting translation in the anesthetized patient; normal posterior translation may be as much as 50% of the glenoid diameter [58]. During EUA the head may be noted to displace to the posterior edge of the glenoid. In contrast, a similar degree of *anterior* translation is unusual in the normal shoulder.

Radiography

Routine radiographic examination of the acutely injured shoulder includes an anteroposterior (AP) view (deviated 30 to 45 degrees from the sagittal

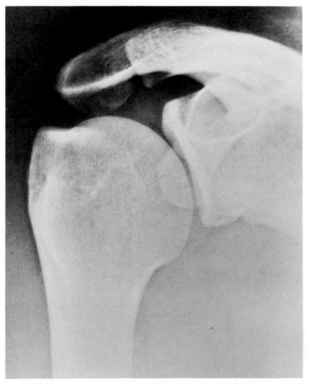

FIGURE 15C–8
True AP view of a normal shoulder. The radiographic beam is directed parallel to the plane of the glenohumeral joint.

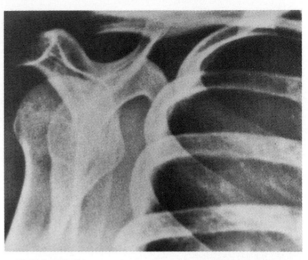

FIGURE 15C–9
Trans-scapular (Y) lateral view of the shoulder. The humeral head is normally located at the intersection of the arms of the scapular Y. (From O'Brien, S. J., Warren, R. F., and Schwartz, E. Anterior shoulder instability. *Orthop Clin North Am* 18(3):395–408, 1987.)

plane of the body in order to parallel the plane of the glenohumeral joint) and a trans-scapular (Y) lateral view (Figs. 15C–8 and 15C–9). The axillary view is extremely valuable for assessing glenoid version, demonstrating humeral head impression fractures, and revealing the position of the humeral head relative to the glenoid. These three views comprise the "trauma series."

In the assessment of more chronic shoulder instability, additional views are helpful to determine bony anatomy and pathology. Internal and external rotation AP radiographs may be ordered to complement the standard AP view. The West Point view [132] often reveals the presence of a fracture or ectopic bone production at the anterior glenoid rim that may not be visualized in other projections (Fig. 15C–10). The West Point view is obtained by placing the patient prone with the arm in 90 degrees of abduction and neutral rotation. The cassette is placed at the superior aspect of the shoulder. The radiographic beam is projected cephalad at an angle of 25 degrees from the horizontal and medially at an angle of 25 degrees. The beam is centered inferomedial to the acromioclavicular joint. The Stryker notch view [52] is especially helpful in demonstrating the Hill-Sachs lesion (the "notch" in the posterolateral humeral head) (Fig. 15C–11). In this radiographic technique, the patient lies supine with the palm of the hand placed on top of the head. The cassette is placed posterior to the shoulder. The arm is aligned in slightly more than 90 degrees of forward flexion, slight internal rotation, and neutral abduction. The beam is then centered upon the coracoid and is tilted 10 degrees from the vertical.

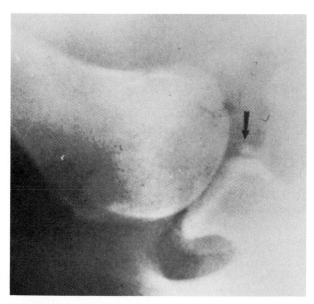

FIGURE 15C–10
West Point view. Arrow demonstrating an osseous Bankart lesion of the anterior glenoid rim. (From O'Brien, S. J., Warren, R. F., and Schwartz, E. Anterior shoulder instability. *Orthop Clin North Am* 18(3):395–408, 1987.)

The "apical oblique projection" described by Garth and colleagues [47] may be helpful in detecting Hill-Sachs and glenoid fractures, especially if the shoulder is acutely injured, because positioning for the West Point or Stryker notch view may be painful in these patients. This is essentially a true

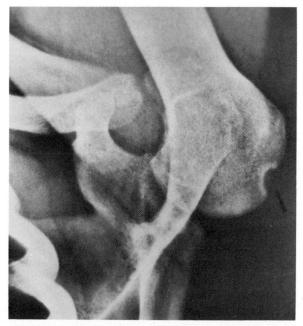

FIGURE 15C–11
Stryker notch view. Arrow revealing a defect in the posterolateral portion of the humeral head (Hill-Sachs lesion). (From O'Brien, S. J., Warren, R. F., and Schwartz, E. Anterior shoulder instability. *Orthop Clin North Am* 18(3):395–408, 1987.)

AP view (in the plane of the joint) modified by aiming the roentgenographic beam from above downward. In the apical oblique view the cassette is placed posterior to the injured shoulder, and the chest is rotated 45 degrees so that the uninvolved shoulder moves further from the cassette. The beam is then directed perpendicular to the cassette and 45 degrees caudally.

An arthrogram may be considered after dislocation of the shoulder in patients older than 45 years of age, especially if the patient seems to be slow to recover after dislocation. Computed tomography (with or without arthrography) is also helpful in selected cases. Finally, magnetic resonance imaging (MRI) is emerging as a potentially helpful tool in the evaluation of labral and capsular lesions, particularly when performed shortly after a dislocation, when the attendant joint effusion permits better visualization of these structures [168]. In the absence of an effusion, intra-articular injection of gadolinium may improve the resolution of these tissues. MRI is also a valuable aid when significant cuff injury is associated with instability.

ANTERIOR DISLOCATION

Anterior instability of the glenohumeral joint is the most common type of shoulder instability. In Rowe's series of 500 shoulder dislocations [133], 98% occurred in the anterior direction. The humeral head is usually displaced to a subcoracoid position after an anterior dislocation (Fig. 15C–12) but rarely may come to rest beneath the clavicle or within the thorax. Indirect forces applied to the upper extremity are the most common cause of anterior dislocation. These forces generally place the shoulder in a position of extreme external rotation combined with abduction or hyperextension [95]. Rarely, a direct force upon the posterior or posterolateral aspect of the shoulder may result in an anterior dislocation. In shoulders of patients with inherent soft tissue laxity, dislocation can occur with little or no injury. In the series of Rowe, 4.4% of dislocations were of this atraumatic subtype [133].

The diagnosis of an anterior dislocation is usually readily apparent on physical examination. The arm is held in a slightly abducted and externally rotated position. Internal rotation and full abduction are not possible. The humeral head may be palpable anteriorly, and the distal acromion may be more prominent than usual. The posterior aspect of the joint appears hollow.

After the diagnosis has been made, the shoulder should be reduced as quickly and as gently as

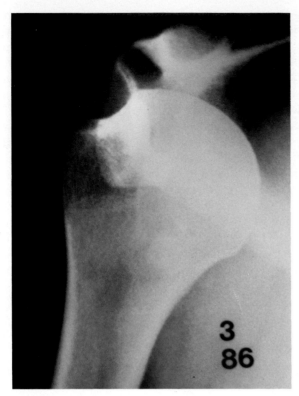

FIGURE 15C–12
Subcoracoid anterior dislocation of the shoulder on an AP radiograph.

possible. If such an injury is encountered in a healthy athlete on the playing field, an immediate reduction maneuver may be performed prior to the onset of muscular spasm, which interferes with reduction. We generally prefer manipulation in slight abduction, forward flexion, and gentle internal rotation in this setting.

If this maneuver is unsuccessful, the patient should be taken to the locker room, and intravenous access should be obtained. The patient is suitably sedated with a narcotic or benzodiazepine. At this point, the reduction method described by Stimson [155] may be successful. The patient is placed prone with the arm hanging free, and a weight of 5 pounds (or more) is attached to the upper extremity. With muscular relaxation the humeral head may spontaneously reduce. If the head fails to reduce spontaneously, a scapular rotation maneuver [4, 111] is attempted. The patient lies prone with the injured arm hanging in a dependent position off the edge of the table. The scapula is then manipulated to "unlock" the anterior aspect of the joint by functionally increasing the amount of glenoid anteversion. If the shoulder remains unreduced after an attempt at scapular rotation, alternative methods of analgesia such as the injection of local anesthetic into the joint or an interscalene nerve block can

facilitate reduction. Adequate muscular relaxation is extremely important in these difficult cases.

The final reduction maneuver is modified from that described by Kocher [86]. The patient is positioned supine, and one sheet is tied around the patient's chest and axilla. This sheet is then tied to the table or held by an assistant to provide countertraction. A second sheet is then wrapped around the physician's waist as he stands in the axillary region. The patient's elbow is flexed to 90 degrees, and the sheet is applied to the forearm just distal to the antecubital fossa. Traction is then applied to the upper extremity by the backward lean of the physician with the arm held in slight abduction. The arm is then gently moved in external and internal rotation until the shoulder reduces. The modified Kocher maneuver carries a risk of iatrogenic damage to bony and neurovascular structures.

In older or debilitated patients, radiographs should be obtained prior to attempting relocation to rule out concomitant fracture. The presence of collagen disease or poor tissue quality raises a concern about vascular injury during reduction. Open reduction may be considered in these cases.

Radiographs of the injured shoulder should always be obtained after reduction. The standard trauma series consists of an anteroposterior, transscapular (Y) lateral, and an axillary view. Evidence of bony damage to the humeral head or glenoid should be noted. Hovelius and colleagues noted Hill-Sachs lesions in 55% of primary dislocators [67], whereas Rowe found this defect in 38% [133]. Greater tuberosity fractures occur in approximately 15% of patients with anterior shoulder dislocation [133] (Fig. 15C–13A and B).

A neurovascular examination should be performed before and after reduction. The axillary nerve is the most frequently damaged neurologic structure after an anterior dislocation. Sensory testing of the lateral shoulder may be unreliable in the assessment of axillary nerve function because the motor component of the nerve appears to be more vulnerable to traction injury.

Several factors may influence the rate of recurrence of an anterior dislocation. The majority of patients who sustain recurrence have an additional dislocation within the first 2 years after the primary event. The age of the patient at the time of the initial dislocation has a major effect on the incidence of recurrence. Rowe and Sakellarides [134] noted a 94% recurrence rate in patients under 20 years of age at primary dislocation but found that only 14% of patients over 40 years of age had a recurrence. In addition, Rowe [133] noted increased recurrence in patients who sustained an atraumatic dislocation

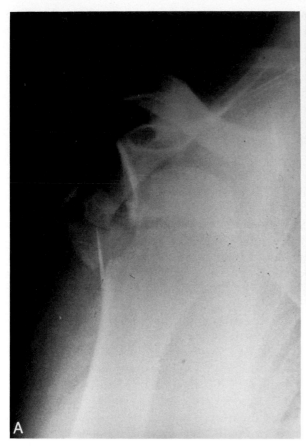

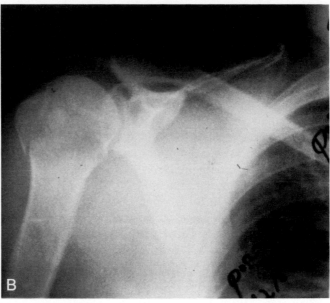

FIGURE 15C–13
Anterior dislocation with greater tuberosity fracture. *A*, AP view revealing dislocation and displaced fracture of the greater tuberosity. *B*, Postreduction view.

compared to those with traumatic dislocation. Patients who were able to reduce the shoulder by themselves or in whom reduction was easily obtained with the help of a nonphysician were also felt to be at increased risk for recurrence. Males may have an increased recurrence rate compared to females [133]. Simonet and Cofield noted that athletes tended to have more frequent recurrences than nonathletes [151]. Concomitant greater tuberosity fractures greatly reduce the risk of recurrence [67, 133]. The effect of a Hill-Sachs lesion on recurrence rates is controversial. Rowe [133] found that the presence of a Hill-Sachs lesion was related to a higher recurrence rate. In contrast, Hovelius and colleagues [67] did not find that the risk of recurrence was increased in the presence of a head defect.

The effect of immobilization and rehabilitation on recurrence rates is also somewhat controversial. Rowe [133] noted a trend toward decreased recurrence in patients immobilized for 3 weeks but no further decrease in patients immobilized for 6 weeks. Simonet and Cofield [151] reported that immobilization had no effect on recurrence, but they found that restricting sports or full activity for

6 weeks did reduce recurrence. Wheeler and associates [181] felt that the method of rehabilitation had no effect on recurrence in army cadets after a primary dislocation.

Recently, Baker and colleagues [12] divided patients with acute anterior dislocations into three groups based on arthroscopic evaluation of intra-articular pathology and on clinical findings:

Type I. Stable on examination under anesthesia. Minimal hemarthrosis (less than 10 mL). Hemorrhage in inferior capsule. No Hill-Sachs lesions or glenoid rim fractures.

Type II. Anterior subluxation on examination under anesthesia. Hemarthrosis of 10 to 15 mL. Mild, partial separation of glenoid labrum and IGHLC. May have bony head or glenoid lesions.

Type III. Gross anterior instability. Moderate hemarthrosis. Complete avulsion of labrum.

It was the hope of these authors that this classification system would prove beneficial in predicting which shoulders are at risk for recurrence. Treatment could then be modified accordingly.

Authors' Recommended Treatment Method

We generally immobilize young patients who suffer a traumatic anterior dislocation for 4 to 6 weeks. The available basic science data [29, 93] suggest that immobilization of this duration is required for optimal capsular healing to occur. This period is followed by a rehabilitative regimen that employs rotator cuff and scapular rotator strengthening. External rotator strengthening receives particular emphasis. Positions of extreme abduction and external rotation are avoided for 3 months after removal of the sling. A full range of motion, complete return of strength, and absence of pain are prerequisites for a return to sports and overhead activities. When the patient returns to athletic activities, a harness that limits abduction may provide additional security against recurrent damage.

In older patients, especially those over 40 years of age, immobilization is continued only until pain subsides. Seven to ten days in a sling are usually required. Because of the diminished risk of recurrence and the difficulty of regaining motion in this age group, the rehabilitation program is started earlier.

If progress with therapy seems slow, an arthrogram should be obtained to rule out a rotator cuff pathology, particularly in the older age group. Evidence of deltoid weakness suggests the need for electromyographic studies to evaluate the axillary nerve.

Open reduction of acute anterior shoulder dislocations is necessary in those rare cases that are irreducible by closed methods and in those that are more than 3 weeks old at the time of presentation. Displaced fractures of the greater tuberosity and large fractures of the glenoid rim may also require surgical treatment in the acute setting.

The role of operative stabilization of acute dislocations in young, scholastic athletes is controversial. These patients appear to be at extremely high risk for recurrence. The acute dislocation often curtails sports activity for the current year, and the development of recurrent instability results in an extended period away from athletics and other activities. For some patients, this long absence from their sport may be extremely undesirable. Wheeler and colleagues [181] reported a dramatic decrease in the recurrence rate in army cadets after arthroscopic staple capsulorrhaphy or after simple arthroscopic debridement of the labrum and capsular insertion site. The treatment of these patients should be pursued individually on a case by case basis. The risk of recurrence should be explained as well as the potential risks of a surgical procedure. In some young athletes engaged in throwing or contact sports, early restoration of the disrupted anatomy appears to provide the best opportunity for continuing in their sport without losing a significant period of time during the following season.

Recurrent Anterior Dislocation

The diagnosis of recurrent anterior dislocation is usually made without difficulty. The patient typically gives a history of a specific initial injury in which the shoulder "popped out," followed by multiple similar episodes that tend to occur with a lesser amount of trauma and are more easily reduced. In some patients, recurrent dislocation occurs with no history of significant trauma. These individuals are frequently noted to have generalized ligamentous laxity of other joints and may have a primary connective tissue disorder.

On physical examination, patients with recurrent anterior dislocation demonstrate increased anterior translation of the humeral head on the glenoid. In many cases, anterior translation is graded as 3+ because the head is completely dislocatable. Translation to this degree may not be possible in the awake, anxious patient. The apprehension and relocation tests are positive. Evidence of excessive inferior translation, as denoted by an abnormal sulcus sign, should be sought to rule out multidirectional instability.

Patients who develop recurrence should have a complete radiographic evaluation for evidence of bony pathology. The "instability series" [121] includes the West Point and Stryker notch views in addition to the more standard projections. Both Pavlov and colleagues [121] and Rowe [133] noted Hill-Sachs lesions in 77% of patients with recurrent anterior dislocation. An osseous Bankart lesion (ectopic bone production or fracture) of the anterior glenoid was found in 15% of these patients in the series of Pavlov and associates [121]. However, Morrey and Janes [103] pointed out that neither lesion is present in many cases. Loose bodies are found in the joints of approximately 10% of operative cases [111, 133, 136].

Rowe and colleagues [136] noted the Bankart lesion in 85% of cases that came to operation. Rowe felt that this lesion occurred gradually as the result of the repeated trauma of dislocation [133]. Patients who did not have a Bankart lesion frequently had dislocation with an atraumatic cause. In Rowe's experience, the great majority of dislocations were intracapsular, only 6% demonstrating perforation of the capsule by the humeral head. Reeves [127] felt that capsular-periosteal disruption was common in younger patients but that capsular rupture was the primary mode of failure in the older age group. On occasion, lateral capsular disruption from the humeral neck may occur with or without a Bankart lesion [8]. This detachment occurs secondary to trauma and should be searched for in those with traumatic dislocation in whom no Bankart lesion is found.

Recurrent Anterior Subluxation

The diagnosis of recurrent anterior subluxation is often more difficult than that of recurrent anterior dislocation. The patient with recurrent subluxation is often unaware that the shoulder has "popped out." The chief complaint may be subtle, such as a sense of movement, pain, or clicking with certain activities. The pain may often be localized posteriorly owing to strain on the posterior capsule and tendons in resisting anterior translation. Rowe and Zarins [137] described the "dead arm syndrome" in patients with anterior subluxation. In this situation, the patient experiences a sharp pain when the arm is placed in extreme external rotation or after a blow has been delivered to the shoulder. The patient may then lose control of the extremity and may drop any object held in the hand. After the acute episode, the severe pain usually subsides quickly, but the shoulder may remain sore and weak. In throwers, pain is often associated with the cocking or acceleration phases of throwing but it may occasionally occur during follow-through. Swimmers commonly experience pain with the backstroke or during turns. Overhead serves and volleys are particularly likely to incite symptoms during racquet sports. Volleyball and water polo are also commonly associated with recurrent anterior subluxation.

A traumatic event may be related to the onset of symptoms. Often such an incident involves extreme external rotation of the arm combined with either abduction or hyperextension. Tackling in football and diving back to a base in baseball may result in this position. Anterior dislocations can lead to recurrent anterior subluxation. In other patients, there is no history of macrotrauma. Instead, repetitive low loading appears to result in microtraumatic changes associated with overuse.

The findings on physical examination are also often quite subtle. Tenderness may be noted over the posterior shoulder. In throwers a 10- to 15-degree loss of internal rotation and a similar gain in external rotation may represent normal physiologic

changes and are not necessarily pathologic. The anterior apprehension test usually causes more pain than apprehension. This pain is relieved upon performance of the relocation test.

Rotator cuff impingement symptoms frequently accompany anterior subluxation, particularly in throwers and overhead athletes [79, 80]. In addition to the physical findings mentioned above, examination may reveal the presence of impingement and rotator cuff signs. Differentiation between pure impingement, pure instability, and mixed pathology may be difficult. Patients with pure impingement may not experience pain relief with the relocation test. Patients with pure instability should have a negative impingement sign. The impingement *test* (which involves injection of a local anesthetic into the subacromial space) localizes the process to the subacromial space and should provide temporary relief of symptoms in patients with pure impingement. Patients with pure instability continue to have pain after this injection. However, the test is nonspecific because the injection may relieve secondary tendinitis in patients who do not have true impingement.

Jobe and Kvitne [80] identified two groups of throwing or overhead athletes with concomitant impingement and instability, based on clinical and operative findings. In one group, instability was thought to result from chronic labral microtrauma with impingement occurring secondarily. These patients had damage to the labrum and attenuation of the IGHLC. In the second group, instability was thought to be due to hyperlaxity. These patients had an intact labrum and demonstrated laxity of the glenohumeral ligaments. The physical findings in the two groups were similar except that patients with hyperlaxity had generalized laxity of other joints as well.

Humeral hypertrophy [175] is another normal physiologic response to throwing that may be noted radiographically. Hill-Sachs lesions are seen in 25% of patients with recurrent anterior subluxation. Calcification of the anteroinferior glenoid margin is found in approximately 50% [111, 121]. Magnetic resonance imaging may be helpful in evaluation of the rotator cuff.

Treatment of Recurrent Anterior Instability

Nonoperative Treatment
The treatment of recurrent anterior instability begins with a period of rest. After an acute event, an arm sling is worn for a few days for comfort.

Nonsteroidal anti-inflammatory medication is administered during this time. Prolonged immobilization after the second dislocation is of no apparent value.

A rehabilitation program emphasizing rotator cuff and periscapular muscle strengthening is then employed. This regimen is generally started at 2 weeks in throwers and overhead athletes. Stretching exercises are instituted in an attempt to regain full motion. Isokinetic testing at this point can help to identify specific muscular weakness and can provide a baseline for comparison during the rehabilitative process [80]. Selective weakness in a particular muscle suggests the need for strengthening. Rotator strengthening begins with spring exercises, progresses to exercises performed with rubber tubing, and advances finally to Nautilus or isokinetic exercises. In the early phase of therapy, especially in patients with concomitant impingement, these activities should be performed with the arm at the side or in the lower ranges of abduction to protect the rotator cuff [186]. Later, these exercises may be performed with the arm in 90 degrees of abduction as well. Muscular endurance should be emphasized in addition to strengthening. The scapular rotators are conditioned by a combination of shoulder shrugs, horizontal adduction exercises, pull-downs, chin-ups, and push-ups [80].

Recently, Townsend and colleagues [162] have described four specific exercises that, based on electromyographic data, specifically strengthen the glenohumeral muscles.

1. Elevation of the arm in the scapular plane with the arm internally rotated and the thumbs down.
2. Elevation of the arm in the sagittal plane.
3. Horizontal adduction from the prone position with the arm externally rotated.
4. The "press-up" exercise: In a seated position, the hands are placed upon the seat, and the body is lifted from the chair by extending the upper extremities.

The first of these four exercises may aggravate rotator cuff symptomatology and should be avoided if rotator cuff pathology is present.

Throwing is not allowed until strength and range of motion are normal. Throwing is slowly advanced in distance, velocity, frequency, and duration. The patient's pitching mechanics should be adjusted to provide efficient energy transfer from the lower extremities and thorax to the shoulder. Specific activities that seemed to incite pain prior to the institution of therapy are withheld for longer periods of time.

In sports that do not require overhead activity,

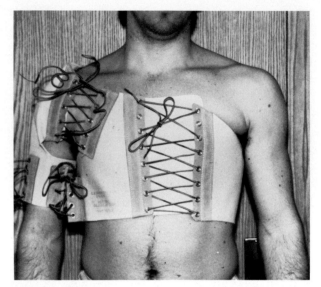

FIGURE 15C–14
Abduction harness. This device may be useful in an athlete who does not require overhand motion.

an abduction harness may be worn to prevent elevation of the arm above 90 degrees. This device is especially useful for selected positions in football (Fig. 15C–14).

A rehabilitative program often succeeds in patients with atraumatic instability. Traumatic dislocators appear to respond less favorably.

Operative Treatment

The indications for surgical treatment of recurrent anterior shoulder instability are highly subjective and include recurrence, pain, or limitation of activity after a thorough trial of nonoperative management. Some surgeons feel that the presence of a Bankart lesion on arthroscopic evaluation is another indication for a stabilization procedure because conservative methods may be likely to fail in the presence of this type of pathology. As discussed earlier, the treatment of acute instability in young athletes is controversial. Patients with voluntary anterior instability are often poor operative candidates and require psychological testing and a rigorous attempt at rehabilitation [138].

Patients with concomitant instability and impingement who do not respond to conservative treatment should undergo surgical intervention to provide stability. Subacromial decompression is not recommended as a primary procedure in these patients. Occasionally, anterior acromioplasty may be required as a secondary part of treatment [61].

The goals of operative treatment for shoulder instability are to stabilize the shoulder and to restore pain-free motion [61]. To accomplish these goals, the pathologic process must be thoroughly investigated by means of the preoperative history, physical examination, and radiography. In general, the surgeon should have gained a clinical sense of the direction and degree of instability from the office evaluation, and this impression should be confirmed by the examination under anesthesia. On occasion, the findings may be equivocal, and the arthroscope may then be used to add objective data to the evaluation.

Arthroscopy

In most cases, arthroscopy of the shoulder does not play a major role in diagnosis of shoulder instability. However, it is highly useful in some cases. Throwers with concomitant instability and rotator cuff symptoms are particularly likely to be good candidates for arthroscopy. Arthroscopy is also valuable in patients in whom the direction or degree of instability is in doubt. Arthroscopic examination may help in identifying intra-articular pathology and in the detailed planning of an open or an arthroscopic stabilization procedure. However, the tendency to overemphasize minor changes noted on arthroscopy should be avoided [6]. Degenerative lesions of the labrum are common in patients older than 40 years of age [5, 36].

Our basic approach is to perform an arthroscopic examination in all patients with traumatic anterior instability because the likelihood of capsular-periosteal disruption of the anterior glenoid is high in these patients. In patients with an atraumatic etiology, an open technique is generally used. Since 1986, we have performed arthroscopy in the modified "beach chair" position [2]. The patient's back is placed at an angle of approximately 75 degrees

from the floor, the hips are flexed to 90 degrees, and the knees are set in 30 degrees of flexion. The thorax is rotated slightly toward the nonoperative shoulder to expose the medial border of the scapula on the operative side. The position is fixed by molding a beanbag to the body and then deflating the beanbag to make it firm (Fig. 15C–15). A special chair (Concepts, Inc., Largo, FL) has been developed that has proved to be a useful aid in positioning the patient.

We feel that there are several advantages to the beach chair position. We perform the great majority of our shoulder procedures under interscalene block anesthesia. The patients are more comfortable in the semisitting position than in the lateral decubitus position and are able to observe the procedure on the arthroscopic monitor. The beach chair position allows the surgeon to examine the shoulder in various positions of abduction and rotation because there is no traction on the arm. There is also a lower risk of neurapraxia [85] from traction, and the anterior structures are not placed in a stretched, nonanatomic orientation. The beach chair position also allows easier access to the anterior shoulder during arthroscopic stabilization. In addition, an arthroscopic procedure can be simply converted to an open procedure without the need for extensive repositioning and redraping. The beanbag is simply deflated, and the head of the operating table is lowered to the appropriate level.

The posterior arthroscopic portal is placed 3 cm inferior and 2 cm medial to the posterolateral corner of the acromion. The anterior portal is situated lateral and superior to the coracoid. Placement of

FIGURE 15C–15
Beach chair position. The patient's back is positioned at an angle of 75 degrees from the floor. The medial border of the scapula is exposed, and the patient is stabilized by deflating a beanbag.

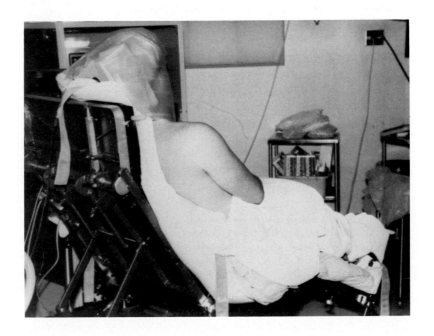

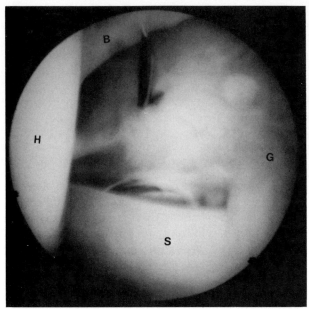

FIGURE 15C–16
Arthroscopic view of anterior portal placement. The tip of a spinal needle is seen in the triangular interval between the glenoid margin (G) at right, the biceps tendon (B) superiorly, and the subscapularis tendon (S) inferiorly. The humeral head (H) is on the left.

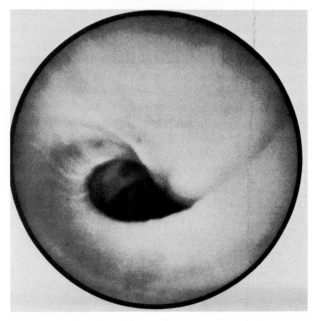

FIGURE 15C–17
Normal "wind-up" of the axillary pouch with rotation of the arm. (From Altchek, D. W., Skyhar, M. J., and Warren, R. F. Shoulder arthroscopy for shoulder instability. *Instr Course Lect* 38:187–198, 1989.)

the anterior portal inferior or medial to the coracoid risks neurovascular injury [94]. The anterior cannula enters the joint in the triangular area bounded by the biceps tendon, the subscapularis tendon, and the anterosuperior glenoid (Fig. 15C–16).

Arthroscopic examination of the shoulder should be performed in a standard systematic fashion, viewing all significant anatomic structures in turn. The first step in our routine is to examine the biceps tendon throughout its intra-articular course. The anterior labrum is then probed for evidence of detachment or tearing. The presence of a "labral sulcus" near the two o'clock position is a normal variant and should not be confused with a pathologic change. The presence and qualities of the SGHL, the MGHL, the subscapularis tendon, and the anterior band of the IGHLC are ascertained. Next, the axillary pouch is inspected to assess excess volume and to search for loose bodies. Internal and external rotation of the arm will reveal whether or not the normal mechanism of reciprocal tightening of the IGHLC components is functional (Fig. 15C–17).

The articular surfaces of the glenoid and humeral head are checked for damage. The undersurface of the rotator cuff is evaluated, including its insertion on the greater tuberosity. The presence or absence of a defect in the posterolateral head is noted. Translatory forces may be delivered in both anterior and posterior directions to assess the relationship between the head and the glenoid.

Pathologic findings during the arthroscopic examination include detachment of the anteroinferior capsulolabral structures (Fig. 15C–18), stretching of the IGHLC, and fraying or tearing of the anteroin-

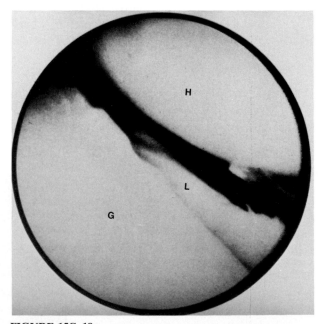

FIGURE 15C–18
Arthroscopic view of the Bankart lesion. Anteroinferior labral detachment and laxity of the inferior glenohumeral ligament complex (IGHLC) are noted. G, glenoid margin; L, labrum IGHLC; H, humeral head. (From Altchek, D. W., Skyhar, M. J., and Warren, R. F. Shoulder arthroscopy for shoulder instability. *Instr Course Lect* 38:187–198, 1989.)

ferior labrum. Laxity of the pouch is present when the arthroscope is easily passed into the anteroinferior joint cavity without the normal restraint of the capsular tissues. This phenomenon is referred to as the "drive-through" sign.

Andrews and colleagues [5] recognized the association between anterosuperior labral lesions and biceps lesions in throwers. Recently, Snyder and associates [152, 153] coined the term "SLAP lesion" to describe "an injury to the superior aspect of the labrum which begins posteriorly and extends anteriorly . . . including the 'anchor' of the biceps tendon to the glenoid." A similar lesion was described earlier by Garth and associates [48]. In the series of Snyder and colleagues [152, 153], the SLAP lesion was most commonly caused by a fall onto the outstretched arm with the shoulder in abduction and slight forward flexion. Traction injuries were also associated with this lesion. In contrast, Andrews and colleagues [5] and Speer and colleagues [154] noted that lesions of the superior labrum are found almost exclusively in throwers. Symptoms include pain that becomes worse with overhead activity and a sensation of "catching" or "popping" in the shoulder. Andrews described a method of eliciting these symptoms by rotating the arm in a position of 100 degrees of elevation. These subtle clinical findings and changes in the superior labrum on magnetic resonance imaging may suggest the diagnosis of a superior labral lesion.

The arthroscopist must be careful not to mistake the normal mobility of the superior portion of the labrum or the labral sulcus for the SLAP lesion [34]. Snyder and associates [152] classified the lesion into four types based on arthroscopic findings:

Type I. Fraying of the superior labrum, but firm attachment of the labrum to the glenoid.

Type II. Stripping of the superior labrum and biceps tendon off the underlying glenoid, resulting in instability of the labral-biceps anchor.

Type III. Bucket-handle tear of the labrum with an intact biceps insertion.

Type IV. Bucket-handle tear of the labrum extending into the biceps tendon.

Patients in the series of Snyder and associates had a high incidence of associated pathology including anterior instability and rotator cuff disease [152, 153]. In our experience, superior labral lesions are not usually associated with overt instability, but the affected patient may have a sense of "looseness" in the shoulder. Superior labral flaps are often associated with posterior rotator cuff pathology [73].

The recommended treatment of SLAP lesions includes debridement of frayed lesions, absorbable tac fixation in cases of stripping from the glenoid, excision of bucket-handle tears, and biceps tenodesis when more than half of the biceps tendon is involved in the process [152, 153]. Speer and associates [154] noted a resolution of the sensation of "looseness" after fixation of hypermobile superior labral lesions with an absorbable tac. Secure fixation of the biceps anchor appears to be the key to successful treatment. Labral debridement of the SLAP lesion has not been highly successful in our experience [3].

Garth and colleagues [48] and Terry and colleagues [159] have theorized that the IGHLC relaxes during the acceleration phase of throwing and that this laxity is a factor in injury to the superior labrum. The eccentric contraction of the biceps during the throwing act functions to decelerate the elbow [5, 50]. Andrews and associates [5] and Rodosky and co-workers [131] have proposed that the biceps may function as a secondary shoulder stabilizer as well. Increased biceps activity has been noted in throwers with anterior instability [49]. The forces incurred by this mechanism may help to explain the pathogenesis of the SLAP lesion in some patients with anterior instability. If this hypothesis is correct, the presence of superior labral hypermobility may worsen the instability already present in these patients.

Anteroinferior labral flaps are often associated with instability. Large labral flaps that appear to be nonfunctional can be debrided, but the surgeon must not destabilize the IGHLC. Altchek and colleagues [3] recently reported poor long-term results in patients who underwent arthroscopic debridement of labral flaps at all sites. This treatment provided temporary pain relief, but symptoms generally recurred upon resumption of normal activities and sports. Patients with evidence of labral detachment and instability responded especially poorly. We do not recommend debridement of anteroinferior labral insufficiency as an isolated treatment.

Open Stabilization Procedures

Many operative procedures have been described for the treatment of anterior shoulder instability. The most popular procedures in North America can be broken down into three basic groups:

1. Procedures designed to repair the avulsed anteroinferior capsulolabral structures to the bony glenoid from which they have been torn. These operations are modifications of the techniques described by Perthes [123] and Bankart [10].

2. Operations designed to limit external rotation of the shoulder. This group includes the Putti-Platt [114] and Magnuson-Stack [91] procedures.

3. Bone block procedures. This category encompasses coracoid transfer procedures [15, 88, 163] such as the Bristow operation [62] and procedures

in which a bone block is obtained from a distant site [39, 74, 115].

In 1956, Carr [28] stated, "I wish to emphasize that completely normal scapulohumeral motion should not be the requirement for an excellent result . . . restriction of movement is the price paid willingly for stability and full confidence in the shoulder." DePalma was of the opinion that limitation of external rotation was the key to preventing recurrence and that the surgeon should adopt the simplest procedure that accomplished this limitation [37]. It was in this spirit that the Putti-Platt and Magnuson-Stack procedures came to the forefront. From the earlier discussion, it is clear that limiting external rotation does not address the true pathology of the anteriorly unstable shoulder. Such functional restraint severely limits the patient unnecessarily. The Putti-Platt procedure, in particular, may cause degenerative arthritis when the subscapularis is shortened excessively [55, 142].

Bone block procedures, when done in isolation, also fail to address the problem of anteroinferior capsulolabral insufficiency. The concept of an anterior bony buttress (upon which these procedures are based) may be invalid, particularly in patients with subluxation [143]. These types of operations also tend to limit motion. When metal hardware is used to fix the bone block, there is a high risk of complications from penetration of the joint or loosening [187]. A revision stabilization procedure is especially difficult after a Bristow procedure due to extensive scarring of the anterior capsule and subscapularis tendon [184]. Regan and colleagues [134] noted limitation and weakness of external rotation in all patients after performance of the Bristow, Magnuson-Stack, and Putti-Platt procedures. Of these three procedures, the Putti-Platt procedure affected external rotation to the greatest degree. Surprisingly, external rotation was better in patients who had undergone a Magnuson-Stack procedure than in those who had been treated with a Bristow operation. Excess capsular laxity is not addressed by these procedures.

The concept of repairing the capsular-periosteal separation at the anterior glenoid neck was first proposed by Perthes [123] and later expounded upon by Bankart [9, 10, 11]. This technique attacks the pathology at its most common site and is directed at reconstitution of the primary static stabilizer of the shoulder, the IGHLC. If abnormal capsular laxity is encountered, the procedure is easily modified to account for this pathologic factor as well. In a long-term review of 50 patients treated by Bankart and his colleagues between 1925 and 1954, recurrence was noted in only two patients [38]. Rowe

and colleagues [136] noted a 3.5% recurrence rate in 145 patients after a Bankart procedure. When properly performed, the procedure results in a superior functional outcome compared to the previously discussed operations. In the series of Rowe and colleagues, 69% of the patients regained full range of motion [136].

Our basic procedure for open surgical treatment of recurrent anterior glenohumeral instability is a modification of the Bankart procedure. Bone blocks are considered only in those rare cases in which there is a severe deficiency (greater than 40%) of the anterior glenoid, and even then the bone block is used in conjunction with a repair of the capsulolabral system. Procedures designed to limit external rotation are not a part of our armamentarium in the initial surgical treatment of anterior instability.

OPERATIVE TECHNIQUE. The patient is positioned supine with the head of the bed raised 30 degrees and the arm abducted 45 degrees on an arm board. A folded towel is placed between the scapulae to rotate the scapula on the involved side laterally. The skin incision is started just lateral to the coracoid and extended approximately 6 cm distally along Langer's lines. (In some cases, when cosmesis is a special concern, the shoulder can be approached from low in the axilla [89]. The axillary approach requires a considerable subcutaneous dissection and offers a limited view when dealing with capsular laxity.) The deltopectoral interval is identified. The cephalic vein is retracted laterally because there are fewer branches on the medial side. The surgeon must take care not to damage the vein as it crosses the superior aspect of the wound.

After dissection through the deltopectoral interval, the coracoid process and the clavipectoral fascia are identified. The fascia is incised lateral to the muscle belly of the short head of the biceps. To facilitate exposure in some cases, a partial oblique incision is made in the conjoined tendon just distal to the coracoid. The musculocutaneous nerve, which may enter the tendon as close as 1 cm distal to the coracoid, is avoided. We do not recommend coracoid osteotomy or complete detachment of the tendon.

The arm is then externally rotated, and the insertion of the subscapularis tendon is revealed. Three small branches of the anterior circumflex vessels lie near the inferior edge of the tendon and may require ligation. A small transverse incision (3 to 4 mm in length) is made at the inferior border of the tendon at a point approximately 3 cm medial to the lesser tuberosity. The anterior capsule is visualized through this incision. A Kelly clamp is then passed from inferomedial to superolateral in the interval between the anterior capsule and the tendon. The medial portion of the tendon is tagged with heavy, nonabsorbable sutures. The tendon is then incised obliquely over this clamp. The medial portion of the tendon is dissected from the capsule with a periosteal elevator (Fig. 15C–19A).

Jobe and colleagues [81] noted that the ability to throw commonly diminishes after operative treatment for shoulder instability. They developed an approach in which the subscapularis tendon is split longitudinally rather than divided. We recommend this method in throwers and overhead athletes who have minimal microinstability. A variation of this approach may be utilized in patients with a greater degree of instability; however, it is difficult to detect and correctly tension an enlarged rotator interval with this method.

The laxity and quality of the capsule are assessed first. If trauma played a significant role in the development of instability, a small transverse capsulotomy is performed, and the joint is explored for evidence of damage to the anteroinferior labrum or to the bony glenoid. The joint is irrigated to remove any loose bodies.

If there is no evidence of capsulolabral separation from the glenoid neck, the shoulder should be reassessed for evidence of multidirectional instability. If there is no abnormal inferior translation in the absence of a Bankart lesion, superior advancement of the capsule may be all that is required. If there is evidence of capsular-periosteal separation, a vertical capsulotomy is created at the glenoid margin. The capsular separation is extended medially to allow placement of a retractor along the glenoid neck. A humeral head retractor is then carefully placed within the joint (Fig. 15C–19B). The glenoid neck is prepared with an osteotome to provide a bleeding surface of bone. Preparation of the glenoid neck is an important step. Zarins and associates blame inadequate attention to this point in their analysis of the causes of failed reconstructions [139, 186].

If the labrum is well attached, the capsule may be sutured to it. If the labrum is detached or attenuated, a dental drill is used to place three to four holes at the edge of the glenoid. The holes are then prepared with an awl, and nonabsorbable sutures are passed through the holes. Recently, we have used suture anchors [129] rather than drill holes to place these sutures. If suture anchors are used, they must be placed at the edge of the glenoid and not along its neck (Fig. 15C–19C). The capsule is then reattached but not overly tightened. The lateral flap is advanced slightly medially and superiorly with the arm in 45 to 60 degrees of external rotation. Each suture is passed through the capsule, and the sutures are tied over the capsule. The

labrum may also be reattached to the glenoid if it is not degenerated. The goal is not to reduce external rotation but to obliterate excess capsular volume and to reattach the IGHLC to the glenoid.

If there is evidence of capsular laxity, the T-plasty modification is used to shift the capsule (Fig. 15C–19D). The small transverse capsulotomy is enlarged, and the capsulotomy is converted to a T shape by the addition of a vertical limb. If a Bankart lesion is present also, this vertical limb should be placed medially at the glenoid margin. The axillary nerve is protected when this limb is created. The capsule is then shifted to correct abnormal anterior or inferior laxity. Specifically, the inferior flap can be taken superiorly and medially to eliminate an abnormally large axillary recess. The same sutures may then be passed through the superior flap for reinforcement and additional tensioning. Again, tension on the capsule should be adjusted with the arm in at least 45 degrees of external rotation to prevent overtightening. If no Bankart lesion is noted in patients with capsular redundancy, the vertical limb of the T-shaped capsulotomy may be placed either medially or laterally. The capsular shift may be technically easier when this limb is placed laterally (Fig. 15C–19E). An arthroscopic examination prior to the open procedure may help in planning the type of capsulorrhaphy to be performed.

It is important to note the presence of an enlarged rotator interval during the assessment of the capsule. If present, an enlarged interval must be closed to create adequate tension in the capsular system. The interval may be closed prior to the creation of a formal capsulotomy or after the capsular tissue has been reattached to the glenoid.

Ten to twenty percent of patients who undergo a stabilization procedure for anterior instability have a significant bony deficiency of the anterior glenoid. In the majority of these cases, stability can be restored by correcting the capsuloligamentous pathology without specifically modifying the procedure to account for the anterior bone loss. A bone block procedure may be indicated in some patients in whom more than one-third of the glenoid is deficient. The capsular pathology should always be addressed in conjunction with this procedure. We have not performed a bone block for anterior instability in more than 10 years.

After the capsule has been satisfactorily addressed, the subscapularis tendon is reapproximated but not tightened. If the conjoined tendon was partially released, it is repaired. A subcutaneous drain is placed, and the wound is closed.

After operation, the shoulder is immobilized for 2 to 3 weeks. Pendulum exercises are instituted at 1 to 2 days. Active rotation and abduction are

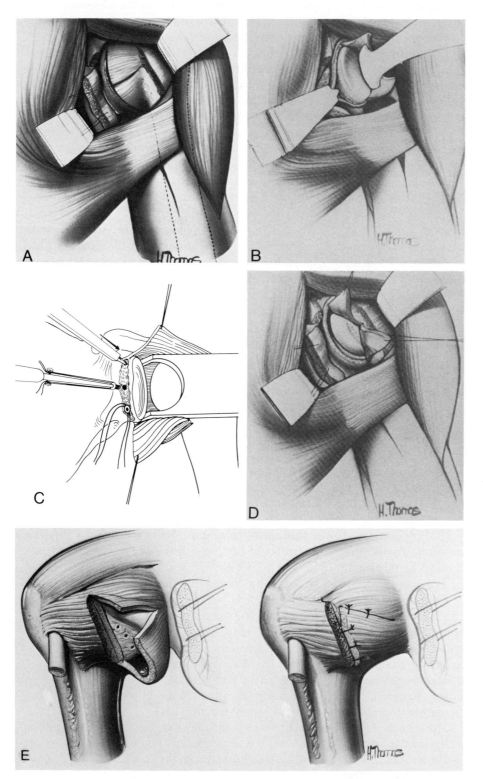

FIGURE 15C–19

Technique of open anterior stabilization. *A,* Capsular exposure in open anterior shoulder stabilization. The subscapularis tendon is divided, exposing the anterior capsule. The size of the "rotator interval" and the degree of capsular laxity are assessed. A small transverse capsulotomy can be performed to determine if a Bankart lesion is present. *B,* If a Bankart lesion is present, a vertical capsular incision is placed at the glenoid margin. Capsular separation is extended medially to allow placement of a retractor along the glenoid neck. A second retractor is then placed to retract the humeral head.

C, Currently, suture anchors are used for suture fixation at the glenoid margin. The anchors should be placed near the glenoid margin rather than medially along the glenoid neck. The capsule is then pulled superomedially and reattached to the bony glenoid so that it obliterates the medial defect in the capsulolabral complex. *D,* A T-plasty is performed if excess capsular laxity is noted. When capsular laxity is accompanied by a Bankart lesion, the vertical limb of the T should be created medially at the glenoid margin. *E,* If no Bankart lesion is present, the capsule can be shifted laterally at the humeral neck. (*A, B,* and *D* from O'Brien, S. J., Warren, R. F., and Schwartz, E. Anterior shoulder instability. *Orthop Clin North Am* 18(3):395–408, 1987.)

allowed after 3 weeks with resistance gradually increased in each subsequent week. Weight training is not allowed in the early healing period. At 6 weeks, aggressive measures are used to increase range of motion. External rotation must be carefully monitored. The exercises outlined in the section on nonoperative treatment are also applicable to postoperative therapy.

Postoperative rehabilitation is more aggressive in throwers. In this group, pendulum exercises are begun in the first postoperative week, and the program is accelerated accordingly. In throwers, the goal is to restore full range of motion by 6 weeks postoperatively. Heavy lifting, contact sports, and throwing are avoided for 6 months after operation.

Arthroscopic Stabilization Procedures

Arthroscopic stabilization techniques have stimulated a great deal of interest in the orthopaedic community since the introduction of arthroscopic staple capsulorraphy by Johnson in the early 1980s [83]. Arthroscopic techniques offer the advantage of less perioperative pain and morbidity, and they can be performed in an outpatient setting. At the present time, however, the risk of recurrence after arthroscopic stabilization is higher than that after an open procedure. Early reports revealed recurrence rates of between 15% and 20% after arthroscopic stabilization procedures [56, 101, 182, 183]. Recently, Morgan [102] reported a recurrence rate of only 5% after 1- to 7-year follow-up of 175 patients who had undergone an anterior stabilization using a transglenoid suture technique.

Techniques of arthroscopic stabilization are still in their infancy. At present, these procedures should be performed only by experienced arthroscopists in selected patients. A well-performed open stabilization is certainly preferable to a failed arthroscopic procedure. Patients treated arthroscopically may require a longer period of postoperative immobilization than patients treated by open techniques [56].

In our view, the ideal candidate for an arthroscopic stabilization is a patient with recurrent anterior subluxation or dislocation who has a demonstrated detachment of a stout, well-defined IGHLC-labral complex on arthroscopic examination. Patients who appear to have an attenuated, patulous anterior capsule are less likely to have a good result [173]. There is great hope that arthroscopic techniques will improve the results of operative treatment of instability in throwers. In the series of Rowe and colleagues [136], 69% of patients treated by an open Bankart procedure regained full motion. Morgan [102] reported recovery of full range of motion in 87% of the first 55 patients that he had treated arthroscopically. Less than one-third of the

throwing athletes treated by Rowe and colleagues were able to return to their premorbid level of pitching. Warner and associates [173] recently reported that 9 of 12 overhead athletes who were stabilized arthroscopically with biodegradable tac were able to return to their preinjury level of function.

Multidirectional instability and voluntary instability are absolute contraindications to the use of an arthroscopic stabilization procedure. The finding of a poorly formed IGHLC or the lack of a Bankart lesion on arthroscopic examination is a relative contraindication. Some surgeons feel that an arthroscopic method should not be used in the presence of a large Hill-Sachs lesion. Morgan [102] recommended against using an arthroscopic technique in collision athletes.

OPERATIVE TECHNIQUE. We have experience with two forms of arthroscopic stabilization procedure. The first type of stabilization employs sutures passed through drill holes in the glenoid neck and tied posteriorly. The second technique involves the use of an absorbable tac. We have recently begun to experiment with the arthroscopic use of suture anchors. We do not recommend the routine use of metal hardware around the shoulder.

SUTURE TECHNIQUE. In 1983, we began using a technique of arthroscopic stabilization that was derived from the pull-out suture techniques described for open Bankart repair by Luckey [90] and by Viek and Bell [169]. A standard arthroscopic examination is performed using the anterior and posterior portals described earlier. The labrum and anteroinferior capsule are carefully assessed to determine the quality of the tissue. If there is neither a functional labrum nor a robust IGHLC, the arthroscopic technique is likely to fail, and an open procedure should be performed.

If the labrum is functional but detached, it is grasped from the anterior portal and brought toward the anterior glenoid neck (Fig. 15C–20A). When tension is applied, the labral-IGHLC complex should tighten. If the labrum is absent or nonfunctional, the anterior band of the IGHLC is advanced with the grasper, and a similar observation is made to assess the quality and tension within the system. The axillary pouch should disappear when these tissues are advanced medially. In some cases, electrocautery is used to extend the separation from the anterior glenoid neck to allow sufficient advancement of the capsulolabral system.

If the tissue quality is felt to permit the performance of an arthroscopic stabilization procedure, the anterior glenoid neck is carefully debrided to bleeding bone. A combination of a rasp, an arthroscopic

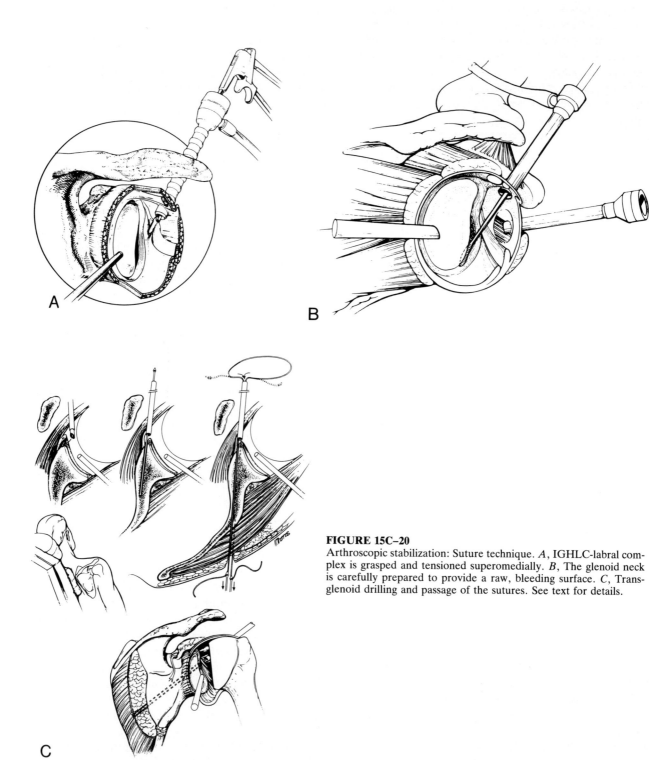

FIGURE 15C–20

Arthroscopic stabilization: Suture technique. *A*, IGHLC-labral complex is grasped and tensioned superomedially. *B*, The glenoid neck is carefully prepared to provide a raw, bleeding surface. *C*, Transglenoid drilling and passage of the sutures. See text for details.

shaver, and an arthroscopic burr is used for this purpose (Fig. 15C–20B). Adequate preparation of the neck is an important part of the procedure.

At this point, a Beath pin is placed through the anterior portal. A second anterior portal is created more superiorly, and the grasper is used to advance the capsulolabral tissue in a medial and superior direction through this accessory portal. The sharp end of the pin is used to spear a robust portion of the tissue, and the pin is placed on the glenoid neck at the two o'clock position. The pin is then drilled in a posterior direction, making sure to avoid the glenoid articular surface (Fig. 15C–20C). The entry point of the pin should be 2 to 3 mm medial to the edge of the glenoid articular surface. The pin is drilled through the posterior cortex of the glenoid and is recovered after exiting from the posterior skin. If the pin exits too far laterally, the suprascapular nerve is at risk for damage.

The pin is identified posteriorly, and a 5-mm incision is made at its exit point. The drill is detached from the tail of the pin, and two size 0 PDS sutures are passed through the eyelet of the pin. One end of each suture is held anteriorly as the pin is advanced completely out of the posterior shoulder. The posterior halves of the sutures are tagged, and the anterior halves are brought out of the accessory portal. The process is repeated a second time with two additional sutures being passed at the four o'clock position.

The anterior sutures are paired together so that each suture is coupled with a suture from the other drill hole. The sutures are then tied, and tension is applied from posterior to bring the capsulolabral structures into position on the neck. A subcutaneous tunnel is then created posteriorly from one group of sutures to the exit point of the other group. This technique allows the posterior halves of the sutures to be tied together in a subcutaneous location. After the sutures have been tied, there should be good tension in the anterior capsulolabral system, and the "drive-through" sign should be eliminated.

BIODEGRADABLE TAC [173, 178]. In an effort to avoid the problems associated with tying sutures over a fascial bridge near the suprascapular nerve as well as the technical difficulty of transglenoid drilling, we have recently used an absorbable tac (Acufex Microsurgical Inc., Mansfield, MA) as a fixation device for arthroscopic stabilization. The tac is cannulated and is made of polyglyconate. Its strength diminishes over a 4-week period. Ribs on the shaft of the tac increase its pull-out strength to approximately 100 newtons. The tac's broad, flat head allows it to capture soft tissue.

The same basic arthroscopic technique is used with the tac as with the suture technique. The glenoid neck is prepared, and an accessory portal is created to grasp and tension the capsulolabral tissue. A cannulated drill bit that contains a guide wire is placed through the anterior portal. The wire is locked so that it protrudes a few millimeters from the end of the drill bit. The wire is used to pierce the capsulolabral tissue, and the tissue is then brought to the two o'clock position on the glenoid neck (Fig. 15C–21A). The drill bit is advanced to a depth of approximately 12 mm into the bony glenoid (Fig. 15C–21B). The wire is then unlocked from the drill bit and gently tapped to free it from the drill bit. The drill bit is removed, leaving the wire in place. A tac is then placed over the wire and forced into place using a cannulated pusher (Fig. 15C–21C). We generally attempt to place a second tac at the four o'clock position as well (Fig. 15C–21D).

In some patients, a combination of the two techniques may be needed for optimal tensioning of the tissue. In this situation, sutures are used superiorly to tension the tissues, and a tac is then placed at the four o'clock position to close the defect.

After an arthroscopic stabilization, patients are maintained in internal rotation in a shoulder immobilizer for 4 weeks postoperatively. Elbow range of motion exercises are encouraged during this period. Shoulder motion is initiated at 4 weeks using active-assisted and passive techniques. When full range of motion is obtained, resistance exercises are instituted. At 4 months, the patient may resume light throwing and underhand racquet sports. After 6 months, contact sports and unrestricted activity are permitted.

Management of Failed Anterior Stabilizations

Recurrence after an anterior shoulder stabilization procedure is uncommon if the pathologic lesions are addressed at the time of the initial procedure. Morrey and Janes [103] reported an increased failure rate in patients with bilateral instability and in those with a family history of instability. These patients may have generalized ligamentous laxity and require careful evaluation for evidence of multidirectional instability. When recurrence does occur after an open operative procedure, Zarins and colleagues [186] noted that special consideration must be given to two key questions:

1. Does the instability occur in the same or in a different direction from that noted preoperatively?

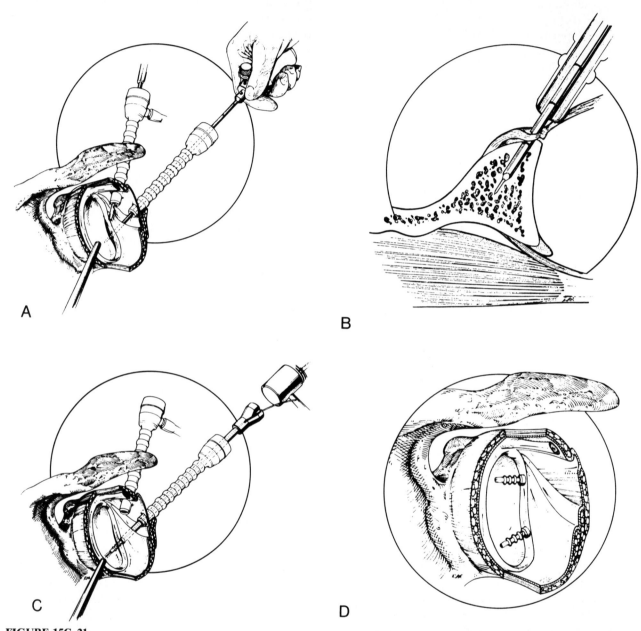

FIGURE 15C–21

Arthroscopic stabilization: Biodegradable tac. *A*, As the IGHLC-labral complex is pulled superomedially, the wire is used to pierce the tissue and is then directed upon the glenoid neck just medial to the glenoid margin. *B*, The drill bit is advanced to a depth of approximately 12 mm into the glenoid neck. *C*, The drill bit is removed, and a tac is placed over the guide wire. A cannulated pusher is then used to force the tac onto the glenoid neck. *D*, Generally, tacs are placed at the two o'clock and four o'clock positions.

2. Was the first postoperative recurrence associated with a significant traumatic event?

If the recurrent instability occurs in the same direction and was secondary to significant trauma, it is likely that the previous operative procedure correctly addressed the pathology but that new trauma disrupted the repair. If instability occurs in the same direction and the first postoperative recurrence was not associated with significant trauma, it is likely that the previous operation did not correct the pathology.

Postoperative recurrence may occur in a different direction from that which the operation was designed to prevent. In cases associated with significant trauma, this may represent a new injury that is unrelated to the original problem. More commonly, the surgeon must question whether or not the previous operation was performed at the correct site. This question is especially salient when preoperative radiographs did not prove the direction of instability.

Atraumatic instability in the opposite direction may indicate that the initial repair was too tight. With shoulder instability in one direction, there is often a degree of increased translation in the opposite direction. If the primary direction of instability is overly constrained, the head will be forced out on the opposite side. Alternatively, atraumatic instability in the opposite direction may occur because the patient actually had posterior or multidirectional instability at the time of the initial procedure.

The evaluation of a patient with recurrent instability after a stabilization procedure begins with a careful review of the history. A search is made for preoperative radiographic evidence of anterior instability. The operative report should be obtained. The degree of trauma that was responsible for the first postoperative recurrence is determined. The shoulder is reexamined, and a special effort is made to rule out multidirectional instability. Voluntary instability and its attendent psychological problems should be ruled out also. New radiographs are obtained to evaluate the bony anatomy and to assess the location of any hardware that may have been placed near the shoulder.

Pathologic findings in a patient with a postoperative recurrence most commonly reveal an unrepaired Bankart lesion [139]. Another common finding is excessive capsular laxity, often associated with a large rotator interval. In some cases, the subscapularis tendon may be disrupted, or the capsular injury may have occurred laterally at the humeral neck. Clinical failure after an anterior shoulder procedure may be due to reasons other than recurrence. Limitation of external rotation may severely compro-mise function. Loosened metal devices or bone grafts can penetrate the joint and may be the source of disabling pain.

Patients with recurrent instability after a stabilization procedure may respond to a rehabilitative program such as the regimen outlined earlier. Arthroscopy may provide helpful diagnostic clues in selected patients who do not respond to conservative measures.

Prior to proceeding with a revision stabilization procedure, the surgeon should consider the manner in which the anatomy was altered by the initial procedure. Careful dissection techniques must be employed. After a failed Bristow procedure, in particular, extensive scarring in the region of the anterior capsule and subscapularis tendon makes it difficult to identify the usual anatomic landmarks [184].

The results of revision stabilization procedures are not as favorable as those reported for primary stabilizations [139, 170, 184]. A modification of the Bankart procedure is used in most cases to address unrepaired lesions of the capsuloligamentous system. In patients who are found to have evidence of multidirectional instability, extensive scarring may make it especially difficult to shift the capsule [60]. In patients with posterior instability after an anterior repair, release of the anterior structures may be required before the procedure continues posteriorly.

The presence of extensive scarring of the anterior structures may necessitate shortening of the subscapularis tendon. The functional limitations of restricted external rotation are unavoidable in these cases. A bone block procedure may have to be considered as well. Rowe and colleagues [139] and Zarins and co-workers [186] have suggested that transfer of the infraspinatus tendon to the defect in the posterolateral humeral head (the Connolly procedure [30]) may be indicated in some cases of postoperative recurrence associated with large Hill-Sachs lesions. Although this procedure is of interest, it has not been required in any case in our experience.

POSTERIOR DISLOCATION

Posterior dislocations constitute only 2% to 4% of all shoulder dislocations. The largest series, 37 patients, was reported by Malgaigne in 1855 [92]. The overwhelming majority of posterior dislocations are of the subacromial type, although occasionally the humeral head may come to rest under the glenoid or under the scapular spine. Probably because of the relative infrequency of posterior dislocation, the diagnosis is initially missed in 50% to 80% of cases [19, 135].

A posterior dislocation may be caused by a direct blow to the anterior shoulder, but indirect forces are much more common causative factors. The classic examples of indirect forces are those associated with electrical shocks and with seizures. In these situations, the powerful internal rotators (latissimus dorsi, pectoralis major, and subscapularis) are thought to overcome the weaker external rotators, forcing the humeral head posteriorly [95]. A less common indirect force that can cause posterior dislocation is a fall on an outstretched hand with the arm in a relatively adducted position. Atraumatic posterior dislocations may occur in patients with congenital laxity.

The patient who presents with an acute posterior dislocation has an internal rotation deformity and limited external rotation of the arm. The arm is also adducted with abduction usually limited to less than 90 degrees. In thin patients, posterior prominence of the shoulder and anterior prominence of the coracoid may be noted [98] (Fig. 15C–22). The patient is unable to supinate the forearm fully with the arm forward flexed [135]. After reduction of the shoulder, only increased posterior translation and posterior apprehension may be evident.

Several radiographic signs have been described to aid in the detection of posterior dislocation on an anteroposterior view; this view is rarely diagnostic, however [19, 145]. Normally, the head of the humerus is superimposed on the glenoid in a smooth-

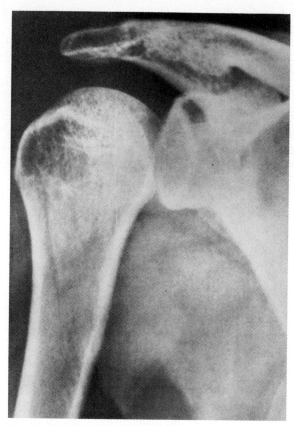

FIGURE 15C–23
The vacant glenoid in a patient with posterior dislocation. (From Schwartz, E., Warren, R. F., O'Brien, S. J. et al. Posterior shoulder instability. *Orthop Clin North Am* 18(3):409–419, 1987.)

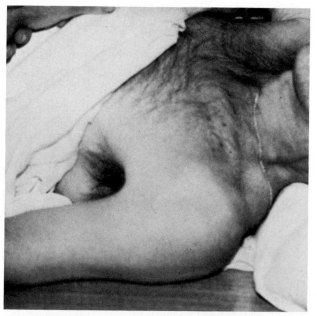

FIGURE 15C–22
Posterior dislocation. Prominence of the coracoid is noted anteriorly, and there is an abnormal fullness in the posterior shoulder.

bordered ellipse. Distortion of this *elliptical overlap shadow* may indicate the presence of a posterior dislocation. In the normal situation, the humeral head appears to fill most of the glenoid on an anteroposterior radiograph. With a posterior dislocation, the glenoid may appear to be vacant (Fig. 15C–23). A positive *rim sign* is present if the space between the anterior rim of the glenoid and the humeral head is greater than 6 mm; a high percentage of patients with this finding have a posterior dislocation [7]. A reverse *Hill-Sachs lesion* is an impaction fracture of the anteromedial humeral head that occurs as the head contacts the posterior rim of the glenoid.

Although the trans-scapular lateral view may show that the humeral head lies posterior to the glenoid, the axillary view is generally the most helpful view if one suspects a posterior dislocation. The presence and assistance of a physician may be required to obtain an adequate study. In addition to showing the position of the humeral head relative to the glenoid and providing information about glenoid version, the axillary view may also reveal impaction fractures, glenoid fractures, and lesser tuberosity fractures. A CT scan may provide additional help in the assessment of glenoid version.

Posterior dislocations may be associated with disruption of the posterior labrum or capsule [180]; however, Warren and associates [176] have shown that detachment of the entire posterior capsule does not cause posterior dislocation unless it is accompanied by disruption of the anterosuperior capsule as well.

Lesser tuberosity fractures are commonly associated with posterior dislocations, and the presence of such a fracture should alert the clinician to the need to rule out a posterior dislocation. Neurovascular injuries and rotator cuff tears are much less common after posterior dislocation compared to anterior dislocations. Recurrence is also more infrequent after a traumatic posterior dislocation than after a traumatic anterior dislocation [95]. An increased recurrence rate after posterior dislocation has been noted in younger patients, in cases with an atraumatic etiology, and when a large bony defect is noted in either the glenoid or the humeral head. These recurrence rates may be of little value, however, because most published series of posterior dislocations consist of relatively small numbers of patients [17, 43, 66, 98, 133, 185].

Treatment of the acute posterior dislocation usually begins with an attempt at closed reduction if no fracture lines are seen in the humeral head [19, 145]. Any associated fracture may extend and result in displacement of the head fragment during reduction. If an attempt at closed reduction is thought to be safe, the patient should be suitably sedated with an intravenous benzodiazepine and a narcotic. Some patients may require general anesthesia. The patient is positioned supine. Lateral traction is applied to the arm, and gentle *internal* rotation is used to unlock the impaction fracture from the glenoid. Use of an excessive *external* rotation force at this point may displace the head fragment (Fig. 15C–24). Posterior pressure is then applied to the head, and a longitudinal traction force is placed on the adducted arm. The head is gently lifted back into the glenoid as the arm is rotated externally.

If the reduction is stable, the arm is immobilized in 0 degrees of abduction and in slight extension. When the shoulder is unstable after reduction, the arm is externally rotated to 20 degrees in a splint or brace. Young patients are immobilized for a period of 4 to 6 weeks and then are started on an aggressive physical therapy program that emphasizes external rotator strengthening. Older patients are immobilized for only 2 to 3 weeks.

Operative treatment with open reduction is indicated if closed reduction fails, if there is a significant risk of further damage to the head with closed reduction, and when there is major displacement of a head or glenoid fragment. A posterior approach to the shoulder is preferred because damage to the posterior capsule can be addressed through this approach.

Postoperatively, the arm is splinted in neutral rotation, slight abduction, and extension for 6 weeks. Rehabilitation is instituted after 6 weeks and initially consists of passive and active-assisted range of motion exercises. Muscle strengthening is begun after the patient has achieved a nearly normal range of motion.

Fixed Posterior Dislocation

Chronic posterior dislocation of the shoulder is not rare, owing largely to the significant number of cases that are missed on initial examination. Rowe and Zarins [135] noted that 14 of 24 chronically unreduced shoulders in their series were of the posterior type. This predominance is especially remarkable when one considers the relative infrequency of posterior dislocation. Samilson and Prieto [142] noted an increased incidence of degenerative arthritis after posterior dislocation, and they attributed this to a delay in diagnosis. Locked posterior dislocations are usually associated with a large humeral head impression fracture [105].

The patient with a fixed posterior dislocation presents with a painful, stiff shoulder that has been unresponsive to physical therapy. The differential diagnosis includes frozen shoulder, adhesive capsulitis, and massive rotator cuff tear. On physical examination, an internal rotation deformity of the arm is noted. This deformity is greater with a larger humeral head defect [59]. No external rotation is present. There is limited supination when the arm is forward flexed (Fig. 15C–25). Abduction is limited but may not be painful.

The axillary radiograph is diagnostic in all cases of chronic posterior dislocation [46]. This view is also helpful in demonstrating the size of the impression fracture. A CT scan of the shoulder is obtained routinely to assess the anatomy of the head and glenoid (Fig. 15C–26).

Nonoperative treatment may be considered in the elderly patient who has minimal pain and little functional loss, but reduction and surgical reconstruction provide improved results in most patients [59]. Gentle closed reduction under anesthesia may be attempted when less than 20% of the humeral head is involved and when the dislocation is less

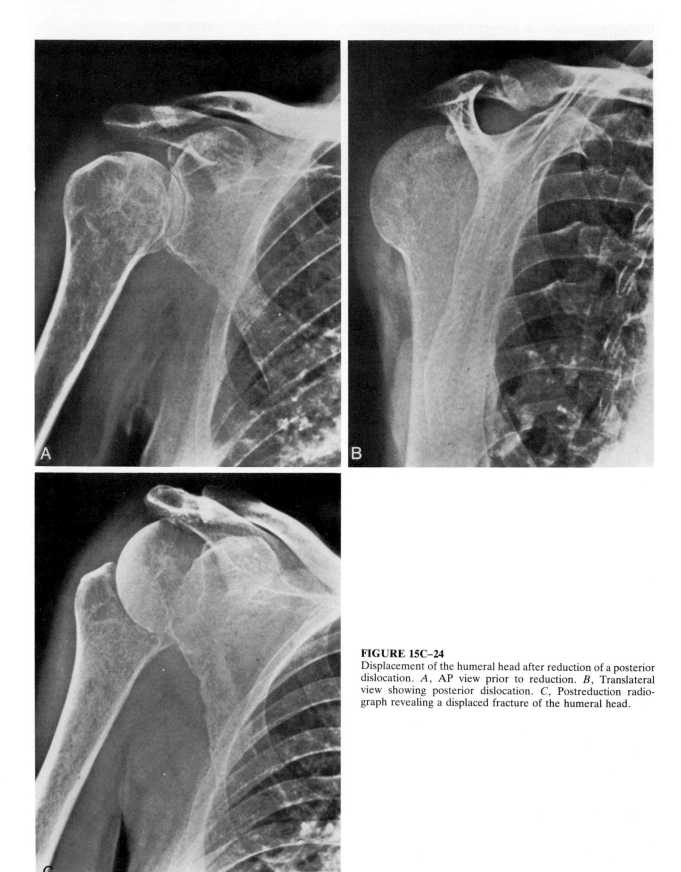

FIGURE 15C–24
Displacement of the humeral head after reduction of a posterior dislocation. *A*, AP view prior to reduction. *B*, Translateral view showing posterior dislocation. *C*, Postreduction radiograph revealing a displaced fracture of the humeral head.

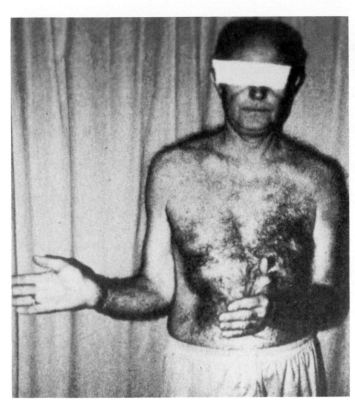
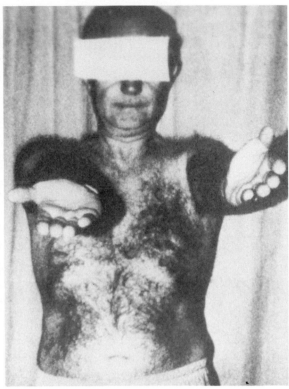

FIGURE 15C–25
The Rowe sign. The patient with a posterior dislocation has limited external rotation and restricted supination with the arm forward flexed. (From Schwartz, E., Warren, R. F., O'Brien, S. J., et al. Posterior shoulder instability. *Orthop Clin North Am* 18(3):409–419, 1987.)

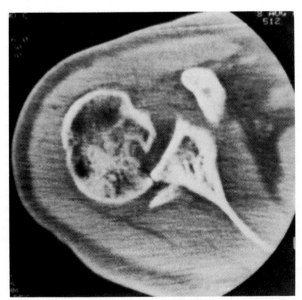

FIGURE 15C–26
CT scan of fixed posterior dislocation. Note the presence of the impression fracture of the humeral head. (From Schwartz, E., Warren, R. F., O'Brien, S. J., et al. Posterior shoulder instability. *Orthop Clin North Am* 18(3):409–419, 1987.)

than 6 weeks old. If closed reduction fails, open reduction is necessary.

When there is 20% to 50% head involvement or when the dislocation is more than 6 weeks old, open reduction with subscapularis transfer is indicated. The shoulder is approached through a deltopectoral incision. The joint is inspected by incising the subscapularis tendon (McLaughlin technique [98]) or by osteotomizing the lesser tuberosity (Neer modification [105]). The long head of the biceps is used as a guide if shoulder anatomy is distorted. The head is reduced, and the subscapularis (or the lesser tuberosity) is advanced into the head defect and secured with sutures or a screw (Fig. 15C–27). If excessive posterior translation is noted following reduction, the stretched posterior capsule is plicated, and the arm is immobilized in 20 degrees of external rotation. We have performed this procedure in patients with dislocations as old as 10 months with acceptable results; viability of the head may be maintained despite the chronic nature of the dislocation.

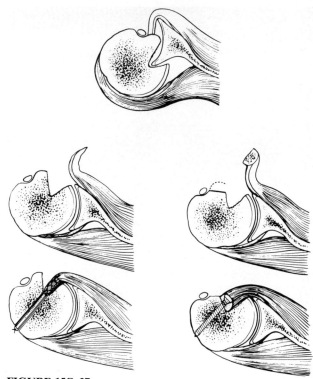

FIGURE 15C–27
Subscapularis transfer for the treatment of a fixed posterior dislocation. Left, McLaughlin technique. Right, Neer modification. See text for details.

Prosthetic replacement is indicated when more than 50% of the head is involved or when articular surface degeneration is present. If the glenoid is degenerated, it should be resurfaced at the time of proximal humeral replacement. Technical hints at the time of prosthetic replacement after a fixed posterior dislocation include placing less retroversion in the humeral component and occasionally considering the use of a bone graft to build up a posteriorly deficient glenoid [59].

Recurrent Posterior Dislocation

Recurrent posterior dislocation, as mentioned earlier, is unusual. Voluntary posterior dislocation has been reported [26, 138]. Boyd and Sisk [17] noted posterior Bankart lesions in four of nine shoulders that were operated upon for recurrent posterior dislocation. Rowe [133] found posterior labral or capsular detachment in three of eight shoulders that were operated upon. Rowe also noted an anteromedial head defect in three of ten patients who had this condition.

Neer [105] reported a series of 23 patients with recurrent posterior dislocation in which no additional recurrences were noted after a posterior cap-

sular shift procedure was performed. Patients with recurrent posterior subluxation may have been included in this series. Operative intervention should be considered if more than one recurrence is noted in a traumatic dislocator. We recommend a posterior capsulorrhaphy procedure as described in the following section.

Recurrent Posterior Subluxation

Recurrent posterior subluxation (RPS) is the most common form of posterior instability. Although there is increased recognition of this entity in athletes, the diagnosis is often missed or delayed. In a recent report, the diagnosis was made an average of 14 months after the onset of symptoms [18]. Patients with RPS do not usually progress to complete dislocation [44]; recurrent dislocation is rare compared to recurrent subluxation.

These patients often have a history of overuse rather than of macrotrauma. They usually present because of *pain*. Pain may be localized anteriorly or posteriorly, or it may be diffuse in nature. Symptoms do not usually limit the activities of daily living or work, but they may interfere with athletic performance. The pain tends to become worse during the follow-through phase of throwing and racquet sports and during the pull-through phase of swimming. The arm is flexed, adducted, and internally rotated during these activities. Symptoms may also increase during the bench press in weightlifters and are not uncommon in offensive linemen in football. Instability is usually a secondary complaint. Instability symptoms tend to increase with time, however, and they appear to be more common in patients who have had an episode of macrotrauma.

Physical examination typically reveals pain or symptoms of instability when the arm is flexed, adducted, and internally rotated. Examination should be performed with the patient supine and the arm abducted 90 degrees and in neutral rotation. A posterior force is exerted while a gentle axial load is provided to the elbow. Posterior subluxation of the head on the glenoid may be noted when the arm is in this position. The examiner should note posterior translation with a click. Painful or marked instability is rare. In contrast to the patient with recurrent dislocation, the patient with RPS may not have a posterior apprehension sign [58]. These patients should be carefully evaluated for concomitant inferior instability to rule out multidirectional instability. Increased ligamentous laxity is not uncommon in patients with RPS.

Many patients with RPS learn to subluxate the

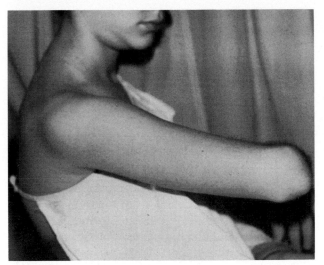

FIGURE 15C–28
Positional voluntary posterior subluxation. Subluxation occurs with forward flexion and internal rotation of the arm.

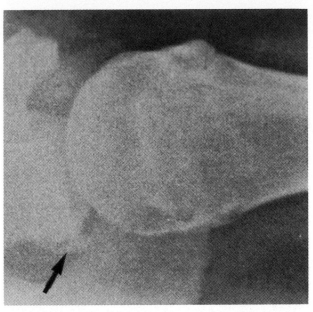

FIGURE 15C–30
Axillary view revealing a posterior osseous Bankart lesion (arrow). (From Schwartz, E., Warren, R. F., O'Brien, S. J., et al. Posterior shoulder instability. *Orthop Clin North Am* 18(3):409–419, 1987.)

shoulder voluntarily [18, 44, 57, 58, 105]. In most of these patients, it is their *involuntary* instability that leads them to seek treatment. A small number of patients subluxate their shoulders habitually to manipulate their environment; these individuals are psychologically disturbed and will not be helped by operative intervention. However, the majority of patients who can voluntarily display posterior subluxation have no underlying psychiatric disease.

Two types of voluntary instability have been described [44]. In the *positional* type, the head subluxates posteriorly as the arm is forward flexed and internally rotated (Fig. 15C–28). The head then reduces with extension of the arm. The second type incorporates selective activation of the internal rotators to cause posterior subluxation (Fig. 15C–29). This *muscular* type of voluntary instability is felt to be indicative of a poor response to surgical management. Hawkins has noted that the diagnosis of RPS may be very difficult in patients who cannot demonstrate instability [57, 58].

The pathoanatomy of RPS is often quite subtle. A reverse Bankart lesion is not usually present, although clefts may be found in the posterior labrum [58]. Articular cartilage degeneration may also be noted. Glenoid version does not appear to play a role in the production of RPS [45, 46]. Radiographic studies are rarely diagnostic. Twenty percent of affected patients exhibit calcification of the posterior glenoid rim and capsule. Another 20% have some erosion of the posterior glenoid [18] (Fig. 15C–30).

Treatment of Recurrent Posterior Subluxation

Most patients with RPS, especially those with an atraumatic or microtraumatic etiology, respond well to an aggressive rehabilitation program that empha-

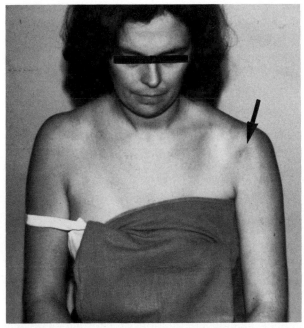

FIGURE 15C–29
Muscular voluntary posterior subluxation. The humeral head subluxates with selective contraction of the internal rotators. (From Schwartz, E., Warren, R. F., O'Brien, S. J., et al. Posterior shoulder instability. *Orthop Clin North Am* 18(3):409–419, 1987.)

sizes rotator cuff strengthening. Patients with complaints of instability and those with a history of macrotrauma appear to do less well after a trial of physical therapy [18].

Operative treatment is indicated in patients with pain or unintentional instability despite rehabilitation, in those with pain during the activities of daily living, and in competitive athletes who develop symptoms during strenuous activity [44]. The early published results of operative treatment for RPS were not encouraging. Hawkins and colleagues [58] noted a 50% failure rate after a variety of stabilization procedures including glenoid osteotomy, biceps transfer, and capsulorrhaphy. There is a substantial risk of glenohumeral arthritis after posterior glenoid osteotomy [84]. Tibone and Ting [160] reported a 30% failure rate following open posterior staple capsulorrhaphy. More recently, Bowen and colleagues [18] reported a more favorable outcome (with a 12% recurrence rate) after posterior T-plasty capsular shift.

Operative Technique [18, 44]. The technique of posterior T-plasty capsular shift is as follows: The patient is placed in either the lateral decubitus or, more recently, the modified beach chair position. A horizontal or vertical skin incision may be used. The horizontal incision, used initially, is placed 1 cm inferior to the scapular spine; it allows the surgeon to obtain a bone graft from the scapular spine if indicated. A vertical incision [24] is presently used in most patients (Fig. 15C–31A). It is made midway between the lateral border of the acromion and the posterior axillary crease. The superficial deltoid fascia is identified, and the deltoid is split to expose the infraspinatus and teres minor tendons. Division of the deltoid begins at the scapular spine and proceeds inferiorly (Fig. 15C–31B). Next, the infraspinatus tendon is incised vertically allowing a large portion of the tendon to remain laterally. (In throwers, the capsule may be exposed by developing an interval within the infraspinatus tendon without dividing the tendon. This option is especially useful if the tissue is of good quality. If tissue quality is suboptimal, the approach may be converted by obliquely incising the tendon.) The surgeon should be mindful of the axillary nerve and the posterior circumflex humeral vessels, which exit the quadrilateral space immediately inferior to the teres minor. A T-shaped capsular incision is then made. The vertical limb of the incision is placed medially, near the glenoid (Fig. 15C–31C). An incision of this type allows easier repair of a Bankart lesion if one is present. The horizontal limb of the capsular incision is extended laterally from the vertical component. With the arm in neutral rotation and 30 to 40 degrees of abduction, the capsule is reattached to the glenoid through drill holes or by means of suture anchors. Prior to reattachment, the inferior limb of the capsule is advanced medially and superiorly to eliminate laxity in the posteroinferior capsule. If marked capsular laxity is noted with the arm in a position of abduction despite the medial T-plication, a second vertical capsular incision is placed laterally to create an H-plasty (Fig. 15C–31D). The inferior portion of the capsule can be advanced further by this method. Inferior laxity with the arm in adduction may indicate that the superior structures need additional tensioning. Tensioning of the superior structures may be technically difficult from a posterior approach. This difficulty has led some surgeons to consider an anterior approach in selected cases.

The capsular repair may be reinforced using the infraspinatus tendon if local tissue is felt to be insufficient. If a large posterior defect is found on the glenoid, consideration can be given to obtaining a tricortical bone block from the scapular spine to compensate for the bony deficiency. When it is used, the graft is placed at the posteroinferior quadrant of the glenoid to increase the articulating surface of the glenoid at a point contiguous with its curvature (Fig. 15C–32). The humeral head should not be allowed to impinge upon the bone graft. The bone block technique has been used infrequently in the past 5 years.

The patient is maintained in an Orthoplast splint with the arm in external rotation for the first 6 weeks after operation. In throwers, early passive motion may be instituted with rotation from neutral to full external rotation. Full internal rotation should be avoided. Flexion is best avoided for 6 weeks, but elevation in the plane of the scapula will allow healing to occur without placing undue stress on the repair.

The anterosuperior capsule has been shown to be an important stabilizer against posterior translation of the humeral head [176]. In our experience, it has not been necessary to address the anterior structures specifically at the time of reconstruction; however, the surgeon may have to consider the role of the anterosuperior capsule in patients with posterior instability. Additional tensioning of this area may be especially applicable in cases with an atraumatic etiology.

The role of arthroscopic Bankart repair in the treatment of RPS remains undefined. Arthroscopic findings include attenuation or capsular-periosteal separation of the posterior capsule, tearing or detachment of the posterior labrum, and the reverse Hill-Sachs lesion. We have used an absorbable tac for arthroscopic stabilization of RPS with good

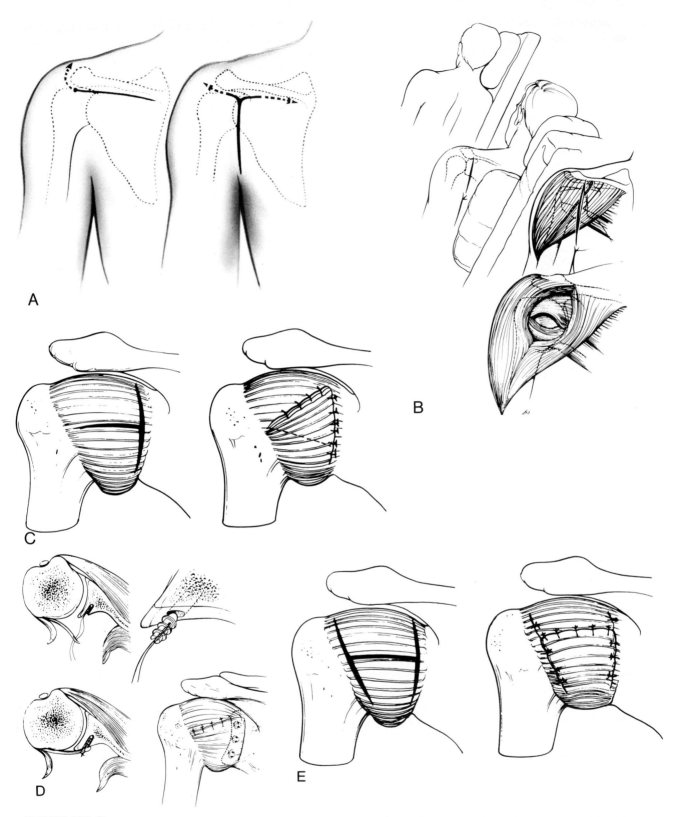

FIGURE 15C–31

Posterior T-plasty capsulorrhaphy. *A*, Horizontal and vertical incisions. (From Fronek, J., Warren, R. F., and Bowen, M. Posterior subluxation of the glenohumeral joint. *J Bone Joint Surg* 71A:205–216, 1989.) *B*, Posterior approach in the beach chair position. A vertical incision is used. The posterior deltoid is split in the direction of its fibers. In this illustration the capsule is exposed by creating a transverse interval within the infraspinatus. *C*, Posterior T-plasty capsular incision and shift. *D*, Posterior T-plasty capsulorrhaphy using suture anchors. *E*, Posterior H-plasty capsular incision and shift.

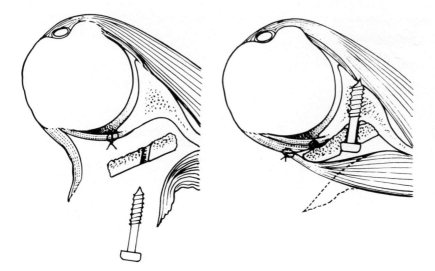

FIGURE 15C–32
Posterior bone block. The graft is placed so that it is contiguous with the curvature of the glenoid. The humeral head should not impinge upon the bone block.

preliminary results, but further follow-up of patients treated in this manner is required [173].

Possible indications for posterior glenoid osteotomy in the treatment of RPS include glenoid retroversion in excess of 20 degrees and congenital hypoplasia of the glenoid [95]. In our opinion, this procedure is rarely indicated.

MULTIDIRECTIONAL INSTABILITY

The patient with multidirectional instability has symptomatic inferior instability in addition to anterior or posterior instability. The presence of inferior instability is a *requirement* for the diagnosis of multidirectional instability [105]. These patients often present with bilateral complaints. Approximately 50% of affected patients have evidence of generalized ligamentous laxity. Classically, either there is no history of trauma or the initial dislocation occurred with trauma that would not dislocate a normal shoulder. In loose-jointed athletes, however, trauma may play a significant role in the development of symptoms.

Physical examination usually reveals little or no apprehension. The sulcus sign is positive. There are three basic types of multidirectional instability, and these can be differentiated on the basis of the direction and degree of abnormal translation. Type I is global instability and dislocation that occurs in all three directions. Type II comprises anterior and inferior dislocation as well as mild posteroinferior subluxation. Patients with posterior and inferior dislocation and mild anteroinferior laxity are classified as type III. Impingement findings occur in approximately 20% of patients with multidirectional instability [1].

The pathology of the multidirectionally unstable shoulder of the *atraumatic* type usually consists of a large inferior capsular pouch that extends both posteriorly and anteriorly [106]. Anterior labral detachment is generally not associated with this capsular redundancy. In contrast, a *traumatic* type of multidirectional instability exists and will be seen with varying frequency depending on the physician's patient population. In loose-jointed athletes, particularly, a traumatic event may damage the shoulder tissue to such an extent that the result is a shoulder with both multidirectional instability and a Bankart lesion of varying size.

To paraphrase Neer [105], "Not all loose shoulders are painful and not all require treatment." Symptomatic patients should be given a thorough trial of internal and external rotator strengthening. Patients with atraumatic multidirectional instability often respond to nonoperative therapy. Intentional dislocators should be recognized. Patients with inferior laxity may fail to respond to standard operative procedures designed for unidirectional instability. In some cases, these standard procedures may cause excessive tightness on one side of the hypermobile shoulder. Subluxation or dislocation will then occur in the opposite direction, and glenohumeral arthritis may occur.

At present, in our opinion, there is no role for arthroscopic stabilization procedures in the treatment of multidirectional instability.

Operative Technique

The type of operative approach used for multidirectional instability should be determined by the patient's history and physical findings. Generally,

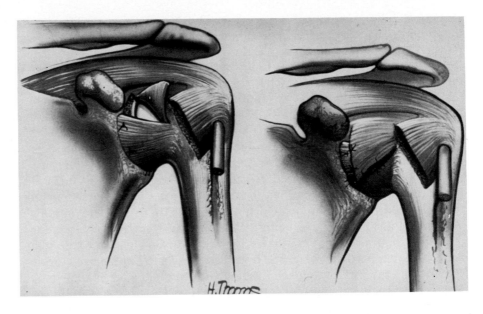

FIGURE 15C–33
Anterior T-plasty capsulorrhaphy for multidirectional instability. Horizontal and medial vertical capsular incisions were created. The inferior flap is shifted superomedially. The superior flap is then tensioned by transposing it inferiorly. An enlarged rotator interval, if present, should be closed in conjunction with the capsulorraphy.

the approach should be made on the side associated with the greatest amount of clinical instability. In patients whose instability is primarily anterior, the anterior deltopectoral approach is used to perform a T-plasty modification of the Bankart procedure [1]. After exposure of the anterior capsule by division of the subscapularis tendon, a search is made for the presence of a rotator interval. If an interval is present, it is closed to increase tension in the capsule. The labrum is viewed through a transverse capsulotomy, which is created obliquely in the midportion of the capsule from lateral to medial. If a Bankart lesion is noted, the capsulotomy is converted to a T by creating a vertical limb at the glenoid margin that extends back to the posterior capsule (Fig. 15C–33). The axillary nerve is vulnerable during this portion of the procedure and must be protected. The inferior flap is then advanced

superiorly to eliminate inferior laxity and medially to a point where 40 degrees of external rotation is just possible. Next, the superior flap is advanced distally, and the repair is buttressed with the medial flap.

If no Bankart lesion is found, the vertical limb of the T is placed laterally near the humeral neck. A medial vertical limb facilitates combining the capsular shift with a Bankart repair when the Bankart lesion is present. On the other hand, a lateral limb allows easier access to the posterior capsule and is preferable if the Bankart lesion is not present. In some cases, an anterior H-plasty may be necessary to allow sufficient mobilization of the capsule (Fig. 15C–34). An arthroscopic examination prior to opening the shoulder may facilitate the decision-making process.

After an anterior capsulorrhaphy for multidirec-

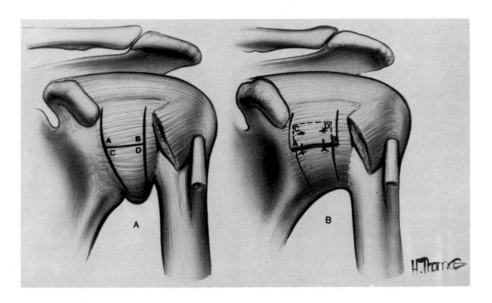

FIGURE 15C–34
Anterior H-plasty for multidirectional instability. In some cases, both medial and lateral capsular incisions may be required to provide sufficient capsular tension.

tional instability, the patient is placed in a shoulder immobilizer, and the arm is kept in adduction and internal rotation for 6 weeks. We formerly placed these patients in an Orthoplast splint in slight abduction and internal rotation, but we have found that this is not necessary if the inferior instability has been eliminated. Gentle passive flexion exercises to 90 degrees are instituted after 3 weeks. At 6 weeks, active-assisted range of motion exercises are begun. When full range of motion is obtained, active resistance exercises are started to strengthen the internal and external rotators and the deltoid. Light throwing and sidearm racquet sports are permitted after 6 months. At 6 to 9 months, patients whose instability had a traumatic etiology may return to contact sports, hard throwing, and overhead racquet sports. Patients with atraumatic instability are protected from these activities for 9 to 12 months.

Although some surgeons prefer to use an anterior approach for all types of multidirectional instability, we feel that a posterior approach should be used when the predominant direction of instability is posterior. The posterior approach for multidirectional instability is similar to the T-plasty capsular shift described for the treatment of RPS. In the patient with multidirectional instability, conversion of the capsulotomy to an H-plasty is often necessary to eliminate inferior redundancy by advancing the inferior capsule superiorly. Hawkins and colleagues [60] noted an increased failure rate after capsular shift in patients whose instability is primarily posterior.

Both anterior and posterior approaches may be required on rare occasions [105]. The combined approaches are indicated when significant anterior labral detachment is noted after a posterior approach, when there is doubt about the adequacy of stabilization after capsular repair of the first side, or when the opposite side of the capsule is tight and requires release before a capsular shift can be performed.

The results of surgical treatment for multidirectional instability have historically been less successful than those for unidirectional anterior instability. Results have improved with current techniques, however. In a recent article, Altchek and colleagues [1] noted only four recurrences in 42 patients and a 95% patient satisfaction rate after T-plasty repair for type II multidirectional instability.

CONCLUSION

Glenohumeral instability is a common cause of disability in the athletic population. The shoulder joint is extremely mobile, and bony stability has been sacrificed to allow this motion. Glenohumeral stability is dependent upon the soft tissues surrounding the joint. These soft tissue stabilizers operate in a complex pattern that varies with shoulder position and activity. If a primary stabilizer is damaged, there appear to be secondary restraints to abnormal translation. However, the high loads incurred at the shoulder during athletic activity can overwhelm both primary and secondary mechanisms of stability. Our understanding of these intricate relationships has increased dramatically in recent years. Posterior and multidirectional instabilities of the glenohumeral joint are important causes of shoulder disability. Although there is an increased recognition of these types of instability, diagnosis is commonly missed or delayed owing to the relative infrequency of presentation and the subtlety of the associated physical findings.

A thorough understanding of the pathomechanics of shoulder instability is required for the proper treatment of these lesions. Nonoperative treatment and rehabilitation are based on the principle of secondary mechanisms of restraint. If operative intervention is employed, the surgeon should carefully define the problem and address the pathologic anatomy accordingly. Operative treatment of anterior instability should correct damage to the anterior capsulolabral system and excess capsular laxity. Standard procedures designed to treat unidirectional anterior instability are likely to fail if used to treat posterior or multidirectional instability and may in fact worsen the problem. A scientific approach directed at the restoration of normal capsular anatomy should be employed to prevent instability while retaining maximal function.

References

1. Altchek, D. W., Warren, R. F., Skyhar, M. J., and Ortiz, G. T-plasty modification of the Bankart procedure for multidirectional instability of the anterior and inferior types. *J Bone Joint Surg* 73A:105–112, 1991.
2. Altchek, D. W., Skyhar, M. J., and Warren, R. F. Shoulder arthroscopy for shoulder instability. *Instr Course Lect* 38:187–198, 1989.
3. Altchek, D. W., Ortiz, G., Warren, R. F., and Wickiewicz, T. L. Arthroscopic labral debridement—A three year follow-up study. *Orthop Trans* 14(2):258, 1990.
4. Anderson, D., Zuirbulis, R., and Ciullo, J. Scapular manipulation for reduction of anterior shoulder dislocation. *Clin Orthop* 164:181–183, 1982.
5. Andrews, J. R., Carson, W. G., and McLeod, W. D. Glenoid labrum tears related to the long head of the biceps. *Am J Sports Med* 13:337–341, 1985.
6. Andrews, J. R. Shoulder arthroscopy: Its role in evaluating shoulder disorders in the athlete (discussion). *Am J Sports Med* 18:483, 1990.
7. Arndt, J. H., and Sears, A. D. Posterior dislocation of the shoulder. *Am J Roentgenol* 94:639–645, 1965.
8. Bach, B. R., Warren, R. F., and Fronek, J. Disruption of

the lateral capsule of the shoulder: A cause of recurrent dislocation. *J Bone Joint Surg* 70B:274–276, 1988.

9. Bankart, A. S. B. Recurrent or habitual dislocation of the shoulder-joint. *Br Med J* 2:1132–1133, 1923.

10. Bankart, A. S. B. The pathology and treatment of recurrent dislocation of the shoulder joint. *Br J Surg* 26:23–29, 1938.

11. Bankart, A. S. B. Recurrent dislocation of the shoulder (discussion). *J Bone Joint Surg* 30B:46–47, 1948.

12. Baker, C. L., Uribe, J. W., and Whitman, C. Arthroscopic evaluation of acute initial anterior shoulder dislocations. *Am J Sports Med* 18:25–28, 1990.

13. Basmajian, J. V., and Bazant, F. J. Factors preventing downward dislocation of the adducted shoulder joint. *J Bone Joint Surg* 41A:1182–1186, 1959.

14. Bestard, E. A., Schuene, H. R., and Bestard, E. H. Glenoplasty in the management of recurrent shoulder dislocation. *Contemp Orthop* 12(5):47–55, 1986.

15. Boicev, B. Solla lussazione abituale della spalla. *Chirurg organi movimento* 23:354, 1938.

16. Bost, F. C., and Inman, V. T. The pathological changes in recurrent dislocation of the shoulder. *J Bone Joint Surg* 24A:595–613, 1942.

17. Boyd, H. B., and Sisk, T. D. Recurrent posterior dislocation of the shoulder. *J Bone Joint Surg* 54A:779–786, 1972.

18. Bowen, M. K., Warren, R. F., Altchek, D. W., and O'Brien, S. J. Posterior subluxation of the glenohumeral joint treated by posterior stabilization. *Orthop Trans* 15(3):764, 1991.

19. Bowen, M. K., and Warren, R. F. Surgical approaches to posterior shoulder instability, 1992.

20. Bowen, M. K., and Warren, R. F. Ligamentous control of shoulder stability based on selective cutting and static translation. *Clin Sports Med* 10(4):757–782, 1991.

21. Bradley, J. P., and Tibone, J. E. Electromyographic analysis of muscle action about the shoulder. *Clin Sports Med* 10(4):789–805, 1991.

22. Brewer, B. J., Wubben, R. G., and Carrera, G. F. Excessive retroversion of the glenoid cavity. *J Bone Joint Surg* 68A:724, 1986.

23. Broca, A., and Hartmann, H. Contribution a l'etude des luxations de l'epaule (luxations dites incompletes, decollements periostiques, luxationes directes et luxations indirectes). *Bull Soc Anat Paris,* 5 me Série, 4:312–336, 1890.

24. Brodsky, J. W., Tullos, H. S., and Gartsman, G. M. Simplified posterior approach to the shoulder joint. *J Bone Joint Surg* 69A:773–774, 1987.

25. Browne, A. O., Hoffmeyer, P., An, K. N., and Morrey, B. F. The influence of atmospheric pressure on shoulder stability. *Orthop Trans* 14(2):259, 1990.

26. Budd, F. W. Voluntary bilateral posterior dislocation of the shoulder joint. Report of a case. *Clin Orthop* 63:181–183, 1969.

27. Cain, P. R., Mutschler, T. A., and Fu, F. H. Anterior stability of the glenohumeral joint: A dynamic model. *Am J Sports Med* 15:144–148, 1987.

28. Carr, C. R. Prognosis in dislocations of the shoulder (discussion). *J Bone Joint Surg* 38A:977, 1956.

29. Clayton, M. L., and Weir, G. J. Experimental investigations of ligamentous healing. *Am J Surg* 98:373–378, 1959.

30. Connolly, J. F. Humeral head defects associated with shoulder dislocations: Their diagnostic and surgical significance. *Instr Course Lect* 21:42–54, 1972.

31. Cooper, A. *A treatise on Fractures and Dislocations of the Joints.* Boston, T. R. Marvin, 1844.

32. Cooper, D. E., Warner, J. J. P., Deng, W., Arnoczky, S. P., Warren, R. F., and O'Brien, S. J. Anatomy and function of the coracohumeral ligament. *Orthop Trans* 15(3):803, 1991.

33. Cooper, D. E., O'Brien, S. J., and Warren, R. F. Supporting layers of the glenohumeral joint: An anatomic study. *Clin Orthop* (in press.)

34. Cooper, D. E., Arnoczky, S. P., O'Brien, S. J., Warren, R. F., DiCarlo, E., and Allen, A. A. Anatomy, histology,

35. and vascularity of the glenoid labrum. *J Bone Joint Surg* 73A:46–52, 1992.

35. Delorme: Die Hemmungsbander des Schultergelenks und ihre Bedeutung fur die Schulterluxation. *Arch Klin Chir* 92:79–101, 1910.

36. DePalma, A. F., Callery, G., and Bennett, G. A. Variational anatomy and degenerative lesions of the shoulder joint. *Instr Course Lect* 6:255–281, 1949.

37. DePalma, A. F. *Surgery of the Shoulder*, Philadelphia, J. B. Lippincott, 1950, p. 236.

38. Dickson, J. W., and Duvas, M. B. Bankart's operation for recurrent dislocation of the shoulder. *J Bone Joint Surg* 39B:114–119, 1957.

39. Eden, R. Zur operativen Behandlung der habituellen Schulterluxation unter Mitteilung, eines neuen Verfahrens bei Abriss am inneren Pfannenrande. *Deutsch Z Chir* 144:269, 1918.

40. Flatow, E. L., Soslowsky, L. J., Ateshian, G. A., Ark, J. W., Pawluk, R. J., Bigliani, L. U., and Mow, V. C. Shoulder joint anatomy and the effect of subluxations and size mismatch on patterns of glenohumeral contact. *Orthop Trans* 15(3):803–804, 1991.

41. Flood, V. Discovery of a new ligament. *Lancet* 671, 1829.

42. Flower, W. On the pathological changes produced in the shoulder-joint by traumatic dislocation; as derived from an examination of all the specimens illustrating this injury in the museums of London. *Trans Pathol Soc London* 12:179–200, 1860–61.

43. Fried, A. Habitual posterior dislocation of the shoulder-joint. *Acta Orthop Scand* 18:329–345, 1949.

44. Fronek, J., Warren, R. F., and Bowen, M. Posterior subluxation of the glenohumeral joint. *J Bone Joint Surg* 71A:205–216, 1989.

45. Galinat, B. J., Howell, S. M., and Kraft, T. A. The glenoid-posterior acromion angle: An accurate method of evaluating glenoid version. *Orthop Trans* 12:727, 1988.

46. Galinat, B. J., and Warren, R. F. Shoulder: Trauma and related instability *In* Poss, R. (Ed.), *Orthopaedic Knowledge Update 3*, Park Ridge, IL, American Academy of Orthopaedic Surgeons, 1990, pp. 303–312.

47. Garth, W. P., Slappey, C. E., and Ochs, C. W. Roentgenographic demonstration of the shoulder: The apical oblique projection. A technical note. *J Bone Joint Surg* 66A:1450–1453, 1984.

48. Garth, W. P., Allman, F. L., and Armstrong, W. S. Occult anterior subluxations of the shoulder in noncontact sports. *Am J Sports Med* 15:579–585, 1987.

49. Glousman, R., Jobe, F., Tibone, J., Moynes, D., Antonelli, D., and Perry, J. Dynamic electromyographic analysis of the throwing shoulder with glenohumeral instability. *J Bone Joint Surg* 70A:220–226, 1988.

50. Gowan, I. D., Jobe, F. W., Tibone, J. E., Perry, J., and Moynes, D. R. A comparative electromyographic analysis of the shoulder during pitching. *Am J Sports Med* 15:586, 1987.

51. Habermeyer, P., Schuller U., and Wiedemann, E. The intra-articular pressure of the shoulder: An experimental study on the role of the glenoid labrum in stabilizing the joint. *Arthroscopy* 8:166–172, 1992.

52. Hall, R. H., Isaac, F., and Booth, C. R. Dislocations of the shoulder with special reference to accompanying small fractures. *J Bone Joint Surg* 41A:489–494, 1959.

53. Harryman, D. T., Sidles, J. A., Clark, J. M., and McQuade, K. J. Translation of the humeral head on the glenoid with passive glenohumeral motion. *J Bone Joint Surg* 72A:1334, 1990.

54. Harryman, D. T., Sidles, J. A., Harris, S. L., and Matsen, F. A. Role of the rotator interval capsule in passive motion and stability of the shoulder. *J Bone Joint Surg* 72A:53–66, 1992.

55. Hawkins, R. B., and Hawkins, R. J. Failed anterior reconstruction for shoulder instability. *J Bone Joint Surg* 67B:709–714, 1985.

56. Hawkins, R. B. Arthroscopic stapling repair for shoulder instability: A retrospective study of 50 cases. *Arthroscopy.* 5:122–128, 1989.

57. Hawkins, R. J., and Belle, R. M. Posterior instability of the shoulder. *Instr Course Lect* 38:211–216, 1989.

58. Hawkins, R. J., Koppert, G., and Johnston, G. Recurrent posterior instability (subluxation) of the shoulder. *J Bone Joint Surg* 66A:169–174, 1984.

59. Hawkins, R. J., Neer, C. S., Pianta, R. M., and Mendoza, F. X. Locked posterior dislocation of the shoulder. *J Bone Joint Surg* 69A:9–18, 1987.

60. Hawkins, R. J., Kunkel, S. S., and Nayak, N. K. Inferior capsular shift for multidirectional instability of the shoulder: 2–5 year follow-up. *Orthop Trans* 15(3):765, 1991.

61. Hawkins, R. J., and Mohtadi, N. G. H. Controversy in anterior shoulder instability. *Clin Orthop* 272:152–161, 1991.

62. Helfet, A. J. Coracoid tranplantation for recurring dislocation of the shoulder. *J Bone Joint Surg* 40B:198–202, 1948.

63. Helmig, P., Sojbjerg, J. E., Kjaersgaard-Andersen, P., Nielsen, S., and Ovesen, J. Distal humeral migration as a component of multidirectional shoulder instability: An anatomical study in autopsy specimens. *Clin Orthop* 252:139–143, 1990.

64. Hermodsson I. Röntgenische Studien über die traumatischen und habituellen Schultergelenkeverrenkungen nach vorn und nach unten. *Acta Radiol* (Suppl.) 20:1–173, 1934.

65. Hill, H. A., and Sachs, M. D. Grooved defect of the humeral head; a frequently unrecognized complication of dislocation of the shoulder. *Radiology* 35:690–700, 1940.

66. Hindenach, J. C. R. Recurrent posterior dislocation of the shoulder. *J Bone Joint Surg* 29B:582–586, 1947.

67. Hovelius, L., Erikkson, K., Fredin, H., Hagberg, G., Hussenrus, A., Lind, B., Thorling, J., and Weckstrom, J. Recurrences after initial dislocation of the shoulder: Results of a prospective study of treatment. *J Bone Joint Surg* 65A:343–349, 1983.

68. Howell, S. M., and Galinat, B. J. The glenoid-labral socket. A constrained articular surface. *Clin Orthop* 243:122, 1989.

69. Howell, S. M., Galinat, B. J., Renzi, A. J., and Marone, P. J. Normal and abnormal mechanics of the glenohumeral joint in the horizontal plane. *J Bone Joint Surg* 70A:227, 1988.

70. Howell, S. M., and Kraft, T. A. The role of the supraspinatus and infraspinatus muscles in glenohumeral kinematics of anterior shoulder instability. *Clin Orthop* 263:128–134, 1991.

71. Howell, S. M., Imobersteg, A. M., Seger O. H., and Marone, P. J. Clarification of the role of the supraspinatus muscle in shoulder function. *J Bone Joint Surg* 68A:398, 1986.

72. Humphry, G. M. *A Treatise on the Human Skeleton (Including the Joints).* Cambridge, Macmillan, 1858.

73. Hurley, J. A., and Anderson, T. E. Shoulder arthroscopy: Its role in evaluating shoulder disorders in the athlete. *Am J Sports Med* 18:480–483, 1990.

74. Hybbinette, S. De la transplatation d'un fragment osseux pour remedier aux luxations recidivantes de l'epaule; constatations et resultats operatiores. *Acta Chir Scand* 71:411–445, 1932.

75. Jens, J. The role of the subscaularis muscle in recurrent dislocation of the shoulder. *J Bone Joint Surg* 46B:780–781, 1964.

76. Jobe, F. W., Tibone, J. E., Perry, J., and Moynes, D. An EMG analysis of the shoulder in throwing and pitching: A preliminary report. *Am J Sports Med* 11:3–5, 1983.

77. Jobe, F. W., Moynes, D. R., Tibone, J. E., and Perry, J. An EMG analysis of the shoulder in pitching: A second report. *Am J Sports Med* 12:218–220, 1984.

78. Jobe, F. W., and Bradley, J. P. Rotator cuff injuries in baseball: Prevention and rehabilitation. *Sports Med* 6:377–386, 1988.

79. Jobe, F. W. Impingement problems in athletes. *Instr Course Lect* 38:205–209, 1989.

80. Jobe, F. W., and Kvitne, R. S. Shoulder pain in the overhead or throwing athlete: The relationship of anterior instability and rotator cuff impingement. *Orthop Rev* 18:963–975, 1989.

81. Jobe, F. W., Giangarra, C. E., Kvitne, R. S., and Glousman, R. E. Anterior capsulolabral reconstruction of the shoulder in athletes in overhead sports. *Am J Sports Med* 19:428–434, 1991.

82. Joessel, D. Ueber die Recidine der Humerus-Luxationen. *Dtsche Z Chir* 13:167–184, 1880.

83. Johnson, L. L. Symposium on arthroscopy. Annual Meeting of Arthroscopy Association of North America, San Francisco, March 1986.

84. Johnston, H. H., Hawkins, R. J., Haddad, R., and Fowler, P. J. A complication of posterior glenoid osteotomy for recurrent posterior shoulder instability. *Clin Orthop* 187:147–149, 1984.

85. Klein, A. H., and France, J. C. Measurement of brachial plexus strain in arthroscopy of the shoulder. *Arthroscopy* 3:45–52, 1983.

86. Kocher, E. T. Eine neue Reductionmethode für Schulterverrenkung. *Berl Klin Wochenschr* 7:101–105, 1870.

87. Kumar, V. P., and Balasubramianium, P. The role of atmospheric pressure in stabilizing the shoulder. An experimental study. *J Bone Joint Surg* 67B:719–721, 1985.

88. Laterjet, M. A propos du traitement des luxations recidivante de l'epaule. *Lyon Chir* 49:994–997, 1954.

89. Leslie, J. T., and Ryan, T. J. Anterior axillary approach to the shoulder joint. *J Bone Joint Surg* 44A:1193–1196, 1962.

90. Luckey, C. A. Recurrent dislocation of the shoulder. Modification of the Bankart capsulorraphy. *Am J Surg* 77:220–222, 1949.

91. Magnuson, P. B., and Stack, J. K. Recurrent dislocation of the shoulder. *JAMA* 123:889–892, 1943.

92. Malgaigne, J. F. *Traite des Fractures et des Luxations.* Paris, J. B. Bulliere, 1855.

93. Mason, M. L., and Allen, H. S. The rate of healing of tendinitis: An experimental study of tensile strength. *Ann Surg* 113:424–456, 1941.

94. Matthews, L. S., Zarins, B., Michael, R. H., and Helfet, D. L. Anterior portal selection for shoulder arthroscopy. *Arthroscopy* 1:33–39, 1985.

95. Matsen, F. A., Thomas, S. C., and Rockwood, C. A. Anterior glenohumeral instability, *In* Rockwood, C. A., and Matsen, F. A. (Eds.), *The Shoulder.* Philadelphia, W. B. Saunders, 1990, pp 526–622.

96. McKernan, D. J., Mutschler, T. A., Rudert, M. J., Klein, A. H., Victorino, G., Harner, C. D., and Fu, F. H. The characterization of rotator cuff muscle forces and their effect on glenohumeral joint stability: A biomechanical study. *Orthop Trans* 14(2):237–238, 1990.

97. McKernan, D. J., Mutschler, T. A., Rudert, M. J., Luo, L., Harner, C. D., and Fu, F. H. Significance of a partial and full Bankart lesion: A biomechanical study. Read at Annual Meeting of Orthopaedic Research Society, Las Vegas, February, 1989.

98. McLaughlin, H. L. Posterior dislocation of the shoulder. *J Bone Joint Surg* 34A:584–590, 1952.

99. McLaughlin, H. L. Posterior dislocation of the shoulder. *J Bone Joint Surg* 44A:1477, 1962.

100. Morgan, C. D., Rames, R. D., and Snyder, S. J. Arthroscopic assessment of anatomic variants of the glenohumeral ligaments associated with recurrent anterior shoulder instability. Read at the Annual Meeting of American Academy of Orthopaedic Surgeons, Washington, D. C., February 1992.

101. Morgan, C. D., and Bodenstab, A. B. Arthroscopic Bankart suture repair: Technique and early results. *Arthroscopy* 3(2):111–122, 1987.

102. Morgan, C. D. Arthroscopic transglenoid Bankart suture repair. *Operative Techniques in Orthopaedics* 1:171–179, 1991.

103. Morrey, B. F., and Janes, J. M. Anterior dislocation of the shoulder: Long-term follow-up of the Putti-Platt and Bankart procedures. *J Bone Joint Surg* 58A:252–256, 1976.

104. Moseley, H. J., and Overgaard, B. The anterior capsular mechanism in recurrent anterior dislocation of the shoulder: Morphological and clinical studies with special reference to the glenoid labrum and glenohumeral ligaments. *J Bone Joint Surg* 44B:913–927, 1962.

105. Neer, C. S. *Shoulder Reconstruction.* Philadelphia, W. B. Saunders, 1990, pp 273–362.

106. Neer, C. S., and Foster, C. R. Inferior capsular shift for involuntary and multidirectional instability of the shoulder. *J Bone Joint Surg* 62A:897–908, 1980.

107. Nicola, T. Anterior dislocation of the shoulder: The role of the anterior capsule. *J Bone Joint Surg* 24A:614–616, 1942.

108. Nobuhara, K., and Ikeda, H. Rotator interval lesion. *Clin Orthop* 223:44–50, 1987.

109. Nuber, G. W., Jobe, F. W., Perry, J., Moyes, D. R., and Antonelli, D. Fine wire electromyographic analysis of muscles of the shoulder during swimming. *Am J Sports Med* 14:7, 1986.

110. O'Brien, S. J., Neves, M. C., Arnoczky, S. P., Rozbruch, S. R., DiCarlo, E. F., Warren, R. F., Schwartz, R., and Wickiewicz, T. L. The anatomy and histology of the inferior glenohumeral ligament complex of the shoulder. *Am J Sports Med* 18:449–456, 1990.

111. O'Brien, S. J., Warren, R. F., and Schwartz, E. Anterior shoulder instability. *Orthop Clin North Am* 18(3):395–408, 1987.

112. O'Brien, S. J., Schwartz, R. E., Warren, R. F., and Torzilli, P. A. Capsular restraints to anterior/posterior motion of the shoulder. *Orthop Trans* 12:143, 1988.

113. O'Connell, P. W., Nuber, G. W., Mileski, R. A., and Lautenslager, E. The contribution of the glenohumeral ligaments to anterior stability of the shoulder joint. *Am J Sports Med* 18:579–584, 1990.

114. Osmond-Clarke, H. Habitual dislocation of the shoulder: The Putti-Platt operation. *J Bone Joint Surg* 30B:19–25, 1948.

115. Oudard, P. La luxation recidivante de l'epaule (variete anterointerne) procede operatoire. *J Chir* 23:13–25, 1924.

116. Oveson, J., and Nielson, S. Experimental distal subluxation in the glenohumeral joint. *Arch Orthop Trauma Surg* 104:78–81, 1985.

117. Oveson, J., and Nielson, S. Anterior and posterior instability: A cadaver study. *Acta Orthop Scand* 57:324–327, 1986.

118. Oveson, J., and Nielson, S. Posterior instability of the shoulder: A cadaver study. *Acta Orthop Scand* 57:436–439, 1986.

119. Pappas, A. M., Zawacki, R. M., and McCarthy, C. F. Rehabilitation of the pitching shoulder. *Am J Sports Med* 13:223–235, 1985.

120. Pappas, A. M., Zawacki, R. M., and Sullivan, T. J. Biomechanics of baseball pitching: A preliminary report. *Am J Sports Med* 13:216–222, 1985.

121. Pavlov, H., Warren, R. F., Weiss, C. B., and Dines, D. M. The roentgenographic evaluation of anterior shoulder instability. *Clin Orthop* 194:153–158, 1985.

122. Perry, J. Anatomy and biomechanics of the shoulder in throwing, swimming, gymnastics, and tennis. *Clin Sports Med* 2:247–270, 1983.

123. Perthes, G. Über Operationen der habituellen Schulterluxation. *Dtsche Z Chir* 85:199, 1906.

124. Pollock, R. G., Bigliani, L. U., Flatow, E. L., Soslowsky, L. J., and Mow, V. C. The mechanical properties of the inferior glenohumeral ligament. *Orthop Trans* 14(2):259–260, 1990.

125. Poppen, N. K., and Walker, P. S. Normal and abnormal motion of the shoulder. *J Bone Joint Surg* 58A:195–201, 1976.

126. Poppen, N. K., and Walker, P. S. Forces at the glenohumeral joint in abduction. *Clin Orthop* 135:165–170, 1978.

127. Reeves, B. Experiments on the tensile strength of the anterior capsular structures of the shoulder in man. *J Bone Joint Surg* 50B:858–865, 1968.

128. Regan, W. D., Webster-Bogaert, S., Hawkins, R. J., and Fowler, P. J. Comparative functional analysis of the Bristow, Magnuson-Stack, and Putti-Platt procedures for recurrent dislocation of the shoulder. *Am J Sports Med* 17:42–48, 1989.

129. Richmond, J. C., Donaldson, W. R., Fu, F. H., and Harner, C. D. Modification of the Bankart reconstruction with a suture anchor. Report of a new technique. *Am J Sports Med* 19:343–346, 1991.

130. Rockwood, C. A. Subluxations and dislocations about the shoulder. *In* Rockwood, C. A., and Green, D. P. (Eds.), *Fractures in Adults.* Philadelphia, J. B. Lippincott, 1984, pp 722–860.

131. Rodosky, M. W., Rudert, M. J., Harner, C. H., Luo, L., and Fu, F. H. Significance of the superior labral lesion of the shoulder: A biomechanical study. *Trans Orthop Res Soc* 15:276, 1990.

132. Rokous, J. R., Feagin, J. A., and Abbott, H. G. Modified axillary roentgenogram: A useful adjunct in the diagnosis of recurrent instability of the shoulder. *Clin Orthop* 82:84–86, 1972.

133. Rowe, C. R. Prognosis in dislocations of the shoulder. *J Bone Joint Surg* 38A:957–977, 1956.

134. Rowe, C. R., and Sakellarides, H. T. Factors related to recurrences of anterior dislocations of the shoulder. *Clin Orthop* 20:40–48, 1961.

135. Rowe, C. R., and Zarins, B. Chronic unreduced dislocations of the shoulder. *J Bone Joint Surg* 64A:494–505, 1982.

136. Rowe, C. R., Patel, D., and Southmayd, W. W. The Bankart procedure: A long-term end-result study. *J Bone Joint Surg* 60A:1–16, 1978.

137. Rowe, C. R., and Zarins, B. Recurrent transient subluxation of the shoulder. *J Bone Joint Surg* 63A:863–872, 1981.

138. Rowe, C. R., Pierce, D. S., and Clark, J. G. Voluntary dislocation of the shoulder: A preliminary report on a clinical, electromyographic, and psychiatric study of twenty-six patients. *J Bone Joint Surg* 55A:445–460, 1973.

139. Rowe, C. R., Zarins, B., and Ciullo, J. V. Recurrent anterior dislocation of the shoulder after surgical repair. *J Bone Joint Surg* 66A:159–168, 1984.

140. Saha, A. K. Dynamic stability of the glenohumeral joint. *Acta Orthop Scand* 42:491–505, 1971.

141. Saha, A. K., and Das, A. K. Anterior recurrent dislocation of the shoulder: Treatment by rotation osteotomy of the upper shaft of the humerus. *Indian J Orthop* 1:132–137, 1967.

142. Samilson, R. L., and Prieto, V. Dislocation arthropathy of the shoulder. *J Bone Joint Surg* 65A:456–460, 1983.

143. Schauder, K. S., and Tullos, H. S. Role of the coracoid bone block in the Bristow procedure. *Am J Sports Med* 20:31–37, 1992.

144. Schlemm, F. Ueber die Verstarkungsbander am Schultergelenk. *Arch Anat* 45, 1853.

145. Schwartz, E., Warren, R. F., O'Brien, S. J., and Fronek, J. Posterior shoulder instability. *Orthop Clin North Am* 18(3):409–419, 1987.

146. Schwartz, R. E., O'Brien, S. J., Warren, R. F., and Torzilli, P. A. Capsular restraints to anterior-posterior motion of the shoulder. *Orthop Trans* 12:727, 1988.

147. Schwartz, R. E., O'Brien, S. J., Warren, R. F., and Torzilli, P. A. Capsular restraints to anterior/posterior motion of the shoulder. *Trans Orthopaedic Research Society* 12:78, 1987.

148. Scott, D. J. Treatment of recurrent posterior dislocations of the shoulder by glenoplasty. Report of three cases. *J Bone Joint Surg* 49A:471–476, 1967.

149. Silliman, J. F., and Hawkins, R. J. Current concepts and recent advances in the athlete's shoulder. *Clin Sports Med* 10(4):693–706, 1991.

150. Simkin, P. A. Structure and function of joints. *In* Schumacher, H. R. (Ed.), *Primer on the Rheumatic Diseases,* 9th ed. Atlanta, Arthritis Foundation.

151. Simonet, W. T., and Cofield, R. A. Prognosis in anterior shoulder dislocation. *Am J Sports Med* 12:19–24, 1984.

152. Snyder, S., Karzel, R., Del Pizzo, W., Ferkel, R., and

Friedman, R. S.L.A.P. lesions of the shoulder (lesions of the superior labrum both anterior and posterior). *Orthop Trans* 14(2):257–258, 1990.

153. Snyder, S. J., Karzel, R. P., and Del Pizzo, W. SLAP lesions of the shoulder. *Arthroscopy* 6:274–279, 1990.

154. Speer, K. P., Altchek, D. W., and Warren, R. F. Displaced lesions of the superior labrum and biceps anchor. A clinical and biomechanical investigation. Unpublished data.

155. Stimson, L. A. An easy method of reducing dislocation of the shoulder and the hip. *NY Med Rec* 57:356–357, 1900.

156. Symeonides, P. P. The significance of the subscapularis muscle in the pathogenesis of recurrent dislocation of the shoulder. *J Bone Joint Surg* 54B:476–483, 1972.

157. Thomas, T. T. Habitual or recurrent anterior dislocation of the shoulder. I. Etiology and pathology. *Am J Med Sci* 137:237–246, 1909.

158. Thomas, T. T. Recurrent anterior dislocation of the shoulder. *JAMA* 54:834, 1910.

159. Terry, G. C., Hammon, D., France, P., and Norwood, L. A. The stabilizing function of passive shoulder restraints. *Am J Sports Med* 19:26–34, 1991.

160. Tibone, J., and Ting, A. Capsulorraphy with a staple for recurrent posterior dislocation of the shoulder. *J Bone Joint Surg* 72A:999–1002, 1990.

161. Townley, C. O. The capsular mechanisms in recurrent dislocation of the shoulder. *J Bone Joint Surg* 32A:370–380, 1950.

162. Townsend, H., Jobe, F. W., Pink, M., and Perry, J. Electromyographic analysis of the glenohumeral muscles during a baseball rehabilitation program. *Am J Sports Med* 19:264–272, 1991.

163. Trillat, A. Traitement de la luxation recidivante de l'epaule. *Lyon Chir* 49:986, 1954.

164. Turkel, S. J., Panio, M. W., Marshall, J. L., and Girgis, F. G. Stabilizing mechanisms preventing anterior dislocation of the glenohumeral joint. *J Bone Joint Surg* 63A:1208–1217, 1981.

165. Uthoff, H., and Piscopo, M. Anterior capsular redundancy of the shoulder: Congenital or traumatic. *J Bone Joint Surg* 67B:363–366, 1985.

166. Vanderhooft, E., Lippitt, S., and Harris, S. Glenohumeral stability from concavity-compression: A quantitative analysis. Read at Annual Meeting of American Shoulder and Elbow Society, Washington, D. C., February 1992.

167. Vangsness, C. T., and Ennis, M. Neural anatomy of the human shoulder ligaments and the glenoid labrum. Read at Annual Meeting of American Academy of Orthopaedic Surgeons, Washington, D. C., February, 1992.

168. Vellet, A. D., Munk, P. L., and Marks, P. Imaging techniques of the shoulder: Present perspectives. *Clin Sports Med* 10(4):721–756, 1991.

169. Viek, P., and Bell, B. T. The Bankart shoulder reconstruction. The use of pull-out wires and other practical details. *J Bone Joint Surg* 41A:236–242, 1959.

170. Walch, G., Charnet, P., Pietro-Paoli, H., and Dejour, H. La luxation recidivante anterieure de l'epaule. Le recidives post-operatoires. *Rev Chir Orthop* 72:541–555, 1986.

171. Warner, J. J. P., Deng, X., Warren, R. F., Torzilli, P. A., O'Brien, S. J., and Altchek, D. W. Static capsuloligamentous restraints to superior-inferior translation of the glenohumeral joint. Presented at Annual Meeting of the Orthopaedic Research Society, Anaheim, 1991.

172. Warner, J. J. P., Deng, X., Warren, R. F., Torzilli, P. A., O'Brien, S. J., and Altchek, D. W. Superior-inferior translation in the intact and vented glenohumeral joint. Presented at Annual Meeting of American Shoulder and Elbow Surgeons, Anaheim, 1991.

173. Warner, J. J. P., Pagnani, M. J., Warren, R. F., Kavanaugh, J., and Montgomery, W. Arthroscopic Bankart repair with an absorbable, cannulated fixation device. *Orthop Trans* 15(3):761–762, 1991.

174. Warner, J. J. P., Paletta, G. J., and Warren, R. F. Biplanar x-ray evaluation of the shoulder in patients with instability and rotator cuff tears. *Orthop Trans* 15(3):763, 1991.

175. Warren, R. F. Instability of the shoulder in throwing athletes. *Instr Course Lect* 34:337–348, 1985.

176. Warren, R. F., Kornblatt, I. B., and Marchand, R. Static factors affecting posterior shoulder stability. *Orthop Trans* 8:89, 1984.

177. Warren, R. F., and O'Brien, S. J. Shoulder pain in the geriatric patient: Part I. Evaluation and pathophysiology. *Orthop Rev* 18:129–135, 1989.

178. Warren, R. F. *Surgical Technique for Suretac.* Mansfield, MA, Acufex Microsurgical, Inc., 1991.

179. Weber, B. G., Simpson, L. A., and Hardegger, F. Rotational humeral osteotomy for anterior dislocation of the shoulder associated with a large Hill-Sachs lesion. *J Bone Joint Surg* 66A:1443–1450, 1984.

180. Weber, S. C., and Caspari, R. B. A biomechanical evaluation of the restraints to posterior shoulder dislocation. *Arthroscopy* 5:115–121, 1989.

181. Wheeler, J. H., Ryan, J. B., Arciero, R. A., and Molinari, R. N. Arthroscopic versus nonoperative treatment of acute shoulder dislocations in young athletes. *Arthroscopy* 5:213–217, 1989.

182. Wiley, A. M. Arthroscopy for shoulder instability and a technique for arthroscopic repair. *Arthroscopy* 4:25–30, 1988.

183. Yahiro, M. A., and Matthews, L. A., Arthroscopic stabilization procedures for recurrent anterior shoulder instability. *Orthop Rev* 11:1161–1168, 1989.

184. Young, D. C., and Rockwood, C. A. Complications of a failed Bristow procedure and their management. *J Bone Joint Surg* 73A:969–981, 1991.

185. Zadik, F. R. Recurrent posterior dislocation of the shoulder joint. *J Bone Joint Surg* 30B:531–532, 1948.

186. Zarins, B., Rowe, C. R., and Stone, J. W. Shoulder instability: Management of failed reconstructions. *Instr Course Lect* 38:217–230, 1989.

187. Zuckerman, J. D., and Matsen, F. A. Complications about the glenohumeral joint related to the use of screws and staples. *J Bone Joint Surg* 66A:175–180, 1984.

Rotator Cuff Problems in Athletes

Richard J. Hawkins, M.D., FRCS(C)
Nicholas Mohtadi, M.D., FRCS(C), M.Sc.

Historical Review

Rotator cuff disorders in athletes have been documented only in recent years. However, our appreciation of such disorders comes from extrapolation of the data on rotator cuff problems dating back to the 1700s. In 1788 Monro illustrated a rotator cuff tear in his thesis entitled "All the Bursae Mucosae of the Human Body" [96]. In 1835 Smith described seven instances of rotator cuff pathology in a cadaver series [90].

The recent history of rotator cuff disorders dates back to 1934, when E. A. Codman published his comprehensive work, *The Shoulder* [25]. In 1909 he performed what was probably the first operative procedure on a torn rotator cuff. He painstakingly described the pathology, proposed the pathophysiology, outlined the clinical syndrome, and recommended and described his surgical management.

Meyer (1937) argued against Codman's theory of a purely traumatic etiology and proposed for the first time the idea that repeated minor trauma caused the pathologic findings [90]. He also described the concept of impingement of the greater tuberosity on the undersurface of the acromion, thus narrowing the interval between the acromion and the humeral head and damaging the intervening soft tissues [90]. There were further reports confirming the clinical and pathologic findings in people with rotator cuff pathology, but no references to rotator cuff involvement as a result of sporting endeavors were made in these early years. The first such report was of biceps tendon rupture associated with significant trauma in pitchers and lifters [46].

Neer, in his classic article in 1972, described the treatment of chronic impingement syndrome with anterior acromioplasty [100]. He subsequently described three stages of impingement and suggested that pitching and overhead sports may be related to the etiology of the first two stages.

Epidemiology

It is difficult to obtain accurate data on the incidence of rotator cuff problems. Cuff tears have been documented in up to 39% of cadaver specimens; however, this figure may not reflect cuff problems in the general population [26]. Shoulder pain in swimmers has been reported in up to 80% of athletes [22, 23, 41, 77, 133]. Football injuries to the shoulder are usually dislocations with a lesser incidence of rotator cuff or bicipital tendinitis [66]. In a series of world class tennis players more than 50% suffered shoulder problems of which the majority involved the rotator cuff and biceps tendon [113]. The shoulder is involved primarily in many other overhead sporting activities such as volleyball, javelin throwing [66], golf [73], and baseball [2, 13, 66]. Although rotator cuff problems were originally described in workers [117], they are of considerable concern in athletes, especially those involved in overhead sports [10, 27, 66, 82, 99].

Current Concepts and Controversies

Concepts relating to the causes of rotator cuff disorders in the athlete are evolving at a rapid rate. Since the 1970s it has been taught that impingement is the primary pathology underlying rotator cuff disorders [100]. This primary impingement results from narrowing of the subacromial space due to edema of the cuff or bony encroachment from the overlying coracoacromial arch [99]. We now appreciate that impingement is a common component of rotator cuff disorders, but frequently and perhaps more commonly it is a secondary phenomenon in athletes. In secondary impingement there is a relative narrowing of the available space that is usually related to eccentric overload of the cuff or glenohumeral instability.

Overuse and fatigue related to eccentric overload resulting in intrinsic fiber failure of the rotator cuff and biceps tendon constitute a common etiology in the young athletic population [42]. This statement applies especially to those involved in repetitive overhead use of the extremity in their sporting endeavors. This initial failure of the fibers may lead to secondary impingement as forces pull the cuff superiorly [62].

Shoulder instability may also be related to both rotator cuff and bicipital disorders [62]. Patients with anterior subluxation [44] are at risk for developing secondary impingement because of the architectural set-up of the subacromial region. Those with loose shoulders or classic multidirectional instability may develop cuff tendinitis due to overwork and overstretching.

Controversies relating to rotator cuff disorders in the athlete are plentiful. There is a lack of appreciation of the underlying pathology described in the above concepts, resulting in confusion and disagreement. It is generally agreed that nonoperative management of these problems should be aggressively pursued, particularly muscular strengthening. The role of stretching remains unclear. Surgical management of these disorders, although rarely applicable, is controversial. The underlying diagnosis should dictate the appropriate surgical management, yet there is considerable debate about stabilization versus decompression, arthroscopic versus open procedures [36], and rotator cuff repair versus no repair.

PERTINENT ANATOMY

The shoulder is a ball and socket joint that provides the link between the trunk and the upper extremity. It is the most mobile joint in the body and as such allows precise positioning of the hand in space for an unlimited number of functions. The shoulder is also the focal point or fulcrum for the long lever arm of the upper limb. As a result, the shoulder absorbs the majority of the forces in sports that require propulsive action of the upper extremity. The rotator cuff is vitally linked to these motions in terms of both precision and propulsion.

The rotator cuff is composed of the supraspinatus, the infraspinatus, the subscapularis, and the teres minor [143] (Fig. 15D–1A). The tendon of the long head of the biceps muscle is closely associated with the cuff and therefore is intimately related.

The cuff surrounds and is united to the shoulder joint capsule on all sides except at the redundant inferior pouch. The biceps tendon originating on the supraglenoid tubercle traverses the glenohumeral joint as an intra-articular structure surrounded by the synovial lining of the joint. It runs deep to the interval between the supraspinatus and subscapularis and exits the joint in the intertubercular sulcus. It is held in this groove by the transverse ligament (Fig. 15D–1B). The groove has a variable shape and depth, and in the supratubercular region an elevation or ridge has been described [34]. The bony anatomy of this area has been implicated in the degenerative processes seen in the groove [34, 96, 128].

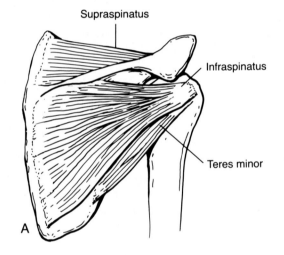

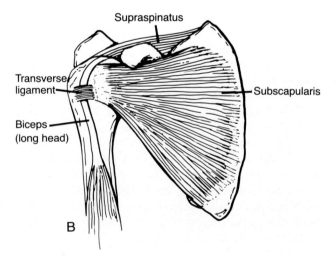

FIGURE 15D–1

A, Posterior view of shoulder. *B*, Anterior view of shoulder.

The vascular supply of the biceps and cuff tendons has been extensively studied [97, 131, 136]. In one study the supply from the axillary artery involved six branches, among which the suprascapular and the anterior and posterior circumflex humeral arteries contributed the most [136]. The pattern of the arterial supply was shown to result in an anastomotic area known as the "critical zone" (the area of the cuff where most degenerative changes occur) [97]. A further study showed areas of hypovascularity in the region of the supraspinatus tendon just proximal to its insertion into the greater tubercle [131]. This was a consistent finding and corresponded to the same critical zone. The biceps tendon also had an area of hypovascularity in its intra-articular portion. These hypovascular regions were shown to be related to tension or pressure from the underlying humeral head when the tendons were in the anatomic position. With abduction of the arm and, presumably, release of this tension, the areas showed complete filling of the vessels [131]. A recent study [21a] demonstrated that there was abundant vascularity adjacent to rotator cuff tears. Another author [83a] has also challenged the theory of vascular deficiency of the rotator cuff.

Any discussion of anatomy related to the rotator cuff includes the structures both superficial and deep to it. The deltoid is the superficial structure completely enclosing the area. Next is the coracoacromial arch. The acromion is an extension of the spine of the scapula that forms the posterolateral bony roof of the arch. It is important to recognize that the acromion can slope inferiorly to the point where it can be hidden under the overlying deltoid [6] (Fig. 15D–2).

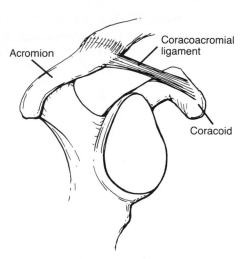

FIGURE 15D–3
Coracoacromial arch.

The acromion provides bony protection to the shoulder but also creates a finite space between its undersurface and the humeral head. The coracoacromial ligament forms the anterior extent of the arch. It extends from the outer edge of the coracoid tapering to insert on the anteromedial aspect of the acromion [141]. It is the anterior third of the acromion, the coracoid, and the attached coracoacromial ligament that is implicated in the impingement syndrome of the rotator cuff [17] (Fig. 15D–3). Deep to this arch lies the subacromial bursa (Fig. 15D–4). It is a filmy synovium-lined sac that is attached at its base on the greater tuberosity with its roof fixed to the undersurface of the acromion and the coracoacromial ligament. The remainder of the roof is loosely applied to the undersurface of the deltoid with the remainder of the base applied to the corresponding surface of the rotator cuff [2]. Although the roof and the base are in intimate contact, the two layers are separated by a thin interface of synovial fluid. It is this mechanism that allows the relatively frictionless motion between the cuff and the overlying deltoid and coracoacromial arch. It is an entirely extra-articular space and structure.

Deep to the bursa lies the rotator cuff and the biceps tendon within the glenohumeral joint. The four short rotators insert as a fused composite tendon into the lesser and greater tuberosities (Fig. 15D–4). There is a recognizable interval superiorly between the subscapularis and supraspinatus tendons. This interval is occupied by the biceps tendon as it enters the joint cavity on the deep surface of the cuff tendons along with the coracohumeral ligament superficially [123] (Fig. 15D–5). This ligament is not only an important anatomic landmark but also requires identification and usually release when mobilization of the subscapularis is required, for example in surgical repair of the cuff. Normally

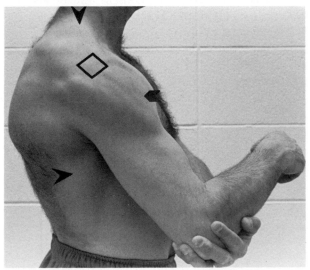

FIGURE 15D–2
Superficial shoulder girdle muscles with deltoid (middle arrowhead) overriding sloping acromion (diamond), trapezius (arrow at top), and latissimus dorsi (lower arrow).

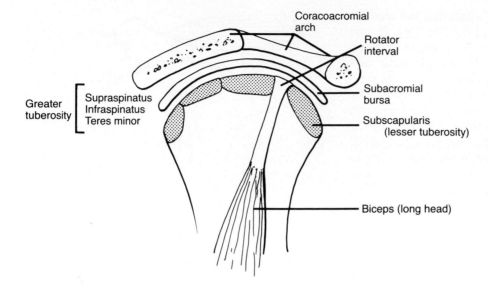

FIGURE 15D–4
Subacromial bursa between overlying coracoacromial arch and underlying cuff tendons attached to greater and lesser tuberosities.

it functions as a suspensory ligament of the humeral head [143].

Recent studies by Aoki and associates [6], Bigliani and colleagues [15], and others suggest that the shape and slope of the acromion vary and may in some circumstances be related to rotator cuff disorders, particularly in the presence of a tear [15, 94]. Because these studies are based primarily on cadaveric specimens, it is not clear whether the variability in shape is the result rather than the cause of cuff degeneration (Fig. 15D–6).

The acromion, when viewed from the lateral perspective, is curved with a beak anteriorly (Fig.

15D–6*B*). Anterior subluxation may result in compromise of the available space in the subacromial region with secondary impingement due to this anatomic set-up [6, 15].

RELEVANT BIOMECHANICS

The rotator cuff provides muscular forces to generate movement in the shoulder and is intimately involved in stabilizing and controlling the humeral head in the glenoid during athletic activity [138]. The effectiveness of the cuff depends upon its force of action, which is related to its size, the type of contraction, and the speed of contraction. Effectiveness is also related to its moment arm or leverage and to its angle of pull [123].

FIGURE 15D–5
Anterior cuff structures with coracohumeral ligament depicted as a superficial structure to the rotator interval between the subscapularis and supraspinatus tendons.

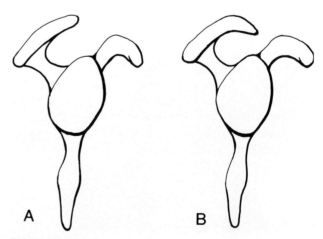

FIGURE 15D–6
A, Normal shaped acromion, lateral view. *B*, Hooked acromion, lateral view.

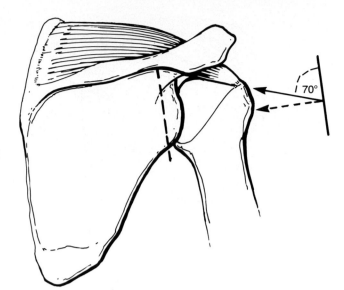

FIGURE 15D–7
Angle of pull of supraspinatus with (solid arrow) direct line of force and (dotted arrow) compressive component of force.

Cuff Function

The supraspinatus, although previously thought to initiate abduction [65], is presently thought to function primarily as a stabilizer of the glenohumeral joint. Its angle of insertion at 70 degrees with respect to the glenoid provides a compressive force [123] (Fig. 15D–7). This allows the powerful deltoid to function more efficiently by maintaining its fulcrum of action at the glenohumeral interface. Without the synergistic action of the supraspinatus the fulcrum would be displaced superiorly, allowing impingement of the humeral head and rotator cuff against the undersurface of the acromion [123, 159]. The infraspinatus and teres minor are primarily external rotators. They also aid in stabilization as humeral head depressors [68]. The subscapularis is an internal rotator; it also depresses the humeral head and provides protection for the anterior capsule. The infraspinatus and subscapularis are important stabilizing muscles of the shoulder, especially during eccentric contraction [74].

The large superficial muscles such as the deltoid, trapezius, latissimus dorsi (see Fig. 15D–2), and pectoralis provide power for movements of the shoulder. The rotator cuff muscles provide the fine tuning. Concentric contraction provides propulsive movement, but eccentric control is crucial to the balance of muscular function about the shoulder and likewise to stabilization of the glenohumeral joint.

Biceps

The long head of the biceps, although implicated as a humeral head depressor [139] (Fig. 15D–8), is

most likely a passive player during shoulder motion [146]. In the presence of a rotator cuff tear, however, the biceps tendon may also be torn or may be enlarged through compensatory functional requirements. It then may become more active in athletic endeavors in older patients, particularly in the presence of a cuff disruption [53]. It may be important in combined shoulder and elbow function as in the deceleration phase of the throwing motion in younger athletes [4, 43].

Static Stabilizers

The static structures of the shoulder such as the glenohumeral ligaments (Fig. 15D–9) are important

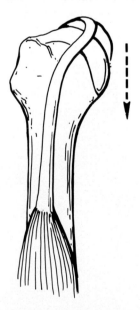

FIGURE 15D–8
Biceps tendon showing potential humeral head depressive action (dotted arrow).

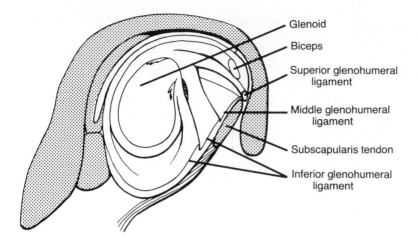

Glenoid
Biceps
Superior glenohumeral ligament
Middle glenohumeral ligament
Subscapularis tendon
Inferior glenohumeral ligament

FIGURE 15D–9
Inside view of the shoulder joint showing the relationship between the rotator cuff biceps tendon and the static stabilizers including the superior, middle, and inferior glenohumeral ligaments.

for stability but also may be implicated in the impingement phenomenon. For example, tight posterior structures cause greater anterior translation of the humeral head with forward elevation and thus may contribute to secondary impingement [52].

The Throwing Motion

Overhead motion, as exemplified by throwing, is the most common motion affecting the shoulder in sports [112]. This action, such as pitching, passing (football), or serving (tennis), can be divided into three phases [119, 120, 150]. The wind-up or cocking phase (Fig. 15D–10) involves abduction, extension, and external rotation. The deltoid is primarily involved during this phase. This results in a levering action anteriorly and can cause superior migration with the natural consequence of secondary impinge-

ment and tension on the anterior joint structures [82, 52]. The supraspinatus provides a stabilizing force through its action in compressing the humeral head in the glenoid. The action of the infraspinatus and teres minor maintains the humeral head within the glenoid fossa through inferiorly directed forces. They also provide external rotation concentrically. The subscapularis, aided by the pectoralis major, restricts the terminal external rotation of this first phase through eccentric control and stabilizes the humeral head within the glenoid fossa through a tethering effect. The acceleration phase then begins instantaneously with a complete reversal of motion (Fig. 15D–11). The internal rotators, the subscapularis and sternal head of the pectoralis major, provide the force [50]. It may be that no synergistic relaxation of the posterior cuff muscles actually occurs but rather eccentric contraction of these muscles, contributing to a stabilizing balance in the

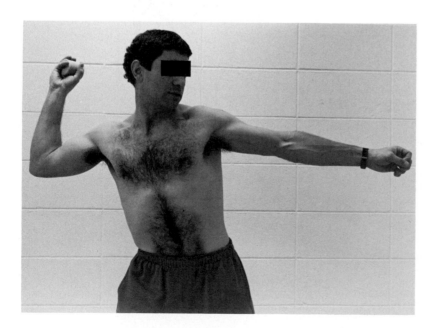

FIGURE 15D–10
Wind-up or cocking phase of throwing motion.

FIGURE 15D–11
Acceleration phase of throwing motion.

joint. The latissimus dorsi, serratus anterior, and triceps provide additional force through concentric activity. The final phase, deceleration or follow-through, involves the entire rotator cuff [75, 120] (Fig. 15D–12). The subscapularis is continuing to rotate internally. The posterior cuff is working to decelerate the arm and maintain the humeral head within the glenoid. The supraspinatus is active through all phases of the throwing motion, but its intensity of activity is only moderate and probably reflects its role in maintaining the humeral and glenoid articulating surfaces [75]. The long head of the biceps tendon, apart from being active during

the cocking phase, in which the elbow is actively bent, also comes into play during deceleration. It is at peak activity following ball release when it acts eccentrically to decelerate terminal elbow extension [71, 74, 120]. This constitutes the active or dynamic description of the throwing motion. Passively, it has been demonstrated that there are obligatory translations of the glenohumeral joint. The humeral head moves posteriorly with extension or external rotation and anteriorly with flexion or cross-body motion. A clear understanding of the interplay between the dynamic and passive motions in the shoulder has yet to be gained. Recognition of these components implies that an even greater complexity of these functions exists.

The actions comprising the overhead throw, the tennis serve, the javelin throw, and the various swimming strokes are all made up of relatively similar mechanisms. Differences exist in the equipment involved, the associated body motions, and the position of the shoulder in each action. Of importance is the degree, repetitiveness, and nature of the forces involved and whether any impact occurs such as in spiking a volleyball. Because of the biomechanical action of the rotator cuff, dysfunction due to injury or disease can easily lead to significant problems, particularly in the athlete's shoulder, in which the stresses are so great.

Concentric Versus Eccentric Muscular Contraction

Recent work in biomechanics suggests that significant interplay between concentric and eccentric

FIGURE 15D–12
Deceleration or follow-through phase of throwing motion.

action of the muscles on opposite sides of a joint occurs, particularly in the rotator cuff. With external rotation of the arm, the infraspinatus contracts concentrically; however, the subscapularis shows significant electromyographic (EMG) action, obviously as a result of eccentric contraction [74, 75]. This action is related to propulsive action on power movements on one side of the joint and perhaps to tethering stabilizing control on the opposite side of the joint. This balance is important biomechanically in fine tuning the movements in the athlete's shoulder. Recent work suggests that emphasis should be placed more on eccentric strengthening.

Scapular Lag

Biomechanically the scapula plays an intimate role in shoulder function. Many pathologic situations such as impingement and various instabilities result in subtle winging through dysfunction of the scapula as it moves on the chest wall. Fatigue of the scapular rotators on the chest wall may lead to secondary impingement because the scapula may not rotate properly and therefore the acromion may fail to clear out of the way when the arm is elevated. This situation, termed scapular lag, may result in secondary impingement [47].

Eccentric and Intrinsic Failure

In many shoulder problems early musculotendinous fiber failure is likely with resultant secondary changes [41, 113]. With time and superior migration of the humeral head, degenerative changes develop, causing wearing and thinning and eventually tears of the rotator cuff. Many years are usually required to develop a rotator cuff tear; however, in the younger throwing athlete, partial tears may result because of the severe stresses placed on the cuff structures. In addition, other mechanisms may come into play such as instability [118] and secondary impingement, causing further stresses on the already weakened musculotendinous cuff. Classic full-thickness tears are not usually seen until the athlete is older, often over the age of 50 years. These tears are related to chronic degeneration and attrition occurring over an extended period of time. This process leads to changes in the biomechanical function of the shoulder with abnormal instant centers of rotation and resultant compensation by the biceps, deltoid, and other shoulder girdle muscles.

CLINICAL EVALUATION

History

An athlete initially diagnosed as having a rotator cuff or biceps tendon problem in the shoulder may present with one or a variety of chief complaints. The majority of patients have pain. They may complain of such things as fatigue, functional catching in the shoulder, stiffness, weakness, and symptoms of instability. On occasion, an athlete may present simply with complaints about deterioration in athletic performance, for example, a pitcher who loses velocity on his fastball.

When considering an athlete's shoulder problem it is important to categorize carefully the sport involved, the intensity of participation (e.g., recreational vs. professional), the offending activity (e.g., throwing vs. blocking in football), and the arm position and activity causing the problem, usually pain.

The chief complaint must be matched with various constitutional factors such as age and sex. The older athlete who has experienced an overhead smash in tennis or a fall on the outstretched arm could sustain a rotator cuff tear, but this would be quite unusual in someone below the age of 40. Female athletes are more likely to demonstrate the generalized hyperlaxity that has been implicated in chronic painful shoulder conditions [18]. Hand dominance also plays a role. In the older nonathletic population, rotator cuff pathology may be distributed equally between the dominant and nondominant shoulders. However, among athletes it is likely to be bilateral in such athletes as swimmers or gymnasts.

Taking these things into consideration, it is usually possible to characterize these complaints into two main categories: (1) the acute or macrotraumatic presentation, in which it is necessary to identify as clearly as possible the mechanism of injury, and (2) the overuse or microtraumatic presentation, in which it is helpful to analyze the pattern of training and competition [70]. It is also important to realize that an acute episode may be superimposed upon a chronic situation.

As previously stated, the most common complaint is that of pain. Although pain can be difficult for the patient to localize accurately, it commonly radiates into the upper arm in the region of the deltoid tuberosity. Biceps tendon pain is usually located anteriorly and can radiate down the belly of the long head of the biceps toward the elbow. Most athletes associate their pain with their sporting activity. It is important to gauge the severity of the complaint. The modified Blazina classification is a useful method of grading severity as follows: grade 1, pain with activity; grade 2, pain during and after activity (not disabling); grade 3, pain during and after activity (disabling); and grade 4, pain with activities of daily living [42]. This classification not only helps to characterize the severity of the pain but also aids in determining the treatment that is required.

The associated symptoms of fatigue, functional catching, stiffness, or weakness must also be characterized carefully with respect to the individual's constitutional factors and how they relate to the particular sporting activity (Table 15D–1). Other symptoms such as clicking in the shoulder, a feeling of instability, numbness, or radiating sharp pain or paresthesias down the arm into the hand, neck, or elbow suggest an alternative or associated diagnosis to rotator cuff and biceps tendon pathology [11, 12, 19, 80, 107].

History taking in the athletic population does not lead to a clear-cut diagnosis as easily as it does in more sedentary people. This is a direct result of the fact that the cause of rotator cuff and biceps tendon pathology can be attributed to a number of different factors. Stresses placed upon the shoulder joint by the athlete can lead to problems such as eccentric overload of the surrounding musculature, impingement either primary or secondary, acute traumatic injury, or uni- or multidirectional instability [71]. This understanding, combined with constitutional factors, the severity of the complaint, and the specific sporting activity involved, constitutes the diagnostic process during the history taking.

The main concern with respect to the differential diagnosis is ruling out glenohumeral instability of one form or another [71]. This can be particularly difficult, and indeed instability may be combined with rotator cuff or biceps tendon pathology. Referred pain from the surrounding joints should always be considered. Radicular pain, numbness, or paresthesias may point to cervical pathology, thoracic outlet syndrome, or a primary neurologic problem such as suprascapular neuropathy [19] (Table 15D–2).

TABLE 15D–1
Rotator Cuff Problems in Athletes

Pain
Difficulty with overhead activities
Night pain
Loss of endurance during activities
Deterioration in sporting performance
Catching, grinding (crepitus)
Weakness
Stiffness

TABLE 15D–2
Differential Diagnosis of Rotator Cuff Problems in Athletes

Rotator cuff or biceps tendon
  Strain
  Tendinitis
  Tear
Glenohumeral instability
Glenohumeral instability with secondary impingement
Primary impingement of the cuff or biceps tendon
Calcific tendinitis
Acromioclavicular joint pathology
Glenohumeral arthritis
Cervical spine pathology
Thoracic outlet syndrome
Adhesive capsulitis
Suprascapular neuropathy
Angina, lung tumor, gallbladder disease

Finally, it should be stated that a diagnosis of rotator cuff or biceps tendon pathology based on the history alone is not as specific or sensitive as is a diagnosis of other athletic injuries made in this way. Confirmation by a physical examination and further investigation is usually required.

Physical Examination

The physical examination should include the necessary evaluation of the general health of the individual in order to consider systemic or regional diseases that can cause shoulder pain. Usually the history is sufficient in this regard; however, a lack of local shoulder findings could point toward conditions such as angina, cervical disc disease, or cholecystitis as the etiology of the shoulder pain. The initial impression considers the patient's age, overall health, and level of specific distress related to the shoulder problem. An organized and comprehensive approach to the physical examination is necessary. (Inspection, palpation, range of motion, strength testing, and neurologic stability assessment and special testing constitute an orderly sequence.)

Inspection considers symmetry (one must appreciate that pitchers or tennis players may have unilateral drooping of their dominant shoulder) [66, 83, 130] (Fig. 15D–13) or deformities such as old acromioclavicular injuries and muscle wasting, which may occur in the infraspinatus or supraspinatus fossa and is characteristic of a rotator cuff tear. A ruptured biceps tendon shows the diagnostic bulge in the arm with muscle contraction [59].

The location and degree of tenderness found on palpation often provide a reliable physical sign leading to an accurate diagnosis. Tenderness in the

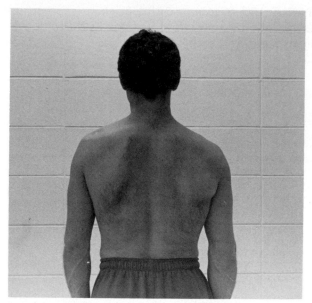

FIGURE 15D–13
Right-sided dominant tennis player showing compensatory droop of the right shoulder girdle.

bicipital groove (2 to 5 cm distal to the anterior acromion, midway between the axilla and the lateral deltoid with the arm in the anatomic position) is a reliable sign of bicipital tendinitis [59] (Fig. 15D–14). The supraspinatus insertion is palpated through the deltoid just distal to the anterolateral border of the acromion with the shoulder extended and inter-

nally rotated. Maximal tenderness over the gleno-humeral or acromioclavicular joint lines may indicate specific joint pathology.

Range of motion requires documentation. A true discrepancy between active and passive ranges of motion is suggestive of a rotator cuff tear. However, in the athlete superior strength and flexibility may easily mitigate against this finding [151]. Many athletes such as swimmers and gymnasts may appear to have a greater range than normal, and it is important to recognize that there is great variation in the normal range [16]. The unilateral overhead athlete may demonstrate an obvious discrepancy with increased external rotation and decreased internal rotation compared to the opposite side [81] (Fig. 15D–15). It is believed that this decreased internal rotation is an adaptive mechanism in athletes engaging in overhead throwing. It may also, however, be a contributing factor in rotator cuff overuse syndromes. It has been shown that limited internal rotation may be associated with tight posterior capsular structures. This could contribute to rotator cuff pathology through excessive obligatory anterior translation and secondary impingement [52]. Stressing the shoulder at the extremes of motion can also provide clues to pathology in the shoulder.

Ranges of motion that should be documented in degrees are active and passive elevation in the scapular plane (Fig. 15D–16), active and passive external rotation with the arm at the side (Fig. 15D–17), and internal rotation where the hitchhiking thumb (or index finger) reaches in reference to the

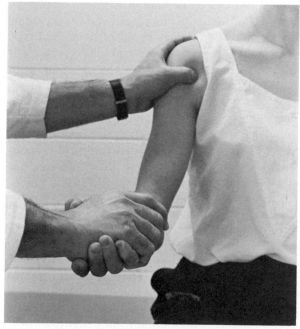

FIGURE 15D–14
Palpation of the biceps tendon in the bicipital groove. Demonstration of Yergason's test with resisted supination and palpation in the bicipital groove.

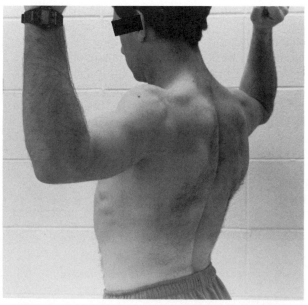

FIGURE 15D–15
Increased external rotation in the right dominant shoulder.

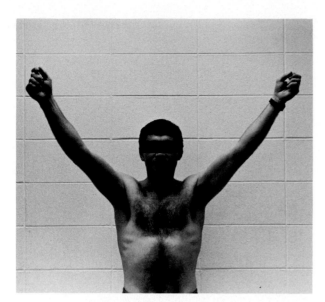

FIGURE 15D–16
Elevation in the scapular plane.

posterior vertebral anatomy (Fig. 15D–18) [58]. It is also important, especially in athletes, to document external rotation, particularly passively in the 90-degree abducted position in the coronal plane. This position represents a more functional measure of external rotation (Fig. 15D–19).

Strength testing is considered along with range of motion, and a clinical grade from 0 to 5 is estimated for each motion involved. Although it is part of the neurologic examination, assessment of strength is particularly important in athletes with rotator cuff pathology. Objective weakness beyond that considered due to pain or a neurologic deficit is a very

specific sign of rotator cuff deficiency. The remainder of the neurologic examination will help to rule out pathology such as a cervical root, brachial plexus, or peripheral nerve lesion.

The assessment of shoulder stability is very important because rotator cuff signs and symptoms are often a secondary manifestation of an underlying problem in instability. It has been suggested in high-profile throwing athletes that shoulder pain is due to instability related to anterior subluxation with secondary impingement until proven otherwise. Stability is assessed by translating the humeral head in the glenoid fossa anteriorly, posteriorly (the load and shift test) (Fig. 15D–20), and inferiorly (the sulcus sign) (Fig. 15D–21). It is also important to assess the presence of an anterior apprehension sign, which is indicative of anterior instability. This is performed by passively placing the arm in increasing degrees of abduction and external rotation (Fig. 15D–22). The relocation test, or Fowler's sign, is a variation of the apprehension sign. The arm is placed in the abducted, externally rotated position until pain or apprehension is elicited. The maneuver is then repeated with the arm supported in a posterior direction. Relief of pain or apprehension with improved external rotation is indicative of anterior subluxation [64] (Fig. 15D–23).

Examination of the regional vascular supply is necessary as a baseline and also for consideration of conditions such as thoracic outlet syndrome.

Finally, there are a number of special tests that should be considered. The signs of impingement are characteristic of rotator cuff tendinitis and tears. These include a painful arc of abduction between

FIGURE 15D–17
External rotation neutral position.

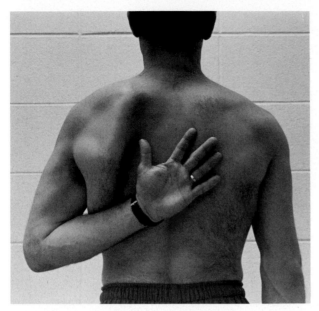

FIGURE 15D–18
Internal rotation.

60 and 120 degrees [79], pain on forced forward flexion in which the greater tuberosity is forced up against the anterior acromion (Fig. 15D–24), and pain on forcible internal rotation of the 90-degree forward flexed arm [59] (Fig. 15D–25). The latter maneuver causes impingement against the coraco-acromial ligament. Biceps tendon involvement is demonstrated by Speed's test, in which pain is reproduced on resisted forward elevation of the humerus against an extended elbow [58] (Fig. 15D–26). Yergason's test is performed with the elbow flexed to 90 degrees and the forearm pronated [160].

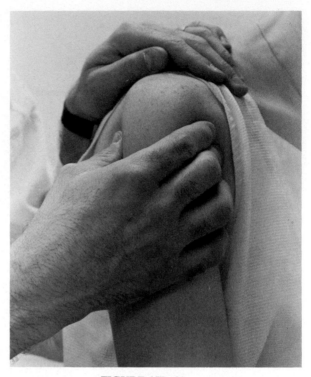

FIGURE 15D–20
The load and shift test.

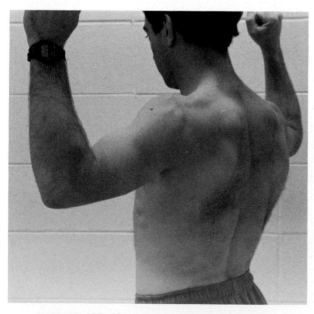

FIGURE 15D–19
External rotation in the 90-degree abducted position.

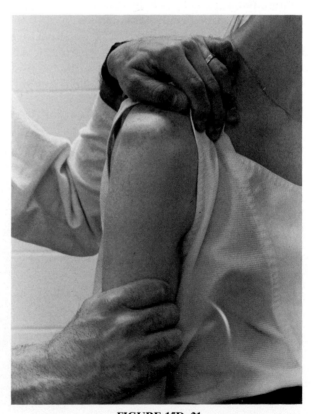

FIGURE 15D–21
The sulcus sign.

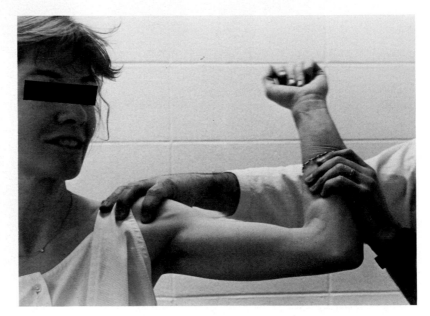

FIGURE 15D–22
Anterior apprehension testing.

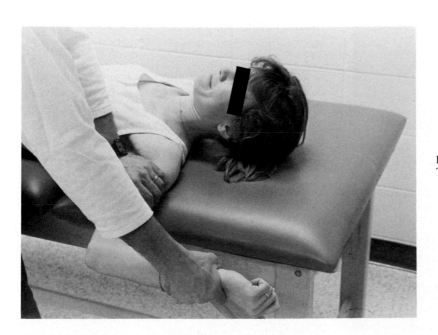

FIGURE 15D–23
The relocation test, or Fowler's sign.

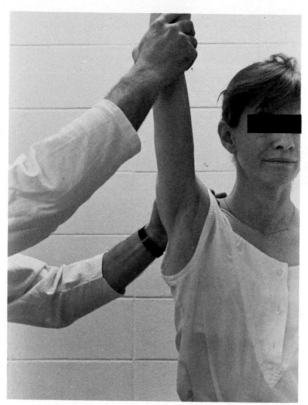

FIGURE 15D–24
Impingement sign: forced forward flexion.

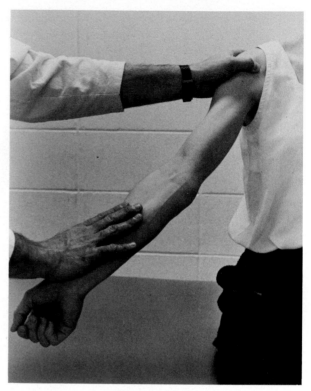

FIGURE 15D–26
Speed's test.

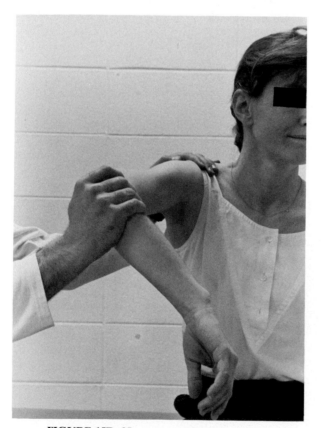

FIGURE 15D–25
Impingement sign: abduction internal rotation.

The examiner grasps the wrist and resists active supination by the patient. Tenderness and pain in the area of the bicipital groove suggest pathology in the long head of the biceps [59, 128] (see Fig. 15D–14).

Biceps tendon instability (i.e., medial subluxation or dislocation) can be determined by passively abducting the shoulder to 80 to 90 degrees and eliciting a palpable snap in the region of the bicipital groove with internal and external rotation, a rare presentation in the majority of cases, especially in the absence of cuff tears [96, 105].

The history and physical examination will lead to an appropriate diagnosis. As previously stated, in athletes the main differential is between instability and primary rotator cuff pathology. Before discussing the relevant diagnostic studies we would like to introduce the concept of the vicious cycle as it applies to the athlete with shoulder pain. This theory is based upon a number of clinical observations, a current appreciation of the biomechanic function of the shoulder, and evaluation of the results of various forms of treatment of painful shoulder conditions. It is now well appreciated that in the athlete there is often an overlap between shoulder instability and rotator cuff tendinitis or impingement (biceps tendinitis included). The primary question is, Does instability cause tendinitis, or does tendinitis cause

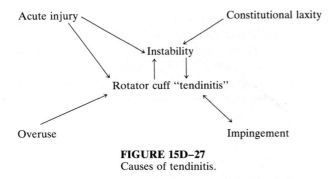

FIGURE 15D-27
Causes of tendinitis.

instability? The diagram in Figure 15D–27 has been proposed to help explain these related problems.

Diagnostic Studies

The use of diagnostic studies to evaluate the athlete with a rotator cuff or proximal biceps tendon problem aids in the confirmation of the clinical findings. These studies are important in ruling out other pathologic entities.

The Impingement Test

This test as described by Neer involves the injection of local anesthetic into the subacromial region [101] (Fig. 15D–28A). It can reasonably be considered an extension of the physical examination in patients presenting with a painful shoulder. The injection is performed under sterile conditions with insertion of the needle anteriorly, laterally (Fig. 15D–28B), or posteriorly into the subacromial space. Impingement signs should be sought as previously described in the section on physical examination. Subjective relief of pain and diminution of the previously present painful signs are helpful signs that impingement may be a component of the problem. It should be emphasized, however, that this is a nonspecific test and can be misleading [20], since it may be positive in patients with either primary impingement or secondary impingement due to instability.

Although not strictly an impingement test, injection of local anesthetic into the acromioclavicular joint or into the bicipital groove can supply additional information about the source of the pain. Subacromial anesthetic can mask or minimize the symptoms from these two areas. It is the clinical examination that is critical in guiding the selections and order of the injection sites.

Radiographic Studies

Plain Radiographs. Plain radiographs of the normal athlete with a rotator cuff complaint are most often normal. These should include an anteroposterior (AP) film at right angles to the scapular plane, a lateral film in the scapular plane with the beam tilted 10 degrees to look at the acromial shape and slope (Fig. 15D–29), and an axillary view. The characteristic changes of advanced rotator cuff disease include sclerosis and cystic changes in the greater tuberosity, osteophyte formation on the acromion [94], a more prominent notch between the

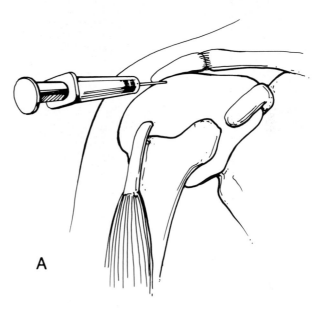

A

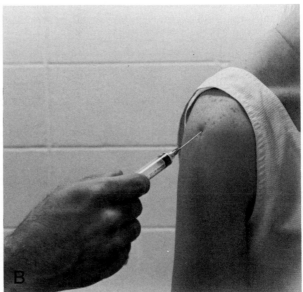

B

FIGURE 15D-28
A, Impingement test showing needle in subacromial region. *B,* Impingement test, lateral approach.

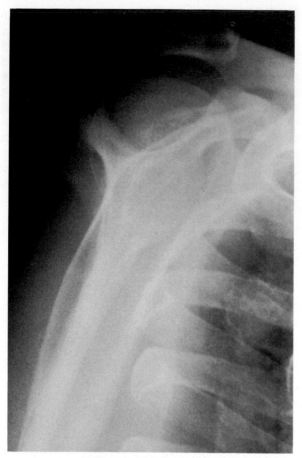

FIGURE 15D–29
Supraspinatus, outlet view.

greater tuberosity and the articular surface, changes in the shape of the acromion, and, in the presence of a cuff tear, sometimes a narrowed acromiohumeral distance [28, 156]. There may be osteophyte formation on the inferior surface of the acromioclavicular joint as part of chronic rotator cuff disease [114]. One value of these films is to rule out other conditions that may present with shoulder pain such as glenohumeral arthritis and calcific tendinitis that would not be considered primary problems in the athlete.

Plain radiographs can also be utilized to evaluate the bicipital groove. A shallow groove may indicate a very rare biceps tendon instability problem [96, 115].

Arthrography. Single or double contrast arthrography is currently considered the gold standard for determining the presence of a full-thickness rotator cuff tears [17] (Fig. 15D–30). It is an extremely sensitive technique for detecting complete cuff tears with reports of almost 100% sensitivity and a correspondingly high positive predictive value [21]. In our series of complete rotator cuff tears, however, we did have an approximately 8% incidence of false-

negative results [63]. Arthrography is only occasionally helpful in diagnosing undersurface partial-thickness cuff tears. Arthrography is an invasive technique and is not without complications. It can be used to diagnose other pathologic conditions such as adhesive capsulitis. Combined with CT scanning it is helpful in diagnosing intra-articular pathology such as labral changes suggestive of instability. However, the arthrogram in the athletic population is most helpful in ruling out a cuff tear [104].

Ultrasonography. Diagnostic ultrasound is a noninvasive form of examination of the rotator cuff [28]. It allows comparison with the other side and, compared with arthrography, can evaluate a greater amount of anatomic detail [83, 145] (Fig. 15D–31). It has a reported 91% sensitivity and specificity [21], with a 100% positive predictive value when it shows nonvisualization or focal thinning [83]. It has also been reported to be very useful in diagnosing bicipital pathology [93] and is helpful in patients who have previously undergone a rotator cuff repair [28, 29]. However, the results are related to the operator's experience, and the technique has inherent limitations because of the surrounding bony anatomy [28, 92].

Magnetic Resonance Imaging. Magnetic resonance imaging (MRI) has recently been used in the investigation of rotator cuff pathology [38] with promising results [144] (Fig. 15D–32). Sensitivities have ranged from 80% to 91% [21, 162] with specificities of 88% to 94% [38, 162]. These studies are operator dependent, and the newer technology has

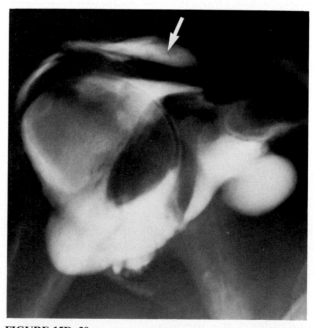

FIGURE 15D–30
Single contrast arthrogram showing contrast in the subacromial bursa. Rotator cuff tear.

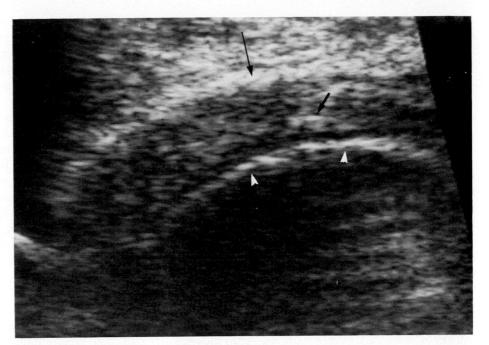

FIGURE 15D–31
Normal suprapinatus, ultrasound. The tendon lies between the cortical surface of the humerus (white arrowheads) and the linear echogenic interface with subacromial fat, bursa, and tendon (long arrow). There is a small irregular focus of increased echogenicity within the tendon (arrow).

improved diagnostic capabilities [91]. MRI can detect the size, location, and characteristics of the pathology in the cuff [21] (Fig. 15D–33). The drawbacks are that the test is time-consuming and costly, and there is an occasional instance of patient intolerance because of the required lack of movement and because of claustrophobia [28].

At present there is a tendency toward using the noninvasive techniques. The arthrogram is very valuable for determining the presence or absence of full-thickness tears. Ultrasound and MRI provide more information about size and other characteris-

tics of full-thickness tears. Ultrasound and MRI also may aid in diagnosing the presence, size, and location (joint, bursal, or in-substance) of partial tears and other pathologies. It is our guess that with time

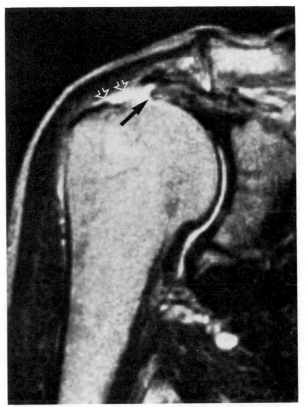

FIGURE 15D–33
MRI demonstrating torn rotator cuff (black arrow). Open white arrows show fluid in the subacromial space.

FIGURE 15D–32
MRI showing a normal suprapinatus tendon between the humeral head and acromion.

MRI will supplant the other modalities. MRI may become particularly helpful in identifying partial cuff tear and labral lesions in the athletic population. Validation of these lesions requires further studies.

Arthroscopy

Arthroscopy can play an important role in the evaluation of a patient with a painful shoulder [4, 88]. Arthroscopy provides direct visualization of the intra-articular structures. The biceps tendon, the articular surfaces, the status of the labrum [88] and ligamentous structures, and the undersurface of the cuff can be observed. Combined with an examination under anesthesia, the clinical picture, and other investigations, a definite diagnosis can usually be made [121]. The subacromial space can be less reliably assessed looking at the cuff on its superficial surface and assessing the bursal tissues. Partial-thickness tears can occasionally be seen, although debridement of the bursa may be required for accurate visualization, thereby introducing an operative component.

In the athletic population the specific lesions to be identified are partial- or full-thickness rotator cuff tears; bicipital lesions of the superior labrum associated with bicipital tendon avulsion or fraying superior labral, anterior, and posterior (SLAP) lesion [36]; anterior-inferior labral avulsions or damage suggestive of associated instability; and chondral changes or even a frank Hill-Sachs lesion. It is surprising how often multiple lesions are identified through the arthroscope, the exact meaning of which requires careful consideration.

Examination Under Anesthesia

Performing an examination under general anesthesia allows assessment of translation of the humeral head within the glenoid fossa and an accurate assessment of the passive range of motion of the shoulder. The "load and shift test" (see Fig. 15D–20) can be applied to document translation clinically and to feel the relationship of the humeral head to the glenoid rim with stressing anteriorly and posteriorly [64, 65]. In patients with multidirectional instability [58] the humeral head can frequently be displaced over the glenoid rim both anteriorly and posteriorly. An inferior sulcus sign (see Fig. 15D–21) is an important indicator of multidirectional instability and is documented under anesthesia by determining in centimeters the amount of inferior translation of the superior part of the humeral head

away from the undersurface of the acromion. More than 2 cm of inferior translation suggests multidirectional instability or certainly indicates excessive inferior translation. Sometimes a diagnosis of adhesive capsulitis is not determined until examination under anesthesia (EUA).

Challenging situations prompt investigations such as arthrography, MRI, EUA, and arthroscopy. Sometimes in the overhead athlete with a painful shoulder the diagnosis of impingement versus instability versus eccentric overload can be confusing, requiring everything available in our investigative armamentarium. The history and physical examination remain, however, the most reliable diagnostic procedures in the majority of situations.

TREATMENT OPTIONS

In the majority of patients, treatment of a rotator cuff and biceps tendon problem is nonoperative. This point is especially important in athletes. Three important items to be considered are (1) the etiology of the condition, (2) the sport and level of performance, and (3) the severity of the problem (based upon the modified Blazina grading).

Etiology

Patients with *primary impingement* [62] fit into Neer's three-stage classification: stage 1, edema and hemorrhage (at any age); stage 2, fibrosis and tendinitis (usually in patients over the age of 25); and stage 3, degeneration, bony changes, and tendon ruptures (patients usually over the age of 40) [103].

Overuse and fatigue of the scapular stabilizers lead to scapular lag and secondary impingement [47, 62].

Eccentric overload [63] of the cuff tendons leads to fatigue and pain. Long-term repetitive overuse combined with the inherent poor blood supply of the tendons can lead to degeneration and tearing, particularly in older individuals [113].

Instability [63] may be due to one of two causes. Loose shoulder or true multidirectional instability can lead to secondary rotator cuff–related pain and pathology. Anterior subluxation can lead to secondary impingement with pain [71]. In the overhead athlete, anterior subluxation is a common cause of secondary impingement, especially in patients with tight posterior structures and limited internal rotation.

Acute trauma to the cuff secondary to either a dislocation of the glenohumeral joint or direct or indirect damage can also be implicated in the etiology [56, 106, 108, 109].

Although the great majority of athletes with these problems can be treated nonoperatively, the surgical management of someone with primary impingement as opposed to instability would be quite different.

Sport and Performance

A number of generalizations can be made with respect to the type of sport. A soccer player with a rotator cuff problem will have only minimal if any disability. The baseball pitcher, on the other hand, may be significantly disabled and may be unable to throw at all. The gymnast can avoid the rings and concentrate on his vaulting, and a swimmer may be able to change from freestyle to breaststroke [132].

The same consideration should be given to the level of performance. Enforcement of therapeutic rest has dramatically different implications to a weekend tennis player than to a professional quarterback in midseason.

Severity

Finally, the severity of the problem, whether measured in terms of the clinical picture (modified Blazina classification) [42] or in terms of the underlying pathology has obvious implications with respect to the proposed treatment.

Types of Treatment

With these principles in mind we can divide the suggested treatments into three categories: (1) preventive, (2) nonoperative, and (3) operative.

Preventive Treatment

First and foremost when dealing with athletes, the sports medicine physician, surgeon, trainer, therapist, or coach should consider prevention of injury and problems [77]. Prevention is of particular importance in relation to the shoulder, in which the majority of injuries are related to overuse. The countless hours of training and practice are to be condemned if faulty technique or improper methods lead to damage of the rotator cuff or biceps. The underlying principle of prevention is applied common sense. A musculotendinous unit is capable of resisting only as much as it has been prepared to resist. A 50-year-old who plays 2 hours of doubles tennis once per week cannot expect his shoulder to tolerate up to 12 hours over a 2-day weekend

tournament. The same argument can be applied to the college or professional level pitcher who arrives in training camp with little or no off-season training.

The basis of prevention is preparation. This involves overall body conditioning [98], flexibility [55], strengthening, and careful attention to technique, recognizing the stresses of training and competition. It may appear from the examples given above that this is possible in the organized sporting situation but unrealistic for the weekend tennis player. Prevention in the latter circumstances refers to avoidance of reinjury and minimization of the current problem as well as to sport-related education.

It has been recommended that the Little League pitcher develop the fastball as his primary pitch [140]. From there the emphasis should be on improving velocity while maintaining consistent mechanics and control before developing the full repertoire of pitches. The off-season is the key to a pitcher's development, a general fitness and weight-training program being essential. It is during this time that the specific adaptation to imposed demand (SAID) principle is applied [140]. For a pitcher this involves alternating between long-toss and short-toss throwing at half speed with enforced rest at least 2 days per week. During the season the same principle applies with a proportionate increase in the frequency and duration of training [140].

The warm-up is also of critical importance [59, 98]. A satisfactory warm-up leads to increases in tissue temperature with improved oxygen uptake, nerve impulse transmission, and increased activity of metabolic enzymes [72, 98].

The influence of shoulder flexibility has been demonstrated in swimmers. Greipp showed a clear correlation between anterior shoulder inflexibility and shoulder pain [51]. It follows that stretching is an important preventive measure [98]. The particular goal in stretching is to try to maintain internal rotation and adduction (Fig. 15D–34). It is not necessarily the intention to normalize internal rotation compared to the opposite side. The intention is to avoid posterior capsular contractures and maladaptive loss of internal rotation and adduction. Stretching is equally important in older athletes, in whom the potential for stiffness is greater.

A well-balanced strong musculature about the shoulder is critically important in the prevention of overuse injuries. The rotator cuff can be strengthened with the use of many aids, for example, free weights, Theraband, or surgical tubing [98]. Isokinetic machines are also recommended [77]. It is evident that any strengthening program must be well controlled, particularly in the young athlete. In addition, the use of some isokinetic equipment does

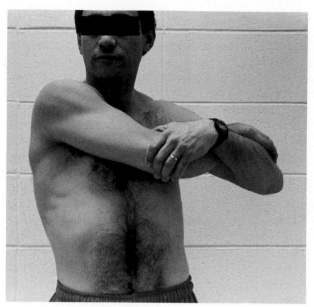

FIGURE 15D–34
Internal rotation and adduction stretching.

FIGURE 15D–36
Resisted internal rotation.

not allow eccentric muscle contraction. This may be of considerable importance given our understanding of the pathophysiology of rotator cuff lesions, which is that of eccentric overload.

A preventive strengthening program for shoulder problems in athletes is critical, and the emphasis should be on eccentric exercises. This can be easily performed by the athlete on a daily basis with some form of resisted Theraband or rubber tubing. The exercises consist of

1. Resisted external rotation exercises with the arm at the side (Fig. 15D–35) and also in the 90-degree abducted position.

2. Resisted internal rotation exercises (Fig. 15D–36)

3. Push-ups with the arms adducted for scapular rotator control (Fig. 15D–37)

4. Sitting rows for serratus and rhomboid scapular control.

5. Shrugs for trapezius strengthening.

6. Latissimus pull-downs for latissimus control, important in the deceleration phase of overhead motion and change.

7. Supraspinatus strengthening exercises: resisted abduction in the scapular plane (Fig. 15D–38) with internal rotation

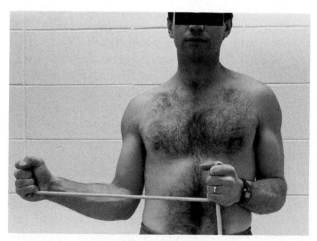

FIGURE 15D–35
Resisted external rotation.

FIGURE 15D–37
Inside push-ups.

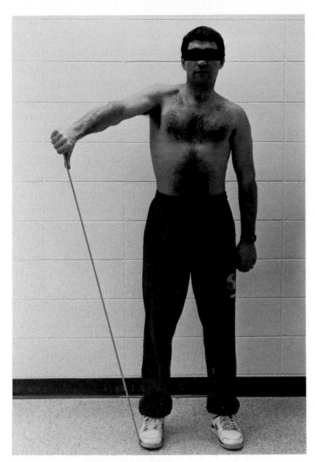

FIGURE 15D–38
Supraspinatus strengthening.

Finally, it is incumbent upon the coaches to teach the correct technique, avoid overtraining, allow for rest periods, and, ideally, recognize the fatigued athlete who is heading for a painful shoulder problem.

Nonoperative Treatment

Overall nonoperative treatment is an extension of preventive management with the addition of specific measures dealing with the injury. It can be divided into four components: (1) modification of activity, (2) local or systemic measures to reduce and relieve the symptoms, (3) stretching and strengthening exercises, and (4) reevaluation and maintenance treatment.

Modification of Activity. In the athlete with mild symptoms, modification of activity means reducing the frequency and duration of the specific activity. It also involves activity substitution or what is sometimes called active rest. In a tennis player, avoiding the service action but still hitting groundstrokes may be all that is required. In a baseball pitcher, it would be helpful to cut back on the daily number of pitches and to slow their velocity. A swimmer should de-

crease yardage or use the kickboard [77]. Other methods of treatment involve specific changes in technique, such as throwing side-arm for a pitcher or, in swimmers, a higher arm entry in the freestyle, ensuring adequate body roll, or even changing from freestyle to breaststroke. Changing equipment may also be of benefit. With time these patients usually improve, but return to sport is a longer term goal, and indeed, more involved treatment including surgery may be required.

Local and Systemic Methods to Relieve Symptoms. Although the athlete experiencing pain with activity that is not disabling may be able to manage with various methods of modified activity, most patients have a more involved problem and require more complicated therapies. The use of nonsteroidal anti-inflammatory medications is ubiquitous. There is no question that they provide symptomatic relief. However, we are dealing with injury, and whether acute or due to overuse the inflammatory response to this injury is a normal part of the healing mechanism. There is also evidence that in chronic overuse type injuries the surgical pathologic tissue is noninflammatory [113]. The value of these medications lies in the initial treatment given to decrease pain and allow rehabilitation. Ice treatment is also a recognized local treatment modality with its action of reducing vascularity, numbing the pain, and possibly reducing swelling and inflammation [87]. Ice is particularly useful after an acute episode or injury. It is commonly advocated after overuse activity to minimize the pain and lessen the immediate postactivity inflammatory response.

Therapeutic ultrasound may be helpful in the treatment of rotator cuff and biceps tendinopathies. It is believed to increase the local vascular response to the injured tissue. This in turn allows the release of the products of injury and the influx of the raw materials needed for repair. A dose of 0.8 to 1.2 w/cm^2 and 1.2 to 1.5 w/cm^2 has been advocated for the biceps and supraspinatus tendons, respectively [77]. Other modalities include high-voltage electrical stimulation, transcutaneous nerve stimulation, electromagnetic field therapy, and the use of lasers. The specific benefits of some of these modalities are not clear, yet many patients seem to gain relief, and side effects are few.

The use of local corticosteroid injections is a more invasive form of therapy. The deleterious effects of steroid injections have been documented [30, 39]. Critical analysis of the literature cannot lead to the conclusion that they are of any long-term benefit [30, 31, 39, 134, 158]. Their use is still advocated and may be of value in a patient with an acutely painful lesion to halt the vicious circle of pain related to overuse. This would then allow earlier institution of corrective rehabilitation.

Stretching. Stretching exercises are not only therapeutic but quite clearly preventive and provide the basis for maintenance treatment. The warm-up including stretching not only helps the athlete to improve his or her timing and control but also allows the muscles to function efficiently. It increases the muscle blood supply and improves contractility [23]. Lack of flexibility has been associated with a higher incidence of shoulder problems in swimmers [51]. An increased range of external rotation compared with internal rotation is also associated, especially in throwers. These factors in the athlete with a painful shoulder require specific attention to the treatment regimen. Stretching should be generalized but should focus on internal rotation and adduction across the chest (Fig. 15D–34) and internal rotation (Fig. 15D–18) and extension behind the back. These exercises should be performed prior to activity in addition to the athlete's usual routine.

Strengthening. Strengthening is the mainstay of treatment for the majority of athletes with rotator cuff or biceps tendon problems. Whether the problem is due to an acute direct injury or eccentric overload the athlete is left with a compromised, weakened musculotendinous unit. This unit is usually contracted either primarily as a result of muscular imbalance or secondarily due to the injury mechanism. A complete tear or disruption of the tendons will obviously need to heal or be approximated through surgery prior to active use, but this is an uncommon circumstance in the young athletic population.

Strengthening should emphasize the external rotators (Fig. 15D–35). The use of rubber tubing is simple and effective for both prophylaxis and treatment. Initially the exercises need to be performed with the arm at the side until the pain has been relieved. When pain is present it is important to avoid the 90-degree abducted position with resisted rotational strengthening. The arm can be brought into the abducted position gradually, initially to 45 degrees until the pain has completely disappeared and then to the functional range. Most recreational and nonthrowing athletes do not require strengthening at 90 degrees of abduction.

The associated musculature, particularly the scapular stabilizers, should be considered. These muscles, when overused, fatigued, or overstretched, lead to scapular lag and secondary impingement [47]. This can be treated with exercises such as shoulder shrugging both upward and backward and with push-ups, arms adducted (Fig. 15D–37).

In addition to the specific components of shoulder strengthening, it is important to understand that the shoulder cannot be viewed in isolation. Equal concern should be paid to the associated joints and muscles to ensure that appropriate body mechanics are utilized in the rehabilitation of the athlete.

Reevaluation and Maintenance. It is during reevaluation that the treatment phase blends into that of prevention of further injury. Throughout the process of treatment the athlete, the coach, the trainer, and the parent should be part of an educational program. This involves teaching these individuals about the clinical problem, its course, and its ultimate prognosis. Most athletes are willing to perform a daily routine of exercises if it means participating in the sports they love. Unfortunately, compliance with this routine is more likely following rather than preceding an injury.

Surgical Management

The surgical management of patients with rotator cuff or biceps tendon pathology should be emphasized less than nonoperative treatment. The great majority of athletes will recover, modify their activities, or even give up their sport before undergoing surgery. The maturing athlete may have not only acute and overuse injuries but also the added concern of degenerative tendon pathology, which involves a far greater incidence of rotator cuff and biceps tears. These athletes have the same symptoms as any individual of a similar age with rotator cuff pathology, but usually they have greater demands and expectations. Greater understanding of the pathophysiology of shoulder pain has led to the realization that sometimes stabilization procedures rather than surgery are required to deal with primary cuff or biceps disorders.

A surgical approach can be divided into open and arthroscopic procedures.

Open Procedures. Historically the procedures designed to correct these problems involved various forms of cuff repair and biceps tenodeses and utilized a variety of approaches [79, 111], including acromionectomy [53, 85, 142]. Even a paraglenoid osteotomy was devised to deal with subacromial bursitis and supraspinatus tendinitis [147]. Today only two procedures are really used: subacromial decompression [15, 57, 154] (and its variants such as debridement) and rotator cuff repair. Neer was the first to describe anterior acromioplasty based on the fact that impingement occurred under the anterior third of the acromion [100]. His initial patient population was nonathletic and included a heterogeneous group of patients with and without cuff tears in whom he reported overall satisfactory results in 80% [100]. His method of anterior acromioplasty has been modified by others to avoid detachment of the deltoid, to include distal clavicle excision, or to release only the coracoacromial ligament. It can now be performed arthroscopically.

The other main open procedure is that of rotator cuff repair. Usually this involves an anterior acromioplasty with either direct side-to-side or, more commonly, tendon-to-bone repair [54, 63, 53]. Many other techniques have been advocated [116] in dealing with an athletic population; however, restoration of normal anatomy is the obvious goal.

Many factors relating to sport, level of activity, and arm dominance influence results. The demands are different for athletes versus nonathletes and throwers versus nonthrowers. Surgery for cuff tendinitis produces different results in a soccer player than in a pitcher.

The use of coracoacromial ligament resection has been successful in 95% of a series of patients who had failed to benefit from nonoperative management for impingement syndrome secondary to coracoacromial ligament entrapment [76]. The use of this procedure, which has a 70% success rate in athletes, has been advocated as an alternative to the more surgically invasive anterior acromioplasty [69].

Comparison of these two procedures in a group of athletes showed similar results [122]. Of the patients with coracoacromial ligament resection 13 of 17 returned to full activity with no further symptoms at an average follow-up of 3½ years. These series did not clearly define the level of activity or throwing capability postoperatively. In a report of anterior acromioplasty in athletes, 89% reported improvement subjectively but only 43% showed good functional results. Of the athletes classified as primarily pitchers and throwers the results are even less satisfactory, with only 22% showing a good functional result [148]. It is very difficult to compare these results. Patient selection is critical to the outcome in this type of operation, and there are no adequate control series.

For athletes with rotator cuff tears the decision to operate is easier in those with a complete cuff lesion but is not so clear with partial-thickness tears. In a series of 45 patients who underwent rotator cuff repair for both full-thickness and partial-thickness tears, 39 (87%) were satisfied and experienced subjective pain relief. Analysis based on sports participation showed that among pitching and throwing athletes performing at the college or professional level only 32% had a good result [149]. The results of surgery in similar athletes with complete cuff tears were slightly better, with 5 of 9 experiencing good results.

Most authors recommend early surgical repair of full-thickness cuff tears that are symptomatic [9, 32, 40, 128]; however, many older athletes can manage by modifying their functional demands provided the pain has resolved. In an athlete with higher demands a prediction of full return to the premorbid level of play is not possible based on our current understanding of the literature. This caution applies both to decompression of the cuff regardless of the method used and to rotator cuff repair. A symptomatic full-thickness tear without surgery yields predictably poor results, and a large tear (i.e., greater than 3 cm) has a poor functional recovery even with surgery [33, 37, 48, 49, 153].

Surgery related to biceps tendon pathology is even less clearly delineated in the literature. It is usually assumed that the biceps is involved along with the rotator cuff based upon similar pathophysiology, mechanisms of injury, and repair [101, 105, 155]. It therefore follows that the treatment involves decompression of the coracoacromial arch. Again, it is very difficult to sort out the results of surgical management, but success similar to that achieved with rotator cuff repair has been reported.

With respect to subluxation or dislocation of the biceps tendon from the bicipital groove, the treatment is relatively simple, tenodesis being the recommended approach. One series reported that 77% of the athletes resumed their sport and could throw satisfactorily [115]. In another report, excellent results were cited; however, no specifics were documented about either the patient population or the actual results [128]. In our experience primary biceps instability in athletes is extremely rare.

Arthroscopy. Arthroscopic management involves initial confirmation of the diagnosis. The intra-articular structures are visualized, and evidence of articular damage or instability is documented. Superior labral, anterior, and posterior (SLAP) lesions with undersurface cuff degeneration may be a manifestation of overuse associated with instability. The undersurface of the cuff is carefully examined. Partial-thickness tears can be judiciously debrided [35], and then an arthroscopic subacromial decompression with coracoacromial (Fig. 15D–39) ligament resection is performed as part of the same procedure. Debridement alone is done by some surgeons, particularly with a diagnosis of eccentric overload, undersurface cuff degeneration, and minimal if any evidence of impingement as part of the process. Massive rotator cuff tears that are the result of previously failed open repair can be treated similarly [7].

The results of subacromial decompression in 24 patients active in sports who had a diagnosis of impingement revealed that 87.5% returned to active participation [7]. The overall success rate in another series of heterogeneous patients, including both sport- and nonsport-related etiology, was 88% [35].

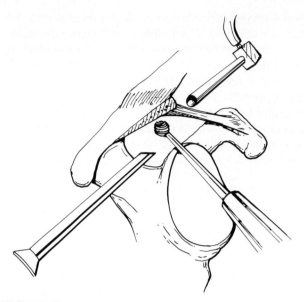

FIGURE 15D–39
Arthroscopic subacromial decompression with removal of anterior acromion and coracoacromial ligament.

The complications of arthroscopic procedures are relatively few. Transient neuropraxias have been reported by a number of authors [4, 35], as has local hematoma [35].

The arthroscope has allowed identification of the intra-articular portion of the biceps tendon. Partial tears can be debrided and tendonesis performed as an associated open procedure if greater than 50% of the tendon is disrupted [4]. A newly reported lesion in athletes involving the superior labrum both anteriorly and posteriorly (SLAP lesion) with respect to the biceps tendon has been attributed to throwing [4, 89]. This lesion involves detachment of the labrum and has been shown to be related to pull off the biceps tendon as it functions as a shunt muscle to stabilize the glenohumeral joint during deceleration of the arm [4]. Early results of treatment of these lesions through arthroscopic debridement have been encouraging in the highly selective patient population of throwing athletes [3].

In some patients there is evidence of a partial undersurface tear of the rotator cuff with associated anterior-inferior quadrant labral pathology indicative of anterior subluxation. Under these circumstances these patients may require an anterior stabilization procedure to prevent the secondary impingement that is likely occurring in this situation.

POSTOPERATIVE MANAGEMENT AND REHABILITATION

The postoperative routine follows principles similar to those governing nonoperative management of cuff and biceps tendon pathology. With arthroscopic procedures immobilization in a sling is not necessary, although it is useful simply for comfort until the inevitable swelling has subsided in 24 to 72 hours. Range of motion exercises can begin in the recovery room and thereafter progress as tolerated with the goal of attaining a full range of motion. Activities of daily living can be resumed as pain relief permits. Strengthening exercises stressing the rotators with the use of rubber tubing is recommended prior to returning to sporting activities, especially sports involving throwing. It has been recommended that strenuous overhead sports be avoided for at least 4 to 6 weeks following simple debridement and longer following decompression, until strength has been restored. This recommendation of course is sport- and activity-related. Throwers require much longer than other athletes.

With open surgical procedures two separate issues determine the postoperative course of management: (1) the status of the deltoid (i.e., whether it was released and, if so, whether it was satisfactorily reattached, and (2) whether or not a rotator cuff repair was performed and the status of the repair.

Postoperative management involves the same considerations in the athlete as in any other individual undergoing these procedures. The main difference relates to the timing and intensity of the postoperative regimen. In the high-performance athlete there is a great deal of pressure to get back to the playing field. There are also considerably higher expectations in regard to ultimate function.

The available literature offers little reason to hope for a full functional recovery regardless of the type of rotator cuff surgery performed [62, 148, 149]. In spite of this there is a consistent and predictable response in terms of pain relief. This dual result leads to the dilemma of wanting to push the rehabilitation as quickly as possible for the sake of improving performance while realizing that repair of tissues after the surgical insult takes time.

Most surgeons advocate a sling postoperatively, with initial rehabilitation involving passive range of motion and pendulum exercises for 6 weeks. This stage is followed by active motion. Once a full range of motion has returned, strengthening and stretching are advocated. It is recommended that athletes not return to sport until 6 months after the operation and pitchers 1 year after [70]. The use of abduction splints has been advocated for patients undergoing cuff repair surgery, [153] especially for large cuff defects.

The use of isokinetic machines in the postoperative period has become very popular, but there are few published guidelines on how to use these devices. It has been shown that strength measured objectively by an isokinetic device improves during the first year following cuff repair. At 1 year the ultimate strength attained is still less than that on the opposite side [152].

Overall, there is little evidence to provide clear recommendations for rehabilitation of an athlete following these surgical options. The most important guideline for postoperative management is to use techniques similar to those used in the nonoperative management of these problems. The bottom line is the need to appreciate the functional demands of the athlete's sport and to ensure that he or she is prepared during the postoperative period to withstand the severe stresses anticipated by return to the sport.

CRITERIA FOR RETURN TO SPORTS PARTICIPATION

The literature provides few recommendations regarding specific criteria governing return to sports participation after either surgery or nonoperative rehabilitation. The determining factors must be individualized for each patient considering the level of participation anticipated, the particular sport, and the specific activity requirement. For example, in a high-profile pitcher following surgical treatment 1 year has been suggested as the time needed for full return of function [71].

There are many guidelines that can be used. The first is pain. To return to sports activity, pain must be relieved sufficiently to allow normal biomechanical function. The range of motion should be within normal limits for that athlete given the specific sport. Strength also needs to be near normal. Endurance is an important component of strength and should be assessed. This, however, is particularly difficult because there is no accurate way to determine a person's endurance. It follows that careful control of return to activity is required. Gradually increasing the intensity and duration of training is the only way to ensure adequate endurance. Finally, performance of the sport must be considered. This can be looked at from two points of view: (1) physical performance such as pitch velocity, or comparison of previous times for swimming events, and (2) mental performance. A high-performance athlete requires a high degree of motivation and confidence before he is able to return to his or her previous level of competition.

It must be clear that these considerations also apply to the weekend tennis player with his sore shoulder. The recreational athlete can often return to sports activity earlier since he has lower demands

and expectations and need not necessarily perform at 100% capability. The major difference is that the physician has less ability to control the circumstances of return to sport in such people. It is here that the principle of applied common sense overrides any specific criteria.

The return to sport is always based upon the relative severity of the original problem. If we use the modified Blazina grading system [2] it is obvious that an athlete with grade 1 pain during activity that is not disabling is going to be able to return to sport at full participatory level far sooner than an athlete with disabling symptoms even at rest (grade 4).

Lastly, there is no way of speeding up the process of recovery. It is imperative to maintain careful control to avoid reinjury in order to determine the optimum time for full return.

AUTHORS' PREFERRED METHOD OF TREATMENT

First and most important in the management of a patient with a rotator cuff or biceps problem is the need to establish the correct diagnosis [57]. A careful history, physical examination, plain radiographs, and the use of diagnostic injections of lidocaine are usually sufficient. Further investigation is reserved for patients who have an atypical presentation, who are older, who have a history of a significant traumatic episode, or who show signs that pose a diagnostic challenge. Magnetic resonance imaging appears to be the investigation of choice but with the proviso that an experienced and competent observer is reading the films. Arthrography is frequently employed to confirm the presence of a full-thickness cuff tear.

The diagnosis of a rotator cuff or biceps tendon problem is based upon the etiology:

1. Acute trauma
2. Primary impingement
3. Instability
4. Overuse: (a) scapular lag with secondary impingement, or (b) eccentric overload
5. Combined etiology

The type of management used follows from this etiologic classification. It should be emphasized that the focus of treatment is nonoperative in the great majority of individuals [135]. It is estimated that within a busy tertiary subspecialty shoulder practice less than 20% of patients with these problems require surgery.

Patients with acute traumatic episode resulting in a strain of these musculotendinous units require rest

until the symptoms have subsided and then a rehabilitation program involving a gradually increasing
regimen of stretching and strengthening. Immobilization should be avoided, and anti-inflammatory
agents are sometimes helpful in this situation. The
prognosis is good, and an early return to sport is
possible depending upon the severity of the injury.
In an older individual (older than 40 years) acute
trauma could suggest a disruption of the rotator
cuff. This should be treated with temporary rest to
allow sufficient healing to take place followed by
range of motion and strengthening exercises. Persistent pain or weakness requires further investigation. Surgical repair should be considered early
(within 2 months) to minimize the chronic effects of
such an injury.

Patients with primary impingement more commonly present at an older age. This diagnosis implies
an anatomic narrowing of the subacromial space.
We advocate surgical decompression at an earlier
date than in someone with secondary impingement
following failure of nonoperative management.

Patients with associated instability are treated
nonoperatively, especially if multidirectional instability is diagnosed. The emphasis is on strengthening
the cuff after symptoms have subsided. Following
failure of a prolonged nonoperative program surgical management involves stabilization in the form
of an inferior capsular shift.

Anterior subluxation causing secondary impingement and pain can produce both a diagnostic and a
therapeutic challenge. In such cases prolonged nonoperative measures are appropriate. If surgery is
considered, the choice between anterior stabilization
and subacromial decompression remains unclear.
Usually a choice can be made, but in questionable
cases arthroscopic decompression is the simplest and
safest diagnostic modality.

Overuse problems should be treated extensively
with nonoperative methods. Surgery is recommended in chronic situations in which nonoperative
management has failed or when the problem has
progressed to the point of a cuff tear.

It must be appreciated that in athletes combined
causes and underlying pathology can lead to these
problems. It is in such cases that treatment is very
difficult. A nonoperative approach is prudent until
the specific components can be sorted out. The
treatment is modified accordingly.

Nonoperative Management

Nonoperative treatment regardless of the level of
performance follows the same generic plan. The
following format is utilized:

1. Activity modification
2. Medications
3. Stretching
4. Strengthening
5. Ice
6. Physiotherapy

Usually a period of *rest* is advised, specifically
avoiding the overhead sporting activity involved. It
should be emphasized that this is an active form of
rest. This means that although the specific activity
causing the symptoms is avoided, substitute activities are used, for example, cycling to maintain
cardiovascular fitness, changing from a butterfly to
a breaststroke in swimming, practicing ground
strokes and avoiding serving and overhead swings
in tennis, and so on. This "active rest" is particularly
important in the high-profile athlete who will not
accept the prescription of total rest.

We then prescribe *nonsteroidal anti-inflammatory
medications* [60]. They are prescribed initially for a
2- to 4-week period with the understanding that they
can be used on a PRN basis thereafter. The use of
a steroid is reserved for patients not responding to
this regimen of anti-inflammatory medications in
addition to the other aspects of the protocol. Subacromial injections of 40 mg of methylprednisolone
or triamcinolone combined with 6 to 8 mL of lidocaine are used as a one-time measure. A peritendinous injection at the level of the transverse humeral
ligament is used for bicipital tendinitis with the same
amount of steroid and 2 to 4 mL of lidocaine. The
injection is rarely repeated.

Stretching is utilized to maintain range of motion
and to correct any obvious discrepancies or contractures, in particular posterior capsular tightness [18,
30] (see Fig. 15D–18). Cross-arm adduction and
overhead adduction stretching are important in the
athlete, particularly one with tight posterior structures.

Strengthening is the hallmark of nonoperative
management of the patient with a biceps or rotator
cuff lesion. The use of rubber tubing or simple free
weights is the most practical method of strengthening the rotator cuff. The scapular muscles cannot be
ignored, however, especially when they are implicated in the etiology. Rotator cuff strengthening
exercises are performed initially at the side using
the rubber tubing. The individual then progresses
to 45 degrees of abduction and then to a more
functional level above 90 degrees in selected, more
competitive athletes, especially throwers. The supraspinatus muscle is isolated by abducting the arm in
the plane of the scapula with the forearm and
shoulder internally rotated [74, 98] (Fig. 15D–40).
The infraspinatus and teres minor are exercised in
external rotation (Fig. 15D–41), the subscapularis

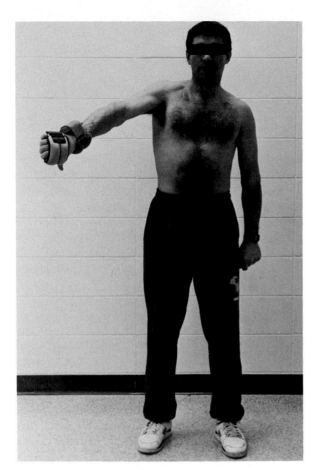

FIGURE 15D–40
Supraspinatus strengthening with weights.

in internal rotation. Biceps function is improved through elbow flexion and forearm supination. Scapular stabilizers are strengthened by resisted scapular elevation, retraction, and protraction. In addition,

the inside push-up with hands placed inside the parasaggital plane of the shoulder has been helpful (Fig. 15D–37). Sitting rows for the serratus and rhomboids, shrugs for the trapezius, and latissimus pull-downs for the latissimus dorsi muscle help control, strengthen, and stabilize the scapula.

Local *ice application* after work-outs and competition can prove beneficial and is frequently employed depending on the athlete's level of participation and response.

Physiotherapy is instituted with specific modalities such as ultrasound [61], transcutaneous nerve stimulation, muscle stimulation, and laser therapy depending on the individual patient's response to these modalities. We employ these modalities only occasionally in resistant cases.

Nonoperative management of the athlete with a rotator cuff or biceps tendon problem initially involves regular visits to the therapist. However, it should be emphasized that long-term benefit will be gained through a regular and almost obsessive home exercise program rather than relying on specific physical therapy modalities or medications. Failure of this regimen over a prolonged period of 6 to 12 months constitutes an indication for surgery, which will be discussed below.

The treatment of the high-profile athlete may differ in subtle ways from the generic program just outlined. Included in this group are individuals who may not necessarily be competing at a professional, national, or international level. Nevertheless, they take their sport very seriously and have the same motivation to compete and perform to capacity within their own level. The main differences between their treatment and the generic program concern the intensity and volume of the exercises

FIGURE 15D–41
External rotation strengthening with weights.

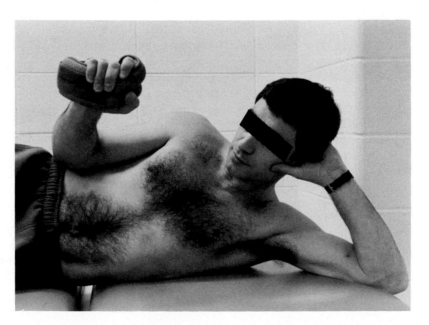

prescribed. Because these individuals place great demands upon their shoulders, the overall management should reflect the stresses involved. Isokinetic machines can be of value, especially those allowing eccentric training of the musculature. Strengthening again is the key to long-term success. Depending upon whether the sport involves primarily aerobic or anaerobic shoulder function, strengthening exercises should be low-intensity, high-volume, or high-intensity, respectively.

The other main difference encountered in treatment of these high-profile athletes is that communication with the coach or team trainer, if applicable, is essential. Communication creates the best possible environment with all the relevant people involved. Specific techniques may need to be changed, equipment modified, and short- and long-term goals of competition identified. This approach helps to focus the athlete and maintain a positive attitude during the rehabilitation treatment.

Surgical Management

The primary indication for surgery is the failure of an adequate nonoperative management program. Inherent in defining failure is the presence of sufficient pain and disability to warrant intervention. This program should be carefully coordinated and the patient followed for a minimum of 1 year before considering surgery.

With an acute traumatic injury, especially in an older athlete, the possibility of a full-thickness rotator cuff tear must be entertained. MRI has been very helpful in providing a diagnosis in this situation, but arthrography remains the gold standard. If a full-thickness tear of the rotator cuff is present, surgical repair is performed. Another situation in which surgery may be considered earlier is the presence of very obvious primary impingement due to bony overgrowth of the acromion or an abnormally angled acromion, usually in an older athlete. In both of these situations the pathologic status of the anatomy dictates a surgical solution.

Overall, the choice of surgical procedure depends upon a number of factors including the underlying etiology and the extent of the abnormality. If a full-thickness tear of the rotator cuff is present, open surgical repair is preferred. The method of repair involves an anterosuperior approach centered between the acromioclavicular (AC) joint medially and the lateral edge of the acromion. The incision is 5 to 7 cm long, with one-third superior and two-thirds inferior to the anterior tip of the acromion (Fig. 15D–42). The skin and subcutaneous tissues

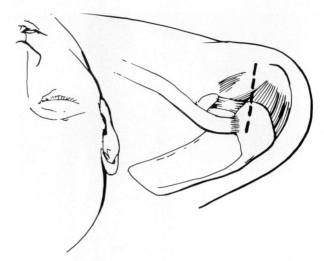

FIGURE 15D–42
Anterosuperior approach to the shoulder.

are incised and blunt dissection performed to allow identification of the AC joint, the lateral acromion, and the muscular attachments overlying the acromion. The deltoid is then detached from the anterior acromion for a distance of no more than 2.5 cm (Fig. 15D–43). It is taken off using cutting cautery, flush with the anterior acromion. Care is taken to leave sufficient tissue to effect reattachment during closure. Access to the cuff can also be achieved without detaching the deltoid. The junction between the anterior and middle parts of the deltoid is identified. The aponeurosis of the muscle is then incised over the anterior acromion in line with the deltoid fibers. The deltoid is then elevated medially and laterally off the acromion. This affords adequate access to perform acromioplasty and repair a small

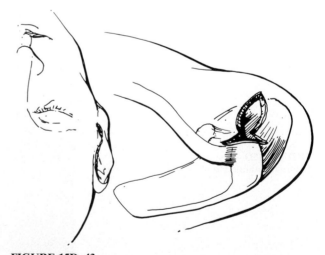

FIGURE 15D–43
Approach through the deltoid with split distally and detachment from anterior acromion.

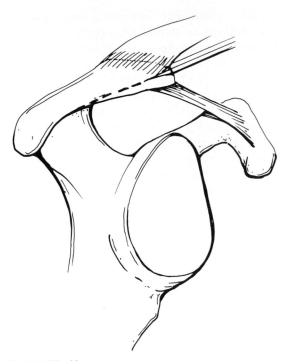

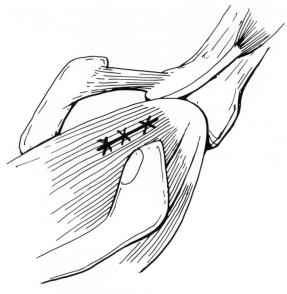

FIGURE 15D–45
Side-to-side repair of cuff defect.

FIGURE 15D–44
Excision of anteroinferior acromion and associated coracoacromial ligament.

to moderate sized cuff tear [54, 63]. Formal detachment is recommended for large or massive tears. The deltoid is then split distally in line with its fibers, taking care to avoid distal extension beyond 5 cm, where the axillary nerve could be in jeopardy [63] (Fig. 15D–43). A straight retractor is placed under the acromion to protect the underlying cuff. An anterior acromioplasty is then performed. A 2.5-cm osteotome is utilized, commencing at the antero-superior edge of the acromion and aiming posteriorly to exit the osteotomy at a distance of 1.5 to 2.0 cm. This results in excision of the anterior inferior portion of the acromion and a portion of the coracoacromial ligament (Fig. 15D–44). Any remaining ligament is excised. Bleeding is frequently encountered in this area due to incising the acromial branch of the thoracoacromial artery. The bony margins are smoothed down using a burr or rasp including any prominence of the inferior surface of the acromioclavicular joint. This completes the subacromial decompression. Although impingement is not considered in the pathophysiology of the acute traumatic tear, residual swelling in the cuff tissues and that resulting from the surgical insult will likely cause impingement. Decompression also allows adequate exposure to perform the necessary repair. Visualization of the cuff is improved by removing the bursa, and the entire extent of the cuff can be brought into view by careful manipulation of the arm. The cuff defect is then repaired as dictated by

the extent of the lesion, either with a side-to-side repair (Fig. 15D–45) or, more commonly, by securing it to a trough in bone (Figs. 15D–46 and 15D–47). The deltoid is reapproximated very carefully to maintain its normal length of attachment to the acromion [48].

Postoperatively a shoulder immobilizer is used. Passive assisted motion is started immediately [56]. The patient progresses to active motion at approximately 6 weeks depending upon the extent of the tear [54, 63]. Active motion is combined with terminal stretching, and resistive motion is added according to each individual's progress, usually at the 8-week mark [62]. The remaining postoperative regimen includes the components outlined in the nonoperative section.

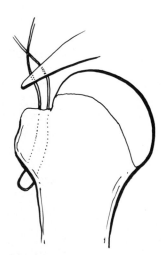

FIGURE 15D–46
Repair of cuff through a trough in bone.

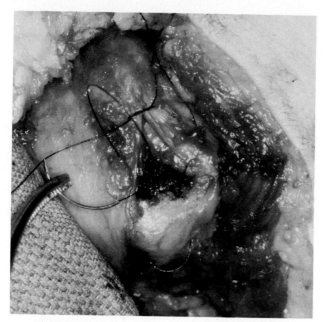

FIGURE 15D–47
Combined side-to-side and repair through trough in bone.

If a full-thickness tear is not present, the arthroscopic method of subacromial decompression is employed and has been for the last few years.

For arthroscopic decompression insertion is initially made from a posterior portal to perform a diagnostic evaluation of the intra-articular structures. An examination under anesthesia (EUA) is also considered an integral part of the procedure. The purpose of the EUA and diagnostic arthroscopy is to rule out alternative or associated pathology. In particular, instability can be diagnosed by evidence of increased translation of the humeral head, associated labral pathology, or even a "Bankart" lesion. If the biceps tendon is implicated as a contributing factor, it can be debrided. If the biceps is not implicated, attention is directed to the undersurface of the rotator cuff, and the appearance of the cuff tendons is documented. A partial-thickness cuff tear can be visualized more easily from the joint than from the bursa.

The arthroscope is then inserted into the subacromial space to pursue the subacromial decompression (Fig. 15D–48). A separate posterior portal may be made to allow a direct approach in line with the longitudinal axis of the acromion. The appearance of the bursa is irregular, and visualization can be very difficult. An anterior portal is then set up to provide flow of fluid. Epinephrine (1:1000 ampule to 3 liters of irrigation fluid) added to the irrigation fluid will help to minimize the inevitable bleeding during the procedure. Initially the bursal tissue is removed using a full radius resector. The landmarks can then be visualized along with the bursal surface

of the rotator cuff, although portal bursal side tears can be difficult to define. A needle can be used percutaneously to help in the identification of the acromioclavicular joint, the anterolateral edge of the acromion, and the coracoacromial ligament. Once these structures have been located, a third portal is made laterally over the deltoid muscle. The portal is located 2 to 3 cm inferior to the lateral edge of the acromion and 1 cm posterior to the anterior border. The flattened oval-shaped or spherical burr is then inserted through this portal into the subacromial space (see Fig. 15D–36). The posterior extent of the acromioplasty is measured directly; it should be 1.5 to 2 cm from the anterior border. The depth of the resection can be difficult to determine; however, we suggest that the anterior border be resected until the soft tissue of the deltoid insertion can be seen. The resection is then tapered to the posterior extent as previously determined. The medial extent is determined by identification of the acromioclavicular joint. Any prominence of the undersurface of the joint is smoothed off using the same burr. Because the anatomic insertion of the coracoacromial ligament is to the anterior border and undersurface of the acromion, adequate bony removal usually renders the ligament nonfunctional. It should be debrided further with a full-radius resector to ensure that it does not reconstitute itself. We have found it unnecessary to utilize cutting cautery for this purpose. However, if significant bleeding is encountered, cautery is used.

Postoperatively, passive assisted motion and stretching are started immediately. The patient progresses rapidly to active and resisted motion as

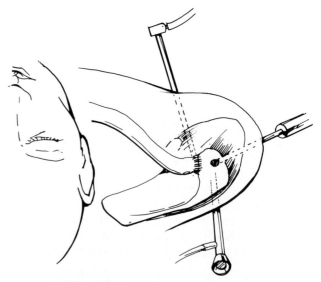

FIGURE 15D–48
Arthroscopic subacromial decompression.

tolerated. The nonoperative routine is then instituted to build strength and regain function.

With respect to surgical management, the biceps tendon must be considered separately from the rotator cuff. Acute rupture of the biceps tendon in an athletic individual is more common at the elbow with severe trauma and requires surgical management [8]. Ruptures at the shoulder are usually due to chronic attrition with or without an acute episode, usually in an older athlete [126]. Although this is a relatively common finding, surgery is only rarely required. A surgical tenodesis within the bicipital groove is indicated in an athlete with a specific need for strong elbow flexion and forearm supination. We have had no experience with this because of its rarity [95]. Occasionally the chronic rupture will become symptomatic, and in those cases exploration and tenodesis have been effective to decrease pain. However, a subacromial decompression for concomitant rotator cuff disease will usually benefit a symptomatic bicipital tendinitis. This surgical procedure is not indicated for the isolated bicipital problem.

We have had little experience with arthroscopic debridement of intra-articular biceps tendon pathology. We consider it a reasonable approach for the athlete with the so-called SLAP lesion [36]. The long-term management and prognosis have yet to be determined with this approach. Our results of intra-articular labral biceps debridement have not proved very satisfactory, but our experience with high-level throwers is small. Postoperative management is geared toward the procedure. A tenodesis will require sufficient healing time before full activity is recommended. With an arthroscopic debridement we progress our patients as quickly as tolerated using the principles of nonoperative management.

Those patients presenting with symptoms and signs of instability of their biceps tendon [115, 124] in our experience are extremely rare in the athletic population. If necessary, we would recommend a biceps tenodesis although we have not ever had to perform this procedure in the athletic population.

It has been our experience that despite our best efforts to make a clear diagnosis, some patients have problems actually related to both instability and rotator cuff pathology. Under these circumstances both a decompression procedure and an anterior stabilization procedure have been performed in the same patient, usually on different occasions. We must emphasize the overlap between instability and rotator cuff and biceps tendon pathology. The vicious cycle concept that was outlined earlier in section on diagnosis attempts to explain this difficulty in diagnosis and the necessity to pay attention to two pathologies.

In summary, it is unlikely for elite overhead athletes to gain full functional recovery after an operation for rotator cuff or biceps tendon pathology. The real benefit comes from pain relief, and, indeed, a secondary improvement in function is likely. However, normality in terms of stressful overhead athletic competition is achieved in the minority. This is not the case with nonoperative management, with which full functional recovery is a reasonable goal. The difference in prognosis is obviously related to the severity, chronicity, and nature of the pathology.

References

1. Ahovuo, J., Paavolainen, P., and Slatis, P. Diagnostic value of sonography in lesions of the biceps tendon. *Clin Orthop* 202:184–188, 1986.
2. Albee, F. H. *Orthopaedic and Reconstructive Surgery*. Philadelphia, W. B. Saunders, 1921, pp. 507–512.
3. Andrews, J. R., and Carson, W. G. The arthroscopic treatment of glenoid labrum tears in the throwing athlete. *Orthop Trans* 8:44, 1984.
4. Andrews, J. R., Carson, W. G., and McLeod, W. D. Glenoid labrum tears related to the long head of the biceps. *Am J Sport Med* 13:337–341, 1985.
5. Andrews, J. R., Carson, W. G., and Ortega, K. Arthroscopy of the shoulder: Technique and normal anatomy. *Am J Sport Med* 12:1–7, 1984.
6. Aoki, M., Ishii, S., and Usui, M. The slope of the acromion and rotator cuff impingement. *Orthop Trans* 10:228, 1986.
7. Azevedo, A. J. Anterior acromioplasty for the shoulder impingement syndrome. *Orthop Trans* 10:505, 1986.
8. Baker, B., and Bierwagen, D. Rupture of the distal tendon of the biceps brachii. *J Bone Joint Surg* 67A:414–417, 1985.
9. Bassett, R. W., and Cofield, R. H. Acute tears of the rotator cuff. *Clin Orthop* 175:18–24, 1983.
10. Bateman, J. E. Cuff tears in athletes. *Orthop Clin North Am* 4:721–745, 1973.
11. Bateman, J. E. *The Shoulder and Neck* (2nd ed.). Philadelphia, W. B. Saunders, 1978, pp. 242–339.
12. Bateman, J. E. Neurological painful conditions affecting the shoulder. *Clin Orthop* 173:44–54, 1983.
13. Bennett, G. E. Elbow and shoulder lesions of baseball players. *Am J Surg* 98:484–492, 1959.
14. Bigliani, L. U., D'Alessandro, D. F., Duralde, X. A., and McIlveen, S. J. Anterior acromioplasty for subacromial impingement in patients younger than 40 years of age. *Clin Orthop* 246:111–116, 1989.
15. Bigliani, L. U., Morrison, D. S., and April E. W. The morphology of the acromion and its relationship to rotator cuff tears. *Orthop Trans* 10:216, 1986.
16. Bonci, C. M., Hensal, F. J., and Torg J. S. A preliminary study on the measurement of static and dynamic motion at the glenohumeral joint. *Am J Sport Med* 14:12–17, 1986.
17. Brems, J. Rotator cuff tear: Evaluation and treatment. *Orthopedics* 11:69–81, 1988.
18. Brewer, B. J. Aging of the rotator cuff. *Am J Sports Med* 7:102–110, 1979.
19. Brown, C. Compressive, invasive referred pain to the shoulder. *Clin Orthop* 173:55–62, 1983.
20. Brown, J. T. Early assessment of supraspinatus tears: Procaine infiltration as a guide to treatment. *J Bone Joint Surg* 31B:423–425, 1949.
21. Burk, D. L., Jr., Darasick, D., Kurtz, A. B., Mitchell, D. G., Rifkin, M. D., Miller, C. L., Levy, D. W., Fenlin, J.

M., and Bartolozzi, A. R. Rotator cuff tears: Prospective comparison of MR imaging with arthrography, sonography and surgery. *Am J Roentgenol* 153:87–92, 1989.

21a. Chansky, H. A., and Iannotti, D. D. The vascularity of the rotator cuff. *Clin Sports Med* 10:807–822, 1991.

22. Cuillo, J. V., and Guise, E. R. Adolescent swimmer's shoulder. *Orthop Trans* 7:171, 1983.

23. Ciullo, J. V., and Stevens, G. G. The prevention and treatment of injuries to the shoulder in swimming. *Sports Med* 7:182–204, 1989.

24. Codman, E. A. Rupture of the supraspinatus tendon. *Surg Gynecol Obstet* 52:579–586, 1931.

25. Codman, E. A. *The Shoulder.* Boston, Thomas Todd, 1934, pp. 1–261.

26. Codman, E. A., and Akerson, I. B. The pathology associated with rupture of the supraspinatus tendon. *Ann Surg* 93:348–359, 1931.

27. Cofield, R. H. Current concepts review: Rotator cuff disease of the shoulder. *J Bone Joint Surg* 67A:974–979, 1985.

28. Crass, J. R., and Craig, E. V. Non-invasive imaging of the rotator cuff. *Orthopedics* 11:57–64, 1988.

29. Crass, J. R., Craig, E. V., and Feinberg, S. R. Sonography of the postoperative rotator cuff. *AJR* 146:561–564, 1986.

30. Crisp, E. J., and Kendall, P. H. Hydrocortisone in lesions of soft tissue. *Lancet* 1:476–479, 1955.

31. Darlington, L. G., and Coomes, E. N. The effects of local steroid injection for supraspinatus tears. *Rheumatol Rehab* 16:172–179, 1977.

32. Davis, T. W., and Sullivan, J. E. Rupture of the supraspinatus tendon. *Ann Surg* 106:1059–1069, 1937.

33. De Orio, J. K., and Cofield, R. H. Results of a second attempt at surgical repair of a failed initial rotator-cuff repair. *J Bone Joint Surg* 66A:563–567, 1984.

34. De Palma, A. F. *Surgery of the Shoulder* (3rd ed.). Philadelphia, J. B. Lippincott, 1983, pp. 242–285.

35. Ellman, H. Arthroscopic subacromial decompression: Analysis of one to three year results. *Arthroscopy* 3:173–181, 1987.

36. Ellman, H. Shoulder arthroscopy: Current indications and techniques. *Orthopedics* 11:45–51, 1988.

37. Ellman, H., Hanker, G., and Bayer, M. Repair of the rotator cuff. *J Bone Joint Surg* 68A:1136–1144, 1986.

38. Evancho, A. M., Stiles, R. G., Fajman, W. A., Flower, S. P., Macha, T., Brunner, M. C., and Fleming, L. MR imaging diagnosis of rotator cuff tears. *AJR* 151:751–754, 1988.

39. Fearnley, M. E., and Vadasz, I. Factors influencing the response of lesions of the rotator cuff of the shoulder to local steroid injection. *Ann Phys Med* 10:53–63, 1969.

40. Fowler, E. B. Stiff, painful shoulders, exclusive of tuberculosis and other infections. *JAMA* 101:2106–2109, 1933.

41. Fowler, P. J., and Webster, S. Shoulder pain in highly competitive swimmers. *Orthop Trans* 7:170, 1983.

42. Fowler, P. J. Shoulder injuries in the mature athlete. *Adv Sports Med Fitness* 1:225–238, 1988.

43. Gainor, B. J., Piotrowski, G., Puhl, J., Allen, W. C., and Hagen, R. The throw: Biomechanics and acute injury. *Am J Sports Med* 8:114–118, 1980.

44. Garth, W. P., Allman, F. L., and Armstrong, W. S. Occult anterior subluxation of the shoulder in non-contact sports. *Orthop Trans* 10:214, 1986.

45. Gerber, C., Terrier, F., and Ganz, R. The role of the coracoid process in the chronic impingement syndrome. *J Bone Joint Surg* 67B:703–708, 1985.

46. Gilcrest, E. L. Rupture of muscles and tendons. *JAMA* 84:1819–1822, 1925.

47. Glousman, R., Jobe, F. W., Tibone, J., Moynes, D. R., Antonelli, D. A., and Perry, J. Dynamic EMG analysis of the throwing shoulder with glenohumeral instability. *Orthop Trans* 11:247, 1987.

48. Godsil, R. D., and Linscheid, R. L. Intratendinous defects of the rotator cuff. *Clin Orthop* 69:181–188, 1970.

49. Gore, D. R., Murray, M. P., and Gardner, G. M. Shoulder

50. Gowan, I. D., Jobe, F. W., Tibone, J. E., Perry, J., and Moynes, D. R. A comparative electromyographic analysis of the shoulder during pitching. *Am J Sports Med* 15:586–590, 1987.

51. Greipp, J. F. Swimmer's shoulder: The influence of flexibility and weight training. *Phys Sportsmed* 13:92–105, 1985.

52. Harryman, D. T., Sidles, J. A., Clark, J. M., McQuade, K. J., Gibb, T. D., and Matsen, F. A. III. Translation of the humeral head on the glenoid with passive glenohumeral motion. *J Bone Joint Surg* 72A:1354, 1990.

53. Hawkins, R. J. *The Rotator Cuff and Biceps Tendon in Surgery of the Musculoskeletal System.* New York, Churchill Livingstone, 1983, pp. 3:5–3:33.

54. Hawkins, R. J. Surgical management of rotator cuff tears. *In* Bateman, J. E., and Welsh, R. P. (Eds.), *Surgery of the Shoulder.* Philadelphia, B. C. Decker, 1984, pp. 161–166.

55. Hawkins, R. J., and Abrams, J. S. Impingement syndrome in the absence of rotator cuff tear (stages 1 and 2). *Orthop Clin North Am* 18:373–382, 1987.

56. Hawkins, R. J., Bell, R. H., Hawkins, R. H., and Koppert, G. J. Anterior dislocation of the shoulder in the older patient. *Clin Orthop* 206:192–195, 1986.

57. Hawkins, R. J., Brock, R. M., Abrams, J. S., and Hobeika, P. Acromioplasty for impingement with an intact rotator cuff. *J Bone Joint Surg* 70B:795–797, 1988.

58. Hawkins, R. J., Chris, T., Bokor, D., and Kiefer, G. N. Failed anterior acromioplasty. *Clin Orthop* 243:106–111, 1989.

59. Hawkins, R. J., and Hobeika, P. E. Physical examination of the shoulder. *Orthopedics* 6:1270–1278, 1983.

60. Hawkins, R. J., and Hobeika, P. E. Impingement syndrome in the athletic shoulder. *Clin Sports Med* 2:391–405, 1983.

61. Hawkins, R. J., and Kennedy, J. C. Impingement syndrome in athletes. *Am J Sports Med* 8:151–158, 1980.

62. Hawkins, R. J., and Kunkel, S. S. Rotator cuff tears. *In* Torg, J. S. (Ed.), *Current Therapy in Sports Medicine.* St. Louis, C. V. Mosby, 1990.

63. Hawkins, R. J., Misamore, G. W., and Hobeika, P. E. Surgery for full thickness rotator-cuff tears. *J Bone Joint Surg* 67A:1349–1355, 1985.

64. Hawkins, R. J., and Mohtadi, N. G. Clinical evaluation of shoulder instability. *Clin J Sports Med* 1:59, 1991.

65. Hawkins, R. J., Schutte, J. P., Huckell, G. H., and Abrams, J. The assessment of glenohumeral translation using manual and fluoroscopic techniques. *Orthop Trans* 12:727, 1988.

66. Hill, J. A. Epidemiologic perspective on shoulder injuries. *Clin Sports Med* 2:241–245, 1983.

67. Howell, S. M., Imobersteg, A. M., Seger, D. H., and Marone, P. J. Clarification of the role of the supraspinatus muscle in shoulder function. *J Bone Joint Surg* 68A:398–404, 1986.

68. Inman, V. T., Saunders, M., and Abbott, L. C. Observations on the function of the shoulder joint. *J Bone Joint Surg* 26:1–30, 1944.

69. Jackson, D. W. Chronic rotator cuff impingement in the throwing athlete. *Orthop Trans* 1:24–25, 1977.

70. Jobe, F. W. Serious rotator cuff injuries. *Clin Sports Med* 2:407–412, 1983.

71. Jobe, F. W., and Bradley, J. P. Rotator cuff injuries in baseball: Prevention and rehabilitation. *Sports Med* 6:378–387, 1988.

72. Jobe, F. W., and Jobe, C. M. Painful athletic injuries of the shoulder. *Clin Orthop* 173:117–124, 1983.

73. Jobe, F. W., Moynes, D. R., and Antonelli, D. J. Rotator cuff function during a golf swing. *Am J Sports Med* 14:388–392, 1986.

74. Jobe, F. W., Moynes, D. R., Tibone, J. E., and Perry, J. An EMG analysis of the shoulder in pitching: A second report. *Am J Sports Med* 12:218–220, 1984.

75. Jobe, F. W., Tibone, J. E., Perry, J., and Moynes, D. R.

An EMG analysis of the shoulder in throwing and pitching. *Am J Sports Med* 11:3–5, 1983.

76. Johansson, J. E., and Barrington, T. W. Coracoacromial ligament division. *Am J Sports Med* 12:138–141, 1984.

77. Kennedy, J. C., Hawkins, R. J., and Krissoff, W. B. Orthopaedic manifestations of swimming. *Am J Sports Med* 6:309–322, 1978.

78. Kennedy, J. C., and Willis, R. B. The effects of local steroid injections on tendons: A biomechanical and microscopic correlative study. *Am J Sports Med* 4:11–21, 1976.

79. Kessel, L., and Watson, M. The painful arc syndrome. *J Bone Joint Surg* 59B:166–172, 1977.

80. King, J. M., and Holmes, G. W. Diagnosis and treatment of 450 painful shoulders. *JAMA* 89:1956–1961, 1927.

81. King, J. W., Brelsford, H. J., and Tullos, H. S. Analysis of the pitching arm of the professional baseball pitcher. *Clin Orthop* 67:116–123, 1969.

82. Lehman, R. C. Shoulder pain in the competitive tennis player. *Clin Sports Med* 7:309–327, 1988.

83. Mack, L. A., Nyberg, D. A., and Matsen, F. A. Sonographic evaluation of the rotator cuff. *Radiol Clin North Am* 26:161–177, 1988.

83a. Matsen, D. D., Matsen, F. A. III, and Arntz, C. T. Rotator cuff tendon failure. *In* Rockwood C. A., Jr., and Matsen, F. A. III (Eds.), *The Shoulder.* Philadelphia, W. B. Saunders, 1990. p. 654. American Shoulder and Elbow Surgeons, 1988.

84. Maylack, F. H. Epidemiology of tennis, squash and racquetball injuries. *Clin Sports Med* 7:233–243, 1988.

85. McLaughlin, H. L. Lesions of the musculotendinous cuff of the shoulder. *J Bone Joint Surg* 26:31–51, 1944.

86. McMaster, P. E. Tendon and muscle ruptures. *J Bone Joint Surg* 15A:705–722, 1933.

87. McMaster, W. C., Liddle, S., and Waugh, T. R. Laboratory evaluation of various cold modalities. *Orthop Trans* 1:23, 1977.

88. McMaster, W. C. Anterior glenoid labrum damage: A painful lesion in swimmers. *Am J Sports Med* 14:383–387, 1986.

89. Mendoza, F. X., Nicholas, J. A., and Reilly, J. P. Anatomic patterns of anterior glenoid labrum tears. *Orthop Trans* 11:246, 1987.

90. Meyer, A. W. Chronic functional lesions of the shoulder. *Arch Surg* 35:646–674, 1937.

91. Middleton, W. D., Kneeland, J. B., Carrera, G. F., Cates, J. D., Kellman, G. M., Campagna, N. G., Jesmanowicz, A., Froncisz, W., and Hyde, J. S. High-resolution MR imaging of the normal rotator cuff. *AJR* 148:559–564, 1987.

92. Middleton, W. D., Reinus, W. R., Melson, G. L., Totty, W. G., and Murphy, W. A. Pitfalls in rotator cuff sonography. *AJR* 146:555–560, 1986.

93. Middleton, W. D., Reinus, W. R., Totty, W. G., Nelson, C. L., and Murphy, W. A. Ultrasonographic evaluation of the rotator cuff and biceps tendon. *J Bone Joint Surg* 68A:440–450, 1986.

94. Morrison, D. S., and Bigliani, L. U. The clinical significance of variations in acromial anatomy. *Orthop Trans* 11:234, 1987.

95. Moseley, H. F. Athletic injuries to the shoulder region. *Am J Surg* 98:401–422, 1959.

96. Moseley, H. F. *Shoulder Lesions* (3rd ed.). Edinburgh, E. & S. Livingstone, 1969, pp. 60–98.

97. Moseley, H. F., and Goldie, I. The arterial pattern of the rotator cuff of the shoulder. *J Bone Joint Surg* 45B:780–789, 1963.

98. Moynes, D. R. Prevention of injury to the shoulder through exercises and therapy. *Clin Sports Med* 2:413–422, 1983.

99. Nash, H. L. Rotator cuff damage: Re-examining the causes and treatments. *Phys Sportsmed* 16:129–135, 1988.

100. Neer, C. S., II. Anterior acromioplasty for the chronic impingement syndrome in the shoulder: A preliminary report. *J Bone Joint Surg* 54A:41–50, 1972.

101. Neer, C. S., II. Impingement lesions. *Clin Orthop* 173:70–77, 1983.

102. Neer, C. S., II, and Poppen, N. K. Supraspinatus outlet. *Orthop Trans* 11:234, 1987.

103. Neer, C. S., II, and Welsh, R. P. The shoulder in sports. *Orthop Clin North Am* 8:583–591, 1977.

104. Nelson, C. L. The use of arthrography in athletic injuries of the shoulder. *Orthop Clin North Am* 4:775–785, 1973.

105. Neviaser, R. J. Lesions of the biceps and tendinitis of the shoulder. *Orthop Clin North Am* 11:343–348, 1980.

106. Neviaser, R. J. Tears of the rotator cuff. *Orthop Clin North Am* 11:295–306, 1980.

107. Neviaser, R. J. Painful conditions affecting the shoulder. *Clin Orthop* 173:63–69, 1983.

108. Neviaser, R. J. Ruptures of the rotator cuff. *Orthop Clin North Am* 18:387–394, 1987.

109. Neviaser, R. J., Neviaser, T. J., and Neviaser, J. S. Concurrent rupture of the rotator cuff and anterior dislocation of the shoulder in the older patient. *J Bone Joint Surg* 70A:1308–1311, 1988.

110. Neviaser, T. J. Arthrography of the shoulder. *Orthop Clin North Am* 11:205–217, 1980.

111. Nevaiser, T. J., and Neviaser, R. J. The anterior superior approach to the shoulder. *Orthop Trans* 11:229, 1987.

112. Nicholas, J. A., Grossman, R. B., and Hershman, E. B. The importance of a simplified classification of motion in sports in relation to performance. *Orthop Clin North Am* 8:499–532, 1977.

113. Nirschl, R. P. Prevention and treatment of elbow and shoulder injuries in the tennis player. *Clin Sports Med* 7:289–308, 1988.

114. Norwood, L. A., Barrack, R., and Jacobson, K. E. Clinical presentation of complete tears of the rotator cuff. *J Bone Joint Surg* 71A:499–505, 1989.

115. O'Donoghue, D. H. Subluxing biceps tendon in the athlete. *Clin Orthop* 164:26–29, 1982.

116. Ozaki, J., Fujimoto, S., Masuhara, K., Tamai, S., and Yoshimoto, S. Reconstruction of chronic massive rotator cuff tears with synthetic materials. *Clin Orthop* 202:173–183, 1986.

117. Outland, T. A., and Shepherd, W. F. Tears of the supraspinatus tendon. *Ann Surg* 107:116–121, 1938.

118. Pappas, A. M., Goss, T. P., and Kleinman, P. K. Symptomatic shoulder instability due to lesions of the glenoid labrum. *Orthop Trans* 10:504, 1986.

119. Pappas, A. M., Zawacki, R. M., and McCarthy, C. F. Rehabilitation of the pitching shoulder. *Am J Sports Med* 13:223–235, 1985.

120. Pappas, A. M., Zawacki, R. M., and Sullivan, T. J. Biomechanics of baseball pitching: A preliminary report. *Am J Sports Med* 13:216–222, 1985.

121. Patel, D., and Sabharwal, S. Correlation of the diagnostic accuracy of double contrast arthrography, computerized arthrotomography and arthroscopy for painful shoulders. *Orthop Trans* 11:235, 1987.

122. Penny, J. N., and Welsh, R. P. Shoulder impingement syndromes in athletes and their surgical management. *Am J Sports Med* 9:11–15, 1981.

123. Perry, J. Anatomy and biomechanics of the shoulder in throwing, swimming, gymnastics and tennis. *Clin Sports Med* 2:247–270, 1983.

124. Petersen, C. J. Spontaneous medial dislocation of the tendon of the long biceps brachii. *Clin Orthop* 211:224–227, 1986.

125. Petri, M., Dobrow, R., Nieman, R., Whiting-O'Keefe, Q., and Seaman, W. E. Randomized, double-blind placebo-controlled study of the treatment of the painful shoulder. *Arthritis Rheum* 30:1040–1045, 1987.

126. Platt, H. Observations on some tendon ruptures. *Br Med J* 1:611–615, April, 1931.

127. Poppen, N. K., and Walker, P. S. Normal and abnormal motion of the shoulder. *J Bone Joint Surg* 58A:195–201, 1976.

128. Post, M., and Benca, P. Primary tendinitis of the long head of the biceps. *Clin Orthop* 246:117–125, 1989.

129. Post, M., Silver, R., and Singh, M. Rotator cuff tear: Diagnosis and treatment. *Clin Orthop* 173:78–91, 1983.
130. Preist, J. D., and Nagel, D. A. Tennis shoulder. *Am J Sports Med* 4:28–41, 1976.
131. Rathburn, J. B., and MacNab, I. The microvascular pattern of the rotator cuff. *J Bone Joint Surg* 52B:540–553, 1970.
132. Richardson, A. B. Overuse syndromes in baseball, tennis, gymnastics and swimming. *Clin Sports Med* 2:379–390, 1983.
133. Richardson, A. B., Jobe, F. W., and Collins, H. F. The shoulder in competitive swimming. *Am J Sports Med* 87:159–163, 1980.
134. Richardson, A. T. The painful shoulder. *Proc R Soc Med* 68:11–16, 1975.
135. Rockwood, C. A., Jr., Burkhead, W. Z., and Brna, J. Subluxation of the glenohumeral joint: Response to rehabilitative exercise traumatic vs. atraumatic instability. *Orthop Trans* 10:220, 1986.
136. Rothman, R. H., and Parke, W. W. The vascular anatomy of the rotator cuff. *Clin Orthop* 41:176–186, 1965.
137. Rowe, C. R. *The Shoulder*. New York, Churchill Livingstone, 1988, pp. 103–155.
138. Saha, A. K. Dynamic stability of the glenohumeral joint. *Acta Orthop Scand* 42:491–505, 1971.
139. Saha, A. K. Mechanism of shoulder movements and a plea for the recognition of "zero position" of glenohumeral joint. *Clin Orthop* 173:3–10, 1983.
140. Sain, J., and Andrews, J. R. Proper pitching techniques. *In* Zarins, B., Andrews, J. R., and Carson, W. G., (Eds.), *Injuries to the Throwing Arm*. Philadelphia, W. B. Saunders, 1985, pp. 30–37.
141. Salter, E. G., Nasca, R. J., and Shelley, B. S. Anatomical observations on the acromioclavicular joint and supporting ligaments. *Am J Sports Med* 15:199–206, 1987.
142. Samilson, R. L., and Binder, W. F. Symptomatic full thickness tears of the rotator cuff. *Orthop Clin North Am* 6:449–466, 1975.
143. Sarrafian, S. K. Gross and functional anatomy of the shoulder. *Clin Orthop* 173:11–19, 1983.
144. Seeger, L. L., Gold, R. H., Bassett, L. W., and Ellman, H. Shoulder impingement syndrome: MR findings in 53 shoulders. *AJR* 150:343–347, 1988.
145. Seitz, W. H., Jr., Abram, L. J., Fromison, A. J., Wiener, S., and Berns, D. Rotator cuff imaging techniques: Comparison of arthrography, ultrasonography and magnetic resonance imaging. *Orthop Trans* 11:235, 1987.

146. Slocum, D. B. The mechanics of some common injuries to the shoulder in sports. *Am J Surg* 98:394–400, 1959.
147. Stamm, T. T., and Crabbe, W. A. Paraglenoid osteotomy of the scapula. *Clin Orthop* 88:39–45, 1972.
148. Tibone, J. E., Jobe, F. W., Kerlan, R. K., Carter, V. S., Shields, C. L., Lombardo, S. J., and Yocum, L. A. Shoulder impingement syndrome in athletes treated by anterior acromioplasty. *Clin Orthop* 198:134–140, 1985.
149. Tibone, J. E., Elrod, B., Jobe, F. W., Kerlan, R. K., Carter, V. S., Shields, C. L., Lombardo, S. J., and Yocum, L. Surgical treatment of tears of the rotator cuff in athletes. *J Bone Joint Surg* 68A:887–891, 1986.
150. Tullos, H. S., and King, J. W. Throwing mechanism in sports. *Orthop Clin North Am* 4:709–720, 1972.
151. Van Linge, B., and Mulder, J. D. Function of the supraspinatus muscle and its relation to the supraspinatus syndrome. *J Bone Joint Surg* 45B:750–754, 1963.
152. Walker, S. W., Couch, W. H., Boester, G. A., and Sprowl, D. W. Isokinetic strength of the shoulder after repair of a torn rotator cuff. *J Bone Joint Surg* 69A:1041–1044, 1987.
153. Watson, M. Major ruptures of the rotator cuff. *J Bone Joint Surg* 67B:618–624, 1985.
154. Watson, M. Rotator cuff function in the impingement syndrome. *J Bone Joint Surg* 71B:361–366, 1989.
155. Watson-Jones, R. *Fractures and Other Bone and Joint Injuries* (2nd ed.). Edinburgh, E & S Livingstone, 1941, pp. 289–297.
156. Weiner, D. S., and MacNab, I. Superior migration of the humeral head. *J Bone Joint Surg* 52B:524–527, 1970.
157. Wilson, P. D. Complete rupture of the supraspinatus tendon. *JAMA* 96:433–439, 1931.
158. Withrington, R. H., Girgis, F. L., and Seifert, M. H. A placebo-controlled trial of steroid injections in the treatment of supraspinatus tendinitis. *Scand J Rheumatol* 14:76–78, 1985.
159. Wolfgang, G. L. Rupture of the musculotendinous cuff of the shoulder. *Clin Orthop* 134:230–243, 1978.
160. Yergason, R. M. Supination sign. *J Bone Joint Surg* 13:160, 1931.
161. Zarins, B., and Prodromos, C. C. Shoulder injuries in sports. *In* Rowe, C. R. (Ed.), *The Shoulder*. New York, Churchill Livingstone, 1988, pp. 411–434.
162. Zlatkin, M. B., Iannotti, J. P., Roberts, M. C., Esterhai, J. L., Dalinka, M. K., Kressel, H. Y., Schwartz, J. S., and Lenkinski, R. E. Rotator cuff tears: Diagnostic performance of MR imaging. *Radiology* 172:223–229, 1989.

Nerve Lesions of the Shoulder

Kenneth P. Butters, M.D.

SUPRASCAPULAR NERVE PALSY

Compression or injury of the suprascapular nerve at the scapular notch can result in pain and motor weakness of the supra- and infraspinatus muscles. Atrophy will later result and is often the clue to the clinical diagnosis. A direct blow or forceful scapular protraction may cause traction on the nerve at Erb's point or kinking at the suprascapular or spinoglenoid notch [47]. Suprascapular neuropathy has been reported in many types of athletes. [7, 19, 44]. Compression of the infraspinatus branch with a ganglion at the spinoglenoid notch has also been reported [49], as have bilateral nerve palsies [21]. Partial infraspinatus palsy may present as a painless posterior shoulder atrophy and weakness in throwers. In the sports setting, spontaneous recovery of nerve function is the usual course, suggesting that the process is a nerve injury rather than entrapment neuropathy.

Anatomy and Biomechanics

The suprascapular nerve arises from C5—C6 at the upper trunk of the brachial plexus, where it passes deep to the trapezius and the omohyoid. It enters the supraspinatus fossa via the suprascapular notch beneath the transverse scapular ligament (Fig. 15E–1). The nerve continues deep to the supraspinatus, innervating it with two branches and sensory branches to the glenohumeral and acromioclavicular joints. There is no cutaneous sensory distribution of the suprascapular nerve. The nerve then reaches the lateral edge of the spine of the scapula and descends through the spinoglenoid notch to enter the infraspinatus fossa. A spinoglenoid ligament is described in 50% of patients [7] passing from the glenoid neck up to the spine of the

scapula and dividing the two spinati muscles. The suprascapular notch may assume various shapes as described by Rengachary and colleagues [41]. It is most commonly U-shaped and varies from wide open to enclosed with bone. The suprascapular nerve, then, is fixed at its origin in the brachial plexus and at its terminal branches into the infraspinatus. In addition to the suprascapular notch and the lateral edge of the spine of the scapula (spinoglenoid notch), several other possible sites of injury have been suggested in anatomic studies [44]. The width of the transverse scapular ligament parallels the size of the notch—that is, a larger bony notch results in a larger foramen [41].

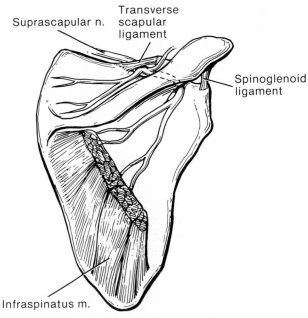

FIGURE 15E–1
Anatomy of the suprascapular nerve. (Redrawn from Black, K. P., and Lombardo, J. A. Suprascapular nerve injuries with isolated paralysis of the infraspinatus. *J Sports Med* 18(3): 225–228, 1990.)

No translational motion of the suprascapular nerve through the foramen has been seen. The nerve, however, forms an angle at the foramen. Nerve contact with the ligament is accentuated with depression retraction or hyperabduction of the shoulder. The mechanism of nerve injury, then, may well result from this "sling effect" [41].

The cadaver studies [42] showing that extremes of scapular motion can render the suprascapular nerve taut and kink it present a concept that is supported by Sunderland [48] and Drez [17].

Clinical Evaluation

The athlete may have a history of trauma, but most often the complaint is vague discomfort or weakness. Pain is usually localized posteriorly. If the infraspinatus alone is involved in the lesion at the spinoglenoid notch distal to the acromioclavicular and glenohumeral branches, the presentation may be one of painless atrophy and external rotation weakness.

Diagnosis of suprascapular nerve compression is often difficult and requires careful shoulder examination including posterior shoulder inspection and testing of external rotation strength as well as complete neurologic evaluation at the neck and proximal extremity and, most important, electrical evaluation. The pain is usually located in the posterior shoulder and radiates to the arm; it may be worse with adduction of the shoulder [39]. Posterior scapular atrophy, especially in the infraspinatus fossa, is an important finding. Supraspinatus atrophy may be difficult to see, and its weakness not as easily exposed as that of the infraspinatus. I have not found suprascapular notch tenderness to be consistently helpful, but Post and Mayer [39] found it in seven out of nine patients.

Nerve Studies

Electrical evaluation should include electromyography (EMG) of the entire shoulder girdle including the paraspinous muscles and nerve conduction studies from Erb's point to the supraspinatus, comparing the values to those from the opposite side. Of course, the supraspinatus may be spared by a lesion at the spinoglenoid notch. Normal latency values in nerve conduction studies to the supraspinatus are 1.7 to 3.7 msec and 2.4 to 4.2 msec to the infraspinatus. Nerve conduction studies should be abnormal to confirm the diagnosis of suprascapular nerve compression. Electromyographic abnormalities oc-

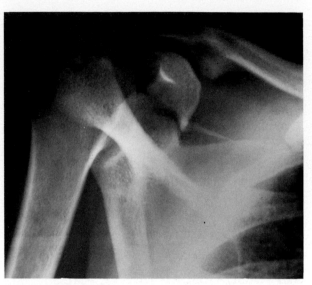

FIGURE 15E–2
Radiograph of suprascapular notch fracture with 30-degree cephalic tilt.

cur with brachial neuritis, cervical root compression, and incomplete brachial plexus stretch. Also, some feel that EMG studies may be normal with an obvious clinical suprascapular nerve deficit [11, 18], confirming the need for the nerve latency examination.

A 30-degree cephalic tilt view x-ray to visualize the suprascapular notch is helpful, especially in patients with fractures (Fig. 15E–2). There are few data on the incidence of fracture of the scapula with associated nerve palsy. Edeland and Zachrisson described 18 scapula fractures, seven with clinical involvement of the suprascapular nerve and only one with positive EMG findings [18]. Treatment in fracture cases should probably include early exploration of the nerve with neurolysis and notch resection.

A work-up for a patient with suprascapular nerve palsy should include shoulder views and probably a cervical spine series, and, if the differential diagnosis includes suspicion of rotator cuff disease, appropriate evaluation of the rotator cuff with ultrasound, (MRI), magnetic resonance imaging or arthrogram. I have also used a local anesthetic block in the suprascapular notch area as part of a series of diagnostic injections.

Treatment

Treatment of a patient with closed, acute suprascapular nerve injury consists of conservative follow-up of the problem with a series of examinations and electrical studies. A patient with a chronic condition with well-established atrophy is a surgical problem, as is suprascapular nerve palsy associated with scapular fracture in the area of the suprascapular notch.

Surgical decompression of the suprascapular nerve is done with the patient in the lateral decubitus position, and the incision is made parallel to the spine of the scapula. Subperiosteal removal of the trapezius attachment to the spine exposes the supraspinatus and its superior border. This upper border of the supraspinatus is carefully retracted inferiorly and posteriorly to expose the superior surface of the scapula and the suprascapular notch and ligament. The suprascapular artery crosses above the ligament and the nerve below. Ligament excision should be done along with appropriate bony resection with a laminectomy rongeur. Rask reported two cases in which a repeat decompression of the nerve with bony resection gave good results; he recommends wide notch resection as primary treatment [40]. Certainly, if there is any question about the nerve being free, notch resection is indicated. If decompression of the nerve at the spinoglenoid notch is necessary, and a surgical approach to the posterior glenoid and base of the scapular spine is attempted, one would try not to detach the infraspinatus, but this has been necessary on occasion [49].

Sports

Suprascapular nerve injury may present after specific trauma, with chronic onset of pain or weakness, or with insidious painless muscle atrophy. Bateman stated that "athletic stress," especially throwing, produces a backward and forward rotation of the scapula and suprascapular nerve compression at the notch [4]. Jobe and colleagues have stated that in the athlete the nerve is often injured as it passes around the lateral spine of the scapula, sparing the supraspinatus [27]. It has been their experience that if the nerve lesion is at the suprascapular notch, the condition does not respond well to surgical management. In patients with spinoglenoid notch lesions, if the infraspinatus alone is not completely denervated, a program of therapy can allow the elite pitcher to return to his high level competition. Jobe and colleagues' EMG studies showed that only 30% to 40% of the maximum strength of the infraspinatus is used during throwing, and, with a partial nerve injury, a return to pitching, at least, is possible [27]. Ferretti and associates studied asymptomatic volleyball players and found that 12 of 96 had isolated partial infraspinatus paralysis, mostly in a dominant shoulder, some electrical abnormalities, muscle atrophy, and a 15% to 30% loss of external rotation

power [19]. They suggested that the cause was nerve tension at the spinoglenoid notch when the arm is cocked in maximum stretch and during follow-through. Suprascapular neuropathy has been reported with acute shoulder dislocation in a cyclist [58] and with sudden onset after a hard throw from center field in a professional baseball player [8]. The literature is confusing, referring to problems with the suprascapular nerve as both a compression syndrome and a nerve stretch injury.

The overall good response to conservative management suggests that nerve injury may be the cause. Rest from sports or other inciting cause may be helpful. Return to activity is permitted according to the judgment of the physician based on factors in the course of follow-up, including the extent of the initial paralysis, electrical studies, symptoms, and improvements in the muscle examination with therapy. Surgical exploration of a well-localized lesion should be performed if conservative management of 3 to 6 months has failed.

LONG THORACIC NERVE

Anatomy and Etiology

Long thoracic nerve palsy causing paralysis of the serratus anterior with winging of the scapula is a rather disabling lesion. The nerve is formed from roots of C5, C6, and C7, which branch shortly after they exit from the intervertebral foramina. Branches of C5 and C6 pass anteriorly through the middle scalene muscle, then fuse and pass over the posterior scalene. A C7 branch joins to form the long thoracic nerve. The nerve courses behind the brachial plexus to perforate the fascia of the proximal serratus anterior. It then passes medial to the coracoid on the frontal view and has an overall length of 30 cm [24] (Fig. 15E–3). The serratus anterior covers much of the lateral thorax and acts with the trapezius to position the scapula for elevation. It arises from the upper nine ribs and attaches at the deep surface of the scapula along the vertebral border. This powerful muscle draws the scapula forward and rotates its inferior angle upward. The serratus anterior also acts as an accessory inspiratory muscle, as seen in runners who fix their scapulae by holding their thighs to catch their breath after a race.

The long thoracic nerve is often affected by the poorly understood syndrome of brachial neuritis. Long thoracic nerve palsy may occur with prolonged recumbency or intraoperative stretch during thoracic surgery. Serratus anterior weakness following transaxillary first rib resection is not uncommon and has

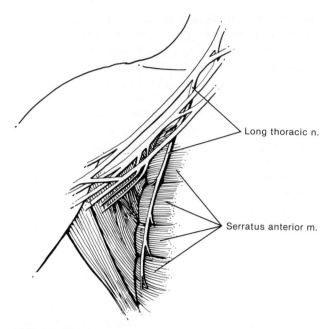

FIGURE 15E–3
The brachial plexus. (Modified from Haymaker, W., and Wood-hall, B. *Peripheral Nerve Injuries*. Philadelphia, W. B. Saunders, 1956.)

a good prognosis although complete paralysis has a poor outlook [27]. Other causes of nerve palsy include backpacking and shoveling. Proposed traumatic mechanisms include crushing of the nerve between the clavicle and the second rib [26], tetanic scalenus medius muscle contraction, and nerve stretch with head flexion or rotation and lateral tilt with ipsilateral arm elevation or backward arm extension [24]. The outcome of acute traction injuries is good [26, 41]. Since the nerve is deeply located, a direct blow seems unlikely to cause isolated palsy. Serratus anterior rupture has been reported in patients with rheumatoid arthritis [35].

Brachial neuritis is a clinical syndrome of unknown etiology and is the most common cause of serratus anterior palsy in the author's experience. Significant pain lasting a variable time—days to weeks—precedes loss of function in one or more shoulder girdle proximal extremity muscles. Sensory loss does not exclude the syndrome. In the literature, there is a good prognosis for recovery, with 36% of patients recovered by the end of the first year and 75% at the end of the second year [50]. Some improvement may occur after 2 years [20]. Recurrent long thoracic nerve palsy is very rare [24]. Parsonage and Turner coined the term "neuralgic amyotrophy" (brachial neuritis) in 136 military personnel, 30 of whom had isolated serratus anterior paralysis. They also noted a right-sided predominance.

Clinical Evaluation

In the clinical syndrome, paralysis of the serratus anterior causes winging and a lack of scapular stabilization, limiting active shoulder elevation to 110 degrees in patients with complete lesions [24]. Winging of the scapula is usually brought out with resisted active arm elevation or by doing a push-up leaning against a wall (Fig. 15E–4). Presenting symptoms in a patient with an early palsy may be subtle changes in the ability to perform his or her sport with findings of decreased active range of motion of the shoulder and altered scapulohumeral rhythm. There are causes of winging other than serratus anterior palsy such as trapezius palsy, painful shoulder conditions resulting in splinting of the glenohumeral joint, winging associated with multidirectional instability, and voluntary winging. The onset of long thoracic nerve palsy may be painful as in brachial neuritis, or it may be more subtle, involving problems with weight lifting or the feeling of pressure from a chair against the winging scapula when sitting. After an acute injury, several weeks may pass before marked scapular winging is evident. Gregg and colleagues believe that time is needed for the trapezius to stretch out and scapular winging to become evident [24].

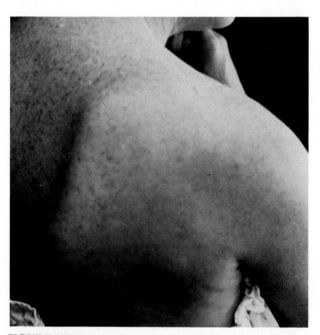

FIGURE 15E–4
Scapular winging is often discovered during weight training as the scapula protrudes with resisted elevation or contacts the flat surface during bench press. If weight lifting is felt to be the cause, resumption of participation should await return of nerve function. Return to sports by patients with long thoracic nerve palsy depends upon the demands placed on the upper extremity by the sport.

Electromyographic studies will confirm the diagnosis of long thoracic nerve palsy. Conduction studies can be performed from Erb's point to the serratus anterior muscle on the anterolateral chest wall.

The appearance of winging with arm elevation due to serratus anterior palsy differs from that of winging due to trapezius palsy. When the serratus anterior muscle does not function, the inferior tip of the scapula is pulled medially and posteriorly. With trapezius paralysis, the scapula body is held in position, and the medial border merely becomes more prominent, a more subtle deformity. In neither type is the scapula rotated laterally to facilitate arm elevation.

Treatment

Cessation of the suspected inciting activity is important. Canvas-reinforced shoulder braces cannot begin to normalize the force couple on the scapula of the serratus anterior and the trapezius. With serratus winging, braces may prevent the stretching out of the trapezius muscle.

Surgically, pectoralis minor transfer [10, 53] to the lateral inferior scapula for dynamic support has been reported. Transfer of the pectoralis major (sternal head) with fascia lata extension to the inferior border of the scapula is the currently favored reconstruction [33].

Sports

Sports have been implicated as a cause of isolated serratus anterior palsy [20, 24, 28] with traction injury—single or repetitive—to the long thoracic nerve being the proposed mechanism. In one series, the repetitive trauma of tennis and archery was thought to be the cause of the lesion in 5 of 20 patients. Other sports implicated in this type of injury are basketball, football, golf, gymnastics, and wrestling [34].

ACCESSORY NERVE

Anatomy

The spinal accessory nerve is a pure motor nerve innervating the trapezius and sternocleidomastoid muscles. The nerve leaves the jugular foramen at the base of the skull, goes through the upper third of the sternocleidomastoid muscle, and crosses the posterior triangle of the neck. It is here that it is superficial and vulnerable to injury. The nerve enters the trapezius and is the predominant motor nerve to that muscle. Root fibers from C3 and C4 also innervate the trapezius and may blend with the accessory nerve; some feel that this C3–C4 contribution is only proprioceptive [36]. The accessory nerve is small—only 1 to 3 mm in diameter [52].

Scapular stabilization and elevation result from the balance of the forces of the trapezius and serratus anterior. The upper trapezius elevates and tilts the scapula, raising the point of the shoulder and assisting in arm elevation respectively. The lower trapezius works with the rhomboids to retract the scapula and balance the pull of the serratus anterior. The nerve may be damaged, as it is most commonly, during a posterior triangle node biopsy or by a direct blow—for example, with a hockey stick or in a traction injury with a cross-face maneuver in wrestling [12]. Stretch injury resulting from distal upper extremity distraction and contralateral head rotation has been reported [32].

Clinical Evaluation

The patient complains of a sagging shoulder and incomplete arm elevation with loss of strength (Fig. 15E–5). The symptoms may be quite severe due to muscle spasm and brachial plexus traction neuritis. Examination does indeed show a drooping of the shoulder or a deepening of the supraclavicular fossa after trapezius atrophy has occurred. Also, winging of the scapula occurs with resisted arm elevation. The levator scapulae is palpable and is seen as a band of muscle in the neck, and rhomboid contraction is also palpable on attempted scapular adduction.

Treatment

Closed injuries should be followed for 6 months before exploration of the nerve is considered and 12 months before reconstruction is done. Accessory nerves that are not functioning clinically or electrically at 6 weeks following open posterior triangle surgery should be explored with neurolysis, repair, or graft. The clinical situation may require earlier exploration. When accessory nerve palsy is judged to be permanent and the patient is symptomatic with upper extremity drooping, aching, and numbness with incomplete active arm elevation, reconstruction may be indicated. Adjacent scapular muscles cannot substitute for a paralyzed trapezius with muscle strengthening alone. The current operation

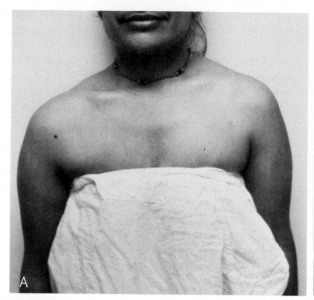
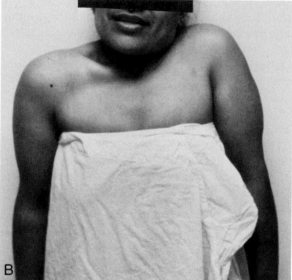

FIGURE 15E–5

A, Drooping of the right shoulder when the patient is relaxed. *B,* There is no voluntary elevation of the shoulder on the right compared with the left.

of choice has been described by Bigliani [5]. The levator scapulae and rhomboids are moved to a more lateral insertion on the scapula to substitute for the upper, middle, and lower trapezius. Other operations described include a scapular suspension with fascial grafts from the vertebral spine to the medial scapula or from the ribs to the scapula, and scapulothoracic fusion [15].

Sports

Cases have been reported of a wrestler and a hockey player [12] with closed accessory nerve palsy, and the author has seen a rugby player with palsy resulting from a direct blow, all of whom were recovering nerve function with observation. Winging is less obvious and often less disabling with trapezius palsy than with serratus anterior palsy. Shoulder function in an athlete with accessory nerve palsy may be inadequate for competition.

References

1. Agre, J. C., Ash, N., Cameron, C., and House, J. Suprascapular neuropathy after intensive progressive resistive exercise: Case report. *Arch Phys Med Rehabil* 67:236–238, 1986.
2. Aiello, I., Serra, G., Traina, G. C., and Tugnoli, V. Entrapment of the suprascapular nerve at the spinoglenoid notch. *Ann Neurol* 12:314–316, 1982.
3. Alon, M., Weiss, S., Fishel, B., and Dekel, S. Bilateral suprascapular nerve entrapment syndrome due to an anomalous transverse scapular ligament. *Clin Orthop* 234:31–33, 1988.
4. Bateman, J. E. Neurologic painful conditions affecting the shoulder. *Clin Orthop* 173:44–54, 1983.
5. Bigliani, L. U. Fracture of the shoulder. Part I: Fractures of the proximal humerus. *In*, Rockwood, C. A., and Green, D. P. (Eds.), *Fractures in Adults*, Vol. 1. Philadelphia, J. B. Lippincott, 1991, pp. 871–927.
6. Bigliani, L. U., Perez-Sanz, J. R., and Wolfe, I. N. Treatment of trapezius paralysis. *J Bone Joint Surg* 67A:871–877, 1985.
7. Black, K. P., and Lombardo, J. A. Suprascapular nerve injuries with isolated paralysis of the infraspinatus. *Am J Sports Med* 18:225–228, 1990.
8. Bryan, W. J., and Wild, J. J. Isolated infraspinatus atrophy—A common cause of posterior shoulder pain and weakness in throwing athletes? *Am J Sports Med* 17:130, 1989.
9. Burge, P., Rushworth, G., and Watson, N. Patterns of injury to the terminal branches of the brachial plexus: The place for early exploration. *J Bone Joint Surg* 67B:630–634, 1985.
10. Chavez, J. P. Pectoralis minor transplanted for paralysis of the serratus anterior. *J Bone Joint Surg* 33B:2128, 1951.
11. Clein, L. J. Suprascapular entrapment neuropathy. *J Neurosurg* 43:337, 1975.
12. Cohn, B. T., Brahms, M. A., and Cohn, M. Injury to the eleventh cranial nerve in a high school wrestler. *Orthop Rev* 15:59–64, 1986.
13. Craig, E. V. Fracture of the shoulder. Part II: Fractures of the clavicle. *In* Rockwood, C. A., and Green, D. P. (Eds.), *Fractures in Adults*, Vol. 1. Philadelphia, J. B. Lippincott, 1991, pp. 928–990.
14. Dewar, F. P., and Harris, R. I. Restoration of the function of the shoulder following paralysis of the trapezius by fascial sling and transplantation of the levator scapulae. *Ann Surg* 132:1111, 1950.
15. Di Benedetto, M. D., and Markey, K. Electrodiagnostic localization of traumatic upper trunk brachial plexopathy. *Arch Phys Med Rehabil* 65:15–17, 1984.
16. Dillin, L., Hoaglund, F. T., and Scheck, M. Brachial neuritis. *J Bone Joint Surg* 67A:878–880, 1985.
17. Drez, D., Jr. Suprascapular neuropathy in the differential diagnosis of rotator cuff injuries. *Am J Sports Med* 4:43–45, 1976.

18. Edeland, H G., and Zachrisson, B. E. Fracture of the scapular notch associated with lesion of the suprascapular nerve. *Acta Orthop Scand* 46:758–763, 1975.

19. Ferretti, A., Cerullo, G., and Russo, G. Suprascapular neuropathy in volleyball players. *J Bone Joint Surg* 69A:260–263, 1987.

20. Foo, C. L., and Swann, M. Isolated paralysis of the serratus anterior. *J Bone Joint Surg* 65B:552–556, 1983.

21. Garcia, G., and McQueen, D. Bilateral suprascapular nerve entrapment syndrome: Case report and review of the literature. *J Bone Joint Surg* 63A:491–492, 1981.

22. Gozna, E. R., and Harris, W. R. Traumatic winging of the scapula. *J Bone Joint Surg* 61A:1230–1233, 1979.

23. Green, R. F., and Brien, M. Accessory nerve latency to the middle and lower trapezius. *Arch Phys Med Rehabil* 66:23–24, 1985.

24. Gregg, J., Jr., Labosky, D., Hartz, M., Lotke, P., Ecker, M., Distefano, V., and Das, M. Serratus anterior paralysis in the young athlete. *J Bone Joint Surg* 61A:825–832, 1979.

25. Hauser, C. U., and Martin, W. F. Two additional cases of traumatic winged scapula occurring in the armed forces. *JAMA* 121:667–668, 1943.

26. Hauser, C. U., and Martin, W. F. Two additional cases of traumatic winged scapula occurring in the armed forces. *JAMA* 121:667–668, 1943.

27. Jobe, F. W., Tibone, J. E., Jobe, C. M., and Kvitne, R. S. The shoulder in sports. *In* Rockwood, C. A., and Matsen, F. A. III (Eds.), *The Shoulder*, Vol. 2. Philadelphia, W. B. Saunders, 1990, pp. 961–990.

28. Johnson, J. T. H., and Kendall, H. Isolated paralysis of the serratus anterior muscle. *J Bone Joint Surg* 37A:567–574, 1955.

29. Kaplan, P. E., and Kernahan, W. T. Jr. Rotator cuff rupture: Management with suprascapular neuropathy. *Arch Phys Med Rehabil* 65:273–275, 1984.

30. Kopell, H., and Thompson, W. Pain and the frozen shoulder. *Surg Gynecol Obstet* 108:92–96, 1959.

31. Leffert, R. D. Neurological problems. *In* Rockwood, C. A., and Matsen, F. A (Eds.), *The Shoulder*, Vol. 2. Philadelphia, W. B. Saunders, 1990, pp. 750–773.

32. Logigian, E. L., McInnes, J. M., Berger, A. R., Busis, N. A., Lehrich, J. R., and Shahani, B. T. Stretch-induced spinal accessory nerve palsy. *Muscle & Nerve* 11:146–150, 1988.

33. Marmar, L., and Bechtal, C. O. Paralysis of the serratus anterior due to electric shock relieved by transplantation of the pectoralis major muscle. *J Bone Joint Surg* 45A:156–160, 1963.

34. Mendoza, F. X., and Main, K. Peripheral nerve injuries of the shoulder in the athlete. *Clin Sports Med* 9:331–341, 1990.

35. Meythaler, J. M., Redd, N. M., and Morris, M. Serratus anterior disruption: A complication of rheumatoid arthritis. *Arch Phys Med Rehabil* 67:770–772, 1986.

36. Olarte, M., and Adams, D. Short report: Accessory nerve palsy. *J Neurol Neurosurg Psychiat* 40:1113–1116, 1977.

37. Overpeck, D. O., and Ghormley, R. K. Paralysis of the serratus magnum caused by lesions of the long thoracic nerve. *JAMA* 114:1994–1996, 1940.

38. Petrera, J. E., and Trojaborg, W. Conduction studies along the accessory nerve and follow-up patients with trapezius palsy. *J Neurol Neurosurg Psychiat* 44:630–636, 1984.

39. Post, M., and Mayer, J. Suprascapular nerve entrapment: Diagnosis and treatment. *Clin Orthop* 223:126–136, 1987.

40. Rask, M. R. Suprascapular nerve entrapment: A report of two cases treated with suprascapular notch resection. *Clin Orthop* 123:73–75, 1977.

41. Rengachary, S. S., Burr, D., Lucas, S., Hassanein, K., Mohn, P., and Matzke, H. Suprascapular entrapment neuropathy: A clinical, anatomical, and comparative study. Part II: Anatomical study. *Neurosurgery* 5:447–451, 1979.

42. Rengachary, S. S., Neff, J. P., Singer, P. A., and Brackett, C. E. Suprascapular entrapment neuropathy: A clinical, anatomical , and comparative study: Part I: Clinical study. *Neurosurgery* 5:441–445, 1979.

43. Richards, R. R., Hudson, A. R., Bertoia, J. T., Urbaniak, J. R., and Waddell, J. P. Injury to the brachial plexus during putti-platt and bristow procedures: A report of eight cases. *Am J Sports Med* 15:374–380, 1987.

44. Ringel, S. P., Treihaft, M., Carry, M., Fisher, R., and Jacobs, P. Suprascapular neuropathy in pitchers. *Am J Sports Med* 18:80–86, 1990.

45. Shabas, D., and Scheiber, M. Case report: Suprascapular neuropathy related to the use of crutches. *Am J Phys Med* 65:298–300, 1986.

46. Solheim, L. F., and Roaas, A. Compression of the suprascapular nerve after fracture of the scapular notch. *Acta Orthop Scand* 49:338–340, 1978.

47. Steiman, I. Painless infraspinatus atrophy due to suprascapular nerve entrapment. *Arch Phys Med Rehabil* 69:641–643, 1988.

48. Sunderland, S. *Nerves and Nerve Injuries.* Edinburgh, Churchill-Livingstone, 1978, p. 1015.

49. Thompson, R. C., Jr., Schneider, W., and Kennedy, T. Entrapment neuropathy of the inferior branch of the suprascapula nerve by ganglia. *Clin Orthop* 166:185–187, 1982.

50. Tsairis, P., Dyck, P. J., and Mulder, D. W. Natural history of brachial plexus neuropathy: Report on 99 patients. *Arch Neurol* 27:109–117, 1972.

51. Valtonen, E. J., and Lilius, H. G. Late sequelae of iatrogenic spinal accessory nerve injury. *Acta Chir Scand* 140:453–455, 1973.

52. Vastamaki, M., and Solonen, K. A. Accessory nerve injury. *Acta Orthop Scand* 55:296–299, 1984.

53. Vastamaki, M. Pectoralis minor transfer in serratus anterior paralysis. *Acta Orthop Scand* 55:293–295, 1984.

54. Weber, L. E. Sport injuries at the peripheral nerve, plexus, and nerve root levels. *Sem Neurol* 1:291, 1981.

55. Woodhead, A. B. III. Paralysis of the serratus anterior in a world class marksman. *Am J Sports Med* 13:359–362, 1985.

56. Wright, T. A. Accessory spinal nerve injury. *Clin Orthop* 108:15–18, 1985.

57. Zibordi, F., Baiocco, F., Bascelli, C., Bini, A., and Canepa, A. Spinal accessory nerve function following neck dissection. *Ann Otol Rhinol Laryngol* 97:83–86, 1988.

58. Zoltan, J. D. Injury to the suprascapular nerve associated with anterior dislocation of the shoulder: Case report and review of the literature. *J Trauma* 19:203–206, 1979.

Injuries of the Proximal Humerus Region

Ralph J. Curtis, Jr., M.D.

FRACTURES OF THE PROXIMAL HUMERUS

Although there is a great deal of information in the historical literature about fractures of the proximal humerus, this is a relatively uncommon sports injury [91, 141, 199]. Fractures of the proximal humerus represent 4% to 5% of all fractures [12, 77]. They are most common in the early adolescent patient who has open physes and in the older patient who has osteoporosis. Fractures of the proximal humerus occur at nearly 70% of the reported rate for proximal femoral fractures in all ages considered [167]. In the most active young and middle-aged groups, the bony structure of the proximal humerus appears to be less vulnerable to injury than the soft tissue structures about the shoulder, i.e., the glenohumeral ligaments, labrum, and rotator cuff. Shoulder instability and rotator cuff injury, therefore, are far more common than proximal humeral fractures in athletes. When proximal humeral fractures do occur in the young athlete, fracture-dislocations, displaced fractures of the greater tuberosity, and nondisplaced fractures due to high-energy trauma are the most common types [141, 172, 196].

Proximal humeral fractures can lead to significant morbidity in the athlete and require aggressive treatment to allow return to competitive sport. Precise anatomic reconstruction is often necessary to restore the shoulder to its normal biomechanical abilities. Precise diagnosis using the four-part classification, described by Neer in 1970, is essential for accurate treatment [135–137]. The judicious use of open reduction and internal fixation techniques for displaced fractures has become an integral part of the treatment program to achieve anatomic restoration [15, 141]. Equally as important is an early aggressive

rehabilitation program designed to prevent the common problems of residual stiffness and dysfunction that can compromise treatment of these fractures [111, 139].

Anatomy

It is extremely important to understand the complex anatomy of the shoulder when dealing with fractures of the proximal humerus. At no other joint is precise restoration of anatomy more important and necessary to restore normal function after injury [76].

Bony Anatomy. The proximal humerus consists of four major bony components: the humeral head, the lesser tuberosity, the greater tuberosity, and the humeral shaft (Fig. 15F–1). The humeral head is a

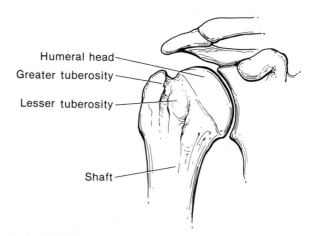

Humeral head
Greater tuberosity
Lesser tuberosity
Shaft

FIGURE 15F–1
Schematic diagram of the bony structure of the proximal humerus and the relationship to the scapula. Note the four major portions of the proximal humerus: (1) head, (2) greater tuberosity, (3) lesser tuberosity, (4) shaft.

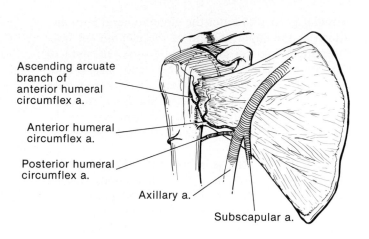

FIGURE 15F–2
Vascular supply to the humeral head. The anterior humeral circumflex artery through the ascending arcuate branch enters the humerus at the intertubercular groove, providing the major source of proximal humeral vascularity.

Labels in figure:
Ascending arcuate branch of anterior humeral circumflex a.
Anterior humeral circumflex a.
Posterior humeral circumflex a.
Axillary a.
Subscapular a.

thin, flattened portion of cancellous bone with a dense subchondral layer that supports the articular surface of the humerus. This articular surface has been described as a flattened hemisphere with a radius of curvature of approximately 2.25 cm. The humeral head is retroverted 30 to 35 degrees and is tilted upward 45 degrees in relation to the humeral shaft. The anatomic neck is at the junction between the head and the greater and lesser tuberosities, whereas the surgical neck is the area below the tuberosities at the junction with the shaft. The anatomic neck is the area between the articular cartilage and the attachment of the rotator cuff that supplies the primary blood supply to the head [99, 131, 144, 160, 168]. Fractures of the anatomic neck can disrupt the blood supply to the humeral head and can lead to avascular necrosis (Fig. 15F–2).

With the arm in the anatomic position, the lesser tuberosity is the small prominence that lies directly anterior and medial to the intertubercular groove. The lesser tuberosity is the attachment area for the subscapularis muscle and anterior capsule of the glenohumeral joint. The bicipital or intertubercular groove lies between the greater and lesser tuberosities and can vary considerably in width and depth. The biceps tendon lies in the bicipital groove as it enters the glenohumeral joint and is covered by the transverse humeral ligament. The greater tuberosity is much larger than the lesser tuberosity and lies posterior and superior on the humeral shaft. It provides attachment for the supraspinatus, infraspinatus, and teres minor muscles.

Closure of the proximal humeral physis leaves an area of dense bone or scar that remains until the sixth or seventh decade. The medullary canal begins distal to the epiphyseal scar, and the entire proximal humeral area is composed of dense cancellous bone surrounded by a cortical shell. In the older patient, osteoporosis leads to a diminution in density of this cancellous bone, and the medullary canal extends all the way to the epiphyseal scar [63]. Mechanical support in the proximal end of the humerus is therefore diminished, and the arm is more vulnerable to fracture in the older patient.

The glenoid is the shallow, concave articular surface of the scapula that provides the "socket" of the glenohumeral joint for the humeral head. It has approximately one-third the surface area of the humeral head itself. Superior to the humeral head lies the acromion. It is a broad, thin shelf of bone extending laterally from the spine of the scapula. The acromion provides a superior origin for the deltoid muscle and a protective bony cover for the glenohumeral joint itself. Together with the coracoacromial ligament and coracoid it forms the coracoacromial arch, which is a rather rigid structure under which the proximal humerus, rotator cuff, and subacromial bursa must pass when the arm is abducted. Residual displacement in fractures of the proximal humerus can result in impingement underneath this coracoacromial arch. The subacromial bursa is a large synovial-lined structure that lies between the proximal humerus and the undersurface of the coracoacromial arch and the deltoid muscle. Adherent scar within this bursa after fracture can cause limitation of motion by reducing the normal ability of the bursa to reduce friction, and therefore not allowing the gliding mechanism to occur.

Muscular Anatomy. The tendons of the muscles of the rotator cuff are closely attached to the capsule of the glenohumeral joint and insert into the tuberosities of the proximal humerus [76]. The subscapularis muscle attaches to the lesser tuberosity and functions as a strong internal rotator of the humerus. The supraspinatus, infraspinatus, and teres minor tendons attach to the greater tuberosity and function in abduction and external rotation. The long head of the biceps sends its tendon through the intertubercular groove to enter the glenohumeral joint and attach at the superior pole of the glenoid. In fractures involving the greater tuberosity, the pull of the supraspinatus, infraspinatus, and teres minor can displace the greater tuberosity fragment both posteriorly and superiorly [22]. Similarly, fractures of the lesser tuberosity will be displaced by the attached subscapularis muscle anteriorly and medially.

The deltoid is the primary motor for the glenohumeral joint, and it is the synchronous activity between the rotator cuff muscles and deltoid that provides the power and stability that permit the wide range of shoulder motion. The deltoid muscle takes origin from the lateral one-third of the clavicle, the acromion, and the spine of the scapula. It inserts into the deltoid tuberosity distally approximately halfway down the lateral shaft of the humerus. The deltoid can cause displacement of fractures of the proximal humerus because of the shearing action imparted to the fracture during muscular contraction (Fig. 15F–3).

The pectoralis major is a large fan-shaped muscle that originates from the clavicle, upper ribs, and sternocostal area. It inserts on the lower portion of the lateral lip of the bicipital groove distal to the tuberosities. It functions as a strong adductor and internal rotator and can displace the proximal shaft of the humerus medially in fractures of the surgical neck.

Blood Supply. The major blood supply to the humeral head is through the intraosseous vessels that cross the anatomic neck from the metaphysis [99]. The anterior humeral circumflex artery is the primary source and continues as the ascending arcuate artery that penetrates the humerus in the area of the intertubercular groove (see Fig. 15F–2). The anterior humeral circumflex also gives branches to the lesser and greater tuberosities through the at-

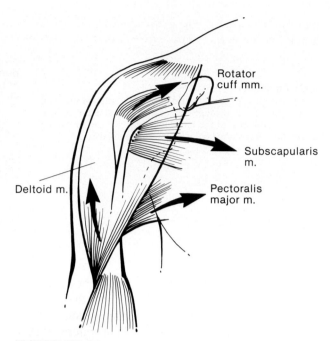

FIGURE 15F–3
The muscular attachments and the direction of their pull can influence displacement in proximal humeral fractures. These forces must be taken into account when attempting to reduce a fracture in this region.

tachments of the rotator cuff [160, 168]. A small contribution to the humeral head circulation is made by branches of the posterior humeral circumflex artery (Fig. 15F–4). Fractures in the region of the anatomic neck therefore can disrupt blood supply to the humeral head, which can lead to avascular necrosis. Knowledge of the delicate vascular supply is also helpful in limiting potential damage by dissection during open reduction and internal fixation.

Nerve Supply. The brachial plexus lies anterior to the scapula, passing below the coracoid to enter the upper arm (Fig. 15F–4). The axillary nerve arises from the posterior cord and courses first along the

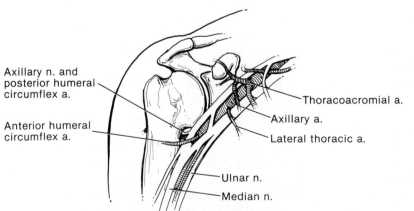

FIGURE 15F–4
The relationship of the proximal humerus to the brachial plexus and axillary artery with prominent branches is demonstrated in this schematic representation.

anterior surface of the subscapularis muscle belly and then below the glenohumeral joint to innervate the deltoid and teres minor. The musculocutaneous nerve arises from the lateral cord, penetrating the coracobrachialis muscle 5 to 7 cm distal to the coracoid [47]. The brachial plexus therefore is tethered in its position anteromedial to the proximal humerus and is vulnerable to injury in displaced fractures. The axillary nerve is the most frequently injured portion of the plexus in proximal humeral fractures.

Biomechanics

The shoulder is the most mobile major joint in the body. The primary function of the shoulder is to position the hand in space to accomplish prehensile activity. The anatomic features of the proximal humerus and glenohumeral joint are well suited to provide this function. The articular surface of the humeral head is two to three times larger than the surface area of the glenoid. The relatively flat glenoid produces very little constraint to humeral motion. The range of motion of the glenohumeral joint is two to one that of scapulothoracic motion, and their combined motion approximates 180 degrees of abduction. The humeral neck shaft angle is approximately 45 degrees, and the head is 30 to 40 degrees retroverted relative to the epicondyles at the elbow. This anatomic configuration aligns the humeral head with the scapula as it lies along the posterolateral thorax in the anatomic position.

The synergy between the stabilizing effect of the rotator cuff and the biceps combined with the power of the deltoid provides normal dynamic shoulder function. The deltoid is the primary motor source for the shoulder but also creates shear stress across the joint. The rotator cuff and biceps provide stability by counterbalancing the humeral head against the deltoid shear. As internal rotation and external rotation components are added, the rotator cuff muscles provide not only humeral head depression but also stability against excessive anterior and posterior translation within the joint. Fractures of the proximal humerus that disrupt or distort this finely tuned anatomy can alter the biomechanics of the shoulder and therefore lead to limitation of motion and function. Restoration of this anatomy along with appropriate muscle strength and coordination are all necessary to return the shoulder to normal function [198].

Classification

The most commonly used classification system for fractures of the proximal humerus is the Neer clas-

sification described in 1970 [135–137]. This classification system was based upon the work of Codman, who proposed that fractures of the proximal humerus could be separated into four distinct types along the anatomic lines of epiphyseal union, including the anatomic head, the greater tuberosity, the lesser tuberosity, and the shaft. Codman's conclusion was that all fractures in this region were some combination of different fracture fragments. The Neer classification is called the "four segment classification" (Fig. 15F–5). It is a comprehensive system that considers anatomy, biomechanical forces, and resultant displacement of fracture fragments and relates these factors to diagnosis and treatment. An adequate knowledge of the anatomy and insertions of the tendons of the rotator cuff are essential to use it properly. Inherent in the use of this classification system is the ability to identify accurately the four major fragments and their relationship to each other. The identification of fragments can be accomplished only with proper x-rays including the "trauma series" [141, 165]. This trauma series consists of the *true anteroposterior view*, the *lateral view* in the *scapular plane*, and the *axillary lateral view*. The essential focus of this fracture classification system is the status of the blood supply to the humeral head and the relationship of the humeral head to the displaced fragments and the glenoid.

When any of the four major segments is displaced more than 1 cm or angulated more than 45 degrees, the fracture is considered displaced. Any other minimally displaced or nondisplaced fractures that do not meet these criteria are not considered separate fragments. Any fracture in which none of the fragments meet the above criteria should be considered a *one-part fracture* or a *minimally displaced fracture*. Eighty percent of all proximal humeral fractures fall into this category. A *two-part fracture* is one in which one fragment is displaced in reference to the other three anatomic units. The displaced fragment most commonly involves the surgical neck but may also involve the greater tuberosity or, less commonly, the lesser tuberosity. Only rarely is the anatomic neck involved in two-part fractures. A *three-part fracture* is one in which two fragments are displaced in relation to each other and in relation to the other two nonfractured or nondisplaced fragments. The surgical neck and one of the tuberosities are most commonly displaced from the intact head, single-tuberosity fragment. These fractures are also known as rotary fracture-subluxations because the head fragment is commonly rotated by the intact tuberosity and rotator cuff attachment but is not dislocated. A *four-part fracture* is one in which all four fracture fragments

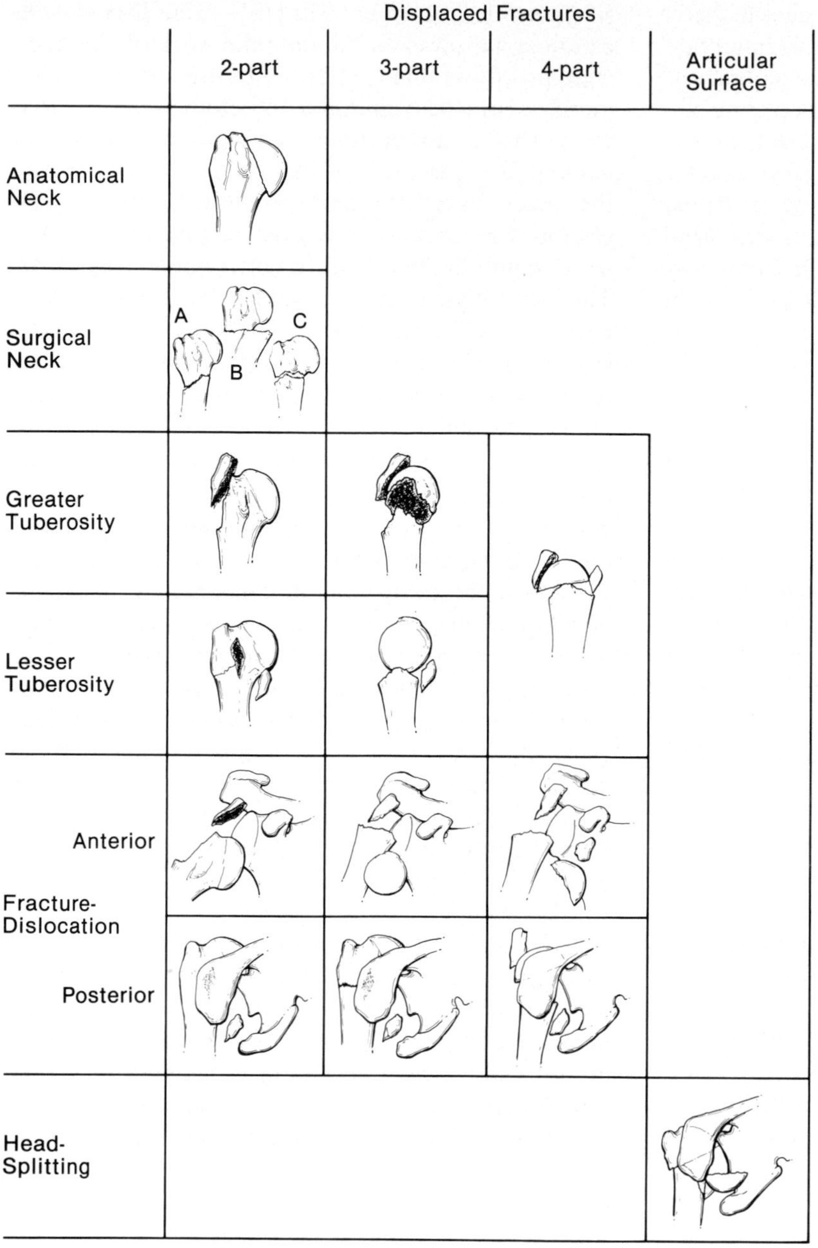

Displaced Fractures

	2-part	3-part	4-part	Articular Surface
Anatomical Neck				
Surgical Neck				
Greater Tuberosity				
Lesser Tuberosity				
Fracture-Dislocation — Anterior				
Fracture-Dislocation — Posterior				
Head-Splitting				

FIGURE 15F–5
The Neer four-segment classification. A fragment is considered displaced when greater than 1 cm displacement or 45 degrees angulation is present. A fracture-dislocation is present only if the articular segment is no longer in contact with the glenoid. (From Neer C. S. Displaced proximal humeral fractures. Part I. Classification and evaluation. *J Bone Joint Surg* 52A:1077–1089, 1970.)

are displaced. In this fracture the head is typically partially out of contact with the glenoid and is angulated or rotated either laterally, anteriorly, posteriorly, inferiorly, or superiorly. Furthermore, it is detached from both tuberosities and therefore from its blood supply. A *fracture-dislocation* exists when the head is displaced completely outside the joint space rather than simply being subluxated or rotated within the joint and there is an additional fracture of the neck or tuberosities. Fracture-dislocations are classified according to the direction of head displacement, either anterior or posterior, as well as by the number of fracture fragments. Isolated fractures of the articular surface are described as *head-splitting*

fractures or *impression fractures*; they are treated as special fractures and will be discussed separately.

Clinical Evaluation

History and Mechanism of Injury. The patient with a proximal humeral fracture typically describes specific trauma and can often outline the mechanism of injury [31, 91, 141, 154, 169]. After the fracture pain, swelling, and inability to use the shoulder are seen immediately. Most patients are not able to continue to participate in the activity at hand. The most common mechanism of fracture is a fall on an outstretched arm, as commonly occurs in both contact and noncontact sports. In the older patient with osteoporotic bone, fracture of the proximal humerus often occurs without significant trauma. In the younger patient, however, significant trauma is necessary, and the resultant fracture is often more serious. These younger patients commonly have displaced fractures or fracture-dislocations with substantial soft tissue disruption. Neurologic or vascular injury can occur and are related to the seriousness of the soft tissue component of the injury.

An additional mechanism of fracture of the proximal humerus occurs with excessive external rotation of the arm, especially in an abducted position. The proximal humerus is wedged against the acromion in a pivotal position, and a proximal humeral fracture can occur through this area.

A direct blow to the lateral arm can also cause injury to the proximal humerus but is less common. This mechanism usually results in a fracture of the greater tuberosity, a minimally displaced fracture, or a fracture involving the articular surface.

Fractures associated with primary dislocation of the shoulder, both anterior and posterior, are usually caused by the forced abducted, externally rotated position (for anterior fracture-dislocations) and forced adduction with posterior displacement (for posterior fracture-dislocations).

Physical Examination. Physical examination reveals in most cases swelling and tenderness to palpation about the shoulder [15, 24, 137]. Commonly, disruption of the normal bony contour of the shoulder is seen in displaced fractures and fracture-dislocations. Crepitus may be present if there is significant displacement of the fracture fragments. Ecchymosis does not occur early, but over 2 or 3 days discoloration may occur in the arm and extends to the elbow and along the chest wall and upper back. These patients often bind the arm to the side to help protect it against motion. It is difficult for the patient to initiate active motion, which usually results in painful muscle spasm. Attention to a detailed neurovascular examination is essential. Brachial plexus and axillary artery injuries are associated with proximal humerus fracture [16, 39, 73, 212]. The axillary nerve is the most commonly injured, but the entire plexus can be involved. The sensory distribution of the axillary nerve over the lateral upper arm should be checked for light touch and pin-prick. Deltoid motor activity is also an axillary nerve function and can be checked by palpating the muscle belly with one hand while supporting the elbow to resist abduction with the other (Fig. 15F–6).

Radiographic Examination. Accurate diagnosis of proximal humeral fractures can be readily made in the majority of cases when proper x-rays of the shoulder are available. The trauma series of plain radiographs has been described as the cornerstone of radiographic evaluation and classification of these fractures [17, 130, 165]. The trauma series includes a *true anteroposterior (AP) view* and a *true lateral view* in the scapular plane as well as an *axillary lateral view*. At least two views of the shoulder at 90 degrees to each other are absolutely essential for accurate diagnosis.

Anatomically the scapula sits obliquely on the chest wall at approximately a 35- to 40-degree angle in the frontal plane. The glenohumeral joint, therefore, does not lie in either the sagittal or coronal plane. To obtain true AP and lateral views in the scapular plane, they must be done in an oblique fashion with regard to the chest. A *true AP view* in the scapular plane is performed with the affected shoulder placed against the x-ray cassette and the body rotated toward that side approximately 40 degrees (Fig. 15F–7). This technique, therefore, gives a true AP view of the glenohumeral joint. This avoids superimposition of the humeral head on the glenoid that occurs with a routine AP view of the shoulder.

The *true lateral view* in the scapular plane is also called the tangential view, Y view, or true scapular lateral view. The lateral view in the scapular plane is accomplished by placing the x-ray cassette laterally against the point of the affected shoulder while the plane of the x-ray beam is directed along the scapular spine (Fig. 15F–8). This can best be accomplished by drawing a line along the scapular spine to help align the x-ray tube with the spine.

The *axillary lateral view* allows evaluation in the axial plane (Fig. 15F–9). The supine position is preferable. The arm can be held in abduction of approximately 60 degrees while the plate is placed along the superior aspect of the shoulder. The tube is angled approximately 20 degrees off the horizontal and vertical axes and is directed toward the axilla. The axillary view often provides the best information about the relationship of the humeral head and glenoid fossa (anterior or posterior dislocation versus rotary subluxation of the head). In addition, posterior displacement of the greater tuberosity and medial displacement of the lesser tuberosity are best delineated in this projection.

A computer tomographic (CT) scan of the shoulder is often helpful for complete evaluation of fractures and fracture-dislocations of the proximal humerus [17, 130] (Fig. 15F–10). Fractures involving the humeral head articular surface such as head-splitting fractures or impaction-type fractures are

Text continued on page 675

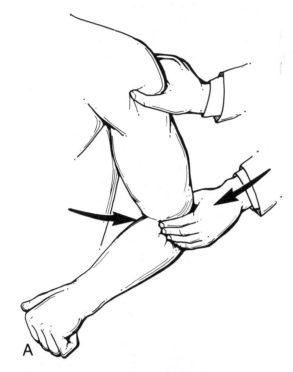

FIGURE 15F–6

A, Schematic demonstrates clinical examination of axillary nerve motor distribution. The deltoid muscle is palpated while an attempt to abduct the arm is made. *B*, This football player has sustained an injury to the axillary nerve. Note atrophy of the deltoid muscle. *C*, The lined area on the upper lateral arm depicts the region of decreased sensation in axillary nerve injury.

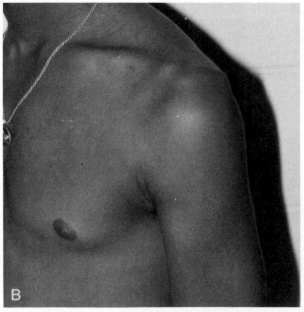

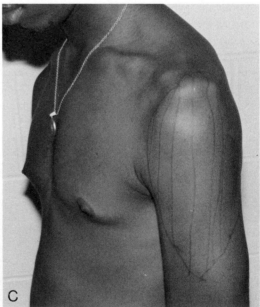

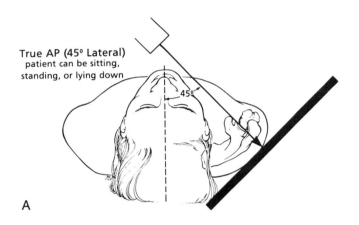

True AP (45° Lateral)
patient can be sitting,
standing, or lying down

45°

A

FIGURE 15F–7
The technique for taking a true AP radiograph of the shoulder is demonstrated. The affected shoulder is rotated approximately 40 degrees toward the cassette. This eliminates the bony overlap between the humeral head and the glenoid. (*A* and *B*, From Rockwood, C. A., and Matsen, D. D. (Eds.), *The Shoulder*. Philadelphia, W. B. Saunders, 1990.)

Illustration continued on following page

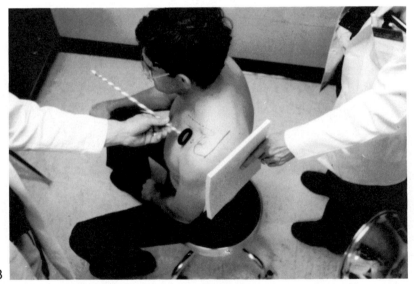

B

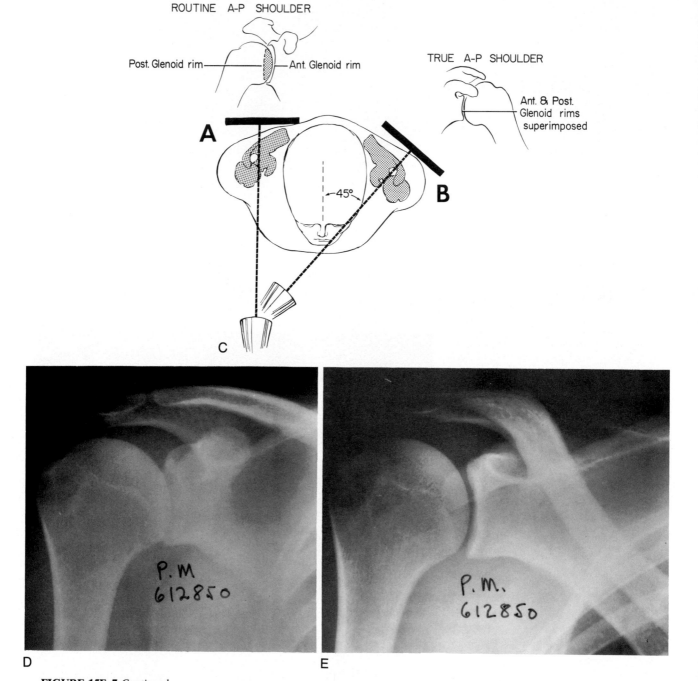

FIGURE 15F–7 *Continued*
C, D and *E*, from Rockwood, C. A., and Green, D. P. (Eds.), Fractures (3 vols), 2nd ed. Philadelphia, J. B. Lippincott, 1984.)

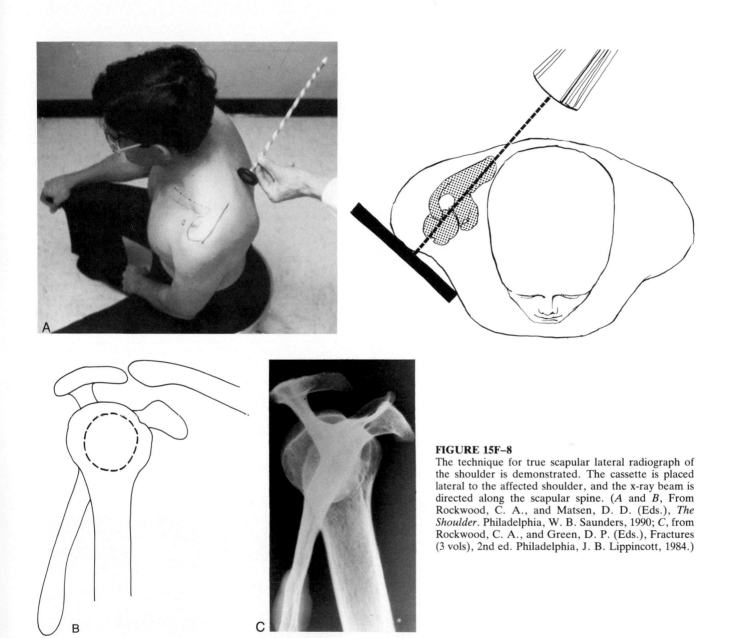

FIGURE 15F–8
The technique for true scapular lateral radiograph of the shoulder is demonstrated. The cassette is placed lateral to the affected shoulder, and the x-ray beam is directed along the scapular spine. (*A* and *B*, From Rockwood, C. A., and Matsen, D. D. (Eds.), *The Shoulder*. Philadelphia, W. B. Saunders, 1990; *C*, from Rockwood, C. A., and Green, D. P. (Eds.), Fractures (3 vols), 2nd ed. Philadelphia, J. B. Lippincott, 1984.)

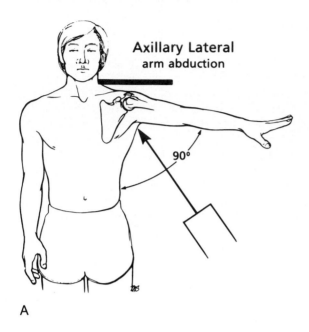

A

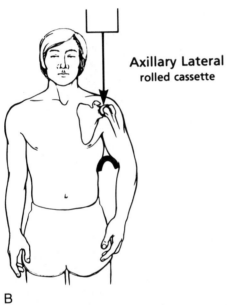

B

FIGURE 15F–9
The technique for taking an axillary lateral radiograph of the shoulder is demonstrated. The tube is angled approximately 20 degrees off the horizontal and vertical axes toward the cassette which is positioned above the shoulder. (From Rockwood, C. A., and Matsen (Eds.), *The Shoulder*. Philadelphia, W. B. Saunders, 1990.)

FIGURE 15F–10
This CT scan of the shoulder and proximal humerus demonstrates excellent bony detail. This technique is useful for evaluating the articular surface of the humeral head and its relationship to the glenoid.

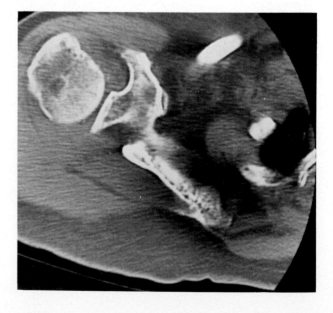

best evaluated with a CT scan. This technique is also useful for evaluating glenoid rim fractures commonly associated with dislocations.

Treatment Options

Historically, many different types of treatment have been described for proximal humeral fractures including a variety of external immobilization devices such as spica and hanging arm casts, traction, and splints as well as internal fixation devices such as percutaneous pinning, plates and screws, rods, staples, and the humeral head prosthesis [1, 2, 21, 29, 40, 42, 50, 75, 78, 85, 86, 87, 110, 127, 129, 156, 161, 177]. Today it is recognized that utilization of appropriate anatomic classification schemes for diagnosis allows development of a consistent plan for treatment that avoids factors that can lead to poor results such as residual scarring, malunion, and aseptic necrosis. Because of the unique anatomy and biomechanics of the shoulder, it is not always necessary to accomplish complete restoration of the anatomy to restore good function. However, in the competitive athlete who requires the arm for throwing or overhead motion (baseball, football, javelin, swimming, and so on), a more rigorous effort should be made to restore anatomy and obtain rigid fixation. Most treatment programs stress early motion to prevent the formation of scarring in the capsule, subacromial space, and other surrounding soft tissues that will often lead to decreased function [51].

Minimally Displaced or Nondisplaced Fractures

The majority (80% to 85%) of proximal humeral fractures belong in this category (Fig. 15F–11). A short period of initial immobilization followed by early motion has been consistently described by most authors as having a high degree of success in treating this fracture [2, 43, 60, 89, 126, 141, 205]. Minimial degrees of displacement and angulation (less than 1 cm or 45 degrees) are well tolerated functionally even in the throwing athlete. Restoration of function depends primarily on restoring the normal range of motion at the shoulder joint, so this goal is given priority in most treatment plans [18]. The arm is initially supported in a sling and swathe, collar and cuff, or Velpeau immobilizer. Range of motion exercises along with isometric exercises usually are started very early. Recommendations for the length of time of immobilization prior to beginning exercise varies from a few days to a few weeks. Some authors feel that a short period of immobilization prior to beginning exercise decreases the risk of displacement and subsequent malunion or nonunion [81, 105, 145]. The fracture should be followed radiographically; however, most minimally displaced fractures are stable. No matter how long the immobilization period lasts, an early aggressive range of motion program is emphasized uniformly. Bertoft and colleagues reported that the greatest improvement in shoulder range of motion

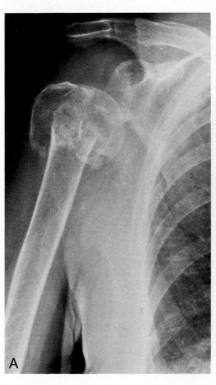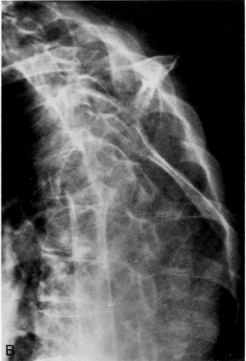

FIGURE 15F–11
Radiograph of a nondisplaced, impacted proximal humeral fracture. This fracture can be treated by a short period of immobilization followed by an aggressive rehabilitation program. *A*, AP view demonstrates impaction of the surgical neck. *B*, Inadequate scapular lateral view of this same fracture.

after proximal humeral fracture occurs between 3 and 8 weeks after injury [14]. Bony healing is usually complete by 6 to 8 weeks in the adult, but return to normal motion and function may require 3 to 4 months. The prognosis for healing and return to normal function is good.

Two-Part Fractures

Two-Part Surgical Neck Fractures. Two-part surgical neck fractures are fractures in which both tuberosities remain attached to the humeral head fragment while the shaft fragment is either impacted and angulated or is displaced. Two-part surgical neck fractures are the most common type of two-part fracture and occur in all age groups.

In the impacted surgical neck fracture with angulation, the posterior periosteum often remains intact with the apex of the angulation anterior (Fig. 15F–12), If anterior angulation of more than 45 degrees is present, union in this position could block motion. If the fracture is otherwise well opposed, correction of this anterior angulation can be obtained with gentle longitudinal traction and posterior pressure at the fracture to improve position [43, 49, 60, 104]. Once reduced, these fractures are immobilized for a short period of 2 to 3 weeks until they are stable, after which motion is instituted. Immo-

bilization of these fractures can be accomplished with a sling and swathe, collar and cuff, or shoulder Velpeau immobilizer. Many authors have described the use of the hanging arm cast in this particular fracture. Because of the risk of nonunion caused by overdistraction, this treatment is less frequently used today [58, 79, 98, 159, 207]. These impacted fractures are usually stable, allowing early motion and therefore carrying a good prognosis [24].

Displaced two-part surgical neck fractures are characterized by anterior and medial displacement of the shaft fragment caused by the strong pull of the pectoralis major muscle. The proximal fragment is typically in a neutral or slightly abducted position through the action of the attached rotator cuff musculature. Occasionally these fractures are comminuted, and severe shortening and instability are present.

In the displaced two-part surgical neck fracture, an attempt at closed reduction is usually warranted. Reduction is best accomplished under general anesthesia. Longitudinal traction is applied to the arm, followed by adduction and flexion of the distal fragment to relax the pectoralis major. Direct pressure from the axilla is then used to engage the fragments [74, 105]. If reduction is accomplished, stability is judged according to whether the shaft and head fragments move as a single unit. If the fracture is stable after reduction, range of motion

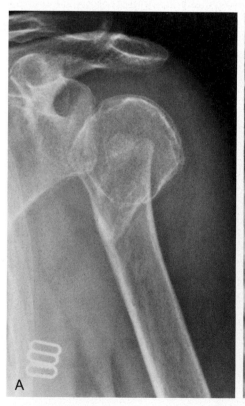
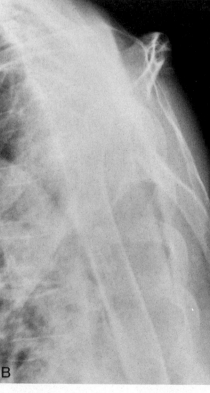

FIGURE 15F–12
This x-ray series demonstrates a two-part angulated fracture involving the surgical neck of the humerus. *A*, Fractures with an anterior angulation of greater than 45 degrees may block shoulder motion if left uncorrected. *B*, Scapular lateral view demonstrates angulation.

and isometric exercises are begun in the first 3 to 7 days after injury. If the fracture is judged to be unstable after reduction, motion is delayed 2 to 3 weeks while the arm is immobilized in a sling and swathe.

Two-part displaced surgical neck fractures may require open reduction with or without internal fixation because of either failure to achieve closed reduction or instability. Soft tissue interposition may block successful closed reduction and usually involves the biceps tendon, deltoid muscle, or periosteum attached to the proximal fragment [12, 103, 169]. If soft tissue interposition is present and closed

reduction fails to align the fracture successfully, surgical intervention is necessary to remove the soft tissue impediment.

Various methods and devices of internal fixation for treatment of the two-part surgical neck fracture have been described. Percutaneous pinning is useful if the fracture is found to be unstable following closed reduction. Jakob and colleagues have nicely outlined the technique, using parallel 2.5-mm AO threaded pins placed in the proximal shaft just above the deltoid insertion and then drilled into the head fragment under C-arm control (Fig. 15F–13) [82]. The pins are removed when adequate stability of

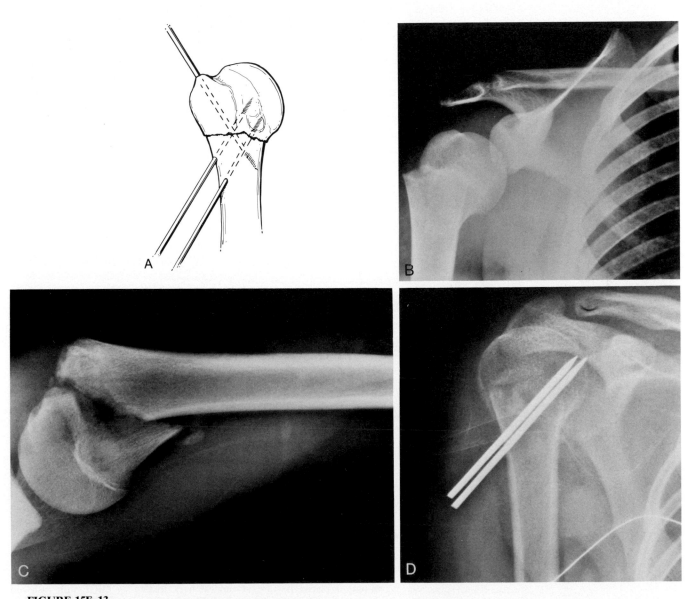

FIGURE 15F–13
A, Schematic depiction of a two-part displaced surgical neck fracture treated with Kirschner wires after technique described by Jakob. *B*, AP view demonstrates surgical neck fracture. *C*, Axillary lateral view demonstrates unacceptable displacement. *D*, Threaded Steinman pins were driven from the distal fragment into the head of the humerus.

the fracture has been achieved at 2 to 4 weeks. This is followed by a vigorous rehabilitation program [82].

Several open techniques of internal fixation have also been described (Fig. 15F–14). The Rush rod technique is very easy and limits the incision to a small split in the deltoid [57, 114, 171]. Good results

have been reported, but drawbacks to this procedure include a relative lack of fixation of this device and its inability to control rotation completely. A figure-of-eight tension band technique with wire or non-absorbable sutures has been described with or without intramedullary fixation and has yielded good results [135, 205]. Use of the AO buttress plate and

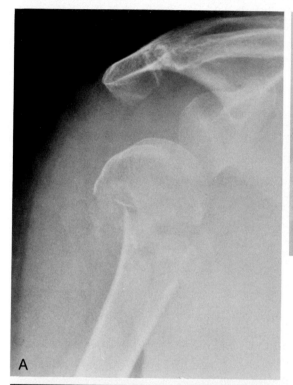

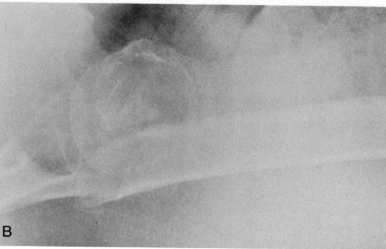

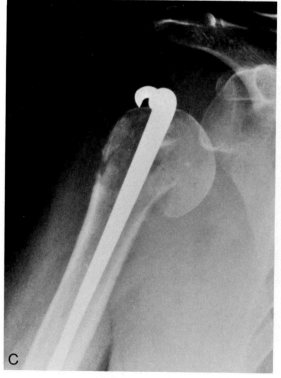

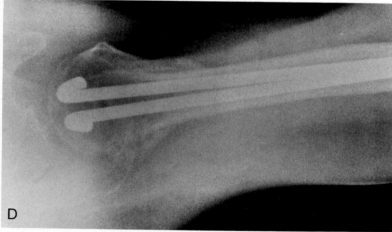

FIGURE 15F–14
Two-part displaced fracture of surgical neck treated by ORIF with rush rod technique. *A*, AP view of injury. *B*, Axillary lateral view of injury. *C*, Postreduction AP view. *D*, Postreduction axillary lateral view.

screws has been associated with good results, especially in the younger patient with good, solid bone [190]. Yamano has described a high success rate with a contoured hooked plate [208]. All of these procedures have the disadvantage of increased soft tissue dissection and the need to remove the implant after healing in most cases.

Skeletal traction has also been described but is not commonly indicated [29, 145]. It can be very useful, however, in the management of comminuted fractures with shortening and instability. It is a difficult and demanding technique that requires attention to detail on the part of the surgeon. A threaded Kirschner wire or Steinman pin is placed transversely across the proximal ulna. The arm should be held flexed and slightly adducted to relax the pectoralis major muscle. The forearm and wrist are suspended in a sling. The hand and elbow should be allowed to undergo range of motion exercise to avoid stiffness. As callus appears on radiographs, traction is progressively lowered to the side until stability in the neutral position is achieved. After removal of the traction pin, vigorous rehabilitation is required to regain motion.

Two-Part Anatomic Neck Fractures. Displaced anatomic neck fractures are rare but are important because of the expected high incidence of avascular necrosis after this type of injury (Fig. 15F–15). The survival of the head fragment is compromised in this fracture owing to the loss of both the intraosseous and the capsular blood supply [99]. There are very few cases in the literature on which to base a discussion concerning treatment. Closed reduction is difficult because the thin, small head fragment is easily rotated or angulated in the joint capsule.

DePalma stated that deviation of up to 25 degrees from the normal head-neck angle of 140 degrees was acceptable [31, 32]. He described a closed reduction technique for displaced anatomic neck fractures that includes the application of longitudinal traction while the head fragment is stabilized with the thumb. The shaft fragment is then abducted and aligned with the head prior to impaction. An attempt at open reduction with internal fixation in younger patients with displaced anatomic neck fractures may be indicated if closed reduction fails [38, 134, 143]. Careful dissection is required to avoid further compromise to the blood supply. Fixation can be accomplished with screws or temporary K-wires. In the older patient with a displaced two-part anatomic neck fracture, use of a prosthesis may be considered [34, 142, 206]. The prognosis in this fracture is guarded because of the high incidence of avascular necrosis.

Two-Part Fractures Involving the Greater Tuberosity. Two-part fractures involving the greater tuberosity are not uncommon and are often displaced (Fig. 15F–16). In these fractures the greater tuberosity is usually retracted posteriorly and superiorly by the pull of the supraspinatus, infraspinatus, and teres minor. These displaced fractures are commonly associated with a split in the rotator cuff that occurs in the rotator interval between the upper border of the subscapularis tendon and the anterior edge of the supraspinatus tendon. Closed reduction of the displaced greater tuberosity fracture is usually difficult [15, 31, 210]. The deforming forces of the rotator cuff are difficult to counterbalance, and the position of the greater tuberosity beneath the acromion makes it difficult to manipulate. Even if suc-

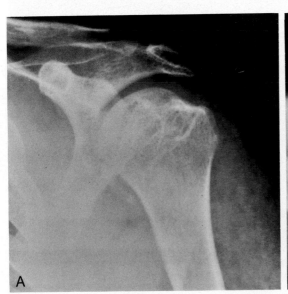
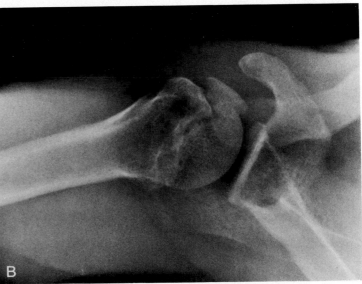

FIGURE 15F–15
Tow-part anatomic neck fracture. *A,* AP view. *B,* Axillary view demonstrates comminution.

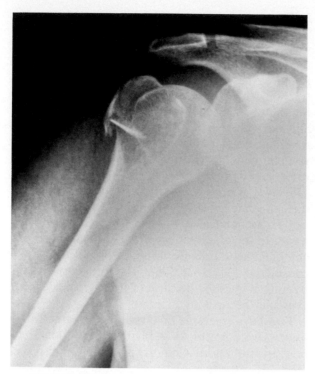

FIGURE 15F–16
Radiograph of displaced fracture of greater tuberosity.

cessful reduction is accomplished, late displacement of the greater tuberosity fragment is not uncommon. If left in a displaced position, these fragments will heal in a malunited position and then will impinge upon the coracoacromial arch (Fig. 15F–17). These fractures are not amenable to percutaneous pinning, plaster splints, or the hanging arm cast. They should be treated by open reduction to improve the posterior and superior displacement of the fragment (Fig. 15F–18). Both a superior approach with splitting of the deltoid and an anterior deltopectoral approach have been utilized. Fixation of larger fragments has been accomplished using screws, wire suture, staples, and heavy nonresorbable suture material [104, 141, 203]. Because of the relatively prominent position of internal fixation in the greater tuberosity, these fragments often require removal to achieve full motion. If the fragment is small, it can be excised and the attached cuff advanced into the defect. The longitudinal rent in the rotator interval should also be repaired in a routine fashion. Healing, requires 4 to 6 weeks, and recovery proceeds slowly, resembling recovery after rotator cuff repair alone.

Two-Part Fractures Involving the Lesser Tuberosity. A two-part fracture involving the lesser tuberosity is rarely reported as an isolated injury [3, 36, 61, 66, 90, 97, 119]. This fracture is more commonly associated with a posterior dislocation or multiple part injury. It is important to recognize

these fractures with good x-ray techniques including an axillary lateral view or CT scan [17, 64, 130]. If the fracture is associated with a posterior dislocation, reduction of the dislocation usually reduces the lesser tuberosity fragment. These fragments are commonly small and heal uneventfully. Displaced fractures that are either large or significantly displaced may block internal rotation if they are allowed to heal in a displaced position. In these large displaced fractures open reduction with internal fixation or excision of the fragment with advancement of the subscapularis tendon into the defect is indicated. Open reduction can be accomplished through a deltopectoral incision using nonabsorbable sutures or screws for internal fixation (Fig. 15F–19). If the fragment is large enough to involve a portion of the articular surface, it should be fixed. Neer and Rockwood have described late excision of a malunited lesser tuberosity fragment if it causes pain or is a block to internal rotation after closed treatment [141].

Three-Part Fractures

Three-part fractures of the proximal humerus are usually quite unstable and are difficult to treat because of this instability (Fig. 15F–20) [15, 45, 170, 173]. A displaced tuberosity fragment is commonly

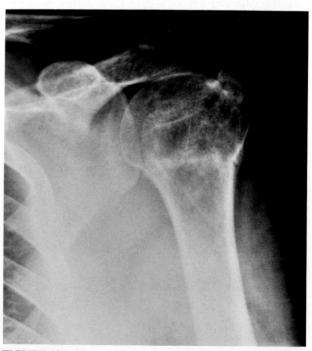

FIGURE 15F–17
Malunited fracture of greater tuberosity. This patient has restricted motion and pain due to impingement of the malunited greater tuberosity on the coracoacromial arch.

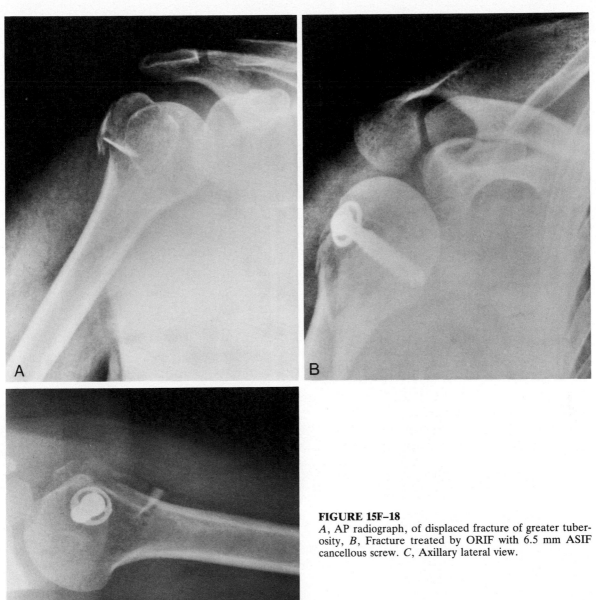

FIGURE 15F–18
A, AP radiograph, of displaced fracture of greater tuberosity, *B*, Fracture treated by ORIF with 6.5 mm ASIF cancellous screw. *C*, Axillary lateral view.

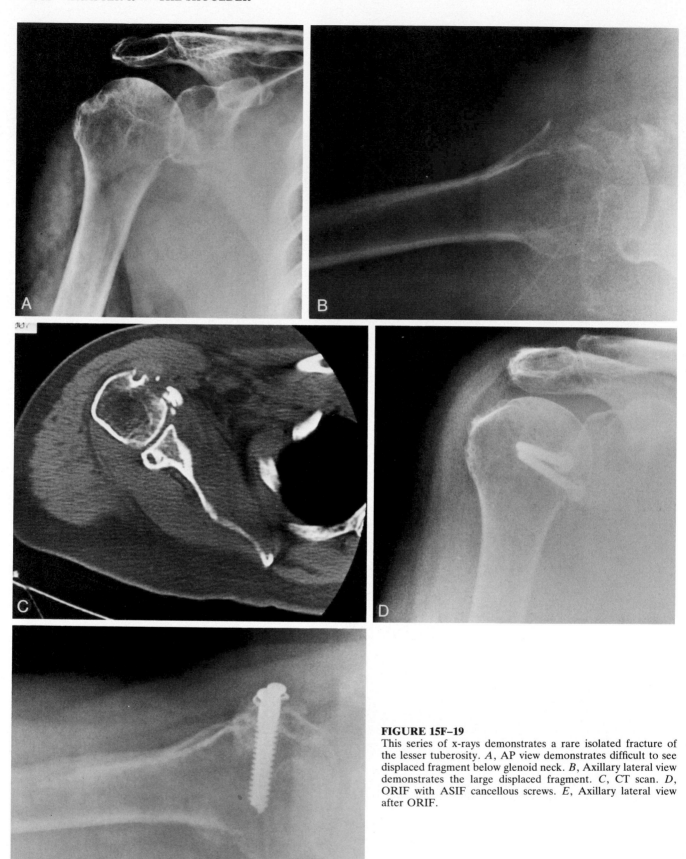

FIGURE 15F–19

This series of x-rays demonstrates a rare isolated fracture of the lesser tuberosity. *A*, AP view demonstrates difficult to see displaced fragment below glenoid neck. *B*, Axillary lateral view demonstrates the large displaced fragment. *C*, CT scan. *D*, ORIF with ASIF cancellous screws. *E*, Axillary lateral view after ORIF.

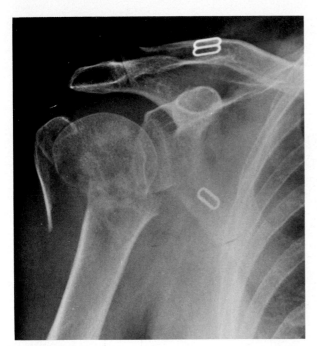

FIGURE 15F–20
Three-part fracture of the proximal humerus with displaced greater tuberosity and surgical neck.

present along with a displaced fracture at the level of the surgical neck. If the lesser tuberosity is fractured and displaced, the greater tuberosity and attached rotator cuff pull the head fragment into external rotation with the articular surface facing anteriorly. If the greater tuberosity is fractured and displaced, then the lesser tuberosity and attached subscapularis pull the head into marked internal rotation with the articular surface facing posteriorly. This rotational displacement of the articular fragment by an intact tuberosity-cuff unit is the reason these fractures are often termed "rotary subluxations." The shaft fragment in these fractures is usually displaced and pulled anteriorly and medially by the pectoralis major. The head fragment remains attached to one tuberosity fragment-rotator cuff until and therefore maintains an adequate blood supply in most cases. Despite the residual soft tissue attachment, avascular necrosis can still be a significant complication of this injury.

Theoretically, closed reduction of a three-part fracture could be accomplished by correction of the rotary subluxation, reapproximation of the displaced tuberosity, and reduction of the shaft fragment. Several factors, however, contribute to the difficulty of obtaining these goals with closed reduction. The position of the tuberosity fragments beneath the acromion as well as the difficulty of stabilizing the proximal head-tuberosity fragment make closed reduction difficult. The deforming muscular forces in three-part fractures including the rotator cuff, deltoid, and pectoralis major are considerable and contribute to the difficulty of closed reduction. Interposition of soft tissues including the long head of the biceps, the rotator cuff, and the periosteum can also cause problems in accomplishing these goals by closed reduction. Manual closed reduction, the hanging arm cast, and skeletal traction have all been used to treat three-part fractures, but most authors report poor results with closed treatment [15, 24, 25, 52, 105]. A high incidence of nonunion and malunion resulting in pain, limitation of motion, and function has been cited.

Open reduction and internal fixation appear to be the treatment of choice for displaced three-part fractures of the proximal humerus, although results are only fair at best [25, 106, 200]. The deltopectoral approach is utilized, and soft tissue dissection is limited to avoid further compromise of blood supply to the bony fragments. The goal of internal fixation is to reduce the rotary subluxation by securing the displaced tuberosity fragment to the head and then to attach the shaft fragment to the head-tuberosity fragment.

Two major types of internal fixation have been described for use in three-part fractures: a combination of wire sutures and intramedullary rods, and the AO buttress plate technique (Fig. 15F–21). Many authors have described the use of wire or heavy nonabsorbable sutures with or without Kirschner wires, Steinman pins, or Rush rods [57]. Neer and associates and Sturzenegger and co-workers have reported good results with internal fixation of three-part fractures utilizing wires and nonabsorbable sutures [141, 136, 189]. In 1986 Hawkins and colleagues reported good results in 14 of 15 patients with three-part fractures using the figure-of-eight wire technique [69]. Two patients later developed avascular necrosis, and one required a humeral head prosthesis, which reinforces the need for careful dissection of soft tissue in these cases. Hagg and Lundberg reported an incidence of avascular necrosis as high as 25% in a series of these fractures treated by open reduction and internal fixation [62].

The use of the AO buttress plate technique for three-part fractures has also been reported [96, 148, 200]. This method has been successful in younger patients with good bone stock and limited comminution. The advantage of this technique is the ability to achieve rigid internal fixation so that early range of motion exercises can be started to help prevent stiffness as a long-term complication. However, several authors including Savoie and associates, Kristiansen and Christensen, and Sturzenegger and colleagues have reported a substantial complication

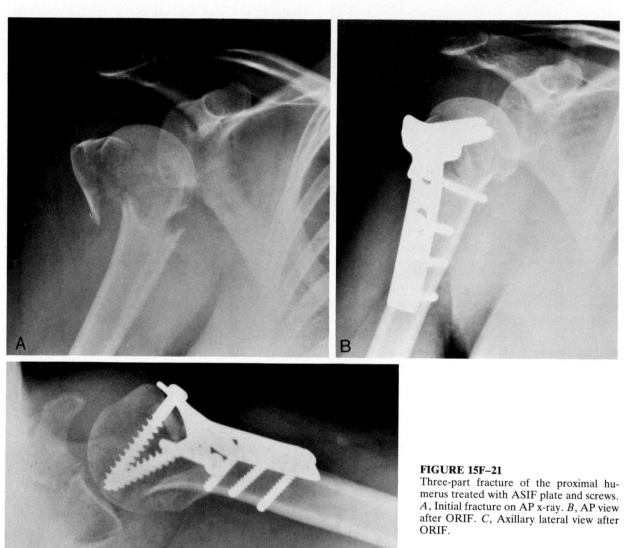

FIGURE 15F–21
Three-part fracture of the proximal humerus treated with ASIF plate and screws. *A*, Initial fracture on AP x-ray. *B*, AP view after ORIF. *C*, Axillary lateral view after ORIF.

rate using this technique, including avascular necrosis, plate impingement requiring removal, and failure of fixation in osteoporotic bone [96, 148, 174].

Four-Part Fractures

In this fracture there is displacement of both tuberosities and a fracture at the level of the anatomic neck. The articular fragment is often displaced laterally and may be completely devoid of soft tissue attachment (Fig. 15F–22). In addition to the severe bony injury, damage to the soft tissue capsule and rotator cuff is present plus a higher incidence of injury to the brachial plexus and axillary vessels. The complications of treatment are similar to those listed for three-part fractures, but there is a higher incidence of avascular necrosis and malunion. Rates of, avascular necrosis of between 10% and 34% have been reported [62, 96]. This is a very difficult fracture to treat, but thankfully it is uncommon in the younger athletic population.

Treatment by closed methods has been unsuccessful as reported by most authors [25, 37, 52, 153]. There is significant difficulty in achieving satisfactory closed reduction, and even greater difficulty in achieving stability when reduction is accomplished. Open reduction with internal fixation has also yielded only limited success [25, 96, 148, 181].

Difficulty in achieving stable internal fixation, a relatively high incidence of avascular necrosis, and residual stiffness due to the severity of the soft tissue injury have all contributed to the high failure rate. Neer and others have reported some limited success with suture techniques but recommend prosthetic hemiarthroplasty in most cases [4, 162]. Mouradian described the use of a modified Zickel supracondylar device with good results in seven cases [132]. Kristiansen and Christensen reported 32 fractures treated with the AO buttress plate technique [96]. They found that they could achieve acceptable reduction in the majority of cases, but at follow-up a 50% unsatisfactory rate was noted as a result of failure of fixation, plate impingement, avascular necrosis, and infection.

Because of the high incidence of failure with either closed treatment or open reduction with internal fixation, many authors recommend primary hemiarthroplasty for treatment of four-part fractures of the proximal humerus (Fig. 15F–23). In the older patient population in which this fracture most commonly occurs, this may be the best alternative. However, in the younger athletic population the functional limitations imposed by a prosthesis in most instances would preclude the athlete's return to competitive sports [80, 94, 140]. Therefore, four-part fractures of the proximal humerus have a very poor prognosis if they occur in the competitive athlete.

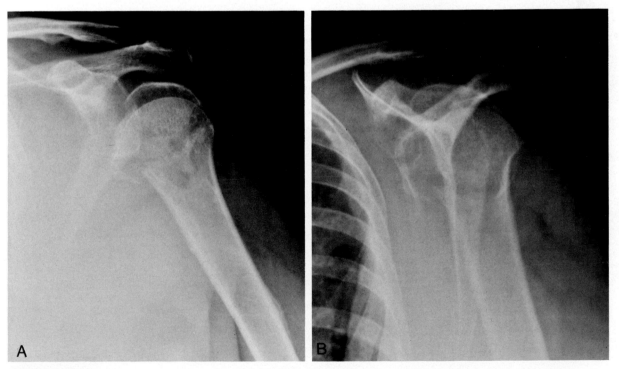

FIGURE 15F–22
Four-part fracture. *A*, AP x-ray demonstrates displacement of both tuberosities and humeral head. *B*, Scapular lateral view.

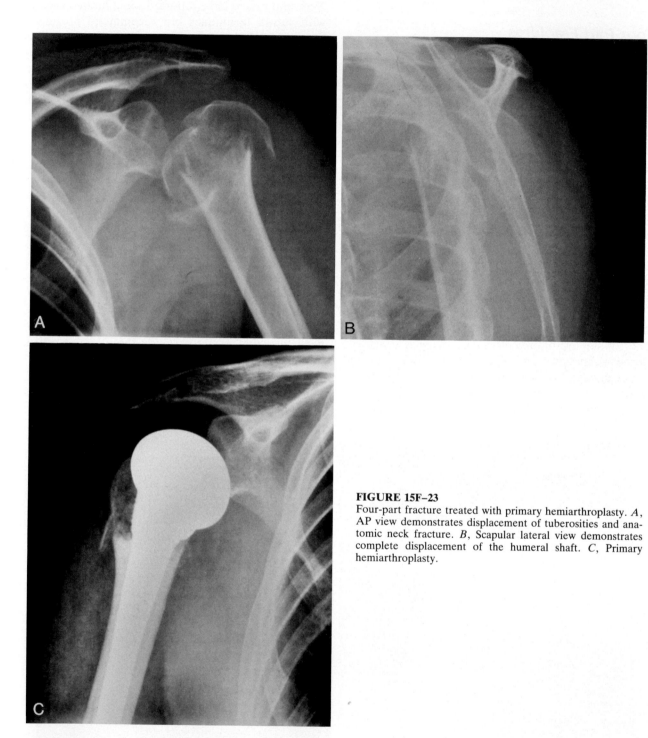

FIGURE 15F–23
Four-part fracture treated with primary hemiarthroplasty. *A*, AP view demonstrates displacement of tuberosities and anatomic neck fracture. *B*, Scapular lateral view demonstrates complete displacement of the humeral shaft. *C*, Primary hemiarthroplasty.

Fracture-Dislocations

These injuries combine fracture of one or more of the four segments with dislocation of the articular fragment out of the glenoid fossa either anteriorly or posteriorly [35–38, 46, 52, 108, 116, 123–125, 128, 133, 138, 155, 178]. Before treatment, adequate radiographs are required to determine the number of fragments involved as well as the direction of the dislocation. These injuries are often associated with severe soft tissue trauma associated with a higher incidence of nerve and vascular injuries. Intratho-

racic dislocation of the proximal humerus has also been reported [55, 65, 151, 204].

Two-Part Fracture-Dislocations. Two-part fracture-dislocations most commonly involve fractures of the greater tuberosity with anterior dislocations and fractures of the lesser tuberosity with posterior dislocations. The greater tuberosity is fractured in as many as 15% of all anterior dislocations (Fig. 15F–24) [5, 56, 146]. In most cases, reduction of the dislocation reduces the greater tuberosity fragment. If this occurs, immobilization for 3 to 4 weeks followed by appropriate rehabilitation usually pro-

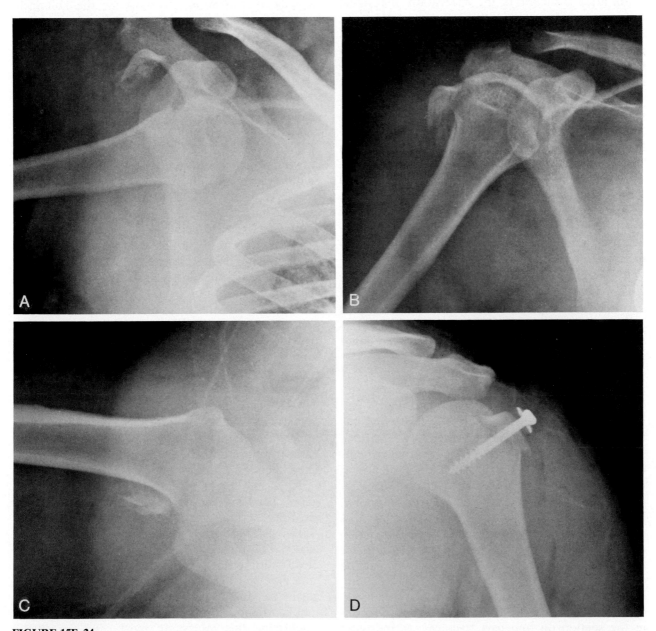

FIGURE 15F–24
A, AP radiograph of anterior fracture-dislocation of the shoulder with displaced fracture of greater tuberosity. *B,* Postreduction AP view demonstrates residual displacement of greater tuberosity. *C,* Axillary lateral view shows the posterior displacement of the greater tuberosity fragment. *D,* ORIF with ASIF screw.

vides a good result with a very low incidence of recurrent dislocation [11, 121]. In patients with greater tuberosity fractures with residual displacement after reduction of the dislocation, the same rules as were described in the section on isolated fractures of the greater tuberosity are applied. Residual displacement should be treated by open reduction with internal fixation or excision of a small fragment and cuff reattachment.

Treatment of lesser tuberosity fractures associated with posterior dislocation is similar to that described for greater tuberosity fractures. Usually the lesser tuberosity fragment demonstrates an adequate position after reduction of the dislocation. If residual displacement is noted, either open reduction and internal fixation or excision of the fragment and reattachment of the subscapularis tendon is performed. In patients with a lesser tuberosity fracture and a large reverse Hill-Sachs lesion, the fragment with the attached subscapularis tendon can be advanced into the defect as described by Neer [8, 141].

Three-Part Fracture-Dislocations. These particular fracture-dislocations represent significant trauma with the additional soft tissue compromise of a dislocated humeral articular surface and a displaced tuberosity fracture [117, 193]. Because of the lack of attachment between the shaft and the humeral head fragment, closed reduction is extremely difficult or impossible. Open reduction with internal fixation is the treatment of choice as in the three-part fracture without dislocation. The risk of avascular necrosis is higher because the degree of soft tissue injury is greater. The methods of internal fixation described in the section on three-part fractures are also utilized in the care of three-part fracture-dislocations.

Four-Part Fracture-Dislocations. This injury represents a major bony and soft tissue injury, with complete disruption of the capsule and rotator cuff because all four segments are displaced and the articular fragment is dislocated (Fig. 15F–25). Blood supply to the articular head fragment is virtually always disrupted. There is a higher incidence of neurovascular injuries with this type of fracture-dislocation if the segments are displaced anteriorly. Late complications regardless of the type of treatment are not uncommon and include nonunion, malunion, stiffness, myositis, and avascular necrosis with collapse. Open reduction with internal fixation can be considered for the young athletic patient, but the risk of problems is high. The favored methods of internal fixation have been described earlier in the section on four-part fractures. In the older patient, primary hemiarthroplasty is the treatment of choice [4].

Articular Surface Fractures

Fractures of the articular surface of the humerus may be associated with a dislocation as an *impression*-type fracture or by direct impaction of the humeral head on the glenoid as a *head-splitting* or *indentation* fracture. Impression fractures, called the Hill-Sachs lesion on the posterior aspect of the humeral head, are commonly found in anterior shoulder dislocations (Fig. 15F–26). They are usually small, the size and depth of the defect correlating with the severity of the original trauma and the number of recurrent dislocations. In most cases no specific treatment is necessary for the Hill-Sachs lesion. In the rare severe head defect that involves enough of the articular surface to be a cause of recurrent dislocation, a rotational osteotomy through the surgical neck with plate fixation may be used [183].

Impression fractures in cases of posterior dislocation can be quite large and carry a poorer prognosis. Often significant trauma is associated with a locked posterior dislocation. Delay in diagnosis is not uncommon and contributes to the size of the anterior impression fracture or "reverse" Hill-Sachs lesion (Fig. 15F–27). The amount of the articular surface involved should be assessed thoroughly with a trauma series of radiographs and in most cases by CT scans because of the treatment implications. Anterior impression fractures involving less than 20% of the articular surface can generally be treated by closed reduction of the posterior dislocation followed by a 3- to 4-week period of immobilization with the arm externally rotated. For fractures involving 20% to 30% of the articular surface, transfer of the subscapularis tendon (McLaughlin procedure) or the lesser tuberosity-subscapularis complex (Neer modification) into the defect has been successful [141]. When the size of the defect reaches 40% of the articular surface or more, poor results have been reported from conservative treatment or subscapularis transfer, and therefore most authors recommend hemiarthroplasty in this situation [4, 9, 191].

Head-splitting or indentation fractures of the humeral head are caused by direct impaction of the articular surface against the glenoid cavity (Fig. 15F–28). These fractures are most commonly associated with three- or four-part fractures. However, head-splitting fractures do occur as isolated injuries and may be difficult to detect on plain films. Once again, it should be emphasized that the trauma series of plain radiographs and CT scan are very useful in confirming this injury. In most cases, prosthetic replacement is recommended. An argument may be made for attempting open reduction and internal

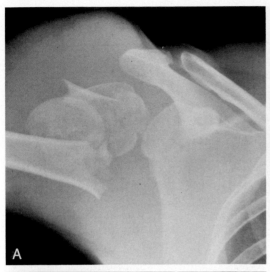

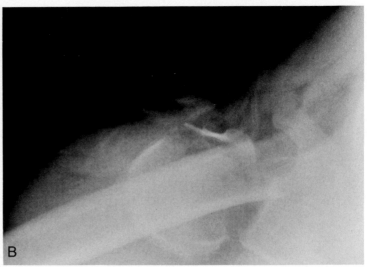

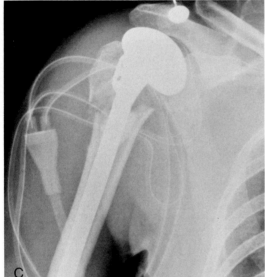

FIGURE 15F–25

Four-part fracture-dislocation of the shoulder. *A*, Note that the articular surface of the humeral head is no longer in continuity with the glenoid. *B*, Axillary view shows the humeral head rotated 180 degrees. *C*, Primary hemiarthroplasty.

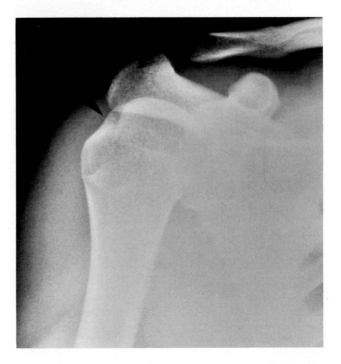

FIGURE 15F–26

AP radiograph of the shoulder with indentation fracture of the posterolateral humeral head after anterior dislocation—the so-called Hill-Sachs lesion.

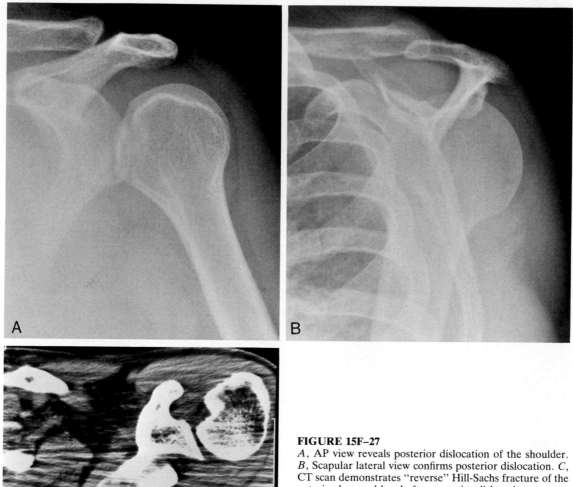

FIGURE 15F–27
A, AP view reveals posterior dislocation of the shoulder. *B*, Scapular lateral view confirms posterior dislocation. *C*, CT scan demonstrates "reverse" Hill-Sachs fracture of the anterior humeral head after posterior dislocation.

fixation in the younger athlete, but there is nothing in the literature that addresses this subject [93].

Complications

Complications in fractures of the proximal humerus are not uncommon. The more severe the initial displacement and associated soft tissue injury, the greater the risk of associated problems. These problems include stiffness, nonunion or malunion, avascular necrosis, myositis ossificans, and neurovascular injury [16, 26, 27, 31, 84]. The same problems may occur in proximal humeral fractures whether treated by open or closed techniques. Surgical treatment adds the risks of infection, hardware failure, and need for hardware removal [27, 174, 185].

Stiffness. Limitation of motion after fracture of the proximal humerus is common. Soft tissue trauma associated with these fractures leads to scarring of the capsule and subacromial region with adhesions involving the rotator cuff and bursa. As a general rule, the more severe the initial displacement of the fracture fragments, the more likely that permanent limitation of motion will result. However, even minimally displaced fractures can be complicated by residual stiffness if motion exercises are not initiated as early as possible [164]. Many patients will continue to improve for 12 to 18 months after this injury. Vigorous physical therapy is the treatment of choice for residual stiffness. Only on rare occasion

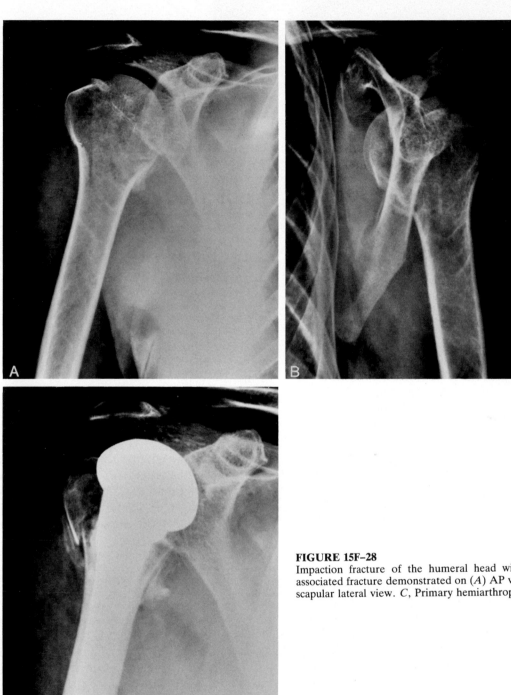

FIGURE 15F–28
Impaction fracture of the humeral head without other associated fracture demonstrated on (*A*) AP view and (*B*) scapular lateral view. *C*, Primary hemiarthroplasty.

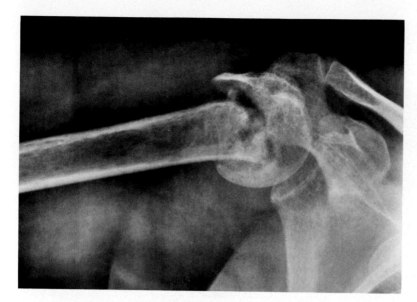

FIGURE 15F–29
Nonunion of humeral neck fracture treated primarily with a hanging arm cast.

does manipulation or open release become necessary.

Nonunion and Malunion. Nonunion has generally been associated with displaced two-part unstable surgical neck fractures; however, it can occur with any type of proximal humeral fracture (Fig. 15F–29). These two-part unstable fractures can result in nonunion because of residual instability, interposition of soft tissues, use of the hanging arm cast, insufficient immobilization, inappropriate physiotherapy, or poor bone quality [10, 26, 63, 101, 166, 176, 184]. Treatment of painful nonunion is accomplished by open reduction and internal fixation with iliac crest bone grafting.

Malunion due to insufficient or inadequate treatment of displaced fractures can occur. Although the shoulder tolerates a considerable amount of deformity without compromise of function owing to its large range of motion, this can become problematic in the athlete who demands maximum motion for his sport. Excessive anterior angulation of surgical neck fractures can restrict forward elevation, and retraction of greater or lesser tuberosity fragments certainly can restrict motion. Greater tuberosity displacement can lead to impingement of the supraspinatus and infraspinatus on the coracoacromial arch, both blocking motion and leading to impingement symptoms. Lesser tuberosity displacement can result in impingement against the coracoid and glenoid with limitation of internal rotation. In three-part and four-part fractures, severe deformity can result in abnormal biomechanics around the joint owing to impingement of displaced tuberosities as well as the varus position of the humeral head. These are very difficult problems to correct with

reconstruction because of fixed malunions of the tuberosities, severe scarring of the soft tissues, and bone loss. Greater tuberosity malunions can be addressed by osteotomy with mobilization of the attached rotator cuff followed by distal advancement and internal fixation or excision and reattachment of the cuff (Fig. 15F–30).

Avascular Necrosis. Avascular necrosis has been well described in proximal humeral fractures and may result from fracture in any classification (Fig. 15F–31) [48, 102]. It is more common after four-part fractures and after three- and four-part fracture-dislocations. The avascular necrosis rate reported by Hagg and Lundberg after closed reduction of three- and four-part fractures is between 3% and 14% [62]. The rate of avascular necrosis in three- and four-part fracture-dislocations may be as high as 60% to 70%. Extensive exposure by stripping of soft tissue during open reduction with internal fixation has been identified as a major contributing factor in avascular necrosis. Sturzenegger and associates reported a 34% incidence in fractures treated with the AO buttress T-plate [189]. Avascular necrosis can occur despite anatomic reduction and internal fixation, but for most athletic individuals this should be attempted despite the risk.

Neurovascular Problems. Vascular complications following proximal humeral fractures are not common but do occur. When they do occur, they are usually serious. Axillary artery injuries are the most common vascular injury secondary to fractures of the proximal humerus and account for approximately 6% of all arterial trauma from all fractures [39, 44, 71, 73, 109, 163, 180, 182, 192, 212]. The risk is increased in older patients because of arterio-

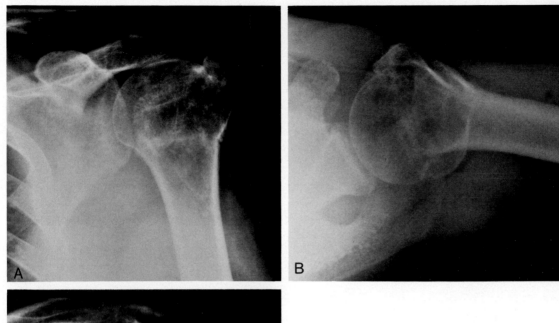

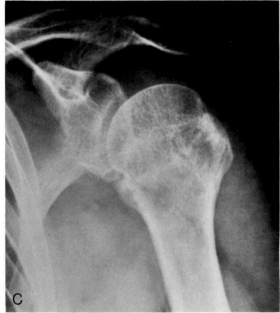

FIGURE 15F–30
Malunited greater tuberosity fracture treated with osteotomy and advancement of the rotator cuff. *A*, AP view demonstrates superior position. *B*, Axillary view. *C*, After excision of the fragment, the rotator cuff is advanced into the defect and anchored with heavy suture.

sclerosis. The most common site of injury is proximal to the take-off of the humeral circumflex arteries (see Fig. 15F–4). The key to successful treatment is early diagnosis and repair. It is important to check the radial pulse in the injured extremity. Other signs of vascular trauma include expanding hematoma, pallor, and paresthesias. If unrecognized, complications are catastrophic including severe infection, possible amputation, and compressive neuropathy due to compartment syndrome.

Brachial plexus injuries occur following fractures of the proximal humerus [16, 141] Stableforth reported an incidence of 6.1% following displaced proximal humeral fractures [185]. Isolated axillary nerve injury is the most common; however, stretch injuries to the brachial plexus of all types have also been described. Injury to the suprascapular and musculocutaneous nerves also occur in an isolated fashion [47]. Clinical testing of the neurologic status is important. The prognosis in most cases is good because these injuries usually represent a neurapraxia. Electromyography (EMG) and nerve conduction velocity studies combined with clinical examination can be utilized to follow the progress of the injury.

Myositis Ossificans. Myositis ossificans can occur after fracture of the proximal humerus. Because the degree of myositis can usually be correlated with the severity of the soft tissue injury, fracture-dislocations have the highest incidence [141]. The het-

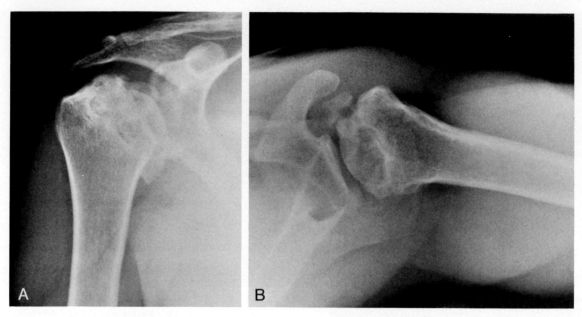

FIGURE 15F–31
Avascular necrosis of the humeral head. *A*, AP view. *B*, Axillary later view.

erotopic calcification is usually not severe, however, and rarely forms a block to motion, although many of these patients have decreased motion owing to soft tissue contracture.

Author's Preferred Method Of Treatment

Minimally Displaced Fractures. Treatment with a sling, ice, and analgesics is instituted initially for comfort. Patients with these minimally displaced fractures are started on pendulum exercises and isometrics early in the first week. When it is determined from follow-up radiographs over the next 10 to 14 days that the fracture is stable, a more aggressive rehabilitation program follows. Healing usually requires 6 to 8 weeks with an additional 4 to 6 weeks required to restore function.

Two-Part Fractures

Two-Part Surgical Neck Fractures. Two-part surgical neck fractures that are impacted but angulated more than 45 degrees should be treated by closed reduction. In the throwing athlete, anterior angulation of more than 15 to 20 degrees will restrict shoulder flexion an unacceptable amount. Reduction is accomplished by applying longitudinal traction to the arm and posterior pressure to the apex of the angulation. In most cases the fracture will be stable after reduction. If instability does exist, percutaneous pinning is carried out under fluoroscopic control with two parallel threaded pins. These pins

are removed after 3 to 4 weeks, and vigorous rehabilitation is begun.

When a two-part surgical neck fracture is displaced, the shaft is pulled anteriorly and medially by the pectoralis major. The proximal head and tuberosity fragment remain within the glenohumeral joint in a neutral or slightly abducted position. An attempt at closed reduction is usually warranted. This reduction is achieved under general anesthesia with fluoroscopic control by applying longitudinal traction to the arm followed by gentle flexion and adduction to relax the pectoralis major. If reduction can be accomplished, gentle impaction is performed, and the arm is immobilized with a sling and swathe as with a nondisplaced fracture. Isometrics are started immediately. Close follow-up radiographs are needed to monitor progress and guard against late displacement. If reduction can be accomplished but the fracture is seen to be unstable, percutaneous pinning is performed using two parallel threaded Steinman pins passed from the distal to the proximal fragment. Additional fixation with a third pin can be accomplished beginning at the greater tuberosity with the pin extending into the distal shaft medially. These pins can be cut short beneath the skin and will remain in place for 3 to 4 weeks. After pin removal, vigorous rehabilitation is begun.

If closed reduction is not possible, open reduction with internal fixation is required. Soft tissue interposition is often found and is removed. I prefer a deltopectoral incision sparing the cephalic vein. Care is taken to palpate and protect the axillary nerve. If the fracture is not comminuted and the bone stock is good, threaded Steinman pins can be used in a fashion similar to that previously described for the percutaneous technique. If mild comminution is present, I prefer to use two Rush rods inserted at the level of the greater tuberosity followed by a tension band technique using a figure-of-eight heavy suture such as No. 5 Ethibond or 1-mm Dacron tape (Fig. 15F–32). This nonabsorbable suture is passed deep to the supraspinatus tendon proximally and through a drill hole an equal distance down the shaft beyond the fracture. Excellent stability can usually be achieved, allowing early range of motion exercise and isometrics. The younger athletic patient demands closer attention to detail in terms of alignment.

Markedly comminuted two-part surgical neck fractures should be treated nonoperatively. These fractures can initially be treated with closed reduction, and if acceptable reduction can be obtained then, sling and swathe immobilization followed by isometrics and early motion is indicated. If excessive shortening or recurrent angulation occurs, skeletal traction through an olecranon pin can be used to maintain length until early callus is present. Open reduction with internal fixation is hazardous in patients with significant comminution.

Two-Part Anatomic Neck Fractures. If this rare fracture is significantly displaced, it should be treated with open reduction and internal fixation in the young athletic patient. Stable anatomic reduction is the goal to allow early rehabilitation to regain motion. Open reduction is achieved through a deltopectoral incision, taking great care to minimize dissection and avoid further vascular compromise to the humeral head. Threaded lag screws or threaded Steinman pins can be chosen for internal fixation depending on the amount of bone remaining within the head fragment. Avascular necrosis is a significant risk in this particular fracture and should be discussed early with the patient. It is hoped that early restoration of anatomy will allow revascularization before collapse and head destruction occur. The prognosis for return to sports demanding heavy use of the affected shoulder is guarded.

Two-Part Greater Tuberosity Fractures. In two-part greater tuberosity fractures that are displaced more than 1 cm or angulated more than 45 degrees, open reduction is indicated to prevent malunion. A superior surgical approach is preferred. A 3- to 4-cm incision is made in Langer's lines overlying the lateral border of the acromion. The deltoid muscle is split in line with its fibers for a short distance, and if additional exposure is required the deltoid can be split off the lateral acromion. The fracture is

identified along with the concomitant split in the rotator interval. The fracture can be stabilized in one of two ways. If good bone stock is present, bicortical screw fixation using 6.5-mm AO screws is utilized. If the fragment of bone is small, repair with heavy suture such as No. 5 Ethibond or 1-mm Dacron tape or excision of the fragment with repair of the cuff into the defect is effective (Fig. 15F–33). The split in the rotator cuff is always repaired as well. Early motion is begun in a controlled fashion, but resistance exercises are delayed 6 to 8 weeks to allow adequate healing. Complete rehabilitation and return to heavy sport may require 4 to 6 months.

Two-Part Lesser Tuberosity Fractures. These fractures are extremely rare as an isolated injury. Most fractures of the lesser tuberosity are associated with posterior dislocations of the shoulder. Recognition of the fracture-dislocation with adequate axillary radiographs or CT scan is important. In cases associated with posterior dislocation, closed reduction is indicated, and in most cases adequate reduction of the fragment is obtained. A short period of immobilization is then required with the arm in the neutral position or slight internal rotation.

In the event of isolated displaced fractures of the lesser tuberosity, the size of the fragment is important. Small fragments that are not excessively displaced can usually be treated with immobilization and early range of motion exercises. If the fragment is large and could potentially block internal rotation, or if it involves a portion of the articular surface, then open reduction with internal fixation is indicated. This can be accomplished either by screw fixation, in the case of a large fragment, or by using nonabsorbable sutures for smaller fragments. If a displaced lesser tuberosity fragment heals with residual pain or restriction of internal rotation, late excision of the fragment is indicated.

Three-Part Fractures. Three-part fractures are usually unstable and are not amenable to closed reduction in the athlete. Open reduction with internal fixation is the treatment of choice in these instances. I prefer a surgical approach through the deltopectoral interval. In the younger athletic patient with good bone stock, I prefer rigid internal fixation using an AO buttress plate with heavy suture to reinforce the displaced tuberosity (Fig. 15F–34). Rigid internal fixation allows early mobilization in an attempt to maximize return of function. In most instances, plate removal will be required after healing and rehabilitation have been accomplished. Attention to detail in placing the plate as distal on the greater tuberosity as possible will decrease the need for plate removal.

In patients with comminution or osteoporotic

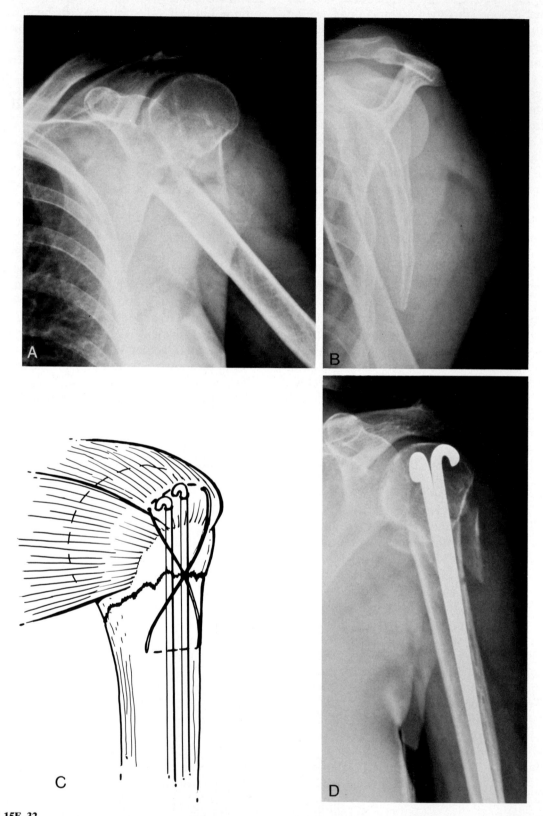

FIGURE 15F–32
Completely displaced surgical neck of humerus demonstrated on *(A)* AP view and *(B)* scapular lateral view. *C,* Schematic illustration of technique using rush rods and figure-of-eight suture beneath the cuff tendons proximally and through a drill hole distally. *D,* X-ray view of the same technique.

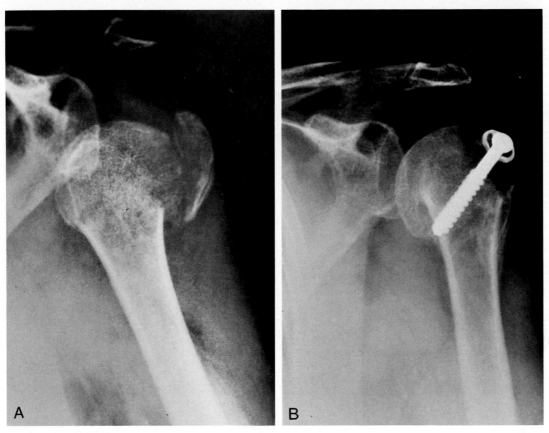

FIGURE 15F–33
A, Injury film shows displaced fracture of the greater tuberosity. *B*, ORIF of greater tuberosity fracture.

bone, I prefer a figure-of-eight wire suture technique. This technique is very successful in maintaining reduction, but early rehabilitation must be protected because fixation is not rigid.

Four-Part Fractures. Four-part fractures of the proximal humerus in the younger patient have a very poor prognosis. The risks of avascular necrosis, malunion, stiffness, and residual pain are substantial. In the younger athletic patient, however, an attempt at open reduction with internal fixation is preferred over primary prosthetic replacement. In the older patient, primary Neer hemiarthroplasty is indicated and provides a more consistent result. Open reduction with internal fixation is performed through an extended deltopectoral approach that avoids detachment of the deltoid muscle. Extra care in dissection should be used to avoid any further stripping of the soft tissues. In patients with adequate bone stock, an AO buttress plate is preferred, but in other patients multiple pins and wires can be utilized. Early motion is achieved through a vigorous rehabilitation program. Diligent follow-up is necessary to discover and prevent avascular necrosis.

Fracture-Dislocations

Two-Part Fracture-Dislocation. Two-part fractures of the greater tuberosity are often associated with anterior dislocation and should be treated by closed reduction of the dislocation. In the majority of cases the greater tuberosity fragment is adequately reduced, the routine nonoperative treatment is then followed. If residual displacement is noted on postreduction radiographs, open reduction with internal fixation of the greater tuberosity is performed. I prefer one or two 6.5-mm AO cancellous screws in patients with large fragments and good bone. In patients with smaller fragments or osteoporotic bone, heavy 1-mm Dacron tape is passed through drill holes to exit laterally. Early rehabilitation is started after open reduction and internal fixation but is delayed 4 weeks if closed reduction only is performed. The prognosis for return to athletics is good.

Posterior dislocations associated with fractures of the lesser tuberosity are treated initially by closed reduction of the dislocation as well. If the lesser tuberosity fragment remains displaced, open reduction and internal fixation with AO screws is performed. If a large reverse Hill-Sachs lesion is present, the lesser tuberosity is transferred into the defect.

Three-Part Fracture-Dislocations. In the athletic individual, this type of fracture-dislocation is best

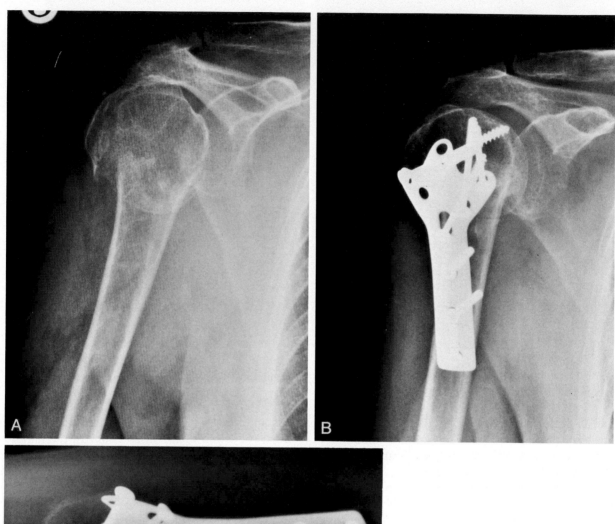

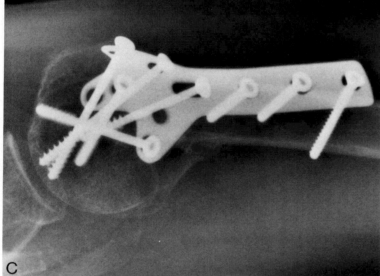

FIGURE 15F–34
Three-part fracture treated with AO-buttress plating. *A*, AP view of injury. *B*, AP view after ORIF. *C*, Axillary lateral view demonstrates reduction.

treated by open reduction and internal fixation. I prefer the extended deltopectoral approach whether the dislocation is anterior or posterior. A combination of 1-mm Dacron tape and the AO buttress plate is favored in athletes with good bone. The tape is used to reattach the displaced tuberosity to the head-tuberosity fragment before application of the buttress plate that fixes the shaft to the head fragment. In most cases plate removal will be necessary, but the advantage of early motion after stable internal fixation outweighs the risks of plate removal.

Four-Part Fracture-Dislocations. Despite the very high incidence of complications with internal fixation of this injury, I still feel that an attempt at open reduction and internal fixation is warranted in the young athletic patient. The technique is described in the section Three-Part Fracture-Dislocations.

Articular Surface Fractures. Impression-type fractures associated with anterior dislocation of the shoulder (Hill-Sachs lesion) are essentially never addressed directly. The reverse Hill-Sachs lesion that occurs in posterior dislocations can be more significant. If this fracture represents 30% or more of the articular surface by CT scan, I feel that the subscapularis tendon and attached lesser tuberosity should be transferred into the defect and securely fixed with AO screw techniques.

Indentation or head-splitting fractures are rare in the young athlete. If this fracture is encountered, an attempt at open reduction and internal fixation is indicated. A high rate of avascular necrosis and late arthritic changes may be expected. However, in the athletic population, primary prosthetic replacement is a less desirable initial treatment due to its inherent limitations for return to athletics.

Postoperative Management and Rehabilitation

Intensive rehabilitation after proximal humeral fractures is essential to maximize the functional result after treatment [139]. Closer attention to detail must be used in the athletic patient to avoid the common complications of stiffness and decreased function associated with this group of injuries.

A four-phase program is utilized for rehabilitation. In the initial phase, passive and assisted range of motion exercises are performed. Isometrics are also begun early to maintain muscle tone. In the second phase, passive and assisted range of motion exercises are continued as active exercises are started. Early resistance exercises, both isometric and isotonic, are begun. The third phase is instituted after fracture healing is complete and moves toward advanced stretching exercises and advanced isotonic and isokinetic strengthening. The final fourth phase is activity-specific exercise, which attempts to return the athlete to his sport.

In the initial phase of rehabilitation, exercises are performed more frequently for a shorter duration. A common protocol consists of exercises performed three to four times a day for 15 to 20 minutes. The use of moist heat prior to exercise helps to relieve discomfort and make the soft tissues more supple. Analgesics are utilized for pain control as necessary. The first phase consists of pendulum exercises along with supine assisted elevation exercises. These can be accomplished with the use of either an overhead pulley mounted to a frame, a 3-foot stick, or the opposite hand. Isometrics are begun along with hand and elbow range of motion exercises early in this initial phase (Fig. 15F–35).

As the fracture begins to consolidate at the 2- or 3-week stage, the second phase of assisted exercises is added. The stick is utilized to perform supine external rotation stretching. The opposite hand is used to help in a combination elevation–external rotation exercise (butterfly). Extension exercises and internal rotation exercises are also performed utilizing a stick (Fig. 15F–36). Later in the same exercise phase, isometrics are continued, and light resistance exercises are performed utilizing a variable resistance Thera-Band in controlled arcs of motion (Fig. 15F–37).

The third phase of exercises includes more advanced stretching exercises such as wall stretch for both elevation and external rotation. Internal rotation stretching is performed using a towel (Fig. 15F–38). Isotonic exercises needed to strengthen the components of the rotator cuff and three components of the deltoid are performed utilizing a pulley and weight system. The trapezius and rhomboids are strengthened using a hand-held weight for shoulder shrugs and scapular retraction exercises. As healing progresses and strengthening advances, isokinetic strengthening can be utilized in both internal and external rotation as well as flexion-extension planes. Attention should be given to the fact that a weak rotator cuff can allow superior shear of the humeral head and therefore can propagate impingement. Care is taken to keep the strengthening exercises out of the impingement planes until adequate cuff strength has been regained (Fig. 15F–39).

The final phase of rehabilitation is sport-specific. Ballistic overhead activities in the throwing plane can be performed using surgical tubing or isokinetic

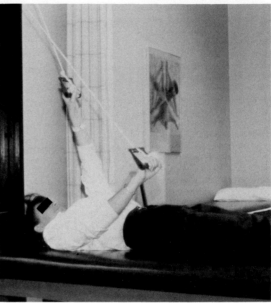

FIGURE 15F–35
Rehabilitation phase I. *A,* Pendulum exercise *B,* Supine elevation using overhead pulley.

equipment (Fig. 15F–40). Subcomponents of the sport-specific activity can be performed until the athlete completes the entire sport-specific motion in a pain-free fashion.

MUSCLE RUPTURES INVOLVING THE PROXIMAL HUMERAL REGION EXCLUDING THE ROTATOR CUFF

Injuries to the musculotendinous unit are quite common in sports. Fortunately, most cases represent only partial injury and result in little long-term disability for the athlete. However, even partial injuries or muscle strains can temporarily cause pain, loss of strength, and decrease in function leading to significant interference with athletic performance. Muscle strains and tendinitis probably represent the most frequent cause of missed playing time in competitive athletes.

Complete avulsion injuries to major tendons such as the Achilles, rotator cuff, distal triceps, pectoralis major, and distal biceps can lead to substantial functional disability. With complete disruption of a musculotendinous unit, significant alteration of joint biomechanics results. Because most complete tendinous avulsion injuries result in retraction of the tendon by muscle contraction and shortening, permanent dysfunction occurs unless appropriate anatomic repair is carried out.

Most significant muscle ruptures occur when an actively contracting muscle group is overloaded by the application of a load or extrinsic force that exceeds tissue tolerance [122]. In the athlete this overload can occur acutely with the application of a single macrotraumatic force, or it can occur chronically with the application of multiple submaximal forces at a rate that exceeds the body's ability to respond. Muscles and tendons clearly do respond to stress and applied demand by varying their strength and dimensions [201]. Tendons become thicker with repeated stresses, and their collagen fibers become more appropriately oriented for function. Repetitive stress can eventually damage tendon tissue if the rate of application of the stress is more rapid than the body's ability to respond. Healing of microscopic fiber failure occurs by the formation of scar and granulation tissue, which leads to an area within the tendon of altered mechanical properties. It is in this area that macrofailure will eventually occur.

In the normal state, it has been shown that tendon appears to be stronger than the muscle belly and the tendon-bone interface [122]. However, in the case of repetitive microtrauma to tendons, as often occurs in sports, this "normal" situation is modified and may lead more commonly to tendinous rupture. Both the rate of force application and the mechanism of injury affect the site of rupture [201].

An additional factor that can affect the normal state of muscle-tendon unit physiology and biomechanics is the use of anabolic steroids. In response to reports in recent years of frequent abuse of steroids by certain groups of athletes, much research has begun to better understand the changes that

Text continued on page 705

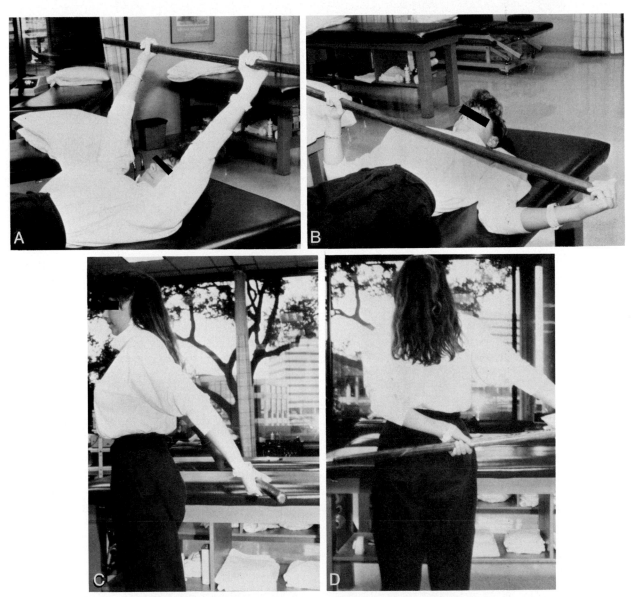

FIGURE 15F–36
Rehabilitation phase II. Range of motion assisted with a stick. *A*, Elevation. *B*, External rotation. *C*, Extension. *D*, Internal rotation.

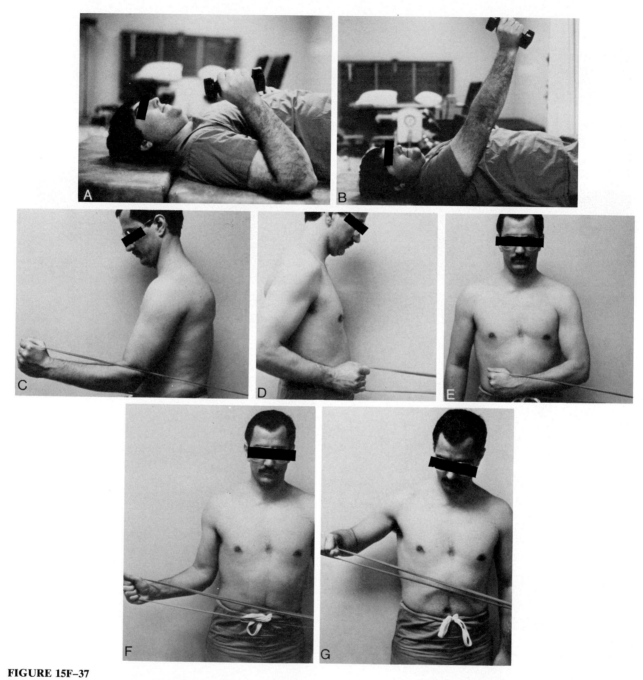

FIGURE 15F–37
Rehabilitation phase II. Strengthening. *A*, Supine deltoid press. *B*, Eccentric deltoid Thera-Band strengthening. *C*, Flexion. *D*, Extension. *E*, Internal rotation. *F*, External rotation. *G*, Abduction.

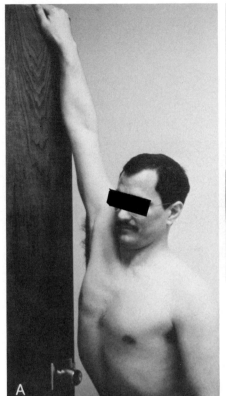

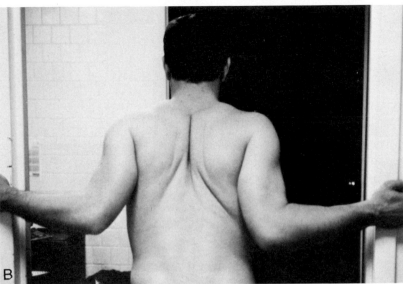

FIGURE 15F–38
Rehabilitation phase III. Range of motion, advanced. *A*, Flexion-abduction. *B*, External rotation. *C*, Internal rotation using a towel.

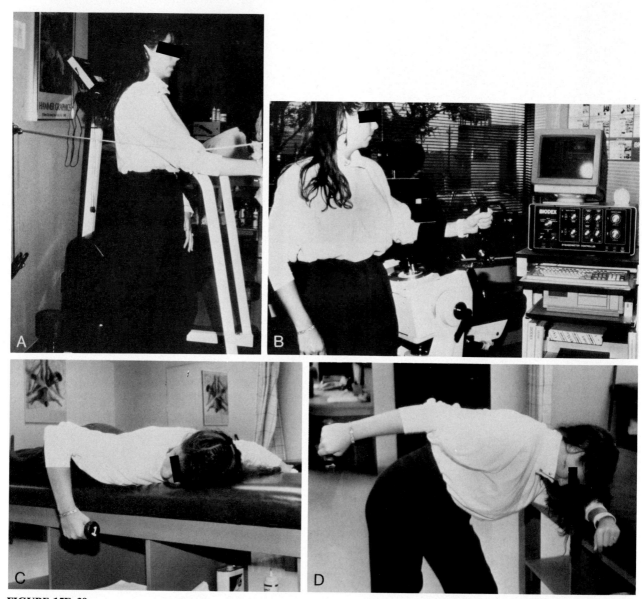

FIGURE 15F–39
Rehabilitation phase III. Strengthening. *A*, Pulley and weight isotonics. *B*, Isokinetics. *C*, Scapular retraction. *D*, Rhomboids and trapezius.

FIGURE 15F-40
Rehabilitation phase IV. Sport specific exercise—ballistics using Thera-Band in throwing position.

develop in the human system [67, 152]. We do understand the deleterious effects that occur as a direct result of steroid use in the treatment of diseases such as rheumatoid arthritis. It is safe to assume, despite a paucity of direct evidence in athletes, that abuse of anabolic steroids may increase the risk of injury to the muscle-tendon unit both by direct effects on structure and physiology and indirectly through changes in biomechanics [67].

Rupture of the Pectoralis Major

Rupture of the pectoralis major is a relatively rare injury. A review of the literature by McEntire and colleagues in 1972 revealed that only 45 cases had been reported prior to the 11 cases noted in their series [118]. Zeman and associates presented nine cases in 1979, and Kretzler and Richardson recently published a review of 19 cases in 1989 [95, 211]. Pectoralis major rupture, therefore, is a relatively rare injury with fewer than 90 reported cases in the literature to date [7, 13, 19, 20, 30, 41, 70, 88, 92, 118, 120, 112, 113, 149, 150, 157, 158, 195, 211].

Of special interest is the fact that most of the patients reported in the literature sustained rupture of the pectoralis major while involved in sports. Weight lifting was by far the most common activity, followed by rugby, wrestling, and other contact sports. This injury has been reported in all age groups, but patients in the third and fourth decades are affected predominantly. There has never been a case reported in a female [13, 30, 41, 95, 147, 175, 211]. These last two facts most likely reflect the fact that men in the early middle age groups are more commonly involved in contact sports and the traditionally greater involvement of men in weight lifting rather than a true sexual predisposition to this injury.

Anatomy. The pectoralis major muscle arises as a broad sheet of muscle from the midclavicle, sternum, ribs, and external oblique fascia (Fig. 15F-41). The muscle is often described as having two parts or "heads," an upper clavicular head and a lower sternocostal head. Muscle fibers, composed of a flat tendon about 5 cm broad, converge toward their insertion distal to the crest of the greater tuberosity of the humerus. The tendon consists of two laminae placed one in front of the other that commonly blend together inferiorly. The fibers from the clavicular head run in line to form the anterior lamina of the tendon. The more distal and deep fibers of the sternocostal head run upward and laterally to form the posterior lamina, those with the lowest origin having the highest insertion, giving a twisted appearance to the muscle [76]. Dissections performed by Kretzler and Richardson showed the tendon to be about 1 cm long on the anterior surface and perhaps 2.5 cm long on the deeper or posterior surface [95]. The tendon is quite thin and appears to be a coalescence of the anterior and posterior vesting fasciae rather than a true tendon or musculotendinous junction. Grossly, the pectoralis major

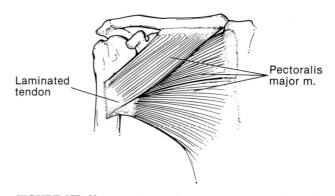

FIGURE 15F-41
This schematic diagram depicts the two "heads" of the pectoralis major muscle arising from a broad area on the chest, converging to attach as a flat laminated tendon into the proximal humerus.

forms the smooth, rounded appearance of the anterior axillary fold. The muscle is innervated by the medial and lateral pectoral nerves, which branch directly from the medial and lateral cords.

The primary function of the pectoralis major muscle is adduction and medial rotation of the humerus. The muscle can also function to flex the humerus if it is extended behind the plane of the body. With the arm at the side, the upper or clavicular portion of the muscle is most effective, but as the shoulder is abducted the lower portion of the muscle provides the bulk of the power [76].

Classification. Pectoralis major ruptures can be classified by the degree and location of the rupture. The degree of rupture can be termed either complete or incomplete. The majority of cases are undoubtedly partial or incomplete and represent strains of the muscle belly or musculotendinous junction. Most cases in the literature that have come to surgery involved complete ruptures [5, 13, 23, 41, 59, 95, 147, 175, 195].

Concerning the location of complete injuries, avulsion-type injuries involving the tendon at or near the humeral insertion predominate. Incomplete and complete ruptures at the musculotendinous junction or in the substance of the muscle belly have also been reported more commonly in association with a direct blow [13, 30, 158, 211].

Clinical Evaluation

History and Mechanism of Injury. Patients with an acute rupture of the pectoralis major present with a definite history of injury. The typical injury occurs when the patient is lifting weight or receives a direct blow in rugby or wrestling. The patient experiences an immediate, severe pain and a tearing sensation with burning at the site of injury. Some patients describe an audible pop or snap associated with a complete rupture of the tendon.

Two major mechanisms have been described in rupture of the pectoralis major muscle. The most common mechanism occurs when excessive force develops in the muscle as a result of indirect application of the force through the upper extremity. Weight lifting and, more specifically, the bench press are examples of sports producing this mechanism [95, 118, 211]. The other described mechanism occurs by direct trauma to the muscle as in wrestling, football, or other contact sports.

In the cases reported by most authors to date, the indirect mechanism occurs most commonly, and the most common activity associated with complete rupture was the bench press weight lift. This indirect mechanism is associated with the development of excessive muscle tension and usually results in an avulsion type of injury at or near the tendinous insertion into the humerus. An attempt to break a fall on an outstretched hand can also apply a severe indirect force to the muscle-tendon unit that can result in rupture.

Direct blows are also associated with injury to the pectoralis major, and many times these injuries are within the substance of the muscle belly. In wrestling there seems to be a propensity to disrupt the muscle at the sternoclavicular head by direct contact. McEntire and colleagues, in their review of the literature, proposed that some of the previously reported cases of pectoralis major rupture may actually represent congenital absence of the muscle [118].

Physical Examination. The patient with acute rupture of the pectoralis major presents with the affected extremity splinted across the chest and often supported by the opposite hand. Early swelling and ecchymosis occur across the chest and upper arm region. Pain with shoulder motion is present. Distal rupture is associated with asymmetry of the anterior axillary fold as the muscle retracts medially and superiorly. There is a prominent bulge involving the retracted muscle belly. A palpable defect is present at the site of injury but may be obscured initially by swelling. Weakness is present with attempted adduction and internal rotation of the arm. In the acute phase it can be difficult to distinguish between complete and partial rupture because of pain and swelling (Fig. 15F–42).

In chronic cases the balled-up, retracted muscle belly is very prominent, and asymmetry from side to side is noticeable. A defect in the anterior axillary fold gives a webbed appearance to the anterior

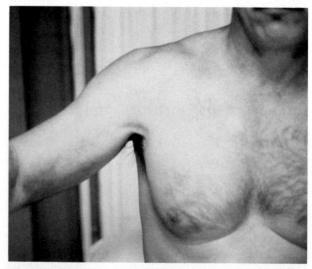

FIGURE 15F–42
Clinical photograph of a patient with an acute rupture of the pectoralis major tendon. Note the ecchymosis on the medial arm and the webbed appearance of the axilla.

axilla. The most reliable clinical examination finding is weakness in adduction and internal rotation. In chronic cases the patient often complains of a dull, aching pain with activity that requires heavy muscle power.

Radiographic Examination. Plain x-rays of the shoulder, chest, and scapula are normal. A loss of the normal pectoralis major shadow has been described but is very subtle [23, 95]. Magnetic resonance imaging (MRI), although not yet reported for this injury, could potentially demonstrate the expected asymmetry and the site and extent of disruption in the muscle.

Treatment Options. There is uniform agreement that partial ruptures of the pectoralis major tendon and incomplete lesions of the muscle belly itself respond to conservative treatment [7, 23, 92, 118]. These partial injuries are characterized by less swelling, ecchymosis, and pain than complete injuries. When adduction is resisted, no defect in the tendon is palpable. Initial treatment begins with ice, rest, and control of the hematoma followed by a program of progressive range of motion and strengthening exercises. Slow return to strength is the rule. These injuries usually heal without major deformity or significant strength deficit. They often require a 6- to 8-week course of recovery prior to return to stressful lifting activities.

Both surgical and nonsurgical treatment methods for complete rupture of the pectoralis major have been described [41, 92, 95, 149, 211]. Three major parameters can be assessed when describing results of treatment after this injury: pain, strength, and cosmetic deformity. Nonoperative treatment is similar to that described for incomplete injuries. Operative treatment for complete tears within the tendon can be carried out by reattaching the tendon through drill holes in the humeral cortex at its anatomic insertion site. If the tear is at or near the musculotendinous junction, repair can be accomplished by an end-to-end technique. In a review of 29 cases by Park and Espiniella [149], only 58% of patients with rupture of the pectoralis major treated by nonoperative means showed good results. In the same series, 90% of patients treated surgically showed good or excellent results [149]. Zeman and colleagues reported that four of four cases treated surgically had excellent results [211]. In five patients treated conservatively, significant strength deficits existed that limited return to athletic competition [211]. Late repair has been shown by Kretzler and Richardson, Lindenbaum, and Zeman and associates to be effective in improving pain, strength, and function in a high percentage of cases [95, 107, 211]. Delayed repair for up to 4 to 6 weeks does not

appear to affect the result. Kretzler and Richardson performed late repair in patients up to 5 years after injury and reported some residual deficiency of strength but good overall results [95].

Author's Preferred Method of Treatment. For partial ruptures involving predominantly the muscle belly itself, I prefer nonoperative treatment. Initially, the patient's arm should be immobilized in a sling and ice should be applied. Pendulum exercises are begun on the second or third day followed by slow, progressive shoulder range of motion exercises. Strengthening exercises are begun at approximately the fourth week, and slow progression to isotonic and isokinetic exercises is allowed. Six to eight weeks are often required for recovery.

Early diagnosis of complete rupture is helpful in advising the patient about treatment options. Diagnosis can usually be made on clinical data, but if there is a question, MRI can be used to obtain more information. Complete ruptures within the substance of the tendon or avulsion from the humerus should be repaired to bone (Fig. 15F–43). The distal portion of a deltopectoral incision is used, and the tendon is isolated. The tendon is surprisingly thin and small and is sometimes difficult to identify because it retracts into the muscle belly and large hematoma. Heavy No. 5 nonabsorbable sutures or 1-mm Dacron tape is woven in a Bunnell-type fashion from the musculotendinous junction out through the end of the tendon. Even if some lateral tendon remnant is found at the humerus, I believe the musculotendinous unit should be anchored back to bone. A bony trough is developed in the cortex of the humerus, and multiple drill holes are made in the adjacent cortex. The tendon is advanced into the trough and anchored by passing the sutures through the adjacent drill holes. If there is residual tendon remaining on the humerus, this can be oversewn for double reinforcement of the repair.

Postoperative Management and Rehabilitation. A sling is worn postoperatively for the first month. Gentle shoulder range of motion exercises are begun early in the first week. Further stretching and range of motion exercises are initiated to obtain a full range of motion by the sixth or eighth week. Light isotonic exercises are begun at about 6 weeks, and slow progression to advanced strengthening and functional activities is made during the first 3 months. Return to full unrestricted weight lifting is often delayed until 6 months after the repair.

Criteria for Return to Sports Participation. The patient with a partial or complete injury who is treated conservatively should regain full shoulder motion and protective strength before return to competition is allowed. The patient treated by sur-

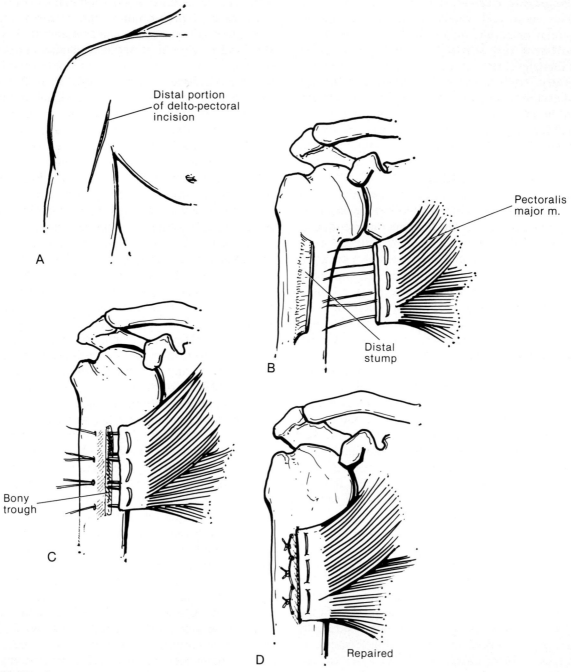

FIGURE 15F–43
Surgical repair of ruptured pectoralis major tendon. *A*, The distal portion of a deltopectoral incision is used. *B*, The tendon is repaired with multiple No. 5 nonresorbable suture or 1-mm Dacron tape. *C*, The suture is passed into a trough burred into the humerus at the anatomic insertion site. *D*, Sutures are tied laterally, pulling the repaired tendon into the bony trough.

gical repair often requires 6 months to return to full strength. Before such a patient can be allowed to return to heavy athletic competition, full range of motion should be achieved. Full strength should be required before the patient returns to contact sports or unrestricted weight lifting.

Rupture of the Deltoid

Complete rupture of the deltoid muscle is a very rare clinical entity, although contusions and strains are not uncommon in both throwing and contact sports. Very little has been written about injury to this muscle, yet even minor injuries can seriously affect athletic performance [23].

Anatomy and Biomechanics. The deltoid is the primary motor for the shoulder girdle. It arises from the anterior clavicle, the acromion, and the spine of the scapula (Fig. 15F–44). There are three major portions of the deltoid that converge upon a common insertion point midway down the humerus at the deltoid tuberosity. The axillary nerve is the primary innervation for the deltoid muscle. It passes beneath the glenohumeral joint to exit posteriorly at the quadrangular space, then branches into anterior and posterior divisions that course on the deep surface of the deltoid approximately 6 to 8 cm distal to its origin. The axillary nerve innervates both the deltoid and the teres minor and provides sensation for the upper lateral arm. Functionally, the deltoid provides power for flexion, extension, and abduction of the glenohumeral joint. It also forms the bulk that covers and protects the underlying rotator cuff and the shoulder joint itself [76].

Clinical Evaluation

History and Mechanism of Injury. Minor injuries to the deltoid such as strains and contusions are common in athletic activities. Usually a direct blow to the upper arm while it is in abduction or forward elevation is the cause. Deltoid muscle strains have also been described in throwing sports. The anterior deltoid can be injured during acceleration, whereas the posterior deltoid can be injured during deceleration. Injury to the origin of the deltoid can occur with grade V acromioclavicular dislocations when the distal clavicle ruptures through the deltotrapezius fascia.

Complete disruption of the deltoid is quite rare and seems to be associated with crushing injuries or severe direct blows received during major trauma from the outside. A few rare examples have been reported in the literature, usually associated with cases of severe trauma such as those described by Gilchrist and Albee in 1939 and McEntire and co-workers in 1972 [54, 118]. The series of Anzel that described over 1000 musculotendinous injuries contained no cases of deltoid rupture [5].

It is important to note that rupture of the deltoid is perhaps most commonly associated with orthopaedic surgery [23, 28]. The deltoid muscle is commonly detached from its origin on the clavicle and acromion in certain shoulder procedures. Inadequate reattachment of the deltoid origin followed by early institution of resistance exercises can result in retraction of the deltoid distally. This is an extremely difficult problem to reconstruct later, and therefore early diagnosis following detachment and retraction of the deltoid is important. Permanent loss of deltoid function due to detachment can result in a severe functional deficit that may preclude return to athletic competition (Fig. 15F–45).

Physical Examination. In the acute deltoid strain without rupture, local tenderness and mild swelling may be the only clinical signs. Ecchymosis may be present in the case of a contusion due to a direct blow. Shoulder motion is often limited, and weakness secondary to pain may be present.

Examination after acute complete rupture will demonstrate massive injury with swelling, deformity, and ecchymosis. With complete avulsion, there is loss of the normal shoulder contour, direct tenderness, and often a palpable defect. Weakness and limited abduction are present. Because this injury usually occurs after massive multiple trauma, masking by the additional injuries may make immediate recognition difficult. In postsurgical dehiscence of the deltoid, the defect is usually palpable in the region of the previous attachment. This late detachment can also be difficult to diagnose because of local swelling.

Treatment. In most strains and contusions involving the deltoid muscle, local conservative treatment

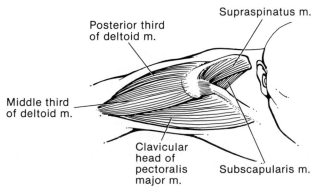

FIGURE 15F–44
Diagram demonstrates the three major portions of the deltoid attaching to the clavicle, acromion, and spine of scapula. Insertion is midway down the humerus at the deltoid tuberosity.

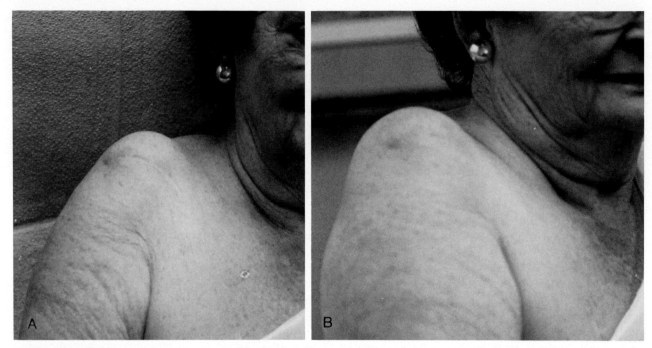

FIGURE 15F–45
This patient has chronic detachment of the anterior deltoid after repair of a rotator cuff tear. Substantial weakness and dysfunction of abduction and forward elevation are present.

is all that is necessary. Ice is applied in the acute phase with immobilization in a sling as symptoms warrant. After several days, heat, gentle mobilization, stretching, and slowly progressive strengthening exercises are usually all that are required to return to full function.

Management of postoperative dehiscence requires early diagnosis followed by surgical reattachment. Prognosis for delayed repair is poor owing to severe scarring and atrophy of the muscle.

In patients with traumatic or complete avulsion, surgical treatment is necessary. Direct repair to bone either at the origin or insertion of the muscle is recommended. With midsubstance injuries, direct repair using a suture technique can be expected to be weak, and the arm requires protection in an abduction splint.

Author's Preferred Treatment. Complete disruption and dehiscence following previous shoulder surgery should be treated surgically. Every attempt should be made to anchor these injuries back to bone. Protection in a sling or abduction splint may be necessary if there is any tension on the repair. Passive motion should be instituted early, but active exercise is delayed for 6 weeks. Prognosis after this injury is guarded.

Partial injuries and contusions should be treated with ice and rest in the acute phase, followed by progressive range of motion exercises, heat, and strengthening. Complete healing and return to sports can be expected.

Rupture of the Subscapularis

This is an extremely rare injury. Hauser, McAuliffe and Dodd, and Gilcreest and Albi have all reported cases of isolated rupture of the subscapularis tendon [54, 68, 115]. Recently, Gerber presented a series of 16 cases of this injury that required surgical treatment [53]. Partial ruptures have been documented by many authors in association with anterior dislocation of the glenohumeral joint [33, 83]. In a traumatic situation, subscapularis tendon avulsions are often associated with avulsion fractures of the lesser tuberosity [97, 179]. See the previous section of this chapter on proximal humeral fractures for a complete discussion of this type of avulsion injury.

Anatomy. The subscapularis muscle arises from the subscapular fossa on the deep surface of the scapula and inserts by a broad tendinous attachment into the lesser tuberosity of the humerus. It is the anterior part of the musculotendinous rotator cuff. The subscapularis forms the upper border of the quadrangular and triangular spaces with the axillary nerve, posterior humeral circumflex vessels, and scapular circumflex vessels passing beneath it. It has an important function in internal rotation of the humerus and as a dynamic humeral head stabilizer. The subscapularis is especially important in preventing anterior instability of the glenohumeral joint [76].

Clinical Presentation. Although poorly described

in the literature, the mechanism of injury is similar to that for an anterior dislocation. For injuries associated with anterior dislocation, diagnosis is usually made at the time of reconstruction. For other isolated injuries, a fall on the outstretched arm in abduction as the patient attempts to bring the arm into an adducted position is the mechanism of injury. Gerber described the cause of injury as forceful hyperextension or traumatic external rotation of the adducted arm [53].

The patient presents with pain, anterior swelling, and decreased mobility about the joint. If the injury is associated with anterior instability, the apprehension test is positive. With an isolated injury, weakness of internal rotation is present. This finding is usually not tested in the acute phase. Gerber has described a new clinical test for evaluation of the integrity of the subscapularis that he terms the "lift-off" test [53]. The patient places the arm in internal rotation with the dorsum of the hand on the back pocket. If the patient is unable to internally rotate the arm any further and lift the hand off the pocket, incompetence of the subscapularis can be suspected (Fig. 15F–46).

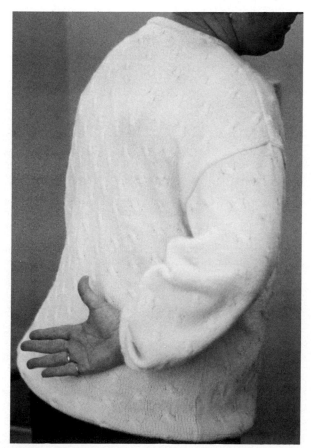

FIGURE 15F–46
The "lift-off" sign was described by Gerber to demonstrate incompetence of the subscapularis muscle. This patient lacks the ability to rotate further the humerus internally, and therefore he cannot "lift off" the hand from the posterior position.

Recommended Treatment. Treatment recommendations in the literature are sparse at best. In cases associated with anterior dislocation, the subscapularis rupture is repaired during anterior reconstruction. Gerber described surgical repair or reconstruction of isolated ruptures and reported good results [53]. McCauliffe and Dowd reported a case of complete avulsion that was repaired by direct reattachment [115]. In cases associated with lesser tuberosity avulsion, treatment by direct reattachment with either sutures or screw fixation has been described [97, 179] (Fig. 15F–47).

Author's Preferred Treatment. I recommend primary repair of injuries to the subscapularis tendon with or without avulsion fracture. If the bone fragment is large, it can be repaired with a screw. If it is a tendinous avulsion or a small portion of bone, it can be repaired through drill holes using multiple strands of heavy suture.

Latissimus Dorsi Injuries

Injuries to the latissimus dorsi are rare. Complete rupture has been described only in conjunction with rupture of the pectoralis major in severe trauma [23, 88]. Pain in the region of the latissimus dorsi does occur in throwing athletes as a result of muscular strain and in contact sports as a result of contusion.

Anatomy. This large muscle arises in the back from the lower six thoracic spinous processes, the thoracolumbar fascia, the posterior part of the iliac crest, the lower three or four ribs and the inferior angle of the scapula. It attaches to the upper medial humerus, forming the posterior axillary fold. It is a powerful adductor of the humerus and acts as an extensor of the humerus when the shoulder is flexed. The thoracodorsal nerve arising from the posterior cord supplies the latissimus [76].

Clinical Presentation. In most cases of muscle strain and contusion of the latissimus dorsi, swelling, ecchymosis, and deformity are notably absent. Tenderness to direct palpation of the muscle belly and pain when adduction of the shoulder is resisted are the major clinical signs. The latissimus is stressed during the throwing motion when the arm is rapidly decelerated during follow-through.

Kawashima and colleagues, have described the only case of complete rupture of the latissimus dorsi tendon [88]. This was associated with a complete rupture of the pectoralis major in a patient who suffered a severe industrial crush injury. Pain, ecchymosis, and swelling of the posterior axillary fold and tenderness to palpation were present.

Recommended Treatment. Surgical repair of complete rupture by direct suture has been reported

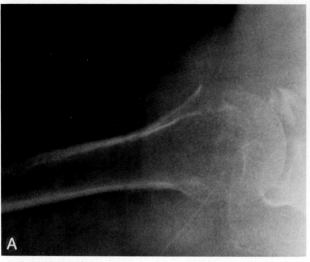
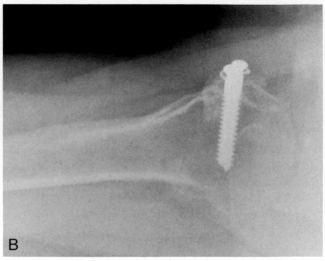

FIGURE 15F–47
Avulsion fracture of the lesser tuberosity. *A*, Axillary lateral x-ray demonstrates displaced lesser tuberosity fracture pulled medially by subscapularis muscle. *B*, Lesser tuberosity internally fixed with two ASIF screws.

with complete return of function [88]. Partial injuries should be treated like other acute strains of the musculotendinous unit. In the acute phase, rest, ice, and stretching are the mainstays, followed by a program of progressive motion, heat, ultrasound, and strengthening exercises.

Author's Preferred Method of Treatment. If diagnosis of a complete rupture is made, primary surgical repair should be performed. Treatment for muscular strain is as described in the previous section.

References

1. Ahlgren, O., and Appel, H. Proximal humeral fractures. *Acta Orthop Scand* 44:124–125, 1973.
2. Aldredge, R. H., and Knight, M. P. Fractures of the upper end of the humerus treated by early relaxed motion and massage. *New Orleans Med Surg J* 92:519–524, 1940.
3. Andreasen, A. T. Avulsion fracture of the lesser tuberosity of the humerus. *Lancet* 1:750, 1941.
4. de Anquin, C. L., and de Anguin, A. Prosthetic replacement in the treatment of serious fractures of the proximal humerus. *In* Bayley, I., and Kessel, L. (Eds.), *Shoulder Surgery*. New York, Springer-Verlag, 1965.
5. Anzel, S. H., Covey, K. W., Weiner, A. D., Disruption of muscles and tendons: an analysis of 1,014 cases. *Surgery* 45 (3):406–414, 1959.
6. Aufranc, O. E., Jones, W. N., and Turner, R. H. Bilateral shoulder fracture-dislocations. *JAMA* 195:1140–1143, 1966.
7. Bakalim, G. Rupture of the pectoralis major muscle. A case report. *Acta Orthop Scand* 36:274–279, 1965.
8. Baker, D. M., and Leach, R. E. Fracture-dislocation of the shoulder. Report of three unusual cases with rotator cuff avulsion. *J Trauma* 5:659–664, 1965.
9. Barrett, W. P., Franklin, J. L., Jackins, S. E., Wyss, C. R., Total shoulder arthroplasty. *J Bone Joint Surg* 69A:865–872, 1987.
10. Baxter, M. P., and Wiley, J. J. Fractures of the proximal

11. Bell, H. M. Posterior fracture-dislocation of the shoulder. A method of closed reduction. A case report. *J Bone Joint Surg* 47A:1521–1524, 1965.
12. Bengner, U. Changes in the incidence of fracture of the upper end of the humerus during a 3-year period. A study of 2125 fractures. *Clin Orthop* 231:179–182, 1988.
13. Berson, B. L. Surgical repair of pectoralis major rupture in an athlete. *Am J Sports Med* 7(6):348–351, 1979.
14. Bertoft, E. S., Lundh, I., and Ringqvist, I. Physiotherapy after fracture of the proximal end of the humerus. *Scand J Rehab Med* 16:11–16, 1984.
15. Bigliani, L. U. Fractures of the proximal humerus. *In* Rockwood, C. A., and Matsen, F. A. (Eds), *The Shoulder*. Philadelphia, W. B. Saunders, 1990.
16. Blom, S., and Dahlback, L. O. Nerve injuries in dislocations of the shoulder joint and fractures of the neck of the humerus. A clinical and electromyographical study. *Acta Chir Scand* 136:461–466, 1970.
17. Bloom, M. H., and Obata, W. Diagnosis of posterior dislocation of the shoulder with use of Velpeau axillary and angle-up roentgenographic views. *J Bone Joint Surg* 49A:943–949, 1967.
18. Brostrom, F. Early mobilization of fractures of the upper end of the humerus. *Arch Surg* 46:614, 1943.
19. Buck, J. E. Rupture of the sternal head of the pectoralis major: A personal description. *J Bone Joint Surg* 45B:224, 1963.
20. Butters, A. G. Traumatic rupture of the pectoralis major. *Br Med J* 2:652–653, 1941.
21. Caldwell, G. A. The treatment of fractures of the upper end of the humerus. *Arch Surg* 46:614, 1943.
22. Callahan, D. J. Anatomic considerations. Closed reduction of proximal humeral fracture. *Orthop Rev* 13(3):79–85, 1984.
23. Caughey, M. A., and Welsh, P. Muscle ruptures affecting the shoulder girdle. *In* Rockwood, C. A., and Matsen, F. A. (Eds), *The Shoulder*. Philadelphia, W. B. Saunders, 1990.
24. Clifford, P. C. Fractures of the neck of the humerus. A review of the late results. *Injury* 12:91–95, 1980.
25. Cofield, R. H. Comminuted fractures of the proximal humerus. *Clin Orthop* 230:49–57, 1988.
26. Coventry, M. A., and Laurnen, E. L. Ununited fractures

of the middle and upper humerus. Special problems in treatment. *Clin Orthop* 69:192–198, 1970.

27. Dameron, T. B. Complications of treatment of injuries to the shoulder. *In* Epps, C. H. (Ed), *Complications in Orthopaedic Surgery* (2nd ed.). Philadelphia, J. B. Lippincott, 1986.
28. Davis, C. B. Plastic repair of the deltoid muscle. *Surg Clin* 3: 287–289, 1919.
29. Dehne, E. Fractures at the upper end of the humerus. *Surg Clin North Am* 25:28–47, 1945.
30. Delport, H. P., and Piper, M. S. Pectoralis major rupture in athletes. *Arch Orthop Trauma Surg* 100:135–137, 1982.
31. DePalma, A. F. *Surgery of the Shoulder* (3rd ed.). Philadelphia, J. B. Lippincott, 1983.
32. DePalma, A. F. and Cautilli, R. A. Fractures of the upper end of the humerus. *Clin Orthop* 20:73–93, 1961.
33. DePalma, A. F., Cooke, A. J., and Prabhaker, M. The role of the subscapularis in recurrent anterior dislocations of the shoulder. *Clin Orthop* 54:35, 1967.
34. Des Marchais, J. E., and Morais, G. Treatment of complex fractures of the proximal humerus by Neer hemiarthroplasty. *In* Bateman, J. E., and Welsh, R. P. (Eds), *Surgery of the Shoulder*. Philadelphia, B. C. Decker, 1984.
35. Dewar, F. B., and Yabsley, R. H. Fracture-dislocation of the shoulder. Report of a case. *J Bone Joint Surg* 49B:540–543, 1967.
36. Dimon, J. H. Posterior dislocation and posterior fracture dislocation of the shoulder. A report of 25 cases. *South Med J* 60:661, 1967.
37. Din, K. M., and Meggitt, B. F. Bilateral four-part fractures with posterior dislocation of the shoulder. A case report. *J Bone Joint Surg* 65B:176–178, 1983.
38. Dingley, A., and Denham, R. Fracture-dislocation of the humeral head. A method of reduction. *J Bone Joint Surg* 55A:1299–1300, 1973.
39. Drapanas, T., Hewitt, R. L., Weichert, R. F., and Smith, A. D. Civilian vascular injuries. A critical appraisal of three decades of management. *Ann Surg* 172:351–360, 1970.
40. Drapanas, T., McDonald, J., and Hale, H. W. A rational approach to classification and treatment of fractures of the surgical neck of the humerus. *Am J Surg* 99:617–624, 1960.
41. Egan, T. M., and Hall, H. Avulsion of the pectoralis major tendon in a weight lifter: Repair using a barbed staple. *Can J Surg* 30:434, 1987.
42. Einarsson, F. Fracture of the upper end of the humerus. *Acta Orthop Scand (Suppl)* 3:10–209, 1958.
43. Ekstrom, T., Lagergren, C., and von Schreeb, T. Procaine injections and early mobilisation for fractures of the neck of the humerus. *Acta Chir Scand* 130:18–24, 1965.
44. Elliot, J. A. Acute arterial occlusion. An unusual cause. *Surgery* 39:825–826, 1956.
45. Fairbank, T. J. Fracture-subluxations of the shoulder. *J Bone Joint Surg* 30B:454–460, 1968.
46. Fellander, M. Fracture-dislocations of the shoulder joint. *Acta Chir Scand* 107:138–145, 1954.
47. Flatow, E. L., Bigliani, L. U., and April, E. W. An anatomic study of the musculocutaneous nerve and its relationship to the coracoid process. *Clin Orthop* 244:166–171, 1989.
48. Fourrier, P., and Martini, M. Post-traumatic avascular necrosis of the humeral head. *Int Orthop* 1:187–190, 1977.
49. Frankau, C. A manipulative method for the reduction of fractures of the surgical neck of the humerus. *Lancet* 2:755, 1933.
50. Funsten, R. V., and Kinser, P. Fractures and dislocations about the shoulder. *J Bone Joint Surg* 18:191–198, 1936.
51. Garceau, G. J., and Cogland, S. Early physical therapy in the treatment of fractures of the surgical neck of the humerus. *J Indiana Med Assoc* 34:293–295, 1941.
52. Geneste, R. Closed treatment of fracture-dislocations of the shoulder joint. *Rev Chir Orthop* 66:383–386, 1980.
53. Gerber, C., and Krushell, R. J. Isolated ruptures of the tendon of the subscapularis muscle. *J Bone Joint Surg* 73B(3):389–394, 1991.
54. Gilcreest, E. L., and Albi, P. Unusual lesions of muscles and tendons of the shoulder girdle and upper arm. *Surg Gynecol Obstet* 68:903–917, 1939.
55. Glessner, J. R. Intrathoracic dislocation of the humeral head. *J Bone Joint Surg* 43A:428–430, 1961.
56. Greeley, P. W., and Magnuson, P. B. Dislocation of the shoulder accompanied by fracture of the greater tuberosity and complicated by spinatus tendon injury. *JAMA* 102:1835–1838, 1934.
57. Grimes, D. W. The use of Rush pin fixation in unstable upper humeral fracture. A method of blind insertion. *Orthop Rev* 9(4):75–79, 1980.
58. Griswold, R. A., Hucherson, D. C., and Strode, E. C. Fractures of the humerus treated with hanging cast. *South Med J* 34:777–778, 1941.
59. Gudmundsson, B. A case of agenesis and a case of rupture of the pectoralis major muscle. *Acta Orthop Scand* 44:213–218, 1973.
60. Gurd, F. B. A simple effective method for the treatment of fractures of the upper part of the humerus. *Am J Surg* 47:433–453, 1940.
61. Haas, S. L. Fracture of the lesser tuberosity of the humerus. *Am J Surg* 63:252, 1944.
62. Hagg, O., and Lundberg, B. Aspects of prognostic factors in comminuted and dislocated proximal humeral fractures. *In* Bateman, J. E., and Welsh, R. P. (Eds), *Surgery of the Shoulder*. Philadelphia, B. C. Decker, 1984.
63. Hall, M. C., and Rosser, M. The structure of the upper end of the humerus, with reference to osteoporotic changes in senescence leading to fractures. *Can Med Assoc J* 88:290–294, 1963.
64. Hall, R. H., Isaac, F., and Booth, C. R. Dislocations of the shoulder with special reference to accompanying small fractures. *J Bone Joint Surg* 41A:489–494, 1959.
65. Hardcastle, P. H., and Fisher, T. R. Intrathoracic displacement of the humeral head with fracture of the surgical neck. *Injury* 12:313–315, 1981.
66. Hartigan, J. W. Separation of the lesser tuberosity of the head of the humerus. *NY Med J* 61:276, 1895.
67. Haupt, H. A., and Rovere, G. D. Anabolic steroids: A review of the literature. *Am J Sports Med* 12(6):469–484, 1984.
68. Hauser, F. D. W. Avulsion of the tendon of subscapularis muscle. *J Bone Joint Surg* 36A:139–141, 1954.
69. Hawkins, R. J., Bell, R. H., and Gurr, K. The three-part fracture of the proximal part of the humerus. Operative treatment. *J Bone Joint Surg* 68A:1410–1414, 1986.
70. Hayes, W. M. Rupture of the pectoralis major muscle. Review of the literature and report of two cases. *J Int Coll Surg* 14:82–88, 1950.
71. Hayes, M. J., and Van Winkle, N. Axillary artery injury with minimally displaced fracture of the neck of the humerus. *J Trauma* 23:431–433, 1983.
72. Henderson, R. S. Fracture-dislocation of the shoulder with interposition of the long head of the biceps. Report of a case. *J Bone Joint Surg* 34B:240–241, 1952.
73. Henson, G. F. Vascular complications of shoulder injuries. A report of two cases. *J Bone Joint Surg* 38B:528, 1956.
74. Heppenstall, R. B. Fractures of the proximal humerus. *Orthop Clin North Am* 6(2):467–475, 1975.
75. Heuget, L. Bone cement in the treatment of certain fractures of the proximal humerus. *Ann Chir* 27:311–313, 1973.
76. Hollinshead, W. H. *Anatomy for Surgeons*, Vol. 3 (3rd ed.) The Back and Limbs. New York, Harper & Row, 1982.
77. Horak, J., and Nilsson, B. Epidemiology of fractures of the upper end of the humerus. *Clin Orthop* 112:250–253, 1975.
78. Howard, N. J., and Eloesser, L. Treatment of the fractures of the upper end of the humerus. An experimental and clinical study. *J Bone Joint Surg* 16:1–29, 1934.
79. Hudson, R. T. The use of the hanging cast in treatment of fractures of the humerus. *South Surgeon* 10:132–134, 1941.
80. Hughes, M., and Neer, C. S. Glenohumeral joint replacement and postoperative rehabilitation. *Phys Ther* 55:850–858, 1975.

81. Hundley, J. M., and Stewart, M. J. Fractures of the humerus. A comparative study in methods of treatment. *J Bone Joint Surg* 37A:681–692, 1955.

82. Jakob, R. P., Kristiansen, T., Mayo, K., Ganz, R., and Müller, M. E. Classification and aspects of treatment of fractures of the proximal humerus. *In* Bateman, J. E., and Welsh, R. P. (Eds), *Surgery of the Shoulder*. Philadelphia, B. C. Decker, 1984.

83. Jens, J. The role of subscapularis muscle in recurring dislocation of the shoulder. *J Bone Joint Surg* 46B:780, 1964.

84. Johnasson, O. Complications and failures of surgery in various fractures of the humerus. *Acta Chir Scand* 120:469–478, 1961.

85. Jones, L. Reconstructive operation for non-reducible fractures of the head of the humerus. *Ann Surg* 97:217–225, 1933.

86. Jones, R. On certain fractures about the shoulder. *Irish J Med Sci* 78:282–291, 1932.

87. Jones, R. Certain injuries commonly associated with displacement of the head of the humerus. *Br Med J* 1:1385–1386, 1906.

88. Kawashima, M., Sato, M., Torisu, T., et al. Rupture of the pectoral major: Report of 2 cases. *Clin Orthop* 109:115–119, 1975.

89. Keene, J. S., Huizenga, R. E., Engber, W. D., and Rogers, S. C. Proximal humeral fractures. A correction of residual deformity with long-term function. *Orthopedics* 6:173–178, 1983.

90. Kelly, J. P. Fractures complicating electroconvulsive therapy and chronic epilepsy. *J Bone Joint Surg* 36B:70–79, 1954.

91. Key, J. A., and Conwell, H. E. *Fractures, Dislocations, and Sprains* (5th ed.). St. Louis, C. V. Mosby, 1951.

92. Kingsley, D. M. Rupture of pectoralis major. Report of a case. *J Bone Joint Surg* 28:644–645, 1946.

93. Knight, R. A., and Mayne, J. A. Comminuted fractures and fracture-dislocations involving the articular surface of the humeral head. *J Bone Joint Surg* 39A:1343–1355, 1957.

94. Kraulis, J., and Hunter, G. The results of prosthetic replacement in fracture-dislocations of the upper end of the humerus. *Injury* 8:129–131, 1976.

95. Kretzler, H. H., and Richardson, A. B. Rupture of the pectoralis major muscle. *Am J Sports Med* 17(4):453–458, 1989.

96. Kristiansen, B., and Christensen, S. W. Plate fixation of proximal humeral fractures. *Acta Orthop Scand* 57:230, 1986.

97. LaBriola, J. H., and Mohaghegh, H. A. Isolated avulsion fracture of the lesser tuberosity of the humerus. A case report and review of the literature. *J Bone Joint Surg* 57A:1011, 1975.

98. LaFerte, A. D., and Nutter, P. D. The treatment of fractures of the humerus by means of hanging plaster cast. "Hanging cast." *Ann Surg* 114:919–930, 1955.

99. Laing, P. G. The arterial supply of the adult humerus. *J Bone Joint Surg* 38A:1105–1116, 1956.

100. Lane, L. B., Villacin, A., and Bullough, P. G. The vascularity and remodelling of subchondral bone and calcified cartilage in adult human femoral and humeral heads. An age- and stress-related phenomenon. *J Bone Joint Surg* 59B:272–278, 1977.

101. Leach, R. E., and Premer, R. F. Nonunion of the surgical neck of the humerus. Method of internal fixation. *Minn Med* 48:318–322, 1965.

102. Lee, C. K., and Hansen, H. R. Post-traumatic avascular necrosis of the humeral head in displaced proximal humeral fractures. *J Trauma* 21:788–791, 1981.

103. Lee, C. K., Hansen, H. T., and Weiss, A. B. Surgical treatment of the difficult humeral neck fracture. Acromial shortening, anterolateral approach. *J Trauma* 20: 67–70, 1980.

104. Lentz, W., and Meuser, P. The treatment of fractures of the proximal humerus. *Arch Orthop Trauma Surg* 96:283–285, 1980.

105. Leyshon, R. L. Closed treatment of fractures of the proximal humerus. *Acta Orthop Scand* 55:48–51, 1984.

106. Lim, T. E., Ochsner, P. E., Marti, R. K., and Holscher, A. A. The results of treatment of comminuted fractures and fracture dislocations of the proximal humerus. *Neth J Surg* 35:139–143, 1983.

107. Lindenbaum, B. L. Delayed repair of a ruptured pectoralis major muscle. *Clin Orthop* 109:120–121, 1975.

108. Lindholm, T. S., and Elmstedt, E. Bilateral posterior dislocation of the shoulder combined with fracture of the proximal humerus. A case report. *Acta Orthop Scand* 51:485–488, 1980.

109. Linson, M. A. Axillary artery thrombosis after fracture of the humerus. A case report. *J Bone Joint Surg* 62A:1214–1215, 1980.

110. Lorenzo, F. T. Osteosynthesis with Blount staples in fracture of the proximal end of the humerus. A preliminary report. *J Bone Joint Surg* 37A:45–48, 1955.

111. Lundberg, B. J., Svenungson-Hartwig, E., and Vikmark, R. Independent exercises versus physiotherapy in nondisplaced proximal humeral fractures. *Scand J Rehab Med* 11:133, 1979.

112. Manjarris, J., Gershuni, D. H., and Moitoza, J. Rupture of the pectoralis major tendon. *J Trauma* 25(8):810–811, 1985.

113. Marmor, L., Bechtol, C. O., and Hall, C. B. Pectoralis major muscle function of sternal portion and mechanism of rupture of normal muscle: Case reports. *J Bone Joint Surg* 43A:81–87, 1961.

114. Mazet, R. Intramedullary fixation in the arm and the forearm. *Clin Orthop* 2:75–92, 1953.

115. McAuliffe, T. B., and Dowd, G. S. Avulsion of the subscapularis tendon: A case report. *J Bone Joint Surg* 69A:1454–1455, 1987.

116. McBurney, C., and Dowd, C. N. Dislocation of the humerus complicated by fracture at or near the surgical neck with a new method of reduction. *Ann Surg* 19:399, 1894.

117. MacDonald, F. R. Intra-articular fractures in recurrent dislocations of the shoulder. *Surg Clin North Am* 43:1635–1645, 1963.

118. McEntire, J. E., Hess, W. E., and Coleman, S. Rupture of the pectoralis major muscle. *J Bone Joint Surg* 54A(5):1040–1046, 1972.

119. McGuinness, J. P. Isolated avulsion fracture of the lesser tuberosity of the humerus. *Lancet* 1:508, 1939.

120. MacKenzie, D. B. Avulsion of the insertion of the pectoralis major muscle. *S Afr Med J* July:147–148, 1981.

121. McLaughlin, H. L. Dislocation of the shoulder with tuberosity fracture. *Surg Clin North Am* 43:1615–1620, 1963.

122. McMaster, P. F. Tendon and muscle ruptures. Clinical and experimental studies and locations of subcutaneous ruptures. *J Bone Joint Surg* 15: 705–22, 1933.

123. Meyerding, H. W. Fracture-dislocation of the shoulder. *Minn Med* 20:717–726, 1937.

124. Michaelis, L. S. Comminuted fracture-dislocation of the shoulder. *J Bone Joint Surg* 26:363–365, 1944.

125. Milch, H. The treatment of recent dislocations and fracture-dislocations of the shoulder. *J Bone Joint Surg* 31A:173–180, 1949.

126. Miller, S. R. Practical points in the diagnosis and treatment of fractures of the upper fourth of the humerus. *Indust Med* 9:458–460, 1940.

127. Mills, K. L. G. Severe injuries of the upper end of the humerus. *Injury* 6:13–21, 1974.

128. Mills, K. L. G. Simultaneous bilateral posterior fracture-dislocation of the shoulder. *Injury* 6:39, 1974.

129. Moriber, L. A., and Patterson, R. L. Fractures of the proximal end of the humerus. *J Bone Joint Surg* 49A:1018, 1967.

130. Morris, M. F., Kilcoyne, R. F., and Shuman, W. Humeral tuberosity fractures: Evaluation by CT scan and management of malunion. *Orthop Trans* 11:242, 1987.

131. Moseley, H. F. The arterial pattern of the rotator cuff of the shoulder. *J Bone Joint Surg* 45B:780–789, 1963.

132. Mouradian, W. H. Displaced proximal humeral fractures. Seven years' experience with a modified Zickel supracondylar device. *Clin Orthop* 212:209–218, 1986.

133. De Mourgues, G., et al. Fracture-dislocations of the shoulder joint. *Rev Chir Orthop* 51:151, 1965.

134. Neer, C. S. Prosthetic replacement of the humeral head. Indications and operative technique. *Surg Clin North Am* 43:1581–1597, 1963.

135. Neer, C. S. Displaced proximal humeral fractures. Part I. Classification and evaluation. *J Bone Joint Surg* 52A:1077–1089, 1970.

136. Neer, C. S. Displaced proximal humeral fractures. Part II. Treatment of three-part and four-part displacement. *J Bone Joint Surg* 52A:1090–1103, 1970.

137. Neer, C. S. Four-segment classification of displaced proximal humeral fractures. *Instr Course Lect* 24:160–168, 1975.

138. Neer, C. S., Brown, T. H., and McLaughlin, H. L. Fractures of the neck of the humerus with dislocation of the head fragment. *Am J Surg* 85:252–258, 1953.

139. Neer, C. S., McCann, P. D., MacFarlane, E. A., and Padilla, N. Earlier passive motion following shoulder arthroplasty and rotator cuff repair. A prospective study. *Orthop Trans* 11:231, 1987.

140. Neer, C. S., and McIlveen, S. J. Recent results and technique of prosthetic replacement for 4-part proximal humeral fractures. *Orthop Trans* 10:475, 1986.

141. Neer, C. S., and Rockwood, C. A. Fractures and dislocations of the shoulder. *In* Rockwood, C. A., and Green, D. P. (Eds), Fractures (2nd ed.). Philadelphia, J. B. Lippincott, 1984.

142. Neer, C. S., Watson, K. C., and Stanton, F. J. Recent experience in total shoulder replacement. *J Bone Joint Surg* 64A:319–337, 1982.

143. Neviaser, J. S. Complicated fractures and dislocations about the shoulder joint. *J Bone Joint Surg* 44A:984–998, 1962.

144. Newton-John, H. F., and Morgan, D. B. The loss of bone with age, osteoporosis and fractures. *Clin Orthop* 71:229, 1970.

145. North, J. P. The conservative treatment of fractures of the humerus. *Surg Clin North Am* 20:1633–1643, 1940.

146. Oni, O. O. Irreducible acute anterior dislocation of the shoulder due to a loose fragment from an associated fracture of the greater tuberosity. *Injury* 15:138, 1984.

147. Orava, S., Sorasto, A., Aalto, K., and Kvist, H. Total rupture of the pectoralis major muscle in athletes. *Int J Sports Med* 5:272–274, 1984.

148. Paavolainen, P., Bjorkenheim, J-M., Slatis, P., and Paukku, P. Operative treatment of severe proximal humeral fractures. *Acta Orthop Scand* 54:374–379, 1983.

149. Park, J. Y., and Espiniella, J. L. Rupture of pectoralis major muscle. A case report and review of literature. *J Bone Joint Surg* 52A:577–581, 1970.

150. Parkes, M. Rupture of the pectoralis major muscle. *Ind Med* 12:226, 1943.

151. Patel, M. R., Pardee, M. L., and Singerman, R. C. Intrathoracic dislocation of the head of the humerus. *J Bone Joint Surg* 45A:1712–1714, 1963.

152. Perry, P. J., Anderson, K. H., and Yates, W. R. Illicit anabolic steroid use in athletes: A case series analysis. *Am J Sports Med* 18(4):422–428, 1990.

153. Pilgaard, S., and Och Oster, A. Four-segment fractures of the humeral neck. *Acta Orthop Scand* 44:124, 1973.

154. Post, M. Fractures of the upper humerus. *Orthop Clin North Am* 11(2):239–252, 1980.

155. Prillaman, H. A., and Thompson, R. C. Bilateral posterior fracture-dislocation of the shoulder. A case report. *J Bone Joint Surg* 51A:1627–1630, 1969.

156. Proximal humeral fractures. What price history? [Editorial]. *Injury* 12:89–90, 1981.

157. Pulaski, E. J., and Chandlee, B. H. Ruptures of the pectoralis major muscle. *Surgery* 10:309–312, 1941.

158. Pulaski, E. J., and Martin, G. W. Rupture of the left pectoralis major muscle. *Surgery* 25:110–111, 1949.

159. Raney, R. B. The treatment of fractures of the humerus with the hanging cast. *North Carolina Med J* 6:88–92, 1945.

160. Rathbun, J. B., and Macnab, I. The microvascular pattern of the rotator cuff. *J Bone Joint Surg* 52B:540–553, 1970.

161. Rechtman, A. M. Open reduction of fracture dislocations of the humerus. *JAMA* 94:1656, 1934.

162. Reckling, F. W. Posterior fracture-dislocation of the shoulder treated by a Neer hemiarthroplasty with a posterior surgical approach. *Clin Orthop* 207:133–137, 1986.

163. Rob, C. G., and Standeven, A. Closed traumatic lesions of the axillary and brachial arteries. *Lancet* 1:597–599, 1956.

164. Roberts, S. M. Fractures of the upper end of the humerus. An end-result study which shows the advantage of early active motion. *JAMA* 98:367–373, 1932.

165. Rockwood, C. A., Szaley, E. A., Curtis, R. J., Young, D. C., and Kay, S. P. X-ray evaluation of shoulder problems. *In* Rockwood, C. A., and Matsen, F. A. (Eds), The Shoulder. Philadelphia, W. B. Saunders, 1990.

166. Rooney, P. J., and Cockshott, W. P. Pseudarthrosis following proximal humeral fractures. A possible mechanism. *Skel Radiol* 15:21–24, 1986.

167. Rose, S. H., Milton, L. J., Morrey, B. F., Ilstrup, D. M., and Riggs, L. B. Epidemiologic features of humeral fractures. *Clin Orthop* 168:24–30, 1982.

168. Rothman, R. H., and Parke, W. W. The vascular anatomy of the rotator cuff. *Clin Orthop* 41:176–186, 1965.

169. Rowe, C. R., and Colville, M. The glenohumeral joint. *In* Rowe, C. R. (Ed.), The Shoulder. New York, Churchill Livingstone, 1988.

170. Rowe, C. R., and Marble, H. Shoulder girdle injuries. *In* Cave, E. F. (Ed.), Fractures and Other Injuries. Chicago, Year Book, 1958.

171. Rush, L. V. *Atlas of Rush Pin Techniques*. Meridian, MI, Beviron, 1959.

172. Salem, I. Bilateral anterior fracture-dislocation of the shoulder joints due to severe electric shock. *Injury* 14:361–363, 1983.

173. Santee, H. E. Fractures about the upper end of the humerus. *Ann Surg* 80:103–114, 1924.

174. Savoie, F. H., Geissler, W. B., and Vader Griend, R. A. Open reduction and internal fixation of three-part fractures of the proximal humerus. *Orthopedics* 12:65–70, 1989.

175. Schechter, R., and Gristina, A. G. Surgical repair of rupture of the pectoralis major muscle. *JAMA* 188:1009, 1964.

176. Scheck, M. Surgical treatment of nonunions of the surgical neck of the humerus. *Clin Orthop* 167:255–259, 1982.

177. Sever, J. W. Fracture of the head of the humerus. Treatment and results. *N Engl J Med* 216:1100–1107, 1937.

178. Shaw, J. L. Bilateral posterior fracture-dislocation of the shoulder and other trauma caused by convulsive seizures. *J Bone Joint Surg* 53A:1437–1440, 1971.

179. Shibuya, S., and Ogawa, K. Isolated avulsion fracture of the lesser tuberosity of the humerus. A case report. *Clin Orthop* 211:215–218, 1986.

180. Shuck, J. M., Omer, G. E., and Lewis, C. E. Arterial obstruction due to intimal disruption in extremity fractures. *J Trauma* 12:481–489, 1972.

181. Silfverskiold, N. On the treatment of fracture-dislocations of the shoulder-joint. With special reference to the capability of the head-fragment, disconnected from capsule and periosteum to enter into bony union. *Acta Chir Scand* 64:227–293, 1928.

182. Smyth, E. H. J. Major arterial injury in closed fracture of the neck of the humerus. Report of a case. *J Bone Joint Surg* 51B:508–510, 1969.

183. Solonen, K. A., and Vastamaki, M. Osteotomy of the neck

of the humerus for traumatic varus deformity. *Acta Orthop Scand* 56:79–80, 1985.

184. Sorensen, K. H. Pseudarthrosis of the surgical neck of the humerus. Two cases. One bilateral. *Acta Orthop Scand* 34:132–138, 1964.

185. Stableforth, P. G. Four-part fractures of the neck of the humerus. *J Bone Joint Surg* 66B:104–108, 1984.

186. Stangl, F. H. Isolated fracture of the lesser tuberosity of the humerus. *Minn Med* 16:435–437, 1933.

187. Stewart, M. J., and Hundley, J. M. Fractures of the humerus. A comparative study in methods of treatment. *J Bone Joint Surg* 37A:681–692, 1955.

188. Stimson, B. B. *A Manual of Fractures and Dislocations* (2nd ed.). Philadelphia, Lea & Febiger, 1947.

189. Sturzenegger, M., Fornaro, E., and Jakob, R. P. Results of surgical treatment of multifragmented fractures of the humeral head. *Arch Orthop Trauma Surg* 100:249–259, 1982.

190. Sven-Hansen, H. Displaced proximal humeral fractures. A review of 49 patients. *Acta Orthop Scand* 45:359–364, 1974.

191. Tanner, M. W., and Cofield, R. H. Prosthetic arthroplasty for fractures and fracture-dislocations of the proximal humerus. *Clin Orthop* 179:116–128, 1983.

192. Theodorides, T., and Dekeizer, G. Injuries of the axillary artery caused by fractures of the neck of the humerus. *Injury* 8:120, 1976.

193. Thompson, F. E., and Winant, E. M. Comminuted fracture of the humeral head with subluxation. *Clin Orthop* 20:94–97, 1961.

194. Tietjen, R. Closed injuries of the pectoralis major muscle. *J Trauma* 20(3):262–264, 1980.

195. Urs, N. D., and Jani, D. M. Surgical repair of rupture of the pectoralis major muscle: A case report. *J Trauma* 16(9):749–750, 1976.

196. Vastamaki, M., and Solonen, K. A. Posterior dislocation and fracture-dislocation of the shoulder. *Acta Orthop Scand* 51:479–484, 1980.

197. Vesely, D. G. Use of the split diamond nail for fractures of the humerus, 1958–1964. *Clin Orthop* 41:145–156, 1965.

198. Wallace, W. A. The dynamic study of shoulder movement. *In* Bayley, I., and Kessel, L. (Eds.), *Shoulder Surgery*. New York, Springer-Verlag, 1982.

199. Watson-Jones, R. *Fractures and Joint Injuries* (4th ed.). Baltimore, Williams & Wilkins, 1955, pp. 473–476.

200. Weise, K., Meeder, P. J., and Wentzensen, A. Indications and operative technique in osteosynthesis of fracture-dislocations of the shoulder joint in adults. *Langenbecks Arch Chir* 351:91–98, 1980.

201. Welsh, R. P., Macnab, I., and Riley, V. Biomechanical studies of rabbit tendon. *Clin Orthop* 81:171–177, 1971.

202. Wentworth, E. T. Fractures involving the shoulder joint. *NY State J Med* 40:1282–1288, 1940.

203. Weseley, M. S. Barenfeld, P. A., and Eisenstein, A. L. Rush pin intramedullary fixation for fractures of the proximal humerus. *J Trauma* 17:29–37, 1977.

204. West, E. F. Intrathoracic dislocation of the humerus. *J Bone Joint Surg* 31B:61, 1949.

205. Whitson, T. B. Fractures of the surgical neck of the humerus. A study in reduction. *J Bone Joint Surg* 36B:423–427, 1954.

206. Willems, W. J., and Lim, T. E. A. Neer arthroplasty for humeral fracture. *Acta Orthop Scand* 56:394–395, 1985.

207. Winfield, J. M., Miller, H., and LaFerte, A. D. Evaluation of the "hanging cast" as a method of treating fractures of the humerus. *Am J Surg* 55:228–249, 1942.

208. Yamano, Y. Comminuted fractures of the proximal humerus treated with hook plate. *Arch Orthop Trauma Surg* 105:359–363, 1986.

209. Yano, S., Takamura, S., and Kobayshi, I. Use of the spiral pin for fractures of the humeral neck. *J Jpn Orthop Assn* 55:1607, 1981.

210. Young, T. B. Conservative treatment of fractures and fracture-dislocations of the upper end of the humerus. *J Bone Joint Surg* 67B:373–377, 1985.

211. Zeman, S. C., Rosenfeld, R. T., and Lipscomb, P. R. Tears of the pectoralis major muscle. *Am J Sports Med* 7(6):343–347, 1979.

212. Zuckerman, J. D., Flugstad, D. L., Teitz, C. C., and King, H. A. Axillary artery injury as a complication of proximal humeral fractures. Two case reports and a review of the literature. *Clin Orthop* 189:234–237, 1984.

Imaging the Glenohumeral Joint

Robert O. Cone, M.D.

Shoulder pain is a frequent complaint of the competitive as well as the "weekend" athlete and provides an important indication for diagnostic imaging procedures. Unfortunately, no universal diagnostic test is available that is reliably sensitive to all of the glenohumeral abnormalities that may be encountered. Thus, a variety of imaging choices are available to evaluate these patients. This chapter will attempt to clarify the selection and utility of glenohumeral joint imaging techniques.

At first glance the shoulder appears to be a deceivingly simple articulation. But with study the true nature of the glenohumeral joint becomes evident: it is a complex mechanism that allows an inherently unstable anatomic arrangement to provide a surprisingly large range of motion as well as powerful force transmission. The only simple aspect of the shoulder is the way in which a variety of abnormalities are manifested by similar symptoms and physical examination findings. The imaging modalities available for the shoulder vary widely in sensitivity to specific abnormalities and are often uncomfortable, invasive, and expensive. Selection of the appropriate test that will allow the correct diagnosis as simply, rapidly, and inexpensively as possible is critical. Plain radiography provides the first step in evaluating most glenohumeral abnormalities. For the most part, the role of plain radiography will be covered in separate sections of this chapter dealing with specific abnormalities. In this section discussion of plain radiographic findings will be limited to those instances when they modify or enhance the secondary and tertiary imaging modalities discussed in detail. Magnetic resonance imaging (MRI) is at present in the process of exerting the same impact on glenohumeral joint imaging that it has demonstrated in the knee. Thus, in this section emphasis will be placed on the evolving role of MRI

in the evaluation of shoulder abnormalities as well as the appropriate role of other modalities.

SHOULDER ARTHROGRAPHY

Shoulder arthrography (Fig. 15G–1) has long provided the gold standard for the preoperative diagnosis of full-thickness rotator cuff tears. When properly performed, shoulder arthrography has a reported accuracy of 98% to 99% [80] in detecting full-thickness rotator cuff tears. Shoulder arthrography is also quite accurate in identifying adhesive capsulitis of the shoulder [38, 92] (see Fig. 15G–29). Partial rotator cuff tears involving the undersurface of the cuff (see Fig. 15G–14) are well demonstrated by shoulder arthrography [36, 61], whereas those involving the superior (bursal) surface or intrasubstance tears are not visualized. The capsular mechanism and glenoid labrum are not well evaluated by simple arthrography, although the addition of polytomography (see Figs. 15G–3 and 15G–38) or computed tomography (see Figs. 15G–2 and 15G–39) to the study rectifies this shortcoming. Shoulder arthrography may be performed as a single contrast [54] or double contrast procedure [34] (see Fig. 15G–1). Most authors recommend the double contrast technique [34, 80] because it is equally sensitive to full-thickness tears and allows better identification of incomplete tears and some cartilage abnormalities. We generally utilize double contrast arthrography with a slightly modified version of the technique described by Goldman [34].

The procedure should begin with appropriate scout radiographs of the affected shoulder, both to ensure appropriate radiographic technique during the arthrogram and to identify abnormalities that may be obscured by radiographic contrast. The scout

717

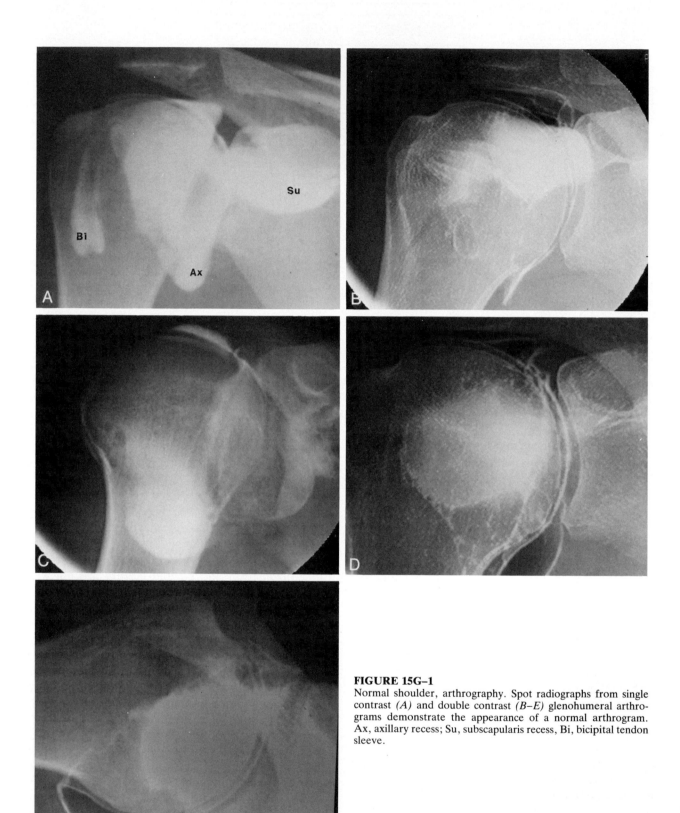

FIGURE 15G–1
Normal shoulder, arthrography. Spot radiographs from single contrast *(A)* and double contrast *(B–E)* glenohumeral arthrograms demonstrate the appearance of a normal arthrogram. Ax, axillary recess; Su, subscapularis recess, Bi, bicipital tendon sleeve.

radiographs consist of anteroposterior (AP) projections of the shoulder in internal and external rotation, an axillary lateral projection, and a bicipital groove view. These radiographs should be a little "light" in relation to normal radiographic techniques because with the double contrast technique air is the major contrast agent. If plain radiographs are not available, another radiograph combining a "true AP" view of the scapula with a 30-degree caudal tilt projection allows identification of subacromial spurs (see Fig. 15G–33) and often gives a more accurate depiction of glenohumeral and humeroacromial relationships than do standard projections. Spot scout films are also obtained to ensure proper radiographic technique. I typically shoot most of my spot films with a one-on-one setting on 9-inch radiographic cassettes with occasional four-on-one spots of the rotator cuff. In most instances spot films may be phototimed at 70 to 80 keV and yield good radiographic quality.

The patient is placed in a supine position on the fluoroscopic table with the arm in external rotation. External rotation of the arm is important because this position maximally "opens" the anterior aspect of the joint to allow easiest joint puncture. The anterior aspect of the shoulder is prepared with a povidone-iodine solution, and the point of joint puncture is located with the fluoroscope. It must be remembered that the radiographically invisible cartilaginous glenoid labrum overlies the joint line and has provided much frustration for novice arthrographers and their patients. A point is chosen within the middle third of the vertical plane of the joint that is approximately 1 to 1.5 cm lateral to the subchondral bone plate of the humeral head. This area is anesthetized with 1% Xylocaine (lidocaine) with a 25-gauge ¾-inch needle that is left in place after infiltration of local anesthetic so that the proper position for joint puncture can be verified by a quick check with the fluoroscope. The 25-gauge needle is then withdrawn and local anesthetic infiltrated to the level of the joint with a 22-gauge 1½-inch needle. In many smaller patients the joint may be entered with this needle as signified by a sudden loss of resistance to injection. In this case the Xylocaine syringe may be removed and the 1½-inch needle used to inject the joint. Otherwise, joint puncture is accomplished with a 22-gauge 3½-inch spinal needle. Correct needle position can be verified by injection of a small amount of air or Xylocaine. If no resistance is felt, a small amount of contrast should be injected. With appropriate needle position the contrast agent will be seen to outline the medial surface of the humeral head and spill

into the subscapularis recess. A mixture of 5 mL of nonionic radiographic contrast agent and 1 mL of 0.5% marcaine is injected followed by approximately 10 cc of air. Luer Lok tubing and syringes are used because "slip tip" devices have an unfortunate habit of blowing apart during injection of relatively thick radiographic contrast through a small caliber needle. A total volume of 12 to 15 mL of (air plus contrast agent) provides distention of a normal shoulder joint without undue discomfort. Injection of a greater volume will often result in decompression of the joint at a weak point of capsular insertion along the medial aspect of the subscapularis recess. This degrades the quality of the examination and should be avoided if at all possible. The precise volume of contrast agent injected varies with the arthrographer. Most texts suggest injection of 2 mL of contrast agent and a slightly larger volume of air, but I find that a slightly larger volume of contrast allows better definition of the outlines of the glenoid labra and occasionally allows better definition of the precise location of a rotator cuff tear. Filling of the intracapsular space should be observed fluoroscopically because large rotator cuff tears often fill immediately, obscuring the location of the tear unless spot films are obtained immediately. If the fluoroscopic unit has a digital subtraction capability an initial digital subtraction film may be useful [137]. In addition, identification of adhesive capsulitis (see Fig. 15G–29) may be made by identifying increasing resistance to contrast injection with a small injected volume as well as by noting an abnormally small axillary pouch and small or absent subscapularis recess [18]. In this instance a normal volume (12 to 15 mL) of contrast and air will overdistend the joint and may extravasate. A smaller volume should be utilized, the precise amount of which is often a matter of "feel." In general, when I encounter adhesive capsulitis and feel increased resistance I relax pressure on the plunger of the syringe; if the pressure is great enough that the plunger backs up, I cease the injection and withdraw the needle.

Following the injection a spot radiograph is obtained before the needle is removed; and this is followed by spot radiographs in maximal internal and external rotation while traction is applied to the arm. Some clinicians prefer to exercise the patient with a sandbag attached to the wrist [33], but the same effect may be obtained with manual traction. It should be emphasized that stressing the joint is important because most false-negative examinations occur because the articulation was not adequately stressed (see Fig. 15G–13*B,C*). A spot radiograph is then obtained with the arm in abduction. The patient is then rotated toward the affected shoulder until the anterior and posterior glenoid processes overlap (true AP scapular projection), which allows a true anteroposterior projection of the glenohumeral joint. In patients with a small supraspinatus tendon tear the site of the tear may in some instances be filled with contrast, and it can be precisely localized in this position if the arm is internally rotated and slowly elevated while the superior aspect of the joint is observed (see Fig. 15G–35). At this point overhead radiographs are obtained using the anteroposterior internal and external rotation projections as well as the axillary lateral, bicipital groove, and true AP scapular projections.

False-negative shoulder arthrograms are probably impossible to eliminate completely, but the great majority can be avoided by careful arthrographic technique. It is not adequate to inject the shoulder with contrast medium and walk away, instructing the radiographic technician to "shoot some films." The process must be a dynamic one with the arthrographer observing a full range of glenohumeral motion and applying stress appropriately in order to demonstrate small tears. Adequate distention of the joint is critical to supply sufficient pressure to displace the hematoma that may be present in small tears, preventing their filling. However, a fine line exists between adequately distending the joint and overinjecting it. Overinjection will result in contrast extravasation, usually from the subscapularis recess or through the bicipital tendon sleeve. When this occurs, the possibility of a false-negative film in the presence of a small tear is undoubtedly increased. Patient discomfort may present another problem. Patients with shoulder pain usually have a limited range of motion and frequently object to attempts by the arthrographer to manipulate the shoulder. However, with the use of intra-articular local anesthetic and slow and careful manipulation the articulation can usually be adequately stressed and mobilized.

At this time glenohumeral arthrography remains the gold standard for the diagnosis of full-thickness rotator cuff tears. However, there are some disadvantages to this technique. Shoulder arthrography is invasive and uncomfortable for the patient. A level of expertise is required on the part of the examiner to minimize discomfort and obtain a useful diagnostic film. Many pathologic conditions affecting the shoulder are not visualized by arthrography, including many incomplete rotator cuff tears. MRI is rapidly gaining ground on shoulder arthrography and will probably replace it for most purposes in the near future [31].

COMPUTED ARTHROTOMOGRAPHY

The technique for joint injection for computed arthrotomography [25, 62, 110, 132] is similar to that used for routine double contrast shoulder arthrography. I usually inject a smaller volume of contrast medium (1 to 2 mL) as well as a smaller total volume (10 mL) because spontaneous decompression of the joint through the subscapularis recess is a disaster that significantly degrades the quality of the examination. With ionic contrast agents the addition of 0.1 to 0.3 mL of epinephrine (1:1000) to 10 mL of contrast may be necessary to prolong retention of the contrast agent within the joint while computed tomography (CT) is performed. Epinephrine does not appear to be necessary with nonionic contrast agents, which seem to persist longer in the joint at high concentrations. Spot radiographs of the shoulder are obtained immediately after injection, but overhead radiographs are not usually obtained because any delay in obtaining the CT scan degrades the quality of the examination. The patient is immediately placed on the CT table, and images are obtained through the joint using 3-mm contiguous sections and a high-resolution, bone-imaging algorithm with the arm placed in a neutral position (Fig. 15G–2). Using this series to determine appropriate levels, the sections are then repeated with the arm internally rotated and then externally rotated, which stresses the anterior and posterior glenoid labra [101].

Using this technique, the size and shape of the cartilaginous glenoid labra as well as labral tears are well demonstrated (see Figs. 15G–2 and 15G–39). Rotator cuff tears can be verified by the presence of contrast within the subacromial and subdeltoid bursae (see Fig. 15G–40) and by the presence of contrast agent within the defect in the superior sections. In addition, the articular cartilage of the humeral head and glenoid are well seen, and other abnormalities such as humeral head impaction fractures (Hill Sachs lesions) or glenoid rim fractures (see Fig. 15G–39E) may also be seen.

ARTHROTOMOGRAPHY

The combination of tomography with arthrography provides a highly sensitive means of evaluating the cartilaginous glenoid labra [8, 27, 63, 100]. The joint is injected, as with computed arthrotomography, and thin-section polytomography performed. Patient positioning is critical and varies with the suspected pathology. Arthrotomography with the

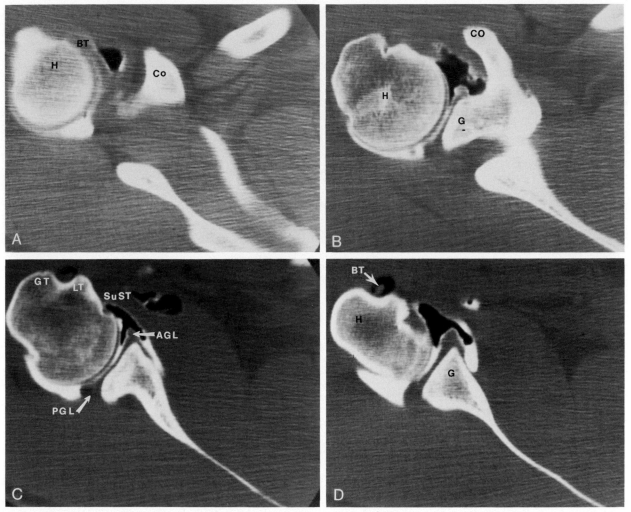

FIGURE 15G–2
Normal shoulder, computed arthrotomography. Normal anatomy is demonstrated by computed arthrotomographic sections at the level of the bicipital tendon origin *(A)*, the coracoid process *(B)*, the subscapularis tendon *(C)*, and the inferior joint level *(D)*. Bt, Bicipital tendon; H, humeral head; Co, coracoid process; G, glenoid process; GT, greater tuberosity; LT, lesser tuberosity; SuST, subscapularis tendon; AGL, anterior glenoid labrum; PGL, posterior glenoid labrum.

patient supine and rotated 40 to 45 degrees toward the affected shoulder demonstrates the superior and inferior glenoid labra as well as the undersurface of the rotator cuff [8]. A more useful examination is performed with the patient lying on his or her side with the arm fully abducted (axillary lateral position), which provides visualization of the anterior and posterior portions of the glenoid labra [27, 63] (Fig. 15G–3). Prior to injection of the joint it is important to obtain scout tomograms to verify appropriate patient position, radiographic technique, and tomographic levels so that the examination can be completed without delay following the injection before the contrast agent is resorbed and diluted. When this examination is properly performed the glenoid labra will be sharply outlined with radio-

graphic contrast medium and surrounded by air. Labral defects can be identified by abnormal morphology or the presence of radiographic contrast agent within the substance of the labrum (see Fig. 15G–38). Arthrotomography has also been suggested for the detailed localization and characterization of rotator cuff tears [60], especially the rare ones involving the infraspinatus, subscapularis, or teres minor tendons. The primary disadvantages of this technique are that it may be extremely uncomfortable for the patient, it is time-consuming, and it is associated with a relatively high radiation dose to the patient. At present the combination of CT with arthrography appears to be as sensitive as or superior to arthrotomography and has replaced it as the gold standard in the imaging of labral defects [25,

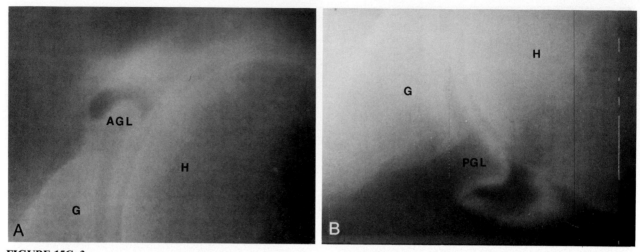

FIGURE 15G–3
Normal shoulder, arthrotomography. The appearance of the normal anterior glenoid labrum *(A)* and posterior glenoid labrum *(B)* is demonstrated in arthrotomography. For abbreviations see Anatomy Key at end of chapter.

39]. Even more recently, MRI shows considerable promise in providing a noninvasive means of imaging labral and capsular abnormalities.

COMPUTED TOMOGRAPHY

Plain computed tomography is primarily useful for evaluating the bony anatomy of the shoulder, especially with fractures of the humeral head or glenoid [21, 26]. The combination of high-resolution bone imaging algorithms with thin (1.5 to 3 mm) contiguous sections through the joint provide exquisite detail of the osseous anatomy of the glenohumeral joint. Plain CT is primarily useful for characterizing fractures and for locating and orienting fracture fragments about the shoulder (see Fig. 15G–44).

SUBACROMIAL BURSOGRAPHY

Subacromial bursography is a relatively simple procedure in which radiographic contrast agent is injected directly into the subacromial bursa. The technical aspects of subacromial bursography are quite simple and consist of introducing a 20-gauge 3½-inch needle immediately inferior to the anterior lip of the acromion and injecting 3 to 4 mL of contrast agent [69, 140]. A normal bursa usually communicates with the subdeltoid bursa and, less frequently, with a subcoracoid extension [140], and is visualized as a hemispherical contrast collection that overlies the superior aspect of the humeral head (Fig. 15G–4). Communication with the glenohu-

meral joint is indicative of a full-thickness rotator cuff tear. The procedure has been described as useful in the identification of small superior surface rotator cuff tears [32, 69, 107, 118]. In cases of subacromial bursitis (see Fig. 15G–11) the margins of the bursa may be small, irregular, or spiculated, whereas in severe cases bursal opacification may not be possible. Shoulder impingement syndrome has been described as being characterized by progressive distention of the bursa when the arm is abducted [17, 140] (see Fig. 15G–36). At present there seem to be few specific indications for this procedure, and we do not currently utilize it.

SONOGRAPHY

In recent years real-time sonography of the rotator cuff has been popularized as a simple and noninvasive way of imaging the rotator cuff [9, 19, 41, 72, 78]. A high-resolution scanner utilizing high-frequency transducers (7.5 to 10 MHz) is mandatory for this procedure. The patient sits facing the sonographer with the arm in the neutral (thumb up) position. In this position the bicipital tendon can be identified at the superolateral margin of the shoulder (Fig. 15G–5C), and the presence or absence of effusion in the subacromial-subdeltoid bursae can be noted [9, 30] (see Fig. 15G–15E). The supraspinatus insertion (Fig. 15G–5) can be identified immediately proximal to the greater tuberosity of the humerus. Only the distal portion of the supraspinatus tendon can be seen in this position before it is obscured by the acromion. From the initial position the arm is internally rotated to displace the

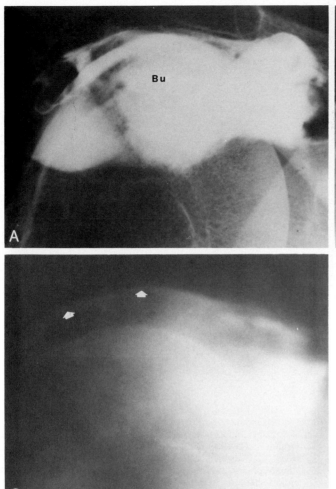

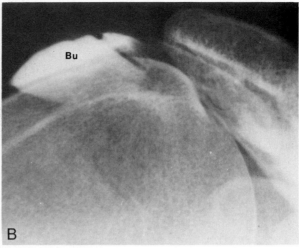

FIGURE 15G–4
Normal shoulder, subacromial bursography. The appearance of normal subacromial bursography in the neutral external rotation position *(A)* and in abduction *(B)* is demonstrated. A normal section from a bursatomogram is illustrated in *C.* Bu, Subacromial-subdeltoid bursa.

supraspinatus anteriorly, allowing a larger viewing window anterior to the acromion process [20]. The normal rotator cuff is sharply defined by a homogeneous internal pattern of intermediate echogenicity and measures 4 to 6 mm in diameter, normally being somewhat thinner posteriorly [19]. The subacromial-subdeltoid bursae are characterized by a thin, highly echogenic band paralleling the upper surface of the cuff [30]. Between the subacromial-subdeltoid bursae and the overlying deltoid muscle the peribursal fat plane [81] can be identified as a thin, sonographically "bright" arcuate band [9] (Fig. 15G–5). The overlying deltoid muscle is characterized by a "speckled" appearance that is quite distinct from that of the normal rotator cuff. This should be a dynamic study with passive shoulder motion used to demonstrate the normal motion of the rotator cuff and the humeral head as well as any potential defects occurring with changes in position. The deltoid muscle, in contrast to the rotator cuff, does not move as the shoulder is mobilized. Examination of both shoulders is performed routinely. Subacro-

mial-subdeltoid effusions are characterized by the presence of a dark bank of fluid density filling the bursae (see Fig. 15G–15E). Sonographic signs of a rotator cuff tear include a focal hypoechoic zone of cuff discontinuity, echogenic foci, central echogenic bands, abnormal cuff morphology, and nonvisualization of the rotator cuff [7, 30, 46, 136]. It should also be noted that recent assessments of the value of rotator cuff sonography have suggested that it is inferior in both sensitivity and specificity to arthrography and MRI [7]. At present we do not regularly use rotator cuff sonography in our practice.

MAGNETIC RESONANCE IMAGING

At this time MRI of the shoulder is still in evolution, but it appears to be well on the way toward replacing shoulder arthrography as the gold standard for imaging internal derangements of the glenohumeral joint. We utilize a 1.5-tesla whole-

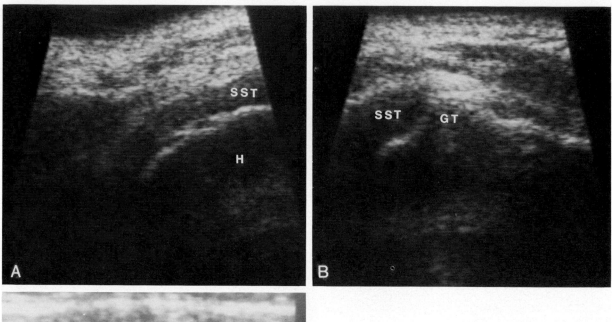

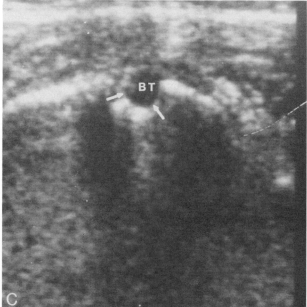

FIGURE 15G–5

Normal shoulder, sonography. Images from a normal shoulder sonogram demonstrate the appearance of the supraspinatus tendon in axial *(A)* and sagittal internal rotation *(B)* projections. *C,* The bicipital groove is demonstrated in an axial sonographic section. For abbreviations see Anatomy Key at end of chapter.

body magnetic resonance scanner.* A "wraparound" surface coil specially designed for the shoulder is utilized, although satisfactory studies may be obtained by using paired (anterior and posterior) 5-inch surface coils. At present our routine shoulder examination consists of multiecho sequences in the coronal oblique (Fig. 15G–6) and transaxial (Fig. 15G–7) imaging planes, which yield an intermediate weighted first echo and T2 weighted second echo images as well as a transaxial MPGR/30 sequence. The MPGR/30 series is T2† weighted and appears

*General Electric, Milwaukee, WI.

†T2 is the time constant for loss of phase coherence among spins oriented at an angle to the static magnetic field.

to provide better definition of the labral and capsular structures of the shoulder (Fig. 15G–8). The coronal oblique sequence is obtained from an axial localizing sequence with the long axis of the supraspinatus muscle utilized as the imaging plane. For both the coronal oblique and transaxial sequences echo times (TE) of 20 and 90 msec are utilized with a repetition time (TR) of 2000 msec. Slice thickness is 4 mm with a 1-mm skip, and a 14-cm field of view (FOV) is utilized. Two averages (NEX) are used, and the image is displayed with a 256 × 128 pixel matrix. Scan time for this sequence is 8:56 minutes. The MPGR/30 sequence utilizes a TE of 15 msec and TR of 500 msec with 3-mm sections at 1-mm skip, a 14-cm FOV, 2 NEX, and the image is displayed on a 256 × 192 pixel matrix. The transaxial images

Text continued on page 730

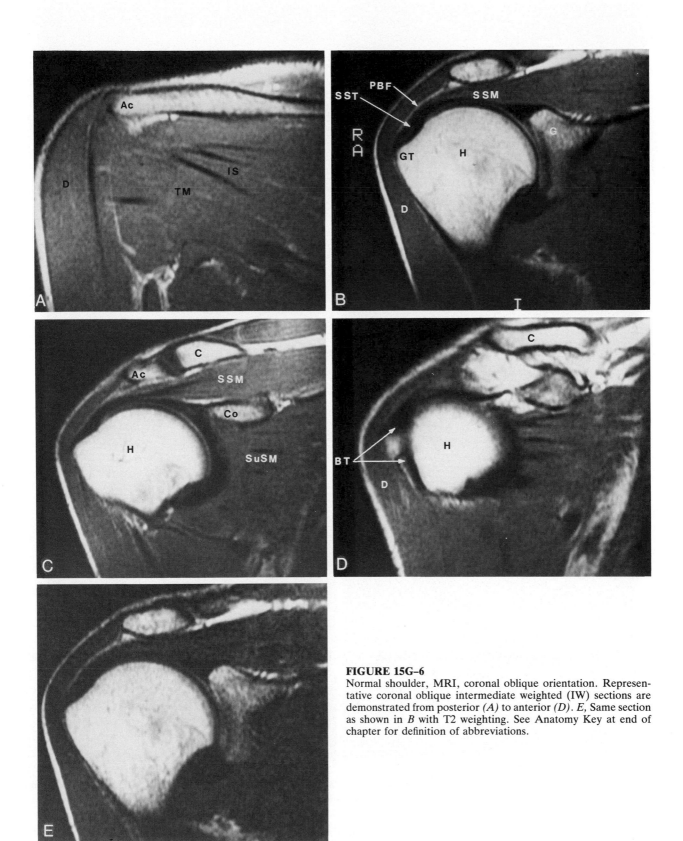

FIGURE 15G–6
Normal shoulder, MRI, coronal oblique orientation. Representative coronal oblique intermediate weighted (IW) sections are demonstrated from posterior *(A)* to anterior *(D)*. *E,* Same section as shown in *B* with T2 weighting. See Anatomy Key at end of chapter for definition of abbreviations.

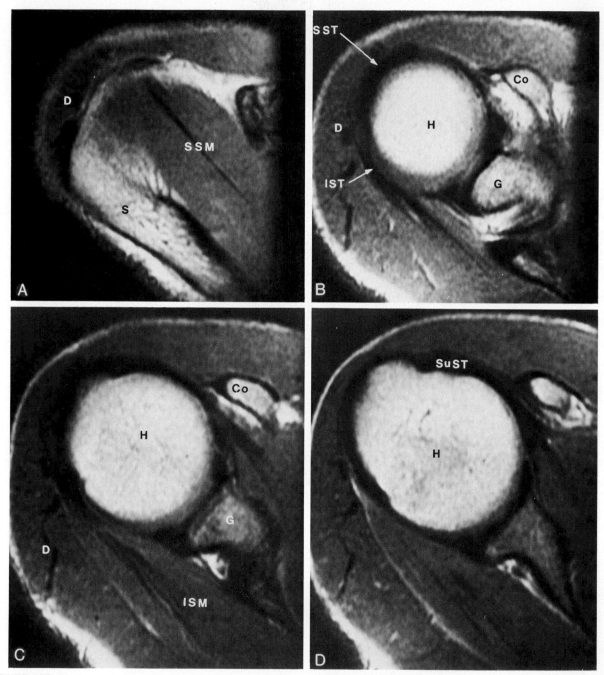

FIGURE 15G–7
Normal shoulder, MRI, transaxial orientation. Representative transaxial IW sections are demonstrated from superior *(A)* to inferior *(H)*. See Anatomy Key at end of chapter for definition of abbreviations.

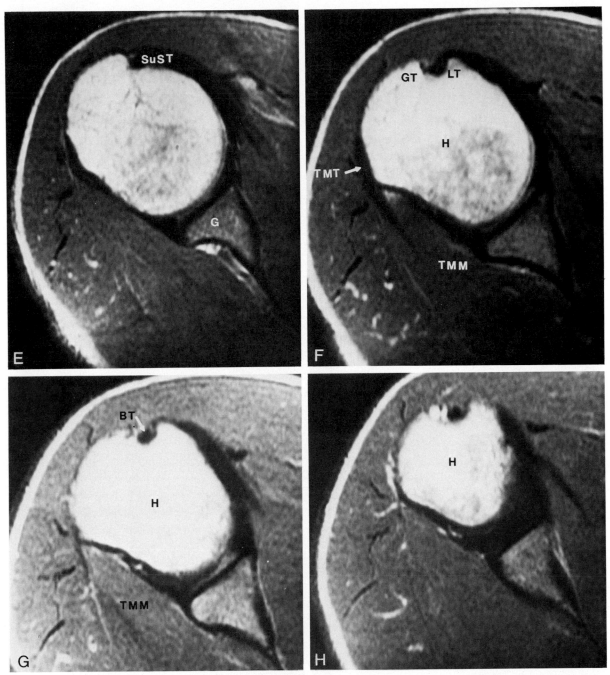

FIGURE 15G–7 *Continued*

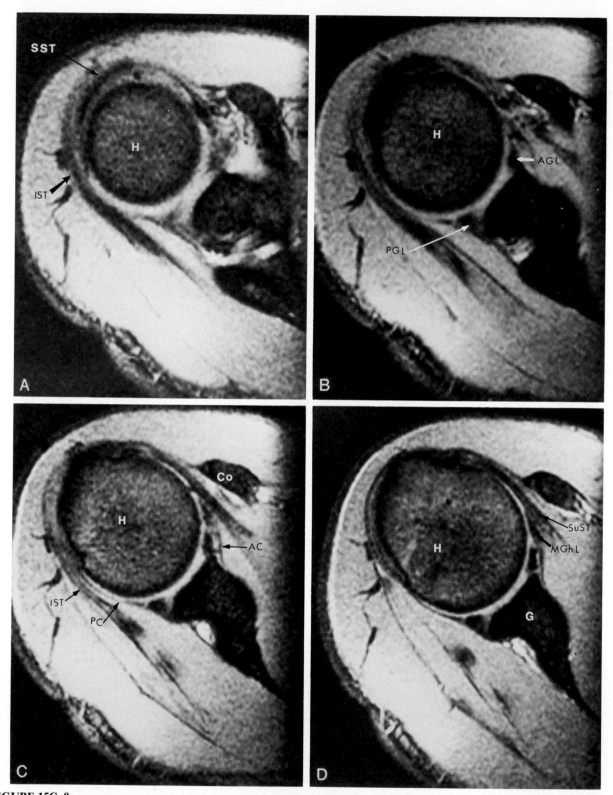

FIGURE 15G–8
Normal shoulder, MRI, transaxial MPGR/30 orientation. Transaxial MPGR/30 MRI sections from superior *(A)* to inferior *(I)* demonstrate superior visualization of the capsular structures and glenoid labrum. See Anatomy Key at end of chapter for definition of abbreviations.

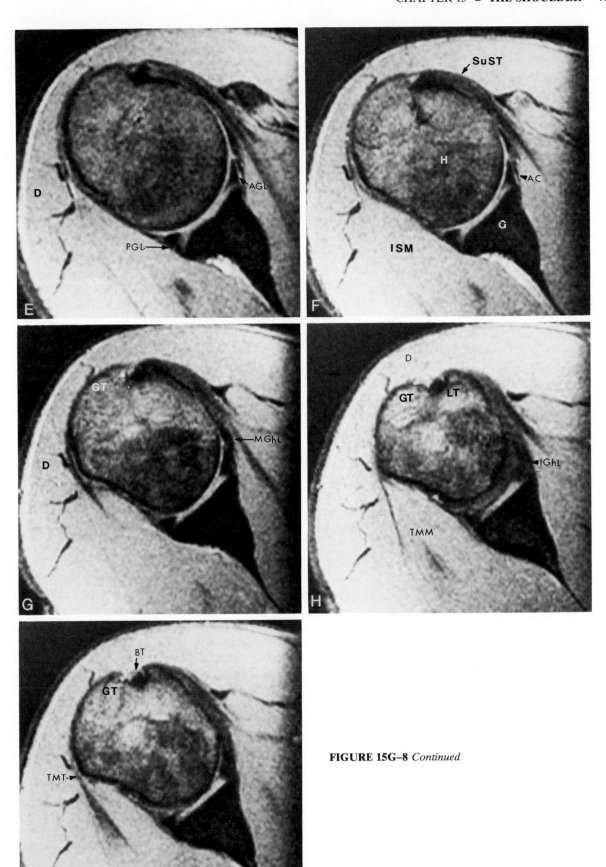

FIGURE 15G–8 *Continued*

are obtained using the coronal oblique sequence as a localizer. Using this technique, a thorough examination is obtained with a total scan time of less than 25 minutes, and the entire examination is easily scheduled in a 30-minute time slot.

As in any MRI examination the relatively long acquisition times require a very cooperative patient because motion artifact is the primary enemy of a good quality scan. Claustrophobia is a frequent problem, and patient sedation is often required. Patient positioning is critical, especially in the coronal oblique sequence. The patient's arm should be in neutral rotation (thumb up) or slight external rotation so that the supraspinatus tendon insertion onto the greater tuberosity is at the lateral border of the shoulder. In this way, the long axis of the muscle and tendon are displayed, and pathologic alterations are easily demonstrated. In internal rotation the supraspinatus insertion rotates anteriorly and is quite difficult to visualize. In the shoulder comparison of T1* or intermediate weighted (IW) sequences with similar T2 weighted sections is critical because identification and differentiation of the signals generated by fat, tendon degeneration, joint fluid, and hemorrhage form the basis of accurate MRI diagnosis (see Fig. 15G–21). In this regard, larger field strength magnets (i.e., 1.5 tesla) that allow true T2 weighted images with reasonable scan times as well as sophisticated software allowing off-center fields of view and oblique imaging offer a distinct advantage. The combination of MRI with intra-articular contrast has been suggested [40], but at present this technique remains investigational.

IMAGING ANATOMY

Detailed descriptions of the gross anatomy of the shoulder girdle and glenohumeral joint are well covered in numerous references [37]. In this section the relevant normal anatomy as it relates to various imaging modalities will be described as a basis for the recognition of pathologic processes [48, 51, 56, 96, 127, 148]. In addition, fine points of anatomic detail as they relate to identification of specific pathologic processes will be introduced in appropriate sections.

Osseous Anatomy [37]

The body of the scapula is a thin, flat sheet of bone that is closely applied to the posterior thorax

along its superior and lateral aspects. The scapula provides sites for muscular and ligamentous attachments as well as the glenoid process and acromion, which form articular components of the glenohumeral and acromioclavicular joints. The scapular spine serves as a site of muscular attachment and provides the means for suspending the acromion process. The coracoid process serves as the site of attachment for the coracoclavicular and coracoacromial ligaments, which are important in the stability of the shoulder girdle. The body of the scapula follows the slope of the posterior thorax and thus does not lie in the coronal anatomic plane but rather forms an angle of approximately 30 degrees to this plane with the lateral margin of the scapula anterior to the medial border. The glenoid process arises from the lateral margin of the scapula to form a shallow articular fossa, which is typically retroverted approximately 6 degrees relative to the scapular body [126]. The vertical plane of the scapula approximates the sagittal anatomic plane. The central cortex of the glenoid fossa is quite thin and thickens peripherally to form the bony glenoid labrum, which slightly deepens the concavity of the glenoid.

The humeral head is roughly hemispherical with a transverse diameter of 4.5 cm and a radius of curvature of approximately 2.25 cm [103], making it much larger than the shallow glenoid fossa (see Fig. 15–2B,C). This allows a nearly unrestricted range of motion but introduces a basic structural instability into the articulation. With the arm in the anatomic plane, the humeral head is retroverted approximately 30 degrees [123] relative to the long axis of the humerus with its articular portion separated from the adjacent metaphysis and tuberosities by a shallow groove that defines the anatomic neck and provides the site for capsular insertion. At the lateral border of the proximal humerus two tubercles arise, separated by a shallow groove (bicipital groove, intertubercular groove), along which passes the tendon of the long head of the biceps brachii muscle [16] (see Fig. 15G–7F,H). The greater tuberosity lies posterior to the bicipital groove and provides insertion points for the tendons of the supraspinatus, infraspinatus, and teres minor muscles (see Fig. 15G–7). The lesser tuberosity is smaller, lies anterior to the bicipital groove, and provides a point of insertion for the subscapularis tendon (see Fig. 15G–7). The bicipital groove is a structure of some importance because it provides the point of egress of the tendon of the long head of the biceps from the shoulder joint and is the spot at which the tendon is most vulnerable to direct trauma as well as forces tending to cause dislocation. The bicipital groove is a shallow semicircular depres-

---

*T1 is the time constant for alignment of spins along the external magnetic field.

sion in the superolateral aspect of the humerus with its posterior wall formed by the greater tuberosity and the anterior wall formed by the lesser tuberosity [16]. There appears to be a correlation between the depth of the groove and the susceptibility of the tendon to chronic tendinitis and rupture. A shallow angle of inclination of the medial wall is said to be associated with a tendency toward subluxation and dislocation of the tendon [45].

Soft Tissue Anatomy

The hyaline cartilage covering of the humeral head is thickest at the center of the articular process and thins progressively toward the periphery, whereas the converse is true for the glenoid, hyaline cartilage being thinnest at the center of the glenoid fossa [23] (see Fig. 15G–2). Along the periphery a fibrocartilaginous rim serves to deepen the glenoid fossa, of which the base is tightly attached to the periphery of the bony glenoid while its free edges are sharp and well defined (see Figs. 15G–2 and 15G–7). The anterior labrum tends to have a triangular shape in cross section while the posterior labrum is frequently more rounded [148]. Superiorly the cartilaginous labrum is contiguous with the insertion of the tendon of the long head of the biceps brachii (see Fig. 15G–9C). As with the menisci of the knee, the normal cartilaginous labrum demonstrates a homogeneously dark appearance on MRI with "crisp" outlines (see Fig. 15G–7). There is considerable individual variation in the size of the glenoid labrum. In some individuals it is a relatively large structure that overlies the peripheral glenoid articular cartilage and resembles the menisci of the knee whereas in others it is virtually absent [95]. There has been some controversy about the true histologic composition of the glenoid rim. In past years the results of Moseley and Overgaard were cited to show that the labrum was composed of fibrous tissue with scattered elastic fibers [83]. More recently, Prodromos and associates [108] have disputed these findings, reporting the labrum to be composed of fibrocartilage similar to the menisci of the knee.

The capsule of the glenohumeral joint arises along the periphery of the glenoid labrum and from the coracoid process (coracohumeral ligament) and passes, loosely, over the articular process of the humeral head to insert in the groove that defines the anatomic neck of the humerus. The capsule is quite loose and demonstrates a prominent inferior area of redundancy, the axillary recess (see Fig. 15G–1A). The capsule is fenestrated anteromedially

to communicate with the subscapularis bursa and superolaterally to communicate with the bicipital sleeve, allowing the tendon of the long head of the biceps to pass out of the articulation. A less constant posterior fenestration may be present to allow communication with the infraspinatus bursa. The capsule is lined with synovium, which envelopes the bicipital tendon and passes into the bicipital sleeve to approximately the level of the anatomic neck of the humerus. Anteriorly the capsule is characterized by three discrete areas of fibrous thickening forming capsular ligaments termed the glenohumeral ligaments (see Fig. 15G–8). The superior glenohumeral ligament arises from the apex of the glenoid with or adjacent to the long head of the biceps and inserts onto the lesser tuberosity of the humerus [22]. The superior glenohumeral ligament is quite variable in size and probably contributes little to glenohumeral stability [95]. The middle glenohumeral ligament passes from the superior aspect of the glenoid labrum or adjacent glenoid neck and inserts onto the base of the lesser tuberosity. It varies somewhat in size but may provide a secondary restraint for anterior translation of the humeral head [127]. The inferior glenohumeral ligament takes origin from the anterior-inferior margin of the labrum or adjacent glenoid neck and inserts onto the anatomic neck of the humerus [23]. The inferior glenohumeral ligament is a complex structure that functions as the primary static stabilizer in the abducted shoulder and thus is quite important in glenohumeral instability. It is a broad structure that may be divided into a thickened anterior band arising from the middle third of the glenoid labrum, a thinner axillary portion, and a thick posterior band that arises from the inferior half of the posterior labrum [95]. The glenohumeral ligaments may be identified on MRI sections as homogeneous intermediate to dark zones of capsular thickening, similar in appearance to tendons (see Figs. 15G–7 and 15G–8).

The next outward layer consists of the tendons of the supraspinatus, infraspinatus, teres minor, and subscapularis muscles, which blend with the outer layer of the capsule to complete a dense fibrous cuff, the rotator cuff, which surrounds the articulation. The intrinsic muscles of the shoulder consist of the supraspinatus, infraspinatus, teres minor, subscapularis, teres major, and deltoid. The first four comprise the rotator cuff group. The supraspinatus muscle arises from the supraspinous fossa of the scapula and passes laterally, deep to the trapezius and under the acromion process, acromioclavicular joint, and coracoacromial ligament to blend with the superior aspect of the glenohumeral joint capsule before inserting on the superior aspect of the greater tuberosity of the humerus (see Fig. 15G–6B,C). The plane of the muscle approximates the plane of the body of the scapula, an important point to bear in mind when selecting MRI planes (see Fig. 15G–7A). The primary role of the supraspinatus is abduction of the humerus. The infraspinatus arises from the infraspinous fossa of the scapula and passes laterally to insert on the greater tuberosity of the humerus slightly lower and posterior to the supraspinatus insertion (see Fig. 15G–7B–E). A bursa (infraspinatus bursa) lies between the belly of the infraspinatus muscle and the scapula and occasionally communicates with the intracapsular space either as a normal variant or with infraspinatus tendon tears (see Fig. 15G–24). The teres minor muscle arises from the middle portion of the lateral scapular border and passes obliquely in a cephalic direction to insert on the posterior aspect of the greater tuberosity of the humerus, posterior and inferior to the infraspinatus insertion (see Fig. 15G–7F–H). Both the infraspinatus and teres minor muscles function as external rotators of the humerus. The subscapularis muscle arises from the subscapular fossa on the anterior surface of the scapula and passes laterally to insert as a broad tendon into the anterior capsule and anterior border of the lesser tuberosity of the humerus. The subscapularis muscle functions as an internal rotator of the shoulder. The tendons and fibrous capsule of the rotator cuff appear as low intensity (dark) on MRI images (see Figs. 15G–6 and 15G–7) and intermediate density, similar to muscle, on computed tomographic images (see Fig. 15G–2). The deltoid is a powerful muscle that arises from the lateral third of the clavicle, the acromion, and the spine of the scapula and passes laterally as a broad band that covers the anterior, posterior, and superior aspects of the glenohumeral articulation (see Figs. 15G–6 and 15G–7). The deltoid inserts on the deltoid tuberosity along the lateral border of the humeral shaft and functions as the major abductor of the shoulder. The teres major arises along the inferior third of the lateral border of the scapula and passes anteriorly to insert on the humeral shaft just below the subscapularis insertion. The primary function of the teres major is internal rotation of the humerus, but, in concert with the latissimus dorsi, it also contributes to extension and adduction of the humerus. Another large group of muscles comprising the extrinsic shoulder muscles are also well seen on MRI. These include the trapezius, serratus anterior, latissimus dorsi, subclavius, rhomboideus major and minor, and pectoralis major and minor. The biceps brachii originates as two heads from the glenoid region. The short head of the biceps originates from the coracoid process and passes over the anterior aspect of the humeral head, superficial to the subscapularis before joining with the belly of the long head of the biceps to form the biceps brachii muscle. The tendon of the long head of the biceps is a unique structure that arises from the poster-superior portion of the glenoid labrum (see Fig. 15G–9C) and adjacent supraglenoid tubercle and passes as an intra-articular structure over the humeral head to exit the joint through the bicipital groove (see Fig. 15G–7). The intra-articular portion of the tendon is extrasynovial, being invested in a thin layer of visceral synovium. At the level of the bicipital groove a synovially lined sheath transmits the tendon out of the joint, terminating at approximately the level of the surgical neck of the humerus (see Fig. 15G–1A). A short dense ligament, the transverse ligament, passes transversely between the humeral tuberosities forming the roof of the bicipital groove (see Fig. 15G–8E,F).

IMAGING EVALUATION OF INJURIES OF THE SHOULDER

Subacromial-Subdeltoid Bursitis

The subacromial and subdeltoid bursae normally are contiguous and form a common hemispherical space extending from the level of the acromioclavicular joint laterally to overlie the greater tuberosity of the humerus (see Fig. 15G–4A). Superiorly the bursa is bounded by the undersurface of the acromion and deep surface of the deltoid muscle. Inferiorly the bursa is in intimate contact with the upper surface of the rotator cuff. Synovitis or effusion in the subacromial-subdeltoid bursa is a common finding in a variety of abnormalities of the shoulder, and as such it forms a worthwhile point for beginning a discussion of imaging pathologic alterations of the shoulder. Subacromial-subdeltoid bursitis can be directly imaged by subacromial bursography (Fig. 15G–11) and can be inferred by sonography, MRI, and plain radiographs [9, 29, 111, 145]. Almost invariably subacromial bursitis is secondary to another process. Primary subacromial-subdeltoid bursitis [125] resulting from a throwing injury has been suggested but has not been identified in imaging studies. Indeed, it has been suggested that primary bursitis is seen only in rheumatoid arthritis, tuberculosis, gout, and pyogenic infections [90]. In sports medicine the presence of subacromial-subdeltoid bursitis should suggest the presence of a rotator cuff tear or a shoulder impingement syndrome [29].

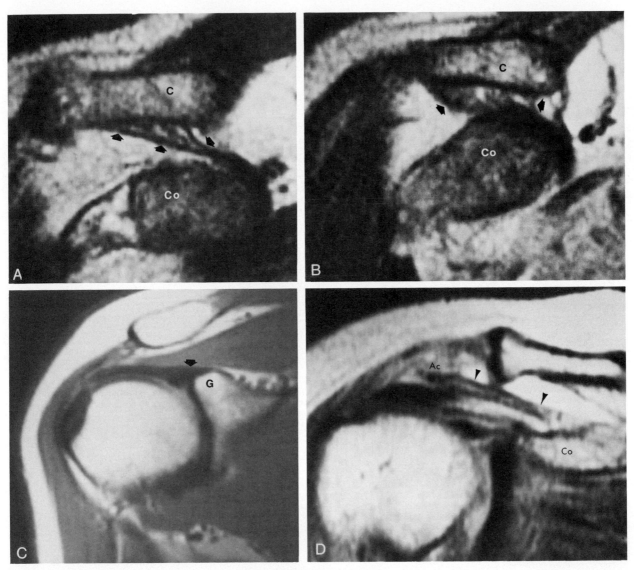

FIGURE 15G–9
Normal shoulder, MRI, ligamentous anatomy. *A* and *B* coronal oblique IW (2000/20) MRI sections demonstrate the trapezoid *(A, arrows)* and conoid *(B, arrows)* ligaments that comprise the coracoclavicular ligamentous complex. *C,* The origin of the tendon of the long head of the biceps (arrow) is seen to arise from the superior aspect of the glenoid (G) process in this coronal oblique IW (2000/ 20) image. *D,* The coracoclavicular ligament (arrowheads) is demonstrated passing from the coracoid process (Co) to the anterior inferior margin of the acromion (Ac) process in a coronal oblique IW (2000/20) MRI section. C, clavicle.

On MRI the normal subacromial-subdeltoid bursa is not directly visualized. However, the location of the bursa may be identified by the presence of the peribursal fat plane [81], which is felt to represent the apposed fatty layers associated with the synovial lining of the subacromial-subdeltoid bursa (see Fig. 15G–6B). There are two MRI criteria that suggest the diagnosis of subacromial-subdeltoid bursitis. These are loss of definition of the normal peribursal fat plane between the bursa and the rotator cuff and the presence of fluid within the bursa (Fig. 15G–10). It has been suggested that in the absence of inflammatory arthritis or infection a visible quantity of fluid can accumulate in the bursa only in the presence of a rotator cuff tear [149]. Loss of the normal peribursal fat plane may be seen with distention of the bursa by fluid or with inflammatory thickening of the bursal walls. In patients with chronic bursitis, usually associated with chronic rotator cuff tendinitis, the peribursal fat plane may also be obliterated in the absence of bursal wall thickening or bursal effusion.

On plain radiographs the peribursal fat plane is best seen in internal rotation views, and its absence is associated with a increased incidence of rotator cuff tears. However, its utility is limited by the fact that it is seen in only 60% of normal individuals, and it may be seen in 21% of patients with rotator

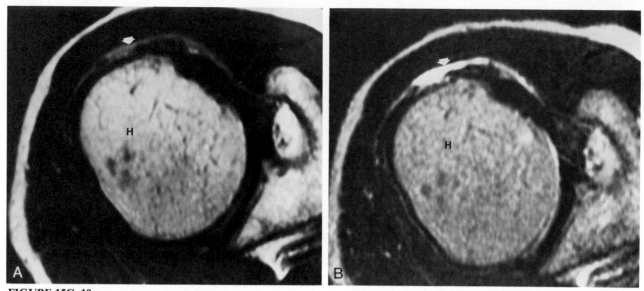

FIGURE 15G–10
Subacromial-subdeltoid bursal effusion, MRI. Transaxial IW 2000/20 *(A)* and transaxial T2 2000/80 *(B)* MRI sections through the superior aspect of the humeral head (H) demonstrate a large effusion (arrows) in the subacromial-subdeltoid bursa. Note the intermediate signal intensity of the fluid on the IW image *(A)* and its marked brightening in the T2 weighted image *(B)*. Coronal oblique images of the same patient are illustrated in Fig. 15G–45*E* and *F*.

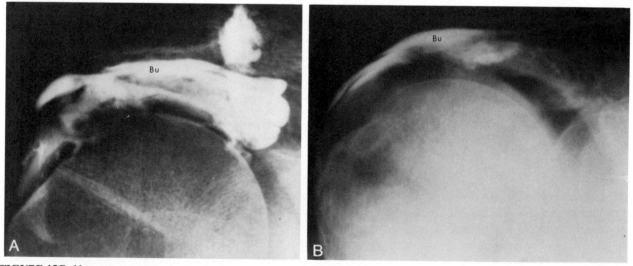

FIGURE 15G–11
Subacromial bursitis, subacromial bursography. Frontal radiographs from two patients with subacromial bursitis demonstrate typical abnormalities. *A*, Subacromial bursa (Bu) demonstrates irregular spiculated margins; nodular defects due to synovitis are also present. Communication with the acromioclavicular joint is seen. *B*, The bursa (Bu) is constricted with spiculated margins.

cuff tears [81]. An enlarged subacromial-subdeltoid bursa, distended by effusion, is occasionally seen on plain radiographs as a water dense mass between the superior aspect of the humeral head and the undersurface of the acromion [145]; however, this is not a reliable finding. Certainly the presence of a "fat-fluid level" (lipohemarthrosis) on upright or cross-table radiographs may be associated with humeral head fractures.

Rotator Cuff Abnormalities

Any review of rotator cuff pathology will rapidly expose the supraspinatus tendon as the weak link of the rotator cuff. A substantial majority of cuff failures occur in the supraspinatus tendon near or at its insertion into the greater tuberosity of the humerus [15]. Most investigators have attributed this tendency for cuff failure to occur within the supraspinatus tendon to its blood supply. The supraspinatus tendon receives its arterial supply from the anterior humeral circumflex, subscapular, suprascapular, and posterior humeral circumflex arteries [70, 123]. A zone of relative avascularity in the tendon proximal to its insertion [70, 123] has been described and may represent a "critical zone" for cuff failure. Other authors have found this area to be richly vascularized by anastomosing vessels from the tendon and humeral tuberosity [82]. It has also been suggested that arterial filling of the cuff vessels in the critical zone depends to a great degree on the position of the arm, poor filling being present when the arm is adducted [112]. Recently, a high association (95%) [86] between rotator cuff tears and subacromial impingement has been suggested, so perhaps it is the location of the supraspinatus tendon and its susceptibility to impingement that is the real cause of this predisposition. Regardless of the precise cause, all of the major abnormalities of the rotator cuff are most common in, though they are not limited to, the critical zone of the supraspinatus tendon.

It is useful, though undoubtedly oversimplistic, to consider rotator cuff pathology on the basis of tendinitis, cuff tears (partial and complete), cuff atrophy, calcific tendinitis, and adhesive capsulitis. A brief overview of these entities will be followed by descriptions of the abnormalities as revealed by the appropriate imaging modalities.

Tendinitis. Rotator cuff tendinitis is a very common clinical diagnosis in all age groups. Some consider rotator cuff tendinitis an example of localized fiber rupture within the cuff manifested as an acute self-limited process [73]. As such, each episode of "cuff tendinitis" results in weakening of the cuff and a predisposition toward further such injury. Others consider a form of rotator cuff tendinitis, especially in younger athletes, to be a primary process without fiber failure [50]. This latter type overlaps with shoulder impingement, which will be discussed in a later section. Sonography has not proved especially useful in identifying rotator cuff tendinitis. It is reasonable to assume that inflammatory changes and local edema may be responsible for the frequent presence of broad bands of increased echogenicity in patients with shoulder pain and without evidence of a rotator cuff tear (Fig. 15G–12). However, the validity of this assumption is seriously limited by the frequent presence of identical findings in the contralateral asymptomatic shoulder in these patients [7]. In the presence of rotator cuff tendinitis (see Fig. 15G–17) MRI demonstrates increased signal intensity in the supraspinatus tendon, near its insertion [150]. These changes are best seen on T1 and IW images and do not increase in T2 weighted images. Presumably, this is a manifestation of an increased concentration of free water in the involved area associated with edema and inflammation. Fluid density, as manifested by very bright T2 weighted signals, is not present in the cuff or in the adjacent subacromial-subdeltoid bursa. The peribursal fat plane is not effaced and remains quite prominent. Shoulder arthrography is normal in these patients [151]. Pathologic evaluation in these patients demonstrates inflammation and mucoid degeneration [57].

Cuff Tears. Rotator cuff tears are relatively rare in individuals under the age of 40 [113] but become progressively more common with advancing age. As previously noted, the supraspinatus tendon is by far the most common site of cuff tears either at or near its insertion into the humeral tuberosity [18], although tears involving other cuff tendons are not rare [18, 141]. Acute full-thickness rotator cuff tears resulting from a single defined injury are relatively rare [15]. In the majority of cases cuff failure represents the end point of a continuum of injury, frequently involving chronic shoulder impingement syndrome [99]. Partial-thickness tears are about twice as common as full-thickness tears and may involve the undersurface or the bursal surface or may occur as an intrasubstance tear. In throwing athletes partial tears consisting of localized areas of avulsion of the supraspinatus insertion may be common [94].

Plain radiography is frequently useful in the initial evaluation of patients with rotator cuff tears, primarily to allow the identification of associated pathology. Elevation of the humeral head with narrow-

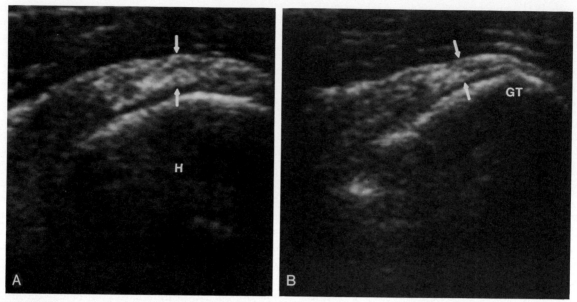

FIGURE 15G–12
Supraspinatus tendinitis, shoulder sonography. Transaxial *(A)* and sagittal internal rotation *(B)* sonographic images demonstrate increased echogenicity within the substance of the supraspinatus tendon (between arrows). Subsequent shoulder arthrography did not demonstrate a rotator cuff tear. H, Humeral head; GT, greater tuberosity.

ing of the normal space between the superior aspect of the humeral head and the undersurface of the acromion to less than 7 mm [28] may be seen in chronic, retracted rotator cuff tears [66] as well as in cuff atrophy. A "scalloped" contour of the undersurface of the acromion with loss of the normal convex shape may also be seen in patients with long-standing rotator cuff tears [38]. Loss of definition of the peribursal fat plane has also been described as suggestive of rotator cuff tears [81]. However, as previously noted, this sign is of limited value.

At the present time shoulder arthrography remains the gold standard for the identification of full-thickness rotator cuff tears with a reported accuracy of 99% [80]. Either single or double contrast techniques may be utilized, as discussed previously. The critical sign in the identification of a full-thickness cuff tear is the appearance of contrast agent(s) in the subacromial-subdeltoid bursa (Fig. 15G–13). With large tears immediate "flooding" of the bursa occurs, and the abnormality is difficult to miss. With smaller tears there may not initially be communication between the joint and the bursa, and in these individuals the shoulder arthrogram must be a "dynamic" examination with the glenohumeral joint taken through a full range of motion, often with external stress applied (Fig. 15G–13*B,C*). Small tears are most likely to fill when the shoulder is maximally rotated internally or fully abducted. Stress may be applied manually with distraction of the patient's arm during shoulder motion or at a

specific shoulder position or by having the patient exercise the extremity with a sandbag in the hand prior to obtaining overhead radiographs. In general, the precise site and size of the cuff tear cannot be reliably identified with arthrography. In some instances, with the patient in a true AP scapular position and the arm internally rotated, the tear can be identified radiographically. Upright radiographs sometimes demonstrate the tear outlined with contrast and distended with air, but unfortunately, the subdeltoid bursa is often flooded with contrast, obscuring the outline of the cuff. In general, a positive shoulder arthrogram demonstrates the result of the cuff tear (i.e., bursal filling) rather than the abnormality itself. With an intact cuff, only the inner surface is visualized on arthrography. A careful examination for focal contrast accumulation within the cuff or at the insertion of the supraspinatus tendon should be performed to identify incomplete undersurface tears or partial avulsions (Fig. 15G–14). It cannot be overemphasized that, as previously noted, the accuracy of shoulder arthrography is dependent on the care with which it is performed.

As previously noted, sonography may also be utilized to evaluate the rotator cuff. Sonographic signs of a cuff tear include focal discontinuity, abnormal morphology, focal or diffuse echogenicity, and fluid in the subacromial-subdeltoid bursa (Fig. 15G–15). Although there is no doubt that sonography can be used to identify rotator cuff tears, the

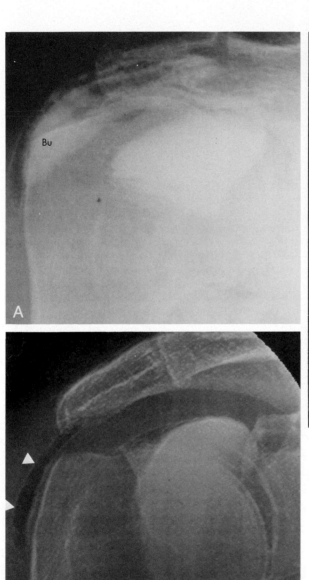

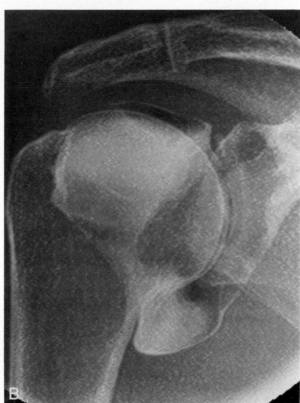

FIGURE 15G–13
Rotator cuff tear, shoulder arthrography. *A,* Frontal external rotation radiograph demonstrates air and radiographic contrast within the subacromial-subdeltoid bursa (Bu) following glenohumeral arthrography. *B,* An initial radiograph after joint injection appears normal, whereas in a postexercise *(C)* radiograph a prominent air collection (arrowheads) within the bursa is readily evident.

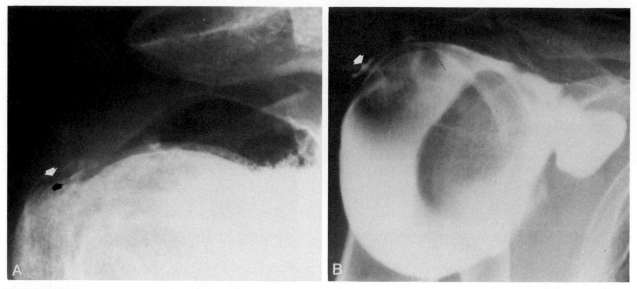

FIGURE 15G–14

Incomplete rotator cuff tear, shoulder arthrography. *A,* Spot radiograph in external rotation demonstrates a small collection of contrast within the supraspinatus tendon (white arrow) near its insertion into the greater tuberosity (black arrow) of the humerus. B, Internal rotation spot radiograph demonstrates radiographic contrast (white arrow) dissecting within the substance of the supraspinatus tendon.

lack of specificity of the procedure as well as the excellent sensitivity and specificity of other modalities place its value in doubt [7, 78].

MRI of the rotator cuff is a new and exciting area of shoulder imaging. Accuracy in the identification of rotator cuff tears (sensitivity 100%, specificity 95%) [49] similar to that of shoulder arthrography has recently been reported. On MRI the tendons of the rotator cuff are characterized by smooth sharp outlines with a homogeneously dark appearance on T1, IW, and T2 sequences that smoothly blends into the normal intermediate signal of muscle at the musculotendinous junction [128, 142, 148] (see Figs. 15G–6 to 15G–8). A rotator cuff tear (Fig. 15G–16) is visualized as a bright zone on a T1 or IW image that increases in intensity on T2 weighted images [11, 29, 59, 64, 65, 76, 77, 102, 149]. This latter point is critical because it is fluid (effusion, hemorrhage) filling the defect that is visualized in the cuff defect. It has already been noted that cuff tendinitis (Fig. 15G–17) also results in lesions that are bright on T1 and IW images [57, 129], but these lesions do not brighten on T2 weighted images and tend to be less prominent. It is the direct visualization of fluid filling a rotator cuff tear that allows the greatest certainty of diagnosis. Unlike arthrography, MRI directly visualizes the cuff defect rather than a consequence of the defect (abnormal joint–bursal communication), and thus the location and size of the tear can be accurately identified [47, 49]. The morphology of the musculotendinous unit must also be evaluated. A normal tendon has smooth, "crisp"

margins with gradual changes in diameter [77] (Fig. 15G–6*B*). Thinning or irregularity of the margins of the tendon should be considered abnormal [111] (Fig. 15G–18). The musculotendinous junction should smoothly blend into the normal intermediate signal of muscle. Abrupt changes in signal intensity and retraction of the musculotendinous junction are indicative of large full thickness tears (Figs. 15G–19 and 15G–20). The muscular portion should be of normal size and homogeneous in texture. In the presence of large long-standing tears the muscle may be atrophic with prominent high-intensity fatty bands noted within and parallel to the long axis of the muscle (see Fig. 15G–20). Identification of fluid within the subacromial-subdeltoid bursa is accomplished by noting a crescentic band of low to moderate intensity paralleling the outer margin of the rotator cuff on T1 or IW images that markedly brightens on T2 weighted images [10, 150] (see Fig. 15G–10). As previously noted, in the absence of direct trauma or infectious or synovial inflammatory processes, the glenohumeral joint must be the source of significant fluid accumulations in this bursa. The peribursal fat plane may be identified as a thin crescentic band of high-intensity signals on T1 and IW images paralleling the upper border of the rotator cuff (see Fig. 15G–6*B*). In contrast to the situation with bursal effusions, the peribursal fat plane does not "brighten" on T2 weighted images but rather becomes less prominent. Absence of the peribursal fat plane suggests severe tendinitis or a rotator cuff tear.

Text continued on page 743

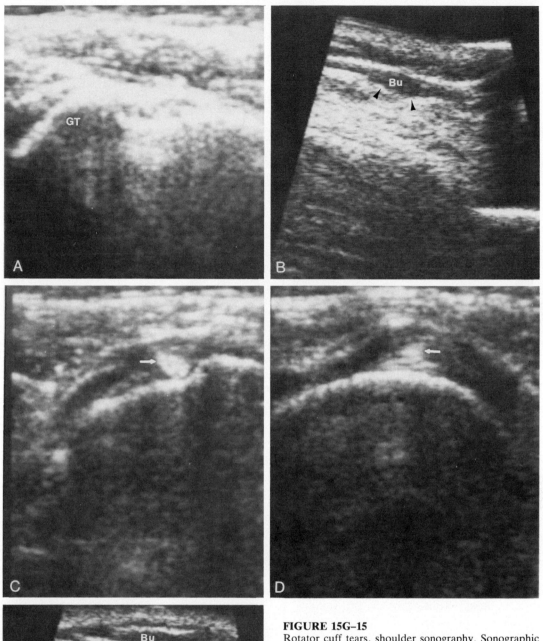

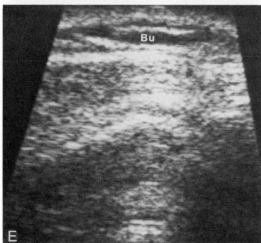

FIGURE 15G–15

Rotator cuff tears, shoulder sonography. Sonographic abnormalities that may be associated with rotator cuff tears are demonstrated. *A,* Absence of the normal supraspinatus insertion into the greater tuberosity (GT) of the humerus. *B,* Morphological alteration of the supraspinatus tendon (arrowheads) consisting of an area of indentation is associated with an overlying bursal effusion (Bu) in this transaxial sonographic section. Sonographic sections in *C* (transaxial) and *D* (sagittal internal rotation) show a "bright" spot (arrows) within the substance of the supraspinatus tendon in patients with subsequently documented rotator cuff tears. *E,* A bursal effusion (Bu) is demonstrated as an elliptical anechoic fluid collection associated with a deeper zone of hyperechogenicity, which is characteristic of a fluid collection.

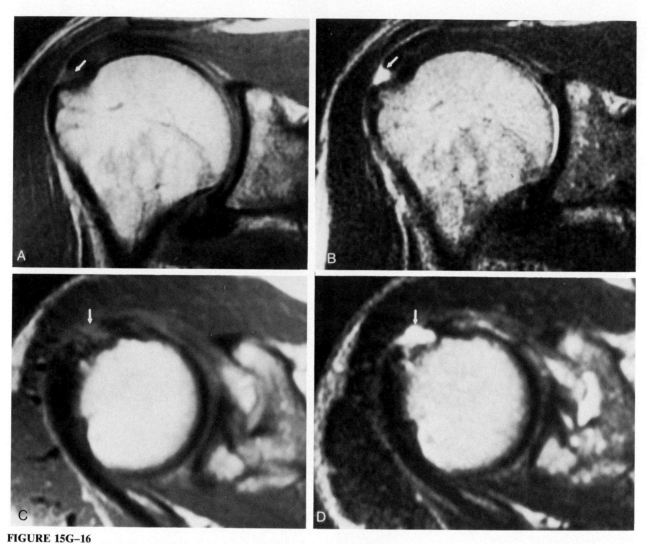

FIGURE 15G–16

Rotator cuff tear, MRI. Coronal oblique *(A)* and transaxial *(C)* IW (2000/20) images demonstrate a zone of intermediate signal intensity (arrow) within the substance of the supraspinatus tendon near its insertion. Corresponding coronal oblique *(B)* and transaxial *(D)* T2 weighted (2000/80) images demonstrate marked brightening of these zones (arrows). These changes are characteristic of fluid and are indicative of a rotator cuff tear. Note the small bursal effusion adjacent to the superior aspect of the supraspinatus tendon in the T2 coronal oblique image *(B)*.

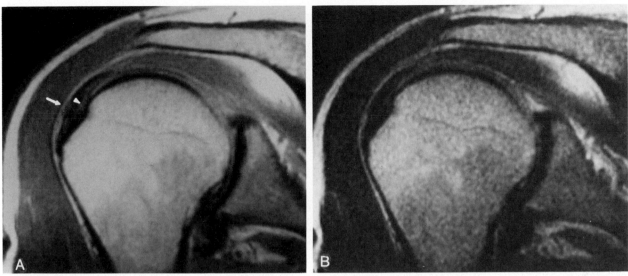

FIGURE 15G–17
Rotator cuff tendinitis, MRI. Coronal oblique IW *(A)* and T2 *(B)* weighted images through the supraspinatus tendon insertion demonstrate a zone of increased intensity (arrowhead) on the IW image that does not brighten on the corresponding T2 image. Also note the presence of a prominent peribursal fat plane (arrow) in the IW image as well as the absence of a bursal effusion; these signs classify this lesion as grade 1 rotator cuff tendinitis. Contrast these changes with those of a rotator cuff tear demonstrated in Figure 15G–16*A* and *B*.

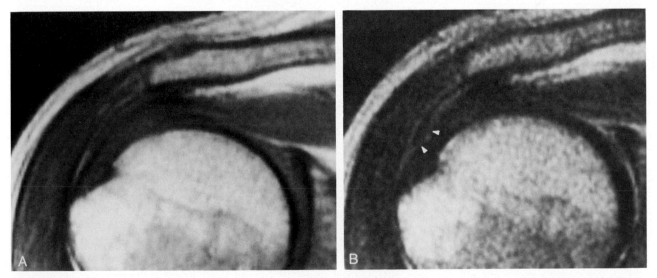

FIGURE 15G–18
Incomplete rotator cuff tear, MRI. Coronal oblique IW *(A)* and T2 *(B)* MRI images in this patient show a morphologic abnormality consisting of an "indentation" (arrowheads) of the superior surface of the supraspinatus tendon, which is most obvious in the T2 weighted image *(B)*. Arthrography was normal in this patient, and the changes are consistent with an incomplete disruption of the superior surface of the tendon, probably secondary to chronic shoulder impingement.

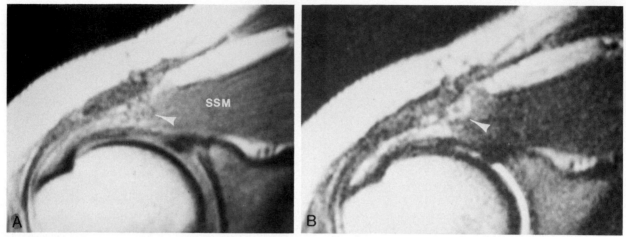

FIGURE 15G–19
Acute supraspinatus tendon rupture, MRI. Coronal oblique IW *(A)* and T2 *(B)* MRI images through the plane of the supraspinatus muscle (SSM) and tendon demonstrate the findings of acute supraspinatus rupture. The normal supraspinatus tendon is absent, being replaced by a zone of increased IW and T2 signal intensity. Small serpiginous densities (arrowheads) are noted adjacent to the musculotendinous junction and represent ruptured fiber bundles. In addition, the humeral head is elevated relative to the glenoid.

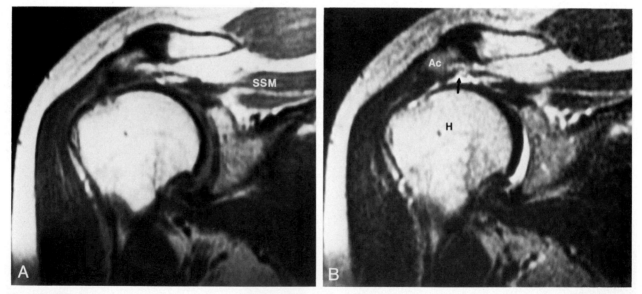

FIGURE 15G–20
Chronic supraspinatus tendon tear with retraction, MRI. Coronal oblique IW *(A)* and T2 *(B)* images in this patient demonstrate the changes of long-standing supraspinatus tendon tear with retraction. Small irregular fragments of residual tendon are identified adjacent to the greater tuberosity of the humerus, and effusion is visible in the subacromial-subdeltoid bursa (arrow) on the T2 weighted image. The supraspinatus muscle (SSM) is quite thin and infiltrated with fat; its musculotendinous junction is retracted medially. The humeral head (H) is elevated and approaches the undersurface of the acromion (Ac).

Zlatkin and colleagues [149] have proposed an MRI grading system (Fig. 15G–21) similar to that utilized for the knee to bring some order to this area. The grading system encompasses four grades (0 to 3) based on MRI observations. Grade 0 (see Fig. 15G–6B) is characterized by a normal tendon signal and morphology with a normal peribursal fat plane. Grade 1 (see Fig. 15G–21A,B) indicates a tendon with increased nonfluid signal intensity but with normal morphology and a normal peribursal fat plane. Grade 2 (Fig. 15G–21C,D) describes tendons with a combination of increased signal intensity and morphologic changes. The peribursal fat plane may be normal or absent with grade 2 changes. Grade 3 (see Fig. 15G–21E,F) is applied when an area of tendon discontinuity, typically associated

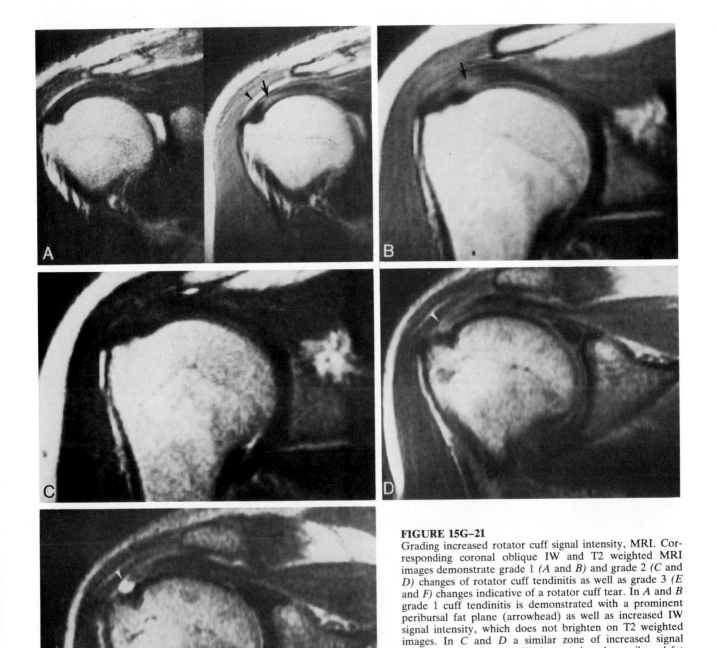

FIGURE 15G–21
Grading increased rotator cuff signal intensity, MRI. Corresponding coronal oblique IW and T2 weighted MRI images demonstrate grade 1 *(A and B)* and grade 2 *(C and D)* changes of rotator cuff tendinitis as well as grade 3 *(E and F)* changes indicative of a rotator cuff tear. In *A* and *B* grade 1 cuff tendinitis is demonstrated with a prominent peribursal fat plane (arrowhead) as well as increased IW signal intensity, which does not brighten on T2 weighted images. In *C* and *D* a similar zone of increased signal (arrow) is noted on the IW images, but the peribursal fat plane is absent indicating grade 2 cuff tendinitis. In *E* and *F* grade 3 changes of rotator cuff tear are evident with marked T2 brightening of the zone of increased signal intensity (arrowheads) noted on the IW image.

with T2 brightening, is identified. The peribursal fat plane is typically absent with grade 3 changes, and fluid is typically present in the subacromial-subdeltoid bursa. Based on these criteria, grade 0 represents a normal rotator cuff, grade 1 or grade 2 changes with a normal peribursal fat plane indicate tendinitis, and grade 2 or grade 3 changes with an absent peribursal fat plane and fluid in the bursa indicate a rotator cuff tear. Evaluation of these and similar criteria has resulted in impressive reports of sensitivity and specificity in the diagnosis of rotator cuff tears with MRI [29, 111, 149]. No system is complete, however, and Rafii and colleagues [111] reported several cases of confirmed tears with tendon defects demonstrating intermediate or low-intensity signals on T2 weighted images. They proposed that the associated presence of gross contour abnormalities, muscle retraction and fatty replacement, and fluid in the adjacent bursae were helpful in suggesting the correct diagnosis in these patients.

The body of experience of MRI in regard to incomplete rotator cuff tears is somewhat limited, and these lesions may be similar in appearance to full-thickness tears [65]. Features that are helpful in identifying incomplete tears include localized contour abnormalities near a localized area of increased signal intensity or localized loss of definition of the peribursal fat plane [149, 150, 151] (Fig. 15G–22). A discrete area of cuff discontinuity with retracted margins has not been reported in the presence of an incomplete tear. Similarly, a tendon defect that is bright on T2 weighted images and is not associated with fluid in the subacromial-subdeltoid bursa has

been suggested as a finding indicative of a partial tear [111].

Tears of cuff tendons other than the supraspinatus can be successfully visualized by arthrography and computed arthrotomography (Figs. 15G–23 and 15G–24). At present, there is a void in the MRI literature concerning these lesions. However, we have visualized infraspinatus and subscapularis tears with MRI on several occasions (Fig. 15G–25), and there seems to be no doubt that MRI will prove to be quite sensitive in this regard.

Cuff Atrophy. Cuff atrophy refers to an abnormal thinning of the rotator cuff, almost always involving and frequently localized to the supraspinatus tendon. This lesion is almost always associated with degenerative arthritic changes of the glenohumeral joint [104]. On appropriately centered plain radiographs narrowing of the acromiohumeral space may be indicative of cuff atrophy or tear. Rotator cuff atrophy may be identified sonographically, although one should be cautious here because a thin band of fibrotic tissue may be present after massive cuff tears with retraction and may simulate an intact though atrophic cuff. On shoulder arthrograms elevation of the humeral head in association with an intact cuff may occasionally be seen, although, as noted, degenerative arthritic changes are frequently present. It has been suggested that in preoperative planning for cuff arthropathy, MRI (Fig. 15G–26) may be useful in determining the integrity of the remaining rotator cuff [12].

Calcific Tendinitis. Calcific tendinitis of the shoulder is a poorly understood disorder characterized

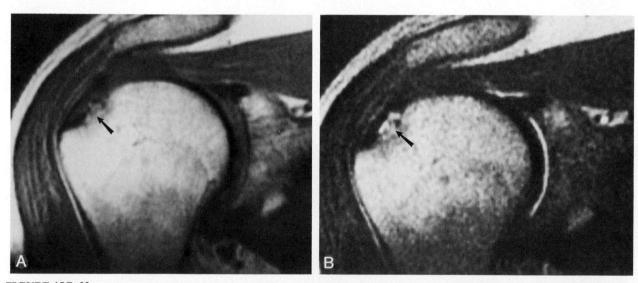

FIGURE 15G–22
Incomplete rotator cuff tear, MRI. Coronal oblique IW *(A)* images through the level of the supraspinatus tendon insertion demonstrates a zone of increased IW signal intensity (arrow) along the undersurface of the tendon that brightens markedly on the T2 weighted image *(B)*. Also note the thinning of the distal portion of the tendon and the absence of both a peribursal fat plane and a bursal effusion.

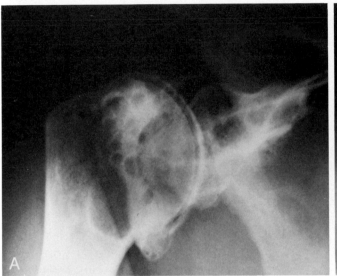

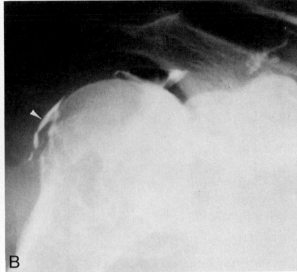

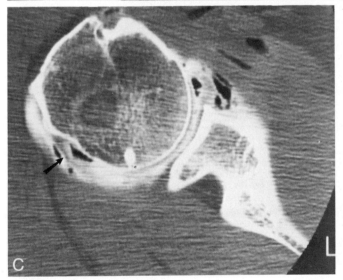

FIGURE 15G–23

Infraspinatus tendon tear, arthrography and CT arthrography. A spot radiograph in the externally rotated position *(A)* appears normal, but in the internally rotated position *(B)* localized extravasation of contrast adjacent to the infraspinatus tendon is evident (arrowhead). A computed arthrotomographic section in the same patient *(C)* demonstrates a localized defect within the infraspinatus tendon (arrow) with air and contrast external to the tendon.

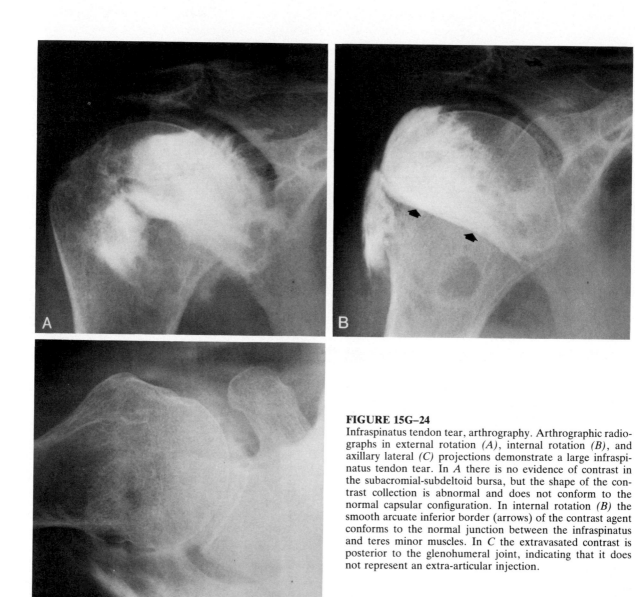

FIGURE 15G–24

Infraspinatus tendon tear, arthrography. Arthrographic radiographs in external rotation *(A)*, internal rotation *(B)*, and axillary lateral *(C)* projections demonstrate a large infraspinatus tendon tear. In *A* there is no evidence of contrast in the subacromial-subdeltoid bursa, but the shape of the contrast collection is abnormal and does not conform to the normal capsular configuration. In internal rotation *(B)* the smooth arcuate inferior border (arrows) of the contrast agent conforms to the normal junction between the infraspinatus and teres minor muscles. In *C* the extravasated contrast is posterior to the glenohumeral joint, indicating that it does not represent an extra-articular injection.

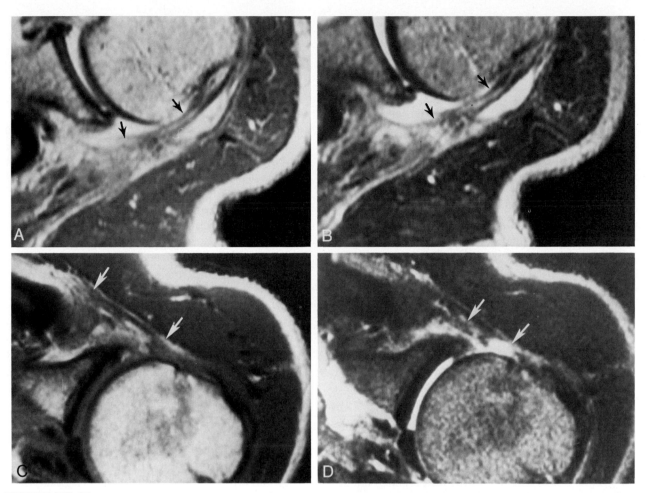

FIGURE 15G–25
Infraspinatus and subscapularis tendon tears, MRI. Axial IW *(A)* and T2 *(B)* weighted MRI images in this patient with an infraspinatus tendon tear show abnormal contours and increased IW and T2 signal intensity within the infraspinatus tendon (arrows) and adjacent muscle as well as an adjacent fluid collection. Axial IW *(C)* and T2 *(D)* weighted MRI images in another patient with a subscapularis tendon tear also demonstrate abnormal tendon morphology (arrows) with thinning and increased IW and T2 signal intensity in the tendon and adjacent muscle belly.

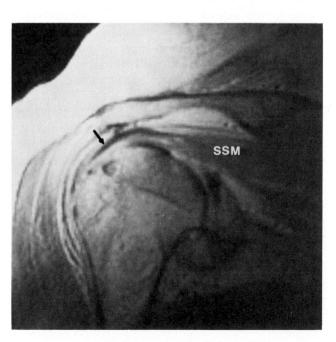

FIGURE 15G–26
Rotator cuff atrophy, MRI. A coronal oblique IW section in this patient with chronic shoulder impingement syndrome demonstrates marked thinning of the supraspinatus tendon (arrow) with atrophic change in the muscle belly (SSM) and elevation of the humeral head. Subsequent surgery demonstrated the cuff to be paper thin but intact.

by the deposition of calcium hydroxyapatite crystals within the substance of one or more tendons of the rotator cuff. The calcific deposits are most common in the supraspinatus tendon, followed in frequency by the infraspinatus, teres minor, and subscapularis tendons [6, 23, 105]. Occasionally the process may also be identified in the tendon of the long head of the biceps brachii muscle [35]. The disorder is most common in the fourth and fifth decades of life and is more often seen in females [6, 23, 144]. There may be a slight predilection for the right shoulder [23, 144], and bilaterality of involvement is relatively common. The pathogenesis of the disorder remains unclear, but primary tendon degeneration as a result of chronic trauma, age, or tissue hypoxia is the most popular theory [144]. The onset of symptoms is often insidious, and asymptomatic deposits are frequently incidental findings on routine radiographs. Acute calcific periarthritis tends to be a self-limiting syndrome characterized by excruciating pain and lasting 1 to 2 weeks. In other instances the symptoms may be considerably milder and may be protracted for weeks to months. In many instances rupture of the calcific deposit into the adjacent subacromial-subdeltoid bursa with subsequent resorption can be observed radiographically. Plain radiographs are adequate to verify the presence of calcific deposits in this disorder. The location of calcific deposits within the rotator cuff can be surmised by evaluating internal rotation, external rotation, and axillary lateral projections and noting the location of the deposit in each projection (Fig. 15G–27). When encountered during shoulder sonography, calcific deposits are characterized by a hypointense deposit associated with acoustic shadowing. Arthrography has little value in the evaluation of calcific tendinitis, and rotator cuff tears are quite unusual in the presence of this disorder. On MRI (Fig. 15G–28) calcific deposits are identified as a low-intensity signal deposit similar to the intensity of the normal rotator cuff [12].

Adhesive Capsulitis. Adhesive capsulitis refers to the process of capsular thickening and contraction about a joint that results in pain and limitation of motion. When it involves the shoulder this disorder is usually referred to as the frozen shoulder syndrome. Clinically this disorder often follows a protracted course, in many cases resulting in a permanent loss of joint motion. Pathologically the disorder is characterized by chronic inflammatory changes with fibrosis localized to the subsynovial layer of the capsule [88]. The frozen shoulder syndrome may be seen as a primary idiopathic disorder, though a variety of predisposing factors have been suggested [84]. The most common predisposing factor is a period of immobility during which normal spontaneous shoulder motion is limited or prevented [92, 134]. The frozen shoulder syndrome is usually a clinical diagnosis, and imaging modalities are not specifically indicated in uncomplicated cases [92]. However, in atypical cases or as incidental findings some imaging procedures may provide useful information. Plain radiographs in patients with adhesive capsulitis are typically normal, although in some patients a mild osteopenia of the humeral head may be identified [117]. This is probably osteoporosis as a manifestation of disuse rather than a specific finding of the disorder. Similarly, radionuclide scintigraphy may demonstrate a mildly increased uptake of radionuclide in the juxta-articular area in these patients [5]. This is probably a manifestation of increased bone turnover with an explanation similar to that of radiographic osteopenia. Shoulder arthrography (Fig. 15G–29) is quite specific in patients with adhesive capsulitis [54, 115, 116, 118]. A generalized constriction of the joint capsule with obliteration of the normal axillary and subscapularis recesses occurs. Typically, the biceps tendon sheath is not affected and fills normally, but absent filling or restriction may occasionally be seen [71, 89, 114]. Total joint volume is reduced to 3 to 10 mL from a normal volume of 12 to 15 mL. Distention arthrography (Brisement procedure) is a mixed therapeutic and diagnostic procedure in which the capsule is progressively distended by serial injections of a mixture of positive contrast agent and local anesthetic in an attempt to break up adhesions and increase joint volume [2, 118].

Bicipital Tendon Abnormalities

As previously described, the tendon of the long head of the biceps brachii is somewhat unique in that its proximal portion is intra-articular and it passes into a synovial sheath through a narrow bony groove. The major pathologic alterations that involve the tendon of the long head of the biceps are tendinitis and subluxation. The former may be divided into impingement and attrition types of tendinitis [91]. Impingement tendinitis is a manifestation of the shoulder impingement syndrome and is frequently associated with full-thickness rotator cuff tears. Following cuff tear and retraction, the bicipital tendon is exposed and chronically traumatized between the humeral head and the coracoacromial arch when the arm is elevated and rotated. Synovitis is usually absent in impingement tendinitis. Attrition tendinitis is associated with intense synovitis. Fraying and thinning of the tendon are frequently prom-

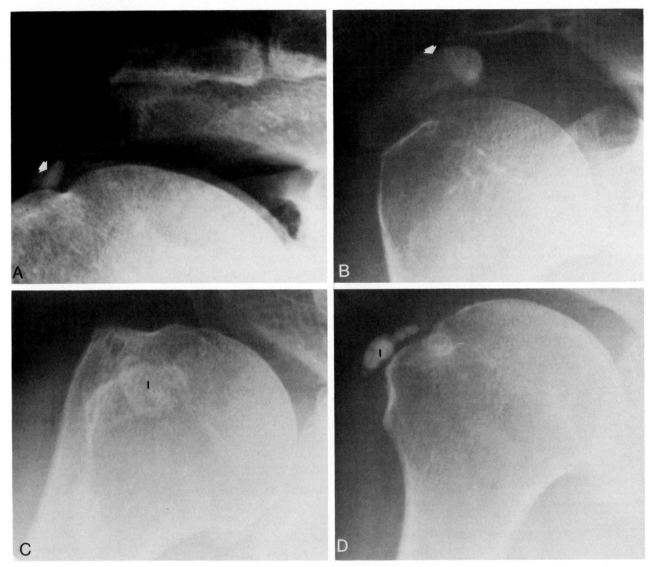

FIGURE 15G–27

Calcific tendinitis, plain radiographs. External rotation radiographs demonstrates typical foci of calcification within the supraspinatus tendon near its insertion *(A, arrow)* and in the subacromial bursa *(B, arrow)*. In another patient, the external rotation position *(C)* demonstrates a focus of calcification (I) overlying the humeral head, which rotates clear in the internal rotation position *(D)*. This is typical of calcifications within the infraspinatus tendon.

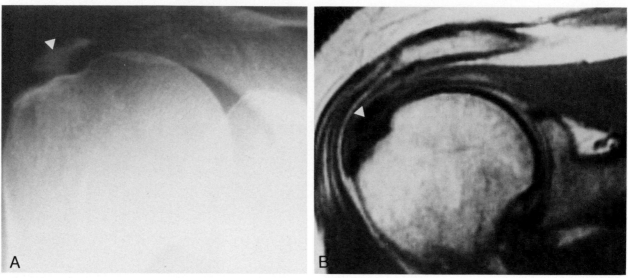

FIGURE 15G–28
Calcific tendinitis, MRI. In this patient an external rotation radiograph *(A)* demonstrates a typical focus of calcification (arrowhead) within the supraspinatus tendon. This calcification is visible but somewhat subtle (arrowhead) on a corresponding coronal oblique IW MRI section *(B)* as an ovoid dark zone within the supraspinatus tendon.

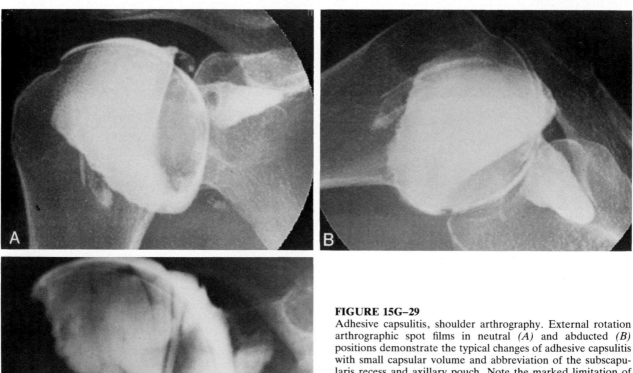

FIGURE 15G–29
Adhesive capsulitis, shoulder arthrography. External rotation arthrographic spot films in neutral *(A)* and abducted *(B)* positions demonstrate the typical changes of adhesive capsulitis with small capsular volume and abbreviation of the subscapularis recess and axillary pouch. Note the marked limitation of motion, virtually all abduction occurring through scapulothoracic rotation with almost no glenohumeral motion. In *C* an external rotation spot radiograph in a patient with a history of seven prior shoulder operations demonstrates similar changes with a "tight" capsule and plication defects. Note in all radiographs the "feathered" appearance of the capsular margins.

inent, usually localized to the intratubercular portion of the tendon where new bone formation and spur formation tend to occur. Attrition tendinitis is the process most commonly associated with bicipital tendon rupture. Subluxation of the biceps tendon occurs with loss of integrity of the transverse intertubercular ligament and medial displacement of the bicipital tendon. Rotator cuff tears are frequently present.

In most instances, bicipital tendon abnormalities are readily identified by clinical examination; however, several imaging modalities may contribute to its evaluation. On plain radiographs the status of the bicipital groove, including depth, medial wall angle, and the presence of spurs may be evaluated. The biceps tendon is well visualized on sonography as a cylindrical structure of uniform intermediate echogenicity (see Fig. 15G–5C). Swelling of the tendon has been associated with bicipital tendinitis as well as the presence of effusion within the bicipital tendon sheath [30]. The latter finding should be viewed with some caution because the bicipital sheath is contiguous with the glenohumeral joint, and effusion may certainly enter the sheath in the absence of tendinitis. In addition, the depth of the bicipital groove can be evaluated sonographically. In the case of bicipital tendon subluxation an empty bicipital groove will be noted with the tendon overlying the anterior aspect of the capsule [79]. On shoulder arthrography the intra-articular portion of the bicipital tendon can be directly visualized. Failure to visualize the intra-articular portion of the bicipital tendon is evidence of tendon rupture. Leakage of contrast through the distal portion of the bicipital sheath may be seen with tendon ruptures (Fig. 15G–30), but caution is advised in such cases because this is a normal weak point at which contrast extravasation may occur due to overdistention of the joint. In a well-positioned bicipital groove projection the intratubercular portion of the tendon can be directly visualized within the bicipital groove. An "empty" bicipital groove may be seen with biceps tendon rupture or with subluxation of the tendon (see Figs. 15G–31C,D and 15G–32). In the latter instance, the tendon may be visualized, in some instances, surrounded by contrast overlying the anterior aspect of the lesser tuberosity [35]. The bicipital tendon is best visualized on MRI in axial sections, where it appears as a round signal void within the intertubercular sulcus (see Fig. 15G–7F,H). The sheath can be identified surrounding the tendon as a thin band of intermediate-intensity signal [48, 56, 130]. With complete tendon rupture, an "empty" fluid-filled sheath (Fig. 15G–31C,D) is seen, whereas with incomplete tears a bright T2

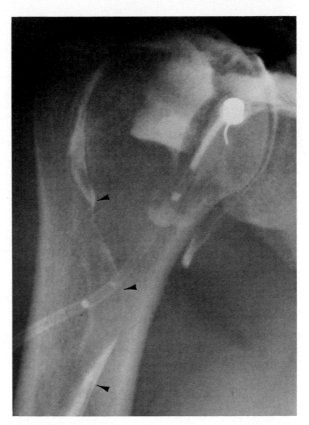

FIGURE 15G–30
Bicipital tendon rupture, shoulder arthrography. An overhead frontal radiograph obtained during shoulder arthrography in this patient demonstrates extravasation of contrast from the tendon sleeve of the long head of the biceps along the muscle belly.

signal may be seen within the central substance of the tendon (see Fig. 15G–31E). The intra-articular portion of the tendon may be visualized in coronal oblique or transaxial images. Loss of the normal dark tendon signal with increased intensity on T1 or IW images may be associated with variable degrees of T2 brightening with bicipital tendon tears (see Fig. 15G–31A,B). Avulsion of the superior glenoid labrum at the biceps origin may also be observed (see Fig. 15G–42A–D).

Shoulder Impingement Syndrome

The shoulder impingement syndrome refers to the impingement of the soft tissues of the superior aspect of the shoulder (rotator cuff, subacromial bursa, and bicipital tendon) between the humeral head and the coracoacromial arch (coracoid process, acromion process, and coracoacromial ligament) [7, 55, 85, 86]. In the normal shoulder the powerful upward pull of the deltoid on the proximal humerus is resisted by an intact rotator cuff so that the

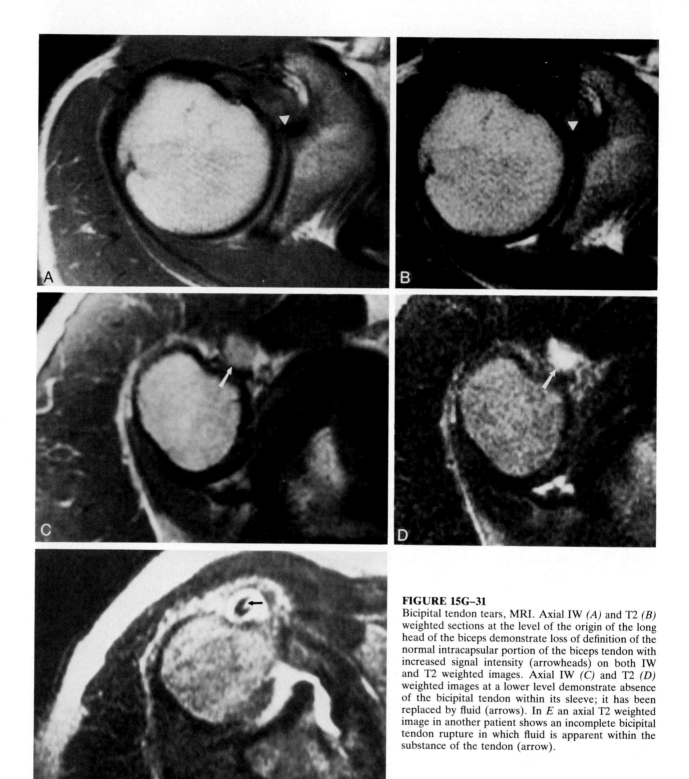

FIGURE 15G–31

Bicipital tendon tears, MRI. Axial IW *(A)* and T2 *(B)* weighted sections at the level of the origin of the long head of the biceps demonstrate loss of definition of the normal intracapsular portion of the biceps tendon with increased signal intensity (arrowheads) on both IW and T2 weighted images. Axial IW *(C)* and T2 *(D)* weighted images at a lower level demonstrate absence of the bicipital tendon within its sleeve; it has been replaced by fluid (arrows). In *E* an axial T2 weighted image in another patient shows an incomplete bicipital tendon rupture in which fluid is apparent within the substance of the tendon (arrow).

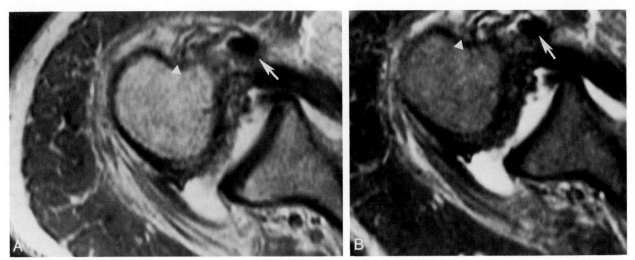

FIGURE 15G–32
Medial dislocation of the bicipital tendon, MRI. Axial IW *(A)* and T2 *(B)* weighted images demonstrate an "empty" bicipital groove (arrowheads), the tendon being medially dislocated (arrows).

humeral head remains centered on the glenoid in all arm positions. If this stabilizing mechanism becomes weakened due to repeated trauma, overuse, or age, the humeral head is pulled upward under the structures of the coracoacromial arch. Initially, impingement is followed by subacromial bursitis and rotator cuff tendinitis. With time, irreversible cuff trauma appears with fibrosis and degeneration. In the latter stages a bony excrescence (subacromial spur) tends to form at the anterior-inferior margin

of the acromion where the coracoacromial ligament (see Fig. 15G–9*D*) inserts (Fig. 15G–33*A,B*). Tears of the rotator cuff are frequent in this stage, undoubtedly in part due to direct cuff trauma from the spur (Fig. 15G–35). In severe cases direct bone-to-bone contact with the greater tuberosity results in sclerosis and proliferative changes in this area [17, 86] (Fig. 15G–34). Tears of the rotator cuff also expose the bicipital tendon to the impingement process [85]. The humeroacromial space is of mini-

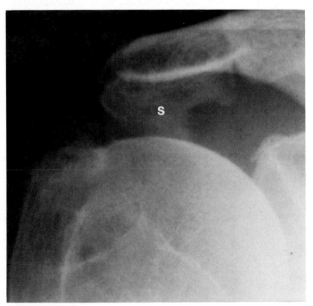

FIGURE 15G–33
Subacromial spur, plain radiograph. A frontal radiograph of the shoulder with a 30-degree caudal tilt demonstrates a large, hooklike, bony excrescence (S) arising from the anterior inferior margin of the acromion process in this patient with shoulder impingement syndrome.

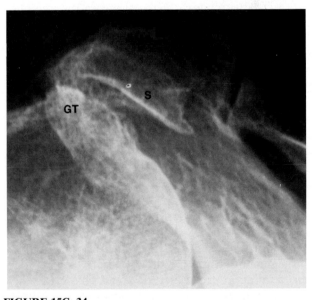

FIGURE 15G–34
Shoulder impingement, fluoroscopy. A fluoroscopic spot radiograph of this patient obtained with the arm in external rotation and abduction demonstrates contact between the subacromial spur and the greater tuberosity (GT) of the humerus. At this point in the examination the patient complained of pain and restriction of abduction. S, Scapula.

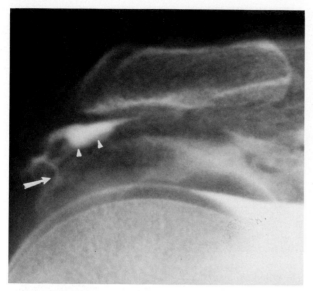

FIGURE 15G–35
Shoulder impingement with rotator cuff tear, shoulder arthrography. A spot radiograph of the shoulder in this patient with shoulder impingement syndrome demonstrates a small, full-thickness supraspinatus tendon tear (arrow). Contrast is present in the subacromial bursa outlining an "ulceration" (arrowheads) in the superior surface of the tendon due to chronic trauma from a subacromial spur.

mal diameter when the arm is elevated in internal rotation. At approximately 80 degrees of elevation the greater tuberosity approaches the anterior-inferior margin of the acromion (see Fig. 15G–34). Thus, the structure exposed to the maximum impingement stress is the area of insertion of the supraspinatus tendon. In recent years the impingement syndrome has been thought to be the most important cause of chronic shoulder pain [86]. The term impingement syndrome was introduced by Neer [86], who described three stages in the disorder in terms of rotator cuff pathology. Stage 1 is characterized by reversible hemorrhage and edema, typically in patients under 25 years of age. In stage 2 cuff fibrosis and tendinitis are seen in patients 25 to 40 years old. In stage 3 bone spurs form along the anterior-inferior margin of the acromion, and rotator cuff tears are common. This latter group is usually over 40 years of age. In younger patients impingement is frequently associated with sports activities, especially throwing, swimming, skiing, and tennis [43, 53, 102]. The diagnosis of shoulder impingement syndrome is usually a clinical one based on appropriate historical and physical examination findings; imaging procedures are useful primarily to assess the extent of abnormalities. Plain radiographs are typically normal in younger patients with shoulder impingement syndrome [87] but often provide important diagnostic information in older patients [17]. The presence of a subacromial spur

(see Figs. 15G–33 and 15G–34) is the single most diagnostic radiographic finding in this disorder. The spur is a bony excrescence arising from the anterior-inferior margin of the acromion process at the site of insertion of the coracoacromial ligament. In routine radiographs the spur is often hidden by the posterior border of the acromion, but it is easily identified in frontal radiographs obtained with the X-ray tube tilted 30 degrees caudally (see Fig. 15G–33). Elevation of the humeral head may be seen in patients with chronic cuff tears and a bony proliferative reaction on the greater tuberosity of the humerus. On sonography the changes of subacromial-subdeltoid bursitis, rotator cuff tendinitis, and rotator cuff tears may be seen as previously described. It has been noted that in impingement syndrome with elevation of the humeral head, rotator cuff tears may be hidden under the acromion and are not visualized on sonography. Shoulder arthrography is useful to demonstrate rotator cuff tears in patients with impingement syndrome [17]. Subacromial bursography (Fig. 15G–36) or fluoroscopy (see Fig. 15G–34) during or before shoulder arthrography can frequently be used to verify the diagnosis [93]. If the subacromial space is observed while the patient is elevating the arm in internal rotation or abducting the arm in external rotation, the point at which motion stops due to pain will be found to be the point at which the greater tuberosity contacts the anterior-inferior margin of the acromion. If the patient can be induced to lean forward during this observation, contact with a subacromial spur can be documented. Occasionally impingement

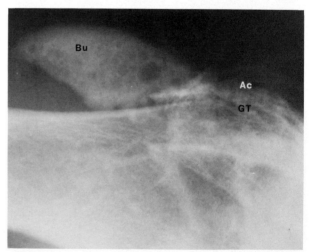

FIGURE 15G–36
Shoulder impingement, subacromial bursography. A spot radiograph from a subacromial bursagram in abduction and external rotation demonstrates distention of an enlarged subdeltoid bursa as contrast is "stripped" from the subacromial bursa as it passes under the large subacromial spur. For abbreviations see Anatomy Key at end of chapter.

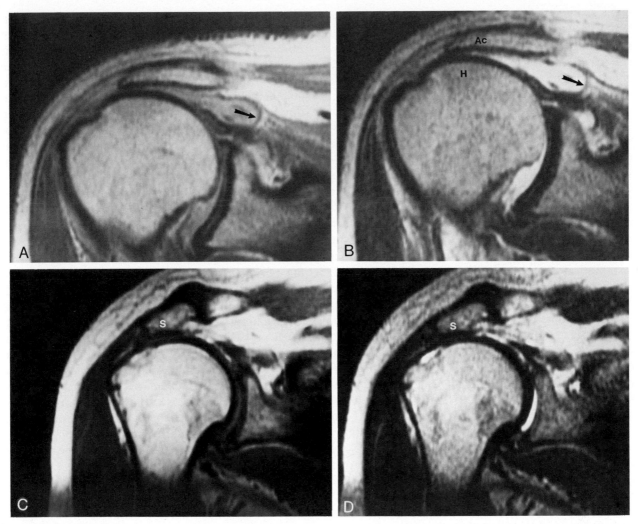

FIGURE 15G–37
Shoulder impingement, MRI. Coronal oblique IW *(A)* and T2 *(B)* weighted images in this patient with chronic shoulder impingement syndrome demonstrate a full-thickness supraspinatus tendon tear with retraction of the musculotendinous junction (arrows). The defect is filled with fluid (effusion), and the humeral head is elevated to contact the undersurface of the acromion (Ac). In another patient, coronal oblique IW *(C)* and T2 *(D)* weighted images also demonstrate a torn and retracted supraspinatus tendon as well as a prominent subacromial spur (S), which is in contact with the humeral head (H).

will be observed not at the acromion but rather at a degenerative acromioclavicular joint. In this latter form of the shoulder impingement syndrome symptoms typically occur at a greater degree of arm elevation (120 degrees). MRI (Fig. 15G–37) provides a noninvasive means of imaging virtually the entire spectrum of impingement-related abnormalities, from subacromial bursitis through chronic rotator cuff tears. The MRI findings of the specific soft tissue impingement lesions have been described in earlier sections of this chapter and include subacromial-subdeltoid bursitis, rotator cuff tendinitis and tears, and bicipital tendon abnormalities [57, 129]. Not infrequently, absence of the supraspinatus tendon will be noted with apposition of the humeral head and acromion as well as retraction of the

supraspinatus muscle. The presence of a subacromial spur may be seen in coronal oblique or sagittal plane images as a dark projection contiguous with the anterior margin of the acromion, pointing toward the coracoid. Similarly, reactive sclerosis of the greater tuberosity is manifested by an irregular increase in dark cortical bone signals in this region.

Shoulder Instability—The Capsular Mechanism

The intrinsic instability of the anatomic configuration of the glenohumeral joint is well known, but, in spite of its morphology, a remarkable degree of stability is present in the normal shoulder. Electromyographic studies [4] have demonstrated that no muscular activity is required to maintain shoulder stability. In cadaver studies [67] as well as in observations of the normal shoulders of patients under general anesthesia, the joint is shown to be quite stable. This intrinsic stability of the glenohumeral joint may be attributed to a combination of active and passive mechanisms. Passive mechanisms [98, 99, 126, 133, 143] include the congruity of the glenohumeral articulation, the normal negative pressure within the capsule, the glenoid labrum, and the shoulder capsule and capsular ligaments. Active mechanisms [14, 52, 106] include the rotator cuff musculature and the long head of the biceps, which come into play during muscular exertion about the shoulder. The great majority of cases of shoulder instability originate in prior trauma [121, 124] with glenohumeral dislocations being by far the most common predisposing factor. Atraumatic shoulder instability comprises approximately 4% of cases of shoulder instability [124] and is frequently associated with congenital abnormalities including abnormal collagen, glenoid malformation, and excessive retroversion of the humeral head. Radiographic identification and classification of glenohumeral dislocations are beyond the scope of this section, but certainly, the evaluation of glenohumeral instability typically involves these patients. Fractures of the anterior glenoid rim [3, 68] with fragments larger than 5 mm in diameter are associated with a 95% incidence of recurrent anterior glenohumeral dislocation [102, 120]. Fractures of the greater tuberosity with anterior glenohumeral dislocation are associated with a lesser (5% to 30%) incidence of recurrent dislocation [120]. Compression fractures of the posterolateral aspect of the humeral head (Hill Sachs defect) [44] following anterior glenohumeral dislocation occur in approximately 60% [122] of patients with anterior dislocation and are seen in 80% [146] of patients with recurrent anterior glenohumeral dislocation. Similarly, compression fractures of the anterior aspect of the humeral head and posterior glenoid rim fractures may be seen in patients with prior posterior shoulder dislocations [13]. Lesser tuberosity fractures are associated with posterior shoulder dislocations, whereas coracoid process, acromion, and acromioclavicular joint fractures are relatively frequent in patients with superior glenohumeral dislocations [138].

Soft tissue injuries are very important in glenohumeral instability. Avulsions or frank tears of the glenoid labra may follow anterior or posterior dislocations and predispose to future instability. As a rule, glenoid labral defects that involve the lower half of the glenoid tend to be associated with instability, whereas those in the superior half are not.

Glenohumeral ligament deficiencies are undoubtedly important in patients with glenohumeral instability. As previously noted, the glenoid labrum is quite variable in size and is unlikely in itself to provide an important component of glenohumeral stability. It is likely that the true significance of anterior labral detachments rests with associated injuries to the anterior band of the inferior glenohumeral ligament or middle glenohumeral ligament. Similarly, posterior labral tears indicate loss of integrity of the posterior band of the inferior glenohumeral ligament. This distinction may explain the difference in significance between supraequatorial and subequatorial labral tears as well as the occurrence of multidirectional instability following a single injury. Stripping of the capsular insertion from the periphery of the labra onto the glenoid neck may be seen anteriorly or posteriorly, and, as with fibrous labral tears, is most significant when the subequatorial portion of the glenoid is involved, again probably due to the loss of inferior glenohumeral ligament integrity. Rotator cuff tears are most common with anterior and inferior shoulder dislocations, whereas subscapularis muscle tears and tendon avulsions may be seen in anterior or posterior dislocations. Massive soft tissue injuries involving the capsule, rotator cuff, and bicipital tendon may be encountered with superior and inferior dislocations.

Plain radiography of the patient with glenohumeral instability is useful in defining the associated fractures and bony abnormalities described above. The radiographic features of these lesions and the projections best demonstrating them are well documented in numerous texts and will not be recounted here.

Arthrography is of little value in the evaluation of glenohumeral instability with the exception of patients in whom an associated rotator cuff tear is suspected. The presence of a prominent medial extension of the capsular space overlying the glenoid neck and adjacent scapular body (dislocation pouch) is occasionally encountered in patients with previous anterior dislocations and may be indicative of capsular laxity. However, this finding is nonspecific and may be seen in patients with no history of prior dislocation and with no symptoms of glenohumeral instability. Arthrotomography (Fig. 15G–38), in which linear tomography or polytomography is combined with arthrography, provided the first means of adequate imaging of the glenoid labra. With proper patient positioning the anterior and posterior glenoid labra are sharply outlined with radiographic contrast medium, and even small labral tears can easily be identified as contrast accumulates within the substance of the labrum [25, 119, 135]. As noted

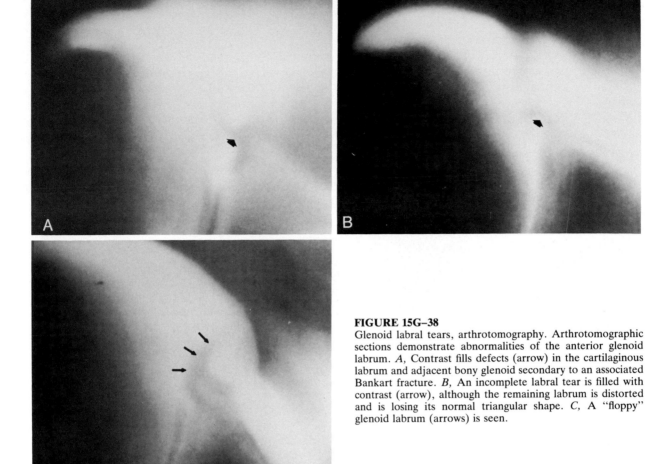

FIGURE 15G–38
Glenoid labral tears, arthrotomography. Arthrotomographic sections demonstrate abnormalities of the anterior glenoid labrum. *A,* Contrast fills defects (arrow) in the cartilaginous labrum and adjacent bony glenoid secondary to an associated Bankart fracture. *B,* An incomplete labral tear is filled with contrast (arrow), although the remaining labrum is distorted and is losing its normal triangular shape. *C,* A "floppy" glenoid labrum (arrows) is seen.

previously, arthrotomography has the disadvantages of being an invasive procedure and being exceedingly uncomfortable for the patient.

Computed arthrotomography [25, 109, 110] has in recent years become the procedure of choice for imaging the glenoid labrum. Utilizing high-resolution algorithms, the anterior and posterior labra can be sharply outlined (see Fig. 15G–2). The normal anterior labrum is somewhat larger than the posterior labrum and frequently has a triangular shape, whereas the posterior labrum typically has a rounded configuration [25]. These shapes are not constant and may be observed to vary on different sections and with different arm positions [75, 128]. This is probably due to the pliability of the labrum as it conforms to pressure exerted by the humeral head or the adjacent capsule. As with meniscal tears in the knee, some form of stress is frequently useful to allow small tears to fill with contrast. We accomplish this by first scanning the entire articulation in

a neutral (thumb up) arm position and then repeating scans through the labrum with the arm in internal and external rotation (Fig. 15G–39*C,D*). As with arthrotomography, labral tears can be identified by identifying contrast agent within the substance of the labrum (see Fig. 15G–39*A*). Detachments of the labrum (see Fig. 15G–39*B*) similar to a "buckethandle" tear of a meniscus may also be seen as well as labral insufficiency with an abnormally shaped and distorted appearance [109]. The anterior and posterior capsular insertions into the glenoid should also be evaluated. The normal capsule inserts along the outer margin of the base of the labrum. "Stripping" of the capsular insertion such that insertion occurs along the glenoid neck or body is suggestive, though not diagnostic, of glenohumeral instability. In sections at midglenoid level a thickened band of capsule, representing the middle glenohumeral ligament, can be observed outlined by contrast within the joint and subscapular recess. Some care must

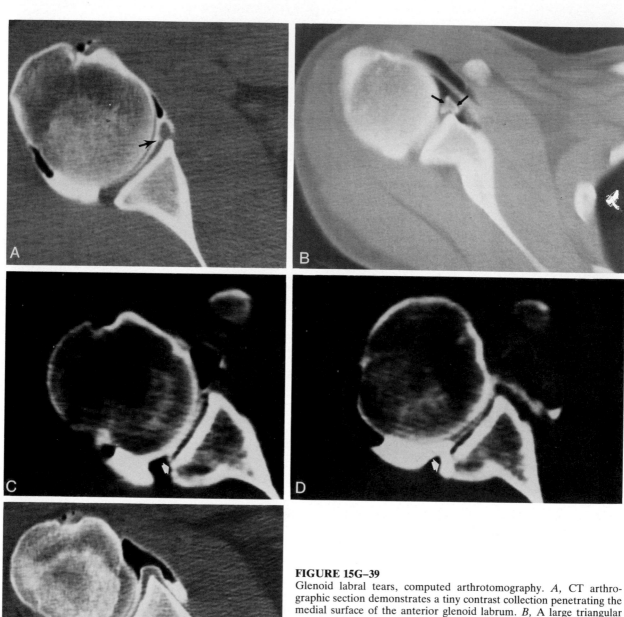

FIGURE 15G–39

Glenoid labral tears, computed arthrotomography. *A,* CT arthrographic section demonstrates a tiny contrast collection penetrating the medial surface of the anterior glenoid labrum. *B,* A large triangular anterior labral fragment (arrows) is surrounded by contrast. In another patient a section in external rotation *(C)* demonstrates a tiny posterior labral tear (arrow), which enlarges when the same level is sectioned in internal rotation *(D). E,* A posterior labral tear is evident as well as a small bony fragment (arrow) related to an associated reverse Bankart fracture.

be used in interpretation because there is substantial individual variation in the size of the glenoid labra. A small labrum does not indicate an abnormal labrum if its contour is normal. Occasionally a torn labral fragment may be displaced resulting in a smaller residual labrum, which invariably has an abnormal contour. Another false-positive finding to be avoided is misinterpreting the small recess that can be seen under the central lip of the labrum, allowing a thin linear collection of contrast to accumulate under the central periphery of the labrum parallel to the glenoid articular surface [25]. Fracture defects involving the glenoid rim (see Fig. 15G–39E) may also be identified and the actual size of the osteocartilaginous fragment defined. Hill Sachs lesions are well delineated with arthrotomography, and full-thickness rotator cuff tears (Fig. 15G–40) are identified by the presence of contrast agent in the subacromial-subdeltoid bursa [147]. The bicipital tendon is also well visualized and sharply outlined by contrast agent in the bicipital sleeve (see Fig. 15G–2D).

MRI is still in its infancy in the evaluation of glenohumeral instability, but early studies suggest that it may become quite important in this regard [59, 130]. The axial imaging plane is preferred because this provides the best cross-sectional view of the anterior and posterior labra. Recently we

have added an axial MPGR/30 sequence to our routine in patients in whom shoulder instability is suspected because this sequence allows better differentiation of the capsule, capsular ligaments, and labrum from the overlying rotator cuff tendons (see Fig. 15G–8). On all imaging sequences (T1, T2, IW), the normal glenoid labrum appears as a signal void [48, 49, 56]. As noted in the earlier section on computed arthrotomography, the anterior labrum is normally larger with a triangular shape, whereas the smaller posterior labrum frequently has a rounded peripheral border [56]. Labral tears (Fig. 15G–41) similar to those seen in the knee demonstrate linear bands of increased signal intensity in contiguity with a free labral margin [49]. If an effusion is present, it frequently serves as a contrast agent on T2 weighted sections to fill labral defects. It has been suggested that poorly localized zones with increased signal intensity within the labra are abnormal and are associated with shoulder instability [128]. However, if recent studies are accurate and the labrum is actually composed of fibrocartilage [108] rather than fibrous tissue, this conclusion should be viewed with caution [49], since similar changes in the fibrocartilage of the knee have been shown to represent myxoid degeneration [139], and degenerative changes in the labrum have been shown to be common in older individuals without associated instability [24, 97]. Labral detachments (see Fig. 15G–41A,B) are identified by the absence of a normal labrum at the margin of the bony glenoid or identification of the detached fragment frequently surrounded by effusion on T2 weighted images. Detachment of the superior labrum at the origin of the tendon of the long head of the biceps (Fig. 15G–42) is not infrequent and tends not to be associated with glenohumeral instability. The capsular insertion is normally well visualized on MRI. The posterior capsule always inserts into the posterior labrum whereas the anterior capsular insertion is more variable [48]. Zlatkin and colleagues have suggested that the anterior capsular insertion be divided into three types, type 1 being insertion into the labrum (Fig. 15G–39A), type 2 being insertion into the glenoid neck adjacent to the labrum (Fig. 15G–39E), and type 3 being insertion into the glenoid neck remote to the labrum (Fig. 15G–43). A type 1 insertion is clearly normal, whereas a type 3 insertion is clearly abnormal. Type 2 capsular insertions are intermediate in significance. Anterior glenohumeral instability has been shown to have a strong association with type 2 and type 3 patterns [49]. Capsular insertion is probably a misnomer in type 3 and some type 2 patterns. The structural importance of the anterior capsule rests with the middle gleno-

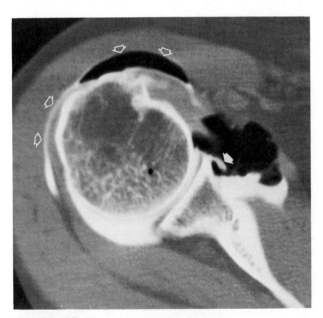

FIGURE 15G–40
Anterior labral and rotator cuff tears, computed arthrotomography. A computed arthrotomographic section in this patient with a history of prior anterior glenohumeral dislocation and recurrent anterior shoulder instability demonstrates a torn anterior glenoid labrum in which the fragment is rotated anteriorly (closed arrow). An associated rotator cuff tear is also present, as seen by contrast and air filling the subdeltoid bursa (open arrows).

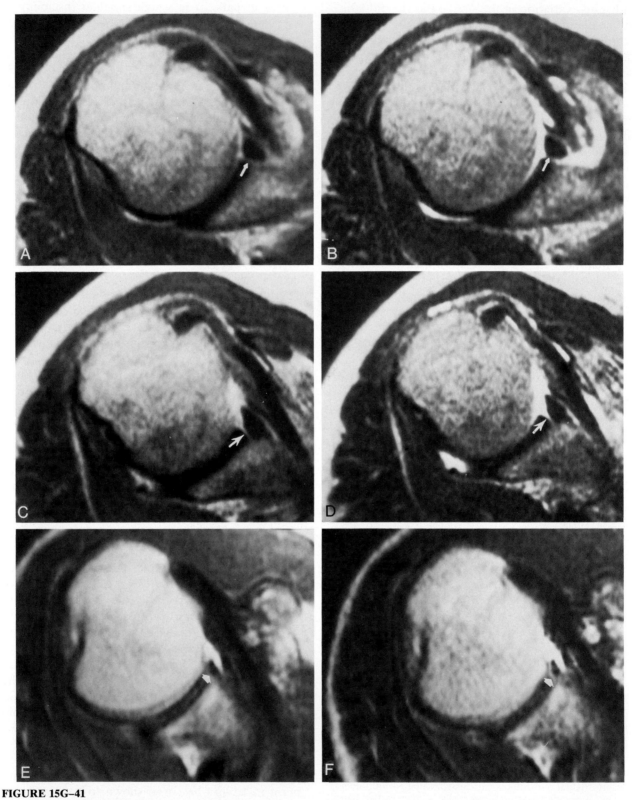

FIGURE 15G–41
Glenoid labral tears, MRI. Axial IW *(A)* and T2 *(B)* MRI sections demonstrate avulsion of the anterior glenoid labrum (arrows), which remains attached to the anterior capsule. Axial IW *(C)* and T2 *(D)* MRI sections in another patient demonstrate an incomplete "split" (arrows) dividing the anterior labrum from the articular fossa of the glenoid. In another patient axial IW *(E)* and axial T2 *(F)* also demonstrate a small labral split (arrows).

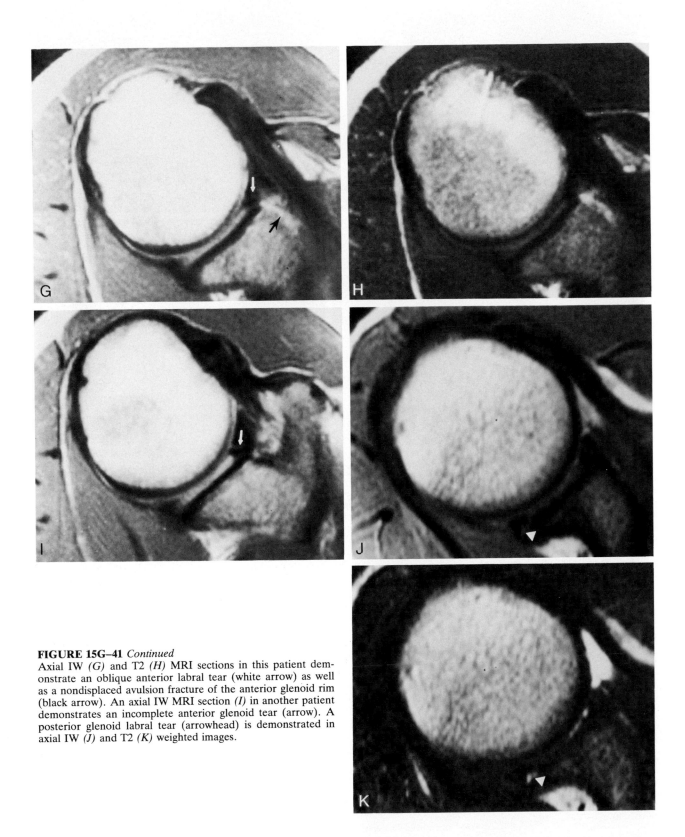

FIGURE 15G–41 *Continued*
Axial IW *(G)* and T2 *(H)* MRI sections in this patient demonstrate an oblique anterior labral tear (white arrow) as well as a nondisplaced avulsion fracture of the anterior glenoid rim (black arrow). An axial IW MRI section *(I)* in another patient demonstrates an incomplete anterior glenoid tear (arrow). A posterior glenoid labral tear (arrowhead) is demonstrated in axial IW *(J)* and T2 *(K)* weighted images.

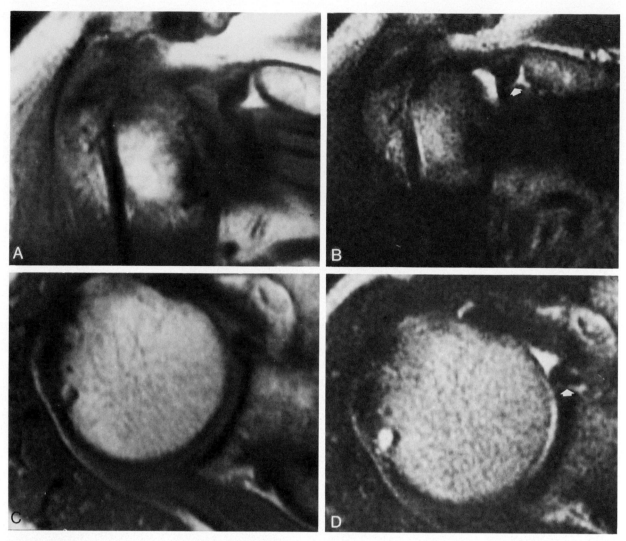

FIGURE 15G–42
Superior glenoid labral avulsion, MRI. Coronal oblique and axial IW *(A* and *C)* and T2 *(B* and *D)* weighted images in this patient demonstrate avulsion of the superior glenoid labrum (arrows) at the origin of the tendon of the long head of the biceps.

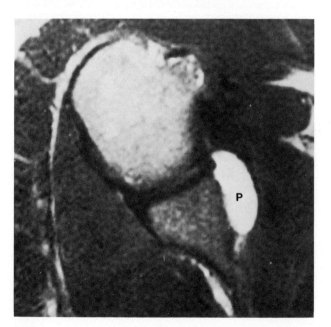

FIGURE 15G–43
Type 3 anterior capsular insertion, MRI. An axial T2 MRI section through the glenoid demonstrates anterior capsular insertion low on the glenoid neck (type 3), outlining a large fluid collection (P) ("dislocation pouch"). This pattern is associated with a high incidence of anterior glenohumeral instability.

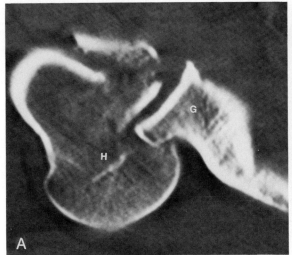
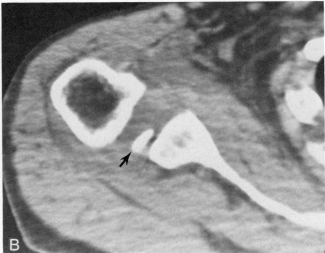
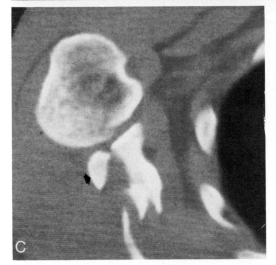

FIGURE 15G–44
Fractures about the shoulder, computed tomography. A computed tomographic section *(A)* in this patient demonstrates a fracture of the surgical neck of the humerus as well as an "irreducible" posterior dislocation of the humeral head (H), which is impacted on the posterior glenoid (G). In another patient a computed tomographic section through the inferior glenoid demonstrates a fracture of the anterior glenoid (Bankart fracture) with the fragment (arrow) displaced posteriorly. In *C* a glenoid neck fracture is accompanied by a displaced fracture involving the posterior third of the glenoid (arrow) as well as an impaction fracture of the anterior aspect of the humeral head (reverse Hill Sachs lesion).

humeral ligament and the anterior band of the inferior glenohumeral ligament. O'Brien and associates [95, 96] have shown that the normal insertion of these structures is either along the free edge of the labrum or at its base with a shallow pouch outlining the outer margin. These patterns conform to type 1 insertions and probably some type 2 insertions. In other types the capsular (glenohumeral ligament) insertion has been "stripped," and the visualized "capsular insertion" is actually the anterior margin of the subscapularis origin from the scapula. The enlarged anterior "recess" ("dislocation pouch") (Fig. 15G–43) is best seen on T2 weighted axial images where it is usually filled with fluid density (bright). Careful evaluation for an associated anterior labral tear or detached glenohumeral ligament remnant should be performed. Soft tissue lesions associated with posterior glenohumeral dislocation include posterior labral tears, subscapularis avulsion and ruptures, and attenuation

of the infraspinatus muscle and tendon [74]. Posterior labral tears have the same appearance as anterior labral tears. Capsular stripping is unusual posteriorly because the posterior capsule is tightly attached to the outer edge of the labrum and the posterior cuff muscles are not attached to the posterior glenoid neck. Subscapularis tendon injuries [24, 42] may be seen in patients with anterior or posterior instability with increased signal and morphologic defects of the tendon as well as medial retraction of the musculotendinous junction and hemorrhage or fatty infiltration of the muscle belly [128]. The infraspinatus is subject to trauma with posterior glenohumeral dislocations [74] but at present, there are no reports of infraspinatus abnormalities visualized by MRI. Glenoid rim fractures that may be associated with anterior glenohumeral dislocations (Bankart lesion) or posterior dislocations are well visualized on MRI, and, as with computed arthrotomography, the actual size and point of ori-

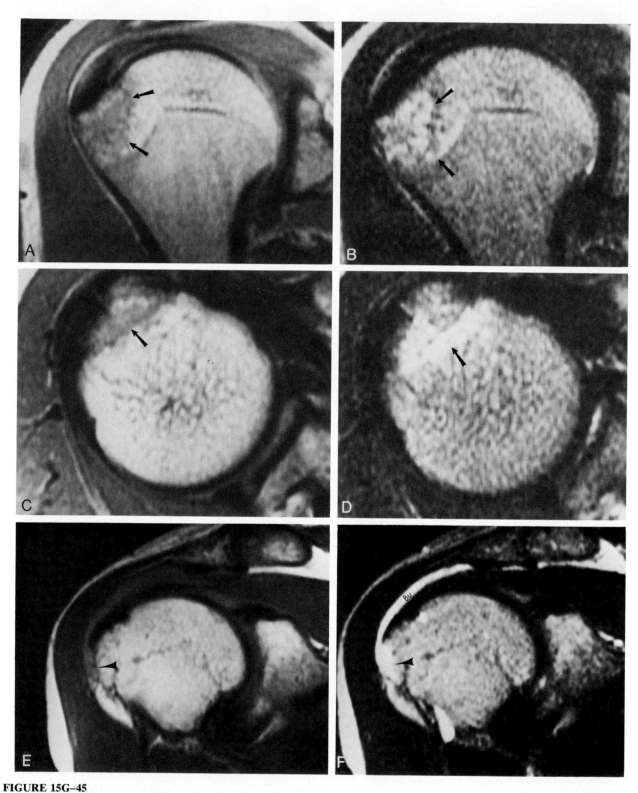

FIGURE 15G–45
Greater tuberosity fractures, MRI. Coronal oblique and axial IW *(A* and *C)* and T2 *(B* and *D)* weighted images in this patient who injured his shoulder in a skiing accident demonstrate a nondisplaced fracture of the greater tuberosity of the humerus (arrows), which was invisible on plain radiographs and clinically mimicked an acute rotator cuff tear. In another patient coronal oblique IW *(E)* and T2 *(F)* weighted images demonstrate the donor site (arrowheads) of a tiny greater tuberosity avulsion fracture as well as a large bursal effusion (Bu).

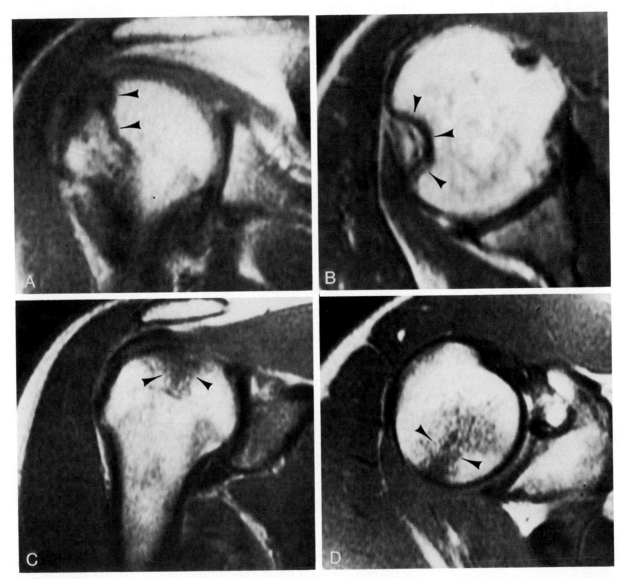

FIGURE 15G–46
Hill-Sachs lesion, MRI. IW coronal oblique *(A)* and transaxial *(B)* images through the shoulder in this patient with a history of prior anterior glenohumeral dislocation demonstrate an impaction fracture (arrowheads) (Hill-Sachs lesion) involving the posterior superior aspect of the humeral head. IW coronal oblique *(C)* and transaxial *(D)* MRI images in another patient who injured his shoulder lifting weights and continued to have pain 3 weeks later demonstrate a subcortical trabecular fracture ("bone bruise") in the typical location of a Hill-Sachs lesion. This finding confirmed the diagnosis of a momentary glenohumeral dislocation.

gin of the osteochondral fragment can be identified. As previously noted, these lesions have an extremely high association with glenohumeral instability. Humeral head impaction fractures (Hill-Sachs fracture, reversed Hill-Sachs fracture), rotator cuff tears, and bicipital tendon abnormalities may also be identified as previously described. A new possibility afforded by MRI is the ability to visualize other soft tissue structures directly such as the glenohumeral ligaments and the subscapularis muscle and tendon, which are known to be important in glenohumeral instability but were not visible by previous imaging procedures.

Osseous Abnormalities

Osseous abnormalities of the glenohumeral joint are usually adequately imaged by plain radiography, although in some circumstances more detailed evaluation is required. In addition, osseous abnormalities are often incidental findings on a variety of imaging modalities.

Computed tomography is the modality of choice for the detailed evaluation of fractures (Fig. 15G–44) about the shoulder [21, 26, 131]. Anterior and posterior glenoid rim fractures can be characterized in regard to the size and displacement of the fragment. CT can also be useful for evaluating osseous union at these fracture sites. Intra-articular extensions of glenoid neck fractures or extension into the

suprascapular notch are other indications for CT. Complex proximal humeral fractures are often evaluated by CT to characterize the size and number of the intra-articular fragments as well as to identify rotated or dislocated fragments. CT is also quite sensitive in the identification of humeral head compression fractures (i.e., Hill-Sachs fracture) associated with prior glenohumeral dislocation.

Fractures are readily visualized on MRI with cortical bone demonstrating a signal void and medullary bone characterized by intermediate to bright intensity on T1 and IW images (Fig. 15G–45). A potential pitfall with MRI is that small cortical fragments may be quite difficult to distinguish from adjacent tendons and ligaments, which are also dark on MRI images [49]. However, the donor site for small fragments is usually readily visualized on MRI and is characterized by a focal loss of cortical integrity, often with hemorrhage in the adjacent medullary bone (see Fig. 15G–45E,F). Lipohemarthroses associated with small fractures have been demonstrated with great sensitivity in the knee joint and undoubtedly can be similarly seen with MRI. We have seen instances of nondisplaced fractures of the greater tuberosity mimicking acute rotator cuff tears that are quite accurately demonstrated on MRI (see Fig. 15G–45A–D). Humeral head impaction fractures (Fig. 15G–46) and glenoid rim fractures are easily identified on MRI [49, 58, 128]. Acromioclavicular joint separations do not typically warrant evaluation on MRI, but the nature of the injury can

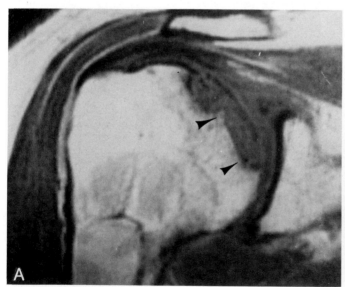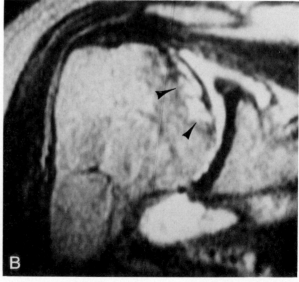

FIGURE 15G–47
Humeral avascular necrosis, MRI. Coronal oblique IW *(A)* and T2 *(B)* weighted MRI images in this patient with a healed fracture of the surgical neck of the humerus demonstrate the MRI characteristics of avascular necrosis of the humeral head. A well-marginated zone of intermediate signal intensity (arrowheads) is identified in the superomedial subchondral region of the humeral head in IW images. On T2 weighted images the defect (arrows) is shown to have the characteristics of fluid, confirming the loss of integrity of the overlying subchondral bone plate.

be readily assessed by this modality [15]. MRI is also quite sensitive to the early changes typical of avascular necrosis and provides important information about cortical integrity (Fig. 15G–47).

ANATOMY KEY

Ac, Acromion process
AC, Anterior capsule
AGL, Anterior glenoid labrum
Ax, Axillary recess
Bi, Bicipital tendon sleeve
BT, Bicipital tendon
Bu, Subacromial-subdeltoid bursa
C, Clavicle
Co, Coracoid process
D, Deltoid muscle
G, Glenoid process
GT, Greater tuberosity
H, Humeral head
IGhL, Inferior glenohumeral ligament
IS, Infraspinatus muscle
IST, Infraspinatus tendon
LT, Lesser tuberosity
MGhL, Middle glenohumeral ligament
PBF, Peribursal fat plane
PC, Posterior capsule
PGL, Posterior glenoid labrum
S, Scapula
SSM, Supraspinatus muscle
SST, Supraspinatus tendon
Su, Subscapularis recess
SuSM, Subscapularis muscle
SuST, Subscapularis tendon
TMM, Teres minor muscle
TMT, Teres minor tendon

References

1. Altchek, D. W., Warren, R. F., and Skyhar, M. J. Shoulder arthroscopy. *In* Rockwood CA, and Matsen FA (Eds.), *The Shoulder*. Philadelphia, W.B. Saunders, pp. 258–277, 1990.
2. Andren, L., and Lundberg, B. J. The treatment of rigid shoulders by joint distension during arthrography. *Acta Orthop Scand* 36:45–53, 1965.
3. Astor, J. W., and Gregory, C. F. Dislocation of the shoulder with significant fracture of the glenoid. *J Bone Joint Surg* 55A:1531, 1973.
4. Basmajian, J. V., and Bazant, F. J. Factors preventing downward dislocation of the adducted shoulder joint. *J Bone Joint Surg* 41A:1182–1186, 1959.
5. Binder, A. I., Bulgen, D. Y, Hazelman, B. L., et al. Frozen shoulder: an arthrographic and radionuclide scan assessment. *Ann Rheum Dis* 43:365–369, 1984.
6. Bosworth, B. M. Examination of the shoulder for calcium deposits. *J Bone Joint Surg* 23:567–577, 1941.
7. Brandt, T. D., Cardone, B. W., Grant, T. H., Post, M. and Weiss, C. A. Rotator cuff sonography: a reassessment. *Radiology* 173:323–327, 1989.
8. Braunstein, E. M., and O'Connor, G. Double-contrast arthrotomography of the shoulder. *J Bone Joint Surg* 64:192, 1982.
9. Bretzke, C. A., Crass, J. R., and Craig, E. V. Ultrasonography of the rotator cuff: Normal and pathologic anatomy. *Invest Radiol* 20:311–315, 1985.
10. Buirski, G. Magnetic resonance imaging in acute and chronic rotator cuff tears. *Skel Radiol* 19:109, 1990.
11. Burk, D. L., Karasick, D. K., and Kurtz, A. B. Rotator cuff tears: prospective study of MR imaging with arthrography, sonography and surgery. *AJR* 153:87–92, 1989.
12. Burk, D. L. J., Karasick, D., and Mitchell, D. G. MR imaging of the shoulder: Correlation with plain radiography. *AJR* 154:549, 1990.
13. Cisternino, S. J., Rogers, L. F., Stufflebaum, B. C., and Kruglik, G. D. The trough line: a radiographic sign of posterior shoulder dislocation. *AJR* 130:951, 1978.
14. Cleland, J. On the action of muscles passing over more than one joint. *J Anat Physiol* 1:85–93, 1986.
15. Colfield, R. H. Current concepts review rotator cuff disease of the shoulder. *J Bone Joint Surg* 67A:974–979, 1985.
16. Cone, R. O., Danzig, L., Resnick, D., and Goldman, A. B. The bicipital groove: radiographic, anatomic, and pathologic study. *AJR* 41:781–788, 1983.
17. Cone, R. O., Resnick, D., and Danzig, L. Shoulder impingement syndrome: Radiographic evaluation. *Radiology* 150:29–33, 1984.
18. Cotton, R. E., and Rideout, D. F. Tears of the humeral rotator cuff: A radiological and pathological necropsy survey. *J Bone Joint Surg* 46B:314–328, 1964.
19. Crass, J. R., Craig, E. V, Bretzke, C. A., and Feinberg, S. B. Ultrasonography of the rotator cuff. *Radiographics* 5:941–953, 1985.
20. Crass, J. R., Craig, E. V., and Feinberg, S. B. The hyperextended internal rotation view in rotator cuff ultrasonography. *Journal of Clinical Ultrasound* 15:416–420, 1987.
21. Danzig, L., Resnick, D., and Greenway, G. Evaluation of unstable shoulders by computed tomography. *Am J Sports Med* 10:138, 1982.
22. DePalma, A. F. *Surgery of the Shoulder* (2nd ed.). Philadelphia, J. B. Lippincott, 1973.
23. DePalma, A. F., Callery, G., and Bennett, G. A. Shoulder joint: variational anatomy and degenerative lesions of the shoulder joint. *AAOS Instr Course Lect* 6:255–281, 1949.
24. DePalma, A. F., and Kruper, J. S. Long term study of shoulder joints afflicted with and treated for calcific tendinitis. *Clin Orthop* 20:61–72, 1961.
25. Deutsch, A. Z., Resnick, D., Mink, J. H. et al. Computed and conventional arthrotomography of the glenohumeral joint: normal anatomy and clinical experience. *Radiology* 153:603–609, 1984.
26. Deutsch, A. L., Resnick, D., and Mink, J. H. Computed tomography of the glenohumeral and sternoclavicular joints. *Orthop Clin North Am* 16:497, 1985.
27. El-Khoury, G. Y., Albright, J. P., Abu Yousef, M. M., Montgomery, W. J., and Tuck, S. L. Arthrotomography of the glenoid labrum. *Radiology* 131:333, 1979.
28. Ellman, H., Hanker, G., and Bayer, M. Repair of the rotator cuff. *J Bone Joint Surg* 68A:1136–1143, 1986.
29. Evancho, A. M., Stiles, R. G., and Fajman, W. A. MR imaging diagnosis of rotator cuff tears. *AJR* 151:751–754, 1988.
30. Farin, P. U. Jaroma, H., Harju, A., and Soimakallio, S. Shoulder impingement syndrome: Sonographic evaluation. *Radiology* 176:845–849, 1990.
31. Flannigan, B., Kursunoglu-Brahme, S., Snyder, S., Karzel, R., Del Pizzo, W., and Resnick, D. MR arthrography of the shoulder: Comparison with MR imaging. *AJR* 155:829–832, 1990.
32. Fukuda, H., Mikasa, M., and Yamanaka, K. Incomplete thickness rotator cuff tears diagnosed by subacromial bursography. *Clin Orthop Rel Res* 223:51–58, 1987.
33. Garcia, J. F. Arthrographic visualization of rotator cuff tears. *Radiology* 150(2):595, 1984.
34. Goldman, A. B. Double contrast shoulder arthrography. *In* Freiberger, R. H., and Kaye, J. J. (Eds.), *Arthrography*. New York, Appleton-Century-Crofts, 1979, pp. 165–188.
35. Goldman, A. B. Calcific tendinitis of the long head of the biceps brachii distal to the glenohumeral joint. *AJR* 153:1011–1016, 1989.

36. Goldman, A. B., and Ghelman, B. The double contrast shoulder arthrogram: A review of 158 studies. *Radiology* 127:655–663, 1978.
37. Gray, H. *Gray's Anatomy* (29th ed.). Philadelphia, Lea & Febiger, 1973, p. 1466.
38. Greenway, G. D., Danzig, L. A., Resnick, D., and Haghighi, P. The painful shoulder. *Med Radiol Photog* 58(2):22–67, 1982.
39. Habibian, A., Stauffer, A., and Resnick, D. Comparison of conventional and computed arthrotomography with MR imaging in the evaluation of the shoulder. *J Comput Assist Tomogr* 13:968, 1989.
40. Hajek, P. C., Sartoris, D. J., and Neumann, C. H. Potential contrast agents for MR arthrography: In vitro evaluation and practical observations. *AJR* 149:97–104, 1987.
41. Hall, F. M. Sonography of the shoulder. *Radiology* 173:310, 1989.
42. Hauser, E. D. W. Avulsion of the tendon of the subscapularis muscle. *J Bone Joint Surg* 36A:139–141, 1954.
43. Hawkings, R. J., and Kennedy, J. C. Impingement syndrome in athletes. *Am J Sports Med* 8:151–157, 1980.
44. Hill, H. A., and Sachs, M. D. The grooved defect of the humeral head—a frequently unrecognized complication of dislocations of the shoulder joint. *Radiology* 35:690, 1940.
45. Hitchcock, H. H., and Bechtol, C. O. Painful shoulder. Observations on the role of the tendon of the long head of the biceps brachii in its causation. *J Bone Joint Surg* 30A:263–273, 1948.
46. Holder, M. D., Fretz, C. J., Terrier, F., and Gerber, C. Rotator cuff tears: Correlation of sonographic and surgical findings. *Radiology* 169:791–794, 1988.
47. Holt, R. G., Helms, C. A., and Steinbach, L. Magnetic resonance imaging of the shoulder: Rationale and current applications. *Skel Radiol* 19:5, 1990.
48. Huber, D. J., Sauter, R. S., and Muller, E. MR Imaging of the normal shoulder. *Radiology* 158:405–408, 1986.
49. Ionnotti, J. P., Zlatkin, M. B., Esterhai, J. L., Kressel, H. Y., Dalinka, M. K., and Spindler, K. P. Magnetic resonance imaging of the shoulder. *J Bone Joint Surg* 73A(1):17–29, 1991.
50. Jackson, D. W. Chronic rotator cuff impingement in the throwing athlete. *Am J Sports Med* 4:231–240, 1976.
51. Jobe, C. M. Gross anatomy of the shoulder. *In* Rockwood, C. A., and Marsen, F. A. I. (Eds.), *The Shoulder*. Philadelphia, W. B. Saunders, 1990, pp. 34–97.
52. Jobe, F. W. Unstable shoulders in the athlete. *AAOS Instr Course Lect* 34:228–231, 1985.
53. Jobe, F. W., and Jobe, C. M. Painful athletic injuries of the shoulder. *Clin Orthop Rel Res* 173:117–124, 1983.
54. Kaye, J. J., and Schneider, R. Positive contrast shoulder arthrography. *In* Freiberger, R. H., and Kaye, J. J. (Eds.), *Arthrography*. New York, Appleton-Century-Crofts, 1979, pp. 137–163.
55. Kessel, L., and Watson, M. The painful arc syndrome: Clinical classification as a guide to management. *J Bone Joint Surg* 59B:166–172, 1977.
56. Kieft, G. J., Bloem, J. L., Obermann, W. R., Verbout, A. J., Rozing, P. M., and Doornbos, J. Normal shoulder: MR imaging. *Radiology* 159:741–745, 1986.
57. Kieft, G. J., Bloem, J. L., and Rozing, P. M. Rotator cuff impingement syndrome: MR Imaging. *Radiology* 166:211–214, 1988.
58. Kieft, G. J., Bloem, J. L., and Rozing, P. M. MR imaging of recurrent anterior dislocation of the shoulder: Comparison with CT. *AJR* 150:1083, 1988.
59. Kieft, G. J., Sartoris, D. J., and Bloem, J. L. Magnetic resonance imaging of gleno-humeral joint diseases. *Skel Radiol* 16:285–290, 1987.
60. Kilcoyne, R. F., and Matsen, F. A. Rotator cuff tear measurement by arthropneumotomography. *AJR* 140:315, 1983.
61. Killoran, P. J., Marcove, R. C., and Freiberger, R. H. Shoulder arthrography. *AJR* 103:658, 1957.
62. Kinnard, P., Tricroie, J. L., Levesque, R. Y., and Bergeron, D. Assessment of the unstable shoulder by computed arthrography. A preliminary report. *Am J Sports Med* 11:157, 1983.
63. Kleinmann, P. K., Kanzaria, P. K., Goss, T. P., and Pappas, A. M. Axillary arthrotomography of the glenoid labrum. *AJR* 141:993, 1984.
64. Kneeland, J. B., Carrera, G. F., and Middleton, W. D. Rotator cuff tears: Preliminary application of high-resolution MR imaging with counter-rotating current loop-gap resonators. *Radiology* 160:695–699, 1986.
65. Kneeland, B. J., Middleton, W. D., and Carrera, G. F. MR imaging of the shoulder: Diagnosis of rotator cuff tears. *AJR* 149:333–337, 1987.
66. Kotzen, L. M. Roentgen diagnosis of rotator cuff tear: Report of 48 surgically proven cases. *AJR* 112:507–511, 1971.
67. Kumar, V. P., and Balasubramaniam, P. The role of atmospheric pressure in stabilizing the shoulder. An experimental study. *J Bone Joint Surg* 67B(5):719–721, 1985.
68. Kummel, B. M. Fractures of the Glenoid causing chronic dislocation of the shoulder. *Clin Orthop* 69:189, 1970.
69. Lie, S., and Mast, W. A. Subacromial bursography. *Radiology* 144:626–630, 1982.
70. Lindblom, K. On pathogenesis of ruptures of the tendon aponeurosis of the shoulder joint. *Acta Radiol* 20:563–567, 1939.
71. Loyd, J. A., and Loyd, H. M. Adhesive capsulitis of the shoulder: Arthrographic diagnosis and treatment. *South Med J* 76(7):879–883, 1983.
72. Mack, L. A., Matsen, F. A., Kilcoyne, R. F., Davies, P. K., and Sickler, M. E., US evaluation of the rotator cuff. *Radiology* 157:205–209, 1985.
73. Matsen, F. A., and Arntz, C. T. Rotator cuff tendon failure. *In* Rockwood, C. A., and Matsen, F. A. (Eds.), *The Shoulder*. Philadelphia, W. B. Saunders, 1990, pp. 647–677.
74. Matsen, F. A. I., Thomas, S. C., and Rockwood, C. A. Anterior glenohumeral instability. *In* Rockwood, C. A., and Matsen, F. A. I. (Eds.), *The Shoulder*. Philadelphia, W. B. Saunders, 1990, pp. 526–622.
75. McNiesh, L. M., and Callaghan, J. J. CT arthrography of the shoulder: Variations of the glenoid labrum. *AJR* 149:963–966, 1987.
76. Meyer, S. J. F., and Dalinka, M. K. Magnetic resonance imaging of the shoulder. *Semin Ultrasound CT MR* 11:253, 1990.
77. Middleton, W. D., Kneeland, J. B., and Carrera, G. F. High resolution MR imaging of the normal rotator cuff. *AJR* 148:559–564, 1987.
78. Middleton, W. D., Reinus, W. R., Melson, G. L., Totty, W. G., and Murphy, W. A. Pitfalls of rotator cuff sonography. *AJR* 146:555–560, 1986.
79. Middleton, W. D., Reinus, W. R., Totty, W. G., Melson, G. L., and Murphy, W. A. Ultrasonographic evaluation of the rotator cuff and biceps tendon. *J Bone Joint Surg* 68A:440–450, 1986.
80. Mink, J. H., Harris, E., and Rappaport, M. Rotator cuff tears: Evaluation using double-contrast shoulder arthrography. *Radiology* 157:621–623, 1985.
81. Mitchell, M. J., Causey, G., and Berhoty, D. P. Peribursal fat plane of the shoulder: Anatomic study and clinical experience. *Radiology* 168:699–704, 1988.
82. Mosely, H. F., and Goldie, I. The arterial pattern of the rotator cuff of the shoulder. *J Bone Joint Surg* 45B(4):780–789, 1963.
83. Mosely, H. F., and Overgaard, B. The anterior capsular mechanism in recurrent anterior dislocation of the shoulder. *J Bone Joint Surg* 44B:913–927, 1962.
84. Murnaghan, J. P. Frozen shoulder. *In* Rockwood, C. A., and Matsen, F. A. (Eds.), *The Shoulder*. Philadelphia, W. B. Saunders, 1990, pp. 837–862.
85. Neer, C. S. Anterior acromioplasty for the chronic impinge-

ment syndrome in the shoulder: A preliminary report. *J Bone Joint Surg* 54A:41–50, 1972.

86. Neer, C. S. Impingement lesions. *Clin Orthop Rel Res* 173:70–77, 1983.

87. Neer, C. S. I., and Welsh, R. P. The shoulder in sports. *Orthop Clin North Am* 8:583–591, 1977.

88. Neviaser, J. S. Adhesive capsulitis of the shoulder: A study of the pathological findings in periarthritis of the shoulder. *J Bone Joint Surg* 27:211–222, 1945.

89. Neviaser, J. S. Arthrography of the shoulder joint. *J Bone Joint Surg* 44:1321–1330, 1962.

90. Neviaser, R. J. Tears of the rotator cuff. *Orthop Clin North Am* 11:295–306, 1980.

91. Neviaser, R. J. Lesions of the biceps and tendinitis of the shoulder. *Orthop Clin North Am* 11:343–348, 1980.

92. Neviaser, R. J., and Nevaiser, T. J. The frozen shoulder diagnosis and management. *Clin Orthop* 223:59–64, 1987.

93. Newhouse, K. E., El-Khoury, G. Y., Nepola, J. V., and Montgomery, W. J. The shoulder impingement view: A fluoroscopic technique for the detection of subacromial spurs. *AJR* 151:539–541, 1988.

94. Nixon, J. E., and DiStefano, V., et al. Ruptures of the rotator cuff. *Orthop Clin North Am* 6:423–447, 1975.

95. O'Brien, S. I., Arnoczky, S. P., Warren, R. F., and Rozbruch, S. R. Developmental anatomy of the shoulder and anatomy of the glenohumeral joint. *In* Rockwood, C. A., and Matsen, F. A. I. (Eds.), *The Shoulder.* Philadelphia, W. B. Saunders, 1990, pp. 1–33.

96. O'Brien, S. J., Neves, M. C., Rozbuck, R. S. The anatomy and histology of the inferior glenohumeral ligament complex of the shoulder. Las Vegas, The Shoulder and Elbow Society, 1989.

97. Olsson, O. Degenerative changes of the shoulder joint and their connection with shoulder pain. *Acta Chir Scand* 181:1–30, 1953.

98. Ovesen, J., and Nielsen, S. Stability of the shoulder joint: Cadaver study of stabilizing structures. *Acta Orthop Scand* 56:149–151, 1985.

99. Ovensen, J., and Nielsen, S. Anterior and posterior shoulder instability. A cadaver study. *Acta Orthop Scand* 57(4):324–327, 1986.

100. Pappas, A. M., Goss, T. P., and Kleinman, P. K. Symptomatic shoulder instability due to lesions of the glenoid labrum. *Am J Sports Med* 11:279, 1983.

101. Pennes, D. R., Jonsson, K., Buckwalter, K., Braunstein, E., Blasier, R., and Wojtys, E. Computed arthrotography of the shoulder: Comparison of examinations made with internal and external rotation of the humerus. *AJR* 153:1017–1019, 1989.

102. Penny, J. N., and Welsh, M. B. Shoulder impingement syndrome in athletes and their surgical management. *Am J Sports Med* 9:11–15, 1981.

103. Perry, J. Biomechanics of the shoulder. *In* Rowe, C. (Ed.), *The Shoulder.* New York, Churchill Livingstone, 1988.

104. Petersson, C. J. Degeneration of the gleno-humeral joint. *Acta Orthop Scand* 54:277–283, 1983.

105. Plenk, H. P. Calcifying tendinitis of the shoulder. *Radiology* 59:384–389, 1952.

106. Poppen, N. K., and Walker, P. S. Normal and abnormal motion of the shoulder. *J Bone Joint Surg* 58A:195, 1976.

107. Preston, B. J., and Jackson, J. P. Investigation of shoulder disability by arthrography. *Br J Radiol* 28:259–266, 1977.

108. Prodromos, C. C., Ferry, J. A., Schiller, A. L., and Zarins, B. Histological studies of the glenoid labrum from fetal life to old age. *J Bone Joint Surg* 72A(9):1344–1348, 1990.

109. Rafii, M., Firooznia, H., Bonamo, J. J., et al. Athlete shoulder injuries: CT arthrographic findings. *Radiology* 162:559–564, 1987.

110. Rafii, M., Firooznia, H., Golimbu, C., Minkoff, J., and Bornamo, J. CT arthrography of capsular structure of the shoulder. *AJR* 146:361–367, 1986.

111. Rafii, M., Firooznia, H., Sherman, O., et al. Rotator cuff lesions: Signal patterns at MR imaging. *Radiology* 177:817–823, 1990.

112. Rathbun, J. B., and MacNab, I. The microvascular pattern of the rotator cuff. *J Bone Joint Surg* 52B:540–553, 1970.

113. Reeves, B. Arthrography of the shoulder. *J Bone Joint Surg* 48B:424–435, 1966.

114. Reeves, B. Arthrographic changes in frozen and post traumatic stiff shoulders. *Proc R Soc Med* 59:827–830, 1966.

115. Resnick, D. Arthrography, tenography and bursography. *In* Resnick, D., and Niwayama, G. (Eds.), *Diagnosis of Bone and Joint Disorders.* Philadelphia, W.B. Saunders, 1989, pp. 544–558.

116. Resnick, D. Shoulder arthrography. *Radiol Clin North Am* 19:243–253, 1981.

117. Resnick, D. Shoulder pain. *Orthop Clin North Am* 14(1):81–97, 1983.

118. Resnick, D. Arthrography, tenography and bursography. *In* Resnick, D., and Niwayama G. (Eds.), *Diagnosis of Bone and Joint Disorders* (2nd ed.), Philadelphia, W.B. Saunders, 1989, pp. 302–340.

119. Resnik, C. S., Deutsch, A. L., and Resnick, D. Arthrotomography of the shoulder. *RadioGraphics* 4:963–976, 1984.

120. Rockwood, C. A. Dislocations about the shoulder. *In* Rockwood, C. A., and Green, D. P. (Eds.), *Fractures.* Philadelphia, J.B. Lippincott, 1975, pp. 624–686.

121. Rockwood, C. A. Subluxation of the shoulder. *Orthop Trans* 4:306, 1979.

122. Rogers, L. F. The shoulder and humeral shaft. *In* Rogers, L. F. (Ed.), *Radiology of Skeletal Trauma.* New York, Churchill Livingstone, 1982, pp. 377–432.

123. Rothman, R. H., and Parke, W. W. The vascular anatomy of the rotator cuff. *Clin Orthop* 41:176–186, 1970.

124. Rowe, C. R. Prognosis in dislocations of the shoulder. *J Bone Joint Surg* 38A:957–977, 1956.

125. Roy, S., and Irvin, R. Throwing and tennis injuries to the shoulder and elbow. *In Sports Medicine.* Englewood Cliffs, Prentice-Hall, 1983, pp. 211–227.

126. Saha, A. K. Dynamic stability of the glenohumeral joint. *Acta Orthop Scand* 42:491, 1971.

127. Sarrafian, S. K. Gross and functional anatomy of the shoulder. *Clin Orthop* 173:11–19, 1983.

128. Seeger, L. L., Gold, R. H., and Bassett, L. W. Shoulder instability: evaluation with MR imaging. *Radiology* 168:695, 1988.

129. Seeger, L. L., Gold, R. H., Bassett, L. W., and Ellman H. Shoulder impingement syndrome: MR findings in 53 shoulders. *AJR* 150:343–347, 1988.

130. Seeger, L. L., Ruszkowski, J. T., and Bassett, L. W. MR imaging of the normal shoulder: Anatomic correlation. *AJR* 148:83–91, 1987.

131. Selzer, S. E., and Weissmann, B. N. CT findings in normal and dislocating shoulders. *J Can Assoc Radiol* 36:41, 1985.

132. Shuman, W. P., Kilcoyne, R. F., Matsen, F. A., Rogers, J. V., and Mack L. A. Double contrast computed tomography of the glenoid labrum. *AJR* 153:581–584, 1983.

133. Simkin, P. A. Structure and function of joints. *In* Schumacher, H. R. (Ed.) *Primer on the Rheumatic Diseases.* Atlanta, Arthritis Foundation, 1988, Chap. 11.

134. Simon, W. H. Soft tissue disorders about the shoulder. *Orthop Clin North Am* 6(2):521–539, 1949.

135. Singson, R. D., Feldman, F., and Bigliani, F. CT arthrographic patterns in recurrent glenohumeral instability. *AJR* 149:749–753, 1987.

136. Soble, M. G., Kaye, A. D., and Guay, R. C. Rotator cuff tear: clinical experience with sonographic detection. *Radiology* 173:319–321, 1989.

137. Stiles, R. G., Resnick, D., Sartoris, D. J., and Andre, M. P. Rotator cuff disruption: Diagnosis with digital arthrography. *Radiology* 168:705–707, 1988.

138. Stimson, L. A. *A Practical Treatise on Fractures and Dislocations* (7th ed.). Philadelphia, Lea and Febiger, 1912.

139. Stoller, D. W., Martin, C., Crues, J. V., Kaplan, L., and Mink, J. H. Meniscal tears: Pathologic correlation with MR imaging. *Radiology* 163:732–735, 1987.

140. Strizak, A. M., Torrance, L. D., and Jackson, D. Subacromial bursography: An anatomic and clinical study. *J Bone Joint Surg* 64A:196–201, 1982.

141. Tamai, K., and Ogawa, K. Intratendinous tear of the supraspinatus tendon exhibiting winging of the scapula. *Clin Orthop Rel Res* 194:159–163, 1985.

142. Tsai, J. C., and Zlatkin, M. B. Magnetic resonance imaging of the shoulder. *Radiol Clin North Am* 28:279, 1990.

143. Turkel, S. J., Panio, M. W., Marshall, J. L., and Girgis, F. G. Stabilizing mechanisms preventing anterior dislocation of the glenohumeral joint. *J Bone Joint Surg* 63A:1208–1217, 1981.

144. Uhthoff, H. K., and Sarkar, K. Calcifying tendinitis. *In* Rockwood, C. A., and Matsen, F. A. (Eds.), *The Shoulder.* Philadelphia, W.B. Saunders, 1990, pp. 774–190.

145. Weston, W. J. The enlarged subdeltoid bursa in rheumatoid arthritis. *Br J Radiol* 27:481–486, 1969.

146. Wiley, A. M. Arthroscopy for shoulder instability and a technique for arthroscopic repair. *Arthroscopy* 4(1):25–30, 1988.

147. Wilson, A. J., Totty, W. G., Murphy, W. A., and Hardy, D. C. Shoulder joint: Arthrographic CT and long-term follow-up. *Radiology* 173:329–333, 1989.

148. Zlatkin, M., Bjorkengren, A. G., Gylys-Morin, V., Resnick, D., and Sartoris, D. J. Cross-sectional imaging of the capsular mechanism of the glenohumeral joint. *AJR* 150:151–158, 1988.

149. Zlatkin, M. B., Ionnati, J. P., and Roberts, M. C. Rotator cuff disease: Diagnostic performance of MR imaging. *Radiology* 172:223–229, 1989.

150. Zlatkin, M. B., and Kneeland, J. B. MR imaging of the glenohumeral joint. *In* Edelman, R. R., and Hesselink, J. R. (Eds.), *Clinical Magnetic Resonance Imaging.* Philadelphia, W.B. Saunders, 1990, pp. 1010–1030.

151. Zlatkin, M. B., Reicher, M. A., and Kellerhouse, L. E. The painful shoulder: MR imaging of the glenohumeral joint. *J Comput Assist Tomog* 12:995–1001, 1988.

THE ARM

Fractures and Soft Tissue Injury

James B. Bennett, M.D.
Thomas L. Mehlhoff, M.D.

ANATOMY

The arm, exclusive of the glenohumeral joint and the elbow joint, consists of the bony architecture of the humerus, the musculotendinous units that originate from the shoulder girdle to insert onto the humerus or pass across the elbow joint, and the associated neurovascular structures, soft tissue, and skin. This anatomic region is also referred to as the brachium. Skeletal support for the arm is provided by the humerus, which is approximately one-fifth of the body height in length [123]. The proximal shaft of the humerus at the upper border of the pectoralis major tendon is initially rounded but becomes more prismatic distally in the region of the supracondylar ridges. A medial torsion of approximately 35 degrees occurs in the humerus during growth and is not complete until nearly 20 years of age [64]. The anterolateral surface of the humerus presents a prominent V-shaped roughening approximately at its midlength, the deltoid tuberosity. The anteromedial surface usually presents no conspicuous surface markings. One exception to this is the supracondylar process. A supracondylar process exceeding 3 mm can be found on the anteromedial surface of the distal humerus, 5 to 7 cm proximal to the elbow, in approximately 1% of anatomic specimens [110]. The posterior surface of the humerus is notable for a shallow groove for the radial nerve,

passing obliquely in a lateral and downward direction.

Strong intermuscular septa arise from the periosteum of the lateral and medial aspects of the humerus, forming a strong tubular investment that divides the musculature of the brachium into anterior and posterior compartments (Fig. 16A–1A). The anterior compartment includes the preaxial muscles, namely, the coracobrachialis, biceps brachii, and brachialis anticus. The posterior compartment includes the postaxial muscles, namely, the triceps brachii and the anconeus. Interfascial spaces provide for the passage of neurovascular structures, including the radial, ulnar, and median nerves, as well as the brachial artery. The brachial artery is a single vessel throughout the brachium in 80% of cases; however, a proximal bifurcation of the brachial artery in the upper arm may result in a high radial artery in 10% of cases, a high ulnar artery in 3% of cases, or both the radial and ulnar arteries in 7% of cases [71].

The biceps brachii (musculocutaneous nerve, C5–C6) is a long, fusiform muscle with two heads proximally and one tendon distally. The long head of the biceps originates from the supraglenoid tubercle and the short head from the coracoid process, which then join as a common muscle belly. The distal tendon proper inserts into the radial tuberosity but also has a flattened attachment to the lacertus

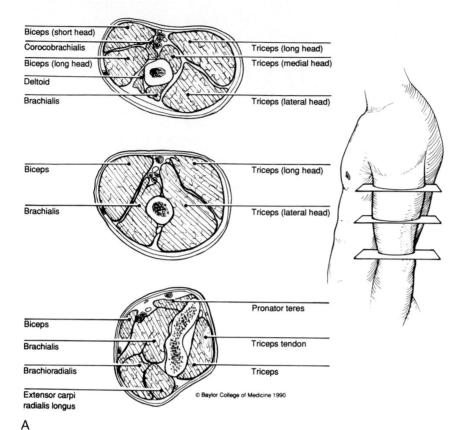

Biceps (short head)
Corocobrachialis
Biceps (long head)
Deltoid
Brachialis

Triceps (long head)
Triceps (medial head)
Triceps (lateral head)

Biceps
Brachialis

Triceps (long head)
Triceps (lateral head)

Biceps
Brachialis
Brachioradialis
Extensor carpi radialis longus

Pronator teres
Triceps tendon
Triceps

© Baylor College of Medicine 1990

A

FIGURE 16A–1
A, Muscles, compartments, and intermuscular septa of the arm. *B,* Neuromuscular anatomy of the upper arm, anterior view.

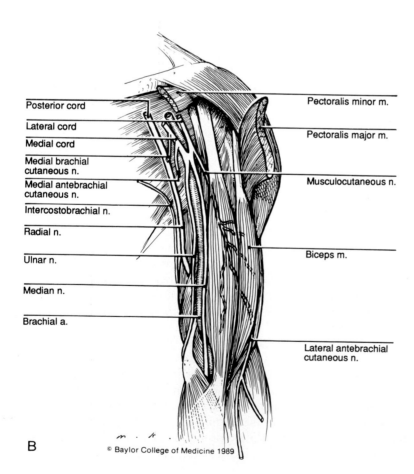

Posterior cord
Lateral cord
Medial cord
Medial brachial cutaneous n.
Medial antebrachial cutaneous n.
Intercostobrachial n.
Radial n.
Ulnar n.
Median n.
Brachial a.

Pectoralis minor m.
Pectoralis major m.
Musculocutaneous n.
Biceps m.
Lateral antebrachial cutaneous n.

B
© Baylor College of Medicine 1989

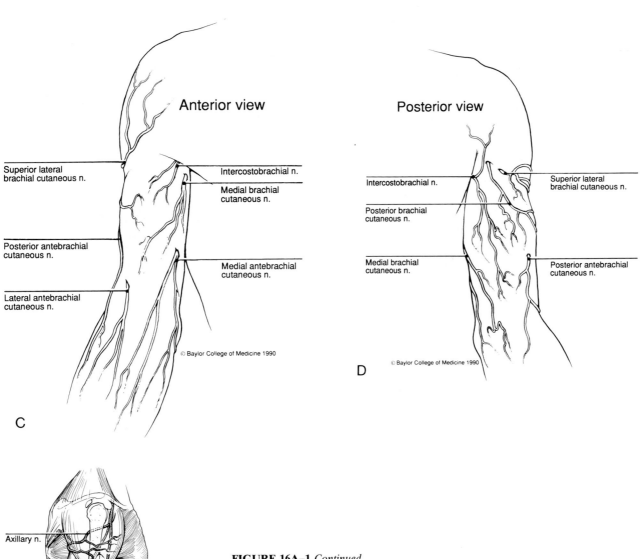

Anterior view

Posterior view

Superior lateral brachial cutaneous n.

Intercostobrachial n.

Medial brachial cutaneous n.

Posterior antebrachial cutaneous n.

Medial antebrachial cutaneous n.

Lateral antebrachial cutaneous n.

© Baylor College of Medicine 1990

C

Intercostobrachial n.

Superior lateral brachial cutaneous n.

Posterior brachial cutaneous n.

Medial brachial cutaneous n.

Posterior antebrachial cutaneous n.

© Baylor College of Medicine 1990

D

Axillary n.

Radial n.

Musculocutaneous n.

© Baylor College of Medicine 1989

E

FIGURE 16A–1 *Continued*
C, Cutaneous nerve supply to the arm, anterior and *D,* posterior. *E,* Neuromuscular anatomy of the upper arm with axillary, radial, and musculocutaneous nerves, lateral view.

fibrosis [85] (Fig. 16A–1*B*). One or more accessory heads of the biceps brachii are not uncommon, and have been reported in 21% of 130 anatomic dissections [43].

The brachialis anticus (musculocutaneous nerve, C5–C6) arises from the anterior surface of the humerus, as do the medial and lateral intermuscular septa. The brachialis insertion cloaks the capsule of the elbow, coronoid process, and the ulnar tuberosity distally.

The triceps brachii (radial nerve, C7–C8) is composed of three heads. The long head of the triceps originates from the infraglenoid tubercle and, with the lateral head, forms a common aponeurosis, which inserts along the proximal olecranon. The third head, the medial head, lies deep to the aponeurosis.

The cutaneous innervation of the medial aspect of the arm in the axilla is provided by the intercostal brachial nerve (T2). The anterior surface of the arm is supplied by the medial brachial cutaneous nerve (C8, T1) arising from the medial cord of the brachial plexus. The posterior surface of the arm is supplied by the superolateral brachial (branch of the axillary nerve), inferior lateral brachial, and posterior brachial cutaneous nerves, derivatives of the radial nerve. The medial antebrachial cutaneous nerve (C8, T1) follows the basilic vein to supply the medial forearm. The lateral forearm is supplied by the lateral antebrachial cutaneous nerve (C5, C6, C7), continuing from the musculocutaneous nerve (Fig. 16A–1*C,D*).

HYPERTROPHY

Athletic participation in competition produces a myriad of developmental changes within the upper arm complex. Hypertrophy results from repetitive muscular activity. In an analysis of 50 professional baseball pitchers, King and colleagues [61] demonstrated striking hypertrophy of the entire shoulder girdle, including the latissimus dorsi, pectoralis major, and serratus, as well as the flexors of the forearm (Fig. 16A–2). The biceps and triceps are hypertrophied as well. Flexion contracture of the elbow was present in over 50% of the pitchers examined. Almost uniformly, hypertrophy of the humerus was noted, with an increase in both cortical size and density of the humerus [61]. Priest and Nagel reported similar hypertrophy of the humerus in tennis players. Cortical thickness of the humerus on the playing side was 35% greater than that on the nonplaying side in males and 28% greater in females [97]. King and colleagues also reported that

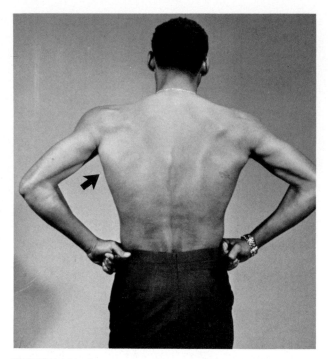

FIGURE 16A–2
Shoulder–upper arm athletic muscular hypertrophy in a professional baseball pitcher.

along with cortical hypertrophy and muscle hypertrophy, the professional throwing athlete usually demonstrated an increase in external rotation with a concomitant decrease in internal rotation in the shoulder of the pitching arm [61]. Hypertrophy is a normal response to prolonged heavy arm exercise and should not be misinterpreted as a pathologic condition (Fig. 16A–3).

King and colleagues [61], in the analysis of the pitching arm of the professional athlete, divided the act of throwing into a cocking phase, an acceleration phase, and a follow-through phase. Each phase of throwing requires a specific interaction of biomechanical forces.

The cocking phase is accomplished during the wind-up, when the shoulder is brought into extreme external rotation, abduction, and extension. Repetition of the cocking phase increases external humeral rotation and decreases internal humeral rotation, as noted in professional baseball pitchers.

The acceleration phase is a two-stage process. The shoulder is brought forward with force while the forearm and hand are left stationary. The elbow is placed in a position of extreme valgus stress as the forces increase in the humeral shaft. The forearm and hand are then rapidly brought forward by the internal rotator of the shoulder. This phase terminates with the release of the ball from the hand.

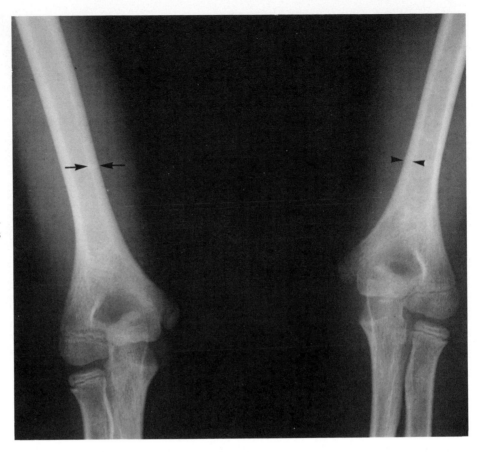

FIGURE 16A–3
Cortical bone thickening and increased bone density in the humerus of a professional athlete.

The follow-through phase is the terminal event of the throwing act. It begins with the release of the ball, applies rotation for ball control, and terminates with the force of deceleration of the arm and forearm rotation [61].

Repetitive throwing, muscle imbalance due to hypertrophy or contracture, and fatigue result in pathologic conditions of the arm without a history of specific trauma in the athlete.

TENDON AND MUSCLE RUPTURES

The first report of a tendon rupture is credited to Petit in 1772 [7]. Two series of 100 tendon ruptures each were reported by Gilcreest [40] in 1934 and by Haldeman and Soto-Hall in 1935 [45]. In each series the authors noted that tears of the supraspinatus, the long head of the biceps, and the finger extensors (extensor digitorum communis and extensor pollicis longus), were the most common tendon ruptures in the upper extremity. Quadriceps and triceps surae tendon ruptures were most common in the lower extremity; disruption of the trunk, such as rupture of the rectus abdominis, composed only a very small percentage of ruptures.

The largest series of tendon injuries was reported in 1959 by Anzel and associates and included lacerations as well as ruptures. Of 1014 tendon injuries in 781 patients, 85% occurred in the upper extremity. Of 106 tendon ruptures in the upper extremity, 54.7% involved the supraspinatus tendon, 26.4% the biceps tendon (predominantly the long head), and 1.9% the triceps tendon [7].

Many factors, both local and systemic, may predispose to tendon ruptures. Degenerative changes within tendons near the osseotendinous insertion, sometimes associated with calcification or even bone spurring, have been implicated [28]. Other intrinsic factors leading to tendon rupture have been reported to include intratendinous tumors such as xanthoma or hemangioendothelioma [126]. Systemic conditions, such as rheumatoid arthritis, systemic lupus erythematosus, hyperparathyroidism, and secondary hyperparathyroidism related to chronic renal failure, have been associated with tendon ruptures [73, 96, 116]. Finally, systemic steroids and local steroids have been implicated with tendon rupture both clinically and experimentally [60, 117].

The pathophysiology of tendon rupture was scientifically investigated and reported by McMaster in 1933 [75]. His work demonstrated that a normal tendon will not rupture when the tendon is under stress. In fact, even when a normal tendon had 75% of its fiber sectioned, normal activity did not result in a rupture. Increasing linear tension across the muscle tendon unit caused failure at one of three locations other than within the tendon itself, namely, the muscle belly, the musculotendinous junction, and the tendo-osseous insertion. Most clinical ruptures occur at the tendo-osseous insertion and are more accurately termed avulsions.

Rupture of the Pectoralis Major Muscle

Patissier first described the rupture of the pectoralis major muscle from its humeral insertion in 1822 [93]. This pectoralis major muscle injury may be incomplete or complete and may occur at the muscle origin from the clavicle or sternum, at the muscle belly, at the musculotendinous junction, or at its insertion into the proximal humerus. This is a relatively uncommon injury but is seen increasingly in sports activities requiring forced contraction against resistance, as in weight lifting, and in those involving forceful adduction and internal rotation against resistance, as in contact sports, football, and wrestling. McEntire and associates reviewed 45 cases in the literature as well as 11 cases from their study in 1972 [74]. Currently, there are approximately 100 reported cases in the literature.

Clinical Evaluation

The pectoralis major muscle rupture presents clinically as swelling and ecchymosis about the muscle insertion to the humerus and the chest wall. Weakness of internal rotation is present. There is a history of severe searing, burning pain with forceful internal rotation or sudden external rotation against a fixed fulcrum by an adducted, internally rotated forearm. The ecchymosis, which may be delayed, will occur at the site of injury and may be located on the chest wall, at the insertion into the proximal humerus, or in the axillary fold (Fig. 16A–4). The axillary fold will be deficient and demonstrate a palpable pectoralis muscle, with a visible defect when the arm is abducted at 90 degrees if insertional avulsion is present. Clinical weakness with adduction and internal rotation is present and is associated with pain and muscular defect.

Plain x-ray films will be negative; however, ultrasound and magnetic resonance imaging will show pectoralis rupture. Hematoma and pseudocyst formation may be shown as well.

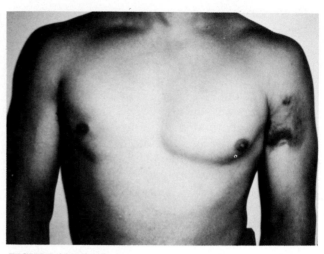

FIGURE 16A–4
Left pectoralis major tendon rupture at insertion into humerus. Note ecchymosis at insertion site and enlargement of pectoralis major over chest wall as a result of muscle retraction.

Treatment Options

Conservative treatment is used in pectoralis major muscle sprain or partial rupture. Also, complete rupture from the chest wall origin and within the muscle substance is treated conservatively. The hematoma is controlled with cooling, ice, and rest, followed by heat and mobilization to restore shoulder function and pectoralis function.

Complete rupture of the pectoralis major muscle in a nonathlete may be treated conservatively as with partial ruptures; however, full strength will not return. The active athlete should be treated with surgical reattachment in cases of tendon avulsion injuries from the proximal humerus for complete restoration of shoulder strength and function.

Orava and colleagues [90] and Berson [14] have demonstrated surgical repair of pectoralis major muscles to the humeral insertion lateral to the long head of the biceps brachii tendon. Musculotendinous tears may be sutured directly, but this may be difficult to accomplish owing to the failure of muscle tissue to hold an appropriate suture repair.

Park and Espiniella report excellent results in 90% of their patients repaired surgically compared with approximately 60% of patients treated nonoperatively [92]. Nonoperative patients had varying degrees of weakness in adduction and internal rotation. However, the function of the pectoralis major was well compensated for by the teres major, subscapularis, and latissimus dorsi.

Zeman and associates report similar excellent results in surgically repaired cases and compare them to the residual weakness in the conservatively treated patient population [127].

Postoperative Management and Rehabilitation

The pectoralis muscle major reattachment to the proximal humerus or repairs to the musculotendinous junction are treated initially in a shoulder immobilizer with the arm adducted to the chest wall and internally rotated on to the abdomen, with the elbow flexed at 90 degrees in a neutral position. Immobilization is maintained full-time for 4 weeks, with a gentle pendulum and circumduction exercise program initiated at 2 weeks following repair. After 4 weeks, immobilization is removed, and active and passive range of motion without resistance is continued for an additional 4 weeks. The arm is maintained in a sling at this time, and full range of motion is obtained. Resistive range of motion and strengthening are begun at 8 weeks and continued for 12 weeks. Gentle restricted throwing activities are begun at this time. The entire upper extremity is involved in the rehabilitation process because of disuse atrophy and weakness. Unprotected heavy lifting is restricted for 6 months during the rehabilitation process.

Criteria for returning to sports participation is maximum functional strength compared with the opposite shoulder and upper arm, with full range of motion. Return to sports activities may be at 3 months, with protection against forceful external rotation. Full sports contact would begin at 6 months if strength and range of motion have plateaued and equal those in the opposite extremity. A shoulder harness, a strap to block full external rotation, may be used if the participant is involved in the contact sport of football. Otherwise, full unrestricted activities are allowed at 6 months.

Authors' Preferred Method of Treatment

Conservative treatment, as outlined above, is used for incomplete tears or tears from the chest at the sternal or clavicular origins.

Surgical repair at the musculotendinous junction with nonabsorbable suture and the rehabilitation program as outlined above is used in cases of high-performance athletes with tears at the musculotendinous junction. In the nonathlete, conservative management is used.

Tendon avulsions at the insertion of the proximal humerus are reattached to surgically created troughs and drill holes into the humerus with nonabsorbable sutures through cortical bone and are reinforced with appropriate soft tissue and deltoid fascia as outlined in the techniques of Orava and co-workers

and Berson [14, 90]. Evacuation of hematoma or pseudocyst with seroma is performed at the same time. Care must be taken to protect the brachial plexus as well as the axillary and brachial vasculature during surgical exposure and reconstruction. The surgical exposure is through a deltopectoral incision, and identification of the tendinous insertion is performed. Mobilization of the pectoralis major muscle from the chest is performed if the muscle is retracted. Repair is lateral to the biceps tendon at the level of the deltoid insertion. In cases of late repair, mobilization is difficult, contraction may be fixed, and reconstruction may be difficult, although Kretzler and Richardson report repair up to 5 years following injury [65]. Rehabilitation in these cases yields less satisfactory results, with restriction of full functional motion and weakness. For the best outcome, early diagnosis and repair (less than 4 weeks following injury) are recommended.

Distal Biceps Avulsion

Avulsion of the distal biceps tendon at the elbow is an uncommon injury [7, 10, 31, 36, 69, 77, 80, 87]. Gilcreest and Albii [39] reported that 97 of 100 biceps ruptures are proximal (long head, 96; short head, 1), whereas only three biceps ruptures occurred at the elbow. The first description of distal biceps avulsion is credited to Starks in 1843 [118]. Operative treatment for a distal rupture was first reported by Acquaviva in 1898 [69, 118]. In 1941, Dobbie reported a collection of 51 cases from 40 surgeons; most reported an experience of only two or three cases each, emphasizing the rarity of this lesion [31]. Less than 200 cases of distal biceps avulsion have been reported in the literature.

Clinical Evaluation

Rupture of the distal biceps brachii tendon typically occurs in men between the fourth and sixth decades of life [36]. All cases reported have been in males, with the possible exception of one case [76]. The average age at rupture is 50 years, but in the literature, the ages range from 21 to 70 years. Approximately 80% of the injuries occur in the dominant extremity [80]. Bilateral avulsion is extremely rare, but at least three cases have been reported [118].

Avulsion typically occurs during a heavy lift with the elbow flexed approximately 90 degrees and is the result of a sudden or prolonged contracture of the biceps against high load resistance [80]. In most instances, a single traumatic event is recalled by the patient [10, 80, 87]. Prodromal or prerupture symptoms are uncommon [69]. The tear usually occurs at the tendo-osseous insertion and notably leaves no distal tendon at the tuberosity [80, 87]. The lacertus fibrosis attachment is damaged to a varying degree, but usually is left intact [31]. In fact, the distal biceps tendon may rupture in stages, first with avulsion of the distal biceps tendon proper from the radial tuberosity, and then by a second tearing from the lacertus fibrosis [85].

Preexisting degenerative changes within the distal tendon at the radial tuberosity are thought to predispose the tendon to rupture [94]. It is unusual to see radiographic changes in the radial tuberosity prior to tendon avulsion [76], but Davis and Yassine [28] did identify degenerative changes at the volar aspect of the radial tuberosity. They postulated that these hypertrophic changes could cause tears in the tendon during pronation and supination of the forearm.

When rupture occurs, the patient usually experiences a popping or tearing sensation and presents with acute pain in the antecubital fossa. He or she may not be aware of a flexion or supination loss of strength following rupture. Clinical findings include tenderness, swelling, and mild to moderate ecchymosis in the antecubital fossa [36, 80]. The distal biceps tendon is not palpable following a complete rupture. Deformity may occur as the muscle belly retracts proximally during an attempted contraction. The patient can still flex the elbow because the brachialis is intact and supinate the forearm owing to an intact supinator, but he will have decreased powers of flexion and especially residual supination [36].

The diagnosis is usually established by the history and physical examination. It is unusual to see radiographic evidence of avulsion fragments from the radial tuberosity. The differential diagnosis for pain in the antecubital fossa includes biceps tendinitis, partial distal biceps avulsion, bicipital bursitis, or the lateral antebrachial cutaneous nerve entrapment syndrome. The absence of a palpable tendon in the antecubital fossa should distinguish a complete rupture from the other entities; however, a partial tendon rupture may be more difficult to diagnose correctly. Magnetic resonance imaging (MRI) may be useful when the diagnosis is unclear.

Treatment Options

Treatment of the distal biceps tendon rupture is surgical. The goal is to restore supination and flexion power to the elbow and forearm through anatomic repair of the tendon to the radial tuberosity [10, 17, 31, 77, 80, 87]. Although only slight functional deficit follows a disruption of the long head of the biceps (provided that the short head is intact), rupture of the distal tendon results in significant loss of flexion and supination strength and endurance [10, 23, 77, 80]. Meherin and Kilgore reported a threefold increase in disability for unrepaired cases [77]. Baker and Bierwagen used Cybex testing to document an 86% decrease in supination endurance in unrepaired cases [10]. Morrey and colleagues reported only 61% flexion strength and 65% supination strength in unrepaired cases [80]. Nonoperative treatment is no longer recommended in the athlete.

In patients undergoing anatomic repair, flexion and supination strength and endurance can be returned to normal. Morrey and colleagues reported 97% flexion strength and 95% supination strength in patients with injuries repaired within 2 weeks after the injury [80]. Late cases (over 4 weeks) with repair of the brachialis resulted in 87% flexion strength but only 43% supination strength.

Operative repair for a complete avulsion can be performed through the anterior extensile exposure of Henry using one incision. In these cases, a pull-out suture is used over a skin button distally [69, 87]. Most authors, however, advocate the two-incision approach proposed by Boyd and Anderson in 1961 [10, 17, 78, 80].

Postoperative Management and Rehabilitation

Following repair of the avulsed distal biceps tendon, the elbow is immobilized in 90 degrees of flexion with either neutral, moderate, or full supination of the forearm, depending on the tension of the tendon repair. Most authors recommend 8 weeks of protection [10, 80, 87]. For the athletic population, 4 weeks of static immobilization in this position can be followed by 4 weeks of dynamic splinting [80]. A gradual range-of-motion and strengthening program is initiated 6 to 8 weeks after surgery, followed by a progressive resistance exercise program for 8 to 12 weeks; unprotected heavy lifting should not be allowed for 6 months. Aerobic conditioning should be emphasized to the athlete for maintaining endurance and performance in uninjured extremities while rehabilitating the arm.

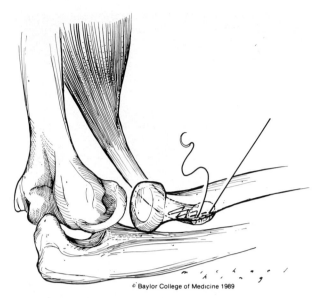

FIGURE 16A–5
Biceps tendon reinsertion technique into the radial tuberosity.

Criteria for Return to Sports Participation

Maximum functional strength return occurs at 4 to 6 months. Maximum range of motion is attained by 6 months. Cybex isokinetic testing may be used as an objective test to determine recovery [10]. Return to sports without contact or maximum resistance forces may be allowed at 3 months. Return to full-contact or maximum resistance sports may be allowed at 6 months after surgery if strength and motion of the involved arm have reached a plateau in the rehabilitation program. An elbow brace or strap to prevent full extension or hyperextension may be required in contact sports during the first few months of active participation.

Authors' Preferred Method of Treatment

Our preference is the use of the two-incision approach described by Boyd and Anderson [17]. A limited antecubital fossa incision is used to retrieve the distal biceps tendon. Care must be taken to identify the lateral antebrachial cutaneous nerve so that it will not be injured. A modified Bunnell suture is placed in the distal aspect of the tendon, using a heavy, nonabsorbable suture. A second incision is placed just lateral to the subcutaneous border of the proximal ulna, and an extraperiosteal dissection along the proximal ulna is then used to expose the radial tuberosity. The forearm must be

kept in a pronated position to minimize risk to the posterior interosseous nerve. The suture in the distal tendon can then be passed through the biceps tunnel with a Kelly clamp, so that it can be seen at the second incision. The tuberosity bed is prepared with a burr, and the suture is then passed through drill holes into the bone to secure the tendon (Fig. 16A–5). The wound is then repaired over a closed Hemovac drainage pump.

In cases of late reconstruction for a complete rupture, the avulsed biceps tendon is difficult to return to the radial tuberosity. Increased risk to the radial nerve is significant. Reconstruction of the biceps tendon may be performed by attaching the tendon to the brachialis muscle or coronoid process of the proximal ulna in repairs done more than 4 weeks following injury [80] (Fig. 16A–6). This will regain most of the flexion strength. If restoration of supination strength is required, use of a fascial graft extension into the radial tuberosity is necessary.

Although the extensile anterior approach of Henry is preferred by some authors, dissection about the radial nerve and recurrent leash vessels is difficult. Radial nerve palsy has been reported with this approach [17]. The use of two incisions avoids dissection of the neurovascular structures. Heterotopic ossification and synostosis of the proximal radial ulnar joint have, however, occasionally been reported with use of the two-incision approach [32, 80]. To decrease this risk, Morrey and colleagues recommend (1) avoiding subperiosteal dissection on

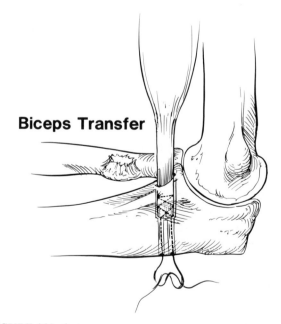

Biceps Transfer

FIGURE 16A–6
Biceps tendon transfer insertion into the coronoid process with or without fascial slip extension for late reconstruction of distal biceps tendon rupture.

the ulna, (2) clearing all bone debris from the radial tuberosity after burring, and (3) draining the wound to prevent hematoma. The lateral antebrachial cutaneous nerve and the posterior interosseous nerve can be at risk with either approach.

Nielson reported a single case of partial rupture of the distal biceps tendon, which resulted in an elongated and redundant biceps and lacertus fibrosis 8 months after injury [85]. The patient presented with weakening of elbow flexion and forearm supination but without significant pain. The tendon was found to be "too long and slack." The problem was addressed by shortening the tendon to normal tension using a step-cut Z-plasty and imbrication, resulting in restoration of normal strength and motion 8 to 12 weeks after surgery.

Distal Triceps Avulsions

The triceps tendon avulsion is also a rare injury and is perhaps the least common of all tendon ruptures [7, 9, 35, 50]. Only four cases of avulsion were reported by Anzel and associates in their review of tendon injuries at the Mayo Clinic [7]. Partridge is credited with the first description of a triceps rupture in 1863 [50]. Tarsney collected seven patients who had sustained tendo-osseous avulsions over a 16-year period [109]. Aso and Torisu noted 35 cases reported in the English literature; only two of these cases had been ruptures of the muscle belly. They added an additional two cases of muscle belly tears [8]. Farrar and Lippert emphasized the importance of distinguishing between complete and incomplete ruptures of the distal triceps tendon [35].

Nearly 75% of the ruptures reported in the literature occurred in males, with a male-to-female ratio of 3:1 [9]. Although the mean age at injury was approximately 26 years, case reports have spanned an age range from 7 to 72 years. Dominant and nondominant extremities appear to be injured with equal frequency [9]. Two cases of bilateral avulsion have been reported [73].

Most avulsions of the distal triceps follow trauma. Indirect trauma is the most common cause of injury and usually involves a fall onto the outstretched upper extremity. This imparts a deceleration stress on an already contracted triceps, resulting in distal avulsion at the tendo-osseous insertion. The tendon usually retracts with bone from the proximal olecranon embedded in it [9, 35, 50, 109]. In exceptional cases, midsubstance belly ruptures and musculotendinous junction ruptures have been reported [41, 78]. Some of these injuries may also result from a direct blow to the elbow. In addition, spontaneous

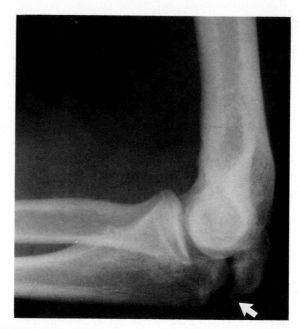

FIGURE 16A–7
Proximal olecranon apophysis separation with triceps insufficiency.

avulsion of the distal triceps tendon has been reported in patients with hyperparathyroidism, renal osteodystrophy, osteogenesis imperfecta, Marfan syndrome, systemic lupus erythematosus, and administration of systemic steroids [35, 73, 96, 103].

The same mechanism of injury responsible for distal triceps tendon avulsions may result in the less common transverse or oblique avulsion fracture through the proximal olecranon. A high incidence of proximal olecranon avulsion fractures was noted in javelin throwers, and proximal olecranon avulsion fractures have also been reported in baseball pitchers [105]. Such fractures are also not uncommon in children, who are more inclined to have a fracture of the proximal olecranon apophysis than a tendo-osseous avulsion of the triceps mechanism (Fig. 16A–7). The radiographic vagaries of the proximal olecranon have previously been addressed in the literature and should be referred to by physicians treating pediatric athletes [104].

Clinical Evaluation

The clinical features of triceps tendon avulsion include pain and swelling of the posterior elbow. A palpable depression just proximal to the olecranon may be noted [35, 109]. These findings may be difficult to appreciate in the face of severe pain and swelling, especially in a well-muscled athlete with large bulk [91, 109]. Ecchymosis may be marked

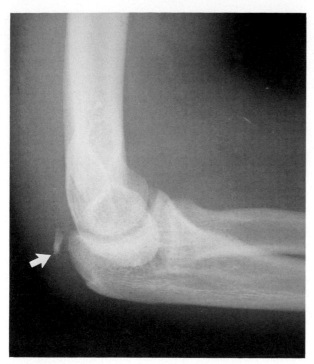

FIGURE 16A–8
Triceps tendon rupture with small avulsion fragment of olecranon.

several days after injury but not immediately. Careful testing of elbow extension strength is important to determine whether the tear is partial or complete [35]. Loss of active extension of the elbow signifies a complete tear of the triceps tendon.

Variations in clinical symptoms and signs after incomplete or complete avulsions of the triceps may lead to a delay in correct diagnosis. Rupture of the triceps tendon may present as a cubital tunnel syndrome [48], snapping elbow [32], "collar stud-shaped" olecranon bursitis [26], or even a posterior compartment syndrome [20].

Radiography should be performed in all suspected cases. Avulsed flecks of bone from the olecranon were demonstrated in approximately 83% of cases reported in the literature [9, 35, 109]. The remainder of the upper extremity should be carefully examined for associated fractures. Levy and associates reported their experience with 16 patients who sustained concomitant triceps ruptures in radial head fractures. The triceps rupture may be overlooked if the posterior aspect of the elbow is not specifically examined. An avulsed bone fleck present on the lateral radiograph of the elbow may be the only clue to the correct diagnosis [67] (Fig. 16A–8).

Treatment Options

Complete avulsion of the triceps should be surgically repaired in order to restore extension

strength [9, 35, 50, 109]. The role of surgery in treating partial triceps tendon ruptures is, however, unclear. Farrar and Lippert have stated that if full active extension can be demonstrated on the physical examination, then these injuries are partial and can be followed closely without surgical repair [35].

For complete tears, the accepted method of repair is reattachment of the avulsed triceps tendon to the olecranon with nonabsorbable sutures through drill holes in bone [35, 91, 109]. If a large fragment (50%) of the olecranon is present and displaced, open reduction and internal fixation of the olecranon is indicated. Otherwise, excision of the bone fragments and repair of the triceps tendon to the articular surface are indicated. Repair of the triceps to the articular surface is necessary to minimize anterior posterior instability of the elbow after excision of the proximal olecranon fragment (Fig. 16A–9).

Other methods of triceps repair have been described, including an inverted tongue of triceps used as a turned-down flap, a periosteal flap from the olecranon, and a posterior forearm fascial flap [6, 50]. These techniques may be useful in cases of delayed reconstruction if primary reattachment failed or initial diagnosis was missed and triceps retraction occurred.

Complications with operative repair may include skin slough, infection, and rerupture. Pantazopoulos and colleagues reported the development of an olecranon bursitis after wire-suture fixation [91]. This resolved after removal of the wire. Generally, good extension strength has been restored [9, 35]. Several authors have noted a mild flexion contrac-

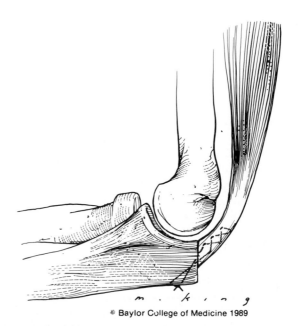

© Baylor College of Medicine 1989

FIGURE 16A–9
Triceps tendon reinsertion approximating the joint surface after olecranon fracture excision.

ture, ranging from 5 to 20 degrees in perhaps 10% of their series [67, 109].

Postoperative Management and Rehabilitation

The elbow is immobilized postoperatively in extension for a variable period of time, ranging from 10 days to 6 weeks in the literature [9]. Generally, the elbow is immobilized at 30 to 45 degrees of flexion for 4 weeks before a graduated range-of-motion and strengthening program is begun. The elbow is immobilized at 30 degrees of flexion for 4 weeks, and then 0 to 45 degrees range of motion is allowed during the next 2 weeks, after which graduated range-of-motion and strengthening exercises are begun. An extension night splint is used for the first 3 months after repair.

Criteria for Return to Sports Participation

Return to contact sports should be restricted until maximum motion and extension strength have been obtained, generally after 6 months. Active resistive strengthening to achieve equal triceps strength is performed from 3 to 6 months after surgery.

Five patients in the literature were treated non-operatively [9]. This treatment has generally been reserved for patients with partial ruptures or muscle belly ruptures that still demonstrated some ability to extend against gravity. Avulsion of the lateral head alone has been stated to result in no significant functional impairment, and in fact may be more frequent than has been recognized [20].

Authors' Preferred Method of Treatment

Avulsions are reattached with nonabsorbable heavy suture through drill holes in the proximal olecranon tip. Mobilization of the triceps tendon may be required, care being taken not to injure the radial nerve with proximal mobilization.

Fracture fragments of less than 50% of the olecranon, nonunion of the olecranon apophysis, or fragments that are severely comminuted are excised, and the triceps tendon is reattached to the olecranon adjacent to the articular surface. Large fragments (50% or greater) are reconstructed with open reduction and internal fixation by the AO/ASIF technique of tension-band wiring.

Delayed or late reconstruction of the triceps is accomplished by triceps mobilization and a turn-down triceps fascia flap or the use of a fascial strip of palmaris longus tendon or tensor fascia lata placed through a drill hole in the olecranon.

FRACTURES

Rotational Stress Fracture of the Proximal Humeral Epiphysis

Rotatory torque stresses to the epiphyseal growth plate of the proximal humerus in adolescent throwers can result in "Little Leaguer's shoulder." Initially described as epiphysitis or osteochondrosis by Adams [1], this lesion is a fatigue fracture or slip of the proximal humeral epiphyseal plate [21, 113] (Fig. 16A–10). Cahill described five cases in Little League pitchers aged 11 to 12 years [21]. Pain and inability to perform were the usual presenting complaints. Decreased range of shoulder motion and anterior glenohumeral tenderness were found on physical examination. Radiographs usually revealed widening of the epiphyseal line with a metaphyseal separation. With rest, the stress fractures healed with periosteal new bone formation secondary to periosteal stripping. The recommended treatment is rest for approximately 6 weeks, or until the lesion is healed [113]. Considering the large number of

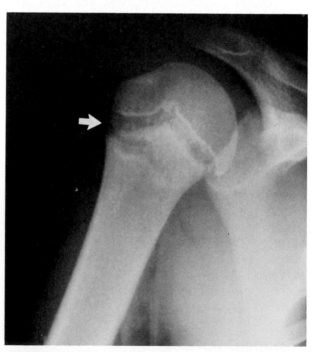

FIGURE 16A–10
"Little Leaguer's shoulder"—proximal humerus epiphyseal stress fracture in the adolescent.

participants in Little League baseball, Little Leaguer's shoulder is uncommon [112]. Resumption of sports activity is allowed when pain is resolved and healing is noted on radiographs.

Humeral Shaft Spiral Fractures

Although the most common cause of humeral fractures is blunt trauma, a spiral fracture of the shaft of the humerus resulting from muscular violence has been reported in various throwing sports, including baseball, javelin, and handball, as well as in arm wrestling [19, 24, 47, 49, 113, 121]. This fracture has become a recognized clinical entity (Fig. 16A–11). One of the first cases of "spontaneous" fracture of the humerus due to muscular violence was reported in 1947 [19]. Herzmark and Klune reported an additional four cases in prison inmates who were throwing baseballs [49]. They postulated

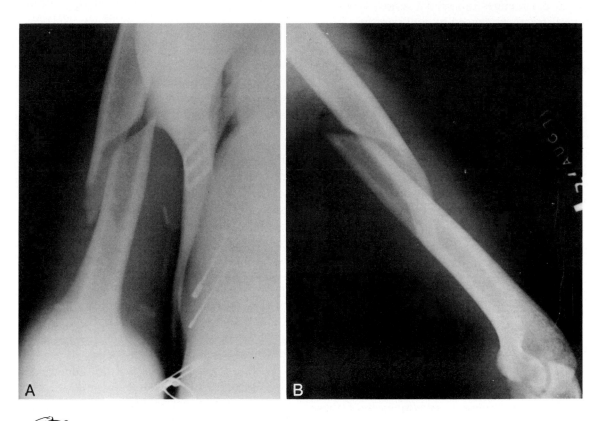

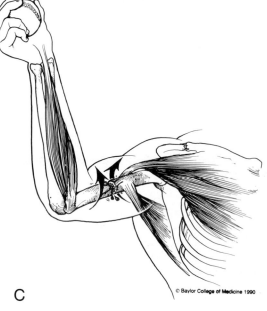

© Baylor College of Medicine 1990

FIGURE 16A–11
Spiral oblique humerus fracture. *A,* Anteroposterior view. *B,* Lateral view. *C,* Muscular rotational and biomechanical forces producing spiral fracture of the humerus.

torsional stress as the cause of this spiral fracture [49]. These fractures were also reported in three soldiers throwing hand grenades, and Chao and colleagues reported 129 cases in the Chinese army between 1959 and 1966; their report also included an experimental study that confirmed the role of torsional stress in causing spiral fractures of the distal humerus and quantitated the necessary force as 7 foot-pounds of torque [24]. A similar fracture was reported in an arm wrestler in 1962, and Heilbronner and associates reported a series of 28 cases of humeral fractures caused by arm wrestling [47].

Clinical Evaluation

The average age of patients sustaining these spiral fractures is 25 years and ranges from 18 to 38 years. The dominant extremity is fractured in over 90% of cases, which is not surprising because usually the stronger arm is used for throwing or arm wrestling. These fractures have been more frequently reported in unorganized sporting events [121].

Typically, a spiral fracture of the humerus results, usually in the middle to distal third of the humerus, and is often associated with a butterfly fragment. Chao and colleagues found these fractures in the upper third of the humerus in 1.6% of cases, in the middle third in 3.1% of cases, and in the distal third in 95.3% of cases [24]. No underlying bone disease has ever been documented in these cases [113].

Radial nerve injury may accompany these fractures. In a review of 42 cases in the English literature, radial nerve palsy was present in five, for an incidence of approximately 12% [47]. The radial nerve injury was a neurapraxia in each of these cases, and recovery occurred by 6 months. Only Chao and colleagues reported laceration of the radial nerve, which necessitated repair at 6 months [24].

Many explanations for the spiral fracture have been offered, most being variations on the theme of torsional stress. Powerful internal rotation is applied to the upper shaft of the humerus by the pectoralis major, subscapularis, teres major, and latissimus dorsi while a force across the forearm imparts external rotation to the distal humerus through the ulnotrochlear joint (see Fig. 16A–11C). Only 7 foot-pounds of torque are necessary for this fracture to occur, even without a stress raiser, as shown experimentally [24]. This degree of force can be attained in the throwing motion during the transition from the cocking phase to the acceleration phase [113]. This force is also easily attained during an arm wrestling match, particularly during a draw or losing match [121].

Treatment Options

The majority of these fractures have been treated successfully with closed management, usually a hanging arm cast [19, 47, 49, 121]. Conservative, nonoperative treatment consists of a coaptation splint followed by functional orthotic bracing with a humeral sleeve. No nonunions or delayed unions have been reported in arm wrestlers with closed treatment [121]. Generally, the fractures have healed within 6 to 10 weeks. Open treatment has been reported in 16 cases in the literature, usually for failure to obtain reduction [47]. Muscle interposition at the fracture site was noted in a number of these open cases. Heilbronner and associates have stated that open reduction and internal fixation should be reserved for markedly displaced segmental fractures, fractures associated with elbow articular injuries requiring early mobilization, fractures associated with vascular injury, radial nerve palsy present after a manipulation of the fracture, and inability to obtain reduction due to soft tissue interposition [47].

There is no series in the literature that demonstrates superiority of open reduction and internal fixation over closed management of these fractures. These fractures are the result of low-energy trauma, which in most cases preserves the periosteal and muscular envelope of the humerus.

With uneventful healing of the humeral fracture at 3 months, a full range of motion should be present, and strengthening exercises from 3 to 6 months are then required.

Criteria for Return to Sports Participation

When strength returns, sports are resumed. Contact sports require a functional humeral brace or a protective pad during the first year of competition.

Postoperative Management and Rehabilitation

A coaptation splint, hanging arm cast, or humeral sleeve orthosis is used until fracture healing occurs with closed treatment. If open reduction and internal fixation are used, a humeral sleeve orthosis is applied.

Authors' Preferred Method of Treatment

We recommend a coaptation splint for 2 weeks, followed by a humeral sleeve orthosis for approximately 6 weeks. These fractures are usually nontender by 8 weeks and demonstrate radiographic callus. A graduated range-of-motion and strengthening program should be resumed at this time. Particular emphasis should be placed on regaining external rotation, which is important to the throwing athlete.

Following apparent healing, refracture of the humerus has been reported [38, 121, 122]. Whitaker reported recurrence of fracture in two of his five cases [121]. Garth and colleagues reported three recurrent fractures in a baseball pitcher over a 21-month span [38]. Each fracture was in a different location of the humerus; the authors postulated that a nonhealed extension of the spiral fracture acted as a stress raiser for each of the new fractures. Because of this, they recommended tomograms to assess healing in all spiral humeral fractures before any torque-producing activity is resumed.

Open reduction and internal fixation is performed for severely displaced fractures that cannot be reduced, combined vascular injuries that require repair, loss of radial nerve function during reduction of fracture fragments, and ipsilateral elbow or forearm fractures.

Supracondylar Process Fracture

A supracondylar process has been noted in approximately 1% of anatomic specimens [11, 71, 110]. Terry noted a supracondylar process exceeding 3 mm in 6 of 515 whites and 1 of 1020 blacks examined [110]. A tendency toward familial occurrence was suspected. A fibrous band, the ligament of Struthers, usually connects the tip of the supracondylar process to the medial epicondyle. The median nerve passes through this arcade, often accompanied by the brachial or radial artery [11, 71]. Frequently, an anomalous pronator teres attaches to the process, and sometimes the process serves as a distal attachment for the coracobrachialis (Fig. 16A–12).

The supracondylar process can usually be palpated on clinical examination if it is carefully sought. Routine anteroposterior and lateral radiographs may fail to catch the process in profile, and oblique views of the anterior medial aspect of the humerus may be necessary [30, 39] (Fig. 16A–13).

Fracture of the supracondylar process has been reported following a direct blow to the process [30,

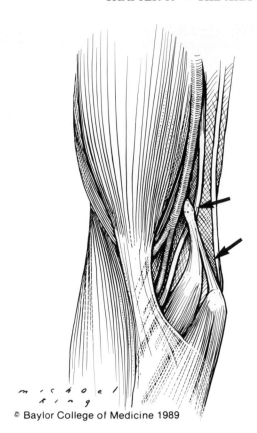

© Baylor College of Medicine 1989

FIGURE 16A–12
Supracondylar process and ligament of Struthers with compression entrapment of the median nerve.

39, 63]. In one case, the healing fracture was misinterpreted as an osteochondroma and excised. Conservative management is recommended for these injuries unless persistent pain or median nerve entrapment symptoms mandate excision and nerve exploration.

Rib Fractures in Throwing Athletes

Fracture of the first rib may occur through the subclavian groove between the attachment of the scalenus anticus and the scalenus medius muscles in throwing athletes, particularly baseball pitchers [113]. This occurs on the side contralateral to the pitching arm. Although it has been postulated that the fracture may result from a sudden muscular contraction [27], most of these fractures are most likely stress fractures [113]. The symptoms of pain are often vague and poorly defined. Technetium bone scanning or tomograms may confirm the diagnosis. The recommended treatment is conservative, and rest is advised until the symptoms subside. Although a high rate of nonunion was reported in

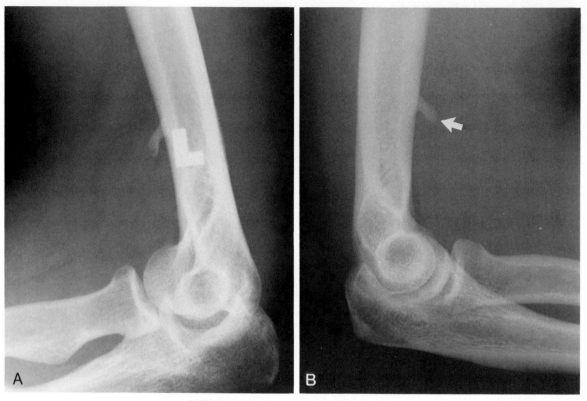

FIGURE 16A–13
Supracondylar process of the distal humerus.

one series, the majority of these fractures are pain-less [15]. Despite the close proximity of this injury to the subclavian artery, no neurovascular injury has been reported [95] (Fig. 16A–14).

Similar fractures in the lower three floating ribs have been described in both pitchers and batters [113]. Sudden or repetitive vigorous ipsilateral con-traction of the external abdominal oblique muscles, interdigitating with the latissimus dorsi, serratus posterior inferior, and internal abdominal oblique muscles at the tips of the tenth, eleventh, and twelfth ribs can result in stress fractures at the tips of these ribs. Multiple oblique radiographs with rib details or bone scan may be required to confirm the diag-nosis. These fractures have also been described in laborers [52]. Conservative treatment is recom-mended, including rest and modification of training activities. Injection with local anesthetic may be necessary to obtain temporary symptomatic relief (Fig. 16A–15).

MYOSITIS OSSIFICANS

Myositis ossificans is a nonmalignant lesion of heterotopic bone localized in the soft tissues and associated with trauma and debilitating illnesses such as poliomyelitis, paraplegia, or tetanus [115].

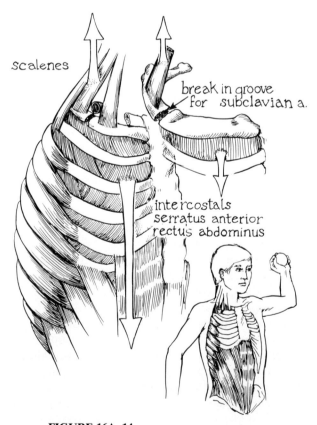

FIGURE 16A–14
First rib stress fracture in the throwing athlete.

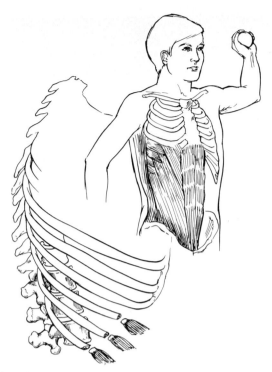

FIGURE 16A–15
Tenth, eleventh, and twelfth avulsion fractures in the throwing athlete.

demonstrate an intact underlying cortex, whereas the cortex is usually violated with osteogenic sarcoma. Biopsy specimens of myositis ossificans after 2 weeks show a definite zonal pattern with the most differentiated tissue at the periphery of the lesion. With osteogenic sarcoma, the most differentiated tissue occurs in the central portion of the lesion [34].

Clinical Evaluation

In the typical history an adolescent football player who has received a direct blow in the upper arm region has persistent swelling and tenderness. A palpable mass may be detected, and a flexion contracture about the elbow or shoulder may occur. The skin may show discoloration or may be normal. Initial radiographs are negative; however, after 2 to 3 weeks radiologic changes are noted [86] (Fig. 16A–16).

The most common cause is trauma, and the lesion occurs particularly in contact sports such as football, rugby, and lacrosse [111]. Although also noted in soft tissues, myositis ossificans has been classified as extraosseous, periosteal, or parosteal [42]. Hait and colleagues related the lesion to trauma of large muscle masses as well as of investing tissues against bone, including avulsion of tendinous origins or fascia from their bony attachments [44].

Related terminology includes heterotopic ossification, myositis ossificans, and ossifying hematoma, all of which appear to be a result of direct trauma, contusion, bleeding, calcification, and subsequent ossification. Heterotopic ossification about the elbow and shoulder joint is well recognized as a sequel to fracture–dislocations of these joints. Myositis ossificans in the muscle of the upper arm is less common than heterotopic ossification about the joints after dislocation; likewise, it is less common than myositis ossificans in the lower extremity [33].

Localized myositis ossificans must be distinguished from osteogenic sarcoma. Myositis ossificans is usually situated over the diaphysis, as opposed to the metaphyseal site of osteogenic sarcoma. With myositis ossificans, the pain and mass decrease with time, whereas the opposite is true with osteogenic sarcoma. In myositis ossificans, radiographs

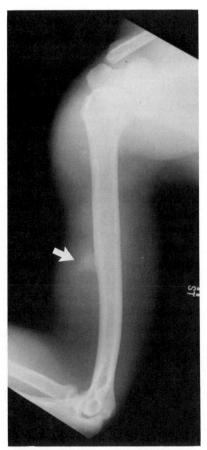

FIGURE 16A–16
Heterotopic bone (myositis ossificans) within the brachialis muscle.

Treatment Options

Rest of the affected extremity, application of protective support such as plaster or foam padding, analgesics, and anti-inflammatory drugs, particularly indomethacin, may be of some benefit. Diamond and McMaster described a form of myositis ossificans traumatica occurring in the area of the lateral humerus in the deltoid insertion in football players, particularly defensive linebackers, and coined the term tackler's exostoses [29]. They described the origin of this lesion as damage to the osseous insertion of the deltoid or the origin of the brachialis, producing a periosteal tear. It may be seen bilaterally and, if treated early with adequate rest and splinting, may undergo spontaneous regression. Huss and Puhl described a triad of pain, the presence of a hard palpable mass in the muscle, and flexion contracture of the elbow [58]. They noted that 70% of these symptoms significantly improved in less than 3 months with conservative, nonoperative management [58]. Thirty percent of these cases required surgery for a persistent painful mass, which recurred postoperatively in two-thirds of the surgical cases, although surgery was delayed until radiologic evidence of maturation was present. Lipscomb and associates likewise recommended conservative treatment consisting of rest, elevation, ice, and immobilization with a compression dressing [68]. They used oral proteolytic enzymes and also recommended aspiration of a fluctuant hematoma mass, if present, under sterile conditions. When all pain and tenderness have subsided and joint motion returns, the athlete is allowed to resume participation in sports with protective padding. If radiographic evidence of residual myositis ossificans persists but no pain or mass is present, no other treatment is recommended. In those cases in which a palpable mass, muscle atrophy, and joint limitation are problems despite conservative treatment and radiographs have revealed mature laminar bone formation within the myositis ossificans traumatic mass, surgical excision is recommended. These masses have been removed at 6 months or later if bone maturation appears to be complete [56]. Bone scans may be used to distinguish active (immature) bone formation from the mature nonactive lesion. The lesion may recur if immature bone is excised. If the lesion is mature, the sedimentation rate and alkaline phosphatase activity should be normal.

Although lesions of heterotopic bone in the biceps and at the deltoid insertion appear to be related to direct trauma, Fulton and associates have described cortical, desmoidlike lesions of the proximal humerus in gymnasts (ringman's shoulder lesion) [37].

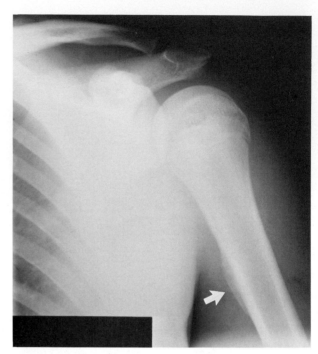

FIGURE 16A–17
Cortical irregularity ("ringman's shoulder") in the proximal humerus of a gymnast.

This is described radiologically as a cortical irregularity of the proximal humerus at the insertion of the pectoralis major muscle into the proximal humerus (Fig. 16A–17). This irregularity may simulate malignancy; however, it is considered a benign reactive lesion secondary to the excessive forces present at the pectoralis major insertion. This lesion is particularly unique to gymnasts because of the strength movements involved in the shoulder girdle and pectoralis muscle group.

Avulsions, either complete or partial, of the pectoralis major insertion or latissimus dorsi insertion onto the humerus may result in heterotopic bone formation and a painful mass. These have been noted particularly in football players (Fig. 16A–18).

Postoperative Management and Rehabilitation

Padded protective splints are worn for contact sports unless pain persists. Sports are terminated until pain resolves and then are resumed with protective equipment.

If the lesion is excised, Hemovac drainage of the wound is used to prevent hematoma formation. Postoperative indomethacin is used to reduce the formation of heterotopic bone. Range-of-motion exercises are begun at 7 to 10 days and are maxi-

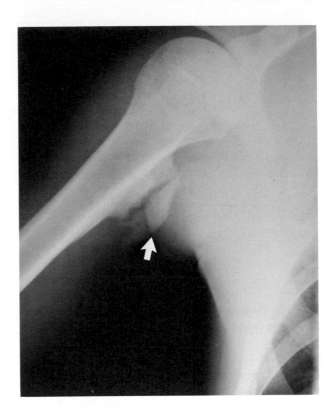

FIGURE 16A–18
Pectoralis major muscle avulsion at the insertion with heterotopic bone formation.

mized at 3 to 4 weeks to prevent contracture formation. A postoperative extension splint is used at night on the elbow to prevent recurrence of a flexion contracture. Strengthening and resistive exercises are begun at 4 weeks.

Criteria for Return to Sports Participation

Painless range of motion and return of strength to a level equal to that in the nonaffected area are the criteria for return to sports participation.

Authors' Preferred Method of Treatment

Small hematomas due to acute direct trauma are treated with ice and cold packs, rest, and nonste-

roidal anti-inflammatory drugs (NSAIDs). Large, fluctuant hematomas are aspirated in a sterile fashion and wrapped for compression, and are then given the same conservative treatment. Painful bruises or hematomas are padded for sports. If heterotopic bone formation is noted on radiographs but the lesion is asymptomatic, sports activity may be continued.

Symptomatic heterotopic bone, if immature on radiographs and characterized by increased uptake on bone scan and increased sedimentation rate and alkaline phosphatase levels, is treated with padding, cast, and cessation of sports activity if this activity is painful. Mature heterotopic bone that remains symptomatic is surgically excised and splinted until skin and soft tissue healing occurs; the patient is then begun on a range-of-motion and strengthening program. NSAIDs, particularly indomethacin, are used postoperatively for 3 to 6 months to retard new bone formation.

Neurovascular Injury

James B. Bennett, M.D.
Thomas L. Mehlhoff, M.D.

NEUROVASCULAR INJURY IN THE UPPER ARM

Neurovascular injuries distal to the glenohumeral joint and proximal to the elbow joint are infrequent and are generally related to direct trauma, particularly fracture–dislocations of either the glenohumeral joint or elbow joint. Spiral oblique fractures of the humerus are well recognized as the cause of neurapraxic injuries to the radial nerve, with predictable recovery in most cases [89]. Axillary nerve neurapraxic injuries with shoulder dislocations, particularly posterior dislocations, are also well documented in the literature [16]. Direct trauma can likewise produce neurapraxic injuries to the axillary nerve as it exits posteriorly at the level of the proximal humerus and enters the deltoid region (Fig. 16B–1), or to the radial nerve as it travels about the spiral groove of the humerus. Motor and sensory deficits are noted in the appropriate nerve distribution.

Specific treatment of nerve injuries is addressed elsewhere in this text. The ulnar nerve is less commonly injured in its course along the medial aspect of the arm. The most common site of ulnar nerve injury about the elbow is in the cubital tunnel; this is likewise discussed elsewhere in this text. Median nerve injury in the humerus may be associated with

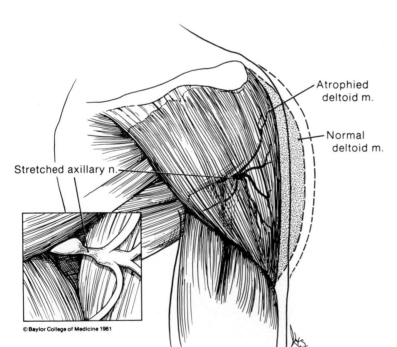

Stretched axillary n.

Atrophied deltoid m.

Normal deltoid m.

© Baylor College of Medicine 1981

FIGURE 16B–1
Axillary nerve compression or stretch injury at the posterior aspect of the deltoid muscle.

injury to the supracondylar process and ligament of Struthers. Decompression is indicated if clinical and electrodiagnostic studies localize compression to this level. Combined or mixed nerve injuries may be associated with major fractures or dislocations of the glenohumeral joint or the elbow joint.

Brachial artery injury in the arm is a rare injury. It may, however, be associated with a high-velocity closed fracture of the humerus, a compound fracture, or lacerations or puncture wounds of the medial aspect of the arm. Arteriographic diagnosis and appropriate surgical repair are indicated. Thrombosis of the brachial artery secondary to trauma is uncommon but, if present, a more proximal lesion should be ruled out arteriographically. Cahill and Palmer have described compression of the circumflex humeral artery as a "quadrilateral space syndrome" [22]. This may also result in compression of the axillary nerve. It is seen in throwing athletes and is produced with forward flexion, abduction, and external rotation of the humerus, aggravating the symptoms of pain, weakness, and fatigability. Documentation is obtained with arteriography that reveals occlusion of the posterior circumflex humeral artery with the arm in abduction and external rotation. Conservative treatment with cessation of throwing activities should relieve the symptoms. If persistent tenderness and pain exist and positive arteriograms are obtained, surgical decompression through a posterior incision is recommended, with exploration of the quadrilateral space and decompression of the posterior circumflex humeral artery and axillary nerve as indicated by pathologic findings in this region.

COMPARTMENT SYNDROME

Compartment syndrome of the upper extremity localized to the upper arm is a rare lesion, as reported by Leguit [66], Holland and associates [51], and Brumback [20]. The patient has a massive, swollen, and tense upper arm with severe pain, loss of active and passive motion, and sensory and motor deficits secondary to a history of direct trauma. Holland and associates described a patient with a normal brachial arteriogram but massive swelling and a secondary hematoma displacing the artery, which necessitated immediate operative decompression, evacuation of the hematoma, and secondary wound closure; treatment was successful with return of function and return to sports activities [51]. Due to the soft tissues and muscles, compartmental pressure measurements are somewhat less reliable in this area than in the forearm and hand, with their constrictive fascia. Pressure measurements are, however, recommended, as is arteriographic evaluation of the axillary and brachial arteries before surgical intervention for compartment syndrome [81].

VENOUS AND ARTERIAL INSUFFICIENCY

The swollen arm in the athlete with no history of significant trauma presents a diagnostic dilemma (Fig. 16B–2). Differential diagnosis of upper extremity pain and swelling includes muscle sprains, tears and contusions, edema, bone or soft tissue infection, cellulitis, and venous thromboses. Ob-

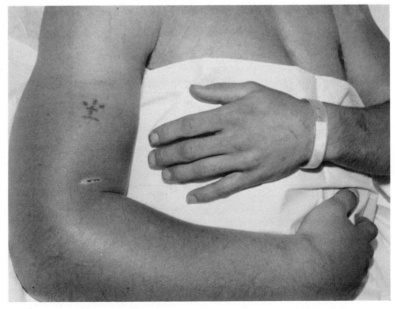

FIGURE 16B–2
Swollen arm-forearm-hand secondary to effort thrombosis of the axillary vein.

viously, a fracture due to direct trauma or a pathologic lesion of bone resulting in fracture must be ruled out.

Upper extremity venous insufficiency results from acute axillary or subclavian vein obstruction, which is due to iatrogenic trauma in the form of central venous catheters in 70% to 80% of cases. Ten to twenty percent of cases are associated with spontaneous thromboses ("effort" thromboses), and the remaining 5% to 10% are secondary to chronic disorders of the venous system. Not infrequently, patients with upper extremity venous thromboses develop complications that include pulmonary embolism [12], septicemia, and chronic postphlebitic syndrome [55]. Sudden, painless arm swelling in a patient with a central venous line is virtually diagnostic of subclavian vein thrombosis.

In the athletic population, spontaneous or effort thrombosis of the axillary vein is more common in men than in women, occurs on the dominant side in the majority of cases, and is most common in the fourth decade of life. The majority of these patients have a history of trauma or recent strenuous exertion [72]. The arm is swollen with unilateral nonpitting edema that becomes cyanotic in appearance with continued exercise.

Idiopathic chronic venous insufficiency without prior symptoms may occur. Von Schroetter described thrombosis of the axillary vein in 1884 and used the term primary or idiopathic thrombosis of the axillary vein [102]. Venous occlusion may be secondary to thoracic outlet syndrome. Vogel and Jensen described effort thrombosis in the competitive swimmer [120]. A similar lesion has also been described in the throwing athlete, particularly baseball pitchers. Direct trauma in contact sports has produced venous thrombosis of the axillary and subclavian veins. The patient has good arterial flow, but the pathology lies in restricted venous outlet flow.

In addition to thoracic outlet syndrome, certain congenital and developmental anomalies may predispose a patient to venous insufficiency. Abnormalities in blood coagulation as well as secondary causes such as malignancy, congestive heart failure, polycythemia vera, and the prior use of any indwelling catheters about the shoulder region must be ruled out [101]. Diagnosis is confirmed with venography [2] (Fig. 16B–3). Doppler ultrasound and plethysmography are less reliable in the upper extremity. Venous pressure measurements have been recommended by Veal and Hussey [119] arteriography shows a patent arterial tree in these cases. A delayed venous phase on the arteriogram shows dilatation and dysfunction of the venous system.

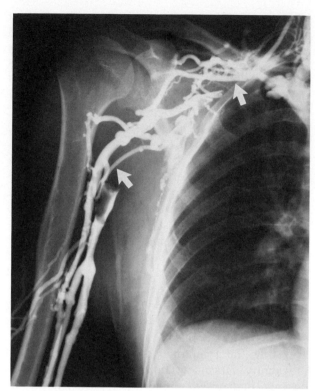

FIGURE 16B–3
Venogram of axillary and subclavian vein occlusion.

Treatment of patients with minimal symptomatology is conservative, with rest and cessation of sports activities. Surgery to reconstitute the vein has not been uniformly successful, and thrombosis tends to recur. Collateral circulation develops over the course of time, and venous decompression results. Historically, anticoagulation with heparin followed by warfarin for 2 to 3 months has been the treatment of choice [70]. More recently, fibrinolytic therapy for the lysis of either intra-arterial or intravenous clots has been used successfully [13]. Streptokinase or urokinase intravenous injections to lyse clot formation may allow restoration of blood flow into the thrombosed segment [79]. Complications are significant with fibrinolytic therapy, and the patient must be monitored closely to detect a bleeding tendency or hemorrhage at other sites within the body, arrhythmias, allergic reaction, or fever secondary to the therapy program. Recannulation of the axillary or subclavian vein, as well as the collateral circulation, develops over the course of 6 months to 1 year. First rib resection to decompress the thoracic outlet and axillary–subclavian vein complex may be required [124].

Acute arterial thrombosis has been described in the pitching or throwing athlete, involving acute occlusion, vascular ischemia, pain, and development of retrograde thrombosis, which may occlude the

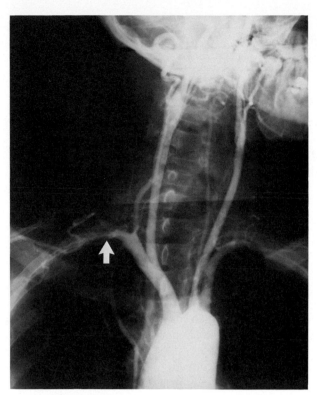

FIGURE 16B–4
Arteriogram with thrombosis of subclavian artery.

proximal carotid and vertebral arteries [113]. Symptoms of severe arm pain, diminished or absent pulse, and a pale extremity suggest acute arterial thrombosis; diagnosis is confirmed through arteriographic studies (Fig. 16B–4). Treatment, depending on the duration of the arterial lesion, may involve surgery, including embolectomy or vascular reconstruction, resolution of the thrombus with fibrinolytic therapy in the form of streptokinase or urokinase, or angioplastic dilatation at the time of catheterization. Laser therapy in the treatment of intra-arterial lesions is under study. With chronic thrombosis, bypass vascular surgery is recommended to restore circulation in the extremity if the collateral circulation is inadequate and evidence of ischemia with exercise or rest pain is present. If the components of a thoracic outlet syndrome or anomalous cervical rib contribute to venous or arterial insufficiency, removal of the pathologic structure is recommended, as well as reconstruction of the arterial supply to the upper extremity.

Postoperative Management and Rehabilitation

Medical management may vary from anticoagulation with heparin and warfarin (Coumadin) to low-dose aspirin, depending on the severity of the occlusion, the collateral blood flow in the extremity, and the extent of surgical reconstruction, if required. Rehabilitation requires range-of-motion and strengthening exercises and medical control of bleeding and clotting factors.

Criteria for Return to Sports Participation

Return to sports may not be possible if arterial or venous reconstruction is required. Medically managed vascular occlusive disease should allow return to competition in athletes who have a patent vascular trunk and good collateral circulation into the extremity.

Authors' Preferred Method of Treatment

Venous occlusion and arterial thrombosis of the upper extremity are managed by the cardiology and cardiovascular departments using the various methods outlined in the preceding section.

Thoracic Outlet Syndrome

James B. Bennett, M.D.
Thomas L. Mehlhoff, M.D.

Thoracic outlet syndrome has various synonyms including Naffziger's syndrome, scalenus anticus syndrome [88], scalenus medius band syndrome, scalenus minimus syndrome, costoclavicular compression syndrome, hyperabduction syndrome, acroparesthesia, cervical rib syndrome and Paget-Schroetter syndrome. Thoracic outlet syndrome was described by Paget as an effort thrombosis of the subclavian vein in 1875 and was similarly discussed by von Schroetter in 1884 [102]. In 1740, Hunauld first described compression of the thoracic outlet secondary to a cervical rib [57]. In 1919, Stopford and Telford demonstrated that the neurovascular structures could be compressed by the first thoracic rib and that surgical removal of this rib would alleviate symptoms of the compression [108].

The thoracic outlet involves the area of the shoulder girdle and thorax in which the subclavian artery and vein exit the chest cavity and combine with the brachial plexus passing through the scalene triangle over the first rib and under the clavicle to enter the axillary region of the shoulder [66A] (Fig. 16C–1A). The anatomic boundary of the thoracic outlet consists of the superior surface of the first rib, the anterior scalene muscle, and the middle scalene muscle, both of which insert into the first rib. The clavicle overrides the neurovascular structures and applies pressure on the thoracic outlet if it is displaced or positioned posteriorly. Neurovascular compression within the thoracic outlet involves the subclavian artery, the subclavian vein, or the brachial plexus within this area (Fig. 16C–1B). Compression may occur within the thorax; may involve a cervical rib abnormality or the scalene anticus or medius muscle; or it may occur at the costoclavicular or subclavian tendon junction, at the level of the first rib, or as far laterally as the pectoralis minor insertion into the coracoid process [84].

Anatomic factors include an inadequate intrascalene triangle (which may be due to scalenus anticus or scalenus medius hypertrophy or displacement), a high first thoracic rib, or descent of the shoulder girdle with age, allowing a sagging effect and compression of the neurovascular structures. Congenital factors such as a cervical rib, a rudimentary first thoracic rib, variant scalene muscles, an elongated transverse process, an anomalous first thoracic rib, or adventitial fibrotic bands may be present [100]. Further considerations include traumatic factors such as fracture of the clavicle (Fig. 16C–2), injuries to the cervical vertebrae, dislocation of the head of the humerus, and atherosclerosis of the major arteries at the isthmus of the neck of the humerus [98]. Wright and Lipscomb describe hyperabduction syndrome or pectoralis minor muscle compression as a cause of arteriovenous or brachial plexus compression beneath the coracoid insertion of the pectoralis minor [125] (Fig. 16C–3). This is made worse when the arm is elevated and externally rotated. Kofoed emphasized the necessity of ruling out cervical disc herniation in the evaluation of thoracic outlet syndrome [62].

Differential diagnoses in thoracic outlet syndrome include any pathology creating pain in the neck, arm, or shoulders, such as bursitis, cervical arthritis, myositis, fibrositis, brachial plexus neuritis, Raynaud's disease, thromboangiitis, neoplasm of the spinal canal, neoplasm of the peripheral nerve, or apical pulmonary neoplasm and cervical disc disease. Neurologic symptoms consist of weakness, fatigability, and numbness and tingling, particularly in the posterior and medial cords of the brachial plexus. Vascular symptoms consist of ischemia, claudication, cold intolerance, swelling with venous congestion, and occasional thromboembolic phenomena with distal arterial occlusion. Symptoms are

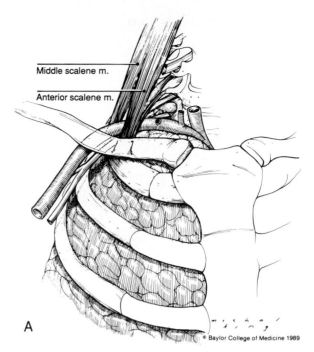

A

FIGURE 16C–1
A, Thoracic outlet anatomy. *B,* Thoracic outlet syndrome with compression of the subclavian artery and brachial plexus by clavicle—first rib—distal aneurysm of the subclavian artery.

Middle scalene m.
Anterior scalene m.

© Baylor College of Medicine 1989

© Baylor College of Medicine 1989

B

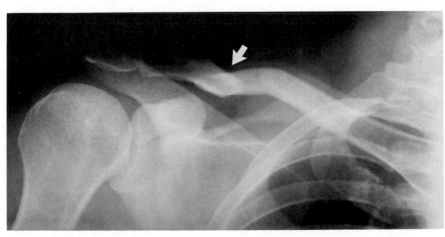

FIGURE 16C–2
Malunion clavicle fracture with thoracic outlet compromise.

795

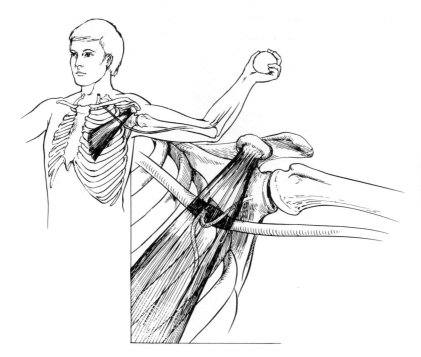

FIGURE 16C-3
Axillary artery compression by the pectoralis minor muscle at coracoid process insertion in the throwing athlete.

especially pronounced with arm elevation above the level of the shoulder, particularly during throwing or activities such as combing the hair or sleeping with the arm above the head. Classically, the patient has one of two typical body types. The first is an individual with heavy shoulders and a muscular neck and upper extremities, which, by sheer muscular bulk, could cause obstruction of the thoracic outlet. The other typical patient is a long, slender-necked individual with sloping shoulders, especially an older person who has lost muscle tone or control and strength in the shoulder girdle. The sloping descent of the shoulders puts traction on the brachial plexus and compromises the neurovascular structures at the first rib level.

CLINICAL EVALUATION

Physical signs of vascular compression are demonstrated by Adson's maneuver, which involves abduction of the arm with elevation, extension of the neck, and rotation of the chin to obliterate the radial pulse [3]. The chin is classically tilted to the side being tested. However, in some patients a greater effect on the radial pulse is noted by tilting the head to the opposite side. Therefore, both positions must be tested (Fig. 16C-4). Allen's maneuver consists in hyperabduction of the hand and forearm straight upward to compress the artery at the coracoid process and obliterate the pulse. Wright's test involves hyperabduction of the arm at 90 degrees and full external rotation of the shoulder,

which obliterates the pulse (Fig. 16C-5) [124]. The military shoulder position (costoclavicular syndrome), with hyperextension of the shoulders and deep inhalation with the arm held at the side, also

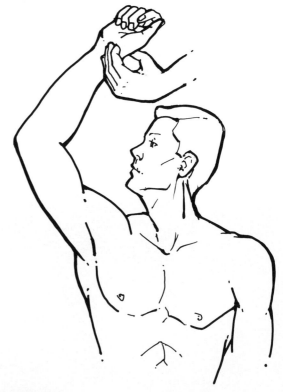

FIGURE 16C-4
Adson's test. The patient takes and holds a deep breath, the arm is adducted, the neck is extended, and the chin is turned to the side being tested.

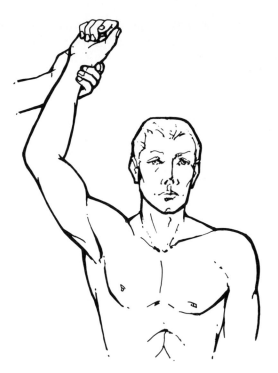

FIGURE 16C–5
Wright's test for hyperabduction syndrome. The arm is held in hyperabduction at 90° with full external rotation of the shoulder.

produces diminution or disappearance of the pulse (Fig. 16C–6). This test is used to detect compression in the costoclavicular interval. This finding may be present in a large percentage of normal individuals who do not have thoracic outlet syndrome. Clavicular compression with the shoulder slumped and the arm held with traction inferiorly causes obliteration of the pulse or neurologic symptomatology. The claudication test, with the hand held above the head for 1 minute, produces pain, fatigue, and numbness.

Although pain, weakness, and neurovascular deficits are associated with thoracic outlet syndrome, often the symptomatology is quite bizarre and may be intermittent. Awareness of the occurrence of thoracic outlet syndrome and the presence of clinical objective findings are necessary to diagnose this syndrome. Repetitive throwing activities in the extended, abducted, externally rotated position of the arm aggravate the symptoms.

Clinical evaluation must include musculoskeletal, neurologic, and vascular examinations. Diagnostic tests consist of routine chest, cervical spine, and shoulder radiographs. Roentgenographic evaluation for cervical ribs (Fig. 16C–7), anomalous first and second ribs, pathologic clavicular fractures, and space-occupying lesions such as tumor or aneurysm must be ruled out.

Arteriography documents arterial compression and possible aneurysmal formation about the first rib. Venography documents venous compression or occlusion. Peripheral vascular studies, including pulses, blood pressure measurements, and Doppler studies, aid in the diagnosis of thoracic outlet compression and occlusion of the arterial supply to the arm. Electromyography evaluates neurologic compression and differentiates cervical radiculopathy from thoracic outlet compression [54].

Thoracic outlet syndrome is seen not only in the throwing athlete [108], but also in heavy muscular athletes such as weight lifters, football players, and athletes who may sustain traction injuries to the upper arm and chest. Direct trauma that results in rib fractures, transverse process fractures, clavicular fractures, and shoulder dislocations may also precipitate thoracic outlet symptomatology [114].

TREATMENT OPTIONS

In most cases nonoperative treatment is the initial form of management once the diagnosis of thoracic outlet syndrome is confirmed. The treatment program includes not only evaluation but also patient education, behavior modification, and joint mobili-

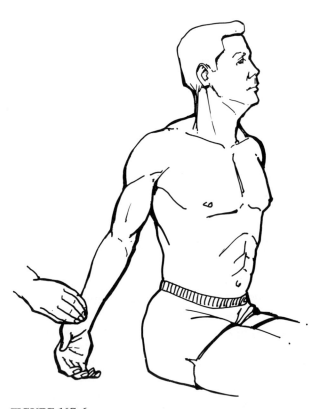

FIGURE 16C–6
Costoclavicular maneuver. Backward and downward thrust of the shoulder combined with deep inhalation results in pressure on the subclavian artery and vein as it narrows the space between the clavicle and first rib.

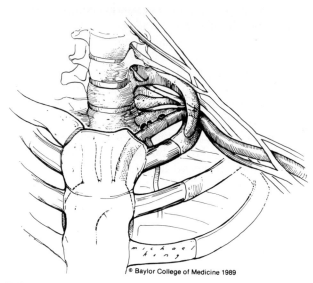

FIGURE 16C–7
Thoracic outlet syndrome produced by synostosis of C7 cervical rib and first thoracic rib.

zation exercises. If successful, this program is followed continually and should not be terminated. Continued evaluation and monitoring of the patient are necessary. Behavior modification consists of altering sleep patterns, working patterns, and driving patterns and taking general precautions for activities that could compromise the thoracic outlet. Faulty posture must be corrected. Shoulder exercises with emphasis on gradual scapular retraction and shoulder range of motion increase joint motion and allow the cervical spine to achieve axial body and extremity extension and open the thoracic outlet space. Diaphragmatic breathing exercises are often indicated to aid in respiration when there is hypertrophy of the accessory muscles of the chest and neck. In particular, elevation of the rib cage by the pectoralis minor and scalene muscles should be eliminated because this tendency decreases the thoracic outlet space. Weak musculature about the neck and shoulder should be strengthened, and shoulder posture should be improved. The upper trunk muscles, such as the serratus anterior, middle and lower trapezius, latissimus dorsi, and rhomboids, must be strengthened. Joint mobilization techniques for the sternoclavicular, acromioclavicular, and scapulothoracic joints improve and increase the costoclavicular space. Likewise, mobilization of the occiput on the atlas facilitates axial extension body movements and improves the symptomatology of the thoracic outlet syndrome. Specific joint mobilization and therapy programs have been outlined by Smith [106]. Conservative management should produce improvement in symptomatology by 1 to 3 months after the onset of symptoms.

When symptoms persist or become worse with conservative management, surgical intervention may be necessary. Patients with severe, intractable pain, disability, arterial or venous compromise, or neurologic compromise fall into this category [46]. Surgery ranges from resection of the scalenus anticus, described by Adson in 1927 [3], to removal of the first thoracic rib, described by Murphy in Australia [82] and Brickner in the United States [18]. Anterior and supraclavicular approaches have been used by these authors for first rib resection. Another approach used by Clagett [25], was a limited posterior thoracotomy incision, and a transaxillary approach was used by Roos [99]. First rib resections are now most commonly done according to the techniques of Roos through a transaxillary approach (Fig. 16C–8). In addition to first rib resection, this approach allows the possibility of performing a thoracic sympathectomy at the same time for pain relief. Scalenectomy by itself does not provide predictable relief of thoracic outlet symptoms. Strict attention must be paid to the pathology, and when other causes appear to be operative, they must be corrected. Such correction may include cervical rib resection or callus formation of a fractured clavicle with thoracic outlet decompression. Vascular changes, particularly aneurysmal dilatation or thromboembolism of the intima of the vessel, must be addressed and surgically corrected at the time of thoracic outlet decompression. In certain instances, a combination of both supraclavicular and transaxillary approaches has been used for decompression of the thoracic

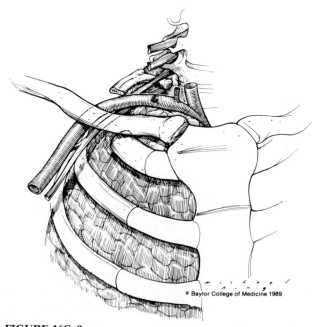

FIGURE 16C–8
Thoracic outlet decompression by first thoracic rib resection—cervical rib removed if present.

outlet, first rib resection, and vascular reconstruction including brachial plexus exploration as indicated.

POSTOPERATIVE MANAGEMENT AND REHABILITATION

Postoperative rehabilitation is similar to the conservative rehabilitation program discussed previously.

CRITERIA FOR RETURN TO SPORTS PARTICIPATION

Resumption of sports depends on return of range of motion, strength, and endurance in the shoulder girdle and upper extremity. A recovery period of 6 months to 1 year is usually required to maximize functional return in competitive athletes.

AUTHORS' PREFERRED METHOD OF TREATMENT

Diagnosis and treatment of thoracic outlet compression syndrome may be difficult and frustrating for both the physician and the patient. Conservative treatment is used as the principal form of treatment but is followed by surgery in cases that do not show clinical resolution. Surgical treatment must be directed at the pathologic lesion that appears to be the most prominent causative factor in the thoracic outlet compression syndrome. Thoracic outlet syndrome that does not respond to conservative management and does not demonstrate arterial or venous occlusion by angiography is treated by surgical removal of the first rib through a transaxillary approach, as described by Roos [99]. If brachial plexus compression is present, as demonstrated by delayed nerve conduction or muscular compromise in electrodiagnostic testing, the brachial plexus is explored, and the first ribs are resected through a supraclavicular approach. The supraclavicular approach also allows cervical rib resection and vascular reconstruction if required.

Sympathectomy may be performed through either approach if severe pain is a component of the thoracic outlet syndrome and the patient has obtained pain relief through a series of sympathetic blocks of the stellate ganglion system.

A guarded prognosis for return to sports activities is reserved for those patients who have chronic thoracic outlet compression syndrome resulting in a neurologic deficit or vascular ischemia with thromboembolic phenomena.

References

1. Adams, J. E. Little League Shoulder. *Calif Med* 105:22–25, 1977.
2. Adams, J. T., and DeWeese, J. A. Primary deep venous thrombosis of the upper extremity. *Arch Surg* 191:29–42, 1965.
3. Adson, A. W. Cervical ribs: Symptoms, differential diagnosis for section of the insertion of the scalenus anticus muscle. *J Int Coll Surg* 16:546, 1951.
4. An, K. N., Hui, F. C., Morrey, B. F., Linscheid, R. L., and Chas, E. Y. Muscles across the elbow joint: A biomechanical analysis. *J Biomech,* 14:659–669, 1981.
5. Anderson, E. Triceps tendon avulsion. *Injury* 17:279–280, 1986.
6. Anderson, K. J., and Le Cocq, J. F. Rupture of the triceps tendon. *J Bone Joint Surg [Am]* 39:444–446, 1957.
7. Anzel, S. H., Covey, K. W., Weiner, A. D., and Lipscomb, P. R. Disruption of muscles and tendons: An analysis of 1,014 cases. *Surgery* 45:406, 1959.
8. Aso, K., and Torisu, T. Muscle belly tear of the triceps. *Am J Sports Med* 12:485–487, 1984.
9. Bach, B. R., Warren, R. F., and Wickiewicz, T. L. Triceps rupture: A case report and literature review. *Am J Sports Med* 15:285–289, 1987.
10. Baker, B. E., and Bierwagen, D. Rupture of the distal tendon of the biceps brachii: Operative versus non-operative treatment. *J Bone Joint Surg [Am]* 67:414–417, 1985.
11. Barnard, L. B., and McCoy, S. M. The supracondyloid process of the humerus. *J Bone Joint Surg* 28:845, 1946.
12. Barnet, T., and Levitt, L. M. "Effort" thrombosis of the axillary vein with pulmonary embolism. *JAMA* 146:1412–1413, 1951.
13. Becker, G. J., and Holden, R. W. Local thrombolytic therapy for subclavian and axillary vein thrombosis. *Radiology* 149:419–423, 1983.
14. Berson, B. L.: Surgical repair of pectoralis major rupture in an athlete. *Am J Sports Med* 7(6):348–351, 1979.
15. Blichert-Toft, M. Fatigue fracture of the first rib. *Acta Chir Scand* 135:675–678, 1969.
16. Blom, S., and Dahlback, L. D. Nerve injuries in dislocation of the shoulder joint and fractures of the neck of the humerus. *Acta Chir Scand* 135:461, 1970.
17. Boyd, H. D., and Anderson, L. D. A method for reinsertion of the distal biceps brachii tendon. *J Bone Joint Surg [Am]* 43:1041–1043, 1961.
18. Brickner, W. M. Brachial plexus pressure by the normal first rib. *Ann Surg* 85:858–872, 1927.
19. Brismar, B., and Spargen, L. Fracture of the humerus from arm wrestling. *Acta Orthop Scand* 46:707, 1975.
20. Brumback, R. J. Compartment syndrome complicating avulsion of the origin of the triceps muscle: A case report. *J Bone Joint Surg [Am]* 69:1445–1446, 1987.
21. Cahill, B. R. Little League shoulder. *Am J Sports Med* 2:150–154, 1974.
22. Cahill, B. R., and Palmer, R. E. Quadrilateral space syndrome. *J Hand Surg [Am]* 8:65–69, 1983.
23. Carroll, R. E., and Hamilton, L. R. Rupture of biceps brachii—A conservative method of treatment. *J Bone Joint Surg [Am]* 49:1016, 1967.
24. Chao, S. L., Miller, M., and Terg, S. W. A mechanism of spiral fracture of the humerus: A report of 129 cases following the throwing of hand grenades. *J Trauma* 11:602–605, 1971.
25. Clagett, O. T. Presidential address, American Association of Thoracic Surgery Research and Prosearch, April 16–18, 1962. *J Thorac Cardiovasc Surg* 44:153–166, 1962.
26. Clayton, M. L., and Thirupathi, R. G. Rupture of the

triceps tendon with olecranon bursitis: A case report with a new method of repair. *Clin Orthop* 184:183–185, 1984.

27. Curran, J. P., and Kelly, D. A. Stress fracture of the first rib. *Am J Orthop* 8:16, 1966.
28. Davis, W. M., and Yassine, Z. An etiological factor in tear of the distal tendon of the biceps brachii: Report of two cases. *J Bone Joint Surg [Am]* 38:1365–1368, 1956.
29. Diamond, P. E., and McMaster, J. H. Tackler's exostosis. *J Sports Med [Am]* 3:238–242, 1975.
30. Doane, C. P. Fractures of the supracondyloid process of the humerus. *J Bone Joint Surg* 18:757–759, 1936.
31. Dobbie, R. P. Avulsion of the lower biceps brachii tendon: Analysis of 51 previously reported cases. *Am J Surg* 51:661, 1941.
32. Dreyfus, U., and Kessler, I. Snapping elbow due to dislocation of the medial head of the triceps: A report of two cases. *J Bone Joint Surg [Br]* 60:56–57, 1978.
33. Ellis, M., and Frank, H. G. Myositis ossificans traumatica: With special reference to the quadriceps femoral muscle. *J Trauma* 6:724–738, 1966.
34. Epps, C. H. Jr. (Ed.). *Complications in Orthopaedic Surgery* (2nd ed.). Philadelphia, J. B. Lippincott, 1986.
35. Farrar, E. L., and Lippert, F. G. Avulsion of the triceps tendon. *Clin Orthop* 161:242–246, 1981.
36. Friedmann, E. Rupture of the distal biceps brachii tendon. *JAMA* 184:60–63, 1963.
37. Fulton, M. N., Albright, J. P., and El-Khoury, G. Y. Cortical desmoid-like lesion of the proximal humerus and its occurrence in gymnasts (ringman's shoulder lesion). *Am J Sports Med* 7:57–61, 1979.
38. Garth, W. P., Leberte, M. A., and Cool, T. A. Recurrent fractures of the humerus in a baseball pitcher: A case report. *J Bone Joint Surg [Am]* 70:305–306, 1988.
39. Genner, B. A. Fracture of the supracondylar process. *J Bone Joint Surg [Am]* 41:1333, 1959.
40. Gilcreest, E. L. Rupture of muscles and tendons. *JAMA* 84:1819, 1925.
41. Gilcreest, E. L. The common syndrome of rupture, dislocation, and elongation of the long head of the biceps brachii: An analysis of 100 cases. *Surg Gynecol Obstet* 58:322–340, 1934.
42. Gilmer, W. S., and Anderson, L. D. Reactions of soft somatic tissue with progress to bone formation. *South Med J* 52:1432–1448, 1959.
43. Greig, H. W., Anson, B. J., and Budinger, J. M. Variations in the form and attachments of the biceps brachii muscle. *Quarterly Bulletin of Northwestern University Medical School* 26:241, 1952.
44. Hait, G., Boswick, J., and Stone, N. Heterotopic bone formation secondary to trauma (myositis ossificans traumatica). *J Trauma* 10:405–411, 1970.
45. Haldeman, K. O., and Soto-Hall, R. Injuries to muscles and tendons. *JAMA* 104:2319, 1935.
46. Hawkes, C. D. Neurosurgical considerations in thoracic outlet syndrome. *Clin Orthop* 207:24–28, 1986.
47. Heilbronner, D. M., Manoli, A., and Morawa, L. G. Fractures of the humerus in arm wrestlers. *Clin Orthop* 149:169–171, 1980.
48. Herrick, R. T., and Herrick, S. Ruptured triceps in a powerlifter presenting as cubital tunnel syndrome. *Am J Sports Med* 15:514–517, 1987.
49. Herzmark, M. H., and Klune, R. F. Ball throwing fracture of the humerus. *Med Ann Dist Columbia* 21:196–199, 1952.
50. Holder, S. F., and Grava, W. A. Complete triceps tendon avulsion. *Orthopedics* 9:1582, 1986.
51. Holland, D. L., Swenson, W. M., Tudor, R. B., and Borge, D. A compartment syndrome of the upper arm—A case report. *Am J Sports Med* 13:363–364, 1985.
52. Horner, D. Lumbar back pain arising from stress fractures of the lower ribs. *J Bone Joint Surg [Am]* 46:1553, 1964.
53. Hovelus, L., and Josefsson, G. Rupture of the distal biceps tendon: report of five cases. *Acta Orthop Scand* 48:280–282, 1977.
54. Huffman, J. D. Electrodiagnostic techniques for and con-

servative treatment of thoracic outlet syndrome. *Clin Orthop* 207:21–23, 1986.
55. Hughes, E. S. R. Venous obstruction in the upper extremity (Paget-Schroetter's syndrome). *International Abstracts of Surgery* 88:89–127, 1949.
56. Hughston, J. C., Whatley, G. S., and Stone, M. N. Myositis ossificans traumatica (myo-osteosis). *South Med J* 55:1167–1170, 1962.
57. Hunauld, F.-J. Sur le nombre des cotes, moindre ou plus grand a l'ordinaire. *Hist Acad roy d sc [de Paris], Amst*, 1740.
58. Huss, C. D., and Puhl, J. J. Myositis ossificans of the upper arm. *Am J Sports Med* 8:419–424, 1980.
59. Jorgenson, U., Hinge, K., and Rye, B. Rupture of the distal biceps brachii tendon. *J Trauma* 26:1061–1062, 1986.
60. Kennedy, J. C., and Willis, R. B. The effect of local steroid injections on tendons: A biomechanical and microscopic correlative study. *Am J Sports Med* 4:4–21, 1976.
61. King, J. W., Brelsford, H. J., and Tullos, H. S. Analysis of the pitching arm of the professional baseball pitcher. *Clin Orthop* 67:116–123, 1969.
62. Kofoed, H. Thoracic outlet syndrome: Diagnostic evaluation by analgesic cervical disk puncture. *Clin Orthop* 146:145–148, 1980.
63. Kolb, L. W., and Moore, R. D. Fractures of the supracondylar process of the humerus: Report of two cases. *J Bone Joint Surg [Am]* 49:532, 1967.
64. Krahl, V. E. The bicipital groove: A visible record of humeral torsion. *Anat Rec* 101:319, 1948.
65. Kretzler, H. H., and Richardson, A. B. Rupture of the pectoralis major muscle. *Am J Sports Med* 17:453–458, 1989.
66. Leguit, P. Compartment syndrome of the upper arm. *Neth J Surg* 34:123–126, 1982.
67. Levy, M., Goldberg, I., and Meir, J. Fracture of the head of the radius with a tear or avulsion of the triceps tendon. *J Bone Joint Surg [Br]* 64:70–72, 1982.
68. Lipscomb, A. B., Thomas E. D., and Johnston, R. K. Treatment of myositis ossificans traumatica in athletes. *Am J Sports Med* 4:111–120, 1976.
68a. Lord, J. W., and Rosati, L. M. Thoracic outlet syndromes. *Clin Symp* 23(2), 1971.
69. Louis, D. S., Hankin, F. M., Eckenrode, J. F., Smith, P. A., and Wovtys, E. M. Distal biceps brachii tendon avulsion: A simplified method of operative repair. *Am J Sports Med* 14:234–236, 1986.
70. Marks, J. Anticoagulant therapy in idiopathic occlusion of the axillary vein. *Br Med J* 1:11–13, 1956.
71. Marquis, J. W., Bruner, A. J., and Keith, H. M. Supracondyloid process of the humerus. *Mayo Clin Proc* 37:691, 1957.
72. Matas, R. Primary thrombosis of the axillary vein caused by strain. *Am J Surg* 24:642–666, 1934.
73. Match, R. M., and Corrylos, E. V. Bilateral avulsion fracture of the triceps tendon insertion from skiing with osteogenesis imperfecta tarda: A case report. *Am J Sports Med* 11:99–101, 1983.
74. McEntire, J. E., Hess, W. E., and Coleman, S. Rupture of the pectoralis major muscle. *J Bone Joint Surg* 54A:1040–1046, 1972.
75. McMaster, P. E. Tendon and muscle ruptures: Clinical and experiment studies on causes and locations of subcutaneous ruptures. *J Bone Joint Surg* 15:705, 1933.
76. McReynolds, I. S. Avulsion of the insertion of the biceps brachii tendon and its surgical treatment. *J Bone Joint Surg [Am]* 45:1780–1781, 1963.
77. Meherin, J. H., and Kilgore, B. S. The treatment of rupture of the distal biceps brachii tendon. *Am J Surg* 99:636, 1960.
78. Montgomery, A. H. Two cases of muscle injury. *Surg Clin Chicago* 4:871, 1920.
79. Mori, K. W., and Bookstein, J. J. Selective streptokinase infusion: Clinical and laboratory correlates. *Radiology* 143:677–682, 1983.
80. Morrey, B. F., Askew, L. J., An, K. N., and Doybns, J.

H. Rupture of the distal tendon of the biceps brachii: A biomechanical study. *J Bone Joint Surg [Am]* 67:418–421, 1985.

81. Mubarak, S. J., and Hargens, A. R. Acute compartment syndromes: Diagnosis and treatment with the aid of the wick catheter. *J Bone Joint Surg [Am]* 60:1091–1095, 1978.

82. Murphy, T. Brachial neuritis caused by pressure of first rib. *Aust Med J* 15:582, 1910.

83. Naffziger, H. C., and Grant, W. T. Neuritis of brachial plexus mechanical in origin, the scalenus syndrome. *Surg Gynecol Obstet* 67:722, 1938.

84. Nichols, H. M. Anatomic structures of the thoracic outlet. *Clin Orthop* 207:13–20, 1986.

85. Nielson, K. Partial rupture of the distal biceps brachii tendon: A case report. *Acta Orthop Scand* 58:287–288, 1987.

86. Norman, A., and Dorfman, H. D. Juxtacortical circumscribed myositis ossificans: Evolution and radiographic features. *Radiology* 96:301, 1970.

87. Norman, W. H. Repair of avulsion of insertion of biceps brachii tendon. *Clin Orthop* 193:189–194, 1985.

88. Ochsner, A., Gage, M., and DeBakey, M. Scalenus anticus (Naffziger) syndrome. *Am J Surg* 28:669, 1935.

89. Omer, G. E. J. Injuries to nerves of the upper extremity. *J Bone Joint Surg [Am]* 56:1615, 1974.

90. Oraua, S., Sorasto, A., Aalto, K., and Kuist, H. Total rupture of the pectoralis major muscle in athletes. *Int J Sports Med* 5:272–274, 1984.

91. Pantazopoulos, T., Exarchow, E. Stavrou, Z., and Hartofilakidis-Garofalidis, G. Avulsion of the triceps tendon. *J Trauma* 15:827, 1975.

92. Park, J. Y., and Espiniella, J. L. Rupture of pectoralis major muscle. *J Bone Joint Surg* 52A:577–581, 1970.

93. Patissier, P. *Traite des Maladies des Artisons*. Paris, 1822, pp. 162–164.

94. Postacchini, F., and Pudda, G. Subcutaneous rupture of the distal biceps brachii tendon. *J Sports Med Phys Fitness* 15:84–90, 1975.

95. Powell, F. I. Fracture of the first rib. *Br Med J* 1:282–285, 1950.

96. Preston, F. S., and Adicoff, A. Hyperparathyroidism with avulsion at three major tendons. *N Engl J Med* 266:968, 1961.

97. Priest, J. D., and Nagel, D. A. Tennis shoulder. *Am J Sports Med* 4:28–42, 1976.

98. Riddell, D. H. Thoracic outlet syndrome: Thoracic and vascular aspects. *Clin Orthop* 51:53–64, 1967.

99. Roos, D. B. Transaxillary approach for first rib resection to relieve thoracic outlet syndrome. *Ann Surg* 163:354–358, 1966.

100. Roos, D. B. Congenital anomalies associated with thoracic outlet syndrome. *Am J Surg* 132:771, 1976.

101. Roos, D. B. Axillary-subclavian vein occlusion. *In* Rutherford, R. D. (Ed.), *Vascular Surgery*. Philadelphia, W. B. Saunders, 1984, pp 1385–1393.

102. Schroetter, L. von. *Erkrankanger der Gefasse: Nothnasel Handbuck der Pathologie und Therapie*. Vienna, Holder, 1884.

103. Searfoss, R., Tripi, J., and Bowers, W. Triceps brachii rupture: A case report. *J Trauma* 16:244–246, 1976.

104. Silbertstein, M. J., Brodeur, A. E., Graviss, E. R., and Atchawee, L. Some vagaries of the olecranon. *J Bone Joint Surg [Am]* 63:722–725, 1981.

105. Slocum, D. B. Classification of elbow injuries from baseball pitching. *Tex Med* 64:48, 1968.

106. Smith, K. F. The thoracic outlet syndrome: A protocol of treatment. *J Orthop Sports Phys Ther* 1:89, 1979.

107. Stopford, J. S. B., and Telford, E. D. Compression of the lower trunk of the brachial plexus by a first dorsal rib with a note on the surgical treatment. *Br J Surg* 7:168, 1919.

108. Strukel, R. J., and Garrick, J. G. Thoracic outlet compression in athletes. *Am J Sports Med* 6:35–39, 1978.

109. Tarsney, F. F. Rupture and avulsion of the triceps. *Clin Orthop* 83:177–183, 1972.

110. Terry, R. J. New data on the incidence of the supracondyloid variation. *Am J Phys Anthropol* 9:265, 1926.

111. Thorndike, A., Thomas, E. D., and Johnston, R. K. Myositis ossificans traumatica. *J Bone Joint Surg* 22:315–323, 1940.

112. Torg, J. S. The effect of competitive pitching on the shoulders and elbows of pre-adolescent baseball players. *Pediatrics* 49:267–272, 1972.

113. Tullos, H. S., Erwin, W. D., Woods, G. W., Wukasch, D. C., Cooley, D. A., and King, J. W. Unusual lesions of the pitching arm. *Clin Orthop* 88:169–182, 1972.

114. Tullos, H. S., and King, J. W. Lesions of the pitching arm in adolescents. *JAMA* 220:264, 1972.

115. Turek, S. L. (Ed.). *Orthopedics: Principles and Their Application* (4th ed.). Philadelphia, J. B. Lippincott, 1984.

116. Twinning, R. H., Marcus, W. Y., and Garez, J. L. Tendon ruptures in systemic lupus erythematous. *JAMA* 187:123–124, 1964.

117. UnVerforth, J. L. The effect of local steroid injections on tendons. *J Sports Med* 11:31–37, 1973.

118. Vastamaki, M., Brummer, H., and Soloren, K. Avulsion of the distal biceps brachii tendon. *Acta Orthop Scand* 52:45–48, 1981.

119. Veal, J. R., and Hussey, H. H. The use of "exercise tests" in connection with venous pressure measurements for the detection of venous obstruction in the upper and lower extremity. *Am Heart J* 20:308–321, 1944.

120. Vogel, C. M., and Jensen, J. E. "Effort" thrombosis of the subclavian vein in a competitive swimmer. *Am J Sports Med* 13:269–272, 1985.

121. Whitaker, J. H. Arm wrestling fractures—A humerus twist. *Am J Sports Med* 5:67–77, 1977.

122. Wilmoth, C. L. Recurrent fracture of the humerus due to sudden extreme muscular violence. *J Bone Joint Surg* 12:168–169, 1930.

123. Woodburne, R. T. *Essentials of Human Anatomy* (3rd ed.). New York, Oxford University Press, 1965, pp 94–96.

124. Wright, I. S. The neurovascular syndrome produced by hyperabduction of the arm. *Am Heart J* 29:1, 1945.

125. Wright, R. S., and Lipscomb, A. B. Acute occlusion of the subclavian vein in an athlete: Diagnosis, etiology and surgical management. *J Sports Med* 2:343–348, 1975.

126. Young, F., and Harris, C. T. Complete excision and reconstruction of both achilles tendons for giant cell xanthoma. *Surg Gynecol Obstet* 61:662–669, 1935.

127. Zeman, S. C., Rosenfeld, R. T., and Lipscomb, P. R. Tears of the pectoralis major muscle. *Am J Sports Med* 7(6):343–347, 1979.

Nerve Lesions of the Arm and Elbow

Kenneth P. Butters, M.D.
Kenneth M. Singer, M.D.

ULNAR NERVE

Entrapment of the ulnar nerve in the region of the elbow is the second most common compressive neuropathy in the upper extremity. Ulnar neuritis at the elbow results from compression, repetitive stretch, or friction. The valgus stress in throwing may cause repeated insult to the nerve, which is properly held behind the medial epicondyle. Subluxation of the nerve is common and usually asymptomatic, but the nerve may be injured when it is resting in the subluxed position. A friction neuritis can also result from nerve movement against a bony irregularity or just the surface of the medial epicondyle. Ulnar nerve compression at the level of the epicondyle or as the nerve enters the flexor carpi ulnaris bridge is commonly called cubital tunnel syndrome. Postoperatively, ulnar nerve compression can result from kinking at the medial intermuscular septum, compression at the arcade of Struthers, or diffuse scarring about the nerve. The diagnosis is made by the history and physical examination, usually supported by electrical testing. Nonsurgical treatment methods are limited, and many patients need operative treatment. Several surgical techniques are available, and surgical failure rates vary from 0 to 75% [13, 25]. Choice of surgical technique based on published reports is clouded by personal bias.

Anatomy

The ulnar nerve enters the distal arm by passing from the anterior to the posterior compartment (Fig. 16D–1). It penetrates the fibrous raphe called the arcade of Struthers, located 8 cm above the medial epicondyle and formed by the medial intramuscular septum, medial head of the triceps, and deep investing fascia of the arm plus the internal brachial ligament. The arcade should be distinguished from the ligament of Struthers, which, when present, connects the medial epicondyle with the bony supracondylar process compressing the median nerve.

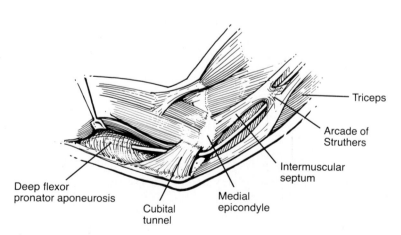

Triceps

Arcade of Struthers

Intermuscular septum

Medial epicondyle

Cubital tunnel

Deep flexor pronator aponeurosis

FIGURE 16D–1
The anatomy of the ulnar nerve is pictured from the arcade of Struthers through the flexor carpi ulnaris. Surgical exploration often includes all potential levels of compression.

An offending arcade of Struthers is suspected when, upon tracing the ulnar nerve proximally from the epicondyle, muscle fibers of the medial triceps are seen crossing obliquely superficial to the nerve [37].

The nerve passes behind the medial epicondyle into the true cubital tunnel, whose medial boundaries are the ulnar collateral ligament of the elbow, the medial edge of the trochlea, and the medial epicondylar groove. The roof of the cubital tunnel is the arcuate ligament, which extends from the medial epicondyle to the olecranon. The ulnar nerve passes, then, between the humeral and ulnar heads of the flexor carpi ulnaris. It courses within the flexor carpi ulnaris and exits by piercing its deep surface to lie between the flexor carpi ulnaris and the flexor digitorum profundus. There are usually no ulnar nerve branches in the arm. At the level of the elbow, there is an articular branch and often a motor branch to the flexor carpi ulnaris.

Biomechanics

The ulnar nerve undergoes stresses with elbow motion. To reach 45 degrees of elbow flexion, the arcuate ligament must stretch nearly 5 mm, and at 90 degrees, the proximal edge of the ligament becomes taut. The ulnar collateral ligament relaxes and bulges with elbow flexion, creating the potential for ulnar nerve compression [37]. The ulnar nerve elongates an average of 4.7 mm with elbow flexion, and the medial head of the triceps can pull the nerve 7 mm medially proximal to the cubital tunnel. The nerve must remain mobile and free to alter its shape or position. Fracture or adhesions may alter the mechanics of the cubital tunnel. The origin of the flexor carpi ulnaris (FCU) can compress the ulnar nerve during elbow flexion [2].

The entrance to the often compressing flexor carpi ulnaris arch is located 3 to 20 mm distal to the medial epicondyle [6]. The exit point of the ulnar nerve from the FCU is 48 mm distal to the epicondyle, another potential point of compression. This correlates with the description by Gabel and Amadio of a "deep flexor pronator aponeurosis," which is 5 cm distal to the epicondyle [25] (see Fig. 16D–1). Miller, using electrical studies, found that the point of compression at the entry to the cubital tunnel was 1.5 to 4 cm distal to the medial epicondyle, indicating, in some cases, a very distal compression site [6].

The ulnar nerve is larger behind the epicondyle [68]. There is an increasing connective tissue component of the nerve, and an increasing number of fascicles are present as the nerve crosses the elbow joint [14].

The topography of the ulnar nerve at the cubital tunnel shows deeply situated flexor profundus and flexor carpi ulnaris fibers and superficially positioned sensory and intrinsic motor fibers.

Pechan and Julis have shown, in fresh cadavers, that elbow flexion and wrist extension cause a threefold increase in ulnar nerve pressure, and with the hand in the cocking position of throwing, pressure is elevated to six times the resting pressure [58]. Chronic nerve compression can certainly result from these pressures.

Etiology

Idiopathic compression of the ulnar nerve within the cubital tunnel is well known. A number of other anatomic causes have been described such as subluxation of the nerve, compression by an anomalous anconeus epitrochlearis muscle, chronic cubital valgus deformity, ganglion, old medial epicondyle fracture, rheumatoid arthritis, or osteoarthritis. Failed surgical decompression may be due to scarring, kinking of the nerve, an unresected medial intramuscular septum or arcade of Struthers, or a fibrous band within the flexor carpi ulnaris [25, 26, 64].

In the athlete, especially the throwing athlete, the syndrome usually results from direct injury or occurs as part of the lateral compression, medial tension forces that accompany the act of throwing. Ulnar nerve compression due to flexor ulnaris and triceps contraction is postulated in cross-country skiers [24]. Post-traumatic elbow changes described by King and colleagues such as humeral hypertrophy, degenerative changes in the medial olecranon fossa, and flexion contracture of the elbow are evidence of the potential for nerve damage [42].

Jobe and associates believe that the thrower may develop a traction neuritis, especially when an attenuated ulnar collateral ligament allows the medial joint to "open up" with a progressive flexion-valgus deformity [37]. Thickening due to scar or calcium in the ulnar collateral ligament and degenerative arthritis within the groove contribute to ulnar nerve compression or tethering, especially with motion. Also, arcuate ligament thickening (Osborne lesion) is important [36].

Ulnar neuritis resulting from recurrent nerve instability anterior to the epicondyle during elbow flexion is common. Sixteen percent of the population have ulnar nerve hypermobility, usually bilateral and secondary to arcuate ligament laxity, which is congenital or developmental [9, 10].

Childress classified ulnar nerve instability [9, 10]. Type A instability occurs when the nerve moves to the tip of the epicondyle when the elbow is flexed to 90 degrees. In type B instability, the nerve has greater excursion and rests anterior to the epicondyle. Childress found that 16% of the population have some nerve instability—12% have type A and 4% type B; in 90% of these people the condition is bilateral. It seems that with type A instability the nerve is more exposed and is likely to receive direct trauma, and in type B the greater excursion of the nerve is more likely to cause a friction neuritis [9, 46]. Lazaro, citing Conway, related the neuritis of recurrent nerve dislocation to progressive fibrosis and a decrease in the elasticity of the nerve [46].

Osteoarthritis with medial trochlear osteophytes diminishes the space available to the ulnar nerve through the cubital tunnel. This is particularly significant with elbow flexion.

Clinical Presentation

Classic symptoms such as ring and small finger numbness are usually seen early, but they are not necessary to make the diagnosis. Medial elbow aching and "hand weakness" after throwing may be the only signs present. In the athlete, physical findings such as decreased two-point discrimination, intrinsic weakness, or atrophy are usually absent. Sensitivity to palpation of the ulnar nerve at the elbow is a nonspecific sign. Lateral elbow compartment compression symptoms are common and should be evaluated appropriately. Pain in the elbow radiating to the medial forearm and painful popping with subluxation of the ulnar nerve with elbow flexion may be the complaint.

Physical Examination

Examination of the patient with small and ring finger numbness and suspected ulnar nerve compression at the elbow extends from the cervical spine to the hand. Especially when there is a history of a certain neck motion or neck position that increases the distal symptoms, a foraminal compression maneuver should be done. Cervical root compression is suggested when a positive Spurling's sign is obtained by rotating the head to the side opposite the symptoms and extending the neck with axial compression, creating the distal symptoms. The thoracic outlet is examined with the arms overhead in the throwing position, opening and closing the hands for 60 seconds, eliciting the distal symptoms, and listening for a supraclavicular bruit. Elevation of the arm with the shoulders back and deep inspiration may produce the distal symptoms (Adson's sign), suggesting thoracic outlet compression.

Tapping along the course of the ulnar nerve through the arm and forearm may give localizing information about ulnar nerve compression. Holding the elbow in maximum flexion for 60 seconds may produce ring and small finger numbness, much as the Phalen's sign does for carpal tunnel syndrome. Palpation of the ulnar nerve through the cubital tunnel with the elbow flexed and extended may show nerve instability, pseudoneuroma, or nonspecific ulnar nerve tenderness. Valgus stability and the presence of cubitus valgus or flexion contracture should be carefully recorded as part of elbow range of motion testing. Specific tenderness over the flexor pronator origin at the epicondyle may be distinguished from that occurring at the ulnar nerve. Guyon's canal, Tinel's sign, and Allen's test are performed next, followed by careful two-point discrimination testing of the fingers. Sensory evaluation of the ulnar dorsum of the hand (dorsal cutaneous branch of the ulnar nerve) and ulnar volar forearm (medial antebrachial cutaneous nerve) helps to distinguish proximal extremity compression from cubital tunnel syndrome and also from ulnar nerve compression at the wrist. Motor examination includes muscle strength testing; the flexor digitorum profundus of the ring and small fingers is compared with the index finger, and the ulnar intrinsics test abduction and adduction of the fingers. Again, careful examination should be done to detect associated elbow joint problems resulting from lateral compression and valgus-extension overload as well as from medial flexor-pronator origin injury or overuse. Subtle intrinsic weakness may be picked up by pinch meter testing and by careful inspection for ulnar intrinsic atrophy and clawing. Jaymar grip testing should also be done.

A Martin-Gruber anastomosis is present in 15% of limbs. In patients with this anomaly, fibers pass from the median to the ulnar nerve in the midforearm; hence, ulnar lesions at the elbow may spare some ulnar intrinsic fibers, confusing the examination.

Diagnostic Studies

X-ray studies of the elbow should be done including a cubital tunnel view, especially if surgery is planned. Other problems may be identified, although they are often asymptomatic and unrelated.

Electrical testing focuses on the nerve conduction velocity delay across the elbow. The examination may include ulnar F wave, cervical paraspinous electromyography (EMG), distal muscle EMG, and wrist conduction studies. Nerve conduction velocity across the elbow of more than 45 meters/second is considered normal. Del Pizzo and colleagues did not rely on nerve conduction velocity changes in the five baseball pitchers they operated on for cubital tunnel syndrome [12]. Three of these patients had normal electrical findings. Nerve conduction velocity across the elbow of less than 45 meters/second and a velocity drop of greater than 10 meters/second compared with the normal side, may be considered abnormal. Provocative testing with the elbow at 90 degrees or more is difficult owing to the difficulty of obtaining accurate surface length measurements from one point on the nerve to another.

With electrical studies, the differential diagnosis of brachial neuritis at the distal roots, mononeuritis multiplex, thoracic outlet syndrome, and cervical root impingement can be sorted out from ulnar nerve compression at the wrist.

Treatment

Nonoperative Management

Nonoperative treatment centers on minimizing the pressure increases around the ulnar nerve with elbow flexion or direct pressure. The application of an anterior splint with the elbow placed at 30 degrees of flexion for several weeks seems to be the best conservative management for cubital tunnel syndrome [15]. Short of this, use of an elbow pad, avoidance of recurrent valgus stress and direct pressure, and an understanding of ulnar nerve subluxation by the athlete are all that can be offered.

Surgery

Basic choices include cubital tunnel release only, anterior transposition of the nerve with a fascial sling, medial epicondylectomy, and submuscular transposition. Beyond simple release, all approaches have in common careful identification of the ulnar nerve proximal to the cubital tunnel, preservation of the medial antebrachial cutaneous nerve branches in the anterior wound, exposure and resection of the medial intermuscular septum, ulnar nerve release through the cubital tunnel, and release of the arch of the flexor carpi ulnaris, allowing mobilization of the nerve. Also, identification of the arcade of Struthers, if present, 8 cm proximal to the epicondyle and release of ulnar nerve tethering are part of this treatment. Transposition of the ulnar nerve may gain as much as 2 cm in length; this is done at the time of micro nerve repair for laceration [63].

Learmonth in 1942 described a submuscular position for the ulnar nerve [47]. Many authors feel that this position offers protection from direct trauma during athletic activity. The flexor pronator muscle group is detached from the medial epicondyle and separated carefully from the ulnar collat-

eral ligament. Elbow arthrotomy can be performed anteriorly or posteriorly with this approach, and the ulnar collateral ligament can be inspected or reconstructed. The ulnar nerve is then placed under the flexor pronator muscle, and the muscle is reattached to the epicondyle. Jobe and Fanton prefer this technique and begin range-of-motion exercises at 2 weeks, light throwing at 2 months, and full activity as early as 6 months. The security of the reattachment of the flexor pronator muscle has not been a problem in throwers or tennis players [37].

Anterior transposition allows predictable ulnar nerve decompression without much dissection. Proximal and distal mobilization without excessive handling or stripping of the nerve is accomplished, and the accompanying veins remain with the nerve if possible. A fasciodermal sling based medially is made from the flexor pronator fascia. A mark is made on the skin over the medial epicondyle before the incision is made, identifying the point of attachment of the free end of the sling as it passes posterior to the transposed nerve. The nerve is not kinked when the elbow is extended; however, only skin and subcutaneous tissue are superficial to the nerve (Fig. 16D–2). Other levels of mechanical compression can be created, and the nerve is placed in a relatively hypovascular bed and may be exposed to additional trauma [25]. In Eaton's series of 16 patients there were seven college or professional pitchers [16]. Simple release has been reported to give results

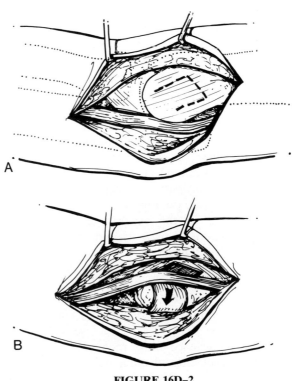

FIGURE 16D–2

comparable with those achieved by anterior transposition [8, 68] if symptoms had been of brief duration and there were no chronic changes in the nerve and no bony deformities, arthritis, subluxation, or nerve fibrosis or scarring.

Medial epicondylectomy probably allows the least amount of dissection of the ulnar nerve. The epicondyle is exposed with a longitudinal subperiosteal approach, allowing its removal with an osteotome while preserving the ulnar collateral ligament. An inverted closure of the flexor pronator origin and periosteum allows a good bed for the nerve and good nerve decompression. The authors are concerned about altering the flexor pronator origin and about potential damage to the ulnar collateral ligament with this approach in athletes. The advantages of medial epicondylectomy compared with simple transposition are that with transposition the nerve continues to overlie the bony prominence of the medial epicondyle and is therefore susceptible to contusion, and it also tends to be irritated when the elbow is in extension; neither situation is present with epicondylectomy.

Results

Jobe and Fanton believe that the prognosis for return to a preoperative level of competition after ulnar nerve surgery at the elbow is good [37]. The degree of ulnar nerve compression is a prognostic indicator, intrinsic motor loss and sensory loss having a less favorable outcome. Associated joint pathology caused by the medial tension and lateral compression that occur in the throwing motion also determines the level of functional recovery. Abnormal electrical studies, which are important in determining the need for surgery in the nonathlete with cubital tunnel syndrome, are not critical in pre- or postoperative assessment of throwers [37]. Del Pizzo and associates performed submuscular transpositions of the ulnar nerve in throwing athletes and returned nine out of ten to their preinjury level [12]. Progression of neuropathy was halted.

In the nonathletic population, the results can be compared based on the existing degree of ulnar nerve compression. In patients with minimal nerve compression, half will recover without surgery, and surgery gives excellent results in 90% of the others [13, 15]. With moderate compression, excellent results occur in 50% to 80% of patients with surgery. With severe compression, the results vary widely, but probably less than 50% recover sensory innervation, 25% experience motor recovery, and around 30% have a recurrence of ulnar nerve compression [13]. Friedman and Chochran's series of patients

with advanced ulnar neuropathy showed a 70% rate of good and fair results; they felt that an advanced stage of disease, long duration of symptoms, diabetes, alcoholism, and muscle atrophy do not necessarily predict a poor result [22]. As stated in the literature, intrinsic muscle atrophy is the problem least likely to improve after ulnar nerve surgery, and a prolonged duration of symptoms tends to indicate a poor prognosis [31, 50].

A lot can be learned from a review of surgical failures. All potential levels of compression along the course of the ulnar nerve must be explored from the arcade of Struthers to the deep pronator aponeurosis at the time of initial surgery. This is necessary even though a preoperative EMG may seem to localize a specific lesion of the ulnar nerve. At primary surgery, one level of compression is usually noted. With reoperation, the average level seen is 2.2 levels of compression [25]. The causes of this discrepancy are failure to release the involved level or creation of a new level of compression including diffuse nerve scarring. During repeat ulnar nerve surgery, all levels should be addressed. The author believes that the location of Tinel's sign and careful nerve inspection at surgery may allow a less extensive procedure in that removal of diffuse scar from the nerve carries a worse prognosis. Gabel and Amadio state that a 10-cm mobilization is necessary to complete nerve exploration and submuscular transposition; they cite experimental data showing that mobilization of 10 to 15 cm of human nerve has no adverse effects on nerve circulation [25].

Postoperative Management and Rehabilitation

Range-of-motion exercises are begun following wound healing, and resistance exercises are begun at 3 weeks. Careful protection of the ulnar nerve from overlying contact is important for 6 weeks, and, depending upon the sport, the subcutaneous transposed nerve should be padded for 6 months after surgery. Associated elbow problems often dictate the proper methods of functional rehabilitation. Again, Jobe and Fanton have stated that elevation and reattachment of the flexor muscle origin have not been a problem. Return to throwing after 6 weeks is based on the specific rehabilitation required for the individual's throwing sport [37].

Treatment in the Athlete

Diagnosis of ulnar nerve compression at the elbow is difficult in the athlete. Sensory loss, motor weakness, electrical abnormalities, and even classic symptoms such as small finger numbness are often absent. Flexor pronator inflammatory problems and associated valgus extension overload elbow joint problems must be carefully differentiated. This often includes differential Xylocaine injection coupled with careful ulnar nerve examination at the elbow. Identification of the etiology is often difficult because the point of compression may be at the arcade of Struthers or the cubital tunnel, or it may be a nerve undergoing friction or contusion as it subluxes out of the groove with elbow flexion. It is important to remember that ulnar nerve instability is not necessarily pathologic. Important in conservative management is the use of an anterior elbow splint with the elbow flexed at 30 degrees, worn for 20 hours a day for a trial period of at least 2 weeks. If the problem becomes chronic—that is, if it lasts 6 months or longer—surgery may be considered. The presence of significant electrical abnormalities (denervation or a significant drop in nerve conduction velocity across the elbow) certainly is a solid indication for surgery, as is sensory loss or intrinsic weakness caused by a lesion at the medial elbow. The author prefers to perform ulnar nerve transposition in the athlete using a sling of flexor pronator fascia as described by Eaton and associates [16]. This approach does not alter the flexor pronator origin to any significant degree; however, it also does not allow arthrotomy or surgery on the medial collateral ligament.

Indications for surgery in the athlete may be symptoms that occur only when the athlete engages in the sport or that only affect playing performance. Most commonly, the symptoms resolve with rest but then recur. The athlete may complain primarily of medial elbow pain with little hand numbness, and electrical studies may be normal. The author believes that symptomatic compression of the ulnar nerve at the elbow in the athlete is a problem that can be corrected surgically with return to full function with the understanding that the disease process is part of a spectrum of problems affecting the medial elbow due to valgus stress. The athlete is more likely to develop ulnar nerve compression that is due to mechanical stretch or friction.

MEDIAN NERVE COMPRESSION

Median nerve compression about the elbow and proximal forearm is often accompanied by vague symptoms of easy fatigability, numbness, and aching in the forearm. The pronator syndrome refers to compression of the median nerve in the anatomic area between the supracondylar process of the distal humerus and the fibrous arch of the sublimis and is separate from the motor-only paralysis characteristic of Kiloh-Nevin (anterior interosseous) syndrome [40].

The etiology is not well defined, but the condition is possibly due to cyclic stress in grasping in pronation, that is, during weight lifting or tennis serves, or to a direct blow on the proximal forearm. Other possible causes are deep space-occupying lesions pushing the nerve up into the flexor sublimis arch and muscle trauma associated with a forearm fracture [55]. Overuse pronator hypertrophy has also been described as a cause.

Anatomy

The median nerve approaches the elbow from the medial arm and crosses the antecubital fossa medial to the brachial artery and biceps tendon; it then courses under the pronator and flexor sublimis arch on the way to the wrist (Fig. 16D–3). Above the elbow, the supracondylar process (present in 1% of limbs) projects distally at the anteromedial humerus

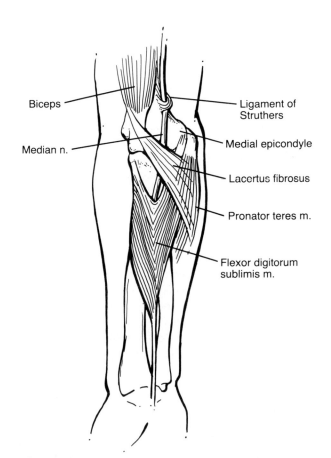

FIGURE 16D–3
The anatomy of the median nerve is pictured from the supracondylar process (ligament of Struthers) through the sublimis arch. Surgical exploration often includes all potential levels of compression.

and continues to the medial epicondyle as the ligament of Struthers, which may be ossified. The median nerve may pass under this seldom-present arch and then rest on the brachialis muscle with the brachial artery. At this point it is deep to the lacertus fibrosus and subject to compression. The median nerve then passes along the lateral edge of the pronator teres and deep to it, passing most commonly between the humeral and ulnar heads (in 80% of cases) [4]. In 15% of arms, the nerve passes below a solitary humeral head, and in 5% it pierces the humeral head. Motor branches, sometimes several, to the pronator muscle arise from the median nerve at the upper part of the muscle with usually a separate ulnar head branch.

The anterior interosseous nerve arises lateral and deep at the level of the ulnar head of the pronator. The median nerve then passes beneath the tendinous arch of the flexor sublimis origin and is adherent to the underside of that muscle. Anatomic variations noted by Spinner and Linscheid [64] include the supracondylar process, Gantzer's muscle and the palmaris profundus, flexor carpi radialis brevis, duplicated lacertus fibrosus, high origin of the pronator teres, and vascular anomalies.

Clinical Syndrome

This is a syndrome of neural compression characterized by proximal forearm aching, easy fatigue, and hand numbness. The nocturnal symptoms seen with carpal tunnel compression are typically absent. Symptoms occur during activity or sport. The diagnosis of pronator teres syndrome is seldom easy.

Examination

Physical examination should begin proximally at the supracondylar level and extend to the fibrous arch of the sublimis. This, of course, is in addition to the examination of the cervical spine, thoracic outlet, and carpal tunnel.

Spinner and Linscheid offer the following observations to help make the often subtle diagnosis of pronator syndrome [64]:

1. Indentation of the pronator teres, suggesting a tight lacertus fibrosus. This may be increased by forearm pronation, active or passive.

2. Tense flexor or pronator musculature, an example being the unilateral muscle hypertrophy seen in throwers and racquet athletes.

3. Reproduction of symptoms with resisted muscle function. This can be elicited in one of four

ways: (1) resisted pronation for 30 to 60 seconds; (2) resisted elbow flexion and supination (tightening of the lacertus); (3) resisted long finger flexion at the PIP joint (tightening the fibrous arch of sublimis); and (4) direct pressure over the proximal part of the pronator during resisted pronation (compare to asymptomatic side).

4. Tapping over the median nerve from above the elbow to the wrist. Specific, localized Tinel's signs are helpful if they are reproducible compared to a similar examination on the other side and with other nerve examinations. There are patients with a diffuse "nerve sensitivity."

5. Passive stretching of the finger and wrist flexors.

6. Median innervated muscle weakness. This is infrequent, and most likely flexor profundus to the index finger and flexor pollicis longus should be distinguished from anterior interosseous syndrome (pure motor syndrome).

Emphasis should be placed on reproduction of the patient's symptoms in a nonsolicited response with careful comparison to the opposite side. Phalen's sign at the wrist is positive in 50% of patients with pronator syndrome who are operated on [4].

Electrical Studies

Electrical studies including EMG and nerve conduction velocity studies are rarely diagnostic. Distal latency testing may diagnose carpal tunnel syndrome, and a median innervated extrinsic abnormality may place the lesion proximal to the carpal tunnel but will not give a specific diagnosis. To confuse matters further, simultaneous occurrence of two-level compression (double-crush syndrome) may occur, or proximal forearm symptoms may occur with carpal tunnel compression alone. In Hartz and colleagues' paper, only 2 of 39 patients operated on had definite electrical evidence of pronator syndrome; 15% to 20% had a nonlocalizable median nerve lesion [32].

The work-up includes a plain x-ray of the elbow that includes the distal humerus to look for a supracondylar process. An oblique x-ray may be necessary to see this.

Differential Diagnosis

The differential diagnosis list is long and comprises causes of median nerve hand numbness and proximal forearm and elbow pain including inflammatory conditions such as distal biceps tendinitis and medial epicondylitis.

Treatment

Conservative management includes rest for the upper extremity, elbow splinting in neutral rotation, nonsteroidal anti-inflammatory agents, and tapering doses of oral steroids in acute cases.

Use of a sterile tourniquet is helpful in the surgical approach to the median nerve proximal to the elbow. The S-shaped incision courses down the medial arm, passing obliquely across the antecubital fossa and then down the volar radial forearm (Fig. 16D–4). The incision design is based on an understanding of surface landmarks; a dot is placed with a marking pen over the structures needing exposure, and then the incision is drawn. The incision crossing the antecubital crease should be drawn as a Z or S shape. Loupe magnification is necessary to identify abnormalities of the median nerve. In the absence of a supracondylar process, the exposure begins with identification of the median nerve above the lacertus fibrosis, preserving branches of the medial antebrachial cutaneous nerve in the medial wound. The lateral antebrachial cutaneous nerve should be protected because it lies just radial to the biceps

FIGURE 16D–4
Return to sports following median nerve compression in the proximal forearm depends on the degree of motor and sensory involvement. Rehabilitation after surgical release may begin after wound healing, progressing from putty grip strengthening to eccentric forearm rotation with mild resistance to progressive resistance exercises. Protected return to sports may occur early in this course. Return to skilled use of the extremity may take many months. Syndromes of pain and numbness without motor loss that resolve may allow quick return to training and competition.

tendon. Lacertus fibrosus tightness should be observed with forearm pronation before and after release of the lacertus. Motor branches come off the median nerve medially. The median nerve is traced under the pronator humeral head to the ulnar head, which may be only a fibrous band or a 1- to 2-cm muscle belly. One to two centimeters distal to this point of possible pronator compression is the arch of the sublimis, which can easily be probed. The ulnar head is resected to allow more distal exposure.

Anterior Interosseous Syndrome

This syndrome classically is a pure motor paralysis syndrome of the flexor pollicis longus and index profundus combined with pronator quadratus muscle weakness, which often goes unnoticed. Flexor pollicis longus or index profundus paralysis alone may occur, and if a Martin-Gruber anastomosis (anterior interosseous to ulnar nerve) is present in the forearm, ulnar intrinsic paralysis or flexor profundus paralysis to other fingers may be involved. The author has seen this syndrome in weight lifters and gymnasts who experienced cumulative trauma and in a football player after a direct blow. The presenting symptom may be pain lasting 12 to 24 hours, leaving a pure motor loss. Electromyography of the affected muscles after 3 weeks shows denervation. To distinguish paralysis from rupture, the forearm squeeze test or wrist tenodesis effect is done to detect passive tendon activity.

Conservative treatment with tapering oral steroid drugs or nonsteroidal anti-inflammatory agents and avoidance of repetitive manual tasks for at least 8 weeks is indicated.

It is not well understood why some patients get this well described specific motor syndrome when more commonly others experience a vague pain and the discomfort of the pronator syndrome. Nerve compression is occurring with both in the same anatomic area. The surgical approach is similar to that used for the pronator syndrome. At surgery, the anterior interosseous nerve, in the author's experience, often feels scarred or indurated, indicating micro-external neurolysis.

References

1. Alvine, F. G., and Schurrer, M. E. Postoperative ulnar nerve palsy: Are there predisposing factors? *J Bone Joint Surg* 69A:255–259, 1987.
2. Apfelberg, D. B., and Larson, S. J. Dynamic anatomy of the ulnar nerve at the elbow. *Plast Reconstr Surg* 51:76–81, 1973.

3. Barss, P. Ulnar compression neuropathy due to an occult post-traumatic synovial cyst. *Med J Aust* 140:428–429, 1984.
4. Beaton, L. E., and Anson, B. J. The relation of the median nerve to the pronator teres muscle. *Anat Rec* 75:23–26, 1939.
5. Brown, W. F., Yates, S. K., and Ferguson, G. G. Cubital tunnel syndrome and ulnar neuropathy. *Ann Neurol* 7:289–290, 1980.
6. Campbell, W., Pridgeion, R. M., and Sahni, S. Entrapment neuropathy of the ulnar nerve at its point of exit from the flexor carpi ulnaris muscle. *Muscle Nerve* 11:467–470, 1988.
7. Campbell, W. W., Sahni, S. K., Pridgeon, R. M., Riaz, G., and Leshner, R. T. Intra-operative electroneurography: Management of ulnar neuropathy at the elbow. *Muscle Nerve* 11:75–81, 1988.
8. Chan, R. C., Paine, K. W. E., and Varughese, G. Ulnar neuropathy at the elbow: Comparison of simple decompression and anterior transposition. *Neurosurgery* 7:545–550, 1980.
9. Childress, H. M. Recurrent ulnar nerve dislocation at the elbow. *Clin Orthop* 180:168–173, 1975.
10. Childress, H. M. Recurrent ulnar nerve dislocations at the elbow. *J Bone Joint Surg* 35A:978, 1956.
11. Dangles, C. J., and Bilos, J. Ulnar nerve neuritis in a world champion weightlifter: A case report. *Am J Sports Med* 8:443–445, 1980.
12. Del Pizzo, W., Jobe, F. W., and Norwood, L. Ulnar nerve entrapment syndrome in baseball players. *Am J Sports Med* 5:182–185, 1977.
13. Dellon, A. L. Review of treatment for ulnar nerve entrapment at the elbow. *J Hand Surg* 14A:688–700, 1989.
14. Dellon, A. L., and MacKinnon, S. E. Human ulnar neuropathy at the elbow: Clinical, electrical and morphometric correlations. *J Reconstr Microsurg* 4:179–184, 1988.
15. Dimond, M. K., and Lister, G. D. Cubital tunnel syndrome treated by long arm splintage (Abstract). *J Hand Surg* 10A:430, 1985.
16. Eaton, R. G., Crowe, J. F., and Parkes, J. C. Anterior Transposition of the ulnar nerve using a non-compressing fasciodermal sling. *J Bone Joint Surg* 62A:820–825, 1980.
17. Eckman, P. B., Peristein, G., and Altrocchi, P. H. Ulnar neuropathy in bicycle riders. *Arch Neurol* 32:130–131, 1975.
18. Eisen, A., Schomer, D., and Melmed, C. The application of F-wave measurements in the differentiation of proximal and distal upper limb entrapments. *Neurology* 27:662–668, 1977.
19. Ekerot, L. Postanesthetic ulnar neuropathy at the elbow. *Scand J Plast Reconstr Surg* 11:225–229, 1977.
20. Eversman, W. W. Entrapment and compression neuropathies. *In* Green, D. P. (Ed.), *Operative Hand Surgery*. New York, Churchill-Livingstone, 1982, p. 957.
21. Foster, R. J., and Edshage, S. Factors related to the outcome of surgically managed compressive ulnar neuropathy at the elbow level. *J Hand Surg* 6:181–192, 1981.
22. Friedman, R. J., and Chochran, T. P. Anterior transposition for advanced ulnar neuropathy at the elbow. *Surg Neurol* 25:446–448, 1986.
23. Froimson, A. I., and Zahrawi, F. Treatment of compression neuropathy of the ulnar nerve at the elbow by epicondylectomy and neurolysis. *J Hand Surg* 5:391–395, 1980.
24. Fulkerson, J. Transient ulnar neuropathy from nordic skiing. *Clin Orthop* 153:230–231, 1980.
25. Gabel, G. T., and Amadio, P. C. Reoperation of the ulnar nerve in the region of the elbow. *J Bone Joint Surg* 72A:213–219, 1990.
26. Gessini, L., Jandolo, B., Pietrangeli, A., and Occhipinti, E. Ulnar nerve entrapment at the elbow by persistent epitrochleoanconeus muscle: Case report. *J Neurosurg* 55:830–831, 1981.
27. Gessini, L., Jandolo, B., and Peitrangeli, A. Entrapment neuropathies of the median nerve at and above the elbow. *Surg Neurol* 19:112–116, 1983.
28. Godshal, R., and Hansen, C. Traumatic ulnar neuropathy in adolescent baseball pitchers. *J Bone Joint Surg* 53A:359–361, 1974.
29. Grevsten, S., Lindsjo, U., and Olerud, S. Recurrent ulnar nerve dislocation at the elbow: Report of a non-traumatic case with ulnar entrapment neuropathy. *Acta Orthop Scand* 49:151–153, 1978.
30. Hang, Y. Tardy ulnar neuritis in a little league baseball player. *Am J Sports Med* 9:244–246, 1981.
31. Harrison, M., and Nurick, S. Results of anterior transposition of the ulnar nerve for ulnar neuritis. *Br Med J* 1:293–301, 1950.
32. Hartz, C. R., Linscheid, R. L., Gramse, R. R., and Danbe, J. R. Pronator teres syndrome: Compressive neuropathy of the median nerve. *J Bone Joint Surg* 63A:885, 1981.
33. Hollinshead, W. H. *Anatomy for Surgeons,* Vol. 3 (3rd ed.). *The Back and Limbs.* New York, Harper & Row, 1982.
34. Holtzman, R. N. N., Mark, M. H., Patel, M. R., and Wiener, L. M. Ulnar nerve entrapment neuropathy in the forearm. *J Hand Surg* 9A:576–578, 1984.
35. Howard, F. M. Controversies in nerve entrapment syndromes in the forearm and wrist. *Orthop Clin North Am* 17:375–381, 1986.
36. Jobe, F. W., Stark, H., and Lombardo, S. J. Reconstruction of the ulnar collateral ligament in athletes. *J Bone Joint Surg* 68A:1158–1163, 1986.
37. Jobe, F. W., and Fanton, G. S. Nerve injuries. *In* Morrey, B. F. (Ed.), *The Elbow and Its Disorders*. Philadelphia, W.B. Saunders, 1985, p. 497.
38. Johnson, R. K., and Shrewsbury, M. M. Median nerve entrapment syndrome in the proximal forearm. *J Hand Surg* 4:48–51, 1979.
39. Kamhin, M., Ganel, A., Rosenburg, B., and Engel, J. Anterior transposition of the ulnar nerve. *Acta Orthop Scand* 51:475–448, 1986.
40. Kiloh, L. G., and Nevin, S. Isolated neuritis of the anterior interosseous nerve. *Br Med J* 1:850, 1952.
41. Kincaid, J. C., Phillips, L. H., and Daube, J. R. The evaluation of suspected ulnar neuropathy at the elbow: Normal conduction study values. *Arch Neurol* 43:44–47, 1986.
42. King, J. W., Brelsford, H. J., and Tullos, H. S. Analysis of the pitching arm of the professional baseball pitcher. *Clin Orthop* 67:116–123, 1969.
43. Koppell, H. P., and Thompson, W. A. L. Pronator syndrome. *N Engl J Med* 259:713, 1958.
44. Larson, R. L., Singer, K. M., Bergstrom, R., and Thomas, S. Little league survey: The Eugene study. *Am J Sports Med* 4:201–209, 1976.
45. Lavyne, M. H., and Bell, W. O. Simple decompression and occasional micro-surgical enineurolysis under local anesthesia as treatment for ulnar neuropathy at the elbow. *Neurosurgery* 11:6–11, 1982.
46. Lazaro, L. III. Ulnar nerve instability: Ulnar nerve injury due to elbow lesion. *So Med J* 70:36–40, 1977.
47. Learmonth, J. R. A technique for transplanting the ulnar nerve. *Surg Gynecol Obstet* 74:792, 1942.
48. Leffert, R. D. Anterior submuscular transposition of the ulnar nerves by the Learmonth technique. *J Hand Surg* 7:147–155, 1982.
49. Loomer, R. Elbow injuries in athletes. *Can J Appl Sport Sci* 7(3):164–166, 1982.
50. Lugnergard, H., Juhlin, L., and Nilsson, B. Ulnar neuropathy at the elbow treated with decompression: A clinical and electrophysiological investigation. *Scand J Plast Reconstr Surg* 16:195–200, 1982.
51. MacNicol, M. The results of operation for ulnar neuritis. *J Bone Joint Surg* 61B:159–164, 1979.
52. MacNicol, M. F. The results of operation for ulnar neuritis. *J Bone Joint Surg* 61B:159–164, 1979.
53. McGowan, A. J. The best results of transposition of the ulnar nerve for traumatic ulnar neuritis. *J Bone Joint Surg* 32B:293–301, 1950.
54. Morris, H. H., and Peters, B. H. Pronator syndrome: Clinical and electrophysiological features in seven cases. *J Neurol Neurosurg Psychiat* 39:461, 1976.

55. Nigst, H., and Dick, W. Syndromes of compressions of the median nerve in the proximal forearm (pronator teres syndrome): Anterior interosseous nerve syndrome. *Arch Orthop Trauma Surg* 93:307–312, 1979.

56. Olney, R. K., and Wilbourn, A. J. Ulnar nerve conduction study of the first dorsal interosseous muscle. *Arch Phys Med Rehabil* 66:16–18, 1985.

57. Overpeck, D. O., and Ghormley, R. K. Paralysis of the serratus magnum: Caused by lesions of the long thoracic nerve. *JAMA* 114:1994–1996, 1940.

58. Pechan, J., and Julis, I. The pressure measurement in the ulnar nerve. A contribution to the pathophysiology of the cubital tunnel syndrome. *J Biomech* 8:75–79, 1975.

59. Rask, M. R. Anterior interosseous serve entrapment (Kiloh-Nevin syndrome). *Clin Orthop* 142:176–181, 1979.

60. Richmond, J. C., and Southmayd, W. W. Superficial anterior transposition of the ulnar nerve at the elbow for ulnar neuritis. *Clin Orthop* 164:42–44, 1982.

61. Robinson, S. C. An anomalous flexor digitorum superficialis muscle-tendon unit associated with ulnar neuropathy: A case report. *Clin Orthop* 194:169–171, 1985.

62. Sisto, D. J., Jobe, F. W., Moynes, D. R., and Antonelli, D. J. An electromyographic analysis of the elbow in pitching. *Am J Sports Med* 15:260–263, 1987.

63. Spinner, M., and Kaplan, E. B. The relationship of the ulnar nerve to the medial intermuscular septum in the arm and its clinical significance. *Hand* 8:239, 1976.

64. Spinner, M., and Linscheid, R. L. Nerve entrapment syndromes. *In* Morrey, B. F. (Ed.), *The Elbow and Its Disorders*. Philadelphia, W.B. Saunders, 1985, p. 691.

65. St. John, J. N., and Palmaz, J. C. The cubital tunnel in ulnar entrapment neuropathy. *Radiology* 158:119–123, 1986.

66. Vanderpool, D. W., Chalmers, J., Lamb, D. W., and Whiston, T. B. Peripheral compression lesions of the ulnar nerve. *J Bone Joint Surg* 50B:792, 1968.

67. Wadsworth, T. G. The external compression syndrome of the ulnar nerve at the cubital tunnel. *Clin Orthop* 124:189–204, 1977.

68. Wertsch, J. J., and Melvin, J. Median nerve anatomy and entrapment syndromes: A review. *Arch Phys Med Rehabil* 63:623–627, 1982.

69. Wilson, D. H., and Krout, R. Surgery of ulnar neuropathy at the elbow: 16 cases treated by decompression without transposition: Technical note. *J Neurosurg* 38:780–785, 1973.

SPORTS INJURIES OF THE UPPER EXTREMITY

Biomechanics of the Elbow and Forearm

Bernard F. Morrey, M.D.

At first glance, the subject of biomechanics may seem complex and unintelligible to many orthopaedic surgeons. If discussed according to the typical joint functions familiar to the clinician's definition, this important body of information may seem more relevant. We will therefore discuss the topic of biomechanics of the elbow along these lines: joint motion, stability, and force-strength. The discussion closes with a consideration of the biomechanical features of the forearm.

JOINT MOTION

The elbow joint is considered a trochleogiglamoid joint; that is, it possesses two degrees of freedom: flexion-extension and forearm rotation [26]. There are certain special features of this articulation that account for the motion attributed to it. The joint surfaces of the humerus, radius, and ulna have specific and discrete characteristic orientations with regard to the long axis of each. Recognition of this orientation is sometimes important for proper treatment of some fractures about the elbow.

Humeral Articulation

The articular surface of the humerus is rotated anteriorly approximately 30 degrees in reference to the long axis of the humerus, as viewed in the lateral projection. Although there is considerable individual variation, a mean of 5 to 7 degrees of valgus tilt of the axis of rotation is thought to be normal. Viewed on end, the distal humeral articulation is also externally rotated approximately 3 to 5 degrees in reference to the posterior surface of the medial and lateral supracondylar columns [20].

Articular Surface

The articular surface of the distal humerus is a complex one, with the trochlea comprising medial and lateral elements. Cartilage covers the trochlea through an arc of approximately 300 degrees. The capitellum is almost a perfect geometric hemisphere, with an arc of curvature of approximately 180 degrees in both the medial lateral and proximal distal meridians. The orientation of this semicircle is directly anterior; thus, a fixation plate can be contoured and applied to the posterior-inferior aspect of the lateral column without violating the articular surface of the capitellum [20]. The thickness of the articular cartilage of the distal humerus is only 2 to 3 mm; as such, it is one of the thinnest articular surfaces of the human body. This is consistent with the observation that the more congruous the joint, the thinner the articular surface [32]; the elbow is considered one of the most congruous articular joints in the body.

Ulna

The articular surface of the ulna makes approximately a 5- to 7-degree valgus angulation with regard to the long axis of the shaft when viewed on the anteroposterior projection. Viewed laterally, the articular surface has a posterior opening that is oriented approximately 30 degrees with regard to the long axis, and, as such, it matches well with the anterior rotation of the distal humeral articulation (Fig. 17A–1). This observation has not been emphasized, but this particular articular orientation allows the elbow to be stable when completely straight, an important function for some athletic requirements.

Articular Surface

The ulnar articulation is composed of two components. The greater sigmoid fossa (often referred to as the olecranon) consists of the coronoid process distally and the olecranon process proximally. The arc of curvature of this articulation is approximately 185 degrees. The articular surface is thin, measuring 2 to 3 mm, as occurs on the humeral counterpart. An important anatomic feature is the lack of artic-

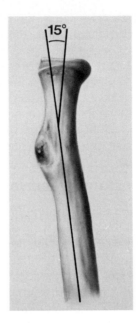

FIGURE 17A–2
The radial head and neck deviate 15 degrees away from the radial tuberosity. This allows a full arc of forearm rotation without translation at the radiohumeral joint. (With permission of the Mayo Foundation.)

ular cartilage at the center of the greater sigmoid fossa [20]. This suggests that the contact area of this joint primarily resides on the anterior coronoid and posterior olecranon surfaces. This has, in fact, been shown to be the case [33]. Recognition of the nonarticular midportion of the greater sigmoid fossa is an important feature of this joint. When investigating the ulnohumeral joint arthroscopically, care must be taken not to interpret this normal variation as articular pathology.

On the lateral side of the greater sigmoid fossa is the lesser fossa, which accommodates the radial head. The arc of curvature of this articulation is about 70 degrees.

Radial Head

The radial head and neck are oriented on an angle of approximately 15 degrees opposite the radial shaft and away from the radial tuberosity (Fig. 17A–2). This angular relationship allows the forearm to undergo an arc of forearm rotation approximating 180 degrees while still maintaining a very precise and constant orientation in regard to the capitellum.

Articular Surface

The articular surface of the radial head consists of a disc-shaped impression with an arc of curvature

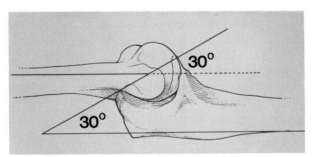

FIGURE 17A–1
The 30-degree anterior rotation of the distal humerus is matched by the 30-degree opening of the greater sigmoid notch, thus providing stability to the elbow in full extension.

of 40 degrees [20]. The margin of the radial head is covered by articular cartilage through an arc of about 240 degrees and articulates with the lesser sigmoid fossa. The remaining 120 degrees that comprise the nonarticular portion of the radial head are thought to be weaker because there is no subchondral bone in this region. This may account for the predilection toward slice fractures of the radial head that typically occur through this region.

Three-Dimensional Passive Motion

The three-dimensional motion in the elbow joint has been carefully studied with a biplanar radiographic technique [15, 26]. This study revealed that the carrying angle typically undergoes a linear change from valgus to varus during elbow flexion (Fig. 17A–3). Furthermore, there is a slight axial rotation of the ulna internally during the initiation of flexion and then externally at the completion of flexion. These patterns are due to the geometric characteristics of the ulnohumeral joint. The subtle axial rotation of the ulna may account for the radial location of the osteophyte that develops on the tip of the olecranon in the throwing athlete due to repetitive elbow extension [34].

Axis of Rotation

The locus of the instant center of rotation of the elbow is less than 3 mm in its widest dimension [10]. Hence, elbow flexion motion may be considered primarily as a spinning motion for all practical purposes. As such, the hinge axis may be approximated by a line that pierces the lateral projection of the center of the capitellum and passes through the center of the trochlea (Fig. 17A–4). This line emerges at the anterior-inferior aspect of the medial epicondyle.

Axis of Forearm Rotation

The forearm rotation axis passes directly through the center of the radial head and thus through the center of the capitellum in both the anteroposterior and lateral projections [19]. This axis then traverses the interosseous membrane and emerges through the base of the styloid process of the ulna, at the center of curvature of the distal ulna.

Normal Elbow Motion

The normal arc of elbow flexion is 0 to 145 degrees, with considerable individual variation [8].

Lax-jointed individuals may hyperextend 10 or more degrees, and well-muscled individuals may flex only 130 degrees.

Pronation and supination arcs are usually not of equal magnitude. Pronation averages about 80 degrees, which is 5 to 10 degrees less than supination, which averages about 85 degrees. Thus, normal forearm rotation is not typically a full arc of 180 degrees but averages approximately 165 to 170 degrees [8].

Measurement of flexion-extension is one of the easiest joint measurements in the body. Hand-held goniometers are accurate to within 5 degrees. To obtain follow-up information about this function, we have found that motion may be accurately estimated by tracing the silhouette in full flexion and full extension. Goniometric measurements of these tracings are comparable to in vivo measurements.

Carrying Angle

The carrying angle varies both as a function of age, being less in children than in adults, and according to sex, with females averaging 3 to 4 degrees greater than males [7, 17]. The carrying angle is formed by the valgus tilt of the axis of rotation (humeral articulation) and the valgus orientation of the ulnar shaft in reference to the axis of the greater sigmoid fossa [35]. The normal distribution of this angle varies greatly and averages approximately 10 degrees in the male and 13 degrees in the female. The proper definition of the carrying angle is the orientation of the forearm in reference to the humerus when the elbow is in full extension. The concept loses its significance as the elbow flexes [3].

Functional Motion

Like many joints in the body, the elbow does not necessarily use the full available arc of motion for routine daily activities. One study has shown that activities involving personal hygiene and daily sustenance are accomplished with an arc of 30 to 130 degrees of flexion and with 50 degrees of pronation and 50 degrees of supination [25]. For the athlete, most functional motion requirements of the various sports have not been previously addressed. It is known, however, that the thrower frequently develops a flexion contracture of up to 10 degrees; this does not impair function. It has also been shown that pitchers tend to develop an increased carrying angle compared to the opposite side. This also is not known to affect performance adversely [34].

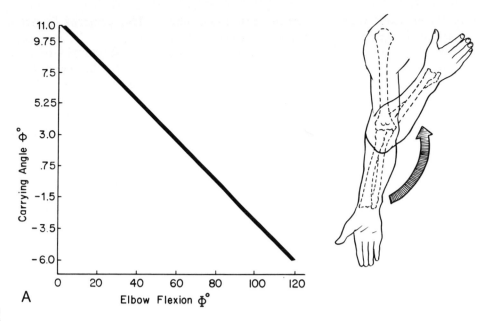

FIGURE 17A–3

A, During elbow flexion, the carrying angle changes in a linear fashion from valgus to neutral or varus. *B,* The ulna undergoes a slight axial rotation during flexion and extension. (With permission of the Mayo Foundation.)

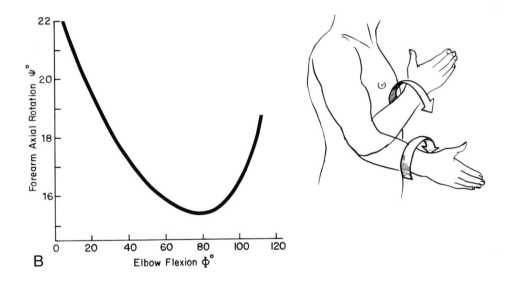

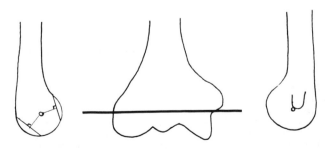

FIGURE 17A–4

The axis of rotation of the elbow is approximated by a line through the middle of the lateral epicondyle, through the center of the trochlea, and emerging in the anterior infra-aspect of the medial epicondyle.

Greater motion might be anticipated from some athletic activities, such as gymnastics. Less motion, however, is associated with increased muscle mass, as might be seen with body builders. Once again, the functional requirements of the elbow motion for the athlete as a function of the particular sport have not been well defined.

ELBOW STABILITY

The stability of any joint is composed of three major elements: the articular surface, the capsular and ligamentous structures, and the muscles. Unlike the shoulder, the dynamic contribution of the musculature to elbow stability under normal circumstances is minimal. The elbow relies on the contributions of the static constraints: the articular surface and the capsular ligamentous complex.

Articular Contribution to Stability

Accurate measurement of the relative contribution of the articular surface to joint stability during normal use is complex and has not been satisfactorily addressed to date.

The Ulnohumeral Joint

Resection of successive portions of the ulnar articulation decreases the stability of that joint in proportion to the amount that has been removed [4]. Hence, a 25% resection of the olecranon causes a 25% reduction in ulnohumeral stability. It must be noted that these data do not account for the presence of the stabilizing effect of the triceps, which, while normally not a major stabilizer, does assume this function after partial resection of the olecranon. Although the subject has not been formally studied, clinical information reveals that gross posterior elbow instability is present with loss of the coronoid (type III fracture), and posterior instability may be present with partial fracture of the coronoid (type II fracture) [31]. The clinical implication of these data is extremely important when managing individuals with coronoid fractures because the goals of such treatment are to ensure stability and provide functional motion. This dilemma is not easily resolved and sometimes requires the use of external fixators and continuous motion machines. Return to high-level competition is, of course, compromised.

The Radial Head

The contribution of the radial head to elbow stability has been investigated indirectly by calculating the amount of force transmitted across this articulation during various loading configurations (see later discussion). These studies have revealed that about 20% to 30% of an applied valgus load is transmitted through and thus resisted by the radiohumeral joint [13, 21, 30]. Realistically, it is difficult to ascribe a specific value of the radial head to resist valgus stress without considering the associated ligamentous structures (Fig. 17A–5) [21]. With space-age technology, we can now study the interrelation between the radial head and the medial collateral ligament to resist valgus stress at the elbow by the use of an electromagnetic, three-dimensional, motion-tracking device. The relative contributions of the radial head and the ligamentous structures were recently demonstrated by a technique of serial release [24]. Contrary to previous reports, these data clearly show that, with simulated flexion and extension, the radial head does not resist physiologic valgus stresses in the presence of an intact medial collateral ligament (Fig. 17A–6). If the medial collateral ligament has been removed or violated, however, the radial head plays a major role in resisting the valgus torque. If both the medial collateral ligament and the radial head are removed, the elbow predictably dislocates.

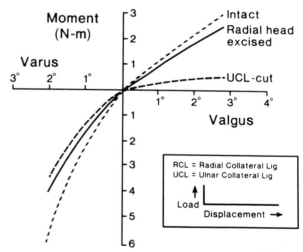

FIGURE 17A–5

For a given amount of displacement, approximately a 25% decreased load is observed in valgus stress after the radial head has been removed. This suggests that, at least with this loading configuration, the radial head contributes approximately 25% to the load-carrying capacity or to the resistance of a valgus load.

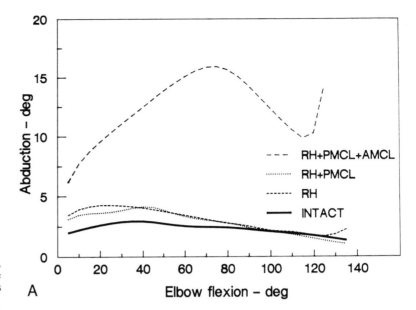

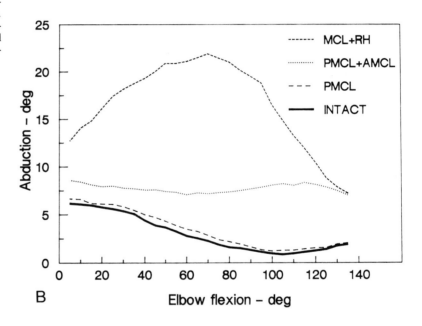

FIGURE 17A–6
A, A quasi-dynamic model with serial sections, removing the radial head (RH) but leaving the medial collateral ligament (MCL) intact, shows no alteration in normal elbow kinematics. *B,* When the medial collateral ligament is first removed, the radial head is observed to provide some resistance to valgus stress. When both constraints are released, the elbow is grossly unstable. (PMCL, posterior bundle medial collateral ligament; AMCL, anterior bundle medial collateral ligament) (With permission of the Mayo Foundation.)

These data may be summarized in a format familiar to the sports medicine physician when considering the relation between the meniscus and the anterior cruciate ligament to knee stability. The radial head has a role similar to the meniscus, and the medial collateral ligament is analogous to the anterior cruciate ligament. In the presence of an anterior cruciate ligament, the menisci play little role in stability. Similarly, the radial head may be considered a secondary stabilizer in valgus elbow instability. Its contribution is seen only if the medial collateral ligament complex is deficient (Fig. 17A–7).

Capsuloligamentous Complex

Medial Collateral Ligament

Experimental studies have emphasized the role of the anterior bundle of the medial collateral ligament in resisting valgus stability [22]. Acute rupture of the medial complex may be, therefore, effectively addressed by reconstructing or repairing the anterior bundle of this complex. The dynamic function of the ligament, however, must be recognized. The anterior and posterior elements of the medial collateral ligament both originate a few millimeters off

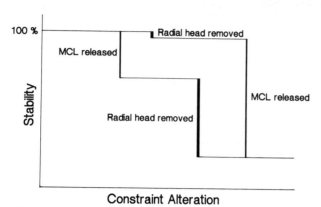

FIGURE 17A–7
The radial head provides a secondary constraint to a valgus stress but is unimportant when the medial collateral ligament (MCL) is intact. (With permission of the Mayo Foundation.)

the axis of rotation for the elbow (Fig. 17A–8). As such, a cam effect is present because the posterior bundle is taut only in flexion. The anterior bundle is somewhat taut throughout the arc of motion; the anterior fibers are most taut in extension, and the posterior bundles become tightened in flexion. Hence, the role of the anterior portion of the medial collateral ligament is not unlike that of the anterior cruciate at the knee. These experiments specifically show that the anterior bundle is the essential com-

ponent of the medial complex. Given this fact, precise restoration of the humeral origin of the ligament must be attained with ligament reconstruction procedures [16].

Lateral Collateral Ligament Complex

A thorough description of the lateral collateral ligament complex has only recently emerged. The traditional concept of the radial collateral ligament originating from the humerus and attaching to the annular ligament has been expanded. A discrete portion of the lateral collateral complex has now been demonstrated to originate from the humerus, to be superficial to the annular ligament, and to attach on the tubercle crista supinatoris of the ulna [22]. This component of the lateral collateral ligament has been termed the lateral ulnar collateral ligament and should be distinguished from the radial collateral ligament (Fig. 17A–9). The existence of this particular structure explains the residual varus stability of the elbow after the radial head has been excised. Because the radial collateral ligament becomes lax as the annular ligament loses its tension, resistance to varus stress is provided by the lateral ulnar collateral ligament. It is, therefore, believed

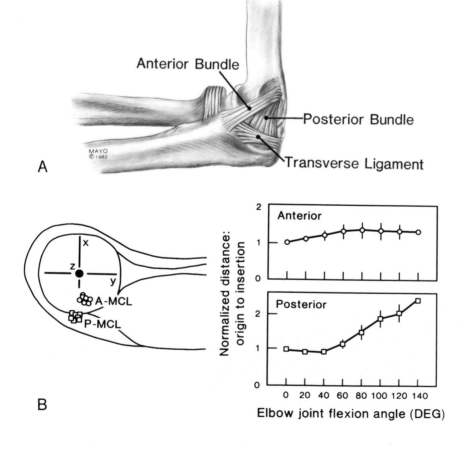

FIGURE 17A–8
A, Anatomic distribution of the medial collateral ligament. *B,* The origins of the anterior and posterior bundles do not lie along the axis of rotation. Thus, some change in length of these ligaments is seen as they function with changes in elbow flexion angle. (With permission of the Mayo Foundation.)

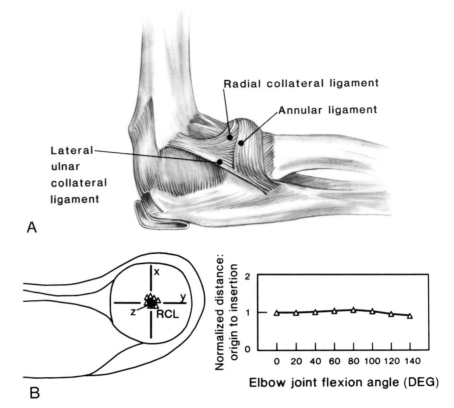

FIGURE 17A–9
A, The medial collateral ligament complex consists not only of the radial collateral ligament but also of a lateral ulnar collateral ligament. *B,* With flexion, there is no change in length of the radial collateral ligament complex. This suggests that the origin is at the axis of rotation. (With permission of the Mayo Foundation.)

that the most appropriate concept of the collateral ligament stability of the elbow is that of a medial and a lateral ulnar collateral ligament complex that functions independently in the presence or absence of a radial head (Fig. 17A–10). This concept is consistent with the anatomic findings as well as with clinical experience.

Unlike the medial collateral complex, the radial collateral ligament originates in the precise center of the axis of rotation of the elbow joint (see Fig. 17A–9). This anatomic and biomechanical fact is important when a lateral collateral ligament reconstruction is undertaken.

Relative Contribution of the Collateral Ligaments

Cadaver specimens have been employed to test the relative contribution of the collateral ligaments and articular surface to elbow stability in flexion and extension [21] (Tables 17A–1 and 17A–2). The data may be simply summarized by stating that the articular surfaces provide approximately 50% of elbow stability, and the collateral ligaments provide the remaining 50%. When the elbow is in full extension, the anterior capsule contributes about 15% of the resistance to varus-valgus stress [21].

FIGURE 17A–10
The ulnohumeral joint is stabilized by both medial and lateral constraints, independent of the presence of the radial head.

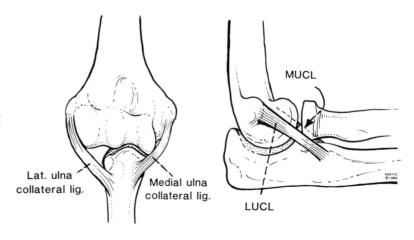

TABLE 17A–1
Elbow Stability: Extension*

	Contribution (%)	
	Varus	*Valgus*
Ligaments		
Lateral collateral	25	—
Medial collateral	—	40
Capsule	15	25
Articulation	60	35

*With permission of the Mayo Foundation.

FORCES ACROSS THE ELBOW JOINT

Accurate determination of forces across the joint is difficult to describe with certainty due to the indeterminate nature of the system; that is, there is more estimation than independent variables. Nonetheless, a rather accurate estimation of the resultant forces across the joint may be predicted with analytical models.

Muscle Contribution

The contribution of the muscles to transmission of force across the joint is essential to estimate or calculate the force at a joint. Muscle function is based on three features: cross-sectional area, orientation or line of action, and specific activity during a given function. For the elbow, all three characteristics of muscle size, orientation, and activity with certain functions have been defined [2, 6, 11, 14]. This information allows one to estimate the resultant force at the elbow in certain positions.

The greatest amount of force generated at the elbow occurs with the initiation of flexion. The flexion moment arm of the muscles is shortened, and the mechanical advantage is decreased, so that greater force of contraction is required. Up to three times body weight can theoretically be transmitted

TABLE 17A–2
Elbow Stability: Flexion of 90 Degrees*

	Contribution (%)	
	Varus	*Valgus*
Ligaments		
Lateral collateral	45	—
Medial collateral	—	55
Articulation	55	45

*With permission of the Mayo Foundation.

to the elbow under maximal loading conditions [1, 23]. The resultant vector is directed axially toward the humeral head. Greater strength of flexion is generated with the elbow in 90 degrees of flexion, however, because the mechanical advantage of the elbow flexors has improved from the fully extended position, so that less force is actually applied to the joint. Thus, the actual force across the elbow is less in this position than in full extension, and the resultant force is deviated posterior and superiorly. It has been estimated that approximately three times body weight may be transmitted across the elbow joint when it is flexed at 90 degrees. It is important to note that with elbow extension and flexion, the resultant vector also undergoes a change in direction (Fig. 17A–11) [29]. Through an arc of 140 degrees of flexion and extension, the direction of the resultant force vector changes by more than 180 degrees. The significance of this finding is that in management following elbow fracture or injuries resulting in elbow instability the physician should consider the significant forces being directed through a wide arc of motion. These large forces are responsible for the frequently observed loss of fixation of distal humeral fractures.

Distributive Forces

Force across the ulnohumeral joint is distributed in a considerably different fashion, depending on

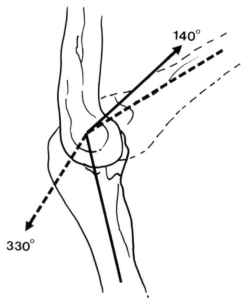

FIGURE 17A–11
With flexion and extension, the resultant force of the elbow undergoes a cyclic change in direction. The order of magnitude can be as much as three times the body weight.

whether the elbow is flexing or extending [1]. During flexion, the biceps and brachialis tend to pull the ulna posteriorly, thus generating contact at the coronoid. The coronoid process is separated into a medial and lateral portion by a central ridge [33]. Thus, the contact forces during flexion occur on three surfaces: the medial and lateral aspects of the coronoid, and the radial head. Although the force is transmitted to three articulations, these comprise a relatively small articular contact area.

During initiation of flexion from an extended position, the contact occurs at the anterior aspect of the articulation. At full flexion, the contact is at the olecranon. With extension from full flexion, most force crosses the anterior articulation. The magnitude of these forces has not been accurately identified. With forced extension, impaction of the olecranon process has been recognized in the throwing athlete, leading to development of an osteophyte at the tip of the olecranon and resulting in a classic impingement syndrome [34].

Radiohumeral Joint

The load-carrying characteristics of the radiohumeral joint have been previously described as they relate to the stabilizing contribution of the radial head. An early study using a direct measurement technique revealed that, with the elbow in full extension, approximately 40% of the applied force is transmitted across the ulnohumeral joint and 60% is transmitted across the radial head [12]. Release of the interosseous membrane did not affect the measurements (Fig. 17A–12). This experiment was repeated by Walker, who used a more refined technique and reported identical findings [36]. Release of the distal radioulnar attachments, however, increased the force transmission at the radial head. Further studies of the load-bearing characteristics of the radiohumeral joint using the MTS machine suggest that, in a valgus-loading configuration, approximately 25% to 30% of the torque occurs at this articulation [13, 30]. Using a more physiologic model of active joint loading in which muscle flexion and extension were simulated, the greatest amount of force across the radiohumeral joint was shown to occur in extension, as previously stated. The maximum amount of force across the radiohumeral articulation, however, approaches but does not exceed body weight [21]. Finally, with varus-valgus stress, the load pivots medially and laterally at a point located in the center of the lateral face of the trochlea.

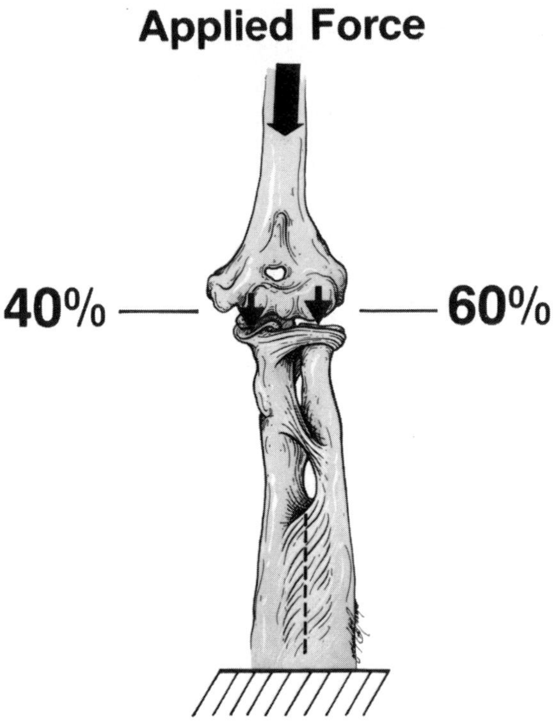

FIGURE 17A–12
The distributive forces with the elbow in full extension and an axial load placed at the wrist show that about 60% of the force goes across the radiohumeral joint and 40% goes across the ulnohumeral joint. This does not change the section of the interosseous membrane. (With permission of the Mayo Foundation.)

ELBOW STRENGTH

The subject of strength function at the elbow joint is particularly important for rehabilitation of the athlete. The normal strength pattern of elbow function has been reported for approximately 100 normal subjects equally divided between males and females [5]. These data are shown in Figure 17A–13.

It is noted that extension strength is only approximately 70% of flexion strength. This is consistent with predictions based on the cross-sectional area and the moment of the flexor and extensor muscle mass. It is further shown that supination strength is approximately 20% to 30% greater than pronation strength, once again owing to the contribution of the biceps muscle to supination. The difference between the dominant and nondominant extremity varies from 4% to 7%, depending on the function studied. The difference between males and females also is rather consistent, females being approximately 60% as strong as males in all functions.

The question of whether static or endurance strength loss should prevent a patient returning to athletics is a very relevant one, but answers are still

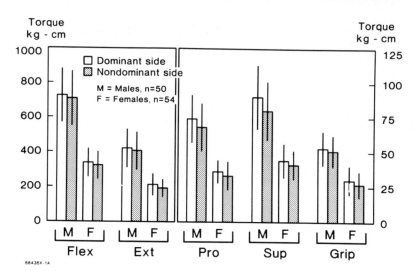

FIGURE 17A–13
Strength variations for different functions of approximately 50 male and 50 female subjects show distinctions between the dominant and nondominant extremities. (With permission of the Mayo Foundation.)

lacking. Unfortunately, no correlation between isometric and isokinetic endurance strength exists, so isokinetic techniques are probably necessary to define endurance. Such testing must be done with proper protocols so that the particular deficiency is accurately identified. Variables such as rate of testing, position of the extremity, motivation, test-retest fatigue, and conditioning are just some of the problems inherent in such testing [5, 9, 18, 27].

References

1. Amis, A. A., Dowson, D., and Wright, V. Elbow joint force predictions for some strenuous isometric actions. *J Biomech* 13:765–775, 1980.
2. An, K. N., Hui, F. C., Morrey, B. F., Linscheid, R. L., and Chao, E. Y. Muscles across the elbow joint: A biomechanical analysis. *J Biomech* 14:659–669, 1981.
3. An, K. N., Morrey, B. F., and Chao, E. Y. S. Carrying angle of the human elbow joint. *J Orthop Res* 1:369–378, 1984.
4. An, K. N., Morrey, B. F., and Chao, E. Y. S. The effect of partial removal of the proximal ulna on elbow constraint. *Clin Orthop* 209:270–279, 1986.
5. Askew, L. J., An, K. N., Morrey, B. F., and Chao, E. Y. S. Isometric elbow strength in normal individuals. *Clin Orthop* 222:261–266, 1987.
6. Basmajian, J. V., and Latif, S. Integrated actions and functions of the chief flexors of the elbow. *J Bone Joint Surg [Am]* 39:1106–1118, 1957.
7. Beals, R. K. The normal carrying angle of the elbow. *Clin Orthop* 119:194, 1976.
8. Boone, D. C., and Azen, S. P. Normal range of motion of joints in male subjects. *J Bone Joint Surg [Am]* 61:756, 1979.
9. Currier, D. P. Maximal isometric tension of the elbow extensors at varied positions, II: Assessment of extensor components at quantitative electromyography. *Phys Ther* 52:1265–1276, 1972.
10. Fischer, G. *Handbuch der Anatomie und Mechanik du Gelenke, unter Berucksichtigung der Bewegenden Muskeln.* Jena, 1911.
11. Funk, D. A., An, K. N., Morrey, B. F., and Daube, J. R.

An EMG analysis of muscles controlling elbow motion. *J Orthop Res* 5:529–538, 1989.
12. Halls, A. A., and Travill, R. Transmission of pressures across the elbow joint. *Anat Rec* 150:243, 1964.
13. Hotchkiss, R. N., and Weiland, A. J. Valgus stability of the elbow. *J Orthop Res* 5:372–377, 1987.
14. Hui, F. C., Chao, E. Y., and An, K. N. Muscle and joint forces at the elbow during isometric lifting (abstr). *Orthop Trans* 2:169, 1978.
15. Ishizuki, M. Functional anatomy of the elbow joint and three-dimensional quantitative motion analysis of the elbow joint. *Journal of the Japanese Orthopedic Association* 53:989–996, 1979.
16. Jobe, F. W., and Fanton, G. S. Nerve injuries. In Morrey, B. F. (Ed.), *The Elbow and Its Disorders.* Philadelphia, W. B. Saunders, 1985, pp. 497–501.
17. Keats, T. E., Tuslink, R., Diamond, A. E., and Williams, J. H. Normal axial relationship of the major joints. *Radiology* 87:904, 1966.
18. Larson, R. F. Forearm positioning on maximal elbow-flexor force. *Phys Ther* 49:748–756, 1969.
19. London, J. T. Kinematics of the elbow. *J Bone Joint Surg [Am]* 63:529–535, 1981.
20. Morrey, B. F. Anatomy of the elbow. In Morrey, B. F. (Ed.), *The Elbow and Its Disorders.* Philadelphia, W. B. Saunders, 1985, pp. 7–42.
21. Morrey, B. F., and An, K. N. Articular and ligamentous contributions to the stability of the elbow joint. *Am J Sports Med* 11:315, 1983.
22. Morrey, B. F., and An, K. N. Functional anatomy of the ligaments of the elbow. *Clin Orthop* 201:84–90, 1985.
23. Morrey, B. F., An, K. N., and Stormont, T. J. Force transmission through the radial head. *J Bone Joint Surg [Am]* 70:250–256, 1988.
24. Morrey, B. F., An, K. N., and Tanaka, S. Valgus stability of the elbow: A definition of primary and secondary constraints. Accepted for publication, *Clin Orthop* 265:187, 1991.
25. Morrey, B. F., Askew, L. J., An, K. N., and Chao, E. Y. A biomechanical study of functional elbow motion. *J Bone Joint Surg [Am]* 63:872–877, 1981.
26. Morrey, B. F., and Chao, E. Y. S. Passive motion of the elbow joint. *J Bone Joint Surg [Am]* 59:501–508, 1976.
27. Nicol, A. C., Berme, N., and Paul, J. P. A biomechanical analysis of elbow joint function. *In Joint Replacement in the Upper Limb.* London, Institute of Mechanical Engineers, 1977, p. 45.

28. O'Driscoll, S. W., Bell, D. F., and Morrey, B. F. Postero-lateral rotatory instability of the elbow: Clinical and radiographic features. Accepted for publication, *J Bone Joint Surg* 73A:440, 1991.

29. Pearson, J. R., McGinley, D. R., and Butzel, L. M. A dynamic analysis of the upper extremity: Planar motions. *Hum Factors* 5:59, 1963.

30. Pribyl, C. R., Kester, M. A., Cook, S. D., Edmunds, J. O., and Brunet, M. E. The effect of the radial head and prosthetic radial head replacement on resisting valgus stress at the elbow. *Orthopedics* 9:723–726, 1986.

31. Regan, W., and Morrey, B. F. Fractures of the coronoid process of the ulna. *J Bone Joint Surg [Am]* 71:1348–1354, 1989.

32. Simon, W. H., Friedenberg, S., and Richardson, S. Joint congruence. *J Bone Joint Surg [Am]* 55:1614–1620, 1973.

33. Stormont, T. J., An, K. N., Morrey, B. F., and Chao, E. Y. Elbow joint contact study: Comparison of techniques. *J Biomech* 18:329–336, 1985.

34. Tullos, H. S., Erwin, W., Woods, G. W., Wukasch, D. C., Cooley, D. A., and King, J. W. Unusual lesions of the throwing arm. *Clin Orthop* 88:169, 1972.

35. von Lanz, T., and Wachsmuth, W. *Praktische Anatomie.* Berlin, Springer-Verlag, 1959.

36. Walker, P. S. *Human Joints and Their Artificial Replacements.* Springfield, IL, Charles C Thomas, 1977, pp. 182–183.

Fractures About the Elbow in Sports and Their Sequelae

FRACTURES

Bernard F. Morrey, M.D.
William D. Regan, M.D.

Fractures of the elbow in sports are classified the same way as fractures in any post-traumatic injury. The majority of the fractures about the elbow in sports are a result of direct trauma, but occasionally they may occur from indirect activities, such as a fall onto an outstretched arm with a twisting motion to the elbow itself. As with any fracture, it is extremely important to manage the soft tissues as delicately as the bone structure and to obtain anatomic restoration of the osseous structures involved. If the fracture and soft tissue are not properly managed, limitation in range of motion will certainly result. It is also important to pay strict attention to both the elbow and wrist joints in the assessment of forearm and radial head fractures. If these structures are not carefully evaluated, a Monteggia or Galeazzi fracture may be overlooked in the course of management.

An open approach with reduction and internal fixation using a compression plate is most typically performed in the mature individual. This allows anatomic restoration of the osseous architecture and early functional activity. The definition of maturity is important and somewhat arbitrary. A closed physis in female patients over age 14 years and in males over age 15 years is considered mature.

The long-term sequelae of fractures about the elbow in an athlete are mainly related to the successful restoration of full function without pain. In the majority of cases of fractures of the long bone, management is carried out such that full function is easily attained. Unfortunately, the characteristics of fractures or soft tissue injury about the elbow often defy even the most meticulous and brilliant surgeon, with a resultant loss in elbow motion. Most serious injuries about the elbow leave a residual flexion deformity, usually ranging between 5 and 15 de-

grees. Although this is fully acceptable for normal activities of daily living, for an athlete who requires absolute flexion and extension of the elbow, this limitation can be career ending. Consequently, it is important for any athlete sustaining a fracture about the elbow to be told specifically about the complications of elbow stiffness.

The classification and management of various fractures about the elbow and forearm are covered in many orthopaedic fracture textbooks. The majority of these are not specifically covered in this chapter, other than those mentioned in the section on elbow dislocations and medial epicondylar fractures found in throwing injuries. As already mentioned, in the context of this presentation, some fractures are worthy of specific discussion.

STRESS FRACTURES OF THE ULNA

Stress fractures in general were first brought to the attention of the orthopaedic world by Evans in 1955 [9]. He described the fracture as a lifting injury, with the ulna flexed to 90 degrees and the forearm either supinated or pronated.

In a more recent report, stress fractures were observed in a softball pitcher throwing underarm and in a volleyball player who experienced pain during underhand maneuvers [13, 17]. Both actions involve movements of the unilateral upper limb with a light load following extreme contraction of the flexor muscles. In the volleyball player, where wrist flexion was more pronounced, the stress was more proximal; in the softball player, it was more distal. Symptoms included pain after strenuous activity in the involved athletic endeavor. In both cases the physical examination revealed swelling, tenderness, local heat over the involved section of the ulna, and pain with movements. Radiographs demonstrated an insignificant crack in the cortex of the ulna, with a thickening of the cortex around it. Radiographic features demonstrated a slight haze of new periosteal bone over the region of the stress fracture. Neither of these patients had a bone scan, but this may be helpful for diagnostic purposes if the physis has closed.

The management of this injury is nonoperative and consists of prohibition of sporting activities until clinical and radiographic evidence of fracture union exists. Activities of daily living are allowed, and complete freedom of pain and resumption of full activities are to be expected.

FRACTURES OF THE HUMERUS IN ARM WRESTLERS

Arm wrestling, also known as Indian or wrist wrestling, has been a popular activity for many years. This sport has gained national and international recognition, and competition is occurring on all levels. With an increase in participation, there have been more cases of patients sustaining fractured humeri during competition. The injury is relatively rare, with only 41 cases reported in the English medical literature [14, 15, 20, 22, 26].

Patient Profile

The usual age of a patient with this injury is the mid-twenties. The dominant extremity is exclusively involved for obvious reasons, and patients are engaged in active competition. In competition, the patient generally is in a winning position, with the arm internally rotated past the vertical position, when the arm suddenly gives way with a snap. By and large, patients are not neurovascularly compromised, but in case reports by Brismar and Spangen [2], one of two patients had a radial nerve paresis. In another series reported by Lihna and Kurttela [14], 3 of 28 patients had a radial nerve paresis. Radiographically, the fracture consists of a spiral fracture in the mid- to distal third of the humerus (Fig. 17B–1).

Patient Management

The treatment of fractures of the humerus usually consists of closed reduction and immobilization in a

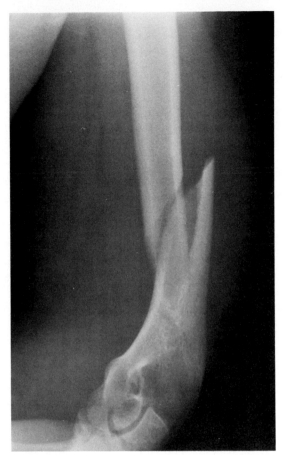

FIGURE 17B–1
Spiral fracture of the distal third of the humerus from arm wrestling.

hanging cast; this remains the optimum treatment for this fracture. In the 41 cases reported in the English literature, 17 had open reduction and internal fixation. Of these 17 patients, the usual reasons accepted for open reduction are compound fractures, fractures with vascular injury, markedly displaced segmental fractures, fractures in which interposed soft tissue prohibits reduction, and fractures associated with elbow injuries requiring early mobilization. As with any such injury, fractures with radial nerve dysfunction that occurred after closed reduction when the radial nerve had been functioning initially are suitable for exploration.

Authors' Preferred Method of Treatment

In the arm wrestler who sustains a spiral fracture of the humerus, the initial attempt is to manage the fracture with a hanging arm cast, provided that none of the conditions for open reduction and internal fixation are present. If indications for surgery do exist, I would proceed to open reduction and internal fixation with a 3.5-mm, eight-hole dynamic compression plate through an anterolateral approach to the humerus, after first identifying and protecting the radial nerve.

It is enough to say that arm wrestling is not a totally benign sport, as evidenced by the occurrence of this fracture.

FRACTURES ABOUT THE FOREARM

Forearm fractures are commonly seen in contact athletics; this seems to be particularly true of high school football. Forearm Monteggia or Galeazzi-type fractures have been observed in the athletes participating in contact sports. The treatment of these fractures follows the basic treatment principles of any forearm fracture. The major concerns relating to athletic injuries are whether the hardware should be removed and what the timing of such an intervention should be. This particular topic deserves specific attention and logically may be considered in three categories. The first is the basic science of defects in bone and the effects of plate on bone. The second has to do with the recommendations for initial treatment. The third is what might be done after the initial treatment to minimize the potential complications.

Incidence of Refracture. The incidence of refracture is variably reported and is a function of a number of variables. This complication was observed in 4 of 80 bones (5%) with plate removal by Rosson and Shearer [24], in 2 of 34 (6%) plate removals by Chapman and associates [5], and in 7 of 62 (11%) plate removals by DeLuca and colleagues [7]. Of the 7 of 333 both-bone forearm fractures (2%) with refracture reported by Anderson and associates [1], the exact number of plate removals is not known with certainty. Two of seven bones (28%) with a Galeazzi fracture had a refracture as reported by Moore and associates [16], and 6 of 23 patients (24%) sustained a refracture with at least one plate removal as reported by Hidaka and Gustilo [11].

Effect of Plates and Screws in Bone

The effects of a circular defect in the bone have been extensively studied in the orthopaedic literature [3, 4, 8]. The difficulty consists of translating basic experimental information to the clinical setting. Thus, the stress-riser effect observed by Bur-

stein and colleagues that resolved so rapidly in their experimental dog model is not applicable to the treatment of athletes [4]. It was believed that any size of drill hole caused a stress-riser effect [3, 4]; this is now known not to be the case [8]. In a careful study performed using sheep bone and a torsional model, Edgerton and associates [8] demonstrated that small defects measuring approximately 10% of the diameter of the bone did not cause any appreciable stress-riser effect. In the clinical setting, this is the type of hole that might be used to pass sutures through the proximal tibia. The size of hole used to apply the forearm plates does constitute greater than 10% of the diameter of the bone; thus, these holes do have the potential to cause a stress fracture.

Furthermore, Edgerton and associates demonstrated that the weakness in bone arising from a circular defect is in direct proportion to the size of the defect (Fig. 17B–2) [8]. Thus, the smaller the hole and plate applied, the less the stress-riser effect. This is consistent with clinical observations with regard to the selection of hardware, described below [5, 11].

Further investigations of the specific effect of the presence of the plate have clearly demonstrated the stress-shielding effect to the bone that is in immediate apposition to the plate [25]. The disuse osteoporosis that develops is in proportion to the duration of the application of the plate [6, 18, 19, 23, 25]. Furthermore, there are vascular changes that are associated with the mechanical effects of stripping the bone and applying the plate [12]. Each of these features, the disuse and the avascular phase, requires a revascularization process that can also potentially cause a weakening effect [12, 25]. Once

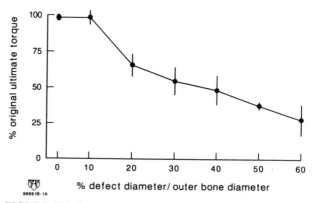

FIGURE 17B–2
Stress-riser effect in circular defects in bone. Note that small, 10% defects have little effect, whereas the remaining size of the defect weakens bone in proportion to the size of the defect. The forearm fracture is approximately a 20% type of defect, typically made to apply the 3.5 DCP plate. (From Edgerton, B. C., An, K. N., and Morrey, B. F. Torsional strength reduction due to cortical defects in bone. *J Orthop Res* 8:851–855, 1990.)

again, these basic investigations are important, relevant, and correlate with clinical observations [10, 11].

Risk Factors for Refracture

The Initial Treatment

The traditional wisdom for the treatment of forearm fractures is immediate plate fixation. Although this is appropriate for the routine case [5], there is some question as to the effect of immediate versus delayed treatment on the intermediate strength of the callus. It has been observed that patients who are treated by plate fixation 2 or fewer days after the initial injury have a higher rate of refracture than those who have delayed fixation at about a week. In fact, all of DeLuca's seven refractures occurred in bones that were plated within the first 2 days [7]. Similarly, no refracture was observed in Anderson's group if an 8-day delay occurred between the injury and the treatment [1]. Unfortunately, the pressures of the practice do not typically allow the arbitrary delay of definitive treatment, so although there may be a theoretical advantage to a delay, from a practical standpoint most of these injuries are dealt with on an immediate basis. This being the case, it is important to recognize other risk factors that may also be controlled and that can be dealt with more acceptably.

The Nature of the Injury

As with any fracture, patients with a high-energy injury, often associated with a compound wound, are known to heal more slowly. It has been demonstrated that these patients are also at risk for refracture [7]. The refracture may occur either through the screw hole or at the fracture site, and the latter two variables (high-energy injury and compound wound) predispose the fracture to recur through the previous fracture site [7].

Technique of Fixation

Compression plate fixation is the accepted mode of treatment for a forearm fracture. The adequacy of the compression, however, is somewhat variable, often relating to the degree of comminution present; an inadequate compression technique has been implicated as predisposing to refracture when the plate is removed, probably due to inadequate fracture

healing [7]. Of particular interest is the selection of the fixation device. Chapman and colleagues demonstrated that of 4 refractures in a series of 34 patients in whom the plates were removed, two occurred in individuals with a 4.5-dynamic compression plate (DCP). Only 4 patients in this series had the 4.5-DCP plate, representing a 50% chance of fracture when this plate was used and removed. Conversely, there were no refractures among 31 patients in whom a 3.5-DCP plate was used for the initial treatment and subsequently removed. This is consistent with the observations of DeLuca and associates and Hidaka and Gustilo, neither of whom reported refractures on removal of a 3.5-DCP plate [7, 11]. Hence, these data are not generally recognized but offer strong support for the use of the 3.5-DCP plate in the fixation of both-bone or single-bone forearm fractures in the athlete in whom plate removal is anticipated.

Postfracture Variables

The duration of application of the plate is probably the most important variable with regard to refracture. This is particularly true because it relates to whether the weakest portion of the system is at the site of the fracture or the screw hole. By far the majority of refractures reported in the literature occur when the plate is removed in less than a year (Fig. 17B–3). Thus, Rosson and Shearer reported that three of six patients in whom the plate was removed in less than 1 year had refracture, compared to only 1 of 42 in whom the plate was removed in greater than 1 year [24]. All seven of Anderson's patients who experienced a refracture were in the group who had plate removal less than 1 year from the time of application [1]. In most instances, the time of plate removal was dictated by social or personal preference, and thus it may range anywhere from less than a year to 5 years [11]. In the athlete, the time of plate removal often revolves around the athletic season or cycle. The logical option is to remove the plate after the time of high-risk contact has passed and a period of time is anticipated and can be allowed before additional contact or stresses take place.

The question of plate removal is a legitimate one, and some recommend leaving the plate intact while an individual continues to play contact sports; this is our preference unless at least a 9-month waiting period can be realized between plate removal and the next stressful event. Lacking this, we prefer to leave the plate in place until the patient has finished his or her contact sport or career (Fig. 17B–4). This

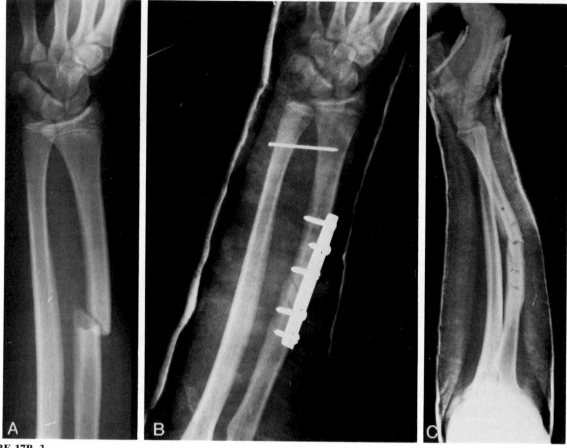

FIGURE 17B–3
A, Galeazzi fracture in a 16-year-old athlete treated by *(B)* DCP plate with distal radioulnar pin. The plate was removed at 11 months, and *(C)* the fracture recurred at the previous site.

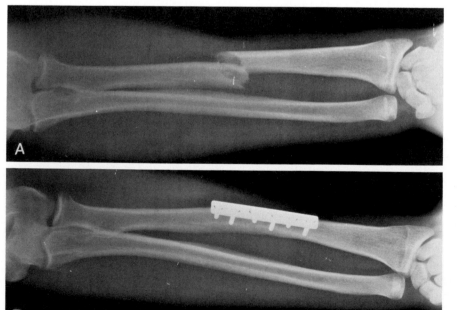

FIGURE 17B–4
A, Typical Galeazzi fracture in a 16-year-old high school football player. *B,* This fracture was treated within 24 hours by a DCP plate. The plate was left intact and the patient was allowed to play baseball 6 months later. The plate was left intact for the subsequent football season and will be removed at the completion of his contact sport career.

concept is in some conflict with the basic science information, which demonstrates continued osteoporosis the longer the plate is left in place; thus, Uhthoff and Finnegan recommend removing the plate as soon as union has taken place [25]. Although this may be acceptable wisdom in the traditional sense, in the athlete this is not, in our judgment, appropriate. The plate should be left in place a minimum of 1 year, regardless of whether union appears to take place; in a patient participating in contact sports, it may be left on until the patient's athletic career has ended, particularly in the younger individual.

The explanation for this approach is based on the causes of refracture. There are two basic time periods during which a patient may be at risk; the first is when the bone undergoes a vascular change immediately under the plate [18, 25]. In this setting, the main risk factor is the bone holes. As the bone remodels after plate removal, however, a revascularization process takes place. This may take up to a year, and during this time the bone and the screw holes are at risk [12]. Some believe that early refracture after plate removal is due to the avascular process and that late refracture is due to the revascularization process [11].

Time of Refracture

If the plate is to be removed, the next logical question concerns when the patient is at risk of refracturing and what might be done to prevent this. Refracture has been reported anywhere from 14 days to more than 1 year after plate or screw removal. Two large recent series have demonstrated that refracture occurs between 40 and 120 days after plate removal in DeLuca and associates' experience of 62 bones, whereas Hidaka and Gustilo report six refractures among 23 patients undergoing at least one plate removal. The time frame of refracture in these six patients was between 14 to 120 days [7, 11].

Site of Refracture

An additional consideration that has not been emphasized to any great extent in the literature relates to the precise site of the refracture. This information places the clinician in a better position to anticipate, protect, and avoid this complication. The previous fracture is the most common site of refracture (see Fig. 17B–3). Thus, combining the data from several series, 75% of the fractures oc-

curred at the fracture site, leaving only one in four occurring at a stress-riser through the screw hole [1, 11, 16, 24].

Authors' Preferred Method of Treatment

Based on our experience and that reported in the literature, we believe that high school athletes with a single- or both-bone forearm fracture should be treated by plate fixation; this includes individuals greater than 15 years of age. The timing of the surgery should be delayed if at all possible, but more than 2 days of delay is typically not logistically possible or socially acceptable. The selection of the plate is most important; the 3.5-DCP plate is the treatment of choice because it is associated with less clinical failure and is consistent with the experimental data. The classic compression plate should be applied with meticulous technique to ensure rigid fixation and compression. The plate should be left in place at least 1 year; if removed, it should be removed in an off-season. Ideally, a minimum of 6 months should be allowed for protection before any contact sport commences, and we recommend a delay of a minimum of 9 months to, ideally, 1 year. In the interim, we do recommend a forearm protective splint for the first 4 to 6 weeks, with weaning from this splint as activity is increased (Fig. 17B–5). In patients with a high school or college career that does not allow the time frame discussed above, we believe leaving the plate in place is the more prudent option.

This general area remains controversial, and additional information would be welcome to help provide more definitive recommendations regarding the management of this difficult injury.

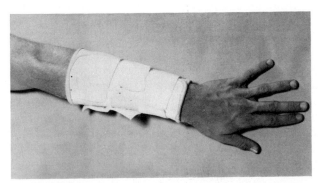

FIGURE 17B–5
In the early period after plating or if the plate has been removed, custom-molded forearm splints are applied. These are maintained for a minimum of 6 months after plate removal.

References

1. Anderson, L. D., Sisk, T. D., Tooms, R. E., and Park, W. I. Compression-plate fixation in acute diaphyseal fractures of the radius and ulna. *J Bone Joint Surg [Am]* 57:287–297, 1975.
2. Brismar, B., and Spangen, L. Fracture of the humerus from arm wrestling. *Acta Orthop Scand* 48:707, 1975.
3. Brooks, D. B., Burstein, A. H., and Frankel, V. H. The biomechanics of torsional fractures: The stress concentration effect of a drill hole. *J Bone Joint Surg [Am]* 52:507–514, 1970.
4. Burstein, A. H., Currey, J., Frankel, V. H., Heiple, K. G., Lunseth, P., and Vessely, J. C. Bone strength: The effect of screw holes. *J Bone Joint Surg [Am]* 54:1143–1156, 1972.
5. Chapman, M. W., Gordon, J. E., and Zissimos, A. G. Compression-plate fixation of acute fractures of the diaphyses of the radius and ulna. *J Bone Joint Surg [Am]* 71:159–169, 1989.
6. Cordey, J., Schwyzer, H. K., Brun, S., Matter, P., and Perren, S. M. Bone loss following plate fixation of fractures: Quantitative determination in human tibiae using computerized tomography. *Helv Chir Acta* 52:181–184, 1985.
7. DeLuca, P. A., Lindsey, R. W., and Ruwe, P. A. Refracture of bones of the forearm after the removal of compression plates. *J Bone Joint Surg [Am]* 70:1372–1376, 1988.
8. Edgerton, B. C., An, K. N., and Morrey, B. F. Torsional strength reduction due to cortical defects in bone. *J Orthop Res* 8:851–855, 1990.
9. Evans, D. L. Fatigue fractures of the ulna. *J Bone Joint Surg [Br]* 37:618–621, 1955.
10. Harkess, J. W., Ramsey, W. C., and Ahmadi, B. Principles of fractures and dislocations. *In* Rockwood, C. A., Jr., and Green, D. P. (Eds.), *Fractures in Adults* (2nd ed.). Philadelphia, J. B. Lippincott, 1984, pp 1–146.
11. Hidaka, S., and Gustilo, R. B. Refracture of bones of the forearm after plate removal. *J Bone Joint Surg [Am]* 66:1241–1243, 1984.
12. Jacobs, R. R., Rahn, B. A., and Perren, S. M. Effect of plates on cortical bone perfusion. *J Trauma* 21:91–95, 1981.
13. Kitchin, I. D. Fatigue fracture of the ulna. *J Bone Joint Surg [Br]* 30:622–623, 1948.
14. Lihna, M., and Kurttela, K. Twenty-eight cases, humerus fractures caused by arm wrestling. *Ann Chir Gynecol* 52:232, 1963.
15. Lofgren, L., and Vigolio, L. Fracture of the upper arm during Indian wrestling. *Acta Chir Scand* 124:36, 1962.
16. Moore, T. M., Klein, J. P., Patzakis, M. J., and Harvey, J. B. Results of compression plating of closed Galeazzi fractures. *J Bone Joint Surg [Am]* 67:1015–1021, 1985.
17. Mutoh, Y., Mori, T., Suzuki, Y., and Sugiura, Y. Stress fractures of the ulna in athletes. *Am J Sports Med* 10:365–367, 1982.
18. Olerud, S., and Danckwardt-Lilliestrom, G. Fracture healing in compression osteosynthesis in the dog. *J Bone Joint Surg [Br]* 50:844–851, 1968.
19. Paavolaimen, P., Karaharju, R., Slatis, P., Shonen, J., and Holmstrom, T. Effect of rigid plate fixation on structure and mineral content of cortical bone. *Clin Orthop* 136:287–293, 1978.
20. Peace, P. K. Fractures of the humerus from arm wrestling. *Injury* 9:162, 1977/78.
21. Perren, S. M., Cordey, J., Rahn, B. A., Gautier, E., and Schneider, E. Early temporary porosis of bone induced by internal fixation implants. *Clin Orthop* 232:139–151, 1988.
22. Rooker, G. D. An unusual cause of a spiral fracture of the humerus. *Injury* 9:62, 1977/78.
23. Rosson, J. W., Petley, G. W., and Shearer, J. R. Bone structure after removal of internal fixation plates. *J Bone Joint Surg [Br]* 73:65–67, 1971.
24. Rosson, J. W., and Shearer, J. R. Refracture after the removal of plates from the forearm: An avoidable complication. *J Bone Joint Surg [Br]* 73:415–417, 1991.
25. Uhthoff, H. K., and Finnegan, M. The effects of metal plates on post-traumatic bone remodelling and bone mass. *J Bone Joint Surg [Br]* 65:66–71, 1983.
26. Whitaker, J. H. Arm wrestling fractures—A humeral twist. *Am J Sports Med* 5:67, 1977.

HETEROTOPIC BONE ABOUT THE ELBOW

Bernard F. Morrey, M.D.

Although ectopic bone is not a common complication in many orthopaedic settings, it can have devastating consequences when it does occur. The elbow is, in fact, one of the anatomic regions of predilection. In the elbow, the involvement has been associated with (1) head or spinal cord trauma [8], (2) extensive burns [3], and (3) elbow trauma, usually fracture-dislocation with the involvement of extensive soft-tissue components [20].

INCIDENCE

The occurrence of this complication is not common. Roberts reported only two patients with ectopic bone after 60 elbow dislocations [18], which is virtually the identical percentage reported by Linscheid and Wheeler in 4 of 110 such injuries [10]. Interestingly, two of the four patients of Linscheid and Wheeler had fracture-dislocations. This confirms the observations of Thompson and Garcia, whose extensive assessment of this complication remains the standard [20]. Among over 1300 elbow injuries, the highest incidence among various injuries was recorded as 3%, and these were associated with fracture-dislocations. In this particular category, radial head or olecranon fracture with dislocation was the single most common circumstance in which this complication developed [21].

DEFINITION

Various authors report ectopic ossification in surprisingly different percentages, from 3% to over 50% [9, 12, 13, 21]. The obvious explanation for the dramatic difference in this reported incidence relates to the definition of what is described.

We have found the following scheme to be helpful in classifying ectopic radiodensity about the elbow [2] (Fig. 17B–6). Radiodensity from calcification is distinguished from true ectopic ossification. Calcification is amorphous material that is poorly circumscribed and has no trabeculation; this is usually found in the ligaments and capsule. Ectopic ossification can occur as an exostosis in the muscle or capsule. The calcification or ossification of the capsule and ligament typically only serves as an indication of the pathoanatomy (Fig. 17B–7). The development of bone in the anterior musculature is of greater clinical significance (Fig. 17B–8). A particularly bothersome problem is the development of proximal radioulnar synostosis. Although this may be due to an injury, the process is sometimes observed when a distal biceps tendon is attached through the Boyd approach (Fig. 17B–9).

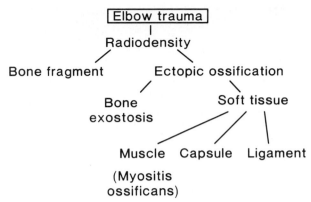

FIGURE 17B–6
Schema of classifying radial density about the elbow after injury. (From Broberg, M. A., and Morrey, B. F. Results of treatment of fracture-dislocation of the elbow. *Clin Orthop* 216:109–119, 1987.)

The development of ectopic bone begins with the initial injury. The process is known to be related somehow to delay in treatment of the initial injury [9, 11]. We believe that the second or third insult is the most serious factor other than the initial injury; thus, the initial injury is the first insult, and surgical treatment is the second. If a definitive procedure

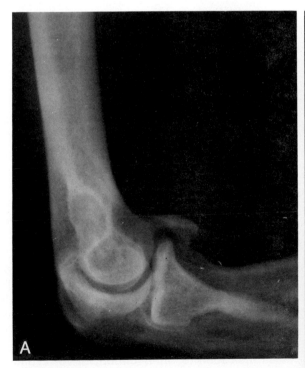

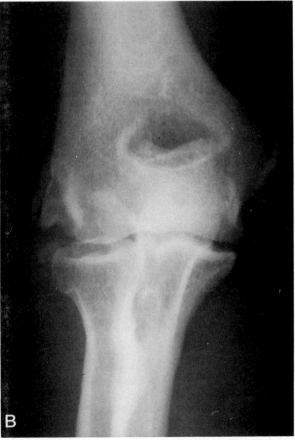

FIGURE 17B–7
Fracture-dislocation of the elbow with development of ectopic bone. Note location *(A)* in the anterior capsule and *(B)* in the collateral ligament.

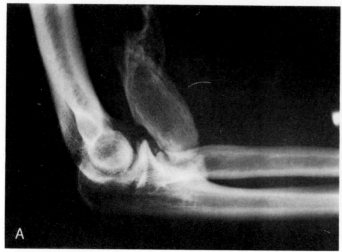

FIGURE 17B–8
(A) Well-defined heterotopic ossification occurring in the brachialis muscle. *(B)* The term myositis ossificans is properly reserved for this process.

cannot be accomplished with a single operation and a second operation is necessary, such as for a compound comminuted fracture, such a patient seems to be at the greatest risk of developing

FIGURE 17B–9
Proximal radioulnar synostosis after biceps rupture due to an approach to the radial tuberosity that exposed the ulna. This approach is thought to predispose to this complication. Excision usually improves motion to a functional but not normal degree.

hyptertrophic ossification. Postsurgical proximal radioulnar synostosis is best treated by avoiding the complication. The Boyd approach has been modified by using a muscle-splitting rather than a muscle-elevating approach, which exposes the ulna, for reattachment of the distal biceps tendon (Fig. 17B–10).

Ectopic bone is usually not observed radiographically until at least 10 days to 2 weeks, and sometimes longer, after the insult (Fig. 17B–11). The clinical presentation is one of increasing pain with decreasing motion. Beware of the patient who loses motion that had been previously gained. The development of ectopic bone is well defined by the technetium bone scan. The staging or maturity is best assessed, in our opinion, by plain radiographs. In our experience, alkaline phosphatase levels and technetium bone scans do correlate with the presence, but not with the maturity, of ectopic bone. Thus, we do not rely on these tests for prognosis, as an indication for intervention, or for estimation of the severity of the process.

TREATMENT

Certainly, the most obvious and desirable mode of treatment is prevention. In this context, we

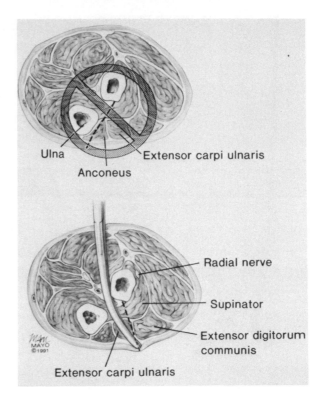

FIGURE 17B–10
The two-incision technique employs a muscle-splitting incision through the extensor muscle mass. The ulna should *not* be exposed. (With permission of the Mayo Foundation.)

assiduously avoid a second surgical procedure. Following surgery, passive manipulation is also strictly avoided. We make use of continuous motion for the severely injured elbow [19] (Fig. 17B–12). The early belief that complete rest was necessary to minimize the likelihood of ectopic bone has not been substantiated in our practice. Although diphosphonates at one time were considered an effective means of inhibiting bone formation [1, 6, 7], current literature

clearly demonstrates that these are not a clinically effective means of preventing ectopic ossification. Our preference is to treat patients with indomethacin or a similar medication that inhibits bone formation through inhibition of prostaglandin activity. Typically, we use 75 mg of sustained-release indomethacin once a day for approximately 2 months after the injury, insult, or surgery [16].

In general, we believe irradiation should be

FIGURE 17B–11
Experimental data correlating laboratory findings as a function of time and the development and maturation of ectopic bone. (From Orzel, J. A., and Rudd, I. G. Heterotopic bone formation: Clinical laboratory and imaging correlations. *J Nucl Med* 26:125, 1985.)

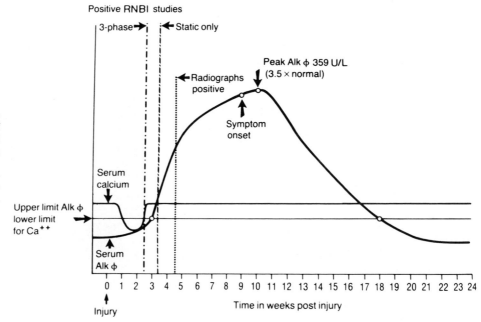

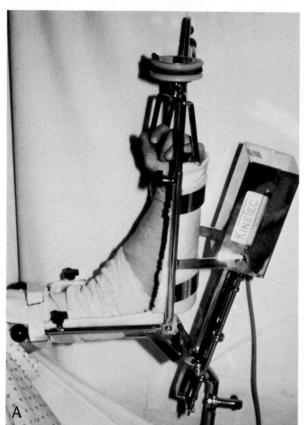
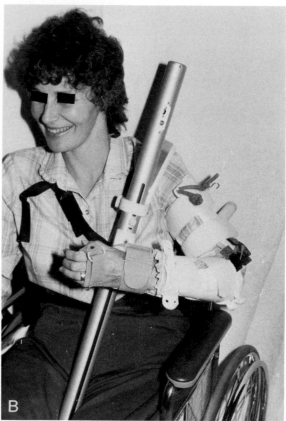

FIGURE 17B–12
Continuous motion is helpful to avoid harmful passive stretch both *(A)* in the hospital and *(B)* sometimes after dismissal with portable units. (With permission of the Mayo Foundation.)

avoided around the elbow. We do not feel comfortable using this treatment modality, particularly in the younger athletic population. If, however, unique circumstances outweigh our reservations, then we use 200 rads in four divided doses delivered with a computer-controlled targeting system to the area of greatest concern. This is started the day after surgery or after the insult, not after the process has begun [4]. The concern about irradiation, of course, relates to its effect on bone healing, as well as on soft tissue healing. The development of sarcomatous degeneration is also a concern. We have not been able to document an instance of sarcomatous degeneration in patients treated with under 3000 rads.

OPERATIVE INDICATIONS

Typically, ectopic bone can be successfully removed if the joint was not initially involved in the insult, such as after a head injury, a burn, or simple fracture-dislocations. The severe injuries may best be left alone [3, 8, 17] (Fig. 17B–13). If surgical intervention is considered, we use plain radiographs until evidence of maturity is demonstrated. In this context, the tomogram may be helpful, but we do not use alkaline phosphatase levels or technetium bone scans. We also find it helpful to make the distinction of intrinsic or extrinsic involvement [14].

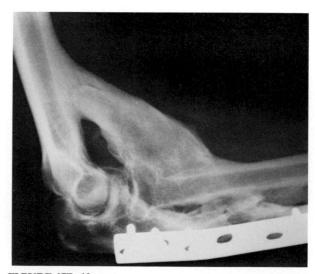

FIGURE 17B–13
Extensive ectopic bone compromises routine function, and athletic competition is out of the question. This lesion is not amenable to surgery.

Extrinsic involvement implies that the joint is uninvolved, and intrinsic involvement is that in which the traumatic event involved the joint and the soft tissue. This has a much poorer prognosis than an injury in which the ectopic bone is primarily part of an extrinsic contracture, with the joint reasonably well intact.

The results of ectopic bone removal are poorly documented. We have reported our experience of removing the synostosis in the proximal radius and ulna after a distal biceps tendon repair [5]. Some return of motion is anticipated, and this is usually in the functional range.

To summarize, formation of true ectopic bone about the elbow after injury does not occur frequently, but when it does, it has grave prognostic implications. Marked joint motion loss may be anticipated, and obviously this is considered a salvage circumstance, with devastating implications for athletic participation.

References

1. Bijoet, O. L. M., Nollen, A. J. G., Slooff, T. J., and Feith, R. Effect of a diphosphonate on para-articular ossification after total hip replacement. *Acta Orthop Scand* 45:926–934, 1974.
2. Broberg, M. A., and Morrey, B. F. Results of treatment of fracture-dislocation of the elbow. *Clin Orthop* 216:109–119, 1987.
3. Cooney, W. P. Treatment of the contracted elbow. *In* Morrey, B. F. (Ed.), *The Elbow and Its Disorders*. Philadelphia, W. B. Saunders, 1985, pp. 433–451.
4. Coventry, M. B. Ectopic ossification about the elbow. *In* Morrey, B. F. (Ed.), *The Elbow and Its Disorders*. Philadelphia, W. B. Saunders, 1985, pp. 464–471.
5. Failla, J. M., Amadio, P. C., and Morrey, B. F. Posttraumatic proximal radioulnar synostosis. *J Bone Joint Surg [Am]* 71:1208, 1989.
6. Fleisch, H., Russel, R. G., Bisaz, S., Muhlbauer, R. C., and Williams, D. A. The inhibiting effect of phosphates on the formation of calcium phosphate crystals in vitro and an aortic kidney calcification in vivo. *Eur J Clin Invest* 1:12, 1970.
7. Francis, M. D., Russell, R. G., and Fleisch, H. Diphosphates inhibit formation of calcium phosphate crystals in vitro and pathological calcifications in vivo. *Science* 165:1264, 1969.
8. Garland, D. E., Hanscom, D. A., Keenan, M. A., Smith, C., and Moore, T. Resection of heterotopic ossification in the adult with head trauma. *J Bone Joint Surg [Am]* 67:1261–1269, 1985.
9. Gaston, S. R., Smith, F. M., and Baab, D. D. Adult injuries of radial head and neck: Importance of time element in treatment. *Am J Surg* 78:631, 1949.
10. Linscheid, R. L., and Wheeler, D. K. Elbow dislocation. *JAMA* 194:1171, 1965.
11. McLaughlin, H. L. Some fractures with a time limit. *Surg Clin North Am* 35:553, 1955.
12. Mikic, Z. D., and Vukadinovic, S. M. Late results in fractures of the radial head treated by excision. *Clin Orthop* 181:220–227, 1983.
13. Mohan, K. Myositis ossificans traumatica of the elbow. *Int Surg* 57:475, 1972.
14. Morrey, B. F. Treatment of post-traumatic elbow stiffness including distraction arthroplasty. *J Bone Joint Surg [Am]* 72:1600–1618, 1990.
15. Orzel, J. A., and Rudd, T. G. Heterotopic bone formation: Clinical laboratory and imaging correlations. *J Nucl Med* 26:125, 1985.
16. Ritter, M. E., and Gide, T. The effect of indomethacin on para-articular ectopic ossification following long-term coma. *Clin Orthop* 167:113, 1982.
17. Roberts, J. B., and Pankratz, D. G. The surgical treatment of heterotopic ossification of the elbow following long-term coma. *J Bone Joint Surg [Am]* 61:760–763, 1979.
18. Roberts, P. H. Dislocation of the elbow. *Br J Surg* 56:806, 1969.
19. Salter, R. B., Hamilton, H. W., Wedge, J. H., Tile, M., Torode, I. P., O'Driscoll, S. W., Murnaghan, J. J., and Saringer, J. H. Clinical application of basic research on continuous passive motion for disorders and injuries of synovial joints: A preliminary report of a feasibility study. *J Orthop Res* 1:325–342, 1983.
20. Thompson, H. C., and Garcia, A. Myositis ossificans: Aftermath of elbow injuries. *Clin Orthop* 50:129, 1967.
21. Wilson, P. D. Fractures and dislocations in the region of the elbow. *Surg Gynecol Obstet* 56:335, 1933.

Elbow Dislocation in the Athlete

Bernard F. Morrey, M.D.

INCIDENCE

Dislocation of the elbow is not a common occurrence, either in the general population or in the athlete. It has been estimated in demographic studies that approximately 6 of every 100,000 individuals will sustain an elbow dislocation [20]. It is of interest that the nondominant extremity is involved in about 60% of such instances [23, 30, 39, 47]. This suggests a protective effect of the dominant side, with the use of the nondominant extremity to break the fall. The mean age of an individual sustaining this injury is 30 years [23, 30, 39, 47].

As suggested above, elbow dislocation is not a common or unique injury in any specific sport. One review of over 2000 downhill skiing injuries in children did not mention elbow dislocation [2]. Any activity associated with contact, such as football or a fall on the outstretched hand, may have an incidence of this injury ranging from 0.1% to 1% [7, 8, 52]. In a recent study of 123 injuries involving rollerskates, no individuals sustained an elbow dislocation [11].

MECHANISM

The mechanism of elbow dislocation has been investigated to some extent, but the precise mechanism has not yet been completely defined. It has been well accepted and reported that a fall on the outstretched hand is the most common cause of this particular injury, which occurs two to two and a half times more frequently in males than in females and has a similar ratio in both adults and children [3, 23]. Investigations by O'Driscoll and associates suggest that extension and a varus stress can disrupt the lateral ligament complex, allowing a perched dislocation; subsequent further forces rotate the

forearm and therefore allow a complete dislocation [40]. A commonly accepted mechanism is a hyperextension injury that occurs with a fall on the outstretched hand [27, 39, 47]. There is little question that, in some, the initial dislocation occurs with the elbow slightly flexed [19, 42]. This mechanism is consistent with the observations of Josefsson [20]. About 40% occur during sports, 10% in play or traffic accidents, and the remaining 50% from poorly defined causes [23].

CLASSIFICATION

The traditional classification divides the injury into anterior and posterior dislocations, with the anterior being quite uncommon, occurring in only 1% to 2% of instances [3, 29]. Posterior dislocations are further divided according to the final resting position of the olecranon in reference to the distal humerus. Thus, posterior, posterolateral, posteromedial, and pure lateral dislocations have been recognized. By far the most common type is the posterolateral position, with a pure lateral dislocation being much less common, and posteromedial being the least common [24].

We use a slightly different classification system, which we think has clinical relevance, particularly for the athlete. The distinction made is simply between a complete or a perched dislocation (Fig. 17C–1). The complete dislocation can have any disposition of the ulna behind the humerus; we believe it makes little difference with regard to treatment or prognosis. Conversely, a perched dislocation is one in which the elbow is actually subluxed but the coronoid appears to be impinged on the trochlea. In this type, the implication is that the ligaments have not been as badly torn, and thus rehabilitation should be more rapid and recovery

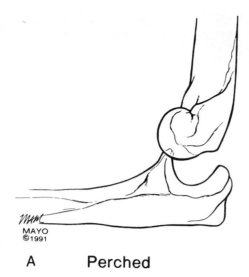

A   Perched

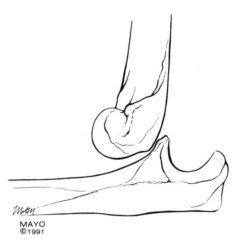

B   Posterior-Complete

FIGURE 17C–1
Simplified classification of elbow dislocation has prognostic implications. *A*, Perched (subluxed); *B*, complete (dislocated). (With permission of the Mayo Foundation.)

more complete. Furthermore, treatment for the complete dislocation requires general anesthesia and a muscle relaxant, particularly if the dislocation persists for several hours before reduction. A perched dislocation can usually be reduced rapidly in the emergency room with intra-articular Xylocaine and muscle relaxation. The implications to the patient, particularly the athlete, are obvious.

The clinical data clearly reflect the fact that an elbow dislocation is a serious injury. When complete dislocation occurs, it is assumed that there has been a complete disruption of the medial collateral ligament [37, 38, 40], probably also indicating a rupture of the lateral collateral ligament structures [10, 21].

For a complete dislocation, the anterior capsule must be disrupted. Furthermore, the brachialis muscle must also be torn or significantly stretched. This has been observed at the time of surgery for the associated fractures and explains the loss of motion so commonly observed with this condition [10, 21].

To clarify this issue further, O'Driscoll and associates have performed laboratory cadaver experiments in which the lateral collateral ligament and anterior capsule are surgically released [40]. By manipulating the forearm with excessive varus and supination, the elbow can be subluxed in a perched position. With further manipulation, a partial stretch of the medial collateral ligament occurs, and the elbow can be completely dislocated, even when most of the medial collateral ligament is intact.

PATHOANATOMY

The pathoanatomy of elbow dislocations has been investigated not only in the laboratory but, more importantly, in clinical studies. There have been few nonsurgical clinical studies to better delineate the pathology of an elbow dislocation, and the use of arthrograms [19] for this purpose has been found to be unreliable [24].

Surgical exploration of the acute injuries in vivo have revealed important data. Although it is possible for the elbow to be dislocated without rupture of the medial collateral ligament, the surgical experience suggests that this entity may exist only theoretically and not in reality. Josefsson and colleagues found that the medial collateral ligament was violated in every instance of elbow dislocation that was explored at surgery; examination at surgery also prompted the conclusion that the lateral ligament is also disrupted in all instances [21]. The anterior bundle of the medial collateral ligament was considered the essential lesion, and disruption at the proximal portion near the humerus was the most common site of pathology [10, 24]. The fact that calcification is seen in one or both of the ligamentous complexes further supports the observation that complete dislocation violates both collateral ligament structures [3, 6, 21, 29].

ASSOCIATED INJURIES

The most common associated injury is a radial head fracture [37]. In general, the radial head fracture occurs in 10% of elbow dislocations; similarly, 10% of patients with a radial head fracture have an associated elbow dislocation (Fig. 17C–2). The other

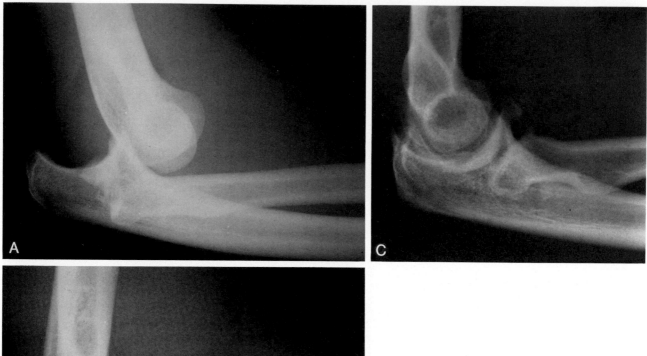

FIGURE 17C–2
A, Fracture of radial head with dislocation; *B,* adequate initial reduction; *C,* type 2 radial head fracture fragment was removed and the radial head has subluxed. The function is now significantly compromised.

associated fractures occur with almost equal frequency, including the coronoid and olecranon, medial and lateral epicondyle, and the capitellum being possibly the least commonly involved [20, 28, 34].

Fractures

It is also well recognized that elbow dislocation is not necessarily an isolated event. Associated fractures have been reported in 25% to 50% of cases [21, 47, 53]. The incidence of associated fractures is also quite high in children, approaching 50% [3]. In this patient group, in whom the physis is still open, a medial epicondyle avulsion is the most common associated injury [12].

Other Injuries

Additional injuries that are uncommon but potentially devastating are those to the neurovascular structures. It is well known that the brachial artery can be injured with posterior displacement of the ulna, and this has been a subject of a number of case reports [1, 14]. Some believe that it is not necessary to explore the brachial artery if the radial pulse is present, even if evidence of arterial injury exists [14]. It is commonly believed, however, that disruptions of the brachial artery should be treated by ligation and vein graft [1].

Various neurologic complications have also been reported with elbow dislocation. The classic Matev sign was first described in association with median nerve entrapment after reduction of a dislocated elbow [32]. Others have also observed that the median nerve may become entrapped in the joint at the time of reduction; thus, a careful neurovascular examination is necessary to ensure that these uncommon but devastating complications are not present [15, 44].

EVALUATION

Assessment of the extremity for neurovascular competence is mandatory before reduction. Initial plain anteroposterior and lateral radiographs are required before reduction. Detailed assessment of associated fractures should be deferred until after reduction. After this, plain films or tomograms are effective in further characterizing any associated injuries of the radial head or coronoid. In our opinion, computed tomography, magnetic resonance imaging, and arthrography are of no value in this injury. The postreduction evaluation is particularly important for the athlete, who is receptive to a quick and aggressive rehabilitation program.

TREATMENT

Treatment of the elbow dislocation consists simply in reducing the joint as soon and as atraumatically as possible. This may require a general anesthetic with adequate muscle relaxation. If the dislocation is complete and has occurred several hours earlier, there can be a great amount of swelling and muscle spasm, making the reduction particularly difficult. Great care must be exercised to avoid multiple attempts at reduction, because this has the potential to tear additional soft tissue and muscle, thus predisposing to ectopic bone [51].

Technique

The technique of reduction is basically one of extending the elbow with countertraction on the brachium and then manipulating the olecranon distally and anteriorly with the thumb, so that the coronoid clears the trochlea [16, 29]. Passive techniques, such as the application of a weight with gravity, have also been recommended, but we do not consider these to be desirable techniques [35]. Repeated attempts are to be avoided, and adequate analgesic is of extreme importance. For the well-muscled athlete, muscle relaxation may also be needed.

The distinction between a perched and a complete dislocation implies different treatment. In a perched dislocation, the use of an analgesic, intra-articular Xylocaine, and manipulation in which direct pressure is applied over the olecranon with the elbow slightly extended with axial distraction is adequate in most instances. It is uncommon to have an associated fracture.

An elbow dislocation that is irreducible by closed means is uncommon [43]. Irreducible dislocations are most frequently seen with associated fracture [9]. When a dislocation is irreducible, the radial head has been shown to be trapped in the soft tissues of the forearm [49] or can buttonhole through the forearm fascia [13]. Both conditions necessitate surgical release for reduction.

The issue of surgical repair of the ligaments, even in the presence of simple dislocations without associated fractures, has been recently studied [10, 21]. Repair followed by early motion has been suggested to be of greater value than treatment by closed means [10]. In a carefully conducted prospective study comparing 14 patients with surgical repair and 14 without surgery, no long-term value in repair of the medial collateral ligament could be identified. Residual pain and stability were similar [21]. There is no need or justification for ligament repair for uncomplicated dislocations, especially in the athlete, because this can only delay rehabilitation.

After Reduction

For simple dislocations, the elbow is immobilized for approximately 10 days, and then early motion is begun. It is particularly important for the athlete to avoid prolonged immobilization. This has been emphasized by Protzman, who has the only series dealing almost exclusively with the athlete, as he reported the West Point experience [45]. At the end of 3 weeks, if a flexion contracture of 45 to 50 degrees is still present, a hinged turnbuckle splint, which both protects the collateral ligaments and aids in eliminating the tendency for flexion contracture, is employed. Perched dislocations are essentially normal within 6 to 8 weeks. After a complete dislocation, most patients should regain 80% to 90% of normal function by 3 months [22, 47].

LONG-TERM RESULTS

The most common sequela of an elbow dislocation is loss of motion, particularly extension [22, 29, 39, 45, 47]. The degree and frequency of this complication has been variably reported. At 10 weeks, a flexion contracture averaging about a 30 degrees is common. After 2 years, a 10-degree flexion contracture is typical, and no change occurs after this period of time [22]. There is some difference in the extent of motion loss between the adult and adolescent population [3]. Only about 5% of patients with a single dislocation have a flexion contracture of greater than 30 degrees, particularly if early motion is initiated.

Some have investigated the loss of motion and correlated it with the type of dislocation [22]. Josefsson thought that the loss of motion was greater with lateral or posterolateral dislocations, and that motion loss was greater in adults than in children [3, 22].

Although gross instability is not commonly observed after elbow dislocation, with careful investigation some have reported mild laxity in approximately one-third of patients in the adolescent or younger age group and in about 20% of adults [3, 34]. Josefsson, however, reported no instance of recurrences among 148 dislocations [23, 24]. Linscheid and Wheeler reported 2 instances of recurrent instability after 110 dislocations [30]. It is generally accepted that recurrent instability occurs in 1% to 2% of simple elbow dislocations.

Recurrent Instability

For the athlete, probably the most devastating complication is recurrent instability. Only 40 patients with recurrent elbow dislocation following the initial event were reported in the English literature up to 1984, according to Lansinger and associates, who reviewed the experience of 17 authors [28]. Although this is a very uncommon event, it is more common when an initial dislocation occurs in children or adolescents, than in adults [28, 31, 54]. The high school athlete is thus at some risk for this complication. In the four patients with recurrent

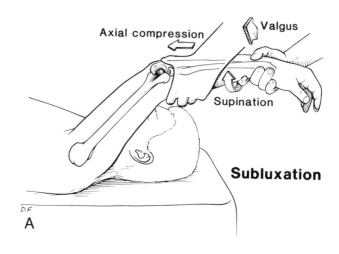

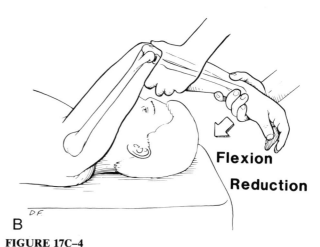

FIGURE 17C–4
Pivot shift test of the elbow. *A,* With extension valgus and supination the elbow subluxes. *B,* Flexion and pronation reduces the subluxation. (With Permission of the Mayo Foundation.)

dislocation reported by Hassmann and colleagues, two were under 16 years of age at the time of the initial injury [17].

Presentation of recurrent instability can be quite variable. Complete recurrent dislocation with an obvious diagnosis is well known and most commonly described [17, 18, 25, 27, 48, 50, 54]. We have found recurrent subluxation to be a more common and subtle presentation. This problem has been recognized as subluxation of the radial head [5] or as a recurrent locked elbow (Fig. 17C–3) [26]. The pathology and mechanism of recurrent subluxation have been described by O'Driscoll and associates [41]. The examination used to replicate the instability is analogous to the pivot shift test of the knee (Fig. 17C–4). When the elbow is extended with valgus stress and supination, the radial head rolls below the capitellum and the ulna externally rotates on the trochlea. With elbow flexion and pronation,

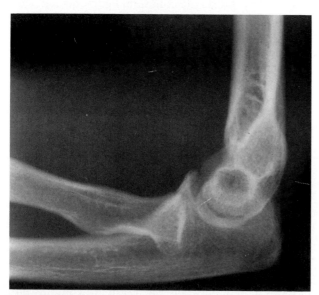

FIGURE 17C–3
Recurrent subluxation in a professional football player. Note inferior position of the radial head.

a clunk or a thud is appreciated as the elbow reduces. This maneuver is truly analogous to the knee pivot shift with regard to both the maneuver demonstrating the pathology and the nature of the pathology itself. Further studies have shown the deficiency to be of the lateral ulnar collateral ligament.

The treatment for recurrent elbow dislocation or subluxation consists of reconstructing the lateral ulnar collateral ligament. Prior reports have clearly shown the value and effectiveness of lateral joint reconstruction [17, 42]. We have had 10 patients with this condition; of these, three were competitive athletes, including one professional football player. Reconstruction of the lateral ulnar collateral ligament has been successful in 80% of this population (Fig. 17C–5).

After surgery, the joint is immobilized in a long-arm cast for 4 weeks. A hinged splint with a 30-degree extension stop but full flexion is employed for an additional month. Activity as tolerated and strength rehabilitation are initiated. Full activity is then allowed. The splint is used without the extension stop for a third month.

Ectopic Bone

Ectopic bone is commonly seen with fracture-dislocations, particularly after radial head fracture [51]. However, in a recent series of elbow dislocations, ectopic bone occurred in only 3 of 142 (2%) instances [22]. In our practice, this problem has not been common but can be a most difficult complication [4]. It is helpful to distinguish ectopic calcification that may occur in the ligaments or capsule from

FIGURE 17C–5
A, The pathology of the recurrent subluxed elbow is stretch of the lateral ligament complex, including the lateral ulnar collateral ligament. The reconstruction must include reestablishment of the integrity of the ulna referrable to the humerus. *B,* We specifically reconstruct the lateral ulnar collateral ligament. (With permission of the Mayo Foundation.)

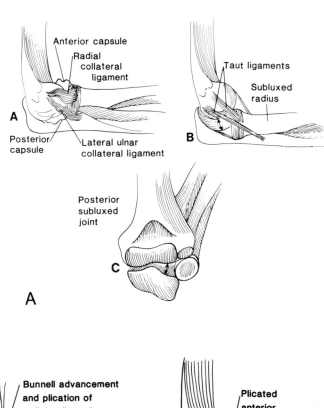

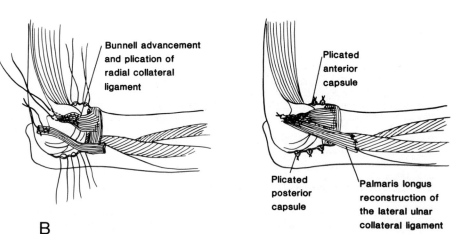

true myositis ossificans. The former is common and innocuous; the latter can be career-ending. Myositis ossificans appears to be associated with multiple attempts at reduction or occurs after several efforts of surgical intervention for a fracture. The time frame of treatment may also be important, as was reported in the early literature [33]. Most fracture-dislocations should be dealt with definitively in the first 24 to 48 hours after the injury to avoid ectopic bone. This topic is dealt with in more detail below.

AUTHOR'S PREFERRED METHOD OF TREATMENT

Our preferred treatment of elbow dislocation is similar to the method that is commonly accepted. A distinction is made between complete and perched dislocations. Postoperative motion is started at 7 to 10 days for complete and at 2 to 3 days for perched dislocations. A hinged splint is sometimes used for those with complete dislocations. Anti-inflammatory medications, heat, and ice are prescribed. Physical therapy is *not* used except for strength exercises. Active assisted extension is encouraged. If more than 50 degrees of contracture are present at 3 weeks, an extension turnbuckle splint is used, especially at night. Gentle strengthening exercises are allowed at 3 weeks and are performed without restriction at 8 to 10 weeks.

Associated fractures are dealt with after ulnohumeral reduction according to the merits of the fracture. Radial head fractures are treated with open reduction and compressive screw fixation when possible. Fracture-dislocations in general are inconsistent with return to competition that requires repetitive stressful use of the involved extremity. Simple dislocations generally do not adversely affect the long-term prognosis of the athlete.

References

1. Amsallem, J. L., Blankstein, A., Bass, A., and Horoszowski, H. Brachial artery injury: A complication of posterior elbow dislocation. *Orthop Rev* 15:61–64, 1986.
2. Blitzer, C. M., Johnson, R. J., Ettlinger, C. F., and Aggeborns, K. Downhill skiing injuries in children. *Am J Sports Med* 12:142–147, 1984.
3. Borris, L. C., Lassen, M. R., and Christensen, C. S. Elbow dislocation in children and adults: Long-term follow-up of conservatively treated patients. *Acta Orthop Scand* 58:649–651, 1987.
4. Broberg, M. A., and Morrey, B. F. Results of treatment of fracture-dislocations of the elbow. *Clin Orthop* 216:109–119, 1987.
5. Burgess, R. C., and Sprague, H. H. Post-traumatic posterior radial head subluxation. *Clin Orthop* 186:192–194, 1984.
6. Buxton, J. D. Ossification in the ligaments of the elbow joint. *J Bone Joint Surg* 20:709–714, 1938.
7. Culpepper, M. I., and Niemann, K. M. High school football injuries in Birmingham, Alabama. *South Med J* 76:873–878, 1983.
8. DeHaven, K. E., and Lintner, D. M. Athletic injuries: Comparison by age, sport, and gender. *Am J Sports Med* 14:218–224, 1986.
9. Devadoss, A. Irreducible posterior dislocation of the elbow. *Br Med J* 3:659, 1967.
10. Durig, M., Muller, W., Ruedi, T. P., and Gauer, E. F. The operative treatment of elbow dislocations in the adult. *J Bone Joint Surg [Am]* 61:239–244, 1979.
11. Esses, S., Zaremba, M., and Langer, F. Roller skating injuries. *Contemp Orthop* 5:99–103, 1982.
12. Fowles, J. V., Kassab, M. T., and Moula, T. Untreated intra-articular entrapment of the medial humeral epicondyle. *J Bone Joint Surg [Br]* 66:562–565, 1984.
13. Greiss, M., and Messias, R. Irreducible posterolateral elbow dislocation: A case report. *Acta Orthop Scand* 58:421–422, 1987.
14. Grimer, R. J., and Brooks, S. Brachial artery damage accompanying closed posterior dislocation of the elbow. *J Bone Joint Surg [Br]* 67:378–381, 1985.
15. Hallett, J. Entrapment of the median nerve after dislocation of the elbow: A case report. *J Bone Joint Surg* 63:408–412, 1981.
16. Hankin, F. M. Posterior dislocation of the elbow: A simplified method of closed reduction. *Clin Orthop* 190:254–256, 1985.
17. Hassmann, G. G., Brunn, F., and Neer, C. Recurrent dislocation of the elbow. *J Bone Joint Surg [Am]* 57:1080–1084, 1975.
18. Jacobs, R. L. Recurrent dislocation of the elbow and review of the literature. *Clin Orthop* 74:151, 1971.
19. Johansson, O. Capsular and ligament injuries of the elbow joint: A clinical and arthrographic study. *Acta Chir Scand (Suppl.)* 287:1, 1962.
20. Josefsson, P. O. The dislocated elbow: With special reference to incidence, ligamentous injuries and stability. Dissertation, Lund University, Malmo, Sweden, 1986.
21. Josefsson, P. O., Gentz, C. F., Johnell, O., and Wendberg, B. Surgical versus non-surgical treatment of ligamentous injuries following dislocation of the elbow joint. *J Bone Joint Surg [Am]* 69:605–608, 1987.
22. Josefsson, P. O., Johnell, O., and Gentz, C. F. Long-term sequelae of simple dislocation of the elbow. *J Bone Joint Surg [Am]* 66:927–930, 1984.
23. Josefsson, P. O., and Nilsson, B. E. Incidence of elbow dislocation. *Acta Orthop Scand* 47:537–538, 1986.
24. Josefsson, P. O., and Wendberg, B. Ligamentous injuries in dislocations of the elbow joint. *Clin Orthop* 221:221–225, 1987.
25. Kepel, A. Operation for habitual traumatic dislocation of the elbow. *J Bone Joint Surg [Am]* 33:707, 1951.
26. Kinast, C., and Jakob, R. P. Differential diagnosis in locking of the elbow joint: The subluxation stress x-ray technique. *Hefte zur Unfallheilkunde* 181:339–341, 1986.
27. King, T. Recurrent dislocation of the elbow. *J Bone Joint Surg [Br]* 35:50, 1953.
28. Lansinger, O., Karlsson, J., Korner, L., and Mare, K. Dislocation of the elbow joint. *Arch Orthop Trauma Surg* 102:183–186, 1984.
29. Linscheid, R. L. Elbow dislocation. *In* Morrey, B. F. (Ed.), *The Elbow and Its Disorders.* Philadelphia, W. B. Saunders, 1985, pp. 414–432.
30. Linscheid, R. L., and Wheeler, D. K. Elbow dislocations. *JAMA* 194:1171–1176, 1965.
31. Malkawi, H. Recurrent dislocation of the elbow accompanied by ulnar neuropathy: A case report and review of the literature. *Clin Orthop* 161:270–274, 1981.
32. Matev, I. A radiological sign of entrapment of the median nerve in the elbow joint after posterior dislocation. *J Bone Joint Surg [Br]* 58:353–355, 1976.

33. McLaughlin, H. L. Some fractures with a time limit. *Surg Clin North Am* 35:553, 1955.

34. Mehlhoff, T. L., Noble, P. C., Bennett, J. B., and Tullos, H. S. Simple dislocation of the elbow in the adult. *J Bone Joint Surg [Am]* 70:244–249, 1988.

35. Meyn, M. A., Jr., and Quigley, J. B. Reduction of posterior dilsocation of the elbow by traction on the dangling arm. *Clin Orthop* 103:106–108, 1974.

36. Milch, H. Bilateral recurrent dislocation of the ulna at the elbow. *J Bone Joint Surg [Br]* 18:777–780, 1936.

37. Morrey, B. F. Fractures of the radial head. *In* Morrey, B. F. (Ed.), *The Elbow and Its Disorders.* Philadelphia, W. B. Saunders, 1985, pp. 355–381.

38. Morrey, B. F., and An, K. N. Articular and ligamentous contribution to the stability of the elbow joint. *Am J Sports Med* 11:315, 1983.

39. Neviaser, J. S., and Wickstrom, J. K. Dislocation of the elbow: A retrospective study of 115 patients. *South Med J* 70:172–173, 1977.

40. O'Driscoll, S. W., Morrey, B. F., and An, K. N. Elbow dislocation and subluxation: A spectrum of instability. *Clin Orthop* 280:186, 1992.

41. O'Driscoll, S., Bell, D., and Morrey, B. F. Pivot shift of the elbow. *J Bone Joint Surg [Am]* 73:440–446, 1991.

42. Osborne, G. V., and Cotterill, P. Recurrent dislocation of the elbow joint. *J Bone Joint Surg [Br]* 48:340, 1966.

43. Pawlowski, R. F., Palumbo, F. C., and Callahan, J. J. Irreducible posterolateral elbow dislocation: Report of a rare case. *J Trauma* 10:260–266, 1970.

44. Pritchard, D. J., Linscheid, R. L., and Svien, H. J. Intra-articular median nerve entrapment with dislocation of the elbow. *Clin Orthop* 90:100–103, 1973.

45. Protzman, R. R. Dislocation of the elbow joint. *J Bone Joint Surg [Am]* 60:539–541, 1978.

46. Reichenheim, P. P. Transplantation of biceps tendon as treatment for recurrent dislocation of the elbow. *Br J Surg* 35:301, 1947.

47. Roberts, P. H. Dislocation of the elbow. *Br J Surg* 56:806–815, 1969.

48. Spring, E. W. Report of a case of recurrent dislocation of the elbow. *J Bone Joint Surg [Br]* 35:55, 1953.

49. Strong, M. L. Irreducible posterolateral dislocation of the elbow without fracture: Report of two cases. *Contemp Orthop* 11:69–70, 1985.

50. Symeonides, P. P., Paschaloglou, C., Stavrou, Z., and Pangalides, T. Recurrent dislocations of the elbow. *J Bone Joint Surg [Am]* 57:1084, 1975.

51. Thompson, H. C., III, and Garcia, A. Myositis ossificans: Aftermath of elbow injuries. *Clin Orthop* 50:129–134, 1967.

52. Watson, A. W. S. Sports injuries during one academic year in 6,799 Irish school children. *Am J Sports Med* 12:65–71, 1984.

53. Wilson, P. D. Fractures and dislocations in the region of the elbow. *Surg Gynecol Obstet* 56:335–359, 1933.

54. Zeier, F. G. Recurrent traumatic elbow dislocation. *Clin Orthop* 169:211–214, 1982.

Entrapment Neuropathies About the Elbow

William D. Regan, M.D.
Bernard F. Morrey, M.D.

Entrapment neuropathies of the elbow basically involve three major mixed nerves, which cross the elbow to innervate the musculature and provide sensation in the forearm and hand. Accordingly, we will discuss neuropathies of the median, radial, and ulnar nerves.

The pathophysiology is said to involve ionic, mechanical, and vascular injury, which together produce an entrapment neuropathy. In recent years, the vascular lesions seem to have become most understandable from the standpoint of the pathophysiologic mechanisms. Obstruction of venous return from the nerve secondary to inflammation initially may cause venous congestion in the epineurial and perineurial vascular plexuses, with a generalized slowing of circulation in the nerve trunk. Resultant anoxia of the nerve leads to dilatation of the small vessels and capillaries within the nerve, and endoneurial edema of the tissue results [42, 46, 73]. The swelling of the nerve attendant on this edema increases the effect of the original compression, further slowing venous return if this is allowed to persist for prolonged periods. With anoxia, fibroblasts proliferate within the nerve, which of course results in permanent scarring within the nerve, further rendering segments of the nerve anoxic because of a barrier of fibroblasts that inhibits circulation within the nerve and the exchange of vital nutrients between the vascular system and the nerve fibers.

When a portion of the axon is rendered ischemic, the axioplasmic transport system within the axon is also affected. This transport system combines transport filaments, microtubules, and neural filaments to transport protein molecules synthesized in the endoplasmic reticulum of the cell body to a location in the axon where they will be chemically active, such as in the wall of the axon or in terminal endings. When there is a 30% to 50% reduction in blood flow, the reduction in oxidated phosphorylation and production of high-energy phosphate decreases the efficiency of the sodium pump and of the axioplasmic transport system, and the integrity of the cell membrane, which in turn eventually leads to a loss of conduction or transmission along the nerve fiber [24, 25]. In summary, the segment of the axon rendered ischemic or relatively ischemic through a change of position or of local external or internal anatomy not only reacts through a series of vascular mechanisms but also, by so doing, alters its ionic relationship to its environment and further aggravates the normal internal pressure of the nerve trunk to such a degree as to account for increased vascular changes and deterioration of normal function of the nerve trunk [24, 25].

PRONATOR SYNDROME

Of the three distinct entrapment neuropathies, the first to be discussed is that of median nerve entrapment at the elbow. This most proximal entrapment neuropathy of the median nerve, which has been widely studied, is known as the pronator syndrome [37, 39]. In order to understand the pathogenesis of the pronator syndrome, it is essential to understand the anatomic relations and functions of the median nerve.

Anatomy

The median nerve arises from the brachial plexus by two heads, a lateral and a medial. The larger lateral head originates from the lateral cord of the brachial plexus. It contains fibers from the anterior division to the C5, C6, and C7 cervical spinal nerves. These spinal nerves are concerned chiefly with innervation controlling the gross movements of forearm flexion and pronation, wrist flexion, and radial deviation. The smaller medial head arises from the medial cord. It contains fibers from the anterior divisions of C8 and T1 spinal nerves. These spinal nerve roots are concerned with the innervation and fine movements of the thumb, including flexion and opposition. They also allow functions of abduction and flexion of the index and middle fingers.

In the arm, the median nerve is intimately related to the brachial artery, first lying lateral to it; then at the elbow it crosses the artery anteriorly, and while in the antecubital fossa it finally comes to lie medial to the artery (Fig. 17D–1) [37, 53].

At the elbow, the median nerve leaves the brachial artery to pass between the two heads of the pronator teres muscle and beneath the tendinous arch of the flexor digitorum superficialis (Fig. 17D–2). From this point, the nerve passes distally be-

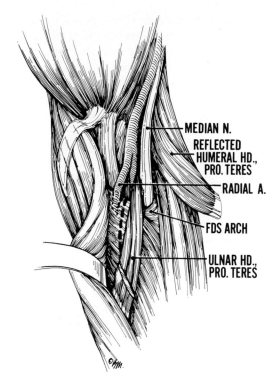

FIGURE 17D–2
The median nerve is identified by reflecting the humeral head of the pronator teres, allowing inspection of the nerve as it passes under the flexor digitorum superficialis (FDS) arch. (Illustration by Elizabeth Roselius, © 1988. Reprinted with permission from Green, D. P. (Ed.), *Operative Hand Surgery*, 2nd ed. New York, Churchill Livingstone, 1988, p. 1429.)

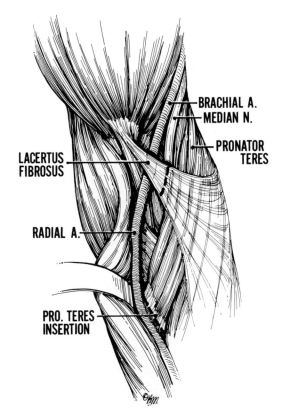

FIGURE 17D–1
Lacertus fibrosus covers the brachial artery and median nerve and may act as a compressive band to these structures, particularly with the forearm in supination. The lacertus fibrosus must be released to adequately explore the structures in this region. (Illustration by Elizabeth Roselius, © 1988. Reprinted with permission from Green, D. P. (Ed.), *Operative Hand Surgery*, 2nd ed. New York, Churchill Livingstone, 1988, p. 1428.)

tween the flexor digitorum superficialis and the flexor digitorum profundus and is loosely adherent to the former muscle by thin fascia [37].

In 85% of cases, the pronator teres muscle arises from two heads, superficial and deep, and the median nerve passes between them as it courses from the antecubital fossa to the forearm. The deep head separates the median nerve from the ulnar artery and vein. The superficial or humeral–ulnar head of the pronator teres arises from the common flexor origin and is attached to the medial epicondyle of the humerus, intermuscular septa, ulnar collateral ligament of the elbow joint, and a slender slip from the medial border of the coronoid process of the ulna. The deep or radial head is smaller and thinner, usually fibromuscular in structure, and arises from the upper two-thirds of the anterior border of the shaft of the radius and, as shown by dissections done by Solnitzky [70], in 15% of cases the median nerve has a different relationship to the heads of the pronator teres muscle. He noted passage of the median nerve deep to both heads of the muscle; passage of the nerve deep to the superficial head, the deep head being absent; and passage of the nerve through a two-layered superficial head [70].

Normally, the two heads of the pronator teres are joined by a strong fibrous band, the tendinous arch of the flexor digitorum superficialis, or "sublimis bridge" (Fig. 17D–3). Proximally, this arch represents a firm and sharp free border, beneath which pass the median nerve and the ulnar artery and vein. Occasionally, an abnormal fibrous band may extend from the deep or ulnar head of the pronator teres to the sublimis bridge. Such bands may be sufficiently strong to exert pressure in the median nerve at this level; this can be an additional source of compression.

Clinical Presentation

A pronator syndrome presents with pain in the proximal volar surface of the forearm that generally increases with activity. The symptoms are often vague, with a fatigue-like pain described in many cases. Repetitive, strenuous motions such as weight training, competitive driving, or underarm fastball pitching often provoke the symptoms. The symptoms usually develop insidiously, but occasionally a specific event triggers a sudden onset of pain in the forearm, which brings the patient to the attention of the sports physician.

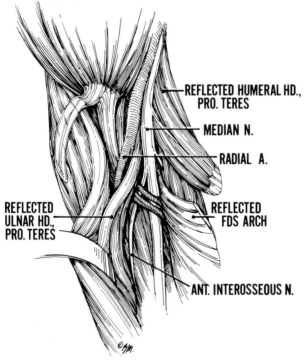

FIGURE 17D–3
The anterior interosseous nerve is identified by reflecting the FDS arch and elevating the ulnar head of the pronator teres. (Illustration by Elizabeth Roselius, © 1988. Reprinted with permission from Green, D. P. (Ed.), *Operative Hand Surgery,* 2nd ed. New York, Churchill Livingstone, 1988, p. 1430.)

In addition, there may be reduced sensibility or at least some sensory abnormality in the radial three and a half digits of the hand [7, 11]. The absence of distal paresthesia does not rule out the condition. There are usually no subjective complaints of hand weakness.

Physical Findings

There are four sites of potential compression, all of which can produce signs and symptoms of pronator syndrome. The first of these is compression of the median nerve in the distal third of the humerus, beneath a supracondyloid process at the ligament of Struthers [76]. If symptoms are aggravated by flexion of the elbow against resistance between 120 and 135 degrees of flexion, the surgeon must be suspicious of a Struthers ligament compression.

A second site of potential compression occurs at the lacertus fibrosis, which courses across the median nerve at the level of the elbow joint. On inspection, an indentation of the pronator muscle mass below the medial epicondyle suggests a lacertus fibrosis constrictive effect at that level (see Fig. 17D–1). The indentation may be increased by active or passive pronation of the forearm [73]. This is thought to occur predominantly when the median nerve is located superficial to and along the lateral edge of the flexor muscle mass, where the lacertus fibrosis crosses from the bicipital tendon across the flexor muscle mass [24, 25].

A third site of potential compression is within the pronator teres muscle, caused by hypertrophy of the pronator teres muscle itself, which can occur with underarm fastball pitching (see Fig. 17D–2). In addition, the aponeurotic fascia on the deep surface of the superficial head or the superficial surface of the deep head of the pronator teres may cause compression. If the patient's symptoms are increased by resistance to pronation of the forearm, usually combined with flexion of the wrist (to relax the flexor digitorum superficialis), the surgeon should be particularly careful to explore the median nerve as it passes through the pronator teres muscle [24, 25, 70]. Spinner and Linscheid suggest that resisted pronation for a period of 60 seconds may initiate the symptoms by prolonged contracture of the flexor pronator muscle. They also believe that direct pressure over the proximal portion of the pronator teres by the examiner's thumb, approximately 4 cm distal to the antebrachial crease, while exerting moderate resistance to pronation has been the most reliable test in their experience [73].

The final area of potential compression occurs at the arch of the flexor digitorum superficialis muscle, where the median nerve passes beneath that muscle to lie immediately deep to it within the muscle fascia (see Fig. 17D–3). If the symptoms of pronator syndrome are aggravated by resisted flexion of the superficialis muscle of the middle finger, the surgeon should be careful to inspect the superficialis arch at the time of exploration [25, 39]. Occasionally, passive stretching of the finger and wrist flexors accentuates the symptoms; this sign is unlikely to be positive if all the other tests have been negative.

Distal neurovascular function is generally intact; however, it is important to carefully document and compare muscle power in the median innervated musculature between the two hands. If there is weakness or paralysis of the flexor digitorum profundus of the index finger, the flexor pollicis longus, or possibly the pronator quadratus muscle, a characteristic posture of pinch is noted in this syndrome. This consists of hyperextension of the distal interphalangeal joint of the index finger and hyperextension of the interphalangeal joint of the thumb as one attempts to pinch the thumb to the index finger [25]. Confirmatory electrodiagnostic studies are necessary to evaluate such patients.

Electromyography

Buchthal and associates [11] have done an exhaustive review of the electrophysiologic findings in pronator syndrome; generally, these have been disappointing. Only 10% of patients had electromyographic findings that supported the clinical diagnosis. Nerve conduction studies are similarly disappointing in that velocity across the median nerve below the elbow is seldom abnormal. Spinner has suggested that the explanation for this relates to the size and complexity of the nerve. In short, the slowed impulses in affected fascicles are blurred and dampened in nerve conduction recordings because the majority of fascicles conduct normally. Accordingly, the clinical history and physical examination are the mainstays to be relied on for an accurate diagnosis.

Treatment

Operative Approach

The operative treatment for pronator syndrome consists of a detailed exploration of the median nerve in the proximal forearm. The exploration for pronator syndrome should begin with an incision at least 5 cm above the elbow joint over the medial neurovascular bundle, zig-zagging across the elbow flexion crease to lie midway between the flexor and extensor muscle masses and extending to the mid-forearm.

The supracondyloid process, an anomalous bony spur on the medial aspect of the lower third of the humerus, is found in approximately 1% of people of European ancestry; it is usually found approximately 5 cm above the medial epicondyle, projecting anteromedially from the surface of the humerus [6]. The apex of the process is often roughened and, in some cases, joined to the medial epicondyle by a band of fibrous tissue known as the ligament of Struthers [6, 76]. The foramen thus formed by the supracondyloid process and its ligamentous band allows passage of the median nerve. Frequently, it is accompanied by the brachial artery or one of its branches, and by the radial or ulnar artery. Following release of the ligament of Struthers, with or without resection of the supracondyloid process, the median nerve can be traced distally to the second area of potential compression, the fascia of the lacertus fibrosis. Laha and associates [42] described a thickening of the lacertus fibrosis that comprised a compression neuropathy of the median nerve. They believed that the compression is only recognized when the forearm is in pronation, which thereby tightens the lacertus fibrosis over the neurovascular bundle and the flexor muscle mass. Accordingly, the lacertus fibrosis should be divided routinely during surgery. The median nerve is then traced distally as it enters the forearm between the two heads of the pronator muscle [37]. After the entrance of the nerve into the pronator teres muscle has been inspected, the superficial head of the muscle can be elevated by dissecting the distal insertion of the head that inserts at the midportion of the radius. The superficial head, having been elevated, is divided along the course of its fibers and reflected ulnarly to explore the median nerve. Division and retraction of the superficial head of the pronator decompress the median nerve and also allow the surgeon to see the fourth and last potential compression site of the pronator syndrome (Fig. 17D–4).

The last site of potential compression is the flexor digitorum superficialis muscle arch. The aponeurotic arch, if compression in the nerve is released, constitutes the distal extent of the exploration for entrapment of the median nerve in the pronator syndrome (Fig. 17D–5).

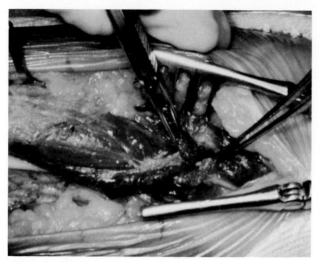

FIGURE 17D–4
The most common location of median nerve compression at the elbow occurs as it passes between the superficial head (forceps covering the nerve) and deep heads of the pronator teres muscle.

Postoperative Management

The superficial head of the pronator teres is loosely sutured to its tendinous insertion in the radial aspect of the wound or, more commonly, is simply laid back over the deep head to heal on its own. After routine closure, a posterior elbow splint is worn with the elbow in neutral rotation at about 90 degrees of flexion. Because there is no detachment of the common flexor origin, the patient can initiate active range-of-motion exercises after routine inspection of the wound at about the third or fourth postoperative day. If indeed the pronator teres muscle has been detached from its insertion from the radius, one would alter the rehabilitation program, keeping the arm maintained at some degree of pronation and allowing only flexion–extension. Forearm rotation is restricted until the pronator teres muscle is sufficiently healed, which takes approximately 3 weeks. This immobilization can be accomplished with a Munster-type cast, applied with the forearm in pronation and limiting full extension of the elbow [25].

Authors' Preferred Method of Treatment

We prefer to explore the median nerve 5 cm proximal to the elbow, making a curvilinear incision and tracing it distally to each of the four areas of compression to ensure that each potential site is divided and confirm that a comprehensive release

of the median nerve has been accomplished. To perform a complete exploration, one must begin from the area of the ligament of Struthers and proceed through the fibrous arch of the flexor digitorum superficialis muscle belly.

NEUROPATHIES OF THE RADIAL NERVE

Compression neuropathies of the radial nerve occur in predictable areas along the course of the nerve. Above the elbow, the most common offending agent appears to be the lateral intermuscular septum, where the nerve appears to be in jeopardy during open reduction and internal fixation of displaced fractures of the humerus. It is rare to release the radial nerve above the elbow for compression neuropathy, although it has been reported [37]. The most common compression neuropathy of the radial nerve at the elbow is known as the radial tunnel syndrome [25, 35, 49, 55].

Anatomic Considerations

The radial nerve pierces the lateral intermuscular septum to proceed from the posterior to the anterior compartment of the humerus [35]. At the level of the radiocapitellar joint, it divides into its major branches, the posterior interosseous and superficial radial nerves. At this level, it enters the radial tunnel. The radial tunnel consists of the structures

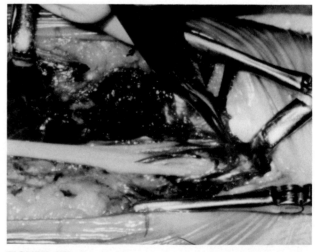

FIGURE 17D–5
The superficial head of the pronator muscle has been divided. Visualization of the aponeurotic arch of the flexor digitorum superficialis muscle is shown ready for division (forceps).

surrounding the radial nerve from the furrow between the brachioradialis and brachialis in the distal arm to the distal edge of the supinator in the proximal forearm. In the radial tunnel, the posterior interosseous nerve passes between the two heads of the supinator muscle. The proximal edge of the supinator forms an arch for the posterior interosseous nerve, the arcade of Frohse. The superficial radial nerve passes superficially to the supinator muscle and is covered anteriorly by the brachioradialis muscle (Fig. 17D–6).

Most authors agree that there are four sites of potential compression within the radial tunnel (Fig. 17D–7) [24, 25, 37, 52, 58, 64, 68, 74]. The first site consists of fibrous bands lying anterior to the radial head at the entrance to the radial tunnel. The second site occurs at a fan-shaped leash of vessels, the so-called leash of Henry, which may lie across the radial nerve to supply the brachioradialis and extensor carpi radialis longus muscles. The third site of potential compression occurs where the radial nerve courses just ulnar to the tendinous margin of the extensor carpi radialis brevis, as the radial nerve enters the supinator muscle. The fourth potential source of compression is the most common and occurs as the radial nerve enters the supinator

muscle through the arcade of Frohse. Recently, a fifth cause of radial tunnel syndrome has been described [71]. This occurs as the radial nerve passes through the supinator muscle and exits along its distal lateral border. A fascial arcade is often present here, lining the superficial head of the supinator muscle just above the exiting posterior interosseous nerve.

Other, less common causes that produce symptoms of radial nerve compression include the proliferating rheumatoid synovium from the radiocapitellar joint [47], fractures [5], vascular aberrations [23], anomalies [47], and tumors [10, 13, 49, 82, 85]. On occasion, the superficial radial nerve can be involved as well or in isolation [20]. There is also a double-entrapment radial tunnel syndrome, in which compression of the nerve occurs at both the arcade of Frohse and the distal border of the supinator [28].

Patient Demographics

Athletes presenting with radial tunnel syndrome most often perform repetitive rotatory movements of the forearm in conjunction with sporting activity, such as rowing, discus, or racquet sports. The ma-

FIGURE 17D–6
The anterolateral approach to the radial nerve provides the best exposure of the radial tunnel when the compressive lesion cannot be localized to the arcade of Frohse. (Illustration by Elizabeth Roselius, © 1988. Reprinted with permission from Eversman, W. W. Entrapment and compression neuropathies. In Green, D. P. (Ed.), *Operative Hand Surgery*. New York, Churchill Livingstone, 1988, p. 1458.)

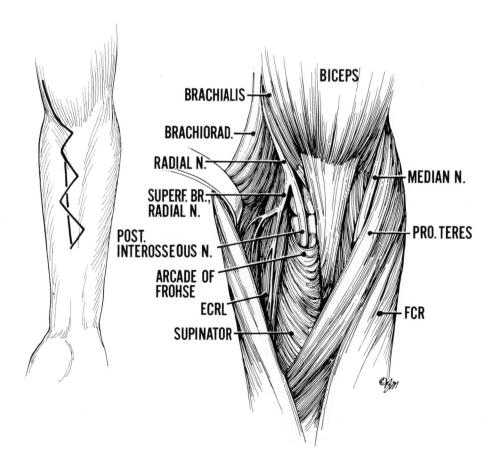

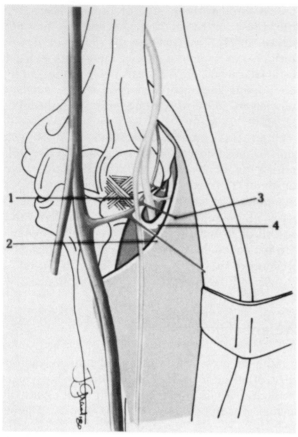

FIGURE 17D–7
Four potentially compressive anatomic elements. (1) Fibrous bands overlying radial head and capsule. (2) Fibrous origin of the extensor carpi radialis brevis. (3) Radial recurrent arterial fan. (4) Arcade of Frohse. (From Moss, S., and Switzer, H. Radial tunnel syndrome: A spectrum of clinical presentations. *J Hand Surg* 8:415, 1983.)

jority of people presenting with this syndrome come from an occupational background requiring heavy manual labor with repetitive motion. For example, in the series of Roles and Maudsley [65], only 4 of 36 patients had symptoms referable to sporting activities. Of those four, three played badminton and one played tennis. Other series, such as that of Werner [82], implicate tennis more often than any other sporting activity. In Werner's entrapment series of 85 patients, many different occupations were represented; however, all performed repetitive rotatory movements such as pronation and supination while at work, and there was a relation between work and increased symptoms of pain.

The patient with radial tunnel syndrome complains primarily of pain. The pain is usually well localized laterally to the extensor mass just below the elbow and is aching in character [45, 64, 82]. Repetitive movements that appear to intensify the pain consist of forearm pronation, often with wrist flexion. There may also be weakness of grip because pain is caused by the use of wrist dorsiflexion. In Werner's series [82], night pain appeared to be a particularly prominent feature in this syndrome, as well as pain after physical exertion.

Clinical Examination

Examination of the patient with radial tunnel syndrome reveals three pathognomonic signs. First, tenderness to palpation is most severe over the radial nerve palpated through the mobile wad muscle mass at and just distal to the radial head. On occasion, pain can be localized to the distal end of the supinator and occasionally down into the forearm [37,71].

Second, resisted extension of the middle finger with the elbow extended produces pain at the site of the previously elicited tenderness [45, 64, 65]. The nerve is compressed by a fascial extension from the extensor carpi radialis brevis muscle during this maneuver. This sign was not confirmed by Werner [82]. In his investigation, there was no relation between preoperative pain on extension of the middle finger and observations at surgery of a fascial extension or arch from the extensor carpi radialis brevis crossing the nerve. Nevertheless, this remains an important diagnostic physical finding.

The third sign is similar pain on resisted supination of the extended forearm. This is distinguished in the material of Lister and associates [45] from pain localized to the lateral epicondyle that is increased on flexion of the wrist and all fingers with the elbow extended. It has been reported by many authors [20, 25, 28, 45, 54] that there are occasional paresthesias in the distribution of the superficial radial nerve; this is the experience of most authors, however, and has come to be known as an unreliable sign.

Electromyographic Studies

Neurophysiologic investigations concerning posterior interosseous nerve entrapment have been scanty. Roles and Maudsley found delay in motor conduction velocity in "some cases" [65]. Two cases showed a revision to normal latency after decompression. Electromyographic (EMG) evidence of a diagnosis was reported in seven of nine cases as well. Werner [82] also demonstrated some decrease in motor conduction velocity in the radial nerve across the entrapment site in 13 of 25 cases examined, and EMG changes in muscles innervated distal to the entrapment site in 8 of 25 cases. Although in this study there was complete relief of pain in most cases with positive electrophysiologic findings, equally good results were obtained where findings were interpreted as normal. Werner concluded that a normal electrophysiologic finding does not exclude the entrapment diagnosis; however, these conduction delays and positive EMG findings may be observed and can be helpful.

This has not been the experience of Van Rossum and associates [78], who found no electrophysiologic evidence of entrapment in 10 cases of resistant tennis elbow. In this series, however, no mention was made of the presence or absence of signs of nerve entrapment. We have to conclude from this review that EMG studies, although not diagnostic, can be helpful in confirming a diagnosis of radial tunnel syndrome.

Resistant Tennis Elbow—Radial Nerve Compression

Werner has done the most extensive work in this regard [82]. In his series of 203 patients treated for lateral epicondylitis, 5% had coincident posterior interosseous nerve entrapment based on the number of cases requiring decompression of the posterior interosseous nerve. Heyse-Moore [35], conversely, believed that in two well-matched groups of patients, results of extensor carpi radialis brevis (ECRB) lengthening and radial tunnel decompression were the same. He thought that this was related to the blended origin of the ECRB and supinator muscles to the lateral epicondyle, capsule of the elbow, and orbicular ligament. Contraction of either of these muscles exerts an equal pull on these structures. A surgical division of the superficial part of the supinator has the same effect as lengthening the tendon of the ECRB. In his study, there was no evidence of radial nerve entrapment in resistant tennis elbow. This study elaborates the work of Van Rossum and associates [78], who found no clinical or electrical evidence of radial nerve entrapment in resistant tennis elbow.

Radial Tunnel Syndrome

Treatment

In the athlete suspected of having radial tunnel syndrome, we have used diagnostic blocks with small amounts of lidocaine administered at various points along the radial nerve. When the instillation of 0.5

to 1 mL of 1% lidocaine four fingerbreadths distal to the lateral epicondyle relieves pain and is accompanied by deep radial palsy, and a complementary injecton more proximal in the region of the lateral epicondyle (usually given 24 to 48 hours later) does not relieve the patient's symptoms, the diagnosis of radial tunnel compression is made.

The basic operation for neuropathies of the radial nerve secondary to entrapment at the level of the radial tunnel is exploration and surgical decompression of this peripheral nerve. Exposure of the radial nerve within the radial tunnel can be accomplished by using either of two surgical approaches.

Transbrachioradialis Approach. This approach has been popularized by Lister and associates and involves an incision beginning 2 cm proximal to the radial head and 3 cm lateral to the bicipital tendon [45]. He first advocated a straight incision, but changed to a lazy-S incision to reduce the unsightly scarring that can result in some patients. The lateral cutaneous nerve of the forearm is preserved if it is encountered, and the fascia of the brachial radialis is incised, revealing the muscle belly of the brachial radialis, which is then split. The muscle belly is thick in this location, and progressively deeper retraction is needed as one proceeds through the brachial radialis muscle belly. The split in this muscle belly should then be lengthened to equal the skin incision and to expose the entire tunnel. The fibrous edge of the extensor carpi radialis brevis can be divided transversely. The arcade of Frohse may be divided as well, until no compression on the nerve is produced by forearm pronation or wrist flexion [45] (Fig. 17D–8). This approach has also been advocated by Roles and Maudsley [65], citing the advantage of limited dissection and thus limited morbidity for the patient in the postoperative period. It is very

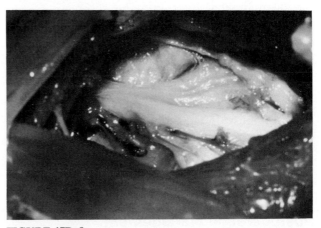

FIGURE 17D–8
The transbrachioradialis approach to the radial nerve allows limited exposure, but is effective for release of the arcade of Frohse and extensor carpi radialis brevis.

important during this approach to pronate the fore-arm passively and volar-flex the wrist to ensure that there is no potential for compression of the radial nerve by the sharp tendinous margin of the extensor carpi radialis brevis. The disadvantage of this approach is the limited anatomic exposure that can be developed through this incision, and, in our hands, inability to localize the lesion to the arcade of Frohse is a contraindication to the use of this limited anatomic approach. This is also the opinion of others [24, 25].

Anterolateral Approach. This is a more generalized approach to decompression of the radial tunnel. It is indicated in any patient with entrapment neuropathy of the radial nerve that cannot be isolated preoperatively to the arcade of Frohse. In this operative procedure, the dissection is carried from the anterior surface just above the elbow joint, tracing the radial nerve distally into and through the radial tunnel, with care taken to débride all the superficial fibers of the supinator muscle and protect the radial nerve in the process (see Fig. 17D–6).

Briefly, the radial nerve is located between the brachial radialis and brachialis muscles; at this point, fibrous bands lying anterior to the radial head are freed from the radial nerve. Likewise, somewhat more distally, the radial recurrent vessels that form a vascular arcade anteriorly across the radial nerve are ligated to free the nerve from compression to the level of the extensor carpi radialis brevis.

Because the surgical dissection is done with the forearm in supination, it is important to pronate the forearm passively and volar-flex the wrist at this point to ensure that the border of the extensor carpi radialis brevis is not causing a dynamic constriction of the radial nerve. This fibrous margin should be excised if compression is noted. Finally, the exploration is continued to the arcade of Frohse, where the deep branch dives into the substance of the supinator muscle beneath the arcade. The arcade should be divided with care, so that the branch of the radial nerve to the superficial head of the muscle is not transected during division of the arcade. This final exploration can be done with the forearm in marked pronation and with the superficial layer of extensor muscles retracted laterally to expose the entire supinator muscle. Complete division of the supinator muscle allows a visualization of the radial nerve to the point where it arborizes as it leaves the distal margin of this muscle (Fig. 17D–9).

Postoperative Management.

A thorough visualization of the entire radial nerve through the course of the radial tunnel is required,

FIGURE 17D–9
Exposure of the radial nerve from the anterolateral approach. The superficial branch of the radial nerve is retracted, and the posterior interosseous nerve is lying beneath the arcade of Frohse *(arrow)*.

and the wound is closed in layers with subcutaneous and skin sutures placed in the usual fashion. A bulky dressing with plaster reinforcement is used for the first 5 to 7 days, with early range of motion beginning 1 week postoperatively. Therapists may fashion a sand splint to be worn for comfort as range of motion progresses back to normal in flexion–extension and pronation–supination. A strengthening program begins after restoration of full range of motion. Resumption of the sport of choice is encouraged when strength of the upper extremity is 80% of the opposite extremity as measured by Cybex testing. Symptoms of paresthesia resolve early. If there has been damage by compressive neuropathy of the radial nerve, the recovery of this function appears to be prolonged over a 3- to 4-month period. Full recovery can, however, be expected to take place eventually [24].

Authors' Preferred Method of Management

Unless we are convinced by accurate physical examination and selective blockage of pain with local anesthesia that the compressive neuropathy is at the arcade of Frohse, we generally decompress the radial tunnel by an anterolateral approach. In this way, our ability to evaluate the entire nerve from above the elbow through the radial tunnel to the distal end of the supinator is complete. This aspect of dealing with this compressive neuropathy cannot be overemphasized. On occasion, if there is a combined lateral epicondylitis and a radial nerve compression, an incision slightly anterior to the standard approach for lateral epicondylitis is made. The skin and subcutaneous tissue are moved forward and backward to deal with the lateral epicondylitis, and then a decompression of the radial tunnel at the arcade of Frohse is performed through a transbrachial radialis approach, as described by Lister and associates [45]. We have not been pleased with the results of this more limited approach to the radial tunnel, and believe that the more extensive exposure is indicated in the majority of cases.

ULNAR NEUROPATHY AT THE ELBOW (CUBITAL TUNNEL SYNDROME)

Ulnar neuropathy at the elbow can be related to a number of etiologic factors. These can range from a muscular source, such as the anconeus epitrochlearis [36, 66], to compression from the aponeurosis of the flexor carpi ulnaris or adhesions within the "cubital tunnel" [27]. Bony abnormalities about the elbow have also been cited as a predisposing cause [56].

Although athletes may indeed fall victim to any of the above problems, they are more likely to develop ulnar nerve irritation on a mechanical basis. This is usually an overuse injury or the result of repetitive tension, compression, or friction, the latter condition being the result of repetitive subluxation of the ulnar nerve across the medial epicondyle. The majority of athletic endeavours that result in the cubital tunnel syndrome involve throwing or racquet sports. It has, however, also been reported as a complication of weight lifting.

Both anatomy and biomechanical forces applied to the elbow joint during the act of pitching must be considered when discussing ulnar neuritis in the athlete.

Anatomy

The ulnar nerve is formed as a continuation of the medial cord of the brachial plexus. It passes from the anterior into the posterior compartment of the arm, where it penetrates a thick fibrous raphe first described by Struthers [75]. The arcade of Struthers is located (Fig. 17D–10A) approximately 8 cm above the medial epicondyle and represents the most proximal source of ulnar nerve entrapment,

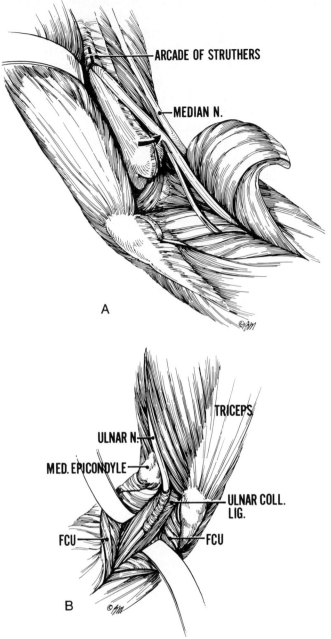

FIGURE 17D–10
The ulnar nerve emerges from the medial margin of the triceps and passes behind the medial epicondyle and under a retinaculum in the cubital tunnel to lie between the two heads of the flexor carpi ulnaris. (Illustration by Elizabeth Roselius, © 1988. Reprinted with permission from Green, D. P. (Ed.), *Operative Hand Surgery*. New York, Churchill Livingstone, 1982, p. 974.)

either primarily or after ulnar nerve transposition about the elbow.

Further distally, the ulnar nerve courses posterior to the medial epicondyle and enters the cubital tunnel. The roof of the tunnel is formed by the triangular arcuate ligament that extends from the medial epicondyle to the medial border of the olecranon process and acts as a tendinous origin of the flexor carpi ulnaris muscle. The remaining boundaries of the cubital tunnel are the ulnar collateral ligament of the elbow, the medial edge of the trochlea, and the medial epicondylar groove (see Fig. 17D–10*B*) [79]. After exiting the cubital tunnel, the ulnar nerve then continues into the forearm between the humeral and ulnar heads of the flexor carpi ulnaris muscle. Ulnar nerve compression within the cubital tunnel is a function of both tunnel boundaries and ulnar nerve changes itself.

The volume in the cubital tunnel is reduced with elbow flexion. Dangles and Bilos [18] have shown that the anterior bundle of the ulnar collateral ligament tends to relax and bulge slightly with elbow flexion, reducing the volume in the cubital tunnel. Wadsworth [79] has also shown that, with increasing angles of flexion, the arcuate ligament stretches at its proximal edge, becoming quite taut at 90 degrees of flexion. Both factors result in external compression of the ulnar nerve.

The ulnar nerve itself is a dynamic structure. It undergoes change throughout the arc of flexion of the elbow. Apfelberg and Larsen [2] have shown that it both elongates and is moved medially by the medial head of the triceps through elbow flexion. Hence, tethering of the nerve by scar tissue interferes with its normal longitudinal and transverse mobility during elbow flexion and adversely affects its function.

Childress [16] has also noted a hypermobility of the ulnar nerve secondary to either congenital or developmental laxity of soft tissue constraints that normally restrain it anatomically in the epicondylar groove. In a study of the normal population, he noted that 16.2% of people demonstrated recurrent dislocation of the ulnar nerve as the elbow was moved into flexion [16]. Friction neuritis was more common in type B patients, or those whose ulnar nerve completely dislocated anterior to the epicondyle, than in type A patients, who had an incomplete dislocation over the medial epicondyle. The latter group is more susceptible to direct trauma to the ulnar nerve. This may explain why athletes such as gymnasts, who have developed expertise in their field not only through ability but through hypermobility, are at increased risk of friction ulnar neuritis from subluxation.

Subluxation ulnar neuritis appears to affect baseball pitchers more than any other athlete. Failure of a baseball pitcher to reach his peak or to maintain a sustained performance appears to be directly related to symptoms arising from the shoulder or elbow. Wilson and associates [84] have shown that considerable tensile forces are generated in the medial side of the elbow and that compressive forces are predominant at the radiocapitellar articulation; these are secondary to the repetitive valgus movements applied to the elbow during the act of pitching. These movements may cause impingement of the medial tip of olecranon articular cartilage in the olecranon fossa, with a resultant area of chondromalacia. This impingement does not occur in all pitchers, but in the professional pitcher a series of adaptive physical changes take place in conjunction with impingement. These have been identified by King and colleagues [40], who analyzed the pitching arms of 50 professional baseball players. Over 50% demonstrated flexor contractures with increased valgus deformities. In addition, bony changes included hypertrophy of the humerus and enlargement of the olecranon and coronoid fossae. These changes can also occur in top-level tennis players [38].

From these various descriptions, one can see that fixed flexion and valgus deformities can increase the possibility of a traction neuritis. Indeed, that has been described by two studies of adolescent pitchers. Godshall and Hansen [32] described medial collateral ligament attenuation and rupture with resultant ulnar irritation and palsy due to subluxation and friction of the ulnar nerve. In a similar age group, Hang [34] described a case report of ulnar neuropathy in a pitcher due to compression of the ulnar nerve between the aponeurosis joining the heads of the flexor carpi ulnaris and the floor of the tunnel formed by the medial ligament of the elbow, the transverse ligament. Hang hypothesized that the pathogenesis was due to repetitive microtear of the arcuate aponeurosis, with subsequent edema and hyperemia producing scarring and thickening of the aponeurosis, which in turn encroached on the ulnar nerve. This entrapment has also been noted in the past by Osbourne [59] and is often seen in conjunction with hypertrophy of the flexor musculature of the forearm.

The pathophysiology of ulnar neuritis in the pitching athlete may also be a consequence of ischemic insult to the ulnar nerve. Pechan and Julius [62] demonstrated that marked elbow flexion with wrist extension and abduction of the shoulder, the attitude of the upper extremity as one initiates a forward pitch, can raise the intraneural pressure up to six times higher than it is in the relaxed state. They believed that the raised intraneural pressure was caused by both the physiologic stretch of the nerve and additional external compression from the overlying aponeurosis of the flexor carpi ulnaris muscle. Indeed, if capillary perfusion pressures are exceeded, this could render the nerve repetitively ischemic with each pitch thrown. This may explain the increase in ulnar neuritis in adolescents as they increase in age and exposure to pitching, as shown by Grana [33].

Clinical Presentation

The clinical findings of ulnar neuritis at the elbow in the athlete involve pain of either a lancinating or an aching quality at the medial side of the proximal forearm, which may radiate proximally or distally and may be accompanied by paresthesias, dysesthesias, or anesthesia in the ulnar one and one-half digits. These paresthesias are encountered early and usually precede any detectable motor weakness of the hand. Weakness of the flexor carpi ulnaris and flexor digitorum profundus muscles is rarely encountered because branches of these muscles occur right at the cubital tunnel and are most deeply situated in the tunnel and therefore usually spared [38]. Muscle wasting of the ulnar innervated intrinsic muscles of the hand is a late finding but is not uncommon. Clumsiness or heaviness of the hand and fingers, especially after pitching a few innings, may be a primary complaint.

The athlete with a recurrent dislocation of the ulnar nerve may complain of a painful snapping or popping sensation when the elbow is rapidly flexed and extended, with sharp pains radiating into the forearm and hand [16, 38].

An important localizing physical finding on examination is a positive percussion test over the ulnar nerve at the elbow. This may send shocklike waves into the ulnar aspect of the hand, and is known as a positive Tinel's sign [24]. The ulnar nerve may be manually subluxated or dislocated from the ulnar groove with abnormal mobility as well [16]. Here, the nerve indeed may feel thickened or doughy and may be quite tender to palpation.

The elbow flexion test as described by Buehler and Thayer [12] either increases or incites symptoms of ulnar nerve compression at the elbow. Both elbows are fully but not forcefully flexed, with full extension of the wrist to maximize both compressive and tensile forces on the nerve. Patients note numbness and tingling through the dermatomal distribution of the ulnar nerve. Pain was not limited to the ulnar nerve distribution exclusively in the series of 13 patients reported by Buehler and Thayer. The symptom complex, its rapid onset, and its rapid resolution demonstrate this to be a useful, reliable, and provocative test for cubital tunnel syndrome.

Finally, a thorough neurologic examination with emphasis on the ulnar nerve should be undertaken to complete the physical examination. A complete series of elbow roentgenograms, including a cubital tunnel view as described by Wadsworth, should be performed [79].

Electrodiagnostic studies of conduction velocities across the elbow are often done as an aid in diagnosis. These studies have been well outlined by Gilliatt and Thomas [30], who have provided some guidelines for clinicians. Loss of conduction velocity of less than 25% compared with other velocities recorded above or below the elbow is said to be insignificant. Nerve conduction velocities that are reduced by more than 33% from those above or below the elbow are, however, always significant and suggestive of the cubital tunnel syndrome.

As in other compressive neuropathies, the use of electrodiagnostic studies in isolation can be fraught with difficulty. It is always important to consider nerve compression at another level, such as the cervical ribs, scalenus anticus syndrome, superior sulcus tumor, cervical rib protrusion with radiculopathy, compression at Guyon's canal, or compression of the deep branch of the ulnar nerve in the hand, which may all produce symptoms along the ulnar nerve distribution and should be specifically ruled out by careful history and physical examination [38].

Treatment

Nonoperative

McGowan's method of grading the degree of ulnar neuropathy has been used by many surgeons [51]. In McGowan's stage 1, the symptoms of ulnar neuropathy are subjective only and consist of minor hypoesthesia and paresthesias. The intermediate stage 2 has accompanying weakness and wasting of the interossei in addition to hypoesthesia. The final stage 3 is marked by weakness and wasting of the interossei, adductor pollicis, and hypothenar muscles, in addition to a complete or partial anesthesia of the ulnar one and one-half digits of the hand.

Nonoperative management is attempted with a minimum degree of compression of the ulnar nerve, or McGowan's stage 1. The nonoperative approach consists of minimizing the pressure increases that occur about the ulnar nerve with elbow flexion or direct contact. This can be accomplished by the use of elbow pads when a patient is on bed rest with a postoperative paresthesia from an unrelated procedure. Icing of the area helps prevent edema and subsequent inflammation about the nerve that could result in scarring. Approximately one-half of the patients with this grade of compression can be expected to recover with this form of treatment [19]. For a moderate degree of compression, one must adopt operative intervention when a nonoperative technique is unsuccessful.

Operative Techniques

There are basically four types of operative procedure that can be performed for ulnar neuropathy at the elbow. They are simple decompression, subcutaneous anterior transposition, submuscular anterior transposition, and medial epicondylectomy.

Decompression. The simplest procedure for ulnar nerve entrapment at the elbow is decompression. This is only applicable if there is a localized compression of the ulnar nerve by the aponeurosis between the heads of the flexor carpi ulnaris or by the muscle itself.

The most useful localizing sign of compression is a positive percussion test over the ulnar nerve because the ulnar nerve lies between the superficial and deep heads of the flexor carpi ulnaris muscle. This preoperative finding, combined with the finding at operation of an indentation of the nerve beneath the stout portion of the aponeurosis overlying the heads of the flexor carpi ulnaris, is the most reliable indication that the site of compression of the nerve is indeed localized and can be relieved by decompression [26]. It has also been known as an Osbourne lesion [59]. Osbourne has described dividing the aponeurotic band that comprises the fibrous arcade of the flexor carpi ulnaris muscle lying over the top of the ulnar nerve as the nerve is traced distally between the two heads of this muscle.

If there is no localized site of compression between the two heads of the flexor carpi ulnaris, with prestenotic swelling of the nerve as described by Gilliatt and Thomas [30], or if the nerve is subluxatable following simple decompression, then the nerve should be transposed anteriorly.

Anterior Transposition of the Ulnar Nerve. Anterior transposition is appropriate when a decompressive procedure has failed, for subluxation of the ulnar nerve, for persistent or progressive valgus deformity at the elbow, and for moderate compressive neuropathy as described by McGowan. The main advantage of anterior transposition is that all pathology is correctable because the nerve is transposed from the cubital tunnel into its new bed anterior to the medial epicondyle in a subcutaneous position or submuscularly deep to the common flexor origin in a well-vascularized muscle bed.

Subcutaneous Anterior Transposition. This operative procedure involves a curvilinear incision posterior to the medial epicondyle and parallel with the ulnar nerve, after identification of the ulnar nerve proximally at the level of the arcade of Struthers, 8 cm above the medial epicondyle [75]. The ulnar nerve lies along the medial head of the triceps, and any encroachment of the muscle on the ulnar nerve [66] or due to hypertrophy of the anconeus epitrochlearis [36] is identified and decompressed. The ulnar nerve is traced distally from this point to the aponeurosis of the flexor carpi ulnaris behind the medial epicondyle, which is released.

The distal exploration is continued to the interval between the heads of the flexor carpi ulnaris musculature, where the arcade and fascia of the flexor carpi ulnaris are divided in the proximal third of the forearm.

The surgeon then elevates the nerve from its bed, taking care to include both the arteries and the veins accompanying the ulnar nerve in a bulk of tissue, which is transposed. Prior to transposition, the medial intermuscular septum is divided and the arcade of Struthers is released so that a new site of compression is not created with the transposition. Following the transposition, a suture from the anterior skin flap to the medial epicondyle is placed so that the nerve will not become relocated posterior to the medial epicondyle.

POSTOPERATIVE MANAGEMENT. A bulky dressing is used after the operative procedure for protection of the extremity and the ulnar nerve in its transposed position. This is used for a week to 10 days. After this, gentle mobilization to regain range of motion is initiated, followed by strengthening activity. Patients are usually back to full athletic endeavors at 3 to 4 weeks after surgery.

Submuscular Transposition. A similar operative procedure is carried out to mobilize the ulnar nerve itself. Particular attention is paid again to preservation of the medial brachial and medial antebrachial cutaneous nerves of the forearm. After immobilization of the ulnar nerve, the flexor muscles are elevated from the medial epicondyle, including the superficial head of the flexor carpi ulnaris, the flexor carpi radialis, the palmaris longus, the pronator teres, and a portion of the flexor digitorum superficialis [25, 44]. After mobilization of these muscle bellies, the origin is reflected from the medial epicondyle using sharp dissection; the muscle origin is reflected distally, preserving innervation of the superficial head of the flexor carpi ulnaris. The nerve is transposed into the submuscular position (Fig. 17D–11); before closing the common flexor origin, the nerve is explored in detail proximally to ensure that there is no entrapment in the muscular septum or arcade of Struthers. The flexor aponeurosis is

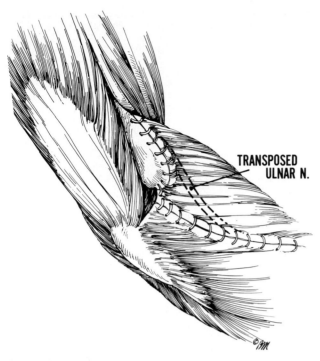

FIGURE 17D–11
Submuscular translocation of the ulnar nerve requires adequate proximal release at the arcade of Struthers. The nerve is brought laterally to lie near the median nerve. (Illustration by Elizabeth Roselius, © 1988. Reprinted with permission from Green, D. P. (Ed.), *Operative Hand Surgery,* 2nd ed. New York, Churchill Livingstone, 1988, p. 1447.)

then undertaken, assuring that there is no compression by the flexor origin across the ulnar nerve.

POSTOPERATIVE MANAGEMENT. After a routine closure, the postoperative management is the same as that in patients undergoing a subcutaneous anterior transposition, except that there is no active strengthening of the common flexor origin for 4 to 6 weeks in order to allow adequate repair of the common flexor origin. According to Learmouth, the submuscular anterior transposition offers the best location for the nerve, placing the nerve back in a "normal intermuscular interval adjacent to the median nerve" [43]. Although this technique forces the surgeon to develop a heightened awareness of the future potential sites of recurrent ulnar compression and their subsequent avoidance, it appears to offer the best results in patients with moderate compression neuropathy, achieving approximately 80% excellent results with the lowest recurrence rate [19].

The reported rates of failure after operative decompression of the ulnar nerve have varied widely, from 0% to as much as 78% [29]. Gobel and Amadio [31] reported on 30 patients who had undergone a revision of failed decompression of the ulnar nerve at the elbow. They found multiple sources of compression of the nerve at repeat surgery and in all cases decompressed the nerve from the arcade of Struthers to the deep flexor pronator aponeurosis. They also believed that after the initial procedure failed to relieve symptoms or when symptoms recurred, a submuscular transposition would be an effective management. They did not believe that internal neurolysis would improve the results of treatment.

Medial Epicondylectomy. King and Morgan thought that removal of the medial epicondyle was the most appropriate form of management for ulnar neuritis [41]. They believed that there was minimum exposure of the nerve, with no damage to its muscular branches or interference with its blood supply. Because tension on the nerve as it passed behind the medial epicondyle was removed with excision of the medial epicondyle, the nerve could slide forward and seek its own position and optimum tension.

The potential disadvantages of this procedure can be the creation of new sources of compression as the nerve slides forward after the medial epicondylectomy, with compression created at the medial intermuscular septum or arcade of Struthers, which is not addressed during the classic description of this procedure. After detachment of the common flexor origin, the anterior band of the medial collateral ligament may indeed be sacrificed by too generous an excision [25].

The operative approach for this is a small skin incision, parallel to the ulnar nerve, approximately 8 cm in length. The epicondyle is exposed by sharp subperiosteal dissection, reflecting off the common flexor pronator origin and part of the anterior bundle of the medial collateral ligament. The adjacent supracondylar ridge is also exposed in the entire medial epicondyle, and the ridge is removed using rongeurs, an osteotome, or bone saw. The natural guide for the proper plane of the osteotomy is the medial border of the trochlea [25]. Bone wax is used prior to periosteal closure. The ulnar nerve is not exposed in the classic description of the operation [41], but should be protected throughout the course of the operation in order to avoid injury to it.

Following the procedure, range-of-motion exercises are initiated almost immediately, followed by a strengthening program.

Authors' Preferred Method of Treatment

Athletes with repetitive flexion–extension and valgus movements applied to the elbow, such as in active pitching, require a careful preoperative evaluation and selection of procedure. Indeed, it is only the rare case that does not require anterior transposition of the ulnar nerve into a subcutaneous location. Decompression of the ulnar nerve should be complete and extend proximally from the arcade of Struthers to a release of the medial muscular septum, the fibrous arcuate ligament behind the medial epicondyle, and the fascial slips of the flexor carpi ulnaris to a distal extent approximately 8 cm between the heads of the flexor carpi ulnaris where the arcade and fascia of the flexor carpi ulnaris are divided. After ulnar nerve transposition, care is taken not to create a new site of constriction along the subcutaneous course of the ulnar nerve. When exposing the ulnar nerve, one must be extremely cautious to avoid injury to the medial brachial or medial antebrachial cutaneous nerves of the forearm, which are localized and protected throughout the exposure. The nerve is placed in a bed of muscle, created by dividing the fascia on the common flexor origin. Subcutaneous tissue is placed over the nerve and sutured to the medial epicondyle to avoid the possibility of relocation of the nerve posterior to the medial epicondyle.

After a routine closure, we initiate early range-of-motion exercises, followed by a strengthening program.

If a throwing athlete has developed medial epicondylitis in addition to ulnar neuritis or neuropathy, we often do a submuscular transposition of the ulnar nerve and resect the diseased and degenerative tendon from the common flexor origin as it is resected off the medial epicondyle. With this technique, the nerve is freed proximally, again to the arcade of Struthers down through the fascial arcade of the flexor carpi ulnaris. After placement of the nerve in its muscular bed, the fascia of the common flexor origin is sutured back to the medial epicondyle, ensuring that there is no tension over the ulnar nerve itself and that no secondary compressive sites have been created by the transposition. After routine closure, we initiate range-of-motion exercises in the immediate postoperative period. We do not initiate any strengthening program for 3 to 4 weeks, to allow the common flexor origin to heal appropriately.

References

1. Ailson, A. E. The surgical treatment of progressive ulnar paralysis. *Collective Papers of the Mayo Clinic* 10:944, 1918.
2. Apfelberg, P. B., and Larson, S. J. Dynamic anatomy of the ulnar nerve at the elbow. *Plast Reconstr Surg* 51:76, 1973.
3. Ashenhurst, E. M. Anatomical factors in the etiology of ulnar neuropathy. *Can Med Assoc J* 87:4, 1962.
4. Atwater, A. E. Biomechanics of overarm throwing movements and of throwing injuries. *Exerc Sport Sci Rev* 7:43, 1979.
5. Austin, R. Tardy palsy of the radial nerve from a Monteggia fracture. *Injury* 7:202–204, 1976.
6. Barnard, L. B., and McCoy, M. S. The supracondyloid process of the humerus. *J Bone Joint Surg* 28:845–950, 1946.
7. Beaton, L. E., and Anson, B. J. The relation of the median nerve to the pronator teres muscle. *Anat Rec* 75:23–26, 1939.
8. Bell, G. E., and Goldner, J. L. Compression neuropathy of the median nerve. *South Med J* 49:966–972, 1956.
9. Blakemore, M. E. Posterior interosseous nerve paralysis caused by lipoma. *J R Coll Surg Edinb* 24:113, 1979.
10. Bowen, T. L., and Stone, K. H. Posterior interosseous nerve paralysis caused by a ganglion at the elbow. *J Bone Joint Surg [Br]* 48:774–776, 1966.
11. Buchthal, F., Rosenfalck, A., and Trojaborg, W. Electrophysiological findings in entrapment of the median nerve to wrist and elbow. *J Neurol Neurosurg Psychiatry* 37:340, 1974.
12. Buehler, M. J., and Thayer, D. T. The elbow flexion test—A clinical test for cubital tunnel syndrome. *Clin Orthop* 233:213, 1988.
13. Campbell, C. S., and Wolf, R. F. Lipoma producing a lesion of the deep branch of the radial nerve. *J Neurosurg* 11:310–311, 1954.
14. Capener, N. Tennis elbow and posterior interosseous nerve. *Br Med J* 2:130, 1960.
15. Capener, N. The vulnerability of the posterior interosseous nerve of the forearm: A case report and anatomical study. *J Bone Joint Surg [Br]* 48:770–773, 1966.
16. Childress, H. M. Recurrent ulnar nerve dislocation at the elbow. *Clin Orthop* 108:168, 1975.
17. Craven, R. P., and Green, D. P. Cubital tunnel syndrome: Treatment by medial or epicondylectomy. *J Bone Joint Surg [Am]* 62:986–989, 1980.
18. Dangles, C. J., and Bilos, Z. J. Ulnar neuritis in a world champion weight lifter. *Am J Sports Med* 8:443, 1980.
19. Dellon, A. L. Review of treatment results for ulnar nerve entrapment at the elbow. *J Hand Surg [Am]* 14:688–700, 1987.
20. Dellon, A. L., and Mackinnon S. E. Radial sensory nerve entrapment in the forearm. *J Hand Surg [Am]* 11:199–205, 1986.
21. Dellon, A. L., and Mackinnon, S. E. Human ulnar neuropathy at the elbow: Clinical, electrical and morphometric correlations. *J Reconstr Microsurg* 4:179–184, 1988.
22. Del Pizzo, W., Jobe, F. W., and Norwood, L. Ulnar nerve entrapment in baseball players. *Am J Sports Med* 5:182, 1977.
23. Dharapak, C., and Nimberg, G. A. Posterior interosseous nerve compression: Report of a case caused by traumatic aneurysm. *Clin Orthop* 101:225–228, 1974.
24. Eversmann, W. W. Compression and entrapment neuropathies of the upper extremity. *J Hand Surg* 8:759–766, 1983.
25. Eversmann, W. W. Entrapment and compression neuropathies. *In* Green, D. P. (Ed.), *Operative Hand Surgery*, vol. 2 (2nd ed.). New York, Churchill, Livingstone, 1988, pp. 1423–1478.
26. Feindel, W., and Stratford, J. The role of the cubital tunnel in tardy ulnar palsy. *Can J Surg* 1:287–300, 1958.
27. Friedman, R. J., and Cochran, T. P. A clinical and electrophysiological investigation of anterior transposition for ulnar neuropathy at the elbow. *Arch Orthop Trauma Surg* 106:375–380, 1987.
28. Gassel, M. M., and Diamantopoulos, E. Pattern of conduction times in the distribution of the radial nerve. *Neurology* 14:222–231, 1964.
29. Gay, J. R., and Love, J. G. Diagnosis and treatment of tardy paralysis of the ulnar nerve based on a study of 100 cases. *J Bone Joint Surg* 29:1087–1097, 1947.
30. Gilliatt, R. W., and Thomas, P. K. Changes in nerve conduction with ulnar nerve lesions at the elbow. *J Neurol Neurosurg Psychiatry* 23:312–320, 1960.

31. Gobel, G. T., and Amadio, P. C. Reoperation for failed decompression of the ulnar nerve in the region of the elbow. *J Bone Joint Surg [Am]* 72:213–219, 1990.

32. Godshall, R. W., and Hansen, C. A. Traumatic ulnar neuropathy in adolescent baseball pitchers. *J Bone Joint Surg [Am]* 53:359–361, 1977.

33. Grana, W. A. Pitcher's elbow in adolescents. *Am J Sports Med* 8:333–336, 1980.

34. Hang, Y. Tardy ulnar neuritis in a little league baseball player. *Am J Sports Med* 9:244, 1981.

35. Heyse-Moore, E. H. Resistant tennis elbow. *J Bone Joint Surg [Br]* 9:64–66, 1984.

36. Hirasawa, Y., Sawamaura, H., and Sakakida, K. Entrapment neuropathy due to bilateral epitrochleoanconeus muscles: A case report. *J Hand Surg* 4:181–184, 1979.

37. Hollingshead, W. H. The back and limbs. *In Anatomy for Surgeons*, Vol. 3. New York, Harper & Row, 1964, pp. 419–422.

38. Jobe, F. W., and Fanton, G. S. Nerve injuries. *In* Morrey, B. F. (Ed.), *The Elbow and Its Disorders*. Philadelphia, W. B. Saunders, 1985, pp. 497–501.

39. Kopell, H. P., and Thompson, W. A. L. Pronator syndrome. *N Engl J Med* 259:713–715, 1958.

40. King, J. W., Bielsford, H. J., and Tullos, H. S. Analysis of the pitching arm of the professional baseball pitcher. *Clin Orthop* 67:116–122, 1969.

41. King, T., and Morgan, F. P. Late results of removing the medial humeral epicondyle for traumatic ulnar neuritis. *J Bone Joint Surg [Br]* 41:51–55, 1959.

42. Laha, R. K., Lunsford, D., and Dujovny, M. Lacertus fibrosis compression of the median nerve: Case report. *J Neurosurg* 48:838–841, 1978.

43. Learmouth, J. R. A technique for transplanting the ulnar nerves. *Surg Gynecol Obstet* 75:792–793, 1942.

44. Leffert, R. D. Anterior submuscular transposition of the ulnar nerves by the Learmouth technique. *J Hand Surg* 7:147–155, 1982.

45. Lister, G. D., Belsole, R. B., and Kleinert, H. E. The radial tunnel syndrome. *J Hand Surg* 4:52–59, 1979.

46. Mannerfelt, L. Median nerve entrapment after dislocation of the elbow—Report of a case. *J Bone Joint Surg [Br]* 50:152–155, 1968.

47. Manske, P. R. Compression of the radial nerve by the triceps muscle: Case report. *J Bone Joint Surg [Am]* 59:835–836, 1977.

48. Marmor, L., Lawrence, J. F., and Dubois, E. L. Posterior interosseous nerve paralysis due to rheumatoid arthritis. *J Bone Joint Surg [Am]* 49:381–383, 1967.

49. Marshall, S. C., and Murray, W. R. Deep radial nerve palsy associated with rheumatoid arthritis. *Clin Orthop* 103:157–162, 1974.

50. Matey, I. A radiological sign of entrapment of the median nerve in the elbow joint after posterior dislocation—A report of two cases. *J Bone Joint Surg [Br]* 58:353–355, 1976.

51. McGowan, A. J. The results of transposition of the ulnar nerves for traumatic ulnar neuritis. *J Bone Joint Surg [Br]* 32:293–301, 1950.

52. Moon, N., and Marmor, L. Parosteal lipoma of the proximal part of the radius. *J Bone Joint Surg [Am]* 55:753–757, 1973.

53. Morris, H. H., and Peters, B. H. Pronator syndrome: Clinical and electrophysiological features in seven cases. *J Neurol Neurosurg Psychiatry* 39:461–464, 1976.

54. Moss, S. H., and Switzer, H. E. Radial tunnel syndrome: A spectrum of clinical presentations. *J Hand Surg* 8:414–420, 1983.

55. Mulholland, R. C. Non-traumatic progressive paralysis of the posterior interosseous nerve. *J Bone Joint Surg [Br]* 48:781–785, 1966.

56. Murakami, Y., and Komiyama, Y. Hypoplasia of the trochlea and medial epicondyle of the humerus associated with ulnar neuropathy. *J Bone Joint Surg [Br]* 60:225–227, 1978.

57. Neblett, C., and Ehni, G. Medial epicondylectomy for ulnar palsy. *J Neurosurg* 32:55–62, 1978.

58. Nielson, H. O. Posterior interosseous nerve paralysis caused by fibrous band compression at the supinator muscle: A report of four cases. *Acta Orthop Scand* 47:304–307, 1976.

59. Osbourne, G. The surgical treatment of tardy ulnar neuritis. *J Bone Joint Surg [Br]* 39:782, 1957.

60. Panas, J. Sur une cause per connue de paralysie du serf cubital. *Arch Generalis Che Med* 2:5, 1978.

61. Pappas, A. M. Elbow problems associated with baseball during childhood and adolescence. *Clin Orthop* 164:30–41, 1982.

62. Pechan, J., and Julius, I. The pressure measurement in the ulnar nerve: A contribution to the pathophysiology of the cubital tunnel syndrome. *J Biomech* 8:75, 1975.

63. Richards, R. L. Traumatic ulnar neuritis: The results of anterior transposition of the ulnar nerve. *Edinburgh Med J* 52:14–21, 1945.

64. Ritts, E. D., Wood, M. B., and Linscheid, R. L. Radial tunnel syndrome: A ten year surgical experience. *Clin Orthop* 279:201–205, 1987.

65. Roles, N. C., and Maudsley, R. H. Radial tunnel syndrome: Resistant tennis elbow as a nerve entrapment. *J Bone Joint Surg [Br]* 54:499–508, 1972.

66. Rolfson, L. Snapping triceps tendon with ulnar neuritis. *Acta Orthop Scand* 41:74–76, 1970.

67. Salsbury, C. R. The nerve to the extensor carpi radialis brevis. *Br J Surg* 26:95–97, 1938.

68. Sharvard, W. J. W. Posterior interosseous neuritis. *J Bone Joint Surg [Br]* 48:777–780, 1966.

69. Sisto, D. J., Jobe, F. W., Moynes, D. R., and Antonelli, D. J. An electromyographic analysis of the elbow in pitching. *Am J Sports Med* 15:260–263, 1987.

70. Solnitzky, O. Pronator syndrome: Compression neuropathy of the median nerve at level of pronator teres muscle. *Georgetown Med Bull* 13:232–238, 1960.

71. Spinner, M. The arcade of Frohse and its relationship to posterior interosseous nerve paralysis. *J Bone Joint Surg [Br]* 50:809–812, 1968.

72. Spinner, M., and Kaplan, E. B. The relationship of the ulnar nerve to the medial intermuscular septum in the arm and its clinical significance. *Hand* 8:239–242, 1976.

73. Spinner, M., and Linscheid, R. L. Nerve entrapment syndrome. *In* Morrey, B. F. (Ed.), *The Elbow and Its Disorders*. Philadelphia, W. B. Saunders, 1985, pp. 691–712.

74. Sponsella, P. D., and Engber, W. D. Double-entrapment radial tunnel syndrome. *J Hand Surg* 8:420–423, 1983.

75. Struthers, J. On some points in the abnormal anatomy of the arm. *British Forensic Medicine and Surgical Review* 14:170, 1854.

76. Struthers, J. *Anatomical and Physiological Observations*. Edinburgh, Sutherland and Knox, 1854.

77. Sunderland, S. Nerve lesion in carpal tunnel syndrome. *J Neurol Neurosurg Psychiatry* 39:615–626, 1976.

78. Van Rossum, J., Buruma, O. J. S., Kamphuisen, H. A. C., and Onvlec, G. S. Tennis elbow—A radial tunnel syndrome? *J Bone Joint Surg [Br]* 60:197–198, 1978.

79. Wadsworth, T. G. The external compression syndrome of the ulnar nerve at the cubital tunnel. *Clin Orthop* 124:189–204, 1977.

80. Wartenberg, R. A sign of ulnar nerve palsy. *JAMA* 112:1688, 1939.

81. Wattys, E. M., Smith, P. A., and Hankin, F. M. A cause of ulnar neuropathy in a baseball pitcher: A case report. *Am J Sports Med* 14:422–424, 1986.

82. Werner, C. O. Lateral elbow pain and posterior interosseous nerve entrapment. *Acta Orthop Scand (Suppl)* 174:1, 1979.

83. Whitely, W. H., and Albers, B. J. Posterior interosseous palsy with spontaneous neuroma formation. *Arch Neurol* 1:226–229, 1959.

84. Wilson, F. D., Andrews, J. R., Blackburn, T. A., and McCluskey, G. Valgus extension overload in the pitching elbow. *Am J Sports Med* 11:83–87, 1983.

85. Wu, K. T., Jordan, F. R., and Eckert, C. Lipoma, a cause of paralysis of deep radial (posterior interosseous) nerve: Report of a case and review of the literature. *Surgery* 75:790–795, 1974.

Tendinopathies About the Elbow

Bernard F. Morrey, M.D.
William D. Regan, M.D.

The symptoms of tennis elbow were originally described in a discussion of lawn tennis by Major in 1883 [25, 55]. Because of the common use of the expression tennis elbow for nearly a century, it has been accepted as a standard diagnosis of a syndrome characterized by epicondylar pain and tenderness and related disability. Although 95% of reported cases occur in other than tennis players [23, 78], it is estimated that from 10% to 50% of people who regularly play tennis experience the symptoms of tennis elbow in varying degrees some time during their tennis lives [57, 58, 65].

Epidemiology

An analysis of 2500 patients at the Vic Braden Tennis Camp revealed a 50% incidence of symptoms of tennis elbow [65]. This study was reaffirmed by another study involving 200 tennis players, which revealed a 50% incidence of players over the age of 30 years experiencing symptoms characteristic of tennis elbow for less than 6 months, with the remaining 50% having major symptoms of an average duration of $2\frac{1}{2}$ years. This malady can affect players involved in any sport involving the use of the upper extremities, such as baseball, gymnastics, or swimming. The incidence of tennis elbow is equal among men and women, but among tennis players it is more common in men [23]. It is rare in Blacks, as evidenced by a series of 1000 patients located in the southern United States in an area of equal racial distribution in which the occurrence was limited to Caucasians only [23]. Although frequently seen in the age group spanning the fourth to sixth decade, it occurs four times more commonly in the fourth decade, with a peak at age 42 years [23, 78]. It

involves the lateral epicondyle approximately seven times more frequently than the medial epicondyle.

Macroscopic Pathology

There has been some confusion regarding the precise gross pathologic anatomy of this condition. In 1922, Osgood related this to a radiohumeral bursitis [59]. Since then, such various etiologies as local inflammation [9, 35, 57], muscular [9, 73] or ligamentous strain [9, 59], synovial fringe inflammation [31, 32, 35, 73], tendoperiosteal tears [24, 34, 55, 56–59], disturbance in local metabolism [70], degenerative changes in the orbicular ligament [11, 12], and cervical root irritations [39] have been held responsible for the symptoms. In 1936, Cyriax [25] found 36 published reports of differing pathologic conditions that were held to be responsible for the entity of lateral epicondylitis. The studies of Coonrad and Hooper [24] confirmed earlier works by Goldie [35], who demonstrated the essential lesion in tennis elbow to be a tear of the common extensor or flexor origin at or near the respective lateral or medial epicondyle. It was believed that these tears were produced in a degenerating tendon fiber by mechanical overload in sports or at work. Tendon tears were found to range from microscopic to macroscopic rupture. The reason for this obscurity of diagnosis is related to the number and complexity of structures originating from the lateral and medial epicondyles.

In the lateral epicondyle, Heyse-Moore [37] has recently shown by anatomic dissection that the origins of the extensor carpi radialis brevis and the superficial part of the supinator are blended and inseparable. Both originate from the lateral epicondyle, the elbow joint capsule, and the orbicular ligament. Furthermore, the extensor carpi radialis longus arises somewhat from the lateral epicondyle, as well as more proximally along the lateral epicondylar ridge. Finally, the extensor digitorum communis takes origin in part from the lateral epicondyle and does indeed contribute to this condition.

Careful gross inspection of the extensor carpi radialis brevis tendon at surgery reveals a characteristic grayish, gelatinous, and friable immature tissue. In Nirschl's series of tennis elbow patients [57, 58], 97% of cases demonstrated various degrees of this pathologic tissue. In his experience, a macroscopic tear of the tendinous origin was found in 35% of cases. In Coonrad and Hooper's work, a tear of the extensor or flexor tendon was demonstrated in 28 of 39 patients, or 72% [24]. In the remaining 11 patients, there was no actual tear, but nine patients demonstrated excessive scar tissue replacement of the tendinous origin. Coonrad states that these macroscopic tears can be superficial or deep [23]. When they are deep, the superficial tendon attachment to bone may completely obscure the pathology beneath it.

Accordingly, a macroscopic tear of the extensor carpi radialis brevis tendon, most commonly with a contribution possibly from the extensor digitorum communis, either superficial or deep, can be singled out as the pathologic anatomy producing the pain of lateral tennis elbow [23, 24, 57, 59].

Microscopic Pathology

To this day, considerable controversy exists regarding the exact etiology and histopathology of the clinical entity of lateral epicondylitis. Although most authors have placed the macroscopic pathology at the extensor carpi radialis brevis origin [23, 24, 35, 56–58], the microscopic features have yet to be precisely defined. The first definitive work was that of Goldie in 1964 [35]. He described subtendinous granulation beneath the origin of the extensor carpi radialis brevis, thus implicating this tendon and the subtendinous space. Previous work by Cyriax had implicated a periostitis [25].

More recently, microscopic studies have implicated the extensor carpi radialis brevis, involved with tendon necrosis, round cell infiltration, foci of calcification, and scar tissue [7]. Others have described the histopathology as an invasion of fibroblasts and vascular granulation tissue and have coined the term angiofibroblastic hyperplasia to describe this pathologic process [58]. Unfortunately, the micropathology is complicated by the fact that

normal tendon injected with steroid shows a somewhat similar histologic feature, including collagen fiber disruption, lymphocyte infiltration, and limpid histiocytes with fibroblastic proliferation [5, 76]. Nirschl believes that cases treated by steroid injection show a clear distinction between the characteristic angiofibroblastic proliferation in the injection site. He further characterizes the injection site as one of nonpolarizable amorphic eosinophilic material, often without any foreign body response and usually without evidence of calcification [56, 58].

To further assess this question, a single-blind, randomized study of control and patient specimens was conducted at the Mayo Clinic [68]. Two questions were addressed: (1) Are there any consistent microscopic pathologic changes characterizing lateral epicondylitis, and (2) Is there an objective means of grading the changes to correlate pathologic features with the clinical presentation in order to increase insight into its etiology? Eleven patients with an unequivocal clinical diagnosis of refractory lateral epicondylitis all experienced relief following surgery. Macroscopic pathology at the origin of the extensor carpi radialis brevis was observed in all, and 40% were also found to have involvement of the common extensor origin [68]. Twelve unembalmed and radiologically normal cadaveric specimens were employed as a control.

Fifteen microscopic pathologic features reported as characteristic of lateral epicondylitis were identified from the literature. All specimens were reviewed, and each histologic feature was graded in a single, randomized manner by a single pathologist. Common features of the surgical specimens included vascular proliferation, hyaline degeneration, fibroblastic proliferation, and calcific debris. The presence of vascular proliferation and hyaline degeneration in the operative group was statistically higher than that in controls (P .001) (Fig. 17E–1). Patients with calcific debris had preoperatively received 2, 3, and 12 steroid injections, respectively (see Fig. 17E–1). None of the 11 surgical specimens demonstrated histologic evidence of subtendinous granulation tissue or inflammation. These findings are consistent with the extensor tendon disruption noted by Cyriax [25], and later by Coonrad and Hooper [24], which microscopically resembled scar tissue.

The findings are somewhat different from those described as angiofibroblastic hyperplasia. The precise pathologic microscopic features characteristic of lateral epicondylitis in our material did include a significant vascular and fibrous proliferation ($P <$ .001), but there was a definite absence of any inflammatory component. Cortisone injections may contribute to the overall microscopic pathology of

FIGURE 17E–1

A, Vascular proliferation is a consistent feature of surgically excised tissue. *B,* Focal hyalin degeneration is also present, but has been infrequently recognized in the literature. *C,* Calcific debris is uncommon and associated with previous cortisone injections.

tennis elbow [5, 76]. The fact that an inflammatory component was not observed in the material allows the following possible explanation of the clinical course of epicondylitis. The initial insult is the disruption of the tendon origin. The response is demonstrated histologically as vascular and fibrous proliferation. The healing response may be incomplete due to a lack of an inflammatory response.

Clinical Evaluation

History

The term tennis elbow has generally become accepted as the standard diagnosis of a syndrome characterized by epicondylar pain and tenderness and related disability. The onset of symptoms can be sudden or gradual. Often, a history of repetitive flexion-extension or pronation-supination activity and overuse is obtained. Unfortunately, in many instances no predisposing event can be determined. Tennis has been implicated in about 50% of cases [9, 58]. The entity has been said by others to affect laborers performing repetitive activities in the majority of cases [24].

Among tennis players and athletes performing other racquet sports, the backhand was the most common swing initiating symptoms [65]. If tennis players are categorized according to their level of play, the incidence of lateral epicondylitis markedly decreases with a superior level of play, suggesting that faulty swing mechanics initiate the process. Nirschl found that the backhand was the most common stroke initiating symptoms (81%), and associated this with the forearm being in full pronation and in a position of anatomic weakness in front of the trunk, with the trunk in turn leaning backward at the point of impact [58]. This position was contrasted with that in the high-quality backhand, which is characteristically initiated with the forearm in midpronation, the front shoulder down, and the trunk leaning forward. In this position, the elbow, forearm, and wrist are in an overall position of anatomic strength, and this allows better stroke mechanics. In Nirschl's series of world-class players, only 13% were affected by lateral epicondylitis [58].

Physical Examination

In lateral epicondylitis, point tenderness characteristically occurs over the extensor carpi radialis brevis origin at the epicondyle. The tenderness may sometimes be present more diffusely over the origin of the extensor longus or the common digital exten-

sors. The pain is often worsened when wrist extension is tested against resistance with the forearm in pronation. The chair test as described by Gardner may be helpful [32]. The patient is asked to lift a chair with one hand, which is pronated. Severe pain in the region of the lateral epicondyle confirms the diagnosis. Coonrad has described the "coffee cup test," which involves picking up a full cup of coffee with associated localized pain at the lateral epicondylar origin [23]. He believes that pain following this test is almost pathognomonic of lateral epicondylitis.

Radiographic Studies

Routine anteroposterior and lateral radiographs are usually of little help in the diagnosis; however, according to Coonrad, a "gun-sight" oblique view of the medial or lateral epicondyle often shows irregularity or punctate calcification [23] (Fig. 17E–2). Nirschl has reported calcification of some degree in 22% of cases in one series [56]; however, gross calcification is very uncommon in this disorder.

Electromyographic Studies

These are of little or no help in a diagnosis of lateral epicondylitis, because they are characteristi-

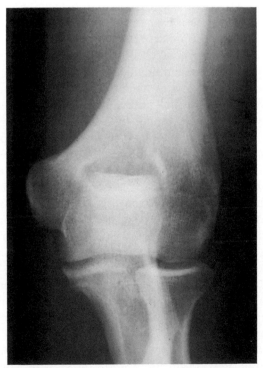

FIGURE 17E–2
The gun-sight oblique view of the lateral epicondyle demonstrates punctate calcification suggestive of lateral epicondylitis.

cally normal. Even if one entertains a diagnosis of radial nerve entrapment, electromyographic (EMG) studies are often normal. Nirschl reported normal EMG findings in 20 unselected patients with findings of lateral epicondylitis [56], and Werner showed that the test was unreliable with documented posterior interosseous nerve entrapment [78].

Differential Diagnosis

Conditions that may mimic tennis elbow include local or distal neurologic entrapment, localized interarticular pathology such as osteochondritis dissecans, radiocapitellar osteoarthrosis, or bony tumors about the elbow. Subtle instability of the radiocapitellar joint with associated pain may also cloud this diagnosis.

Localized pathology within the elbow joint or localized tumor processes are usually ruled out with routine anteroposterior, lateral, and oblique radiographs of the elbow. A localized cervical root irritation is best diagnosed by a careful history and examination of the cervical spine for point tenderness and a distal neurologic examination for cervical radiculopathy.

Pain from localized instability of the radiocapitellar joint can also be diagnosed by a careful history of either a previous posterolateral dislocation of the elbow, a severe varus stress, or prior surgery for release of the common extensor origin.

The posterior interosseous or radial nerve proper is vulnerable to compression in the radial tunnel. The tunnel originates as the radial nerve courses between the brachial radialis and brachialis proximally through the tunnel to the arcade of Frohse, which is its distal extent. In this condition, the pain is located directly over the point of nerve compression, which is generally 3 or 4 cm distal to the lateral epicondyle. Werner [78] has shown that the most common site of entrapment of the posterior interosseous nerve occurs as the nerve courses through the arcade of Frohse. Other common causes of compression in this area include adhesions of the nerve to vessels at the leash of Henry overlying the nerve, adhesions of the extensor carpi radialis muscle as it passes anterior medial to the nerve, and adhesions to the supinator muscle itself. This subject is discussed further in Section D of this chapter. It is helpful to know, however, that radial nerve compression may occur in 5% of those with lateral epicondylitis [78].

Treatment Options

Nonsurgical Treatment

In the nonathlete, elimination of activities that are painful is of the utmost importance. Usually, repetitive pronation-supination motions plus occupational lifting of heavy weights can be modified. For example, avoidance of grasping in pronation and substituting controlled supination lifting occasionally relieves much of the symptomatology. With rest and the use of anti-inflammatory medications, most of the pain can often be eliminated. If the pain is persistent, relief of acute inflammation with physiotherapy consists of ultrasound and galvanic stimulation for 3 minutes daily for a maximum of 5 days. Once the acute inflammatory phase has been relieved, an exercise program is initiated.

In the athlete, specifically the tennis player who develops this syndrome, one must first obtain relief of acute pain and then increase the forearm extensor power, flexibility, and endurance. In combination with these efforts, the athlete must attempt to decrease the moment of force placed against the elbow by altering sport mechanics or changing his equipment. Proper stroke techniques are emphasized for the club tennis player, particularly for backhand strokes, so that the forearm is not placed in the fully pronated position while swinging at the tennis ball. Avoidance of ball impact without the proper forward body weight transference is also stressed.

Nirschl has devised a technique for determining the proper raquet handle size (Fig. 17E–3). He suggests that the distance from the midpalmar crease to the ring finger is helpful in selecting the proper handle size, and we have used this technique effectively in treating patients with this condition.

Isotonic eccentric hand exercises using graduated weights of usually no more than 5 pounds may also be helpful. Progressive repetitions on a daily basis are increased, limited by symptoms. If pain recurs, return to a lower level of exercise in combination with anti-inflammatory medications or rest is mandatory. Transverse friction massage is also implemented in conjunction with the physiotherapist in an attempt to lyse scar tissue.

If this baseline program fails, we prescribe phonophoresis [42]. In this technique 10% hydrocortisone cream is administered with an ultrasonic beam through the skin into underlying tissues. This is done in a continuous fashion at 1.5 to 2 W/cm^2 every other day over a course of 10 days. This treatment is not recommended except in a patient

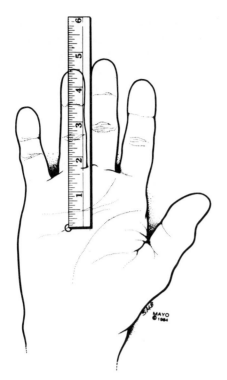

FIGURE 17E–3
Nirschl technique for proper handle size measured from proximal palmar crease to tip of ring finger. Place measuring rule between ring and long fingers for proper ruler placement on palmar crease. The measurement obtained is the proper handle size— that is, if this distance is $4\frac{1}{2}$ inches, the proper grip size is $4\frac{1}{2}$ inches. (With permission of the Mayo Foundation.)

with a chronic, refractory elbow condition, with signs similar to those initating recommendation of an injectable steroid. If local injections are performed, we do not recommend more than three over a period of a year, because the injections themselves can produce microscopic tendon disorientation and rupture [76]. If one prefers a local injection, we use 0.5 mL of betamethasone (Celestone) to 2 mL of Xylocaine. A separate syringe of corticosteroid is preferred to prevent skin infiltration with corticosteroid and subsequent skin atrophy. This is injected directly over the point of maximal tenderness at the origin of the extensor carpi radialis brevis tendon, staying just off the bone and not entering the tendon substance itself.

Both Froimson [30] and Nirschl [56, 58] advocate the use of counterforce braces. These are approximately 5 to 6 cm wide and consist of a band of heavy-duty, nonelastic fabric lined with foam rubber padding to prevent slipping (Fig. 17E–4). Velcro fasteners allow easy application of the band, which encircles the forearm just below the elbow. Tension is adjusted to a comfortable degree with the muscles

relaxed, so that maximum contraction of the wrist and finger extensors is inhibited by the band. The patient is advised to use the support only during actual play, to avoid excessive tightness, and to remove it during periods of inactivity to avoid venous congestion and edema.

It is believed that such bracing provides gentle compression of the muscle–tendon areas and partially decreases muscle expansion at the time of intrinsic muscle contraction. This was investigated by Groppel and Nirschl [36] with EMG studies using counterforce bracing. A lateral counterforce brace demonstrated lower muscular activity in the two extensor muscles across all skill levels with the serve and one-handed backhand. The three muscles involved were the extensor carpi ulnaris, the extensor carpi radialis brevis, and the flexor carpi radialis. This objective evidence provides more credibility to counterforce bracing in tennis players.

Finally, tennis players should also consider reducing racquet string tension, enlarging the grip handle, and avoiding the use of heavy tennis balls, as advocated by Coonrad [23].

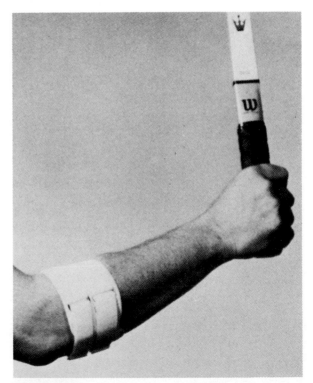

FIGURE 17E–4
Lateral elbow counterforce brace. Note that the wide, nonelastic support is curved to fit the conical forearm shape. This concept does not allow for full muscular expansion, thereby diminishing intrinsic muscular force on the lateral epicondyle. (From Morrey, B. F. (Ed.), *The Elbow and Its Disorders.* Philadelphia, W. B. Saunders, 1985.)

Surgical Treatment

Surgery is not usually recommended unless patients have had more than a year of symptoms referrable to lateral epicondylitis. If the symptoms are disabling and recurrent in spite of faithful compliance to the nonoperative program, surgical intervention may be initiated sooner than 1 year. Various operations have been described for tennis elbow and may be classified into four basic types: (1) repair of the extensor origin after excision of torn tendon, granulation tissue, and part of the epicondyle [13, 24, 56, 57, 59] (Fig. 17E–5); (2) relief of tension of the common extensor origin by fasciotomy, direct release of the extensor origin, or lengthening of the extensor carpi radialis brevis tendon distally [31]; (3) treatment by decompressing the radial or posterior interosseous nerves [39, 78]; and (4) intra-articular procedures, such as division of the orbicular ligament [11, 12] or synovectomy [59].

It is now generally accepted that the macroscopic lesion of tennis elbow is at the extensor carpi radialis brevis origin. Because this is an extra-articular structure, an intra-articular procedure for relief of pain is not necessary. Occasionally, as reported by Coonrad, synovitis is present with effusion, which can resolve after extensor origin repair [24].

According to Froimson [30], fasciotomy and complete extensor release done openly or percutaneously has resulted in loss of strength, particularly in skilled tennis players. Because the fasciotomy extends across normal fibers as well as scarred ones, the proper extent of the fasciotomy is unknown. Pain relief resulting from additional resection of a partially torn tendon such as the extensor carpi radialis brevis, by creating a larger defect in the tendon and fascia, seems illogical. Extensor tendon lengthening at the wrist, as described by Garden [31], appears to relieve pressure from the extensor carpi radialis brevis tendon proximally but does not address the pathology directly. Although Garden believed this technique was universally effective, this was not Froimson's experience. In his hands, pain relief occurred in only about half of the cases. Others found it effective in only about 20% of cases [19]. Accordingly, we do not perform this procedure.

In our opinion, the treatment of choice is to excise the torn, scarred origin of the extensor carpi radialis brevis and extensor digitorum communis if involved, remove the granulation tissue, and drill the subchondral bone of the lateral epicondyle to enhance blood supply. The elbow joint capsule is not intentionally violated unless an intra-articular problem coexists, which is rare.

Authors' Preferred Operative Technique

Although this operation can be performed under regional anesthesia such as intravenous lidocaine, we do prefer a general anesthetic. We also use a pneumatic tourniquet to obtain a bloodless field, in order to recognize more precisely pathologic tissue for its subsequent removal.

Under general anesthesia, the patient is placed supine and the affected upper extremity is supported on an armboard with the forearm in a pronated position. A short, oblique incision is made, extending approximately 2 cm proximal to the lateral epicondyle and 4 or 5 cm distal to it. Through this incision, the common extensor origin is exposed after the deep fascia, which lies immediately over the extensor aponeurosis, is excised and gently retracted. The fibers of the common extensor origin beneath the lateral epicondyle and the inferior and superior margins of the epicondyle are then inspected. Occasionally, a tear or scar tissue is evident. The fibers of the common extensor tendon are separated, and the superficial fibers of the exensor brevis, which usually lie partially covered by the fibers of the extensor longus and the extensor digitorum communis, are observed. If a tear or scar tissue is not identified by observing the superficial

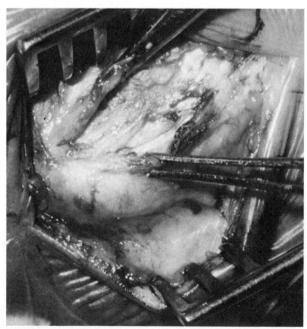

FIGURE 17E–5
The interval between the extensor digitorum communis and the extensor carpi radialis longus tendons is opened, exposing the degenerative tendon of the extensor carpi radialis brevis beneath.

fibers of the extensor brevis, the common extensor origin should be elevated more deeply. This is accomplished by splitting the tendon longitudinally and then sharply dissecting it off the lateral epicondyle, approximately 0.5 cm both anteriorly and posteriorly. The tear in the substance of the tendon or subaponeurotic space is typically identified at this time (see Fig. 17E–5). At this point, necrotic and torn tendon fibers are excised, along with any granulation tissue in the tendon. When actual tearing or scar tissue replacement is not identified, calcification or other pathology may be found adjacent to these osseous attachments, and usually the area of dissection is expanded to include more of the extensor origin elevated from the epicondyle.

Usually the extensor carpi radialis brevis is involved; however, the extensor digitorum communis aponeurosis may also be involved. It is important when excising all fibers and granulation tissue that this be done sharply with a 15-blade knife, dissecting particularly the extensor carpi radialis brevis cleanly from the fibers of the radial collateral ligament and ensuring that this ligament is left in continuity. Complete removal of abnormal granulation tissue generally encompasses the total origin of the extensor brevis (Fig. 17E–6). The remaining part of the extensor brevis tendon does not retract because of

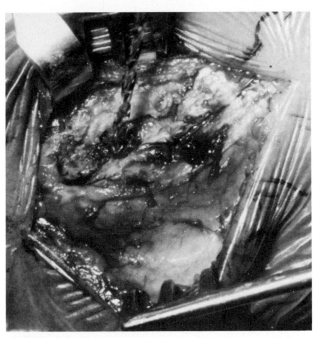

FIGURE 17E–7
To ensure improved blood supply, the exposed lateral epicondyle is decorticated and drilled with multiple 2-mm drill holes, using a hand-held drill.

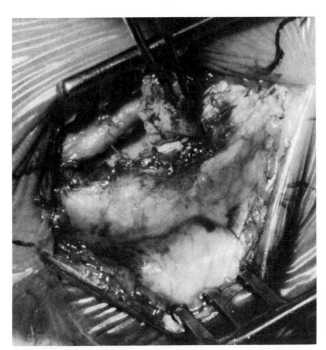

FIGURE 17E–6
The extensor carpi radialis brevis tendon is removed in its entirety with a 15-blade knife, dissecting cleanly this tendon origin from the fibers of the radial collateral ligament and ensuring that this ligament is left in continuity. The remaining part of the extensor brevis tendon does not retract because of the closed fascial adherence to the extensor longus muscle.

the closed fascial adherence to the extensor longus muscle.

To ensure improved blood supply, the exposed lateral epicondyle is decorticated by drilling multiple small 2.0-mm drill holes under controlled conditions with a hand-held drill (Fig. 17E–7). Alternatively, the bone may be stimulated by curettage. This area of decortication is slightly anterior and distal to the lateral epicondyle, in the area of insertion of the scarred extensor carpi radialis brevis tendon and part of the extensor digitorum communis tendons.

The interface between the extensor carpi radialis longus and the anterior edge of the extensor aponeurosis is repaired with a running 1–0 Dexon suture. The subcutaneous and skin layers are closed with 2–0 and 3–0 Dexon subcuticular suture and Steri-strips (Fig. 17E–8).

A posterior plaster splint is worn for approximately 7 days, with the elbow placed at 90 degrees of flexion and the forearm in neutral rotation. After removal of this bulky splint, we continue to protect the elbow in a posterior splint for 2 more weeks before range-of-motion exercises are begun. Once full range of motion is achieved, a gentle, graduated exercise program is initiated. For the first 6 weeks, the patient is encouraged to regain grip strength by use of a hand-held rubber ball. After this, gentle resistive elbow and wrist eccentric loading exercises are performed, using no more than a 5-pound

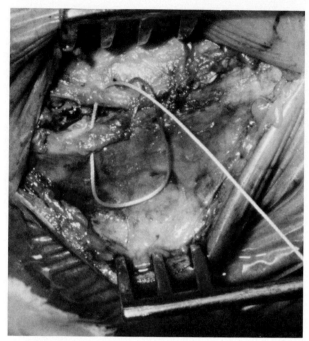

FIGURE 17E–8
The interface between the extensor carpi radialis longus and the anterior edge of the extensor aponeurosis is repaired with a running 1-0 absorbable suture.

weight. Residual pain is generally treated with decrease in activity, nonsteroidal anti-inflammatory medication, and ultrasound in conjunction with transverse massage performed gently on the affected area.

Tennis usually can be gradually resumed at about 12 weeks when adequate strength has returned and little or no pain remains. A counterforce support band should be worn while working or playing tennis for several months to help prevent any recurrence of injury.

MEDIAL TENNIS ELBOW

Medial epicondylitis is an example of medial tension overload of the elbow. This results primarily from indirect trauma, as in young throwing athletes, in whom repetitive valgus stress and flexor forearm muscle pull can produce an overuse syndrome in the common flexor origin. Sporting activities producing this condition include squash, racquetball, and particularly tennis. The activities of tennis or any racquet sport most likely to initiate difficulty are the serve and forearm strokes [57].

Symptoms and Signs

Patients complaining of medial epicondylitis describe aching pain in the flexor musculature, arising

with the medial epicondyle. There may also be weakness of grip strength specifically related to pain. Because patients may have either an acute disruption of the common flexor origin with a throwing injury or, on the contrary, rupture of the medial collateral ligament, it is important to differentiate between these two conditions. In the former, forearm flexor muscle pain or medial tendinitis is enhanced by flexing and pronating the wrist against resistance (Fig. 17E–9). In the latter, the valgus stress test localizes injury to the ligaments (Fig. 17E–10). In either condition, there may be associated symptoms of mild ulnar neuropathy. Although this does not occur in all instances, forearm pain or radicular symptoms must be sought and considered if surgical intervention is undertaken. Nirschl [57] found ulnar nerve dysfunction in 60% of cases undergoing surgery for medial epicondylitis. Concomitant osteocartilaginous loose bodies and triceps tendinitis were also found in 1% of his cases [57].

Nirschl [56] has subsequently divided the medial epicondylar groove into three zones, zone 1 being proximal to the epicondyle, zone 2 at the medial epicondyle, and zone 3 distal to the medial epicondyle (Fig. 17E–11). The symptoms of ulnar neuritis usually imply involvement in zone 3, at the level of the medial epicondylar groove. Symptoms are generally mild, intermittent, and primarily sensory, occurring after prolonged use and with heavy forearm activity. Tinel's sign may be positive. This is presumably secondary to local inflammation and edema, causing a compressive neuropathy of the ulnar nerve in the region of the cubital tunnel. In the majority of cases, EMG conduction studies are

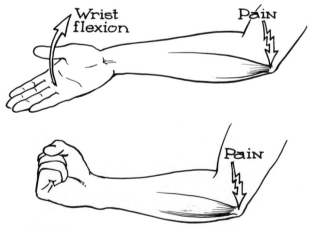

FIGURE 17E–9
Medial epicondylitis may be diagnosed clinically by pain localized to the medial epicondyle during wrist flexion and pronation against resistance. There is often pain elicited after making a tight fist, and grip strength is usually diminished on the affected side. (From Morrey, B. F. *The Elbow and Its Disorders.* Philadelphia, W. B. Saunders, 1985.)

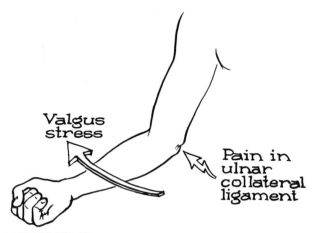

FIGURE 17E–10
Medial joint line pain elicited by placing a valgus stress to the elbow identifies injury to the ulnar collateral ligament. (From Morrey, B. F. *The Elbow and Its Disorders*. Philadelphia, W. B. Saunders, 1985.)

normal. Decompression of this zone by release of the flexor carpi ulnaris arcade generally resolves these symptoms [56].

Treatment

Nonsurgical Treatment

The treatment of medial epicondylitis is quite similar to that of lateral epicondylitis; the principles center on the relief of the acute or chronic inflammatory process. This involves the use of nonsteroidal anti-inflammatory medication, in combination with ice and a decrease or modification in the activities that produce the tension overload. Finally, a therapy program aimed at gradually increasing flexibility, power, and endurance is initiated. A counterforce elbow splint has been devised that is similar to the lateral counterforce brace but provides an additional support just distal to the medial epicondyle. This has not proved to be of much help in our hands. A gradual resumption in play is usually recommended when symptoms have subsided, generally between 6 and 12 weeks after injury.

For the most refractory cases, we have reserved a local deposition of 0.5 mL of betamethasone (Celestone) as a corticosteroid injection. When performing this injection, one must be careful to remain anterior to the medial epicondyle and to be aware that the ulnar nerve lies just posterior to the injection site. We have sometimes noted that ulnar neuropathy subsides following an injection of the medial epicondyle. We have also, however, observed instances of a subluxing ulnar nerve masquerading as medial epicondylitis, so a careful examination must be conducted before the injection or surgery.

Operative Treatment

The lack of reported series of surgical management of medial epicondylitis is interesting. It is usually discussed in association with lateral epicondylitis [23, 56]. It is believed that intrinsic musculotendinous overload results in the same sequence of pathologic changes that occurs in lateral epicondylitis. It is important to obtain radiographs prior to surgery to ensure that there is no evidence of degenerative changes in the posterior medial aspect of the olecranon, mimicking medial epicondylitis. Such changes can be visualized radiologically by simple radiographic assessment.

FIGURE 17E–11
Ulnar nerve zones at the cubital tunnel. Zone 1, proximal to medial epicondyle; zone 2, at medial epicondyle; zone 3, distal to medial epicondyle. Zone 3 includes penetration of the nerve through the flexor ulnaris arcade and is the most common site for compression neuropraxia of the ulnar nerve. (From Morrey, B. F. *The Elbow and Its Disorders*. Philadelphia, W. B. Saunders, 1985.)

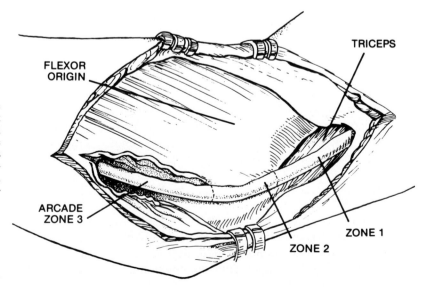

If the diagnosis is certain, localization of the maximal point of tenderness with a skin marker, such as a 25-gauge needle, may be helpful [80]. The medial epicondyle is exposed while the needle is in position. The tendon origin of the pronator teres or a portion of the flexor carpi radialis is generally involved and is elevated similarly, with the exposure directed toward the synovium of the elbow joint. Torn or scarred tissue is excised, and repair is done in a manner similar to that described for the lateral epicondyle. Coonrad identifies the ulnar nerve in each case with this exposure, although he does not transpose the ulnar nerve or perform a neurolysis unless evidence of associated neuropathy is present [23].

An alternative approach is a longitudinal incision extending from the medial epicondyle distally for about 2 inches. The tendons are spread longitudinally, and the lesion comes clearly into view in appropriately selected cases [56]. All pathologic tissue is excised longitudinally and elliptically, including resection to the joint in occasional cases. All normal tissue is left attached to the medial epicondyle for fear that total excision of the common flexor origin may include a portion of the origin of the medial collateral ligament, which may then lead to subtle posteromedial instability.

Management of the ulnar nerve usually involves decompression in zone 3, just distal to the medial epicondylar groove [57] (see Fig. 17E–11). Decompression of this zone by release of the flexor ulnaris arcade generally resolves these symptoms in Nirschl's experience [57]. His indications for anterior ulnar nerve transposition include (1) nerve subluxation or dislocation from the epicondylar groove; (2) skeletal valgus, with symptomatic tension neurapraxia, which is common in throwers; (3) a hostile environment, such as scarring from prior surgery; and (4) necessity of transfer to facilitate surgical exposure to the medial aspect of the elbow.

Following a routine closure, the elbow is protected for approximately 3 weeks in 90 degrees of flexion and neutral rotation of the forearm. Limbering exercises are then performed for 2 weeks, followed by strength and endurance-resistance exercises, which begin about 6 weeks after surgery. This progresses to strength-resistance exercises, including isometrics, isotonics, and isokinetics in proper sequence and intensity. These resistance exercises are continued until full strength returns. In Nirschl's hands [57], full strength in the dominant arm is 10% greater in the average person and 20% greater in the competitive racquet or throwing sport athlete. Most strength returns about 6 months after surgery. Modified sporting technique patterns beginning 6

weeks after surgery and no competitive sports are recommended until full strength returns at approximately 6 months.

It appears that there are few complications except a loss of elbow extension of up to 5 degrees in about 1% of patients, and superficial infection rate in 0.6%. In our experience, the major problem has been the overall lack of predictability of the procedure in reliably relieving symptoms.

Authors' Preferred Method of Treatment

We prefer to have the patient lie supine with the arm abducted and flexed to 90 degrees and resting on an armboard, with an extra sheet beneath the lateral humerus. A gentle, curved incision is made, beginning just posterior to the medial epicondyle. After excising down to the fascia overlying the medial epicondyle, the ulnar nerve is identified and protected proximally. If there is any sign of compression of the nerve, or if symptoms have suggested that this is the case, the forearm fascia is split in zone 3, where the nerve enters the forearm between the two heads of origin of the flexor carpi ulnaris. We avoid a detailed dissection of the nerve unless intrinsic pathology is found.

Following exposure of the medial epicondyle by excising the deep fascia, the fibers between the pronator teres and the flexor carpi radialis are excised longitudinally. The tendinous origin of these muscles is sharply excised from the medial epicondyle in the direction of the joint. The fascia and fleshy muscle fibers are sutured back to the cuff of tissue left on the medial epicondyle, while the scarred and degenerative tendon is excised in an elliptical and longitudinal fashion. Little attempt is made to close the defect, but normal tissue, if detached, is repaired. The longitudinal incision made through the common flexor origin is then closed with 2–0 absorbable suture.

When the ulnar nerve is extremely irritable or subluxed, or if there is objective clinical or EMG evidence of ulnar compromise, this is addressed by a submuscular transposition of the nerve. The entire origin of the medial epicondyle is resected, leaving a cuff of normal tendon (Fig. 17E–12), and the ulnar nerve is transposed submuscularly. The tendinous origin of the common flexors is removed, and the fascia plus muscular fibers are then reattached to the cuff of tissue at the origin of the medial epicondyle (Fig. 17E–13). Prior to reattachment, the degenerative tendinous portions of the common flexor muscle mass are excised, and the

FIGURE 17E–12
The common flexor origin is detached, leaving a cuff of tendon. The degenerative tendon is resected and the ulnar nerve, if involved, is transposed beneath the common flexor origin. Drill holes are placed in the medial epicondyle to enhance blood supply.

medial epicondyle is drilled with a 2-mm drill or freshened with a curette. The latter step is performed to ensure that adequate vascularity is present to enhance healing.

After a routine closure, the patient is kept in a bulky arm dressing, which is well padded with the elbow at 90 degrees of flexion in the form of neutral rotation. This is maintained for approximately 3 weeks. The patient begins gentle range-of-motion exercises without resistance until range of motion has been restored. At the end of 6 weeks, we encourage the restoration of grip strength by having the patient squeeze a rubber ball or use exercise putty. Between 10 and 12 weeks postoperatively, resisted elbow and wrist flexion exercises are initiated with the use of light dumbbell weights or, for those patients who cannot attend formal physiotherapy classes, such household items as a can of soup.

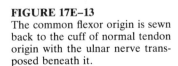

FIGURE 17E–13
The common flexor origin is sewn back to the cuff of normal tendon origin with the ulnar nerve transposed beneath it.

At the end of 4 to 6 months, patients have usually regained adequate strength to resume light duties or sporting activities. We recommend no return to their level of sport or work until we have evidence by objective testing that they have gained 80% or more strength compared to the opposite extremity. In our experience, this usually takes approximately 6 months. We recommend the use of a forearm support band while working or playing any overhand sports for up to 6 months after surgery [56].

We recognize that the above program is more prolonged than that which is often followed. This regime is, however, based on our observation that a number of failed procedures appear to be the result of too early rehabilitation and return to sport.

TRICEPS TENDINITIS AND RUPTURE

Anatomy

The triceps muscle arises anatomically from the dorsal arm and scapula and is composed of the long, lateral, and medial heads, which coalesce to form the triceps tendon. The tendon of the triceps brachii muscle consists of two aponeurotic lamellae that join together above the elbow and insert into the posterior portion of the dorsal surface of the olecranon. A lateral band extends over the anconeus muscle to attach to the dorsal fascia of the forearm.

Triceps Tendinitis

Triceps tendinitis has not been described in the literature, other than in an allusion by Nirschl [57]. It is thought most often to be an isolated entity and is usually associated with loose bodies in the posterior compartment of the elbow or with lateral tennis elbow. We have observed this condition in baseball players and those with occupations involving repetitive elbow extension, such as carpenters. The same conservative management program is used for triceps tendinitis as is used for lateral or medial tennis elbow. We have never operated on a case of triceps tendinitis and doubt the value of operative treatment.

In cases that have been refractory to all conservative treatment for more than 1 year, Nirschl has elected to excise surgically a small portion of the triceps insertion. The site of resection coincides with the area of maximum point tenderness, previously eliminated by local Xylocaine injection.

Triceps Rupture

A rupture or avulsion of the triceps tendon is a rare injury. In 1868, Partridge reported the first case, which was managed by rest and graduated exercise [61]. Clinically, the site of disruption was at the tendo-osseous junction.

Since that time, there have been 40 cases reported in series available in the English literature to describe this condition. Ages affected ranged from 7 to 72 years, with a mean age of 26 years. The majority of patients are men aged 29 to 41 years. The right and left triceps are equally affected. There is no correlation of the side of rupture with the dominant or nondominant side.

The first extensive series of triceps rupture was compiled by Anzel and associates [3]. They reported a 9-year experience of consecutive tendon ruptures in 781 patients, involving 1014 tendon ruptures. Eight triceps ruptures represented 1.9% of the tendon injuries. Although the mechanism of injury or mode of treatment was not described, this percentage does attest to the rarity of this injury. We have collected three instances of triceps rupture in a series of 100 fractures of the olecranon reviewed at the Mayo Clinic. In this experience, the patients were older, and the triceps was ruptured as an avulsion at its insertion with a portion of osteoporotic olecranon.

Mechanism of Injury

The mechanism of injury is nearly always a fall onto an outstretched hand. A direct blow to the olecranon resulted in two triceps ruptures in a series of seven patients [74]. We have treated patients with both mechanisms, as have others [63, 64]. Disruption of the triceps may also occur spontaneously with minimal trauma, in individuals who have been compromised by a systemic disease process such as hyperparathyroidism [64], and in those receiving steroid treatments for lupus erythematosus [75]. A weakening of tendons in general has become increasingly associated with local deposition of steroid [75].

The injury location is almost always in the tendo-osseous region and represents avulsion of the insertion, usually with a fleck of bone. Sherman and colleagues [71] and Bach and associates [4] have noted musculotendinous injuries as well.

The association of additional injuries cannot be overlooked. Levy and coworkers [46, 47] reported 16 patients who had incurred triceps avulsion associated with radial head fractures, and termed this a new syndrome. Clayton and Thirupathi [21] found an association with chronic olecranon bursitis. Accordingly, additional injuries and systemic conditions should be considered at the time of clinical examination.

Diagnosis

The history is commonly that of a fall onto the outstretched hand. The most common mechanism of injury is a deceleration stress superimposed on a contracted triceps muscle, with or without a concomitant blow to the posterior aspect of the elbow. A diagnosis of triceps tendon avulsion is usually evident in patients who present with a characteristic history, including pain, swelling, and a palpable depression just proximal to the olecranon. Considering the 16 patients reported by Levy and coworkers [46, 47] and others who have detected radial head fractures, an increased valgus load against a contracted triceps muscle is a viable causative mechanism. On examination, features usually include pain, swelling, and a palpable depression just proximal to the olecranon and posterior ecchymosis. Radiographs are particularly helpful, with flecks of osseous material visible on the lateral radiograph in 80% of patients [26, 71] (Fig. 17E–14).

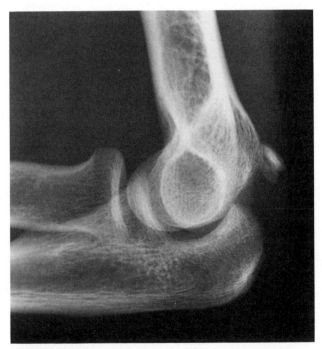

FIGURE 17E–14
Anteroposterior and lateral roentgenogram illustrating the small avulsion fracture off the tip of the olecranon, indicative of a triceps tendon rupture.

Treatment

Immediate surgery has been recommended as the treatment of choice for complete ruptures. At the time of exploration, one of two different situations may be evident. The first is that the tendinous portion of the triceps may be found to be retracted within the muscle, usually accompanied by a bony avulsion from the olecranon. The second is a palpable thinning and partial defect of the tendon encountered proximal to the olecranon process; this is due to a partial disruption of the triceps tendon, usually at its central third [1].

In the acute rupture, several techniques may be employed to reattach the triceps tendon to the olecranon. A nonabsorbable suture, placed in drill holes located transversely in the olecranon or through a subperiosteal flap, is usually quite effective in this situation [46, 47] (Fig. 17E–15). Others have used stainless steel wire to facilitate better fixation [7]. We do not recommend this, because the wire fatigues and fragments.

If treatment has been delayed or if the patient has a systemic disease that compromises the quality of tissue and hence the quality of repair, a reconstructive technique may be considered. Farrar and Lippert [26] suggested using a periosteal flap from the olecranon to reinforce the repair. Bennett [7] used a flap of fascia taken from the posterior aspect of the forearm with its base attached to the medial and lateral epicondyles to the humeral anterior olecranon process. Clayton and Thirupathi [21] achieved successful results by splitting the triceps tendon into a partial-thickness flap at the base 2 cm proximal to the cut end. This flap was rotated on its proximal base and pulled down to the olecranon with appropriate tension applied to help reinforce the repair.

For most sedentary older patients, a nonoperative approach is indicated if a partial tear is present and some elbow extension is still present. Farrar and Lippert [26] stressed close clinical follow-up in these cases. Surgery is indicated in this type of patient if there is significant weakness. In the athlete, we explore and repair all partial and complete ruptures.

Postoperative Rehabilitation

Postoperative immobilization varies from 10 days to 6 weeks, but most patients are generally immobilized 3 to 4 weeks in a cast or posterior splint at 35 to 40 degrees of elbow flexion. Active range-of-motion exercises are then begun. In a single report,

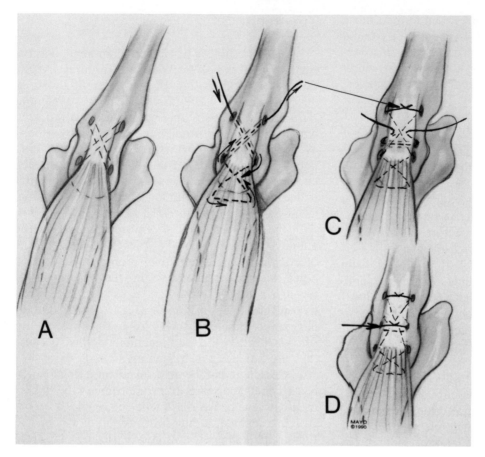

FIGURE 17E–15
A criss-cross suture is secured to the olecranon by a cruciate attachment through drill holes in the olecranon. A second transverse suture is also employed to further stabilize the fixation. (With permission of the Mayo Foundation.)

an Olympic weight lifter began progressive active and active-assisted exercises at 2 months; by 4 months after the procedure, he had resumed weight training [4].

Results

In most instances of acute or delayed repair, normal strength and full motion are restored with no pain. Levy and associates [46, 47] noted a 10% to 20% limitation of range of motion in two patients. Anderson and LeCoco [1] and Bennett [7] have noted 5-degree flexion-contraction in two patients; this seems to be the norm in most reports. Sherman and colleagues [71] noted that a professional body builder, at reconstruction 3 months after injury, lacked 3 degrees of extension but had full power and was able to bench-press 370 pounds without difficulty. Most of these reports are without the benefit of precise objective strength measurements.

Complications

Pantagopoulos and associates [60] noted olecranon bursitis in a patient in whom repairs were made with wire suture; this has been the only complication of note. There have been no wound infections reported, and no reruptures have been described. Return of strength and range of motion has been consistently reported as near normal. The relationship between the use of high-dose oral steroids and tendon and ligament strength has previously been unproved; however, a report by Bach and associates [4] suggests that high-dose oral steroids may contribute to rupture of the triceps tendon. A competitive weight lifter had taken large doses of oral and injectable anabolic steroids prior to and after a triceps tendon rupture.

Authors' Preferred Method of Treatment

In our opinion, the treatment of choice is immediate repair using a posterior base incision just lateral to the midline. We prefer two nonabsorbable 5–0 Mersilene sutures. The first is placed in a Bunnell fashion through the proximal torn triceps tendon and then through crossed holes drilled through the olecranon (see Fig. 17E–15). A second transverse suture is used to secure the precise site of attachment. The sutures are tied with the elbow at 90 degrees of flexion. This is the same technique that is used to reattach the triceps after elective reconstructive procedures.

If the repair requires additional support, the tendon is split in partial thickness proximal to the cut end and this flap is turned down, anchoring it to the olecranon through these same drill holes, as advocated by Clayton and Thirupathi [21]. If possible, however, we prefer to mobilize the anconeus and lateral triceps extension and relocate this over the olecranon rather than create free tissue flaps.

We advocate protected immobilization for 4 weeks, with the arm held in 30 degrees of flexion. We then use a dynamic extension splint with a 100-degree flexion stop for 4 additional weeks. This is followed by active range-of-motion to active resisted exercises. Training for avid athletes is not recommended until Cybex testing indicates that 80% strength has returned; this may take 6 months. We allow full participation in all active sporting events, and expect 90% to 100% return of strength if the injury has occurred at the site of attachment.

OLECRANON BURSITIS

There is an abundance of superficial and deep bursae in the human body; however, none is more frequently involved in athletic trauma that the prepatellar bursa at the knee and the olecranon bursa at the elbow. Football and ice hockey have been the chief athletic events implicated in the development of olecranon bursitis at the elbow.

Anatomy

In 1934, Black [10] examined a large number of fetuses and showed that only the subacromial bursa was present at birth. He was able to identify subacromial bursae in 72.5% of the specimens. Neither the commonly found subcutaneous olecranon nor the prepatellar bursa was present at birth in his dissections. A more recent anatomic study by Chen and colleagues [20] analyzed the incidence of bursae at autopsy in three groups: those under 7 years of age, those from 7 to 10 years of age, and adults. Dissection of both elbows in 63 autopsy specimens demonstrated no bursae in those under 7 years of age. In the 7- to 10-year age group, a minute unilateral bursa was found in four of six cases. Above age 10, the bursae were noted to increase in size with age and were present in all specimens. In both these studies and in others [79], the presence

of these bursae is variable and is probably developmental; thus, one is forced to the conclusion that they develop after birth in response to movement and function.

The bursae in the olecranon region exist in three locations: (1) the subcutaneous bursa, so commonly seen clinically; (2) an intratendinous bursa, in the substance of the triceps tendon near its insertion; and (3) the subtendinous bursa between the tendon and capsule (Fig. 17E–16), as described by Morrey [53].

Although there have been no recognizable clinically documented presentations concerning the two deep bursae about the olecranon, the intratendinous bursa may indeed be involved with tears of the triceps tendon as described earlier in this chapter, or with cases of triceps tendinitis. The amorphous gray, white gelatinous material described by Nirschl as removed in triceps tendinitis may be intratendinous bursal tissue. The only subtendinous bursa has been described by Vizkelety [77] and was idiopathic in nature; there has been no other clinical material on this subject. We have not encountered this bursa in our practice.

Inflammation of the superficial olecranon bursa is most commonly encountered in clinical practice. The bursitis may be acute or chronic, septic or nonseptic and is associated most commonly with occupational or sports trauma. It has been called a miner's elbow or student's elbow. In sports injuries, Larsen and Osternig [43] have shown that this superficial olecranon bursitis is a common football

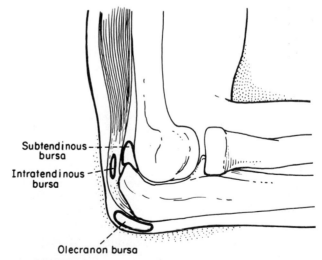

FIGURE 17E–16
Lateral illustration of the elbow demonstrating the superficial olecranon bursa, the intratendinous bursa found in the substance of the tendon, and the subtendinous bursa lying between the tip of the olecranon and the triceps tendon. (With permission of the Mayo Foundation.)

injury, almost exclusively associated with artificial turf. Larsen and Osternig collected 16 cases of olecranon bursitis over the course of one regular football season. Fourteen of the sixteen cases, or 87.5%, were sustained on artificial turf, and the remaining two cases occurred on a grass field. We have encountered this lesion regularly in hockey players more frequently than in football players.

The relation between artificial turf and the mechanism of injury centers on the construction features of the artificial turf. This turf is composed of an upper layer of very durable, synthetic grass, which is applied over a layer of padding of varying thickness and resiliency. These two surfaces are usually applied over a hard surface, such as asphalt. Consequently, artificial turf gives a very even and consistent upper playing surface, but has a relatively unyielding quality at the base compared to actual grass. It is this latter aspect that appears to be the culprit in the increased incidence of olecranon bursitis and similar problems. Repeated falling on the partially flexed elbow can result in trauma to the olecranon bursa, and sometimes an olecranon spur develops (Fig. 17E–17). Such traumatic episodes may produce an acute inflammatory response, with excessive production of bursal fluid and subsequent bursal sac distention. Occasionally, with severe trauma vascular disruption with hemorrhagic distension of the bursal sac may occur. There may also be an inflammatory response following this effusion in the bursal sac. The initial episode is usually one of a heme bursitis. After resolution, recurrent episodes occur with less trauma and are not associated with intrabursal hemorrhage.

Olecranon bursitis may be classified into three types: acute, chronic, and suppurative. The acute type usually results either from a direct blow or from acute repeated insults to the superficial bursa, which may also be due to pressure for prolonged periods. Chronic bursitis develops as a sequel to recurrent acute episodes when the trauma is relatively mild but occurs frequently when the resorptive phase of an acute bursitis is repeatedly interrupted by further trauma. In this situation, the bursae lining is replaced by fibrous tissue, which becomes the predominant characteristic. Suppurative bursitis results from an infective process developing in an acute or chronic bursitis after contamination through a skin wound or dermatitis, or, most commonly, it may be of hematogenous origin.

Acute Bursitis

Clinical Presentation

The presentation is characteristic, with tenderness and distention of the bursa after a direct blow. It may be difficult to differentiate the condition from cellulitis as a result of abrasion over the injured area, and indeed, one may have a superficial infection of an abrasion with an underlying aseptic bursitis. Larsen and Osternig [43] suggest a differential diagnosis including acute arthritis, ligamentous injury, and tendinitis.

Joint motion is usually not limited in patients with subcutaneous olecranon bursitis except when flexion produces skin tension and increased pressure over the tender, distended bursa. This is a very important point because septic bursae are painful. Fluctuation of the distended bursa in the absence of true joint findings further localizes the injury to the bursa. Canoso [16, 17] has characterized the clinical features of 30 patients with acute traumatic olecranon bursitis. Repetitive trauma was cited as the most important factor in 14 cases. When symptoms were present more than 14 days, the bursa was discretely swollen; but if symptoms were observed in less than 2 weeks, parabursal edema was present, and swelling was also observed in the arm and forearm in 50% of cases. We have not observed this associated swelling except with septic bursitis.

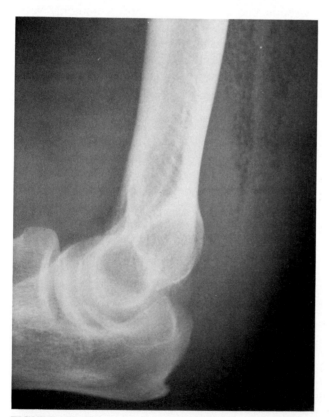

FIGURE 17E–17
Lateral roentgenogram demonstrating spur off tip of olecranon, indicative of olecranon bursitis.

Aspiration of the acute entity shows evidence of recent hemorrhage. The bursal fluid was characterized by a low white blood cell count with a high percentage (80%) of monocytes [16].

Treatment

In the athletic population, olecranon bursitis is most commonly a post-traumatic disorder, but care must be taken to make the correct diagnosis. One must be careful to rule out systemic inflammatory processes such as calcium pyrophosphate, dihydrate crystal deposition disease [33], and other disorders. Control of the underlying disease process is an obvious prerequisite in any treatment program.

In the post-traumatic condition, the best treatment, of course, is prevention. A decrease in the energy of impact by better dissipation of forces lessens the likelihood of the occurrence of olecranon bursitis; this requires the use of elbow pads. Natural turf is desirable. A high-quality elbow pad helps prevent the injury in ice hockey. Once a bursitis has occurred, recurrence becomes more frequent with less trauma, resulting in the need to provide additional protective covering to lessen the forces of impact.

In the acute situation, if the bursa is distended and uncomfortable or interferes with the use of the joint, aspiration is recommended. This, of course, is done under strictly sterile conditions and is followed by the application of continuous compressive bandages for 48 hours and cold packs for 15 minutes every hour for 24 to 48 hours.

If the distention is not too severe, there is no need to aspirate the bursa. Compression and cold packs at the first sign of swelling help to minimize bleeding into the sac, which usually reaches a maximum 24 hours after injury. After this, warm packs can be used to hasten absorption of bursal fluid. The bursal fluid often resorbs if it is not subjected to recurrent trauma. If continued trauma occurs, however, accumulation of fluid recurs. This can result in fibrous tissue formation becoming the predominant characteristic and then leads to subacute and chronic bursitis, which may require surgery.

Chronic Bursitis

Presentation

This follows recurrent traumatic episodes and, in this situation, the bursal walls usually have become much thickened. Trabeculae and villi form, increasing in number and density and filling the bursal space [41]. The villi that arise represent granulation in the floor of the bursa, consisting of central blood vessels surrounded by fibrous tissue cells. This granulation tends to grow into the cavity of the bursa, so that eventually a villous projection forms consisting of a central vessel surrounded by a stalk of fibrous tissue. Clinically, in a very mild subacute form, this may be evident as a slight, palpable thickening of the bursa; in the chronic form, a large, rubbery mass occurs in the subcutaneous tissues containing numerous hard moveable bodies (Fig. 17E–18).

Frequently, the causal factor in the development of a chronic bursitis is not sports but the patient's work, which necessitates exposure of the bursa to continued trauma. Occurrences have been so common in certain characteristic occupations that they have been given names such as miner's elbow for chronic olecranon bursitis.

By far the largest number of cases of chronic olecranon bursitis develop from an acute traumatic episode such as a football injury. In such cases, the acute injury is superimposed on a chronic or recurrent bursitis. The bursal sac is distended and fluctuant, but on palpation the indurated walls of the bursa are easily noted, and the fibrous trabeculation and villi can be rolled under the finger. Definitive

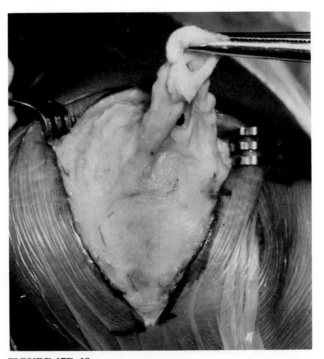

FIGURE 17E–18
Trabeculae and villi fill the bursal sac to produce a scarred, thickened bursa. This is excised through an incision centered lateral to the midline.

measures are usually required in this situation. Using sterile technique, a 16-gauge needle is inserted into the bursal sac to allow decompression, followed by application of a compressive dressing for about 3 days. This is recommended as an effective means of lessening the tendency for recurrence of bursal swelling [27]. Unfortunately, aspiration of fluid from such a bursa usually results in rapid and almost complete refilling of the bursa in a period of 24 to 48 hours, and applications of heat or pressure have little effect, even when they are maintained for several weeks [53]. The surgeon should not be surprised to find little free fluid during aspiration of the chronic bursa.

Steroid injections have no place in the treatment of acute traumatic bursitis. They usually interfere with the normal protective mechanism against infection. In subacute and chronic bursitis, steroid injections have occasionally been used if recurrences follow minimal trauma, but they have had limited success [43]. We do not use steroid injections for acute, subacute, or chronic bursitis.

Prevention and Protective Equipment

The mainstay in prevention of olecranon bursitis is equipment design that helps to dissipate the energy of impact on the elbow. The difficulty comes when players must trade restriction and movement for additional padding around the vulnerable area, such as the elbow and specifically the olecranon. It is essential that these pads, which are layered with soft and firm foam, be comfortable but have minimal restriction; they should not slip away from the elbow when subjected to severe contact and joint movement. The majority of sleevelike pads have a doughnut-shaped elevation on the inside of the pad encircling the olecranon, which distributes the force of impact to the peripheral areas. An additional layer of material over the prominence of the olecranon provides further protection. These are currently worn within the sleeves of all players participating in active contact sports, such as football and hockey and some noncontact sports, such as volleyball.

Operative Intervention

Surgery is indicated if the process is refractory to nonoperative intervention and bothers the athlete to the point where he or she cannot participate in the sport of choice or if a septic episode has been imposed on the chronic process.

A longitudinal incision just medial to the midline [40] or centered directly over the olecranon bursa has been recommended [14]. The bursa is carefully dissected out, preferably by means of a 15-blade knife, and without opening into it if at all possible (Fig. 17E–19). Occasionally, a bursa must be removed piecemeal; if this is necessary, all pieces of the bursal sac should be carefully removed. Because freeing the bursa from the skin can devitalize the skin over the olecranon process or cause problems with healing, Morrey [53] recommends a compressive dressing with the elbow held in less than 45 degrees of flexion. Breck and Higinbotham [14] have suggested placing mattress sutures on either side of the incision centered over the appropriate half of the dead space under the skin flaps. These sutures are passed down through the skin and then into the underlying deep tissues of the muscle and fascia of the triceps. The suture is then brought back up to the skin and snugly tied over a button. This brings the skin into firm contact with the underlying structures and tends to prevent the formation of a hematoma and recurrence of a bursa or skin breakdown.

Quayle and Robinson [66] have avoided the problem of wound healing secondary to the subdermal dissection of the bursa by simply reflecting the skin with the bursal tissue from a medial to a lateral direction. After this, the tip of the olecranon is obliquely osteotomized, leaving the bursal tissue intact. The subcutaneous tissue plus bursa is then

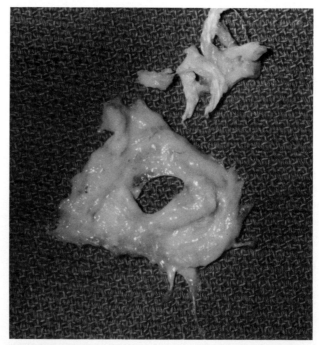

FIGURE 17E–19
The olecranon bursa is resected without opening the bursal sac.

reflected back over the olecranon, and the wound is closed with a drain. In a series of 11 patients, there have been no recurrences.

Authors' Preferred Method of Treatment

We prefer a skin incision over the lateral aspect of the olecranon. The skin and subcutaneous tissue are carefully incised, exposing the bursal sac. If possible, the bursal sac is enucleated in total, without violating its contents. The minimal amount of subdermal dissection required to accomplish this is performed equidistant on the medial and lateral aspects of the olecranon. Any olecranon spur or prominence is removed. The tourniquet is deflated to assess the viability of the skin, and meticulous hemostasis is attained.

Closure is accomplished with absorbable sutures placed in the medial and lateral skin flaps and sewn down to the deep fascia. In this way, the dead space is obliterated, and the chance of subdermal hematoma is minimized. The wound is then closed in the usual fashion, and the elbow is immobilized from 0 to 45 degrees of flexion to further prevent recurrence. A compressive dressing is maintained for 2 to 3 weeks. Normal elbow motion is easily attained.

Septic Bursitis

A large number of patients developing a septic bursa give a history of prior idiopathic or traumatic bursitis. The presentation varies widely from acute onset of localized cellulitis, generalized cellulitis involving the forearm, a low-grade subacute process of 10 to 14 days duration, to a fulminating process with systemic symptoms.

In general, the patient with septic bursitis is unlikely to be febrile, but, unlike the patient with an aseptic process, pain is present over the olecranon bursa and with motion. The infected bursa is often tender to palpation, and sometimes there is an abrasion or a skin lesion. These findings are not, however, diagnostic of infection; some nonseptic cases also show these features. In addition, their absence does not rule out an infected bursa.

In all cases of a painful bursa, the diagnosis is made by needle aspiration of the bursal contents. With aspiration, a cell count, crystal determination, and Gram stain should be done in all cases. Crystals are determined because gout may coexist with or even predispose the patient to olecranon bursitis [33].

The septic fluid drawn generally appears purulent as either frank or bloody pus [16, 40]. Bursal fluid analysis in septic cases shows a predominance of polymorphonuclear cells, and nonseptic fluids have a predominance of mononuclear cells. In septic bursitis, the fluid-to-serum glucose ratio is less than 50%. Gram stain smears of bursal fluid demonstrated organisms in all 10 septic cases reported by Ho and Tice [38]. The commonest organism was *Staphylococcus aureus* in 94% of cases in their series [38]; this is in agreement with our experience.

Following an accurate diagnosis, initial treatment should include adequate drainage coupled with antibiotic therapy. Occasionally, aspiration and oral antibiotics may suffice, but most authors recommend formal incision and drainage of the bursa. Intravenous antibiotics are usually required if systemic symptoms are present as well.

The length of treatment with antibiotics is usually based on the medical response of the patient and of the olecranon bursa itself. Certainly, return to any athletic endeavor is attempted only after full resolution of all symptoms from the septic bursitis, both systemically and locally. Special efforts are made to pad the elbow on return to sport.

Authors' Preferred Method of Treatment

All painful bursae are aspirated. If the aspirate is purulent or cloudy, or if the patient is febrile, we inject 0.5 g of methicillin in 10 mL of saline into the bursa after it has been lavaged. Oral antibiotics effective against *S. aureus* are started. If the process recurs, a second aspiration is performed. The elbow is elevated and splinted in 45 degrees of flexion with an anterior splint. There is no reason for an incision in the great majority of cases. If a septic bursa occurs or if the septic process has followed a course of recurrent aseptic bursitis, the bursa is excised only after the septic process has resolved.

References

1. Anderson, K. J., and LeCoco, J. F. Rupture of the triceps tendon. *J Bone Joint Surg [Am]* 39:441–446, 1957.
2. Anderson, R. L. Traumatic rupture of the triceps tendon. *J Trauma* 19:134, 1979.
3. Anzel, S. H., Coney, K. W., Weiner, A. D., and Lipscomb, P. R. Disruption of muscles and tendons—An analysis of 1014 cases. *Surgery* 45:406–414, 1959.
4. Bach, B. R., Warren, R. F., and Wickiewicz, T. L. Triceps rupture: A case report and literature review. *Am J Sports Med* 15:285–289, 1987.
5. Balasubramanian, P., Prathap, K., and Zumpur, K. The

effects of injection of hydrocortisone into rabbit calcaneal tendons. *J Bone Joint Surg [Br]* 54:729–734, 1972.

6. Bauman, B. S. Triceps tendon rupture. *J Bone Joint Surg [Am]* 44:741, 1962.
7. Bennett, B. S. Triceps tendon ruptures. *J Bone Joint Surg [Am]* 44:741–744, 1961.
8. Bennett, J. B., and Tullos, H. S. Ligamentous and articular injuries in the athlete. *In* Morrey, B. F. (Ed.), *The Elbow and Its Disorders*. Philadelphia, W. B. Saunders, 1985, pp. 502–522.
9. Bernhang, A. M. The many causes of tennis elbow. *NY State J Med* 79:1363–1366, August 1979.
10. Black, B. Development of human synovial bursae. *Anat Rec* 60:333, 1934.
11. Bosworth, D. M. The role of the orbicular ligament in tennis elbow. *J Bone Joint Surg [Am]* 37:527–533, 1955.
12. Bosworth, D. M. Surgical treatment of tennis elbow—A follow-up study. *J Bone Joint Surg [Am]* 47:1533–1536, 1965.
13. Boyd, H. B., and McLeod, A. C. Tennis elbow. *J Bone Joint Surg [Am]* 55:1183–1187, 1973.
14. Breck, L. W., and Higinbotham, W. L. Patellar and olecranon bursitis: With a report of an improved operative procedure. *Milit Surg* 98:386–399, 1946.
15. Bywaters, E. G. The bursae of the body. *Ann Rheum Dis* 24:215, 1965.
16. Canoso, J. J. Idiopathic or traumatic olecranon bursitis: Clinical features and bursal fluid analysis. *Arthritis Rheum* 20:1213, 1977.
17. Canoso, J. J. Intrabursal pressures in the olecranon and prepatellar bursae. *J Rheumatol* 7:570, 1980.
18. Canoso, J. J., and Sheikman, P. R. Septic subcutaneous bursitis: Report of sixteen cases. *J Rheumatol* 6:1, 1979.
19. Carroll, R. E., and Jorgensen, E. C. Evaluation of the Garden procedure for lateral epicondylitis. *Clin Orthop* 60:201–204, 1968.
20. Chen, J., Alk, D., Eventov, I., and Weintroub, S. Development of the olecranon bursa: An anatomic cadaveric study. *Acta Orthop Scand* 58:408–409, 1987.
21. Clayton, B. S., and Thirupathi, R. G. Rupture of the triceps tendon with olecranon bursitis. *Clin Orthop* 184:183–185, 1984.
22. Conwell, H. R., and Alldredge, R. H. Ruptures and tears of muscles and tendons. *Am J Surg* 35:22, 1937.
23. Coonrad, R. W. Tennis elbow. *Instr Course Lect* 35:94–101, 1986.
24. Coonrad, R. W., and Hooper, W. R. Tennis elbow: Its course, natural history, conservative and surgical management. *J Bone Joint Surg [Am]* 55:1177–1182, 1973.
25. Cyriax, J. H. The pathology and treatment of tennis elbow. *J Bone Joint Surg [Am]* 18:921–940, 1936.
26. Farrar, E. L., and Lippert, F. G. Avulsion of the triceps tendon. *Clin Orthop* 161:242–246, 1981.
27. Fisher, R. H. Conservative treatment of disturbed patellae and olecranon bursae. *Clin Orthop* 123:98, 1977.
28. Friedlander, H. L., Reid, R. L., and Cape, R. F. Tennis elbow. *Clin Orthop* 51:109–116, 1967.
29. Froimson, A. I. Treatment of tennis elbow with forearm support band. *J Bone Joint Surg [Am]* 53:183–184, 1971.
30. Froimson, A. I. Tenosynovitis and tennis elbow. *In* Green, D. P. (Ed.), *Operative Hand Surgery*, Vol. 3 (2nd ed.), New York, Churchill Livingstone, 1989, pp. 2127–2133.
31. Garden, R. S. Tennis elbow. *J Bone Joint Surg [Br]* 43:100–106, 1961.
32. Gardner, R. C. Tennis elbow: Diagnosis, pathology and treatment. *Clin Orthop* 72:248–253, 1970.
33. Gerster, J. C., Lagier, R., and Voivan, G. Olecranon bursitis related to calcium pyrophosphate dihydrate crystal deposition disease. *Arthritis Rheum* 25:989, 1982.
34. Goldberg, E. J., Abraham, E., and Siegel, I. The surgical treatment of chronic lateral humeral epicondylitis by common extensor release. *Clin Orthop* 223:208–212, 1988.
35. Goldie, I. Epicondylitis lateralis humeri. *Acta Chir Scand (Suppl)* 339, 1964.

36. Groppel, J. L., and Nirschl, R. P. A mechanical and electromyographical analysis of the effects of various joint: Counterforce braces on the tennis player. *Am J Sports Med* 14:195–200, 1986.
37. Heyse-Moore, G. H. Resistant tennis elbow. *J Hand Surg [Br]* 9:64–66, 1984.
38. Ho, G. and Tice, A. D. Comparison of nonseptic and septic bursitis. *Arch Intern Med* 139:1269–1273, 1979.
39. Hohl, M. Epicondylitis—Tennis elbow. *Clin Orthop* 19:232–238, 1961.
40. Justis, E. J. Affection of fascia and bursae. *In* Crenshaw, A. H. (Ed.), *Campbells Operative Orthopaedics* Vol. 3. 7th ed. St. Louis, C. V. Mosby, 1987, pp. 247–261.
41. Kaplan, L., and Ferguson, L. K. Bursitis. *Am J Surg* 37:455–465, 1937.
42. Kleinkort, J. A., and Wood, F. Phonophoresis with 1 percent versus 10 percent hydrocortisone. *Phys Ther* 55:1320–1324, 1975.
43. Larson, R. L., and Osternig, L. R. Traumatic bursitis and artificial turf. *J Sports Med* 2:183, 1974.
44. Lasher, H. W., and Mathewson, L. M. Olecranon bursitis. *JAMA* 90:1030, 1928.
45. Lee, M. L. H. Rupture of the triceps tendon. *Br Med J* 2:197, 1960.
46. Levy, M., Fisher, R. E., and Stern, E. M. Triceps tendon avulsion with or withut fracture of the radial head—A rare injury? *J Trauma* 18:677, 1978.
47. Levy, M., Goldberg, I., and Meir, I. Fracture of the head of the radius with a tear or avulsion of the triceps tendon. *J Bone Joint Surg [Br]* 64:70–72, 1982.
48. Major, H. P. Lawn tennis elbow. *Br Med J* 2:557, 1883.
49. Mathews, R. E., Gould, J. S., and Kashlan, M. B. Diffuse pigmented villomodular tenosynovitis of the ulnar bursa—A case report. *J Hand Surg* 6:64, 1981.
50. McCarty, D. J., and Gatter, R. A. Recurrent acute inflammation associated with focal crystal deposition. *Arthritis Rheum* 9:84, 1966.
51. Montgomery, A. H. Two cases of muscle injury. *Surg Clin Chicago* 4:871, 1920.
52. Morrey, B. F. Tendon injuries about the elbow. *In* Morrey, B. F. (Ed.), *Disorders of the Elbow*. Philadelphia, W. B. Saunders, 1985, pp. 452–463.
53. Morrey, B. F. Bursitis. *In* Morrey, B. F. (Ed.), *The Elbow and Its Disorders* Philadelphia, W. B. Saunders, 1985, pp. 745–751.
54. Morris, M., Jobe, F. W., Perry, J., Pink, M., and Healy, B. S. Electromyographic analysis of elbow function in tennis players. *Am J Sports Med* 17:241–247, 1989.
55. Nirschl, R. P. The etiology and treatment of tennis elbow. *J Sports Med Phys Fitness* 2:308–323, 1974.
56. Nirschl, R. P. Muscle and tendon trauma: Tennis elbow. *In* Morrey, B. F. (Ed.), *The Elbow and Its Disorders*. Philadelphia, W. B. Saunders, 1985, pp. 481–496.
57. Nirschl, R. P. Prevention and treatment of elbow and shoulder injuries in the tennis player. *Clin Sports Med* 7:289–308, 1988.
58. Nirschl, R. P., and Pettrone, F. A. Tennis elbow—The surgical treatment of lateral epicondylitis. *J Bone Joint Surg [Am]* 61:832–839, 1979.
59. Osgood, R. B. Radiohumeral bursitis, epicondylitis, epicondylalgia (tennis elbow). *Arch Surg* 4:420–433, 1922.
60. Pantagopoulos, T., Exarchow, E., Stavrou, Z., and Hartofilakidis-Garofalidis, G. Avulsion of the triceps tendon. *J Trauma* 15:827–829, 1975.
61. Partridge, N. A case report of rupture of the triceps cubiti. *Medical Times and Gazette* 1:175, 1868.
62. Patterson, R. L., and Darrach, W. Treatment of acute bursitis by needle irrigation. *J Bone Joint Surg [Am]* 19:993–1002, 1937.
63. Penhallow, D. P. Report of a case of ruptured triceps due to direct violence. *NY Med J* 91:76, 1910.
64. Preston, F. S., and Adicoll, A. Hyperparathyroidism with avulsion at three major tendons. *N Engl J Med* 266:968, 1961.

65. Priest, J. D., Braden, V., and Gerberich, J. G. The elbow and tennis (part 1). *Physician Sports Med* 8:80, 1980.
66. Quayle, J. B., and Robinson, M. P. A useful procedure in the treatment of chronic olecranon bursitis. *Injury* 9:299, 1976.
67. Quillin, W. S. Ultrasonic phonophoresis. *Physician Sports Med* 10:211, 1982.
68. Regan, W., Wold, L., Coonrad, R., and Morrey, B. F. Microscopic pathology of lateral epicondylitis. *Am J Sports Med* (accepted for publication).
69. Rennie, W. R. J. The elbow. *In* Cruess, R. L., Rennie, W. R. J. (Eds.), *Adult Orthopaedics,* Vol. 2. New York, Churchill Livingstone, 1984, pp. 1055–1076.
70. Sarkar, K., and Uhthoff, H. Ultrastructure of the common extensor tendon in tennis elbow. *Virchows Arch [A]* 386:317–330, 1980.
71. Sherman, O. H., Snyder, S. J., and Fox, J. M. Triceps tendon avulsion in a professional body builder. *Am J Sports Med* 12:328–329, 1988.
72. Spencer, G. E., and Herndon, C. H. Surgical treatment of epicondylitis. *J Bone Joint Surg [Am]* 35:421–424, 1953.
73. Stack, J. K. Acute and chronic bursitis in the region of the elbow joint. *Surg Clin North Am* 29:155–162, 1949.
74. Tarsney, F. F. Rupture and avulsion of the triceps. *Clin Orthop* 83:177–183, 1972.
75. Twinning, R. H., Marcus, W. Y., and Garey, J. L. Tendon ruptures in systemic lupus erythematosis. *JAMA* 187:123–124, 1964.
76. Unverferth, J. L., and Olix, M. L. The effects of local steroid injections on tendon. *J Sports Med* 1:3–37, 1973.
77. Vizkelety, T., and Aszodi, K. Bilateral calcareous bursitis at the elbow. *J Bone Joint Surg [Br]* 58:644, 1968.
78. Werner, C. O. Lateral elbow pain and posterior interosseous nerve entrapment. *Acta Orthop Scand Suppl* 174:1–62, 1979.
79. Whittaker, C. R. The arrangement of the bursae in the superior extremities of the full term fetus. *J Anat Physiol* 44:133, 1910.
80. Wilson, F. D., Andrews, J. R., Blackburn, T. A., and McClusky, G. Valgus extension overload in the pitching elbow. *Am J Sports Med* 11:83–88, 1983.

Throwing Injuries

Bernard F. Morrey, M.D.
William D. Regan, M.D.

LITTLE LEAGUER'S ELBOW

The elbow is the most frequent area of complaint in children and adolescent baseball players [11, 24]. Activities that involve throwing or catching or pushing or hammering may produce significant stresses about the elbow joint, involving the ligaments, capsule, and articular surfaces of both the ulnohumeral and radial capitellar joint [7, 12, 26]. These forces may impair elbow function and render the athlete unable to perform.

In some instances, presentation is associated with a specific episode, and acute traumatic conditions about the elbow joint are considered. More commonly, however, chronic elbow pain and instability in the young athlete is specifically referred to as little leaguer's elbow [8, 20]. At the adult level and in competitive throwing sports for the college or professional athlete, it is more common to find acute ligamentous rupture and chronic valgus insufficiency. In order to more fully appreciate the injury associated with throwing sports, one must study the functional anatomy of the elbow ligaments, specifically the medial collateral ligament.

The Ligaments

The radiocollateral ligament arises from the lateral epicondyle and inserts onto the annular ligament, which surrounds the head of the radius. This ligamentous complex offers very little varus stability and is rarely stressed in the athlete. More recently, an accessory collateral ligament has been described, arising from the lateral epicondyle and inserting across the radial head and into the ulna [18]. This ligament is termed the lateral ulnar collateral liga-

ment, and it offers varus stability in both flexion and extension [5, 18, 21].

The anconeus muscle that traverses the radiocapitellar articulation laterally may serve as a dynamic stabilizer [4, 5]. The medial collateral ligament is composed of three parts: an anterior oblique, a posterior oblique, and a transverse ligament. A transverse band is of little significance. The anterior oblique portion of the medial collateral ligament arises from the medial epicondyle and inserts into the coronoid process. The anterior fibers become taut in extension; in flexion, the posterior fibers become tight. If viewed from within the elbow joint, however, the medial collateral ligamentous complex is a continuum along the greater sigmoid fossa of the ulna. Fibers sequentially become tighter from anterior to posterior and as the elbow is brought from extension into flexion [21]. In conjunction with this, the posterior oblique ligament inserting onto the olecranon is tight in flexion only and is lax in extension.

Recently, the medial collateral ligamentous complex has been studied biomechanically. Through this work, it has been determined that the anterior bundle of the medial collateral ligamentous complex is both stronger and stiffer than its posterior counterpart [21]. This work was done by splitting the medial collateral ligamentous complex into two distinct bundles, anterior and posterior, and mounting these into a bone–ligament–bone complex for measurement of displacement curves. In addition, it was found that the radiocollateral ligament has intermediate strength and stiffness, compared to the anterior and posterior bundles of the medial collateral ligament [21]. The anterior bundle of the medial collateral ligament is important in throwing sports, because this is the primary stabilizer to valgus stress.

Sixty-five percent of the load to valgus strain is taken up by the medial collateral ligament, according to Morrey and An [17]. Repetitive valgus stress, which is associated with the act of throwing, may result in microtrauma to the anterior oblique ligament. This repetitive insult can result in attenuation and stretching of the anterior bundle of the medial collateral ligament and subsequent compression of the radiocapitellar articulations. The effect has been analyzed in the pitching arm of professional baseball pitchers [16, 25]. It has been shown that valgus deformity of the elbow is very common secondary to attenuation of the medial collateral ligament. In addition, in response to this injury, there are often flexion contractures, which were present in over 50% of the pitchers examined [16].

Because valgus elbow stress and strain are associated with the act of pitching, it is important to study this condition in both the adolescent and the mature thrower. In the acute situation, the abnormal medial stress applied to the elbow in the adult can produce a medial collateral ligament tear or a muscular strain to the common flexor origin. In a growing child, bony epicondylar avulsion may occur with the acute valgus strain. In addition, the adolescent may develop a stress fracture of the medial epicondyle. In the adolescent or adult, valgus instability may result.

Medial Epicondylar Stress Fracture

Skeletal maturation factors in children or adolescents are related to the pattern of ossification centers about the distal humerus. The medial epicondyle appears at approximately 5 years of age in females and at 7 years of age in males; it enlarges until it fuses to the humerus at approximately age 14 years in females and age 17 years in males [20].

In the little league group of pitchers, the most frequent complaints are related to sensitivity about the medial epicondyle. Adams [1] drew attention to the potential risk of injury in a study of 80 little league pitchers, ages 9 to 14 years. He found that 45% of these players gave a history of pain about the elbow; in all 80 pitchers, some degree of radiographic abnormality was present in the dominant elbow when compared to the opposite side. Although he did not document his findings statistically, he states that changes were in direct proportion to the amount and type of throwing [1].

During the process of pitching, particularly in the cocking phase, repetitive forces secondary to contracture of the flexor muscle origin at the medial epicondyle may stress the normal chondro-osseous

transformation and result in a series of microinjuries [1, 6, 20]. This may result in the characteristic pain that is localized to the common flexor origin, swelling localized to the same area with pain on performance, and an irregular ossification pattern on roentgenograms. This pain pattern and radiographic appearance are usually referred to as an overstress syndrome and are classically called little leaguer's elbow [6].

When the athlete is entering adolescence, muscle mass and strength increase rapidly. In this situation, the valgus stress and sudden contracture of the common flexor origin may be strong enough to result in an avulsion fracture of the entire medial epicondyle, requiring surgical relocation and fixation [1, 2].

On approaching the final stages of fusion of the medial epicondyle to the humerus, the forces are not generally distributed to the entire medial epicondyle but to one point of muscle origin. Here, instead of the entire epicondyle becoming avulsed, a fragment may be separated from the bone, which may go on to heal or leave a mild residual spur on the inferior surface of the medial epicondyle. In addition, it may not unite, and it may become a painful loose osteocartilaginous fragment in the origin of the common flexor mass [20].

Woods and Tullos have classified medial epicondylar fractures into two types [28]. Type I occurs in younger children, in whom a large fragment involving the entire epicondyle is often displaced and malrotated. In the adolescent pitcher, a type II fracture occurs. Here, the medial epicondyle is fragmented, and the fracture fragment is small. The small fracture fragment may indicate that the anterior oblique ligament has been ruptured.

On examination, a markedly swollen and painful elbow is typical, as is ecchymosis along the medial epicondylar region. Typically, limited range of motion is present secondary to pain and sometimes to an effusion. Point tenderness over the common flexor origin and medial epicondylar region are consistent features that are associated with marked instability to valgus stressing of the elbow.

Radiographic review often shows a displaced medial epicondylar fragment with or without entrapment of this fragment within the joint (Fig. 17F–1). A simple radiograph, using gravity to impart valgus stress to the elbow, may be helpful in difficult cases (Fig. 17F–2) [22, 27]. In this method the patient lies supine with the shoulder in maximal external rotation. The sagittal plane of the elbow is now parallel to the floor, with only the flexor forearm mass and a medial collateral ligament providing resistance to the weight of the forearm (see Fig. 17F–2).

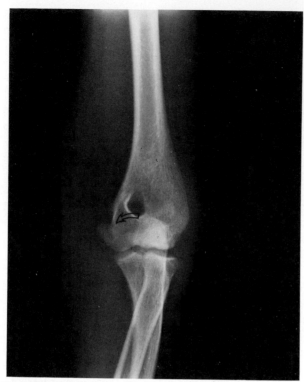

FIGURE 17F–1
Displaced medial epicondyle fracture in a 14-year-old gymnast after falling during a floor routine.

Treatment

If the radiograph of a little league elbow reveals some widening of the epiphyseal lines compared to the opposite normal elbow, this patient should be rested for 2 to 3 weeks, with gradual resumption of a throwing sport over the next 6 weeks [5].

Violent, single-event stress, such as throwing a pitch, falling in gymnastics, or even throwing a javelin, can cause avulsion of the medial epicondyle or a larger fragment. Treatment usually depends on the amount of displacement that has occurred. Patients sustaining fracture with displacement of more than 5 mm or with marked rotation of the fragments are managed with open reduction and internal fixation with smooth Kirschner wires. The Kirschner wires are removed after 3 weeks, and active motion is begun. After injury, the physis of the epicondyle usually closes, but because the physis does not contribute to longitudinal length or articulate with the ulna, there are no long-term sequelae of this injury or treatment [4].

If there is only mild displacement of less than 5 mm, the patient is usually managed with immobilization for 10 to 14 days, followed by early mobilization with active assisted and active range-of-mo-

tion exercises. It is important to reexamine the elbow at approximately 6 weeks to assess the ligamentous stability. Because the ulnar collateral ligament may be attached to the medial epicondyle fragment, the stability of the joint should be assessed following union of the medial epicondylar fracture.

CHRONIC VALGUS OVERLOAD SYNDROME IN ADULTS

Bennett was the first to direct attention to the problems that occur in the pitching elbow of professionals [4]. Since then, a great deal of work has gone into describing the injuries about the elbow during pitching. Because there are unusual stresses applied to the elbow joint during the pitching maneuver, it is essential to understand the basic aspects of the act of pitching in order to recognize the different areas of stress concentration about the elbow. The throwing mechanism may be divided into various phases, which include the wind-up or cocking phase, the acceleration phase, release, and follow-through [16, 23]. The combination of medial elbow stress, cubitus valgus, and a narrow bony fossa secondary to hypertrophy can cause painful impingement of the olecranon process against the medial wall of the olecranon fossa. This problem has been particularly troublesome in throwing athletes and is related to the valgus overload during the acceleration phase of pitching [14]. The development of a posterior osteophyte occurs in time. Wilson and associates believe that the mechanism of injury in the valgus extension overload syndrome occurs in the early acceleration phase of pitching [27]. In this early phase of acceleration, excessive valgus stress is applied to the elbow, causing a wedging effect of the olecranon into the olecranon fossa [2, 27].

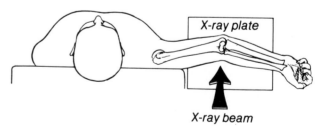

Gravity Test

FIGURE 17F–2
Gravity valgus stress view of elbow demonstrates medial collateral ligament insufficiency. (From Schwab, G. H., Bennett, J. B., and Woods, G. W. Biomechanics of elbow instability. *Clin Orthop* 146:42, 1980.)

Subjective Complaints

This condition occurs predominantly in pitchers but is also observed in other sports, such as javelin throwing. The pain is usually bothersome with forced extension and valgus strain, as occurs during the acceleration phase of pitching. Due to the painful inhibition, the pitcher is usually effective for two to three innings and then suffers a gradual loss of control, especially with early release, which causes him to throw high. Often, a pitcher compensates by snapping the elbow in an attempt to gain speed, but unfortunately this often causes more loss of control. Throwers generally localize pain over the olecranon process posteriorly and medially.

Examination

Because about 50% of throwers have flexion contractures and 30% exhibit cubitus valgus deformities, this syndrome is fairly common [17]. Local pain is present over the posterior aspect of the olecranon and is made worse by stressing the elbow in valgus and in extension. Marked pain and occasionally palpable osteophytes may be noted in the posteromedial aspect of the olecranon. Occasionally, individuals demonstrate some ulnar neuritis due to impingement on the posteromedial olecranon, and the ulnar nerve should be examined carefully for this associated problem.

The roentgenogram is important in this condition. The lateral radiograph occasionally demonstrates a posteromedial osteophyte. This is unfortunately difficult to localize, and axial projections at various degrees of flexion can demonstrate a posteromedial osteophyte more readily (Fig. 17F–3). Tomograms are also most helpful in this circumstance. Wilson and associates found that, with the elbow flexed at 110 degrees and lying on a cassette, the lesion can be demonstrated if the beam is angled 45 degrees to the ulna [27]. This gives the best view of the olecranon as it articulates with the trochlea and puts the medial aspect of the olecranon on profile (see Fig. 17F–3). It is important to realize that it is possible not to see the obvious bony osteophytes if the lesion is purely cartilaginous.

Treatment

Conservative Management

A conservative physiotherapy program is initially the mainstay of treatment of this condition. One

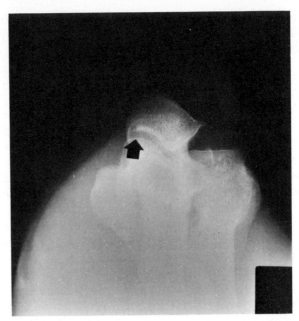

FIGURE 17F–3
Radiograph demonstrating articulation of olecranon and trochlea with posteromedial osteophyte on profile.

must try to achieve an improved arc of flexion-extension without pain.

Limitation of range of motion may be secondary to a synovitis; to assist in its resolution, a nonsteroidal anti-inflammatory drug (NSAID) may be prescribed along with the physiotherapy program. This program involves stretching the elbow with the wrist held in supination; a slow stretching is held for a slow count of 10, released, and repeated 25 times, 3 times per day. In addition, the wrist is actively extended as one is stretching the elbow. In this way, the extensor musculature is also stretched.

Wilson and associates [27] recommend an isotonic regime of light weights and high repetitions, which results in less stress about the elbow. Wrist curls are performed, and flexion moving to extension of the elbow with the forearm in pronation and supination. Usually, a 1-pound dumbbell is used to begin, and the weights are gradually increased to 5 pounds.

In addition, other physiotherapy modalities include heat and ultrasound. Occasionally, 10% hydrocortisone given by phonophoresis through the skin may be effective [27]. Deep transverse friction massage may indeed increase circulation and flexibility of the scar tissue building up in the posteromedial aspect of the elbow. If the athlete continues to throw easily through this period of rehabilitation, ice packs are applied to lessen the inflammatory component and synovitis that accompany this condition.

Following an adequate course of physiotherapy

and NSAIDs, if the athlete is unable to return to his prefunctional level of performance surgical intervention may be considered.

Surgical Management

Exposure is through a straight posterolateral approach to the posterior compartment of the elbow, with the triceps tendon identified and sharply elevated off the epicondylar ridge of the humerus. Following various degrees of flexion and extension with variable stress, a straight osteotomy is made 1 cm proximal to the tip of the olecranon. The second excision of bone is made with a curved osteotome that transects the posteromedial olecranon, removing the posteromedial osteophyte. One must be careful to remove enough of the posteromedial olecranon tip to clear this area of chondromalacia and osteophytes and thereby prevent further impingement (Fig. 17F–4).

The ulnar nerve is more posterior than the osteotomy incision; extreme caution must be taken at all times to avoid injury to the nerve.

Following a routine closure, active range of motion is begun. Because little has been done to cause instability, rapid return to full function can be expected following restoration of full range of motion.

In a series of five baseball pitchers, three college and two professional, following this program of operative intervention and gradual rehabilitation with an average follow-up of 12 months, all athletes returned to their preoperative sport at full strength for at least one full season. The average time to return was 11 weeks. Since the advent of elbow arthroscopy, this problem can be addressed using arthroscopic techniques. The posterolateral portal is

employed for this approach, and osteophytes can be removed using the high-speed bur. It is important when undertaking of this operative procedure that the hood of the abrader lie in the medial direction to avoid injury to the ulnar nerve.

The advantages of this technique are that it avoids incisions in the elbow itself, quickens the rehabilitation of the extremity, and allows more rapid return to sporting activity. The incisions may cause rapid periarticular fluid extravasation, necessitating a limited incision and osteotomy under direct vision.

MEDIAL COLLATERAL LIGAMENT RUPTURES

In skeletally mature pitchers and javelin throwers, excessive valgus forces on the medial compartment in the late cocking and acceleration phases of throwing lead to flexor-pronator muscle tears, medial ligament ruptures, and ulnar neuritis [7, 10]. The incidence of these injuries is well described [1–4, 11, 13–15].

The main structure involved in throwing sport is the anterior bundle of the ulnar collateral ligament. This structure has recently been biomechanically tested with the bone-ligament-bone preparation, and has been found to be both the strongest and stiffest ligament of the elbow [21]. It is also a major stabilizing structure of the elbow [17, 22].

Those who present with problems referable to the medial collateral ligament are usually baseball pitchers or javelin throwers. The patients presenting with these collateral ligament ruptures have often had months to years of recurrent pain and tenderness along the medial aspect of the elbow, associated with the throwing sport. In some, a sudden catastrophic event occurs that heralds the rupture and forces the athlete to stop playing immediately. In the majority, however, a slow deterioration in function of the elbow occurs, accompanied by increasing pain and loss of control [Frank Jobe, personal communication, 1991].

On clinical examination, valgus instability is present in all cases. This is best assessed by flexing the elbow 25 degrees to unlock the olecranon from its fossa, externally rotating the humerus to lock the shoulder, and gently stressing the medial aspect of the elbow with a valgus stress [17] (Fig. 17F–5). If there is an acute rupture of the anterior bundle of the ulnar collateral ligament, the elbow often is ecchymotic and tender along the medial aspect. The tenderness is usually localized to the ulnar aspect of the medial collateral ligament compared to the humeral side because the ligament usually tears off the

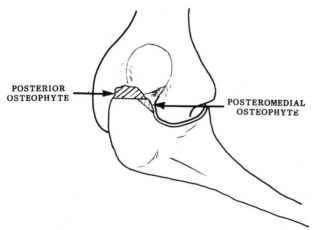

FIGURE 17F–4
Surgical management of valgus overload syndrome in adults.

POSTERIOR OSTEOPHYTE

POSTEROMEDIAL OSTEOPHYTE

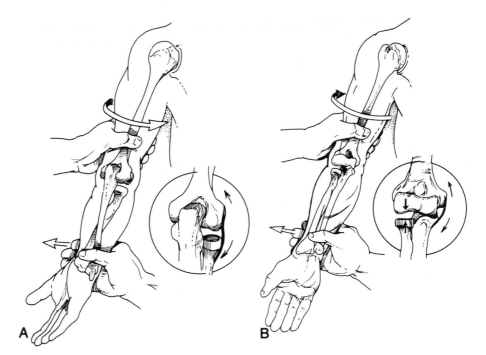

FIGURE 17F–5
Valgus instability is assessed by loading the humerus in full external rotation and stressing the elbow in about 10 degrees of flexion. (With permission of the Mayo Foundation.)

medial aspect of the coronoid process. In addition, the ulnar nerve may be tender, with or without paresthesias.

Radiographic review may demonstrate calcification in the ligament. This may be more common in those in whom a chronic pain syndrome has led to one or more courses of steroid injection into the medial epicondylar region.

Treatment for most patients is conservative, consisting of rest until the swelling settles, usually with icing and NSAIDs, followed by restoration of range of motion. Following a complete resolution of pain and restoration of preinjury level of range of motion, the athlete is allowed to resume a throwing program, usually beginning at 3 months after injury. Progressive velocity and endurance are allowed, with careful physician supervision. Patients who continue to have ongoing pain or cannot throw effectively after 6 months of this conservative management are considered candidates for operative intervention.

Operative Treatment

Little has been written regarding the operative management of acute medial elbow ruptures. A direct repair of torn medial structures without ligamentous reconstruction has been reported for acute injuries diagnosed by abduction stress tests at 15 degrees of flexion [19]. Three elbows had associated ulnar nerve compression with neurologic dysfunction; all were decompressed, and one had an anterior transposition. The medial collateral ligaments were repaired in each case; one that was torn from the medial epicondyle was repaired by direct suture. In the second, the ligament was repaired to the humerus by passing suture through drill holes. The posterior bundle of the medial collateral ligament was repaired with direct suture to the olecranon.

In this series, all elbows were placed in a posterior splint, and three of the four were removed at 10 days following surgery and placed in a single-axis elbow orthosis that allowed protected elbow flexion and extension and prohibited valgus stress, pronation, and supination. The last elbow was in a splint for 4 weeks and then was started on an active range-of-motion program without additional orthotic support.

The results at 20 months demonstrated full range of motion compared to the uninjured elbow, with no evidence of myositis ossificans. There was no sign of instability.

The issue of repair versus reconstruction has been addressed more recently by Jobe and associates [15]. They reported on 16 athletes, using a free tendon graft of either palmaris longus, plantaris, or the lateral aspect of the Achilles tendon. All had ruptured the anterior bundle of the medial collateral ligament. Subsequently, Jobe has had experience with more than 150 of these operative procedures (personal communication, 1991).

Technique

The surgical technique uses a posteromedial approach to the elbow, with isolation of the anterior bundle of the ulnar collateral ligament by elevating the common flexor origin. The ulnar nerve is then identified and prepared for transposition anteriorly.

Using a slow-speed drill with guide, 3.2-mm holes are placed in the medial epicondyle and in the ulna, corresponding to the anatomic sites of attachment of the ligament (Fig. 17F–6). A tendon graft, usually the palmaris longus, is passed through the holes to form a figure-of-eight that acts as a functional substitute for the anterior bundle of the medial collateral ligament. Following reconstruction, the flexor and pronator muscles are placed over the transferred ulnar nerve and reattached to the fringe of tendon left on the epicondyle.

The postoperative regimen for these patients is initiated immediately. A soft rubber ball is squeezed to maintain grip strength, and then immobilization is discontinued and active range-of-motion exercises are begun. Beginning 1 month after surgery, a muscle-strengthening program is initiated. This consists of tossing a baseball 30 feet while standing flat-footed at 3 months after surgery. At 4 months, throwing a baseball 60 feet while standing flat-footed is allowed. At 6 months after surgery, pitchers may throw three-quarter speed. At 7 to 8 months, throwing at full speed is expected. Typically, 1 year is required to regain the ability to participate in competitive sport.

Results

In the initial experience of Jobe, 10 of the 16 patients returned to their previous level of participation, 1 returned to a lower level of participation, and 5 retired from professional athletics.

Due to the success of ulnar collateral ligament reconstruction using the technique popularized by Jobe and associates [15], we now advocate reconstruction of the anterior bundle of the ulnar collateral ligament for those patients who have been unsuccessful in conservative treatment for at least 6 months, or who have been diagnosed with either a complete rupture of the anterior bundle of the medial collateral ligament or chronic laxity secondary to repetitive microtrauma and tearing. These patients all undergo the reconstructive procedure, followed by the detailed rehabilitation program described for return to active pitching. In the competitive throwing athlete with a complete active medial collateral ligament rupture, we have not had good success with primary repair of the anteroposterior bundle of the medial collateral ligament.

References

1. Adams, J. E. Injury of the throwing arm: A study of traumatic changes in the elbow joints of boy baseball players. *Calif Med* 102:127, 1965.
2. Andrews, J. R., McCluskey, G. M., and McLeod, W. D. Musculo-tendinous injuries of the shoulder and elbow in athletes, Schering symposium. *Athletic Training* 11(2):68–71, 1976.
3. Barnes, D. A., and Tullos, H. S. An analysis of 100 symptomatic baseball players. *Am J Sports Med* 6:62–67, 1978.
4. Bennett, G. E. Shoulder and elbow lesions of the professional baseball pitcher. *JAMA* 117:510–514, 1941.
5. Bennett, J. B., and Tullos, H. S. Ligamentous and articular injuries in the athlete. *In* Morrey, B. F. (Ed.), *The Elbow and Its Disorders*. Philadelphia, W. B. Saunders, 1985, pp. 502–522.
6. Brodgon, B. E., and Crow, W. F. Little leaguer's elbow. *AJR* 8:671, 1960.
7. Childress, H. M. Recurrent ulnar nerve dislocation at the elbow. *J Bone Joint Surg [Am]* 38:978–984, 1956.
8. Dehaven, K. E., and Evarts, C. M. Throwing injuries of the elbow in athletes. *Orthop Clin North Am* 1:801, 1973.
9. Gainor, B. J., Piotrowski, G., Puhl, J., Allen, W. C., and Hagen, R. The throw: Biomechanics and acute injury. *Am J Sports Med* 8:114, 1980.
10. Godshall, R. W., and Hansen, C. A. Traumatic ulnar neuropathy in adolescent baseball pitchers. *J Bone Joint Surg [Am]* 53:359–361, 1971.
11. Guggenheim, J. J., Stanley, R. F., Woods, G. W., and Tullos, H. S. Little league survey: The Houston study. *Am J Sports Med* 4:189, 1976.

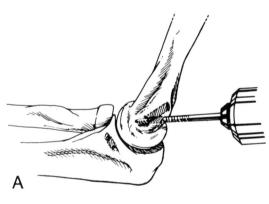

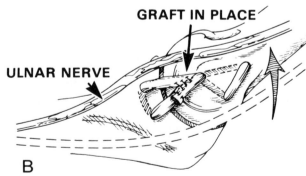

GRAFT IN PLACE

ULNAR NERVE

FIGURE 17F–6

A, Holes are placed in the anatomic site of origin and insertion of the anterior bundle of the medial ulnar collateral ligament. *B,* The free tendon graft forms a figure-of-eight repair. (From Jobe, F. W., Stark, H., and Lombardo, S. J. Reconstruction of the ulnar collateral ligament in athletes. *J Bone Joint Surg [Am]* 68:1158–1163, 1986.)

12. Hang, V. S., Lippert, F. G., Spolek, G. A., Frankel, V. H., and Harrington, R. M. Biomechanical study of the pitching elbow. *Int Orthop* 3:217, 1979.
13. Hane, Y. S. Tardy ulnar neuritis in a little league baseball player. *Am J Sports Med* 9:244–246, 1981.
14. Indelicato, P. A., Jobe, F. W., Kerlin, R. K., Carter, V. S., Shields, C. L., and Lombardo, S. I. Correctable elbow lesions in professional baseball players. *Am J Sports Med* 7:72, 1979.
15. Jobe, F. W., Stark, H., and Lombardo, S. J. Reconstruction of the ulnar collateral ligament in athletes. *J Bone Joint Surg [Am]* 68:1158–1163, 1986.
16. King, J. W., Brelsford, J. H., and Tullos, H. S. Analysis of the pitching arm of the professional baseball pitcher. *Clin Orthop* 67:116, 1969.
17. Morrey, B. F., and An, K. N. Stability of the elbow joint: A biomechanical assessment. *Am J Sports Med* 12:315–319, 1984.
18. Morrey, B. F., and An, K. N. Functional anatomy of the ligament of the elbows. *Clin Orthop* 201:84–90, 1985.
19. Norwood, L. A., Shork, J. A., and Andrews, J. R. Acute medial elbow ruptures. *Am J Sports Med* 9:16, 1981.
20. Pappas, A. M. Elbow problems associated with baseball during childhood and adolescence. *Clin Orthop* 163:30, 1982.
21. Regan, W. D., Korinek, S., An, K., and Morrey, B. F. Biomechanical study of ligaments around the elbow joint. *Clin Orthop* 271:170, 1991.
22. Schwab, G. H., Bennett, J. B., and Woods, G. W. Biomechanics of elbow instability. *Clin Orthop* 146:42, 1980.
23. Sisto, D. J., Jobe, F. W., Moynes, D. R., and Antonelli, D. J. An electromyographic analysis of the elbow in pitching. *Am J Sports Med* 15:260–263, 1987.
24. Torg, J. S., Pollack, H., and Sweterlitsch, P. The effect of competitive pitching on the shoulders and elbows of preadolescent baseball players. *Pediatrics* 49:267, 1972.
25. Tullos, H. S., Erwin, W., Woods, G. W., Wukasch, D. C., Caoley, D. A., and King, J. W. Unusual lesions of the pitching arm. *Clin Orthop* 88:169, 1972.
26. Tullos, H. S., and King, J. W. Throwing mechanism in sports. *Orthop Clin North Am* 4:709–721, 1973.
27. Wilson, F. D., Andrews, J. R., Blackburn, T. A., and McCluskey, G. Valgus extension overload in the pitching elbow. *Am J Sports Med* 11:83–87, 1983.
28. Woods, G. W., and Tullos, H. S. Elbow instability and medial epicondylar fractures. *Am J Sports Med* 5:23, 1977.

Olecranon Bursitis

Kenneth M. Singer, M.D.
Kenneth P. Butters, M.D.

The olecranon bursa is one of the many superficial bursae in the body, and its function is to allow the skin to glide freely over the bony prominence of the olecranon (Fig. 17G–1). It is a closed sac, with no traversing fibers, and is lined by opposing surfaces of synovium, which allow the surfaces to glide freely over each other. The synovium is similar to that of any other joint and secretes high-viscosity lubrication fluid. Although most of the deeper bursae are present at birth, the olecranon bursa, like most other subcutaneous bursae, probably does not form until after birth [2].

Bursal problems in athletes may present in three forms—acute hemorrhagic bursitis, chronic bursitis, or septic bursitis. The superficial location of the olecranon bursa exposes it to trauma with both athletic and nonathletic activities. Repetitive bursal traumas often result in small tears that heal spontaneously, leaving behind an enlarged bursa containing intrabursal bands and thickened walls. These are rarely symptomatic, but with additional repetitive trauma can become chronically inflamed. The bursa is also subjected to direct blows to the tip of the elbow, resulting in hemorrhage into the bursa. It may present as an acute or subacute inflammatory process, and most likely is associated with major or minor repetitive trauma. If the overlying skin has been broken, an acute infectious or septic bursitis may result.

The blood supply to the overlying skin comes in part from the bursa; therefore, in performing surgery through the bursa, such as repairing olecranon fractures, it is important not to excise the bursa as an incidental part of the procedure.

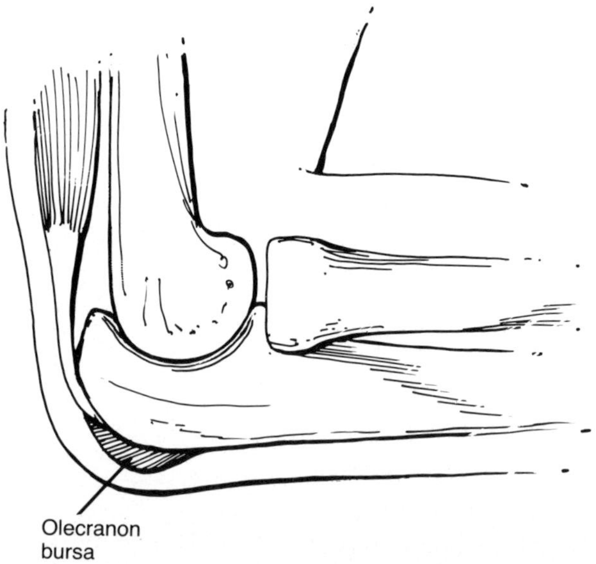

PERTINENT ANATOMY AND PATHOLOGY

The olecranon bursa is a closed sac interspersed between the skin and the triceps tendon and olecranon process (see Fig. 17G–1). In an anatomic cadaver study, no bursa was found in children below the age of 7 years; in the 7- to 10-year age group, the bursa, when present, was very small and usually unilateral. The size of the bursa increased with age and was usually larger on the dominant side [4]. The floor of the bursa is intimately adherent to the olecranon, but when the bursa is abnormal, it may be partially absent or inseparable from the olecranon.

The normal bursa is lined by smooth, glistening synovial layers. When the bursa is pathologic, the

Olecranon
bursa

FIGURE 17G–1
Schematic drawing showing the relation of the olecranon bursa to the skin and olecranon.

lining contains tufts of inflamed tissue, crossing strands and bands of abnormal, thickened synovium, and movable, often partially attached pieces of organized fibrinous material. These mobile pieces are often palpable and may be quite symptomatic.

The bursa has never been shown to communicate with the elbow joint except in rheumatoid arthritis, where the bursa may be involved in the inflammatory process. In fact, we have had this experience with two recreational athletes who came to surgery because of persistent chronic olecranon bursitis but were otherwise asymptomatic. Pathologic examination of the tissue revealed typical rheumatoid nodules. One patient developed other symptoms 8 years later, and the other patient remains asymptomatic.

Traumatic olecranon bursitis is the most common condition affecting the olecranon bursa [3]. The olecranon bursa in the athlete is subject to trauma in a variety of circumstances. These traumatic episodes may result in an acute inflammatory response; the bursal walls become thickened and edematous, and the bursal lining cells produce excess fluid. If the trauma is sufficiently severe to disrupt vessels, the bursa will contain either frank blood or bloody fluid. Repeated episodes of lesser trauma give rise to a chronic inflammatory process of the bursa with persistent effusions.

Septic olecranon bursitis occurs more commonly in athletes than the literature suggests [7]. The source of infection may be from skin breaks, often quite superficial and seemingly innocuous; from coexisting dermatitis or acneiform lesions that are colonized with bacteria; or from a hematogenous source. Trauma seems to be the most frequently implicated predisposing factor. Special note should be taken of the finding that steroid injections have preceded infection in approximately 10% of the infections studied [14, 15].

The most common organism by far is *Staphylococcus aureus*, but beta-hemolytic *Streptococcus* and other *Staphylococcus* species have been isolated [7, 14] as well. There are several case reports of septic olecranon bursitis caused by algae from *Prototheca*, the only algae pathogenic in humans; none of these has been seen in an athlete [1]. In the single case of olecranon bursitis reported in a child, *Streptococcus pyogenes* was isolated from the aspirate in a 3-year-old child [11].

CLINICAL EVALUATION

The most common cause of olecranon bursitis is trauma [3, 13], and a history of either a single event or multiple, less extensive trauma to the tip of the

elbow is frequently obtained. Soft tissue swelling is always present, and careful examination determines whether the swelling is the thickened bursa, fluid within the bursa, or both. Often inclusion nodules, which consist of fibrin masses covered by bursal lining cells, can be felt within the bursa if the swelling has persisted. If the process is of recent onset, there may be pitting edema in the skin overlying the olecranon bursa.

Radiographic evaluation consists of routine elbow views and may show soft tissue swelling or may be normal. In older individuals, olecranon spurs or calcium deposits may be seen, and the incidence of these findings seems to be higher in affected elbows than on the opposite side [13] (Fig. 17G–2). Bursography may be performed by first distending the olecranon bursa with air and then obtaining radiographs with the elbow both flexed and extended. The visible bubble changes shape with elbow flexion, and films may reveal nodules or septa [13] if they are present. Although bursography may be of interest, it is unlikely to influence the diagnostic or therapeutic decision-making process.

The patient with painless swelling of the olecranon bursa without redness or warmth is easy to evaluate (Fig. 17G–3). The problem arises in evaluation of the athlete with a history of either acute or repetitive

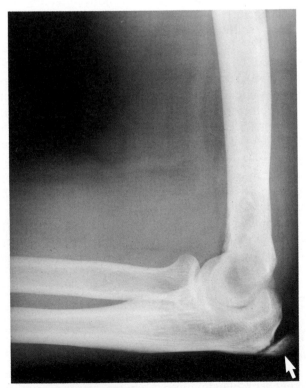

FIGURE 17G–2
Radiograph showing an olecranon spur.

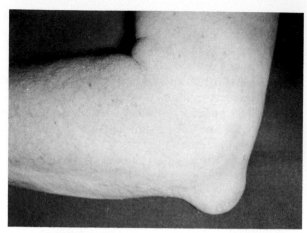

FIGURE 17G–3
Photograph of an enlarged olecranon bursa.

trauma, perhaps with a small abrasion over the elbow and with a tender, swollen olecranon bursa with overlying skin that is red, warm, and edematous. This is a common presentation in athletes, such as football players who play on an artificial surface without elbow pads [9], or wrestlers who have mat trauma. The physician must determine whether or not this is an infectious process.

Septic bursitis occurs in athletes, and usually there is a history of trauma either with or without skin breakage. There may be an abrupt onset and progression of symptoms that are quite disabling and are associated with systemic signs such as fever (35%), cellulitis (75%), or regional adenopathy (25%) [7]. In other instances, mild local trauma is followed by gradually increasing subacute symptoms, progressing to redness, warmth, and subcutaneous edema. This may result in a delay of several days before the athlete seeks medical attention. In such instances, if the athlete had been seen early in the course of the process, the symptoms would have suggested a nonseptic inflammatory process, and the athlete may have already been started on nonsteroidal anti-inflammatory drugs (NSAIDs), which may mask the real diagnosis.

Whenever there is concern that septic bursitis is a possibility, sterile aspiration of the bursal fluid for analysis by Gram stain and culture is essential.

TREATMENT OPTIONS

Traumatic Bursitis

In the acute traumatic event, in which there has likely been bleeding into the bursa, sterile aspiration for the purpose of evacuation of the blood, followed by a compression dressing and frequent icing, decreases the chances of progression to chronic bursitis [10]. Two or three aspirations may be necessary if the blood reaccumulates.

In most instances, the bursa is enlarged, minimally tender, and nontense. Treatment is symptomatic, and NSAIDs are the most common form of treatment. If there is fluid in the bursa, it may be aspirated, a small amount of steroid instilled, and a compression dressing applied. It is not necessary to interrupt athletic participation. If the bursa is enlarged, nontender, and nontense, it is probably best to protect it with an elbow pad but otherwise leave it alone.

If the bursa is tender and inclusion bodies can be felt, aspiration and steroid injection followed by a 3- to 5-day period of relative immobilization in a splint allow the most rapid recovery. NSAIDs may be used for 7 to 10 days. If the fluid continues to reaccumulate, insertion of a No. 16 angiocatheter into the bursa for several days, protected by a compression dressing, has been recommended [5].

When the bursitis is chronic and symptomatic, has recurred numerous times, and thickened bursal tissue or inclusion bodies are present, it is unlikely that resolution will occur without surgery. The surgical procedure then depends on the underlying pathology.

Surgical Procedure

If the bursal wall is not particularly thickened but has nodules and areas of inflamed tissue, it may be necessary only to open the bursa and debride it. Usually, however, it is advisable to excise the bursa itself and a small portion of the underlying bone at the point of the olecranon.

Either a transverse or longitudinal incision may be used, and the plane between the skin and the bursa developed to expose the bursa, keeping the skin flap as thick as possible. If the plane is difficult to identify, the bursa may be opened and the sac dissected free, using the same technique as one would with excision of an inguinal hernia. The skin flap must be treated gently and the ulnar nerve avoided. Usually, it is not necessary to expose the nerve, but if there is any extension of the bursa medially, the safest course is to identify and retract the nerve. The floor of the bursa is intimately adherent to the olecranon, and the tip of the bursa or any spurs that are present may be removed with a rongeur or an osteotome (Fig. 17G–4).

Quayle and Robinson [12] have suggested that the bursa itself not be excised but rather that the

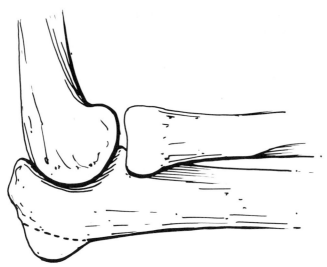

FIGURE 17G–4
Illustration of excision of olecranon bursa with decompression of the underlying bone.

triceps be retracted, a generous osteotomy of the olecranon performed, and the triceps repaired over the denuded olecranon (Fig. 17G–5). We have had no experience with this procedure but believe it may be useful, except perhaps in unusual circumstances where there has been significant bursal hypertrophy or where the skin overlying the bursa has become thin and healing might be compromised.

Postoperative Care and Rehabilitation

The wound should be drained for 24 to 48 hours with either a rubber or suction drain, and the elbow splinted in 45 to 60 degrees of flexion until the sutures are removed. Immobilization in 90 degrees of flexion increases tension over the olecranon and might compromise wound healing. Range of motion exercises are then begun while the patient is still partially immobilized and combined with biceps- and triceps-strengthening exercises. Athletic participation can begin in 4 weeks, but contact sports are avoided for 6 weeks. Soft elbow pads should be used for several months or until all tenderness has subsided.

Septic Bursitis

The significant incidence of previous steroid injections into the bursa in individuals who have developed septic olecranon bursitis is disconcerting, and a history of previous injection in a patient with persistent symptoms should arouse suspicion that a septic process may be present.

When septic bursitis is suspected and if fluid can be palpated in the bursa, it should be aspirated and cultures obtained. The elbow should be splinted, and frequent heat treatments instituted. Initially, if there are no systemic signs of infection and little cellulitis, the bursa can be decompressed by aspiration, and oral broad-spectrum antibiotics are begun in reasonably high doses. Daily observation is necessary, and repeat aspiration is accomplished. If significant clinical improvement ensues, the treatment is continued; if not, the bursa should be opened and drained and intravenous antibiotics used.

Knight and coworkers [8] reported 10 cases of septic olecranon bursitis treated with catheter drainage and antibiotics with good results. A 3-mm polyethylene tube was inserted into the bursa, antibiotic solution of 1% kanamycin and 0.1% polymyxin instilled for 3 hours, and the tube connected to suction for 9 hours. Intravenous antibiotics were used. When the cultures from the tube were negative for 3 consecutive days, the tube was removed, and

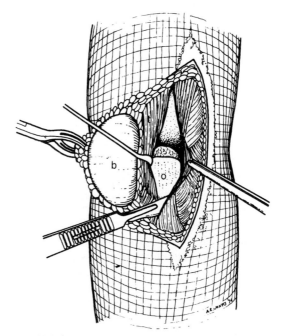

FIGURE 17G–5
Technique for excision of bursa. (*b,* bursa; *o,* olecranon process; *t,* triceps, musculotendinous part.) (From Quayle, J. B., and Robinson, M. P. A useful procedure in the treatment of chronic olecranon bursitis. *Injury* 9:299, 1976.)

the antibiotics were changed to oral drugs the following day and continued on an outpatient basis for 1 week. The average length of hospitalization was 12 days, and in all cases the septic bursitis resolved with no recurrences and no sinus tract formation.

Ho and associates [7] reported 20 cases of septic olecranon bursitis treated with bursal fluid drainage and antibiotics, usually parenteral. None of their patients was less than 37 years old, and only one, a gymnastics instructor, was involved in athletics. The remainder were engaged in work that predisposed them to frequent trauma to the elbows, such as gardening or plumbing. It was their belief that early and prompt recognition, often not the case in their series, was the most important factor in decreasing morbidity.

The duration of antibiotic therapy has not been clearly delineated, but the length of treatment necessary to sterilize the bursal fluid seems to be proportional to the length of time infection has existed and averages approximately 12 days [6].

CRITERIA FOR RETURN TO SPORTS PARTICIPATION

In instances of traumatic bursitis, usually the problem can be managed with minimal or brief interruption of sports activity. In the individual with acute hemorrhagic bursitis, several days of sports interruption are needed to prevent rebleeding into the bursa.

In the more chronic cases, symptoms are the deciding factor. If pain is not prevalent, most athletes can continue in their sport, and aspiration and steroid injection do not require cessation of athletics. Protection of the involved area to prevent or minimize the recurrent trauma is quite important, and this can be done with protective padding.

Should surgery be required, it can wait until the end of the athlete's season. Six weeks are required after surgery before the elbow can be subjected to significant trauma; however, sports such as swimming, golfing, or others that do not subject the elbow to significant insult can be allowed in 3 weeks, provided that there is full range of motion, normal strength, and no swelling.

Septic bursitis is a more serious situation, and return to sports will probably not be possible for several weeks and often not until the next playing season. Complete resolution of the infection, with no residual bursal symptoms and normal elbow function, likely takes several weeks, except in the mildest cases.

AUTHORS' PREFERRED METHOD OF TREATMENT

We believe that acute hemorrhagic bursitis should be treated aggressively with aspiration and the techniques mentioned earlier. The more chronic forms are usually treated with oral NSAIDs and padding, and sports participation is allowed. If inclusion bodies are present but asymptomatic, no treatment is recommended. Most patients respond and do not require more aggressive treatment.

Whenever a needle is inserted into a bursa for any purpose, it should be done under sterile conditions, including cleansing of the skin with an iodophor solution, use of sterile gloves by the physician, and single-dose or previously unused containers of local anesthetic and steroid.

If surgery is required, we prefer to use the longitudinal incision, lateral to the midline, so that in the normal sitting position the elbow does not rest on the incisional scar. The plane just superficial to the bursa is identified and, using sharp and blunt dissection, the bursa is exposed. The bursal tissue, particularly any inclusion bodies or granulation tissue, is removed and the ulna decompressed. Suction drainage is used for 12 to 24 hours and is then replaced with a compression dressing; the elbow is splinted until the sutures are removed or for 3 weeks, depending on the extent of the surgery. No recurrences have been noted.

In any instance where infection is suspect or cellulitis of the overlying skin of the elbow is present, cultures are taken of any open areas, and antibiotics effective against penicillin-resistant *S. aureus* are begun at once. A wait-and-see attitude should not be used; the morbidity associated with a few days of unnecessary antibiotics is much less than that with a few days of an untreated infection.

References

1. Ahbel, D. E., Alexander, A. H., Kleine, M. L., and Lichtman, D. M. Protothecal olecranon bursitis. *J Bone Joint Surg [Am]* 62:835–836, 1980.
2. Black, B. M. The prenatal incidence, structure and development of some human synovial bursae. *Anat Rec* 60:333, 1934.
3. Canoso, J. J. Idiopathic or traumatic olecranon bursitis. *Arthritis Rheum* 20:1213–1216, 1977.
4. Chen, J., Alk, D., Eventov, I., and Wientroub, S. Development of the olecranon bursa: An anatomic cadaver study. *Acta Orthop Scand* 58:408–409, 1987.
5. Fisher, R. H. Conservative treatment of distended patellar and olecranon bursae. *Clin Orthop* 123:98, 1977.

6. Ho, G. Jr., and Tice, A. D. Comparison of non-septic and septic bursitis: Further observations on the treatment of septic bursitis. *Arch Intern Med* 139:1269–1273, 1979.

7. Ho, G. Jr., Tice, A. D., and Kaplan, S. R. Septic bursitis in the prepatellar and olecranon bursae. *Ann Intern Med* 89:21–27, 1978.

8. Knight, J. M., Thomas, J. C., and Maurer, R. C. Treatment of septic olecranon and prepatellar bursitis with percutaneous placement of a suction-irrigation system. *Clin Orthop* 206:90–93, 1986.

9. Larson, R. L., and Osternig, L. R. Traumatic bursitis and artificial turf. *J Sports Med* 2:183–188, 1974.

10. O'Donoghue, D. H. *Treatment of Injuries to Athletes* (4th ed.). Philadelphia, W. B. Saunders, 1984, p. 242.

11. Paisley, J. W. Septic bursitis in childhood. *J Pediatr Orthop* 2:57–61, 1982.

12. Quayle, J. B., and Robinson, M. P. A useful procedure in the treatment of chronic olecranon bursitis. *Injury* 9:299, 1976.

13. Saini, M., and Canoso, J. J. Traumatic olecranon bursitis: Radiologic observations. *Acta Radiol Diagn* 23:255–258, 1982.

14. Soderquist, B., and Hedstrom, S. A. Predisposing factors, bacteriology, and antibiotic therapy in 35 cases of septic bursitis. *Scand J Infect Dis* 18:305–311, 1986.

15. Weinstein, P. S., Canoso, J. J., and Wohlgethan, J. R. Long-term follow-up of corticosteroid injection for traumatic olecranon bursitis. *Ann Rheum Dis* 43:44–46, 1984.

Osteochondroses of the Elbow

Kenneth M. Singer, M.D.
Kenneth P. Butters, M.D.

Osteochondroses about the elbow have become much more common as organized sports activities have involved younger individuals. Nearly all of the cases reported prior to 1980 have been in individuals participating in organized sports activities, primarily baseball, and have occurred in the dominant arm [6]. Although initially seen almost exclusively in prepubescent and adolescent boys, the entities have lost their sex specificity as girls have become more athletically involved. It has become apparent that these entities are related to the specific demands of the sport, not the gender of the athlete.

Osteochondrosis is defined as an alteration in an area of endochondral ossification, and at the elbow it may affect a traction apophysis, such as the medial epicondyle or olecranon, or an intra-articular epiphysis, specifically the capitellum.

PERTINENT ANATOMY

The medial epicondylar, lateral epicondylar, and trochlear epiphyses are really apophyses because they function as attachment sites for strong muscles. Their vulnerability, therefore, is associated with either overuse of the involved muscle groups or the stresses that can be applied to the apophyses by those muscle forces via violent contractions that cause the apophysis to fracture, or by the repetitive forces of sports that may cause osteochondroses.

The capitellum and radial head, being intra-articular structures, are vulnerable to compression and shear stresses, which can permanently damage the structural integrity of the articular surfaces.

The other important anatomic aspects of this entity relate to the radiographic appearance of the various ossification centers about the elbow, including the ages of their appearance and fusion (Table 17H–1).

LITTLE LEAGUE ELBOW

The entity termed Little League (or Little Leaguer's) elbow was initially used in 1960 by Brodgon and Crow [5] to describe the clinical and radiographic changes on the medial side of the elbow in young baseball pitchers. Adams [1] expanded the entity to include all of the problems that he associated with pitching, including lateral injuries, and correctly focused attention on the damage occurring to the immature elbow. The sports medicine community rightly became concerned with what appeared to be an epidemic of potentially harmful injuries occurring as a result of an activity that was intended to be fun and safe for children.

In 1968, Slocum [37] described the throwing motion as consisting of four distinct phases: wind-up, cocking, acceleration, and follow-through (Fig. 17H–1). He and others classified the types of injuries likely to occur by correlating the phases of the throwing motion with the forces occurring to the various soft tissue, osseous, and articular components of the elbow during each phase of throwing.

The early reports raised the concern that pitching or throwing at a young age caused changes in the growing elbow that would cause major problems in adulthood [22]. Subsequent surveys of the elbows of young baseball pitchers [10, 13, 20] confirmed the existence of radiographic changes on the medial side but failed to demonstrate that these were detrimental and found no increased incidence of the more serious lateral compression injuries. Pitching certainly causes elbow symptoms, the incidence

TABLE 17H–1
Ages of Appearance and Fusion for Secondary Ossification Centers About the Elbow*

	Approximate Age of Appearance (Years)		Approximate Age of Fusion (Years)	
	Males	*Females*	*Males*	*Females*
Capitellum	2	2	14.5	13
Radial head	5	4.5	16	14
Medial epicondyle	7	5	17	14
Lateral epicondyle	11	10	15	12.5
Trochlea	9	8	13	11.5
Olecranon	10	8	16	14

*From Pappas, A. M. *Clin Orthop* 164:30–41, 1982. Used by permission.

varying from 20% [13, 20], 45% [1], 58% [11], to 70% [42]. Fortunately, most elbow problems occur on the medial side, are not severe, and, if treated reasonably, cause no permanent problems. Whether rule changes restricting the number of innings a Little League player may pitch or simply an increased awareness of the fragility of the immature elbow by coaches and parents is responsible for the lack of the more serious intra-articular injuries is not apparent.

This, however, is not the case in Japan [39]. In a study of over 2500 baseball players, almost half of those between 9 and 12 years of age had elbow symptoms. Radiographic abnormalities were seen in one-fifth, and of those 89% had medial changes, 15% lateral abnormalities, and 5% posterior abnormalities. In the entire group, 2% had osteochondritis dissecans and 18% had medial radiographic changes. The radiographic changes correlated directly with the amount of throwing: 38% in pitchers, 32% in catchers, 13% in infielders, and 8% in outfielders.

Little League rules now limit the number of innings a child may pitch each week and the length of time between appearances. What the child does on his own cannot however, be controlled. Most existing non-Little League organized community programs have a minimum age for pitching, limit the amount of pitching allowed, and disallow breaking pitches, all of which may contribute to the reduced injury rate.

Osteochondritis dissecans involving the elbow is now quite uncommon in the Little League age group (9 to 12 years) in the United States but is more common in older adolescents and young adults. The average age of individuals coming to surgery in this country because of capitellar osteochondritis is 16 years of age or older [45], and therefore the low incidence in the Little League age groups, while reassuring, may impart a false sense of security. It is possible that the repetitive throwing activities at the younger ages sets the stage for the development of osteochondritis dissecans, but the entity does not fully evolve or become manifest until later. The true incidence of this problem in baseball in the United States remains unknown.

Biomechanics of the Throwing Motion

A complex series of biomechanical events must occur in a precisely orchestrated and timed sequence

Stages of Pitching

FIGURE 17H–1
The four phases of throwing: cocking, wind-up, acceleration, and follow-through.

in order to propel an object forcefully, be it for speed, distance, or accuracy. The prototypic overhand throwing motion is seen in baseball, but the same general principles apply to similar motions occurring in tennis [12], volleyball, and javelin throwing.

The throwing motion derives much of its energy from the powerful muscles of the legs and trunk, and then, in sequence, involves the muscles of the shoulder, arm, forearm, and finally the wrist and hand.

The throwing motion can be divided into distinct phases (see Fig. 17H–1). The initial wind-up and cocking phases place the arm in a preparatory position for throwing. These motions bring the shoulder into a position of external rotation, abduction, and extension, and bring the elbow into flexion. They are essentially static phases, and place no major stresses on the elbow. By the end of the cocking phase, the body is turned 90 degrees to the direction of the throw, the weight transfer from the back to the front foot has begun, and the ball is poised behind the body ready to begin its forward propulsion. The shoulder adductors, internal rotators, and flexors are all preloaded, ready to begin their strong contractions.

The acceleration phase is the most forceful part of throwing, and the phase during which most elbow injuries occur. Early in acceleration, the shoulder propels the arm forward, and because of the inertia of the aftercoming arm, the elbow is thrust forward ahead of the wrist, forcing it first into valgus and then into extension.

The injuries that occur from throwing may be divided into medial tension injuries, lateral compression injuries, and posterior injuries [9,43,44]. During the acceleration phase, the medial aspect of the elbow is placed under tension and the lateral side is subjected to high compression forces.

On the medial side, the injuries occur primarily during the acceleration phase, as the arm is forcefully brought forward. The strong contraction of the flexor-pronator muscle attachments as the arm is started on its forward journey can cause injury to the medial ligaments, the flexor tendon attachment, and the medial epicondyle. The valgus stress that results causes tension on the medial ligament, which also attaches to the medial epicondyle, and thus strong static forces are also transmitted to the medial epicondyle. Traumatic avulsion of the medial epicondyle through the epiphyseal plate can occur with a single throw (Fig. 17H–2), or damage can be more gradually caused by repetitive throwing maneuvers, as in the classic Little League elbow (Fig. 17H–3).

As the arm is forced into valgus at the elbow,

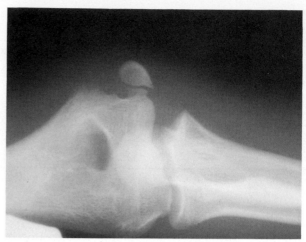

FIGURE 17H–2
Radiograph of a young athlete's elbow, showing avulsion of the medial epicondyle.

high compression and shear stresses are created between the radial head and the capitellum. The capitellar side is more susceptible to injury, and osteochondrosis or osteochondritis dissecans of the capitellum (Fig. 17H–4) and, rarely, the radial head occur.

Just before and during the release of the ball, the wrist flexors contract forcefully, adding to the stresses on the medial ligament, the forearm flexor muscles, and the medial epicondyle.

Also during the acceleration phase, the triceps contracts strongly to forcibly extend the elbow,

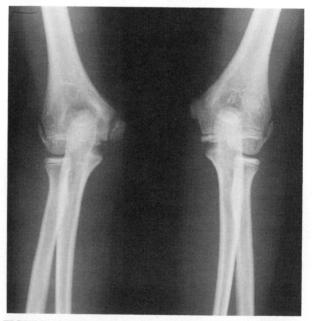

FIGURE 17H–3
Radiograph showing enlargement and fragmentation of the medial epicondyle of the throwing arm *(right)*.

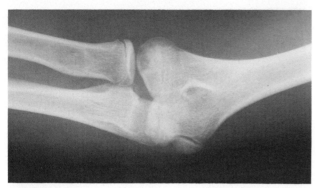

FIGURE 17H–4
Osteochondrosis of the capitellum in the dominant arm of a young athlete.

exposing the muscle, its tendon, and its olecranon attachment site to injury. Very large forces occur during this phase; angular velocities averaging over 4000 degrees per second have been reported in professional pitchers [29].

As the motion continues into the late acceleration phase, the ball is released, the arm continues into the follow-through phase, and the arm is rapidly decelerated. The elbow flexors contract to brake the rapidly extending elbow, and the ligamentous and osseous components must absorb the forces that are not dampened by the eccentric muscle contractions. As the elbow reaches full extension, the olecranon strikes against supracondylar fossa of the humerus, causing additional trauma to the olecranon at impact.

The forces applied by the routine throwing motion to the various aspects of the elbow described above are shown schematically in Figure 17H–5. These forces may be further enhanced. If the pitcher is throwing a fast ball or a screw ball, the elbow is snapped into pronation and radial deviation; if he throws a curve ball, forceful supination and ulnar deviation occur, adding additional stresses to the elbow. Often, when attempting to learn how to throw a breaking pitch, youngsters snap the wrist forcefully, increasing the muscle forces on the medial side of the elbow and the compression forces laterally.

Clinical Evaluation

It is important to obtain a careful history of the nature, duration, and location of the symptoms and, if possible, to relate the pain to the phase of the throwing motion. The history is usually that of a gradual onset of pain, and its localization is helpful. The pain of capitellar osteochondritis is deep or

lateral, whereas most medial tension problems cause medial elbow pain, either in the soft tissues of the proximal forearm or directly over the medial epicondyle.

The abrupt onset of pain medially with a single throw suggests an avulsion of the medial epicondyle, an epiphyseal fracture, or a significant soft tissue tear. Sudden pain laterally, associated with catching or true locking, occurs in osteochondritis dissecans when a piece has broken free.

Correlation of pain with the activity or the phase of throwing may suggest a specific diagnosis. A history of swelling may be obtained; loss of joint motion can be related to either an effusion or an interposed piece.

The physical examination should identify the site or sites of specific tenderness. Joint tenderness medially may be either from ligament or tendon injury. The range of motion of the elbow should be compared with the normal side. The differentiation between effusion or extra-articular swelling helps to separate intra-articular from extra-articular pathology. Pain may be elicited by active wrist flexion or passive wrist extension on the medial side in soft tissue injuries of the medial flexors. A careful search for tenderness posteriorly should be able to distinguish between injury to the triceps insertion and the olecranon. One must always do a complete neurologic and vascular examination, and muscle strength of the various muscle groups should be assessed. Laxity of the medial and lateral ligaments of the elbow is examined by varus and valgus stress tests in extension and comparing them to the opposite side. Pain elicited by valgus stress localized to the medial side is suggestive of ligament irritation and pain localized to the lateral side is suggestive of

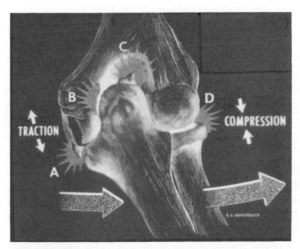

FIGURE 17H–5
Forces occurring on the elbow with the throwing motion.

radial head or capitellar involvement. Similarly, rotational maneuvers to examine pronation and supination, and to seek pain with either active or passive rotation, is helpful in localizing the pathology.

Radiographic examination should initially include anteroposterior, lateral, and oblique views. When the epiphyses are open, comparison views of the normal side are often necessary. The radiologic findings in typical Little League elbow with pain and tenderness medially are enlargement, fragmentation, or separation of the medial epicondylar epiphysis (see Fig. 17H–3). Avulsion of the medial epicondyle (see Fig. 17H–2) can usually be demonstrated on routine views, but if radiographs are normal and suspicion of an epiphyseal fracture exists, it may be necessary to stress the elbow gently into valgus and obtain additional stress views. Unless specifically searching for osteochondritis of the capitellum (see below), further studies are usually not necessary, and treatment may be instituted.

Authors' Preferred Method of Treatment

Treatment of medial elbow pain in adolescent throwing athletes, once intra-articular pathology has been excluded, is primarily symptomatic. Mild muscle strains are treated with rest until pain is absent, and then the patient is returned to activity. In the more serious muscle injuries, where swelling and limitation of motion are noted, the elbow should be splinted for 7 to 10 days, or until swelling and pain have largely subsided, and then the elbow is rehabilitated prior to return to sports.

Similarly, in the young athlete with medial ligament or medial epicondyle tenderness, symptoms dictate the treatment. The amount and duration of the pain determines whether activities should be decreased or ceased. Radiographic appearance of the medial epicondyle is not helpful in the absence of a fracture. Neither the type nor the duration of symptoms seems to correlate with the medial epicondylar radiographic findings of enlargement or fragmentation [20].

Avulsion of the medial epicondyle is treated like any other epiphyseal fracture. If there is minimal displacement, cast immobilization for 4 weeks and limitation of activities for another 4 weeks is advised. Return to sports must not be allowed until there is full motion, no symptoms, and normal strength. Usually, this requires sitting out the entire sports season.

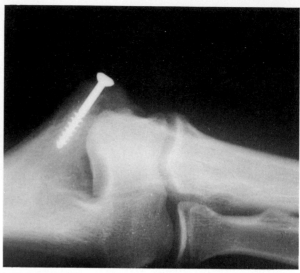

FIGURE 17H–6
Avulsion of the medial epicondyle in an immature young athlete, which has been reduced and internally stabilized.

If the medial epicondyle is avulsed more than 5 mm, replacement should be accomplished surgically using internal fixation with pins or a small screw (Fig. 17H–6). Displacements greater than 5 mm heal but have a higher incidence of future problems, such as delayed ulnar neuropathy or persistent medial elbow discomfort with throwing. Because these are youngsters who are going to continue to be very active in sports, we would advise internal fixation of displaced fractures.

OSTEOCHONDROSIS AND OSTEOCHONDRITIS DISSECANS OF THE CAPITELLUM

Capitellar injuries in young athletes are a cause of considerable concern because they involve the articular surface and can lead to traumatic arthritis and permanent joint impairment. There are varying opinions as to whether osteochondrosis and osteochondritis dissecans are two separate entities [3, 21] or a single entity composed possibly of different manifestations [28].

Osteochondrosis is defined as a "disease of the growth or ossification centers in children that begins as a degeneration or necrosis followed by regeneration or recalcification," and osteochondritis is defined as "inflammation of both bone and cartilage . . . resulting in the splitting of pieces of cartilage into the joint" [3]. However, as Pappas [28] correctly points out, histopathologic studies of osteochondritis dissecans have never documented signs of inflammation, and therefore osteochondrosis dissecans more accurately describes what is occurring. Although the terms osteochondrosis, osteochondritis, and osteochondritis dissecans are often used interchangeably, it is likely that the different manifestations that are seen clinically represent different stages of the alteration of endochondral ossification [25, 26], and whether the ultimate fate is healing, nonhealing, or loose body formation depends on the age of the patient, the type and level of activities, and the severity of the original lesion [36]. Thus, the so-called Panner's disease [27], or osteochondrosis of the capitellum, is seen in younger children, and osteochondritis dissecans (OCD), with a true cleavage line or loose body, occurs in the older child or adolescent [21].

History

Osteochondritis dissecans was probably first described in the orthopaedic literature by Paré in 1558. In 1870, Paget named the process "quiet necrosis," and it was in 1888 that Konig, who is often credited with the original description, first used the term osteochondritis dissecans. Subsequently, an extensive body of literature has accumulated regarding this entity, describing its elusive etiology and numerous theories about its pathology, natural history, incidence, and treatment.

Numerous causes have been suggested such as cartilage rests, bacterial embolism, vascular insufficiency, primary fracture with separation, constitutional factors, heredity, and a reactive dissecting process. As one reviews the debate with a historical perspective, however, the trail invariably returns to the entity as being some form of disordered endochondral ossification associated either with vascular impairment or with trauma.

It should be noted that although these elements seem the most logical, there has been no experimental evidence either proving or disproving on a scientific basis the cause or true pathogenesis of this entity because the published information regarding the osteochondroses has been essentially descriptive. Chiroff and Cooke [7] studied the histopathology of a group of patients with osteochondritis dissecans involving the knee and found focal avascular necrosis of the subchondral epiphyseal bone at the cleavage line and normal histology of the overlying articular cartilage. They found evidence of revascularization, which they interpreted as early repair. Interestingly, the subchondral bone and the loose fragment were alive, and there was no evidence of inflammation [2, 26].

The first involvement of the elbow is attributed to Panner [27], who described osteochondrosis of the capitellum and called the entity "Perthes disease of the elbow." By far the most common site of osteochondroses involving the elbow is the capitellum, and yet as late as 1950 Roberts and Hughes [33] were able to find only 44 cases in the literature. Haraldsson [15] surveyed 22 Swedish hospitals in 1959 and reported 27 cases; the average age of onset was 7.5 years, and all but three cases were in boys. He described the evolutionary cycle of destruction and regeneration as determined by radiograph. There were no loose bodies, and whereas trauma was postulated as an etiologic factor, there was no clear association with athletic activity.

Naguro [24] reported a series of 59 patients with OCD, of whom 50 had disease located in the elbow. Of these, 30 cases involved the capitellum, 12 involved other specific sites in the elbow, and in 8 the site of the lesion was not identified. Of the 30 with OCD of the capitellum, 16 were baseball players, 11 laborers, and 4 participated in jujitsu. All of his patients were males, with an age range of 11 to 61 years, and only three were under age 15 years. In 1964, Smith [38] reported another 28 cases involving the capitellum. Lindholm and associates [21] were the first to suggest that osteochondrosis and OCD were different entities, but this distinction received very little attention in the literature.

Although the incidence of capitellar involvement appears to be decreasing in baseball players, it is now being seen in increasing numbers in female gymnasts. Priest and Weise [31] reported two instances of osteochondrosis of the capitellum in a series of 32 elbow injuries in gymnastics. Singer and Roy [36] described seven cases of osteochondrosis and osteochondritis dissecans in young female gymnasts and proposed that it was caused by valgus compression overload as a result of using the arms as weight-bearing extremities and transferring high compressive and shear stresses to the capitellum (Fig. 17H–7). They cited as support of this theory Haraldsson's work [15] on the vascular anatomy of the epiphysis of the capitellum, which suggests that the epiphysis undergoes avascular necrosis as a result of the repetitive valgus compressive stresses placed on the epiphysis during a vulnerable time in its development when, because of its vascular supply, it is particularly susceptible to injury.

Gymnasts, in many of their routines, use their arms as weight-bearing extremities. Because of the increased carrying angle, the elbows of female gymnasts are in valgus in full extension and therefore concentrate the large forces generated by such maneuvers as round-offs, handstands, or vaulting to

FIGURE 17H–7
A young gymnast in action. Note the valgus position of the elbows.

the lateral side of the elbow. It is likely that these compressive forces are probably much larger than those that occur with throwing and are transmitted to the relatively small cross-sectional area of the radial-capitellar joint, causing large pressures there. Female gymnasts begin young and, as they excel, spend many hours daily in their sport. The true incidence of lateral compression injuries in gymnasts is unknown; when the number of participants is considered, the injury rate may well be higher than that seen in the throwing sports. Subsequently, Jackson and associates [17] presented 10 similar cases in young female gymnasts and indicated that this entity did not have a particularly good prognosis.

Certain common features have become apparent when the cases from all sources are combined. The vast majority of patients were males in baseball and females in gymnastics, and less than 5% of cases were bilateral; when unilateral and a throwing activity was involved, the dominant arm was virtually always involved. In children below age 8 years, the lesions are usually similar to those described by Panner, but in older adolescents loose bodies were more frequently found. It seems apparent that the entity is related to the specific activity, not the sex of the athlete or any anatomic predisposition.

Etiology

The pathologic entity of osteochondrosis contains an area of avascular necrosis. This conclusion is based on the microscopic appearance, which is that of infarction in the subchondral bone, followed by resorption of the necrotic bone and changes suggestive of repair [3, 26, 30]. The vascular contribution to the etiology of the entity has been strongly suggested by the vascular studies of Haraldsson [15], described below. The implication of trauma in the etiology is apparent by the fact that whenever it occurs in sports involving a dominant arm, that arm is the involved side, and by the fact that the entity occurs rarely, if at all, in inactive children or adolescents.

The probable etiology, therefore, involves a contribution from all three of those factors. Repetitive trauma to an epiphysis made vulnerable by dependence on a fragile blood supply may interrupt that blood supply and result in avascular necrosis [19]. This results in the subchondral changes described. If shear and compression stresses continue, a cleavage plane develops through the weakened subchondral support, eventually involving previously intact articular cartilage with the formation and eventual separation of the avascular osteochondral fragment.

Pertinent Anatomy

Haraldsson [15] used injection studies to demonstrate the vascular anatomy of the capitellum. Early in life, the large chondroepiphysis has a rich vascular supply that then diminishes with growth. At the age of 3 years, when the ossific nucleus of the capitellum has appeared, there are only one or two large vessels entering posteriorly, and no vessels crossing the epiphyseal plate. By the age of 8 years, when the ossification center is much larger, the blood supply still comes only from the posterior vessels, with no contribution from the metaphysis. The vascularity of the cartilaginous portion of the epiphysis has diminished significantly, particularly in the areas directly adjoining the ossification center of the capitellum. Once complete ossification and plate closure have occurred, the primary blood supply still comes from the discrete posterior vessels, but abundant anastomoses have occurred between the blood supplies of the epiphysis, metaphysis, and diaphysis. Therefore, there seems to be a period of several years when the blood supply through the posterior vessels is the only source of blood to the capitellum, and injury to that blood supply could render at least a portion the epiphysis avascular.

Clinical Evaluation

The athlete presenting with a history of lateral elbow pain must be taken seriously. The location of the pain and its relation to the sport must be evaluated. Careful history and physical examination should be able to determine whether it is related to the soft tissues or the articular components.

A history of catching or locking presumes an intra-articular cause. A history of swelling or findings of effusion are of particular importance. Limitation of motion may occur as a result of effusion, pain, or an interposed loose fragment.

Radiographic examination should begin with routine views. Osteochondrosis appears on the radiographs as a lucent area in the capitellum (see Fig. 17H–4). If present, a careful search must follow for a loose body, particularly if symptoms are suggestive or if an effusion is present. The loose piece may be entirely chondral and may not show up on routine radiographs. Lateral tomograms may be helpful, and we prefer them to computed tomography scans at present (Fig. 17H–8).

With normal radiographs and persistent findings, radionucleotide scans may be helpful, particularly with respect to suspected OCD. Magnetic resonance imaging should be particularly helpful in patients with OCD, but we have not had sufficient experience to determine its exact role. Arthrography of the elbow has not been helpful in our experience.

Treatment Options

Treatment options depend on the diagnosis. If the patient presents with pain, little or no effusion, and radiographic evidence of lucency in the capitellum, but with no clinical or radiographic evidence of loose body, he or she is considered to have osteochondrosis. This presentation is most common in younger individuals, likely does not progress to osteochondritis dissecans, and heals without surgery [19, 28, 45].

Nonsurgical treatment consists of resting the elbow by immobilizing it in either a splint or a cast until the irritability has resolved, and then instituting active range-of-motion exercises without applying forceful stresses across the elbow. Throwing, gymnastics, or strenuous physical activities and contact sports are not allowed for a minimum of 6 weeks [19]. Gradual return to activities is allowed once full range of motion, strength, power, and endurance have returned, and progression is continued as long as the patient remains asymptomatic. Sequential radiographs are used to follow the evolution of the lesion; in many instances, the radiographic abnormalities continue (Fig. 17H–9).

Although surgery has been recommended in osteochondrosis without signs or symptoms of loose bodies, the results of drilling the lesion have not been shown, and although bone grafting the capitellar lesion from a supracondylar window has been suggested, the results have not been encouraging [40].

In individuals with osteochondritis dissecans, various treatment regimens have been advocated, depending on the stage of the lesion. Tivnon and associates [40] reported removal of loose bodies and curettage of the crater to bleeding bone in six patients, with two good and four fair results. All patients reported subjective improvement, but normal motion was not likely. In those individuals who had fragmented lesions of the capitellum but no loose bodies, debridement of the capitellum and curettage resulted in improvement in all patients, but all remained symptomatic. Shaving of the articular cartilage did not help.

If a loose body is present but is relatively asymptomatic, some [32] advocate leaving it alone, and others advocate removing it [15, 19, 23, 35]. Woodward and Bianco [45] stated that no procedure other than removal of the loose body was warranted, even

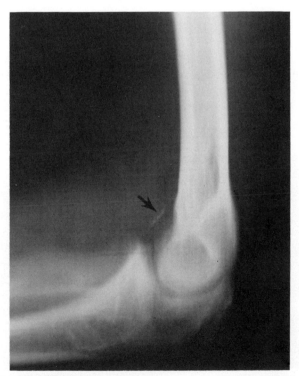

FIGURE 17H–8
Lateral tomograms of the elbow showing a loose body that was not apparent on routine radiographs.

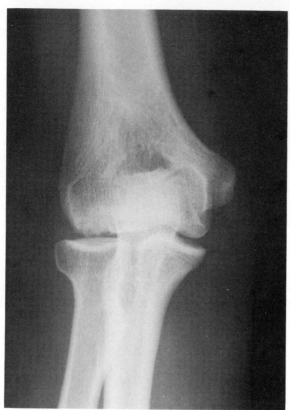

FIGURE 17H–9
Residual radiographic appearance of a young athlete 3 years after an episode of osteochondrosis. She is currently asymptomatic, and the radiographic appearance has not changed.

curetting the crater. McManama and associates [23] believe that the lesion should be curetted to bleeding bone, and that pain relief and improvement of motion is to be expected. All of their patients returned to competitive activities.

Applications of arthroscopic surgical techniques to the elbow have, as in other joints, increased the apparent indications for surgery. Arthroscopic evaluation, removal of loose bodies, drilling or abrasion of osteochondrotic craters, and resection of inflamed and presumably symptomatic synovium have all been reported [2, 14], and convalescent times and morbidity have been decreased, as occurs with arthroscopic surgery elsewhere. Arthroscopic surgery offers advantages in the treatment of loose bodies, but its success in the treatment of lesions of the articular cartilage has yet to be demonstrated. The technique of elbow arthroscopy is discussed elsewhere.

Replacement of osteochondritic fragments is usually not advocated [15, 19, 23], but since the advent of arthroscopic surgery, techniques for in situ screw fixation have been developed [21]. Long-term success of replacing the loose fragment has not been demonstrated in the elbow, as it has in the knee.

OSTEOCHONDROSIS OF THE OLECRANON

Osteochondrosis of the apophysis of the olecranon, although quite rare, can occur in the child, much as Osgood-Schlatter disease occurs in the apophysis of the tibial tubercle [35]. There is a case report of a competitive diver [25], a series of four including one field hockey player and one gymnast [8], and we have had two patients, both competitive gymnasts.

The clinical features are similar to those of Osgood-Schlatter disease because during childhood the growth center of the olecranon, like the tibial tubercle, is nonarticular. The child is usually involved in an organized sports activity and presents with activity-related pain, swelling, and tenderness involving the olecranon process. Although we have seen two cases in gymnasts, we have seen none in throwing athletes.

Radiographic examination shows the typical fragmentation and irregularity of the olecranon apophysis (Fig. 17H–10), and comparison views of the opposite elbow are often helpful.

Treatment is entirely symptomatic; usually, activity restriction or brief immobilization allows the symptoms to subside. It may take several months before the symptoms allow return to the sport.

Olecranon Epiphyseal Fractures

In the adolescent, posterior impingement can elicit disturbances in the olecranon epiphysis. Although not truly an osteochondrosis, stress fractures or nonunion of stress injuries of the olecranon

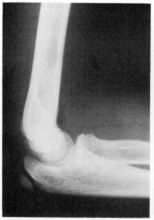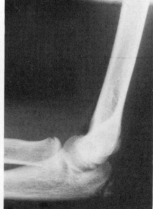

FIGURE 17H–10
Osteochondrosis of the olecranon in a 9-year-old girl, with comparison view of the opposite side.

epiphysis can and do occur in athletics [19, 41]. Pavlov and associates [30] described two instances of nonunion of the olecranon epiphysis in adolescent baseball pitchers; Retrum and colleagues [32] described two instances in tennis players. The olecranon epiphyseal plate usually closes at approximately age 16 years, but repetitive stress may cause it to fuse incompletely, rendering it susceptible to later traumatic disruption [18].

The epiphysis of the olecranon is usually thought to be nonarticular. Confusion may result, however, because as the olecranon matures, the epiphyseal line often migrates distally so that it moves into the joint, the so-called wandering physeal line of the olecranon [4], and thus may simulate a fracture [34].

The patient presents with elbow pain, and radiographic examination shows persistence of the epiphyseal line, with perhaps some widening. It is imperative to obtain comparison views of the opposite side, and nuclear bone scans may be helpful if the diagnosis is not apparent on routine radiographs.

The treatment should initially be a trial of immobilization in an attempt to secure union. If the elbow remains symptomatic, or if nonunion is well established, surgical correction is required. Internal fixation with tension band wires and axial compression screw fixation [35] have been advocated, and placing an inlay bone graft across the nonunion site has been recommended [41].

Authors' Preferred Method of Treatment

Our recommended treatment for capitellar lesions depends on the symptoms and radiographic appearance. If the individual presents with pain, tenderness in the lateral compartment, and radiographs showing a lucent area in the capitellum, but there is no evidence clinically or radiographically of a loose body, our approach is nonsurgical. The elbow should be placed at rest until symptoms subside, and then active motion and strengthening are instituted. Return to sports activities is not allowed for at least 6 weeks, and then only if all symptoms have subsided. Once it has been ascertained that there is no tenderness and that normal range of motion and normal strength are present, cautious return to sports is allowed except for pitchers or predominantly throwing athletes; those we do not allow to return that season. Serial radiographs are taken, but often in our experience the radiographic appearance does not change. Should more fragmentation or

lucency evolve radiographically, then continued observation and sports limitation are advised.

If there is an effusion or symptoms of catching or locking, an extensive search is made for a loose body. If one is present, we recommend removing it, and our preferred method is to do so by means of arthroscopic surgery. Lesions of the capitellum are debrided and gently curetted or drilled, leaving the structural integrity of the subchondral plate as minimally disturbed as possible. It is not always possible to remove the loose body or treat the capitellar lesion arthroscopically, and we do not hesitate to open the elbow to accomplish the desired procedure.

Following arthroscopic removal of loose bodies, with or without treatment of the capitellar lesions, postoperative management consists of a splint for 7 to 10 days and then institution of motion out of a sling. We protect the elbow in a sling for 3 weeks and then start strengthening exercises. If an arthrotomy has been performed, postoperative management is essentially the same, with 7 to 10 days of splint immobilization. Then we allow the splint to be removed three or four times daily to begin active and active assisted flexion, extension, pronation, and supination of the elbow. At 3 weeks, the splint is removed, and exercises are performed out of a sling, which is worn for additional 3 weeks. It is not unusual for it to take 10 to 12 weeks to regain the last 10 degrees of extension. Active strengthening exercises are begun as soon as comfort allows and are progressed until full strength is achieved.

We strongly recommend that athletes with OCD in their dominant arm never return to pitching; if they choose to continue playing competitive baseball, we recommend that they change their position. If they are gymnasts, they may return approximately 4 months after the surgery if symptoms are gone and full motion and normal strength have been achieved. We are concerned about the long-term effects and do not yet know whether these young athletes should or should not be allowed to return to gymnastics.

Individuals with osteochondrosis of the olecranon are treated symptomatically, and the outcome is uniformly good. They miss perhaps 6 weeks of sports, often less, depending on symptoms. In those who still have symptoms after 2 or 3 weeks at rest, we recommend a splint for an additional 3 weeks and then a rehabilitation program.

We recommend an aggressive approach to the treatment of stress fractures or nonunion of the olecranon epiphysis. Our preferred method of treatment of nonunion of stress fractures of the olecranon is stabilization by means of an axial transolecranon

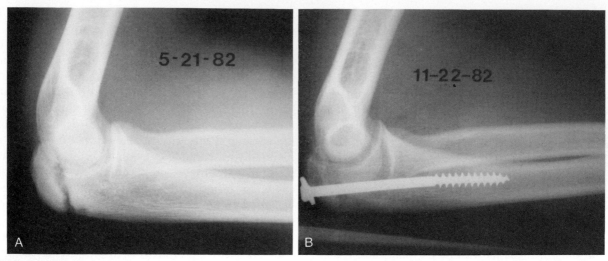

FIGURE 17H–11
Nonunion of a stress fracture of the olecranon apophysis. *A,* Appearance prior to treatment. *B,* After in situ axial screw fixation.

compression screw without supplemental bone graft (Fig. 17H–11). At the time of insertion of the axial screw, several small holes are drilled across the nonunion site adjacent to where the screw is placed to allow additional blood access to the fracture site. The elbow is immobilized for 6 weeks, and a sling is worn until union is judged to be complete clinically and radiographically. If the nonunion line is wide or if the olecranon has displaced, an inlay slot graft is placed across the nonunion site in addition to axial fixation. Postoperative management is the same. The athlete is able to return to sports in approximately 3 to 4 months.

References

1. Adams, J. E. Injury to the throwing arm: A study of traumatic changes in the elbow joints of boy baseball players. *Calif Med* 102:127–132, 1965.
2. Andrews, J. R., and Carson, W. G. Arthroscopy of the elbow. *Arthroscopy* 1:92–107, 1985.
3. Bianco, A. J. Osteochondritis dissecans. *In* Morrey, B. F. (Ed.), *The Elbow and its Disorders.* Philadelphia, WB Saunders, 1985, pp. 254–259.
4. Bodeur, A. E., Silberstein, M. J., and Gaviss, E. R. *Radiology of the Pediatric Elbow.* Boston, GK Hall, 1981.
5. Brodgon, M. D., and Crow, N. E. Little leaguers elbow. *AJR* 83:671–675, 1960.
6. Brown, R., Blazina, M. E., Kerlan, R. K., Carter, V. S., Jobe, F. W., and Carlson, G. J. Osteochondritis of the capitellum. *J Sports Med* 2:27–46, 1974.
7. Chiroff, R. T., and Cooke, C. P. Osteochondritis dissecans: A histological and microradiographic analysis of surgically excised lesions. *J Trauma* 15:689–696, 1975.
8. Danielsson, L. G., Hedlund, S. T., and Henricson, A. S. Apophysitis of the olecranon. *Acta Orthop Scand* 54:777–778, 1983.
9. DeHaven, K. E., and Evarts, C. M. Throwing injuries of the elbow in athletes. *Orthop Clin North Am* 4:301, 1973.
10. Frances, R., Bunch, T., and Chandler, B. Little league elbow: A decade later. *Physician Sports Med* 6:88–94, 1978.
11. Grana, W. A., and Rashkin, A. Pitcher's elbow in adolescents. *Am J Sports Med* 8:333–336, 1980.
12. Gregg, J. R., and Torg, J. E. Upper extremity injuries in adolescent tennis players. *Clin Sports Med* 7:359–370, 1988.
13. Guggenheim, J. J. Jr., Stanley, R. F., Woods, G. W., and Tullas, H. S. Little league survey: The Houston study. *Am J Sports Med* 4:189–219, 1976.
14. Guhl, J. F. Arthroscopy and arthroscopic surgery of the elbow. *Orthopedics* 8:1290–1296, 1985.
15. Haraldsson, S. On osteochondritis deformans juveniles capituli humeri including investigation of intra-osseous vasculature in distal humerus. *Acta Orthop Scand (Suppl)* 38:1–232, 1959.
16. Hunter, L. Y., and O'Connor, G. A. Traction apophysitis of the olecranon. *Am J Sports Med* 8:51–52, 1980.
17. Jackson, D. W., Silvino, N., and Reiman, P. Osteochondritis in the female gymnast's elbow. *Arthroscopy* 5:129–136, 1989.
18. Kovach, J. II, Baker, B. E., and Mosher, J. F. Fracture separation of the olecranon ossification center in adults. *Am J Sports Med* 13:105–111, 1985.
19. Kvidera, A., Madera, D., and Pedegano, A. L. Stress fracture of the olecranon: A report of two cases and review of the literature. *Orthop Rev* 12:113–116, 1983.
20. Larson, R. L., Singer, K. M., Bergstrom, R., and Thomas, S. Little league survey: The Eugene study. *Am J Sports Med* 4:201–209, 1976.
21. Lindholm, T. S., Osterman, K., and Vankka, E. Osteochondritis dissecans of the elbow, ankle and hip: A comparison survey. *Clin Orthop* 148:245–253, 1980.
22. Lipscomb, A. B. Baseball pitching injuries in growing athletes. *J Sports Med* 3:25–34, 1975.
23. McManama, G. B., Micheli, L. J., Berry, M. V., and Sohn, R. S. The surgical treatment of osteochondritis dissecans of the capitellum. *Am J Sports Med* 13:11–21, 1985.
24. Naguro, S. The so-called osteochondritis dissecans of Konig. *Clin Orthop* 18:100–122, 1960.
25. Ogden, J. A. *Skeletal Injury in the Child.* Philadelphia, Lea & Febiger, 1982.
26. Omer, G. E. Primary articular osteochondroses. *Clin Orthop* 158:33–40, 1981.
27. Panner, H. J. A peculiar affection of the capitellum humeri resembling Calvé-Perthes disease of the hip. *Acta Radiol* 10:234, 1929.
28. Pappas, A. M. Osteochondritis dissecans. *Clin Orthop* 158:59–69, 1981.

29. Pappas, A. M., Zawacki, R. M., and Sullivan, T. J. Biomechanics of baseball pitching. A preliminary report. *Am J Sports Med* 13:216–222, 1985.

30. Pavlov, H., Torg, J. S., Jacobs, B., and Vigorita, V. Nonunion of olecranon epiphysis: Two cases in adolescent baseball pitchers. *AJR* 136:819–820, 1981.

31. Priest, J. D., and Weise, D. J. Elbow injury in women's gymnastics. *Am J Sports Med* 9:288–295, 1981.

32. Retrum, R. K., Wepfer, J. F., Olen, D. W., and Laney, W. H. Case report 355. *Skeletal Radiol* 15:185–187, 1985.

33. Roberts, N., and Hughes, R. Osteochondritis dissecans of the elbow joint. *J Bone Joint Surg [Br]* 32:348–360, 1950.

34. Silberstein, M. J., Brodeur, A. E., Graveiss, E. R., and Luisiri, A. Some vagaries of the olecranon. *J Bone Joint Surg [Am]* 63:722–725, 1981.

35. Singer, K. M., and O'Neill, D. Update on elbow injuries in the young athlete. *In* Grana, W. B. (Ed.), *Advances in Sports Medicine*. Vol. 3. Chicago, Year Book Medical Publishers, 1990, pp. 147–167.

36. Singer, K. M., and Roy, S. P. Osteochondritis of the humeral capitellum. *Am J Sports Med* 12:351–360, 1984.

37. Slocum, D. B. Classification of elbow injuries from baseball pitching. *Texas Med* 64:48–53, 1968.

38. Smith, M. G. H. Osteochondritis of the humeral capitellum. *J Bone Joint Surg [Br]* 46:50–54, 1964.

39. Takenoabu, I., and Ikata, T. Baseball elbow of young players. *Tokushima J Exp Med* 32:57–64, 1985.

40. Tivnon, M. C., Anzel, S. H., and Waugh, T. R. Surgical management of osteochondritis dissecans of the capitellum. *Am J Sports Med* 4:121–128, 1976.

41. Torg, J. S., and Moyer, A. Nonunion of a stress fracture through the olecranon epiphyseal plate observed in an adolescent baseball pitcher. *J Bone Joint Surg [Am]* 59:264–265, 1977.

42. Torg, J. S., Pollack, H., and Swetersitsch, P. The effect of competitive pitching on the shoulders and elbows of preadolescent baseball players. *Pediatrics* 49:267–272, 1972.

43. Tullos, H. S., and King, J. W. Throwing mechanism in sports. *Orthop Clin North Am* 4:709–721, 1973.

44. Woods, C. W., Tullos, H. S., and King, J. W. The throwing arm: Elbow joint injuries. *J Sports Med (Suppl)*:43–47, 1973.

45. Woodward, A. H., and Bianco, A. J. Osteochondritis dissecans of the elbow. *Clin Orthop* 110:35–41, 1975.

Osteochondritis Dissecans

Bernard F. Morrey, M.D.

Osteochondritis dissecans of the elbow is an uncommon disorder in the general population. Among young adolescent athletes, however, particularly baseball pitchers or those engaged in any other athletic endeavor that involves throwing, the condition is neither unusual or rare.

It is important to differentiate between osteochondrosis of the capitellum and osteochondritis dissecans of the capitellum. The differences are related to age and degree of involvement of the capitellar secondary ossification center [2]. Osteochondrosis is defined as "a disease of the growth or ossification centers in children that begins as a degeneration or necrosis followed by regeneration or recalcification" [2]. This condition is also known as Panner's disease [13], but its treatment is entirely different from that of osteochondritis dissecans of the capitellum.

Osteochondritis dissecans of the capitellum, in its simplest terminology, indicates fragmentation and possible separation of a portion of the articular surface [15]. It has also been described as an inflammation of bone and cartilage, resulting in the splitting of pieces of cartilage into the joint [2].

It is important to realize that the capitellum usually appears in males at age 2 years and fuses at age 14.5 years; hence, skeletal maturation is nearly complete when the condition occurs in adolescence. Accordingly, although the physis is still actively functioning in this age group, this condition should not be confused with osteochondrosis of the capitellum, otherwise known as Panner's disease.

Of interest is the fact that osteochondritis dissecans of the capitellum is the same condition that can occur in the ankle and hip regions, although in the series of Lindholm and associates [10] it comprised only 6% of all patients with osteochondritis dissecans reviewed in a 20-year period. In that series as well as others [3, 4, 9, 11, 16, 18, 21], the process involved localized areas of subchondral bone with the overlying adjacent articular cartilage. In these cases, one of three events can occur: (1) the subchondral bone with overlying cartilage can remain in situ, (2) it can eventually heal, or (3) it can partially separate. The elastic can completely separate from the capitellum and become a loose body within the elbow joint itself.

ETIOLOGY

The cause of osteochondritis dissecans is controversial, although the role of trauma is undeniable in most instances [1, 3, 9, 10, 11, 21]. There have been reports of a familial tendency [5, 19] and of multiple joint involvement in the same patient, which clouds the issue of whether trauma or ischemia is the cause. In spite of these reports, there is no convincing evidence that osteochondritis dissecans of the capitellum is an inheritable disease.

A traumatic etiology was popular in the past, but there is no conclusive evidence that a single traumatic episode has produced osteochondritis dissecans of the capitellum [2, 10, 14]. On the contrary, a history of repetitive use of the elbow on a constant basis in sport has commonly been associated with this condition. Pitching frequency and repetitive trauma have been known to be major factors in predicting the frequency of elbow problems [14]. When one studies the biomechanical forces of the throwing motion, considerable distractive force is noted concentrated on the medial aspect of the elbow in the cocking phase, with resultant compressive forces in the lateral side of the elbow and a translatory force across the humeral olecranon articulation [14]. As the arm is brought forward in the acceleration phases, these forces neutralize. As the follow-through is initiated, two major forces act on the posterior and anterior aspects of the elbow: triceps contraction through its insertion on the olecranon, and the force delivered by a rapid rotational pronation of the forearm, associated with a combined compressive and shearing force of the radial head on the capitellum. In baseball pitchers it is repetitive trauma, particularly forceful extension and pronation of the elbow, that creates the most severe compressive shearing forces transmitted by the radius to the adjacent articular surface of the capitellum. This may result in infarction of an area of subchondral bone in the overlying articular cartilage, which is not yet securely bonded to the rest of the epiphysis by normal bony maturation [2, 21].

There is a third theory of the etiology of osteochondritis dissecans of the capitellum—namely, ischemia. The ischemic theory is based primarily on the fact that the capitellum is the recipient of end arterials that terminate in the subchondral plate. In addition, the histopathologic characteristics of the involved area are typical of infarction of bone due to interruption of the subchondral terminal arterial vessels [2]. If the cartilage remains intact following the ischemic episode, absorption of the avascular segment occurs, being replaced by viable osseous tissue. Because this is not a weight-bearing articulation, the normal architecture of the articular surface is preserved. If, however, the articular cartilage is fractured during the initial stage of disease, the osteochondral fragment may become detached and may subsequently represent an intra-articular loose body.

CLINICAL CHARACTERISTICS

Epidemiology

There are few large series of osteochondritis dissecans affecting various joints. In the series of Smillie, 30% of 165 cases of osteochondritis dissecans affected the elbow [17]. In Green and Banks' [6] series, slightly less than 10% of 27 cases in patients under the age of 15 years affected the capitellum. All authors agree that males are more frequently affected than females. Bilateral disease is not rare, affecting 8 of 42 patients in Woodward and Bianco's series [21] and 6 of 38 patients described by Roberts and Hughes [16].

The most common subjective finding in athletes suffering from this condition is pain. The pain is usually aggravated by activity and relieved by rest. For many pitchers, the first symptom is pain in the elbow after a season of baseball pitching. The elbow pain can, however, occur with any other type of

activity. These complaints may be accompanied by intermittent swelling. These symptoms usually occur early in the course of the disease, when there are no intra-articular loose fragments.

As the disease progresses, the next most frequent complaint is loss of motion, usually an inability to fully extend the elbow. This can be accompanied by locking, clicking, and intermittent severe pain. These latter complaints are generally encountered when loosely attached or free bodies are discharged into the joint. Less common complaints, such as limitation of flexion of the elbow and pronation and supination of the forearm, also may occur.

The physical examination of an athlete with this condition demonstrates some decrease in the range of motion of the elbow, most often a loss of extension. Loss of full flexion of the elbow is a less frequent finding. Crepitus is often noted through the range of motion; to direct palpation, there is often pain located over the lateral aspect of the elbow. An uncommon but positive physical finding is the locking of the elbow through flexion and extension, the result of interposed osteochondral fragments.

Roentgenographic Findings

Before intra-articular loose bodies are found, all radiographic changes are confined to the capitellum or radial head. There is rarefaction in the capitellum, usually surrounded by a sclerotic rim of subchondral bone adjacent to the articular surface (Fig. 17I–1). This is most commonly seen on the anteroposterior view. The lateral view usually shows flattening and irregularity of the capitellar articular surface. If the disease is more advanced, there are multiple loose fragments within the joint (Fig. 17I–2). Tomograms, often using contrast medium, can be an accurate method of establishing a diagnosis that is difficult to see on plain radiographs.

Later radiographic findings include an enlarged radial head compared to the opposite elbow on comparison views [2]. Premature skeletal maturity has been evident in the affected elbow compared with the unaffected normal elbow [2]. The late sequelae of this condition may include degenerative changes of both the capitellar and radial head articular surfaces, characterized by irregularity and incongruity of these surfaces.

If sequestration does occur, multiple loose fragments can be seen within the elbow joint (see Fig. 17I–2). These are accompanied by irregularity of the articular surface of the capitellum. If sequestration does not occur, the central sclerotic fragment

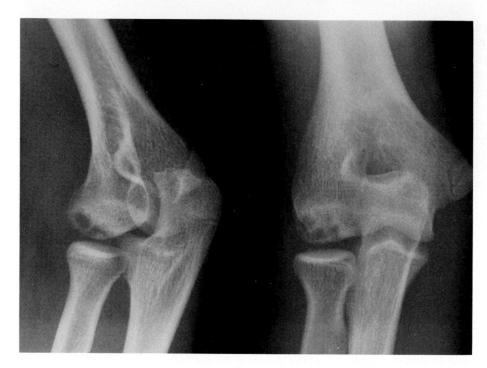

FIGURE 17I–1
Oblique and anteroposterior views of a 15-year-old boy with osteochondritis dissecans, demonstrating rarefaction and irregular ossification (anteroposterior view). The crater has a sclerotic rim (oblique view). The fragments have not separated (type I lesion).

gradually becomes less distinct and opaque with time, and the surrounding area of rarefaction gradually ossifies as the lesion heals. This may indeed take several years to occur, long after objective complaints have subsided.

TREATMENT

The treatment of osteochondritis dissecans depends on the stage of the disease; the stage of the disease is dictated by the clinical findings and the radiographic appearance. In addition, arthroscopy is occasionally necessary to determine whether or

not the osteochondral fragment has separated. Bianco has described three types of osteochondritis dissecans lesions and their treatment [2].

Type I Lesions

These are lesions with no evidence of subchondral displacement or fracture of the articular cartilage as noted on contrast tomography. All authors manage this entity nonoperatively [2, 14, 17, 21]. The nonoperative regimen involves rest of the elbow and avoidance of all activity requiring the use of the elbow. The elbow is splinted or casted if pain

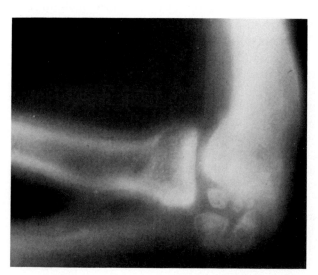

FIGURE 17I–2
Lateral tomogram of a 14-year-old boy with osteochondritis dissecans. Multiple loose bodies are visible, and the capitellar surface is irregular and flattened.

continues to be a significant complaint after rest. With splinting, active range-of-motion exercises are begun almost immediately to preserve motion. Serial radiographs are performed every 3 to 6 months to confirm the maintenance of the ostechondritic portion of the capitellum. Rest of the elbow is continued indefinitely until the affected area revascularizes. In addition, cessation of the offending sporting activity is crucial to control pain and elbow stiffness.

Type II Lesions

If, on radiographic or arthroscopic review, there is evidence of subchondral detachment or fracturing of the articular cartilage, the surgeon is left with two options: (1) to try to reattach the area of avascular bone surgically, or (2) to excise the fragment to prevent loose body formation. Pappas advocates removal of the loosened fragment [14]. Roberts and Hughes [16] do not recommend any surgical curettage, drilling, or trimming of the crater after removal of the loose body; however, Smillie [17] has advocated the latter debridement procedure. There has been no conclusive evidence that surgical procedures on the crater either help or hinder the overall prognosis of this stage II disease. In Woodward and Bianco's series, the poorest results were associated with bone grafting of the defect or replacement of the loose fragment [21].

If one is to pin the partially detached lesion in situ, several unthreaded Kirschner wires may be used under direct vision. The wires are inserted following exposure of the capitellum through a Kocher approach. The tips of the pins are embedded deep into the articular cartilage so that they do not protrude into the joint. They are brought out through the skin in the lateral epicondyle and may be removed under local anesthesia at 6 to 8 weeks.

One hopes for evidence of healing, demonstrated by resorption of the sclerotic bone at the base of the lesion with preservation of the articular surface. This, however, is not guaranteed, even with a successful reattachment. Occasionally, collapse and deformation of the capitellar articular surface may occur [2].

Type III Lesions

If loose bodies have formed and are causing symptoms, their removal is indicated. Smillie recommends removal to relieve the symptoms of lock-ing [17]. This does not improve range of motion. This has also been demonstrated by Woodward and Bianco [21]. The most effective form of removal is by arthroscopy [2]; however, loose bodies may be removed by open arthrotomy. There does not appear to be any evidence that either debridement or drilling of the crater leads to improved results. Unfortunately, late degenerative arthritis still may be a sequel of this condition, even with removal of intra-articular loose bodies.

If significant radiocapitellar degenerative changes are noted, excision of the radial head may improve the lateral elbow pain and restore some pronation and supination to the forearm. This is contraindicated in a skeletally immature patient.

AUTHOR'S PREFERRED METHOD OF TREATMENT

Conservative treatment is the mainstay of treatment for this condition. The only indication for operative intervention is a symptomatic loose body within the radiocapitellar joint. Loose bodies are arthroscopically excised; the crater is not drilled but may be debrided to smooth the crater edges. Debridement of a synovitis of the radiocapitellar articulation is carried out at that time, as indicated by its arthroscopic appearance. Postoperatively, a rehabilitation program stressing active range of motion is instituted to maintain the patient's arc of flexion-extension and forearm rotation. Return to an active sport that requires vigorous elbow function is discouraged. Certainly, all competitive throwing sports are terminated.

References

1. Adams, J. E. Injury to the throwing arm: A study of traumatic changes in elbow joints by boy baseball players. *Calif Med J* 102:127–132, 1965.
2. Bianco, A. J. Osteochondritis dissecans. *In* Morrey, B. F. (Ed.), *The Elbow and Its Disorders.* Philadelphia, W. B. Saunders, 1985, pp. 254–259.
3. Claxton, T. O., and DeLee, J. E. Osteochondritis dissecans. *Clin Orthop* 167:50, 1982.
4. Dixit, R. K., and Gilula, L. A. Roentgen rounds '82. *Orthop Rev* 15:92–95, 1986.
5. Gardiner, J. B. Osteochondritis dissecans in three members of one family. *J Bone Joint Surg [Br]* 37:139, 1955.
6. Green, W. T., and Banks, H. H. Osteochondritis dissecans in children. *J Bone Joint Surg [Am]* 35:26, 1953.
7. Haraldson, S. On osteochondrosis deformans juvenitis, capitali humeri including investigation of intraosseous vasculature in distal humerus. *Acta Orthop Scand (Suppl.)* 38:1, 1959.
8. Heller, C. J., and Wietse, L. Avascular necrosis of the

capitellum humeri (Panner's disease): A report of a case. *J Bone Joint Surg [Am]* 42:513–516, 1960.

9. King, D. Osteochondritis dissecans: A clinical study of twenty-four cases. *J Bone Joint Surg* 14:535, 1932.

10. Lindholm, I. S., Osterman, K., and Vankka, E. Osteochondritis dissecans of elbow, ankle and hip. *Clin Orthop* 148:245, 1980.

11. Mitsuwaga, M. M., Adishian, D. O., and Bianco, A. J. Jr. Osteochondritis dissecans of the capitellum. *J Trauma* 22:53, 1982.

12. Omer, G. E. J. Primary articular osteochondrosis. *Clin Orthop* 158:133, 1981.

13. Panner, H. J. A peculiar affection of the capitellum humeri resembling Calvé-Perthes disease of the hip. *Acta Radiol* 8:1617, 1927.

14. Pappas, A. M. Osteochondritis dissecans. *Clin Orthop* 158:59, 1981.

15. Resnick, D., and Niwayama, G. Physical injury. *In* Resnick, D. *Diagnosis of Bone and Joint Disorders.* Philadelphia, W. B. Saunders, 1981, p. 2893.

16. Roberts, W., and Hughes, R. Osteochondritis dissecans of the elbow joint: A clinical study. *J Bone Joint Surg [Br]* 32:348, 1950.

17. Smillie, I. S. *Osteochondritis Dissecans: Loose Bodies in Joints. Etiology, Pathology, Treatment.* Edinburgh, E. and S. Livingstone, 1960.

18. Smith, M. G. H. Osteochondritis of the humeral capitellum. *J Bone Joint Surg [Br]* 46:50–54, 1964.

19. Stougaard, J. Familial occurrence of osteochondritis dissecans. *J Bone Joint Surg [Br]* 46:542, 1964.

20. Wilson, F. D., Andrews, J. R., Blackburn, T. A., and McCluskey, G. Valgus extension overload in the pitching elbow. *Am J Sports Med* 11:83–87, 1983.

21. Woodward, A. H., and Bianco, A. J. Jr. Osteochondritis dissecans of the elbow. *Clin Orthop* 110:35, 1975.

HAND AND WRIST

The Wrist

Frank C. McCue, III, M.D.
James F. Bruce, Jr., M.D.

For many years the wrist joint has been an enigma. Even today our understanding of problems concerning the wrist remains incomplete. Questions persist in regard to basic function and the

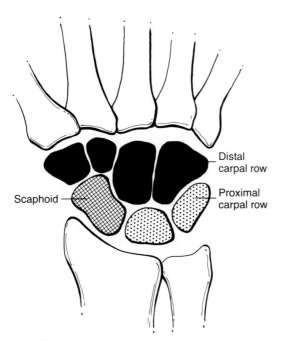

FIGURE 18A–1
The carpal bones are aligned in two transverse rows, with the scaphoid bridging the proximal and distal carpal rows. (Redrawn from Green, D.P. *Operative Hand Surgery* (2nd ed.). New York, Churchill Livingstone, 1988.)

effect of trauma and disease on that function. Methods of treatment in many cases have been derived from trial and error rather than from a basic understanding of this unique collection of joints. The long-term effect of some treatment modalities has not been fully determined.

The wrist is a complex structure with multiple parts, and an infinite combination of positions and motions is possible. Understanding these complexities requires a thorough knowledge of the anatomy and kinematics of the normal wrist. The effect of trauma and disease on the wrist must be appreciated to deliver optimal care to the injured athlete.

It is beyond the scope of this chapter to include an exhaustive review of all the contributors to our understanding of the wrist. References will be given in each section to those pioneers whose work was pertinent. It is enough to say that great strides have been made that now permit us to diagnose and treat disorders of this complex articulation more accurately.

BIOMECHANICS

Anatomy

Osseous Anatomy

Eight bones make up the carpus. Some authors consider the pisiform a sesamoid rather than one of the carpal bones. The other seven carpal bones are aligned in two transverse rows, with the scaphoid bridging the proximal and distal carpal row (Fig. 18A–1).

The carpal bones each have multiple articulations. The carpus as a unit basically functions as two articulations: (1) the radiocarpal joint between the distal radius and the proximal carpal row and (2) the midcarpal joint between the proximal and distal carpal rows. The carpometacarpal articulation is relatively fixed and does not contribute significantly to carpal motion.

There are no tendon insertions on the seven carpal bones except the flexor carpi ulnaris, which inserts on the pisiform. Motion of the wrist therefore results from a complex combination of indirect forces exerted by muscle units attaching proximal or distal to the carpus and from the complex osseous and ligamentous orientation.

Ligamentous Anatomy

There are two excellent studies of the ligamentous anatomy of the wrist, one by Taleisnik [118] and one by Mayfield and associates [90]. Although there are some minor discrepancies between the two studies, they agree in general. The major points have been summarized well by Green [55] as follows:

1. The major ligaments of the wrist are intracapsular and are covered by the wrist capsule.

2. The volar ligaments are much more substantial than the dorsal ligaments (Fig. 18A–2).

3. The prime stabilizer of the proximal pole of the scaphoid is the volar radioscapholunate ligament.

4. The collateral ligaments do not exist as static stabilizers of the wrist.

Vascular Anatomy

The wrist has an excellent vascular supply with multiple anastomoses between the radial, ulnar, and anterior interosseous arteries [120]. These vessels supply three pairs of transverse arches, which are located volarly and dorsally [45] (Fig. 18A–3). Despite their excellent collateral blood supply, there is an increased incidence of avascular necrosis in the scaphoid, lunate, and capitate [108]. It has been found that each of these bones is dependent on a single external vessel or group of vessels for its blood supply [100]. The proximal poles of both the scaphoid and the capitate are supplied by intraosseous vessels coursing retrograde from the waist [44] (Fig. 18A–4). These bones are therefore more susceptible to vascular disruption and avascular necrosis subsequent to trauma.

Function

Biomechanical Concepts

Our understanding of the biomechanics of the wrist has evolved greatly since Navarro described

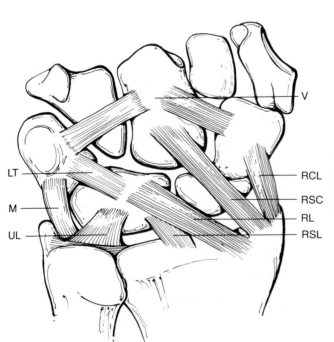

FIGURE 18A–2
The volar carpal ligaments from Taleisnik. RCL, radial collateral ligament; RSC, radioscaphocapitate ligament; RL, radiolunate ligament; RSL, radioscapholunate ligament; UL, ulnolunate ligament; M, ulnocarpal meniscus homologue; LT, lunotriquetral ligament; V, deltoid ligament. (Redrawn from Taleisnik, J. *The Wrist*, New York, Churchill Livingstone, 1985.)

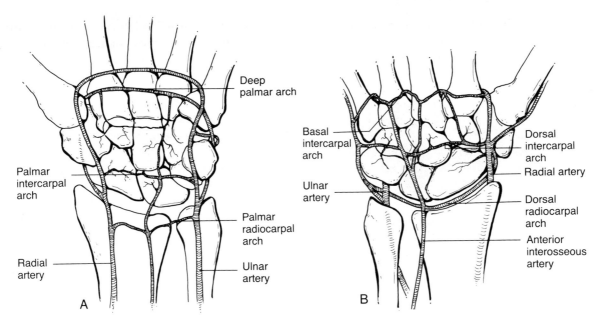

FIGURE 18A–3
The arterial supply of the wrist. Anastomoses from the radial ulnar and anterior interosseous arteries form three transverse arches. *A*, The arterial supply of palmar aspect of the wrist. *B*, Arterial supply of the dorsal aspect of the wrist.

carpal mechanics in 1921 [94]. Several different models have been developed to help explain the complex pattern of motion that exists in the wrist. Lambrinudi [49] proposed the link mechanism of the wrist joint: the distal radius, the proximal carpal row, and the distal carpal row function as links in a chain that is stable in tension but inherently unstable in compression. The scaphoid bridges the proximal and distal carpal rows, provides stability, and helps to control motion (Fig. 18A–5). Taleisnik [119] modified Navarro's columnar column concept to explain patterns of carpal instability (Fig. 18A–6). The scaphoid is the mobile lateral column, the triquetrum is the rotary medial column, and the lunate and entire distal carpal row function as a flexion-extension column. Weber [126] proposed a

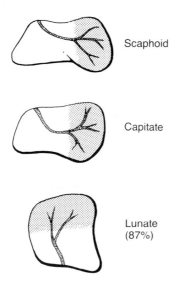

FIGURE 18A–4
The scaphoid, capitate, and occasionally the lunate are each dependent on a single intraosseous vessel for their blood supply. (Redrawn from Gelberman, R. H., and Gross, M. S. The vascularity of the wrist. *Clin Orthop* 202:42, 1986.)

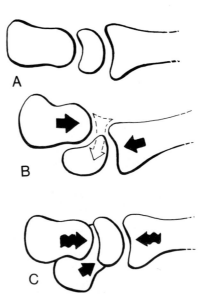

FIGURE 18A–5
A–C, The link mechanism of Lambrinudi, which is inherently unstable in compression. The scaphoid bridges the proximal and distal carpal rows, providing stability in compression. (Redrawn from Gilford, W. W., Bolton, R. H., and Lambrinudi, C. The mechanism of the wrist joint. *Guy's Hosp Rep* 92:52–55, 1943.)

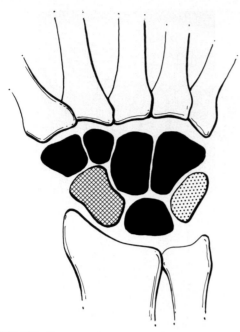

FIGURE 18A–6
The columnar column concept of Taleisnik. (Redrawn from Lichtman, D. M., Schneider, J. R., Swafford, A. R., and Mack, G. R. Ulnar midcarpal instability—Clinical and laboratory analysis. *J Hand Surg* 6:522, 1981.)

slightly different concept, dividing the carpus into two longitudinal columns (Fig. 18A–7). The radial side of the carpus functions as a force-bearing column while the ulnar aspect of the carpus acts to control the position of the lunate relative to the capitate. Lichtman [81] has proposed an oval ring concept (Fig. 18A–8). The carpus is viewed as two mobile links that allow reciprocal motion between the proximal and distal carpal rows. The mobile links are at the scaphotrapezial joint and the triquetrohamate joint. Although these models differ in their concepts of wrist mechanics, each serves to explain carpal motion and helps to increase our understanding of carpal injury patterns.

Patterns of Injury

Mayfield [90] has demonstrated the perilunate pattern of injury to the wrist (Fig. 18A–9). With increased force there is a progression of ligamentous injury from the scapholunate joint continuing around the lunate until all ligamentous support is disrupted. Stage I corresponds to a scapholunate dissociation, whereas stage IV results in a lunate or perilunate dislocation. Various fracture patterns can accompany the ligamentous injury. Varying the position of the hand can change the pattern of injury [128]. With progressive radial deviation the path of injury moves from a lesser arc about the lunate to

a greater arc in a more peripheral course through the carpus.

Kinematics

Flexion and extension of the wrist occur through both the radiocarpal joint and the midcarpal joint [131], although the relative contribution of each joint has been disputed. The radiocarpal joint participates more prominently in dorsiflexion, whereas the midcarpal joint is the major contributor to palmar flexion. Ulnar and radial deviation involve more complex intercarpal motion. In radial deviation the scaphoid palmar-flexes and becomes more vertical relative to the long axis of the forearm. This in effect shortens the radial column of the wrist. As the wrist moves into ulnar deviation, the scaphoid orients in a more longitudinal position, which lengthens the radial column (Fig. 18A–10). This has the effect of locking the midcarpal joint, preventing further flexion or extension from occurring through the midcarpal joint. On the ulnar side of the wrist, the helicoid triquetrohamate joint allows the triquetrum to move proximal and distally relative to its position on the hamate (Fig. 18A–11). This allows a similar shortening and lengthening of the ulnar column in ulnar and radial deviation respectively. The position of the lunate is influenced by the

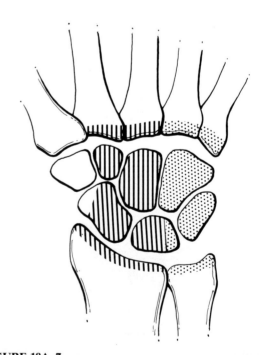

FIGURE 18A–7
The longitudinal column concept of Weber. (Redrawn from Weber, E. R. Concepts governing the rotational shift of the intercalated segment of the carpus. *Orthop Clin North Am* 15:196, 1984.)

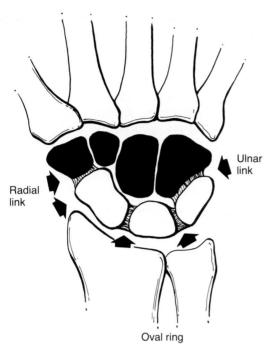

FIGURE 18A–8
The oval ring concept of Lichtman. (Redrawn from Lichtman, D. M., Schneider, J. R., Swafford, A. R., and Mack, G. R. Ulnar midcarpal instability—Clinical and laboratory analysis. *J Hand Surg* 6:522, 1981.)

scaphoid and triquetrum; this results in dorsiflexion of the lunate in ulnar deviation and palmar flexion in radial deviation. Understanding these relationships is important in evaluating x-rays and appreciating patterns of ligamentous instability.

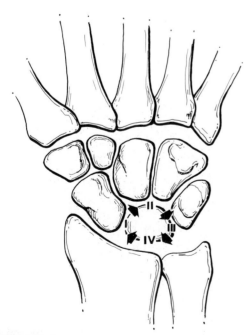

FIGURE 18A–9
The perilunate pattern of injury from Mayfield. (Redrawn from Mayfield, J. K. Mechanism of carpal injuries. *Clin Orthop* 149:50, 1980.)

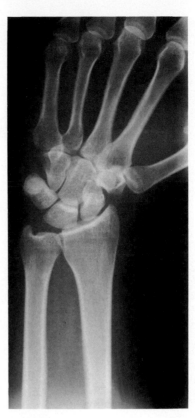

FIGURE 18A–10
In radial deviation the scaphoid appears shortened owing to its more vertical alignment, and the triquetrum has moved to a proximal position on the hamate.

Radiography of the Wrist

Although a thorough history and physical examination are critical in diagnosis of injuries to the carpus, radiography of the wrist is also an essential element in diagnosis. The two basic views are the anteroposterior (AP) view and the lateral view, which should each be taken with the wrist in a neutral position. In cases of trauma, we usually add an anteroposterior view with the wrist in ulnar deviation and another with an active "fist" compression by the patient in a neutral anteroposterior alignment. These views are helpful in identifying fractures of the scaphoid and accentuating a scapholunate gap that may not be apparent in the standard views. Additional views include the carpal tunnel view, which allows visualization of the hook of the hamate and demonstrates any calcific deposits in the carpal tunnel. On occasion, additional views including oblique and anteroposterior views in maximum radial and ulnar deviation and lateral views in maximum flexion and extension are helpful in diagnosing fractures or dislocations of the carpus.

Instability patterns can be identified on the lateral x-ray. In the normal wrist in the neutral position

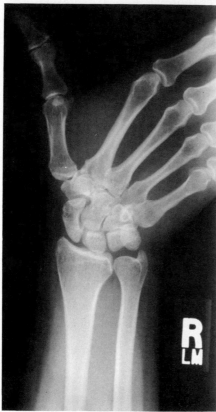

FIGURE 18A–11
In ulnar deviation the scaphoid has assumed a more horizontal position and appears lengthened whereas the triquetrum has moved to a more distal position on its articulation with the hamate.

the axes of the radius, lunate, and capitate are co-aligned. Dorsal intercalary segmental instability (DISI) is present when the lunate lies palmar to the capitate but is flexed dorsally (Fig. 18A–12). Volar intercalary segmental instability (VISI) is present if the lunate lies dorsal to the capitate and is palmar-flexed (Fig. 18A–12). Either of these collapsed patterns can be associated with carpal instabilities, and each is discussed later in this section.

Various other radiographic techniques can be used to evaluate further injuries that are not completely defined by routine radiographs. A bone scan performed 48 to 72 hours after an injury can be helpful in confirming an occult fracture and may be helpful in chronic cases in defining an area of arthrosis. Trispiral tomography is especially helpful for documenting an occult fracture such as a fracture of the hook of the hamate. Motion studies with an imaging intensifier or formal cineradiography will help to document abnormal carpal kinematics that are not apparent on static views. We have found it helpful to use cineradiography to create a permanent record in these cases. We routinely study the uninvolved wrist to have it as a normal comparison. Our

standard study includes an anteroposterior view with the wrist moving from full radial deviation to full ulnar deviation and a lateral view with the wrist moving from full dorsiflexion to full volar flexion. In difficult cases we frequently follow this study with an arthrogram, in which contrast material is injected into the radiocarpal joint. We observe to see if there is any abnormal filling from the radiocarpal joint into the midcarpal joint or the distal radial ulnar joint. To complete the study we then ask the patient to go through the range of motion again and record this later for review.

Additional methods of study include computed tomography (CT) scan, magnetic resonance imaging (MRI), and, more recently, the three-dimensional scanning techniques. We have occasionally found these methods useful in difficult cases in which a definitive diagnosis could not be made with the preceding studies. However, this has been the exception rather than the rule, and we do not routinely employ these techniques.

WRIST DISORDERS

Carpal Instabilities

Scapholunate Instability

Scapholunate instability or rotary subluxation of the scaphoid is the most common form of carpal instability. It is frequently diagnosed late owing to the patient's own delay in presenting for treatment.

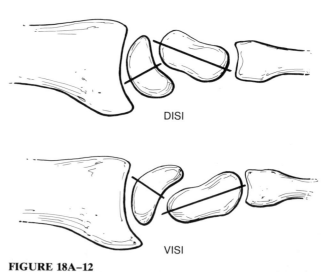

FIGURE 18A–12
Patterns of carpal instability. Dorsal intercalated segmental instability (DISI) is present when the lunate lies volar to the capitate but is flexed dorsally. Volar intercalated segmental instability (VISI) is present if the lunate lies dorsal to the capitate and is flexed volarly.

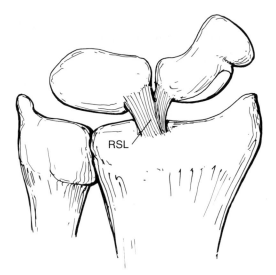

FIGURE 18A–13
Volar ligaments. The most important stabilizer of the proximal pole of the scaphoid is the radioscapholunate ligament (RSL). (Redrawn from Taleisnik, J. *The Wrist.* New York, Churchill Livingstone, 1985.)

Diagnosis can be difficult because of the subtle findings in this condition.

Three ligaments are involved in scapholunate instability: the interosseous scapholunate ligament, the dorsal scapholunate ligament, and the volar radioscapholunate ligament (Fig. 18A–13), which is the strongest and most important of the three. Anatomic studies have demonstrated that two of the three ligaments must be disrupted for dissociation to occur, and the volar radioscapholunate ligament must be ruptured before complete subluxation can occur [77, 119].

Clinical Evaluation

The usual history involves a fall or direct blow that causes hyperextension of the wrist. Pain, swelling, and tenderness are noted over the dorsoradial aspect of the wrist, particularly in the anatomic snuffbox. Watson [124] has described a provocative test in which the examiner stabilizes the scaphoid by placing his thumb over the volar pole of the scaphoid while the wrist is held in ulnar deviation. As the hand is brought into radial deviation, pain is produced as the force is transmitted to the injured scapholunate ligaments (Fig. 18A–14).

X-ray findings may be subtle (Fig. 18A–15). An AP view in full supination may demonstrate widening of the scapholunate interval. A gap of more than 2 mm is abnormal. The scaphoid will be shortened because it is subluxated into a more vertical position, and a ring sign may be present representing the cortical projection of the distal pole in its more vertical position. The lateral x-ray may reveal a DISI pattern. A scapholunate angle of more than

70 degrees (normal 30 to 60 degrees) is also suggestive of scapholunate dissociation [37] (Fig. 18A–16). The scaphoid axis is drawn longitudinally through the long axis of the scaphoid transecting the proximal (dorsal) and distal (volar) convexities at each end of the bone. The lunate axis is drawn perpendicular to the proximal and distal articular surfaces of the lunate and should transect the bone into dorsal and volar halves. The angle formed by these two lines is the scapholunate angle, and its normal values range from 30 to 60 degrees [82].

Special radiographic studies usually are not necessary. If the diagnosis remains in doubt, cineradiography can be helpful in demonstrating the instability [96]. Bone scans and arthrography can aid in localizing the area of injury [63] but are not specifically indicated for scapholunate instability.

Treatment of Acute Injuries

Early treatment of ligament injuries yields the best results. Injuries diagnosed within 3 weeks can be treated as acute [13], although Linscheid has suggested that acute treatment may be successful for up to 3 months [82].

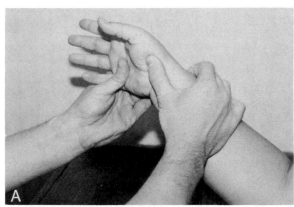
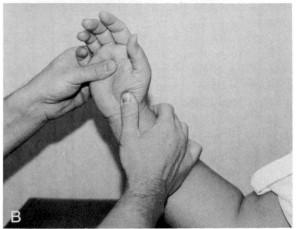

FIGURE 18A–14
The Watson test for scapholunate instability. *A,* The scaphoid is stabilized with the thumb over the volar pole. *B,* When the hand is brought from ulnar to radial deviation, pain results.

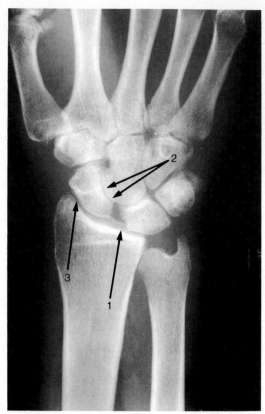

FIGURE 18A–15
Radiographic findings of the scapholunate dissociation: (1) widening of the scapholunate interval, (2) shortened appearance of the scaphoid, and (3) the "ring sign" of the cortical projection of the distal pole.

Cast immobilization has been recommended by King [71] with the wrist in full supination, mild dorsiflexion, and ulnar deviation. Most authors feel that it is difficult to maintain reduction in a cast. This treatment is probably applicable in cases of suspected scapholunate injury in which actual dissociation has not occurred and x-ray studies are negative.

Closed reduction and percutaneous pinning may be successful, but care must be taken that a complete reduction has been obtained [97]. The wrist should be maintained in dorsiflexion to reduce the proximal pole of the scaphoid while the scapholunate joint is pinned. A Kirschner wire (K-wire) is passed percutaneously through the snuffbox and across the scapholunate joint. A second K-wire may be used to provide additional stability across the scapholunate joint. The wrist should then be gently palmarflexed to approximate the volar ligaments [34]. A K-wire is then passed across the scaphoid into the capitate for final stabilization. A K-wire across the radiocarpal joint is not necessary. The wires can be removed after 8 to 10 weeks, and a protective splint is then utilized an additional 4 weeks.

Open reduction and internal fixation offer the obvious advantage that the scapholunate joint is reduced, and direct repair of the ligaments can be accomplished. Most authors recommend a dorsal approach [97]. Repair of the volar ligaments does not appear to be necessary for a good result. Blatt [15] has utilized a portion of the dorsal capsule as a tenodesis to reinforce acute repairs.

It is particularly important to remember that in patients with acute or chronic scapholunate instability, treatment of any type rarely gives completely satisfactory results. In patients with partial injuries, closed reduction and cast immobilization for 6 to 10 weeks may allow healing. However, in complete tears with significant displacement, we have found it difficult to obtain a satisfactory reduction by closed means. In certain cases, percutaneous pinning with image intensification may allow satisfactory reduction. To allow proper approximation of the torn ends of the ligaments, the scaphoid should be reduced initially with the wrist in dorsiflexion and pinned to the carpals to stabilize the proximal pole of the scaphoid. Then the wrist is brought into volar flexion to approximate the torn ends of the volar ligaments, and the radioscaphoid joint is pinned in this position. The wrist is then immobilized in a short-arm thumb spica cast. The wires can be removed at 8 weeks and the wrist immobilized for an additional 4 weeks in a short-arm cast. The patient is then fitted with a protective splint, and an active rehabilitation program is started.

Authors' Preferred Method of Treatment. In the majority of cases we have not felt that a satisfactory

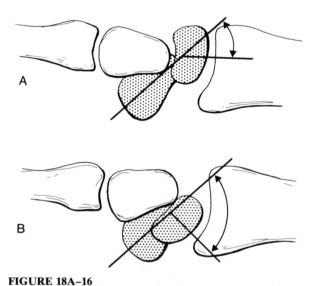

FIGURE 18A–16
The normal scapholunate angle *(A)* is 30 to 60 degrees. A scapholunate angle of more than 70 degrees *(B)* is suggestive of scapholunate dissociation. (Redrawn from Green, D. P. *Operative Hand Surgery* (2nd ed.). New York, Churchill Livingstone, 1988.)

reduction could be attained by closed methods. We prefer to treat most patients with open techniques. In our experience, open reduction is best performed through a dorsal approach with repair of the dorsal scapholunate interosseous ligaments. When the ligament cannot be adequately repaired, reinforcement with local soft tissue is carried out. This can be accomplished by utilizing a portion of the dorsal capsule. In addition, a second incision is made on the volar aspect of the wrist to expose and repair the avulsed volar ligaments, which are usually detached en masse. Decompression of the medial nerve can also be accomplished at the same time because associated carpal tunnel syndrome is often present. It is important to remember that the fresh injury may be superimposed on a chronic instability, and soft tissue as well as articular cartilage changes may be encountered. Fixation with K-wires is indicated for temporary stabilization of the scaphoid to the capitate and lunate as well as for stabilization of the radiocarpal joint. The wrist can be immobilized in a neutral or slightly dorsiflexed position in a thumb spica cast for 8 weeks, at which time the pins are removed. A short-arm cast is utilized for an additional 4 weeks, and then the patient is fitted with a protective splint, and a rehabilitation program is begun. Our rehabilitation program is discussed at the end of this chapter.

Criteria for Return to Sports Participation. The athlete may be allowed to return to sports activity after he demonstrates progress in a supervised rehabilitation program. Normally, 1 to 2 months are necessary to demonstrate significant progress in strength and range of motion. The athlete should be protected by an orthosis such as a silicone or synthetic fiberglass cast during participation in sports. At 6 months the athlete may be allowed to participate without an orthosis if he demonstrates maximum strength and range of motion.

Treatment of Chronic Injuries Without Degenerative Changes

Chronic scapholunate instability presents a more complex problem and a greater challenge in management. Chronic instability is defined as an injury that has lasted longer than 3 months. The diagnostic criteria and clinical evaluation for chronic injuries are the same as described previously for acute injuries. Additionally, degenerative changes may be noted on x-ray with loss of the interosseous space at the radiocarpal and intercarpal joints as well as osteophyte formation in the late stages. There is still no consistently reliable and predictable procedure for treatment of chronic injuries.

Blatt [15] utilizes a dorsal tenodesis technique in which the dorsal capsule acts as a check rein to prevent volar rotation of the distal pole of the scaphoid. Other authors [37, 119] have used tendon grafts to reconstruct the scapholunate ligament or the radioscapholunate ligament. These methods are complex and not consistently reliable.

A variety of intercarpal arthrodeses have been performed to correct late scapholunate instability. These limited wrist arthrodeses attempt to control the subluxation of the scaphoid. Scapholunate joint arthrodesis intuitively seems to be the most appropriate procedure; however, this is a difficult joint in which to attain successful fusion [122] owing to its small size and the significant forces directed across it. The capitate can be included in a scapholunate arthrodesis to increase success in attaining solid fusion by increasing the area of fusion. However, this results in further loss of wrist motion [55], primarily palmar flexion. Triscaphe (trapezium-trapezoid-scaphoid) fusion in which the distal pole is stabilized has been popularized by Watson [124]. This prevents subluxation of the proximal pole, but the technique is demanding, and care must be taken that the scaphoid is properly reduced.

The delay in diagnosis may lessen the chance of a successful result, and it is important to remember that a completely successful result almost never occurs. In chronic cases without degenerative changes, various methods of reconstruction of the ligaments may be carried out. We initially used a variation of the procedure described by Dobyns and colleagues [37] to reconstruct the interosseous ligament with a strip of the extensor carpi radialis. We have been relatively pleased with the initial results, although stretching of the structures with widening of the scapholunate interval often occurs. The presence of an interpositional ligament, however, seems to prevent proximal migration of the capitate and the subsequent changes caused by migration. Motion, although decreased in most cases, is usually within a functional range. However, the primary drawback to this procedure is the technical difficulty in performing the reconstruction of the interosseous ligaments.

In our experience, attempts to perform a scapholunate arthrodesis to stabilize the scapholunate joint are functionally gratifying. Although bony arthrodesis may not occur, it appears that a fibrous union does occur to prevent migration of the capitate. Including the capitate in the fusion increases the chances of obtaining a bony fusion but does result in further restriction of motion. The scapho-trapezial-trapezoid arthrodesis described by Peterson and Lipscomb [103A] and popularized by Watson [124] has met with limited success in our hands. We are concerned about the increased stress on the other

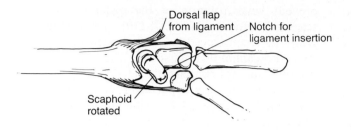

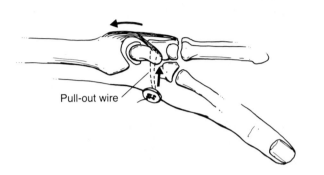

FIGURE 18A–17
Blatt's dorsal capsulodesis for chronic scapholunate dissociation. (Redrawn from Green, D. P. *Operative Hand Surgery* (2nd ed.). New York, Churchill Livingstone, 1988.)

joints, particularly the radioscaphoid joint and the lack of control of the scapholunate interval. We have used a number of these procedures with varied success.

Authors' Preferred Method of Treatment. At present we are using the procedure described by Blatt, the dorsal capsulodesis, and we have been gratified by the functional results and lack of problems with healing compared with the other procedures described. The goal of the procedure is to reduce the scaphoid from its rotary subluxation and secure it in a reduced position with a strip of the dorsal capsule (Fig. 18A–17). A dorsal approach to the wrist is made, and a 1-cm strip of dorsal capsule is fashioned based on the ulnar aspect of the distal radius. This provides a flap of the capsule that will be oriented obliquely across the carpus in line with the long axis of the scaphoid. The scaphoid is reduced by manipulating the scaphoid out of its volar rotated position. The reduction is maintained by passing a K-wire through the distal pole of the scaphoid and into the capitate. An additional K-wire may be passed across the radioscaphoid joint for additional stability. A trough is created in the distal pole of the scaphoid, taking care to make the trough distal to the mid axis. The dorsal flap of tissue is then secured to the trough with a pull-out wire over the button on the volar aspect of the wrist. The wrist is immobilized in a thumb spica cast. The K-wires and pull-out wire are removed at 8 to 12 weeks. Total cast immobilization lasts for 12 weeks followed by active rehabilitation. Our rehabilitation program is discussed at the end of the chapter.

Criteria for Return to Sports Participation. The athlete may be allowed to return to sports activity after he demonstrates progress in a supervised rehabilitation program. Normally 1 to 2 months are necessary to demonstrate significant progress in strength and range of motion. The athlete should be protected by an orthosis such as a silicone or synthetic fiberglass cast during participation in sports. At 6 months the athlete may be allowed to participate without an orthosis if he demonstrates maximum strength and range of motion.

Salvage Procedures

In patients with scapholunate instability and established arthrosis, pain will not be relieved by the preceding procedures. Athletes with an advanced stage of arthrosis [1] will probably have already experienced significant limitation of sports activity that is dependent on wrist function. Even with a salvage procedure they would not be expected to regain their previous level of activity.

A variety of salvage procedures have been suggested. Styloidectomy may be applicable in patients with arthrosis localized to the radioscaphoid joint, but it is often only a temporizing procedure and can compromise later procedures. Occasionally, it can cause increased instability if too much of the styloid is removed. Proximal row carpectomy appears to have relatively good results but does result in decreased grip strength [56]. The procedure must be done before degenerative changes involve the proximal pole of the capitate or the lunate fossa of the radius. Silastic scaphoid replacement carries the risk of silicone synovitis. The implications of this complication are not clear at this time. If this technique is to be used it should probably be combined with a midcarpal arthrodesis [123]. One should avoid placing K-wires or sutures through the implant if possible because this process may be related to later silicone synovitis. Total wrist arthroplasty is applicable only in persons who are not involved in activities that place heavy stress on the wrist. Wrist arthrodesis is more applicable as a salvage procedure in patients who are athletically active.

Salvage procedures by definition are probably not consistent with maintaining athletic activities. However, for completeness we will briefly cover procedures that can be considered. We have found that the proximal row carpectomy, despite its controversial aspects, has been very useful in allowing preservation of functional wrist motion. It should be performed before degenerative changes are present in the proximal pole of the capitate. This procedure does result in some weakness of grip strength. If the procedure is unsuccessful owing to persistent pain, an arthrodesis can be carried out at a later date. If degenerative changes in the carpus are significant, particularly in individuals desiring to return to heavy activity, wrist arthrodesis is the procedure of choice. Arthroplasty can be carried out to maintain motion but does not hold up with heavy use. Implants with associated intercarpal fusion may be helpful in certain individuals, but again these do not do so well in people requiring heavy or repetitious use of the wrist.

Medial (Ulnar) Carpal Instability

Our appreciation of wrist instabilities on the ulnar aspect of the wrist has increased tremendously over the last decade. In the past these problems have been more difficult to diagnose and treat because of our lack of knowledge. Recent advances have resulted in an increased awareness and more appropriate management. Patients present with wrist pain and often complain of a painful click, which may be audible or palpable. There may be a history of trauma, but frequently the patient does not recall a specific inciting episode.

Triquetrohamate Instability
The most common ulnar instability is triquetrohamate dissociation. These patients have a characteristic "clunk" that can be audible and palpable. It is reproduced by active motion. Alexander and Lichtman state that the ulnar arm of the arcuate ligament extending from the capitate to the triquetrum is torn, causing the instability [2]. This clunk represents the triquetrum suddenly snapping back and forth over the lunate articulation with ulnar and radial deviation rather than following its normal smooth synchronous course.

Clinical Evaluation. Physical examination may elicit point tenderness over the triquetrum. Plain x-rays usually offer no clues to the pathology. In advanced stages a DISI pattern may be present, although a VISI pattern has rarely been noted [115]. Cineradiographs are most useful and demonstrate abnormal motion well [2]. In particular, the lunate can be seen to snap suddenly at the moment the triquetrum reduces on the lunate. An arthrogram should be negative in isolated triquetrohamate instability because the lesion does not involve communication between the radiocarpal joint and the midcarpal joint. However, when a triquetrolunate lesion coexists with a triquetrohamate instability, the arthrogram will be positive owing to flow of dye through the incompetent triquetrolunate joint. In these cases, the clinical effect of triquetrohamate instability will take precedence over the less severe triquetrolunate lesion [78].

Triquetrolunate Instability
Clinical Evaluation. Triquetrolunate instability is not as common as triquetrohamate instability. Patients may or may not have a wrist click. Local tenderness can be elicited over the triquetrolunate joint. Reagan and colleagues [106] have described a ballottement test (Fig. 18A–18) in which the lunate is stabilized by one hand while the other hand shifts the triquetrum in a palmar and dorsal direction. This reproduces the pain the patient has experienced. Plain x-rays are usually not helpful in our experience, although a VISI pattern has occasionally been noted [78]. Measurement of the lunate-triquetral angle has been described [98], but we have found it difficult to consistently measure this angle accurately. Cineradiography is usually normal, unlike in triquetrohamate instability, in which it is notably abnormal. Arthrograms can be helpful but should be observed closely to determine whether dye flows through the triquetrolunate joint to confirm the site of pathology [22].

Treatment
All patients should have a trial of conservative treatment consisting of immobilization in a splint or cast and an anti-inflammatory medication. Local steroid injection into the involved joint can be helpful. Modification of activities may be necessary after acute symptoms resolve. Athletes who are symptomatic primarily during sports activity can be fitted with a splint to correct triquetrohamate instability [78]. The splint reduces the VISI sag of the proximal row by pushing dorsally on the pisiform (Fig. 18A–19).

Authors' Preferred Method of Treatment. Surgery

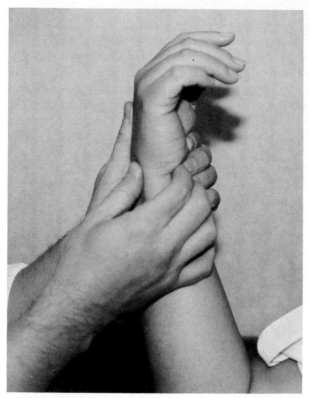

FIGURE 18A–18
The ballottement test described by Reagan and colleagues [106]. The lunate is stabilized by one hand while the triquetrum is shifted in a palmar and dorsal direction with the examiner's opposite hand.

can be considered in patients who do not respond to conservative treatment. Reconstruction of the ligaments is technically demanding. We prefer to perform an intercarpal arthrodesis of the involved joint. In this procedure we use a dorsal approach between the fourth and fifth dorsal compartments. The triquetrohamate and triquetrolunate joints are inspected, and the wrist is stressed by axial compression and deviation to confirm the location of the instability. The triquetrum is then reduced and fixed internally by inserting K-wires across the involved joint. The joint surfaces are denuded, and a trough is created across the joint. The spatial relationship of the joint is maintained by filling the joint space with cancellous bone chips. A corticocancellous graft is taken from the distal radius to fill the trough (Fig. 18A–20). The wrist is immobilized for 8 weeks in a short-arm cast, after which the pins are removed and a rehabilitation program is initiated. Patients should expect to lose some wrist motion. Usually range of motion will improve during the first year postoperatively. Proportionally more dorsiflexion than palmar flexion can be expected to be regained.

Criteria for Return to Sports Participation
The athlete may be allowed to return to sports activity after he demonstrates progress in a supervised rehabilitation program. Normally 1 to 2 months are necessary to demonstrate significant progress in strength and range of motion. The athlete should be protected by an orthosis such as a silicone or synthetic fiberglass cast during participation in sports. At 6 months the athlete may be allowed to participate without an orthosis if he demonstrates maximum strength and functional range of motion.

Fractures

Scaphoid Fractures

Fractures of the scaphoid are the most common fractures involving the carpal bones [19]. The incidence of this fracture in college-level football players has been reported to be as high as 1 in 100 players per year [132]. The trauma causing the injury is frequently minor and leads many physicians to discount it as a sprain, resulting in delayed diagnosis. The mechanism of injury is probably forced hyperextension with the wrist in ulnar deviation [89].

The vascular supply of the scaphoid plays a prominent role in the high incidence of nonunion associ-

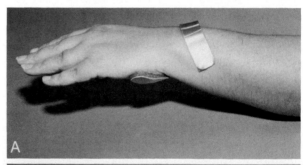

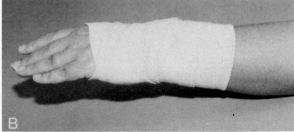

FIGURE 18A–19
A splint supporting a wrist with triquetral hamate instability helps to reduce the proximal row by putting pressure dorsally on the pisiform.

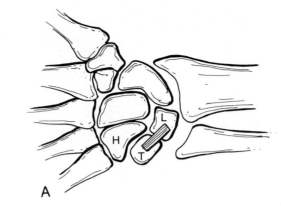

FIGURE 18A–20
A, Triquetrolunate arthrodesis. *B,* Triquetrohamate arthrodesis. H, hamate; T, triquetrum; L, lunate.

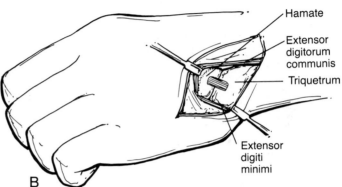

ated with this fracture. Extraosseous vessels enter the middle and distal portions of the scaphoid [42]. The proximal pole (see Fig. 18A–3) receives no direct vascular supply but derives its supply from the intraosseous vessels passing retrograde from the waist. This arrangement provides an anatomic basis for the notably high rate of nonunion and avascular necrosis characteristic of proximal pole fractures.

Clinical Evaluation

Patients present with wrist pain localized to the snuffbox. This finding alone should be enough to justify immobilization until a definitive diagnosis is made. X-rays should include a posteroanterior (PA) view in neutral and ulnar deviation, a lateral view, and a closed fist view (Fig. 18A–21). Polytomography is helpful in defining a nondisplaced fracture [83]. A bone scan after 72 hours can also identify a fracture, and a negative bone scan effectively rules out the diagnosis [96] (Fig. 18A–22).

Factors Affecting Healing

Several factors have been identified as having an adverse effect on healing. Fractures of the proximal pole clearly have a higher incidence of delayed union [36], nonunion, and avascular necrosis [31]. Russe [110] states that vertical and oblique fractures are potentially unstable and require prolonged immobilization. Uncorrected displacement of more than 1 mm and delay in diagnosis for longer than 4 weeks affect healing adversely [39]. Uncomplicated scaphoid fractures have a union rate of 95% when they are diagnosed early and immobilized [97].

FIGURE 18A–21
Scaphoid fracture.

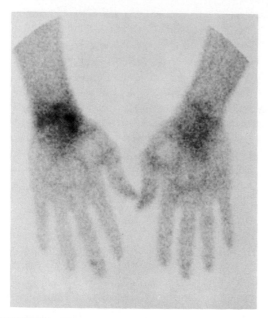

FIGURE 18A–22
Bone scan demonstrating increased uptake over the scaphoid.

Treatment

Nondisplaced Fractures. Nondisplaced fractures can be treated by immobilization. There is significant disagreement about the position and type of immobilization. A review of the literature [89, 67, 97] supports placing the wrist in slight (10 degrees) palmar flexion and radial deviation. A recent prospective study gives strong support to using a long-arm thumb spica cast for 6 weeks followed by a short-arm spica cast [46]. This study indicates that healing time is shortened by utilizing a long arm spica cast initially. In athletes, nondisplaced fractures of the wrist have been effectively immobilized for competition in contact sports in plaster, synthetic, or Silastic casts [107]. If serial x-rays taken monthly demonstrate no progress toward union after 3 to 4 months, consideration can be given to bone grafting or electrical stimulation [16].

Displaced Fractures. Displaced fractures of more than 1 mm of the scaphoid require accurate reduction by either closed or open methods. Failure to reduce a displaced fracture accurately results in an unacceptably high rate of nonunion [39]. Several studies [72, 106] have reported successful treatment of displaced fractures by closed methods. However, attaining and maintaining accurate reduction of a displaced fracture is quite difficult, and open reduction is usually necessary [97]. Open reduction and internal fixation (ORIF) can be accomplished by volar approach and fixation with K-wires. Internal fixation by means of the Herbert screw [60] is technically demanding but has the advantage of providing compression and allowing early mobilization. When a displaced fracture of the scaphoid is associated with a dorsiflexion instability, a dorsal approach is necessary to reduce and pin the lunate along with the scaphoid fracture.

Authors' Preferred Method of Treatment. Nondisplaced fractures (less than 1 mm displacement) are treated in a short-arm thumb spica cast using synthetic (fiberglass) casting material. Based on individual factors, athletes may be allowed to participate in their sport if immobilization does not prevent sport-specific function of the extremity. Immobilization is continued until healing is demonstrated by x-ray, usually within 3 months. The wrist is protected during strenuous use with a rigid splint for an additional 2 months.

If there is no evidence of healing after 6 months, we prefer to proceed with bone grafting and internal fixation as needed. The scaphoid is approached through a volar incision, and accurate reduction is attained. A trough is developed longitudinally across the nonunion with osteotomes and a power burr. A corticocancellous bone graft is harvested from the distal radius through the same incision and placed in the trough. If the fracture is not thought to be sufficiently stable, internal fixation is attained with two smooth K-wires. The extremity is immobilized for 6 weeks in a long-arm thumb spica cast followed by a short-arm thumb spica until healing has occurred. The K-wires are removed prior to rehabilitation.

Proximal third fractures of the scaphoid are treated similarly but tend to have a higher incidence of avascular necrosis and nonunion. If the proximal fragment is large enough, cases of nonunion are treated by bone grafting as described above. However, when the proximal fragment is too small for adequate bone grafting and internal fixation, we have excised the proximal fragment and inserted a hand-fashioned spacer of silicone or a soft tissue spacer of capsule or tendon. The wrist is immobilized for 4 weeks in a short-arm cast, and rehabilitation is then begun.

Displaced fractures are treated by open reduction through a volar Russe approach. Any rotary deformity is corrected, and the fracture is stabilized with at least two smooth K-wires. In isolated scaphoid fractures, the K-wires are not placed across the intercarpal joints. The wires are cut off beneath the skin. The extremity is immobilized in a long-arm thumb spica cast for 6 weeks followed by a short-arm thumb spica cast, which is continued until x-rays demonstrate healing. The K-wires are removed, and rehabilitation is initiated.

Criteria for Return to Sports Participation

Following removal of the cast, the athlete is placed on a supervised program of strengthening and range of motion exercises. The wrist is protected with a rigid splint for athletic activities until strength approaches that of the opposite wrist and motion approaches functional range. The wrist is protected for a minimum of 3 months following cast removal on an empirical basis.

Hamate Fractures

Fractures of the hamate usually occur at the hook, and diagnosis is commonly missed [104]. In athletes this injury commonly results from the direct force of a bat, club, or racket [114]. Pain can be localized to the hamate hook over the hypothenar eminence. A carpal tunnel view is necessary to demonstrate the fracture radiographically (Fig. 18A–23). Tomograms and CT scans can aid in diagnosis in questionable cases. Chronic fractures have been associated with rupture of the flexor tendons to the ring and little fingers [3] and with neuropathy of the ulnar nerve [18].

Treatment. Healing of acute cases is rare even with ORIF [14]. Excision of the hook through the fracture site is an effective treatment and allows early return to sports in 6 to 8 weeks [101]. Care must be taken to protect the deep branch of the ulnar nerve during excision.

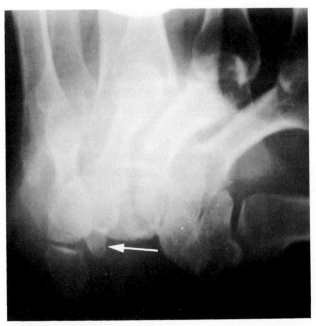

FIGURE 18A–23
Carpal tunnel view isolating the hook of the hamate to allow visualization of a fracture.

Fractures of the Body of the Hamate

Fractures of the body of the hamate are less common than hook fractures. They may be associated with dorsal dislocation of the fourth and fifth metacarpals [86]. They are appreciated best on oblique x-rays of the carpus.

Treatment. Nondisplaced fractures heal with cast immobilization for 4 to 6 weeks. Displaced fractures should be reduced and pinned with K-wires.

Authors' Preferred Method of Treatment. Fractures of the hook of the hamate are treated by excision through a palmar incision over the hook. The motor branch of the ulnar nerve is adjacent to the hook and must be identified and protected. The hook is excised through the fracture site, and the base is smoothed. The hand can be splinted until the incisional pain has resolved; vigorous mobilization follows.

See the following discussion on fractures of the body of the hamate for criteria on return to sports participation.

Nondisplaced fractures of the body of the hamate are immobilized in a short-arm cast for 4 to 6 weeks; this is followed by a mobilization program. Displaced fractures of more than 2 cm are treated by open reduction and internal fixation with K-wires. The wrist is immobilized for 4 to 6 weeks in a short-arm cast and then mobilized. The K-wires can be removed at 4 to 6 weeks or left in place until a later date if they are not causing discomfort.

Criteria for Return to Sports Participation. Athletes who have fractures that are treated nonsurgically may return to play immediately. The wrist is protected in a semirigid synthetic cast. Athletes with surgically treated fractures are usually restricted from active participation in sports until after 4 to 6 weeks of mobilization. Protective splinting is continued for 3 months and thereafter until the wrist has normal strength and range of motion.

Capitate Fractures

Fractures of the capitate resemble fractures of the scaphoid in several ways. The mechanism of injury is variable and can result in different patterns of injury [1]. Fractures can result from a direct blow to the dorsum of the wrist or from forced dorsiflexion or palmar flexion. The wrist is usually found to be swollen with point tenderness over the dorsum at the base of the third and fourth rays. Detection on initial plain x-rays may be difficult, and follow-up x-rays in 7 to 10 days, tomograms, or bone scans may be needed to confirm the diagnosis [1]. Like the scaphoid, the blood supply to the proximal pole of the capitate is retrograde from vessels entering at the waist [100]. This explains the higher incidence of avascular necrosis seen in capitate fractures. Capitate fractures may be associated with fractures of the scaphoid and perilunate dislocations [105]; care should be taken not to overlook the capitate injury in these situations.

Treatment. Nondisplaced fractures can be managed by immobilization. Displaced fractures of more than 2 mm require ORIF with K-wires. Attention should be given to possible avascular necrosis of the proximal pole, although collapse is uncommon. An increased incidence of arthrosis has been noted on late follow-up [105].

Authors' Preferred Method of Treatment. Nondisplaced fractures are evaluated carefully to ensure the absence of associated fractures of the scaphoid or dislocations of the carpometacarpal joints or perilunate dislocations. Isolated nondisplaced fractures of the capitate are immobilized in a short arm cast for 6 weeks. The athlete may compete in a semirigid orthosis after the acute pain has resolved. At 6 weeks the wrist is mobilized and strengthened.

Displaced fractures of more than 2 mm are treated with ORIF with K-wires. The capitate is approached by a dorsal incision, and reduction of the carpometacarpal joints is ensured. These can also be pinned with K-wires if necessary. The wrist is immobilized for 6 weeks in a short-arm cast. The K-wires can be removed at that time or following rehabilitation at the surgeon's discretion. Therapy to regain strength and range of motion is instituted as described later in this chapter.

Criteria for Return to Sports Participation. Athletes with fractures treated nonsurgically may return to play immediately. The wrist is protected in a semirigid synthetic cast. Athletes with surgically treated fractures are usually restricted from active participation in sports until after 4 to 6 weeks of mobilization. Protective splinting is continued for 3 months or until the wrist has normal strength and range of motion.

Pisiform Fractures

Fractures of the pisiform are uncommon and are usually nondisplaced. They usually result from direct trauma to the palm. Tenderness to palpation over the base of the hypothenar eminence suggests a pisiform fracture. These fractures are visualized best on 30-degree oblique AP views or on carpal tunnel views [43]. Pisiform fractures usually heal with immobilization lasting 3 to 6 weeks. Nonunion or malunion that is symptomatic is an indication for excision [58].

Criteria for Return to Sports Participation. Taping the wrist or using a semirigid synthetic cast usually allows an athlete to resume sports as soon as the acute pain subsides. In rare cases of nonunion, the pisiform can be excised at the end of the athletic season.

Triquetrum Fractures

Triquetrum fractures can result from hyperextension with impingement of the distal ulna on the triquetrum. This typically causes an avulsion fracture from the dorsal cortex, which is demonstrated on a lateral or oblique x-ray of the wrist. The dorsal chip fracture routinely becomes painless after immobilization in a short-arm cast for 3 weeks. The athlete can participate in sports wearing a semirigid cast. Nonunion of a dorsal chip fracture has been noted but rarely causes persistent symptoms [10].

Criteria for Return to Sports Participation. Athletes can usually participate in sports wearing a semirigid cast. When participation is precluded by a cast, the athlete can resume activity whenever the discomfort lessens to an extent that it does not interfere with sport-specific activity.

Osteochondroses

Lunate—Kienbock Disease

Kienbock in 1910 published an article titled Traumatic Malacia of the Semilunar Bone [70] in which he postulated that traumatic lesions of the lunate occurred more frequently than was believed at that time. The cause of this lesion and an accepted treatment have yet to be well established. Kienbock attributed the progressive collapse of the lunate to avascular necrosis, and although this theory is generally accepted, the cause of the lesion remains undetermined. Trauma has been implicated by Peste [103] and Kienbock. Many subsequent authors have shared this view [4, 13, 102, 41]. Hulten [64] noted the increased incidence of avascular necrosis of the lunate in wrists that have an ulnar minus variant in which the articular surface of the ulna is proximal to the articular surface of the radius (Fig. 18A–24). Axelson [8] and Gelberman and colleagues [43] have verified this finding. The shape of the lunate was implicated by Zapico [5], who suggested that a lunate with an apical proximal articular surface is most susceptible to avascular necrosis (Fig. 18A–25).

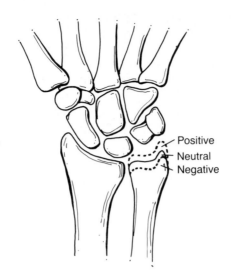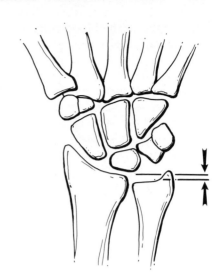

FIGURE 18A–24
Ulnar variance. Ulnar minus variant is associated with an increased incidence of avascular necrosis of the lunate.

Positive
Neutral
Negative

Although no single factor can be blamed for Kienbock's disease, avascular necrosis probably results from repetitive compression forces that cause microfractures of the cancellous bone in patients with a vascular or mechanical predisposition [118]. Continued stress prevents healing and leads ultimately to progressive collapse of the lunate and subsequent degenerative changes in the wrist.

Clinical Evaluation. Patients may complain of pain or stiffness. On occasion, symptoms of carpal tunnel syndrome have been noted [11]. Routine radiographs in the early stages may reveal no changes in the lunate, but the ulnar minus variant may be noted. Tomography is sometimes helpful to

demonstrate early lunate changes. Bone scans may demonstrate increased local uptake of contrast material [12]. Progressive x-ray changes noted include sclerosis, cyst formation, fragmentation and collapse with loss of carpal height, and finally degenerative joint changes throughout [79]. Stahl developed a classification of Kienbock's disease based on the radiographic changes [26] (Fig. 18A–26).

Treatment. Multiple treatment methods have been suggested for Kienbock's disease. Basically they can be categorized into two types. The first is an attempt to allow revascularization of the lunate by relieving the compression forces. This is accomplished by lengthening the ulna or shortening the radius in stage I prior to collapse of the lunate. In stage II, intercarpal arthrodesis may be appropriate. In the more advanced stages III and IV, treatment methods include excision of the lunate with either soft tissue interposition or silicone replacement arthroplasty, or various types of intercarpal arthrodesis. Salvage procedures used in end-stage disease include proximal row carpectomy and wrist arthrodesis.

Immobilization in stage I was recommended by Stahl [113], but other authors have noted progression of the disease when this method alone is used [80, 121]. Immobilization with an external fixator has been suggested to unload as well as immobilize the lunate [79], but no results have yet been reported. Attempts to revascularize the lunate directly with pedicle flaps have been described [21, 26], but the results so far have been inconclusive.

Ulnar shortening or radial lengthening to correct ulnar minus variance is based on the theory that correction of this predisposing factor will relieve stress on the lunate and allow healing. Studies have demonstrated good results from these methods [4,

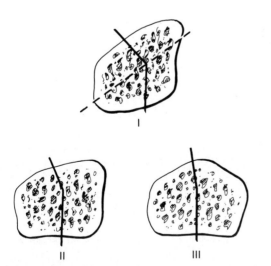

I

II   III

FIGURE 18A–25
A lunate with an apical proximal surface (type I) has been implicated as being more susceptible to avascular necrosis and is associated with an ulnar minus variant. (Redrawn from Green, D. P. *Operative Hand Surgery* (2nd ed.). New York, Churchill Livingstone, 1988.)

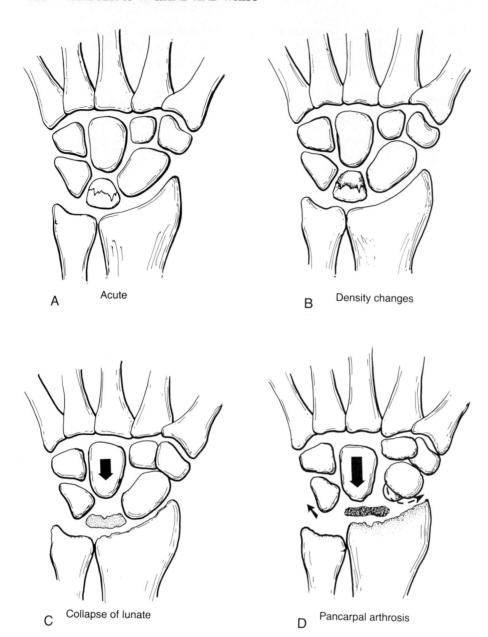

A   Acute

B   Density changes

C   Collapse of lunate

D   Pancarpal arthrosis

FIGURE 18A–26
Modified Stahl's classification of Kienbock's disease. *A*, Stage I: Normal-appearing lunate with a compression fracture demonstrated by a radiolucent line. *B*, Stage II: Sclerosis of the lunate. *C*, Stage III: Collapse of the lunate. *D*, Stage IV: Pancarpal arthrodesis. (Redrawn from Lichtman, D. M., Alexander, A. H., Mack, G. R., and Gunther, S. F. Kienbock's disease—Uptake on silicone replacement arthroplasty. *J Hand Surg* 7:343–347, 1982.)

6]. This treatment is logical if it is applied prior to collapse of the lunate in stage III. Both methods require osteotomy and internal fixation. Ulnar lengthening in addition requires harvesting bone graft from a donor site.

Limited intercarpal fusion in theory will relieve compression on the lunate and prevent further distortion of the normal carpal alignment. This method also has the advantage of preserving some carpal motion, unlike complete carpal arthrodesis. Several different methods have been advocated including arthrodesis of the lunate to the adjacent carpal bones [87], capitohamate fusion [91], and triscaphe fusion [79]. These methods may be combined with lunate arthroplasty to obtain improved results. Lunate excision with or without interposition arthro-

plasty has been utilized as a treatment method in patients with stage III lesions. Various tendons and capsules have been used as interposition material with fair results [7], but this technique has been criticized by others [54] because proximal migration of the capitate is probable. Silicone replacement arthroplasty has been advocated for this reason in patients with stage II or III disease [29]. Newer prosthesis design has reduced the incidence of dislocation and silicone synovitis [125]. In the young, active patient, silicone replacement arthroplasty should probably be combined with either a lengthening or a shortening procedure or with a limited intercarpal fusion to reduce stress on the prosthesis.

In end-stage Kienbock's disease with diffuse arthrosis, the previously mentioned procedures will not be adequate to relieve pain. Salvage procedures such as proximal row carpectomy [33] or complete wrist arthrodesis [80] have been utilized successfully in these instances. However, these methods obviously sacrifice function to accomplish pain relief.

Authors' Preferred Method of Treatment. Lichtman and colleagues [80] recommended correlating treatment with the radiographic stages of Kienbock's disease as modified from Stahl's original classification (see Fig. 18A–26). In stage I, the lunate is "decompressed" by procedures to prevent further trauma. This can be accomplished by using a leveling procedure (ulnar lengthening or radial shortening for ulnar minus variant) or by a scapho-trapezial-trapezoid arthrodesis to decrease the compressive forces on the lunate. In stage II, collapse of the lunate exists, and we have used excision of the lunate and replacement with a silicone prosthesis. Despite the recent concern about silicone synovitis, we have not found this a significant problem in lunate replacement arthroplasty. It is important to avoid placing K-wires or suture through the implant. Also, sufficient stability of the implant must be attained at the time of surgery. The arthroplasty may also be combined with methods to reduce shear stress such as radial shortening, ulnar lengthening, or intercarpal arthrodesis.

The procedure is done through a dorsal incision over the lunate. A dorsal flap of capsule based distally is developed, and the lunate is excised with osteotomes and rongeur. The palmar cortex of the lunate is retained to provide volar support. Trials are used to obtain the appropriate size for the implant. There should be a full range of motion of the wrist without subluxation of the implant. The dorsal capsule is closed to provide additional support. If satisfactory stability cannot be attained, silicone arthroplasty should be abandoned, and a tendon interposition should be substituted for the implant [79]. The wrist is immobilized in a short-arm cast for 6 weeks.

With further collapse and associated carpal instability in stage III, we prefer to combine silicone replacement arthroplasty with scapho-trapezial-trapezoid arthrodesis. In stage IV, when advanced perilunate degenerative changes are present, a proximal row carpectomy is useful to maintain wrist motion. A complete wrist arthrodesis is an alternative if extensive arthrosis exists.

Criteria for Return to Sports Participation. Following removal of the cast, the athlete is placed on a supervised program of strengthening and range of motion exercises. The wrist is protected with a rigid splint for athletic activities until its strength approaches that of the opposite wrist and motion approaches the functional range. The wrist is protected for a minimum of 3 months following cast removal on an empirical basis.

Osteochondrosis of the Capitate

Avascular necrosis of the capitate is a rare disease. It was first reported by Destot of France in 1924 [93]. In 1942 Jonsson recognized that avascular necrosis of the capitate was similar to Kienbock's disease [69]. The number of reported cases had been less than 10 until the 1980s, when this condition began to be recognized more frequently [1, 17, 20, 61]. There are no distinct differences in regard to age, sex, or occupation. Among athletes this condition is seen more frequently in gymnasts [93]. It is postulated that chronic or repeated trauma is the cause.

Anatomically, the proximal portion of the capitate is supplied by intraosseous vessels flowing in a retrograde direction. Case reports describe the main lesion located anywhere from the isthmus to the proximal portion of the capitate. It is postulated that a microfracture may take place at the isthmus, thereby blocking blood flow to the more proximal portion and causing avascular necrosis.

Clinical Evaluation. Patients complain of gradual onset of pain without a specific episode of trauma. The pain is worse with exercise. Swelling and point tenderness are noted over the dorsum of the wrist, and range of motion is limited primarily in dorsiflexion. Grip strength is typically decreased. Radiographs show avascular changes in the capitate with sclerosis and absorption of bone.

Treatment. Treatment methods are varied and include intercarpal fusion and partial resection of the capitate. Intercarpal fusion is effective in relieving pain but does result in significant loss of motion. Partial resection will preserve motion but may not be as effective in the long term in relieving pain [93]. However, the limited number of cases reported does not allow any definite conclusions about the appropriate management of these rare lesions.

Authors' Preferred Method of Treatment. We have not treated an athlete for this condition. Murakami and Nakajiwa [93] reviewed eight patients who were treated by partial resection of the capitate. This was combined with intercarpal arthrodesis in some patients. We agree with their recommendation that in athletes intercarpal arthrodesis should be avoided if full range of motion is required for athletic activity.

Criteria for Return to Sports Participation. Reports indicate that recovery in these cases may be prolonged. Return to sports must be judged individually based on reduction in pain and recovery of motion and strength. This appears to be particularly applicable to gymnasts [93].

Entrapment Neuropathies

Median Nerve Compression (Carpal Tunnel Syndrome)

Carpal tunnel syndrome is the most common entrapment neuropathy associated with sports activities. The carpal tunnel is a fibro-osseous canal bound volarly by the transverse carpal ligament; the floor and walls are formed by the carpal bones (Fig. 18A–27). The transverse carpal ligament attaches radially to the scaphoid and trapezium and ulnarly to the pisiform and hamate. The flexor pollicis longus, flexor digitorum profundus, and flexor digitorum superficialis tendons traverse the carpal tunnel along with the median nerve. Prior to entering the carpal tunnel, the median nerve gives off the palmar cutaneous branch, which passes superficial to the transverse carpal ligament and radial to the course of the median nerve (Fig. 18A–28). The motor branch to the thenar muscles can vary its anatomic pattern from the median nerve [73] (Fig. 18A–29). Usually the motor branch comes off distal to the transverse carpal ligament, but occasionally it can branch proximal to it and pass below the ligament or over the ligament. Usually the motor branch comes off on the radial side of the median nerve, but rarely it branches on the ulnar side and passes superficial to the transverse carpal ligament. Appreciation of these anatomic variants by the surgeon is critical to ensure protection of the motor branch during surgical exposure. The distal branches

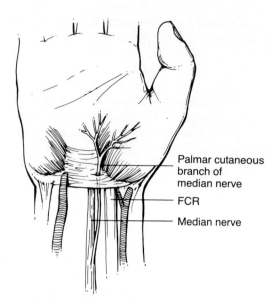

FIGURE 18A–28
Palmar cutaneous branch of the median nerve.

of the median nerve supply sensation to the thumb, index, and long fingers and to the radial aspect of the ring finger.

Any condition that reduces space in the carpal tunnel can cause compression of the median nerve. Usually compression is caused by flexor tenosynovitis. Any sports activity that involves repetitive flexion or grasping can provoke symptoms [74], which include numbness or tingling in the median distribution and aching pain radiating to the forearm, elbow, or shoulder. Pain or paresthesia often awakens the patient at night [27]. Symptoms can be triggered by position (wrist flexion) or repetitive grasping. Frequently, patients complain of clumsiness and loss of dexterity [51].

Clinical Evaluation. The physical examination should include sensibility, motor, and provocative testing. Two-point discrimination is abnormal (>5 mm) in more advanced cases but can be normal in mild cases. Vibratory sensation may be a more sensitive test for detection of early carpal tunnel syndrome [47]. Motor testing should include observing for thenar atrophy and testing abduction against resistance (abductor pollicis brevis). Provocative tests include those to elicit Phalen's sign and Tinel's sign. Electromyography (EMG) can be used to confirm the diagnosis by demonstrating prolonged motor or sensory latency across the wrist. EMG testing is not necessary for patients who give a typical history and have appropriate physical findings. It is more useful for differentiating carpal tunnel syndrome from cervical radiculopathy [92].

Treatment

Treatment is discussed in the following section on ulnar nerve entrapment.

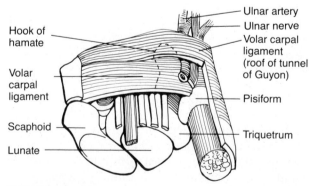

FIGURE 18A–27
The carpal tunnel is formed by the transverse carpal ligament volarly and the carpal bones on the floor and sides. Guyon's canal is formed by the volar carpal ligament (roof), the hamate (lateral wall), and the pisiform (medial wall).

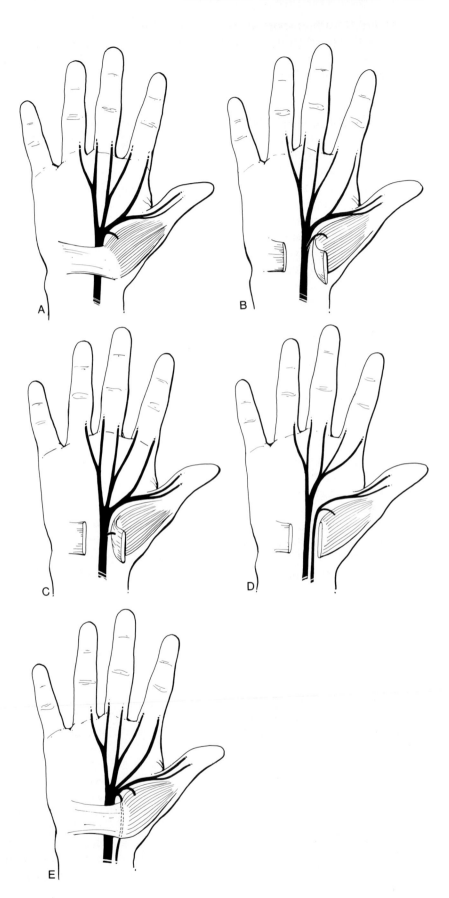

FIGURE 18A–29
Variations of the motor branch of the median nerve. *A,* The most common pattern is extraligamentous and recurrent followed by *(B)* subligamentous or *(C)* transligamentous. *D–E,* Proximal branching is rare.

Ulnar Nerve Compression (Guyon's Canal Syndrome)

Entrapment neuropathy of the ulnar nerve in Guyon's canal is much less common than carpal tunnel syndrome, but it is particularly apt to occur in cyclists and has been called "handle-bar palsy" [98]. The cause in this case is direct pressure, and the syndrome is characterized by paresthesia in the ulnar one and a half digits.

Guyon's canal is triangular and is formed by the volar carpal ligament (roof), the hamate (lateral wall), and the pisiform and pisohamate ligament (medial wall) [35] (see Fig. 18A–27). The ulnar nerve bifurcates within the canal into superficial and deep branches. The deep branch of the ulnar nerve, along with a branch of the ulnar artery, passes deep to the hypothenar muscles and is the motor branch to the interosseous muscles and adductor pollicis (Fig. 18A–30). The superficial branch of the ulnar nerve passes superficial to the hypothenar muscles and supplies sensation to the ring and little fingers [111].

Clinical Evaluation. Symptoms due to compression vary with the location of the lesion. Pressure proximal to the bifurcation of the ulnar nerve will result in both sensory and motor findings. Sensation is decreased in the ring and little fingers. Intrinsic muscle weakness or atrophy may be noted. Tinel's test may cause paresthesias into the little and ring fingers. Conversely, a lesion distal to the bifurcation of the nerve will result in findings limited to either motor or sensory changes depending on which branch is affected, the deep motor branch or the superficial sensory branch.

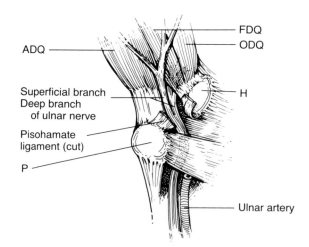

FIGURE 18A–30
Branches of the ulnar nerve. ADQ, abductor digiti quinti; FDQ, flexor digiti quinti; ODQ, opponens digiti quinti; H, hamate; P, pisiform.

Electromyography is useful in confirming the diagnosis. EMG will demonstrate denervation potentials in the interosseous muscles. Nerve conduction velocity studies will demonstrate prolongation of the motor latency to the first dorsal interosseous. A difference of over 1 msec is considered significant. These changes aid in localizing the lesion to Guyon's canal as opposed to a more proximal area of compression.

Compression of the ulnar nerve usually has a definite temporal onset related to trauma, which can include acute fractures of the hamate [52] or distal radius [62] as well as the chronic insult seen in bicycle riders. Other causes include anomalous muscles, lipomas, or ganglia in Guyon's canal [59]. Ulnar artery aneurysm or thrombosis can also result in pressure on the ulnar nerve.

Treatment. Conservative treatment consists of splinting, anti-inflammatory medication, and modification of activity. We use an off-the-shelf splint that maintains the wrist in a position of function in slight dorsiflexion. We recommend wearing the splint at night and as tolerated during the day. The splint may be removed as needed for essential activity. Local injection of a corticosteroid into the carpal tunnel or Guyon's canal may be effective. The appropriate technique of injection must be followed to avoid further injury to the nerve [130]. Injection of corticosteroids is frequently transient in effect, and 60% to 90% of patients experience recurrent symptoms [68]. Surgical treatment is indicated in patients who fail to respond appropriately to conservative treatment or when symptoms recur after an initial response.

Authors' Preferred Method of Treatment. Surgical release of the carpal tunnel and Guyon's canal must take into account the anatomic variations in the motor branch and the position of the palmar cutaneous branch and must provide for adequate release. Many approaches have been recommended. We use a slightly curved incision over the axis of the ring finger ray extending from the distal palmar crease to the volar wrist crease. The palmar fascia is exposed and partially excised to provide good exposure of the transverse carpal tunnel, which is then incised with care to protect the underlying median nerve. The antebrachial fascia proximal to the incision can be released subcutaneously with scissors to ensure that no residual compression remains. The motor branch of the nerve should be identified and avoided. This approach also allows easy access to the volar carpal ligament over Guyon's canal, permitting decompression of the ulnar nerve as well.

We do not routinely release Guyon's canal at the

time of carpal tunnel release. In patients with symptoms of ulnar neuropathy or changes in the ulnar nerve documented on EMG, we also release Guyon's canal, but these cases are rare. Epineurectomy is not routinely necessary but may be considered if there is evidence of significant epineural fibrosis. A strip of the transverse carpal ligament is routinely removed to prevent recurrent compression. The skin is closed, and a bulky hand dressing is applied to compress the palm.

Endoscopic carpal tunnel release offers the potential for reduced morbidity and an earlier return to activity. However, this technique requires much more skill from the surgeon, and the risk of neurovascular injury is much greater in the hands of an inexperienced surgeon. Long-term studies are needed to determine the efficacy of this procedure compared with an open carpal tunnel release. We have limited experience with this technique and do not use it on a routine basis.

The wrist is immobilized for 10 to 14 days, during which time the patient is encouraged to use the hand as tolerated and to flex and extend the fingers. The sutures are removed at 10 to 14 days, and the patient is started on a program comprising gentle range of motion of the wrist and massage of the incision to decrease scar formation.

Criteria for Return to Sports Participation. After removal of the sutures the athlete can return to sport-specific activities as tolerated. Pain in the palm will limit activities that depend on fine manipulation and gripping. A volar splint should be worn or taping of the hand and wrist should be done for athletes who return to sport activity before 8 weeks postoperatively.

Vascular Thrombosis

Thrombosis occurring in the ulnar artery in Guyon's canal [32] and in a persistent median artery in the carpal tunnel [76] will precipitate an acute neuropathy that frequently requires early surgical intervention. Blunt trauma to the palm causing this problem is usually associated with activity in which the hand is used as a hammer. Generally, one or more sensory branches of the ulnar nerve are affected; the median nerve is affected uncommonly.

Clinical Evaluation. Thrombosis of the ulnar artery can present as a tender mass in the hypothenar area. The patient may experience ischemic symptoms of pain and pallor as well as numbness due to compression of the involved nerve. The symptoms may mimic those of carpal tunnel syndrome or ulnar tunnel syndrome depending on the nerve involved.

These thromboses can result from blunt trauma sustained during athletic activity localized to the hypothenar area. Occasionally, such thrombosis occurs with a fracture of the hamate [25]. The diagnosis is suggested by the history. Allen's test will be abnormal with poor filling through the ulnar artery. Doppler ultrasound can help to confirm obliteration of the superficial palmar arch pulse when the radial artery is compressed [9].

Treatment. The most widely accepted treatment is surgical resection to remove the mass and to perform a sympathectomy [75]. Indirect methods include vasodilators [132], stellate ganglion block [30], and chemical clot lysis [28]. Some authors have recommended reanastomosis or vein grafting to reestablish blood flow [50]. However, numerous studies have demonstrated that simple resection is curative [53].

Authors' Preferred Method of Treatment. We prefer surgical resection of the thrombus and ligation of the involved artery. The approach is made through a curved incision parallel to the thenar crease over the fourth metacarpal. This provides access to Guyon's canal or the carpal tunnel as needed. The ulnar nerve is identified and carefully protected. The thrombosed section of the artery is resected, and the proximal and distal stumps of the artery are ligated with nonabsorbable suture. The tourniquet should be deflated to ensure adequate hemostasis and also adequate perfusion of the hand from the radial artery prior to closure. Ischemia of the hand at this point is the only indication for arterial reanastomosis or vein grafting. Postoperative care involves a standard bulky dressing for 2 weeks before the sutures are removed.

Criteria for Return to Sports Participation. After the sutures are removed, a standard rehabilitation program is begun emphasizing range of motion and strengthening. We restrict participation in vigorous activity or contact sports for 6 weeks. The athlete may then return to play if he demonstrates satisfactory motion and strength. The wrist is protected in a splint or by standard taping of the hand and wrist for 3 months after surgery.

Tendinitis

Tendinitis of the wrist is the most frequent problem in sports that requires medical attention. Symptoms often begin some time after an inciting event, which is usually overuse of some kind [84]. The athlete complains of localized pain that is made worse by activity. Findings include local swelling and pain that is intensified by movement of or tension on the affected tendon. X-rays are usually negative except in chronic cases, in which calcification rarely may be present in the soft tissues [129]. Calcification is much more common in the supraspinatus tendon of the shoulder.

Flexor Carpi Ulnaris–Flexor Carpi Radialis Tendinitis

Tendinitis of the two wrist flexors is relatively common [38]. It is caused by chronic repetitive trauma. Localized tenderness and swelling are present over the tendon. Pain is increased with passive dorsiflexion or resisted palmar flexion. Crepitus may be noted with movement.

Treatment. Treatment with splinting, anti-inflammatory agents, and temporary restriction of activity is usually successful. Unless the splint prohibits essential activity-related motion, most athletes can continue with their sport. The wrist is splinted in a neutral position or in slight dorsiflexion. At the end of the athletic season, restriction of the inciting activity usually results in resolution of persistent symptoms. In recalcitrant cases, surgery may be necessary to excise calcific deposits or to lyse adhesions. The flexor carpi ulnaris (FCU) may require excision of the pisiform in rare cases [99]. The flexor carpi radialis (FCR) may require decompression of its fibro-osseous tunnel.

Authors' Preferred Method of Treatment. We splint the wrist and, if necessary, restrict the activity causing stress on the tendons. Local heat and anti-inflammatory agents are utilized as indicated. In difficult cases we apply a short-arm cast for 2 weeks. Most cases respond to conservative treatment. In rare patients who have recurrent episodes we perform a tenolysis through a standard volar approach. Following surgery, the wrist is immobilized for 2 weeks. After suture removal a standard rehabilitation program is begun to allow the patient to regain range of motion and strength. The athlete can be expected to progress rapidly.

Criteria for Return to Sports Participation. The athlete can return to sports as soon as full range of motion and strength return, usually within 1 month after surgery. In individual cases, we allow earlier return to sports if the athlete wears a protective splint. Splinting should not be necessary after full motion and normal strength are attained.

Subluxation of the Extensor Carpi Ulnaris

Subluxation of the extensor carpi ulnaris (ECU) has been noted in athletes [23]. A painful snap over the dorsoulnar aspect of the wrist occurs with pronation and supination. This injury results when the ulnar septum of the sixth dorsal compartment is ruptured allowing the extensor carpi ulnaris tendon to subluxate in supination and to reduce in pronation.

Treatment. Initial treatment of an acute episode requires immobilization in a long-arm cast with the wrist in pronation and slight dorsiflexion to reduce the tendon [129]. In acute cases the wrist should be immobilized for 6 weeks. Chronic cases may respond to standard taping of the wrist and hand. If this is unsuccessful, the fibro-osseous tunnel can be reconstructed using a flap of the extensor retinaculum [112].

Authors' Preferred Method of Treatment. In acute cases (less than 2 weeks since the initial injury), the injury is treated with a long-arm cast, with the wrist in full pronation and slight dorsiflexion. The cast is worn for 6 weeks. Following removal of the cast, a standard rehabilitation program is begun to regain range of motion and strength.

In chronic cases and when treatment of acute cases has failed, we recommend reconstruction of the fibro-osseous tunnel as described by Spinner and Kaplan [112]. The distal ulna is approached by a dorsal longitudinal incision. The dorsal sensory branch of the ulnar nerve should be identified and protected. The extensor retinaculum is exposed, and a flap one-half inch wide is elevated from Lister's tubercle and based adjacent to the extensor digiti minimi (Fig. 18A–31). The vertical septum forming the dorsal compartment of the extensor carpi ulnaris

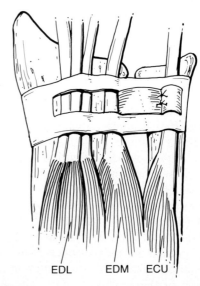

EDL   EDM   ECU

FIGURE 18A–31
Extensor carpi ulnaris (ECU) stabilized with a flap of retinaculum. EDM, extensor digiti minimi; EDC, extensor digitorum communis. (Redrawn from Spinner, M., and Kaplan, E. B. Extensor carpi ulnaris. *Clin Orthop* 68:124–129, 1970.)

is released radially and medially. The flap is then passed beneath the tendon from the radial side and back over the tendon with the synovial side of the flap facing the tendon. The flap is sutured to itself. The tendon is centralized over the ulna. The wrist is immobilized for 3 weeks in a short-arm cast, after which a splint is worn for 3 weeks. The athlete follows the standard rehabilitation program to regain motion and strength.

Criteria for Return to Sports Participation. Whether treatment is surgical or nonsurgical, we allow the athlete to resume active sports when range of motion and strength reach approximately 70% of the levels of the opposite extremity. Protective splinting or a semirigid orthosis is utilized for 8 weeks or until the athlete attains normal motion and strength.

De Quervain's Disease

De Quervain's disease is a tenosynovitis of the abductor pollicis longus and extensor pollicis brevis at the first dorsal compartment [24]. Sports activities that require repetitive ulnar deviation place an athlete at risk for this syndrome.

Clinical Evaluation. Findings include local swelling and tenderness at the radial styloid. Positive results on the Finkelstein test [40], in which ulnar deviation of the wrist with the thumb fully adducted causes marked pain, is considered pathognomonic (Fig. 18A–32). Differential diagnosis includes osteoarthritis of the first carpometacarpal joint and intersection syndrome.

Treatment. Treatment consists of splinting, anti-

inflammatory medication, and avoidance of the inciting activity. Local corticosteroid injection at the first dorsal compartment can aid in resolution of the symptoms. In patients in whom inflammation has resulted in thickening and stenosis of the fibro-osseous tunnel, conservative measures may fail.

Resistant cases of de Quervain's disease may require surgical release of the first dorsal compartment. Adequate decompression requires complete longitudinal release on the dorsal aspect. Frequently, there are multiple slips of the abductor pollicis longus, and the extensor pollicis brevis may be in a separate tunnel. Care should be taken that all slips of both tendons are released [85]. Particular attention should be paid to avoid trauma to the sensory branches of the radial nerve either by cutting or by applying undue traction to prevent formation of a painful neuroma. These neuromas can be particularly resistant to treatment and have a poor prognosis.

Authors' Preferred Method of Treatment. Initial treatment consists of splinting or taping to restrict ulnar deviation of the wrist. An oral anti-inflammatory medication is used for 2 weeks. If the symptoms persist, a local corticosteroid injection into the first dorsal compartment is utilized. The patient should be cautioned that local depigmentation may result from the injection.

Cases that do not respond satisfactorily after 6 weeks of conservative treatment are considered for surgical release of the first dorsal compartment. We perform this procedure through a transverse incision, usually under local anesthesia. Care is taken to release all slips of the abductor pollicis longus and extensor pollicis brevis. The sensory branches of the radial nerve are carefully protected during the procedure. Postoperatively, the hand and wrist are placed in a bulky compressive dressing for 1 week. The thumb is left free, and the patient is encouraged to begin range of motion exercises immediately. Sutures are removed at 10 to 14 days. Range of motion and strengthening exercises are begun.

Criteria for Return to Sports Participation. These patients usually regain strength and range of motion quickly. Taping or splinting can be utilized as needed for comfort. Athletes can return to their sport as soon as discomfort allows.

Intersection Syndrome

The intersection syndrome is characterized by pain in the area where the abductor pollicis longus and extensor pollicis brevis cross over the underlying

FIGURE 18A–32
Finkelstein test.

wrist extensors. This syndrome is common in oarsmen, canoeists, and weight lifters secondary to the repetitive wrist activity required.

Clinical Evaluation. Clinically, there is swelling and tenderness over the dorsoradial aspect of the forearm about 6 cm proximal to Lister's tubercle [38]. Crepitus is the distinguishing finding of this syndrome and may result in an audible squeak.

Treatment. Initial treatment includes splinting in neutral or slight dorsiflexion, anti-inflammatory medication, local heat, and restriction of activity. If symptoms persist, a local injection of a corticosteroid may be helpful. Usually, conservative treatment is successful. In recalcitrant cases surgical exploration for lysis of adhesions could be considered, but this has not been necessary in our experience.

Protective Splinting

The wrist is exposed in almost all sporting activities and is highly susceptible to injury. Such injuries occur even in sports such as lacrosse and hockey in which padded gloves are worn. The major objective in treating injuries to the wrist is to return the athlete to competition as early as possible while preventing permanent disability. Initially, an athlete's injury must be evaluated by a physician or trainer who is qualified to diagnose and treat the injury. A decision must be made about whether the athlete's injury can be safely protected. Through proper protective splinting it is possible to allow earlier return to participation in sports while minimizing the risk of recurrent injury. It is the responsibility of the therapist or trainer to fabricate a splint that allows adequate protection from reinjury combined with minimal interference with the functional skills required of each athlete.

Prophylactic taping of a joint has most commonly been done for the ankle. Taping can likewise be used for the wrist in cases of mild injury when only slight restriction of motion is required. Wrist taping does not completely restrict motion but does give support to the wrist for comfort and provides some protection from further injury. A tape adherent is utilized. An underwrap is optional. The methods of taping vary but usually consist of a figure-of-eight taping technique. The wrist is usually positioned in neutral while it is being taped. However, the wrist can be positioned in hyperflexion, hyperextension, radial deviation, or ulnar deviation depending on which plane of motion is to be restricted. In general, the wrist is positioned to relax the involved tendons. Radial deviation for de Quervain's disease relaxes the affected extensor pollicis brevis and abductor pollicis longus. In patients with intersection syndrome the wrist is positioned in hyperextension and radial deviation for taping to relax the wrist extensors and the abductor pollicis longus and extensor pollicis brevis. With subluxation of the extensor carpi ulnaris the wrist is positioned in ulnar deviation.

Commercial wrist splints that incorporate the distal forearm and wrist can also be utilized for support (Fig. 18A–33). Molded orthoplast splints with Velcro fasteners can also be effective as both functional and prophylactic braces in the athletic population (Fig. 18A–34). These splints are easily removed and provide about the same support as taping.

If more effective splinting is required for protection of the wrist, a semirigid material such as silicone or one of the newer synthetic fiberglass materials provide a safer means of protection. Regulations regarding acceptable protective materials may vary between states as well as between high schools. College and professional teams may wear a hard cast of fiberglass covered by a ½-inch slow rebound rubber. Fiberglass casts are desirable in that the same cast may be used for competition and for daily use. A silicone cast must be alternated with a bivalved fiberglass cast because silicone casts are nonporous and can only be worn for several hours at a time. If the silicone cast is worn for longer

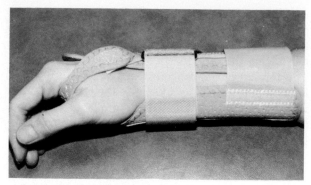

FIGURE 18A–33
Commercial wrist splint.

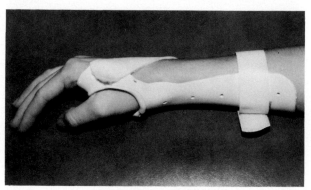

FIGURE 18A–34
Orthoplast wrist splint.

FIGURE 18A–35
Silicone cast materials.

periods, skin maceration may occur. The silicone cast may be worn for competition in which the hard cast is prohibited. A bivalved fiberglass cast is then applied for periods other than athletic competition.

Silicone cast materials include (Fig. 18A–35) 1 lb RTV silicone with catalysts (General Electric, Silicone Products Department, Waterford, NY), 1-inch adhesive tape, skin lubricant, 3-inch Kling (Johnson & Johnson, New Brunswick, NJ; approximately three rolls), 4 × 4 inch gauze pads, roller gauze, and plastic sheeting. Preparation directions are as follows:

Mix silicone and catalyst.
Tape area to be protected.
Coat affected area with lubricant to prevent silicone from adhering to skin and tape.
Apply an even layer of silicone to the area.
Wrap Kling over silicone.
Soak 4 × 4 inch gauze pads in silicone and apply to areas that require protection.
Alternate Kling and silicone until all silicone is used.
Wrap splint in plastic sheeting and secure with roller gauze until silicone is set.
When silicone has adequately hardened (3 to 4 hours), remove with bandage scissors, cutting along ulnar area.
Trim splint and check for fit.
Remove inside tape [48].

Scotchrap is a semirigid support wrap that was recently introduced by 3M (Fig. 18A–36). This material is a silicone product, but it is very similar to fiberglass casting material. It is applied over stockinette and cast padding after a brief submersion in water. The product appears very similar to a synthetic fiberglass cast, but it is semirigid in nature and meets the requirements of regulations regarding acceptable protective materials.

It should be emphasized that regulations for acceptable protective materials vary between high schools and colleges and between school districts. Athletic trainers and attending physicians should familiarize themselves with the local regulations governing these protective materials before using any of the specific products mentioned here.

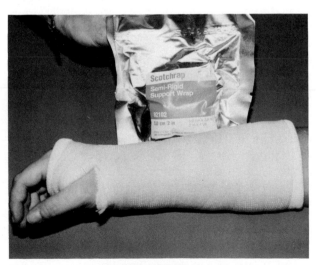

FIGURE 18A–36
Semirigid support wrap.

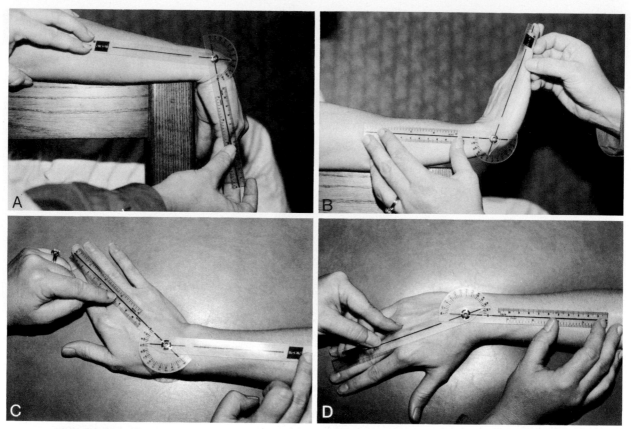

FIGURE 18A–37
Range of motion measurements. *A,* Palmar flexion. *B,* Dorsiflexion. *C,* Ulnar deviation. *D,* Radial deviation.

Rehabilitation

All athletic injuries should be evaluated by a physician, therapist, or athletic trainer before rehabilitation is started. The athlete's status should be documented before the rehabilitation program is started to provide an objective method of determining progress and to assist in the evaluation of the treatment program. Range of motion should be assessed frequently to monitor progress as well as to determine the need for changing the treatment plan (Fig. 18A–37). Manual muscle testing and sensation testing should be performed [48]. Edema should be assessed by recording circumferential measurements at the wrist. Early mobilization of the injured wrist should be initiated as pain subsides. Active motion is the only modality of hand therapy that has been shown to be of lasting benefit [88]. Active motion should be within pain tolerance.

The use of modalities such as heat, cold, electrical stimulation, and ultrasound may also aid in improving function. Cold is initially used for 12 to 14 hours to reduce the effects of swelling. It may be used in conjunction with active exercises to improve range of motion. Cold may also be used after exercise for 30 minutes to an hour to reduce the resultant edema. Following the postacute phase, warm whirlpool baths (102°F) can be used. The whirlpool aids in early wrist motion and is used while the athlete performs active exercise. Duration of the treatment is about 15 minutes. Care must be taken to avoid allowing the wrist to assume a dependent position in the whirlpool, which may induce edema to increase. To prevent this, the elbow should be flexed as close to 90 degrees as possible, and the individual should be encouraged to exercise actively. If edema occurs, the temperature of the water may need to be reduced [48].

Prevention and control of edema are important in the rehabilitation of wrist injuries. Failure to control edema may lead to stiff and painful joints. Methods of controlling edema include:

1. Elevation of the extremity above heart level.
2. Active exercises.
3. Retrograde massage.
4. ICE.
5. Elasticized tape (Coban, 3M Company). This is available in various widths.

6. Compression gloves, which may be off-the-shelf or custom-fitted (Jobst, Toledo, OH).

7. Air splints or Jobst intermittent air compression splints.

8. Static splints [48].

Passive motion is another adjunct used to regain motion. The benefits of passive exercises in the wrist are probably transient. Overzealous passive exercises may cause increased pain and may be detrimental [66]. Acceptable methods of passive motion include gentle active-assistive exercises, slow deliberate stretching, and joint mobilization techniques after application of heat or cold. Ice should be used until motion is free. Heat should not be used until progress with the use of ice reaches a plateau and there is no increased pain or edema [48].

Grip strength depends on wrist mobility, stability, and a combination of many muscle groups. Grip is divided into two types: power grip and precision grip [94]. Power grip requires the wrist to be dorsiflexed, thereby providing a mechanical advantage for the long digital flexors. Carpal pathology may produce pain that considerably limits functional grip

strength. Precision grip is performed with the wrist in any position of volar flexion and dorsiflexion. The thumb is opposed to the fingers, and the intrinsic muscles provide most of the finger movement.

A dynamometer can be utilized to measure grip strength accurately (Fig. 18A–38). Grip strength varies according to the size of the object that is grasped and should be evaluated and recorded in five grip spans. Grip strengthening programs should be modified according to the needs of the individual athlete. Strengthening should occur within pain-free limits. Initially, it may be necessary to begin with isometric gripping exercises, gradually increasing these to gentle resistive exercises.

A comprehensive home program is vital to the success of any rehabilitation protocol. Exercises are best done frequently throughout the day with a small number of repetitions. Oral, written, and visual instructions can be utilized. Exercise aids, splints, and modalities can be utilized as well. Home programs should be reviewed intermittently and altered as needed.

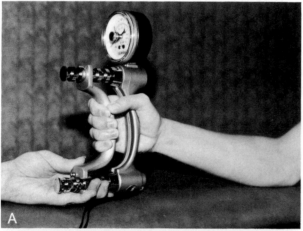

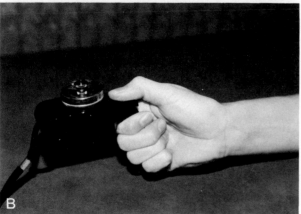

FIGURE 18A–38
Dynamometer to measure grip strength.

References

1. Adler, J. B., and Shaftan, G. W. Fractures of the capitate. *J Bone Joint Surg* 44A:1537–1547, 1962.
2. Alexander, C. E., and Lichtman, D. M. Ulnar carpal instabilities. *Orthop Clin North Am* 15:307–320, 1984.
3. Alho, A., and Kanhaanjaa, U. Management of fractured scaphoid bones. A prospective study of 100 fractures. *Acta Orthop Scand* 46:737–743, 1975.
4. Almquist, E. E., and Burns, J. F. Radial shortening for the treatment of Kienbock's disease—a 5- to 10- year follow-up. *J Hand Surg* 7:348–352, 1982.
5. Antuna Zapico, J. M. Malacia del semilunar. Thesis, Universidad de Valladolid. Valladolid (Spain), Industrias y Editorial Sever Cuesta, 1966.
6. Armstead, R. B., Linscheid, R. L., Dobyns, J. H., et al. Ulnar lengthening in the treatment of Kienbock's disease. *J Bone Joint Surg* 64A:170–178, 1982.
7. Atkinson, R. E., Smith, R. J., and Jupiter, J. B. Silicone synovitis of the wrist. Presented at the 40th Annual Meeting of the American Society for Surgery of the Hand, January, 1985.
8. Axelsson, R. Behandling av lunatomalaci. Gotenborg (Sweden), Elanders Boktrycker, Aktieboalag, 1971.
9. Barker, N. W., and Fines, F. H., Jr. Arterial occlusion in the hands and fingers associated with repeated occupational trauma. *Mayo Clin Proc* 19:345–349, 1944.
10. Bartone, N. F., and Grieco, R. V. Fracture of the triquetrum. *J Bone Joint Surg* 38A:353–356, 1956.
11. Beckenbaugh, R. D., Shives, T. C., Dobyns, J. H., and Linscheid, R. L. Kienbock's disease: The natural history of Kienbock's disease and considerations of lunate fractures. *Clin Orthop* 149:98, 1980.
12. Bellinghausen, H. W., Weeks, P. M., Young, L. V., et al. Roentgen rounds No. 62. *Orthop Rev* 11:73, 1982.
13. Blaine, E. S. Lunate osteomalacia. *JAMA* 96:492, 1931.
14. Blair, W. F., Kilpatrick, W. C., and Over, G. E. Open fracture of the hook of the hamate: A case report. *Clin Orthop* 163:180–184, 1982.
15. Blatt, G. Capsulodesis in reconstructive hand surgery: Dorsal capsulodesis for the unstable scaphoid and volar capsu-

lodesis following excision of the distal ulna. *Hand Clinics* 3:81–102, 1987.

16. Bora, F. W., Jr., Osterman, A. L., Woodbury, D. F., et al. Treatment of non-union of the scaphoid by direct current. *Orthop Clin North Am* 15:107–112, 1984.
17. Borgeskov, S., Christiansen, B., Kjaer, A., et al. Fractures of the carpal bones. *Acta Orthop Scand* 37:276, 1966.
18. Bowen, T. L. Injuries of the hamate bone. *Hand* 5:235–238, 1973.
19. Boyes, J. H. *Bunnell's Surgery of the Hand* (5th ed.). Philadelphia, J. B. Lippincott, 1970.
20. Brainard, C. W. *J Bone Joint Surg* 20:486, 1958.
21. Braun, R. M. Viable pedicle bone grafts. *What's New and What's True Orthopaedic Symposium* (abstract). University of California at Davis, March, 1985, pp. 28–29.
22. Brown, D. E., and Lichtman, D. M. The evaluation of chronic wrist pain. *Orthop Clin North Am* 15:183–192, 1984.
23. Burkhart, S. S., Wood, M. B., and Linscheid, R. L. Posttraumatic recurrent subluxation of the extensor carpi ulnaris tendon. *J Hand Surg* 7:1, 1982.
24. Burman, M. Stenosing tendovaginitis of the dorsal and volar compartments of the wrist. *Arch Surg* 65:752, 1952.
25. Butsch, J. L., and Janes, J. M. Injuries of the superficial palmar arch. *J Trauma* 3:505–516, 1963.
26. Chacha, P. B. Vascularized pedicular bone grafts. *Int Orthop* 8:117–138, 1984.
27. Cherington, M. Proximal pain in the carpal tunnel syndrome. *Arch Surg* 108:69, 1974.
28. Cho, K. O. Entrapment occlusion of the ulnar artery in the hand. *J Bone Joint Surg* 60A:841–843, 1978.
29. Chuinard, R. G., and Zeman, S. C. Kienbock's disease: An analysis and rationale for treatment by capitate-hamate fusion. *Orthop Trans* 4:18, 1980.
30. Conn, J., Jr., Bergan, J. J., and Bell, J. L. Hypothenar hammer syndrome: Post traumatic digital ischemia. *Surgery* 68:1122–1128, 1970.
31. Cooney, W. P., Dobyns, J. H., and Linscheid, R. L. Nonunion of the scaphoid. Analysis of the results from bone grafting. *J Hand Surg* 8:343–354, 1980.
32. Costigan, D. G., Riley, J. M., Jr., and Coy, F. E., Jr. Thrombofibrosis of the ulnar artery in the palm. *J Bone Joint Surg* 41A:702–704, 1959.
33. Crabbe, W. A. Excision of the proximal row of the carpus. *J Bone Joint Surg* 46B:708–711, 1964.
34. Crawford, G. P., and Taleisnik, J. Rotary subluxation of the scaphoid after excision of dorsal carpal ganglion and wrist manipulation—A case report. *J Hand Surg* 8:921–925, 1983.
35. Denman, E. E. The anatomy of the space of Guyon. *Hand* 10:69–76, 1978.
36. Dickinson, J. C., and Shannon, J. G. Fractures of the carpal scaphoid in the Canadian Army. *Surg Gynecol Obstet* 79:225–239, 1944.
37. Dobyns, J. H., Linscheid, R. L., Chao, E. Y. S., Weber, E. R., and Swanson, G. E. Traumatic instability of the wrist. *Instr Course Lect* 24:182–199, 1975.
38. Dobyns, J. H., Sim, F. H., and Linscheid, R. L. Sports stress syndromes of the hand and wrist. *Am J Sports Med* 6:236, 1978.
39. Eddeland, A., Eiken, O., Hellgren, E., et al. Fractures of the scaphoid. *Scand J Plast Reconstr Surg* 9:234–239, 1975.
40. Finkelstein, H. Stenosing tendovaginitis at the radial styloid process. *J Bone Joint Surg* 12A:509, 1930.
41. Gelberman, R. H., Bauman, T. D., and Menon, J. The vascularity of the lunate bone and Kienbock's disease. *J Hand Surg* 5:272–278, 1980.
42. Gelberman, R. H., and Menon, J. The vascularity of the scaphoid bone. *J Bone Surg* 5:508–513, 1980.
43. Gelberman, R. H., Salamon, P. B., Jurist, J. M., and Posch, J. L. Ulnar variance in Kienbock's disease. *J Bone Joint Surg* 57A:674–676, 1975.

44. Gelberman, R. H., and Gross, M. S. The vascularity of the wrist. *Clin Orthop* 202:43, 1986.
45. Gelberman, R. H., Panagis, J. S., Taleisnik, J., and Baumgartner, M. The arterial anatomy of the human carpus. Part I: The extraosseous vascularity. *J Hand Surg* 8:367, 1983.
46. Gellman, H., Caputo, R. J., Carter, V., Aboulafia, A., and McKay, M. A comparison of short and long thumb-spica casts for nondisplaced fractures of the carpal scaphoid. *J Bone Joint Surg* 71A:354–357, 1989.
47. Gerstner, A. L., and Omer, G. E., Jr. Peripheral entrapment neuropathies in the upper extremity. *J Musculoskel Med* 14–29, 1988.
48. Gieck, J. H., and Mayer, V. Protective splinting for the hand and wrist. *Clin Sports Med* 5(4):795, 1986.
49. Gilford, W., Baltar, R., and Lambrinudi, C. The mechanics of the wrist joint. *Guy's Hosp Rep* 92:52–59, 1943.
50. Given, K. S., Puckett, C. L., and Kleinert, H. E. Ulnar artery thrombosis. *Plast Reconstr Surg* 61:405–411, 1978.
51. Goldner, J. L. Median nerve compression lesions: Anatomic and clinical analysis. *Bull Hosp Joint Dis* 44:199–223, 1984.
52. Gore, D. R. Carpometacarpal dislocation producing compression of the deep branch of the ulnar nerve. *J Bone Joint Surg* 53A:1387–1390, 1971.
53. Goren, M. L. Palmar intramural thrombosis in the ulnar artery. *Calif Med* 89:424–425, 1958.
54. Graner, O., Lopes, E. I., Carvalho, B. C., et al. Arthrodesis of the carpal bones in the treatment of Kienbock's disease, painful ununited fractures of the navicular and lunate bones with avascular necrosis, and old fracture-dislocations of carpal bones. *J Bone Joint Surg* 48:767–774, 1966.
55. Green, D. P. *Operative Hand Surgery* (2nd ed.). New York, Churchill Livingstone, 1988.
56. Green, D. P. Proximal row carpectomy. *Hand Clinics* 3:163–168, 1987.
57. Green, D. P. The sore wrist without a fracture. *Instr Course Lect* 34:300–313, 1985.
58. Grundy, M. Fractures of the carpal scaphoid in children. *Br J Surg* 56:523–524, 1969.
59. Hayes, C. W. Ulnar tunnel syndrome from giant cell tumor of tendon sheath: A case report. *J Hand Surg* 3A:187–188, 1978.
60. Herbert, T. J., and Fisher, W. E. Management of the fractured scaphoid using a new bone screw. *J Bone Joint Surg* 66B:114–123, 1984.
61. Hoshi, S., et al. *Orthop Surg (Jpn)* 21:860, 1970.
62. Howard, F. M. Ulnar nerve palsy in wrist fractures. *J Bone Joint Surg* 43A:1197–1201, 1961.
63. Hudson, R. M., Caragol, W. J., and Faye, J. J. Isolated rotary subluxation of the carpal navicular. *Am J Roentgenol* 126:601, 1976.
64. Hulten, O. Über anatomische Variatonen de Handgelenk-knochen. *Acta Radiol Scand* 9:155–168, 1928.
65. Hunter, J., Schneider, J., Mackin, F., and Bell, J. (Eds.). *Rehabilitation of the Hand.* St. Louis, C. V. Mosby, 1978.
66. Hunter, J., Schneider, J., Mackin, F., and Bell, J. (Eds.). *Rehabilitation of the Hand* (2nd ed.). St. Louis, C. V. Mosby, 1984.
67. Johnson, R. P. The acutely injured wrist and its residuals. *Clin Orthop* 149:33–44, 1980.
68. Jones, K. G. Carpal tunnel syndrome. *J Arkansas Med Soc* 75:58, 1978.
69. Jonsson, C. *Acta Radiol Scand* 23:562, 1942.
70. Kienbock, R. Über traumatishe Malazie des Mondbeins, und ihre Folgezustaude: Eufartungsformen und Kompressions Frakturen. *Fortschr Roengenstr* 16:77, 1910.
71. King, R. J. Scapholunate diastasis associated with a Barton fracture treated by manipulation, or Terry-Thomas and the wine waiter. *J R Soc Med* 76:421–423, 1983.
72. King, R. J., Machenna, R. P., and Elnur, S. Suggested method for closed treatment of fractures of the carpoid

scaphoid: Hypothesis supported by dissection and clinical practice. *J R Soc Med* 75:860–867, 1982.

73. Lanz, U. Anatomical variations of the median nerve in the carpal tunnel. *J Hand Surg* 2:44–53, 1977.

74. Layfer, L. F., and Jones, J. V. Hand paresthesias after racquetball. *Ill Med J* 152:190, 1977.

75. Leriche, R., Fontaine, R., and Dupertuis, S. M. Arterectomy with follow-up studies on 78 operations. *Surg Gynecol Obstet* 64:149–155, 1937.

76. Levy, M., and Pauker, M. Carpal tunnel syndrome due to thrombosed persisting median artery. A case report. *Hand* 10:65–68, 1978.

77. Lewis, O. J., Hamshere, R. J., and Bucknill, T. M. The anatomy of the wrist joint. *J Anat* 106:539, 1970.

78. Lichtman, D. M. *The Wrist and its Disorders.* Philadelphia, W. B. Saunders, 1988.

79. Lichtman, D. M., Alexander, A. H., Mack, G. R., and Gunther, S. F. Kienbock's disease—Update on silicone replacement arthroplasty. *J Hand Surg* 7:343–347, 1982.

80. Lichtman, D. M., Mack, G. R., MacDonald, R. I., et al. Kienbock's disease: The role of silicone replacement arthroplasty. *J Bone Joint Surg* 59A:899–908, 1977.

81. Lichtman, D. M., Schneider, J. R., Mack, G. R., and Swafford, A. R. Ulnar midcarpal instability. *J Hand Surg* 6:515–523, 1981.

82. Linscheid, R. L. Scapholunate ligamentous instabilities (dissociations, subdislocations, dislocations). *Ann Chir Main* 3:323–330, 1984.

83. Linscheid, R. L., Dobyns, J. H., and Younge, D. K. Trispiral tomography in the evaluation of wrist injury. *Bull Hosp J Dis Orthop Inst* 44:297–308, 1984.

84. Lipscomb, P. R. Chronic nonspecific tenosynovitis and peritendinitis. *Surg Clin North Am* 24:780, 1944.

85. Loomis, K. L. Variations of stenosing tenosynovitis at the radial styloid process. *J Bone Joint Surg* 33A:340, 1951.

86. Marck, K. W., and Klasen, H. J. Fracture-dislocation of the hamato-metacarpal joint: A case report. *J Hand Surg* 11A:128–130, 1986.

87. Marek, R. M. Avascular necrosis of the carpal lunate. *Clin Orthop* 10:96–107, 1957.

88. Mayer, V., and Gieck, J. H. Rehabilitation of hand injuries in athletes. *Clin Sports Med* 5(4):783–793, 1986.

89. Mayfield, J. K. Mechanism of carpal injuries. *Clin Orthop* 149:45–54, 1980.

90. Mayfield, J. K., Johnson, R. P., and Kilcoyne, R. F. The ligaments of the human wrist and their functional significance. *Anat Rec* 186:417–428, 1976.

91. McMurty, R. Y., Youm, Y., Flatt, A. E., et al. Kinematics of the wrist. II. Clinical applications. *J Bone Joint Surg* 60A:955–961, 1978.

92. Melvin, J. L., Schuckmann, J. A., and Lanese, R. R. Diagnostic specificity of motor and sensory nerve conduction variables in the carpal tunnel syndrome. *Arch Phys Med Rehabil* 54:69, 1973.

93. Murakami, S., and Nakajima, H. Aseptic necrosis of the capitate bone. *Am J Sports Med* 12(2):170–173, 1984.

94. Napier. The prehensile movements of the human hand. *J Bone Joint Surg* 38B:902, 1956.

95. Navarro, A. Luxaciones del carpo. *An Fac Med Montevideo* 6:113, 1921.

96. Nielsen, P. T., and Hedeboe, J. Posttraumatic scapholunate dissociation detected by wrist cineradiography. *J Hand Surg* 9A:135–138, 1984.

97. O'Brien, E. T. Acute fractures and dislocations of the carpus. *Orthop Clin North Am* 15:237–258, 1984.

98. Palmer, A. K., Levinsohn, E. M., and Kuzma, G. R. Arthrography of the wrist. *J Hand Surg* 8:15–23, 1983.

99. Palmieri, T. J. Pisiform area pain treatment by pisiform excision. *J Hand Surg* 7:477, 1982.

100. Panagis, J. S., Gelberman, R. H., Taleisnik, J., and Baumgartner, M. The arterial anatomy of the human carpus. Part II: The interosseous vascularity. *J Hand Surg* 8:35, 1983.

101. Parker, R. D., Berkowitz, M. S., Brahms, M. A., and Bohl, W. R. Hook of the hamate fractures in athletes. *Am J Sports Med* 14:517–523, 1986.

102. Perrson, M. Casual treatment of lunatomalacia. *Acta Chir Scand* 100:531–544, 1950.

103. Peste. Discussion. *Bull Soc Anat* 18:169–170, 1843.

103a. Peterson H. A., and Lipscomb P. R. Intercarpal arthrodesis. *Arch Surg* 95:127, 1967.

104. Polivy, K. D., Millender, L. H., Newberg, A., et al. Fractures of the hook of the hamate: A failure of clinical diagnosis. *J Hand Surg* 10A:101–104, 1985.

105. Rand, J., Linscheid, R. L., and Dobyns, J. H. Capitate fractures. A long term follow-up. *Clin Orthop* 165:209–216, 1982.

106. Reagan, D. S., Linscheid, R. L., and Dobyns, J. H. Lunotriquetral sprains. *J Hand Surg* 9A:502–514, 1984.

107. Reister, J. N., Baker, B. E., Mosher, J. F., and Lowe, D. A review of scaphoid fracture healing in competitive athletes. *Am J Sports Med* 13:159–161, 1985.

108. Rockwood, C. A., and Green, D. P. *Fractures.* Vol. 1. Philadelphia, J. B. Lippincott, 1975, p. 421.

109. Ruby, L. K. Common hand injuries in the athlete. *Orthop Clin North Am* 11:819, 1980.

110. Russe, O. Fracture of the carpal navicular. *J Bone Joint Surg* 42A:759–768, 1960.

111. Shea, J. D., and McClain, E. J. Ulnar nerve compression syndromes at and below the wrist. *J Bone Joint Surg* 51A:1095–1103, 1969.

112. Spinner, M., and Kaplan, E. B. Extensor carpi ulnaris. Its relationship to stability of the distal radio-ulnar joint. *Clin Orthop* 68:124–129, 1970.

113. Stahl, F. On lunatomalacia (Kienbock's disease), a clinical and roentgenological study, especially on its pathogenesis and the late results of immobilization treatment. *Acta Chir Scand* [Suppl.] 126:1–133, 1947.

114. Stark, H. H., Jobe, F. W., Boyes, J. H., et al. Fracture of the hook of the hamate in athletes. *J Bone Joint Surg* 59A:575–582, 1977.

115. Sutro, C. J. Bilateral recurrent intercarpal subluxation. *Am J Surg* 72:110–113, 1946.

116. Taleisnik, J. The ligaments of the wrist. *J Hand Surg* 1:110–118, 1976.

117. Taleisnik, J. Scapholunate dissociation. *In* Strickland, J. W., and Steichen, J. B. (Eds.), *Difficult Problems in Hand Surgery.* St. Louis, C. V. Mosby, 1982.

118. Taleisnik, J. *The Wrist.* New York, Churchill Livingstone, 1985, p. 174.

119. Taleisnik, J. Wrist: Anatomy, function, and injury. *Instr Course Lect* 27:61–87, 1978.

120. Taleisnik, J., and Kelley, P. J. The extraosseous and intraosseous blood supply of the scaphoid bone. *J Bone Joint Surg* 48(6)A:1125–1137, 1966.

121. Viernstein, K., and Weigert, M. Die Radiusverkurzungsosteotomie bei der Lunatummalazie. *Münch Med Wochenschr* 109:1992, 1967.

122. Watson, H. K. Limited wrist arthrodesis. *Clin Orthop* 149:126–136, 1980.

123. Watson, H. K., and Ballet, F. L. The SLAC wrist: Scapholunate advanced collapse pattern of degenerative arthritis. *J Hand Surg* 9A:358–365, 1984.

124. Watson, H. K., and Hempton, R. F. Limited wrist arthrodesis I: The triscaphoid joint. *J Hand Surg* 5:320–327, 1980.

125. Watson, H. K., Ryu, J., and DiBella, A. An approach to Kienbock's disease: Triscaphe arthrodesis. *J Hand Surg* 10A:179–187, 1985.

126. Weber, E. R. Biomechanical implications of scaphoid wrist fractures. *Clin Orthop* 149:83–89, 1980.

127. Weber, E. R. Concepts governing the rotational shift of the intercalated segment of the carpus. *Orthop Clin North Am* 15:193–207, 1984.

128. Weber, E. R., and Chao, E. Y. An experimental approach

to the mechanism of scaphoid wrist fractures. *J Hand Surg* 3:142–148, 1978.

129. Wood, M. B., and Dobyns, J. H. Sports-related extra-articular wrist syndromes. *Clin Orthop* 202:93–102, 1986.

130. Wood, M. R. Hydrocortisone injections for carpal tunnel syndrome. *Hand* 12:62–64, 1980.

131. Yorun, Y., McMurtry, R. Y., Flatt, A. E., and Gillespie, T. E. Kinematics of the wrist. I. An experimental study of radioulnar deviation and flexion-extension. *J Bone Joint Surg* 60A:423–431, 1978.

132. Zemel, N. P., and Stark, H. H. Fractures and dislocations of the carpal bones. *Clin Sports Med* 5:709–724, 1986.

133. Zweig, J., Lie, K. K., Posch, J. L., and Larsen, R. D. Thrombosis of the ulnar artery following blunt trauma to the hand. *J Bone Joint Surg* 51A:1191–1198, 1969.

The Hand

David P. Green, M.D.
James W. Strickland, M.D.

There is a great tendency to minimize the importance of hand injuries in athletes. When compared with the obvious disability resulting from ligamentous injuries and fractures in the lower extremity, great pressure is frequently brought to bear on the athlete to continue to play after similar injuries in the hand or wrist. Unfortunately, this attitude may lead to a compromise in the quality of medical care for athletes who are permitted or even pressured into returning to competition without proper diagnosis or treatment.

It is surprising how often athletes with wrist and hand problems are managed by orthopaedic surgeons or team physicians with little or no training in hand surgery. It is by no means axiomatic that a well-trained sports medicine physician, highly skilled in the diagnosis and treatment of knee injuries, can provide the same expert care for hand and wrist injuries, some of which have the potential for permanent functional impairment if not treated appropriately.

Parents, coaches, fans, teammates, school officials, and (sometimes most of all) the involved athlete all want to minimize the "down time" following an injury. It is the responsibility of the treating physician to isolate himself from the pressures imparted by these individuals and to ensure that a hastened return to competition does not violate the time-honored tenets of sound orthopaedic management. Before initiating a particular course of treatment for an athletic injury, the physician must pose several pertinent questions to himself and answer those questions with absolute honesty. He must be unbiased by a zeal to demonstrate his reparative abilities and unaffected by the persuasive efforts of others whose primary concern is to return the athlete to the team and to competition. These questions include:

1. Is the method of treatment that I have selected the most appropriate for this injury, and can it be expected to provide the best long-term result?

2. If this injury had occurred in a nonathlete or during the off season, would I manage it in basically the same manner?

3. Are the potential complications of my anticipated treatment method significantly greater than might be expected from a more conservative approach?

4. Will the treatment allow the athlete to return to competition with little or no risk of a reinjury?

5. If reinjury does occur, would it unfavorably influence the prognosis for a satisfactory recovery?

If the physician can sincerely answer these questions in a manner that is consistent with the best possible outcome for the patient, regardless of his competitive desires or demands, then he is justified in proceeding with that method of treatment. It is understandable and appropriate that occasional compromises may be made, depending on the level of competition and the specific circumstances and requests of a given athlete. Considering the significant financial impact that upper extremity injuries may have on those who are reimbursed for their sports participation, it would seem appropriate that the physician might accede to somewhat greater treatment risks in the professional athlete than in the amateur. Inasmuch as the overwhelming number of injuries treated by readers of this text will be for amateur athletes in the pre-high school or high school age range, the basic axiom must be that the patient's athletic career is almost always of secondary importance to the implementation of the best method of treatment. One should, therefore, employ those therapeutic measures that most reliably ensure the best long-term function, regardless of the obligatory loss of playing time which that method may mandate.

The age of the athlete and his or her level of performance are important factors that must enter into the decision concerning return to competition. The delicate status of the musculoskeletal system in the preadolescent athlete makes him or her vulnerable to injury, and there should be absolutely no reason to cut any corners in the management of injuries in this age group. If the college or professional athlete chooses to "play hurt," it is the responsibility of the treating physician to provide a candid and thorough explanation of the injury, as well as the implications of compromising treatment so that he or she can return to competition sooner. The athlete and all interested parties (trainers, coaches, parents, and friends) must clearly understand the potential consequences of what the surgeon considers to be premature return to competition or less than ideal treatment for a given injury.

The use of local anesthetic agents to alleviate discomfort emanating from an injured part is subject to considerable debate. One must weigh the potential damage that could occur during competition if the athlete fails to recognize pain in the affected area. It would seem that little harm could arise from a single injection into an area of bruising, strain, or inflammation if there is little or no likelihood that additional damage could occur during practice or a game. If, conversely, the injection is to be used to arrest pain in an area of recent fracture or ligamentous injury, where reinjury would likely increase the severity of the condition, it should be considered an unacceptable treatment alternative. In actual practice, the indications for pregame injections of local anesthetic medications in the hand and wrist are extremely limited.

As physicians dealing with athletic injuries of the hand and wrist, we have consummate responsibility to fully appreciate the magnitude and severity of the condition, and we must make treatment decisions that are totally consistent with the traditional principles of managing musculoskeletal injuries. Accommodations cannot be made as a result of the inevitable pressures that are generated to get the athlete back into competition as soon as possible. Unfortunately, one of those mitigating pressures often emanates from the physician himself as he strives to devise a treatment alternative that will publicly demonstrate his abilities to rapidly deal with sports injuries and to contribute to the success of a team with which he may have strong personal loyalties. Although not always easy, it is far better for the physician to detach himself from such pressures and to isolate his considerations to the best management of a particular injury for a specific

athlete. Treatment must be that which provides the most predictable healing of the injured tissues with the fewest complications and the best long-term functional prognosis. Although athletic participation is sometimes not inconsistent with those treatment principles, the surgeon must in all cases recognize when such a potential conflict exists.

TENDON INJURIES

Flexor Digitorum Profundus Avulsion

Of all the injuries discussed in this chapter, avulsion of the flexor digitorum profundus (FDP) from its insertion into the base of the distal phalanx is the one that is most consistently incurred by athletes. Two classic mechanisms of injury are (1) grabbing the jersey of a ball carrier in football and (2) catching a finger on the rim of the goal while slam-dunking a basketball. The ring finger is the most frequently involved and, although the reason for this is not totally understood, several theories have been suggested [12, 327, 368].

Anatomy

The FDP tendon inserts into the volar base of distal phalanx just distal to the last major annular pulley (A4) (Fig. 18B–1). The tendon flares out

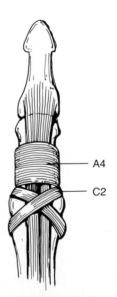

FIGURE 18B–1
The flexor digitorum profundus (FDP) tendon flares out after it passes through the A4 pulley and is wider at its insertion than it is more proximally. This may make it difficult to pass the tendon through the pulley when it is reinserted surgically, but it is very important to preserve the A4 pulley.

after it passes through the pulley, and is wider at its insertion than it is more proximally. This creates some technical difficulty at the time of repair, because it may be difficult to retrieve the tendon and pass it beneath the intact distal components of the pulley system.

The flexor tendons receive their blood supply through the long and short vincula, which remain at least partially intact if the tendon retracts no further proximally than the proximal interphalangeal (PIP) joint (type II, below). If, however, the tendon retracts into the palm (type I), both vincula are ruptured and the blood supply of the tendon is more severely impaired. At the time of FDP avulsion, the tendon sheath remains intact and is usually filled with blood, potentially setting the stage for extensive fibrosis and scarring, with the obligatory compromise of tendon gliding and digital performance.

Classification

Leddy and Packer [51] identified three different types of FDP avulsion injuries that have differing implications with regard to treatment:

Type I: the tendon retracts into the palm, severing all blood supply and frequently creating extensive scarring in the tendon sheath (Fig. 18B–2A). Repair

within 7 to 10 days is required, or the distal end of the retracted tendon softens and deteriorates.

Type II: the tendon retracts to the level of the PIP joint, where it becomes entrapped at the chiasma of the flexor digitorum superficialis (Fig. 18B–2B). Occasionally, a small fleck of avulsed bone can be seen on a true lateral radiograph and serves as a marker for the position of the tendon. Early treatment is advised, but successful repair can be done as late as 3 months after injury.

Type III: a large bony fragment is avulsed by the tendon and is usually lodged at the level of the distal pulley (A4) (Fig. 18B–3).

Smith [354] and others [11, 49] have described yet another type of flexor tendon avulsion that cannot be classified in Leddy and Packer's system, and they suggested that this be called type IV. In addition to avulsion of the profundus insertion (with or without a small osseous fragment), there is a separate, concomitant intra-articular fracture at the base of the distal phalanx. Early fixation of the intra-articular fracture and reinsertion of the avulsed tendon are both required to restore function.

Clinical Evaluation

A high index of suspicion and awareness of this injury are imperative, because in most cases early

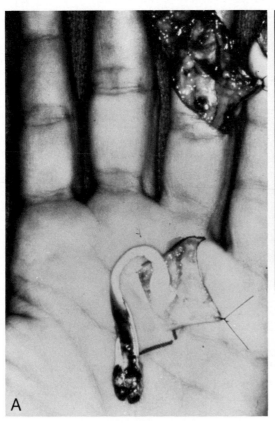

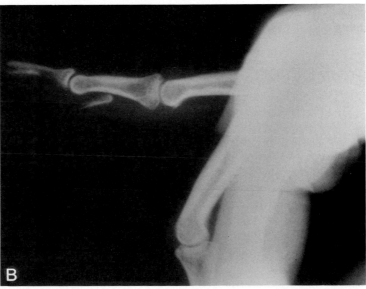

FIGURE 18B–2
A, In a type I FDP avulsion, the tendon retracts into the palm. Attempted repair after 7 to 10 days may be difficult or impossible. *B,* In a type II FDP avulsion, the tendon gets hung up at the chiasma of the flexor digitorum superficialis. Occasionally, the tendon may avulse a fragment of bone off the volar cortex of distal phalanx (usually the fragment is smaller and lies more proximal than that seen in this patient).

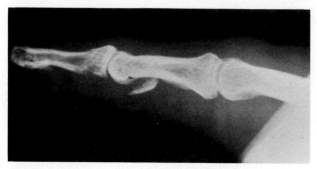

FIGURE 18B–3
In a type III FDP avulsion, a large bony fragment is avulsed by the tendon, preventing more proximal retraction because the fragment gets hung up at the level of the A4 pulley.

operative treatment is required in order to achieve a good result. Immediately after injury, the finger is characteristically swollen and somewhat ecchymotic. The diagnosis is established by demonstrating the absence of FDP function [4, 51, 341], but this must be specifically tested for by stabilizing the PIP joint and having the patient try to actively flex the distal interphalangeal (DIP) joint. Local tenderness is usually more marked at the site where the FDP stump has come to rest; that is, over the PIP joint or at the level of the A1 pulley in the distal palm. Radiographs should be taken to look for an avulsed fragment, but in most cases the tendon ruptures directly from the distal phalanx and no fragment is seen radiographically. Therefore, a correct diagnosis usually depends on the clinical examination to establish the absence of FDP function.

Treatment Options

In most cases, early operative reinsertion of the avulsed tendon is mandatory to restore active flexion of the distal joint. The success of repair is directly related to the length of delay after injury, and the most satisfactory results are usually obtained with immediate operative treatment. Because the tendon is retained in the sheath and usually does not undergo substantial deterioration or shortening, type II injuries often can be treated successfully on a delayed or secondary basis. When the tendon has retracted into the palm (type I), reinsertion is difficult or impossible after 7 to 10 days. The problem is that one cannot always be absolutely certain which type of avulsion is present unless a fragment of bone is seen on radiograph. Therefore, FDP avulsion is not an injury that can wait until the end of the season for repair; proper management demands surgical treatment as soon after injury as possible.

Certainly, in young or amateur athletes, there

should be no question regarding the decision to proceed with surgical reinsertion of the avulsed flexor tendon. Although the loss of terminal (DIP) joint flexion may not seem to result in a major functional deficit, it does compromise digital facility and strength. In addition, not repairing the tendon may lead to finger flexion deformities, and the patient may experience tenderness if the proximal tendon stump remains retracted in the palm. For those reasons, it is always preferable to proceed with immediate repair in young athletes, despite the profound influence that the surgery might have on the player's athletic career or the win–loss record of the team. Professional athletes, conversely, may take a different view of the need for surgical reconstitution of a ruptured flexor tendon, at least when it is contemplated on an emergent basis. Many athletes observe the strong flexion of the PIP joint produced by the intact flexor digitorum superficialis (FDS) and refuse surgical repair, opting to continue to play largely because of the financial impact that the obligatory postoperative disability would impose.

Technically, the repair may be difficult because the avulsed end of the tendon is usually wider than the annular pulleys (A3, A4, A5); it is extremely important that these pulleys be completely preserved during reinsertion. It is often necessary to narrow the FDP stump to allow passage beneath the distal pulleys before reinserting it directly into the base of the distal phalanx. During the operative repair, it is also imperative to avoid inflicting any damage to the flexor tendon sheath or to the intact and usually well-functioning FDS tendon (Fig. 18B–4). As with any flexor tendon surgery, meticulous, atraumatic technique is required to prevent intense scarring and adhesion formation with an obligatory compromise of tendon excursion and digital motion. The loss of FDS function after an effort to repair or reconstruct the FDP substantially adds to the patient's disability, often requiring further surgery and time lost from athletic participation.

Authors' Preferred Method of Treatment

For the reasons stated above, we prefer operative repair of the acutely avulsed tendon when the diagnosis is made, preferably within a few days of injury and certainly within 7 to 10 days. The standard postoperative flexor tendon protocol as outlined below is begun a few days postoperatively. Return to athletic competition is not allowed until the tendon is securely healed (8 weeks), and even

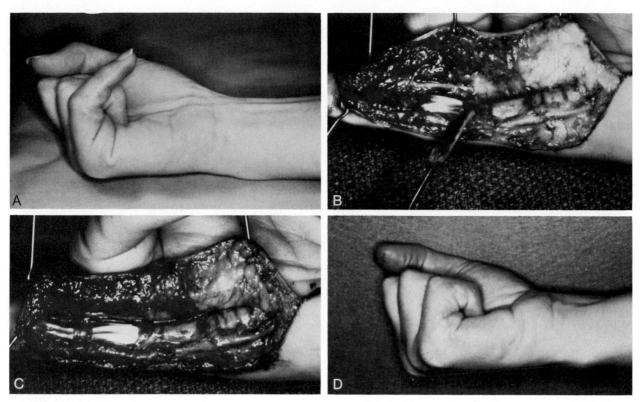

FIGURE 18B–4

A, Appearance of the small finger after a type II FDP avulsion. There is absence of flexion in the DIP joint and limited flexion in the PIP joint. *B,* Note the hemorrhagic bed at the time of digital exploration and tendon retrieval. *C,* The tendon has been reinserted, preserving the A4 pulley. *D,* The patient had full recovery of active flexion 3 months postoperatively.

then some type of splint that prevents hyperextension of the finger should be worn for an additional month.

Postoperative Management and Rehabilitation

Flexor digitorum profundus avulsion is unfortunately one of the few hand injuries that almost always eliminates an athlete from competition until complete healing has occurred. The repaired tendon must be protected against disruptive forces for at least 6 to 8 weeks, and it is almost impossible to devise a splint that safely allows an athlete to return to competition during this period of time. If the tendon is not repaired primarily and it is a type I injury, functional recovery is likely to be impaired.

As with any repaired flexor tendons in zones I and II, the recovery of digital function is long and difficult. Programs have been devised to impart early protected stress to repaired tendons in order to improve gliding and the tensile strength at the suture site [22, 53]. The effectiveness of these techniques have now been well demonstrated in both the laboratory and clinical settings, and there are many modifications of the original protocols that are now recommended by various authors [13, 15, 16, 29–33, 46, 75, 84–89, 91]. The post-repair regimens most applicable to flexor profundus avulsions are those for zone I interruption and may be instituted within the first few days if the reinsertion is thought by the surgeon to be strong and secure. Similar early motion programs are advised by some after free tendon grafting of unreparable or late avulsions, although some surgeons are reluctant to impart immediate stress to grafts for fear of rupture. In those instances, the wrist and digits are immobilized for 3 to 4 weeks and the mobilization programs are picked up at that point in their progression. Almost all of these protocols recommend that some protection of the repair be continued for at least 3 months. This precaution mandates restricted athletic participation for the majority of that time period and protective splinting for the remainder.

Late Treatment

It is not unusual for athletes to appear weeks or even months after injury with an untreated FDP avulsion; in fact, such patients are frequently seen at the end of each football season. At this point in time, the condition is a far more difficult problem to treat than the acute injury. The restorative possibilities are substantially altered, and mature clinical judgment based on many factors unique to that injury and that patient must be exercised by the treating physician. Secondary repair is usually not possible, and the options include doing nothing, tenodesing the distal profundus stump to the middle phalanx or arthrodesing the DIP joint, free tendon grafting, or staged tendon grafting using a silicone implant. The decision as to which approach is most appropriate for a given athlete depends on his or her needs and requests, the status of the digit, and the philosophy and experience of the treating physician.

The first priority in such patients is to try to regain full, painless PIP motion with a supervised exercise and splinting program. If this is successful, the patient may be able to get along reasonably well without FDP function. In some patients, arthrodesis of the DIP joint may be the appropriate treatment. If, however, active DIP flexion is required, tendon grafting can be done through or around the intact superficialis, but not without significant risk of limiting PIP motion. This is an operation that should be done only by a surgeon who is well versed and experienced in tendon surgery.

Mallet Finger of Tendon Origin

By common usage, the term mallet finger has come to mean a flexion deformity of the DIP joint resulting from loss of extensor tendon continuity [81]. The same clinical picture can be caused by an intra-articular avulsion fracture involving one-third or more of the dorsal lip of the distal phalanx. This troublesome fracture is discussed under Mallet Finger of Bony Origin.

Anatomy

The extensor tendon inserts into the dorsal base of the distal phalanx just proximal to the germinal nail matrix. There is virtually no subcutaneous fat between the tendon and the overlying skin. These two anatomic facts make operative procedures at this level difficult and fraught with potential complications, such as permanent nail deformity and skin slough.

Mechanism of Injury and Classification

Forcible flexion of the DIP joint is thought to be the mechanism of injury resulting in the mallet or "baseball" finger. Bunnell [9] preferred the term drop finger. The injury usually occurs when the extensor tendon is taut, as in catching a ball or striking an object with the finger extended. The deformity may result from relatively trivial trauma, and it has been suggested that a familial predisposition [43] or an area of relative avascularity in the tendon [97] may play a role in such patients.

Several patterns of injury may be seen [100]:

1. The extensor tendon fibers over the distal joint may be stretched without completely dividing the tendon. The degree of drop of the distal phalanx is usually not pronounced—a loss of perhaps 5 to 20 degrees of extension—and the patient retains some weak active extension.

2. The extensor tendon may be ruptured or torn from its insertion into the distal phalanx (Fig. 18B–5A) [6, 44]. In this instance, there is a greater flexion deformity (up to 60 degrees), and the patient has complete loss of active extension at the distal joint.

3. A small fragment of the distal phalanx may be avulsed with the extensor tendon (Fig. 18B–5A). In this instance, the injury has the same characteristics as the tendon injuries mentioned above, and the degree of drop depends not on the small fragment, but rather on the amount of loss of continuity of the tendon mechanism. These injuries should be treated as tendon injuries rather than fractures. In the series of Stark and co-workers [81], 24% of the injuries were associated with small bony avulsion fractures, but this finding had no effect on the final result.

In all of these injuries, if the flexion deformity of the distal joint is severe, a secondary hyperextension deformity of the PIP joint may develop because of imbalance of the extensor mechanism, the so-called swan-neck deformity (Fig. 18B–5B).

Treatment Options

Considerable controversy has existed in the past as to whether or not the PIP joint need be immobilized in the treatment of mallet finger. Probably in large part because such authorities as Bunnell [9] and Watson-Jones [100] recommended immobilization of the PIP joint in 60 degrees of flexion, many splints made by therapists with experience in hand splinting may be the most comfortable and best accepted by the patient (Fig. 18B–7). Regardless of the type of splint used, precise instructions must be given to the patient to achieve optimal results and avoid minor problems [18] (see Authors' Preferred Method, below).

Poor results after conservative treatment prompted some surgeons to carry out immediate operative repair of the torn tendon. As early as 1930, Mason [55] had recommended early operative repair of the ruptured terminal extensor expansion. Operative treatment is aimed at freshening and approximating the edges of the tear; it is followed by immobilization of the distal and middle joints. An avulsed bone fragment can be reattached with a pull-out wire or, if too small, it may be excised and the tendon directly reinserted by a pull-out wire through tiny drill holes. The results of operative repair of tendinous mallet fingers are not always satisfactory. Although they may give improved cosmetic results, some flexion is often lost due to scarring on the dorsal aspect of the joint. Assess-

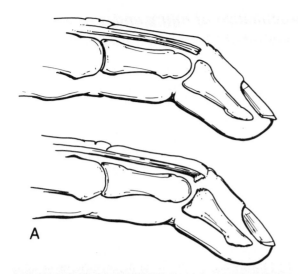

FIGURE 18B–5
A, The two most common types of mallet finger. The tendon may be avulsed cleanly from the base of distal phalanx *(top),* or the tendon may pull a small fragment of bone with it *(bottom).* Treatment for these two types is identical (see text). *B,* Severe flexion deformity of the DIP joint secondary to a mallet finger may lead to swan-neck deformity. This is particularly likely to occur in patients with lax PIP joints.

authors advised such treatment. However, most authors in the current literature believe that only the DIP joint need be immobilized [1, 8, 18, 23, 25, 39, 45, 58, 59, 79, 80, 83, 98]. Several types of DIP splints have been described (Fig. 18B–6), including at least two modifications of the original Stack splint, one with perforations [45] and another with larger windows [80], both designed to minimize skin problems. The results of treatment with these various types of splints do not appear to be substantially different [74], but minor complications were reported in one series with the dorsal aluminum splint [83]; in another series, patients preferred the Stack splint over the Abouna splint [98]. For many patients, simple, conforming synthetic rubber or plastic

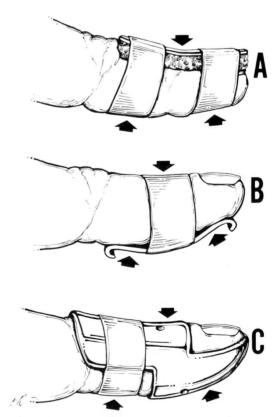

FIGURE 18B–6
Several different types of splints have been designed to immobilize the DIP joint in full extension. *A,* Dorsal padded aluminum splint. *B,* Volar unpadded aluminum splint. *C,* Stack splint. All three employ a three-point fixation principle.

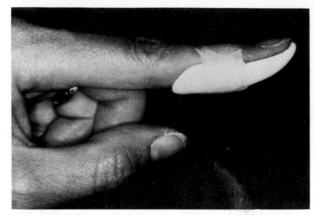

FIGURE 18B–7
A custom-made mallet finger splint molded from Orthoplast by a hand therapist.

ment of the late results in Robb's series [70] suggested that early operative repair was unnecessary and even undesirable, a viewpoint shared by most authors.

Stark and colleagues [81] analyzed their results of 163 mallet fingers and concluded that not all patients with mallet finger should be treated in the same way. Treatment should depend on the elapsed time following injury, the previous treatment, the degree of loss of extension, the degree of functional disability, and the age of the patient. In their series, only 4 patients were treated surgically and none by Kirschner wire fixation.

The results of treatment of mallet fingers are not universally good by any method. Burke [10] emphasized that assessment at the time of splint removal is valueless because subsequent recurrence of at least some of the extensor lag is likely. Flinchum [26] probably said it best when he noted that "this is a difficult injury to treat without some residual cosmetic or functional loss." Cold weather intolerance has been noted to be the most common late symptom [96], and is not directly related to the amount of residual deformity [62]. Mikic and Helal [59] reported that although most patients are improved, only about 30% to 40% regain normal function in the DIP joint after treatment. In a very comprehensive study of mallet fingers, Abouna and Brown [1] discovered that the following factors are likely to lead to a poor prognosis: age over 60 years; delay in treatment of more than 4 weeks; an initial active extensor lag greater than 50 degrees; less than 4 weeks of immobilization; and patients with short, stubby fingers. Treatment of patients with associated arthritis (rheumatoid or degenerative) or peripheral vascular disease was particularly unrewarding in their experience.

Authors' Preferred Method of Treatment: Acute Mallet Finger

The authors' preferred treatment for acute mallet finger is continuous extension splinting of the DIP joint only for 6 to 10 weeks. We do not immobilize the PIP joint, and have never opted for operative repair of the extensor tendon. Percutaneous, transarticular Kirschner wire fixation is used in selected patients who have unique needs or requests. In some patients, we prefer a dorsal padded alumafoam splint, which tends to interfere less with the tactile pad of the finger (see Fig. 18B–6A). An individually contoured splint fabricated from thermoplastic materials is frequently used when the dorsum of the DIP joint is tender or swollen, because these splints tend to cause less pressure over the joint. A prefabricated Stack plastic splint may also be employed, but these splints often do not hold the DIP joint in complete extension. In some instances, we may alternate several types of splints to accommodate and relieve local areas of pressure or tape irritation or to satisfy the particular needs or wishes of the patient.

We usually attempt to position the DIP joint in slight hyperextension, but the degree depends on the mobility of the patient's joints and the level of discomfort. The splint should never cause pain, and the amount of hyperextension should never produce blanching of the skin over the DIP joint. Rayan and Mullins [69] showed that skin blanching occurs in most patients at 50% of total passive hyperextension. The patient is shown how to take the splint off and reapply it occasionally for skin care. He or she must appreciate that splint removal and reapplication require assistance, because both hands are necessary to hold the DIP joint in extension and properly apply the splint [18]. At no point in the healing process is the DIP joint allowed to drop into flexion. If it does, the splinting process must begin again just as though the injury had just occurred. With the Stack splint, it is important to keep the tip of the finger in contact with the end of the splint, which may require an additional strip of tape placed longitudinally. If the finger does not fit any Stack splint well, Crawford [18] suggests using the smaller of the two sizes and cutting the splint longitudinally in the midline. To minimize skin irritation with the dorsal aluminum splint, Stern and Kastrup [83] suggested a layer of moleskin or gauze beneath the splint.

This method of treatment is successful only for patients who are reliable and reasonably intelligent, because it requires compliance and cooperation. It is advisable to see these patients after the initial 1 to 2 weeks of treatment to be certain that they understood the instructions and are using the splint properly.

Postoperative Splinting, Management, Rehabilitation, and Criteria for Return to Play

Athletes may continue to participate, provided the splint is secure and worn continuously. This is usually no problem for most football players, especially nonball handlers, for whom the splint can be covered with a layer of foam rubber. However, a basketball player or a quarterback may object to the restricted motion of the DIP joint and the cumbersome nature of the splint. Therefore, in those sports or positions requiring full mobility of the fingers, a decision must be reached as to whether treatment can be deferred until after the season, which is probably not unreasonable if only 4 to 6 weeks of playing time remain (see Late Mallet Finger, below).

A minimum of 6 weeks of continuous DIP immobilization is required [101], and some authors recommend 8 weeks [18]. Some recurrence of the extensor lag is virtually inevitable, and at least 2 to 4 weeks of additional nighttime extension splinting of the DIP joint is advisable after the continuous splint has been discontinued; we prefer a minimum of 4 weeks. Careful follow-up is required, and this treatment must be individualized. Some patients show a greater tendency for recurrence, and the splinting must be extended for more nights and additional weeks. If the recurrent extensor lag is severe, Crawford [18] suggests a second course of full-time splinting for 8 weeks. After the removal of the continuous splint, the recovery of DIP flexion may come slowly. Patients should be cautioned against exerting strong passive pressure on the distal phalanx in an effort to speed up flexion. A gradual progression of improved flexion almost always occurs without passive assistance, which may risk additional damage to the extensor insertion. Occasionally, a patient has difficulty in regaining flexion, and in this situation, less part-time splinting is indicated.

Late Mallet Finger

A drop finger seen within 3 to 4 weeks should be treated as an acute injury, although the longer the delay, the less successful the result [1, 14, 81]. Mallet deformities that were not seen until 2 to 3 months after injury have also been improved with prolonged (at least 8 weeks) splinting of the distal joint [18, 45, 64]. Auchincloss [2] has suggested that Kirschner wire splintage may possibly be more effective than external splinting in these patients with late recognition of the injury.

Many patients with late, untreated mallet fingers have no functional problems and require no treatment. Mild discomfort such as cold weather intolerance is not uncommon, but surgical treatment is unlikely to affect this and is rarely indicated. If the flexion deformity is severe (i.e., greater than 45 degrees), the bent finger may get in the way and cause some functional impairment. For those few patients who do have such problems, a variety of operative procedures have been proposed, including plication [79] or reefing [14, 23] of the scarred tendon, tenodermodesis [42, 48, 96], arthrodesis (more likely to be necessary in the patient with post-traumatic arthritis secondary to a mallet finger of bony origin; discussed below) [42, 81], and even DIP disarticulation [71], although few, if any, authors endorse Rosenzweig's recommendation for this.

The patients most likely to be symptomatic with late mallet finger are those who develop a swan-neck deformity, with a supple hyperextension posture of the PIP joint that accentuates the DIP extensor lag. Certain carefully selected patients with this problem may be candidates for Fowler's central slip release [38]. However, as Bowers and Hurst [5] have noted, the indications for this procedure are very strict, and the operation must be done with meticulous care to avoid significant potential complications, including weakness of PIP joint extension. Grundberg and Reagan [35] also reported satisfactory results with this procedure, but stressed that the swan-neck deformity must be due to a mallet finger and not to pathology at the PIP joint if the operation is to be successful.

Satisfactory correction of such deformities can also be achieved with the spiral oblique retinacular ligament (SORL) reconstruction [47, 92], although this operation is much more technically demanding and requires a thorough understanding of the extensor mechanism. In most instances, the magnitude and potential pitfalls of this procedure outweigh the probable functional benefit.

Authors' Preferred Method of Treatment

We have used both of the above-mentioned operations for late correction of mallet finger. In general, the Fowler procedure is preferred for supple swan-neck deformity, and the SORL is preferred reconstruction for more severe deformity. We rarely do procedures directly on the DIP joint, except for the occasional arthrodesis for painful post-traumatic arthritis, which is decidedly uncommon following mallet fractures, especially in the young athlete.

Mallet Thumb

Disruption of the extensor pollicis longus insertion into the base of distal phalanx can result in a flexion deformity of the thumb similar to that seen in the finger. Both splinting [65, 67] and operative repair [20] have been shown to yield satisfactory results, although a review of the cases in these small series suggests that more limited motion in the interphalangeal joint may follow surgical treatment. In reporting the largest series of mallet thumbs, Miura and colleagues [60] recommended splinting (dorsal padded aluminum splint to immobilize only the interphalangeal joint in slight hyperextension) for closed injuries and operative repair for lacerations of the tendon.

Authors' Preferred Method of Treatment

Our experience with this relatively uncommon injury is limited, but we tend to agree that Miura's method of splinting is appropriate treatment for closed injuries, but that in an open wound it would be reasonable to repair the tendon at the time of wound closure [60].

Mallet Finger of Bony Origin (Mallet Fractures)

A fracture involving the dorsal articular surface of the distal phalanx produces a mallet deformity because the extensor tendon is attached to the avulsed fragment. Commonly, the fragment includes one-third or more of the articular surface (Fig. 18B–8). Displacement ranges from minimal separation to a wide gap with tilting and malrotation of the fragment, and the remaining distal phalanx may subluxate volar to the condyles of the middle phalanx (Fig. 18B–9). Several authors [17, 50] believe that the mallet fracture with volar subluxation is caused by a different mechanism than the usual hyperflexion injury. They suggest that these fractures, involving 50% or more of the articular sur-

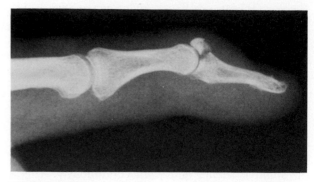

FIGURE 18B–8
A mallet finger of bony origin (mallet fracture), typically involving approximately one-third of the articular surface.

face, are caused by a hyperextension force, and that attempted reduction by extension of the DIP joint will fail. They believe that open reduction and internal fixation (ORIF) is necessary for these fractures.

Treatment Options

The management of mallet fractures is quite controversial. Several authors have cited specific indications for open reduction, including more than 2 mm displacement [321], 3 mm displacement [41], more than 30% involvement of the articular surface [61, 63, 82], and volar subluxation of the distal phalanx [1, 18, 58, 79, 81]. However, Schneider [73, 101] concluded from a large series of patients with mallet fractures that there is no difference in the results from operative treatment and splinting, and that in both groups the fractures heal with a "bump" over the DIP joint. Stern and Kastrup [83] noted a higher complication rate and more limited flexion of the DIP joint in their operated group, and even

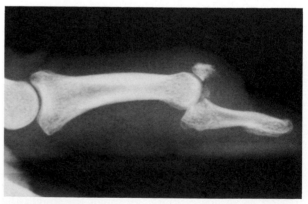

FIGURE 18B–9
Volar fracture-subluxation of the DIP joint secondary to a mallet fracture.

the staunchest advocates of open reduction [82] concede that "very few had normal motion" in the DIP joint after open reduction and internal fixation.

Operative treatment of a mallet fracture is a deceptively difficult procedure [101], fraught with many potential problems, including fragmentation of the small dorsal lip fracture, difficulty in exposing and reducing the fragment anatomically, skin slough, loss of fixation postoperatively, and subsequent limited motion in the DIP joint. Stark and associates [81, 82] emphasized the importance of exact anatomic reduction, but noted that one collateral ligament must be divided to provide adequate exposure in order to visualize the articular surface. The technique described by Hamas and colleagues [36], in which the extensor tendon is divided 5 mm proximal to the bone, provides good exposure, although at least partial division of one or both collateral ligaments is still necessary to see the articular surface. Our experience with this technique led us to conclude that it may cause proximal adherence of the extensor tendon and thus limit the recovery of DIP motion.

Several methods [36, 82, 49] have been described for operative fixation of the fracture, each with its advantages and disadvantages. Small Kirschner wires provide better purchase and can usually be left in place longer than a pull-out wire, but even the smallest (0.028 inches) Kirschner wire may shatter the small dorsal fragment.

Authors' Preferred Method of Treatment

Having treated many mallet fractures in the past with open reduction, we have now come to the conclusion that virtually all mallet fractures should be treated with splinting of the DIP joint, in much the same fashion as the mallet finger of tendon origin. The reason for this is that nonoperative treatment avoids all of the potential complications noted above, and even fractures with rather significant displacement remodel into a remarkably smooth articular surface (Fig. 18B–10). Perhaps most importantly, in our experience, patients treated with splinting tend to regain a better range of motion (especially flexion) of the DIP joint than those treated with open reduction and internal fixation. Our only indication for open reduction is the mallet fracture with volar subluxation of the distal phalanx.

Postoperative Management and Rehabilitation

Because our treatment of mallet fractures in most instances does not differ from the treatment of mallet finger of tendon origin, athletes with mallet fractures can generally continue to compete while wearing a splint, providing it can be reliably main-

FIGURE 18B–10
In our opinion, most mallet fractures should be treated nonoperatively. *A,* This patient was seen 1 month following injury, having had no treatment. He was treated with simple splint immobilization of the DIP joint for 6 weeks. *B,* Six months later, there is excellent remodeling of the articular surface. Range of motion in the DIP joint was 10 to 70 degrees.

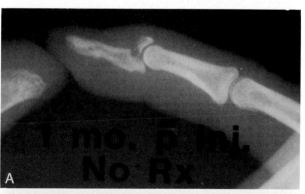
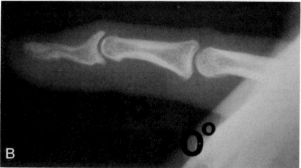

tained in position throughout competition. The only major difference between the management of tendon and bony mallet injuries is that, in those rare situations requiring open reduction, surgery should not be deferred until after the season.

Boutonnière Injuries

Boutonnière or buttonhole deformity is caused by disruption of the central slip of the extensor tendon combined with tearing of the triangular ligament on the dorsum of the middle phalanx (Fig. 18B–11) which, in time, allows the lateral bands to slip below the axis of the PIP joint. Rupture of the central slip may not immediately lead to a significant loss of PIP extension, as long as the lateral bands remain in position dorsal to the axis of joint rotation. After several days or perhaps even weeks, the lateral bands, untethered by the triangular ligament, displace volarly beneath the axis, and full active joint extension is lost. The deformity is further aggravated by the pull of the lateral bands, which have now become flexors of the joint. The PIP deformity is compounded by the limitation of flexion, both passive and active, which develops in the DIP joint. This results from the tenodesis effect of the displaced lateral bands and, in time, may lead to the fixed hyperextension deformity of the distal joint seen in the classic boutonnière lesion (Fig. 18B–12).

Clinical Evaluation

Most boutonnière deformities are caused by rupture of the central slip directly from its bony insertion; the diagnosis must be made by clinical examination. The patient with acute injury usually does not present with a typical boutonnière deformity, but simply with a swollen, painful PIP joint. Rupture of the central slip is differentiated from the more common injury to the collateral ligament by the location of the maximum area of tenderness on the dorsum rather than on the sides of the joint and, most importantly, by the patient's inability to strongly extend the PIP joint. If pain prevents active extension, it must be eliminated by digital or metacarpal nerve block so that active motion can be tested. On some occasions, one can palpate a depression directly over the joint, resulting from the loss of tendon continuity. If a patient is suspected of having a ruptured central slip, radiographs of the involved finger should always be made; in some instances, there is an avulsion fracture arising from the dorsum of the base of the middle phalanx (Fig.

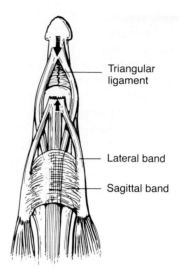

FIGURE 18B–11
Boutonnière deformity is caused by disruption of the central slip of the extensor tendon *(small arrow)* and tearing of the triangular ligament *(large arrow)*, which in time allows the lateral bands to slip below the axis of the PIP joint.

18B–13). Patients with suspected central slip injuries should always be reexamined 7 to 10 days after the injury, when the boutonnière lesion may be more apparent.

Treatment Options—Acute Injuries

A boutonnière lesion without a fracture should be treated closed, by splinting the PIP joint in full extension for at least 6 weeks [57, 76] (Fig. 18B–14). Souter [76] has shown that successful results can be achieved in closed boutonnière lesions where treatment is delayed up to 6 weeks after injury. It is extremely important that the distal joint not be immobilized; in fact, it must be actively and passively flexed during the period of treatment. After the initial period of immobilization, a removable

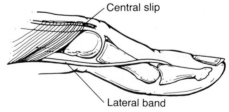

FIGURE 18B–12
In an established boutonnière deformity, flexion deformity of the PIP joint is caused by loss of active extension through the central slip, plus the deforming force of the lateral bands, which now pass volar to the axis of the PIP joint. Intrinsic muscle pull through the lateral bands leads to hyperextension of the DIP joint.

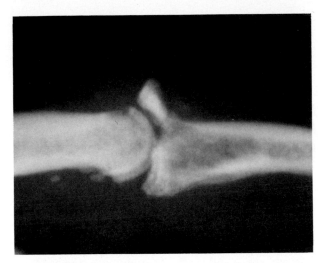

FIGURE 18B–13
Occasionally, the central slip may avulse a small fragment of bone off the base of middle phalanx. More commonly, however, the tendon is pulled cleanly off the bone, and the radiograph of a patient with a boutonnière lesion usually shows no abnormality.

splint that holds the PIP joint in extension should be worn at night for at least an additional 4 weeks. From the onset of treatment, the patient must be taught to actively and passively bend the distal joint while holding the PIP joint in full extension. Treatment should not be considered complete until active flexion of the distal joint is equal to that in the opposite, normal finger.

A boutonnière injury with a displaced avulsion fracture of significant size demands open reduction and internal fixation. At the time of operation, it is important also to repair the triangular ligament to correct the volar subluxation of the lateral bands, although care must be taken not to reef these tendons dorsally under excessive tension, because this limits subsequent flexion of the joint. Postoperative immobilization should be limited to the PIP joint only (in full extension, by means of an external splint or with a small, smooth Kirschner wire passed obliquely across the joint). Again, the patient should be carefully instructed in active assisted flexion of the distal joint.

Postoperative Management, Rehabilitation, and Criteria for Return to Play

If a boutonnière injury is recognized early, before a fixed flexion contracture develops in the PIP joint, the only immobilization required is continuous splinting of the PIP joint in full extension for 6 weeks. Thus, most athletes with this problem can

continue to compete, although ball handlers, basketball players, and others requiring dexterity in the hands may object to the splint. If the injury occurs late in the season and no more than 6 weeks will transpire before instituting continuous treatment, intermittent splinting can be alternated with non-splinting during competition. It should be emphasized, however, that the earlier treatment is begun, the easier and more effective it is likely to be.

Chronic Boutonnière Deformity

The long-standing boutonnière deformity represents a very difficult reconstructive challenge. Unfortunately, some athletes will, for varying reasons, elect not to seek medical management until after the conclusion of a season or perhaps until the end of the school year. By that time, the deformity may have progressed to severe, fixed PIP flexion and DIP extension contractures that cannot be appreciably improved by passive manipulation. These athletes must be told that the deformity will be extremely difficult to correct, that treatment will be prolonged, and that the final result is unpredictable.

The mechanism of the chronic boutonnière deformity has been well described by many authors and is appreciated by most surgeons dealing with upper extremity injuries [7, 34, 77, 93, 103]. In most instances, the deformity is fixed and not amenable to conservative treatment. If the contractures are mild and represent little functional disability and are not progressing, the athlete may be better off simply accepting the status of the digit with no further treatment. Particularly in the ring and little

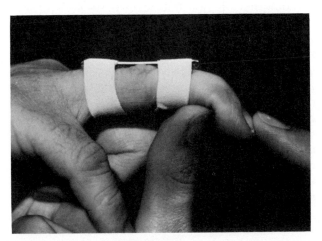

FIGURE 18B–14
Appropriate treatment of a boutonnière deformity involves two elements: (1) continuous immobilization of the PIP joint in full extension for 6 weeks, and (2) passive stretching of the DIP joint into flexion.

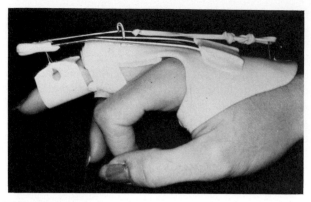

FIGURE 18B–15
In long-standing boutonnière deformity, there is frequently a fixed flexion contracture of the PIP joint, which should be corrected with dynamic splinting before surgical repair is undertaken.

fingers, a reasonable (about 30 to 40 degrees) amount of PIP flexion deformity can be tolerated.

Prior to any effort at surgical reconstruction of a long-standing boutonniere deformity, it is imperative that dynamic extension splinting with a safety pin splint, Joint Jack, Capener splint, serial digital casts, or an outrigger apparatus be used to maximize passive PIP extension (Fig. 18B–15). Recommended methods of surgical repair have included extensor tenotomy for mild cases where the recovery of DIP flexion is the major consideration [21, 28], transposition and repositioning of the lateral bands [54, 56], and shortening and imbrication of the attenuated and scarred central slip [24, 72]. Interpositional tendon grafts have also been recommended for some severe cases with substance loss [95], and arthrodesis may be the best option in some instances, where the deformity is not amenable to surgical restoration. Curtis and associates [19] have described a rational sequence of steps for the systematic surgical management of the chronic boutonnière deformity secondary to trauma. We have found this approach to be practical and effective in the treatment of these difficult disorders. Their staged protocol includes the following: stage I, tenolysis of the extensor tendon and freeing of the transverse retinacular ligament; stage II, sectioning of the transverse retinacular ligaments; stage III, tenotomy or lengthening of the lateral bands over the middle phalanx; and stage IV, repair of the central extensor tendon.

Full passive extension is a prerequisite for surgery, and the procedure is performed under local anesthesia, consisting of a digital nerve block. After each stage, the patient is asked to demonstrate digital function and, if adequate extension has been reestablished, the operation can be terminated. Seven-

teen of 23 patients were successfully managed by some combination of stages I, II, and III, and 6 patients received adequate treatment with stages I, II, and IV.

Postoperative Management, Rehabilitation, and Criteria for Return to Play

Following any of these treatment methods, the PIP joint is splinted in extension, and the DIP joint is either placed in flexion or allowed to actively flex and extend without the constraints of a splint. The length of time that extension splinting is continued depends on whether the tendon was repaired or grafted or if it was lysed or tenotomized. Tendon repairs require 4 to 6 weeks postoperative immobilization, whereas early active and passive range-of-motion exercises can be initiated after tenotomy or lysis. Night extension splinting may be continued for many weeks after any of these methods. As far as athletic participation is concerned, there is usually no treatment urgency once the fixed boutonnière deformity has occurred, and surgery can be delayed until season's end in most cases. Dynamic extension splinting can be initiated during nonparticipating hours so that the athlete can achieve maximum improvement by the time the season is over.

LIGAMENTOUS INJURIES AND DISLOCATIONS

Ligament injuries and dislocations of the small joints of the hand are probably the most common athletic injuries of the upper extremity. Many are relatively benign injuries that require only minimal protection and do not preclude continued participation in sports, but a few have the potential for serious impairment of function if not treated properly in the early stages.

Proximal Interphalangeal Joint

Anatomy

Although sharing some common characteristics, the PIP and metacarpophalangeal (MP) joints have numerous important anatomic differences, several of which have been clearly identified by Kuczynski [172]. The PIP joint is essentially a ginglymus or hinge joint, allowing only flexion and extension. It

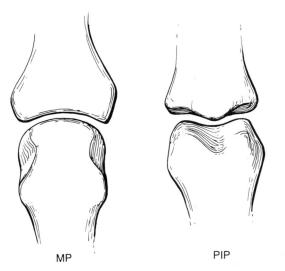

MP   PIP

FIGURE 18B–16
The MP and PIP joints are quite different structurally. The PIP joint has a bicondylar configuration, making it inherently more stable than the globular MP joint.

is inherently more stable than the MP joint, by virtue of its bicondylar configuration, which gives it a modified tongue-in-groove appearance (Fig. 18B–16). The shape of the head of the proximal phalanx is less eccentric than that of the metacarpal as seen in the lateral view, and therefore the cam effect is less significant (Fig. 18B–17).

The collateral ligaments of the PIP joint are similar to those in the MP joint, with a cordlike collateral ligament and a fan-shaped palmar component—the accessory ligament (Fig. 18B–18*A*). The major difference between the ligaments of the

MP and PIP joints is that the collateral ligaments in the PIP joint are more parallel than those in the MP joint (Fig. 18B–18*B*). The PIP collaterals are, therefore, tight throughout the entire range of motion, whereas those of the MP joint are tightest in flexion. The clinical importance of these differences is that, in order to minimize the likelihood of developing contractures after joint injuries, the preferred position of immobilization of the PIP joints is in extension, whereas the MP joints are more properly placed in flexion.

Clinical Evaluation

Every injury of the PIP joint, regardless of how seemingly insignificant it might appear at first glance, must be examined in a systematic and thorough fashion that should include at least the following three major points.

Determine the point of maximum tenderness. Although this may sometimes be difficult, the examiner should attempt to ascertain if the most tender area is over the central slip (dorsal), the collateral ligaments (radial and ulnar), or the volar plate (volar).

Test for stability of the supporting structures of the joint. Both collateral ligaments must be tested by radial and ulnar deviation stress, and the volar plate by hyperextension stress. Partial tears (sprains) of the collateral ligaments do not allow the joint to open up in excess of physiologic limits. In complete tears (ruptures), there is little or no resistance to lateral deviation stress, and the joint opens up on

FIGURE 18B–17
The shape of the metacarpal head is eccentric, resulting in a cam effect that makes the collateral ligaments more taut in flexion than in extension. The cam effect is not present in the PIP joint.

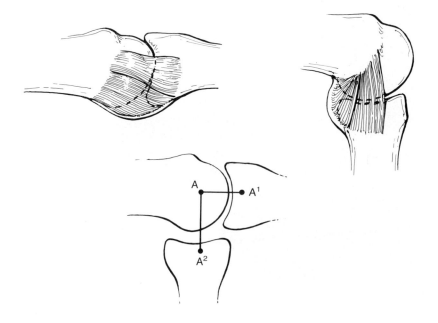

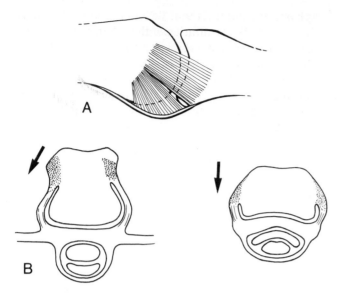

FIGURE 18B–18

A, In both the MP and PIP joints, the collateral ligaments have a cord-like dorsal component and a fan-shaped volar component. *B,* As seen in cross-section, the collateral ligaments of the PIP joint *(right)* are more parallel than those of the MP joint *(left)*. (Redrawn from Eaton, R.G. *Joint Injuries of the Hand.* Springfield, Charles C Thomas, 1971.)

the injured side. Complete tears that are suspected on clinical examination can be documented by stress radiographs, comparing these with stress films of the same digit in the uninjured hand.

Test both active and passive range of motion, with specific emphasis on the patient's ability to actively extend the PIP joint completely and with strength equal to other digits. If he is unable to do so because of pain, the examination must be repeated under digital block anesthesia. Inability to actively extend the PIP joint against resistance is diagnostic of rupture of the central slip, and a boutonnière deformity may develop if the patient is not treated appropriately.

Radiographic Examination

All injuries of the finger should be evaluated with at least two (anteroposterior and lateral) radiographic views of the involved digit. Of these, the most important is a true lateral view of the individual finger. More errors are made in the diagnosis of PIP joint injuries because of failure to obtain this radiograph than any other single reason. A lateral view of the entire hand is unacceptable, because important details are obscured by superimposition of the other three fingers (Fig. 18B–19*A*). An oblique view, although often helpful in delineating other injuries such as condylar fractures, often fails to

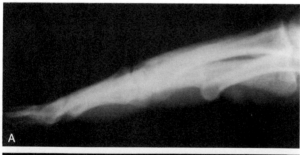
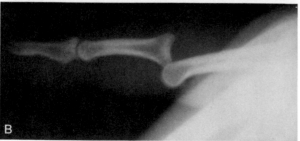

FIGURE 18B–19

A lateral radiograph of the entire hand is unacceptable for evaluating injuries of the fingers, and a true lateral radiograph of the *individual* digit is mandatory. *A,* In this lateral view of the entire hand, superimposition of the other fingers obscures dislocation of the PIP joint, which is clearly seen in *B,* the true lateral view of the involved finger.

demonstrate the presence of a small chip fracture at the base of the middle phalanx.

Stress Radiographs

Partial tears of the collateral ligaments can be difficult to differentiate from complete ruptures on clinical examination alone. Therefore, when instability is suspected, stress radiographs (Fig. 18B–20) should be obtained under local anesthesia. Kiefhaber and associates [169] performed extensive laboratory studies on collateral ligament instability and concluded that angulation greater than 20 degrees is diagnostic of complete rupture.

Collateral Ligament Injuries

Mechanism of Injury

Athletes are particularly prone to injuries of the collateral ligaments of the digital joints. Perhaps the most common mechanism is a nearly direct longitudinal force created by catching a ball on the end of a finger or "jamming" the digit into an opposing player. Basketball and baseball players are particularly vulnerable to the former cause, whereas football and soccer players often incur the latter mechanism. If the longitudinal force of this injury is applied in a slightly off-center manner, a strong bending moment is placed on one or more of the supporting ligaments along the phalangeal chain. If a lateral force is applied to the digit, the collateral ligaments of the DIP, PIP, or MP joint are stressed, and partial or complete ruptures may result. Also, the force that places strong pressure on the volar surface of the tip of the extended digit may result in hyperextension of the joints and tearing or avulsing of the palmar plates. Similar injuries can result from lateral forces applied to a digit during athletic participation, particularly if some object or body part serves as a fulcrum over which the digit is bent. Dislocation is the end stage of some of these injuries and results in an obligatory disruption of several components of the supporting ligamentous complex of the joint.

Some pain, swelling, and loss of motion result from these injuries, and, not infrequently, the athlete simply discounts the problem as a "jammed finger." In most instances, participation is continued with or without supportive splinting. Reinjury is quite possible during continued play, and a partial tear of a ligament may be converted to a complete tear. For that reason, it is imperative that team physicians and trainers examine all players who have sustained digital injuries with a high level of suspicion concerning possible ligament injuries. When in doubt, the digit should be securely splinted or taped before allowing the resumption of competition. Following the game, a thorough reexamination should be carried out, including routine and stress radiographs.

Partial Tears (Sprains)

Most authors agree that incomplete tears of a collateral ligament should be treated with 2 to 5 weeks of buddy taping or immobilization [183, 193], depending on the degree of severity. We prefer to treat all partial tears (sprains) of the PIP collateral ligaments with buddy taping. If possible, the injured finger is taped to the adjacent digit that is directly opposite the injured ligament in order to prevent any additional deviation forces. Active motion is encouraged from the outset, and the tape is worn continuously for 3 weeks and during participation in sports for the remainder of the season. Usually, the athlete is allowed to play during the entire period of treatment.

Complete Tears (Ruptures)

Treatment Options

The treatment of complete tears of the collateral ligaments of the PIP joints is somewhat controversial. Some authors advocate simple immobilization of the finger, although the recommended position and duration of immobilization are also subject to disagreement. Moberg [193] advised immobilization of the PIP joint in the position of function (flexion) for as long as 5 weeks. Eaton and Littler [131] recommended immobilization of the joint in 25 to 30 degrees of flexion for 2 to 3 weeks, followed by buddy taping for further protection as active motion is begun. In discussing McCue's paper, Coonrad [181] suggested that the finger be immobilized for 3 to 4 weeks and that operative treatment be considered only if the joint is still unstable after that time. At one time, Milford [186] suggested splinting the PIP joint in 60 degrees of flexion for 2 to 3 weeks for complete ruptures, but he subsequently advocated surgical repair in young adults, especially for a torn radial collateral ligament of the index finger [187–189].

A few authors [107, 160, 181, 216] have advised surgical repair of all complete ruptures, but others

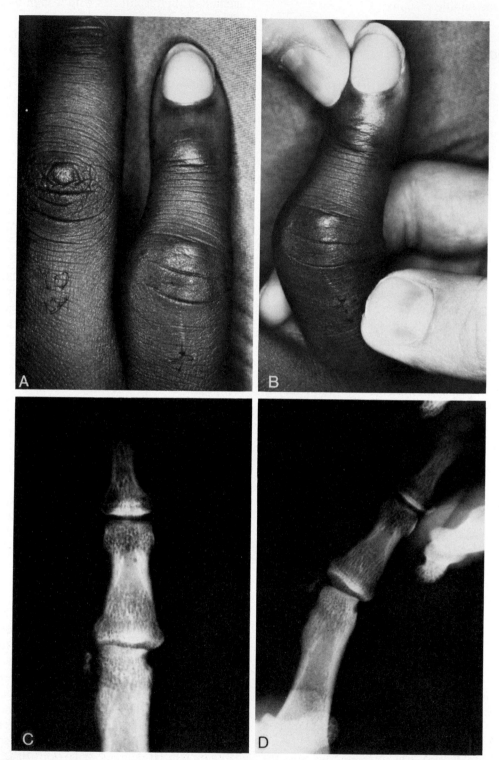

FIGURE 18B–20

A, The appearance of the small finger in a college basketball player with a 3-year history of chronic problems in the PIP joint. *B*, Instability of the radial collateral ligament was easily demonstrated clinically. *C*, Radiographs showed a small avulsed bony fragment originating from the neck of the proximal phalanx. *D*, Stress radiographs demonstrated distal movement of the avulsed fragment. Late reconstruction of the ligament was done by reinserting the avulsed fragment, and the patient now earns more than $2 million per year in professional basketball.

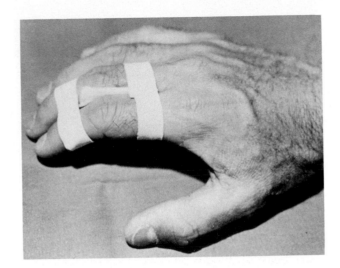

FIGURE 18B–21
Our preferred treatment for acute complete ruptures of the PIP collateral ligament is continuous buddy taping for 6 weeks.

have provided more specific indications. Bowers [119] suggested two indications for operative repair: (1) inability to achieve a perfectly congruent reduction, and (2) "unstressed instability." Wilson and Liechty [256] advised surgical repair for (1) instability demonstrated on active range of motion, (2) tissue interposition preventing joint motion, or (3) lack of joint congruity on radiographic evaluation.

Authors' Preferred Method of Treatment

It is our opinion that most complete ruptures of the collateral ligaments of the PIP joints do not require operative treatment. We prefer continuous buddy taping for 6 weeks (Fig. 18B–21). The major reason for not advocating primary repair for all collateral ligament ruptures is that the potential for disability resulting from motion loss or persisting instability after surgical intervention may exceed that which is seen in the few nonoperatively managed injuries that still have some slight ligamentous laxity after treatment. As Flatt [138] pointed out, surgical repair of the collateral ligament is delicate and difficult work. In our experience, it is unusual to obtain a completely full range of motion after any type of surgical procedure on the PIP joint. In our opinion, the only indication for surgical repair is a total rupture of the radial collateral ligament of the index finger of a young adult, where stability is more important than a full range of motion.

Postoperative Management and Rehabilitation

If buddy taping is used as the treatment method, there should be no playing restrictions in virtually any athlete. If operative repair of the collateral ligament is performed, playing can probably be resumed within a few days of surgery, provided that the finger is protected with very secure buddy taping. If the surgeon is concerned about the quality of the repair, playing should probably not be allowed for 3 weeks unless the finger is splinted, which precludes athletic participation by skilled position players such as ball handlers and basketball players.

DISLOCATIONS

There are three distinctly different types of dislocations that occur in the PIP joint: dorsal, volar, and rotatory.

Dorsal Dislocation of the PIP Joint (Volar Plate Injuries)

Clinical Evaluation

Dorsal dislocation (Fig. 18B–22) is by far the most common type of dislocation of the PIP joint. Physicians rarely have the opportunity to see the finger when it is dislocated, unless they happen to be on the sidelines. Reduction is usually performed by a coach, trainer, observer, or even the athlete himself. If the reduction has been performed before the physician examines the athlete, it is important to try to ascertain whether the initial displacement of the finger was dorsal or volar, because the treatment implications for these two types of dislocations are fundamentally different.

In order for a dorsal dislocation to occur, the volar plate must be torn either directly from bone or by avulsing a tiny chip of bone off the volar base

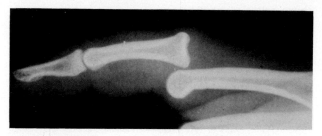

FIGURE 18B–22
Dorsal dislocation is by far the most common type of dislocation of the PIP joint.

of middle phalanx (Fig. 18B–23). This small volar plate avulsion fracture is rarely displaced more than 1 mm and generally heals with slight spurring of the volar beak of middle phalanx. Occasionally, sufficient scarring occurs at the site of injury to produce a flexion contracture, which McCue has referred to as a pseudoboutonnière deformity. Collateral ligament rupture sometimes occurs concomitant with dorsal dislocation [120, 193], and should always be tested for by the examiner.

Complex (irreducible) dorsal dislocations of the PIP joint are relatively uncommon, but have been reported [142, 148, 171, 204]. If the patient is seen with an unreduced dislocation, radiographs should be taken prior to reduction. More commonly, the reduction has been done elsewhere and the digit is swollen and painful, with some limitation of motion. A careful examination of the joint must be carried out in the manner previously described. Radiographs in two planes (including a true lateral view of the involved finger) are indicated.

Treatment Options

Many authors advocate immobilization of the PIP joint for 2 to 4 weeks following closed reduction

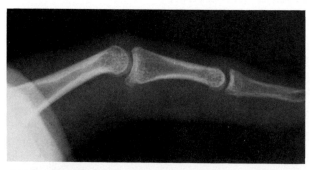

FIGURE 18B–23
Small chip fractures avulsed off the base of middle phalanx by the volar plate are not uncommon following dorsal PIP dislocations. These do not affect the routine management of the PIP dislocation, and should not be confused with the far more serious fracture-dislocation of the PIP joint shown in Figure 18C–28.

[197, 234, 251]. There is not much support for the suggestion of Zook and colleagues [263] that better results can be achieved with primary surgical repair of the volar plate. Stern and Lee [239] studied open dislocation of the PIP joint and concluded that the severity of these injuries is usually underestimated. They emphasized the importance of open reduction in the operating room and they advocated the use of parenteral antibiotics. Kjeldal [171] pointed out that cleansing and debridement should precede reduction of the dislocation, which has obvious significance if the injury occurs on the playing field.

Authors' Preferred Method of Treatment

Following closed reduction, we prefer to treat all acute volar plate injuries or dislocations with 3 to 6 weeks of buddy taping. This allows early active motion and prevents hyperextension, which is, of course, to be avoided in these patients.

Postoperative Management, Rehabilitation, and Criteria for Return to Play

Because these patients are treated with buddy taping, there are usually no playing restrictions. Joint swelling can be expected to persist for many months and is usually associated with some stiffness and, occasionally, mild crepitance. Vigorous active and passive range-of-motion exercise programs should be initiated early and monitored closely after these injuries. Dynamic splinting may be employed for recalcitrant joints that appear to be developing chronic contractures, and intra-articular steroid injections are infrequently used for joints that remain inflamed and painful for an inordinate period of time.

Volar Dislocation of the PIP Joint

Clinical Evaluation

Volar dislocations of the PIP joint (Fig. 18B–24) are relatively uncommon injuries. They may be pure dislocations or fracture–dislocations, but for both types the implications are the same: for this type of dislocation to occur, the central slip must be disrupted, and the potential for boutonnière deformity is present. If the dislocation is reduced before the physician sees the patient, and no fracture is seen

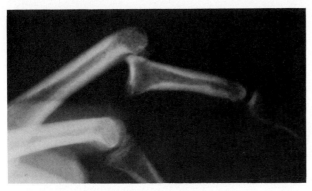

FIGURE 18B–24
Volar dislocation of the PIP joint is a relatively uncommon injury. It should be obvious from this radiograph that in order for this injury to occur, the central slip must be torn; therefore, these patients should be treated in the same manner as a patient with a boutonnière injury.

radiographically at the dorsal base of the middle phalanx, the pitfall is that it will be treated as if it were the much more common dorsal dislocation. Unfortunately, if the finger is treated with splinting in flexion or by buddy taping (both appropriate for a dorsal dislocation), a boutonnière deformity inevitably develops.

Treatment Options

Very little is to be found in the literature regarding this uncommon injury. Spinner and Choi [233] reported the largest series (five cases), and their conclusion that open reduction and repair of the central slip is mandatory has been widely quoted since that time. Conversely, Thompson and Eaton [246] advocated splinting of the joint in extension for 3 weeks, and reserved primary operative treatment for situations where the dislocation is irreducible, the postreduction joint surfaces are incongruent, or the active extensor lag is in excess of 30 degrees.

Authors' Preferred Method of Treatment

Unless the dislocation is irreducible by manipulation, we prefer to treat this injury closed. After closed reduction of the dislocation, anteroposterior and true lateral radiographs of the finger should be taken to ensure normal congruity of the joint. The presence of a small avulsion fracture at the dorsal base of middle phalanx is not an indication for open reduction unless it is displaced and does not reduce adequately with the PIP joint in full extension. If

these criteria are met, the treatment is the same as for a closed boutonnière lesion; that is, immobilization of only the PIP joint in full extension with a padded dorsal splint for 6 weeks, combined with active and passive flexion exercises of the DIP joint.

Postoperative Management and Rehabilitation

Because most of these patients are treated with splinting of the PIP joint, there are usually no playing restrictions, although ball handlers may have some problems with the splint. As with dorsal dislocations, vigorous postsplinting active and passive range-of-motion exercises are prescribed, and the recovery of joint motion is carefully monitored.

Rotatory Subluxation of the PIP Joint

Clinical Evaluation

Rotatory subluxation of the PIP joint is even more uncommon than volar dislocation but, in our opinion, the two are distinctly different entities, even though previous reports in the literature have not always made this distinction [162, 170, 199, 201, 212, 229]. The mechanism of injury in most cases is a twisting injury [207], and the resulting pathologic anatomy is buttonholing of one condyle of the head of the proximal phalanx through a longitudinal rent in the extensor hood between the central slip and the lateral band, both of which remain intact (Fig. 18B–25). The key radiographic feature of rotatory subluxation is seen in the lateral view, where there

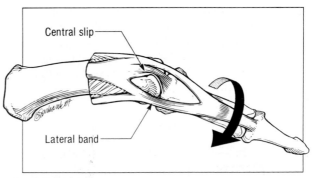

FIGURE 18B–25
Rotatory subluxation of the PIP joint results from a twisting injury of the finger. One condyle of the head of proximal phalanx is buttonholed between the central slip and lateral band, both of which remain intact.

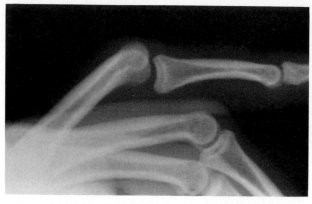

FIGURE 18B–26
A lateral radiograph of the finger with rotatory subluxation of the PIP joint shows a true lateral view of the middle phalanx and a slightly obliqued view of the proximal phalanx (as seen here), or vice versa.

is a true lateral profile of the proximal phalanx and an oblique appearance of the middle phalanx (Fig. 18B–26), or vice versa. In addition, a subtle widening of one side of the joint, as seen on the anteroposterior view, may be a further indication that a lateral band is impaled in the joint, preventing complete reduction.

Treatment Options

Rotatory subluxation of the PIP joint is sometimes irreducible by closed manipulation, but Eaton [130, 246] described a closed reduction maneuver that should be tried before resorting to open reduction. Under digital block anesthesia, gentle traction is applied to the finger with both the MP and PIP joints flexed to 90 degrees. This relaxes the volarly displaced lateral band and allows the band to be disengaged and slip dorsally when a gentle rotatory and traction force is applied. Further relaxation of the extensor mechanism can be achieved by dorsiflexion of the wrist. A gentle pop may be felt as the lateral band reduces to its dorsal position. Successful reduction is followed by full active and passive motion of the PIP joint, and must of course be confirmed by postreduction radiographs. After successful closed reduction, no static immobilization is required, and early active motion can be started immediately with buddy taping. In such cases, there are no playing restrictions.

Failure of closed reduction is an indication for open reduction. The joint is exposed through a curved dorsal incision, which reveals the condyle of the proximal phalanx protruding through the longitudinal rent between the central slip and lateral band (Fig. 18B–27). It is relatively easy to reduce

the condyle under direct vision by retracting the lateral band and lifting it from its volarly displaced position beneath the condyle. The collateral ligament should be inspected; if torn, it is repaired. Immediately after reduction, the joint should glide through a full range of passive motion.

Postoperative Management, Rehabilitation, and Criteria for Return to Play

Because the central slip is intact, minimal postoperative immobilization is required, and the athlete can return to competition within a few days, protecting the finger with buddy taping.

Dorsal Fracture–Dislocation of the PIP Joint

By far the most potentially disabling injury of the PIP joint is a dorsal fracture–dislocation (Fig. 18B–28). Usually the result of a longitudinal compression type injury, the volar articular surface of the base of middle phalanx is fractured (almost always comminuted and involving up to 75% of the joint surface), and the remaining intact portion of the middle phalanx is subluxated dorsal to the head of the proximal phalanx. Unfortunately, these deformities are often seen many weeks after the initial injury, after having been written off as a "jammed finger" or having been reduced but improperly splinted (usually in extension). Treatment at that point in time may be very difficult.

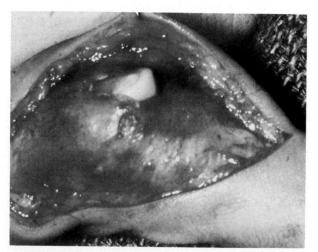

FIGURE 18B–27
Operative view of a rotatory subluxation of the PIP joint. The condyle of the proximal phalanx is buttonholed up between the intact central slip and lateral band.

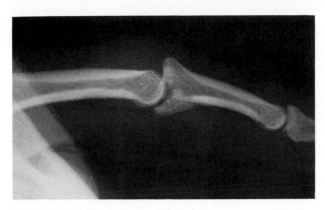

FIGURE 18B–28
Fracture-dislocation (more correctly called fracture-subluxation) of the PIP joint. The volar base of middle phalanx is crushed, usually comminuted, and generally involves 30% or more of the articular surface. The remaining base of middle phalanx is subluxated dorsally.

Clinical Evaluation

The PIP joint is usually quite swollen, and range of motion, both active and passive, is severely limited and very painful. The common pitfall seen in athletes is for the assumption to be made that this is simply a "jammed" or "sprained" finger and to allow continued athletic participation. This is an injury in which prompt diagnosis and treatment is critical to the outcome; delay in treatment can result in marked impairment of PIP joint function.

A true lateral radiograph is mandatory to confirm the diagnosis. This film demonstrates the comminuted fracture of the volar base of middle phalanx, with the intact portion of the middle phalanx having subluxated dorsally (see Fig. 18B–28). Superimposition of the base of the middle phalanx and the distal articular surface of the proximal phalanx is a more subtle finding, seen on the anteroposterior radiographic view.

Treatment Options

The best method for the management of PIP fracture–dislocation is rather controversial. Closed techniques, traction, and operative methods have all been recommended. In the older literature, various forms of simple taping or strapping of the finger were described [228, 235], and Trojan [247] advocated closed reduction and percutaneous pin fixation of the joint with a Kirschner wire. Complicated, tridirectional traction devices have also been described, first by Robertson and associates [221], and later by Agee [104, 105]. Schenck [226] devised a technique that combines skeletal traction and pas-

sive motion, and Hastings and Carroll [155] have devised a custom-made external fixator that allows immediate active range of motion.

McElfresh and colleagues [183] first described the extension block splint method, and Strong [244] suggested a much simpler but perhaps less secure method that employs the dorsal extension block principle.

There has been little enthusiasm in the literature for open reduction and internal fixation (ORIF) since Wilson and Rowland [255] in 1966 reported the largest series of patients treated in this manner. The major objection to open reduction is that the volar lip fragment is invariably comminuted (not unlike a tibial plateau fracture), making operative restoration technically difficult.

Authors' Preferred Method of Treatment

Our best results with this difficult injury have by and large been obtained with the dorsal extension block splinting method described by McElfresh and colleagues [183]. The technique requires careful attention to detail, but can result in restoration of a full range of stable motion if properly applied.

It must be emphasized that this technique should not be used if a satisfactory closed reduction cannot be achieved. Adequacy of reduction must be judged by congruity between the intact dorsal portion of the articular surface of middle phalanx and the distal surface of the proximal phalanx (Fig. 18B–29*A*). The V sign (Fig. 18B–29*B*) described by Light [175] is indicative of an inadequately reduced joint. Although it is not necessary for the comminuted volar lip fragment to be anatomically reduced (usually this fragment remains slightly depressed), it is imperative that the subluxation of the joint be precisely repositioned.

Technique of Dorsal Extension Block Splinting

Under digital block anesthesia, the PIP joint is reduced by longitudinal traction on the digit with downward pressure exerted on the dorsally subluxated base of middle phalanx. A lateral radiograph is taken to be certain that the joint can be reduced satisfactorily. This is usually fairly easy to accomplish within a few days of injury, but becomes increasingly difficult with the passage of time and may be impossible as early as 2 to 3 weeks after injury. If an adequate reduction cannot be documented radiographically, open reduction or volar

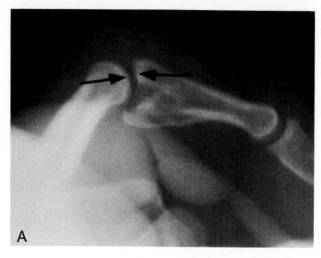

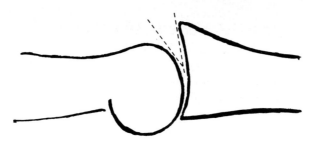

FIGURE 18B–29

A, Adequacy of reduction of a fracture-dislocation is judged by congruency between the intact articular portion of base of middle phalanx and the head of proximal phalanx. The volar fragments can be disregarded in a closed reduction if the dorsal aspect of the joint is congruent. *B,* Light's V sign *(dotted lines)* is a clear indication that the joint has not been satisfactorily reduced. (*B* from Light, T.R. Buttress pinning techniques. *Orthop Rev* 10:49–55, 1981.)

plate arthroplasty is indicated. McElfresh and co-workers [183] reported success with this method in patients with involvement of 30% to 50% of the articular surface, and we have been able to use it effectively in some patients with involvement of up to 70%.

If the initial test radiograph reveals a satisfactory reduction (Fig. 18B–30), a short-arm cast is applied, incorporating a 1-inch-wide padded aluminum splint over the dorsum of the involved finger. The splint is fabricated so that it extends approximately one-half inch beyond the tip of the digit. The PIP joint is then reduced and the splint bent to conform to the amount of flexion required to maintain the reduction; usually this is about 60 degrees, but occasionally a bit more is necessary. A strip of adhesive tape is applied over the full extent of the aluminum splint and secured to the volar aspect of

the plaster cast in order to prevent inadvertent straightening of the splint (Fig. 18B–31). An essential part of the technique is that the proximal segment of the finger must be held firmly to the splint with a strip of half-inch adhesive (see Fig. 18B–31); if this is not done, MP joint flexion causes the proximal phalanx to pull away from the splint, allowing extension of the PIP joint and loss of reduction. Immediately after application of the splint, true lateral radiographs are taken with the PIP joint in both extension and flexion to ensure that the reduction has been maintained.

Active flexion of the involved finger is encouraged from the outset, with PIP extension blocked by the splint. The patient is seen at weekly intervals, and if a lateral radiograph shows continued maintenance of the reduction, the adhesive tape holding the splint to the cast is cut, the splint is extended to reduce

FIGURE 18B–30
These two radiographs show the difference between acceptable *(bottom)* and unacceptable *(top)* reductions of a fracture-dislocation of the PIP joint.

the amount of flexion in the PIP joint, and the tape is reapplied (Fig. 18B–32). Usually, the flexion can be reduced approximately 15 degrees each week, so that full extension is achieved by 4 to 6 weeks (Fig. 18B–33). At that time, the splint is removed and buddy taping is applied for an additional 2 to 3 weeks. Most patients regain full active flexion from the beginning and achieve extension gradually over

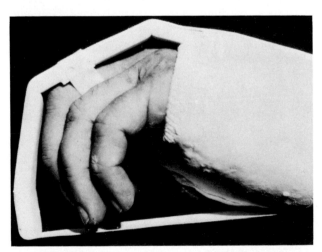

FIGURE 18B–31
Dorsal extension block splinting for treatment of fracture-dislocation of the PIP joint. An essential part of the technique is that the proximal phalanx must be secured to the splint with tape; if it is not, the PIP joint extends when the patient flexes the MP joint. Another important element is that the aluminum outrigger must be secured to the cast with tape to prevent inadvertent straightening of the splint, thereby allowing excessive extension of the PIP joint.

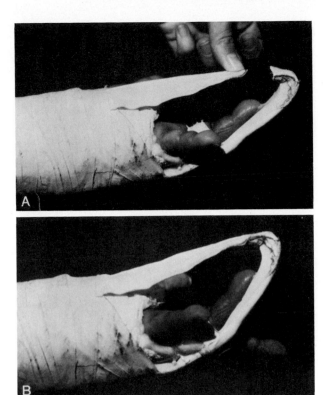

FIGURE 18B–32
When the patient being treated with dorsal extension block splinting is seen weekly, *A,* the tape holding the splint is cut, the splint is straightened a bit, and *B,* the tape is reapplied. Active flexion of the PIP and DIP joints is encouraged from the outset.

a 4- to 6-week period as the splint is extended. If the patient has not regained full extension of the PIP joint by 6 weeks, dynamic splinting may be initiated at that time.

Strong's [244] simpler type of extension block splint fashioned from two pieces of padded aluminum splint (Fig. 18B–34) is a bit less secure but is probably a reasonable method to use in reliable patients.

It must be reemphasized that if a satisfactory closed reduction cannot be achieved, dorsal exten-

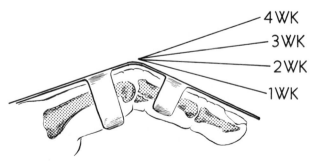

FIGURE 18B–33
The dorsal extension splint is gradually straightened at weekly intervals and may be discontinued at 4 weeks.

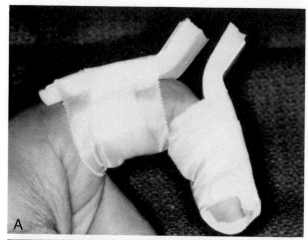

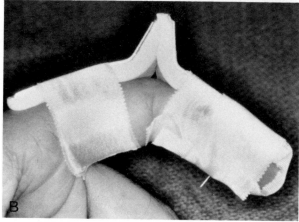

FIGURE 18B–34

A and *B,* Strong's simpler type of extension block splinting [244] is a bit less secure but is probably a reasonable method to use in reliable patients. However, an athlete should certainly not return to competition wearing this type of splint.

sion block splinting should not be used, and the treatment options are either open reduction or volar plate arthroplasty. If the latter procedure is chosen, the surgeon should follow precisely the excellent description of the operation by Eaton and Malerich [132].

Postoperative Management, Rehabilitation, and Criteria for Return to Play

PIP fracture–dislocation is an injury that almost always interrupts participation in athletics. As noted above, this injury demands early treatment, and the functional end result is likely to be significantly impaired if treatment is deferred until the end of the season. If dorsal extension block splinting is used, the minimum time required before the return to participation is 3 to 4 weeks; that is, that period

in which the finger is actually in the splint. If the above protocol has been followed and the patient has maintained good flexion of the PIP joint while gradually regaining extension, he should have a reasonably good range of motion when the splint is removed. At that time, he can probably begin to play with the finger protected by full-time buddy taping, which he should wear for the remainder of the season. The time out of competition is probably closer to 6 weeks if open reduction or volar plate arthroplasty has been performed.

Metacarpophalangeal Joints of the Fingers

Anatomy

The MP joint is a condyloid joint that allows flexion, extension, abduction, adduction, and a limited amount of circumduction [145]. The globular head of the metacarpal articulates with the reciprocally concave base of the proximal phalanx, although the surface of the latter has a slightly less acute curve than the metacarpal head. The articular surface of the head is broader volarly than dorsally, allowing for the recesses in the dorsolateral aspects of both sides of the head to accommodate the origin of the collateral ligaments. The stability of the joint depends on the collateral ligaments and volar plate, which together form a snug, boxlike configuration, as noted by Eaton [130] (Fig. 18B–35).

Each collateral ligament has essentially two parts: an upper, cordlike metacarpophalangeal ligament and a lower, fan-shaped accessory, or metacarpoglenoidal ligament (see Fig. 18B–18). The latter, which

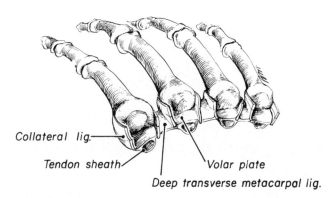

FIGURE 18B–35

Eaton has described the snug, boxlike configuration formed by the collateral ligaments and volar plates of the MP joints. Note also the deep transverse metacarpal ligaments, which Eaton has called the intervolar plate ligaments. (From Eaton, R.G. *Joint Injuries of the Hand.* Springfield, Charles C Thomas, 1971.)

attaches directly into the volar plate, is less rigid, so that it can fold on itself when the joint flexes. The metacarpal head has an eccentric configuration: the distance from the center of rotation to the articular surface is greater in a volar direction than it is dorsally (see Fig. 18C–17). This produces a camlike effect on the collateral ligaments, making them tight in flexion and lax in extension. One can readily demonstrate this in his own hand by noting that passive abduction and adduction are much more restricted with the joint held in maximum flexion than in full extension. The tendency for loose or relaxed ligaments to tighten following injury and immobilization combined with the eccentricity of the metacarpal head creates a situation that favors extension contractures of the MP joint.

The volar plate of the MP joint is a relatively thick, fibrocartilaginous condensation of the joint capsule that forms the anterior wall of the joint. Distally, it is firmly attached to the base of the phalanx, but its proximal attachment to the neck of the metacarpal is more areolar and flexible, allowing passive hyperextension of the joint and permitting the volar plate to fold on itself in flexion. The volar plates of the four palmar metacarpals are firmly held together by the continuous deep transverse metacarpal ligament (see Fig. 18B–35). Perhaps more appropriately, Eaton [130] has called this structure the intervolar plate ligament.

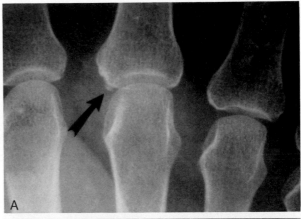

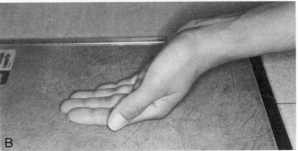

FIGURE 18B–36
A, MP joint collateral ligament injury is occasionally (but not usually) associated with a small avulsion fracture *(arrow). B,* The Brewerton view may be helpful in delineating obscure avulsion fractures of the MP joint collateral ligaments. The x-ray beam is centered over the MP joints and directed 30 degrees from the ulnar side.

Lateral MP Dislocations (Collateral Ligament Injuries)

Isolated injuries of the collateral ligaments of the finger MP joints are relatively uncommon in the general population, presumably because of their shielded proximal position within the web space and the protection provided by the adjacent digits. They are probably more common in athletes than generally recognized, because of the unusual forces inflicted on the hand during contact, and even during so-called noncontact sports such as basketball. The diagnosis may be missed early, and the patient may present later with rather vague pain in the region of the MP joint. Most of these injuries appear to involve the radial collateral ligament, and the mechanism of the injury is usually a strong, ulnarly directed force on the MP joint.

Clinical Evaluation

The diagnosis may be suggested by local tenderness and subtle swelling in the valley between the two metacarpal heads directly over the involved collateral ligament. The most specific and significant clinical sign is pain in response to a lateral stress applied to the digit with the MP joint held in full flexion. Because of the normal laxity of the collateral ligaments with the joint in full extension, stress in this position is usually not painful. Radiographs frequently demonstrate no abnormality, but they may reveal a small fragment that has been avulsed from the metacarpal head (Fig. 18B–36). Even relatively large collateral ligament avulsion fractures may be difficult to see on routine radiographs [143], and McElfresh and Dobyns [182] have suggested the use of the Brewerton view (Fig. 18B–36A) in order to better see the lesion. Other associated bony injuries include an intra-articular corner fracture at the base of the proximal phalanx (Fig. 18B–37) in the acute injury, or an area of cortical irregularity at the site of attachment of the collateral ligament in chronic injuries (Fig. 18B–38).

Ishizuki [161] reported the largest series of MP joint collateral ligament injuries: 22 patients, all of whom had arthrograms. The technique of MP joint arthrography is described in detail in his article.

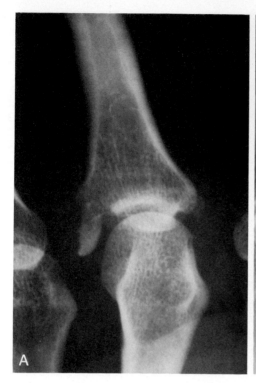
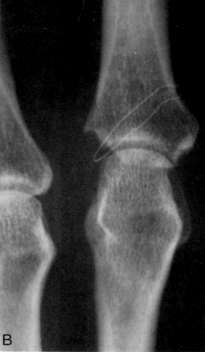

FIGURE 18B–37
A, An intra-articular corner fracture of the base of proximal phalanx is a variant of MP joint collateral ligament injury. *B,* This patient was treated with excision of the fragment and reinsertion of the ligament.

Treatment Options for Acute Injuries

If the diagnosis is made early (within 10 to 14 days), and if a significant avulsion fracture is not present radiographically, splinting of the joint in 50 degrees of flexion for 3 weeks is the recommended treatment [130], although Ishizuki [161] recommended that grossly unstable joints should be repaired surgically. If the radiograph shows wide dis-placement of a tiny avulsion chip (greater than 2 to 3 mm), or if the fragment involves more than 20% of the articular surface and is displaced or rotated, primary operative treatment is indicated.

Postoperative Management, Rehabilitation, and Criteria for Return to Play

Following repair of a pure ligamentous avulsion or rupture, immobilization of the MP joint in 50 to 70 degrees of flexion is indicated for 3 weeks. Football linemen or linebackers could return to competition with the fingers splinted in this position, but ball handlers and other skill position players will probably be out for at least 3 to 4 weeks. After healing, gentle active and passive motion exercises should be commenced to ensure the recovery of full MP extension, and protection against lateral stresses should be continued for at least another month.

Treatment Options for Chronic Injuries

In reality, most of these injuries are probably not recognized acutely, and many present with chronic symptoms late in the season or, more frequently, after the season is over. The major complaint is

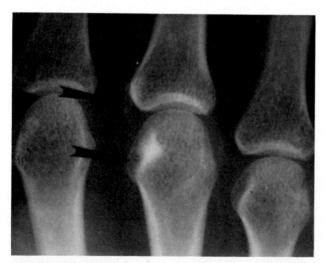

FIGURE 18B–38
Patients with chronic MP collateral ligament injuries may show some cortical irregularity or ectopic calcification on radiograph. These injuries cause discomfort for a long period of time and are difficult to treat.

generally mild to moderate pain, especially with the MP joint in full flexion; instability is usually not a problem. A trial of a steroid injection and splinting may alleviate the symptoms and preclude the need for operative treatment. If pain persists, surgical repair is indicated. There are insufficient reports in the literature to document the best treatment, but the choice is between ligament reconstruction (which may lead to limitation of joint motion) or ligament resection (which may increase the instability, but may be acceptable on the ulnar side of the joints).

Dorsal MP Dislocations

Two types of dorsal dislocations occur in the MP joint: simple and complex. It is important to differentiate between these two, because simple dislocations can be reduced by closed manipulation, whereas complex dislocations are irreducible by closed methods and require open reduction. Both result from hyperextension injuries, and in both the volar plate is torn from its proximal insertion into the neck of the metacarpal. Distinction between the two can usually be made fairly easily by clinical and radiographic examination, and the surgeon should not have to resort to multiple unsuccessful attempts at closed reduction to conclude that he is dealing with an irreducible dislocation.

Simple Dorsal Dislocation (Subluxation)

Eaton [130] correctly refers to simple dorsal dislocation as subluxation of the joint, because the articular surfaces are still in partial contact, with the proximal phalanx resting in 60 to 90 degrees of hyperextension on the dorsum of the metacarpal head. Closed reduction is relatively simple, although McLaughlin [184] cautioned that it is possible to convert a simple dislocation into a complex dislocation if the reduction is performed by traction alone.

Method of Reduction. Closed reduction of MP subluxation is usually successful using the following technique (modified from McLaughlin [184] and Malerich and associates [179]): first, the wrist and the IP joints are flexed in order to relax the flexor tendons. The proximal phalanx is hyperextended to a position almost perpendicular to the metacarpal head. The base of the phalanx is then *pushed* volarly across the articular surface of the metacarpal in an effort to push the volar plate out of the joint. The

phalanx should maintain contact with the head of the metacarpal throughout the entire manuever in an effort to prevent reentrapment of the volar plate in the joint. When the proximal phalangeal base is across the metacarpal articular surface, the thumb or digit is straightened and flexed; an audible "clunk" may be heard if the reduction is successful. Gentle active and passive flexion and extension confirm the repositioning of the joint, which is usually found to be stable in all directions. Radiographs should be taken to confirm the restoration of articular congruency and the absence of associated fractures.

Although the reduction can frequently be done without anesthesia, we prefer to do it under wrist or axillary block. Otherwise, muscle contracture resulting from pain may preclude a successful reduction and lead the surgeon to the erroneous conclusion that he is dealing with a complex dislocation. If the team physician on the sideline sees such an injury immediately after it has occurred, a single attempt at closed reduction without anesthesia is indicated, unless the signs of a complex dislocation (see below) are present.

Postreduction Management, Rehabilitation, and Criteria for Return to Play. Although a short (7 to 10 day) period of immobilization with the MP joint in 50 to 70 degrees of flexion is acceptable treatment, we generally prefer to permit immediate active motion, preventing hyperextension by buddy taping the involved digit to one or two adjacent digits. No loss of time from athletic participation is required. Recurrent dislocations or chronic symptoms following simple MP dislocations are rare.

Complex (Irreducible) MP Dislocation

Although Farabeuf [137] first coined the term "complex dislocation" in 1876, irreducible dislocation of the MP joints is said to have been described by Malgaigne in his 1855 text [141, 257]. Isolated case reports were reported in the early English literature [111, 113, 220], but it was not until 1957 that the pathologic anatomy became widely appreciated. In that year, Kaplan [164] published the classic article in which he described the buttonholing of the metacarpal head into the palm and the constricting anatomic structures that prevent reduction by closed methods. Although not all authors have agreed with Kaplan regarding the role of these various structures, most agree that the single most important factor is the interposition of the volar plate between the base of the proximal phalanx and

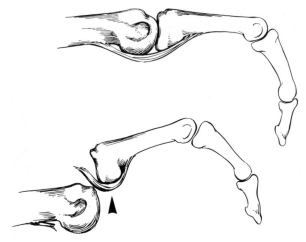

FIGURE 18B–39
The single most important element preventing reduction in a complex MP dislocation is interposition of the volar plate between the base of the middle phalanx and the metacarpal head.

the head of the metacarpal [112, 122, 147, 150, 190] (Fig. 18B–39).

Numerous case reports in the earlier literature implied that this is a rare injury, but more recently reported larger series [109, 115, 147, 157, 184, 190, 200, 248] suggest that it is not as uncommon as formerly believed.

Clinical Evaluation. Recognition of a complex dislocation should be relatively simple, because there are three clinical and radiographic clues to the diagnosis.

The complex dislocation does not present as dramatic an appearance as the simple dislocation (subluxation) described above. The joint is only slightly hyperextended, with the proximal phalanx lying on the dorsum of the metacarpal head and the finger partially overlapping the adjacent digit. The IP joints are slightly flexed. Radiographically, the proximal phalanx and metacarpal are nearly parallel, with only slight angulation.

A consistent finding of complex dislocations is a puckering of the volar skin. This is more difficult to see when the index finger is involved, because the skin dimple lies within the proximal palmar crease (Fig. 18B–40). It is more readily apparent when the dislocation occurs in the thumb, and the dimple is present in the thenar eminence.

A pathognomonic radiographic sign of complex dislocation is the presence of a sesamoid within a widened joint space [147, 203, 230, 245]. Because the sesamoids reside within the volar plate, the presence of a sesamoid in the joint space should be considered an unequivocal sign of a complex dislocation (Fig. 18B–41). This finding should not be confused with a chip fracture of the metacarpal head (Fig. 18B–42), which may also occur but does not

necessarily carry with it the same diagnostic significance. Tsuge and Watari [248] noted that this frequent concomitant injury represents an avulsion fracture of the ulnar tuberosity by the collateral ligament.

Complex dislocations occur most commonly in the index finger, followed in incidence by the thumb and small finger and, rarely, the long and ring fingers. Many other combinations have been described [158, 159, 254], including simultaneous dislocation of all four MP joints [260, 194].

Treatment Options. An attempt at closed reduction should be made in all dislocations of the MP joint. Even if the pathognomonic signs of a complex dislocation are present, the surgeon is justified in making a single attempt at gentle reduction under adequate anesthesia, but he should be prepared to follow this with an immediate open reduction if the manipulation is unsuccessful. The preferred method of closed reduction is that described by McLaughlin [184].

If operative treatment is required, there is some

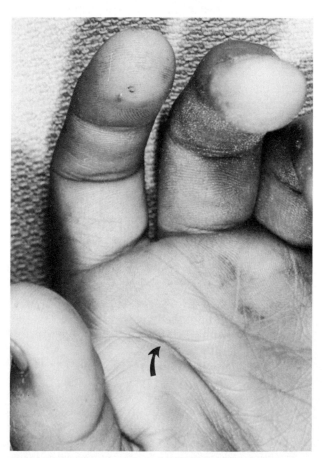

FIGURE 18B–40
A consistent finding in patients with complex dislocation of the index MP joint is puckering of the palmar skin, which may be difficult to see because it lies directly in the distal palmar crease *(arrow)*.

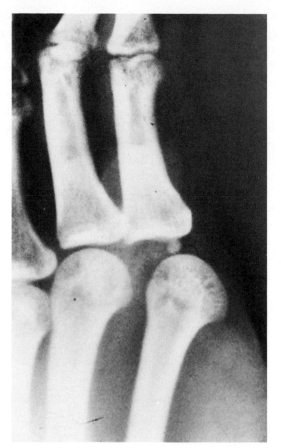

FIGURE 18B–41
A pathognomonic radiographic sign of complex MP dislocation is the presence of a sesamoid in the widened joint space, indicative of interposition of the entrapped volar plate.

period of time, and a therapist-assisted exercise program should be initiated.

If the surgeon is concerned about possible instability in hyperextension, a dorsal extension block splint can be used, in which case return to competition is delayed for 3 to 4 weeks. Hubbard [156] has suggested the use of continuous passive motion postoperatively, but we have not found this to be necessary.

Volar MP Dislocations

Volar dislocation of the MP joint is quite rare [116, 118, 218, 258]. Perhaps because it is such an uncommon injury, the pathologic anatomy of volar MP dislocations is not as well understood as in the more common complex dorsal dislocation. The volar plate [116, 218], collateral ligament [116, 198], and dorsal capsule [218, 614] have all been implicated as a cause of irreducibility. As a result, some authors [116, 198] have suggested that both dorsal and volar approaches may be necessary to achieve a reduction.

An attempt at closed reduction should always be made under adequate anesthesia (wrist block should suffice). The principles of closed reduction should be followed, reversing the maneuver used for dorsal MP dislocation described on page 973; that is, hyperflexing the MP joint and then *pushing* the proximal phalanx back into position rather than applying traction. Failure to achieve a congruous reduction (following which the patient should have

controversy about whether the preferred exposure is dorsal [114, 137, 153, 248] or volar [164, 184]. The dorsal approach is thought by its advocates to be safer because it avoids jeopardizing the neurovascular bundle, which lies just beneath the volar skin and is, therefore, more vulnerable with the palmar approach. The dorsal approach has the disadvantage of necessitating that the volar plate be split longitudinally if it cannot be pushed out of the joint with a flat, blunt elevator. If there is a concomitant fracture of the metacarpal head, the dorsal approach generally provides better access.

Postoperative Management, Rehabilitation, and Criteria for Return to Play. The joint is invariably stable after reduction, a fact usually confirmed at the operating table by passively moving the joint through a full range of motion. Thus, no immobilization is necessary postoperatively, and early active motion is encouraged, protecting the finger with buddy taping. Therefore, in most cases, the athlete can return to competition as soon as he has regained reasonably good range of motion. In some instances, the joint remains swollen and stiff for a protracted

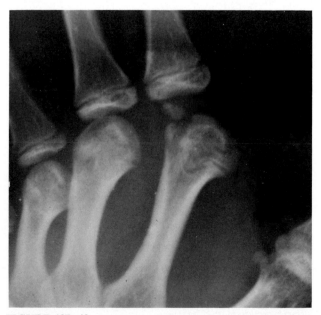

FIGURE 18B–42
A concomitant injury sometimes seen with a complex MP dislocation is a chip fracture of the metacarpal head. This should not be confused with the entrapped sesamoid shown in Figure 18C–41.

unrestricted active motion of the joint under wrist block anesthesia) is an indication for open reduction. We have very limited experience with this injury but would probably make the initial approach from the palmar side, adding a dorsal exposure if the joint could not be reduced under direct vision from the volar aspect.

MP Joint of the Thumb

Anatomy

The MP joint of the thumb is basically a condyloid joint, allowing flexion, extension, abduction, adduction, and a very limited amount of rotation [145]. The range of normal motion in the MP joint of the thumb varies widely [126] and appears to be related to the contour of the metacarpal head [163]. Harris and Joseph [154] noted that motion in joints with flat or flattish metacarpal heads tends to be considerably limited. Palmer and Louis [208] recorded that the normal arc of MP motion ranges from 5 to 115 degrees. Coonrad and Goldner's studies [126] revealed that the normal range of abduction–adduction varies from 0 to 20 degrees, with an average of 10 degrees (measured with the joint in 15 degrees of flexion). Mediolateral stability is provided mainly by the collateral ligaments [236].

On the ulnar aspect of the joint, the adductor pollicis muscle is inserted partly through the ulnar sesamoid bone into the volar plate and partly through a strong tendon directly into the proximal phalanx. Additional fibers fuse with the ulnar expansion of the dorsal aponeurosis [231, 236]. This part of the dorsal aponeurosis is called the adductor aponeurosis, and it plays an important role in the pathomechanics of injuries of the ulnar collateral ligament. The adductor aponeurosis directly overlies the ligament and must be divided in order to provide operative exposure. Despite some contradictory studies by Kaplan [165], most authors now accept Stener's findings that passive stability of the joint is provided by the collateral ligament, with the adductor aponeurosis providing active stabilization against violent abduction forces the thumb.

Ulnar Collateral Ligament Injury (Gamekeeper's Thumb, Skier's Thumb)

Mechanism of Injury

A sudden valgus (abduction) stress (probably combined with hyperextension [232]) applied to the MP joint of the thumb results in partial or complete disruption of the ulnar collateral ligament and volar plate. In 1955, Campbell [124] reported that chronic laxity of this ligament can develop without a specific incident of acute trauma, and he found this to be an occupational deformity in the hands of British gamekeepers. Their customary method of killing wounded rabbits was such that, over a period of time, attenuation of the ligament resulted in chronic instability of the joint. Through common usage, the term gamekeeper's thumb has come to include any injury of the ulnar collateral ligament, although most of these injuries seen currently are acute injuries, rather than the chronic stretching of the ligament reported by Campbell.

Several authors [127, 144, 195] have suggested

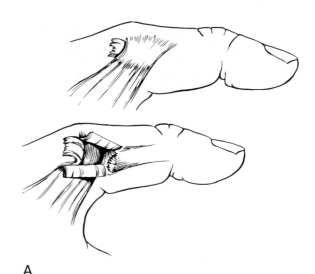

A

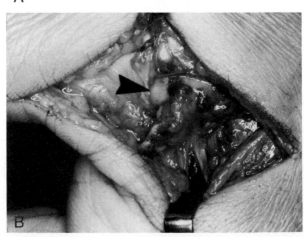

B

FIGURE 18B–43

The Stener lesion seen in gamekeeper's (skier's) thumb. *A,* The distal end of the ulnar collateral ligament has been ruptured off the base of proximal phalanx and is turned 180 degrees, facing proximally. The adductor tendon then becomes interposed between the torn end of the ligament and its site of insertion. *B,* At operation, the torn end of the avulsed ligament is seen end-on, just proximal to the adductor aponeurosis (beneath the *arrow*). Healing cannot occur until the tendon is repositioned into the base of proximal phalanx.

that acute injuries of the ulnar collateral ligament of the thumb should more properly be called "skier's thumb," because this is probably the most common mechanism of injury and, in fact, is one of the most common ski injuries [117, 125, 135, 151]. Indeed, it has been noted that this injury was reported in the German literature in 1939–40 [144, 240, 249], and skier's thumb is the preferred appellation in Europe [144]. The ski pole has been implicated as the causative factor, and there apparently has been no decrease in incidence of this injury with the use of newer types of strapless poles [125, 127, 135, 191]. Primiano [213] suggested that if a strapless pole is used, injury is less likely to occur if the pole does not block full flexion of the IP joint. A special glove has been designed [136] in an effort to protect the ulnar collateral ligament.

Stener Lesion

Stener's important contribution [236] to our understanding of this injury was the recognition that the adductor aponeurosis frequently becomes interposed between the two ends of the torn ligament, thereby preventing adequate healing (Fig. 18B–43). The most frequent site of rupture is directly from the distal attachment of the ligament into the proximal phalanx, although interposition may occur even if the ligament is torn through its substance. Stener found this interposition in 25 of his 39 cases, and other authors [173, 196, 210, 262] have reported an incidence of the Stener lesion ranging from 14% to 83%. Our own impression is that the Stener lesion is present in more than 50% of acute ruptures of the ulnar collateral ligament.

Clinical Evaluation

The patient presents with a painful, swollen MP joint of the thumb. Usually, the point of maximum tenderness can be localized to the ulnar aspect of the joint. A particularly important point in the evaluation of these patients is to differentiate a sprain (partial tear) from a rupture (complete tear). The obvious way to do this is to determine the amount of radial deviation produced by abduction stress of the MP joint, but this may be more difficult than it sounds. If the joint opens up easily (Fig. 18B–44), the diagnosis of rupture is obvious, but pain and muscle spasm may limit passive abduction of the thumb and give a false-negative impression. Unless there is easily demonstrable gross instability, we believe that the abduction stress test should be done under some type of anesthesia. Most authors

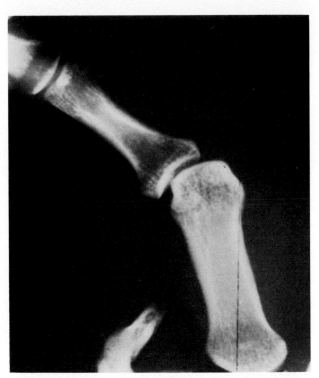

FIGURE 18B–44
Stress radiographs in patients with ulnar collateral ligament injuries of the thumb are indicated if complete rupture is not obvious on clinical examination.

[121, 166, 168, 180, 208, 210, 225, 227] advocate local infiltration of the ligament; others prefer a block of the median and radial nerves at the wrist [139, 196, 208]. Testing should be done with the MP joint in both flexion and extension, and the opposite thumb should be used for comparison. What constitutes an abnormal or positive stress test remains somewhat controversial: Smith [232] said 45 degrees of abduction; Frank and Dobyns [139] and Bowers and Hurst [121] suggested more than 10 degrees greater than the opposite side, and Palmer and Louis [208] indicated that 35 degrees (tested in full flexion) represented a complete tear. It may be rather difficult to clinically measure the angle of abduction, and Bowers and Hurst [121] noted that clinical estimates consistently were 5 to 15 degrees greater than radiographic measurements in their studies. Therefore, if there is any question about the diagnosis, we believe that the stress test should be measured radiographically.

Radiographic Examination

Several authors [139, 238] have suggested that routine radiographs should be made before stressing the joint, in order to prevent possible displacement of an undisplaced fracture. Three types of avulsion

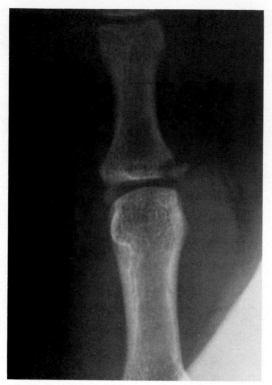

FIGURE 18B–45
A typical fracture seen in gamekeeper's thumb is a small fragment avulsed by the ulnar collateral ligament.

fractures may be seen in the initial films; the most common types are a small fragment pulled away from the base of the proximal phalanx (Fig. 18B–45) and a large, intra-articular fracture involving one-fourth or more of the articular surface of the base of the proximal phalanx (Fig. 18B–46). Louis and associates [178] have identified a third type of avulsion fracture that is not attached to the ulnar collateral ligament but rather to the volar plate. Stothard and Caird's arthrographic studies [241] showed two fractures that probably represented this type of injury, which is not unstable and can be treated with cast immobilization alone.

Stener and Stener [238] have stressed the importance of differentiating avulsion and shear fractures (Fig. 18B–47). The latter may originate from the *radial* side of the head of the metacarpal and are decidedly less common in our experience than avulsion fractures, which arise from the ulnar aspect of the base of the proximal phalanx.

If no fracture is seen on the initial radiographs and a clear distinction cannot be made on clinical examination between sprain and rupture, or if documentation of the rupture is desired, stress radiographs are indicated. We believe that stress radiographs should be taken, even though Mogan and Davis [195] advised against their use. We agree with Engel, who stated that stress views without anesthe-

sia are of questionable value [134]. Bowers and Hurst [121] believe that the incomplete relief of pain provided by local infiltration anesthesia is a safeguard against further disruption of any of the torn structures. Nonetheless, we prefer to perform the stress radiographs under median and radial nerve wrist block. Films are taken with the MP joint in full extension, with the radial abduction stress applied to the thumb by the surgeon, who wears leaded gloves. To our knowledge, no satisfactory method has yet been described to record radiographically the stress test done with the MP joint in flexion.

Arthrography

Several authors have attempted to evaluate ruptures of the MP collateral ligaments with arthrograms [121, 134, 176, 219, 222, 252]. This is done by injecting 1 to 2 mL of contrast material (1.2 mL 60% Renografin mixed with 0.8 mL 1% lidocaine) into the joint with a tuberculin syringe in the interval between the extensor pollicis brevis and radial collateral ligament [121, 176]. In our opinion, arthrography offers little additional information, unless it can be used to successfully identify the Stener lesion, as Bowers and Hurst [121] attempted to do. We

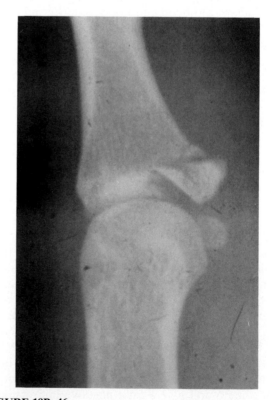

FIGURE 18B–46
A variant of gamekeeper's thumb is a large fracture fragment disrupting a significant portion of the articular surface. Open reduction and anatomic restoration is indicated.

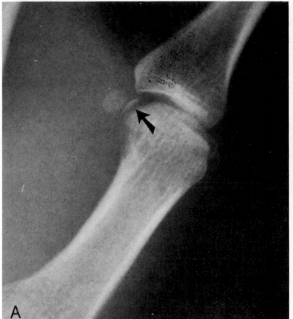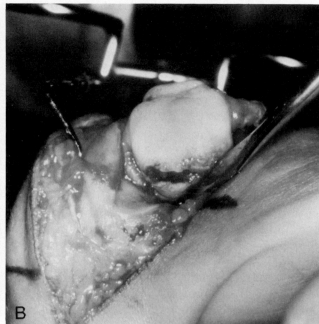

FIGURE 18B–47
A, A less common fracture seen in gamekeeper's thumb is a shear fracture, seen radiographically on the volar side of the joint. It is important to always explore the joint at the time of operation to look for *B,* shear fractures or osteochondral defects.

agree with them that this specific use of the arthrogram has great merit, because it can aid the surgeon in deciding between operative and nonoperative treatment. Mogan and Davis [195] reported that they were able to identify the Stener lesion by arthrography, but they did not correlate their radiographic impressions with operative findings. We agree with other authors [191, 241] who believe that it is difficult to identify the Stener lesion by arthrography with any degree of confidence.

Arthroscopy

Arthroscopy does not at this time appear to have any value in the diagnosis or treatment of MP joint collateral ligament injuries, but Vaupel and Andrews [250] reported its use in one case to treat a chondral defect in the base of proximal phalanx.

Treatment Options for Acute Injuries

Partial Tears. There is general agreement that partial tears (sprains) of the ulnar collateral ligament should be treated nonoperatively. We prefer a well-molded thumb spica cast with the MP joint in slight flexion (hyperextension must be avoided) for 3 to 6 weeks, depending on the severity of the injury. Rovere and associates [223] described a fiberglass

"mini" thumb spica that they used in hockey players, allowing continued participation in the sport. Primiano [214] also advocated the use of a modified thumb spica cast that allows full flexion and extension of the wrist.

Complete Tears. Since the publication of Stener's classic article in 1962 [236], there has been increasingly strong support for the operative treatment of all acute ruptures of the ulnar collateral ligament of the thumb [139, 166, 177, 180, 206, 232], even in the rare case of complete rupture in a child with an open epiphyseal plate [253]. Theoretically, a complete rupture without a Stener lesion can be treated satisfactorily nonoperatively, and Coonrad and Goldner [126] have advocated cast immobilization. There are, however, several reasons why we prefer early operative repair of all acute complete ulnar collateral ligament injuries: (1) the Stener lesion is definitely present in a significant number of cases, and we are not confident that we can differentiate preoperatively the presence or absence of that lesion with any of the currently available modalities; (2) operative repair of the acute tear is a relatively uncomplicated procedure with minimal morbidity; and (3) as is true in ligamentous injuries in other joints, the results of primary repair are better than with any of the late reconstructive operations [206].

Associated fractures that are also indications for operative treatment, in our opinion, are (1) a displaced intra-articular corner fracture involving 25%

or more of the articular surface at the base of the proximal phalanx, and (2) a small avulsion fracture displaced more than 5 mm. Smith [232] suggested that volar subluxation of the proximal phalanx is also an indication for surgical treatment.

Authors' Preferred Method of Repair

Acute Tear of the Ulnar Collateral Ligament of the Thumb. An oblique or chevron-shaped incision is used across the dorsum of the MP joint and extended proximally along the radial border of the first metacarpal. Branches of the superficial radial nerve should be identified and protected in the deep subcutaneous tissue. The edematous stump of the ruptured ligament is usually the first structure encountered if a Stener lesion is present. It usually lies on top of the proximal edge of the adductor aponeurosis, which must be split longitudinally and retracted volarly to expose the ulnar aspect of the MP joint and the base of the proximal phalanx. The ligament is then replaced and repaired by direct suture (if ruptured in the middle two-thirds) or by a pull-out technique (Fig. 18B–48) (if avulsed completely from the proximal phalangeal bone). If a fragment of bone is attached to the avulsed ligament, the method of repair depends on the size of the ligament. If it is small, it may be excised and reattachment is accomplished by the pull-out technique. If the fragment is large, an anatomic repositioning into the defect is carried out and is secured by the pull-out method or by Kirschner wires or small mini-screws.

Dray and Eaton [128] emphasized that, at the time of closure, four additional steps should be taken: (1) a suture is placed between the distal volar portion of the repaired ligament and the volar plate to secure the "vital three-dimensional complex"; (2) the dorsal ulnar capsule is carefully repaired to prevent volar subluxation or rotation of the joint; (3) the stability of the repair is gently stressed; and (4) the joint is flexed to 20 degrees and transfixed in an overcorrected ulnar-deviated position with an oblique transarticular Kirschner wire. (We do *not* routinely use a Kirschner wire.)

Postoperative Management, Rehabilitation, and Criteria for Return to Play

If the ulnar collateral ligament is repaired surgically, it must be protected postoperatively. At 10 days, the splint and sutures are removed and a

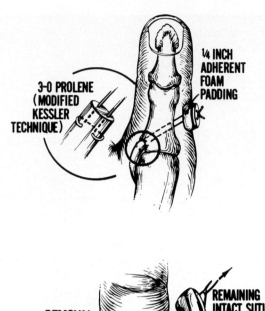

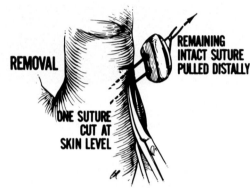

FIGURE 18B–48
The preferred pull-out technique. Prolene suture is used because it is much easier to use and less painful than wire for the patient at the time of removal, which is accomplished by cutting one end of the suture and pulling through on the other. This type of pull-out suture can be used in the hand for any situation in which a ligament or tendon is sutured into bone.

thumb spica cast is applied, leaving the IP joint free. The Kirschner wire is carefully protected, or it can be removed through a window in the cast. If the athlete's position does not require dexterous use of the hand, he may return to competition at this time wearing the thumb spica cast or splint. At 4 weeks, the cast is removed (as well as the Kirschner wire if it is still in place), and flexion-extension exercises are instituted for the MP joint. A removable splint is worn to protect the thumb for an additional 2 to 3 weeks, and forceful stress and participation in strenuous athletic activities without splint protection are not allowed until 10 to 12 weeks after the operation, depending on the mobility of the MP joint and the strength of the thumb.

Treatment of Chronic Injuries

If the patient is seen within 2 to 3 weeks of injury, operative repair can usually be accomplished with a reasonable chance of success. Unfortunately, chronic untreated injuries of the ulnar collateral ligament of the thumb are commonly seen for the first time weeks or months following injury, especially in athletes who "jam" the thumb and are not referred for evaluation until after the season.

Treatment of Partial Tears

Untreated sprains of the MP collateral ligaments can produce long-term refractory symptoms, which may be difficult to fully alleviate. If there is no evidence of instability on stress examinations, we generally treat these patients with cast immobilization (with or without a local steroid injection) for 3 weeks, followed by an intensive course of physical therapy. Perhaps the most important aspect of management of these patients is to make them aware of the chronicity of the problem, emphasizing that it is likely to take several months for their symptoms to subside.

Treatment of Complete Tears

The patient with an untreated or inadequately treated rupture of the ulnar collateral ligament with pain and demonstrable instability of the joint is a candidate for reconstruction, because post-traumatic arthritis is likely to develop unless stability is restored. It must be emphasized, however, that no type of soft-tissue reconstruction will relieve the patients' pain if post-traumatic arthritis is already established. If preoperative radiographs demonstrate significant arthritic changes, or if significant articular damage is seen at operation, arthrodesis of the MP joint is the treatment of choice [126, 139, 152, 202].

Many different types of operative procedures have been described for treatment of chronic laxity of the ulnar collateral ligament of the thumb [106, 108, 140, 174, 202, 224, 225, 232, 242]. However, it should be underscored that, as with most ligament injuries, the result of reconstruction is generally not as good as the results of primary repair. For this reason, if a complete rupture of the ulnar collateral ligament is recognized acutely, primary repair is the treatment of choice, and it is not appropriate to defer surgical treatment until the end of the season.

Radial Collateral Ligament Injury

Injury of the radial collateral ligament of the MP joint of the thumb is less common than that of the

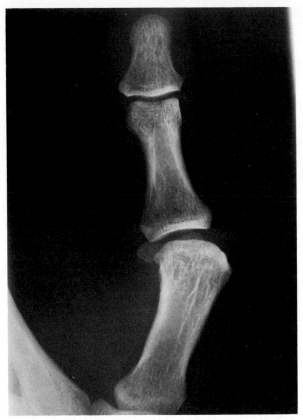

FIGURE 18B–49
Patients with chronic instability of the radial collateral ligament of the thumb MP joint usually present with a tender prominence of the radial aspect of the metacarpal head, which on radiograph is seen to be due to volar and ulnar subluxation of the proximal phalanx on the metacarpal.

ulnar collateral ligament. Perhaps partly for this reason, the injury is frequently not diagnosed early, and our experience has been that these patients are almost always seen late. Such patients usually present with a tender prominence of the radial aspect of the metacarpal head (Fig. 18B–49) and often have pain with activities such as opening a large jar lid or a car door [123].

There is scanty information in the literature regarding the acute management of such injuries, but anatomically, a situation analogous to the Stener lesion cannot exist on the radial side of the thumb. Theoretically, then, nonoperative treatment (cast immobilization) of a complete rupture of the radial collateral ligament is more likely to be successful than similar treatment of an ulnar collateral ligament rupture. However, in one of the few collected series of such patients, Woods and associates [259] advised primary repair for complete ruptures. In either case, immobilization in a thumb spica cast for 3 to 4 weeks is required.

Dislocations of the Thumb MP Joint

As in the fingers, dorsal MP joint dislocations are more common than volar dislocations. These injuries may be simple (reducible) or complex (irreducible), usually due to interposition of the volar plate [129], although entrapment of the flexor pollicis longus has also been reported [205]. Volar dislocations seem to be almost always associated with collateral ligament ruptures and appear to be more commonly irreducible, due to interposition of the dorsal capsule and one or both extensor tendons [149, 192, 198]. Less severe dorsal capsular tears, with or without rupture of the extensor pollicis brevis, have also been reported [209].

Clinical Evaluation

The most important initial step in management of the dorsal dislocation is to differentiate between a simple and a complex dislocation. As with MP dislocations in the fingers, it is usually possible to make this distinction on the basis of clinical findings rather than after repeated unsuccessful attempts at closed reduction. The attitude of the finger offers the examiner the first clue as to the condition of the joint. In a simple dislocation (subluxation), the phalanx usually rests on the head of the metacarpal in nearly 90 degrees of hyperextension. In the complex dislocation, the proximal phalanx is more nearly parallel to the metacarpal, with only slight hyperextension. The most important diagnostic clinical sign of a complex dislocation is a skin dimple found on the volar aspect of the thenar eminence. Demonstration on radiographs of a sesamoid within the widened joint space is also pathognomonic of a complex dislocation.

Treatment Options

Simple dislocations of the MP joint can be reduced easily by closed manipulation; complex dislocations cannot. Even in the presence of the pathognomonic skin dimple, however, we believe that a single attempt at closed reduction under adequate anesthesia should be made in all dorsal dislocations of this joint. If the skin dimple is not present, reduction of a simple dislocation (subluxation) can usually be accomplished fairly easily, and it is reasonable to attempt to carry out the manipulation on the sideline or in the locker room. McLaughlin [184] emphasized the point originally made by Farabeuf [137], that a simple dislocation can be converted into a complex dislocation by improper reduction.

Attempting to reduce the deformity by traction and pulling the phalanx back into position may not only fail, but it may cause the volar plate to become interposed in the joint, greatly decreasing the likelihood of successfully reducing the joint by closed means. Closed reduction should be performed with the wrist flexed to relax the tension on the flexor tendons [179]. The proximal phalanx is hyperextended as far as possible on the first metacarpal, and then its base rotated slightly and pushed against and volarly across the dorsal articular surface of the metacarpal head with the IP joint in flexion. The phalanx is then straightened and flexed, and reduction is usually achieved. Even when the volar plate is entrapped over the metacarpal head, this maneuver may dislodge it and avoid the necessity of open reduction. Very few authors advocate immediate repair of the volar plate after a successful closed reduction, because satisfactory healing usually occurs with adequate immobilization. If closed manipulation of the dorsal dislocation fails, the surgeon should be prepared to follow this with an immediate open reduction.

Following a successful closed reduction, lateral stability of the joint should be carefully tested to check the integrity of the radial and ulnar collateral ligaments. If there is evidence of complete rupture of one of the collateral ligaments, treatment should be directed toward management of this injury, as outlined in the previous section on injuries of the collateral ligaments.

Postreduction Management, Rehabilitation, and Criteria for Return to Play

Usually, the joint is stable after closed reduction, and protected motion can be started almost immediately. However, 3 to 6 weeks of protection with a removable thumb spica splint may minimize the prolonged MP joint soreness that these patients frequently experience. Athletic participation may be permitted within a few days of a successful closed reduction, providing the joint is taped or splinted in such a manner that it cannot hyperextend.

The patient must realize that the dislocation has injured several of the important supporting structures of the MP joint and that the joint will remain swollen, stiff, and somewhat tender for many weeks or even months. Vigorous active and passive motion exercises should be encouraged, although forceful hyperextension or lateral manipulation may actually reinjure the joint and prolong the recovery process. In rare cases, improper healing of the plate may result in chronic pain, with either instability in hyperextension or a flexion contracture necessitating late reconstruction (see Volar Plate Injuries, below).

Volar Plate Injuries

Relatively little has been written about hyperextension injuries of the MP joint of the thumb [194]. Perhaps the best article is that by Stener [237] in 1963, in which a distinction is made between the passive (volar plate and accessory collateral ligaments) and active (adductor pollicis and flexor pollicis brevis) restraints to hyperextension of the joint.

Clinical Evaluation

The diagnosis is made by a history of a hyperextension injury and well-localized tenderness over the volar aspect of the joint. Excessive hyperextension of the joint may be present, and pain on passive hyperextension is a good diagnostic sign. If pain limits mobility of the joint, the examination must be performed under anesthesia (local infiltration or, preferably, median plus radial wrist block) to obtain a reliable test. Hyperextension should be compared with the opposite uninjured thumb and tested with the IP joint flexed to relax the flexor pollicis longus.

Stener has suggested that the position of the sesamoids may aid in differentiating between volar plate rupture and tear of the intrinsic muscles. If the sesamoids are displaced distally with the proximal phalanx on hyperextension stress, muscle tear should be suspected. He suggested that muscle rupture should be treated operatively.

Treatment of Acute Injuries

Despite Stener's studies, there appears to be little support for his recommendation of primary repair of acute hyperextension injuries of the thumb. Most of these can probably be treated satisfactorily with thumb spica immobilization of the MP joint in 15 to 20 degrees of flexion for 3 to 4 weeks.

Criteria for Return to Play. Sports activity can be resumed immediately, but taping or splinting to protect the joint against extension forces should be carried out for 4 to 6 weeks, or until the volar plate is no longer palpably tender.

Treatment of Chronic Injuries

Chronic hyperextension instability of the MP joint of the thumb can cause pain with grasping and pinching. Posner and colleagues [211] have pointed out that it is important to differentiate passive instability due to chronic volar plate injury from a patient who can actively hyperextend the MP joint. The latter individual can stabilize the thumb in flexion when grasping or pinching, but the patient with passive instability cannot.

Although this is a relatively uncommon entity, we have treated several patients in whom this created a painful and functional problem. Operative treatment is required if the patient does not respond to nonoperative modalities (splinting, anti-inflammatory medications, and local injections). Several different types of operative procedures have been described for late reconstruction of the volar plate [133, 167, 185, 217, 261].

FRACTURES OF THE PHALANGES AND METACARPALS

Fractures of the phalanges and metacarpals are relatively common injuries in athletes. Fracture patterns are no different from those in nonathletes, but treatment of the athlete may be altered somewhat to allow him or her to return to competition sooner. Two very important principles must be kept in mind, however: (1) there is no way to speed up the natural healing time of a fracture, and (2) a treatment plan must not jeopardize the eventual outcome of a fracture just to allow an athlete to play sooner.

Anatomy

Phalanges

As in other long bone fractures, displacement and angulation of fractures of the phalanges and metacarpals are influenced by two factors: the mechanism of injury and the muscles acting as deforming forces on the fractured bone. The type of injury often determines the nature of the fracture; for example, a direct blow is more likely to cause a transverse or comminuted fracture, whereas a twisting injury more often results in an oblique or spiral fracture. The direction of angulation seen in fractures of the phalanges and metacarpals primarily depends on the muscles acting on the bone.

Unstable transverse fractures of the proximal phalanx typically present with volar (recurvatum) angulation (Fig. 18B–50). The proximal fragment is flexed by the bony insertions of the interossei into the base of the proximal phalanx. Although there are no tendons inserting on the distal fragment, it tends to be pulled into hyperextension by the central

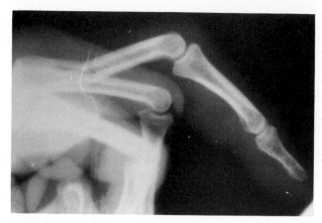

FIGURE 18B–50
Unstable transverse fractures of the proximal phalanx typically present with volar (recurvatum) angulation.

slip acting on the base of the middle phalanx. Once the stability of the proximal phalanx is lost, there is an accordion-like collapse at the fracture site, aggravated by further pull on the extensor hood by the intrinsic muscles.

The middle phalanx is much less commonly fractured than the proximal phalanx, and muscle forces acting at that level are different. The important deforming forces to be considered are the insertion of the central slip into the dorsum of the base of the middle phalanx and the insertion of the flexor digitorum superficialis volarly. The central slip has a well-defined area of insertion, and its action is to extend the middle phalanx. Although the action of the flexor superficialis is to flex the middle phalanx, its insertion is rather complex and is not confined to a short segment of the phalanx. Kaplan [322] described in detail the decussation of the superficialis tendon to allow the profundus tendon to pass through its two slips, and the reader is referred to Kaplan's text for this detailed description. In short, the superficialis divides into halves, each half turning 90 degrees to allow the profundus to pass through and then completing another 90-degree rotation to insert into nearly the entire volar surface of the middle phalanx. Thus, angulation of middle phalanx fractures cannot always be predicted by theoretical analysis of the muscle insertions.

Malrotation at the fracture site is one of the most frequent complications of phalangeal fractures, and one that can be avoided only by careful attention to anatomic detail. When the fingers are individually flexed, they point toward the region of the scaphoid tubercle just inside the thenar crease at the base of the thumb. They do not actually converge on a single fixed point, as is sometimes depicted (Fig. 18B–51). It is relatively easy to detect malrotation when the fingers are in full flexion, but with the

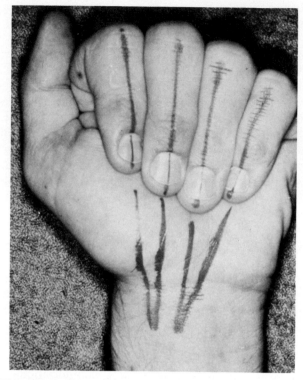

FIGURE 18B–51
When the fingers are flexed, they tend to converge, but not toward a single fixed point, as is sometimes depicted.

fingers only semi-flexed, malalignment is more subtle, and it is helpful to use the planes of the fingernails as an additional guide to correct rotation (Fig. 18B–52). The opposite hand must be checked for comparison, because often the border fingers lie in a slightly different plane of rotation when viewed end-on.

Metacarpals

The metacarpals are small, long bones that are slightly arched in the long axis and concave on the

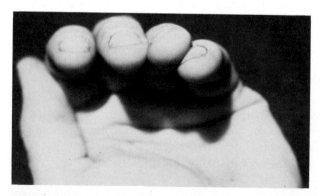

FIGURE 18B–52
Maintenance of correct rotational alignment with the fingers semi-flexed (as they are in a cast or splint) is best done by noting the planes of the fingernails as seen end-on. Comparison with the opposite hand is helpful.

palmar surface [287, 322]. Their weakest point is just behind the head [322].

The proximal ends of the index and long finger metacarpals articulate with the distal carpal row in practically immobile articulations, whereas those of the ring and little fingers have approximately 15 degrees and 30 degrees of anteroposterior motion, respectively [306]. The metacarpal shafts radiate like spokes of a wheel, terminating in bulbous articular heads, which are weakly joined by the transverse metacarpal ligaments. The collateral ligaments that join the metacarpal head to the proximal phalanx are relaxed in extension, permitting lateral motion, but become taut when the joint is fully flexed (see Fig. 18B–17). This occurs because of the unique shape of the metacarpal head, which acts as a cam. The distance in extension from the pivot point of the metacarpal to the phalanx is less than the distance in flexion, so that the collateral ligament is tightened with flexion of the MP joint. This anatomic point explains why the MP joints stiffen if the collateral ligaments are allowed to shorten, as they do when the MP joints are immobilized in extension [316, 344, 346].

The dorsal and volar interosseous muscles arise from the shafts of the metacarpals and act as strong flexors of the MP joint. It is thought that these muscles account for the dorsal angulation of metacarpal neck and shaft fractures.

Clinical Evaluation

The examining physician should carefully observe the resting posture of the injured hand and note deformities resulting from disruption of the skeleton. Some fractures and dislocations may be immediately obvious because of local swelling and deformity. Angulation and displacement are often readily apparent, but in fractures of the metacarpals and phalanges it is even more important to recognize rotational malalignment.

One of the greatest pitfalls in treating injuries of the hand is to focus on the obvious fracture and overlook more subtle, but often more significant, damage to soft tissues. One must always determine the precise area of tenderness in order to accurately appraise the damage to soft tissues as well as to bone. Careful assessment must be made of an open wound with regard to its precise location, its relationship to skin creases, the direction and viability of skin flaps, the extent of actual skin loss, and the degree of wound contamination. An open wound should not be probed or handled excessively, especially on the sidelines of an athletic field. Gentle

inspection with sterile instruments and gloves in the emergency room gives sufficient preliminary information until a thorough exploration can be done in the operating room. Damage to nerves and tendons should be determined by careful motor and sensory testing rather than by probing the wound.

Both open and closed injuries must be examined meticulously for injury to adjacent tendons, nerves, and blood vessels. Precarious circulation may be particularly subtle in closed injuries and must be assessed by noting color and temperature, capillary filling, and patency of the collateral circulation by Allen's test at the wrist and in the digit itself [307]. Satisfactory physical examination may not be possible without local or regional anesthesia, but the block should be postponed until an initial assessment of nerves and blood vessels has been made in the unanesthetized hand.

Radiographic Examination

Radiographs are essential in most injuries of the hand, even if no bone injury is obvious on clinical examination. Many significant fractures and joint injuries are missed simply because adequate radiographs were not taken on the day of injury. Three views are necessary: posteroanterior (PA), lateral, and oblique. Oblique films are particularly helpful in accurately assessing intra-articular fractures. For injuries involving a finger, it is absolutely mandatory that a true lateral of the individual digit be obtained. Superimposition of the other fingers on a lateral view of the entire hand obscures significant details that are easily seen on a lateral view of the single digit. Angulation of metacarpal fractures may be difficult to assess accurately on a true lateral film. We have found that added information can be obtained by including a lateral view of 10 degrees of supination (for the fourth and fifth metacarpals) or in 10 degrees of pronation (for the second and third metacarpals) (Fig. 18B–53).

Again, the temptation is great to concentrate on obvious abnormalities in the radiograph and overlook important but more obscure injuries. For example, one or more carpometacarpal (CMC) joints may be subluxated or dislocated when there is a displaced or angulated fracture of an adjacent metacarpal.

Treatment Options

There are many ways to treat fractures of the phalanges and metacarpals, and the decision to use

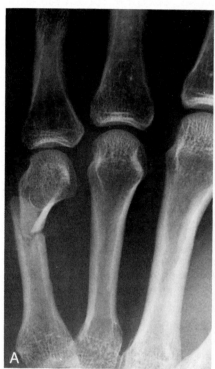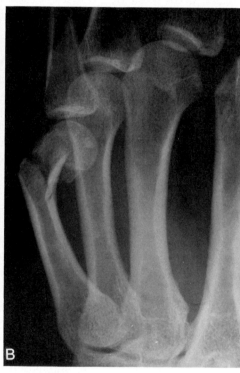

FIGURE 18B–53
A, Anteroposterior view of a fracture through the neck of the fifth metacarpal with apparent modest angulation. *B*, The same metacarpal fracture seen in a lateral view in 10 degrees of supination; the true amount of angulation is much better appreciated in this view.

a particular method results from the distillation of many factors. The experience, expertise, and personal prejudices of the surgeon will, in the final analysis, dictate the treatment selected for each patient. No single method of treatment can be applied to all fractures of the phalanges and metacarpals, and the surgeon must have in his armamentarium several different techniques with which he is experienced and comfortable. He must then choose that method which in his hands is most likely to restore maximum functional recovery for that specific fracture. For the athlete, this choice may be influenced by a desire to minimize the time lost from competition, but, in each case, potential complications must be weighed against the benefits of the particular method chosen for treatment.

A comprehensive discussion of all the various methods of treating fractures in the hand is beyond the scope of this chapter, and for this the reader is referred to fracture texts such as Rockwood and Green [301]. However, it is pertinent to point out that the treatment options for phalangeal and metacarpal fractures fall into three major categories, as outlined below.

Closed Reduction and Immobilization

Many fractures of the phalanges can be satisfactorily managed by closed reduction and external immobilization, but the fracture must be stable after reduction in order for the splint or cast to maintain the reduction. Transverse fractures generally fall into this category, but spiral or oblique fractures are inherently unstable, and it may be impossible to hold them reduced with any sort of splint or cast. Likewise, excessive swelling, such as is frequently seen in metacarpal fractures, may make it impossible to hold the reduction of even a transverse fracture.

A wide variety of splints and casts have been used to immobilize hand fractures, including circular casts, casts with outriggers, gutter splints, functional braces [362, 363], and a combination of anterior and posterior splints. The generally preferred position of immobilization is the intrinsic plus position [317]; that is, with the MP joints in at least 70 degrees flexion and the PIP joints in nearly full extension. More PIP flexion may, however, be required to prevent the recurvatum type angulation typically seen in transverse fractures of the proximal phalanx.

A modification of this method is that described by Burkhalter and associates [280, 343], in which early active motion is permitted in the splint, preventing volar angulation by keeping the MP joints

in maximum flexion and thereby tightening up the extensor hood to act as a tension band and, at the same time, lessening the deforming influence of the intrinsic muscles (Fig. 18B–54).

Closed Reduction and Percutaneous Pin Fixation

The theoretical advantage of closed reduction and percutaneous pinning is that an unstable fracture can be rendered sufficiently stable to allow early motion without subjecting the hand to the surgical trauma of open reduction. Three distinctly different types of percutaneous pinning can be used (Fig. 18B–55).

Intramedullary Pin Fixation. As described by Belsky and colleagues [266], this method is particularly applicable to transverse and short oblique fractures. A smooth Kirschner wire is passed either retrograde from the head of the proximal phalanx (Fig. 18B–55B) or, more commonly, from the metacarpal neck through the flexed MP joint (Fig. 18B–55C). With this method, motion cannot be started early and the MP joint must be immobilized with a cast or splint for 3 weeks. It is a useful way, however, to prevent the recurvatum angulation that may be difficult to control with a cast alone.

Direct Pinning of the Fracture. Suited primarily for long oblique or spiral fractures of the phalanges, this technique (Fig. 18B–55A) is considerably more difficult than intramedullary fixation, because there is a narrow margin for error. A power-driven drill is essential, and the image intensifier facilitates the procedure (except in intra-articular fractures, where the images produced on the monitor are not sharp enough to ensure precise anatomic reduction) [326]. Several authors [282, 349] have suggested the use of hypodermic needles to serve as guides for more precise pin placement, and special fracture-reduction clamps [272, 297] (Fig. 18B–56) may also be useful. A few millimeters of error in pin placement can easily result in inadequate fixation, but with proper pin placement, buddy taping and active range of motion can be started immediately (Fig. 18B–57).

Fixation to an Adjacent Intact Bone. Especially in metacarpal shaft fractures, transverse pin fixation to an adjacent intact metacarpal [269, 275, 323, 352, 367] can provide sufficiently stable fixation to allow early motion, leaving the pins in place for 4 weeks (Fig. 18B–55D).

Each of these techniques for closed reduction and percutaneous pinning can achieve good results if it is properly applied. It must be emphasized, how-

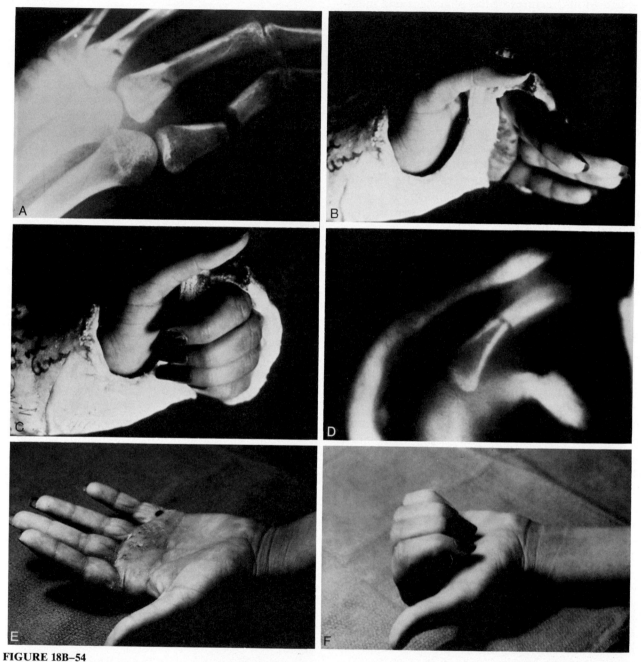

FIGURE 18B–54
Burkhalter's technique of functional casting for fractures of the proximal phalanx [280,343]. *A*, A 30-year-old woman sustained fractures of the bases of all four proximal phalanges, with particularly wide displacement and angulation in the index finger. *B*, A short-arm cast immobilizes the MP joints in marked flexion, *C*, while allowing full flexion of the PIP and DIP joints. *D*, A lateral tomogram in the splint demonstrates satisfactory alignment of the fracture. The patient's hand 8 weeks post-injury showing satisfactory *E*, extension and *F*, flexion of the fingers. (Courtesy of Dr. William E. Burkhalter.)

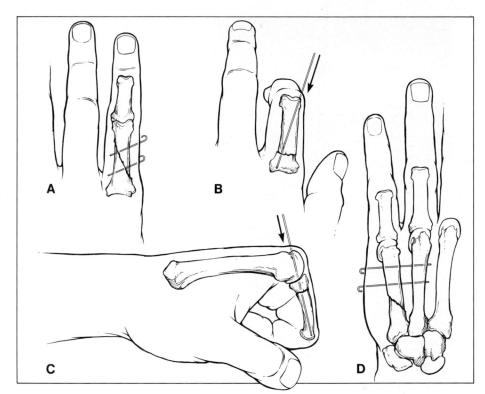

FIGURE 18B–55
Four methods of closed reduction and percutaneous pin fixation (see text for details).

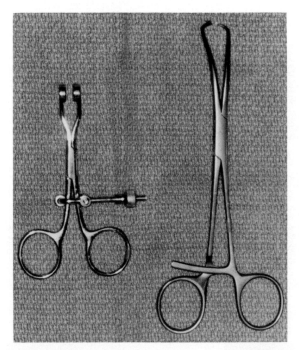

FIGURE 18B–56
Commercially available fracture reduction clamps, such as the Blalock (Weck) *(right)* and Aesculap *(left)* are useful in percutaneous pin fixation of fractures, especially for the direct pinning technique shown in Figure 18C–55A.

ever, that they are contraindicated if a satisfactory reduction of the fracture cannot be accomplished by closed manipulation. Furthermore, we are reluctant to advise any return to athletic participation as long as temporary fixation pins are in place. Losing tenuous reductions, damaging adjacent or remote tissues, and creating local factors favoring pin-tract infections are all pitfalls of trying to combine percutaneous fixation with an early return to competition.

Open Reduction and Internal Fixation

Specific indications for open reduction and internal fixation vary widely, depending on the experience, technical skills, and judgment of the surgeon. The desire to get an athlete back into competition sooner should in itself rarely if ever be the sole indication for choosing open reduction and internal fixation. In every instance, there must also be sound and appropriate medical/orthopaedic reasons for selecting that method of treatment over other, more

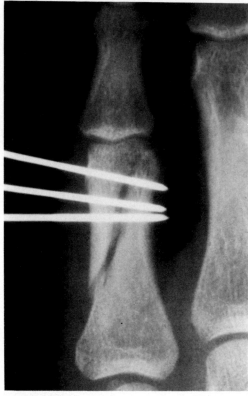

FIGURE 18B–57
Closed reduction and percutaneous pin fixation of a spiral fracture of the proximal phalanx. Reduction is not perfect, but overall alignment is satisfactory, there is no rotational deformity, and the pin placement is good. Active range of motion with buddy taping can therefore be started immediately.

conservative techniques. If, however, internal fixation is selected for an athlete, a prime goal should be to achieve as solid and rigid fixation as can be accomplished.

The choices of internal fixation techniques include Kirschner wires (Fig. 18B–58) [289, 336, 338, 349], intramedullary fixation [305, 308, 309, 314, 324, 337], intraosseous wiring [296, 325, 331, 353, 361], tension band wiring [302, 321, 350, 335], cerclage wiring [304, 320], and the AO (ASIF) system of mini-fragment plates and screws [265, 278, 286, 288, 292, 311, 312, 319, 332, 333, 356, 357, 370] (Fig. 18B–59). The relative strengths of these different types of internal fixation constructs have been studied in the laboratory by many investigators [270, 271, 298, 328–330, 342, 361]. Unfortunately, the results of these studies have frequently yielded conflicting conclusions, and true comparisons among the various studies are virtually impossible because of the differing testing methods used. Most have agreed, however, that plates provide the most rigid fixation, with tension band wiring and intraosseous wiring also being superior to Kirschner wires in holding capacity. No one has ever answered the question of how strong does internal fixation in the hand have to be. Although plates and screws may provide more rigid fixation, the AO system is technically difficult, with a very narrow margin for error, and the surgeon must have the experience and expertise to use it effectively and safely.

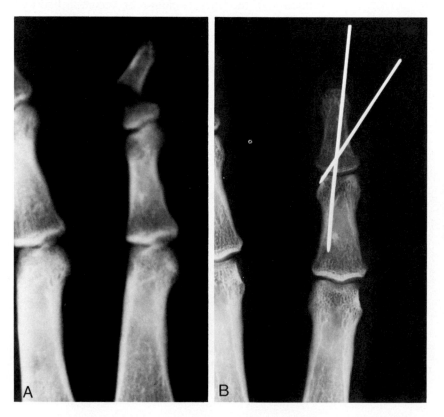

FIGURE 18B–58
A, A widely displaced fracture through the base of the distal phalanx in a weight lifter. *B,* This was treated by open reduction and internal fixation with Kirschner wires.

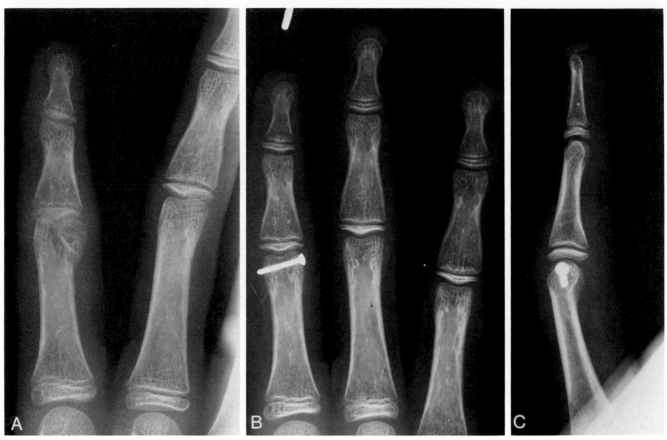

FIGURE 18B–59
Displaced condylar fractures virtually always demand internal fixation. *A,* This displaced and rotated fracture of the condyle in a young football player was, *B and C,* treated with open reduction and anatomic restoration of the articular surface with a single 1.5 mm AO screw.

Postoperative Management, Rehabilitation, and Criteria for Return to Play

The decision as to when to allow the athlete to return to competition is often a difficult one. The surgeon must be quite confident that there is little or no likelihood that the fracture and its fixation device can be in any way disrupted by the vigorous stresses that occur during a hard-fought game. The designers of fracture-fixation devices have not intended or recommended that these materials be subjected to the tremendous forces that are incurred in many individual and team sports. Although they have created screws, plates, and wiring techniques that can be shown to impart considerably greater resistance to deformity than with no fixation or the use of one or more Kirschner wires, the main advantage of these devices has been to provide accurate reduction and stability sufficient to allow early, protected joint motion in an effort to minimize stiffness. To apply these methods with the expectation that the extremity can perform as though there were no fracture at all is wishful thinking, totally unsupported by any scientific investigation. Even the provision of additional external support to the part after open reduction and internal fixation may be of little protection against the strong insults that occur during competition.

The surgeon must have a clear understanding of the stresses to which the wrist and hand of the patient/athlete is subjected during the course of competition. The specific performance demands at the particular level of competition, the forces that are generated by the competitive activities of a certain position, and the peculiarities of the individual player's method of participation are all important considerations. The stability of the fracture and its ability to withstand the forces applied to it during sports participation are the primary considerations in deciding when an athlete can safely return to competition. In some cases, the fracture itself may be inherently stable, and in other situations the surgeon may render stability to an unstable fracture by some means of external or internal fixation. In the final analysis, however, the type of sport and the player's position are major factors in determining when an athlete can actually resume competition without undue concern for endangering fracture stability and healing. For example, an intra-articular fracture involving the PIP joint might keep a quarterback out of competition for 4 to 6 weeks, but a lineman could conceivably be playing within a few days as long as adequate external protection can be provided. Unrestricted participation in a contact sport without any form of external protection should probably not be allowed until there is solid, clinical (but not necessarily radiographic) healing. This usually means 4 to 6 weeks, and it is important that the athlete should have regained practically a full range of motion in the involved digit and hand before returning to competition.

Authors' Preferred Method of Treatment

There are so many different types of fracture combinations in the hand that it would be impossible to try to include all of them in this discussion. Therefore, in this section, we will address the most common fractures that are likely to be seen in athletes.

Intra-articular Fractures

Truly undisplaced intra-articular fractures are uncommon, but they do occur. They are best treated by carefully guarded and protected early range of motion with buddy taping. Frequent (weekly) clinical and radiographic reevaluations are mandatory to be sure that the fracture does not displace, especially in the athlete who continues to play.

The goal of treatment in displaced intra-articular fractures is anatomic restoration of the joint surface, and in most cases this is best accomplished by open reduction and internal fixation.

Condylar Fractures

Fractures involving the condyles of the proximal or middle phalanges are inherently unstable and virtually always require internal fixation. Occasionally, this can be accomplished with closed reduction and percutaneous pin fixation with the aid of a fracture-reduction clamp, but more often open reduction is required (see Fig. 18B–59). The major

pitfall in this operation is preservation of the integrity of the extensor tendon insertion just distal to the PIP joint. Large fragments may be fixed with AO mini-fragment screws, but it is safer, especially for the less experienced surgeon, to transfix the repositioned fracture with small Kirschner wires.

Avulsion Fractures at the Base of Proximal Phalanx

Marginal fractures at the base of the proximal phalanx usually represent avulsion fractures of the collateral ligament. Small or nondisplaced fragments can be satisfactorily managed with buddy taping, but larger displaced fragments should be fixed as anatomically as possible by open reduction. Although precise indications for operative treatment are difficult to define, we would consider open reduction for fractures involving more than 20% of the articular surface, fractures displaced more than 2 to 3 mm away from the base of proximal phalanx, and fracture fragments with obvious rotation of the articular surface.

Comminuted Intra-articular Fractures

Severely comminuted intra-articular fractures pose one of the most difficult problems in the hand, and they most frequently involve the PIP joint, the result of a so-called jamming injury. Sometimes these are the typical fracture-dislocation, but more difficult problems are those with bony comminution that leave virtually no part of the articular surface intact. Open reduction and internal fixation of such fractures should be attempted only by surgeons with extensive experience and expertise; in some cases, the most efficacious treatment is traction or external fixation in an effort to restore some semblance of articular congruity by distraction across the joint.

Bennett's Fracture

More properly called a fracture-subluxation, Bennett's fracture was first described by Bennett in 1882 [268]. The volar lip of the thumb metacarpal is avulsed by the anterior oblique ligament, and the remaining shaft of metacarpal is subluxated dorsally by the pull of the abductor pollicis longus.

At least 20 different methods of treatment have been advocated for this fracture [273, 279, 283, 285,

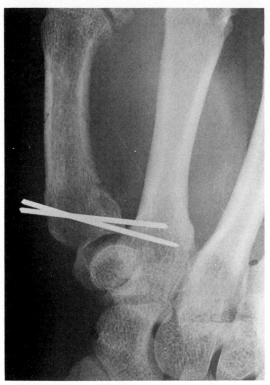

FIGURE 18B–60

A reasonable treatment alternative for Bennett's fracture is closed reduction and percutaneous pin fixation. The pins do not necessarily enter the small volar fracture fragment; rather the intent is to reduce the fracture-subluxation by closed manipulation and then secure the base of the thumb metacarpal to the carpal bones or to the adjacent second metacarpal, as shown here.

310, 315, 340, 345, 347, 348, 359, 360, 366], but there is fairly universal agreement that it is difficult or impossible to hold an anatomic reduction with any sort of cast. This then leaves two reasonable alternative methods of treatment: closed reduction and percutaneous pin fixation (Fig. 18B–60) [318, 351, 364, 367, 369], or open reduction and internal fixation [264, 276, 290, 291, 293–295, 313, 339, 358]. In the athlete, our preferred method is open reduction and internal fixation, because this is likely to minimize the time out of competition. If stable internal fixation is achieved with Kirschner wires, a Herbert screw [313], or an AO screw [293, 358], the player can usually return to competition within 7 to 10 days, provided the thumb is protected in a spica splint. Return to competition is of course longer (4 to 6 weeks) if his sport or position precludes the use of a thumb spica splint.

Extra-articular Fractures

It is the treatment of displaced extra-articular fractures of the phalanges and metacarpals where the most diverse differences of opinion exist. Many of these fractures can be very satisfactorily managed with closed methods, but in almost all such situations, nonoperative treatment precludes participation in athletics for at least 3 to 4 weeks.

There seems to be an increasing tendency to allow continued participation in athletics to assume a more important role in the treatment decision process than it deserves. The surgeon must keep foremost in mind the best interests of the injured athlete and not necessarily the desires of the player, the team, the coaches, or the parents. Stretching indications for operative treatment or compromising what is truly best for the individual may occasionally be acceptable for the professional athlete, but surgeons must resist the temptation to allow this mentality to drift down into the college, high school, and junior high school ranks. In general, the younger the athlete and the less proficient the level of competition, the less the consideration that should be given to compromising what is, in the opinion of the treating surgeon, strictly best for the individual. Exceptions are made from time to time, depending on the specific circumstances. For example, some compromises may be made in the case of a high school athlete whose chances for attending college are totally dependent on his receiving an athletic scholarship or the college stand-out whose chance of making the professional ranks is contingent on his senior year performance. With these considerations in mind, we have briefly outlined below what would be our preferred methods of treatment for specific fractures in the average high school athlete.

Transverse Fracture of Base of Proximal Phalanx. As noted previously, the typical angulation in this fracture is recurvatum (apex volar), and our preferred method is either Burkhalter's technique (see Fig. 18B–54) or immobilization of the hand in the intrinsic plus position with anterior and posterior splints. Eaton's intramedullary fixation with a Kirschner wire passed across the flexed MP joint (see Fig. 18C–55*B*) is an acceptable alternative, but all of these methods require external protection for at least 3 weeks, during which time the athlete cannot play. Rarely do we treat this fracture with open reduction and internal fixation.

Spiral or Long Oblique Fracture of Proximal Phalanx. If a satisfactory reduction can be achieved with closed manipulation, we prefer percutaneous pin fixation with smooth Kirschner wires (Fig. 18B–61). Immediate active range of motion may be commenced with protective buddy taping to one or more adjacent digits. Nonetheless, these patients usually should not return to aggressive physical contact sports participation until the pins are re-

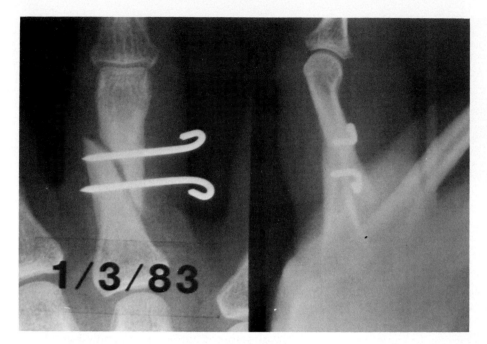

FIGURE 18B–61
Unsatisfactory closed reduction and percutaneous pin fixation technique. The fracture was not adequately reduced, and pin fixation in this position assures a malunion. The closed method should have been abandoned in this patient and the fracture fixed with open reduction and internal fixation.

moved. Less demanding competition may be permissible if one can be certain that no additional damage to the fracture or the digital soft tissues is likely to occur. If a satisfactory closed reduction cannot be achieved, then open reduction and internal fixation, preferably with AO mini-fragment lag screws, is our choice. If stable fixation is achieved with such internal fixation, and strong support can be provided by splinting or buddy taping, early return to competition is often acceptable.

Transverse Fracture of Neck of Proximal or Middle Phalanx. This fracture, which more commonly occurs in children, is a classic booby-trap fracture because reduction is easy to achieve but difficult to hold (Fig. 18B–62). For this reason, we generally prefer closed reduction and percutaneous pinning. As previously mentioned, the presence of Kirschner wires almost always precludes the return to vigorous contact sports, although less demanding participation may sometimes be resumed shortly after fixation has been achieved. If it is necessary to cross the PIP joint with the Kirschner wire, return to any competition should be prohibited until the Kirschner wire is removed at 3 weeks (Fig. 18B–63).

Fracture of Metacarpal Neck (Fighter's Fracture). Because it is virtually impossible to hold this fracture well reduced with any type of external splint [301], it is probably appropriate treatment to accept the deformity (if angulation is less than 30 degrees), apply a splint for comfort, and allow the athlete to return to competition as his pain subsides. He must understand, however, that he will have a bump on the dorsum of the hand and some loss of prominence

of the knuckle; if this is unacceptable to him, then closed reduction and percutaneous pinning or open reduction and internal fixation should be considered.

Transverse Fracture of Metacarpal Shaft. The further away from the MP joint, the more obvious and potentially unacceptable is the typical dorsal angulation deformity. For this reason, and because

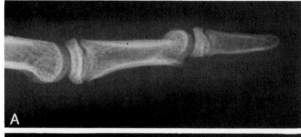

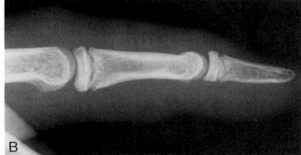

FIGURE 18B–62
Fractures through the neck of the proximal or middle phalanges are classic booby trap fractures because they are unstable. *A,* In this transcondylar fracture of the middle phalanx in a soccer player, *B,* near anatomic alignment was achieved with closed reduction, but the fracture redisplaced twice before a percutaneous pin was used to provide stable fixation.

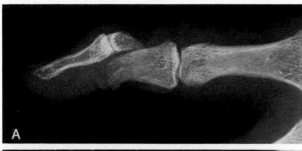

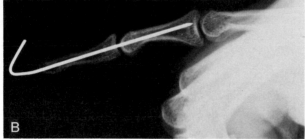

FIGURE 18B–63

A, This widely displaced fracture through the neck of the middle phalanx was treated with open reduction and internal fixation using a single, longitudinally oriented Kirschner wire. *B,* Because the Kirschner wire crosses the DIP joint, return to athletic competition should not be permitted until the pin is removed at 3 weeks.

many of these patients have marked swelling that precludes satisfactory maintenance of reduction in a cast or splint, open reduction and internal fixation is probably the best treatment, usually with an AO mini-fragment plate and screws.

Spiral Fracture of Metacarpal Shaft. These fractures usually do not angulate, and shortening of at least 5 mm can be easily accepted, but care must be taken to ensure correct rotational alignment. If there is malrotation, open reduction and internal fixation with AO mini-fragment lag screws is indicated; if stable fixation is achieved, early return to competition with a good protective splint is usually permissible. Any one of the various protective splints described in the section on splinting is acceptable, but for this particular injury the MP joints should be included (preferably in at least 50 degrees of flexion). It is probably advisable to wait at least 7 to 10 days until the postoperative swelling has subsided and the sutures are removed before fitting the splint and allowing the athlete to return to play.

Multiple Fractures. In general, the more fractures and the greater the associated soft tissue injury, the greater the indication for stable internal fixation. However, such situations are not seen nearly so commonly in athletics as they are in other forms of more violent trauma.

Fractures of the Distal Phalanx. Fractures of the distal phalanx deserve special mention primarily because of the associated soft-tissue damage. Except for mallet fractures (see page 954) and flexor tendon avulsion (see page 946), most fractures of the distal phalanx are crush injuries that result in comminution of the tuft portion of the bone. In sports, these injuries often result from having a digit stepped on by an opposing player's athletic shoe, such as in football. The fracture itself rarely requires anything more than minimal splinting or padding for comfort, but the associated soft-tissue damage does demand early treatment. Many of these injuries result in open fractures with lacerations of the nailbed that require special treatment consideration.

Nailbed Injuries. If the nail plate is intact and there is a subungual hematoma, evacuation of the blood gives marked relief of pain. This can be done with the time-honored paper clip heated to a red-hot temperature, but a small, battery-operated disposable cautery (Accu-Temp, Concept) is even better.

Late nail deformity is often an unavoidable complication following crush injuries of the distal phalanx, but a meticulous repair of the nailbed is the best means of minimizing such a deformity [277, 326]. We recommend that the nailbed be repaired as carefully as possible with fine (6–0 or 7–0) absorbable sutures, using loupe magnification. If there is loss of nailbed tissue, various types of local and distant nailbed grafts can be considered [371].

If the nail plate has been avulsed at its base, it is carefully removed, scrubbed with povidone-iodine (Betadine), and sutured back into place following repair of the nailbed [334, 372]. This not only serves as a protective splint but also minimizes local tenderness. The old nail is pushed off gradually by growth of the new nail. The patient should be told that some deformity of the nail is to be expected, although the full extent of the deformity will not be known until the new nail has fully regrown, a process that usually takes about 4 to 5 months.

Occasionally, the base of the nailbed may become entrapped in the fracture site, a phenomenon most commonly seen when the root of the nail has been avulsed from beneath the proximal nail fold. Failure to recognize this complication and extricate the nailbed from the fracture site at the time of the wound debridement may result in nonunion of the fracture.

NERVE COMPRESSION INJURIES

Fortunately, upper extremity compression neuropathies are relatively rare in athletes. Although it is worthwhile to consider some of the nerve disorders that may produce motor or sensory dysfunction in the hand or digits, an extensive discussion is beyond the scope of this section. In this section, we briefly discuss the mechanics, symptoms, physical findings, diagnostic tests, and treatment of the most common neuropathies that affect the hands of athletes.

Most of these conditions follow direct contusion of the tissues that overlie peripheral nerves, particularly in anatomic areas, where they are most vulnerable. Vigorous, repetitious athletic activity that produces inflammation and tissue swelling can also lead to nerve compression; usually in closed passages such as the carpal, cubital, or ulnar tunnels.

The physician examining an athlete whose particular activity places undue or repeated strain on the muscles, tendons, or ligaments of the upper extremity must be particularly wary of the potential for compression neuropathy, and a thorough neurologic examination should be carried out. Nerve conduction studies and electromyographic evaluations may add to the diagnostic information. In most instances, these conditions can be effectively managed by conservative means, and rarely will symptoms be severe enough to preclude continued athletic participation. The most frequent nerve compression syndromes involving athletes are discussed below.

Median Nerve

Pronator Syndrome

Anatomy. The median nerve at the elbow is covered by the bicipital aponeurosis (lacertus fibrosis). It crosses the brachial artery in the midarm and lies medial to the artery in the cubital fossa (Fig. 18B–64). Although there are some variations, the median nerve usually passes between the two heads of the pronator teres at the distal end of the fossa and proceeds down the forearm beneath the deep surface of the flexor digitorum superficialis. In the proximal one-third of the forearm, the nerve gives off a deep (posterior) branch, the anterior interosseous branch, which passes inferiorly to lie between and innervate the flexor pollicis longus (FPL), the flexor digitorum profundus (FDP) to the index and, occasionally, a portion of the FDP to the middle (long) finger.

Mechanism of Compression and Presenting Symptoms. The median nerve may be compressed proximally by the ligament of Struthers, which usually connects a supracondylar process of the humerus to the medial epicondyle, the lacertus fibrosus, the pronator teres, or the proximal arch of the flexor

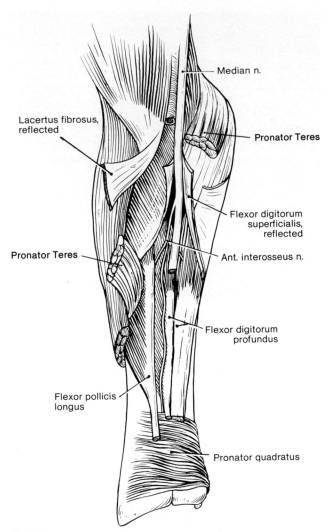

FIGURE 18B–64
The course of the median nerve in the antecubital fossa and forearm. Note the deep position of the anterior interosseous nerve. Any of the structures that cross or surround either the median or anterior interosseous nerves may contribute to a compression neuropathy.

digitorum superficialis. In sports such as weight lifting, arm-wrestling, or rowing, which are characterized by repetitive, forceful pronation, the athlete may experience aching discomfort in the proximal forearm and paresthesias in the median nerve distribution. The dominant side is likely to be involved, and nocturnal symptoms are infrequent.

Physical Findings. On examination, the symptoms can most reliably be reproduced by firm, deep compression directly over the pronator muscle for about 1 minute. Symptoms may also be elicited by resisted forearm pronation and elbow flexion. If the discomfort is increased by resisted elbow flexion of 120 to 135 degrees, one should suspect compression at either the ligament of Struthers or the lacertus fibrosus, and pain resulting from resisted long finger flexion may indicate compression at the superficialis arch. Tinel's sign (sensitivity to tapping or percussion directly over the site of compression with "tingling" or "electric" sensations radiating distally from the site of percussion in the distribution of the nerve) is usually positive at the site of compression. Electrodiagnostic studies are seldom helpful in this condition, and radiographs should be taken to rule out the presence of a supracondylar process.

Treatment Options. This condition should be managed conservatively for a period of at least 6 weeks, with a regimen of rest, splinting, and anti-inflammatory medication. In particular, weight-lifting activities should be curtailed until improvement is achieved. Surgical decompression is reserved for severe or refractory cases, and includes a complete release of all offending structures.

Anterior Interosseous Nerve Syndrome

This condition is similar to the pronator syndrome in that it also involves the median nerve in the proximal forearm and may produce vague activity-aggravated myalgia in the proximal forearm [382]. It differs from pronator syndrome in that it exclusively involves the anterior interosseous branch of the nerve and, as such, is an entirely motor condition without any sensory deficit.

Mechanism of Compression and Presenting Symptoms. Anterior interosseous syndrome is caused by a single violent muscle contracture, by repetitive trauma to the forearm, or by constricting bands from the deep head of the pronator teres, anomalous proximal forearm muscles or vessels, or the origins of the flexor superficialis (FDS), FDP, or flexor carpi radialis (FCR). It is occasionally seen in "junk" baseball pitchers or tennis players who attempt to impart a great deal of topspin to their shots.

There is often a history of a single episode of strong contracture of the elbow, wrist, and finger flexors accompanied by pain and, subsequently, by motor loss. The athlete notes the inability to end pinch between the index finger and the thumb, and there is a hyperextension attitude to the terminal joints of those two digits (Fig. 18B–65). Weakness or paralysis of the FPL, FDP of the index finger, and pronator quadratus characterize this disorder, and there are no sensory alterations in the median nerve distribution. Electrodiagnostic studies are occasionally of value if the appropriate muscles are tested by an experienced electromyographer. Treatment is essentially identical to that described for the pronator syndrome.

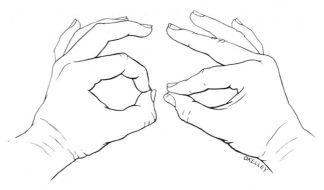

FIGURE 18B–65
Normal *(left)* and impaired *(right)* tip pinch in a patient with anterior interosseous nerve syndrome. Absence of active flexion in the index DIP and thumb IP joints is due to loss of the flexor digitorum profundus to index and the flexor pollicis longus, respectively.

Carpal Tunnel Syndrome

Anatomy. The carpal tunnel is an anatomic passageway bounded dorsally and laterally by the hemicircular carpal bones, and on the palmar surface by the transverse carpal ligament (Fig. 18B–66). Through this tunnel pass the nine digital and thumb flexor tendons and the median nerve. It is important to appreciate the fact that the radial and ulnar arteries and the ulnar nerve do not pass through the tunnel. The passageway is relatively unyielding, and pressure within the tunnel may be increased by anything that increases the volume of its contents. The median nerve is particularly susceptible to pressure increases, and the symptoms characteristic of this disorder result.

Mechanism of Compression and Presenting Symptoms. Closed wrist trauma with or without carpal

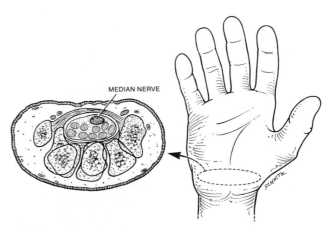

FIGURE 18B–66
Cross-sectional anatomy of the carpal tunnel, bounded on three sides by the carpal bones and volarly by the transverse carpal ligament. Nine flexor tendons and the median nerve pass through the tunnel. Anything that causes increased pressure within this canal can produce the symptoms of carpal tunnel syndrome.

fracture may produce hemorrhage or edema in the carpal tunnel, with the resultant development of the familiar symptoms of median nerve compression. Athletes may complain of activity-aggravated weakness or clumsiness of the hand and hypesthesia, anesthesia, dysesthesia, or paresthesias of the thumb, index, middle, and radial one-half of the ring finger. Nocturnal discomfort is frequent, and the hand may fall asleep during athletic participation or when driving a car or carrying out other activities that require prolonged or repetitious wrist flexion or extension. Aching of the forearm or upper arm in association with more distal numbness or discomfort is a common presenting complaint.

Physical Findings. The most important diagnostic test for carpal tunnel syndrome is the wrist flexion test described by Phalen in 1951 [380]. This maneuver consists of placing the wrist in unforced complete flexion for as long 60 seconds. Athletes with this condition usually describe hypesthesia or paresthesias in the median nerve distribution that closely simulate their symptoms. Tinel's tapping test may also contribute useful information and is elicited by tapping proximal to, directly over, and distal to the palmar wrist crease. A positive response occurs when the athlete describes tingling or an electric shock sensation in all or part of the median nerve distribution. Sensory evaluation may show a deficiency in patients with advanced or long-standing carpal tunnel syndrome, and electrodiagnostic studies may give supporting diagnostic information.

Treatment Options. The conservative treatment of carpal tunnel syndrome consists of splinting the wrist (particularly at night) and the use of nonsteroidal anti-inflammatory drugs (NSAIDs). Athletes should, if possible, wear their splints during practice and avoid those activities known to produce their symptoms. On occasion, the injection of a corticosteroid preparation directly into the carpal tunnel may be carried out for both diagnostic and therapeutic reasons. Only when the athlete fails to respond to conservative treatment and has a demonstrable neurologic deficit should surgical decompression be considered.

Bowler's Thumb

Anatomy, Mechanism, and Presenting Symptoms. This condition occurs almost exclusively in bowlers, who develop a traumatic neuroma of the ulnar digital nerve to the thumb. The cause of this entity appears to be repeated friction between the thumb and the edge of the thumb hole in the bowling ball. In the proximal thumb, the ulnar digital nerve lies superficial to the sesamoid bone and can be trapped during the repeated insertion and withdrawal of the thumb during bowling. The condition may also result from repetitive pressure on the ulnar digital nerve during racket sports. Presenting complaints usually include sensitivity over the ulnar digital nerve and hypesthesia in the area of irritation. Atrophy of the overlying skin or callus formation is usually present and the indurated area of the nerve can almost always be palpated. "Rolling" of the nerve over the metacarpophalangeal joint prominence reproduces the patient's complaints.

Treatment Options. Early recognition and treatment of this condition is important, because permanent disability may result if the lesion is allowed to persist too long. Conservative treatment may include adjustment of the size and spacing of the bowling ball hole, and the racket player must shift his grip position in order to eliminate the repetitive pressure against the inner aspect of the proximal thumb. Protective guards may be tried, but athlete compliance is poor because of the bulk of the devices usually used.

Surgery consisting of neurolysis of the involved nerve may be beneficial if the symptoms are severe [373]. Transposition of the nerve posterior to the adductor pollicis tendon has been recommended [379] and may be indicated for recalcitrant patients. Unfortunately, surgery rarely completely eliminates the condition, and athletic participation that produces pressure on the involved area may have to be terminated.

Repetitive trauma to other digital neurovascular bundles may create similar clinical conditions in other areas of the hand. Handball players are frequent offenders, with at least transient damage produced to one or more digital nerves from repetitive percussion of the digits against the ball.

Ulnar Nerve

Ulnar compression neuropathy in athletes is usually produced by compression or stretching of the nerve at the elbow, proximal forearm, or at the proximal ulnar palm. The nerve is particularly vulnerable as it passes subcutaneously under a thick band of fascia around the medial epicondyle and in the area where it passes beneath the aponeurotic connection between the two heads of origin of the flexor carpi ulnaris muscle (Fig. 18B–67).

Cubital Tunnel Syndrome

Mechanism and Presenting Complaints. Neuropathy at the elbow may be produced by direct trauma,

FIGURE 18B–67
The anatomy of the ulnar nerve in the cubital tunnel, where the nerve passes behind the medial epicondyle and beneath the arcade of the flexor carpi ulnaris (FCU).

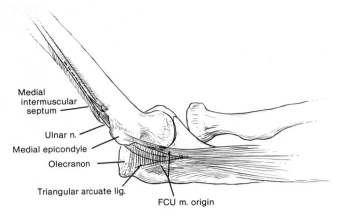

Medial intermuscular septum

Ulnar n.

Medial epicondyle

Olecranon

Triangular arcuate lig.

FCU m. origin

stretching due to a cubitus valgus deformity, or repetitive subluxation of the nerve over the epicondyle after rupture of the transverse ligament. The latter condition has been described in pitchers and is presumably related to microscopic tears in the ligament caused by the throwing mechanism. Compression secondary to hypertrophy of the confluence of the flexor carpi ulnaris heads can also occur in baseball pitchers [376]. The condition has also been reported in cross-country skiers and appears to be related to forceful repetitious poling. Extension and pronation of the elbow with simultaneous contracture of the flexor carpi ulnaris and triceps narrows the cubital tunnel, causing nerve irritation during this maneuver [374]. Patients with ulnar neuropathy at the elbow present with pain and paresthesias radiating into the ring and little fingers. Weakness of grip usually accompanies these symptoms and may be the major complaint in more subtle cases.

Physical Findings. The diagnosis can be made by reproducing the symptoms by asking the patient to perform the inciting activity or by hyperflexing the elbow for at least 1 minute. This test usually produces ulnar nerve dysesthesia or paresthesia in the ring and little fingers. The nerve may also be percussed in the tunnel and particularly over the origin of the two heads of the flexor carpi ulnaris. In cases where the nerve tends to subluxate, it may be readily palpated moving from its posterior position to a position on or anterior to the medial epicondyle during elbow flexion. Motor involvement or objective sensory loss are infrequently seen and usually indicate prolonged compression. Electrodiagnostic information may be valuable in diagnosing ulnar nerve conditions and establishing the level of embarrassment.

Treatment Options. Early treatment consists of rest and avoidance of the causative activity. A splint with the elbow in midposition between extension and flexion and in full supination may be tried for 4 to 6 weeks, and NSAIDs may be used in conjunc-

tion with the splint. Chronic ulnar nerve embarrassment at the elbow unrelieved by conservative measures may require surgical decompression when there is evidence of motor weakness, intrinsic atrophy, or a significant loss of sensation. Anterior transposition of the nerve has been recommended when the condition is severe, when exploration fails to reveal a compressed nerve, when there is a positive elbow flexion test, and when the elbow is in cubitus valgus.

Ulnar Tunnel Syndrome (Guyon's Canal)

Mechanism of Compression and Presenting Symptoms. Ulnar neuropathy at the wrist is much less common than median neuropathy, because the nerve passes outside the confines of the carpal tunnel (Fig. 18B–68). The condition occurs frequently in bicycle riders and has been termed cyclist's palsy. It typically develops following a prolonged road trip or heavy training, and nerve compression is probably caused by the direct pressure over the hypothenar eminence as riders support the weight of their trunk on the handlebar. Hyperextension of the wrist is said to contribute to the neuropathy, which almost always includes both motor and sensory involvement, although sometimes the deficit is purely motor. Athletes such as football linemen who repeatedly strike the ulnar aspect of their proximal palm may also develop this condition. It has also been seen in basketball players who repeatedly strike their proximal palms against the rim while slam dunking. Occasionally, ganglions or lipomas may produce ulnar compression, and a thorough examination is necessary to distinguish the entity from more proximal lesions.

Patients with ulnar nerve embarrassment at the wrist differ from those with higher ulnar nerve pathology in that they have normal sensation on the dorsal medial side of the hand in the distribution of

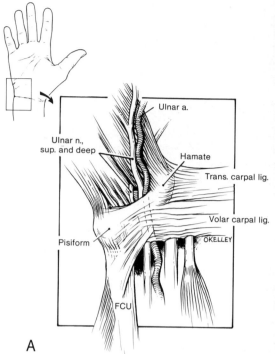

FIGURE 18B–68

A, The anatomy of the ulnar nerve at the wrist. The ulnar artery and nerve both pass beneath the piso-hamate ligament into Guyon's canal, with the nerve lying ulnar to the artery. *B,* The carpal tunnel view is useful in depicting fractures of the hook of the hamate and *C,* pisiform, which are frequently not seen on routine views of the hand.

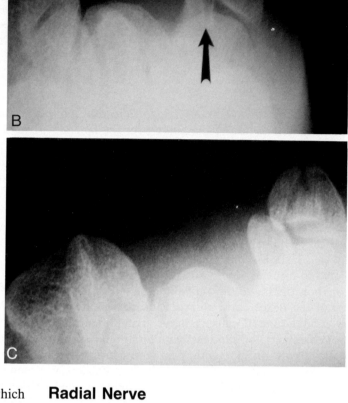

the dorsal sensory branch or the ulnar nerve, which is involved in cubital tunnel disorders. Weakness of the flexor carpi ulnaris or flexor profundi to the ring and small fingers also dictates a more proximal injury. Electromyography and nerve conduction studies may be of considerable value in confirming the diagnosis, and radiographs, including a carpal tunnel view (Fig. 18B–68*B, C*) should be obtained to diagnose fractures of the hamate and pisiform. Calcific tendinitis of the flexor carpi ulnaris, arthropathy of the pisotriquetral joint, and ligamentous wrist injuries may be other causes of pain in the region, but are seldom associated with ulnar neuropathy. Thrombosis of the ulnar artery may occur in combination with ulnar neuritis and should be determined by an Allen's test or, when in doubt, by noninvasive vascular studies or arteriography.

Treatment Options. Conservative treatment is recommended, and avoidance of the inciting trauma is paramount. Splinting or padding the proximal ulnar palm may be helpful, and anti-inflammatory drugs may contribute to the resolution of the problem. Surgery is rarely required except when there is a neurologic deficit.

Radial Nerve

Athletic injuries to the radial nerve are quite rare. Symptoms are usually those of vague weakness and pain, and the diagnosis is often difficult to make. Several conditions involving the radial nerve (Fig. 18B–69) may become a problem for athletes and are discussed here.

Radial Tunnel Syndrome

Mechanism of Compression and Presenting Symptoms. This condition has been referred to as resistant tennis elbow because the patient has often been unsuccessfully treated for lateral epicondylitis. Differentiation of the two conditions may be quite difficult, because tenderness over the lateral epicondyle is almost always present. Careful examination usually reveals tenderness over the extensor muscle mass directly overlying the radial nerve about four finger breadths distal to the lateral epicondyle. In radial tunnel syndrome, there is usually pain with resisted supination of the extended forearm; a pos-

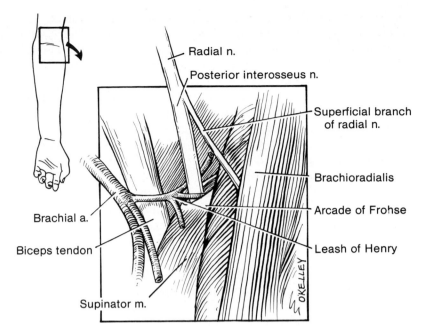

FIGURE 18B–69
The anatomy of the radial nerve in the proximal forearm. The area where the posterior interosseous nerve passes beneath the arcade of Frohse at the proximal edge of the supinator muscle has been implicated as a cause of radial tunnel syndrome.

itive middle finger sign (resisted extension of the middle finger that elicits pain in the proximal forearm) is said to be diagnostic of the condition. The condition may be seen in athletes who use their extensor muscles vigorously. In particular, weight lifters and bowlers occasionally present with the symptoms characteristic of this disorder.

In addition to lateral epicondylitis, other clinical conditions that should be included in the differential diagnosis of radial tunnel syndrome include wrist extensor tendinitis, stenosis of the orbicular ligament, fibrillation of the radial head, and cervical spine disorders. An unusual radial neuropathy, called runner's radial palsy, is distinctly different from radial tunnel neuritis and is manifested by numbness in the sensory distribution of the forearm and hand in runners who keep their arms acutely flexed at the elbow. This is thought to be caused by compression of the radial nerve between the humerus and triceps, and symptoms resolve by simply having the patient run with his elbow in a more extended posture [381].

Treatment Options. Conservative management, including rest and splinting with the forearm in neutral may be helpful, and anti-inflammatory medication should be tried. Surgery should be reserved for refractory cases not responding to conservative management.

Posterior Interosseous Nerve Syndrome

Mechanism of Compression and Presenting Symptoms. This condition is also occasionally seen in athletes and apparently results from overuse of the extensor muscle mass. Again, weight lifters and bowlers have been known to develop this syndrome, which usually begins with soreness and aching of the proximal portion of the extensor muscle group just distal to the lateral epicondyle. The radial nerve is usually compressed under the inverted fibrous arch at the proximal edge of the supinator muscle (arcade of Frohse). At this point, the nerve carries almost entirely motor fibers, and weakness or paralysis of the thumb abductor and thumb and digital extensors as well as extensor carpi ulnaris may result. A feature that distinguishes this condition from high radial nerve compression is that the extensor carpi radialis longus and brevis are still innervated (having received their motor fibers from the radial nerve proximal to the area of compression). As a result, wrist extension is still present, albeit with radial deviation because of the absence of the balancing effect of the extensor carpi ulnaris.

Treatment Options. When the condition presents in the acute stage, it should be treated by rest, splinting, and anti-inflammatory medications for at least 2 months to determine if spontaneous recovery will occur. If it does not, then surgical decompression should probably be carried out.

Distal Posterior Interosseous Nerve Syndrome

Dorsal wrist pain is a fairly frequent complaint among athletes, particularly when they are engaged in activities that produce repetitive forceful hyperextension. These injuries are frequently seen in gymnasts doing floor activities [377], and must be differentiated from scaphoid or radial impaction syndromes, tears of the scapholunate ligament, occult ganglia, and carpometacarpal joint sprains.

Distal posterior interosseous nerve syndrome is a condition in which the terminal sensory division of the posterior interosseous nerve is thought to be compressed or otherwise irritated as it passes over the distal radius and enters the dorsal wrist capsule. It is a difficult diagnosis to make with certainty and is generally considered when other, more obvious causes of dorsal wrist pain have been ruled out. The diagnosis of distal posterior interosseous impingement is therefore made by exclusion of the more common conditions and a clinical examination in which no masses are identified and there is an absence of clicking or snapping of the wrist. Provocative tests for carpal instability are negative and pain is elicited by forceful wrist extension or deep palpation of the dorsal wrist while it is in flexion. In some instances, an injection of a local anesthetic material just proximal to the wrist joint may aid with the diagnosis. If distal posterior interosseous embarrassment is felt to exist, rest, splinting, and anti-inflammatory medications should be tried initially. Only in rare instances is surgical exploration of the wrist indicated and, in the absence of any other pathology, the posterior interosseous nerve should be transected.

Superficial Radial Nerve Compression (Cheiralgia Paresthetica or Wartenberg's Disease)

Although this condition is not usually related to any specific sports activity, it may be seen in athletes wearing a tight wristwatch or wristband, such as those worn by weight lifters. It is frequently confused with other problems, including de Quervain's stenosing tenosynovitis, sprain, or arthropathy of the thumb carpometacarpal joint and gamekeeper's thumb. Patients with this condition usually present with numbness or decreased sensation over the dorsal radial hand and the dorsum of the thumb and index finger. The clinical findings often include a negative Finkelstein's sign; however, there will be a positive Tinel's test over the superficial branch of the radial nerve in the mid-forearm and decreased sensation over the dorsal radial hand and thumb when compared to the opposite side. Thumb extension or abduction is usually normal, indicating the absence of coexisting first dorsal extensor tenosynovitis. A history of the use of a wristband, tight

watch, or leather support can usually be obtained, and symptoms usually improve with the removal of the constricting influence of these devices. Direct trauma can produce the condition and is usually obvious to the involved athlete. Treatment of direct injury to the nerve should be rest and splinting.

SPLINTING AND REHABILITATION

An integral part of effectively managing athletic injuries of the hand is the employment of splinting and rehabilitation techniques that most efficiently protect the injured part during practice and competition and restore maximum joint mobility and hand strength as quickly as possible. In this chapter we have tried to emphasize the appropriate splinting and exercise methods for each injury, and an overview of the general principles of splinting and rehabilitation is provided here.

With the development of therapists trained in the post-injury and post-surgery management of hand conditions, there has been a tremendous improvement in the ability of the team managing athletic injuries (trainer, physician, and therapist) to expeditiously assure tissue healing while protecting injured parts during competition. The members of this team must work together with frequent communication to be sure that the best interests of the athlete are respected. Throughout the treatment period techniques must be employed that conform to good medical management and to the rules of the governing bodies for the particular sport. Specific rehabilitation protocols must be coordinated so that an effective game plan is implemented without contradictions between various members of the treating team.

Splinting

General Principles

The principles of splinting athletic injuries of the hand are really quite simple. During the healing phase, the injured part is usually rested with static splints used to position the involved tissues and joints in a manner that facilitates prompt healing with little additional stress. When motion of adjacent uninjured joints does not jeopardize tissue healing, gentle range-of-motion exercises should be commenced as quickly as possible, and it may be appropriate to alternate static splints with periods of exercise. It is often possible to design splints that allow motion within certain planes but limit motion in a direction that might reinjure healing tissues.

For example, following a fracture dislocation of the PIP joint, where extension might lead to disruption of the healing volar tissues and resubluxation of the joint, dorsal blocking splints are often employed to permit flexion but limit extension. Similarly, collateral ligaments can often be protected by cleverly designed splints that permit motion only in the anterior-posterior plane. When joint stiffness is a problem, dynamic splinting may be instituted in either the extension or flexion plane or both in order to facilitate a rapid recovery of active and passive motion.

Rules and Regulations Concerning Splints and Braces

The physician treating high school or college athletes must be familiar with the rules governing competition in his state or conference. This is not always easy, because considerable confusion exists about what types of splints are allowed to be worn by athletes in competition, especially in football. One of the major reasons for this is that there are at least two major organizations that publish football rules: (1) the National Collegiate Athletic Association (NCAA), which governs most college athletics; and (2) the National Federation of State High School Associations (NFSHSA), which oversees most, but not all, high school competition. For example, Texas does not belong to the NFSHSA, but instead uses the NCAA rules. The confusion stems from two diametrically opposing statements regarding the use of hard splints, as follows.

The NCAA rules [388] state that illegal equipment includes "hard, abrasive or unyielding substances on the hand, wrist, forearm or elbow of any player *unless* covered on all sides with closed cell, slow-recovery foam padding no less than one-half-inch thick or an alternate material of the same minimum thickness and similar physical properties."

The NFSHSA rules [391], conversely, prohibit any "elbow, hand, wrist, or forearm guard or brace made of sole leather, plaster, metal or other hard substance in its final form, and *even though* covered with soft padding."

Therefore, strict interpretation and enforcement of these rules would make it illegal for a high school player in most states (excluding Texas) to wear any type of hard splint (plaster, fiberglass, or plastic), even if it is covered with foam padding.

Further confusion arises because, in many jurisdictions, the final determination as to what type of splint a high school or college player can wear is left to the discretion of the game officials.

If the athlete is permitted to wear a hard splint,

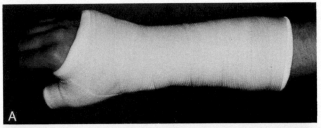
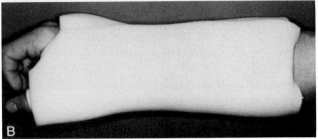
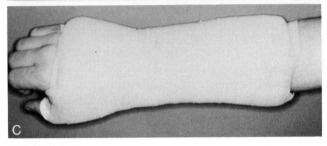

FIGURE 18B–70
Padding of hard casts for competition. *A,* A standard thumb spica cast is padded with *B,* thick rubber foam, and *C,* wrapped with Coban.

we prefer a standard fiberglass cast for injuries requiring full-time immobilization, and a plastic orthosis for those requiring only protection during practice and games. Both of these must of course be padded during competition (Fig. 18B–70). If the athlete is playing under rules that forbid a hard cast, we then prefer the silicone rubber splint originally described by Bassett and co-workers [385] and subsequently popularized by Bergfeld and associates [386] (Fig. 18B–71) (see description of RTV-11 on page 1005). These rubber splints are heavy and do not allow the skin to breathe, and therefore are worn only during practice and games. A plastic orthosis is worn at all other times.

Types of Splints

Adhesive Tape. Athletic trainers are masters with adhesive tape, and it is a delight to watch the speed and efficiency with which a skilled trainer can tape an ankle or wrist. Adhesive tape alone, however, provides very little support, and its functional use in the hand is primarily in the form of buddy taping, in which an injured finger is securely taped to an

FIGURE 18B–71
The RTV-11 silicone cast can be worn in competition without additional padding (see Figure 18C–73 for details of fabrication).

adjacent normal digit. For the thumb, minimal protection but perhaps some psychological effect can be gained from the use of an adhesive tape thumb spica splint. When taping the thumb or wrist, care must be taken to avoid wrapping too tightly to avoid swelling of the hand distal to the circumferential tape about the wrist.

Padded Aluminum Splints. These readily available and versatile splints are useful particularly when only a single joint needs to be splinted; for example, the DIP joint for mallet fingers and the PIP joint for boutonniere lesions. The metal part of the splint must be trimmed to eliminate sharp edges, the splint securely taped to the digit, and the entire splint covered with a thin layer of soft padding to prevent injury to other players.

Synthetic Rubber and Plastic Splints. When an athlete is returned to competition with a healing injury that requires protection, this is best provided with a custom-made orthosis designed to protect the injured part. There are many different types of splinting materials available, the various properties and qualities of which are described below. It is neither feasible nor necessary to have all of these materials on hand, for many of the commercially available products have similar characteristics. However, the therapist must be familiar with the various types of splinting materials and have available materials representing each type.

Materials. An appreciation of the unique characteristics and properties of each material allows the splint maker to determine the optimum type to be used for the support and protection of a particular injured part. Table 18B–1 lists the material type, trade name, characteristics, and manufacturer of some of the most commonly used splinting materials.

It can be seen that many products with differing characteristics are available and, when used correctly, they can prove valuable and effective in splinting the injured hand of the athlete. Most athletic hand splinting uses low-temperature thermoplastic materials, the majority of which are plastic combined with varying amounts of rubber. A high content of plastic generally gives the material good drapeability and contour, whereas the amount of rubber determines the elasticity and memory. Most low-temperature thermoplastics may be softened at about 160°F by the use of wet or dry heat. Working times vary according to manufacturer, but usually fall between 3 and 6 minutes.

We have found the use of two materials, Polyform and Orthoplast, to be versatile enough for most athletic hand-splinting situations (Fig. 18B–72). Polyform, with its high plastic content, has excellent stretching and molding capabilities, producing an exact fit with good contour. When cured, Polyform is quite rigid, with little flexibility. Orthoplast, in

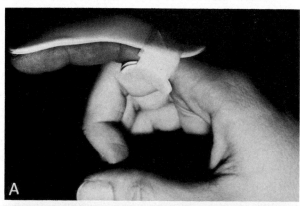

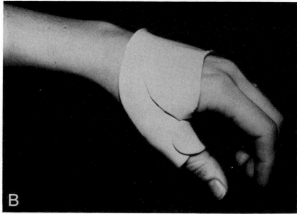

FIGURE 18B–72
The use of Orthoplast for protective (noncompetition) splinting. *A,* A dorsal extension block splint that allows flexion of the finger but prevents full extension, for patients recovering from PIP fracture-dislocation or volar plate injuries. *B,* A hand-based thumb spica splint to protect the MP joint of the thumb. Although this splint could be padded and used in competition, the silicone RTV-11 splint provides better protection for the thumb.

TABLE 18B–1
Splint Materials

Material	Features	Manufacturer
Splint Materials		
Low-Temperature Thermoplastics		
Polyform (plastic polymer)	Good drapeability with excellent stretch No memory Excellent contourability Excellent rigidity and impact strength	Smith and Nephew Rolyan, Inc. N93 W14475 Whitaker Way Menomonee Falls, WI 53051
Polyflex II (plastic polymer and rubber)	Good draping and contourability Good resilient stretch with slight memory Excellent rigidity—resists fatigue Ideal for larger splints	Smith and Nephew Rolyan, Inc. N93 W14475 Whitaker Way Menomonee Falls, WI 53051
Orthoplast (transpolyisoprene)	High rubber content Offers some stretch with good memory Good rigidity yet lightweight Ideal for circumferential splinting due to flexibility	Johnson & Johnson Products New Brunswick, NJ 08903
Aquaplast-T nonsticky (polycaprolactone) with surface coating	Clear when heated enabling control of pressure points over small areas Good conforming qualities Excellent elasticity when heated; good memory with repeated heating Excellent rigidity with small splints	WFR Aquaplast Corporation Post Office Box 327 Ramsey, NJ 07446

Others: Ezeform (plastic polymer), Orthoform (transpolyisoprene), Hexcelite (polymers, filler, and pigment impregnated onto cotton netting), Aquaplast (original sticky) (poly [epsilon] caprolactone), Monorthos (polycaprolactone, other polymers, and pigment), Green Stripe Aquaplast (polycaprolactone, with or without surface coating), Burns Quality Aquaplast (blue stripe), Kay-Splint Series 3 (plastic polymer), K-Splint Isoprene (plastic polymer), Kay Splint I (plastic polymer), JU 1000 Splinting Compound (plastic polymer), MR 2000 Splinting Compound (plastic polymer), RS 3000 Splinting compound (plastic polymer)

High-Temperature Thermoplastics

Vinyl (polyvinyl chloride), Kydex (copolymer of polyvinyl chloride and acrylic), W-Clear (transparent polyester-based orthotic material (copolyester of polyethylene terephthalate)

Playing Cast Materials		
RTV-11	Liquid-based silicone rubber compound, cured with a catalyst. Impregnated with gauze during application Usually padded externally (often with Temperfoam) Requires 3–4 hours to completely cure	General Electric Co. Silicone Rubber Products Div. Waterford, NY 12188 (518) 237-3330
3M Scotchrap	Semi-rigid support wrap External padding Can be used pending specific regulations for competition	3M Orthopedic Products Div. 6850 S. Harlem Ave. Bedford Park, IL 60501

Shock-Absorbing Materials		
Neoprene	Closed cell rubber available in variety of thicknesses Retains maximum body heat Provides uniform compression	Local distributor of divers' supplies.
Sorbothane	Viscoelastic polymer Elasticity and flowability Can absorb 94% of energy in dynamic load displacement tests High compression resistance	IEM Orthopaedics 251 West Garfield Road Aurora, OH 44202

Table continued on following page

TABLE 18B–1
Splint Materials *Continued*

Material	Features	Manufacturer
Viscolis	Viscoelastic polymer Has elasticity and flowability Excellent fine-vibration shock absorber High resistance to compression	Chattanooga Corporation Visco-Elastic Products 101 Memorial Drive Post Office Box 4287 Chattanooga, TN 37405
Foam Materials		
Polycushion	Soft, closed cell foam with adhesive backing Resists compression well Available in ¼- and ⅛-thicknesses	North Coast Medical, Inc. 450 Salmar Avenue Campbell, CA 95008
Temperfoam	Soft, viscoelastic foam product with adhesive back Fluidlike support conforms to hand Resists impact, shock, and vibration	North Coast Medical, Inc. 450 Salmar Avenue Campbell, CA 95008
Splint Inserts		
Otoform-K (transpolyisoprene)	Paste form of silicone elastomer cured with catalyst Amount of catalyst used will affect firmness of finished rubber mold Good contourability for interfacing splints Good shock-absorbing quality	WFR Aquaplast Corporation Post Office Box 327 Ramsey, NJ 07446
Elastomer (Silastic 382, medical grade)	Liquid-based silicone elastomer curved with a catalyst Produces firm, closed cell rubber mold Excellent contouring for interfacing of splints Good shock-absorbing quality	WFR Aquaplast Corporation Post Office Box 327 Ramsey, NJ 07446
Padded Gloves		
Antivibration gloves	Leather or jersey glove with Viscolis insert Available in a variety of sizes and styles	Steel Grip Inc. 700 Garfield Street Danville, IL 61832

contrast, has a higher rubber component, which provides better flexibility and elasticity. These characteristics make Orthoplast the preferred material for larger, more circumferential types of splints, where exacting contour is not so important. Polyflex II and Aquaplast-T are two other splinting materials that may prove useful in specific instances. Polyflex II provides flexibility for circumferential splints, and is more rigid when cured than Orthoplast. It does not, however, have as good a memory. Aquaplast-T has very good elasticity and conformability, in addition to being transparent when heated. This may be beneficial when attempting to splint over small bony areas of the hand that are prone to the development of pressure points. The splint materials should be chosen on the basis of their ability to meet the basic requirements of the injury for which the splint is being fabricated and the experience and preferences of the splint maker.

Some recently introduced products designed to provide reasonable support to the injured hand and wrist are RTV-11 and 3M Scotchrap. These materials attempt to solve the dilemma of effectively splinting and protecting the injured part while overcoming the restriction on competition resulting from the use of hard devices. RTV-11 is a silicone rubber compound that conforms well and provides better support than some of the previous elastomers (Fig. 18B–73). Splints made from RTV-11 can be worn without external padding (Fig. 18B–73). 3M Scotchrap is applied much like a cast and provides semirigid support. It remains soft enough to be acceptable for competition in most instances, but should be supplemented by an additional layer of padding.

Early protective splinting for return to competition usually requires some type of padding or lining for comfort and shock absorption. Again, a wide variety of materials is available for customizing the splint to the athlete's injury and the needs of the sport. Foam padding has value for lining areas of

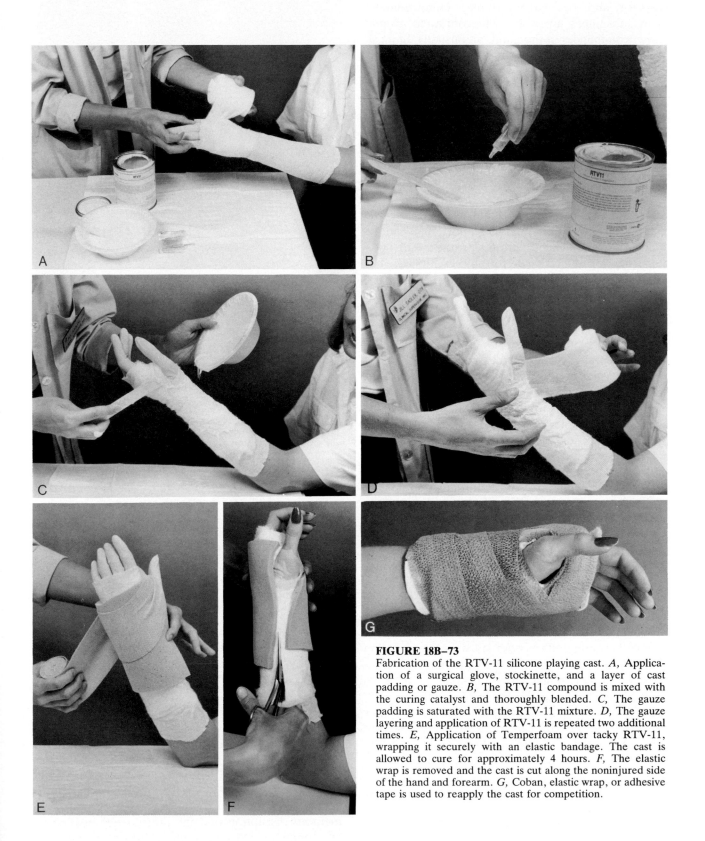

FIGURE 18B–73

Fabrication of the RTV-11 silicone playing cast. *A*, Application of a surgical glove, stockinette, and a layer of cast padding or gauze. *B*, The RTV-11 compound is mixed with the curing catalyst and thoroughly blended. *C*, The gauze padding is saturated with the RTV-11 mixture. *D*, The gauze layering and application of RTV-11 is repeated two additional times. *E*, Application of Temperfoam over tacky RTV-11, wrapping it securely with an elastic bandage. The cast is allowed to cure for approximately 4 hours. *F*, The elastic wrap is removed and the cast is cut along the noninjured side of the hand and forearm. *G*, Coban, elastic wrap, or adhesive tape is used to reapply the cast for competition.

splints where firm materials may produce excessive pressure and be uncomfortable. Unfortunately, foam substances are not easily contoured around bony prominences in the hand, and usually diminish the protective support provided by the splint. Several types of foam are available in various densities, and most have a peel-off adhesive backing for easy application. We have found the most useful foam linings to be Polycushion, which is supplied in both ¼- and ⅛-inch, and Temperfoam, ¼-inch. Polycushion is an excellent, all-purpose, closed-cell foam, used primarily as a splint interface for patient comfort. It can be applied to the splint material before or after fabrication, and the material resists compression well. Temperfoam is a viscoelastic foam that provides fluidlike support and can be contoured to the hand through warmth and body heat, making it quite comfortable. Although it is soft to the touch, Temperfoam is firm on impact, making it a good shock absorber. With the advent of playing cast materials such as RTV-11, Temperfoam has become popular as an external coating that provides the final cushioning that makes the splint acceptable for athletic competition.

Medical-grade silicone elastomers such as Otoform-K and Elastomer also provide protective padding and may be useful in the preparation of splints for the injured hands of athletes. Otoform-K has a putty-like base and catalyst, whereas Elastomer has a liquid base and catalyst. Both materials may be molded and contoured to a specific area, which can provide an exacting fit. Both materials form flexible rubber molds when they cure, resulting in excellent shock-absorbing capabilities. Elastomer has also been used for playing casts, although it is not as strong and supportive as RTV-11.

Other materials that have gained popularity for their shock-absorbing qualities are Neoprene, Sorbothane, and Viscolis. Neoprene, which has long been used by wetsuit manufacturers, is a closed-cell rubber material available in various thicknesses. It is usually laminated on both sides to stretch nylon or terrylike materials that provide uniform compressive support while retaining body heat for increased comfort, a quality especially useful in managing compression syndromes and tendinitis. It may be used as an interface for rigid splints, or by itself for custom-made soft splints. Sorbothane and Viscolis are both classified as viscoelastic polymer products. They possess the flowability of a viscous liquid that enables the material to absorb and dissipate impact energies very effectively. Studies have shown Sorbothane to absorb as much as 94% of impact energy during dynamic load-displacement testing. Viscolis is most noted for its ability to absorb

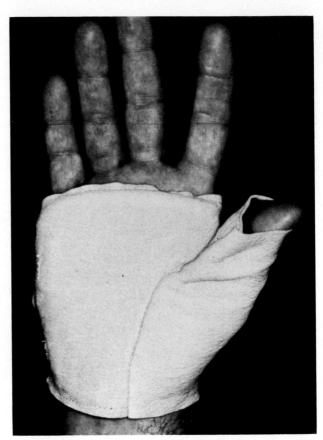

FIGURE 18B–74
A prefabricated antivibration glove (Steel Grip) may be useful for some athletes, such as cyclists.

vibrations between 90 and 400 cycles/second, the range in which most soft tissue damage occurs. These materials are available in sheets that can be cut to the desired shapes for lining of gloves and splints. A tacky exterior has been created for securing the inserts in position. Both of these products represent important advances in the ability to produce effective splints for the management of athletic injuries as well as for the protection of vulnerable areas of the hand during practice and competition.

Many types of prefabricated antivibration gloves or pads are currently manufactured (Steel-grip), and may be useful for some athletes, such as cyclists (Fig. 18B–74).

Choice of Splints

Static splints used as protective devices may be constructed from rigid or flexible materials. Firm splints are employed for protection following ligamentous disruption or bony injury when healing is not complete. However, rigid splints that cross the wrist are often impractical for competitive use and may only be appropriate for the athlete whose sport requires minimal use of the hand. Rigid, low-profile splints that are confined to the digits may be suitable for competitive use following ligament injuries of the PIP or DIP joints, but do not offer adequate protection for newly healed fractures. Flexible splints are obviously preferred by athletes because they are less restrictive. Although this type of splint may not provide much true protection, they generally provide enough support to help minimize pain and prevent excessive movement.

With contact sports, a wider area of application or more circumferential splints may be indicated in order to provide more support and distribute pressure forces more evenly. Regardless of the splinting program employed for competition, the physician, therapist, and coach must all agree that it will not in any way jeopardize the healing process for a particular injury. Further, it must not risk reinjury that might unfavorably alter the prognosis for the long-term performance of the injured part. Continued protective splinting may be indicated in those situations where the athlete is prone to reinjury.

Rehabilitation

Although minor sprains or strains or even undisplaced stable fractures can be managed by simple protective splinting with the initiation of early motion programs and the expectation of a rapid full recovery, more severe injuries and those that require a protracted recovery often necessitate formal rehabilitation programs. For the most part, these programs should be directed by well-trained sports and hand therapists in an effort to restore maximum joint mobility and extremity strength. Active and passive range-of-motion programs may be augmented by dynamic splinting or even the use of continuous passive motion machines in a few recalcitrant cases. Specific strengthening programs that employ simple devices such as hand grippers and Silly Putty may be helpful, and exercises designed to maximize recovery in specific muscle groups may be taught to the patient and supervised by the therapist. Throughout this process, serial recordings of joint motion and measurements of grip and pinch strength help the therapists assess recovery and, when necessary, alter the rehabilitation program. Modalities such as transcutaneous electrical nerve stimulation (TENS) to modify pain, or functional electrical stimulation (FES) in order to augment and retrain injured or unused muscles, may be of value in the rehabilitation of the athlete's hand.

SUMMARY

An assortment of injuries may occur in the hands of athletes during competition or practice. There are no shortcuts to the effective management of these injuries, and sound hand surgical and orthopaedic procedures must be employed. The managing physician should be cognizant of the unique demands and reinjury potential of each injured athlete's sport and playing position before allowing the return to competition. There may be certain calculated alterations of traditional conservative treatment methods to return the athlete to full participation, but these departures should never be made if they run any appreciable risk of compromising the healing or functional result of a given injury. No player's career, no team's record, and no athletic contest is worth permanent deformity or dysfunction of the hand.

References

Tendon Injuries

1. Abouna, J.M., and Brown, H. The treatment of mallet finger: The results in a series of 148 consecutive cases and a review of the literature. *Br J Surg* 55:653–667, 1968.
2. Auchincloss, J.M. Mallet-finger injuries: A prospective, controlled trial of internal and external splintage. *Hand* 2:168–173, 1982.
3. Backdahl, M. Ruptures of the extensor aponeurosis at the distal digital joints. *Acta Chir Scand* 111:151–157, 1956.
4. Blazina, M.E., and Lane, C. Rupture of the flexor digitorum profundus tendon in student athletes. *J Am Coll Health* 14:248–249, 1966.
5. Bowers, W.H., and Hurst, L.C. Chronic mallet finger: The use of Fowler's central slip release. *J Hand Surg* 3:373–376, 1978.
6. Boyes, J.H. *Bunnell's Surgery of the Hand* (3rd ed.). Philadelphia, J.B. Lippincott, 1956.
7. Boyes, J.H. *Bunnell's Surgery of the Hand* (5th ed.). Philadelphia, J.B. Lippincott, 1970.
8. Brooks, D. Splint for mallet fingers. *Br Med J* 1:1238, 1964.
9. Bunnell, S. *Surgery of the Hand.* Philadelphia, J.B. Lippincott, 1944, pp. 490–493.
10. Burke, F. Editorial: Mallet finger. *J Hand Surg [Br]* 13:115–117, 1988.
11. Buscemi, M.J., and Page, B.J. Flexor digitorum profundus avulsions with associated distal phalanx fractures. *Am J Sports Med* 15:366–370, 1987.
12. Bynum, D.K., and Gilbert, J.A. Avulsion of the flexor digitorum profundus: Anatomic and biomechanical considerations. *J Hand Surg [Am]* 13:222–227, 1988.
13. Cannon, N.M., and Strickland, J.W. Therapy following flexor tendon surgery. *Hand Clin* 1:147–166, 1985.
14. Casscells, S.W., and Strange, T.B. Intramedullary wire

We would like to thank Jill Sadler, OTR (The Hand Rehabilitation Center of Indiana, Inc.) and Jan Kinnunen Lopez, OTR (Hand Rehabilitation Associates of San Antonio) for assisting us with their knowledge and expertise in the writing of the section on splinting.

fixation of mallet finger. *J Bone Joint Surg [Am]* 39:521–526, 1957.

15. Chow, J.A., Thomes, L.J., Dovelle, S.W., et al. A combined regimen of controlled motion following flexor tendon repair in "no man's land." *Plast Reconstr Surg* 79:447–453, 1987.

16. Chow, J.A., Thomes, L.J., Dovelle, S.W., Monsisvias, J., Milnor, W.H., and Jackson, J.P. Controlled motion rehabilitation after flexor tendon repair and grafting: A multicentre study. *J Bone Joint Surg [Am]* 70:591–595, 1988.

17. Cohn, B.T., and Froimson, A.I. Case report of a rare mallet finger injury. *Orthopedics* 9:529–531, 1986.

18. Crawford, G.P. The molded polythene splint for mallet finger deformities. *J Hand Surg [Am]* 9:231–237, 1984.

19. Curtis, R.M., Reid, R.L., and Provost, J.M. A staged technique for the repair of the traumatic boutonniere deformity. *J Hand Surg* 8:167–171, 1983.

20. Din, K.M., and Meggitt, B.F. Mallet thumb. *J Bone Joint Surg [Br]* 65:606–607, 1983.

21. Dolphin, J.A. Extensor tenotomy for chronic boutonniere deformity of the finger. *J Bone Joint Surg [Am]* 47:161–164, 1965.

22. Duran, R.J., Houser, R.G., Coleman, C.R., and Postlewaite, D.S. A preliminary report in the use of controlled passive motion following flexor tendon repair in zones II and III (abstr). *J Hand Surg* 1:79, 1976.

23. Elliott, R.A. Injuries to the extensor mechanism of the hand. *Orthop Clin North Am* 1:335–354, 1970.

24. Elliott, R.A. Boutonniere deformity. *In* Cramer, L.M., and Chase, R.A. (Eds.). *American Academy of Orthopaedic Surgeons Symposium on the Hand.* St. Louis, C.V. Mosby, 1971, pp. 42–54.

25. Elliott, R.A. Splints for mallet and boutonniere deformities. *Plast Reconstr Surg* 52:282–285, 1973.

26. Flinchum, D. Mallet finger. *J Med Assoc Ga* 48:601–603, 1959.

27. Fowler, F.D. New splint for treatment of mallet finger. *JAMA* 170:945, 1959.

28. Fowler, S.B. Quoted by Littler, J.W., *In* Converse, J.M. (Ed.). *The Hand and Upper Extremity in Reconstructive Plastic Surgery.* Philadelphia, W.B. Saunders, 1964, p. 1630.

29. Gelberman, R.H., Amifl, D., Gonsalves, M., and Akeson, W.H. The influence of protected passive mobilization on the healing of flexor tendons: A biochemical and microangiographic study. *Hand* 13:120–128, 1981.

30. Gelberman, R.H., Botte, M.J., Spiegelman, J.H., and Akeson, W.H. The excursion and deformation of repaired flexor tendons treated with protected early motion. *J Hand Surg [Am]* 11:106–110, 1986.

31. Gelberman, R.H., Menon, J., Gonsalves, M., and Akeson, W.H. The effects of mobilization on the vascularization of healing flexor tendons in dogs. *Clin Orthop* 153:283–289, 1980.

32. Gelberman, R.H., Van De Berg, J.S., Lundborg, G.N., and Akeson, W.H. Flexor tendon healing and restoration of the gliding surface. *J Bone Joint Surg [Am]* 65:70–80, 1983.

33. Gelberman, R.H., Woo, S.L.Y., Lothringer, K., Akeson, W.H., and Amiel, D. Effects of early intermittent passive mobilization on healing canine flexor tendons. *J Hand Surg* 7:170–175, 1982.

34. Goldner, J.L. Deformities of the hand incidental to pathological changes of the extensor and intrinsic muscle mechanisms. *J Bone Joint Surg [Am]* 35:115–131, 1953.

35. Grundberg, A.B., and Reagan, D.S. Central slip tenotomy for chronic mallet finger deformity. *J Hand Surg [Am]* 12:545–547, 1987.

36. Hamas, R.S., Horrell, E.D., and Pierret, G.P. Treatment of mallet finger due to intra-articular fracture of the distal phalanx. *J Hand Surg* 3:361–363, 1978.

37. Harris, C., and Rutledge, G.L. The functional anatomy of the extensor mechanism of the finger. *J Bone Joint Surg [Am]* 54:713–726, 1972.

38. Harris, C. Jr. The Fowler operation for mallet finger. *J Bone Joint Surg [Am]* 48:613, 1966.

39. Hillman, F.E. New technique for treatment of mallet fingers and fractures of distal phalanx. *JAMA* 161:1135–1138, 1956.

40. Howie, H. The treatment of mallet finger: A modified plaster technique. *N Z Med J* 46:513, 1947.

41. Isani, A. Small joint injuries requiring surgical treatment. *Orthop Clin North Am* 17:407–419, 1986.

42. Iselin, F., Levame, J., and Godoy, J. A simplified technique for treating mallet fingers: Tenodermodesis. *J Hand Surg* 2:118–121, 1977.

43. Jones, N.F., and Peterson, J. Epidemiologic study of the mallet finger deformity. *J Hand Surg* 13A:334–338, 1988.

44. Kaplan, E.B. Mallet or baseball finger. *Surgery* 7:784–791, 1940.

45. Kinninmonth, A.W.G., and Holburn, F. A comparative controlled trial of a new perforated splint and a traditional splint in the treatment of mallet finger. *J Hand Surg [Br]* 11:261–262, 1986.

46. Kleinert, H.E., and Verdan, C. Report of the Committee on Tendon Injuries. *J Hand Surg* 8:794–798, 1983.

47. Kleinman, W.B., and Petersen, D.P. Oblique retinacular ligament reconstruction for chronic mallet finger deformity. *J Hand Surg [Am]* 9:399–404, 1984.

48. Kon, M., and Bloem, J.J.A.M. Treatment of mallet fingers by tenodermodesis. *Hand* 14:174–176, 1982.

49. Langa, V., and Posner, M.A. Unusual rupture of a flexor profundus tendon. *J Hand Surg [Am]* 11:227–229, 1986.

50. Lange, R.H., and Engber, W.D. Hyperextension mallet finger. *Orthopedics* 6:1426–1431, 1983.

51. Leddy, J.P., and Packer, J.W. Avulsion of the profundus tendon insertion in athletes. *J Hand Surg* 2:66–69, 1977.

52. Lewin, P. A simple splint for baseball finger. *JAMA* 85:1059, 1925.

53. Lister, G.D., Kleinert, H.E., Kutz, J.E., and Atasoy, E. Primary flexor tendon repair followed by immediate controlled mobilization. *J Hand Surg* 2:441–451, 1977.

54. Littler, J.W., and Eaton, R.G. Redistribution of forces in the correction of the boutonniere deformity. *J Bone Joint Surg [Am]* 49:1267–1274, 1967.

55. Mason, M.L. Rupture of tendons of the hand: With a study of the extensor tendon insertions in the fingers. *Surg Gynecol Obstet* 50:611–624, 1930.

56. Matev, I. Transposition of the lateral slips of the aponeurosis in treatment of long-standing "boutonniere deformity" of the finger. *Br J Plast Surg* 17:281–286, 1964.

57. McCue, F.C., and Abbott, J.L. The treatment of mallet finger and boutonniere deformities. *Va Med* 94:623–628, 1967.

58. McFarlane, R.M., and Hampole, M.K. Treatment of extensor tendon injuries of the hand. *Can J Surg* 16:366–375, 1973.

59. Mikic, Z., and Helal, B. The treatment of the mallet finger by Oakley splint. *Hand* 6:76–81, 1974.

60. Miura, T., Nakamura, R., and Torii, S. Conservative treatment for a ruptured extensor tendon on the dorsum of the proximal phalanges of the thumb (mallet thumb). *J Hand Surg [Am]* 11:229–233, 1986.

61. Mixa, T.M., Blair, S.J., and Dvonch, V.M. Acute and chronic management of mallet finger: A case study. *Orthopedics* 8:1044–1046, 1985.

62. Moss, J.G., and Steingold, R.F. The long term results of mallet finger injury: A retrospective study of one hundred cases. *Hand* 15:151–154, 1983.

63. Niechajev, I.A. Conservative and operative treatment of mallet finger. *Plast Reconstr Surg* 76:580–585, 1985.

64. Patel, M.R., Desai, S.S., and Lipson, L.B. Conservative management of chronic mallet finger. *J Hand Surg [Am]* 11:570–573, 1986.

65. Patel, M.R., Lipson, L.B., and Desai, S.S. Conservative treatment of mallet thumb. *J Hand Surg [Am]* 11:45–47, 1986.

66. Pratt, D.R. Internal splint for closed and open treatment

of injuries of the extensor tendon at the distal joint of the finger. *J Bone Joint Surg [Am]* 34:785–788, 1952.

67. Primiano, G.A. Conservative treatment of two cases of mallet thumb. *J Hand Surg [Am]* 11:233–235, 1986.

68. Ratliff, A.H.C. Mallet finger: A review of forty-five cases. *Manchester Medical Gazette* 26:4, 1947.

69. Rayan, G.M., and Mullins, P.T. Skin necrosis complicating mallet finger splinting and vascularity of the distal interphalangeal joint overlying skin. *J Hand Surg [Am]* 12:548–552, 1987.

70. Robb, W.A.T. The results of treatment of mallet finger. *J Bone Joint Surg [Br]* 41:546–549, 1959.

71. Rosenzweig, N. Management of the mallet finger. *S Afr Med J* 24:831–832, 1950.

72. Rothwell, A.G. Repair of the established post-traumatic boutonniere deformity. *Hand* 10:241–145, 1978.

73. Schneider, L.H. Fractures of the distal phalanx. *Hand Clin* 4:537–547, 1988.

74. Smillie, I.S. Mallet finger. *Br J Surg* 24:439–445, 1937.

75. Sourmelis, S.G., and McGrouther, D.A. Retrieval of the retracted flexor tendon. *J Hand Surg [Br]* 12:109–111, 1987.

76. Souter, W.A. The boutonniere deformity: A review of 101 patients with division of the central slip of the extensor expansion of the fingers. *J Bone Joint Surg [Br]* 49:710–721, 1967.

77. Souter, W.A. The problem of boutonniere deformity. *Clin Orthop* 104:116–133, 1974.

78. Spigelman, L. New splint for management of mallet finger. *JAMA* 153:1362, 1953.

79. Stack, H.G. Mallet finger. *Hand* 1:83–89, 1969.

80. Stack, H.G. A modified splint for mallet finger. *J Hand Surg [Br]* 11:263, 1986.

81. Stark, H.H., Boyes, J.H., and Wilson, J.N. Mallet finger. *J Bone Joint Surg [Am]* 44:1061–1068, 1962.

82. Stark, H.H., Gainor, B.J., Ashworth, C.R., Zemel, N.P., and Rickard, T.A. Operative treatment of intra-articular fractures of the dorsal aspect of the distal phalanx of digits. *J Bone Joint Surg [Am]* 69:892–896, 1987.

83. Stern, P.J., and Kastrup, J.J. Complications and prognosis of treatment of mallet finger. *J Hand Surg [Am]* 13:329–334, 1988.

84. Strickland, J.W. Management of acute flexor tendon injuries. *Orthop Clin North Am* 14:827–849, 1983.

85. Strickland, J.W. Flexor tendon repair. *Hand Clin* 1:55–68, 1985.

86. Strickland, J.W. Results of flexor tendon surgery in zone II. *Hand Clin* 1:167–179, 1985.

87. Strickland, J.W. Opinions and preferences in flexor tendon surgery. *Hand Clin* 1:187–191, 1985.

88. Strickland, J.W. Flexor tenolysis. *Hand Clin* 1:121–132, 1985.

89. Strickland, J.W. Biologic rationale, clinical application, and results of early motion following flexor tendon repair. *J Hand Ther* 10:71–83, 1989.

90. Strickland, J.W. Flexor tendon surgery: A review article. Part 1: Primary flexor tendon repair. *J Hand Surg [Br]* 14:261–272, 1989.

91. Strickland, J.W. Flexor tendon surgery: A review article. Part 2: Free tendon grafts and tenolysis. *J Hand Surg [Br]* 14:368–382, 1989.

92. Thompson, J.S., Littler, J.W., and Upton, J. The spiral oblique retinacular ligament (SORL). *J Hand Surg* 3:482–487, 1978.

93. Tubiana, R. Surgical repair of the extensor apparatus of the fingers. *Surg Clin North Am* 48:1015–1031, 1968.

94. Van DeMark, R.E. A simple method of treatment for recent mallet finger. *Milit Surg* 107:385–386, 1950.

95. Van der Meulen, J.C. The treatment of prolapse and collapse of the proximal interphalangel joint. *Hand* 4:154–162, 1972.

96. Warren, R.A., Kay, N.R.M., and Ferguson, D.G. Mallet finger: Comparison between operative and conservative management in those cases failing to be cured by splintage. *J Hand Surg [Br]* 13:159–160, 1988.

97. Warren, R.A., Kay, N.R.M., and Norris, S.H. The microvascular anatomy of the distal digital extensor tendon. *J Hand Surg [Br]* 13:161–163, 1988.

98. Warren, R.A., Norris, S.H., and Ferguson, D.G. Mallet finger: A trial of two splints. *J Hand Surg [Br]* 13:151–153, 1988.

99. Watson, F.M. *Am Soc Surg Hand Correspondence Newsletter* 104, 1983.

100. Watson-Jones, R. *Fractures and Joint Injuries* (4th ed.). Edinburgh, E & S Livingstone, 1956, pp. 645–646.

101. Wehbe, M.A., and Schneider, L.H. Mallet fractures. *J Bone Joint Surg [Am]* 66:658–669, 1984.

102. Williams, E.G. Treatment of mallet finger. *Can Med Assoc J* 57:582, 1947.

103. Zancolli, E. Structural and Dynamic Bases of Hand Surgery (2nd ed.). Philadelphia, J.B. Lippincott, 1979, p. 79.

Ligamentous Injuries and Dislocations

104. Agee, J.M. Unstable fracture dislocations of the proximal interphalangeal joint of the fingers: A preliminary report of a new treatment technique. *J Hand Surg* 3:386–389, 1978.

105. Agee, J.M. Unstable fracture dislocations of the proximal interphalangeal joint: Treatment with the force couple splint. *Clin Orthop* 214:101–112, 1987.

106. Ahmad, I., and DePalma, A.F. Treatment of gamekeeper's thumb by a new operation. *Clin Orthop* 103:167–169, 1974.

107. Ali, M.S. Complete disruption of collateral mechanism of proximal interphalangeal joint of fingers. *J Hand Surg* 9:191–193, 1984.

108. Alldred, A.J. Rupture of the collateral ligament of the metacarpo-phalangeal joint of the thumb. *J Bone Joint Surg [Br]* 37:443–445, 1955.

109. Andersen, J.A., and Gjerloff, C.C. Complex dislocation of the metacarpophalangeal joint of the little finger. *J Hand Surg [Br]* 12:264–266, 1987.

110. Araki, S., Ohtani, T., and Tanaka, T. Open dorsal metacarpophalangeal dislocations of the index, long, and ring fingers. *J Hand Surg [Am]* 12:458–460, 1987.

111. Barnard, H.L. Dorsal dislocation of the first phalanx of the little finger: Reduction by Farabeuf's dorsal incision. *Lancet* 1:88–90, 1901.

112. Barry, K., McGee, H., and Curtin J. Complex dislocation of the metacarpo-phalangeal joint of the index finger: A comparison of the surgical approaches. *J Hand Surg [Br]* 13:466–468, 1988.

113. Battle, W.H. Backward dislocation of the fingers upon the metacarpus. *Lancet* 1:1223–1224, 1888.

114. Becton, J.L., and Carswell, A.S. The natural history of an unreduced dislocated index finger metacarpophalangeal joint in a child. *J Med Assoc Ga* 64:413–415, 1975.

115. Becton, J.L., Christian J.D. Jr., Goodwin, H.N., and Jackson, J.G. III. A simplified technique for treating the complex dislocation of the index metacarpophalangeal joint. *J Bone Joint Surg [Am]* 57:698–700, 1975.

116. Betz, R.R., Browne, E.Z., Perry, G.B., and Resnick, E.J. The complex volar metacarpophalangeal-joint dislocation: A case report and review of the literature. *J Bone Joint Surg [Am]* 64:1374–1375, 1982.

117. Bezes, P.M.H. Severe metacarpophalangeal sprain of the thumb in ski accidents. *Ann Chir Main* 3:101–112, 1984.

118. Boland, D. Volar dislocation of the ring finger metacarpophalangeal joint. *Orthop Rev* 13:69–72, 1984.

119. Bowers, W.H. Management of small joint injuries in the hand. *Orthop Clin North Am* 14:793–810, 1983.

120. Bowers, W.H., and Fajgenbaum, D.M. Closed rupture of the volar plate of the distal interphalangeal joint. *J Bone Joint Surg [Am]* 61:146, 1979.

121. Bowers, W.H., and Hurst, L.C. Gamekeeper's thumb. Evaluation by arthrography and stress roentgenography. *J Bone Joint Surg [Am]* 59:519–524, 1977.

122. Burman, M. Irreducible hyperextension dislocation of the metacarpophalangeal joint of a finger. *Bull Hosp Jt Dis* 14:290–291, 1953.

123. Camp, R.A., Weatherwax, R.J., and Miller, E.B. Chronic posttraumatic radial instability of the thumb metacarpophalangeal joint. *J Hand Surg* 5:221–225, 1980.

124. Campbell, C.S. Gamekeeper's thumb. *J Bone Joint Surg [Br]* 37:148–149, 1955.

125. Carr, D., Johnson, R.J., and Pope, M.H. Upper extremity injuries in skiing. *Am J Sports Med* 9:378–383, 1981.

126. Coonrad, R.W., and Goldner, J.L. A study of the pathological findings and treatment in soft-tissue injury of the thumb metacarpophalangeal joint. *J Bone Joint Surg [Am]* 50:439–451, 1968.

127. Derkash, R.S., Matyas, J.R., Weaver, J.K., Oden, R.R., Kirk, R.E., Freeman, J.R., and Cipriano, F.J. Acute surgical repair of the skier's thumb. *Clin Orthop* 216:29–33, 1987.

128. Dray, G.J., and Eaton, R.G. Dislocations and ligament injuries in the digits. *In* Green, D.P. (Ed.). *Operative Hand Surgery* (2nd ed.). New York, Churchill, 1988, pp. 795–796.

129. Dutton, R.O., and Meals, R.A. Complex dorsal dislocation of the thumb metacarpophalangeal joint. *Clin Orthop* 164:160–164, 1982.

130. Eaton, R.G. *Joint Injuries of the Hand.* Springfield, Charles C Thomas, 1971.

131. Eaton, R.G., and Littler, J.W. Joint injuries and their sequelae. *Clin Plast Surg* 3:85–98, 1976.

132. Eaton, R.G., and Malerich, M.M. Volar plate arthroplasty of the proximal interphalangeal joint: A review of ten years' experience. *J Hand Surg* 5:260–268, 1980.

133. Eiken, O. Palmaris longus-tenodesis for hyperextension of the thumb metacarpophalangeal joint. *Scand J Plast Reconstr Surg* 15:149–152, 1981.

134. Engel, J., Ganel, A., Ditzian, R., and Militeanu, J. Arthrography as a method of diagnosing tear of the ulnar collateral ligament of the metacarpophalangeal joint of the thumb ("Gamekeeper's Thumb"). *J Trauma* 19:106–109, 1979.

135. Engkvist, O., Balkfors, B., and Lindsjo, U. Thumb injuries in downhill skiing. *Int J Sports Med* 3:50–55, 1982.

136. Fairclough, J.A., and Mintowt-Czyz, W.J. Skier's thumb—A method of prevention. *Injury* 17:203–204, 1986.

137. Farabeuf, L.H.F. De la luxation du ponce en arriere. *Bull Soc Chir* 11:21–62, 1876.

138. Flatt, A.E. *The Care of Minor Hand Injuries.* St. Louis, C.V. Mosby, 1959, pp. 188–189.

139. Frank, W.E., and Dobyns, J. Surgical pathology of collateral ligamentous injuries of the thumb. *Clin Orthop* 83:102–114, 1972.

140. Frykman, G., and Johansson, O. Surgical repair of rupture of the ulnar collateral ligament of the metacarpophalangeal joint of the thumb. *Acta Chir Scand* 112:58–64, 1956.

141. Fultz, C.W., and Buchanan, J.R. Complex fracture-dislocation of the metacarpophalangeal joint: Case report. *Clin Orthop* 227:255–260, 1988.

142. Garroway, R.Y., Hurst, L.C., Leppard, J. III, and Dick, H.M. Complex dislocations of the proximal interphalangeal joint: A pathoanatomic classification of the injury. *Orthop Rev* 13:21–28, 1984.

143. Gee, T.C., and Pho, R.W.H. Avulsion-fracture at the proximal attachment of the radial collateral ligament of the fifth metacarpophalangeal joint—A case report. *J Hand Surg* 7:526–527, 1982.

144. Gerber, C., Senn, E., and Matter, P. Skier's thumb: Surgical treatment of recent injuries to the ulnar collateral ligament of the thumb's metacarpophalangeal joint. *Am J Sports Med* 9:171–177, 1981.

145. Goss, C.M. *Gray's Anatomy of the Human Body* (26th ed.). Philadelphia, Lee & Febiger, 1954, pp. 324, 372–373.

146. Green, D.P., and Rowland, S.A. Fractures and dislocations in the hand. *In* Rockwood, C.A., and Green, D.P. (Eds.). *Fractures in Adults* (3rd ed.). Philadelphia, J.B. Lippincott, 1991.

147. Green, D.P., and Terry, G.C. Complex dislocation of the

148. Green, S.M., and Posner, M.A. Irreducible dorsal dislocations of the proximal interphalangeal joint. *J Hand Surg [Am]* 10:85–87, 1985.

149. Gunther, S.E., and Zielinski, C.J. Irreducible palmar dislocation of the proximal phalanx of the thumb—Case report. *J Hand Surg* 7:515–517, 1982.

150. Gustilo, R.B. Dislocation of the metacarpophalangeal joint of the index finger. *Minn Med* 1119–1121, 1966.

151. Gutman, J., Weisbuch, J., and Wolf, M. Ski injuries in 1972–1973: A repeat analysis of a major health problem. *JAMA* 230:1423–1425, 1974.

152. Hagan, H.J., and Hastings, H. Fusion of the thumb metacarpophalangeal joint to treat posttraumatic arthritis. *J Hand Surg [Am]* 13:750–753, 1988.

153. Hall, R.F. Jr., Gleason, T.F., and Kasa, R.F. Simultaneous closed dislocations of the metacarpophalangeal joints of the index, long, and ring fingers: A case report. *J Hand Surg [Am]* 10:81–85, 1985.

154. Harris, H., and Joseph, J. Variation in extension of the metacarpo-phalangeal and interphalangeal joints of the thumb. *J Bone Joint Surg [Br]* 31:547–559, 1949.

155. Hastings, H., and Carroll, C. Treatment of closed articular fractures of the metacarpophalangeal and proximal interphalangeal joints. *Hand Clin* 4:503–527, 1988.

156. Hubbard, L.F. Metacarpophalangeal dislocations. *Hand Clin* 4:39–44, 1988.

157. Hunt, J.C., Watts, H.B., and Glasgow, J.D. Dorsal dislocation of the metacarpophalangeal joint of the index finger with particular reference to open dislocation. *J Bone Joint Surg [Am]* 49:1572–1578, 1967.

158. Iftikhar, T.B., and Kaminski, R.S. Simultaneous dorsal dislocation of MP joints of long and ring fingers: A case report. *Orthop Rev* 10:71–72, 1981.

159. Imbriglia, J.E., and Sciulli, R. Open complex metacarpophalangeal joint dislocation. Two cases: Index finger and long finger. *J Hand Surg* 4:72–75, 1979.

160. Isani, A., and Melone, C.P. Jr. Ligamentous injuries of the hand in athletes. *Clin Sports Med* 5:757–772, 1986.

161. Ishizuki, M. Injury to collateral ligament of metacarpophalangeal joint of a finger. *J Hand Surg [Am]* 13:456–460, 1988.

162. Johnson, F.G., and Greene, M.H. Another cause of irreducible dislocation of the proximal interphalangeal joint of a finger: A case report. *J Bone Joint Surg [Am]* 48:542–544, 1966.

163. Joseph, J. Further studies of the metacarp-phalangeal and interphalangeal joints of the thumb. *J Anat* 85:221–229, 1951.

164. Kaplan, E.B. Dorsal dislocation of the metacarpophalangeal joint of the index finger. *J Bone Joint Surg [Am]* 39:1081–1086, 1957.

165. Kaplan, E.B. The pathology and treatment of radial subluxation of the thumb with ulnar displacement of the head of the first metacarpal. *J Bone Joint Surg [Am]* 43:541–546, 1961.

166. Kessler, I. Complex avulsion of the ulnar collateral ligament of the metacarpophalangeal joint of the thumb. *Clin Orthop* 29:196–200, 1961.

167. Kessler, I. A simplified technique to correct hyperextension deformity of the metacarpophalangeal joint of the thumb. *J Bone Joint Surg [Am]* 61:903–905, 1979.

168. Kessler, I., and Heller, J. Complete avulsion of the ligamentous apparatus of the metacarpophalangeal joint of the thumb. *Surg Gynecol Obstet* 116:95–98, 1963.

169. Kiefhaber, T.R., Stern, P.J., and Grood, E.S. Lateral stability of the proximal interphalangeal joint. *J Hand Surg [Am]* 11:661–669, 1986.

170. Kilgore, E.S., Newmeyer, W.L., and Brown, L.G. Posttraumatic trapped dislocations of the proximal interphalangeal joint. *J Trauma* 16:481–487, 1976.

171. Kjeldal, I. Irreducible compound dorsal dislocations of the

proximal interphalangeal joint of the finger. *J Hand Surg [Br]* 11:49–50, 1986.

172. Kuczynski, K. The proximal interphalangeal joint: Anatomy and causes of stiffness in the fingers. *J Bone Joint Surg [Br]* 50:656–663, 1968.

173. Lamb, D.W., Abernethy, P.J., and Fragiadakis, E. Injuries of the metacarpophalangeal joint of the thumb. *Hand* 3:164–168, 1971.

174. Lamb, D.W., and Angarita, G. Ulnar instability of the metacarpophalangeal joint of thumb. *J Hand Surg [Br]* 10:113–114, 1985.

175. Light, T.R. Buttress pinning techniques. *Orthop Rev* 10:49–55, 1981.

176. Linscheid, R.L. Arthrography of the metacarpophalangeal joint. *Clin Orthop* 103:91, 1974.

177. Linscheid, R.L., Grainger, R.W., and Johnson, E.W.: The thumb metacarpophalangeal joint: Injuries. *Minn Med* 55:1037–1040, 1972.

178. Louis, D.S., Huebner, J.J. Jr., and Hankin, F.M. Rupture and displacement of the ulnar collateral ligament of the metacarpophalangeal joint of the thumb: Preoperative diagnosis. *J Bone Joint Surg [Am]* 68:1320–1326, 1986.

179. Malerich, M.M., Eaton, R.G., and Upton, J. Complete dislocation of a little finger metacarpal phalangeal joint treated by closed technique. *J Trauma* 20:424–425, 1980.

180. McCue, F.C., Hakala, M.W., Andrews, J.R., and Gieck, J.H. Ulnar collateral ligament injuries of the thumb in athletes. *J Sports Med* 2:70–80, 1974.

181. McCue, F.C., Honner, R., Johnson, M.C., and Gieck, J.H. Athletic injuries of the proximal interphalangeal joint requiring surgical treatment. *J Bone Joint Surg [Am]* 52:937–956, 1970.

182. McElfresh, E.C., and Dobyns, J.H. Intra-articular metacarpal head fractures. *J Hand Surg* 8:383–393, 1983.

183. McElfresh, E.C., Dobyns, J.H., and O'Brien, E.T. Management of fracture-dislocation of the proximal interphalangeal joints by extension-block splinting. *J Bone Joint Surg [Am]* 54:1705–1711, 1972.

184. McLaughlin, H.L. Complex "locked" dislocation of the metacarpophalangeal joints. *J Trauma* 5:683–688, 1965.

185. Milch, H. Recurrent dislocation of thumb: Capsulorrhaphy. *Am J Surg* 6:237–239, 1929.

186. Milford, L. The hand. *In* Crenshaw, A.H. (Ed.), *Campbell's Operative Orthopaedics* (4th ed.). St. Louis, C.V. Mosby, 1963, p. 166.

187. Milford, L. The hand. *In* Crenshaw, A.H. (Ed.), *Campbell's Operative Orthopaedics* (5th ed.). St. Louis, C.V. Mosby, 1971, pp. 188–189.

188. Milford, L. The hand. *In* Edmonson, A.S., and Crenshaw, A.H. (Eds.), *Campbell's Operative Orthopaedics* (6th ed.). St. Louis, C.V. Mosby, 1980, p. 160.

189. Milford, L. Interphalangeal dislocations. *In* Crenshaw, A.H. (Ed.), *Campbell's Operative Orthopaedics* (7th ed.). St. Louis, C.V. Mosby, 1987, pp. 249–253.

190. Miller, P.R., Evans, B.W., and Glazer, D.A. Locked dislocation of the metacarpophalangeal joint of the index finger. *JAMA* 203:138–139, 1968.

191. Miller, R.J. Dislocations and fracture dislocations of the metacarpophalangeal joint of the thumb. *Hand Clin* 4:45–65, 1988.

192. Miyamoto, M., Hirayama, T., and Uchida, M. Volar dislocation of the metacarpophalangeal joint of the thumb—A case report. *J Hand Surg [Br]* 11:51–54, 1986.

193. Moberg, E. Fractures and ligamentous injuries of the thumb and fingers. *Surg Clin North Am* 40:297–309, 1960.

194. Moberg, E., and Stener, B. Injuries to the ligaments of the thumb and fingers: Diagnosis, treatment and prognosis. *Acta Chir Scand* 106:166–186, 1953.

195. Mogan, J.V., and Davis, P.H. Upper extremity injuries in skiing. *Clin Sports Med* 1:295–308, 1982.

196. Mogensen, B.A., and Mattsson, H.S. Post-traumatic instability of the metacarpophalangeal joint of the thumb. *Hand* 12:85–90, 1980.

197. Moller, J.T. Lesions of the volar fibrocartilago in finger

198. Moneim, M.S. Volar dislocation of the metacarpophalangeal joint: Pathological anatomy and report of two cases. *Clin Orthop* 176:186–189, 1983.

199. Murakami, Y. Irreducible volar dislocation of the proximal interphalangeal joint of the finger. *Hand* 6:87–90, 1974.

200. Murphy, A.F., and Stark, H.H. Closed dislocation of the metacarpophalangeal joint of the index finger. *J Bone Joint Surg [Am]* 49:1579–1586, 1967.

201. Neviaser, R.J., and Wilson, J.N. Interposition of the extensor tendon resulting in persistent subluxation of the proximal interphalangeal joint of the finger. *Clin Orthop* 83:118–120, 1972.

202. Neviaser, R.J., Wilson, J.N., and Lievano, A. Rupture of the ulnar collateral ligament of the thumb (gamekeeper's thumb). *J Bone Joint Surg [Am]* 53:1357–1364, 1971.

203. Nutter, P.D. Interposition of sesamoids in metacarpophalangeal dislocations. *J Bone Joint Surg* 22:730–734, 1940.

204. Oni, O.O.A. Irreducible buttonhole dislocation of the proximal interphalangeal joint of the finger (a case report). *J Hand Surg [Br]* 10:100, 1985.

205. Onuba, O., and Essiet, A. Irreducible dislocation of the metacarpophalangeal joint of the thumb due to tendon interposition. *J Hand Surg [Br]* 12:60–61, 1987.

206. Osterman, A.L., Hayken, G.D., and Bora, F.W. A quantitative evaluation of thumb function after ulnar collateral repair and reconstruction. *J Trauma* 21:854–861, 1981.

207. Ostrowski, D.M., and Neimkin, R.J. Irreducible palmar dislocation of the proximal interphalangeal joint: A case report. *Orthopedics* 8:84–86, 1985.

208. Palmer, A.K., and Louis, D.S. Assessing ulnar instability of the metacarpophalangeal joint of the thumb. *J Hand Surg* 3:542–546, 1978.

209. Palmer, R.E. Injury to dorsal MCP joint of the thumb. *Orthop Rev* 11:127–129, 1982.

210. Parikh, M., Nahigian, S., and Froimson, A. Gamekeeper's thumb. *Plast Reconstr Surg* 58:24–31, 1976.

211. Posner, M.A., Langa, V., and Ambrose, L. Intrinsic muscle advancement to treat chronic palmar instability of the metacarpophalangeal joint of the thumb. *J Hand Surg [Am]* 13:110–115, 1988.

212. Posner, M.A., and Wilenski, M. Irreducible volar dislocation of the proximal interphalangeal joint of a finger caused by interposition of an intact central slip: A case report. *J Bone Joint Surg [Am]* 60:133–134, 1978.

213. Primiano, G.A. Skiers' thumb injuries associated with flared ski pole handles. *Am J Sports Med* 13:425–427, 1985.

214. Primiano, G.A. Functional cast immobilization of thumb metacarpophalangeal joint injuries. *Am J Sports Med* 14:335–339, 1986.

215. Quinton, D.N. Dorsal locking of the metacarpophalangeal joint. *J Hand Surg [Br]* 12:62–63, 1987.

216. Redler, I., and Williams, J.T. Rupture of a collateral ligament of the proximal interphalangeal joint of the fingers: Analysis of eighteen cases. *J Bone Joint Surg [Am]* 49:322–326, 1967.

217. Reikeras, O., and Kvarnes, L. Rupture of the ulnar ligament of the metacarpophalangeal joint of the thumb. *Arch Orthop Trauma Surg* 100:175–177, 1982.

218. Renshaw, T.S., and Louis, D.S. Complex volar dislocation of the metacarpophalangeal joint: A case report. *J Trauma* 13:1086–1088, 1971.

219. Resnick, D., and Danzig, L.A. Arthrographic evaluation of injuries of the first metacarpophalangeal joint: Gamekeeper's thumb. *AJR* 126:1046–1052, 1976.

220. Ridge, E.M. Dorsal dislocation of the first phalanx of the little finger. *Lancet* 1:781, 1901.

221. Robertson, R.C., Cawley, J.J. Jr., and Faris, A.M. Treatment of fracture-dislocation of the interphalangeal joints of the hand. *J Bone Joint Surg* 28:68–70, 1946.

222. Rosenthal, D.I., Murray, W.T., and Smith, R.J. Finger arthrography. *Radiology* 137:647–651, 1980.

223. Rovere, G.D., Gristina, A.G., Stolzer, W.A., and Garver,

joints: A 2-year material. *Acta Orthop Scand* 45:673–682, 1974.

E.M. Treatment of "gamekeeper's thumb" in hockey players. *J Sports Med* 3:147–151, 1975.

224. Sakellarides, H.T. Treatment of recent and old injuries of the ulnar collateral ligament of the MP joint of the thumb. *Am J Sports Med* 6:255–262, 1978.

225. Sakellarides, H.T., and DeWeese, J.W. Instability of the metacarpophalangeal joint of the thumb. Reconstruction of the collateral ligaments using the extensor pollicis brevis tendon. *J Bone Joint Surg [Am]* 58:106–112, 1976.

226. Schenck, R.R. Dynamic traction and early passive movement for fractures of the proximal interphalangeal joint. *J Hand Surg [Am]* 11:850–858, 1986.

227. Schultz, R.J., and Fox, J.M. Gamekeeper's thumb. *NY State J Med* 73:2329–2331, 1973.

228. Schulze, H.A. Treatment of fracture-dislocations of the proximal interphalangeal joints of the fingers. *Milit Surg* 99:190–191, 1946.

229. Selig, S., and Schein, A. Irreducible buttonhole dislocation of the fingers. *J Bone Joint Surg* 22:436–441, 1940.

230. Silberman, W.W. Clear view of the index sesamoid: A sign of irreducible metacarpophalangeal joint dislocation. *J Am Coll Emerg Phys* 8:371–373, 1979.

231. Smith, M.A. The mechanism of acute ulnar instability of the metacarpophalangeal joint of the thumb. *Hand* 12:225–230, 1980.

232. Smith, R.J. Post-traumatic instability of the metacarpophalangeal joint of the thumb. *J Bone Joint Surg [Am]* 59:14–21, 1977.

233. Spinner, M., and Choi, B.Y. Anterior dislocation of the proximal interphalangeal joint: A cause of rupture of the central slip of the extensor mechanism. *J Bone Joint Surg [Am]* 52:1329–1336, 1970.

234. Sprague, B.L. Proximal interphalangeal joint injuries and their initial treatment. *J Trauma* 15:380–385, 1975.

235. Spray, P. Finger fracture-dislocation proximal at the interphalangeal joint. *J Tenn Med Assoc* 59:765–766, 1966.

236. Stener, B. Displacement of the ruptured ulnar collateral ligament of the metacarpophalangeal joint of the thumb: A clinical and anatomical study. *J Bone Joint Surg [Br]* 44:869–879, 1962.

237. Stener, B. Hyperextension injuries to the metacarpophalangeal joint of the thumb—Rupture of ligaments, fracture of sesamoid bones, rupture of flexor pollicis brevis: An anatomical and clinical study. *Acta Chir Scand* 125:275–293, 1963.

238. Stener, B., and Stener, I. Shearing fractures associated with rupture of ulnar collateral ligament of metacarpophalangeal joint of thumb. *Injury* 1:12–16, 1969.

239. Stern, P.J., and Lee, A.F. Open dorsal dislocations of the proximal interphalangeal joint. *J Hand Surg [Am]* 10:364–370, 1985.

240. Sternbach, G. C.S. Campbell: Gamekeeper's thumb. *J Emerg Med* 1:345–347, 1984.

241. Stothard, J., and Caird, D.M. Experience with arthrography of the first metacarpophalangeal joint. *Hand* 13:257–266, 1981.

242. Strandell, G. Total rupture of the ulnar collateral ligament of the metacarpophalangeal joint of the thumb: Results of surgery in 35 cases. *Acta Chir Scand* 118:72–80, 1959.

243. Stripling, W.D. Displaced intra-articular osteochondral fracture—Cause for irreducible dislocation of the distal interphalangeal joint. *J Hand Surg* 7:77–78, 1982.

244. Strong, M.L. A new method of extension-block splinting for the proximal interphalangeal joint—Preliminary report. *J Hand Surg* 5:606–607, 1980.

245. Sweterlitsch, P.R., Torg, J.S., and Pollack, H. Entrapment of a sesamoid in the index metacarpophalangeal joint. *J Bone Joint Surg [Am]* 51:995–998, 1969.

246. Thompson, J.S., and Eaton, R.G. Volar dislocation of the proximal interphalangeal joint (abstr). *J Hand Surg* 2:232, 1977.

247. Trojan, E. Fracture dislocation of the bases of the proximal and middle phalanges of the fingers. *Hand* 4:60–61, 1972.

248. Tsuge, K., and Watari, S. Dorsal dislocation of the metacarpophalangeal joint of the index finger. *Hiroshima J Med Sci* 22:65–81, 1973.

249. Van Der Kloot, J.F.V. Injury to the ulnar ligament of the thumb. *Arch Chir Neerl* 17:179–185, 1965.

250. Vaupel, G.L., and Andrews, J.R. Diagnostic and operative arthroscopy of the thumb metacarpophalangeal joint: A case report. *Am J Sports Med* 13:139–141, 1985.

251. Vicar, A.J. Proximal interphalangeal joint dislocations without fractures. *Hand Clin* 4:5–13, 1988.

252. Weston, W.J. The normal anthrograms of the metacarpophalangeal, metatarso-phalangeal and inter-phalangeal joints. *Aust Radiol* 13:211–218, 1969.

253. White, G.M. Ligamentous avulsion of the ulnar collateral ligament of the thumb of a child. *J Hand Surg [Am]* 11:669–672, 1986.

254. Wilhelmy, J., and Hay, R.L. Dual dislocation of metacarpophalangeal joints. *Hand* 4:168–170, 1972.

255. Wilson, J.N., and Rowland, S.A. Fracture-dislocation of the proximal interphalangeal joint of the finger: Treatment by open reduction and internal fixation. *J Bone Joint Surg [Am]* 48:493–502, 1966.

256. Wilson, R.L., and Liechty, B.W. Complications following small joint injuries. *Hand Clin* 2:329–345, 1986.

257. Wolov, R.B. Complex dislocations of the metacarpophalangeal joints. *Orthop Rev* 17:770–775, 1988.

258. Wood, M.B., and Dobyns, J.H. Chronic, complex volar dislocation of the metacarpophalangeal joint: Report of three cases. *J Hand Surg* 6:73–76, 1981.

259. Woods, D.W., Mudge, M.K., and Wood, V.E. Radial instability of the thumb metacarpophalangeal joint: A clinical and cadaveric study. Scientific Exhibit, AAOS Annual Meeting, San Francisco, 1987.

260. Wright, C.S. Compound dislocations of four metacarpophalangeal joints. *J Hand Surg [Br]* 10:233–235, 1985.

261. Zancolli, E. *Structural and Dynamic Bases of Hand Surgery* (2nd ed.). Philadelphia, J.B. Lippincott, 1979, pp. 212–213.

262. Zilberman, Z., Rotschild, E., and Krauss, L. Rupture of the ulnar collateral ligament of the thumb. *J Trauma* 5:447–481, 1965.

263. Zook, E.G., Van Beek, A.L., and Wavak, P. Transverse volar skin laceration of the finger: A sign of volar plate injury. *Hand* 11:213–216, 1979.

Fractures of the Phalanges and Metacarpals

264. Badger, F.C. Internal fixation in the treatment of Bennett's fractures. *J Bone Joint Surg [Br]* 38:771, 1956.

265. Barton, N.J. Fractures of the hand (review article). *J Bone Joint Surg [Br]* 66:159–167, 1984.

266. Belsky, M.R., Eaton, R.G., and Lane, L.B. Closed reduction and internal fixation of proximal phalangeal fractures. *J Hand Surg [Am]* 9:725–729, 1984.

267. Belsole, R. Physiological fixation of displaced and unstable fractures of the hand. *Orthop Clin North Am* 11:393–404, 1980.

268. Bennett, E.H. Fractures of the metacarpal bones. *Dublin J Med Sci* 73:72–75, 1882.

269. Berkman, E.F., and Miles G.H. Internal fixation of metacarpal fractures exclusive of the thumb. *J Bone Joint Surg* 25:816–821, 1943.

270. Black, D.M., Mann, R.J., Constine, R., and Daniels, A.U. Comparison of internal fixation techniques in metacarpal fractures. *J Hand Surg [Am]* 10:466–472, 1985.

271. Black, D.M., Mann, R.J., Constine, R.M., and Daniels, A.U. The stability of internal fixation in the proximal phalanx. *J Hand Surg [Am]* 11:672–677, 1986.

272. Blalock, H.S., Pearce, H.L., Kleinert, H., and Kutz, J. An instrument designed to help reduce and percutaneously pin fractured phalanges. *J Bone Joint Surg [Am]* 57:792–794, 1975.

273. Blum, L. The treatment of Bennett's fracture-dislocation of the first metacarpal bone. *J Bone Joint Surg* 23:578–580, 1941.

274. Borden, J. Complications of fractures and ligamentous injuries of the hand. *Orthop Rev* 1:29–38, 1972.

275. Bosworth, D.M. Internal splinting of fractures of the fifth metacarpal. *J Bone Joint Surg* 19:826–827, 1937.

276. Breen, T.F., Gelberman, R.H., and Jupiter, J.B. Intra-articular fractures of the basilar joint of the thumb. *Hand Clin* 4:491–501, 1988.

277. Brunet, M.E., and Haddad, R.J. Fractures and dislocations of the metacarpals and phalanges. *Clin Sports Med* 5:773–781, 1986.

278. Buchler, U., and Fischer, T. Use of a minicondylar plate for metacarpal and phalangeal periarticular injuries. *Clin Orthop* 214:53–58, 1987.

279. Bunnell, S. *Surgery of the Hand.* Philadelphia, J.B. Lippincott, 1944.

280. Burkhalter, W.E., and Reyes, F.A. Closed treatment of fractures of the hand. *Bull Hosp Jt Dis Orthop Inst* 44:145–162, 1984.

281. Burton, R.I., and Eaton, R.G. Common hand injuries in the athlete. *Orthop Clin North Am* 4:809–838, 1973.

282. Caffee, H.H. Atraumatic placement of Kirschner wires. *Plast Reconstr Surg* 63:433, 1979.

283. Cannon, S.R., Dowd, G.S.E., Williams, D.H., and Scott, J.M. A long-term study following Bennett's fracture. *J Hand Surg [Br]* 11:426–431, 1986.

284. Charnley, J. *The Closed Treatment of Common Fractures.* Edinburgh, E & S Livingstone, 1961.

285. Cotton, F.J. *Dislocations and Joint Fractures.* Philadelphia, W.B. Saunders, 1910.

286. Crawford, G.P. Screw fixation for certain fractures of the phalanges and metacarpals. *J Bone Joint Surg [Am]* 58:487–492, 1976.

287. Culver, J.E. *Am Soc Surg Hand Correspondence Newsletter.* 1981, p. 69.

288. Dabezies, E.J., and Schutte, J.P. Fixation of metacarpal and phalangeal fractures with miniature plates and screws. *J Hand Surg [Am]* 11:283–288, 1986.

289. Edwards, G.S., O'Brien, E.T., and Heckman, M.M. Retrograde cross-pinning of transverse metacarpal and phalangeal fractures. *Hand* 14:141–148, 1982.

290. Ellis, V.H. A method of treating Bennett's fracture. *Proc R Soc Med* 39:21, 1946.

291. Fisher, E. Bennett's fracture in general practice. *Med J Aust* 1:434–438, 1976.

292. Ford, D.J., El-Hadidi, S., Lunn, P.G., and Burke, F.D. Fractures of the phalanges: Results of internal fixation using 1.5 mm and 2 mm A.O. screws. *J Hand Surg [Br]* 12:28–33, 1987.

293. Foster, R.J., and Hastings, H. Treatment of Bennett, Rolando, and vertical intraarticular trapezial fractures. *Clin Orthop* 214:121–129, 1987.

294. Gedda, K.O. Studies on Bennett's fracture: Anatomy, roentgenology, and therapy. *Acta Chir Scand (Suppl)* 193, 1954.

295. Gedda, K.O., and Moberg, E. Open reduction and osteosynthesis of the so-called Bennett's fracture in the carpometacarpal joint of the thumb. *Acta Orthop Scand* 22:249–256, 1953.

296. Gingrass, R.P., Fehring, B., and Matloub, H. Intraosseous wiring of complex hand fractures. *Plast Reconstr Surg* 66:383–394, 1980.

297. Glasgow, M., and Lloyd, G.J. The use of modified A.O. reduction forceps in percutaneous fracture fixation. *Hand* 13:214–216, 1981.

298. Gould, W.L., Belsole, R.J., and Skelton, W.H. Tension-band stabilization of transverse fractures: An experimental analysis. *Plast Reconstr Surg* 73:111–115, 1984.

299. Green, D.P., and Anderson, J.R. Closed reduction and percutaneous pin fixation of fracture phalanges. *J Bone Joint Surg [Am]* 55:1651–1654, 1973.

300. Green, D.P., and O'Brien, E.T. Fractures of the thumb metacarpal. *South Med J* 65:807–814, 1972.

301. Green, D.P., and Rowland, S.A. Fractures in and dislocations in the hand. *In* Rockwood, C.A. Jr., and Green, D.P. (Eds.), *Fractures in Adults* (3rd ed.). Philadelphia, J.B. Lippincott, 1991.

302. Greene, T.L., Noellert, R.C., and Belsole, R.J. Treatment of unstable metacarpal and phalangeal fractures with tension band wiring techniques. *Clin Orthop* 214:78–84, 1987.

303. Griffiths, J.C. Fractures at the base of the first metacarpal bone. *J Bone Joint Surg [Br]* 46:712–719, 1964.

304. Gropper, P.T., and Bowen, V. Cerclage wiring of metacarpal fractures. *Clin Orthop* 188:203–207, 1984.

305. Grundberg, A.B. Intramedullary fixation for fractures of the hand. *J Hand Surg* 6:568–573, 1981.

306. Gunther, S.F. The carpometacarpal joints. *Orthop Clin North Am* 15:259–277, 1984.

307. Haggart, G.E. Fractures of the metacarpal, metatarsal bones, and phalanges treated by skeletal traction. *Surg Clin North Am* 14:1203–1210, 1934.

308. Hall, R.F. Closed flexible intramedullary rodding of metacarpal and phalangeal fractures. *Orthop Trans* 8:187–188, 1984.

309. Hall, R.F. Treatment of metacarpal and phalangeal fractures in noncompliant patients. *Clin Orthop* 214:31–36, 1987.

310. Harvey, F.J., and Bye, W.D. Bennett's fracture. *Hand* 8:48–53, 1976.

311. Hastings, H. Unstable metacarpal and phalangeal fracture treatment with screws and plates. *Clin Orthop* 214:37–52, 1987.

312. Heim, U., and Pfeiffer, K.M. *Small Fragment Set Manual: Technique Recommended by the ASIF Group.* New York, Springer-Verlag, 1982.

313. Howard, F.M. Fractures of the basal joint of the thumb. *Clin Orthop* 220:46–51, 1987.

314. Iselin, F., and Thevenin, R. Fixation of fractures of the digits with intramedullary flexible screws. *J Bone Joint Surg [Am]* 56:1096, 1974.

315. James, E.S., and Gibson, A. Fractures of the first metacarpal bone. *Can Med Assoc J* 43:153–155, 1940.

316. James, J.I.P. Fractures of the proximal and middle phalanges of the fingers. *Acta Orthop Scand* 32:401–412, 1962.

317. James, J.I.P. Common, single errors in the management of hand injuries. *Proc R Soc Med* 63:69–71, 1970.

318. Johnson, E.C. Fracture of the base of the thumb: A new method of fixation. *JAMA* 126:27–28, 1944.

319. Jones, W.W. Biomechanics of small bone fixation. *Clin Orthop* 214:11–18, 1987.

320. Jordan, S.E., and Greider, J.L. The angiocatheter in the management of hand injuries. *J Hand Surg [Am]* 11:446, 1986.

321. Jupiter, J.B., and Sheppard, J.E. Tension wire fixation of avulsion fractures in the hand. *Clin Orthop* 214:113–120, 1987.

322. Kaplan, E.B. *Functional and Surgical Anatomy of the Hand* (2nd ed.). Philadelphia, J.B. Lippincott, 1965.

323. Lamb, D.W., Abernethy, P.A., and Raine, P.A.M. Unstable fractures of the metacarpals: A method of treatment by transverse wire fixation to intact metacarpals. *Hand* 5:43–48, 1973.

324. Lewis, R.C., Nordyke, M., and Duncan, K. Expandable intramedullary device for treatment of fractures in the hand. *Clin Orthop* 214:85–92, 1987.

325. Lister, G. Intraosseous wiring of the digital skeleton. *J Hand Surg* 3:427–435, 1978.

326. Lucas, G.L. Internal fixation in the hand: A review of indications and methods. *Orthopedics* 3:1083–1089, 1980.

327. Manske, P.R., and Lesker, P.A. Avulsion of the ring finger flexor digitorum profundus tendon: An experimental study. *Hand* 10:52–55, 1978.

328. Mason, S.M., and Fyfe, I.S. Comparison of rigidity of whole tubular bones. *J Biomech* 12:367–372, 1979.

329. Massengill, J.B., Alexander, H., Langrana, N., and Mylod, A. A phalangeal fracture model—quantitative analysis of rigidity and failure. *J Hand Surg* 7:264–270, 1982.
330. Massengill, J.B., Alexander, H., Parson, J.R., and Schecter, M.J. Mechanical analysis of Kirschner wire fixation in a phalangeal model. *J Hand Surg* 4:351–356, 1979.
331. Meals, R.A. *Amer Soc Surg Hand Correspondence Newsletter* 1986, p. 30.
332. Melone, C.P. Rigid fixation of phalangeal and metacarpal fractures. *Orthop Clin North Am* 17:421–435, 1986.
333. Meyer, V.E., Chiu, D.T., and Beasley, R.W. The place of internal skeletal fixation in surgery of the hand. *Clin Plast Surg* 8:51–64, 1981.
334. Michon, J., and Delagoutte, J.P. Crush injuries of the digital extremities. *In* Pierre M. (Ed.), The Nail (G.E.M. Monograph). Edinburgh, Churchill Livingstone, 1981.
335. Muller, M.E., Allgower, M., Schneider, R., and Willenegger, H. *Manual of Internal Fixation* (2nd ed.). New York, Springer-Verlag, 1979, pp. 42–43.
336. Namba, R.S., Kabo, J.M., and Meals, R.A. Biomechanical effects of point configuration in Kirschner-wire fixation. *Clin Orthop* 214:19–22, 1987.
337. Nordyke, M.D., Lewis, R.C., Janssen, H.F., and Duncan, K.H. Biomechanical and clinical evaluation of expandable intramedullary fixation device. *J Hand Surg [Am]* 13:129–134, 1988.
338. O'Brien, E.T. Fractures of the metacarpals and phalanges. *In* Green, D.P. (Ed.), *Operative Hand Surgery.* New York, Churchill Livingstone, 1982.
339. Pellegrini, V.D. Fractures at the base of the thumb. *Hand Clin* 4:87–101, 1988.
340. Pollen, A.G. The conservative treatment of Bennett's fracture-subluxation of the thumb metacarpal. *J Bone Joint Surg [Br]* 50:91–101, 1968.
341. Posch, J.L., Walker, P.J., and Miller, H. Treatment of ruptured tendons of the hand and wrist. *Am J Surg* 91:669–681, 1956.
342. Rayhack, J.M., Belsole, R.J., and Skelton, W.H. A strain recording model: Analysis of transverse osteotomy fixation in small bones. *J Hand Surg [Am]* 9:383–387, 1984.
343. Reyes, F.A., and Latta, L.L. Conservative management of difficult phalangeal fractures. *Clin Orthop* 214:23–30, 1987.
344. Riordan, D.C. Fractures about the hand. *South Med J* 50:637–640, 1957.
345. Roberts, J.B., and Kelly, J.A. *Treatise on Fractures.* Philadelphia, J.B. Lippincott, 1916.
346. Roberts, N. Fractures of the phalanges of the hand and metacarpals. *Proc R Soc Med* 31:793–798, 1938.
347. Robinson, S. The Bennett fracture of the first metacarpal bone: Diagnosis and treatment. *Boston Med Surg J* 158:275–276, 1908.
348. Ross, J.W., and Sinclair, A.B. The treatment of Bennett's fracture with the Stader splint. *J Can Med Serv* 3:507–511, 1946.
349. Ruggeri, S., Osterman, A.L., and Bora, F.W. Stabilization of metacarpal and phalangeal fractures in the hand. *Orthop Rev* 9:107–110, 1980.
350. Russotti, G.M., and Sim, F.H. Missile wounds of the extremities: A current concepts review. *Orthopedics* 8:1106–1116, 1985.
351. Salgeback, S., Eiken, O., Carstam, N., and Ohlsson, N.M. A study of Bennett's fracture. Special reference to fixation by percutaneous pinning. *Scand J Plast Reconstr Surg* 5:142–148, 1971.
352. Saypol, G.M., and Slattery, L.R. Observations on displaced fractures of the hand. *Surg Gynecol Obstet* 79:522–525, 1944.
353. Scheker, L.R. Department of technique: A technique to facilitate drilling and passing intraosseous wiring in the hand. *J Hand Surg* 5:629–630, 1982.
354. Smith, J.H. Avulsion of a profundus tendon with simultaneous intra-articular fracture of the distal phalanx—case report. *J Hand Surg* 6:600–601, 1981.

355. Spangberg, O., and Thoren, L. Bennett's fracture: A method of treatment with oblique traction. *J Bone Joint Surg [Br]* 45:732–739, 1963.
356. Steel, W.M. The A.O. small fragment set in hand fractures. *Hand* 10:246–253, 1978.
357. Stern, P.J., Wieser, M.J., and Reilly, D.G. Complications of plate fixation in the hand skeleton. *Clin Orthop* 214:59–65, 1987.
358. Stromberg, L. Compression fixation of Bennett's fracture. *Acta Orthop Scand* 48:586–591, 1977.
359. Thoren, L. A new method of extension treatment in Bennett's fracture. *Acta Chir Scand* 110:485–493, 1956.
360. Thoren, L. Basal fractures of the first metacarpal bone—A method of treatment by excision. *Acta Orthop Scand* 27:40–48, 1957.
361. Vanik, R.K., Weber, R.C., Matloub, H.S., Sanger, J.R., and Gingrass, R. The comparative strengths of internal fixation techniques. *J Hand Surg [Am]* 9:216–221, 1984.
362. Viegas, S.F. New method and brace for metacarpal fractures. *Surg Rounds Orthop* March:47–55, 1987.
363. Viegas, S.F., Tencer, A., Woodard, P., and Williams, C.R. Functional bracing of fractures of the second through fifth metacarpals. *J Hand Surg [Am]* 12:139–143, 1987.
364. Wagner, C.J. Method of treatment of Bennett's fracture dislocation. *Am J Surg* 80:230–231, 1950.
365. Wagner, C.J. Transarticular fixation of fracture-dislocations of the first metacarpal-carpal joint. *West J Obstet Gynecol* 59:362–365, 1951.
366. Watson-Jones, R. *Fractures and Joint Injuries* (4th ed.). Edinburgh, E & S Livingstone, 1956.
367. Waugh, R.L., and Ferrazzano, G.P. Fractures of the metacarpals exclusive of the thumb: A new method of treatment. *Am J Surg* 59:186–194, 1943.
368. Wenger, D.R. Avulsion of the profundus tendon insertion in football players. *Arch Surg* 106:145–149, 1973.
369. Wiggins, H.E., Bundens, W.D. Jr., and Park, B.J. A method of treatment of fracture-dislocations of the first metacarpal bone. *J Bone Joint Surg [Am]* 36:810–819, 1954.
370. Woods, G.L. Troublesome shaft fractures of the proximal phalanx. Early treatment to avoid late problems at the metacarpophalangeal and proximal phalangeal joints. *Hand Clin* 4:75–85, 1988.
371. Zacher, J.B. Management of injuries of the distal phalanx. *Surg Clin North Am* 64:747–760, 1984.
372. Zook, E.G. Care of nail bed injuries. *Surg Rounds* September: 44–61, 1985.

Nerve Compression Injuries

373. Dobyns, J.H., O'Brien, E.T., Linscheid, R.L., and Farrow, G.M. Bowler's thumb: Diagnosis and treatment. A review of seventeen cases. *J Bone Joint Surg [Am]* 54:751–755, 1972.
374. Fulkerson, J.P. Transient ulnar neuropathy from Nordic skiing. *Clin Orthop* 153:230–231, 1980.
375. Green, D.P. Diagnostic and therapeutic value of carpal tunnel injection. *J Hand Surg [Am]* 9:850–854, 1984.
376. Hang, Y.S. Tardy ulnar neuritis in a little league baseball player. *Am J Sports Med* 9:244–246, 1981.
377. Linscheid, R.L., and Dobyns, J.H. Athletic injuries of the wrist. *Clin Orthop* 198:141–151, 1985.
378. Lister, G.D., Belsole, R.B., and Kleinert, H.E. The radial tunnel syndrome. *J Hand Surg* 4:52–59, 1979.
379. Minkow, F.V., and Bassett, F.H. III. Bowler's thumb. *Clin Orthop* 83:112, 1972.
380. Phalen, G.S. Reflections on 21 years' experience with carpal tunnel syndrome. *JAMA* 212:1365–1367, 1970.
381. Pickering, T.G. Runner's radial palsy. *N Engl J Med* 10:768, 1981.
382. Stern, P.J., and Kutz, J.E. An unusual variant of anterior interosseous tunnel syndrome. *J Hand Surg* 5:32–34, 1983.
383. Struthers, J. On a peculiarity of the humerus and humeral artery. *Monthly Medical Science* 8:264, 1848.

384. Struthers, J. *Anatomical and Physiological Observations: Part I.* Edinburgh, Sutherland and Knox, 1854.

Splinting and Rehabilitation

385. Bassett, F.H. III, Malone, T., and Gilchrist, R.A. A protective splint of silicone rubber. *Am J Sports Med* 7:358–360, 1979.
386. Bergfeld, J.A., Weiker, G.G., Andrish, J.T., and Hall, R. Soft playing splint for protection of significant hand and wrist injuries in sports. *Am J Sports Med* 10:293–296, 1982.
387. Ditmars, D.M., and Zawlocki, J.M. The Mickey Stanley splint for functional thumb support. *Orthop Rev* 12:107–111, 1983.
388. Nelson, D.M. (Ed.). *Illegal Equipment. 1988 NCAA Football.* Mission, KS, National Collegiate Athletic Association, 1988.
389. Peppard, A. Thumb taping. *Physician Sports Med* 10:139, 1982.
390. Rovere, G.D. How I manage skier's thumb. *Physician Sports Med* 11:73–83, 1983.
391. Schindler, D. (Ed.). *Player Equipment. 1988 Official Football Rules.* Kansas City, MI, National Federation of State High School Associations, 1988.

INDEX

Note: Page numbers in *italics* refer to illustrations;
numbers followed by (t) indicate tables.

3